THE OXFORD HANDBOOK OF

NUMERICAL COGNITION

Roi Cohen Kadosh is a Professor of Cognitive Neuroscience at the University of Oxford, UK. His work combines basic and applied science, with a focus on high level cognitive abilities, such as numerical and mathematical cognition and learning. At the theoretical level, his work challenges and revises previous theories in mathematical cognition, with implications to psychology, neuroscience and education. At the translational level, he works at the forefront, integrating brain stimulation with enhancement of high-level and complex cognitive functions, including mathematical abilities. His work not only focuses on research, but also on ethical implications. He is actively involved in policy making. His pioneering work has received prestigious awards in the fields of neuroscience, psychology, and education, and coverage by leading media channels (e.g., BBC, CNN, Science Magazine, Nature, Scientific American, Time Magazine).

Ann Dowker is a University Research Lecturer at the Department of Experimental Psychology, University of Oxford, UK. She has carried out extensive research on developmental psychology and individual differences, especially with regard to mathematical learning. Her interests include the effects of culture and language on mathematics; mathematics anxiety; links between neuroscience and education; and the development of intervention programs for children with mathematical difficulties. She is the lead researcher on the Catch Up Numeracy Intervention project. She is author of *Individual Differences in Arithmetic: Implications for Psychology, Neuroscience, and Education* (Psychology Press); has edited one book and co-edited another; has produced two reports for the British government on mathematical difficulties and interventions; and has published many articles in academic journals.

THE OXFORD HANDBOOK OF

NUMERICAL COGNITION

Edited by

ROI COHEN KADOSH

and

ANN DOWKER

OXFORD
UNIVERSITY PRESS

OXFORD
UNIVERSITY PRESS

Great Clarendon Street, Oxford, OX2 6DP,
United Kingdom

Oxford University Press is a department of the University of Oxford.
It furthers the University's objective of excellence in research, scholarship,
and education by publishing worldwide. Oxford is a registered trade mark of
Oxford University Press in the UK and in certain other countries

First published 2015
First published in paperback 2016

Published in the United States of America by Oxford University Press
198 Madison Avenue, New York, NY 10016, United States of America

British Library Cataloguing in Publication Data
Data available

Library of Congress Cataloging in Publication Data
Data available

ISBN 978-0-19-964234-2 (Hbk.)
ISBN 978-0-19-879575-9 (Pbk.)

Printed in Great Britain by
CPI Group (UK) Ltd, Croydon, CR0 4YY

Roi Cohen Kadosh would like to dedicate this book to Avishai Henik for introducing him to this fascinating field, his ongoing support and friendship.

Ann Dowker dedicates this book to her late but never forgotten supervisor Neil O'Connor (1917–1997)

Preface

Numbers are crucial and pervasive in many areas: science, economics, sports, education, and everyday life. The human ability to represent and understand numbers is a puzzling paradox: in some ways universal; in other ways subject to huge individual and cultural differences, with many people experiencing difficulties that seriously affect them in everyday life. Numerical cognition, which in this book refers to basic numerical skills as well as mathematical abilities and learning, has in recent years become a vibrant area that brings together scientists from different and diverse research areas (e.g. neuropsychology, cognitive psychology, neuroscience, developmental psychology, comparative psychology, anthropology, education, computer science, and philosophy) using different methodological approaches (e.g. behavioral studies of healthy children and adults and of patients; electrophysiology and brain imaging studies in humans; physiology and lesion studies in non-human primates, habituation studies in human infants and animals, and computational modeling).

Despite its importance, the study of numerical cognition had been relatively neglected for a long time. However, especially during the last decade there has been an explosion of studies and new findings, due partly to technological advances enabling sophisticated human neuroimaging techniques and neurophysiological studies of monkeys, and partly to advances in more traditional psychological and educational research. This has resulted in an enormous advance in our understanding of the neural and cognitive mechanisms of numerical cognition. In addition, there has recently been increasing interest and concern about pupils' mathematical achievement, resulting in attempts to use research to guide mathematics instruction in schools, and to develop interventions for children with mathematical difficulties.

The primary goal of this handbook is to bring together the different research areas that constitute contemporary research in the field of numerical cognition to one comprehensive and authoritative volume. This provides a broad and extensive review of this field that is written in an accessible form for scholars and students, as well as educationists and policy makers. The book covers the most important aspects of numerical cognition research from the areas of development psychology, cognitive psychology, neuropsychology and rehabilitation, learning disabilities, human and animal neuroscience, computational modeling, education and individual differences, and philosophy in fifty-three chapters and seven overview chapters that provide a state-of-the-art review of the current literature.

The book is divided into eight sections. The first section provides a general introduction including a chapter on how mathematics can be communicated to a general audience and a chapter that discusses numerical cognition from a philosophical perspective.

The section after the introduction introduces the major theoretical accounts that have been formulated in the field of cognitive psychology based primarily on behavioral research in humans. The following two sections cover empirical findings on the ontogeny and phylogeny of numerical understanding, and the effect of culture and language on numerical competence and mathematical understanding. These sections are followed by a section on neuroscience and mathematics, which deals with neuronal and neural mechanisms that are involved in mathematical understanding in monkeys and humans, and a chapter that relates some of these empirical findings to computational models.

After the sections on typical numerical understanding with regard to its development, cognitive mechanisms, and neural architecture, we provide a section that deals with numerical disabilities of congenital or acquired origin, and methods for rehabilitation and intervention.

The last two sections discuss mathematics from the perspective of education and individual differences. These sections provide a useful way to link the research discussed up to that point with its practical and social applications, e.g. improving classroom mathematics education.

A special feature of this book is the 'navigator' chapters. The role of these chapters, which appear at the beginning of each section (aside from the Introduction) is to provide an integration of the section, together with the future direction that the field needs to go (therefore *navigating* the field). We chose to include this type of chapter in order that the non-specialist could read an overview on the field, and decide if they would like to read part or the whole of the section, or just use the navigator for a quick update. This will make it more likely that readers will gain a comprehensive overview of the area, even if they choose not to engage in a detailed reading of sections that do not relate to their work (albeit they are encouraged!). We hope that these navigator chapters will also be useful to people who are actively seeking brief reviews of specific fields.

This book could have not been produced by ourselves alone, and we are honored and privileged to have engaged scientists and researchers whose valuable contributions have advanced and elucidated these areas of research. These authors have paved the way to several of the most relevant discoveries in the field, and have provided excellent chapters that provide state-of-the-art knowledge. Undoubtedly, as much as we wanted to, we could not cover all topics or involve all researchers in the field in this volume, and some of those who did not contribute to this book as authors were courteous enough to serve as reviewers and provide excellent feedback that improved the chapters in this book.

We wish to thank all the authors for their contributions, which have also increased our own knowledge of the field. We would also like to thank the reviewers who gave time to providing such useful feedback for the various chapters. We are grateful to Martin Baum, the acquisitions editor, and to Charlotte Green, the Senior Assistant Commissioning Editor, for all their support during the various stages of this project. Thanks are due to the Wellcome Trust for their support of Roi, and to the Esmee Fairbairn Foundation, the Caxton Trust and the Education Endowment Fund for financial support to Ann, which enabled them to complete this book. Last, Roi would like to thank Kathrin, his wife, Jonathan and Itamar, his children, and his students for their patience and support. Ann would like to thank all who have helped and supported her over the years.

We trust that this book will provide a valuable contribution in promoting future discoveries, providing material for critical discussions, and educating readers about this fascinating and promising field.

Ann Dowker
Roi Cohen Kadosh

Contents

PART III: PHYLOGENY AND ONTOGENY OF MATHEMATICAL AND NUMERICAL UNDERSTANDING

PART IV: CULTURE AND LANGUAGE

PART V: NEUROSCIENCE OF MATHEMATICS

PART VI: NUMERICAL IMPAIRMENTS, CO-MORBIDITY, AND REHABILITATION

PART VII: INDIVIDUAL DIFFERENCES

PART VIII: EDUCATION

List of Figures

List of Tables

List of Contributors

Dor Abrahamson
Assistant Professor, Graduate School of Education
University of California
Berkeley
California
USA

Christian Agrillo
Dipartimento di Psicologia Generale "Vittorio Benussi"
Università degli Studi di Padova
Padova
Italy

Michael E. Andres
Professor
Université Catholique de Louvain
Louvain-la-Neuve
Belgium

Daniel Ansari
Department of Psychology
University of Western Ontario
London
Ontario
Canada

Sarit Ashkenazi
School of Medicine
Stanford University
Stanford
California
USA

Mark H. Ashcraft
University of Nevada
Las Vegas
Nevada
USA

David Barner
Assistant Professor, Department of Psychology
University of California
La Jolla
California
USA

Pierre Barrouillet
Professor, University Medical Center (CMU) Dept of Neuroscience
University of Geneva
Geneva
Switzerland

Michael J. Beran
Language Research Center
Georgia State University
Atlanta
Georgia
USA

Wesley Birdsall
Barnard College
Columbia University
New York
USA

Elizabeth Brannon
Professor
Duke University
USA

Christopher Budd
Professor, Mathematical Sciences
University of Bath
Bath
UK

Brian Butterworth
Professor, Institute of Cognitive Neuroscience and Department of Psychology
University College London
London
UK

Jamie Campbell
Department of Psychology
University of Saskatchewan
Saskatoon
Canada

Marinella Cappelletti
Institute of Cognitive Neuroscience
University College London
London
UK

Kara Carpenter
Teachers College
New York
USA

Julie Castronovo
Department of Psychology
University of Hull
Hull
UK

Roi Cohen Kadosh
Department of Experimental Psychology
University of Oxford
Oxford
UK

Richard Cowan
Professor, Institute of Education
University of London
London
UK

Bert De Smedt
Assistant Professor, Parenting and Special Education Research Unit
Leuven
Belgium

Margarete Delazer
Professor, Clinical Department of Neurology
Innsbruck Medical University
Innsbruck
Austria

Fien Depaepe
Katholieke Universiteit
Leuven
Belgium

Annemie Desoete
Department of Experimental Clinical and Health Psychology
Ghent University
Ghent
Belgium

Chris Donlan
Division of Psychology and Language Sciences
University College London
London
UK

Ann Dowker
University of Oxford
Oxford
UK

Nick Dowrick
Edge Hill University
Lancashire
UK

Nancy Dyson
Assistant Professor
University of Delaware
Delaware
USA

Theodore A. Evans
Language Research Center
Georgia State University
Atlanta
Georgia
USA

Wim Fias
Department of Experimental Psychology
Ghent University
Ghent
Belgium

Martin H. Fischer
Professor,Division of Cognitive Science
University of Potsdam
Potsdam
Germany

Lynn S. Fuchs
Professor
Vanderbilt University's Peabody College
Nashville
Tennessee
USA

Karen C. Fuson
Northwestern University Institute of the Learning
Evanston
Illinois
USA

Dana Ganor-Stern
Achva Acedemic College
University of the Negev
Beer Sheva
Israel

David C. Geary
Professor, Department of Psychological Sciences
University of Missouri
Columbia
Missouri
USA

Titia Gebuis
Department of Experimental Psychology
Katholieke Universiteit
Leuven
Belgium

Wim Gevers
Faculty of Psychology
Université Libre de Bruxelles
Brussels
Belgium

Marcus Giaquinto
Professor, Department of Philosophy
University College London
London
UK

Camilla K. Gilmore
Mathematics Education Centre
Loughborough University
Loughborough
UK

Herbert P. Ginsburg
Professor, Department of Human Development
Teachers College Columbia University
New York
USA

Veronique Ginsburg
Université Libre de Bruxelles
Brussels
Belgium

Luisa Girelli
Professor
Università degli Studi di Milano – Bicocca
Milan
Italy

Silke Göbel
Centre for Reading and Language
Department of Psychology
University of York
York
UK

Meghan Goldman
University of California Irvine
Costa Mesa
California
USA

Roland H. Grabner
Professor, Institute for Behavioral Sciences
ETH Zürich
Zürich
Switzerland

Minna Hannula-Sormunen
Department of Teacher Education
University of Turku
Turku
Finland

Claire M.A. Haworth
Professor, Institute of Psychiatry
King's College London
UK

Avishai Henik
Professor, Department of Psychology
Ben-Gurion University of the Negev
Beer-Sheva
Israel

Jo Van Herwegen
Kingston University
London
UK

Ian D. Holloway
Department of Psychology
University of Western Ontario
London
Ontario
Canada

Julie-Ann Jordan
School of Education
Queen's University Belfast
Belfast
Ireland

Nancy C. Jordan
Professor, School of Education
University of Delaware
Newark
Delaware
USA

Adam Tauman Kallai
University of Pittsburgh
Pittsburgh
Pennsylvania
USA

Annette Karmiloff-Smith
Professor, Birkbeck College
University of London
London
UK

Liane Kaufmann
Professor, Department of Psychiatry and Psychotherapy A
General Hospital
Hall in Tyrol
Austria

Yulia Kovas
Department of Psychology
Goldsmiths College
University of London
London
UK

Karin Kucian
Center for MR-Research
University Children's Hospital
Zurich
Switzerland

Rachael Labrecque
Teachers College
Columbia University
New York
USA

Diana Laurillard
Professor
London Knowledge Lab
London
UK

Jo-Anne Le Fevre
Institute of Cognitive Science and Department of Psychology
Carleton University
Ottawa
Ontario
Canada

Patrick Lemaire
Professor
Université de Provence
Marseille
France

Oliver Lindemann
Radboud Universiteit Nijmegen
Nijmegen
The Netherlands

Koleen McCrink
Assistant Professor
Barnard Cognitive Development Center
New York
USA

Tyler Marghetis
Department of Cognitive Science
University of California, San Diego
USA

Tetsuro Matsuzawa
Primate Research Institute
Kyoto University
Kyoto
Japan

Michèle Mazzocco
Professor, Institute of Child Development
Director, Early Math and Numeracy Research Lab
Research Director, Center for Early Education and Development
University of Minnesota
Minneapolis
USA

Vinod Menon
Professor
Stanford School of Medicine
Stanford
California
USA

Korbinian Moeller
Knowledge Media Research Center
Tübingen
Germany

Alex M. Moore
Professor, Department of Psychology
University of Columbia-Missouri
Missouri
USA

Kinga Morsanyi
School of Psychology
Queen's University Belfast
Belfast
Northern Ireland

Aki Murata
Assistant Professor, Stanford University School of Education
Stanford
California
USA

Kevin Muldoon
School of Life Sciences
Heriot-Watt University
Edinburgh
UK

Gerry Mulhern
School of Psychology
Queen's University Belfast
Belfast
Northern Ireland

Andreas Nieder
Professor, Department of Animal Physiology
University of Tübingen
Tübingen
Germany

Marie-Pascale Noël
Université Catholique de Louvain
Louvain-la-Neuve
Belgium

Hans-Christoph Nuerk
Professor
University of Tübingen
Tübingen
Germany

Rafael Nunez
University of California, San Diego
La Jolla
California
USA

Yukari Okamoto
Professor, Department of Education
University of California
Santa Barbara
California
USA

Dana Pagar
Teachers College, Columbia University
Columbia University
New York
USA

Joonkoo Park
Assistant Professor
Department of Psychological and Brain Sciences
University of Michigan
USA

Bonnie M. Perdue
Assistant Professor, Department of Psychology
Agnes Scott College
Decatur
Georgia
USA

Mauro Pesenti
Professor
Université Catholique de Louvain
Louvain-la-Neuve
Belgium

Michal Pinhas
Duke University
Durham
North Carolina
USA

Pekka Räsänen
Niilo Mäki Institute
University of Jyväskylä
Jyväskylä
Finland

Geetha B. Ramani
Department of Human Development
University of Maryland
Baltimore
Maryland
USA

Bert Reynvoet
Faculty of Psychology and Educational Sciences
Katholieke Universiteit Leuven
Leuven
Belgium

Chantal Roggeman
Department of Experimental Psychology
Universiteit Gent
Belgium

Bethany Rittle-Johnson
Professor, Department of Psychology
Vanderbilt University
Nashville
Tennessee
USA

Orly Rubinsten
Department of Learning Disabilities
University of Haifa
Haifa
Israel

Nathan O. Rudig
University of Nevada at Las Vegas
Las Vegas
Nevada
USA

Elena Salillas
Basque Center on Cognition, Brain and Language
Donostia
Spain

Barbara Sarnecka
University of California at Irvine
Irvine
California
USA

Geoffrey Saxe
Graduate School of Education
University of California
Berkeley
California

USA **Michael Schneider**
Professor, Faculty I - Psychology
University of Trier
Trier
Germany

Carlo Semenza
Professor, Department of Neurosciences
University of Padova
Padova
Italy

Robert Siegler
Professor, Psychiatry Department
Carnegie Mellon University
Pittsburgh
Pennsylvania
USA

Victoria Simms
Department of Health Sciences,
University of Leicester
Leicester
UK

Emily Slusser
Professor, Department of Child and Adolescent Development
San Jose State University
San Jose
California
USA

Carla Sowinski
Department of Psychology
Carleton University
Ottawa
Ontario
Canada

Linda Sturman
National Foundation for Educational Research
Slough
UK

Dénes Szücs
Faculty of Education
University of Cambridge
Cambridge
UK

Catherine Thevenot
University Medical Center (CMU)
University of Geneva
Geneva
Switzerland

Maria Grazia Tosto
Department of Psychology, Goldsmiths College
University of London
London
UK

John Towse
Department of Psychology
University of Lancaster
Lancaster
UK

Joseph Tzelgov
Professor, Department of Behavioral Sciences
Ben Gurion University of the Negev
Beer Sheva
Israel

Kim Uittenhove
School of Psychology
University of Geneva
Geneva
Switzerland

Wim van Dooren
Education and Training
Katholieke Universiteit
Leuven
Belgium

Jean Philippe van Dijck
Ghent University
Ghent
Belgium

Sashank Varma
Assistant Professor, Educational Psychology
University of Minnesota
Minneapolis
Minnesota
USA

Tom Verguts
Department of Experimental Psychology
Ghent University
Ghent
Belgium

Lieven Verschaffel
Faculty of Psychology and Educational Sciences
Center for Instructional Psychology & Technology
Leuven
Belgium

Michael von Aster
Professor, Department of Child and Adolescent Psychiatry
University Children's Hospital Zurich
Zurich
Switzerland

Vincent Walsh
Professor, Institute of Cognitive Neuroscience
University College London
London
UK

Emma Wells
Department of Psychology
Carleton University
Ottawa
Ontario
Canada

Rebecca Wheater
National Foundation for Educational Research
Slough
UK

Klaus Willmes
Professor, University Hospital of the RWTH Aachen
Aachen
Germany

Judith Wylie
School of Psychology
Queen's University Belfast
Belfast
Northern Ireland

Laura Zamarian
Clinical Department of Neurology
Medical University Innsbruck
Innsbruck
Austria

PART I

INTRODUCTION

CHAPTER 1

PROMOTING MATHS TO THE GENERAL PUBLIC

CHRIS J. BUDD

What's It All About?

Mathematics is all around us, it plays a vital role in much of modern technology from Google to the Internet and from space travel to the mobile phone. It is central to every school student's education, and anyone needing to get a mortgage, buy a car, sort out their household bills, or just understand the vast amount of information now thrown at them, needs to know some maths. Maths is even used to help us understand, and image, the complex networks and patterns in the brain and many of the processes of perception. However, like the air around us, the importance of mathematics is often invisible and poorly understood, and as a result many people are left unaware of the vital role that it could, and does, play in their lives. In an increasingly technology and information driven world this is potentially a major problem.

However, we have to be honest, mathematics and its relevance, is a difficult subject to communicate to the general public. It certainly doesn't have the instant appeal of sex and violence that we find in other areas (although it does have applications to these) and there is a proud cultural tradition in the UK that it is good to be bad at maths. For example when I appeared once on the One Show, both presenters were very keen to tell me that they were rubbish at maths and that it didn't seem to have done them any harm! (I do wonder whether they would have said the same to a famous author, artist, or actor.) Maths is also perceived as a dry subject without any applications (this is also very untrue and I will discuss this later) and this perception does put a lot of school students (and indeed their teachers) off. Finally, and (perhaps this is what makes it especially hard to communicate), maths is a linear subject, and a lot of background knowledge, and indeed investment of time, is required of any audience to whom you might want to communicate its beauty and effectiveness. For example, one of the most important way that maths affects all of our lives is through the application of the methods of calculus. But very few people have heard of calculus, and those that have are generally scared by the very name. It also takes time and energy to communicate maths well and (to be honest), most mathematicians are not born communicators (in fact rather the opposite). However, it is a pleasure to say that there are some gifted maths communicators out there who are making a very positive impact, as well as university courses teaching maths communication skills. Indeed, the popularization of mathematics has become an increasingly respectable and widespread activity, and I will describe some of this work in this chapter.

So why do we bother communicating maths in the first place, and what we hope to achieve when we attempt to communicate maths to any audience, whether it is a primary school class, bouncing off the walls with enthusiasm, or a bored class of teenagers on the last lesson of the afternoon? Well, the reason is that maths is insanely important to everyone's lives whether they realize it directly (for example through trying to understand what a mortgage percentage on an APR actually means) or indirectly through the vital role that maths plays in the Internet, Google, and mobile phones to name only three technologies that rely on maths. Modern technology is an increasingly mathematical technology and unless we inspire the next generation then we will rapidly fall behind our competitors.[1] However, when communicating maths we always have to tread a narrow line between boring our audience with technicalities at one end, and watering maths down to the extent of dumbing down the message at the other. Ideally, in communicating maths we want to get the message across that maths is important, fun, beautiful, powerful, challenging, all around us and central to civilization, to entertain and inspire our audience and to leave the audience wanting to learn more maths (and more about maths) in the future, and not to be put off it for life. Rather than dumbing down maths, public engagement should be about making mathematics come alive to people. This is certainly a tall order, but is it possible? While the answer is certainly YES, there are a number of pitfalls to trap the unwary along the way.

In this chapter I will explore some of the reasons that maths has a bad image and/or is difficult to communicate to the general public. I will then discuss some general techniques which have worked for myself, and others, in the context of communicating maths to a general audience. I will then go on to describe some initiatives which are currently under way to do this. Finally I will give some case studies of what works and what does not.

What's the Problem with Maths?

Let's be honest, we do have a problem in conveying the joy and beauty of mathematics to a lay audience, and maths has a terrible popular image. A lot of important maths is built on concepts well beyond what a general audience has studied. Also mathematical notation can be completely baffling, even for other mathematicians working in a different field. Here for example is a short quote from a paper, authored by myself, about the equations describing the (on the face of it very interesting) mathematics related to how things combust and then explode:

Let $-\Delta\phi = f(\phi)$. *A weak solution of this PDF satisfies the identity*
$\int \nabla\phi\nabla\psi dx = \int f(\phi)\psi dx \forall\ \psi \in H_0^1(\Omega)$.
Assume that $f(\phi)$ *grows sub* − *critically it is clear from Sobolev embedding that* $\exists\phi \in H_0^1$

This quote is meaningless to any other than a highly specialist audience. Trying to talk about (say in this example) the detailed theory and processes involved in solving differential equations with an audience which (in general) doesn't know any calculus, is a waste of everyone's time and energy. As a result it is extremely easy to kill off even a quite knowledgeable audience when giving a maths presentation or even talking about maths in general. The same problem extends to all levels of society. Maths is perceived by the greater majority of the country as a boring, uncreative, irrelevant subject, only for (white, male) geeks. All mathematicians know this to be untrue. Maths is an extraordinarily creative subject, with mathematical ideas taking us well beyond our imagination. It is also a subject with limitless applications without which the modern world would simply not function. Not being able to do maths (or at least being numerate) costs the UK an estimated £2.4B every year according to a recent Confederation of British Industry report (CBI Report, 2010). Uniquely amongst all (abstract) subjects, mathematicians and mathematics teachers are asked to justify why their subject is useful. Not only is this unfair (why is maths asked to justify itself in this way, and not music or history), it is also ridiculous given that without maths the world would starve, we would have no mobile phones and the Internet would not function.

I have thought very hard about why the popular image and perception of maths is so different from reality and why it is culturally fine to say that you are bad at maths. There are many possible reasons for this.

Firstly, the obvious. Maths is really hard, and not everyone can do it. Fair enough. However, so is learning a foreign language or taking a free kick, or playing a musical instrument, and none of these carry the same stigma that maths does.

Secondly, maths is often taught in a very abstract way at school with little emphasis on its extraordinary range of applications. This can easily turn an average student off 'what's the use of this Miss' is an often heard question to teachers. Don't get me wrong, I'm all in favor of maths being taught as an abstract subject in its own right. It is the abstractness of maths that underlies its real power, and even quite young students can be captivated by the puzzles and patterns in maths. However, I am also strongly in favor of all teaching of maths being infused with examples and applications. Mathematicians often go much too far in glorifying in the 'uselessness' of their subject (witness the often quoted remarks by Hardy in '*A Mathematicians Apology*' (Hardy 1940) see for example his concluding remark in that book,[2] which was certainly not true, given Hardy's huge impact on many fields of science). However this is sheer nonsense. Nothing in maths is ever useless. I think that it is the duty of all mathematicians to understand, and convey, the importance and applications of the subject to as broad an audience as possible, and to teachers in particular.

Thirdly, we have structural problems in the way that we teach maths in English schools (less so in Scotland). Most UK students give up maths at the age of fifteen or sixteen and never see it again. These students include future leaders in government and in the media. What makes this worse is that the huge majority of primary school teachers also fall into this category. The result is that primary level maths is taught by teachers who are often not very confident in it themselves, and who certainly cannot challenge the brightest pupils. They certainly cannot appreciate its creative and useful aspects. (Indeed when I was at primary school in the 1960s maths lessons were actually banned by the headmistress as 'not being creative'.) Students at school are thus being put off maths far too early, and are given no incentive to take it on past GCSE. Even scientists (such as psychologists!) who need mathematics (and especially statistics) are giving up maths far too early. Perhaps most seriously of all, those in government or positions of power, may themselves have had no exposure to maths after the age of 15, and indeed there is a woeful lack of MPs with any form of scientific training. How are these policy makers then able to cope with the complex mathematical issues which arise (for example) in the problems associated with climate change (see the example at the end of this chapter). We urgently need to rectify this situation, and the solution is for every student to study some form of maths up to the age of 18, with different pathways for students with different abilities and motivation. (See the *Report on Mathematical Pathways post 16* (ACME Report 2011 and also Vorderman 2011).

Finally, and I know that this is a soft and obvious target, but I really do blame the media. With notable (and glorious) exceptions, maths hardly ever makes it onto TV, the radio, or the papers. When it does it is often either extremely wrong (such as the report in the Daily Express about the chance of getting six double-yoked eggs in one box) or it is treated as a complete joke (the local TV reports of the huge International Conference in Industrial and Applied Mathematics at Vancouver in 2011 are a good example of this, see <www.youtube.com/watch?v=M4beANEdl4A&lr=1&feature=mhee>).

Sadly this type of report is the rule rather than the exception, or is given such little airtime that if you blink then you miss it. Contrast this with the acres of time given to the arts or even to natural history, and the reverence that is given to a famous author when they appear on the media. Part of this can be explained by the ignorance of the reporters (again a feature of the stopping of mathematics at the age of 15), but nothing I feel can excuse the antagonistic way in which reporters treat both mathematicians and mathematics. I have often been faced by an interviewer who has said that they couldn't do maths when they were at school, or they never use maths in real life, and that they have done really well. To which my answers are that they are not at school anymore and that if they can understand their mortgage or inflation or APR without maths then they are doing well. Worst of all are those journalists that ask you tough mental arithmetic questions live on air to make you look a fool (believe me your mind turns to jelly in this situation). It is clearly vital to work with the media (see later), but the media also needs to put its own house in order to undo the damage that it has done to the public's perception of mathematics.

How can Maths be Given a Better Image?

As with all things there is no one solution to the problem of how to communicate to the broader public that maths isn't the irrelevant and scary monster that they (and the media) often make it out to be. Many different maths presenters have adopted different (and equally successful) styles. However some techniques that I have found to have worked with many audiences (both young and old) include the following.

- Starting with an application of maths relevant to the lives of the audience, for example Google, iPods, crime fighting, music, code breaking, dancing (yes, dancing). Hook them with this and then show, and develop, the maths involved (such as in the examples above, network theory, matrix theory, and group theory). Science presenters can often be accused of 'dumbing down' their subject, and it is certainly true that it is impossible to present higher level maths to a general audience for the reasons discussed above. However, a good application can often lead to many fascinating mathematical investigations

- Being proud not defensive of the subject. Maths really DOES make a difference to the world. If mathematicians can't be proud and passionate of it then who will be? Be very positive when asked by any interviewer 'what's the point of that'.
- Showing the audience the surprise and wonder of mathematics. It is the counter-intuitive side of maths, often found in puzzles or 'tricks', that often grabs attention, and can be used to reveal some of the beauty of maths. The public loves puzzles, witness the success of Sudoku, and many of these (such as Griddler, Killer Sudoku, and problems in code breaking) have a strong mathematical basis. (Those that say that Sudoku has nothing to do with maths simply don't understand what maths really is all about!) There are also many links between maths and magic (as we shall see later); many good magic tricks are based on theorems (such as fixed point theorems in card shuffling and number theorems in mind-reading tricks). Indeed a good mathematical theorem itself has many of the aspects of a magic trick about it, in that it is amazing, surprising, remarkable, and when the proof is revealed, you become part of the magic too.
- Linking maths to real people. Many of our potential audiences think that maths either comes out of a book, or was carved in stone somewhere. Nothing could be further from the truth. One of the problems with the image of maths in the eyes of the general public is that it does not seem to connect to people. Indeed a recent letter in *Oxford Today* (<http://www.oxfordtoday.ox.ac.uk/>) the Oxford alumni magazine (which really should have known better!) said that the humanities were about people and that science was about things (and that as a consequence the humanities were more important). What rubbish! All maths at some point was created by a real person, often with a lot of emotional struggle involved or with argument and passion. No one who has seen Andrew Wiles overcome with emotion at the start of the BBC film *Fermat's Last Theorem* produced by Simon Singh and described in his wonderful book (Singh 1997), can fail to be moved when he describes the moment that he completed his proof. Also stories such as the life and violent death of Galois, the recent solution of the Poincare Conjecture by a brilliant, but very secretive Russian mathematician, or even the famous punch up surrounding the solution of the cubic equation or the factorization of matrices on a computer, cannot fail to move even the most stony-faced of audiences.
- Not being afraid to show your audience a real equation. Stephen Hawking famously claimed that the value of a maths book diminishes with every formula. This is partly true as my earlier example showed. There are, however, many exceptions to this. Even an audience that lacks mathematical training can appreciate the elegance of a formula that can convey big ideas so concisely. Some formulae indeed have an eternal quality that very few other aspects of human endeavor can ever achieve. Mind you, it may be a good idea to warn your audience in advance that a formula is coming so that they can brace themselves. So here goes:

$$\frac{\pi}{4} = 1 - \frac{1}{3} + \frac{1}{5} - \frac{1}{7} + \frac{1}{9} - \frac{1}{11} + \frac{1}{13} - \ldots$$

Isn't that sheer magic. You can easily spend an entire lecture, or popular article, talking about that formula alone. If I am ever asked to 'define mathematics' then that is my answer. Anyone who does not appreciate that formula simply has no soul! You can find out more in my article (Budd 2013). Whole (and bestselling) books (Nahin 2006) have been written on arguably the most important and beautiful formula of all time

$$e^{i\pi} = -1$$

which was discovered by Euler and lies behind the technology of the mobile phone and also the electricity supply industry. For more fabulous formulae see the book *17 Equations that Changed the World*, by Ian Stewart (2012).

- Above all, be extremely enthusiastic. If you enjoy yourself then there is a good chance that your audience will too.

So, What's Going On?

As I said earlier, we have seen a rapid increase in the amount of work being done to popularize maths. Partly this is a direct result of the realization that we do need to justify the amount of money being spent on maths, and to increase the number of students both studying maths and also using it in their working lives. I also like to think that more people are popularizing maths because it is an exciting thing to do which brings its own rewards, in much the same way that playing an instrument or acting in a play does. Maths communication activities range from high profile work with the media, to writing books and articles, running web-based activities, public lectures, engaging with schools, busking, stand up events, outreach by undergraduates, and science fairs. In all these activities we are trying to reach three groups; young people, the general public, and those who control the purse strings.

The Media. As I described above, the media is a very hard nut to crack, with a lot of resistance to putting good maths in the spotlight. However, having said that we are very fortunate to have a number of high profile mathematicians currently working with the media in general and TV/radio in particular. Of these I mention in particular Ian Stewart, Simon Singh, Matt Parker, Marcus du Sautoy, and Sir David Spiegelhalter, but there are many others. The recent BBC4 series by Marcus du Sautoy on the history of maths was a triumph and hopefully the DVD version of this will end up in many schools) and we mustn't also forget the pioneering work of Sir Christopher Zeeman and Robin Wilson. Marcus du Sautoy, Matt, and Steve Humble (aka Dr Maths) also show us all how it can be done, by writing regular columns for the newspapers. It is hard to underestimate the impact of this media work, with its ability to reach millions, although it is a long way to go before maths is as popular in the media as cooking, gardening and even archaeology.

Popular Books. Ian Stewart, Robin Wilson, Simon Singh, and Marcus du Sautoy are also well-known for their popular maths books and are in excellent company with John Barrow, David Acheson, and Rob Eastaway, but I think the most 'popular' maths author by quite a wide margin is Kjartan Poskitt. If you haven't read any of his Murderous Maths series then do so. They are ostensively aimed at relatively young people and are full of cartoons, but every time I read them I learn something new. Certainly my son has learnt (and become very enthusiastic about maths) from devouring many of these books.

The Internet. Mathematics, as a highly visual subject, is very well-suited to being presented on the Internet and this gives us a very powerful tool for not only bringing maths into peoples homes but also being able to have a dialogue between them and experienced mathematicians via blog sites and social media. The (Cambridge-based) Mathematics Millennium Project (the MMP) has produced a truly wonderful set of Internet resources through the NRICH and PLUS websites and the STIMULUS interactive project. Do have a look at these if you have time. I have personally found the PLUS website to be a really fantastic way of publishing popular articles which reach a very large audience. The Combined mathematical Societies (CMS) have also set up the Maths Careers website, <http://www.mathscareers.org.uk/>, showcasing the careers available to mathematicians. I mustn't also forget the very popular Cipher Challenge website run by the University of Southampton.

Direct engagement with the public. There is no substitute for going into schools or engaging directly with the public. A number of mechanisms exist to link professional mathematicians to schools, of which the most prominent are the Royal Institution Mathematics Masterclasses. I am biased here, as I am the chair of maths at the Royal Institution, but the masterclasses have an enormous impact. Every week many schools in over 50 regions around the country will send young people to take part in Saturday morning masterclasses on topics as various as the maths of deep sea diving to the Fibbonacci sequence. These masterclasses are often run (and are based in) the university local to the region and are a really good way for university staff to engage with young people. Of course it is impossible to get to every school in the country and it is much more efficient to bring lots of schools to really good events. One way to do this is through the LMS Popular Lectures, the Training Partnership Lectures, and the *Maths Inspiration* series (<http://www.mathsinspiration.com/index.jsp>). The latter (of which I'm proud to be a part) are run by Rob Eastaway and deliver maths lectures in a theatre setting, often with a very interactive question and answer session. A recent development has been the growth of 'Maths Busking' (<http://mathsbusking.com/>). This is really busking where maths itself is the gimmick and reaches out to a new audience who would otherwise not engage with mathematics or mathematicians. Closely related are various stand up shows linked to maths such as the *Festival of the Spoken Nerd* or *Your Days are Numbered.* These link maths to comedy and reach out to a very non-traditional maths audience, appearing, for example, at the Edinburgh Fringe.

Science fairs are a popular way of communicating science to the public. Examples range from the huge, such as the British Science Association annual festival, the Big Bang Fair, and the Cheltenham festival of science, to smaller local activities such as Bath Taps Into Science and Maths in the Malls (Newcastle). I visit and take part in a lot of science fairs and it is fair to say that in general maths has traditionally been very much under-represented. Amongst the vast number of talks/shows on biology, astronomy, archaeology, and psychology you may be lucky to find one talk on maths. The problems we referred to earlier of a resistance to communicate maths in the media often seem to extend to science communicators as well. Fortunately things are improving, and the maths section of the British Science Association has in recent years been very active in ensuring that the annual festival of the BSA has a strong maths presence. Similarly, the maths contribution towards the Big Bang has grown significantly, with the IMA running large events since 2011, attended by approximately

50 000 participants. Hopefully mathematics will have a similar high profile presence at future such events. Indeed 2014 marks the launch of the very first Festival of Mathematics in the UK. A related topic is the presence of mathematics exhibits in science museums. It is sad to say that the maths gallery in the Science Museum, London, is very old and is far from satisfactory as an exhibition of modern mathematics. Fortunately it is now in a process of redesign. Similarly the greater majority of exhibits in science museums around the UK have no maths in them at all. There seems to be a surprising reluctance from museum organizers to include maths in their exhibits. However, our experience of putting maths into science fairs shows that maths can be presented in an exciting and hands on way, well-suited to a museum exhibition. It is certainly much cheaper to display maths than most other examples of STEM (Science, Technology, and Mathematics) disciplines. The situation is rather better in Germany where they have the 'Mathemtikum' (<http://www.mathematikum.de/>) which contains many hands-on maths exhibits as well as organizing popular maths lectures, and in New York with the Museum of Maths. Plans are underway to create 'MathsWorld UK' which will be a UK-based museum of maths.

Maths Communicators. Finally, my favorite form of outreach are ambassador schemes in which undergraduates go into the community to talk about mathematics. They can do this for degree credit (as in the Undergraduate Ambassador Scheme (<http://www.uas.ac.uk/>) or the Bath 'Maths Communicators' scheme), for payment as in the Student Associate Scheme, or they can act as volunteers such as in the Cambridge STIMULUS programme which encourages undergraduates to work with school students through the Internet. The undergraduates can be mainly based in schools, or can have a broader spectrum of activities. Whatever the mechanism Student Ambassador Schemes have been identified as one of the most effective activities in terms of Widening Participation and Outreach. They combine the enthusiasm and creative brilliance of the pool of maths undergraduates that we have in the UK, with the very need no only to communicate maths but to teach these undergraduates communication skills which will be invaluable for their subsequent careers. Everybody wins in this arrangement. The students often describe these courses as the best thing that they do in the degree, and they create a lasting legacy of resources and a lasting impression amongst the young people and general public who they work with. The recent IMA report on Maths Student ambassador Case studies (<http://www.hestem.ac.uk/sites/default/files/mark_inner.pdf>) gives details on a number of these schemes.

What Doesn't Work

I repeat the fact that maths can remain hard to communicate, and it is very easy to fall into a number of traps. For the sake of a balanced chapter (and to warn the unwary) here are a few examples of these.

Too much or too little. We have already seen an example of where too much maths in a talk can blow your audience away. It is incredibly easy to be too technical in a talk, to assume too much knowledge and to fail to define your notation. We've all been there, either on the giving or the receiving end. The key to what level of mathematics to include is to find out about your audience in advance. In the case of school audiences this is relatively easy—knowing the year group and whether you are talking to top or bottom sets should give you a good idea of how much maths they are likely to know. Yet too often I have seen speakers standing in front of a mixed GCSE group talking about topics like dot products and differentiation and assuming that these concepts will be familiar. It is equally dangerous to put in too little maths and to water down the mathematical content so that it becomes completely invisible, or (as is often the case) to talk only about arithmetic and to miss out maths all together. With a few notable exceptions, most producers and presenters in the media, think that any maths is too much maths and that their audience cannot be expected to cope with it at all. But this only highlights the real challenge of presenting maths in the media where time and production constraints make it very hard indeed to present a mathematical argument. In his Royal Institution Christmas Lectures in 1978, Prof Christopher Zeeman spent 12 minutes proving that the square root of two was irrational. It is hard to think of any mainstream prime time broadcast today where a mathematical idea could be investigated in such depth. A couple of minutes would probably be the limit, far too short a time to build a proof. Perhaps at some point in the future this will change, but for the time being, maths communicators have to accept that television is a very limited medium for dealing with many accessible mathematical ideas.

The curse of the 'formula'. As I have said, one of the ways of engaging audiences in maths is by relating it to everyday life and done correctly this can be very effective. This can, however, be taken too far. Taking a topic that is of general interest—romance, for example—and attempting to 'mathematize' it in the hope that the interest of the topic will rub off on the maths, can backfire badly. Much of the maths that gets reported in the press is like this. Although we love the use of formulae when they are relevant, the use of irrelevant formulae in a talk or an article can make maths appear trivial. For example, I was once rung up by the press just before Christmas and asked for the 'formula for the best way to stack a fridge for the Christmas Dinner'. The correct answer to this question is that there is no such formula, and an even better answer is that if anyone was able to come up with one they would (by the process of solving the NP-hard Knapsack problem) pocket $1 000 000 from the Clay Foundation. However the journalist concerned seemed disappointed with the answer. No such reluctance however got in the way of the person that came up with

$$K = \frac{F(T+C)-L}{S}$$

Which is apparently the formula for the perfect kiss. All I can say is: whatever you do, don't drop your brackets. For the mathematician collaborating with the press this might seem like a great opportunity to get maths into the public eye. To the journalist and the reading public, however, more often it is simply a chance to demonstrate the irrelevance of the work done by 'boffins'. Such things are best avoided.

And What Does Work

I will conclude this chapter with some examples of topics that contain higher level of maths in them than might be anticipated and communicate maths in a very effective way. More examples of case studies can be found in my article (Budd and Eastaway 2010), or on my website <http://people.bath.ac.uk/mascjb/>, or on the Plus maths website <http://plus.maths.org>.

Example 1. Asperger's Syndrome. In the book *The Curious Incident of the Dog in the Night-time* by Mark Haddon (2004) the reader was invited to find an example of a right-angled triangle in which its sides could not be written in the form $n^2 + 1$, $n^2 - 1$ and $2n$ (where $n > 1$). On the face of it this was quite a high level of mathematics for a popular book (which has now been turned into a play). *The Curious Incident* is a book about Asperger's Syndrome, written from a personal perspective. Millions of people have read this book, and many of these (who are not in any sense mathematicians) have read this part of it and have actually enjoyed, and learned something, from this. The reason this worked was twofold. First, the maths was put into the context of a human story, which made it easier for the reader to empathize with it. The second was that the author used a clever device whereby he allowed the lead character to speak for maths, while his friend spoke for the baffled unmathematical reader. As a result, Haddon (a keen mathematician) managed to sneak a lot of maths into the book without coming across as a geek himself.

Example 2. Maths Magic. Everyone (well nearly everyone) likes the mystery and surprise that is associated with magic. To a mathematician, mathematics has the same qualities, but they are less well appreciated by the general public. One way to bring them together is to devise magic tricks based on maths. I have already alluded to some of these. The general idea is to translate some amazing mathematical theorem into a situation which everyone can appreciate and enjoy. These may involve cards, or ropes, or even mind-reading. As an example, it is a well-known theorem that if any number is multiplied by nine, then the sum of the digits of the answer is itself a multiple of nine. Similarly, if you take any number and subtract from it the sum of its digits then you get a multiple of nine. Put like this these results sound rather boring, but in the context of a magic trick they are wonderful ambassadors for mathematics. The first leads to a lovely mind-reading trick. Ask your audience to think of a whole number between one and nine and then multiply it by nine. They should then sum the digits and subtract five from their answer. If they have a one they should think A, two think B, three think C, etc. Now take the letter they have and think of a country beginning with that letter. Take the last letter of that country and think of an animal beginning with that letter. Now take the last letter of the animal and think of a color beginning with that letter. Got that. Well hopefully you are now all thinking of an *Orange Kangaroo from Denmark.*

The reason that this trick works, is that from the first of the above theorems, the sum of the digits of the number that they get must be nine. Subtract five to give four, and the rest is forced. This trick works nearly every time and I was delighted to once use it for a group of blind students, who loved anything to do with mental arithmetic. For a second trick, take a pack of cards and put the Joker in as card number nine. Ask a volunteer for a number between 10 and 19 and deal put that number of cards from the top. Pick this new pack up and ask for the sum of the digits of the volunteer's number. Deal that number of cards from the top. Then turn over the next card. It will always be the Joker. This is because if you take any number between 10 and 19 and subtract the sum of the digits then you always get nine.

With a collection of magic tricks you can introduce many mathematical concepts, from primary age maths to advanced level university maths. The best was to do this, is to first show the trick, then explain the maths behind it, then get the audience to practice the trick, and then (and best of all) get them to devise new tricks using the maths that they have just learned. You never knew that maths could be so much fun!

Example 3. How Maths Won the Battle of Britain. It may be unlikely to think of mathematicians as heroes, but without the work of teams of mathematicians the Allies would probably have lost the Second World War. Part of this story is well-known. The extraordinary work of the mathematical code breakers, especially Alan Turing and Bill Tutte, at Bletchley Park has been the subject of many documentaries and books (and this is one area where the media has got it right). This has been described very well in the Code Book (also) by Simon Singh (1999). However, mathematics played an equally vital role in the Battle of Britain and beyond. One of the main problems faced by the RAF during the Battle of Britain was that of detecting the incoming bombers and in guiding the defending fighters to meet them. The procedure set up by Air Vice Marshall Dowding to do this, was to collect as much data as possible about the likely location of the aircraft from a number of sources, such as radar stations and the Royal Observer Corps, and to then pass this to the 'Filter Room' where it was combined to find the actual aircraft position. The Filter Room was staffed by mathematicians whose job was to determine the location of the aircraft by using a combination of (three-dimensional) trigonometry to predict their height, number, and location from their previous known locations, combined with a statistical assessment of their most likely position given the less than reliable data coming from the radar stations and other sources. Once the location of the aircraft was known further trigonometry was required to guide the fighters on the correct interception path (using a flight direction often called the 'Tizzy' angle after the scientific civil servant Tizard). An excellent account of this and related applications of maths is given in Korner (1996). In a classroom setting this makes for a fascinating and interactive workshop in which the conditions in the filter room are recreated and the students have to do the same calculations under extreme time pressure. One of the real secrets to popularizing maths is to get the audience really involved in a hands-on manner! (It is worth saying that the same ideas of comparing predictions with unreliable data to determine what is actually going on are used today both in Air Traffic Control, meteorology and robotics.)

Whilst it might be thought that this is a rather 'male oriented' view of applied mathematics, it is well worth saying that the majority of the mathematicians employed in the filter rooms were relatively young women in the WAAF, often recruited directly from school for their mathematical abilities. In a remarkable book, Eileen Younghusband (2011) recounts how she had to do complex three-dimensional trigonometric under extreme pressure, both in time and also knowing how many lives depended on her getting the calculations right. After the Battle of Britain she 'graduated' to the even harder problem of tracking the V2 rockets being fired at Brussels. When I tell this story to teenagers, they get incredibly involved and there is not a dry eye in the house. No one can ever accuse trigonometry of not being useful or interesting!

Example 4. Weather and Climate. One of the most important challenges facing the human race is that of climate change. It is described all the time in the media and young people especially are very involved with issues related to it. The debates about climate change are very heated. From the perspective of promoting mathematics, climate change gives a perfect example of how powerful mathematics can be brought to bear on a vitally important problem, and in particular gives presenters a chance to talk about the way that equations can not only model the world, but are used to make predictions about it. Much of the mathematical modeling process can be described and explained through the example of predicting the climate and the audience led through the basic steps of:

(1) Making lots of observations of pressure, temperature, wind speed, moisture, etc.
(2) Writing down the (partial differential) equations, which tell you how these variables are related.
(3) Solving the equations on a computer.
(4) Constantly updating and checking the computer simulations with new data.
(5) Assessing the reliability of the prediction.
(6) Informing policy bodies about the results of the simulations.

There are plenty of mathematics and human elements to this story, starting from Euler's derivation of the first laws of fluid motion, the work of the mathematicians Navier and Stokes on fluids or Kelvin in thermodynamics (the latter was a real character), the pioneering work of Richardson (another great character) in numerical weather forecasting, and the modern day achievements and work of climate change scientists and meteorologists. However, the real climax of talking about the climate should be the maths itself which comes across well as being an impartial factor in the debate, far removed from the hot air of the politicians. As a simple example, if 'T' is the temperature of the Earth, 'e' is its emmisivity (which decreases as the carbon dioxide levels in the atmosphere increase), 'a' is its albedo (which decreases as the ice melts), and 'S' is the energy from the sun (which is about ⅓kW per m^2 on average) then:

$$e\sigma T^4 = (1-a)S$$

This formula can be solved using techniques taught in A level mathematics, and allows you to calculate the average temperature of the Earth. The nice thing about this formula is that unlike the formula for the perfect kiss, this one can be easily checked against actual data. From the perspective of climate science its true importance is that it clearly shows the effects on the Earth's temperature (and therefore on the rest of the climate) of reducing the emmisivity 'e' (by increasing the amount of Carbon Dioxide in the atmosphere) or of reducing the albedo 'a' (by reducing the size of the ice sheets. This leads to a frightening prediction. The hotter it is the less ice we have as the ice sheets melt. As a consequence the albedo, 'a', decreases, so the Earth reflects less of the Sun's radiation. Our formula then predicts that the Earth will get hotter, and so more ice melts and the cycle continues. Thus we can see the possible effects of a positive feedback loop leading to the climate spiraling out of control. This is something that any audience can connect with, and leads to fierce debates! It may come as a surprise, but I have always found that audiences generally like the 'unveiling' of this equation, and seeing how it can be used to make predictions. A talk about mathematics can be exactly that, i.e. 'about' mathematics. If the audience gains the impression that maths is important, and that the world really can be described in terms of mathematical equations and that a lot of mathematics has to be (and still is being done) to make sense of these equations, then the talk to a certain extent has achieved its purpose. Talks on climate change often lead to a furious email (and other) correspondence, which goes against the implicit assumption in the media that no one is really interested in a mathematical problem.

At another level, climate change is exactly the sort of area where mathematicians and policy makers need to communicate with each other as clearly as possible, with each side understanding the language (and modus operandi) of the other.

Example 5. Maths and Art. One of the aspects of mathematics which tends to put people off is that it is perceived as a dry subject, far removed from 'creative' subjects such as art and music. Of course this is nonsense, as maths is as creative a subject as it is possible to get (I spend my life creating new mathematics), but it is worth making very explicit the wonderful links between mathematics and art. (When faced with the question: is maths an art or a science? The correct answer is simply 'Yes'.) Some of these links run very deep, for example the musical scale is the product of many centuries of mathematical thought (started by Pythagoras). The subject of origami was for many years treated simply as an art form. However, working out the folding pattern to create a three-dimensional object (such as a beetle) from a single sheet of paper is fundamentally a mathematical problem. This was realized recently by Robert Lang <http://www.langorigami.com/> amongst others, and the fusion of mathematics with Origami leads to sublime artistic creations. Another area where art meets maths in a multicultural setting is in Celtic Knots and the related Sona drawings from Africa. Examples of both of these are illustrated in Figure 1.1a, b, with Figure 1.1a showing a circular Celtic Knot created by a school student, and Figure 1.1b a Sona design called the 'Chased Chicken'.

Celtic Knots are drawn on a grid according to certain rules. These rules can be translated into algebraic structures and manipulated using mathematics. By doing this, students can explore various combinations of the rules, and then turn them into patterns of art. This is an incredibly powerful experience for them as they see the direct relation between quite deep symmetry patterns in mathematics and beautiful art work. Usually when I do Celtic Art workshops I have two sessions, one where I describe the maths and then I wait for a month whilst the students work with an art department. By doing this they learn both maths and art at the same time. As I said, a very powerful experience all round. A nice spin-off is the related question of investigating African Sona patterns. Mathematically these are very similar to Celtic Knots, and in fact the ideas behind them predate those of Celtic Knots. An excellent account of these patterns along with many other examples of the fusion of African mathematics and art, is given in Gerdes (1999). Doing a workshop on Celtic Knots and Sona

(a) (b)

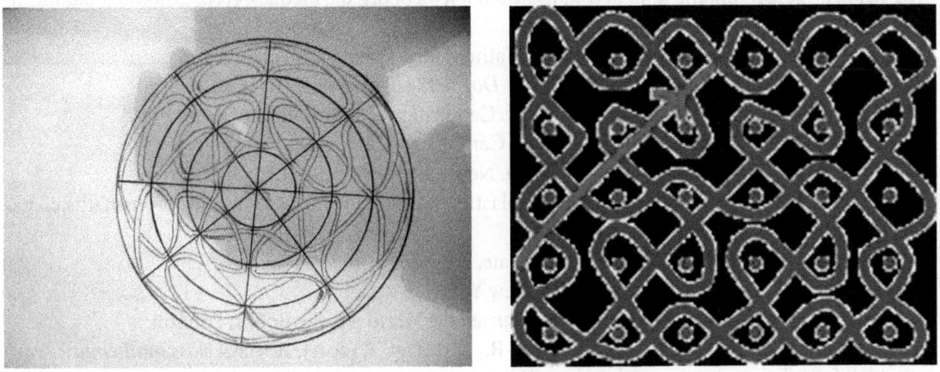

FIGURE 1.1 (a) A Circular Celtic Knot. (b) The Chased Chicken Sona pattern.

patterns, demonstrates the fact that maths is not a creation of the Western World, but is a truly international and multi-cultural activity.

And Finally

I hope that I have demonstrated in this chapter that although maths is hard and has a terrible public image, it is a subject that can be presented in a very engaging and hands on way to the general public. Indeed it can be used to bring many ideas together from art to engineering and from music to multi-culturalism. By doing so, everyone can both enjoy, and see the relevance, of maths. There is still a long way to go before maths has the same popularity (and image) on the media as (say) cooking or gardening (or even astronomy or archaeology), but significant progress is being made (as mathematicians say it 'has a positive gradient') and I am very optimistic that in ten years time, maths will have a very much better public image than it does at the present.

Notes

1. A lot of these issues were explored in the 'Vorderman report' *A world class mathematics education for all our young people* (2011), of which I was a co-author.
2. I have never done anything 'useful'. No discovery of mine has made, or is likely to make, directly or indirectly, for good or ill, the least difference to the amenity of the world.

References

ACME Report (2011). *Mathematical needs in the workplace and in higher education.* <http://www.acme-uk.org/media/7624/acme_theme_a_final%20(2).pdf>.

Budd, C.J. (2013). 'How to add up quickly'. *Plus Maths Magazine*, <http://plus.maths.org/content/how-add-quickly>.

Budd, C.J. and Eastaway, R. (2010). 'How much maths is too much maths?' *Maths Today*, (The journal of the IMA), October 2010.

CBI Report. (2010). *Making it all add up.* <http://www.cbi.org.uk/media/935352/2010.08-\making-it-all-add-up.pdf>.

Gerdes, P. (1999). *Geometry from Africa.* The Mathematical Association of America.

Haddon, M. (2004). *The Curious Incident of the Dog in the Night time.* Vintage, London.

Hardy, G.H. (1940). *A Mathematicians Apology.* Cambridge: Cambridge University Press.

Korner, T.W. (1996). *The Pleasures of Counting.* Cambridge: Cambridge University Press.

Nahin, P.J. (2006). *Dr Euler's Fabulous Formula.* New Jersey, Princeton University Press.

Poskitt, K. *Murderous Maths (series).* Scholastic, London. <http://www.murderousmaths.co.uk/BOOKS/books.htm>.

Singh, S. (1997). *Fermat's Last Theorem.* 4th Estate, London.

Singh, S. (1999). *The Code Book.* Doubleday, New York.

Stewart, I. (2012). *Seventeen Equations that Changed the World.* Profile Books, London.

Vorderman, C., Porkess, R., Budd, C., Dunne, R., and Hart, P. (2011). *A world class mathematics education for all our young people.* <http://www.conservatives.com/News/News_stories/2011/08/~/media/Files/Downloadable%20Files/Vorderman%20maths%20report.ashx>.

Younghusband, E. (2011). *One Woman's War.* Candy Jar Books, Cardiff.

CHAPTER 2

PHILOSOPHY OF NUMBER

MARCUS GIAQUINTO

THERE are many kinds of number: natural numbers, integers, rational numbers, real numbers, complex numbers and others. Moreover, the system of natural numbers is instantiated by both the finite cardinal numbers and the finite ordinal numbers. We cannot deal properly with all of these number kinds here. This chapter concentrates on the finite cardinal numbers. These are the numbers which are answers to questions of the form 'How many Fs are there?' In what follows, an unqualified use of the word 'number' abbreviates 'cardinal number'.

Numbers cannot be seen, heard, touched, tasted, or smelled; they do not emit or reflect signals; they leave no traces. So what kind of things are they? How can we have knowledge of them? These are the central philosophical questions about numbers. Plausible combinations of answers have proved elusive. The aim of this chapter is to present and assess the main views – classical and neo-classical, nominalism, mentalism, fictionalism, logicism, and the set-size view. All views are disputed, including the view I will argue for – the set-size view. The final section relates finite cardinal numbers to natural numbers.

The Classical View: Multitudes of Units

At the start of Book VII of *Euclid's Elements*, having defined a 'unit' to be a single individual thing, a number (arithmos) is defined thus:

> A number is a multitude of units.

(Euclid 2002, p. 157, Book VII, definitions 1 and 2). On this view, any pair of items is a 2 and so there are many 2s; any trio is a 3 and so there are many 3s. In general, any plurality of *k* things is a *k* and there are many *k*s. There was no notion of zero; a 1 is a unit, not a plurality of units and therefore not a number. We retain a corresponding use of the word 'number', as when we say that a number of authors were late with their submissions. We can apply arithmetic perfectly well taking numbers to be pluralities: as the number of authors who were late is a 9 and the number who were on time is a 14, the number of authors in total is a 23. Multiplication entails that pluralities themselves may constitute units, as 6 multiplied by 3 is the aggregate of a trio of 6s; the potential for confusion seems not to have been a problem.

While the truths of arithmetic are specific numerical facts (such as: $19 \times 19 = 361$), number theory consists of general truths about numbers (such as: there is no greatest prime number) and proofs of those truths. A tradition of number theory was one of the impressive intellectual achievements of the ancient Greek speaking world. It was studied for its own sake and regarded as a body of unqualified and unchanging truths. Many are proved in *Euclid's Elements* (2002), Books VII–IX. Here, for example, is Proposition 30 of Book VII:

> If a prime number measures [divides without remainder] the product of two numbers, it also measures one of those two numbers.

Number theory, in contrast to applied arithmetic, seems not to be about pluralities of the ordinary things we count (such as sheep, votes or chimes). On the contrary, the mathematicians seemed to have in mind pluralities of units without distinguishing characteristics. Plato (1997a, 56d) proposed exactly this: the numbers of number theory are pluralities of pure units, where a pure unit is a single entity lacking any characteristic distinguishing it from any other pure unit. This account has two advantages. It avoids a problem about the perceived inequality of ordinary units: How could one ship be equal to one plank? And it avoids making the subject matter of number theory contingent.

But there are also several problems with Plato's proposal. Pure units are mysterious. How can there be two or more entities whose only difference is that they are different? Two distinct things may have all the same qualities. But would they not have to be in different positions? A pure unit, however, lacks position. And there are further questions. What is the origin of pure units? Are they internal or external to the mind? Is there an inexhaustible supply of them? How can we know of them? How can we have a cognitive grasp of a plurality of them if they are indistinguishable from each other? Plato (1997b, 526a) says that these pure numbers can be grasped only in thought, but does not elaborate.

Although Plato's account has an echo in a mentalist view put forward by Georg Cantor over two millennia later, it was too fraught with difficulties to have much staying power. While the classical view continued to be accepted for applied arithmetic by some later thinkers, other accounts of number in pure arithmetic and number theory were sought, and it is to these that we now turn.

Nominalism

The decimal place system of numerals, originating in India, reached Europe in the 13th century via Arab mathematicians. By the 17th century, symbol-manipulation algorithms using the decimal place system had superseded calculation by abacus. This was the backdrop for the view proposed by the philosopher George Berkeley (Berkeley, 1956, p. 25; Berkeley, 1989, entry 763) that the numbers of pure arithmetic (i.e. when not applied to pluralities of physical objects) are 'nothing but names,' meaning that, for example, the number 26 is nothing over and above the numeral '26.' The mathematician David Hilbert (1967, p. 377) suggested that a number is a horizontal string of short vertical strokes; arabic numerals abbreviate the corresponding strings, e.g. '3' is short for 'III'. More recently, the philosopher Saul Kripke has suggested, in unpublished lectures, that numbers are numerals in a place system of numerals. These views are versions of nominalism, by which I refer to the identification of numbers with numerals. This is to be sharply distinguished from the claim that there is nothing abstract, also sometimes called 'nominalism'. A numeral, as opposed to its particular occurrences, has to be abstract, being a type of mark.

Why think that numbers are numerals? Berkeley (1956, p. 25; 1989, entry 761) noted that large numbers within the range of performable calculations defy precise sensory representation. So, when we think of 201, what is present to the mind is not a representation of 201 items, but just the numeral. Although empirical studies indicate that we cannot, without counting, tell the precise number of any large collection of things presented to us, it does not follow that our idea of a large number is just a representation of its numeral. An alternative is that we have a descriptive way of mentally designating the number in terms of smaller numbers, such as 'two tens of ten, plus one'; the number ten is known as the number of one's fingers and the numbers one and two have precise representations in what cognitive scientists have called the 'number sense', which I say more about in the section 'Numbers as Set-sizes'.

Berkeley's other reason for nominalism is that when we make arithmetical calculations we seek an answer given by a numeral in the decimal place system (1989, entry 766). An answer in any other format is not what is wanted (e.g. abacus display, Roman or binary numerals). That is right. If you ask 'What is eight to the power of six' I can answer right away that it is 1,000,000 in base 8 notation – a trivial and unhelpful answer. We want answers in the decimal place format. Is that fact best explained, however, by claiming that numbers in arithmetic *are* the decimal place numerals? Here is an alternative explanation. We want answers in the decimal place format because (a) we want to be able to use answers as inputs for other calculations, and our calculation algorithms require inputs in decimal place format; (b) our sense of number size is tied to the numerals we are most familiar with, the arabic decimal place numerals. Consider the following number here presented in binary notation: 1010101. Is it larger or smaller than seventy? You will probably have to convert this into a verbal number expression or decimal notation in order to be sure of the answer, but if the number is presented as a decimal numeral you will know immediately that it is larger than seventy: it is 85. Assuming that our sense of number size is well-linked to our verbal number expressions, this is evidence that our number sense is more strongly associated with decimal than with binary numerals, even though we understand both.

A merit of nominalism is that it says clearly what numbers are and brings them within the bounds of human cognition. But all versions of nominalism face serious objections. A decisive objection to Berkeley's nominalism is that the same common core of arithmetical information can be expressed using different numeral systems or even ordinary words: 'XII et IX fit XXI'; '12 + 9 = 21'; '1100 + 1001 = 10101'; 'twelve plus nine is twenty-one'. This can be met by Hilbert's prescription (1967) that the numerals of customary systems be regarded as abbreviations for rows of short vertical strokes, but this is unconvincing. Why rows of strokes, as opposed to columns of dots? Any choice of canonical numerals will be arbitrary – we can have no good reason for thinking that the chosen symbols are what mathematicians are *really* referring to.

A second objection, decisive against any version of nominalism when extended to the objects of number theory, is that truths of number theory are independent of numeral systems. Consider the theorem that any plural number is a prime or product of primes. This is a consequence of the fact that there is no infinite decreasing sequence of smaller positive numbers, so that if one factors a non-prime into two smaller numbers and continues factoring the factors, the process is bound to terminate after finitely many steps in primes.

Mentalism

Mentalism is the view that a number is a mental entity, an innately supplied representation or a product of intellectual activity. Within the mentalist camp views diverge. The mathematician Georg Cantor (1955, p. 86) claimed that the number of things in a given class is an image or mental projection that results when we abstract from the nature of members of the class and the order in which they are given. The mathematician Luitzen Brouwer (1983, p. 80), founder of 'intuitionist' philosophy of mathematics, regarded numbers as resulting from the mental splitting of an experience of a temporal period into two, and the simultaneous representation of this into a remembered 'then' and a current 'now' as constructions of the first two numbers 1 and 2. Then with the passage of time, what was the 'now' extends into a new 'then' and 'now', to give us 3, and so on. An alternative idea is that numbers have their origin in the practice of counting. More recently, the cognitive scientist Stanislas Dehaene (1997) has suggested that numbers are just our mental number representations, an internal version of nominalism.

The advantage of mentalism is that the knowledge of numbers becomes less mysterious. As Dehaene (1997, p. 242) puts it, 'If these objects are real but immaterial, in what extrasensory ways does a mathematician perceive them?' But if numbers are just mental items, they may be knowable by inner awareness and reflection.

Let us put aside the question of the plausibility of the various cognitive hypotheses proposed by mentalists. The big problem for any version of mentalism is that only finitely many brain states have actually been realised; hence, there are only finitely many mental entities, whether innately given or produced by intellectual activity or a combination of the two. So the idea that numbers are mental entities conflicts with the fact that, for any number, there is a yet greater number. How can we know this fact? There are many ways. For example, any number *n* is the number of preceding numbers, as we start with 0; so the number of numbers up to and including *n* is greater than *n* by one.

There are two responses to this. One is 'strict finitism', the view that despite accepted number theory, the numbers run out at some finite point. This view has little plausibility and commands scant support, although it has been investigated by some logicians. The other response is to concede and weaken the claim: numbers are *possible* mental entities. The philosopher Michael Dummett (1977, p. 58) has suggested that a number *n* is the possibility of counting up to *n*. The immediate problem with this response is that what we *could* mentally represent or construct, as well as what we *actually do* mentally represent or construct, has finite limitations. The reason is that there are only finitely many possible brain states – take an upper bound on the number of neurons in a human brain and multiply it by an upper bound on the number of possible states of a neuron; the result will be a finite upper bound on the number of possible human brain states. So the numbers outstrip our possible mental constructions. One might seek to escape this by holding that there could be ever more powerful minds, so that any limitation on one possible mind could be surpassed by another. But this is just a metaphysical speculation. How do we know that there could be such an intellectual hierarchy? It falsifies our real epistemic situation – our knowledge that there are more than finitely many numbers does not depend on our knowing that this metaphysical speculation is true.

Fictionalism

Having reviewed several answers to the question 'What kind of things are cardinal numbers?' and found them wanting, what options are left? One answer is fictionalism: there are no numbers, and so accepted arithmetical claims such as '2 + 3 = 5' or '3 is prime' are untrue, as they entail that there are numbers; but accepted arithmetic is useful and mathematical practice should continue as if it were true.

How might one reach this desperate conclusion? Here is how the main line of thought goes. Numbers are not material or mental. If numbers are not material or mental, they must be abstract. But if abstract, they must be unknowable, it is argued, as abstracta are unperceivable, leave no traces, and do not influence the behaviour of perceptible things. So our numerals do not refer to anything.

Over recent decades fictionalism has been advocated by several philosophers and taken very seriously by others (Balaguer, 2011; Field, 1980; Leng, 2010; Yablo, 2005). But it involves a serious methodological flaw. Opting for one philosophical solution over others may be fine if one is denying nothing but a bunch of other philosophical views, but not if one is denying both rival philosophical views and propositions of independent standing that are generally regarded by rational thinkers as among the most certain things that we know. No metaphysical or epistemological doctrine has greater rational credibility than basic arithmetic. Our confidence in basic arithmetic is not an article of faith; our belief that 2 + 3 = 5, for example, is well supported by our counting experience. In the sections to come I will argue that there are credible non-fictionalist responses to our questions about number and I will pinpoint an error that may prevent fictionalists from appreciating this.

Neo-Classical Views

The classical view that cardinal numbers are multitudes of units was taken up by the philosopher John Stuart Mill, with one modification. The change is that units, or ones, count as numbers too. He avoids problems about pure numbers by denying that there are such things and he avoids denying arithmetical theorems by construing them as general statements (Mill, 1974, II. vi. 2):

> All numbers must be numbers of something; there are no such things as numbers in the abstract. *Ten* must mean ten bodies, or ten sounds, or ten beatings of the pulse. But though numbers must be numbers of something, they may be numbers of anything. Propositions, therefore, concerning numbers, have the remarkable peculiarity that they are propositions concerning all things whatever.

Outside of any application, an equation such as '3 = 2 + 1' means that any parcel of three things can be arranged to form one parcel of two things and one other thing. More complicated equations, such as 'the cube of 12 is 1728' can be dealt with in a similar way (Mill, 1974, III, xxiv, 5). On this reading numerical terms do not designate pure numbers, but merely signify differing ways in which a plurality of things can be arranged. Mill also held that the things in question, the units, are perceptible; thus, the equations of arithmetic are general claims with empirical content.

A major problem with Mill's account of equations is that it has restricted application. Referring to Mill's claim that '3 = 2 + 1' means that any parcel of three things can be separated into a parcel of two things and one other thing, the logician Gottlob Frege (1980, p. 9) says:

> What a mercy, then, that not everything in the world is nailed down; for if it were, we should not be able to bring off this separation and 2 + 1 would not be 3!

We certainly want to be able to apply arithmetic to things that cannot be re-arranged, such as lunar eclipses or solutions of an equation. Another problem for Mill: What fact about separating and re-arranging objects is expressed by the equation '$2^0 = 1$'?

The way out of these problems is to throw off empiricist constraints and understand arithmetic as a body of general truths about sets of any kind (including 1-membered sets and the empty set), and to interpret numerical equations in term of 1-1 correlations, as is done in standard set theory. Precisely this is proposed by the mathematical logician John Mayberry (2000). On this view '3 = 2 + 1' means that there is a 1-1 correlation between any 3-membered set and the union of any pair set with any single-membered set not included in the pair set; '2 × 3 = 6' means that there is a 1-1 correlation between the set of units in any pair of disjoint triples and any sextet; even '$2^0 = 1$' comes out right, though this is not immediately obvious. In fact, the whole of cardinal arithmetic is preserved this way.

Any cardinal number on Mayberry's view is a set and any set is a cardinal number. Accordingly, for any number *k* apart from zero, there are many *k*s: many pairs, many trios, and so on. There are several advantages to this account:

(1) Frege's objections do not apply: the account caters for numbers of eclipses and solutions as well as for numbers of apples and pebbles.
(2) It allows for the arithmetic of 0 and 1, as in '$2^0 = 1$'.
(3) The problem of determining which particular mathematical object is named by a numeral, say '5', does not arise, as any quintet is a 5.
(4) If numbers are abstract objects, we need to explain how we can know of them; if numbers are sets, we can know of them by knowing sets.

In many cases knowing a set, such as the set of your siblings, is far less mysterious than knowing an abstract denizen of some special mathematical realm.

But Mayberry's account runs into difficulty with general theorems about cardinal numbers, precisely because it allows that there are many numbers of each size. For example, it is a theorem that any number has a unique successor, but any given pair can be extended to many different trios, by adding different objects to the pair. Another theorem is that exactly one positive number has a square that is equal to its double. That number is 2, but on the neo-classical account there are many 2s. Moreover, we count numbers themselves. For example, we say that there are exactly four primes less than 10, namely, 2, 3, 5, and 7. This makes no sense unless there is just one number per numeral. The possibility of enumerating numbers gives rise to important functions in number theory, such as the number of primes less than or equal to n. So the account stands in *prima facie* conflict with number theory.

The neo-classicist can make two kinds of response, concessive, or aggressive. The concession says that the account does not apply to number theory in general, only to arithmetical equations. While I agree that modern number theory is not to be construed as a theory of finite cardinal numbers (as I will explain in the final section), the theorems of number theory surely apply to the finite cardinal numbers. So the concession does not save the neo-classical view from conflict with number theory. The aggressive response is to claim that propositions of number theory must be interpreted to have a hidden prefix, meaning: 'in any omega-sequence of numbers (sets) starting with the empty set, each later set extending its predecessor by one member ...', where an omega-sequence is a sequence of elements conforming to the Dedekind axioms (given in the section 'The Finite Cardinal Numbers and the Natural Numbers'). While that would eliminate the conflict, it is implausible as an interpretation of the claims of number theory made by mathematicians before the 19th century, as they did not yet have the concept of an omega sequence. Other responses are possible, but those I know about are difficult and no less contentious. Time to look at other views.

Logicism: Numbers as Special Sets

On the neo-classical view that numbers are just sets, there are *many* twos, *many* threes, and so on, instead of just *one* number two, *one* number three, etc. In the previous section, we found that this view runs into trouble with number theory. An alternative that avoids that problem is to take a cardinal number k to be something like a species to which all the k-membered sets belong.

Views of this kind were put forward by Gottlob Frege (1980, pp. 79–80) and Bertrand Russell (1919), as part of their philosophy that the mathematics of number is part of pure logic. With wrinkles ironed out, the claim is:

> The number k is the set of k-membered sets.

This runs into two problems – one mathematical, the other metaphysical. The mathematical problem is that there is no set of sets with exactly 1 member; hence there would be no cardinal number 1. From the assumption that there is such a set, two uncontroversial principles about sets (union and separation) lead straight to Russell's paradox. Russell (1908) evades the paradox by means of his theory of types. The details of his theory of types need not detain us, but an essential element is that things are regarded as falling into exclusive layers or 'types' – ordinary individual items are of type 0, *sets* of individuals are of type 1, sets of sets of individuals are of type 2, and in general sets of things of type n are of type $n + 1$. Then each number k splits into many, the set of all k-membered sets of things of type 0, the set of all k-membered sets of things of type 1, and so on.

Russell's many-types view faces several problems. First, it takes us back to the disadvantage of the neo-classical view, having many ones, many twos, etc. Secondly, it conflicts with mathematical practice, which allows that some sets of different type have the same cardinal number. Finally, to establish the correctness of the principles of number theory, Russell had to assume that there are infinitely many individuals, but this is clearly awry (and contrary to his logicist outlook), because we know that the principles of number theory are true without knowing that there are infinitely many individuals.

Both versions of logicism face the metaphysical objection. The argument here is for Frege's version; the same argument applies to Russell's version for numbers of individuals (things of type 0.) Call a two-membered set a *pair*, for short. The proposition to be challenged entails that the number 2 is the set of all pairs. Call the actual set of pairs 'P'. The set of Charles Windsor's sons, {William, Harry}, is a member of P. Now consider the possible circumstance that Harry had never been conceived: the set {William, Harry} would not have existed; so P would not have existed; so the set that would have been the set of pairs is not P, but some other set. In general, which set is the set of pairs depends on contingent events, just as the identity of the 43rd US president – it would have been Gore not Bush, had the Supreme Court ordered a rerun of the Florida ballot. But does the identity of the number 2 depend on contingent events, such as the results of royal mating? Surely not. The number of protons in the nucleus of a helium atom is the same in all possible circumstances; and that number is 2 in all possible circumstances. So the identity of the number 2 is not dependent on circumstances, whereas the identity of the set of pairs is dependent on circumstances. So 2 is not the set of pairs. A parallel argument works for any other positive cardinal number.

Although the logicist proposal is wrong, there are many mathematically adequate ways of representing the finite cardinal numbers as sets. Set theory has settled on one of these systems of set representations as the most convenient, but we should not take the set representations of cardinal numbers to be the things represented. The point is cogently argued for in Benacerraf (1965). We should be no more tempted to think that cardinal numbers really are sets of a certain kind than that spatial points really are ordered triples of real numbers.

It is clear, however, that there is an intimate relation between a cardinal number *n* and sets with exactly *n* members. Any satisfactory answer to our question must make this relation clear. A satisfactory answer, however, must also make it possible to account for our cognitive grasp of some cardinal numbers, which in all of us antedates knowledge of even moderately sophisticated set theory. So we need to turn away from set theory (and, for the same reason, from any mathematical theory) and look in another direction.

Numbers as Set-Sizes

Cardinal numbers are answers to questions of the form 'How many Fs are there?'. This gives us a big clue. Answers to questions beginning 'How much', 'How far', 'How long', 'How heavy', 'How loud' and others of this form are sizes, quantities or magnitudes. So are answers to questions beginning 'How many': cardinal numbers are sizes of sets. This is the set-size view. There is disagreement about whether every definite plurality of things constitutes a set; the set-size view allows that a definite plurality has a numerical size, that is, a cardinal number, even if the plurality is not a set.

Let me say up-front that the set-size view of cardinal numbers is the view I judge to be correct. An immediate advantage of the set-size view is that it is consonant with the way we ordinarily think and talk. When we talk of family sizes or class sizes, we refer to the number of family members or the number of pupils in a class. The set-size view also reveals the connection between the number *n* and sets with exactly *n* members: the number *n* is what all and only sets with exactly *n* members are bound to have in common, namely, their size. This view was expressed by the philosopher John Locke (1975) and more recently supported in (Giaquinto 2001).

A cardinal number, on the set-size view, is not an object, but a *property* of sets. This has prompted two major philosophical objections:

(1) Properties are not real; talk of properties is a mere façon de parler.
(2) If numbers were properties, they would be causally inert; so we could not have knowledge of them.

I will respond to these questions in turn. In responding to the second, I will argue that empirical work on numerical cognition reveals how knowledge of cardinal size is possible.

Debate about the reality of properties (or 'universals') stretches from at least mediaeval times to the late 20th century (Mellor & Oliver, 1997). However, we can cut through these scholastic thickets by noting that empirical science quite often delivers properly substantiated judgments about the reality or unreality of properties. For example, Joseph Priestley thought that all combustible material contained phlogiston, a substance that is liberated in combustion from the material, with the dephlogisticated substance left as an ash or residue. On this theory, a candle flame in an enclosed lantern will go out because the contained air will become saturated with phlogiston. Antoine Lavoisier held that there is no such substance as phlogiston and no such property as phlogiston saturation. Combustion involves absorption of oxygen, rather than release of phlogiston, and a candle flame in an enclosed lantern will go out because the contained air will become depleted of oxygen. Eventually, the judgments of Lavoisier were substantiated – nothing could be phlogiston saturated because there is no such property as phlogiston saturation, but oxygen depletion is real. Medicine and psychiatry make similar judgments: possession by demons is not a real condition; but multiple personality disorder may be real and bipolar disorder definitely is. The conclusion must be that some putative properties are real and some are unreal.

Are set sizes real? Locke (1975, II.VIII.17) included number in his list of real properties, in contrast to sensory qualities such as flavours, which he took to be in us, rather than in the substances to which we attribute them. About number Locke was right. Scientists appeal to the number of protons in the nucleus of an atom to explain properties of the atom; they do not explain the number as a merely subjective phenomenon like a rainbow. This is reason to believe that the cardinal number of the set of protons in a helium nucleus, for example, is a real property of that set, a property that does not depend on us or our mental life. Moreover, the fact that the number of electrons in a neutral atom and the number of protons in its nucleus are the same is a significant objective fact. The number of legs on a normal spider, the number of major branches of a snowflake – these surely are real properties of the relevant sets, not to be explained away as illusory phenomena or mere ways of talking.

The second objection to the set-size view is that if numbers are set sizes we could not have knowledge of them, for the following reason. Set sizes, being properties, cannot have any causal effect on us: they emit no signals, leave no traces, and have no influence on perceptible things; therefore they cannot be known.

This last inferential step is the main error. It may arise from using as a general model of knowing things a model that is appropriate for physical objects, especially Spelke-objects; but it is not appropriate for more abstract kinds of thing, such as properties and relations. As properties do not causally interact with other things, we cannot have knowledge of them in the way that we have knowledge of planets and protons. Yet we often know such things as Beethoven's pastoral symphony, the letters of the Greek alphabet, or other things that do not *themselves* have causal effects on other things. However, their *instances*, the sounds of an actual performance or the actual inscriptions of Greek letters, do have causal effects on us: we perceive them. We can come to know a musical composition or a letter type through repeated exposure to their physical instances.

To know Beethoven's pastoral symphony it is enough than we can recognise performances *as* performances of Beethoven's pastoral symphony and to tell them apart from performances of other music. To know the Greek lower case alpha it is enough than we can recognise inscriptions of it *as* inscriptions of the lower case alpha and tell them apart from inscriptions of other letters. The parallel holds for cardinal numbers. To know the number *n* it is enough that one can recognise sets of things as *n*-membered and discriminate *n*-membered sets from sets with fewer or more than *n* members.

How is it possible to acquire this capacity for number recognition and discrimination? On this matter, philosophers must attend to the findings of cognitive science. First, the data provide evidence that we have an innately given number sense, that is, a system of mental magnitude representations of rough cardinal size, with a neural basis in the intraparietal sulcus (Butterworth & Walsh, 2011). The evidence comes from a variety of sources: experiments on healthy adults and children, clinical tests on brain damaged patients, brain imaging, and studies on animals from parrots to primates (Butterworth, 1999, chapters 3–6; Dehaene, 1997, Chapters 1–3, 7, 8; and several book chapters in this handbook that describe the most recent lines of research).

There are no dedicated exteroreceptors and there is no specialised organ for number detection. So why a number *sense*? One reason is that our capacity for detecting number does not involve applying a procedure (such as counting); it is subjectively immediate. Another reason is that number detection has the signature features of other quantity senses, such as sense of duration (Cohen Kadosh, Lammertyn, & Izard, 2008; Walsh, 2003). One of these features is the 'distance effect': the smaller the distance between two levels of a quantity (for fixed mean), the harder it is to distinguish them. It takes longer to distinguish 7 from 9 than 4 from 12, whatever the stimulus format (e.g. random dot arrays, arabic numerals, sequences of knocks). The other feature is the 'magnitude effect': the greater the mean of two levels of a quantity (for fixed distance), the harder it is to distinguish them. It takes longer to distinguish 8 from 10 than 2 from 4. These effects follow from the kind of formula (Welford, 1960) to which the reaction time data for single and double digit number comparison conform:

$$RT = a + k \cdot log[L/(L-S)]$$

where *L* is the larger number, *S* is the smaller, and 'a' and 'k' denote constants. (Butterworth, Zorzi, Girelli, & Jonckheere, 2001; Dehaene, 1989; Hinrichs, Yurko, &, Hu, 1981; Moyer & Landauer, 1967, 1973). This is typical of response data for comparison of other physical magnitudes, such as line-length, loudness, and duration. Finally, experiments with pre-linguistic children and with animals lacking language or symbol system show that they too have a capacity for number discrimination (see Beran et al., Agrillo, and McCrink, this volume). So it is reasonable to posit an innate number sense.

This number sense decreases in precision as numbers increase, enabling us to gauge approximate size for larger numbers, but for numbers 1, 2, and 3 it is precise. This is predicted by a neural network model for the number sense (Dehaene & Changeux, 1993), although it may be due to a second system of mental representation (Feigenson, Dehaene, & Spelke, 2004).

The number sense is just one of the resources of numerical cognition. There is also the culturally supplied instrument of verbal counting, which enables us to determine cardinal size precisely for sets too great to be gauged with precision by the number sense. Counting also helps us appreciate a feature that seems to distinguish set-size from other magnitudes, namely discreteness. Between any two lengths there is (or seems to be) an intermediate length, but each set-size has an immediate successor, with no set-size in between.

Practice with verbal counting may produce in us an association of number sense representations with our representations of number words, thence with our representations of numerals (see Sarnecka, this volume); and it may help sharpen our number sense representations, so that the range of numbers represented with precision extends beyond 3 (though perhaps not very far). Familiarity with counting also supplies us with uniquely identifying positional information about numbers within our counting range. Thus 1 is the first number and 2 is the next, 3 the next after 2, and so on.

All this is surely enough for possession of concepts for the first few positive cardinals. With these resources it is possible not merely to discriminate between 3-membered sets and sets with more or fewer than 3 members, but also to recognise such a set *as* 3-membered. The same for other very small positive cardinal numbers. So we have the framework at least for an account of how we can have cognitive grasp of these numbers, without appealing to modes of cognition not recognised by cognitive scientists.

The number 0 is a special case (see Tzelgov et al., this volume). We probably do not have any number sense representation for zero (Wynn & Chang, 1998). It is the cardinal size of the empty set. The empty set can seem to be an artificial posit, but it does not seem so artificial when one considers all the sets of possible winnings in a two-person zero-sum game played for valuable items; moreover, the existence of a unique empty set is provable from the established axioms of set theory. So we can know the cardinal number 0 by description as the cardinal size of the empty set. We do not grasp 0 as we grasp the small positive numbers; we cannot literally recognise that a set has zero members, though we may deduce it.

What about larger numbers? When we know an identifying description of a number in terms of smaller numbers, we can know the number descriptively (assuming we already know the smaller numbers). Often we have more than one identifying description of a number, giving us a better grasp of it. For example, we know five identifying descriptions of 10 as the sum of two positive numbers and one identifying description of it as the product of two smaller numbers (treating the order of operands as irrelevant.) By 'knowing' an identifying description of a number I mean that we can retrieve the relevant number fact from memory. We can, of course, *figure out* many more than five identifying descriptions of 53 as the sum of two numbers, but before having learnt those addition facts they cannot make us more familiar with the number. Contrast 53 with 60, of which numerate adults know five identifying descriptions as a product of two smaller numbers (2 × 30, 3 × 20, etc.) and three as sums of two decades (10 + 50, 20 + 40, etc.)

These then are ways in which we can properly be said to know a number: by means of our number sense, by a variety of identifying descriptions in terms of smaller numbers, and by a combination of these two. Beyond numbers knowable in those ways are numbers still small enough to refer to by means of their decimal notation: viewing decimal numerals in blocks of three digits (from the right) gives us some relative awareness of size. Still further out are numbers which we can designate, but not transcode into their decimal numerals and which utterly defeat our number sense, such as 9^(9^(9^9)), where '*n*^*p*' denotes '*n* to the power of *p*'. Of course, most numbers will lie totally beyond our ability to refer to them, using whatever is our currently most compact notation; even if we allow for ever improving means of reference (using symbols for faster growing functions), most will remain outside the light cone of human intellect.

My claims, in summary, are these. Cardinal numbers are size properties of sets (or of definite collections or definite pluralities). Some cardinal numbers can be known. Very small numbers can be known by means of the number sense and the practice of counting. This knowledge is not a quasi-perception of the number *n* itself, but a capacity for recognising *n*-membered sets as *n*-membered and for discriminating sets of *n* items from sets with fewer or more items. Some larger numbers are knowable in a different way, as the cardinal number designated by one or more identifying descriptions in terms of smaller numbers, when these descriptions are stored in memory.

This account of the nature of cardinal numbers has these crucial advantages. It makes clear the necessary relation of a cardinal number *n* to sets with exactly *n* members, it avoids any conflict with number theory, and it permits an account of our knowledge of cardinal numbers within the framework of cognitive science.

THE FINITE CARDINAL NUMBERS AND THE NATURAL NUMBERS

Should we take the natural numbers of modern number theory to be the finite cardinal numbers? Number theory became a highly abstract subject in the 19th century, with the work of the mathematicians Kummer, Kronecker, and especially Dedekind. There was a focus on structural properties and relations that has remained ever since. The subject matter of modern number theory can be described as the structure of natural numbers and its extensions to other number structures. The sequence of finite cardinal numbers has the structure of natural numbers, but indefinitely many other mathematical sequences also have that structure. So we should not take the natural numbers to be the finite cardinal numbers.

To make the point a bit clearer, let us go into a little detail. The finite cardinal numbers have a natural ordering: starting with zero, each number *n* is immediately succeeded by *n* + 1 (the size of sets with *n* + 1 members), and each number is reached by finitely many applications of this successor operation starting with zero. This ordering is shared with any sequence of things satisfying the following conditions given by Dedekind (1996):

(1) There is a single distinguished element – call it 'the zero'.
(2) There is a unary operation **s** – call it 'the successor operation' – such that every element *x* has a unique immediate successor element **s**(*x*).

Table 2.1 Instances of the structure of the natural numbers

	Starting element	Successor operation
Cardinal numbers	Zero	$\alpha \mapsto \alpha+1$
Hilbert numerals	I	$\alpha \mapsto \alpha^\frown \mathrm{I}$
Von Neumann ordinals	Ø (the empty set)	$\alpha \mapsto \alpha \cup \{\alpha\}$
Zermelo ordinals	Ø	$\alpha \mapsto \{\alpha\}$

(3) The zero is not the successor of any element.
(4) No two elements have the same successor.
(5) Any set that contains the zero and that contains s(x) if it contains x is a set that contains all the elements.

Some examples are given in Table 2.1.

Any two of these sequences, known collectively as omega sequences, are isomorphic: their elements can be paired off one-to-one in a way that preserves order, i.e. without any crossovers. In other words, these ordered sets have the same structure, the structure of the natural numbers.

What then are the natural numbers? It is a mistake to look for some mathematical entities specifiable independently of the structure. This leaves two options. One is to deny that the question has an absolute answer; the elements of any omega sequence can serve as natural numbers, but none is privileged as the real sequence of natural numbers, as Benacerraf (1965) argues. On this view, the full content of a theorem of Dedekind–Peano number theory is a proposition tacitly about all omega sequences. The other option is to take the natural numbers to be the *positions* in the natural number structure: 0 is the initial position, and for any position x, s(x) is the next position along (Shapiro, 1997).

My aim here is merely to relate the finite cardinal numbers to the natural numbers and, for that purpose, it is not necessary to adjudicate between the two views of the natural numbers just presented. On the first view, as the finite cardinals in their natural ordering constitute one of the many sequences which instantiate the natural number structure, it is permissible in a suitable context to think and talk as though they are the natural numbers. This would be a manner of speaking or thinking, not an expression of metaphysical fact. On the alternative view, the relationship between the finite cardinals and the natural numbers is (again) not identity; it is occupation. The finite cardinals in their natural sequence occupy the positions of the structure of natural numbers.

Either way, the theory of natural numbers has greater generality and abstractness than the theory of finite cardinal numbers, being about features of the structure common to all omega sequences. Here, then, is an area for future cognitive research. What cognitive resources are involved, and how are they involved, in the development of one's grasp of number theory?

References

Agrillo, C. (This volume). Numerical and Arithmetic abilities in non-primate species. In Roi Cohen Kadosh & Ann Dowker (Eds.) The Oxford Handbook of Mathematical Cognition. Oxford: Oxford University Press.

Balaguer, M. (2011). Fictionalism in the Philosophy of mathematics. In E.N. Zalta (Ed.), The Stanford Encyclopedia of Philosophy (Autumn 2011 edn). Available at http://plato.stanford.edu/archives/fall2011/entries/fictionalism-mathematics/

Benacerraf, P. (1965). What numbers could not be. Philosophical Review 74, 47–73.

Beran, M.J., Perdue, B.M., & Evans, T.A. (This volume). Monkey Mathematical Abilities. In Roi Cohen Kadosh & Ann Dowker (Eds.) The Oxford Handbook of Mathematical Cognition. Oxford: Oxford University Press.

Berkeley, G. (1956). Letter to Molyneux [1709]. In A. Luce & T. Jessop (Eds), The Works of George Berkeley, Bishop of Cloyne, Volume 8. London: Nelson.

Berkeley, G. (1989). Philosophical Commentaries, G. Thomas (ed.). London: Garland.

Brouwer, L. (1983). Intuitionism and Formalism [1913]. In P. Benacerraf & H. Putnam (Eds). Philosophy of Mathematics, 2nd edn. Cambridge: Cambridge University Press.

Butterworth, B. (1999). The Mathematical Brain. London: Macmillan.

Butterworth, B., & Walsh, V. (2011). Neural basis of mathematical cognition. Current Biology, 21, 1337–1420.

Butterworth, B., Zorzi, M., Girelli, L., & Jonckheere, A. (2001). Storage and retrieval of addition facts: the role of number comparison. Quarterly Journal of Experimental Psychology, 54A, 1005–1029.

Cantor, G. (1955). Contributions to the Founding of the Theory of Transfinite Numbers [1895], transl. P. Jourdain. New York: Dover Publications.

Cohen Kadosh, R., Lammertyn, J., & Izard, V. (2008). Are numbers special? An overview of chronometric, neuroimaging, developmental and comparative studies of magnitude representation. Progress in Neurobiology, 84, 132–147.

Dedekind, R. (1996). Was sind und was sollen die Zahlen? Vieweg, Braunschweig. In W. Ewald (Ed.), From Kant to Hilbert. A Source Book in the Foundations of Mathematics, Vol. 2. Oxford: Oxford University Press (original published 1888).

Dehaene, S. (1989). The psychophysics of numerical comparison: a re-examination of apparently incompatible data. Perception and Psychophysics, 45, 557–566.

Dehaene, S. (1997). The Number Sense. Oxford: Oxford University Press.

Dehaene, S., & Changeux, J. (1993) Development of elementary numerical abilities: a neuronal model. Journal of Cognitive Neuroscience, 5, 390–407.

Dummett, M. (1977). Elements of Intuitionism. Oxford: Oxford University Press.

Euclid (2002). Euclid's Elements: All Thirteen Books Complete in One Volume, transl. T. Heath. Santa Fe, NM: Green Lion Press.

Feigenson, L., Dehaene, S., & Spelke, E. (2004). Core systems of number. Trends in Cognitive Sciences, 8, 307–314.

Field, H. (1980). Science Without Numbers. Princeton, NJ: Princeton University Press.

Frege, G. (1980). The Foundations of Arithmetic [1884], transl. J. Austin. Evanston, IL: Northwestern University Press.

Giaquinto, M. (2001). Knowing numbers. Journal of Philosophy, 98, 5–18.

Hilbert, D. (1967). On the Infinite. In J. van Heijenoort (Ed.), From Frege to Gödel: a source book in mathematical logic, transl. S. Bauer-Mengelberg. Cambridge, MA.: Harvard University Press, 367–392.

Hinrichs, J., Yurko, D., & Hu, J. (1981). Two-digit number comparison: use of place information. Journal of Experimental Psychology: Human Perception and Performance, 7, 890–901.

Kripke, S. (1992). Logicism, Wittgenstein, and *De Re* Beliefs About Numbers. Unpublished lectures.

Leng, M. (2010). Mathematics and Reality. Oxford: Oxford University Press.

Locke, J. (1975). An Essay Concerning Human Understanding [1689]. Oxford: Clarendon Press.
Mayberry, J. (2000). The Foundations of Mathematics in the Theory of Sets. Cambridge: Cambridge University Press.
McCrink, K., & Birdsall, W. (This volume). Numerical Abilities and Arithmetic in Infancy. In Roi Cohen Kadosh & Ann Dowker (Eds.) The Oxford Handbook of Mathematical Cognition. Oxford: Oxford University Press.
Mellor, D., & Oliver, A. (1997). Properties. Oxford: Oxford University Press.
Mill, J.S. (1974). A System of Logic Ratiocinative and Inductive [1843]. In J. Robson (Ed.) The Collected Works of John Stuart Mill, Vol. *VII*. London: Routledge and Kegan Paul.
Moyer, R., & Landauer, T. (1967). Time required for judgements of numerical inequality. Nature, *215*, 1519–1520.
Moyer, R., & Landauer, T. (1973). Determinants of reaction time for digit inequality judgements. Bulletin of the Psychonomics Society, *1*, 167–168.
Plato (1997a). Philebus. J. Cooper (ed.), In Plato: complete works, transl. D. Frede. Indianapolis, IN: Hackett Publishing Company, 399–456.
Plato (1997b). Republic. In J. Cooper (Ed.), Plato: complete works, transl. G. Grube Indianapolis, IN: Hackett Publishing Company, 972–1223.
Russell, B. (1908). Mathematical logic as based on the theory of types. American Journal of Mathematics, *30*, 222–262. Reprinted in R. Marsh (Ed.) Logic and Knowledge: essays 1901–1950. London: George Allen and Unwin, 59–102.Russell, B. (1919). Introduction to Mathematical Philosophy. London: George Allen and Unwin.
Sarnecka, B.W., Goldman, M.C., & Slusser, E.B. (This volume). How Counting Leads to Children's First Representations of Exact, Large Numbers. In Roi Cohen Kadosh & Ann Dowker (Eds.) The Oxford Handbook of Mathematical Cognition. Oxford: Oxford University Press.
Shapiro, S. (1997). Philosophy of Mathematics: Structure and Ontology. Oxford: Oxford University Press.
Tzelgov, J., Ganor-Sterm, D., Kallai, A., & Pinhas, M. (This volume). Primitives and non-primitives of numerical representations. In Roi Cohen Kadosh & Ann Dowker (Eds.) The Oxford Handbook of Mathematical Cognition. Oxford: Oxford University Press.
Walsh, V. (2003). A theory of magnitude: common cortical metrics of time, space and quantity. Trends in Cognitive Sciences, *7*, 483–488.
Welford, A. (1960). The measurement of sensory-motor performance: survey and reappraisal of twelve years progress. Ergonomics, 3, 189–230.
Wynn, K., & Chang, W. (1998). Limits to infants' knowledge of objects: the case of magical appearance. Psychological Science, *9*, 448–455.
Yablo, S. (2005). The Myth of the Seven. In M. Kalderon (Ed.), Fictionalism in Metaphysics. New York: Oxford University Press, 88–115.

PART II

HUMAN COGNITION

CHAPTER 3

COGNITIVE FOUNDATIONS OF HUMAN NUMBER REPRESENTATIONS AND MENTAL ARITHMETIC

OLIVER LINDEMANN AND MARTIN H. FISCHER

Representation of Numerical Knowledge

THE philosopher Henri Poincaré (1854–1912) stated that intuitions, and not formal logic, are the foundation upon which humans base their understanding of mathematics (McLarty, 1997). Interestingly, modern psychological research provides empirical support for Poincaré's notion and shows that a "sense of numbers" is part of a human's core knowledge that is already present early on in infancy. The origin and the underlying cognitive codes on which this number sense is grounded have, however, so far not been fully understood. This is where the chapters collected in this volume deliver significant advances in our understanding of the component processes and representations involved in numerical cognition and arithmetic.

Primitives of Number Representation

The chapter by Tzelgov et al. (this volume) discusses several basic cognitive mechanisms underlying the processing of Arabic digits. While we know that humans and non-human animals share the ability to process approximate magnitudes and numerosity information (see also the chapters by Agrillo and by Beran Perdue & Evans, this volume), only humans possess the ability to generate numerical notation systems that allow for a symbolic representation of exact quantities of natural numbers. Throughout civilisation, these notational systems became more and more sophisticated (e.g., Ifrah, 1981): the progressive introduction of syntactic features, such as the place-value principle to code magnitudes with multi-digit numbers, the polarity sign to denote negative values, or fraction symbols to denote non-natural numbers, made it possible to generate compound expressions to represent magnitudes that do not correspond to simple single-digit numbers (see also Nuerk et al., this volume).

In their chapter in this volume, Tzelgov and colleagues report their long-running research program aimed at identifying elementary entities – called primitives – for cognitive numerical representations. While mathematicians often consider the prime numbers to be such elementary units (since they make up all other natural numbers), the authors take a psychological view and define primitives as numbers whose meanings are holistically retrieved from memory without further processing. In contrast, the semantics of non-primitive numbers are generated on-line from primitives in order to perform a specific task. In other words, the direct and automatic meaning retrieval from memory is, according to Tzelgov and colleagues, the central processing criterion by which a numerical primitive can be identified. One approach to investigate such automaticity of number processing is to determine the "size congruity effect," that is, the interaction between numerical magnitude meaning and physical size of the number symbol being processed: more efficient processing in congruent conditions (e.g., "1" printed in small font or "9" printed in large font) establishes such automaticity (Henik & Tzelgov, 1982). It is noteworthy that this interaction points to an inescapable link between sensory experience and conceptual representation of magnitudes, which is a core aspect of the embodied cognition approach (cf. Barsalou, 2008). The chapter by Tzelgov and colleagues provides a detailed review of studies on numerical primitives and also encompasses work on multi-digit numbers (cf. Nuerk et al., this volume), fractions, negative numbers, and the number zero. The authors conclude that not only natural single-digit numbers but also some double-digit numbers and certain types of fractions seem to be holistically represented. Together with the basic concept of place-value, they should, therefore, be conceived as primitives of number representation.

In the modern numerical cognition literature, the concept of a holistic representation of number meaning is often linked to the notion of a mental number line, that is, the hypothesis that numbers are systematically associated to spatial codes, as if magnitudes were represented along a spatial continuum with small numbers typically to the left of larger numbers (Fias & Fischer, 2005; Hubbard et al., 2005). Tzelgov and colleagues acknowledge the mental number line not only as a culturally shaped representational medium to efficiently code and compare the meaning of natural numbers but also as a cognitive scaffolding mechanism that might help us to utilise syntactic processes to derive the meaning of multi-digit integers, negative numbers, or fractions.

Numbers and Space

The idea of a spatially oriented mental number line is further elaborated by van Dijk et al. (this volume), who provide a detailed introduction into the research on the association between numbers and space. While the association of spatial and numerical information is probably one of the most investigated phenomena in numerical cognition research of the last two decades, the cognitive mechanisms underlying this phenomenon, as well as its theoretical and practical implications, are still heavily debated. The chapter by van Dijk and colleagues captures this controversy and proposes a new theoretical development: the working memory account. The authors especially discuss, in this context, the effect of Spatial Numerical Association of Response Codes (SNARC; Dehaene et al., 1993). This effect reflects the tendency of participants, in a wide range of tasks, to respond faster with the left hand to relatively small numbers, while right-handed responses are faster for relatively large numbers (for a recent review, see Wood et al., 2008). This observation has typically been conceived of as the key evidence in favor of the mental number line hypothesis. However, although the SNARC effect seems to emerge in an automatic fashion, it has been shown that the spatial reference frame that is used to arrange numbers in space (e.g. from left to right, or right to left) is affected both by long-term cultural habits (such as the person's habitual reading direction) and current task demands (for a review, see Göbel et al., 2011). Besides presenting several behavioral paradigms for the assessment of spatial–numerical associations, such as number interval bisection and random number generation, the chapter also reviews neuropsychological research showing how selective brain damage disturbs this cognitive process.

Although spatial–numerical associations have been traditionally explained with the mental number line hypothesis, van Dijk and colleagues (this volume) remark critically that a growing number of studies have recently challenged this interpretation. First of all, several authors have shown that number associations strongly depend on short-term contextual influences and not merely on cultural preferences about a particular number line orientation (e.g. Bächthold et al., 1998; Fischer et al., 2009). In the same vein, recent studies demonstrated that requiring participants to hold number sequences in working memory disrupts (Lindemann et al., 2008) or modulates their spatial associations (van Dijk & Fias, 2011). That is, under conditions of memory load, the associated location of a particular digit in a number classification task seems no longer to depend on its numerical magnitude but is instead determined by its ordinal position in the memorised sequence.

Van Dijck and colleagues discuss different theoretical explanations for the ubiquitous spatial–numerical association and propose a hybrid account that assumes multiple sources for the cognitive mapping between numbers and space: the number line seems to play an important role for the learning of visual associations, while a general bipolar distinction between small and large magnitudes (so-called polarity coding; cf. Proctor & Cho, 2006) accounts for the additional impact of abstract language-driven categorisations on numerical cognition. Finally, the order of information in working memory nicely captures the short-term contextual modulations of SNARC and SNARC-like effects. Thus, the authors underline the important influence of contextual constraints on number representation (see also Morsanyi & Szucs, this volume).

The authors' main conclusion, namely that the degree of involvement of these three different processes in spatial–numerical associations will depend on the task at hand, is consistent with a recent proposal according to which number representations possess situated, embodied, and grounded features (Fischer, 2012; Fischer & Brugger, 2011). Specifically, grounding refers to universal constraints imposed by the physical world, such as the accumulation of objects along a vertical dimension or the influence of gravity; embodiment refers to an individual's sensory–motor learning history and, thus, also encompasses culture-specific directional spatial habits; finally, situated representations are set up flexibly and rapidly in response to current task-specific processing constraints.

Numbers and The Body

Andres and Pesenti (this volume) elaborate in their chapter on the idea of embodied number representations and highlight the impact of finger-counting habits on number representations in adults. Specifically, they remind us that the acquisition of number concepts in early childhood typically occurs while mapping number words to finger postures during counting. The authors convincingly show that these habits have long-lasting implications for the way adults comprehend and represent numerical information. In contrast to the other chapters in this section of the volume, which give overviews of specific cognitive mechanisms underlying quantity processing within different domains of numerical cognition research, Andres and Pesenti address the broader issue of the nature of cognitive codes involved in numerical cognition.

The approach of embodied cognitive representations provides readers with an alternative account of numerical cognition, which holds that symbols and abstract concepts become meaningful only when they are somehow mapped onto sensorimotor experiences (Barsalou, 2008; Glenberg et al., 2013). The chapter of Andres and Pesenti examines the types of sensorimotor experiences that might provide a basis for the development of an intuitive understanding of numbers and other magnitude-related information. The authors report cross-cultural, behavioral, and neuroscientific evidence in support of the notion that finger use during counting and while manipulating objects of different sizes leads to embodied representations of magnitudes in the adult brain. This evidence highlights the important contribution of finger counting to the development of numerical knowledge in general, and to establishing a one-to-one correspondence between an ordered series of count words and the available objects in particular. As a consequence of such systematic and contiguous sensorimotor and linguistic activity, the bodily experiences become associated with the number concepts, consistent with the well-established neuroscientific principles of Hebbian association learning (cf. Pulvermüller, 2013). During knowledge retrieval, in turn, these associated sensorimotor features become (re-)activated whenever adults process numbers and semantic information about numerosities and magnitudes. In this way, sensory and motor mechanisms are co-opted to assist numerical cognition.

Multi-Digit Numbers and Language

The contribution by Nuerk and colleagues (this volume) extends the investigation of number representation beyond the limits of single digits and into the realm of multi-digit number processing. It begins with a thorough review of the group's work on the unit-decade compatibility effect, which refers to a performance penalty in multi-digit number comparison when the magnitude ordering for decades and units of the two numbers goes in opposite directions. This observation is taken as evidence against a holistic representation of number meaning and, thereby, conflicts with the notion that multi-digit numbers may be considered as primitives of the numeration system (see Tzelgov et. al., this volume, for more details on this discussion). Nuerk et al. move on to present the basic assumptions of the triple-code model of number representation, probably the most influential cognitive model of number processing to date (cf. Dehaene, 1992, 2011). This model postulates three distinct but interconnected representations of number knowledge: a visual–Arabic code, an auditory–verbal word form, and an approximate analogue representation of the numerical magnitude (see previous section in this chapter). The model is, thus, already committed to a multi-modal representation of number knowledge, a feature that deserves further elaboration in the light of the hypothesized sensory components of all embodied knowledge. Nuerk et al.'s chapter is particularly remarkable for its focus on the detailed cognitive representation of place-value knowledge. Their novel proposal of a distinction between identification, activation, and computation processes is aimed at a more detailed understanding of place-value coding and might help to better pinpoint the origin of transcoding errors in children and in some neuropsychological patients, who might read a number such as "201" mistakenly as "two hundred and ten."

Mental Arithmetic

The three remaining chapters in this section of the volume go beyond the mere representation of numerical information and address the cognitive manipulation of magnitudes, i.e. mental arithmetic.

Cognitive Architecture of Mental Arithmetic

A reasonable assumption would be that our mental arithmetic operations are independent of the specific format of the numbers, be they written as digits or number words, spoken or merely imagined. In conflict with this assumption, the chapter by Campbell points to effects of number format on calculation and strategy choices, and also highlights several other effects of the surface form of arithmetic tasks on performance. In a detailed comparison of the strengths and weaknesses of various cognitive models of number representations involved in mental arithmetic, Campbell reviews Dehaene's (2011) "triple code model" (see "Multi-digit Numbers and Language" in this chapter) as well as McCloskey's (1992) "abstract code model" which claims that all number information converges on a single abstract mental representation. Finally, he introduces the "encoding complex model" (Campbell, 1994) which accounts for format-specific activations in mental arithmetic, and provides several examples in support of this proposal. For example, reading number words instead of digits not only prolongs the encoding of number information but also affects the subsequent calculation strategies used.

Interestingly, while Tzelgov et al. (this volume) attributed processing difficulties with larger compared to small numbers (the so-called "problem size effect") to their syntactically more complex representation, Campbell points to the simple fact that we encounter larger problems less frequently as an alternative explanation for this effect. Eventually, the author concludes that neither an abstract processing model nor the perhaps dominant triple code model of number knowledge adequately captures performance signatures of mental arithmetic. The cognitive arithmetic research rather supports an "encoding complex" model and points, in line with the chapter of Andres and Pesenti, toward integrated multi-modal or embodied representations of quantities that are closely coupled with concrete sensorimotor experiences, such as visual perception (e.g. Landy & Goldstone 2007) or finger counting (e.g. Badets et al., 2010; Andres & Pesenti, this volume).

Arithmetic Word Problem-Solving

Finally, it should be emphasized that a full understanding of human mathematical problem-solving is not possible without taking into account everyday mathematical cognition in the form of situated calculations. For example, instead of solving well-formed mathematical equations, we often encounter everyday situations that require us to transform numerosities as part of social exchanges, such as shopping or itinerary coordination. The chapter by Thevenot and Barrouillet in this volume discusses these processes in great detail and reviews research on arithmetic word problems (also called verbal or story problems). Word problem-solving is not only an important field for research on real-life numerical cognition; word problems are of particular relevance also in formal schooling because they are typically used as diagnostic test situations where students are expected to show their understanding of mathematical concepts. The chapter discusses both developmental and educational studies and demonstrates convincingly that a large part of the difficulty that children encounter when solving word problems arises from difficulties in understanding the described situation and constructing the adequate mental model. We learn from Thevenot and Barrouillet's review that the most important predictor of successful word problem-solving is the nature of the semantic situation described in the problem. These contextual influences do not only prime the strategies used by young children (cf. De Corte & Verschaffel, 1987) but also determine the structure and complexity of the mental representation that needs to be constructed to solve the problem correctly. In other words, a situation that cannot be represented by a straightforward mental model will impede the child's success at solving that problem. In contrast, problems that can be easily mentally simulated are solved more often. For instance, if children are not yet familiar with the concepts "more than" or "less than", they will face difficulties in solving the following arithmetic word problem: *"There are five birds and three worms. How many more birds than worms are there?"* (De Corte & Verschaffel, 1985). However, reformulating the question and asking the same children *"How many birds won't get a worm?"* results in almost perfect performance (Hudson, 1983).

Thevenot and Barrouillet distinguish between different theoretical accounts, such as schema-based approaches that focus on the semantic structure and the relations between elements involved in the story, and situation model approaches that highlight, instead, the functional and temporal features of mental models that are constructed in working memory (Reusser, 1989). The chapter also provides a brief but important discussion of the role of individual differences in word problem-solving, such as working memory capacity and reading abilities. These parameters affect the construction of mental models and, thus, the performance in arithmetic word problem-solving (see also Morsanyi & Szucs, this volume). The authors end their chapter with a description of techniques for enhancing word problem-solving performance. Successful interventions emphasize the conceptual characteristics of a particular situation and, thereby, support the construction of an adequate mental model. This can be accomplished, for instance, by simply rewording the problem (as seen in the example already given). Interestingly, the most efficient interventions imply either an activation of sensory or motor experiences, such as visualizations in the form of pictures and graphs, or the active exploration of the situation with the aid of manipulable materials (e.g. Cuisenaire rods). It seems that these observations reinforce the importance of sensorimotor experiences for successful mathematical performance and hark back to longstanding pedagogical traditions (e.g., Montessori, 1906).

Intuitive Reasoning in Mental Arithmetic

Complementing the chapter on arithmetic word problem-solving, the contribution by Morsanyi and Szucs (this volume) addresses the effects of rather general cognitive biases and overlearned strategies on arithmetic problem-solving. The authors demonstrate, with several empirical examples, that mathematical reasoning is often guided by intuitive heuristics instead of normative rules. An illustrative example of such overlearned but sometimes misleading cognitive strategies are the heuristics "multiplication makes bigger" and "division makes smaller" (Greer, 1994) – two assumptions that only hold for natural numbers. This so-called natural number bias is, therefore, misleading when inappropriately applied in problems comprising rational numbers (e.g., 5×0.5 or $16 \div 0.3$). A further example is the observation that problems with miscalculated results are more likely to be judged as correct when they can be more fluently processed due to superficial perceptual characteristics, such as their visual symmetry (Kahneman, 2011; Reber et al., 2008) or their temporal contiguity (Topolinski & Reber, 2010). Not surprisingly, these examples are consistent with the proposed role of sensory and motor activation in knowledge retrieval.

The research on intuitive rules and heuristic problem-solving has a rather long tradition compared to most other lines of mathematical cognition research. Some of the earliest studies on problem-solving heuristics have used mathematical word problems and pointed out that judgment biases become especially evident when people deal with uncertainties and problems that require an understanding of probabilities and randomness (e.g., Fischbein, 1975; Kahneman & Tversky, 1972; Kahneman et al., 1982). The chapter by Morsanyi and Szucs conveniently summarizes this classic research and gives a brief overview of the different types of cognitive biases, heuristics, and intuitions that are relevant for our understanding of numerical cognition. The authors introduce, in this context, the distinction between primary and secondary intuitions in problem-solving: primary intuitions are experience-based perceptual regularities that are characterized as bottom-up influences on decision making, whereas secondary intuitions are rule-based abstractions that reflect formal education. Interestingly, Morsanyi and Szucs not only provide an overview of the different kinds of misleading cues that can lead people astray in mathematical reasoning but also, as do Campbell and Thevenot and Barrouillet in their chapters (this volume), highlight the importance of individual differences for arithmetic problem-solving. They provide evidence supporting the notion that a person's cognitive capacity, mathematical education, and thinking dispositions all affect their mathematical reasoning. Taken together, the complex interactions between long-lasting personality traits (e.g., Krause et al., 2014) and short-lived situational factors (e.g., Werner & Raab, 2013) are key to our understanding why certain intuitive but misleading tendencies can be tackled more easily than others, and why performance on certain tasks is especially resistant to educational interventions as compared with other tasks.

Summary

This section of this volume provides a detailed and up-to-date introduction to the cognitive foundations of human number processing and mental arithmetic. The seven chapters identify a large number of cognitive signatures and biases that contribute to the representation and processing of numerical information; their authors discuss these empirical observations in the light of current theories and developments in the field. Despite this diversity of cognitive mechanisms and processes involved in mental arithmetic and the considerable differences in the theoretical approaches taken by the authors of the chapters, the section as a whole points to some of the over-arching principles that provide the scaffolding for human number knowledge. Given the strong evidence for a fundamental sense of number and at least partially automatic coding of numerical information (Tzelgov et al., this volume), it becomes clear that human numerical cognition cannot be fully understood if we conceive it as a specialised, domain-specific mechanism that operates in isolation from other non-numerical cognitive representations and processes. First, the section nicely illustrates that number processing depends on non-numerical contextual information (Thevenot & Barrouillet, this volume) as well as current working memory resources (van Dijk et al., this volume); it is strongly affected by individual biases such as the influences of one's native language (Nuerk et al., this volume) and intuitive rules and heuristics (Morsanyi & Szucs, this volume). Second, it becomes clear that each representation of numerical information is coupled with associated sensorimotor experiences. Besides the mental number line hypothesis and the association of numerical information with positions in space, the chapters also demonstrate that the coding of numerical magnitude information is modulated by motor representation of hand postures and fingers (Andres & Pesenti, this volume) and by visual information such as perceptual size (Tzelgov et al., this volume), perceptual fluency (Morsanyi & Szucs, this volume), or the visual format of the number (Campbell, this volume).

Together, these general principles of an interrelation of numerical and modality-specific sensory or motor codes support the idea that a complete perspective on human numerical cognition must take into account the body as a representational medium. The idea that the development of number knowledge is grounded in sensorimotor experiences will help us to gain new insights into the nature of human numerical cognition.

References

Agrillo, C. (this volume). Numerical and arithmetic abilities in non-primate species. In R. Cohen Kadosh & A. Dowker (Eds.), *The Oxford Handbook of Numerical Cognition*. Oxford: Oxford University Press.

Andres, M. & Pesenti, M. (this volume). Finger-based representation of mental arithmetic. In R. Cohen Kadosh & A. Dowker (Eds.), *The Oxford Handbook of Numerical Cognition*. Oxford: Oxford University Press.

Bächtold, D., Baumüller, M., & Brugger, P. (1998). Stimulus-response compatibility in representational space. *Neuropsychologia*, 36(8), 731–735.

Badets, A., Pesenti, M., & Olivier, E. (2010). Response-effect compatibility of finger-numeral configurations in arithmetical context. *Quarterly Journal of Experimental Psychology*, 63(1), 16–22.

Barsalou, L.W. (2008). Grounded cognition. *Annual Review of Psychology*, 59, 617–645.

Beran, M.J., Perdue, B.M., & Evans, T.A. (this volume). Monkey mathematical abilities. In R. Cohen Kadosh & A. Dowker (Eds.), *The Oxford Handbook of Numerical Cognition*. Oxford: Oxford University Press.

Campbell, J.I.D. (1994). Architectures for numerical cognition. *Cognition*, 53(1), 1–44.

Campbell, J.I.D. (this volume). How abstract is arithmetic? In R. Cohen Kadosh & A. Dowker (Eds.), *The Oxford Handbook of Numerical Cognition*. Oxford: Oxford University Press.

De Corte, E. & Verschaffel, L. (1985). Beginning first graders' initial representation of arithmetic word problems. *The Journal of Mathematical Behavior*, 4, 3–21.

De Corte, E. & Verschaffel, L. (1987). The effect of semantic structure on first graders' strategies for solving addition and subtraction word problems. *Journal for Research in Mathematics Education*, 18(5), 363–381.

Dehaene, S. (1992). Varieties of numerical abilities. *Cognition*, 44(1–2), 1–42.

Dehaene, S. (2011). *The Number Sense: How the Mind Creates Mathematics*. New York: Oxford University Press.

Dehaene, S., Bossini, S., & Giraux, P. (1993). The mental representation of parity and number magnitude. *Journal of Experimental Psychology: General*, 122(3), 371–396.

Fias, W. & Fischer, M. (2005). Spatial representation of numbers. In J.I.D. Campbell (Ed.), *Handbook of Mathematical Cognition* (pp. 43–54). New York: Psychology Press.

Fischbein, E. (1975). *The Intuitive Sources of Probabilistic Thinking in Children*. Doordrecht: Reidel.

Fischer, M.H. (2012). A hierarchical view of grounded, embodied, and situated numerical cognition. *Cognitive Processing*, 13(1), S161–S164.

Fischer, M.H. & Brugger, P. (2011). When digits help digits: spatial-numerical associations point to finger counting as prime example of embodied cognition. *Frontiers in Psychology*, 2, 260.

Fischer, M.H., Shaki, S., & Cruise, A. (2009). It takes just one word to quash a SNARC. *Experimental Psychology*, 56(5), 361–366.

Glenberg, A.M., Witt, J.K., & Metcalfe, J. (2013). From the revolution to embodiment: 25 years of cognitive psychology. *Perspectives on Psychological Science*, 8(5), 573–585.

Göbel, S.M., Shaki, S., & Fischer, M.H. (2011). The cultural number line: a review of cultural and linguistic influences on the development of number processing. *Journal of Cross-Cultural Psychology*, 42(4), 543–565.

Greer, B. (1994). Extending the meaning of multiplication and division. In G. Harel & J. Confre (Eds.), *The Development of Multiplicative Reasoning in the Learning of Mathematics* (pp. 61–85). Albany: State University of New York Press.

Henik, A. & Tzelgov, J. (1982). Is three greater than five: the relation between physical and semantic size in comparison tasks. *Memory & Cognition*, 10(4), 389–395.

Hubbard, E.M., Piazza, M., Pinel, P., & Dehaene, S. (2005). Interactions between number and space in parietal cortex. *Nature Reviews Neuroscience*, 6(6), 435–448.

Hudson, T. (1983). Correspondences and numerical differences between disjoint sets. *Child Development*, 54(1), 84–90.

Ifrah, G. (1981). *The Universal History of Numbers: From Prehistory to the Invention of the Computer*. London: The Harvill Press.

Kahneman, D. (2011). *Thinking, Fast and Slow*. New York: Farrar, Strauss, Giroux.

Kahneman, D. & Tversky, A. (1972). Subjective probability: a judgment of representativeness. *Cognitive Psychology*, 3, 430–454.

Kahneman, D., Slovic, P., & Tversky, A. (1982). Judgment under uncertainty. *Science*, 185, 1124–1131.

Krause, F., Lindemann, O., Toni, I., & Bekkering, H. (2014). Different brains process numbers differently: structural bases of individual differences in spatial and non-spatial number representations. *Journal of Cognitive Neuroscience*, 26(4), 768-776.

Landy, D. & Goldstone, R.L. (2007). How abstract is symbolic thought? *Journal of Experimental Psychology: Learning, Memory and Cognition*, 33(4), 720–733.

Lindemann, O., Abolafia, J.M., Pratt, J., & Bekkering, H. (2008). Coding strategies in number space: memory requirements influence spatial-numerical associations. *Quarterly Journal of Experimental Psychology*, 61(4), 515–524.

McCloskey, M. (1992). Cognitive mechanisms in numerical processing: evidence from acquired dyscalculia. *Cognition*, 44(1–2), 107–157.

McLarty, C. (1997). Poincare: mathematics & logic & intuition. *Philosophia Mathematica*, 5(2), 97–115.

Montessori, M. (1906). *The Montessori Method*. New York: F.A. Stokes.

Morsanyi, K. & Szucs, D. (this volume). Intuition in mathematical and probabilistic reasoning. In R. Cohen Kadosh & A. Dowker (Eds.), *The Oxford Handbook of Numerical Cognition*. Oxford: Oxford University Press.

Nuerk, H.-C., Moeller, K., & Willmes, K. (this volume). Multi-digit number processing – overview, conceptual clarifications, and language influences. In R. Cohen Kadosh & A. Dowker (Eds.), *The Oxford Handbook of Numerical Cognition*. Oxford: Oxford University Press.

Proctor, R.W. & Cho, Y.S. (2006). Polarity correspondence: a general principle for performance of speeded binary classification tasks. *Psychological Bulletin*, 132(3), 416–442.

Pulvermüller, F. (2013). How neurons make meaning: brain mechanisms for embodied and abstract-symbolic semantics. *Trends in Cognitive Sciences*, 17(9), 458–470.

Reber, R., Brun, M., & Mitterndorfer, K. (2008). The use of heuristics in intuitive mathematical judgment. *Psychonomic Bulletin & Review*, 15(6), 1174–1178.

Reusser, K. (1989). *Textual and Situational Factors in Solving Mathematical Word Problems*. Bern: University of Bern.

Thevenot, C. & Barrouillet, P. (this volume). Arithmetic word problem solving and mental representations. In R. Cohen Kadosh & A. Dowker (Eds.), *The Oxford Handbook of Numerical Cognition*. Oxford: Oxford University Press.

Topolinski, S. & Reber, R. (2010). Immediate truth – temporal contiguity between a cognitive problem and its solution determines experienced veracity of the solution. *Cognition*, 114(1), 117–122.

Tzelgov, J., Ganor-Stern, D., Kallai, A., & Pinhas, M. (this volume). Primitives and non-primitives of numerical representations. In R. Cohen Kadosh & A. Dowker (Eds.), *The Oxford Handbook of Numerical Cognition*. Oxford: Oxford University Press.

van Dijck, J., Ginsburg, V., Girelli L., & Gevers, W. (this volume). Linking numbers to space: from the mental number line towards a hybrid account. In R. Cohen Kadosh & A. Dowker (Eds.), *The Oxford Handbook of Numerical Cognition*. Oxford: Oxford University Press.

Van Dijck, J.-P. & Fias, W. (2011). A working memory account for spatial-numerical associations. *Cognition*, 119(1), 114–119.

Werner, K. & Raab, M. (2013). Moving to solution. *Experimental Psychology*, 60(6), 403–409.

Wood, G., Willmes, K., Nuerk, H., & Fischer, M.H. (2008). On the cognitive link between space and number: a meta-analysis of the SNARC effect. *Psychology Science Quarterly*, 50(4), 489–525.

CHAPTER 4

PRIMITIVES AND NON-PRIMITIVES OF NUMERICAL REPRESENTATIONS

JOSEPH TZELGOV, DANA GANOR-STERN, ARAVA Y. KALLAI, AND MICHAL PINHAS

Introduction

It is well documented that humans share with other non-human animals an analogue system that represents magnitudes or quantities (for a review see Feigenson, Dehaene, & Spelke, 2004; see Agrillo, and Beran et al., this volume). The use of natural (counting) numbers – symbols intended to represent exact magnitudes – emerges on the basis of this ancient system (Gallistel & Gelman, 2000). Numbers are symbolically represented by numerals that are single symbols or symbol combinations used to externally represent numbers. For the sake of simplicity, in most cases we will use the term 'number' to refer to the mathematical object that corresponds to a specific magnitude and to the numeral corresponding to it. We use the term 'numeral' whenever the distinction between numbers and numerals seems absolutely necessary.

Throughout history, different cultures used different symbolic numeration systems. Zhang and Norman (1995) made a distinction between external symbolic numeration systems and internal numeration systems that refer to the way numbers are mentally represented. Some of the external numeration systems are unidimensional, and they use quantity as the only dimension; however, such systems are limited to representing only small magnitudes. In the process of enculturation, humans extended the repertoire of numbers they used. Consequently, most external numeration systems are at least two-dimensional. In particular, the Arabic numeration system uses shapes (digits) as one dimension representing quantity, and digit position representing power with a base ten as the other dimension. Thus, for example, the number (specific magnitude) 74 is represented as $7 \times 10^1 + 4 \times 10^0$.

We focus on internal numeration systems. The mental number line (MNL; Dehaene, 1997; Restle, 1970) is the most frequent metaphor in discussions of the internal representation of numbers as symbolic representation of magnitudes. We are interested in the MNL as the representation of numbers in our long-term memory (LTM), in contrast to what humans may generate when required to perform a specific task. According to the MNL metaphor, numbers are mentally represented along a line, which at least in some cultures spreads in space from left to right (Dehaene, Bossini, & Giraux, 1993). Such a representation is, in essence, unidimensional, with magnitude or quantity being translated into spatial locations along this line (see van Dijck et al., this volume). This unidimensional representation is in contrast to the at least two-dimensional representation of external numeration systems in general, and the Arabic numeration system in particular. A unidimensional internal representation of quantity implies that humans understand the quantity represented by a number by extraction of its meaning from the MNL. This readout mechanism seems, in the Arabic numeration system, natural for single-digit (1D) numbers, where the value of the power dimension equals zero, and no additional processing is needed to access the represented magnitude. This becomes more complicated if the representation extends beyond 1D natural numbers, to include numbers such as 1789 or 35/42.

Verguts and Fias (2004, 2008) described a model that shows how a system can learn to use symbols as representations of magnitudes when provided with input from both a summation field that codes non-symbolic inputs of magnitude (Zorzi & Butterworth, 1999) and a symbolic input field; input that is passed to a common net of number detectors (see Roggeman et al., this volume). This implies a place coding representation as proposed by Dehaene and Changeux (1993), which is consistent with coding, at least by humans, of small magnitudes and the corresponding digits in specific brain areas of the intraparietal sulcus and the prefrontal cortex (Piazza, Pinel, Le Bihan, & Dehaene, 2007). Place coding is viewed by many as the neural realization of the MNL (Ansari, 2008).

Identifying Primitives and The Case of 1-Digit Numbers

We differentiate between numbers whose meaning can be holistically retrieved from memory and those whose meaning is generated online in order to perform a specific task. We refer to the first kind as the *primitives* of the numeration system. Such primitives are stored as distinctive nodes in LTM and humans are able to access their meaning without further processing. Because 1-digit (1D) numbers map into place coding (i.e. onto their location on the MNL) without additional preprocessing, they are natural candidates to serve as the basic set of numerical primitives. Thus, if 5 is a primitive, we retrieve its meaning from memory once we look at it. Other numbers (e.g. 3574) are stored in terms of their component dimensions; if 3574 is not a primitive, we have to generate its meaning from its components by applying the relevant algorithm to combine its component primitives.

Note that we make no assumption that these primitives are unaffected by culture, or that they are the same for all numerical notations. We primarily investigated people from Western culture using the Arabic notation and the decimal system.

The empirical identification of primitives is based on a distinction between two modes of processing – intentional and automatic. Bargh (1992) pointed out that processing without conscious monitoring is common to all automatic processes. This led Tzelgov (1997) to use such processing as the defining feature of automaticity. Tzelgov also suggested that processing without conscious monitoring can be diagnosed once the process in question is not part of the task requirements. By contrast, representations resulting from intentional processing of numerical values in most cases reflect the task requirements (Bonato, Fabbri, Umilta, & Zorzi, 2007; Ganor-Stern, Pinhas, Kallai, & Tzelgov, 2010; Shaki & Petrusic, 2005). The notion of a primitive as a mental entity that can be accessed without further processing is consistent with the view of automatic processing being memory retrieval, rather than mental computation (see, e.g. Logan, 1988; Perruchet & Vinter, 2002). Accordingly, numbers that are primitives have to be shown to be retrieved from LTM automatically, that is, as whole units. As such, they differ from numbers that are not primitives and are generated online if needed.

In the domain of numerical cognition two markers of automatic processing of numerical size are frequently used. One of them is the size congruity effect (SiCE). The SiCE is obtained when participants perform *physical size comparisons*[1] on stimuli varying also in their numerical magnitude. It refers to the increased latency in the incongruent (e.g. 3; 5) as compared with congruent (e.g. 3; 5) condition (Henik & Tzelgov, 1982). The SiCE was obtained in many studies when 1D numbers were used as stimuli, thereby validating the claim that 1D numbers are represented as primitives. Furthermore, consistent with the metaphor of the MNL, the SiCE increases with the intrapair distance along the irrelevant numerical dimension (Cohen Kadosh & Henik, 2006; Schwarz & Ischebeck, 2003; Tzelgov, Yehene, Kotler, & Alon, 2000). Consistent with the assumption of a compressed mental representation of numbers as implied by a model presuming the scalar variability principle in humans and animals alike (Gallistel, Gelman, & Cordes, 2005) or a logarithmic mapping function (Dehaene & Changeux, 1993), the SiCE is smaller for pairs of larger numbers (Pinhas, Tzelgov, & Guata-Yaakobi, 2010). Thus, this suggests that the spatial representation of a MNL is stored in memory, rather than constructed to meet the requirement of a specific task.

The other marker for automatic processing in the numerical domain is the SNARC (Spatial Numerical Association of Response Codes) effect (Dehaene et al., 1993). This effect is obtained when participants respond manually to number classification tasks (e.g. parity judgments). The effect is indicated by a negative correlation between the difference of the right- and left-hand response latencies when responding to a given number and the numerical magnitude of that number. Such a correlation is consistent with the assumption that the MNL spreads from left to right. Yet, the SNARC effect was hypothesized to reflect not only the representation of numerical magnitude but also reading habits (Shaki, Fischer, & Petrusic, 2009), thus questioning its usefulness as a tool to diagnose numerical primitives. Hence, in the present chapter we focus on the SiCE as the marker of automaticity of numerical processing.

The SiCE is a quite general and robust phenomenon that has been reported in children (e.g. Rubinsten, Henik, Berger, & Shahar-Shalev, 2002) and in adults of different mathematical expertise (Kallai & Tzelgov 2009), and in those speaking differing langauges from Chinese (e.g. Zhou, Chen, Chen, Jiang, Zhang, & Dong, 2007) and Hebrew (e.g. Henik & Tzelgov, 1982) to German (e.g. Schwarz & Ischenbeck, 2003) and English (Ansari, Fugelsang,

Bibek, & Venkatraman, 2006). Most of the studies on the SiCE have been perfomed on the Arabic numeration system. Yet the SiCE has been shown to work not just on Arabic numerals but also on number words (Cohen Kadosh, Henik, & Rubinsten, 2008), on Indian numerals (Ganor-Stern & Tzelgov 2011), on artificial numbers (Tzelgov et al., 2000), and in synesthetes on colours that representing numbers (Cohen Kadosh, Tzelgov, & Henik, 2008). In any case, in the last part of the article, we will return to the issue of the generality of the finding reported in the present chapter.

Assuming that 1D natural numbers are primitives, the question to be answered is: what happens once we move beyond the first decade? Consider natural numbers outside of the first decade. A holistic (unidimensional) representation of such numbers would imply that they are represented on the same continuum with 1D numbers, and therefore, their meaning is retrieved from the MNL. In contrast, a components (multi-dimensional) representation implies that the magnitudes of such numbers are constructed from their components, and they are not represented on the same continuum with 1D numbers. A similar question applies to numbers that are not natural, such as fractions, negative numbers, and even zero.

It might be that not all numbers outside of the first decade are represented in the same way. Some multi-digit numbers can apparently acquire a status of primitives given enough practice (see Logan, 1988; Palmeri, 1997; Rickard, 1997, for possible models) that could cause multi-digit numbers to become unitized by and retrieved from memory without additional processing. This would allow representing them mentally together with 1D numbers on the MNL in LTM and make them primitives. In this chapter, we present a series of studies, conducted in our laboratory, aimed at revealing the primitives of the (Arabic) numeration system. This series included an investigation of double-digit (2D) numbers, fractions, negative fractions, and zero. Thus, in order to show that a certain types of numbers are primitives, one has to show the SiCE for that type of numbers. Furthermore, if primitives are aligned along the MNL, the SiCE should be moderated by intrapair distance (Figure 4.1).

Are Double-Digit Numbers Primitives?

The first study in our exploration of the primitives of a numeration system dealt with 2D numbers (Ganor-Stern, Tzelgov, & Ellenbogen, 2007). Mathematically, 2D numbers, as all natural numbers, are written according to the rules of two-dimensional representation. The case of how 2D numbers are represented mentally is especially interesting as, on the one hand, they are compounds constructed from basic units, but on the other hand, they are the simplest numbers outside the first decade in that they are natural numbers, including only two elements, and they are highly familiar as they represent numerical magnitudes that are often used in daily life.

It is not at all obvious, however, that the mental representation of such numbers in LTM would mimic their mathematical characteristics. It is still debated whether 2D numbers are represented holistically along the MNL or compositionally (e.g. Brysbaert, 1995; Dehaene, 1997; Nuerk, Weger, & Willmes, 2001, see Nuerk, Moeller, & Willmes, this volume). A holistic representation means that 2D numbers are represented as integrated values, much like 1D numbers (e.g. Brysbaert, 1995; Dehaene, Dupoux, & Mehler, 1990) and accordingly,

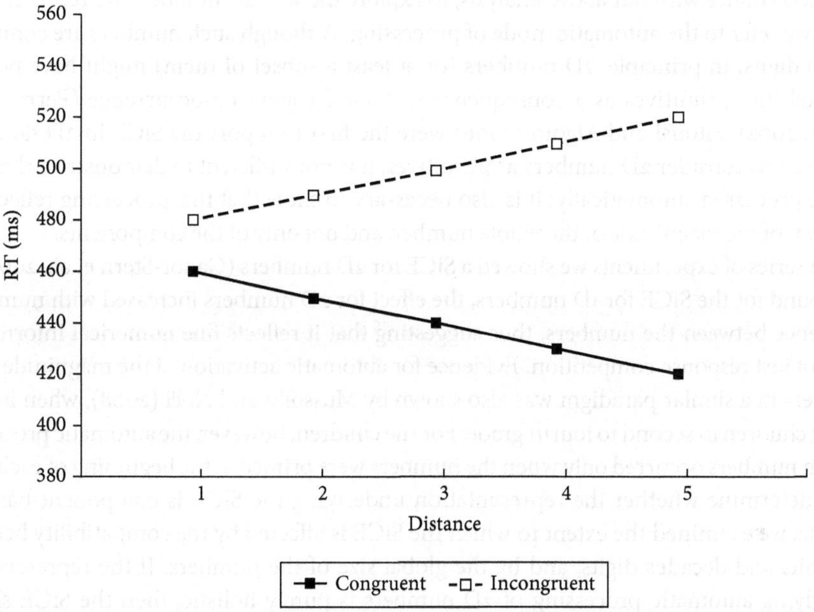

FIGURE 4.1 An illustration of the typically found modulation of the SiCE by the task-irrelevant intrapair numerical distance. The SiCE is larger for larger compared with smaller numerical distances. For example, the SiCE is expected to be larger for physical comparisons of 2 with 8 (i.e. distance = 6) than of 2 with 3 (i.e. distance = 1).

their meaning is accessed by direct retrieval from memory. This also implies that 1D and two-digit (2D) numbers alike are aligned mentally along one dimension. In contrast, a components representation implies that the internal representation of 2D numbers mimics the base-10 structure, such that the magnitudes of the different digits are stored separately, with an indication of their syntactic roles (e.g. Barrouillet, Camos, Perruchet, & Seron, 2004; McCloskey, 1992; Verguts & De Moor, 2005). In this case, the magnitude of a 2D number is produced by the application of an algorithm on the components' magnitudes according to their syntactic roles (i.e. decades, units).

When processed intentionally in the context of the numerical comparison task, 2D numbers seem to be mentally represented both in terms of their holistic values and the values of their components. Support for the holistic representation is provided by the distance and size effects. The distance effect (i.e. better performance for larger numerical differences) suggests the representation of 2D numbers is along a mental continuum, while the size effect (i.e. worse performance for larger numbers) suggests that the resolution of this representation decreases for larger magnitudes, consistent with Weber's law (Moyer & Landauer, 1967).

Support for the components representation was provided by the unit-decade compatibility effect, which shows that response latency to a numerical comparison task is influenced by the compatibility between the units and decades digits across the two numbers (e.g. Nuerk et al., 2001). Specifically, performance is faster for pairs that are unit-decade-compatible (e.g. for the pair 42; 57, both 4 < 5 and 2 < 7) than for pairs that are unit-decade-incompatible (e.g. for 47; 62, 4 < 6, but 7 > 2).

In accordance with our above analysis, to explore the way 2D numbers are represented in LTM, we refer to the automatic mode of processing. Although such numbers are composed of two digits, in principle, 2D numbers (or at least a subset of them) might have become unitized into primitives as a consequence of their frequent co-occurrence (Perruchet & Vinter, 2002). Fitousi and Algom (2006) were the first to report the SiCE in 2D numbers. However, to consider 2D numbers as primitives, it is not sufficient to demonstrate that they can be processed automatically; it is also necessary to show that this processing reflects the retrieval of the magnitude of the whole number, and not only of the components.

In a series of experiments we showed a SiCE for 2D numbers (Ganor-Stern et al., 2007). As was found for the SiCE for 1D numbers, the effect for 2D numbers increased with numerical difference between the numbers, thus suggesting that it reflects fine numerical information and not just response competition. Evidence for automatic activation of the magnitude of 2D numbers in a similar paradigm was also shown by Mussolin and Noël (2008), when investigating children in second to fourth grade. For the children, however, the automatic processing of such numbers occurred only when the numbers were primed at the beginning of each trial.

To determine whether the representation underlying the SiCE is component-based or holistic, we examined the extent to which the SiCE is affected by the compatibility between the units and decades digits, and by the global size of the numbers. If the representation underlying automatic processing of 2D numbers is purely holistic, then the SiCE should be unaffected by the compatibility between decades and units digits within the pair, but it should be affected by the global size of the numbers, with a smaller SiCE for larger numbers. The latter pattern was found for 1D numbers (Pinhas et al. 2010).

In contrast, if the representation underlying automatic processing of 2D numbers is a pure components representation, then the SiCE should be affected by the compatibility between decades and units digits (i.e. it should be reduced for pairs that are unit-decade incompatible compared to compatible), but unaffected by the global magnitude of the numbers. The results support the latter, as they showed that the SiCE was reduced for pairs that were unit-decade incompatible compared to compatible (see Figure 4.2, left panel), but it was unaffected by the pairs' global size (Ganor-Stern et al., 2007). That is, the SiCE was similar for 2D pairs in the range of 11-20 and in the range of 71-80.

These results were corroborated by Mussolin and Noël's (2008) study on children showing a similar pattern according to which the SiCE for 2D numbers was unaffected by the global size. A more recent study replicated the interaction between the SiCE and the unit-decade compatibility effect using both Arabic and Indian notation (Ganor-Stern & Tzelgov, 2011).

These findings show that the SiCE found for 2D numbers was mainly a product of the numerical magnitudes of the components and not of the magnitude of the whole numeral, thus suggesting that 2D numbers may not be considered primitives of the numeration system. We refer to this possibility as the Components Model, according to which the two digits are integrated into a representation of the whole 2D number only when numerical magnitude is processed intentionally, and when this integration is needed to fulfill the task. We further distinguish between two variants of this model. In the first variant, the Components with Syntactic Structure Model, the representation of the components also includes their syntactic roles, and therefore more weight is given to the decades compared to the units digits. Grossberg and Reppin's (2003) ESpaN model, according to which the magnitudes corresponding to the numbers 1–9 in each decade are represented by a different neural strip, belongs to this variant. In such a case, the SiCE should be more affected by the numerical

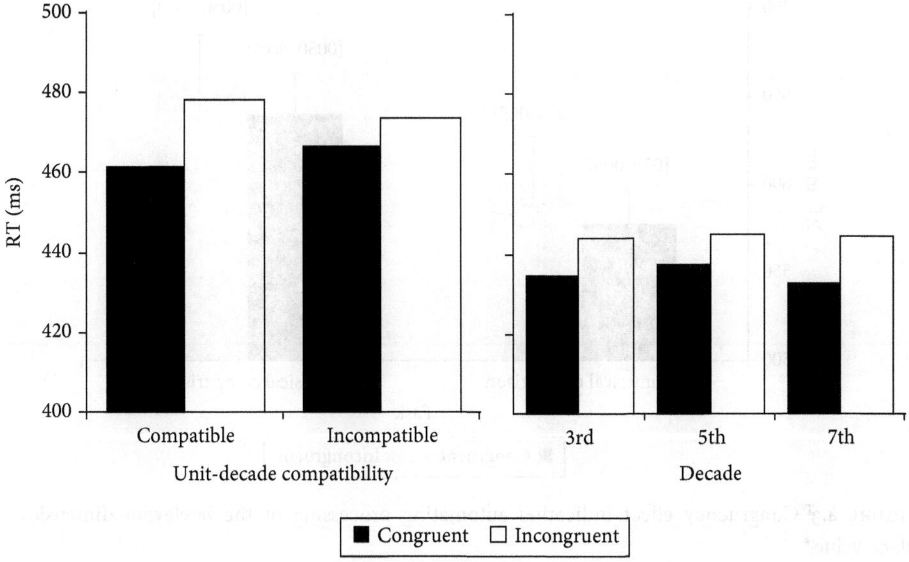

FIGURE 4.2 SiCE is larger for unit-decade compatible versus incompatible trials, but is unaffected by the decade size.

Data taken from Ganor-Stern, D., Tzelgov, J., and Ellenbogen, R., Automaticity and two-digit numbers, *Journal of Experimental Psychology: Human Perception and Performance*, 33, pp. 483–96, 2007.

difference between the decades digits than by the difference between the units digits, and it should be larger in pairs with a large compared to small numerical difference between the numbers' decade digits. The effect of numerical difference between the numbers' units digits should be reduced or absent. In the second variant, the Components Without Syntactic Structure Model, the representation of the components includes no information about the components' syntactic roles and, therefore, equal weight is given to the decades compared with the units digits. In such a case, the SiCE should be similarly affected by a numerical difference between the decades digits and by a numerical difference between the units digits.

The results show that the SiCE was more affected by the numerical difference between the decades digits than between the units digits, thus supporting the Components with Syntactic Structure representation. Although the SiCE for 2D numbers emerged from the magnitude of the 2D numbers' components, and not from the magnitude of the whole number, the syntactic roles of the different digits were processed automatically. Further support for the role of the syntactic structure of 2D numbers came from a study that tested the automatic processing of the place-value in multi-digit numbers (Kallai & Tzelgov, 2012a). In this study, summarized in Figure 4.3, the irrelevant dimension of place-value (i.e. units, decades, hundreds, etc.) was automatically processed in both numerical and physical comparison tasks. Thus, in the numerical comparison task, when selecting the larger non-zero digit in incongruent pairs (e.g. 030; 005) participants were slower and had more errors than in the case of congruent pairs (e.g. 060; 004). Similarly, in the physical comparison task, when selecting the physical larger string, incongruent pairs (e.g. 0200; 0020) were processed slower and with more errors than were congruent pairs (e.g. 0050; 0005).

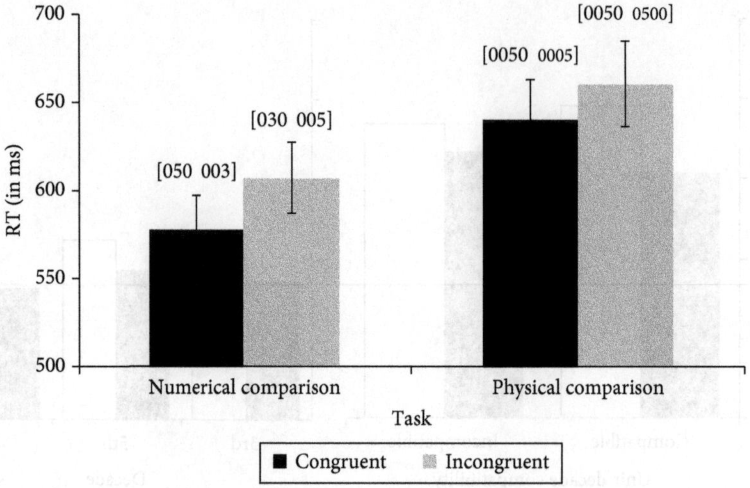

FIGURE 4.3 Congruency effect indicating automating processing of the irrelevant dimension of place-value.

Data taken from Kallai, A.Y. and Tzelgov, J., The place-value of a digit in multi-digit numbers is processed automatically, *Journal of Experimental Psychology: Learning, Memory, and Cognition*, 35(8), pp. 1221–33, 2012.

These results suggest that the decimal place-value of a digit within a numerical string is a primitive of the numeration system. Since place-value is a product of the decimal numeration system, this is a clear example of the culture-dependent nature of at least some of the primitives of the numeration system.

Are Fractions Primitives?

Fractions are rational numbers that can be described as any natural number divided by any other natural number except for zero. They can be represented as a ratio of two numbers (*x*/*y*) or as decimals (*x.y*), and they can be smaller (proper fractions) or larger (improper fractions) than 1. In this chapter we focus on common fractions (i.e. *x*/*y*).

Education researchers found that children have a hard time understanding fractions (Cramer, Post, & delMas, 2002; Gelman, 2000; Hartnett & Gelman, 1998; Miura, Okamoto, Vlahovic-Stetic, Kim, & Han, 1999; Moss & Case, 1999; Ni & Zhou, 2005; Robert, 2003; Smith, Solomon, & Carey, 2005). For instance, Ni and Zhou (2005) termed this the 'whole number bias', which is 'the tendency in children to use the single-unit counting scheme applied to whole numbers to interpret instructional data on fractions' (p. 27). Recently, cognitive and neurocognitive researchers started to study the mental representation of fractions, seeking to reveal the representational and neuronal basis for these difficulties.

Bonato et al. (2007) showed that when psychology and engineering students compared fractions to a standard of 1/5 or 1, the resultant distance effect (Moyer and Landauer 1967) and SNARC effect (Dehaene et al. 1993) were determined by the components of the fraction and not by the fraction value. This led the authors to conclude that the holistic magnitudes of the fractions were not accessed. However, the results of Bonato et al.'s (2007) study could have been a consequence of the stimuli set they used, which might have encouraged the use of component-based strategies. Indeed, when fractions differed in their denominators (Meert, Grégoire, & Noël, 2009) or shared no common component (Meert, Grégoire, & Noël, 2010), a higher correlation was found between response latency and distance between the values of the fractions than with the distance between the components (see similar results in Schneider & Siegler, 2010). These findings suggest that access to the holistic values of fractions is not automatic, but rather it occurs under conditions when componential strategies are difficult to implement. Further evidence for the holistic representation of unit fractions is provided by Ganor-Stern, Karasik-Rivkin, & Tzelgov (2011), who demonstrated such evidence when the fractions were compared with the digits 0 and 1. Thus, it might be concluded that although in some conditions the holistic values of fractions are accessed, this access is not automatic.

Common to the studies described above is their employment of tasks requiring intentional numerical processing. Automatic processing of fractions was tested by Kallai and Tzelgov (2009) using the SiCE. The presence of the SiCE was tested for comparisons of pairs of fractions and for comparisons of a fraction and a natural number. For unit fractions, the larger the denominator was, the smaller the whole fraction was. A SiCE was present in comparisons of unit fractions (i.e. $1/x$); however, it reflected the values of the denominators and not that of the fractions. Since the components' values in this case are in an inverse relationship with the fractions' values, the SiCE found was reversed. However, as can be seen in Figure 4.4, comparing fractions with natural numbers produced a SiCE that increased with the magnitudes of the natural numbers (see the black lines), but was insensitive to the magnitude of the fractions (see the gray lines)[2]. The same results were found for error rates.

These two results suggested the existence of a primitive representation of a fraction as an entity 'smaller than one'. We referred to this entity as a 'generalized fraction' (GeF) and assumed it was based on the general structure of the fraction (i.e. a ratio of two integers). The automatic processing of the GeF was found robust across different size relations between the digits comprising the fraction and the to-be compared natural number. Since the GeF showed numerical effects (i.e. SiCE and distance effect) in comparisons to natural numbers, we suggested the GeF was processed as a numerical entity and was represented on the same MNL as natural numbers (see also Ganor-Stern, 2012, for evidence for the representation of a GeF on the same MNL as natural numbers). In accordance with the notion of the GeF, improper fractions (i.e. fractions larger than one) were found to be automatically processed as smaller than one (Kallai and Tzelgov, 2009). Thus, Kallai and Tzelgov's results showed that when numerical processing was not required by the task, individual fractional values were *not* processed automatically. Instead, the components of the fractions were processed together with the notion of a GeF.

In a training study, Kallai and Tzelgov (2012b) demonstrated the crucial role of the symbolic representation of fractions on the way the fractions are represented. Participants were trained to map unit fractions (1/2 to 1/8) to arbitrary figures. With this notation each fraction was mapped to one figure, thus eliminating the components' external representation. The automatic processing of the magnitude of the newly acquired figures was tested following training. The SiCE for comparisons of pairs of fractions presented in this unfamiliar notation indicated that the participants automatically processed the fractions' holistic values. Taken together, the results of Kallai and Tzelgov (2009, 2012b) suggest that although specific fractions, and not only GeFs, *can* be represented as unique units in LTM, the more

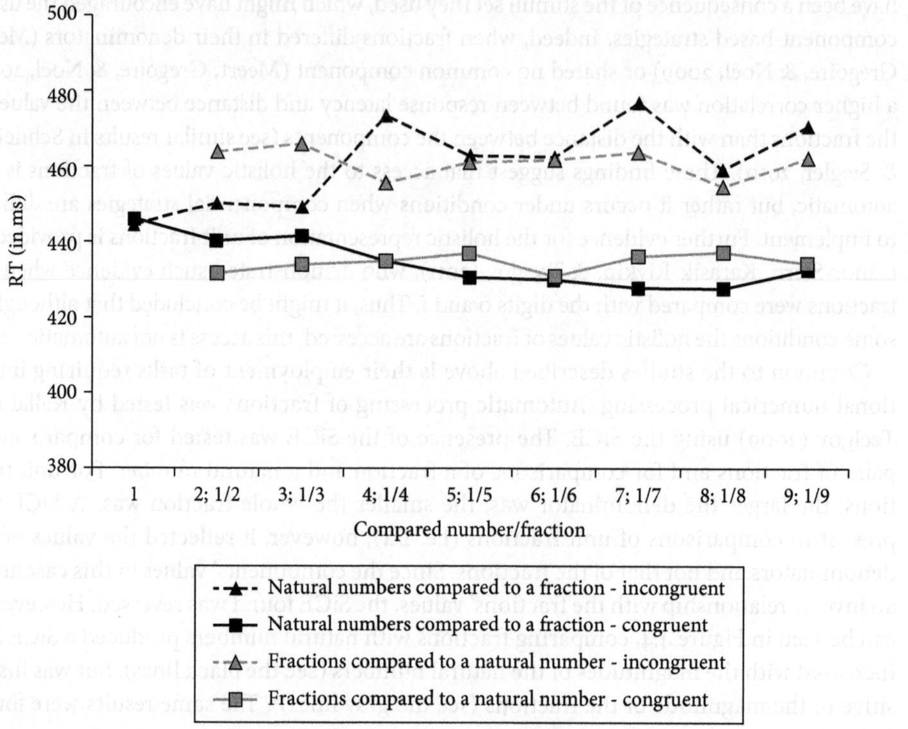

FIGURE 4.4 SiCE in comparisons of a set of natural numbers to a given fraction (black lines) increases with the magnitudes of the natural numbers. SiCE in comparisons of a set of fractions to given natural number (grey lines) is insensitive to the magnitudes of the fractions.

Data taken from Kallai, A.Y. and Tzelgov, J., A generalized fraction: An entity smaller than one on the mental number line, *Journal of Experimental Psychology: Human Perception and Performance*, 35, pp. 1845–64, 2009.

basic components they are composed of (i.e. natural numbers) interfere with their processing and, in fact, dominate automatic processing.

The training procedure used by Kallai and Tzelgov (2012b) to map unit fractions to unfamiliar figures consisted of a comparison task in which in each trial an unfamiliar figure appeared with a fraction, and the participant's task was to indicate which member of the pair represented a larger magnitude, with feedback provided after each trial. Thus, the knowledge acquired during practice was about the ordinal relationships between the unfamiliar figures and fractions. During testing, Kallai and Tzelgov showed that, unlike in previous studies (Bonato et al., 2007; Kallai & Tzelgov, 2009; Meert, et al. 2009), the distance between the fractional values predicted reaction time better than the componential distance did, thus suggesting that numerical meanings of fractions were assigned to the new symbols. The fractional values assigned to the figures by the participants might have been drawn from the participants' past knowledge. This, in turn, suggests that fractional values (or at least unit fractions) might have been represented in LTM before training and that the mapping to the new figures only eliminated the interference of the fractional components. Moreover, it might also be the case that fractions (as ratios between integers) do serve as primitives of the numeration system, but the effects of their processing are overshadowed by the stronger activation of the fractional components. This possibility is in line with the conclusion of Meert et al. (2010) suggesting hybrid representations for fractions (i.e. a combination of componential and holistic representations).

The possibility that fractions do serve as primitives of the numeration system is in line with a series of imaging studies that were used to find out whether mental representations of fractions exist independently of their components. Recent studies with primates and humans showed some evidence that the brain can process ratios and fractions (see also Nieder, this volume). Vallentin and Nieder (2008) found that monkeys can process the relation of two line lengths. The authors further found that different populations of neurons were sensitive to different ratios. This finding suggests that ratios might be represented as such in the primate brain. In an adaptation fMRI (functional magnetic resonance imaging) experiment with human subjects, Jacob and Nieder (2009) presented their participants with a stream of different fractions (e.g. 2/12, 3/18, 4/24) that all corresponded to the quantity of 1/6 for passive viewing. After the adaptation period, deviants were introduced as fractions of different value (e.g. 1/4). A distance effect was found in the bilateral horizontal section of the intra-parietal sulcus (hIPS), an area that was found to be sensitive to analogue numerical knowledge (Dehaene, Piazza, Pinel, & Cohen, 2003), and this effect was not sensitive to notation. The authors concluded that humans have access to an 'automatic representation of relative quantity'. Similar results were obtained by Ischebeck, Schocke, & Delazer (2009).

While the results described by Jacob and Nieder (2009) are impressive, we, in contrast with the authors, do not think that the participants automatically derived ratios of numbers. At least some of the fractions used by Jacob and Nieder were unfamiliar and rarely used (e.g. 4/24), and thus it raises the question whether such fractions could be unitized by experience, and stored in memory. In addition, it should be clear that passive viewing instructions do not specify what should be done by the participant in terms of the cognitive processing of whatever is presented. We believe that the participants did just what they were told to do – they fixated on and attended to features of the visual display. Thus, when presented with a fraction and not told to do anything specific, they did what they usually do under similar situations. For instance, when presented with the fraction 4/24, they processed its meaning because that is what humans do while looking at numbers. This indicates not that unfamiliar numbers (like 4/24) are retrieved from memory, but that when participants 'passively look' at numbers without specific instructions, they apply the default activity applied by humans when looking at numbers; they compute their value. This corresponds to the 'mental set' alternative for automatic processing proposed by (Besner, Stoltz, & Boutilier, 1997), which suggests that automaticity can be explained in terms of context effects (i.e. the meaning of the numbers can be ignored, but the context encourages participants to process them). In contrast, we propose that primitives are retrieved from memory (e.g. Logan, 1988) rather than computed. Thus, we believe that in order to argue that the processing of a number is automatic, a stronger test of automaticity than passive looking is needed.

To summarize, our findings suggest that when a fraction is compared with an integer, since the structure common to all fractions is enough to decide that the fraction is smaller, a GeF (which is the structure common to the whole set of fractions) dominates behaviour. In addition, when very familiar unit fractions are compared with one another, although their fractional values might be represented as numerical primitives, when using numerals that are composed of more basic components (i.e. numerals corresponding to natural numbers), the result is that the more basic components interfere with the processing of the fraction and overshadow it.

Are Negative Numbers Primitives?

Negative numbers are formally defined as real numbers smaller than zero. Their external representation is composed of a digit and a polarity minus sign in front of the digit. Similar to unit fractions, for negative numbers the larger the digit is, the smaller the value of the negative number is. In this chapter we focus on 1D negative integers. Fischer (2003) has taken as a starting point the assumption that, at least in English readers, numbers on the MNL are arranged in order of their magnitudes from left to right. Consequently, and in line with the SNARC effect (Dehaene et al. 1993), if negative and positive numbers are represented on the MNL and participants are presented in each trial with a pair of numbers and asked to select the number that is smaller (or larger) in its real value, responses should be faster when the smaller number appears on the left. This prediction was supported for positive, negative, and mixed (including a positive and a negative number) pairs alike. This study supports the conclusion that at least under some conditions, negative and positive numbers can be mentally aligned. However, since the participants had to perform numerical comparisons, it can argued that the alignment of negative and positive integers on a single number line reflects the task requirements rather than the representation stored in LTM (Bonato et al., 2007; Ganor-Stern et al., 2010). Fischer and Rottmann (2005) tested this hypothesis by asking participants to perform a speeded parity judgement task on positive and negative numbers in three different blocks – a block of positive numbers, a block of negative numbers, and a mixed block. As expected, independent of block type, a negative correlation between the difference in latencies of the right- and left-hand responses, and the (real) values of the positive numbers was found. However, in the case of negative integers the correlation was positive, indicating that participants ignored the polarity sign, consistent with the assumption of componential representation. Thus, if the SNARC effect taps into automatic processing of numerical values, negative integers are not primitives.

Tzelgov, Ganor-Stern, and Maymon-Schreiber (2009) used the SiCE to address this issue. They presented participants with pairs of numbers for a physical size decision. The pairs were generated from the set of 1D positive and negative integers, and differed in their physical and numerical size. As expected, a regular SiCE was found for pairs of positive integers (e.g. pairs made of the numbers 3; 5). A similar pattern was obtained in the case of mixed pairs in which the positive number was larger in value than the absolute value of the negative number (e.g. –3; 5). Note that in such pairs, ignoring polarity does not affect the magnitude relation between the two numbers. However, when pairs of negative numbers were presented (e.g. –3; –5), an inverse SiCE was obtained. Thus, the latency of congruent pairs like –3; –5 was *longer* than that of incongruent pairs like –3; –5 (see also Pinhas & Tzelgov, 2012). A similar pattern was found for mixed pairs in which the negative number was larger in its absolute value than the positive number (e.g. –5; 3). These results indicate that under the condition of automatic processing, negative numbers are perceived as the positive numbers corresponding to them. Thus the pair –3; –5 is perceived as 3; 5 and the same happens in the cased of 3; –5.

Recently, Parnes, Berger, & Tzelgov, 2012) have shown that the inverse relation between congruity effects in positive and negative integers is not constrained to the behavioural level, by showing a similar phenomenon using ERP. In the P300 peak, an inverse effect of congruity in the negative pairs as compared with the positive pairs was found. This implies that the polarity signs were ignored, and supports the hypothesis of a componential representation of negative numbers in LTM.

Tzelgov et al. (2009) tested whether the componential representation is due to the fact that magnitude and polarity are represented by different symbols. They trained participants to code polarity by colour (e.g. red numerals represented positive numbers and blue numerals represented negative numbers) and then asked them to perform the physical comparison task. Once again, the SiCE effect for comparisons of negative numbers was inversed. However, one could still argue that although the colour-coding method enabled the presentation of polarity and magnitude as part of the same integrated stimulus, they could still be represented as two separate dimensions of the same stimulus. This might encourage participants to represent polarity and magnitude separately. Therefore, in another experiment, Tzelgov et al. (2009) used unfamiliar figures to represent positive and negative integers. They mapped each of the integers in the range from –5 to 5 (zero excluded) into Japanese letters and trained participants by presenting them with pairs of stimuli that included two of the figures or one of the figures and zero, and asking them to decide which symbol corresponded to a larger magnitude. In the test phase, participants performed both physical and numerical comparison tasks. In the numerical comparison task, the distance effect was obtained even in the case of mixed pairs, indicating that under intentional processing conditions participants aligned the positive and negative integers along a single line. The picture was different under conditions of automatic processing. When performing physical comparisons of the unfamiliar Japanese letters, they showed a normal SiCE for positive and for mixed polarity insensitive pairs, but an inverse effect for negative and for mixed polarity sensitive pairs. Thus, even after extensive training using unfamiliar figures that eliminated the influence of the components of the negative numbers, no indication for automatic processing of negative integers was found. This leads us to conclude that negative numbers are resistant to automatization. We are skeptic whether they are potentially changeable to numerical primitives.

Is Zero a Primitive?

The idea of a number that represents a null quantity was born many years after people were already dealing with numbers. Similarly, infants who exhibit enumeration and computation abilities from the first few months of life show no expectation for a result of zero items (Wynn, 1998; Wynn & Chiang, 1998). Yet, like chimpanzees, 12-month-old infants can point to an empty location in which a desired object is missing, as a way for communicating about absent objects (Liszkowski, Schäfer, Carpenter, & Tomasselo, 2009). Later on, and with formal education, children gradually develop the understanding of the number zero (Wellman & Miller, 1986). Studies on non-human animals show that primates are able to associate a symbol with a null quantity (squirrel monkeys – Olthof, Iden, & Roberts, 1997; chimpanzees – Biro & Matsuzawa, 2001; Boysen & Berntson, 1989), as well as to perceive empty sets as representing a smaller magnitude than nonempty sets (rhesus monkeys – Merritt, Rugani, & Brannon, 2009). Furthermore, a series of studies on a grey parrot demonstrated the comprehension of a zero-like concept (Pepperberg, 1988, 2006;

Pepperberg & Gordon, 2005). Taken together, it seems that the cognitive tools necessary for the understanding of zero as a number exist in humans even before the learning of symbolic numerical representations (for further discussion see Merritt et al. 2009).

While children and non-human animals can represent a null quantity, the question is whether zero, that is, a symbolic representation of a null quantity acquired through schooling, is represented in LTM as a primitive. Furthermore, even if zero is a primitive, given its special status as representing a null quantity, it is still an open question whether it is represented as part of the MNL. As aforementioned, previous studies have shown that negative numbers are not stored as such in LTM (e.g. Tzelgov et al., 2009), and if needed, they are generated to fulfill task requirements (Ganor-Stern et al., 2010). If only positive integers are represented as primitives on the MNL, it follows that if zero is stored in LTM as a primitive, it is not represented as a polarity-neutral midpoint. Furthermore, given that zero is not part of the counting sequence, it is learned much later on in life, and it applies unique mathematical rules, it might be the case that, even if it is represented as a primitive, zero is stored separately from the other numerical primitives. Alternatively, despite zero's dissimilarities to the counting numbers, it is possible that it has acquired the status of the smallest primitive on the MNL through an extensive learning process (for detailed discussions see Kallai & Tzelgov, 2009; Perruchet & Vinter, 2002; Pinhas & Tzelgov, 2012). Such learning is possible considering zero's re-occurring appearances with positive single digits (presumably our numerical primitives), both as a number that represents a null quantity and as a placeholder that signifies an empty place in the decimal system.

Previous studies that tested the representation of zero in adults found inconsistent results. In a larger/smaller than zero comparison task, Fischer and Rottmann (2005) argued that the MNL starts with zero, while other studies (Brysbaert, 1995; Nuerk et al., 2001) suggested that zero is represented separately from other integers. However, all of these studies used tasks in which the processing of the numbers was intentional and was thereby affected by the task requirements. As such, they are less revealing with regard to the underlying representation of zero in LTM.

Pinhas and Tzelgov (2012) used the SiCE as a signature of automatic processing to investigate the representation of zero as a number that represents a null quantity. They found a regular SiCE in comparisons of zero with positive integers, and an inverted SiCE in comparisons of zero with negative integers (see also Kallai & Tzelgov, 2009; Tzelgov et al., 2009). These results confirmed that zero is stored in LTM as a primitive and that it is automatically perceived as smaller than the other 1D integers. To test whether zero is represented as part of the MNL stored in LTM, Pinhas and Tzelgov (2012) compared whether when processed automatically, zero behaves similarly to other numbers known to be members of the MNL. In particular, they tested the modulations of the SiCE by two *task-irrelevant* factors: the numerical intrapair distance and the presence of an end-stimulus (e.g. the numerically smallest number in the range) within the pair. *Numerical* comparisons to pairs that involve end-stimuli are responded to faster than pairs that involve numbers from the middle of the range (i.e. the end effect, e.g. Banks, 1977; Leth-Steensen & Marley, 2000). Furthermore, distance effects are attenuated for pairs that involve end-stimuli (for a possible model see Leth-Steensen & Marley, 2000).

Accordingly, Pinhas and Tzelgov (2012) examined the presence of an end effect in automatic processing while manipulating the numbers that constitute the end numbers. The stimulus set was manipulated such that for different groups of participants, 0, 1, or 2 were used as the smallest numbers in terms of their absolute values. When 0 was the smallest number in the set, the SiCE found in comparisons of the other numbers to 0 was enlarged and, minimally, if at all, modulated by the intrapair distance. This was in contrast to comparisons of non-end numbers, which showed a smaller SiCE that increased with numerical distance. The pattern of an enlarged SiCE, minimally affected by numerical distance, was termed the 'automatic end effect'. Its presence in comparisons of other numbers with 0 suggests that 0 was perceived automatically as an end value. To examine whether this status of being perceived automatically as an end value is unique to zero, Pinhas and Tzelgov looked at the SiCE found in comparisons of integers with 1 in blocks where 1, and not 0, was the smallest number. Interestingly, the same pattern of results was found when 1 was the smallest number in the stimulus set in terms of its absolute value. However, it was not found when 2 was the smallest number in the stimulus set in terms of its absolute value, suggesting that only 0, or 1 in the absence of 0, are automatically perceived as end values (see Figure 4.5).

Thus, with respect to the main question addressed in this series of experiments, when processed automatically, 0 behaves as the smallest number on the MNL. Therefore, although it might be considered a primitive, it has a special status. Interestingly, in the absence of 0, 1 fulfills this function and behaves as the smallest number. Larger numbers, such as 2, do not

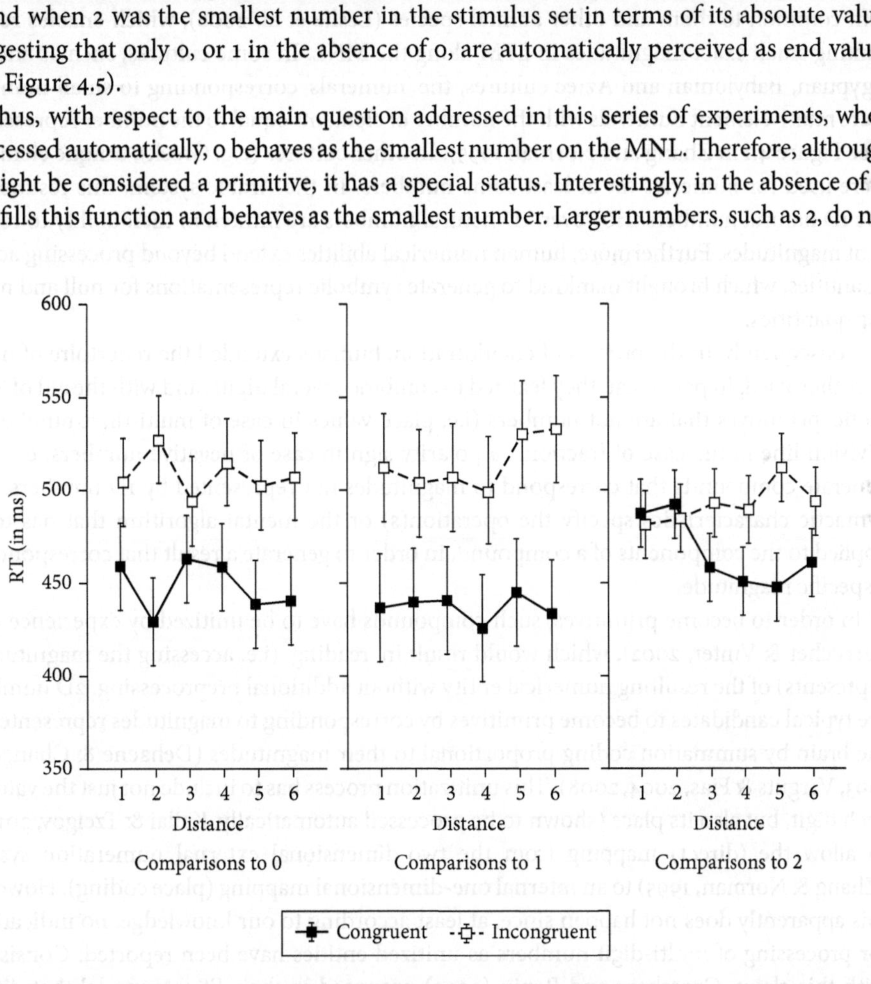

FIGURE 4.5 Mean RTs as a function of comparison type, congruency and (absolute) distance. Comparisons to 0 and 1 as the smallest number in the set resulted in the 'automatic end effect', whereas comparisons to 2 as the smallest number resulted in the increase of the SiCE with the increase of intrapair distance. Vertical bars denote 0.95 confidence intervals.

Data taken from Pinhas, M. and Tzelgov, J., Expanding on the mental number line: Zero is perceived as the "smallest", *Journal of Experimental Psychology: Learning, Memory, and Cognition*, 38(5), pp. 1187–205, 2012.

elicit the same pattern of responses, even when they are, in fact, the smallest in their absolute values in the number set, thus suggesting that the special role of zero reflects semantic, rather than episodic memory (Pinhas & Tzelgov, 2012).

General Discussion

Humans share with other non-human animals the ability to process magnitudes, and yet only humans use symbols to represent magnitudes. Humans are apparently born with an ability to generate a set of symbols on the basis of the ancient approximate number system common to them and other animal species (Leslie et al. 2008), which enables representing small exact magnitudes in LTM along the MNL. In some cultures, such as ancient Egyptian, Babylonian and Aztec cultures, the 'numerals' corresponding to small numbers were multi-element numerals with the number of elements equal to the number represented (see Figure 4.1 in Zhang and Norman 1995). In other cultures (e.g. Greek), unique elements were used for numerals for at least some numbers. In the Arabic system, this resulted in the 1D natural numbers. However, 1D natural numbers are limited in their ability to represent magnitudes. Furthermore, human numerical abilities extend beyond processing actual quantities, which brought mankind to generate symbolic representations for null and missing quantities.

Consequently, in the process of enculturation, humans extended the repertoire of numbers they used. In particular, they learned to combine several digits, and with the aid of 'syntactic' primitives that are not numbers (i.e. place values in case of multi-digit numbers, a division line in the case of fractions, a polarity sign in case of negative numbers, etc.), to generate compounds that correspond to magnitudes not represented by 1D numbers. The syntactic characteristics specify the operation(s) or the mental algorithm that has to be applied to the components of a compound, in order to generate a result that corresponds to a specific magnitude.

In order to become primitives, such compounds have to be unitized by experience (e.g. Perruchet & Vinter, 2002), which would result in 'reading' (i.e. accessing the magnitude it represents) of the resulting numerical entity without additional preprocessing. 2D numbers are typical candidates to become primitives by corresponding to magnitudes represented in the brain by summation coding proportional to their magnitudes (Dehaene & Changeux, 1993, Verguts & Fias, 2004, 2008). This unitization process has to include not just the value of each digit, but also its place (shown to be processed automatically, Kallai & Tzelgov, 2012a), to allow the (direct) mapping from the two-dimensional external numeration system (Zhang & Norman, 1995) to an internal one-dimensional mapping (place coding). However, this apparently does not happen since, at least according to our knowledge, no indications for processing of multi-digit numbers as unitized entities have been reported. Consistent with this claim, Grossberg and Repin (2003) proposed in their ESpaN model that different strips of neural structure in the brain correspond to parallel values representing units, decades, hundreds, etc., which implies a 'number plane,' rather than a number line. Yet, it still may be that some of the more familiar 2D numbers do become primitives, but this is difficult to detect due to 'components bias' that the well documented 'whole numbers bias,' reported for fractions (Ni & Zhou, 2005), is a special case. Kallai and Tzelgov (2012b) have shown that once such bias is eliminated, familiar fractions can be shown to behave as primitives. A parallel study still waits to be carried out on 2D numbers.

Common fractions, similar to 2D numbers, include (at least) two numerical components (digits) and a syntactic non-numerical component (the division line), which have to become unitized. The compound of two natural numbers and a division line has a unique notational structure, which Kallai and Tzelgov (2009) have shown behaves like a 'generalized fraction'; that is, it is automatically perceived to be smaller than 1. Because fractions are written as ratios of natural numbers, which apparently results in a components bias, referred to by Ni and Zhou (2005) as the 'whole numbers bias,' there is no direct evidence that even very familiar fractions are unitized and processed as primitives. Once this bias is eliminated by training participants to denote fractions by single (unfamiliar) figures, the typical behavioural signature of numerical primitives emerges: a SiCE reflecting the holistic magnitudes of fractions emerges (Kallai & Tzelgov, 2012b). Consistent with these findings, Jacob and Nieder (2009), using an adaptation fMRI paradigm under conditions of passive viewing, found a distance effect for deviations from the presented fraction. However, as we have pointed out above, the results of Jacob and Nieder, while very promising, should be interpreted with caution. Thus, at this point we may conclude there is some behavioural evidence that some of the more familiar fractions are primitives and there is also promising imaging data pointing in this direction; however, additional research that will apply a more rigorous, rather than 'passive viewing' definition of automaticity is needed.

Zero and negative numbers represent 'missing magnitudes'; zero represents a null magnitude and negative numbers represent missing magnitudes of specific value. In contrast to natural numbers, neither zero nor negative numbers have parallel analogue counterparts and consequently, such numbers have no summation representations. Yet, humans use the symbol '0,' which may also be acquired by primates (e.g. Merritt et al., 2009) and gray parrots (e.g. Pepperberg & Gordon, 2005), as representing a null magnitude. Zero is not composed of familiar components and apparently, due to the fact that a null magnitude is simpler to comprehend than a missing magnitude of a specific value, zero is represented as being part of the MNL. The inverse SiCE for negative numbers indicates that such numbers are not primitives. In contrast to fractions (and maybe to 2D numbers), this is not just due to the 'components bias'. Ganor-Stern (2012) has shown that even under conditions of intentional processing, 1D natural numbers and unit fractions, but not negative integers, are aligned. Yet, when the components bias is eliminated by using novel symbols, after extensive training, a distance effect in mixed pairs made of positive and negative numbers is evident under conditions of intentional processing. However, under conditions of automatic processing, an inverse SiCE is still evident (Tzelgov et al., 2009). Thus, it may be that the intrinsic complexity of the notion of 'missing magnitude of a specific value' makes it very difficult, if not impossible, to automatize specific negative numbers.

To conclude, if one uses the SiCE as the behavioural signature that a given numerical entity has a status of a primitive, the set of primitives includes, in addition to 1D numbers, at the very least also zero and the GeF. Place-value as well should be considered a primitive. Zero serves as the smallest member of the set of primitives and the GeF is apparently located in-between zero and 1. The automatic perception of fractions as smaller than 1, and the distance effect found in intentional comparisons of fractions and natural numbers (Kallai & Tzelgov, 2009), are consistent with this claim. In addition, the training study by Kallai & Tzelgov (2012b) raises the possibility that at least familiar fractions are represented as primitives, but this is overshadowed by the stronger representation of their components. Thus, further research is needed on this issue, as well as on the issue of emergence of 2D numbers as primitives.

The set of studies described in this chapter was performed on the Arabic numeration system and, thus, one could ask how general the reported findings are. The notion of primitives implies that for some numbers there is a one-to-one mapping between their external representation in terms of specific numerals and their internal representation on the MNL. Leslie et al. (2008) proposed a model according to which the concept of 1 and the function next are enough to acquire a set of consecutive natural numbers. Repetitive joint appearance of a specific magnitude with a unique symbol makes this symbol a numeral and we have shown that in the Arabic numeration system 1D numerals are primitives. More generally, we propose that consecutive natural numbers starting with 1 have the best chances to become primitives. We believe that exactly which numbers (and when) become primitives is culture – or numeration-system-specific. On the basis of our results, we would predict that given an equal amount of practice (i.e. co-occurrence with the corresponding magnitude), it will be easier for a given number to become a primitive if the numeral corresponding to it is made of a single symbol, because no unitization is needed. Thus, for example, because in the Greek numeration system the numeral corresponding to eight is a single-symbol, while in the Roman numeration system it is collection of four symbols of two different types (i.e. VIII), eight will be harder to become a primitive in Greek. Another prediction that follows from our results is that in the case of multi-symbol numbers, the process of primitivization will be slower if the components themselves serve as numerals. These predictions have still to be tested.

While the focus of the present review is behavioural, one could ask where in the brain the primitives are represented. As pointed out by Nieder (2009, see also Nieder, this volume), the symbolic representation of numbers starts apparently in the prefrontal cortex, but acquisition and automatization of numerical processing is characterized by a frontal-to-parietal shift in neural representation (for a similar argument see Cohen Kadosh & Walsh, 2009). This leads to the hypothesis that at the neural level, the representations of primitives should be looked for in the parietal lobe and, in particular, in the IPS. The test of this hypothesis requires further research.

Acknowledgments

This research was supported by grant No. 1664/08 for the Center for the Study of the Neurocognitive Basis of Numerical Cognition from the Israel Science Foundation.

Notes

1. Physical size comparisons refer to the font size of the numerals presented.
2. The two sets of lines display the same data (of mixed pairs of a natural number and a fraction) from two different points of view. *Note:* the double X-axis legend indicating both comparisons of a fraction to a set of natural numbers (as depicted by the black lines) and comparisons of a natural number to a set of fractions (as depicted by the gray lines).

References

Agrillo, C. (this volume). Numerical and arithmetic abilities in non-primate species. In: R. Cohen Kadosh, and A. Dowker (Eds), The Oxford Handbook of Numerical Cognition. Oxford University Press.

Ansari, D. (2008). Effects of development and enculturation on number representation in the brain. Nature Reviews Neuroscience, 9(4), 278–291.

Ansari, D., Fugelsang, J.A., Bibek, D., & Venkatraman, V. (2006). Dissociating response conflict from numerical magnitude processing in the brain: an event-related fMRI study. Neuroimage, 32, 799–805.

Beran, M.J., Perdue, B.M., and Evans, T.A. (this volume). Monkey mathematical abilities. In: R. Cohen Kadosh, and A. Dowker (Eds), The Oxford Handbook of Numerical Cognition. Oxford University Press.

Banks, W. (1977). Encoding and processing of symbolic information in comparative judgements. In: G. Bower (Ed.), The Psychology of Learning and Motivation, 11 (pp. 101–159). New York: Academic Press.

Bargh, J. (1992). The ecology of automaticity: towards establishing the conditions needed to produce automatic processing effect. *American Journal of Psychology*, 105, 181–199.

Barrouillet, P., Camos, P., Perruchet, P., & Seron, X. (2004). ADAPT: a developmental, asemantic, and procedural model for transcoding from verbal to Arabic numerals. Psychological Review, 111, 368–394.

Besner, D., Stoltz, J., & Boutilier, C. (1997). The Stroop effect and the myth of automaticity. Psychonomic Bulletin & Review, 4, 221–225.

Biro, D., & Matsuzawa, T. (2001). Use of numerical symbols by the chimpanzee (*Pan troglodytes*): cardinals, ordinals, and the introduction of zero. Animal Cognition, 4, 193–199.

Bonato, M., Fabbri, S., Umilta, C., & Zorzi, M. (2007). The mental representation of numerical fractions: real or integer? *Journal of Experimental Psychology: Human Perception and Performance*, 33, 1410–1419.

Boysen, S.T., & Berntson, G.G. (1989). Numerical competence in a chimpanzee (*Pan troglodytes*). Journal of Comparative Psychology, 103, 23–31.

Brysbaert, M. (1995). Arabic number reading: on the nature of the number scale and the origin of the phonological recoding. Journal of Experimental Psychology: General, 124, 434–452.

Cohen Kadosh, R., & Henik, A. (2006). A common representation for semantic and physical properties: a cognitive-anatomical approach. Experimental Psychology, 53, 87–94.

Cohen Kadosh, R., Henik, A., & Rubinsten, O. (2008). Are Arabic and verbal numbers processed in different ways? Journal of Experimental Psychology: Learning, Memory and Cognition, 34, 1377–1391.

Cohen Kadosh, R., Tzelgov, J., & Henik, A. (2008). A synesthetic walk on the mental number line: the size effect. Cognition, 106, 548–557.

Cohen Kadosh, R., & Walsh, V. (2009). Numerical representation in the parietal lobes: abstract or not abstract? Behavioral and Brain Sciences, 32, 313–328.

Cramer, K.A., Post, T.R., & delMas, R.C. (2002). Initial fraction learning by fourth- and fifth-grade students: a comparison of the effects of using commercial curricula with the effects of using the Rational Number Project Curriculum. Journal for Research in Mathematics Education, 33, 111–144.

Dehaene, S. (1997). The Number Sense. New York: Oxford University Press.

Dehaene, S., Bossini, S., & Giraux, P. (1993). The mental representation of parity and number magnitude. Journal of Experimental Psychology: General, 122, 371–396.

Dehaene, S., & Changeux, J.P. (1993). Development of elementary numerical abilities—a neuronal model. Journal of Cognitive Neuroscience, 5, 390–407.

Dehaene, S., Dupoux, E., & Mehler, J. (1990). Is numerical comparison digital? Analogical and symbolic effects in two-digit number comparison. Journal of Experimental Psychology: Human Perception and Performance, 16, 626–641.

Dehaene, S., Piazza, M., Pinel, P., & Cohen, L. (2003). Three parietal circuits for number processing. Cognitive Neuropsychology, 20, 487–506.

Feigenson, L., Dehaene, S., & Spelke, E. (2004). Core systems of number. Trends in Cognitive Science, 8, 307–314.

Fischer, M.H. (2003). Cognitive representation of negative numbers. Psychological Science, 14, 278–282.

Fischer, M.H., & Rottmann, J. (2005). Do negative numbers have a place on the mental number line? Psychology Science, 47, 22–32.

Fitousi, D., & Algom, D. (2006). Size congruency effects with two-digit numbers: expanding the number line. Memory & Cognition, 34, 445–457.

Gallistel, C.R., & Gelman, R. (2000). Non-verbal numerical cognition: from real to integers. Trends in Cognitive Science, 4, 59–65.

Gallistel, C.R., Gelman, R., & Cordes, S. (2005). The cultural and evolutionary history of the real numbers. In: S. Levinson & P. Jaisson (Eds), Culture and Evolution (pp. 247–274). Oxford: Oxford University Press.

Ganor-Stern, D. (2012). Fractions but not negative numbers are represented on the mental number line. Acta Psychologica, 139, 350–357.

Ganor-Stern, D., Karasik-Rivkin, I., & Tzelgov, J. (2011). Holistic representation of unit fractions. Experimental Psychology, 58, 201–206.

Ganor-Stern, D., Pinhas, M., Kallai, A., & Tzelgov, J. (2010). A holistic representation of negative numbers will be formed to meet the task requirement. Quarterly Journal of Experimental Psychology, 63, 1969–1981.

Ganor-Stern, D. & Tzelgov, J. (2011). Across-notation automatic processing of two-digit numbers. Experimental Psychology, 58, 147–153.

Ganor-Stern, D., Tzelgov, J., & Ellenbogen, R. (2007). Automaticity and two-digit numbers. Journal of Experimental Psychology: Human Perception and Performance, 33, 483–496.

Gelman, R. (2000). The epigenesis of mathematical thinking. Journal of Applied Developmental Psychology, 21(1), 27–37.

Grossberg, S., & Repin, D.V. (2003). A neural model of how the brain represents and compares multi-digit numbers: spatial and categorical processes. Neural Networks, 16, 1107–1140.

Hartnett, P., & Gelman, R. (1998). Early understandings of numbers: paths or barriers to the construction of new understandings? Learning and Instruction, 8, 341–374.

Henik, A., & Tzelgov, J. (1982). Is three greater than five: the relation between physical and semantic size in comparison tasks. Memory & Cognition, 10, 389–395.

Ischebeck, A., Schocke, M., & Delazer, M. (2009). The processing and representation of fractions within the brain: an fMRI investigation. NeuroImage, 47, 403–413.

Jacob, S.N., & Nieder, A. (2009). Notation-independent representation of fractions in the human parietal cortex. Journal of Neuroscience, 29, 4652–4657.

Kallai, A.Y., & Tzelgov, J. (2009). A generalized fraction: an entity smaller than one on the mental number line. *Journal of Experimental Psychology: Human Perception and Performance*, 35, 1845–1864.

Kallai, A.Y., & Tzelgov, J. (2012a). The place-value of a digit in multi-digit numbers is processed automatically. Journal of Experimental Psychology: Learning, Memory, and Cognition, 35(8), 1221–1233.

Kallai, A.Y., & Tzelgov, J. (2012b). When meaningful components interrupt the processing of the whole: the case of fractions. Acta Psychologica, 139, 358–369.

Leslie, A.M., Gelman, R., & Gallistel, C.R. (2008). The generative basis of natural number concepts. Trends in Cognitive Sciences, 12, 213–218.

Leth-Steensen, C., & Marley, A. (2000). A model of response-time effects in symbolic comparison. Psychological Review, 107, 62–100.

Liszkowski, U., Schäfer, M., Carpenter, M., & Tomasselo, M. (2009). Prelinguistic infants, but not chimpanzees, communicate about absent entities. Psychological Science, 20, 654–660.

Logan, G. (1988). Toward an instance theory of automatization. Psychological Review, 91, 295–327.

McCloskey, M. (1992). Cognitive mechanisms in numerical processing: evidence from acquired dyscalculia. Cognition, 44, 107–157.

Meert, G., Grégoire, J., & Noël, M.-P. (2009). Rational numbers: componential versus holistic representation of fractions in a magnitude comparison task. Quarterly Journal of Experimental Psychology, 62, 1598–1616.

Meert, G., Grégoire, J., & Noël, M.P. (2010). Comparing 5/7 and 2/9: adults can do it by accessing the magnitude of the whole fractions. Acta Psychologica, 135, 284–292.

Merritt, D.J., Rugani, R., & Brannon, E.M. (2009). Empty sets as part of the numerical continuum: conceptual precursors to the zero concept in rhesus monkeys. Journal of Experimental Psychology: General, 138, 258–269.

Miura, I.T., Okamoto, Y., Vlahovic-Stetic, V., Kim, C.C., & Han, J.H. (1999). Language supports for children's understanding of numerical fractions: cross-national comparisons. Journal of Experimental Child Psychology, 74, 356–365.

Moss, J., & Case, R. (1999). Developing children's understanding of the rational numbers: a new model and an experimental curriculum. Journal for Research in Mathematics Education, 30, 122–147.

Moyer, R.S., & Landauer, T.K. (1967). Time required for judgements of numerical inequality. Nature, 215, 1519–1520.

Mussolin, C., & Noël, M.-P. (2008). Automaticity for numerical magnitude of two-digit Arabic numbers in children. Acta Psychologica, 129, 264–272.

Ni, Y., & Zhou, Y.-D. (2005). Teaching and learning fraction and rational numbers: the origins and implications of whole number bias. Educational Psychologist, 40, 27–52.

Nieder, A. (2009). Prefrontal cortex and the evolution of symbolic reference. Current Opinion in Neurobiology, 19, 99–108.

Nieder, A. (this volume). Neuronal correlates of nonverbal numerical competence in primates. In: R. Cohen Kadosh, and A. Dowker (Eds), The Oxford Handbook of Numerical Cognition. Oxford University Press.

Nuerk, H.-C., Moeller, K., and Willmes, K. (this volume). Multi-digit number processing: overview, conceptual clarifications and language influences. In: R. Cohen Kadosh, and A. Dowker (Eds), The Oxford Handbook of Numerical Cognition. Oxford University Press.

Nuerk, H.C., Weger, U., & Willmes, K. (2001). Decade breaks in the mental number line? Putting the tens and units back in different bins. Cognition, 82, B25–B33.

Olthof, A., Iden, C.M., & Roberts, W.A. (1997). Judgments of ordinality and summation of number symbols by squirrel monkeys (*Saimiri sciureus*). Journal of Experimental Psychology: Animal Behavior Processes, 23, 325–339.

Palmeri, T. (1997). Exemplar similarity and the development of automaticity. Journal of Experimental Psychology: Learning, Memory, and Cognition, 23, 324–354.

Parnes, M., Berger, A., & Tzelgov, J. (2012). Brain representations of negative numbers. Canadian Journal of Experimental Psychology, 66(4), 251–258.

Pepperberg, I.M. (1988). Comprehension of 'absence' by an African grey parrot: learning with respect to questions of same/different. Journal of the Experimental Analysis of Behavior, 50, 553–564.

Pepperberg, I.M. (2006). Grey parrot (*Psittacus erithacus*) numerical abilities: addition and further experiments on a zero-like concept. Journal of Comparative Psychology, 120, 1–11.

Pepperberg, I.M. & Gordon, J.D. (2005). Number comprehension by a grey parrot (*Psittacus erithacus*), including a zero-like concept. Journal of Comparative Psychology, 119, 197–209.

Perruchet, P., & Vinter, A. (2002). The self-organizing consciousness. Behavioral and Brain Sciences, 25, 297–330.

Piazza, M., Pinel, P., Le Bihan, D., & Dehaene, S. (2007). A magnitude code common to numerosities and number symbols in human intraparietal cortex. Neuron, 53, 293–305.

Pinhas, M., & Tzelgov, J. (2012). Expanding on the mental number line: zero is perceived as the 'smallest'. Journal of Experimental Psychology: Learning, Memory, and Cognition, 38(5), 1187–1205.

Pinhas, M., Tzelgov, J., & Guata-Yaakobi, I. (2010). Exploring the mental number line via the size congruity effect. Canadian Journal of Experimental Psychology, 64, 221–225.

Restle, F. (1970). Speed of adding and comparing numbers. Journal of Experimental Psychology, 83, 274–278.

Rickard, T.C. (1997). Bending the power law: a CMPL theory of strategy shift and the automatization of cognitive skill. Journal of Experimental Psychology: General, 126, 288–311.

Robert, P.H. (2003). Part-whole number knowledge in preschool children. Journal of Mathematical Behavior, 22, 217–235.

Roggeman, C., Fias, W., and Verguts, T. (this volume). Basic number representation and beyond: neuroimaging and computational modeling. In: R. Cohen Kadosh, and A. Dowker (Eds), The Oxford Handbook of Numerical Cognition. Oxford University Press.

Rubinsten, O., Henik, A., Berger, A., & Shahar-Shalev, S. (2002). The development of internal representations of magnitude and their association with Arabic numerals. Journal of Experimental Child Psychology, 81, 74–92.

Schneider, M. and Siegler, R.S. (2010). Representations of the magnitudes of fractions. Journal of Experimental Psychology: Human Perception and Performance, 36, 1227–1238.

Schwarz, W., & Ischebeck, A. (2003). On the relative speed account of number-size interference in comparative judgments of numerals. Journal of Experimental Psychology: Human Perception and Performance, 29, 507–522.

Shaki, S., Fischer, M.H., & Petrusic, W.M. (2009). Reading habits for both words and numbers contribute to the SNARC effect. Psychonomic Bulletin & Review, 16, 328–331.

Shaki, S., & Petrusic, W.M. (2005). On the mental representation of negative numbers: context dependent SNARC effects with comparative judgments. Psychonomic Bulletin & Review, 12, 931–937.

Smith, C.L., Solomon, G.E.A., & Carey, S. (2005). Never getting to zero: elementary school students' understanding of the infinite divisibility of number and matter. Cognitive Psychology, 51, 101–140.

Tzelgov, J. (1997). Specifying the relations between automaticity and consciousness: a theoretical note. Consciousness and Cognition, 6, 441–451.

Tzelgov, J., Ganor-Stern, D., & Maymon-Schreiber, K. (2009). The representation of negative numbers: exploring the effects of mode of processing and notation. Quarterly Journal of Experimental Psychology, 62, 602–624.

Tzelgov J., Yehene, V., Kotler, L., & Alon, A. (2000). Automatic comparisons of artificial digits never compared: learning linear ordering relations. Journal of Experimental Psychology: Learning, Memory, and Cognition, 26, 103–120.

Vallentin, D., & Nieder, A. (2008). Behavioral and prefrontal representation of spatial proportions in the monkey. Current Biology, 18, 1420–1425.

van Dijck, J.-P., Ginsburg, V., Girelli, L., and Gevers, W. (this volume). Linking numbers to space: from the mental number line towards a hybrid account. In: R. Cohen Kadosh, and A. Dowker (Eds), The Oxford Handbook of Numerical Cognition. Oxford University Press.

Verguts, T., & De Moor, W. (2005). Two-digit comparisons: decomposed, holistic or hybrid? Experimental Psychology, 52, 195–200.

Verguts, T., & Fias, W. (2004). Representation of number in animals and humans: a neural model. Journal of Cognitive Neuroscience, 16, 1493–1504.

Verguts, T., & Fias, W. (2008). Symbolic and nonsymbolic pathways of number processing. Philosophical Psychology, 21, 539–554.

Wellman, H.M., & Miller, K.F. (1986). Thinking about nothing: development of concepts of zero. British Journal of Developmental Psychology, 4, 31–42.

Wynn, K. (1998). Psychological foundations of number: numerical competence in human infants. Trends in Cognitive Sciences, 2, 296–303.

Wynn, K., & Chiang, W-C. (1998). Limits to infants' knowledge of objects: the case of magical appearance. Psychological Science, 9, 448–455.

Zhang, J., & Norman, D.A. (1995). A representational analysis of numeration systems. Cognition, 57, 271–295.

Zhou, X., Chen, Y., Chen, C., Jiang, T., Zhang, H., & Dong, Q. (2007). Chinese kindergartners' automatic processing of numerical magnitude in Stroop-like tasks. Memory & Cognition, 35, 464–470.

Zorzi, M., & Butterworth, B. (1999). A computational model of number comparison. In: M. Hahn & S. C. Stoness (Eds), Proceedings of the Twenty-First Annual Meeting of the Cognitive Science Society, August 19–21 (pp. 778–78). Mahwah, NJ: Lawrence Erlbaum Associates.

CHAPTER 5

FINGER-BASED REPRESENTATION OF MENTAL ARITHMETIC

MICHAEL ANDRES AND MAURO PESENTI

Introduction

We live in a mathematical world. All day long, we calculate prices, quantities (e.g. amount of food needed to prepare the dinner, the number of people in a supermarket queue, etc.), delays (e.g. the time we need to leave for work before we miss the next train or the time needed to get home before the beginning of our favourite TV show), or distances. We are able to calculate areas (e.g. to buy the right volume of paint to cover a wall), ratios (e.g. to compare the price of two items with different weights), rates (e.g. to discriminate between good and bad investments), and much more. Yet most of these calculation abilities are not innate and, before being mastered, require rather long, boring, sometimes painful, study during childhood. In fact, although the primary school cycle, during which the bases of arithmetic knowledge are formally taught, lasts 5–7 years depending on country, nearly 40% of pupils end this cycle with insufficient arithmetic knowledge (Brun & Paster, 2010), and 5–10% of them even present learning disabilities (Shalev, 2007) from which they might still suffer into adulthood (Barbaresi, Katusic, Colligan, Weaver, & Jacobsen, 2005; Shalev, Manor, & Gross-Tsur, 2005). However, while complex mathematics remain abstract for most of us and seem distant from our everyday needs, basic arithmetic does speak to our mind because it is useful in everyday life, hence, the need to find the right teaching methods to help children learn. There are various teaching strategies, from intense learning by drill, without concrete mediation, to learning through concrete manipulation of actual objects. Using the fingers to keep track of a count or back up mental calculation is one of these concrete means.

In this chapter, we will argue that the way we express numerical concepts physically, by raising fingers while counting or using grip aperture to describe magnitude, leads to embodied representations of numbers and calculation procedures in the adult brain. By *embodied*, we mean here that sensory and motor mechanisms are co-opted to assist and possibly develop mental activities, whose referents are distant in time and space or even imaginary (Wilson, 2002); the co-opted sensory-motor mechanisms are thus decoupled from their initial function in bodily experience in order to support off-line conceptual processing (Anderson, 2010). This view must be distinguished from other views suggesting that cognition aims to serve action, so every mental activity is situated in the context of on-going interactions between the body and the environment (e.g. Glenberg, 1997; Walsh, 2003). As a result, the structure and content of 'embodied' concepts should be influenced by the sensory-motor mechanisms they rely on (Barsalou, 1999; Lakoff & Johnson, 1999), and their implementation in the brain should be accompanied by the colonization of the brain areas underlying these sensory-motor mechanisms (Martin, 2007). This functional and brain overlap does not imply that numerical cognition depends exclusively on sensory-motor mechanisms, as predicted by some sensory-motor theories of conceptual processing (e.g. Gallese & Lakoff, 2005). We will specify the properties of number representations and the calculation procedures assumed to involve a redeployment of sensory and motor resources for off-line conceptual processing. We will also stress the importance of brain lesion data to assess the causal relationship between sensory-motor mechanisms and conceptual processing (Mahon & Caramazza, 2009). To illustrate the concept of embodied numerical cognition, we will focus on number and finger interactions in the context of simple arithmetic operations. As we will show, the fixed order of fingers on the hand provides human beings with unique facilities to increment numerical changes or represent a cardinal value, while solving arithmetic problems. Number processing was also found to interact with the execution (Andres, Davare, Pesenti, Olivier, & Seron, 2004; Andres, Ostry, Nicol, & Paus, 2008c; Chiou, Wu, Tzeng, Hung, & Chang, 2012; Lindemann, Abolafia, Girardi, & Bekkering, 2007;Moretto & di Pellegrino, 2008) or perception of grasping movements (Badets, Andres, Di Luca, & Pesenti, 2007; Badets & Pesenti, 2010, 2011; Chiou, Chang, Tzeng, & Wu, 2009; Ranzini, Lugli, Anelli, Carbone, Nicoletti, & Borghi, 2011), indicating that the adjustment of the hand grip to object size shares processes with the computation of number magnitude estimates. However, these interactions will not be discussed here (but see Box 5.1), since they might be part of a magnitude system whose role extends beyond arithmetic problem solving (Walsh, 2003). In order to specify the influence of finger representation on mental arithmetic, both at the cognitive and neural level, we will review the findings of anthropological, behavioural, electrophysiological, and brain imaging studies. We will start with anthropological and developmental data showing the role of fingers in the acquisition of arithmetic knowledge. We will then address the issue of whether number and finger interactions are also observed in adults used to solve arithmetic problems mentally. We will suggest arithmetic performance depends on the integrity of finger representations in children and adults. Finally, we will provide an overview of the results of recent functional magnetic resonance imaging (fMRI) studies showing a common brain substrate for finger and number representations in children and adults.

Box 5.1 Grasping Number Magnitude

Using the fingers to count or calculate is only one type of motor behaviour that could mediate the functional link between fingers and numbers. Number magnitude and the representation of the finger movements involved in grasping actions also influence each other. This suggests that the sensory-motor mechanisms of object grasping may underlie number processing, presumably because adjusting grip aperture to object size requires computing internal magnitude estimates.

Effect Of Number Magnitude On Grip Adjustment

Recent findings demonstrate the interference of processing number magnitude on adjusting grip aperture. Small numbers are shown to delay the initiation of finger movements when the response requires opening the grip, whereas large numbers slow down movement start when the response requires closing the grip (Andres et al., 2004). Similarly, small numbers facilitate precision grip actions (i.e. the fingertips of the thumb and index finger are pressed against each other to grasp small objects) whereas large numbers facilitate power grip actions (i.e. the thumb makes opposition with the other fingers to grasp large objects; Moretto & di Pellegrino, 2008; Lindemann et al., 2007). A similar effect is observed when numbers are displayed in close synchrony with pictures of objects affording precision or power grip actions (Chiou et al., 2009). Kinematics recordings during reach-to-grasp actions further reveal that processing large compared to small numbers increases grip aperture as the hand starts to move towards the object. The time course of grip aperture shows the effect of number is maximal during the first stage of the movement and decreases progressively before object contact, suggesting that number magnitude influences movement planning whereas visual feedback about the object leads to on-line adjustment during movement execution (Box 5.1 Figure 5.1; Andres et al., 2008c; Chiou et al., 2012; Lindemann et al., 2007). Number magnitude interference is also observed when imagining a grasping movement without actually performing it. When participants are asked to make

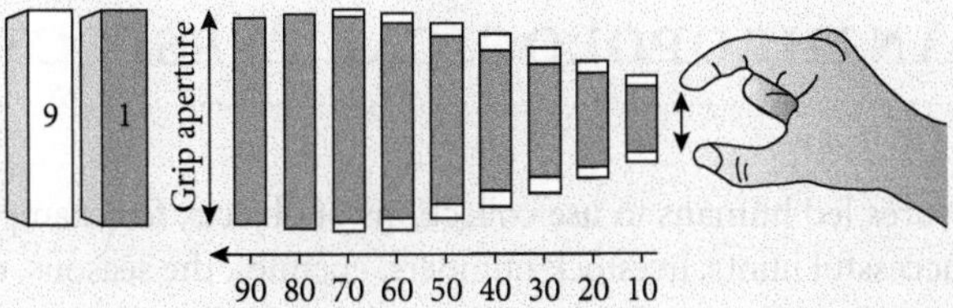

BOX FIGURE 5.1 Grip aperture during the grasping of wooden blocks with small or large numbers printed on their visible face. Large numbers induce an increase of grip aperture during the first stages of the movement (illustrated by the grey–white ratio of the rectangles); this effect decreases progressively as the hand approaches the object

(Adapted from Cortex, 44 (4), Michael Andres, David J. Ostry, Florence Nicol, and Tomas Paus, Time course of number magnitude interference during grasping, pp. 414–9, doi: 10.1016/j.cortex.2007.08.007, Copyright (2008), with permission from Elsevier).

subjective judgements about their capacity to grasp objects of various lengths, the average length they find impossible to grasp is longer when small numbers are presented before the objects, and shorter when large numbers are presented (Badets et al., 2007). Importantly, numerical magnitude does not influence perceptual size judgements on the same objects, suggesting that the mechanisms shared by number magnitude and grip aperture are specific to action. However, these mechanisms might not share the same metric, as the effect of numbers on grip aperture seems to be dependent on the relative magnitude of the numbers used in a specific task set, rather than on their absolute value (Chiou et al., 2012). Finally, number magnitude is found to interfere with the perception of grip movements: responding to closing versus opening grips is facilitated by the prior display of small *versus* large numbers, respectively (Badets & Pesenti, 2010).

Effect Of Grasping Movements On Number Magnitude Processing

Activation of magnitude estimates also flows from the perception of grasping movements to the processing of number magnitude. Movements mimicking a grip closure facilitate the processing of small numbers (Badets & Pesenti, 2010, 2011), and merely viewing graspable objects speeds up numerical judgements compared to ungraspable objects, suggesting that the representations of object affordances and numbers share internal magnitude estimates (Ranzini et al., 2011). In random number generation tasks, observing irrelevant pictures mimicking grip closure or aperture influences number production: more small numbers than large numbers are produced after grip closure observation whereas small and large numbers are produced as frequently after observing grip aperture (Badets, Bouquet, Ric, & Pesenti, 2012). The effect is observed for closing grips, perhaps because grip closure evokes the final step of reach-to-grasp actions, and is thus more predictive of the perceived object size. Critically, parity judgement and random number generation are not affected by viewing closing or opening movements of non-biological geometric shapes (Badets et al., 2010, 2012), indicating that the aforementioned effects do not result from a general system processing motion amplitude, but rather from the activation of a magnitude code shared by numbers and grasping movements.

An Anthropological Perspective

Evolutionary pressures led humans to use collections of objects, for example, shells or rods, to keep track of successful hunts, livestock numbers, enemies, the seasons, or the stars in the sky. This use of external counting and calculation tools can be found in the numerous archaeological traces left by ancient cultures (e.g. pieces of wood or bones with notches, artefacts such as reliefs and mosaics, etc.; Boyer, 1968; Ifrah, 1981). Body parts were also used in such calculations at different times and by different cultures. The use of fingers to count or calculate has, indeed, been reported by several authors from ancient times (e.g. Cicero, Epistole ad Atticum, V, 21, 13, 106–43 BCN) and is further evidenced in famous counting systems (e.g. Beda Venerabilis, De temporum ratione, 672–735 ACN) commonly used across Europe in the Middle Ages (Williams & Williams, 1995), as well as in many Oriental cultures. Box 5.2 provides two examples of such finger-based calculation methods. An elaborate use of body parts for counting has also been observed in remote cultures, such as the Oksapmin and Torres tribes from Papua New Guinea, who have had little contact with occidental culture (Butterworth, 1999; Lancy, 1978; Saxe, 1982). In North America, it has been estimated that

Box 5.2 Arithmetic and Algebra On Our Fingers

There are dozens of finger-based calculation methods, all invented to help solving arithmetical problems with reduced time and effort. Some are rudimentary, others quite complex; all are mathematically correct. Here are some of them. (Fuller descriptions and other examples of such finger-based calculation methods can be found in Ifrah, 1981.)

Finger Multiplication

This very ancient method allows one to multiply any pair of numbers between 5 and 10 using the hands. For example, to multiply 6 by 7 (Box 5.2 Figure 5.1), lower as many fingers as there are extra-units compared to 5 on one hand for the first operand, and on the other hand for the second operand (i.e. $6-5=1$: lower 1 finger on the left hand; $7-5=2$: lower 2 fingers on the right hand). Then, multiply by 10 the total number of lowered fingers (i.e. $10\times 3=30$), multiply the number of raised fingers of each hand (i.e. $4\times 3=12$), and add the two answers (i.e. $30+12=42$). This method is based on the fact that the product of two numbers, say x and y, equals $10[(x-5)+(y-5)]+[5-(x-5)]\times[5-(y-5)]$. Similar methods can be used to multiply numbers between 10 and 15, 15 and 20, 20 and 25, etc.

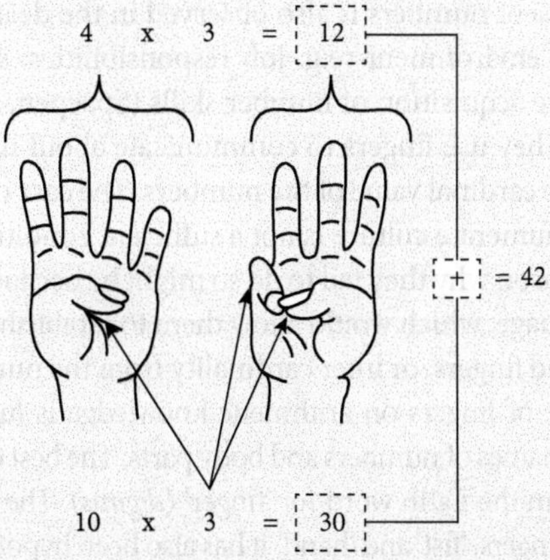

BOX FIGURE 5.2 A finger-based method to multiply any pair of numbers between 5 and 10, here to multiply 6 by 7 (see text for a description of the method).

The 9-Table Case

You think your fingers would not help much proving that '$9\times n=10(n-1)+(10-n)$'? If you think so, you are wrong! Your fingers will actually help you – in just one move and two glances. To prove that this equation holds true for '$n=5$', put your hands in front of you, then, starting from the left end, move the *5th* finger (i.e. your left thumb). Look to the left of this finger: the number of fingers gives you the decade value of the answer; look to the right: the number of fingers gives you the unit value of the answer. Put them together: 9×5 equals 45 (Box 5.2 Figure 5.2). Do it for any $1<n<10$ and see that it always works. By the way, you have just learned an error-free method to retrieve the 9-table on your fingers at any time …

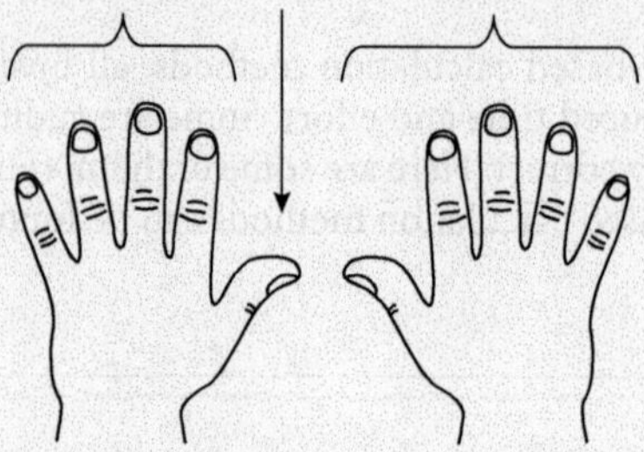

BOX FIGURE 5.3 A finger-based method to prove that $9 \times n = 10(n-1) + (10-n)$. For $n = 5$, $9 \times 5 = 10(5-1) + (10-5) = 45$ (see text for a description of the method).

around two-thirds of the several hundred Native American tribes used base-5 or base-10 systems derived from finger counting (Eels, 1913; cited in Boyer, 1968). More recent research indicates that Amazonian tribes with a limited number vocabulary (i.e. fewer than five words) only make rudimentary use of their fingers for counting (Gordon, 2004; Pica, Lemer, Izard, & Dehaene, 2004). In these populations, the failure to develop efficient finger-counting strategies could be related to the infrequent use of numbers or to a lack of education. An inappropriate use of fingers to represent numbers is also observed in the deaf population of Nicaragua, even though their direct environment (e.g. job responsibilities, social interactions, money exchanges, etc.) fosters the acquisition of number skills (Spaepen, Coppola, Spelke, Carey, & Goldin-Meadow, 2010). They use fingers to communicate about numbers, but their gestures are not consistent with the cardinal value of the numbers. The case of deaf people in Nicaragua indicates that living in a numerate culture is not a sufficient condition for developing flawless counting routines. The reason why they fail to do so might be because they do not benefit from a spoken (or signed) language, which would allow them to establish a one-to-one correspondence between numbers and fingers, or infer cardinality from the number of raised fingers.

The pervasive influence of fingers on arithmetic knowledge is further corroborated by diachronic links between the names of numbers and body parts. The best example is the English word 'digit', which originates from the Latin word for 'finger' (*digitus*). The word 'five' shares common ancestors with the words 'fingers', 'fist' and 'hand'. It has also been hypothesized the Indo-European root of the word 'four' is an expression meaning 'the sequence of fingertips', presumably from one hand, excluding the thumb (Menninger, 1969). Moreover, most cultures shared a base-5 or base-10 system, although the base-12 system occupies an important place in the story of mathematics (e.g. a dozen) and presents mathematical advantages (Ifrah, 1981). The preference for base-5 or base-10 systems underlines the influence of finger and hand configurations on the symbolic, numerical representations used in arithmetic tasks. A recent study capitalized on cultural differences to show that even the comparison of Arabic digits is biased by the inherent representation of a sub-base-5 system in the counting habits of Western participants (Domahs, Moeller, Huber, Willmes, & Nuerk, 2010) or deaf Korean signers (Domahs, Klein, Moeller, & Nuerk, 2012). Moreover, as recently suggested, the variety of finger-counting systems used throughout history and across cultures may have led to different representational systems as a function of the specific structure and properties of each system (e.g. base and sub-base values, extent and extendibility of the count, etc.; Bender & Beller, 2012).

In summary, this short overview has shown that fingers endowed human beings with unique facilities to keep track of a count or represent intermediary results during arithmetic problem solving. Because the domain of exact arithmetic is very recent in the evolution of the human species, the transmission of finger-based calculation abilities is likely to result from cultural conventions. However, the success of cultural conventions over time depends on their capacity to pass through the bottleneck of pre-existing representations in each individual's brain (De Cruz, 2006; Dehaene & Cohen, 2007). From this perspective, fingers constitute a useful means for acquiring and transmitting arithmetic knowledge, because they offer a permanent physical support for mental operations and rely on pre-existing representations in the sensory-motor system (Andres, Di Luca, & Pesenti, 2008a; Michaux, Pesenti, Badets, Di Luca, & Andres, 2010).

Finger Counting in Children

Whereas infants exhibit preverbal numerical abilities that allow them to approximate the result of arithmetic operations, the learning of a stable and flexible counting routine to perform exact calculation is a long process lasting four years, between the ages of two and six (Fuson, Richards, & Briars, 1982). Finger-counting strategies occur in the course of this process. Systematic observations taken from mothers' diaries indicate the use of fingers for counting is influenced by observation and imitation of others' behaviour (Fuson, 1988), but finger counting soon becomes a self-initiated strategy observed in most counting and arithmetic tasks, even when there is no explicit instruction to use it (Siegler & Shrager, 1984). Group studies show 2-year-old children satisfy the one-to-one correspondence principle in gestures prior to speech (Graham, 1999) and commit fewer violations when they perform the gestures themselves than when the gestures are performed by a puppet manipulated by the experimenter (Alibali & DiRusso, 1999). Between the ages of three and five, children use fingers in very different ways to keep track of the count, for example by lifting fingers sequentially or by repeatedly pressing the table (Fuson et al., 1982). In order to enumerate the result of large problems, such as 8 + 5, older children may also represent their count by raising five fingers successively while reciting the sequence from 8 to 13 (for other examples, see Box 5.3).

The development of finger-counting strategies may also contribute to the assimilation of the stable-order principle (Gallistel & Gelman, 1992; Gelman & Gallistel, 1978). Indeed, the omnipresence of finger-counting systems, relative to other iconic systems, is not just related to the fact we can carry our fingers around in full view, all the time (Wiese, 2003). Fingers also have the advantage of occupying fixed positions on the hand. This feature opens the way for the emergence of routines linking fingers to objects in a sequential, stable order. Although multiple combinations of finger movements could be used to count a set of objects, the order in which fingers are lifted or folded generally respects the relative positions of the four fingers on the hand. The emergence of this routine is a crucial step in the development of a child's arithmetic skills, because it allows the transfer of the stable-order principle to the sequence of counting words and it creates associations between finger configurations and cardinal values (Di Luca, Lefèvre, & Pesenti, 2010; Di Luca & Pesenti, 2008).

The influence of finger knowledge on the development of arithmetic skills is corroborated by longitudinal studies showing that the ability to discriminate fingers is a better predictor of mathematical achievement in 5- and 6-year-old children than standard developmental or visuospatial tests (Fayol, Barrouillet, & Marinthe, 1998; Noël, 2005;

Box 5.3 When the Verbal Chain Meets Finger Counting

The learning of the conventional sequence of number words can be divided into five levels (Fuson et al., 1982): (a) at the *string* level, number words are produced as a single expression and they are not understood by the child as separate words (e.g. 'one-two-three-four-five ...'); (b) at the *unbreakable chain* level, the mental representation of the sequence is split into single units that correspond to separate words in recitation (e.g. 'one, two, three, four, five ...'); (c) the *breakable chain* level indicates that children are able to start reciting the sequence from arbitrary entry points rather than always starting at the beginning (e.g. 'four, five, six, seven ...'); (d) at the *numerable chain* level, number words acquire a mathematical meaning and children get used to counting and keeping track of segments of connected words (e.g. the child is able to solve $5 + 3$ by starting to count with 5 and keeping track of the 3 following words: 'five ... six, seven, eight'); (e) the *bidirectional chain* level reflects a flexible use of number words in either direction (e.g. 'five, four, three, two, one').

Finger movements have been related to the procedures for keeping track of the counting words uttered while counting up or down. Because keeping track implies mathematical operations on segments of counting words, the contribution of finger movements is likely to occur at the numerable chain level. Small addition problems are usually solved by representing the operands separately on the left and right hands and moving each raised finger while reciting the counting words (Geary, 1994).

In 5- and 6-year-old children, finger-counting strategies are also observed during the resolution of subtraction problems. Children may perform subtractions by lifting a number of fingers corresponding to the larger operand and then bending a number of fingers corresponding to the smaller operand while counting down (Box 5.3 Figure 5.1). Alternatively, children may represent the smaller operand on raised fingers and count up while lifting the other fingers, until the number of fingers corresponding to the larger operand is reached.

Anecdotic reports reveal that fingers may even be used to keep track of intermediary results when multiplying numbers between 5 and 10 (Ifrah, 1981; see Box 5.2).

$5 - 1 = 4$

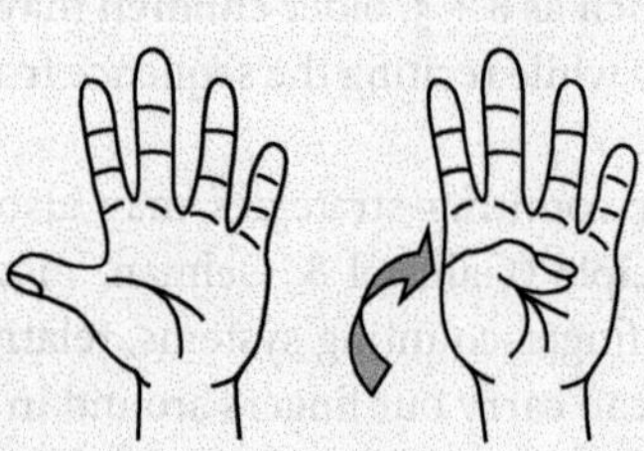

BOX FIGURE 5.4 Finger-based method to subtract 1 from 5 (see text for a description of the method).

Penner-Wilger, 2007). Moreover, the performance of 6-year-old children in finger discrimination is a good predictor of their numerical skills, but not of their reading skills one year later, meaning that finger knowledge specifically enriches mathematical competence (Noël, 2005). Two recent surveys show that finger gnosis helps to differentiate good and bad calculators in large samples of children aged between five and seven (Gracia-Baffaluy & Noël, 2008; Reeve & Humberstone, 2011), and between 10 and 11 years old (Costa, Silva, & Chagas, 2011). Children with normal finger gnosia use their fingers more frequently than those with poor finger gnosia, and are faster and more accurate in arithmetic tasks (Reeve & Humberstone, 2011). Moreover, this relationship remains significant when general intelligence and visuospatial working memory are taken into account (Costa et al., 2011). Subtraction is assumed to rely more heavily on finger counting in childhood because it benefits less than other single-digit operations from the direct retrieval of the answer in long-term memory (Fuson & Kwon, 1992) and because the answers to most basic problems are less than 10, which makes finger-based calculation strategies more efficient than for other operations. For example, in 9-year-old children, the percentage of problems solved by memory retrieval is only 19% for subtraction as opposed to 65% for addition (Barrouillet & Lépine, 2005; Barrouillet, Mignon, & Thevenot, 2008), whereas from the age of 10, most multiplication problems are solved by memory retrieval (Cooney, Swanson, & Ladd, 1988).

Moreover, an analysis of the errors made by 6- to 8-year-old children during mental arithmetic revealed that incorrect answers were most often five units off the correct answer, which is reminiscent of the use of a fully-opened hand to keep track of the count (Domahs, Krinzinger, & Willmes, 2008). For example, children often produce *6* rather than *11* in response to *18 – 7*, as if the mental representation of one hand was 'lost' during the count. This example illustrates how the use of fingers to keep track of the count progressively evolves towards the use of finger configurations to represent the cardinal value of intermediary results. These split-5 errors tend to decrease over time, from ages 6–9, but a transient increase is observed at the end of this period for simple addition problems. The occurrence of split-5 errors at this age seems to be sensitive to the automatic retrieval of competitive answers from long-term memory, as indicated by operation-related errors (e.g. *14* is given as an answer to *7 + 2*). In other words, the mental representation of finger patterns might play a critical role in the transition from calculation procedures to memory retrieval for simple problems. This could also explain why the correlation between finger gnosia and mental arithmetic does not always differ across operations, even in older children (Costa et al., 2011; Noël, 2005). Finally, the use of hands and fingers to keep track of calculation steps was also found to enhance arithmetic reasoning. For example, nine- to 10-year-old children learned better how to solve complex addition problems when they were trained to indicate their grouping strategy by making pointing gestures (Goldin-Meadow, Cook, & Mitchell, 2009).

In summary, current findings suggest finger movements may reflect one-to-one correspondence and stable-order principles before counting words take on a mathematical meaning. The time frame of the shift from finger to verbal counting may be extremely variable, especially assuming that the conceptual understanding of counting is primarily shaped by the child's experience. Children may rely on fingers in some contexts, whereas they may perform number word assignments without any digital support in others. Some other characteristics of fingers may justify their use, parallel to counting words, in arithmetic problem solving. Although fingers provide an iconic representation of numbers, their use is completely non-referential. It is remarkable that finger movements, although they usually serve to designate a specific object or individual, do not refer to object or people identity in the context of number processing. Moreover, the joint use of the fingers and hands may prompt the acquisition of the number base concept. Indeed, once children have discovered that the number 6, for example, can be represented with one hand raised and one finger lifted on the other hand, they acquire a concrete means of representing higher-order units while counting.

Influence of Finger Representation on Arithmetic in Adulthood

Although adults mostly perform numerical tasks mentally or using computing tools, it has been shown that the processes and/or representations underlying finger control continue interacting with number processing into adulthood. Initial evidence for this is provided by a study in which the performance of Italian adults in a choice reaction time (RT) task revealed a systematic association of the digits one to five with the fingers of the right hand and the digits six to 10 with the fingers of the left hand, in accordance with the counting strategy of the participants (Di Luca, Granà, Semenza, Seron, & Pesenti, 2006). This preferred association between small digits and the right hand, and large digits and the left hand was observed whether the choice RT task was performed palms up or palms down, indicating that space representations have little influence on finger-number mappings. Nevertheless, another study showed these mappings can be flexibly adjusted to task demands, as revealed by a simple RT task in which tactile stimulation of the thumb and little finger of the right hand was primed by small and large numbers respectively, regardless of palm orientation (Brozzoli, Ishihara, Göbel, Salemme, Rossetti, & Farnè, 2008). In order to disentangle the effects of palm orientation, allocentric spatial codes and counting habits, participants were asked to perform a choice RT number task, either with their right or left hand in a prone or supine posture (Riello & Rusconi, 2011). The results showed that the mapping of numbers onto a left–right representation of the fingers was observed only if this mapping preserved the direction of counting from thumb to little finger, that is when the right hand was in a prone posture or when the left hand was in supine posture. These results corroborate the view that, in adults, the mental frame of reference for number representation is indeed a posture-invariant representation of the finger sequence.

Another line of evidence can be found in electrophysiological studies showing increased corticospinal excitability (CSE) in the hand motor circuits during simple counting tasks. Because the tasks were performed with both hands at rest, the increase was attributed to the mental activation of finger representations while manipulating numbers (Andres et al., 2008b). This increase was found to be specific to hand muscles, since it was not observed in foot muscles (Andres, Seron, & Olivier, 2007). Moreover, a selective increase of CSE was observed in the right hand in response to numbers between one and five when compared to numbers between six and ten, in line with the use of this hand by all participants to this study to begin counting (Sato, Cattaneo, Rizzolatti, & Gallese, 2007). It should be noted that hand preference in counting habits has garnered more attention in recent research (Lindemann, Alipour, & Fischer, 2011). Individual preference for one counting direction was found to be much more variable than expected across countries, with people from Anglo-Saxon countries using mostly their left hand to begin counting, while people from the Middle East use their right hand. Mixed patterns were observed in Mediterranean and other European countries. In contrast to Westerners, who start with the thumb and end with the little finger of each hand successively, Iranians were found to count from the little finger to the thumb. Other results in a sample of more than 100 native French-speaking children and adults, with the same cultural and educational backgrounds, suggest that the direction of counting could also be influenced by handedness, with a trend for a higher proportion of left starters in left-handed rather than right-handed individuals (Sato & Lalain, 2008). The investigation of finger-number mappings in groups showing different counting habits might help to specify the embodied status of these mappings in future studies. Indeed, if number representations are constrained by finger-counting strategies, these representations should capture specific features of the finger movement sequence used by participants. Further research is required, however, to evidence the consistency of counting behaviours in repeated trials or in contexts where fingers are used in the absence of explicit instructions (e.g. count the number of letters between K and V in the alphabet). Moreover, several behavioural data suggest that the finger-number mappings observed in adults also correspond to 'finger-montring' configurations (i.e. finger configurations used to show cardinal values to someone else; Di Luca & Pesenti, 2008), and these might differ from finger-counting configurations (e.g. the index finger is mostly used to show number one whereas the thumb is the first finger lifted when counting for oneself).

The finding of number and finger interactions in adults highlights an intrinsic relationship between body knowledge and numerical cognition. The question arises as to whether this embodied representation of numbers influences the performance of adults in arithmetic tasks. In line with the performance of children, crossing a sub-base-5 (e.g. 4 + 3) or base-10 boundary (e.g. 8 + 3) delays the solving of addition problems in adults (Klein, Moeller, Willmes, Nuerk, & Domahs, 2011). Crossing the boundary between two decades takes longer presumably because it requires a carry operation, but the increase in RT observed while crossing a sub-base-5 boundary is difficult to explain, unless one considers that mental arithmetic is constrained by non-mathematical boundaries arising from finger counting or finger-based calculation strategies. This finding provides evidence that finger use is not just a transitory step in the development of arithmetic skills. Another inventive experiment used the so-called *response effect paradigm* to show the embodied status of the representation of arithmetic results (Badets, Pesenti, & Olivier, 2010). Addition problems were solved faster in blocks where the result (e.g. 2 + 2 = 4) triggered the display of a congruent finger configuration (i.e. four lifted fingers) than in blocks where the result was followed by the digital representation of an incorrect result (e.g. three lifted fingers), whereas no facilitation of arithmetic reasoning was observed at all when the correct result was represented by a set of rods. These behavioural data suggest that mental arithmetic induces concomitant activation of finger representations in adults, but they do not allow one to infer that finger knowledge is causally involved in arithmetic operations. A direct test of this assumption comes from dual-task experiments where participants had to answer arithmetic problems while moving their fingers. Michaux and colleagues (Michaux, Masson, Pesenti, & Andres, 2013) found that sequential movements from the thumb to the little finger slowed down the solving of subtraction and addition, but not multiplication problems, matched for difficulty. Such a selective interference was not observed when participants were asked to move their foot in successive locations from left to right. Since the rate of use of memory retrieval strategies is much higher in multiplication (97%) than in subtraction (57%) or addition (76%; Campbell & Xue, 2001), it has been proposed that finger movements interfere only with those operations less frequently solved through memory retrieval. This view was confirmed by another study where participants received explicit instructions to solve addition and subtraction problems either by retrieving the answer from long-term memory, working out the answer through problem decomposition or counting one by one (Imbo, Vandierendonck, & Fias, 2011). During arithmetic problem solving, participants' index finger was moved by the experimenter across four locations arranged in a square. The results showed that finger movements interfered with both arithmetic operations, but only when participants relied on counting to solve the problems. In this study, interference cannot be linked to the intention to act, since the participants' index finger was moved passively. The results, therefore, strengthen the view that, in adults, the processes and/or representations underlying finger movements are recruited to assist calculation procedures during arithmetic operations.

The strongest evidence for a causal relationship between finger and arithmetic knowledge comes from the study of brain-lesioned patients. In the first half of the 20th century, Gerstmann reported a series of patients with parietal damage who were unable to calculate, write by hand, discriminate the left and right sides of their body, name the fingers moved by the experimenter or move the fingers named by the experimenter (Gerstmann, 1930, 1940, 1957). In order to explain this cluster of symptoms, Gerstmann advanced the necessary role of hands and fingers in learning arithmetic, performing highly differentiated handwriting movements, or recognizing left and right. He argued all symptoms may originate from the disturbance of a single process underlying finger differentiation and their association indicates selective damage in the lower part of the left parietal lobe (see also Kinsbourne & Warrington, 1962). Further neuropsychological studies have replicated the pattern of deficits observed by Gerstmann, in the absence of any other cognitive deficit (Cipolotti, Butterworth, & Denes, 1991; Delazer & Benke, 1997; Mayer, Martory, Pegna, Landis, Delavelle, & Annoni, 1999; Mazzoni, Pardossi, Cantini, Giorgetti, & Arena, 1990; Tucha, Steup, Smely, & Lange, 1997; Turconi & Seron, 2002; Varney, 1984). An alternative hypothesis suggests that the cluster of symptoms might result from damage to a convergence zone in the white matter of the posterior parietal cortex (Rusconi, Pinel, Dehaene & Kleinschmidt, 2009a). Indeed, fMRI data gathered in six healthy participants showed no common region when overlaying the four activation maps measured during calculation, finger discrimination, left–right orientation and writing tasks, whereas diffusion tensor imaging (DTI) revealed that the fibre tracts reconstructed from the seed of each activation map converged to a white matter locus underlying the posterior parietal cortex (Rusconi, Pinel, Eger, Lebihan, Thirion, Dehaene & Kleinschmidt, 2009b). Although this finding questions the functional relevance of the four symptoms taken as a whole, it is important to stress that fMRI activations depend highly on the sensitivity of the task used to assess symptom-related abilities, and that the specific overlap between areas involved in finger discrimination and arithmetic was not investigated. Indeed, the use of electrical stimulation to map patients' posterior parietal cortex during awake open-brain surgery revealed that acalculia and finger agnosia could occur in isolation of other symptoms (Morris, Luders, Lesser, Dinner, & Hahn, 1984; Roux, Boetto, Sacko, Chollet, & Trémoulet, 2003). As we will see in the next section, other brain imaging and transcranial magnetic stimulation (TMS) studies suggest that the neural networks underlying calculation and finger discrimination do overlap in the human brain.

In summary, the occurrence of deficits in both number processing and finger discrimination after brain damage supports the view finger counting and finger-based calculation procedures form the basis of arithmetic performance in adults. A more extensive examination of these deficits is now required, independently of their relationship to Gerstmann syndrome, to clarify the specific shared processes.

Neural Substrate of Finger Representation and Mental Arithmetic

The posterior parietal cortex is subdivided in the inferior and superior parietal lobules by the intraparietal sulcus (IPS), a sulcus orientated in an antero-posterior direction with a downward convexity. Electrophysiological recordings in the posterior parietal cortex of the monkey indicate that neurons coding an upcoming movement are capable of keeping track of the number of times this movement has been repeated previously (Sawamura, Shima, & Tanji, 2002). Neurons showing numerosity-selective activity ahead of movement execution were found in the ascending branch of the IPS starting at the junction between the anterior end of the IPS and the inferior part of the post-central sulcus. They were found to discharge in response to a preferred ordinal position in the motor sequence, with a selective increase when compared with previous and following positions. Interestingly, transient chemical inactivation of these numerosity-selective neurons using muscimol, a γ-aminobutyric acid receptor agonist, showed they were involved in numerical aspects of an action (i.e. how many times an action had to be performed in a specific learning context), while leaving the motor aspects of this action unaffected (Sawamura, Shima, & Tanji, 2010). Anterior to the post-central sulcus, in the primary somatosensory cortex, numerosity-selective neurons were relatively rare. It is worth noting that visual numerosity-selective neurons were also found, in the fundus of the IPS, in a slightly more posterior region than sensory-motor numerosity-selective neurons (Diester & Nieder, 2007; Nieder, 2006; Nieder & Miller, 2004). This anterior-to-posterior gradient fits with the functional organization of sensory-motor and visuospatial functions in monkey IPS (Grefkes & Fink, 2005; Lewis & Van Essen, 2000). These findings suggest the sensory-motor system of non-human primates is endowed with precursor abilities for keeping track of incremental changes useful during arithmetic operations, and these abilities seem to be distinct from those involved in visuospatial processing.

In humans, several brain imaging studies showed that number processing and calculation are associated with increased activity in the premotor and posterior parietal cortex (Andres, Pelgrims, Michaux, Olivier, & Pesenti, 2011; Dehaene, Spelke, Pinel, Stanescu, & Tsivkin, 1999; Ischebeck, Zamarian, Siedentopf, Koppelstätter, Benke, Felber, et al., 2006; Pesenti, Thioux, Seron, & De Volder, 2000; Piazza, Mechelli, Price, & Butterworth, 2006; Pinel, Dehaene, Riviere, & Lebihan, 2001; Simon, Mangin, Cohen, Le Bihan, & Dehaene, 2002; Thioux, Pesenti, Costes, De Volder, & Seron, 2005; Zago, Pesenti, Mellet, Crivello, Mazoyer, & Tzourio-Mazoyer, 2001), two regions known for their contribution to finger movement control (Pelgrims et al., 2009; Hamilton & Grafton, 2009; de Lange et al., 2005; Haaland et al., 2004; Harrington et al., 2000). Interestingly, these regions were used not only when performing finger movements but also when imaging the same movements, which underlines their potential role in tasks where finger representations are manipulated without overt motor output (de Lange, Hagoort, & Toni, 2005; Pelgrims, Andres, & Olivier, 2009; Pelgrims, Michaux, Olivier, & Andres, 2011; Willems, Toni, Hagoort, & Casasanto, 2009). The mapping of parietal functions in humans conforms with monkey data to reveal a specific organization of the IPS, with calculation areas in the anterior part of the IPS dedicated to sensory-motor functions and the posterior part dedicated to visuospatial functions (Simon et al., 2002). A similar gradient was observed along the inferior-to-superior axis in the ventral (PMv) and dorsal premotor cortex (PMd; Simon, Kherif, Flandin, Poline, Riviere, Mangin et al., 2004). No overlap between sensory-motor and calculation areas was observed in these studies, but the motor tasks involved grasping and pointing movements, which were unrelated to the use of fingers for calculation.

In order to isolate the brain regions involved in digital representations of numbers, adult participants were asked to identify a number from films showing finger configurations associated with counting or lip movements associated with number naming (Thompson, Abbott, Wheaton, Syngeniotis, & Puce, 2004). Increased activity was found along the left IPS in response to finger movements and along the superior temporal sulcus in response to lip movements, with common activation in the right IPS. These results indicate that the left IPS is specifically involved in decoding number from finger movements. In another fMRI study, 8-year-old children and adults had to discriminate two hand pictures as a function of palm orientation, thumb colour or number of raised fingers in each picture (Kaufmann, Vogel, Wood, Kremser, Schocke, Zimmerhackl, et al., 2008). Finger counting led to a selective increase of activity in the left parietal lobe, extending from the IPS to the post-central gyrus. When compared with adults, children also showed higher activity in the left and right inferior parietal lobules during finger counting. Inferior parietal activations were centred on the supramarginal gyrus, and extended to the post- and pre-central gyri in the right hemisphere. These fMRI data suggest that number processing activates the brain circuits underlying the observation of finger configurations, especially in the left hemisphere. However, because numerical judgements were performed on body images, these results cannot be taken as evidence that finger-related circuits are spontaneously engaged during counting or calculation.

More recent fMRI studies used distinct tasks to assess the overlap between the brain areas involved in numerical and motor tasks. The similarities and differences between the neural networks involved in number comparison or addition and those involved in saccades or finger movements were first studied in nine-year-old children (Krinzinger, Koten, Horoufchin, Kohn, Arndt, Sahr, et al., 2011). This study showed adding numbers and making saccades led to overlapping activations in the posterior part of the IPS, whereas common activations were found in the anterior part of the IPS during addition and finger movements. Other parietal and premotor areas underlying finger movements were also activated during number comparison and addition, with a larger increase of activity for non-symbolic than symbolic stimuli. These results led the authors to conclude that, in children, this parieto-premotor network could be particularly important to matching non-symbolic and symbolic numerical knowledge via somatosensory integration.

In adults, fMRI was used to reveal the brain regions showing overlapping activity during mental arithmetic and finger judgements that required participants to discriminate between flexed and extended fingers without looking at the imposed hand posture (see Figure 5.1(a) and 5.1(b); Andres, Michaux, & Pesenti, 2012). Solving basic subtraction and multiplication problems was found to increase activity bilaterally in the anterior part of the IPS and in the superior parietal lobule, with additional activations in the middle and superior temporal gyri during multiplication only (see Figure 5.1(c)). Finger discrimination was associated with increased activity in a bilateral occipito-parieto-precentral network extending from the extrastriate body area to the primary somatosensory (S1) and motor (M1) cortices contralateral to the tested hand. A conjunction analysis showed common areas of activity for mental arithmetic and finger representation in the IPS and superior parietal lobule bilaterally. In order to ensure these results were not due to spatial blurring, smoothing procedures or distinct, but intermingled neural networks, voxel-wise correlations were computed between the patterns of decreased and increased activity observed in each region during finger discrimination and mental arithmetic tasks. In all areas, but the right hIPS, the patterns of activity observed during finger discrimination and arithmetic were highly correlated, meaning that brain activity was distributed similarly across voxels during these tasks. The finding of larger between-task correlations for the left than the right hIPS conforms with other fMRI results to suggest a left-hemispheric specialization for precise numerical

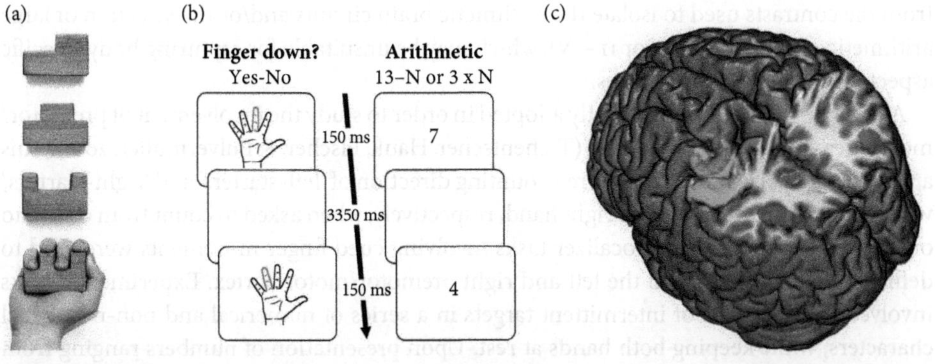

FIGURE 5.1 (a) In this fMRI experiment a wooden block was placed in each hand of the participant, with half of the fingers extended over the bumps of the block and the other half flexed in the holes. (b) In the finger discrimination task, participants viewed hand drawings with one finger coloured red. They had to decide whether their matching finger was flexed or not, without visual feedback about hand posture. In arithmetic tasks, they had to multiply the Arabic digit by three or four, or subtract it from 11 or 13. (c) Arithmetic tasks induced increased activity in the same intraparietal areas (IPS) as those involved in finger discrimination (red). A selective increase of activity was found in the middle and superior temporal gyri during multiplication only (blue). No overlap was observed in the primary motor (M1) and somatosensory (S1) cortex contralateral to the tested hand (green). Another fMRI

(Data from Andres, M., Di Luca, S. & Pesenti, M., Finger Counting: the Missing Tool? *Behavioral and Brain Sciences*, 31(6), pp. 642–643, 2012)

representations during the development of finger-counting abilities (Andres et al., 2008a; Andres, Seron, & Olivier, 2005; Piazza et al., 2006; Piazza, Pinel, Le Bihan, & Dehaene, 2007). Pattern similarity was more important for subtraction than multiplication, in line with behavioural results indicating that interactions between finger and number representations are modulated by the involvement of retrieval strategies in arithmetic problem solving (Imbo, Vandierendonck, & Fias, 2011; Campbell & Xue, 2001; Michaux et al., 2013). Since neural overlap was found in those parietal areas with a contribution to finger discrimination not specific to the left or right hand, it is unlikely that mental arithmetic involves somatic representations. In the finger discrimination task, the absence of visual feedback put strong demands on the ability to represent the relative position of the fingers, as evidenced by increased response latencies for the middle fingers, which have more neighbours than the thumb or little finger. The same pattern of activation was observed in other fMRI studies comparing complex and simple finger movements (Haaland, Elsinger, Mayer, Durgerian, & Rao, 2004; Harrington, Rao, Haaland, Bobholz, Mayer, Binder, et al., 2000). These studies showed increased activity along the IPS and superior parietal lobule as a function of the number of transitions between fingers, irrespective of handedness, whereas finger transitions did not affect the activity observed in the M1 and S1 contralateral to the tested hand. We therefore suggest that mental arithmetic uses the parietal areas underlying finger discrimination to keep track of incremental changes while calculating. The exact nature of the shared mechanisms requires further research, especially with regard to the respective role of motor and spatial mechanisms. Since increased activity in premotor/motor areas was previously observed in several other studies, its absence in this study is likely to originate from the contrasts used to isolate the arithmetic brain circuits and/or the selection of large arithmetic problems (13 – *N* or 11 – *N*), which may be unsuitable for capturing body-specific aspects of calculation procedures.

A different approach was recently adopted in order to study the involvement of premotor/motor areas in number processing (Tschentscher, Hauk, Fischer, & Pulvermüller, 2012). This approach capitalized on the preferred counting direction of 'left-starters' and 'right-starters,' who start moving their left and right hand, respectively, when asked to count from one to 10 on their fingers. Functional localizer tasks involving cued finger movements were used to define regions of interest in the left and right premotor/motor cortex. Experimental tasks involved the detection of intermittent targets in a series of numerical and non-numerical characters, while keeping both hands at rest. Upon presentation of numbers ranging from one to five, left-starters showed a larger increase of the blood-oxygen-level-dependent (BOLD) signal than right-starters in the right premotor/motor cortex, suggesting that the preferred starting hand influenced the activation induced by small numbers in the contralateral hemisphere. No difference was observed for numbers ranging from six to 10. Other studies showed that cognitive tasks assumed to involve covert hand movements, such as motor imagery or action verb understanding (Willems et al., 2009), are indeed associated with increased activity in the motor cortex contralateral to the hand preferentially used by participants to perform the related action. This paradigm could, therefore, be exploited to investigate the continuing influence of finger counting and finger-related calculation procedures in adults performing mental arithmetic tasks.

TMS studies also revealed a contribution of parietal and premotor areas to arithmetic and counting tasks. TMS over the inferior parietal cortex disrupted finger movements in response to tactile stimulation of the same fingers on the other hand and also increased response times for numerical magnitude judgments (Rusconi, Walsh, & Butterworth, 2005). TMS over the inferior part of the PMv delayed overt and covert counting of the number of syllables in a word (Aziz-Zadeh, Cattaneo, Rochat, & Rizzolatti, 2005), whereas counting large – but not small – collections of visual and auditory stimuli was biased after TMS over the superior part of the PMv (Kansaku, Carver, Johnson, Matsuda, Sadato, & Hallett, 2007). However, further TMS studies are needed, to disentangle the role of the aforementioned parietal and premotor areas in motor control from a potential role in spatial cognition, language, or working memory in relation to calculation procedures.

Conclusion

Human beings are constantly required to process the world numerically and perform computations to adapt their behaviour. Consequently, they have developed various calculation strategies, some of which are based on specific manipulation of the fingers. Among the various finger-number interactions reported so far, we focused in this chapter on those underlying simple arithmetic operations. We have presented anthropological and developmental data showing the long-term role of such finger-based calculation strategies in the acquisition and development of numerical and arithmetic knowledge throughout human history and individual childhood. We have also shown that this finger-number link is still observed in educated adults, with interactions between number processing and control of finger movement reported in various numerical and arithmetical tasks. Finally, we have reviewed lesional, electrophysiological, and brain imaging studies demonstrating some commonalities between the brain substrates for finger and number representations. In human adults, areas around the IPS not only support both finger representations and calculation procedures, but also show correlated patterns of activity distribution in finger discrimination and basic arithmetic problem solving.

Where does this finger-number link in mental arithmetic originate? We argue that the way numerical concepts and calculation are physically expressed by raising fingers, while counting or calculating leads to embodied representations of numbers and calculation procedures in the adult brain. As we have shown above, the fixed order of fingers on the hands provides human beings with unique facilities to increment numerical changes or represent a cardinal value while solving arithmetic problems. For this reason, it is sensible to assume that motor representations end up supporting mental arithmetic during ontogenetic development and continue to assist some arithmetic procedures in adults, even when overt finger-based calculation strategies are no longer used in most daily situations. In his authoritative Mathematical Brain, Butterworth (1999) hypothesized that 'without the ability to attach number representations to the neural representations of fingers and hands […] numbers themselves will never have a normal representation in the brain' (p. 250). He made this risky prediction on the basis of the kind of anthropological, developmental and neuropsychological data we reviewed in this chapter. More than a decade of novel behavioural and brain-imaging studies has now provided enough evidence to support the idea that finger and number representations are not only functionally, but also neuro-anatomically linked. However, the strongest part of the prediction—that number representations do need finger representation to properly develop—has yet to be demonstrated. We are confident this next step will be taken soon.

Acknowledgments

Some of the studies reported in this chapter were supported by grant FSR09-CWS/09.285 from the Fonds Spéciaux de Recherche of the Université catholique de Louvain (Belgium). MA was a post-doctoral researcher and MP is a research associate at the National Fund for Scientific Research (Belgium). We thank Nicolas Michaux (Université catholique de Louvain) for his comments and his help preparing the figures.

References

Alibali, M.W., & DiRusso, A.A. (1999). The function of gesture in learning to count: more than keeping track. Cognitive Development, 14, 37–56.

Anderson, M.L. (2010). Neural reuse: a fundamental organizational principle of the brain. Behavioral and Brain Sciences, 33(4), 245–266.

Andres, M., Davare, M., Pesenti, M., Olivier, E., & Seron, X. (2004). Number magnitude and grip aperture interaction. NeuroReport, 15(18), 2773–2777.

Andres, M., Di Luca, S., & Pesenti, M. (2008a). Finger counting: the missing tool? Behavioral and Brain Sciences, 31(6), 642–643.

Andres, M., Michaux, N., & Pesenti, M. (2012). Common substrate for mental arithmetic and finger representation in the parietal cortex. NeuroImage, 62, 1520–1528.

Andres, M., Olivier, E., & Badets, A. (2008b). Actions, words, and numbers. Current Directions in Psychological Science, 17(5), 313.

Andres, M., Ostry, D. J., Nicol, F., & Paus, T. (2008c). Time course of number magnitude interference during grasping. Cortex, 44(4), 414–419.

Andres, M., Pelgrims, B., Michaux, N., Olivier, E., & Pesenti, M. (2011). Role of distinct parietal areas in arithmetic: an fMRI-guided TMS study. NeuroImage, 54(4), 3048–3056.

Andres, M., Seron, X., & Olivier, E. (2005). Hemispheric lateralization of number comparison. Brain Research Cognitive Brain Research, 25(1), 283–290.

Andres, M., Seron, X., & Olivier, E. (2007). Contribution of hand motor circuits to counting. Journal of Cognitive Neuroscience, 19(4), 563–576.

Aziz-Zadeh, L., Cattaneo, L., Rochat, M., & Rizzolatti, G. (2005). Covert speech arrest induced by rTMS over both motor and nonmotor left hemisphere frontal sites. Journal of Cognitive Neuroscience, 17(6), 928–938.

Badets, A., Andres, M., Di Luca, S., & Pesenti, M. (2007). Number magnitude potentiates action judgements. Experimental Brain Research, 180(3), 525–534.

Badets, A., Bouquet, C., Ric, F., & Pesenti, M. (2012). Number-generation bias after movement perception. Experimental Brain Research, 221, 43–49.

Badets, A., & Pesenti, M. (2010). Creating number semantics through finger movement perception. Cognition, 115(1), 46–53.

Badets, A., & Pesenti, M. (2011). Finger-number interaction: an ideomotor account. Experimental Psychology, 58(4), 287–292.

Badets, A., Pesenti, M., & Olivier, E. (2010). Response-effect compatibility of finger-numeral configurations in arithmetical context. Quarterly Journal of Experimental Psychology, 63(1), 16–22.

Barbaresi, W.J., Katusic, S.K., Colligan, R.C., Weaver, A.L., & Jacobsen, S.J. (2005). Math learning disorder: incidence in a population-based birth cohort, 1976–82, Rochester, Minn. Ambulatory Pedriatrics, 5(5), 281–289.

Barrouillet, P., & Lépine, R. (2005). Working memory and children's use of retrieval to solve addition problems. Journal of Experimental Child Psychology, 91, 183–204.

Barrouillet, P., Mignon, M., & Thevenot, C. (2008). Strategies in subtraction problem solving in children. Journal of Experimental Child Psychology, 99(4), 233–251.

Barsalou, L.W. (1999) Perceptual symbol systems. Behavioral and Brain Sciences, 22, 577–660.

Bender, A., & Beller, S., (2012). Nature and culture of finger counting: diversity and representational effects of an embodied cognitive tool. Cognition, 124, 156–182.

Boyer, C.B. (1968). A History of Mathematics. New York: John Wiley and Sons.

Brozzoli, C., Ishihara, M., Göbel, S., Salemme, R., Rossetti, Y., & Farnè, A. (2008). Touch perception reveals the dominance of spatial over digital representation of numbers. Proceedings of the National Academy of Sciences, 105(14), 5644.

Brun, A., & Paster, J-M. (2010). Les compétences en mathématiques des élèves en fin d'école primaire. Report #10.17 of the Direction de l'Evaluation, de la Prospective et de la Performance, French National Education Ministry (pp. 6).

Butterworth, B. (1999). The Mathematical Brain. London: MacMillan.

Campbell, J.I.D., & Xue, Q. (2001). Cognitive arithmetic across cultures. Journal of Experimental Psychology General, 130(2), 299–315.

Chiou, R.Y., Chang, E.C., Tzeng, O.J., & Wu, D.H. (2009). The common magnitude code underlying numerical and size processing for action but not for perception. Experimental Brain Research, 194, 553–562.

Chiou, R.Y-C., Wu, D.H., Tzeng, O.J-L., Hung, D.L., & Chang, E.C. (2012). Relative size of numerical magnitude induces a size-contrast effect on the grip scaling of reach-to-grasp movements. Cortex, 48(8), 1043–1051.

Cipolotti, L., Butterworth, B., & Denes, G. (1991). A specific deficit for numbers in a case of dense acalculia. Brain, 114(6), 2619–2637.

Cooney, J.B., Swanson, H.L., & Ladd, S.F. (1988). Acquisition of mental multiplication skill: evidence for the transition between counting and retrieval strategies. Cognition & Instruction, 5, 323–345.
Costa, A., Silva, J., & Chagas, P. (2011). A hand full of numbers: a role for offloading in arithmetics learning? Frontiers in Psychology, 2, Article 368.
De Cruz, H. (2006). Why are some numerical concepts more successful than others? An evolutionary perspective on the history of number concepts. Evolution and Human Behavior, 27(4), 306–323.
de Lange, F., Hagoort, P., & Toni, I. (2005). Neural topography and content of movement representations. Journal of Cognitive Neuroscience, 17(1), 97–112.
Dehaene, S., & Cohen, L. (2007). Cultural recycling of cortical maps. Neuron, 56(2), 384–398.
Dehaene, S., Spelke, E., Pinel, P., Stanescu, R., & Tsivkin, S. (1999). Sources of mathematical thinking: behavioral and brain-imaging evidence. Science, 284(5416), 970–974.
Delazer, M., & Benke, T. (1997). Arithmetic facts without meaning. Cortex, 33(4), 697–710.
Di Luca, S., Granà, A., Semenza, C., Seron, X., & Pesenti, M. (2006). Finger-digit compatibility in arabic numeral processing. Quarterly Journal of Experimental Psychology, 59(9), 1648–1663.
Di Luca, S., Lefèvre, N., & Pesenti, M. (2010). Place and summation coding for canonical and non-canonical finger numeral representations. Cognition, 117(1), 95–100.
Di Luca, S., & Pesenti, M. (2008). Masked priming effect with canonical finger numeral configurations. Experimental Brain Research, 185(1), 27–39.
Diester, I., & Nieder, A. (2007). Semantic associations between signs and numerical categories in the prefrontal cortex. PLoS Biology, 5(11), e294.
Domahs, F., Klein, E., Moeller, K., & Nuerk, H. (2012). Multimodal semantic quantity representations: further evidence from Korean sign language. Frontiers in Psychology, 2, Article 389.
Domahs, F., Krinzinger, H., & Willmes, K. (2008). Mind the gap between both hands: evidence for internal finger-based number representations in children's mental calculation. Cortex, 44(4), 359–367.
Domahs, F., Moeller, K., Huber, S., Willmes, K., & Nuerk, H.-C. (2010). Embodied numerosity: implicit hand-based representations influence symbolic number processing across cultures. Cognition, 1–16.
Eels, W.C. (1913). Number systems of North America Indians. American Mathematical Monthly, 20, 293.
Fayol, M., Barrouillet, P., & Marinthe, C. (1998). Predicting arithmetical achievement from neuro-psychological performance: a longitudinal study. Cognition, 68, B63–70.
Fuson, K.C. (1988). Children's Counting and Concepts of Number. New York: Springer-Verlag.
Fuson, K.C., & Kwon, Y. (1992). Learning addition and subtraction: effects of number word and other cultural tools. In J. Bideau, C. Meljac, & J. P. Fisher (eds), Pathways to Number (pp. 351–374). Hillsdale, NJ: Erlbaum.
Fuson, K.C., Richards, J., & Briars, D.J. (1982). The acquisition and elaboration of the number word sequence. In C. J. Brainerd (ed.), Children's Logical and Mathematical Cognition (pp. 33–92). New York: Springer-Verlag.
Gallese, V., & Lakoff, G. (2005). The brain's concepts: the role of the sensory-motor system in conceptual knowledge. Cognitive Neuropsychology, 22, 455–479.
Gallistel, C.R., & Gelman, R. (1992). Preverbal and verbal counting and computation. Cognition, 44(1–2), 43–74.
Geary, D.C. (1994). Developing arithmetical skills. In D.C. Geary (ed.), Children's mathematical development: Research and practical applications. Washington, DC: American Psychological Association.
Gelman, R., & Gallistel, C.R. (1978). The Child's Understanding of Number. Cambridge, MA: Harvard University Press.
Gerstmann, J. (1930). Zur Symptomatologie Der Hirnläsionen Im Übergangsgebiet Der Unteren Parietal-Und Mittleren Occipitalwindung. Nervenarzt, 3, 691–695.
Gerstmann, J. (1940). Syndrome of finger agnosia, disorientation for right and left, agraphia, acalculia. Archives of Neurology and Psychology, 44, 398–408.
Gerstmann, J. (1957). Some notes on the Gerstmann syndrome. Neurology, 7(12), 866–869.
Glenberg, A.M. (1997). What memory is for. Behavioral & Brain Sciences, 20, 1–55.
Goldin-Meadow, S., Cook, S.W., & Mitchell, Z.A. (2009). Gesturing gives children new ideas about math. Psychological Science, 20(3), 267–272.

Gordon, P. (2004). Numerical cognition without words: evidence from Amazonia. Science, 306(5695), 496–499.

Gracia-Baffaluy, M., & Noël, M-P. (2008). Does finger training increase young children's numerical performance? Cortex, 44(4), 368–375.

Graham, T.A. (1999). The role of gesture in children's learning to count. Journal of Experimental Child Psychology, 74, 333–355.

Grefkes, C., & Fink, G.R. (2005). The functional organization of the intraparietal sulcus in humans and monkeys. Journal of Anatomy, 207(1), 3–17.

Haaland, K., Elsinger, C., Mayer, A., Durgerian, S., & Rao, S. (2004). Motor sequence complexity and performing hand produce differential patterns of hemispheric lateralization. Journal of Cognitive Neuroscience, 16(4), 621–636.

Hamilton, A., & Grafton, S. (2009). Repetition suppression for performed hand gestures revealed by fMRI. Human Brain Mapping, 30, 2898–2906.

Harrington, D., Rao, S., Haaland, K., Bobholz, J., Mayer, A., Binder, J., & Cox, R. (2000). Specialized neural systems underlying representations of sequential movements. Journal of Cognitive Neuroscience, 12(1), 56–77.

Ifrah, G. (1981). Histoire Universelle Des Chiffres. Paris: Robert Laffont.

Imbo, I., Vandierendonck, A., & Fias, W. (2011). Passive hand movements disrupt adults' counting strategies. Frontiers in Psychology, 2, Article 201.

Ischebeck, A., Zamarian, L., Siedentopf, C., Koppelstätter, F., Benke, T., Felber, S., & Delazer, M. (2006). How specifically do we learn? Imaging the learning of multiplication and subtraction. NeuroImage, 30(4), 1365–1375.

Kansaku, K., Carver, B., Johnson, A., Matsuda, K., Sadato, N., & Hallett, M. (2007). The Role of the human ventral premotor cortex in counting successive stimuli. Experimental Brain Research, 178(3), 339–350.

Kaufmann, L., Vogel, S., Wood, G., Kremser, C., Schocke, M., Zimmerhackl, L., & Koten, J. (2008). A developmental fMRI study of nonsymbolic numerical and spatial processing. Cortex, 44(4), 376–385.

Kinsbourne, M., & Warrington, E.K. (1962). A study of finger agnosia. Brain, 85, 47–66.

Klein, E., Moeller, K., Willmes, K., Nuerk, H-C., & Domahs, F. (2011). The Influence of implicit hand-based representations on mental arithmetic. Frontiers in Psychology, 2, Article 197.

Krinzinger, H., Koten, J.W., Horoufchin, H., Kohn, N., Arndt, D., Sahr, K., Konrad, K., & Willmes, K. (2011). The role of finger representations and saccades for number processing: an fMRI study in children. Frontiers in Psychology, 2, Article 373.

Lakoff, G., & Johnson, M. (1999). Philosophy in the Flesh: The Embodied Mind and its Challenge to Western Thought. NY: Basic Books.

Lancy, D.F. (1978). Cognitive testing in the Indigenous Mathematics Project. Papua New Guinean Journal of Education, 14, 114–142.

Lewis, J., & Van Essen, D. (2000). Corticocortical connections of visual, sensorimotor, and multimodal processing areas in the parietal lobe of the macaque monkey. Journal of Comparative Neurology, 428(1), 112–137.

Lindemann, O., Abolafia, J.M., Girardi, G., & Bekkering, H. (2007). Getting a grip on numbers: numerical magnitude priming in object grasping. Journal of Experimental Psychology: Human Perception and Performance, 33(6), 1400–1409.

Lindemann, O., Alipour, A., & Fischer, M.H. (2011). Finger counting habits in middle-eastern and western individuals: an on-line survey. Journal of Cross-Cultural Psychology, 42, 566–578.

Mahon, B.Z., & Caramazza, A. (2009). Concepts and categories: a cognitive neuropsychological perspective. Annual Review of Psychology, 60, 27–51.

Mayer, E., Martory, M., Pegna, A.J., Landis, T., Delavelle, J., & Annoni, J. (1999). A pure case of Gerstmann syndrome with a subangular lesion. Brain, 122, 1107–1120.

Martin, A. (2007). The representation of object concepts in the brain. Annual Review of Psychology, 58, 25–45.

Mazzoni, M., Pardossi, L., Cantini, R., Giorgetti, V., & Arena, R. (1990). Gerstmann syndrome: a case report. Cortex, 26(3), 459–467.
Menninger, K. (1969). Number Words and Number Symbols. Cambridge, MA: MIT Press.
Michaux, N., Masson, N., Pesenti, M., & Andres, M. (2013). Selective interference of finger movements on basic addition and subtraction problem solving. Experimental Psychology, 60(3), 197–205.
Michaux, N., Pesenti, M., Badets, A., Di Luca, S., & Andres, M. (2010). Let us redeploy attention to sensorimotor experience. Behavioral and Brain Sciences, 33(4), 283–284.
Moretto, G., & di Pellegrino, G. (2008). Grasping numbers. Experimental Brain Research, 188(4), 505–515.
Morris, H.H., Luders, H., Lesser, R.P., Dinner, D.S., & Hahn, J. (1984). Transient neuropsychological abnormalities (including Gerstmann's syndrome) during cortical stimulation. Neurology, 34(7), 877–883.
Nieder, A. (2006). Temporal and spatial enumeration processes in the primate parietal cortex. Science, 313(5792), 1431–1435.
Nieder, A., & Miller, E. (2004). A parieto-frontal network for visual numerical information in the monkey. Proceedings of the National Academy of Sciences, USA, 101(19), 7457–7462.
Noël, M-P. (2005). Finger gnosia: a predictor of numerical abilities in children? Child Neuropsychology, 11(5), 413–430.
Pelgrims, B., Andres, M., & Olivier, E. (2009). Double dissociation between motor and visual imagery in the posterior parietal cortex. Cerebral Cortex, 19(10), 2298–2307.
Pelgrims, B., Michaux, N., Olivier, E., & Andres, M. (2011). Contribution of the primary motor cortex to motor imagery: a subthreshold TMS study. Human Brain Mapping, 32(9), 1471–82.
Penner-Wilger, M. (2007). The foundations of numeracy: subitizing, finger gnosia, and fine motor ability. Poster presented at the 29th Annual Cognitive Science Society, Nashville, TN, August 1–4, 2007.
Pesenti, M., Thioux, M., Seron, X., & De Volder, A. (2000). Neuroanatomical substrates of arabic number processing, numerical comparison, and simple addition: a PET study. Journal of Cognitive Neuroscience, 12(3), 461–479.
Piazza, M., Mechelli, A., Price, C., & Butterworth, B. (2006). Exact and approximate judgements of visual and auditory numerosity: an fMRI study. Brain Research, 1106(1), 177–188.
Piazza, M., Pinel, P., Le Bihan, D., & Dehaene, S. (2007). A magnitude code common to numerosities and number symbols in human intraparietal cortex. Neuron, 53(2), 293–305.
Pica, P., Lemer, C., Izard, V., & Dehaene, S. (2004). Exact and approximate arithmetic in an Amazonian indigene group. Science, 306(5695), 499.
Pinel, P., Dehaene, S., Riviere, D., & Lebihan, D. (2001). Modulation of Parietal activation by semantic distance in a number comparison task. NeuroImage, 14(5), 1013–1026.
Ranzini, M., Lugli, L., Anelli, F., Carbone, R., Nicoletti, R., & Borghi, A.M. (2011). Graspable objects shape number processing. Frontiers in Human Neuroscience, 5, Article 147.
Reeve, R., & Humberstone, J. (2011). Five-to 7-year-olds' finger gnosia and calculation abilities. Frontiers in Psychology, 2, Article 359.
Riello, M., & Rusconi, E. (2011). Unimanual SNARC effect: hand matters. Frontiers in Psychology, 2, Article 372.
Roux, F.-E., Boetto, S., Sacko, O., Chollet, F., & Trémoulet, M. (2003). Writing, calculating, and finger recognition in the region of the angular gyrus: a cortical stimulation study of Gerstmann syndrome. Journal of Neurosurgery, 99, 716–727.
Rusconi, E., Pinel, P., Dehaene, S., & Kleinschmidt, A. (2009a). The enigma of Gerstmann's syndrome revisited: a telling tale of the vicissitudes of neuropsychology. Brain, 1–13.
Rusconi, E., Pinel, P., Eger, E., Lebihan, D., Thirion, B., Dehaene, S., & Kleinschmidt, A. (2009b). A disconnection account of Gerstmann syndrome: functional neuroanatomy evidence. Annals of Neurology, 66(5), 654–662.
Rusconi, E., Walsh, V., & Butterworth, B. (2005). Dexterity with numbers: rTMS over left angular gyrus disrupts finger gnosis and number processing. Neuropsychologia, 43(11), 1609–1624.

Sato, M., Cattaneo, L., Rizzolatti, G., & Gallese, V. (2007). Numbers within our hands: modulation of corticospinal excitability of hand muscles during numerical judgment. Journal of Cognitive Neuroscience, 19(4), 684–693

Sato, M., & Lalain, M. (2008). On the relationship between handedness and hand-digit mapping in finger counting. Cortex, 44(4), 393–399.

Sawamura, H., Shima, K., & Tanji, J. (2002). Numerical representation for action in the parietal cortex of the monkey. Nature, 415(6874), 918–922.

Sawamura, H., Shima, K., & Tanji, J. (2010). Deficits in action selection based on numerical information after inactivation of the posterior parietal cortex in monkeys. Journal of Neurophysiology, 104, 902–910.

Saxe, G.B. (1982). Culture and the development of numerical cognition: studies among the Oksapmin of Papua New Guinea. In C. J. Brainerd (ed.), Children's Logical and Mathematical Cognition (pp. 157–176). New York: Springer-Verlag.

Shalev, R.S. (2007). Prevalence of developmental dyscalculia. In D. B. Berch & M. M.M. Mazzocco (eds), Why is Math so Hard for Some Children? The Nature and Origins of Mathematical Learning Difficulties and Disabilities (pp. 49–60). Baltimore: Brookes Publishing Co.

Shalev, R.S., Manor, O., & Gross-Tsur, V. (2005). Developmental dyscalculia: a six-year follow-up. Developmental Medicine and Child Neurology, 47(2), 121–125.

Siegler, R.S., & Shrager, J. (1984). Strategy choice in addition and subtraction: How do children know what to do? In C. Sophian (ed.), Origins of Cognitive Skills (pp. 229–293). Hillsdale: LEA.

Simon, O., Kherif, F., Flandin, G., Poline, J., Riviere, D., Mangin, J., Le Bihan, D., & Dehaene, S. (2004). Automatized clustering and functional geometry of human parietofrontal networks for language, space, and number. NeuroImage, 23(3), 1192–1202.

Simon, O., Mangin, J.F., Cohen, L., Le Bihan, D., & Dehaene, S. (2002). Topographical layout of hand, eye, calculation, and language-related areas in the human parietal lobe. Neuron, 33(3), 475–487.

Spaepen, E., Coppola, M., Spelke, E.S., Carey, S.E., & Goldin-Meadow, S. (2010). Number without a language model. Proceedings of the National Academy of Science, 108(8), 3163–8.

Thioux, M., Pesenti, M., Costes, N., De Volder, A., & Seron, X. (2005). Task-independent semantic activation for numbers and animals. Cognitive Brain Research, 24(2), 284–290.

Thompson, J., Abbott, D., Wheaton, K., Syngeniotis, A., & Puce, A. (2004). Digit representation is more than just hand waving. Cognitive Brain Research, 21(3), 412–417.

Tschentscher, N., Hauk, O., Fischer, M.H., & Pulvermüller, F. (2012). You can count on the motor cortex: finger counting habits modulate motor cortex activation evoked by numbers. NeuroImage, 59(4), 1–10.

Tucha, O., Steup, A., Smely, C., & Lange, K.W. (1997). Toe agnosia in Gerstmann syndrome. Journal of Neurology, Neurosurgery & Psychiatry, 63(3), 399–403.

Turconi, E., & Seron, X. (2002). Dissociation between quantity and order meanings in a patient with Gerstmann syndrome. Cortex, 38, 911–914.

Varney, N.R. (1984). Gerstmann syndrome without aphasia: a longitudinal study. Brain and Cognition, 31(1), 1–9.

Walsh, V. (2003). A theory of magnitude: common cortical metrics of time, space and quantity. Trends in Cognitive Sciences, 7(11), 483–488.

Wiese, H. (2003). Numbers, Language, and the Human Mind. Cambridge, MA: Cambridge University Press.

Willems, R.M., Toni, Y., Hagoort, P., & Casasanto, D. (2009). Body-specific motor imagery of hand actions: neural evidence from right- and left-handers. Frontiers in Human Neuroscience, 3, Article 39.

Williams, B.P., & Williams, R.S. (1995). Finger numbers in the Greco-Roman world and the Early Middle Ages. Isis, 86, 587–608.

Wilson, M. (2002). Six views on embodied cognition. Psychonomic Bulletin & Review, 9(4), 625–636.

Zago, L., Pesenti, M., Mellet, E., Crivello, F., Mazoyer, B. & Tzourio-Mazoyer, N. (2001). Neural correlates of simple and complex mental calculation. NeuroImage, 13(2), 314–327.

CHAPTER 6

LINKING NUMBERS TO SPACE

from the mental number line towards a hybrid account

JEAN-PHILIPPE VAN DIJCK, VÉRONIQUE GINSBURG, LUISA GIRELLI, AND WIM GEVERS

Introduction

Because of its relevance in daily life and its contribution to the expansion of our technological human societies, mathematics and the way in which the human mind deals with such abstract information has triggered the curiosity of many cognitive neuroscientists. Numbers form the cornerstone of this intellectual achievement. Therefore, it is not surprising that much research has been devoted to understanding how our brain is able to represent and process this type of information. Both introspection and more formal research approaches converge on the idea that the processing of numbers and space are highly related, both at a functional and anatomical level.

The first scientific articles, illustrating the link between numbers and space, date back to the 19th century. Sir Francis Galton published two papers in which he reported people who experienced vivid spatial images when processing numbers that he termed 'natural lines of thought' (Galton, 1880a, b). Later, large-scale screening studies showed that about 15% of normal adults report such vivid visuospatial experiences when processing numbers (Seron, Pesenti, Noël, Deloche, & Cornet, 1992). Although these observations support the current conception, the fact that part of the population reports such an explicit experience is, in itself, not sufficient to conclude that common or similar processing mechanisms underlie both of them. To illustrate, a certain part of the population (e.g. 2%, Simner, Mulvenna, Sagiv, Tsakanikos, Witherby, Fraser, et al., 2006) reports that they always 'see' Arabic numbers in a specific colour, a condition called colour-number synaesthesia. Not surprisingly, colour-number synaesthesia is never used as evidence for the existence of a close link between the processing of numbers and the processing of colour. Nevertheless, numerous other investigations using different research techniques in both healthy participants and in neurologically impaired patients have repeatedly demonstrated the close link between the processing of numbers and the processing of space (for overviews see Fias & Fischer, 2005; Gevers & Lammertyn, 2005; de Hevia, Vallar, & Girelli, 2008; Hubbard, Piazza, Pinel, & Dehaene et al., 2005).

Although recently several alternative explanations have been formulated, the most popular account for the interaction between numbers and space argues that this association has its origin in the underlying mental representation of numbers taking the form of a horizontally-orientated mental number line (Dehaene, Bossini, & Giraux, 1993). However, as we tried to illustrate on the basis of Galton's study, it is an easy pitfall to describe empirical observations in line with current popular theoretical accounts. In the attempt to avoid this pitfall, we start with a descriptive overview of three key behavioural observations, i.e. the SNARC effect, the number interval bisection bias and the asymmetry of the distance effect in neglect, which are currently used to illustrate the association between numbers and space. In doing so, we first describe the empirical observations, as much as possible, free of imposed theoretical interpretations. In a second step, those theories that provide a unified framework for these observations are described. By using this approach, it will become clear that none of these accounts is able to fully explain the available evidence by itself. To go beyond this impasse, a hybrid account is proposed in which emphasis is put on both the mental representation and the processing mechanisms that operate on it.

Evidence From The Snarc Effect

The interaction between numbers and space has a strong empirical foundation and has been demonstrated in a variety of experimental designs and contexts. One of the most convincing and robust demonstrations for this intrinsic link is the Spatial Numerical Associations of Response Codes (SNARC) effect. This effect was originally reported by Dehaene et al. (1993; Dehaene, Dupoux, & Mehler, 1990). It was observed that participants' left-handed responses were faster when judging (relatively) small numbers while right-handed responses were faster when judging (relatively) large numbers (see Figure 6.1).

Automatic Nature of The Snarc Effect

Following this seminal report, this phenomenon was of inspiration for several researchers who tested the effect with different tasks and in several experimental settings. The SNARC effect was observed in tasks that do not require the explicit processing of numerical magnitude, like parity judgment (e.g. is the presented number odd or even; Dehaene et al., 1993), phoneme monitoring (e.g. does the name of the presented number contain the phoneme/e/; Fias, Brysbaert, Geypens, & D'Ydewalle, 1996), or even when numerical stimuli were completely task irrelevant (e.g. is a superimposed triangle pointing up or downward? Fias, Lauwereyns, & Lammertyn, 2001; Lammertyn, Fias, & Lauwereyns, 2002). A related observation was made in tasks where participants were required to detect a lateralized target preceded by a centrally displayed number. Even though this number was completely task irrelevant, left-sided targets were detected faster when preceded by a small number, whereas right-sided targets were detected faster when preceded by a large number (Fischer, Castel,

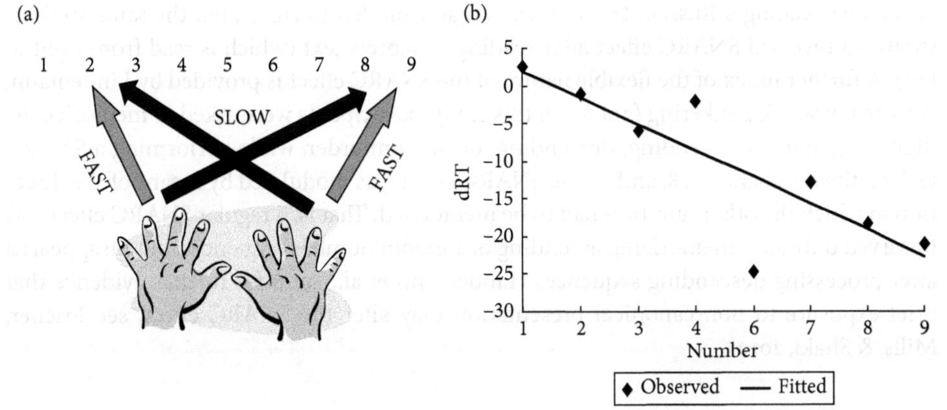

FIGURE 6.1 Schematic depiction of the SNARC effect. (a) Participants are required to judge whether a number is odd or even, or whether it is smaller or larger than a reference number. In both conditions, left-hand responses are faster for smaller numbers compared to larger numbers, whereas the opposite is observed for right-hand responses. (b) The results of a magnitude comparison task. The reaction time difference (dRT) between right-hand and left-hand responses. Values greater than zero indicate that the response is given faster with the left hand, whereas values under zero indicate that the response is given faster with the right.

Adapted from W. Gevers, T. Verguts, B. Reynvoet, B. Caessens, and W. Fias, Numbers and space: A computational model of the SNARC effect, Journal of Experimental Psychology: Human Perception and Performance, 32 (1), pp. 32–44, doi: 10.1037/0096-1523.32.1.32 © 2006, American Psychological Association.

Dodd, & Pratt, 2003; see Ruiz Fernández, Rahona, Hervás, Vázquez, & Ulrich, 2011, for similar observations in free vision). A similar congruency was observed when participants had to judge the parity status (e.g. say 'Ti' if odd, 'To' if even) or the magnitude of a centrally presented number immediately followed (but not preceded) by a lateralized prime (Kramer, Stoianov, Umiltà, & Zorzi, 2011; Stoianov, Kramer, Umilta, & Zorzi, 2008).

Flexibility of The Snarc Effect

Critical to the present review, is the observation that the relative, rather than the absolute magnitude of a number drives the SNARC effect. That is, for example, the digits 4 and 5 elicited faster left than right responses when the digits ranged from 4 to 9, but elicited faster right than left responses when digits ranged from 0 to 5 (Dehaene et al., 1993; Fias et al., 1996). Recently, Ben Nathan, Shaki, Salti, & Algom (2009; see also Ren, Nicholls, Ma, & Chen, 2011) refined the range effect in a magnitude comparison task by dynamically changing the standard reference from trial to trial. They found that the SNARC effect was driven by the relative instead of the absolute magnitude as, for example, '7' was associated with 'left' when the referent was '8,' but with 'right' when the referent was '6'. Bächtold, Baumuller, & Brugger (1998; see also Vuilleumier, Ortigue, & Brugger, 2004) extended this idea to mental imagery by showing that inducing participants to imagine the numbers on a clock face, is sufficient to observe a reversed spatial congruency effect (i.e. the smaller hours were associated with right, and the larger ones with left). Moreover, such a reversed congruency effect is observed even when no explicit reference to mental imagery occurs. For instance, Shaki and Fischer (2008) reported that Russian–Hebrew bilinguals showed a normal SNARC effect after reading a Russian text (which is read from left to right), but the same subjects showed a reversed SNARC effect after reading a Hebrew text (which is read from right to left). A further index of the flexible nature of the SNARC effect is provided by Lindemann, Abolafia, Pratt, & Bekkering (2008). In this study, participants were asked to memorize the digits 3, 4, and 5 in ascending, descending, or random order, while performing a SNARC task on the numbers 1, 2, 8, and 9. The SNARC effect was modulated by means of the direction in which the other numbers had to be memorised. That is, a regular SNARC effect was observed only after memorizing ascending or random number sequences, but disappeared after processing descending sequences (Lindemann et al., 2008 for further evidence that brief exposure to non-canonical presentation may alter the SNARC effect, see Fischer, Mills, & Shaki, 2010).

Nature of The Spatial Code Driving The Snarc Effect

Another line of research deals with the nature of the spatial codes underlying the SNARC effect, by means of methodological manipulations that controlled for on-line number-space associations. Departing from the standard bimanual left/right response paradigm, Santens and Gevers (2008) required participants to perform a magnitude comparison task adopting uni-manual responses in the close/far dimension (response locations were placed physically close or far from a starting position). They observed that 'small' numbers were associated with 'close' responses, whereas 'large' numbers were associated with 'far' responses, regardless of the lateralized movement direction. Following this observation, Gevers, Santens, Dhooge, Chen, Fias, & Verguts (2010) further investigated the possibility that the SNARC effect mainly reflects an association between number magnitude and spatial concepts associated with the responses such as 'close'/'far' and 'left'/'right'. In their design, conceptual labels and physical response locations were directly pitted against one another. The verbal labels 'left' and 'right' were presented either in their canonical position (left right) or not (right left) on the screen. In a series of four experiments, it was shown that the verbal-spatial coding of magnitude was dominant over the visuospatial one: participants associated numbers more strongly with the verbal labels 'left' and 'right' regardless of their physical position on the screen.

Ordinal Information And The Snarc Effect

Numbers convey not only magnitude meaning, but also ordinal meaning. Several studies showed that ordinal information is also associated with lateralized responses. Gevers, Reynvoet, & Fias (2003) first reported that the SNARC effect holds when either letters or names of the months are used as stimuli. Later research demonstrated that lateralized responses to newly-learned ordinal sequences could also lead to a SNARC effect. In two separate studies, participants either had to extensively learn new arbitrary sequences of words (Previtali, de Hevia, & Girelli, 2010) or of arbitrary visual elements (Van Opstal, Fias, Peigneux, & Verguts, 2009). Regardless of the type of stimuli, elements early in the sequence were responded to faster with the left-hand side, whereas elements later in the sequence were responded to faster with the right-hand side, even when order was irrelevant to the task (Previtali, de Hevia, & Girelli, 2010). Critically, ordinal meaning does not need to be internalized on a long-term basis to be associated with a lateralized response. Recently, van Dijck and Fias (2011) directly investigated the role of working memory in associating ordinal position to lateralized response. In their study, participants were asked to memorize a sequence of (centrally presented) numbers. Subsequently, using a go–no go procedure, participants had to make a parity judgment, but only if the presented number belonged to the memorized sequence. Regardless of their magnitude (i.e. no SNARC effect was observed), numbers presented at the beginning of the memorized sequence were responded to faster with the left-hand side, whereas items presented at the end of the sequence were responded to faster with the right-hand side. The same observation was made when words (e.g. fruits and vegetables) instead of numbers were used, again strongly suggesting that it is the ordinal position in the sequence and not the cardinal meaning of the stimuli that interacts with the response side. More recent findings demonstrate that the association between the ordinal position in the sequence and space can be observed without lateralized responses. Similar to the design introduced by Fischer et al. (2003), van Dijck, Abrahamse, Majerus, & Fias (2013) asked participants to memorize a sequence of centrally presented digits in the order of presentation, and found that when centrally presented as a cue, these memorized numbers can modulate spatial attention according to their position in the working memory sequence, irrespective of their magnitude. That is, after retrieving these numbers from memory, subjects were faster to detect left-sided dots (vocally or with a central key press) when the number was from the beginning of the sequence, and faster to detect right-sided dots when the number was from the end of the sequence.

Evidence From The Number Interval Bisection Task

Besides investigating the behaviour of neurologically intact participants, more insights into normal cognitive functioning can be obtained by studying how selective brain damage disturbs these functions. In this respect, valuable signatures of the association between numbers and space are provided by neuropsychological studies on neglect patients. Patients suffering from neglect after (mainly) right-side brain damage struggle with difficulties to report, respond, and orientate to stimuli in the left side of space (e.g. Driver & Mattingley, 1998). Neglect not only manifests itself in perception. Since the seminal study of Bisiach and Luzzatti (1978), it is known that neglect can also affect the contra-lesional side of mental representations, like familiar places or objects (Grossi, Modafferi, Pelosi, & Trojano, 1989). This observation led to the hypothesis that, if numbers are mapped in the representational space along a left-to-right orientation, the left-sided attentional deficit specific to neglect should also affect the processing of small numbers (Zorzi, Priftis, & Umilta, 2002). To test this idea, neglect patients with intact arithmetic skills were recruited and orally presented with two numbers that identified the extremes of a numerical interval. The task required them to indicate, without calculation and with no explicit reference to spatial imagery, the midpoint of the interval. Neglect patients systematically over-estimated the objective midpoint of the intervals (e.g. they indicated 7 as being the midpoint of the interval 1–9) as if they ignored the smallest numbers, and based their decision on the item-range entering their focus of attention (see Figure 6.2). Interestingly, the bias increased as a function of the number interval length (with a cross over-effect for the smallest intervals), similarly to that observed in neglect patients' bisection of physical lines (e.g. Marshall & Halligan, 1989). Moreover, Rossetti, Jacquin-Courtois, Rode, Ota, Michel, & Boisson (2004) showed that,

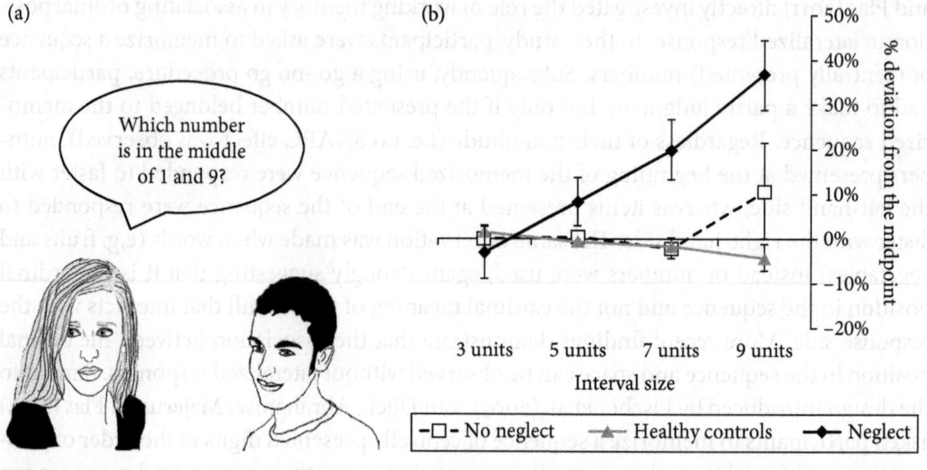

FIGURE 6.2 Schematic depiction of the number interval bisection task. (a) Several pairs of numbers are orally presented and subjects are asked to verbally indicate the numerical midpoint between two different numbers. (b) This figure shows the percentage of deviation from the numerical midpoint. The zero corresponds to a correct response whereas positive values indicate an overestimation of the midpoint and negative values indicate an underestimation of this midpoint.

Reprinted from van Dijck J-P, Gevers W, Lafosse C and Fias W, The heterogeneous nature of number–space interactions, *Frontiers in Human Neuroscence*, 5(182), Figure 1b, doi: 10.3389/fnhum.2011.00182 Copyright: © 2012, van Dijck, Gevers, Lafosse and Fias. Reproduced under the terms of the Creative Commons Attribution License.

similarly to that observed in visuospatial tasks (Rossetti, Rode, Pisella, Farné, Li, Boisson, et al., 1998), their performance on the number bisection task benefit from prism adaptation. Also optokinetic stimulation, known to ameliorate perceptual neglect (for a review see Kerkhoff, 2003), exerts a positive influence on number interval bisection (Priftis, Pitteri, Meneghello, Umiltà, & Zorzi, 2012).

A bisection bias towards the smaller numbers was observed in a left-side brain damaged patient suffering from right-sided neglect (Pia, Corazzini, Folegatti, Gindri, & Cauda, 2009). Similarly, schizophrenic patients, having difficulties in orienting attention to the right side of space, showed a number bisection bias towards the smaller numbers (Cavezian, Rossetti, Danckert, D'Amato, Dalery, & Saoud, 2007). Moreover, according to the slight leftward bias generally shown by healthy participants when bisecting physical lines, (i.e. pseudo-neglect; for a review see Jewell & McCourt, 2000) a slight bias towards the smaller numbers was also observed when they bisected number intervals (Longo & Lourenco, 2007, 2010).

However, defective attentional orientating towards the left side of physical space in neglect is not always associated with a number interval bisection bias towards the larger numbers. A double dissociation was reported between physical line bisection and interval bisection tasks (Doricchi, Guariglia, Gasparini, & Tomaiuolo, 2005). Specifically, neglect patients showing a strong bias in number bisection did not necessarily demonstrate a similar bias in line bisection, and vice versa. This dissociation was further corroborated by anatomical data, since the number bisection bias was found to be associated with damage to prefrontal areas, known to be involved in working memory, whereas the line bisection bias was associated with more posterior damage. Accordingly, the size of the number interval bisection bias was associated with a reduction of working memory capacity, both in the verbal and in the spatial modality (Doricchi, Merola, Aiello, Guariglia, Bruschini, Gevers, et al., 2009). Similarly, in a single-case study, van Dijck, Gevers, Lafosse, Doricchi, & Fias (2011) reported a patient showing right-sided extra personal and representational neglect, as marked by a left bias in line bisection, but a number bisection bias that was orientated towards the larger numbers. Evaluation of the patient's working memory resources confirmed a non-spatial origin of the number bisection bias, revealing that the bias was associated with a reduced working memory capacity mainly affecting the first elements within verbal sequences (van Dijck et al., 2011). The observed dissociation between number interval and line bisection was confirmed in several subsequent studies entailing neglect (e.g. Rossetti et al., 2004; van Dijck, Gevers, Lafosse, & Fias, 2012), schizophrenic (Tian, Wei, Wang, Chen, Jin, Wang, et al., 2011) and developmental dyscalculic (Ashkenazi & Henik, 2010) populations (see Rossetti, Jacquin-Courtois, Aiello, Ishihara, Brozzolli, & Doricchi, 2011, for an overview). Importantly, number interval bisection is a task that taps on a mental representation, while physical line bisection does not. Since it has been shown that representational and extra personal neglect can doubly dissociate (e.g. Guariglia, Padovani, Pantano, & Pizzamiglio, 1993), the above reports can be an instantiation of this dissociation. This alternative interpretation, however, does not hold for the patient reported by van Dijck et al. (2011) who showed right-sided representational neglect together with a bias towards the larger numbers in a number interval bisection task. Given the limited generalizability of single case reports, larger group studies are needed on this debate (see for instance Aiello, Jacquin-Courtois, Merola, Ottaviani, Tomaiuola, Bueti, et al., 2012).

Evidence From The Asymmetrical Distance Effect

Additional studies on neglect patients have shown that the impact of the attentional deficit on number processing is not limited to bisection tasks. Vuilleumier, Ortigue, & Brugger (2004) asked neglect patients to perform a series of magnitude comparison tasks in which numbers from 1 to 9 were to be compared against different referents. For this purpose, numbers were visually presented in random order and participants were required to compare them against 5 or 7 in different blocks. Typically, in such comparison tasks, reaction times to the targets increase as an inverse function of the distance between this number and the referent. This observation, also known as the distance effect (e.g. Moyer & Landauer, 1967), is symmetrical in healthy subjects, meaning that the increase in reaction times is similar for numbers either smaller or larger than the referent. Vuilleumier et al. (2004) showed that this is not the case for neglect patients as their distance effect was much more marked for small (associated to the left) than for large (associated to the right) numbers. Importantly, this distance asymmetry in neglect patients was similar for both referents, suggesting that the asymmetry is driven by the relative, and not the absolute size of the numbers (see Figure 6.3).

Evidence From Other Tasks

Beside what has been reviewed so far, there is abundant additional evidence in favour of the idea that numbers and space are closely related. For example, another demonstration comes from a study using the random digit generation task (Loetscher, Schwarz, Schubiger, &

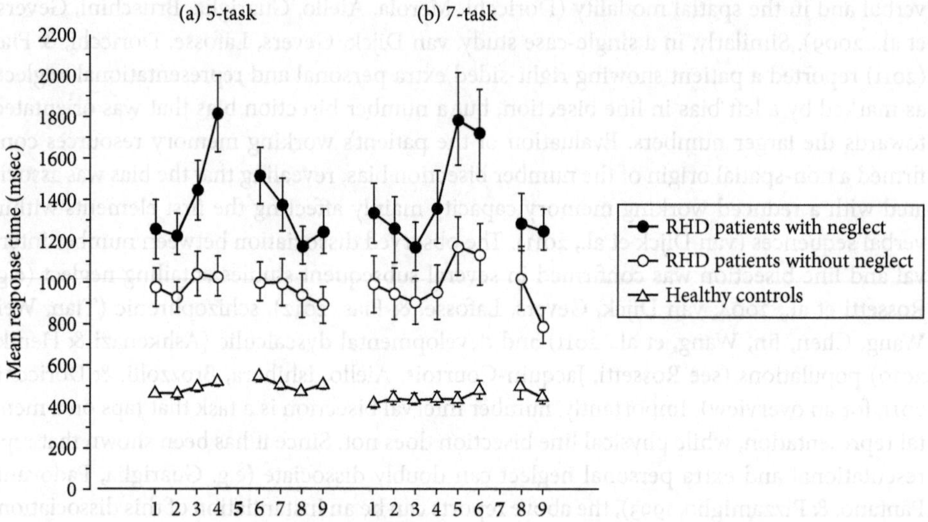

FIGURE 6.3 Depiction of the magnitude comparison distance effect in neglect patients. Participants have to judge whether numbers are smaller or larger than the reference number (5 or 7). Neglect patients show increasing RTs for those numbers immediately preceding the referent (e.g. in the condition with number 5 as reference, neglect patients have particularly slower RTs for the number « 4 »).

Brugger, 2008). In this study, subjects were asked to name, as randomly as possible, numbers between 1 and 30 in a sequence. Data were collected in two runs: a baseline condition in which the subjects had to look straight ahead, and a condition in which left- or right-sided head turns had to be made. When looking straight ahead, the participants generated a higher amount of small numbers (i.e. numbers smaller than 16) compared with larger ones (i.e. numbers larger than 15). This small number bias, however, changed when subjects moved their head during the generation process, i.e. small numbers were more frequently produced when the head was rotated leftwards, while the small number bias decreased when rotating the head towards the right. Interestingly, when looking straight ahead, the direction in which the eye moves just before a number is said, betrays whether this will be a small or a large one (Loetscher, Bockisch, Nicholls, & Brugger, 2010).

Further evidence comes from a variant of the bisection task. When neurologically healthy participants bisect digit strings, their performance is biased towards the left when the line is composed of small numbers (e.g. 1 or 2) and towards the right when the line is composed of large numbers (e.g. 8 or 9; Fischer, 2001). Moreover, when the line to be bisected is flanked by irrelevant digits, participants systematically shift the subjective centre toward the larger digit irrespective of its position (De Hevia, Girelli, & Vallar, 2006; Ranzini & Girelli, 2012). More recently, a similar association has been observed when writing digits by hand. Analyses of the spatial properties of handwritten digits revealed that they were dislocated as a function of their magnitude with small numbers being written more leftwards relative to large numbers (Perrone, de Hevia, Bricolo, & Girelli, 2010). Both findings indicate that, in healthy participants, the automatic spatial coding of numbers not only interacts with spatially defined response buttons (like in the SNARC tasks described above), but also with more spontaneous hand movements.

Finally, beyond the mental representation of numbers itself, there is also evidence that spatial coding contributes to mental arithmetic. It has been shown that when participants are asked to point to number locations (1–9) on a visually-presented number line after computing them from addition or subtraction problems, pointing was biased leftward after subtracting and rightward after adding (Pinhas and Fischer, 2008; McCrink, Dehaene, & Dehaene-Lambertz, 2007). Moreover, by using multivariate classification techniques of brain imaging data, these spatial biases were associated with the brain circuitry involved in making left or right eye movements. This suggests that performing mental arithmetic operations, like the mental representation of numbers, co-opts neural circuitry and cognitive mechanisms associated with spatial coding (Knops, Thirion, Hubbard, Michel, & Dehaene, 2009).

Theories Explaining The Associations between Numbers and Space

The Mental Number Line

A popular account for the empirical findings described above is that numbers are represented on a horizontal mental number line (MNL) with numbers orientated in ascending order from left-to-right or from right-to-left (e.g. Dehaene et al., 1993). The orientation of the number line is thought to reflect a long-term association built up gradually across development according to the culturally dominant reading/writing direction. People educated in Western societies would have a long-term representation of numbers going from small-left to large-right, whereas a reversed long-term association would be observed in societies reading from right-to-left. Processing of a number results in the automatic activation of the corresponding spatial location along the MNL. Activating such a location or shifting from one location to another on the MNL is supposed to involve spatial attentional processes (e.g. Fischer et al., 2003). Critically, it is believed that this MNL is based on the same coordinate system used to represent physical space (Umilta, Priftis, & Zorzi, 2009).

Polarity/Conceptual Coding

This alternative view suggests that number-space associations are an instance of the many associations that can exist between categorical conceptual dimensions. In particular, Proctor and Cho (2006) developed a polarity coding account in which the relationship between numbers and space derives from a systematic association between the verbal concepts that are linked to stimulus and response properties (e.g. the size of the number being small or large, and the side of the response being left or right). Based on the assumption that such categorical conceptual dimensions have a specific polarity (e.g. 'left' being negative and 'right' being positive; and 'small' being negative and 'large' being positive), it is suggested that the congruency between polarities drives the association between numbers and space. Importantly, such conceptual coding is not limited to numbers as it applies to other kinds of binary categories as well (e.g. good/bad, odd/even, yin/yang, early/late are also coded in terms of positive and negative polarities). In that way, this account can also explain the SNARC-like effects observed with other ordinal information, like days of the week or letters from the alphabet.

According to a similar view, the conceptual coding account, numbers are considered to be first conceptually categorized as either 'small' or 'large,' and then linked in an associative network to other dichotomous (output) dimensions, such as left-right, (Gevers, Santens, Dhooge, Chen, Fias, & Verguts, 2010; Gevers, Verguts, Reynvoet, Caessens, & Fias, 2006; Santens & Gevers, 2008). Unlike the MNL it is assumed that this network is verbal-spatial rather than visuospatial in nature.

Working Memory

In a final account it is argued that the ordinal position of information in working memory is spatially coded (Fias, van Dijck, & Gevers, 2011; van Dijck & Fias, 2011). The processing of numerical magnitude would use the same underlying architecture, resulting in an association between numbers and space. In itself, this working memory account does not specify whether the coding of ordinal position is visuospatial or verbal-spatial in nature. In contrast to the MNL account, it supposes that the association between numbers and space is not long-term, but built up during task execution, probably as a strategy to facilitate task execution (for a similar strategic interpretation of the SNARC effect see Fischer, 2006, Fischer, Mills, & Shaki, 2010). The link between serial order coding in working memory and spatial attention (Van Dijck, Abrahamse, Majerus, & Fias, 2013) would associate the working memory account to theories which assume that working memory and attention are closely related and partly overlapping constructs (e.g. Cowan, 1995; Awh & Jonides, 2001).

Strengths and Weaknesses of The Described Accounts

The Mental Number Line

Because of the visuospatial nature of the representation, the MNL can well account for the observation of the SNARC effect: it reflects a congruency between the activated location on the MNL and the location of the response buttons. In addition, the close link to spatial attention explains why numbers can act as directional cues, so that the mere perception of them gives rise to lateralized shifts of spatial attention (e.g. Fischer et al., 2003) or, the other way around, that attentional modulations can influence the processing of numbers (e.g. Stoianov et al., 2008). Assuming such spatial attention as processing mechanism, the MNL can easily account for the number bisection bias and the distance asymmetry observed in neglect. The deficit in orientating attention towards the left affects the access to the left side of the

MNL and, thereby, the processing of small numbers. Despite this explanatory strength, the MNL account cannot explain several of the observations described above. For example, it remains unclear why the association between numerical magnitude and space (as reflected in the SNARC effect) is mediated by verbal-conceptual processes (e.g. Gevers et al., 2010). Moreover, without making additional assumptions, the MNL account cannot explain why the severity of neglect and the size of the number bisection bias do not consistently correlate (e.g. Doricchi, Merola, Aiello, Guariglia, Bruschini, Gevers, et al., 2009; Rossetti et al., 2011; van Dijck et al., 2012). Because the MNL account assumes long-term associations between numbers and space, it may hardly justify the flexible nature of the SNARC effect. To this aim, it requires the extra assumption that short-term associations can easily and rapidly overrule the existing long-term ones.

Polarity/Conceptual Coding

The conceptual coding account, on the other hand, provides a parsimonious explanation for the conceptual nature of the SNARC effect, its automatic emergence, its flexibility and for the observations of spatial coding for ordinal information. After all, the categorical concepts (e.g. small/large; early/late; close/far) or the polarity codes (negative/positive) associated to a specific number or ordinal position are determined spontaneously by contextual factors. For instance, in a magnitude-related context, the number 5 can be either categorized as large (or positive polarity), when it occurs in the range from 0 to 5, or as small (negative polarity) when the range goes from 4 to 9. Similarly, in an ordinal context, the letter E can be categorized as late (positive polarity) within the A to E range, but as early (negative polarity) when the range goes from E to I. However, it is unclear how this theory can be applied to the interval bisection and distance asymmetry effects or to the observations that number-space interactions are observed in situations where no lateralized responses were required (e.g. Fischer et al., 2003).

Working Memory

Finally, by assuming that ordinal information in working memory is spatially coded and that during task execution, numbers are spontaneously and strategically stored in their canonical order in working memory, the working memory account can well explain the SNARC effect. Along the same line, a working memory deficit for the initial or final elements of ordered sequences, including number intervals, would result in a mental bisection task, in a directional bias towards the end or the beginning of the list. Due to the involvement of working memory, this account can also explain the flexibility of the effects and why spatial coding is not limited to numbers. It is unclear, however, how working memory may account for the asymmetry in the distance effect observed in neglect patients. After all, in magnitude comparison (in which the asymmetry of the distance effect is observed) the need to store information in working memory is only minimal. In addition, the working memory account can only explain the beneficial effect of prism adaptation and optokinetic stimulation in neglect patients' interval bisection (Priftis et al., 2012; Rossetti et al., 2004), assuming that this improvement would extend to any serial order in working memory.

Towards a Hybrid Account for Spatial Numerical Associations

In the previous section we described several theories that can account for most of the observed number–space interactions. Apparently, none of them is able to explain the full range of observations. This suggests that the relationship between numbers and space is not due to one single underlying processing mechanism or representation. This idea is well supported by a dual-task study where participants were asked to perform a parity judgement and magnitude comparison task while memorizing spatial and verbal information (van Dijck, Gevers, & Fias, 2009). A double dissociation was observed between the type of task and the type of working memory load. The SNARC effect in parity judgment disappeared when verbal working memory was taxed, and no SNARC effect in magnitude comparison was found when memorizing spatial information. These findings provide evidence against the view that all behavioural signatures of the association between numbers and space have their origin in a single underlying mental representation. Besides, they show that the nature of these associations and the type of working memory resources involved, depend on the task at hand. Elaborating upon this idea, van Dijck et al. (2012) investigated the functional relationship between the several tasks and effects typically used to illustrate the link between numbers and space. Therefore, a group of right-sided brain damaged patients (with and without neglect) and healthy participants performed physical line bisection, number interval bisection, parity judgment, and magnitude comparison. After replicating the previously reported ANOVA patterns and the dissociations at the level of the individual subjects, the data were submitted to a principal component analyses (PCA) to unravel the 'internal structure of number-space'. Further confirming the heterogeneous nature of the association between numbers and space, the results showed that a three-component solution provided the best fit of the data pattern (see Figure 6.4). The first component was loaded by the magnitude comparison SNARC effect and number interval bisection, the second component by physical line bisection and the distance asymmetry, and the third component by the parity judgment SNARC effect and number interval bisection, and the distance asymmetry. These results clearly refute a single mechanism account by showing that multiple factors are needed to capture the correlations between the tasks. Furthermore, the observed component structure fitted the dissociations previously described in the literature. This led to the suggestion that the three components reflect the involvement of spatial attention, verbal and spatial working memory, and that the different effects draw differently upon these processing mechanisms. In an attempt to explain the component loadings, it was proposed that the parity judgment SNARC effect emerges by using verbal working memory while the magnitude comparison SNARC effect emerges by using visuospatial working memory. The number interval bisection bias and distance asymmetry would arise from the interplay of different processes. Both effects draw upon verbal working memory (probably for the recollection of the numbers defining the interval and the encoding the ordinal relations with regard the comparison referent, respectively), but whereas the distance asymmetry is further related to spatial attention, the number interval bisection recruits spatial working memory (probably to construct a spatial representation in mental imagery in searching for the numerical midpoint).

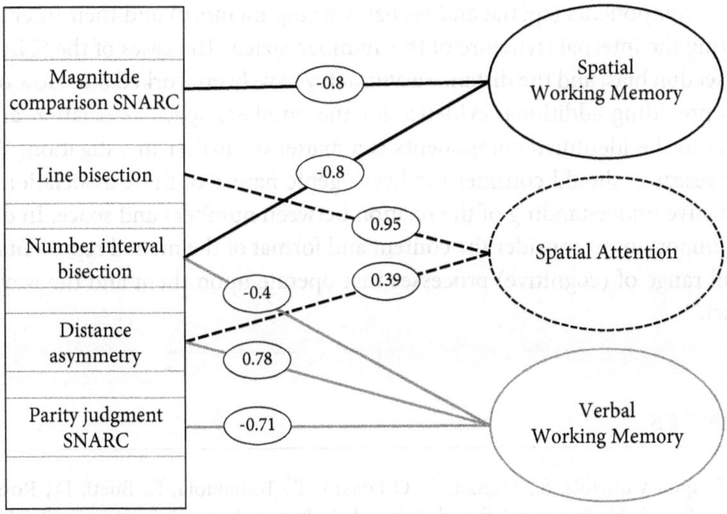

FIGURE 6.4 Results of the Principal Component Analyses (PCA) as described by van Dijck et al. (2012). To unravel the « internal structure » of the association between numbers and space, a PCA was used. Instead of 1 single underlying component (spatial attention), this analysis revealed that at least two additional factors (verbal and spatial working memory) are involved.

In the PCA analyses, no direct measures of working memory capacity were included. This makes it difficult to determine exactly the functional details of these underlying mechanisms. Therefore, we acknowledge that the current proposal should be consider a starting point for an alternative framework since a fully elaborated functional elaboration is still to be reached. The observations from the dual-task paradigm and the group-study, however, clearly indicate that the relation between numbers and space is more complex than so far proposed by each theory in isolation. Specifically, they indicate that the association between numbers and space results from the interplay of different processing mechanisms and that the degree of involvement of those mechanisms depends on the task at hand (Chen & Verguts, 2010; see also the chapter by Roggeman, Fias, & Verguts, 2014).

CONCLUSION

In the present review, we directly opposed the existing theories to explain the association between numbers and space. Although we acknowledge that not everyone will fully agree with our interpretation of the described theoretical account, we aimed at illustrating their individual strengths and weaknesses. In doing so, we concluded that none of the described theoretical accounts by itself can explain the full range of empirical findings and we proposed a hybrid account to overcome this impasse. Here, it is argued that instead of one single underlying representation associated with attentional processes; at least two additional independent components (spatial and verbal working memory) and their interactions are characterizing the internal structure of the 'number-space.' The cases of the SNARC effect, number bisection bias, and the distance asymmetry have been worked out. How other tasks and effects providing additional evidence for the numbers-space association are situated with respect to the identified components is a matter of further investigation. We believe that future research should consider the heterogenic nature of these associations to come to an exhaustive understanding of the relation between numbers and space. In doing so, it is not only important to consider the content and format of the mental representations, but also the full range of (cognitive) processes that operate upon them and the way in which they interact.

References

Aiello, M., Jacquin-Courtois, S., Merola, S., Ottaviani, T., Tomaiuola, F., Bueti, D., Rossetti, Y., & Doricchi, F. (2012). No inherent left and right side in human 'mental number line': evidence from right brain damage. *Brain*, 135, 2492–2505.

Ashkenazi, S., & Henik, A. (2010). A disassociation between physical and mental number bisection in developmental dyscalculia. *Neuropsychologia*, 48, 2861–2868.

Awh, E., & Jonides, J. (2001). Overlapping mechanisms of attention and spatial working memory. *Trends in Cognitive Sciences*, 5, 119–126.

Bächtold, D., Baumuller, M., & Brugger, P. (1998). Stimulus-response compatibility in representational space. *Neuropsychologia*, 36, 731–735.

Ben Nathan, M., Shaki, S., Salti, M., & Algom, D. (2009). Numbers and space: associations and dissociations. *Psychonomic Bulletin & Review*, 16, 578–582.

Bisiach, E., & Luzzatti, C. (1978). Unilateral neglect of representational space. *Cortex*, 14, 129–133.

Cavezian, C., Rossetti, Y., Danckert, J., D'Amato, T., Dalery, J., & Saoud, M. (2007). Exaggerated leftward bias in the mental number line of patients with schizophrenia. *Brain and Cognition*, 63, 85–90.

Chen, Q., & Verguts, T. (2010). Beyond the mental number line: a neural network model of number-space interactions. *Cognitive Psychology*, 60, 218–240.

Cowan, N. (1995). *Attention and memory: an integrated framework*. Oxford Psychology Series, No. 26. New York: Oxford University Press.

De Hevia, M.D., Girelli, L., & Vallar, G. (2006). Numbers and space: a cognitive illusion? *Experimental Brain Research*, 168, 256–264.

De Hevia, M.D., Vallar, G., & Girelli, L. (2008). Visualizing numbers in the mind's eye: the role of visuo-spatial processes in numerical abilities. *Neuroscience and Biobehavioral Reviews*, 32, 1361–1372.

Dehaene, S., Bossini, S., & Giraux, P. (1993). The mental representation of parity and number magnitude. *Journal of Experimental Psychology-General*, 122, 371–396.

Dehaene, S., Dupoux, E., & Mehler, J. (1990). Is numerical comparison digital – analogical and symbolic effects in 2-digit number comparison. *Journal of Experimental Psychology-Human Perception and Performance*, 16, 626–641.

Doricchi, F., Guariglia, P., Gasparini, M. & Tomaiuolo, F. (2005). Dissociation between physical and mental number line bisection in right hemisphere brain damage. *Nature Neuroscience*, 8, 1663–1665.

Doricchi, F., Merola, S., Aiello, M., Guariglia, P., Bruschini, M., Gevers, W., Gasparini, M., & Tomaiuola, F. (2009). Spatial orienting biases in the decimal numeral system. *Current Biology*, 19, 682–687.

Driver, J., & Mattingley, J.B. (1998). Parietal neglect and visual awareness. *Nature Neuroscience*, 1, 17–22.

Fias, W., Brysbaert, M., Geypens, F. & D'Ydewalle, G. (1996). The importance of magnitude information in numerical processing: evidence from the SNARC effect. *Mathematical Cognition*, 2, 95–110.

Fias, W., & Fischer, M.H. (2005). Spatial representation of number In: J. I. D. Campbell (Ed.) *Handbook of Mathematical Cognition*. New York: Psychology Press, 43–54.

Fias, W., Lauwereyns, J., & Lammertyn, J. (2001). Irrelevant digits affect feature-based attention depending on the overlap of neural circuits. *Cognitive Brain Research*, 12, 415–423.

Fias, W., van Dijck, J.-P. & Gevers, W. (2011). How number is associated with space? The role of working memory In: S. Dehaene & E. M. Brannon (Eds), *Space, time and number in the brain: searching for the foundations of mathematical thought*. Amsterdam: Elsevier Science, 133–148.

Fischer, M.H. (2001). Number processing induces spatial performance biases. *Neurology*, 57, 822–826.

Fischer, M.H. (2006). The future for SNARC could be stark. *Cortex*, 42, 1066–1068.

Fischer, M.H., Castel, A.D., Dodd, M.D., & Pratt, J. (2003). Perceiving numbers causes spatial shifts of attention. *Nature Neuroscience*, 6, 555–556.

Fischer, M.H., Mills, R.A. & Shaki, S. (2010). How to cook a SNARC: number placement in text rapidly changes spatial-numerical associations. *Brain and Cognition*, 72, 333–336.

Galton, F. (1880a). Visualised numerals. *Nature*, 21, 252–256.

Galton, F. (1880b). Visualized Numerals. *Nature*, 21, 494–495.

Gevers, W., & Lammertyn, J. (2005). The hunt for SNARC. *Psychology Science*, 47, 10–21.

Gevers, W., Reynvoet, B., & Fias, W. (2003). The mental representation of ordinal sequences is spatially organized. *Cognition*, 87, B87–B95.

Gevers, W., Santens, S., Dhooge, E., Chen, Q., Fias, W., & Verguts, T. (2010). Verbal-spatial and visuo-spatial coding of number-space interactions. *Journal of Experimental Psychology: General*, 139, 180–190.

Gevers, W., Verguts, T., Reynvoet, B., Caessens, B., & Fias, W. (2006). Numbers and space: a computational model of the SNARC effect. *Journal of Experimental Psychology: Human Perception and Performance*, 32, 32–44.

Grossi, D., Modafferi, A., Pelosi, L., & Trojano, L. (1989). On the different roles of the cerebral hemispheres in mental-imagery: The o'clock test in 2 clinical cases. *Brain and Cognition*, 10, 18–27.

Guariglia, C., Padovani, A., Pantano, P., & Pizzamiglio, L. (1993). Unilateral neglect restricted to visual-imagery. *Nature*, 364, 235–237.

Hubbard, E.M., Piazza, M., Pinel, P., & Dehaene, S. (2005). Interactions between number and space in parietal cortex. *Nature Reviews Neuroscience*, 6, 435–448.

Jewell, G., & McCourt, M.E. (2000). Pseudoneglect: a review and meta-analysis of performance factors in line bisection tasks. *Neuropsychologia*, 38, 93–110.

Kerkhoff, G. 2003. Modulation and rehabilitation of spatial neglect by sensory stimulation. In: C. Prablanc, D. Pélisson, & Rossetti, Y. (Eds), *Neural Control of Space Coding and Action Production*. Amsterdam: Elsevier, 257–271.

Knops, A., Thirion, B., Hubbard, E.M., Michel, V., & Dehaene, S. (2009). Recruitment of an area involved in eye movements during mental arithmetic. *Science*, 324, 1583–1585.

Kramer, P., Stoianov, I., Umiltà, C., & Zorzi, M. (2011). Interactions between perceptual and numerical space. *Psychonomic Bulletin & Review*, 18, 722–728.

Lammertyn, J., Fias, W., & Lauwereyns, J. (2002). Semantic influences on feature-based attention due to overlap of neural circuits. *Cortex*, 38, 878–882.

Lindemann, O., Abolafia, J.M., Pratt, J., & Bekkering, H. (2008). Coding strategies in number space: memory requirements influence spatial-numerical associations. *Quarterly Journal of Experimental Psychology*, 61, 515–524.

Loetscher, T., Bockisch, C.J., Nicholls, M.E.R., & Brugger, P. (2010). Eye position predicts what number you have in mind. *Current Biology*, 20, R264–R265.

Loetscher, T., Schwarz, U., Schubiger, M., & Brugger, P. (2008). Head turns bias the brain's internal random generator. *Current Biology*, 18, R60–R62.

Longo, M.R., & Lourenco, S.F. (2007). Spatial attention and the mental number line: evidence for characteristic biases and compression. *Neuropsychologia*, 45, 1400–1407.

Longo, M.R., & Lourenco, S.F. (2010). Bisecting the mental number line in near and far space. *Brain and Cognition*, 72, 362–367.

Marshall, J.C., & Halligan, P.W. (1989). When Right Goes Left—An Investigation Of Line Bisection In A Case Of Visual Neglect. *Cortex*, 25, 503–515.

McCrink, K., Dehaene, S., & Dehaene-Lambertz, G. (2007). Moving along the number line: operational momentum in nonsymbolic arithmetic. *Perception & Psychophysics*, 69, 1324–1333.

Moyer, R.S., & Landauer, T.K. (1967). Time Required For Judgements Of Numerical Inequality. *Nature*, 215, 1519–20.

Perrone, G., De Hevia, M.D., Bricolo, E., & Girelli, L. (2010). Numbers can move our hands: a spatial representation effect in digits handwriting. *Experimental Brain Research*, 205, 479–487.

Pia, L., Corazzini, L.L., Folegatti, A., Gindri, P., & Cauda, F. (2009). Mental number line disruption in a right-neglect patient after a left-hemisphere stroke. *Brain and Cognition*, 69, 81–88.

Pinhas, M., & Fischer, M.H. (2008). Mental movements without magnitude? A study of spatial biases in symbolic arithmetic. *Cognition*, 109, 408–415.

Previtali, P., De Hevia, M.D., & Girelli, L. (2010). Placing order in space: the SNARC effect in serial learning. *Experimental Brain Research*, 201, 599–605.

Priftis, K., Pitteri, M., Meneghello, F., Umiltà, C., & Zorzi, M. (2012). Optokinetic stimulation modulates neglect for the number space: evidence from mental number interval bisection. *Frontiers in Human Neuroscience*, 6, 1–7.

Proctor, R.W., & Cho, Y.S. (2006). Polarity correspondence: a general principle for performance of speeded binary classification tasks. *Psychological Bulletin*, 132, 416–442.

Ranzini, M., & Girelli, L. (2012). Exploiting illusory effects to disclose similarities in numerical and luminance processing. *Attention Perception & Psychophysics*, 75, 1001–1008.

Ren, P., Nicholls, M.E.R., Ma, Y.-Y., & Chen, L. (2011). Size matters: non-numerical magnitude affects the saptial coding of response. *PLoS One*, 6, e23553.

Roggeman, C., Fias, W., & Verguts, T. (2014). Basic number representation and beyond: neuroimaging and computational modeling. In: R. Cohen Kadosh, & A. Dowker (Eds), *Oxford Handbook of Mathematical Cognition*. Oxford: Oxford University Press.

Rossetti, Y., Jacquin-Courtois, S., Aiello, M., Ishihara, M., Brozzolli, C., & Doricchi, F. (2011). Neglect 'around the clock': dissociating number and spatial neglect in right brain damage. In: S. Dehaene, & E. M. Brannon (Eds), Space, time and number in the brain: searching for the foundations of mathematical thought. Amsterdam: Elsevier Science, 149–171.

Rossetti, Y., Jacquin-Courtois, S., Rode, G., Ota, H., Michel, C., & Boisson, D. (2004). Does action make the link between number and space representation? Visuo-manual adaptation improves number bisection in unilateral neglect. *Psychological Science*, 15, 426–430.

Rossetti, Y., Rode, G., Pisella, L., Farné, A., Li, L., Boisson, D., & Perenin, M.T. (1998). Prism adaptation to a rightward optical deviation rehabilitates left hemispatial neglect. *Nature*, 395, 166–169.

Ruiz Fernández, S., Rahona, J., Hervás, G., Vázquez, C., & Ulrich, R. (2011). Number magnitude determines gaze direction: spatial – numerical association in a free-hoice task. *Cortex*, 47, 617–620.

Santens, S., & Gevers, W. (2008). The SNARC effect does not imply a mental number line. *Cognition*, 108, 263–270.

Seron, X., Pesenti, M., Noel, M.P., Deloche, G., & Cornet, J.A. (1992). Images of numbers, or when 98 is upper left and 6 sky blue. *Cognition*, 44, 159–196.

Shaki, S., & Fischer, M.H. (2008). Reading space into numbers – a cross-linguistic comparison of the SNARC effect. *Cognition*, 108, 590–599.

Simner, J., Mulvenna, C., Sagiv, N., Tsakanikos, E., Witherby, S.A., Fraser, C., Scott, K., & Ward, J. (2006). Synaesthesia: the prevalence of atypical cross-modal experiences. *Perception*, 35, 1024–1033.

Stoianov, I., Kramer, P., Umilta, C., & Zorzi, M. (2008). Visuospatial priming of the mental number line. *Cognition*, 106, 770–779.

Tian, Y., Wei, L., Wang, C., Chen, H., Jin, S., Wang, Y., & Wang, K. (2011). Dissociation between visual line bisection and mental number line bisection in schizophrenia. *Neuroscience Letters*, 491, 192–195.

Umilta, C., Priftis, K., & Zorzi, M. (2009). The spatial representation of numbers: evidence from neglect and pseudoneglect. *Experimental Brain Research*, 192, 561–569.

van Dijck, J.-P., Abrahamse, E.L., Majerus, S., & Fias, W. (2013). Spatial attention drives serial order retrieval in verbal working memory. *Psychological Science*, 24(9), 1854–1859.

Van Dijck, J.-P., & Fias, W. (2011). A working memory account for spatial numerical associations. *Cognition*, 119, 114–119.

van Dijck, J.-P., Gevers, W., & Fias, W. (2009). Numbers are associated with different types of spatial information depending on the task. *Cognition*, 113, 248–253.

van Dijck, J.-P., Gevers, W., Lafosse, C., Doricchi, F., & Fias, W. (2011). Non-spatial neglect for the mental number line. *Neuropsychologia*, 49, 2570–2583.

van Dijck, J.-P., Gevers, W., Lafosse, C., & Fias, W. (2012). The heterogeneous nature of number-space interactions. *Frontiers in Human Neuroscience*, 5, 182, 1–13.

van Opstal, F., Fias, W., Peigneux, P., & Verguts, T. (2009). The neural representation of extensively trained ordered sequences. *Neuroimage*, 47, 367–375.

Vuilleumier, P., Ortigue, S., & Brugger, P. (2004). The number space and neglect. *Cortex*, 40, 399–410.

Zorzi, M., Priftis, K. & Umilta, C. (2002). Brain damage – neglect disrupts the mental number line. *Nature*, 417, 138–139.

CHAPTER 7

MULTI-DIGIT NUMBER PROCESSING

overview, conceptual clarifications, and language influences

HANS-CHRISTOPH NUERK, KORBINIAN MOELLER, AND KLAUS WILLMES

Prologue

When travelling by train, most numbers one encounters, time schedules, prices, train numbers, seat number of a reservation and—luckily—percentage of battery load on one's laptop, require multi-digit number understanding. Likewise, in our everyday life, when we encounter numbers when dealing with money, time and many other matters, we have to deal with multi-digit numbers. Therefore, multi-digit number understanding is essential for managing life in a modern information society. Failing with numbers or math produces substantial individual risks (Bynner and Parsons 1997; Parsons and Bynner 2005) and societal costs (Beddington et al. 2008; Gross et al. 2009). Such failure which is measured with standardized mathematical achievement tests in schools and elsewhere usually involves deficits in multi-digit number processing as well (e.g., HRT, Haffner et al. 2005; Tedi-Math, Kaufmann et al. 2009; for tests in German, e.g., TEMA-3, Ginsburg and Baroody 2003; TOMA-2, Brown et al. 1994, for examples in English).

Against this background, it is surprising that the majority of cognitive-numerical research has dealt with single-digit numbers, for instance when the distance effect (Moyer and Landauer 1967) or the SNARC effect (Dehaene et al. 1993) were studied. And even when research has used multi-digit numbers as stimuli, the specific properties of multi-digit numbers often have not been taken into account. Instead, multi-digit numbers have been (sometimes implicitly) been dealt with like an extrapolation of single-digit numbers into the two-digit and multi-digit number range.

In the first part of this chapter, we will argue that this assumption may only hold partially. Results obtained and—consequently—conclusions drawn from single-digit number research cannot simply be generalized to multi-digit number research. Even findings related to some multi-digit numbers (e.g., two-digit numbers) cannot easily be generalized to all multi-digit numbers (e.g., four-to-six-digit integer numbers or fractions). We outline that this is not only of theoretical interest, but also has practical implications. In the second part of the chapter, we suggest a taxonomy to categorize specific multi-digit number effects. These effects—with only one exception—have already been described in more detail by Nuerk et al. (2011) in more detail. However, in the current chapter we will further categorize them with regard to the tasks used and the representations involved; and we will propose a new framework for the cognitive mechanisms recruited by multi-digit integer processing in the third part of the chapter, differentiating place-value identification, place-value activation, and place-value computation. In the fourth part we will directly apply this framework by indicating how recent research provides accumulating evidence that multi-digit numbers are strongly and specifically influenced by language. The word 'specific' is of major importance here because we are particularly concerned with studies showing that specific linguistic properties influence particular multi-digit number processes in a specific way. In the final part we address the longitudinal developmental aspects of multi-digit number processing in general and the importance of early place-value understanding for children's later arithmetic development in particular. We will then conclude the chapter with a summary and an outlook on future perspectives.

Why Studying Single-Digit Numbers is Not Sufficient for Understanding Multi-Digit Numbers

The vast majority of numerical research is concerned with effects related to single-digit numbers (e.g. Dehaene et al. 1993; Moyer and Landauer 1967). Generally, results from single-digit number research are helpful and perhaps even *necessary* as a first step towards a deeper understand multi-digit number processing. However, single-digit number research is not *sufficient* to understand multi-digit number processing because the latter draws on additional representations, mental manipulations, and processing of inter-digit relations beyond those which can be studied with single-digit numbers (see Nuerk et al. 2011, for a review). Nevertheless, the influence of such specific multi-digit number representations is often considerable. In some studies (e.g., Moeller et al. 2011a), multi-digit number effects indexing such specific representations accounted for more variance of overall performance than did single-digit number effects. So there are both conceptual and empirical reasons not to ignore the specificities of multi-digit number processing. This will be discussed in detail in the following.

Numerical Distance for Multi-Digit Numbers

Multi-Digit Integers

The numerical distance effect describes that two numbers are compared faster when they have a larger numerical distance (Moyer and Landauer 1967). The distance effect for single-digit numbers has been generalized to various forms of multi-digit numbers, e.g. two-digit integer numbers (Dehaene, 1989; Dehaene et al. 1990) as well as fractions (Ischebeck et al. 2009; Jacob and Nieder 2009). It has been claimed that comparison behavior for multi-digit integers or fractions can be described sufficiently by their dependency on their overall magnitude. This is theoretically important because, as a consequence, it has been postulated that the magnitude of multi-digit numbers is represented in a way comparable to the magnitude representation of single-digit numbers, namely in a holistic analogue fashion (Dehaene et al. 1990), suggesting that multi-digit integers or fractions are represented on some kind of an analogue mental number line. However, whenever distance effects were examined more carefully and stimulus attributes were considered more thoroughly, the notion that multi-digit number processing is sufficiently explained by overall magnitude is no longer consistent with the data.

For multi-digit integers, Verguts and de Moor (2005) examined the distance effects for between-decade items (58_61 vs. 58_64) and within-decade items (51_54 vs. 51_57). If the magnitude of two-digit numbers were represented holistically in an analogue fashion, the distance effects should be comparable. However, they were not. Stable within-decade distance effects were observed, whilst no such distance effects could be obtained for between-decades stimuli, although overall distances were the same for both conditions. This suggests that depending on the decade unit structure, different distances and magnitudes are represented and processed mentally. It is important to note that even Dehaene and colleagues (1990), who argued for a holistic, analogue representation, found a regression predictor 'within-decade comparison' against between-decade comparison in two of three conditions (comparison standards 55 and 66, but not with standard 65). Another, more specific effect, indicating specific processing of tens and units, is the so-called unit-decade compatibility effect.

The Unit-Decade Compatibility Effect

In a two-digit number comparison task, Nuerk et al. (2001) observed that number pairs such as 42_57, in which comparing tens and units separately leads to the same decision bias (4 < 5 and 2 < 7), are responded to faster and are less error-prone than number pairs such as 47_62, in which separate comparisons of tens and units leads to opposing decision biases (4 < 6, but 7 > 2)[1]. As overall distance was matched (i.e., 15 in both examples above), no holistic or overall distance effect can account for the compatibility effect.

This compatibility effect is not only important because it suggests that the magnitude of tens and units is processed in a decomposed fashion with separate magnitude representations of tens and units (Nuerk and Willmes 2005, for a descriptive model; Moeller et al. 2011b, for computational modelling). It is also important because the compatibility effect is modulated by distinct decomposed decade and unit distances. In different experiments, this unit-decade compatibility effect was more pronounced for small decade distances (in particular for error rates) and larger unit distances (in particular for RT; e.g. Nuerk et al. 2001; Nuerk and Willmes 2005; Nuerk et al. 2011, for reviews). Moreover, multiple regression analyses revealed that overall distance is not a sufficient predictor when aiming at identifying the best fitting model. Rather, distinct unit (e.g., Nuerk et al. 2001) as well as decade distance predictors (e.g., Nuerk et al. 2005a) explain additional variance of mean RT, suggesting that the distance effect for two-digit numbers can be conceptualized as the joint impact of distance effects of tens and units. This seems reasonable because overall distance reflects 10 times unit distance plus one time decade distance (Nuerk and Willmes 2005; Nuerk et al. 2011). So, even the existence of an overall distance effect does not necessarily imply that there is an overall magnitude representation determining responses for symbolic number comparison. Indeed, Verguts and de Moor (2005) argued for fully decomposed processing of decades and units because of the lack of a distance effect for between-decade trials. Similarly, Moeller et al. (2011b) found that a decomposed computational model of two-digit number comparison (with strictly decomposed representations of tens and units) can account for the data pattern usually obtained in two-digit number comparison better or at least equally as well as compared to a hybrid model which assumes an additional representation of overall magnitude besides distinct representations of tens and units. Finally, the simulations of Moeller et al. (2011b) also indicated that an influence of overall distance as identified in the regression analyses by for instance Nuerk et al. (2001) can also be obtained for the data produced by a model assuming no specific representation of overall magnitude.

To summarize, overall distance does not account for all the variance in two-digit magnitude comparison; therefore overall distance effects, which are simply generalized from single-digit distance effects which are not *sufficient* to explain multi-digit number understanding exhaustively. Computational modelling (Moeller et al. 2011b) as well as empirical evidence (Verguts and de Moor 2005) suggest that overall distance might not even be *necessary* for understanding multi-digit number processing because the underlying representation of multi-digit integers may be different from the analogue representation which is commonly assumed to underlie the overall distance effect. Importantly, this argument is not restricted to multi-digit integers but also holds for other multi-digit numbers such as fractions.

Numerical Distance For Fractions

In the preceding paragraph we claimed that simply generalizing the distance effect from single-digit research to multi-digit research is not sufficient because of the findings from research conducted with multi-digit integers. This argument can be easily transferred to the case of fraction processing.

In research on fraction processing, the majority of studies reported overall analogue distance to be the most important predictor of, for instance, comparison performance (Ischebeck et al. 2009; Jacob and Nieder 2009; Schneider and Siegler 2010; Siegler et al. 2011; Sprute and Temple 2011). In different variants of fraction comparison tasks when stimulus characteristics were manipulated (i.e., comparison with a fixed standard, comparing two fractions with a common numerators), overall distance predicted performance best. This led various authors to suggest a holistic magnitude representation of fractions, which in turn determines the overall fraction distance effect comparable to the distance effect in single-digit number research.

However, the results of these studies contrast with other empirical evidence (e.g., Bonato et al. 2007; Meert et al. 2009 for common denominators), corroborating the notion of decomposed processing of the magnitudes of numerators and denominators of fractions. In these experiments, different fraction components, which were the relevant ones for the actual comparison (e.g., the numerators in 3/4 vs. 2/4 or the denominators in 2/3 vs. 2/5), were processed in a decomposed fashion.

As a consequence of these inconsistent findings, Meert et al. (2010) suggested that fraction processing style may depend on fraction types and proposed a hybrid model of magnitude representation for fractions similar to the model of Nuerk and colleagues (2001) for two-digit numbers. Meert and colleagues suggested that magnitude representations of the fraction as an integrated entity and decomposed magnitude representations of numerators and denominators exist in parallel. For fractions without common components, the overall magnitude representation was suggested to be more important, because in that case an overall distance effect predicts performance best. In contrast, for fractions with common components, those components relevant for the task itself may be predominantly processed because distance effects for the relevant components predict task performance best. However, to date it is important to note that what one gets from a regression study depends on what predictors one enters into the model and maybe also on which dependent variables one uses. Faulkenberry and Pierce (2011) suggested the cross-multiplication strategy[2] (amongst other strategies) as a predictor variable to explain performance in a fraction magnitude comparison task. This is the strategy which is often taught in school. Strategy reports suggest that some people use it, and the related predictor (average cross-product) explained a significant part of performance variance in addition to overall distance. Previous studies on fraction processing could not find any variance explained by cross-multiplication simply because they had not included this predictor. However, the fact that Faulkenberry and Pierce (2011) identified this predictor as relevant in the regression model and the fact that participants repeatedly reported such strategies, suggest that component-based cross-multiplication may explain fraction magnitude comparison performance in addition to or even instead of other (correlated) predictors.

In summary, the exact processes and distance measures which determine fraction performance are still controversial and under investigation because fraction research is a very young field in numerical cognition research. However, even from the research conducted so far, it has become evident that a single overall distance predictor cannot explain fraction processing sufficiently for all fractions.

Commonalities and Differences Between Multi-Digit Numbers

In the previous sections, we argued that multi-digit number processing differs from single-digit number processing. But, this does not mean that we assume all multi-digit numbers (e.g., multi-digit integers vs. fractions) are one homogeneous category. The place-value relation for multi-digit integers is of course different from the division relation for fractions and possibly also from multi-digit decimal numbers. However, as pointed out above, there are also obvious commonalities in the research questions for different types of multi-digit numbers. For instance, theoretical issues such as whether the magnitude of any multi-digit number is processed either holistically or in a decomposed manner are investigated and discussed controversially for different types of multi-digit numbers (i.e., fractions and multi-digit integers). We would even suggest that the issue of holistic vs. decomposed representations is relevant for any type of multi-digit numbers requiring the integration of multiple digits into one entity during the course of the numerical processing.

Nevertheless, because the vast majority of research has either been restricted to single-digit numbers or has not considered the specifics of multi-digit number attributes, we argue in this chapter that single-digit number research is not sufficient to understand multi-digit number representations. However, we do not wish to imply that all different types of multi-digit numbers are processed in the same way. In fact, recent research suggests that they are important differences even within the same type. For instance, the processing of multi-digit integers seems to be moderated by the number of digits involved (e.g., Meyerhoff et al. 2012).

Practical and Diagnostic Implications

So, far the stated arguments have largely been about theoretical developments in multi-digit number processing. However, distance effects for single-digit and multi-digit numbers may not only differ theoretically, they may also differ in diagnostic applications. Ashkenazi et al. (2009) observed a larger numerical distance effect for dyscalculic children as compared to typically developing children for two-digit numbers, but not for single-digit numbers. Although individual differences in single-digit number distance effects may be observed (e.g., Delazer et al. 2006; Holloway and Ansari 2009), impairments in arithmetic in the study by Ashkenazi et al. (2009) would have been overlooked if only single-digit distance effects had been studied. In particular, it might be the case that single-digit numbers may be highly overlearned, and thus the complexity of single-digit number processing may be too limited to detect reliable deficits in magnitude representations between dyscalculic and normally developing children.

Understanding Multi-Digit Number Processing: A Taxonomy of Studies and Conceptual Clarifications

Definition and Clarifications

Nuerk and colleagues (2011, p. 6) gave the following definition of multi-digit number processing: 'Multi-digit number processing relies on the integration or computation of multiple (at least two) digits to realize a numerical entity.' Thus, as introduced in the discussion of the distance effect, multi-digit number processing is: (i) not only about integer numbers only, but may also include decimal numbers and fractions; (ii) it is also not specific for a particular representation, because different effects of multi-digit number processing have been observed not only for the magnitude representation, but also for verbal and other representations (see Table 7.2); (iii) it is also not about a particular task, such as the magnitude comparison task, since multi-digit number processing also leads to specific effects in parity processing, number matching, number naming and addition, subtraction, or multiplication tasks (see Table 7.3); and (iv), the definition given above (namely 'entity') allows for different conceptual positions with respect to the representations (e.g., decomposed, holistic, hybrid), which may underlie a multi-digit number entity in a particular task. Finally, (v) the definition is not only concerned with a particular effect, as for instance, the compatibility effect. Rather, Nuerk and colleagues (2011) summarized 16 effects which are specific for multi-digit number processing in the way that they cannot be studied in single-digit numbers (such as the distance effect)[3], and which were observed in basic and complex number processing studies (cf. Table 7.1[4]). The list shows that over recent years, multi-digit number processing has become a research field of its own covering different effects, tasks, and representations. In this chapter about multi-digit numbers, we wish to go beyond only listing these effects. Before suggesting a taxonomy of representations involved in multi-digit number processing, we will first summarize which tasks have been used so far for investigating multi-digit number processing.

Tasks Used for Multi-Digit Number Processing Research

Multi-digit number processing has been studied using different tasks. As can be seen in Table 2, the magnitude comparison task has been used predominantly for studying a variety of different effects, while other tasks have been used much less frequently. Whether this dominance relates to the importance and validity of the magnitude comparison task for answering theoretical questions remains to be seen. Additionally, it becomes obvious that there are effects that so far seem to be specific for a task (e.g., consistency effect, digit repetition effects, unit-decade compatibility effect, etc.). Further research is needed here to clarify whether structurally corresponding effects can be observed in other tasks as well. For instance, one might argue for such a correspondence for the compatibility effect found in magnitude comparison and the borrowing effect observed in subtraction: For all incompatible number pairs a borrowing operation is needed when subtracting the smaller from the larger number, whereas no borrowing operation is needed for subtraction made from compatible pairs. Investigating the generalizability of multi-digit number effects over different tasks may be a suitable possibility in order to identify systematic influences in multi-digit number processing. The references given in Table 7.2 indicate exemplary studies investigating the respective effect using a particular task.

Representations and Processes Involved In Multi-Digit Number Processing

The taxonomy of basic numerical processes which we wish to use in this chapter is adapted from Nuerk et al. (2006; see also Moeller et al. 2009a, for similar suggestions). The taxonomy is based on the Triple Code Model of Dehaene (1992; see also Dehaene and Cohen 1995) and its latest amendments (Dehaene et al. 2003; Hubbard et al. 2005), but it also adds a specific place-value representation to the processes and representations (see Table 7.3 for a categorization of which multi-digit number effect is associated with which underlying representation).

Visual Number Form

Whenever we process symbolic numbers in the Arabic or any other symbolic number system, we first have to identify and represent the visual forms to convey numerical information. Therefore, Dehaene (1992) suggested the representation of a so-called visual

Table 7.1 Effects specific to multi-digit number processing which cannot be investigated using single-digit numbers (as described by Nuerk et al. 2011). The list has been augmented with a new effect—the place-value congruency effect reported only recently by Kallai and Tzelgov (2012). The references are not comprehensive due to space limitation but represent typical examples.

	Name	Description	Example	Exemplary References
Basic number processing				
1	Unit-Decade Compatibility Effect	When comparing two two-digit numbers in a magnitude comparison task, incompatible number pairs are processed more slowly and with more errors than compatible number pairs.	compatible pair: 42_57 (4<5 and 2<7) Incompatible pair: 47_62 (4<6 but 7>2)	Nuerk et al. (2001) Nuerk and Willmes (2005)
2	Parity Effects	Odd numbers are processed more slowly than even numbers. Both parity of the whole number and parity of the constituents influence processing.	Parity Congruency Effect: The parity decision for 64 is faster than for 74 because both constituting digits of 64 are even.	Reynvoet et al. (2011) Zhou et al. (2008)
3	Cross-Place Congruency Effect	When comparing two two-digit numbers in a magnitude comparison task, number pairs for which the within-number comparison and the overall comparison lead to the same conclusion are processed more easily.	Congruent: 34_76 (3<7, and 6<7 and 3<4) Incongruent: 32_79 (3<7, but 9>7 and 3>2)	Wood et al. (2005)
4	Transcoding Effects	Transcoding describes the transfer of numbers from one code (e.g., verbal number words) to another. (e.g. Arabic numbers). During the transcoding process typical syntactic errors can be observed that originate from the place-value-structure of the Arabic number system.	*Inversion Error*: The order of units and decades is inverted (e.g., the German word for 27 is 'sieben-und-zwanzig' which means 'seven-and-twenty'). *Additive Composition Error*: 'Four-hundred-and-twenty-seven' is written as 40027. *Multiplicative Composition Error*: 'Four-hundred' is written a 4100.	Deloche and Seron (1982) Zuber et al. (2009)
5	Positional Digit Repetition Effect	When priming a digit at the wrong position, the naming of the number takes longer than in neutral conditions.	28 primed by 86 (8 in the unit-position primed by 8 in the decade-position) vs. 28 primed by 46 (neutral condition)	Ratinckx et al. (2005) Gazzellini and Laudanna (2011)

(*Continued*)

Table 7.1 Continued

	Name	Description	Example	Exemplary References
6	Serial Order Effect	In a perceptual number matching task probes that continue previously seen serially ordered cues are rejected more slowly than probes with no such computational relation.	Cues: 45 67, probe: 89 (serial order: 4 5 6 7 8 9) vs. Cues: 45 67, probe: 92	García-Orza and Damas (2011)
7	Place-value congruency effect	In a single-digit magnitude comparison task, responses are slower and more error prone when the single number is at a position (place) in a digit string which is typically associated with a smaller value due to automatic but irrelevant processing of place-value information (incongruent condition).	congruent 070 vs. 005 (Task $7 > 5$ congruent to $70 > 5$) incongruent 050 vs 007 (Task $5 < 7$ incongruent to $50 > 7$)	Kallai and Tzelgov (2012)
Complex number processing				
8	Carrying	A carry in the addition task needs to be conducted when the digit sum of the operands at any position in the place-value system reaches or exceeds 10. Solving problems requiring a carry operation leads to higher reaction times and error rates as compared to problems that do not require a carry operation.	Carry: 37 + 28 (the unit sum is equal to/larger than 10; executed by adding 1 to the decade digits of the addends)	Ashcraft and Stazyk (1981)
9	Borrowing	Borrowing in a subtraction task needs to be conducted when the digit sum of the subtrahends exceeds the digit sum of the minuend at any position in the place value system. Solving problems requiring a borrowing operation leads to higher reaction times and error rates as compared to problems that do not require a borrowing operation.	*Borrowing*: 65–28 (the unit sum is smaller than 0; executed by subtracting 1 from the difference of the decade digits)	Sandrini, Miozzo, Cotelli, and Cappa (2003)
10	Consistency Effect in Multiplication	Errors in multiplication tasks are more likely when the correct result and the error have the same digit at the same place-value position in common.	The error $7 \times 3 = 24$ is more likely than the error $7 \times 3 = 18$ (The correct result 21 and the error 24 have the decade digit in common).	Verguts and Fias (2005) Domahs et al. (2006, 2007)

11	Decade Crossing Effect	When determining the midpoint of a numerical interval in a number bisection task, number triplets where no decade boundary has to be crossed are easier to process.	21_25_29 (without decade crossing) vs. 23_27_31 (with decade crossing from 20 to 30)	Nuerk et al. (2002) Hoeckner et al. (2008) Wood et al. (2008)
12	Decade Number Effects	Decade numbers occur more often than other numbers and are used as anchors. Decade numbers are named faster than one would expect by their magnitude alone. It is easier to add a single-digit number to a decade number as well as to bisect correctly a number triplet in a number bisection task when it involves a decade number.	30 + 4 vs. 32 + 4 21_30_35 vs. 23_32_37	Brysbaert (1995) Brysbaert et al. (1998) Nuerk et al. (2002)
13	Fraction Effects	When comparing two fractions in a number comparison task both the type of fraction and the strategy used to compare the fraction influence reaction times as well as error rates. *Distance Effects*: When comparing fractions with common denominators participants prefer to use componential strategies leading to a distance effect of the numerators. When comparing fractions with common numerators participants prefer to use holistic strategies leading to a distance effect of the whole fraction. *Problem-Size Effect:* When comparing fractions without common components participants use component based strategies like for example the cross product, whereby larger average cross products would result in longer reaction times.	5/8_6/8 takes longer than 3/8_7/8 because the distance between numerators is smaller. 3/4_3/5 takes longer than 3/4_3/8 because the distance between the whole fractions is smaller. Two fractions a/b and c/d can be compared by the products a×d and b×c: 8/9_2/3 takes longer than 4/5_2/3 because the average cross product is larger (21 vs. 11).	Meert et al. (2009) Schneider and Siegler (2010) Faulkenberry and Pierce (2011)

(*Continued*)

Table 7.1 Continued

	Name	Description	Example	Exemplary References
14	Decimal Effects	As decimal digits differ from integer digits concerning the rules of the place-value system, specific effects of length and value of decimal numbers can be observed.	*Decimal Length Effect*: 3.6 is erroneously said to be smaller than 3.24. (6 < 24). *Decimal Value Effect*: 5.25 vs. 5.258: The 2 does not change its value, irrespective of the digits to the right.	Sackur-Grisvard and Leonard (1985) Desmet et al. (2010)
15	Multi-Linear Effects in the Number Line Estimation Task	The spatial representation of number magnitude on the mental number line changes with a more elaborated representation of the place value structure of Arabic digits. While young and unexperienced children show two separate linear number lines with different slopes for one—and two-digit numbers, older and more experienced children and adults show just one linear number line slope across all number ranges.	Adults and older children have only one linear number line, which means the distance between 0 and 60 is 10 times as large as the distance between 0 and 6. Children overestimate the numerical intervals for single-digit numbers (e.g., between 0 and 6) relative to the numerically 10 times larger intervals between corresponding two-digit numbers (e.g., between 0 and 60), which leads to a representational bias (two linear number lines).	Ebersbach et al. (2008) Moeller et al. (2009d)
Other effects also observed for single digits				
16	Multi-Digit Modulation of Other Effects Previously Observed for Single-Digit Numbers	*Operational Momentum Effect*: In the number line estimation task with multi-digit numbers addition leads to a right and subtraction to a left bias. For two-digit number the effect was modulated by the carry operation. *Size-Congruity Effect*: For single-digit numbers comparisons are faster when a number is numerically and physically larger/smaller. For two-digit numbers the effect was modulated by compatibility and digit magnitude but not by holistic magnitude.	21 + 14 --> non-symbolic dot number response: > 35 49 – 14 --> non-symbolic dot number response < 35 Congruent: 4 8 Incongruent: 4 8	Knops et al. (2009) Ganor-Stern et al. (2007) Lindemann et al. (2011)
17	Probability and Cumulative Frequency Effects	People tend to misjudge probabilities, particularly conditional probabilities. For example, when trying to estimate a certain probability, base-rates are neglected and instead specific cases are used for judgements (base-rate neglect).	Taking hormones for ten years increases the risk for breast cancer from 60 to 66 out of 1000 women (0.6%). vs. taking hormones for ten years increases the risk of breast cancer by 10%.	Barbey and Sloman (2007) Gigerenzer and Hoffrage (1995)

Table 7.2 Overview of tasks used to study multi-digit number processing with exemplary studies employing the respective task. Note that the references given are typical examples and cannot be exhaustive due to space limitations.

	Tasks							
Effects	Magnitude Comparison	Parity judgement	Number naming	Addition/ Subtraction	Multiplication	Number line estimation	Number bisection	Other
Unit-decade compatibility effect	Macizo and Herrera 2011; Nuerk et al. 2001			Guillaume et al. (2012)				
Priming effects		Reynvoet et al. 2011; Zhou et al. 2008						
Cross-place congruity effect	Wood et al. 2005							
Transcoding effects			Power and Dal Martello 1997; Seron and Fayol 1994					
Positional digit repetition effects			Gazellini and Laudanna 2011; Ratinckx et al. 2005					
Serial order effects								García-Orza and Damas, 2011
Place-value congruency effect	Kallai and Tzelgov (2012)							
Carry and borrowing				Ashcraft and Stazyk 1981; Groen and Parkman 1970				
Multiplication consistency effect					Verguts and Fias 2005; Domahs et al. 2007			

(Continued)

Table 7.2 Continued

Effects	Tasks							
	Magnitude Comparison	Parity judgement	Number naming	Addition/ Subtraction	Multiplication	Number line estimation	Number bisection	Other
Decade crossing effects							Nuerk et al., 2002; Wood et al. 2008	
Decade number effects			Brysbaert 1995	Brysbaert et al. 1998			Nuerk et al, 2002; Wood et al. 2008	
Fraction effects	Meert et al. 2009; Schneider and Siegler 2010							
Decimal effects	Desmet et al. 2010							
Multilinear number line effects						Ebersbach et al. 2008; Moeller et al. 2009d		
Operational momentum effect				Knops et al. 2009; McCrink et al. 2007		Lindemann et al., 2011		
Size congruity effect	Ganor-Stern et al. 2007							
Probability / cumulative effects								Barbey and Sloman, 2007

Table 7.3 Categorization indicating in what way effects described in the text are directly associated to (marked 'x') or influenced by [marked '(x)'] the assumed underlying basic numerical representations.

	Representations					
Effects	Number Form	Magnitude	Spatial Magnitude	Verbal/Arithmetic facts	Place-value	Procedures, rules, etc.
Unit-decade compatibility effect		(x)		(x)	x	x
Priming effects	x					
Cross-place congruity effect		x			x	
Transcoding effects				X	x	
Positional digit repetition effects	x				x	
Serial order effects	x			X		
Place-value congruency effect		x			x	
Carry and borrowing		x			x	x
Multiplication consistency effect				X	x	
Decade crossing effects		x			x	
Decade number effects					x	
Fraction effects		x			(x)	(x)
Decimal effects		x			x	
Multilinear number line effects		x	x	(x)	x	
Operational momentum effect		x	x			x
Size congruity effect		x				
Probability/cumulative effects		x				x

number form (in the fusiform gyrus bilaterally) much in analogy to the visual word form area (Dehaene and Cohen, 2011 for a review). For multi-digit number processing in particular, the successful identification of a string of symbols is essential. Impairments of processing the visual number form are rare but have nevertheless been reported (e.g. Cohen and Dehaene 2000; Dehaene and Cohen 1995; Temple 1989).

Semantic Representation of Numerical Magnitude

Numerical magnitude is probably the most basic and most prominent semantic information to be represented by numbers (cf. Dehaene and Cohen 1995; Dehaene 2003) and is supposed to be automatically activated whenever we encounter a number (e.g. Dehaene and Akhavein 1995; Nuerk et al. 2005b; but see Cohen 2009). Therefore, the representation of number magnitude is often seen as the core numerical representation which underlies symbolic and non-symbolic magnitude alike (Cohen Kadosh and Walsh 2009, for a discussion; Piazza et al. 2007). A total loss of the representation of number magnitude is unlikely as it is assumed to be subserved by structures in the intraparietal sulcus bilaterally (but see Cipolotti et al. 1991; Delazer et al. 2006; Dehaene and Cohen 1997, for case studies). However, deficits in magnitude representation, be it symbolic, non-symbolic, single-digit or multi-digit, representations have often been associated with difficulties in numerical development (e.g., Ashkenazi et al. 2008; Halberda et al. 2008; Holloway and Ansari 2009; Kaufmann and Nuerk 2008; Moeller et al. 2009b; Rousselle and Noël 2007; Rubinsten and Henik 2005). Accordingly, this has led some authors to consider this representation to be the best candidate for a core number sense and its related developmental deficits (Wilson and Dehaene 2007). For multi-digit number processing, the nature of the magnitude representation is one of the most controversially debated topics in numerical cognition research; the debate on holistic, strictly decomposed, or hybrid, representations of multi-digit integers and fractions is actually a debate about multi-digit *magnitude* representations.

Verbal Numerical Representations

The verbal representation of numbers is the third representation originally proposed by the Triple Code Model of Dehaene and colleagues (Dehaene 1992; Dehaene and Cohen 1995; 1997). In the first place it refers to the written or spoken number words associated with a given number. Indeed, mapping the verbal representation of numbers to the corresponding symbolic Arabic number system presents quite a challenge in numerical development in general and particularly when the correspondence between symbolic and number word notation is not transparent (e.g., Pixner et al. 2011a; Seron and Fayol 1994; Zuber et al. 2009). Moreover, it has been suggested that arithmetic fact knowledge, in particular multiplication facts, are stored in a verbal format in long-term memory (e.g., Cohen and Dehaene 2000; Delazer et al. 2003; Ischebeck et al. 2006; Lee and Kang 2002; Moeller et al. 2011c). The presence of specific verbal deficits in children has been reported as well (e.g., Dehaene and Cohen 1995, 1997; Geary 1993; Temple 1991).

Generally, deficits associated with the verbal representation of numbers are almost always deficits of processing multi-digit numbers. So-called lexical transcoding errors (e.g., naming '5' as 'six') are actually very rare, whereas syntactic transcoding errors (e.g., naming 31 as thirteen) are much more frequent (e.g. Zuber et al. 2009).

Spatial Representation of Numbers

The spatial representation of number magnitude is often referred to by using the metaphor of a *mental number line,* upon which numbers are represented in ascending order from left to right (Dehaene et al. 1993; Restle 1970). In the earliest version of the Triple Code Model by Dehaene and Cohen (1995), the spatial representation of numbers was considered part of the semantic magnitude representation of numbers. However, Dehaene et al. (2003) suggested that semantic magnitude representation and spatial magnitude representation may have different underlying neural correlates (namely, the horizontal part of the intraparietal sulcus for the semantic magnitude representation, and the posterior superior parietal lobule for the spatial representation). Interestingly, Nuerk et al. (2005b) observed behavioral effects of semantic number magnitude in the absence of spatial-numerical effects of the same magnitudes indicating that spatial and semantic representations of magnitude can differ. Based on this, we suggest a spatial representation of number magnitude is a separate representation, which may or may not be co-activated with semantic magnitude representations.

Interestingly, explicit spatial representations of numbers, when people report an explicit and reliable mental spatial arrangement of numbers, are a rather frequent phenomenon (also termed number form synaesthesia, cf. Sagiv et al. 2006, see also Cohen Kadosh and Henick 2007, for a discussion, Gertner et al., in press, for implications). Seron et al. (1992, see Galton, 1880, for earlier data) reported that up to 26 per cent of their female and 14 per cent of their male participants reported interindividually highly varying but intraindividually highly reliable mental number lines. Given that such explicit spatial-numerical representations are quite frequent, it is astonishing that they are studied so rarely. However, even without such conscious spatial number lines, a spatial representation of numbers seems to be quite universal. In left-to-right reading cultures, smaller numbers are responded to faster with the left hand and larger numbers with the right hand (the so-called SNARC-effect; Dehaene et al., 1993; Wood et al. 2008, for a meta-analysis; but see Gevers et al. 2010, for a different account, see also van Dijck et al., this volume).

The association of numbers and space is observed in elementary school children (Berch et al. 1999; van Galen and Reitsma 2008) but can also be found even earlier in life (e.g., de Hevia and Spelke 2009, 2010; McCrink and Wynn 2009; Opfer and Furlong 2011; Patro and Haman 2012). Specific deficits of the spatial representation of numbers have been observed both in patient studies (e.g., Hoeckner et al. 2008; Zorzi et al. 2002) and in developmental studies (e.g., Bachot et al. 2005).

Strategic, Conceptual, and Procedural Components

Already in an early version of the Triple Code Model by Dehaene and Cohen (1995) it was postulated that strategic, conceptual, and procedural components are important for numerical cognition, in particular for the case of complex which means predominantly multi-digit tasks. We do not argue that these latter numbers reflect a unitary representation
of their own, rather a category of still to-be-specified representations and processes needed for complex numerical tasks.

It is important to note that we suggest the relationship between the components of this category and the other more basic representations described above depends on the respective task at hand and the actual availability of these other more basic representations. To illustrate this argument consider the following example: In the verification version of the number bisection task, participants have to decide whether the middle number of a given triplet also represents the numerical mean of the two outer numbers (e.g., 12_15_18 vs. 12_16_18). This task is generally considered to tap on the semantic magnitude representation because it requires a magnitude decision (e.g., Dehaene and Cohen 1997, but see also Zorzi et al. 2002; Hoeckner et al. 2008, for spatial-numerical components). Nevertheless, it was also observed that triplets taken from a multiplication table (e.g., 21_24_27) were responded to faster than triplets which were not part of a multiplication table (e.g., 22_25_28, e.g., Nuerk et al. 2002; Moeller et al. 2009e). We have attributed this to strategic or procedural knowledge indicating that the middle number of a multiplication table is always the numerical mean of the outer numbers (Moeller et al. 2009e). In this example, strategies and procedures in a magnitude decision task systematically recruit (verbal) representations of number facts. Thus, strategic, conceptual, and procedural components complement the processing of basic representations to ensure the most economic way to deal with the task at hand.

Future research may lead to a further differentiation of this category because automatic (possibly rule-based) procedures (e.g., x*0 = 0), the understanding of number concepts (e.g., the relation between addition and multiplication) and the choice of adaptive strategies (e.g., back-up strategies like x*9 = x*10—x) may be distinguishable conceptually and/or with respect to their neuro-cognitive correlates. Nevertheless, specific deficits for procedural components of arithmetic concepts have been reported (Geary 1993; Temple 1991) and the use of procedural and strategic components refers almost exclusively to multi-digit numbers because procedures and strategies are usually not needed when only single-digit numbers are involved as operands and results of an arithmetic task.

Structural Representation of The Symbolic Number System (Place-Value Representation)

Last but not least we suggest a structural representation, serving as some kind of framework for our understanding of multi-digit integer numbers. As we view this representation to be of particular importance, we will introduce it in more detail and discuss its implications in the remainder of this chapter.

A cognitive taxonomy of number systems (Zhang and Norman 1995) shows that the Arabic symbolic system has one quite unique attribute: its place-value structure. Rather than number values being associated with certain symbols like in the Roman number system (X for 10, C for 100), the number values (0,1, …,9) are associated with certain positions in a string of digits.

Considering this, it is obvious that place-value understanding plays a crucial role for virtually all multi-digit number processing that has been studied in the literature. But is it justified to postulate this external structural representation as an internal mental representation of its own?

We argue that this is the case. To begin with, the place-value representation is fast and automatic. For two-digit numbers, we automatically activate the values of the constituting digits as shown by the compatibility effect and its interaction with the magnitudes of units and decade distances (Nuerk et al. 2011, for a review). Like spatial representations, the place-value magnitude activation is automatically activated even when it is irrelevant for the task. Ganor-Stern et al. (2007) examined the size congruity effect. It describes that it is harder to decide which is the physically larger number when this number is numerically smaller and vice versa. This size congruity effect is affected more by the magnitude of the decade digits than by the unit digits, indicating that the value of one digit in the place-value system is processed automatically as has been found for other representations. Recently, Kallai and Tzelgov (2012) provided further evidence for the influence of the place-value structure of the Arabic number system. In their study, participants had to compare items like, for instance 050_vs. 003 or 030 vs._005 (among other neutral trials such as 030 vs. 050). Importantly, participants had to indicate which digit string contained the larger non-zero digit while ignoring the zeros. Nevertheless, participants were slower and (at least for small distances) more error-prone for items in which the larger number was at the position with the lower place-value (i.e., 030 vs._005, considering place-value 30 vs. 5, whereas participants simply had to respond to 5 as being larger than 3). This is an illustrative example that shows although the place-value of the digit was irrelevant for the task at hand, it was automatically processed by the participants and could not be ignored (see also Moeller et al. 2009c, for similar evidence). Thus, place-value knowledge is not something that is cognitively constructed when necessary but rather automatically activated (also see Chan et al. 2011, for data observed in children).

Second, one could argue that the place-value representation is a cultural convention and not an innate representation like semantic magnitude itself. The latter may be true for the magnitude representation, but all the other representations, number form, verbal representation of numbers and even the spatial representation of numbers are also cultural conventions because they depend on the individual number forms used in a culture, the number words given for particular numbers, and the reading direction influencing particular number-space associations (see e.g., Chen and Verguts 2012, for possible origins of spatial numerical associations). In fact, most problems children and patients have with verbal number representations refer to the syntactic structure of the verbal number representation (Blanken et al. 1997; Pixner et al. 2011b; Zuber et al. 2009). Similarly, for symbolic numbers, many problems children have also refer to the syntactic structure of the symbolic magnitude representation; and early place-value understanding is a better predictor for later arithmetic achievement than indices of number magnitude representation (Moeller et al., 2011a).

Third, place-value integration of multi-digit numbers seems to be subserved by a particular neuro-cognitive structure, namely, posterior portions of the intraparietal sulcus (Knops et al. 2006; Wood et al. 2006; Wood et al. 2008) in which a purely semantic representation of magnitude (value) and a spatial representation (place) may converge.

Theoretical Considerations on Multi-Digit Place-Value Processing

In this book chapter, we wish to present another theoretical distinction for multi-digit number processing, which to our knowledge has not been introduced so far: The distinction between place identification, place-value activation, and place-value computation.

3.1 *Place identification* refers to the fact that in order to understand multi-digit numbers, one first has to correctly identify the positions at which the individual digits are located. This holds for multi-digit integers (e.g. hundred position, decade position, unit position), decimals as well as fractions (e.g. nominator, denominator position). A place identity representation may be activated in some paradigms without the corresponding value being automatically activated (e.g., priming effects, Ratinckx et al. 2005; but see Gazzellini and Laudanna 2011).

3.2 *Place-value activation* refers to the fact that in order to understand multi-digit numbers, one does not only have to identify the places at which particular numbers are located, but also the values which are associated with these positions. The interaction of the compatibility effect with unit magnitude distances and decade magnitude distances (Nuerk et al. 2001) shows that such magnitudes can be automatically activated in number magnitude comparison (see also Kallai and Tzelgov 2012). However, the differential size congruity effect for two-digit numbers (Ganor-Stern et al. 2007) shows that place-values for multi-digit integers are even automatically activated when no numerical response is required.

3.3 *Place-value computation* refers to the fact that in multi-digit number processing it is sometimes no sufficient to identify the places for digits and to activate the magnitudes associated with those places. Sometimes computations across place-values are necessary to solve a multi-digit arithmetic task. An example is the carry operation: In trials like 47 + 38, the added magnitudes of the unit digits have to be carried over to the decade digits, i.e., magnitudes of units and digits are no longer only separately presented but have to be computed in relation to each other. This process can be rather complex—Moeller et al. (2011d) observed at least three subprocesses involved in the carry operation.

In the remainder of this chapter, we will discuss this differentiation of place identification, activation, and computation with reference to a domain of research particularly suited to investigate characteristics of multi-digit number processing: language differences in numerical cognition. The intention for doing so is twofold: On the one hand, we wish to illustrate those theoretical considerations with real world instances. On the other hand, we aim to indicate that language influences multi-digit number processing in a more essential way than single-digit number processing.

Language Modulations

Language modulates all aspects of multi-digit number processing and a variety of language related effects have been observed (see Nuerk et al. 2011 for a review). In the following, we will elaborate on specific language related differences regarding (i) the verbal representation of numbers, (ii) the number magnitude representation and its spatial representation, and (iii) calculation competencies.[5]

Place Identification

The most basic process of place identification primarily involves perceptual processes of encoding the fact that the relevant symbols (in our case Arabic digits) are spatially ordered in a very specific manner: with each position step to the left, the value of the respective digits increases by a power of ten. Therefore, we assume that on a first processing level the single digits of a multi-digit number need to be differentiated and assigned to their positional identity (i.e., units, tens, hundreds, etc). This process is in some way comparable to the visual number form representation suggested by Dehaene and Cohen in the Triple Code Model (1995; 1997). Yet, the visual number form representation is assumed to subserve identification of the single digits and digit strings, whereas place identification is supposed to specifically assign these digits in their place-value information. Importantly, it has to be noted that we assume place identification to be a rather low level perceptual process differentiating and assigning the place-value stack information of the single digits. Therefore, the semantic numerical, which means quantity information conveyed by the individual digits, is not yet processed at this early stage of place identification.

Not surprisingly, correct identification of units, tens, hundreds, and so on, based on the verbal representations of numbers, differs across languages. Languages are more or less transparent with respect to the construction of their number words. In Japanese '327' is verbally coded as 'three hundred—two ten—seven'. Thus, the place-value structure of multi-digit numbers is coded explicitly in the Japanese number word system. In contrast, many European languages are less transparent. In many Western languages, the first multi-digit numbers whose place-value attributes have to be captured (i.e., 11, 12, etc.) are not transparent regarding their place-value properties. For instance, in English, these numbers are often coded with unique lexical terms like 'eleven' or 'twelve' and not as ten-one or ten-two reflecting their place-value structure. The latter system would allow a child to acquire more transparent knowledge about the composition of 'eleven' to be composed of ten and one. And indeed, this seems to affect children's understanding of the verbal representation of numbers. Nuerk et al. (2005) showed that German first grade children committed many more transcoding errors (both in writing and reading) than comparable first grade Japanese children. More importantly, the errors committed were not arbitrary—the type of errors corresponded to the inversion property of the German number word system. In German, Dutch, Maltese, Arabic, and other languages, the number word order of two-digit numbers is inverted. This means that '54' is spoken as 'fünfundvierzig' <four-and-fifty>. German children have trouble with this inversion and, consequently, when hearing or reading '54', they may write 45 and thus misidentify units and tens.

Interestingly, also most of the (altogether very few) errors in Japanese corresponded to a peculiarity of the Japanese number word system. For instance, '316' in Japanese is 'three-hundred—ten—six' but not 'three-hundred—one-ten—six'. No value at the tens position is named when there is a '1' at the tens position. Consequently, some Japanese children would write 306 instead of 316 because no (i.e., possibly null) value is verbally assigned to the tens position (Nuerk et al. 2005).

However, even among the European languages, the number word systems differ considerably (see also Pixner et al. 2011a; Zuber et al. 2009). For the verbal representation, Seron and Fayol (1994) showed that these differences influence performance in transcoding. They compared Belgian French and French as it is spoken in France. The difference is that in Belgium there is a decade-based number word system like in English with septante, octante, nonante for the numbers 70, 80, 90, respectively. In French, there are two differences compared to the Belgian system. First, the decade number words are not transparent; e.g., 80 is spoken as quatre-vingt (literally translated four-twenty). Second, from 60 onwards, there is a number word system in French which follows a base-20-system; So '98' would be 'quatre-vingt-dix-huit' (literally translated four-twenty-ten-eight). Therefore, when French children have to write down '98' to dictation, they would probably write 42018. Again, this might be interpreted as a failure to correctly specify the place-value identities of the respective number words and/or digits. Obviously, such an error would usually not happen in Belgium or any other country without a base-20 number word system.

Taken together, errors in transcoding from and/or into the verbal representation of numbers can be attributed to misidentification of the place-value identities of the digits involved. This is further corroborated by the fact that such inversion transcoding errors have not been reported in other languages without inversion (e.g., French: Barrouillet et al. 2004; Camos 2008; Italian: Power and Dal Martello 1990; 1997).

Nevertheless, it is important to note that while transcoding may influence or may be influenced by the assignment of semantic magnitude information, there is also evidence that transcoding can be performed without such magnitude assignment along an asemantic route both by children and patients (see, Cipolotti and Butterworth 1995, for patients). In fact, the typical syntactical transcoding errors described reflect correct identification of the constituent elements but erroneous identification of the place of the correctly encoded digits. Additionally, it does not seem plausible to assume place identification to involve a semantic transcoding route because then children should be aware that 29 and 92 are different numbers (at least regarding their semantic values). Therefore, such syntactic transcoding errors do not seem to be magnitude-related in the sense that place-value has already been activated. However, the influence of the verbal representation of numbers is not limited to the case of place identification but also generalizes to processes of place-value activation.

Place-Value Activation

The fact that we have affects transcoding may not seem too surprising because the number word system is directly involved. However, inversion also influences multi-digit number processing, even when number words are not directly tapped and processing of the digits' magnitude information is necessary beyond their place-value identities. For instance, Nuerk et al. (2005a) examined the compatibility effect in two-digit number magnitude comparison for different languages with (German) and without inversion (English). The hypothesis regarding inversion is straightforward. When number words influence unit interference in the compatibility effect, this interference should be larger in a language with inverted number words because the order of the Arabic digits is incongruent with the number word order. Therefore, these unit and decade digits at different positions of the place-value system should be harder to discern, resulting in a more pronounced compatibility effect. Importantly, this hypothesis was corroborated by the data. For Arabic number comparison, the compatibility effect was larger in German adults than in English adults (see also Macizo and Herrera 2010, for differences in number words; but see Ganor-Stern and Tzelgov 2011).

However, processing of two-digit numbers is highly automatic and parallel in adults and differs from that of young children. Nuerk et al. (2004) observed that multi-digit number integer processing in young children seemed to be sequential in a left-to-right order.

Interestingly, this sequential processing is—comparable to adults—influenced by processes of cognitive control. When more within-decade trials are included in the stimulus set, making the units more salient (see above), even first-grade children seem to focus more prominently on the units and as a consequence exhibit more parallel processing (see Mann et al. 2011; Pixner et al. 2009).

Nevertheless, the question in this section is whether language influences on multi-digit number processing generalize to children in general and whether such influences can be observed for Arabic number magnitude processing in particular. Pixner and colleagues (2011a) studied such language influences on two-digit magnitude comparison in first grade children from Austria (German language with inversion), Italy (without inversion), and Czech (with both inverted and non-inverted number forms). The results for unit interference in the compatibility effect study with only Arabic numbers were straightforward. The units interfered most in the language with inversion (German), least in the language without inversion (Italian) and intermediately in the language with inverted and non-inverted number word forms (Czech). Taken together, the results showed language influences even when no number words were involved and not only place identification but also place-value activation were required as in number magnitude comparison. Nevertheless, there is also evidence for language influences on place-value activation for the spatial representation of numbers.

Like the magnitude representation itself, the spatial representation of multi-digit numbers has long been regarded as an analogue holistic spatial magnitude representation. It has been argued that children start with a logarithmic spatial magnitude representation and later with increasing age and experience they acquire a linear spatial magnitude representation (Opfer and Siegler 2007; Siegler and Opfer 2003). However, one fundamental question remains: when the non-spatial representation of multi-digit number magnitude is decomposed (see above, Nuerk et al. 2001; Nuerk and Willmes 2005): Why should the spatial representation of number magnitude be holistic? And indeed, the idea that the spatial magnitude representation of multi-digit numbers is holistic has been questioned in a series of studies and opinion papers (Barth and Paladino 2011; Barth et al. 2011; Ebersbach et al. 2008; Helmreich et al. 2011; Moeller and Nuerk 2011; Moeller et al. 2009d; Slusser et al. 2013; Sullivan et al. 2011).

For instance, according to the bilinear account, at a certain point of development different linear space-to-number mappings exist for single—and two-digit numbers. Consider that a child is somewhat familiar with the place-value system but not yet with its base-10 structure. As a consequence, this child understands that the distance between 30 and 40 may be larger than between 3 and 4. Yet, it may be less than 10 times as large, for example only two-and-a-half or five times as large. If this is the underlying assumption of the child, a four times ($4 = 10/2.5$) or twice ($2 = 10/5$) flatter regression line would result for two-digit as compared to single-digit numbers. Moeller and colleagues (2009d) have outlined this argument in detail and related it to regression RT data in adults as follows.

However, when older children and/or adults may have learned to apply this linear decimal relation, does this necessarily mean that they represent number magnitude in such a way? Perhaps not. In regression analyses of two-digit number comparison data, we usually do not observe the beta weight of linear decade distance to be 10 times as large as the beta weight of linear unit distance (see Nuerk and Willmes 2005, for a review). In fact, when reanalyzing the original data of Nuerk et al. (2001) by incorporating only the predictors linear decade and unit distance, the beta weight for linear decade distance was only 4.39 times larger than that for linear unit distance. At the least, these results open up the possibility that even in adults tens might not be represented as being 10 times as large as units. Nevertheless, adults may have learned to overcome this representational bias when they solve numerical tasks that require processing of place value information.

However, if this assumption of decomposed processing in the number line task is true, decomposed spatial magnitude processing should be similarly affected by language as magnitude comparison. Indeed, cultural effects have been reported. Scottish children were—surprisingly—relatively better than Chinese children (Muldoon et al. 2011), while in another study Chinese children were relatively better than North American children (Siegler and Mu 2008), and the Italian children were relatively better than Austrian (German-speaking) children (Helmreich et al. 2011). However, general differences may not be due to specific language characteristics such as the transparency of the number word system, but rather caused by general educational and curricular differences between cultures. Therefore, more specific differences corresponding to a language's number word system should be examined. Helmreich and colleagues (2011) investigated whether inversion affected performance in the number line task (range: 0–100) in a specific way.

Two specific influences of inversion were observed.

(i) The authors compared items with a large inter-digit distance (e.g. 28 or 93 have a distance of six between decade and unit digit) to items with a smaller inter-digit distance (e.g., 45 or 87 have a distance of one between decades and units). If inversion led to systematic errors in marking a two-digit number on a multi-digit number line, estimation errors for numbers with a large inter-digit distance should be much more affected by inversion because marking 82 instead of 28 produces a large deviation on the number line. In contrast, numbers with a small inter-digit distance should be less affected by inversion because marking 45 instead of 54 does not result in a large estimation error. Indeed, this was corroborated by the data. The performance of the Italian children did not differ much from that of Austrian children for numbers with a small inter-digit distance but they were reliably more accurate for numbers with a large inter-digit distance.

(ii) The influence of inversion would predict another systematic error in the number line task in languages with inversion is not related to the overall error per se but to the direction of the error. When an inversion error is committed, numbers like 49 should rather be overestimated more strongly (because they might be confused with 94 due to inversion), while numbers like 51 should rather be underestimated less because they may be mixed up with 15. And indeed, Helmreich and colleagues (2011) observed such directional errors in the number line task to be more pronounced in German-speaking children than in Italian-speaking children, underlining a possible influence of the inversion property and thereby of language on place-value activation for the spatial representation of numbers.

Place-Value Computation

Finally, there is now first evidence indicating that calculation and in particular place-value computation as required in carry operations is specifically influenced by language as well.

While general cross-cultural differences in arithmetic have often been reported, such specific language effects have been noted less often. Brysbaert et al. (1998) studied two-digit addition in Dutch (with inversion) and Walloon adults (without inversion) and observed language-specific effects with oral but not with manual responses. Moreover, addition performance also seems to be influenced by the inversion property in children (Göbel et al. 2014). In a choice reaction task, German-speaking children (number words with inversion) and Italian children (without inversion) had to respond to addition problems with and without carry. As argued above, carry problems (e.g., 27 + 48 = 75) not only require an understanding of the place-value system but also explicit place-value computation because different sums of units and tens have to be related to each other in line with the constraints of the place-value base-10-system. This means that the tens digit of the unit sum must be carried to the tens stack to compute the correct tens digit of the result (see Moeller et al. 2011d, for a systematic analysis with eye-tracking). Furthermore, this carry has to be kept in working memory as an intermediate result in mental addition. Again, this should be much more difficult when the corresponding language system is incongruent with regard to the order of number words for tens and units because this order has to be constantly monitored and manipulated. Therefore, carry problems should not only be more difficult in general but also more difficult in languages with inversion. Göbel and colleagues (2014) observed in fact that the carry effect was larger in German—as compared to Italian-speaking children, although this effect was mediated by working memory performance.

However, inversion is not the only property which influences place-value computation. Recently, Colomé et al. (2010) investigated the role of the base-20 system in calculation in Basque. The Basque number word system has another language specificity. It is a full base-20 system. 35 is spoken as 20 + 15. And indeed, such addition problems were solved faster than comparable problems like 25 +10. In contrast to Brysbaert et al. (1998), this result held for both oral and manual production. The study is important and instructive because so far, most studies about specific influences of the number word system so far have been about an impact of the inversion property[6].

In summary, it can be said that comparable to place identification and place-value activation, there are language related differences for place-value computation. In particular, inverted number words not only complicate place identification and place-value activation but also computation along place-value constraints. Thereby, it becomes evident, that the place-value representation is a necessary prerequisite for successful multi-digit (integer) number processing. As a consequence, an early understanding of the place-value structure of the Arabic number system seems essential for the numerical development of children. In the following, we will address this issue in more detail.

Long-Term Consequences of Early Multi-Digit Number Understanding

We have reviewed evidence that verbal and non-verbal basic number and arithmetic processing in general and place identification, place-value activation and computation in particular, are influenced by language properties. For instance, children in languages with non-transparent number words due to inversion commit more errors and in particular, many more inversion and thus place-value related transcoding errors. The important question now is whether this is relevant for later arithmetic performance or whether it is just a transient developmental phase? A recent longitudinal study by Moeller and colleagues (2011a) provides a first answer to this question. For single-digit numbers, it is well known that basic representations influence later arithmetic performance (e.g., Moeller et al. 2011a), although causal inferences need to be made with care because most data on the relation of basic and later arithmetic abilities are correlational (e.g. Holloway and Ansari 2009). For multi-digit numbers, Moeller et al. (2011a) observed three longitudinal effects.

First, successful basic multi-digit number processing in first grade, as indexed by successful two-digit comparison, predicts better multi-digit addition performance in third grade. This prediction was specific for numerical capabilities because non-numerical measures of IQ as well as verbal and visual working memory were also included in this regression but did not explain any additional variance.

Second, and this is important with regard to language modulation, Moeller et al. (2011a) followed what they called an *effect-based approach* instead of a *task-based approach*. The task-based approach is so far predominantly used in longitudinal research and assumes that a particular representation can be indexed by the performance in a corresponding task. For instance, magnitude representation may be indexed by performance in a magnitude comparison task. Although this approach is commonly used, it is at odds with an approach using (multi-digit) number processing effects as indices for particular representations such as the distance effect for the magnitude representation. Following an effect-based approach, Moeller and colleagues (2011a) investigated not only overall task performance but also specific effects which relate to multi-digit number processing and are specific to or more pronounced in the inverted number word system of their German speaking sample. They used the number of inversion errors in transcoding as a predictor in first grade; such errors only occur in languages with inversion (cf. Pixner et al. 2011a). They also used the compatibility effect as a first predictor; the effect is larger in German than in other languages due to the inversion property of German number words which seems to make incompatible trials in particular more difficult. The authors observed that besides the distance effect (thereby replicating Holloway and Ansari 2009, for two-digit number stimuli), inversion errors and the compatibility effect explained most of the variance of later addition performance; fewer inversion errors and a smaller compatibility effect were associated with better addition performance in the future.

Finally, specific numerical effects have been used as predictors previously (e.g., Holloway and Ansari 2009) but rarely as a criterion variable. Using them as a criterion variable means that we do not only predict overall performance but also the indices of specific representations underlying that performance. Moeller and colleagues (2011a) therefore used the carry effect as a criterion for place-value processing in carry and non-carry trials separately. Inversion errors predicted the carry effect in particular and also the performance in carry trials but not in non-carry trials. The higher the number of inversion transcoding errors in grade 1, the larger the carry effect/the more errors occurred in carry problems in grade three. We have argued above that carry trials present a particularly difficult case of place-value computation which is heavily influenced by language. This place-value computation in third grade is heavily influenced by inversion errors in first grade although various other numerical and non-numerical variables were controlled for in the regression to partial out general effects. In non-carry trials, when there are fewer demands on place-value integration, the distance effect was the only numerical predictor. The compatibility effect followed the same pattern as the inversion errors being a significant predictor for carry trials but not for the less complex non-carry trials, for which distance was the strongest predictor.

Why is this longitudinal study important? It is important because it shows that early place-value understanding (i.e., place identification (transcoding) and place-value activation (compatibility effect)), which is strongly related to the language system, predicts later arithmetic performance (i.e., place-value computation). Importantly, for multi-digit numbers, the variance explained by such specific multi-digit number effects is larger than the variance explained by overall magnitude effects (distance) and by non-numerical variables (IQ, working memory). In sum, there is first evidence suggesting that for complex multi-digit numbers, specific multi-digit number effects predict later performance better than simple distance effects or non-numerical variables.

As regards the developmental aspect, early place-value understanding is not only a hurdle which might be higher in non-transparent number systems, rather it continues to influence calculation performance with Arabic numbers later. And it does so in a very specific way. Those trials (carry trials), which require more sophisticated and effortful manipulation of the place-values of the numbers involved, are predicted by those predictors which index the place-value system and possible problems with it (inversion errors in transcoding). This influence also generalizes to the mathematics marks at the end of grade three. The more inversion related transcoding errors a child committed in grade one, the worse its mathematics grade was two years later. Thereby, early place-value understanding is an externally valid predictor of later arithmetic performance.

Summary, Conclusions, and Perspectives

In this chapter on multi-digit number processing, we addressed five issues: (i) we outlined why single-digit number processing may not be sufficient for understanding multi-digit number processing. (ii) We argued that even though there are numerous studies specifically examining multi-digit number processing, our overview shows that there is a research bias towards the properties of certain representations (namely magnitude representation). While this may reflect the theoretical importance of the magnitude representation, it also indicates that in our view other representations (e.g., place-value representation) have been less well studied. (iii) We introduced a new conceptual distinction for multi-digit numbers between place-identification, place-value activation, and place-value computation. Future studies are needed to evaluate whether this conceptual framework will prove empirically helpful. (iv) We devoted a large part of this chapter to language influences which have never been described in such detail in line with that new conceptualization. We argued that language influences are probably most pronounced for multi-digit numbers because the differences between number word systems are more pronounced for multi—than for single-digit numbers. Thus, the focus on single-digit numbers in number processing research may have led our field to underestimate language and cultural influences. (v) Finally, regarding numerical development, we indicated that predictors specific to multi-digit number processing (e.g., inversion transcoding errors) may predict later arithmetic performance better than predictors reflecting single-digit number processing or non-numerical cognitive functioning.

Taken together, we are confident that the current review not only showed that multi-digit number processing cannot be understood by studying single-digit number processing alone, but also that we may understand many questions like language influences on numerical cognition, predictions of later arithmetic performance, etc. more thoroughly when specifically considering multi-digit numbers as tools and means to examine these questions. In number processing research, there has been a long and fruitful tradition in studying single-digit numbers and their properties. However, sometimes traditions may also keep the scientific community on well charted roads, while new pathways to less-studied domains are not taken. We suggest that multi-digit number processing may be such an understudied domain. As a consequence, our understanding not only of multi-digit integers but also of fractions, proportions, decimals, etc. is still scarce and may deserve future research.

Acknowledgments

We wish to thank Marielle Borsche and Joshua Schmidt for their assistance with the reference list and Jennifer Pröhl for proofreading the English grammar and spelling.

Notes

1. Please note that decomposed processing as indicated by the unit-decade compatibility effect seems to be a general characteristic of multi-digit number processing rather than a peculiarity only occurring under particular circumstances. This notion is corroborated by recent observations indicating that the compatibility effect is (i) not restricted to a particular external perceptual presentation format (e.g. Ganor-Stern and Tzelgov 2011; Nuerk et al. 2004; Macizo and Herrera 2008; Moeller et al. 2009c) (ii) not specific to two-digit numbers but extends to multi-digit numbers (Korvorst and Damian 2008; Mann et al. 2012; Meyerhoff et al. 2012). (iii) not restricted to adult participants but can already be observed in children (e.g. Mann et al. 2011; Nuerk et al. 2004; Pixner et al. 2011a).
2. Following the cross-multiplication strategy, the larger one of two fractions m/n and x/y can be identified by comparing the products of opposite numerators and denominators, that is, the products my and xn with $m/n > x/y$ if $my > xn$ and vice versa.
3. See Reynvoet et al. (2011) for distance effects different from the so-called comparison distance effect referred to here.
4. Note that the place-value congruency effect published by Kallai and Tzelgov 2012, has been added to the 16 effects summarized by Nuerk et al. (2011), so that there are 17 effects in the Table.
5. It should be noted that so far language relations have mostly been investigated for multi-digit integers and not for other multi-digit numbers such as unit fractions or decimal fractions. The fact that language properties influence multi-digit number processing in a specific way for both children and adults will hopefully become clear from the following. However, if - and if so - how language influences processing of other types of multi-digit numbers remains a topic for future investigation.
6. We are well aware that there are numerous studies which compare different calculation tasks across different cultures and henceforth, different languages (e.g. Campbell and Epp 2004; Campbell and Xue 2001; Imbo and LeFevre 2009, 2011; Wang et al. 2007 and many others). However, such studies were not considered here because they are either not about the specifics of multi-digit number processing in particular or do not investigate specific effects related to the specificity of a particular

language or number word system so that eventual differences may be also be due to culture, curriculum or education. In this section, only differences are presented which specifically correspond to a particular language property or peculiarity and therefore just show up under specific circumstances or as one particular corresponding error type (e.g., inversion error). Such effects cannot be explained easily by general differences in culture or curriculum because then one would have to account for the fact that only specific conditions are affected. Nevertheless, these specific effects can be explained best and most parsimoniously by language properties. Other explanations cannot be ruled out logically in transcultural studies due to their inevitable quasi-experimental nature.

References

Ashcraft, M.H., and Stazyk, E.H. (1981). Mental addition: A test of three verification models. *Memory & Cognition* 9: 185–96.

Ashkenazi, S., Mark-Zigdon, N., and Henik, A. (2009). Numerical distance effect in developmental dyscalculia. *Cognitive Development* 24: 387–400.

Ashkenazi, A., Henik, H., Ifergane, G., and Shelef, I. (2008). Basic numerical processing in left intraparietal sulcus (IPS) acalculia. *Cortex* 44: 439–48.

Bachot, J., Gevers, W., Fias, W., and Roeyers, H. (2005). Number sense in children with visuospatial disabilities: orientation of the mental number line. *Psychology Science* 47: 172–83.

Barbey, A.K., and Sloman, S.A. (2007). Base-rate respect: From ecological rationality to dual processes. *Behavioral and Brain Sciences* 30: 241–54.

Barrouillet, P., Camos, V., Perruchet, P., and Seron, X. (2004). ADAPT: a developmental, asemantic, and procedural model for transcoding from verbal to Arabic numerals. *Psychological Review* 111: 368–94.

Barth, H. and Paladino, A.M. (2011). The development of numerical estimation: Evidence against a representational shift. *Developmental Science* 14: 125–35.

Barth, H., Slusser, E., Cohen, D., and Paladino, A.M. (2011). A sense of proportion: Commentary on Opfer, Siegler, and Young. *Developmental Science* 14: 1205–6.

Beddington, J., Cooper, C.L., Field, J., Goswami, U., Huppert, F., Jenkins, R., Jones, H., Kirkwood, T.B. L., Sahakian, B.J., and Thomas, S.M. (2008). The Mental Wealth of Nations. *Science* 455: 1057–60.

Berch, D.B., Foley, E.J., Hill, R.J., and McDonough Ryan, P. (1999). Extracting Parity and Magnitude from Arabic Numerals: Developmental Changes in Number Processing and Mental Representation. *Journal of Experimental Child Psychology* 74: 286–308.

Blanken, G., Dorn, M., and Sinn, H. (1997). Inversion errors in Arabic number reading: is there a non-semantic route? *Brain & Cognition* 34: 404–23.

Bonato, M., Fabbri, S., Umilta, C., and Zorzi, M. (2007). The Mental Representation of Numerical Fractions: Real or Integer? *Journal of Experimental Psychology: Human Perception and Performance* 33: 1410–19.

Brown, V., Cronin, M.E., and McEntire, E. (1994). *TOMA-2 test of mathematical abilities*. Austin, TX: Pro-Ed.

Brysbaert, M. (1995). Arabic number reading: On the nature of the numerical scale and the origin of phonological recoding. *Journal of Experimental psychology: General* 124: 434–52.

Brysbaert, M., Fias, W., and Noël, M.P. (1998). The Whorfian hypothesis and numerical cognition: Is “twenty-four” processed in the same way as “four-and-twenty”? *Cognition* 66: 51–77.

Bynner, J. and Parsons, S. (1997). *Does numeracy matter?* London: The Basic Skills Agency.

Camos, V. (2008). Low working memory capacity impedes both efficiency and learning of number transcoding in children. *Journal of Experimental Child Psychology* 99 1: 37–57.

Campbell, J.I. D. and Epp, L.J. (2004). An encoding-complex approach to numerical cognition in Chinese-English bilinguals. *Canadian Journal of Experimental Psychology* 58: 229–44.

Campbell, J.I. D. and Xue, Q. (2001). Cognitive arithmetic across cultures. *Journal of Experimental Psychology: General* 130: 299–315.

Chan, W.W. L., Au, T.K., and Tang, J. (2011). Exploring the developmental changes in automatic two-digit number processing. *Journal of Experimental Child Psychology* 109: 263–74.

Chen, Q. and Verguts, T. (2012). Spatial intuition in elementary arithmetic: A neurocomputational account. *PLoS:ONE* 7: 2.

Cipolotti, L. and Butterworth, B. (1995). Toward a multiroute model of number processing: Impaired number transcoding with preserved calculation skills. *Journal of Experimental Psychology: General* 124: 375–90.

Cipolotti, L., Butterworth, B., and Denes, G. (1991). A specific deficit for numbers in a case of dense acalculia. *Brain* 114: 2619–37.

Cohen, D.J. (2009). Integers do not automatically activate their quantity representation. *Psychonomic Bulletin and Review* 16: 332–6.

Cohen, L. and Dehaene, S. (2000). Calculating without reading: Unsuspected residual abilities in pure alexia. *Cognitive Neuropsychology* 17: 563–83.

Cohen Kadosh R., and Henik A. (2007). Can synaesthesia research inform cognitive science? *Trends in the Cognitive Sciences* 11: 177–84.

Cohen Kadosh, R., and Walsh, V. (2009). Numerical representations in the parietal lobes: Abstract or not abstract? *Behavioral and Brain Sciences* 32: 313–28.

Colomé, A., Laka, I., and Sebastián-Gallés, N. (2010). Language effects in addition: How you say it counts. *The Quarterly Journal of Experimental Psychology* 63: 965–83.

Dehaene, S. (1989). The psychophysics of numerical comparison: A re-examination of apparently incompatible data. *Perception and Psychophysics* 45: 557–66.

Dehaene, S. (1992). Varieties of numerical abilities. *Cognition* 44: 1–42.

Dehaene, S. and Akhavein, R. (1995). Attention, automaticity, and levels of representation in number processing. *Journal of Experimental Psychology: Learning, Memory, and Cognition* 21: 314–26.

Dehaene, S. and Cohen, L. (1995). Towards an anatomical and functional model of number processing. *Mathematical Cognition* 1: 83–120.

Dehaene, S. and Cohen, L. (1997). Cerebral pathways for calculation: Double dissociation between rote verbal and quantitative knowledge of arithmetic. *Cortex* 33: 219–50.

Dehaene, S. and Cohen, L. (2011). The unique role of the visual word form area in reading. *Trends in Cognitive Sciences* 15: 254–62.

Dehaene, S., Dupoux, E., and Mehler, J. (1990). Is numerical comparison digital? Analogical and symbolic effects in two-digit number comparison. *Journal of Experimental Psychology: Human Perception and Performance* 16: 626–41.

Dehaene, S., Bossini, S., and Giraux, P. (1993). The mental representation of parity and number magnitude. *Journal of Experimental Psychology: General* 122: 371–96.

Dehaene, S., Piazza, M., Pinel, P., and Cohen, L. (2003). Three parietal circuits for number processing. *Cognitive Neuropsychology* 20: 487–506.

de Hevia, M.D. and Spelke, E.S. (2009). Spontaneous mapping of number and space in adults and young children. *Cognition* 110: 198–207.

de Hevia, M.D. and Spelke, E.S. (2010). Number-space mapping in human infants. *Psychological Science* 21: 653–60.

Delazer, M., Domahs, F., Bartha, L., Brenneis, C., Lochy, A., Trieb, T., and Benke, T. (2003). Learning complex arithmetic—an fMRI study. *Cognitive Brain Research* 18: 76–88.

Delazer, M., Karner, E., Zamarian, L., Donnemiller, E., and Benke, T. (2006). Number processing in posterior cortical atrophy—A neuropsychological case study. *Neuropsychologia* 44: 36–51.

Deloche, G., and Seron, X. (1982). From three to 3: A differential analysis of skills in transcoding quantities between patients with Broca's and Wernicke's aphasia. *Brain* 105: 719–33.

Desmet, L., Grégoire, J., and Mussolin, C. (2010). Developmental changes in the comparison of decimal fractions. *Learning and instruction* 20: 521–32.

Domahs, F., Delazer, M., and Nuerk, H.C. (2006). What makes multiplication facts difficult. *Experimental Psychology* 53: 275–82.

Domahs, F., Domahs, U., Schlesewsky, M., Ratinckx, E., Verguts, T., Willmes, K., and Nuerk, H.C. (2007). Neighborhood consistency in mental arithmetic: Behavioral and ERP evidence. *Behavioral and Brain Functions* 3: 66.

Ebersbach, M., Luwel, K., Frick, A., Onghena, P., and Verschaffel, L. (2008). The relationship between the shape of the mental number line and familiarity with numbers in 5- to 9-year old children: evidence for a segmented linear model. *Journal of Experimental Child Psychology 99*: 1–17.

Faulkenberry, T.J., and Pierce, B.H. (2011). Mental representations in fraction comparison: Holistic versus component-based strategies. *Experimental Psychology 58*: 480–9.

Galton, F. (1880). Visualised numerals. *Nature 21*: 252–6.

Ganor-Stern, D. and Tzelgov, J. (2011). Across-Notation Automatic Processing of Two-Digit Numbers. *Experimental Psychology 58*: 147–53.

Ganor-Stern, D., Tzelgov, J., and Ellenbogen, R. (2007). Automaticity of two-digit numbers. *Journal of Experimental Psychology: Human Perception and Performance 33*: 483–96.

García-Orza, J., and Damas, J. (2011). Sequential Processing of Two-Digit Numbers. *Journal of Psychology 219*: 23–9.

Gazzellini, S. and Laudanna, A. (2011). Digit repetition effect in two-digit number comparison. *Journal of Psychology 219*: 30–6.

Geary, D.C. (1993). Mathematical disabilities: Cognitive, neuropsychological, and genetic components. *Psychological Bulletin 114*: 345–62.

Gertner, L., Henik, A., Reznik, D., Cohen Kadosh, R. (2013). Implications of number-space synesthesia on the automaticity of number processing. *Cortex* 49(5): 1352–1362. doi: 10.10.16/j.cortex,2012.03.019.

Gevers, W., Santens, S., Dhooge, E., Chen, Q., Van den Bossche, L., Fias, W., and Verguts, T. (2010). Verbal-spatial and visuo-spatial coding of number-space interactions. *Journal of Experimental Psychology: General 139*: 180–90.

Gigerenzer, G., and Hoffrage, U. (1995). How to improve Bayesian reasoning without instruction: Frequency formats. *Psychological Review 102*: 684–704.

Ginsburg, H.P. and Baroody, A.J. (2003). *Test of Early Mathematics Ability* (3rd edition). Austin, TX: Pro-Ed, Incorporated.

Göbel, S., Moeller, K., Pixner, S., Kauffmann, L., and Nuerk, H.-C. (2014). Language affects symbolic arithmetic in children: the case of number word inversion. *Journal of Experimental Child Psychology 119*, 17–25.

Gross, J., Hudson, C., and Price, D. (2009). *The Long Term Costs of Numeracy Difficulties*. Every Child a Chance Trust and KPMG: London.

Guillaume, M., Nys, J., and Content, A. (2012). Is adding 48 + 25 and 45 + 28 the same? How addend compatibility influences the strategy execution in mental addition. *Journal of Cognitive Psychology 24*: 836–43.

Haffner, J., Baro, K., Parzer, P., and Resch, F. (2005). *Heidelberger Rechentest (HRT)*. Göttingen: Hogrefe.

Halberda, J., Mazzocco, M., and Feigenson, L. (2008). Individual differences in nonverbal number acuity predict maths achievement. *Nature 455*: 665–8.

Helmreich, I., Zuber, J., Pixner, S., Kaufmann, L., Nuerk, H.-C., and Moeller, K. (2011). Language effects on children's mental number line: How cross-cultural differences in number word systems affect spatial mappings of numbers in a non-verbal task. *Journal of Cross-Cultural Psychology 42*: 598–613.

Hoeckner, S., Moeller, K., Zauner, H., Wood, G., Haider, C., Gaßner, A., and Nuerk, H.-C. (2008). Impairments of the mental number line for two-digit numbers in neglect. *Cortex 44*: 429–38.

Holloway, I.D. and Ansari, D. (2009). Mapping numerical magnitudes onto symbols: The numerical distance effect and individual differences in children's mathematics and achievement. *Journal of Experimental Child Psychology 103*: 17–29.

Hubbard, E.M., Piazza, M., Pinel, P., and Dehaene, S. (2005). Interactions between number and space in parietal cortex. *Nature Reviews Neuroscience 6*: 435–48.

Imbo, I. and LeFevre, J. (2009). Cultural differences in complex addition: Efficient Chinese versus adaptive Belgians and Canadians. *Journal of Experimental Psychology: Learning, Memory, & Cognition 35*: 1465–76.

Imbo, I. and LeFevre, J.-A. (2011). Cultural differences in strategic behavior: A study in computational estimation. *Journal of Experimental Psychology: Learning, Memory, and Cognition 37*: 1294–301.

Ischebeck, A., Zamarian, L., Siedentopf, C., Koppelstätter, F., Benke, T., Felber, S., and Delazer, M., (2006). How specifically do we learn? Imaging the learning of multiplication and subtraction. *NeuroImage 30*: 1365–75.

Ischebeck, A., Schocke, M., and Delazer, M. (2009). The processing and representation of fractions within the brain: an fMRI investigation. *Neuroimage 47*: 403–13.

Jacob, S.N. and Nieder, A. (2009). Notation-independent representation of fractions in the human parietal cortex. *Journal of Neuroscience 29*; 4652–7.

Kallai, A., and Tzelgov, J. (2012). When meaningful components interrupt the processing of the whole: the case of fractions. *Acta Psychologica 129*: 358–69.

Kaufmann, L., and Nuerk, H.-C. (2008). Basic number processing deficits in ADHD: A broad examination of elementary and complex number processing skills in 9 to 12 year-old children with ADHD-C. *Developmental Science 11*: 692–9.

Kaufmann, L., Nuerk, H.-C., Graf, M., Krinzinger, H., Delazer, M., and Willmes, K. (2009). *TEDI-MATH: Test zur Erfassung numerisch-rechnerischer Fertigkeiten für 4–8 Jährige.* Bern: Hans-Huber-Verlag.

Knops, A., Nuerk, H.-C., Sparing, R., Foltys, H., and Willmes, K. (2006). On the Functional Role of Human Parietal Cortex in Number Processing: How Gender Mediates the Impact of a Virtual Lesion Induced by rTMS. *Neuropsychologia 44*: 2270–83.

Knops, A., Viarouge, A., and Dehaene, S. (2009). Dynamic representations underlying symbolic and nonsymbolic calculation: Evidence from the operational momentum effect. *Attention, Perception, & Psychophysics 71*: 803–21.

Korvorst, M., and Damian, M. F. (2008). The differential influence of decades and units on multi-digit number comparison. *The Quarterly Journal of Experimental Psychology 61*: 1250–64.

Lee, K., and Kang, S. (2002). Arithmetic operation and working memory: differential suppression in dual tasks. *Cognition 83*: B63–B68.

Macizo and Herrera, (2008). The effect of number codes in the comparison task of two-digit numbers. *Psicologica, 29*: 1–34.

Macizo, P., and Herrera, A. (2010). Two-digit number comparison: Decade-unit and unit-decade produce the same compatibility effect with number words. *Canadian Journal of Experimental Psychology 64*: 17–24.

Macizo, P., and Herrera, A. (2011). Cognitive control in number processing: Evidence from the unit–decade compatibility effect. *Acta Psychologica 136*: 112–8.

Mann, A., Moeller, K., Pixner, S., Kaufmann, L., and Nuerk, H.-C. (2011). Attentional strategies in place–value integration: a longitudinal study on two-digit number comparison. *Journal of Psychology 219*(1): 42–9.

Mann, A., Moeller, K., Pixner, S., Kaufmann, L. and Nuerk, H.-C. (2012). On the development of Arabic three-digit number processing in primary school children. *Journal of Experimental Child Psychology 113*: 594–601.

McCrink, K., Dehaene, S., and Dehaene-Lambertz, G. (2007). Moving along the number line: Operational momentum in nonsymbolic arithmetic. *Perception & Psychophysics 69*: 1324–33.

McCrink, K. and Wynn, K. (2009). Operational momentum in large-number addition and subtraction by 9-month-old infants. *Journal of Experimental Child Psychology 103*: 400–8.

Meert, G., Gregoire, J, and Noël, M.P. (2009). Rational numbers: Componential versus holistic representation of fractions in a magnitude comparison task. *The Quarterly Journal of Experimental Psychology 62*: 1598–616.

Meert, G., Grégoire, J., and Noël, M.P. (2010). Comparing 5/7 and 2/9: Adults can do it by accessing the magnitude of the whole fractions. *Acta Psychologica 135*: 284–92.

Meyerhoff, H.S., Moeller, K., Debus, K., and Nuerk, H.-C. (2012). Multi-digit number processing beyond the two-digit number range: A combination of sequential and parallel processes. *Acta Psychologica 140*: 81–90.

Moeller, K., Pixner, S., Klein, E., Cress, U., and Nuerk, H.-C. (2009a). Zahlenverarbeitung ist nicht gleich Rechnen—Eine Beschreibung basisnumerischer Repräsentationen und spezifischer Interventionsansätze. *Prävention und Rehabilitation 21*: 121–36.

Moeller, K., Neuburger, S., Kaufmann, L., Landerl, K., and Nuerk, H.-C. (2009b). Basic number processing deficits in developmental dyscalculia. Evidence from eye-tracking. *Cognitive Development 24*: 371–86.

Moeller, K., Nuerk, H.-C., and Willmes, K. (2009c). Internal number magnitude representation is not holistic, either. *The European Journal of Cognitive Psychology 21*: 672–85.

Moeller, K., Pixner, S., Kaufmann, L., and Nuerk, H.-C. (2009d). Children's early mental number line: Logarithmic or rather decomposed linear? *Journal of Experimental Child Psychology 103*: 503–15.

Moeller, K., Fischer, M.H., Nuerk H.-C., and Willmes, K. (2009e). Eye fixation behaviour in the number bisection task: Evidence for temporal specificity. *Acta Psychologica 131*: 209–20.

Moeller, K., and Nuerk, H.-C. (2011). Psychophysics of numerical representation: Why seemingly logarithmic representations may rather be multi-linear. *Journal of Psychology 219*: 64–70.

Moeller, K., Pixner, S., Zuber, J., Kaufmann, L., and Nuerk, H.-C. (2011a). Early place-value understanding as a precursor for later arithmetic performance—a longitudinal study on numerical development. *Research in Developmental Disabilities 32*: 1837–51.

Moeller, K., Huber, S., Nuerk, H.-C., and Willmes, K. (2011b). Two-digit number processing—holistic, decomposed or hybrid? A computational modelling approach. *Psychological Research 75*: 290–306.

Moeller, K., Klein, E., Fischer, M.H., Nuerk H.-C., and Willmes, K. (2011c). Representation of multiplication facts—Evidence for partial verbal coding. *Behavioral and Brain Functions 7*: 25.

Moeller, K., Klein, E., and Nuerk, H.-C. (2011d). Three processes underlying the carry effect in addition—Evidence from eye-tracking. *British Journal of Psychology 102*: 623–45.

Moyer, R.S. and Landauer, T.K. (1967). Time required for judgements of numerical inequality. *Nature 215*: 1519–20.

Muldoon, K., Simms, V., Towse, J., Menzies, V., and Yue, G. (2011). Cross-Cultural Comparisons of 5-Year-Olds' Estimating and Mathematical Ability. *Journal of Cross-Cultural Psychology 42*: 669–81.

Nuerk, H.-C. and Willmes, K. (2005). On the magnitude representations of two-digit numbers. *Psychology Science 47*: 52–72.

Nuerk, H.-C., Weger, U., and Willmes, K. (2001). Decade breaks in the mental number line? Putting the tens and units back in different bins. *Cognition 82*: B25–B33.

Nuerk, H.-C., Geppert, B.E., van Herten, M., and Willmes, K. (2002). On the impact of different number representations in the number bisection task. *Cortex 38*: 691–715.

Nuerk, H.-C., Kaufmann, L., Zoppoth, S., and Willmes, K. (2004). On the development of the mental number line. More, less or never holistic with increasing age. *Developmental Psychology 40*: 1199–211.

Nuerk, H.-C., Olsen, N., and Willmes, K. (2005). *Better teach your children Japanese number words: how transparent number structure helps number acquisition.* Poster presented at the 23rd Workshop of Cognitive Neuropsychology, Bressanone, January 23–8.

Nuerk, H.-C., Weger, U., and Willmes, K. (2005a). Language effects in magnitude comparison: Small, but not irrelevant. *Brain and Language 92*: 262–77.

Nuerk, H.-C., Bauer, F., Krummenacher, J., Heller, D., and Willmes, K. (2005b). The power of the mental number line: How the magnitude of unattended numbers affects performance in an Eriksen task. *Psychology Science 47*: 34–50.

Nuerk, H.-C., Graf, M., and Willmes, K. (2006). Grundlagen der Zahlenverarbeitung und des Rechnens. *In Sprache, Stimme, Gehör: Zeitschrift für Kommunikationsstörungen 30*: 147–53.

Nuerk, H.-C., Moeller, K., Klein, E., Willmes, K., and Fischer, M.H. (2011). Extending the Mental Number Line—A Review of Multi-Digit Number Processing. *Journal of Psychology 219*: 3–22.

Opfer, J.E. and Siegler, R.S. (2007). Representational change and children's numerical estimation. *Cognitive Psychology 55*: 169–95.

Opfer, J.E. and Furlong, E.E. (2011). How numbers bias preschoolers' spatial search. *Journal of Cross-Cultural Psychology 42*: 682–95.

Parsons, S. and Bynner, J. (2005). *Does Numeracy Matter More?* London: National Research and Development Centre for Adult Literacy and Numeracy, Institute of Education.

Patro, K and Haman, M. (2012). The spatial–numerical congruity effect in preschoolers. *Journal of Experimental Child Psychology 111*: 534–42.

Piazza, M., Pinel, P., Le Bihan, D., and Dehaene, S., 2007. A Magnitude Code Common to Numerosities and Number Symbols in Human Intraparietal Cortex. *Neuron 53*: 293–305.

Pixner, S., Moeller, K., Zuber, J., and Nuerk, H.-C. (2009). Decomposed but parallel processing of two-digit numbers in 1st graders. *The Open Psychology Journal 2*: 40–8.

Pixner, S., Kaufmann, L., Moeller, K., Hermanova, V., and Nuerk, H.-C. (2011a). Whorf reloaded: Language effects on non-verbal number processing in 1st grade—a trilingual study. *Journal of Experimental Child Psychology 108*: 371–82.

Pixner, S., Zuber, J., Hermanova, V., Kaufmann, L., Nuerk, H.-C., and Moeller, K. (2011b). One language, two number-word systems and many problems: Numerical cognition in the Czech language. *Research in Developmental Disabilities 32*: 2683–9.

Power, R.J. D. and Dal Martello, M.F. (1990). The dictation of Italian numerals. *Language and Cognitive Processes 5*: 237–54.

Power, R.J. D. and Dal Martello, M.F. (1997). From 834 to eighty thirty four: The reading of Arabic numerals by seven year old children. *Mathematical Cognition 3*: 63–85.

Ratinckx, E., Brysbaert, M., and Fias, W. (2005). Naming two-digit arabic numerals: evidence from masked priming studies. *Journal of Experimental Psychology: Human Perception Performance 31*: 1150–63.

Restle, F. (1970). Speed of adding and comparing numbers. *Journal of Experimental Psychology 83*: 32–45.

Reynvoet, B., Notebaert, K., and Van den Bussche, E. (2011). The processing of two-digit numbers depends on task instructions. *Journal of Psychology 219*: 37–41.

Rousselle, L. and Noël, M.P. (2007). Basic numerical skills in children with mathematics learning disabilities: A comparison of symbolic vs. non-symbolic number magnitude processing. *Cognition 102*: 361–95.

Rubinsten, O. and Henik, A. (2005). Automatic activation of internal magnitude: A study of developmental dyscalculia. *Neuropsychology 19*: 641–8.

Sackur-Grisvard, C., and Léonard, F. (1985). Intermediate cognitive organizations in the process of learning a mathematical concept: The order of positive decimal numbers. *Cognition and Instruction 2*: 157–74.

Sagiv, N., Simner, J., Collins, J., Butterworth, B., and Ward, J. (2006). What is the relationship between synaesthesia and visuo-spatial number forms? *Cognition 101*: 114–28.

Sandrini, M, Miozzo, A, Cotelli, M, and Cappa, S.F. (2003) The residual calculation abilities of a patient with severe aphasia: Evidence for a selective deficit of subtraction procedures. Cortex *39*: 85–96.

Schneider, M. and Siegler, R.S. (2010). Representations of the magnitudes of fractions. *Journal of Experimental Psychology: Human Perception and Performance 36*: 1227–38.

Seron, X. and Fayol, M. (1994). Number transcoding in children: A functional analysis. *British Journal of Developmental Psychology 12*(3): 281–300.

Seron, X., Pesenti, M., Noel, M.P., Deloche, G., and Cornet, J.A. (1992). Images of numbers, or "When 98 is upper left and 6 sky blue." *Cognition 44*: 159–96.

Siegler, R.S. and Mu, Y. (2008). Chinese children excel on novel mathematics problems even before elementary school. *Psychological Science 19*: 759–63.

Siegler, R.S. and Opfer, J. (2003). The development of numerical estimation: Evidence for multiple representations of numerical quantity. *Psychological Science 14*: 237–43.

Siegler, R.S., Thompson, C.A., and Schneider, M. (2011). An integrated theory of whole number and fractions development. *Cognitive Psychology 62*: 273–96.

Slusser, E., Santiago, R., and Barth, H. (2013). Developmental change in numerical estimation. *Journal of Experimental Psychology: General 142*: 193–208.

Sprute, L. and Temple, E. (2011). Representations of Fractions: Evidence for Accessing the Whole Magnitude in Adults. *Mind, Brain, and Education 5*: 42–7.

Sullivan, J., Juhasz, B., Slattery, T., and Barth, H. (2011). Adults' number-line estimation strategies: evidence from eye movements. *Psychonomic Bulletin and Review 18*: 557–63.

Temple, C.M. (1989). Digit dyslexia: A category-specific disorder in developmental dyscalculia. *Cognitive Neuropsychology 6*: 93–116.

Temple, C.M. (1991). Procedural dyscalculia and number fact dyscalculia: Double dissociation in developmental dyscalculia. *Cognitive Neuropsychology 8*: 155–76.

van Dijck, J.-P., Ginsburg, V., Girelli, L., and Gevers, W. (this volume). Linking Numbers to Space: From the Mental Number Line towards a Hybrid Account. In R. Cohen Kadosh, and A. Dowker, (eds.) *The Oxford Handbook of Numerical Cognition*, Oxford University Press.

van Galen, M., and Reitsma, P. (2008). Developing access to number magnitude: a study of the SNARC effect in 7- to 9-year-olds. *Journal of Experimental Child Psychology 101*: 99–113.

Verguts, T., and De Moor, W. (2005). Two-digit comparison decomposed, holistic or hybrid? *Experimental Psychology 52*: 195–200.

Verguts, T., and Fias, W. (2005). Neighborhood effects in mental arithmetic. *Psychology Science 47*: 132–140.

Wang, Y., Lin, L., Kuhl, P., and Hirsch, J. (2007). Mathematical and linguistic processing differs between native and second languages: An fMRI study. *Brain Imaging and Behavior 1*: 68–82.

Wilson, A.J. and Dehaene, S. (2007). *Number sense and developmental dyscalculia.* In: Coch, D., Fischer, K, and Dawson, G. (eds). *Human Behavior and the Developing Brain: Atypical Development.* 212–38. New York: Guilford Press.

Wood, G., Mahr, M. and Nuerk, H.-C. (2005). Deconstructing and reconstructing the 10-Base structure of Arabic numbers. *Psychology Science,* 42, 84–95.

Wood, G., Nuerk, H.-C., and Willmes, K. (2006). Neural representations of two-digit numbers: A parametric fMRI study. *NeuroImage 46*: 358–67.

Wood, G., Nuerk, H.-C., Moeller, K., Geppert, B., Schnitker, R., Weber, J., and Willmes, K. (2008). All for one but not one for all: How multiple number representations are recruited in one numerical task. *Brain Research 1187*: 154–66.

Zhang, J. and Norman, D.A. (1995). A representational analysis of numeration systems. *Cognition 57*: 271–95.

Zhou, X., Chen, C., Chen, L., and Dong, Q. (2008). Holistic or compositional representation of two-digit numbers? Evidence from the distance, magnitude, and SNARC effects in a number-matching task. *Cognition 106*: 1525-36.

Zorzi, M., Priftis, K., and Umiltà, C. (2002). Neglect disrupts the mental number line. *Nature 417*: 138–9.

Zuber, J., Pixner, S., Moeller, K., and Nuerk, H.-C. (2009). On the language-specificity of basic number processing: Transcoding in a language with inversion and its relation to working memory capacity. *Journal of Experimental Child Psychology 102*: 60–77.

CHAPTER 8

HOW ABSTRACT IS ARITHMETIC?

JAMIE I. D. CAMPBELL

THE abstractness of cognitive representations and the extent to which mental processes depend on the modality or format of stimulus inputs are central issues in cognition and neuroscience (Barsalou, 2008; Cohen Kadosh, Henik, & Rubinsten, 2008). One view is that modal inputs (e.g. auditory, visual) are recoded into amodal symbols that represent concepts in abstract form. An alternative view is that modal representations participate in multimodal interactions that form integrated multimodal concepts, rather than abstract codes. The cognitive domains of number processing and arithmetic seem to be obvious candidates for abstract representation. The cardinality of a set of elements is an abstract property because it is invariant for all possible kinds of elements. Arithmetic is similarly abstract because the results of arithmetic operations are invariant regardless of numeral format (e.g. 3, three, ...) or the referents of problem operands (e.g. $3x + 2x = 5x$, regardless of whether x refers to aardvarks or zebras). Thus, number and arithmetic are abstract when viewed as formal mathematical concepts, but does this extend to the cognitive domain?

This chapter reviews experimental evidence about the abstractness of elementary cognitive arithmetic when skilled adults solve single-digit problems, such as $4 + 5 = 9$ and $6 \times 2 = 12$. The issue of abstractness is important because it potentially discriminates prominent theoretical views of the cognitive architecture for arithmetic. For many years there has been a quite general consensus that answering a simple arithmetic problem entails cognitive subsystems for encoding the problem, remembering or calculating an answer, and producing the appropriate answer or response (Ashcraft, 1992). How these systems are interrelated, however, has been the focus of much empirical analysis and theoretical debate. One fundamental question is whether number processing stages are additive or interactive. If two stages are additive, this implies that each completes its component operations independently. For example, in simple arithmetic, if problem encoding and calculation stages are additive, this implies that calculation processes largely operate independently of the conditions of encoding, which would imply abstract processes for arithmetic. In contrast, if encoding and calculation are interactive, then the conditions of encoding could directly affect how calculation proceeds. If cognitive processes for arithmetic are context dependent, this implies at least some degree of non-abstraction.

First, three prominent recent proposals about cognitive number processing will be reviewed. These models make different assumptions about the basic cognitive architecture of numerical skills and the abstractness of cognitive arithmetic. Then evidence that calculation processes depend upon the conditions of encoding, including numeral surface format (e.g. 4 + 5 versus four + five) is reviewed. A diverse variety of experimental studies demonstrate that calculation is not abstracted away from surface form or the encoding context. Moreover, there is quite direct evidence that cognitive processes for elementary calculation are not inherently abstract. Additionally, evidence that memory for arithmetic facts often is based on language-specific representations as opposed to abstract codes will be reviewed.

Architectures For Cognitive Arithmetic

The *abstract code model*, introduced by McCloskey and colleagues (e.g. McCloskey, 1992; McCloskey & Macaruso, 1994, 1995; Sokol, McCloskey, Cohen, & Aliminosa, 1991), assumed a strictly additive architecture. According to this view (see Figure 8.1), number processing is based on comprehension, calculation, and response production systems that communicate via an abstract, semantic quantity code. Comprehension systems, specialized for different numerical input formats (e.g. Arabic digits, written, or spoken number words, etc.) convert that input into an abstract format that is independent of the original surface form. This abstract code is the representation upon which subsequent calculation and production systems operate. Calculation systems include long-term memory representations for some arithmetic facts (e.g. 5 + 2 = 7), stored arithmetic rules such $0 \times N = 0$, $0 + N = N$, as well as procedures to solve more complex arithmetic problems (e.g. multi-digit addition or multiplication). The model assumes that the numerical output of calculation processes is in abstract code format. Production subsystems convert the abstract output from the comprehension and calculation systems into Arabic, written, or spoken verbal number form as required. This model is supported largely on the basis of number processing deficits

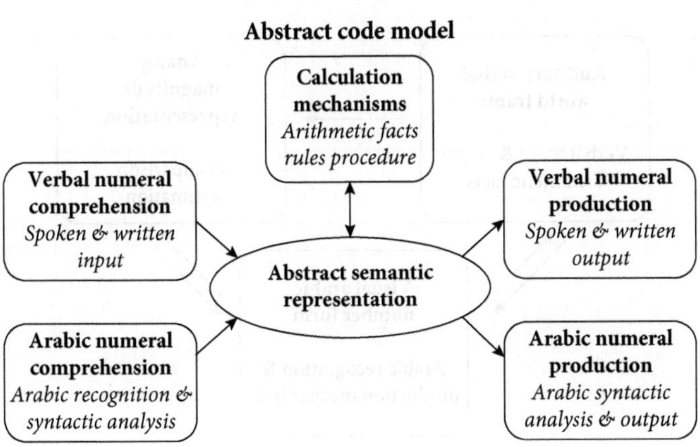

FIGURE 8.1 Illustration of the abstract-code model.

Reprinted from M. McCloskey and P. Macaruso, Representing and using numerical information, *American Psychologist*, 50(5), pp. 351–363 © 1995, American Psychological Association.

following brain injury, and can account for a variety of specific patterns of impaired and spared number processing functions (see McCloskey & Macaruso, 1995, for a review).

An alternative architecture the, *triple-code model*, was proposed by Dehaene (Dehaene, 1992; Dehaene & Cohen, 1995; see also Dehaene, Molko, Cohen, & Wilson, 2004; Dehaene, Piazza, Pinel, & Cohen, 2005). In contrast to the abstract-code model, which proposed a *functional* modularity (i.e. the architecture is primarily modular in relation to what one does with numbers) the triple-code model emphasized *representational* modularity (i.e. the architecture is primarily modular with respect to distinct types of representational codes for numbers). As depicted in Figure 8.2, the triple-code theory posits three primary representational formats for number processing: a semantic quantity code (i.e. numerical magnitude), a visual-Arabic number form, and an auditory-verbal code system. The three types of codes differentially participate in different number processing tasks. The quantity code, which is often conceptualized as a visual-spatial number line, mediates numerical size comparisons, approximate calculation, and supports semantic elaboration strategies, such as those commonly used for subtraction. The Arabic number form system processes Arabic input and output, and mediates multi-digit operations. The auditory-verbal system includes components for written and spoken numeral input and output, and also includes verbal representations of memorized addition and multiplication facts. The triple-code model thereby assumes language-based representation of arithmetic facts, as opposed to the language-independent representation proposed in the abstract code model. As in the abstract-code model, however, the triple-code model assumes additive, not interactive communication between representational systems. Once numerical input is transcoded into the appropriate internal code, subsequent processing proceeds the same regardless of the original input format. Thus, any format-related differences in performance for a given numerical operation (e.g. arithmetic-fact retrieval) presumably reflects differences in the efficiency of transcoding from the stimulus code to the type of internal code required for that operation (Dehaene, 1996; Dehaene & Akhavein, 1995; Dehaene, Bossini, & Giraux, 1993). Consequently, encoding and calculation processes in the triple code model are generally expected to be independent and additive.

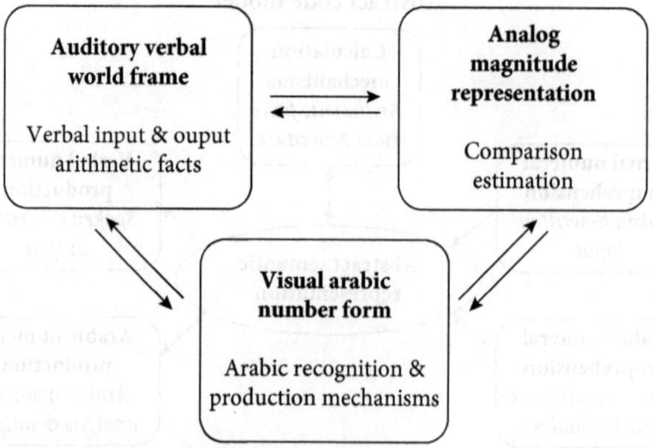

FIGURE 8.2 Illustration of triple-code theory.

Reprinted from *Current Opinion in Neurobiology*, 14(2), Stanislas Dehaene, Nicolas Molko, Laurent Cohen, and Anna J Wilson, Arithmetic and the brain, pp. 218–24, Copyright (2004), with permission from Elsevier.

Campbell and colleagues (e.g. Campbell, 1994; Campbell & Clark, 1988; Campbell & Epp, 2004) argued that strictly additive architectures cannot account for many basic phenomena of cognitive number processing. Instead, an *encoding complex hypothesis* is suggested by evidence that resolution among competing numerical responses and operations is pervasive in numerical cognition. Numerals are associated with many functions (e.g. number reading, comparison, estimation, arithmetic operations, etc.); consequently, numerals encoded for a given task activate a complex of associated representations and processes that include both relevant and irrelevant information. Successful performance requires extraction of relevant information, while overcoming interference from irrelevant information. This is clearly demonstrated by the characteristic errors produced when skilled adults answer simple arithmetic problems under speed pressure. Errors are usually semantic neighbours in the same (3 × 6 = 21) or a related operation (3 × 6 = 9), and often involve intrusions by one of the problem operands (e.g. 2 + 9 = 9) or an answer retrieved previously. Furthermore, errors often reflect a combination of these influences (e.g. 8 × 4 = 24). Problem error rate is highly correlated with time for a correct response, suggesting that resolution of interference from competing facts or operations is a global factor governing arithmetic performance (Campbell, 1995). In the encoding-complex view, such phenomena demonstrate that the subsystems for number processing often are interactive, rather than strictly additive.

Figure 8.3 depicts assumptions of the encoding-complex view (see Campbell & Epp, 2004, for an elaborated scheme for Chinese–English bilinguals). Input via written number words and Arabic numerals are emphasized because these two formats have received the lion's share of experimental research discussed later, but input via other modalities or formats (e.g. spoken numbers) must also be represented. As in the triple-code model, the Arabic visual code is assumed to subserve Arabic recognition and production processes, but the encoding-complex view also posits a visual form of number-fact memory (e.g. Kashiwagi, Kashiwagi, & Hasegawa, 1987; McNeil & Warrington, 1994). The magnitude code provides a semantic representation of quantity and also provides an approximate quantity in the context of arithmetic. Verbal-number codes represent lemma, phonological, and syntactic processes for verbal-number production

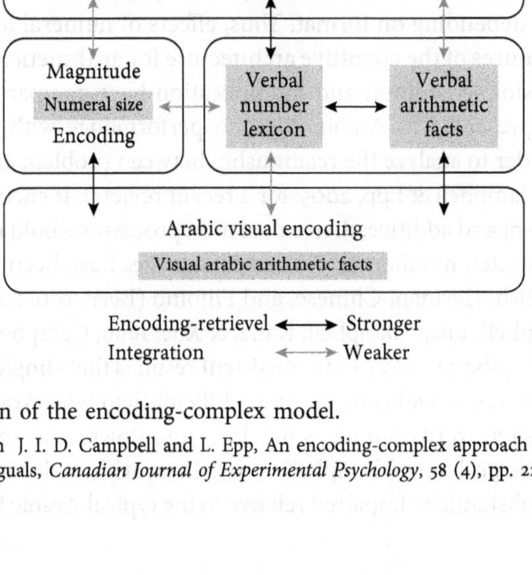

FIGURE 8.3 Illustration of the encoding-complex model.

Reprinted from J. I. D. Campbell and L. Epp, An encoding-complex approach to numerical cognition in English-Chinese bilinguals, *Canadian Journal of Experimental Psychology*, 58 (4), pp. 229–244 © 2004, American Psychological Association.

and language-based memory for number facts. Thus, Arabic, magnitude, and verbal codes all potentially participate in memory and retrieval of basic arithmetic facts.

In the encoding-complex view, communication between representational systems often involves interactive, rather than strictly additive processes. The assumption of format- and task-specific encoding-retrieval interactions constitutes the most fundamental theoretical departure from Dehaene and Cohen's triple-code model. These interactions reflect the degree of encoding-retrieval integration between systems, which is expected to vary with the amount of task- and code-specific practice or experience. The bi-directional arrows in Figure 8.3 represent this assumption. Darker arrows correspond to greater encoding-retrieval integration, which means more automaticity and a stronger capacity to coordinate task-specific processing associated with each code. The specific pattern of stronger and weaker links in Figure 8.3 correspond to assumptions about how the history of task-specific experience strengthens encoding-retrieval pathways. For present purposes, it is sufficient to emphasize certain task-related asymmetries that exist for Arabic digits and written number words (see Campbell & Epp, 2004, for more details). Arabic numerals are used regularly to represent numbers for calculation and become associated with skilled retrieval processes for calculation. In contrast, written number words are rarely used to represent numbers for the purpose of calculation; consequently, it is more difficult to activate or utilize calculation-relevant information given written number words than Arabic digits. Effects of these asymmetries are implicated in numerous phenomena of cognitive arithmetic.

Numeral Format and Arithmetic

Effects of numerical surface form (e.g. 3, three, or ...) on performance of elementary arithmetic have held an important place in cognitive arithmetic research. This is because effects of format are potentially diagnostic of whether calculation processes are abstracted away from the modality of input. Format might affect only operand encoding or response output processes and have no direct effect on calculation. Alternatively, calculation processes for arithmetic could vary both in kind (memory retrieval versus procedural strategies) and efficiency depending on format. Thus, effects of numeral format on calculation may reflect basic features of the cognitive architecture for arithmetic.

Many studies of simple addition and multiplication have compared performance with problem operands presented as Arabic digits to performance with operands as written number words in order to analyze the relationship between problem encoding and calculation processes (see Campbell & Epp, 2005, for a recent review). If encoding and calculation stages are independent and additive, then calculation processes would not vary with format. Performance with written numbers in a variety languages have been examined, including French, Dutch, English, German, Chinese, and Filipino (Bernardo, 2001; Blankenberger & Vorberg, 1997; Campbell, 1994; Campbell, Kanz, & Xue, 1999; Campbell, Parker, & Doetzel, 2004; Noël, Fias, & Brysbaert, 1997). One consistent result is that simple arithmetic with visual number words (seven × five) is much more difficult than with Arabic numerals (7 × 5). Arithmetic with written words can be as much as 30% slower and 30% more error prone compared with Arabic problems (Campbell, 1994). Thus, arithmetic for the less typical written word format is substantially impaired relative to the typical Arabic format.

Are these large word-format costs necessarily incurred during calculation or could they arise at another stage of processing? One possibility is that it simply takes longer to encode written number words than Arabic numerals, which could reduce time devoted to calculation and promote errors (McCloskey, Macaruso, & Whetstone, 1992; Noël et al. 1997). There is clear evidence contrary to this, however. Campbell et al. (2004) compared simple addition, multiplication, and parity comparison tasks on pairs of Arabic digits or pairs of English number words. For addition and multiplication (e.g. two + three, 7 × 6), participants spoke the answer as quickly as possible. For the parity task, participants saw two numerals separated by a horizontal line (e.g. 7 | 9 or seven | nine) and stated 'yes' or 'no' to indicate whether or not both numbers had the same parity (i.e. odd-even status). There was a Format × Task RT interaction that reflected much larger word-format costs for addition (264 ms) and multiplication (241 ms) than comparison (123 ms). Similarly, Campbell et al. (1999) compared Chinese–English bilinguals' performance with Arabic numerals and Chinese symbols. The Chinese symbols are effectively number words and rarely used for calculation. Word-format RT costs for addition and multiplication (114 ms) were large compared with word format costs for number comparison (choose the numerically larger or smaller of two Arabic or two Chinese symbols; 43 ms). These results demonstrate that word-format costs to add or multiply two single-digit numbers are much larger than format-specific differences in the time required to identify or compare semantic properties of those two numbers. This makes it unlikely that word format costs can be attributed simply to encoding differences.

Format and Problem Difficulty

Evidence that surface format can directly affect calculation comes from the *problem-size effect*. This is the nearly-ubiquitous finding that RT and errors for simple arithmetic problems increase with the numerical value of problem operands; for example, the problem 2 + 3 usually is answered more quickly and accurately than 6 + 8. Small-number problems (e.g. sum ≤ 10) are encountered more often in everyday arithmetic and therefore develop strong memory representations compared with larger problems (Zbrodoff & Logan, 2005). Consequently, large problems are more likely to be solved by procedural strategies (e.g. counting, transformation) that are relatively slow and error prone compared with direct memory retrieval (Campbell & Xue, 2001). Most important here, many studies have shown that the problem-size effect is larger for problems in written-word format (e.g. three + eight) than digit format (3 + 8; e.g. Campbell, 1994; Campbell & Fugelsang, 2001; Campbell & Penner-Wilger, 2006). The finding that word-format costs increase with problem difficulty suggests that format directly affects calculation.

For example, Campbell and Fugelsang (2001) examined adults' performance of simple addition problems in a true/false verification task with equations presented either in digit format (3 + 4 = 8) or written English format (three + four = eight). The Format × Size effect on RT (i.e. greater word-format RT costs for large than for small problems) was approximately twice as large for true as for false equations. Error rates also presented this pattern. Given that encoding conditions and response requirements were practically equivalent for true and false equations, the triple interaction cannot have occurred at these stages. Rather, surface format evidently affected the efficiency of retrieval or calculation processes differently for true and false equations. This would occur because true verification trials provide a more sensitive measure of arithmetic efficiency than do false trials. Specifically, Campbell and Fugelsang observed that the problem-size effect was substantially larger for true (+368 ms) than for false equations (+281 ms). Arithmetic processing may be terminated relatively quickly for false equations because substantial discrepancies between the presented false answer and the correct answer (e.g. more than ±2 from correct) may be detected while retrieval or calculation is in progress. Consequently, the difference in retrieval difficulty between small and large problems is more fully expressed on true trials, which makes the problem-size effect on true trials more sensitive to manipulations that affect retrieval efficiency. Thus, the fact that the Format × Size effect was larger for true than for false equations would occur if the word format reduced the efficiency of arithmetic retrieval processes.

Format can also interact with the difficulty of applying basic arithmetic rules. Campbell and Metcalfe (2007) examined adults' performance of simple addition and multiplication involving rule-based zero and one problems ($N + 0 = N$, $N \times 1 = N$, $N \times 0 = 0$) presented as Arabic digits (6 × 1; 0 + 5) or written English number words (six × one; zero + five). The results showed that digits and number word format provided differential access to the zero and one rules. With Arabic numerals, $N \times 0 = 0$ problems were relatively difficult compared with the other zero and one problems. This likely reflects interference from activation of the competing $N + 0 = N$ and $N \times 1 = N$ rules, an interpretation supported by a very high rate of $N \times 0 = N$ errors with Arabic problems. In contrast, with word stimuli, $N \times 0$ items were among the fastest items and yielded a relatively low rate of $N \times 0 = N$ errors. These format-related differences in the relative difficulty of zero and one problems are consistent with the view that calculation-related associations are differentially activated by different surface forms of numbers.

Format and Strategy Choice

Campbell and Epp (2005) proposed that poor arithmetic performance with written-word problems occurs because arithmetic fact retrieval processes are less efficient with word than digit operands. One type of evidence for this is that people report direct memory retrieval (rather than procedures, such as counting) for simple addition less often for word than digit format. For example, Campbell and Fugelsang found that non-retrieval strategies were reported nearly 60% more often for simple addition verification problems presented in word format (four + seven = eleven) than in digit format (4 + 7 = 11). Moreover, this word-format cost on retrieval usage increases with problem size (Campbell & Alberts, 2009; Campbell & Fugelsang, 2001; Campbell & Penner-Wilger, 2006). In theory, the unfamiliar associative activation produced by word operands in the context of arithmetic (see also Metcalfe & Campbell, 2007) reduces retrieval efficiency but impacts large problems more because of their already-low memory strength.

Campbell and Alberts (2009) argued that if the written word format reduces retrieval efficiency then word-format costs on retrieval efficiency should be closely linked to the probability of retrieval. This is because strategy choice often is adaptive, in that it reflects the relative efficiency (i.e. speed and accuracy) of alternative strategies (Lemaire & Reder, 1999; Schunn & Reder, 2001; Shrager & Siegler, 1998; Steel & Funnell, 2001). Consequently, we would expect reduced retrieval efficiency to promote a switch to procedural strategies (e.g. counting). Given this, Campbell and Alberts predicted different effects of format on retrieval usage for simple addition and subtraction. Indeed, addition but not subtraction presented a Format × Size interaction in retrieval usage that reflected greater word-format retrieval costs (i.e. increased use of procedures) for the larger additions, which possess relatively weak memory strength. In contrast, large and small subtractions, which both have relatively weak retrieval strength, presented similar word-format costs on retrieval usage. Indeed, reported use of direct memory retrieval was closely linked to retrieval efficiency. The mean rate of reported retrieval varied from 40 to 82% across combinations of operation (addition or subtraction), problem size (small or large), and format (digits or words). This variability was closely tied to mean condition retrieval RT and error rate, which together in a multiple regression analysis accounted for 98% of mean retrieval usage across experimental conditions. The close calibration of retrieval usage with retrieval performance reinforces the view that strategy choice is often precisely adapted to strategy efficiency (Schunn & Reder, 2001; Siegler & Lemaire, 1997). The observation that the effect of format on retrieval usage was precisely proportional to the effects of problem size and operation supports the conclusion that these diverse factors all affected retrieval efficiency.

What mechanisms mediate the effect of digit versus word format on strategy choice for arithmetic? A shift from retrieval to an alternative strategy may be preceded by a judgment of 'feeling of knowing' (FOK; Cary & Reder, 2002). If the FOK is sufficiently strong then a retrieval attempt is made. The unfamiliar written word format (i.e. adults rarely perform arithmetic on written words) might activate a weaker FOK compared with problems in the more-familiar digit format, thereby reducing the frequency of retrieval attempts with word operands. Indeed, Metcalfe and Campbell (2007) demonstrated that practicing specific operands in the context of one set of problems (six + seven, four + nine) increased reported use of direct memory retrieval for unpracticed problems composed of the familiarized operands (six + nine). Thus, operand-format familiarity in the context of arithmetic likely plays a role in format effects on strategy choice, but the effect size of the familiarity manipulation observed by Metcalfe and Campbell was small compared with the overall effects of digit versus word format on strategy choice typically observed.

Another possibility is that written word operands, compared with digits, make it relatively difficult to initiate a retrieval task set. For example, the written word format more strongly activates a tendency to read the presented problem as a two-digit number (e.g. to process 'four × eight' as ' forty-eight'; Campbell, 1994). A strong reading task set would interfere with initiating the retrieval task set, possibly promoting a shift to procedural strategies. Strong reading-based activation with the word format could also promote a shift after retrieval is initiated by activating associative competitors that reduce the efficiency of retrieval (Campbell, 1994; Campbell & Timm, 2000). Word format might also promote specific procedural strategies directly by priming verbal working memory representations implicated in verbal counting (Carlson & Cassenti, 2004) and procedural strategies for arithmetic (Trbovich & LeFevre, 2003). Thus, there are several mechanisms by which the written word format could reduce arithmetic retrieval and promote use of procedural strategies.

Format-induced strategy shifts confirm that different formats can recruit different neural processes for elementary arithmetic. Imaging research suggests that retrieval of arithmetic facts is associated with linguistic representations in the left angular gyrus, whereas procedural strategies requiring semantic quantity processing recruit bilateral components of the intraparietal sulcus (Dehaene et al., 2004). As direct retrieval and procedural strategies activate distinct brain regions (see also Dehaene et al., 2003) the effects of format on strategy choice for elementary arithmetic imply that calculation is not generally abstracted away from surface form.

Format and Arithmetic Errors

Analysis of specific arithmetic errors has provided much information about the representation and retrieval of arithmetic facts (Campbell, 1995). To investigate whether different combinations of input-output formats involve a common arithmetic system, Sokol, McCloskey, Cohen, and Aliminosa (1991) examined the multiplication errors of patient PS, who had left hemisphere cerebral vascular damage. Multiplication problems from 0×0 to 9×9 were tested in combinations of three formats, including Arabic digits, written number words, and dots, producing nine stimulus by response format combinations. They found no differences in error rates across formats (stimulus: 12.4% Arabic, 13.0% Verbal, 13.0% Dots; response: 12.8% Arabic, 13.8% Verbal, 11.7% Dots). Furthermore, PS made several different types of errors, and Sokol et al. found that the proportions of each type of error did not differ across stimulus and response formats. They concluded that the errors produced by PS support the conclusion that arithmetic facts are stored in a format independent code.

Nonetheless, there is ample evidence that format can have large effects on the specific errors that people produce (e.g. Campbell, 1994; Campbell et al., 2004; LeFevre & Liu, 1997; Noël et al., 1997). Intrusions errors, in which one of the operands appears in the error (e.g. 2 + 9 = 'nine' or 9 × 6 = 'thirty six') are much more common and account for a higher percentage of errors with written words than digit stimuli. Words have a stronger capacity to activate numeral reading processes that interfere with answer retrieval. In contrast, operation errors (e.g. 2 + 4 = 'eight' or 2 × 4 = 'six') are more common with Arabic than word stimuli (Campbell, 1994; Campbell et al., 1999, 2004), suggesting that Arabic operands produce relatively stronger activation of arithmetic associations. Similarly, most errors involve answers that are correct to a neighbour (i.e. a problem created if one of the operands was changed by ±1), but the word format promotes a higher percentage of 'far off' errors relative to digit stimuli (Campbell, 1994; Campbell et al., 1999). Thus, errors with written word stimuli appear to be less constrained by semantic distance, suggesting that numerical magnitude information is utilized less efficiently with word than Arabic stimuli.

LeFevre, Lei, Smith-Chant, and Mullins (2001) tested Chinese and English speakers who provided answers to simple multiplication problems with operands presented as Arabic numerals or auditory words in their native language. Performance with auditory problems is relatively good, probably because arithmetic often is practiced in silent verbal rehearsal. Indeed, LeFevre et al. found only subtle differences in RT (see also Metcalfe & Campbell, 2008). English speakers made more errors than Chinese speakers, but with Arabic format, the Chinese speakers produced more operand intrusions (e.g. 4 × *8* = 'twenty *eight*') than the English. LeFevre et al. suggested that this occurred because the Chinese emphasize verbal rote memory drills during learning; consequently, their arithmetic memory may be especially susceptible to phonological interference from the operands. As is discussed later, there is diverse evidence for a linguistic basis of number-fact representation.

Different Arithmetic Memory Networks For Arabic and Word Formats?

As the preceding paragraphs demonstrate, there is much evidence that calculation processes can vary depending on surface form, but does this imply distinct memory networks? Blankenberger and Vorberg (1997) concluded that different surface forms may or may not access a common number-fact representation depending upon the involvement of working memory. If solving a problem requires holding one or both operands in working memory (e.g. in phonological format), this will force conversion to the common working-memory code regardless of surface format. In contrast, if a problem is solved by direct retrieval and does not depend upon holding numerical operands in working memory, then the retrieval processes engaged may be format specific. Their participants produced answers to simple addition and multiplication problems presented in Arabic digits, German number words, or dice patterns. Participants either were shown both operands (visual display condition), or one operand was presented and the other held in working memory (memory display condition). In the memory condition, addition was faster than multiplication, but multiplication was faster in the visual condition. This difference was due entirely to multiplication number-word problems in the visual condition; specifically, multiplication was faster than addition with simultaneously displayed word pairs, but was slower than addition in all other conditions. Blankenberger and Vorberg suggested that addition performance may be especially susceptible to phonological interference when the two operands appeared simultaneously in word format.

Campbell and Thompson (2012) obtained results that clearly support a common retrieval network for digit and written-word format arithmetic. They found that retrieval practice of multiplication problems in word format (two × four) produced inhibition (e.g. slowed response times) for the corresponding addition fact tested in digit format (2 + 4). This implies that the addition retrieval structure inhibited in the context of word-format multiplication was the same structure subsequently accessed via digit-format addition. Cross-format inhibition cannot be easily reconciled with digits and written number words activating entirely different long-term memory structures, and instead supports the conclusion that they access a common network of addition and multiplication facts. This coincides with the results of Zhou, Chen, Qiao, Chen, Chen, Lu, et al. (2009), who found no differences in the event-related potentials produced by single-digit addition and multiplication problems in Arabic digits and Chinese number words.

Nonetheless, the results do not rule out the possibility that format-specific problem-encoding processes can interact with the long-term retrieval structures that subserve arithmetic memory (e.g. Jackson & Coney, 2007). Number facts may be stored as linguistic structures that are associated with semantic quantity codes that represent the numerical magnitude of the operands and answers (e.g. Campbell & Epp, 2004; Dehaene & Cohen, 1997; Dehaene et al., 2005). Number-fact retrieval entails the coordinated activation of this collection of associated memory codes and generation of the appropriate response output. Different numeral surface forms, however, may differentially activate components of the semantic and associative structures representing number facts. As discussed previously, characteristics of arithmetic errors observed with digit and word operands suggest that digits more effectively recruit magnitude information for number facts and that written words activate irrelevant reading processes that interfere with retrieving a problem's answer. These types of interactions between problem encoding and arithmetic retrieval processes would give rise to robust format effects on number fact-retrieval, despite an overlapping representational network for retrieval with digit and word formats.

Semantic Alignment Effects

Format effects have provided one line of evidence that problem encoding conditions and number-fact retrieval processes are interactive. In this sense, cognitive processes for arithmetic processes are not abstracted way from surface form or the encoding context. Bassok and colleagues (e.g. Bassok, Chase, & Martin, 1998; Bassok, Pedigo, & Oskarsson, 2008; Wisniewski & Bassok, 1999) provided another type of evidence, demonstrating *semantic alignment* effects in adults' elementary arithmetic. Although arithmetic is formally abstract, people routinely apply arithmetic operations in combination with their knowledge about specific referents (e.g. 3 cats + 2 dogs = 5 pets; 12 roses/2 vases = 6 roses per vase). This involves analogical coordination of semantic relations and arithmetic relations (Bassok, 2001). For example, determining the total set of categorically related objects (e.g. the total number of pets) is aligned with addition because both addends and related objects (e.g. cats and dogs) bear the same commutative relation to their sum (i.e. $a + b = b + a = c$). Similarly, functional object relations that are asymmetric (e.g. containers and their contents) align with division because such objects (e.g. vases and flowers) bear an identical non-commutative relation to the dividend and the divisor ($a/b \neq b/a$; Bassok et al., 2008, p. 344).

Bassok and her colleagues found that alignments between semantic and arithmetic knowledge influenced people's performance in both obvious and subtle ways. For example, Bassok et al. (1998) presented university students with word pairs under instructions to construct addition word problems. With categorically related words (e.g. tulips and roses) participants typically made simple additive equations (e.g. 3 roses + 2 tulips = 5 flowers). In contrast, with functionally-related words (e.g. plums and baskets) participants avoided simple addition equations and, instead, produced elaborated equations that preserved the semantic alignment with addition. When asked to construct division problems, participants produced the reverse pattern. Thus, people's use of arithmetic operations was biased to preserve semantic alignment between words as conceptual references and as arithmetic operands.

A more subtle, but theoretically telling, phenomenon was investigated by Bassok et al. (2008). They demonstrated that memory activation of simple addition facts (e.g. 3 + 5 = 8) can depend on semantic alignment with word primes. Bassok et al. adapted the number-matching task developed by LeFevre, Bisanz, and Mrkonjic (1988). In this task, a probe digit pair (e.g. 3, 4) is presented followed by a target number that either matches (e.g. 3) or does not match one of the probe pair (e.g. 6). Participants press a button to indicate whether there is a match or not. The critical manipulation occurs within the no-match trials: On some no-match trials the target is the sum of the probe pair (e.g. 7). Compared with neutral no-match targets, RT is slower when the target on a no-match trial is the sum of the probe pair. This *sum interference* effect suggests that the probe pair can automatically activate its sum, and a match between the target and sum interferes with a 'no match' response. Bassok et al. modified this paradigm by displaying word-pair primes 480 ms before the probe pair. The primes were either semantically aligned with addition (e.g. tulips daisies) or misaligned with addition (e.g. hens radios). When primes were semantically aligned, the sum-interference effect occurred, but the effect was eliminated when primes were semantically misaligned with addition. The activation of addition facts in long-term memory, thereby depended on the analogical consistency of the semantic context activated by the primes. This clearly is contrary to the concept of arithmetic operations and number-fact representations being encoded abstractly or that their access is context independent. In contrast, the results directly demonstrate that the kinds and referents of problem operands are relevant to cognitive arithmetic, despite being irrelevant to arithmetic as a formal operation. Thus, representation and retrieval of arithmetic facts is integrated with knowledge of non-numerical concepts and their relations; therefore, processes for cognitive arithmetic are not inherently abstract.

The non-abstractness of arithmetic is not restricted to elementary number facts. Landy and Goldstone (2007) investigated adults' judgements of algebraic equality ($b^*c + a = a + c^*b$?). The purpose was to determine if algebraic analysis is amodal or bound to the perceptual structure of the equations. If the representations are amodal, then performance should be unaffected by perceptual grouping manipulations that are congruent or incongruent with the correct order of processing the operations (i.e. multiplication takes precedence over addition). They found that perceptual groupings that were inconsistent with the correct order of operations (e.g. $w + n^*r + k = r + k^*w+ n$) disrupted performance compared with perceptually consistent groupings (e.g. $w + n^*r + k = r + k^*w + n$). This indicates that the algebraic analysis was grounded in spatial-perceptual cues, rather than based on amodal or abstract representations. These effects, like the semantic alignment effects and effects of surface format, demonstrate that activation of processes for cognitive arithmetic is context dependent, rather than abstract and independent of the conditions of problem encoding.

Language and Arithmetic

Taken together, the semantic alignment and surface format effects argue against a purely abstract representation of arithmetic facts. An emerging view is that representation of exact arithmetic facts like 3 + 2 = 5, often is based on linguistics codes (Dehaene, Spelke, Pinal, Stanescu, & Tsivkin, 1999; Hodent, Bryant, & Houdé, 2005; Rusconi, Galfano, & Job, 2007; Venkatraman, Siong, Chee, & Ansari, 2006). Memorized arithmetic facts may be stored as a sequence of lemma-level word representations (Dehaene et al., 2005) or perhaps as phonological codes (De Smedt & Boets, 2010; De Smedt, Taylor, Archibald, & Ansari, 2010; but see Whalen, McCloskey, Lindemann, & Bouton, 2002).

Language-Specific Transfer and Simple Arithmetic

One striking source of evidence for linguistic representation of memorized arithmetic facts comes from transfer of practice effects observed in the arithmetic performance of multilingual speakers. For example, Dehaene et al. (1999; see also Spelke & Tsivkin, 2001) presented evidence that language-based processes mediate exact arithmetic, whereas a language independent magnitude representation mediates approximate arithmetic. In the Dehaene et al. study, Russian–English bilinguals were trained on sets of exact and approximate arithmetic problems in their two languages. At test, exact calculations showed large RT costs of language switching and poor generalization to novel problems for both languages. In contrast, approximate arithmetic showed language independence at test, and training transferred to novel problems without cost. Brain imaging with both ERP and fMRI revealed activation in bilateral parietal areas for approximation tasks, whereas exact calculation tasks activated the left inferior frontal lobe. The parietal area is associated with visuospatial tasks, whereas the area activated during exact calculation is implicated in language processes. Dehaene at al. concluded that exact arithmetic is language dependent, whereas approximate arithmetic depends upon a language independent, visuospatial magnitude representation.

Campbell and Dowd (2012) examined inter-operation transfer of practice in adult Chinese–English bilinguals' memory for simple multiplication ($6 \times 8 = 48$) and addition facts ($6 + 8 = 14$). The purpose was to determine if they possessed distinct number-fact representations in both Chinese (L1) and English (L2). Participants repeatedly practiced multiplication problems (e.g. $4 \times 5 =?$), answering a subset in L1 and another subset in L2. Then separate groups answered corresponding addition problems ($4 + 5 =?$) and control addition problems in either L1 ($N = 24$) or L2 ($N = 24$). The results demonstrated language-specific negative transfer of multiplication practice to corresponding addition problems. Specifically, large simple addition problems (sum > 10) presented a significant response time cost (i.e. inhibition) after their multiplication counterparts were practiced in the same language relative to practice in the other language. The results indicate that their Chinese–English bilinguals had multiplication and addition facts represented in distinct language-specific memory stores (see also Imbo & LeFevre, 2011). This converges with other evidence of bilingual number-fact representation. Frenck-Mestre and Vaid (1993) reported evidence from French–English bilinguals that verification of simple equations (e.g. $3 \times 4 = 7$, true or false?) was more susceptible to associative interference when stimuli were presented visually in their first-language than in their second language. If arithmetic facts are weakly represented in the second language then interference from associatively related facts in the second language would be weak (see also Bernardo, 2001).

Other Language-Related Evidence

There is also other diverse evidence that language contributes to knowledge and performance of exact arithmetic. Several studies have found evidence that exact arithmetic is very limited in cultures with a language that has few number words, although approximate arithmetic performance is comparable with groups with a complete number vocabulary (Pica, Lemer, Izard, & Dehaene, 2004; but see Gelman & Butterworth, 2005). Furthermore, linguistic structure for number words can affect arithmetic performance. Colomé, Laka, and Sebastián-Gallés (2009) tested speakers with a base-10 language for numbers to Basque speakers for whom number words are constructed by combining multiples of 20 and units or teens (e.g. '35' is said 'twenty and fifteen'). Basque participants answered additions composed of a multiple of 20 and a teen (e.g. 20 + 15) faster than control problems with identical answers (e.g. 25 + 10), whereas this difference was not apparent with a base-10 language for arithmetic. The linguistic structure of number words evidently can affect performance on corresponding arithmetic problems (but see Brysbaert, Fias, & Noel, 1998). Cohen, Dehaene, Chochon, Lehericy, and Naccache (2000) examined the simple arithmetic performance of patient ATH, who had a left parietal injury in the area typically implicated in language processing, namely the perisylvian language cortex. Cohen et al. argued that simple multiplication and division usually are solved by direct retrieval of verbal associations, whereas addition and subtraction problems often are solved by procedure and semantic elaboration, which do not necessitate the use of verbal processes. With Arabic input–output, ATH displayed a pattern of impairment consistent with this hypothesis; specifically, multiplication and division were quite severely impaired, while addition and subtraction remained relatively intact (see also Grabner, Ansari, Ebner, Koschutnig, Neuper, & Reishofer, 2009). Thus, various types of evidence converge to the conclusion that representation of exact arithmetic facts can include a linguistic component.

Despite evidence favouring a linguistic representation, it is unlikely that arithmetic facts are always stored in a linguistic code or that they are stored exclusively in linguistic format. For example, the possibility of an Arabic visual code for number facts is suggested by the results of Kashiwagi et al. (1987), who successfully retrained patients on simple multiplication with Arabic input–output, while these patients could not relearn the multiplication facts with verbal input–output (see also McNeil & Warrington, 1994). Additionally, retrieval of arithmetic facts appears to involve magnitude processing that provides an approximate range for the correct answer (Campbell, 1995; Dehaene & Cohen, 1991). Finally, Butterworth, Reeve, & Reynolds (2011) demonstrated effective use of spatial strategies in a nonverbal exact-addition task performed by children whose native language did not have corresponding number words. Thus, although the evidence is substantial that linguistic codes have a key role to play, they probably are neither necessary nor exclusive codes for the representation of exact number facts.

Summary and Conclusions

Accessing arithmetic memory from different surface forms appears to involve overlapping representations, but calculation efficiency often is format specific. This would occur if problem encoding processes and calculation processes were interactive, rather than strictly additive. Indeed, format-related strategy shifts demonstrate that calculation performance sometimes involves discrete, format-specific processes (Campbell & Alberts, 2009). Nonetheless, when procedural strategy trials are removed from analysis and only retrieval trials are analysed, there remain substantial word-format costs relative to digit format, and word-format retrieval costs tend to increase with problem difficulty (Campbell & Penner-Wilger, 2006). This reinforces the conclusion that arithmetic retrieval processes, per se, are not abstracted away from surface format. The non-abstractness of elementary arithmetic is demonstrated further by semantic context effects on the automatic activation of arithmetic facts (Bassok et al., 2008) and substantial evidence for language-specific representations for exact arithmetic. Calculation also entails embodied cognition as perceptual-motor processes mediate spatial grouping effects on algebra performance (Landy & Goldstone, 2007), and link finger representation to counting and quantity representation (Badets & Pesenti, 2010; Sato, Cattaneo, Rizzolatti & Gallese, 2007). Collectively, these observations imply that the kinds and referents of problem operands are fundamentally relevant to cognitive arithmetic, despite being irrelevant to arithmetic as a formal operation. Processes for cognitive arithmetic reflect a variable complex of interacting representations, and processes that depend directly upon context and the format of problem encoding. Cognitive arithmetic research thereby points toward integrated, multimodal mechanisms as proposed in the encoding-complex model, and aligns with approaches that emphasize concrete representations and embodied cognitive processes (e.g. Barsalou, 2008; Landy & Goldstone, 2007), rather than abstract or amodal processes.

Acknowledgments

This research was supported by a grant from the Natural Sciences and Engineering Research Council of Canada.

References

Ashcraft, M.H. (1992). Cognitive arithmetic: a review of data and theory. Cognition, 44, 75–106.

Badets, A., & Pesenti, M. (2010). Creating number semantics through finger movement perception. Cognition, 115, 46–53.

Barsalou, L.W. (2008). Grounded cognition. Annual Review of Psychology, 59, 617–645.

Bassok, M. (2001). Semantic alignments in mathematical word problems. In D. Gentner, K. J. Holyoak, & B.N. Kokinov (eds), The Analogical Mind: Perspectives From Cognitive Science. Cambridge, MA: MIT Press.

Bassok, M., Chase, V., & Martin, S. (1998). Adding apples and oranges: alignment of semantic and formal knowledge. Cognitive Psychology, 35, 99–134.

Bassok, M., Pedigo, M., & Oskarsson, F. (2008). Priming addition facts with semantic relations. Journal of Experimental Psychology: Learning, Memory, and Cognition, 34(2), 343–352.

Bernardo, A.B.I. (2001). Asymmetric activation of number codes in bilinguals: further evidence for the encoding-complex model of number processing. Memory & Cognition, 29, 968–976.

Blankenberger, S., & Vorberg, D. (1997). The single-format assumption in arithmetic fact retrieval. Journal of Experimental Psychology: Learning, Memory, and Cognition, 23, 721–738.

Brysbaert, M., Fias, W., & Noel, M.P. (1998). The Whorfian hypothesis and numerical cognition: is 'twenty-four' processed in the same way as 'four-and-twenty'? Cognition, 66, 51–77.

Butterworth, B., Reeve, R., & Reynolds, F. (2011). Using mental representations of space when words are unavailable: studies of enumeration and arithmetic in indigenous Australia. Journal of Cross-Cultural Psychology, 42, 630–638.

Campbell, J.I.D. (1994). Architectures for numerical cognition. Cognition, 53, 1–44.

Campbell, J.I.D. (1995). Mechanisms of simple addition and multiplication: a modified network-interference theory and simulation. Mathematical Cognition, 1, 121–164.

Campbell, J.I.D., & Alberts, N.A. (2009). Operation-specific effects of numerical surface form on elementary calculation. Journal of Experimental Psychology: Learning, Memory, and Cognition, 35, 999–1011.

Campbell, J.I.D., & Clark, J.M. (1988). An encoding-complex view of cognitive number processing: comment on McCloskey, Sokol, & Goodman (1986). Journal of Experimental Psychology: General, 117, 204–214.

Campbell, J.I.D., & Dowd, R. (2012). Interoperation transfer in Chinese-English bilinguals' arithmetic. Psychonomic Bulletin and Review, 19, 948–954.

Campbell, J.I.D., & Epp, L. (2004). An encoding-complex approach to numerical cognition in English–Chinese bilinguals. Canadian Journal of Experimental Psychology, 58, 229–244.

Campbell, J.I.D., & Epp, L.J. (2005). Architectures for arithmetic. In J. I. Campbell (ed.), Handbook of Mathematical Cognition. New York: Psychology Press.

Campbell, J.I.D., & Fugelsang, J. (2001). Strategy choice for arithmetic verification: effects of numerical surface form. Cognition, 80, B21–B30.

Campbell, J.I.D., Kanz, C.L., & Xue, Q. (1999). Number processing in Chinese–English bilinguals. Mathematical Cognition, 5, 1–39.

Campbell, J.I.D., & Metcalfe, A.W.S. (2007). Arithmetic rules and numerical format. European Journal of Cognitive Psychology, 19, 335–355.

Campbell, J.I.D., Parker, H.R., & Doetzel, N.L. (2004). Interactive effects of numerical surface form and operand parity in cognitive arithmetic. Journal of Experimental Psychology: Learning, Memory, and Cognition, 30, 51–64.

Campbell, J.I.D., & Penner-Wilger, M. (2006). Calculation latency: the μ of memory and the τ of transformation. Memory & Cognition, 34, 217–226.

Campbell, J.I.D., & Thompson, V.A.T. (2012). Retrieval-induced forgetting of arithmetic facts. Journal of Experimental Psychology: Learning, Memory, and Cognition, 38, 118–129.

Campbell, J.I.D., & Timm, J.C. (2000). Adults' strategy choices for simple addition: effects of retrieval interference. Psychonomic Bulletin & Review, 7, 692–699.

Campbell, J.I.D., & Xue, Q. (2001). Cognitive arithmetic across cultures. Journal of Experimental Psychology: General, 130, 299–315.

Carlson, R.A. & Cassenti, D.N. (2004). Intentional control of event counting. Journal of Experimental Psychology: Learning, Memory, and Cognition, 30, 1235–1251.

Cary, M. & Reder, L.M. (2002). Metacognition in strategy selection: giving consciousness too much credit. In P. Chambres, M. Izaute, & P. J. Marescaux (eds), Metacognition: Process, Function, and Use (pp. 63–78). New York, NY: Kluwer.

Cohen, L., Dehaene, S., Chochon, F., Lehericy, S., & Naccache, L. (2000). Language and calculation within the parietal lobe: a combined cognitive, anatomical, and fMRI study. Neuropsychologia, 38, 1426–1440.

Cohen Kadosh., R., Henik., A., & Rubinsten., O. (2008). Are Arabic and verbal numbers processed in different ways? Journal of Experimental Psychology: Learning, Memory and Cognition, 34(6), 1377–1391.

Colomé, Á., Laka, I., & Sebastián-Gallés, N. (2009). Language effects in addition: how you say it counts. Quarterly Journal of Experimental Psychology, 1, 1–19.

Dehaene, S. (1992). Varieties of numerical abilities. Cognition, 44, 1–42.

Dehaene, S. (1996). The organization of brain activations in number comparison: event-related potentials and the additive-factors method. Journal of Cognitive Neuroscience, 8, 47–68.

Dehaene, S., & Akhavein, R. (1995). Attention, automaticity, and levels of representation in number processing. Journal of Experimental Psychology: Learning, Memory and Cognition, 21, 314–326.

Dehaene, S., Bossini, S., & Giraux, P. (1993). The mental representation of parity and number magnitude. Journal of Experimental Psychology: General, 122, 371–396.

Dehaene, S., & Cohen, L. (1991). Two mental calculation systems: a case study of severe acalculia with preserved approximation. Neuropsychologia, 29, 1045–1074.

Dehaene, S., & Cohen, L. (1995). Toward an anatomical and functional model of number processing. Mathematical Cognition, 1, 83–120.

Dehaene, S., & Cohen, L. (1997). Cerebral pathways for calculation: double dissociation between rote verbal and quantitative knowledge of arithmetic. Cortex, 33, 219–250.

Dehaene, S., Molko, N., Cohen, L., & Wilson, A.J. (2004). Arithmetic and the brain. Current Opinion in Neurobiology, 14, 218–224.

Dehaene, S., Piazza, M., Pinel, P., & Cohen, L. (2003). Three parietal circuits for number processing. Cognitive Neuropsychology, 20, 487–506.

Dehaene, S., Piazza, M., Pinel, P., & Cohen, L. (2005). Three parietal circuits for number processing. In J. I. D. Campbell (ed.), Handbook of Mathematical Cognition (pp. 433–453). New York: Psychology Press.

Dehaene, S., Spelke, E., Pinal, P., Stanescu, R., & Tsivkin, S. (1999). Sources of mathematical thinking: behavioral and brain-imaging evidence. Science, 284, 970–974.

De Smedt, B., & Boets, B. (2010). Phonological processing and arithmetic fact retrieval: evidence from developmental dyslexia. Neuropsychologia, 48, 3973–3981.

De Smedt, B., Taylor, J., Archibald, L., & Ansari, D. (2010). How is phonological processing related to individual differences in children's arithmetic skills? Developmental Science, 13, 508–520.

Frenck-Mestre, C., & Vaid, J. (1993). Activation of number facts in bilinguals. Memory & Cognition, 21, 809–818.

Gelman, R., & Butterworth, B. (2005). Number and language: How are they related? Trends in Cognitive Science, 9, 1–10.

Grabner, R.H., Ansari, D., Ebner, F., Koschutnig, K., Neuper, C., & Reishofer, G. (2009). To retrieve or to calculate? Left angular gyrus mediates the retrieval of arithmetic facts during problem solving. Neuropsychologia, 47, 604–608.

Hodent, C., Bryant, P., & Houdé, L. (2005). Language-specific effects on number computation in toddlers. Developmental Sciences, 8(5), 420–423.

Imbo, I., & LeFevre, J-A. (2011). Cultural differences in strategic behavior: a study in computational estimation. Journal of Experimental Psychology: Learning, Memory, & Cognition, 37, 1294–1301.

Jackson, N., & Coney, J. (2007). Simple arithmetic processing: surface form effects in a priming task. Acta Psychologica, 125, 1–19.

Kashiwagi, A., Kashiwagi, T., & Hasegawa, T. (1987). Improvement of deficits in mnemonic rhyme for multiplication in Japanese aphasics. Neuropsychologia, 25, 443–447.

Landy, D., & Goldstone, R.L. (2007). How abstract is symbolic thought? Journal of Experimental Psychology: Learning, Memory and Cognition, 33(4), 720–733.

LeFevre, J., Bisanz, J., & Mrkonjic, L. (1988). Cognitive arithmetic: evidence for obligatory activation of arithmetic facts. Memory & Cognition, 16, 45–53.

LeFevre, J., Lei, Q., Smith-Chant, B.L., & Mullins, D.B. (2001). Multiplication by eye and by ear for Chinese-speaking and English-speaking adults. Canadian Journal of Experimental Psychology, 55, 285–295.

LeFevre, J., & Liu, J. (1997). The role of experience in numerical skill: multiplication performance in adults from China and Canada. Mathematical Cognition, 3, 31–62.

Lemaire, P., & Reder, L.M. (1999) What affects strategy selection in arithmetic? An examination of parity and five effects on product verification. Memory & Cognition, 27(2), 364–382.

McCloskey, M. (1992). Cognitive mechanisms in numerical processing: evidence from acquired discalculia. Cognition, 44, 107–157.

McCloskey, M., & Macaruso, P. (1994). Architecture of cognitive numerical processing mechanisms: contrasting perspectives on theory development and evaluation. Cahiers de Psychologie Cognitive/Current Psychology of Cognition, 13, 275–295.

McCloskey, M., & Macaruso, P. (1995). Representing and using numerical information, American Psychologist, 50, 351–363.

McCloskey, M., Macaruso, P., & Whetstone, T. (1992). The functional architecture of numerical processing mechanisms: Defending the modular model. In J. I. D. Campbell (Ed.) The Nature and Origins of Mathematical Skills (pp. 493–537). Amsterdam: Elsevier.

McNeil, J.E., & Warrington, E.K. (1994). A dissociation between addition and subtraction with written calculation. Neuropsychologia, 32, 717–728.

Metcalfe, A.W.S., & Campbell, J.I.D. (2007). The role of cue familiarity in adult's strategy choices for simple addition. European Journal of Cognitive Psychology, 19, 356–373.
Metcalfe, A.W.S., & Campbell, J.I.D. (2008). Spoken numbers vs. Arabic numerals: differential effects on adults' multiplication and addition. Canadian Journal of Experimental Psychology, 62, 56–61.
Noël, M-P., Fias, W., & Brysbaert, M. (1997). About the influence of the presentation format on arithmetic-fact retrieval processes. Cognition, 63, 335–374.
Pica, P., Lemer, C., Izard, V., & Dehaene, S. (2004). Exact and approximate arithmetic in an Amazonian indigene group. Science, 306, 499–503.
Rusconi, E., Galfano. G., & Job, R. (2007). Bilingualism and cognitive arithmetic. In I. Kecskes and L. Albertazzi (eds), Cognitive Aspects of Bilingualism (pp. 153–174). New York, NY: Springer.
Sato, M., Cattaneo, L., Rizzolatti, G., & Gallese, V. (2007). Numbers within our hands: modulation of corticospinal excitability of hands muscles during numerical judgment. Journal of Cognitive Neuroscience, 19, 684–693.
Schunn, C.D., & Reder, L.M. (2001). Another source of individual differences: strategy adaptivity to changing rates of success. Journal of Experimental Psychology: General, 130(1), 59–76.
Shrager, J., & Siegler, R.S. (1998). SCADS: a model of children's strategy choices and strategy discoveries. Psychological Science, 9, 405–410.
Siegler, R.S., & Lemaire, P. (1997). Older and younger adults' strategy choices in multiplication: Testing predictions of ASCM via the choice/no-choice method. Journal of Experimental Psychology: General, 126, 71–92.
Sokol, S.M., McCloskey, M., Cohen, N.J., & Aliminosa, D. (1991). Cognitive representations and processes in arithmetic: inferences from the performance of brain-damaged subjects. Journal of Experimental Psychology: Learning, Memory, and Cognition, 17, 355–376.
Spelke, E.S., & Tsivkin, S. (2001). Language and number: a bilingual training study. Cognition, 78, 45–88.
Steel, S., & Funnell, E. (2001). Learning multiplication facts: a study of children taught by discovery methods in England. Journal of Experimental Child Psychology, 79, 37–55.
Trbovich, P., & LeFevre, J. (2003). Phonological and visual working memory in mental addition. Memory & Cognition, 31, 738–745.
Venkatraman, V., Siong, A.C. Chee, M.W.L., & Ansari, D. (2006). Effect of language switching on arithmetic: a bilingual fMRI study. Journal of Cognitive Neuroscience, 18, 64–74.
Whalen, J., McCloskey, M., Lindemann, M., & Bouton, G. (2002). Representing arithmetic table facts in memory: evidence from acquired impairments. Cognitive Neuropsychology, 19, 505–522.
Wisniewski, E., & Bassok, M. (1999). What makes a man similar to a tie? Stimulus compatibility with comparison and integration. Cognitive Psychology, 39, 208–238.
Zbrodoff, N.J., & Logan, G.D. (2005). What everyone finds: the problem size effect. In J. I. Campbell (ed.), Handbook of Mathematical Cognition (pp. 331–346). New York: Psychology Press.
Zhou, X., Chen, C., Qiao, S., Chen, C., Chen, L., Lu, N., & Dong, Q. (2009). Event-related potentials for simple arithmetic in Arabic digits and Chinese number words: a study of the mental representation of arithmetic facts through notation and operation effects. Brain Research 1302, 212–224.

CHAPTER 9

ARITHMETIC WORD PROBLEM SOLVING AND MENTAL REPRESENTATIONS

CATHERINE THEVENOT AND
PIERRE BARROUILLET

INTRODUCTION

ARITHMETIC word problems, also called verbal or story problems, are verbal descriptions of numerically quantified situations involving one or several explicit questions. It is tacitly assumed that the answer to these questions must correspond to a numerical value that has to be reached after one or a series of arithmetic operations has been carried out. Arithmetic word problems are of particular importance because they are often considered and used as a privileged pedagogical situation in which students are expected to show their understanding and mastery of previously learned mathematical procedures and concepts. In some sense, world problem solving is seen as the testing ground of mathematical achievement. However, as testified by several international surveys (see for example Fayol, Barrouillet, & Camos, 1997), this school exercise remains the area of mathematics in which students experience the greatest difficulties, word problem solving lagging far behind others domains such as the mastery of computational algorithms. In this chapter, we review recent theoretical and empirical studies that have been conducted to classify the different additive and multiplicative problems that can be constructed, to describe the representations and processes involved in their solution, to identify the characteristics of the problems, as well as the individual factors that modulate performance, and to design intervention tools for enhancing performance. As we will see, these studies allow researchers to better understand the difficulties encountered by children and provide practitioners with tools to help children in overcoming them.

To anticipate, a consensus has developed among researchers that a large part of the difficulties children encounter in solving word problems arise from an underlying difficulty in understanding the situation that the text describes. Indeed, solving a word problem does not simply consist in successfully performing arithmetic calculations, but anteriorly in understanding the relational structure in which the numerical values are embedded. In this respect, we will see that the way the text is formulated and presented has a strong impact on the ease with which children and adults grasp this structure and solve the problems. It is the reason why word problems have been traditionally classified according to their semantic characteristics, and more precisely to the relational structure of the situation they refer to. According to this view, even when they require the same arithmetic operation, solving problems that refer to quantities that undergo some transformation would not involve the same cognitive processes as solving problems that refer to comparisons between unchanged quantities. In the first part of this chapter, we describe the most frequently used classifications, and we report the empirical evidence that have been gathered in order to attest for their psychological relevance. The second part of the chapter is devoted to the question of the nature of the representations involved in word problem solving. Do problem solvers progressively extract through practice schematic abstract representations that match the most frequent relational structures and then instantiate these representations with the numerical values of the problem at hand, or do they construct in working memory ad hoc transient mental representations for each word problem encountered? This is of importance because the strategies used probably depend on the nature of the representation constructed. Empirical evidence from developmental and adult studies that favor these opposite conceptions will be reviewed. Of course, performance in arithmetic word problem solving does not entirely depend on the semantic characteristics of the problems, but also on the characteristics of the problem solvers themselves. In a third part, we focus on the impact on performance of individual differences in those abilities that are directly involved in word problem solving, such as calculation and text comprehension, but also in more general cognitive resources, such as working memory capacity. Finally, we address the studies that have focused on methods for enhancing performance in word problem solving, either by making clearer the semantic aspects of the text or by providing students with more appropriate knowledge and strategies. We conclude by pointing what we consider as the main pending questions and suggesting some leads for future research.

Semantic Characteristics of Word Problems

The most famous classification of addition problems has been proposed by Riley, Greeno and Heller (1983). It is based on the type of semantic relations described in the text. Thus, according to the authors, many problems can be classified into three semantic types: Change, Combine, and Comparison problems. Note that, as acknowledge by Riley et al. (1983), these categories have been used by several other investigators sometimes using different names (Nesher, 1981; Vergnaud, 1982). Change problems describe actions that cause increases or decreases in quantities and the unknown is either the resulting amount (John has 6 marbles. Tom gives him 3 more marbles. How many marbles does John have now?), the amount of the change (John had 4 marbles. Tom gave him some more marbles. Now, John has 6 marbles. How many marbles did Tom give him?), or the starting amount (John had some marbles. Tom gave him 5 marbles. Now, John has 7 marbles. How many marbles did John have at the beginning?). In contrast to such dynamic problems, Combine and Comparison problems are relative to static situations. Combine problems describe two quantities that have to be considered in combination to produce the answer. The question can be relative to the total amount (John has 2 marbles. Peter has 4 marbles. How many marbles do John and Tom have altogether?) or to the amount in one of the subsets (Joe and Tom have 8 marbles altogether. Joe has 3 marbles. How many marbles does Tom have?). Finally, in Compare problems, a Comparison is made between two quantities using the expressions 'more than' and 'less than'. The unknown is either the difference between the two quantities (John has 3 marbles. Tom has 5 marbles. How many marbles does John have less than Tom?), the compared quantity (John has 3 marbles. Tom has 5 more marbles than John. How many marbles does Tom have?), or the referent (John has 3 marbles. He has 5 more marbles less than Tom. How many marbles does Tom have?).

The ecological relevance of this classification is attested by the fact that different problem types are not of equal difficulty, even when they require the same arithmetic operation for solution (Riley et al., 1983). Besides, an improvement in performance for each category of problem is observed through development and education. More concretely, Change problems are easier for Grade 2 and 3 children than for kindergarteners and Grade 1 children. When the question of the problem is relative to the resulting state, problems are quite easy, even for young children. In contrast, when the question is about the starting quantity, children experience more difficulties and successful performance for this category of problems only reach a satisfactory rate in Grade 2 (75%). Similarly, Combine problems are easy for young children when the unknown is the sum of the sets but, when the unknown is one of these sets, children experience great difficulties (only 30% of successful solving in kindergarten and Grade 1). Finally, Compare problems are more difficult than Combine and Change problems.

If Riley et al.'s classification is limited to addition problems, Vergnaud (1983) conducted semantic analyses of multiplication problems. Three different forms of relations between the problem elements are described by the author. The first form (or structure) is called 'isomorphism of measures' and consists in direct and simple proportions between two measures or quantities (e.g. Linda wants to share her 12 cakes with her two friends. How many cakes each of the three little girls have?). The second form called 'product of measures' describes situations where three measures are involved and the third measure is the product of the first two, as in calculating surfaces or volumes (e.g. What is the surface of a room that is 27 feet long and 12 feet large?). The third form of relations is called 'multiple proportions' and corresponds to situations that are essentially similar to the isomorphism of measures structure, but in which several variables are involved (e.g. One cow produces an average of 30 liters of milk per day, each liter of milk costs $1.05. How much will the farmer earn over a 2-week period?).

Greer (1992) also proposed a classification of problems wherein a distinction is made between situations that are psychologically commutative and non-commutative. In non-commutative situations, the multiplier and the multiplicand can be distinguished. In other words, one of the quantities involved in multiplication (i.e. multiplier) is conceptualized as operating on the other (i.e. the multiplicand) to produce the result. Concerning division problems, this implies that two types of divisions can be distinguished: division by multiplier (e.g. A college passed the top 3/5 of its students in an exam. If 48 passed, how many students sat the exam?) and division by multiplicand (e.g. A college passed the top 48 out of 80 students who sat the exam. What fraction of the students passed?). In commutative situations, it is impossible to distinguish between multiplier and multiplicand and, consequently, between the two types of division.

Contrary to addition problems, the ecological validity of multiplication problem classifications has not yet been systematically investigated. This is probably due to the dissimilarities in the classification schemata used by different researchers. However, Vergnaud (1983) showed that children master scalar problems, wherein there is a number of groups or objects having the same number in each group (e.g. 3 children have 4 cookies each. How many cookies do they have altogether?) before Cartesian-product problems (e.g. In the ballroom there are 5 boys and 4 girls. If all the boys dance with all the girls, how many couples will be constituted?) or problems involving conversion of measures (e.g. A foot is 0.3 meters. How many meters are 3 feet?). Therefore, as addition problems, the difficulty of multiplication problems does not simply depend on the nature of the operation that has to be performed. The nature and the structure of the semantic situation in which the problem is embedded play a crucial role in individuals' performance. This is due to the fact that these factors determine the structure, nature, and complexity of the mental representation that has to be constructed to reach the solution.

The Nature of The Mental Representations in Word Problem Solving

It is generally admitted that children struggle with arithmetic word problems when they experience difficulties in constructing the adequate mental representation of the situation that the text describes (De Corte & Verschaffel, 1985). For instance, Change problems are relatively straightforward because they describe dynamic situations that can be easily mentally simulated. What is represented by children is a sequence of events ordered in time in a unidirectional manner (Nesher, Greeno, & Riley, 1982). In contrast, Compare problems are difficult for children because they are not familiar with the expressions 'more than' and 'less than' (De Corte & Verschaffel, 1985). Interestingly, if the problem 'There are 5 birds and 3 worms. How many more birds than worms are there?' leads to only 25% of correct responses in kindergarten children, reformulating the question into 'How many birds won't get a worm?' leads to 96% of correct responses in children at the same level (Hudson, 1983). Again, compared with the first version, the second version of the problem is easier to represent because the birds and the worms can be concretely or mentally matched to form a correspondence between the two sets. However, as stated above, some non-semantic factors modulate these patterns of results. The situation is more difficult to represent when the amount of change is unknown (Ibarra & Lindvall, 1982) and even more when the start set is unknown (Garcia, Jimenez, & Hess, 2006; Riley et al., 1983). The lack of knowledge of the first or intermediate quantity would severely hinder the solving process because problems cannot be represented directly by models.

Interestingly, the semantic characteristics of the problems play a role on children's performance before and after formal instruction. Indeed, it has been shown that some preschool children can solve arithmetic word problem before any explicit instruction by using informal strategies based on their understanding of the situation described in the text of the problem (Carpenter & Moser, 1982; De Corte & Verschaffel, 1987; Ibarra & Lindvall, 1982). Children who are successful in solving verbal problems have this ability to develop physical models to serve as an aid to solution (Lindvall & Ibarra, 1980). Carpenter and Moser (1984) showed that the semantic characteristics of subtraction problems strongly shape the way the models are constructed. More precisely, problems such as 'John has 8 cookies. He gives 3 cookies to his friend Leo. How many cookies does John have left?' is likely to be solved by young children by a counting down strategy, which consists, in our example, in counting backwards three times from the starting amount of 8. This strategy perfectly models the decrease Change situation described in the problem. In contrast, a problem such as 'Trudy has 3 cookies. How many more cookies does she have to get so that she has 8 cookies altogether?' is likely to be solved by a counting up strategy from 3 to 8, which is derived from a model that mimics the situation of joining the missing addend. The answer to the problem is the number of counting words in the sequence. Whereas Carpenter and Moser did not find evidence for a relationship between the semantic structure of addition problems and children's strategies, it was done by De Corte and Verschaffel (1987). The authors showed that Change problems such as 'Lynn has 2 cinema tickets. Her friend Sara gives her 2 more tickets. How many tickets does Lynn have now?' is likely to be solved through an adding procedure. In this case, the child constructs a set of blocks corresponding to the first number in the problem, then adds to this set a number of blocks corresponding to the second number, and finally counts the total number of blocks. In contrast the Combine problem 'John has 3 bottles of soda. Tom has 4 bottles. How many bottles do they have altogether' is more likely to be solved by a joining or a no move strategy, where the child constructs two sets corresponding to the two given numbers and either counts the total number of blocks without physically moving the sets (i.e. no move strategy) or moves the sets together with both hands and finally counts the total number of blocks (i.e. joining strategy).

These studies clearly showed that the semantic of the problems determines the strategy used by young children. Nonetheless, even after instruction and practice, semantic factors still determine strategies. Indeed, it is assumed that through practice, problem schemata would be progressively abstracted from and associated with different semantic categories and would determine the procedures used by individuals to solve the problems.

Schemata and Word Problems

Riley et al. (1983) suggest that schemata have the form of semantic network structures that consist of elements and relations between those elements. For example, a Change problem schema would contain an initial quantity that represents a start set, an event that changes the start set (increase or decrease), and an unknown final quantity or result set. The empty slots in the schema are filled in progressively as soon as the relevant information is read in the text of the problem. When all the empty slots are filled in, specific expressions or propositions triggers a second type of schema (i.e. higher order schemata) that contains the procedure required to solve the problem (Kintsch & Greeno, 1985). For example, in Compare problems, 'more than' and 'less than' propositions would trigger the More-than and Less-than schemata, respectively, which include a large-set, a small-set and a difference variables to be instantiated by numerical values. Both schemata also include a Difference strategy triggered by the same propositions that allows the comparisons of the two sets by the appropriate arithmetic operation of subtraction.

Arguments in favor of the use of arithmetic word problem schemata have been provided by Lewis and Mayer (1987). These authors focused on the study of two forms of Compare problems, namely consistent and inconsistent problems. In consistent problems, the unknown variable is the subject of the second sentence that contains a relational term consistent with the necessary arithmetic operation. For example, in the consistent problem 'John has 3 marbles. Tom has 5 more marbles than Joe. How many marbles does Tom have?', the expression 'has more' is consistent with the fact that an addition is required to solve the problem. In contrast, in inconsistent problems, the unknown variable is the object of the second sentence and the relational term conflicts with the necessary arithmetic operation. In the problem 'Joe has 3 marbles. He has 5 marbles less than Tom. How many marbles does Tom have?' the term 'less' is inconsistent with the addition that has to be performed to solve the problem. According to the authors, problem solvers come to the task with a set of schemata associated to Compare problems. When the form of the given relational sentence is not consistent with the problem solver's schema, she or he must mentally rearrange the information. The fact that inconsistent problems are more error prone and take longer to solve than consistent problems is coherent with the author's interpretation. These findings were extended by Fuson, Carroll, and Landis (1996) who showed that the capacity to solve inconsistent problems represents the fourth level of a progression in conceptualizing and solving Compare problems in first and second graders. In a first level, children can only identify who has more or less without being able to quantify the difference. In a second level, they strongly rely on language cues and solve consistent problems ('John has 3 marbles. Tom has 5 more marbles than Joe. How many marbles does Tom have?') better than inconsistent problems and even problems requiring to calculate the difference ('John has 9 marbles. Tom has 7 marbles. How many marbles does John have more than Tom?'). In a third level, children solve these two kinds of problems, but not yet the tricky inconsistent problems. This slow progression indicates how problems inducing the wrong schemata from misleading language cues are difficult to solve.

This sketchy presentation of the schemata-based approach makes clear that schemas are abstract general frames in which the quantitative elements of the problem (numerical values and relations) constitute the only relevant information needed for successful problem solving. However, the abstract nature of these representational frames makes that the schema approach encounters difficulties in accounting for content and context effects ubiquitous in word problem solving. This limitation prompted the emergence of theories that postulate the construction of representations, such as situation models or mental models, which bring to bear a rich real-word experience that makes them more appropriate to grasp semantic content and context.

Situation Model and Word Problem

Reusser (1989; Staub & Reusser, 1995) postulates the existence of an Episodic Situation Model that includes functional and temporal features of the story problem. This qualitative model would be activated by the solver before a more logico-mathematical representation of the problem corresponding to the schemas described above. Evidence for the construction of such qualitative model during the solving process has been collected through several studies. Coquin-Viennot and Moreau (2003) showed that a problem such as 'For a prize-giving the florist prepares for each of the 14 candidates 5 roses and 7 tulips. How many flowers does the florist use in total?' is likely to be solved by children with a distributed strategy consisting in performing (14 × 5) + (14 × 7). However, when formulated as 'For a prize-giving, the florist prepares for each of the 14 candidates a bouquet composed of 5 roses and 7 tulips. How many flowers does the florist use in total?' the problem is more likely to be solved with a factorized strategy, by performing 14 × (5 + 7). In the second formulation of the problem, the term 'bouquet' plays a structuring role in the organization of the elements contained in the mental representation constructed by children to solve the problem. This result is difficult to interpret within the schema framework, and Coquin-Viennot and Moreau concluded that less formal representations based on real-world knowledge necessarily come into play during the resolution. Similar results were obtained by Bassok, Chase, and Martin (1998) who showed that, when college students construct mathematical problems, they align the mathematical relations between arguments of arithmetic operations with the semantic relations that are evoked by the pairs of objects they are asked to include in the problems. More concretely, division involves an asymmetric mathematical relation between dividend and divisor ($a/b \neq b/a$). Bassok et al.'s participants applied this asymmetric operation to object sets that readily evoked functionally asymmetric semantic relations (e.g. *a* apples and *b* baskets), but refrained from dividing object sets that did not evoke functionally asymmetric relations (e.g. *a* apples and *b* oranges). Instead, they usually related such functionally symmetric sets with the mathematically symmetric operation of addition ($a + b = b + a$).

In addition to real-world knowledge, it has also been shown that characteristics, such as the size of numbers used in the problem can affect the strategies selected by individuals. Thevenot and Oakhill (2005) presented adults with segmented problems such as 'How many marbles does John have more than Tom and Paul altogether?/John has 49 marbles./Tom has 24 marbles./Paul has 17 marbles./(the '/'s represent the boundaries of the segment)'. When a segment appeared, the previous one disappeared. The strategy preferred by adults to solve the problem was to calculate: 49 – (24 + 17), whereas an alternative sequential strategy that is far less demanding could have been possible to implement, namely (49 – 24) – 17. Nevertheless, when three-digit instead or two-digit numbers were used in the text (How many marbles does John have more than Tom and Paul altogether? / John has 649 marbles./ Tom has 324 marbles./ Paul has 217 marbles./'), the alternative strategy consisting in two successive subtractions was preferred by participants. Obviously, the difficulty of the task pushed individuals to construct an alternative representation to the one directly induced by the wording of the problem. Again, these results are difficult to reconcile with a pure schema-based approach in which the use of different strategies for a specific type of problem is difficult to interpret. Still, as stated above, it is not to say that schemata did not play any role in the course of the solution process, but rather that a transitional representation between the text problem and a pure mathematical mental representation comes into play.

Mental Models and Word Problems

Nevertheless, some studies might more fundamentally question the systematic resort to problem schema in children and adults. It was shown that placing the question before a story problem rather than classically presenting it at the end facilitates problem solution in children. In the seminal studies, this finding was first attributed to mobilization by problem solvers of the appropriate schema (Devidal, Fayol, & Barrouillet, 1997; Fayol, Abdi, & Gombert, 1987). In a standard word problem, the question often contains all the information needed to describe the relations between the different elements of the text as well as specific linguistic expressions. Therefore, according to Devidal et al. (1997), the question placed at the beginning of the text, via the specific information it contains, directly triggers the appropriate schema (e.g. the expression 'less than' will trigger a Comparison schema). This early activation allows the numerical data to be integrated into the schema as soon as they are encountered, and the calculations can be achieved during reading. The resulting release of working memory resources would explain the improvement in performance. However, Thevenot, Barrouillet, and Fayol (2004) confirmed the facilitatory effect of the question at the beginning, but showed that this effect is observable whether or not the calculations are achieved during reading. Consequently, the increase in the number of successful resolutions when the question is placed before, rather than after, the text cannot be attributable only to the on-line integration of the numerical data into the activated schema. The authors proposed that the aid provided by the position of the question before the text would only be due to its facilitatory effect on the construction of the representation required to solve the problem. Just like a title before a narrative text (Rawson & Kintsch, 2002), the question before an arithmetic problem would facilitate the subsequent encoding and integration of the information in the representation. This interpretation is alternative to the one formulated within the Schema framework and is compatible with an approach that considers the representation as a specific and temporary mental structure constructed in working memory, such as the Situation Model approach described above (Reusser, 1989) or the Mental Model theoretical framework (Johnson-Laird, 1983).

This interpretation was confirmed by Thevenot, Devidal, Barrouillet, and Fayol (2007) who showed that the facilitatory effect of the question before the text was more pronounced for children with poor abilities in mathematics. These results are not compatible with the schema framework. If the facilitatory effect of the question before the text was due to the early activation of the appropriate schema, high-skilled children would be those who are more likely to benefit from this effect. Indeed, these children are more often successful in solving problems and, consequently, are more likely to have extracted schemas. As stated above, the reverse result was obtained: it were the low-skilled children who benefited more from the position of the question before the problem. While the simple fact that high-skilled children are not those who benefit more from the position of the question before the text is sufficient to rule out the Schema interpretation, the fact that the inverse result is obtained supports the Mental Model theory. We know that low-skilled children are those children who experience the highest difficulty in the construction of the Mental Model of the situation (Cain & Oakhill, 1999; Oakhill, 1996; Oakhill, Cain, & Yuill, 1998; Yuill, Oakhill, & Parkin, 1989). Therefore, it seems natural that they are the same children who benefit more from an aid to the construction of the representation. Moreover, Thevenot et al. (2007) showed that the facilitatory effect of the question before the text was more pronounced for difficult than easier problems. Again, this result was not compatible with the Schema theory. Difficult problems, which by definition are often failed, are less likely to be associated with the appropriate schema than easier problems, which are often correctly solved. Therefore, the higher facilitatory effect for difficult problems cannot be attributable to the early activation of a problem schema.

The relative relevance of Mental Model and Schemas in the domain of story problems was also directly tested by Thevenot (2010). In her experiment, adults were asked to solve Compare problems wherein the question took a form such as 'How many marbles does Louis have more than Jean?' After the solving phase, an unexpected task of problem recognition was proposed to participants. They were presented either with the original problems, with inconsistent problems that had never been solved, or with paraphrases that respected the relational structure of the original problems, but not their exact wording. These paraphrases were constructed by inverting the terms and the linguistic expressions in the original problems (our example became 'How many marbles does Jean have less than Louis?). Inconsistent problems were constructed by keeping the same relational terms as in the original problem, but by inverting the names of the protagonists (i.e. 'How many marbles does Jean have more than Louis'). Whereas the literal form of paraphrastic problems bore the least resemblance to original problems, paraphrastic problems were associated to higher recognition rates than inconsistent problems. These results provided strong evidence that it is the structure of the arithmetic word problem and not its exact wording or its propositional level of representation that is mentally represented by individuals engaged in a resolution task. This is perfectly coherent with the mental model theory proposed by Johnson-Laird (1983; see Mani & Johnson-Laird, 1982, for a similar demonstration in the domain of spatial reasoning), and at odds with the approach postulating that the relational expression ('have more' in our example) plays a crucial role in triggering the appropriate schema or procedure.

These results also fit nicely with the Situation Strategy First framework proposed by Brissiaud and Sander (2010), which posits that the initial mental representation constructed from a problem text activates a situation-based strategy even after instruction (see also Brissiaud, 1994). It would be only when it is not efficient for providing the solution that the representation would be modified so that the relevant arithmetic knowledge might be used. For example, the situation-based strategy for a problem such as 'Luc is playing with 42 marbles. During the game, he loses 39 marbles. How many marbles does Luc have now?' would consist in counting backward from 42, which is cognitively demanding. In order to gain in efficiency, children would have to 're-represent' the problem in order to apply a mental arithmetic strategy consisting in counting forward from 39 to 42. In accordance with their Situation Strategy First model, the authors show that such problems are indeed more difficult for 7 – and 8-year-old children than problems where the situation-based strategy directly corresponds to the most economic strategy (i.e. 'Luc is playing with 42 marbles. During the game he loses 3 marbles. How many marbles does Luc have now?').

This series of works, which provides evidence that schemata do not always constitute the core of the mental representation constructed by individuals, questions assumptions relative to the source of age-related increase in performance. According to Judd and Bilsky (1989), young children experience difficulties with word problems because they do not yet have a repertoire of highly automatized schemata for representing the different problem types. However, it seems that schemata are not always necessarily convoked to solve familiar problems. Thus, other sources of difficulties have to be considered, such as limited working memory capacities, reading and text comprehension difficulties, or insufficient mastery of number facts and calculation procedures.

Individuals' Factors Impacting Arithmetic Word Problem Performance

We have seen that arithmetic word problem solving is a complex activity that requires a series of cognitive processes involving the comprehension of a text, the instantiation of abstract representations with current information or the construction of ad hoc mental models, as well as reasoning activities to select appropriate arithmetic operations. Once these operations have been identified, calculation of the answers often requires selection, running, and control of algorithmic procedures that necessitate retrieval of relevant arithmetic facts from long-term memory and temporary storage of intermediary results. These activities thus involve manipulation and maintenance of verbal, but also visuospatial information, as well as the selection and control of procedures. Thus, arithmetic problem solving is a complex cognitive activity that would involve all the components of working memory. Accordingly, several studies have investigated the role of working memory in word problem solving and how individual differences in working memory capacity affect performance.

Working Memory

The relationship between mathematical cognition and working memory has been extensively studied, especially in the domain of arithmetic problem solving (see DeStefano & LeFevre, 2004; LeFevre, DeStefano, Coleman, & Shanahan, 2005, for reviews). In the same way, several studies have provided support for the importance of working memory (WM) in word problem solving. Most of them used the theoretical framework provided by Baddeley's (1986) multi-component model and aimed at identifying the specific contribution of the central executive, which is considered to be primarily responsible for coordinating activity within the cognitive system, the phonological loop that assumes the storage of verbal material, and the visuospatial sketchpad devoted to the maintenance of visual and spatial information. Among these components, the phonological loop can be expected to play a crucial role in solving arithmetic problems presented through text. Accordingly, Passolunghi and Siegel (2001) showed that 9-year-old children (Grade 4) who are poor at problem solving have difficulties with digit span, which is often used as a measure of the phonological loop. Rasmussen and Bisanz (2005) further supported the notion that, for school-age children, phonological WM is what differentiates between poor and good verbal arithmetic performers. However, Passolunghi and Siegel (2001) also identified difficulties with central executive tests of WM in children who are poor at problem solving. Several other studies by Passolunghi shed light on the role of the central executive. For example, Passolunghi, Cornoldi, and De Liberto (1999) found that Grade 4 children who present difficulties in arithmetic word problem solving also present difficulties in working memory tasks that require the inhibition of irrelevant information. Passolunghi and Pazzaglia (2004) investigated the role of memory-updating, a central process needed to active the relevant information while removing no longer relevant information from WM (this capacity is, for example, assessed by asking children to recall the five smallest animals at the end of the following list of items presented successively: *hare*, tiger, game, type, *mole*, *fly*, fox, return, *mosquito*, idea, answer, and *spider*). They observed that a group of fourth-graders with high memory-updating ability performed better in problem solving compared with a group of low memory-updating ability. Andersson (2007) extended these results by demonstrating that three different measures related with the central executive contributed unique variance to word problem solving in children in grades 2, 3, and 4 – the capacity to coordinate processing and storage, to shift attention from one task to another, and to access knowledge from long-term memory. This relation holds even when the influence of reading, age, and IQ were controlled. The involvement of central executive does not seem to be restricted to arithmetic word problems and several studies have found that this component of WM also contributes significant variance in predicting algebraic word problem solving in older children (Lee, Ng, & Ng, 2009; Lee, Ng, Ng, & Lim, 2004).

Studies that have addressed the possible implication of the visuospatial sketchpad are rarer. Nonetheless, Passolunghi and Mammarella (2010) observed that fourth graders who are poor problem solvers exhibited specific difficulties in spatial WM assessed through a Corsi block and a spatial matrix tasks, but not visual working memory. It is worth noting that, in this study, the authors did not found any relation with verbal memory. A more extensive investigation has recently been conducted by Zheng, Swanson, and Marcoulides (2011) who assessed the three components of WM along with reading and mathematical achievement in a large sample of 310 children in grades 2, 3, and 4. The results indicated that the three components predicted word-problem solving performance. Interestingly, reading and calculation proficiency mediated the effect of verbal WM and central executive, whereas academic achievement did not mediate the relationship between visuospatial WM and word problem solving, a result that echoes Passolunghi and Mammarella's (2010) observation.

In summary, there is a strong consensus that WM plays a crucial role in word problem solving and that individual differences in WM capacity underlie at least a part of individual differences in performance. Importantly, the three components of WM have been identified as having a specific impact, though reading abilities could act as a moderating factor, a point that will be addressed below. Although the studies reported here do not leave many doubts about the involvement of WM in the complex activity that word problem solving is, they have also the common characteristic of assessing correlations between abilities and capacities measured at a given moment in development. As such, they provide only indirect evidence concerning the question of the role of WM in the development of problem solving abilities. Interestingly, Swanson (2011) addressed the question of whether the development in problem solving abilities is related to growth in WM capacity. The results showed that WM performance in Grade 1 contributes approximately 26% of the variance to problem-solving accuracy in Grade 3. The significant contribution of WM to word problem-solving accuracy was maintained when measures of attention and long-term memory capacities were entered into the regression analysis. Swanson's results also showed that Grade 1 performance on measures of naming speed (see also Kail, 2007, for similar results) and inhibition (see Passolunghi et al., 1999) contribute unique variance to Grade 3 problem-solving performance. Finally, the growth curve analysis showed that growth on the executive component of WM was related to growth in word problem solving. Thus, the executive system (controlled attention) and not measures of storage (phonological loop) were related to growth in word problem solving, suggesting that this skill and its development strongly solicits the central cognitive system in charge of coordinating cognitive activities.

Reading and Text Comprehension Abilities

When thinking about the capacities needed to solve arithmetic word problems, reading comprehension and calculation skills are among the first abilities that come to mind. Indeed, solving word problems requires, as a minimum, comprehending the text, storing facts during solution, and performing calculations on the stored numbers. Thus, along with WM, reading comprehension and calculation skills should be among the main determinants of word problem-solving performance. Accordingly, several studies have reported a role of these factors. For example, Kail and Hall (1999) measured reading skills, arithmetic knowledge, and short-term memory span in 8–12-year-old children who were asked to solve Change, Equalize, Combine, and Compare problems. Not surprisingly, knowledge of basic arithmetic facts had a strong impact on word problem performance, but reading abilities contributed also consistently and strongly, although these abilities were assessed through a simple word recognition task, whereas memory was a less consistent predictor of word-problem performance. This latter finding has been reported in several studies. For example Swanson, Cooney, and Brok (1993) observed that the correlation between WM and word-problem performance became non significant when reading comprehension was partialed out. This suggests that reading comprehension takes precedence over memory, a phenomenon also observed by Fuchs and colleagues (Fuchs, Fuchs, Compton, Powell, Seethaler, Capizzi, et al., 2006) who found that, when word-reading proficiency is entered in the regression model, WM no longer plays a significant role in predicting children's word problem accuracy (see Lee et al., 2004, for similar findings in the domain of algebraic word problems).

A number of following studies confirmed the impact of reading comprehension skills on problem solving. We mentioned above the study by Andersson (2007) who assessed a variety of WM capacities in relation with word problem-solving performance, but also arithmetic calculation and reading skills. Although the main conclusion of the authors were related with the unique contribution of WM capacities, the best predictor of arithmetic word-problem solving was, as Kail and Hall (1999) observed, arithmetic calculation ($r = 0.69$) immediately followed by reading ability ($r = 0.56$). In the same way, Zheng et al., (2011) investigated the specific role of several components of WM as we reported above, but they also assessed calculation capacities and reading comprehension. Once more, arithmetic calculation was the best predictor of word problem solving, followed by reading comprehension. Interestingly, this study also revealed that all the components of WM predict word problem performance, but that reading abilities can compensate for some of the influence of WM in children. It is worth to note that comprehension per se is the key aspect of the relationship between reading abilities and word problem solving. Pinperton and Nation (2010) studied the mathematical profile of 7–8-year-old poor comprehenders, who are children who show significant deficits in reading comprehension despite average, or above-average, word-reading ability. When compared with controls showing age-appropriate reading comprehension abilities, poor comprehenders showed a deficit in mathematical reasoning assessed through verbally presented problems, whereas they did not differ from their controls in solving numerical operations.

Although the relationship between reading comprehension and word problem solving seems straightforward, it could be more complex than expected. A first interpretation could be that better reading comprehension facilitates the construction of a correct representation of the situation the problem describes, a correct representation which is necessary for solving the problem. However, as Kail and Hall (1999, p. 667) noted, another interpretation could be that because reading and mathematics achievement are correlated (Geary, 1993; Jordan, Kaplan, & Hanish, 2002), reading comprehension may simply be a proxy for general mathematical skills, including word problem solving. This latter interpretation is reinforced by Lerkkanen, Rasku-Puttonen, Aunola, and Nurmi (2005). They observed in a longitudinal study that mathematics and reading comprehension are highly associated during the first and second years of primary school, but, surprisingly, it appeared that mathematical performance predicted reading comprehension rather than vice versa. Though Lerkkanen et al. (2005) did not assess word problem solving performance, their finding suggests that the correlation observed by Kail and Hall (1999) does not necessarily reveal a causal relationship from reading comprehension to word problem solving performance.

Surprisingly, some studies focusing on groups of children suffering from learning disabilities have reached the same temperate conclusions. Vukovic, Lesaux, and Siegel (2010) studied mathematical skills of third graders who present two types of reading difficulties, either a dyslexia or specific reading comprehension difficulties. When contrasted with a large control group, the two groups presenting reading difficulties performed lower than the control group on word problems. However, once mastery of arithmetic facts was controlled, children with reading difficulties performed in the average range, suggesting that reading difficulties have no effect per se in word problem-solving performance. Nonetheless, as the authors note, the problems were read to the children, eliminating the reading demand of the task and, despite this aid, children with reading difficulties performed lower than the control group. In fact, several studies have shown that children who have difficulties in both mathematics and reading perform worse in word problems than children who have difficulties in mathematics only (Jordan & Hanish, 2000; Jordan & Montani, 1997).

To summarize, many studies have established that, after fluency in arithmetic fact and calculation abilities, reading comprehension is a strong predictor of word problem-solving performance. Even if the directionality of the causal link remains uncertain and the specific contribution of reading comprehension is difficult to isolate, reading difficulties contribute to a lower achievement in word problem solving. Alternatively, it seems that reading abilities can compensate for the influence of more general cognitive traits like WM capacity. In the next section, we address the studies that have explicitly focused on those factors susceptible to improve performance.

Enhancing Word Problem Performance

Schema-Based Instruction (Sbi)

As noted by Jitendra, Griffin, Deatline-Buchman, and Sczesniak (2007), traditional instruction teaches students to use keywords in order to solve verbal problems. For example, 'all' suggests addition, whereas 'share' suggests division. This approach is limited because many problems do not have keywords. Moreover, as already noted when commenting on inconsistent problems, basing the solving process on keywords can be misleading and does not help students to make sense of problem situations. In order to better promote elementary students' mathematical problem solving skills, Jitendra and her colleagues (e.g. Jitendra et al., 2007; Jitendra & Hoff, 1996; Jitendra & Star, 2011) designed an intervention tool called the schema-based instruction (SBI). SBI promotes the explicit analysis of the problem schema (e.g. part-part-whole) and the relationships between its different elements (e.g. parts make up the whole). The rationale behind SBI is that understanding these relationships is crucial in selecting appropriate operations needed to solve the problems. In fact, 'SBI allows students to approach the problem by focusing on the underlying problem structure, thus facilitating conceptual understanding and adequate word-problem-solving skills' (Jitendra et al., 2007, p. 286). The instruction program consists of asking children to read and retell the problem to discover its type (Change, Combine, or Comparison, for example, for addition problems), to underline and map important information onto a schematic diagram, to decide which operation has to be performed, to write the mathematical sentence and the answer, and finally to check it. Such program conducted in low – and high-ability, as well as special education classrooms revealed performance improvement from pretest to posttest on word problem solving and computation fluency measures.

Using Manipulatives

As seen above, capacities in WM can be related to word problem performance. Therefore, decreasing the WM demand of a problem by using external aids such as manipulatives could help children to solve it. Manipulative materials, such as Cuisenaire rods or craft sticks, are concrete models that incorporate mathematical concepts, and that can be touched and moved around by students. They can be used by children in order to represent the objects and their relationships described in the text. Then, using manipulatives may be beneficial in constructing an adequate situation or mental model of the problem. Indeed, providing kindergarten children with manipulatives enable them to solve simple word problem (Carpenter, Hiebert, & Moser, 1983). However, Stellingwerf and Van Lieshout (1999) showed that 11-year-old children from schools for special education do not improve their performance on word problem after a training in which only manipulatives were used. The authors conclude that, whereas young children are helped by the use of manipulatives (Carpenter et al., 1983), older ones do not benefit any longer from such instructional tools. Stellingwerf and Van Lieshout also demonstrated the existence of a transitional stage (i.e. between the age of 6 and 11) in which manipulatives are helpful when used in combination with other educational tools such as the use of number sentences (e.g. $a + ? = c$).

Rewording The Problems

As already stated above, reformulating the problem can dramatically improve children performance. We saw that whereas the problem 'There are 5 birds and 3 worms. How many more birds than worms are there?' leads to only 25% of correct response in kindergarten children, a reformulation into 'How many birds won't get a worm?' leads to 96% of correct response in children at the same level (Hudson, 1983). According to Staub and Reusser (1995), the two versions of the problem refer to different episodic situation models. Whereas the first version describes a static and abstract situation, the second version describes a dynamic and more concrete situation. As a consequence, everyday and real-world knowledge can easily be applied during the course of the comprehension and solving processes, hence the improvement in performance. Likewise, Stern and Lehrndofer (1992) recorded better results on Compare problems when they added qualitative information to the text. For example, in a problem such as 'How many fewer pencils does Laura have than Peter?', adding that Peter is the older brother and, as such, enjoys a number of advantage, helps children in their understanding of the problem. However, other studies failed to obtain such effects. Cummins, Kintsch, Reusser, & Weimer (1988) compared second graders' comprehension and solution of standard word problems with enriched problems embedded into little stories showing plausible and realistic situations. For example, the problem 'Joe has 8 dollars. He has 5 more dollars than Tom. How many dollars does Tom have?' was reformulated into 'Joe and Tom play tennis together twice a week. They both always try hard to beat each other. Both of them decided to buy new tennis balls. So far Joe has saved 8 dollars for his balls. He has saved 5 dollars more than Tom. How many dollars has Tom saved?' Contrary to the author's expectations, the reworded problems did not elicit significantly higher solution performance than the standard ones.

As a matter of fact, Vicente, Orrantia, and Verschaffel (2007) have shown recently that such situational rewordings are, indeed, not always very efficient. In contrast, conceptual rewording, which consists of the enhancement of the underlying semantic structure of the problem, more systematically elicits positive effects. It is the case in De Corte, Verschaffel, and de Win's study (1985), where the semantic relationship between the sets implied in the problem are stated more transparently. First and second graders were asked to solve Change, Combine, and Compare problems, either in their classical form, or in a reworded way. The problems were reformulated in such a way that the relations between the given and the unknown sets were made more explicit. However, the underlying mathematical structure of the problems was not affected. For example, a sentence relative to the initial situation was added at the beginning of Change problems ('Joe had some marbles. He won 3 more marbles. Now he has 5 marbles. How many marbles did Joe have at the beginning?') or the part-whole relation was made explicit in Combine problems ('Tom and Ann have 9 nuts altogether. Three nuts belong to Tom. The rest belong to Ann. How many marbles does Ann have?'). Such conceptual rewordings facilitated the solving process in both age groups, but especially in younger children. The interpretation of Verschaffel, De Corte, and De Win (1985) was that their reformulations helped children in constructing the adequate mental representation of the problem, especially in children whose canonical schemata are not fully developed and, therefore, depend more on text-driven (bottom-up) processing to construct their mental representation.

Using Pictures, Graphs, and Figures

Using pictures or self-generated drawings tends to facilitate the solution of word problems. Willis and Fuson (1988) showed that asking children to choose a schematic drawing that matches the situational structure of a given problem, then to fill the problem numbers into appropriate locations in the drawing and, finally, to use the drawing in order to decide the solution procedure is an efficient mean to improve performance in second-graders. Children from high and average-achieving classrooms were generally accurate in choosing the correct drawing for Combine, Change, and Compare problems. According to the authors, this technique is useful because it helps children to grasp the semantic structure of the problem situation. Sprinthall and Nolan (1991) added that the structure of Combine problems is made even more salient when different objects are introduced within the pictures. For example, it is more efficient to present first-grade children with pictures representing 5 tulips and 2 roses than representing the same objects: 5 tulips and 2 tulips. The pictures or drawing representing different objects would present clearer visual arrays, which would allow students to form contrasting perceptions of the quantities in each set and, consequently, would help them to grasp more easily the semantic structure of the problem.

However, as for rewording, Vicente, Orrantia, and Verschaffel (2008) showed that only conceptual (mathematical) drawings and not situational ones are efficient. Conceptual drawings make salient the mathematical organization, whereas situational drawings highlight the components of the situation such as the temporal sequence of events or the context in which the story is embedded. Vicente et al. (2008) asked children with high and low abilities in arithmetic to solve easy and difficult problems. The former could be solved following the time sequence described in the text, whereas for the latter it was necessary to solve the second part of the problem before its first part in order to reach the solution (e.g. 'Laura had 47 beads. She bought some more beads. She used 126 beads and she was left with 11 beads. How many beads did she buy?'). The problems could be presented in their standard form (as in the example), with extra mathematical information that highlighted the part-whole relation described in the problem ('Laura had 47 beads. She bought some more beads *and put them together with the beads she already had. From the total amount of beads she had after buying some,* she used 126 beads and she was left with 11 beads. How many beads did she buy?'), or with extra situational information that emphasized the temporal sequence of events described in the problem ('*Two days ago*, Laura had 47 beads. *Yesterday*, she bought some more beads. This morning she decided to make a collar for her mother. *To do so*, she used 126 beads and, *when she finished*, she was left with 11 beads. How many beads did she buy?'). The problems could be presented by themselves or along with a drawing. Drawings that were presented with standard problems did not contribute to the understanding of the conceptual or mathematical situation of the problem but merely represented the protagonist(s) and the objects described in the problem. The mathematical drawing represented the part-whole relation described in the problem. It was constituted of two pictures, one for each part of the Change situation. The first picture depicted the reunion of the two parts into the whole, for example, Figure 9.1.

The second picture showed that the initial total amount of objects can be portioned into two parts, for example, Figure 9.2.

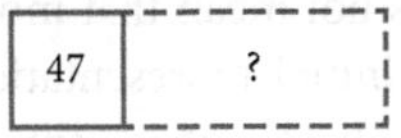

FIGURE 9.1 Each part of the Change situation.

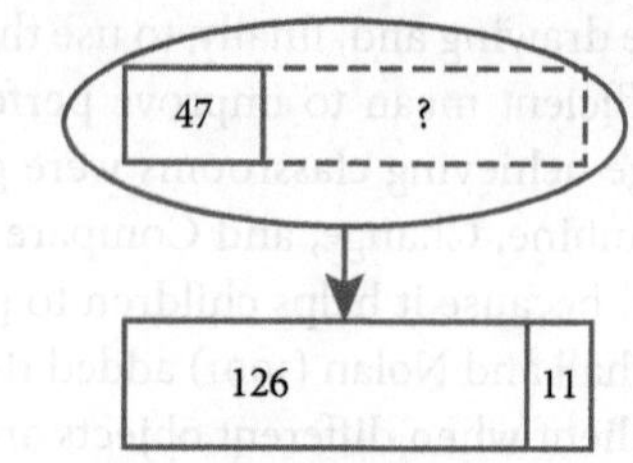

FIGURE 9.2 Initial total amount of objects portioned into two parts.

Finally, situational drawings corresponded to a kind of comic strip in which each of the frames represented a temporal window described in the problem. In line with previous research (see the above section about rewording), the authors showed that when no drawing was presented with the problems, the situational and the mathematical versions of the problems elicited more correct answers than the standard versions. This was especially true for difficult problems. More importantly for our purpose, when the problem was accompanied with a mathematical drawing, children's rate of correct answers was greater compared with a situation in which a standard or situational drawing was presented. However, and interestingly, only children with higher abilities in mathematics benefited from the presentation of mathematical drawings. The interpretation of the authors concerning this last finding was that only children with high abilities have the necessary skills and cognitive resources to process the mathematical drawing efficiently.

Conclusions and Directions

The main conclusion that could be drawn from this rapid survey of the literature on arithmetic word problem solving during the last decades is that this research domain has progressively moved towards an increasing emphasis of the role of semantically rich and complex representations. If the seminal works on this domain and the first semantic classifications could lend credence to a word problem-solving activity relying on a restricted and finite number of definite schemas progressively abstracted through repeated exposition with problems, the difficulties that this view encountered in accounting for ubiquitous effects of content and context brought about the necessity of envision more complex representations and associated cognitive processes. Instead of retrieving abstract schemata from long-term memory that would be readily instantiated with the available numerical values before running the associated calculation, problems solvers seem to construct ad hoc transient representations (mental models or situation models) that involve a fairly rich real-world knowledge. However, the level of abstractness of representations and procedures remains a controversial issue in research on numerical and arithmetic processes (see Cohen Kadosh & Walsh, 2009; Landy & Goldstone, 2007), and the widespread use of context-bound representations does not mean that more abstract mathematical representations cannot emerge from these initial representations and direct further calculation. Nonetheless, empirical evidence indicates that semantic and pragmatic aspects influence the entire process of problem solving. The difficulty of constructing these complex representations from text reading, and the necessity to maintain them in an active state, while selecting and performing appropriate calculations, probably explains why, along with trivial factors, such as calculation proficiency and reading comprehension, working memory capacities are so good predictors of performance. This is also probably why, among the variety of interventions and tutorials that have been imagined and tested, such as rewordings or use of drawings, those that emphasize the conceptual characteristics of the problems are the most efficient. Arithmetic word problem solving thus appears as a highly complex cognitive activity that mobilizes a variety of conceptual knowledge and procedural skills, and it does not come as a surprise that it constitutes the domain in which pupils experience their main difficulties.

As other high-level cognitive activities, such as reasoning, decision making, or expertise acquisition, the complexity of the activity of arithmetic word problem solving makes its study especially arduous for cognitive and educational psychologists. However, we would like to conclude by suggesting some leads for future research. The present review of the literature made probably clear that developmental studies have up to now focused on relatively restricted ranges of ages and degrees. Nonetheless, studies that would encompass the complete developmental era or the educational curriculum, say from kindergarten age to adulthood, would be of interest. Indeed, very little is known about how strategies and their underlying representations evolve in the long run. It is possible that the use of manipulatives to model problem situations in young children is progressively interiorized into some mental schemas that would coexist with the use of semantic representations like mental or situation models, enriching children's toolbox. It is also possible that these semantic representations evolve themselves into more abstract representations that could direct problem solving, in the same way as experts in several domains have memorized an impressive list of strategies in response to a variety of recurrent situations. It cannot be excluded that these developmental changes interact with content and contexts, rendering developmental trends even more complex. For example, contents that facilitate the construction of mental images, such as problems involving count nouns, could more easily elicit mental models than problems involving mass nouns that could more efficiently trigger schema retrieval due to their low level of imageability. In other words, it might be that problem solvers mobilize at each developmental level a variety of strategies varying in representational format and level of abstraction, in the same way as children and adults use a variety of strategies, even for solving very simple arithmetic problems such as elementary additions or subtractions (Siegler, 1996). Another necessary extension of the current research concerns the variety of the problems studied. As the present review made clear, and apart from rare attempts, the range of the problems studied has up to now remained restricted to the simplest forms of the additive problems. The ecological and psychological validity of more extended and complex classifications requires to be established. More complex problems might reveal even more elaborated strategies involving an increased variety of representational formats. Finally, psychologists should not forget that arithmetic word problem solving is not only a complex cognitive activity that constitutes a privileged tool for exploring high-level cognitive processes, but also a school activity and one of the main pedagogical tools for mathematics teaching. In this respect, a better understanding of the underlying factors that make this activity so difficult, and of the determinants of individual difficulties is needed that could help to promote more rational teaching curricula and more efficient intervention programs.

References

Andersson, U. (2007). The contribution of working memory to children's mathematical word problem solving. Applied Cognitive Psychology, 21, 1201–1216.

Baddeley, A. (1986). Working Memory. New York: Oxford University Press.

Bassok, M., Chase, V.M., & Martin, S.A. (1998). Adding apples and oranges: alignment of semantic and formal knowledge. Cognitive Psychology, 35, 99–134.

Brissiaud, R. (1994). Teaching and development: solving 'missing addend' problems using subtraction. European Journal of Psychology of Education, 4, 343–365.

Brissiaud, R., & Sander, E. (2010). Arithmetic word problem solving: a Situation Strategy First Framework. Developmental Science, 13(1), 92–107.

Cain, K., & Oakhill, J. (1999). Inference making ability and its relation to comprehension failure in young children. Reading and Writing, 11, 489–503.

Carpenter, T.P., Hiebert, J., & Moser, J.M. (1983). Problem structure and first-grade-children's initial solution processes for simple addition and subtraction problems. Journal for Research in Mathematics Education, 12, 27–39.

Carpenter, T.P., & Moser, J.M. (1982). The development of addition and subtraction problem-solving skills. In T. P Carpenter, J. M. Moser, & T. A. Romberg (Eds), Addition and Subtraction: A Cognitive Perspective (pp. 9–25). Hillsdale: Erlbaum.

Carpenter, T.P., & Moser, J.M. (1984). The acquisition of addition and subtraction concepts in grades one through three. Journal for Research in Mathematics Education, 15, 179–202.

Cohen Kadosh, R., & Walsh, V (2009). Numerical representation in the parietal lobes: abstract or not abstract? Behavioral and Brain Sciences, 32(3–4): 313–28.

Coquin-Viennot, D., & Moreau, S. (2003). Highlighting the role of the episodic situation model in the solving of arithmetical problems. European Journal of Psychology of Education, 18, 267–279.

Cummins, D., Kintsch, W., Reusser, K., & Weimer, R. (1988). The role of understanding in solving word problems. Cognitive Psychology, 20, 405–438.

De Corte, E., & Verschaffel, L. (1985). Beginning first graders' initial representation of arithmetic word problems. Journal of Mathematical Behavior, 4, 3–21.

De Corte, E., & Verschaffel, L. (1987). The effect of semantic structure on first grader's strategies for solving addition and subtraction word problems. Journal for Research in Mathematics Education, 18, 363–381.

De Corte, E., Verschaffel, L., & De Win, L. (1985). Influence of rewording verbal problems on children's problem representations and solutions. Journal of Educational Psychology, 77, 460–470.

DeStefano, D., & LeFevre, J. (2004). The role of working memory in mental arithmetic. European Journal of Cognitive Psychology, 16, 353–386.

Devidal, M., Fayol, M. & Barrouillet, P. (1997). Stratégies de lecture et résolution de problèmes arithmétiques. L'Année Psychologique, 97, 9–31.

Fayol, M., Abdi, H., & Gombert, J.E. (1987). Arithmetic problem formulation and working memory load. Cognition and Instruction, 4, 183–202.

Fayol, M., Barrouillet, P., & Camos, V. (1997). Early mathematics learning: what can research tell us, Report for the DG 22 of the European Community, Brussels. Brussels: EU.

Fuchs, L.S., Fuchs, D., Compton, D.L., Powell, S.R., Seethaler, P.M., Capizzi, A.M., Schatschneider, C., & Fletcher, J.M. (2006). The cognitive correlates of third-grade skill in arithmetic, algorithmic computation, and arithmetic word problems. Journal of Educational Psychology, 98, 29–43.

Fuson, K.C., Carroll, W.M., & Landis, J. (1996). Levels in conceptualizing and solving addition and subtraction compare word problems. Cognition and Instruction, 14, 345–371.

Garcia, A.I., Jimenez, J.E., & Hess, S. (2006). Solving arithmetic word problems: an analysis of classification as a function of difficulty in children with and without arithmetic LD. Journal of Learning Disabilities, 39, 270–281.

Garcia, A.I., Jimenez, J.E., & Hess, S. (2006). Solving arithmetic word problems: an analysis of classification as a function of difficulty in children with and without arithmetic LD. Journal of Learning Disabilities, 39, 270–281.

Geary, D.C. (1993). Mathematical disabilities: cognitive, neuropsychological, and genetic components. Psychological Bulletin, 114, 345–362.

Greer, B. (1992). Multiplication and division as models of situations. In D. A. Grouws (Ed.) Handbook of Research in Mathematics Teaching and Learning (pp. 276–295). New York: Macmillan.

Hudson, T. (1983). Correspondences and numerical differences between disjoint set. Child Development, 54, 84–90.

Ibarra, C., & Lindvall, C. (1982). Factors associated with the ability of kindergarten children to solve simple arithmetic story problems. Journal of Educational Research, 75, 149–155.

Jitendra, A, K., Griffin, C. C., Deatline-Buchman, A., & Sczesniak, E. (2007). Mathematical word problem solving in third-grade classrooms. Journal of Educational Research, 100, 283–302.

Jitendra, A.K., & Hoff, K. (1996). The effects of schema-based instruction on the mathematical word-problem solving performance of students with learning disabilities. Journal of Learning Disabilities, 29, 422–431.

Jitendra, A.K., & Star, J.R. (2011). Meeting the needs of students with learning disabilities in inclusive mathematics classrooms: the role of schema-based instruction on mathematical problem-solving. Theory into Practice, 50, 12–19.

Johnson-Laird, P.N. (1983). Mental Models: Towards a Cognitive Science of Language, Inference, and Consciousness. Cambridge, MA: Harvard University Press/Cambridge: Harvard University Press.

Jordan, N.C., & Hanish, L.B. (2000). Mathematical thinking in second-grade children with different forms of LD. Journal of Learning Disabilities, 33, 567–578.

Jordan, N.C., Kaplan, D., & Hanish, L.B. (2002). Achievement growth in children with learning difficulties in mathematics: findings of a two-year longitudinal study. Journal of Educational Psychology, 94, 586–597.

Jordan, N.C., & Montani, T.O. (1997). Cognitive arithmetic and problem solving: a comparison of children with specific and general mathematics difficulties. Journal of Learning Disabilities, 30, 624–634.

Judd, T.P., & Bilsky, L.H. (1989). Comprehension and memory in the solution of verbal arithmetic problems by mentally retarded and nonretarded individuals. Journal of Educational Psychology, 81, 541–546.

Kail, R.V. (2007). Longitudinal evidence that increases in processing speed and working memory enhance children's reasoning. Psychological Science, 18, 312–313.

Kail, R., & Hall, L.K. (1999). Sources of developmental change in children's word problem performance. Journal of Educational Psychology, 91, 660–668.

Kintsch, W., & Greeno, J.G. (1985). Understanding and solving word arithmetic problems. Psychological Review, 92, 109–129.

Landy, D., & Goldstone, R.L. (2007). How abstract is symbolic thought? Journal of Experimental Psychology: Learning, Memory and Cognition, 33(4), 720–733.

Lee, K., Ng, S.F., & Ng, E.L. (2009). The contributions of working memory and executive functioning to problem representation and solution generation in algebraic word problems. Journal of Educational Psychology, 101, 373–387.

Lee, K., Ng, S.F., Ng, E.L., & Lim, Z.Y. (2004). Working memory and literacy as predictors of performance on algebraic word problem. Journal of Experimental Child Psychology, 89, 140–158.

LeFevre, J., DeStefano, D., Coleman, B., & Shanahan, T. (2005). Mathematical cognition and working memory. In J. I. D. Campbell (Ed.), *Handbook of Mathematical Cognition* (pp. 361–377). New York: Psychology Press.

Lerkkanen, M.-K., Rasku-Puttonen, H., Aunola, K., & Nurmi, J-E. (2005). Mathematical performance predicts progress in reading comprehension among 7-year olds. European Journal of the Psychology of Education, 20, 121–137.

Lewis, A., & Mayer, R.E. (1987). Students' miscomprehension of relational statements in arithmetic word problems. Journal of Educational Psychology, 79, 363–371.

Lindvall, C.M., & Ibarra, C.G. (1980). Incorrect procedure used by primary grade pupils in solving open addition and subtraction sentences. Journal for Research in Mathematics Education, 11, 50–62.

Mani, K., & Johnson-Laird, P.N. (1982). The mental representation of spatial description. Memory & Cognition, 10, 181–187.

Nesher, P. (1981). Levels of description in the analysis of addition and subtraction. In T. P. Carpenter, J. M. Moser, & T. A. Romberg (Eds), Addition and Subtraction: Developmental Perspective (pp. 25–38). Hillsdale, NJ: Lawrence Erlbaum.

Nesher, P., Greeno, J.G., & Riley, M.S. (1982). The development of semantics categories for addition and subtraction. Educational Studies in Mathematics, 13, 373–394.

Oakhill, J. (1996). Mental models in children's text comprehension. In J. Oakhill & A. Garnham (Eds), Mental Models in Cognitive Science (pp. 77–94). Hove: Psychology Press.

Oakhill, J., Cain, K., & Yuill, N. (1998). Individual differences in children's comprehension skill: Toward an integrated model. In C. Hulme & R. M. Joshi (Eds), Reading and Spelling: Development and Disorders (pp. 343–367). Mahwah, NJ: Lawrence Erlbaum Associates, Publishers.

Passolunghi, M.C., Cornoldi, C., & De Liberto, S. (1999). Working memory and intrusions of irrelevant information in a group of specific poor problem solvers. Memory & Cognition, 27, 779–790.

Passolunghi, M.C., & Mammarella, I.C. (2010). Spatial working memory impairments in children with difficulties in arithmetic word problem-solving. European Journal of Cognitive Psychology, 22, 944–963.

Passolunghi, M.C., & Pazzaglia, F. (2004). Individual differences in memory updating in relation to arithmetic problem solving. Learning and Individual Differences, 14, 219–230.

Passolunghi, M.C., & Siegel, L.S. (2001). Short-term memory, working memory, and inhibitory control in children with difficulties in arithmetic problem solving. Journal of Experimental Child Psychology, 80, 44–57.

Pinperton, H., & Nation, K. (2010). Understanding words, understanding numbers: an exploration of the mathematical profiles of poor comprehenders. British Journal of Educational Psychology, 80, 255–268.

Rasmussen, C., & Bisanz, J. (2005). Representation and working memory in early arithmetic. Journal of Experimental Child Psychology, 91, 137–157.

Rawson, K.A., & Kintsch, W. (2002). How does background information improve memory for test content? Memory & Cognition, 30 (5), 768–778.

Reusser, K. (1989). Textual and Situational Factors in Solving Mathematical Word Problems. Bern: University of Bern.

Riley, M.S., Greeno, J.G., & Heller, J.I. (1983). Development of children's problem solving ability in arithmetic. In H. P. Ginsburg (Ed.), The Development of Mathematical Thinking. New York: Academic Press.

Siegler, R.S. (1996). Emerging Minds: The Process of Change in Children's Thinking. Oxford: Oxford University Press.

Sprinthall, R.C., & Nolan, T.E. (1991). Efficacy of representing quantities with different pictures in solving arithmetic word problems. Perceptual and Motor Skills, 72, 274–274.

Staub, F.C., & Reusser, K. (1995). The role of presentational structures in understanding and solving mathematical word problems. In C. A. Weaver, S. Mannes, & C. R. Fletcher (Eds), Discourse Comprehension. Essays in Honor of Walter Kintsch (pp. 285–305). Hillsdale, NJ: Lawrence Erlbaum.

Stellingwerf, B.P., & Van Lieshout, E.C.D.M. (1999). Manipulatives and number sentences in computer aided arithmetic word problem solving. Instructional Science, 27, 459–476.

Stern, E., & Lehrndorfer, A. (1992). The role of situational context in solving word problems. Cognitive Development, 2, 259–268.

Swanson, H.L. (2011). Working memory, attention, and mathematical problem solving: a longitudinal study of elementary school children. Journal of Educational Psychology, 103, 821–837.

Swanson, H.L., Cooney, J.B., & Brock, S. (1993). The influence of working memory and classification ability on children's word problem solution. Journal of Experimental Child Psychology, 55, 374–395.

Thevenot, C. (2010). Arithmetic word problem solving: evidence for the construction of a mental model. Acta Psychologica, 133, 90–95.

Thevenot, C., Barrouillet, P., & Fayol, M. (2004). Mental representation and procedures in arithmetic word problems: The effect of the position of the question. L'Année Psychologique, 104 (4), 683–699.

Thevenot, C., Devidal, M., Barrouillet, P., & Fayol, M. (2007). Why does placing the question before an arithmetic word problem improve performance? A situation model account. Quarterly Journal of Experimental Psychology, 60, 43–56.

Thevenot, C., & Oakhill, J. (2005). The strategic use of alternative representation in arithmetic word problem solving. Quarterly Journal of Experimental Psychology – A, 58, 1311–1323.
Verschaffel, L., De Corte, E., & De Win, L. (1985). Influence of rewording verbal problems on children's problem representations and solutions. Journal of Educational Psychology, 77, 460–470.
Vergnaud, G. (1982). A classification of cognitive tasks and operations of thought involved in addition and subtraction problems. In T. P. Carpenter, J. M. Moser, & T. A Romberg (Eds), Addition and Subtraction: A Cognitive Perspective (pp. 39–59). Hillsdale: Erlbaum.
Vergnaud, G. (1983). Multiplicative structures. In R. Lesh, & M. Landau (Eds), Acquisition of Mathematic Concepts and Processes. New York: Academic Press.
Vicente, S., Orrantia, J., & Verschaffel, L. (2007). Influence of situational and conceptual rewording on word problem solving. British Journal of Educational Psychology, 77, 829–848.
Vicente, S., Orrantia, J., & Verschaffel, L. (2008). Influence of mathematical and situational knowledge on arithmetic word problem solving: textual and graphical aids. Infancia y Aprendizare, 31, 463–483.
Vukovic, R.K., Lesaux, N.K., & Siegel, L.S. (2010). The mathematics skills of children with reading difficulties. Learning and Individual Differences, 20, 639–643.
Willis, G.B., & Fuson, K.C. (1988). Teaching children to use schematic drawings to solve addition and subtraction word problems. Journal of Educational Psychology, 80, 192–201.
Yuill, N., Oakhill, J., & Parkin, A. (1989). Working memory, comprehension ability and the resolution of text anomaly. British Journal of Psychology, 80, 351–61.
Zheng, X., Swanson, H.L., & Marcoulides, G.A. (2011). Working memory components as predictors of children's mathematical word problem solving. Journal of Experimental Child Psychology, 110, 481–498.

CHAPTER 10

INTUITION IN MATHEMATICAL AND PROBABILISTIC REASONING

KINGA MORSANYI AND DENES SZUCS

INTRODUCTION

MATHEMATICS, similarly to other scientific subjects, is based on a complex system of interrelated concepts, principles, and procedures. Although most people study mathematics at school for many years, it is only a small minority who feel comfortable with the subject. Indeed, mathematics is notorious for eliciting feelings of anxiety and 'threat' (see, e.g., Moore, Rudig, & Ashcraft (this volume) for a review). In order to explore the origins of people's difficulty with understanding different forms of numerical information, we will first give an overview of some typical errors that people make when they try to solve mathematical or probabilistic reasoning problems. We will argue that most people have fragmented knowledge of the rules, procedures, and concepts of mathematics, and, as a result, they are susceptible to use inappropriate strategies to approach problems. In particular, we will focus on the inappropriate use of intuitive strategies (i.e. responses or procedures that come quickly and easily to mind, and which are usually accompanied by feelings of ease and confidence—see, e.g., Fischbein, 1987; Thompson & Morsanyi, 2012). Intuitive strategies are usually very simple, and can be applied automatically, without much conscious reflection. These strategies might be based on personal experiences in real-life settings (i.e. primary intuitions—cf. Fischbein, 1987) or they might develop as a result of formal education (i.e. secondary intuitions). Although intuitive responses are often correct or, at least, satisfactory, here we will focus on the cases where intuition leads to systematic error and bias. Understanding the intuitive sources of typical errors is essential for educators working in the areas of mathematics and probability, as mistaken intuitions can make it hard for students to grasp certain concepts. After reviewing some of the typical intuitive strategies that people tend to adopt, we will explore the cognitive and contextual factors that affect the choice between different strategies to solve problems and, ultimately, the likelihood that a person will be able to find the correct solution. Finally, we will discuss the relevance of these findings for educators and researchers.

As described above, mathematics is considered to be a difficult subject, and people are often confused when they are confronted with numerical information in everyday settings. For example, Gigerenzer, Hertwig, Van den Broek, Fasolo, and Katsikopoulos (2005) asked pedestrians in a number of big cities in Europe and the USA what it means if the weather forecast says for tomorrow that 'there is a 30% chance of rain'. The standard meteorological interpretation would be that 'when the weather conditions are like today, in 3 out of 10 cases there will be rain the next day'. Nevertheless, most people judged this interpretation as the least appropriate. Instead they understood the statement as an indication that 'it will rain tomorrow 30% of the time' or 'it will rain in 30% of the area'.

Similar misconceptions have also been identified in classroom settings. Richland, Stigler, and Holyoak (2012) investigated US secondary-school students' typical approaches to mathematics. These researchers proposed that a potential reason for students' problems is that they do not perceive mathematics as a coherent body of knowledge, which can be used as a basis of reasoning. Instead, they see the subject as an ad-hoc collection of rules and procedures which have to be memorized. Consequently, as students progress through the curriculum, and the body of knowledge grows, mathematics appears to be an increasingly challenging test of memory.

Given the fragmented nature of mathematical knowledge, and in order to reduce memory load, students often rely on simplifying heuristics or intuitive solutions when they try to select from available strategies and procedures to approach problems that include numerical information. For example, they choose the operations to solve a mathematical word problem based on the title of the chapter that the problem appears in (Van Dooren, De Bock, Evers, & Verschaffel, 2009), or they might develop the expectation that a correct solution for a textbook problem always has to be an integer. Intuitive mistakes can also be based on overgeneralizing well-practised strategies to novel contexts. In the following sections we give an overview of some of the typical intuitive approaches that students adopt. Some of these strategies might be specific to the domain of mathematics or science, whereas others could be applied in various domains.

Intuitive Tendencies in Mathematics

Using Overlearned Strategies

The Natural Number Bias

A prominent intuitive trend in mathematics (just as in other domains) is to use overlearned strategies without giving much consideration to whether these are appropriate or not in the context of a specific task. One example of this is the natural number bias (or whole number bias—cf. Ni & Zhou, 2005) whereby students apply natural number rules inappropriately in the case of rational numbers (e.g. fractions). For example, although in the case of natural numbers (with the exception of multiplying with/dividing by 1) 'multiplication makes bigger' and 'division makes smaller' (Greer, 1994), this is not necessarily true for rational numbers (e.g. $6 \times 0.5 = 3$). There is also evidence that students associate addition with 'more' and subtraction with 'less' (De Corte, Verschaffel, & Pauwels, 1990), although this does not hold in the case of negative numbers. In a recent study Vamvakoussi, Van Dooren, and Verschaffel (2012) investigated participants' response patterns and reaction times when judging the correctness (i.e. True/False) of mathematical statements. These authors aimed to demonstrate that the biases based on natural number rules are still present in the case of adults, and that overcoming these intuitive tendencies takes substantial cognitive effort (i.e. it is a time-consuming and error-prone process). Their findings were mostly in line with this prediction. That is, statements that were congruent (as opposed to incongruent) with natural number rules (e.g. $2/3 < 4/3$ vs. $3/4 < 3/2$) generally took less time to judge, and participants' judgements were also more likely to be correct.

Overusing Proportionality

Another typical example of using overlearned strategies in mathematics is the overuse of proportionality (e.g. Fischbein & Gazit, 1984; Fischbein & Schnarch, 1997; Kahneman & Tversky, 1972, Van Dooren, De Bock, Hessels, Janssens, & Verschaffel, 2005). For example, consider the following problem (based on Van Dooren et al., 2005):

> *Ellen and Kim are running around a track. They run equally fast but Ellen started later. When Ellen has run 5 laps, Kim has run 15 laps. When Ellen has run 30 laps, how many has Kim run?*

Van Dooren et al. (2005) administered this and other similar problems to children from grades 4 to 6. They found that the proportion of children who gave the correct (additive) response of 40 decreased with grade levels, whereas the likelihood of giving the incorrect, proportional response of 90 increased. Indeed, in an earlier study Cramer, Post, and Currier (1993) found that 32 out of 33 preservice elementary teachers made the proportional error when faced with a similar problem, which suggests that the tendency to use proportionality inappropriately increases further with education. Arguably, the arithmetical operations involved in applying the proportional strategy (i.e. $30 \times 3 = 90$) are not any less complex than the computations required for the use of the additive strategy (i.e. $30 + 10 = 40$). Thus, when applying the proportional strategy, students do not simply want to save the effort of engaging in complex computations. Rather, their preferred strategy is cued by the superficial features of the problem. Indeed, Van Dooren et al. (2005) demonstrated that changing the numbers in the task made a significant difference in children's tendency to commit the proportional error. In particular, if the ratios of the numbers in the problem were non-integers (for example, if the problem stated that Ellen has run 5 laps and Kim has run 8 laps), children were much less likely to use the proportional strategy.

The Effect of Fluency

The Role of Numbers Included in A Problem

Now consider the following problems (based on Frederick, 2005) as further examples for how the particular numbers included in problems can cue incorrect responses:

(1) *A bat and a ball cost \$1.10 in total. The bat costs \$1.00 more than the ball. How much does the ball cost? _____ cents*

(2) *If it takes 5 machines 5 minutes to make 5 widgets, how long would it take 100 machines to make 100 widgets?* _____ *minutes*
(3) *In a lake, there is a patch of lily pads. Every day, the patch doubles in size. If it takes 48 days for the patch to cover the entire lake, how long would it take for the patch to cover half of the lake?* _____ *days*

Frederick (2005) administered the above problems (which he called the Cognitive Reflection Test—CRT) to students from elite US universities (such as Harvard, MIT, and Princeton). Although these problems are very simple, Frederick found that most of the students in his samples got at least one or two of the solutions wrong. What is more interesting is that the participants who gave an incorrect response typically made the same mistake (i.e. they answered 10 to the first, 100 to the second, and 24 to the third problem). In a recent paper, Thompson and Morsanyi (2012) proposed that the reason for these errors is that these incorrect responses come very easily (i.e. fluently) to mind, which then leads to a feeling of confidence that the answer is correct (i.e. a *feeling of rightness*—see, e.g., Thompson, Prowse-Turner, & Pennycook, 2011). For example, in the case of the first problem, the text implies that participants have to use addition to find a solution, and the sum of $1.10 naturally breaks down to $1.00 (which also appears in the text) + 10 cents. Indeed, there is direct evidence that fluency affects the judgement of correctness of mathematical solutions, as we will demonstrate below.

We should note that the experience of fluency (i.e. ease of processing) can arise as a result of various factors, such as familiarity, symmetry, the perceptual or semantic clarity of presentation, or the speed with which an item can be recalled or recognized (see, e.g., Alter & Oppenheimer, 2009; Thompson & Morsanyi, 2012; Topolinski & Strack, 2009). Nevertheless, all of these properties lead to similar experiences of correctness, liking, truth, or, even, beauty (e.g. Topolinski & Reber, 2010; Whittlesea, 1993; Winkielman, Halberstadt, Fazendeiro, & Catty, 2006). Thompson and colleagues (e.g. Thompson & Morsanyi, 2012; Thompson et al., 2011) argued that once a solution is reached that feels right, this serves as a signal for reasoners that no more investment of cognitive effort is necessary. This is a potential explanation for why errors stemming from intuitive solutions often remain uncorrected.

The Effect of Symmetry

As a demonstration of the effect of fluency on judged correctness of mathematical solutions, Reber, Brun, and Mitterndorfer (2008) administered symmetric and asymmetric arithmetic verification tasks (more specifically, additions of dot patterns) to their participants. The symmetry of the patterns was manipulated in the following way (see Figure 10.1). Apart from the visual properties of the patterns, the additions were equivalent across the symmetric and asymmetric problems. In Experiment 1, the additions were shown for 600 ms, after which participants had to indicate within a 600-ms time window if the addition was correct or incorrect. Although participants performed at chance level, there was a significant effect of symmetry on their judgements of correctness (i.e. symmetric patterns were more likely to be judged as correct). In Experiment 2, the same additions were presented for 1800 ms, after which participants had to generate a response within 600 ms. The purpose of increasing presentation time was to give participants a chance to engage in exact calculations (and, thus, to improve performance). Whereas symmetric patterns were expected to facilitate calculations (and to increase

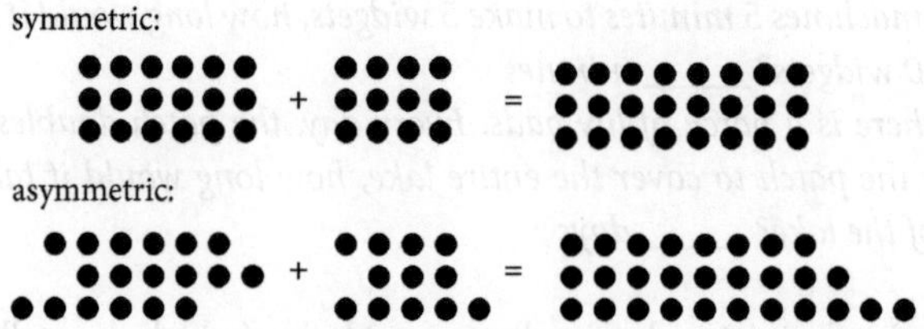

FIGURE 10.1 Examples for symmetric and asymmetric dot patterns used by Reber et al. (2008).

Reproduced from *Psychonomic Bulletin & Review*, 15(6), 2008, pp. 1174–1178, The use of heuristics in intuitive mathematical judgment, Rolf Reber, Morten Brun, and Karoline Mitterndorfer, © 2008, Springer Science and Business Media. With kind permission from Springer Science and Business Media.

endorsement rates for correct and reduce endorsement rates for incorrect solutions), the authors also predicted, and found, that symmetric patterns would be more likely to be judged as correct. That is, even when participants had the opportunity to engage in exact calculations, and they were able to judge the correctness of the solutions above chance level, the ease of processing of the patterns still affected their judgements of correctness.

The Effect of Temporal Contiguity

In another study Topolinski and Reber (2010, Experiment 2) used 40 correct and 40 incorrect algebraic equations, with two addends being higher than 200, and their sum being lower than 1000 (e.g. 256 + 573 = 829). The incorrect equations contained the same addends as the correct ones, but 55 or 100 was added to or subtracted from the solutions. The solution of the problem (e.g. 829) was presented for 500 ms, followed by the addends (e.g. 256 + 573). In high-fluency trials the addends appeared immediately after the presentation of the solution, whereas in low-fluency trials there was a 50-ms delay before the presentation of the addends. Besides accepting a higher number of correct than incorrect additions, participants were more likely to judge the addition to be correct if the addition appeared immediately after the solution, although they were unaware that two different presentation times were used. This suggests that temporal immediacy created the illusion of self-evidence or straightforwardness (see also Hertwig, Herzog, Schooler, & Reimer, 2008).

Arguably, these findings are very similar to the links between fluency and judgements of truth that have been demonstrated in other domains, such as that rhyming versions of aphorisms seem to be truer than non-rhyming aphorisms with the same meaning (McGlone & Tofighbakhsh, 2000), or that answers to trivia questions that come quickly and easily to mind are more likely to be judged as correct by the individuals who retrieve them, regardless of whether they are actually correct (e.g. Koriat, 2007). Although the link between fluency and judgements of correctness is not specific to mathematics, the impact of perceived fluency in the mathematical domain should not be underestimated. Indeed, this cannot only be demonstrated in psychology labs with participants who have limited knowledge of mathematics, or through using speeded procedures. There is anecdotal evidence that prominent mathematicians and scientists used symmetry and beauty as indicators of the correctness of mathematical solutions (e.g. Chandrasekhar, 1987). Furthermore, mathematical theories that are difficult to understand are usually received more negatively than simpler theories (McColm, 2007).

Intuitive Rules in Science and Mathematics

The Intuitive Rules Approach

At the beginning of the previous section we presented evidence that the structural characteristics of problems can invoke certain highly familiar and well-practised procedures and schemas to solve problems, regardless of whether these are appropriate or not. Then we described the ubiquitous effect of fluency on judgements of correctness. In the following, we introduce another tendency, which is to base judgements on 'intuitive rules' which students assume to be universally applicable in different areas of science. These rules include 'more A—more B', 'same A—same B', and 'everything can be divided endlessly' (see, e.g., Babai, Brecher, Stavy, & Tirosh, 2006). We will describe these rules together with a framework that is aimed at explaining the cognitive bases of how these rules are activated and applied.

The intuitive rules approach (Stavy & Tirosh, 2000) deals with the development of certain heuristics (i.e. intuitive rules) which are defined as self-evident and self-consistent cognitions (based on Fischbein, 1987). These rules are retrieved without conscious intention and they are activated by certain characteristics of tasks. Which aspects of a task will trigger a particular response depends on saliency, which can be determined by bottom-up or top-down processes (see below).

Implicit, Explicit, and Automatic Reasoning

An interesting aspect of the intuitive rules approach is the distinction between implicit, explicit, and automatic forms of reasoning (see Osman, 2004). Although both implicit and automatic processes can be considered as intuitive, and they both operate outside of conscious control, according to the intuitive rules approach, they correspond to the two extremes of a continuum. In this view, implicit reasoning is based on weak representations that are not stable, and, as a result, they influence participants' behaviour without arousing the feelings of intention and conscious awareness. However, these representations are still capable of influencing explicit processes (an example of this could be the effect of fluency on judgements of truth or correctness). Explicit representations are stable, strong, and distinctive, and, as a result, they can be consciously controlled and manipulated in working memory, and are available for declarative knowledge (for example, these are the processes that are activated when students solve equations that contain large numbers). Finally, automatic representations are also strong and stable and, thus, accessible for consciousness. However, they became so strong through repeated activation that they are hard to control or modify (see Cleeremans & Jimenez, 2002; Osman & Stavy, 2006). For example, the solutions of highly familiar multiplications (e.g. $4 \times 4 = 16$) are retrieved from long-term memory, rather than being computed. Automatic representations are considered to be the final products of cognitive development, and they form the basis of skill-based reasoning and mental flexibility.

It is important to note that the status of representations can change over time. Specifically, representations gain strength with repeated exposure, and the procedures based on them can become conscious and, eventually, automatic. As the strength of representations increases, reasoners become more aware of them, and they are also more able to control them (up to the point of automaticity, when representations gain a certain level of autonomy and become hard to control).

Bottom-Up and Top-Down Saliency

Salient stimuli are arousing, they capture attention, and behavioural resources are preferentially directed toward them (cf. Osman & Stavy, 2006). *Bottom-up saliency* is usually based on the similarity between items in a task. The more an item differs from other elements of a task, and the more easily a response based on a stimulus can be generated, the more salient it is (consider the 'widget problem' from the CRT as an example, or the effect of symmetry on judgements of correctness in the case of additions of dots). By contrast, *top-down saliency* is based on the number of times a particular stimulus has been experienced in a learning situation. If a stimulus is strongly associated with a response, it will generate a response in novel situations, regardless of its relationship with other information provided in the task. An example of this could be when students see a mathematical word problem where they have to find a missing number based on information about three other numbers, and they apply the proportionality rule (e.g. Van Dooren et al., 2005).

Osman and Stavy (2006) give the example of the *ratio bias problem* (e.g. Brecher, 2005) for a task that elicits bottom-up saliency effects. According to Osman and Stavy (2006), in this task the intuitive rule 'more A—more B' is invoked (for other conceptualizations of this problem see, e.g., Denes-Raj & Epstein, 1994; Reyna & Brainerd, 2008). In Brecher's experiment participants were presented with a probability task in which they had to decide which one out of two boxes would give them a better chance of drawing a black counter from a mix of black and white counters (see Figure 10.2). In this task the salient feature is the number of black counters. In congruent tasks the box that contains more black counters also contains black counters in a higher proportion than white counters. By contrast, in the case of incongruent tasks the box that contains more black counters contains black counters in a lower proportion. Task difficulty was also manipulated based on the difference between the number of black and white counters in the two boxes.

Given that the salient feature is the number of black counters, this task activates the 'more A—more B' intuitive rule (cf. Osman & Stavy, 2006). Babai et al. (2006) found that participants responded to congruent tasks more quickly and more accurately than to incongruent tasks. However, response speed and accuracy was also affected by the perceptual discriminability between the number of black and white counters (that is, the ratios of black and white counters within each box). This indicates that the status of bottom-up salient task stimuli is dependent on their relationship with other presented stimuli. In the case of bottom-up

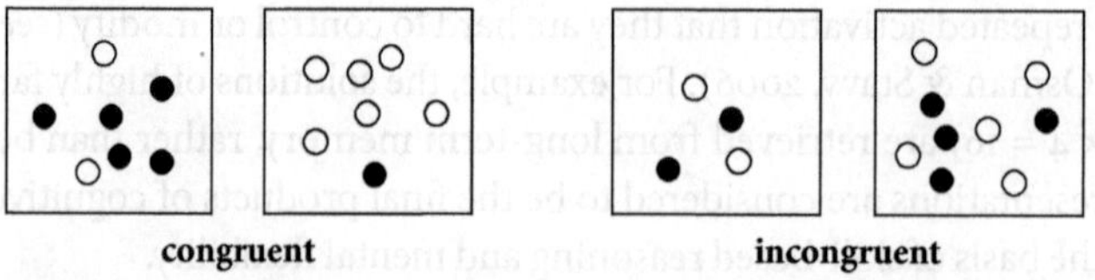

FIGURE 10.2 Examples for Brecher's (2005) ratio bias task.

Reproduced from T. Brecher, *Application of the intuitive rule: "More A-more B" in a probability task.* A reaction time study, Master's thesis, Tel Aviv University, Israel, 2005.

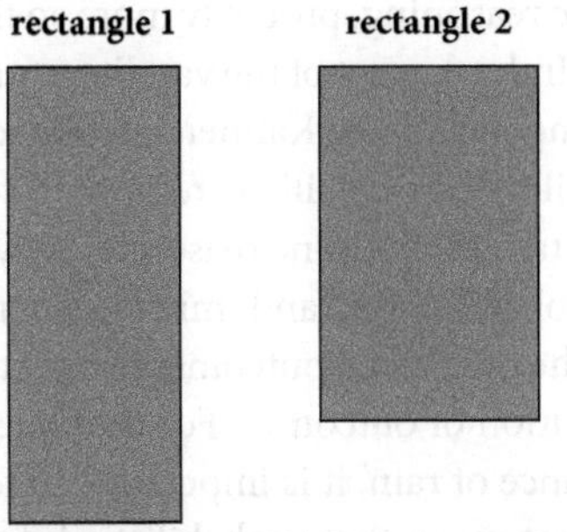

FIGURE 10.3 Illustration of Mendel's (1998) rectangle task.

Reproduced from N. Mendel, *The intuitive rule "same of A, same of B": The case of comparison of rectangles*, Master's thesis, Tel Aviv University, Israel, 1998.

saliency, participants are unaware of how the relevant intuitive rule got activated, and, thus, they are unable to control it. This is not to say that they are unaware of the perceptual stimuli that invoked the rule, as these are salient and, as such, are in the focus of attention.

By contrast, top-down salient stimuli activate relevant rules automatically, which means that individuals are consciously aware of them, and they also possess meta-knowledge about them. Osman and Stavy (2006) describe the 'same A—same B' rule as an example of an intuitive rule based on top-down saliency (the application of this rule is related to the acquisition and stabilization of the proportionality schema). For example, Mendel (1998) presented students with a problem in which two rectangles were shown. Students were told that the second rectangle is a modified version of the first one, where the length of the rectangle was decreased by 20% and the width was increased by 20% (see Figure 10.3). Students were asked about the perimeters of the two rectangles (i.e. whether they were equal, or whether the perimeter was longer in the case of one of the rectangles). In this experiment participants had knowledge about the relevant rule (i.e. how to compute the perimeter of a rectangle), and the available perceptual information was also in line with the rule (i.e. that the perimeter of rectangle 1 was longer). However, instead of generating a solution based on either of these, over 70% of the students applied the 'same A—same B' rule. That is, they claimed that the perimeters of the two rectangles were equal, because 'adding 20% and removing 20% equals to no change.' The fact that the justification in itself makes sense explains why it is so hard to eliminate biases based on top-down saliency. By contrast, young children tend to respond to similar questions correctly, because they do not possess relevant knowledge, and, thus, they rely on perceptual information (and apply the 'more A—more B' rule in the present case). This example shows how saliency is dependent on individuals' knowledge. That is, different aspects of the same input might appear salient for different individuals, and, thus, the same stimulus does not always evoke the same type of response or the same rule. Instead, it depends on reasoners' experiences and knowledge of what sort of response they will generate based on given stimuli.

Problems With Understanding Probability and Randomness

If understanding mathematics is difficult, these difficulties are magnified in the case of probabilistic reasoning. Thus, it is not surprising that intuition plays a very important role in the domain of probabilistic reasoning, probably more so than in other domains of mathematics (cf. Fischbein, 1975). Indeed, most of the vast literature on heuristics and biases (see, e.g., Gilovich, Griffin, & Kahneman, 2002; Kahneman, Slovic & Tversky, 1982) is concerned with reasoning about probabilities. Probabilistic reasoning consists of drawing conclusions about the likelihood of uncertain events. One reason for why this might be particularly difficult is that the concept of probability (or randomness) incorporates two seemingly contradictory ideas: that although the individual outcomes of events are unpredictable, in the long run there is a regular distribution of outcomes. For example, based on the knowledge that tomorrow 'there is a 30% chance of rain', it is impossible to tell whether it will actually rain or not. This, however, does not mean that probabilistic information is completely useless. Indeed, based on this statement, we can be reasonably confident that it will not be raining tomorrow. A failure to integrate these two aspects of probability leads to either interpreting probabilistic events in an overly deterministic way or disregarding the pattern and focusing entirely on the uncertainty aspect (Metz, 1998).

Intuitively, it might seem surprising that people are not very skilled in reasoning about probabilistic events. Indeed, most outcomes that we encounter during our daily lives are probabilistic, and even very young children and babies seem to have some basic grasp of certain laws of probability (e.g. Girotto & Gonzalez, 2008; Xu & Denison, 2009). Moreover, we have to make decisions on the basis of uncertain outcomes all the time. Nevertheless, we do not necessarily perceive random events as random, and, indeed, short sequences of random events do not necessarily exhibit a 'random pattern' (see, e.g., Hahn & Warren, 2009). For example, whether we find a parking space in front of the house of a friend who lives in the city centre depends purely on chance. However, most people will attribute this to being lucky or unlucky. By contrast, we might see patterns in otherwise random events, as illustrated by the survivorship bias. For example, stock market millionaires are usually seen as very skilled rather than very lucky, although their success might be based purely on chance (cf. Taleb, 2001). Indeed, if we consider the large number of people who follow similar investment strategies as the winners, but nevertheless lose money, the picture is not so different from the national lottery. Although the number of stock market (and lottery) millionaires is relatively small, what we encounter much more often in our daily lives is examples of short sequences of random events which show a regular pattern (as we might see in the case of coin tosses or, for example, when out of two sports teams of similar standing in the championship, one wins over the other in a number of consecutive games). As the above examples illustrate, there are various ways that we can misunderstand probability, and a large number of heuristics and biases have been identified which reflect these problems. Nevertheless, in the following sections we will just focus on two typical examples: the representativeness heuristic (Kahneman & Tversky, 1972) and the equiprobability bias (Lecoutre, 1992). Not only do these heuristics and biases demonstrate two typical erroneous tendencies in probability (i.e. overrating/disregarding information), but also we will use these examples to illustrate the notion of primary and secondary intuitions (Fischbein, 1987).

The Representativeness Heuristic

The representativeness heuristic (an example of a primary intuition) is a tendency for people to base their judgement of the probability of a particular event on how much it represents the essential features of the parent population or of its generating process. The representativeness heuristic often manifests itself in the belief that small samples will contain the same proportion of outcomes as the parent population, and that they will also 'look' similar to the sample that they were drawn from. A typical example is that, when tossing a fair coin, after a series of heads people have the feeling that a tails should follow, because this corresponds more to their expectation of having a mix of heads and tails, which they would expect from a random sequence. This is called the negative recency effect (or the gambler's fallacy). Or consider this problem (where people often prefer response *a*, as that sequence looks 'more random'):

> *A fair coin is flipped six times. Which of the following sequence of outcomes is the most likely result of six flips of the fair coin? (H: Heads, T: Tails)*
>
> a. *THHTHT*
> b. *HTHTHT*
> c. *Both sequences are equally likely.*

The Equiprobability Bias

The equiprobability bias (a secondary intuition) was described by Lecoutre (1992) as a tendency for individuals to think of probabilistic events as 'equiprobable' by nature, and, as a result, to judge outcomes that occur with different probabilities as equally likely (see also Callaert, 2004). This bias appears to be related to formal education in probability, and especially to the misunderstanding of the concept of randomness. Indeed, a number of researchers found that this bias increased with formal education (see, e.g., Batanero, Serrano, & Garfield, 1996; Chiesi & Primi, 2009; Lecoutre, 1985; Morsanyi, Primi, Chiesi, & Handley, 2009). For example, when they have to predict the sum of two dice, students often think that no total is harder or easier to obtain than any other. This notion might be based on the fact that the dice are individually fair and they cannot be controlled (Pratt, 2000). More surprisingly, many students also exhibit the equiprobability bias in the case of the following problem (based on Green, 1982), although in this case information about base rates is explicitly presented:

> *A mathematics class has 13 boys and 16 girls in it. The teacher does a raffle. Each pupil's name is written on a slip of paper. All the slips are put in a hat. The teacher picks one slip without looking. Which of the following results is most likely?*
>
> a. *The winner is more likely to be a boy than a girl.*
> b. *The winner is more likely to be a girl than a boy.*
> c. *The winner is just as likely to be a girl as a boy.*

It is interesting to note that the incorrect equiprobability response (i.e. to select option *c*) is based on reasoning (although this reasoning is misguided). In fact, when thinking about this problem, it seems very simple to integrate the knowledge that either a girl or a boy will win with the information about base rates. Moreover, students without much knowledge of random processes will rarely give an incorrect response. Nevertheless, as Gigerenzer, Todd, & The ABC Research Group (1999) noted, base rates are irrelevant when making predictions about a single, individual outcome (although it is not incorrect to make predictions about the *likelihood* of a single outcome). This awareness of uncertainty might contribute to the increases in this type of error with education in probability.

In a recent review of the literature on probabilistic reasoning Bryant and Nunes (2012) argued that understanding that probabilistic events are uncertain is an essential first step in solving probability problems correctly. Nevertheless, having a grasp of the notion of uncertainty in itself is not sufficient to solve probability problems. Indeed, students also need to be able to work out the sample space (i.e. what all the possible events are) for each problem, and then quantify and compare the probabilities of each possible event. Thus, the equiprobability error demonstrates the grasp of an important aspect of probability (i.e. uncertainty), and participants making this error also work out the sample space correctly (i.e. that, in the above example, either a boy or a girl will win). Nevertheless, they are unable to correctly assign probabilities to the different possible events, and default to assigning equal probabilities.

In summary, whereas errors based on the representativeness heuristic arise as a result of personal experiences with short sequences of probabilistic outcomes (and are not based on explicit rules), the equiprobability bias is more likely to affect people with explicit instruction in probabilistic reasoning. Indeed, in one of our studies (Morsanyi et al., 2009, Experiment 1) we found that whereas the tendency to rely on representativeness decreased with grade level amongst students who had education in statistics and critical thinking (in this particular case, psychology students), the equiprobability bias increased in the same population.

Besides Morsanyi et al. (2009), various authors found evidence for simultaneous increases and decreases with age and/or education in different biases within the same population (e.g. Chiesi, Gronchi, & Primi, 2008; Fischbein & Schnarch, 1997), or for an increase in normative performance which was accompanied by a simultaneous increase in a certain bias (e.g. Jacobs & Potenza, 1991; Van Dooren et al., 2005). Fischbein and Schnarch (1997) proposed that the reason for these paradox findings is that, in general, the impact of intuitions increases with age and with education. Nevertheless, students' relevant normative knowledge also increases. When a problem is easy to conceptualize, and the relevant normative rule is readily available, misconceptions will diminish with age. However, if the task is hard to conceptualize, the effects of misconceptions or irrelevant strategies activated by superficial cues will increase (cf. Fischbein & Schnarch, 1997).

Primary and Secondary Intuitions

Earlier we described the representativeness heuristic as a primary intuition and the equiprobability bias as a secondary intuition (cf. Fischbein, 1987). Here we discuss these concepts in more detail. *Primary intuitions* are based on individual experiences and vivid personal memories (such as when a child thinks that throwing six with a dice is harder than throwing two, based on vivid memories of playing a game and having to wait for a long time for a six—cf. Callaert, 2004). By contrast, *secondary intuitions* (such as the equiprobability bias, the 'same A—same B' rule, or the proportionality schema) are formed by scientific education at school. Both primary and secondary intuitions can lead to correct as well as incorrect responses. Thus, relying on intuition in itself does not necessarily lead to non-normative solutions. In fact, some authors noted that students often get worse in some areas of mathematical or probabilistic reasoning with education, as a result of a general tendency to rely on rigid rules and to give up common-sense explanations (e.g. Fischbein, 1975; Sedlmeier, 2007).

A similar distinction between two different types of intuition/heuristics has been made by Fong, Krantz, and Nisbett (1986) who used the term *statistical heuristics* (see also Nisbett, Krantz, Jepson, & Kunda, 1983) to refer to rudimentary but abstract versions of statistical principles, such as the law of large numbers. They contrasted these with *non-statistical heuristics*, such as representativeness and availability (Kahneman & Tversky, 1972, 1973; Tversky & Kahneman, 1974) which have no underlying abstract representations. Finally, Betsch (2008) also noted that *intuitions* (which are quick responses associated with a feeling of confidence) should be distinguished from *heuristics* (which are based on a conscious implementation of simple rules). These concepts are not unlike the distinction between *implicit* and *automatic* processes made by Osman (2004).

As we described earlier, intuitions or heuristics are self-evident cognitions, which get activated by certain task characteristics, and this might happen outside of conscious awareness (in the case of implicit processes/primary intuitions), or these processes might be hard to control, even if the reasoner is aware of them (in the case of automatic processes/secondary intuitions). It should also be kept in mind that primary and secondary intuitions have different origins, and they might be present in the case of different individuals, or, if they are present in the same individual, they might be activated differentially by certain task components or experimental manipulations. Indeed, whereas primary intuitions are usually based on the perceptual characteristics of problems (i.e. bottom-up processes), secondary intuitions are based on the explicit rules activated by a particular task format (i.e. top-down processes).

Earlier we introduced the representativeness heuristic and the equiprobability bias as examples of primary and secondary intuitions, and as demonstrations of two potential ways of how people might misunderstand probability (i.e. as examples for under-/overestimating information). Nevertheless, as the two tendencies have different origins, they are expected to be independent (rather than being negatively correlated), and interventions that target one of them will not necessarily affect the other. Having said this, primary and secondary intuitions share some of their core characteristics, such as that they emerge quickly and effortlessly. Given the effortless nature of intuitive reasoning, these processes can be executed in parallel (e.g. Evans, 2006). That is, performance on a task can be simultaneously affected by a primary and secondary intuition. For example, in proportionality problems, the ease/fluency with which a proportional response can be generated has a great effect on error rates in proportional problems (e.g. Van Dooren et al., 2005).

In two recent papers (Morsanyi et al., 2009; Morsanyi, Handley, & Serpell, 2012) we argued that distinguishing between primary and secondary intuitions could contribute to a better understanding of the cognitive underpinnings of intuitive reasoning, and could help with predicting performance on various tasks. We will discuss this issue in the next section, together with the question of how the tendency to rely on intuition and heuristics can be predicted on the basis of some individual differences variables, as well as contextual factors.

Predicting Performance in Mathematical and Probabilistic Reasoning

The Role of Individual Differences Variables

There are many factors that have been found to influence mathematical and probabilistic reasoning performance. Providing a complete list, and giving an overview of how each of these factors affect reasoning performance, and all the possible ways these factors might interact is beyond the scope of this chapter. Here we will just focus on some of the best researched phenomena regarding how these factors affect the use of intuitive strategies in mathematics and probability. We think that this overview might help to address some of the 'mysteries' of research into intuitive reasoning in mathematics and probability, and, in particular, the question of why certain intuitive tendencies can be tackled more easily than others, and why performance on certain tasks is especially resistant to educational interventions.

The Effects of Cognitive Capacity and Cognitive Effort

It is a general finding in reasoning research that using greater cognitive resources to solve a problem leads to better performance, as well as a reduced tendency to solve problems by using heuristic shortcuts (e.g. Kokis et al., 2002; Stanovich & West, 2000). This has been demonstrated through correlations between reasoning performance and cognitive ability, as well as reasoning performance and need for cognition (i.e. a measure of the tendency to spontaneously engage in and to enjoy effortful cognitive activities—Cacioppo & Petty, 1982), through the application of working memory load (e.g. Gillard, Van Dooren, Schaeken, & Verschaffel, 2009, Experiment 2; De Neys, 2006) or time constraints (which makes it hard or impossible for participants to engage in complex reasoning processes—e.g. Gillard et al., 2009, Experiment 1), and through instructing participants to reason effortfully and/or logically (e.g. Ferreira, Garcia-Marques, Sherman, & Sherman, 2006). With regard to primary and secondary intuitions, whereas these factors and manipulations are correlated with or have an effect on secondary intuitions, they are mostly unrelated to primary intuitions. More specifically, primary intuitions might possibly be reduced or eliminated when participants are able to engage in effortful thinking, but only if they have knowledge of the correct strategies to solve problems that are typically affected by these primary intuitions (cf. Morsanyi et al., 2013). Without such instruction, even participants with high cognitive ability will be susceptible to using primary intuitions. By contrast, participants with high cognitive ability are generally less likely to rely on secondary intuitions when these lead to incorrect solutions, because they have a better understanding of the situations where particular procedures (such as the proportionality schema) might be applied. As participants with higher cognitive ability have meta-knowledge of secondary intuitions, they will be less likely to use them inappropriately.

The Effects of Education and Training

We will describe a curious finding from one of our recent studies (Morsanyi et al., 2013) to demonstrate how knowledge and cognitive ability interact in shaping reasoning performance (see also Stanovich & West, 2008, for a theoretical analysis). In order to reduce the equiprobability bias, we provided psychology students with a 40-minute training session, where they had the opportunity to experiment with random generators (i.e. they were invited to toss coins and to throw dice). We used these exercises as demonstrations of certain characteristics of random sequences, and also as illustrations of the law of large numbers. The probabilistic reasoning performance of participants was measured immediately after the training session, and it was compared to the performance of another group who did not take part in the training.

Whereas the training successfully reduced the equiprobability bias, at the same time it *increased* students' susceptibility to the representativeness heuristic. Nevertheless, these two effects were unrelated, which was evidenced by the fact that susceptibility to the two biases was uncorrelated. Furthermore, whereas the training was especially useful in reducing the equiprobability bias in the case of students with high cognitive ability, susceptibility to the representativeness heuristic was unrelated to cognitive ability. By contrast, the effect of training on the equiprobability bias was unrelated to students' achievement in critical thinking (which they studied as part of their psychology course). Nevertheless, education in critical thinking was an important factor in preventing the inappropriate application of the representativeness heuristic (but only in the case of participants in the training group). These findings demonstrate very clearly how educational interventions can lead to counterintuitive results, such as simultaneous increases and decreases in certain intuitive tendencies, even as a result of a brief training intervention. Furthermore, these effects were moderated by factors, such as participants' cognitive ability and their knowledge about some typical biases in probability. However, if we keep in mind that different aspects of the same input might simultaneously weaken and strengthen certain intuitive tendencies, the above findings seem less puzzling. Indeed, as we described earlier, it is very often the case that a reduction in certain intuitive tendencies is accompanied by increases in other intuitive tendencies (e.g. Chiesi et al., 2008; Fischbein & Schnarch, 1997; Morsanyi et al., 2009).

The Effects of Thinking Styles and Thinking Dispositions

Besides cognitive ability or effort and relevant knowledge or education, another important group of factors that contribute to mathematical and probabilistic reasoning performance is thinking styles and thinking dispositions. We have already mentioned the role of need for cognition in increasing the tendency for individuals to engage in effortful cognitive activity. As a result, individual differences in need for cognition can eliminate some of the differences in reasoning performance between participants with higher and lower cognitive ability. It has also been shown that individuals who preferentially rely on spatial strategies when they solve mathematical problems, outperform other participants who prefer verbal or visual strategies (e.g. Blazhenkova & Kozhevnikov, 2009). Other factors, such as positive attitudes towards mathematics and statistics, and high motivation and self-efficacy (e.g. Galli,

Chiesi, & Primi, 2011; Morsanyi, Primi, Handley, Chiesi, & Galli, 2012), can lead to similar effects, whereas anxiety and low self-efficacy, or testing conditions which induce stereotype threat (i.e. a situation in which a member of a group fears that their performance will validate an existing negative performance stereotype) decrease performance (see, e.g., Moore et al., this volume, for a review). These effects were shown to operate through moderating the availability or the use of cognitive resources (e.g. Ashcraft & Kirk, 2001). Additionally, these factors can have a direct effect on learning new skills in the areas of mathematics and probability, and, thus, on the quality of individuals' knowledge in these areas. Although, to the best of our knowledge, the effect of these factors on intuitive reasoning have never been studied, we can expect that intuitive response tendencies will increase in the case of low motivation, negative attitudes, and high anxiety or threat, whereas high motivation and self-efficacy, together with positive attitudes and low anxiety, should increase normative performance, and decrease the tendency to rely on shortcuts and to use inappropriate strategies. Nevertheless, these claims should be tested in future studies.

Finally, certain thinking dispositions (which were termed 'contaminated mindware' by Stanovich & West, 2008) can replace relevant knowledge, and can be used as a basis of generating erroneous responses, which sometimes correspond to the typical intuitive tendencies in the domains of mathematics and probability. These thinking dispositions are made up of collections of mistaken beliefs about probability, chance, luck, or gambling. For example, consider these two representative items from the Gambling Related Cognitions Scale (Raylu & Oei, 2004): 'Specific numbers and colours can help increase my chances of winning' and 'I have some control over predicting my gambling wins'. Whereas the first statement expresses a belief that it is possible to control random outcomes by creating a 'lucky environment', the second statement indicates that the person thinks it is possible to predict the outcome of single random events. Thus, in the first case the person disregards the laws of probability and attributes the outcomes to situational factors, whereas in the second case they try to employ a rule to make predictions (for example, they might commit the gambler's fallacy). Although these tendencies seem contradictory, they both express a belief that random outcomes can be predicted or even controlled, and, indeed, responses to these items are significantly correlated (e.g. Raylu & Oei, 2004). Such mistaken beliefs about random events have been found to be positively related to real-life gambling behaviour, and negatively related to probabilistic reasoning ability (e.g. Michalczuk, Bowden-Jones, Verdejo-Garcia, & Clark, 2011; Toplak, Liu, Macpherson, Toneatto, & Stanovich, 2007). There is also evidence that these thinking dispositions are negatively related to cognitive ability and need for cognition (e.g. Chiesi, Primi & Morsanyi, 2011). This suggests that low cognitive ability or 'lazy' reasoners might employ heuristics or inappropriate (intuitive) rules when they reason about probabilistic events. Furthermore, this tendency might also prevent the acquisition of relevant normative rules, and the retrieval and application of probability knowledge when reasoning about random events.

Implications For Research and Education

Although predicting actual reasoning performance in general might be complicated, our knowledge about intuitive tendencies in mathematics and probability has some important implications for researchers and educators. First of all, it is possible to create ideal circumstances for reasoning by reducing anxiety and threat, giving enough time for individuals to work on problems, and by employing instructions to reason carefully and to check the correctness of responses which seem obviously right. Nevertheless, all of these factors might be ineffective if a person does not have knowledge of the relevant normative rule, or if they show a tendency to rely on mistaken beliefs (see also Stanovich & West, 2008). In terms of teaching relevant rules and procedures, educators should present the examples where certain rules apply together with problems with similar formats where the same rules and procedures lead to incorrect responses, in order to help students with discriminating between these problems. This has to be done explicitly, as students tend not to make comparisons when they are presented with multiple problems (cf. Richland et al., 2012). Finally, besides improving knowledge, educators might also focus on interventions to reduce anxiety, improve students' attitudes towards mathematics and probabilistic reasoning, and to highlight the relevance of this knowledge to their professional and everyday lives. Indeed, this kind of intervention should start early on, as it seems that whereas the majority of children have very positive attitudes towards mathematics at the beginning of their primary education, and they feel that they are highly competent (e.g. Moore & Ashcraft, 2012), their attitudes become increasingly negative with age and education (cf. Moore et al., in press).

Regarding future directions of research, an interesting issue is the interactions between the factors that affect the use of intuitive strategies. For example, in recent studies we found that participants with high cognitive ability were more responsive to instructions to rely on effortful reasoning strategies (e.g. Chiesi et al., 2011; Morsanyi et al., 2009), and they benefited more from training (Morsanyi et al., 2013) than lower-ability participants. Other researchers found that students with high anxiety, or when placed in situations with high stakes, were less able to retrieve and apply relevant knowledge (e.g. Beilock, Kulp, Holt, & Carr, 2004), which might render efforts to teach normative rules ineffectual.

Additionally, although it is assumed that intuitive rules and heuristic strategies are activated implicitly or automatically (e.g. Osman & Stavy, 2006), and that there is a causal link between feelings of reduced effort (i.e. when producing a quick, intuitive response), confidence, and judgements of correctness (e.g. Thompson & Morsanyi, 2012; Topolinski & Reber, 2010), there is virtually no research to examine the physiological processes that underlie the production of intuitive responses in the areas of mathematics and probability, apart from a small number of studies on gambling behaviour and risky choice (e.g. De Martino, Kumaran, Seymour & Dolan, 2006; Studer & Clark, 2011).

One method which could provide particularly interesting results would be to measure the activation of facial muscles that are associated with smiling and frowning, using electromyography (see Topolinski, Likowski, Weyers, & Strack, 2009), while participants evaluate mathematical/probabilistic statements (e.g. $3/4 < 3/2$). Whereas fluent processing (i.e. intuitive appeal) is expected to lead to incipient smiles, reduced fluency would be expected to activate the 'frowning muscles'. Electrophysiological measures could also be used to detect conflicts between intuitions and normative rules. This could be done using electromyography whereby the competing responses are associated with the left and right hands, respectively (e.g. Szucs, Soltész, & White, 2009) or by measuring skin conductance responses (e.g. De Neys, Moyens, & Vansteenwegen, 2010). Such studies could provide fruitful links between research into educational and neuroscience approaches to learning about mathematics and probability.

Summary and Conclusion

We started this chapter by describing mathematics as a subject that is composed of a large set of interrelated rules. Thus, when people attempt to solve mathematical problems, they have to face the challenge of selecting the appropriate rule(s) from this large set. In mathematics classrooms this process might be supported by certain cues, such as that problems that can be solved by applying the same procedure appear in the same chapter in the textbook, or that they are presented in a similar format (for example, as a missing value problem in the case of tasks where the proportionality schema needs to be applied). Indeed, when they select a strategy, students are often guided by superficial problem features, such as the familiarity of the problem format, the numbers that appear in the problem, and the ease with which a particular rule can be applied in the context of a task. The activation of certain rules by salient cues in a problem might happen implicitly or automatically, which provides reasoners with quick responses that often appear self-evident and might be accepted with high confidence. Although these cues are indeed valid sometimes, in other cases they lead to systematic error. As intuitive responses are based on learning (through either personal experiences or formal education), the tendency to produce intuitive responses increases with age and education, which might result in a counterintuitive increase in certain errors with level of education. The purpose of this chapter was to give an overview of what these misleading cues might be, and in what kind of situations they will be especially likely to lead reasoners astray.

We also discussed the effect of cognitive ability, thinking dispositions, and the context of task administration on the tendency to rely on intuitive strategies. In general, factors that prevent the use of effortful and time-consuming strategies (e.g. time constraints, the application of working memory load) will lead to a higher proportion of intuitive responses. An increase in intuitive responses might also be observed in the case of participants with lower cognitive ability or with a reduced tendency to spontaneously rely on effortful reasoning, as well as in the case of those participants who experience threat or anxiety, or who generally have negative attitudes towards mathematics. Additionally, students' knowledge about the relevant rules of probability, as well as some related, mistaken beliefs will have a great impact on reasoning performance, as knowledge and beliefs constrain the set of rules that might be applied by an individual to solve problems. Finally, the way certain rules are activated and retrieved (i.e. whether this happens implicitly, explicitly, or automatically) will be an important determinant of how and to what extent reasoners will be able to control the implementation of these rules.

Although the concept of intuition might be associated with something puzzling, uncontrollable, and unconscious, and students might be unaware of the sources of their intuitive preferences, this does not mean that intuition has to be a mystery for researchers and educators. In fact, intuitive tendencies are based on students' personal experiences and their (explicit and implicit) learning history. Intuitive tendencies also tend to be similar across individuals. Whereas teaching students about the normative rules of mathematics and probability is essential, educational and training efforts should also aim to develop the right intuitions, together with an awareness of when and how these intuitions can be employed successfully.

References

Alter, A.L. & Oppenheimer, D.M. (2009). Uniting the tribes of fluency to form a metacognitive nation. *Personality and Social Psychology Review, 13*, 219–235.

Ashcraft, M.H. & Kirk, E.P. (2001). The relationships among working memory, math anxiety, and performance. *Journal of Experimental Psychology: General, 130*, 224–237.

Babai, R., Brecher, T., Stavy, R., & Tirosh, D. (2006). Intuitive interference in probabilistic reasoning. *International Journal of Science & Mathematics Education, 4*, 627–639.

Batanero, C., Serrano. L., & Garfield, J.B. (1996). Heuristics and biases in secondary school students' reasoning about probability. In L. Puig & A. Gutiérrez (Eds.), *Proceedings of the 20th Conference on the International Group for the Psychology of Mathematics Education* (Vol. 2, pp. 51–59), University of Valencia.

Beilock, S.L., Kulp, C.A., Holt, L.E., & Carr, T.H. (2004). More on the fragility of performance: choking under pressure in mathematical problem solving. *Journal of Experimental Psychology: General, 133*, 584–600.

Betsch, T. (2008). The nature of intuition and its neglect in research on judgment and decision making. In H. Plessner, C. Betsch, & T. Betsch (Eds.), *Intuition in Judgment and Decision Making* (pp. 3–22). New York: Erlbaum.

Blazhenkova, O., & Kozhevnikov, M. (2009). The new object-spatial-verbal cognitive style model: Theory and measurement. *Applied Cognitive Psychology, 23*, 638–663.

Brecher, T. (2005). *Application of the intuitive rule: 'more A–more B' in a probability task. A reaction time study*. Unpublished master's thesis, Tel Aviv University, Israel (in Hebrew).

Bryant, P. & Nunes, T. (2012). *Children's understanding of probability. A literature review* (Full report). Retrieved 9 August 2012, <http://www.nuffieldfoundation.org/sites/default/files/files/nuffield_cup_full_reportv_final.pdf>.

Cacioppo, J.T. & Petty, R.E. (1982). The need for cognition. *Journal of Personality and Social Psychology, 42*, 116–131.

Callaert, H. (2004). In search of the specificity and the identifiability of stochastic thinking and reasoning. In M.A. Mariotti (Ed.), *Proceedings of the Third Conference of the European Society for Research in Mathematics Education*. Pisa: Pisa University Press.

Chandrasekhar, S. (1987). *Truth and Beauty. Aesthetics and Motivations in Science*. Chicago: University of Chicago Press.

Chiesi, F., Gronchi, G., &, Primi, C. (2008). Age-trend related differences in task involving conjunctive probabilistic reasoning. *Canadian Journal of Experimental Psychology, 62*, 188–191.

Chiesi, F. & Primi, C. (2009). Recency effects in primary-age children and college students using a gaming situation. *International Electronic Journal of Mathematics Education, Special issue on 'Research and Development in Probability Education', 4*, <www.iejme.com>.

Chiesi, F., Primi, C. & Morsanyi, K. (2011). Developmental changes in probabilistic reasoning: the role of cognitive capacity, instructions, thinking styles and relevant knowledge. *Thinking & Reasoning, 17*, 315–350.

Cleeremans, A. & Jimenez, L. (2002). Implicit learning and consciousness: a graded, dynamic perspective. In R. M. French & A. Cleeremans (Eds), *Implicit Learning and Consciousness* (pp.1–40). Hove: Psychology Press.

Cramer, K., Post, T., & Currier, S. (1993). Learning and teaching ratio and proportion: research implications. In D. Owens (Ed.), *Research Ideas for the Classroom* (pp. 159–178). New York: Macmillan.

De Corte, E., Verschaffel, L., & Pauwels, A. (1990). Influence of the semantic structure of word problems on second graders' eye movements. *Journal of Educational Psychology, 82*, 359–365.

De Martino, B., Kumaran, D., Seymour, B., & Dolan, R.J. (2006). Frames, biases, and rational decision-making in the human brain. *Science, 313*, 684–687.

Denes-Raj, V. & Epstein, S. (1994). Conflict between intuitive and rational processing: when people behave against their better judgment. *Journal of Personality and Social Psychology, 66*, 819–829.

De Neys, W. (2006). Dual processing in reasoning: two systems but one reasoner. *Psychological Science, 17*, 428–433.

De Neys, W., Moyens, E., & Vansteenwegen, D. (2010). Feeling we're biased: autonomic arousal and reasoning conflict. *Cognitive, Affective, and Behavioral Neuroscience, 10*, 208–216.

Evans, J. St. B.T (2006). The heuristic-analytic theory of reasoning: extension and evaluation. *Psychnomic Bulletin and Review, 13*, 378–395.

Ferreira, M.B., Garcia-Marques, L., Sherman, S.J., & Sherman, J.W. (2006). Automatic and controlled components of judgment and decision making. *Journal of Personality and Social Psychology, 91*, 797–813.

Fischbein, E. (1975). *The Intuitive Sources of Probabilistic Thinking in Children*. Dordrecht: Reidel.

Fischbein, E. (1987). *Intuition in Science and Mathematics*. Dordrecht: Reidel.

Fischbein, E. & Gazit, A. (1984). Does the teaching of probability improve probabilistic intuitions? *Educational Studies in Mathematics, 15*, 1–24.

Fischbein, E. & Schnarch, D. (1997). The evolution with age of probabilistic intuitively based misconceptions. *Journal for Research in Mathematics Education, 28*, 96–105.

Fong, G.T., Krantz, D.H., & Nisbett, R.E. (1986). The effects of statistical training on thinking about everyday problems. *Cognitive Psychology, 18*, 253–292.

Frederick, S. (2005). Cognitive reflection and decision making. *Journal of Economical Perspectives, 19*, 25–42.

Galli, S., Chiesi, F., & Primi, F. (2011). Measuring mathematical ability needed for 'non-mathematical' majors: the construction of a scale applying IRT and differential item functioning across educational contexts. *Learning and Individual Differences, 21*, 392–402.

Gigerenzer, G., Hertwig, R., Van den Broek, E., Fasolo, B., & Katsikopoulos, K.V. (2005). A 30% chance of rain tomorrow: how does the public understand probabilistic weather forecasts? *Risk Analysis, 25*, 623–629.

Gigerenzer, G., Todd, P.M., & The ABC Research Group (1999). *Simple Heuristics That Make Us Smart*. New York: Oxford University Press.

Gillard, E., Van Dooren, W., Schaeken, W., & Verschaffel, L. (2009). Proportional reasoning as a heuristic-based process: time constraint and dual-task considerations. *Experimental Psychology, 56*, 92–99.

Gilovich, T., Griffin D., & Kahneman, D. (Eds). (2002). *Heuristics and Biases: The Psychology of Intuitive Judgment*. Cambridge: Cambridge University Press.

Girotto, V. & Gonzalez, M. (2008) Children's understanding of posterior probability. *Cognition, 106*, 325–344.

Green, D.R. (1982). *Probability Concepts in 11–16 Year Old Pupils*, 2nd edn. Loughborough: Centre for Advancement of Mathematical Education in Technology, College of Technology.

Greer, B. (1994). Extending the meaning of multiplication and division. In G. Harel & J. Confrey (Eds), *The Development of Multiplicative Reasoning in the Learning of Mathematics* (pp. 61–85). Albany: State University of New York Press.

Hahn, U. & Warren, P.A. (2009). Perceptions of randomness: why three heads are better than four. *Psychological Review, 116*, 454–461.

Hertwig, R., Herzog, S.M., Schooler, L.J, & Reimer, T. (2008) Fluency heuristic: a model of how the mind exploits a by-product of information retrieval. *Journal of Experimental Psychology: Learning, Memory and Cognition, 34*, 1191–1206.

Jacobs, J.E. & Potenza, M. (1991). The use of judgment heuristics to make social and object decision: a developmental perspective. *Child Development, 62*, 166–178.

Kahneman, D., Slovic, P., & Tversky, A. (Eds). (1982). *Judgment Under Uncertainty: Heuristics and Biases*. Cambridge: Cambridge University Press.

Kahneman, D. & Tversky, A. (1972). Subjective probability: a judgment of representativeness. *Cognitive Psychology, 3*, 430–454.

Kahneman, D. & Tversky, A. (1973). On the psychology of prediction. *Psychological Bulletin, 80*, 237–251.

Kokis, J., Macpherson, R., Toplak, M.E., West, R.F., & Stanovich, K.E. (2002). Heuristic and analytic processing: age trends and associations with cognitive ability and cognitive styles. *Journal of Experimental Child Psychology*, *83*, 26–52.

Koriat, A. (2007). Metacognition and consciousness. In P.D. Zelazo, M. Moscovitch, E. Thompson (Eds) *The Cambridge Handbook of Consciousness* (pp. 289–326). Cambridge, NY: Cambridge University Press.

Lecoutre, M.P. (1985). Effect d'informations de nature combinatoire et de nature fréquentielle sur les judgements probabilistes. *Recherches en Didactique des Mathématiques*, *6*, 193–213.

Lecoutre, M.P. (1992). Cognitive models and problem spaces in purely random situations. *Educational Studies in Mathematics*, *23*, 557–568.

McColm, G. (2007). A metaphor for mathematics education. *Notices of the American Mathematical Association*, *54*, 499–502.

McGlone, M.S. & Tofighbakhsh, J. (2000). Birds of a feather flock conjointly(?): rhyme as reason in aphorisms. *Psychological Science*, *11*, 424–428.

Mendel, N. (1998). *The Intuitive Rule 'Same of A, Same of B': The Case of Comparison of Rectangles.* Unpublished master's thesis, Tel Aviv University, Israel (in Hebrew).

Metz, K.E. (1998). Emergent understanding and attribution of randomness: comparative analysis of the reasoning of primary grade children and undergraduates. *Cognition and Instruction*, *16*, 285–365.

Michalczuk, R., Bowden-Jones, H., Verdejo-Garcia, A., & Clark, L. (2011). Impulsivity and cognitive distortions in pathological gamblers attending the UK National Problem Gambling Clinic: a preliminary report. *Psychological Medicine*, *41*, 2625–2636.

Moore, A.M. & Ashcraft, M.H. (2012). Relationships across mathematics tasks in elementary school children. Unpublished manuscript, University of Nevada, Las Vegas.

Moore, A.M., Rudig, N.O., & Ashcraft, M.H. (2015). Affect, motivation, working memory and mathematics. In R. Cohen-Kadosh & A. Dowker (Eds), *The Oxford Handbook of Numerical Cognition.* Oxford: Oxford University Press.

Morsanyi, K., Handley, S.J., & Serpell, S. (2013). Making heads or tails of probability: an experiment with random generators. *British Journal of Educational Psychology*, *83* (Part 3), 379–395.

Morsanyi, K., Primi, C., Chiesi, F., & Handley, S. (2009). The effects and side-effects of statistic education. Psychology students' (mis-)conceptions of probability. *Contemporary Educational Psychology*, *34*, 210–220.

Morsanyi, K., Primi, C., Handley, S.J., Chiesi, F., & Galli, S. (2012). Are systemizing and autistic traits related to talent and interest in mathematics and engineering? Testing some of the central claims of the empathizing-systemizing theory. *British Journal of Psychology*, *103*, 472–496.

Ni, Y. & Zhou, Y.-D. (2005). Teaching and learning fraction and rational numbers: the origins and implications of whole number bias. *Educational Psychologist*, *40*, 27–52.

Nisbett, R.E., Krantz, D.H., Jepson, C., & Kunda, Z. (1983). The use of statistical heuristics in everyday inductive reasoning. *Psychological Review*, *90*, 339–363.

Osman, M. (2004). An evaluation of dual-process theories of reasoning. *Psychonomic Bulletin and Review*, *11*, 988–1010.

Osman, M. & Stavy, R. (2006). Development of intuitive rules: evaluating the application of the dual-system framework to understanding children's intuitive reasoning. *Psychonomic Bulletin & Review*, *13*, 935–953.

Pratt, D. (2000). Making sense of the total of two dice. *Journal for Research in Mathematics Education*, *31*, 602–625.

Raylu, N. & Oei, T.P. S. (2004). The Gambling Related Cognitions Scale (GRCS): development, confirmatory factor validation and psychometric properties. *Addiction*, *99*, 757–769.

Reber, R., Brun, M., & Mitterndorfer, K. (2008). The use of heuristics in intuitive mathematical judgment. *Psychonomic Bulletin & Review*, *15*, 1174–1178.

Reyna, V.F. & Brainerd, C.J. (2008). Numeracy, ratio bias, and denominator neglect in judgments of risk and probability. *Learning and Individual Differences*, *18*, 89–107.

Richland, L.E., Stigler, J.W., & Holyoak, K.J. (2012). Teaching the meaningful structure of mathematics. *Educational Psychologist, 47* (3), 189–203.

Sedlmeier, P. (2007). Statistical reasoning: valid intuitions put to use. In M. Lovett & P. Shah (Eds), *Thinking with Data* (pp. 389–419). New York: Erlbaum.

Stanovich, K.E. & West, R.F. (2000). Individual differences in reasoning: implications for the rationality debate. *Behavioral and Brain Sciences, 23*, 645–726.

Stanovich, K.E. & West, R.F. (2008). On the relative independence of thinking biases and cognitive ability. *Journal of Personality and Social Psychology, 94*, 672–695.

Stavy, R. & Tirosh, D. (2000). *How students (mis-)understand science and mathematics: intuitive rules.* New York: Teachers College Press.

Studer, B. & Clark, L. (2011). Place your bets: psychophysiological correlates of decision-making under risk. *Cognitive, Affective and Behavioural Neuroscience, 11*, 144–158.

Szucs, D., Soltész, F., & White, S. (2009). Motor conflict in Stroop tasks: direct evidence from single-trial electro-myography and electro-encephalography. *NeuroImage, 47*, 1960–1973.

Taleb, N.N. (2001). *Fooled by Randomness*. London: Penguin.

Thompson, V.A. & Morsanyi, K. (2012). Analytic thinking: do you feel like it? *Mind & Society, 11*, 93–105.

Thompson, V.A., Prowse-Turner, J., & Pennycook, G. (2011). Intuition, reason and metacognition. *Cognitive Psychology, 63*, 107–140.

Toplak, M., Liu, E., Macpherson, R., Toneatto, T., & Stanovich, K.E. (2007). The reasoning skills and thinking dispositions of problem gamblers: a dual-process taxonomy. *Journal of Behavioral Decision Making, 20*, 103–124.

Topolinski, S., Likowski, K.U., Weyers, P., & Strack, F. (2009). The face of fluency: semantic coherence automatically elicits a specific pattern of facial muscle reactions. *Cognition and Emotion, 23*, 260–271.

Topolinski, S. & Reber, R. (2010). Immediate truth: temporal contiguity between a cognitive problem and its solution determines experienced veracity of the solution. *Cognition, 114*, 117–122.

Topolinski, S. & Strack, F. (2009). The architecture of intuition: fluency and affect determine intuitive judgments of semantic and visual coherence, and of grammaticality in artificial grammar learning. *Journal of Experimental Psychology: General, 138*, 39–63.

Tversky, A. & Kahneman, D. (1974). Judgment under uncertainty: heuristics and biases. *Science, 185*, 1124–1131.

Vamvakoussi, X., Van Dooren, W., & Verschaffel, L. (2012). Naturally biased? In search for reaction time evidence for a natural number bias in adults. *Journal of Mathematical Behaviour, 31*, 344–355.

Van Dooren, W., De Bock, D., Evers, M., & Verschaffel, L. (2009). Students' overuse of proportionality on missing-value problems: how numbers may change solutions. *Journal for Research in Mathematics Education, 40*, 187–211.

Van Dooren, W., De Bock, D., Hessels, A., Janssens, D., & Verschaffel, L. (2005). Remedying secondary school students' illusion of linearity: developing and evaluating a powerful learning environment. In L. Verschaffel et al. (Eds), *Powerful Environments for Promoting Deep Conceptual and Strategic Learning (Studia Paedagogica, 41)* (pp. 115–132). Leuven: Universitaire Pers.

Whittlesea, B.W.A. (1993). Illusions of familiarity. *Journal of Experimental Psychology: Learning, Memory, & Cognition, 19*, 1235–1253.

Winkielman, P., Halberstadt, J., Fazendeiro, T., & Catty, S. (2006). Prototypes are attractive because they are easy on the mind. *Psychological Science, 17*, 799–806.

Xu, F. & Denison, S. (2009). Statistical inference and sensitivity to sampling in 11-month-old infants. *Cognition, 112*, 97–104.

PART III

PHYLOGENY AND ONTOGENY OF MATHEMATICAL AND NUMERICAL UNDERSTANDING

CHAPTER 11

PHYLOGENY AND ONTOGENY OF MATHEMATICAL AND NUMERICAL UNDERSTANDING

ELIZABETH M. BRANNON AND JOONKOO PARK

INTRODUCTION

THE case study of numerical cognition has proven to be a remarkable test bed for understanding the evolution and development of the mind. Comparative psychologists seek to understand what differentiates human minds from animal minds, and importantly the study of how the mind represents number illustrates both evolutionary continuity and discontinuity. On the one hand, the continuity is evident in a shared system for making approximate numerical judgments (i.e. the approximate number system or ANS). On the other hand, there is a clear evolutionary discontinuity in that only the human species has invented arbitrary symbols for number. This unique capacity for representing number symbolically allows humans to mentally manipulate precise numerical values and permits complex and abstract mathematics. In parallel, developmental psychologists seek to understand the origins of human conceptual abilities. The study of how number is represented from infancy into adulthood reveals profound continuity whereby infants appear to enter the world able to represent and compare numerical values approximately. While this system for representing number approximately dramatically improves in its precision over early childhood and into middle age, its fundamental signatures remain constant. At the same time, numerical development reveals a paradigmatic example of discontinuity and conceptual change whereby language transforms a child's ability to represent number (Carey 2009). The chapters in this section represent new and exciting research on both the development and the evolution of numerical thinking and highlight cutting edge questions in cognitive science that emerge from this field.

Evolutionary and Developmental Continuity

Species throughout the animal kingdom are sensitive to the numerical attributes of the world around them (Agrillo, this volume; Beran, this volume). The primary characteristic of animal number representations is that they are imprecise. Discrimination follows Weber's Law whereby the ability to discriminate between two values is determined by their ratio, not their absolute difference. Figure 1 illustrates that accuracy in a numerical comparison task is predicted by the ratio of the two numerical values being compared. In this task the animal is presented with two numerosities on a touch sensitive screen and is required to choose the array with the larger number of dots. In fact, Figure 11.1 shows that when adult humans are tested in the same task that prohibits verbal counting, their performance is indistinguishable from that of nonhuman animals (see also Agrillo et al. 2012; Beran et al. 2011a; Beran et al. 2008, 2011b; Cantlon & Brannon, 2007; Jordan & Brannon, 2006a,b). This pattern of results suggests that we share with many diverse species a system for representing number approximately, which is referred to as the approximate number system (ANS).

Sensitivity to number is part of the human repertoire even at birth (McCrink, this volume; Izard et al. 2009). A recent set of studies shows parametric ratio dependence in the time infants spend looking at visual arrays as predicted by Weber's Law. As shown in Figure 11.2, infants were shown two simultaneous streams of visual arrays. One stream maintained a constant numerosity and the other stream alternated between two numerosities. Infants'

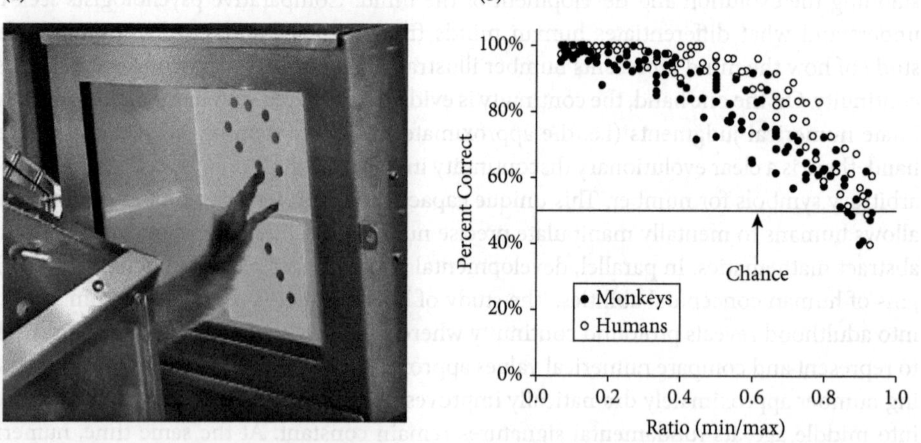

FIGURE 11.1 Monkeys show ratio dependence when ordering numerosities. (a): A photograph of a monkey engaging in a numerosity ordering touch-screen task. (b): Accuracy as a function of the ratio between the two numerical values. Open symbols are data from college students and closed symbols are data from rhesus monkeys. Ratio refers to small/large numerosity.

Reproduced from Jessica F. Cantlon and Elizabeth M. Brannon, *Psychological Science*, 17(5), Shared System for Ordering Small and Large Numbers in Monkeys and Humans, pp. 401–406, doi: 10.1111/j.1467-9280.2006.01719.x,

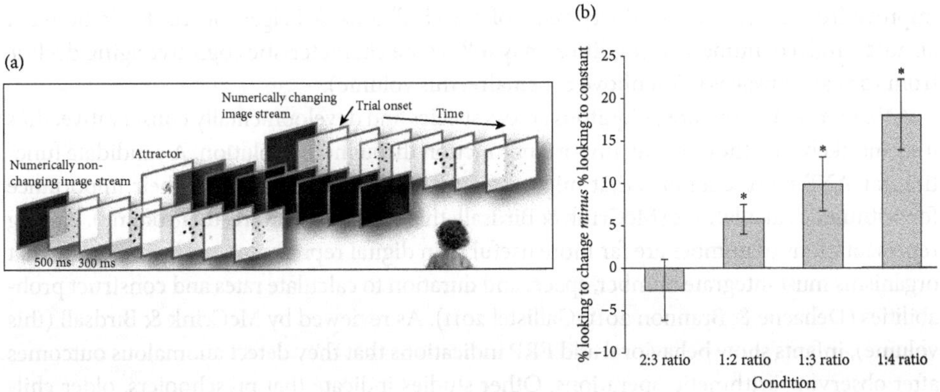

FIGURE 11.2 Infant numerical change detection tasks demonstrates ratio dependence. (a): Experimental design of numerical change detection task. During each trial, infants were presented with two image streams simultaneously on two peripheral screens. In the numerically changing image stream, images contained two different numerosities in alternation (here: 10 and 20), while the numerically non-changing image stream contained only images with the same numerosity (here: 10). Infants' looking time to each of the image streams was measured. Reprinted from Libertus & Brannon 2010. (b): six-month-old infants' preference scores for numerically changing image streams in four different ratio conditions. Significantly positive preference scores were found for the 1:2, 1:3, and 1:4 ratio conditions. Furthermore, preference scores increased with increasing relative numerical disparity. Error bars reflect standard errors.

Reproduced from Stable individual differences in number discrimination in infancy, Melissa E. Libertus and Elizabeth M. Brannon, *Developmental Science*, 13(6), pp. 900–906, DOI: 10.1111/j.1467-7687.2009.00948.x, © 2010 Blackwell Publishing Ltd, with permission.

preference for the changing stream depended on the ratio between the two values in the changing stream (Libertus & Brannon 2010; Starr et al. 2013a; see also Xu & Spelke 2000).

Particularly compelling evidence for the existence of this primitive number system in human infants and animals comes from studies that require cross-modal matching of arrays based on number. Newborn infants preferentially look at a visual array that numerically matches the number of phonemes to which they have been familiarized (Izard et al. 2009). Rhesus monkeys and human infants presented with arrays of two and three conspecifics vocalizing allocate more visual attention to the array that numerically matches the number of conspecific calls or voices they hear (Jordan et al., 2005; Jordan & Brannon, 2006c). Unfortunately, those studies did not address ratio-dependence and instead were limited to the small values 2 and 3. However, rhesus monkeys were tested with an operant task and a wide range of numerical values and were able to reliably choose a visual array that numerically matched an auditory sample sequence (Jordan et al. 2008). These cross-modal findings are one of the strongest arguments against the alternative explanation that humans and non-human animals are sensitive not to number, but to perceptual variables such as surface area, density, or duration that are typically confounded with number.

Although the characteristic signatures of the ANS appear to be present very early in human development, the system undergoes a great deal of change over the lifespan. While newborns require a 1:3 ratio to detect a change in numerosity (Izard et al. 2009), by six months of age infants are capable of differentiating a 1:2 ratio change, and by nine months they succeed with a 2:3 ratio change (Xu & Spelke 2000; Lipton & Spelke 2004). Numerical comparison tasks indicate that precision in numerosity discrimination continues to improve from three years to thirty years of age (Halberda & Feigenson 2008; Halberda et al. 2012). Indeed, numerical cognition may follow the characteristic cognitive aging decline from 30 years onwards (Uittenhove & Lemaire, this volume).

ANS representations are ubiquitous across species and developmentally conservative, thus it seems likely that they have an important function throughout evolution. A candidate function for ANS representations that might account for such universality is their importance for arithmetic calculations (McCrink & Birdsall, this volume; Gilmore, this volume). Analog representations of number are far more useful than digital representations if we assume that organisms must integrate number, space, and duration to calculate rates and construct probabilities (Dehaene & Brannon 2011; Gallistel 2011). As reviewed by McCrink & Birdsall (this volume), infants show behavioral and ERP indications that they detect anomalous outcomes after observing arithmetic operations. Other studies indicate that preschoolers, older children, adults, and even monkeys appreciate ordinal relations between numerical values and perform approximate arithmetic on ANS representations (Gilmore, this volume).

As Agrillo describes (this volume), the ANS may not be the only way number can be represented without language. Infants and adults appear to use an alternative route for keeping track of small numbers of individual items, referred to as the object-file system (Uller et al. 1999; See Feigenson et al. 2004 for review). In adults this presents as exceedingly fast and accurate enumeration or subitizing of one to four items in contrast to the serial increase in reaction-time that is observed with each additional item to-be-enumerated for arrays of five or greater. Evidence for an object-file system in infancy comes from a few different experimental paradigms. In each of these nonverbal procedures infants seem only capable of tracking small numbers of objects and display a remarkable inability to compare small with large sets. Nevertheless, it must be emphasized that while these findings are robust and replicable, they are context specific. In other experimental contexts infants appear to be capable of representing small and large sets with a single cognitive system and even show ratio dependence when discriminating small sets (e.g. Starr et al. 2013a; vanMarle & Wynn 2011).

Agrillo describes research with a wide range of species including elephants, dogs, parrots, beetles, and primates. In some cases animals exhibit ratio dependence throughout the entire numerical range and in other cases they show the pattern described above for infants whereby they are either unable to represent large values or exhibit differences in the ratio required for small and large number discrimination. These data raise interesting questions about whether species differences or experimental parameters explain whether behavioral patterns are controlled by an object-file system or the ANS.

Conceptual Foundations For Uniquely Human Mathematical Cognition

While the above section illustrates significant continuities in numerical cognition throughout evolution and development, there are also important discontinuities (Carey 2009; Spelke 2011; Sarnecka et al., this volume). A major discontinuity is that only humans develop a concept of natural number. As children acquire the verbal count list, it provides an avenue for distinguishing between sets that cannot be accommodated by the ANS. In other words, the ANS cannot support our ability to distinguish between 18 and 19 apples, however by verbally counting we can easily do so. Without the verbal count list and the successor principle, full-blown mathematics as observed in human societies would not be possible (Carey 2009; Spelke 2011).

At first blush, the cases reviewed by Beran (this volume) in which animals appear to have learned a mapping between numerosities and arbitrary symbols may suggest that there is more continuity than discontinuity between animal and human minds in the domain of numerical cognition. However, we argue that there remains a fundamental difference that sets all nonhuman animals apart from human children. As reviewed by Sarnecka & Goldman (this volume) somewhere around the age of four years children internalize the successor principle. They grasp that verbal counting is a means by which they can determine the cardinality of a set. When they do this it allows a mapping that transcends arduous learning of each symbol-referent pairing and lays the foundation for uniquely human mathematics (Carey 2009). In contrast, the few studies that have trained animals to match a symbol to a numerosity reveal no evidence that the animals show a learning curve or make an inference such as 'this new symbol represents the previous numerosity plus one'. Instead non-human animals seem to learn each new pairing as a new problem. Furthermore there is no compelling evidence that acquisition of numerical symbols allows nonhuman animals to represent large values precisely (although see Pepperberg & Carey 2012 for a different position). Most importantly, while there are many examples of animals attending to numerosity spontaneously it is inarguable that no species other than humans has independently devised systems for representing number symbolically.

How are the evolutionary and developmental building blocks of numerical cognition transformed into uniquely human mathematical cognition? We have discussed two systems for representing number without language: the ANS which is ratio-dependent and can represent number independently from other stimulus attributes (Feigenson et al. 2004), and the object-file system which is almost accidently numerical and is potentially more bound to continuous variables (Feigenson et al. 2004). An important question for this field is whether either of these systems provides the conceptual basis for symbolic math. As children learn the meaning of number words and symbols they must ground these symbols with nonverbal representations. What work does each system do in this important process?

Sarnecka & Goldman give an account of how children come to have natural number concepts. They offer a compelling account of the 'conceptual-role bootstrapping' hypothesis proposed by Carey (2009) which maintains that natural number concepts are constructed. The procedural counting routine that children master at an early age serves as a placeholder structure whereby the symbols are ordered but do not yet represent natural numbers for the child. Over time the symbols take on more meaning and become fully defined natural number concepts. Under this scenario the ANS is not part of this construction but is mapped onto natural number concepts later in development only after the child has become a cardinal principal knower and has internalized the successor principle. However as reviewed by Gilmore (this volume) and Gebuis & Reynvoet (this volume) recent data linking individual differences in the acuity of the ANS with individual differences in symbolic math ability suggest that the ANS may play a more important role than the bootstrapping hypothesis allows. There is agreement that children have mapped symbols onto ANS representations by the age of five, as suggested by the distance effect when they compare Arabic numerals, (Duncan & McFarland 1980; Sekuler & Mierkiewicz 1977), however the specific role if any the ANS plays in grounding children's symbolic understanding of number is not understood.

The Quest for The Cognitive Foundations of Mathematics

As reviewed by Gilmore (this volume) and Gebuis & Reynvoet (this volume) there are a growing number of recent datasets that demonstrate that some of the variance in symbolic math performance, as measured by SATs or standardized math achievement tests, can be explained by ANS precision as measured by Weber fraction (DeWind & Brannon 2012; Halberda & Feigenson 2008, Halberda et al. 2012; Libertus et al. 2011, 2012; Lyons & Beilock 2011; Piazza et al. 2010; Gilmore et al. 2010; Mazzocco et al. 2011). However other studies have failed to find any relationship between ANS and symbolic math achievement (Sasanguie et al. 2013; Holloway & Ansari 2009; Inglis et al. 2011; Iuculano et al. 2008; Nosworthy et al. 2013; Price et al. 2012; Soltesz et al. 2010; Wei et al. 2012), or have found that the relationship is mediated by symbolic number knowledge (Lyons & Beilock 2011), executive function (Fuhs and McNeil 2013), or only holds for children with low math ability (Bonny and Lourenco 2013).

An alternative possibility, as reviewed in Gebuis & Reynvoet, (this volume), is that facility with symbol ordering may play a more critical role in mathematical cognition and competence. For example, facility with ordering Arabic numerals is predictive of group differences between those with and without math deficiencies (Rousselle & Noel 2007; De Smedt & Gilmore 2011; Iuculano et al. 2008) and explains individual differences in typically developing children and healthy adults (Castronovo & Gobel 2012; Holloway & Ansari 2009; Bugden & Ansari 2011; Sasanguie et al. 2012; De Smedt et al. 2009; Landerl et al. 2004; Lyons & Beilock 2011). Clearly a great deal of future research is needed to uncover the mechanisms underlying the interactive links between ANS, symbolic number system, and math. Thus the relationship between ANS acuity and symbolic math may not be clear cut and further research is needed to determine whether it is mediated by other cognitive factors.

Most of the research exploring the relationship between the ANS and math achievement has been limited to a correlational approach, and very little work has yet to investigate the causal relationship between ANS and math. Two recent studies attempt to probe a possible causal relationship. We used a pre- and post-test training paradigm in adult participants and found that approximate arithmetic training improved symbolic math performance in two experiments (Park & Brannon 2013). Specifically participants were given six to ten days of training in which they solved approximate non-symbolic addition and subtraction problems. A second group of participants either received no intervening training or were trained on other tasks. Prior to and after the training sessions, all participants solved two- and three-digit symbolic arithmetic problems. The participants who were trained on non-symbolic approximate arithmetic showed stark improvement in symbolic math compared to participants in other groups. While the exact mechanisms underlying these transfer effects are not yet understood, these results suggest a promising avenue for future research and provide the first evidence that at least part of the relationship between ANS and math may indeed reflect a causal relationship. In another study, Hyde & colleagues (in press) asked first-grade children to complete an exact symbolic arithmetic test. Critically, prior to this test, some of the children received a small set of problems that engaged representation and operation of approximate numerical quantities while others received other kinds of non-numerical problems. Interestingly, those children who received the ANS-based problems prior to the symbolic arithmetic test performed better at the symbolic arithmetic test compared to other children. There are a few limitations of the study such as an absence of a pre- and post-test design, brief exposure of the training problems, and immediate testing, which restricts the interpretations of these results. Nevertheless, they support the hypothesis that there is a causal relationship between ANS and symbolic math and suggest the exciting possibility that training aspects of the ANS may be useful in improving symbolic mathematics in children.

If there is in fact a causal relationship between ANS and symbolic math this still leaves open the question of why. What role does ANS play in scaffolding symbolic math? On the one hand, ANS enables numerical quantity estimations and operations, so it is plausible to expect that formal mathematics, which involves the representations and operations of numbers, is rooted in the ANS. On the other hand, it is challenging to understand how such a primitive system that is not capable of representing exact large numbers could give rise to formal mathematics, which is uniquely human.

One possibility is that young children with high ANS acuity have more distinct internal representation of large numerosity that may facilitate better symbol to numerosity mapping, which in turn may facilitate the learning of symbolic number system. This possibility is supported by recent findings indicating that children's ANS acuity measured prior to their formal math education at the age of three to four years correlates with their performance in standardized math achievement tests measured at the age of five to six (Mazzocco et al. 2011). Furthermore, six-month-old infants' ANS acuity measured from a change detection paradigm correlates with their standardized math scores three years later (Starr et al. 2013b). However, these findings do not explain why the relationship between ANS representations and math achievement appears to hold even into adulthood (e.g. Lyons & Beilock 2011; DeWind & Brannon 2012; Halberda et al. 2012). Thus another possibility is that ANS representations allow people to reject incorrect calculations that are off by orders of magnitude (Feigenson et al. 2013).

Yet a third possibility is that the causal arrow is reversed and that mathematical ability instead functions to sharpen ANS acuity. For example, children who are particularly adept at math or enjoy math will spend more time playing with numbers, which in turn may facilitate better performance in ANS tasks. In this way, the link between ANS and math may be bidirectional into adulthood. A recent study by Piazza & colleagues (2013) supports this idea in that indigenous Amazonian people with access to schooling have higher ANS acuity than those without access to schooling. Similarly, Hannula-Sormunen (this volume) argues that a domain-selective attentional system for number, namely the spontaneous focus on number, is another important factor that gives rise to mathematical competence in such an interactive way throughout development.

A final caveat is in order. Even if the answer turns out to be that ANS precision predicts some of the variance in math ability and that this reflects a causal relationship, it is notable that in most samples only a very small proportion of the variability in math competence appears to be explained by ANS acuity. Thus, even if there is a direct causal relationship between ANS and math there are likely many other more important predictors of math aptitude. Understanding the causal relationship may do more for the epistemology of numerical cognition than it does for practical interventions. In other words, understanding the relationship between the ANS and symbolic math is critical for uncovering the developmental roots of uniquely human numerical cognition; however, whether this link can be harnessed to meaningfully improve symbolic math is an open and distinct question.

Conclusions and Outstanding Questions

Why has the study of the evolution and development of number concepts received so much attention? A primary reason is that it provides one of the best examples of abstract thought. Even more so than categorical distinctions such as vehicle vs animal, number is abstract in that it can be divorced from perceptual variables. Three giraffes and three cars look, smell, feel, and taste nothing alike and yet they are both sets of three. The study of number therefore provides a case study for examining how complex cognition emerges over evolution and within the human lifespan. The chapters in this section illustrate the remarkable primitives that allow a wide range of species and prelinguistic humans to represent the quantitative aspect of the world within which they live. The chapters also raise compelling questions about how these primitives give rise to uniquely human mathematical cognition and should foster many questions for future research such as:

(1) What are the factors that determine whether animals and babies rely on the ANS or object-file system? Are there species differences in the presence or limits of the object-file system?
(2) What are the fundamental cognitive differences that allow the human mind to transcend the ANS that are lacking in non-human animals?
(3) How do children acquire natural numbers? Is there a causal role for both object-files and ANS in this process?
(4) What is the relationship between the ANS and symbolic math throughout the lifespan? What are the causal mechanisms that drive this relationship?
(5) How can the knowledge we gain about the primitives of numerical cognition be used to improve school-based math?

References

Agrillo, C., Piffer, L., Bisazza, A., & Butterworth, B. (2012). Evidence for two numerical systems that are similar in humans and guppies. *PLoS ONE* 7(2): e31923.

Beran, M.J., Johnson-Pynn, J.S., & Ready, C. (2008). Quantity representation in children and rhesus monkeys: Linear versus logarithmic scales. *Journal of Experimental Child Psychology* 100(3): 225–233.

Beran, M.J., Decker, S., Schwartz, A., & Schultz, N. (2011a). Monkeys (*Macaca mulatta* and *Cebus apella*) and human adults and children (Homo sapiens) compare subsets of moving stimuli based on numerosity. *Frontiers in Psychology* 2: 61, 1–7.

Beran, M.J., Johnson-Pynn, J.S., & Ready, C. (2011b). Comparing children, *Homo sapiens* and chimpanzees, *Pan troglodytes* quantity judgments of sequentially presented sets of items. *Current Zoology* 57(4): 419–428.

Bonny, J.W., & Lourenco, S.F. (2013). The approximate number system and its relation to early math achievement: evidence from the preschool years. *Journal of Experimental Child Psychology* 114(3): 375–388.

Bugden, S., & Ansari, D. (2011). Individual differences in children's mathematical competence are related to the intentional but not automatic processing of Arabic numerals. *Cognition* 118(1): 32–44.

Cantlon, J.F., & Brannon, E.M. (2006). Shared system for ordering small and large numbers in monkeys and humans. *Psychological Science* 17(5): 401–406.

Cantlon, J.F., & Brannon, E.M. (2007). Basic math in monkeys and college students. *PLoS Biology*, 5(12): e328.

Carey, S. (2009). *The Origin of Concepts*. Oxford: Oxford University Press.

Castronovo, J., & Göbel, S. M. (2012). Impact of high mathematics education on the number sense. *PLoS One* 7(4): e33832.

De Smedt, B., & Gilmore, C.K. (2011). Defective number module or impaired access? Numerical magnitude processing in first graders with mathematical difficulties. *Journal of Experimental Child Psychology* 108(2): 278–292.

De Smedt, B., Verschaffel, L., & Ghesquiere, P. (2009). The predictive value of numerical magnitude comparison for individual differences in mathematics achievement. *Journal of Experimental Child Psychology* 103(4): 469–479.

Dehaene, S., & Brannon, E.M. (2011). Space, time and number in the brain: Searching for the foundations of mathematical thought. Academic Press.

DeWind, N.K., & Brannon, E.M. (2012). Malleability of the approximate number system: effects of feedback and training. *Frontiers in Human Neuroscience* 6: 68.

Duncan, E., & McFarland, C. (1980). Isolating the effects of symbolic distance, and semantic congruity in comparative judgments: an additive-factors analysis. *Memory and Cognition* 8(6): 612–622.

Feigenson, L., Dehaene, S., & Spelke, E. (2004). Core systems of number. *Trends in Cognitive Sciences* 8(7): 307–314.

Feigenson, L., Libertus, M.E., & Halberda, J. (2013). Links Between the Intuitive Sense of Number and Formal Mathematics Ability. *Child Development Perspectives* 7(2): 74–79.

Fuhs, M.W., & McNeil, N.M. (2013). ANS acuity and mathematics ability in preschoolers from low-income homes: contributions of inhibitory control. *Developmental Science* 16(1): 136–148.

Gallistel, C. (2011). Mental magnitudes. In: *Space, Time, and Number in the Brain: Searching for the Foundations of Mathematical Thought*, edited by S, Dehaene & E.M. Brannon (pp. 3–12). Academic Press.

Gilmore, C.K., McCarthy, S.E., & Spelke, E.S. (2010). Non-symbolic arithmetic abilities and mathematics achievement in the first year of formal schooling. *Cognition* 115(3): 394–406.

Halberda, J., & Feigenson, L. (2008). Developmental change in the acuity of the "Number Sense": The Approximate Number System in 3-, 4-, 5-, and 6-year-olds and adults. *Developmental Psychology* 44(5): 1457–1465.

Halberda, J., Ly, R., Wilmer, J.B., Naiman, D.Q., & Germine, L. (2012). Number sense across the lifespan as revealed by a massive Internet-based sample. *Proceedings of the National Academy of Sciences*, 109(28): 11116–11120.

Holloway, I.D., & Ansari, D. (2009). Mapping numerical magnitudes onto symbols: the numerical distance effect and individual differences in children's mathematics achievement. *Journal of Experimental Child Psychology* 103(1): 17–29.

Hyde D.C., Khanum S., & Spelke E.S. (2014). Brief non-symbolic approximate number practice enhances subsequent exact symbolic arithmetic in children. *Cognition* 131(1): 92–107.

Inglis, M., Attridge, N., Batchelor, S., and Gilmore, C. (2011). Non-verbal number acuity correlates with symbolic mathematics achievement: but only in children. *Psychonomic Bulletin and Review* 18(6): 1222–1229.

Iuculano, T., Tang, J., Hall, C. W., & Butterworth, B. (2008). Core information processing deficits in developmental dyscalculia and low numeracy. *Developmental Science* 11(5): 669–680.

Izard, V., Sann, C., Spelke, E.S., & Streri, A. (2009). Newborn infants perceive abstract numbers. *Proceedings of the National Academy of Sciences of the United States of America* 106(25): 10382–10385.

Jordan, K.E., & Brannon, E.M. (2006a). A common representational system governed by Weber's law: Nonverbal numerical similarity judgments in 6-year-olds and rhesus macaques. *Journal of Experimental Child Psychology* 95(3): 215–229.

Jordan, K.E., & Brannon, E.M. (2006b). Weber's Law influences numerical representations in rhesus macaques (Macaca mulatta). *Animal Cognition* 9(3): 159–172.

Jordan, K.E., & Brannon, E.M. (2006c). The multisensory representation of number in infancy. *Proceedings of the National Academy of Sciences of the United States of America* 103(9): 3486–3489.

Jordan, K.E., Brannon, E.M., Logothetis, N.K., & Ghazanfar, A.A. (2005). Monkeys match the number of voices they hear to the number of faces they see. *Current Biology* 15(11): 1034–1038.

Jordan, K.E., MacLean, E.L., & Brannon, E.M. (2008). Monkeys match and tally quantities across senses. *Cognition* 108(3): 617–625.

Landerl, K., Bevan, A., & Butterworth, B. (2004). Developmental dyscalculia and basic numerical capacities: A study of 8-9-year-old students. *Cognition* 93(2): 99–125.

Libertus, M.E., & Brannon, E.M. (2010). Stable individual differences in number discrimination in infancy. *Developmental Science* 13(6): 900–906.

Libertus, M.E., Feigenson, L., & Halberda, J. (2011). Preschool acuity of the approximate number system correlates with school math ability. *Developmental Science* 14(6): 1292–1300.

Libertus, M.E., Odic, D., & Halberda, J. (2012). Intuitive sense of number correlates with math scores on college-entrance examination. *Acta Psychologica* 141(3): 373–379.

Lipton, J.S., & Spelke, E.S. (2004). Discrimination of large and small numerosities by human infants. *Infancy* 5(3): 271–290.

Lyons, I.M., & Beilock, S.L. (2011). Numerical ordering ability mediates the relation between number-sense and arithmetic competence. *Cognition* 121(2): 256–261.

Mazzocco, M.M.M., Feigenson, L., & Halberda, J. (2011). Preschoolers' precision of the approximate number system predicts later school mathematics performance. *PLoS ONE* 6(9): 1–8.

Nosworthy, N., Bugden, S., Archibald, L., Evans, B., & Ansari, D. (2013). A two-minute paper-and-pencil test of symbolic and nonsymbolic numerical magnitude processing explains variability in primary school children's arithmetic competence. *PLoS ONE* 8(7): e67918.

Park, J., & Brannon, E.M. (2013). Training the approximate number system improves math proficiency. *Psychological Science* 24(10): 1013–1019.

Pepperberg, I.M., & Carey, S. (2012). Grey parrot number acquisition: The inference of cardinal value from ordinal position on the numeral list. *Cognition*, 125(2): 219–232.

Piazza, M., Facoetti, A., Trussardi, A.N., Berteletti, I., Conte, S., Lucangeli, D., et al. (2010). Developmental trajectory of number acuity reveals a severe impairment in developmental dyscalculia. *Cognition* 116(1): 33–41.

Piazza, M., Pica, P., Izard, V., Spelke, E.S., & Dehaene, S. (2013). Education enhances the acuity of the nonverbal approximate number system. *Psychological Science* 24(6): 1037–1043.

Price, G.R., Palmer, D., Battista, C., & Ansari, D. (2012). Nonsymbolic numerical magnitude comparison: Reliability and validity of different task variants and outcome measures, and their relationship to arithmetic achievement in adults. *Acta Psychologica* 140(1): 50–57.

Rousselle, L., & Noel, M.P. (2007). Basic numerical skills in children with mathematics learning disabilities: a comparison of symbolic vs non-symbolic number magnitude processing. *Cognition* 102(3): 361–395.

Sasanguie, D., De Smedt, B., Defever, E., & Reynvoet, B. (2012). Association between basic numerical abilities and mathematics achievement. *The British Journal of Developmental Psychology* 30(Pt 2): 344–357.

Sasanguie, D., Gobel, S. M., Moll, K., Smets, K., & Reynvoet, B. (2013). Approximate number sense, symbolic number processing, or number-space mappings: what underlies mathematics achievement? *Journal of Experimental Child Psychology* 114(3): 418–431.

Sekuler, R., & Mierkiewicz, D. (1977). Children's judgments of numerical inequality. *Child Development* 48(2): 630–633.

Soltesz, F., Szucs, D., & Szucs, L. (2010). Relationships between magnitude representation, counting and memory in 4- to 7-year-old children: a developmental study. *Behavioral and Brain Functions: BBF* 6: 13.

Spelke, E. (2011) Natural number and natural geometry. In: *Space, Time, and Number in the Brain: Searching for the Foundations of Mathematical Thought*, edited by S, Dehaene & E.M. Brannon (pp. 287–317). Academic Press.

Starr, A.B., Libertus, M.E., & Brannon, E.M. (2013a). Infants show ratio dependent number discrimination regardless of set size. *Infancy*, 18(6): 1–15.

Starr, A.B., Libertus, M.E., & Brannon, E.M, (2013b) *Proceedings of the National Academy of Sciences*, 110(45): 18116–18120.

Uller, C., Carey, S., Huntley-Fenner, G., & Klatt, L. (1999). What representations might underlie infant numerical knowledge? *Cognitive Development* 14(1): 1–36.

vanMarle, K., & Wynn, K. (2011). Tracking and quantifying objects and non-cohesive substances. *Developmental Science* 14(3): 502–515.

Wei, W., Yuan, H., Chen, C., & Zhou, X. (2012). Cognitive correlates of performance in advanced mathematics. *The British Journal of Educational Psychology* 82(Pt 1): 157–181.

Xu, F., & Spelke, E.S. (2000). Large number discrimination in 6-month-old infants. *Cognition* 74(1): B1–B11.

CHAPTER 12

NUMERICAL AND ARITHMETIC ABILITIES IN NON-PRIMATE SPECIES

CHRISTIAN AGRILLO

INTRODUCTION

ALTHOUGH mathematical abilities are mainly cultural achievements related to language, comparative research has widely demonstrated that non-human primates possess basic numerical abilities, such as learning to use Arabic numbers, adding or subtracting small sets of items, and using ordinal information (for a recent review of these topics, see the chapter by Beran and colleagues). More recently, an increasing number of studies suggest that other mammals, birds, amphibians, fish, and even some invertebrates display at least a rudimental capacity to discriminate among quantities.

There are many real-life situations in which having numerical abilities can be useful, and there is no reason to believe that selective pressures in favour of processing quantitative information have acted on primates only. For instance, the ability to decide whether to attack or retreat, based on the assessment of the number of individuals, seems to be a crucial cognitive skill, widespread among species. This is particularly clear in the spontaneous behaviour of lionesses at the Serengeti National Park. In a study by McComb, Packer, and Pusey (1994), recordings of single female lions roaring and groups of three females roaring together were played back to lionesses to simulate the presence of unfamiliar intruders. Interestingly, defending females were less likely to approach playbacks of three intruders than playbacks of a single intruder; furthermore, when the observed female approached three intruders, she did so more cautiously. In other words, lionesses decide to approach intruders aggressively only if they outnumber the latter, which shows their ability to take into account quantitative information. This strongly resembles what has been observed in primates (Wilson, Hauser, & Wrangham, 2001), with chimpanzees being more willing to enter contests if they outnumber potential opponents. More recently, using a playback technique similar to that of McComb et al. (1994), numerical abilities have been reported in hyenas at the Masai Mara National Reserve. Hyenas in larger groups (hence, with better numerical odds) appeared to be more proactive by approaching more often when hearing the calls of intruding individuals (Benson-Amram, Heinen, Dryer, & Holekamp, 2011). Numerical assessment in social contexts is not a vertebrates' prerogative, however, as Tanner (2006) reported that even ants are somehow capable of assessing their group's numerical size before a competitive encounter and are more likely to attack when they perceive themselves as being part of a larger group.

Numerical information could be useful in nature in other ecological contexts. Lyon (2003) reported an example of the ecological and evolutionary context of numerical abilities as a strategy to reduce the costs of avian conspecific brood parasitism in American coots. It is fundamental that these birds raise their own offspring to ensure that they can feed them all. In this sense, egg recognition and rejection is a particularly useful defence and, when making their clutch-size decisions, American coots seem to be able to avoid brood parasitism by estimating the number of eggs perceived as their own and ignoring parasitic eggs, thus increasing their own fitness. Animals are also likely to rely on quantitative information to guide their foraging decisions, as it is expected to be more advantageous to select and defend larger amounts of food when competitors are in the vicinity. Social animals can also take advantage by joining the larger group when being chased by predators. For instance, numerous studies have shown that many fish species prefer to join the larger shoal when exploring an unfamiliar and potentially dangerous environment, to reduce the probability of being spotted by predators, thus showing their capacity to discriminate the larger quantity of conspecifics (Hager & Helfman, 1991; Buckingham, Wong, & Rosenthal, 2007).

Most of the above-mentioned studies have been conducted in nature, an important condition to consider for grasping the ecological relevance of numerical information. However, it is difficult to control non-numerical cues in the field, and controlled laboratory studies (Figure 12.1) are required to investigate the cognitive mechanisms underlying the numerical abilities of animals.

As in many other research fields, psychologists have used rats and pigeons as model species. For instance, Fernandes and Church (1982) trained rats to press a lever on the right after two sounds and to press a lever on the left after four sounds. Subjects were able to discriminate the number of sounds, even when non-numerical cues (such as total sound duration, or interval between each sound) were controlled for. However, it was argued that rats might have simply learned to respond on the lever on the right whenever they heard 'few' sounds and to respond on the left whenever they heard 'many' sounds. To verify this hypothesis, Davis and Albert (1987) set up a new study. Rats were trained under a discrimination procedure in which responding was reinforced only following the repeated presentation of three bursts of white noise; on the contrary, responding in the presence of either two or four bursts of noise was not reinforced. Subjects successfully solved the task, showing themselves able to make intermediate numerical discriminations based upon something more than a simple many-versus-few dichotomy.

As concerns pigeons, Alsop and Honig (1991) trained four subjects through a matching-to-sample procedure. The sample stimuli were made up of sequences of red and blue flashes of light. If more blue flashes appeared in a sequence, then one side key was correct and provided reinforcement, while the other side key produced a short blackout. The opposite contingencies occurred if there were more red flashes. Pigeons were found to be able to make relative numerical judgments. By using a similar procedure, Roberts (2010) asked whether pigeons show distance and magnitude effects[1] when two groups of visual stimuli are sequentially presented, finding evidence of the presence of both cognitive effects in pigeons. In this sense, the birds' performance parallels that observed in experiments involving non-human and human

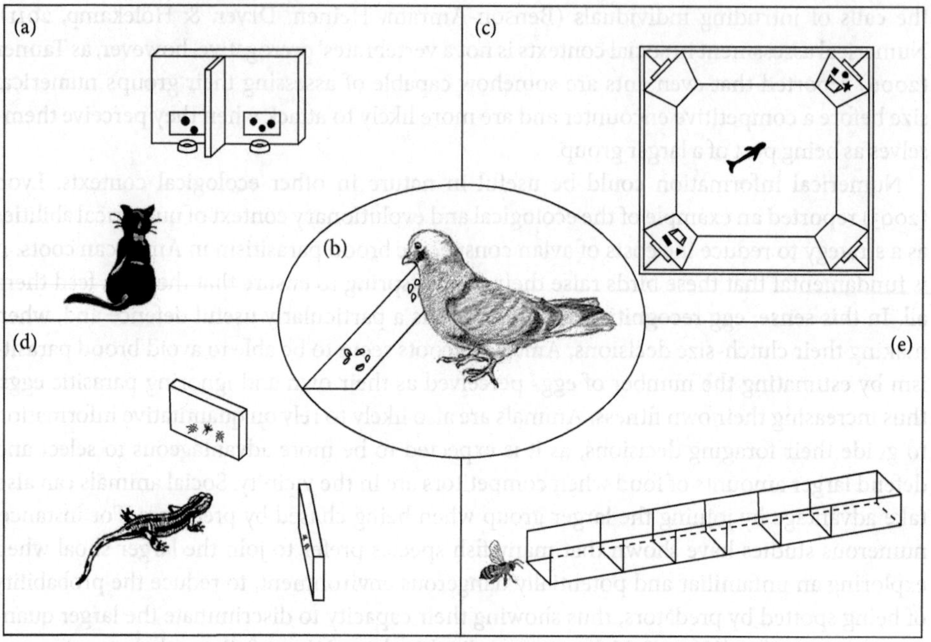

FIGURE 12.1 Laboratory studies have shown the existence of quantity abilities in mammals, birds, basal vertebrates, and invertebrates. (a) Cats can be trained to discriminate between two quantities in order to get a food reward (b) pigeons can spontaneously select the larger quantity of food items (c) fish can be trained to discriminate between two groups of figures to rejoin their shoal mates (d) salamanders spontaneously select the larger group of potential prey and (e) bees are able to enumerate the number of landmarks on the floor to reach a food source . All of this evidence encourages researchers to consider relative numerical judgments as being based on low-level mechanisms, just as was recently proposed in regard to humans.

(Data from *Journal of Ethology*, 27(2), Paola Etel Pisa and Christian Agrillo, Quantity discrimination in felines: a preliminary investigation of the domestic cat (Felis silvestris catus), pp. 289–293, DOI: 10.1007/s10164-008-0121-0, 2009);

(Data from J. Emmerton, *Birds' judgments of number and quantity*, in 'Avian Visual Cognition', R.G. Cook (ed), Boston, USA, Comparative Cognition Press, 2001);

(Data from *Cognition*, 119(2), Christian Agrillo, Laura Piffer, and Angelo Bisazza, Number versus continuous quantity in numerosity judgments by fish, pp. 281–7, doi: 10.1016/j.cognition.2010.10.022, 2011);

(Data from *Animal Cognition*, 6(2), Claudia Uller, Robert Jaeger, Gena Guidry, and Carolyn Martin, Salamanders (Plethodon cinereus) go for more: rudiments of number in an amphibian, pp. 105–112, DOI: 10.1007/s10071-003-0167-x, 2003);

(Data from *Animal Cognition*, 11(4), Marie Dacke and Mandyam V. Srinivasan, Evidence for counting in insects, pp. 683–689, DOI: 10.1007/s10071-008-0159-y, 2008)

primates. Having said that, probably the most famous 'counting' bird in literature was not a pigeon, but an African Grey parrot called 'Alex.' This parrot was the subject of numerous cognitive studies for approximately 30 years, including those involving numerical competence. Taken together, these studies demonstrated that he was able to enumerate sets of up to six items, even using vocal English labels (Pepperberg, 2006); Alex was also able to use both cardinal and ordinal information (Pepperberg, 2012). The astonishing abilities of this parrot have received a lot of media coverage because they have undoubtedly opened a window into the cognitive systems of a bird's brain. We can only speculate about his outstanding abilities; it is possible, however, that Alex's training on human number labels may have enabled him to use representational abilities that would otherwise be inaccessible to other birds.

With few exceptions, all laboratory studies reported in the literature cover one of these main issues:

1) Primates can discriminate between two quantities without attending numerical information: several non-numerical continuous quantities co-vary with numbers, and human and non-human primates have sometimes been found to use this information instead of numbers (Gebuis & Reynvoet, 2012; Stevens, Wood, & Hauser, 2007). However, recent evidence suggests that number processing is not more cognitively demanding for primates than continuous quantities (Cantlon & Brannon, 2007). It is unclear whether other mammals, birds, and basal vertebrates use numerical information as a 'last resort strategy' or, like primates, can make spontaneous use of numerical information without showing more effort than that required to process continuous quantities.
2) There is compelling evidence that primates display an approximate number system (Beran, 2004; Cantlon & Brannon, 2006), while there is an open debate as to whether human and non-human primates also possess a precise number system for a small number of items, usually ≤4 items (Hauser, Carey, & Hauser, 2000; Tomonaga & Matsuzawa, 2002). This debate has been recently enlarged to encompass non-primate species as well. The question is: Do mammals, birds, and basal vertebrates display a subitizing-like process or only one numerical system for the whole numerical range?
3) With the exception of human infants (Xu & Arriaga, 2007; Xu & Spelke, 2000), the ontogeny of non-verbal numerical abilities has been rarely investigated in primates. Primates are alcitrial species, incapable of moving independently when born, and fed by the parents for a long while. The use of precocial species (such as domestic chicken and fish) as model species can permit us to identify the development trajectories of non-verbal numerical abilities using the same experimental paradigms for newborn, juveniles, and adults, thus allowing for a fine comparison of numerical abilities across the ages. The questions are as follows: Is there an inborn number sense in animals? Does numerical discrimination increase in precision across development? Are there differences in the developmental trajectories of small and large number discrimination?

Numbers versus Continuous Quantities

As we have all experienced in everyday life when looking at fruit baskets or a group of beans, we can discriminate between two quantities without necessarily counting the number of objects. Numerosity covaries with other physical attributes (i.e. cumulative surface area, brightness, density, or the overall space occupied by the sets) and organisms can use the relative magnitude of continuous variables to estimate which group is larger/smaller. Discriminations based on number or continuous extent often yield comparable results; therefore, carefully controlled experiments are necessary to understand whether an animal is really using numerical information. To this end, in this chapter I refer to 'quantity discrimination' as the general condition in which animals proved able to discriminate between groups differing in numerosity, but the exact mechanism is unclear (or no evidence of a use of numbers was reported). 'Numerical discrimination,' on the contrary, is adopted to refer to the specific use of discrete numerical information (when continuous quantities are controlled for).

For instance, Uller, Jaeger, Guidry, & Martin (2003) set up the very first study to explore quantity abilities in basal vertebrates. In their work, a group of salamanders were given a choice between tubes containing fruit flies (potential prey) differing in numerosity (2 versus 3); they found that the subjects spontaneously selected the larger group. However, as outlined by the authors themselves, the nature of this ability was unclear. It is possible that the salamanders were selecting the larger group on the basis of cumulative surface area, density, or quantity of movement. In addition, the group containing more flies had a higher probability of having at least one insect active, thus increasing the probability of it being detected by the amphibians' eyes. In this sense, we can only conclude that salamanders spontaneously discriminate between quantities. This also highlights one of the main problems of studying numerical abilities using spontaneous choice tests—namely, the control of continuous quantities with biologically relevant stimuli (such as living organisms).

Only recently, an attempt to determine the cognitive mechanisms underlying amphibians' quantity discrimination has been made. Using a two-alternative choice task, salamanders were required to discriminate between 8 and 16 potential prey. Stimuli were live crickets, videos of live crickets, or images animated by a computer program. Salamanders were able to discriminate the larger group, but the performance dropped to chance level when stimuli were controlled for the total movement of the stimuli, which shows that this non-numerical cue was a dominant feature for their quantity discrimination (Krusche, Uller, & Ursula, 2010).

Several authors have tried to understand whether animals preferentially attend numerical or continuous information. In a seminal paper, Davis and Perusse (1988) argued that numerical discrimination may be difficult for animals, and that they rely on number only as a 'last resort' strategy when no alternative solution is available to accomplish the task. The traditional explanation for the 'last resort' strategy is that number would require more effort to process compared with continuous quantities; in this sense, animal species would preferentially or exclusively exploit continuous quantities when confronted with two magnitudes. Numerical information would be used eventually only after extensive training.

Some studies have supported this view, reporting a spontaneous (or exclusive) use of continuous quantities. In a recent study by Pisa and Agrillo (2009), four cats were trained to discriminate between two groups of dots differing in numerosity (2 versus 3) in order to get a food reward. There were two identical bowls and only the one associated with the reinforced quantity presented commercial wet cat food. During the initial training, stimuli were not controlled for continuous quantities, and both numbers and continuous quantities could be used to select the larger/smaller group. After 100 trials, the cats easily learnt to associate food and stimuli, preferentially selecting the bowl placed below the reinforced quantity. However, in the subsequent test phase, stimuli were controlled for cumulative surface area by enlarging the size of the dots in the smaller groups and reducing those included in the large one. The cats' performance dropped to chance level, suggesting that they can learn to discriminate between quantities using spontaneously non-numerical continuous quantities instead of numbers. Unfortunately, there is a lack of further studies on felines to help us understand whether or not they can also process discrete numerical information.

Such an issue was previously investigated in aquatic mammals by Kilian, Yaman, Fersen, & Güntürkün (2003). A bottlenose dolphin was trained to discriminate two stimuli differing in numerosity (2 versus 5 objects submerged beneath the water's surface) in order to obtain a food reward. The subject successfully learned to select the larger group, also proving able to generalize the response in presence of novel objects. However, when the exact mechanism underlying quantity discrimination was investigated, it was demonstrated that some non-numerical cues, such as element configuration and overall brightness, primarily determined the dolphin's choice. Only after all continuous quantities were controlled for was the subject able to discriminate the two numerosities, which suggests that aquatic mammals attend to number as a 'last resort' only if they cannot use other attributes of the stimuli.

The 'last resort' hypothesis was recently challenged by studies on primates showing that both monkeys and infants can automatically process numerical information. Cantlon and Brannon (2007) have demonstrated that macaques spontaneously encode numerical information, despite the fact that continuous quantities are also available. When number and continuous quantities were contrasted, macaques were more likely to use number if the numerical ratio was favourable. Similarly, Cordes and Brannon (2008a) found that infants do not use numerical information as a last resort when distinguishing between groups of dots differing in continuous quantities, and there is even evidence that it may be harder for 8- or 9-year-old children to compute continuous quantities than to compute numerosity when the objects differ in size (Iuculano, Tang, Hall, & Butterworth, 2008).

The capacity of primates to spontaneously process numerical information initially brought researchers to believe in the existence of a sharp discontinuity in cognitive skills between primates and other animal species. However, a spontaneous use of numerical information was recently observed in non-primate species as well. For instance, West and Young (2002) investigated numerical competence in dogs without the need for any training. The authors adapted a well-known procedure used with primates (Wynn, 1992; Hauser, MacNeilage, & Ware, 1996)—what is known as 'preferential look time'—which works on the principle that subjects should look longer at an event that appears unexpected. Two food treats were sequentially hidden in front of dogs behind an opaque barrier (1 + 1); the screen was then lowered to allow subjects to see three possible outputs: two treats (expected calculation), or one or three treats (unexpected calculation). The dogs looked significantly longer at the unexpected calculations, suggesting that a spontaneous use of numbers characterizes this species (Figure 12.2). In particular, the dogs are likely to have mentally added the presented discrete information (1 item + 1 item), even though no firm conclusion can be drawn on this latter point.

A spontaneous use of numbers has been also demonstrated in birds (Hunt, Low, & Burns, 2008). Different numbers of mealworms were presented sequentially to New Zealand robins in a pair of artificial cache sites, and were then obscured from view. In this way, the subjects could not see each stimulus group as a whole, preventing the possibility of attending continuous quantities. Even in this condition, subjects chose the site containing more prey. Interestingly, in one of these experiments some prey items were hidden behind a trapdoor after being shown to the birds (for instance, 1+1+1 = 3(− 1)); if the subjects were choosing cache sites on the basis of number alone, then they should have searched for longer when they were allowed to retrieve only a fraction of the prey they were initially shown. Actually, the birds searched for longer when they expected to retrieve more prey. The results are particularly interesting since the study was conducted in nature with naïve subjects, thus emphasizing the salience of numerical information in birds' food choices.

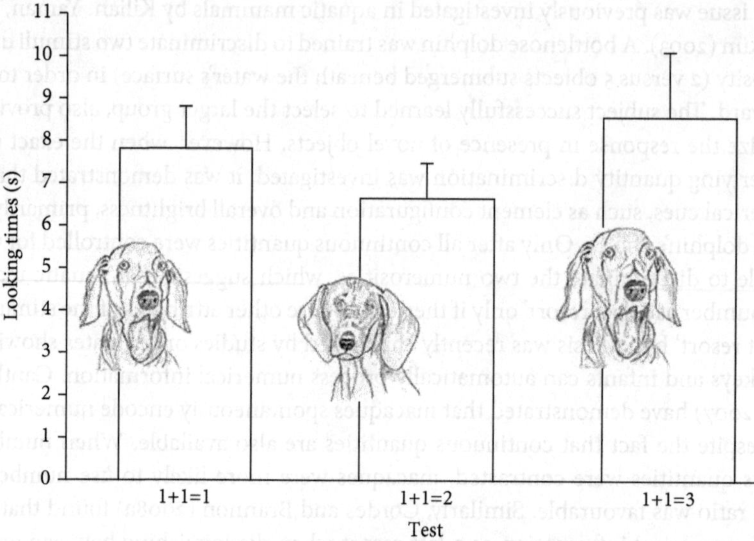

FIGURE 12.2 Preferential looking time was recorded to assess whether dogs can spontaneously add numerical information. Two food treats were hidden behind an opaque barrier. When the screen was lowered, three outputs were presented: 1, 2, or 3 treats. Subjects looked significantly longer at both unexpected calculations, suggesting a use of numerical information in the absence of training. Bars in this and in the following graphs represent the standard error.

(Data from *Animal Cognition*, 5(3), Rebecca E. West and Robert J. Young, Do domestic dogs show any evidence of being able to count?, pp. 183–186, DOI: 10.1007/s10071-002-0140-0, 2002)

To directly test whether numerical competence is more cognitively demanding than continuous quantities, a sample of mosquitofish was trained using an operant conditioning procedure (Agrillo, Piffer, & Bisazza, 2011). Fish were inserted into an unfamiliar tank and the only way to reach other conspecifics was by selecting one of two identical tunnels. Above each tunnel stimuli were presented, groups of black geometrical figures on a white background. Only the tunnel associated with the reinforced stimuli permitted fish to leave and re-join other conspecifics. In particular, subjects were required to discriminate two items from three in three different conditions. In one condition, continuous quantities were controlled, while numerical information was available (they could use only numerical information); in another, the number was kept constant (1 versus 1) and information relating to continuous variables was available (only continuous quantities could be used); in the third condition, stimuli differed for both numbers and continuous quantities (they could use both sets of information). Fish learn to discriminate more quickly when both numbers and continuous quantities were available compared with when they could use continuous quantities or numbers only; interestingly, no difference in the learning rate between the two latter conditions was found. Hence, processing numbers does not seem to impose a greater cognitive load than processing continuous quantities, not only in primates, but even in a fish brain, which is less than a thousandth the size of a primate brain.

This might be surprising at first, but it is in line with an increasing number of studies suggesting that numbers per se are not as cognitively demanding as previously thought.

First, organisms with a nervous system much simpler than that of vertebrates are able to use numerical information after non-numerical variables are controlled for (for a discussion on brain size and cognitive abilities, see Chittka & Niven, 2009). Nelson and Jackson (2012), for instance, studied specialized predatory strategy in spiders (*Portia africana*). In this species, when practicing communal predation, two or more juveniles spiders typically settle by the same oecobiid nest (the prey is another spider), in order to share the prey as soon as one of the two individuals captures it. The authors observed nest preference when different numbers of conspecifics were present: the results showed that settling decisions are based on the specific number of already settled conspecifics, with spiders showing a preference for settling when the number is 1 (instead of 0, 2 or 3 conspecifics). Control tests for continuous quantities apparently suggest that numerical information is a salient cue for this species.

Numerical abilities have been advanced for other invertebrates too. Ants reportedly pass numerical information to other ants when transferring information about which branch of a maze contains food (Reznikova & Ryabko, 2011). Bees were found to reach a food reward by enumerating the number of landmarks (yellow strips of paper decorating the walls and floor of the experimental apparatus) encountered sequentially during flight. The bees were also able to identify the food reward when the distance to it was controlled for, thereby showing their use of numerical information instead of the absolute distance from the starting point (Dacke & Srinivasan, 2008). In a subsequent study, bees made use of numbers in a sequential matching to sample task in which continuous quantities were controlled for (Gross, Pahl, Si, Zhu, Tautz, & Zhang, 2009).

Data from artificial neural networks also support the idea according to which numerical discrimination would not be as cognitively demanding as previously thought. Hope, Stoianov, & Zorzi (2010) found that fewer than 25 units may be sufficient for a system to represent numerical information with a performance comparable to that observed in fish (Agrillo et al., 2011) and salamanders (Uller et al., 2003). Similar conclusions have recently been drawn in another neural network study by Stoianov and Zorzi (2012), and a psychophysical study suggests that adult humans' processing of numbers might, indeed, be based on low level mechanisms (Burr & Ross, 2008). After being exposed for 30 seconds to an array of dots in one portion of their visual field, participants showed a strong after-effect, underestimating the number of dots in arrays of dots subsequently presented in the same region of the retina. The presence of a retinotopic adaptation would indicate that the visual system is able to extract the numerical information, indicating that numerosity might simply be another primary visual property of a stimulus, such as colour, brightness, and contrast.

To summarize the issue, the picture of non-primate species is not clear. While some authors provided evidence of a preferential (or exclusive) use of continuous quantities, others have reported a spontaneous use of numerical information or a lack of difference in the learning rate between numbers and continuous quantities. To explain this inconsistency, three different hypotheses may be formulated:

1) *Inter-species variation:* it is possible that the relative salience of numerical information over continuous quantities varies among species. For instance, for the visual system of an amphibian (highly sensitive to detect any movement in the visual field), the quantity of movement can be the main cue to determine the choice. Indeed, salamanders (Krusche et al., 2010) proved to be sensitive to the quantity of movement

when choosing between groups of crickets differing in numerosity. On the contrary, this continuous quantity may be less relevant in other species. An inter-specific variation in the mechanism used to discriminate between quantities has been also suggested by Emmerton and Renner (2009). The cognitive systems of the species might be shaped by evolution on the basis of different pressure selections, with some species preferentially attending a range of continuous quantities and others not doing so.

2) *Salience of numerical information is stimulus-dependent:* we often assume that if subjects discriminate when number and continuous quantities are both available, and they do not subsequently select the larger/smaller group when continuous information is controlled for, they cannot process numerical information. This reasoning, although sound in theory, does not take into account the possibility that the relative salience of numerical information can vary as a function of the type of stimuli. For instance, in a study by Feigenson, Carey, and Hauser (2002), after seeing crackers placed sequentially into two containers, 12-month-old infants were allowed to crawl and choose one of the containers. The infants successfully discriminated between 1 versus 2 and 2 versus 3. However, when the crackers were of different sizes, the choice was determined by total surface area or total volume. The authors concluded that infants discriminate small quantities by using continuous quantities. However, the reward was food and it is not unexpected that natural selection had shaped the quantification systems in order to maximize the amount of food (calories) retrieved, rather than the number of pieces. We should expect the choice of the larger volume of crackers regardless of whether infants can or cannot discriminate using numerical information. If total area and volume are the most salient information in a food choice task, numerical information may be the most salient information in other contexts. For instance, it is known that, when chased by predators, fish tend to join the larger shoal. In this circumstance, for a dilution effect, the number of conspecifics (not their overall brightness or area) is the main information that may enhance the probability of survival. Similarly, when searching for potential mates, a male may gain advantage by joining the larger number of females (and not the group of females encompassing the larger space). The differences reported in the literature may be partially due to these factors.

3) *Salience of numerical information is task-dependent:* experiments on infants using a habituation paradigm suggest that they can attend to number and continuous quantities simultaneously. The relative salience of these two dimensions would depend on the type of task, with infants preferentially attending to number over continuous quantities when large sets of objects were presented (reviewed in Cordes & Brannon, 2008b). It is possible that the relative salience of numerical information in animal species varies depending on the experimental conditions (numerical range, ratio, type of paradigm, etc.), thus explaining part of the inconsistency reported in the literature. Unfortunately, a fine comparison across the species is difficult, since different studies use different stimuli, numerical contrasts, and procedures. If we report a spontaneous use of numbers in birds by using the item-by-item procedure, and then a preferential use of continuous quantities in a dolphin after an extensive procedure, can we finally say that the latter species undoubtedly differs from the former in its spontaneous capacity to process numerical information? As a partial confirmation, it was shown recently that the performance in a quantity discrimination task

in Goldbelly topminnows is influenced by the type of procedure, with fish able to discriminate 3 versus 2 only in one of two different procedures (Agrillo & Dadda, 2007). This also highlights the importance of replication in comparative psychology. In peer-reviewed journals, scientists are strongly encouraged to make original studies with novel conceptual approaches, while attention should also be devoted to replicating previous findings, either by using the same methodology or by using different methods. With the former, we might eliminate the possibility that the previous results were an accident; with the latter, we might determine whether the results were an artefact of the previous methodology (for a recent discussion on this subject, see Agrillo & Miletto Petrazzini, 2012).

Do Non-Primate Species Subitize?

Non-verbal numerical abilities in primates (human adults, infants, and non-human primates) appear to be rooted in two distinct mechanisms that operate over different parts of the number range. One is a system for representing approximate numerosities, and is usually called the analogue magnitude system. This mechanism has virtually no upper limit, but is subject to a ratio limit in accordance with Weber's Law, which states that the just-noticeable difference between two stimuli is proportional to the magnitude of the stimuli. As a consequence, the capacity to discriminate between two quantities becomes increasingly accurate as the ratio between them increases. For instance, using the habituation-dishabituation paradigm, Xu and Spelke (2000) demonstrated that 6-month-old infants can distinguish between 8 and 16 dots (1:2 ratio), while they cannot distinguish between 8 and 12 dots (2:3). Similarly, Flombaum, Junge, and Hauser (2005) found that rhesus monkeys successfully discriminated between four and eight lemons (1:2), but not between four and six (2:3).

The second system is referred to as the object-file system and depends on a mechanism for representing and tracking small numbers of individual objects. Each item would be represented by a symbol (a file) and numerical equivalence would be established indirectly by evaluating a one-to-one correspondence between files and objects (Uller, Huntley-Fenner, Carey, & Klatt, 1999). This mechanism would be precise, but would allow for the parallel representation of a small number of objects (usually 3–4 elements). It is assumed to support 'subitizing,' the accurate reporting of small sets of items without serial counting. For instance, it has been shown that 12-month-old infants are able to select the larger group of crackers when the paired numbers are 1 versus 2 and 2 versus 3, but they fail with 3 versus 4 and 3 versus 6 (Feigenson et al., 2002). Similarly, rhesus monkeys, confronted with two quantities of apple slices, successfully choose the greater number with comparison of 1 versus 2, 2 versus 3, and 3 versus 4, but fail with 4 versus 5 and 4 versus 6 (Hauser et al., 2000). In chimpanzees, the error rate and reaction time are constant in the range of 1–4, while they tend to increase monotonically for larger numbers (Tomonaga & Matsuzawa, 2002).

The two mechanisms (object file and analogue magnitude) appear to differ in many respects, including speed, accuracy, and cognitive load (Kahneman, Treisman, & Gibbs, 1992); however, the lack of a ratio effect is the main signature to experimentally differentiate the object-file from the analogue magnitude system (Feigenson, Dehaene, & Spelke,

2004). In other words, we are quicker and more accurate when discriminating 6 versus 24 (1:4 ratio) than 6 versus 9 (2:3), while we need approximately the same amount of time (with similar accuracy) to distinguish 1 versus 4 and 2 versus 3.

However, not all empirical studies support the existence of separate cognitive mechanisms in primates for small and large numbers. vanMarle and Wynn (2008) found that infants' discrimination of auditory events was ratio-dependent even for small numbers, suggesting that infants can use analogue magnitudes for both small and large quantities in the auditory domain. Another study comparing rhesus monkeys and adult humans showed that accuracy and reaction time were similarly affected by numerical ratios in the large and small number range (Cantlon & Brannon, 2006). Other studies do not support the two-system hypothesis (Beran, Taglialatela, Flemming, James, & Washburn, 2006; Cordes, Gelman, Gallistel, & Whalen, 2001); they report an influence of ratio—the typical signature of the analogue magnitude system—also in the small number range.

The debate has enlarged in the last 5 years to encompass non-primate species. Indeed, the study of other mammals, birds, and basal vertebrates may help us to add further evidence supporting (or against) the two-system hypothesis, and to better understand the evolutionary roots of the cognitive mechanisms supporting small and large number discrimination. One possibility is that the supposed object-file system is a more recent evolutionary development that affords more precise comparisons of smaller quantities; therefore, ratio insensitivity in the small number range should not be seen in species more distantly related to primates. Another possibility is that the object-file system is more evolutionarily pervasive and not restricted to primates.

Evidence supporting an analogue magnitude system for the whole numerical range has been collected in mammals, birds, and in vertebrates as well. Ward and Smuts (2007) studied dogs' ability to select the larger quantity of food items in the range 1–5 (1 versus 5, 1 versus 4, 1 versus 3, 1 versus 2, 2 versus 4, 2 versus 5, 3 versus 5, 2 versus 3, 3 versus 4), and found that their accuracy was a function of ratio between quantities. The authors concluded that dogs rely on an approximate mechanism of quantity representation that fits the analogue magnitude system. In addition, if the dogs had been using an object file mechanism, they would have successfully discriminated from chance level 1 versus 2 and 2 versus 3, a result not found in that study. Other mammals were investigated with regard to this topic: by using an item-by-item presentation, Perdue, Talbot, Stone, & Beran (2012) observed the spontaneous ability of African elephants to discriminate between numbers in the 1–10 range. The performance of elephants was ratio-dependent, with accuracy decreasing when increasing the numerical ratio between the small and the large numbers, supporting the idea of an analogue magnitude system in the whole numerical range.

Al Aïn, Giret, Grand, Kreutzer, and Bovet (2009) approached the same question by testing parrots' ability to discriminate numerical and continuous information. Birds were trained to select the larger number of food items (all the numerical discrimination in the range 1–5) or the larger volume of a food substance (0.1–1.0 mL); numerical ratio was the best predictor of the performance for both number and continuous quantities. The authors concluded that parrots use an analogue magnitude, rather than an object-file mechanism. The typical signature of the analogue magnitude system in the small quantity range has also been reported in beetles. Using a spontaneous two-choice procedure, male mealworm beetles were presented substrates bearing odours from different numbers of females (< 5) in increasing numerosity ratios (1/4, 1/3, and 1/2). The subjects proved able to discriminate sources of odours reflecting 1 versus 4 and 1 versus 3 females, while no choice was observed in 1 versus 2 or 2 versus 4. Numerical ratio was again the best predictor of performance (Carazo, Font, Forteza-Behrendt, & Desfilis, 2009).

Nonetheless, we also see an increasing number of studies supporting the existence of a precise mechanism limited up to 3–4 units and insensitive to the numerical ratio. New Zealand robins were found to successfully discriminate 1 versus 2, 2 versus 3, and 3 versus 4, showing a very similar performance in the three numerical contrasts (Hunt et al., 2008). On the contrary, larger numerosities (>4) could be discriminated when the numerical ratio was 1:2 (4 versus 8). This is in contrast with the existence of one general system based on the ratio; the capacity to discriminate small quantities seems to be more precise, with very few differences in the ability to discriminate 2 versus 3 (0.67 ratio) and 3 versus 4. In the large number range, birds can only discriminate when the numerical distance between the groups is increased. This indirectly suggests a different mechanism within and beyond the subitizing range.

Additional evidence comes from a semi-naturalistic study by Bonanni, Natoli, Cafazzo, and Valsecchi (2011) on feral dogs. As outlined at the beginning of this chapter, the decision whether to attack/retreat in conflicts between social groups is based on the assessment of the quantity of individuals in one's own and the opposing group. The authors observed the attack/retreat decisions in free-ranging dogs, suggesting that the dogs spontaneously assessed large numbers of conspecifics as noisy magnitudes (with their capacity to estimate between numerosities strongly affected by the ratio). In contrast, in the small number range, dogs approached aggressively with the same probability when they outnumbered opponents by a 1:2, 2:3, or 3:4 ratio, which suggests that dogs discriminate two small quantities using an object file mechanism, or at least a mechanism poorly affected by the ratio.

Basal vertebrates were also involved in the debate. Uller et al. (2003) found that salamanders can discriminate 1 versus 2 (0.50) and 2 versus 3 (0.67) fruit flies, while their performance dropped to chance level for 3 versus 4. Interestingly, they could not discriminate 4 versus 6, even though the same ratio (0.67) was successfully discriminated in the small number range. If numerical ratio is the best predictor of animals' performance, we should expect the same discrimination capability in both small and large numbers. The subsequent study by Krusche et al. (2010) shed new light on the cognitive mechanism underlying salamanders' quantity discrimination, showing that they can also discriminate larger numerosities provided that numerical ratio is 1:2 (8 versus 16). Similar results have been reported in other amphibians (Stancher, Regolin, & Vallortigara, 2007), with bombinas able to discriminate 1 versus 2, 2 versus 3, and 4 versus 8 larvae.

Recently, the problem has been tackled using fish as an experimental model. There is substantial evidence that individual social fish that happen to be in an unknown environment tend to join other conspecifics and, if choosing between two shoals, they exhibit a preference for the larger one (Hager & Helfman, 1991). Such behaviour is thought to be an anti-predatory strategy allowing them to reduce the chance of being spotted by predators. As a consequence, some research groups took advantage of this spontaneous tendency to go to the larger shoal to study the limits of fish quantity discrimination (Binoy and Thomas, 2004; Buckingham et al., 2007; Hager & Helfman, 1991). In these studies, subjects were singly inserted into the middle of the experimental tank and two groups of conspecifics differing in numerosities were visible at the two bottoms. The experimental tank represented an unfamiliar environment without any shelter, hence inducing subjects to join the larger shoal. As a measure of accuracy, the researchers recorded the proportion of time spent near the larger shoal.

Through this procedure, it has been shown that mosquitofish discriminate between shoals that differ in numerosity when the paired numbers are 1 versus 2, 2 versus 3, and 3 versus 4, but they succeed with only up to a 1:2 numerical ratio (4 versus 8 or 8 versus 16) when they have to discriminate between large numbers (Agrillo, Dadda, Serena, & Bisazza, 2008). Similarly, two different studies (Gòmez-Laplaza & Gerlai, 2011a,b) showed that angelfish can discriminate 1 versus 2 (0.50) and 2 versus 3 (0.67), while their capacity to discriminate large numbers is limited up to a ratio of 0.56 (5 versus 9). In other words, both mosquitofish and angelfish are more accurate in the so-called subitizing range, and show a different ratio sensitivity in the small and large number ranges. Interestingly, the set size limit in small number discrimination seems to be different between the two species (4 for mosquitofish, 3 for angelfish). These differences raise the question of why different results are sometimes observed across species. One possibility is that the differences observed between species may have been shaped by evolution on the basis on different pressure selections (i.e., high/low predator density, different shoal composition, etc.). In this sense, some species might use object-tracking mechanisms more than others, or have a more precise system of this kind. Another possibility is that different results may depend on the different methodologies adopted among the studies. A previous work has indeed shown that the limit of precise number discrimination in topminnows varies depending on the type of procedure (Agrillo & Dadda, 2007). Interestingly, a similar issue about task dependence has been also raised in numerical representation of humans (Cohen Kadosh & Walsh, 2009).

To shed light on this topic, the performance of fish was directly compared to that of humans (Agrillo, Piffer, Bisazza, & Butterworth, 2012). Five numerical ratios were presented (0.25, 0.33, 0.50, 0.67, 0.75) for small (1 versus 4, 1 versus 3, 1 versus 2, 2 versus 3, and 3 versus 4) and large (6 versus 24, 6 versus 18, 6 versus 12, 6 versus 9, 6 versus 8) numerical contrasts. For the human participants, a procedure commonly used to measure non-verbal numerical abilities in adults was adopted—namely, a computerized numerical judgement with sequential presentation of the stimuli. Undergraduates were required to estimate the larger of two groups of dots, while being prevented from using verbal counting. The fish were tested by observing their spontaneous tendency to join the larger group of conspecifics when inserted into an unfamiliar environment. Interestingly, the performance of the fish aligns with that of humans (Figure 12.3); in both species, the ability to discriminate between large numbers (>4) was approximate and strongly dependent on the ratio between the numerosities. In contrast, in both fish and humans, discrimination in the small number range was not dependent on ratio and discriminating 3 from 4 was as easy as discriminating 1 from 4. To date, this represents the very first study comparing human and basal vertebrates in proto-numerical skills. These results open the possibility of common mechanisms between primates and basal vertebrates, suggesting that the evolutionary emergence of numerical abilities (both object file and analogue magnitude systems) may be much more ancient than previously thought, possibly dating back to before the separation of land vertebrates from bony fish.

It seems we are coming to an impasse in the literature; some studies on mammals, birds, and invertebrates support the existence of an analogue magnitude system throughout the whole numerical range, while other works on mammals, birds, and basal vertebrates suggest the existence of a precise mechanism devoted to small number discrimination. Future

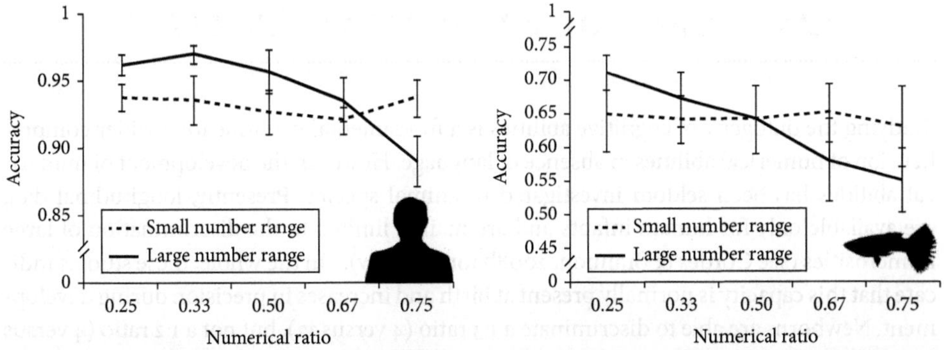

FIGURE 12.3 Agrillo et al. (2012) compared guppies and humans in small and large number discrimination. Accuracy is plotted against the numerical ratio of the contrasts for both large and small number ranges. The performance of humans and fish showed ratio sensitivity for large numbers and ratio insensitivity in the small number range, suggesting the existence of different cognitive systems devoted to small and large numbers.

Adapted from Agrillo C, Piffer L, Bisazza A, Butterworth B, Evidence for Two Numerical Systems That Are Similar in Humans and Guppies, *PLoS ONE*, 7(2), e31923, Figure 3, doi: 10.1371/journal.pone.0031923 © 2012, The Authors.

studies are needed to clarify the issue. In the meantime, two different hypotheses can be advanced to explain these differences:

1) *Inter-species variation:* some species may be equipped with an object-file system, while others are not; it depends on the selective pressures involved. It would be interesting to delineate which selective pressures may have caused the emergence of an object-file mechanism.
2) *Small numbers can be discriminated through object-file and analogue magnitude systems depending on the experimental procedure:* this hypothesis has been advanced regarding primates. In particular, it has been argued that small quantities may be represented by both analogue magnitudes and object-files, and the context in which the representation is elicited determines which of the two systems is employed (Cordes & Brannon, 2008b; Hyde, 2011; vanMarle & Wynn, 2009). In other words, we can always discriminate numerosities by using analogue magnitudes, but in some circumstances we may adopt the object-file system originally devoted for tracking and store moving objects in our memory. Confirmation of this can be found in a recent study on undergraduates, which shows that the ratio effect in the small number range is determined by the type of task (Piffer et al., in preparation).

At the moment, it is difficult to find a clear answer in the animal kingdom, given that the different studies on mammals, birds, amphibians, fish, and invertebrates use different methodologies. Future studies using identical stimuli (for instance, black dots on a white background) and a similar procedure (training procedure) may help us to understand whether animal species, just like humans, display a specific mechanism for small numbers. This would increase the correspondence in numerical abilities between humans and animal species, thus supporting the intriguing idea of a potential common origin of vertebrate numerical abilities.

Ontogeny of Numerical Abilities

Studying the ontogeny of cognitive abilities is a fundamental keystone for a wider comprehension of numerical abilities in absence of language. However, the development of numerical abilities has been seldom investigated in animal species. Presently, longitudinal data are available only for human infants and are mainly limited to the discrimination of large numerosities (see Cordes & Brannon, 2008b for a review). On the whole, these studies indicate that this capacity is normally present at birth and increases in precision during development. Newborns are able to discriminate a 1:3 ratio (4 versus 12), but not a 1:2 ratio (4 versus 8; Izard, Sann, Spelke, & Streri, 2009); 6-month-old infants can discriminate numerosities with a 1:2 ratio (such as 8 versus 16), but not a 2:3 ratio (Xu & Spelke, 2000), whereas 10-month-old infants are able to discriminate numerosities with a 2:3, but not a 4:5 ratio (Xu & Arriaga, 2007). The resolution of this system continues to increase throughout childhood, with 6-year-olds being able to discriminate a 5:6 ratio and adults a ratio of 9:10 (Halberda, Mazzocco, & Feigenson, 2008). However, very different paradigms were used across the different ages, so the results cannot be easily compared. This highlights one of the problems of studying the development trajectories of numerical competence in humans—namely, the difficulty of devising experimental paradigms that are simultaneously applicable to newborns, infants, toddlers, and adults. A second important limit of the research on humans and non-human primates is that, for practical and ethical reasons, it is very difficult to manipulate experience during development, which therefore precludes the possibility of disentangling the relative contribution of maturation and experience. The recent discovery that even simple organisms, such as birds and fish, are capable of numerical abilities similar to primates has paved the way for the use of new animal models in developmental research.

Domestic chicks in particular have become a pre-eminent model, given the possibility they present to set controlled-rearing experiments on newborn subjects. Rugani, Regolin, & Vallortigara (2007) asked whether ordinal abilities are inborn or depend on maturation/experience. The term *ordinal abilities* refers to the capacity to identify an object on the basis of its position in a series of identical objects. Such a skill was previously reported in adult rats by Davis and Bradford (1986), who required subjects to enter one box of a defined ordinal number among a series of boxes. A food reward could be obtained only by entering into one of the boxes (for instance, the fourth one); the rats proved able to take into account ordinal information by entering the fourth of six boxes arrayed in the experimental apparatus. However, the research employed adult rats and was not aimed to test the role of maturation/experience. To specifically address this question, Rugani et al. (2007) trained groups of chicks to peck either at the 3rd, 4th, or 6th position in a series of 10 identically-spaced locations, sagittally aligned in front of the subjects' starting point. The chicks successfully learned the task and were able to select the correct location above chance level. The authors also controlled for non-numerical information, such as spatial information provided by the experimental apparatus or the absolute distance necessary to reach the reinforced location. The subjects were initially trained to peck at the fourth position, and were then required to generalize the learned response to a new sequence in which the distance from the starting point to each position was manipulated. For instance, in one condition, the distance among each location was doubled. In this way, had the chicks used the absolute distance from the starting point, they would have pecked at the second position of the series, a circumstance not observed here. Indeed, the chicks generalized their response in the novel apparatus by pecking at the correct serial position, even though that position was located much further away than before.

In a subsequent series of studies (Rugani, Kelly, Szelest, Regolin, & Vallortigara, 2010a), once the subjects had learned the task (pecking at the 4th location), the authors rotated the whole series of locations by 90°. In this way, the chicks could see the array from a different perspective (frontoparallel to their starting position) and were required to spontaneously choose the side from which they would start to order the items. Again, the chicks generalized the learned response to a new sequence without relying on the absolute distance from the starting point or other geometrical information. Interestingly, most of the chicks spontaneously selected the 4th position starting from the left end, and not the 4th position from the right end, even though they were totally free to start ordering in either direction (Figure 12.4). The authors suggest that a sort of disposition to map the number line from left to right would also exist in non-human species, indirectly evoking the intriguing idea of a mental number line in a bird's brain. However, as acknowledged by the authors themselves, this spontaneous preference could be simple due to a bias in the allocation of attention, a sort of 'pseudoneglect' similar to that described in humans. Based on this evidence, the capacity to take into account ordinal information seems to be present at birth, and might have been selected due to its potential adaptive value. We can only speculate about it, but it is easy to imagine ecological contexts in which displaying such a skill might be useful from the first days of life—for instance, when searching for a source of food ('after three rocks') or a shelter ('the third cave') for predator avoidance.

Simple arithmetic seems also to be processed by a newborn bird's brain. Using a paradigm previously adopted for infants, it was demonstrated that 5-day-old chicks are able to perform addition and subtraction (Rugani, Fontanari, Simoni, Regolin, & Vallortigara, 2009). Chicks were initially reared for 3 days with five identical objects that served as imprinting objects. Subsequently, the subjects were confined to a holding box behind a transparent barrier where two opaque screens were visible in front of them. The chicks could see two sets of objects—for instance, one composed by three of the five imprinting objects and the other made of the remaining two. Each set was hidden sequentially or simultaneously behind one of the two screens. Only after the disappearance of the sets were the subjects free to search for their imprinting objects within the experimental apparatus. There is evidence that chicks spontaneously prefer to join the larger number of imprinting objects (Rugani, Regolin, & Vallortigara, 2010b), probably because a larger group of them represents a sort of 'super-stimulus.'[2] Interestingly, the subjects inspected the screen occluding the larger set even when continuous quantities were controlled for, proving that they were able to keep track of single items and of which screen hid the larger number of objects. After the initial disappearance of the two sets (for instance, three objects behind the left screen and two behind the right one), the researchers transferred some of the objects visibly, one by one, from one screen to the other before releasing the chicks into the arena. The chicks again selected the group containing the larger number of objects. In another test (4 versus 1), the chicks were required to choose against the potential directional cue provided by the final visible displacement (indeed, in the 3[−1] versus 2[+1], the chicks could have simply followed the last moving objects) by presenting the following comparison: 4(−1) versus 1(+1). Even in this case, the chicks spontaneously chose the screen hiding the larger number of

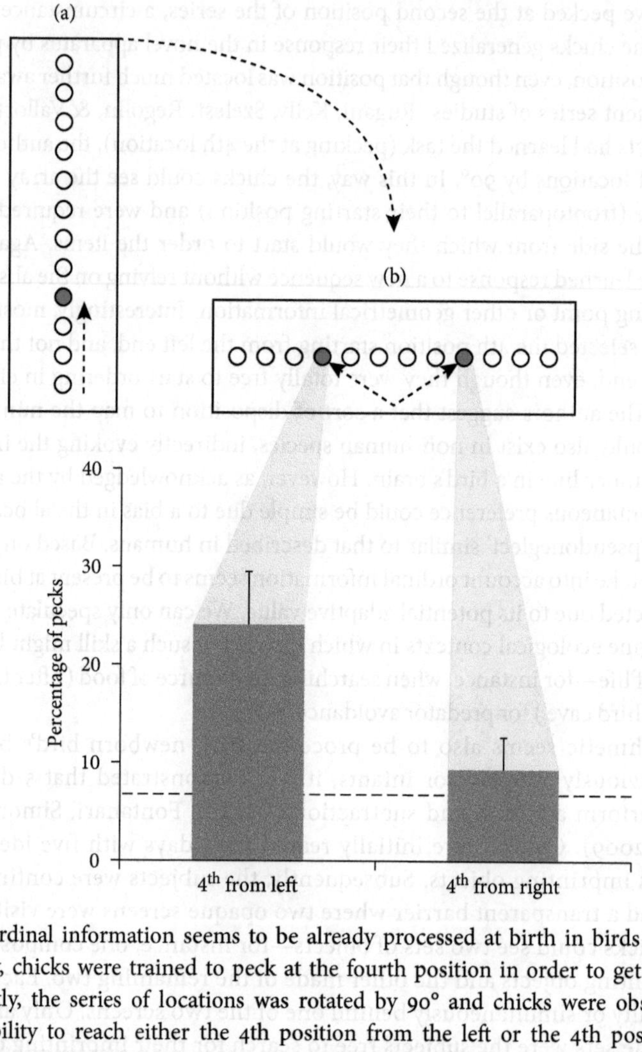

FIGURE 12.4 Ordinal information seems to be already processed at birth in birds (Rugani et al., 2010a). Initially, chicks were trained to peck at the fourth position in order to get a food reward (a). Subsequently, the series of locations was rotated by 90° and chicks were observed in their spontaneous ability to reach either the 4th position from the left or the 4th position from the right (b). Chicks significantly pecked the 4th position from the left more frequently, suggesting that the disposition to map the number line from left to right might not be exclusive to humans.

Adapted from Rosa Rugani, Debbie M. Kelly, Izabela Szelest, Lucia Regolin, and Giorgio Vallortigara, Is it only humans that count from left to right?, *Biology Letters*, 6(3), pp. 290–2, Figure 1, DOI: 10.1098/rsbl.2009.0960 © 2010 The Royal Society.

objects, thus selecting the larger group irrespective of the directional cues provided by the final displacement.

The study of developmental trajectories of numerical abilities has also been used to verify one prediction of the one-system hypothesis according to which there would be an approximate number system underlying small and large number processing (see previous section). Indeed, if a single analogue system underlies numerical discrimination, one would expect the same developmental trajectories for the discrimination of numbers in the small and large ranges. Conversely, a developmental dissociation, either in terms of onset timing or age-related change in performance, would indicate that different systems are probably at work. To this end, Bisazza, Serena, Piffer, and Agrillo (2010) have investigated the ontogeny of numerical competence in a freshwater fish, the guppy. Guppies are viviparous with a relatively short lifespan; they give birth to fully developed offspring that are completely independent and display a full social repertoire. In this sense, they represent a suitable model for studying the developmental trajectories of small–large number discrimination. The authors used a small-scale version of the experimental apparatuses adopted to study spontaneous numerical discrimination in adult fish (Agrillo et al., 2008; Gòmez-Laplaza and Gerlai, 2011a,b), and the spontaneous tendency to join the larger shoal was recorded as a measure of their capacity to discriminate between the two quantities. At birth, the capacity of guppies to discriminate between sets differing by one unit includes all numerical contrasts in the range 1–4 (1 versus 2, 2 versus 3, 3 versus 4), but not contrasts involving larger numbers, such as 4 versus 5 and 5 versus 6. In this sense, the capacity of newborn guppies to discriminate between small numerosities appears to be the same as that shown by adult individuals (Agrillo et al., 2012). At the same time, 1-day-old fish were unable to find a difference in a 1:2 (4 versus 8) and 1:3 (4 versus 12) ratio, which suggests that the capacity to discriminate large numbers is absent at birth (or highly approximate). Because adult fish can easily discriminate these ratios, the development of large quantity discrimination (4 versus 8) was investigated in a subsequent experiment by testing guppies at three different ages (1, 20, or 40 days old). The role of experience was also assessed, since one group was reared in large shoals with the possibility of seeing sub-groups of variable numerosities, while another group was reared in pairs without the possibility of seeing more than one fish at a time. The authors found that, at 40 days, fish from both treatments were able to discriminate 4 from 8 fish, while at 20 days this was only observed in fish with normal social experience. These results again point toward the existence of two separate quantificational systems in species distantly related to primates. One, the small number system, would be innate and displayed immediately at birth; the other, the large number system, would emerge later as a consequence of both maturation and environment.

Obviously, the choice of a species that is only distantly related to primates indubitably imposes some limitations on the debate surrounding the object-file versus analogue magnitude system; nonetheless, this study represents an almost unique opportunity to compare newborns, juveniles, and adults performing the same task, as well as manipulating experience and social influences during ontogeny. More generally, the literature on birds and fish shows that some proto-counting abilities appear quite early, suggesting the existence of a 'number sense,' a core system of numbers that is inborn not only in primates (Feigenson et al., 2004; Izard et al., 2009), but also in other vertebrates.

Conclusions

Numerical competence is a field in which considerable progress has been made during the last decade. Nowadays, we know that numbers are more than a cultural invention, and our numerical abilities seem to have their roots in vertebrates' cognitive evolution. In this chapter, I have reviewed most of the studies surrounding three main issues of numerical cognition in animal species. As I presume the reader has already understood, no firm conclusion is possible at this stage, but it is undeniable that non-primate species have contributed to increase the body of experimental evidence on these issues.

As outlined throughout the chapter, more comparative studies with reduced methodological variability are needed. Most of the current studies have used different stimuli, procedures, and numerical contrasts, making any comparison difficult. To understand the origin of numerical systems of vertebrates, it is necessary to minimize all of the potential biases related to different methodologies. To this purpose, Scarf, Hayne, and Colombo (2011) recently investigated pigeons' numerical ability by using the same stimuli adopted by Brannon and Terrace with rhesus monkeys (1998): the performance of birds adhered to that of primates, highlighting the importance of using the same experimental material in cross-species comparisons.

However, it must be pointed out that, even when using the same stimuli and procedures, similar performance between two species may not necessarily reflect a common origin. As with most comparative data, it is always possible that a strong similarity in cognitive abilities is the product of convergent evolution and that similar performance may reflect very dissimilar underlying mechanisms. However, if numerical abilities have evolved independently in many different taxa, it would be challenging to understand which selective constraints have shaped them in a converging way.

In addition, the research needs to be broadened by encompassing other vertebrates. Several mammal species have not yet been investigated, such as wolves, giraffes, goats, squirrels, bats, and mice. For some of organisms, we now have full knowledge of the genome (i.e. zebrafish, drosophila). Recent evidence suggests that dyscalculia might be due to a deficit in non-verbal numerical systems (Butterworth & Laurillard, 2010) and the genetic origin of this disorder has been commonly invoked. As in many other sciences, the advancement of research is crucially dependent upon the availability of suitable animal models. In this sense, the use of animals already adopted as model systems for genetic investigation will permit us to shed light on the genetic mechanisms underlying non-verbal numerical abilities.

As a final note, it is worth mentioning the curious lack of interest in reptiles; nobody has tried to investigate whether snakes, turtles, or crocodiles display at least the capacity to discriminate different amounts of food. To date, this is the only class of vertebrates for which no data has been collected. Only as we collect information on most vertebrate species we will presumably come to understand better whether our numerical abilities derive from the cognitive system of a common ancestor of mammals, birds, and basal vertebrates that lived more than 450 million years ago.

Acknowledgments

I would like to thank two anonymous referees for their useful comments. I am profoundly indebted to Maria Elena Miletto Petrazzini to have provided me the drawings included in the chapter and to Angelo Bisazza for the productive discussion of the issues here addressed. Financial support was provided by research grant 'FIRB 2013' entitled: 'Development of animal models for the study of pre-verbal numerical abilities' (prot. N: RBFR13KHFS) from '*Ministero dell'Istruzione, dell'Università e della Ricerca*' (MIUR).

Notes

1. Distance effect refers to the empirical finding according to which the ability to make relative numerosity judgments improves as the numerical distance between the two numerosity increases. Magnitude effects refers to the finding that, for equal numerical distance, relative numerosity judgments worsen as the numerical size increases (see Dehaene, Dehaene-Lambertz, & Cohen, 1998).
2. Super stimulus is commonly defined by ethologists as an exaggerated version of a natural stimulus to which there is a spontaneous response, or any stimulus eliciting a response more strongly than the stimulus for which it evolved.

References

Agrillo, C., & Dadda, M. (2007). Discrimination of the larger shoal in the poeciliid fish *Girardinus falcatus*. Ethology, Ecology and Evolution, 19, 145–157.

Agrillo, C., Dadda, M., Serena, G., & Bisazza, A. (2008). Do fish count? Spontaneous discrimination of quantity in female mosquitofish. Animal Cognition, 11, 495–503.

Agrillo, C., & Miletto Petrazzini, M.E. (2012). The importance of replication in comparative psychology: the lesson of elephant quantity judgments. Frontiers in Comparative Psychology, 3, 181.

Agrillo, C., Piffer, L., & Bisazza, A. (2011). Number versus continuous quantity in numerosity judgments by fish. Cognition, 119, 281–287.

Agrillo, C., Piffer, L., Bisazza, A., & Butterworth, B. (2012). Evidence for two numerical systems that are similar in humans and guppies. PLoS ONE, 7(2), e31923.

Al Aïn, S., Giret, N., Grand, M., Kreutzer, M., & Bovet, D. (2009). The discrimination of discrete and continuous amounts in African grey parrots (*Psittacus erithacus*). Animal Cognition, 12, 145–154.

Alsop, B., & Honig, W. K. (1991). Sequential stimuli and relative numerosity discriminations in pigeons. Journal of Experimental Psychology: Animal Behavior Processes, 17, 386–395.

Benson-Amram, S., Heinen, V.K., Dryer, S. L., & Holekamp, K.E. (2011). Numerical assessment and individual call discrimination by wild spotted hyaenas, *Crocuta crocuta*. Animal Behaviour, 82, 743–752.

Beran, M.J. (2004). Chimpanzees (Pan troglodytes) respond to nonvisible sets after one-by-one addition and removal of items. Journal of Comparative Psychology, 118, 25–36.

Beran, M.J., Perdue, B.M., and Evans, T.A. (this volume). Monkey mathematical abilities. In The Oxford Handbook of Numerical Cognition, R. Cohen Kadosh, and A. Dowker, eds. (Oxford University Press).

Beran, M.J., Taglialatela, L.A., Flemming, T.M., James, F.M., & Washburn, D.A. (2006). Nonverbal estimation during numerosity judgements by adult humans. Quarterly Journal of Experimental Psychology, 59, 2065–2082.

Binoy, V.V., & Thomas, K.J. (2004) The climbing perch (*Anabas testudineus Bloch*), a freshwater fish, prefers larger unfamiliar shoals to smaller familiar shoals. Current Science, 86, 207–211.

Bisazza, A., Serena, G., Piffer, L., & Agrillo, C. (2010). Ontogeny of numerical abilities in guppies. PLoS ONE, 5(11), e15516.

Bonanni, R., Natoli, E., Cafazzo, S., & Valsecchi, P. (2011). Free-ranging dogs assess the quantity of opponents in intergroup conflicts. Animal Cognition, 14, 103–115.

Brannon, E.M., & Terrace, H.S. (1998). Ordering of the numerosities 1–9 by monkeys. Science, 282, 746–749.

Buckingham, J.N., Wong, B.B.M., & Rosenthal, G.G. (2007). Shoaling decision in female swordtails: how do fish gauge group size? Behaviour, 144, 1333–1346.

Burr, D., & Ross, J. (2008). A visual sense of number. Current Biology, 6(18), 425–428.

Butterworth, B., & Laurillard, D. (2010). Low numeracy and dyscalculia: identification and intervention. ZDM Mathematics Education, 42(6), 527–539.

Cantlon, J.F., & Brannon, E.M. (2006). Shared system for ordering small and large numbers in monkeys and humans. Psychological Science, 17, 401–406.

Cantlon, J.F., & Brannon, E.M. (2007). How much does number matter to a monkey (*Macaca mulatta*)? Journal of Experimental Psychology: Animal Behavior Processes, 33(1), 32–41.

Carazo, P., Font, E., Forteza-Behrendt, E., & Desfilis, E. (2009). Quantity discrimination in *Tenebrio molitor*: evidence of numerosity discrimination in an invertebrate? Animal Cognition, 12, 463–470.

Chittka, L., & Niven, J. (2009). Are bigger brains better? Current Biology, 19, R995–1008.

Cohen Kadosh, R., & Walsh, V. (2009). Numerical representation in the parietal lobes: Abstract or not abstract? Behavioral and Brain Sciences, 32, 313–373.

Cordes, S., & Brannon, E.M. (2008a). The difficulties of representing continuous extent in infancy: using number is just easier. Child Development, 79(2), 476–489.

Cordes, S., & Brannon, E.M. (2008b). Quantitative competencies in infancy. Developmental Science, 11(6), 803–808.

Cordes, S., Gelman, R., Gallistel, C.R., & Whalen, J. (2001). Variability signatures distinguish verbal from nonverbal counting for both large and small numbers. Psychonomic Bulletin and Review, 8, 698–707.

Dacke, M., & Srinivasan, M.V. (2008). Evidence for counting in insects. Animal Cognition, 11(4), 683–689.

Davis, H., & Albert, M. (1987). Failure to transfer or train a numerical discrimination using sequential visual stimuli in rats. Bulletin of the Psychonomic Society, 25, 472–474.

Davis, H., & Bradford, S.A. (1986). Counting behaviour by rats in a simulated natural environment. Ethology, 73, 265–280.

Davis, H., & Perusse, R. (1988). Numerical competence in animals: Definitional issues, current evidence, and a new research agenda. Behavioural Brain Science, 11, 561–579.

Dehaene, S., Dehaene-Lambertz, G., & Cohen L. (1998). Abstract representations of numbers in the animal and human brain. Trends in Neuroscience, 21, 355–361.

Emmerton, J., & Renner, J.C. (2009). Local rather than global processing of visual arrays in numerosity discrimination by pigeons (*Columbia livia*). Animal Cognition, 12, 511–526.

Emmerton, J. (2001). Birds' judgments of number and quantity. In 'Avian Visual Cognition', ed. R. G. Cook. Boston: Comparative Cognition Press.

Feigenson, L., Carey, S., & Hauser, M.D. (2002). The representations underlying infants' choice of more: object-files versus analog magnitudes. Psychological Science, 13, 150–156.

Feigenson, L., Dehaene, S., & Spelke, E.S. (2004). Core systems of number. Trends in Cognitive Sciences, 8, 307–314.

Fernandes, D.M., & Church, R.M. (1982). Discrimination of the number of sequential events by rats. Animal Learning & Behavior, 10, 171–176.

Flombaum, J.I., Junge, J.A., & Hauser, M.D. (2005). Rhesus monkeys (*Macaca mulatta*) spontaneously compute addition operations over large numbers. Cognition, 97, 315–325.

Gebuis, T., & Reynvoet, B. (2012). The role of visual information in numerosity estimation. PLoS ONE, 7(5), e37426.

Gòmez-Laplaza, L.M., & Gerlai, R. (2011a). Can angelfish (*Pterophyllum scalare*) count? Discrimination between different shoal sizes follows Weber's law. Animal Cognition, 14(1), 1–9.

Gòmez-Laplaza, L.M., & Gerlai, R. (2011b) Spontaneous discrimination of small quantities: shoaling preferences in angelfish (*Pterophyllum scalare*). Animal Cognition, 14(4), 565–574.

Gross, H.J., Pahl, M., Si, A., Zhu, H., Tautz, J., & Zhang S. (2009). Number-based visual generalisation in the honeybee. PLoS ONE, 4(1), e4263.

Hager, M.C., & Helfman, G.S. (1991). Safety in numbers: shoal size choice by minnows under predatory treat. Behavioral Ecology, 29, 271–276.

Halberda, J., Mazzocco, M.M.M., & Feigenson, L. (2008). Individual differences in nonverbal number acuity correlate with maths achievement. Nature, 455, 665–668.

Hauser, M.D., Carey, S., & Hauser, L.B. (2000). Spontaneous number representation in semi-free-ranging rhesus monkeys. Proceedings of the Royal Society of London: Biological Science, 267, 829–833.

Hauser, M.D., MacNeilage, P., & Ware, M. (1996). Numerical representations in primates. Proceedings of the National Academy of Sciences USA, 93, 1514–1517.

Hope, T., Stoianov, I., & Zorzi, M. (2010). Through neural stimulation to behavior manipulation: A novel method for analyzing dynamical cognitive models. Cognitive Science, 34(3), 406–433.

Hunt, S., Low, J., & Burns, K.C. (2008). Adaptive numerical competency in a food-hoarding songbird. Proceedings of the Royal Society of London: Biological Science, 267, 2373–2379.

Hyde, D.C. (2011). Two systems of non-symbolic numerical cognition. Frontiers in Human Neuroscience, 5, 150.

Iuculano, T., Tang, J., Hall, C.W.B., & Butterworth, B. (2008). Core information processing deficits in developmental dyscalculia and low numeracy. Developmental Science, 11, 669–680.

Izard, V., Sann, C., Spelke, E.S., & Streri, A. (2009). Newborn infants perceive abstract numbers. Proceedings of the National Academy of Sciences USA, 106, 10382–10385.

Kahneman, D., Treisman, A., & Gibbs, B. J. (1992). The reviewing of object files: object-specific integration of information. Cognitive Psychology, 24, 175–219.

Kilian, A., Yaman, S., Fersen, L., & Güntürkün, O. (2003). A bottlenose dolphin discriminates visual stimuli differing in numerosity. Learning and Behaviour, 31, 133–42.

Krusche, P., Uller, C., & Ursula, D. (2010). Quantity discrimination in salamanders. Journal of Experimental Biology, 213, 1822–1828.

Lyon, B.E. (2003). Egg recognition and counting reduce costs of avian conspecific brood parasitism. Nature, 422, 495–499.

McComb, K., Packer, C., & Pusey, A. (1994). Roaring and numerical assessment in the contests between groups of female lions, Panther leo. Animal Behaviour, 47, 379–387.

Nelson, X.J., & Jackson, R.R. (2012). The role of numerical competence in a specialized predatory strategy of an araneophagic spider. Animal Cognition, 15, 699–710.

Pepperberg, I.M. (2006). Grey parrot numerical competence: a review. Animal Cognition, 9, 377–391.

Pepperberg, I.M. (2012). Further evidence for addition and numerical competence by a Grey parrot (*Psittacus erithacus*). Animal Cognition, 15(4), 711–717.

Perdue, B.M., Talbot, C.F., Stone, A., & Beran, M.J. (2012). Putting the elephant back in the herd: elephant relative quantity judgments match those of other species. Animal Cognition, 15(5), 955–961.

Pisa, P.E., & Agrillo, C. (2009). Quantity discrimination in felines: a preliminary investigation of the domestic cat (Felis silvestris catus). Journal of Ethology, 27, 289–293.

Reznikova, Z., & Ryabko, B. (2011). Numerical competence in animals, with an insight from ants. Behaviour, 148, 405–434.

Roberts, W.A. (2010). Distance and magnitude effects in sequential number discrimination by pigeons. Journal of Experimental Psychology: Animal Behavior Processes, 36, 206–216.

Rugani, R., Fontanari, L., Simoni, E., Regolin, L., & Vallortigara, G. (2009). Arithmetic in newborn chicks. Proceedings of the Royal Society of London: Biological Sciences, 276, 2451–2460.

Rugani, R., Kelly, D.M., Szelest, I., Regolin, L., & Vallortigara, G. (2010a). Is it only humans that count from left to right? Biology Letters, 6, 290–292.

Rugani, R., Regolin, L., & Vallortigara, G. (2007). Rudimental numerical competence in 5-day-old domestic chicks (Gallus gallus): identification of ordinal position. Journal of Experimental Psychology: Animal Behavior Processes, 33, 21–31.

Rugani, R., Regolin, L., & Vallortigara, G. (2010b). Imprinted numbers: newborn chicks' sensitivity to number vs. continuous extent of objects they have been reared with. Developmental Science, 13, 790–797.

Scarf, D., Hayne, H., & Colombo, M. (2011). Pigeons on par with primates in numerical competence. Science, 334, 1664.

Stancher, G., Regolin, L., & Vallortigara, G. (2007). Bombinas (*Bombina orientalis*) go for more. Neural Plasticity, 23250, 60.

Stevens, J.R., Wood, J.N., & Hauser, M.D. (2007). When quantity trumps number: discrimination experiments in cotton-top tamarins (*Saguinus oedipus*) and common marmosets (*Callithrix jacchus*). Animal Cognition, 10, 429–437.

Stoianov, I., & Zorzi, M. (2012). Emergence of a 'visual number sense' in hierarchical generative models. Nature Neuroscience, 15, 194–196.

Tanner, C.J. (2006). Numerical assessment affects aggression and competitive ability: a team-fighting strategy for the ant Formica xerophila. Proceedings of the Royal Society of London: Biological Sciences, 273, 22737–22742.

Tomonaga, M., & Matsuzawa, T. (2002). Enumeration of briefly presented items by the chimpanzee (*Pan troglodytes*) and humans (*Homo sapiens*). Animal Learning and Behavior, 30, 143–157.

Uller, C., Huntley-Fenner, G., Carey, S., & Klatt, L. (1999). What representations might underlie infant numerical knowledge? Cognitive Development, 14, 1–36.

Uller, C., Jaeger, R., Guidry. G., & Martin, C. (2003). Salamanders (*Plethodon cinereus*) go for more: rudiments of number in an amphibian. Animal Cognition, 6, 105–112.

vanMarle, K., & Wynn, K. (2009). Infants' auditory enumeration: evidence for analog magnitudes in the small number range. Cognition, 111, 302–316.

Ward, C., & Smuts, B.B. (2007). Quantity-based judgments in the domestic dog (*Canis lupus familiaris*). Animal Cognition, 10, 71–80.

West, R.E., & Young, R.J. (2002). Do domestic dogs show any evidence of being able to count? Animal Cognition, 5, 183–186.

Wilson, M.L., Hauser, M.D., & Wrangham, R.W. (2001). Does participation in intergroup conflict depend on numerical assessment, range location, or rank for wild chimpanzees? Animal Behaviour, 61, 1203–1216.

Wynn, K. (1992). Addition and subtraction in infants. Nature, 358, 749–750.

Xu, F., & Arriaga, R.I. (2007). Number discrimination in 10-month-old infants. British Journal of Developmental Psychology, 25, 103–108.

Xu, F., & Spelke, E.S. (2000). Large number discrimination in 6-month-old infants. Cognition, 74, B1–B11.

CHAPTER 13

MONKEY MATHEMATICAL ABILITIES

MICHAEL J. BERAN, BONNIE M. PERDUE, AND THEODORE A. EVANS

'Ah-ah, I know what you're thinking punk. You're thinking did he fire six shots or only five? And to tell you the truth I've forgotten myself in all this excitement. But being this is a .44 Magnum – the most powerful handgun in the world and will blow your head clean off, you've got to ask yourself a question – Do I feel lucky? Well, do ya punk?'

Dirty Harry

A Brief History

The above quote nicely illustrates one situation in which the ability to accurately and perfectly remember and represent a counted array of items can provide clear survival advantages for a human (in this case, the 'bad guy' in the movie who has to decide whether Clint Eastwood's character still has a bullet left for him or not). This example and many others make clear that, for adult humans, number is a highly salient property of our environment, and it is hard to find situations in which humans are not counting, estimating, adding, subtracting, or otherwise manipulating and calculating numerosities. A longstanding question in comparative psychology, however, is whether this suite of abilities is unique to our species, or something we share with other animals.

Perhaps no area of comparative investigation has had as long and controversial a history as the assessment of the numerical skills of non-human animals. This area of inquiry has progressed tremendously in the past few decades, recovering from its early ignominy (see Boysen & Capaldi, 1993; Davis, 1993; Davis & Perusse, 1988). At the turn of the 20th century, there was already great interest in whether animals showed anything like the skill that adult humans showed in quantifying things in the world around them. This early work focused primarily on the question of counting by animals, driven in large part by the performances of the so-called clever animals (see Candland, 1995; Davis, 1993). The most famous of these animals, Hans the horse, reportedly could calculate and perform any number of mathematical computations, from arithmetic to much more complicated word problems involving maths. However, the debunking of this animal as being much more astute about reading human behaviour than reporting mathematical outcomes put the field into disarray (Pfungst, 1911), and for many decades after there was little or no trust of reports of mathematical performances by animals (see Candland, 1995). These concerns about cuing remain today for all areas of animal cognition research (e.g. Lit, Schweitzer, & Oberbauer, 2011; Sebeoke & Rosenthal, 1981), although many researchers have made great strides in minimizing or eliminating these concerns through the use of adequate controls over cuing.

Despite the reservations of many because of concerns about cuing, a number of interesting and important findings occurred in the first half of the 20th century (see Boysen & Capaldi, 1993; Davis & Perusse, 1988; Wesley, 1961). A few of these early studies are worth noting, particularly given that they occurred from using tests with monkeys. The present chapter will constrain itself to discussing numerical cognition in monkeys, in part because monkey species are the most often used non-human animals in this area of research, and also because other chapters in this volume will cover the important findings with other

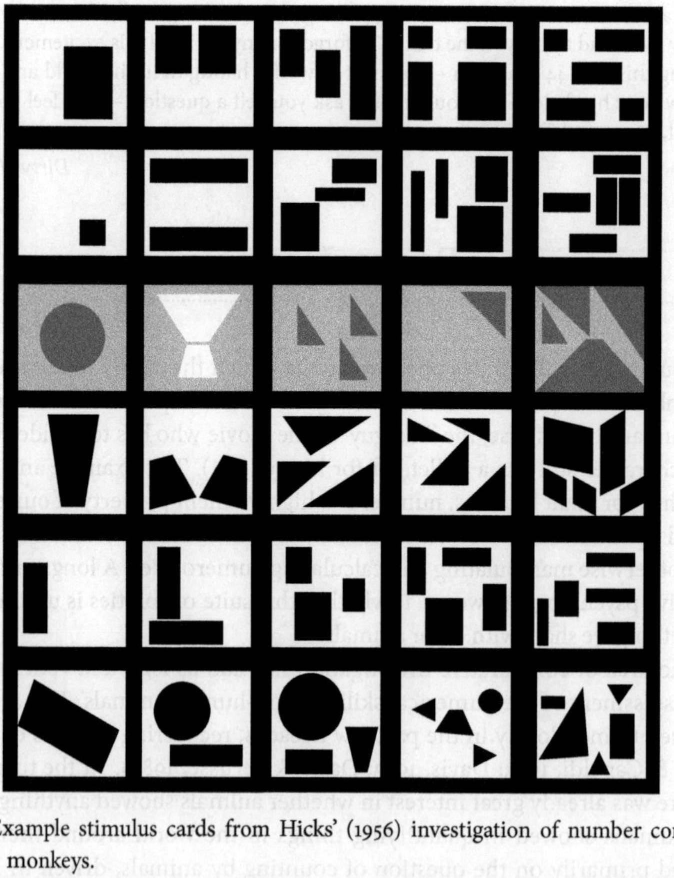

FIGURE 13.1 Example stimulus cards from Hicks' (1956) investigation of number concept formation in rhesus monkeys.

Reprinted from L. H. Hicks, An analysis of number-concept formation in the rhesus monkey, *Journal of Comparative and Physiological Psychology*, 49(3), pp. 212–218, doi: 10.1037/h0046304 © 1956, American Psychological Association.

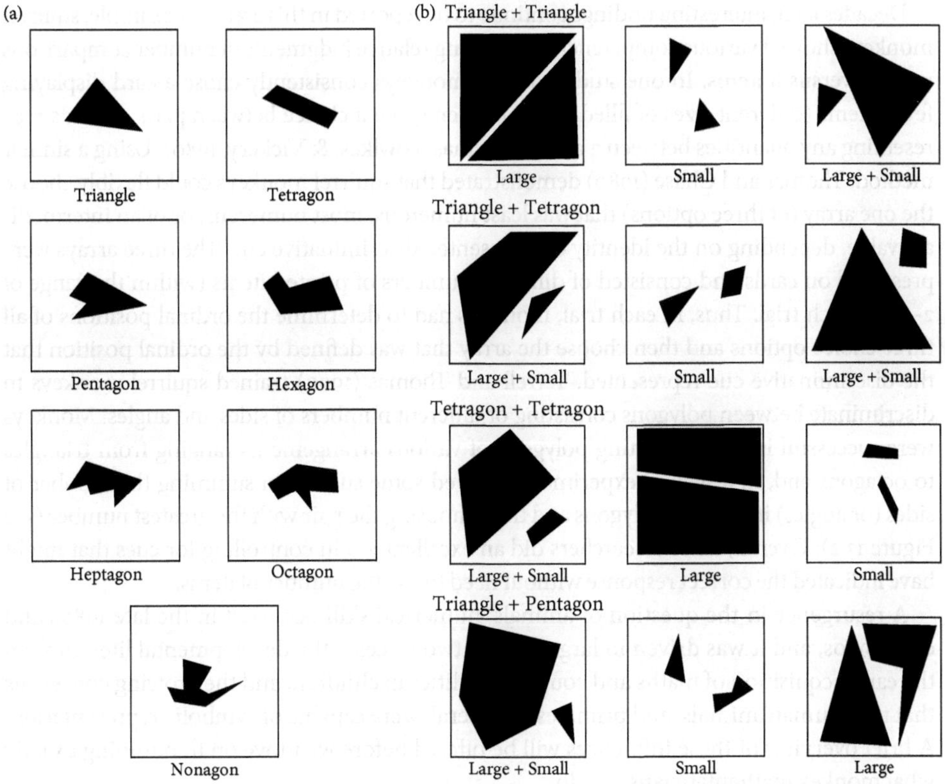

FIGURE 13.2 Stimuli used by Terrell and Thomas (1990) to show that squirrel monkeys could make relative numerousness judgments between polygons of various arrangements (a), as well as sum the number of sides (or angles) in pairs of polygons (b).

Reprinted from D. F. Terrell and R. K. Thomas, Number-related discrimination and summation by squirrel monkeys (Saimiri sciureus sciureus and S. boliviensus boliviensus) on the basis of the number of sides of polygons, *Journal of Comparative and Physiological Psychology*, 104(3) pp. 238–47, doi: 10.1037/0735-7036.104.3.238 © 1990, American Psychological Association.

species (e.g. the chapter by Agrillo, this volume). This is not to diminish the importance of work with other species, but to provide a focus on one type of animal and the diverse manner in which that type of animal can process and deal with numerical and quantitative information.

Kuroda (1931) attempted one of the early systematic investigations of numerical discrimination, asking whether a macaque monkey could discriminate between one and two, and two and three sounds. Performance was mixed, but suggested the monkey might have some faculty for numerical discriminations. Hicks (1956) trained rhesus monkeys to select a card with three figures on it rather than 1, 2, 4, 5, or 6 figures. Background area was controlled as a possible cue. Many different stimulus types, and many different configurations were presented to control for memorization of the correct stimuli and area cues, both important manipulations (see Figure 13.1). The monkeys still did better than chance when all of these potential cues were controlled or eliminated except for the concept of 'threeness.'

Decades later, interesting findings continue to be reported in this area. For example, squirrel monkeys showed various competencies in making relative judgments of number comparisons up to 9 versus 8 items. In one study, squirrel monkeys consistently chose a card displaying fewer items (different sizes of filled circles), when given a choice between pairs of cards representing any quantities between 2 and 9 (Thomas, Fowlkes, & Vickery, 1980). Using a similar method, Thomas and Chase (1980) demonstrated that squirrel monkeys could flexibly choose the one array (of three options) that was least numerous, most numerous, or of an intermediate value, depending on the identity of a presented discriminative cue. The three arrays were presented on cards and consisted of different numbers of printed items (within the range of 2–7) on each trial. Thus, in each trial, monkeys had to determine the ordinal positions of all three choice options and then choose the array that was defined by the ordinal position that the discriminative cue represented. Terrell and Thomas (1990) trained squirrel monkeys to discriminate between polygons consisting of different numbers of sides and angles. Monkeys were successful in discriminating polygons of various arrangements ranging from triangles to octagons and, in a second experiment, showed some success in summing the number of sides (or angles) in pairs of polygons and then choosing the pair with the greatest number (see Figure 13.2). Overall, these researchers did an excellent job in controlling for cues that might have indicated the correct response without need to use the number of items.

A resurgence in the question of animals' numerical skill occurred in the late 1980s and early 1990s, and it was driven in large part by two forces – the developmental literature on the early acquisition of maths and counting abilities in children, and the growing consensus that non-human animals, and primates in general, were capable of symbolic representation. A brief overview of these influences will be offered before we move on to outlining exactly what monkey mathematicians can do.

Counting Principles and Symbolic Representation

As anyone who has been around a growing human child can attest, counting proficiency occurs rapidly somewhere between the ages of 3 and 6 years, and as it does, it can be difficult to *stop* children from counting just about everything around them. This proficiency, however, is not *all or none* in the sense that the necessary principles for becoming a proficient counter can take time to emerge, and can wax or wane in the daily attempts by the young child to count. This issue has been covered in much detail elsewhere, and for our purposes the main point is that there appear to be a number of critical principles that the 'counting' child must possess (see Gelman & Gallistel, 1978). These principles, outlined below, helped set the stage for new assessments of the numerical competence of non-human animals. The counting principles include:

Stable order principle: the numeric tags used to enumerate and represent items within a counted array must be applied in the same order across counting events, so that the labels consistently represent the same number of items.

One to one correspondence principle: the numeric tags are applied to each item once, and only once, within a counting event.

Ordinality principle: the numeric tags, when applied in a stable order, represent each item's sequential place within the array (e.g. first, third, fifth). This principle affords the knowledge for making relational judgments ('more,' 'less,' 'bigger,' 'smaller).

Cardinality principle: the final numeric tag applied when counting a set of items represents not only the last item's ordinal position, but also the sum total of the set. This principle allows one to answer the question 'how many?'

Abstraction principle: this principle simply states that the counting routine can be applied to any array of things, so that one can count pennies, dogs, finger taps, and seasons of the year using all of the principles above in the same way.

The question, then, for comparative psychologists was the extent to which non-human animals showed evidence of these principles in their performances on tasks designed to mimic the studies used with children, or with new methods that were more animal friendly, but were also designed to require these principles for proficient performance. These early studies mainly assessed the question of animal counting and, specifically, the cardinality principle, and much of the evidence was promising. In many cases, the positive evidence came from chimpanzees and parrots (e.g. Boysen & Berntson, 1989; Matsuzawa, 1985; Pepperberg, 1987, 1994; Rumbaugh et al., 1989), or from more traditional laboratory subjects such as rats (Burns, Goettl, & Burt, 1995; Capaldi & Miller, 1988) or pigeons (e.g. Roberts & Mitchell, 1994).

At the same time, it was becoming clear that animals could learn associations between arbitrary stimuli, such as Arabic numerals and specific quantities. This research complemented the ongoing demonstrations that great apes and other animals could use various symbol systems to map symbols to real world referents (e.g. Rumbaugh, 1977; Savage-Rumbaugh, 1986), and so it was a natural extension of this research to the numerical domain. It was clear that chimpanzees (e.g. Beran & Rumbaugh, 2001; Boysen & Berntson, 1989; Matsuzawa, 1985) and an African Gray Parrot (Pepperberg, 1987, 1994) could associate Arabic numerals with specific quantities, and they did so in a way that allowed them to make sets to match numerals or label sets of quantities with the correct numerals. For example, Matsuzawa (1985) trained a chimpanzee named Ai to assess the number of stimuli presented to her and then select, on a computer screen, the Arabic numeral that matched that array quantity (see also Tomonaga & Matsuzawa, 2002). Another chimpanzee, Sheba, showed some of the most impressive abilities yet found in a non-human animal, including the ability to label arrays of items with the correct numeral, even when she had to travel to two locations and sum together the total number of items seen in those locations (Boysen & Berntson, 1989). In another test paradigm, chimpanzees were trained to collect sets of items, on a computer screen, to match a presented Arabic numeral, and performance was above chance for numbers up to 7 (Beran, 2004a; Beran & Rumbaugh, 2001; Beran, Rumbaugh, & Savage-Rumbaugh, 1998; Rumbaugh et al., 1989). These two kinds of tests, constructing sets of items to match numerals or labelling sets of items with numerals, were also performed successfully by other species, such as pigeons (e.g. Xia, Emmerton, Siemann, & Delius, 2000; Xia, Siemann, & Delius, 2001). Thus, by the mid-1980s and early 1990s, evidence re-emerged that showed that, with more proper controls, and intensive training, some species did seem to exhibit mathematical abilities and even approached the counting performances of humans.

However, as the evidence accumulated, it became clearer and clearer that non-human animals did not really show the same ability to count that human children show by 6 or

7 years of age. Instead, the evidence suggested that non-human animals had a more 'approximate' representation of number (Brannon & Roitman, 2003; Gallistel & Gelman, 2000), albeit one that still allowed for a number of competencies in responding to numerical tasks. Extensive research then followed, much of it with monkey species that proved capable of performing well in both manual and computerized tests, making them excellent subjects for this area of research. We will now outline some of the evidence that has accumulated so that the reader will understand what monkeys can do when it comes to numerical competence. Then, we will return to the question of *how* animals, and specifically monkeys, represent quantities, and what mathematical abilities are within their grasp.

Ordinality Judgments

Perhaps the most clearly demonstrated capacity in non-human animals is a sense of ordinality that allows for ordered judgments of stimuli. Many species encode the ordinal properties of stimulus sets and make judgments between two or more sets of items on the basis of the quantity or number of items in those sets (e.g. Beran et al., 2005; Biro & Matsuzawa, 1999; Boysen, Bernston, Shreyer, & Quigley, 1993; Brannon, Cantlon, & Terrace, 2006; Brannon & Terrace, 1998, 2000; Matsuzawa, 1985; Pepperberg, 2006; Terrace, Son, & Brannon, 2003; Washburn & Rumbaugh, 1991). Monkeys, in particular, have been shown to do well on these tasks, not only with visible quantities of items, but also with respect to symbolic representations of number. For example, Washburn and Rumbaugh (1991) trained rhesus monkeys to select among 2, 3, 4, or 5 Arabic numerals within the range of the numerals 0–9 presented on a computer screen (see Figure 13.3). Selection of each numeral led to the presentation of a number of food pellets equal to the value of that numeral. Washburn and Rumbaugh first trained monkeys to select the larger numeral of a presented pair. The amount of reward affected performance in that the monkeys had more difficulty discriminating pairs of symbols that were associated with similar numbers of rewards (e.g. 6 and 7). Importantly, these researchers withheld multiple pairings of numerals for later use as novel probe trials. Monkeys' successful performance with these novel pairs suggested that they had learned the ordinal sequence for all of the numerals, rather than learning a matrix of independent two-choice discrimination problems. A subsequent test involving sets of 3, 4, or 5 numerals provided additional evidence to support the ordinality explanation. This experiment did not address the role of cardinality, because monkeys did not have to match numerals to specific quantities, but this was an important demonstration of ordinal knowledge that was subsequently confirmed through a number of other test paradigms.

In one of the most well-known studies of this kind, rhesus monkeys demonstrated the capacity to touch, in ascending numerical order, pairs of computer generated visual stimuli representing the quantities one to nine (Brannon & Terrace, 1998, 2000). Critically, Brannon and Terrace tested the monkeys with multiple stimulus categories that controlled for the contributions of non-numeric stimulus properties (e.g. item size, shape, colour) to judgments between digital arrays (see Figure 13.4). This ensured that the monkeys could only use the number of items to guide their responses, making this truly a numerical judgment. This study led to multiple replications with other monkey and non-monkey species (e.g. baboon and squirrel monkey: Smith, Piel, & Candland, 2003; capuchin monkeys: Judge, Evans, & Vyas, 2005; pigeons: Scarf, Hayne, & Colombo, 2011). Also, Brannon

FIGURE 13.3 A macaque monkey performing the task used in Washburn and Rumbaugh (1991) and Beran et al. (2008a). The monkey is given an array of numerals and must select them in descending order using a cursor to contact the numerals in sequence.

and colleagues used this method to produce some extensions of their original ordinality study, including investigations of the influence of stimulus heterogeneity (Cantlon & Brannon, 2006a) or reference points (Brannon et al., 2006) on rhesus monkey ordinal performance, as well as studies outlining the mechanisms underlying such judgments (e.g. Cantlon & Brannon, 2006b).

In a manual test for ordinal knowledge, Olthof, Iden, and Roberts (1997) trained and tested squirrel monkeys to order Arabic numerals presented as physical tokens on a version of the Wisconsin General Test Apparatus (WGTA). As in Washburn and Rumbaugh (1991), Olthof and colleagues rewarded monkeys' choices with a number of food items equal to the numeral that had been selected, and they reserved certain pairs of numerals for a final test to show that monkeys understood the ordinality relations underlying the set of numerals. In a subsequent experiment, these researchers presented monkeys with choices between sets of multiple numerals (i.e. 2 numerals versus 2 numerals, 1 numeral versus 2 numerals, and 3 numerals versus 3 numerals). Monkeys tended to choose the set of numerals representing the largest overall quantity (e.g. 3 + 3 over 5 + 0) and were not biased towards the largest individual

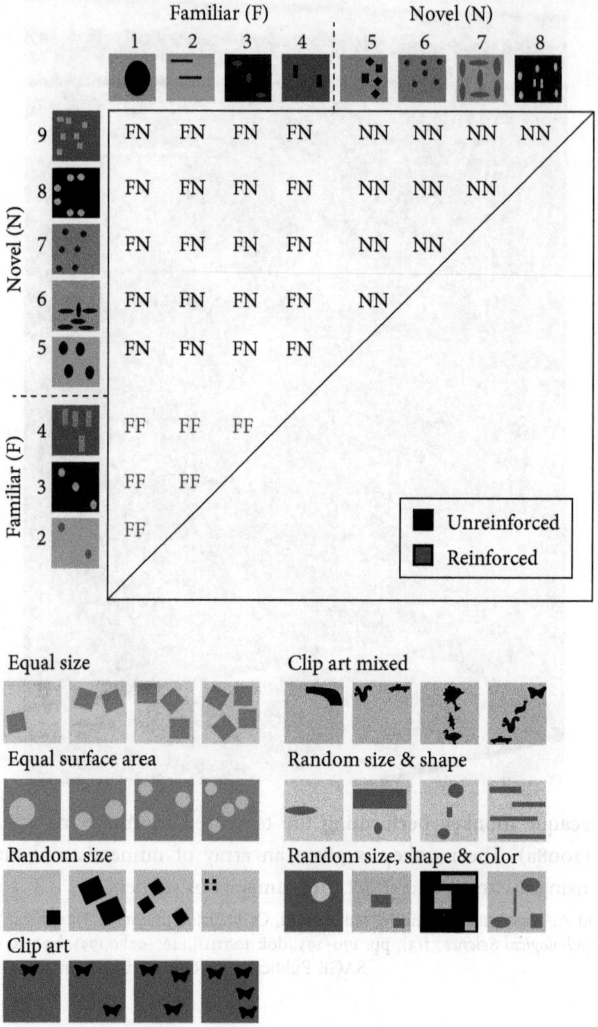

FIGURE 13.4 Brannon and Terrace (2000) conducted one of the most systematically controlled tests of relative numerousness judgment in rhesus monkeys. Their experiments controlled for stimulus familiarity (top panel) as well as several other non-numeric stimulus properties (bottom panel).

Reprinted from D. F. Terrell and R. K. Thomas, Number-related discrimination and summation by squirrel monkeys (Saimiri sciureus sciureus and S. boliviensus boliviensus) on the basis of the number of sides of polygons, *Journal of Comparative and Physiological Psychology*, 104(3) pp. 238–47, doi: 10.1037/0735-7036.104.3.238 © 1990, American Psychological Association.

numeral or against the smaller individual numeral. This suggested that monkeys were able to sum and then compare the different combinations of numerals presented in each set.

In another related study, Beran et al. (2008a) compared ordinality judgments in rhesus monkeys and capuchin monkeys. This study followed the methods of Washburn and Rumbaugh (1991) quite closely, but included two important differences that were introduced for the sake of investigating alternative explanations for monkeys' performance in prior studies. First, the authors considered that presenting different numbers of food rewards based on the differing stimuli (as in Olthof et al., 1997; Washburn & Rumbaugh, 1991) allowed for the possibility that animals had come to assign different hedonic values to each numeral. In other words, each different stimulus, because it led to a different number of food items, may have acquired different response strength on the basis of its hedonic value rather than some quantitative property. To test this possibility, Beran and colleagues rewarded half of the monkeys of each species by presenting a single food reward for choosing the larger of two numerals. The remaining participants of each species were presented with a number of food rewards equal to the selected numeral (as in the above studies). All monkeys learned to choose the larger of two numerals in training trials, but only rhesus monkeys that were rewarded with food amounts equal to the value of the numerals succeeded in novel probe trials consisting of numeral pairs not presented during training. Beran and colleagues also considered the possibility that, in the final experiment of the Washburn and Rumbaugh (1991) study that involved responding to sets of 3, 4, or 5 numerals, monkeys' performance may have been the result of rapid learning of a new ordinal sequence without knowledge of the ordinal relations of the stimuli established through prior pairing of those stimuli. To test this idea, after monkeys succeeded in ordering sets of 5 numerals in their study, Beran and colleagues presented monkeys with sets of novel stimuli (letters of the alphabet). Monkeys failed to learn to order the letters within the same number of trials presented to test their ordinal ability with previously trained Arabic numerals, and so the rapid ordinal learning hypothesis was not supported. Instead, Beran and colleagues concluded that monkeys did learn the ordinal sequence of Arabic numerals rather than a matrix of two-choice discriminations that did not exhibit ordinality.

Despite the popularity of the use of Arabic numerals for tests with monkeys (i.e. Beran et al., 2008a; Olthof et al., 1997; Washburn & Rumbaugh, 1991) and also with apes (e.g. Beran et al., 1998; Boysen & Berntson, 1989; Matsuzawa, 1985), other methods have been devised to illustrate the ordinal knowledge of non-human primates. For example, Beran, Beran, Harris, and Washburn (2005) employed a set of five coloured food containers to test whether a rhesus monkey (and two chimpanzees) could learn the ordinal relations of these arbitrary objects based on the food items they contained. These animals learned to choose the more valuable of two presented containers and later transferred their ordinal knowledge of the container values to a test involving choices between containers and visible (uncontained) food amounts. Performance in the latter test demonstrated that monkeys had learned the approximate value of each individual container in addition the relative values of paired containers. This study has since been replicated with capuchin monkeys (and chimpanzees) using both the original container method as well as a similar method involving solid tokens to represent different quantities of food items (Evans, Beran, & Addessi, 2010).

Quantity Judgments of Food Items

For any animal, being able to choose the larger of two quantities of food items leads to clear survival advantages, and a large variety of organisms have shown this capability in experimental tests (e.g. Hunt, Low, & Burns, 2008; Krusche, Uller, & Dicke, 2010; Perdue, Talbot,

Stone, & Beran, 2012). Monkey species are particularly well-represented in this kind of experiment. Capuchin monkeys, in particular, have participated in several such tests and have shown performance levels akin to great apes (as in Beran, 2001, 2004b). In one study, experimenters tested the ability of capuchin monkeys to choose the larger of two sets of discrete food items, or the greater of two quantities of a continuous substance (i.e. banana puree; vanMarle, Aw, McCrink, & Santos, 2006). Monkeys succeeded in both conditions and their performance was dependent on the ratio of the two quantities being compared. In another study, capuchins were presented with two sets of food items, identical in food type but differing in number from 1 to 6 (Beran, Evans, Leighty, Harris, & Rice, 2007). Monkeys accurately chose the greater of two quantities when each quantity was presented one at a time (by uncovering and recovering each in succession). Monkeys continued to choose the larger choice option when the covered food sets were combined with additional visible food items, which required summation of the visible and covered food items in each set. The same research group conducted a second study with capuchin monkeys in which food sets were presented only one item at a time. Monkeys succeeded in choosing the more numerous set when one quantity was presented item by item and the alternative set was entirely visible throughout the trial. Some monkeys even accurately compared two sets of food items, each presented item by item at different rates (which controlled for temporal cues; Evans, Beran, Harris, & Rice, 2009).

Semi-free-ranging rhesus monkeys have been presented with similar tests. In one of the first tests of this kind presented to monkeys, Hauser, Carey, and Hauser (2000) reported that rhesus monkeys could make discrete quantity judgments without explicit training prior to the test. Researchers presented a single trial to each of 225 monkeys, in which they dropped different numbers of apple slices into two opaque containers, 5–10 metres away from the target monkey, and then walked away from the containers giving the monkey an opportunity to make a selection. Monkeys were successful overall for the majority of quantity comparisons, and control conditions ruled out the use of temporal or volume cues by monkeys to make these judgments. This research group followed this study with multiple extensions of this paradigm, for example, assessing judgments between continuous quantities (Wood, Hauser, Glynn, & Barner, 2008), and evaluating spontaneous addition of discrete items (Flombaum, Junge, & Hauser, 2005).

Numerousness Judgments of Non-Food Items

Beyond tests involving quantities of food items, monkeys also have shown the ability to make relative quantity judgments between arrays of artificial items (e.g. digital dot arrays, cards with printed arrays; Terrell & Thomas, 1990; Thomas & Chase, 1980; Thomas et al., 1980). Interesting relative numerousness tasks have been presented to monkeys using a computerized method. For instance, Beran (2007) created a computerized analogue of the item-by-item quantity presentation method (as used in Beran, 2001, 2004b; Hauser et al., 2000; vanMarle et al., 2006) and presented the task to rhesus monkeys. Monkeys succeeded in choosing the digital container into which the largest number of digital items was dropped, despite precise control of non-numeric stimulus properties, including presentation duration and the total amount of digital item volume presented in each set. In a related study, Beran (2008) assessed the ability of rhesus monkeys to compare arrays of moving digital items. Monkeys in this study accurately tracked, enumerated, and compared sets of moving items, and these judgments were numerical in that they were not controlled by non-numeric stimulus properties, such as cumulative item area. Beran, Decker, Schwartz, and Schultz (2011) used this task to make a broader comparative assessment of this ability, and found similar performances in rhesus monkeys, capuchin monkeys, and adult humans and children. This capacity is not unique to primates, as even black bears show some competencies in this kind of test (Vonk & Beran, 2012).

Although it is most common to assess numerical cognition by having individuals choose the larger (or smaller) of a pair (or set) of quantities, there are other paradigms that require assessing numerousness. For example, in an absolute numerousness judgment, one must select a target stimulus consisting of a particular quantity of items when given a choice between the target quantity and one or more non-target quantities. Squirrel monkeys have shown this ability by consistently responding to a stimulus panel containing four dots when given the choice between this target quantity and other panels containing different dot quantities (Noble, 1986; see also Hicks, 1956). Moreover, rhesus monkeys performed a computerized delayed matching-to-sample task in which the sample and choice stimuli were arrays of digital elements, and the correct choice matched the number of elements shown in the sample stimulus while other non-numeric stimulus dimensions were controlled (Jordan & Brannon, 2006). This study indicated that monkeys could discern the absolute number of items in the sample and then match that array to another containing the same absolute number of items. This is an impressive ability that would seem to require the same sensitivity to cardinality shown in some of the studies outlined earlier in which animals created sets equal to an Arabic numeral's value or labelled arrays with a numeral. Only through representing and matching exact numerosities (i.e. using cardinal representations) could the animals perform at such high levels, as ordinal knowledge would not be sufficient to perform this well. Careful control over non-numerical features of the stimulus sets also meant that monkeys could not match on the basis of amount or other non-numerical cues, further supporting their necessary use of cardinal features of the arrays.

Numerosity and Self-Control

The assumption that monkeys (and other species) will always prefer the larger of two food amounts is so prevalent that it has been used to build other psychological tests. These include the discrete-trial self-control test (originally used with non-primate participants, e.g. Ainslie, 1974), which is an assessment of preferences for larger, more delayed food amounts or smaller, more immediately available food amounts (see Logue, 1988, for a review). In this test, self-control is assumed when the participant consistently prefers the larger amount, even when it is associated with a longer delay to food consumption. A variety of monkeys have been tested in versions of this paradigm and in many cases exhibit self-control for delayed rewards (e.g. Addessi, Paglieri, & Focaroli, 2011; Amici, Aureli, & Call, 2008; Stevens, Hallinan, & Hauser, 2005; Szalda-Petree, Craft, Martin, & Deditius-Island, 2004;

Tobin, Logue, Chelonis, Ackerman, & May, 1996). A related test is the reversed-reward contingency task (originally devised for use with chimpanzees; Boysen & Berntson, 1995), which involves having to choose the smaller of two food amounts in order to receive the larger (see Shifferman, 2009, for a review). Because the preference for larger reward amounts is assumed to be inherent in the participant, overcoming prepotent responses to those rewards in order to choose the lesser reward is considered to be another form of self-control. Monkeys have been tested in this task as well, but like most other species, they require additional training or task modifications to exhibit consistent success (Addessi & Rossi, 2011; Anderson, Awazu, & Fujita, 2000, 2004; Kralik, 2005). Thus, monkeys' sensitivity to numerosity and their ability to make numerousness judgments have been central features in other assessments of their behaviour in choice situations.

Mechanisms of Monkey Mathematics

One major question that has driven research into monkey mathematics is the nature of the mechanism or mechanisms that underlie numerical competence in non-human animals. In humans, there is an exactness in numerical representation that can only occur with a symbolic system in place for representing, perfectly, the numerousness of stimuli. Adult humans can count, perform arithmetic, calculate, and use higher mathematics to deal with complex problems of geometry, algebra, and calculus. Hopefully, by now the reader will agree that, although animals cannot do all of these things, they share with humans some of the core processes that support high level mathematical reasoning, such as some understanding of the counting principles outlined earlier. That said, the question that must be answered is what system or systems are available to animals for numerical representation, and what do the presence or absence of these systems in non-human animals tell us about the evolution of numerical cognition in humans.

This debate has, in recent years, centred around two systems. The first is a system that represents number inexactly by representing quantities as analogue magnitudes. This system, sometimes called the Approximate Number System (ANS), is inherently inexact, and this system conforms to Weber's Law in its ability to provide an organism with comparisons between quantities. There are two signature behavioural effects of the approximate number system. Specifically, performance improves in judging between quantities as the difference between those quantities increases. Also, in situations in which the difference between sets is held constant, performance is better when comparisons are between smaller sets rather than larger sets (for reviews, see Brannon & Roitman, 2003; Dehaene, 1997; Gallistel & Gelman, 2000). These two effects, called the *distance effect* and the *size effect* (sometimes called the *magnitude effect*), have been found in a large number of species. This analogue magnitude estimation capacity seems to underlie many aspects of performance by animals, including those that involve estimating continuous and discrete amounts (for overviews, see Cantlon, Platt, & Brannon, 2009; Gallistel & Gelman, 2000). Humans also show evidence of analogue magnitude estimation. For example, when they are prevented from counting stimuli, they show greater variability as a function of increasing set size (e.g. Beran, Taglialatela, Flemming, James, & Washburn, 2006; Cordes, Gelman, Gallistel, & Whalen, 2001; Huntley-Fenner, 2001; Huntley-Fenner & Cannon, 2000; Whalen, Gallistel, & Gelman, 1999).

However, a second system for numerical representation has been proposed. The *object file* model states that discriminations can be made for small numerosities, but not for large numerosities. The idea is that there are memory limits for the individuation of items (typically, there are not more than 3 files available in short-term memory for individual items in a set). Thus, when one counts or otherwise quantifies arrays of items, separate items are encoded in terms of separate object files, and these are limited in number (Feigenson, Carey, & Hauser, 2002). Object files then operate as representations of items within the array (Simon, 1997; Uller, Carey, Huntley-Fenner, & Klatt, 1999).

The object file model draws support from the data reported by Hauser et al. (2000). To date, that is the only study with monkeys that has shown set size limits on discrimination performance with food items. In the Hauser et al. (2000) study, monkeys watched experimenters place pieces of apple, one-at-a-time, into two opaque containers. Monkeys then moved to one of the containers and received the apple pieces. Some comparisons involved all smaller numbers of apple pieces (one versus two, two versus three, three versus four, and three versus five). On these, the monkeys almost always chose the container with the larger number of apple pieces. However, for comparisons with larger numbers (four versus five, four versus six, four versus eight, and three versus eight), the monkeys had more trouble. Thus, there appeared to be a sharp difference in performance between trials with small sets only, and trials with small sets and larger sets of items. This led to the suggestion that rhesus monkeys may be accessing a limited-capacity mechanism (the object file model) for representing quantities that are presented sequentially. Some researchers have suggested that monkeys (and young human children) have two distinct systems for representing number that include the object file system and the analogue magnitude system, and this idea is the basis for the 'two core number systems' hypothesis (Feigenson, Dehaene, & Spelke, 2004; Xu, 2003).

Contrary to the Hauser et al. (2000) study, a large number of studies with monkeys found no evidence of set size limits. For example, Beran (2007) trained two monkeys to make judgments between two sets of sequentially presented arrays of items. In addition to determining that the monkeys used the number of items, and not the amount or size of the elements in each array to guide responding, there was no evidence that monkeys had difficulty discriminating small arrays from large ones. In fact, many of the problematic comparisons for the monkeys in Hauser et al. (2000) were easily performed by the monkeys in Beran et al. study (2007; see also Nieder & Miller, 2004). These data also matched those from manual tests given to all four great ape species (gorillas, bonobos, chimpanzees, and orangutans; e.g. Beran, 2004b; Hanus & Call, 2007) capuchin monkeys (Evans et al., 2009), and many other species ranging from sea lions (Abramson, Hernandez-Lloreda, Call, & Colmenares, 2011) to parrots (Aïn, Giret, Grand, Kreutzer, & Bovet, 2009) and pigeons (Roberts, 2010). There have been very few comparative studies that have found evidence in support of the object file model with different species. Some evidence has been reported for fish (e.g. Gomez-Laplaza & Gerlai, 2011; Piffer, Agrillo, & Hyde, 2012), birds (e.g. Garland, Low, & Burns, 2012) and Asian elephants (Irie-Sugimoto, Kobayashi, Sato, & Hasegawa, 2009), but the latter outcome was not replicated in a subsequent study with African elephants (Perdue et al., 2012).

Thus, nearly all studies with non-human species show the signature effects of the approximate number system, and there is very limited evidence of restrictions on performance as might be expected if two core systems of number were at work. This is a critical point, and one that bears repeating because the implications of this debate have been wide reaching, particularly in the developmental literature, where the comparative data suggesting there are two core systems have been used to outline the evolutionary foundations of two core systems for humans. For example, Feigenson et al. (2004) stated that 'Monkeys' restriction to the numerosities 1–4 in situations involving small arrays, coupled with their capacity to create noisy representations of large sets, suggests that monkeys, like humans, have two distinct systems for representing number. The two core number systems therefore offer a strong case of representational continuity across development and across species.' This seems far less likely to be the case as the comparative database grows, and cases in which animals struggle to discriminate 3 items from 8, or 2 items from 6 become less and less evident. Rather, it appears that the true nature of numerical representation by animals is through use of a continuous, approximate scaling of numerosity for which the exactness of the representation fades more and more as magnitude increases.

Even if one discards the object file model, and focuses solely on one core system of number (ANS), there is still necessary work in the search for understanding the mechanism that supports numerical (and other quantitative) representations. Several important details remain to be worked out regarding this approximate number system. Ongoing debates are important, not only for understanding the numerical abilities of animals, but also for understanding the nature of non-verbal numerical representation in humans. One such debate regards whether the scaling that occurs in numerical representation is linear, but with increasing variability that accompanies increasing magnitude, or if it occurs through logarithmic compression of the mental number line that is used to represent numerosity (e.g. Beran et al., 2008b; Dehaene, 2003; Roberts, 2006; Siegler & Opfer, 2003). Another important research avenue is to understand the neural processes that support mathematical and numerical competencies in non-human animals, and progress has already begun on this issue.

The Neuroscience of Monkey Mathematics

Given recent advances in neuroimaging techniques, we can identify the underlying neural structures involved in processing numerical information and uncover similarities in human and non-human neural structures and pathways involved in this process. Although a wide network of brain regions is likely involved in numerical processing (Dehaene & Cohen, 1995; see also the chapters by Menon and Kaufmann et al., this volume), research in humans suggests that the parietal cortex plays a critical role. In particular, the left and right intraparietal cortices have been reported as a critical substrate in humans, with some lateralization of function across hemispheres (Dehaene, Dehaene-Lambertz, & Cohen, 1998). Several lines of evidence support the idea that this specific brain substrate is dedicated to numerical processing:

- lesions to this area of the brain will impair numerical representations;
- functional MRI studies reveal activation in this region during numerical processing (for reviews, see Cohen Kadosh, Lammertyn & Izard, 2008; Cohen Kadosh & Walsh,

2009; Dehaene et al., 1998; Kaufmann, Wood, Rubinsten, & Henik, 2011; Nieder & Dehaene, 2009).

Furthermore, research suggests that the intraparietal cortex is involved in both symbolic and non-symbolic numerical representation (Cohen Kadosh, Henik, Rubinsten, Mohr, Dori, Van de Ven, et al., 2005; Fias, Lammertyn, Reynvoet, Dupont, & Orban, 2003), which supports the possibility that this substrate is common across species. Although this topic is covered in much greater detail in the chapter by Nieder (this volume), a few points are worth making here. Recent research has confirmed that similar brain regions may be involved in numerical processing in monkeys, such as the prefrontal cortex, posterior parietal cortex, and the fundus of the intraparietal sulcus (Nieder, Freedman, & Miller, 2002; Nieder & Miller, 2004).

The 'scaling problem' refers to the question of how a sensation such as a sense of number that occurs on a continuum is represented in the brain, and neurological research has provided insights into the question of how, specifically, numerical information is scaled in the brain (Dehaene, 2003). Nieder and Miller (2003) trained monkeys to match a numerical stimulus after a brief delay (delayed match-to-numerosity task), while recording neurons in the prefrontal cortex. Performance followed Weber's Law, and the distributions of responses were plotted on a linear and logarithmic scale. The logarithmic scale produced more symmetrical distributions than the linear scale, suggesting that, at the neural level, numerical information may be compressed in a logarithmic manner in accordance with Fechner's Law (Nieder & Miller, 2003).

Summary and The Future of Monkey Mathematics

Monkeys clearly respond to the numerical properties of the world around them. Their performances across tasks show many of the capacities necessary for formal counting to occur, although to date there has been no clear demonstration of 'true counting' in a monkey species, but that should not be the goal. Rather, research into the numerical competence of monkeys (and other non-human animals, see Agrillo, this volume) has provided great insight into the competencies (and mechanisms underlying those competencies) that monkeys share with humans. These data have improved our understanding of the evolutionary emergence of numerical competence, and the constraints on that emergence. Monkeys represent number imperfectly, but with an imperfection that makes sense in an evolutionary perspective. Being able to tell the difference between a food source with 10 items versus one with only 5 items (or telling 6 predators apart from 3 predators) offers a clear survival advantage. It seems much less obvious that telling 15 food items from 16 offers such an advantage. So, the approximate number system seems more than adequate to provide monkeys (and other non-human species) with all that they need in order to survive and thrive. At the same time, monkeys appear quite capable of learning ordinal relationships among stimuli, as would be necessary to learn and remember dominance hierarchies, and they seem capable of abstracting from arrays the numerical properties independent of other properties such as colour, amount, shape, and density. Monkeys even seem capable of performing 'fuzzy' calculations as when they sum amounts across arrays or keep track of accumulating sets of items. Although these calculations and responses to arithmetic operations are not exact, they again serve their purpose, which is to provide enough information to make good decisions in most cases, such as not foregoing large differences in food choices or failing to notice large differences in group sizes. Thus far, higher-order mathematical operations have appeared to be beyond the capacity of monkeys (and other animals), but monkeys do share with humans many of the same foundational mechanisms for the representation of simple arithmetic operations, and for the representation of numerosity itself. Future studies will certainly highlight new capacities that are at present unknown, and these studies, along with those outlined here, will continue to help us outline the evolutionary foundations of mathematics.

Acknowledgments

Preparation of this chapter was supported by funding from the National Institutes of Health (HD060563) and the Rumbaugh Fellowship of Georgia State University.

References

Abramson, J.Z., Hernandez-Lloreda, V., Call, J., & Colmenares, F. (2011). Relative quantity judgments in South American sea lions (*Otaria flavescens*). Animal Cognition, 14, 695–706

Addessi, E., Paglieri, F., & Focaroli, V. (2011). The ecological rationality of delay tolerance: insights from capuchin monkeys. Cognition, 119, 142–147

Addessi, E., & Rossi, S. (2011). Tokens improve capuchin performance in the reverse–reward contingency task. Proceedings of the Royal Society B: Biological Sciences, 278, 849–854

Agrillo, C. (this volume). Numerical and arithmetic abilities in non-primate species. In R. Cohen Kadosh, and A. Dowker (eds), The Oxford Handbook of Numerical Cognition. Oxford University Press

Aïn, S.A., Giret, N., Grand, M., Kreutzer, M., & Bovet, D. (2009). The discrimination of discrete and continuous amounts in African grey parrots (*Psittacus erithacus*). Animal Cognition, 12, 145–154

Ainslie, G. (1974). Impulse control in pigeons. Journal of the Experimental Analysis of Behavior, 21, 485–489

Amici, F., Aureli, F., & Call, J. (2008). Fission-fusion dynamics, behavioral flexibility, and inhibitory control in primates. Current Biology, 18, 1415–1419

Anderson, J.R., Awazu, S., & Fujita, K. (2000). Can squirrel monkeys (*Saimir sciureus*) learn self-control? A study using food array selection tests and reverse-reward contingency. Journal of Experimental Psychology: Animal Behavior Processes, 26, 87–97

Anderson, J.R., Awazu, S., & Fujita, K. (2004). Squirrel monkeys (*Saimiri sciureus*) choose smaller food arrays: long-term retention, choice with nonpreferred food, and transposition. Journal of Comparative Psychology, 118, 58–64

Beran, M.J. (2001). Summation and numerousness judgments of sequentially presented sets of items by chimpanzees (*Pan troglodytes*). Journal of Comparative Psychology, 115, 181–191

Beran, M.J. (2004a). Long-term retention of the differential values of Arabic numerals by chimpanzees (*Pan troglodytes*). Animal Cognition, 7, 86–92

Beran, M.J. (2004b). Chimpanzees (*Pan troglodytes*) respond to nonvisible sets after one-by-one addition and removal of items. Journal of Comparative Psychology, 118, 25–36

Beran, M.J. (2007). Rhesus monkeys (*Macaca mulatta*) enumerate large and small sequentially presented sets of items using analog numerical representations. Journal of Experimental Psychology: Animal Behavior Processes, 33, 42–54

Beran, M.J. (2008). Monkeys (*Macaca mulatta* and *Cebus apella*) track, enumerate, and compare multiple sets of moving items. Journal of Experimental Psychology: Animal Behavior Processes, 34, 63–74

Beran, M.J., Beran, M.M., Harris, E.H., & Washburn, D.A. (2005). Ordinal judgments and summation of nonvisible sets of food items by two chimpanzees and a rhesus macaque. Journal of Experimental Psychology: Animal Behavior Processes, 31, 351–362

Beran, M.J., Decker, S., Schwartz, A., and Schultz, N. (2011). Monkeys (*Macaca mulatta* and *Cebus apella*) and human adults and children (*Homo sapiens*) enumerate and compare subsets of moving stimuli based on numerosity. Frontiers in Comparative Psychology, 2, Article 61

Beran, M.J., Evans, T.A., Leighty, K.A., Harris, E.H., & Rice, D. (2007). summation and quantity judgments of sequentially presented sets by capuchin monkeys (*Cebus apella*). American Journal of Primatology, 70, 191–194

Beran, M.J., Harris, E.H., Evans, T.A., Klein, E.D., Chan, B., Flemming, T.M., & Washburn, D.A. (2008a). Ordinal judgments of symbolic stimuli by capuchin monkeys (*Cebus apella*) and rhesus monkeys (*Macaca mulatta*): the effects of differential and nondifferential reward. Journal of Comparative Psychology, 122, 52–61

Beran, M.J., Johnson-Pynn, J.S., & Ready, C. (2008b). Quantity representation in children and rhesus monkeys: linear versus logarithmic scales. Journal of Experimental Child Psychology, 100, 225–233

Beran, M.J., & Rumbaugh, D.M. (2001). 'Constructive' enumeration by chimpanzees (*Pan troglodytes*) on a computerized task. Animal Cognition, 4, 81–89

Beran, M.J. Rumbaugh. D.M., & Savage-Rumbaugh. E.S. (1998). Chimpanzee (*Pan troglodytes*) counting in a computerized testing paradigm. Psychological Record, 48, 3–20

Beran, M.J., Taglialatela, L.A., Flemming, T.J., James, F.M., & Washburn, D.A. (2006). Nonverbal estimation during numerosity judgements by adult humans. Quarterly Journal of Experimental Psychology, 59, 2065–2082

Biro, D., & Matsuzawa, T. (1999). Numerical ordering in a chimpanzee (*Pan troglodytes*): planning, executing, and monitoring. Journal of Comparative Psychology, 113, 178–185

Boysen, S.T., & Bernston, G.G. (1989). Numerical competence in a chimpanzee (*Pan troglodytes*). Journal of Comparative Psychology, 103, 23–31

Boysen, S.T., & Bernston, G.G. (1995). Responses to quantity: perceptual versus cognitive mechanisms in chimpanzees. Journal of Experimental Psychology: Animal Behavior Processes, 21, 82–86

Boysen, S.T., Bernston, G.G., Shreyer, T.A., & Quigley, K.S. (1993). Processing of ordinality and transitivity by chimpanzees (*Pan troglodytes*). Journal of Comparative Psychology, 107, 208–215

Boysen, S.T., & Capaldi, E.J. (eds) (1993). The development of numerical competence: animal and human models. Comparative cognition and neuroscience. Hillsdale, NJ: Erlbaum

Brannon, E.M., Cantlon, J.F., & Terrace, H.S. (2006). The role of reference points in ordinal numerical comparisons by rhesus macaques (*Macaca mulatta*). Journal of Experimental Psychology: Animal Behavior Processes, 32, 120–134

Brannon, E.M., & Roitman, J.D. (2003). Nonverbal representations of time and number in animals and human infants. In W. H. Meck (ed.), Functional and Neural Mechanisms of Interval Timing (pp. 143–182). Boca Raton, FL: CRC Press

Brannon, E.M., & Terrace, H.S. (1998). Ordering of the numerosities 1 to 9 by monkeys. Science, 282, 746–749

Brannon, E.M., & Terrace, H.S. (2000). Representation of the numerosities 1–9 by rhesus macaques (*Macaca mulatta*). Journal of Experimental Psychology: Animal Behavior Processes, 26, 31–49

Burns, R.A., Goettl, M.E., & Burt, S.T. (1995). Numerical discriminations with arrhythmic serial presentations. Psychological Record, 45, 95–104

Candland, D.K. (1995). Feral children and clever animals: reflections on human nature. Oxford, MA: Oxford University Press

Cantlon, J.F. & Brannon, E.M. (2006a). The effect of heterogeneity on numerical ordering in rhesus monkeys. Infancy, 9, 173–189

Cantlon, J.F., & Brannon, E.M. (2006b). Shared system for ordering small and large numbers in monkeys and humans. Psychological Science, 17, 401–406

Cantlon, J.F., Platt, M.L., & Brannon, E.M. (2009). Beyond the number domain. Trends in Cognitive Sciences, 13, 83–91

Capaldi, E.J., & Miller, D.J. (1988). Counting in rats: its functional significance and the independent cognitive processes that constitute it. Journal of Experimental Psychology: Animal Behavior Processes, 14, 3–17

Cohen Kadosh, R., Henik, A., Rubinsten, O., Mohr, H., Dori, H., Van de Ven, V., Zorzi, M., Hendler, T., Goebel, R., & Linden, D.E.J. (2005). Are numbers special? The comparison systems of the human brain investigated by fMRI. Neuropsychologia, 43, 1238–1248

Cohen Kadosh, R., Lammertyn, J., & Izard, V. (2008). Are numbers special? An overview of chronometric, neuroimaging, developmental and comparative studies of magnitude representation. Progress in Neurobiology, 84, 132–147

Cohen Kadosh R., & Walsh, V. (2009). Numerical representation in the parietal lobes: abstract or not abstract? Behavioral and Brain Sciences, 32, 313–373

Cordes, S., Gelman, R., Gallistel, C.R., & Whalen, J. (2001). Variability signatures distinguish verbal from nonverbal counting for both large and small numbers. Psychonomic Bulletin and Review, 8, 698–707

Davis, H. (1993). Numerical competence in animals: life beyond Clever Hans. In S. T. Boysen & E. J. Capaldi (eds), The Development of Numerical Competence: Animal and Human Models (pp. 109–125). Hillsdale, NJ: Erlbaum

Davis, H., & Perusse, R. (1988). Numerical competence in animals: definitional issues, current evidence, and a new research agenda. Behavioral and Brain Sciences, 11, 561–615

Dehaene, S. (1997). The number sense. New York: Oxford University Press

Dehaene, S. (2003). The neural basis of the Weber-Fechner law: a logarithmic mental number line. Trends in Cognitive Sciences, 7, 145–147

Dehaene, S., & Cohen, L. (1995) Towards an anatomical and functional model of number processing. Mathematical Cognition, 1, 83–120

Dehaene, S., Dehaene-Lambertz, G., & Cohen, L. (1998). Abstract representations of numbers in the animal and human brain, Trends in Neuroscience, 28, 355–361

Evans, T.A., Beran, M.J., Harris, E.H., & Rice, D. (2009). Quantity judgments of sequentially presented food items by capuchin monkeys (*Cebus apella*). Animal Cognition, 12, 97–105

Evans, T.A., Beran, M.J., & Addessi, E. (2010). Can nonhuman primates use tokens to represent and sum quantities? Journal of Comparative Psychology, 129, 369–380

Feigenson, L., Carey, S., & Hauser, M.D. (2002). The representations underlying infants' choice of more: object files versus analog magnitudes. Psychological Science, 13, 150–156

Feigenson, L., Dehaene, S., & Spelke, E. (2004). Core systems of number. Trends in Cognitive Sciences, 8, 307–314

Fias, W., Lammertyn, J., Reynvoet, B., Dupont, P., & Orban, G.A. (2003). Parietal representation of symbolic and nonsymbolic magnitude. Journal of Cognitive Neuroscience, 15, 47–56

Flombaum, J.I., Junge, J.A., & Hauser, M.D. (2005). Rhesus monkeys (*Macaca mulatta*) spontaneously compute addition operations over large numbers. Cognition, 97, 315–325

Gallistel, C.R., & Gelman, R. (2000). Non-verbal numerical cognition: from reals to integers. Trends in Cognitive Sciences, 4, 59–65

Garland, A., Low, J., & Burns, K.C. (2012). Large quantity discrimination by North Island robins (*Petroica longipes*). Animal Cognition, 15, 1129–1140

Gelman, R., & Gallistel, C.R. (1978). The Child's Understanding of Number. Cambridge, MA: Harvard University Press

Gomez-Laplaza, L.M., & Gerlai, R. (2011). Spontaneous discrimination of small quantities: shoaling preferences in angelfish (*Pterophyllum scalare*). Animal Cognition, 14, 565–574

Hanus, D., & Call, J. (2007). Discrete quantity judgments in the great apes (*Pan paniscus, Pan troglodytes, Gorilla gorilla, Pongo pygmaeus*): the effect of presenting whole sets versus item-by-item. Journal of Comparative Psychology, 121, 241–249

Hauser, M.D., Carey, S., & Hauser, L.B. (2000). Spontaneous number representation in semi-free-ranging rhesus monkeys. Proceedings of the Royal Society of London. Series B: Biological Sciences, 267, 829–833

Hicks, L.H. (1956). An analysis of number-concept formation in the rhesus monkey. Journal of Comparative and Physiological Psychology, 49, 212–218

Hunt, S., Low, J., & Burns, K.C. (2008). Adaptive numerical competency in a food-hoarding songbird. Proceedings of the Royal Society B: Biological Sciences, 275, 2373–2379

Huntley-Fenner, G. (2001). Children's understanding of number is similar to adults' and rats': numerical estimation by 5–7-year-olds. Cognition, 78, B27-B40

Huntley-Fenner, G., & Cannon, E. (2000). Preschoolers' magnitude comparisons are mediated by a preverbal analog mechanism. Psychological Science, 11, 147–152

Irie-Sugimoto, N., Kobayashi, T., Sato, T., & Hasegawa, T. (2009). Relative quantity judgment by Asian elephants (*Elephas maximus*). Animal Cognition, 12, 193–199

Jordan, K.E. & Brannon, E.M. (2006). Weber's law influences numerical representations in rhesus macaques (*Macaca mulatta*). Animal Cognition, 9, 159–172

Judge, P.G., Evans, T.A., & Vyas, D.K. (2005). Ordinal representation of numeric quantities by brown capuchin monkeys (*Cebus apella*). Journal of Experimental Psychology: Animal Behavior Processes, 31, 79–94

Kaufmann, L., Kucian, K., and von Aster, M. (this volume). Development of the numerical brain. In R. Cohen Kadosh, and A. Dowker (eds), The Oxford Handbook of Numerical Cognition. Oxford University Press

Kaufmann, L., Wood, G., Rubinsten, O., & Henik, A. (2011). Meta-analyses of developmental fMRI studies investigating typical and atypical trajectories of number processing and calculation. Developmental Neuropsychology, 36, 763–787

Kralik, J.D. (2005). Inhibitory control and response selection in problem solving: how cotton-top tamarins (*Saguinus oedipus*) overcome a bias for selecting the larger quantity of food. Journal of Comparative Psychology, 119, 78–89

Krusche, P., Uller, C., & Dicke, U. (2010). Quantity discrimination in salamanders. Journal of Experimental Biology, 213, 1822–1828

Kuroda, R. (1931). On the counting ability of a monkey (*Macacus cynomolgus*). Journal of Comparative Psychology, 12, 171–180

Lit, L., Schweitzer, J.B., & Oberbauer, A.M. (2011). Handler beliefs affect scent detection dog outcomes. Animal Cognition, 14, 387–394

Logue, A. W. (1988). Research on self-control: An integrating framework. Behavioral and Brain Sciences, 11, 665–709

Matsuzawa, T. (1985). Use of numbers by a chimpanzee. Nature, 315, 57–59

Menon, V. (this volume). Arithmetic in the child and adult brain. In R. Cohen Kadosh, and A. Dowker (eds.), The Oxford Handbook of Numerical Cognition. Oxford University Press

Nieder, A. (this volume). Neuronal correlates of nonverbal numerical competence in primates. In R. Cohen Kadosh, and A. Dowker (eds), The Oxford Handbook of Numerical Cognition. Oxford University Press

Nieder, A., & Dehaene, S. (2009). Representation of number in the brain. Annual Review of Neuroscience, 32, 185–208

Nieder, A., Freedman, D.J., & Miller, E.K. (2002). Representation of the quantity of visual items in the primate prefrontal cortex. Science, 297, 1708–1711

Nieder, A., & Miller, E.K. (2003). Coding of cognitive magnitude: compressed scaling of numerical information in the primate prefrontal cortex. Neuron, 37, 149–157

Nieder, A., & Miller, E.K. (2004). Analog numerical representations in rhesus monkeys: evidence for parallel processing. Journal of Cognitive Neuroscience, 16, 889–901

Noble, L.M. (1986). Absolute versus relative conceptual numerousness judgments in squirrel monkeys (*Saimiri sciureus*). Dissertation Abstracts International, 46, 3257–3258
Olthof, A., Iden, C.M., & Roberts, W.A. (1997). Judgments of ordinality and summation of number symbols by squirrel monkeys (*Saimiri sciureus*). Journal of Experimental Psychology: Animal Behavior Processes, 23, 325–339
Pepperberg, I.M. (1987). Evidence for conceptual quantitative abilities in the African grey parrot: labeling of cardinal sets. Ethology, 75, 37–61
Pepperberg, I.M. (1994). Numerical competence in an African grey parrot (*Psittacus erithacus*). Journal of Comparative Psychology, 108, 36–44
Pepperberg, I.M. (2006). Ordinality and inferential abilities of a grey parrot (*Psittacus erithacus*). Journal of Comparative Psychology, 120, 205–216
Perdue, B.M., Talbot, K.F., Stone, A., & Beran, M.J. (2012). Putting the elephant back in the herd: elephant relative quantity judgments match those of other species. Animal Cognition, 15, 955–961
Pfungst, O. (1911). Clever Hans (*The horse of Mr von Osten*): a contribution to experimental animal and human psychology, transl. by C. L. Rahn. New York: Henry Holt. (Originally published in German, 1907.)
Piffer, L., Agrillo, C., & Hyde, D.C. (2012). Small and large number discrimination in guppies. Animal Cognition, 15, 215–221
Roberts, W.A. (2006). Evidence that pigeons represent both time and number on a logarithmic scale. Behavioural Processes, 72, 207–214
Roberts, W.A. (2010). Distance and magnitude effects in sequential number discrimination by pigeons. Journal of Experimental Psychology: Animal Behavior Processes, 36, 206–216
Roberts, W.A., & Mitchell, S. (1994). Can a pigeon simultaneously process temporal and numerical information? Journal of Experimental Psychology: Animal Behavior Processes, 20, 66–78
Rumbaugh, D.M. (1977). Language learning by a chimpanzee: the LANA Project. New York: Academic Press
Rumbaugh, D. M., Hopkins, W. D., Washburn, D. A., & Savage-Rumbaugh, E. S. (1989). Lana chimpanzee learns to count by "NUMATH": A summary of a videotaped experimental report. Psychological Record, 39, 459–470
Savage-Rumbaugh, E.S. (1986). Ape language: from conditioned response to symbol. New York: Columbia University Press
Scarf, D., Hayne, H., & Colombo, M. (2011). Pigeons on par with primates in numerical competence. Science, 334, 1664
Sebeoke, T., & Rosenthal, L. (eds) (1981). The Clever Hans phenomenon: communication with horses, whales, apes, and people. New York: New York Academy of Sciences
Siegler, R.S., & Opfer, J.E. (2003). The development of numerical estimation: evidence for multiple representations of numerical quantity. Psychological Science, 14, 237–243
Smith, B.R., Piel, A.K., & Candland, D.K. (2003). Numerity of a socially housed hamadryas baboon (*Papio hamadryas*) and a socially housed squirrel monkey (*Saimiri sciureus*). Journal of Comparative Psychology, 117, 217–225
Stevens, J.R., Hallinan, E.V., & Hauser, M.D. (2005). The ecology and evolution of patience in two New World monkeys. Biology Letters, 1, 223–226
Shifferman, E.M. (2009). Its own reward: lessons to be drawn from the reversed-reward contingency paradigm. Animal Cognition, 12, 547–558
Simon, T.J. (1997). Reconceptualizing the origins of number knowledge: a 'non-numerical' account. Cognitive Development, 12, 349–372
Szalda-Petree, A.D., Craft, B.B., Martin, L.M., & Deditius-Island, H.K. (2004). Self-control in rhesus macaques (*Macaca mulatta*): controlling for differential stimulus exposure. Perceptual and Motor Skills, 98, 141–146
Terrace, H.S., Son, L.K., & Brannon, E.M. (2003). Serial expertise of rhesus macaques. Psychological Science, 14, 66–73

Terrell, D.F., & Thomas, R.K. (1990). Number-related discrimination and summation by squirrel monkeys (*Saimiri sciureus sciureus* and *S. boliviensus boliviensus*) on the basis of the number of sides of polygons. Journal of Comparative Psychology, 104, 238–247

Thomas, R.K., & Chase, L. (1980). Relative numerousness judgments by squirrel monkeys. Bulletin of the Psychonomic Society, 16, 79–82

Thomas, R.K., Fowlkes, D., & Vickery, J.D. (1980). Conceptual numerousness judgments by squirrel monkeys. American Journal of Psychology, 93, 247–257

Tobin, H., Logue, A.W., Chelonis, J.J., Ackerman, K.T., & May, J.G.I. (1996). Self-control in the monkey *Macaca fascicularis*. Animal Learning & Behavior, 24, 168–174

Tomonaga, M., & Matsuzawa, T. (2002). Enumeration of briefly presented items by the chimpanzee (*Pan troglodytes*) and humans (*Homo sapiens*). Animal Learning and Behavior, 30, 143–157

Uller, C., Carey, S., Huntley-Fenner, G., & Klatt, L. (1999). What representations might underlie infant numerical knowledge? Cognitive Development, 14, 1–36

vanMarle, K., Aw, J., McCrink, K., & Santos, L.R. (2006). How capuchin monkeys (*Cebus apella*) quantify objects and substances. Journal of Comparative Psychology, 120, 416–426

Vonk, J., and Beran, M.J. (2012). Bears 'count' too: quantity estimation and comparison in black bears (*Ursus americanus*). Animal Behaviour, 84, 231–138

Washburn, D.A., & Rumbaugh, D.M. (1991). Ordinal judgments of numerical symbols by macaques. Psychological Science, 2, 190–193

Wesley, F. (1961). The number concept: a phylogenetic review. Psychological Bulletin, 58, 420–428

Whalen, J., Gallistel, C.R., & Gelman, R. (1999). Nonverbal counting in humans: the psychophysics of number representation. Psychological Science, 10, 130–137

Wood, J.N., Hauser, M.D., Glynn, D.D., & Barner, D. (2008). Free-ranging rhesus monkeys spontaneously individuate and enumerate small numbers of non-solid portions. Cognition, 106, 207–221

Xia, L., Emmerton, J., Siemann, M., & Delius, J.D. (2001). Pigeons (*Columba livia*) learn to link numerosities with symbols. Journal of Comparative Psychology, 115, 83–91

Xia, L., Siemann, M., & Delius, J.D. (2000). Matching of numerical symbols with number of responses by pigeons. Animal Cognition, 3, 35–43

Xu, F. (2003). Numerosity discrimination in infants: evidence for two systems of representations. Cognition, 89, B15-B25

CHAPTER 14

NUMERICAL ABILITIES AND ARITHMETIC IN INFANCY

KOLEEN MCCRINK AND WESLEY BIRDSALL

The 'Number Sense' in Infancy

The field of infant cognition has revealed a surprising prowess for representing numerical quantities before any formal education or mathematical training on the topic. Preverbal infants have been found to represent the number of both small and large numbers of objects (Antell and Keating 1983; Bijeljac-Babic et al. 1993; Cheries et al. 2006; vanLoosbroek and Smitsman 1990; Cooper 1984; Starkey and Cooper 1980; Strauss and Curtis 1981; Wood and Spelke 2005a; Xu and Spelke 2000; Xu et al. 2005). They can also enumerate the number of events in a scene (Sharon and Wynn 1998; Wood and Spelke 2005b; Wynn 1996), discriminate small numbers of collections of objects (Chiang and Wynn 2000; Wynn et al. 2001), enumerate tones as well as objects (vanMarle and Wynn 2009; Lipton and Spelke 2003, 2004), and match number across visual and auditory modalities (Izard et al. 2009; Jordan & Brannon 2006; Kobayashi et al. 2005; Starkey et al. 1983, 1990; though c.f. Mix et al. 1997; Moore et al. 1987). There is also evidence for infants' sensitivity to other, co-occurring perceptual variables such as area and contour length, especially in small-number arrays (Clearfield and Mix 1999, 2001; Feigenson et al. 2002a, Feigenson et al. 2002b.)

Several prominent theorists have proposed that there are actually two distinct systems for keeping track of number. The first system, commonly referred to as the 'object-file' system (after Kahneman et al. 1992), tracks the spatiotemporal and featural characteristics of precise *small* numbers of individuals, with an upper representation limit of three to four objects (see Scholl 2001 for a review). The second system represents the approximate numerical magnitude of *large* numbers of objects, with variability in representation that increases as the numbers increase (see Feigenson et al. 2004; McCrink and Wynn 2008; Xu 2003 for overviews of this theory and supporting evidence for the two-systems view.) This increasing variability with increasing amount results in the 'mental' difference between two amounts being determined by their ratio, and not the absolute difference between them (a basic physiological phenomenon also known as Weber's law; ten and twenty objects are as psychologically distinct as 100 and 200). The large-number system, referred to as the 'number sense' (after Dehaene 1997) or the Approximate Number System (hereafter ANS), is thought to be one of the few *core knowledge* systems (Spelke 2000) that is ontogenetically and phylogenetically continuous, as it is found in many non-human animal species (see Gallistel 1990, Nieder 2005 for a review) as well as throughout the human lifespan (Cordes et al. 2001; Barth et al. 2003; Barth et al. 2005), irrespective of formal education (Pica et al. 2004.)

Although there has been a great deal of work on representation and discrimination of number and quantity, as can be seen above, the main goal of this chapter is to review evidence for productive arithmetic operations and calculations in infancy, using one or both of these numerical systems. For the purposes of this review, we will consider the evidence for addition, subtraction, ordering, division, and multiplication (as the implied inverse of division) in infancy (after Gallistel and Gelman's (1992) seminal conception of the components of a foundational number system), and ask whether and when infants can perform simple and complex arithmetic calculations.

Addition and Subtraction

In one of the first studies on the topic, Wynn (1992) examined whether the process of learning to add and subtract was inculcated through extensive experience and training, or if there might be an underlying set of numerical concepts existing as early as a few months after birth. The author utilized a classic violation-of-expectation paradigm, in which infants were shown several events, some of which were physically possible, and some of which were impossible. (Infants tend to look longer to events which are physically impossible, suggesting that these particular events violate the infants' underlying expectations of how the world works.) Five-month-old infants were randomly placed into 'addition' conditions or 'subtraction' conditions, in which they saw arithmetic problems acted out by a series of placements of dolls on a stage. Infants who saw addition problems viewed a second doll being added to an initial single doll that had been occluded; those who saw subtraction scenarios viewed an initial display of two dolls become occluded and one doll subtracted from the array. In both cases, the occluder was taken away to reveal either one or two dolls. Infants who saw an addition scenario looked longer when only one doll was revealed, and those who saw a subtraction scenario looked longer when two were revealed. A follow-up experiment was conducted to determine if infants were truly calculating the outcomes to these problems, or if they were instead deploying generic arithmetic or logical principles. For example, infants may possess a simple rubric to accept any outcome that is not the same as the initial amount, or, when adding accept any number greater than the initial amount and when subtracting accept any number less than the initial amount. To test this possibility, a new group of infants was presented with the same addition problem of one plus one, and outcomes of either two or three dolls. Infants in this experiment continued to look significantly longer at incorrect outcomes relative to correct ones, indicating that they were estimating an actual outcome and not applying a generic mathematical principle. Although Wynn (1992) herself acknowledged the possibility that infants were calculating not discrete numbers but amount of substance (e.g., expecting a certain amount of blue behind the screen, as the dolls are wearing blue pants), the author pointed to the finding that infants are predisposed to interpret their physical surroundings in terms of discrete entities (Spelke 1990) and suggests that this is the likely interpretation of the scene by the infants in her study.

A follow up study by Simon et al. (1995) examined whether understanding of object principles in infancy provided a more straightforward explanation of Wynn's finding than a purely numerical account that invoked addition and subtraction calculations. Work by Baillargeon (1993), Leslie (1984), and Spelke (1990) illustrated that infants have a basic understanding of the physical world, such as the fact that objects are 'substantial, permanent, and occupy space' (Simon et al. 1995, p. 254). Additionally, Strauss and Curtis (1984), vanLoosbroek and Smitsman (1990), and Antell and Keating (1983) all illustrated that infants can reliably and repeatedly distinguish between one to three discrete objects without an appeal to mathematical addition or subtraction. Taken together, these studies offer the alternative that a relatively sophisticated understanding of physical knowledge and individuation of objects in infancy was the driving factor behind infants' performance in Wynn (1992). To test this, Simon et al. (1995) modified Wynn's (1992) paradigm by including dolls with unique identities: Elmo and Ernie, from a popular children's television show. The authors presented three different types of possible incorrect outcomes in addition to the correct outcome: arithmetically impossible (e.g., Ernie + Ernie = Ernie), physically impossible (Ernie + Ernie = Ernie and Elmo), or both (Ernie + Ernie = Elmo). Results from both addition and subtraction versions of this study found that infants looked longer at those conditions with incorrect arithmetic outcomes, regardless of the object identities. Simon et al. (1995) interpreted this finding in light of work done by Xu (1993; as cited in Simon et al. 1995) which found that ten-month-old infants preferentially attend to spatiotemporal information in a scene, and not information about individual identity, to track the existence of objects. The authors suggest that this spatiotemporal preference, combined with their result that infants violate a detection of the numerical outcome, supports a link between numerical and spatiotemporal knowledge in infancy. This link would later be elucidated and appear as a central component to the small-number, object-tracking system for numerical arrays up to a magnitude of three or four (as outlined in the adult literature by Trick and Pylyshyn 1994, and Scholl and Pylyshyn 1999, and expanded to a developmental theory in Carey and Xu 2001.)

Koechlin et al. (1997) explored whether infants in Wynn (1992) may have developed an expectation as to what should be behind the screen based solely on the spatial location of specific objects, since each item was repeatedly placed in the same location on the stage and behind the screen. This hypothesis was particularly likely given Xu's (1993) and Simon et al.'s (1995) findings of the importance of spatiotemporal object tracking, and studies in which infants exhibited the ability to relate locations in space with the objects that had previously existed in those locations (Baillargeon 1993; Baillargeon and DeVos 1991.) Koechlin et al. (1997) randomly placed five-month-old infants into either a 'fixed' or 'rotating' conditions (in which the objects were always moving) using the same values as the original Wynn (1992) study. In both conditions the infants discriminated the mathematically-correct from incorrect outcomes. As with Simon et al. (1995), the authors discuss the possibility that a system that is built for tracking the existence (and not just the placement) of a small number of objects is underlying competence in tasks such as these, and that although infants do indeed possess the ability to detect numerically-based violations of expectation, there may be no actively arithmetic factor at play. It might be that infants are not actually manipulating number per se, but simply detecting a mismatch between what they see and what 'open' object-file records they are holding in memory. Taken together, the results from Koechlin et al. (1997) and Simon et al. (1995) illustrate that in the domain of small numbers, it is extremely difficult to dissociate early knowledge of physics and object persistence from early knowledge of arithmetic.

Along these lines, Baillargeon et al (1994) (as cited in Uller et al. (1999)) found that 10-month-olds infants shown a 1 + 1 + 1 = 3 or 2 addition problem failed to discriminate the possible from impossible outcomes, despite being relatively old compared to Wynn's (1992) infants. Uller et al. (1999) explain that a purely arithmetic model provides no real explanation as to why this task would be more difficult to compute than 2 + 1 = 3 or 2, an 'equation' the ten-month-old infants were able to comprehend. However, in the proposed object-file system, the mental tags assigned to each object require updating each time the imagined set of objects behind the screen change, and this process taxes the immature short-term memory of the infants. Therefore, a 1 + 1 + 1 problem—which requires more mental updates than a 2 + 1 problem—leads to an increased margin of error in following and comprehending the experiment. To specifically test this hypothesis, Uller et al. (1999) tested eight-month-olds infants in two conditions which manipulated the level of mental updating needed to successfully keep track of the number of objects in the scene. In the object-first conditions, they saw one object become occluded and another added; in the screen-first they saw a blank stage, a screen rotated upward, and two objects added serially behind the screen. The latter condition requires more-taxing mental imagery and memory updates. Infants succeeded to a greater degree in the object-first condition, an outcome suggested by the object-file representation model. Follow-up studies illustrated that by the time they reached ten-months-old, infants were able to succeed in both conditions (likely due to increased memory as they developed); even eight-month-olds who are given extra 'help' in being shown a unique screen for each object (and therefore are given explicit location information) can succeed in both conditions. This set of experiments suggests a clear role for memory and spatiotemporal considerations when considering infants' ability to perform addition and subtraction over small sets of objects.

The studies outlined thus far have shown that infants do indeed possess some amount of numerical operational ability, even when other object reasoning-based explanations are experimentally accounted for. Simon et al (1995), Koechlin et al. (1997), and Uller et al. (1999) all replicated Wynn's original study to some extent, and found results generally consistent with hers. However, Wakeley et al. (2000) found no success by young infants on a straightforward replication of Wynn's (1992) designs, as well as an additional design in which infants were presented with an incorrect problem of 3-1=1 vs. a correct problem of 3-1=2. They suggest that the findings from addition and subtraction experiments were inconsistent and variable (a claim that Wynn (2000), strongly rebutted in the same journal issue.) Researchers such as Clearfield and Mix (1999, 2001) and Cohen and Marks (2002) also doubt the existence of calculation abilities, instead pointing to alternate non-numerical possibilities which lie more in the domain of perception than cognition. Clearfield and Mix (1999, 2001) found that infants readily attend to the area and contour length of a set of small objects, an ability that could explain performance in Wynn's (1992) study (but importantly, not subsequent studies like Simon et al. 1995, which manipulate the identity of the objects at test and still find correct calculations of outcomes.) Cohen and Marks (2002) appeal to a perceptual, non-numerical, dual-process model in which infants faced with addition and subtraction tasks succeed only by exhibiting (1) a preference for a familiar number of objects and (2) a preference for more objects. The authors find evidence for both of these preferences and suggest that in tandem they could account for the data in Wynn's (1992) paper as well as subsequent replications. Additionally, Moore and Cocas (2006) explored the role of familiarity preferences rather than true numerical computation by thoroughly familiarizing infants to the test displays prior to the actual testing sessions, and found mixed evidence for competence following this manipulation (with more cognitively-mature female infants succeeding while the less cognitively-mature males did not.)

One way to explore the impact of visual perceptual factors as the driving force behind calculation abilities is to design an addition and subtraction study *without* the use of visible objects. Kobayashi et al. (2004) performed such an experiment by first familiarizing five-month-old infants to a stage in which dolls appeared and made a tone-like sound when they landed. They then presented the infants with test scenarios in which they saw one object + one tone resulting in two objects vs. one object + two tones resulting in two objects (the two-outcome condition), or one object + two tones resulting in three objects vs. one object + one tone resulting in three objects (the three-outcome condition). The infants looked significantly longer to the incorrect outcomes, a finding which circumvents perceptual accounts such as those proposed by Clearfield and Mix (1999, 2001) or outcome-preference or familiarity accounts such as Cohen and Marks (2002).

It is important to note that this does not mean that infants *cannot* add and subtract over continuous perceptual quantities such as area or contour length. In particular, work by Feigenson and colleagues (Feigenson et al. 2002a; Feigenson et al. 2002b) suggests that they actively do pay attention to these variables and can, for example, respond preferentially to larger perceptual total amounts over larger numerical amounts (in the case of food) or detect when objects' additive amount of surface area is tabulated incorrectly. In so-called 'crawling paradigms.' infants between ten and fourteen months of age selectively crawl towards locations in which they saw a greater sum total of food substance surface area after an addition scenario (involving crackers, or sometimes cheerios; Cheries et al. 2008; Feigenson et al. 2002a,b; vanMarle 2013; vanMarle and Wynn 2011). Further, the discrimination ratio of this addition function over continuous extent is distinct from that found for discrete number, lending support to the two-systems view in which small numbers of objects are reasoned about differently than large numbers of objects (Feigenson et al. 2002a, vanMarle and Wynn 2011).

All of the studies reviewed thus far have utilized infants' gaze as a measure of their level of surprise at the presented outcomes. Looking time is an indirect and oftentimes contradictory, or inconsistent, measure of cognition; sometimes comprehension of a particular stimulus is indicated by a familiarity preference and sometimes a novelty preference, depending on the complexity of the stimuli and the age of the infant (Hunter et al. 1983). With this in mind, Berger et al. (2006) sought to provide physiological, neurological methods to gain further insight into the question of numerical knowledge in infancy. The authors presented seven-month-old infants with stimuli based on Wynn's (1992) study, while recording brain activity via EEG/ERP (electroencephalogram/event-related potential) technology. They observed behavioral responses that supported Wynn's findings; the infants looked longer to the incorrect outcomes. The EEG/ERP results showed activation in the portion of the brain that has been repeatedly shown in existing literature as the site of localized responses to error in adult brains. Adults tested in this paradigm concurrently also showed similar localization of response to the infants. This result provides physiological evidence to support the claim that there is an acknowledgment in the infant brain of an error occurring in the incorrect-outcome conditions of these studies. However, as with any study that looks at calculation within the realm of small numbers, this research does not illuminate if the error detect is arithmetic or simply that of a one-to-one object correspondence between an imagined set and a perceived set gone awry.

If infants do possess the ability to understand small numbers and manipulate them in an operational manner, the literature at this point is unable to dissociate between a calculation-based, arithmetic account and one in which each object is mentally represented by a single object file and then tracked spatiotemporally. To some degree, this distinction is unclear; because the object-tracking mechanism provides for an exact numerical calculation, it can be, and is often, considered a type of number system itself. However, the object-tracking mechanism is very different from a large-number mechanism in that it is not thought to yield any cardinal representations, or representations of the whole set in a numerical fashion. Each object is thought of as an individual, with its own defining characteristics (e.g., the type and amount of spatial extent it exhibits, its location in the world.) To many theorists the engagement of a system that is specifically for number, and not one that is co-opted in order to provide an idea of what is present in a scene, would provide a more satisfying answer to the question of whether infants possess purely arithmetic calculation abilities. Such a system is the aforementioned Approximate Number System (ANS), which yields mental representations of approximate quantities of objects. This system has no upper limit, unlike the object-tracking system, which can only keep track of about four objects (Scholl 2001). Xu and Spelke (2000) established that this system is also present in infancy, with six-month-old infants able to successfully discriminate between eight and sixteen objects, but not eight and twelve, indicating a relatively high level of noise in their representations. Further research by Lipton and Spelke (2003) shows a similar discriminatory ability in the auditory system, suggesting that the ANS can estimate magnitudes in multiple modalities.

With this research in mind, McCrink and Wynn (2004), attempted to create a study that would address the question of infant operational capabilities beyond object-tracking system limits. In this study, nine-month-old infants were assigned to either an addition (5 + 5 = 5 or 10), or subtraction (10 − 5 = 5 or 10) condition. In a departure from earlier infant number studies, the stimuli used here were presented entirely on video, rather than a live action version of events. This allowed the authors to equate the perceptual variables of area and contour length between the two presented outcomes of five and ten objects, with the express purpose of eliminating those variables as drivers of infants' performance at test. Infants in both conditions looked longer to the incorrect outcomes than correct outcomes, lending credence to the idea that there may not be only one process at work, but instead, a parallel set of mechanisms for number: object-tracking for small number sets, and magnitude estimation for use with larger number sets. McCrink and Wynn (2009) replicated this ability to add and subtract over large numbers, and found that infants were prone to overestimating addition problems and underestimating subtraction problems, a tendency that is also found in adults who perform non-symbolic addition/subtraction calculations (Cordes et al. 2007; McCrink et al. 2007). The results of McCrink and Wynn (2004, 2009), taken with the large body of evidence for addition/subtraction calculations in the small-number realm, strongly support the idea that infants possess an intuitive sense of adding and subtracting quantities.

Ordering

The process of ordering numbers is also considered part of the suite of arithmetic operations underlain by the ANS (alongside addition, subtraction, multiplication, and division; see Gallistel and Gelman (1992) for a review). In one of the first studies on the topic, Cooper (1984) measured ordinal processing in the small-number realm by ten to twelve-month-olds and fourteen to sixteen-month-olds. The infants were habituated to a stimulus pair that exhibited either a greater-than or less-than relationship of two small amounts (from one to four). They were then tested with four different trial type pairings: one drawn directly from habituation, one that had the same ordinal relationship as habituation, one that had a reversed relationship, and one that exhibited an equality relationship. Older infants (> fourteen months of age) exhibited longer looking to any new ordinal relationship, while younger infants (<twelve months of age) looked longer solely to the test item which displayed a pair of equal-number stimuli, indicating that they encoded only a notion of equal/unequal instead of a notion of increasing/decreasing.

The ability to appreciate ordinal relationships in infancy has also been studied more recently by Brannon and colleagues (Brannon 2002; Suanda et al. 2008), and as with many abilities associated with the ANS, has been found in non-human animals using a variety of convergent methods (Brannon and Terrace 2000; Pepperberg 2006; Rugani et al. 2007). To test whether infants perceive numerical amounts such as ten and twenty as distinct but related values along the same continuum (or if they are, instead, conceived of as two different entities '... like a blender is to a chair,' p. 224) Brannon (2002) habituated nine and eleven-month-old infants to displays of arrays of dots presented in either an ascending or descending fashion. The displays contained variable numbers of objects (e.g., 1-2-4, or 2-4-8), with the only constant being that each sequence contained the same type of ordering (ascending, or descending).

They were then presented with test items that were new values, arranged in either a familiar order relative to habituation, or an unfamiliar order. It is worth noting that this design tests for ordinal processing across both the small—and large number range, and the infants must utilize their ANS to represent the larger numbers in some of the arrays. Eleven-month-old infants, but not nine-month-old infants, looked longer to the new ordinal pattern at test, and did so regardless of whether non-numerical cues (such as area and contour length) largely correlated with number (exp. 1) or were controlled for (exp. 2). Interestingly, in a follow-up experiment, nine-month-old infants *were* able to successfully detect the ordinality of a non-numerical stimulus set of a square that either increased or decreased in area, leading the author to suggest that the developmental change between nine and eleven months of age is specific to understanding of numerical ordinality per se. The relative prowess of the eleven-month-old infants in this study is in contrast to the poor performance by a similarly-aged population in Cooper (1984). This is likely due to several stimulus-driven factors; the infants saw values that bridged the small and large number realm and thus readily activated the ANS, the values were in triplicate across the exemplar instead of just a pair, and there was more than one type of habituation trial.

Suanda et al. (2008) further explored the nature of this change during late infancy. The authors replicated the finding that eleven-month-olds could detect a change in ordinality of number per se, and could do so even when the initial numerosity presented was equally-novel across the novel and familiar test stimuli relative to habituation, something that was not controlled in Brannon (2002). Nine-month-olds continue to fail at detecting a change in numerical ordinality, even when given extra time to process the stimuli (up to three seconds per stimulus, compared to one second in Brannon 2002). In contrast to the initial findings of Brannon (2002), they found that infants younger than eleven months of age could not detect a change in the ordinal direction of a non-numerical stimulus of a square increasing/decreasing in area. This is likely because the previous stimulus set in Brannon (2002), of a square that was fixed on the center of the screen while increasing or decreasing in size, yielded a 'looming' (growing) or 'zooming' (shrinking) percept that the children coded as such. Thus the response by nine-month-olds in Brannon (2002) was likely based on not on the concept of quantity ascension or descension, but whether an object was coming at them or receding from them, a percept to which even very young infants are attuned (Yonas et al. 1979). Suanda at al. (2008) also found that, unlike eleven-month-olds, nine-month-olds could not detect ordinality in *any* single dimension (with cumulative area, item size, and number all being tested). These younger infants needed redundant cues across all three dimensions in order to detect a change in ordinality, which dovetails with work by Bahrick (Bahrick et al. 2002; Bahrick and Lickliter 2000) that shows enhanced learning when presented with multiple convergent sensory cues.

Furthering this hypothesis, Picozzi et al. (2010) found that infants as young as seven months of age could detect ordinal numerical sequences, provided that each of the exemplar sequences had its own unique featural information (e.g., specific color), and that the relationship *between* each of the habituation sequences was consistent with the ordering relationship showed *within* the sequence (e.g., the ascending condition was presented 6-12-24, then 9-18-36, then 12-24-48.) Additionally, there is evidence that the infants can abstract an ascension/descension concept and apply it across numerical and non-numerical domains. de Hevia and Spelke (2010) habituated eight-month-old infants to successive arrays that either increased or decreased in number, while controlling for other non-numerical cues. At test, the infants saw sequences of horizontal lines which either increased or decreased in length. Infants looked longer at test trials which presented a new type of order relative to habituation, indicating an amodal transfer of ordinality across different quantity dimensions. (Interestingly, the competence shown by these eight-month-olds in de Hevia and Spelke (2010)—but not by the nine-month-olds tested by Suanda et al. (2008)—suggests that the ability to amodally transfer ordinality across quantity dimensions comes in earlier than the ability to reason about ordinality within one tightly-controlled set of numerical stimuli.) Taken as a whole, the results illustrate an early, non-verbal, and non-symbolic understanding of more and less in both the numerical and non-numerical domains of quantity in the first year of life.

Multiplication and Division

Classic work by psychologists such as Piaget (1954) as well as work by current educational psychologists (such as Kelly Mix and Arthur Baroody) on the development of arithmetic reasoning in early childhood suggests that an explicit, symbolic understanding of multiplication and division—and perhaps even number in general—is best characterized as a prolonged and effortful developmental process (see Baroody and Dowker (2003), or Mix et al. (2002) for a volume on and review of the complex relationship between infant and child concepts of number, respectively; also see Cordes and Gelman (2005) for a rebuttal to Mix et al. (2002).) The challenging nature of complex symbolic operations is evident in the many errors children make when questioned on concepts like fractions, such as not understanding the part-whole relationship, or adding numerators and denominators incorrectly (Hecht 1998), and the poor performance on standardized test questions involving fractional calculations deep into adolescence (Brown et al. 1988; Hope and Owens 1987; Kouba et al. 1997). This is in stark contrast to work on proportional understanding in infancy, which capitalizes on non-symbolic, approximate representations generated by the ANS and/or a related system for tabulating non-numerical quantity. The following studies illustrate an implicit, indirect prowess in preverbal infants for calculating proportions and extracting statistics from an event or scene.

Perhaps the first hint that infants are excellent statisticians came from Saffran et al. (1996). In this series of experiments, eight-month-old infants were presented with two minutes of a continuous synthetic speech stream consisting of four different 'words,' or three syllables that always occurred together (e.g., bi-da-ku or pa-do-ti), with the words themselves occurring randomly after each other. This stream was presented without pause or emphasis, and the infants were then presented with either the words themselves (bi-da-ku), non-words in which the word-forming syllables were arranged incorrectly (da-bi-ku), or part-words which spanned a word boundary (da-ku-pa). Infants exhibited a novelty response to both non-words and part-words, indicating that they were calculating the transitional probability between syllables (which is 1.0, or 100 per cent, in words, .33 in part-words, and 0 in non-words) via a divisory process. This process has been found even when units are not adjacent (Creel et al. 2004; Gebhart et al. 2009; Gomez 2002; Newport and Aslin 2004), when speech is natural instead of synthetic (Hay et al. 2011), and for non-speech units such as tones (Saffran et al. 1999). Although this body of work is conventionally labeled as 'statistical learning,' and is not often thought of in a numerical fashion, the infants in these studies may in fact be responding on the basis of an arithmetic calculation that involves a process of division—a calculation of relative frequency of occurrence and probability of a relationship between distinct units. Many theorists who study the process of statistical learning posit that it is domain-general across many aspects of cognition (see Saffran and Thiessen 2007 for a review of these theories). However, some developmentalists have proposed that the probability information guiding infant behavior in many domains is itself computed by a specific system for reasoning about quantity in an abstract fashion, such as the ANS (see Xu and Denison (2009) for such a proposal). As detailed below, recent research on the juncture between statistical learning and the ANS is beginning to illuminate what, if any, connection exists between the two systems.

Teglas et al. (2007) examined whether infants can spontaneously calculate probabilities when making predictions about novel events. The authors presented twelve-month-olds with computerized movies in which four objects moved randomly inside a container that had a single opening at the bottom. Three of the four objects were yellow, and one blue. The container was then occluded, and a single object emerged that was either yellow (a probable event) or blue (an improbable event.) The infants looked reliably longer to the improbable event, implicating a process which calculated a .25 vs. a .75 probability—an inherently division-based computation. A follow-up experiment established that this was not due to low-level perceptual preferences for singleton objects or a less-common color. Follow-up work using this paradigm by Teglas et al. (2011) found that these computations over the quantity domain were readily integrated with other considerations such as how close the objects were to the opening when the test trial started, or how fast the objects were traveling as they bounced around the enclosed area. Interestingly, and in accordance with work on symbolic proportional reasoning in later childhood, Teglas et al. (2007) found in a final set of experiments with preschoolers that it is not until five years of age that children can explicitly talk about which event was most probable, despite its clear impact on behavior well before then.

Xu and Garcia (2008; see also Xu and Denison 2009, and Denison and Xu 2009) designed convergent violation-of-expectation paradigms to investigate whether infants were 'intuitive statisticians'(p. 5012) in the realm of large numbers as well as small numbers. Xu and Garcia (2008) first familiarized eight-month-old infants to large boxes with an opaque front panel that was opened to reveal a mix of many ping-pong balls that was either overwhelmingly red with some white (for half the familiarization trials) or overwhelmingly white with some red (for the other half of the familiarization trials). The infants were then shown test trials in which the experimenter removed a subset of balls that was largely red, or largely white from a large box, and then opened the box up to reveal its contents. The infants' looking time to the contents of the large box was measured. When infants saw a mismatch between the proportional distribution of the sample relative to the revealed composition of the box, they looked longer than when the sample matched the composition. This indicates they were matching proportions across the two amounts; the only manner in which one red: four white (drawn sample) is comparable to ten red: forty white (composition of balls in the box) is through a division calculation that reduces the larger sample to an equivalent ratio of 1:4. The infants also looked longer when a non-representative sample of balls was drawn from the box, illustrating that these intuitive statistics can be employed both when inferring a larger population from a small sample, and a small sample from a large population. Additionally, infants as young as twelve months of age can integrate this computation with their knowledge of a person's preference (Xu and Denison 2009). These studies have been conceptually replicated using a different paradigm in which infants as young as twelve months of age calculate the probability of a favored-color lollipop being drawn from a container to guide their crawling behavior (Denison and Xu 2009).

As Xu and Garcia (2008) themselves point out, it is unclear if these computations are being done over continuous variables such as color patches, or the number of items themselves. However, the fact that infants in these experiments were given familiarization experience with the items as individual objects makes it unlikely that they were viewing the contents of the box as that of an undifferentiated mass of substance, and points to the ANS as the likely driver behind this ability. Also adding to the likelihood of ANS-supported proportional reasoning over abstract number *per se* is work by McCrink and Wynn (2007). In this set of experiments, six-month-old infants were habituated to slides of arrays of yellow pacmen-like objects and blue balls. Half of the infants saw slides which conveyed an exemplar of a particular ratio (either one pacman: four balls or one pacman: two balls), while the absolute number of each object type and overall number of objects differed between the slides (e.g., for the 1:2 condition—four pacmen: eight pellets, then ten pacmen: twenty pellets, etc.) At test, infants saw either new exemplars of the habituated ratio, or an exemplar of an entirely new ratio. Perceptual factors such as area and contour length were strictly controlled for in both habituation and test movies, meaning that the calculation must be performed over the variable of numerical quantity and not non-numerical quantity. Infants who were habituated to a 1:2 ratio looked significantly longer at test when presented with a 1:4 ratio vs. a 1:2 ratio, while infants who were habituated to a 1:4 ratio showed the opposite pattern. A second experiment examined whether this ratio-extraction ability exhibited poorer performance when the ratio values to be discriminated were arithmetically closer (a version of Weber's law, one well-known signature of representations generated by the ANS). Infants were presented with exemplars of 1:2 or 1:3 and then tested with new versions of these ratios at test. In this experiment, the infants did not distinguish between the habituated and new ratio, indicating that the system underlying this process is 'noisy' and cannot discriminate ratios that differ only by a factor of 1.5 (but can do so for ratios that differ by a factor of 2.0.) Taken as a whole, the body of literature on infant statistics and ratio-based reasoning provides evidence for a subtle but complex ability to relate two quantities in an inherently proportional fashion.

Conclusion

As with any particular field, it is important to consider these findings in tandem with the limitations of the research used to establish them—especially since the conclusions the one draws from these studies are so theoretically telling. The 'drop-out' rate for infant studies is relatively high compared to other populations, which often creates a selection bias and can limit what one concludes about the generality of any found competence. Second, the looking-time measure that infancy researchers rely on so readily is not a straightforward one. Depending on the age of the infant, complexity of the stimuli, and looking time paradigm used, the exhibition of a novelty preference *or* a familiarity preference can be taken as evidence that stimulus discrimination has taken place (for a recent review see Houston-Price and Nakai (2004) and accompanying commentary in the same issue). The reliance on changing levels of attention, as measured by a single factor of looking time, makes it difficult to know if there are multiple representations that are being encoded—but only one 'strongest' representation that drives performance. Finally, as with any comparative psychology program (such as studies on the cognition of non-human animals), these participants cannot speak for themselves, and the designs of the studies must account for multiple alternate strategies that the participants may engage in. This is reflected in the logic of the literature summarized here, and the extensive back-and-forth as to how to interpret the infants' responses to particular experimental designs. However, it is not always a drawback; the alternate interpretations of the studies have led to a deeper understanding of the multiple systems which must work together to achieve involved mathematical understanding.

Indeed, knowledge on the network of systems that support math reasoning—two of which are detailed here—has the potential to be of great benefit to special populations, such as those suffering from dyscalculia (a severe and selective impairment in math skills.) Most notably, Butterworth (Butterworth 1999, 2005) has proposed that a biological deficit in special neural circuits which support reasoning about cardinal numbers and discrete amounts—a 'number module'—underlies this learning disability. The broader question of how typically-developing populations are performing these operations forms a burgeoning neuroscientific literature. In particular, the parietal lobe has been extensively linked to representations of number in adults, children, and infants (Izard et al. 2008; Cantlon et al. 2006; Piazza et al. 2004; see Dehaene 2009 for a review), with the horizontal segment of the intraparietal sulcus (hIPS) reliably showing activation during simple operations such as comparison, addition, and subtraction (Chochon et al. 1999). (In adults, activation during calculation frequently involves linguistic areas of the brain as well, such as the left angular gyrus (cf. Dehaene et al. 1999), reflecting the symbolic nature of number as adults become educated as well as the symbolic format of the testing materials themselves.)

In non-human primates, ventral intraparietal area (VIP) hosts 'tuned neurons' for specific numerosities of whole displays, up to values in the 30s; this tuning is approximate as these neurons show a response peak at the specialized value, with a present-but-lesser response as the stimulus moves away from the optimal value (Nieder and Merten 2007; Nieder and Miller 2003, 2004.) Roitman et al. (2007) also found another type of neuron which encodes number; found in lateral intraparietal area (LIP), these neurons fire only to the information present in their limited receptive field and not the whole display, increasing or decreasing firing sharply as a function of the log of the number of objects in their receptive fields. Models of number representation (Dehaene and Changeux 1993; Verguts and Fias 2004; Verguts et al. 2005) suggest that these two representations act in concert to extract approximate number from a scene. There is also evidence for neurons sensitive to the rank order, found in both the usual number regions as well as the prefrontal, cingulated, and caudate areas (as reviewed in Nieder 2005); in tandem with the approximate number representations these two types of neurons have the potential to yield a sense of ordinality that underlies performance in ordering tasks. The potential neural processes that give rise to outcomes for more complex calculations such as addition and subtraction, or multiplication and division, are less straightforward. Work by Gallistel and colleagues (Balci and Gallistel 2006; Kheifets & Gallistel 2012; see Gallistel 1990 for a broad theory of learning which uses magnitude representations as a central component, and Gallistel 2011 for an overview of his view of 'mental magnitudes' as they enter into operations) on calculating risk and making decisions, and the arithmetic processes implicit in these studies, provide at minimum a computational framework for ways in which organisms operate over numerical (and non-numerical) variables.

Overall, the field of infant cognition has revealed a surprising set of abilities by preverbal infants to compute arithmetic operations. Infants tap into two systems when performing these calculations; a system which allows for tracking of individuals, and a system which yields an imprecise notion of 'how many.' Using one or both of these systems, even very young infants can intuit that five plus five equals roughly ten, ten is greater than five, and that ten is to five in the same way that two is to one. These remarkable abilities set the stage for later math learning during formalized education (Gilmore et al. 2007; Halberda et al. 2008), and suggest a level of deep continuity between the infant and adult mind.

REFERENCES

Antell, S.E., and Keating, D.P. (1983). Perception of numerical invariance in neonates. *Child Development* 54: 695–701.

Bahrick, L.E., Flom, R., and Lickliter, R. (2002). Intersensory redundancy facilitates discrimination of tempo in 3-month-old infants. *Developmental Psychobiology* 41: 352–63.

Bahrick, L.E. and Lickliter, R. (2000). Intersensory redundancy guides attentional selectivity and perceptual learning in infancy. *Developmental Psychology 36*: 190–201.

Baroody, A.J. and Dowker, A. (2003). *The development of arithmetic concepts and skills: Constructing adaptive expertise*. Mahwah, NJ: Lawrence Erlbaum Associates.

Baillargeon, R. (1993). The object concept revisited: New direction in the investigation of infants' physical knowledge. In: C. Granrud (ed.), *Visual perception and cognition in infancy*. London: Psychology Press.

Baillargeon, R. and DeVos, J. (1991). Object permanence in young infants: Further evidence. *Child Development 62*: 1227–48.

Baillargeon, R., Miller, K., and Constantino, J. (1994). Ten-month-old infants intuitions about addition. Unpublished manuscript, University of Illinois at Urbana Champaign.

Balci, F. and Gallistel, C.R. (2006). Cross-domain transfer of quantitative discriminations: Is it all a matter of proportion? *Psychonomic Bulletin and Review 13*: 636–42.

Barth, H., La Mont, K., Lipton, J., and Spelke, E. (2005). Abstract number and arithmetic in preschool children. *Proceedings of the National Academy of Sciences 102*: 14116–21.

Barth, H., Kanwisher, N., and Spelke, E. (2003). The construction of large number representations in adults. *Cognition 86*: 201–21.

Berger, A., Tzur, G., and Posner, M.I. (2006). Infant brains detect arithmetic errors. *Proceedings of the National Academy of Sciences 103*: 12649–53.

Bijeljac-Babic, R., Bertoncini, J., and Mehler, J. (1993). How do 4-day-old infants categorize multisyllabic utterances? *Developmental Psychology 29*: 711–21.

Brannon, E. (2002). The development of ordinal numerical knowledge in infancy. *Cognition 83*: 223–40.

Brannon, E. and Terrace, H.S. (2000). Representation of the numerosities 1–9 by rhesus macaques (Macaca mulatta). *Journal of Experimental Psychology: Animal Behavior Processes 26*: 31–49.

Brown, C., Carpenter, T., Kouba, V., Lindquist, M., Silver, E., and Swafford, J. (1988). Secondary school results for the fourth NAEP mathematics assessment: Algebra, geometry, mathematical methods and attitudes. *Mathematics Teacher 81*: 337–47.

Butterworth, B. (1999). *The Mathematical Brain*. London: Macmillan.

Butterworth, B. (2005). Developmental Dyscalculia. In: J. Campbell (ed.), *The Handbook of Mathematical Cognition*, pp. 455–67. New York, NY, USA: Psychology Press.

Cantlon, J. F., Brannon, E. M., Carter, E. J., et al. (2006). Functional imaging of numerical processing in adults and 4-year-old children. *PLoS Biogy*, *4*(5): e125.

Carey, S. & Xu, F. (2001). Infants' knowledge of objects: beyond object files and object tracking. *Cognition 80*: 179–213.

Cheries, E.W., Wynn, K., and Scholl, B.J. (2006). Interrupting infants' persisting object representations: an object-based limit? *Developmental Science 9*: F50–F58.

Cheries, E.W., Mitroff, S., Wynn, K., and Scholl, B.J. (2008). Cohesion as a constraint on object persistence in infancy. *Developmental Science 11*(3): 427–32.

Chiang, W. and Wynn, K. (2000). Infants' tracking of objects and collections. *Cognition 77*: 169–95.

Chochon, F., Cohen, L., van de Moortele, P. F., et al. (1999). Differential contributions of the left and right inferior parietal lobules to number processing. *Journal of Cognitive Neuroscience 11*: 617–30.

Clearfield, M.W. and Mix, K.S. (1999). Number versus contour length in infants' discrimination of small visual sets. *Psychological Science 10*: 408–11.

Clearfield, M.W. and Mix, K.S. (2001). Amount versus number: Infants' use of area and contour length to discriminate small sets. *Journal of Cognition and Development 2*: 243–60.

Cohen, L.B., and Marks, K.S. (2002). How infants process addition and subtraction events. *Developmental Science 5*: 186–201.

Cooper, R.G. (1984). Early number development: Discovering number space with addition and subtraction. In: C. Sophian (ed.), *The Origins of Cognitive Skill*, pp. 157–92. Hillsdale: Erlbaum.

Cordes, S., Gelman, R., Gallistel, C. R., and Whalen, J. (2001). Variability signatures distinguish verbal from nonverbal counting for both large and small numbers. *Psychonomic Bulletin and Review 8*: 698–707.

Cordes, S. and Gelman, R. (2005). The young numerical mind: When does it count? In: J. Campbell (ed.), *Handbook of Mathematical Cognition*, pp. 127–42. New York: Psychology Press.

Cordes, S., Gallistel, C. R., Gelman, R., and Latham, P. (2007). Nonverbal arithmetic in humans: Light from noise. *Perception and Psychophysics 69*(7): 1185–203.

Creel, S.C., Newport, E.L., and Aslin, R.N. (2004). Distant melodies: statistical learning of nonadjacent dependencies in tone sequences. *Journal of Experimental Psychology: Learning, Memory, and Cognition 30*: 1119–30.

de Hevia, M.D. and Spelke, E.S. (2010) Number-space mapping in human infants. *Psychological Science 21*: 653–60.

Dehaene, S. (1997). *The Number Sense*. New York: Penguin Press.

Dehaene, S. (2009). Origins of mathematical intuitions: the case of arithmetic. *Annals of the New York Academy of Sciences 1156*: 232–59.

Dehaene, S. and Changeux, J. P. (1993). Development of elementary numerical abilities: A neuronal model. *Journal of Cognitive Neuroscience 5*: 390–407.

Dehaene, S., Spelke, E., Pinel, P., et al. (1999). Sources of mathematical thinking: behavioral and brain imaging evidence. *Science 284*(5416): 970–4.

Denison, S. and Xu, F. (2009). Twelve—to fourteen-month-old infants can predict single-event probability with large set sizes. *Developmental Science 13*: 798–803.

Feigenson, L., Carey, S., and Hauser, M. (2002a). The representations underlying infants' choice of more: Object files versus analog magnitudes. *Psychological Science 13*: 150–6.

Feigenson, L., Carey, S., and Spelke, E. (2002b). Infants' discrimination of number vs. continuous extent. *Cognitive Psychology 44*: 33–66.

Feigenson, L., Dehaene S., and Spelke, E. (2004). Core systems of number. *Trends in Cognitive Science 8*: 307–14.

Gallistel, C.R. (1990). *The Organization of Learning: Learning, Development, and Conceptual Change*. Cambridge: The MIT Press.

Gallistel, C.R. (2011). Mental Magnitudes. In S. Dehaene and E. Brannon (eds.) Space, Time, and Number in the Brain. London, UK: Academic Press.

Gallistel, C.R. and Gelman, R. (1992). Preverbal and verbal counting and computation. *Cognition 44*: 43–74.

Gebhart, A.L., Newport, E.L., and Aslin, R.N. (2009). Statistical learning of adjacent and nonadjacent dependencies among nonlinguistic sounds. *Psychonomic Bulletin & Review 16*: 486–90.

Gilmore, C.K., McCarthy, S.E., and Spelke, E. (2007). Symbolic arithmetic knowledge without instruction. *Nature 447*: 589–91.

Gomez, R.L. (2002). Variability and detection of invariant structure. *Psychological Science 13*: 431–6.

Halberda, J., Mazzocco, M., and Feigenson, L. (2008). Individual differences in nonverbal number acuity predict maths achievement. *Nature 455*: 665–8.

Hay, J.F., Pelucchi, B., Estes, K.G., and Saffran, J.R. (2011). Linking sounds to meaning: Infant statistical learning in a natural language. *Cognitive Psychology 63*: 93–106.

Hecht, S.A. (1998). Toward an information-processing account of individual differences in fraction skills. *Journal of Educational Psychology 90*: 545–59.

Hope, J.A. and Owens, D.T. (1987). An analysis of the difficulty of learning fractions. *Focus on Learning Problems in Mathematics 9*: 25–40.

Houston-Price, C. and Nakai, S. (2004). Distinguishing novelty and familiarity effects in infant preference procedures. *Infant and Child Development 13*: 341–8.

Hunter, M., Ames, E., and Koopman, R. (1983). Effects of stimulus complexity and familiarization time on infant preferences for novel and familiar stimuli. *Developmental Psychology 19*: 338–52.

Izard, V., Dehaene-Lambertz, G., and Dehaene, S. (2008). Distinct cerebral pathways for object identity and number in human infants. *PLoS Biology 6*: e11.

Izard, V., Sann, C., Spelke, E.S., and Streri, A. (2009). Newborn infants perceive abstract numbers. *Proceedings of the National Academy of the Sciences 106*: 10382–5.

Jordan, K. and Brannon, E.M. (2006). The multisensory representation of number in infancy. *Proceedings of the National Academy of Sciences, 103*(9): 3486–9.

Kahneman, D., Treisman, A., and Gibbs, B.J. (1992). The reviewing of object files: Object-specific integration of information. *Cognitive Psychology 24*: 175–219.

Kheifets A. and Gallistel C.R. (2012). Mice take calculated risks. *Proceedings of the National Academy of Sciences 109*(22): 8776–9.

Kobayashi, T., Hiraki, K., Mugitani, R., and Hasegawa, T. (2004). Baby arithmetic: one object plus one tone. *Cognition 91*: B23–B34.

Kobayashi, T., Hiraki, K., and Hasegawa, T. (2005). Auditory-visual intermodal matching of small numerosities in 6-month-old infants. *Developmental Science 8*(5): 409–19.

Koechlin, E., Dehaene, S., and Mehler, J. (1997). Numerical transformations in five month old human infants. *Cognition 3*: 89–104.

Kouba, V., Zawojewski, J., and Strutchens, M. (1997). What do students know about numbers and operations? In: P. A. Kenney and E. A. Silver (eds.), *Results from the Sixth Mathematics Assessment of the National Assessment of Educational Progress*, pp. 87–140. Reston, VA: National Council of Teachers of Mathematics.

Leslie, A. (1984). Spatiotemporal continuity and the perception of causality in infants. *Perception I3*: 287–305.

Lipton, J.S. and Spelke, E.S. (2003). Origins of number sense: Large-number discrimination in human infants. *Psychological Science 14*: 396–401.

Lipton, J.S. and Spelke, E.S. (2004). Discrimination of large and small numerosities by human infants. *Infancy 5*: 271–90.

McCrink, K. and Wynn, K. (2004). Large-number addition and subtraction by 9-month-old infants. *Psychological Science 15*: 776–81.

McCrink, K. and Wynn, K. (2007). Ratio abstraction by 6-month-old infants. *Psychological Science 18*: 740–5.

McCrink, K. and Wynn, K. (2008) Mathematical Reasoning. In: M. Haith and J. Benson (eds.), *Encyclopedia of Infant and Early Childhood Development* Vol. 2, pp. 280–9. San Diego, CA, USA: Academic Press.

McCrink, K. and Wynn, K. (2009). Operational momentum in large-number addition and subtraction by 9-month-olds. *Journal of Experimental Child Psychology 103*: 400–8.

McCrink, K., Dehaene, S., and Dehaene-Lambertz, G. (2007) Moving along the number line: The case for operational momentum. *Perception and Psychophysics 69*(8): 1324–33.

Mix, K.S., Huttenlocher, J., and Levine, S.C. (2002). *Quantitative development in infancy and early childhood.* Oxford: Oxford University Press.

Mix, K.S., Levine, S.C., and Huttenlocher, J. (1997). Numerical abstraction in infants: Another look. *Developmental Psychology 33*: 423–8.

Moore, D.S. and Cocas, L.A. (2006). Perception precedes computation: Can familiarity preferences explain apparent calculation by human babies? *Developmental Psychology 42*: 666–78.

Moore, D., Benenson, J., Reznick, J.S., and Kagan, J. (1987). Effect of auditory numerical information on infants' looking behavior: Contradictory evidence. *Developmental Psychology 23*: 665–70.

Nieder, A. (2005) Counting on neurons: The neurobiology of numerical competence (Review). *Nature Reviews Neuroscience 6*: 177–90.

Nieder, A. and Merten, K. (2007). A labeled-line code for small and large numerosities in the monkey prefrontal cortex. *Journal of Neuroscience 27*(22): 5986–93.

Nieder, A. and Miller, E. K. (2003). Coding of cognitive magnitude. Compressed scaling of numerical information in the primate prefrontal cortex. *Neuron 37*(1): 149–57.

Nieder, A. and Miller, E. K. (2004). A parieto-frontal network for visual numerical information in the monkey. *Proceedings of the National Academy of Sciences 101*(19): 7457–62.

Newport, E.L. and Aslin, R.N. (2004). Learning at a distance I. Statistical learning of non-adjacent dependencies. *Cognitive Psychology 48*: 127–62.

Pepperberg, I.M. (2006). Cognitive and communicative abilities of Grey parrots. *Applied Animal Behaviour Science 100*: 77–86.

Piaget, J. (1954). *The Construction of Reality in the Child.* New York: Ballentine.

Piazza, M., Izard, V., Pinel, P., et al. (2004). Tuning curves for approximate numerosity in the human intraparietal sulcus. *Neuron 44*(3): 547–55.

Pica, P., Lemer, C., Izard, V., and Dehaene, S. (2004). Exact and approximate arithmetic in an Amazonion idigene group. *Science 306*: 499–503.

Picozzi, M., de Hevia, M.D., Girelli, L., and Cassia, V.M. (2010). Seven-month-olds detect ordinal numerical relationships within temporal sequences. *Journal of Experimental Child Psychology 107*: 359–67.

Roitman, J. D., Brannon, E. M., and Platt, M. L. (2007). Monotonic coding of numerosity in macaque lateral intraparietal area. *PLoS Biology 5*(8): e208.

Rugani, R., Regolin, L., Vallortigara, G. (2007). Rudimental numerical competence in 5-day-old domestic chicks (Gallus gallus): Identification of ordinal position. *Journal of Eperimental Psychology: Animal Behavior Processes 33*: 21–31.

Saffran, J.R. and Thiessen, E.D. (2007). Domain-general learning capacities. In: E. Hoff and M. Shatz (eds.), *Handbook of Language Development*, pp. 68–86. Cambridge: Blackwell.

Saffran, J.R., Aslin, R.N., and Newport, E.L. (1996). Statistical learning by 8-month-olds. *Science 274*: 1926–8.

Saffran, J., Johnson, E., Aslin, R., and Newport, E. (1999). Statistical learning of tone sequences by human infants and adults. *Cognition 70*: 27–52.

Scholl, B.J. (2001). Objects and attention: The state of the art. *Cognition 80 ½*: 1–46.

Scholl, B.J. and Pylyshyn, Z.W. (1999). Tracking multiple items through occlusion: Clues to visual objecthood. *Cognitive Psychology 38*: 259–90.

Sharon, T. and Wynn, K. (1998). Individuation of actions from continuous motion. *Psychological Science 9*: 357–62.

Simon, T.J., Hespos, S.J., and Rochat, P. (1995). Do infants understand simple arithmetic? A replication of Wynn (1992). *Cognitive Development 10*: 253–69.

Spelke, E.S. (1990). Principles of object perception. *Cognitive Science 14*: 29–56.

Spelke, E.S. (2000). Core knowledge. *American Psychologist 55*: 1223–43.

Starkey, P. and Cooper, R.G. (1980). Perception of numbers by human infants. *Science 210*: 1033–5.

Starkey, P., Spelke, E.S., and Gelman, R. (1983). Detection of intermodal numerical correspondences by human infants. *Science 222*: 179–81.

Starkey, P., Spelke, E.S., and Gelman, R. (1990). Numerical abstraction by human infants. *Cognition 36*: 97–127.

Strauss, M.S. and Curtis, L.E. (1981). Infant perception of numerosity. *Child Development 52*: 1146–52.

Strauss, M.S. and Curtis, L.E. (1984). Development of numerical concepts in infancy. In: C. Sophian (ed.), *The Origins of Cognitive Skills*. pp. 131–155. Hillsdale: Erlbaum.

Suanda, S.H., Tompson, W., and Brannon, E.M. (2008). Changes in the ability to detect ordinal numerical relationships between 9 and 11 months of age. *Infancy 13*: 308–37.

Teglas, E., Girotto, V., Gonzalez, M., and Bonatti, L.L. (2007). Intuitions of probabilities shape expectations about the future at 12 months and beyond. *Proceedings of the National Academy of Sciences 104*: 19156–9.

Teglas, E., Vul, E., Girotto, V., Gonzalez, M., Tenenbaum, J.B., and Bonatti, L.L. (2011). Pure reasoning in 12-month-old infants as probabilistic inference. *Science 332*: 1054–9.

Trick, L.M. and Pylyshyn, Z.W. (1994). Why are small and large numbers enumerated differently? A limited-capacity preattentive stage in vision. *Psychological Review 101*: 80–102.

Uller, C., Carey, S., Huntley-Fenner, G., and Klatt, L. (1999). What representations might underlie infant numerical knowledge? *Cognitive Development 14*: 1–36.

vanLoosbroek, E. and Smitsman, A.W. (1990). Visual perception of numerosity in infancy. *Developmental Psychology 26*: 916–22.

vanMarle, K. (2013). Infants use different mechanisms to make small and large number ordinal judgments. *Journal of Experimental Child Psychology 114*(1): 102–10.

vanMarle, K. and Wynn, K. (2011). Tracking and quantifying objects and non-cohesive substances. *Developmental Science 14*(3): 302–16.

vanMarle, K. and Wynn, K. (2009). Infants' auditory enumeration: Evidence for analog magnitudes in the small number range. *Cognition 111*: 302–16.

Verguts, T. and Fias, W. (2004). Representation of number in animals and humans: a neural model. *Journal of Cognitive Neuroscience 16*(9): 1493–504.

Verguts, T., Fias, W, and Stevens, M. (2005). A model of exact small-number representation. *Psychonomic Bulletin and Review 12*(1): 66–80.

Wakeley, A., Rivera, S., and Langer, J. (2000). Can young infants add and subtract? *Child Development 71*: 1525–34.

Wynn, K. (1992). Addition and subtraction by human infants. *Nature 358*: 749–50.

Wynn, K. (1996). Infants' individuation and enumeration of physical actions. *Psychological Science 7*: 164–9.

Wynn, K. (2000). Findings from addition and subtraction in infants are robust and consistent: Reply to Wakely, Rivera, and Langer. *Child Development 71*: 1535–6.

Wood, J.N. and Spelke, E.S. (2005a). Infants' enumeration of actions: Numerical discrimination and its signature limits. *Developmental Science 8*: 173–81.

Wood, J.N. and Spelke, E.S. (2005b). Chronometric studies of numerical cognition in five-month-old infants. *Cognition 97*: 23–39.

Wynn, K., Bloom, P., and Chiang, W. (2001). Enumeration of collective entities by 5-month-olds. *Cognition 83*: B55–B62.

Xu, E (1993, March). *The concept of object identity.* Paper presented at the biennial meeting of the Society for Research in Child Development, New Orleans, LA.

Xu, F. (2003) Numerosity discrimination in infants: Evidence for two systems of representations. *Cognition 89*: B15–B25.

Xu, F. and Garcia, V. (2008). Intuitive statistics by 8-month-old infants. *Proceedings of the National Academy of Sciences 105*: 5012–15.

Xu, F. and Spelke, E.S. (2000). Large number discrimination in 6-month-old infants. *Cognition 74*: B1–B11.

Xu, F. and Denison, S. (2009). Statistical inference and sensitivity to sampling in 11-month-old infants. *Cognition 112*: 97–104.

Xu, F., Spelke, E.S., and Goddard, S. (2005). Number sense in human infants. *Developmental Science 8*: 88–101.

Yonas, A., Pettersen, L., and Lockman, J.J. (1979). Young infants' sensitivity to optical information for collision. *Canadian Journal of Psychology 33*: 268–76.

CHAPTER 15

SPONTANEOUS FOCUSING ON NUMEROSITY AND ITS RELATION TO COUNTING AND ARITHMETIC

MINNA M. HANNULA-SORMUNEN

Introduction

In this chapter, both the theoretical and methodological issues of the formation of individual differences in Spontaneous Focusing On Numerosity (SFON) as a part of numerical development will be reviewed. In one of the first tests in the set of these studies, the experimenter introduced a toy parrot and his favourite berries to the child and said: 'Watch carefully what I do, and then you do just like I did.' Then the experimenter put two berries, one at a time into the parrot's mouth, and asked the child to do exactly like she had done. It appears that in this test there are 3- to 5-year-old children who immediately, without any guidance, notice the exact number of berries the experimenter gave and systematically give the same amount of berries. However, some children of the same age do not pay attention to the number of berries, but instead focus on other aspects, such as the way in which the experimenter held the berries and gave them to the bird. These children do not give the exact number of berries the experimenter give but many more, or just one berry to the parrot. These differences between children appear despite the fact that they do not differ from the other group on measures of attention or motivation to complete the task, and they are able to recognize and produce sets of two when asked.

SFON has been defined as a process of spontaneously (i.e. in a self-initiated way, not prompted by others in a certain situation) focusing attention on the aspect of exact number of a set of items or incidents (Hannula, 2005; Hannula & Lehtinen, 2001, 2005). Because exact number recognition is not a totally automatic process that would take place every time a person faces something to enumerate (Railo, Koivisto, Revonsuo, & Hannula, 2006; Trick & Pylyshyn, 1994), focusing of attention is needed for triggering exact number recognition processes and using the recognized exact number in action. This proposal is based on the notion that number is not a property of the physical world itself, but is rather determined as a result of how we—based on our goals—choose to carve up the physical world into individual elements (e.g. Wynn, 1998). Sometimes sets of objects are readily distinct from other sets, but often not. In our natural surroundings, we have to focus our attention on the aspect of number in the set of items in question, and define the set of items (i.e. which kind of targets belong to the set) before we can determine the exact number of items in a set. This is different from most experimental settings in enumeration studies, which typically present sets of items in such a way that the set to be enumerated is clearly distinct and already defined. The concept of a 'set of individuals' is central to counting, simple addition and subtraction, and all natural number concepts (Spelke, 2003). The development of this concept is crucial for a young child's development in understanding what oneness, twoness, and threeness mean (Spelke, 2003). It is also essential for focusing on the aspect of numerosity, because numerosity (i.e. cardinality) is a quality of a set requiring focusing on the set of individuals, not only on the individuals.

SFON tendency refers to a generalized tendency to spontaneously focus attention on exact number across different contexts and time (Hannula, 2005; Hannula & Lehtinen, 2001, 2005). The measures of SFON tendency are an indicator of the amount of a child's self-initiated practice in using exact enumeration in his or her natural surroundings (Hannula, Mattinen, & Lehtinen, 2005).

SFON tendency is strongly related to the development of verbal-counting skills, which are the building blocks for the natural number concept and numeracy and have been reviewed, for example, in Fuson (1988) and Gelman and Gallistel (1978). According to Gelman and Gallistel five principles govern and define counting, and even form the innate basis for counting. The first three deal with rules of how to count, the fourth with the definition of what to count, and finally the fifth involves a composite of the features of the other four principles. Mastering these five principles provides accurate counting skills: one to one, the stable-order, cardinality, abstraction, and order-irrelevance principle. In addition to these principles, Hannula (2005) suggested that an additional relevant aspect of cardinality recognition based on counting—SFON—needs to be taken into account if we want to understand the formation of developmental differences in early counting skills. Namely, before a set of items can be enumerated, attention needs to be focused on the aspect of exact numerosity in the set of items. Focusing of attention on the aspect of numerosity is thus needed for exact number recognition. As well, practice with enumeration skills produced by children's SFON is needed for the normal development of counting skills before school age.

The research of SFON originated in research questions aimed at discovering the reasons for large individual differences in young children's mathematical skills and mathematical learning difficulties emerging already at the beginning of the school career, which suggest that there are kindergartners who lack the basic counting skills and arithmetic knowledge necessary for later development in mathematics (Aunola, Leskinen, Lerkkanen, & Nurmi, 2004; Hanich, Jordan, Kaplan, & Dick, 2001; Krajewski & Schneider, 2009ab). There are ample research projects investigating individual differences in domain-specific and domain-general aspects of early numeracy and counting (e.g. Dowker, 2008; Lepola et al., 2005; Sarnecka & Carey, 2008). Typically most studies of early numeracy either use tasks in which a child's attention is specifically guided toward numerosity (e.g. Huttenlocher, Jordan, & Levine, 1994; Jordan et al., 2006) or the tasks do not differentiate between mathematical skills and attentional processes needed to trigger mathematical thinking within the task (e.g. Aunola et al., 2004; Baroody, Li, & Lai, 2008). Thus, these studies are not able to capture the individual differences in children's mathematically relevant attentional focusing processes that may trigger the use of early numerical skills across different early learning environments. This results in a lack of knowledge of all the necessary and mathematically meaningful learning processes, particularly regarding individual differences in the amount to which these activities take place in everyday surroundings. Children's everyday surroundings are a potentially rich arena for their mathematical thinking and embedded informal mathematical knowledge (e.g. Ginsburg, Inoue, & Seo, 1999; Nunes & Bryant, 1996; Tudge & Doucet, 2004). Indeed, preschool and home learning environments have been shown to influence mathematics achievement up to the age of 10 years (Melhuish et al., 2008). Likewise, the amount and quality of parents' number-related talk has also been shown to be related to early numerical development (Gunderson & Levine, 2011; Levine, Suriyakham, Rowe, Huttenlocher, & Gunderson, 2010).

Individual Differences and Stability in SFON

Results of previous SFON studies suggest that it is possible to distinguish, within a person's existing mathematical competence, a distinct SFON process and also measure individual differences in this attentional tendency. Results show that there are significant individual differences in SFON tendency in children without diagnosed learning impairments or developmental disorders at the ages of 3–12 years in Finland and in the USA (Edens & Potter 2013; Hannula & Lehtinen, 2005; Hannula-Sormunen, Lehtinen, & Räsänen, in press; Hannula, Lepola, & Lehtinen, 2010; Hannula, Räsänen, & Lehtinen, 2007; Potter, 2009). Likewise, in a large sample of 2-year-old children, substantially prematurely born children and full-term children did not differ from each other in their SFON tendency, yet there were individual differences overall in children's SFON tendency (Takila, Hannula, & Pipari Study Group, 2004). In a study conducted in New York City, primary-school children (aged from 6–12 years) showed substantial individual differences in two SFON tasks (Choudhury, McCandliss, & Hannula, 2007). Dyscalculic children had weaker SFON tendency than the control children at the age of 7-11 years (Kucian et al., 2012). In these studies, there has been within-subject stability either across two or three different SFON tasks or even across years of time, like in Hannula and Lehtinen (2005) from the age of 4 to the age of 5 years, and in Hannula-Sormunen and colleagues (in press) from the age of 4 to the age of 12 years.

SFON Assessments

Over the course of conducting the research studies on SFON, more than two dozen behavioural tasks and observational methods have been designed as SFON assessments. In the following section the principles of SFON assessments will be described. The aim of SFON assessments is to obtain a reliable indicator of a child's general SFON tendency across different task contexts. This kind of indicator is aimed at capturing the extent of a child's self-initiated focusing on numerosity, and thus the amount of practice acquired in utilizing enumeration skills in his or her surroundings. In order to enable the measuring of children's spontaneous behaviour the tasks must be novel and not be explicitly mathematical. Furthermore, when presenting a SFON task, no use can be made of any phrases that could suggest that the task is somehow mathematical or quantitative. Neither can the experimenter give any feedback during the testing of SFON that could help the child figure out which are the relevant aspects of the task[1].

The consenting procedures and practical arrangements of the testing situations need to take account that participants of SFON studies do not expect to be faced (only) with mathematical or counting tasks. In practice, using larger projects including both mathematical and non-mathematical tasks is a way to avoid participants expecting to be tested on their numerical skills, especially, if only non-mathematical tasks are presented before SFON tasks.

Each SFON task trial is presented only when the experimenter has got the child's full attention on the task, so that the child's general attentional state, or task motivation, does not explain individual differences in SFON. In order to hinder the confounding effect of number recognition skills on the measure of SFON, the SFON task may include only such small numbers of items or incidents that all children should be able to recognize. Moreover, the SFON task should not exceed children's memory capacity, visuo-motorical, or verbal comprehension skills, so that if the child focuses on number in the task, she or he is capable of proceeding in the task in accordance with his or her numerical focusing target.

One way of directly testing whether the prerequisites of SFON tasks are at a low enough level is to provide a guided focusing version of the SFON task to those children who did not produce accurate numbers of items or incidents involved in the SFON task, and who did not make any observable enumeration attempts during the SFON task. This was done in the study by Choudhury and collegues (2007), Hannula and Lehtinen (2005), and in Hannula-Sormunen, Lehtinen, and Räsänen (in press). This was done to ensure that all children had the enumeration and cognitive skills needed for the SFON task, and thus the individual differences in the SFON tasks would be a result of differences in SFON, and not other skills. The guided numerical focusing task versions included the same materials and settings as the SFON tasks except that children were given explicit instructions to focus on numerosity. Results of the guided versions of the SFON tasks show that children who had earlier not focused on numerosity in the SFON task got substantially higher scores on the guided numerical focusing task version than their previous 0 scores from the SFON task version. This suggests that the results of SFON tasks reflect individual differences in SFON, not in enumeration or other cognitive skills required for completing the tasks.

Whether SFON tasks capture domain-specific, numerical focusing differences, or if they just measure more general ability to focus on a task-relevant aspect remained an open question to our team until the Hannula and colleagues (2010) study. In this study, it was tested whether children's focusing on some other non-numerical aspect (i.e. focusing on spatial locations) would iron out individual differences in their SFON tendency and whether the relationship between SFON and mathematical skills is explained by children's individual differences in spontaneous focusing on spatial locations (SFOL).

For this purpose, the SFON Model task of Hannula and Lehtinen (2005) was used. In the task, similar pictures of dinosaurs were placed on the table. The child was told that in this task the experimenter would make her dinosaur into a model and would then turn the model upside down. After this, the child would be asked to make his or her dinosaur look exactly like the model one. After introducing the task in this way, the experimenter said, 'Now, watch carefully. I am making this dinosaur into a model'. After stamping six nodes going from the head of the dinosaur to the dinosaur's back and pausing for 5 seconds to let the child memorize the model, the experimenter turned her model upside down, gave the stamp to the child, and said, 'Now, make your dinosaur look like the model dinosaur'. The procedure was repeated with seven spikes placed onto the back of the dinosaur as the second item and five spikes placed onto the tail of the dinosaur as the third item.

For the SFOL measure, it was analysed how precise the location of the child's set of stamps was by measuring the location of the first and last stamps of the row of stamps compared to the ones on the experimenter's model sheet. Results showed marked individual differences in children's SFOL, but in a way that did not negate the relationship between counting skills and SFON (Hannula et al., 2010).

When using the Model task as a SFON measure, the scoring has been based on analyses of video-recorded task situations or in structured observations following the same criteria as the analyses of video-recordings. All the child's (1) utterances including number words (e.g. 'I'll give him two candies'), (2) use of fingers to express numbers, (3) counting acts, such as a whispered number word sequence and indicating acts by fingers and/or head, (4) other comments referring to either quantities or counting (e.g. 'Oh, I miscounted them'), or, (5) interpretation of the goal of the task as quantitative (e.g. 'I gave an exactly accurate number of them') were identified. The child was scored as focusing on numbers, if she or he produced the correct numerosity, and/or if she or he was observed presenting any of the aforementioned (1–5) quantifying acts. Due to the tasks being within all children's enumeration range, more than 90% of all SFON scores have resulted from the children's accurate production of the same number of objects as the target set. Observations of quantifying acts work as additional criteria for making sure that if the child makes an enumeration mistake, his or her focusing on numerosity is noticed. Furthermore, in some SFON tasks for children over 6 years of age, a stimulated recall interview about the child's focusing targets and solution strategies immediately after the task was used to evaluate if the child had considered number in the task (Choudhury et al., 2007; Hannula & Lehtinen, 2005).

In order to explore whether SFON assessments are a valid indicator of a child's more general SFON tendency, observational data on children's SFON in day-care settings, in addition to the SFON assessments, were gathered as part of a SFON enhancement study of 3-year-old children by Hannula and colleagues (2005). The analyses showed a positive correlation ($r = 0.55$) between children's spontaneous, i.e. self-initiated, focusing on numerosity observed by day-care personnel in day-care settings such as free play, lunch, outdoor activities, and dressing up and their SFON scores in the experimental SFON tasks. This suggests that the proposal of SFON tasks being indicators of the amount of children's spontaneous focusing on numerosity in their everyday life is justifiable.

The Need For More Specific Measures of SFON: Brain Imaging Studies

Many SFON tasks, such as the bird task and the dinosaur task described in this chapter, used imitation as the main context for activity. Because of this, the interpretations of SFON in these assessments were based on the child's performance that takes place after the experimenter's model performance; thus it remained unclear whether the differences in SFON are due to the encoding of the number of stimuli or the recall and use of the number of stimuli in action. Therefore these purely behavioural SFON tasks could not exhaustively solve the question of whether or not there are perceptual differences during encoding of the stimuli. This led our research team to look for methods that would reveal whether the differences in SFON are due to differences in the phase of perception and encoding of the stimuli in the task and/or due to differences in the phase of recall and utilizing of numerosities in action. Therefore we explored electroencephalography (EEG) responses during encoding and perception of photos of natural scenes as a part of our 9-year longitudinal study with children of 12 years of age (Hannula-Sormunen, Grabner & Lehtinen, in preparation). Similarly, we investigated adults' neural correlates of SFON with functional Magnetic Resonance Imaging (fMRI) while the participants were memorizing photos ((Hannula, Grabner, Lehtinen, Laine, Parkkola, & Ansari, 2009; Hannula-Sormunen, Grabner, Lehtinen, Laine, Parkkola, & Ansari, in preparation). The results show that the distinct nature of SFON can also be captured by brain imaging methods, such as EEG and fMRI, revealing that SFON specifically engages the temporal-parietal cortex. The oscillatory EEG activity while looking at photographs differs in trials from which participants either did not, spontaneously did (SFON trials), or were asked to (Guided Focusing On Numerosity GFON trials) report an exact number of something in the photograph (Hannula, Grabner & Lehtinen, 2009; Hannula-Sormunen, Grabner, & Lehtinen, in preparation). Correspondingly, the fMRI results indicate that participants engage a distinctive frontoparietal attentional network for encoding numerosity as opposed to encoding colour from photographs (Hannula-Sormunen et al., in preparation). These brain imaging methods show that individual differences in SFON are due to differences in the encoding of stimuli.

Positive Relationship of SFON and Numerical Skills

A set of longitudinal studies covering the development of numerical skills from the age of 3 to the age of 12 years have shown that SFON is positively related to the development of cardinality recognition, subitizing, number sequence, and arithmetic skills (Hannula, 2005; Hannula & Lehtinen, 2001, 2005; Hannula et al., 2007, 2010; Hannula-Sormunen et al., in press). Follow-up data of 39 Finnish children from the age of 3.5 years to the age of 6 years were analysed with path models of the development of SFON and counting skills (see Figure 15.1), which indicate a reciprocal relationship between SFON and counting skills (Hannula & Lehtinen, 2005). Thus, better early mathematical skills would be associated with a stronger SFON tendency, which in turn would be related to subsequent stronger development in mathematical skills.

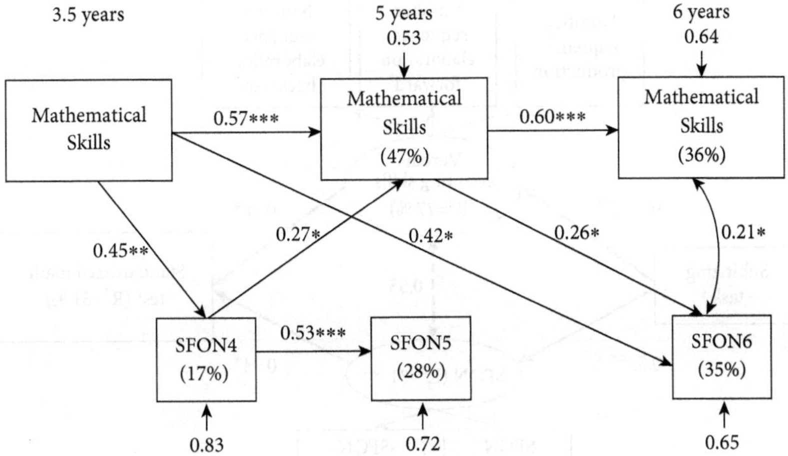

FIGURE 15.1 Path model of development of SFON and mathematical skills from the age of 3.5 years to the age of 6 years (Hannula and Lehtinen, 2005). Percentages shown refer to the amount of variance explained.

Reprinted from *Learning and Instruction*, 15(3), Minna M. vHannula and Erno Lehtinen, Spontaneous focusing on numerosity and mathematical skills of young children, pp. 237–56, Copyright (2005), with permission from Elsevier.

Hannula and colleagues (2007) showed that children with a strong general long-term tendency to focus on numerosity tended to enumerate by subitizing larger numbers of items and had better verbal-counting skills at the age of 5 years. Mediation analyses based on separate sets of regression analyses showed that children's better skills to produce number sequence were directly related to a strong SFON tendency also when subitizing-based enumeration was controlled for, while the association between SFON and object-counting skills was significantly mediated by subitizing-based enumeration. A significant Sobel's test value ($z = 2.33$ $p < 0.05$) revealed that subitizing-based enumeration serves as a mediator between SFON and object-counting skill. These results indicate that the associations between the child's SFON tendency and subskills of verbal counting may differ on the basis of how significant a role the understanding of the cardinal meanings of number words plays in learning these skills. LeFevre and colleagues (2010) found that 5-year-old children's latency in subitizing is related to object-counting skills, but not to number naming. The skills for object counting develop as a result of the integration of the ability to recognize the first small cardinal numbers and the learning of object-counting procedures (Bermejo, Morales, & deOsuna, 2004; Wynn, 1990). More longitudinal studies on the role of subitizing-based enumeration for substantially later numerical skills are still needed, even though there is already some evidence of a positive correlation between subitizing and numerical skills before school age (Fischer, Gebhardt, & Hartnegg, 2008; Landerl, Bevan, & Butterworth, 2004).

One such study is a further follow-up of Hannula and Lehtinen (2005) with 36 Finnish children from the age of 5 years to the age of 12 years (Hannula-Sormunen et al., in press). It provides unique evidence on how these children's subitizing-based enumeration, SFON, and counting skills assessed at the ages of 5 or 6 years predict their school mathematics achievement at the age of 12 years (see Figure 15.2). The results based on partial least squares modelling demonstrate that SFON and verbal-counting skills before school age predict mathematical performance on a standardized test for typical school mathematics in grade

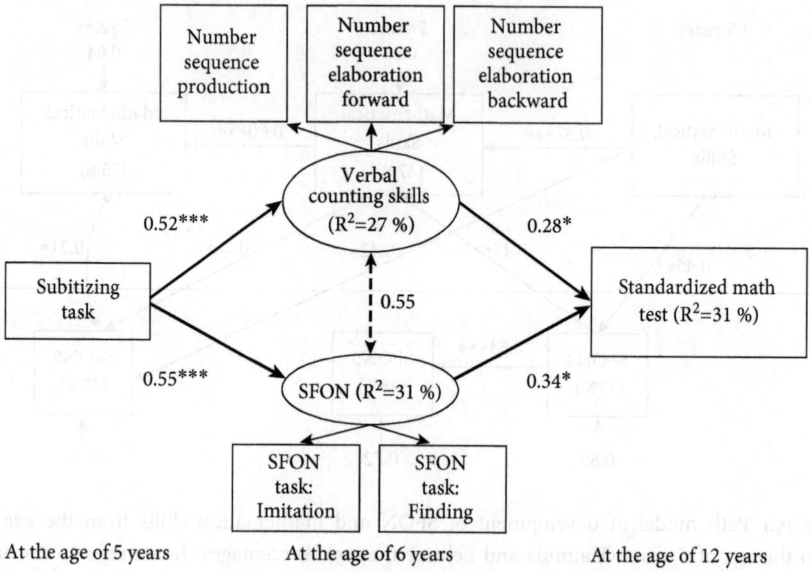

FIGURE 15.2 The partial least squares (PLS) model illustrating significant direct effects (solid lines) and correlations (dashed lines) between subitizing-based enumeration, verbal-counting skills, SFON, and mathematics achievement measured by a standardized test.

(Data from *Mathematical Thinking and Learning* (accepted), M.M. Hannula-Sormunen, E. Lehtinen, and P. Räsänen, Children's preschool subitizing, spontaneous focusing on numerosity and counting skills as predictors of mathematical performance 6-7 years later at school).

5, even after non-verbal IQ is controlled for. Subitizing-based enumeration skill has an indirect effect, via number sequence skills and SFON, on mathematical performance at the age of 12 years. Non-verbal IQ measured at the age of 12 years did not add any additional variance to the model.

Domain Specificity of The Relationship between SFON and Numerical Skills

In a longitudinal follow-up of 139 kindergarteners at the end of 2nd grade, the associations of SFON and contemporaneous cognitive, attentional, linguistic, and numerical skills as well as their spontaneous focusing on location SFOL were analysed (Hannula et al., 2010). By using a set of hierarchical regression analyses it was shown that SFON predicts academic skills, such as arithmetical skills but not reading skills, 2 years later in primary school after controlling for the effects of other predictor variables. Number sequence skills are strongly related to arithmetical skills and share a significant portion of variance with SFON, yet SFON accounts for additional variance even after controlling for not only number sequence skills but also non-verbal IQ, linguistic skills, and focusing on spatial locations. Even though the variance explained by SFON after controlling for cognitive, attentional, and linguistic skills in later arithmetical skills was not large, it is a significant finding. This result corresponds with the results of Hannula-Sormunen and colleagues (submitted) showing that SFON, subitizing, and verbal-counting skills before school-age predict mathematics achievement but not reading skills at the age of 12 years. Controlling for non-verbal IQ did not significantly add explained variance of the model. SFOL was associated with non-verbal IQ, phonological awareness, listening comprehension, as well as with later reading comprehension, in addition to mathematical skills. It did not iron out the significant association of SFON and contemporary numerical skills. This suggests that the numerical practice resulted by children's own focusing on numerical aspects of their surroundings—unlike more general focusing of attention on details such as spatial locations—does not relate to children's reading skills, but to arithmetical skills.

The Reciprocal Nature of The Development of SFON and Numerical Skills

The studies investigating SFON suggest that children's own spontaneous focusing on numerosity frequently leads them to perceive different numbers of objects or events in their surroundings, and thus they get practice in recognizing and producing numerosity. This, in turn, develops their quantifying skills in several ways: not only by increasing their counting range, but also, along with more developed counting (quantifying) skills, a larger quantity may appear as a possible subject for counting. Moreover, knowledge about the use of enumeration skills in different tasks will increase with practice, so the child may tend to focus on numerosity more often in new, more demanding tasks. It is not only the sociocultural mediation of numerical cognition that develops the child's skill to focus on the aspect of numerosity and to utilize innate and cultural tools for enumeration, but also focusing on the aspect of numerosity that develops the child's enumeration skills by activating the enumeration process and thus producing practice in enumeration (Hannula, 2005; Hannula & Lehtinen, 2005).

According to Sophian (1988) children's conceptual knowledge about numbers is dynamically related to their goal-based numerical activities: conceptual advances facilitate new goals and corresponding activities, which in turn provide the input for further advances in numeracy. Saxe, Guberman, and Gearhart (1987) and Sophian (1988) describe counting development as a reciprocal activity, in which socially structured goals of quantification change along with the development of skills and direct children's attention to different aspects of numbers and the ways in which others use them. In concert with this, as described earlier in this chapter, the results of the longitudinal study indicate a reciprocal relationship between counting skills and SFON during the age when children learn to enumerate sets by counting, in other words, from the age of 3 to the age of 6 years (Hannula & Lehtinen, 2005). When comparing performance of 3- and 4-year-old children in the same SFON task in the studies of Hannula and colleagues (2005) and Hannula and Lehtinen

(2005), it becomes evident that when children become older they focus more on numerosity in a certain task situation.

Hannula and colleagues (2010) showed that SFON is a domain-specific predictor of arithmetical skills. What could be the developmental mechanisms of the learning of arithmetical skills with which SFON is associated? One potential mechanism could be that a stronger SFON tendency during kindergarten may be associated with children's stronger tendency to focus on other mathematical aspects, not just on numerosity, in their learning environment. Thus, the children with a strong SFON tendency may acquire more practice also in other numerical tasks than direct enumeration-related mathematical ones, such as the learning of arithmetical operations and number symbols. Thus, SFON tasks may capture a more general mathematically focused tendency than just focusing on numerosity of items or events in the surroundings. Potter (2009) investigated SFON, counting skills, and motivational state and mathematical interest of 4- to 5-year-old children. The correlation of SFON with counting skills but not with motivational state and interest measures suggests that SFON is more strongly associated with cognitive rather than affective factors (Potter, 2009).

It could also be that a large amount of practice in cardinality recognition and counting could lead to learning more sophisticated enumeration strategies, such as grouping and adding up the subtotals of the groups. This set of skills could be defined as an arithmetical skill, and it is tightly linked with number sequence elaboration skills. A cross-sectional study of 6- to 11-year-old children (Yun, Hannula, & McCandliss, 2008) showed that during primary school children learn to spontaneously use grouping and adding-up strategies when enumerating items. Children's spontaneous creation and use of grouping was related to standardized mathematical tests even after controlling for arithmetical fluency (Martin, Hannula, & Schwartz, 2008).

SFON as A Mechanism of Self-Initiated Practice in Early Number Skills

The present study addresses the role of children's spontaneous, i.e. self-initiated, practice in utilizing exact enumeration in the development of counting. Thus, those children who more readily recognize and utilize exact number without guidance to do so in situations that are not explicitly numerical gain more practice in enumerating than their peers who less readily do so. These individual differences in enumeration experiences may explain SFON tendency's positive association with developmental trajectories of early numerical skills. Ericsson and Lehman (1996) show that experts seem to be capable of 'seeing' multiple possibilities to practise their skills in everyday situations, and this has been an essential part of their development from a very early age. In both formal and informal learning situations natural number is not always the most mathematically relevant aspect that a person could utilize. More mathematically advanced aspects, such as quantitative relations, may often be more appropriate for a situation. Those who become experts in a domain seem to search for opportunities to master more and more demanding skills (Ericsson, 1996). This suggests that increased practice with enumeration, through a stronger SFON tendency, may be only one step in the progression of skills needed to be practised for the long-term development of numerical skills. At later points in development, it may not be sufficient or even productive to solely focus on exact number in some situations. Therefore, the investigation of children's tendency to focus on more mathematically advanced aspects of a task, such as quantitative relations, is warranted.

Recent cross-sectional investigations of 5- to 8-year-old children's spontaneous recognition and focusing on quantitative relations (McMullen, Hannula-Sormunen, & Lehtinen, 2011,2013, in press) have started to shed light on the role of spontaneous focusing of attention in the development of quantitative relations, such as relations of numbers. The first results of the project show substantial individual differences in children's spontaneous recognition of quantitative relations in tasks that can be completed by focusing on exact numerosity, non-numerical aspects of the tasks, and on relations of quantities in the task. Furthermore, the results of McMullen and colleagues (in press) indicate that it is possible to isolate an attentional component of quantitative thinking, referred to as Spontaneous Focusing On quantitative Relations (SFOR), and that SFOR is positively related to both symbolic and non-symbolic arithmetical skills.

Possibilities of Promoting SFON Tendency

It is tempting to speculate on the origins of SFON tendency, whether it is innate, or purely or partially culturally based. No matter what the nature of the very first appearance of focusing on numerosity is, what seems to hold true in later phases of numerical development is that culturally based activities and support of more experienced others play a crucial role in supporting children's development in SFON. The joint processes of children and adults directing children's attention to relevant aspects of tasks helps children acquire the culturally based numerical tools necessary for living in a society. These joint processes also help children understand the purposes of the tasks and certain cognitive strategies embedded in a variety of everyday activities (Gauvain, 2001; Gibson & Spelke, 1983).

Our quasi-experimental study on SFON (Hannula, Mattinen, & Lehtinen, 2005; Mattinen, 2006) was aimed at encouraging day-care personnel to support children's numerical focusing by deliberately directing children's attention towards variation in small numbers of objects or incidents. The idea of making children more aware of the aspect of number by using deliberate variation was based on the proposal of Marton and Booth (1997). According to them, to learn something means to become capable of experiencing various aspects of the set of items in a certain, specific way. Thus, 'twoness' is experienced against the background of a potential variation in the aspect of number, against 'oneness' and 'threeness', for instance (Marton & Booth, 1997). In this way, children's implicit knowledge about small numbers can become a more explicit target of intentional focusing, and children could learn the affordances of numerical aspects in a variety of everyday activities. During a 4-week training period, day-care personnel observed and kept a record of incidents when each child spontaneously focused their attention on numerosity while also purposefully guiding children's attention to exact numbers involved in everyday behaviour such as during eating, picking up toys, and outdoor activities. In addition, they played numerosity matching games with children and had a board of variable numbers of animal figures on the wall of the play room at the day-care centre. Adults secretly changed the number of animals on the board when children were away several times a day, which was deliberately used to promote children to look at how many animals there were on the board. All materials that were used, as well as all numerosities of everyday actions that were used for adult-guided numerical focusing, included only very small numbers of items, ranging from 0 to 3. Counting was not practised, just noticing small numbers of items everywhere in the surroundings of the children, being it two fire trucks driving on the street next to the playground, three similar T-shirts, or spoonfuls of cereal. It appeared that similar objects or colours in the set of items made it easier for the children to focus on numerosity. The moment of placing ready-made cookies on the plate, for example, made a 3-year-old girl notice the aspect of exact number spontaneously for the very first time. 'Two cookies' she spontaneously reported when seeing the two next to each other. Another child herself noticed that exactly two papers fell on the floor from a copy machine.

The results of the experimental study show that it is possible to enhance children's SFON tendency by means of social interaction and it even can be an enjoyable, fun activity for 3-year-old children. There was an experimental long-term effect on SFON tendency and subsequent development in cardinality related skills from pretest to delayed post-test 6 months after the 1-month enhancement programme for the children with some initial SFON tendency in the experimental group. Hence, it is possible to increase children's tendency to transfer their existing number skills to new situations. Focusing attention on the aspect of numerosity, wearing 'mathematical lenses', and having the abstract idea of numerosity seem to be important factors in producing the transferable number skills. Learning to focus on numerosity may be one significant step for young children on their way to learning to adopt mathematically meaningful perspectives on perceiving the world around them. Related to this, Goldin-Meadow, Alibali, and Church (1993) have proposed that children's first counting attempts are one of the first significant, easily noticeable signals for adults to start providing guidance in quantitative skills. SFON enhancement could be a necessary part of programmes for preventing and overcoming of mathematical difficulties.

Conclusion

The studies reviewed in this chapter suggest that the use of number skills, such as exact number recognition, in natural surroundings is not an automatic act—the amount of practice young children acquire in using their early number skills may differ substantially according to how frequently they focus on numerical properties (Hannula, 2005; Hannula & Lehtinen, 2005). The resulting differences in the amount of self-initiated practice children acquire in using their numerical skills may help explain developmental differences in numeracy from early childhood to the end of primary school. Differentiating specific attentional processes that trigger the use of number skills allows us to capture more exclusively all of the relevant subprocesses that are needed for the exact number recognition taking place in everyday surroundings. These number-related, domain-specific attentional processes are related to more general attention, such as focusing on tasks or inhibition of off-task behaviour, yet they are a part of number-related activities. Substantial individual differences in SFON tendency during the early childhood years suggests that the uncovering and modelling of focusing of attention on numerical aspects at an early age can make children more apt to practise and use their numerical skills in their everyday surroundings. This could be an efficient tool for promoting young children's mathematical development. Future studies should investigate whether the enhancement of SFON tendency together with other numerical skills could prevent later learning difficulties in mathematics.

Note

1. Huttenlocher, Jordan, and Levine (1994) presented a child with a non-verbal number recognition task within the number range 1–5 with instructions like 'Make your mat like mine', which may remind the reader of SFON imitation tasks. However, in their task, if the child did not respond or placed the wrong number of disks on the mat, when the experimenter placed one disk on her mat, the child was corrected and the item was repeated one more time. The same demonstration procedure was used with two disks. Only after this guidance to focus on the numerosity of disks on the mat, the testing of quantity matching skills started, ranging in numerosity from 1–5.

References

Aunola, K., Leskinen, E., Lerkkanen, M.-L., & Nurmi, J.-E. (2004). Developmental dynamics of math performance from pre-school to grade 2. *Journal of Educational Psychology*, *96*, 699–713.

Baroody, A. J., Li, X., & Lai, M.-L. (2008). Toddlers' spontaneous attention to number. *Mathematical Thinking and Learning*, *10*, 1–31.

Bermejo, V., Morales, S., & deOsuna, J. G. (2004). Supporting children's development of cardinality understanding. *Learning and Instruction*, *14*, 381–398.

Choudhury, A., McCandliss, B, & Hannula, M. M. (2007). *Individual differences in spontaneous focusing on numerosity are related to numerical skills at school age*. Paper presented in B. McCandliss (Chair), Symposium 'Developmental Changes in Number Skills Reflect the Emergence of Children's Ability to Focus Attention on Different Aspects of Number', 19th Annual Convention for Psychological Science, Washington, DC, 27 May.

Dowker, A. (2008). Individual differences in numerical abilities in preschoolers. *Developmental Science*, *11*, 650–654.

Edens, K.M. & Potter, E.F. (2013). An Exploratory Look at the Relationships Among Math Skills, Motivational Factors and Activity Choice. *Early Childhood Education Journal* 41(3), pp 235–243.

Ericsson, K. A. (1996). *The Road to Excellence: The Acquisition of Expert Performance in the Arts and Sciences, Sports, and Games*. Hillsdale, NJ: Erlbaum.

Ericsson, K. A. & Lehman, A. C. (1996). Expert and exceptional performance: evidence of maximal adaptation to task constraints. *Annual Review of Psychology*, *47*, 273–305.

Fischer, B., Gebhardt, C., & Hartnegg, K. (2008). Subitizing and visual counting in children with problems in acquiring basic arithmetic skills. *Optometry & Vision Development*, *39* (1), 24–29.

Fuson, K. (1988). *Children's Counting and Concepts of Number*. New York: Springer.

Gauvain, M. (2001). *The Social Context of Cognitive Development*. New York: Guilford Press.

Gelman, R. & Gallistel, C. R. (1978). *The Child's Understanding of Number*. Cambridge, MA: Harvard University Press.

Gibson, E. J. & Spelke, E. (1983). The development of perception. In J. H. Flavell & E. M. Markman (Eds), *Handbook of Child Psychology: Vol 3. Cognitive Development* (pp. 1–76). New York: Wiley.

Ginsburg, H. P., Inoue, N., & Seo, K.-H. (1999). Preschoolers doing mathematics: observations of everyday activities. In J. Copley (Ed.), *Mathematics in the Early Years* (pp. 88–99). Reston, VA: National Council of Teachers of Mathematics.

Goldin-Meadow, S., Alibali, M. W., & Church, R. B. (1993). Transitions in concept acquisition: using the hand to read the mind. *Psychological Review*, *100*, 279–297.

Gunderson, E. A. & Levine, S. C. (2011). Some types of parent number talk count more than others: Relations between parents' input and children's number knowledge. *Developmental Science*, *14* (5), 1021–1032.

Hanich, L. B., Jordan, N. C., Kaplan, D., & Dick, J. (2001). Performance across different areas of mathematical cognition in children with learning difficulties. *Journal of Educational Psychology*, *93*, 615–626.

Hannula, M. M. (2005). *Spontaneous Focusing On Numerosity in the Development of Early Mathematical Skills*. Annales Universitatis Turkuensis B, 282. Turku: Painosalama.

Hannula, M. M., Grabner, R. H., Lehtinen, E., Laine, T., Parkkola, R. & Ansari, D. (2009). *Neural correlates of Spontaneous Focusing On Numerosity (SFON). NeuroImage*, *47*, S44.

Hannula, M. M., Grabner, R. H., & Lehtinen, E. (2009b). *Neural correlates of Spontaneous Focusing On Numerosity (SFON) in a 9-year longitudinal study of children's mathematical skills*. Paper presented at the EARLI Advanced Study Colloqium 'Cognitive neuroscience meets mathematics education' in Brugge, Belgium.

Hannula, M. M., Lepola, J. & Lehtinen, E. (2010). Spontaneous focusing on numerosity as a domain-specific predictor of arithmetical skills. *Journal for Experimental Child Psychology*. *107*, 394–406.

Hannula, M. M., Räsänen, P. & Lehtinen E. (2007). Development of counting skills: Role of spontaneous focusing on numerosity and subitizing-based enumeration. *Mathematical Thinking and Learning 9 (1)*, 51–57.

Hannula, M. M., & Lehtinen, E. (2001). Spontaneous tendency to focus on numerosities in the development of cardinality. In M. Panhuizen-Van Heuvel (Ed.), Proceedings of the 25th Conference of the International Group for the Psychology of Mathematics Education, 3, 113–120. Drukkerij Wilco, The Netherlands: Amersfoort.

Hannula, M. M. & Lehtinen, E. (2005). Spontaneous Focusing On Numerosity and mathematical skills of young children. *Learning and Instruction*, *15*, 237–256.

Hannula, M. M., Lepola, J. & Lehtinen, E. (2010). Spontaneous Focusing On Numerosity as a domain-specific predictor of arithmetical skills. *Journal for Experimental Child Psychology*. *107*, 394–406.

Hannula, M. M., Mattinen, A., & Lehtinen, E. (2005). Does social interaction influence 3-year-old children's tendency to focus on numerosity? A quasi-experimental study in day-care. In L. Verschaffel, E. De Corte, G. Kanselaar, & M. Valcke (Eds), Powerful learning environments for promoting deep conceptual and strategic learning. *Studia Paedagogica*, *41*, 63–80.

Hannula, M. M., Räsänen, P., & Lehtinen, E. (2007). Development of counting skills: role of Spontaneous Focusing On Numerosity and subitizing-based enumeration. *Mathematical Thinking and Learning*, *9* (1), 51–57.

Hannula-Sormunen, M. M., Grabner, R., & Lehtinen, E. (in preparation). *Neural correlates of Spontaneous Focusing On Numerosity (SFON) in a 9-year-longitudinal study of children's mathematical skills*.

Hannula-Sormunen, M. M., Grabner, R., Lehtinen, E., Laine, T., Parkkola, R., & Ansari, D. (in preparation). *Spontaneous focusing on numerosity: an fMRI study on adults*.

Hannula-Sormunen, M. M., Lehtinen, E., & Räsänen, P. (in press). *Children's preschool subitizing, spontaneous focusing on numerosity and counting skills as predictors of mathematical performance 6-7 years later at school.. Mathematical Thinking and Learning*.

Huttenlocher, J., Jordan, N. C., & Levine, S. C. (1994). A mental model for early arithmetic. *Journal of Experimental Psychology: General*, *123*, 284–296.

Jordan, N. C., Kaplan, D., Nabors Olah, L., & Locuniak, M. N. (2006). Number sense growth in kindergarten: a longitudinal investigation of children at risk for mathematics difficulties. *Child Development*, *77*, 153–175.

Krajewski, K. & Schneider, W. (2009a). Early development of quantity to number-word linkage as a precursor of mathematical school achievement and mathematical difficulties: findings from a four-year longitudinal study. *Learning and Instruction, 19*, 513–526.

Krajewski, K. & Schneider, W. (2009b). Exploring the impact of phonological awareness, visual-spatial working memory, and preschool quantity-number competencies on mathematics achievement in elementary school: findings from a 3 year longitudinal study. *Journal of Experimental Child Psychology, 103*, 516–531.

Kucian, K., Kohn, J., Hannula-Sormunen, M.M., et al. (2012). Kinder mit Dyskalkulie fokussieren spontan weniger auf Anzahligkeit [Children with Developmental Dyscalculia Focus Spontaneously Less on Numerosities]. *Lernen und Lernstörungen,1*(4), pp 241–253.

Landerl, K., Bevan, A., & Butterworth, B. (2004). Developmental dyscalculia and basic numerical capacities: a study of 8- to 9-year-old students. *Cognition, 93* (2), 99–125.

LeFevre, J.-A., Fast, L., Skwarchuk, S.-L., et al. (2010). Pathways to mathematics: longitudinal predictors of performance. *Child Development, 81*, 1753–1767.

Lepola, J., Niemi, P., Kuikka, M., & Hannula, M. M. (2005). Linguistic skills and motivation as predictors of children's difficulties in reading and arithmetics: a follow-up study from 6 to 8 year of age. *International Journal for Educational Research 43*, 250–271.

Levine, S. C., Suriyakham, L. W., Rowe, M. L., Huttenlocher, J., & Gunderson, E. A. (2010). What counts in the development of young children's number knowledge? *Developmental Psychology, 46* (5), 1309–1319.

McMullen, J. A., Hannula-Sormunen, M. M., & Lehtinen, E. (2011). Young children's spontaneous focusing on quantitative aspects and verbalizations of their quantitative reasoning. In B. Ubuz (Ed.). *Proceedings of the 35th Conference of the International Group for the Psychology of Mathematics Education, Vol. 3*, pp. 217–224. Ankara, Turkey: PME.

McMullen, J., Hannula-Sormunen, M.M., & Lehtinen, E. (2013). Young children's recognition of quantitative relations in mathematically unspecified settings. The Journal of Mathematical Behavior, 32 (3), 450–460.

McMullen, J., Hannula-Sormunen, M.M., & Lehtinen, E. (in press). Spontaneous focusing on quantitative relations in the development of children's fraction knowledge. Cognition and Instruction.

Martin, L., Hannula, M. M., & Schwartz, D. (2008). Spontaneous creation and use of grouping structure by elementary school students. In K. Pilner (Chair), *Group Think: Measuring and Improving Set Semantics in Elementary School Mathematics*. Symposium conducted at meeting of American Educational Research Association, 24–28 March, New York.

Marton, F. & Booth, S. (1997). *Learning and Awareness*. Mahwah, NJ: Erlbaum.

Mattinen, A. (2006). *Huomio lukumääriin: Tutkimus 3-vuotiaiden lasten matemaattisten taitojen tukemisesta päiväkodissa* [Focus on Numerosities: A Study on Supporting 3 Year-Old Children's Mathematical Development in Day Care]. Turku, Finland: Painosalama.

Melhuish, E. C., Sylva, K., Sammons, P., et al. (2008). Preschool influences on mathematics achievement. *Science, 321*, 1161–1162.

Nunes, T. & Bryant, P. (1996). *Children Doing Mathematics*. Oxford: Blackwell.

Potter, E. (2009). Spontaneous Focusing On Numerosity: motivational and skill correlates in young children in a public preschool and kindergarten program. In S. L. Swars, D. W. Stinson, & S. Lemons-Smith (Eds), *Proceedings of the 31st Annual Meeting of the North American Chapter of the International Group for the Psychology of Mathematics Education, 5*. (pp. 152–155). Atlanta, GA: Georgia State.

Railo, H. M, Koivisto, M., Revonsuo, A., & Hannula, M. M. (2008). Role of attention in subitizing. *Cognition, 107*, 82–104.

Sarnecka, B.W. & Carey, S. (2008). How counting represents number: what children must learn and when they learn it. *Cognition, 108*. 662–674.

Saxe, G. B., Guberman, S. R., & Gearhart, M. (1987). Social processes in early number development. *Monographs of the Society for Research in Child Development, 52*, Serial No. 216.

Sophian, C. (1988). Limitations on preschool children's knowledge about counting: using counting to compare sets. *Developmental Psychology, 24*, 634–640.

Spelke, E. (2003). What makes us smart? Core knowledge and natural language. In D. Gentner & S. Goldin-Meadow (Eds), *Language in Mind*. Cambridge, MA: MIT Press.

Takila, P., Hannula, M. M., & the PIPARI Study Group (2004). *Spontaneous tendency to focus on numerosity (SFON)—comparison of premature and full term infants at the age of two years*. Poster presented at the 8th Nordic Meeting in Neuropsychology, Turku, Finland.

Trick, L. & Pylyshyn, Z. W. (1994). Why are small and large numbers enumerated differently? A limited-capacity preattentive stage in vision. *Psychological Review*, *101*, 80–102.

Tudge, J. R. H. & Doucet, F. (2004). Early mathematical experiences: observing young black and white children's everyday activities. *Early Childhood Research Quarterly*, *19*, 21–39.

Wynn, K. (1990). Children's understanding of counting. *Cognition*, *36*, 144–193.

Wynn, K. (1998). Numerical competence in infants. In C. Donlan (Ed.), *The Development of Mathematical Skills* (pp. 3–25). Hove: Psychology Press.

Yun, C., Hannula M. M., & McCandliss, B. (2008). Children's use of grouping information for rapid enumeration of items: a cross-sectional study. In K. Pilner (Chair), *Group Think: Measuring and Improving Set Semantics in Elementary School Mathematics*. Symposium conducted at meeting of American Educational Research Association, 24–28 March, New York.

CHAPTER 16

HOW COUNTING LEADS TO CHILDREN'S FIRST REPRESENTATIONS OF EXACT, LARGE NUMBERS

BARBARA W. SARNECKA, MEGHAN C. GOLDMAN, AND EMILY B. SLUSSER

COUNTING is the first overtly numerical activity that most children do. However, young children's counting behaviour is often difficult to interpret, because what they *do* is not a perfect indicator of what they *know*. Of course, this is true for all behaviour, but in the case of counting, the gap between doing it right and knowing what it means has been of particular interest to developmental scientists.

The gap is important because over the past several decades, children's counting behaviour has been the main evidence used to argue that children have (or do not have) natural-number concepts. (By natural-number concepts, we mean the mental representation of exact numerical quantities, such as six or seven, not explicit ideas about entities called 'natural numbers.') Interest in these topics dates back to at least the 1970s, when a few researchers (e.g. Schaeffer, Eggleston, & Scott, 1974; Gelman & Gallistel, 1978) shifted their focus away from Piagetian conservation-of-number tasks, and started to look instead at the number-concept knowledge demonstrated by younger children, especially in their counting behaviour.

As these researchers observed, correct counting follows several rules. For example, you must recite the counting list in the same order every time; you must point to one and only one object with each number word you say; you must point to all the objects without skipping or double-counting any of them; and crucially, you must understand that the last number word you say tells the number of objects in the whole set. Note that all but the last rule is about something you must *do*. The last rule, which Gelman and Gallistel (1978) called the *cardinal principle* and Schaeffer et al. (1974) called the *cardinality rule*, is about something you must *know*. It is easy to measure what children do, but hard to measure what they know. So, naturally, this is the rule that scientists have been arguing about ever since.

Gelman and Gallistel (1978) originally proposed that children have an innate or early-developing knowledge of how to count, and that this is itself evidence that children have innate or early-developing number concepts. These claims inspired many researchers in the 1980s and 1990s to try to figure out which came first: procedural knowledge of counting (knowing that you have to say the counting list in the same order every time, that you have to point to one object for every number word, etc.) versus conceptual knowledge of counting, especially the cardinal principle. Although space precludes a complete review of that literature here, the general finding was that there is a gap between procedural and conceptual knowledge, with children acquiring the procedural knowledge first. In other words, children learn to count (and they appear to be doing everything right) long before they understand that counting reveals the cardinal number of items in the set (Baroody & Price, 1983; Briars & Siegler, 1984; Freeman, Antonucci, & Lewis, 2000; Frye, Braisby, Lowe, Maroudas, & Nicholls, 1989; Fuson, 1988; Miller, Smith, Zhu, & Zhang, 1995; Slaughter, Itakura, Kutsuki, & Siegal, 2011; Wagner & Walters, 1982).

We will take time here to describe just one task that demonstrates this gap, because the task re-appears throughout this chapter. This task is called 'Give-N' or 'Give-A-Number.' In this task, the child is given a set of objects (e.g. a bowl of 15 small plastic apples) and is asked to give a certain number of the objects to a puppet. For example, the child might be asked to 'Give *five* apples to Kermit.' The oft-replicated result is that, while many 2–4-year-olds can count perfectly well (e.g. they count a row of 10 apples without error), these same children are unable to give the right number of objects in the Give-N task. Instead of using counting to generate the right number of items, they just grab a handful. Even when they are explicitly told to count (e.g. 'Can you count and make sure you gave Kermit five apples?' or, 'But Kermit wanted five apples – can you fix it so there are five?') they fail to use the results of their counting to solve the problem (Le Corre, Van de Walle, Brannon, & Carey, 2006).

Studies using the Give-N task have shown that children move through a predictable series of performance levels, often called number-knower levels (e.g. Condry & Spelke, 2008; Le Corre & Carey, 2007; Le Corre et al., 2006; Lee & Sarnecka, 2010, 2011; Negen & Sarnecka, 2012; Sarnecka & Gelman, 2004; Sarnecka & Lee, 2009; Slusser & Sarnecka, 2011; Wynn, 1990). These number-knower levels are found not only in child speakers of English, but also in Japanese (Sarnecka, Kamenskaya, Yamana, Ogura, & Yudovina, 2007), Mandarin Chinese (Le Corre, Li, & Jia, 2003; Li, Le Corre, Shui, Jia, & Carey, 2003) and Russian (Sarnecka et al., 2007).

The number-knower levels are as follows. At the earliest (i.e. the 'pre-number-knower') level, the child makes no distinctions among the meanings of different number words. On the Give-N task, pre-number knowers might always give one object, or might always give a handful, but the number given is unrelated to the number requested. At the next level (called the 'one-knower' level), the child knows that 'one' means 1. On the Give-N task, this child gives exactly 1 object when asked for one, and gives 2 or more objects when asked for any other number. After this comes the 'two-knower' level, when the child knows that 'two' means 2. Two-knowers give 1 object when asked for 'one,' and 2 objects when asked for 'two,' but they do not reliably produce the right answers for any higher number words. The two-knower level is followed by a 'three-knower', then a 'four-knower' level.

Mathieu Le Corre and colleagues (2006) coined the term 'subset-knowers' to describe children at the one-, two-, three-, and four-knower levels, because even though they can typically count to 10 or higher, they only know the exact meanings of a subset of the words in their counting list. Subset-knowers are distinct from both pre-number knowers (who do not yet know the exact meanings of any number words), and 'cardinal-principle-knowers' (abbreviated 'CP-knowers') who know the exact meanings of all the number words, as high as they can count. More precisely, CP-knowers understand that the set size associated with any number word is generated by counting up to that number word from 'one,' and adding one object to the set for every word in the counting list. In this way, counting links each number word to an exact set size.

For children, the rules of counting (the ordered list of words, to be recited while gesturing to objects, one at a time) form a placeholder structure. This is interesting because it suggests that number and counting *concepts* are acquired through a special kind of learning called conceptual-role bootstrapping (Carey, 2009; see also Block, 1986; Quine, 1960). Conceptual-role bootstrapping is a process where the learner first learns a placeholder structure – a set of symbols that have some fixed relation to each other. At first, the individual symbols are just placeholders, because they are defined just in terms of their place in the structure. However, over time, the individual symbols may be filled in with meaning, and become fully defined concepts.

As an example, consider the periodic table of the elements. As I (BWS) write this, I am looking at a laminated copy of the periodic table that serves as a placemat at my house. Many of the element names on the periodic table are mere placeholders for me. The only meaning I attach to them comes from their position in the table itself.

Take the word *beryllium,* for example. I know that beryllium is an element, because it is listed on this table. I can read that it has an atomic number of 4 and an average atomic weight of 9.0122, but that is meaningless to me. The table legend tells me that beryllium is an alkaline earth metal, which sounds vaguely threatening, like something that might poison the well water, but then I notice that magnesium and calcium are in the same group, which makes beryllium seem friendlier.

Magnesium and calcium are examples of concepts that are more than placeholders – they have some meaning outside the periodic table; they are defined in terms of other knowledge I have. Specifically, I know that magnesium and calcium are vitamins; that calcium is found in dairy products and is important for bones and teeth; and so forth.

Of course, there are people who have full (not just placeholder) concepts for all the elements – chemists. Chemists have completed the conceptual-role bootstrapping process that I started when I first encountered the periodic table, but never finished. (That is, many of the element names are still mere placeholders to me.) I was reminded of this when I consulted with an actual chemist about writing this article. 'I'd like to use the periodic table as an example of bootstrapping.' I said. 'What's a good example of an element that nobody knows anything about? You know – an element that you see in the periodic table, but otherwise no one has ever heard of it? How about beryllium?'

'Oh, no,' he said. 'Beryllium is very useful. It's used in computer chips.'

'Oh.' I said, 'How about boron?'

'Boron is used a lot,' he said. 'Boron nitride is almost as hard as diamond. Because it's one element either side of carbon, see?'

Here, he pointed to the positions of boron, carbon, and nitrogen on an imaginary periodic table. Then he sketched molecules in the air, saying,

'So instead of carbon-carbon-carbon-carbon, you go boron-nitrogen-boron-nitrogen. Get it?'

I did not get it.

I said, 'OK, forget those. I'm going to use this one because it has a cool name: *molybdenum*.'

'Oh, people will have heard of that,' he said. 'Molybdenum disulphide is a high-quality lubricant for ball bearings...'

That was when I realized that I was asking precisely the wrong person. For chemists, the elements in the periodic table are not placeholders – they are fully elaborated concepts. The words *beryllium, boron* and *molybdenum* actually mean something to a chemist.

In the case of exact-number concepts, young children are like me with the periodic table; older children and adults are like chemists. In this case, the structured set of symbols is the counting list. The list is structured in the sense that it has an order – one cannot count *three, five, two, four, one*; the order must always be *one, two, three, four five*. However, the individual words (e.g. *three*) are not initially defined in terms of any other knowledge the child has. In the beginning, *three* is just the counting word that comes after *two* and before *four*. (Just as for me, beryllium is merely the element after lithium and above magnesium.)

Under Carey's (2009) conceptual-role bootstrapping account, the process of learning what the words mean *is* the process of acquiring natural-number concepts. This is where natural-number concepts come from. If this proposal is correct, then we should be able to find behavioural evidence for it. Specifically, there should be an intermediate state of knowledge where children have acquired the symbol structure (i.e. memorized the counting list), but do not yet understand what each number word means. The children in this intermediate state are the subset-knowers.

Remember that what separates subset-knowers from CP-knowers is that only CP-knowers understand the cardinal principle (Gelman and Gallistel, 1978), which is much more than a rule about how to count – it is the principle that makes the meaning of any cardinal number word a function of that word's ordinal position in the counting list. In other words, the cardinal principle guarantees that for any counting list, in any language, the sixth word in the list must mean 6, the twentieth word must mean 20, and the thousandth word must mean 1000. To understand the cardinal principle is to understand how counting represents number.

An Aside: Why are There No 'Five-Knowers'?

People sometimes ask why the knower-levels only go up to 'four-knower.' Why do we not find five-knowers, six-knowers, or seven-knowers? Is there no period where children know what the numbers *one* through *seven* mean, but not what *eight* means? According to the bootstrapping account, the answer is *no*. This is because only the meanings of the smallest number words (1–4) can be perceived accurately without counting (Antell & Keating, 1983; Butterworth, 1999; Feigenson & Carey, 2003, 2005; Feigenson, Carey, & Hauser, 2002a; Scholl & Leslie, 1999; Simon, 1997; Starkey & Cooper, 1980; Uller, Carey, Huntley-Fenner, & Klatt, 1999; Wynn, 1992a). Humans have innate, non-verbal cognitive systems that allow them to represent small, exact numbers (up to 3 or 4), as well as large, approximate numbers (see Feigenson, Dehaene, & Spelke, 2004, for review), but these non-verbal systems cannot distinguish, for example, 7 from 8 items. Thus, the largest exact number that children can recognize directly seems to be 4. Numbers higher than 4 must be represented through counting.

Studies of Number-Knower Levels and Number-Concept Development

If this account of where exact-number concepts come from is right, then subset-knowers present a wonderful opportunity to study number-concept construction in progress. The meanings that subset-knowers assign to number words should be partial and incomplete, including some, but not all aspects of the exact-number concepts that adults and older children have. Describing this process, and figuring out what meanings subset-knowers assign to higher numbers and when, is the general motivation for the studies discussed below.

Number-Knower Levels versus Counting or Estimation

First, it is necessary to confirm that number-knower levels are a real phenomenon. As explained above, the knower-levels framework describes a pattern where each child either succeeds up to a given number (4 or below), and then fails at all the higher numbers (in which case the child is a subset-knower), or uses counting and succeeds at the higher numbers as well (in which case the child is a CP-knower). However, this description is idealized. In real life, even subset-knowers sometimes grab (by lucky accident) 6 items when asked for 'six'; and CP-knowers sometimes make counting errors, and so end up giving the wrong number of items, even though they understand how counting works. Also, subset-knowers sometimes do count the objects, even though they do not know how counting solves the problem. So how can we be sure that the 'number-knower levels' framework is actually the right way to think about all this?

An alternative way of thinking about these data would be to assume that children do have exact-number concepts from the beginning. So when they count, we might assume that they do know what all of the number words mean, and that they use either estimation or counting to do the Give-N task. In either case (estimation or counting), smaller sets are easier to produce than larger ones, so something like the knower-levels pattern might emerge. That is, each child would still perform correctly up to a given number, and fail at higher numbers. When children give the right answer (e.g. they give 4 items when asked for 'four'), it's difficult to know what strategy they used – estimation, counting, or simple recognition as in the knower-levels account. However, when they give the wrong answer for higher number words, we can learn something from their mistakes. In particular, both estimation and counting produce different patterns of errors than simple guessing – and simple guessing on the higher numbers is what the knower-levels account predicts.

In a recent meta-analysis (Sarnecka & Lee, 2009), we looked at Give-N data from 280 children, aged 24–48 months (about 4500 Give-N trials in all), and found that most wrong answers were simply guesses, not counting or estimation errors. We could see this in several ways: first, the mean of the errors was unrelated to the number asked for. If children were either estimating or counting, then errors that fell close to the target should be more common than errors that fell far away from it. (For example, if you are trying to estimate or count 10 items, you are more likely to produce 9 or 11 by mistake than 5 or 15.) However, in the data we reviewed, the wrong guesses produced by subset-knowers were unrelated to the number word asked for, indicating that they really did not know what number they were trying to produce.

Second, children's errors were asymmetrical and lower-bounded by the numbers they knew. Both counting and estimation should produce symmetrical patterns of error, because people are equally likely to undercount/underestimate as to overcount/overestimate. However, in the data we reviewed, children's errors were asymmetrical, in exactly the way predicted by the knower-levels account.

Take, for example, children who correctly gave 1, 2, and 3 items when asked (these are three-knowers in the knower-levels account). These children did *not* give 1, 2, or 3 items when asked for 'four' or any higher number. If they were asked for 'four', they were much more likely to give 5 items (as a wrong answer) than 3 items, even though these errors are both the same distance from the correct answer of 4.[1] If children were counting or estimating, this asymmetry would not make sense. Counting and estimation errors are just as likely to produce too few items as too many. Only the knower-levels framework explains why errors would be asymmetrical: three-knowers actually know what 'one,' 'two,' and 'three' mean, and they know that the number-word meanings are exclusive (Wynn, 1990 1992b), so they restrict their guesses about the meaning of 'four' to higher set sizes.

The analysis verified other aspects of the knower-levels account as well – children learned the number-word meanings in order (i.e. there was no evidence of any child learning the meaning of a higher number word before a lower one), and once they figured out the cardinal principle of counting, they generalized this principle to the rest of their count list (i.e. there was no evidence of any child knowing what some, but not all of the higher number words meant). Overall, these analyses strongly supported the number-knower levels/bootstrapping account of number-concept development.

Age Ranges for Each Knower Level

An obvious question to ask about the number-knower levels is, 'How old are the children at each level?' The answer depends on the individual child, and differs dramatically for children in different socioeconomic environments. Most studies to date have involved children from relatively privileged socioeconomic backgrounds. Aggregate results from many studies with high-SES children show that these children typically reach an understanding of cardinality sometime between 34 and 51 months old (see Figure 16.1). That is, any time from a few months before their third birthday to a few months after their fourth birthday. (Because these data are cross-sectional, the fact that many CP-knowers were above age 51 months is

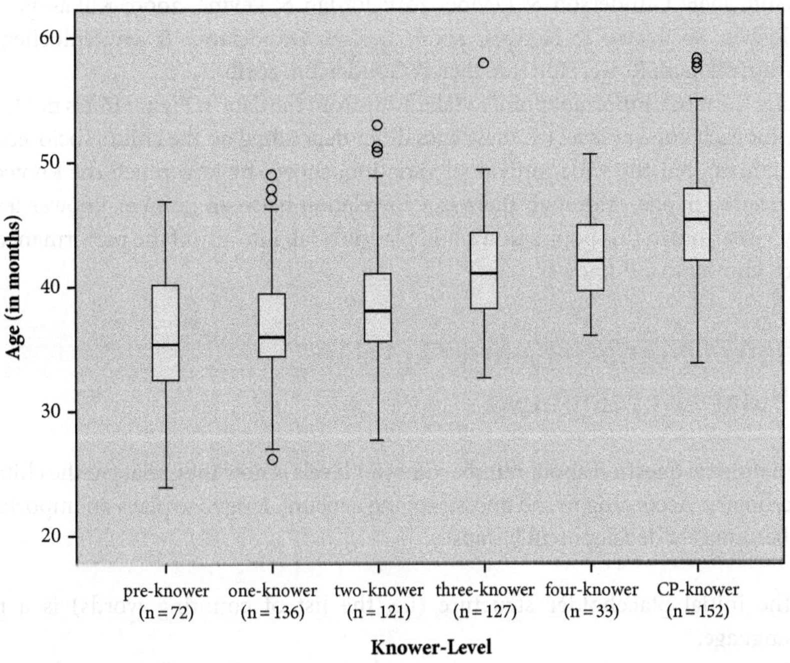

FIGURE 16.1 Aggregate data from 641 2–4-year-old children tested on the Give-N task.

Data from James Negen and Barbara W. Sarnecka, *Young children's number-word knowledge predicts their performance on a nonlinguistic number task*. In N. A. Taatgen and H. van Rijn (eds.), 'Proceedings of the 31st Annual Conference of the Cognitive Science Society'. Austin: Cognitive Science Society, pp. 2998–3003, 2009, James Negen and Barbara W. Sarnecka, Number-concept acquisition and general vocabulary development, *Child Development*, 83(6) pp. 2019–2027, DOI: 10.1111/j.1467-8624.2012.01815.x, 2012. Barbara W. Sarnecka and Susan Carey, How counting represents number: What children must learn and when they learn it, *Cognition*, 108(3), pp. 662–674, DOI: 10.1016/j.cognition.2008.05.007, 2008. Barbara W. Sarnecka and Susan A. Gelman, Six does not just mean a lot: Preschoolers see number words as specific, *Cognition*, 92(3), pp. 329–352, DOI: 10.1016/j.cognition.2003.10.001, 2004, Barbara W. Sarneckaa, Valentina G. Kamenskayab, Yuko Yamanac, Tamiko Ogurad, and Yulia. B. Yudovinae, From grammatical number to exact numbers: Early meanings of "one," "two," and "three" in English, Russian, and Japanese, *Cognitive Psychology*, 55(2), pp. 136–168, DOI: 10.1016/j.cogpsych.2006.09.001, 2007, Barbara W. Sarnecka and Michael D. Lee, Levels of number knowledge during early childhood, *Journal of Experimental Child Psychology*, 103(3), pp. 325–337, DOI: 10.1016/j.jecp.2009.02.007, 2009, Emily B. Slusser and Barbara W. Sarnecka, Find the picture of eight turtles: A link between children's counting and their knowledge of number-word semantics, *Journal of Experimental Child Psychology*, 110(1), pp. 38–51, DOI: 10.1016/j.jecp.2011.03.006, 2011, and Emily Slusser, Annie Ditta, and Barbara W. Sarnecka, Connecting numbers to discrete quantification: A step in the child's construction of integer concepts, *Cognition*, 129(1), pp. 31–41, DOI: 10.1016/j.cognition.2013.05.011, 2013.

not informative; we do not know how long those children were CP-knowers before they were tested. To estimate the range of ages at which children in this population typically become CP-knowers, we looked at the youngest CP-knower and the oldest subset-knower, excluding outliers.)

These data were collected from children attending private preschools in the relatively affluent and educated communities of Ann Arbor, Michigan; Cambridge, Massachusetts; Irvine, California; and Taipei, Taiwan. Children from less privileged backgrounds come to understand cardinality significantly later, at age 4 or older (e.g. Dowker, 2008; Fluck &

Henderson, 1996; Gunderson & Levine, 2011; Jordan & Levine, 2009; Klibanoff, Levine, Huttenlocher, Vasilyeva, & Hedges, 2006; Levine, Gunderson, & Huttenlocher, 2011a; Levine, Suriyakham, Rowe, Huttenlocher, & Gunderson, 2011b).

Perhaps the most important point to take away from the data in Figure 16.1 is not the absolute age for each knower level (as these ages differ depending on the child's socio-economic circumstances), but the wide individual variation, shown by how much the knower-level groups overlap in age. Although there is a correlation between number-knower level and age, any 3-year-old in this population could plausibly fall into any of the performance levels, from pre-knower to CP-knower.

Relation of Number-Knower Level to Vocabulary Development

A basic empirical question about number-knower levels is how they relate to the child's general vocabulary. According to the bootstrapping account, language plays an important role in number-concept development because:

1. The initial placeholder structure (i.e. the list of counting words) is a part of language.
2. The processes used to construct the number concepts (i.e. to figure out the meanings of these symbols) likely subserve word learning in other domains as well.

Thus, it is reasonable to expect that controlling for age, children who know more word meanings in general will also know more number-word meanings. (Note that this is not a direct test of the bootstrapping account itself – it is something that should be true if number-concept development is tied in any way to language.)

To answer this question, we gave 59 children, aged 30–60 months, the Give-N task, as well as tests of expressive and receptive vocabulary (Negen & Sarnecka 2012). As predicted, strong correlations were found between number-knower level and vocabulary, independent of the child's age (see Figure 16.2). The study was very preliminary. It did not control for IQ or SES (although all of the children were from high-income and highly educated families), and it used standardized measures of general vocabulary. The Woodcock Johnson II-R was used for expressive vocabulary (Woodcock & Johnson, 1985) and the Peabody Picture Vocabulary Test III (Dunn & Dunn, 1997) for receptive vocabulary. Because these tests measure vocabulary development as a whole, the analysis also could not tell us whether some domains of vocabulary (e.g. spatial terms or quantifiers) are more highly predictive of number knowledge than others. (Although one study did report a connection between number knowledge and knowledge of quantifiers such as *a*, *some*, and *all*; Barner, Chow, & Yang, 2009a; see also Barner, Libenson, Cheung, & Takasaki, 2009b).

Despite being very preliminary, the Negen and Sarnecka (2012) study does provide indirect evidence for theories of number-concept development that assign an important role to language. A correlation between language and number development is necessary (although certainly not sufficient) for any of these theories to be true.

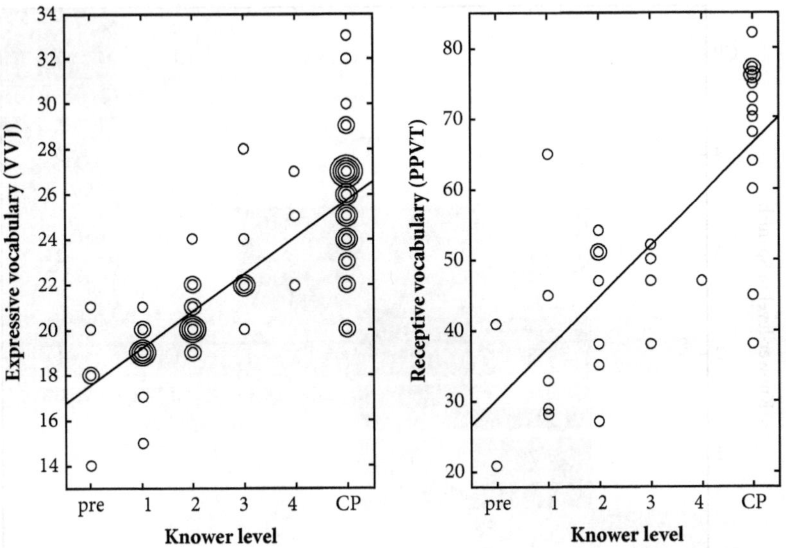

FIGURE 16.2 Relations among expressive vocabulary, receptive vocabulary, and knower-level variables. Concentric markers are used when data points overlap. Correlations were significant at $p < 0.01$, even when partialled for age.

Reproduced from James Negen and Barbara W. Sarnecka, Number-Concept Acquisition and General Vocabulary Development, *Child Development*, 83(6) pp. 2019–2027, DOI: 10.1111/j.1467-8624.2012.01815.x © 2012 The Authors. Child Development © 2012 Society for Research in Child Development, Inc.

Within-Child, Cross-Linguistic Consistency of Number-Knower Levels

A more specific prediction of the bootstrapping account is that if a child speaks two languages (e.g. Spanish and English), their number-knower level should be the same (or nearly the same) in both of them. Here is why. Subset-knowers are in an intermediate state of knowledge. They have memorized a counting list (i.e. they have acquired a placeholder structure), but they have not yet learned what each word means. Bilingual children have ample time to memorize the counting lists in both of their languages before they learn the exact meanings of the words in either language. So when a bilingual child learns that, for example, 'three' means 3, it should be a short step to realizing that *tres* (the third word in the Spanish counting list) also means 3. If the Spanish and English placeholder structures are both known, and are waiting to be filled in with meaning, then the newly-acquired concept '3' can take its place in both of them.

Thus, the prediction of the bootstrapping account is that children should know the same (or nearly the same) set of number-word meanings in both of their languages. To test whether this was true, we looked at data from 65 bilingual children, aged 41–66 months (Goldman & Sarnecka, 2011). These children used English in their Head Start programme, which most of them started at age 4, and either a mixture of Spanish and English, or only Spanish at home. Results are illustrated in Figure 16.3. In general, the data conform to

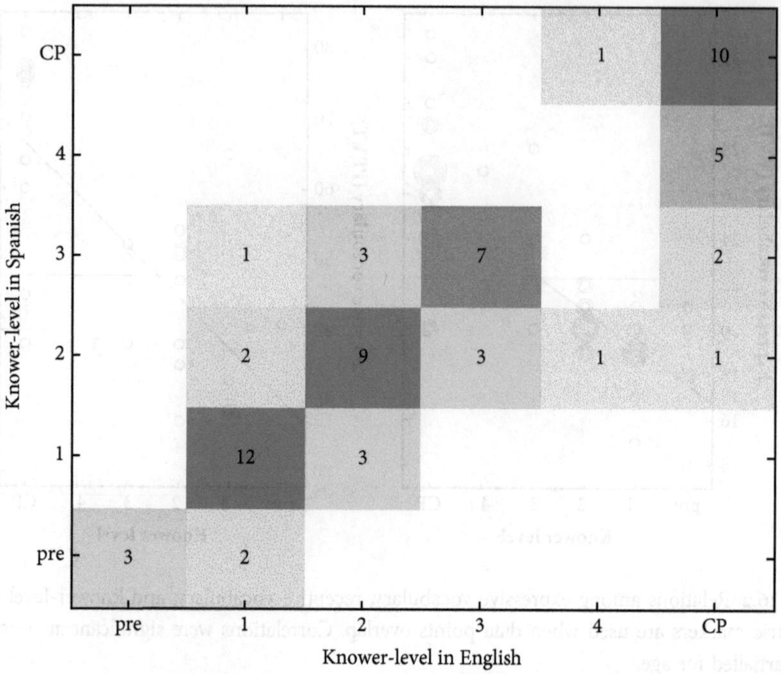

FIGURE 16.3 Heat map of data from 65 bilingual children, aged 41–66 months, tested on the Give-N task in each of their languages separately. Numbers in each cell indicate the number of children with that combination of English and Spanish knower-levels. Cells on the diagonal represent children with the same knower-level in English and Spanish. For all participants, the overall correlation between their knower-levels in English and Spanish was highly significant at $p < 0.0001$.

Reproduced from Goldman, M. C. and Sarnecka, B. W., Cross-linguistic consistency within bilingual children: Are number words special? Paper presented at the Symposium on Cognitive and Language Development, Los Angeles, CA © 2011, The Authors.

the bootstrapping prediction – that is, even though there was wide individual variation in how much time the children spent in Spanish-language versus English-language environments, most children showed the same or similar knower-levels in both Spanish and English. In Figure 16.3, this is illustrated by the fact that the cells on the diagonal (which correspond to having the same knower-level in both languages) had the highest number of children, and children who did not fall on the diagonal usually fell close to it (indicating that their knower-levels were similar, but not the same in both languages). Data from Korean/English and Mandarin/English bilingual preschoolers from higher-income households showed the same pattern (Sarnecka, Wright, & Goldman, 2011).

Importantly for the bootstrapping account, this within-child, cross-language consistency is more true for numbers than for other kinds of words. We looked at bilingual children's knowledge (in both their languages) of three types of words: numbers, colours (*red, yellow*, etc.) and common nouns for animals and vehicles (*tiger, bus*, etc.). Children's number-word knowledge was significantly more correlated (across their two languages) than their colour-word or common-noun knowledge (Sarnecka et al., 2011), again confirming the predictions of the bootstrapping account.

Describing The Number Concepts Under Construction

According to the bootstrapping account, subset-knowers should have a partial and incomplete understanding of what the number words mean – these are the number concepts still under construction. So, many of our studies are aimed at figuring out exactly what subset-knowers understand about numbers (or operationally, what meanings they assign to number words) along the way.

Number is About Discrete Things, Not Continuous Substances

Part of understanding numbers is knowing that number is a property of sets – and that sets are comprised of discrete individuals. We reasoned that this aspect of number knowledge might plausibly be acquired during the bootstrapping process, and that it might be understood by subset-knowers before they acquire the cardinal principle. In other words, 'discreteness' might be a piece of the number-concept 'puzzle' that children discover relatively early.

Language may provide a clue to this aspect of numbers (Bloom & Wynn, 1997), because number words quantify over count nouns (e.g. *five blocks*), but not over mass nouns (e.g. **five water* is ungrammatical). To find out whether subset-knowers understand this about number words, we tested 170 children, aged 30–54 months, on the Blocks and Water task (Slusser, Ditta, & Sarnecka, 2013). This task asked whether young children – even those who cannot yet produce or identify a set of exactly 5 or 6 objects – already know that *five* and *six* refer to discontinuous quantities (i.e. to sets of discrete individuals).

In the first experiment, children were presented with two empty cups. The experimenter then placed five (or six) objects in one cup and five (or six) scoops of a continuous substance in the other cup. Four trials asked children about a number word (e.g. 'Which cup has five?'). The other four trials asked about a quantifier (e.g. 'Which cup has more?'). For half of the trials, the cup with discrete objects was full; for the other half, the cup with the continuous substance was full (see Figure 16.4).

Results showed that, while children correctly chose the full cup when asked which cup has *more* or *a lot*, only children at the three-knower level or higher consistently chose the blocks over the water as examples of *five* and *six*. One- and two-knowers, on the other hand, were as likely to extend *five* or *six* to continuous substances as to sets of discrete objects.

A second experiment used a slightly modified version of the Blocks and Water task, and asked children about low number words (*one* and *two*), as well as high number words (*five* and *six*). Results from this experiment confirmed that even one- and two-knowers understand that *one* and *two* refer to sets of discrete objects. Moreover, in this experiment one- and two-knowers chose the cup with discrete objects when asked about *five* and *six*, but only if they were asked about the low numbers first. This finding suggests that even one- and two-knowers have at least a fragile grasp of the fact that *five* and *six* (and perhaps number words as a class) are about discontinuous, rather than continuous quantities; by the time children have learned the meanings of the first three number words, this knowledge is quite robust.

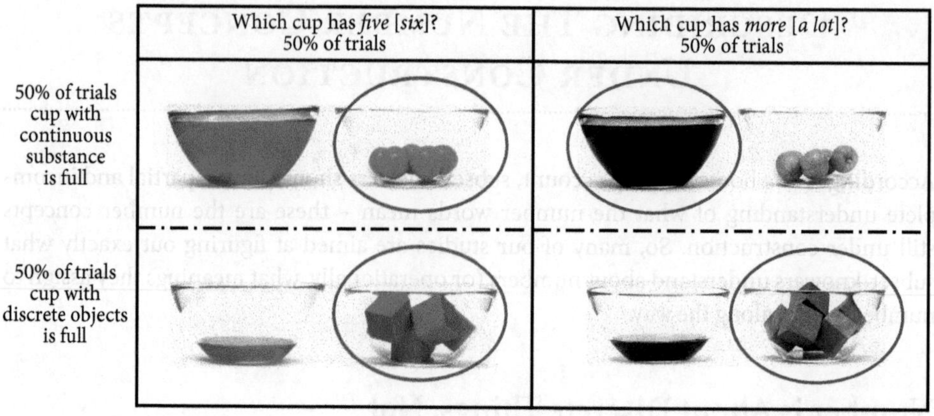

FIGURE 16.4 Blocks and Water task design. The correct answer for each trial type is circled.

Connecting number words to discrete quantification is only one step in natural-number-concept construction. Another thing that children must understand is what types of changes to a set are relevant to number (or operationally, to number words).

Some Kinds of Actions Affect Number, Others Do Not

To test whether children understand that only changing the numerosity of a set will also change its number word, we gave 54 children, aged 34–49 months, the Transform Sets task (Sarnecka & Gelman, 2004). In this task, children were shown a set of objects labelled with a number word (e.g. 'I'm putting *six* buttons in this box.'). Then some action was performed on the set (shaking the whole box, rotating the whole box, adding or removing an object), and the children were asked, 'Now how many buttons? Is it five, or six?' Both subset- and CP-knowers judged that the original number word should still apply on trials where the box had been shaken or rotated, but that the number word should change on trials where an item had been added or removed from the set. Like the results from the Blocks and Water task described above, this result also shows how subset-knowers' understanding of higher numbers (literally, of higher number-word meanings) includes some, but not all aspects of the full concept. A three-knower, for example, seems to understand that 'six' is a word that pertains to sets of discrete items (rather than continuous substances), and that 'six' will no longer apply to the set if items are added or removed. However, that three-knower still does not know exactly how many items 'six' is, or how counting is related to the exact meaning of the word.

Number Words are about Numerosity, Not Total Area or Other Properties of a Set

The Transform-Sets task showed that children recognize some actions (i.e. adding or removing items) as changing the number of items in a set, and other actions (i.e. shaking or rotating the whole box) as not changing the number of items. However, the Transform-Sets task did not distinguish number from the broader dimension of quantity, and there is a difference between the two. An undergraduate assigned to write a 15-page paper is tempted to use a large font size to fill up the space; a researcher writing a 15-page grant proposal looks for ever smaller and narrower fonts to cram in more information. At the end, the essays will each occupy the same quantity of space (15 pages) and may even use the same quantity of ink, but the number of words will be very different.

Even very young infants can detect changes in both numerosity and continuous spatial extent (such as summed surface area or contour length, e.g. Clearfield & Mix, 1999; Feigenson, Carey, & Spelke, 2002b). In the Transform-Sets task described above, the number of items in the box changed, but so did the quantity of stuff. After all, six plastic apples minus one plastic apple equals, not only fewer apples, but also less plastic. Thus, the fact that subset-knowers succeed on the Transform-Sets task, does not necessarily demonstrate that they know number words are about *numerosity per se* – they might just know that number words are about *quantity*.

To answer this question, we tested 116 children, aged 30–48 months, on the Match-to-Sample task (Slusser & Sarnecka, 2011). For this task, the experimenter showed the child a sample picture while saying, for example, 'This picture has *four* turtles.' The experimenter then placed two more pictures on the table, saying, 'Find another picture with *four* turtles.' One picture had the same number of items as the sample. The other had a different number of items (either half or twice as many), but matched the sample in either total summed area, or total summed contour length of the items. (See Figure 16.5.)

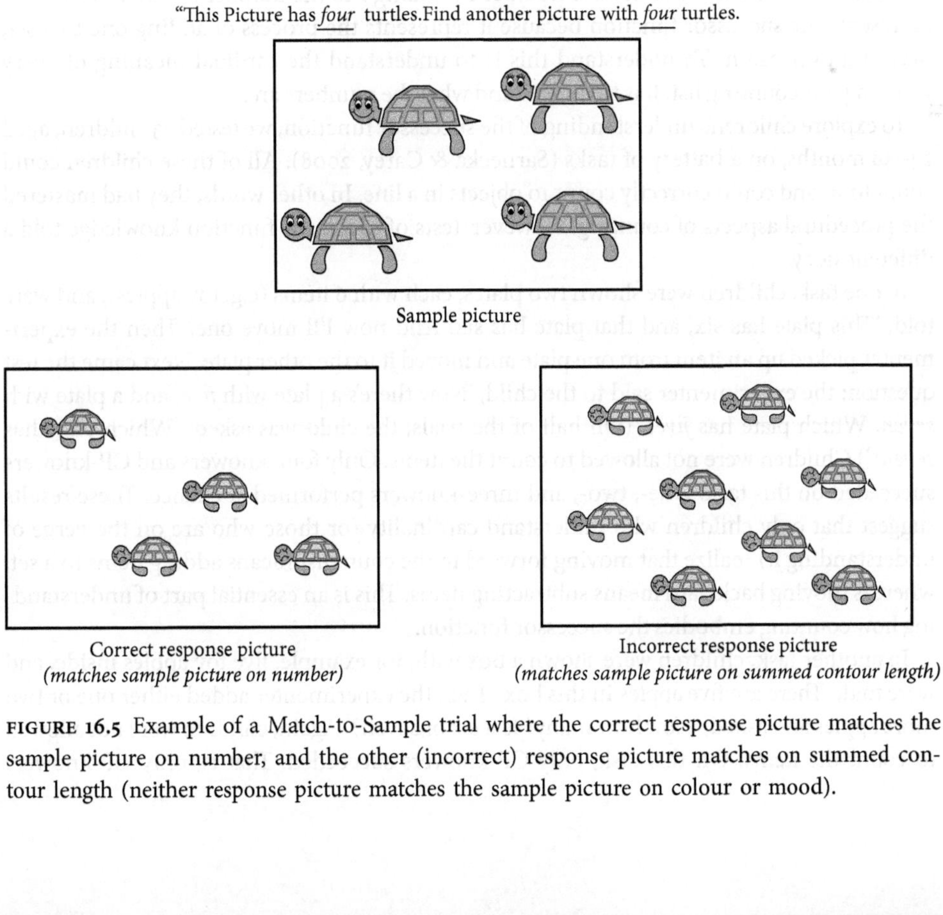

FIGURE 16.5 Example of a Match-to-Sample trial where the correct response picture matches the sample picture on number, and the other (incorrect) response picture matches on summed contour length (neither response picture matches the sample picture on colour or mood).

The experiment also included control trials, where children were asked to match pictures according to mood (happy or sad) or colour. For example, 'This picture has *happy* turtles. Find another picture with *happy* turtles.'

Results from these experiments showed that subset-knowers generally failed to extend number words (*four, five, eight,* and *ten*) based on numerosity. (These children did just fine on the mood and colour trials, so we know that they understood the task.) CP-knowers, on the other hand, succeeded robustly. We did not allow children to count the items, but even without counting, the CP-knowers understood that two sets of the same numerosity should be labelled by the same number word, whereas sets of clearly different numerosities (they were always different by a ratio of 1:2) must have different number words.

As in the Blocks and Water task described above, there was also evidence for progressively better performance across the subset-knower levels, suggesting that children gradually identify numerosity as the dimension of experience that number words refer to. Again, this result supports the bootstrapping account, which posits step-by-step construction of natural-number concepts.

Cardinality and The Successor Function

If the bootstrapping story is correct, then the shift from subset-knower to CP-knower is more profound than it appears. It is not just about learning a counting rule; it is also about understanding how counting instantiates the successor function. The successor function is the rule that generates each natural number by adding 1 to the number before it. Counting represents the successor function because it represents the process of adding one to a set, over and over again. To understand this is to understand the cardinal meaning of every word in your counting list. It is to understand what the numbers are.

To explore children's understanding of the successor function, we tested 73 children, aged 24–48 months, on a battery of tasks (Sarnecka & Carey, 2008). All of these children could count to 10 and could correctly count 10 objects in a line. In other words, they had mastered the procedural aspects of counting. However, tests of successor-function knowledge told a different story.

In one task, children were shown two plates, each with 6 items (e.g. toy apples), and were told, 'This plate has six, and that plate has six. And now I'll move one.' Then the experimenter picked up an item from one plate and moved it to the other plate. Next came the test question: the experimenter said to the child, 'Now there's a plate with *five*, and a plate with *seven*. Which plate has *five*?' (On half of the trials, the child was asked, 'Which plate has *seven*?') Children were not allowed to count the items. Only four-knowers and CP-knowers succeeded on this task. One-, two-, and three-knowers performed at chance. These results suggest that only children who understand cardinality (or those who are on the verge of understanding it) realize that moving forward in the count list means adding items to a set, whereas moving backward means subtracting items. This is an essential part of understanding how counting embodies the successor function.

In another task, children were shown a box with, for example, five toy apples inside, and were told, 'There are five apples in this box.' Then the experimenter added either one or two more apples and asked, 'Now how many? Is it six or seven?' Again, children were not allowed to count the items. On this task, only CP-knowers succeeded. That is, only CP-knowers understood that adding 1 item to the set meant moving forward one word in the counting list, whereas adding 2 items meant moving forward two words. This may be the final piece of the puzzle – the last thing that children figure out as they come to understand how counting embodies the successor function, which is effectively to understand what numbers are.

Cardinality and Equinumerosity

Another profound difference between the knowledge of subset-knowers and CP-knowers is that only CP-knowers appear to robustly understand *equinumerosity* (also called *exact equality*) – the notion that any set of N can be put into one-to-one correspondence with any other set of N (Muldoon, Lewis, & Freeman, 2009). For example, any set of ten things can be matched up one-to-one with any other set of ten things: If you have ten flowers and ten vases, you will have exactly one flower for each vase.

We (Sarnecka & Gelman, 2004; Sarnecka & Wright, 2013) tested children's understanding of equinumerosity, by presenting preschoolers with a scenario in which two stuffed animals (a frog and a lion) were each given 'snacks' (laminated cards with pictures of food). On half of the trials, the snacks were equal (e.g., the Frog and Lion each got 6 peaches); the other half of the trials featured unequal snacks (e.g., the Frog got 5 muffins; the Lion got 6). The pictures were designed to line up clearly, so that the one-to-one correspondence (or lack thereof) between the sets was visually obvious.

At the beginning of each trial, the child was asked whether the animals' snacks were 'just the same.' If children did not correctly identify the sets as 'the same' or 'not the same,' the experimenter corrected the error, and drew the child's attention to the sameness or difference between the sets.

For the test question, the experimenter told the child the number of items in one snack, and then asked about the other (e.g. 'Frog has *six* peaches. Do you think Lion has *five* or *six*?'). Children were not allowed to count the items.

Only CP-knowers succeeded robustly on this task. In other words, CP-knowers knew (without counting) that if Frog had six, and the sets were the same, then Lion must also have six. On the other hand, if Frog had six and the sets were not the same, then Lion must have some other number. This understanding of *equinumerosity* as an aspect of number-word meanings is not a rule about counting, but is understood by children who understand the cardinal principle of counting. This provides convergent evidence that the cardinal-principle induction (i.e. the conceptual achievement of figuring out the cardinal principle) is not just an insight about counting – it is a deeper insight about what numbers are.

Conclusions

The studies described here flesh out Carey's (2009) bootstrapping story of where natural-number concepts come from. These concepts do not appear to be innate – rather, they are constructed. Specifically, they are constructed piece by piece, as meanings for a set of initially meaningless placeholder symbols – the counting words. Children initially learn to say these words in order, and to point to objects while saying them, much as one might point to objects while chanting *eenie, meenie, minie, mo.* Gradually, over a period of months or years, the number words are assigned progressively more sophisticated meanings. Children work out that these words have something to do with discrete objects, quantification, and numerosity, and eventually that the counting list represents numerosities in increasing order, with each word in the list indicating the addition of one item to the set. This one-to-one correspondence (between words in the counting list and items in the set) is then generalizable to relations among sets – leading to the knowledge that any set of six can be put into one-to-one correspondence with all and only other sets of six.

Thus, cardinality seems to be the marker of a profound conceptual achievement, involving an implicit understanding of the successor function and of equinumerosity, as well as of how counting works. In this way, the process of learning to count, which starts when the child memorizes a few words of the counting list, and continues to the point of understanding the cardinal principle, is actually the process by which children develop their first mental representations of exact, large numbers.

Note

1. Note that the model assumes only one act of estimation per trial. That is, it assumes that children estimate the target number and hand over some number of items without looking at their answer and reflecting on it, which could involve a second act of estimation. A model that did include self-correction behaviour could explore a different explanation for why responses of 5 might be more common than responses of 3, for the target number 4. The explanation is that 4 and 5 are more confusable than 3 and 4 in the approximate number system (i.e. 4/5 is a smaller ratio than ¾). To our knowledge, no existing model investigates this possibility.

References

Antell, S.E., & Keating, D.P. (1983). Perception of numerical invariance in neonates. Child Development, 54, 695–701.

Barner, D., Chow, K., & Yang, S. (2009a). Finding one's meaning: a test of the relation between quantifiers and integers in language development. Cognitive Psychology, 58, 195–219.

Barner, D., Libenson, A., Cheung, P., & Takasaki, M. (2009b). Cross-linguistic relations between quantifiers and numerals in language acquisition: evidence from Japanese. Journal of Experimental Child Psychology, 103, 421–440.

Baroody, A.J., & Price, J. (1983). The development of the number-word sequence in the counting of three-year-olds. Journal for Research in Mathematics Education, 14, 361–368.

Block, N.J. (1986). Advertisement for a semantics for psychology. Midwest Studies in Philosophy, 10, 615–678.

Bloom, P., & Wynn, K. (1997). Linguistic cues in the acquisition of number words. Journal of Child Language, 24, 511–533.

Briars, D., & Siegler, R.S. (1984). A featural analysis of preschoolers' counting knowledge. Developmental Psychology 20, 607–618.

Butterworth, B. (1999). What Counts: How Every Brain is Hardwired for Math. New York: Free Press.

Carey, S. (2009). The Origin of Concepts. New York: Oxford University Press.

Clearfield, M.W., & Mix, K.S. (1999). Number versus contour length in infants' discrimination of small visual sets. Psychological Science, 10, 408–411.

Condry, K.F., & Spelke, E. (2008). The development of language and abstract concepts: the case of natural number. Journal of Experimental Psychology: General, 137, 22–38.

Dowker, A. (2008). Individual differences in numerical abilities in preschoolers. Developmental Science, 11, 650–654.

Dunn, L.M., & Dunn, L.M. (1997). Peabody Picture Vocabulary Test III. New York: American Guidance Service.

Feigenson, L. & Carey, S. (2003). Tracking individuals via object-files: evidence from infants' manual search. *Developmental Science*, 5, 568–584.

Feigenson, L., & Carey, S. (2005). On the limits of infants' quantification of small object arrays. Cognition, 97, 295–313.

Feigenson, L., Carey, S., & Hauser, M. (2002a). The representations underlying infants' choice of more: object files versus analog magnitudes. Psychological Science, 13, 150–156.

Feigenson, L., Carey, S., & Spelke, E. (2002b). Infants' discrimination of number vs. continuous extent. Cognitive Psychology, 44, 33–66.

Feigenson, L., Dehaene, S., & Spelke, E. (2004). Core systems of number. Trends in Cognitive Sciences, 8, 307–314.

Fluck, M., & Henderson, L. (1996). Counting and cardinality in English nursery pupils. British Journal of Educational Psychology, 66, 501–517.

Freeman, N.H., Antonucci, C., & Lewis, C. (2000). Representation of the cardinality principle: early conception of error in a counterfactual test. Cognition, 74, 71–89.

Frye, D., Braisby, N., Lowe, J., Maroudas, C., & Nicholls, J. (1989). Young children's understanding of counting and cardinality. Child Development, 60, 1158–1171.

Fuson, K.C. (1988). Children's Counting and Concepts of Number. New York: Springer-Verlag Publishing.

Gelman, R., & Gallistel, C.R. (1978). The Child's Understanding of Number. Cambridge: Harvard University Press.

Goldman, M.C., & Sarnecka, B.W. (2011, May). Cross-linguistic consistency within bilingual children: are number words special? Paper presented at the Symposium on Cognitive and Language Development, Los Angeles, CA.

Gunderson, E.A., & Levine, S.C. (2011). Some types of parent number talk count more than others: relation between parents' input and children's number knowledge. Developmental Science, 14, 1021–1032.

Jordan, N.C., & Levine, S.C. (2009). Socioeconomic variation, number competence, and mathematics learning difficulties in young children. Developmental Disabilities Research Reviews, 15, 60–68.

Klibanoff, R.S., Levine, S.C., Huttenlocher, J., Vasilyeva, M., & Hedges, L.V. (2006). Preschool children's mathematical knowledge: the effect of teacher 'math talk.' Developmental Psychology, 42, 59–69.

Le Corre, M., & Carey, S. (2007). One, two, three, four, nothing more: an investigation of the conceptual sources of the verbal counting principles. Cognition, 105, 395–438.

Le Corre, M., Li., P., & Jia, G. (2003). On the role of singular/plural in number word learning. Paper presented at the biennial meeting at Society for Research in Child Development, Tampa, FL, April 24–27, 2003.

Le Corre, M., Van de Walle, G., Brannon, E.M., & Carey, S. (2006). Re-visiting the competence/performance debate in the acquisition of the counting principles. Cognitive Psychology, 52, 130–169.

Lee, M.D., & Sarnecka, B.W. (2010). A model of knower-level behaviour in number-concept development. Cognitive Science, 34, 51–67.

Lee, M.D., & Sarnecka, B.W. (2011). Number-knower levels in young children: insights from Bayesian modeling. Cognition, 120, 391–402.

Levine, S.C., Gunderson, E.A., & Huttenlocher, J. (2011a). Mathematical development during the preschool years in context: home and school input variations. In N. L. Stein & S. W. Raudenbush (Eds), Developmental Science Goes to School: Implications for Education and Public Policy Research (pp. 189–202). New York: Taylor and Francis.

Levine, S.C., Suriyakham, L., Rowe, M., Huttenlocher, J., & Gunderson, E.A. (2011b). What counts in the development of children's number knowledge? Developmental Psychology, 46, 1309–1313.

Li, P., Le Corre, M., Shui, R., Jia, G., & Carey, S. (2003). Effects of plural syntax on number word learning: a cross-linguistic study. Paper presented at the 28th Boston University Conference on Language Development, Boston, MA, October 31–November 2, 2003.

Miller, K.F., Smith, C.M., Zhu, J., & Zhang, H. (1995). Preschool origins of cross-national differences in mathematical competence: the role of number-naming systems. Psychological Science, 6, 56–60.

Muldoon, K., Lewis, C., & Freeman, N. (2009). Why set-comparison is vital in early number learning. Trends in Cognitive Sciences, 13 203–208.

Negen, J., & Sarnecka, B.W. (2009). Young children's number-word knowledge predicts their performance on a nonlinguistic number task. In N. A. Taatgen & H. van Rijn (Eds), Proceedings of the 31th Annual Conference of the Cognitive Science Society (pp. 2998–3003). Austin: Cognitive Science Society.

Negen, J., & Sarnecka, B.W. (2012). Number-concept acquisition and general vocabulary development. Child Development, 83, 2019–2027.

Quine, W.V.O. (1960). Word and Object. Cambridge: MIT Press.

Sarnecka, B.W. & Carey, S. (2008). How counting represents number: what children must learn and when they learn it. Cognition, 108, 662–674.

Sarnecka, B.W. & Gelman, S.A. (2004). Six does not just mean a lot: preschoolers see number words as specific. Cognition, 92, 329–352.

Sarnecka, B.W., Kamenskaya, V.G., Yamana, Y., Ogura, T., & Yudovina, J.B. (2007). From grammatical number to exact numbers: early meanings of 'one,' 'two,' and 'three' in English, Russian, and Japanese. Cognitive Psychology, 55, 136–168.

Sarnecka, B.W., & Lee, M.D. (2009). Levels of number knowledge during early childhood. Journal of Experimental Child Psychology, 103, 325–337.

Sarnecka, B.W. & Wright, C.E. (2013). The exact-numbers idea: Children's understanding of cardinality and equinumerosity. Cognitive Science. doi: 10.1111/cogs.12043

Sarnecka, B.W., Wright, C.E., & Goldman, M.C. (2011, March). Cross-linguistic associations in the vocabularies of bilingual children: Number words vs. color words and common nouns. Poster presented at the biennial meeting of the Society for Research in Child Development, Montreal, Canada.

Schaeffer, B., Eggleston, V.H., & Scott, J.L. (1974). Number development in young children. Cognitive Psychology, 6, 357–379.

Scholl, B.J., & Leslie, A. (1999). Explaining the infant's object concept: beyond the perception/cognition dichotomy. In E. Lepore & Z. Pylyshyn (Eds), What is Cognitive Science? (pp. 26–73). Oxford: Blackwell.

Simon, T.J. (1997). Reconceptualizing the origins of number knowledge: a 'non-numerical' account. Cognitive Development, 12, 349–372.

Slaughter, V., Itakura, S., Kutsuki, A., & Siegal, M. (2011). Learning to count begins in infancy: evidence from 18 month olds' visual preferences. Proceedings of the Royal Society Biological Sciences, 278, 2979–2984.

Slusser, E., Ditta, A., and Sarnecka, B. W. (2013). Connecting numbers to discrete quantification: A step in the child's construction of integer concepts. Cognition, 129, 31–41.

Slusser, E., & Sarnecka, B.W. (2011). Find the picture of eight turtles: a link between children's counting and their knowledge of number-word semantics. Journal of Experimental Child Psychology, 110, 38–51.

Slusser, E.B., & Sarnecka, B. W. (2010, March). Children's use of morpho-syntactic information to connect number words to discrete quantification. Paper given as part of symposium Early links among number, plural, and discrete objects (Lisa Cantrell, Chair), International Conference on Infant Studies, Baltimore, MD.

Starkey, P., & Cooper, R.G. (1980). Perception of numbers by human infants. Science, 28, 1033–1035.

Uller, C., Carey, S., Huntley-Fenner, G., & Klatt, L. (1999). What representations might underlie infant numerical knowledge? Cognitive Development, 14, 1–36.

Wagner, S.H. & Walters, J. (1982). A longitudinal analysis of early number concepts: from numbers to number. In G. Forman (Ed.), Action and Thought: From Sensorimotor Schemes to Symbolic Operations (pp. 137–161). New York: Academic Press,.

Woodcock, R., & Johnson, M.B. (1985). Woodcock-Johnson II, revised. Itasca, IL: Riverside Publishing.

Wynn, K. (1990). Children's understanding of counting. Cognition, 36, 155–193.

Wynn, K. (1992a). Addition and subtraction by human infants. Nature, 358, 749–750.

Wynn, K. (1992b). Children's acquisition of the number words and the counting system. Cognitive Psychology, 24, 220–251.

CHAPTER 17

APPROXIMATE ARITHMETIC ABILITIES IN CHILDHOOD

CAMILLA GILMORE

Young children draw on a variety of skills and experiences in the early stages of learning arithmetic. One of the first numerical activities that they appear to engage in is to learn the sequence of digits and counting words, and to use their counting skills to understand and perform simple arithmetic with objects. As well as these symbolic skills, it has been suggested that children also possess an approximate number system (ANS), which supports the non-symbolic representation of quantities. The ANS allows children to compare the numerosities of sets of items without counting them. It has also been suggested that young children can use their ANS to perform approximate arithmetic, and new research continues to reveal the impressive extent of their early arithmetical abilities. In this article, I will review studies that have explored children's non-symbolic and symbolic approximate arithmetic skills, consider various alternative explanations of this ability and discuss its potential role in the learning of symbolic mathematics.

The Approximate Number System and Approximate Arithmetic

There is now a wealth of evidence that infants, children and adults can use their ANS to represent numerical information about the world about them (Dehaene, 1997). Research with adults, children, and animals has demonstrated that the ANS can support abstract non-symbolic representations of quantity generated from a range of different input stimuli, e.g. static or dynamic visual displays or sound sequences (e.g. Barth, Kanwisher, & Spelke, 2003; Brannon, 2006; Feigenson, Dehaene, & Spelke, 2004; Halberda & Feigenson 2008). Representations of quantity within the ANS are approximate and grow more approximate as the to-be-represented quantity increases. If an individual is asked to compare the numerosities of two stimuli, the accuracy of this comparison will be affected by the level of overlap of the two representations, which will depend upon the ratio between the two quantities.

It has been suggested that the ANS can be used not only to compare approximate representations of quantities, but also to manipulate numerosity representations by performing arithmetic. Approximate arithmetic in this context refers to the ability to perform arithmetic with approximate representations, rather than estimating or approximating the result of exact calculations.[1] If true, this suggests that individuals can add and subtract quantities not only through learned formal or informal calculation strategies used with digits or number words, but also by recourse to unlearned manipulation of non-symbolic representations. Barth and colleagues (Barth, Beckmann, & Spelke, 2008) suggested that there are four properties that would be observed if children or adults use the ANS to solve approximate arithmetic problems. First, performance on arithmetic problems should be ratio dependent. Accuracy would decline as the ratio between the quantities being compared approaches one, due to the level of overlap of the underlying non-symbolic representations. Secondly, performance on addition problems should be equivalent to performance on matched comparison problems. This is because errors arise from comparison of the noisy representations, rather than the addition process.[2] Thirdly, performance on subtraction problems should be lower than comparison or addition problems. This arises because when subtraction problems are matched to addition problems with the same result, the subtraction problems will necessarily involve larger magnitude operands with greater associated noise in the representations. Fourthly, performance should be as accurate across modalities as within modalities, because ANS representations are abstract and not modality-specific. Below I review studies that have explored children's approximate arithmetic ability and consider to what extent it provides evidence for these performance signatures.

Evidence for Preschool Approximate Arithmetic

One of the first sets of studies to explore preschool children's abilities to perform approximate arithmetic was by Hilary Barth and colleagues (Barth, La Mont, Lipton, Dehaene, Kanwisher, & Spelke, 2006; Barth, La Mont, Lipton, & Spelke, 2005). In a series of experiments they demonstrated that 5-year-old children, who had not yet received formal training in arithmetic at school, were nevertheless able to perform approximate addition using non-symbolic stimuli. The children were shown animated displays involving arrays of coloured dots. Initially, an array of blue dots appeared on the screen and this was then covered by an occluder. A second array of blue dots then appeared and also moved behind the occluder. Finally, an array of red dots appeared and the children were asked to decide whether there were more blue dots in total or more red dots (see Figure 17.1).

During the animation, only one dot array was visible at a time. Each of the blue dot arrays were smaller than the red dot array and the sum of the blue dot arrays was larger than the red dot array on half of the trials. The children were therefore unable to solve the problems simply by comparing single arrays. The arrays were also too numerous (sum totals ranged from 10 to 58) and presented too briefly to allow counting. Instead, addition of the unseen blue arrays was

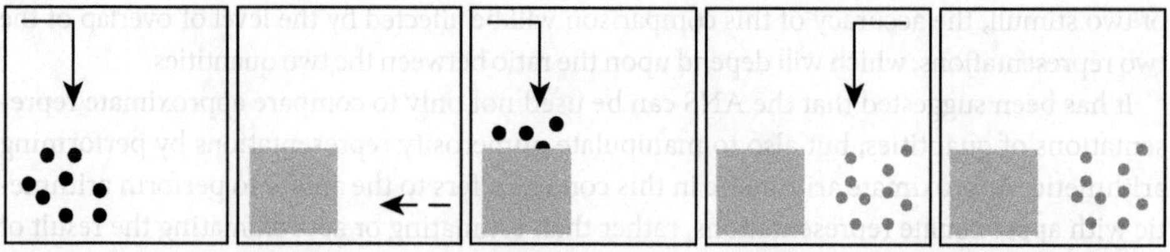
"Look, here come some blue dots."
"Now they're being covered up."
"Here come some more blue dots. Now they're *all* back there."
"And here come some red dots."
"Are there more blue dots, or more red dots?"

FIGURE 17.1 Example of an animated sequence showing addition of dot arrays. Children watch the sequence on a computer while the experimenter narrates it.

Adapted from *Cognition*, 98(3) Hilary Barth, Kristen La Mont, and Jennifer Lipto, Non-symbolic arithmetic in adults and young children, pp. 199–222, Copyright (2006), with permission from Elsevier.

necessary to reliably select the correct answer. Across a series of experiments the children were found to perform reliably above chance on this task, and their accuracy varied with the ratio between the total number of blue dots and the total number of red dots. As would be expected if the children were using ANS representations to solve these problems, accuracy decreased as the ratio approached one. This suggests that preschool children are able to form accurate expectations of the outcomes of additions involving approximate representations of number.

In many ways, we should not be surprised that preschool-age children are able to solve these approximate arithmetic problems. We have known for some time that even babies show evidence of understanding addition and subtraction with small sets of objects. In a classic study, Karen Wynn (1992) demonstrated that 4-month-old babies were able to create expectations of the outcomes of simple calculations, such as 1 + 1 or 2 + 1. When these simple problems were demonstrated with objects, the babies showed surprise if they were shown an incorrect outcome. While these studies demonstrate that very young children appear to have some understanding of arithmetical transformations, they differ in many ways from more recent work aimed to tap into the ANS. Wynn's studies involved very small numbers of objects and required an exact outcome to be calculated. As a result it is likely that babies were using an object tracking system, rather than the ANS to perform these calculations.

The method developed by Barth and colleagues allows an investigation of young children's approximate arithmetic skills involving large sets and has formed the basis of a number of subsequent studies. Three key questions have been addressed:

1. What is the extent of children's approximate arithmetic ability?
2. What are the processes that underpin this ability?
3. What are the influences on children's performance of approximate arithmetic?

How Far Does this Ability Extend?

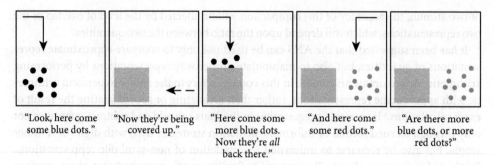

Children's approximate arithmetical ability is not restricted to addition with dot arrays. Studies have explored whether accurate judgements can be made about other operations, and the nature of the stimuli upon which these operations can be performed. Iuculano, Tang, Hall, and Butterworth (2008) tested children's ability to perform approximate subtraction using an animated sequence involving dot arrays based on those developed by Barth and colleagues. The children watched a blue dot array being covered by an occluder and then a subset of dots moved out from behind the occluder and off screen. They were asked to determine whether the number of blue dots remaining behind the occluder was more or less than the number of dots in a comparison red array. The children were able to solve these problems and their performance was correlated with performance on approximate comparison and addition tasks. The children in this study were aged 8–9 years old and, therefore, would have received some years of arithmetic instruction in school. Using a different paradigm, Slaughter, Kamppi, and Paynter (2006) demonstrated that young preschoolers (aged 3-and-a-half to 4 years old) can also make accurate judgements of the outcomes of subtraction operations on large and very large sets when an approximate result, but not an exact result, is required. In this study children watched the experimenter place objects inside two boxes and remove different numbers of items from each box. Children were able to predict which box contained more items at the end of the sequence.

Within the realm of exact symbolic arithmetic, multiplicative relations have consistently been found to be more difficult than additive relations and take longer for children to learn (Dowker, 2005). Is a similar pattern observed with approximate non-symbolic arithmetic? Some recent evidence has begun to suggest that young children may be able to perform simple approximate multiplication and division operations; however, this ability appears to be less reliable than approximate addition and subtraction. Barth, Starr, and Sullivan (2009) explored whether 6 and 7-year-old children could make accurate predictions about the doubling and halving of non-symbolic quantities (i.e. arrays of dots). In a set of training trials, children were initially familiarized with the outcome of either a doubling or halving transformation. The children saw an initial array of dots that were then covered by an occluder, and then they heard a sound to tell them that a 'magic' transformation had occurred. The occluder was removed and the children saw the outcome of the transformation. In the experimental trials the children did not see the outcome of the transformation, instead they were asked to decide whether the outcome was more or less than a comparison quantity. The children performed above chance on these trials and their performance varied with the ratio of the outcome of the transformation and the comparison array. This suggests that they were able to make accurate predictions about the outcome of doubling and halving operations and that they were employing approximate ANS representations to do so. Is this evidence of the ability to perform multiplicative relations? For the doubling operation in particular this is not clear. The children could have solved these problems by employing a strategy of repeated addition in which they predicted the outcome of adding the initial dot array to itself. In order to test for the ability to perform true multiplicative relations, McCrink and Spelke (2010) challenged children to perform multiplicative relations that could not be solved using repeated addition; they also modified the procedure to limit the use of learnt strategies based on the training trials. Children aged 5–7 years old watched animated sequences in which they saw arrays of rectangles undergo transformations involving multiplication by 2, 4 or 2.5. In training trials the children were shown the outcome of the transformation only after they had made a prediction about its numerosity, and in experimental trials they received no feedback. The children were reliably accurate in judging whether the outcome of the transformation was more or less than a comparison array and they could do this from the earliest training trials. Two findings suggested that the children were employing ANS representations to solve these problems. First, performance was related to the ratio between the outcome of the transformation and the comparison array. Secondly, performance was related to the multiplicative factor. Both group and individual level results showed that accuracy decreased as the multiplicative factor (×2, ×2.5, ×4) increased.

These studies appear to demonstrate that the ANS is capable of supporting a range of arithmetical operations, including additive and multiplicative relations. These operations involve comparing and manipulating quantities. Understanding relationships among quantities forms a major part of arithmetical reasoning; however, understanding relationships among operations is also a key element (Gilmore, 2006). Is there any evidence that the ANS supports reasoning about relationships between operations? Gilmore and Spelke (2008) explored this question by testing whether 5–6-year-old children could understand the inverse relationship between addition and subtraction using approximate non-symbolic representations of number. The children were shown animated displays depicting three-term arithmetic problems, involving an addition operation followed by a subtraction operation. A set of dots appeared on screen and was covered by an occluder, a second array of dots appeared and also moved behind the occluder, then a different array of dots moved out from behind the occluder and off screen. In some trials (inversion trials), the number of dots in the arrays that were added and subtracted was the same, while on other trials (control trials) they were different. The children were asked to decide if the number of dots left behind the occluder at the end of the sequence was more or less than a comparison array. Children were able to make this judgement with above chance accuracy for the inversion trials and, crucially, they were more accurate on the inversion trials than the control trials. This suggests that they were able to recognize that the quantities added and subtracted were the same and that these transformations meant that the initial quantity was left unchanged. In a subsequent experiment the authors demonstrated that children of a similar age and background were unable to solve these problems when they involved Arabic symbols and required exact numerical judgements (Gilmore & Spelke, 2008). This suggests that the ANS was the basis for the children's successful performance in the initial experiment.

This series of studies suggests that young children, who have received little formal instruction in symbolic arithmetic, are able to add, subtract, multiply, and divide quantities represented as arrays of dots. Characteristics of children's performance in these experiments indicate that the ANS may underlie this performance. Research has shown that adults are able to represent and compare numerical information about a wide range of stimuli presented in different modalities (e.g. sequences of tones). If children are using ANS representations to perform approximate arithmetic, then their ability to do so should extend beyond the use of dot array stimuli and, moreover, performance should be equivalent whether they are dealing with a single or multiple modalities of input stimuli. A small number of studies have addressed this question and largely confirmed this prediction. Barth et al. (2005) modified the animated sequences that had been used in previous studies of dot array addition and asked 5–6-year-old children to add two arrays of dots and compare the result to a sequence of tones. As in previous experiments the two dot arrays were never visible at the same time, and were presented too briefly to allow counting. Similarly, the tone sequence was presented too rapidly for children to be able to count the tones. The children were, nevertheless, able to make accurate judgements as to whether the sum of the dot arrays or the number of tones was greater, and their performance on this task was equivalent to previous studies when the comparison set was also a dot array. This study suggests that children can compare the results of addition operations to quantities presented in a different modality, but it doesn't demonstrate the ability to add quantities presented in different modalities. A later study by Barth and colleagues (Barth et al. 2008), however, did demonstrate that children are able to add across modalities. In this study 5–6-year-old children were asked to add a dot array to a sound sequence and compare the result to a dot array. Again, children were able to solve these problems with above-chance levels of accuracy and their performance was no less accurate than in previous experiments involving only visual stimuli.

These studies suggest that children's ability to perform approximate arithmetic extends beyond the use of dot array stimuli to include non-symbolic stimuli presented in other modalities. A final study explored whether children could use their ANS to add quantities presented using symbols, even though they had not yet received formal instruction in exact symbolic arithmetic. It has been established that when children learn symbolic representations of number (e.g. number words, Arabic digits) they map these representations onto pre-existing ANS representations. Evidence for this comes from studies demonstrating that children show a numerical distance effect when they compare symbolic numbers (e.g. Holloway & Ansari, 2008; Temple & Posner, 1998), and are able to map between non-symbolic and symbolic representations (Mundy & Gilmore, 2009). Given that children can perform arithmetic using their ANS and that they map symbolic representations onto ANS representations, Gilmore and colleagues (Gilmore, McCarthy, & Spelke, 2007) explored whether preschool children could perform approximate addition and subtraction using symbolic representations. They gave 5–6-year-old children animated displays similar to those used in previous experiments involving non-symbolic stimuli, but the quantities were represented using Arabic digits. The experimenter also narrated the problem to children, for example 'Sarah has 21 candies. She gets 30 more. John has 34 candies. Who has more?' (Gilmore et al., 2007, p. 589). The children were able to solve these problems with above chance levels of accuracy, and their performance showed the ratio effect that is characteristic of the ANS—accuracy decreased as the ratio between the sum of the addends and the comparison quantity approached one. There was no evidence that children were solving these problems using learnt exact symbolic arithmetic and they performed at chance levels on addition problems when they were required to make an exact, rather than an approximate comparison. This study therefore appears to provide evidence that young children are able to use their ANS to perform approximate arithmetic when the input stimuli are exact symbolic representations. Children's ability to do so appears to be fragile, however, since children failed to solve symbolic approximate addition problems with above-chance accuracy when given a small number of symbolic problems presented verbally with no supporting visual displays (Barth et al., 2005, 2006).

How Do Children Perform Approximate Arithmetic?

The studies outlined above seem to show that children can add, subtract, multiply, and divide approximate quantities represented in different ways. Do these studies provide convincing evidence that children can perform approximate arithmetic of numerosity information? Children appear to be able to make reliable predictions of the outcomes of transformations, and consistently perform above chance levels on these arithmetic problems. However, before we accept this conclusion we must explore whether there are alternative explanations of how children solve these problems. Could they be drawing on non-numerical information in the stimuli or using non-arithmetical strategies, which nevertheless allow above-chance performance? One such strategy that children might make use of is to use information derived from non-numerical aspects of the non-symbolic stimuli, rather than numerosity information. With all types of non-symbolic stimuli (dot arrays, tone sequences, etc.) there is information about continuous variables that may be related to numerosity, for example, the total surface area, density, or size of dots in dot arrays, and the duration or rate in tone sequences. In order to conclude that children and adults solve non-symbolic approximate arithmetic problems using their ANS, then it is important to demonstrate that individuals use numerosity information, rather than continuous quantity information, to solve these problems. This is especially important, given evidence that suggests very young children may base their decisions on continuous quantity information, rather than numerosity information when making comparison judgements involving small numerosities (Feigenson, Carey, & Hauser, 2002).

Researchers have taken different approaches to controlling for these continuous quantity variables in the construction of their stimuli and in testing for evidence of this type of strategy use; however, some general conclusions can be drawn. The most common approach to testing for these strategies has been to generate subsets of stimuli that have varying relationships between numerosity information and continuous quantity information, then to compare performance on the different subsets of problems. For example, Barth et al. (2005) generated two subsets of problems, on half of the problems, dot size, total contour length, summed dot area, and density were negatively correlated with number and, therefore, the envelope area (the area taken up by the array) was positively correlated with number. On the other half of the problems these correlations were reversed. If children were basing their answers solely on continuous quantity variables, then performance would be significantly above chance for one subset of problems, and significantly below chance for the other subset of problems. No published study of approximate non-symbolic arithmetic has found this pattern, which suggests that children do take account of numerosity, rather than simply basing their responses solely on correlated continuous quantity variables. However, a number of studies have found that while children's performance is above chance on all subsets of trials with different continuous quantity characteristics, performance is better on trials in which envelope area is positively correlated with numerosity compared with trials in which envelope area is negatively correlated with numerosity (Barth et al., 2005, 2006; Gilmore & Spelke, 2008). This suggests that envelope area is used as a cue to the numerosity of dot array stimuli, which is in keeping with research examining adults' numerosity estimates (Shuman & Spelke, 2006). Studies involving cross-modal arithmetic are less susceptible to the influences of continuous quantity variables, since participants are required to add or compare non-symbolic stimuli presented in different modalities (for example, adding a dot array to a sequence of tones). Barth et al. (2008) investigated whether children performing cross-modal addition and subtraction drew on a spatiotemporal strategy in which they integrated information about the area of a dot array with information about the duration of the tone sequence. Children performed above chance on subsets of trials in which this strategy either did or did not provide a cue to numerosity, and there was no difference in performance on these two subsets of trials, suggesting that children did not generally make use of this strategy.

Another approach to investigating the role of continuous quantity information was taken by Barth et al. (2009) in their studies of doubling and halving operations. In these studies children performed approximate multiplications and divisions on either discrete stimuli (dot arrays) or continuous stimuli (bars of varying lengths). When continuous stimuli are used, variables such as area and length are direct indicators of the outcome of transformations; in contrast, when discrete stimuli are used, continuous quantity variables are only correlated with numerosity. Thus, if children were using continuous quantity variables to solve problems we would expect performance to be higher on tasks involving continuous stimuli compared with equivalent tasks involving discrete stimuli. There was no evidence for this pattern of performance, with both doubling and halving operations children performed as accurately when the problems involved discrete stimuli as they did with continuous stimuli. Moreover, evidence that children made use of alternative strategies (e.g. choosing the comparison quantity, if it is particularly large, without considering the outcome of the transformation) for continuous stimuli, but not for discrete stimuli, suggests that children were not solving continuous and discrete problems the same way.

An alternative strategy that children might make use of in these types of arithmetical studies involves being sensitive to the range of values involved in the experiment. For example, if children take note of the range of comparison values (i.e. the quantity against which they are asked to compare the result of an arithmetical operation) then they might be able to use this as a cue to the correct answer. In most of the studies described above, when the comparison quantity is particularly large, then it is correct to conclude that it is the larger amount, and when the comparison quantity is particularly small, then it is correct to conclude that it is the smaller amount. Children could use this information to select the correct answer for these types of trials without performing any arithmetical operation and this might be sufficient to result in above-chance performance for the task as a whole. Two approaches have been taken to control for this strategy. The problems can be designed so that there are subsets of trials on which this strategy cues the incorrect answer. Alternatively, performance can be examined on trials where the comparison quantity lies in the middle of the range of comparison quantities and thus does not cue either answer. Examining children's performance only on the trials where the size of the comparison number does not cue the correct answer can reveal whether children are relying on this strategy. This approach has ruled out the range strategy as an explanation of children's performance on non-symbolic and symbolic approximate addition problems (Barth et al., 2005, 2006; Gilmore et al., 2007). However, there is some evidence that children do use this strategy to solve three-term addition and subtraction problems, and cross-modal subtraction problems (Barth et al., 2008; Gilmore & Spelke, 2008). In one study, investigating children's performance of multiplicative operations there was evidence that children used this strategy to solve doubling, but not halving, operations (Barth et al., 2009). However, using a redesigned method McCrink and Spelke (2010) demonstrated that children could perform doubling and other multiplicative operations without recourse to this strategy. This highlights that minor details of the experimental procedure employed can make the use of this strategy more or less likely.

Finally, if children are not solving these problems by basing their answers on continuous quantity variables, or using non-arithmetical strategies, could they draw on learnt knowledge of exact symbolic arithmetic? Although children in the majority of these studies were in preschool or kindergarten, and therefore had not received much formal arithmetic instruction, it is possible that children had learnt some early symbolic arithmetic either at home or in childcare settings. Studies have shown that, before beginning school, children can use counting and objects to perform simple exact additions and subtractions (Geary, 2000). Could children have used learnt knowledge of exact symbolic arithmetic to solve approximate symbolic or non-symbolic arithmetic problems? For studies that involved approximate symbolic arithmetic (e.g. Gilmore et al., 2007; Gilmore & Spelke, 2008), children could have directly applied any knowledge of exact symbolic arithmetic to the approximate problems they were solving. To use this strategy for studies involving approximate non-symbolic arithmetic (e.g. Barth et al., 2005, 2006, 2008, 2009; McCrink & Spelke, 2010), children would need to assign a symbolic label to each non-symbolic array, draw on knowledge of exact symbolic arithmetic to determine the outcome of the operation, and then compare this to the non-symbolic comparison quantity. Three lines of evidence suggest that this is not how children solve these problems. First, when given symbolic problems requiring exact comparisons in the format of approximate arithmetic tasks, children performed at chance level (Gilmore et al., 2007; Gilmore & Spelke 2008), indicating that they could not generate exact answers to approximate symbolic arithmetic problems. Secondly, young children struggle to assign accurate symbolic labels to non-symbolic stimuli (Lipton & Spelke 2005), indicating that they would be unable to apply exact symbolic arithmetic knowledge to problems involving non-symbolic stimuli. Finally, there is no difference in performance on approximate addition problems with non-symbolic or symbolic stimuli (McNeil, Fuhs, Keultjes, & Gibson, 2011), which is inconsistent with the hypothesis that children used a method that required additional processing steps for non-symbolic compared with symbolic stimuli. Thus, it is highly unlikely that children solve approximate arithmetic problems by drawing on learnt exact symbolic arithmetic.

Children's performance on approximate arithmetic problems does not appear to result from judgements based on continuous quantity variables, the use of non-arithmetical strategies, or the knowledge of exact symbolic arithmetic. What evidence exists to directly support the thesis that children solve these problems by drawing on ANS representations? Does children's performance match the four signatures of the ANS suggested by Barth et al. (2008)? First, there is evidence across numerous studies that performance is ratio-dependent. Children's accuracy levels decrease as the ratio between the outcome of an operation (e.g. the sum of two dot arrays) and the comparison quantity approaches one. This has been found for studies involving both non-symbolic (Barth et al., 2005, 2008, 2009; McCrink & Spelke, 2010; Pica, Lemer, Izard, & Dehaene, 2004; Slaughter et al., 2006) and symbolic (Gilmore et al., 2007) stimuli. Secondly, children do solve approximate addition problems as accurately as they solve approximate comparison problems when dot array (Barth et al., 2005), cross-modal (Barth et al., 2005, 2008) or symbolic (Gilmore et al., 2007) stimuli were involved. Thirdly, a smaller number of studies have included addition and subtraction approximate arithmetic problems, but there is some evidence to suggest that performance is less accurate for subtraction than addition for both non-symbolic (Barth et al., 2008; Shinskey, Chan, Coleman, Moxom, & Yamamoto, 2009) and symbolic stimuli (Gilmore et al., 2007). Finally, studies involving cross-modal stimuli have demonstrated that the nature of the stimuli has no impact on children's accuracy for approximate addition problems (Barth et al., 2005).

The evidence to date suggests that children can perform approximate arithmetic problems, and that they draw on ANS representations and process to do so. Although children do appear to be influenced to some extent by features of the problem set and they show evidence of drawing on alternative strategies in some situations, these tendencies cannot fully account for their performance. Factors such as sensitivity to the range of numerosities in a problem set and the area of dot array stimuli may influence children's performance of arithmetic, but are not used instead of arithmetic. Children's use of these types of cues may be able to tell us more about the processes that underpin their successful performance on approximate arithmetic tasks. Further research is needed to understand how these strategies influence children's performance, e.g. to discover at what stage of solving an arithmetic problem stimuli characteristics, such as density or envelope area have an impact on children's solutions. Most of the research conducted to date has used dot array stimuli. To gain a better understanding of the way that the ANS is used it would also be informative to use a wider range of stimuli, including those with greater ecological validity.

What Influences Performance?

The studies outlined above have focused on discovering the upper limits of young children's approximate arithmetical abilities and understanding the processes that underlie this. A question that has received less attention concerns the cognitive and environmental factors that influence children's ability to solve these types of problems. A small number of studies have investigated how language, education level and socio-economic status (SES) influence performance on approximate addition problems, with mixed results.

Pica et al. (2004), found no difference in the performance of non-symbolic approximate addition between Munduruku speaking participants drawn from a community in Brazil and French control participants. The Munduruku language has no consistent words for exact numerosities above 5 and thus monolingual Munduruku-speakers do not have symbolic representations for exact numbers. The Munduruku participants in this study varied in age, educational experience, and languages spoken, and included a subset who were monolingual and had received no mathematics instruction. The control participants were French adults who had received many years of mathematics instruction. Despite these environmental differences, the authors found that all groups of participants were able to perform non-symbolic approximate addition and there were no differences between the groups, either among the different groups of Munduruku participants, or between them and the French controls. Similarly, Gilmore, McCarthy, and Spelke (2010) also failed to find environmental effects on approximate arithmetic ability. In this study, child participants recruited from two distinct communities were given a test of non-symbolic approximate addition, as well as a range of other measures. There was no difference in performance on the approximate addition task between children recruited from an area of high SES and those recruited from an area of low SES, although the two groups did differ on tests of exact symbolic arithmetic. These studies suggest that non-symbolic approximate arithmetic skills are robust to the effects of different environmental and educational experiences.

In contrast to these findings, McNeil and colleagues (McNeil, Fuhs, Keultjes, & Gibson, 2011) recently demonstrated evidence of SES effects on young children's approximate arithmetic skills. As well as examining the influence of SES on performance the authors also explored the effect of varying problem format on children's ability to perform non-symbolic and symbolic approximate addition. Children were presented with animated sequences demonstrating approximate addition problems based on those used by Barth et al. (2005) and Gilmore et al. (2007) with one modification. Half of the children observed sequences identical to those used in the previous studies in which the addends were presented first and appeared on the left-hand side of the screen and the comparison quantity appeared later on the right-hand side of the screen, these were termed canonical sequences as they mirror the canonical a + b = c format of formal symbolic arithmetic. The other children observed sequences in which the order and side of presentation were reversed so that the comparison quantity was presented first and appeared on the left-hand-side of the screen and the addends appeared later on the right-hand-side of the screen (see Figure 17.2).

Children observed sequences involving either non-symbolic stimuli (dot arrays) or symbolic stimuli (Arabic numerals). Two experiments were conducted, in the first children were recruited from childcare centres on private college campuses described as mid-high SES, while in the second experiment children were recruited from Head Start classrooms and drawn from a lower SES population. Two key differences in performance between these experiments were observed. First, children from the lower SES population performed less accurately overall than the mid-high SES sample. Secondly, there was an interaction between problem format and SES. Children from the mid-high SES population performed more accurately on problems presented in canonical format than on problems presented in non-canonical format, but there was no difference for children from the lower SES population. In both experiments, performance was equivalent for non-symbolic and symbolic stimuli. The authors suggest that one explanation for these effects was that children from the higher SES background may have received more exposure to the early stages of formal symbolic arithmetic through games and early educational experience than the children from the lower SES background. This additional exposure may have led to a familiarity with canonical arithmetical forms, resulting in the problem format effect for this population, and may have also served to raise their performance on approximate addition problems in general. However, further research is needed to explore the underlying reasons for these effects.

This small set of studies highlight that the influences on children's approximate arithmetic are poorly understood. While the systems supporting approximate arithmetic appear to be robust to the effects of environmental influences in some studies, there is also evidence that these factors play a role in determining performance. The fact that educational

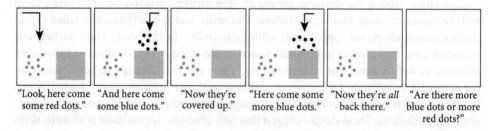

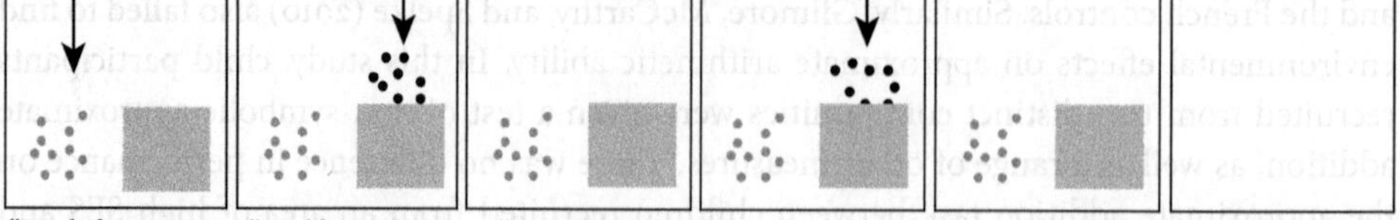

FIGURE 17.2 Example of an animated sequence for a non-canonical problem used by McNeil et al. (2011).

Reprinted from *Cognitive Development*, 26(1), Nicole M. McNeil, Mary Wagner Fuhs, M. Claire Keultjes, and Matthew H. Gibson, Influences of problem format and SES on preschoolers' understanding of approximate addition, pp, 57–71, Copyright (2011), with permission from Elsevier.

experiences may influence the performance of approximate arithmetic can help to inform our understanding of how children solve these problems. While the patterns of children's performance shows the characteristics that would be expected if children were using their ANS, learnt experiences may also play a role in how children solve these problems. Further work is needed to explore how so-called 'core knowledge' (Feigenson et al., 2004) and learnt skills interact in the performance of approximate arithmetic. In particular, a focus on individual differences is overdue. To date, most studies have concentrated on group-level performance and have failed to explore the factors that explain the differences in children's abilities to solve these problems. A range of factors may contribute to children's accuracy on approximate arithmetic problems, including domain-general skills, such as working memory, learnt experience with symbolic numerical representations, as well as the precision of their underlying ANS. An attempt to disentangle these factors will not only allow us to clarify the extent to which the ANS drives performance on these tasks, but will also highlight the way in which such domain-specific knowledge interacts with domain-general skills and knowledge.

Approximate Arithmetic and School Mathematics

The studies discussed above have demonstrated that children have an impressive range of approximate arithmetic abilities, including additive and multiplicative relations, and that they can perform these operations on a wide range of non-symbolic and symbolic stimuli. Do these abilities have any relationship with formal arithmetic that children are taught in school? On the one hand, approximate arithmetic abilities may appear to have little in common with exact symbolic arithmetic. First, unlike symbolic arithmetic, the ANS is unable to represent exact quantities. Secondly, performing exact symbolic arithmetic requires the acquisition of slow effortful strategies, while ANS arithmetic is apparently available from an early age and employs fast unconscious processing. Finally, ANS arithmetic can be applied to a range of stimuli with ease, while exact symbolic arithmetic performance is very context dependent (Bryant, 1995). However, the two arithmetic systems do share some common characteristics, for example, subtraction is more difficult than addition for both ANS arithmetic and exact symbolic arithmetic, and as outlined above there is some evidence that both ANS and exact symbolic arithmetic abilities are influenced by a child's early educational environment. Thus, the nature of the relationship between these two abilities is a question that has attracted recent attention by researchers.

Only a small number of studies have examined the relationship between formal or school mathematics achievement and approximate arithmetic (addition or subtraction) abilities, although more studies have explored the relationship with approximate comparison performance.

Table 17.1 summarizes recent studies that have used either approximate arithmetic or approximate comparison tasks. Those studies that have used an approximate addition task have found mixed results. Iuculano et al. (2008) found no significant correlation between performance of approximate addition and exact symbolic addition. In contrast, Gilmore, McCarthy, and Spelke (2010) found a significant correlation between approximate addition

Table 17.1 Summary of studies investigating the relationship between ans measures and mathematics achievement

Study	Age group	Design	ANS task[a]	Measure[b]	Math task[c]	Relationship between ANS and mathematics?
Bugden & Ansari (2011)	6–8	Concurrent	S comp (9)	ratio effect	WJ-III	Yes (*r* ranges from −0.22 to −.42)
De Smedt et al. (2009)	6	Predictive	S comp (9)	acc, RT, ratio effect	National mathematics test	Yes (*r* range: 0.38–0.40)
Desoete et al. (2012)	5–6	Predictive	NS & S comp (9)	acc	TEDI-Math; calculation; fact retrieval	Yes (*r* range: 0.11–0.20)
Durand et al. (2005)	7–10	Concurrent	S comp (9)	acc	WOND	Yes ($r = 0.49$)
Gilmore et al. (2010)	5–6	Predictive	NS add (65)	acc	School mathematics test	Yes ($r = 0.38$–0.58)
Halberda et al. (2008)	5(math) and 14 (ANS)	Retrospective	NS comp (16)	*w*	TEMA-2 WJ-R	Yes (*r* range: 0.34–0.57)
Halberda et al. (2012)	11–85	Concurrent/ retrospective	NS comp (20)	*w*, RT	Self-report of school performance	Yes ($r = -0.22$ overall)
Holloway & Ansari (2009)	6–8	Concurrent	NS & S comp (9)	NDE	WJ-III	No for NS NDE Yes for S NDE ($r = -0.22$ to -0.34)
Inglis et al. (2011)	7–9 and adults	Concurrent	NS comp (22 children 70 adults)	*w*	WJ-III	Yes for children ($r = -0.55$) No for adults.
Iuculano et al. (2008)	8–9	Concurrent	NS add (58)	efficiency	Verification of single digit addition	No
Libertus et al. (2011)	2–6	Concurrent	NS comp (15)	acc, RT, *w*	TEMA-3	Yes (*r* range: 0.26–0.42)
Lyons & Beilock (2011)	Adults	Concurrent	NS comp (9)	*w*	Mental arithmetic test	Yes ($r = -0.34$)

(continued)

Table 17.1 Continued

Study	Age group	Design	ANS task[a]	Measure[b]	Math task[c]	Relationship between ANS and mathematics?
Lonnemann et al. (2011)	8–10	Concurrent	NS & S comp (6)	NDE	Addition and subtraction problems	No for addition Yes for symbolic NDE and subtraction ($r = -.35$)
Mazzocco et al. (2011)	3–4	Predictive	NS comp (14)	acc	TEMA-3	Yes ($r = 0.53$).
Mundy & Gilmore (2009)	6–7	Concurrent	NS & S comp (9)	acc, RT, NDE	Curriculum mathematics test	No for NS NDE Yes for S NDE, S, & NS accuracy (r range: 0.35–0.52)
Price et al. (2012)	Adults	Concurrent	NS comp (40)	ratio effect, w	WJ-III	No

[a] NS = non-symbolic; S = symbolic; comp = comparison task; add = addition task. The number in parentheses refers to the maximum numerosity used in the task.

[b] acc = accuracy, RT = response time, efficiency = median RT ÷ accuracy, w = estimates of Weber fraction, NDE = numerical distance effect, ratio effect = slope (& intercept) of ratio regressed against RT (smaller numbers indicate higher precision).

[c] WJ = Woodcock-Johnson Tests of Achievement (calculation and/or math fluency subtests), TEMA = Test of Early Mathematics Ability, WOND = Wechsler Objective Numerical Dimensions (numerical operations subtest).

and school mathematics achievement across two experiments. This relationship remained significant after controlling for IQ and literacy. One explanation for the contrast between these findings may be due to the nature of the exact symbolic task and the measures of performance used. Gilmore et al. (2010) used a measure of school mathematics achievement that assessed a range of early symbolic number knowledge, while children in the Iuculano et al. (2008) study were asked to verify the result of single digit addition problems. Because verification and production of arithmetic solutions involve different processes (Campbell & Tarling, 1996), these are not equivalent measures of symbolic mathematics ability. The studies also differed in the measure of approximate addition performance used. Iuculano et al. (2008) used an efficiency measure for all tasks, which incorporates both accuracy and response time, while Gilmore and colleagues used accuracy measures. It is likely that these different measures are tapping into different aspects of the ANS, with accuracy a more direct measure of ANS precision, whilst efficiency measures also incorporate speed of comparison processes. These differences may account for the contrasting findings of these two studies.

Therefore, while there is some limited evidence of the relationship between approximate arithmetic and school mathematics performance, a larger number of studies have explored the relationship between approximate comparison tasks and school mathematics performance. Again, these studies have found mixed results. As can be ascertained from Table 1, these studies have employed different tasks to measure the ANS (ANS acuity estimates from non-symbolic or symbolic comparison tasks), across different number ranges (maximum numerosity ranges from 9 to 70), differing measures of performance (accuracy, RT, estimates of the precision of representations, numerical distance effect), differing measures of exact symbolic arithmetic (various standardized tests or curriculum mathematics tests) and have looked at this relationship across a wide age range. Evidence for a relationship between ANS tasks and exact symbolic arithmetic appears to depend on a combination of these factors. When non-symbolic comparison is employed to measure ANS acuity, significant concurrent relationships have been found between ANS measures and mathematics performance for children aged 2–9 years old (Inglis, Attridge, Batchelor, & Gilmore, 2011; Libertus, Feigenson, & Halberda, 2011). Children's ANS acuity measure on non-symbolic tasks is also correlated with mathematics performance measured two years later (Desoete, Ceulemans, De Weerdt, & Pieters, 2012; Mazzocco, Feigenson, & Halberda, 2011) or 3–9 years earlier (Halberda & Feigensen, 2008). However, the nature of the measure from ANS tasks may be important as no relationship between ANS ability and mathematics performance was found when numerical distance effect measures were used (Holloway & Ansari, 2009; Mundy & Gilmore, 2009). Additionally, numerosity range may also be a factor as no relationship between non-symbolic comparison and symbolic addition performance was found when the number range of the non-symbolic task was restricted (Lonnemann, Linkersdörfer, Hasselhorn, & Lindberg, 2011). For adult participants the evidence is mixed (Halberda, Ly, Wilmer, Naiman, & Germine, 2012; Inglis et al., 2011; Lyons & Beilock, 2011; Price, Palmer, Battista, & Ansari, 2012), and the presence of a relationship between ANS acuity and mathematics performance may depend on the particular mathematical task used, or the numerosity range of the non-symbolic task. When symbolic comparison is used to measure ANS acuity, e.g. by measuring distance effects, the relationship with mathematics performance appears to be more consistent. Significant concurrent relationships have been found in children aged from 6 to 10 years (Bugden & Ansari 2011; Durand, Hulme, Larkin, & Snowling, 2005; Holloway & Ansari, 2009; Mundy & Gilmore, 2009) and predictive relationships over

1 or 2 years have been found for children aged 5–6 years old (Desoete et al., 2012; De Smedt, Verschaffel, & Ghesquière, 2009). However, when symbolic ANS tasks are used it is more difficult to pinpoint whether relationships with mathematics performance are due to ANS acuity itself, or to mappings between the ANS and symbolic representations.

The studies conducted to date do not, however, allow us to adequately determine the nature of the relationship between ANS abilities and achievement with exact symbolic mathematics for three key reasons. First, a number of studies do not control adequately for confounding variables that may provide alternative explanations of the relationships that have been found. For example, measures of IQ are either not collected, or controlled for in a number of studies (e.g. Bugden & Ansari, 2011; Libertus et al., 2011; Mundy & Gilmore, 2009). Similarly, when symbolic ANS tasks are used, it is important to control for symbolic number knowledge to ensure that any relationship observed is specific to ANS processing. This is particularly the case when accuracy or efficiency measures are reported, rather than numerical distance effects or estimates of the precision of representations (e.g. Durand et al., 2005; Iuculano et al., 2008; Mazzocco et al., 2011). Accuracy measures are likely to be influenced by general abilities, as well as ANS acuity to a greater extent than estimates of the precision of representations, which should be more direct measures of ANS acuity. Furthermore, employing alternative outcome variables (such as performance on standardized measures of reading) would allow the specificity of relationships between ANS performance and academic achievement to be tested. However, few studies include such measures.

Secondly, there is a need for systematic studies to investigate the effect of different ANS tasks, and measures on the nature of the relationship between ANS and symbolic mathematics performance. It is unclear, at this time, the extent to which performing different ANS tasks across different number ranges draw on the same underlying system and processes. Indeed, recent evidence suggests that performance on tasks that may have been assumed to be equivalent measures of ANS performance are, in fact, unrelated. Maloney, Risko, Preston, Ansari, and Fugelsang (2010) found that the numerical distance effect obtained from a symbolic comparison task was unrelated to the numerical distance effect from a non-symbolic comparison task with adult participants. These effects had often previously been assumed to arise from the same mechanism of overlapping approximate numerical representations (however, see Verguts & Fias, 2004, or Van Opstal, Gevers, De Moor, & Verguts, 2008, for alternative explanations). Similarly, Gilmore, Attridge, and Inglis (2011) found that accuracy or precision estimates derived from non-symbolic addition and comparison tasks were unrelated in adult participants. Gebuis and van der Smagt (2011) also found that adult performance on a non-symbolic discrimination task was not equivalent to performance on a non-symbolic detection task (i.e. identifying a stimulus of a different numerosity amongst a series of identical numerosities). Thus, different tasks employed to measure ANS performance are not equivalent, at least in adults. If performance on one ANS task is not related to performance on another task then it is not surprising that evidence for the relationship between ANS performance and mathematics achievement has been mixed.

Finally, there is currently no evidence to suggest the causal nature of any relationship between ANS performance and symbolic mathematics achievement. To date, all studies have used a correlational design. The majority of these have involved concurrent testing of ANS abilities and mathematics achievement (see Table 1), although a few studies have tested ANS ability in advance of mathematics achievement and one study has tested mathematics achievement in advance of ANS ability (e.g. Desoete et al., 2012; De Smedt et al. 2009; Halberda, Mazzocco, & Feigenson, 2008; Mazzocco et al., 2011). Longitudinal studies have found that ANS ability predicts mathematics achievement 1 or 2 years later (De Smedt et al. 2009; Desoete et al. 2012; Mazzocco et al. 2011); however, there have been insufficient controls for IQ and educational experience to determine causality from these studies. Because both ANS acuity and mathematics ability are developmental, it is difficult to determine the role of age and educational experience in any relationship observed. However, recent data has begun to tease apart these factors. Piazza, Izard, Spelke, Pica, and Dehaene (2013) studied ANS performance and exact symbolic arithmetic ability in a population of child and adult participants who had received differing amounts of educational instruction and at different ages. They found that years of educational experience was strongly associated with performance on ANS tasks. Thus, the direction of any causal relationship is far from clear. Detailed studies, with appropriate controls, across a wide age range are needed to examine the strength of the relationship between ANS ability and mathematics achievement, and also how this changes with age and educational experience. Moreover, intervention studies are required to determine whether this is a causal relationship.

What mechanisms could account for a causal relationship in either direction between ANS ability and mathematics achievement? A causal relationship between ANS ability and mathematics performance could arise because more precise non-symbolic representations may support the acquisition of symbolic representations. For example, Gilmore et al. (2010) found that the relationship between performance on a non-symbolic approximate addition task and mathematics achievement at age 5 years was no longer significant when symbolic number knowledge was taken into account. Thus, the ANS may play a significant role at the earliest stages of mathematics learning. The ANS may also support later stages of learning arithmetic, e.g. by helping learners to predict and check the plausibility of answers to exact symbolic arithmetic problems. Children with more accurate ANS representations may thus be at an advantage in learning symbolic mathematics (De Smedt et al., 2009). Recently, Lyons and Beilock (2011) have demonstrated that knowledge of ordinal relations of symbolic number mediates the relationship between ANS acuity and arithmetic skill in adults. Thus, the ANS may support learning of symbolic mathematics via better knowledge of ordinal relations. On the other hand, exposure to symbolic stimuli could serve to refine the ANS. Children who are more skilled at symbolic arithmetic may spend more time dealing with symbolic numerical representations and this could, in time, increase the precision of their ANS representations. It is possible that the causal relationship between ANS ability and mathematics achievement could be bi-directional, or the direction of causality may change at different ages. In the early years, the ANS may help a young child to acquire symbolic representations, but later exposure to symbolic arithmetic could serve to refine the ANS.

One way to better understand how the ANS is related to mathematics achievement would be to explore how the ANS is related to numerical and arithmetical estimation skills. A number of studies have explored children's ability to estimate the size of numerical quantities. When children are asked to place a given number on a number line (e.g. place 35 on a line that is labelled 0 at one end and 100 at the other), the accuracy of their estimates increase with age. The estimates that they make also develop from a logarithmic to a linear distribution. This is found to occur at different ages for different number ranges (see, e.g. Booth & Siegler, 2006, 2008). Importantly, the ability to correctly estimate the position of numbers on a number line has been found to be predictive of current and future mathematics achievement. Laski and Siegler (2007) also found that the linearity of children's estimates on a number line task were strongly correlated with performance on a symbolic comparison task. This suggests that children were using the same underlying representations to solve each of these tasks. A better understanding of how magnitude comparison tasks are related to different types of estimation skills, such as numerical or arithmetical estimation would help to clarify how the ANS might be involved in learning mathematics more generally.

If the ANS does play a role in predicting mathematics achievement it is crucial to consider how important this factor is in explaining performance. As can be seen in Table 1, ANS performance only accounted for a small proportion of the variance in mathematics achievement scores. In all studies, less than a third of the variance in outcome measures was explained by ANS score and, in many cases, less than 15% of the variance was accounted for. ANS ability must therefore be considered alongside other predictors of mathematics achievement. One area that has yet to be explored is the nature of the interactions between the ANS and other domain-specific and domain-general predictors of mathematics achievement. It is likely that these abilities interact in a complex fashion, so that individuals with strengths and weaknesses in different areas can use a range of compensatory strategies. Moreover, the importance of the ANS is likely to change over development. During the early years of mathematics education the relationship between the ANS and mathematics achievement may be strong, but over time, ANS precision reaches a peak and other factors come to play a greater role in accounting for individual differences in mathematics achievement, so the relationship between the ANS and mathematics achievement will be weaker. Thus, the ANS is just one element of the range of skills, abilities and experiences that combine to give rise to individual differences in mathematics achievement. Some researchers have suggested that ANS type measures should be used to monitor students' progress in school mathematics. In view of the mixed results surrounding the relationship between the ANS and formal mathematical achievement, the low R^2 values for those studies which have found a relationship, and the uncertainty about the direction of causation (if any), these suggestions are premature.

Conclusions

Research has shown that children have an impressive range of approximate arithmetic skills. Their performance on non-symbolic and symbolic approximate arithmetic tasks cannot easily be explained by the use of strategies or learnt mathematics but appears to reflect the processing of the ANS. Currently, however, we know little about the factors that affect performance on approximate arithmetic tasks, and the skills, knowledge, and experiences, which children draw on when solving these problems. Research in this area is in its infancy and, to date, there has been much emphasis on simply discovering and demonstrating what children are capable of. There is now a need to move on to exploring and understanding the mechanisms by which children solve approximate arithmetic problems, and the way that approximate arithmetic may play a role in learning exact symbolic arithmetic. There are a wealth of questions concerning the factors that influence performance of approximate arithmetic and the way in which the ANS interacts with other cognitive systems that clearly require further study.

Acknowledgments

C.G. is supported by a British Academy Postdoctoral Fellowship.

Notes

1. Performing arithmetic with approximate representations is not the same skill as arithmetical estimation – the ability to estimate the approximate outcome of an exact computation. A substantial body of research has explored the development of children's arithmetical estimation skills. A thorough review of this work is beyond the scope of this article, but it is important to note that children's accuracy and use of arithmetical estimation is related to their mathematical achievement (see Dowker, 2003, for a review). To date, the relationship between arithmetical estimation and the approximate number system has not been explored. A better understanding of this relationship might help to clarify the role of the ANS in learning mathematics.
2. Barth et al. (2006) proposed a mathematical model to account for performance on non-symbolic comparison and addition problems. In this model representations of a numerosity *n* follow a normal distribution with mean *n* and standard deviation *wn*. This model, in fact, predicts that error rates will be lower for addition than for matched comparison problems. This is because the standard deviation of the sum of addends can be less than the standard deviation of the original addends. Since this effect was not observed in the data, Barth and colleagues added an additional variance term to the model, to account for noise introduced by the temporary storage of the operation result. This modified model predicts that addition will be as accurate as comparison.

References

Barth, H., Beckmann, L., & Spelke, E.S. (2008). Nonsymbolic, approximate arithmetic in children: abstract addition prior to instruction. Developmental Psychology, 44, 1466–1477.

Barth, H., Kanwisher, N., & Spelke, E. (2003). The construction of large number representations in adults. Cognition, 86, 201–221.

Barth, H., La Mont, K., Lipton, J., Dehaene, S., Kanwisher, N., & Spelke, E.S. (2006). Non-symbolic arithmetic in adults and young children. Cognition, 98, 199–222.

Barth, H., La Mont, K., Lipton, J., & Spelke, E.S. (2005). Abstract number and arithmetic in preschool children. Proceedings of the National Academy of Sciences of the United States of America, 102, 14116–14121.

Barth, H., Starr, A., & Sullivan, J. (2009). Children's multiplicative transformations of discrete and continuous quantities. Journal of Experimental Child Psychology, 103, 441–454.

Booth, J.L., & Siegler, R.S. (2006). Developmental and individual differences in pure numerical estimation. Developmental Psychology, 42, 189–201.

Booth, J.L., & Siegler, R.S. (2008). Numerical magnitude representations influence arithmetic learning. Child Development, 79, 1016–1031.

Brannon, E.M. (2006). The representation of numerical magnitude. Current Opinion in Neurobiology, 16, 222–229.

Bryant, P. (1995). Children and arithmetic. Journal of Child Psychology and Psychiatry and Allied Disciplines, 36, 3–32.

Bugden, S. & Ansari, D. (2011). Individual differences in children's mathematical competence are related to the intentional but not automatic processing of Arabic numerals. Cognition, 118, 32–44.

Campbell, J.I.D. & Tarling, D.P.M. (1996). Retrieval processes in arithmetic production and verification. Memory & Cognition, 24, 156–172.

Dehaene, S. (1997). The Number Sense: How the Mind Creates Mathematics. New York: Oxford University Press.

De Smedt, B., Verschaffel, L., & Ghesquière, P. (2009). The predictive value of numerical magnitude comparison for individual differences in mathematics achievement. Journal of Experimental Child Psychology, 103, 469–479.

Desoete, A., Ceulemans, A., De Weerdt, F., & Pieters, S. (2012). Can we predict mathematical learning disabilities from symbolic and non-symbolic comparison tasks in kindergarten? Findings from a longitudinal study. British Journal of Educational Psychology, 82, 64–81.

Dowker, A. (2003). Young children's estimates for addition: the zone of partial knowledge and understanding. In A. J. Baroody and A. Dowker (Eds), The Development of Arithmetic Concepts and Skills: Constructing Adaptive Expertise (pp. 243–265). Mahwah, NJ: Erlbaum.

Dowker, A. (2005). Individual Differences in Arithmetic: Implications for Psychology, Neuroscience and Education. Hove: Psychology Press.

Durand, M., Hulme, C., Larkin, R., & Snowling, M. (2005). The cognitive foundations of reading and arithmetic skills in 7- to 10-year-olds. Journal of Experimental Child Psychology, 91, 113–136.

Feigenson, L., Carey, S., & Hauser, M. (2002). The representations underlying infants' choice of more: object files versus analog magnitudes. Psychological Science, 13, 150–156.

Feigenson, L., Dehaene, S., & Spelke, E. (2004). Core systems of number. Trends in Cognitive Sciences, 8, 307–314.

Geary, D.C. (2000). From infancy to adulthood: the development of numerical abilities. European Child & Adolescent Psychiatry, 9, S11–S16.

Gebuis, T., & van der Smagt, M.J. (2011). False approximations of the Approximate Number System? PLoS ONE, 6, e25405.

Gilmore, C.K. (2006). Investigating children's understanding of inversion using the missing number paradigm. Cognitive Development, 21, 301–316.

Gilmore, C., Attridge, N., & Inglis, M. (2011). Measuring the approximate number system. Quarterly Journal of Experimental Psychology, 64, 2099–2109.

Gilmore, C.K., McCarthy, S.E., & Spelke, E.S. (2007). Symbolic arithmetic knowledge without instruction. Nature, 447, 589–591.

Gilmore, C.K., McCarthy, S.E., and Spelke, E.S. (2010). Non-symbolic arithmetic abilities and mathematics achievement in the first year of formal schooling. Cognition, 115, 394–406.

Gilmore, C.K., & Spelke, E.S. (2008). Children's understanding of the relationship between addition and subtraction. Cognition, 107, 932–945.

Halberda, J., & Feigenson, L. (2008). Developmental change in the acuity of the 'number sense': the approximate number system in 3-, 4-, 5-, and 6-year-olds and adults. Developmental Psychology, 44, 1457–1465.

Halberda, J., Ly, R., Wilmer, J. B., Naiman, D.Q., & Germine, L. (2012). Number sense across the lifespan as revealed by a massive Internet-based sample. Proceedings of the National Academy of Sciences, 109, 11116–1120.

Halberda, J., Mazzocco, M.M.M., & Feigenson, L. (2008). Individual differences in non-verbal number acuity correlate with maths achievement. Nature, 455, 665–668.

Holloway, I.D., & Ansari, D. (2008). Domain-specific and domain-general changes in children's development of number comparison. Developmental Science, 11, 644–649.

Holloway, I.D., & Ansari, D. (2009). Mapping numerical magnitudes onto symbols: the numerical distance effect and individual differences in children's mathematics achievement. Journal of Experimental Child Psychology, 103, 17–29.

Inglis, M., Attridge, N., Batchelor, S., & Gilmore, C. (2011). Non-verbal number acuity correlates with symbolic mathematics achievement: but only in children. Psychonomic Bulletin & Review, 18, 1222–1229.

Iuculano, T., Tang, J., Hall, C.W.B., & Butterworth, B. (2008). Core information processing deficits in developmental dyscalculia and low numeracy. Developmental Science, 11, 669–680.

Laski, E.V., & Siegler, R.S. (2007). Is 27 a big number? Correlational and causal connections among numerical categorization, number line estimation, and numerical magnitude comparison. Child Development, 78, 1723–1743.

Libertus, M.E., Feigenson, L., & Halberda, J. (2011). Preschool acuity of the approximate number system correlates with school math ability. Developmental Science, 14, 1292–1300.

Lipton, J.S., & Spelke, E.S. (2005). Preschool children's mapping of number words to nonsymbolic numerosities. Child Development, 76, 978–988.

Lonnemann, J., Linkersdörfer, J., Hasselhorn, M., & Lindberg, S. (2011). Symbolic and non-symbolic distance effects in children and their connection with arithmetic skills. Journal of Neurolinguistics, 24, 583–591.

Lyons, I.M., & Beilock, S.L. (2011). Numerical ordering ability mediates the relation between number-sense and arithmetic competence. Cognition, 121, 256–261.

Maloney, E.A., Risko, E.F., Preston, F., Ansari, D., & Fugelsang, J. (2010). Challenging the reliability and validity of cognitive measures: the case of the numerical distance effect. Acta Psychologica, 134, 154–161.

Mazzocco, M.M.M., Feigenson, L., & Halberda, J. (2011). Preschoolers' precision of the approximate number system predicts later school mathematics performance. PLoS ONE, 6, e23749.

McCrink, K., & Spelke, E.S. (2010). Core multiplication in childhood. Cognition, 116, 204–216.

McNeil, N.M., Fuhs, M.W., Keultjes, M.C., & Gibson, M.H. (2011). Influences of problem format and SES on preschoolers' understanding of approximate addition. Cognitive Development, 26, 57–71.

Mundy, E., & Gilmore, C.K. (2009). Children's mapping between symbolic and nonsymbolic representations of number. Journal of Experimental Child Psychology, 103, 490–502.

Piazza, M., Izard, S., Spelke, E., Pica, P., & Dehaene, S. (2013) Education increases the acuity of the non-verbal approximate number system. Psychological Science, 24, 1037–1043.

Pica, P., Lemer, C., Izard, V., & Dehaene, S. (2004). Exact and approximate arithmetic in an Amazonian indigene group. Science, 306, 499–503.

Price, G.R., Palmer, D., Battista, C., & Ansari, D. (2012). Nonsymbolic numerical magnitude comparison: reliability and validity of different task variants and outcome measures, and their relationship to arithmetic achievement in adults. Acta Psychologica, 140, 50–57.

Shinskey, J.L., Chan, C.H., Coleman, R., Moxom, L., & Yamamoto, E. (2009). Preschoolers' nonsymbolic arithmetic with large sets: is addition more accurate than subtraction? Journal of Experimental Child Psychology, 103, 409–420.

Shuman, M., & Spelke, E. (2006). Area and element size bias numerosity perception. Journal of Vision, 6, 777a.

Slaughter, V., Kamppi, D., & Paynter, J. (2006). Toddler subtraction with large sets: further evidence for an analog-magnitude representation of number. Developmental Science, 9, 33–39.

Temple, E., & Posner, M.I. (1998). Brain mechanisms of quantity are similar in 5-year-old children and adults. Proceedings of the National Academy of Sciences, 95, 7836–7841.

Van Opstal, F., Gevers, W., De Moor, W., & Verguts, T. (2008). Dissecting the symbolic distance effect: comparison and priming effects in numerical and nonnumerical orders. Psychonomic Bulletin & Review, 15, 419–425.

Verguts, T., & Fias, W. (2004). Representation of number in animals and humans: a neural model. Journal of Cognitive Neuroscience, 16, 1493–1504.

Wynn, K. (1992). Addition and subtraction by human infants. Nature, 358, 749–750.

CHAPTER 18

NUMBER REPRESENTATIONS AND THEIR RELATION WITH MATHEMATICAL ABILITY

TITIA GEBUIS AND BERT REYNVOET

The Development of Numerical Abilities

The approximate number system (ANS) is suggested to support our understanding of ordinal relations as well as the ability to approximate and manipulate non-symbolic number. This ability to detect changes in large non-symbolic number is already present at infancy (for a review see Libertus and Brannon 2009). The acuity of the ANS in infants is generally investigated using habituation and violation of expectation paradigms. In these studies, an image with a number of items (e.g. 12 dots) is presented and the time the infant explores this image is measured. The duration of exploration is referred to as looking time. This procedure is repeated with images that represent the same number of items but in different configurations and sizes to account for visual confounds. After a number of repetitions looking time significantly decreases. At this point it is assumed that the infant detected the regularity in number and therefore lost interest in the image. Next a novel image is presented, an image that consists of a different number of items (e.g. 24 dots). If the infant detects the change in number, the infant will regain its interest in the image and looking time will increase.

Infants at six months of age can detect these number changes when the relative difference between the numbers has a ratio of 1:2 (e.g. 12 versus 24 dots) or larger (Lipton and Spelke 2003; Wood and Spelke 2005a, 2005b; Xu and Spelke 2000). The precision in detecting numerical changes increases with increasing age. While six months olds can only detect changes of ratio 1:2, infants of ten months of age can detect changes of ratio 1:2 as well as 2:3 (Lipton and Spelke 2003; Wood and Spelke 2005b). The precision with which we can compare sets of items increases during development until a ratio between 7:8 and 11:12 is reached in adulthood (Gebuis and Van der Smagt 2011; Halberda and Feigenson 2008; van

Oeffelen and Vos 1982). Contrary to infants, the acuity of the ANS in children or adults is assessed using the number discrimination task (see Figure 18.1). In this task, the participant has to decide which of two dot arrays contains more dots. Although it is difficult to compare results from different designs, an increase in ANS acuity from infancy onwards seems apparent (Gebuis and Van der Smagt 2011).

On the basis of these studies investigating (the development of) non-symbolic number processes it is concluded that we are equipped with a system that supports non-symbolic number processes (Cantlon et al. 2009; Feigenson et al. 2004). Some researchers suggested that two separate systems for non-symbolic number processes should be dissociated: an exact system for small numbers that can readily be identified, which is referred to as subitizing (e.g. numbers smaller than four or five), and an inexact system for large numbers that can only be approximated (four or five and larger) (Feigenson et al. 2004; Hyde 2011). This dissociation derived from the observation that human adults can quickly and without making many mistakes name numbers smaller than four or five. In contrast, for numbers larger than four or five, reaction time and the number of mistakes drastically increases (Trick and Pylyshyn 1994). Similarly, six-month-old infants that can dissociate small (one and two items) and large (12 and 24 items) numbers differing with a ratio of 1:2 fail to dissociate numbers that cross the subitizing range (three and six) (e.g. Cordes and Brannon 2009; Feigenson et al. 2002; Xu et al. 2005; for a similar dissociation in nonhuman primates see Beran et al. of this volume). Whether these behavioral differences imply that different neural processes are used for numbers that can be subitized versus approximated is still debated (Feigenson et al. 2004; Hyde 2011).

ANS acuity is suggested to form the basis for later acquired symbolic number abilities (e.g. Libertus et al. 2011; Mazzocco et al. 2011b). From two or three years onwards children are confronted with symbolic numbers at home and later at school as a part of their education. Initially, the child learns counting routines without having the knowledge that each number word refers to a specific number of items. The development of the understanding that a number symbol refers to a specific number of items is suggested to occur via connections that are established between the arbitrary number symbols and the approximate (Piazza 2010), or alternatively with the exact (Carey 2001) non-symbolic number system.

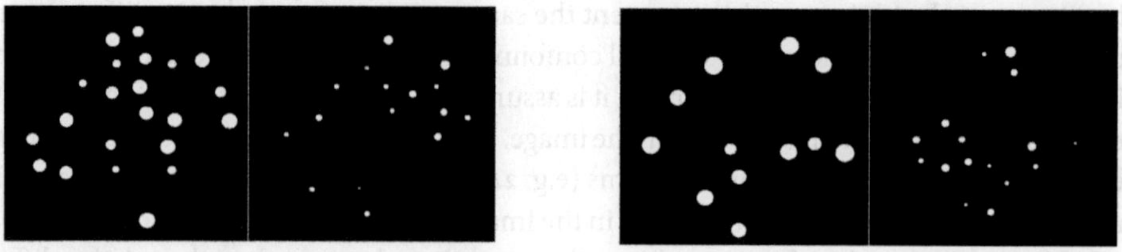

FIGURE 18.1 *Two examples of the stimuli used in the non-symbolic number comparison task.* In the non-symbolic number comparison task, participants have to decide whether the left or the right dot pattern contains more dots. To control for the correlation between number and its visual properties (e.g. when more dots are present, the total surface of the dots is larger) researchers manipulate the visual properties of the stimuli. For example, in half of the trials the numerically larger number also consists of larger visual cues (left example) and in the other half of the trials the numerically larger number consists of smaller visual cues (right image). Consequently, participants cannot rely on a single visual variable throughout the experiment to judge number.

After sufficient exposure to number symbols the relation between number and number meaning becomes automatized, meaning that number can be accessed unintentionally (Gebuis et al. 2009a; Gebuis et al. 2009b; Girelli et al. 2000; Reynvoet et al. 2009; Rubinsten et al. 2002). The age at which symbolic number is automatically processed ranges from five to nine years (Berch et al. 1999; Gebuis et al. 2009b; Girelli et al. 2000; Rubinsten et al. 2002). This large diversity in onset of automatic symbolic number processing is most likely dependent on the amount of exposure to symbolic numbers. The shift from attentional to automatic processing was nicely shown in recent imaging studies (Gebuis et al. 2009b; Kaufmann et al. 2006). In these studies researchers observed a change in the areas recruited during number processing. Initially more frontal areas were involved in number processing but at a later stage, when the access to number meaning became automatized, the parietal areas supporting number processing in adults, were recruited.

The symbolic number system forms the basis for complex mathematics. Being able to represent number in an exact manner is necessary to perform complex mathematical procedures. At elementary school we are all expected to reach a certain level of mathematical proficiency. Unfortunately, around 15–20% of children and adults do not manage to reach this level (Butterworth 2010; Shalev et al. 2000; Shalev et al. 2005). Children with severe mathematical deficiencies generally perform comparable to children that are two years younger and often rely on inefficient strategies, such as finger counting (Ginsburg 1997; Jordan and Montani 1997). Early detection of children at risk for mathematical deficiencies is necessary for the effectiveness of intervention programs. It is therefore of great value to identify the processes that support our (future) mathematical abilities at a young age.

Measuring Non-Symbolic and Symbolic Number Processes

Electrophysiology studies in monkeys showed that groups of neurons exist that are maximally responsive to a specific number (Nieder et al. 2002; see Nieder, this volume). The percept of a number induces an activation pattern that is maximal for neurons that respond to that number and decreases with increasing distance between the number presented and the number the neurons respond to. The overlap in response activation between neurons coding for neighboring numbers increases with increasing number size (Nieder and Merten 2007; Nieder and Miller 2003). Similar results were obtained in neuroimaging studies with adult humans for large non-symbolic numbers (Piazza et al. 2004; but see for an alternative explanation Gebuis and Reynvoet 2012a, 2013) as well as symbolic numbers (Notebaert et al. 2011).

The overlap in neural representations of different numbers is also apparent at the behavioral level (Moyer and Landauer 1967). A behavioral measure of the neural overlap is the *numerical distance effect*. When participants decide which of two non-symbolic or symbolic numbers is numerically larger, reaction time increases and accuracy decreases with decreasing numerical distance (see Figure 2 a,b). Thus faster and more accurate responses can be expected when participants have to compare four and eight (i.e. a distance of four) than four and five (i.e. a distance of one). At the same time, when the numerical distance is kept

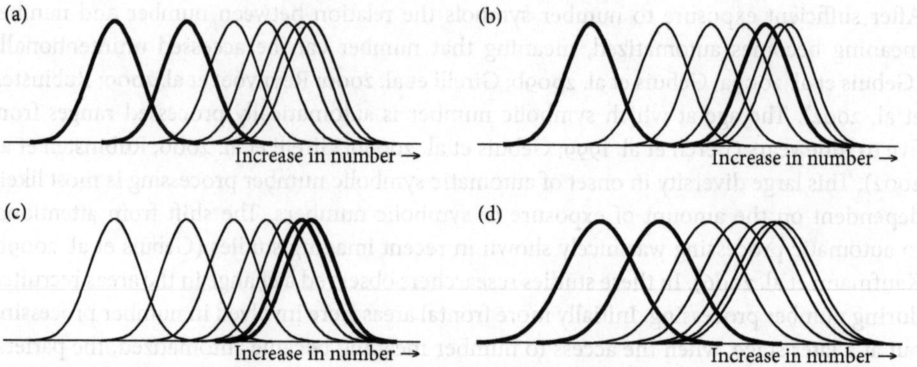

FIGURE 18.2 *A schematic representation of the neural representation of number.* The processing of a number results in the activation of the neural representation of that number as well as its neighboring numbers. This activation of neighboring numbers decreases with increasing distance from the number that is processed. Consequently, a larger neural overlap exists between two neighboring numbers (a) than two numbers further apart (b). Also, the overlap in neural representation increases with increasing number. It is therefore easier to compare two relatively small numbers than two relatively large numbers, even when their number distance is kept constant (a versus c). The approximate number system hypothesis suggests that neural representation of numbers in humans that are less skilled in mathematics will be less accurate (the width of the Gaussian curve is larger) (a versus d).

constant but the size of the numbers is changed (e.g. two and four versus eight and ten), participants are faster and make fewer errors for the numerically smaller numbers. This is called the *size effect* and is a result of the increase in neural overlap of the Gaussian distributions when number increases (see Figure 18.2 a,c). Because of this size effect, differences in large non-symbolic numbers are often expressed as a relative (ratio) instead of an absolute variable (the numerical distance).

The number comparison task is often used to estimate the neural overlap of the underlying number representations. Most non-symbolic number studies use numbers above the subitizing range and include *relative* distance conditions (e.g. ratio 1:2, 1:3, 2:3) to account for the logarithmic compression. From the accuracy data, the Weber fraction can be calculated which is the minimum change in number that is needed to perceive a difference in number. For instance, infants at six months of age have a Weber fraction around two, as they need a twofold difference in number to perceive a change in number (e.g. 12 to 24 or 8 to 16). Contrary to non-symbolic number, for symbolic number comparison tasks, researchers typically include single digits (one to nine) and create *absolute* numerical distance conditions (e.g. a small distance condition of one or two and a large distance condition of three or four). The difference in reaction time or accuracy between two numerical distance conditions is used as a measure of the precision of the underlying neural representation. When the distance effect is small, there is less overlap in the neural representation and thus a more precise representation. When symbolic and non-symbolic processes are measured in the same experiment, the design that is used for symbolic number is also used for non-symbolic number (absolute distances and numbers one to nine) to keep the two tasks as much comparable as possible. The downside of this procedure is that numbers from below and above the subitizing range are included in the non-symbolic number condition. If indeed differences exist between the processes supporting small and large non-symbolic numbers, it can be argued that different neural responses contributed to the observed effects. Together, important differences exist between research designs created to study small symbolic (and non-symbolic) number and large non-symbolic number in children and adults. This should be taken into account when studies investigating the mechanisms underlying mathematical abilities are discussed.

The Impaired Approximate Number System and The Access Deficit Hypothesis

Initially, mathematical deficiencies were related to domain general problems such as phonological skills, working memory, or visuospatial processing (Geary 1993). Later, humans that have problems in mathematics without deficits in these general cognitive abilities were identified (Butterworth 2010; Wilson and Dehaene 2007). This suggests that the fundamental elements of future mathematical abilities might be domain specific and not domain general. The basis for proficiency in mathematics might therefore be the direct result of the establishment of proper elementary number processes. It should be noted that the factor causing the deficiencies observed in humans with mathematical deficiencies does not necessarily have to be the same as the factor that causes the variance in mathematical abilities observed in humans without mathematical deficiencies. We will therefore discuss the studies investigating the mechanisms underlying mathematical abilities in humans with and without mathematical deficiencies separately.

On the basis of the developmental trajectory of symbolic and non-symbolic number processes different domain specific hypotheses about the possible mechanisms that underlie our mathematical abilities are proposed. The most prominent theories are the impaired approximate number system (Butterworth 1999, 2005; Halberda et al. 2008; Wilson and Dehaene 2007) and the access deficit hypothesis (Rousselle and Noel 2007). According to the theory of the impaired approximate number system, problems in mathematics arise when the ANS (the reasoning about non-symbolic number) is inaccurate (see Figure 18.2d). The ANS is suggested to be fundamental for symbolic number learning; impairments at the level of the ANS would therefore result in problems at the symbolic level. In contrast, the access deficit hypothesis suggests that it is not the ANS that is impaired but access to non-symbolic number meaning when a symbolic number has to be processed. These two hypotheses can be disentangled. Namely, the approximate number system hypothesis would predict a deficit in non-symbolic number processing and consequently also in symbolic number processing. In contrast, the access deficit hypothesis would not predict an impairment of non-symbolic number abilities but only for symbolic number processing. To this end, researchers explored possible relations between non-symbolic and symbolic number abilities in typical and atypical developing children and related those measures to their mathematical abilities.

The Approximate Number System Hypothesis

According to the approximate number system hypothesis, ANS acuity plays a key role in the development of mathematical abilities. The precision of the underlying neural representation of non-symbolic number is expected to directly relate to mathematical abilities. The approximate number system hypothesis predicts that children with poor mathematical skills have a less accurate representation of non-symbolic number. Inglis et al. (2011) investigated this relation between ANS acuity and mathematical abilities in children around seven to nine years. The children had to compare sets of dots that could differ with a ratio of 1:2, 2:3, 3:4, or 4:5. They manipulated the visual properties of the number stimuli to account for the visual cues that are confounded with number (e.g. if number increases also its visual properties such as surface and density increase). In addition to the non-symbolic number comparison task, a standardized mathematics and a non-verbal intelligence test were administered to the children. Now, the role that the approximate number system plays in the development of mathematical abilities could be assessed independent of general cognitive factors such as age and IQ. In the number comparison task, the children responded above chance level and a strong correlation between the Weber fraction and performance on the mathematics test was visible. Approximately 30% of the variance in the mathematics test could be explained by the Weber fraction, even after controlling for age and non-verbal IQ.

The ANS acuity appears to correlate with mathematical abilities when both are measured concurrently. In order to make a stronger claim that ANS acuity predicts mathematical abilities; the mathematical abilities should be assessed at a later point in time than ANS acuity. Mazzocco et al. (2011b) therefore measured children's performance on a non-symbolic number comparison task when they were around four years. At this age, children did not start formal (mathematics) education yet. Receiving formal education might not only improve knowledge about mathematics but could also increase the acuity of the approximate number system. Examining children at a young age therefore excludes the possibility that mathematics education influenced the development of ANS acuity. The mathematical abilities were assessed two years later when the children were around six years of age. Again, approximately 30% of the variance on the mathematics test could be explained by ANS acuity suggesting a relatively strong relation between both measures. This was even apparent after correcting for age and grade at the moment of testing.

The above-mentioned studies show that the relatively small variance in mathematical abilities of normal achieving children can be explained by ANS acuity. These results however do not necessarily imply that ANS acuity also underlies the differences between children with mathematical deficiencies and normal achieving children. Several studies therefore included children of varying levels of mathematics. For instance, Mazzocco et al. (2011a) investigated whether differences can be found in ANS acuity between children with a mathematical learning deficiency (mathematics scores below 10th percentile), low achieving children (mathematics scores between 10th and 25th percentile), typical (mathematics scores between 25th and 95th percentile), and high achieving children (mathematics scores > 95th percentile). This study was a longitudinal study following children from kindergarten to the eighth grade. The results showed that the Weber fraction of children with a mathematics learning deficiency was significantly larger than that of the other children. In a related study, Piazza et al. (2010) specifically investigated the acuity of the

ANS in children diagnosed with dyscalculia relative to that of typical achieving children. They tested ten-year-old dyscalculic children as well as typical achieving kindergartners, ten-year-olds and adults on a non-symbolic number comparison task. The results showed that the Weber fraction increased from kindergartners to ten-year olds and adults, which was compatible to the developmental patterns observed in related studies (Halberda and Feigenson 2008). However, the Weber fraction of the dyscalculic children was larger than that of the age-matched controls. The performance of the ten-year old dyscalculic children was comparable to that of the five-year-old kindergartners.

Although the above-described studies indicate that ANS acuity might relate to (future) mathematical abilities in children with and without proficiency in mathematics, contradictory findings exist as well. For instance there are studies investigating the relation between both non-symbolic and symbolic number processes and mathematical abilities that found a non-significant relation between non-symbolic number comparison and (future) mathematical abilities (De Smedt and Gilmore 2011; Holloway and Ansari 2009; Rousselle and Noel 2007; Sasanguie et al. 2012a; Soltesz et al. 2010). An important difference between these and the above-mentioned studies is that these studies did not include large non-symbolic numbers but small numbers from around the subitizing range (one to nine). This was done to allow a direct comparison with the symbolic number condition. Because small numbers were included, absolute numerical distance instead of a relative variable like the Weber fraction was used as a measure for ANS acuity. Only a single study found a relation between small non-symbolic number and mathematical abilities (Price et al. 2007). Price et al. (2007) showed that children with developmental dyscalculia have a larger distance effect for accuracy than typical developing children. Together, it could be that the differences in task designs and/or the number range chosen can explain the lack of a relation between performance on small non-symbolic number tasks and mathematical abilities in these studies.

Still, studies remain that did not observe a (strong) correlation between mathematical abilities and ANS acuity in children with (Landerl and Kolle 2009; Rousselle and Noel 2007) and without mathematical deficiencies (Halberda et al. 2012; Inglis et al. 2011; Libertus et al. 2011; Sasanguie et al. 2012b; Soltesz et al. 2010) even though large non-symbolic number stimuli were included. Because an absence of a correlation was present for children with and without mathematical deficiencies, an explanation on the basis of a deficient ANS seems therefore not sufficient. Alternatively, Soltesz et al. (2010) suggested that the visual cues of the number stimuli (e.g. surface and diameter) might play a significant role in the variance observed in the outcomes of non-symbolic number studies. Soltesz et al. (2010) showed that four-year-old children have the tendency to respond to the visual properties of the stimuli instead of number when they have to decide which stimulus contains more items. In everyday life, such behavior would be accurate as number and its visual cues are confounded: with increasing number the visual cues also increase. However, to measure non-symbolic number processes researchers manipulate the stimuli to account for this confound. They create trials where the numerically larger stimulus consists of larger visual cues in some trials while in other trials the numerically larger stimulus consists of smaller visual cues (see Figure 18.1). The children in the Soltesz et al. (2010) study were strongly biased by the visual cues and made significantly more errors when the numerically larger number consisted of smaller visual cues. Such biases were also observed in older children (Inglis et al. 2011; approximately 30% of the eight-year old children responded to visual cues) and adults (Gebuis and Reynvoet 2012b, 2012c; Gilmore et al. 2011). This is surprising because adults and eight-year-old children can be expected to have the necessary knowledge about cardinal meaning to respond to numerosity instead of its visual properties. Apparently, adults and children rely on the sensory cues even though they are not informative about the sensory cues. Different methods to control the visual cues can affect overall performance (Gebuis and Reynvoet 2012b). As each research group uses their own methods to control the visual cues, the discrepancies in results (i.e. the relation between the performance on non-symbolic number comparison and mathematical abilities) could result from the differences in the stimuli used.

The Access Deficit Hypothesis

The access deficit hypothesis holds that the problems faced by some children when performing mathematics are not induced by deficiencies at the level of the approximate number system but from an inability to access the non-symbolic meaning when symbolic number has to be processed (Rousselle and Noel 2007). If this is indeed the case, a relationship between performance on a symbolic number processing task and a mathematics test can be expected while at the same time non-symbolic number processes are unimpaired. To investigate this hypothesis, Holloway and Ansari (2009) used a symbolic and a non-symbolic number comparison task in which magnitudes between one to nine had to be compared. Children between six and eight years of age were tested on both tasks and their mathematics proficiency and general cognitive skills were assessed. Two tasks were used to measure mathematics performance: the Woodcock Johnson subtest for mathematics fluency and the subtest mathematics calculation. These tasks are similar except that mathematics fluency is timed. A significant correlation was present between symbolic number comparison and the mathematics fluency task. No significant effects were present for non-symbolic number, which is in accordance with the access deficit hypothesis. However, this relation between the symbolic number comparison and mathematics fluency results was only present for the six year olds, not for the older children. This finding led to the conclusion that a critical period might exist. The impairment in accessing symbolic number meaning is only observed when children have to intentionally process symbolic number. When automatic access to number meaning was measured via a numerical Stroop task (Gebuis et al. 2009a; Rubinsten et al. 2002), no significant relation in performance on the Stroop task and mathematics test was present (Bugden and Ansari 2011; Heine et al. 2010), while a significant correlation between the symbolic distance effect and mathematical abilities was shown (Bugden and Ansari 2011).

Similarly as for large non-symbolic number, stronger evidence for a link between symbolic number processes and mathematical abilities comes from longitudinal designs. De Smedt et al. (2009) tested children on a symbolic number comparison task and assessed their mathematical skills one year later. These children just entered formal education and consequently, the influence of their acquired mathematics skills on number comparison processes should be negligible. Results revealed a significant correlation between the symbolic numerical distance effect and mathematical abilities. Unfortunately, the researchers did not correct for average reaction time on the number comparison task. Reaction time strongly correlated with mathematical abilities and it therefore cannot be excluded that response time was the causal factor for the results.

Strong correlations between reaction time on the symbolic comparison task and mathematical abilities are also present in related studies (e.g. Landerl et al. 2004; Rousselle and Noel 2007; Sasanguie, De Smedt et al., 2012a; Sasanguie, Gobel et al. 2012b). Generally, these effects are even stronger than the correlation between the numerical distance effect and mathematical abilities. The speed with which numerical meaning can be accessed might therefore be a better predictor of mathematical abilities. Furthermore, it should be noted that in most studies only a small portion of the variance on mathematics tests could be explained by the size of the numerical distance effect (effect sizes varied between 3% and 8%) (e.g. Bugden and Ansari 2011; De Smedt et al. 2009). Also inconsistent results were derived from studies investigating the numerical distance effect in typical and atypical achieving children. Some researchers showed that children with mathematical deficiencies have larger distance effects for symbolic number comparison than typical developing children (Mussolin, et al., 2010) while others found the opposite (Rousselle and Noel 2007) or no differences at all (De Smedt and Gilmore, 2011). Cohen Kadosh et al. recently suggested that the distance effect might not necessarily reflect proficiency in basic number processes (2012). Such reasoning could explain the observed inconsistencies regarding to the number distance effect and mathematical abilities. For example, fast as well as slow responses to both small and large number distance conditions result in a small number distance effect. Now, on the basis of this small distance effect it can be concluded that the participants had good mathematical abilities but this would not be concluded on the basis of the overall speed in responding. Namely, the relatively fast responses to number trials are indicative of good mathematical abilities while the relatively slow responses to number trials are indicative of poor mathematical abilities.

Together, these studies suggest that the numerical distance effect might not play a role in predicting future mathematical abilities but instead reaction time to symbolic number stimuli. Indeed, children with a mathematical learning disorder, when compared to typical and/or low achieving children, responded significantly slower on a symbolic but not on a non-symbolic number comparison task (De Smedt and Gilmore 2011; Landerl and Kolle 2009; Rousselle and Noel 2007). These results suggest that the problem lies in the speed of accessing numerical meaning, not an inaccurate activation of numerical meaning.

Implications for Developmental Models

From this review, it is clear that there is some evidence for the approximate number system hypothesis as well as the access deficit hypothesis. The Weber fraction obtained in large non-symbolic number comparison and the speed of responding in small symbolic number comparison are related to individual differences on a mathematics test in typically and atypically achieving children. Children with poor mathematical skills have a larger Weber fraction, indicating a less precise representation or they are slower in comparing symbolic numbers.

In a recent review paper by Noel and Rouselle (2012), two different theoretical positions for developmental models of mathematical abilities were proposed. A first group of developmental models assumes that the innate approximate number system is fundamental for later acquired symbolic number processes as well as more complex mathematical skills. Children acquire meaning for the later learned number symbols through connections with the approximate representation (Piazza 2010; Verguts and Fias 2004). Following this assumption, it can be expected that a less precise representation of non-symbolic number already observed in young children (e.g. Mazzocco et al. 2011b) not only affects non-symbolic number processes but also the development of associations between symbols and their corresponding meaning. In line with this idea, Piazza et al. (2010) showed that the Weber fraction in dyscalculic children is related to the number of errors they make in manipulating large symbolic numbers: a less precise representation results in more errors. However, a problematic observation is that many studies reported no relation between mathematical skills and non-symbolic number comparison with small dot collections (i.e. one to nine; e.g. Holloway and Ansari 2009). In contrast, the same numbers lead to difficulties when presented symbolically. If the approximate number system serves as the basis for later acquired symbolic number, this asymmetric performance in small non-symbolic and symbolic conditions is difficult to account for.

A second group of developmental models assumes two distinct magnitude representations (Carey 2001; Noel and Rousselle 2012). Contrary to the previous views on the development of semantic representations, this view suggests that an exact magnitude representation co-exists with the approximate representation. This exact representation originates from our ability to represent small sets exactly (Trick and Pylyshyn 1994) and symbolic numbers are mapped onto it (Sarnecka and Carey 2008). This representation is initially limited to three or four but as soon as the child learns the analogy between the order of the counting list and the order of the quantities, a way to represent number exactly is established. According to this view, it should be possible to find dissociations between large non-symbolic tasks and small symbolic tasks in typically achieving children and children with mathematical learning deficits: the severity of both problems (i.e. a larger Weber fraction and slower symbolic reaction times) may differ. However, to date, only a few studies looked at the performance with symbolic and large non-symbolic number in the same sample (but see Rouselle and Noel 2007).

Although evidence in favor of either model exists clear methodological problems in both lines of research are apparent. The research regarding non-symbolic numerosities should question the role that the sensory cues play in numerosity processing (Gebuis and Reynvoet 2011) as the sensory cues that comprise the numerosity stimuli influence performance independent of the method used to generate the stimuli. Different methods to generate the stimuli could cause differences in performance and thus the possible presence or absence of a relation with mathematical abilities. In contrast, studies concerning symbolic number processes should question whether the number distance effect is a good indicator for mathematical abilities (Cohen Kadosh et al. 2012).

In sum, this review shows shortcomings to disentangle whether one or more numerical representations are necessary to account for the pattern of observed numerical deficiencies. Besides the lack of studies comparing the performance of subjects on (small) symbolic and (large) non-symbolic tasks, also methodological differences between studies make it difficult to favor one of the type of developmental models (e.g. the different methods to control the visual properties of the non-symbolic number stimuli, the tests to measure mathematical performance, the different measures for acuity and the different ranges of numbers included). This opens venues towards new research directly contrasting both theoretical positions.

References

Berch, D.B., Foley, E.J., Hill, R.J., and Ryan, P.M. (1999). Extracting parity and magnitude from Arabic numerals: developmental changes in number processing and mental representation. *J Exp Child Psychol* 74(4): 286–308.

Bugden, S. and Ansari, D. (2011). Individual differences in children's mathematical competence are related to the intentional but not automatic processing of Arabic numerals. *Cognition* 118(1): 32–44. doi: S0010-0277(10)00218-0 [pii] 10.1016/j.cognition.2010.09.005.

Butterworth, B. (1999). *The Mathematical Brain*. London: Macmillan.

Butterworth, B. (2005). The development of arithmetical abilities. *J Child Psychol Psychiatry* 46(1): 3–18. doi: JCPP374 [pii] 10.1111/j.1469-7610.2004.00374.x.

Butterworth, B. (2010). Foundational numerical capacities and the origins of dyscalculia. *Trends Cogn Sci* 14(12): 534–41. doi: S1364-6613(10)00214-7 [pii] 10.1016/j.tics.2010.09.007.

Cantlon, J.F., Platt, M.L., and Brannon, E.M. (2009). Beyond the number domain. *Trends Cogn Sci* 13(2): 83–91. doi: S1364-6613(08)00259-3 [pii] 10.1016/j.tics.2008.11.007.

Carey, S. (2001). Cognitive Foundations of Arithmetic: Evolution and Ontogenisis. *Mind and Language* 16(1): 37–55.

Cohen Kadosh, R., Bien, N., and Sack, A.T. (2012). Automatic and intentional number processing both rely on intact right parietal cortex: a combined FMRI and neuronavigated TMS study. *Front Hum Neurosci* 6: 2. doi: 10.3389/fnhum.2012.00002.

Cordes, S. and Brannon, E.M. (2009). Crossing the divide: infants discriminate small from large numerosities. *Dev Psychol* 45(6): 1583–94. doi: 2009-19928-008 [pii] 10.1037/a0015666.

De Smedt, B. and Gilmore, C.K. (2011). Defective number module or impaired access? Numerical magnitude processing in first graders with mathematical difficulties. *J Exp Child Psychol* 108(2): 278–92. doi: S0022-0965(10)00177-3 [pii] 10.1016/j.jecp.2010.09.003.

De Smedt, B., Verschaffel, L., and Ghesquiere, P. (2009). The predictive value of numerical magnitude comparison for individual differences in mathematics achievement. *J Exp Child Psychol* 103(4): 469–79. doi: S0022-0965(09)00026-5 [pii] 10.1016/j.jecp.2009.01.010.

Feigenson, L., Carey, S., and Hauser, M. (2002). The representations underlying infants' choice of more: object files versus analog magnitudes. *Psychol Sci* 13(2): 150–6.

Feigenson, L., Dehaene, S., and Spelke, E. (2004). Core systems of number. *Trends Cogn Sci*, 8(7), 307–314. doi: 10.1016/j.tics.2004.05.002 S1364661304001317 [pii]

Geary, D.C. (1993). Mathematical disabilities: cognitive, neuropsychological, and genetic components. *Psychol Bull* 114(2): 345–62.

Gebuis, T. and Reynvoet, B. (2011). Generating non-symbolic number stimuli. *Behav Res Methods* 43(4): 981–6. doi: 10.3758/s13428-011-0097-5.

Gebuis, T. and Reynvoet, B. (2012a). Continuous visual properties explain neural responses to nonsymbolic number. *Psychophysiology* 49(11): 1649–59. doi: 10.1111/j.1469-8986.2012.01461.x.

Gebuis, T. and Reynvoet, B. (2012b). The interplay between nonsymbolic number and its continuous visual properties. *J Exp Psychol Gen.* doi: 2011-25898-001 [pii] 10.1037/a0026218.

Gebuis, T. and Reynvoet, B. (2012c). The role of visual information in numerosity estimation. *PLoS One* 7(5): e37426. doi: 10.1371/journal.pone.0037426 PONE-D-11-22563 [pii].

Gebuis, T. and Reynvoet, B. (2013). The neural mechanisms underlying passive and active processing of numerosity. *Neuroimage* 70C: 301–7. doi: S1053-8119(12)01234-7 [pii] 10.1016/j.neuroimage.2012.12.048.

Gebuis, T. and Van der Smagt, M.J. (2011). False approximations of the approximate number system? *PLoS ONE.*

Gebuis, T., Cohen Kadosh, R., de Haan, E., and Henik, A. (2009a). Automatic quantity processing in 5-year olds and adults. *Cogn Process* 10(2): 133–42. doi: 10.1007/s10339-008-0219-x.

Gebuis, T., Herfs, I.K., Kenemans, J.L., de Haan, E.H., and van der Smagt, M.J. (2009b). The development of automated access to symbolic and non-symbolic number knowledge in children: an ERP study. *Eur J Neurosci*. doi: EJN6994 [pii] 10.1111/j.1460-9568.2009.06994.x.

Gilmore, C., Attridge, N., and Inglis, M. (2011). Measuring the approximate number system. *Q J Exp Psychol (Colchester)* 64(11): 2099–109. doi: 10.1080/17470218.2011.574710.

Ginsburg, H.P. (1997). Mathematics learning disabilities: a view from developmental psychology. *J Learn Disabil* 30(1): 20–33.

Girelli, L., Lucangeli, D., and Butterworth, B. (2000). The development of automaticity in accessing number magnitude. *J Exp Child Psychol* 76(2): 104–22. doi: 10.1006/jecp.2000.2564 S0022-0965(00)92564-5 [pii].

Halberda, J. and Feigenson, L. (2008). Developmental change in the acuity of the 'Number Sense': The Approximate Number System in 3-, 4-, 5-, and 6-year-olds and adults. *Dev Psychol* 44(5): 1457–65. doi: 2008-12114-023 [pii] 10.1037/a0012682.

Halberda, J., Mazzocco, M.M., and Feigenson, L. (2008). Individual differences in non-verbal number acuity correlate with maths achievement. *Nature* 455(7213): 665–8. doi: nature07246 [pii] 10.1038/nature07246.

Halberda, J., Ly, R., Wilmer, J.B., Naiman, D.Q., and Germine, L. (2012). Number sense across the lifespan as revealed by a massive Internet-based sample. *Proc Natl Acad Sci USA* 109(28): 11116–20. doi: 1200196109 [pii] 10.1073/pnas.1200196109.

Heine, A., Tamm, S., De Smedt, B., Schneider, M., Thaler, V., and Torbeyns, J. (2010). The numerical stroop effect in primary school children: a comparison of low, normal, and high achievers. *Child Neuropsychol* 16(5): 461–77. doi: 921780905 [pii] 10.1080/09297041003689780.

Holloway, I.D. and Ansari, D. (2009). Mapping numerical magnitudes onto symbols: the numerical distance effect and individual differences in children's mathematics achievement. *J Exp Child Psychol* 103(1): 17–29. doi: S0022-0965(08)00052-0 [pii] 10.1016/j.jecp.2008.04.001.

Hyde, D.C. (2011). Two systems of non-symbolic numerical cognition. *Front Hum Neurosci* 5: 150. doi: 10.3389/fnhum.2011.00150.

Inglis, M., Attridge, N., Batchelor, S., and Gilmore, C. (2011). Non-verbal number acuity correlates with symbolic mathematics achievement: But only in children. *Psychon Bull Rev* 18(6): 1222–9. doi: 10.3758/s13423-011-0154-1.

Jordan, N.C. and Montani, T.O. (1997). Cognitive arithmetic and problem solving: a comparison of children with specific and general mathematics difficulties. *J Learn Disabil* 30(6): 624–34, 684.

Kaufmann, L., Koppelstaetter, F., Siedentopf, C., Haala, I., Haberlandt, E., Zimmerhackl, L.B. (2006). Neural correlates of the number-size interference task in children. *Neuroreport* 17(6): 587–91.

Landerl, K. and Kolle, C. (2009). Typical and atypical development of basic numerical skills in elementary school. *J Exp Child Psychol* 103(4) 546–65. doi: S0022-0965(08)00196-3 [pii] 10.1016/j.jecp.2008.12.006.

Landerl, K., Bevan, A., and Butterworth, B. (2004). Developmental dyscalculia and basic numerical capacities: a study of 8-9-year-old students. *Cognition* 93(2): 99–125. doi: 10.1016/j.cognition.2003.11.004 S0010027704000149 [pii].

Libertus, M.E. and Brannon, E.M. (2009). Behavioral and Neural Basis of Number Sense in Infancy. *Curr Dir Psychol Sci* 18(6): 346–51. doi: 10.1111/j.1467-8721.2009.01665.x.

Libertus, M.E., Feigenson, L., and Halberda, J. (2011). Preschool acuity of the approximate number system correlates with school math ability. *Dev Sci* 14(6): 1292–300. doi: 10.1111/j.1467-7687.2011.01080.x.

Lipton, J.S. and Spelke, E.S. (2003). Origins of number sense. Large-number discrimination in human infants. *Psychol Sci* 14(5): 396–401.

Mazzocco, M.M., Feigenson, L., and Halberda, J. (2011a). Impaired acuity of the approximate number system underlies mathematical learning disability (dyscalculia). *Child Dev* 82(4): 1224–37. doi: 10.1111/j.1467-8624.2011.01608.x.

Mazzocco, M.M., Feigenson, L., and Halberda, J. (2011b). Preschoolers' precision of the approximate number system predicts later school mathematics performance. *PLoS One* 6(9): e23749. doi: 10.1371/journal.pone.0023749 PONE-D-11-09239 [pii].

Moyer, R.S. and Landauer, T.K. (1967). Time required for judgements of numerical inequality. *Nature* 215(109): 1519–20.

Mussolin, C., De Volder, A., Grandin, C., Schlogel, X., Nassogne, M.C., and Noel, M.P. (2009). Neural Correlates of Symbolic Number Comparison in Developmental Dyscalculia. *J Cogn Neurosci*. doi: 10.1162/jocn.2009.21237.

Mussolin, C., Mejias, S., and Noel, M.P. (2010). Symbolic and nonsymbolic number comparison in children with and without dyscalculia. *Cognition, 115*(1), 10–25. doi: S0010-0277(09)00258-3 [pii]

Nieder, A. and Merten, K. (2007). A labeled-line code for small and large numerosities in the monkey prefrontal cortex. *J Neurosci* 27(22): 5986–93. doi: 27/22/5986 [pii] 10.1523/JNEUROSCI.1056-07.2007.

Nieder, A. and Miller, E.K. (2003). Coding of cognitive magnitude: compressed scaling of numerical information in the primate prefrontal cortex. *Neuron* 37(1): 149–57. doi: S0896627302011443 [pii].

Nieder, A., Freedman, D.J., and Miller, E.K. (2002). Representation of the quantity of visual items in the primate prefrontal cortex. *Science* 297(5587): 1708–11. doi: 10.1126/science.1072493 297/5587/1708 [pii].

Noel, M.P. and Rousselle, L. (2012). Developmental Changes in the Profiles of Dyscalculia: An Explanation Based on a Double Exact-and-Approximate Number Representation Model. *Front Hum Neurosci* 5: 165. doi: 10.3389/fnhum.2011.00165.

Notebaert, K., Nelis, S., and Reynvoet, B. (2011). The magnitude representation of small and large symbolic numbers in the left and right hemisphere: an event-related fMRI study. *J Cogn Neurosci* 23(3): 622–30. doi: 10.1162/jocn.2010.21445.

Piazza, M. (2010). Neurocognitive start-up tools for symbolic number representations. *Trends Cogn Sci* 14(12): 542–51. doi: S1364-6613(10)00215-9 [pii] 10.1016/j.tics.2010.09.008.

Piazza, M., Izard, V., Pinel, P., Le Bihan, D., and Dehaene, S. (2004). Tuning curves for approximate numerosity in the human intraparietal sulcus. *Neuron* 44(3): 547–55. doi: S0896627304006786 [pii] 10.1016/j.neuron.2004.10.014.

Piazza, M., Facoetti, A., Trussardi, A.N., Berteletti, I., Conte, S., and Lucangeli, D. (2010). Developmental trajectory of number acuity reveals a severe impairment in developmental dyscalculia. *Cognition* 116(1): 33–41. doi: S0010-0277(10)00076-4 [pii] 10.1016/j.cognition.2010.03.012.

Price, G.R., Holloway, I., Rasanen, P., Vesterinen, M., and Ansari, D. (2007). Impaired parietal magnitude processing in developmental dyscalculia. *Curr Biol* 17(24): R1042–3. doi: S0960-9822(07)02072-6 [pii] 10.1016/j.cub.2007.10.013.

Reynvoet, B., De Smedt, B., and Van den Bussche, E. (2009). Children's representation of symbolic magnitude: the development of the priming distance effect. *J Exp Child Psychol* 103(4): 480–9. doi: S0022-0965(09)00005-8 [pii] 10.1016/j.jecp.2009.01.007.

Rousselle, L. and Noel, M.P. (2007). Basic numerical skills in children with mathematics learning disabilities: a comparison of symbolic vs non-symbolic number magnitude processing. *Cognition* 102(3): 361–95.

Rubinsten, O., Henik, A., Berger, A., and Shahar-Shalev, S. (2002). The development of internal representations of magnitude and their association with Arabic numerals. *J Exp Child Psychol* 81(1): 74–92.

Sarnecka, B.W. and Carey, S. (2008). How counting represents number: what children must learn and when they learn it. *Cognition* 108(3): 662–74. doi: S0010-0277(08)00114-5 [pii] 10.1016/j.cognition.2008.05.007.

Sasanguie, D., De Smedt, B., Defever, E., and Reynvoet, B. (2012a). Association between basic numerical abilities and mathematics achievement.. *British Journal of Developmental Psychology*.

Sasanguie, D., Gobel, S.M., Moll, K., Smets, K., and Reynvoet, B. (2012b). Approximate number sense, symbolic number processing, or number-space mappings: What underlies mathematics achievement? *J Exp Child Psychol* 114(3): 418–31. doi: S0022-0965(12)00202-0 [pii] 10.1016/j.jecp.2012.10.012.

Shalev, R.S., Auerbach, J., Manor, O., and Gross-Tsur, V. (2000). Developmental dyscalculia: prevalence and prognosis. *Eur Child Adolesc Psychiatry* 9 Suppl. 2: II58–64.

Shalev, R.S., Manor, O., and Gross-Tsur, V. (2005). Developmental dyscalculia: a prospective six-year follow-up. *Dev Med Child Neurol* 47(2): 121–5.

Soltesz, F., Szucs, D., and Szucs, L. (2010). Relationships between magnitude representation, counting and memory in 4- to 7-year-old children: A developmental study. *Behav Brain Funct* 6: 13. doi: 1744-9081-6-13 [pii] 10.1186/1744-9081-6-13

Trick, L.M. and Pylyshyn, Z.W. (1994). Why are small and large numbers enumerated differently? A limited-capacity preattentive stage in vision. *Psychol Rev* 101(1): 80–102.

van Oeffelen, M.P. and Vos, P.G. (1982). A probabilistic model for the discrimination of visual number. *Percept Psychophys* 32(2): 163–70.

Verguts, T. and Fias, W. (2004). Representation of number in animals and humans: a neural model. *J Cogn Neurosci* 16(9): 1493–504. doi: 10.1162/0898929042568497.

Wilson, A.J. and Dehaene, S. (2007). Number sense and developmental dyscalculia. *Human behavior, learning and the developing brain: Atypical development*. New York: Guilford Press.

Wood, J.N. and Spelke, E.S. (2005a). Chronometric studies of numerical cognition in five-month-old infants. *Cognition* 97(1): 23–39. doi: S0010-0277(04)00173-8 [pii] 10.1016/j.cognition.2004.06.007.

Wood, J.N. and Spelke, E.S. (2005b). Infants' enumeration of actions: numerical discrimination and its signature limits. *Dev Sci* 8(2): 173–81.

Xu, F. and Spelke, E.S. (2000). Large number discrimination in 6-month-old infants. *Cognition* 74(1): B1–B11.

Xu, F., Spelke, E.S., and Goddard, S. (2005). Number sense in human infants. *Dev Sci* 8(1): 88–101.

CHAPTER 19

NUMERICAL COGNITION DURING COGNITIVE AGING

KIM UITTENHOVE AND PATRICK LEMAIRE

LIKE many other cognitive domains, some aspects of numerical cognition change with age; other aspects remain stable. Understanding how numerical cognition evolves with aging is not trivial. First, it involves both biologically primary and secondary processes, which are differently affected by aging (Geary and Lin 1998). Biologically primary processes are competencies that emerge naturally (e.g. the ability to distinguish two objects from one object), whereas secondary processes are dependent upon training (e.g. acquisition of number words). Most primary processes will suffer from aging due to physiological brain changes (e.g. decreased processing speed through white matter disintegration, Reuter-Lorenz and Park 2010). Secondary processes can sometimes benefit from age through expertise (e.g. base-10 arithmetic, Saxe, 1982). Moreover, older adults often show behaviors similar to young adults in numerical cognition, but supported by different brain activation patterns or different cognitive strategies (e.g. El Yagoubi et al. 2005; see also Reuter-Lorenz and Park 2010, for similar results in other domains). This demonstrates that although young and older adults' numerical cognition can seem similar, it may arise quite differently. In this chapter, we first discuss aging effects on numerical performance. Second, we discuss age-related changes in strategic variations. Finally, we discuss promising avenues for future research on aging and numerical cognition.

THE EFFECTS OF AGING ON NUMERICAL COGNITION

Research has documented age-related changes in several components of numerical cognition, approximate and exact number system, quantification (i.e. subitizing and counting), and arithmetic (i.e. arithmetic fact retrieval and computation). These components rely on biologically primary as well as on biologically secondary processes, resulting in declines, maintenance, or improvements during cognitive aging.

Approximate Number System. The approximate number system refers to number sense, which is the ability to grasp quantity. It is the ability to determine that a set of *N* items contains more than a set of *N*-1 items. It relies on activation of the horizontal intraparietal sulcus to specific numerosities (e.g. Eger et al. 2003). This property has been found in infants and in animals (Dehaene et al. 1998) and is thus close to a biologically primary process. Acuity of approximate number sense can be assessed by the difference in quantities that can be distinguished by an individual, expressed in Weber fractions. In an online study, Halberda and colleagues (2012) recently collected data from 10 000 individuals ranging from 11 to 85 years of age on a non-symbolic number sense task. Participants had to indicate the largest of two collections of dots. The researchers found a decrease in Weber fractions from 11 to 30 years of age, supposedly reflecting a refining of number sense. From age 30 to 85 however, Weber fractions steadily increased, with the Weber fraction of an 85-year-old adult being similar to that of an 11-year old child (see Figure 19.1). This could suggest decreased approximate number sense in older adults (some caution is necessary when interpreting these results however because they could also arise from decreased visual acuity in older adults, or other cognitive functions).

Exact Number System. The exact number system contains symbolic expressions of underlying quantities. Interestingly, when subjects are asked to compare pairs of numbers like 3 and 4 or like 2 and 7, reaction times are longer when adjacent numerical representations (e.g. 3 and 4) are compared than when distant numerical representations (e.g. 2 and 7) are compared, an effect known as the numerical distance effect (Moyer and Landauer 1967). The presence of numerical distance effects suggests that underlying magnitudes are evoked by symbolic numbers. The size of the numerical distance effect informs us on the accuracy of the numerical representations. Larger numerical distance effects suggest less accurate quantity representations (But see Cohen Kadosh et al. 2012). Wood et al. (2009) conducted one of those rare studies investigating the magnitude of numerical distance effects in children, young adults, and older adults. Participants had to decide which of two presented symbolic numbers was larger (for example 7 or 8). Young and older adults presented similar numerical distance effects with

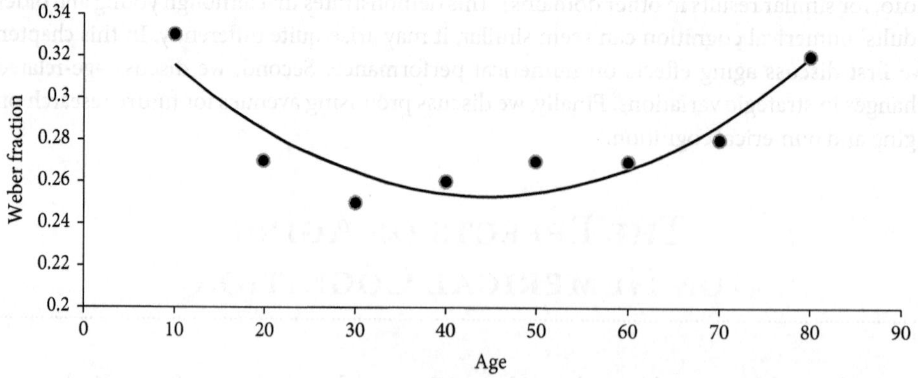

FIGURE 19.1 Age-related changes in number sense acuity with age. Smaller Weber fractions reflect better number sense acuity.

Data from Halberda, J., Ly, R., Wilmer, J.B., Naiman, D.Q., and Germine, L., Number sense across the lifespan as revealed by a massive Internet-based sample, *Proceedings of the national academy of sciences of the United States of America*, 109(28), pp. 11116–20, 2012.

error rates and RT, suggesting access to underlying magnitudes in both groups alike, but older adults showed overall higher error rates for small and large distances (see Figure 19.2), suggesting some differences in number representations nonetheless. This is supported by the findings of different brain activation patterns during number processing in older adults compared to younger adults (e.g. more activation in parietal regions in older adults than in younger adults, possibly suggesting more effortful retrieval of magnitude information).

In older adults presenting pathology such as mild cognitive impairment (MCI) patients, Kaufmann et al. (2008), using the same number-comparison task as Wood et al. (2009), also found numerical distance effects, suggesting access to the magnitudes underlying symbolic numbers in MCI patients (see Kaufmann et al. 2002 for the same results in early Alzheimer's disease (AD)). However MCI patients' numerical distance effects were larger than those of healthy older adults; they committed significantly more errors on close-distance trials (see Figure 19.2). When presented with close-distance trials, MCI patients would more often than on large-distance trials respond very fast (<200 ms), leading to high error rates. It is unclear whether changes in the accuracy of MCI patients' numerical representations underlie this tendency to respond too fast on close-distance trials. Perhaps more informative are the fMRI data obtained by Kaufmann et al. (2008). These authors found more brain activation in MCI patients upon processing magnitude, and most interestingly stronger distance effects in parietal regions adjacent to the intraparietal sulcus. According to the authors this may reflect more effortful retrieval of magnitude information in MCI patients.

More than the approximate number system, the exact number system is sensitive to environmental input (Castronovo and Göbel 2012). The 'mental number line' (Dehaene et al. 1993) is the appealing notion that symbolic numbers are presented on a scale with small numbers on the left and large numbers on the right, at least in Western cultures (for a different representation of numbers in other cultures, see Nunez et al. 2012). The SNARC (*S*patial *N*umerical

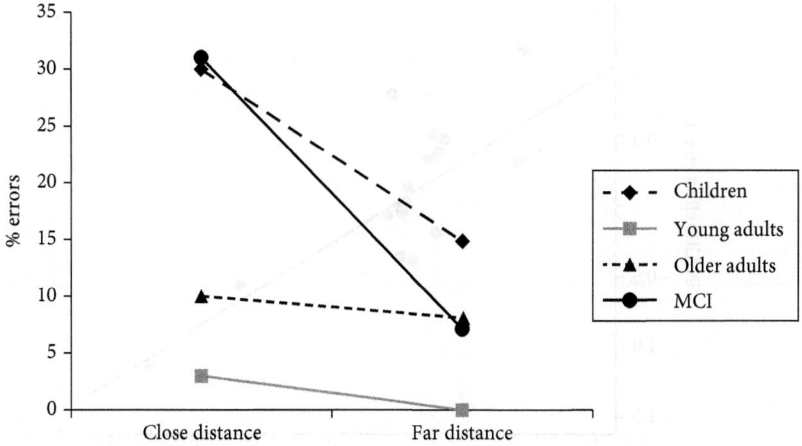

FIGURE 19.2 Numerical distance effects in children, young and older adults, and in MCI patients.

Data from Guilherme Wood, Anja Ischebeck, Florian Koppelstaetter, Thaddaeus Gotwald, Liane Kaufmann, Developmental Trajectories of Magnitude Processing and Interference Control: An fMRI Study, *Cerebral Cortex*, 19 (11), pp. 2755–2765, doi: 10.1093/cercor/bhp056, 2009 and Liane Kaufmann, Anja Ischebeck, Elisabeth Weiss, Florian Koppelstaetter, Christian Siedentopf, Stephan E. Vogel, Thaddaeus Gotwald, Josef Marksteiner, Guilherme Wood, An fMRI study of the numerical Stroop task in individuals with and without minimal cognitive impairment, *Cortex*, 44(9), pp. 1248–1255, DOI: 10.1016/j.cortex.2007.11.009, 2008.

Association of Response Codes) effect is the finding of faster responding to small numbers with the left hand and to large numbers with the right hand (see also van Dijck et al. this volume). This organizational aspect of number representations seems to be spared during aging. Wood et al. (2008) even found linearly increasing SNARC effects with age in a meta-analysis of seventeen studies with ages ranging from 10 to 55 (see Figure 19.3). A later empirical study submitted by Wood et al. (cited in Wood et al. 2008) confirmed these findings in a large sample of individuals aged between 9 and 79 years. SNARC effects were present in every age group, but were significantly larger in older participants than in younger individuals.

Subitizing. Subitizing is the rapid (40–100 ms/item), effortless, and accurate judgment of numbers up to four or five items (Mandler and Shebo 1982). Subitizing supposedly reflects a visuospatial attention mechanism that allows the simultaneous processing of a small number of visual items (Piazza et al. 2011; Trick and Pylyshyn 1993, 1994). Animal-comparative and developmental studies support the notion that subitizing is independent from experience and exists in animals and human infants (Davis and Pérusse 1988). Sliwinski (1997) found that subitizing speed decreased linearly across 20- to 86-year-old adults (See Figure 19.4; see also Geary and Lin 1998; Kotary and Hoyer 1995; Watson et al. 2002, 2005). For numerosities up to four items, these researchers found generally longer reaction times, as well as larger increases in reaction times with each additional item in older adults (see Figure 19.4). Sliwinski (1997) interpreted this as reflecting slowing of the perceptual processes involved in subitizing.

In AD patients, Maylor et al. (2005) (see also Maylor et al. 2008 for similar deficits in AD and vascular dementia) found even larger declines in subitizing. Besides a general increase in reaction times, AD patients seem to subitize items up to two only, and to resort to counting thereafter (see Figure 19.4). Decreased subitizing spans suggest that the capacity

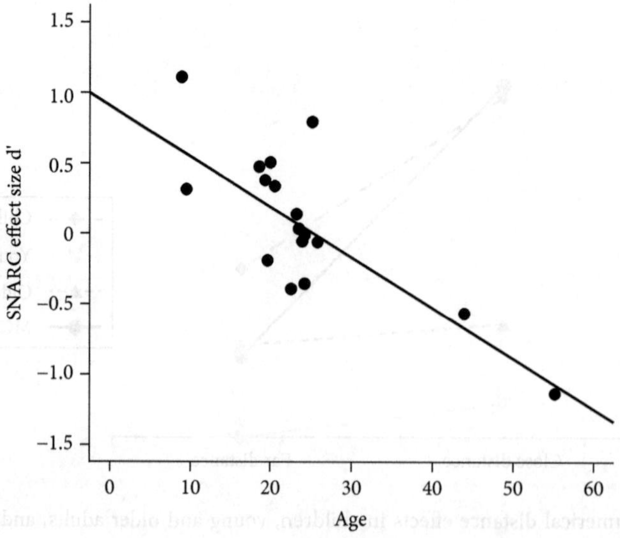

FIGURE 19.3 Age-related changes in SNARC effects. Negative effect sizes represent larger SNARC effects.

Data from Guilherme Wood, Klaus Willmes, Hans-Christoph Nuerk, and Martin H, Fischer, On the cognitive link between space and number: A meta-analysis of the SNARC effects, *Psychology Science Quarterly*, 50(4), pp. 489–525, 2008.

to simultaneously process a limited number of visual items is impaired in AD (Trick and Pylyshyn 1993, 1994; see also Piazza et al. 2011).

Counting. When item sets exceed five items, participants resort to counting, a more consuming process (250–350 ms/item). Counting is largely dependent on enumeration proficiency and the coordination between enumeration and keeping track of to-be-counted items; it is based on experience more than subitizing (Camos et al. 1999). Sliwinski (1997) found that counting speed remains stable across 20- to 86-year-old adults (see Figure 4, see also Geary and Lin 1998; Kotary and Hoyer 1995; Watson et al. 2002, 2005). Older adults were generally slower than young adults, perhaps due to perceptual slowing, but the increase in reaction times with each additional item from four items onwards was similar in all ages. This suggests that the central counting processes themselves were unaffected in healthy older adults.

In AD patients however, counting speed was significantly slowed down (see Figure 19.4; see also Nebes et al. 1992; see also Maylor et al. 2008 for similar deficits in AD and vascular dementia). The reason for this decline in AD patients could be difficulties coordinating the multiple procedures involved in counting. To count, we need to keep track of the to-be-counted items, which may require attentional capacities and we need to combine this process with enumeration, which may require additional executive resources, known to be impaired in AD patients (Baddeley 1986, 1991).

Arithmetic Fact Retrieval. Arithmetic fact retrieval is the retrieval of a solution from the arithmetic facts database when presented with a problem. The learning of these arithmetic facts depends on experience (Geary et al. 1996; Koshmider and Ashcraft 1991; Lemaire

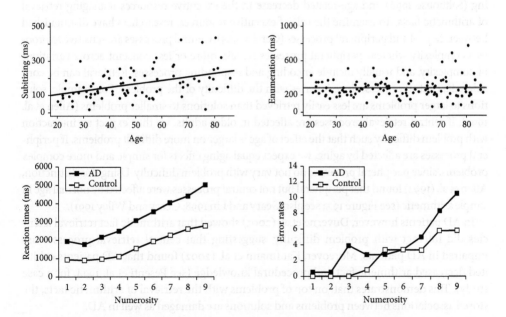

FIGURE 19.4 Top: Age-related changes in counting and subitizing speed per item.

Data from Martin Sliwinski, Aging and counting speed: evidence for process-specific slowing, *Psychology and aging*, 12(1), pp. 38–49 doi: 10.1037/0882-7974.12.1.38, 1997.

Bottom: reaction times for counting up to nine items in healthy older adults and AD patients.

Data from Elizabeth A. Maylor, Derrick G. Watson, and Zoe Muller Effects of Alzheimer's disease on visual enumeration, *The journals of Gerontology, Series B*, 60(3), pp. 129–135, doi: 10.1093/geronb/60.3.P129, 2005.

et al. 1994). This type of rapid retrieval puts minimal demands on executive resources, and is crucial to numerical cognition. To investigate the effect of healthy and pathological aging on arithmetic fact retrieval, researchers usually present individuals with simple arithmetic problems to which the answer is likely to be obtained by retrieval (e.g. 3 × 4) (however, see Fayol and Thevenot 2012, for a recent account on the use of procedures in simple addition problems, e.g. 3 + 4). The speed and accuracy with which answers to such simple problems are retrieved, informs us on the quality of the underlying arithmetic associations. Geary and Wiley (1991) found that older adults were slower than young adults (930 vs. 833 ms) when retrieving arithmetic facts in addition and subtraction production tasks (e.g. 9 + 8 =?). Similarly, Allen et al. (1992) found that older adults were slower than young adults when verifying multiplications (e.g. 4 × 7 = 28, TRUE). However, in spite of being slower, older adults were not less accurate than their younger counterparts when retrieving arithmetic facts. Longer solution latencies in older adults may come as somewhat of a surprise, given that experience is very important in the acquisition of arithmetic facts, and older adults have had more time to acquire this experience than younger adults. Based on these facts, we could expect healthy older adults to be faster in arithmetic fact retrieval. Moreover, Schaie (1996) has demonstrated that arithmetic skills have been better acquired in older adults than in contemporary young adults (see also Geary and Lin 1998; Geary et al., 1996, 1997; Green et al. 2007), constituting another reason for expecting better arithmetic fact retrieval in older adults.

However, even though the content of the arithmetic facts database indeed relies on experience, executive resources are needed for retrieving these facts and maintaining them. The fact that older adults were slower than young adults could be due to general cognitive slowing (Salthouse 1996) and age-related decrease in the executive resources managing retrieval of arithmetic facts. To examine the role of executive resources, researchers have distinguished between central and peripheral processes (Cerella 1985). Central processes are sensitive to problem complexity whereas peripheral processes remain more or less constant across variations of complexity, and usually include encoding and responding processes. Retrieval can be considered a central process, since it varies with the difficulty of the problem. For example, solutions to larger problems are less easily retrieved than solutions to smaller problems (Allen et al. 1992). If central retrieval processes are affected in older adults, we thus expect an interaction with problem difficulty such that the effect of age is larger on more difficult problems. If peripheral processes are affected by aging, we expect equal aging effects for simple and more complex problems, since peripheral processes do not vary with problem difficulty. Using this distinction, Allen et al. (1992) found that peripheral but not central processes were affected by normal age in simple arithmetic (see Figure 19.5; see also Geary and Lin 1998; Geary and Wiley 1991).

In AD patients however, Duverne et al. (2003) showed that arithmetic fact retrieval latencies did interact with problem difficulty, suggesting that central retrieval processes are impaired in AD patients. Moreover, Kaufmann et al. (2002) found that AD patients exhibited decreased arithmetic fact and procedural knowledge (see Pesenti et al. 1994, for a case study). This demonstrates that, on top of problems with the retrieval of arithmetic facts, the stored associations between problems and solutions are damaged as well in AD.

Computation. Computation typically requires multiple retrievals, maintenance of the problem structure, and integration of partial results. Consequently, such activities will depend relatively more on executive resources (Trbovich and LeFevre 2003). Given the declines of executive resources in normal and pathological aging, computation could suffer more from cognitive aging than arithmetic fact retrieval. To investigate the effect of healthy and pathological aging on arithmetic computation, researchers can present individuals with complex

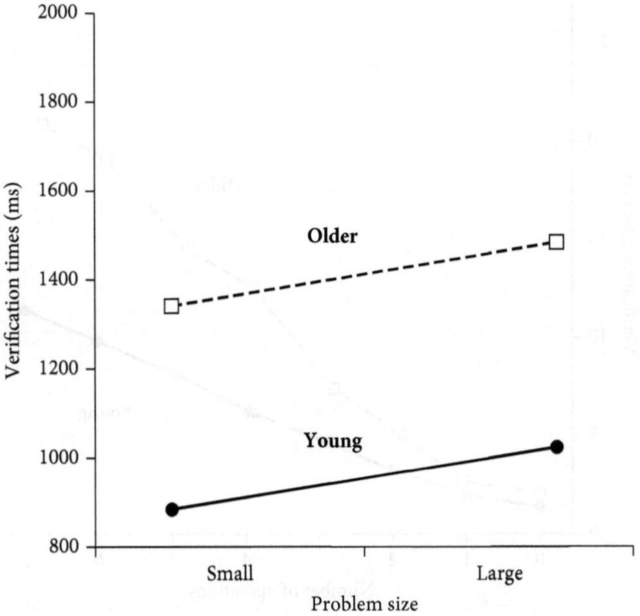

FIGURE 19.5 Age-related changes in arithmetic problem size effects. Data from Allen et al. (1992) showing response latencies to verify small and large single-digit addition problems.

Data from Philip A Allen, Mark H. Ashcraft, and Timothy A. Weber, On mental multiplication and age, *Psychology and aging*, 7(4), pp. 536–545, doi: 10.1037/0882-7974.7.4.536,1992.

arithmetic problems to which the answer can not be obtained via arithmetic fact retrieval. Salthouse and Coon (1994) found that older adults were increasingly slowed relative to young adults when the number of arithmetic operations required to verify an equation was increased in a hierarchical problem format (e.g. 5 + 3 − 1 − (3 + 4) − 1 = 6, FALSE; see Figure 19.6). For example, when three operations were required, young adults would verify a problem in approximately four seconds, whereas older adults would need six seconds. When seven operations were required, this difference would increase from 10 to 21 seconds. Duverne and Lemaire (2005) interpreted this as impairment in the ability to coordinate several procedures in older adults, who have more limited executive capacities than young adults.

AD patients are more impaired in complex arithmetic than healthy older adults (Mantovan et al. 1999). For example, an AD patient tested by McGlinchey-Berroth et al. (1989) showed impairments when asked to apply a procedure to square two-digit numbers. His impairments were in particular due to his difficulties in combining and integrating the multiple steps of the procedure.

In conclusion, age-related changes have been documented in basic components of numerical cognition. First, it seems that the ability to discriminate numerosity (see Halberda et al. 2012) and numerical representations (see Wood et al. 2009) decreases during aging. However, since currently available results leave room for interpretations (e.g. contributions of general, non-specific numerical processes, such as processing speed, inhibition, executive control mechanisms), more evidence supporting this conclusion is needed. Other aspects of numerical representations seem to benefit from aging, such as the association of number

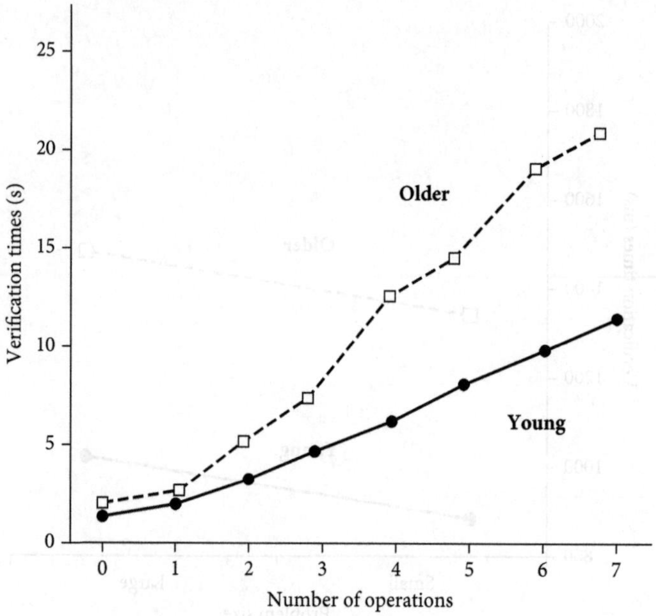

FIGURE 19.6 Effects of arithmetic problem complexity with age. Data from Salthouse and Coon (1994) showing larger increased verification times with problem complexity in older than in young adults.

Data from Timothy A. Salthouse and Vicky E. Coon, Interpretation of differential deficits: The case of aging and mental arithmetic, *Journal of Experimental Psychology*, 20(5), pp. 1172–1182, doi: 10.1037/0278-7393.20.5.1172, 1994.

to space, which had time to strengthen with age (Wood et al. 2008). Here again, future research will discriminate between increased specific spatial-numerical associations and general age-related dedifferentiation of modalities. Second, whereas rapid subitizing, reflecting a combination of perceptual and visuospatial attention core processes, is slowed down in older adults, counting is not impaired. This could be due to counting processes being highly automatized with age/life experience and practice, making them less susceptible to age-related cognitive declines. Finally, older adults' arithmetic fact retrieval seems to be as efficient as that of younger adults, although peripheral processes (e.g. encoding, responding) may be less efficient as we age. When computation is required to solve a problem, however, older adults are less efficient, and more so when problem complexity increases.

Aging and Strategies in Numerical Cognition

An important aspect of numerical cognition is the strategies that we use to accomplish numerical processing tasks. A cognitive strategy can be generally defined as a 'procedure or a set of procedures used to achieve a higher level goal' (Lemaire and Reder 1999, p. 365). In arithmetic, strategies are procedures that permit solution of particular arithmetic problems; they can vary between individuals and between problems. For example, when trying to calculate 38 + 97, one individual may decide to calculate 38 + 100 - 3 whereas another (or even the same individual on another occasion) may calculate 30 + 90 + 8 + 7. Arithmetic performance depends on the selected strategy. Therefore, to understand numerical cognition in older adults as compared to young adults, we need a thorough understanding of which and how strategies are used. Do older adults select strategies as adequately and do they execute these strategies as efficiently as younger adults?

Following Lemaire and Siegler (1995), to investigate strategic variations in aging, Lemaire (2010) proposed a conceptual framework that distinguishes among four strategy dimensions: Strategy repertoire, selection, execution, and distribution. These strategy dimensions are differently affected by aging in many cognitive domains, numerical cognition included. Moreover, older adults can use strategies in ways that compensate for age-related deficits. In this section, we discuss age-related changes in strategy use in numerical cognition.

Strategy repertoire refers to the set of strategies that people use to accomplish a given task. Lemaire and Arnaud (2008) collected verbal protocols of participants while they were solving two-digit additions (e.g. 37 + 58) and observed that, at group level, both young and older adults used the same nine strategies. However, at individual levels, Lemaire and Arnaud (2008) found that young individuals used 5.5 strategies on average whereas older individuals used only 3.2 strategies. This suggests a reduction with age in the number of strategies employed during a cognitive task (see also Hodzik and Lemaire 2011; Geary, French, and Wiley, 1993). Furthermore, Gandini et al. (2008a) found that there were more older than young adults who used a single strategy when assessing the number of dots in a collection (see also Duverne et al. 2008; El Yagoubi et al. 2005; Lemaire and Arnaud 2008; Hodzik and Lemaire 2011). Monostrategic individuals are even more numerous in the AD population. Arnaud et al. (2008) observed that 15% of older adults and 17% of AD patients compared to 5% of young adults used only retrieval strategies on all arithmetic problems. Similarly, in an approximate quantification task, Gandini et al. (2009) found that 7% of young adults compared to 35% of healthy older adults, and an astonishing 49% of AD patients, used a visual estimation strategy on more than 94% of problems.

Thus, although healthy older adults and AD patients use the same set of strategies as young adults, they generally use fewer strategies in numerical cognition. A smaller strategy repertoire may lead to decreased ability to adaptively choose the best strategy on each problem. The issue then concerns the reasons for older adults to use fewer strategies than young adults, despite both groups knowing the same set of strategies. Recent data suggest that age-related decrease in executive functions may lead older adults to intentionally use fewer strategies. This would make sense as maintaining multiple strategies simultaneously active for switching dynamically between strategies as a function of problem characteristics places important demands on working memory and executive control resources (Kray and Lindenberger 2000). In numerical cognition, Ardiale et al. (2012) found increased costs when switching between strategies in older adults as compared to young adults when the number of strategies to switch between increased from two to three strategies. Moreover, Hodzik and Lemaire (2011) assessed young and older participants' strategy repertoire and executive functions while they solved two-digit addition problems. They found that older adults' smaller strategy repertoire was entirely predicted by reductions in executive functions (inhibition and flexibility) in older adults. In other words, age did no longer have a significant effect on strategy repertoire once EFs had been accounted for (Figure 19.7).

Strategy selection is the ability to choose a strategy from multiple available strategies to solve each problem. In healthy older adults, we could expect improvements of adaptive strategy selection, since older adults may have more experience relating strategies to problems and thus may

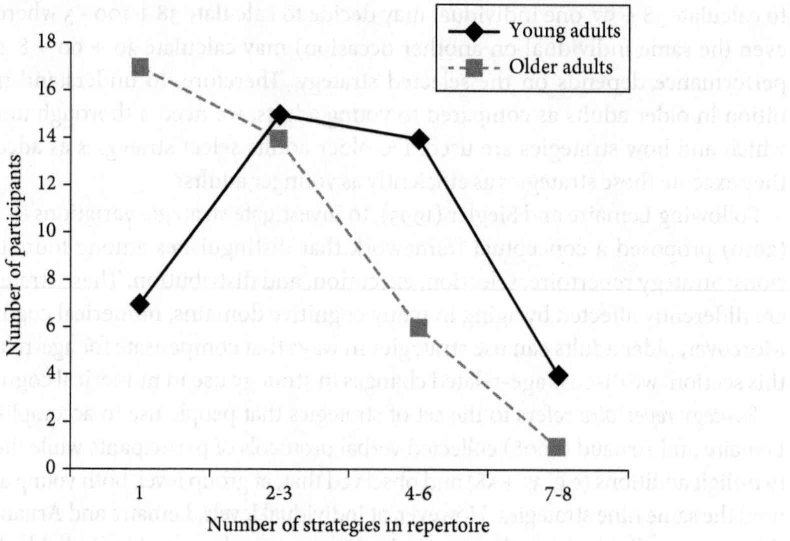

FIGURE 19.7 Age-related changes in strategy repertoire.

Data from Hodzik, S. and Lemaire, P., Inhibition and shifting capacities mediate adults' age-related differences in strategy selection and repertoire, *Acta Psychologica*, 137(3), pp. 335–44, 2011.

have better problem-strategy associations than younger adults. Indeed, formal computational models of strategy selection (Lovett and Andersons (1996) ACT–R; Lovett and Schunns (1999) RCCL; Rieskamp and Otto's (2006) SSL; Siegler and Araya's (2005) SCADS*) assume that the primary mechanism by which a strategy gets chosen on a problem is the association strength between the problem and the strategy, which is shaped by experience. However, some research suggests that the ability to select a strategy requires inhibitory control to suppress competing strategies, and to flexibly alternate between strategies (Lemaire and Lecacheur 2010; Luwel et al. 2009). Given general decrease of executive functions with age (see Jurado and Rosseli 2007, for a review), we could expect declines with age of adaptive strategy selection.

Empirical results show that older adults choose the best strategy less systematically than young adults on each problem (e.g. Duverne and Lemaire 2005; El Yagoubi et al. 2005; Geary and Lin 1998; Lemaire and Arnaud 2008; Siegler and Lemaire 1997; Lemaire et al. 2004). For example, Geary and Lin (1998) found that young adults were able to avoid columnar retrieval (i.e. to solve 29 - 13, participants first subtract the tens and then the units, 2 – 1 = 1 and 9 – 3 = 6, 16) on problems where it is inefficient (i.e. problems requiring a borrow procedure, e.g. 23 – 19) whereas older adults did not. Duverne and Lemaire (2005) found that younger adults more consistently selected plausibility-checking strategies on large-split problems (i.e. arithmetic verification problems with a large difference between the left and right side of an inequality, e.g. 12 + 9 < 30, FALSE) and exhaustive-verification strategies on small-split problems (i.e. arithmetic verification problems with a small difference between the left and right side of an inequality, e.g. 12 + 9 < 21, TRUE) than older adults. Lemaire et al. (2004) showed that in a computational estimation task (i.e. providing estimates to two-digit multiplication problems such as 43 × 78), older adults calibrated their choices to problem characteristics less systematically than young adults. For example, older adults were less able than young adults to choose the best rounding strategy on a multiplication problem like 43 ×

74, when they could choose between rounding down (i.e. rounding both operands down to the nearest decades, e.g. doing 40 × 70 to estimate 43 × 74) and rounding up (i.e. rounding both operands up to the nearest decades, e.g. doing 50 × 80 to estimate 43 × 74). Older adults were less efficient at making these strategy choices, especially on difficult problems such as 34 × 79, for which the best strategy is to round both operands up.

Older adults' decreased selection of the best strategy on each problem has negative performance outcomes. For example, selection of exhaustive-verification strategies to verify an arithmetic equation, where plausibility checking would have been possible, leads to longer overall solution latencies in older adults. As another example, in the computational estimation task of Lemaire et al. (2004), rounding both operands of 41 × 62 up instead of down, leads to overall poorer estimates.

Arnaud et al. (2008) found that AD patients chose retrieval and non-retrieval strategies as systematically during an arithmetic task than did healthy older adults (see also Duverne and Lemaire 2004). These findings suggest that pathological aging did not affect strategy selection mechanisms above and beyond normal aging.

Thus, it seems that older adults' strategy selection suffers from declines in executive resources rather than benefit from accrued experience. Consistent with this, Hodzik and Lemaire (2011) showed that decreased best strategy selection in older adults (see Figure 19.8) was partly mediated by decreased efficiency of executive functions (inhibition and flexibility).

To compensate for, or as a result of, their deficiencies in inhibition and flexibility, older adults could repeat strategies across trials more often, leading to less adaptive strategy selection. Ardiale and Lemaire (2012; see also Lemaire and Lecacheur 2010) found that, during within-item strategy switching (changing strategy on the same item after starting to execute the poorest strategy), older adults tended to continue executing the same strategy more often than young adults, even when it would have been more efficient to switch strategy. Recent work by Lemaire and Leclere (submitted) also found that older adults tended to repeat strategies more often between items than did younger adults. In extreme cases, older adults will consistently select only one strategy, to avoid switching (e.g. Duverne et al. 2008; El Yagoubi et al. 2005; Geary et al. 1993; Geary and Wiley 1991; Lemaire and Arnaud 2008; Gandini et al. 2008a). Indeed, when forced to switch, older adults have larger costs than young adults (Sliwinski et al. (1994; see Figure 19.8 see also Ardiale et al. 2012; Lemaire and Lecacheur 2010).

Strategy execution is the implementation of a strategy. It is characterized by speed and precision. The fact that healthy older adults and AD patients have difficulties managing many different strategies or procedures at the same time (which may cause their reduced strategy repertoire) may have repercussions on strategy speed and precision as well. The difficulties maintaining multiple procedures could be found within numerical strategies, when these require a large degree of coordination of different processes (e.g. procedural strategies). Such strategies may be executed less well in healthy older adults and AD patients.

Gandini et al. (2008a) tested execution of numerosity estimation strategies such as the visual estimation strategy and the anchoring strategy. When using a visual estimation strategy to determine the numerosity of a set of items, we retrieve a numerosity representation from long-term memory. When using an anchoring strategy, we take a more precise approach, decomposing the item sets in smaller groups, subitizing these, and adding the number of similar groups. When comparing young and older adults with these strategies, Gandini et al.

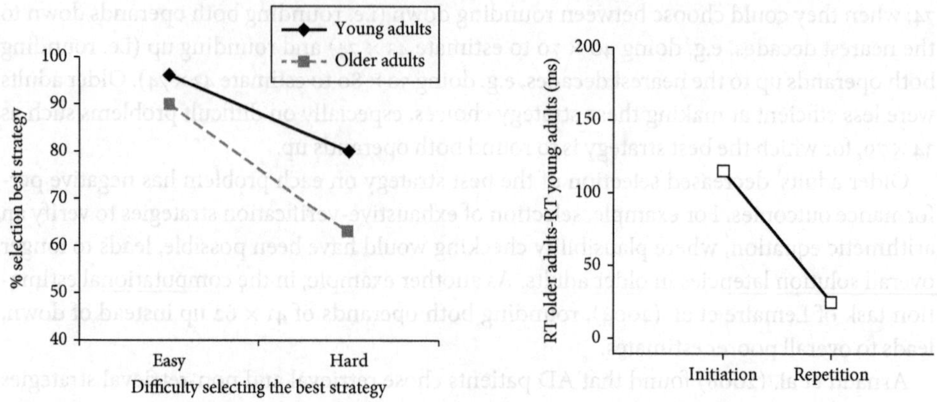

FIGURE 19.8 Strategy selection. *Left:* Showing young and older adults' strategy selection adaptivity in a computational estimation task. *Right:* Showing the difference in response latencies between old and young adults when they had to initiate a new arithmetic operation and when they had to repeat the previous arithmetic operation.

Left: Data from Hodzik, S. and Lemaire, P., Inhibition and shifting capacities mediate adults' age-related differences in strategy selection and repertoire, *Acta Psychologica*, 137(3), pp. 335–44, 2011.

Right: Data from Martin Sliwinski, Herman Buschke, Gail Kuslansky, Graeme Senior, and David Scarisbrick, Proportional slowing and addition speed in old and young adults, *Psychology and aging*, 9(1), pp. 72–80, 1994.

(2008a) found that both populations were equally efficient with the visual estimation strategy. However, when using the anchoring strategy, older adults were slower than young adults (see Figure 19.9; see also Gandini et al. 2009). In arithmetic, healthy older adults have also been found to be slower and more error-prone in the application of strategies, especially of complex arithmetic strategies involving computation (Geary and Wiley 1991; Lemaire and Lecacheur 2001; Lemaire and Arnaud 2008; Salthouse and Coon 1994; Duverne and Lemaire 2005; Hodzik and Lemaire 2011; Siegler and Lemaire 1997). For example, Lemaire et al. (2004) found that older adults made more errors and were slower with computational estimation strategies, especially when using the harder rounding-up strategies or when solving difficult problems.

In AD patients, strategy execution is further impaired (e.g. Grafman et al. 1989; McGlinchey-Berroth et al. 1989; Arnaud et al. 2008; Kaufmann et al. 2002; Duverne et al. 2003), especially for more procedural strategies (Mantovan et al. 1999). For example, in numerosity estimation, Gandini et al. (2009) found that specific impairment of the anchoring strategy was stronger in AD patients than in healthy older adults. AD patients were 5.5 seconds slower when using anchoring than when using visual estimation, compared to 4.2 seconds in healthy older adults. In another example, Arnaud et al. (2008) showed that AD patients were slower than healthy older adults when using a direct retrieval strategy on simple subtraction problems (e.g. 12 – 8 = ?), but were relatively more impaired when using a simple yet procedural strategy such as counting.

Duverne et al. (2008) found that age-related impairments in the execution of strategies on difficult problems further increased when working-memory executive components were taxed by a secondary task. This suggests that executive resources play an important role in the age-related deficits in strategy execution.

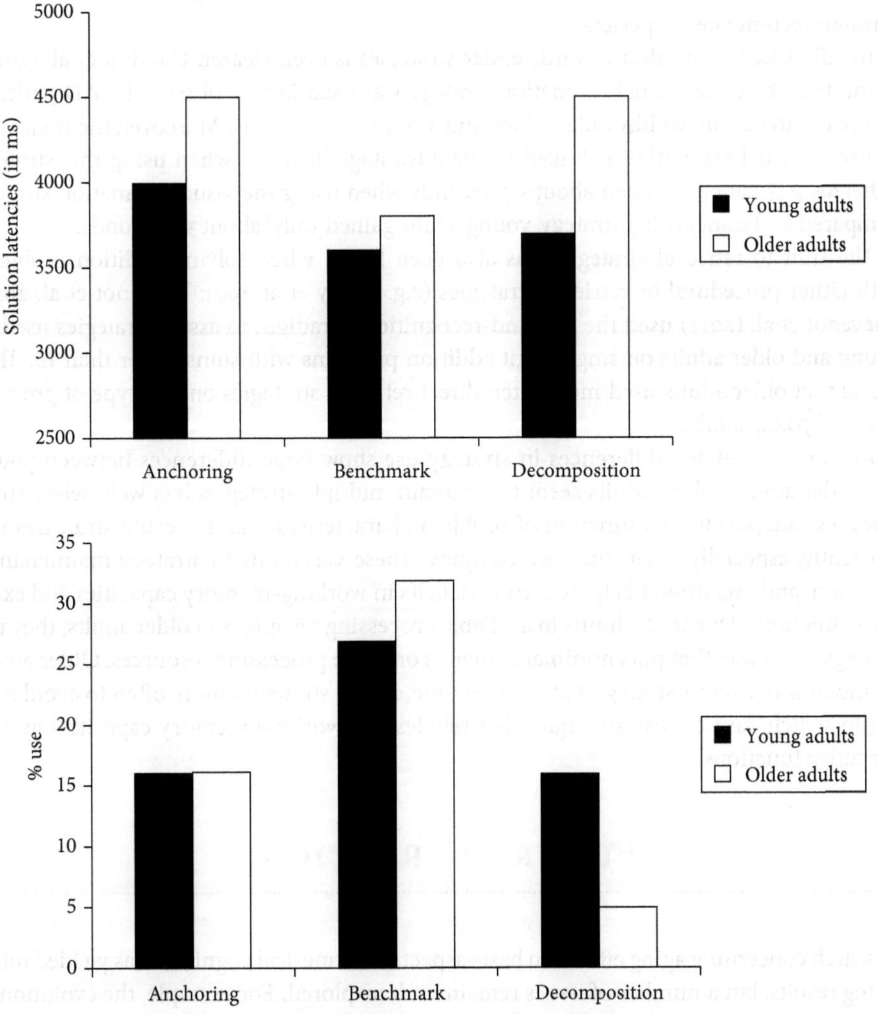

FIGURE 19.9 Top: strategy execution (solution latencies with three numerosity estimation strategies). Bottom: strategy distribution (percent use of each of the three numerosity estimation strategies). We see that older adults use most often the strategy for which their execution is least impaired compared to younger adults (i.e. benchmark or visual estimation).

Data from Gandini, D., Lemaire, P., and Dufau, S., Older and younger adults' strategies in approximate quantification, *Acta Psychologica*, 129(1), pp. 175–89, 2008.

Strategy distribution is the proportions which available strategies are used by an individual. With fewer resources, older adults tend to have a bias for easier strategies, even if those strategies do not yield the best performance. For example, in a numerosity estimation task, older adults more often used the visual estimation strategy than younger adults (68% vs. 46%) (Gandini et al. 2008b; see Figure 19.9). Visual estimation is the numerosity estimation strategy older adults execute most efficiently (Gandini et al. 2008b) because it relies in large part on retrieval from long-term memory, which is rich and well-organized in older adults through accumulated experience.

In AD patients, the shift towards easier strategies is even clearer. Gandini et al. (2009) found that the easier visual estimation strategy was used in 75% of trials by AD patients, compared to 62% in healthy older adults and 59% in young adults. Moreover, the frequency of use of visual estimation is linked to the advantage in speed when using this strategy. Whereas AD patients gained about 5.5 seconds when using the visual estimation strategy compared to the anchoring strategy, young adults gained 'only' about 3.9 seconds.

The shift to retrieval strategies has also been found when solving addition problems with either procedural or retrieval strategies (e.g. Geary et al. 1993; Thevenot et al. 2013). Thevenot et al. (2012) used the operand-recognition paradigm to assess strategies used by young and older adults on single-digit addition problems with sums larger than 10. They found that older adults used more often direct retrieval strategies on this type of problem than did young adults.

In sum, age-related differences in strategy use show large differences between young and older adults. Older adults seem to maintain multiple strategies less well, select strategies less adaptively as a function of problem characteristics, and execute strategies less efficiently, especially when they are complex. These variations in strategy maintenance, selection, and execution likely stem from deficits in working-memory capacities and executive functions. Due to the limits in available processing resources in older adults, they use strategies in a way that puts minimal demands on these processing resources. Older adults maintain active fewer strategies at any one time, repeat strategies more often to avoid having to switch, and choose strategies that rely less on working memory capacities and/or executive functions.

Future Directions

Research concerning aging effects on basic aspects of numerical cognition has yielded interesting results, but a number of issues remain underexplored. For example, the evolution of exact and approximate number sense acuity deserve more attention. The recent study of Halberda and colleagues (2012) suggests that approximate number acuity decreases with age. However, alternative interpretations of their data are possible (e.g. decreased visual acuity), so that more research is needed to validate this result. Moreover, it would be interesting to know whether the potential decreased efficiency of approximate number sense during aging has repercussions for the exact number system.

Concerning the evolution of quantification and arithmetic problem solving, the strategy approach has proven to be fruitful. Some research indicates a role for strategies in more basic components of numerical cognition as well. Siegler and Booth (2005) have shown the importance of strategies in the exact number system in children. They linked more accurate linear representations of quantities (i.e. each quantity is separated by the same distance on a mental number line) to the use of proportioning strategies for placing numbers on a line, whereas they associated less accurate logarithmic representations of quantities (i.e. the distance between numbers grows increasingly smaller as the numbers get larger) to counting strategies for placing numbers on a line. For example, to place 50 on a line with endpoints 0 and 100, a proportioning strategy would consist of placing 50 roughly in the middle of the line, whereas a counting strategy would consist of counting from 0 onwards and marking numbers until one reaches 50. Proportioning strategies are associated to much more accurate number line performance than counting strategies.

Camos (2003) has also shown strategic variations that account for counting performance in children. Children can count items one by one, by groups, or by multiplication. The use of strategies for quantity representations and counting in children varies with their age, with older children using more sophisticated strategies. The use of more sophisticated strategies depends on learning but also on the availability of necessary processing resources. For example, counting items by groups puts higher demands on working memory capacities than counting items one by one. Future research should investigate whether, in older adults, the decline of processing resources could change strategy distributions compared to young adults in basic components of numerical cognition.

A strategy approach human cognition aims at understanding how people accomplish cognitive tasks, or the series of mental processes they use to accomplish whichever tasks we ask them to do. This approach is helpful in determining how age-related differences in performance are best explained by age-related differences in cognitive mechanisms. Future studies may further pursue this overarching goal, in general as well as to understand aging effects in numerical cognition. Several achievements will concur to this goal. For example, future research will uncover specific mechanisms involved in strategy selection and how these are affected by aging. Several models have attempted to formalize the mechanisms by which people can choose a strategy: Lovett and Andersons (1996) ACT–R, Lovett and Schunns (1999) RCCL, Rieskamp and Otto's (2006) SSL model, and Siegler and Araya's (2005) SCADS*. These models have in common that strategies are associated to problems as a function of the success of the strategies for a particular problem or type of problems. The strategy with the strongest association to the problem has the largest probability of being chosen. Models for strategy choices provide a fruitful avenue for investigating the effects of age on strategy use in numerical cognition. For example, do the associations between numerical problems and strategies become more fine-tuned and stronger with age?

Another issue that has received little research is sequential effects on strategy use. In computational estimation, Uittenhove and Lemaire (2012) have shown that the execution of a strategy is harder when it follows execution of a difficult strategy (see also Uittenhove et al. 2013). They suggest that this is due to temporarily reduced working-memory resources following difficult strategies. In their view, working-memory capacities dynamically fluctuate as a function of current but also of immediately prior demands, on a trial-to-trial basis. Populations with less efficient working memory, such as older adults and Alzheimer patients, are expected to suffer more from such sequential difficulty effects. Given the potential age sensitivity of these effects, it is an important factor to take into account when examining age-related differences in numerical cognition. Moreover, sequential difficulty effects could apply not only to arithmetic strategies, but to sequential numerical problem solving in general.

A final issue in need of more research is age-related compensation in numerical cognition. Older adults will try to match young adults' performance (e.g. Geary and Lin 1998, Geary and Wiley 1991) by using strategies differently than younger adults (i.e. functional compensations). However, a successful older adult can also compensate by recruiting alternative neural circuitry to perform the same strategies as do younger adults (e.g. Scaffolding Theory of Aging and Cognition, Park and Reuter-Lorenz 2009). The additional neural circuitry recruited by older adults usually comprises extra prefrontal and contra-lateral activations.

For example, El Yagoubi et al. (2005) found that successful arithmetic performance in solving addition problems in older adults was accompanied by older adults' recruiting contra-lateral brain regions to match young adults' performance. Similarly, in a functional MRI study, Gandini et al. (2008a) found that older adults' strategies in numerosity estimation were supported by different cortical networks than in young adults. Wood et al. (2009) found enhanced parietal activations in older adults in response to numerical distance in a magnitude comparison task, which may have reflected more effortful retrieval of magnitude information.

When task load increases, young adults will show a similar shift in brain activation, to meet the rising task demands. However, older adults will already have depleted their possibilities at lower loads, so they cannot match young adults in meeting the increased task demands. This leads to performance declines at higher task loads in older adults (Reuter-Lorenz and Park 2010). Accordingly, older adults obtain relatively poorer performance than young adults when they are asked to execute more complex strategies compared to simpler strategies. For example, Gandini et al. (2008b) found that older adults took longer than young adults when executing the decomposition/recomposition strategy in numerosity estimation tasks, but were equally fast with using approximate counting. Future studies should investigate whether spared performance during easier strategies such as approximate counting is accompanied by neural compensations in older adults.

Moreover, patterns of neural compensations need to be better understood. In some older adults neural compensations permit matching of younger adults' performance whereas in other older adults it does not. This could depend on inter-individual differences in the intactness of the additionally recruited neural circuitry. All in all, understanding age-related differences in brain activations with a strategy perspective may be fruitful to further our understanding of age-related changes during adulthood in numerical cognition (and in other cognitive domains as well).

Summary

Healthy older adults seem to show good performance with more basic components of numerical cognition. Although a recent study suggests they have poorer number sense acuity (Halberda et al. 2012), and some age differences exist when processing symbolic numbers (Wood et al. 2008), they have good counting and arithmetic fact retrieval skills (Sliwinski 1997; Geary and Wiley 1991). When asked to solve difficult problems that require a high degree of coordination and integration (Salthouse and Coon 1994), older adults are more impaired.

Such deficits seem to have repercussions on the type of strategies older adults select (Gandini et al. 2008b), how many different strategies they use (Hodzik and Lemaire 2011), how adaptively they choose between strategies (Hodzik and Lemaire 2011), and how efficiently they execute strategies (Gandini et al. 2008b). Older adults choose between fewer strategies, repeat strategies more often, and use strategies that they execute most efficiently even if these are not the best strategies. Also, older adults sometime activate more brain regions than young adults to boost their own performance (El Yagoubi et al. 2005; Gandini et al. 2008a). However, when tasks become too complex, compensations are no longer possible or less efficient, and older adults show impaired performance. When aging effects are exacerbated through pathology such as AD and MCI, the damage to the neural substrates of numerical cognition becomes too important, and deficits in numerical cognition start to become obvious at lower levels of complexity. Older adults suffering from AD and MCI may have a less accurate exact number system (Kaufmann et al. 2008), poorer abilities to retrieve arithmetic facts (Duverne et al. 2003), poorer subitizing capacities and counting skills (Maylor et al. 2005).

Authors' Note

This research was supported in part by the CNRS (French NSF) and by a grant from the Agence Nationale de la Recherche (Grant # BLAN-1912-01). Correspondence about this paper should be directed to Patrick Lemaire, Aix-Marseille Université and CNRS, 3 Place Victor Hugo, Case D, 13331 Marseille, France (email: patrick.lemaire@univ-amu.fr).

References

Allen, P.A., Ashcraft, M.H., and Weber, T.A. (1992). On mental multiplication and age. *Psychology and Aging* 7(4): 536–45.

Ardiale, E. and Lemaire, P. (2012). Within-item strategy switching: An age-comparative study in adults. *Psychology and Aging* 27: 1138–51.

Ardiale, E., Hodzik, S., and Lemaire, P. (2012). Aging and Strategy Switch Costs: A study in Arithmetic Problem Solving. *L'Année Psychologique* 112: 345–60.

Arnaud, L., Lemaire, P., Allen, P., and Michel, B.F. (2008). Strategic aspects of young, healthy older adults', and Alzheimer patients' arithmetic performance. *Cortex* 44(2): 119–30.

Baddeley, A., Logie, R., Bressi, S., Della Sala, S., and Spinnler, H. (1986). Dementia and working memory. *Quarterly Journal of Experimental Psychology* 38(4): 603–18.

Baddeley, A., Bressi, S., Della Sala, S., Logie, R., and Spinnler, H. (1991). The decline of working memory in Alzheimer's disease. A longitudinal study. *Brain* 114(6): 2521–42.

Camos, V. (2003). Counting strategies from 5 years to adulthood: Adaptation to structural features. *European Journal of Psychology of Education* 18(3): 251–65.

Camos, V. Fayol, M., and Barrouillet, P. (1999). L'activité de dénombrement chez l' enfant: Double tâche ou procédure? *L'Année Psychologique* 99: 623–45.

Castronovo, J. and Göbel, S.M. (2012). Impact of high mathematics education on the number sense. *Plos One* 7(4): Epub.

Cerella, J. (1985). Information processing rates in the elderly. *Psychological Bulletin* 98(1): 67–83.

Cohen Kadosh, R., Bien, N., Sack, A.T. (2012). Automatic and intentional number processing both rely on intact right parietal cortex: a combined FMRI and neuronavigated TMS study. *Frontiers in human neuroscience* 6(2).

Davis, H. and Perusse, R. (1988). Numerical competence in animals: definitional issues, current evidence, and a new research agenda. *Behavioral and Brain Sciences* 11: 561–615.

Dehaene, S. Dehaene-Lambertz, G., and Cohen, L. (1998). Abstract representations of numbers in the animal and human brain. *Trends in Neuroscience* 21(8): 355–61.

Dehaene, S., Bossini, S., and Giraux, P. (1993). The mental representation of Parity and Number Magnitude. *Journal of Experimental Psychology* 122(3): 371–96.

Duverne, S. and Lemaire, P. (2004). Age-related differences in arithmetic problem-verification strategies. *Journals of Gerontology: Psychological Sciences* 59: 135–42.
Duverne, S., and Lemaire, S. (2005). Arithmetic split effects reflect strategy selection: An adult age comparative study in addition comparison and verification tasks. *Canadian Journal of Experimental Psychology* 59(4): 262–78.
Duverne, S., Lemaire, P., and Michel, B.F. (2003). Alzheimer's disease disrupts arithmetic fact retrieval processes but not arithmetic strategy selection. *Brain and Cognition* 52(3): 302–18.
Duverne, S., Lemaire, P., and Vandierendonck, A. (2008). Do working memory executive components mediate the effects of age on strategy selection or on strategy execution? Insights from arithmetic problem solving. *Psychological Research* 72(1): 27–38.
Eger, E., Sterzer, P., Russ, M.O., Giraud, A.L., and Kleinschmidt, A. (2003). A supramodal number representation in human intraparietal cortex. *Neuron* 37: 719–25.
El Yagoubi, R., Lemaire, P., and Besson, M. (2005). Effects of aging on arithmetic problem solving: An event-related brain potential study. *Journal of Cognitive Neuroscience* 17(1): 37–50.
Fayol, M., & Thevenot, C. (2012). The use of procedural knowledge in simple addition and subtraction problems. *Cognition*, 123, 392–403.
Gandini, D., Lemaire, P., Anton, J.L., and Nazarian, B. (2008a). Neural correlates of approximate quantification strategies in young and older adults: An fMRI study. *Brain Research*, 1246, 144–57.
Gandini, D., Lemaire, P., and Dufau, S. (2008b). Older and younger adults' strategies in approximate quantification. *Acta Psychologica* 129(1): 175–89.
Gandini, D., Lemaire, P., and Michel, B.F. (2009). Approximate quantification in young, healthy older adults', and Alzheimer patients. *Brain and Cognition* 70(1): 53–61.
Geary, D.C. and Wiley, J.G. (1991). Cognitive addition: Strategy choice and speed-of-processing differences in young and elderly adults. *Psychology and Aging*, 6(3), 474–483.
Geary, D.C. and Lin, J. (1998). Numerical cognition: age-related differences in the speed of executing biologically primary and biologically secondary processes. *Experimental Aging Research* 24(2): 101–37.
Geary, D. C., Frensch, P. A., and Wiley, J. G. (1993). Simple and complex mental ubtraction: Strategy choice and speed-of-processing differences in younger and older adults. *Psychology and Aging* 8: 242–56.
Geary, D.C., Bow-Thomas, C.C., Liu, F., and Siegler, R.S. (1996). Development of arithmetical competencies in Chinese and American children: Influence of age, language, and schooling. *Child Development* 67(5): 2022–44.
Geary, D. C., Hamson, C. O., Chen, G. P., Liu, F., Hoard, M. K., and Salthouse, T. A. (1997). Computational and reasoning abilities in arithmetic: Cross-generational change in China and the United States. *Psychonomic Bulletin & Review* 4: 425–30.
Grafman, J., Kampen, D., Rosenberg, J., Salazar, A.M., and Boller, F. (1989). The progressive breakdown of number processing and calculation ability: A case study. *Cortex* 25(1): 121–33.
Green, H., Lemaire, P., and Dufau S. (2007). Eye movement correlates of younger and older adults' strategies for complex addition. *Acta Psychologica* 125: 257–78.
Halberda, J., Ly, R., Wilmer, J.B., Naiman, D.Q., and Germine, L. (2012). Number sense across the lifespan as revealed by a massive Internet-based sample. *Proceedings of the national academy of sciences of the United States of America* 109(28): 11116–20.
Hodzik, S. and Lemaire, P. (2011). Inhibition and shifting capacities mediate adults' age-related differences in strategy selection and repertoire. *Acta Psychologica* 137(3): 335–44.
Jurado, M. B. and Rosselli, M. (2007). The elusive nature of executive functions: A review of our current understanding. *Neuropsychological Review* 17: 213–33.
Kaufmann, L., Montanes, P., Jacquier, M., Matallana, D., Eibl, G., and Delazer, M. (2002). About the relationship between basic numerical processing and arithmetics in early Alzheimer's disease: A follow-up study. *Brain and Cognition* 48(2–3): 398–405.
Kaufmann, L., Ischebeck, A., Weiss, E., Koppelstaetter, F., Siedentopf, C., Vogel, S.E. et al. (2008). An fMRI study of the numerical Stroop task in individuals with and without minimal cognitive impairment. *Cortex* 44(9): 1248–55.

Koshmider, J. W. and Ashcraft, M. H. (1991). The development of children's mental multiplication skills. *Journal of Experimental Child Psychology* 51: 53–89.
Kotary, L. and Hoyer, W.J. (1995). Age and the ability to inhibit distractor information in visual selective attention. *Experimental Aging Research* 21(2): 159–171.
Kray, J. and Lindenberger, U. (2000). Adult age differences in task switching. *Psychology and Aging* 15(1): 126–47.
Lemaire, P. (2010). Cognitive strategy variations during aging. *Current Directions in Psychological Science* 19(6): 363–9.
Lemaire, P. and Arnaud, L. (2008). Young and older adults' strategies in complex arithmetic. *The American Journal of Psychology* 121(1): 1–16.
Lemaire, P., & Lecacheur, M. (2001). Older and younger adults' strategy use and execution in currency conversion tasks: insights from French franc to euro and euro to French franc conversions. *Journal of Experimental Psychology*, 7(3), 195–206.
Lemaire, P. and Lecacheur, M. (2010). Strategy switch costs in arithmetic problem solving. *Memory & Cognition* 38(3): 322–32.
Lemaire P. and Leclère M. Strategy repetition in young and older adults (revision submitted).
Lemaire, P. and Reder, L. (1999). What affects strategy selection in arithmetic? An example of parity and five effects on product verification. *Memory & Cognition* 27(2): 364–382.
Lemaire, P. and Siegler, R.S. (1995). Four aspects of strategic change: contributions to children's learning of multiplication. *Journal of Experimental Psychology* 124(1): 83–97.
Lemaire, P., Barrett, S.E., Fayol, M. and Abdi, H. (1994). Automatic activation of addition and multiplication facts in elementary school children. *Journal of Experimental Child Psychology* 57: 224–58.
Lemaire, P., Arnaud, L., and Lecacheur, M. (2004). Adults' age-related differences in adaptivity of strategy choices: Evidence from computational estimation. *Psychology and Aging* 10(3): 467–81.
Lovett, M.C. and Anderson, J.R. (1996). History of success and current context in problem solving: Combined influences on operator selection. *Cognitive Psychology* 31: 168–217.
Lovett, M. C. and Schunn, C. D. (1999). Task representations, strategy variability, and base-rate neglect. *Journal of Experimental Psychology: General* 128(2): 107–30.
Luwel, K., Schillemans, V., Onghena, P., and Verschaffel, L. (2009). Does switching between strategies within the same task involve a cost? *British Journal of Psychology* 100(4): 753–71.
McGlinchey-Berroth, R., Milberg, W.P., and Charness, N. (1989). Learning of a complex arithmetic skill in dementia: Further evidence for a dissociation between compilation and production. *Cortex* 25(4): 697–705.
Mandler, G. and Shebo, B.J. (1982). Subitizing: An analysis of its component processes. *Journal of Experimental Psychology* 111: 1–22.
Mantovan, M.C., Delazer, M., Ermani, M., and Denes, G. (1999). The breakdown of calculation procedures in Alzheimer's disease. *Cortex* 35(1): 21–38.
Maylor, E.A., Watson, D.G., and Muller, Z. (2005). Effects of Alzheimer's disease on visual enumeration. *The journals of Gerontology* 60(3): 129–35.
Maylor, E.A., Sheehan, B., Watson, D.G., and Henderson, E.L. (2008). Enumeration in Alzheimer's disease and other late life psychiatric syndromes. *Neuropsychologia* 46(11): 2696–708.
Moyer, R.S. and Landauer, T.K. (1967). Time required for judgements of numerical inequality. *Nature* 215(5109): 1519–20.
Nebes, R.D., Brady, C.B., and Reynolds, C.F. (1992). Cognitive slowing in Alzheimer's disease and geriatric depression. *The Journals of Gerontology* 47(5): 331–6.
Nunez, R., Cooperrider, K., and Wassmann, J. (2012). Number concepts without number lines in an indigenous group of Papua New Guinea. *Plos One* 7(4): Epub.
Park, D.C., and Reuter-Lorenz, P. (2009). The Adaptive Brain: Aging and Neurocognitive Scaffolding. *Annual review of psychology*, 60, 173–196.
Pesenti, M., Seron, X., and Van der Linden, M. (1994). Selective impairment as evidence for mental organisation of arithmetical facts: BB, a case of preserved subtraction? *Cortex* 30(4): 661–71.

Piazza, M., Fumarola, A., Chinello, A., and Melcher, D. (2011). Subitizing reflects visuo-spatial object individuation capacity. *Cognition* 121(1): 147–53.

Reuter-Lorenz, P.A. and Park, D.C. (2010). Human neuroscience and the aging mind: A new look at old problems. *The journals of Gerontology: Psychological sciences* 65(4): 405–15.

Rieskamp, J. and Otto, P.E. (2006). SSL: A theory of how people learn to select strategies. *Journal of Experimental Psychology: General* 135(2): 207–36.

Salthouse, T.A. (1996). The processing-speed theory of adult age differences in cognition. *Psychological Review* 103(3): 403–28.

Salthouse, T.A. and Coon, V.E. (1994). Interpretation of differential deficits: The case of aging and mental arithmetic. *Journal of Experimental Psychology* 20(5): 1172–82.

Saxe, G. B. (1982). Developing forms of arithmetical thought among the Oksapmin of Papua New Guinea. *Developmental Psychology* 18: 583–94.

Schaie, K.W. (1996). Intellectual development in adulthood: The Seattle longitudinal study. New York: Cambridge University Press.

Siegler, R.S. and Arraya, R. (2005). *A computational model of conscious and unconscious strategy discovery*. In: Advances in child development and behaviour, edited by R.V. Kail, pp. 1–42. Oxford: Elsevier.

Siegler, R.S. and Booth, J.L. (2005). *Development of numerical estimation: A review*. In Handbook of Mathematical Cognition, edited by J.I.D. Campbell, pp. 197–212. New York: Psychology Press.

Siegler, R.S. and Lemaire, P. (1997). Older and younger adults' strategy choices in multiplication: testing predictions of ASCM using the choice/no-choice method. *Journal of Experimental Psychology* 126(1): 71–92.

Sliwinski, M. (1997). Aging and counting speed: evidence for process-specific slowing. *Psychology and aging* 12(1): 38–49.

Sliwinski, M., Buschke, H., Kuslansky, G., Senior, G., and Scarisbrick, D. (1994). *Psychology and aging* 9(1): 72–80.

Thevenot, C., Castel, C., Danjon, J., Fanget, M., and Fayol, M. (2013). The Use of the Operand-Recognition Paradigm for the Study of Mental Addition in Older Adults. *The Journals of Gerontology* 68(1): 64–7.

Trbovich, P.L. and Lefevre, J.A. (2003). Phonological and visual working memory in mental addition. *Memory and cognition* 31(5): 738–45.

Trick, L.M. and Pylyshyn, Z.W. (1993). What enumeration studies can show us about spatial attention: evidence for limited capacity preattentive processing. *Journal of Experimental Psychology: Human Perception and Performance* 19(2): 331–51.

Trick, L.M. and Pylyshyn, Z.W. (1994). Why are small and large numbers enumerated differently? A limited-capacity preattentive stage in vision. *Psychological Review* 101(1): 80–102.

Uittenhove, K. and Lemaire, P. (2012). Strategy sequential difficulty effects on strategy execution: A study in arithmetic. *Experimental Psychology* 59(5): 295–301.

Uittenhove, K., Poletti, C., Lemaire, P., and Dufau, S. (2013). The time course of strategy sequential difficulty effects: An ERP study in arithmetic. *Experimental Brain Research*, 227(1), 1–8.

Watson, D.G., Maylor, E.A., and Manson, N.J. (2002). Aging and enumeration: a selective deficit for the subitization of targets among distractors. *Psychology and Aging* 17(3): 496–504.

Watson, D.G., Maylor, E.A., and Bruce, L.A. (2005). Effects of age on searching for and enumerating targets that cannot be detected efficiently. *Quarterly Journal of Experimental Psychology* 58(6): 1119–42.

Wood, G., Willmes, K., Nuerk, H-C., and Fischer, M.H. (2008). On the cognitive link between space and number: A meta-analysis of the SNARC effects. *Psychology Science Quarterly* 50(4): 489–525.

Wood, G., Ischebeck, A., Koppelstaetter, F., Gotwald, T., and Kaufmann, L. (2009). Developmental trajectories of magnitude processing and interference control: an fMRI study. *Cerebral Cortex* 19(11): 2755–65.

Wood, G., Stiglbauer, B., Kaufmann, L., Fischer, M., and Nuerk, H.C. (submitted). Age effect on the spatial numerical associations.

PART IV

CULTURE AND LANGUAGE

CHAPTER 20

CULTURE, LANGUAGE, AND NUMBER

GEOFFREY B. SAXE

Vygotsky and Piaget: Seminal Formulations

Some time ago, Bryant (1997) remarked, 'Piaget and Vygotsky set the scene for much of the work that has been done over the last twenty years or so on children's mathematical understanding.' (p. 142). Today, Piaget and Vygotsky's conceptual and empirical frameworks still define principal contours of contemporary work on cognitive development. In introducing this section on culture, language, and number with chapters by Okamoto, Towse, Núñez and Marghetis, and Sturman, I situate the authors' contributions in relation to Vygotsky's and Piaget's seminal writings and some contemporary strands of empirical and conceptual inquiry. I am particularly attentive to the way the authors extend ideas that were core to Piaget's and Vygotsky's writings.

In their treatments, Piaget and Vygotsky each addressed questions about the origins and development of cognition. Although each used a developmental method, their principal foci differed. Vygotsky's treatment is geared for understanding the situatedness of cognitive development in socio-historical processes; Piaget's contributions focused largely on the individuals' constructions of cognitive structures across varied knowledge domains, including number. Themes developed in Vygotsky's and Piaget's writings are reflected across most of the chapters in this section, and in each chapter, the Piagetian and Vygotskian themes appear interwoven with one another in interesting ways.

Vygotsky's Cultural-Historical Perspective

As a part of his general treatment of cognition, Vygotsky distinguishes between 'natural' and 'cultural' lines in children's cognitive development (Vygotsky 1986). For Vygotsky, the *natural line* begins with native capacities that include involuntary memory and a host of reflexive processes that organized the infants' world, whether visual, tactile, or auditory. By their nature, these cognitive and perceptual processes are not under conscious control; rather they are elicited by environmental stimuli or generated by impulse.

The *cultural line* of development is captured in Vygotsky's treatment of mediation. For human cognition to emerge, young children have to break the link between the stimulus world and immediate responding. Vygotsky argues that children begin to interrupt the stimulus-response link by consciously drawing upon aspects of their environments (social, physical, linguistic) to mediate their interaction with their immediate worlds. In the early mediational act, the child begins to take control and organize their own responses via historically elaborated representational and knowledge systems. An example is children's early use of speech to talk through a problem or to plan.

In Vygotsky's treatment, the child's early use of mediational forms is the beginning of a long and complex developmental process in the 'cultural line' of development. In the case of language, for example, Vygotsky argues that early speech initially serves to support social contact and gradually becomes incorporated as speech used to describe, and then to organize their activity. This 'egocentric speech' used to organize activity undergoes further transformations, becoming deeply interwoven with properties of thought and action, developing into 'inner speech'. Vygotsky contributed little to empirical research on children's developing understandings of number In the chapters to follow, investigators, in their treatments of number, bring forward a central construct in Vygoptsky's work – mediation. Foci include linguistic group's number word systems (Okamoto, Towse) and conceptual metaphors that serve as a basis to make sense of and make inferences about a numerical world (Nunez and Marghetis).

Piaget's Treatment of Number Development

Piaget's theoretical and empirical work on cognition sought to illuminate the origins of and qualitative changes in the organization of knowledge over development, and one central focus was on numerical understandings. Piaget's principal monograph on number documents continuities and discontinuities in the transformation of these fundamental numerical ideas like a numerical unit and one-to-one correspondence, and articulates a dialectical treatment of developmental change.

To illustrate areas of complementarity in Piaget's and Vygotsky's writings, which are unified in several of the chapters to follow, consider an anecdote that Piaget presents in a published lecture (Piaget 1970). Though the anecdote and its interpretation is Piaget's, the anecdote is also interpretable by Vygotsky's framework. In the anecdote, Piaget conveys an account told to him by a mathematician friend.

> When he was a small child, he (the mathematician) was counting pebbles one day; he lined them up in a row, counted them from left to right, and got ten. Then, just for fun, he counted them from right to left to see what number he would get, and was astonished that he got ten again. He put the pebbles in a circle and counted them, and once again there were ten. He went around the circle in the other way and got ten again. And no matter how he put the pebbles down, when he counted them, the number came to ten. He discovered here what is known in mathematics as commutativity, that is, the sum is independent of the order. But how did he discover this? Is this commutativity a property of the pebbles? It is true that the pebbles, as it were, let him arrange them in various ways; he could not have done the same thing with drops of water. So in this sense there was a physical aspect to his knowledge. But the order was not in the pebbles; it was he, the subject, who put the pebbles in a line and then in a circle. Moreover, the sum was not in the pebbles themselves; it was he who united them. The knowledge that this

future mathematician discovered that day was drawn, then, not from the physical properties of the pebbles, but from the actions that he carried out on the pebbles.

(Piaget 1970)

Piaget's use of the narrative illustrates his constructive perspective about the origins of number. For Piaget, fundamental ideas like commutativity are not native, nor are they contained in a linguistic system like number words, for it was the child who generated the idea through his own actions. The narrative clearly conveys Piaget's thesis that number is not 'in the environment' or 'in the head' but emerges in an interaction between properties of the world and the actions of the subject.

At the same time, the anecdote also points to the utility of Vygotsky's treatment of a cultural line of development. After all, the boy is a participant in a cultural world in which a number word system has its roots, a number word system that both enables and constrains his activity. Further, number word systems are not 'immaculately conceived' but rather have developed over particular cultural histories (Saxe and Esmonde 2005; Saxe and Posner 1982). Further, the problems with which developing children engage emerge in a socially organized world, and the kinds of solutions that they generate are hence historically situated (Saxe 2012).

Some Contemporary Strands of Research That Engage Culture, Language, and The Development of Numerical Cognition

In the chapters that follow, the reader will find that the authors extend the themes of mediation (akin to Vygotsky) and developing structures of numerical thought (akin to Piaget). At the same time, the authors also contribute to contemporary strands of research on relations between culture, language, and the development of numerical cognition. By way of introduction to the chapters, I provide a brief overview of these areas of contemporary work.

Research On Language and Number Development

A first strand of contemporary work focuses explicitly on the way the number word registers of linguistic groups may mediate the development of numerical thinking, a broad area of theorizing and empirical research (cf. Bowerman 1996; Everett 2005; Frank et al. 2008; Lancy 1983; Whorf 1956). Some languages, like many of the Papua New Guinea highlands, for example, support body part counting systems (cf. Laycock 1975; Saxe 2012). Other linguistic groups use linguistic registers, but these groups vary in the structure of the lexicon (e.g. whether they use a base or multi-unit structure) as well as the magnitude that can be counted with a system (Dehaene et al. 2008; Everett 2005; Saxe and Posner 1982). A general issue in this literature is whether and/or in what way the number system children acquire influences the character of their numerical thought.

In a series of studies to which a number of chapters that follow make reference, Miura and her associates (Miura 1987; Miura et al. 1988; Miura and Okamoto 2003; Miura et al. 1999) investigated the role of number word systems on children's developing conceptual understanding of number representations. Miura's focus picks up Vygotsky's treatment of relations between language and thought; at the same time, it extends Piaget's treatment of numerical unit to a child's developing conceptual differentiations and coordinations of units and multiunits.

Miura began by noting that languages that are derived from Ancient Chinese (like Japanese, Chinese, and Korean) have a numerical register organized by base-10 that is entirely regular and maps directly to the Hindu-Arabic written base-10 system. Other non-Asian systems, like English also make use of a base-10 register but with many irregularities that make the mapping to the written system less transparent. In a series of studies, Miura asked, does the number word system that children are acquiring influence their developing conceptualization of number? To address the question, Miura created tasks in which children in the early elementary grades from different language backgrounds would be required to show an interviewer different numbers using base-10 blocks, unit blocks (cubes), and 10-blocks (ten times longer than the unit blocks). Miura and her associates' principal finding was that children speaking languages rooted in Ancient Chinese much more frequently use multi-unit representations to show two-digit numbers, whereas the children from comparison language groups tended to produce non-multiunit representations, simply counting unit blocks.

Subsequent work has built upon or questioned Miura's interpretation of her findings about the role of a number word system in conceptual understanding; some of the authors of this work are also are contributors to chapters in this section. Among the issues that are raised include: how robust is the effect that Miura reports? Can task variation or particular treatments reduce the effect easily? If the effect is robust, how general is it? Is the effect limited, for example to performance on Miura's type of task? Does it extend to place value understandings? Does it affect a range of conceptual developments in domain of mathematics? Finally, does the effect of language account in any way for the results of international assessments of mathematics achievement (TIMMS, PISA), which document higher levels of achievement among students from Asian countries than students from non-Asian countries?

Research on The 'Mental Number Line'

A second strand of research focuses on what some have referred to as the 'mental number line'. Dehaene (1997) posited that people naturally represent numbers spatially on a linear dimension, with the spacing of larger numbers compressed. In a subsequent study, Siegler and Opfer (2003) generated evidence for a developmental process, beginning with a logarithmic models and shifting to a linear ('accumulator') model, in which numbers progress as a linear function (Gibbon and Church 1981). In one of Seigler's tasks (2003), participants were presented with a number line marked with 0 at one end and 100 at the other, and asked their participants to estimate the positions of numbers presented to them. When he tested 2nd, 4th, 6th graders, and adults, he found a shifting prominence of how numbers were positioned on the line. For younger populations, numbers were compressed at the higher ends, resulting in a function that appeared logarithmic (like original Dehaene et al. 2008 findings). For the older participants, individuals produced linear functions in spacing numbers.

In some important respects, the literature on the mental number line echoes themes in Vygotsky's and Piaget's developmental approaches. On the one hand, Dehaene argues that the logarithmically scaled number line is a 'natural' or 'intuitive' native cognitive capacity with which we organize the quantitative world, much akin to the way Vygotsky discusses native capacities and a natural line of development. Further, it may be, following Vygotsky in part, that the shift to a line that is organized by a linear function may be understood as a shift in mediational processes. Consider, for example that Dehaene and his associates investigated the use of modified number estimation number line tasks with a remote Amazonian group. The investigators found that adults in the Amazonian group perform like children in the US, their estimates conforming to a logarithmic model. Dehaene concludes from this that the logarithmic function is the intuitive one and perhaps universal (Dehaene et al. 2008). One might imagine that the mediational shift is linked to participation in schooling. From Vygotsky's perspective, one might conjecture that children may be drawing upon artifacts like rulers to mediate their estimates of positions in a one-dimensional space.

The shift from logarithmic to linear functions is also, in some respects, in accord with Piaget's early findings on the development of spatial cognition. Early in development, children's spatial constructions have properties of a topological space, with order and continuity respected, but not measured distances. With age, Piaget produced a great deal of data to support the argument that children construct a Euclidean space, with dimensional frames of reference and metric properties (Piaget and Inhelder 1956).

Assessment

Assessing cognition is a dominant concern in cultural and linguistic approaches to numerical cognition, and much has been written about the topic (a useful reference on this is Ginsburg 1997). For their part, Vygotsky and Piaget each elaborated a variety of assessment methods keyed to their conceptual frameworks.

To explore the shifting properties of mediational activity, Vygotsky (1978, 1986) used a technique that he referred to as the 'method of double stimulation'. With this technique, the child was presented a primary problem (primary stimulus) and artifacts that could be drawn into and mediate the solution in order to restructure the problem to achieve an adequate solution. An example is Vygotsky's study of mediated memory—children of different ages were engaged with a game in which they could not say particular colors and could not say any color twice. The interviewer would proceed to ask questions like, 'What color is the sky?' and 'What color are your shoes?' To support their efforts, children were presented, as auxiliary stimuli, color chips that could be drawn into their efforts to support their play. What Vygotsky found in this study, as in others, was that with age, participants shifted in the way that they used the chips to mediate their memory for what colors they could or couldn't say. Younger children typically ignored or were confused by the chips; older children used the chips productively; and the still older children and adult group did not rely on the physical chips to mediate their solutions.

Piaget also developed a wide range of methodological techniques. One for which he is particularly well known was the clinical interview (Piaget 1979). The clinical interview is used to explore the child's thinking much as an anthropologist might explore the phenomenological worlds of people from a different and remote culture. In one approach to the interview, the child is presented with a cognitive task and the interviewer explores the approach that the child takes to the task, probing their thinking in ways that maintain rapport and at the same time illuminate the way the task is conceptualized and solved by the interviewee. The interviewer uses non-leading questions and probe quests that help to disambiguate different interpretations of the way the child is conceptualizing and solving a task.

The chapters to follow each present perspectives and approaches to assessment keyed to understand issues of mediation and children's numerical thinking and achievement. Some are 'high stakes' assessments used to evaluate national differences in students' achievements in the TIMMS and PISA studies. Others are modification of Miura's tasks and study designs; still others involve new designs and approaches to assessing participants' positional estimates for numbers on a number line. Across studies, there are some important methodological issues. Consider one: Claims about a 'mental number line' stem from assessment tasks in which children are presented with a linear representation and asked to place numbers on it. Are claims about a 'mental number line' warranted from such findings? Do such findings from linear tasks warrant claims about processes native to the central nervous system? Alternatively, is this 'mental number line' an artifact that results from presenting participants with a linear display and asking them to produce numerical estimates? In the latter case, there is no 'mental number line', merely an approach to solving a particularly kind of task requiring linear representations.

Some Notes on Research on Number Development in Cultural Practices

A third strand of research does not enter in any direct way the chapters in this section, but from my perspective is important in interpreting studies on mediation and numerical understanding. The work examines the interplay between children's participation in cultural (or collective) practices and the development of numerical understanding (cf. Saxe 2012). It is in cultural practices that numerical problems emerge and are conceptualized, and mediational forms are acquired or invented. So, a focus on practices would seem an important part of the puzzle in making sense of the role of mediational processes in the development of numerical understandings.

Some cultural practices targeted by researchers include family interaction related to the play of number games with toddlers (Saxe et al. 1987), children running errands to purchase goods for families living in Brazilian shantytowns (Guberman 1996), children purchasing goods after school at small markets/liquor stores in the United States (Taylor 2009), and children selling trinkets to tourists in the streets of Oaxaca (Sitabkhan 2012). In these practices, children acquire and invent mediational forms, and they construct numerical ideas in communication about number and in numerical problem solving.

To illustrate Vygotskian and Piagetian themes in research that address the issue of the situatedness of mediation in practices, I consider some of my own work conducted with the Oksapmin, a remote cultural group in Papua New Guinea (Saxe 2012). The Oksapmin traditionally use a 27-body part counting system for number representation (see Figure 20.1). I was interested in both the functions of the number system in traditional practices, as well

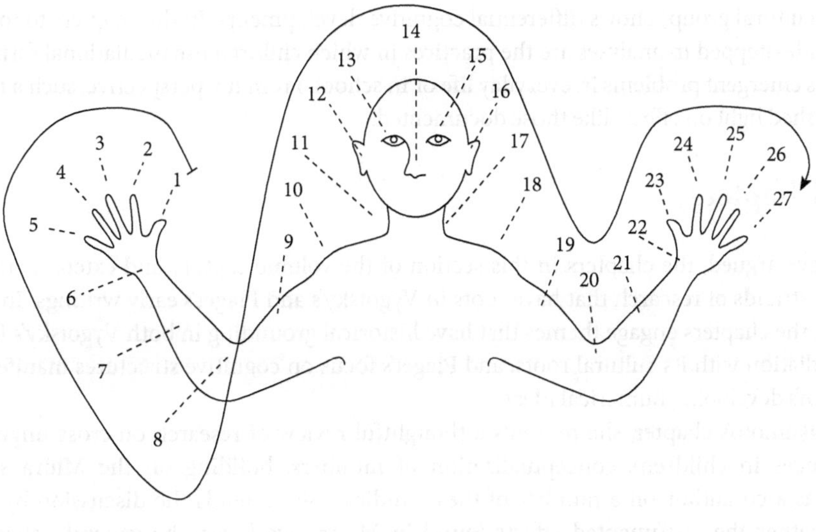

FIGURE 20.1 Oksapmin's 27-body part counting system.

as shifts in its functions over historical time, with the introduction of a money-economy and Western-style schooling.

First, some context: In traditional life in Oksapmin communities, people did not use the system for the purpose of arithmetical computations; in fact, there was no evidence that people engaged in such computations at all. With the introduction of schooling, however, children were engaged with a Western-styled curriculum of which arithmetic was a part. My interest was how children might be representing and making sense of the problems, given that instruction was in English, a language that children were only just learning. I expected that children might be using the 27-body part counting system. Observation and interview studies revealed that children were creating new ways of using the body system to solve arithmetical problems as a function of participating in mathematics classrooms. The new mediational approaches were not taught in school—teachers were foreigners and didn't know the system. Rather, children invented the new forms of mediation. An example of a new mediation function for the body system is evident in a child's solution of 16-7. To solve the problem, the first enumerates body parts to ear (16) and then counts down from the forearm (7) through the thumb (1) as he counts down from the ear (16) to the shoulder (10), leaving biceps (9) as the remainder. A video of this procedure is available at the link: <http://www.culturecognition.com/fourth-grader-solving-16-7-body-system>.

In the Oksapmin children's solutions we can find themes reflected in both Vygotsky's treatment of mediation. From a Vygotskian perspective, one can note the use of the cultural form of the body system to conceptualize and solve numerical problems. At the same time, from a Piagetian perspective, one can note children's construction of new kinds of one-to-one correspondences between body parts as they produce novel solutions to arithmetic problems.

What is both important and intriguing about the practice perspective on mediation is that one should attend to the functions that cultural forms serve for individuals as they structure solutions to recurring problems in daily life. From this perspective, without considering such practices it is problematic to draw conclusions about why one group (whether age or cultural group) shows differential cognitive developments. In the chapters to follow, often side-stepped in analyses are the practices in which children use mediational forms to address emergent problems in everyday life or in school. From my perspective, such a focus would shed light on effects like those documented.

The Chapters

As I have argued, the chapters in this section of the volume capture and extend contemporary strands of research that have roots in Vygotsky's and Piaget's early writings. In particular, the chapters engage themes that have historical grounding in both Vygotsky's focus on mediation with its cultural roots, and Piaget's focus on cognitive structures manifest in children's developing numerical ideas.

In Okamoto's chapter, she presents a thoughtful review of research on cross-linguistic differences in children's conceptualization of numbers, building on the Miura study (she was a co-author on a number of these studies). She extends the discussion by asking whether the documented effects found in Miura's work may be general—affecting mathematical development across many mathematical ideas and domains—or whether the language mediation effects are specific to closely related ideas, limited for example to the multiunit structure of number systems and place value. She reviews some of the contemporary thinking about numerical estimation without counting, and considers these capacities as resources that all children bring to emergent problems of quantification. She ends with a reasoned position that the linguistic effects are specific to a particular array of mathematical ideas.

In Towse's chapter, he takes up two issues related to mediation and numerical understanding. The first is a critical examination of Miura and colleagues original findings related to differences in the way that languages encode number and their implications for children's developing conceptualizations of number. Towse points out that Miura admirably crafted her study as an experiment of nature: She noted that languages differ in their alignment between the structure of their number word registers and the structural features of the written system of Hindu-Arabic numerals. One of Miura's insights was to take advantage of this 'natural experiment' by contrasting different linguistic groups on a clever task that captured students' multiunit interpretations of number words. But Towse also raises cautionary notes. He is concerned about attributing a causal link from language to children's conceptualizations of number. To support his argumentation, he describes some well-crafted experimental studies that probe the causal link, showing that with an experimental paradigm, English-only speaking children can be induced into performing like their peers in East Asia.

In a second contribution to issues of culture and number, Towse considers the ongoing strand of research on the 'mental number line' and ways that children position numbers on the line or label indicated positions with numbers. Among the questions that Towse brings forward are whether and in what way children's representation of numerical distances between numbers can be described with logarithmic or linear functions. In particular, his concern is with how children of the same ages or numerical skill levels differ with respect to their number representations. Again, the designs are clever. Building upon cross-cultural research by Siegler and others, Towse both documents cross-cultural differences in the shift to a linear representation, as well as evidence that is inconsistent with the claim that the shift drives early arithmetical developments.

In the chapter by Nunez and Marghetis, the authors focus on mediational shifts in development but of a different sort than number word systems alone. They argue for a conceptualization similar to Vygotsky's two lines of development, one 'natural' and the other 'cultural'. Among the processes subsumed under a natural or biological line, they include the apprehension of small numbers (through 'subitizing') and varied kinds of numerical discriminations through perceptual mechanisms. But in their mediational focus they point to some fascinating possibilities previously explicated in Nunez's book with Lakoff (Lakoff and Nuñez 2000). Their focus is on forms of mediation involved conceptual metaphor and fictive motion: two important cognitive mechanisms that, in their view, mediate children's construction of a numerical world. The chapter presents some compelling argumentation that extends Vygotsky's cultural line of development in some very interesting and thought provoking ways.

The chapters close with a contribution from Sturman. The chapter has a different focus from the others. The concern is with two well-known international assessments of mathematics achievement, TIMSS and PISA. Sturman presents a thoughtful review of interpretive issues in making sense of cross-national differences in mathematical performance. In doing so, Sturman examines both the cultural and educational factors that can affect mathematical performance, and the methodological and conceptual difficulties in assessing international differences in attainment and in interpreting the results of such assessments. Thus, the section concludes appropriately with an emphasis on the complexity of assessing national differences in mathematics, and the dangers of overly simplistic interpretations.

REFERENCES

Bowerman, M. (1996). Learning how to structure space for language: A crosslinguistic perspective. In P. Bloom, M.A. Peeterson, L. Nadel & M.F. Garrett (Eds.), *Language and space* (pp. 385–436). Cambridge, MA: MIT Press.

Bryant, P. (1997). Piaget, mathematics, and Vygotsky. In: L. Smith, J. Dockrell, and P. Tomlinson (eds.), *Piaget, Vygotsky and Beyond: Future Issues for Developmental Psychology and Education*. (pp. 131–144). New York, NY: Routledge.

Dehaene, S. (1997). *The Number Sense: How the Mind Creates Mathematics*. New York: Oxford.

Dehaene, S., Izard, V., Spelke, E., and Pica, P. (2008). Log or linear? Distinct intuitions of the number scale in Western and Amazonian Indigene cultures. *Science* 320(5880): 1217–1220.

Everett, D.L. (2005). Cultural constraints on grammar and cognition in Pirahã: Another look at the design features of human language. *Current Anthropology* 46(4): 621–646.

Frank, M.C., Everett, D.L., Fedorenko, E., and Gibson, E. (2008). Number as a cognitive technology: Evidence from Pirahã language and cognition. *Cognition* 108(3): 819–824.

Gibbon, J. and Church, R.M. (1981). Time left: Linear versus logarithmic subjective time. *Journal of the Experimental Analysis of Behavior* 7: 87–107.

Ginsburg, H.P. (1997). *Entering the Child's Mind: The Clinical Interview in Psychological Research and Practice*. New York: Cambridge University Press.

Guberman, S.R. (1996). The development of everyday mathematics in Brazilian children with limited formal education. *Child Development* 67(4): 1609–1623.

Lakoff, G. and Nuñez, R.E. (2000). *Where Mathematics Comes From*. New York: Basic Books.

Lancy, D. (1983). *Cross-cultural Studies in Cognition and Mathematics*. New York: Academic Press.

Laycock, D. (1975). Observations on number systems and semantics. *New Guinea Area Languages and Language Study* 1 Pacific Linguistics, series C, no. 38: 219–33.

Miura, I.T. (1987). Mathematics achievement as a function of language. *Journal of Educational Psychology* 79: 79–82.

Miura, I.T., Kim, C.C., Chang, C.-M., and Okamoto, Y. (1988). Effects of language characteristics on children's cognitive representation of number: Cross-national comparisons. *Child Development* 59(6): 1445–1450.

Miura, I.T., Okamoto, Y., Vlahovic-Stetic, V., Kim, C.C., and Han, J.H. (1999). Language supports for children's understanding of numerical fractions: Cross-national comparisons. *Journal of Experimental Child Psychology* 74(4): 356–365.

Miura, I.T. and Okamoto, Y. (2003). *Language Supports for Mathematics Understanding and Performance*. Mahwah, NJ: Lawrence Erlbaum Associates, Publishers.

Piaget, J. (1952). *The Child's Conception of Number*. New York: Norton.

Piaget, J. (1970). *Genetic Epistemology*. New York: Columbia University Press.

Piaget, J. (1979). *The Child's Conception of the World*. Translated by J. Tomlinson and A. Tomlinson. New Jersey: Littlefield, Adams and Co.

Piaget, J. and Inhelder, B. (1956). *The Child's Conception of Space*. London: Routledge & K. Paul.

Saxe, G.B. (2012). *Cultural Development of Mathematical Ideas: Papua New Guinea Studies*. New York: Cambridge University Press.

Saxe, G.B., & Esmonde, I. (2005). Studying cognition in flux: A historical treatment of fu in the shifting structure of Oksapmin mathematics. *Mind, Culture, and Activity Special Issue: Combining longitudinal, cross-historical, and cross-cultural methods to study culture and cognition*, 12(3–4): 171–225.

Saxe, G.B., Guberman, S.R., and Gearhart, M. (1987). Social processes in early number development. *Monographs of the Society for Research in Child Development* 52, Serial No. 162.

Saxe, G.B. and Posner, J. (1982). The development of numerical cognition: Cross-cultural perspectives. In: H.P. Ginsburg (ed.), *The Development of Mathematical Thinking*, (pp. 291–317). New York: Academic Press.

Siegler, R.S. and Opfer, J.E. (2003). The development of numerical estimation: Evidence for multiple representations of numerical quantity. *Psychological Science* 14(3): 237–243.

Sitabkhan, Y. (2012). *Economic Ideas Developed through Young Sellers' Successful Selling Strategies in Oaxaca, Mexico*. University of California, Berkeley, Berkeley, CA.

Taylor, E.V. (2009). The purchasing practice of low-income students: The relationship to mathematical development. *The Journal of the Learning Sciences* 18: 370–415.

Vygotsky, L.S. (1978). *Mind in society: The development of higher psychological processes*. Cambridge, MA: Harvard University Press.

Vygotsky, L.S. (1986). *Thought and Language*. Cambridge, MA: MIT Press.

Whorf, B.L. (1956). *Language, Thought, and Reality*. Cambridge, MA: MIT Press.

CHAPTER 21

COGNITIVE LINGUISTICS AND THE CONCEPT(S) OF NUMBER

RAFAEL NÚÑEZ AND TYLER MARGHETIS

WHAT'S IN A NAME? THAT WHICH WE CALL A NUMBER

THE study of numerical cognition, as the name suggests, investigates the cognitive, psychological, developmental, and neural bases of *numbers*. The term 'number,' however, is highly polysemous. The meaning of 'number' in expressions such as 'seven is a prime *number*,' 'I forgot my passport *number*,' 'not all languages mark *number* grammatically,' or 'an infinitesimal *number* has non-Archimedean properties,' are related but resist capture by a single definition – in fact, some seem to differ radically from others. To make things more confusing, even in scholarly articles the term 'number' is sometimes used to mean 'numeral' (i.e. a sign for a specific number) as in 'the big *number* 2 on the shirt with the giraffe' (McMullen, 2010), and, more importantly, often it is used in lieu of 'numerosity' – the numerical magnitude of stimuli arrays or designated collections – as in the title 'spontaneous number representation in mosquitofish' (Dadda, Piffer, Agrillo, & Bisazza, 2009). In the latter case, there is a risk of unwittingly ascribing 'numerical' properties, such as order or operativity to a far simpler ability to discriminate between stimuli. The mosquitofish is no mathematician. Yet such an ascription leads to the teleological argument that thousands of species, from fish to humans, have full-fledged 'number' representations as a result of biological evolution. Confusion must be avoided. So, what, then, is meant by *number* in numerical cognition?

In mathematics, numbers are abstract entities in their own right, governed by precise properties (e.g. for the real numbers, completeness). They are represented by specific signs – numerals, spoken or written words – and in prototypical cases, can be used to perform calculations. Depending on their properties, specific collections of these numbers are designated as *natural* numbers, *negative* numbers, *whole* numbers, *rational* numbers, *irrational* numbers, *real* numbers, *complex* numbers, *infinitesimal* numbers, *hyper-real* numbers, *surreal* numbers, *transfinite* cardinal numbers, and so on, all of which designate sets of 'numbers.' Some of these are strict subsets of each other; others don't intersect at all. Within this vast universe of mathematical numbers, the field of 'numerical cognition,' has, however, been concerned almost exclusively with *prototypical counting numbers* (PCNs) – 'small' and 'common' natural numbers, often less than 10 – not, as one might think, with *natural numbers* proper. Indeed, no study of numerical cognition has investigated natural numbers such as 745,987,345,316,472,995 ,084,821,410,488,458,390 – even though in mathematics, this number is as 'natural' as the number 4. The study of numerical cognition has, thus, taken 'number' to *primarily* mean PCNs[1], and from there has often, and inappropriately, we believe, advanced claims about the totality of natural numbers, integers, arithmetic, and even mathematical cognition in general. For instance, focusing on small counting numbers has led developmental psychologists to attribute knowledge of the natural numbers when children's knowledge was actually piecemeal, restricted to PCNs and even further to particular small numbers (Davidson, Eng, & Barner, 2012). It just so happens that PCNs are the most frequently manifested, observed, and (cognitively) investigated numbers, and happen to be the smallest natural numbers. However, since 'number' is polysemous in both everyday language and mathematics, this raises the question of how the various mathematical meanings of 'number' relate to each other and, in particular, of what cognitive mechanisms make possible the passage from PCNs to actual arithmetic – and eventually to mathematics. In this chapter, we will address these questions from the perspective of cognitive linguistics, a subdiscipline of linguistics, and cognitive science.

Cognitive linguistics developed in the 1980s with the work of Charles Fillmore (1982), Ron Langacker (1987), Len Talmy (1988), George Lakoff (Lakoff & Johnson, 1980), Gilles Fauconnier (1985), and others. Based on analyses of everyday language, cognitive linguistics has argued that language, rather than emerging from an innate domain-specific language faculty (e.g. Universal Grammar; Chomsky, 2007), is the product of domain-general cognitive mechanisms. Two well-studied cognitive linguistic phenomena are *conceptual metaphor* (Lakoff & Johnson, 1980) and *fictive motion* (Talmy, 1996). Conceptual metaphor is manifested in linguistic expressions where one domain is discussed in terms of another, as in the English expressions, 'Send her my warm helloes,' or 'He is a cold person' – expressions which discuss affection as having thermic properties, 'warm' when there is affection, 'cold' when there is a lack of it. Lakoff and Johnson (1980) argued that these systematic metaphorical expressions were the linguistic manifestation of underlying conceptualization – and thus, that metaphor was a phenomenon of *thought* and not just of *language*. Fictive motion is manifested in expressions such as, 'The fence *runs* along the river,' or 'The Equator *passes through* many countries,' in which real or imaginary static entities are discussed as if they were dynamic. Talmy (1996) argued that fictive motion in *language* may actually reflect a process of dynamic *conceptualization*. Over the last two decades, the cognitive reality of both conceptual metaphor and fictive motion has been supported by psychological experiments, gesture studies, EEG/ERP studies, and neuroimaging (for conceptual metaphor see Gibbs, 2008, for an overview; for fictive motion see Marghetis & Núñez, 2013; Matlock, 2010).

Inspired by the possibility that these two phenomena might be central to human imagination, Lakoff and Núñez (2000) argued that they play a role in the creation and conceptualization of number concepts, arithmetic, and mathematics in general. In this chapter, we review the proposal that cognitive linguistic mechanisms structure numerical cognition – including PCNs, but also the positive integers and arithmetic – and survey the empirical evidence that supports this proposal. The focus throughout is on reasoning, understanding, and conceptualization – the way we think about numbers themselves, as abstract objects – rather than the psychological processing of particular numbers.

From Counting to Arithmetic (And Still Far from Mathematics)

Counting is an apparently simple practice that many children around the world learn at an early age, usually based on a lexicon of 'counting' numbers, which, varying across the languages of the world (Beller & Bender, 2008), refer to prototypical quantities ('one,' 'two,' 'three,' 'four,' etc.). Two questions then immediately come up. First, is this apparently simple act a human universal? If not, what does that tell us about the nature of numbers as a whole, and about the nature of the simplest PCNs, in particular? Secondly, how do we get from 'counting' to PCNs, to natural numbers, to arithmetic, and to mathematics proper?

It is well established that many languages around the world do not have words for exact numerosities beyond 'four' or 'five' (Strömer, 1932), and that native speakers of these languages seem to reason only approximately about numerosities above that range (Pica, Lemer, Izard, & Dehaene, 2004). The apparently simple act of counting beyond, say half a dozen items, is not part of our biological endowment. That is, numbers of all sorts – PCNs, but also natural, negative, irrational, real, complex, infinitesimal, transfinite – have not been selected for by natural selection, but rather emerged out of specific cultural practices that developed over historical time. To get beyond our basic cognitive machinery, therefore, we need an account of how abstract concepts can acquire complex structure – a process that, we suggest, is realized via mechanisms like conceptual metaphor and fictive motion. However, there is more. All languages around the world seem to have a precise number lexicon for at least the most basic numerosities 'one,' 'two,' and 'three' (or at least 'one' and 'two;' Gordon, 2004). What this suggests is that count practices and PCNs seem to build on basic *cognitive pre-conditions for numerical abilities* (Núñez, 2009) that, although not being *numerical* as such, are part of the human biological endowment and are, therefore, universal. One such pre-condition is *subitizing*, the capacity to make quick, error-free, and precise judgments of the numerosity of small collections of items (Kaufmann, Lord, Reese, & Volkmann, 1949), which seems to be based on domain general mechanisms that deal with multiple object individuation (Piazza, Fumarola, Chinello, & Melcher, 2011). Humans can subitize up to about three to four items. Findings suggest that the ability to subitize is innate, that it exists even in some amphibians (Uller, Jaeger, Guidry, & Martin, 2003; see also chapter by Agrillo, this volume), that it is not merely a pattern-recognition process, and that it is not restricted to the visual modality but also operates on, e.g. sequences of knocks or beeps (Davis & Pérusse, 1988). Moreover, deficits in subitizing seem to be associated with developmental dyscalculia (Moeller, Neuburger, Kaufmann, Landerl, & Nuerk, 2009), suggesting an influence on the consolidation of numerical capabilities. Another innate pre-condition is the ability to discriminate, approximately, between large collections on the basis of numerosity – an ability that emerges in humans as early as 6 months, provided the collections' numerosities differ by a sufficiently large ratio (e.g. 8 versus 16, but not 8 versus 12; Xu & Spelke, 2000). Although these inborn abilities exist independently of culture, education, and linguistic practices

(Wiese, 2003), the findings are consistent with the fact that all human languages have lexical items that capture the meaningful experience of discriminating small, precise numerosities (provided by subitizing), and others that express the approximate relations (greater/smaller than) involved in large numerosity discrimination. Language, in a uniquely human form, consolidates at a higher symbolic level these inborn cognitive pre-conditions for numerical abilities, which we share with many other species of the animal kingdom. Subitizing and large numerosity discrimination, however, are not numerical proper, let alone arithmetical or mathematical. They are simply capacities for discriminating numerosities, which lack essential components of *number* concepts. The essential notion of *order*, for instance, does not appear in humans until as late as 11 months (Brannon, 2002). Most importantly, these basic capacities also lack compositionality, the core of the creation of proper numbers, arithmetic, and mathematics. Thus, these innate cognitive preconditions for PCNs are:

(1) insufficient for number at the level of complexity of natural numbers, and
(2) not 'proto' or 'early' *numerical* any more than the vestibular system supporting an infant's first steps is proto – or early snowboarding (Núñez, 2009).

Innate cognitive preconditions cannot provide foundations for the precision, richness, and range of PCNs, let alone for more sophisticated number concepts, arithmetic, and complex mathematical concepts.

Any account of *number cognition* must account for the unique features of PCNs, natural numbers and arithmetic – things like precision, objectivity, rigor, generalizability, stability, and symbolism. These concepts are highly sophisticated and developed culturally only in recent human history. How to get from our basic cognitive toolbox to these rich cultural accomplishments? One possibility is that PCNs, natural numbers and arithmetic are realized through precise combinations of domain-general (i.e. non-mathematical) everyday cognitive mechanisms that make human imagination and abstraction possible, such as conceptual metaphor and fictive motion (Lakoff & Núñez, 2000; Núñez, 2009). According to this proposal, these cognitive mechanisms – mediated through language and other cultural practices – support the conceptualization of precise abstract entities like number and arithmetic, grounding them in our embodied experience[2].

Conceptualizing Number: Conceptual Metaphor and Fictive Motion

In this section we analyse how conceptual mappings – conceptual metaphor and fictive motion in particular – provide the inferential structure of number concepts. We begin with how these mechanisms work in everyday language.

Introduction to Conceptual Mappings and Fictive Motion

Conceptual Metaphor

Consider the following two everyday linguistic expressions: 'The election is *ahead* of us,' and, 'Winter is now *behind* us.' Even though 'an election' is not physically 'ahead' of us, and

'Winter' is not literally 'behind' us, these expressions are, nevertheless, easily understood to convey a precise meaning – namely, a *temporal* meaning. Countless such expressions, whose meaning is not literal, but *metaphorical*, can be observed in human everyday language in many domains. 'Metaphor,' in this sense, is not just a figure of speech, or a rhetorical tool reserved for poets and politicians. It is a mechanism of thought, usually operating unconsciously and effortlessly, but ubiquitous in everyday (and technical) language (Lakoff & Johnson, 1980).

Cognitive linguistics has shown that these metaphorical expressions are systematic, such that the countless metaphorical expressions can be modelled by a relatively small number of conceptual metaphors. By 'conceptual metaphor' we mean both a particular inference-preserving cross-domain mapping, and also the cognitive mechanism that enables such mappings. Conceptual metaphor involves projecting the inferential structure of a grounded *source domain* (e.g. spatial experience) to a different *target domain*, usually more abstract (e.g. time). In the above examples, specific notions related to sagittal bodily space like 'ahead' and 'behind' get mapped onto 'future' and 'past,' respectively, and open up an entire world of inferences where the relatively abstract domain of 'time' is conceived in terms of the more concrete domain of spatial experience. Crucially, conceptual metaphor can account for abstract domains' inferential organization: the network of inferences[3] that is generated via the mappings (see Table 21.1).

Notice that, although the expressions in our initial example use completely different *words* (i.e. 'ahead,' 'behind'), they are both linguistic manifestations of a single general conceptual metaphor, namely, TIME EVENTS ARE LOCATIONS IN SAGITTAL UNI-DIMENSIONAL SPACE[4]. As in any conceptual metaphor, the inferential structure of target domain concepts (time, in this case) is created via a precise mapping from the source domain (in this case sagittal uni-dimensional space – the linear space in front and behind an observer). The inferential structure of this mapping accounts for a number of linguistic expressions, such as 'The summer is still *far away*,' 'The end of the world is *near*,' and 'Election day is *here*.' Many important entailments follow from the mapping. For instance, transitive properties applying to spatial relations between the observer and the objects in the source domain are preserved in the target domain of time: if, relative to the front of the observer, object A is further away than object B, and object B is further away than object C, then object C is closer than object A. Via the mapping, this implies that time C is in a 'nearer' future than time A. The same

Table 21.1 The TIME EVENTS ARE LOCATIONS IN SAGITTAL UNI-DIMENSIONAL SPACE metaphor

Source Domain		Target domain
Sagittal uni-dimensional space relative to ego		**Time**
Locations in front of ego	→	Future times
Locations behind ego	→	Past times
Co-location with ego	→	Present time
Farther away in front of ego	→	'Farther away' in the future
Farther away behind ego	→	'Farther away' in the past

relationships hold for objects behind the observer and times in the past. Furthermore, the conceptual mapping can inject novel structure into the target domain. Time, for instance, is seen as having measurable 'extension' and as extending like a segment of a path, and is thus conceivable as a linear bounded region.

Fictive Motion

Fictive motion is a cognitive mechanism through which we unconsciously (and effortlessly) conceptualize static entities in dynamic terms, as when we say, 'The road goes along the coast.' The road itself doesn't actually move anywhere. It is simply standing still. But we may conceive it as moving 'along the coast.' Len Talmy (1996) analysed linguistic expressions taken from everyday language in which static scenes are described in dynamic terms. Consider:

- The fence *stops* right after the tree.
- The Equator *passes through* many countries.
- The border between Switzerland and Germany *runs along* the Rhine.
- The California coast *goes all the way down* to San Diego

Motion, in these cases, is fictive, imaginary, not real in any literal sense. Not only do these expressions recruit verbs of action, but they provide precise descriptions of the quality, manner, and form of motion. In all cases of fictive motion there is a *trajector* (the moving agent) and a *landscape* (the construed space in which the trajector moves). The trajector can be a physical object (e.g. 'the road goes;' 'the fence stops'), or a social, imagined, or metaphorical object (e.g. 'the Equator passes through;' 'the border runs'). In many cases of fictive motion, real world trajectories don't move, but are associated with something that could; we can talk about a 'road running from Berkeley to San Diego,' and roads are associated with cars, which *can* move. As we will see, this is never the case in mathematics, where the motion of the trajector is always metaphorical: we talk of *oscillating* functions or *approaching* a limit, in the absence of any concrete objects that could move (Marghetis & Núñez, 2013; Núñez, 2006).

Conceptual Mappings and Fictive Motion in The Conceptualization of Numbers And Arithmetic

As discussed in the previous section, number systems and arithmetic are qualitatively more complex than subitizing or judgments of approximate numerosity, the preconditions for numerical abilities that are found in monkeys and human infants. Subitizing is exact within its range, but natural numbers are always exact; approximate judgments of numerosity apply to collections of all sizes, but are inherently *approximate*. From where, then, do PCNs (let alone natural numbers) get their properties? What cognitive mechanisms are needed to go from simple innate abilities to full-blown arithmetic?

Humans seem capable of transcending the limitations of these simple abilities. When sociocultural demands are present, counting emerges. In order to count (e.g. finger-counting), several additional capacities are required – capacities like grouping objects into real or imagined collections, ordering these collections for serially counting, assigning number-labels to individual objects, and using memory to keep track of this process, amongst others (Lakoff & Núñez, 2000)[5]. When these capacities are used within the subitizing range (between 1 and 4), stable results are obtained because cardinal-number assignment is co-extensive with outcomes obtained via subitizing. To count beyond four – the range of the subitizing capacity – requires the above cognitive capacities, but also additional capacities that allow for putting together perceived or imagined groups to form larger groups, and a capacity to associate physical symbols (or words) with the resulting exact quantities. However, subitizing and counting only provide some of the cognitive preconditions for mature numerical abilities, and additional mechanisms are required to go beyond this foundation – imaginative mechanisms like conceptual metaphor and fictive motion. These mechanisms allow for the conceptualization of cardinal numbers and arithmetic operations in terms of basic ordinary experiences of various kinds – experiences with groups of objects, with the part–whole structure of objects, with distances, with movement and locations, and so on. Conceptual metaphor and fictive motion are among the most basic domain-general everyday cognitive mechanisms that take us beyond minimal early abilities and simple counting to the elementary arithmetic of PCNs and natural numbers.

Since conceptual metaphors preserve inferential organization, they make possible the conceptualization of arithmetic in terms of the prior understanding of commonplace physical activities. The understanding of elementary arithmetic seems to be based on a systematic correlation between (1) the most basic literal aspects of arithmetic, such as subitizing and counting, and (2) everyday activities, such as collecting objects into groups or piles, taking objects apart and putting them together, taking steps on a path, and so on. Such correlations allow humans – unlike other animals – to form conceptual mappings by which they greatly extend their subitizing and counting capacities. Thus, if we conceptualize numbers as collections or as locations in space, we can project the logic of collections and of spatial locations, respectively, onto numbers thus providing their inferential structure.

On the basis of these considerations, Lakoff and Núñez (2000) suggested that two fundamental conceptual metaphors are responsible for number concepts and arithmetic: ARITHMETIC IS OBJECT COLLECTION and ARITHMETIC IS MOTION ALONG A PATH. These metaphors are a way to ground our conceptualization of arithmetic in shared, precise bodily experiences, and thus provide the necessary inferential organization. The detailed analysis of these mappings can be found elsewhere (Lakoff and Núñez, 2000, chapters 3 and 4). In order to give a flavour of the robustness and inferential richness of these conceptual mappings, let us point to some of their crucial components.

The ARITHMETIC IS OBJECT COLLECTION metaphor is a mapping from the domain of physical objects to the domain of numbers. The metaphorical mapping consists of:

(1) The source domain of object collection (based on our experiences with grouping objects).
(2) The target domain of arithmetic of PCNs (structured non-metaphorically by subitizing and counting).

Table 21.2 The ARITHMETIC IS OBJECT COLLECTION metaphor

Source Domain		Target domain
Object collection		**Arithmetic of PCNs**
Collections of objects of the same size	→	Numbers
The size of the collection	→	The 'size' of the number
The smallest collection	→	The unit (one)
Bigger	→	Greater than
Smaller	→	Less than
Putting collections together	→	Addition
Taking a smaller collection from a larger collection	→	Subtraction

(3) A mapping across the domains (based on our experience subitizing and counting objects in groups). The basic mapping of this conceptual metaphor is the following (Table 21.2):

Evidence of this metaphor shows up in everyday language. The word 'add' has the physical meaning of placing a substance or a number of objects into a container (or group of objects), as in 'add sugar to my coffee,' 'add some logs to the fire,' and 'add onions and carrots to the soup.' Similarly, we can 'take 7 from 10,' 'take 3 out of 4,' or 'take away 2' – but 'take … from,' 'take … out of,' and 'take away … ' have the physical meaning of removing a substance, an object, or a number of objects from some container or collection. Linguistic examples include 'take some water from this pot,' 'take some books out of the box,' and 'take away some of these logs.' On our proposal, these linguistic regularities are neither random nor superficial. They reflect the recycling of inferences about real-world actions, such as 'to add' and 'to take away,' recruited via the metaphor to conceptualize abstract arithmetic facts. It follows from the metaphor that adding yields something bigger (more) and subtracting yields something smaller (less). Accordingly, lexical items like 'big' and 'small,' which indicate literal size for objects and collections of objects, are metaphorically extended so they apply to numbers, as in 'which is bigger, 5 or 7?' and 'two is smaller than four.' However, this isn't the only way to conceptualize arithmetic. Complementing the OBJECT COLLECTION metaphor is the ARITHMETIC IS MOTION ALONG A PATH metaphor. The metaphorical mapping consists of:

(1) The source domain of paths (based on shared experiences with motion through linear space).
(2) The target domain of arithmetic of PCNs plus Zero (structured non-metaphorically by subitizing and counting).
(3) A mapping across the domains (based on our experience subitizing and counting steps on a path). The basic mapping is the following (Table 21.3):

Table 21.3 The ARITHMETIC IS MOTION ALONG A PATH metaphor

Source domain		Target domain
Motion along a path		**Arithmetic of PCNs (plus 0)**
Point-locations on a path	→	Numbers
The origin, the beginning of the path	→	Zero
A point-location on the path	→	One
Farther from the origin than	→	Greater than
Closer to the origin than	→	Less than
The smallest collection	→	The unit (one)
Acts of moving along the path	→	Arithmetic operations
Moving from a point-location *A* away from the origin, a distance that is the same as the distance from the origin to a point-location *B*	→	Addition of *B* to *A*
Moving toward the origin from *A*, a distance that is the same as the distance from the origin to *B*	→	Subtraction of *B* from *A*

This understanding of numbers as point-locations is expressed linguistically in a variety of ways:

- 'How *close* are these two numbers?'
- '37 is *far away* from 189,712.'
- '8.9 is *near* 9.'
- 'The result is *around* 50.'
- 'Count *up to* 20, without *skipping* any numbers.'
- 'Count *backward* from 20.'
- 'Count *to* 100, *starting at* 20.'
- 'Name all the numbers *from* 2 *to* 10.'

Note that this metaphor relies additionally on fictive motion, with arithmetic operations conceptualized dynamically as individual acts of motion along a path.

These conceptual metaphors are rich in their inferential entailments, derived from basic facts about collections of physical objects or locations on paths. The result is a set of inferences about PCNs, only later extended to the natural numbers (Davidson et al., 2012). For example, suppose we have two collections, *A* and *B*, of physical objects, with *A* bigger than *B*. What happens if we add the same collection *C* to each? Well, *A* plus *C* will be a bigger collection of physical objects than *B* plus *C*. This is a fact about collections of physical objects of the same size. Using the mapping ARITHMETIC IS OBJECT COLLECTION, this physical fact that we experience in grouping objects can now be conceptualized as an arithmetical truth about *numbers*: if a number *A* is greater than number *B*, then *A* plus number *C* is greater than *B* plus *C*. Similar metaphorical analyses are available for various 'truths' of arithmetic, including such things as stability of results for addition and subtraction, closure for addition, the existence of inverse operations, the possibility of unlimited iteration for addition but not for subtraction, and so on (Lakoff & Núñez, 2000, chapters 3 and 4). In each case, the conceptual metaphor maps a property from the source domain of object collections or motion along a path, to the target domain of numbers and arithmetic. On this account, the laws and truths of arithmetic are, in fact, metaphorical entailments of the conceptual mapping we operate with. The truths seem certain because the metaphors are invisible, but formative of our number cognition; we all agree because these metaphors are shared.

The inferential organization of these metaphorical mappings supplies a wide range of essential arithmetic properties. For instance, several *equational properties* that apply to object collections and motion along a path, respectively, get metaphorically extended to numbers:

- *Equality of result:* different operations produce the same resulting object collection or landing location on a path.
- *Preservation of equality:* combining equals to equals yields equals; likewise for subtraction
- *Commutativity:* combining collections is order-independent, as are forward motions along a path.

Moreover, several *relationship properties* are also preserved by the metaphorical mappings:

- *Duality of ordering:* if *A* is greater than *B*, then *B* is less than *A*.
- *Trichotomous ordering:* if *A* and *B* are two collection/numbers or locations on a path, then either *A* is greater than *B*, or *B* is greater than *A*, or *A* and *B* are of the same magnitude.
- *Symmetry of equality:* if *A* is the same size as *B*, then *B* is the same size as *A*.
- *Transitivity:* if *A* is greater than *B* and *B* is greater than *C*, then *A* is greater than *C*.

These powerful conceptual metaphors have many more entailments, and lend themselves to further extensions. Within the domain of object collection, for instance, we can engage in pooling and repeated addition of object collections, as well as splitting and repeated subtraction. It is possible then to state precisely how, via the metaphorical mappings, these entailments in the realm of object collection and motions along a path give rise to arithmetical laws for multiplication and division of natural numbers, including commutativity and associativity for multiplication, distributivity for multiplication over addition, and the existence of multiplicative identity and inverse. The details of this analysis, however, exceed the space available here (for details, see Lakoff & Núñez 2000, chapter 3).

In this brief analysis, one can see the extent to which the two conceptual metaphors ARITHMETIC IS OBJECT COLLECTION and ARITHMETIC IS MOTION ALONG A PATH ground the rich inferential structure of PCN concepts. A crucial component of the robustness of the grounded inferential structure comes from the close agreement in the inferential structure of the two source domains – collections and motion along a path. Many actions in one source domain (e.g. collecting objects) have a corresponding action in the other source domain (e.g. moving along a path), with both actions mapping to identical results in the target domain of numbers. In fact, in the realm of PCNs (without the concept ZERO) and basic arithmetic operations, there is an *isomorphism* between the source domains, thus securing solid foundations of numbers and arithmetic in everyday embodied experience.

These two conceptual metaphors, however, differ in inferential and cognitive complexity. Inferentially, the ARITHMETIC IS OBJECT COLLECTION metaphor does not have a number zero, for instance. Strictly speaking, the result of taking away collection *A* from collection *A* yields a *lack* of collection. It takes extra metaphorical extensions to make the lack of a collection to 'be' a 'collection with no elements.' ARITHMETIC IS MOTION ALONG A PATH, on the other hand, comes with a natural 'built-in' number zero (the origin of motion), which, via the mapping, gets the same ontological status as the rest of PCNs. Zero is a location on the path, and thus a number like any other. Moreover, this metaphor readily affords a further extension to negative numbers, which the object collection metaphor cannot provide. And as we'll argue in the next section, there is now strong evidence suggesting that ARITHMETIC IS OBJECT COLLECTION is more easily acquired and culturally universal than ARITHMETIC IS MOTION ALONG A PATH (Núñez, 2009; Núñez, Cooperrider, & Wassmann, 2012).

A metaphorical analysis suggests that many of the basic properties of arithmetic can be derived from shared, embodied experiences – collecting discrete objects and moving along a path. However, this theoretical account – based primarily on evidence from linguistic practices – says nothing about when and to what end these metaphors are used during real-time reasoning. For that, we must leave the armchair and seek empirical evidence.

Empirical Support for The Embodied Conceptualization of Number

The foregoing analysis grew out of a linguistic tradition that relied primarily on regularities in figurative language. This linguistic evidence, however, cannot tell us when or in what contexts embodied reasoning makes its contribution, if at all. Linguistic evidence on its own does not provide a transparent window into conceptual structure (Murphy, 1996) and is, therefore, insufficient to demonstrate the cognitive reality of metaphorical thought; that we speak of death as 'kicking the bucket' is weak evidence that we conceptualize death in terms of boots and buckets (Casasanto, 2009). We can reason about arithmetic, moreover, solely by manipulating abstract notations, thus taking advantage of external space to structure the intelligent manipulation of inscriptions (Kirsh, 1995; de Cruz, 2008; Landy & Goldstone, 2009; cf. Rumelhart, Smolensky, McClelland, & Hinton, 1986). So, metaphorical thought cannot be the whole story of how we reason about numbers and arithmetic. It's an empirical project, therefore, to determine exactly when, why, and for what purposes we engage in metaphorical thought about number and arithmetic.

In the last decade, research on embodied cognition has begun that project, using an increasingly diverse toolkit of empirical methods to document the functional role of embodied thought in language and cognition more generally (Gibbs, 2006). There is now converging evidence that number cognition is, indeed, embodied and metaphorical, at least sometimes and under certain circumstances. This section surveys that evidence.

What Roles for Imagination and Conceptualization?

An embodied approach to PCNs, and to number cognition in general, needs to distinguish between the many possible contributions that metaphorical thought could make to arithmetic reasoning (cf. Gibbs, 1994, p. 18). Consider the following hypotheses:

H1: Embodied metaphors show up in language, but have no cognitive reality.
H2: Embodied metaphors play a role in learning.
H3: Embodied metaphors are activated during real-time reasoning.

The first hypothesis was the prevailing view for most of history, dating back to Aristotle, and is a reasonable working hypothesis (Lakoff, 1993). Indeed, there are certainly linguistic metaphors that no longer reflect metaphorical conceptualization, linguistic metaphors that have become *dead* metaphors. The term '*until*,' for instance, has lost the spatial meaning it had in Middle English, inherited from Old Norse (Traugott, 1985). This could be true of mathematics, so that expressions like '*bigger* and *smaller* numbers' and 'keep adding numbers until you *pass* one hundred' are merely the sedimentation of mathematical history and uninformative about contemporary number cognition.

The last few decades of research in cognitive linguistics, however, has demonstrated that where there's linguistic smoke, there's often cognitive fire. According to the second hypothesis, metaphorical thought may show up during concept acquisition and in the classroom, but not necessarily outside of that pedagogical context. After all, physicist and Nobel Laureate Richard Feynman claimed that, 'Light is something like raindrops' (1985, p. 14), but presumably did not conceptualize electromagnetic radiation as small wet blobs.

The third hypothesis goes further to suggest that embodied conceptual resources may play a functional role in mature reasoning about number and arithmetic. Consider the task of understanding why the sum of two even integers is itself even. There exist technical proofs of this fact – proofs that rely on algebraic notation – but conceptualizing arithmetic as either *Object Collection* or as *Motion Along A Path* might afford a more grounded, experiential insight into the behaviour of PCNs, and other numbers in general. In cases like these, we might expect embodied thought to guide reasoning.

These latter two hypotheses are not meant to be mutually exclusive, but focus attention on two compelling proposals for how conceptual metaphor and fictive motion might actually influence number cognition. In what follows, we address these two hypotheses in turn.

Embodied Learning

The first possibility is that metaphorical reasoning appears during learning, supporting the early acquisition and later elaboration of number concepts. Right from the very beginning, acquiring the concept of EXACT NUMBER seems to depend on manipulating collections of objects. Developmental psychologists disagree about the details, but most agree that the acquisition process relies on the child's ability to apply a counting routine to collections of discrete objects (Carey, 2009; Spelke & Tviskin, 2001; but see Rips, Bloomfield, & Asmuth, 2008). According to Carey (2009), for instance, learning the meaning of the number lexicon ('one,' 'two,' etc.) relies on early experiences with small collections of objects, extended via analogy to all the positive counting numbers as a result of applying a count routine to varied sets of objects (Carey, 2009). The child's first concept of number, therefore, is intimately tied to experiences of manipulating collections, the source domain of the OBJECT COLLECTION metaphor. The formative influence of this source domain is reflected in children's early difficulties with arithmetic. Hughes (1986) reports that 3- and 4-year old children can accurately add and subtract small numbers when the numbers are used to describe collections of perceived or imagined objects, such as bricks being placed in a box. However, when the number words are used in isolation, without a concrete context, children reliably fail to respond correctly. The ARITHMETIC IS OBJECT COLLECTION metaphor, therefore, seems central to the acquisition and early use of the concept of exact number.

Conceptualizing arithmetic as MOTION ALONG A PATH, on the other hand, does not seem to emerge until later in development, and requires extensive cultural scaffolding. Earlier research suggested that mapping numbers to space might be a cultural universal. Dehaene, Izard, Spelke, & Pica (2008) reported that an Amazonian indigenous group, the Munduruсu, were able to reliably and systematically map numbers to locations along a physical line segment, even though their language has a limited number lexicon, they lack formal education, and make little or no use of cultural artifacts like rulers, graphs, or maps. On the basis of this surprising result, they concluded that 'the mapping of numbers onto space is a universal intuition' (p. 1217). Recent research suggests that this conclusion may have been premature. Núñez, et al. (2012) found that the Yupno of Papua New Guinea – who, unlike the Munduruсu, have an elaborate number lexicon, including words for exact numbers beyond twenty – did not map numbers to locations along a physical line segment. Instead, they systematically mapped small numbers to one endpoint and mid-size and large numbers to the other, ignoring the extent of the line segment. This categorical response pattern persisted even when the instructions explicitly described the line segment as a path, and explicitly demonstrated that a mid-sized numerosity was associated with a location around the mid-point of the path. These Yupno findings suggest that mapping between number and continuous space is not innate, does not emerge spontaneously, varies across cultures, and may require extensive cultural scaffolding (Núñez, 2011).

That scaffolding could be provided by rulers, calendars, graphs, and countless other cultural artifacts that map numbers to locations in space. It is a common pedagogical practice to introduce a material representation of a path-based construal of arithmetic: a literal 'number-line' with locations along a line segment labeled with the integers (Herbst, 1997). The ubiquity of such representations creates a cultural milieu in which associations between number and space are unavoidable, thus supporting the conceptualization of numbers and arithmetic as MOTION ALONG A PATH (Núñez, 2011). As expected, children raised in such a milieu gradually learn to map numbers to locations along a line. If North American children are shown a physical line segment with '0' at one endpoint and a numeral representing a larger number on the other (e.g. '100'), as early as kindergarten they are able to place intermediate numbers along the line segment in a systematic way (Siegler & Booth, 2004) – much like the reported Munduruсu, but unlike the Yupno. North American children's facility with this number-to-line mapping increases across development, and children gradually transition from a logarithmic mapping – allocating more space to smaller numbers – to a linear mapping, so that by the sixth grade children use a linear mapping to map numbers between 0 and 1000 to a line segment (Siegler & Opfer, 2003). Importantly, however, the logarithmic mapping persists even up to college age if participants report quantities with unconventional non-spatial methods (e.g. by squeezing or vocalizing), suggesting that linear mappings are the result of training and education with culturally-developed notational devices like the physical number-line (Núñez, Doan, & Nikoulina, 2011). However, does acquiring these metaphors affect long-term learning outcomes? It seems plausible that it could. Conceptual metaphors ground the understanding of arithmetic in more concrete, experiential domains, and thus are a way to ground mathematical reasoning in shared, embodied intuitions – intuitions about motion through space, for instance, or about manipulating objects (Núñez, Edwards, & Matos, 1999). Inspired by this possibility, Danesi (2003, 2007) used *Conceptual Metaphor Theory* as a pedagogical tool for teaching mathematical reasoning to eighth-grade students with difficulties in mathematics. In one exploratory study, eighth-grade students were explicitly taught to identify conceptual metaphors in algebra word problems. At the beginning of the school year, the students had 'severe difficulties in solving word problems'; by the end of the year, these at-risk students had closed the gap with their peers. While these preliminary studies require follow-up with fully controlled studies, nevertheless, they reinforce the promise of metaphor-based interventions in education – an area ripe for investigation.

Furthermore, long-term mathematical success depends on the acquisition of a linear spatial construal of number. In a longitudinal study of mathematics learning, Gunderson, Ramirez, Beilock, and Levine (2012) followed first – and second – grade students and, for five 4-month intervals, measuring their spatial abilities, symbolic mathematical abilities, and ability to associate numbers with locations along a line segment using a linear mapping. Students' early spatial abilities were predictive of their end-of-year abilities in symbolic arithmetic, replicating previous findings (e.g. Kurdek & Sinclair, 2001; Lachance & Mazzocco, 2006). Crucially, however, this was mediated by their ability to map numbers to a line using a linear mapping. Learning to think of numbers as evenly distributed locations along a path (i.e. preserving magnitude and, therefore, spatial invariance of the increment between the successor and the predecessor of any given PCN) facilitates the acquisition of symbolic mathematical proficiency.

Both PCN concept acquisition and the transition to more advanced mathematics, including algebra word problems, seems to rely on learning to map between PCNs and more concrete domains, either collections of objects or locations along a linear path. These two conceptual metaphors, therefore, seem to play a role in mathematics learning, as suggested by Hypothesis 2. Once acquired, though, do these conceptual metaphors remain active, or are they discarded in favour of more abstract modes of reasoning?

Embodied Reasoning

When Lakoff and Núñez (2000) first suggested that numbers and arithmetic were conceptualized as OBJECT COLLECTION or as MOTION ALONG A PATH, there was little evidence that these metaphors were active during real-time number cognition. The last decade, however, has begun to produce converging evidence that mathematical metaphors are not mere pedagogical scaffolds, but retain their psychological reality and are activated during mathematical thought.

If the ARITHMETIC IS MOTION ALONG A PATH metaphor is active during mature number cognition, then thinking about numbers should systematically activate linear spatial schemas, and thinking about arithmetic, motion through space. This is precisely what has been observed. Processing PCNs can bias the spatial trajectory of subsequent responses. In a visual search task, task-irrelevant numbers prime subsequent eye movements, with smaller numbers priming left eye movements and larger numbers priming right eye movements (Fischer, Castel, Dodd, & Pratt, 2003). These effects also go in the other direction, from space to number. When participants were asked to generate random numbers while shaking their head back and forth, head position systematically biased the magnitude of the 'random' number, with smaller numbers generated when facing leftward, larger numbers when facing rightward (Loetscher, Schwarz, Schubiger, & Brugger, 2008); an analogous effect exists for shifts in eye gaze (Loetscher, Bockisch, Nicholls, & Brugger, 2010). So, task-irrelevant numbers and spatial orientation can bias each other in both directions.

Perhaps the best-known behavioural phenomenon is the Spatial-Numerical Association of Response Codes, or *SNARC* effect. Dehaene, Bossini, & Giraux (1993) had participants judge the relative magnitude ('Greater or less than 5?') or parity ('Even or odd?') of numbers between 1 and 9, and respond by pressing one of two buttons. Participants were faster to respond to smaller numbers when they were responding on the left, and to larger numbers on the right, as if participants were automatically activating a mental 'number-line.' Crucially, the SNARC shows up for non-manual responses (Schwarz & Keus, 2004; Schwarz & Müller, 2006), so it is not effector-specific – it is not, for instance, merely the result of experience with keyboards that display numbers from left to right.

On the basis of this phenomenon, and other evidence of number-space associations (see Hubbard, Piazza, Pinel, & Dehaene, 2005; van Dijck et al., this volume), some authors have concluded that 'the representation of [...] number magnitude [...] is spatial in nature' (Treccani & Umiltà, 2010). Note that these authors are *not* claiming that number magnitude is *mapped* to space, guided by culture and mechanisms of creative cognition. Instead, their claim is that number magnitude just *is* spatial, at least as represented in the brain. There are reasons to doubt this stronger claim. For one, the SNARC effect is shaped by lifelong participation in cultural practices; the orientation of the SNARC effect – left-to-right or right-to-left – is influenced by reading direction for words (Shaki & Fischer, 2008) and numbers (Shaki, Fischer, & Petrusic, 2009), and perhaps finger-counting practices (Fischer, 2008; Lindemann, Alipour, & Fischer, 2011). Moreover, we recently found that number magnitude is just as easily associated with *pitch* as it is with space (Marghetis, Walker, Bergen, & Núñez, 2011). Following Dehaene et al (1993), we had participants make relative magnitude judgments for numbers between 1 and 9, but with one difference: instead of responding spatially by pressing a left or right button, participants responded vocally by producing high- or low-pitched vocalizations ('Ahhh'). Much like the SNARC effect, we found an interaction between number magnitude and pitch – participants responded faster to 'lower' numbers if they had to respond with a *low pitch*, but faster to 'higher' numbers with a *high pitch*. It seems, then, that participants were *Seeing Number As Pitch* – a *SNAP* effect. Perhaps number magnitude is both spatial *and* pitch-based, but we find that unlikely. Rather, the SNAP effect seems to reflect the human capacity to rapidly and unconsciously map between conceptual domains[6] (cf. Fauconnier & Turner, 2002; Lakoff & Johnson, 1980). These considerations suggest that the SNARC and other effects are not evidence that 'number magnitude is spatial,' but reflect a learned mapping from numbers to space, shaped by culture, context, and embodied experience.

Much like number, arithmetic interacts with space in ways consistent with a metaphorical analysis. Recent studies suggest that arithmetic may involve shifts in attention along a spatial representation of number. In one study (McCrink, Dehaene, & Dehaene-Lambertz, 2007), subjects saw two collections of dots that were added (i.e. combined) or subtracted behind an occluding screen. The screen then disappeared to reveal a third collection, and subjects had to judge whether it had the correct number of dots. Responses on this verification task were systematically biased by the arithmetic operation, such that subjects were more likely to accept a collection with too many dots after addition, but more likely to accept a collection with too *few* dots after subtraction. Although space was not directly implicated in their study, McCrink and colleagues interpreted this 'Operational Momentum' effect as evidence that participants were 'overshooting' as they shifted their attention along a mental number-line. Operational Momentum is not restricted to approximate arithmetic, but has also been observed with symbolic arithmetic, such as adding and subtracting single digit numbers (Pinhas & Fischer, 2008), which, moreover, prompts systematic spatial processing (Marghetis, Núñez, & Bergen, 2014). Finally, it appears that arithmetic calculation shares a neural substrate with brain areas responsible for shifts in spatial attention, with leftward shifts in attention associated with subtraction and rightward shifts in attention associated with addition (Knops, Thirion, Hubbard, Michel, & Dehaene, 2009). Taken together, the SNARC and Operational Momentum effects demonstrate that real-time reasoning about number and arithmetic automatically activates spatial representations, exactly as expected if the ARITHMETIC IS MOTION ALONG A PATH metaphor plays an enduring role in number cognition.

More recently, we have turned to the study of gesture – the spontaneous bodily actions that we all co-produce while speaking or thinking, most often involving the hands, but sometimes also the heads, eyes, and other body parts. Gestures are spontaneous and largely unmonitored. Crucially, they are not mere echoes of speech. Temporally, gestures reliably *precede* the associated speech; semantically, they often *complement*, rather than duplicate the content of accompanying speech (McNeil, 1992). Gestures, therefore, offer a glimpse into real-time cognitive processing, a window into thought (Goldin-Meadow, 2003; Goldin-Meadow, Alibali, & Church, 1993). Moreover, gestures are often metaphorical (Cienki & Müller, 2008). We recently observed a friend saying, 'My mood has really been improving,' while simultaneously tracing an upwards trajectory with her finger. Note that, in this case, the speech contains no hint of a spatial metaphor, but the accompanying gesture reflects the HAPPY IS UP metaphor in which mood is conceptualized as vertical space (e.g. 'I'm *down* in the dumps,' 'Things are looking *up*;' Lakoff & Johnson, 1980). Analysing spontaneous co-speech gesture, therefore, can reveal the presence of embodied, metaphorical thought in the absence of metaphorical speech (Núñez, Cooperrider, Doan, & Wassmann, 2012).

The spontaneous gestures of non-expert undergraduate students suggest that they conceptualize arithmetic using OBJECT COLLECTION or MOTION ALONG A PATH metaphors (Marghetis, Bergen, & Núñez, in preparation). Undergraduate volunteers were asked, 'Can you explain why the sum of an odd number and an even number is always odd?' and their responses were video-recorded as they reasoned aloud. Just previously, they had read a proof of a related theorem, and completed one of two mental imagery tasks. For participants in the 'Path' condition, the mental imagery task involved memorizing a picture of a bead on a wire and then imagining sliding the bead back and forth along the wire. For those in the 'Collecting' condition, the picture was of collections of beads, and participants had to imagine combining the different collections. If reasoning about arithmetic relies on an embodied understanding of arithmetic, then participants' gestures should reflect their use of these metaphors. Moreover, if metaphorical reasoning involves activating the associated source domain, then the mental imagery task should systematically bias participants to reason in a particular way: using a MOTION ALONG A PATH metaphor after imagining sliding beads along a path, or an OBJECT COLLECTION metaphor after imagining manipulating collections of beads.

This was exactly what was found. Overall, participants deployed two recurring types of gesture – *Collecting* and *Path* gestures – that evoked the two complementary metaphorical construals of arithmetic. In its canonical form, the Collecting gesture consisted of both hands moving inward, with the hands shaped as if they were grasping, pinching, or holding. Morphologically, this type of gesture used volume to evoke numerical magnitude; kinematically, inward motion to evoke addition. Collecting gestures, therefore, suggested an ARITHMETIC IS OBJECT COLLECTION conceptualization (Figure 21.1).

Path gestures, meanwhile, used one hand in a canonical pointing handshape to trace motion along a horizontal axis placed slightly in front of the body. Locations along the horizontal axis indicated object identity and magnitude, and motion along the axis indicated arithmetic operations, suggesting a conceptualization of arithmetic using fictive motion and the ARITHMETIC IS MOTION ALONG A PATH metaphor (Figure 21.2).

Participants' gesture suggested that they were conceptualizing arithmetic metaphorically, grounding their understanding in primitive embodied experiences: *Object collection* and *motion along a path*. Moreover, mental imagery had a significant effect on conceptualization, as indexed by gesture. After imagining combining sets of beads, participants were significantly more likely to produce grasping, bimanual *Collection* gestures

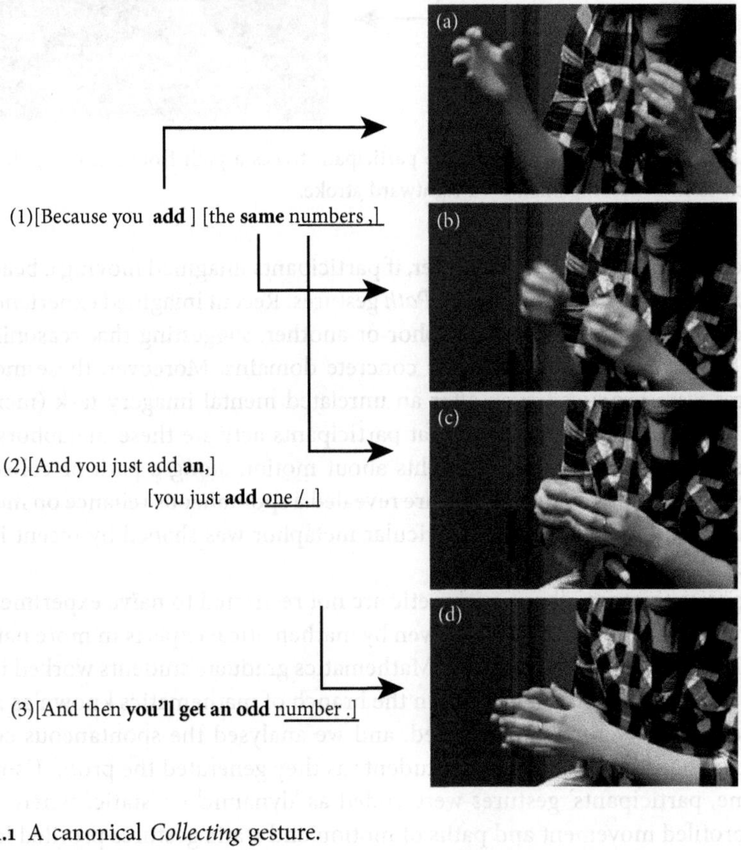
(a)
(b)
(c)
(d)
(1)[Because you add] [the same numbers ,]
(2)[And you just add an,]
[you just add one /.]
(3)[And then you'll get an odd number .]

FIGURE 21.1 A canonical *Collecting* gesture.

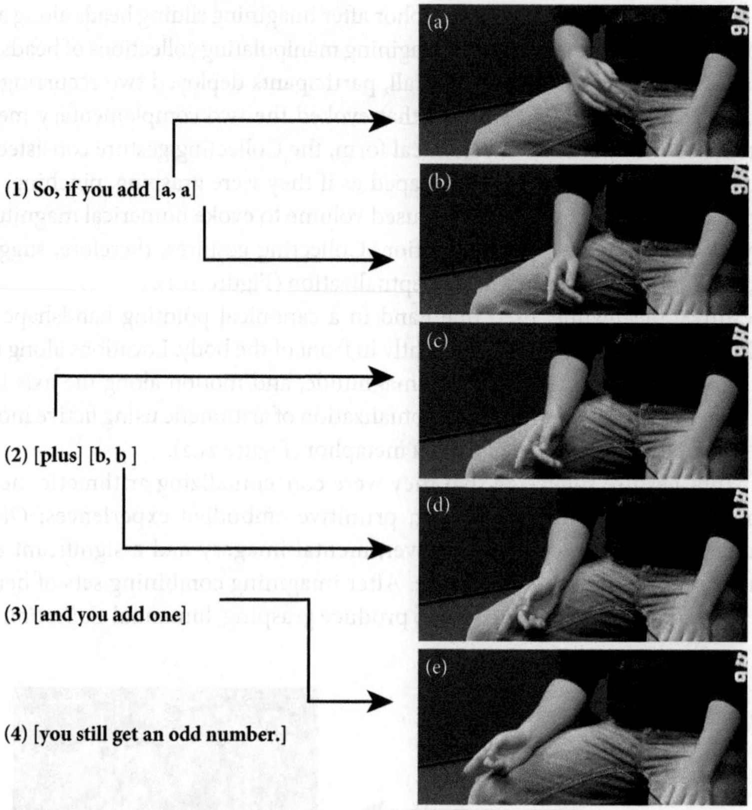
(1) So, if you add [a, a]
(2) [plus] [b, b]
(3) [and you add one]
(4) [you still get an odd number.]
(a)
(b)
(c)
(d)
(e)

FIGURE 21.2 A canonical *Path* gesture. The participant traces a path from left to right. The addition of each term is accompanied by a rightward stroke.

while talking about arithmetic. However, if participants imagined moving a bead along a wire, they were more likely to produce *Path* gestures. Recent imagined experience, therefore, systematically primed one metaphor or another, suggesting that reasoning about arithmetic remains grounded in these concrete domains. Moreover, these metaphorical gestures were produced even after an unrelated mental imagery task (memorizing a picture of animals), which shows that participants activate these metaphors spontaneously, even in the absence of thoughts about motion along a path or collections of objects (Figure 21.3). In summary, gesture revealed a spontaneous reliance on metaphorical reasoning, and the choice of a particular metaphor was shaped by recent imagined experience.

Metaphorical gestures about arithmetic are not restricted to naïve experimental participants in the lab, but are produced even by mathematical experts in more naturalistic settings (Marghetis and Núñez, 2013). Mathematics graduate students worked in groups of two to prove a non-trivial theorem in the branch of mathematics known as Analysis. Their interactions were video-recorded, and we analysed the spontaneous co-speech gestures produced by these graduate students as they generated the proof. Using a coding scheme, participants' gestures were coded as 'dynamic' or 'static,' where dynamic gestures profiled movement and paths of motion and static gestures profiled locations, motionless figures, etc. Independently, the content of speech was searched for talk of

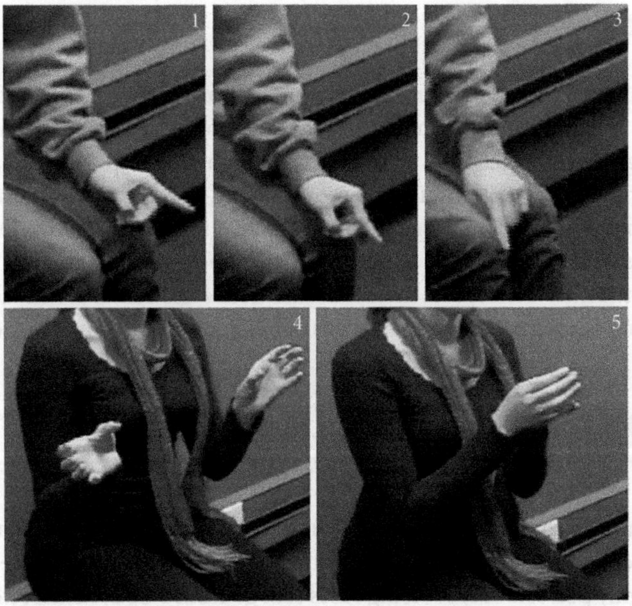

FIGURE 21.3 Participants spontaneously produced *Path* (frames 1–3) and *Collecting* gestures (frames 4–5) while discussing arithmetic, even without first performing related mental imagery. In the top frames, the participant is saying, 'Because seven is three plus three plus one.' In the bottom frames, the participant is saying, 'And you add your two new even numbers and they make another even number.'

concepts that, on a cognitive linguistic analysis, should be conceptualized either statically or dynamically – that is, concepts where fictive motion is thought to play a central role. Examples of dynamic concepts included the concepts of INCREASING, FUNCTION INTERSECTION and LIMIT. Crucially, however, none of these concepts are *technically* dynamic; according to their mathematical definitions, they are entirely static, defined in terms of static quantifiers, inequalities, and set notation. For instance, technically, a series of numbers is *increasing* if subsequent numbers are *greater than* preceding numbers – a static notion, which only becomes dynamic if one conceptualizes the numerical change as a [fictive] movement. As predicted, the graduate students were more likely to gesture dynamically when talking about dynamic concepts. For instance, one student was reasoning aloud about the behaviour of a particular function, out of sight of his collaborator, when he noted that the function was increasing. As he said, 'increasing,' his right hand shot up from his waist, index finger extended, and traced a leftward and upward trajectory. At the time, the blackboard was covered only in formal symbols, mostly set notation, and his speech was notably absent of metaphorical expressions. This graduate student spontaneously used gesture to spatialize a numerical increase as a motion along a path (for details, see Marghetis & Núñez, 2013). Even the gestures of mathematics graduates students, therefore, reflect the spontaneous use of metaphorical reasoning, construing the real numbers – even more sophisticated than PCNs or the natural numbers – as locations in space, and numerical change as motion through space.

Discussion

Broad claims about the 'embodiment' of mathematics can gloss over the sundry, but specific ways in which numbers and arithmetic might be embodied. In this section, we canvassed the converging evidence that early acquisition of number concepts – PCNs, but also more elaborated concepts like arithmetic – rely on domain-general mechanisms that support human imagination: conceptual metaphor and fictive motion. Moreover, we argued that these mechanisms remain active in mature reasoning about arithmetic, even among experts like graduate students. Conceptual metaphor and fictive motion, it seems, allow us to conceptualize the most abstract concepts in terms of basic, shared human experiences.

Conclusion and Future Prospects

The field of number cognition investigates the concept(s) of 'number' to gain insight into mathematical cognition, and perhaps the nature of mathematics itself. The term number, however, is highly polysemous, covering cases that go from PCNs, which often can be written with single digits, to natural, negative, rational, real, complex, infinitesimal, or transfinite ones. The vast majority of research on number cognition has focused on PCNs and the cognitive pre-conditions for numerical abilities, such as subitizing and approximate discrimination of large numerosity. These pre-conditions, however, are not sufficient to account for the rich inferential organization of PCN concepts, let alone natural numbers and arithmetic. It is crucial that we understand that these pre-conditions are just that, 'pre-conditions,' not 'proto' or 'early' *numerical* abilities – not any more than the vestibular system supporting an infant's first steps is proto – or early snowboarding (Núñez, 2011).

Humans build on these pre-conditions to move onto counting and PCNs, elaborate those first concepts to get to arithmetic, and eventually construct the astonishing edifice of mathematics. To do so, we have argued, humans use every day cognitive mechanisms for imagination, such as conceptual metaphor and fictive motion. While these mechanisms were first studied by cognitive linguists concerned primarily with language, they appear to be domain-general and ubiquitous in thought. There is accumulating evidence that these metaphors and mapping mechanisms have, beyond the linguistic expressions, a clear psychological and neurological reality – evidence from a variety of domains, ranging from psychological experiments, to gesture studies, to neuroimaging. Of course, much more needs to be understood about the psychological and neural underpinnings of these powerful conceptual mechanisms, but it is now increasingly clear that when it comes to number cognition, they play a role in early acquisition, support more advanced learning, and remain active during mature mathematical cognition.

Our analysis has not even touched the richness and grandeur of the number concepts found in contemporary mathematics: real, complex, infinitesimal, transfinite. We do ourselves a disservice by equating simple concepts of numerosity with counting numbers, or even with natural numbers, let alone the more complex concepts that populate the mathematical universe. Once all the relevant distinctions have been made clear, the numerical cognition of the future must not shy away from investigating the role of language, culture, and history in the consolidation of these rich, magnificent, and diverse human accomplishment – numbers.

NOTES

1. As other contributions in this Handbook show, some processing research has investigated multi-digit integers, fractions, and negative numbers as well (see Nuerk et al. this volume). Still, the bulk of number cognition research – shaping the development of the field – has been devoted to PCNs.
2. Like 'number,' the term embodied is polysemous. In this chapter, we are primarily interested in the use of 'embodied' as the grounding of the inferential structure in schematized bodily experience as when one moves along generic paths or gather generic collections of objects. We will put less emphasis on concrete specific bodily-supported actions such as finger-counting or on the use of 'embodied' brain areas during number processing.
3. Spatial construals of time have, of course, many other complexities, which go beyond the scope of this chapter (for a review, see Núñez & Cooperrider (2013).
4. Throughout, we reserve SMALL CAPITALS to denote concepts or conceptual metaphors, to distinguish them from particular linguistic manifestations of these mappings (e.g. 'She has a great future in front of her').
5. Note that these differ from the counting principles identified by Gelman and Gallistel (1978), which identify the abstract principles that must be acquired to use a count routine. Here we are concerned with the cognitive prerequisites for applying such a count routine.
6. These results are also consistent with the ATOM proposal (Walsh, 2003), which posits a domain-general representation of magnitude, shared between space, time, number, and even luminance (Cohen Kadosh & Henik, 2006). It's unclear, however, that pitch is naturally understood as in a single, reliable way: While higher pitches are sometimes associated with higher space (Rusconi, Kwan, Giordano, Umiltà, & Butterworth, 2006), they are also associated with smaller objects or animals (e.g. Ohala, 1994), making it difficult with the ATOM framework to predict a particular association between pitch and number.

REFERENCES

Agrillo, C. (this volume). *Numerical and Arithmetic abilities in non-primate species*. In: R. Cohen Kadosh & A. Dowker (Eds), The Oxford Handbook of Numerical Cognition, Oxford University Press.

Beller, S., & Bender, A. (2008). The limits of counting: numerical cognition between evolution and culture. *Science*, *319*, 213–215.

Brannon, E. (2002). The development of ordinal numerical knowledge in infancy. *Cognition*, *83*, 223–240.

Carey, S. (2009). *The Origin of Concepts*. New York: Oxford University Press.

Casasanto, D. (2009). When is a linguistic metaphor a conceptual metaphor? In V. Evans & S. Pourcel (Eds), *New Directions in Cognitive Linguistics* (pp. 127–145). Amsterdam: John Benjamins.

Chomsky, N. (2007). Of minds and language. *Biolinguistics*, *1*, 9–27.

Cienki, A., & Müller, C. (2008). Metaphor, gesture, and thought. In: R. W. Gibbs, Jr. (Ed.), *The Cambridge Handbook of Metaphor and Thought* (483–501). Cambridge: Cambridge University Press.

Cohen Kadosh, R., & Hanik, A. (2006). A common representation for semantic and physical properties: a cognitive-anatomical approach. *Experimental Psychology*, *53*(2), 87–94.

Dadda, M., Piffer, L., Agrillo, C., & Bisazza, A. (2009). Spontaneous number representation in mosquitofish. *Cognition*, *112*, 343–348.

de Cruz, H. (2008). An extended mind perspective on natural number representation. *Philosophical Psychology*, *21*, 475–490.

Danesi, M. (2003). Conceptual metaphor theory and the teaching of mathematics: findings of a pilot project. *Semiotica*, *145*, 71–83.

Danesi, M. (2007). A conceptual metaphor framework for the teaching of mathematics. *Studies in Philosophy and Education*, *26*, 225–236.

Davidson, K., Eng, K., & Barner, D. (2012). Does learning to count involve a semantic induction? *Cognition*, *123*, 162–173.

Davis, H., & Pérusse, R. (1988). Numerical competence in animals: definitional issues, cur- rent evidence, and new research agenda. *Behavioral and Brain Sciences*, *11*: 561–615.

Dehaene, S., Bossini, S., & Giraux, P. (1993). The mental representation of parity and number magnitude. *Journal of Experimental Psychology: General* *122*, 371–396.

Dehaene, S., Izard, V., Spelke, E., & Pica, P. (2008). Log or linear? Distinct intuitions of the number scale in Western and Amazonian indigene cultures. *Science*, *320*: 1217–1220.

Fauconnier, G. (1985). *Mental Spaces: Aspects of Meaning Construction in Natural Language*. Cambridge: MIT Press.

Fauconnier, G., & Turner, M. (2002). The Way We Think: Conceptual Blending and the Mind's Hidden Complexities. New York: Basic Books.

Feynman, R. (1985). *QED: The Strange Theory of Light and Matter*. Princeton: Princeton University Press.

Fillmore, C. (1982). Frame semantics. In: Linguistic Society of Korea (Eds), *Linguistics in the Morning Calm*. Seoul: Hanshin.

Fischer, M.H. (2008). Finger counting habits modulate spatial-numerical associations. *Cortex*, *44*, 386–392.

Fischer, M.H., Castel, A.D., Dodd, M.D., & Pratt, J. (2003). Perceiving numbers causes spatial shifts of attention. *Nature Neuroscience*, *6*, 555–556.

Gelman, R., & Gallistel, C. (1978). *The Child's Understanding of Number*. Cambridge, MA.: Harvard University Press.

Gibbs, R. (1994). *The Poetics of Mind: Figurative Thought, Language, and Understanding*. New York: Cambridge University Press.

Gibbs, R. (2006). *Embodiment and Cognitive Science*. New York: Cambridge University Press.

Gibbs, R. (Ed.) (2008). *Cambridge Handbook of Metaphor and Thought*. Cambridge, MA: Cambridge University Press.

Goldin-Meadow, S. (2003). *Hearing gesture: how our hands help us think*. Cambridge, MA: Harvard University Press.

Goldin-Meadow, S., Alibali, M.W., & Church, R.B. (1993). Transitions in concept acquisition: using the hand to read the mind. *Psychological Review*, *100*, 2.

Gordon, P. (2004). Numerical cognition without words. *Science*, *306*, 496–499.

Gunderson, E.A., Ramirez, G., Beilock, S.L., & Levine, S.C. (2012). The relation between spatial skill and early number knowledge: the role of the linear number line. *Developmental Psychology*, *48*, 1229–1241.

Herbst, P. (1997). The number-line metaphor in the discourse of a textbook series. *For the Learning of Mathematics*, *17*, 3.

Hubbard, E.M., Piazza, M., Pinel, P., & Dehaene, S. (2005). Interactions between number and space in parietal cortex. *Nature Reviews Neuroscience* *6*, 435–448.

Hughes, M. (1986). *Children and number: difficulties in learning mathematics*. Oxford: Basil Blackwell.

Kaufmann, E.L., Lord, M.W., Reese, T.W., & Volkmann, J. (1949). The discrimination of visual number. *American Journal of Psychology*, *62*, 498–525.

Kirsh, D. (1995). The intelligent use of space. *Artificial Intelligence* *73*, 31–68.

Knops, A., Thirion, B., Hubbard, E., Michel, V., & Dehaene, S. (2009). Recruitment of an area involved in eye movements during mental arithmetic. *Science*, *324*, 1583–1585.

Kurdek, L.A., & Sinclair, R.J. (2001). Predicting reading and mathematics achievement in fourth-grade children from kindergarten readiness scores. *Journal of Educational Psychology, 93*, 451–455.

Lachance, J.A., & Mazzocco, M.M.M. (2006). A longitudinal analysis of sex differences in math and spatial skills in primary school age children. *Learning and Individual Differences, 16*, 195–216.

Lakoff, G. (1993). The contemporary theory of metaphor. In: A. Ortony (Ed.), *Metaphor and Thought*, 2nd edn. Cambridge: Cambridge.

Lakoff, G., & Johnson, M. (1980). *Metaphors We Live By*. Chicago: Chicago University Press.

Lakoff, G., & Núñez, R. (2000). *Where Mathematics Comes From: How the Embodied Mind Brings Mathematics into Being*. New York: Basic Books.

Landy, D., & Goldstone, R.L. (2009). *Pushing Symbols. The 31st Annual Conference of the Cognitive Science Society*, Amsterdam.

Langacker, R. (1987). *Foundations of Cognitive Grammar*. Stanford: Stanford University Press.

Lindemann, O., Alipour, A., & Fischer, M. H. (2011). Finger counting habits in Middle-Eastern and Western individuals: an online survey. *Journal of Cross-Cultural Psychology, 42*, 566–578.

Loetscher, T., Bockisch, C., Nicholls, M.E.R., & Brugger, P. (2010). Eye position predicts what number you have in mind. *Current Biology, 20*, R264–R265.

Loetscher, T., Schwarz, U., Schubiger, M., & Brugger, P. (2008). Head turns bias the brain's random number generator. *Current Biology. 18*, R60–R62.

Marghetis, T., Bergen, B., & Núñez, R. (in preparation). Abstract mathematical reasoning involves spatial metaphors: Evidence from spontaneous gesture.

Marghetis, T., Bergen, B., & Núñez, R. (2014). Doing arithmetic by hand: hand movements during exact arithmetic reveal systematic, dynamic spatial processing. *Quarterly Journal of Experimental Psychology, 67*, 1579–1596.

Marghetis, T., & Núñez, R. (2013). The motion behind the symbols: a vital role for dynamism in the conceptualization of limits and continuity in expert mathematics. *Topics in Cognitive Science 5*(2), 299–316.

Marghetis, T., Walker, E., Bergen, B., & Núñez, R. (2011). Making SNAP judgments: Rethinking the spatial representation of number. In: L. Carlson, C. Hölscher, & T. Shipley (Eds), *Proceedings of the 33rd Annual Conference of the Cognitive Science Society* (pp. 1781–1786). Austin, TX: Cognitive Science Society.

Matlock, T. (2010). Abstract motion is no longer abstract. *Language and Cognition, 2*, 243–260.

McCrink, K., Dehaene, S., & Dehaene-Lambertz, G. (2007). Moving along the number line: operational momentum in nonsymbolic arithmetic. *Perception & Psychophysics, 69*, 1324–1333.

McMullen, M.B. (2010). Confronting the baby blues: a social constructivist reflects on time spent in a behaviorist infant classroom. *Early Childhood Research and Practice, 12*(1).

McNeill, D. (1992). Hand and Mind: What Gestures Reveal About Thought. Chicago: Chicago University Press.

Moeller, K., Neuburger, S., Kaufmann, L., Landerl, K., Nuerk, H.-C. (2009). Basic number processing deficits in developmental dyscalculia: evidence from eye tracking. *Cognitive Development, 24*, 371–386.

Murphy, G.L. (1996). On metaphoric representation. *Cognition, 60*, 173–204.

Nuerk, H.-C., Moeller, K., and Willmes, K. (this volume). Multi-digit number processing: Overview, conceptual clarifications and language influences. In: R. Cohen Kadosh, and A. Dowker (Eds), *The Oxford Handbook of Numerical Cognition*. Oxford University Press.

Núñez, R. (2006). Do Real Numbers Really Move? Language, Thought, and Gesture: The Embodied Cognitive Foundations of Mathematics. Reprinted in: R. Hersh (Ed.), *18 Unconventional Essays on the Nature of Mathematics* (pp. 160–181). New York: Springer.

Núñez, R. (2009). Numbers and arithmetic: neither hard-wired nor out there. *Biological Theory, 4*(1), 68–83.

Núñez, R. (2011). No innate number line in the human brain. *Journal of Cross-Cultural Psychology, 45*(4), 651–668.

Núñez, R. & Cooperrider, K. (2013). The tangle of space and time in human cognition. *Trends in Cognitive Sciences, 17*, 220–229.

Núñez, R., Cooperrider, K., & Wassmann, J. (2012). Number concepts without number lines in an indigenous group of Papua New Guinea. PLoS ONE.

Núñez, R., Cooperrider, K., Doan, D, & Wassmann, J. (2012). Contours of time: topographic construals of past, present, and future in the Yupno Valley of Papua New Guinea. *Cognition, 124*, 25–35.

Núñez, R., Doan, D, & Nikoulina, A. (2011). Squeezing, striking, and vocalizing: is number representation fundamentally spatial? *Cognition, 120*, 225–235.

Núñez, R., Edwards, L., & Matos, J.F. (1999). Embodied cognition as grounding for situated—ness and context in mathematics education. *Educational Studies in Mathematics, 39* (1–3): 45–65.

Ohala, J.J. (1994). The frequency codes underlies the sound symbolic use of voice pitch. In: L. Hinton, J. Nichols, & J. J. Ohala (Eds), *Sound Symbolism*. Cambridge: Cambridge University Press.

Piazza, M., Fumarola, A., Chinello, A., & Melcher, D. (2011). Subitizing reflects visuo-spatial object individuation capacity. *Cognition, 121*, 147–153.

Pica, P., Lemer, C., Izard, V., & Dehaene, S. (2004). Exact and approximate arithmetic in an Amazonian indigene group. *Science, 306*, 499–503.

Pinhas, M., & Fischer, M.H. (2008). Mental movements without magnitude? A study of spatial biases in symbolic arithmetic. *Cognition, 109*, 408–415.

Rips, L.J., Bloomfield, A., Asmuth, J. (2018). From numerical concepts to concepts of number. *Behavioral and BrainSciences, 31*, 623-642.

Rumelhart, D.E., Smolensky, P., McClelland, J.L., & Hinton, G.E. (1986). Schemata and sequential thought processes in PDP models. In: J. L. McClelland, D. E. Rumelhart, & the PDP Research Group (Eds), Parallel Distributed Processing: Explorations in the Microstructure of Cognition, Vol. 2.

Rusconi, E., Kwan, B., Giordano, B., Umiltà, C., & Butterworth, B. (2006). Spatial representation of pitch height: the SMARC effect. *Cognition, 99*, 113–129.

Schwarz, W., & Keus, I.M., (2004). Moving the eyes along the mental number line: comparing SNARC effects with saccadic and manual responses. *Perception & Psychophysics, 66*, 651–664.

Schwarz, W., & Müller, D. (2006). Spatial associations in number-related tasks: a comparison of manual and pedal responses. *Experimental Psychology, 53*, 4–15.

Shaki, S., & Fischer, M.H. (2008). Reading space into numbers – a cross-linguistic comparison of the SNARC effect. *Cognition, 108*, 590–599.

Shaki, S., Fischer, M., & Petrusic, W. (2009). Reading habits for both words and numbers contribute to the SNARC effect. *Psychonomic Bulletin and Review, 16*, 328–331.

Siegler, R.S., & Booth, J.L. (2004). Development of numerical estimation in young children. *Child Development, 75*, 428–444.

Siegler, R.S., & Opfer, J.E. (2003). The development of numerical estimation: evidence for multiple representations of numerical quantity. *Psychological Science, 14*, 237–243.

Spelke, E.S., & Tsivkin, S. (2001). Initial knowledge and conceptual change: space and number. In: M. Bowerman & S. Levinson (Eds), *Language Acquisition and Conceptual Development*. Cambridge: Cambridge University Press.

Strömer, C. (1932). *Die Sprache der Mundurukú (The language of the Munduruku) Verlag der Internationalen Zeitschrift 'Anthropos', 11*, 1–146.

Talmy, L. (1988). Force dynamics in language and cognition. *Cognitive Science, 12*: 49–100.

Talmy, L. (1996). Fictive motion in language and 'ception.' In: P. Bloom, M. Peterson, L. Nadel, & M. Garrett (Eds), *Language and Space*. Cambridge: MIT Press.

Traugott, E.S. (1985). 'Conventional' and 'dead' metaphors revisited: In: W. Paprotté & R. Dirven (Eds), *The Ubiquity of Metaphor: Metaphor in Language and Thought* (pp. 17–56). Amsterdam: John Benjamins.

Treccani, B., & Umilta, C. (2010). How to cook a SNARC? Space may be the critical ingredient, after all: a comment on Fischer, Mills, and Shaki (2010). *Brain and Cognition, 75*(3), 310–315; discussion 316–318.

van Dijck, J.-P., Ginsburg, V., Girelli, L., and Gevers, W. (this volume). Linking Numbers to Space: From the Mental Number Line towards a Hybrid Account. In: R. Cohen Kadosh & A. Dowker, (Eds), *The Oxford Handbook of Numerical Cognition*. Oxford University Press.

Uller, C., Jaeger, R., Guidry, G., & Martin, C. (2003). Salamanders (*Plethodon cinereus*) go for more: rudiments of number in an amphibian. *Animal Cognition, 6*, 105–112.

Walsh, V. (2003). A theory of magnitude: common cortical metrics of time, space and time. *Trends in Cognitive Science, 7*, 483–488.

Wiese, H. (2003). Iconic and non-iconic stages in number development: the role of language. *Trends in Cognitive Sciences, 7*(9), 385–390.

Xu, F., & Spelke, E. (2000). Large number discrimination in 6-month-old infants. *Cognition 74*: B1–B11.

CHAPTER 22

FIGURING OUT CHILDREN'S NUMBER REPRESENTATIONS

lessons from cross-cultural work

JOHN N. TOWSE, KEVIN MULDOON, AND VICTORIA SIMMS

INTRODUCTION

THIS chapter is concerned with cross-cultural research into the development of number representations (i.e. abstract mental representations of number) during childhood. We begin by posing a key question, one that perhaps you may be thinking right now: why is this an important topic to focus on? Partly embedded within this question is the subtext: why choose an *abstract* topic such as number representations, when there are many real world issues dealing with concrete numerical performance? We argue that this key question has at least two central components, and that it is valuable to make these components explicit from the outset. That way, the empirical studies that we subsequently review have an appropriate context and interpretative framework. So, we begin with a brief consideration of two components of our analytic focus. What is the purpose of studying number representations? And why might cross-cultural perspectives be relevant to this?

In introducing a focus on number representations, it is necessary to outline what we mean when we refer to numerical 'representations'. Some representations of number are clearly observable, such as the tokens, symbols, or marks that children produce in response to a quantitative question. For example, when asked, 'get me five sheep', the set of toys that a child groups together is a concrete representation of a target number ('5'). Similarly, a mark on a line that is bound by specific endpoints (e.g. '0' and '100') is an observable indicator of a number on a scale. Of course, such physical indicators can be accurate or inaccurate, depending on the child creating them. But this is because sets of objects (and marks along a scale) are indices of a different type of representation: those of a child's mental representation of number. We show how these two meanings of 'representation' have come together in much recent work on early numeracy.

We also need to outline why there should be an emphasis on cross-cultural comparisons of number representations. The principal force behind this is the body of work that consistently indicates that some cultures (especially from Far-East Asia) appear to have an early advantage in the realm of numerical and mathematical understanding over other cultures (e.g. the USA and Europe). For example, it has been suggested that Asian children have an advantage over US and European children both when they start formal education at age 5 and when they finish school around the age of 15–16 (Geary, Salthouse, Chen, & Fan, 1996; Huntsinger, Jose, Liaw, & Ching, 1997). Confirmatory evidence of important gains in numeracy before children engage in mathematical calculations has come in our work studying 5 year olds (Muldoon, Simms, Towse, Burns, & Yue, 2011) where Chinese children were roughly 9 months ahead of children from Scotland in terms of performance (using a standardized set of number problems).

A core principle of the work to be considered in this chapter is that examples of external number representations, which offer clues to or windows into children's mental representations, are important to understand mathematical cognition. Mathematical activities, for example mental addition, as well as estimation and graphing, etc., are procedures that are heavily influenced by the underlying conceptual relationships that children have constructed about number. That is, if children do not grasp that there is a compositional structure to numbers that reflects a base-ten system (at least for modern, Western number systems), or that numbers posses a linear scale, then many of the mathematical activities that children might try to accomplish will be compromised.

Numbers, Mathematics, and Representations

We agree with the position that, for children, mathematics is embedded in (and outside) the classroom as a form of cognitive performance. That is, mathematics is something that children *do*, for example when establishing the numerosity of a group of items or discovering what happens when objects are added to, or taken away from, a set. As children get older, they also increasingly engage in symbol manipulation, undertaking mathematical operations on numbers and carrying out mental arithmetic. Of course, it is vital that all these skills, and others, are assessed; we need to know about mathematics attainment, and many contributions to this volume are concerned directly with performance on standardized mathematics tasks. Yet to reiterate, if we examine only the outcome of mathematics problem solving, we risk missing out on a full and proper understanding of numerical processing. Since many aspects of mathematical cognition rely on the way that individuals conceptualize numerical relationships between symbols (including numbers), we need to ensure that we understand the basis on which mathematical cognition is built. And because children are being introduced to new, formal, mathematical concepts and procedures, we need to establish some of the potential sources of difficulty, confusion, and error in their learning. One possible source is the way children mentally represent number, and the influences on the formation of such representations (see Greeno, Riley, & Gelman, 1983, for an analysis of the distinction between conceptual and procedural development among children).

Many mathematical notions, skills, and procedures are universal. For example, it is necessary that 2 + 2 = 4 regardless of the often subtly different ways in which this relationship may be expressed (e.g. across language). Nonetheless, mathematics can be communicated and trained in different ways, and because there is no universal 'gold standard' for learning mathematics, we have a situation whereby children in different countries potentially become exposed to mathematics in different learning contexts. For example, there are cultural differences in teaching styles (Perry, van der Stoep, & Yu, 1993) and in the volume of mandated teaching (Stigler, Lee, & Stevenson, 1987). Furthermore, there is cross-cultural variation in the expectations of mathematics development in the home and in the classroom (Hess, Chang, & McDevitt, 1987). Indeed, what is taught varies too, with the curriculum and its emphasis on certain topics (see Brown, 1996).

All of these differences represent both an enormous opportunity and a challenge to the research psychologist. The variations between groups can be approached as a natural experiment in mathematical cognition, in which independent variables are manipulated in ways that would be nigh on impossible to achieve in experimental projects. At the same time, because there are *so many* factors at play, some covarying, the opportunity to identify causal relationships between one or even a few independent variables and mathematical cognition is very difficult. Moreover, these varied factors are not always under the direct control of the experimenter (i.e. they are quasi-experimental or natural variables). As a consequence, it can difficult to exert the degree of experimental control that one might wish for. Nonetheless, if we are to try to uncover the underlying reasons for cultural differences, we need to start somewhere, and an obvious starting point is the language of number.

Language and Number

One of the potentially important ways in which numerical representations might differ across cultures originates from the way in which languages code number. Obviously, different languages have different names for number values. However, beyond this, languages also differ in how number relationships and concepts are expressed. Thus, a numerical value such as '25' is rendered as being 'twenty-five' in English, which gives some clue—but only a partial one—to the base-ten nature of the number (i.e. 'twenty' is derived from 'two-ten'). In French, the number would be 'vingt-cinq', where 'vingt' means twenty, and thus this is even less transparent as an indicator of a base-ten system since there is no obvious link between this word and the words for 2 and 10 ('deux' and 'dix'). Moreover, if we take '25' as comprising two numbers ('20' and '5') there is no logical case for referring to one before the other when expressing the sum; in Dutch, '25' is 'vijfentwintig' or five-and-twenty, thus placing the unit value first (see Brysbaert, Fias, & Noel, 1998, for an interesting analysis of this reversed ordering). However, some languages appear to support much greater transparency in the way numbers are expressed. In Japanese, '25' would be spoken as 'ni jyu go' which translates as 'two-ten-five', which, as a means of expression is much more clearly a 'literal base-ten' term, and this is common to many Pacific Rim languages, including Chinese (Göbel, Shaki, & Fischer, 2011, provide a recent overview of some of the more exotic number-language systems).

Linguistic differences in themselves, of course, do not necessarily reflect differing conceptualizations. This begs another important question: psychologically, does this matter, or does this just reflect interesting linguistic idiosyncrasies? After all, different languages have different lexicons and there are more words in some languages for certain concepts than are found in other languages. This need not *necessarily* mean that there are differences in underlying cognitions and, indeed, in many cases there is no reason to suspect that the concepts tied to those words differ (see also below). Yet empirical evidence has been offered to suggest that number words may matter for children's development of number concepts.

Irene Miura and colleagues (e.g. Miura, Kim, Chang, & Okamoto, 1988; Miura & Okamoto, 1989) developed a highly influential account of how language-based differences in labelling multi-digit numbers may affect the development of number representations and conceptualizations of place value within those representations. They presented approximately 6-year-old children from a range of cultures and languages with the task of 'showing' the experimenter particular numbers. They provided a large set of physical 'tokens' for this, with unit blocks (i.e. unit cubes) and ten-blocks (a bar marked into ten segments, each segment equivalent to a unit). Children were told that the blocks could be used for showing (i.e. physically representing) the numbers and, following two single digit practice trials where children were 'coached' on how to respond if necessary, they were then presented with a range of numerals to represent with the tokens.

Their principal finding was that children from Pacific Rim countries such as Japan and Taiwan were more likely to employ base-ten, multi-digit representations for the values. That is, these children would often use the appropriate, most 'efficient' combination of ten blocks along with unit blocks to represent the numeral, in contrast with US children who were much more likely to use just the unit blocks. For example, the most efficient way of using tokens to represent '23' is to use two '10' blocks (i.e. 2 × 10) and add three 'unit' blocks; in contrast, the least efficient is to count out 23 unit blocks. Why should there be national differences in 'efficiency'? One explanation is that in Asia, the '2 × 10' cue to the more efficient solution is made explicit in the language, whereas in English it is not. By extension, this raises the possibility that the two groups were drawing on different underlying representations of numbers brought about via linguistic differences (see also Miura et al., 1994, for similar findings with overlapping data).

It is worth noting that these ideas resonate with the more general (and older) issue as to whether thought and language are linked, with the hypothesis that the latter is constrained by the ways in which the former can divide or partition up the world (e.g. Whorf, 1956). The history of the linguistic relativity hypothesis is long and contested, and this chapter is not the place to provide a comprehensive analysis of the Whorfian hypothesis in all its variations. Suffice to recognize that it has been considered in several domains including colour perception (Rosch, 1972; see also Davidoff, Davies, & Roberson, 1999; Kay & Regier, 2006, for recent perspectives), grammatical structure (Au, 1983; Bloom, 1981), and size-comparison terms (Cole, Gay, Glick, & Sharp, 1969). Thus, the possible link between language and numerical cognition is part of a broader issue concerning language and thought.

The cross-cultural data on number representations from Miura and colleagues are certainly striking. That is, they explored an apparently simple and basic task, and probed behaviour that might be expected to draw on a universal understanding of numbers. That they obtained large differences in the strategies for using number tokens between US children and children from several Pacific Rim countries suggests non-universality, even in the first 1 or 2 years of schooling where children begin to encounter formal curricula to help them grapple with cultural number systems. And these differences map onto the degree of transparency evident in how their languages code multi-digit numbers. Nonetheless, is it necessarily the case that these two dimensions are causally linked? That is, while the data are consistent with the idea that number language might shape conceptual ideas about place value, is this the only plausible explanation?

Towse and Saxton (1997) attempted to examine this hypothesis in several ways. They used English-speaking children to assess whether their language was really a constraint on their multi-digit thinking (in work that was later generalized to cross-cultural comparisons by Saxton & Towse, 1998). First, they sought to establish the *specificity* of the idea that number names affect place value understanding. That is, is it the case that certain number words give better cues to children about their underlying origins, as a function of the transparency of the words involved, than others? Although it is hard to vary the language for a number system, it is more straightforward to adapt the experimental protocol so as to induce a closer and more transparent match between numbers and the token referents in the number representation paradigm. In particular, Towse and Saxton employed a standard condition essentially replicating the original methods of Miura and others but with the blocks colour-coded (i.e. 'unit' cubes were green, '10' blocks were blue). Towse and Saxton also employed an extra condition where '20' (orange) blocks were available too. We asked children to explain to a soft toy what the different written number values meant, using the available materials (under the cover story that the toy knew all about the cubes, but nothing about written number).

In the condition where 20-unit blocks were present, target numbers between 20 and 29 had a more linguistically transparent and obvious link to these particular blocks. Two interesting questions, therefore, are (1) did children use the 20-unit blocks only for this particular range, or also for other multi-digit target numbers (i.e. less than 20 or more than 29), and (2) did the introduction of more multi-digit blocks reduce the likelihood that children would laboriously count out the required number of unit blocks? Towse and Saxton found that the introduction of the larger blocks did indeed induce greater use of the multi-digit blocks (both '10' and '20' blocks) and not just for numbers in the twenties. Moreover, they observed fewer children adopting the less-efficient 'unit block only' strategy in this condition. In other words, children's problem-solving strategy was sensitive to the materials they were supplied with, and was not simply an index of their mental representation of numbers. This in turn undermines the idea that English-speaking children are being substantially impaired or delayed in the acquisition of thinking about place values because of an opaque structure for linking numbers to a base-ten system. It seems harder to argue that (the English) language is holding back conceptual development of place value when it can be induced fairly simply.

In the light of these data, Towse and Saxton also sought to evaluate an alternative account of the language–number link, according to which the exact details of the experimental discourse (rather than the materials) may be important in biasing children one way or another when asked to create target sets using blocks of different magnitude. That is, do some children believe that the experimenter is implicitly asking them to use only unit-value blocks, whilst other children recognize that the experimenter legitimizes the use of multi-digit blocks as and where appropriate? Of course, this might interact or overlay any language-based differences in number conceptions. One basis for taking this account seriously is that the example or practice trials given to children in the original task involved single digits. As a result, the practice trials allowed only unit blocks to be used. Whilst this might help children understand what to do, it could also lead some children to think that the experimenter wanted them to use unit blocks only.

Thus, Towse and Saxton (1997) attempted to examine the impact of prompting children towards a multi-unit representation by using an example of a teen number prior to the experimental trials. This was found to have a substantial and significant effect on how 7 year olds (Experiment 2) and 6 year olds (Experiment 3) gave responses to a whole range of multi-digit representation trials. It greatly increased the use of 10-block strategies amongst the children for the full set of trials with different numerals. Moreover, it also led to the emergence of a 'mixed-unit' strategy among some children, particularly 6-year-olds. This is where children represented numbers with a single 10-block and supplemented this with as many unit blocks as were required to make the number that was shown. For a teen number, this produces a 10-block strategy, but with a number such as 24, it involves one 10-block and 14 units. As a result it is clearly not a canonical base-ten response. Children adopting this mixed-unit strategy appeared to take the teen number example trial very literally as 'show me one 10-block and the remainder as units'. This further emphasizes the impact of the experimental scenario and the instructions, over and above linguistic coding of numbers.

But how might this emphasis on the interpretation of instructions explain or lead to cross-national differences? We suggest that children who differ in both their baseline experience of and practice in mathematics might be differentially inclined to infer that *they* could complete the task in an efficient way. In other words, it may be that experiential differences in numerical activities, along with changes in confidence in exploring number tasks even where these are subtle, affect how children interpret the number representation task and thus the strategies that they adopt to produce collections of numbers.

Thus it is important to recognize that the data from Towse and Saxton (1997; see also Saxton & Towse, 1998) do not challenge the very presence of cross-cultural differences in number representations. Far from it. Also, insofar as there is strong empirical evidence for cross-cultural differences in mathematical attainment (e.g. Reynolds & Farrell, 1996), we would expect that children might differ in the handling and processing of numbers. And indeed there is converging evidence that particular tasks such as number reversal (reading '17' as 'seventy-one') can be affected by language structure (Miller & Zhu, 1991).

These data encourage alternative accounts for why there are differences in multi-digit number use across nations. In other words, they speak to the question of whether language is necessarily responsible for conceptual development. They also suggest we need to draw a careful distinction between manipulation of numeric material, likely to involve a variety of cognitive systems including working memory (Baddeley, 1986), and conceptions of numbers. In the former case, variables such as the articulatory lengths of words can affect their memorability and the way they may be manipulated in different formats (e.g. Campbell, 1994; Hoosain & Salili, 1987). In the latter case, the question is whether there are differences in the sophistication of children's understanding as a result of language differences.

Subsequent research has provided further evidence in favour of local effects of language on number comparison tasks; identifying which is the numerically larger of two numbers. Dowker, Bala, and Lloyd (2008) report an advantage for children speaking Welsh (which has a more transparent base-ten number system) over children speaking English in number comparison, though they did not find a similar task advantage for Tamil speakers (for whom teen numbers are coded in a relatively transparent way). We note that such findings speak mostly to the first issue above (language effect on number manipulation) rather than the second (language effect on number concepts). We believe they feed into a general conclusion that can be drawn from studies that span a range of numerical tasks and language groups, and which reiterates the point made earlier. The interpretation of cross-cultural differences in mathematics is not straightforward since there may be both multiple cognitive processes at play and multiple potential causal factors that are responsible. We turn next to the detailed consideration of a separate paradigm, but one that raises some overlapping issues with regards to numeric representations.

The Mental Number Line and The Number Estimation Task

What if the cross-cultural differences found in 6- to 7-year-olds above reflect differences in mental (i.e. conceptual) representations of abstract number, differences that are evident at an earlier age? An increasingly popular and important method used for accessing young children's representation of abstract number is the number-line estimation task. Siegler and Opfer (2003) brought the number estimation task to the attention of many psychologists through the presentation of some highly interesting data. Essentially, they asked children to mark the position on a line where a particular number would fall, given that the line was bounded by 0 as the lower value and 100 as the higher (the number to position task). They also asked children to suggest what number might be denoted when an existing mark was present between the values on a range (the position to number task).

Their data suggested that young children showed a highly compressed scaling of numbers. A 'compressed' number scale is one where, for example, the difference between 0 and 50 is greater than that between 50 and 100 because numbers are not 'spaced' linearly along the scale (see Figure 22.1a). They also showed that older children exhibited a more linear (i.e. appropriate) scaling for the same ranges. In subsequent work that followed up the original demonstration of highly non-linear estimation, Booth and Siegler (2006) reported that the accuracy of children's number estimation was strongly associated with attainment on mathematics tasks, underlining the exciting possibility that the scaling of the mental number line, as indexed by the number estimation task, might index conceptual development in mathematics.

Looking again at Figure 22.1a, when plotted on a graph these 'compressed' estimations of numbers look more like a logarithmic curve than a straight line (which would result

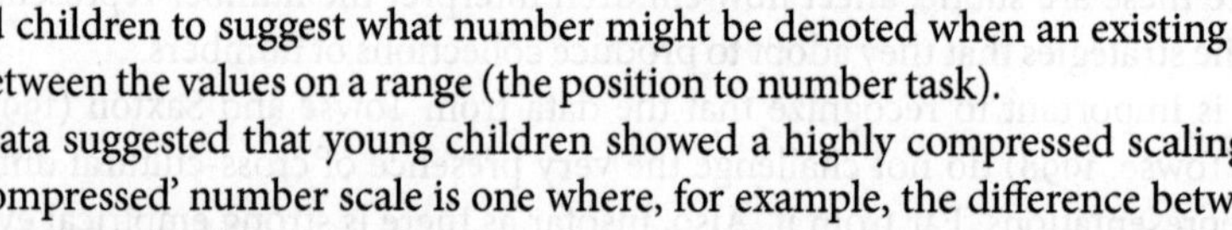

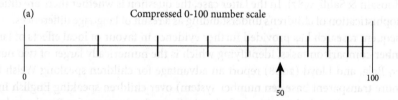

FIGURE 22.1A Compressed 0–100 number scale.

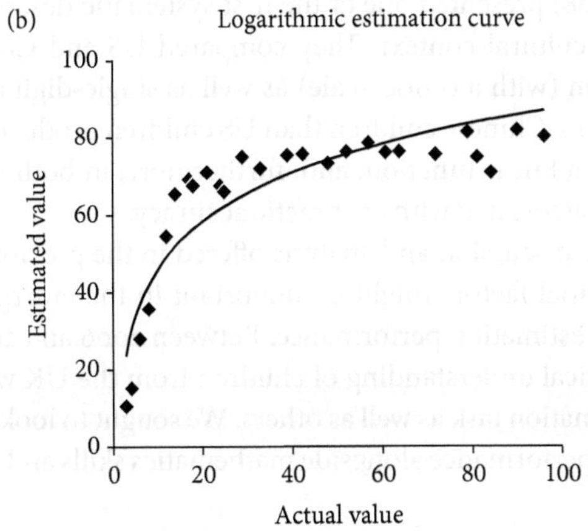

FIGURE 22.1B Logarithmic estimation curve.

from 1–1 correspondence between the target numbers and the child's estimated value—see Figure 22.1b). Notwithstanding the contention by some that children's early data should be described not as logarithmic but more accurately as 'bilinear' (Ebersbach, Luwel, Frick Onghena, & Verschaffel, 2008; Helmreich et al., 2011), there are claims that this 'mental compression' of numbers within a given scale may affect mathematics learning during, and even prior to, schooling. Such claims are founded on the discovery that the quality of children's estimations predicts how well they perform on mathematics problems (Booth & Siegler, 2006; Geary, Bailey, & Hoard, 2009; Siegler & Mu, 2008). According to the account developed by Siegler and Opfer (2003), our mental 'number scale' becomes less compressed with age. However, while the ability to estimate accurately improves steadily during early school years, some children are slower to develop than others, whilst others show incomplete development in this domain, exemplifying an overlapping waves model of development (Siegler & Booth, 2004). The shift from logarithmic to linear patterns of numerical estimates *typically* occurs sometime between the ages of 5 and 8, at least for numbers between 0 and 100 (Booth & Siegler, 2008). Of course, it is also around the age of 5 or 6 that children are introduced to formal mathematics schooling.

The suspicion that non-linear number representations might be a source of difficulty in mathematical learning has been borne out with the more recent finding that the accuracy—and linearity—of young children's estimations correlate with the ability to carry out simple addition problems (Siegler & Mu, 2008). If, as one might reasonably suppose, children self-assess their answers to number problems, then the ability to anticipate when a calculation is reasonable or 'sensible', rather than obviously inaccurate or 'silly', is a key metacognitive attribute. This account is consistent with the association between number-line estimation and mathematics achievement (Siegler & Booth, 2004; see also Dowker, 1998, for an analysis of estimation in mathematics). If it is the case that there is a developmental association between number estimation and other mathematical abilities, it raises the possibility that cross-national differences in mathematics might be explained, at least in part, by differences in the quality of children's mental, conceptual, representations of positive integers.

Siegler and Mu (2008) presented one of the first systematic descriptions of number-line estimation in a cross-cultural context. They compared US and Chinese preschoolers on number-line estimation (with a 0–100 scale) as well as single-digit arithmetic tasks. Their findings were clear; more Chinese children than US children produced number estimations that could be fitted by a linear function, and, furthermore, in both samples that quality of number estimates was associated with arithmetic accuracy.

In the context of the discussion and analysis offered in the previous section, it is important to ask what potential factors might be important in the emergence of cross-cultural differences in number estimation performance. Between 2006 and 2009 we have been able to compare the numerical understanding of children from the UK with children in China, using the number estimation task as well as others. We sought to look at the detailed pattern of number estimation performance alongside mathematics skills and discuss this in the next section.

Estimation as an Index of Conceptual Representation

In 2006 Muldoon (unpublished) conducted an exploratory pilot study that examined cross-cultural differences in the then-recently emerging topic of children's accuracy when estimating numbers along a number line. How might the underlying number representations be stored? Sometimes numbers are encountered as on a vertical straight line (e.g. numbers denoting floors in a tower block); at other times they are presented in a circle (e.g. numbers on a clock face). However, the belief that numbers are stored on a horizontal plane (resulting in the use of a horizontal scale in number-line estimation tasks) is a natural development in light of the evidence for correspondence between numerical magnitude and the 'sidedness' of space. Within a fixed number range (e.g. 0–10) we respond quicker to smaller numbers (e.g. 1, 2, 3) with the left hand than with the right hand. and vice versa, producing a SNARC (Small Numerical Association of Response Codes) effect (Dehaene, 1997). Thus, it appears that numerals (visual and auditory) evoke a magnitude code in the neural system that is effectively an analogue of an integer from the base-ten number system. Importantly, the neural system appears to organize these representations along a left-to-right scale, with zero lying at the leftmost point. Not coincidentally, the area of the brain involved in processing quantity is the same as that employed when estimating magnitudes on a left-to-right axis (Zorzi, Priftis, & Umiltá, 2002).

In our pilot study we restricted the analysis to children's performance on number-line estimating and counting tasks. Thus we considered a precursor skill to arithmetic that was the focus of Siegler and Mu (2008). Six year olds in the UK displayed a linear set of estimations on a small (0–10) scale but a non-linear set on a larger (0–100) scale, a finding that corresponds with Siegler and Opfer's (2003) discovery that even 2 years after starting school, US children apparently refer to a logarithmic-based mental number scale when estimating quantity. In contrast, Chinese children who were between 1 and 2 years *younger* produced estimations that were also linear on the 0–10 scale, but significantly more linear on the 0–100 scale. Crucially, the more accurate mental representations of the younger Chinese children were evident in a population that had yet to start school; if accurate, and abstract representations of number are formed via cultural number practices, then such practices are in play before formal educational curricula are introduced. This begs the question of what 'cultural practices' might be responsible.

Our data revealed that the 'compression' of mental representations of number seemed to be associated with counting proficiency, as the Chinese children outperformed the UK children on all counting tasks given to them. This led us to explore further the developmental associations between estimating skill and a wider battery of number skills. This more extensive study recruited nearly 200 children and confirmed the relative mathematical precociousness of Chinese children (Muldoon, Simms, Towse, Menzies, & Yue, 2011). That is, even though the Chinese children were approximately 10 months younger than the Scottish sample, they were performing nearly as well on a standardized assessment of mathematics attainment, the British Ability Scales (BAS). The BAS number tasks involve a 24-item battery that assesses skills such as reciting number words, counting objects, classifying and matching sets by number, recognizing ordinal relationships, and solving basic addition and subtraction problems

The Chinese sample performed similarly on BAS number tasks despite their younger chronological age. However, the Chinese and UK children were not exactly matched. Therefore we next identified a subsample of both the Chinese and UK children so as to obtain a closer ability-match group with comparable group sizes. Thus, we specifically sought an ability-match design, in contrast with the chronological age-match design from Siegler and Mu (2008). We wanted to determine whether number estimation skill would differ for groups who had reached the same milestones of mathematical development. An ability-match design is valuable in this respect because it can be used to tease apart the trajectory of simultaneously developing skills. If the two samples differed in estimation performance despite equivalent number skills, this would represent good convergent evidence that culturally led differences in the mental number line were helping to drive changes in mathematics. However, we did not find evidence that the Chinese sample were precocious in number estimation. Instead, the effects were in the direction of the (older) Scottish children showing more linear and accurate number estimation.

Recently, we have been able to explore the number estimation task with both cross-sectional and longitudinal approaches, the latter involving repeated testing over the course of a school year (Muldoon, Towse, Simms, Perra, & Menzies, 2013) using on this occasion a sample of British children. This has enabled us to address a range of research issues. For example, the data show that it is feasible to administer the number estimation task to children repeatedly, and that there are consistencies in performance with respect to individual differences from one time point to another. Thus, we found that children's estimation accuracy on the same task correlated over several months, whilst they did not show reliable correlations when estimating with different number ranges even though these were administered at the same time. It appears that although there are stable task strategies, they are quite specific since they vary across estimation ranges. Importantly, this dataset also showed that change in mathematics attainment over time was not directly yoked to changes in estimation performance, casting doubt on whether there is a direct causal link between these skills. That is, whilst number estimation did correlate with mathematics performance, changes in the former did not correspond to changes in the latter.

To summarize this section, we have attempted, first, to show that the number estimation task offers some interesting indicators and insights into the structure of the mental number line. There are some quite dramatic performance changes in the way in which children respond to the apparently simple request to mark where numbers would lie on a bounded scale. These changes are consistent with the idea that children begin with a logarithmic or compressed representation of numbers, which gradually becomes more linear and accurate with age and numerical experience. At the same time, we note that the initial conception of a logarithmic scale is not the only potential account of number estimation performance. Second, we have documented that there is now good evidence from various sources that the number estimation task can and does correlate with mathematics tasks and attainment. Third, we have reported and interpreted evidence that the number estimation task can be studied from a variety of cross-cultural perspectives, and that doing so can throw light on both the nature of number estimations and the emergence of cross-cultural differences in mathematics. The evidence, at present, for the idea that the linearity of the mental number line drives mathematical cognition is not compelling. At the same time, there is much scope to learn more about how the many different facts of numerical thinking, including estimation and computation, are linked.

Conclusions

We look upon the field of cross-cultural research into numerical representations and its development with intrigue, excitement, and caution. As we have already noted, there are fantastic opportunities afforded by the many differences in the way that children experience number and are taught number skills across cultures and concomitant educational settings. Some of these differences arise from the way in which mathematics is taught in schools, while some of these differences are inherent consequences of how languages code number systems.

Researchers therefore have the chance to explore research questions that would otherwise be likely to be seen as being beyond the realms of practicability, whilst at the same time dealing with extant individual differences in a key cognitive skill. However, the degree of experimental control that can be leveraged in cross-cultural research may be sometimes limited, and the interpretive force of conclusions can be blunted by the potential confounds that might be present. Nonetheless, mathematical cognition is much enhanced by the insights that can be gained from cross-cultural studies, and we expect that this contribution will continue for many years to come.

References

Au, T.K. (1983). Chinese and English counterfactuals: the Sapir-Whorf hypothesis revisited. *Cognition*, 15, 155–187.

Baddeley, A.D. (1986). *Working Memory*. Oxford: Clarendon Press.

Bloom, A.H. (1981). *The Linguistic Shaping of Thought: A Study in the Impact of Language on Thinking in China and the West*. Hillsdale, NJ: Lawrence Erlbaum.

Booth, J.L. & Siegler, R.S. (2006). Developmental and individual differences in pure numerical estimation. *Developmental Psychology*, 41, 189–201.

Brown, M. (1996). FIMS and SIMS: the first two IEA international mathematics surveys. *Assessment in Education*, 3, 181–200.

Campbell, J.I.D. (1994). Architectures for numerical cognition. *Cognition*, 53, 1–44.

Cole, M., Gay, J., Glick, J. & Sharp, D. (1969). Linguistic structure and transposition. *Science*, 164, 90–91.

Davidoff, J., Davies, I. & Roberson, D. (1999). Colour categories of a stone-age tribe. *Nature*, 398, 203–204.

Dehaene, S. (1997). *The number sense*. London: Allen Lane The Penguin Press.

Dowker, A. (1998). Individual differences in normal arithmetical development. In C. Donlan (Ed.), *The Development of Mathematical Skills*. Hove, UK: Psychology Press.

Dowker, A., Bala, S. & Lloyd, D. (2008). Linguistic influences on mathematical development: how important is the transparency of the counting system? *Philosophical Psychology*, 21, 523–538.

Ebersbach, M., Luwel, K., Frick, A., Onghena, P. & Verschaffel, L. (2008). The relationship between the shape of the mental number line and familiarity with numbers in 5-to 9-year old children: Evidence for a segmented line. *Journal of Experimental Child Psychology*, 99, 1–17.

Geary, D.C., Bailey, D.H. & Hoard, M.K. (2009). Predicting mathematical achievement and mathematical learning disability with a simple screening tool: the number sets test. *Journal of Psychoeducational Assessment*, 27, 265–279.

Geary, D.C., Salthouse, T. A., Chen, G. & Fan, L. (1996). Are East Asian versus American differences in arithmetical ability a recent phenomenon? *Developmental Psychology*, 32, 254–262.

Göbel, S.M., Shaki, S. & Fischer, M.H. (2011) The cultural number line: a review of cultural and linguistic influences on the development of number processing. *Journal of Cross-Cultural Psychology*, 42, 543–565.

Greeno, J.G., Riley, M.S. & Gelman, R. (1984). Conceptual competence and children's counting. *Cognitive Psychology*, 16, 94–143.

Helmreich, I., Zuber, J., Pixner, S., Kaufman, L., Nuerk, H.-C. & Moeller, K. (2011). Language effects on children's nonverbal number line estimations. *Journal of Cross-Cultural Psychology*, 42, 598–613.

Hess, R. D., Chang, C.M. & Mcdevitt, T.M. (1987). Cultural variations in family beliefs about children's performance in mathematics: comparisons among People's Republic of China, Chinese American, and Caucasian-American Families. *Journal of Educational Psychology*, 79, 179–188.

Hoosain, R. & Salili, F. (1987). Language differences in pronunciation speed for numbers, digit span, and mathematical ability. *Psychologia*, 30, 34–38.

Huntsinger, C.S., Jose, P.E., Liaw, F. & Ching, W. 1997. Cultural differences in early mathematics learning: a comparison of Euro-American, Chinese-American, and Taiwan-Chinese families. *International Journal of Behavioral development*, 21, 371–388.

Kay, P. & Regier, T. (2006). Language, thought, and color: recent developments. *Trends in Cognitive Sciences*, 10, 51–54.

Miller, K. F. & Zhu, J. (1991). The trouble with teens: accessing the structure of number names. *Journal of Memory and Language*, 30, 48–68.

Miura & Okamoto, Y. (1989). Comparisons of US and Japanese first graders cognitive representation of number and understanding of place value. *Journal of Educational Psychology*, 81, 109–114.

Miura, I.T., Kim, C.C., Chang, C. & Okamoto, Y. (1988). Effects of language characteristics on children's cognitive representation of number: cross-national comparisons. *Child Development*, 59, 1445–1450.

Miura, I.T., Okamoto, Y., Kim, C.C., Chang, C.M., Steere, M. & Fayol, M. (1994). Comparisons of children's cognitive representation of number: China, France, Japan, Korea, Sweden, and the United States. *International Journal of Behavioral Development*, 17, 401–411.

Muldoon, K., Simms, V., Towse, J., Burns, V. & Yue, G. (2011). Cross-cultural comparisons of 5-year-olds' estimating and mathematical ability. *Journal of Cross-Cultural Psychology*, 42, 669–681.

Muldoon, K., Towse, J.N., Simms, V., Perra, O. & Menzies, V. (2013). A longitudinal analysis of number estimation, counting skills and mathematical ability across the first school year. *Developmental Psychology*, 49(2), 250–257.

Perry, M., Vanderstoep, S.W. & Yu, S.L. (1993). Asking questions in first grade mathematics classes: potential influences on mathematical thought. *Journal of Educational Psychology*, 85, 31–40.

Rosch Heider, E. (1972). Universals in color naming and memory. *Journal of Experimental Psychology*, 93, 10–20.

Saxton, M. & Towse, J.N. (1998). Linguistic relativity: the case of place value in multi-digit numbers. *Journal of Experimental Child Psychology*, 69, 66–79.

Siegler, R.S. & Booth, J.L. (2004). Development of numerical estimation in young children. *Child Development*, 75, 428–444.

Siegler, R. S. & Mu, Y. (2008) Chinese children excel on novel mathematics problems even before elementary school. *Psychological Science*, 19, 759–763.

Siegler, R.S. & Opfer, J.E. (2003). The development of numerical estimation: evidence for multiple representations of numerical quantity. *Psychological Science*, 14, 237–243.

Stigler, J.W., Lee, S. & Stevenson, H.W. (1987). Mathematics classrooms in Japan, Taiwan, and the United States. *Child Development*, 58, 1272–1285.

Towse, J.N. & Saxton, M. (1997). Linguistic influences on children's number concepts: methodological and theoretical considerations. *Journal of Experimental Child Psychology*, 66, 362–375.

Towse, J.N. & Saxton, M. (1998). Mathematics across national boundaries: cultural and linguistic perspectives on numerical competence. In C. Donlan (Ed.), *The development of mathematical skills*. Hove, England: Psychology Press.

Whorf, B.L. (1956). *Language, Thought and Reality: Selected Writings of Benjamin Lee Whorf* (ed JB Carroll). Cambridge, MA: MIT Press.

Zorzi, M., Priftis, K. & Umiltà, C. (2002). Brain damage: neglect disrupts the mental number line. *Nature*, 417, 138–139.

CHAPTER 23

MATHEMATICS LEARNING IN THE USA AND JAPAN

influences of language

YUKARI OKAMOTO

Mathematical competence is crucial to educational and economic success in contemporary society. Yet, US students do not fare well in mathematics achievement in comparison to their peers in East Asian nations (e.g. Mullis, Martin, & Foy, 2008; OECD, 2010). What can account for achievement differences in mathematics between these high-performing East Asian students and US students? Potential sources of differences are many and multifaceted, and include cultural and contextual factors. For example, varying home and school experiences account for part of achievement differences (see Ng & Rao, 2010 for a recent review). No one factor could explain all of the variances in achievement differences, and it is difficult to disentangle unique contributions of these factors. This chapter suggests yet another factor that may account for part of this variation, namely, spoken language of mathematics.

Throughout the world, mathematics is taught using Arabic numerals. Although the written forms of Arabic numerals are universal, spoken names associated with these numerals vary from language to language. This has implications for how young children acquire numerical skills and knowledge. In particular, the meaning young children assign to numbers – as they learn to count through oral language – may differ depending on the language they speak. Thus, differences in numerical language characteristics may lead to variations in later number processing when attempting numerical reasoning tasks in school.

The goal of the chapter is to examine how spoken number names influence children's attempts at making sense of the Arabic numeration system. A particular focus is placed on variations in numerical language characteristics between East Asian languages, such as Chinese, Japanese, and Korean, and non-East Asian languages, such as English, French, and German. The chapter examines if variations in number naming systems provide different mathematical experiences to the speakers of those languages. It also addresses the question of generality versus specificity of language influences on mathematics competence.

Effects of Number Naming Systems

The number naming systems of East Asian languages, such as Chinese, Japanese, and Korean, have their roots in ancient Chinese. Common among these systems is the structure of the number names. The number names in these languages are organized so that they are congruent with the traditional base-ten numeration system. In this system, the value of a given digit in a multi-digit numeral depends on the face value of the digit (0–9) and on its position in the numeral, with the value of each position increasing by powers of ten from right to left. Number names in these East Asian languages can be generated from a set of base-ten rules and a base sequence of number names. Table 23.1 shows number names for Chinese, Japanese, and Korean, as well as English for comparison purposes. As can be seen, the number names in these East Asian languages follow the base-ten pattern. For example, 11, 12, and 13 are spoken as *ten one, ten two* and *ten three*, respectively. Similarly, no new names are used for decades, allowing one to generate number names. For example, 25 is spoken as *two-tens five* and 65 is spoken as *six-tens five*. In contrast, the number naming systems in western languages such as English, French, and German lack a clear base-ten structure. For example, 11 is spoken as *eleven* in English as opposed to *ten one,* and 80 is *four-twenty* in French as opposed to *eight-tens*. In addition, the order of spoken words and that of Arabic numerals in multi-digit numbers do not necessarily agree. For example, 13 is spoken as *thirteen* in English as opposed to *ten three* and 43 is *three and forty* in German as opposed to *four-tens three*. These variations may have significant impact on primary grade students' acquisition of number concepts within the base-ten system.

Potential effects of East Asian number naming systems were first examined in Miura's (1987) study of Japanese- and English-speaking first-graders who resided in the US, with whom Miura met individually. She first introduced base-ten blocks to the children. These blocks included small cubes (i.e. one blocks), each of which standing for a unit of one, and three-dimensional rectangles (i.e. ten blocks), each of which standing for a unit of ten. No numbers were written on the blocks. Miura showed children how these blocks could be used to count or *show* numbers. Children first worked on two practice items involving single-digit numbers. They were then asked to show a set of five two-digit numbers, one at a time, using the blocks. Once the initial set of five representations were made, children were reminded of their representations and prompted to make alternative representations.

The results showed that Japanese-speaking children's initial preference to represent two-digit numbers was a combination of the precise number of ten and one blocks (e.g. one ten block and three one blocks for 13). This was in stark contrast to English-speaking children who rarely chose to use ten blocks to represent two-digit numbers. That is, they preferred to use one blocks only (e.g. thirteen one blocks for 13). Across the initial and alternative representations, three-quarters of the Japanese-speaking children showed all five numbers using correct combinations of ten and one blocks. In comparison, only half of English-speaking children did so.

Miura's initial effort was not without limitations. For instance, the Japanese participants all resided in the USA. Thus, they were not exposed to oral Japanese language (in particular, number names in everyday conversations, such as during shopping) to the same degree as those who lived in Japan. Another limitation was that the Japanese participants were recruited from a Saturday Japanese School, where they were placed in the first grade

Table 23.1 Number naming systems of English, Chinese, Japanese, and Korean

Number	English	Chinese	Japanese[1]	Korean[2]
1	one	yi	ichi	il
2	two	er	ni	ee
3	three	san	san	sam
4	four	si	shi	sah
5	five	wu	go	oh
6	six	liu	roku	yook
7	seven	qi	shichi	chil
8	eight	ba	hachi	pal
9	nine	jiu	kyuu	goo
10	ten	shi	juu	shib
11	eleven	shi-yi	juu-ichi	shib-il
12	twelve	shi-er	juu-ni	shib-ee
13	thirteen	shi-san	juu-san	shib-sam
14	fourteen	shi-si	juu-shi	shib-sah
15	fifteen	shi-wu	juu-go	shib-oh
16	sixteen	shi-liu	juu-roku	shib-yook
17	seventeen	shi-qi	juu-shichi	shib-chil
18	eighteen	shi-ba	juu-hachi	shib-pal
19	nineteen	shi-jiu	juu-kyuu	shib-goo
20	twenty	er-shi	ni-juu	ee-shib
25	twenty-five	er-shi-wu	ni-juu-go	ee-shib-oh
35	thirty-five	san-shi-wu	san-juu-go	sam-shib-oh
45	forty-five	si-shi-wu	shi-juu-go	sah-shib-oh
55	fifty-five	wu-shi-wu	go-juu-go	oh-shib-oh
65	sixty-five	liu-shi-wu	roku-juu-go	yook-shib-oh
75	seven-five	qi-shi-wu	shichi-juu-go	chil-shib-oh
85	eighty-five	ba-shi-wu	hachi-juu-go	pal-shib-oh
95	ninety-five	jiu-shi-wu	kyuu-juu-go	goo-shib-oh
100	one hundred	bai	hyaku	baek

[1] Sino-Japanese

[2] Sino-Korean.

classrooms according to the age cut-off rules of the Japanese school system. The majority of these children, however, were placed in second grade classrooms in local schools in the USA. Thus, these Japanese participants might have been exposed to more advanced mathematics, in particular, arithmetic with two-digit numbers.

Subsequent studies by Miura and colleagues recruited participants who resided in the target nation and spoke only the official language of that nation. The studies were conducted within the first half of first grade, that is, before students were taught two-digit numbers. The results of those studies were remarkably similar to Miura's original findings. When asked to show the numeral 28, for example, Chinese-, Japanese-, and Korean-speaking first-graders chose to use two ten blocks and eight one blocks, whereas English-speaking first-graders (Miura, Kim, Chang, & Okamoto, 1988) and French- and Swedish-speaking first-graders (Miura, Okamoto, Kim, Chang, Steere, & Fayol, 1994) chose to use 28 one blocks. Across their initial and alternative number representations, all of the East Asian first-graders were able to show at least one of the five numbers using the correct numbers of tens and ones. In contrast, half of the English speakers did not use this type of number representation at all (Miura et al., 1988).

Across the three studies, the common finding is that East Asian speakers differed significantly from non-East Asian speakers in the kinds of representations they made for two-digit numbers. A counter-example to this general finding is seen in the study of English- and Japanese-speaking children who resided in the UK (Saxton & Towse, 1998). Saxton and Towse reported that less than half of Japanese 6-year-olds preferred to use one blocks. Similar to Miura's (1987) initial study, however, this study used Japanese children residing outside of Japan. Thus, these children's exposure to oral numerical language in Japanese might have been limited in comparison to those who experience oral language in Japanese every day and all day long. A significant contribution of Saxton and Towse's study was the experimental manipulation in which half the participants were shown how to construct a two-digit number using a correct combination of ten and one blocks prior to the test phase. This is in contrast to the method used in the previous studies in which children were shown how to construct a one-digit number using one blocks. Saxton and Towse's instructional manipulation proved successful. That is, the majority of English- and Japanese-speaking participants in the experimental condition used the correct numbers of ten and one blocks to show two-digit numbers. Their findings point out how instruction could affect children's choice of base-ten blocks. It remains unanswered, however, whether this instructional manipulation has a lasting influence on children's representation of two-digit numbers. It also remains unanswered if performance resulting from their instruction has any relation to children's numerical competence, such as place value.

Place Value Understanding

The ability to represent two-digit numbers as consisting of tens and ones (as opposed to a collection of ones) facilitates children's understanding of place value – the idea that the value of a digit depends on its relative position in a number. Some children struggle to develop an accurate understanding of multi-digit numbers well into middle grades (Hunter, Turner, Russell, Trew, & Curry, 1994; Kamii, 1986). This difficulty in developing a functional understanding of place value has been linked to later difficulties in learning and performing multi-digit addition and subtraction computations with regrouping (e.g. Ho & Cheng, 1997). The reverse was also found to be true. That is, early mastery of base-ten knowledge in first grade was found to be a reliable predictor of later arithmetic performance in third grade (Moeller, Pixner, Zuber, Kaufmann, & Nuerk, 2011). The question of interest then was whether or not the number naming system depicting the rules of base ten, such as Chinese, Japanese, and Korean, might provide an effective tool for those who use that system.

To address this question, Miura and colleagues (Miura, Okamoto, Kim, Steere, & Fayol, 1993; Miura & Okamoto, 1989) gave first-graders a set of five problems about place value of two-digit numbers. An example of an easier problem in the set was to explain the relation between the written numeral, 44, and a representation of 44 as four ten blocks and four one blocks. An example of a more difficult problem was to explain a relation between the written numeral 13 and a representation of 13 as four one blocks in each of three clear plastic cups (a total of 12 one blocks) and an additional one block set next to the plastic cups (see Figure 23.1). Miura et al. (1993) reported that overall Japanese- and Korean-speaking first-graders showed significantly greater understanding of place-value than did English-, French-, and Swedish-speaking counterparts. This was particularly so for the easier problems in which over 90% of the Japanese- and Korean-speaking first-graders were able to explain that the 4 in the tens column stands for four ten blocks and the 4 in the ones column stands for four one blocks. In contrast, the majority of the English-, French-, and

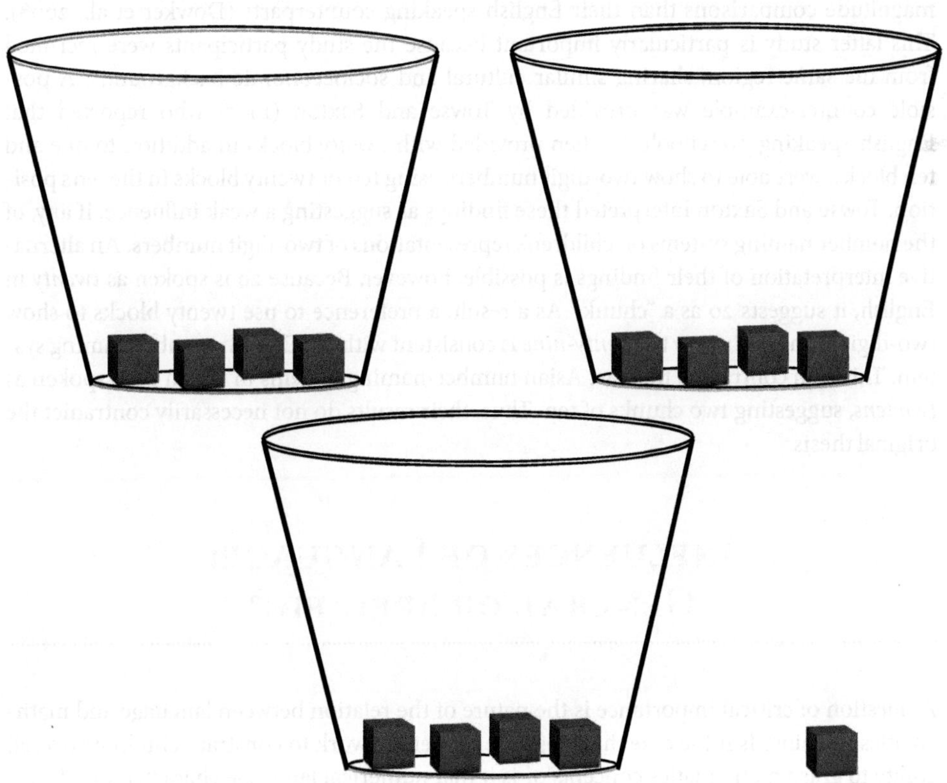

FIGURE 23.1 A test item assessing children's place value understanding.

Swedish-speaking first-graders failed to demonstrate this understanding. On more difficult problems, East Asian language speakers showed stronger understanding of place value than their non-East Asian language speakers, although the majority of children were not able to explain for example, why three cups each with four one blocks and an additional one block would represent numeral 13. Their studies showed clear evidence for differences in first-graders' developing understanding of place value between East Asian and non-East Asian groups. It requires further evidence, however, to make a claim that this early advantage of East Asian speakers leads to more robust understanding of place value beyond first grade.

Support for Miura and colleagues' thesis is seen in other studies. Evidence from these other studies suggests that the compatibility of number names with written notations influences the rate at which children form an understanding of multi-digit numbers such as that of grouping, part-whole relations, and place-value (Alsawaie, 2004; Dowker, Bala, & Lloyd, 2008; Ho & Fuson, 1998). That is, children's mental connections among multi-digit numbers and place value could be facilitated if the structure of number names and the written numerals are closely related. On the other hand, this process could be hindered if there is a disconnect between the number naming system and the written numerals that convolutes the relations between these two representations. For example, Arabic-speaking first-graders, who use a number naming system similar to that of English, showed initial preference for using one blocks to represent two-digit numbers (Alsawaie, 2004). Welsh-speaking children, who use a number naming system similar to that of Japanese, did better on two-digit magnitude comparisons than their English-speaking counterparts (Dowker et al., 2008). This latter study is particularly important because the study participants were recruited from the same region, sharing similar cultural and socioeconomic backgrounds. A possible counter-example was provided by Towse and Saxton (1997) who reported that English-speaking preschoolers, when provided with *twenty* blocks in addition to one and ten blocks, were able to show two-digit numbers using ten or twenty blocks in the ten's position. Towse and Saxton interpreted these findings as suggesting a weak influence, if any, of the number naming systems on children's representations of two-digit numbers. An alternative interpretation of their findings is possible, however. Because 20 is spoken as *twenty* in English, it suggests 20 as a "chunk." As a result, a preference to use twenty blocks to show two-digit numbers *twenty* to *twenty-nine* is consistent with the English number naming system. This is in contrast to the East Asian number-naming systems in which 20 is spoken as *two tens*, suggesting two chunks of ten. Thus, their results do not necessarily contradict the original thesis.

Influences of Language: General or Specific?

A question of critical importance is the nature of the relation between language and mathematics learning. Is it the case that language influences work to constrain children's overall ability to grasp mathematics concepts or is it that numerical language characteristics facilitate or hinder specific aspects of mathematics learning? The studies examining influences of

East Asian number naming systems suggest that the systematic nature of their number naming systems works as an effective tool to think about multi-digit numbers in the base-ten system (e.g. Miura & Okamoto, 1999, 2003). Children speaking these East Asian languages are exposed to this linguistic tool from a young age. This early exposure may result in East Asian children's early mastery of place value. As part of a larger study testing Case's developmental theory (1992; Case & Okamoto, 1996), Okamoto and colleagues (Okamoto, 2010; Okamoto, Case, Bleiker, & Henderson, 1996) reported that children in the USA developed a good understanding of place value by 8 years old. In contrast, Japanese children showed this understanding as early as 6 years old. The effectiveness of the number naming systems of East Asian languages may be specific to the early mastery of place value. However, because mastery of place value is foundational to multi-digit arithmetic (Moeller et al., 2011), speakers of East Asian languages might enjoy this linguistic advantage into the middle grades.

This advantage is seen not only in how children think about multi-digit numbers, but also how solutions to such problems are taught in school. Consider the following subtraction problem: 37 – 29. Once children realize that there are not enough units in the ones column (7) from which to subtract 9, children are taught to trade or regroup one ten from the tens column for ten ones. After regrouping, children in the USA are then taught to subtract 9 from 17 to come up with an answer of 8. This solution method fits well with English number names that encourage a collection of ones. In contrast, East Asian languages afford another solution method: 17 spoken as *ten-seven* allows one to think of 17 as consisting of not seventeen ones but one ten and seven ones. This invites the solution of subtracting 9 from 10 and adding the resulting 1 to 7 to reach the answer 8. As Miura and Okamoto (2003) pointed out, the latter solution takes advantage of the ten-based number name system. By eliminating the need to subtract from a two-digit number, the second solution places less demand on working memory and is thus less prone to error (Fuson, 1992). Murata (2004) also pointed out the importance of the Japanese number naming system in first-graders' acquisition of addition solutions methods.

Studies of preschoolers paint a different picture. Miller, Smith, Zhu, and Zhang (1995) asked Chinese and US preschoolers (3–5 years old) to recite number names as high as they could without counting any objects or on fingers (i.e. abstract counting). Miller and colleagues found that Chinese children at ages 4 and 5 (but not at age 3) were able to recite more number names and did so more accurately than their US counterparts. The authors explained these differences in terms of the base-ten organization of the Chinese number naming system, as well as short syllables involved in pronouncing Chinese number names. When the same children were asked to count small sets of objects or to solve simple numerical problems, however, there was little difference between Chinese and US children. Thus, the Chinese advantage was seen only in the accuracy and amount of number names recalled. This led Miller et al. to conclude that the effects of the Chinese number naming system over that of English at the preschool level are limited to specific aspects of mathematics competence.

Chinese kindergartners (5–6 years old), however, were found to outperform their US counterparts. Geary, Bow-Thomas, Fan, and Siegler (1993), for example, reported that Chinese kindergartners solved more addition problems correctly and used more advanced solution strategies in doing so than US kindergartners. These results per se could have been due to a host of factors, such as more time spent on mathematics at home and school. What is intriguing among the findings is that these same Chinese children also showed superior numerical memory span than their US counterparts. Geary et al. (1993) attributed Chinese children's superior performance on memory span to Chinese number names, which, in turn, enabled them to make use of available memory space to acquire strategies that produced accurate solutions. Chinese number names are not only systematically designed, but also use short syllables, both of which may lead to greater memory space for further processing of numerical information. It should be noted that syllables in Japanese and Korean number names are longer than those in Chinese number names. Thus, it is conceivable that the developmental timetable in the numerical domain during preschool years may not be the same for Japanese and Korean children as for Chinese children.

Effects of Single, Dual, or No Number Name Systems

Use of culturally developed symbol systems may restructure mental activities, but not basic human facilities, such as memory and numerical reasoning. Current thinking suggests that humans have access to two innate systems of numerical representations – an object-file system to keep track of a small number of up to three or four items and an analogue-magnitude system to estimate large numbers (Carey, 2009; Feigenson & Carey, 2005; Feigenson, Carey, & Spelke, 2002). These systems do not require the use of number names to represent quantities. Speakers of any language possess abilities to estimate large numbers and represent small sets of quantities. However, language appears to make a difference beyond these two functions. Pica, Lemer, Izard, and Dehaene's (2004) study clearly demonstrated this point. The authors worked with speakers of Mundurukú, an Amazonian language. Mundurukú has words that roughly correspond to numerical values of 1 through 5, with no apparent verbal counting system. When asked to estimate relative magnitudes of quantities, performance of Mundurukú children and adults was similar to that found in other cultures with elaborate number naming systems. When asked to state cardinal values of sets with 1 to 15 dots, however, their responses were accurate for set sizes of four and fewer, but not for larger set sizes. In the absence of cultural demands for enumeration skills, Mundurukú speakers did not (or have yet to) develop a language-based counting system for exact arithmetic.

Number naming systems of East Asian languages, such as Japanese, provide advantages to the speakers of those languages. These advantages, however, could be limited to abstract counting and base-ten knowledge (and possibly greater use of working memory space by young Chinese speakers). Potential disadvantages have also been reported. For example, Song and Ginsburg (1987) found that Korean preschoolers lagged behind their peers in the USA in informal mathematics, although primary grade children in Korea outperformed their US counterparts on both formal and informal mathematics. The authors speculated that, among various social/cultural factors, there are potential negative effects of the Korean number naming system. In Korean, two number naming systems are used in everyday life (Song & Ginsburg, 1988). One has a structure of base ten, similar to the Chinese system, referred to as Sino Korean. Mathematic instruction is typically done using this system. The other system uses traditional Korean words (referred to as indigenous Korean). This latter system organizes the words for one through nineteen in a base-ten manner and

Table 23.2 Two number naming systems in Korean

Number	English	Indigenous-Korean	Sino-Korean
1	one	hana	il
2	two	dool	ee
3	three	set	sam
4	four	net	sah
5	five	dasut	oh
6	six	yusut	yook
7	seven	ilgop	chil
8	eight	yudulp	pal
9	nine	ahop	goo
10	ten	yul	shib
11	eleven	yul-hana	shib-il
12	twelve	yul-dool	shib-ee
13	thirteen	yul-set	shib-sam
14	fourteen	yul-net	shib-sah
15	fifteen	yul-dasut	shib-oh
16	sixteen	yul-yusut	shib-yook
17	seventeen	yul-ilgop	shib-chil
18	eighteen	yul-yudulp	shib-pal
19	nineteen	yul-ahop	shib-goo
20	twenty	sumul	ee-shib
25	twenty-five	sumul-dasut	ee-shib-oh
35	thirty-five	sulheun-dasut	sam-shib-oh
45	forty-five	maheun-dasut	sah-shib-oh
55	fifty-five	shiheun-dasut	oh-shib-oh
65	sixty-five	yesun-dasut	yook-shib-oh
75	seven-five	ilheun-dasut	chil-shib-oh
85	eighty-five	yudeun-dasut	pal-shib-oh
95	ninety-five	aheun-dasut	goo-shib-oh
100	one hundred	baek	baek

the words for twenty to ninety-nine much like the English number naming system (see Table 23.2).

To examine the effects of this dual number naming system, Song and Ginsburg (1988) asked Korean children (3–9 years old) to recite number names as high as they could (i.e.

abstract counting). They found that young Korean children showed difficulty reciting number names and their performance was worse than that of US children.[1] But by 6 or 7 years old, Korean children's performance surpassed that of their US counterparts and was at the level of their Chinese peers.[2] Young Korean children may initially suffer from this *numerical bilingualism* (Song & Ginsburg, 1988), which might produce a cognitive overload that, in turn, would restrict working memory space to practice arithmetic skills. However, once the dual system is mastered by age 6 or 7, Korean children appear to benefit from the base-ten structure of the Sino Korean number naming system (Song & Ginsburg, 1987). The initial difficulty experienced by young Korean children appears to stem from their attempt to relate two ways of expressing the same quantity – an experience similar to what bilingual children go through (e.g. Bialystok, 2011; Hakuta, 1986; Hoff, Core, Place, Rumiche, Senor, & Parra, 2012).

The Japanese language also includes two number naming systems. Similar to the Korean dual system, one is rooted in ancient Chinese and follows the base-ten structure (referred to as Sino Japanese). The other system uses traditional number names (referred to as indigenous Japanese). In this latter system, only those words that correspond to one to ten are typically used in everyday life (see Table 23.3). Similar to the Korean number naming systems, this indigenous system is rarely used in school mathematics instruction. Thus the potential impact of the Japanese dual system on children's acquisition of base-ten knowledge appears to be minimal.

Children of all languages must learn the number names specified by the language they speak, and the correspondence between the proper number names and the quantities to which those names refer (e.g. Baroody, Lai, & Mix, 2006; Carey, 2004; Spelke, 2003). Japanese and Korean toddlers must do so using not one, but two sets of number names that refer to single concepts of *oneness, twoness, threeness*, and beyond. Thus, the dual number naming systems of Japanese and Korean may delay the acquisition of the cardinal meaning of number. The data from Sarnecka, Kamenskaya, Yamana, Ogura, and Yudovina's (2007)

Table 23.3 Two number naming systems in Japanese

Number	English	Sino-Japanese	Indigenous -Japanese
1	one	ichi	hi
2	two	ni	hu
3	three	san	mi
4	four	shi	yo
5	five	go	itsu
6	six	roku	mu
7	seven	shichi	nana
8	eight	hachi	ya
9	nine	kyuu	kokono
10	ten	juu	tou

study provide support for this speculation. Sarnecka et al. asked English-, Japanese-, and Russian-speaking children 2–3 years old[3] to count small sets of objects. They also asked the children to select the precise number of items from a larger set (i.e. Give-a-Number task; Wynn, 1992). The authors reported that young Russian-speaking children were able to count three or four objects, whereas English- and Japanese-speaking children counted up to four or five objects. When asked to give a specific number of items, however, Japanese children could only give one item, whereas their US and Russian counterparts gave two items. When individual children were classified as knowing the cardinal meaning of one, two, or three, the authors reported that almost all of their US and Russian children knew the meaning of one (i.e. they are '*one knowers*'), whereas only half of the Japanese children were 'one knowers'. In addition, more than half of the US and Russian children knew the meaning of two (i.e. they are '*two knowers*'), whereas only about one-fifth of the Japanese children were classified as such. Japanese children appeared to be delayed in numeral comprehension (i.e. cardinal meaning of numbers). More recently, Barner, Libenson, Cheung, and Takasaki (2009) confirmed this finding for Japanese 2-year-olds. When the performance of 2-, 3-, and 4-year-old children was examined, they found that Japanese 2-year-olds were delayed in numeral comprehension in comparison to their English-speaking counterparts. This delay, however, disappeared when 3- and 4-year-old children's performance was compared. The results of these two studies suggest that the dual number names of Japanese may hinder the onset of numerical comprehension.[4] However, the influence of language appears to be minimal and is quickly overcome in the early years.

Studies of toddlers, as well as preschoolers, suggest that the dual number naming systems of Japanese and Korean appear to affect the onset of numerical comprehension and continue to affect Korean preschoolers' mastery of the base-ten system. Once mastered, however, the Chinese-based number naming systems become a useful tool to reason with natural numbers, which in turn, enable East Asian children to excel in specific areas of mathematics learning.

Summary and Conclusion

This chapter examined the idea that spoken language of mathematics could be a potential factor in explaining part of variations in mathematics achievement between high-performing East Asian students and US students. A particular focus was placed on differences in numerical language characteristics between East Asian (e.g. Japanese) and non-East Asian languages (e.g. English). The chapter traced various findings in roughly chronological order in each section. This section briefly summarizes these findings in developmental order and comes to tentative conclusions about the nature of the relation between language and mathematics, i.e. whether numerical language characteristics influence children's overall mathematics learning or only specific aspects of it.

Japanese 2-year-olds are delayed in acquiring the meaning of oneness, twoness and threeness. Japanese toddlers who hear two sets of words that refer to oneness, twoness, and threeness are delayed in numerical comprehension in comparison to English- or Russian-speaking toddlers (Sarnecka et al., 2007). The initial delay, however, disappears by 4, if not 3 years old (Barner et al., 2009).

Chinese preschoolers and kindergartners excel in numerical competence. During preschool years, Chinese-speaking children show superior performance in the accuracy and amount of number names recalled in comparison to their US counterparts (Miller et al., 1995). Chinese kindergartners also show superior performance in numerical memory span (Geary et al., 1993). Both these results are explained in terms of Chinese number names containing shorter syllables than English names and are organized according to the rules of base ten.

Korean preschoolers are delayed in their mastery of numerical competence. The dual number naming systems used in Korean may initially delay their mastery of cardinal meanings of numbers. Korean preschoolers who must learn two different words for each cardinal value had difficulty reciting number names (Song & Ginsburg, 1988).

Chinese, Japanese, and Korean primary grade children show earlier mastery of place value. Korean children's initial difficulty with the dual number naming systems disappears by the time they begin kindergarten or first grade (Song & Ginsburg, 1988). By early first grade, these children appear to think of two-digit numbers in terms of tens and ones, and begin to show a more mature understanding of place value than do their non-Asian language speaking counterparts (e.g. Miura et al., 1993).

In short, these findings point to the advantages and disadvantages of East Asian languages. During the toddler and preschool years, Japanese- and Korean-speakers face challenges that are rooted in the characteristics of the dual number naming systems. These disadvantages, however, are quickly overcome. By the time they enter kindergarten or first grade, Chinese-, Japanese-, and Korean-speakers develop a preference to think of two-digit numbers as units of tens and ones, which, in turn, contribute to their early mastery of place value.

Reflecting on the findings emerged from the literature, I tentatively reach a conclusion that, in general, influences of spoken language, be it positive or negative, are confined to specific aspects of mathematics learning. One aspect of language influences that has more general impact on children's learning is place value. Because understanding of the base-ten system is foundational to building children's ability to reason with multi-digit numbers, early mastery of place value has broader implications for children's learning of a wide range of arithmetic problems than place value itself (e.g. Miura & Okamoto, 1999). It is thus reasonable to speculate that differences in numerical language characteristics lead to varying degrees of understanding of base-ten knowledge, which, in turn, contribute to the frequently reported, widening achievement gaps between East Asian and US children at later grades (e.g. Mullis, Martin, & Foy, 2008; OECD, 2010). Language alone, of course, cannot account for all of the variance in achievement differences. Together with children's varying home and school experiences, language is an important factor in explaining mathematics achievement differences between the USA and East Asia.

Notes

1. US data were from Fuson (1982).
2. Chinese data were from Miller and Stigler (1987).
3. The authors did not distinguish 2-year-olds from 3-year-olds for analysis. Children were between 2 years 9 months and 3 years 7 months old.
4. Japanese, as well as Chinese, lacks singular-plural markers. This may affect children's acquisition of singular/plural distinction. This possibility was examined in the study of Japanese- and Mandarin-speaking 2-year-olds (Li, Ogura, Barner, Yang, & Carey, 2009). The authors reported

that both groups of children were able to distinguish between singular and plural sets despite the lack of singular-plural knowledge

References

Alsawaie, O.N. (2004). Language influence on children's cognitive representation of number. School Science and Mathematics, 104, 105–111.

Barner, D., Libenson, A., Cheung, P., & Takasaki, M. (2009). Cross-linguistics relations between quantifiers and numerals in language acquisition: evidence from Japanese. Journal of Experimental Child Psychology, 103, 421–440.

Baroody, A.J., Lai, M-L., & Mix, K.S. (2006). The development of young children's number and operation sense and its implications for early childhood education. In O. Saracho & B. Spodek (Eds), Handbook of Research on the Education of Young Children (pp. 187–221). Mahwah, NJ: Lawrence Erlbaum Associates.

Bialystok, E. (2011). Reshaping the mind: the benefits of bilingualism. Canadian Journal of Experimental Psychology, 65, 229–235.

Carey, S. (2004). Bootstrapping and the origin of concepts. Daedalus, 59–68.

Carey, S. (2009). The Origin of Concepts. New York: Oxford University Press.

Case, R. (1992). The role of central conceptual structures in the development of children's scientific and mathematical thought. In. A. Demetriou, R. Shayer, & A. Elfklides (Eds), Neo-Piagetian Theories of Cognitive Development: Implications and Applications for Education (pp. 52–64). London: Routledge.

Case, R., & Okamoto, Y. (1996). The role of central conceptual structures in the development of children's thought. Monographs of the Society for Research in Child Development, 61 (1–2, Serial No. 246).

Dowker, A., Bala, S., & Lloyd, D. (2008). Linguistics influence of mathematical development: how important is the transparency of the counting system? Philosophical Psychology, 21, 523–538.

Feigenson, L., & Carey, S. (2005). On the limits of infants' quantification of small object arrays. Cognition, 97, B13–B23.

Feigenson, L., Carey, S., & Spelke, E. (2002). Infants' discrimination of number vs. continuous extent. Cognitive Psychology, 44, 33–66.

Fuson, K.C. (1982). An analysis of the counting-on solution procedure in addition. In T.P. Carpenter, J.M. Moser, & T.A. Romberg (Eds), Addition and Subtraction: A Cognitive Perspective (pp. 67–82). Hillsdale, NJ: Erlbaum.

Fuson, K.C. (1992). Research on whole number addition and subtraction. In D.A. Grouws (Ed.), Handbook of Research on Mathematics Teaching and Learning: a Project of the National Council of Teachers of Mathematics (pp. 243–275). New York: Macmillan.

Geary, D.C., Bow-Thomas, C.C., Fan, L., & Siegler, R.S. (1993). Even before formal instruction, Chinese children outperform American children in mental addition. Cognitive Development, 8, 517–529.

Hakuta, K. (1986). Mirror of Language: The Debate on Bilingualism. New York: Basic Books.

Ho, C.S., & Cheng, F.S. (1997). Training in place-value concepts improves children's addition skills. Contemporary Educational Psychology, 22, 495–506.

Ho, C.S., & Fuson, K.C. (1998). Children's knowledge of teen quantities as tens and ones: comparisons of Chinese, British, and American kindergarteners. Journal of Educational Psychology, 90, 536–544.

Hoff, E., Core, C., Place, S., Rumiche, R., Senor, M., & Parra, M. (2012). Dual language exposure and early bilingual development. Journal of Child Language, 39, 1–27.

Hunter, J., Turner, I., Russell, C., Trew, K., & Curry, C. (1994). Learning multi-unit number concepts and understanding decimal place value. Educational Psychology, 14, 269–282.

Kamii, C. (1986). Place value: an explanation of its difficulty and educational implications for the primary grades. Journal of Research in Childhood Education, 1, 75–86.

Li, P., Ogura, T., Barner. D., Yang, S-J., & Carey, S. (2009). Does the conceptual distinction between singular and plural sets depend on language? Developmental Psychology, 45(6), 1644–1653.

Miller, K.F., Smith, C.M., Zhu, J., & Zhang, H. (1995). Preschool origins of cross-national differences in mathematical competence: the role of number-naming systems. Psychological Science, 6, 56–60.

Miller, K.F., & Stigler, J.W. (1987). Computing in Chinese: cultural variation in a basic cognitive skills. Cognitive Development, 2, 279–305.

Miura, I.T. (1987). Mathematics achievement as a function of language. Journal of Educational Psychology, 79, 79–82.

Miura, I.T., Kim, C.C., Chang, C.-M., & Okamoto, Y. (1988). Effects of language characteristics on children's cognitive representation of number: cross-national comparisons. Child Development, 59, 1445–1450.

Miura, I.T., & Okamoto, Y. (1989). Comparisons of U.S. and Japanese first graders' cognitive representation of number and understanding of place value. Journal of Educational Psychology, 81, 109–113.

Miura, I.T., & Okamoto, Y. (1999). Counting in Chinese, Japanese, and Korean: Support for number understanding. In W. G. Secada & C. A. Edwards (Eds.), Changing the Faces of Mathematics: Perspectives on Asian Americans and Pacific Islanders (pp. 29–36). Reston, VA: National Council of Teachers of Mathematics.

Miura, I.T., & Okamoto, Y. (2003). Language supports for mathematics understanding and performance. In A.J. Baroody & A. Dowker (Eds), The Development of Arithmetic Concepts and Skills: The Construction of Adaptive Expertise (pp. 229–242). Mahwah, NJ: Lawrence Erlbaum Associates.

Miura, I.T., Okamoto, Y., Kim, C.C., Chang, C.-M., Steere, M., & Fayol, M. (1994). Comparisons of children's cognitive representation of number: China, France, Japan, Korea, Sweden, and the United States. International Journal of Behavioral Development, 17(3), 401–411.

Miura, I.T., Okamoto, Y., Kim, C.C., Steere, M., & Fayol, M. (1993). First graders' cognitive representation of number and understanding of place value: cross-national comparisons – France, Japan, Korea, Sweden, and the United States. Journal of Educational Psychology, 85(1), 24–30.

Moeller, K., Pixner, S., Zuber, J., Kaufmann, L., & Nuerk, H.C. (2011). Early place-value understanding as a precursor for later arithmetic performance – a longitudinal study on numerical development. Research in Developmental Disabilities, 32(5), 1837–1851.

Mullis, I.V.S., Martin, M.O., & Foy, P. (with Olson, J.F., Preuschoff, C., Erberber, E., Arora, A., & Galia, J.). (2008). TIMSS 2007 International Mathematics Report: findings from IEA's Trends in International Mathematics and Science Study at the Fourth and Eighth Grades. Chestnut Hill, MA: TIMSS & PIRLS International Study Center, Boston College

Murata, A. (2004). Paths to learning ten-structured understanding of teen sums: addition solution methods of Japanese Grade 1 students. Cognition and Instruction, 22, 185–218.

Ng, S.S.N., & Rao, N. (2010). Chinese number words, culture, and mathematics learning. Review of Educational Research, 80, 180–206.

OECD (2010). PISA 2009 Results: What Students Know and Can Do: Student Performance in Reading, Mathematics and Science (Volume I). Available at: M http://dx.doi.org/10.1787/9789264091450-en

Okamoto, Y. (2010). Children's developing understanding of number: mind, brain, and culture. In M. Ferrari, & L. Vuletic (Eds), Developmental Relations Among Mind, Brain, and Education: Essays in Honor of Robbie Case (pp. 129–148). New York: Springer.

Okamoto, Y., Case, R., Bleiker, C., & Henderson, B. (1996). Cross cultural investigations. In R. Case & Y. Okamoto (Eds), The Role of Central Conceptual Structures in the Development of Children's Thought. Monographs of the Society for Research in Child Development, 61 (1–2, Serial No. 246), 131–155.

Pica, P., Lemer, C., Izard, V., & Dehaene, S. (2004). Exact and approximate arithmetic in an Amazonian indigene group. Science, 306, 499–503.

Sarnecka, B.W., Kamenskaya, V.G., Yamana, Y., Ogura, T., & Yudovina, Y.B. (2007). From grammatical numbers to exact numbers: early meanings of 'one, 'two', and 'three' in English, Russian, and Japanese. Cognitive Psychology, 55, 136–168.

Saxton, M., & Towse, J.N. (1998). Linguistic relativity: the case of place value in multi-digit numbers. Journal of Experimental Child Psychology, 69, 66–79.

Song, M.J., & Ginsburg, H.P. (1987). The development of informal and formal mathematical thinking in Korean and U.S. children. Child Development, 58, 1286–1296.
Song, M.J., & Ginsburg, H.P. (1988). The effect of the Korean number system on young children's counting: a natural experiment in numerical bilingualism. International Journal of Psychology, 23, 319–332.
Spelke, E.S. (2003). What makes us smart? Core knowledge and natural language. In D. Gentner & S. Goldin-Meadow (Eds), Language in Mind: Advances in the Study of Language and Thought (pp. 277–311). Cambridge, MA: MIT Press.
Towse, J.N., & Saxton, M. (1997). Linguistic influences on children's number concepts: methodological and theoretical considerations. Journal of Experimental Child Psychology, 66, 362–375.
Wynn, K. (1992). Children's acquisition of number words and the counting system. Cognitive Psychology, 24, 220–251.

CHAPTER 24

WHAT IS THERE TO LEARN FROM INTERNATIONAL SURVEYS OF MATHEMATICAL ACHIEVEMENT?

LINDA STURMAN[1]

Introduction

TIMSS and PISA are two high-profile international surveys evaluating the achievement of students in mathematics or mathematical literacy, among other subjects. Since their inception, these large scale surveys have had an impact on education, either by prompting questions about policy (Askew, Hodgen, Hossain, & Bretscher, 2010; Jerrim & Choi, 2013; Stevenson & Stigler, 1992) or by prompting nations directly to change policy in response to survey outcomes (for example, Germany founding the Institute for Educational Quality Improvement, (IQB, 2013) in 2004. Surveys such as these are of increasing importance to policy-makers in the UK and further afield, allowing participants to measure their national achievement at a given point in time and over time, to explore factors related to that achievement and to benchmark their achievement against that of other countries.

At the most basic level, the findings from such international surveys are used to identify the highest performing countries or jurisdictions,[2] with a view to learning from their experience and improving performance in other countries. While this ultimate objective of learning from others' experience is admirable, it cannot be achieved solely through looking at performance data, since the factors affecting performance are complex and, in many cases, interdependent. As a result, further analysis is necessary in order to make informed use of performance outcomes. As well as assessing students, therefore, international surveys also collect background information from key stakeholders, including students, teachers and, in some cases, parents, in order to investigate factors that underlie achievement in the target domains.

International surveys of this type collect a wealth of data relating to achievement and provide enormous scope for analysis. The organisation running each survey produces international reports outlining the main findings of international interest and relevance. These tend to be supplemented by national reports and, in many cases, further national analysis carried out by national centres, investigating questions of interest in their own particular contexts. In addition, the survey databases are released into the public domain, allowing further interrogation by researchers and academics more widely. The databases are accompanied by technical guidance and the offer of training, to support those using the complex datasets and to enable them to do so in an informed manner.

So, potentially, it would appear that there is a lot to be learned from international surveys. Yet, on occasion, doubts are raised about their findings and/or their relevance. This chapter explores some recent headline findings from the mathematical surveys (TIMSS and PISA), the possible uses to which those findings could be put, and some potential abuses of the data. It also looks at issues related to apparently different findings from the two surveys of mathematical achievement/literacy within a country.

As well as direct involvement in the international coordination of some of these international studies, the NFER has long-standing experience as the National Centre coordinating and administering international surveys on behalf of the UK education departments. These studies include TIMSS (mathematics and science) and PISA (reading, mathematical literacy, scientific literacy and problem solving), as well as other international surveys such as PIRLS (reading literacy), ICCS (civics and citizenship), PIAAC (adult competences), and ESLC (European languages). This perspective of experience gives the NFER's international comparisons team strong insight into the nature of the surveys, the richness of the datasets produced, the lessons that can, in theory, be learned from them, and key limitations of the lessons that can be drawn from them.

The Surveys

The two mathematical studies differ in several key aspects. While this means that their outcomes cannot be compared directly, it has the advantage of rendering them complementary. Their different findings reflect their different contexts and help to build a broader national picture for each participating country. Furthermore, where related outcomes from one survey do not correspond exactly to outcomes from the other, the differences can potentially raise interesting questions about mathematical education or mathematical literacy in a given country.

The structure and cycles of the TIMSS and PISA surveys are summarised in Table 24.1, to provide background for the discussion that follows.

Table 24.1 Structure and cycles of TIMSS and PISA

TIMSS	Survey	PISA
Trends in International Mathematics and Science Study	**Title**	Programme for International Student Assessment
International Association for the Evaluation of Educational Achievement (IEA)	**Run by**	Organisation for Economic Co-operation and Development (OECD)

(*continued*)

Table 24.1 Continued

TIMSS	Survey	PISA
Every 4 years	**Frequency**	Every 3 years
9–10-year-olds and 13–14-year-olds	**Student age range**	15-year-olds
Is curriculum-based: assesses mathematics and science achievement	**Focus of survey**	Is not curriculum-based: assesses mathematical literacy, scientific literacy and reading. Other options are problem solving, financial literacy and computer-based assessment of reading and mathematics.
Both subjects in each cycle		All core areas in each cycle, with one as the 'major domain' and the others as 'minor domains' (mathematics was last the major domain in PISA 2003)
TIMSS 2007	**Most recently published cycle***	PISA 2009 (major domain, reading literacy)
TIMSS 2011, due for publication in Dec 2012	**Next cycle to be published**	PISA 2012, due for publication in Dec 2013 (major domain, mathematics)
TIMSS 2007: over 60 countries or jurisdictions, covering all continents	**Participants in most recently published cycle**	PISA 2009: 65 countries participated and a further 10 countries carried out the same assessment in 2010, covering all continents
Collects relevant background data from students, teachers and head teachers	**Background data**	Collects relevant background data from students, teachers and head teachers. Option to also collect data from parents.
Contextualised test items, mostly one or two items (occasionally more) to each context, each item worth one or two marks	**Assessment contexts**	Contextualised test items, usually more than two related items per context, each item worth one or two marks
Mixture of multiple choice and constructed response: at least half of the items to be multiple-choice (Mullis *et al.*, 2009)	**Test items**	Approximately equal numbers of open constructed-response, closed constructed-response† and multiple-choice (OECD, 2009)
Assessments: 2 × 36 minutes (age 9–10 years), 2 × 45 minutes (age 13–14 years). Questionnaires: approximately 30–40 minutes each.	**Administration time**	Paper-based assessment of mathematics, reading and science: 2 hours. Student questionnaire: approximately 35 minutes.

*At the time of writing.

†Open constructed-response items require an extended written response; closed constructed-response items give a more structured context.

Table 24.1 shows the scope of the differences between the two surveys. Each survey addresses a specific context in a specific way. Some potential issues raised by these differences are addressed later in the chapter.

Achievement: Headline Findings

Outcomes from TIMSS 2007, the most recently reported cycle of TIMSS at the time of writing, show that a group of Asian Pacific Rim countries dominates in terms of performance. Similarly, the top performers in PISA 2003, the most recently reported cycle of PISA in which mathematics was the major domain, included several Pacific Rim countries as well as some European countries. Key findings from the PISA 2003 and TIMSS 2007 surveys are summarised below (full details can be found in the international and national reports, which also give details of the within-country distributions).

Pisa 2003 (15-Year-Olds)

- The scale average was set at 500. The scores of participating countries ranged from 550 to 356.
- Hong Kong and Korea were in the top three, along with Finland. Their scores were 550, 542, and 544, respectively. They were closely followed by the Netherlands, Liechtenstein, and Japan (538, 536, and 534, respectively).
- Eleven other countries scored above the average, with scale scores between 532 and 509. These were the remaining participating Pacific Rim country, Macao (527), as well as Canada, Australia, New Zealand, and several European countries, including Denmark and Sweden.
- Four other European countries, including Austria and Germany, scored around the average.[3]
- The remaining 19 countries scored below the scale average. These included Norway, Hungary, Latvia, Russian Federation, United States, and Italy.

Timss 2007, Eighth Grade (13–14-Year-Olds)

- Scores ranged from 598 to 307, with the scale average at 500.
- The three top performers were Chinese Taipei, Korea, and Singapore, with scale scores from 598 to 593. They were followed by Hong Kong and Japan, with scores of 572 and 570, respectively.
- The next highest-scoring group comprised four countries: Hungary, England, Russian Federation and United States, with scale scores from 517 to 508. These four countries, with Lithuania, scored above the international scale average of 500, but were a long way behind the top performers.

- Four countries performed at the scale average: Czech Republic, Slovenia, Armenia, and Australia. The remaining 35, including Sweden, Scotland, Italy, and Norway, were below the scale average.
- Five benchmarking states performed above the average (Massachusetts, Minnesota, Quebec, Ontario, and British Columbia, with scores between 547 and 509). The Basque Country in Spain scored at the average, while one region (Dubai) was below it.

Timss 2007, Fourth Grade (9–10-Year-Olds)

- Scores ranged from 607 to 224, with the scale average at 500.
- Hong Kong and Singapore were the top performers, significantly outscoring all other countries with scale scores of 607 and 599, respectively. They were closely followed by Chinese Taipei and Japan with scale scores of 576 and 568.
- Kazakhstan, Russian Federation, England, Latvia, and the Netherlands formed the next highest-scoring group, with scale scores from 549 to 535, some way below the scores of the top performers.
- The remaining 27 countries performed less well. Of these, eight scored significantly above the TIMSS scale average of 500: Lithuania, United States, Germany, Denmark, Australia, Hungary, Italy, and Austria. Four were not significantly different from the average: Sweden, Slovenia, Armenia, and the Slovak Republic. The remaining 15, including Scotland, New Zealand, Czech Republic, and Norway, were below the international scale average.
- Four benchmarking states performed above the average (Massachusetts and Minnesota in the USA, and Quebec and Ontario in Canada, with scores of 572 to 512). Two Canadian provinces scored at the average (Alberta and British Columbia) and Dubai scored below average.

These findings highlight the variability in performance from highest to lowest performers. At both TIMSS grades and for PISA, there was a wide range of attainment. It is worth noting that the scores cannot be compared directly across grades or across surveys. That is, the average scale score of 500 in one survey does not imply the same level of attainment as the scale average of 500 in another survey. The mean is standardised at 500 in order to allow comparisons within the scale only.

For TIMSS 2007, the highest performers were Asian Pacific Rim countries. This was true for PISA 2003, too, although their performance in this survey was matched by a handful of European countries. For TIMSS, the European countries showed a wide range of attainment, including above and below the international scale average. This was also true of the remaining European countries in PISA.

The Asian Pacific Rim countries differ culturally from the other countries performing above the average in these surveys. For TIMSS, whilst it is true that the least wealthy countries tend to congregate at the lower end of the achievement table (Kazakhstan and Armenia are exceptions), the highest achievers are not necessarily the wealthiest participating countries. They are also not among those spending the largest proportion of their GDP on their public education system or those with the lowest teacher/pupil ratio at primary school level (see Exhibit 3, in Mullis, Martin, Olson, Berger, Milne, & Stanco (Eds), 2008).

In order to understand these findings, therefore, it is useful to know more. The data presented in each survey can provide more detail regarding students' achievement and its context. For example, the international reports for PISA and/or TIMSS include a breakdown of achievement data by gender, by levels of attainment ('benchmarks'), by subject domain and by cognitive process. They also describe trends in attainment over time, and look for associations in the data. These compare achievement with factors related to students' views of their learning, their educational experience, their attitudes towards mathematics, and their home environment. PISA also uses breakdowns by socio-economic status and school type as the basis of much of its analysis of education systems.

Context of Achievement

This contextual information supports the process of learning lessons from complex datasets. In order to draw valid conclusions, it is vital to understand how any given dataset was derived, how to interpret the findings from it, and the context of those findings. The PISA and TIMSS datasets contain achievement data and a wealth of contextual data. This contextual data can suggest variables that might be associated with achievement. This is potentially where the greatest value lies in any international dataset: in suggesting elements of an educational system or home environment that can impact on achievement when changed.

However, causality is an issue when interpreting these aspects of international datasets: it is easy to misinterpret such data. For instance, a reader of an international PISA or TIMSS report might notice that achievement seems to vary with a particular factor, such as number of books in the home. If the trend indicates that achievement increases as the number of books in the home increases, it might then seem logical to conclude that making more books available to all students will raise achievement. But, is it that simple?? It is important to consider the direction of causality. Are students higher achievers because they have more books at home? Or is it the other way round: do their families buy them more books because they see that their young people are capable and interested? There is also a third possibility, that another factor might be implicated. In this example, the number of books in the home may be seen as a proxy measure of socio-economic status or cultural capital. Therefore, it might mask other factors that impact on achievement, such as having a family that is able to provide a supportive environment for learning. In this scenario, a country that decides simply to make more books available to all its students might not achieve its aims of improving its education system, because it is not necessarily the number of books in the home that matters, but possibly the wider environment that supports the provision and usefulness of those books. This example highlights the potential dangers of drawing simplistic solutions from complex data.

Some of the contextual factors that appear to be associated with mathematics achievement internationally, according to PISA and/or TIMSS results, are listed below.

- The more positively students feel towards mathematics, the more highly they achieve.
- The smaller the students' secondary school class for mathematics, the more highly they achieve.

- The more emphasis there is on mathematics homework at secondary school, the more highly students achieve.
- Mathematics achievement is higher where more constructed response items are used in secondary mathematics tests or exams.
- Mathematics achievement is higher in schools with fewer economically disadvantaged students.
- Students achieve more highly in mathematics if they are in schools with more resources.
- Their mathematics achievement is higher in schools with a more positive educational climate.
- Mathematics achievement is higher in schools where teachers feel safe; the same applies in schools where students feel safe.

While many of these findings appear intuitive, they nevertheless hide some variation. For example, in both PISA and TIMSS, positive attitudes towards mathematics were not universal among high achieving students. In some countries, the association between positive attitudes and achievement was stronger than in other countries. Additionally, in some high-performing countries, the proportion of students stating that they enjoyed mathematics was relatively low, compared with the proportions claiming enjoyment in lower-performing countries. This example indicates that a general trend can hide variation and this must be taken into account when making comparisons and drawing conclusions from survey outcomes.

It is also important to bear in mind the caveat mentioned earlier regarding causality and the fact that related variables can act together to influence achievement, or that the effect of one variable might mask the effect of another. Another consideration is that a factor that appears to be associated with achievement might not actually be statistically significant when investigated in a model alongside other variables.

Why Context Matters: Countries of Interest

Three groups of countries are of particular interest to policy-makers. Firstly, those performing especially highly can offer potential insights for other countries into how to improve their mathematics education. Examples of such countries for PISA and TIMSS include Finland, Singapore and Chinese Taipei, high-performing countries, which have taken part in both surveys.

However, high-performing countries might operate in a very different educational or social context from those countries looking towards them for lessons to learn. This might limit the transferability of any lessons to be learned, resulting in 'policy tourism,' rather than true learning. Potential policy lessons should be evaluated and introduced with care. They should take account of the context in which a country's high attainment was achieved, and evaluate whether the factors that appear to underlie its success can, in fact, transfer successfully to another context.

For this reason, the second group of potential interest to policy-makers comprises countries deemed similar to their own. The basis on which policy-makers might make such comparisons may vary over time. For example, depending on what they wish to learn, policy-makers might look to countries with a similar education system to their own, or a similar socio-economic or linguistic profile, or countries deemed to be effective economic competitors. This second group is not necessarily as widely referenced for comparison as the high-performing group, but it might be equally or more useful to policy-makers than high-performing yet dissimilar countries. The same caveat applies to this second group: any country trying to learn lessons from another country needs to consider whether transferability is feasible.

A third group could also be of some interest, although comparison with this group may be even less widespread. This group comprises countries which perform less well than might be expected and which may reform their education systems as a result (for example, Germany in PISA 2000). The outcomes of any reforms or other responses to surprising results are potentially informative for comparator countries.

Drawing Intelligent Conclusions

Achievement, Trends And Ranking

The most obvious lesson that any country will want to learn from an international survey relates to its achievement, specifically, its position in the international rankings and its progress over time. Because the results are reported on a standardised scale, a country's achievement at different time points can be compared directly. Thus, for example, the TIMSS 2007 fourth grade mathematics results are able to indicate that England's performance rose significantly by ten scale points between 2003 and 2007, whereas Scotland's performance rose by four scale points, but this gain was not sufficient to be statistically significant.

However, the situation is less straightforward regarding ranking. A country's ranking is dependent on several factors. Key amongst these is the other countries participating. The list of participating countries tends to change between survey cycles so that, over time, like is not compared with like. For example, the increasing popularity of PISA means that the survey grew from 43 participants in the 2000 cycle to 75 participants in the 2009/2010 cycle.

Even if the list of other participating countries were to remain stable, each country's performance cannot be ranked in isolation. Its ranking will be affected by the trend performance of other countries; whether they improved, declined or remained at the same level of attainment as in the previous cycle. The combined effect of each country's individual trend can render the rankings volatile over time. Thus, rankings should be interpreted with some caution and, when evaluating the progress of any given country, emphasis should be placed on the change in that country's standardised score over time.

For example, in TIMSS 2003, England would have been[4] ranked sixth in terms of its eighth grade mathematics score. In TIMSS 2007, it was ranked seventh on score, suggesting a decline in performance. However, based on the rise in scale points, we know that England's achievement actually increased significantly over that period. The change in England's ranking is not absolute: it will have been affected by the addition of new countries into the list of participants, and by the trends of those which had also participated in earlier cycles.

It is also important to remember that rankings are complex and can be interpreted in different ways. For example, as noted above, for eighth grade mathematics in TIMSS 2007,

England was ranked seventh in terms of score. Its score, however, was not significantly different from that of four other countries, and the scores of the countries outperforming England were similarly banded into groups that were not significantly different from each other. As a result, an alternative conclusion might be that England was ranked in the third band of achievement.

An additional complication for making comparisons over time is that, for PISA specifically, each subject is a major domain only once in every three cycles (i.e. once every nine years). Whilst this design feature enables PISA to report on three subject areas in each cycle, it limits the amount of data that can be collected on two of the domains, whilst maximising the third, in order to maintain manageability.

Causality and Further Analysis

Earlier discussion highlighted that contextual data can be informative for countries, although its interpretation raises potential issues. The difficulty of evaluating causality should not be under-estimated. It is important to recognise which factors might matter when making international comparisons and which might not, to place those hypotheses in context, and to test them further.

There is also the issue that international reports tend to identify factors that show an association with achievement at international level. However, there might be other variables associated with achievement in a subset of countries only. This might be useful comparative information for some countries, but is unlikely to be included in the international reports. Secondary analysis can, therefore, be required. As an example, further analysis was conducted of England's eighth grade mathematics data and that of the five Pacific Rim countries which outperformed England in TIMSS 2007 (Sturman & Lin, 2010). This identified several factors that were significantly associated with achievement in one or more of these six countries, some of which had not been reported in the international report as they had not shown an association with achievement overall. The analysis also showed that, where the same variables had been entered into parallel statistical models for each country, the *combination* of factors significantly associated with achievement differed across countries. The results could not have been divined from the international reports, but raised some interesting questions about how to consider changing the educational context in England in order to raise attainment from good to the excellent levels seen in the Pacific Rim countries.

Same Country, Different Outcomes

A further – and very important – element to take into account when drawing conclusions from international surveys is that different surveys in the same country can appear to give different outcomes, often at different time points but, sometimes, consistently over time. For example, in TIMSS 2007, Australia performed at an average level in eighth grade mathematics, but was above average in PISA 2003. In contrast, the Russian Federation was above average in TIMSS 2007 but below average in PISA 2003. England is another example, with different results in each survey over time. TIMSS results tend to show that England performs well in mathematics, while PISA results tend to suggest that performance in England (and the UK as a whole) is rather more average.

Such differences should be borne in mind when attempting to learn lessons for one country from an international survey: the outcomes of any given survey are a snapshot of performance and background factors in the context of the survey in question and different outcomes might have been obtained under different circumstances. In the case of England, several possibilities have been considered, to try and explain the apparent differences in PISA and TIMSS outcomes over time. In 2006, the NFER conducted a Validation Study of PISA and TIMSS test items on behalf of the Department for Education and Skills (DfES) (Ruddock, Clausen-May, Purple, & Ager, 2006). This explored the extent to which a selection of test items from each survey would be sufficiently familiar to England's students to enable comparison, or whether any differences in relative familiarity might be a reason for difference in performance. The study concluded that test items from both surveys would, on the whole, have been familiar to students. However, it noted that reading demands between the two surveys differed and hypothesised that this might be implicated in some of the differences in achievement.[5]

Of course, another factor that must be considered is that, as noted earlier, the results of surveys on different scales cannot be compared directly. Since scale scores from PISA and TIMSS cannot be compared directly, comparisons of their relative outcomes in England can only be made based on their rankings. However, as also outlined earlier, rankings are a product of the group of countries involved in any given comparison and are potentially volatile. The groups of countries involved in PISA and TIMSS are very different and, thus, rankings within each group are not comparable. This point has been discussed further by Maughan (2011). In addition, and as highlighted in Table 24.1, the contexts of the two surveys differ. Jerrim (2011) highlights several structural and contextual factors that differ between the two surveys, which make it impossible for their findings to be compared directly. He also argues that different cycles within a survey (in this case, PISA) cannot necessarily be compared directly, if the context of the administration of the survey has changed. These factors combined mean that the results of the PISA and TIMSS surveys must be interrogated separately.

Implications

Do caveats such as these mean, therefore, that countries cannot learn from, and perhaps should not try to learn from, international surveys? This question brings us full-circle: this chapter opened with the observation that international surveys, such as these, are increasingly important to policy-makers, allowing them to measure national achievement, explore factors related to that achievement, and benchmark their achievement against other countries. If, however, there are limitations on the interpretation of outcomes based on trends, ranking, causality, and differential outcomes, then the fundamental question becomes: 'Is there, in fact, any purpose to the surveys at all?'.

The survey organisers clearly believe that there is purpose to their endeavours. Regarding PISA, the OECD (2009) states:

> Parents, students, teachers, governments and the general public – all stakeholders – need to know how well their education systems prepare students for real-life situations. Many countries monitor students' learning to evaluate this. Comparative international assessments can extend and enrich the national picture by providing a larger context within which to interpret national performance. They can show what is possible in education, in terms of the quality of educational outcomes as well as in terms of equity in the distribution of learning opportunities. They can support setting policy targets by establishing measurable goals achieved by other systems and help to build trajectories for reform. They can also help countries work out their relative strengths and weaknesses and monitor progress. (p. 9)

Similarly, the IEA (2013) states that:

> Fundamental to IEA's vision is the notion that the diversity of educational philosophies, models, and approaches that characterize the world's education systems constitute a natural laboratory in which each country can learn from the experiences of others. TIMSS participants share the conviction that comparing education systems in terms of their organization, curricula, and instructional practices in relation to their corresponding student achievement provides information crucial for effective education policy-making.

Thus, the explicitly stated purpose of both surveys is to provide information that can help nations to monitor progress and outcomes, and thereby to improve their educational decision-making and support their policy-making. These are ambitious aims and it seems reasonable that countries *should* seek to learn from each other. However, the discussion in this chapter highlights that conclusions from large-scale international datasets should be drawn with extreme care. Any indications of lessons that might be learned should be examined critically before being adopted or acted upon, with consideration given to what the dataset cannot say, in addition to what it appears to say.

For decision-makers, there are two main reasons for taking such a critical and cautious approach. Firstly, policy-making affects learners directly – and learners' education is too valuable to risk unwarranted experimentation. Secondly, and more materially, changing policy can be costly. In the current economic climate, policy decisions with significant financial implications cannot be taken lightly.

Given the difficulty of assigning causality when reporting outcomes from international surveys and the risks inherent in taking performance rankings at face value, it would seem wise to treat international survey outcomes not as end points, but as useful starting points for further investigation of the data in the national context. Survey outcomes might suggest factors that can impact on achievement, but they cannot indicate the direction of the effect and there might be other factors not associated with achievement internationally that, nevertheless, are implicated at national level. As such, the international reports are perhaps most useful in prompting questions that can be investigated through additional analysis of the dataset. Both PISA and TIMSS datasets are released so that they can be interrogated further in this way (OECD, 2010 and IEA, 2008). Clearly, any further analysis needs to be conducted from a position of understanding the data and the context in which it was gathered, and both the IEA and the OECD provide technical guidance with their survey datasets, in order to help researchers to make effective use of the data. As with any research, the surveys will only provide meaningful insight if the data is used appropriately to answer carefully constructed questions, which are tailored to the context and capable of being answered using the data available.

Conclusion

Despite the apparent differences between the surveys and the limitations on interpretation discussed in this chapter, each of the TIMSS and PISA surveys has valid comments to make on the state of mathematics education and mathematical literacy around the world. Their findings can complement each other, even though they might sometimes appear to contradict. In order to use them effectively, however, it is important that their outcomes are not over-interpreted or compared in ways that cannot be justified.

Note

Those interested in learning more about TIMSS, PISA and other international surveys can visit http://www.nfer.ac.uk/what-we-do/international-comparisons. Those interested in learning more on the specific national results and the assessments referred to above can visit http://www.nfer.ac.uk/timss/ and http://www.nfer.ac.uk/pisa/.

Notes

1. Linda Sturman is a Research Director in the Centre for International Comparisons at the NFER and National Research Coordinator for TIMSS (email: l.sturman@nfer.ac.uk). The author wishes to thank NFER colleagues in the Centre for International Comparisons for their contributions to this chapter, in particular Rebecca Wheater (Research Manager and PISA 2012 National Project Manager).
2. The term 'countries' is used throughout the rest of this chapter to refer both to countries and to other participating jurisdictions.
3. The UK did not meet the PISA response rate standards in 2003 and was not included in the international report.
4. In TIMSS 2003, England's sample met the participation guidelines only after the inclusion of a number of replacement schools. Further analysis concluded that its sample was sufficiently robust to be included in analysis, but it was reported 'below the line' and, hence, not included directly in the ranked results.
5. At the time of writing, the NFER is planning an updated version of the Validation Study.

References

Askew, M., Hodgen, J., Hossain, S., & Bretscher, N. (2010). *Values and Variables: Mathematics Education in High-Performing Countries*. London: Nuffield Foundation [online]. Available: http://www.nuffieldfoundation.org/sites/default/files/Values_and_Variables_Nuffield_Foundation_v__web_FINAL.pdf [15 May, 2013].

IEA (2008). *TIMSS 2007 International Database and User Guide* [online]. Available : http://timssandpirls.bc.edu/TIMSS2007/idb_ug.html [15 May, 2013]. (or to access datasets from other cycles, go to http://timss.bc.edu

IEA (2013). *About TIMSS 2011* [online]. Available: http://timssandpirls.bc.edu/timss2011/index.html [15 May, 2013].

IQB (2013). *Competence in Education* [online]. Available at: www.iqb.hu-berlin.de/institut (IQB-Flyer Englisch, accessed 15 May, 2013).

Jerrim, J. (2011). *England's 'Plummeting' PISA Test Scores Between 2000 and 2009: Is the Performance of Our Secondary School Pupils Really in Relative Decline?* DoQSS Working Paper No. 11-09. London: Institute of Education, University of London [online]. Available: http://www.ioe.ac.uk/Study_Departments/J_Jerrim_qsswp1109.pdf [15 May, 2013].

Jerrim, J., & Choi, A. (2013). *The Mathematics Skills of School Children: How Does England Compare to the High Performing East Asian Jurisdictions?* (DoQSS Working Paper No. 13-03). London: Institute of Education, University of London [online]. Available: http://repec.ioe.ac.uk/REPEc/pdf/qsswp1303.pdf [15 May, 2013].

Maughan, S. (2011). '7th or 28th: Why does England Perform so Differently in the International Surveys for Mathematics?' Paper presented at the 37th IAEA Annual Conference, Manila, 23–28 October.

Mullis, I.V.S., Martin, M.O., Olson, J.F., Berger, D.R., Milne, D., & Stanco, G.M. (Eds) (2008). *TIMSS 2007 Encyclopedia: a Guide to Mathematics and Science Education Around the World, Vols 1 and 2*. Chestnut Hill, MA: TIMSS & PIRLS International Study Center, Boston College [online]. Available: http://timssandpirls.bc.edu/TIMSS2007/encyclopedia.html [15 May, 2013].

Mullis, I.V.S., Martin, M.O., Ruddock, G.J., O'Sullivan, C.Y., & Preuschoff, C. (2009). *TIMSS 2011 Assessment Frameworks*. Chestnut Hill, MA: TIMSS & PIRLS International Study Center, Lynch School of Education, Boston College [online]. Available: http://timssandpirls.bc.edu/timss2011/frameworks.html [15 May, 2013].

OECD (2009). *PISA 2009 Assessment Framework: Key Competencies in Reading, Mathematics and Science*. Paris: OECD [online]. Available: http://www.oecd.org/document/44/0,3746,en_2649_35845621_44455276_1_1_1_1,00.html [15 May, 2013].

OECD (2010) *Database PISA 2009* [online]. Available: http://pisa2009.acer.edu.au/ (or to access datasets from other cycles, go to http://www.oecd.org/pisa/pisaproducts/)

Ruddock, G., Clausen-May, T., Purple, C., & Ager, R. (2006). *Validation Study of the PISA 2000, PISA 2003 and TIMSS 2003 International Studies of Pupil Attainment* (DfES Research Report 772). London: DfES [online]. Available: http://webarchive.nationalarchives.gov.uk/20130401151715/https://www.education.gov.uk/publications/eOrderingDownload/RR772.pdf [15 May, 2013].

Stevenson, H.W., & Stigler, J.W. (1992). *The Learning Gap: Why Our Schools are Failing, and What We Can Learn from Japanese and Chinese Education*. New York, NY: Summit Books.

Sturman, L., & Lin, Y. (2010). 'Exploring the Mathematics Gap: TIMSS 2007', *RicercAzione*, 3(1), 43–58.

PART V

NEUROSCIENCE OF MATHEMATICS

CHAPTER 25

FROM SINGLE-CELL NEUROPATHOLOGY TO MATHEMATICS EDUCATION

navigator chapter

ROI COHEN KADOSH

PREVIOUS studies in the field of numerical cognition based their observations mainly on reaction time and accuracy in healthy subjects and in neurological patients (see for example Sections 2, 3, and 6 in this handbook). However, with the advent of non-invasive neuroimaging methods, studies have started to examine the neural correlates, as in the case of functional magnetic resonance imaging (fMRI), and the causal effect of these brain structures, as in the case of brain stimulation, of numerical cognition. Such studies date back to the original study by Roland and Friberg (1985) who used intracarotid ^{133}Xe injection technique to assess changes in regional cerebral blood flow, which indicated increases in metabolism and neuronal activity, during subtraction. Later work has used other methods such as positron emission tomography (PET) (Dehaene et al. 1996), electroencephalography (EEG) (Ruchkin et al. 1991), and fMRI (Dehaene et al. 1998b), and later, single-cell neurophysiology in monkeys (Orlov et al. 2000).

Such a revolution in the way we examine numerical cognition has changed our knowledge of the subject dramatically. For example, different cognitive models of numerical cognition had been based until then mostly on neurological patients (McCloskey et al. 1985) as opposed to later neurocognitive models (Dehaene et al. 2003). As I will discuss in this navigator studies in the field of neuroscience that have extended much of our knowledge, and which have consequences for those interested in cognition, development, evolution, neuropsychology, and education.

This navigator will follow the same order that the chapters are presented in this section, from single-cell neurophysiology in monkeys to human data, and eventually to implications as illustrated by the combination of neuroscience and education.

MONKEY THINKS, MONKEY DO (MATHS?)

Nieder (this volume) in his excellent review provides an integrative knowledge of the advancement that has been made in the last few years.

In several studies Nieder and others have examined how behavioral effects in the field of numerical cognition, which are observed in different species including humans, are represented at the single neuron level in the monkey brain. Such effects include the distance and size effect, processing of continuous and discrete quantity, proportions, as well as symbolic and nonsymbolic numbers.

Aside from providing a better understanding of the relationship between behavior and the brain, with excellent spatial and temporal resolutions (neuronal level, and milliseconds, respectively), this line of research has provided important insights to theories in numerical cognitions that span from developmental psychology to cognitive neuroscience.

An example is the discovery of accumulator neurons with response functions that systematically increased or decreased with increase of stimulus set size (Roitman et al. 2007) corroborated earlier behavioral findings (Meck and Church 1983). The discovery of such neurons in the lateral intraparietal areas LIP by Roitman et al. (2007) has also influenced recent neuroimaging and computational modelling (Roggeman et al. this volume). Nieder and Miller (2003) finding of numerical coding that is best characterized by a logarithmic scale at the neuronal level, has shed further light on the ways in which numerical information can be represented (Dehaene 2003). The discovery of a dissociation between numerosity and numerical symbols (Diester and Nieder 2007) also provided another important contribution to the field, by demonstrating the existence of format-dependent representation at the neuronal level (Cohen Kadosh and Walsh 2009).

In addition, these studies have provided us with some results that changed the focus of the field. For example, most of the studies and discussions in the early days of the neuroscience research were focused on the human parietal lobes. Other brain regions have been largely ignored or have not been discussed. Nieder and colleagues were the first to draw attention to other regions, such as the prefrontal cortex (Nieder et al. 2002), which has since been examined in a more thorough fashion, in adults and especially in children (Kaufmann et al. this volume; Menon, this volume). This led to a shift in attention from a more localist approach that is focused on the parietal lobes to a more distributed approach that examines not just a single brain region, but the network between the parietal, prefrontal, and other brain regions (Menon, this volume). There are many more findings that Nieder describes in his chapter including discrete and continuous quantity processing, ordinal processing, and comparing proportions. Together the different studies that Nieder described in his chapter suggest the phylogenetic foundations for more elaborate numerical skills in humans, which might occur very early on in human infants. Therefore, in the next section I will discuss the neuroscientific knowledge gained from developmental studies of numerical processing.

If Monkeys Can, Human Should: The Developing Brain

Kaufman, Kucian, and von Aster (this volume) open their chapter by noting that, despite the advancement in our understanding of the human brain and the neural correlates of numerical cognition, developmental studies of this sort are relatively scarce. This is unfortunately true, and might be due to a number factors including the relatively greater difficulties in recruiting children (as compared with the comparative ease of recruiting university students) for neuroimaging studies; parental concerns due to unfamiliarity with the research tool and its safety; increased undesired movements by children in comparison to adults inside the magnetic scanner in the case of magnetic resonance imaging (MRI) studies; and other methodological issues that they describe in their chapter. Nevertheless, as neurocognitive models are currently the building blocks upon which the findings of brain imaging studies may be adequately interpreted, and as it is problematic to extrapolate from the adult to the child brain, the need for developmental cognitive neuroscience to inform and revise neurocognitive models is clear (e.g. von Aster and Shalev 2007).

In this respect, some findings in infants support cognitive models in infants and adults. For example Hyde and Spelke (2011) have suggested that six- to seven-and-a-half-month-olds process numerosity in two distinct neural pathways as a function of item set size, which is similar to findings with adults (but see Izard et al. 2008).

One inconsistent finding, as mentioned by Kaufmann et al. (this volume) relates to the similarity between the developing brain and the monkey brain as described in the previous section. While some studies found relative larger involvement of anterior brain regions (the prefrontal cortex), in comparison to more posterior brain region (the parietal lobes) (Rivera et al. 2005), others did not (Cantlon et al. 2006), which might have led to a failure to note prefrontal activation as a function of age in their meta-analysis (Kaufmann et al. 2011). However, Kaufmann et al. (2011, this volume) do report a shift from the anterior part of the intraparietal sulcus (IPS) in children to the posterior part of the IPS in adults, which might reflect a shift in calculation strategies.

Another interesting finding concerns the effect of numerical format on brain activation. In contrast to previous theories of format independent processing (Dehaene et al. 1998a), Kaufmann et al. describe results that indicate that symbolic and non-symbolic number magnitudes are supported by distinct and partly overlapping parietal brain regions, which fits with more recent theories (Cohen Kadosh and Walsh 2009).

In addition, Kaufmann et al. describe the neural correlates in children with and without developmental dyscalculia (see also Kucian et al., this volume), and conclude that children with typical mathematical abilities and children with low maths proficiency differ in their parietal activations, both with respect to lateralization and anterior-posterior location within the parietal cortices.

Lastly, Kaufmann and colleagues identify the neural correlates supporting calculation skills. Findings have disclosed symmetric but rather distributed networks including frontoparietal and occipital brain regions bilaterally. They further discussed how the type of mental operation (e.g. subtraction, addition) and task difficulty modulates these brain networks.

In the third chapter, Menon (this volume) takes into account both the contribution of local neural circuits as well as their distributed patterns of neural connectivity (Bressler and Menon 2010). This chapter provides very important insights that go beyond the current dominant view of the mathematical brain. Menon fractionates arithmetic into several basic neurocognitive processes (see Figure 1 in his chapter) that involves domain-specific processes. These include number forms (inferior temporal cortex) and quantity representations (dorsal parietal lobes). He also describes domain-general processes that include procedural and working memory systems, which are subserved by subcortical and cortical structures (the basal ganglia and frontoparietal circuits, respectively), episodic and semantic memory systems (medial and lateral temporal cortex), and attention allocation, memory retrieval, and goal-directed problem solving (the prefrontal cortex). According to Menon's view, which is supported by previous cognitive and neuroimaging studies, arithmetic requires the orchestration of these neurocognitive functions, which is achieved as a function of brain and cognitive development. After placing the foundations for arithmetic, including the subdivision of the parietal cortex and its functional heterogeneity, Menon examines this mental process in the developing brain both as a function of developmental age (children vs adults), and during sensitive stages of skill acquisition. One of the interesting points is that Menon highlights neurodevelopmental models that go beyond parietal cortex regions involved in number processing. In addition, he highlights different regions including not only the prefrontal cortex, which I mentioned previously, but also the ventral visual stream and the hippocampus. Such brain regions had been previously neglected as most of the research has been focused on adults. Studying the child brain and its development can reveal brain regions that might be critical for scaffolding of the adult mathematical brain.

Lastly, Menon provides a good example of how state-of-the-art neuroimaging methods such as multivariate analysis, diffusion tensor imaging, and resting-state functional connectivity, can provide important insights that cannot be revealed by a traditional univariate analysis of fMRI. Adapting such an integrative approach can foster our understanding of the child mathematical brain and its development.

Holloway and Ansari (this volume) continue to discuss the bridge between child and adults brains by combining the current scientific understanding of numerical symbol processing at the behavioral and neural level. First they discuss the issue of numerical symbols and their semantic representation. They suggest that comparing symbolic and nonsymbolic numbers can help to uncover how the representation of numerical information is changed through the process of learning numerical symbols. Such a change, in my view, could be quantitative (involving changes to an existing single, some might say abstract, representation), which may eventually cause a qualitative change (the emergence of a new representation for symbolic numbers), or vice versa (albeit the preponderance of empirical evidence suggests the former), with different levels of interactions between the representations (Cohen Kadosh and Walsh 2009). Indeed, Holloway and Ansari provide convincing evidence for the dissociation between symbolic and non-symbolic numbers at the behavioral level during childhood.

Future research that is described in much greater details in this clear chapter leads Holloway and Ansari to conclude that the semantic representation of numerical symbols is more complicated than originally thought, that the link between symbolic numbers and non-symbolic numbers does not follow a 1:1 correspondence, and that developmental changes to the numerical representations exist during childhood and adulthood.

One of the issues that in my view should be flagged with this kind of research is the processing of non-symbolic numbers, which according to recent views confound non-numerical sensory cues that by and large do not provide a clear measurement of numerical representation (Gebuis and Reynvoet 2012). Moreover, recent studies show that performance in this task is influenced by inhibitory control mechanisms (Fuhs and McNeil 2013; Gilmore et al. 2013). Such confounds can explain to some degree the lack of 1:1 correspondence between numerical and non-numerical symbols, but also cast doubt on whether we can examine numerical abilities accurately using non-symbolic numbers.

Numerical Quantity or Magnitude

Walsh (this volume) aims in his chapter to elaborate his original A Theory of Magnitude (ATOM) (Walsh 2003), and to discuss why the processing of important information such as numerical quantity is subserved by the parietal cortex, a region that is mainly involved in automatic and motoric processing of which we are seldom aware. According to Walsh, one should have expected the temporal lobe, a region that is associated with more sophisticated processing like language, faces, episodic memory and object recognition to be involved. Walsh suggests that this is due to the fact that the parietal lobe is a central hub for other magnitude functions, such as time and space, leading numerical information to be based in the parietal lobe. In his chapter he elaborates further on his ATOM. He discusses issues that range from the development of magnitude and metaphorical theories, to asymmetrical interference between numbers and other magnitude that has been proposed to challenge the ATOM. While this chapter has been written in an accessible fashion, the ATOM in my view is based solely on the localist view. The parietal cortex is indeed an area that plays a fundamental role in numerical cognition, but it is not the only one, and as has been demonstrate very convincingly by Menon (this volume), we need to go beyond a mere focus on a single region to understanding how our cognitive abilities are subserved (and shared) by brain networks.

Stimulating The Brain

Studies using fMRI and to some degree EEG have dominated the way in which we examine the connections between brain and behavior in the field of numerical cognition and in other fields as well. Notably, these neuroimaging techniques provide the neural correlates of numerical cognition. Salillas and Semenza (this volume) review findings from studies using brain stimulation, which allows one to draw causal inferences about the relationship between brain functions and cognition (Walsh and Cowey 2000).

Most of the studies reviewed in this chapter used transcranial magnetic stimulation (TMS), a method that allows noninvasive stimulation of neurons to trigger action potentials. Salillas and Semenza in their chapter review findings from numerical comparison tasks, using different formats, the interaction between number, space, and time, and calculation (including the usage of fingers in arithmetic). While most of the studies have suggested involvement of the bilateral parietal lobes, and in some cases the prefrontal cortex, Salillas and Semenza suggest that number semantics is based mainly on the left hemisphere, which is in line with neuropsychological findings (Cappelletti, this volume). Nevertheless, it should be noted that the research in this field is relatively sparse and more studies are needed. One of the interesting points that Salillas and Semenza noted is that the results obtained with fMRI are not always in agreement with TMS findings. This might indicate that some of the results obtained using fMRI do not give an accurate account of numerical cognition (see also Price and Friston 2002). However, the reverse is also true; contrary to the left lateralization of calculation that would be expected from fMRI studies or patient data, TMS studies have shown the necessities of both parietal lobes.

The other noninvasive method that Salillas and Semenza discuss is transcranial electric stimulation (tES), which allows painless stimulation of the brain using low direct or alternated current. In contrast to TMS that is used mainly as a virtual lesion approach by impairing the stimulated brain functions for a brief period of time, tES has often been shown to impair or improve performance in numerical skills including symbolic and arithmetic learning (Cappelletti et al. 2013; Cohen Kadosh et al. 2010; Hauser et al. 2013; Iuculano and Cohen Kadosh 2013; Klein et al. 2013; Snowball et al. 2013). Notably, there are even fewer tES studies than TMS, but in my view, tES due to its possibility to improve cognitive performance holds great promise for future research (see below).

Another method that Salillas and Semenza are discussing is intraoperative direct cortical electrostimulation (DCE). DCE is a very powerful method for investigating the location of brain functions. DCE is used to map sensory processing, motor functions, and speech, and more recently calculation skills, mainly when epileptic patients are operated on and diseased brain tissue is removed.

Given the importance of language (some will say even more than numerical abilities) most of the studies have focused on the left hemisphere, and more specifically the left angular gyrus and IPS, and examined the commonalities and differences between arithmetic operations. Nevertheless, it is important to note that studies that examined the right parietal lobe were also able to dissociate between arithmetic operations (Della Puppa et al. 2013). Few studies that examined both left and right hemispheres have been able to dissociate the roles of left and right parietal lobes in calculation. It is important to note that knowledge derived from intraoperative studies is based on the atypical brain, and caution is needed before drawing inferences to the healthy brain.

Salillas and Semenza suggest that future combination of results from TMS and DCE would allow stronger converging evidence of the necessity of different brain regions in numerical skills in different populations. Whether such converging evidence is actually obtainable, given the different types of populations, is a topic for further research.

Computational Modelling

I have discussed so far findings from single-cell neurophysiology in monkeys, neuroimaging in children and adults, and brain stimulation in healthy and atypical population. Roggeman, Fias, and Verguts (this volume) extend the area of research and combine research that incorporates their influential computational modelling (Verguts and Fias 2004) and research with neuroimaging and single-cell neurophysiology. After a short introduction of numerical processing in the animal and human brain they present their original model, which includes number-sensitive and number-selective coding. In number-sensitive coding, sets of objects are represented on a spatial neuronal map in which each neuron responds to a specific location independent of the size of other visual properties of the object (object location map). This map is not specifically dedicated to number processing, but has more general functions such as visuospatial short term memory. In this map each object is represented by only one location neuron. In a number-selective map, neurons are more sensitive to a given number and their sensitivity to other numbers is decreased as the numerical distance from the preferred number increases, similar to what Nieder discussed in his chapter on single-cell neurophysiology in monkeys (Nieder, this volume). Roggeman et al. review different studies including single-unit in monkeys, and behavioral and neuroimaging findings in humans that their model can explain. They discuss how symbolic and non-symbolic numbers recruit different brain networks that provide similar computational processes to those outlined in their model. At the end of their chapter they discuss different issues including the relationship between object location maps and visual short term memory, and the role of working memory, task demands, and the suitability of their model to represent multi-digit numbers, which in this case assumes decomposed-multi-digit architecture (see Nuerk et al. this volume). They also discuss the future directions that should be taken by future models: in particular the need to go beyond basic numerical processing to complex numerical competencies, such as counting and natural number system, and explain how humans can work with concepts of number and its derivations, such as infinity, interest rates, and integral calculus. In addition, in their view, future work should aim to link number processing to other cognitive domains, such as working memory, attention, and language, as well as explaining typical and atypical number development.

Translational Neuroscience: The Case of Education?

In the last chapter, De Smedt and Grabner (this volume) discuss how neuroscientific research can be translated into mathematics education. They offer three pathways: (1) neuro-understanding, (2) neuro-prediction, and (3) neuro-intervention. In *neuro-understanding* they refer to the knowledge that is gained by neuroscientific research on how the numerical skill acquisition is implemented in the brain, research that echoes some of the findings presented in this section by Kaufmann et al. (this volume) and Menon (this volume). However, they go a step forward and suggest that this knowledge might allow a better understanding of success and failure in numerical skill acquisition in schools. By *neuro-prediction* De Smedt and Grabner refer to the potential of neuroimaging data not to only explain existing differences but more importantly to predict future mathematical skill acquisition as well as the success of educational interventions. This is at the moment a futuristic field as neuroscientific tools are quite expensive and time consuming, and it is controversial whether they can at this stage provide more accurate information than cognitive assessment. Success in this pathway can nevertheless have a very important impact on the field of education and medicine. At the same time it is also important to note that such research has neuroethical implications as it is unlikely to yield 100% success and as such has the potential to incorrectly tag infants or children at risk. The last pathway, *neuro-intervention,* refers to how brain imaging data have been used to ground interventions targeted at mathematics learning and how education shapes the neural circuitry that underlies school-taught mathematics. The authors discuss different strands including computer and board games, and the use of tES to affect mathematical learning.

While these pathways offer exciting applications of neuroscientific knowledge to the field of education, De Smedt and Grabner mark important caveats that should be considered and addressed in future research. These include issues such as low ecological validity, as many neuroscientific studies that have educational implications are based on adults, rather than children, and scientific research that does not mimic the learning process that occurs in the classroom. Other issues are the relative lack of knowledge of neuroscientists on how mathematics is learned and taught at school, which might lead to naïve experiments with little or no relevance to educational practice, and unrealistic expectations that stems from several factors that the authors discuss.

With this in mind, De Smedt and Grabner conclude that future translational research in this field will allow a stronger link between neuroscience and education. This in turn will allow findings from the behavioral and neural levels to be integrated in the field of education, which may lead to the development of effective mathematics education.

Adding Up The Findings and Finding Out What Can be Added Up

This chapter outlines the basic research described in this section. I will also outline several points that in my view are currently missing in the field of numerical cognition, but that have important implications for numerical cognition and beyond.

(1) The first major implication concerns single-cell neurophysiology. Assuming that the human mathematical brain represent a quantitative, rather than qualitative, difference from monkeys, it would be interesting to investigate what factors at the neuronal level reflect better numerical skill acquisition and performance in monkeys. While this would require larger samples than in previous studies, accumulation of this data across studies would allow important neuroscientific developments with implications to education.

(2) Developmental studies at the moment rarely provide longitudinal information. More importantly they are usually focused on fMRI. Just as knowledge of human behavior cannot be drawn solely from reaction time data, understanding of the human brain cannot be by referring only to fMRI. Other methods such as EEG, magnetoencephalography, near infrared spectroscopy, magnetic resonance spectroscopy, which can provide understanding at the neurochemical level that might proceed anatomical and function changes, and analysis of gray matter and white matter, including fiber tracts are sparse and critically needed.

(3) Research has been focused on the parietal lobe, but has failed to examine how it subserves the numerical brain network. Focusing on one particular brain region will provide maybe more knowledge, but will hinder understanding of the full picture, which will bear consequence to atypical and typical brain and cognitive development.

(4) Aside from the need to combine them with other neuroscientific tools, brain stimulation studies should be carried out in more naturalistic settings. In this respect tES, which allows greater mobility, can fit nicely to fulfill this aim.

(5) As stated by Roggeman et al. (this volume), computational models should be extended to involve more sophisticated numerical skills that go beyond simple

numerical processes such as number comparison, and should integrate them with models of other cognitive skills that are linked to numerical cognition (see Figure 3 of Kaufmann et al. (this volume)), as well as seeking to explain individual differences in performance. This could be an important step toward a better understanding of learning difficulties and intervention.

(6) Educational neuroscience research should aim to be more applied. This however will come on the expense of basic research. It is hard to conceive how ecological settings would not require a multifactorial design, which in turn will not allow a careful examination of the impact of each factor on education achievement. However, such research will have real potential to translate basic neuroscience findings, which has been described here, into the public benefit (Cohen Kadosh et al. 2013).

To sum up, this section includes state-of-the-art research, starting with basic research in single-cell neurophysiology in monkeys and ending with translational research in the field of neuroscience and education. Progress in the different components that I described here, and the directions offered by the authors of the various chapters will allow greater understanding of the biological and cognitive mechanisms involved in numerical cognition, with potential translation to improve numerical cognition.

References

Bressler, S.L. and Menon, V. (2010). Large-scale brain networks in cognition: emerging methods and principles. *Trends in Cognitive Sciences* 14: 277–290.

Cantlon, J.F., Brannon, E.M., Carter, E.J., and Pelphrey, K.A. (2006). Functional imaging of numerical processing in adults and 4-y-old children. *PLoS Biology* 4: e125.

Cappelletti, M. (this volume). The neuropsychology of acquired calculation disorders. In: *The Oxford Handbook of Numerical Cognition*, edited by R. Cohen Kadosh and A. Dowker. Oxford: Oxford University Press.

Cappelletti, M., Gessaroli, E., Hithersay, R., Mitolo, M., Didino, D., Kanai, R., et al. (2013). Transfer of Cognitive Training across Magnitude Dimensions Achieved with Concurrent Brain Stimulation of the Parietal Lobe. *The Journal of Neuroscience* 33: 14899–14907.

Cohen Kadosh, R. and Walsh, V. (2009). Numerical representation in the parietal lobes: Abstract or not abstract? *Behavioral and Brain Sciences* 32: 313–373.

Cohen Kadosh, R., Soskic, S., Iuculano, T., Kanai, R., and Walsh, V. (2010). Modulating neuronal activity produces specific and long lasting changes in numerical competence. *Current Biology* 20: 2016–2020.

Cohen Kadosh, R., Dowker, A., Heine, A., Kaufmann, L., and Kucian, K. (2013). Interventions for improving numerical abilities: Present and future. *Trends in Neuroscience and Education* 2: 85–93.

De Smedt, B. and Grabner, R.H. (this volume). Applications of Neuroscience to Mathematics Education. In: *The Oxford Handbook of Numerical Cognition*, edited by R. Cohen Kadosh and A. Dowker. Oxford: Oxford University Press.

Dehaene, S. (2003). The neural basis of the Weber–Fechner law: A logarithmic mental number line. *Trends in Cognitive Sciences* 7: 145–147.

Dehaene, S., Dehaene-Lambertz, G., and Cohen, L. (1998a). Abstract representations of numbers in the animal and human brain. *Trends in Neurosciences* 21: 355–61.

Dehaene, S., Naccache, L., Le Clec'H, G., Koechlin, E., Mueller, M., Dehaene-Lambertz, G., et al. (1998b). Imaging unconscious semantic priming. *Nature* 395: 597–600.

Dehaene, S., Tzourio, N., Frak, V., Raynaud, L., Cohen, L., Mehler, J., et al. (1996). Cerebral activations during number multiplication and comparison: a PET study. *Neuropsychologia* 34: 1097–1106.

Dehaene, S., Piazza, M., Pinel, P., and Cohen, L. (2003). Three parietal circuits for number processing. *Cognitive Neuropsychology* 20: 487–506.

Della Puppa, A., De Pellegrin, S., d'Avella, E., Gioffrè, G., Munari, M., Saladini, M., et al. (2013). Right parietal cortex and calculation processing: intraoperative functional mapping of multiplication and addition in patients affected by a brain tumor. *Journal of Neurosurgery* 119: 1107–1111.

Diester, I. and Nieder, A. (2007). Semantic associations between signs and numerical categories in the prefrontal cortex. *PLoS Biology* 5: e294.

Fuhs, M.W. and McNeil, N.M. (2013). ANS acuity and mathematics ability in preschoolers from low-income homes: contributions of inhibitory control. *Developmental Science* 16: 136–148.

Gebuis, T. and Reynvoet, B. (2012). The interplay between nonsymbolic number and its continuous visual properties. *Journal of Experimental Psychology: General* 141: 642–648.

Gilmore, C., Attridge, N., Clayton, S., Cragg, L., Johnson, S., Marlow, N., et al. (2013). Individual differences in inhibitory control, not non-verbal number acuity, correlate with mathematics achievement. *PLoS One* 8.

Hauser, T.U., Rotzer, S., Grabner, R.H., Mérillat, S., and Jäncke, L. (2013). Enhancing performance in numerical magnitude processing and mental arithmetic using transcranial Direct Current Stimulation (tDCS). *Frontiers in Human Neuroscience* 7.

Holloway, I. and Ansari, D. (this volume). Numerical symbols: an overview of their cognitive and neural underpinnings. In: *The Oxford Handbook of Numerical Cognition*, edited by R. Cohen Kadosh and A. Dowker. Oxford: Oxford University Press.

Hyde, D.C. and Spelke, E.S. (2011). Neural signatures of number processing in human infants: Evidence for two core systems underlying numerical cognition. *Developmental Science* 14: 360–371.

Iuculano, T. and Cohen Kadosh, R. (2013). The mental cost of cognitive enhancement. *The Journal of Neuroscience* 33: 4482–4486.

Izard, V., Dehaene-Lambertz, G., and Dehaene, S. (2008). Distinct cerebral pathways for object identity and number in human infants. *PLoS Biology* 6: e11.

Kaufmann, L., Kucian, K., and von Aster, M. (this volume). Development of the numerical brain. In: *The Oxford Handbook of Numerical Cognition*, edited by R. Cohen Kadosh and A. Dowker. Oxford: Oxford University Press.

Kaufmann, L., Wood, G., Rubinsten, O., and Henik, A. (2011). Meta-Analyses of Developmental fMRI Studies Investigating Typical and Atypical Trajectories of Number Processing and Calculation. *Developmental Neuropsychology* 36: 763–787.

Klein, E., Mann, A., Huber, S., Bloechle, J., Willmes, K., Karim, A.A., et al. (2013). Bilateral Bi-Cephalic Tdcs with Two Active Electrodes of the Same Polarity Modulates Bilateral Cognitive Processes Differentially. *PLoS One* 8: e71607.

Kucian, K., Kaufmann, L., and Von Aster, M. (this volume). Brain Correlates of Numerical Disabilities. In: *The Oxford Handbook of Numerical Cognition*, edited by R. Cohen Kadosh and A. Dowker. Oxford: Oxford University Press.

McCloskey, M., Caramazza, A., and Basili, A. (1985). Cognitive mechanisms in number processing and calculation: Evidence from dyscalculia. *Brain and Cognition* 4: 171–196.

Meck, W.H. and Church, R.M. (1983). A mode control model of counting and timing process. *Journal of Experimental Psychology: Animal Behavior Processes* 9: 320–334.

Menon, V. (this volume). Arithmetic in the child and adult brain. In: The Oxford Handbook of Numerical Cognition, edited by R. Cohen Kadosh and A. Dowker. Oxford: Oxford University Press.

Nieder, A. (this volume). Neuronal correlates of nonverbal numerical competence in primates. In: *The Oxford Handbook of Numerical Cognition*, edited by R. Cohen Kadosh and A. Dowker. Oxford: Oxford University Press.

Nieder, A. and Miller, E.K. (2003). Coding of cognitive magnitude: Compressed scaling of numerical information in the primate prefrontal cortex. *Neuron* 37: 149–157.

Nieder, A., Freedman, D.J., and Miller, E.K. (2002). Representation of the quantity of visual items in the primate prefrontal cortex. *Science* 297: 1708–1711.

Nuerk, H.-C., Moeller, K., and Willmes, K. (this volume). Multi-digit number processing: Overview, conceptual clarifications and language influences. In: *The Oxford Handbook of Numerical Cognition*, edited by R. Cohen Kadosh and A. Dowker. Oxford: Oxford University Press.

Orlov, T., Yakovlev, V., Hochstein, S., and Zohary, E. (2000). Macaque monkeys categorize images by their ordinal number. *Nature* 404, 77–80.

Price, C.J. and Friston, K.J. (2002). Degeneracy and cognitive anatomy. *Trends in Cognitive Sciences* 6: 416–421.

Rivera, S.M., Reiss, A.L., Eckert, M.A., and Menon, V. (2005). Developmental changes in mental arithmetic: Evidence for increased functional specialization in the left inferior parietal cortex. *Cerebral Cortex* 25: 1779–1790.

Roggeman, C., Fias, W., and Verguts, T. (this volume). Basic number representation and beyond: Neuroimaging and computational modeling. In: *The Oxford Handbook of Numerical Cognition*, edited by R. Cohen Kadosh and A. Dowker. Oxford: Oxford University Press.

Roitman, J.D., Brannon, E.M., and Platt, M.L. (2007). Monotonic coding of numerosity in macaque lateral intraparietal area. *PLoS Biology* 5: e208.

Roland, P.E. and Friberg, L. (1985). Localization of cortical areas activated by thinking. *Journal of Neurophysiology* 53: 1219–1243.

Ruchkin, D.S., Johnson, R., Jr, Canoune, H., and Ritter, W. (1991). Event-related potentials during arithmetic and mental rotation. *Electroencephalography and Clinical Neurophysiology* 79: 473–487.

Salillas, E. and Semenza, C. (this volume). Mapping the Brain for Math: Reversible inactivation by direct cortical electrostimulation and transcranial magnetic stimulation. In: *The Oxford Handbook of Numerical Cognition*, edited by R. Cohen Kadosh and A. Dowker. Oxford: Oxford University Press.

Snowball, A., Tachtsidis, I., Popescu, T., Thompson, J., Delazer, M., Zamarian, L., et al. (2013). Long-Term Enhancement of Brain Function and Cognition Using Cognitive Training and Brain Stimulation. *Current Biology* 23: 987–992.

Verguts, T. and Fias, W. (2004). Representation of number in animals and humans: A neural model. *Journal of Cognitive Neuroscience* 16: 1493–1504.

von Aster, M.G. and Shalev, R.S. (2007). Number development and developmental dyscalculia. *Developmental Medicine & Child Neurology* 49: 868–873.

Walsh, V. (2003). A theory of magnitude: Common cortical metrics of time, space and quantity. *Trends in Cognitive Sciences* 7: 483–488.

Walsh, V. (this volume). A theory of magnitude: The parts that sum to number. In: *The Oxford Handbook of Numerical Cognition*, edited by R. Cohen Kadosh and A. Dowker. Oxford: Oxford University Press.

Walsh, V. and Cowey, A. (2000). Transcranial magnetic stimulation and cognitive neuroscience. *Nature Reviews Neuroscience* 1: 73–79.

CHAPTER 26

NEURONAL CORRELATES OF NON-VERBAL NUMERICAL COMPETENCE IN PRIMATES

ANDREAS NIEDER

INTRODUCTION

SYMBOLIC number representations, such as numerals and number words, and the infinite mathematical manipulations they enable are uniquely human cultural achievements and shape our technologically advanced and scientific culture. However, research over the last decades has shown that basic numerical competence does not depend on language; rather it is rooted in biological primitives that can already be found in animals (Nieder & Dehaene, 2009). Animals possess impressive numerical capabilities and are able to non-verbally and approximately grasp the numerical properties of objects and events (Nieder, 2005). Such a numerical estimation system for representing number as language-independent mental magnitudes (analogue magnitude system) is a precursor on which verbal numerical representations build (Halberda, Mazzocco, & Feigenson, 2008), and their neural foundations can be studied in animal models.

Neurophysiological experiments in awake, behaving animals are an essential approach to understand the brain and its cognitive abilities. One of the most powerful preparations for studying cognitive functions in an alert organism is the awake, behaving monkey (Evarts 1966; Jasper, Ricci & Doane, 1960). Monkeys can be trained with operant conditioning techniques to perform numerical tasks, such as discriminating and memorizing numerosities, or processing numerical information according to behavioural principles. During the animals' performance, the electrical activity of individual nerve cells can be monitored by means of microelectrodes positioned at known locations within the brain; no other method shows this high spatial and temporal resolution needed for the investigation of single electrical nerve signals. The number of action potentials a single neurons and groups of neurons generate per time interval is a reliable code for the neuronal representations of information, such as the number of items. Recording neuronal activity simultaneously with behavioural performance presents a rich opportunity for experimental analysis of the neuronal foundations of mental functions, such as numerical competence.

The current chapter reviews the progress that has been made in our understanding of the neuronal substrates and mechanisms of non-verbal numerical competence in non-human primates. It is structured according to the two major concepts numerical cognition encompasses: numerical quantity and numerical rank (Wiese, 2003). Numerical quantity refers to the empirical property *cardinality* ('*numerosity*', the size of a set) of objects and events. It pertains to the question '*How many?*' for numerable quantity, and '*How much?*' for innumerable quantity. Numerical rank refers to the empirical property *serial order* and is sought after by the question '*Which position?*'

Numerical Quantity (Cardinality)

Neurons Encoding Numerical Quantity

Investigations of cognitive processing and it neuronal underpinning require subjects that are engaged in controlled behavioural tasks. To that aim, monkeys are trained to perform discrimination tasks based on numerical information. In the basic layout of the delayed match-to-numerosity task, monkeys viewed a sequence of two displays separated by a memory delay and were required to judge whether the second display (test) matches the first (sample) with respect to the number of items shown on it. Thus, the monkeys are required to discriminate matching from non-matching numerosities to solve this task (Figure 26.1a). To ensure that the monkeys solved the task by judging number per se, rather than simply memorizing sequences of visual patterns or exploiting low-level visual features that correlate with number, sensory cues (such as position, shape, overall area, circumference, and density) were varied considerably and controlled for (Nieder, Freedman & Miller, 2002).

In such discrimination tasks, monkeys and other animals can discriminate numerosities, but they do so in an approximate way (Nieder & Miller, 2003). Unlike counting human adults who represent the number of items in a precise way, animals can only non-verbally estimate numerosity. Thus, when animals discriminate a sample numerosity from smaller or larger (non-match) numerosities, it is difficult for them to discriminate numerosities close to the target numerosity (e.g. 5 versus 6), but progressively easier to dissociate numerically remote quantities (such as 2 versus 6). Because of this *numerical distance effect*, the discrimination performances result in peak distributions centred around the target (sample) numerosity (Figure 26.1b). In addition, the behavioural discrimination distributions also grow broader with increasing target numerosities, an indication of the *numerical size effect*. This size effect captures the finding that pairs of numerosities of a constant numerical distance are easier to discriminate if the quantities are small (e.g. 2 versus 3), but more difficult if large (e.g. 5 versus 6). This size effect is in accord with Weber's Law, stating that the just noticeable difference between two magnitudes increases in proportion to the size of the target magnitude.

Where in the brain might neurons be able to represent numerical information? As classical association cortices, the prefrontal and posterior parietal cortices are ideal brain

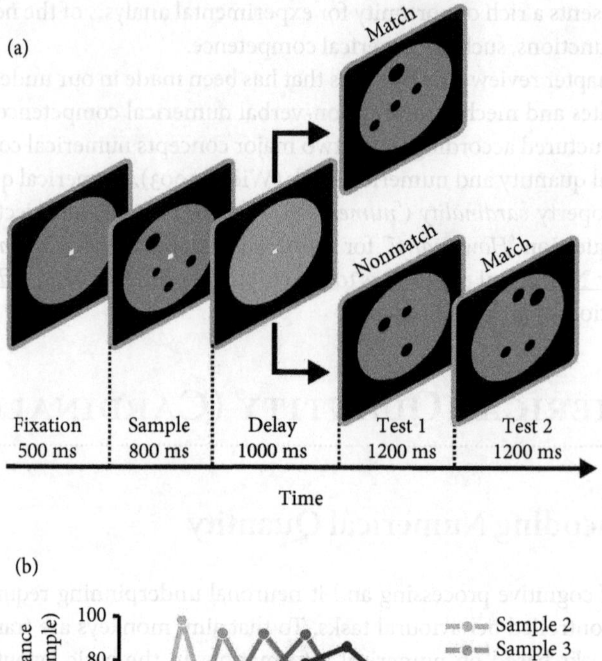

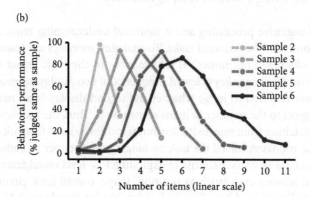

FIGURE 26.1 Representation of visual cardinality in rhesus monkeys. (a) Delayed match-to-sample task with visually presented numerosity as the stimulus dimension of interest. A trial started when the monkey grasped a lever and fixated at a central target. After 500 ms of pure fixation, the sample stimulus (800 ms) cued the monkey for a certain numerosity it had to memorize during a 1000-ms delay period. Then, the test1 stimulus was presented, which in 50% of cases was a match showing the same number of dots as cued during the sample period. In the other 50% of cases the test1 display was a non-match, which showed a different numerosity as the sample display. After a non-match test stimulus, a second test stimulus (test2) appeared that was always a match. To receive a fluid reward, monkeys were required to release the lever as soon as a match appeared. Trials were pseudo-randomized and balanced across all relevant features. Monkeys were required to maintain fixation throughout the sample and delay period. (b) Behavioural numerosity discrimination functions of two monkeys. The curves indicate whether they judged the first test stimulus (after a delay) as containing the same number of items as the sample display. The function peaks (and the colour legend) indicate the sample numerosity at which each curve was derived.

structures for an abstract encoding of quantity. They receive highly processed multimodal input (Bremmer, Schlack, Shah, Zafiris, Kubischik, Hoffmann, et al., 2001; Lewis & Van Essen 2000; Miller & Cohen 2001) – a prerequisite for numerical competence because the number concept applies equally well to all sensory modalities. Both are key processing stages for executive functions (e.g. categorization, working memory, decision-making, goal-directed behaviour, etc.) and play an important role in maintaining information 'on line' (Freedman, Riesenhuber, Poggio, & Miller, 2001; Stoet & Snyder, 2004; Wallis, Anderson & Miller, 2001), and the PPC, in particular, also hosts neural circuitry dedicated to the representation of abstract spatial information (Colby & Goldberg, 1999). Moreover, functional imaging in the human primate points towards the prefrontal and posterior parietal cortices as key structures (Dehaene, Spelke, Pinel, Stanescu, & Tsivkin, 1999).

Recordings in monkeys actively discriminating numerosity demonstrated the capacity of single neurons to encode cardinality (Nieder et al., 2002; Nieder & Miller 2004a; Nieder, Diester, & Tudusciuc, 2006). Numerosity-selective neurons were tuned to the number of items in a visual display, that is, they showed maximum activity to one of the presented quantities – a neuron's 'preferred numerosity' – and a progressive drop off as the quantity became more remote from the preferred number (Nieder et al., 2002; Nieder & Merten, 2007). Importantly, changes in the physical appearance of the displays had no effect on the activity of numerosity-selective neurons (Nieder et al., 2002; Nieder & Miller, 2004a; Nieder et al., 2006). A high proportion of numerosity detectors (Figure 26.2) was found in the lateral prefrontal cortex (PFC) (Nieder et al., 2002). In the posterior parietal cortex (PPC), numerosity-selective neurons were sparsely distributed in several areas, but relatively abundant in the fundus of the intraparietal sulcus (IPS), termed VIP (Colby, Duhamel & Goldberg, 1993). There were few such cells in the anterior inferior temporal cortex (aITC; Nieder & Miller 2004a) (Figure 26.2a).

Item numbers can be determined in two fundamentally different spatiotemporal presentation formats. When presented simultaneously as in multiple-item patterns, numerosity can be estimated at a single glance in a direct, perceptual-like way from a spatial arrangement. On a behavioural level, constant reaction times and equal numbers of scanning eye movements to individual items (Nieder & Miller, 2004b) indicate parallel processing mechanisms for quantity assessments from multiple-dot patterns. Moreover, the response latencies of single neurons are the same across numerosities (Nieder et al., 2002), which suggests that individual elements of a set are processed in parallel to form the neuronal representation of different cardinalities at the same time. In contrast to a simultaneous presentation, the elements of a set can be presented one by one and, thus, need to be enumerated successively across time (Cordes, Gelman, Gallistel, & Whalen, 2001; Meck and Church, 1983; Whalen, Gallistel & Gelman, 1999). Sequential enumeration is cognitively more demanding; it incorporates multiple encoding, memory, and updating stages; it may even be regarded as a form of addition of one. Sequential enumeration is particularly interesting in that it constitutes a non-verbal precursor of real counting; after all, verbal counting is a sequential enumeration process using number symbols (i.e. 1-2-3, etc.).

To address the neuronal representation of an abstract counting-like accumulation of sensory events and to compare it to the encoding of numerosity in simultaneous displays, Nieder et al. (2006) recorded single-cell activity in the fundus of the IPS, while monkeys performed a delayed match-to-sample task in which sample numerosity was specified either by single dots appearing one-by-one to indicate the number of items in sequence ('sequential protocol',

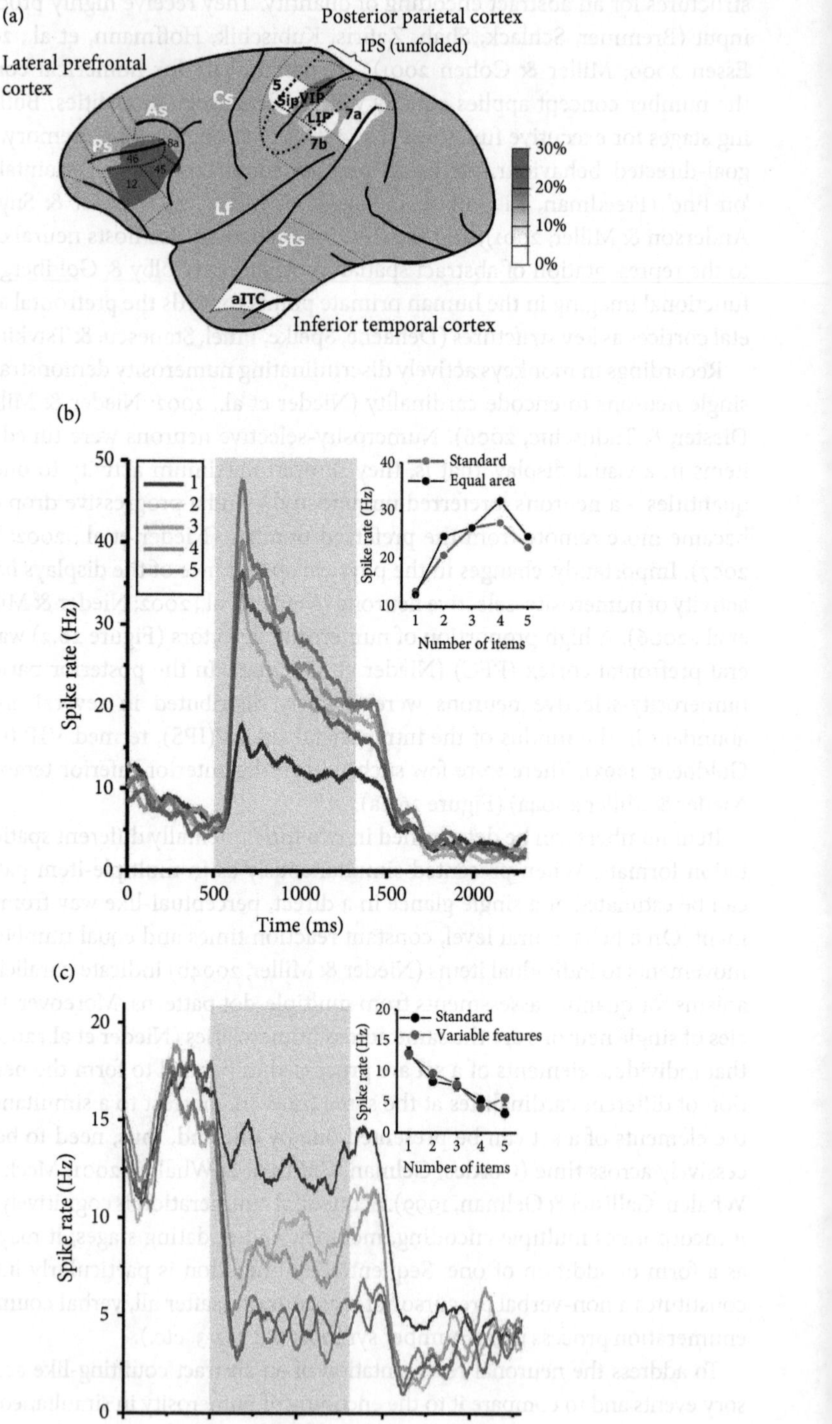

FIGURE 26.2 (Continued)

Figure 26.3a) or by multiple-dot patterns ('simultaneous protocol', Figure 1a). It was ensured that temporal or spatial cues could not be used by the animals to solve the task. In addition to the previously described neurons selective to numerosity in multiple-dot patterns, roughly 25% of the neurons in the fundus of the IPS also encoded sequentially presented numerical quantity (Figure 26.3b). However, numerical quantity was represented by distinct populations of neurons during the ongoing spatial or temporal enumeration process (i.e. in the sample phase); cells encoding the number of sequential items were not tuned to numerosity in multiple-item displays, and vice versa. Once the enumeration process was completed, however, and the monkeys had to store information in mind, a third population of neurons coded numerosity both in the sequential and simultaneous protocol; about 20% of the cells were tuned to numerosity irrespective of whether is was cued simultaneously or in sequence. This argues for segregated processing of numerosity during the actual encoding stage in the parietal lobe, but also for a final convergence of the segregated information to form most abstract quantity representations. The intermediate numerosity of an ongoing quantification process and the storage of the final cardinality are accomplished by different neuronal populations.

In contrast to visual stimuli used in the studies above mentioned, Jun Tanji's group used a different sensory-motor system to investigate numerical representations: cells in the superior parietal lobule (SPL) have been reported to keep track of the number of movements (Sawamura, Shima, & Tanji, 2002, 2010). The authors trained monkeys to alternate between five arm movements of one type ('push' and 'turn') and five of another. They found neurons in a somatosensory-responsive region (part of area 5) of the SPL that maintained the number of movements. Relatively few such neurons were found in the same lateral PFC regions where other perceptual categories were investigated. One possibility for the difference between these studies may be modality (touch vs. vision), but another may be the level of abstraction. Most movement-number representations (85%) found by Sawamura and colleagues (2002) were not abstract; number-selective activity depended on whether the monkey's movement was 'push' or 'turn'. By contrast, the visual numerosity representations found in the PFC and fundus of the IPS were abstract and generalized (Nieder et al., 2002, 2006; Nieder & Miller, 2004a).

FIGURE 26.2 (CONTINUED) Numerosity-selective neurons in the monkey. (a) Lateral view of a monkey brain showing the recording sites in LPFC, PPC, and aITC. The proportion of numerosity-selective neurons in each area is colour coded according to the colour scale. The IPS is unfolded to show the different areas in the lateral and medial walls. Numbers on PFC indicate anatomical areas. (As, arcuate sulcus; Cs, central sulcus; IPS, intraparietal sulcus; LF, lateral fissure; LS, lunate sulcus; Ps, principal sulcus; Sts, Superior temporal sulcus). (b,c) Responses of single neurons that were recorded from the PFC (b) and the IPS. (c) Both neurons show graded discharge during sample presentation (interval shaded in gray, 500–1300 ms) as a function of numerosities 1 to 5 (colour coded averaged discharge functions). The insets in the upper right corner show the tuning of both neurons and their responses to different control stimuli. The preferred numerosity was 4 for the PFC neuron (B), and 1 for the IPS neuron (c).

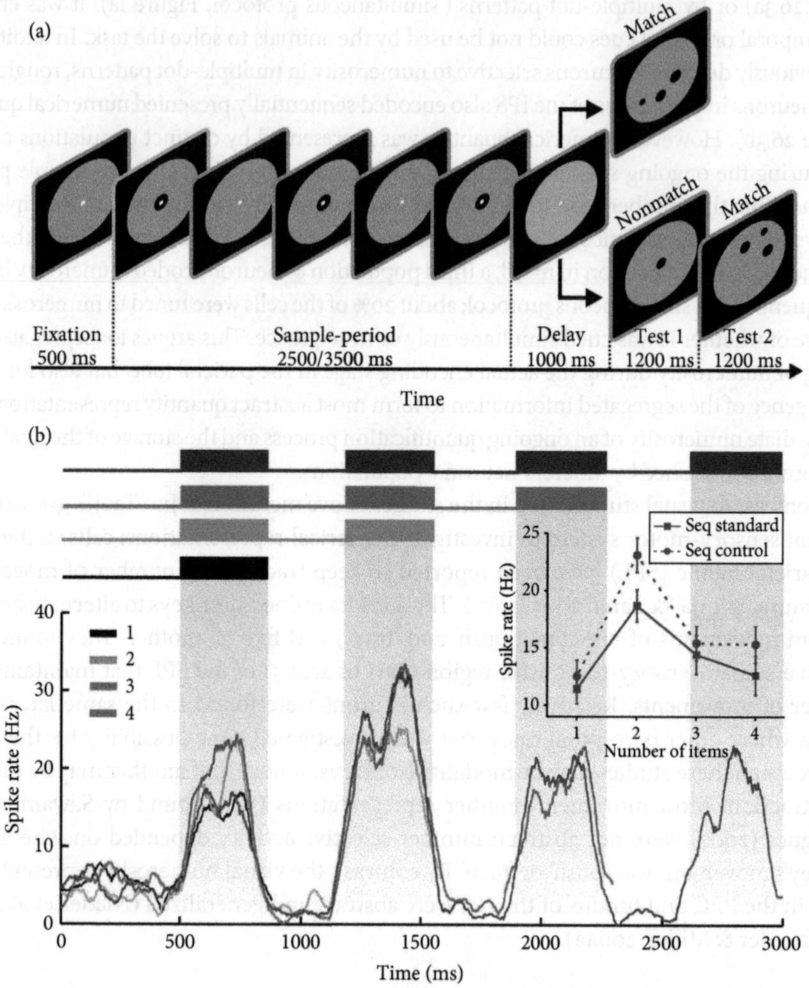

FIGURE 26.3 Coding of sequentially presented numerosity. (a) Sequential delayed match-to-numerosity task (here for numerosity 3). The sample numerosity was cued by sequentially presented items temporally separated by pauses containing no items. The temporal succession and duration of individual items were varied within and across quantities. (b) Responses of an example neuron selective to the sequential quantity 2 (only one condition shown for clarity). Top panel illustrates the temporal succession of individual items (square pulses represent single items). The corresponding latency-corrected discharges for many repetitions of the protocol are plotted as averaged spike density functions. The first 500 ms represent the fixation period. Corresponding colors were used for the stimulation illustration and the plotting of the neural data. Gray shaded areas denote item presentation The inset shows the tuning functions of the neuron to the standard and a control protocol (error bars represent SEM) for four sequential dots. In both protocols, the neuron was tuned to numerosity 2.

From Andreas Nieder, Ilka Diester, and Oana Tudusciuc, Temporal and Spatial Enumeration Processes in the Primate Parietal Cortex, *Science*, 313(5792), pp. 1431–1435, Figures 1a and 2b, DOI: 10.1126/science.1130308 (c) 2006, The American Association for the Advancement of Science. Reprinted with permission from AAAS.

Parietal neurons exhibit selectivity for numerosity with shorter latencies than prefrontal neurons, which suggests that it might be the first cortical stage to extract visual numerical information.

The parietal cortex, and the IPS in particular, might be the first cortical stage that extracts visual numerical information because its neurons require shorter latencies to become numerosity selective than PFC neurons (Nieder & Miller, 2004a). As PPC and PFC are functionally interconnected (Quintana, Fuster, & Yajeya, 1989; Chafee & Goldman-Rakic, 2000), that information might be conveyed directly or indirectly to the PFC where a larger proportion and more selective neurons represent numerosity, particularly during the working memory phase, to gain control over behaviour.

Behavioural Significance of Numerosity-Selective Neurons

The activity of all numerosity-selective neurons, each tuned to a specific preferred numerosity, formed an array of numerically-overlapping tuning functions. Because such tuning functions represent the range of numerosities encoded, or filtered out, by the neurons, these neurons can be said in technical terms to form a bank of overlapping numerosity filters (Figure 26.4). The neuronal tuning functions mirror the animals' behavioural performance functions (Figure 26.4a). Interestingly, the neurons' sequentially-arranged overlapping tuning curves preserved an inherent order of cardinalities. This is important because numerosities are not isolated categories, but exist in relation to one another (for example, 3 is greater than 2 and less than 4); they need to be sequentially ordered to allow meaningful quantity assignments.

The response properties of numerosity-selective cells in both parietal and prefrontal cortices can explain basic psychophysical phenomena in monkeys, such as the numerical distance and size effect (Figure 26.1b). The *numerical distance effect* results from the fact that the neural filter functions that are engaged in the discrimination of adjacent numerosities heavily overlap (Nieder & Miller, 2003). As a consequence, the signal-to-noise ratio of the neural signal detection process is low and the monkeys make many errors. On the other hand, the filter functions of neurons that are tuned to remote numerosities barely overlap, which results in a high signal-to-noise ratio and, therefore, good performance in cases where the animal has to discriminate sets of a larger numerical distance. The behavioural consequences of the numerical size effect therefore accord to Weber's Law. The *numerical size effect* is directly related to the precision of the neuronal numerosity filters – the widths of the tuning curves (or neuronal numerical representations) increase linearly with preferred numerosities (that is, on average, tuning precision deteriorates as the preferred quantity increases). Hence, more selective neural filters that do not overlap extensively are engaged if a monkey has to discriminate small numerosities (say, 1 and 2), which results in high signal-to-noise ratios and few errors in the discrimination. Conversely, if a monkey has to discriminate large numerosities (such as 4 and 5), the filter functions would overlap considerably. Therefore, the discrimination has a low signal–to–noise ratio, which leads to poor performance.

Weber's Law predicts that the behavioural performance functions – the monkeys' behavioural numerical representations – are best described (i.e. symmetrical) on a non-linear, possibly logarithmically compressed scale, or 'number line'. This finding is formally captured by Fechner's Law which states that the perceived magnitude (*S*) is a logarithmic function of stimulus intensity (*I*) multiplied by a modality and dimension specific constant (*k*). In fact, both the behavioural performance functions (Figure 26.4a) and the neuronal tuning functions (Figure 26.4b) are better described by a compressed, as opposed to a linear scale (Nieder & Miller, 2003). Therefore, single-neuron representations of numerical quantity in monkeys obey the Weber–Fechner Law, just as the behavioural discrimination performance does.

An important piece of evidence for the contribution of numerosity-selective neurons to behavioural performance came from the examination of error trials. When the monkeys made judgment errors, the neural activity for the preferred quantity was significantly reduced as compared to correct trials (Nieder et al., 2002, 2006; Nieder & Miller, 2004a; Nieder & Merten, 2007). As a result of this (and the ordered representation of quantity), the activity to a given preferred numerosity on error trials was more similar to that elicited by adjacent non-preferred quantities on correct trials. In other words, if the neurons did not encode the numerosity properly, the monkeys were prone to mistakes.

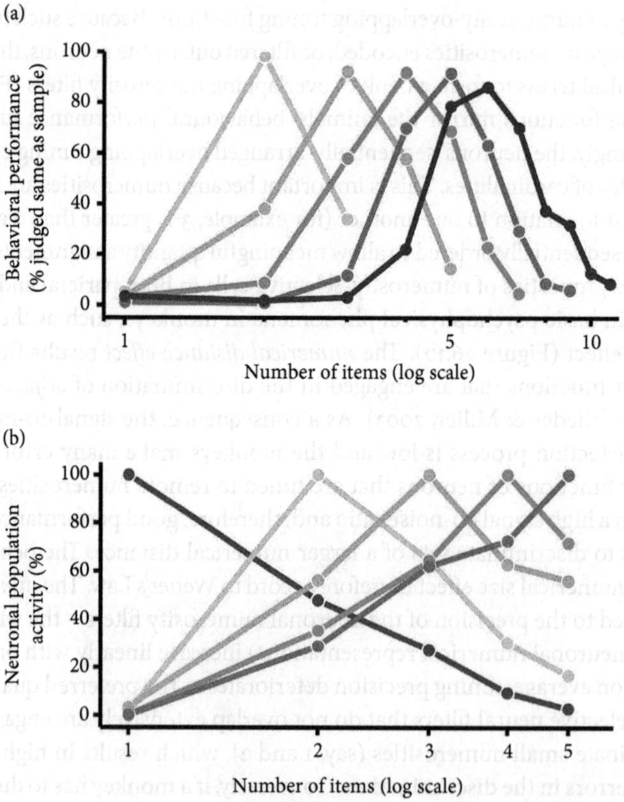

FIGURE 26.4 Relation between monkey behaviour and numerosity-selective neurons. (a) Behavioural filter functions (same layout as in Figure 26.1b) are symmetric on a logarithmic scale. (b) The averaged single-cell numerosity-tuning functions (from PFC) are also symmetric after logarithmic transformation

Reprinted from *Neuron*, 37(1), Andreas Nieder and Earl K. Miller, Coding of Cognitive Magnitude Compressed Scaling of Numerical Information in the Primate Prefrontal Cortex, pp. 149–157, Copyright (2003), with permission from Elsevier.

The most direct evidence for the importance of parietal numerosity-selective cells in representing quantity information was recently collected by Sawamura et al. (2010). These authors transiently and focally inactivated parietal area 5 (by application of muscimol) to test its functional contributions to numerosity-based action selection in monkeys trained to push or turn a handle a variable number of times in response to a visual stimulus. As a consequence of chemical inactivation, the error rate in the numerical task increased significantly, mainly based on omissions. A control task showed that the errors were not caused by motor deficits or impaired ability to select between actions. These results indicate that parietal area 5 is crucial for selecting actions on the basis of numerical information.

Implementing Numerosity Detectors

How may numerosity-selective neurons tuned to preferred numerosities arise in the course of cortical processing? Purely sensory, non-numerical properties (such as binocular disparity, wavelength and contrast in the visual system) are encoded already at the earliest processing stages of the sensory epithelia (Van Essen & DeYoe, 1993). Number, on the other hand, is a most abstract category devoid of specific sensory features; two cats and two calls have nothing in common, except that the size of their sets is 'two'. How then may the cardinality of objects or events, the pure number of entities, be derived in terms of neuronal information processing?

Two main models have been proposed to explain the implementation of quantity information. The *mode-control model* by Meck and Church (1983) works in series and suggests that each item is encoded by an impulse from a pacemaker, which is added to an accumulator (Figure 26.5a). The magnitude in the accumulator at the end of the count is then read into memory, forming a representation of the number of a set. Thus, it is assumed that quantity is encoded by 'summation coding', i.e. the monotonically increasing and decreasing response functions of the neurons (see also network model by Zorzi & Butterworth, 1999; Zorzi, Di Bono, & Fias, 2005).

Another model, the *neural filtering model* by Dehaene and Changeux (1993) implements numerosity in parallel (Figure 26.5b). First, each (visual) stimulus is coded as a local Gaussian distribution of activation by topographically organized input clusters (simulating the retina). Next, items of different sizes are normalized to a size-independent code. At that stage, item size, which was initially coded by the number of active clusters on the retina (quantity code) is now encoded by the position of active clusters on a location map (position code). Clusters in the location map project to every unit of downstream 'summation clusters', whose thresholds increase with increasing number and pool the total activity of the location map. The summation clusters finally project to 'numerosity clusters'. Numerosity clusters are characterized by central excitation and lateral inhibition so that each numerosity cluster responds only to a selected range of values of the total normalized activity, i.e. their preferred numerosity. Since the numerosity of a stimulus is encoded by peaked tuning functions with a preferred numerosity (causing maximum discharge) this mechanism is termed 'labelled-line code'. A similar architecture was proposed by Verguts and Fias (2004) using a back propagation network. Interestingly, summation units developed spontaneously in the second processing stage (the 'hidden units') after tuned numerosity detectors were determined as output stage. More recently, Stoianov and Zorzi (2012) proposed a model architecture in which units in a hidden layer accomplish numerosity discrimination by behaving like accumulators.

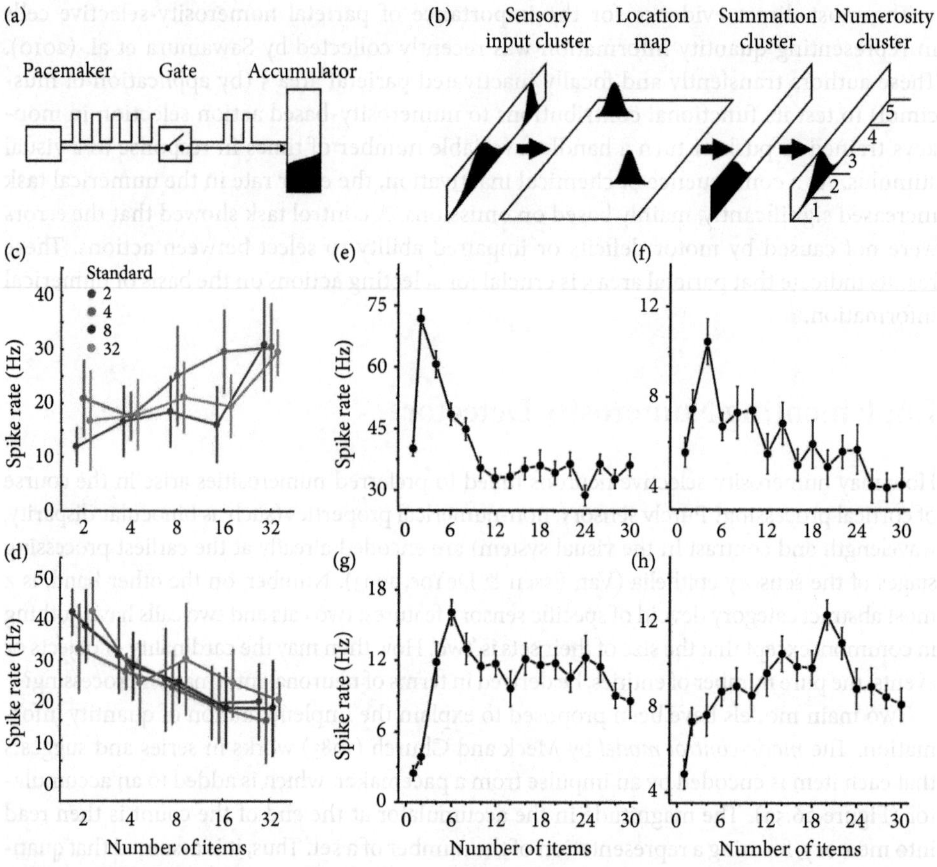

FIGURE 26.5 Implementation of numerosity detectors. (a) Mode-control model. (b) Neural filtering model. (c, d) Neurons in LIP discharge monotonically as a function of set size during an implicit numerosity task. Two single cells are depicted that show an increase (c) or decrease (d) of discharge rate, respectively, with increasing numerosity. Each neuron was tested with different standard (colour code) and deviant numerosities (see text for explanation of the task). (e–h) PFC neurons tuned to preferred numerosities in monkeys performing a delayed match-to-numerosity task. Preferred numerosity was 2 (e), 4 (f), 6 (g) and 20 (h).

Even though numerosity representations derived with both models are noisy (approximate) and obey Weber's Law, the two models differ in important aspects. The mode-control model by Meck and Church (1983) operates serially and assumes representation of cardinality on a linear scale, whereas in the neural filtering model by Dehaene and Changeux (1993) numerosity is encoded in parallel and represented on a logarithmic scale (the same holds for the back-propagation model by Verguts and Fias (2004)). Both models, however, have summation units implemented that accumulate number in a graded fashion prior to feeding into numerosity detectors at the output.

As a putative physiological reflection of this computational stage in models of number processing, Roitman, Brannon, and Platt (2007) recorded neurons in the parietal lobe, the lateral intraparietal areas LIP, of monkeys performing a delayed saccade task in which sets of dots provided implicit information about the reward magnitude a monkey was to receive. The responses of many neurons resembled the output of accumulator neurons with response functions that systematically increased or decreased with increase of stimulus set size (Figure 26.5c,d). The authors suggested that these two classes of number-selective neurons might constitute the physiological instantiation of the summation units and numerosity units proposed in neural network models of numerical representation; monotonic magnitude coding of LIP neurons may provide input to neurons in the PPC and PFC that compute cardinal numerical representations via tuning to preferred numerosities (Figure 26.5e–h; Nieder et al., 2002; Nieder & Miller, 2004a; Piazza, Izard, Pinel, Le Bihan, & Dehaene, 2004; Nieder et al., 2006; Nieder & Merten, 2007; Sawamura et al., 2010). Alternatively, the neuronal representation may change if quantity is encoded as an explicit category (Cohen Kadosh & Walsh, 2009).

Coding of Continuous and Discrete Quantity

Cardinality, the number of items in a set, is a most abstract magnitude category. However, not only enumerable, but also continuous abstract magnitudes, such as size (spatial magnitude), need to be represented and processed. Interestingly, the representation of numerical magnitudes and spatial magnitudes share many features and are not completely segregated. In a number comparison task, for example, choosing the numerically larger number takes significantly longer if the numeral is smaller in size compared to the numerically smaller number (e.g. in the comparison 2 versus 7; Henik & Tzelgov, 1982; Pinel, Piazza, Le Bihan, & Dehaene, 2004; Washburn, 1994). Functional imaging studies in humans suggest that anatomical vicinity (Castelli, Glaser, & Butterworth, 2006; Fias, Lammertyn, Reynvoet, Dupont, & Orban, 2003; Pinel et al., 2004) or even a common magnitude system (Walsh, 2003) for the representation of numerical (discrete) and spatial (continuous) quantity in the parietal cortex might be responsible for behavioural interference phenomena between numerical and spatial quantity.

To investigate how continuous quantity is encoded by single nerve cells and how it relates to numerosity representations, Tudusciuc and Nieder (2007) trained two rhesus monkeys in a delayed match-to-sample task to discriminate different types of quantity randomly alternating within each session. In the '*length protocol*', the length of a line (out of four different lengths) needed to be discriminated (continuous-spatial quantity). In the '*numerosity protocol*' (Figure 26.1a), the number of (one to four) items in multiple-dot displays (discrete-numerical quantity) was the relevant stimulus dimension. To ensure that the monkeys solved the task, based on the relevant quantitative information, other co-varying features of the stimuli were again controlled, and the positions of the dots and lines were greatly varied.

After the monkeys solved more than 81% of the trials correctly for both the length and the numerosity protocols, single unit activity from the depth of the IPS was analysed while the animals performed the task. About 20% of anatomically-intermingled single neurons in the monkey IPS each encoded discrete-numerical (Figure 26.6a), continuous-spatial (Figure 26.6b), or both types of quantities (Figure 26.6c), suggesting that two partly overlapping populations of neurons within this area may give rise to as a generalized magnitude system (Walsh 2003, see also Walsh, this volume). Subsequent analyses using a population decoding technique (Laubach, 2004) based on an artificial neuronal network (Kohonen, 1997) showed that the relatively small population of quantity-selective neurons carried most of the categorical information. By exploiting the classical spike-rate measure, which contributes to the monkeys' quantity discrimination performance (Nieder et al., 2002, 2006; Nieder & Miller, 2004a), the classifier was able to accurately and robustly discriminate both continuous and discrete quantity classes in a behaviourally-relevant way.

Coding of Proportions

Neurons in the prefrontal and posterior parietal cortices are selectively tuned to abstract quantity, and the cellular response characteristics can explain basic psychophysical phenomena in dealing with them. However, many vital decisions in animals require an estimation of the relation between two quantities, or proportion. Vallentin and Nieder (2008, 2010) demonstrated that rhesus monkeys were able to grasp proportionality in a delayed match-to-sample test. They trained two rhesus monkeys to judge the length ratio (proportion) between two lines – a reference and a test line. The length ratios between the test and reference lines were 1:4, 2:4, 3:4 and 4:4 (Figure 26.7). After demonstrating that the monkeys could discriminate spatial proportions, Vallentin and Nieder (2010) investigated this capacity's neuronal underpinning and recorded from neurons of the PFC while the animals performed the proportion discrimination task. Both during the sample and delay presentation, 25% of the tested neurons were significantly tuned only to proportion, irrespective of the absolute lengths of the test and reference bars. Each of the selective neurons preferred one of the four tested proportions. Just as with numerosities or lines, a labelled-line code was found for the coding of proportions, with neurons exhibiting peaked tuning curves and preferred proportions. The areas where such proportion-selective neurons were found coincided with PFC regions that also house numerosity-selective neurons. These data suggest that the perception of relational quantity is represented by the same frontal network and magnitude code as absolute quantity in the primate brain.

These single-cell data provide a neurophysiological explanation for recent functional imaging studies describing selectivity to quantity relations in a parietofrontal network. Using an fMRI adaptation protocol to investigate automatic quantity processing, a recovery from repetition suppression was detected both for line and numerosity proportions in lateral PFC and posterior parietal cortex (Jacob & Nieder, 2009b). Because recovery from blood-oxygenation-level-dependent (BOLD)-adaptation was a function of ratio distance, tuning in the human cortex also seems to be present to preferred proportions. Moreover, both numerosity and proportion seem to be processed by the same dedicated brain areas, as witnessed by a strong overlap of the distance effect for numerosity and proportions stimuli. Using the same methodology but presenting fractions in symbolic notation, Jacob and Nieder (2009a) could show tuning in human parietal cortex to preferred fractions that even generalize across the format of presentation. The distance effect was invariant to changes in

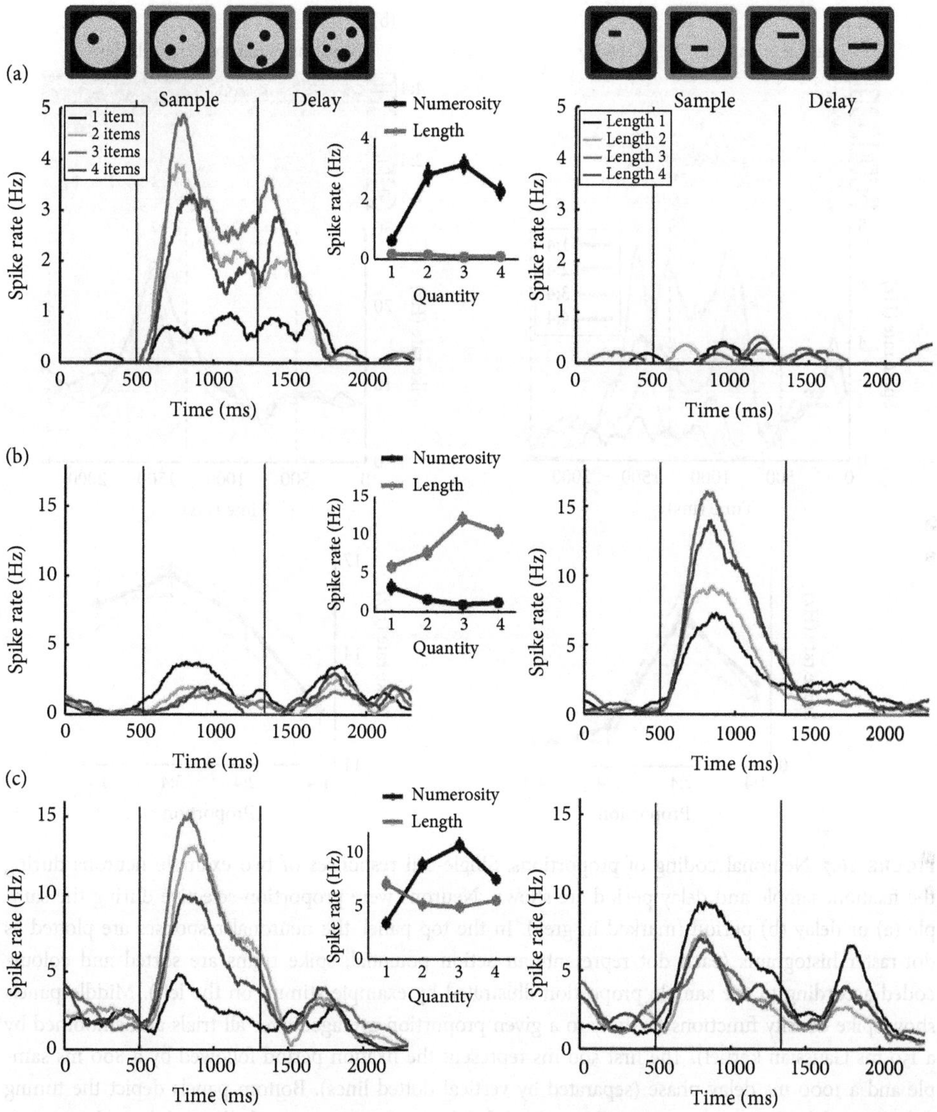

FIGURE 26.6 Neuronal coding of continuous and discrete quantity. (a–c) Three example neurons exhibiting selectivity for quantity. Top panels in (a) illustrate the four different numerosities (*left*) and four different line lengths (*right*) used as stimuli. Left and right graphs illustrate the discharge rates (displayed as smoothed spike density histograms) of the same neuron in the numerosity and length protocol, respectively. The first 500 ms represent the fixation period. The area between the two black vertical bars represents the sample presentation, the following 1000 ms indicate the delay phase. Colours correspond to the quantity dimensions. The insets between two histograms depict the tuning functions of each of the three neurons to numerosity and length. (a) Neuron tuned to numerosity 3, but not to length. (b) Neuron tuned to the third longest line, but not to any tested numerosity. (c) Neuron encoding both discrete and continuous quantity.

Adapted from Oana Tudusciuc and Andreas Nieder, Neuronal population coding of continuous and discrete quantity in the primate posterior parietal cortex, Proceedings of the Academy of Sciences of the United States of America, 104 (36), pp. 14513–14518 figure 3, doi: 10.1073/pnas.0705495104, Copyright (2007) National Academy of Sciences, U.S.A.

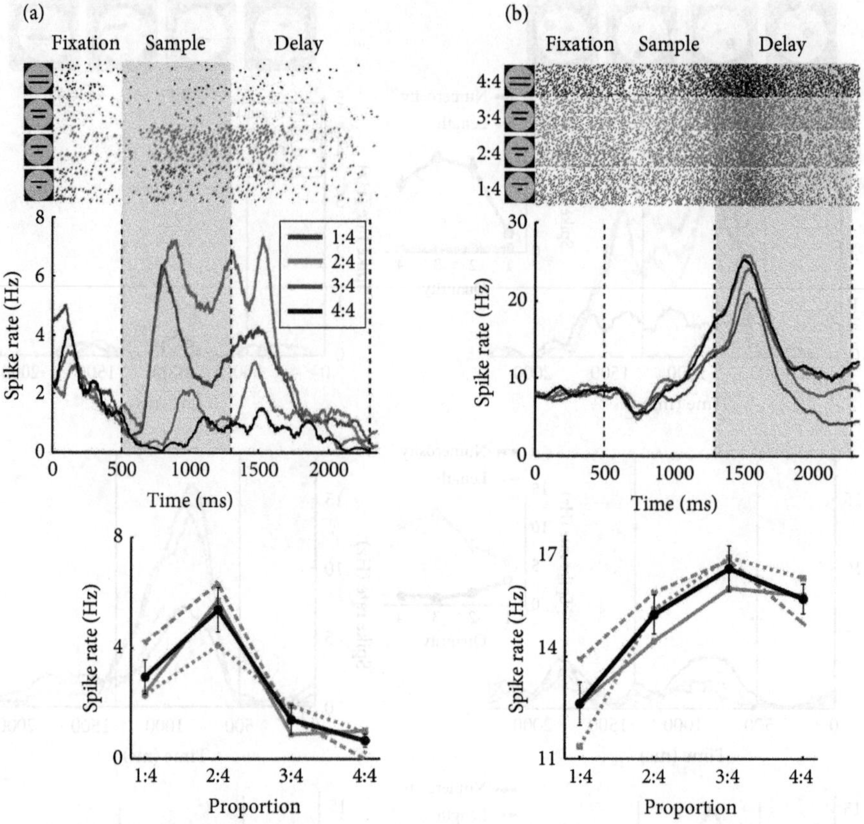

FIGURE 26.7 Neuronal coding of proportions. Single-cell responses of two example neurons during the fixation, sample and delay period are shown. Neurons were proportion-selective during the sample (a) or delay (b) period (marked in grey). In the top panel, the neuronal responses are plotted as dot-raster histograms (each dot represents an action potential, spike trains are sorted and colour-coded according to the sample proportion illustrated by example stimuli on the left). Middle panels show spike density functions (activity to a given proportion averaged over all trials and smoothed by a 150 ms Gaussian kernel). The first 500 ms represent the fixation period followed by a 800 ms sample and a 1000 ms delay phase (separated by vertical dotted lines). Bottom panels depict the tuning functions of the respective neurons for each of the three stimulus protocols derived from the periods of maximum proportion selectivity (error bars represent SEM).

notation from number to word fractions and strongest in the anterior intraparietal sulcus, a key region for the processing of whole numbers. The intraparietal cortex was also active in adults solving a fraction comparison problem (Ischebeck, Schocke, & Delazer, 2009). These findings demonstrate that the primate brain uses the same analogue magnitude code to represent both absolute and relative quantity. A fraction might be represented by its numerical value as a whole, rather than by the numerical values of its numerator and denominator. Together with previous studies in the numerical domain, the current findings indicate close similarities between non-symbolic quantity processing in the human and monkey brain.

Towards Symbolic Number Representations

As shown previously, humans and animals share an evolutionarily old quantity representation system, which allows the estimation of set sizes. Non-verbal numerical cognition, however, is limited to approximate quantity representations and rudimentary arithmetic operations. Language-endowed humans, on the other hand, invented number symbols (numerals and number words) during cultural evolution. These mental tools enable us to create precise quantity representations and perform exact calculation that is beyond the reach of any animal species.

Even though number symbols are of paramount importance in today's scientifically and technologically-advanced culture, their invention dates back only a couple of thousand years (Ifrah, 2000). Given the time scale of brain evolution, a *de novo* development of brain areas with distinct, culturally-dependent number symbol functions is more than unlikely (Dehaene, 2005). Rather, it is conceivable that brain structures that originally evolved for other purposes are built upon in the course of continuing evolutionary development (Gould, 1982). According to the 'redeployment hypothesis' (Anderson, 2007) or 'recycling hypothesis' (Dehaene, 2005; Dehaene & Cohen, 2007), already existing simpler cell assemblies are largely preserved, extended, and combined as networks become more complex (Sporns & Kotter, 2004).

In the number domain, existing neuronal components in PFC and IPS subserving non-verbal quantity representations could be used for the new purpose of number symbol encoding, without disrupting their participation in existing cognitive processes (Piazza, Pinel, Le Bihan, & Dehaene, 2007). Guided by the faculty of language, children learn to use number symbols as mental tools during childhood. During this learning process, and as a prerequisite for the utilization of signs as numerical symbols, long-term associations between initially meaningless shapes (that later become numerals) and inherent semantic numerical categories must be established. This necessary, but by no means sufficient step towards the utilization of number symbols in humans can also be mastered by different animal species (Boysen & Berntson, 1989; Washburn & Rumbaugh, 1991; Matsuzawa, 1985; Xia, Emmerton, Siemann, & Delius, 2001).

To investigate the single-neuron mechanisms of associating the quantity meaning of a set to an arbitrary visual shape (semantic association), Diester and Nieder (2007) trained two monkeys to associate the a priori meaningless visual shapes of Arabic numerals (that became 'signs', or more precisely, 'indices' (Wiese, 2003)) with the inherently meaningful numerosity of multiple-dot displays. After this long-term learning process was completed, a relatively large proportion of PFC neurons (24%) encoded plain numerical values, irrespective of whether they had been presented as a specific number of dots or as a visual sign (Figure 26.8a,b). Such 'association neurons' showed similar tuning during the course of the trial to both the direct numerosity in dot stimuli and the associated numerical values of signs (Figure 26.8c). Interestingly, the tuning functions of association neurons showed a distance effect for the discrimination of both dot patterns and shapes associated with numerical values, i.e. a decrease of neuronal activity with increasing numerical distance from a neuron's preferred numerical value (Figure 26.8d). This distance effect found in the shape protocol indicates that association neurons responded as a function of numerical value rather than visual shape per se. Most cells coded the (direct and associated) numerical values during specific time phases in the trial (e.g. only at sample onset or towards the end of

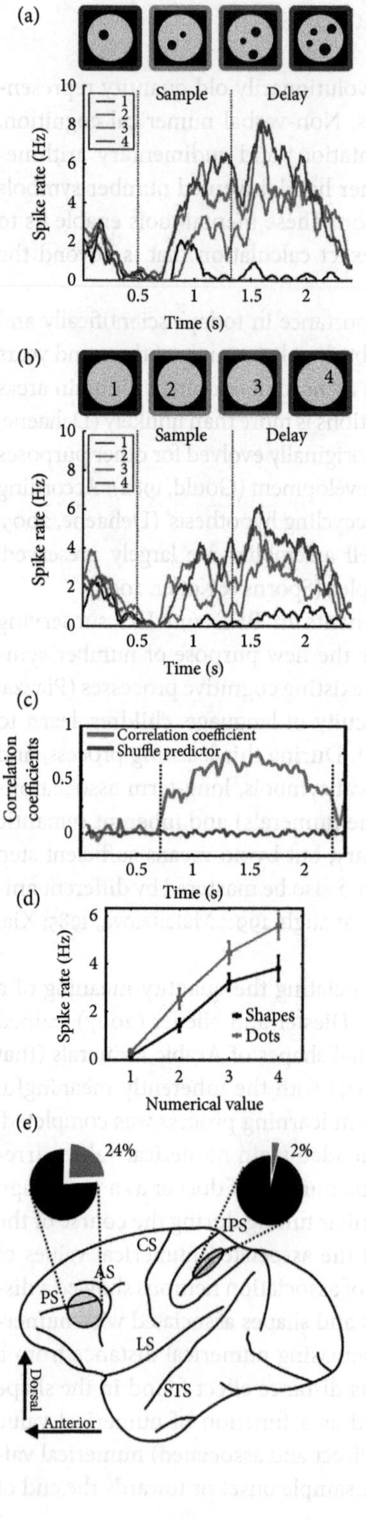

FIGURE 26.8 Semantic associations between signs and numerical categories by single neurons. (a,b) The responses of the same single PFC neuron to both direct numerosities and associated numeral shapes (top panels in (a) and (b) illustrate sample stimuli). Neuronal responses in (a) and (b) are shown as smoothed spike density histograms (colors denote discharge to the corresponding sample numerical value 1 to 4). The first 500 ms indicate the fixation period. Black vertical lines mark sample-onset (500 ms) and offset (1300 ms). This neuron's preferred numerical value in the sample and delay period was 4. Note the similarity in the association neurons' temporal discharge profiles in response to the multiple-dot displays and the shape of Arabic numerals. (c) Time course of original cross correlation coefficients (red) and chance cross correlation coefficients (shuffle predictor, blue). The interval bordered by vertical dotted lines indicates the time phase of significant cross correlation (as determined by measures from Signal Detection Theory) between tuning to numerical values in the multiple-dot displays and Arabic numerals; in this period, the neuron associated numerical values in the two protocols. (d) Tuning functions to numerical values in the multiple-dot displays and Arabic numerals. (e) Lateral view of a monkey brain indicating the recording sites in the PFC and IPS. Proportions of association neurons in the PFC and IPS are displayed as pie charts.

Adapted from Ilka Diester and Andreas Nieder, Semantic associations between signs and numerical categories in the prefrontal cortex, *PLoS Biology*, 5, e294, figure 1, DOI: 10.1371/journal.pbio.0050294. Copyright © 2007, Diester and Nieder. Reproduced under the terms of the Creative Commons Attribution License.

the delay period). The neuronal population as a whole, however, represented the numerical association throughout the entire trial and thus provided crucial associative information over time. The activity of association neurons predicted the monkeys' judgement performance; if the monkeys failed to match the correct number of dots to the learned signs, the tuning behaviour of a given neuron to numerosities and their associated visual shapes were severely disrupted. These findings argue for association neurons as a neuronal substrate for the semantic mapping processes between signs and categories. In the same study (Dieser and Nieder 2007), the activity of neurons in the fundus of the IPS was also recorded. In contrast to PFC, only 2% of all recorded IPS neurons associated signs with numerosities (Figure 26.8e). Moreover, the quality of neuronal association in the IPS was weak and occurred much later during the trial.

The conclusion drawn from these results is that – even though monkeys use the PFC and IPS for non-symbolic quantity representations – only the prefrontal part of this network is engaged in semantic cardinality-to-shape associations. Interestingly, the prefrontal cortex initially also seems to be more involved in semantic associations in children (Ansari, Garcia, Lucas, Hamon, & Dhital, 2005; Kaufmann, Koppelstaetter, Siedentopf, Haala, Haberlandt, et al., 2006; Rivera, Reiss, Eckert, & Menon, 2005). In contrast to adults, pre-scholars lacking ample exposure to number symbols show elevated PFC activity when dealing with symbolic cardinalities. With age and proficiency, however, parietal areas are becoming more and more involved in representing number symbols. This frontal-to-parietal shift has been interpreted as being a result of increasing automaticity in number tasks. This shift of symbolic associations to the parietal lobe could release the limited cognitive resources of PFC for new demanding tasks. The PFC could, thus, be ontogenetically and phylogenetically the first cortical area establishing semantic associations, which might be relocated to the parietal cortex in human adolescents in parallel with maturing language capabilities that endow our species with a sophisticated symbolic system (Deacon, 1997; Nieder, 2009). Similar findings have been made for other stimulus dimensions (e.g. faces), suggesting a general learning effect (Johnson, Grossman, & Cohen Kadosh, 2009).

The prefrontal region is strategically situated to establish semantic associations (Miller & Cohen, 2001); it receives input from both the anterior inferotemporal cortex encoding shape information (Tanaka, 1996), as well as the posterior parietal cortex that contains numerosity-selective neurons (Nieder & Miller, 2004a; Nieder et al., 2006). Previous studies showed that neurons in the PFC encode learned associations between two purely sensory stimuli without intrinsic meaning (e.g. the association of a certain colour with a specific sound, or pairs of pictures; Fuster, Bodner, & Kroger, 2000; Miller, Erickson, & Desimone, 1996; Rainer, Rao, & Miller, 1999). In the anterior inferotemporal cortex, Miyashita and co-workers found 'pair-coding neurons,' that responded to arbitrary pairs of images monkeys learned to match in a pair-association task (Sakai & Miyashita, 1991); the same group found evidence that the PFC is important for active retrieval of these associative representations (Tomita, Ohbayashi, Nakahara, Hasegawa, & Miyashita, 1999). Our findings (Diester & Nieder, 2007) demonstrate that neurons in the PFC represent semantic long-term associations not only between pairs of pictures, but also between arbitrary shapes and systematically arranged categories with inherent meaning (i.e. the ordered cardinalities of sets). In that respect, the PFC of primates may not only control the retrieval of long-term associations, but may in fact constitute a crucial processing stage for abstract semantic associations.

Numerical Rank (Ordinality)

Besides numerical quantity which refers to the empirical property cardinality of objects and events, numerical rank is the second major concept of numerical cognition. Numerical rank refers to the empirical property *serial order* (first, second, third, …) and is sought after by the question '*Which position?*' Animals are also able to represent numerical rank. An elegant study by Brannon and Terrace (1998) demonstrated that rhesus monkeys can be trained to choose numerosities (dot patterns) in ascending or descending order. Importantly, the monkeys also succeeded to rank novel numerosities they had not been trained on, suggesting a conceptual understanding of numerical rank.

The frontal lobe is an ideal region in the brain to encode both sensory object properties and rank-order information because the PFC receives massive sensory input from the temporal and parietal lobes, and projects to pre-motor and motor areas of the frontal lobe (Miller & Cohen, 2001). Thus, neurons that encode the ordinal position of task-related hand or eye movements have been found frequently in prefrontal (Funahashi, Inoue, & Kubota, 1997) and a subset of motor-related cortical areas in trained monkeys. Joseph and co-workers (Barone & Joseph, 1989; Kermadi & Joseph, 1995; Procyk & Joseph, 2001; Procyk, Tanaka, & Joseph, 2000) identified order-selective neurons in the frontal eye field (FEF), caudate nucleus, and anterior cingulate cortex of monkeys that had been trained to sequentially order spatially arranged items. These neurons were only active when the monkeys reached for the 'first,' 'second,' or 'third' target, irrespective of the targets' location and the precise type of hand movement. Clower and Alexander (1998) trained a monkey to position a cursor on a video display by moving a joystick clockwise or counterclockwise along a spatially-arranged four-item path. In the pre-supplementary motor areas (pre-SMA), more than two-thirds of the recorded neurons showed significant effects of numerical order, but only about one-third of the neurons displayed an effect of rank order in the supplementary motor area (SMA). Rank-order selectivity was also identified in the pre-SMA of monkeys that had been trained to sequentially perform three different hand movements ('push,' 'pull,' or 'turn') in four to six different orders separated by waiting times (Shima & Tanji, 2000; Tanji & Shima, 1994). In the pre-SMA, the activity differed selectively in the process of preparing the first, second, or third movements in individual trials. The SMA, on the other hand, was more involved in linking the occurrence of two different movements and, therefore, in determining the order of the component movements in the sequence (relational order). Ordinal position of movements in pre-SMA seems to be encoded in an effector-independent manner, i.e. regardless of effector of movements (eyes or arms; Isoda & Tanji, 2004). In two motor areas that are specialized in processing eye movements, activity that reflected saccade sequence or the numerical position of a saccade within a sequence (rank) was more common in the supplementary eye field (SEF), whereas activity that reflected saccade direction was more dominant in the FEF (Isoda & Tanji, 2003).

Interestingly, encoding of numerical order has not only been observed in these pre-motor and supplementary motor areas, but also in the primary motor cortex. Carpenter and colleagues (Carpenter, Georgopoulos, & Pellizzer, 1999) showed monkeys five spatially-arranged visual targets that appeared successively on a screen. After the target sequence was complete, one of the items changed its colour. The monkeys needed to memorize the order in which the targets appeared and point to the item that appeared just after the one that had changed its colour at the end of the list presentation. In approximately one-third of the neurons recorded from the arm region of the topographically organized primary motor cortex (M1), the ordinal position of the targets was the only factor that co-varied with neuronal activity. Therefore, the motor cortex – an area that is traditionally regarded as purely motor executive – also participates in the processing of cognitive information about serial order within the context of a motor task. The authors pointed out, however, that the motor cortex most likely is just one component in a distributed network that encodes, stores, and recalls a sequence. Motor-related areas, such as M1, SMA, pre-SMA, and FEF may receive numerical information that has already been computed in earlier stages of the cortical hierarchy to perform appropriate serial-order actions.

Ordinal categorization of items requires both information about the rank of an item (for example, based on temporal order) and its identity. Neuropsychological studies emphasize the importance of the lateral prefrontal cortex in maintaining temporal order information (McAndrews & Milner, 1991; Milner, 1971). In monkeys, lesioning the dorsolateral frontal cortex causes impairments in tasks that require recall of the temporal order of events and stimuli (Petrides, 1995). In two recent elegant studies (Ninokura, Mushiake, & Tanji, 2003, 2004), the single-neuron correlate of temporal rank order information in visual lists was addressed. Monkeys were trained to observe and remember the order in which three visual objects appeared. Subsequently, the animals planned and executed a triple-reaching movement in the same order as previously seen (Figure 26.9). Neurons in the ventrolateral PFC selectively encoded visual object properties (26% of the total sample), whereas neurons in the dorso-lateral PFC (44%) were selectively tuned to the rank order of the objects irrespective of the sensory properties of objects. For example, a rank-order selective neuron would be active whenever the second item of a shuffled list appears. A third class of neurons (30%), found in the ventrolateral PFC, showed the most complex responses, integrating the objects' sensory and order information. Such neurons would only discharge whenever a certain object appeared at a given position in the sequence. Similar results have also been reported by Inoue and Mikami (2006).

The representational formats of non-verbal serial order information are still poorly understood. However, the behavioural and neuronal data indicate an imprecise representation of discrete numerical rank, which is reminiscent of the analogue-magnitude mechanism that has been proposed for cardinality. To elaborate a computational model of working memory for serial order Botvinick and Watanabe (2007) recently wove item, numerosity and rank information together. Their network combined graded neuronal responses to different items (not yet verified experimentally) and tuning functions for sequential enumeration processes (Nieder et al., 2006) with the data showing that neurons in the PFC code the rank of items within a sequence (Ninokura et al., 2004; Figure 26.9). The model's output, a recalled multi-item sequence, replicated many behavioural characteristics of working memory such as the primacy effect (a recall advantage for initial items) and the recent effect (advantage for the last one or two items). Furthermore, changing the width of the model rank tuning curve simulated the developmental finding of improved recall accuracy with age. Thus, this model integrates across several neurophysiological studies to demonstrate how higher cognitive functions may exploit both quantity and rank (Jacob & Nieder, 2008). It posits that working

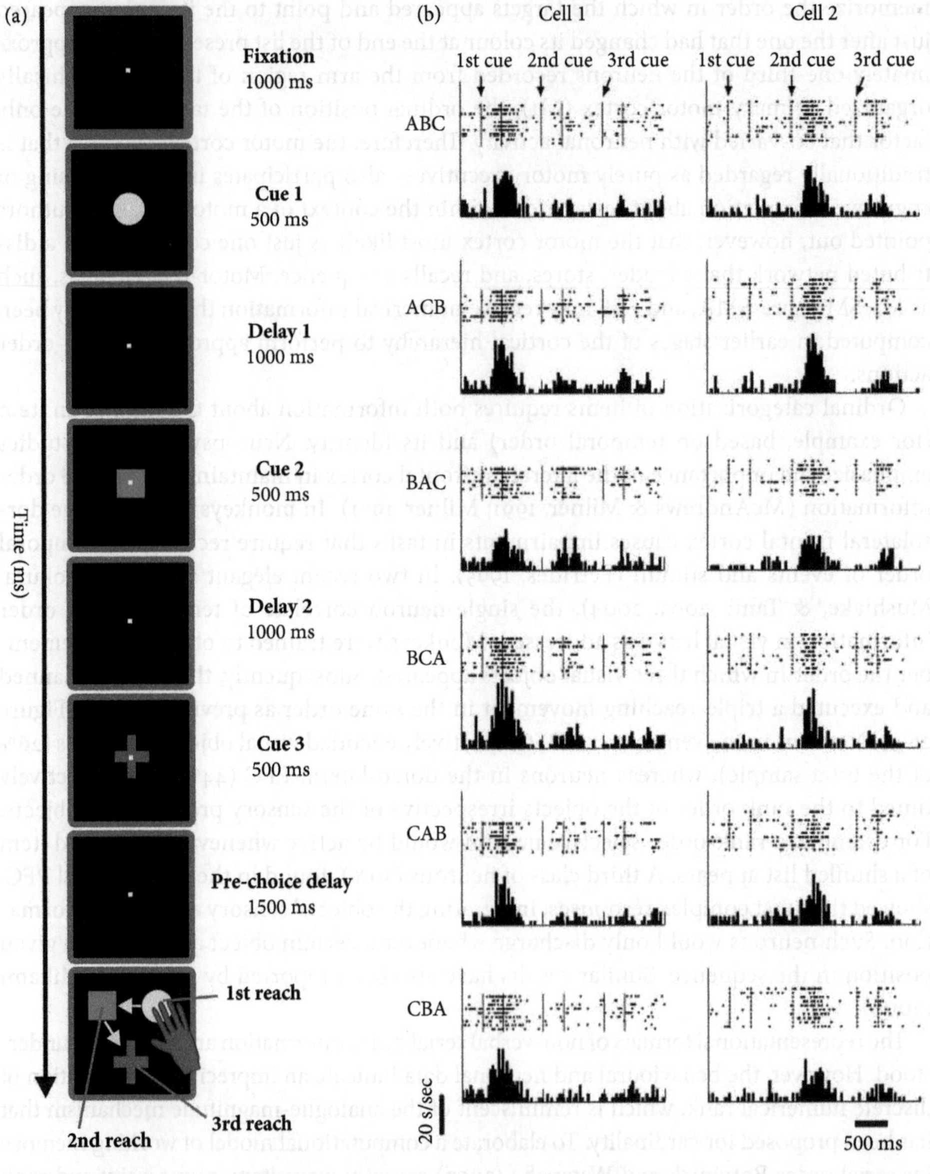

FIGURE 26.9 Temporal ordering task and single cell responses from the PFC. (a) Monkeys were required to observe and remember the order in which three visual objects appeared, so that the animals could plan a subsequent triple-reaching movement in the same order. (b) Two single neurons encoding the first (cell 1) and the second rank (cell 2), irrespective of the order in which the three items (symbolized by letters ABC) appeared. Neural responses are shown in a dot-raster histogram (top panels, each dot represents an action potential) and averaged as peri-stimulus time histograms (bottom panels).

Adapted from Yoshihisa Ninokura, Hajime Mushiake, and Jun Tanji, Integration of Temporal Order and Object Information in the Monkey Lateral Prefrontal Cortex, *Journal of Neurophysiology*, 91(1), pp. 555–560, figure 1, DOI: 10.1152/jn.00694.2003 © The American Physiological Society (APS).

memory of ordered sequences at least in part relies on neuronal assemblies in the parietal and prefrontal cortices that also contribute to numerosity representations.

Numerical Rules

Mathematics is based on highly abstract principles, or rules, of how to structure, process and evaluate numerical information. Also in our everyday life, we obey to quantitative rules. For instance, we typically adopt a 'less than' strategy when shopping for a product to pay the smallest amount of money. When searching for a job, on the other hand, our plan of action is 'greater than,' and we strive to earn the largest sum of money. In such pragmatic situations, our decisions on quantities are guided by mathematical rules applied to them. However, mathematical principles also play a major role in our scientifically and technologically advanced culture (Danzig, 1954). They allow us to structure and process numerical information in the most sophisticated ways, and with the most impressive results, as witnessed by the discovery of laws of nature and their technological applications.

The PFC is intensively engaged during the processing of numbers and arithmetic operations requiring mathematical rules (Roland & Friberg, 1985; Rueckert, Lange, Partiot, Appollonio, Litvan, Le Bihan, et al., 1996; Dehaene, Tzourio, Frak, Raynaud, Cohen, Mehler, et al., 1996; Gruber, Indefrey, Steinmetz, & Kleinschmidt, 2001). Consistent with these findings, damage to the PFC impairs quantity representations (Luria, 1966; Shallice & Evans, 1978; Smith & Milner, 1984). Recently, Bongard and Nieder (2010) recorded the activity of individual prefrontal cortex neurons in rhesus monkeys required to flexibly switch between 'greater than/less than'-rules (Figure 26.10). The monkeys performed this task with different numerical quantities and generalized to novel set sizes, indicating that they had learned an abstract mathematical principle. Roughly 20% of randomly selected PFC neurons encode basic mathematical rules required to process most abstract numerical information. For these neurons, the most prevalent neuronal activity reflected the 'greater than/less than' rules. Approximately one half of the selective neurons preferred the 'greater than'-rule, whereas the other half preferred the 'less than' rule. Purely sensory- and memory-related activity was almost absent. These data show that single PFC neurons have the capacity to represent flexible operations on most abstract numerical quantities. It seems the brain operates with specific 'rule-coding' units that control the flow of information between sensory, memory and motor stages.

The cognitive capacity we traced in a non-human primate is reminiscent of young children's understanding of simple mathematical operations. Based on an intuitive knowledge of set size, they learn, for instance, that three pieces of candy are 'less than' five candies, or five candies are 'greater than' three candies. This comparative numerical relationship is formalized later in education to read '$3 < 5$' or '$5 > 3$', with the symbolic relational operators (<, >) governing how numerical information is to be evaluated. We propose that the brain mechanism we describe serves as an evolutionary precursor for higher mathematics in adult humans. Symbolic mathematical operations may co-opt or 'recycle' prefrontal circuits (Dehaene & Cohen, 2007; Nieder, 2009) to dramatically enrich and enhance our symbolic mathematical skills. Comparative work in non-human primates, children, and adults should test this hypothesis.

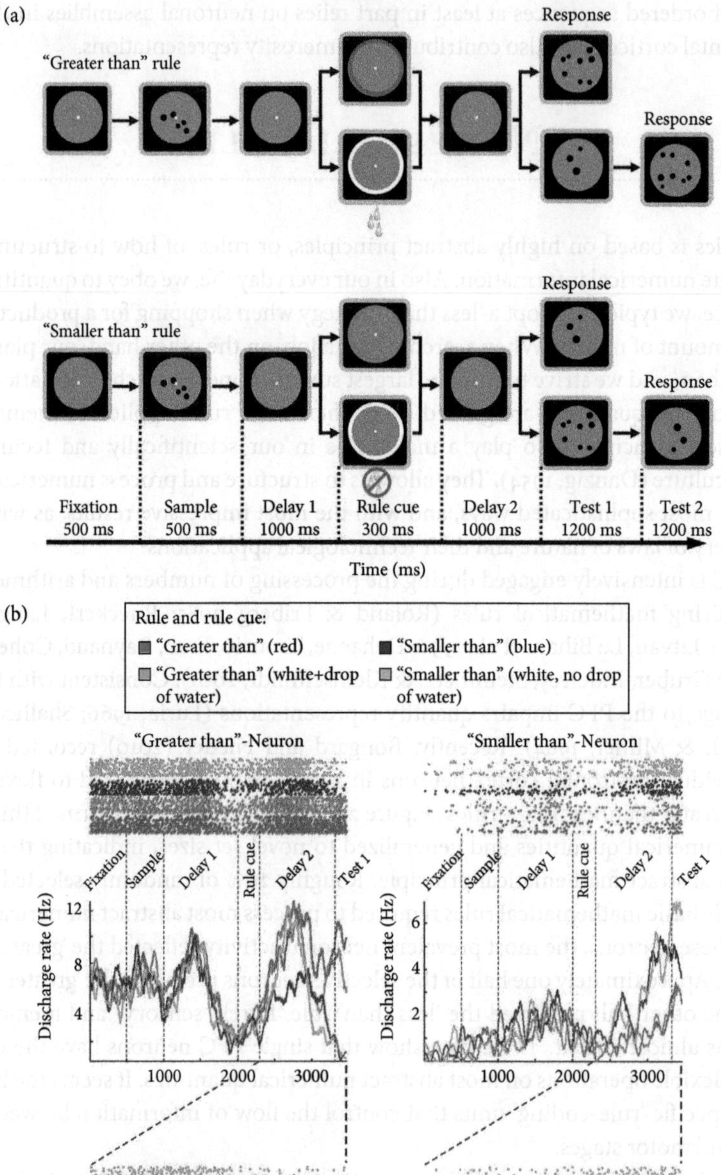
(a)
"Greater than" rule
Response
Response
"Smaller than" rule
Response
Response
Fixation 500 ms
Sample 500 ms
Delay 1 1000 ms
Rule cue 300 ms
Delay 2 1000 ms
Test 1 1200 ms
Test 2 1200 ms
Time (ms)
(b)
Rule and rule cue:
"Greater than" (red)
"Smaller than" (blue)
"Greater than" (white+drop of water)
"Smaller than" (white, no drop of water)
"Greater than"-Neuron
"Smaller than"-Neuron
Fixation
Sample
Delay 1
Rule cue
Delay 2
Test 1
Discharge rate (Hz)
Delay 2
Test 1
Time (ms)

FIGURE 26.10 (Continued)

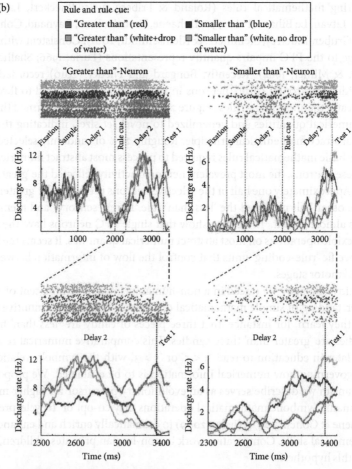

FIGURE 26.10 (CONTINUED) Numerical rule coding. (a) Behavioural protocol. Monkeys grasped a lever and maintained central fixation. A sample numerosity was followed by a brief working memory delay (Delay 1). Next, a cue indicated either the 'greater than' – or the 'less than'-rule. Each rule was signified by two different sensory cues (red and water for the 'greater than'-rule, blue or no-water for the 'less than'-rule; first bifurcating arrows), followed by a rule delay (Delay 2) requiring the monkeys to assess the rule at hand for the subsequent choice. For each rule, two trial types are illustrated (second bifurcating arrows). For the 'greater than'-rule (*top*), the monkeys released the lever if more dots were shown in the first test display than in the sample display; otherwise, they held the lever until appearance of a second test display that always required a response. For the 'less than'-rule (*bottom*), the lever had to be released if the numerosity in the first test display was smaller than in the sample display. Thus, only test 1 required a decision; test 2 was used so that a Behavioural response was required on each trial, ensuring that the monkeys were paying attention at all trials. (b) Example neuron selective for the 'greater than'-rule towards the end of the Delay 2 phase (*left panels*). Responses across the entire trial (*top left*) and magnified during the Delay 2 period (*bottom left*) are shown. In the top panels, the neuronal responses are plotted as dot-raster histograms (each dot represents an action potential, spike trains are sorted and colour-coded according to the rules and rule cues). Bottom panels show spike density functions (activity averaged over all trials and smoothed by a 150 ms Gaussian kernel). Rule-selectivity was regardless of which cue signified the rule. *Right panels:* Example neuron selective for the 'less than'-rule.

Conclusion

This chapter illustrates that non-verbal numerical representations can engage a wide cortical network, with the PFC and the IPS, in particular, as key structures. Neurons in these areas are characterized by response properties that underlie basic psychophysical phenomena during cardinality or serial order judgments. So far, single-cell studies in monkeys have concentrated on either cardinal or ordinal aspects, but never both. In the human primate, a recent study showed that the cortical network for rank coincides with the areas processing numerical quantity (Fias, Lammertyn, Caessens, & Orban, 2007), even though this does not necessarily mean that single neurons may encode both cardinality and ordinality (Zorzi et al., 2011). Based on the findings that discrete and continuous quantities and even simultaneously and sequentially presented numerosities are encoded by largely distinct neuronal populations in the IPS, a cellular segregation also for cardinal and ordinal stimulus aspects is likely. Single-cell electrophysiology is well poised to answer this and other important questions about the neural basis of numerical cognition in the years to come.

References

Anderson, M.L. (2007). Evolution of cognitive function via redeployment of brain areas. Neuroscientist, 13, 13–21.

Ansari, D., Garcia, N., Lucas, E., Hamon, K., & Dhital, B. (2005). Neural correlates of symbolic number processing in children and adults. NeuroReport, 16, 1769–1773.

Barone, P., & Joseph, J.P. (1989). Prefrontal cortex and spatial sequencing in macaque monkey. Experimental Brain Research, 78, 447–464.

Bongard, S., & Nieder, A. (2010). Basic mathematical rules are encoded by primate prefrontal cortex neurons. Proceedings National Academy Sciences USA, 107, 2277–2282.

Botvinick, M., & Watanabe, T. (2007). From numerosity to ordinal rank: a gain-field model of serial order representation in cortical working memory. Journal of Neuroscience, 27, 8636–8642.

Boysen, S.T., & Berntson, G.G. (1989). Numerical competence in a chimpanzee (Pan troglodytes). Journal of Comparative Psychology, 103, 23–31.

Brannon, E.M., & Terrace, H.S. (1998). Ordering of the numerosities 1 to 9 by monkeys. Science, 282, 746–749.

Bremmer, F., Schlack, A., Shah, N.J., Zafiris, O., Kubischik, M., Hoffmann, K., Zilles, K., & Fink, G.R. (2001). Polymodal motion processing in posterior parietal and premotor cortex: a human fMRI study strongly implies equivalencies between humans and monkeys. Neuron, 29, 287–296.

Carpenter, A.F., Georgopoulos, A.P., & Pellizzer, G. (1999). Motor cortical encoding of serial order in a context-recall task. Science, 283, 1752–1757.

Castelli, F., Glaser, D.E., & Butterworth, B. (2006). Discrete and analogue quantity processing in the parietal lobe: a functional MRI study. Proceedings of the National Academy of Sciences USA, 103, 4693–4698.

Chafee, M.V., & Goldman-Rakic, P.S. (2000). Inactivation of parietal and prefrontal cortex reveals interdependence of neural activity during memory-guided saccades. Journal of Neurophysiology, 83, 1550–1566.

Clower, W.T., & Alexander, G.E. (1998). Movement sequence-related activity reflecting numerical order of components in supplementary and presupplementary motor areas. Journal of Neurophysiology, 80, 1562–1566.

Cohen Kadosh, R., & Walsh, V. (2009). Numerical representation in the parietal lobes: abstract or not abstract? Behavioural Brain Sciences, 32, 313–328.

Colby, C.L., Duhamel, J.R., & Goldberg, M.E. (1993). Ventral intraparietal area of the macaque – anatomical location and visual response properties. Journal of Neurophysiology, 69, 902–914.

Colby, C.L., & Goldberg, M.E. (1999). Space and attention in parietal cortex. Annual Review of Neuroscience, 22, 319–349.

Cordes, S., Gelman, R., Gallistel, C.R., & Whalen, J. (2001).Variability signatures distinguish verbal from nonverbal counting for both large and small numbers. Psychonomic Bulletin & Review, 8, 698–707.

Danzig, T. (1954). Number, the Language of Science. New York: Free.

Deacon, T. (1997). The Symbolic Species: The Co-evolution of Language and the Human Brain. London: Norton.

Dehaene, S. (2005). Evolution of human cortical circuits for reading and arithmetic: the 'neuronal recycling' hypothesis. In S. Dehaene, J. R. Duhamel, M. D. Hauser, & G. Rizzolatti (eds), From Monkey Brain to Human Brain (pp. 137–157). Cambridge, MA: MIT Press.

Dehaene, S., & Changeux, J.P. (1993). Development of elementary numerical abilities: a neural model. Journal of Cognitive Neuroscience, 5, 390–407.

Dehaene, S., & Cohen, L. (2007). Cultural recycling of cortical maps. Neuron, 56, 384–398.

Dehaene, S., Spelke, E., Pinel, P., Stanescu, R., & Tsivkin, S. (1999) Sources of mathematical thinking: behavioral and brain-imaging evidence. Science, 284, 970–974.

Dehaene, S., Tzourio, N., Frak, V., Raynaud, L., Cohen, L., Mehler, J., & Mazoyer, B. (1996). Cerebral activations during number multiplication and comparison: a PET study. Neuropsychologia, 34, 1097–1106.

Diester, I., & Nieder, A. (2007). Semantic associations between signs and numerical categories in the prefrontal cortex. PLoS Biology, 5, e294.

Evarts, E.V. (1966). Methods for recording activity of individual neurons in moving animals. In R. F. Rushmer (ed.), *Methods in Medical Research* (pp. 241–250). Chicago: Year Book.

Fias, W., Lammertyn, J., Caessens, B., & Orban, G.A. (2007). Processing of abstract ordinal knowledge in the horizontal segment of the intraparietal sulcus. Journal of Neuroscience, 27, 8952–8956.

Fias, W., Lammertyn, J., Reynvoet, B., Dupont, P., & Orban, G.A. (2003). Parietal representation of symbolic and nonsymbolic magnitude. Journal of Cognitive Neuroscience, 15, 47–56.

Freedman, D.J., Riesenhuber, M., Poggio, T., & Miller, E.K. (2001). Categorical representation of visual stimuli in the primate prefrontal cortex. Science, 291, 312–316.

Funahashi, S., Inoue, M., & Kubota, K. (1997). Delay-period activity in the primate prefrontal cortex encoding multiple spatial positions and their order of presentation. Behavioural Brain Research, 84, 203–223.

Fuster, J.M., Bodner, M., & Kroger, J.K. (2000). Cross-modal and cross-temporal association in neurons of frontal cortex. Nature, 405, 347–351.

Gould, S.J., & Vrba, E.S. (1982). Exaptation: a missing term in the science of form. Paleobiology, 8, 4–15.

Gruber, O., Indefrey, P., Steinmetz, H., & Kleinschmidt, A. (2001). Dissociating neural correlates of cognitive components in mental calculation. Cerebral Cortex, 11, 350–359.

Halberda, J., Mazzocco, M.M., & Feigenson, L. (2008). Individual differences in non-verbal number acuity correlate with maths achievement. Nature, 455, 665–668.

Henik, A., & Tzelgov, J. (1982). Is three greater than five: the relation between physical and semantic size in comparison tasks. Memory and Cognition, 10, 389–395.

Ifrah, G. (2000). The Universal History of Numbers: From Prehistory to the Invention of the Computer. New York: Wiley.

Inoue, M., & Mikami, A. (2006). Prefrontal activity during serial probe reproduction task: encoding, mnemonic, and retrieval processes. Journal of Neurophysiology, 95, 1008–1041.

Ischebeck, A., Schocke, M., & Delazer, M. (2009). The processing and representation of fractions within the brain: an fMRI investigation. NeuroImage, 47, 403–413.

Isoda, M., & Tanji, J. (2003). Contrasting neuronal activity in the supplementary and frontal eye fields during temporal organization of multiple saccades. Journal of Neurophysiology, 90, 3054–3065.

Isoda, M., & Tanji, J. (2004). Participation of the primate presupplementary motor are in sequencing multiple saccades. Journal of Neurophysiology, 92, 653–659.

Jacob, S.N., & Nieder, A. (2008). The ABC of cardinal and ordinal number representations. Trends in Cognitive Sciences, 12, 41–43.

Jacob, S.N., & Nieder, A. (2009a). Notation-independent representation of fractions in the human parietal cortex. Journal of Neuroscience, 29, 4652–4657.

Jacob, S.N., & Nieder, A. (2009b). Tuning to non-symbolic proportions in the human frontoparietal cortex. European Journal of Neuroscience, 30, 1432–1442.

Jasper, H.H., Ricci G., Doane, B. (1960). Microelectrode analysis of cortical cell discharge during avoidance conditioning in the monkey. International Journal of Electroencephalography & Clinical Neurophysiology, Suppl 131, 137–156.

Johnson, M.H., Grossman, T., & Cohen Kadosh, K. (2009). Mapping functional brain development: building a social brain through interactive specialization. Developmental Psychology, 45, 151–159.

Kaufmann, L., Koppelstaetter, F., Siedentopf, C., Haala, I., Haberlandt, E., Zimmerhackl, L.B., Felber, S., & Ischebeck, A. (2006). Neural correlates of the number-size interference task in children. NeuroReport, 17, 587–591.

Kermadi, I., & Joseph, J.P. (1995). Activity in the caudate nucleus of monkey during spatial sequencing. Journal of Neurophysiology, 74, 911–933.

Kohonen, T. (1997). Self-organizing Maps, 2nd edn. Berlin: Springer-Verlag.

Laubach, M. (2004). Wavelet-based processing of neuronal spike trains prior to discriminant analysis. Journal of Neuroscience Methods, 134, 159–168.

Lewis, J.W., & Van Essen, D.C. (2000). Corticocortical connections of visual, sensorimotor, and multimodal processing areas in the parietal lobe of the macaque monkey. Journal of Comparative Neurology, 428, 112–137.

Luria, A.R. (1966). Higher Cortical Functions in Man. London: Tavistock.

Matsuzawa, T. (1985). Use of numbers by a chimpanzee. Nature, 315, 57–59.

McAndrews, M.P., & Milner, B. (1991). The frontal cortex and memory for temporal order. Neuropsychologia, 29, 849–859.

Meck, W.H., & Church, R.M. (1983). A mode control model of counting and timing processes. Journal of Experimental Psychology: Animal Behaviour Proceedings, 9, 320–334.
Miller, E.K., & Cohen, J.D. (2001). An integrative theory of prefrontal cortex function. Annual Review of Neuroscience, 24, 167–202.
Miller, E.K., Erickson, C.A., & Desimone, R. (1996). Neural mechanisms of visual working memory in prefrontal cortex of the macaque. Journal of Neuroscience, 16, 5154–5167.
Milner, B. (1971). Interhemispheric differences in the localization of psychological processes in man. British Medical Bulletin, 27, 272–277.
Nieder, A. (2005). Counting on neurons: the neurobiology of numerical competence. Nature Reviews Neuroscience, 6, 177–190.
Nieder, A. (2009). Prefrontal cortex and the evolution of symbolic reference. Current Opinions in Neurobiology, 19, 1–10.
Nieder, A., & Dehaene, S. (2009). Representation of number in the brain. Annual Reviews of Neuroscience, 32, 185–208.
Nieder, A., Diester, I., & Tudusciuc, O. (2006). Temporal and spatial enumeration processes in the primate parietal cortex. Science, 313, 1431–1435.
Nieder, A., Freedman, D.J., & Miller, E.K. (2002). Representation of the quantity of visual items in the primate prefrontal cortex. Science, 297, 1708–1711.
Nieder, A., & Merten, K. (2007). A labeled-line code for small and large numerosities in the monkey prefrontal cortex. Journal of Neuroscience, 27, 5986–5993.
Nieder, A., & Miller, E.K. (2003). Coding of cognitive magnitude: compressed scaling of numerical information in the primate prefrontal cortex. Neuron, 37, 149–157.
Nieder, A., & Miller, E.K. (2004a). A parieto-frontal network for visual numerical information in the monkey. Proceedings of the National Academy of Sciences USA, 101, 7457–7462.
Nieder, A., & Miller, E.K. (2004b). Analog numerical representations in rhesus monkeys: evidence for parallel processing. Journal of Cognitive Neuroscience, 16, 889–901.
Ninokura, Y., Mushiake, H., & Tanji, J. (2003). Representation of the temporal order of visual objects in the primate lateral prefrontal cortex. Journal of Neurophysiology, 89, 2868–2873.
Ninokura, Y., Mushiake, H., & Tanji, J. (2004). Integration of temporal order and object information in the monkey lateral prefrontal cortex. Journal of Neurophysiology, 91, 555–560.
Petrides, M. (1995). Impairments on nonspatial self-ordered and externally ordered working memory tasks after lesions of the mid-dorsal part of the lateral frontal cortex in the monkey. Journal of Neuroscience, 15, 359–375.
Piazza, M., Izard, V., Pinel, P., Le Bihan, D., & Dehaene, S. (2004). Tuning curves for approximate numerosity in the human intraparietal sulcus. Neuron, 44, 547–555.
Piazza, M., Pinel., P, Le Bihan, D., & Dehaene, S. (2007). A Magnitude code common to numerosities and number symbols in human intraparietal cortex. Neuron, 53, 293–305.
Pinel, P., Piazza, M., Le Bihan, D., & Dehaene, S. (2004). Distributed and overlapping cerebral representations of number, size, and luminance during comparative judgments. Neuron, 41, 983–993.
Procyk, E., & Joseph, J.P. (2001). Characterization of serial order encoding in the monkey anterior cingulate sulcus. European Journal of Neuroscience, 14, 1041–1046.
Procyk, E., Tanaka, Y.L., & Joseph, J.P. (2000). Anterior cingulate activity during routine and non-routine sequential behaviors in macaques Nature Neuroscience, 3, 502–508.
Quintana, J., Fuster, J.M., & Yajeya, J. (1989). Effects of cooling parietal cortex on prefrontal units in delay tasks. Brain Research, 503, 100–110.
Rainer, G., Rao, S.C., & Miller, E.K. (1999). Prospective coding for objects in primate prefrontal cortex. Journal of Neuroscience, 19, 5493–5505.
Rivera, S.M., Reiss, A.L., Eckert, M.A., & Menon, V. (2005). Developmental changes in mental arithmetic: evidence for increased functional specialization in the left inferior parietal cortex. Cerebral Cortex, 15, 1779–1790.
Roitman, J.D., Brannon, E.M., & Platt, M.L. (2007). PLoS Biology, 8, e208.

Roland, P.E., & Friberg, L. (1985). Localization of cortical areas activated by thinking. Journal of Neurophysiology, 53, 1219–1243.
Rueckert, L., Lange, N., Partiot, A., Appollonio, I., Litvan, I., Le Bihan, D., & Grafman, J. (1996). Visualizing cortical activation during mental calculation with functional MRI. NeuroImage, 3, 97–103.
Sakai, K., & Miyashita, Y. (1991). Neural organization for the long-term memory of paired associates. Nature, 354, 152–155.
Sawamura, H., Shima, K., & Tanji, J. (2002). Numerical representation for action in the parietal cortex of the monkey. Nature, 415, 918–922.
Sawamura, H., Shima, K., & Tanji, J. (2010). Deficits in action selection based on numerical information after inactivation of the posterior parietal cortex in monkeys. Journal of Neurophysiology, 104, 902–910.
Shallice, T., & Evans, M.E. (1978). The involvement of the frontal lobes in cognitive estimation. Cortex, 14, 294–303.
Shima, K., & Tanji, J. (2000). Neuronal activity in the supplementary and presupplementary motor areas for temporal organization of multiple movements. Journal of Neurophysiology, 84, 2148–2160.
Smith, M.L., & Milner, B. (1984). Differential effects of frontal-lobe lesions on cognitive estimation and spatial memory. Neuropsychologia, 22, 697–705.
Sporns, O., & Kotter, R. (2004). Motifs in brain networks. PLoS Biology, 2, e369.
Stoet, G., & Snyder, L.H. (2004). Single neurons in posterior parietal cortex of monkeys encode cognitive set. Neuron, 42, 1003–1012.
Stoianov, I., & Zorzi, M. (2012). Emergence of a 'visual number sense' in hierarchical generative models. Nature Neuroscience, 15, 194–196.
Tanaka, K. (1996). Inferotemporal cortex and object vision. Annual Reviews in Neuroscience, 19, 109–139.
Tanji, J., & Shima, K. (1994). Role for supplementary motor area cells in planning several movements ahead. Nature, 371, 413–416.
Tomita, H., Ohbayashi, M., Nakahara, K., Hasegawa, I., & Miyashita, Y. (1999). Top-down signal from prefrontal cortex in executive control of memory retrieval. Nature, 401, 699–703.
Tudusciuc, O., & Nieder, A. (2007). Neuronal population coding of continuous and discrete quantity in the primate posterior parietal cortex. Proceedings of the National Academy of Sciences USA, 104, 14513–14518.
Vallentin, D., & Nieder, A. (2008). Behavioural and prefrontal representation of spatial proportions in the monkey. Current Biology, 18, 1420–1425.
Vallentin, D., & Nieder, A. (2010). Representations of visual proportions in the primate posterior parietal and prefrontal cortices. European Journal of Neuroscience, 32, 1380–1387.
Van Essen, D.C., & DeYoe, E.A. (1993). Concurrent processing in the primate visual cortex. In M. S. Gazzaniga (ed.), *The Cognitive Neurosciences* (pp. 383–400). Cambridge, MA: MIT Press.
Verguts, T., & Fias, W (2004). Representation of number in animals and humans: a neural model. Journal of Cognitive Neuroscience, 16, 1493–1504.
Wallis, J.D., Anderson, K.C., & Miller, E.K. (2001). Single neurons in prefrontal cortex encode abstract rules. Nature, 411, 953–956.
Walsh, V. (2003). A theory of magnitude: common cortical metrics of time, space and quantity. Trends in Cognitive Sciences, 7, 483–488.
Walsh, V. (this volume). A theory of magnitude: The parts that sum to number. In R. Cohen Kadosh, & A. Dowker, (eds.), *The Oxford Handbook of Numerical Cognition*. Oxford University Press.
Washburn, D.A. (1994). Stroop-like effects for monkeys and humans: processing speed or strength of association? Psychological Science, 5, 375–379.
Washburn, D.A., & Rumbaugh, D.M. (1991). Ordinal judgments of numerical symbols by macaques (Macaca mulatta). Psychological Science, 2, 190–193.
Whalen, J., Gallistel, C.R., & Gelman, R. (1999). Nonverbal counting in humans: the psychophysics of number representation. Psychological Science, 10, 130–137.

Wiese, H. (2003). Numbers, Language and the Human Mind. Cambridge, MA: Cambridge University Press.

Xia, L., Emmerton, J., Siemann, M., & Delius, J.D. (2001). Pigeons (Columba livia) learn to link numerosities with symbols. Journal of Comparative Psychology, 115, 83–91.

Zorzi, M., & Butterworth, B. (1999). A computational model of number comparison. In: M. Hahn & S. C. Stoness (eds), Proceedings of the Twenty First Annual Conference of the Cognitive Science Society (pp. 778–783). Mahwah, NJ: Erlbaum.

Zorzi, M., Di Bono, M.G., & Fias, W. (2011). Distinct representations of numerical and non-numerical order in the human intraparietal sulcus revealed by multivariate pattern recognition. NeuroImage, 56, 674–680.

Zorzi, M., Stoianov, I., & Umiltà, C. (2005). Computational modeling of numerical cognition. In J. Campbell (ed.), Handbook of Mathematical Cognition (pp. 67–84). New York: Psychology Press.

CHAPTER 27

DEVELOPMENT OF THE NUMERICAL BRAIN

LIANE KAUFMANN, KARIN KUCIAN, AND MICHAEL VON ASTER

INTRODUCTION

DESPITE the increasing scientific and public interest into numerical cognition, in general, and developmental calculation disorders, in particular, developmental brain imaging studies of numerical cognition are still scarce. Compared with attention or reading research scientific endeavours targeted at numerical cognition are rather in their infancy. Thus, it is not surprising to learn that the majority of functional magnetic resonance imaging (fMRI) studies in the field are devoted to adults and, thus, mature brain systems. Among others, developmental fMRI studies may be limited because of several challenges accompanying brain imaging research with children. A rather trivial, but nevertheless crucial arduousness for researchers examining young children is the noisy and narrow scanning environment. Furthermore, experimental paradigms need to be adapted to children's cognitive and physical capacities. For example, upon designing experimental paradigms to be solved in the scanner, researchers need to take into account children's skill levels (e.g. receptive language, reading skills, working memory, abstract thinking) and ensure that children are able to meet the specific experimental requirements such as response demands and task duration. Obviously, paradigms developed for adults may not be adequate to test children. If, for instance, a task requires children to read task instructions or to memorize complex sequelae of stimuli-response associations, the observed neurofunctional activity will most likely be considerably confounded by task-irrelevant supporting processing mechanisms. Depending on a child's developmental status, which might vary across functional domains, the same experimental task will elicit quite different demands on supporting mechanisms, even in children of the same age. In other words, recruiting children of the same age range is necessary, but maybe not sufficient to ensure comparable cognitive skill levels. Hence, developmental fMRI studies are much more resource-intensive because the neuropsychological background examination needs to be much more comprehensive than is the case if investigating healthy adults (in whom the functional cerebral specialization is already established). Likewise, young children may find it difficult to comply with manual response requirements, either because they have not yet mastered right/left discrimination (this is important, e.g. if the task at hand requires the participant to 'press the right key in case of X' and 'the left key in case of Y') or because response requirements place heavy demands on fine-motor skills (e.g. if more than one response key needs to be pressed by one hand). Please note that verbal responses are highly unusual in brain imaging studies because the orofacial movements associated with speech production require a large language-related network that may partially overlap with task-relevant brain regions and thus mask or diminish part or all of the neurofunctional activity under investigation. Recently, passive viewing paradigms have been employed to circumvent response-related neural activity pertaining to either mouth or hand movements (adults: Ansari, Dhital, & Siong, 2006; Piazza, Izard, Pinel, Le Bihan, & Dehaene, 2004; 4-year-old children: Cantlon, Brannon, Carter, & Pelphrey, 2006). Importantly, findings of the latter studies disclosed that number-relevant brain regions are recruited even in passive viewing paradigms that do not require participants to respond explicitly. Passive viewing fMRI paradigms rest on neural (dis)habituation and in numerical cognition research, participants are habituated to specific numerosities (i.e. set sizes). Once habituation has taken place, the neural responsivity to a specific numerosity becomes negligible and only after the neural system detects a deviancy (here, the deviancy would be a change in numerosity) the system dishabituates which in turn is reflected in a change (i.e. increase or decrease) of the fMRI response. Nonetheless, while passive viewing paradigms have been successfully used in studies investigating numerosity processing, they have not yet been employed in studies targeted at investigating more complex aspects of numerical cognition such as arithmetic skills. Furthermore, upon examining processing mechanisms habituation paradigms may not be as informative as paradigms utilizing explicit response requirements because the latter enable researchers to collect response accuracies and latencies alike upon solving the experimental tasks in the scanner and, thus, may elucidate the underlying processing mechanisms. Thus, both types of paradigms (i.e. passive viewing paradigms and those with explicit response requirements) have specific peculiarities that need to be taken into account when planning and interpreting fMRI data.

Before elaborating on the cerebral correlates of typical pathways of numerical development we will briefly sketch current developmental calculation models, and how they converge and diverge with adult calculation models. Please note that neurocognitive models are necessary building blocks upon which the findings of brain imaging studies may be adequately interpreted (and even before that, neurocognitive models are essential to develop sensible research questions and experimental paradigms, see Berl, Vaidya,& Gaillard, 2006; Cohen Kadosh, Lammertyn, & Izard, 2008).

From Adult Calculation Models to Developmental Models of Numerical Cognition

Most of what we know about number processing and calculation is based upon findings from adults. Adult calculation models mainly rest on reports from neuropsychological patients with acquired calculation disorders. These early studies were informative with respect to both the cognitive architecture of number processing and the neural correlates subserving specific aspects of numerical cognition. In particular, the presence of double dissociations in two patients with diverging lesions (e.g. preserved skill A in the presence of impaired skill B in patient X, while patient Y exhibits the opposite performance pattern) has been interpreted as evidence for the modular architecture of the neurocognitive system (Shallice, 1988). In neuropsychology, the term 'modular architecture' is used to describe how neurocognitive (sub)systems are related and how they communicate with each other. It is assumed that brain systems are modularly organized if their functions (and structures) are dissociable, but yet interrelated to each other. According to Shallice (1988), the presence of a double dissociation between two functions may be regarded as evidence for the modular organization of these functions. With respect to numerical cognition, double dissociations have been reported repeatedly (e.g. Hittmair-Delazer et al., 1995; Dehaene and Cohen, 1995; McCloskey, 1992) and have been interpreted as evidencing the existence of distinct but yet interrelated components of arithmetical processing. For instance, Hittmair-Delazer, Sailer, & Benke (1995) have shown that number fact retrieval (simple mental calculations such as $3 \times 5 = 15$) and arithmetical conceptual knowledge (i.e. basic understanding of operations and arithmetic principles) are dissociable from each other, both at a behavioural and neural level. Likewise, a double dissociation between number fact knowledge and procedural arithmetic knowledge (i.e. knowledge how to perform multi-step operations) has been reported previously by Temple (1991) in developmental dyscalculia[1] and, thus, in the absence of overt brain damage. Nonetheless, it has been argued that during development, neurocognitive systems are still immature (i.e. characterized by undifferentiated neural networks) and thus, the functional specialization of specific brain regions has not yet taken place or is not yet completed in children (Berl et al., 2006; Karmiloff-Smith, 1998). Hence, double dissociations in mature and developing brain systems may reflect distinct underlying processes. In particular, double dissociations observed in immature brain systems might not be indicative of a modular cognitive architecture (Karmiloff-Smith, Scerif, & Ansari, 2003).

Beyond providing converging empirical evidence supporting the distinction between number fact, procedural and conceptual arithmetic knowledge (e.g. Hittmair-Delazer et al., 1995; McCloskey, 1992), the adult literature on numerical cognition has proposed three different number representations that are mediated by distinct neural networks (Dehaene & Cohen, 2005; for an extension, see Dehaene, Piazza, Pinel, & Cohen, 2003). The so-called triple-code model proposed by Dehaene & Cohen (1995) comprises an analogue magnitude representation mediating semantic number processing (i.e. numerosity), a verbal-phonological number representation supporting verbal counting and number fact retrieval, and a visual-Arabic number representation that comes into play upon solving written arithmetical problems. Indeed, neuropsychological findings in adults with acquired calculation disorders (as a consequence of traumatic brain injury) have shown that, depending on the lesion location, one or any combination of the three afore-mentioned number representations may be affected (e.g. Dehaene & Cohen, 1995; Hittmair-Delazer et al., 1995; McCloskey et al., 1992). During the last decade, the triple-code

[1] The term developmental dyscalculia denotes severe difficulties to perform simple arithmetics despite average intellectual abilities and adequate schooling (and in the absence of gross structural brain lesion; American Psychiatric Association, 1994). Nonetheless, in clinical practice, developmental dyscalculia may be diagnosed in the presence of low achievement on a standardized mathematical achievement solely (i.e. without requiring a discrepancy between low maths achievement and average intellectual abilities; Shalev & Gross-Tsur, 2001; see also the fifth version of the Diagnostic and Statistical Manual of Mental Disorders/DSM-V published by the American Psychiatric Association/APA 2013).

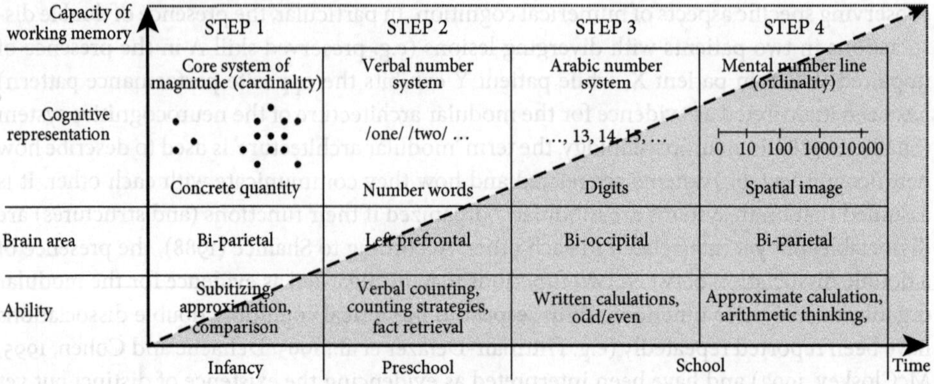
Capacity of working memory
STEP 1
STEP 2
STEP 3
STEP 4
Core system of magnitude (cardinality)
Verbal number system
Arabic number system
Mental number line (ordinality)
Cognitive representation
/one/ /two/ …
…, 13, 14, 15, …
0 10 100 1000 10000
Concrete quantity
Number words
Digits
Spatial image
Brain area
Bi-parietal
Left prefrontal
Bi-occipital
Bi-parietal
Ability
Subitizing, approximation, comparison
Verbal counting, counting strategies, fact retrieval
Written calulations, odd/even
Approximate calulation, arithmetic thinking,
Infancy
Preschool
School
Time

FIGURE 27.1 Four-step-developmental model of numerical cognition as proposed by von Aster and Shalev (2007). Please note that the latter model emphasizes the dynamic nature of development and acknowledges the age- and skill-dependent interplay between numerical and non-numerical processing demands.

Reproduced from Number development and developmental dyscalculia, Michael G Von Aster and Ruth S Shalev, *Developmental Medicine & Child Neurology*, 49(11), pp. 868–73, figure 1, DOI: 10.1111/j.1469-8749.2007.00878.x Copyright © 2007, John Wiley and Sons.

model has been widely employed in both adult and developmental studies of numerical cognition. However, due to striking developmental differences in brain function, developmental and acquired calculation disorders are not readily comparable (Ansari, 2010; see also Kaufmann & Nuerk, 2005; Kaufmann et al., 2013). Therefore, a direct application of findings from adult neuropsychological patients to developmental disorders without gross brain lesions is inappropriate and requires a truly developmental perspective (Karmiloff-Smith, 1998, 2010).

Recently, several developmental calculation models have been launched, most of which remain hypothetical thus far. Von Aster and Shalev (2007) propose a developmental model that partially rests on Dehaene's adult calculation model (Dehaene & Cohen, 2005). Like Dehaene & Cohen (2005), von Aster and Shalev (2007) differentiate semantic and symbolic (verbal and Arabic) number representations (see Figure 27.1). Notably, and with reference to the theory of 'representational redescriptions' (RR;[2] Karmiloff-Smith, 1992) von Aster and Shalev further subdivide the semantic number system into two components: an early, implicit 'core system of magnitude' (step 1 as depicted in Figure 27.1) and a later, explicit representation of a 'mental number line' (step 4 shown in Figure 27.1). Importantly, according to von Aster and Shalev (2007) the mental number line is thought to develop successively, thereby depending on previous processes of representing numerical magnitudes by verbal and Arabic symbols as well as on growing capacities of domain general abilities like working memory.

Unlike Dehaene who proposed distinct yet interrelated number representations in adults (and hence mature brain systems) that may work – dependent on task requirements –in a quasi-parallel fashion, von Aster and Shalev (2007) assume that during development the different types of cognitive representations are acquired in a quasi-hierarchical fashion

[2] Karmiloff-Smith (1992) coined the term 'representational redescriptions' (RR) to emphasize that children's mental (number) representations undergo qualitative changes that are age– and experience-dependent.

and consequently, introduce their model as a *four-step-developmental model*. In addition to the afore-mentioned cognitive representations, von Aster and Shalev (2007) outline additional processing levels (i.e. brain areas and abilities) and, furthermore, stress the impact of increasing working memory demands on the development of numerical cognition. Overall, the four-step developmental model of numerical cognition may be regarded as a first and valuable attempt to conceptualize the dynamic nature of developing maths proficiency. A further asset of the developmental model is that it acknowledges the interplay of numerical and non-numerical skills (i.e. working memory). Indeed, considerable behavioural evidence suggests that working memory – like many other non-numerical (cognitive and non-cognitive) factors – influence the development of numerical thinking. For instance, language (e.g. Pixner, Moeller, Nuerk, Hermanova, & Kaufmann, 2011), spatial skills (e.g. Geary, 1996) and attention (e.g. Kaufmann & Nuerk, 2006; see also Bugden & Ansari, 2011) have been found to be tightly linked to numerical development.

Furthermore, a developmental calculation model based on empirical data derived from functional Magnetic Resonance Imaging (fMRI) has been formulated by Kaufmann and colleagues (2011). As depicted in Figure 27.2, the two model components 'number

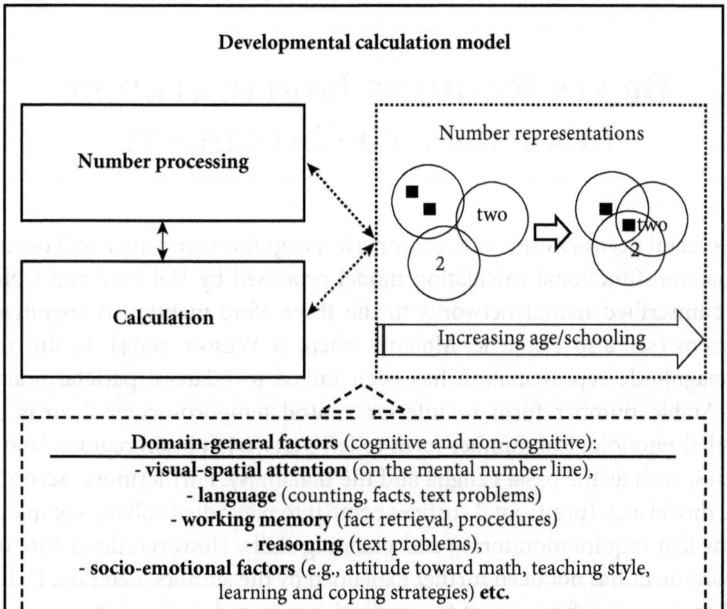

FIGURE 27.2 Tentative developmental calculation model emphasizing the dynamic nature of numerical development on the one side and the interplay of domain-specific and domain-general factors during arithmetical development on the other side. Please note that the partitioning in a number processing and a calculation component rests on the findings of a meta-analyses of developmental fMRI studies on numerical cognition (Kaufmann et al., 2011), while the assumption of domain-general factors influencing the development of numerical cognition is primarily based upon behavioral findings.

processing' and 'calculation' are linked to domain-specific (numerical) and domain-general (non-numerical and supporting) factors alike.

With respect to the domain-specific factor influencing the two afore-mentioned components, the authors argue that both number processing and calculation are fed by cognitive representations of number magnitude that are more or less well established and interlinked (as reflected by cross-notational access). The proposal that with increasing age and experience, different types of mental number representations (i.e. analogue magnitude representations, visual-Arabic number representations and verbal-phonological number representations) overlap, which in turn facilitates cross-notational access has been adopted from Kucian and Kaufmann (2009). Importantly, the notion of developmental and experience-dependent changes affecting the cognitive representations of number magnitude knowledge is an attempt to acknowledge the dynamic nature characterizing developmental processes (Ansari, 2010; Karmiloff-Smith, 1998; Kaufmann & Nuerk, 2005) and thus, may be considered an extension of the triple-code model proposed by Dehaene and Cohen (1995).

The next section discusses further details on the crucial differences between mature and developing neurocognitive systems that come into play upon acquiring arithmetical knowledge.

Brain Regions Implicated in Learning to Calculate

In adults, cerebral networks subserving numerical cognition are rather well described. The popular anatomo-functional calculation model proposed by Dehaene and Cohen (1995) assigns circumscribed neural networks to the three afore-mentioned cognitive number representations (see also Dehaene, Molko, Cohen, & Wilson. 2004). In this model, the analogue magnitude representation has been linked to bilateral parietal brain regions, the visual-Arabic number form to inferior ventral temporo-occipital areas bilaterally, and the verbal-phonological number form to left perisylvian brain regions (and subcortical structures, such as the basal ganglia and the thalamus). Furthermore, according to the triple-code model also (pre)frontal cortices come into play when solving complex arithmetic problems that require monitoring and planning skills. However, the nature of the prefrontal involvement has not been further explained by the authors. Later on, Dehaene et al. (2003) elaborate on the function of the parietal cortex and propose three distinct neural circuits within the parietal cortex. In particular, the intraparietal sulcus (IPS) is thought to mediate numerosity processing per se (i.e. analogue magnitude representation), while adjacent regions such as the posterior superior parietal sulcus (PSPL) and the angular gyrus support the orientating of spatial attention on the mental number line and the verbally-mediated rote retrieval of number knowledge, respectively. Importantly, the initial assumption of an amodal (i.e. notation-independent) numerosity representation in the IPS as suggested by the triple-code-model has been seriously challenged by adult data (Cohen Kadosh, Cohen Kadosh, Kaas, Henik, & Goebel, 2007; Piazza, Pinel, Le Bihan, & Dehaene, 2007). Rather, it has been suggested that adjacent neuronal populations in the IPS support both notation-dependent and abstract number representations, while the left IPS possibly plays a particular role '... in the representation of enculturated symbols of numerical magnitudes, such as Arabic numerals and number words' (Ansari, 2007, p. 166).

Recent Modifications and Extensions of Early Neurofunctional Calculation Models

Although the triple-code model had a considerable impact on numerical cognition research, it has become insufficient to explain some of the more recent brain imaging findings in adults. One possibility to identify common and robust areas of activation across studies is to systematically compare data sets across these studies by conducting fMRI meta-analyses. Significant activation clusters obtained by fMRI meta-analyses depict those brain areas that – across studies – have been reported to be significantly involved in task-related processes more often than chance. fMRI meta-analyses on *adult* imaging findings were conducted recently by Arsalidou and Taylor (2011). Based on the latter findings the authors suggest a modification and extension of Dehaene's triple-code calculation model (see Figure 27.3 in Arsalidou & Taylor, 2011). Notably, the findings of the meta-analyses enable the authors to draw a refined picture of (pre)frontal functions and, thus, of supporting and domain-general functions implicated in solving arithmetic tasks. Beyond differentiating between the monitoring of simple and of complex rules (inferior frontal regions versus middle frontal regions), the findings of the meta-analyses suggest that frontal brain regions also are likely to be implicated in formulating and following goals (superior frontal regions) and in navigating eye movements (required upon visual processing of number magnitudes; precentral gyrus). Further extensions of the triple-code model as suggested by Arsalidou and Taylor (2011) concern brain regions that have been repeatedly reported to be involved in numerical tasks, but thus far have not been studied systematically in numerical cognition research, namely the cerebellum (suggested to come into play whenever calculation

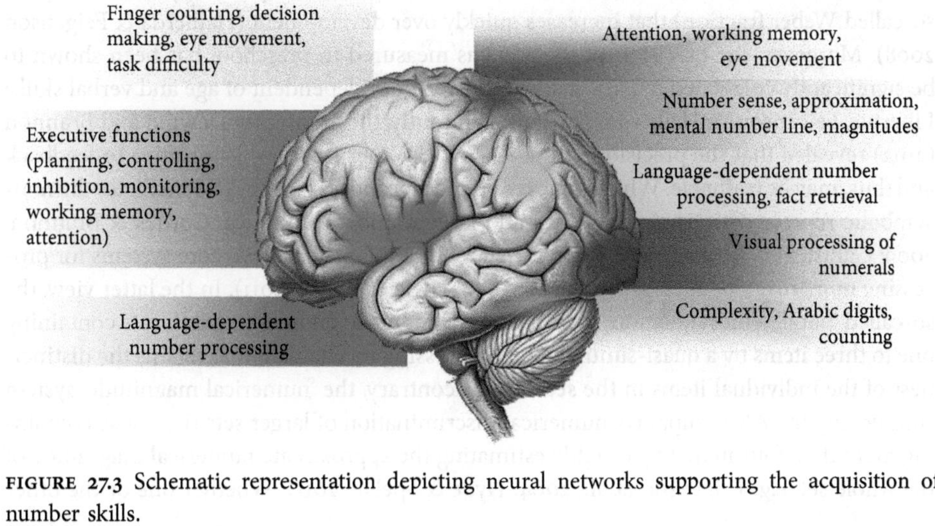

FIGURE 27.3 Schematic representation depicting neural networks supporting the acquisition of number skills.

activities require goal-directed visual-motor sequencing) and the insula (thought to switch between task-relevant behavior and resting state default mode).

Development of The Numerical Brain

Neurofunctional calculation models based on adult data are not adequate to explain developmental data derived from immature brain systems because the latter are less likely to be modularly organized (for an elaboration, see Kaufmann et al., 2013). Rather, developmental data are thought to be characterized by more widespread – and depending on developmental age also more undifferentiated – neural networks (Berl et al, 2006; Johnson, 2001; Johnson, Grossman, & Cohen Kadosh, 2009; Karmiloff-Smith, 2010). With respect to the developmental trajectories of numerical cognition it is important to note that, to the present, developmental imaging studies are scarce and, moreover, findings are partially inconsistent. Inconsistent findings may, in part, be attributable to considerable methodological differences across studies. For instance, studies diverge with respect to the populations under investigation (e.g. age range, grade levels, school type, intellectual level and inclusion/exclusion criteria), experimental and control tasks employed in the scanner, as well as data analyses (preprocessing, templates, threshold selection, etc.). Below, we will provide relevant developmental studies targeted at identifying and characterizing the neurocognitive underpinnings of numerical cognition in both non-verbal infants and school-age children.

Neural Correlates of Number Processing in Non-Verbal Infants

Converging behavioural evidence suggests that already non-verbal infants are capable to process numerosities. The so-called 'approximate number system' (ANS: e.g. Feigenson et al., 2004) denotes infants' ability to discriminate number magnitudes. The ANS is thought to underlie numerical estimation in infants and animals alike (for reviews, see Hyde & Spelke, 2011; Cohen Kadosh et al., 2008) and is characterized by a ratio-dependency (indexed by the so-called Weber fraction) that increases quickly over development (Halberda & Feigenson 2008). Moreover, the precision of the ANS as measured in preschool has been shown to be significantly correlated to later maths proficiency (independent of age and verbal skills: Libertus, Feigenson, & Halberda, 2011). Most recently, the findings of Dewind and Brannon (2012) revealed that the precision of the ANS (i.e. Weber fraction) is sensitive to feedback and thus, may be trainable. While some researchers propose that the ANS mediates the non-symbolic representations of small (<4) and large numbers alike (e.g. Cordes & Brannon, 2009; Gallistel & Gelman, 2000), others believe that infants have two core systems for processing numerosity (e.g. Feigenson et al., 2004; Hyde & Spelke, 2011). In the latter view, the so-called 'parallel individuation' system enables infants to enumerate small sets containing one to three items by a quasi-simultaneous processing mechanism that retains the distinctness of the individual items in the set. On the contrary, the 'numerical magnitude' system (mimicking the ANS) supports numerical discrimination of larger sets (i.e. those containing more than four items) by roughly estimating the approximate numerical magnitude of the whole set; e.g. Feigenson et al., 2004; Hyde & Spelke, 2011). Whether one or the other core system comes into play seems to be determined by task demands or stated differently, by non-numerical factors, such as attentional load (Hyde & Spelke, 2012). Most interestingly, the latter two core systems may be differentiated from each other at both the behavioral and neural level (Hyde & Spelke, 2011, 2012).

Studies targeted at identifying the neural correlates of the two afore-mentioned core systems of number representation in human infants are scarce. This is due to the fact that – as already mentioned in the introduction – brain imaging studies with young children and even more so with infants are challenging because infants are not able to respond actively to the presented stimuli. Thus, respective infant studies need to employ passive viewing paradigms incorporating habituation techniques. The probably most intriguing finding of these studies is that infants' non-verbal number processing may be two-partitioned and dissociable at a neural level (e.g. Hyde & Spelke, 2011, 2012; Izard, Dehaene-Lambertz, & Dehaene, 2008; Libertus et al., 2011). Upon utilizing event-related potentials (ERP) Hyde and Spelke (2011) investigated 6–7.5-month-old infants by having them view either small- (1–3) or large-item sets (8–32). Results revealed two distinct neural pathways underlying the processing of small and large item sets. While discrimination of small item sets was found to evoke earlier electrophysiological positivity (P400) over occipital-temporal brain regions, large item sets elicited a ratio-dependent electrophysiological positivity somewhat later (P500) over parietal areas (for different findings, see Izard et al., 2008). According to Hyde and Spelke (2011), the neural signatures of the two non-verbal number systems as displayed by infants are functionally comparable with those reported previously in adults. Accordingly, the latter authors argue that the two number systems develop early (and independent of language processing) and continue to be functional irrespective of experience and formal schooling.

Neural Correlates of Number Processing in Childhood

Two meta-analyses targeted at *developmental* fMRI studies on numerical cognition have been published recently (Houdé, Rossi, Lubin, & Joliot, 2010; Kaufmann, Wood, Rubinsten, & Henik, 2011). While the meta-analyses reported by Houdé and colleagues (2010) encompass several functional domains, including number processing, reading, and executive function, and with respect to number processing, merge studies investigating different aspects of numerical cognition (i.e. number processing, calculation, algebra equation learning), the meta-analyses reported by Kaufmann et al. (2011) was targeted at drawing a more differentiated picture of numerical cognition. In particular, the latter authors aimed at disentangling effects of notation (symbolic versus non-symbolic number processing), age (children versus adults), task complexity (i.e. number processing and calculation) and competency (i.e. with and without developmental dyscalculia) on cerebral activation extents. In the numerical cognition literature the term non-symbolic is used to depict concrete representations of entities, such as dots or finger patterns. On the contrary, number representations are considered to be symbolic when the symbols used to represent magnitudes are abstract in nature. Although Arabic digits are probably the most common symbolic number representations, other symbolic number formats do exist but have rarely been the target of neurocognitive research (e.g. Roman numerals). In the following, we will briefly present the main findings of the four meta-analyses in Kaufmann et al. (2011).

Age Effects

First, *age effects* were investigated by contrasting children's and adult's activation foci upon processing non-symbolic number magnitudes. Interestingly, within parietal cortex, findings revealed an age-dependent activation shift from anterior to posterior brain regions. The more anterior activations in the IPS observed in children may reflect children's stronger dependency on finger-based calculation strategies (Kaufmann et al., 2008, 2011). The latter interpretation is plausible because the observed activations in the anterior IPS are in close vicinity to finger areas of the sensory homunculus located in the post-central gyrus. The task used by Kaufmann, Vogel, Wood, Kremser, Schocke, Zimmerhackl, et al. (2008) may have provoked finger-based strategies, because finger patterns were used as stimuli. An alternative interpretation of the latter findings is that solely the viewing of finger patterns may have elicited automatic mirroring processes (known to activate the so-called mirror neuron network incorporating parietal regions, among others).

The previously reported developmental shift from frontal to parietal activations could not be replicated in this meta-analysis, possibly due to methodological differences between studies. While the meta-analysis conducted by Kaufmann et al. (2011) was restricted to studies utilizing non-symbolic number comparison tasks, several other studies employed symbolic number tasks requiring participants to process Arabic digits (number comparison tasks: Ansari et al., 2005; Ansari & Dhital, 2006; Cantlon, Libertus, Pinel, Dehaene, Brannon, & Pelphrey, 2009; Holloway & Ansari, 2010; Kaufmann et al., 2008; Kaufmann, Koppelstaetter, Siedentopf, Haala, Haberlandt, Zimmerhackl et al., 2006 calculation tasks: Kucian, Grond, Rotzer, Henzi, Schönmann, Plangger, et al., 2008; Rivera, Reis, Eckert, & Menon, 2005). Indeed, findings of virtually all afore-mentioned studies disclosed a frontoparietal shift of number-related activations revealing that compared with children adults produce stronger (intra-)parietal fMRI responses upon processing symbolic number magnitudes (i.e. Arabic digits), which has been interpreted as reflecting increasing functional specialization of (intra-)parietal regions for number processing. Interestingly, despite comparable behavioural performance, children were reported to stronger recruit frontal supporting brain regions (e.g. Ansari et al., 2005; Kaufmann et al., 2006; Kucian et al., 2008).

Notation Effects

Secondly, *effects of notation* were investigated by contrasting number comparison tasks utilizing symbolic (Arabic digits) versus non-symbolic stimuli (dot or finger patterns). Generally, non-symbolic number processing is mastered well before children are able to process magnitudes symbolically. Moreover, accumulating evidence suggests that the successful mapping between non-symbolic and symbolic number processing is a crucial prerequisite to acquire formal school mathematics (typically developing children: e.g. Mundy & Gilmore, 2009; children with mild intellectual impairments: Brankaer, Ghesquiére, & De Smedt, 2011). The findings of the meta-analyses ought to elucidate whether in children (intra-)parietal regions host an amodal (abstract) number representation as suggested by the triple-code model (Dehaene & Cohen 1995; for respective animal data, see Nieder, 2012). Findings are clearly incompatible with the assumption of abstract number representations in the IPS, but rather reveal that symbolic and non-symbolic number magnitudes are supported by distinct and partly overlapping parietal brain regions. While the processing of non-symbolic number magnitudes yielded significant activations in the right parietal cortex (adjacent to the right IPS), symbolic number magnitudes were found to be supported by bilateral parietal regions including the right IPS and the left PSPL (see Cohen Kadosh & Walsh, 2009's target article and commentaries for the discussion of notation-dependent effects on lateralization of IPS activation in adults). Most interestingly, notation effects were visible even within the right IPS, where activations pertaining to non-symbolic number magnitudes were situated in the anterior portion of the IPS while those pertaining to symbolic processing were located more posterior and thus, closely resemble intraparietal activation foci reported by Dehaene et al. (2003). According to Kaufmann et al. (2011) a potential explanation for the observed link between anterior intraparietal regions and non-symbolic number processing may have to do with finger-based calculation strategies that might more easily be elicited when processing non-symbolic magnitudes (see also Kaufmann et al., 2008).

Effects of Maths Competency

The third meta-analyses was targeted at investigating neural responsivity to number tasks in children with and without developmental dyscalculia. We would like to refer the reader to 'Brain Correlates of Numerical Disabilities' by Kucian, Kaufmann and von Aster (this volume) for a more detailed discussion. Overall, compared with average calculators children with dyscalculia and those with low maths proficiency seem to need more cognitive and neural resources to solve simple number tasks (for similar findings see Houdé et al., 2010). Interestingly, children with good and poor maths performance are distinguishable regarding parietal activations, both with respect to lateralization and anterior–posterior location within the parietal cortices (Kaufmann et al., 2011).

Neural Correlates of Number tasks That Go Beyond Number Comparison

Finally, the fourth meta-analyses aimed at identifying the neural correlates supporting calculation skills. Findings disclosed symmetric, but rather distributed networks including frontoparietal and occipital brain regions bilaterally. In parietal cortex, strong and consistent activations were found in inferior and superior parietal cortices (including bilateral IPS and right supramarginal gyrus). Note that, to the present, developmental studies aiming at identifying whether the four arithmetic operations (addition, subtraction, multiplication, division) are differentially modulated by parietal and extra-parietal regions are scarce (for a respective study in young adults, see Rosenberg-Lee, Ting Chang, Young, Wu, & Menon, 2011b). To the best of our knowledge, the study of De Smedt and colleagues (De Smedt, Holloway, & Ansari, 2011) is the first – and at the time of writing this article the only one – that examined the effects of arithmetic operation on children's fMRI responses. Interestingly, the findings of the latter study revealed two dissociable neural systems that are differentially involved in single-digit addition and subtraction and, furthermore, are sensitive to task difficulty (i.e. problem size). First, (intra-)parietal regions were activated upon solving operations that were not yet automatized and thus, were solved by quantity-based and procedural solution strategies (large problems and subtraction). On the contrary, problems that were mastered easily (small problems and addition) were found to be supported by left-lateralized hippocampal structures, possibly reflecting verbal processing and retrieval. Consequently, De Smedt and coauthors (2011) suggest that, in children, the hippocampus may play a key role upon acquiring arithmetic fact knowledge. The latter interpretation is compatible with the assumptions purported by Dehaene's adult calculation model (Dehaene and Cohen, 1995) and furthermore, extends respective adult findings to children (e.g. Delazer, Ischebeck, Domahs, Zamarian, Koppelstaetter, Siedentopf, et al. 2005; Zamarian, Ischebeck, & Delazer, 2009). The proposed link between arithmetic fact retrieval and hippocampal activity (De Smedt et al., 2011) is readily explained upon considering that arithmetic facts are encoded and retrieved directly from memory. Importantly, arithmetic facts are considered to be part of the semantic memory network and alike other semantic memory aspects, arithmetic fact knowledge seems to be mediated by hippocampal structures known to be involved in memory functions. However, once automatized, arithmetic fact retrieval seems to be mainly supported by inferior parietal regions and, in particular, by the angular gyrus (Grabner, Ansari, Reishofer, Stern, Ebner, & Neuper, 2007; for a review, cf. Zamarian et al., 2009).

SUMMARY

Taken together, in children and adults alike, arithmetical learning is accompanied by adaptive changes in neurofunctional systems (Houdé et al., 2010; Kaufmann et al., 2011; see also a review by Zamarian et al., 2009). Most interestingly, non-verbal number processing may be traced neurally already in infants (e.g. Hyde & Spelke, 2011, 2012; Izard et al., 2008; Libertus et al., 2011). With increasing age and maths proficiency cerebral activations seem to shift from domain-general and supporting predominantly frontal to number-relevant (fronto-) parietal brain regions. Nonetheless, it has to be noted that the latter interpretation rests on cross-sectional fMRI studies solely that included children of varying age-ranges and skill-levels (however, see Rosenberg-Lee, Barth, & Menon, 2011a). In the field of numerical cognition, true longitudinal fMRI studies have not been published thus far, but are urgently needed to elucidate the dynamic nature of neurofunctional development (Karmiloff-Smith, 2010). Furthermore, in adults, competency-dependent activation shifts are observable also within the parietal cortices (i.e. from the intraparietal sulci mediating number magnitude processing to the left angular gyrus supporting automatized, probably verbally-based arithmetical processing) and compared with non-experts neurofunctional activations of adult individuals with exceptional calculation skills seem to be more distributed (Zamarian et al., 2009). It is important to note that various aspects of number processing and calculation skills also rely on supporting domain-general skills. An example of a so far rather neglected supporting function in numerical cognition research concerns visually-guided eye movements upon solving visually presented tasks in the scanner. The findings of Krinzinger and colleagues (Krinzinger, Koten, Horoufchin, Kohn, Arndt, Sahr, et al., 2011) suggest that saccadic movements are supported by frontoparietal networks that are partly overlapping those reported to be involved in number processing (see also Anderson, Jones, O'Gorman, Leemans, Catani, & Husain, 2011). Upon acknowledging similar results from the attention literature disclosing a key role of inferior and superior parietal cortices for the top-down modulation of attention (Cabeza, Ciaramelli, Olson, & Moscovitch, 2008), it becomes readily apparent that number-relevant regions are not specific to number processing and, hence, brain imaging studies should be planned and interpreted with great care. Most importantly, future research endeavours are needed that aim:

1. At disentangling activations pertaining to numerical and non-numerical functions.
2. At characterizing how developmental changes affect these neurofunctional response patterns.

Overall, proficient calculation skills depend on the integrity of a distributed neural network. A tentative, but yet incomplete schematic representation of brain regions implicated in learning to count and calculate is provided in Figure 27.3. Please note that Figure 27.3 depicts number-relevant brain regions, but does not show fibre bundles connecting these brain regions. Respective research is still in its infancy and there is a clear need for future research targeted at systematically investigating neurofunctional connectivity related to the developmental of numerical cognition.

Upon considering the continuing technological advances it is plausible to assume that within the near future, new findings will emerge that are based on refined analyses of brain structure and function. With respect to brain function, traditional fMRI methods provide valuable information regarding the spatial distribution of cerebral activations, but were not able to reveal temporal dynamics of fMRI responses (e.g. whether task-relevant activations emerged first in parietal or frontal regions). Nowadays, functional connectivity studies may describe fMRI responses with respect to their spatial and temporal dynamics alike (e.g. Cohen Kadosh, Bahrami, Walsh, Butterworth, Popescu, & Price, 2011). Another aspect of connectivity concerns their structural correlates. An increasingly popular method to study and quantify white matter structures, and in particular, fibre bundles connecting regions of interest is the so-called diffusion tensor imaging (DTI) (Tsang, Dougherty, Deutsch, Wandell, & Ben-Shachar, 2009; van Eimeren, Niogi, McCandliss, Holloway, & Ansari, 2008). Exciting outlooks into the possibilities of these novel technologies provide two recent developmental studies, one of which was targeted at investigating whether maths proficiency is associated with differences in cerebral white matter in children (van Eimeren et al., 2008), and the other study aimed at identifying neurofunctional connectivity upon processing number tasks (Rosenberg-Lee et al., 2011a). Upon subjecting 7–9-year-old children to a DTI study, van Eimeren and collaborators were able to show that measures of fractional anisotropy (indexing the integrity of white matter microstructure) in the inferior longitudinal fasciculus and the superior corona radiata were correlated to maths competency. In particular, the authors report that the left inferior longitudinal fasciculus was specifically associated to performance on numerical operations (but not to performance on mathematical reasoning), while the superior corona radiata was found to be related to both number processing components. However, it has to be noted that the left superior corona radiata has been reported to be implicated in reading as well (e.g. Ben-Shachar, Dougherty, & Wandell, 2007), which in our view challenges the authors interpretation that the link between the superior corona radiata and numerical cognition is a specific one. Finally, upon studying functional connectivity within parietal as well as frontoparietal brain regions in second- and third-grade children, Rosenberg-Lee and colleagues (2011a) report that considerable changes in neural responsivity emerge during this narrow developmental period. Upon considering that most earlier developmental fMRI studies encompass wider age ranges, the latter findings are important and at the least, call for modest and careful data interpretation of findings based on merging children of different age and grade levels.

Synopsis

Neuroscientific research delineating developmental pathways of numerical cognition is still in its infancy. Thus, it is not surprising that, to the present, many developmental brain imaging studies were conceptualized and interpreted based on adult calculation models. Nonetheless, due to considerable differences regarding brain structure and function a direct comparison of mature and developing brain systems may not be feasible. With respect to the development of numerical cognition, it has been argued that with increasing age and experience, number representations and their interconnections undergo qualitative changes (Kucian & Kaufmann, 2009). Likewise, evidence suggests that modulations of neural connectivity are considerable, are sensitive to age, competency and notation (Kaufmann et al., 2011), and can emerge during relatively narrow developmental periods (Rosenberg-Lee et al., 2011a). Recent attempts to acknowledge the dynamic nature of numerical development (Ansari, 2010) have culminated in the formulation of *developmental* calculation models (e.g. Kaufmann et al., 2011; von Aster & Shalev, 2007). With respect to the brain systems implicated in learning to count and calculate, converging evidence suggests that with increasing age and expertise an anterior-posterior shift of neurofunctional activity takes places that is thought to reflect decreasing reliance on domain-general supporting (frontal) processing mechanisms, on the one hand, and increasing functional specialization of number-relevant (frontoparietal) brain regions on the other hand. Nonetheless, future research is needed:

1. To elaborate on these first model sketches.
2. To identify further cognitive and non-cognitive factors that impact on the development of numerical thinking.
3. To better characterize the neurofunctional underpinnings of the dynamics of numerical development.

In order to achieve these research goals, we need to adopt true developmental approaches acknowledging that experience-dependent neuroplastic changes in numerical processing are substantial and qualitative, occur during narrow time windows (i.e. 1 year of schooling), and manifest at multiple processing levels (for an elaboration, see Kaufmann et al., 2013).

References

American Psychiatric Association (2013). *Diagnostic and statistical manual of mental disorders (5th edn)*. Washington, DC: APA.

Anderson, E.J., Jones, D.K., O'Gorman, R.L., Leemans, A., Catani, M., & Husain, M. (2011). Cortical network for gaze control in humans revealed using multimodal MRI. Cerebral Cortex, 22, 765–775.

Ansari, D. (2007). Does the parietal cortex distinguish between '10', 'ten', and ten dots=? Neuron, 53, 165–167.

Ansari, D. (2010). Neurocognitive approaches to developmental disorders of numerical and mathematical cognition: the perils of neglecting the role of development. Learning and Individual Differences, 20, 123–129.

Ansari, D., & Dhital, B. (2006). Age-related changes in the activation of the intraparietal sulcus during non-symbolic magnitude processing: An event-related functional magnetic resonance imaging study. Journal of Cognitive Neuroscience, 18, 1820–1828.

Ansari, D., Dhital, B., & Siong, S.C. (2006). Parametric effects of numerical distance on the intraparietal sulcus during passive viewing of rapid numerosity changes. Brain Research, 1067, 181–188.

Ansari, D., Garcia, N., Lucas, E., Hamon K., & Dhital, B. (2005). Neural correlates of symbolic number processing in children and adults. Neuroreport, 16, 1769–1773.

Arsalidou, M., & Taylor, M.J. (2011). Is 2 + 2 = 4? Meta-analyses of brain areas needed for numbers and calculations. NeuroImage, 54, 2382–2393.

Ben-Shachar, M., Dougherty, R.F., & Wandell, B.A. (2007). White matter pathways in reading. Current Opinion in Neurobiology, 17, 258–270.

Berl, M.M., Vaidya, C.J., & Gaillard, W.D. (2006). Functional imaging of developmental and adaptive changes in neurocognition. NeuroImage, 30, 679–691.

Brankaer, C., Ghesquière, P., & De Smedt, B. (2011). Numerical magnitude processing in children with mild intellectual disabilities. Research in Developmental Disabilities, 32, 2853–2859.

Bugden, S., & Ansari, D. (2011). Individual differences in children's mathematical competencies are related to the intentional but not automatic processing of Arabic numerals. Cognition, 118, 32–44.

Cabeza, R., Ciaramelli, E., Olson, I.R., & Moscovitch, M. (2008). The parietal cortex and episodic memory: an attentional account. Nature Reviews Neuroscience, 9, 613–625.

Cantlon, J., Brannon, E.M., Carter, E.J., & Pelphrey, K.A. (2006). Functional imaging of numerical processing in adults and 4-y-old children. PLoS Biology, 4, e125.

Cantlon, J.F., Libertus, M.E., Pinel, P., Dehaene, S., Brannon, E.M., & Pelphrey, K.A. (2009). The neural development of an abstract concept of number. Journal of Cognitive Neuroscience, 21, 2217–2229.

Cohen Kadosh, R., Bahrami, B., Walsh, V., Butterworth, B., Popescu, T., & Price, C. (2011). Specialization in the human brain: the case of numbers. Frontiers in Human Neuroscience, 5: 62.

Cohen Kadosh, R., Cohen Kadosh, K., Kaas, A., Henik, A., & Goebel, R. (2007). Notation-dependent and – independent representations of numbers in the parietal lobes. Neuron, 53, 307–314.

Cohen Kadosh, R., Lammertyn, J., & Izard, V. (2008). Are numbers special? An overview of chronometric, neuroimaging, developmental and comparative studies of magnitude representation. Progress in Neurobiology, 84, 132–147.

Cohen Kadosh, R., & Walsh, V. (2009). Numerical representation in the parietal lobes: abstract or not abstract? Behavioral and Brain Sciences, 32, 313–373.

Cordes, S., & Brannon, E.M. (2009). Crossing the divide: infants discriminate small from large numerosities. Developmental Psychology, 45, 1583–1594.

Dehaene, S., & Cohen, L. (1995). Towards an anatomical and functional model of number processing. Mathematical Cognition, 1, 82–120.

Dehaene, S., Molko, N., Cohen, L., & Wilson, A.J. (2004). Arithmetic and the brain. Current Opinion in Neurobiology, 14, 218–224.

Dehaene, S., Piazza, M., Pinel, P., & Cohen, L. (2003). Three parietal circuits for number processing. Cognitive Neuropsychology, 20, 487–506.

Delazer, M., Ischebeck, A., Domahs, F., Zamarian L, Koppelstaetter F, Siedentopf CM, et al. (2005). Learning by strategies and learning by drill: Evidence from an fMRI study. NeuroImage, 25, 838–849.

De Smedt, B., Holloway, I.D., & Ansari, D. (2011). Effects of problem size and arithmetic operation on brain activation during calculation in children with varying levels of arithmetic fluency. NeuroImage, 57, 771–781.

Dewind, N.K., & Brannon, E.M. (2012). Malleability of the approximate number system: effects of feedback and training. Frontiers in Human Neurosciences, 6: 68.

Feigenson, L., Dehaene, S., & Spelke, E. (2004). Core systems of number. Trends in Cognitive Sciences, 8, 307–314.

Gallistel, C.R., & Gelman, R. (2000). Non-verbal numerical cognition: from reals to integers. Trends in Cognitive Sciences, 4, 59–65.

Geary, D.C. (1996). Sexual selection and sex differences in mathematical abilities. Behavioral and Brain Sciences, 19, 229–247.

Grabner, R.H., Ansari, D. Reishofer, G., Stern, E., Ebner, F., & Neuper, C. (2007). Individual differences in mathematical competence predicts parietal brain activation during mental calculation. NeuroImage, 38, 346–356.

Halberda, J., & Feigenson, L. (2008). Developmental changes in the acuity of the 'number sense': the approximate number system in 3-, 4-, 5-, and 6-year-olds and adults. Developmental Psychology, 44, 1457–1465.

Hittmair-Delazer, M., Sailer, U., & Benke, T. (1995). Impaired arithmetic facts but intact conceptual knowledge—a single case study of dyscalculia. Cortex, 31, 139–148.

Holloway, I.D., & Ansari, D. (2010). Developmental specialization in the right intraparietal sulcus for the abstract representation of numerical magnitude. Journal of Cognitive Neuroscience, 22, 2627–2637.

Houdé, O., Rossi, S., Lubin, A., & Joliot, M. (2010). Mapping numerical processing, reading, and executive functions in the developing brain: an fMRI meta-analysis of 52 studies including 842 children. Developmental Science, 13, 876–885.

Hyde, D.C., & Spelke, E.S. (2011). Neural signatures of number processing in human infants: evidence for two core systems underlying non-verbal numerical cognition. Developmental Science, 14, 360–371.

Hyde, D.C., & Spelke, E.S. (2012). Spatiotemporal dynamics of processing nonsymbolic number: an event-related potential source localization study. Human Brain Mapping, 33, 2189–2203.

Izard, V., Dehaene-Lambertz, G., & Dehaene, S. (2008). Distinct cerebral pathways for object identity and number in human infants. PloS Biology, 6(2): e11.

Johnson, M.H. (2001). Functional brain development in humans. Nature Reviews Neuroscience, 2, 475–483.

Johnson, M.H., Grossman, T., & Cohen Kadosh, K. (2009). Mapping functional brain development: building a social brain through interactive specialization. Developmental Psychology, 45, 151–159.

Karmiloff-Smith, A. (1992). *Beyond Modularity*. Cambridge: MIT Press.

Karmiloff-Smith, A. (1998). Development itself is the key to understanding developmental disorders. Trends in Cognitive Sciences, 2, 389–398.

Karmiloff-Smith, A. (2010). Neuroimaging of the developing brain: taking 'developing' seriously. Human Brain Mapping, 31, 934–941.

Karmiloff-Smith, A., Scerif, G., & Ansari, D. (2003). Double dissociations in developmental disorders? Theoretically misconceived, empirically dubious. Cortex, 39, 1–7.

Kaufmann, L., Koppelstaetter, F., Siedentopf, C., Haala, I., Haberlandt, E., Zimmerhackl, L.B., et al. (2006). Neural correlates of a number-size interference task in children. Neuroreport, 17, 587–591.

Kaufmann, L., & Nuerk, H-C. (2005). Numerical development: current issues and future perspectives. Psychology Science, 47, 142–170.

Kaufmann, L., & Nuerk, H.-C. (2006). Interference effects of a numerical Stroop paradigm in 9 to 12 year-old ADHD children. Child Neuropsychology, 12, 223–243.

Kaufmann, L., Vogel, S., Wood, G., Kremser, C., Schocke, M., Zimmerhackl, L.B., et al. (2008). A developmental fMRI study of nonsymbolic numerical and spatial processing. Cortex, 44, 376–385.

Kaufmann, L., Wood, G., Rubinsten, O., & Henik, A. (2011). Meta-analysis of developmental fMRI studies investigating typical and atypical trajectories of number processing and calculation. Developmental Neuropsychology, 36, 763–787.

Kaufmann, L., Mazzocco, M.M., Dowker, A., von Aster, M., Göbel, S.M., Grabner, R.H., Henik, A., Jordan, N.C., Karmiloff-Smith, A.D., Kucian, K., Rubinsten, O., Szucs, D., Shalev, R., & Nuerk, H.-C. (2013). Dyscalculia from a developmental and differential perspective. *Frontiers in Psychology*, 4: 516.Krinzinger, H., Koten, J.W., Horoufchin, H., Kohn, N., Arndt, D., Sahr, K., et al. (2011). The role of finger representations and saccades for number processing: an fMRI study in children. Frontiers in Psychology, 2, 373.

Kucian, K., Grond, U., Rotzer, S., Henzi, B., Schönmann, C., Plangger, F., et al. (2011). Mental number line training in children with developmental dyscalculia. NeuroImage, 57, 782–795.

Kucian, K., & Kaufmann, L. (2009). A developmental model of number representations. Behavioral and Brain Sciences, 32, 340–341. A commentary to Cohen Kadosh, R., & Walsh, V. (2009). Numerical representations in the parietal lobes: abstract or not abstract? Behavioral and Brain Sciences, 32, 313–373.

Kucian, K., Kaufmann, L., & von Aster, M. (in preparation). *Brain correlates of numerical disabilities*. In: R. Cohen Kadosh & A. Dowker (Eds.), Oxford Handbook of Numerical Cognition

Kucian, K., von Aster, M., Loenneker, T., Dietrich, T., & Martin, E. (2008). Development of neural networks for exact and approximate calculation: a fMRI study. Developmental Neuropsychology, 33, 447–473.

Libertus, M.E., Feigenson, L., & Halberda, J. (2011). Preschool acuity of the approximate number system correlates with school math ability. Developmental Science, 14, 1292–1300.

McCloskey, M. (1992). Cognitive mechanisms in numerical processing: evidence from acquired dyscalculia. Cognition, 44, 107–157.

Moeller, K., Pixner, S., Zuber, J., Kaufmann, L., & Nuerk, H.-C. (2011). Early place-value understanding as a precursor for later arithmetic performance – a longitudinal study on numerical development. Research in Developmental Disabilities, 32, 1837–1851.

Mundy, E., & Gilmore, C.K. (2009). Children's mapping between symbolic and nonsymbolic representations of number. Journal of Experimental Child Psychology, 103, 490–502.

Nieder, A. (2012). Supramodal numerosity selectivity of neurons in primate prefrontal and posterior parietalcortices. Proceedings of the National Academy of Sciences USA, 109, 11860–11865.

Piazza, M., Izard, V., Pinel, P., Le Bihan, D., & Dehaene, S. (2004). Tuning curves for approximate numerosity in the human intraparietal sulcus. Neuron, 44, 547–555.

Piazza, M., Pinel, P., Le Bihan, D., & Dehaene, S. (2007). A magnitude code common to numerosities and number symbols in the human intraparietal cortex. Neuron, 53, 293–305.

Pixner, S., Moeller, K., Nuerk, H-C., Hermanova, V., & Kaufmann, L. (2011). Whorf reloaded: language effects on non-verbal number processing in first grade – a trilingual study. Journal of Experimental Child Psychology, 108, 371–382.

Rivera, S.M., Reis, A.L., Eckert, M.A., & Menon, V. (2005). Developmental changes in mental arithmetic: evidence for increased functional specialization in the left inferior parietal cortex. Cerebral Cortex, 15, 1779–1790.

Rosenberg-Lee, M., Barth, M., & Menon, V. (2011a). What difference does a year of schooling make? Maturation of brain response and connectivity between 2nd and 3rd grades during arithmetic problem solving. NeuroImage, 57, 796–808.

Rosenberg-Lee, M., Ting Chang, T., Young, C.B., Wu, S., & Menon, V. (2011b). Functional dissociations between four basic arithmetic operations in the human posterior parietal cortex: a cytoarchitectonic mapping study. NeuroImage, 49, 2592–2608.

Shalev, R., & Gross-Tsur, O. (2001). Developmental dyscalculia. Pediatric Neurology, 24, 337–342.

Shallice, T. (1988). *From Neuropsychology to Mental Structure*. Cambridge: Cambridge University Press.

Temple, C.M. (1991). Procedural dyscalculia and number fact dyscalculia: double dissociation in developmental dyscalculia. Cognitive Neuropsychology, 8, 155–176.

Tsang, J.M., Dougherty, R.F., Deutsch, G.K., Wandell, B.A., & Ben-Shachar, M. (2009). Frontoparietal white matter diffusion properties predict mental arithmetic skills in children. Proceedings of the National Academy of Sciences of the United States of America, 106, 22546-22551.

Van Eimeren, L., Niogi, S.N., McCandliss, B.D., Holloway, I.D., & Ansari, D. (2008). White matter microstructure underlying mathematical abilities in children. Neuroreport, 19, 1117–1121.

Von Aster, M., & Shalev, R. (2007). Number development and developmental dyscalculia. Developmental Medicine and Child Neurology, 49, 868–873.

Zamarian, L., Ischebeck, A., & Delazer, M. (2009). Neuroscience of learning arithmetic—evidence from brain imaging studies. Neuroscience and Biobehavioral Reviews, 33, 909–925.

CHAPTER 28

ARITHMETIC IN THE CHILD AND ADULT BRAIN

VINOD MENON

INTRODUCTION

ARITHMETIC skills build on a core number knowledge system for representing numerical quantity using abstract symbols that is typically in place by the age of 5 (Barth, La Mont, Lipton, & Spelke, 2005). This is as true of brain processes as it is of cognitive processes described in previous chapters. In addition to core number processing systems in the parietal and inferior temporal cortex (Ansari, 2008), arithmetic also involves distributed brain systems mediating different memory processes, including working, episodic, and sematic memories, as well as cognitive control and decision making. With practice, learning, and development, these systems together help to build rich visuospatial, phonological, and mnemonic representations that result in proficiencies that are a hallmark of human cognition.

In this chapter I use a cognitive neuroscience approach to examine brain systems involved in arithmetic problem solving, highlighting their developmental origins. This review takes a distinctly developmental perspective, because neither the cognitive nor the brain processes involved in arithmetic can be adequately understood outside the framework of how ontogenetic processes unfold. This chapter first highlights major findings related to key cognitive component processes involved in arithmetic problem solving and reasoning. I review core cognitive and brain processes involved in arithmetic processing and discuss the implications of relevant studies in adults for understanding the neural basis of arithmetic skill development. It then discusses recent brain imaging studies of arithmetic in children and examine how they inform our understanding of typical and atypical skill development. This review does not focus on developmental dyscalculia and related math learning disabilities – the interested reader is referred to other reviews in this volume (Butterworth, Varma, & Laurillard, this volume; Henik, Rubinsten, & Askenzai, this volume; Kucian, Kaufmann, & von Aster, this volume).

Arithmetic: Core Neurocognitive Processes

Arithmetic skills rely on four basic neurocognitive processes (Figure 28.1). First, basic number sense, including number magnitude and cardinality, and manipulations of numerical quantity are the basic building blocks from which arithmetic is constructed in the brain. These basic building blocks require the integrity of visual and auditory association cortex, which help decode the visual form and phonological features of the stimulus, and the parietal attention system (Dehaene, Piazza, Pinel, & Cohen, 2003) which helps to build semantic representations of quantity (Ansari, 2008) from multiple low level visuospatial primitives such as eye gaze and pointing (Simon, Mangin, Cohen, Le Bihan, & Dehaene,

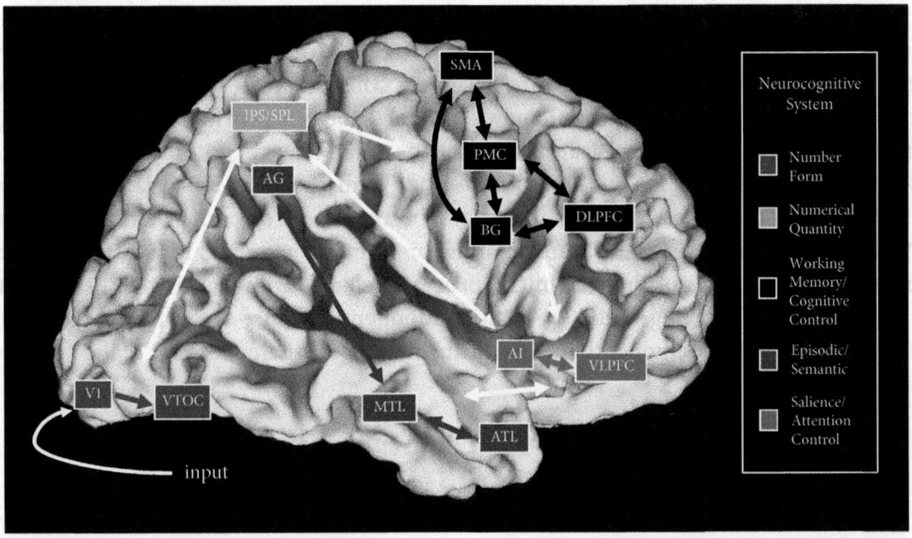

FIGURE 28.1 Schematic circuit diagram of basic neurocognitive processes involved in arithmetic. The ventral temporal-occipital cortex (shown in brown) decodes number form and together with the intra-parietal sulcus (IPS) in the parietal cortex helps builds visuospatial representations of numerical quantity. Procedural and working memory systems anchored in frontoparietal circuits involving the IPS and supra-marginal gyrus in the parietal cortex and the pre-motor cortex (PMC), supplementary motor area (SMA), and the dorsolateral prefrontal cortex (DLPFC) in the prefrontal cortex together with the basal ganglia (BG) create a hierarchy of short-term representations that allow manipulation of multiple discrete quantities over several seconds. Episodic and semantic memory systems anchored in the medial temporal cortex (MTL) and anterior temporal lobe (ATL), and the angular gyrus (AG) within the parietal cortex, play an important role in long-term memory formation and generalization beyond individual problem attributes. Finally, prefrontal control processes anchored in the saliency network encompassing the anterior insula (AI) and ventrolateral prefrontal cortex (VLPFC) guide, and maintain attention in the service of goal-directed problem solving and decision making. Relative transparency for BG and MTL indicates sub-surface cortical structures.

2002). Secondly, procedural and working memory systems anchored in the basal ganglia and frontoparietal circuits create short-term representations that support the manipulation of multiple discrete quantities over several seconds. Thirdly, episodic and semantic memory systems play an important role in long-term memory formation and generalization beyond individual problem attributes. Fourthly, prefrontal control processes guide and maintain attention in the service of goal-directed decision making.

The manner in which these neurocognitive processes are engaged depends critically on both problem complexity and stage of an individual's cognitive development. This is most clearly evident in developmental studies that show that children's gains in problem solving skills are characterized by shifts in the mix of problem solving strategies used, with inefficient procedural strategies being gradually replaced with direct retrieval of domain relevant facts (Geary, Bow-Thomas, & Yao, 1992; Geary, Hoard, Byrd-Craven, & DeSoto, 2004). Over time and with development, episodic and semantic memory systems build representations in long-term memory that allow for fast access of learned arithmetic facts. Working memory resources, including temporary storage, sequencing, and manipulation of information, are needed when problem solutions cannot be directly retrieved from memory – in this case, reliance on different strategies, such as decomposition or more elaborate sequential computations, are necessary. These domain-general cognitive processes are as vital as core numerical knowledge; they not only provide the scaffold for the development of more efficient strategies during the initial stages of arithmetic learning and skill development (Bull, Epsy, & Wiebe, 2008), but, as discussed by Zamarian and Delazer (this volume), they also facilitate learning of new and more complex materials in adults. An important component of arithmetic in the brain therefore relates to how the involvement and interactions of these cognitive processes change with learning and the maturation of problem-solving skills. The multi-componential nature of arithmetic reasoning and how individual components emerge during development are the focus of the next two sections.

Arithmetic in The Adult Brain

Brain imaging studies of arithmetic have used a variety of manipulations including problem size, type of operations, format and modality of operands, self- and experiment-paced tasks, verification, and production, in order to address specific aspects of information processing (Menon, Rivera, White, Glover, & Reiss, 2000b; Menon, Mackenzie, Rivera, & Reiss, 2002; Zago, Petit, Turbelin, Andersson, Vigneau, & Tzourio-Mazoyer, 2008). In adults, previous imaging studies have manipulated problem complexity by varying the number of operations in a problem (Menon et al., 2000b) or the number of digits in the operands (Rosenberg-Lee, Lovett, & Anderson, 2009a; Zago, Pesenti, Mellet, Crivello, Mazoyer, & Tzourio-Mazoyer, 2001). Closely-matched control tasks are needed for precise information about brain responses in relation to arithmetic complexity, independent of basic number processing, decision making, and motor response. These manipulations, together with comparisons with other related visuospatial and working memory tasks (Gruber, Indefrey, Steinmetz, & Kleinschmidt, 2001; Simon et al., 2002), as well as evaluation of individual differences in problem solving, have been a mainstay of arithmetic studies in the adult brain.

A Canonical Circuit for Arithmetic

Figure 28.2 shows canonical brain areas involved in arithmetic problem solving, identified using a Bayesian meta-analytic model (Yarkoni, Poldrack, Nichols, Van Essen, & Wager, 2011). These regions are largely localized to the dorsal aspects of the posterior parietal cortex (PPC), ventral temporal-occipital (VTOC), and the premotor cortex in the prefrontal cortex (PFC). Depending on task complexity, other brain areas, most notably the inferior and middle frontal gyri in the PFC, basal ganglia and cerebellum, are also engaged as working memory requirements increase. The interested reader can browse www.neurosynth.org (Yarkoni et al., 2011) to examine canonical circuits involved in arithmetic and calculation tasks to gain insights into the underlying anatomy and distributed systems for themselves. It should be noted that these maps are all derived from studies in adults and neurodevelopmental studies are not represented at this point. As we discuss below, in addition to these canonical regions, medial and lateral temporal lobe areas crucial for episodic and semantic memory also need to be included when we consider development, learning, and skill acquisition.

Uniqueness and Specificity of Brain Areas

Despite the anatomical specificity of brain areas highlighted in these canonical maps, closer examination with multiple other search terms exposes considerable overlap with brain areas engaged by tasks involving visuospatial attention and working memory. Even a superficial visualization of these overlapping maps highlights the distributed and non-unique nature of brain systems engaged by arithmetic. Each of the brain areas identified above is also involved in other non-arithmetic and non-numerical operations. Indeed, considerable overlap exists between front-parietal regions implicated in arithmetic and working memory (Dumontheil & Klingberg, 2012; Kaufmann, Wood, Rubinsten, & Henik, 2011). Research has also shown overlap in bilateral PPC regions across multiple mathematical and non-mathematical task manipulations involving different types of visuospatial information (Gruber et al., 2001; Simon et al., 2002; Venkatraman, Ansari, & Chee, 2005). Thus, the brain draws upon multiple basic visuospatial functions in the service of arithmetic. Disentangling these processes is a topic of ongoing research; it requires careful synthesis and integration with other domains of cognitive neuroscience, taking into account both the contribution of local neural circuits as well as their distributed patterns of neural connectivity, as reviewed elsewhere (Bressler & Menon, 2010).

Dissecting Functional Subdivisions

Up until a decade ago, lesion studies formed the basis of much of our understanding of the brain bases of arithmetic. Lesions in the parietal cortex (Figure 28.3) have historically been implicated in the classical model of acalculia, a specific disorder of numerical competence and arithmetic skill. Within the PPC, the intraparietal sulcus, angular gyrus, supramarginal gyrus, and perisylvian cortex have all been implicated in acalculia (Henschen, 1920; McCarthy & Warrington, 1988; Takayama, Sugishita, Akiguchi, & Kimura, 1994; Warrington, 1982). A number of functional dissociations between brain regions

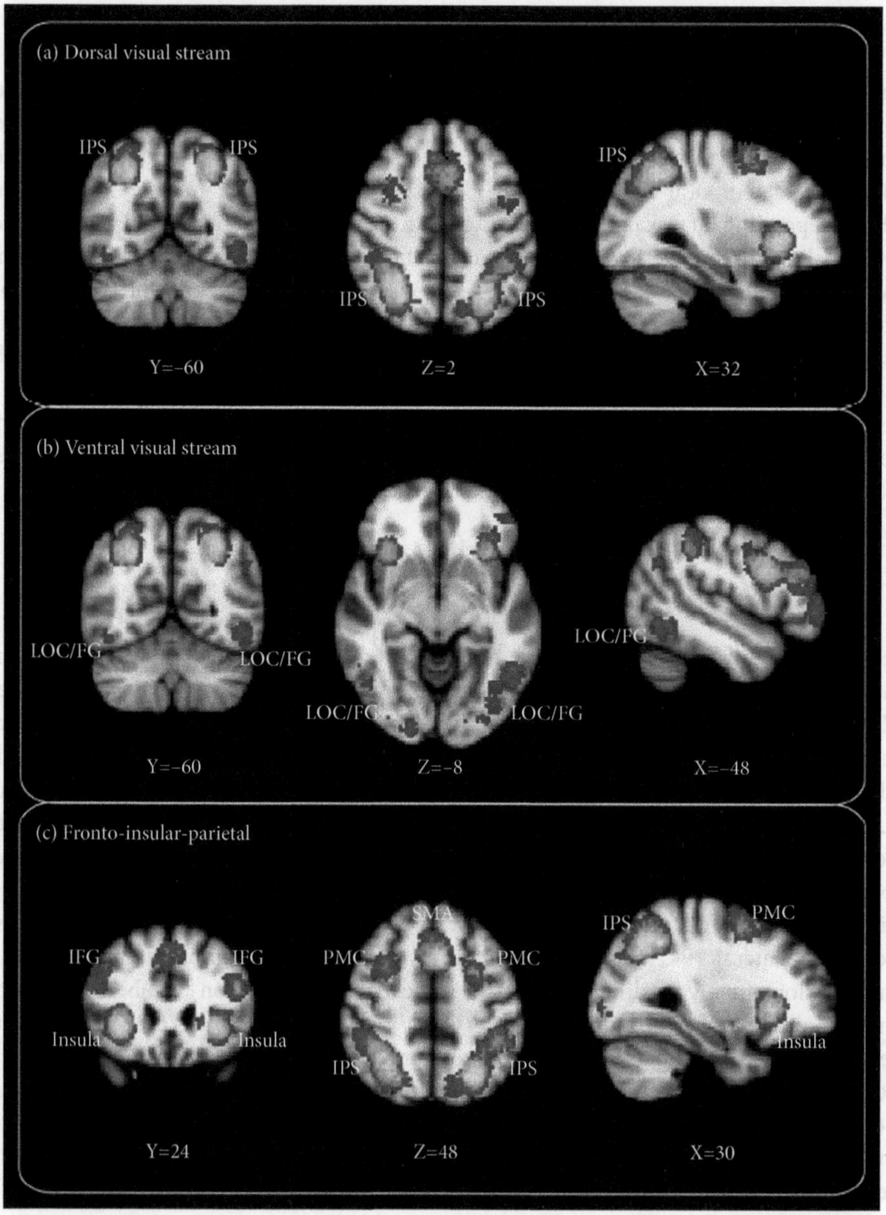

FIGURE 28.2 Canonical brain areas involved in arithmetic problem solving. (a) Dorsal visual stream anchored in the intraparietal sulcus within the posterior parietal cortex. (b) Ventral visual stream anchored in the lateral occipital cortex (LOC) and fusiform gyrus (FG). (c) Prefrontal cortex control system anchored in the anterior insula, inferior frontal gyrus (IFG) and premotor cortex. Maps are based on meta-analysis of 44 studies of arithmetic in neurosynth.org. In this and all subsequent figures the left hemisphere is on the left side of each coronal and axial brain slice.

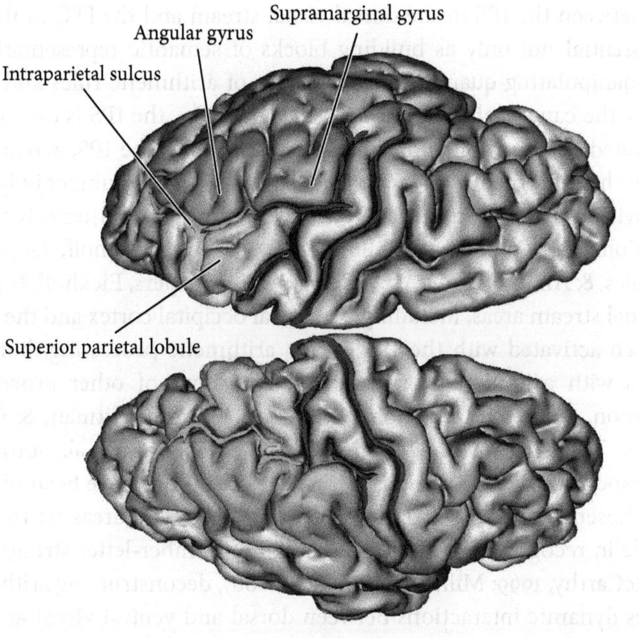

FIGURE 28.3 Neuroanatomy of posterior parietal cortex regions involved in arithmetic. Left and right hemisphere and dorsal views of the posterior parietal cortex regions that are typically activated during arithmetic tasks, including the superior parietal lobule (SPL: yellow), angular gyrus (ANG: blue), and supramarginal gyrus (SMG: orange) delineated by the intraparietal sulcus (IPS: green) and the post central sulcus (pink). The central sulcus (blue) is included as a reference point.

differentially involved in specific operations such as addition, subtraction and multiplication have been suggested in the literature (Chochon, Cohen, van de Moortele, & Dehaene, 1999; McNeil & Warrington, 1994; van Harskamp & Cipolotti, 2001). While lesions in the PPC often have dramatic consequences for mathematical information processing, they appear to be variable in the specific type of arithmetic deficits that are seen across individual patients (Delazer & Benke, 1997; Kahn & Whitaker, 1991; McCloskey, Harley, & Sokol, 1991). Beyond this, in spite of an array of dissociations reported in the literature, lesion studies have lacked adequate anatomical specificity and have yielded limited knowledge about the functional role of specific brain regions in arithmetic. In the following sections we summarize findings, mainly based on functional magnetic resonance imaging (fMRI) studies in healthy adults, which address this gap.

Dorsal Versus Ventral Visual Streams

The dorsal and ventral visual streams play an important role in number representation and manipulation, two key components of any arithmetic task. Normative functional neuroimaging studies in adults have consistently implicated the intraparietal sulcus (IPS) within the PPC as a region specifically involved in the representation and manipulation of numerical quantity (Cohen Kadosh, Lammertyn, & Izard, 2008; Dehaene et al., 2003). Dynamic interactions between the IPS in the dorsal visual stream and the ITC in the ventral visual stream are essential not only as building blocks of semantic representation of quantity, but also for manipulating quantity in the context of arithmetic rules and procedures. As highlighted by the canonical circuit shown in Figure 28.1, the IPS is one of the most consistently activated brain regions during arithmetic. Within the IPS, activations have been observed in the hIPS segment, which is particularly sensitive to number judgment, as well as the IPS subdivisions hIP1, hIP2, and hIP3 more anteriorly and posteriorly (Caspers, Geyer, Schleicher, Mohlberg, Amunts, & Zilles, 2006; Caspers, Eickhoff, Geyer, Scheperjans, Mohlberg, Zilles, & Amunts, 2008; Wu, Chang, Majid, Caspers, Eickhoff, & Menon, 2009).

Ventral visual stream areas, including the lateral occipital cortex and the fusiform gyrus, are not only co-activated with the IPS during arithmetic processing, but their response also increases with arithmetic complexity, independent of other processing demands (Keller & Menon, 2009; Rickard, Romero, Basso, Wharton, Flitman, & Grafman, 2000; Rosenberg-Lee, Tsang, & Menon, 2009b; Wu et al., 2009; Zago et al., 2001). Furthermore, functional dissociations across various arithmetic operations have been observed in these ITC regions (Rosenberg-Lee et al., 2011b). Although these ITC areas are thought to play an important role in recognition and discrimination of number-letter strings (Allison, Puce, Spencer, & McCarthy, 1999; Milner & Goodale, 2008), deconstructing arithmetic problems likely requires dynamic interactions between dorsal and ventral visual areas, particularly when the problem format is less familiar and problem solving routines less well automatized, as is often the case with children (Rosenberg-Lee et al., 2011b).

IPS Versus Angular Gyrus

The inferior parietal lobule, encompassing the lateral aspects of the IPS, shows significant heterogeneity in its response to arithmetic. In contrast to the IPS, which shows greater response to arithmetic problems compared with 'rest' baseline, the angular gyrus shows prominent reductions in activation ('deactivation'), relative to both passive 'rest' and other low-level baselines involving simple arithmetic problems or magnitude judgment tasks. Although Rickard and colleagues first reported this robust effect in 2000, their findings were largely ignored until recently. In their study, they found relative decreases or deactivation in the left and right angular gyrus during a simple well-automated multiplication task, compared with a magnitude judgment task (Rickard et al., 2000). Since then, several studies have consistently found deactivation relative to rest baseline in the angular gyrus during arithmetic problem solving (Grabner, Ansari, Reishofer, Stern, Ebner, & Neuper, 2007; Mizuhara, Wang, Kobayashi, & Yamaguchi, 2005; Wu et al., 2009; Zhou, Chen, Zang, Dong, Chen, Qiao, & Gong, 2007). More recently, Wu and colleagues used precise cyto-architectonic mapping, which permitted systematic investigations of both activation and deactivation in specific subdivisions of the PPC as a function of task difficulty and arithmetic operations (Wu et al., 2009). In each case, the angular gyrus showed task-related deactivation, with greater deactivation relative to a number identification control task. Furthermore, greater angular gyrus deactivation is associated with poorer performance on the arithmetic task (Wu et al., 2009); as well as lower overall math abilities (Grabner et al., 2007). More broadly, these findings suggest a close link between the anatomy and functions within the parietal cortex and they provide new evidence for striking functional heterogeneity within the inferior aspects of the PPC during arithmetic. These findings are noteworthy because they suggest that the IPS and angular gyrus play completely different roles in arithmetic. The functional role of the angular gyrus in arithmetic remains an open question with respect to both the episodic (Cabeza, Ciaramelli, & Moscovitch, 2012; Cabeza, Ciaramelli, Olson, & Moscovitch, 2008) and the semantic (Binder & Desai, 2011) processes subserved by this region and its large-scale interconnected functional networks (Greicius, Krasnow, Reiss, & Menon, 2003; Raichle, MacLeod, Snyder, Powers, Gusnard, & Shulman, 2001; Uddin, Supekar, Amin, Rykhlevskaia, Nguyen, Greicius, & Menon, 2010). For additional discussion of the differential roles of the IPS and angular gyrus in arithmetic fact learning, see 'Training, Learning, and Plasticity'.

IPS Versus Superior Parietal Lobule

Arithmetic problem solving also shows consistent activation of the superior parietal lobule (Dehaene et al., 2003; Grabner, Ansari, Koschutnig, Reishofer, Ebner, & Neuper, 2009; Ischebeck, Zamarian, Siedentopf, Koppelstatter, Benke, Felber, & Delazer, 2006; Menon et al., 2000b; Rickard et al., 2000; Wu et al., 2009; Zago et al., 2001), but a recent virtual lesion study suggests that its role is likely more supportive than obligatory (Andres, Pelgrims, Michaux, Olivier, & Pesenti, 2011). Andres and colleagues used fMRI to first localize parietal areas involved in subtraction and multiplication. As with many prior studies, they found increased activation, bilaterally, in the IPS and superior parietal lobule during both arithmetic operations. In order to examine the causal involvement of these brain areas, transcranial magnetic stimulation (TMS) was used to create a virtual lesion of either the IPS or superior parietal lobule in each participant, over the sites corresponding to the peaks of activation gathered in fMRI. An increase in response latencies was found for both operations after a virtual lesion of either the left or right IPS, but not of the superior parietal lobule. TMS over the IPS also increased error rates in the multiplication task. These results are consistent with fMRI studies comparing brain responses to multiple arithmetic operations (Rosenberg-Lee et al., 2011b) and suggest that even operations solved by memory retrieval, such as multiplication, depend on the IPS (see also Salillas & Semenza, this volume). The superior parietal lobule does not appear to be as critical for solving basic subtraction and multiplication problems (Andres et al., 2011). These functional dissociations might be related to the type of attentional processes engaged by the IPS and superior parietal lobule. Whereas the IPS is primarily involved in visuomotor functions, such as pointing and finger counting, the superior parietal lobule is primarily involved in visuospatial functions, such as saccades and covert shifts in attention. How these two parietal systems interact and share attentional resources during arithmetic problem solving is at present unclear, and is an important topic for future investigations.

PPC Versus PFC

Prefrontal and parietal cortices are often co-activated during arithmetic tasks. Prefrontal areas involved in arithmetic include the premotor cortex, and ventrolateral and dorsolateral PFC, although the specific roles of these PFC regions in arithmetic have not been yet delineated using experimental manipulations [see however (Cho, Metcalfe, Young, Ryali, Geary, & Menon, 2012; Supekar & Menon, 2012)]. Despite co-activation of PPC and the PFC in most arithmetic tasks, their functional roles can be dissociated. One study examined the effects of cognitive load on arithmetic by varying the number of operands and the rate of stimulus presentation in a factorial design (Menon et al., 2000b). This study found quantitative differences in activation of the parietal and prefrontal cortices in relation to increasing task difficulty. The main effect of arithmetic complexity was observed in the left and right IPS, while the main effect of domain-general task difficulty was observed in the left ventrolateral PFC. These findings are consistent with the view that the IPS plays a more crucial and obligatory role in arithmetic processing, independent of general cognitive demands. In contrast to the PFC, the PPC also shows stronger relation to arithmetic performance (Grabner et al., 2007; Menon, Rivera, White, Eliez, Glover, & Reiss, 2000c). Additional support for similar dissociations comes from developmental studies in children and learning studies in adults, which point to a decreasing role of the prefrontal cortex with proficiency and skill acquisition, as reviewed below. Finally, analysis of differential brain responses to incorrect (e.g. '2 + 3 = 5') versus correct (e.g. '2 + 3 = 4') problems shows that the PFC, but not the PPC is sensitive to incorrect arithmetic expressions (Menon et al., 2002). The pattern of brain response observed is consistent with the hypothesis that processing incorrect equations involves detection and resolution of the interference between the internally computed and externally presented incorrect answer. Electrophysiological studies further demonstrate that this process is similar to the semantic incongruity effect as indexed by the 'N400' component of the event-related potential (Niedeggen, Rosler, F. & Jost, 1999).

Anterior Temporal Cortex

Outside the parietal cortex, the most novel evidence from patients with neurological disorders has emerged from studies of semantic dementia. Semantic dementia impacts the anterior aspects of the temporal lobe, and patients perform generally well compared with healthy controls on tests of addition and subtraction. However, they use increasingly basic, inflexible strategies to retrieve multiplication 'facts', and in multi-digit calculations they made procedural errors that pointed to a failure to understand the differential weighting of left and right hand columns (Julien, Thompson, Neary, & Snowden, 2008). The findings are noteworthy because they challenge the notion that arithmetic knowledge is a totally separate semantic domain, and instead suggest that anterior lateral temporal cortex, which is known to be impacted in semantic dementia also plays an important role in arithmetic understanding (Cappelletti, Butterworth, & Kopelman, 2012). Paradoxically, although this brain system has not yet been directly examined in neurotypical adults using functional neuroimaging techniques, recent studies (reviewed below) in children point to its differential engagement in relation to different types of strategies used to solve arithmetic problems (Cho, Ryali, Geary, & Menon, 2011).

Functional Dissociations Across Arithmetic Operations

Addition, subtraction, multiplication, and division constitute the four basic arithmetic operations. There has been considerable behavioral research on the cognitive processes associated with these operations over the past several decades. Surprisingly, even adults use a variety of strategies when solving simple arithmetic problems (Campbell & Timm, 2000; Hecht, 1999; LeFevre, Bisanz, Daley, Buffone, Greenham, & Sadesky, 1996; Siegler & Shrager, 1984). In particular, the rate of retrieval versus alternate calculation strategies differs widely across operations (Campbell, 2008). Retrieval is the dominant method for addition and multiplication, whereas subtraction and division rely more on alternate strategies such as counting and inversion (Campbell & Xue, 2001). Problem-solving using a related fact from previously learned inverse operations is clearly a more parsimonious strategy than memorizing facts for all four operations (Campbell & Alberts, 2009). Although lesion studies over the past several decades have focused on functional dissociations in PPC during arithmetic, no consistent view has emerged of its differential involvement in addition, subtraction, multiplication, and division. To circumvent problems with poor anatomical localization, Rosenberg-Lee and colleagues examined functional overlap and dissociations in cyto-architectonically-defined subdivisions of the IPS, superior parietal lobule, and angular gyrus, across these four operations (Rosenberg-Lee et al. 2011b). Compared with a number identification task, all operations except addition, showed a consistent profile of left posterior IPS activation and deactivation in the right posterior angular gyrus (Figure 28.4). Multiplication and subtraction differed significantly in right, but not in left, angular gyrus activity, challenging the view that the left angular gyrus differentially subserves retrieval during multiplication (Rosenberg-Lee et al., 2011b). Although addition and multiplication both rely on retrieval, multiplication evoked significantly greater activation in right posterior IPS, as well as the prefrontal cortex, lingual, and fusiform gyri, demonstrating that addition and multiplication engage different retrieval processes. The results further suggest that brain responses associated with calculation and retrieval processes cannot be uniquely mapped to specific PPC regions. Rather, these findings point to distributed representation of arithmetic processes in the human PPC, which explains why lesion studies have yielded inconsistent findings. Taken together, these results lead to five key conclusions: first, the PPC shows considerable functional heterogeneity across basic arithmetic operations; secondly, operations matched for task difficulty differ in PPC activation; thirdly, PPC subdivisions are variably activated and deactivated; fourthly, activation is strongest in IPS and deactivation in angular gyrus; and finally, individual differences in performance are related to both PPC activation and deactivation. The manner in which such distributed representations contribute to operation specific information processing remains unknown at this time.

Strategy Use and Individual Differences

Only one brain imaging study, to date, has examined retrieval strategy use during arithmetic problem-solving in adults. Grabner and colleagues used trial-by-trial strategy self-reports to identify brain regions underlying retrieval and calculation strategies in arithmetic problem solving (Grabner et al., 2009). They found stronger activation of the left angular gyrus, while solving arithmetic problems for which participants reported fact retrieval, whereas the application of procedural strategies was accompanied by greater activation in other PFC and PPC regions. Related research has examined individual differences in brain responses associated with different levels of performance and overall mathematical skills. One study examined regional differences in brain activation between perfect and imperfect performers and found a relationship between left PPC activity and mental calculation expertise (Menon et al., 2000c). Perfect performers had an accuracy of 100% and were significantly faster than the rest of the subjects. They also showed significantly less activation only in the left supramarginal gyrus and IPS, leading the authors to suggest that such reduction may be related to functional optimization of performance associated with mastery of arithmetic skills. Adults with higher mathematical competence, as assessed using standardized measures, have also been reported to display stronger activation of the left angular gyrus,

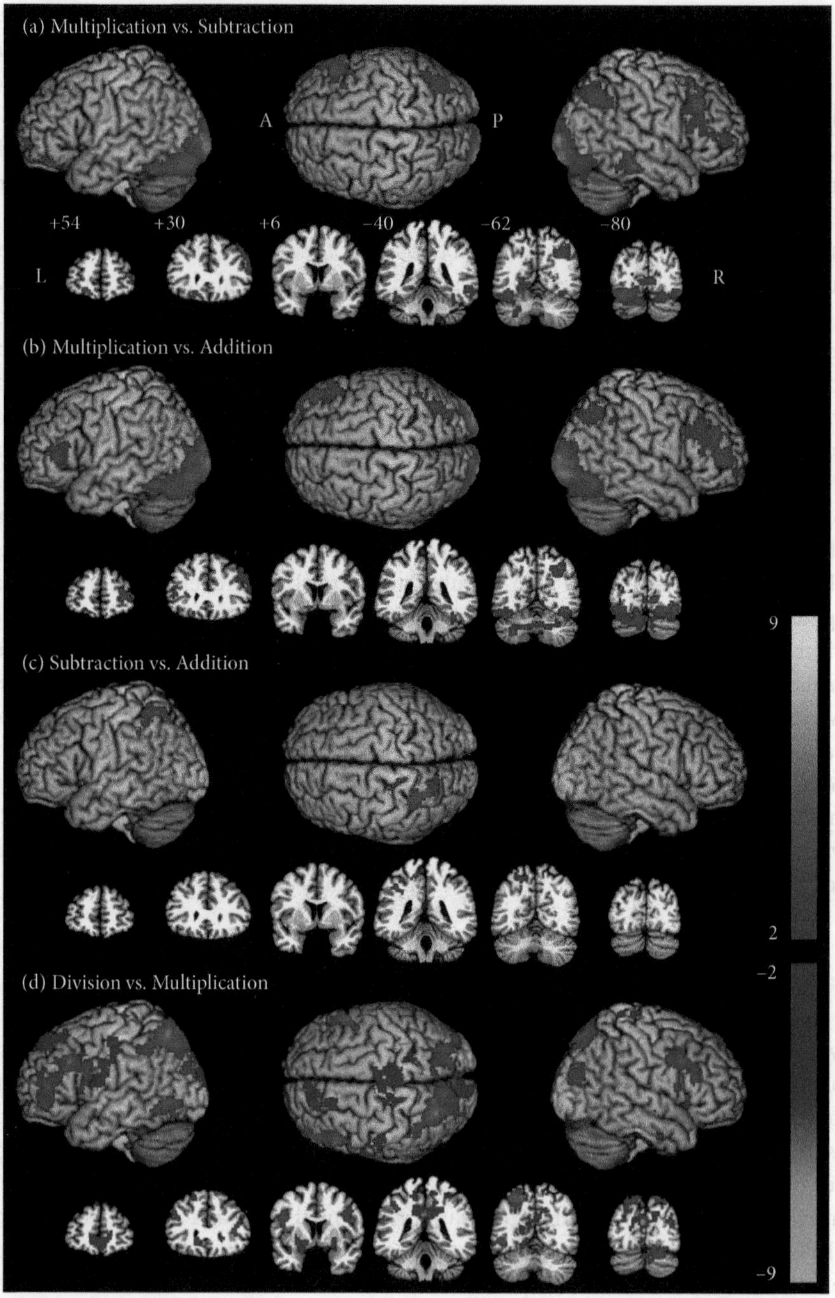

FIGURE 28.4 Brain regions that showed significant differences in activation between arithmetic operations. Surface rendering and coronal sections of brain regions that showed significant activation (red) and deactivations (blue) are shown for (a) multiplication versus subtraction, (b) multiplication versus addition, (c) subtraction versus addition and (d) division versus multiplication.

Adapted from *Neuropsychologia*, 49(9), Miriam Rosenberg-Lee, Ting Ting Chang, Christina B. Young, Sarah Wu, and Vinod Menon, Functional dissociations between four basic arithmetic operations in the human posterior parietal cortex: A cytoarchitectonic mapping study, pp. 2592–608, Copyright (2011), with permission from Elsevier.

while solving single-digit and multi-digit multiplication problems (Grabner et al., 2007). However, based on the discussion above, it would be more accurate to say that adults with low mathematical competence displayed stronger deactivation in this region. Similarly, Wu and colleagues found that lower accuracy during addition and subtraction tasks was associated with lower bilateral angular gyrus response (Wu et al., 2009). However, these differences arose from lower levels of deactivation in poorly performing individuals, rather than activation above baseline in high performers. One interpretation of these findings is that more competent adults have a stronger reliance on automatic, language-mediated efficient retrieval processes, another is that angular gyrus responses reflect subjective task difficulty. Further research on the core semantic and memory functions subserved by the angular gyrus are necessary to disambiguate these two processes and to shed more light on whether the angular gyrus does, indeed, facilitate efficient fact retrieval.

Training, Learning, and Plasticity

Training studies are beginning to provide new insights into how arithmetic is organized in the brain (see Zamarian and Delazer, this volume). Most importantly, this research offers a new conceptual framework for examining how multiple PPC, PFC, and ITC areas come to play the roles they do in the adult brain. A particular focus on these studies has been how the functional organization of the IPS, angular gyrus and the superior parietal lobe in the PPC changes with learning. One paradigm for short-term arithmetic training, lasting about a week, typically consists of repetition for one set of complex problems and a lower frequency for the other novel set. Training effects are rapid, and typically become significant after approximately eight repetitions of a problem and remain stable over the course of the experiment (Ischebeck, Zamarian, Egger, Schocke, & Delazer, 2007). Comparison of brain responses between repeated and novel problems was used to examine the effects of training; implicit in this design is the assumption that brain responses to untrained and trained problems are otherwise well-matched prior to training. Training typically decreases response in the ventrolateral PFC, IPS as well as the caudate nucleus of the basal ganglia (Delazer, Ischebeck, Domahs, Zamarian, Koppelstaetter, Siedentopf et al., 2005; Ischebeck et al., 2006, 2007). In parallel, relative increases in activation of the left angular gyrus have been reported, but this most likely reflects decreases in deactivation relative to baseline as the problems become more automatized. Collectively, a handful of these studies have converged on similar results, and they suggest that learning arithmetic is associated with major functional reorganization within the PPC, such that the load on the IPS is reduced, while angular gyrus responses show lower levels of deactivation (Delazer et al., 2005; Ischebeck et al., 2007; Zamarian, Ischebeck, & Delazer, 2009). These and similar studies in children are likely to guide research into learning in children with dyscalculia.

Rapid learning of arithmetic has also been examined using repetition priming paradigms (Salimpoor, Chang, & Menon, 2009). Although repetition priming has been widely used to examine the neural basis of behavioral facilitation, these studies have focused primarily on word and object identification tasks. In the first such brain imaging study of repetition priming during arithmetic problem solving, repeated stimulus presentation of three-operand arithmetic problems was associated with widespread and robust bilateral suppression of responses in the ventrolateral PFC, extrastriate cortex and middle occipital gyrus, fusiform gyrus, dorsal striatum, and the thalamus (Salimpoor et al., 2009). Repeated trials were associated with mean reaction time improvements of about 100 milliseconds. Across individuals, reaction time improvements were also directly correlated with repetition enhancement (rather than suppression) in the hippocampus and the posteromedial cortex (posterior cingulate cortex, precuneus, retrosplenial cortex), regions known to support memory formation and retrieval, and in the supplementary motor area and the dorsal mid-cingulate cortex, regions known to be important for motor learning. Furthermore, improvements in reaction time were also correlated with increased functional connectivity of the hippocampus with both the supplementary motor area and the dorsal mid-cingulate cortex. These findings provided novel support for the hypothesis that repetition enhancement and associated stimulus-response learning facilitates behavioral performance during arithmetic problem solving.

Summary

Normative functional neuroimaging studies have implicated the IPS within the dorsal aspects of the parietal cortex as a region critically involved in the representation and manipulation of numerical quantity (Ansari, 2008; Cantlon, Brannon, Carter, & Pelphrey, 2006; Cantlon, Libertus, Pinel, Dehaene, Brannon, & Pelphrey, 2009; Cohen Kadosh et al., 2008; Cohen Kadosh & Walsh, 2009; Dehaene et al., 2003). In addition to the dorsal PPC, ventral visual stream areas, encompassing the lateral occipital cortex and fusiform gyri in the ITC, also play an important, although often under-appreciated, role in arithmetic processing (Delazer, Domahs, Bartha, Brenneis, Lochy, Trieb, & Benke, 2003; Grabner et al., 2009; Menon et al., 2000b; Rickard et al., 2000; Wu et al., 2009; Zago et al., 2001). A recent meta-analysis found that the left fusiform gyrus is consistently activated across a wide range of numerical tasks (Arsalidou & Taylor, 2011), consistent with its hypothesized role in processing orthographic structure (Binder, Medler, Westbury, Liebenthal, & Buchanan, 2006). Beyond this, more complex calculation abilities place demands on multiple cognitive systems involving working memory and cognitive control (Menon et al., 2000b). This involves multiple regions including the anterior insula and anterior cingulate cortex which are involved in directing attentional resources (Arsalidou & Taylor, 2011; Supekar & Menon, 2012), the ventrolateral PFC, which is engaged by tasks that involve effortful maintenance and retrieval, and the dorsolateral PFC, which is engaged by tasks that require manipulation of information in working memory during multi-stage calculation tasks (Menon et al., 2000b; Menon et al., 2002; Zago et al., 2008).

Arithmetic in The Developing Brain: Mapping Changes From Childhood to Adulthood

The first phase of neurodevelopmental studies focused on maturation of arithmetic skills over an extended age range, often spanning one or more decades from childhood to adulthood. These studies have shown that, like adults, children reliably engage frontal, parietal, and ventral temporal cortex in response to arithmetic tasks and, more importantly, that activations in these regions are modulated by notation, task complexity, and competence level (Kaufmann et al., 2011). However, despite superficial similarities, arithmetic in the developing brain is characterized by major, protracted, developmental changes in brain activation patterns. Key findings from neurodevelopmental studies are summarized below.

Shift From Frontal to Parietal and Ventrotemporal Cortex

Rivera and colleagues examined the neural correlates of arithmetic skill development in children, adolescents, and adults ranging in age from 8 to 22 (Rivera, Reiss, Eckert, & Menon, 2005). Participants viewed arithmetic equations in the form 'a + b = c' and were asked to judge whether the results were correct or not. Matching for accuracy allowed the researchers to examine dissimilar trajectories of functional maturation independent of performance differences. Children showed less activation in the left supramarginal gyrus and IPS within the PPC, encompassing both the dorsal visual stream areas highlighted above. Increased activation in the left lateral occipital temporal cortex, an area thought to be important for visual word and symbol recognition (Cohen & Dehaene, 2004; Hart, Kraut, Kremen, Soher, & Gordon, 2000; Kronbichler, Hutzler, Wimmer, Mair, Staffen, & Ladurner, 2004; Price & Devlin 2004) was also observed in adults. On the other hand, children showed greater activation in the PFC, including the ventrolateral and dorsolateral PFC, as well as anterior cingulate cortex (Figure 28.5). In general agreement with these findings, Kucian and colleagues (Kucian, von Aster, Loenneker, Dietrich, & Martin, 2008) found greater left IPS activity in 22–32-year-old adults and greater right anterior cingulate cortex activity in 9- and 12-year-old children during approximate addition. These findings suggest a process of increased functional specialization of the PPC and ITC with age, with decreased dependence on PFC working memory and attention resources.

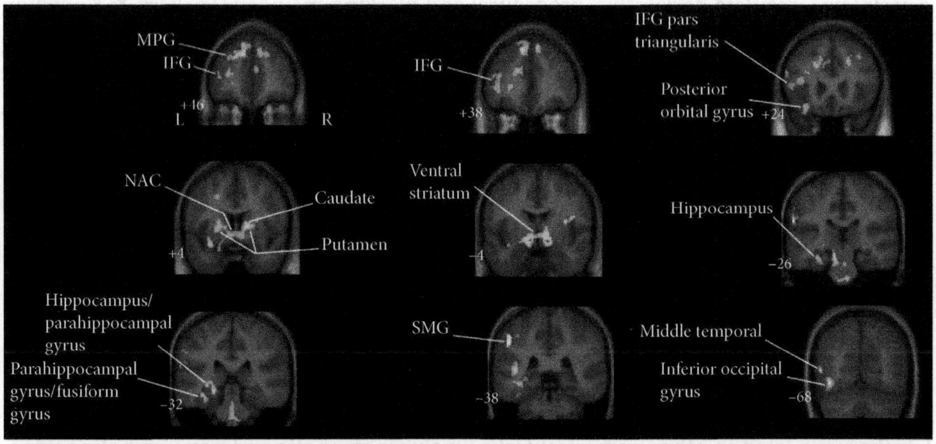

FIGURE 28.5 Neurodevelopmental changes in arithmetic. Compared with adults, children showed greater activation in the prefrontal cortex, basal ganglia and the hippocampus (cyan-blue scale) during two operand arithmetic tasks. Adults showed greater activation in the supramarginal gyrus and the lateral occipital cortex (yellow-red scale). Task accuracy was matched across the groups. IFG = inferior frontal gyrus; MFG = middle frontal gyrus, NAC = nucleus accumbens, SMG = supramarginal gyrus.

Adapted from *Cerebral Cortex*, 15(11), S.M. Rivera, A.L. Reiss, M.A. Eckert, and V. Menon, Developmental Changes in Mental Arithmetic: Evidence for Increased Functional Specialization in the Left Inferior Parietal Cortex, pp. 1779–1790, doi: 10.1093/cercor/bhi055 © 2005, Oxford University Press.

Children Engage Medial Temporal Lobe and Basal Ganglia Memory Systems More

The study by Rivera and colleagues also highlighted the differential role of the medial temporal lobe memory system in children for the first time. Younger children exhibited significantly greater engagement of the left hippocampal and parahippocampal gyrus. Both these brain structures are known to play a major role in encoding and retrieval of facts (Squire, Stark, & Clark, 2004). The parahippocampal gyrus mediates convergence of high-level input from the visual association cortex into the hippocampus (Suzuki & Amaral, 1994), thereby facilitating the persistence of representations in short-term memory (Eichenbaum, 2000). The greater activation seen in this region in younger subjects may reflect the greater recruitment of processing resources to sustain appropriate memory representations and may also reflect generalized novelty effects. In the same vein, De Smedt and colleagues (De Smedt, Holloway, Ansari, 2011) found greater hippocampal response in children when solving addition problems, but not when solving subtraction problems that are less well rehearsed and more difficult to memorize, because subtraction problems are not commutative (e.g., $5 - 3 \neq 3 - 5$). Thus, with increased experience and exposure, medial temporal lobe activations decrease as initially unfamiliar stimuli become less novel (Menon, White, Eliez, Glover, & Reiss, 2000a).

Children also showed greater activation in the dorsal basal ganglia, including the caudate and putamen. The basal ganglia are known to be critical for procedural memory (Graybiel, 2005), i.e. memory for procedures and habits, and it plays a supportive role in the maintenance of information in working memory (Chang, Crottaz-Herbette, & Menon, 2007). Furthermore, the PFC, in concert with medial temporal lobe and dorsal basal ganglia memory systems, regulate declarative, procedural and working memory (Packard & Knowlton 2002). All of these three regions showed greater activation in children. Parallel increases in hippocampus and basal ganglia activation in children have also been reported in a task involving overriding a learned action in favor of a new one (Casey, Thomas, Davidson, Kunz, & Franzen, 2002). These findings provide evidence for greater reliance on multiple memory systems, subserved by the hippocampus and the basal ganglia, in children. Taken together, these studies highlight the important role of episodic and procedural memory systems in children during initial stages of arithmetic learning.

Similarities with Development of Number Processing

The pattern of developmental shifts from the PFC to the PPC observed in arithmetic parallels those reported in other numerical tasks involving symbolic and non-symbolic magnitude comparisons. For example, in a non-symbolic magnitude discrimination task Cantlon and colleagues (Cantlon et al., 2009) found that while 6–7-year-old children engaged the bilateral inferior frontal gyrus and adjoining insular cortex, these PFC areas were not significantly activated in 24-year-old adults. In contrast, both groups showed activation of the left IPS, although the spatial extent of activity was greater in adults. Furthermore, numerical distance effects (greater activity for comparisons involving smaller ratios) were correlated with left IPS activity in adults (Pinel, Dehaene, Riviere, & LeBihan, 2001), whereas children displayed this effect in the left inferior frontal gyrus (Ansari, Garcia, Lucas, Hamon, &

Dhital, 2005; Ansari & Dhital 2006; Cantlon et al., 2009). In a non-symbolic comparison task, Ansari and Dhital (2006) found that only 9–11-year-old children displayed a distance effect in the right dorsal lateral PFC, whereas 19–21-year-old adults had stronger distance effect in the left anterior IPS. Similarly, in a symbolic number comparison task using the same age groups, Ansari and colleagues (Ansari et al., 2005) found that adults showed sensitivity to numerical distance bilaterally in the IPS, whereas in children, the right precentral gyrus and right inferior frontal gyrus were the regions sensitive to numerical distance. Thus, although the precise neural locus of development varies with cognitive process and stimulus, a consistent profile of decreased reliance on the PFC and increased reliance on the PPC has been found in a wide range of studies of numerical cognition involving comparisons of children and adults. The extent to which basic numerosity and arithmetic processes co-develop is currently unknown and remains an important topic for future investigation.

Role of Cognitive Control Processes

A common recurring theme in arithmetic problem solving is the engagement of PFC processes, which are needed both for controlled retrieval of facts from memory, as well as allocation of attention resources. Like adults, children as young as 7 show reliable, and consistent, patterns of brain activity during arithmetic problem-solving in multiple PFC regions (Houde, Rossi, Lubin, & Joliot, 2010). Importantly, as noted by Houde and colleagues in their review, these same PFC regions, most notably the fronto-insular cortex, and anterior insula in particular (see also canonical circuit in Figure 28.1), have also been implicated in reading and executive control tasks in young children. This profile of anatomical overlap suggests a common mechanism by which maturation of basic cognitive control can influence skill development in multiple cognitive and academic domains.

The ability to solve arithmetic problems relies on the ability to control attention and successfully direct cognitive efforts. Using a novel multi-pronged neuroimaging approach, Supekar and Menon identified for the first time the dynamic control processes underlying the maturation of arithmetic problem solving abilities (Supekar & Menon, 2012). They used a novel multimodal neurocognitive network-based approach combining task-related fMRI, resting-state fMRI and diffusion tensor imaging to investigate the maturation of control processes underlying problem solving skills in 7–9-year-old children. Their analysis focused on two key neurocognitive networks implicated in a wide range of cognitive control tasks: the insula-cingulate salience network, anchored in anterior insula, ventrolateral PFC and anterior cingulate cortex, and the frontoparietal central executive network, anchored in dorsolateral PFC and PPC. They found that, by age 9, the anterior insula node of the salience network is a major causal hub initiating control signals during arithmetic problem solving (Figure 28.6). The anterior insula, part of a larger network of regions previously shown to be important for salience processing and generating influential control signals, shows weaker influence over the ventrolateral and dorsolateral PFC, and anterior cingulate cortex in children compared with adults. Furthermore, structural connections between the anterior insula and other key regions were found to be weaker in children compared with adults. Importantly, measures of causal influences between key regions could be used to predict individual differences in behavioral performance on the arithmetic task. In this manner, maturing dynamic causal influences from the anterior insula play an important role in the development of arithmetic abilities.

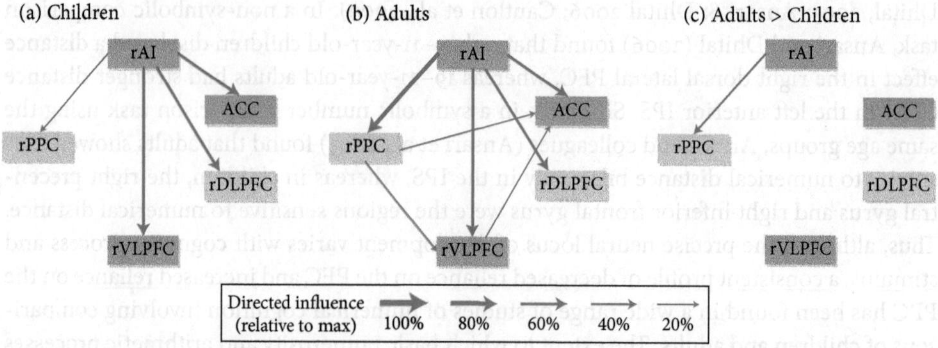

FIGURE 28.6 Developmental changes in causal network interactions during arithmetic problem solving. Casual interactions between five key nodes of the salience network (blue rectangles), and Central Executive network (green rectangles) are shown in (a) children and (b) adults. (c) Weaker causal interactions in children, compared with adults. ACC = anterior cingulate cortex; rAI = right anterior insula; rDLPFC = right dorsolateral prefrontal cortex; rPPC = right posterior parietal cortex; rVLPFC= right ventrolateral prefrontal cortex.

ARITHMETIC IN THE CHILD BRAIN

Arithmetic from A Developmental Perspective

Cross-sectional studies, such as those reviewed above, do not adequately capture neurodevelopment processes involved in acquisition of math skills. Cognitive developmental change should not be conflated with age-related change. To capture the mechanisms driving cognitive change it is critical to focus on the age ranges in which rate of change is the steepest. Examining age-related differences between children, adolescents, and adults alone is not sufficient; for example, comparison of 7- and 12-year-olds will be insensitive to change between the ages of 7 and 9. Critically, it cannot be assumed that learning in adults or contrasts between children and adults is comparable to learning in the developing brain (Karmiloff-Smith, 2010).

It has recently been shown that developmental changes cannot be inferred from, or characterized by a gross comparison between adults and children or by examining the effects of training on novel problems in adults (Cho, et al., 2011). Indeed, most behavioral studies of cognition in children focus on developmental questions; comparisons with adults have rarely been the focus of these studies, as it is well understood that children's emerging memory and representational and logical thinking abilities are not analogous to those of a skilled adult. Approaches to the study of cognitive development that differ fundamentally from the dominant paradigm in brain imaging studies based on 'child versus adult' comparisons are critically needed, to parallel the more sensitive models of behavioral studies (Geary, 1994; Karmiloff-Smith, 2010; Siegler, 1996).

Development during Critical Learning Periods

Early elementary school represents an important period for the acquisition and mastery of arithmetic fact knowledge. Behavioral research has characterized a progression of increasingly sophisticated calculation procedures leading to the greater use of direct retrieval over time (Siegler & Shrager, 1984). By 2nd grade, children are typically able to answer single-digit addition problems, although rapid fact retrieval is still not mature in most children of that age (Jordan, Hanich, & Kaplan, 2003). Between 2nd and 3rd grade, problem-solving abilities generally progress from effortful counting strategies to more automatic retrieval strategies, although the extent and magnitude of skill maturation, and the sources of individual variability are less well understood. In a school-based study of typically developing children, growth curve modeling revealed that from the beginning of 2nd to the end of 3rd grade there was only a modest increase in the number of correctly answered problems (less than one item); however, there was a large decrease in the use of finger counting and an increase in the number of correctly retrieved items (Jordan et al., 2003). Although previous behavioral and classroom-based research studies have shown that arithmetic proficiency undergoes significant improvement in elementary school, surprisingly little is known about its neurodevelopmental underpinnings.

A recent study examined the neural correlates of the maturation of arithmetic problem solving skills over a narrow developmental window spanning a one year interval between 2nd and 3rd grades (Rosenberg-Lee, Barth, & Menon, 2011a). In both 2nd and 3rd graders, arithmetic complexity was associated with increased responses in the right inferior frontal sulcus and anterior insula, regions implicated in domain-general cognitive control, and in left IPS and superior parietal lobule regions important for numerical and arithmetic processing. Compared with 2nd graders, 3rd graders showed greater activity in dorsal stream parietal areas, right IPS, superior parietal lobule and angular gyrus, as well as prominent differences in ventral visual stream areas, bilateral lingual gyrus, right lateral occipital cortex and right parahippocampal gyrus (Figure 28.7). Significant differences were also observed in the PFC, with 3rd graders showing greater activation in left dorsal lateral PFC and greater deactivation in the ventromedial PFC. Third graders also showed greater functional connectivity between the left dorsolateral PFC and multiple posterior brain areas, with larger differences in superior parietal lobule and angular gyrus, compared with ventral stream visual areas lateral occipital gyrus and parahippocampal gyrus. These results suggest that even the narrow one-year interval spanning grades 2 and 3 is characterized by significant arithmetic task-related changes in brain response and connectivity, and argue that pooling data across wide age ranges and grades can miss important neurodevelopmental changes that occur during key stages of academic learning.

Between childhood and adulthood, number and arithmetic processing have highlighted a consistent shift from the PFC to posterior brain areas (Ansari et al., 2005; Ansari & Dhital, 2006; Cantlon et al., 2009; Rivera et al., 2005). In contrast, dorsolateral PFC responses increased in the narrow time window between 2nd and 3rd grades. This data provides new evidence that the initial stages of learning may be accompanied by increases, rather than decreases, in PFC response. There are several possible reasons for this. One possibility is that learning may reduce variability in PFC response. A second possibility is that PFC responses may become more focal with learning and functional maturation (Durston, Davidson, Tottenham, Spicer, Galvan, Fossella, & Casey, 2006). In either case, these findings suggest

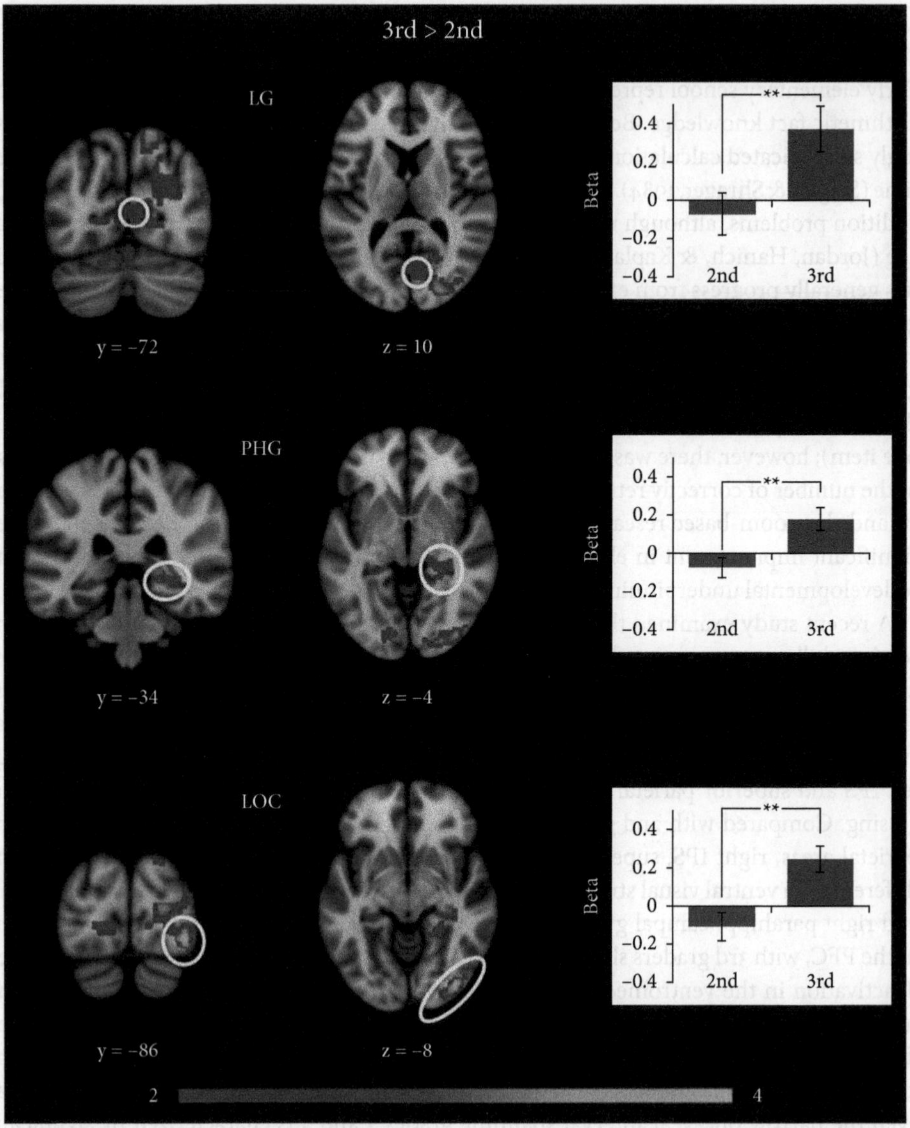

FIGURE 28.7 Comparison of ventral visual stream responses in 2nd and 3rd grade children. Between-grade differences in ventral visual stream response related to arithmetic complexity. Children in the 3rd grade (blue) showed greater activity than 2nd grade children (red) in the left and right parahippocampal gyrus (PHG), left and right lingual gyrus (LG) and right lateral occipital cortex (LOC). The graph on the right shows average beta-values, contrasting the Complex and Simple addition tasks, in each group. The bar graphs on the right show parameter estimated for 2nd and 3rd graders in red and blue, respectively.

a nonlinear trajectory of developmental changes characterized by an initial increase in dorsolateral PFC engagement during the early stages of learning, followed by more protracted decreases in response between childhood and adulthood (Rivera et al., 2005). Precise knowledge of this trajectory is important not only for understanding the effects of various types of instruction in typically developing children, but also for assessing and remediating abnormal developmental patterns in children with dyscalculia and math learning disabilities at an early age (Geary, Bailey, Littlefield, Wood, Hoard, & Nugent, 2009b; Geary, Bailey, & Hoard, 2009a; Rykhlevskaia, Uddin, Kondos, & Menon, 2009).

Development of Memory-Based Strategies

The ability to efficiently retrieve basic facts from memory is a key characteristic of mature problem solving in children (Siegler, 1996). During development, children's math problem solving skills become gradually more dependent on memory-based strategies, such as direct retrieval and less dependent on effortful procedures such as counting (Geary et al., 2004). Despite considerable advances in our understanding of the behavioral and cognitive mechanisms characterizing these shifts (Siegler & Svetina, 2006) little is known about the underlying brain mechanisms. Children's arithmetic problem solving provides an ideal domain for studying the brain mechanisms that underlie this cardinal feature of children's cognitive development because the underlying behavioral characteristics and cognitive processes are particularly well known (Geary, 1994). Little, however, is known about neurodevelopmental mechanisms underlying these transitions in children. As described below, recent fMRI studies have begun to make progress in addressing this gap.

Cognitive development and learning are characterized by diminished reliance on effortful procedures and increased use of memory-based problem solving. Cho and colleagues identified the neural correlates of this strategy shift in 7–9-year-old children at an important developmental period for arithmetic skill acquisition (Cho et al., 2011). Univariate and multivariate approaches were used to contrast brain responses between two groups of children who relied primarily on either retrieval or procedural counting strategies. Children who used retrieval strategies showed greater responses in the left ventrolateral PFC; notably, this was the only brain region that showed univariate differences in signal intensity between the two groups. In contrast, multivariate analysis revealed distinct multivoxel activity patterns in bilateral hippocampus, PPC and left ventrolateral PFC regions between the two groups (Figure 28.8). These results demonstrate that retrieval and counting strategies during early learning are characterized by distinct patterns of activity in a distributed network of brain regions involved in arithmetic problem solving and controlled retrieval of arithmetic facts. Findings suggest that even when the same brain areas are engaged to a similar degree in the two groups of children, compared with procedural strategies, memory-based arithmetic problem solving evokes distinct fine-scale neural representations. Critically, such differences were highly prominent in medial temporal lobe regions important for memory formation. These findings suggest that the reorganization and refinement of neural activity patterns in multiple brain regions plays a dominant role in the transition to memory-based arithmetic problem solving.

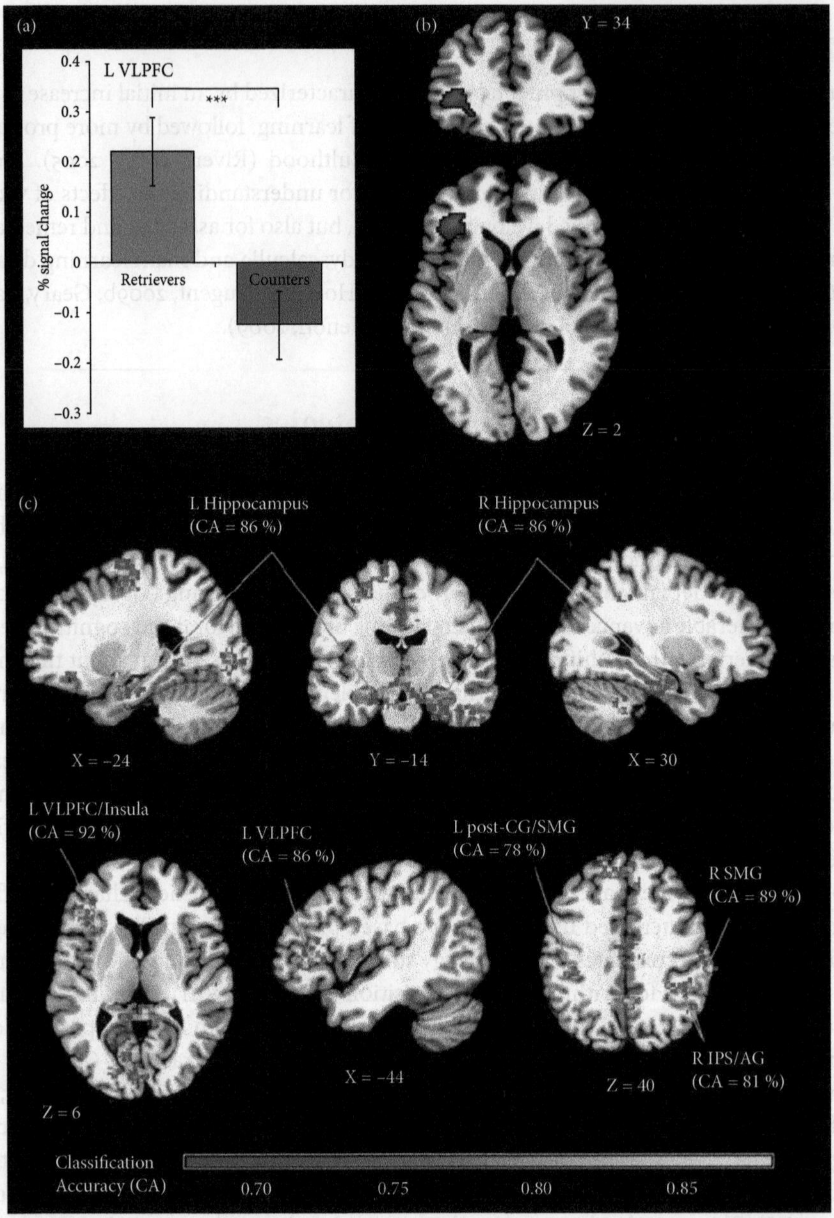

FIGURE 28.8 Strategy differences in arithmetic problem solving. (a) Percentage signal change in the left ventrolateral prefrontal cortex (VLPFC) across the two groups. (b) Brain areas that showed significant differences in voxel-wise activation levels during the Addition task between retrievers and counters. The left ventrolateral prefrontal cortex (VLPFC) showed greater activation in retrievers compared with counters. No brain regions showed greater activation in counters, compared with retrievers. ***$p < .001$. (c) Multivariate pattern analysis (MPA) revealed significant differences in spatial activation patterns between Retrievers and Counters in three brain areas: (1) bilateral medial temporal lobe – hippocampi and parahippocampal gyri; (2) left prefrontal cortex – ventrolateral prefrontal cortex (VLPFC) and adjoining anterior insula; (3) bilateral posterior parietal cortex – left and right supramarginal gyrus (SMG), right intra-parietal sulcus (IPS) and angular gyrus (AG). Peak classification accuracies (CA) in each brain region are shown in parentheses.

Adapted from How does a child solve 7 + 8? Decoding brain activity patterns associated with counting and retrieval strategies, Soohyun Cho, Srikanth Ryali, David C. Geary, and Vinod Menon, *Developmental Science*, 15(5), pp. 989–1001, DOI: 10.1111/j.1467-7687.2011.01055.x Copyright © 2117, John Wiley and Sons.

Summary

Although there are similarities in the brain areas that are engaged by arithmetic tasks in adults and children, there are major differences as well. The development of arithmetic problem solving skills between childhood and adulthood is characterized by decreased engagement of the PFC and increased engagement and functional specialization of the PPC (Ansari & Dhital, 2006; Cantlon et al., 2006; Rivera et al., 2005). Furthermore, even the seemingly brief 1-year interval spanning grades 2 and 3, for example, is characterized by significant task-related changes in brain response and connectivity, a finding that suggests that pooling data across wide age ranges and grades can miss important neurodevelopmental changes (Rosenberg-Lee et al., 2011a). Over the long term there is a shift from more controlled and effortful to more automatic processing of both numerical magnitude and arithmetic problem solving, but even over shorter time periods there is a process of continued refinement of brain responses. From a developmental and learning perspective, emerging data suggests that it is no longer appropriate to focus solely on parietal circuits as the basis of arithmetic skill development. A number of scaffolding systems are called into place during development, and development studies highlight brain systems otherwise missed in studies involving adults or even those that compare children with adults. One of the major emerging contributions of developmental studies is the crucial role played by the medial temporal lobe memory system in the development of fluid fact retrieval. Thus, brain systems and circuits in the developing child brain are clearly not the same as those seen in more mature adult brains sculpted by years of learning.

Conclusions

Converging evidence from studies of infants (Feigenson, Dehaene, & Spelke, 2004), preschool children (Cantlon et al., 2006) and adults (Ansari, 2008), as well as non-human primates (Cantlon & Brannon, 2006), indicates that the representation of approximate quantities is supported by the IPS in the dorsal aspects of the PPC. In addition to the dorsal PPC, the VTOC also plays an important, though often underappreciated, role in number processing. The brain builds arithmetic skills with the support of these systems, but this is only one part of the necessary circuitry. Arithmetic relies on and requires multiple cognitive systems involving working memory, episodic and semantic memory systems and executive control functions. Both developmental studies in young children and training studies in adults are beginning to highlight the important role of these systems in building new representations in the dorsal parietal, as well as the VTOC. Furthermore, as task manipulations become more sophisticated with better matching of control tasks on multiple dimensions, multivariate approaches are likely to be more useful for examining distinct neural representations in these brain areas.

We are still in the initial stages of understanding how these interacting systems unfold with development. It is, nevertheless, clear that the exclusive focus on activity levels in a small set of brain regions identified in highly skilled adults can miss important changes in functional organization that accompany learning and development associated with schooling. Increasingly, the focus has also shifted to multivariate analyses, as it is evident that similar signal levels across task conditions does not necessarily imply similar kind of information processing (Blair Rosenberg-Lee, Tsang, Schwartz, & Menon, 2012; Prado, Mutreja, Zhang, Mehta, Desroches, Minas, & Booth, 2011; Raizada, Tsao, Liu, Holloway, Ansari, & Kuhl, 2010). These types of fine-grained analyses clearly have important implications for understanding brain mechanisms mediating the formation of unique stimulus representations and how they mature with learning and development (Ashkenazi, Rosenberg-Lee, Tenison, & Menon, 2012).

Most previous normative adult and developmental studies of arithmetic have mainly focused on localization of activation and age-related changes, but it is becoming increasingly clear that cognition depends on interactions within and between large-scale brain networks (Bressler & Menon, 2010). Anatomical and physiological connectivity analysis provides insights into the neural processing subserved by these regions. We have recently suggested that IPS acts as an intermediate station for relaying salient visual information into the dorsal attention and working memory network (Uddin et al., 2010). Combined DTI and resting-state functional connectivity analysis uncovered differential connectivity patterns for these regions (Uddin et al., 2010). For example, posterior IPS regions have greater connectivity to striate and extrastriate cortex and the anterior IPS has greater connectivity with inferior and middle frontal gyri. New research is beginning to highlight the significant and specific changes in frontal-posterior functional connectivity that take place during a time period important for arithmetic skill development. In the long run a systems neuroscience approach, with its emphasis on networks and connectivity, rather than a pure localization approach, is better-suited to further understanding how complex skills such as arithmetic develop and are expressed in adulthood. Arithmetic requires the integration of multiple cognitive processes, which rely on the engagement of distributed brain areas subserved by long-range connections that undergo significant changes with development (Fair, Cohen, Dosenbach, Church, Miezin, Barch, et al. 2008; Supekar, Uddin, Prater, Amin, Greicius, & Menon, 2010). The recruitment of brain circuits anchored in the parietal cortex changes dynamically as a function of training and development. In this context, and the many other ways highlighted above, arithmetic serves as a model neurocognitive system for studying the ontogenesis of human cognitive and problem solving skills.

Acknowledgments

It is a pleasure to thank Dr Arron Metcalfe for his valuable feedback on the chapter. Supported by grants from the NIH (HD047520, HD059205) and NSF (DRL-0750340).

References

Allison, T., Puce, A., Spencer, D.D., & McCarthy, G. (1999). Electrophysiological studies of human face perception. I: Potentials generated in occipitotemporal cortex by face and non-face stimuli. Cerebral Cortex 9(5), 415–430.

Andres, M., Pelgrims, B., Michaux, N., Olivier, E., & Pesenti, M. (2011). Role of distinct parietal areas in arithmetic: an fMRI-guided TMS study. NeuroImage 54(4), 3048–3056.

Ansari, D. (2008). Effects of development and enculturation on number representation in the brain. Nature Reviews Neuroscience 9 (4), 278–291.
Ansari, D., & Dhital, B. (2006). Age-related changes in the activation of the intraparietal sulcus during nonsymbolic magnitude processing: an event-related functional magnetic resonance imaging study. Journal of Cognitive Neuroscience 18 (11), 1820–1828.
Ansari, D., Garcia, N., Lucas, E., Hamon, K., & Dhital, B. (2005). Neural correlates of symbolic number processing in children and adults. Neuroreport, 16 (16), 1769–1773.
Arsalidou, M., & Taylor, M.J. (2011). Is 2 + 2 = 4? Meta-analyses of brain areas needed for numbers and calculations. NeuroImage 54 (3), 2382–2393.
Ashkenazi, S., Rosenberg-Lee, M., Tenison, C., & Menon, V. (2012). Weak task-related modulation and stimulus representations during arithmetic problem solving in children with developmental dyscalculia. Developmental Cognitive Neuroscience 2(1), 152–166.
Barth, H., La Mont, K., Lipton, J., & Spelke, E.S. (2005). Abstract number and arithmetic in preschool children. Proceedings National Academy Science USA 102(39), 14116–14121.
Binder, J.R., & Desai, R.H. (2011). The neurobiology of semantic memory. Trends in Cognitive Science 15(11), 527–536.
Binder, J.R., Medler, D.A., Westbury, C.F., Liebenthal, E., & Buchanan, L. (2006). Tuning of the human left fusiform gyrus to sublexical orthographic structure. NeuroImage 33(2), 739–748.
Blair, K.P., Rosenberg-Lee, M., Tsang, J.M., Schwartz, D.L., & Menon, V. (2012). Beyond natural numbers: negative number representation in parietal cortex. Frontiers in Human Neuroscience 6, 7.
Bressler, S.L., & Menon, V. (2010). Large-scale brain networks in cognition: emerging methods and principles. Trends in Cognitive Science 14(6), 277–290.
Bull, R., Epsy, K., & Wiebe, S. (2008). Short-term memory, working memory, and executive functioning in preschoolers: longitudinal predictors of mathematical achievement at age 7 years. Developmental Neuropsychology 33(3), 205–208.
Butterworth, B., Varma, S., & Laurillard, D. (this volume). Dyscalculia: from brain to education. In R. Cohen Kadosh, & A. Dowker, (Eds.), The Oxford Handbook of Numerical Cognition, (Oxford University Press).
Cabeza, R., Ciaramelli, E., & Moscovitch, M. (2012). Cognitive contributions of the ventral parietal cortex: an integrative theoretical account. Trends in Cognitive Science 16(6), 338–352.
Cabeza, R., Ciaramelli, E., Olson, I.R., & Moscovitch, M. (2008). The parietal cortex and episodic memory: an attentional account. Nature Review Neuroscience 9(8), 613–625.
Campbell, J.I.D. (2008). Subtraction by addition. Memory & Cognition, 36(6), 1094–1102.
Campbell, J.I.D., & Alberts, N.M. (2009). Operation-specific effects of numerical surface form on arithmetic strategy. Journal of Experimental Psychology and Learning Memory and Cognition 35(4), 999–1011.
Campbell, J.I.D., & Timm, J.C. (2000). Adults' strategy choices for simple addition: effects of retrieval interference. Psychonomic Bulletin & Review 7(4), 692–699.
Campbell, J.I.D., & Xue, Q. (2001). Cognitiveitive arithmetic across cultures. Journal of Experimental Psychology: General 130(2), 299–315.
Cantlon, J.F., & Brannon, E.M. (2006). Shared system for ordering small and large numbers in monkeys and humans. Psychological Science 17(5), 401–406.
Cantlon, J.F., Brannon, E. M., Carter, E. J., & Pelphrey K. A. (2006). Functional imaging of numerical processing in adults and 4-y-old children. PLoS Biology 4(5), e125.
Cantlon, J.F., Libertus, M.E., Pinel, P., Dehaene, S., Brannon, E.M., & Pelphrey, K.A. (2009). The neural development of an abstract concept of number. Journal of Cognitive Neuroscience 21(11), 2217–2229.
Cappelletti, M., Butterworth, B., & Kopelman, M. (2012). Numeracy skills in patients with degenerative disorders and focal brain lesions: a neuropsychological investigation. Neuropsychology 26(1), 1–19.
Casey, B.J., Thomas, K.M., Davidson, M.C., Kunz, K., & Franzen, P.L. (2002). Dissociating striatal and hippocampal function developmentally with a stimulus-response compatibility task. Journal of Neuroscience 22(19), 8647–8652.
Caspers, S., Geyer, S., Schleicher, A., Mohlberg, H., Amunts, K., & Zilles, K. (2006). The human inferior parietal cortex: cytoarchitectonic parcellation and interindividual variability. NeuroImage 33(2), 430–448.

Caspers, S., Eickhoff, S.B., Geyer, S., Scheperjans, F., Mohlberg, H., Zilles, K., & Amunts, K. (2008). The human inferior parietal lobule in stereotaxic space. Brain Structure and Function 212(6), 481–495.
Chang, C., Crottaz-Herbette, S., & Menon, V. (2007). Temporal dynamics of basal ganglia response and connectivity during verbal working memory. NeuroImage 34(3), 1253–1269.
Cho, S., Ryali, S., Geary, D.C., & Menon, V. (2011). How does a child solve 7 + 8? Decoding brain activity patterns associated with counting and retrieval strategies. Developmental Science 14(5), 989–1001.
Cho, S., Metcalfe, A.W., Young, C.B., Ryali, S., Geary, D.C., & Menon, V. (2012). Hippocampal-prefrontal engagement and dynamic causal interactions in the maturation of children's fact retrieval. Journal of Cognitive Neuroscience 24(9), 1849–1866.
Chochon, F., Cohen, L., van de Moortele, P.F., & Dehaene, S. (1999). Differential contributions of the left and right inferior parietal lobules to number processing. Journal of Cognitive Neuroscience 11(6), 617–630.
Cohen, L., & Dehaene, S. (2004). Specialization within the ventral stream: the case for the visual word form area. NeuroImage 22(1), 466–476.
Cohen Kadosh, R., Lammertyn, J., & Izard, V. (2008). Are numbers special? An overview of chronometric, neuroimaging, developmental and comparative studies of magnitude representation. Progress in Neurobiology 84(2), 132–147.
Cohen Kadosh, R., & Walsh, V. (2009). Numerical representation in the parietal lobes: abstract or not abstract? Behavioural Brain Science 32(3–4), 313–28; discussion 28–73.
De Smedt B., Holloway, I. D., & Ansari, D. (2011). Effects of problem size and arithmetic operation on brain activation during calculation in children with varying levels of arithmetical fluency. Neuroimage 57(3), 771–781.
Dehaene, S., Piazza, M., Pinel, P., & Cohen, L. (2003). Three parietal circuits for number processing. Cognitive Neuropsychology 20(3–6), 487–506.
Delazer, M., & Benke, T. (1997). Arithmetic facts without meaning. Cortex 33(4), 697–710.
Delazer, M., Domahs, F., Bartha, L., Brenneis, C., Lochy, A., Trieb, T., & Benke, T. (2003). Learning complex arithmetic – an fMRI study. Brain Research: Cognitive Brain Research 18(1), 76–88.
Delazer, M., Ischebeck, A., Domahs, F., Zamarian, L., Koppelstaetter, F., Siedentopf, C.M., Kaufmann, L., Benke, T., & Felber, S. (2005). Learning by strategies and learning by drill – evidence from an fMRI study. NeuroImage 25(3), 838–849.
Dumontheil, I., & Klingberg, T. (2012). Brain activity during a visuospatial working memory task predicts arithmetical performance 2 years later. Cerebral Cortex 22(5), 1078–1085.
Durston, S., Davidson, M.C., Tottenham, N., Spicer, J., Galvan, A., Fossella, J.A., & Casey, B.J. (2006). A shift from diffuse to focal cortical activity with development. Dev Sci. 9(1), 1–8.
Eichenbaum, H. (2000). Hippocampus: mapping or memory? Current Biology 10, R785–R787.
Fair, D.A., Cohen, A.L., Dosenbach, N.U., Church, J.A., Miezin, F.M., Barch, D.M., Raichle, M.E., Petersen, S.E., & Schlaggar, B.L. (2008). The maturing architecture of the brain's default network. Proceedings of the National Academy of Science USA, 105(10), 4028–4032.
Feigenson, L., Dehaene, S., & Spelke, E. (2004). Core systems of number. Trends in Cognitive Sciences 8(7), 307–314.
Geary, D.C. (1994). Children's mathematical development: research and practical applications. Washington, DC: American Psychological Association.
Geary, D.C., Bailey, D.H., & Hoard, M.K. (2009a). Predicting mathematical achievement and mathematical learning disability with a simple screening tool the number sets test. Journal of Psychoeducational Assessment 27(3), 265–279.
Geary, D.C., Bow-Thomas, C.C., & Yao, Y. (1992). Counting knowledge and skill in cognitive addition: a comparison of normal and mathematically disabled children. Journal of Experimental Child Psychology 54(3), 372–391.
Geary, D.C., Hoard, M.K., Byrd-Craven, J., & DeSoto, M.C. (2004). Strategy choices in simple and complex addition: contributions of working memory and counting knowledge for children with mathematical disability. Journal of Experimental Child Psychology 88(2), 121–151.

Geary, D.C., Bailey, D.H., Littlefield, A., Wood, P., Hoard, M.K., & Nugent, L. (2009b). First-grade predictors of mathematical learning disability: a latent class trajectory analysis. Cognitive Development 24(4), 411–429.

Grabner, R.H., Ansari, D., Reishofer, G., Stern, E., Ebner, F., & Neuper, C. (2007). Individual differences in mathematical competence predict parietal brain activation during mental calculation. NeuroImage 38(2), 346–356.

Grabner, R.H., Ansari, D., Koschutnig, K., Reishofer, G., Ebner, F., & Neuper, C. (2009). To retrieve or to calculate? Left angular gyrus mediates the retrieval of arithmetic facts during problem solving. Neuropsychologia 47(2), 604–608.

Graybiel, A. M. (2005). The basal ganglia: learning new tricks and loving it. Current Opinion in Neurobiology 15(6), 638–644.

Greicius, M.D., Krasnow, B., Reiss, A.L., & Menon, V. (2003). Functional connectivity in the resting brain: a network analysis of the default mode hypothesis. Proceedings of the National Academy of Sciences of the USA 100(1), 253–258.

Gruber, O., Indefrey, P., Steinmetz, H., & Kleinschmidt, A. (2001). Dissociating neural correlates of cognitive components in mental calculation. Cerebral Cortex 11(4), 350–359.

Hart, J., Jr., Kraut, M.A., Kremen, S., Soher, B., & Gordon, B. (2000). Neural substrates of orthographic lexical access as demonstrated by functional brain imaging. Neuropsychiatry, Neuropsychology, and Behavioral Neurology 13(1), 1–7.

Hecht, S.A. (1999). Individual solution processes while solving addition and multiplication math facts in adults. Memory & Cognitive 27(6), 1097–1107.

Henik, A., Rubinsten, O., & Askenazi, S. (this volume). Developmental dyscalculia as a heterogeneous disability. In R. Cohen Kadosh, & A. Dowker, (Eds.) The Oxford Handbook of Numerical Cognition. Oxford University Press.

Henschen, S. (1920). Klinische und anatomische beitraege sur pathologie des Gehirns. Stockholm: Nordiska Bokhandeln.

Houde, O., Rossi, S., Lubin, A., & Joliot, M. (2010). Mapping numerical processing, reading, and executive functions in the developing brain: an fMRI meta-analysis of 52 studies including 842 children. Developmental Science 13(6), 876–885.

Ischebeck, A., Zamarian, L., Siedentopf, C., Koppelstatter, F., Benke, T., Felber, S., & Delazer, M. (2006). How specifically do we learn? Imaging the learning of multiplication and subtraction. NeuroImage 30(4), 1365–1375.

Ischebeck, A., Zamarian, L., Egger, K., Schocke, M., & Delazer, M. (2007). Imaging early practice effects in arithmetic. NeuroImage 36(3), 993–1003.

Jordan, N.C., Hanich, L.B., & Kaplan, D. (2003). Arithmetic fact mastery in young children: a longitudinal investigation. Journal of Experimental Child Psychology 85(2), 103–119.

Julien, C.L., Thompson, J.C., Neary, D., & Snowden, J.S. (2008). Arithmetic knowledge in semantic dementia: is it invariably preserved? Neuropsychologia 46(11), 2732–2744.

Kahn, H.J., & Whitaker, H.A. (1991). Acalculia: an historical review of localization. Brain and Cognition 17, 102–115.

Karmiloff-Smith, A. (2010). Neuroimaging of the developing brain: taking 'developing' seriously. Human Brain Mapping 31(6), 934–941.

Kaufmann, L., Wood, G., Rubinsten, O., & Henik, A. (2011). Meta-analyses of developmental fMRI studies investigating typical and atypical trajectories of number processing and calculation. Developmental Neuropsychology 36(6), 763–787.

Keller, K., & Menon, V. (2009). Gender differences in the functional and structural neuroanatomy of mathematical cognition. NeuroImage 47(1), 342–352.

Kronbichler, M., Hutzler, F., Wimmer, H., Mair, A., Staffen, W., & Ladurner, G. (2004). The visual word form area and the frequency with which words are encountered: evidence from a parametric fMRI study. NeuroImage 21(3), 946–953.

Kucian, K., von Aster, M., Loenneker, T., Dietrich, T., & Martin, E. (2008). Development of neural networks for exact and approximate calculation: a fMRI study. Developmental Neuropsychology 33(4), 447–473.

Kucian, K., Kaufmann, L., & Von Aster, M. (this volume). Brain Correlates of Numerical Disabilities. In R. Cohen Kadosh, & A. Dowker, (Eds.) The Oxford Handbook of Numerical Cognition. Oxford University Press.

LeFevre, J., Bisanz, J., Daley, K.E., Buffone, L., Greenham, S.L., & Sadesky, G.S. (1996). Multiple routes to solution of single-digit multiplication problems. Journal of Experimental Psychology: General 125(3), 23.

McCarthy, R.A., & Warrington, E.K. (1988). Evidence for modality-specific meaning systems in the brain. Nature 334(6181), 428–430.

McCloskey, M., Harley, W., & Sokol, S.M. (1991). Models of arithmetic fact retrieval: an evaluation in light of findings from normal and brain-damaged subjects. Journal of Experimental Psychology: Learning, Memory and Cognitiveition 17(3), 377–397.

McNeil, J.E., & Warrington, E.K. (1994). A dissociation between addition and subtraction with written calculation. Neuropsychologia 32(6), 717–728.

Menon, V., White, C.D., Eliez, S., Glover, G.H., & Reiss, A.L. (2000a). Analysis of a distributed neural system involved in spatial information, novelty, and memory processing. Human Brain Mapping 11(2), 117–129.

Menon, V., Rivera, S.M., White, C.D., Glover, G.H., & Reiss, A.L. (2000b). Dissociating prefrontal and parietal cortex activation during arithmetic processing. NeuroImage 12(4), 357–365.

Menon, V., Rivera, S.M., White, C.D., Eliez, S., Glover, G.H., & Reiss, A.L. (2000c). Functional optimization of arithmetic processing in perfect performers. Cognitive Brain Research 9(3), 343–345.

Menon, V., Mackenzie, K., Rivera, S.M., & Reiss, A.L. (2002). Prefrontal cortex involvement in processing incorrect arithmetic equations: evidence from event-related fMRI. Human Brain Mapping 16(2), 119–130.

Milner, A.D., & Goodale, M.A. (2008). Two visual systems re-viewed. Neuropsychologia 46(3), 774–785.

Mizuhara, H., Wang, L., Kobayashi, K., & Yamaguchi, Y. (2005). Long-range EEG phase synchronization during an arithmetic task indexes a coherent cortical network simultaneously measured by fMRI. NeuroImage 27(3), 553–563.

Niedeggen, M., Rosler, F., & Jost, K. (1999). Processing of incongruous mental calculation problems: evidence for an arithmetic N400 effect. Psychophysiology 36(3), 307–324.

Packard, M.G., & Knowlton, B.J. (2002). Learning and memory functions of the basal ganglia. Annual Review of Neuroscience 25, 563–593.

Pinel, P., Dehaene, S., Riviere, D., & LeBihan, D. (2001). Modulation of parietal activation by semantic distance in a number comparison task. NeuroImage 14(5), 1013–1026.

Prado, J., Mutreja, R., Zhang, H., Mehta, R., Desroches, A.S., Minas, J.E., & Booth, J.R. (2011). Distinct representations of subtraction and multiplication in the neural systems for numerosity and language. Human Brain Mapping 32(11), 1932–1947.

Price, C.J., & Devlin, J.T. (2004). The pro and cons of labelling a left occipitotemporal region: 'the visual word form area'. NeuroImage 22(1), 477–479.

Raichle, M.E., MacLeod, A.M., Snyder, A.Z., Powers, W.J., Gusnard, D.A., & Shulman, G.L. (2001). A default mode of brain function. Proceedings of the National Academy of Sciences, USA 98(2), 676–682.

Raizada, R.D., Tsao, F.M., Liu, H.M., Holloway, I.D., Ansari, D., & Kuhl, P.K. (2010). Linking brain-wide multivoxel activation patterns to behaviour: examples from language and math. NeuroImage 51(1), 462–471.

Rickard, T.C., Romero, S.G., Basso, G., Wharton, C., Flitman, S., & Grafman, J. (2000). The calculating brain: an fMRI study. Neuropsychologia 38(3), 325–335.

Rivera, S.M., Reiss, A.L., Eckert, M.A., and Menon, V. (2005). Developmental changes in mental arithmetic: evidence for increased functional specialization in the left inferior parietal cortex. Cerebral Cortex 15(11), 1779–1790.

Rosenberg-Lee, M., Barth, M., & Menon, V. (2011a). What difference does a year of schooling make? Maturation of brain response and connectivity between 2nd and 3rd grades during arithmetic problem solving. NeuroImage 57(3), 796–808.

Rosenberg-Lee, M., Lovett, M.C., & Anderson, J.R. (2009a). Neural correlates of arithmetic calculation strategies. Cognitive Affective Behavioural Neuroscience 9(3), 270–285.
Rosenberg-Lee, M., Tsang, J.M., & Menon, V. (2009b). Symbolic, numeric, and magnitude representations in the parietal cortex. Behavioral and Brain Sciences 32(3–4), 350–351.
Rosenberg-Lee, M., Chang, T.T., Young, C.B., Wu, S., & Menon, V. (2011b). Functional dissociations between four basic arithmetic operations in the human posterior parietal cortex: a cytoarchitectonic mapping study. Neuropsychologia 49(9), 2592–2608.
Rykhlevskaia, E., Uddin, L.Q., Kondos, L., and Menon, V. (2009). Neuroanatomical correlates of developmental dyscalculia: combined evidence from morphometry and tractography. Frontiers in Human Neuroscience 3, 51.
Salimpoor, V.N., Chang, C., & Menon, V. (2009). Neural basis of repetition priming during mathematical cognitiveition: repetition suppression or repetition enhancement? Journal Cognitive Neuroscience 22(4), 790–805.
Salillas, E., & Semenza, C. (this volume). Mapping the Brain for Math: Reversible inactivation by direct cortical electrostimulation and transcranial magnetic stimulation. In R. Cohen Kadosh, & A. Dowker, (Eds.), The Oxford Handbook of Numerical Cognition. Oxford University Press.
Siegler, R.S. (1996). Emerging minds: the process of change in children's thinking. New York: Oxford University Press.
Siegler, R.S., & Shrager, J. (1984). Strategy choices in addition and subtraction: how do children know what to do? In C. Sophian (Ed.), The Origins of Cognitive Skills (pp. 229–293). Hillsdale: Erlbaum.
Siegler, R.S., & Svetina, M. (2006). What leads children to adopt new strategies? A microgenetic/cross-sectional study of class inclusion. Child Develpoment 77(4), 997–1015.
Simon, O., Mangin, J.F., Cohen, L., Le Bihan, D., & Dehaene, S. (2002). Topographical layout of hand, eye, calculation, and language-related areas in the human parietal lobe. Neuron 33(3), 475–487.
Squire, L.R., Stark, C.E., & Clark, R.E. (2004). The medial temporal lobe. Annual Review of Neuroscience 27, 279–306.
Supekar, K., & Menon, V. (2012). Developmental maturation of dynamic causal control signals in higher-order cognition: a neurocognitive network model. PLoS Computational Biology 8(2), e1002374.
Supekar, K., Uddin, L.Q., Prater, K., Amin, H., Greicius, M.D., & Menon, V. (2010). Development of functional and structural connectivity within the default mode network in young children. NeuroImage 52(1), 290–301.
Suzuki, W.A., & Amaral, D.G. (1994). Perirhinal and parahippocampal cortices of the macaque monkey: cortical afferents. Journal of Comparative Neurology 350(4), 497–533.
Takayama, Y., Sugishita, M., Akiguchi, I., & Kimura, J. (1994). Isolated acalculia due to left parietal lesion. Archives of Neurology 51(3), 286–291.
Uddin, L.Q., Supekar, K., Amin, H., Rykhlevskaia, E., Nguyen, D.A., Greicius, M.D., & Menon, V. (2010). Dissociable connectivity within human angular gyrus and intraparietal sulcus: evidence from functional and structural connectivity. Cerebral Cortex 11, 2636–2646.
van Harskamp, N.J., & Cipolotti, L. (2001). Selective impairments for addition, subtraction and multiplication. Implications for the organisation of arithmetical facts. Cortex 37(3), 363–388.
Venkatraman, V., Ansari, D., & Chee, M.W. (2005). Neural correlates of symbolic and non-symbolic arithmetic. Neuropsychologia 43(5), 744–753.
Warrington, E.K. (1982). The fractionation of arithmetical skills: a single case study. Quarterly Journal of Experimental Psychology 34, 31–51.
Wu, S., Chang, T.T., Majid, A., Caspers, S., Eickhoff, S.B., & Menon, V. (2009). Functional heterogeneity of inferior parietal cortex during mathematical cognition assessed with cytoarchitectonic probability maps. Cerebral Cortex 19(12), 2930–2945.
Yarkoni, T., Poldrack, R.A., Nichols, T.E., Van Essen, D.C., & Wager, T.D. (2011). Large-scale automated synthesis of human functional neuroimaging data. Nat Methods, 8 (8), 665–670.
Zago, L., Pesenti, M., Mellet, E., Crivello, F., Mazoyer, B., & Tzourio-Mazoyer, N. (2001). Neural correlates of simple and complex mental calculation. NeuroImage 13(2), 314–327.

Zago, L., Petit, L., Turbelin, M.R., Andersson, F., Vigneau, M., & Tzourio-Mazoyer, N. (2008). How verbal and spatial manipulation networks contribute to calculation: an fMRI study. Neuropsychologia 46(9), 2403–2414.

Zamarian, L., Ischebeck, A., & Delazer, M. (2009). Neuroscience of learning arithmetic – evidence from brain imaging studies. Neuroscience Biobehaviour Review 33(6), 909–925.

Zamarian, L., & Delazer, M. (this volume). Arithmetic learning in adults–Evidence from brain imaging. In R. Cohen Kadosh, & A. Dowker, (Eds.), The Oxford Handbook of Numerical Cognition. Oxford University Press.

Zhou, Xinlin, Chen, C., Zang, Y., Dong, Q., Chen, C., Qiao, S., & Gong, Q. (2007). Dissociated brain organization for single-digit addition and multiplication. NeuroImage 35(2), 871–880.

CHAPTER 29

NUMERICAL SYMBOLS
an overview of their cognitive and neural underpinnings

IAN D. HOLLOWAY AND DANIEL ANSARI

INTRODUCTION

Introduction and Plan of Chapter

FROM quotidian uses, such as deciding how much food to prepare, to more rarefied applications, like landing a rover on Mars, numbers can be found in almost every activity across every human culture on Earth. Empirical research has repeatedly demonstrated that human numerical competencies are built upon an evolutionarily-preserved ability to understand approximate numerical magnitude, such as which of two trees contains more apples. Human cultures have invented ways of representing numerical magnitude as abstract symbols, such as spoken number words and written numerical notations, which has allowed us to extend our abilities far beyond the simple perception of less and more.

In the following chapter, we will outline what is currently known about the representation and use of numerical symbols. This narrative is divided into three sections. The first section following the introduction provides an overview of the nature of numerical symbols and the variety of types of information that these symbols can be used to represent. The second section characterizes current research into the cognitive and neural underpinnings of numerical symbols and their use. The third and final section summarizes the corpus of open questions that can and should be addressed by future research.

Definitions

Before beginning this discussion, we offer the following definitions:

(1) In the following chapter, we distinguish between two uses of the word 'representation'. '***Mental representations***' refer to the semantic knowledge of a given quantity.

Mental representations exist within the mind of individuals and can only be shared between individuals through the use of '***expressed representations***'. Although expressed representations of number can occur in any sensory modality, they are most often either auditory or visual representations.

(2) For the sake of variety, we use the terms ***number***, ***quantity***, and ***numerical magnitude*** interchangeably. Each of these words refers to the semantic meaning (mental representation) of number.

(3) We use the phrases '***numerical symbol***' and '***symbolic number***' to refer to an expressed representation of number that is associated with a mental representation of number in a purely abstract manner. That is, the sound of the word 'seven' and the shape of the written numeral '4' give no direct information about the quantity that they represent. Rather, the association between the numerical symbol and its quantitative referent is conventional and must be learned by members of the society. The information represented by a symbol is called its '***referent***.' Numerical symbols are most often associated with semantic referents (numerical meaning) and auditory referents (name of the number).

(4) We use the term '***non-symbolic***' to describe expressed representations of quantities that encode the quantity in a direct, 1-to-1 fashion. A visual array of 23 triangles or a series of 19 tones are both examples of non-symbolic representations of quantity.

(5) ***Numerals*** are a specialized class of numerical symbol. They can be distinguished from number words. The most common form of numerals are the Hindu-Arabic numerals that are used in most cultures of the world, but other writing systems have their own forms of numerals.

Numerals and their Referents

In Figure 29.1, nine symbols are presented. To anyone unfamiliar with these symbols, it is impossible to know much about them. One could perhaps guess that they come from an East Asian script. However, to anyone who reads contemporary Thai, two pieces of information expressed by these symbols would be obvious. Each symbol has a name, an auditory referent that is associated with the visual form. Secondly, these symbols express representations of numerical magnitude and, therefore, are associated with a specific quantity. This information (the name and the meaning of the symbols) is culturally constructed and transmitted, both formally and informally, through various cultural forces.

While each example of numerical symbols is bound to a particular culture, it is possible to generalize these characteristics to all numerical symbols. Understanding the general nature of numerical symbols requires an understanding of the nature of the information expressed

๑ ๒ ๓ ๔ ๕ ๖ ๗ ๘ ๙

FIGURE 29.1 Numerals 1–9 in the Thai script.

by these symbols. Accordingly, the following sections detail what is known of the semantic and auditory processing of numerals.

2.1 Semantic Referents of Numerals

2.1.1 *Background*

While the association between a numeral and its semantic referent requires some type of formal or informal training, the understanding of the semantic referent (numerical magnitude) has its roots in a phylogenetically continuous capacity for using numerical information found in the environment. Evidence of sensitivity to numerical magnitude has been reported in many varieties of animal species (Brannon 2006; Nieder and Dehaene 2009), human infants (Antell and Keating 1983; Feigenson et al. 2002; Libertus and Brannon 2009; Lipton and Spelke 2004; Wynn 1992; Xu and Spelke 2000; Xu et al. 2004), and in human cultures who use extremely few or no number words (Butterworth et al. 2008; Gordon 2004; Pica et al. 2004). Taken together, these studies provide strong evidence that the capacity to perceive numerical magnitude is a predisposition of the human mind adapted for by evolution rather than something that needs to be instructed.

One of the central questions in the study of numerical symbol processing, then, asks about the nature of the relationship between numerical symbols and the biologically-mediated mental representation of number described above. Clues about this relationship can be garnered from the psychophysical study of numerical processing. In one psychophysical task called the numerical comparison task, human participants are asked to choose which of two simultaneously presented quantities is numerically larger. Data collected from literate adults show that numerical decisions are slower and more inaccurate as numerical ratio between the to-be-compared numerals increases (Buckley and Gillman 1974; Dehaene et al. 1999; Moyer and Landauer 1967; Sekuler and Mierkiewicz 1977). In other words, the symbolic representation of numerical magnitude suffers from imprecision. For example, Moyer and Landauer (1967) demonstrated that when adults compared which of two simultaneously presented numerals was numerically larger, the reaction time and error rate of the judgments decreased as numerical distance (or difference) increased. This pattern, dubbed the distance effect, demonstrates that the semantic representations underlying the numerical symbols are imprecise in nature. If, instead, the representations of the numerals were digital and precise, there would be no reason why the comparison of 7 vs. 9 should take significantly longer or be more inaccurate than the comparison of 3 vs. 9, although for another viewpoint see Link (1990). Considering this evidence, it seems clear that the semantic referents of numerical symbols are built upon the approximate representations of numerical magnitude seen in non-human animals and illiterate humans.

While the distance effect in the comparison of symbolic numbers points to a coupling between the symbolic external representations of number and the imprecise internal representation of numerical magnitude, it must still be acknowledged that numerical symbols allow for the discrimination of quantities far greater than those that can be discriminated non-symbolically. This implies that when a symbol is associated with the inherently imprecise mental representations of numerical magnitude, the 'symbolization' of the number adds a degree of precision to the representation. Thus, when the meaning of numerical symbols is learned, the discrete nature of the symbols interacts with the imprecision of the referents with which they are associated. This interaction results in a representation of numerical symbols that is neither purely digital nor as imprecise as the corresponding non-symbolic representation of numerical magnitude. Against the background of these types of considerations, a growing number of studies has begun to elucidate the nature of the relationship between numerical symbols and their quantitative referents (Holloway and Ansari 2009; Lyons et al. 2012; Maloney et al. 2010; Piazza et al. 2007; Verguts and Fias 2004). These studies hinge on an underlying assumption that comparing the representation of symbolic number with the representation of non-symbolic number, one can elucidate characteristics of quantitative representation that are specific to numerical symbols. In other words, one can see how the representation of numerical information is changed through the process of learning numerical symbols by looking at the differences between symbolic and non-symbolic numerical magnitude representation. This logic assumes that both types of external representations of quantity (symbolic and non-symbolic) reference a common internal representation of numerical magnitude. Although this assumption has not been (and likely cannot be) verified, the logic of Ockham's Razor (all things being equal, the simplest explanation should be taken as true) dictates that a single internal representation is more likely than multiple internal representations. Moreover, human beings have been using numerical symbols for only about 5000 years, which is not enough evolutionary time for the adaptation of a distinct system of representation for symbolic numerical magnitude (Dehaene and Cohen 2007). For these reasons, and because the debate regarding single vs. multiple representations has been reviewed and discussed quite thoroughly elsewhere (Cohen and Walsh 2009) we, for the purposes of this discussion, eschew debating the precise nature of the underlying representation(s) of quantity. Instead, we focus on summarizing important differences between the symbolic and non-symbolic representations of number as these differences provide some important clues as to what is unique about the representation of numerical symbols.

Symbolic Vs. Non-Symbolic Representations

In a seminal paper, Verguts and Fias (2004) used a neural network to demonstrate one possibility of how small quantities (1–5) could be represented. The neural network was trained first to 'represent' non-symbolic quantities. In this case, 'represent' is used to describe the process by which every possible input of a given quantity, say five, yields the same internal state in the network. Furthermore, the internal state for five objects would be distinguishable from the internal state for any other number of objects. This way of understanding representation is analogous to how humans abstract the quantity 'five' from any set of five objects, while simultaneously understanding that five is different from any other quantity of objects. After being trained to represent non-symbolic quantity, the model learned to associate symbolic quantities with the previously learned non-symbolic quantities. The resulting model of numerical representation proved plausible as it accounted for the distance effect. Moreover, the model showed an interesting, format-dependent difference in the representation of symbolic and non-symbolic quantities that has guided the generation of hypotheses for the study of numerical symbol processing.

In the first phase of their modeling, Verguts and Fias (2004) hypothesized that the neural network would require three layers to successfully learn to represent numerical quantity: an input layer, an output layer, and a hidden layer. As seen in Figure 29.2, the input layer is called the location field and corresponds to the location of objects, i.e. each node corresponds to a

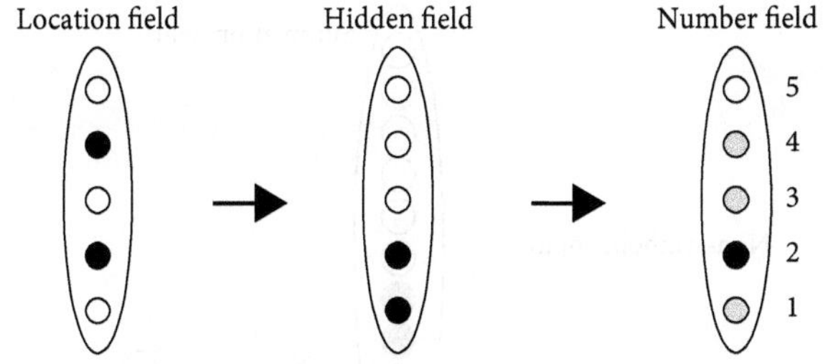

FIGURE 29.2 The three fields in the first phase of the experiment conducted by Verguts and Fias (2004).

Adapted from Tom Verguts and Wim Fias, Representation of Number in Animals and Humans: A Neural Model, *Journal of Cognitive Neuroscience*, 16(9), pp. 1493–1504, doi: 10.1162/0898929042568497 Copyright © 2004, Massachusetts Institute of Technology.

different location in the 'environment'. The output layer is called the number field and corresponds to the representation of number learned by the neural network. Each node in this layer corresponds to a specific number that is place coded, i.e. has a specific place on the number line. The hidden layer, called the hidden field, corresponds to the process by which the location field is transformed into the number field. This layer is summation coded, meaning that each node corresponds to the presence of an additional object in the location field, regardless of the location of that object. In the example presented in Figure 29.2, two objects are encoded in the location field. This causes two nodes in the hidden field to become activated, which, in turn, activates the node corresponding to the quantity '2' in the number field. Interestingly, the presence of two objects also partially activated (as illustrated by grey rather than black coloring) the nodes for quantities '1, 3, and 4'. One can extrapolate this to account for the distance effect as numbers that are closer to 2 show more partial activation than numbers which are further from 2. The presence of this partial activation would make the discrimination of 2 from 3 more challenging than the discrimination of 2 from 5.

In the second phase of their experiment, the authors asked whether the neural network could learn to associate discrete and arbitrary numerical symbols with numerical information. This question was operationalized as depicted in Figure 29.3. The summation field is similar to the hidden field in Figure 29.2, but each quantity is encoded with the activation of two, rather than one, nodes. This resulted in ten nodes in the summation field. Each node in the symbolic field represented a discrete numeral; the coding of this field was arbitrary, i.e., the nodes were not ordered in numerical sequence. The nodes in the number field, which was place coded, were activated through a linear combination of the summation field and the symbolic field. The results demonstrated that the model could learn to link an arbitrary input with numerical information. In addition, as seen in the number field, the resulting activation was less noisy. In other words, when two objects were encoded, the node for 2 in the number field was most strongly activated. Just like in the first phase, the nodes surrounded 2 were also partially activated. However, in this case, the amount of partial activation was noticeably attenuated. Thus, the representation of the number field was sharpened (less spread out) by the association with a discrete symbolic input.

Although neural network models do not provide results that are directly applicable to human learning, the evidence implies that associating a discrete and arbitrary symbol with a given semantic referent could result in a representation that is more discrete than similar representations developed in the absence of symbolic information. As we alluded to above,

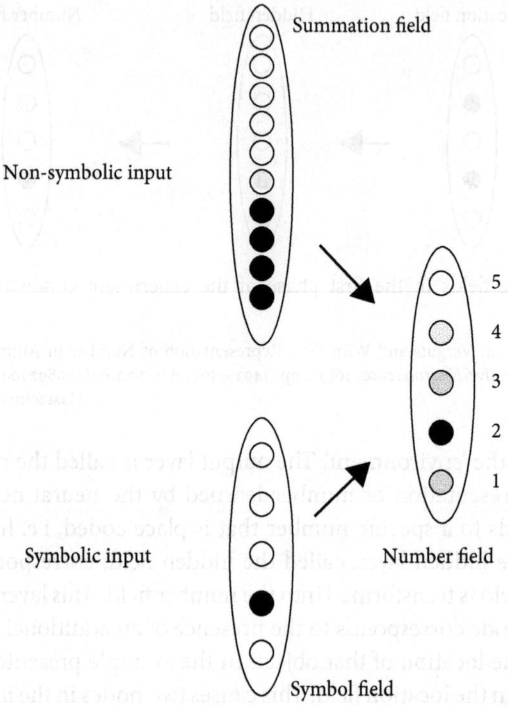

FIGURE 29.3 Illustration of the second phase of the experiment. The symbolic and non-symbolic inputs were presented simultaneously to train the neural network to associate a symbolic input with a non-symbolic representation.

Adapted from Tom Verguts and Wim Fias, Representation of Number in Animals and Humans: A Neural Model, *Journal of Cognitive Neuroscience*, 16(9), pp. 1493–1504, doi: 10.1162/0898929042568497 Copyright © 2004, Massachusetts Institute of Technology.

this idea became the point of departure for more contemporary research investigating and interpreting differences in symbolic and non-symbolic representation in humans.

For example, Holloway and Ansari (2009) tested symbolic and non-symbolic numerical comparison abilities in children aged six to nine years. They found that children's responses to both types of numerical comparison were characterized by the distance effect. The effect that numerical distance had on reaction time and accuracy decreased similarly over developmental time for both numerical formats. Similar developmental changes in the distance effect, the authors suggested, likely represent the fine-tuning of the representations underlying both symbolic and non-symbolic numerical formats. However, the authors also quantified the numerical distance effect for each format into an individual difference variable for each individual. Thus, the numerical comparison performance for each child was redescribed as a single number for each numerical format. The authors failed to find a correlation between the symbolic and non-symbolic distance effect. In other words, although the symbolic distance effect and the non-symbolic distance effect undergo similar developmental stages at the group level, these two effects do not correlate within individuals. This suggests that, insofar as the numerical distance effect can be used as an effective metric of numerical representation, the similarity of representations of symbolic and non-symbolic formats of number depends upon the level of description at which one examines them.

A similar pattern of findings was reported by Maloney and colleagues (2010), who tested the validity and reliability of the distance effect using either symbolic or non-symbolic representations of number. In addition to showing a much stronger degree of reliability in the non-symbolic distance effect relative to the symbolic distance effect, the authors also showed that the symbolic and non-symbolic distance effects are not correlated. Thus, like Holloway and Ansari (2009), these authors call into question the tight coupling of symbolic and non-symbolic representations of number. Taken together, the studies conducted by Maloney et al. (2010) and Holloway and Ansari (2009) showed that while the distance and/or ratio effects elicited during the numerical comparison of symbolic and non-symbolic representations of numerical magnitude are superficially similar, they are not correlated with each other within individuals. In other words, both the comparison of symbolic numbers and the comparison of non-symbolic magnitudes produces a distance effect (smaller numerical distance results in larger reaction time and lower accuracy). If the distance effect were indexing similar processes and/or representations, one would expect that, within individuals, a tight coupling would be evident in quantifications of the symbolic and non-symbolic distance effects. While null effects must be interpreted with care, neither Holloway and Ansari (2009) nor Maloney and colleagues (2010) found evidence of such coupling. Two possible conclusions could be drawn from these data. One possibility is that the representations of numerical magnitude are quantitatively different, which results in their uncorrelated nature. A second possibility is that differences in the symbolic and non-symbolic distance effects reflect different processes required to perform the tasks that are used to generate the effects. It is possible that these two theoretically distinct sources of variation in the distance effect could interact to create differences between the symbolic and non-symbolic distance effects.

Recent research seems to support the second possibility (Van Opstal et al. 2008; Defever et al. 2011). Van Opstal and colleagues (2008) called into question the utility of numerical comparison to create a measure of numerical representation. In brief, when two quantities are compared, the distance effect is reliably observed. However, the distance effect could, logically, depend upon either overlap in the numerical representations used to make the judgments, on other, non-numerical, factors such as response-selection, or on a combination of both (see Verguts et al. 2005 for a related model of numerical comparison). Van Opstal et al. (2008), proposed that a different task, called the numerical priming task, should be used in order to measure the distance effect, as this task is not subject to the same limitations as numerical comparison. Against this background, Defevre and colleagues (2011) measured the priming distance effect of kindergarteners, first-graders, second-graders, and sixth-graders in response to both symbolic and non-symbolic numerical stimuli. These authors showed that the priming distance effect was present in all four age groups in response to both symbolic and non-symbolic stimuli. Moreover, in contrast to the findings reported by Holloway and Ansari (2009), the authors showed that the priming distance effect shows no significant developmental changes. Furthermore, both the priming distance effect elicited by symbolic stimuli and the priming distance effect elicited by non-symbolic stimuli showed correlations with mathematics achievement. Future studies will be required to further elucidate how the distance effect elicited by different tasks is related to the underlying representation of numerical magnitude.

Thus far, the discussion has focused on comparing symbolic and non-symbolic representations of numerical magnitude as represented by the distance effect. A small, but growing number of studies have begun to analyze the behavioral correlates when symbolic and non-symbolic quantities are mapped onto one another. For example, Mundy and Gilmore (2009) demonstrated that children as young as six years of age can successfully map symbols onto dot arrays and vice versa. Moreover, children found it easier to map a numerical symbol onto a dot array than to map a dot array onto a symbol. This pattern of unequal performance in symbol-to-array vs. array-to-symbol mapping extends into adulthood (Crollen et al. 2010) and suggests that symbolic and non-symbolic representations are not mapped onto one another in a 1:1 manner. Against this background, Lyons et al. (2012) proposed the 'symbolic estrangement hypothesis'. According to this hypothesis, the relationship between numerical symbols and the approximate representation of numerical magnitude, while initially (early in development) strong, grows weaker as individuals gain more experience and fluency with numerical symbols (Lyons et al. 2012). In their paper, Lyons and colleagues, tested this hypothesis by asking adults, in addition to comparing symbolic and non-symbolic magnitudes, to engage in mixed-format comparisons. In other words, participants were asked to compare an array of dots with an Arabic numeral or number word or vice a versa. The researchers hypothesized that if number symbols and non-symbolic numerical magnitudes are closely linked/mapped to another, then mixed comparisons should be at least as efficient as symbol-symbol and dot-dot comparisons. The results, however, revealed that mixed comparisons were by far the most difficult for participants and greater than would be expected from the symbol-symbol and dot-dot comparisons. These data therefore suggest that symbolic and non-symbolic representations of number may not be as closely mapped as is assumed by many of the studies as well as models discussed above.

In addition to this evidence from adults, a recent study with children also provides evidence that speaks against a strong association between non-symbolic and symbolic representations of numerical magnitude. Specifically Sasanguie et al. (2013) examined the extent to which kindergartners' non-symbolic numerical magnitude comparison abilities correlated with their performance on symbolic comparison six months later. If the representations tapped by non-symbolic comparison provide the foundations for the semantic processing of number symbols, then early non-symbolic magnitude processing skills should correlate with later symbolic abilities. However, Sasanguie and colleagues did not find a correlation between kindergartner's non-symbolic comparison performance and their symbolic number processing abilities six months later. Therefore these findings, like those of Lyons and colleagues, speak against a strong mapping between symbolic and non-symbolic numerical magnitudes and, contrary to the prediction of the symbolic estrangement hypothesis, show that such a mapping is also absent early in development.

The bottom line of this research is that the semantic representation of numerical symbols is more complicated than originally thought. Rather than a static and transparent 1:1 correspondence between numerals and their referents, there exists a nuanced interaction between the mental representation of numerical magnitude and the symbols that are used to express it. Moreover, this interaction appears to undergo developmental changes, not only in childhood (Holloway and Ansari 2008, 2009; Siegler and Booth 2004) but throughout the lifespan (Lyons et al. 2012).

In summary, the central question of how numerical symbols are grounded in their semantic referents remains unresolved. The data surveyed above each imply the existence of important differences between the symbolic and non-symbolic representation of magnitude. Although the precise nature of these differences must be clarified with future data, one intriguing possibility is that the process of mapping a symbol onto a mental representation of quantity appears to affect the underlying representation in a way that results in a higher degree of precision relative to non-symbolic expressions of number (Ansari 2008; Verguts and Fias 2004). While the exact mechanisms of this increase in representational precision remains opaque, it is likely related to other aspects of numerical symbol processing, such as verbal processing, that have, relative to semantic processing, remained largely ignored by the scientific literature. In other words, our understanding of the semantic processing of numerical symbols suffers from a tendency to view semantic processing as independent from other aspects of numerical symbols. Although it is a minority, some research has been conducted to examine the asemantic processing of numerical symbols. The following section surveys what is currently known about this topic.

Auditory Referents of Numerals

Relative to the semantic referents of numerals, very little is known about the nature of their auditory referents. The study of the cognitive and neural correlates of numerals has focused almost exclusively on the semantic level of representation. The dominant theory guiding the study of numerical cognition considers the auditory processing of numerals as a purely linguistic process (Dehaene 1992). Yet, the degree to which the auditory processing of numerals is comparable to the auditory process of other types of verbal representations has not been empirically tested.

Although the auditory referents of numerals have been largely ignored, they nonetheless play a crucial role in the acquisition of the symbolic representation of number. This fact has been demonstrated clearly by studies of the acquisition of counting. Typically developing children know the names of numbers (what becomes the auditory referents of numerals) well before they understand what those numbers mean (Le Corre and Carey 2007; Le Corre et al. 2006; Wynn 1990). Wynn (1990), for example, used the so-called 'Give-a-Number' task with children two to four years of age. In this task, children are first asked to demonstrate their ability to count, typically by counting as high as they can. Following this, children are introduced to a stuffed animal that has forgotten how to count. They are asked to give the stuffed animal a certain quantity of objects from a large bowl of those objects placed near the child. Children younger than three-and-a-half or four years of age show a peculiar pattern of performance such that they are able to recite the number sequence quite well, but are not able to utilize counting to enumerate the proper quantity of objects for the stuffed animal. For example, a child of three years might be able to count to 25, but when asked to give a stuffed toy five marbles, the child does not understand that the auditory words in the counting sequence can be used to count out five marbles. Through a process that is still poorly understood, learning to count assists young children in associating the auditory number words with a semantic meaning. In other words, an asemantic auditory representation becomes associated with the representation of numerical magnitude. As children begin formal education, children learn the written numerical symbols that become connected with both the auditory and the numerical magnitude representations. Although the auditory referents of numerical symbols could potentially play an important scaffolding role in the acquisition of number symbols, little research has been conducted to specifically study them (Ansari 2008).

When the auditory number words are linked to visual representations such as numerals, an additional level of processing emerges—one in which the auditory and visual representations are integrated into a bimodal audio-visual percept. This audio-visual level of processing is crucial in reading alphabetic languages where the phonological information of speech sounds is associated with the visual information of letters (for review see Blomert 2011). It can be assumed that a similar process of audio-visual integration is important for reading numerals. Unlike letters, however, numerals have an additional semantic content embedded in them. It, therefore, remains unclear whether comparable audio-visual processing exists in the reading of letters and numerals. Moreover, the interaction between the audio-visual and semantic representations of numerals has, to the best of our knowledge, not been studied.

In summary, number words are important for the initial acquisition of symbolic representations of numerical magnitude. Children learn number words and associate them with their pre-existing representations of numerical magnitude. In early elementary school, children then learn to read a special set of symbols (numerals) that can be used to stand in for the number words. The degree to which the verbal and audio-visual processing of numerals is similar to and divergent from the processing of words and letters remains to be determined.

In the next section, we turn to a brief overview of what is currently known about how numerical symbols are processed in the brain both in the semantic and the asemantic (auditory, visual, and audio-visual) domains.

Neural Substrates of Symbolic Number Processing

Semantic Processing of Numerals

The above discussion introduced the notion that the semantic representation of numerals is constructed from the intrinsic, approximate, non-symbolic representation of numerical magnitude. The basic evidence for this is the effect of numerical ratio on the processing of numerical symbols, which originates from the imprecise nature of numerical magnitude representations. Almost all of the neuroimaging research that has been conducted on the processing of numerical symbols has focused on testing this hypothesis. A growing number of studies have demonstrated a neural correlate that is common to the semantic processing of both symbolic numerals (Ansari et al. 2005; Cantlon et al. 2009; Chochon et al. 1999; Eger et al. 2003; Fias et al. 2003; Notebaert et al. 2011; Pesenti et al. 2000; Pinel et al. 1999, 2001) and the quantitative processing of non-symbolic stimuli, such as arrays of dots (Ansari and Dhital 2006; Ansari et al. 2006; Cantlon et al. 2006, 2009; Cappelletti et al. 2010; Fias et al. 2003; Piazza et al. 2004, 2007). This body of literature has converged on the intraparietal sulcus (IPS), which runs between the inferior and superior parietal lobes, as the brain region that houses the representation of numerical magnitude (see Figure 29.4).

The data outlined immediately above present a clear link between activation in the IPS and the processing of numerical information. However, until recently, no studies

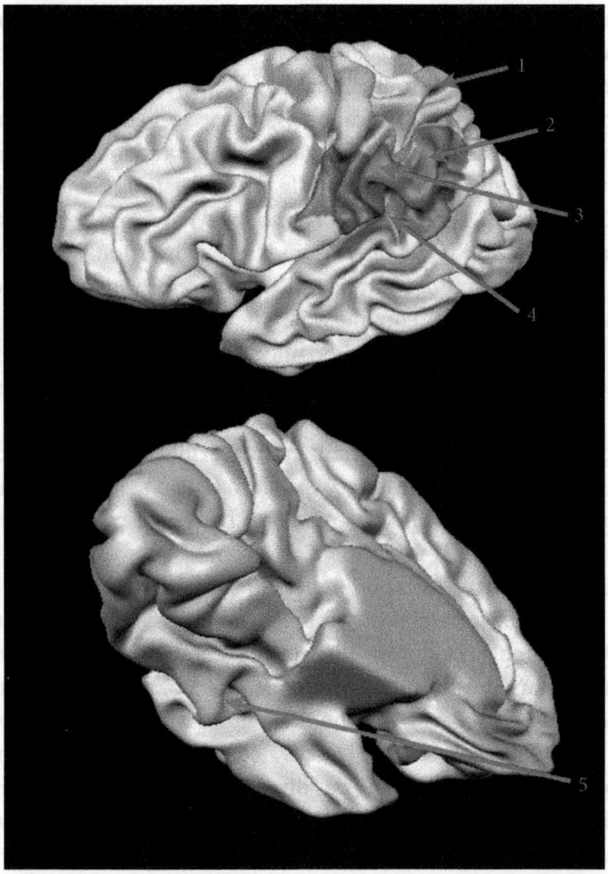

FIGURE 29.4 Brain regions commonly involved in number processing and referred to throughout this chapter: (1) superior parietal lobe Brodmann Area 7 (yellow); (2) posterior portion of inferior parietal lobe including angular gyrus Brodmann Area 39 (blue); (3) anterior portion of inferior parietal lobe including supramarginal gyrus Brodmann Area 40 (brown), (4) posterior superior temporal gyrus (green) Brodmann Area 42; Note that the intraparietal sulcus runs between the superior and inferior parietal lobes.

specifically examined the relationship between IPS activity and the processing of numerical symbols. Recent studies have begun to address this question. Methodologically, these studies have typically used one of two different paradigms to elicit the neural correlates of symbolically-presented magnitude representation. The first is the numerical comparison task, which was discussed above. In this task, participants are asked to compare two numerical magnitudes. The numerical ratio between these magnitudes is varied so that experimenters can identify brain regions whose response is modulated by numerical information. The IPS responds more strongly when numerical ratio is large (magnitudes are more similar) than when numerical ratio is small (magnitudes are less similar). This response is thought to reflect the greater reliance on the neural system of semantic representation to disambiguate two quantities whose representations are highly overlapping. In other words, the IPS must be brought online much more strongly to distinguish 32 from

42 dots relative to when it is asked to distinguish 32 from 12 dots. As aforementioned, this neural ratio effect is seen in the IPS regardless of whether the numerical comparison task uses symbolic or non-symbolic stimuli. The second experimental paradigm used to study the semantic processing of numerals is called the numerical adaptation paradigm. In such paradigms, the repeated presentation of the same quantity (adaptation stimulus) will cause the neurons that encode numerical magnitude representation to reduce in the extent of their response (Cantlon et al. 2006; Piazza et al. 2004). When one presents a novel quantity (deviant stimulus), the size of the rebound is inversely proportional to the amount of overlap between the adaptation and deviant quantities. Therefore, as the IPS is adapted to the quantity '23', it will show a larger rebound when the deviant is 11 than when the deviant is 22. This is due to the fact that the quantities 22 and 23 are highly overlapping in their respective representations and, therefore, utilize many of the same neurons to encode the quantity. Thus, when deviant 22 is presented, most of the neurons that encode it have already been adapted. Presenting 22 would not result in the recruitment of many new neurons that were not involved in the previous response to 23. Presenting 11, on the other hand, would result in a much larger rebound effect as many of the representational neurons were not previously adapted to 23.

Recently published data have used both types of paradigm to investigate similarities and/or differences in the neural correlates of symbolic and non-symbolic numerical magnitude representation. For example, Holloway et al. (2010) compared and contrasted the neural correlates of symbolic and non-symbolic numerical comparison. The authors first looked for commonalities in the correlates of symbolic and non-symbolic number processing. They showed that the right IPS was commonly activated for both types of comparison. When symbolic and non-symbolic comparison was directly contrasted, the authors demonstrated that non-symbolic comparison was associated with activation in the right superior parietal lobe—a region associated with visuospatial processing. In contrast, the symbolic comparison was associated with the left superior temporal gyrus (STG) and the left angular gyrus (AG), regions both associated with verbal and linguistic processing. Similarly to Santens and colleagues (2010), the authors argued that the data could represent distinct encoding pathways that converge upon a common semantic representation housed in the IPS (but also see Cohen Kadosh, et al. 2011 for a different viewpoint regarding whether similar common activation in a given region implies a common semantic representation). Asking a slightly different question, Piazza et al. (2007) used a numerical adaptation paradigm to examine the representation of symbolic and non-symbolic numerical magnitude in the parietal lobe. These authors demonstrated that the left IPS showed evidence of greater representational precision for numerical symbols relative to non-symbolic magnitudes. The right IPS showed no differences between formats. Taken together, these two studies show compelling commensurability with the model developed by Verguts and Fias (2004). It is possible that the increased precision in the left IPS for the representation of numerical symbols reported by Piazza et al. (2007) is a mirror of the increase in precision demonstrated by Verguts and Fias (2004) in their neural network model described above. In addition, the data reported by Holloway et al. (2010) could correspond with the two distinct types of encoding (summation and place coding) described in the Verguts and Fias (2004) model.

In addition to being commensurate with the model put forth by Verguts and Fias (2004), the data reported by Piazza and colleagues (2007) also suggested that the left IPS might specialize in the representation of numerical symbols. Further evidence along this line of reasoning came from Notebaert and colleagues (2011) who used a numerical adaptation task to show that activity in the left IPS, but not the right, is correlated with the numerical ratio of both single- and double-digit Arabic numerals. However, this claim must be balanced by evidence reported by Cohen Kadosh and colleagues (2007), who used an adaptation paradigm to demonstrate that the left IPS showed ratio-dependent adaptation to both numerals and number words, whereas the right IPS showed quantity adaptation only to numerals. These data demonstrated that the left IPS might specialize in the representation of verbal representations of number, such as written words, while both the right and left IPS are involved in representing numerical symbols. The role of the left and right IPS in numerical symbol processing was further clarified by Holloway and colleagues (2012). Instead of comparing symbolic with non-symbolic processing, the authors compared two different types of numerical symbols: the Hindu-Arabic numerals used by most cultures around the world and numerical ideographs used in the Chinese languages. Using an approach similar to that taken by Notebaert et al. (2011), the authors measured the adaptation response to both types of symbols in a group of bilingual Chinese-English participants and compared that response with a group of bilingual English control participants who had no knowledge of Chinese ideographs. In response to the numerals, both groups showed activation isolated to the left IPS. In response to the ideographs, the Chinese-English group activated regions in the bilateral IPS, whilst the control group only activated areas in visual regions. In addition to providing strong corroborating evidence for the role of the IPS in semantic processing of numerical symbols, the data provided more nuance to the link between symbolic number processing and the laterality of activation in the IPS. In the Chinese-English participants, the ideographs were correlated with bilateral activation in the IPS and the numerals were correlated with only the left IPS. The authors proposed that the lateralization difference in response to the two types of numerical symbol reflected differences in the fluency with which the two symbol types are used in Chinese culture. Specifically, Hindu-Arabic numerals are used much more broadly in the culture and serve as the symbols with which mathematics is taught. Ideographs on the other hand, have very limited use. Perhaps, the authors argued, increased left lateralization in IPS activation is related to increased fluency and automaticity in a given numerical symbol. To date, this hypothesis remains untested, but other data shows convergence with the idea. Specifically, Bugden et al. (2012) correlated mathematical fluency scores from a standardized test with the distance effect in the IPS. Only the left IPS showed a significant correlation, which suggests that the extent to which this region is activated during numerical tasks is related to the fluency with which individuals process numerical symbols.

Taken together, these data are broadly commensurate with a recently proposed hypothesis suggesting specialization of the left IPS for the representation of symbolic number (Ansari 2007, 2008). Although Hindu-Arabic numerals are associated with activation in both the left and right IPS, more nuanced research has demonstrated that the left IPS houses a representation of symbolic number that is more finely tuned that its representation of non-symbolic number. Additionally, symbols that are arguably used less often and less fluently (Chinese ideographs) recruit both the left and right IPS, symbols that are used more often and more fluently (Hindu-Arabic numerals) show a greater degree of left-lateralization. This theory of left IPS specialization must be tempered with the findings of Cohen Kadosh et al. (2007), who published results that do not converge with this interpretation. Future studies will undoubtedly further clarify the role of the left IPS in the semantic representation of numerical symbols.

Auditory Processing of Numerals

In this section, we turn to a discussion of what is known about the asemantic processing of numerals in the brain, which can include processing numerals as an asemantic visual form, an asemantic auditory representation or an asemantic audio-visual percept. The overwhelming majority of research studying the neurocognitive correlates of number processing have focused on understanding the representation of numerical magnitude, whether the representations are encoded in a non-symbolic or symbolic manner. This emphasis on numerical representation has resulted in a marked lack of research investigating any aspect of the asemantic processing of numerals in the human brain. Studies that have measured the neural responses to asemantic tasks such as Hindu-Arabic number naming have used these responses as a comparison with semantic processing (Chochon et al. 1999; Pesenti et al. 2000; Zago et al. 2001). The specific neural responses to these asemantic tasks were never reported.

One recent study has examined the neural correlates involved in the asemantic visual processing of Hindu-Arabic numerals (Price and Ansari 2011). These authors presented participants with numerals, letters, and scrambled versions of the same. Relative to scrambled stimuli, a dorsal region of the left angular gyrus showed more activation for whole numerals and letters. In addition, a ventral region of the left angular gyrus showed more activation for numerals relative to letters. These results suggest that the left angular gyrus is involved in some aspect of the asemantic visual processing of numerals.

The findings by Price and Ansari (2011) converge with a previous study conducted by Pesenti et al. (2000) which suggest that, contrary to the predictions for a visual number form area in the triple code model by Dehaene and Cohen, there is no category-specific representation for numerals in the ventral stream. A recent study using a far more anatomically fine-grained methodology, however, has revealed clear evidence for a visual number form area in the ventral visual stream. Specifically, Shum et al. (2013) used intracranial electrophysiological recordings to record from areas within the ventral stream while participants viewed numerals, letters, false fonts, number words, non-number words, foreign numerals as well as scrambled letters and numerals. When contrasting the electrophysiological response to these different conditions against one another along the ventral stream, Shum et al. found a region in the inferior temporal gyrus that responded more to numerals than any other condition. These data provide clear evidence for a region in the ventral stream that responds to visual numerals in a category-specific way. Interestingly, the region revealed by electrophysiology to respond preferentially to the visual numerals is a region that lies with an fMRI signal-dropout zone associated with the auditory canal and venous sinus artifact. This finding may explain why Price and Ansari, using fMRI, were unable to find a region that responded in a category-specific way to numerals in the ventral stream.

More clues about the neural correlates of number name processing were provided in a recent doctoral thesis (Holloway 2012). In this study, the author used fMRI to examine the asemantic visual, auditory, and audio-visual processing of eight numerals (1–9, excluding 7). These correlates were compared with those of the visual, auditory, and audio-visual processing of eight English letters (b, h, j, k, l, p, r, v). Neural responses to both the names of the letters and the sounds of the letters were recorded in separate conditions. Participants passively attended to two unimodal conditions in which either numerals were presented visually or number words were presented aurally. As hypothesized, these conditions were correlated with activity in the primary visual and auditory cortices, respectively. The analysis failed to reveal any differences between the neural correlates of visual processing of numerals and letters. However, the angular gyrus showed greater activation in response to auditory letters than auditory number names. In addition to the unimodal conditions, two bimodal conditions were also recorded. Bimodal (audio-visual) congruent trials were ones in which the visual information was presented simultaneously with matching auditory information. For example, the numeral 8 was presented on the screen while the word 'eight' was presented over headphones. Bimodal incongruent trials differed from the congruent trials solely in that the auditory and the visual information did not match, e.g. 4 was presented with 'nine'. Collapsed across congruency, audio-visual numerals showed more activity in the right supramarginal gyrus, relative to audio-visual letters. In contrast, relative to numeral-number name pairs, audio-visual letters showed more activation in left-lateralized inferior junction of the occipital and temporal lobes including the left fusiform gyrus. The key analysis in this study was the comparison of congruent and incongruent conditions of the audio-visual numerals. As Goebel and van Atteveldt (2009) recently argued, one can draw an important distinction between audio-visual processing in general, which occurs whenever any visual and auditory information is simultaneously present, and audio-visual integration, which occurs in response to an over-learned and automatized pairing of a specific visual and auditory stimuli. Put in the context of this chapter, the neural correlates associated with the linkage between numerals and their number names can be illuminated by identifying brain regions whose activity to congruent pairings of numerals and number names is significantly different from incongruent pairings. This analysis revealed bilateral regions of the superior temporal gyrus and the left supramarginal gyrus were more active during the presentation of congruent numeral-number name pairs than during the presentation of incongruent numeral-number name pairs.

In summary, the results of the studies by Holloway (2012) and Price and Ansari (2011) provide some of the first evidence of the brain regions whose activity is correlated with the asemantic processing of number. Price and Ansari (2011) demonstrated that a portion of the left angular gyrus showed some specialization for the processing of numerals relative to letters. In a similar comparison of letters and numerals, Holloway did not find a similar effect in the left angular gyrus. This difference in results is likely due to the fact that Holloway's study was not optimized for testing the differences between the visual processing of numerals and letters, but rather designed to canvass the brain for the audio-visual correlates of the audio-visual processing of numerals. Holloway (2012) revealed that regions of the superior temporal gyrus appear to be central to processing the learned associations between the visual form of numerals and their auditory referents. The regions correlated with the presentation of audio-visual number pairs show strong convergence with recent results describing the correlates of audio-visual letter processing (Blomert and Froyen 2010; van Atteveldt et al. 2004). Taken together, the results of these two studies provide data that can further nuance our understanding of the verbal, i.e. audio-visual processing of numerals. In the seminal 'Triple Code Model', Dehaene (1992) reviewed evidence from lesion patients and proposed that numerals can be processed asemantically using solely their verbal characteristics. Dehaene and his colleagues later proposed that the primary neural circuit involved in the verbal processing of number would be found in the left angular gyrus (Dehaene et al. 2003). The two studies reviewed above tell a more complicated story. While the left angular gyrus does, indeed, show some specialization for the asemantic processing of the visual form of numerals (relative to letters), the audio-visual processing appears to be highly localized to portions of the superior temporal gyrus. As future studies are conducted, the precise roles of these two regions of the brain in the asemantic visual, auditory, and audio-visual processing of numerals will become better understood.

General Summary

In the previous two sections, we have described what is currently understood about the neurocognitive correlates of numerical symbol representation and processing. Put broadly, numerical symbols can be processed in either a semantic or an asemantic manner that is task dependent. Accordingly, the neural correlates of semantic numerical symbol processing can be distinguished from those of asemantic symbol processing. One way to consolidate the information presented above is to imagine a map. The map defined by the current state of the science is relatively coarse. The major landmarks (brain regions) have been plotted. What is missing is finer detail about each region. Moreover, the roads connecting these regions have yet to be discovered. In the final section, which immediately follows this one, we outline some of the most important questions that future researchers can address to help fill in the details of the map of numerical symbol processing.

Open Questions

The past twenty years of brain imaging research has painted an unprecedented portrait of the functions and complexity of the human brain. This research has provided data that help to constrain theories of cognitive and emotional processing in a wide range of topics, including numerical processing. As we reviewed above, the fundamental neural correlates of the semantic, and more recently asemantic, processing of numerical symbols have been identified. Yet several key areas of inquiry remain largely unexplored. Below, we propose some ideas for future research into the cognitive and neural underpinnings of numerals and other numerical symbols.

One of the most important topics to address is the interaction between the semantic and asemantic levels of processing of numerical symbols. Although a substantial amount of information has been provided about these two levels, almost nothing is known about what regions in the brain connect the asemantic knowledge of numerals with the evolutionarily-conserved mental representations of numerical magnitude and exactly how they are connected. While current theory and computational models (e.g. Verguts and Fias 2004) suggest a strong connection between symbolic and non-symbolic representations of numerical magnitude, many of the empirical data reviewed above suggest that the way in which symbols are grounded and linked with non-symbolic representations of numerical magnitude is far from clear.

This topic is particularly important when trying to understand the developmental trajectory of numerical knowledge. As discussed above, preschool children typically know number words and can recite the counting sequence before they understand that the words are associated with a specific cardinal quantity (Le Corre and Carey 2007; Wagner and

Johnson 2011). Future research could investigate what experiences help children develop this understanding—an understanding shown to be important for future mathematical achievement (Halberda et al. 2008; Holloway and Ansari 2009). Moreover, researchers interested in this topic could conduct neuroimaging studies on younger children to gain insight into what changes in brain structure and function may link numerals to their meanings. Another option is the use of cross-linguistic neuroimaging, such as comparing individuals who understand numerical symbols in a culturally-specific notation (such as the Thai numerals in the introduction) with individuals who do not understand the symbols (Holloway et al. 2013).

One of the most likely bridges between the semantic and asemantic processing of numerals is ordinality. Although not discussed specifically in this chapter, previous research has demonstrated a connection between the cardinal understanding of numerals and the successor function, which refers to the fact that each progressive number in the counting list implies an increase of one in cardinal value (Sarnecka and Carey 2008). Moreover, other researchers have shown that the linkages between the mental representation of numerical magnitude and mathematical achievement are mediated by ordinality knowledge (Lyons and Beilock 2011). Previous neuroimaging studies of ordinality processing have revealed regions that are highly overlapping with those correlated with the semantic representation of number (Zorzi et al. 2011). This implies a variety of possibilities, which could be clarified by future research. For example, do populations of neurons that encode order and numerical magnitude happen to be intermixed in the same regions of the neocortex? Or, does the ordered way in which humans learn number change the innate representations of numerical magnitude? Here again, developmental neuroimaging could be used to help clarify how the brain adapts a general sense of less and more into a specific ordered sequence.

Another important and unresolved issue is the interaction between numerical symbol processing and processes that underpin reading ability. For example, although some work has already been conducted on the topic, the precise interaction between phonological processing and the processing of numerical symbols in the form of mathematical facts is currently unknown (Simmons and Singleton 2008). Moreover, the left angular gyrus appears to be involved in both reading processes (Pinel and Dehaene 2010; Price 2002) and the processing of numerical symbols (Price and Ansari 2011) and mathematical facts (Grabner et al. 2009). Similar to the questions that could be asked of numerical order, future research could attempt to clarify whether this overlap between numerical symbol processing and reading is due to biological happenstance, learning, or a fundamental similarity in the processing of both types of stimuli. As neuroimaging methodologies, particularly MRI, continue to be improved, these and other questions will become increasingly tractable (Mamin et al. 2013; Staudacher et al. 2013).

On a broader level, other interesting questions could be asked of numerical symbols. One such question is how the processing of numerical symbols changes as the referents of the symbols becomes increasingly more abstract. For example, the numeral '5' is relatively concrete; there are many instances in an average life in which an individual may interact with five items. The same could not be said of '555' or even '55'. Although we understand what these numerals mean on an abstract level, concrete examples of precisely 777 items are rare. Moreover, some numerical symbols, such as pi or Euler's number are irrational and, therefore, impossible to represent in a concrete way. Another interesting question emerges when one broadens the definition of a numeral's semantic representation. Certainly, the most common semantic representation of a numeral is quantitative in nature. But there are many examples in modern life in which numerals are used in a nominal manner, such as the numerals on a sports jersey or the numbers used to differentiate different types of produce in the grocery store. How are these types of numerical symbols processed differently from those with quantitative referents? Should they even be grouped into the same category? More importantly, how does the exposure to different types of numeral-referent pairs affect the way our understanding of number is constructed?

In summary, the invention of numerical notation can be considered one of the key intellectual accomplishments of humanity. Such notation served as the fundament for innumerable mathematical and scientific discoveries. Despite their importance in every aspect of human life, modern cognitive science and cognitive neuroscience are just beginning to describe how our understanding of numerical symbols develops and is manifested in the brain. Both behavioral and neuroimaging evidence suggests that numerical symbols rely on the interaction of various processes, including but not limited to visual, auditory, audio-visual, ordinal, and semantic. While the specific processes and their correlated brain regions have become relatively well understood in isolation from each other, the ways in which these processes interact to support the human understanding of numerical symbols remains largely unknown. Investigating the nature of these interactions on both the behavioral and the neural levels will undoubtedly provide many important and interesting insights into this uniquely human capacity.

References

Ansari, D. (2007). Does the parietal cortex distinguish between '10,' 'ten,' and ten dots? *Neuron* 53(2): 165–7. doi:10.1016/j.neuron.2007.01.001.

Ansari, D. (2008). Effects of development and enculturation on number representation in the brain. *Nature Reviews Neuroscience* 9(4): 278–91. doi:10.1038/nrn2334.

Ansari, D., and Dhital, B. (2006). Age-related changes in the activation of the intraparietal sulcus during nonsymbolic magnitude processing: an event-related functional magnetic resonance imaging study. *Journal of Cognitive Neuroscience* 18(11): 1820–8.

Ansari, D., Garcia, N., Lucas, E., Hamon, K., and Dhital, B. (2005). Neural correlates of symbolic number processing in children and adults. *Neuroreport* 16(16): 1769–73.

Ansari, D., Dhital, B., and Siong, S.C. (2006). Parametric effects of numerical distance on the intraparietal sulcus during passive viewing of rapid numerosity changes. *Brain Research*, 1067(1): 181–8. doi:10.1016/j.brainres.2005.10.083.

Antell, S.E. and Keating, D.P. (1983). Perception of numerical invariance in neonates. *Child Development* 695–701.

Blomert, L. (2011). The neural signature of orthographic–phonological binding in successful and failing reading development. *Neuroimage* 57(3): 695–703. doi:10.1016/j.neuroimage.2010.11.003.

Blomert, L. and Froyen, D. (2010). Multi-sensory learning and learning to read. *International Journal of Psychophysiology*. 77(3): 195–204. doi:10.1016/j.ijpsycho.2010.06.025.

Brannon, E.M. (2006). The representation of numerical magnitude. *Current Opinion in Neurobiology* 16(2): 222–9. doi:10.1016/j.conb.2006.03.002.

Buckley, P.B. and Gillman, C.B. (1974). Comparisons of digits and dot patterns. *Journal of Experimental Psychology* 103(6): 1131.

Bugden, S., Price, G.R., McLean, D.A., and Ansari, D. (2012). The role of the left intraparietal sulcus in the relationship between symbolic number processing and children's arithmetic competence. *Developmental Cognitive Neuroscience* 2(4): 448–57. doi:10.1016/j.dcn.2012.04.001.

Butterworth, B., Reeve, R., Reynolds, F., and Lloyd, D. (2008). Numerical thought with and without words: Evidence from indigenous Australian children. *Proceedings of the National Academy of Sciences* 105(35): 13179–84.

Cantlon, J.F., Brannon, E.M., Carter, E.J., and Pelphrey, K.A. (2006). Functional imaging of numerical processing in adults and 4-year-old children. *PLoS Biology* 4(5): e125. doi:10.1371/journal.pbio.0040125.

Cantlon, J.F., Libertus, M.E., Pinel, P., Dehaene, S., Brannon, E.M., and Pelphrey, K.A. (2009). The neural development of an abstract concept of number. *Journal of Cognitive Neuroscience* 21(11): 2217–29.

Cappelletti, M., Lee, H.L., Freeman, E.D., and Price, C.J. (2010). The role of right and left parietal lobes in the conceptual processing of numbers. *Journal of Cognitive Neuroscience* 22(2): 331–46.

Chochon, F., Cohen, L., Moortele, P.F., and Dehaene, S. (1999). Differential contributions of the left and right inferior parietal lobules to number processing. *Journal of Cognitive Neuroscience* 11(6): 617–30.

Cohen Kadosh, R. and Walsh, V. (2009). Numerical representation in the parietal lobes: Abstract or not abstract? *Behavioral and Brain Sciences* 32(3): 313.

Cohen-Kadosh, R., Bahrami, B., Walsh, V., Butterworth, B., Popescu, T. & Price, C.J. (2011). Specialization in the human brain: the case of numbers. Frontiers in Human Neuroscience, 5: 62.

Cohen Kadosh, R., Cohen Kadosh, K., Kaas, A., Henik, A., and Goebel, R. (2007). Notation-dependent and-independent representations of numbers in the parietal lobes. *Neuron* 53(2): 307–14. doi:10.1016/j.neuron.2006.12.025.

Crollen, V., Castronovo, J., and Seron, X. (2010). Under- and Over-Estimation. *Experimental Psychology* (formerly *Zeitschrift für Experimentelle Psychologie*) 58(1): 39–49. doi:10.1027/1618-3169/a000064.

Defever, E., Sasanguie, D., Gebuis, T., and Reynvoet, B. (2011). Children's representation of symbolic and nonsymbolic magnitude examined with the priming paradigm. *Journal of Experimental Child Psychology* 109(2): 174–86. doi:10.1016/j.jecp.2011.01.002.

Dehaene, S. (1992). Varieties of numerical abilities. *Cognition* 44(1): 1–42.

Dehaene, S. & Cohen, L. (2007). Cultural recycling of cortical maps. *Neuron* 56(2): 384–98.

Dehaene, S., Spelke, E., Pinel, P., Stanescu, R., and Tsivkin, S. (1999). Sources of mathematical thinking: Behavioral and brain-imaging evidence. *Science* 284(5416): 970–4.

Dehaene, S., Piazza, M., Pinel, P., and Cohen, L. (2003). Three parietal circuits for number processing. *Cognitive Neuropsychology* 20(3–6): 487–506. doi:10.1080/02643290244000239.

Eger, E., Sterzer, P., Russ, M.O., Giraud, A.L., and Kleinschmidt, A. (2003). A supramodal number representation in human intraparietal cortex. *Neuron* 37(4): 719–26.

Feigenson, L., Carey, S., and Spelke, E. (2002). Infants' discrimination of number vs. continuous extent. *Cognitive Psychology* 44(1): 33–66. doi:10.1006/cogp.2001.0760.

Fias, W., Lammertyn, J., Reynvoet, B., Dupont, P., and Orban, G. A. (2003). Parietal representation of symbolic and nonsymbolic magnitude. *Journal of Cognitive Neuroscience* 15(1): 47–56.

Goebel, R. and van Atteveldt, N. (2009). Multisensory functional magnetic resonance imaging: a future perspective. *Experimental Brain Research* 198(2): 153–64. doi:10.1007/s00221-009-1881-7.

Gordon, P. (2004). Numerical Cognition Without Words: Evidence from Amazonia. *Science* 306(5695): 496–9. doi:10.1126/science.1094492.

Grabner, R.H., Ansari, D., Koschutnig, K., Reishofer, G., Ebner, F., and Neuper, C. (2009). To retrieve or to calculate? Left angular gyrus mediates the retrieval of arithmetic facts during problem solving. *Neuropsychologia* 47(2): 604–8. doi:10.1016/j.neuropsychologia.2008.10.013.

Halberda, J., Mazzocco, M.M.M., and Feigenson, L. (2008). Individual differences in non-verbal number acuity correlate with maths achievement. *Nature* 455(7213): 665–8. doi:10.1038/nature07246.

Holloway, I.D. (2012, July 19). Symbolizing Number: fMRI investigations of the semantic, auditory, and visual correlates of Hindu-Arabic numerals, edited byD. Ansari. London: University of Western Ontario.

Holloway, I.D. and Ansari, D. (2008). Domain—specific and domain—general changes in children's development of number comparison. *Developmental Science* 11(5): 644–9. doi:10.1111/j.1467-7687.2008.00712.x.

Holloway, I.D. and Ansari, D. (2009). Mapping numerical magnitudes onto symbols: The numerical distance effect and individual differences in children's mathematics achievement. *Journal of Experimental Child Psychology*. 103(1): 17-29. doi:10.1016/j.jecp.2008.04.001.

Holloway, I.D., Battista, C., Vogel, S.E., and Ansari, D. (2013). Semantic and perceptual processing of number symbols: Evidence from a cross-linguistic fMRI adaptation study. *Journal of Cognitive Neuroscience* 25(3): 388–400.

Holloway, I.D., Price, G.R. & Ansari, D. (2010). Common and segregated neural pathways for the processing of symbolic and nonsymbolic numerical magnitude: an fMRI study. *Neuroimage*, 48(1): 1006-17.

Le Corre, M. and Carey, S. (2007). One, two, three, four, nothing more: An investigation of the conceptual sources of the verbal counting principles. *Cognition* 105(2): 395–438. doi:10.1016/j.cognition.2006.10.005.

Le Corre, M., Van de Walle, G., Brannon, E.M., and Carey, S. (2006). Re-visiting the competence/performance debate in the acquisition of the counting principles. *Cognitive Psychology* 52(2): 130–69. doi:10.1016/j.cogpsych.2005.07.002.

Libertus, M.E. and Brannon, E.M. (2009). Behavioral and neural basis of number sense in infancy. *Current Directions in Psychological Science* 18(6): 346–51. doi:10.1111/j.1467-8721.2009.01665.x.

Lipton, J.S. and Spelke, E.S. (2004). Discrimination of large and small numerosities by human infants. *Infancy* 5(3): 271–90.

Lyons, I.M. and Beilock, S.L. (2011). Numerical ordering ability mediates the relation between number-sense and arithmetic competence. *Cognition* 121(2): 256–61. doi:10.1016/j.cognition.2011.07.009.

Lyons, I.M., Ansari, D., and Beilock, S.L. (2012). Symbolic Estrangement: Evidence Against a Strong Association Between Numerical Symbols and the Quantities They Represent. Journal of Experimental Psychology: General, 141(4): 635–41.

Link, S. (1990). Modeling imageless thought: The relative judgment theory of numerical comparisons. *Journal of Mathematical Psychology* 34(1): 2–41.

Maloney, E.A., Risko, E.F., Preston, F., Ansari, D., and Fugelsang, J. (2010). Challenging the reliability and validity of cognitive measures: The case of the numerical distance effect. *Acta Psychologica* 134(2): 154–61. doi:10.1016/j.actpsy.2010.01.006.

Mamin, H.J., Kim, M., Sherwood, M.H., Rettner, C.T., Ohno, K., Awschalom, D.D., and Rugar, D. (2013). Nanoscale Nuclear Magnetic Resonance with a Nitrogen-Vacancy Spin Sensor. *Science* 339(6119): 557–60. doi:10.1126/science.1231540.

Moyer, R.S. and Landauer, T.K. (1967). Time required for judgements of numerical inequality, 215(5109): 1519-20.

Mundy, E. and Gilmore, C.K. (2009). Children's mapping between symbolic and nonsymbolic representations of number. *Journal of Experimental Child Psychology* 103(4): 490–502. doi:10.1016/j.jecp.2009.02.003.

Nieder, A. and Dehaene, S. (2009). Representation of number in the brain. *Annual Review of Neuroscience* 32: 185–208. doi:10.1146/annurev.neuro.051508.135550.

Notebaert, K., Nelis, S., and Reynvoet, B. (2011). The magnitude representation of small and large symbolic numbers in the left and right hemisphere: an event-related fMRI study. *Journal of Cognitive Neuroscience* 23(3): 622–30.

Pesenti, M., Thioux, M., Seron, X., and Volder, A.D. (2000). Neuroanatomical substrates of Arabic number processing, numerical comparison, and simple addition: A PET study. *Journal of Cognitive Neuroscience* 12(3): 461–79.

Piazza, M., Izard, V., Pinel, P., Le Bihan, D., and Dehaene, S. (2004). Tuning curves for approximate numerosity in the human intraparietal sulcus. *Neuron* 44(3): 547–55.

Piazza, M., Pinel, P., Le Bihan, D., and Dehaene, S. (2007). A magnitude code common to numerosities and number symbols in human intraparietal cortex. *Neuron* 53(2): 293–305. doi:10.1016/j.neuron.2006.11.022.

Pica, P., Lemer, C., Izard, V., and Dehaene, S. (2004). Exact and approximate arithmetic in an Amazonian indigene group. *Science* 306(5695): 499–503.

Pinel, P. and Dehaene, S. (2010). Beyond hemispheric dominance: brain regions underlying the joint lateralization of language and arithmetic to the left hemisphere. *Journal of Cognitive Neuroscience* 22(1): 48–66.

Pinel, P., Dehaene, S., Rivière, D., and LeBihan, D. (2001). Modulation of parietal activation by semantic distance in a number comparison task. *Neuroimage* 14(5): 1013–26. doi:10.1006/nimg.2001.0913.

Pinel, P., Le Clech, G., van de Moortele, P. F., Naccache, L., Le Bihan, D., and Dehaene, S. (1999). Event-related fMRI analysis of the cerebral circuit for number comparison. *Neuroreport* 10(7): 1473–9.

Price, C.J. (2002). The anatomy of language: contributions from functional neuroimaging. *Journal of Anatomy* 197(3): 335–59.

Price, G.R. and Ansari, D. (2011). Symbol processing in the left angular gyrus: evidence from passive perception of digits. *Neuroimage* 57(3): 1205–11. doi:10.1016/j.neuroimage.2011.05.035.

Santens, S., Roggeman, C., Fias, W., and Verguts, T. (2010). Number Processing Pathways in Human Parietal Cortex. *Cerebral Cortex* 20(1): 77–88. doi:10.1093/cercor/bhp080.

Sarnecka, B.W., and Carey, S. (2008). How counting represents number: What children must learn and when they learn it. *Cognition* 108(3): 662–74. doi:10.1016/j.cognition.2008.05.007.

Sasanguie, D., Defever, E., Maertens, B., and Reynvoet, B. (2013). The approximate number system is not predictive for symbolic number processing in kindergartners. *Quarterly Journal of Experimental Psychology*, 67(2), 271–80.

Sekuler, R. and Mierkiewicz, D. (1977). Children's judgments of numerical inequality. *Child Development*, 48(2): 630 3.

Shum, J., Hermes, D., Foster, B.L., Dastjerdi, M., Rangarajan, V., Vinawer, J., et al. (2013). A brain area for visual numerals. *Journal of Neuroscience* 17: 6709–15.

Siegler, R.S. and Booth, J.L. (2004). Development of numerical estimation in young children. *Child Development* 75(2): 428–44.

Simmons, F.R. and Singleton, C. (2008). Do weak phonological representations impact on arithmetic development? A review of research into arithmetic and dyslexia. *Dyslexia* 14(2): 77–94. doi:10.1002/dys.341.

Staudacher, T., Shi, F., Pezzagna, S., Meijer, J., Du, J., Meriles, C.A., et al. (2013). Nuclear Magnetic Resonance Spectroscopy on a (5-Nanometer)3 Sample Volume. *Science* 339(6119): 561–3. doi:10.1126/science.1231675.

van Atteveldt, N., Formisano, E., Goebel, R., and Blomert, L. (2004). Integration of letters and speech sounds in the human brain. *Neuron* 43(2): 271–82.

van Opstal, F., Gevers, W., De Moor, W., and Verguts, T. (2008). Dissecting the symbolic distance effect: Comparison and priming effects in numerical and nonnumerical orders. *Psychonomic Bulletin & Review* 15(2): 419–25. doi:10.3758/PBR.15.2.419.

Verguts, T. and Fias, W. (2004). Representation of number in animals and humans: a neural model. *Journal of Cognitive Neuroscience* 16(9): 1493–504.

Verguts, T., Fias, W. & Stevens, M. (2005). A model of exact small-number representation. *Psychonomic Bulletin and Review*, 12(10): 66-80.

Wagner, J.B. and Johnson, S.C. (2011). An association between understanding cardinality and analog magnitude representations in preschoolers. *Cognition* 119(1): 10–22. doi:10.1016/j.cognition.2010.11.014.

Wynn, K. (1990). Children's understanding of counting. *Cognition* 36(2): 155–93.

Wynn, K. (1992). Addition and subtraction by human infants. *Nature* 358(6389): 749–50.

Xu, F. and Spelke, E.S. (2000). Large number discrimination in 6-month-old infants. *Cognition* 74(1): B1–11.

Xu, F., Spelke, E.S., and Goddard, S. (2004). Number sense in human infants. *Developmental Science* 8(1): 88–101.

Zago, L., Pesenti, M., Mellet, E., Crivello, F., Mazoyer, B., and Tzourio-Mazoyer, N. (2001). Neural Correlates of Simple and Complex Mental Calculation. *Neuroimage* 13(2): 314–27. doi:10.1006/nimg.2000.0697.

Zorzi, M., Bono, D., Grazia, M., and Fias, W. (2011). Distinct representations of numerical and non-numerical order in the human intraparietal sulcus revealed by multivariate pattern recognition. *Neuroimage* 56(2): 674–80. doi:10.1016/j.neuroimage.2010.06.035.

CHAPTER 30

A THEORY OF MAGNITUDE

the parts that sum to number

VINCENT WALSH

INTRODUCTION

IN cognitive neuroscience, there is a debate, actually the same debate for over a century, about whether functions are localized in the brain. Irrespective of what the positions in the debate might be, scientists work in conceptually modular frameworks. We have literatures on language, vision (color, form, motion, faces, eye movements), learning, emotion, attention, movement, time, space, number, reading, memory, sleep, and many other fractionated components of cognition. There are several assumptions behind this approach. The main one, undoubtedly incorrect, is that we have already parsed the major functional divisions of mental life. This is a dangerous and consequential position. The consequences are easiest to see in the field of attention and how it interfaces with other modules. Here, there is agreement that *something* (though no one seems to know precisely what) is important. It is a concept with many degrees of freedom, often small effects, and is closely associated with parietal cortex function. In studies of faces, number, color, time, space, vision, audition, etc., the assumption is that the attentional component is directed by the parietal cortex. This is a constant and deep assumption. So much so that when interpreting imaging data or modeling vision, it is assumed that the parietal cortex holds a position from which it exerts "top down control" over sensory areas. If we persist with this mode of operation, we will continue to address the literature, rather than nature, and construct a very detailed picture of how the brain operates under peculiar circumstances that have little or nothing to do with integrated brain operation (the operation of a brain that does not know it has to conform to chapter headings in a textbook) or the brain in the real world (the parietal cortex seems to be less important in attentional tasks when, as in the real world, subjects are familiar with target and distractors; Walsh et al., 1998).

Against this background of dissatisfaction with modularity and, in particular, lazy thinking regarding attention, A Theory of Magnitude (ATOM) (Walsh, 2003a) was proposed to address the problem of what a smart thing like number is doing in a region of the human cortex associated with automatic and motoric processing of which we are seldom aware, while all the other smart things we do and are aware of (such as language, faces, episodic memory, and object recognition) are associated with regions of the temporal lobe. In doing so, ATOM had to consider what other functions were associated with the intraparietal sulcus and surrounding cortex and I noted that, although the historical accidents of psychology and physiology had segregated time, space, and number into entirely separate functions and domains of study, the brain itself had taken no notice of psychology textbooks and single unit recording papers. It was clear that these three functions were, at the very least, related, quite possibly anatomically overlapping, and may even share the same basic spatiotemporal metrics upon which numerical understanding is built.

ATOM is, in part, an attempt to present a challenge to simplistic views of modularity in human perception and cognition. It is also an exercise in thinking "what could one say about parietal cortex without reaching for the crutch of attention when things get complex?" Our discipline is replete with examples of modularization that underpin and severely limit our conceptualization of the language of the brain. Sometimes we use the terminology as shorthand, and in my own work, for example, I sometimes write of "the visual motion area". Others write about "the face area", "the color area", "the body area", "the locus of attention", "the location of episodic memory", "the number area", "time neurons", "space neurons", or "number neurons". At best, these formulations are useful shorthand, but they are also over-simplified and, at worst, misleading or even hideaways of wilful ignorance.

By embracing these designations of modularity, we close the door to new discoveries and new ways of thinking. Indeed, we sometimes undergo "retrothink" in order to make interpretations fit what we know to be true. The clearest example of this is in the attention and brain imaging literature, wherein activations in the parietal or prefrontal cortex are often considered de facto evidence of attentional processing even in the absence of attention being manipulated in the experiment. It is difficult to develop a new idea based on the premise that the old one *must* be right, and this is perhaps why the field of attention, rather than advancing conceptually, seems to be in a phase of saying the same thing and changing only the name of the technique and the modality being studied.

What underlies ATOM is that whenever we look hard at a problem, simplistic views of modularity break down, and to understand the development of relative specialization, and our ability to display modular-like behavior, we need to understand the ontogeny of the apparent modules and our experience of categories. If we consider early vision, the textbooks still teach a relative segregation of color, motion, and form – and it is useful as shorthand. But it does not survive inspection, even at the level of the retina or V1. Although we emphasize modularity, many cells display double duty or even triple duty responses. Higher up the system, if one considers area V4, the so-called color area (as strong a candidate for a module as any visual area), even here we see responses to color, wavelength, orientation, contrast. and even motion (Ferrera & Maunsell, 2005).

One of the reasons we cling to simple modularity is that it is difficult to study too many things – mastering the psychophysics of color is hard enough – and disciplines therefore emerge as psychologists and physiologists specialize[1]. It is also hard to examine a neuron for all its possible response properties and single unit studies usually have a target stimulus in which they are interested. These are all fair reasons to limit experimentation, but they do not change the fact that when we examine sensory cortex, we usually do so from a perspective based on the somewhat accidental sub-disciplines of psychology and physiology and on the exigency of getting any data at all.

What is true for the sensory cortex is true for the association cortex too. It is no more than an historical accident that time, space, and number have been studied for a century as three separate subjects, and the difficulties of examining responses of cells in the parietal lobe for more than one stimulus group are just as great as in the sensory cortices (and one can always wave a hand at attentional modulation, so why work hard?).

A Recap: What ATOM did and did Not Predict

The contention of ATOM is that to understand why the parietal cortex is organized as it is, and to understand why number-related processes occur in it, we need to consider numerical processing and development in the context of sensorimotor integration and action. The basis of this contention is that the parietal cortex surrounding the intraparietal sulcus is a major hub for these functions (see Bueti & Walsh, 2009). The principles of cortical organization seem to suggest that nearby areas will not differ markedly in their functions, and even when they do, they seem not to differ markedly in their organization or the way they perform their functions (Shamma, 2001; Sur & Leamey, 2001). I think this is a better starting point than trying to shoehorn the anomaly of number in the parietal cortex into an existing framework simply because it is the framework we have. For example, it has been suggested that because we spatially attend to number, that number therefore needs to be represented in or close to the areas associated with spatial attention. However, one also spatially attends to faces, cars, text, sound, memories, color, and a whole host of other objects and attributes that are not primarily represented in the same areas. Indeed, if everything we attended to spatially had to be represented in the parietal cortex, it would be difficult to imagine what the occipital and temporal lobes might be for.

ATOM made several specific predictions. Based on some transcranial magnetic stimulation (TMS) studies, behavioral data, and reinterpretations of imaging and single-unit studies, it was suggested that our experience of time, space, number, weight, and other prothetic[2] magnitudes originate from a single metric early in child development, and that we learn about these dimensions through motor interaction with the environment: the basis of ATOM is prelinguistic. This is a strength, because it constrains the idea; and a weakness, because an understanding of magnitudes that does not take into account language and metaphor is somewhat limited.

The development of magnitude processing proceeds by interactions with the environment and is, therefore, closely linked with motor reaching, grasping, and manipulating objects. It was further suggested that the emergence of our ability to manipulate discrete quantities evolved from our abilities with continuous quantities. Among the predictions made were that different magnitudes should show interference and priming effects (e.g., Cohen Kadosh et al., 2008; Droit-Volet, 2010; Herrera et al., 2008; Vicario, 2011; Xuan et al., 2007; see also Bueti and Walsh, 2009); that other brain areas associated with magnitude processing (such as V5 for motion processing) should also display some evidence of involvement in other magnitudes (in the case of V5, time; see Bueti et al., 2008a,b); and that the SNARC effect, in which small number judgments are associated with response codes in left space and large numbers with response codes in right space, should prove to be a SQUARC effect, in which any spatially or action-coded magnitude will yield a relationship between magnitude and space (cf. Notebaert et al., 2006; Ishihara et al., 2008). To date, all of these predictions have been confirmed.

There are several other interference studies relevant to the common cortical processing of magnitudes: number can influence spatial orientation (Fischer et al., 2003; Salillas et al. 2009); and temporal judgments are susceptible to spatial–numerical association of response code (SNARC)-like effects, consistent with a generalized spatial quantity association of response code (SQUARC) effect predicted in Walsh (2003a,b) (see Brugger, 2008; Ishihara et al., 2008; and Muller & Schwarz, 2008.)

An oversimplistic view of a generalized magnitude system might expect all interference effects to be symmetrical – that temporal cues, number, space, luminance, and action cues would all impinge on each other equally (an impression ATOM seems to have given unintentionally and one for which the picture is complicated by development – see the following). This is clearly not the case. Brown (1997), for example, found that number interfered with time but not vice versa, and Dormal & Pesenti (2007), for example, found that in a modified Stroop paradigm, spatial cues interfered with number processing but number did not interfere with spatial processing. Hurewitz et al. (2006) suggested a possible hierarchy of magnitudes from continuous to discrete variables following their finding that an amount of "stuff" interfered with numerosity judgments more than numerosity interfered with "stuff". Whether these findings are evidence of constant asymmetries or are task- dependent remains to be established (cf. Göbel et al., 2006). From the point of view of cortical loci, it is clear that some activation sites for time, space, and number overlap, and a few do not. This should not be surprising: the architecture activated in any given experiment is highly dependent on the task and one should, therefore, not expect a single locus to account for all instances of magnitude processing (Cohen Kadosh & Walsh, 2009).

There are some things that ATOM did not predict. The two most important misconceptions are (1) that ATOM proposes that all prothetic dimensions are created equal: they are not; and (2) that prothetic magnitudes will always interfere with each other. During development, the dimensions will not require equal effort to master, they will be associated with different levels of awareness and will receive different amounts and types of feedback. Different magnitudes will also be differently affected by the onset of language. If one considers time versus size, for example, we learn about timing relatively implicitly by interacting with moving objects or playing expectation games such as peek-a-boo. But size is emphasized in language, in games (big Ted, little Ted), and is made explicit when dealing with food, toys, brothers and sisters, etc. It should be no surprise that time is often the weaker stimulus in interference experiments. Similarly, number is learned explicitly through language, and we need to consider this in experiments. Where stimuli can be coded linguistically, number will be likely to dominate other magnitudes. So, ATOM is consistent with asymmetrical interactions. Interactions between magnitudes are a function of the history of the dimensions and the tasks being used to test interactions.

Number and Action

One of ATOM's counter-intuitive predictions is that numerical information should influence action. A number–action link is not as well established as the number–space link, of which there are many examples. Fias et al. (2001) carried out a series of experiments in which subjects performed a judgment on a stimulus attribute that was more (orientation) or less (color and shape) associated with parietal cortex processes. The stimuli were presented along with digits that were irrelevant to the task. The orientation judgments, but not the color or shape judgments, were influenced by the presence of an irrelevant number. A less intuitive interference study, but an important one in the context of our suggestion that discrete number evolved on the back of an analog quantity system necessary for computing the metrics of action, by Andres et al. (2004), required subjects to perform either a grip opening or closing movement to digit stimuli. Closure was initiated more quickly for small digits and opening more quickly for large digits. Andres et al. (2008) later established that as the hand neared the object, the interaction between digit magnitude and grip aperture decreased, and therefore concluded that magnitude influences action at the planning or programing stage of grasping movements (see also, Badets et al., 2007; Badets & Pesenti, 2010; Ishihara et al., 2006). Both Lindemann et al. (2007) and Moretto & di Pellegrino (2008) found evidence that mere exposure to magnitude information automatically primes grasping actions. Subjects were presented with numerical stimuli to which they made grip responses according to the semantic (parity) or surface (color) properties of the stimulus. Although the value of the digit was irrelevant, lower numerical values facilitated precision grip responses (associated with grasping smaller objects) and larger numerical values facilitated power grip responses (associated with grasping larger objects). Lindemann et al. (2007) additionally found that larger numbers were associated with a larger initial power grip.

These studies show that magnitude information influences the selection of action type, but at least one study (Fischer & Miller, 2008) suggested that the influence of magnitude information does not extend to the dynamics of action, such as force (see also, Taylor-Cooke et al., 2006). One other study places the origin of interactions between space and number earlier in the chain of processing. Stoianov et al. (2008) conducted a spatial–numerical priming experiment in which they assessed forward and backward priming with verbal responses. They observed greater effects when the spatial prime followed a number target both for number comparisons and parity judgments, and concluded that the effects could not be ascribed to spatial–numerical response codes. It is a hypothesis and experiment that deserves further exploration.

Space and Time

One of the predictions of ATOM was that time perception should change as a function of the distance of the events being judged. Spatial judgments are affected as a function of being made in "near space" or "far space" (e.g., Bjoertomt et al., 2002). Another way of conceptualizing near and far space is as being in or outside of "action space". Action space is within arms' reach, but if we use tools, such a broom, we can bring objects from far space into our action space. If magnitude systems originate in the need to compute space, time, and size for action, they should behave differently towards stimuli that are within or outside action space because we have learned about them in different ways. Zach & Brugger (2008) tested this by requiring subjects to make duration estimates of clock movement imagined at two distances. Subjects reported time to run faster for the near clock than for the far clock. There is a possibility, however, that this experiment tested the relationship between size and time rather than distance and time.

Development of Magnitude Processing

One prediction from ATOM is that there should be some monotonic mapping of quantities: bigger, faster, brighter, further in one domain should correlate with bigger, faster, brighter, further in another. This kind of intuitive "more A–more B" mapping has been noted in developmental contexts and described in some detail by Stavy & Tirosh (2000) who give many examples of such mapping and reinterpreted several classical findings from developmental psychology (also cf. Kaufmann & Nuerk, 2006; Rousselle & Noel, 2007, 2008). Stavy and Tirosh suggested that children will often base magnitude judgments on irrelevant dimensions. In one of their studies demonstrating this, children were shown two trains running along a track. The children were given all the information necessary to know that the trains ran at the same rate. When the trains differed in size, however, the subjects stated that the larger train was faster. In this case, size is affecting a judgment (speed) in which time is implicit, but children make the same class of error when making explicit temporal judgments. Levin (1977, 1979, 1982) asked children in kindergarten to judge which of two lights was presented for the longest time. The lights differed in brightness and size, and the children consistently judged the larger or brighter stimuli to have persisted for more time.

A more pressing issue is how magnitudes develop pre-linguistically, and this is best left until I have discussed ATOM in the context of metaphorical theories of time and space.

ATOM and Metaphorical Theories

The territory of ATOM was well defined:

> In the context of this article, then, the 'important decision variables' are short 'action-time' durations in the millisecond-to-seconds range, spatial information used for action, and co-ordinate transformations for action or predictions about the immediate sensorimotor consequences of action.
>
> Walsh, 2003a, p. 483

So, there is no necessary prediction that ATOM should extend to episodic memory, planning, higher mathematical operations, or allocentric spatial tasks such as navigation. The proposal is that we learn about space and time through action and that associations between space, time, and magnitudes relevant for action (such as size, speed, and, under some conditions, luminance and contrast) will be made through action. When we later learn about number, the neurons with capacity to represent quantity are those that have information about the continuous variables learned about motorically. Thus, the neuronal scaling mechanisms used for dimensions with action-relevant magnitude information will be co-opted in development for the scaling of number. Psychophysically, this has been shown to be the case (Burr & Ross, 2008); the relevant neurons are found, as predicted, in the parietal cortex and dual-task experiments show interference between number and action (as previously discussed). The single-unit recording literature emphasizes neurons specialized for time (cf. Walsh, 2003a,b), space, or number, but there are good reasons to doubt the simplistic modularity proposed. Similar overemphasis of specializations in the visual system led to concentrating on such things as "the color area", "the form area", or "the motion area", but it is now clear that many neurons at every level of visual analysis are double or even triple duty for form, motion, and chromatic content. Given the similarity of rules observed by sensory processing within and between modalities (Shamma, 2001; Sur & Leamey, 2001), we predict that similar multi-duty neurons will be reported in the magnitude scaling system in due course.

However, one line of work has made me consider whether ATOM is itself too modular and narrow. Work on metaphorical representations of time and quantity offers challenges and extensions (Boroditsky, 2000; Casasanto, 2008; Casasanto & Boroditsky, 2008). The issue of asymmetry of magnitudes was addressed by Casasanto et al., (2010). They investigated the relationship between space and time in children by showing them movies of animals traveling for different distances or different times. The children were between four and ten years old and they were asked to judge whether one of two animals stopped in the same place, whether one traveled further than the other, whether the animals stopped at the same time, and whether one animal moved for more time. In all conditions, there was some effect of interactions between time and space judgments but temporal judgments were disproportionately affected by space information (and space relatively lightly disrupted by temporal information). These results were not caused by the space task being easier than the time tasks. So, in line with Brown (1997), Dormal & Pesenti (2007), and Hurewitz et al. (2006), we see that all magnitudes are not created equal.

There is a challenge to be met here. I do not think there is any conflict between metaphorical theories and ATOM: they represent different levels of analysis, ATOM being prelinguistic and motoric, metaphorical theories addressing linguistic and conceptual representation. The challenge is how we get from one level to the next. So, my question for the Casasanto et al. (2010) experiment would be "is there a way of testing these potential asymmetries in prelinguistic children?" Laurenco et al., (2010) trained nine-month-old infants to associate patterns with magnitudes of time, size, and numerosity. The preference for patterns associated with larger/smaller amounts of time, size, and numerosity transferred bidirectionally between magnitudes, and the transfer was symmetrical. This is an important study because it establishes prelinguistic associations between magnitudes and it also, together with the work of Casasanto et al. (2010), sets the trajectory any explanation has to follow: from prelinguistic and symmetrical, to linguistic and asymmetrical. Another study also emphasizes that the division is not between ATOM and metaphor but between prelinguistic and linguistic processes. Merrit et al. (2010) tested for interference between space and time and found that adults showed the asymmetry of space affecting time more than vice versa, while rhesus monkeys, like Laurenco et al's children, showed symmetrical interference. There are many open questions. When does the asymmetry begin? Is it perhaps best to track the decline (if that is the right word) of time? Does it become the "weaker" dimension because, unlike space and size, it is abstract and not made explicit in every interaction? What is the frequency of time and space words used with children? Can the asymmetries be reversed? How does the acquisition of explicit quantities affect the (a)symmetry of magnitudes?

The work of Laurenco et al. and Casasanto et al. provides a developmental line between infancy and adulthood. The ATOM hypothesis provides a way of testing the links directly, to describe where our experience of magnitudes in the world shifts from equality to, perhaps, hierarchy. One goal of ATOM was to ask "why is the parietal cortex organized as it is?" These studies generate the question "how does the parietal cortex change over development?"

ATOM, Modularity, and Eight Problems to Solve

One challenge I have is that of finding a framework within which we can describe the developmental trajectories of sensory development, describe interactions between the modalities in the adult brain, and describe the experience and expression of those modalities in language and higher cognition. This sentence could refer to magnitude processing or it could refer to sensory processing – both fields present us with the challenge of a general explanation. We can tweak away for ever and generate another thousand manipulations, because a parameter can always be modified, but we will not sneak up on new conceptual approaches by adding more and more data points in a cognitive vacuum. I think it is worth considering whether magnitude processing helps to identify common principles and, if they do, how these principles can guide us.

The first problem is that of modularity. ATOM delivers a fatal blow to simple views of modularity. We never see "pure" color or "pure" motion. They, and other attributes, exist for us only in the context of other attributes. Similarly, we never experience "pure" time (whatever that might be), space, or quantity. There is always a context, a number of something, time since or time to, and a space in which we act. It is true that some brain areas show specialization for color more than motion, motion more than color, etc., but as I noted in the introduction, these are descriptions of the limits of single-unit recording as much as they are descriptions of the brain. The brain would have three possible strategies for integrating these relative specializations: (i) the putative specializations might not be as great as we imagine and, notwithstanding the sampling bias of single units and the analysis bias of brain imaging experiments, there may be many more double or even triple duty cells in so- called specialized areas; (ii) the relatively specialized areas integrate by direct cortico–cortical connections; (iii) the relatively specialized areas are under the control of a master integration area, usually the parietal cortex.

The second problem is that of symmetry/directionality. In synaesthesia, as a comparison, the experience is mostly unidirectional (stimulus A triggers experience X), but as Cohen Kadosh and others have shown, it can be bidirectional (Cohen Kadosh et al., 2005). In the magnitude domain, time is usually the poor relation of the magnitudes, at least after language acquisition. In both cases, more effort is needed to establish the conditions of directionality and also the development of directionality. The answer may well address principles of development that extend beyond magnitudes or synaesthesia.

The third problem is that of mapping/integration. How much space is "7"? How much time is "30 centimeters"? If magnitudes share a common metric, then we need to know how that metric is developed and how it changes from task to task. One could waive these questions away and argue that they are random associations that stick in memory, but the cortex does not map randomly. Even when the visual cortex reorganizes in the blind, the new map preserves the topography of the attribute being remapped (Kupers et al., 2006). As Cohen Kadosh et al. (2007) have shown, the mappings of size and luminance are not random. I think it would even be worth stretching the idea to ask whether the rules for attribute mapping in synaesthesia are the same cortical rules as used for magnitude matching à la ATOM and remapping following acquired blindness (see the sixth problem).

The fourth problem is that of the shift from prelinguistic to linguistic associations. As Laurenco et al. (2010) and Casasanto et al.'s (2010) findings suggest, the co-registration of attributes or magnitudes may differ as a function of language. One prediction, then, would be that prelinguistic patterns will be preserved for some functions (rapid motor responses, for example) and that these might prevail in the kinds of experiments published by Andres (2004, 2008) and Lindemann et al. (2007).

The fifth problem is that of sensory vs. metaphorical processing. If I can use synaesthesia as a comparison again, one of the reasons that synaesthesia has had such a hard time gaining its credibility badges is that it disturbs our habit, in perceptual science, of asking subjects for "yes/no", "same/different" answers within a single modality: to have to consider people's impressions of "a kind of blue that is in front of the object" is a leap for many (I know it was for me). It may be that every sensation triggers a number of other sensations. The real question might be, not why do synaesthetes see/hear/feel these things, but rather, how the brain manages to keep all the secondary associations out of awareness. Both synaesthesia and our magnitude estimations are subject to language at some point. The question is, how? Casasanto (2008), in his essay "Who's afraid of the big bad Whorf?" (do you not wish you had thought of that title first?), reminds us that whether we think in language, and whether language shapes thought, are two very different questions. There is no doubt about the latter: language shapes thought (Boroditsky, 2000, 2001; Boroditsky & Ramscar, 2002). I would not want to try to morph ATOM to include these influences, but I would like to know how language shapes the representations underlying thought. What happens to the columnar representations of vision, the topographic maps of sight, the somatotopy of touch, and the tonotopy of sound when language reshapes our experience of them? This may be a more interesting avenue by which to study consciousness[3] than pitting seen vs. unseen stimulus representations against each other in binocular rivalry experiments.

The sixth problem is how to think about the functions of the parietal cortex. Indeed, one of the initial motivations behind ATOM was to ask why the parietal cortex is organized as it is. The role of the parietal in magnitude processing is indisputable and I have made several predictions about such functions based on ATOM (see previous section). Interference with parietal function clearly impairs time (Alexander et al., 2005), space (Bjoertomt et al., 2002), and number (Göbel et al., 2006; see the chapter by Salillas & Semenza in this volume) functions, and the usual explanations are attentional or based on the role of the parietal cortex in sensory integration.

The seventh problem is that ATOM raises similar questions for development. When does language begin to influence our sensory experience and, of course, how? Understanding the transition between prelinguistic and linguistic thought and experience may reveal common rules. Which brings me to the most important problem.

The principles of cortical organization are often ignored in our domain-specific view of cortical specialization, but it seems to be a principle of brain mechanisms that when one solution to a problem is found, that solution is recycled over and over again. Perhaps "solution to a problem" is too active and it might be better to say that neurons have only a limited way of doing "stuff", and that what we get out of them is determined by a few strategies. The most important example for this chapter is that of cortical mapping. We have somatotopy, tonotopy, and retinotopy, and which area represents touch, sound, and vision is not due to any intrinsic properties of the somatosensory, auditory, or visual cortex. It is determined by the inputs. The visual cortex could just as well map sound or touch as the auditory and somatosensory cortex could represent vision, as we have seen in animal rewiring studies and studies of short- and long-term cortical reorganization in the blind

(Kupers et al., 2006; Shamma, 2001; Sur & Leamey, 2001). What this means is that there should be common rules underlying different modalities of perception (spatiotopic representation, relational coding (e.g. Weber's law), adaptation effects, for example). The principles of wiring between areas also seem to be universal (most connections are with the next nearest neighbor and the further one travels from the source, the smaller the number of connections, Barone et al., 2000). These cortical rules (and even metaphors need the cortex) remind us that modularity, directional symmetry, mapping and integration, differences between prelinguistic and linguistic thinking, sensory and metaphorical approaches to experience, and the functions of the parietal cortex are likely to yield to a small number of rules not only across magnitudes, but across other sensory and cognitive functions.

Conclusions, Future Directions (And A Final Warning About Your Favorite Module)

Many challenges remain, but some tough questions have already been faced. Making the link between action and metaphor (linking Laurenco to Casasanto; linking ATOM to metaphor) is perhaps one especially difficult enterprise. Sell and Kaschack. (2011) suggest some link between the two, but Srinisivan and Carey (2010) posit a break between "lower" and metaphoric processing and. in doing so. suggest a "general purpose" spatial mechanism. Oliveri et al. (2009) remind us that the parietal cortex is not everything. By interfering with cerebellar function, they selectively disrupted the mapping of time-related words for the past, in left space, and for the future, in right space. Yamakawa et al. (2009) extend the issue to social space: we have "close" friends, "distant" relatives, people who are "out of reach", people of "high" status, "low" profile, and a host of social relationships expressed in spatial metaphor. Yamakawa et al. reported overlapping regions of parietal activation when subjects were asked to judge social compatibility and physical distance. What does this tell us about magnitude processing and modularity? I think there are two things we can take from the work and ideas I have reviewed in this chapter. First, the brain is an addictive comparator. "More than" and "less than" are the foundations of our understanding of almost everything. I would not propose a single mechanism for all comparisons, but we need to ask where this learning begins. My suggestion was, and still is, that it begins with motor development: action is the beginning of all cognition. The second message to be taken away is that, whatever your favorite module, it is less of a module than you think. Our continual use of shared maps between time, space, number, and other magnitudes is a good reminder of this and perhaps a place to make a start.

Acknowledgments

Vincent Walsh is supported by a Royal Society Wolfson Award.

Notes

1. *We* are undoubtedly modular in our specializations, but we should remember that the brain has a right to ignore us.
2. Prothetic dimensions are those that can be "more than" or "less than". For example, we can have more or less light, noise, heat, "stuff". Time, space, and number fall into this category because you can have more or less of them. Speed is also in this category because things can move more or less fast. Pitch is not in this category because we do not say that a frequency is more or less than: we identify it as a different category of sound, say a C instead of a D (but pitch differences may be associated with finger movements of different sizes and there are differences between identification and discrimination – cf Schwenzer et al., 2011). Color is a troublesome one: something can be more or less red but, at some point, these differences cross categorical boundaries and we have pink, red, and scarlet. The category of "stuff" that changes identity with changes in the amount of stuff in this way is termed metathetic. The distinction does not always work, but it is a useful guide (Stevens, 1957).
3. But please count me out of that one.

References

Alexander, I., Cowey, A., Walsh, V. (2005). The right parietal cortex and time perception: back to Critchley and the Zeitraffer phenomenon. *Cognitive Neuropsychology*, 22, 306–315.

Andres, M., Davare, M., Pesenti, M., Olivier, E., & Seron, X. (2004). Number magnitude and grip aperture interaction. *NeuroReport*, 15, 2773–2777.

Andres, M., Ostry, D.J., Nicol, F., & Paus, T. (2008). Time course of number magnitude interference during grasping. *Cortex*, 44, 414–419. doi: 10.1016/j.cortex.2007.08.007

Badets, A. & Pesenti, M. (2010). Creating number semantics through finger movement perception. *Cognition*, 115, 46–53.

Badets, A., Andres, M., Di Luca, S., & Pesenti, M. (2007). Number magnitude potentiates action judgements. *Experimental Brain Research*, 180(3), 525–534. doi: 10.1007/s00221-007-0870-y

Barone, P., Batardiere, A., Knoblauch, K., & Kennedy, H. (2000). Laminar distribution of neurons in extrastriate areas projecting to visual areas V1 and V4 correlates with the hierarchical rank and indicates the operation of a distance rule. *Journal of Neuroscience*, 20(9), 3263–3281.

Bjoertomt, O., Cowey, A., &Walsh, V. (2002). Spatial neglect in near and far space investigated by repetitive transcranial magnetic stimulation. *Brain*, 125, 2012–2022. doi: 10. 1093/brain/awf211

Boroditsky, L. (2000). Metaphoric restructuring: understanding time through spatial metaphors. *Cognition*, 75, 1–28.

Boroditsky, L. (2001). Does language shape thought: Mandarin and English speakers' conceptions of time. *Cognitive Psychology*, 43, 1–22.

Boroditsky, L. & Ramscar, M. (2002). The roles of body and mind in abstract thought. *Psychological Science*, 13, 185–189.

Brown, S.W. (1997). Attentional resources in timing: interference effects in concurrent temporal and non-temporal working memory tasks. *Perception & Psychophysics*, 59(7), 1118–1140.

Brugger, P. (2008). SNARC, SCARC, SMARC and SPARC. Are there non spatial magnitudes? *Zeitschrift fur Neuropsychologie*, 19, 271–274.

Bueti, D. & Walsh, V. (2009). The parietal cortex and the representation of time, space, number and other magnitudes. *Philosophical Transactions of the Royal Society B*, 364, 1831–1840.

Bueti, D., Bahrami, B., & Walsh, V. (2008b). Sensory and association cortex in time perception. *Journal of Cognitive Neuroscience*, 20, 1054–1062. doi: 10.1162/jocn.2008.20060

Burr, D.C. & Ross, J. (2008). A visual sense of number. *Current Biology*, 18, 425–428. doi: 10.1016/j.cub.2008.02.052

Casasanto, D. (2008). Who's afraid of the big bad Whorf?: crosslinguistic differences in temporal language and thought. *Language Learning*, 58, 63–79.

Casasanto, D. & Boroditsky, L. (2008). Time in the mind: using space to think about time. *Cognition*, 106, 579–593.

Casasanto, D., Fotakoupoulou, O., & Boroditsky, L. (2010). Space and Time in the Child's Mind: Evidence for a Cross-Dimensional Asymmetry. *Cognitive Science*, 34, 387–405.

Cohen Kadosh, R. & Walsh, V. (2009). Numerical representation in the paroetal cortex. Abstract or not abstract? *Behavioural & Brain Sciences*, 32, 313–328.

Cohen Kadosh, R., Henik, A., & Walsh, V. (2007). Small is bright and big is dark in synaesthesia. *Current Biology*, 17, R834–R835. doi: 10.1016/j.cub.2007.07.048

Cohen Kadosh, R., Cohen Kadosh, K., & Henik, A. (2008). When brightness counts: the neuronal correlate of numerical-luminance interference. *Cerebral Cortex*, 18(2), 337–343. doi: 10.1093/cercor/bhm058

Cohen Kadosh, R., Sagiv, N., Linden, D. E. J., Robertson, L. C., Elinger, G., & Henik, A. (2005). When blue is larger than red: Colors influence numerical cognition in synesthesia. *Journal of Cognitive Neuroscience*, 17, 1766–1773.

Dormal, V. & Pesenti, M. (2007). Numerosity-length interference – a stroop experiment. *Experimental Psychology*, 54(4), 289–297. doi: 10.1027/1618-3169.54.4.289

Droit-Volet, S. (2010). Speeding up a master clock common to time, number and length? *Behavioural Processes*, 85, 126–134.

Ferrera, V. P., & Maunsell, J. H. R. (2005). Motion processing in macaque V4. *Nature Neuroscience*, *8*(9), 1125–1125.

Fias, W., Lauwereyns, J., & Lammertyn, J. (2001). Irrelevant digits affect feature-based attention depending on the overlap of neural circuits. *Cognitive Brain Research*, 12(3), 415–423. doi: 10.1016/S0926-6410(01)00078-7

Fischer, R. & Miller, J. (2008). Does the semantic activation of quantity representations influence motor parameters? *Experimental Brain Research*, 189(4), 379–391. doi: 10.1007/s00221-008-1434-5

Fischer, M.H., Castel, A.D., Dodd, M.D., & Pratt, J. (2003). Perceiving numbers causes spatial shifts of attention. *Nature Neuroscience*, 6(6), 555–556. doi: 10.1038/nn1066

Göbel, S.M., Rushworth, M.F.S., & Walsh, V. (2006). Inferior parietal rTMS affects performance in an addition task. *Cortex*, 42, 774–781. doi: 10.1016/S0010-9452(08)70416-7

Herrera, A., Macizo, P., & Semenza, C. (2008). The role of working memory in the association between number magnitude and space. *Acta Psychologica*, 128, 225–237. doi: 10.1016/j.actpsy.2008.01.002

Hurewitz, F., Gelman, R., & Schnitzer, B. (2006). Sometimes area counts more than number. *Proceedings of the National Academy of Sciences of the United States of America*, 103, 19 599–19 604. doi: 10.1073/pnas.0609485103

Ishihara, M., Keller, P.E., Rossetti, Y., & Prinz, W. (2008). Horizontal spatial representations of time: evidence for the STEARC effect. *Cortex*, 44, 454–461. doi: 10.1016/j.cortex.2007.08.010

Ishihara, M., Jacquin-Courtois, S., Flory, V., Salemme, R., Imanaka, K., & Rossetti, Y. (2006). Interaction between space and number representations during motor preparation in manual aiming. *Neuropsychologia*, 44, 1009–1016.

Kaufmann, L. & Nuerk, H.C. (2006). Interference effects in a numerical stroop paradigm in 9-12 year old children with ADHD-C. *Child Neuropsychology*, 12, 223–243.

Kupers, R., Fumal, A., Maertens de Noordhout, A., Gedde, A., Scoenen, J., & Ptito, M. (2006). Transcranial magnetic stimulation of the visual cortex induces somatotopically organized qualia in blind subjects. *Proceedings of the National Academy of Sciences of the United States of America*, 103(35), 13256–13260.

Laurenco, S.F. & Longo, M.R. (2010). General magnitude representation in human infants. *Psychological Science*, 21, 873–881.

Levin, I. (1977). The development of time concepts in young children. Reasoning about duration. *Child Development*, 48(2), 435–444. doi: 10.2307/1128636

Levin, I. (1979). Interference of time related and unrelated cues with duration comparisons of young children: analysis of Piaget's formulation of the relation of time and speed. *Child Development*, 50(2), 469–477. doi: 10.2307/1129425

Levin, I. (1982). The nature and development of time concepts in children. The effects of interfering cues. In W.J. Friedman (Ed.), *The Developmental Psychology of Time* (pp. 47–85). New York: Academic Press.

Lindemann, O., Abolafia, J.A., Girardi, G., & Bekkering, H. (2007). Getting a grip on numbers: numerical magnitude priming in object grasping. *Journal of Experimental Psychology: Human Perception and Performance*, 33(6), 1400–1409. doi: 10.1037/0096-1523.33.6.1400

Merrit, D.J., Casasanto, D., & Brannon, E. (2010). Do monkeys think in metaphors? Representations of space and time in monkeys and humans. *Cognition*, 117, 191–202.

Moretto, G. & di Pellegrino, G. (2008). Grasping numbers. *Experimental Brain Research*, 188(4), 505–515. doi: 10.1007/s00221-008-1386-9

Muller, D. & Schwarz, W. (2008). '1-2-3': is there a temporal number line? Evidence from a serial comparison task. *Experimental Psychology*, 55(3), 143–150. doi: 10.1027/1618-3169.55.3.143

Notebaert, W., Gevers, W., Verguts, T., & Fias, W. (2006). Shared spatial representation for numbers and space: the reversal of the SNARC and Simon effects. *Journal of Experimental Psychology: Human Perception and Performance*, 32(5), 1197–1207.

Oliveri, M., Bonni, S., Turriziani, P., et al. (2009). Motor and linguistic linking of space and time in the cerebellum. *PloS One*, 4(11), e7933.

Rousselle, L. & Noel, M.P. (2007). Basic numerical skills in children with mathematics learning disabilities: a comparison of symbolic vs non-symbolic number magnitude processing. *Cognition*, 102, 361–395. doi: 10.1016/j.cognition.2006.01.005

Rousselle, L. & Noel, M.P. (2008). The development of automatic numerosity processing in preschoolers: evidence for numerosity-perceptual interference. *Developmental Psychology*, 44(2), 544–560. doi: 10.1037/0012-1649.44.2.544

Salillas, E., Basso, D., Baldi M., Semenza, C., & Vecchi, T. (2009). Motion on numbers. Transcranial magnetic stimulation on the ventral intraparietal sulcus alters both numerical and motion processes. *Journal of Cognitive Neuroscience*, 21(11), 2129–2138.

Schwenzer, M. & Mathiak, K. (2011). Numeric aspects in pitch identification: an fMRI study. *BMC Neuroscience*, 12(1), 26.

Sell, A.J. & Kaschack, M.P. (2011). Processing time shift affects the execution of motor responses. *Brain & Language*, 117, 39–44.

Shamma, S. (2001). On the role of space and time in auditory processing. *Trends in Cognitive Sciences*, 5(8), 340–348. doi: 10.1016/S1364-6613(00)01704-6

Srinisivan, M. & Carey, S. (2010). The long and short of it: on the nature and origin of functional overlap of representations of space and time. *Cognition*, 116, 217–241.

Stavy, R. & Tirosh, D. (2000). *How Students (Mis-)understand Science, Mathematics: Intuitive Rules*. New York/London: Teachers College Press, Columbia University.

Stevens, S.S. (1957). On the psychophysical law. *Psychological Review*, 64(3), 153–181.

Stoianov, I., Kramer, P., Umilta, C., & Zorzi, M. (2008). Visuospatial priming of the mental number line. *Cognition*, 106, 770–779. doi: 10.1016/j.cognition.2007.04.013

Sur, M. & Leamey, C.A. (2001). Development, plasticity of cortical areas and networks. *Nature Reviews Neuroscience*, 2(4), 251–262. doi: 10.1038/35067562

Taylor-Cooke, P.A., Ricci, R., Baños, J.H., Zhou, X., Woods, A.J., & Mennemeier, M.S. (2006). Perception of motor strength and stimulus magnitude are correlated in stroke patients. *Neurology*, 66, 1444–1456.

Vicario, C.M. (2011). Perceiving numbers affects the subjective temporal midpoint. *Perception*, 40, 23–29.

Walsh, V. (2003a). A theory of magnitude: common cortical metrics of time, space and quantity. *Trends in Cognitive Sciences*, 7(11), 483–488. doi: 10.1016/j.tics.2003.09.002

Walsh, V. (2003b). Time: the back-door of perception. *Trends in Cognitive Sciences*, 7(8), 335–338. doi: 10.1016/S1364-6613(03)00166-9

Walsh, V., Ashbridge, E., & Cowey, A. (1998). Cortical plasticity in perceptual learning demonstrated by transcranial magnetic stimulation. *Neuropsychologia*, 36(1), 45-49

Xuan, B., Zhang, D., He, S., & Chen, X.C. (2007). Larger stimuli are judged to last longer. *Journal of Vision*, 7(10), 1–5. doi: 10.1167/7.10.2

Yamakawa, Y., Kanai, R., Matsumura, M., & Naito, E. (2009). Social distance evaluation in human parietal cortex. *PloS One*, 4(2), e4360.

Zach, P. & Brugger, P. (2008). Subjective time in near and far representational space. *Cognitive and Behavioral Neurology*, 21(1), 8–13. doi: 10.1097/WNN.0b013e31815f237c

CHAPTER 31

BASIC NUMBER REPRESENTATION AND BEYOND

neuroimaging and computational modeling

CHANTAL ROGGEMAN, WIM FIAS, AND TOM VERGUTS

Introduction: Number in The Human and Animal Brain

Early in the morning, we open our eyes, and read off the time on our alarm clock. Late in the evening, we go to sleep, perhaps counting how many nights are left before the holidays. On both occasions, and many occasions in between, we are confronted with numbers. Numbers allow calculation, ordering and planning ahead, and the ability to work with numbers is considered one of the principal accomplishments of humanity.

Nevertheless, the ability to process numerical quantities in a non-symbolic format is shared with many animal species (e.g. Cantlon and Brannon 2006; Hubbard et al. 2005). Carefully controlled experiments have shown repeatedly that many animal species can indeed represent number in an abstract way (e.g. Brannon 2006). Human infants have likewise been shown to possess numerical abilities long before language develops (Feigenson et al. 2002; Xu and Spelke 2000; Xu et al. 2005). Remarkably, many properties emerge consistently across species, when humans and animals are engaged in numerical tasks (Dehaene et al. 1998; Roitman et al. 2007; Whalen et al. 1999). Two of these omnipresent properties are the distance and the size effect. The distance effect refers to the observation that it is easier to discriminate between two numbers as the numerical distance between them increases (e.g. sets of two and nine dots are easier to discriminate than sets of eight and nine dots; Moyer and Landauer 1967). The size effect refers to the observation that for an equal numerical distance, small numbers are easier to discriminate than large numbers (e.g. sets of two and three are easier to discriminate than sets of eight and nine). The distance and size effects appear in humans when numbers are presented as Arabic digits (Dehaene et al. 1990), as verbal number words (Koechlin et al. 1999) and as non-symbolic numerosities (patterns of dots) (Buckley and Gillman 1974, see also Gebuis and Reynvoet, this volume). This close correspondence between humans and non-human animals suggests that we share common structures for number processing (Feigenson et al. 2004), the characteristics of which have now begun to become uncovered.

The first neural findings describing how number is processed in the brain were provided by Nieder and colleagues (Nieder et al. 2002; Nieder and Miller 2004). These authors trained macaque monkeys to perform a delayed match-to-numerosity task. The monkeys were presented with two consecutive dot patterns, each containing one to five dots, and were asked to indicate if the second display contained the same number of dots as the first one. Other features in the displays such as individual dot size and spatial configuration were varied randomly. The authors recorded the neuronal activity of single cells during both the sample and the test phase. They observed neurons in the prefrontal cortex and intraparietal sulcus that are tuned to numerosity: The neuronal response is maximal when that neuron's preferred quantity is presented, and decreases systematically when the presented number of dots is numerically more distant from the preferred numerosity (cf. number-selective code in Figure 31.1). Hence, these neurons were called number-selective neurons. When the number of dots presented is not the neuron's preferred quantity, but numerically close, the neuron still responds, albeit at less than maximum. Hence, neuronal activity shows overlap for two numerosities. This overlap is larger for two numerosities with a small numerical distance, making it harder to discriminate between these numerosities. For two numerosities with a large numerical distance, there is little neural overlap, making discrimination easier. In this way, the tuning properties of number-selective neurons can explain the distance effect in the match-to-numerosity task. Nieder and colleagues also observed that the tuning curves of the number-selective neurons broaden as number increases (Nieder et al. 2002). Consequently, larger numerosities are more coarsely represented and thus harder to discriminate, generating a size effect.

Also in humans, there is evidence for a number-selective coding system. In behavioral experiments, Reynvoet and colleagues (e.g. Reynvoet and Brysbaert 2004; Reynvoet et al. 2002) presented two consecutive numbers in each trial, but only the second stimulus (the target) had to be named. By varying the numerical distance between the first stimulus (the prime) and the target, the influence of the prime on target processing was investigated. The priming effects were distance-dependent, meaning that the target is named faster when the numerical distance between the prime and the target is small. This can be explained by number-selective coding: when the prime activates a number in a number-selective system, neurons that prefer numbers close to the prime number will also be somewhat pre-activated by the prime (cf. number-selective code in Figure 31.1), thereby facilitating the naming of a subsequent numerically close target.

Neural evidence for number-selective neurons in humans has been found using the fMRI adaptation paradigm. This method is based on the fact that, when the same visual stimulus is repeated, the activity of neurons responsive to this stimulus is reduced. Therefore, the activity of number-selective neurons responsive for a specific quantity should decrease when this quantity is repeatedly presented. This is known as adaptation of the neuronal response. Piazza et al. (2004) performed an fMRI adaptation study in which they showed adaptation of the neuronal response in the anterior part of the intraparietal sulcus after repeated

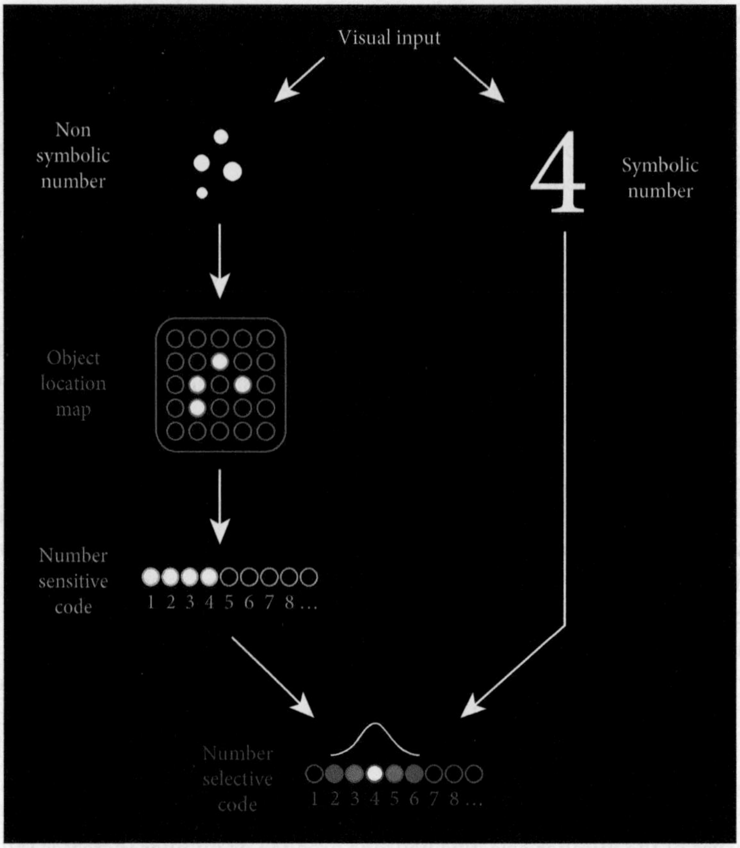

FIGURE 31.1 Illustration of the pathway for the processing of number from visual input to a number-selective code. Left: processing pathway for non-symbolic number; the non-symbolic numerosity is represented in an object location map, which is transformed to a number-sensitive code (the hidden layer in our model). Number-sensitive coding is then transformed to a number-selective code to yield the final representation. Right: no nonlinearity is involved in the transformation from symbols to number-selective neurons, hence symbolic input is directly connected to the number-selective neurons.

presentation of the same numerosity. The response recovered from adaptation when a different numerosity was occasionally presented, but not when the same numerosity was shown with different shapes, consistent with adaptation of number-selective neurons. Moreover, Piazza et al. (2004) showed that the recovery of the response was larger for numerosities with a large distance from the adapted numerosity than for numerosities with a small distance from the adapted numerosity. This provides evidence for an activation profile of the underlying quantity neurons in terms of number-selective coding. Indeed, the response profile of this coding scheme predicts that neighboring numbers will also be activated, and thus also adapted. By plotting the recovery of the BOLD signal as a function of the presented numerosity, the authors obtained tuning curves similar to those obtained by Nieder et al. (2002). Like the tuning curves obtained with single-unit recording in monkeys, the tuning curves obtained with fMRI adaptation in humans were also broader with increasing numerosity.

Model: Number-Sensitive and Number-Selective Coding

The characteristics of the number-selective neurons account for many aspects of behavior (Nieder and Miller 2004). The question remains, however, how visual input is converted into a number-selective coding system. This conversion is most challenging for the transformation from a non-symbolic number, which consists of a number of objects (Figure 31.1). In an attempt to confront this issue, the systems required for this conversion have been investigated by computational modeling (Dehaene and Changeux 1993; Verguts and Fias 2004).

The models start with the representation of the visual input (i.e. a set of objects) on an object location map (see Figure 31.1). An object location map is a spatial neuronal map in which each neuron responds to a specific location. If an object is presented at this location, the neuron detects it and is activated, independent of the size or other visual properties of the object (Goldberg et al. 2002). Rather than being specifically devoted to number processing, this map has a more general function (e.g. supporting visuo-spatial short-term memory, Awh and Jonides 2001; Srimal and Curtis 2008; Roggeman et al. 2010). This map is already a highly abstracted transformation of primary visual cortex, because it has to generalize across different sizes and shapes of the physical appearance of individual objects (e.g. Dehaene and Changeux 1993). These visual transformation processes are beyond the scope of the current discussion. As a result of these processes, each object is represented as 'one' by only one location neuron.

The information in the object location map must then be converted into a number-selective code. For example, the number neuron coding for '1', should be activated if only one object is presented in the object location map; the neuron coding for '2' should be activated if two objects are present in the object location map, and so on. However, because objects are not always presented in the same spatial configuration, a direct transformation from the neurons in the object location map to the number-selective neurons is not possible. The reason for this is that an activated object location neuron should (for example) activate the number 3 neuron, only if exactly two other neurons in the object location map are activated. Unfortunately, the neuron in the object location map has no way of knowing this. It does not know how many other objects are represented, and hence cannot know whether to activate the number 3 neuron or another number neuron. Technically, this corresponds to an instance of the XOR (exclusive OR) rule in logic (Minsky and Papert 1969). These problems require a nonlinear transformation, which cannot be achieved in a single step. The most straightforward way to solve this problem is to implement an intermediate processing step, where the object location map information from different object location neurons can be combined. Therefore, a hidden layer was introduced in the models between the object location map and the number-selective coding system.

Verguts and Fias (2004) trained a neural network with an object location map as input. The network was trained with backpropagation to transform the object location representation of the numerosity at input via the hidden layer to a number-selective coding representation at output. Because of the backpropagation training algorithm, the network was allowed to come up with the computationally optimal solution. After training, it was found that the neurons in the hidden layer displayed a monotonically varying activation pattern (i.e. monotonously increasing or decreasing) when more objects were presented. Hence, the intermediate step between the object location map and the number-selective coding system consisted of neurons accumulating (in a positive or negative way) the number of objects that was presented at input. In other words, the neurons were sensitive to number but, importantly, were not number-selective, since they did not selectively respond to a specific number (cf. number-sensitive code in Figure 31.1). This way of representing number is known as number-sensitive (in contrast to number-selective). This model demonstrates that number-sensitive coding is a biologically plausible way of solving the (generalized) XOR problem implied in mapping from an object location map to number-selective coding.

Number-sensitive coding also emerged spontaneously from visual input in an unsupervised model by Stoianov and Zorzi (2012). Rather than normalize object size, these authors presented (unnormalized) visual input representing visual scenes with different numbers of objects to a so-called deep network model (Hinton 2007). The visual layer projected to a first hidden layer, which projected to a second hidden layer. Crucially, the model training was unsupervised, being required only to reconstruct the visual layer activation based on the hidden layer activation. After training, neurons in the first hidden layer were sensitive to various visual properties. Crucially, the second hidden layer contained a subset of neurons that responded more strongly (or more weakly) to larger numbers of objects (similar to the number-sensitive code in Figure 31.1), but were not sensitive to total area. In this sense, such neurons can be labeled number-sensitive.

In summary, a number-sensitive coding system may be a necessary preceding step in the transformation from a non-symbolic number to an abstract number-selective representation. The models thus suggested that the cardinality of a set of objects is represented differently in different stages of the processing stream: visual input is first transformed into an object location map, which activates a number-sensitive coding system, which subsequently generates a number-selective code (see Figure 31.1).

Data: Number-Sensitive and Number-Selective Coding

The evidence for number-selective coding was discussed in the introduction. Inspired by the computational models, researchers have recently started to look for evidence for number-sensitive coding too.

Single-Unit Evidence in The Monkey

The biological reality of number-sensitive coding was demonstrated by means of single-unit recording. Roitman et al. (2007) recorded neurons in the lateral intraparietal area (LIP) of the macaque monkey. The monkeys were asked to plan an eye movement to a target. At the same time, visual arrays of 2, 4, 8, 16, or 32 dots were displayed at a distal location from the eye-movement target. The numerosity of the array predicted the amount of reward the monkey would receive when he performed the eye movement, but was task-irrelevant otherwise. Activity was recorded from neurons having the distal location of the numerical display in their receptive field. More than half of the neurons recorded in LIP displayed a monotonic response to the numerosity of the numerical arrays: the activity increased or decreased monotonically with increasing numerosity, indicating that these neurons summated (in a positive or negative way) the number of elements displayed. This finding supports the existence of number-sensitive coding in the monkey brain.

Behavioral Evidence in Humans

In humans, support for number-sensitive coding was found in a priming study by Roggeman et al. (2007). Here, the effect of a briefly presented prime on the naming of a subsequently presented target number was evaluated. Both primes and targets could be either symbolic (Arabic digits) or non-symbolic (dot patterns) number stimuli. When primes were symbolic stimuli, naming times increased with increasing distance between prime and target. This is the distance-dependent priming effect (see Reynvoet et al. 2002) reported above. In contrast, when primes were non-symbolic stimuli, naming the target value was faster whenever the value of the prime was larger than or equal to the value of the target. This step-like priming pattern is consistent with number-sensitive coding. Because large numbers lead to more activation in number-sensitive coding, a prime that is larger than the target will activate the representation of the target, facilitating the naming of the target. If the prime is smaller than the target, the representation of the target will be only partially activated, and additional neurons will have to be activated to name the target, increasing response time.

Neuroimaging Evidence in Humans

Neural evidence for number-sensitive coding in humans was provided by Santens et al. (2010, Experiment 1). Participants were presented with dot displays containing one to five dots, and neural activity was measured for each numerosity separately with event-related fMRI. Number-sensitive areas were localized as areas showing increasing activation with increasing number. In order to be sure that we actually detected number-sensitive areas rather than areas that are sensitive to physical parameters that correlate with numerosity (such as total luminance or object size), stimuli were constructed such that confounds of these non-numerical parameters were eliminated. The results revealed a network of bilateral occipital and parietal areas and an area in the medial frontal gyrus. Given that Roitman et al. (2007) found number-sensitive coding neurons in monkey area LIP, we assessed the correspondence of non-symbolic number processing in the monkey and the human brain. We therefore identified areas that functionally correspond to monkey LIP, as assessed with a saccade localizer task. In the area obtained by this localizer task (bilateral posterior superior parietal cortex), number-sensitivity was observed.

Networks for Number Representation

Having established the existence of number-sensitive and number-selective in the human brain, we investigated how these different stages are located relative to one another in the human brain, and how non-symbolic and symbolic number representations are related.

Stages of Non-Symbolic Number Processing

The computational models proposed three different stages for numerosity processing. In the first stage, the spatial locations of the to-be-enumerated elements are stored in an object location map (Goldberg et al. 2002). This information is then transformed into a number-sensitive code in the second stage, which is subsequently transformed into a number-selective code in the third stage. In Santens et al. (2010, Experiment 1), we observed a larger BOLD signal for larger numerosities in the same area. These areas could correspond either to the object location map or to the number-sensitive code, as both stages show increasing activity for increasing number on the population level. Hence, these studies did not allow distinguishing between the first two stages.

To resolve this issue, we used fMRI adaptation to identify the three postulated stages of numerosity processing and their anatomical location relative to one another (Roggeman et al. 2011). We repeatedly presented the same non-symbolic numerosity (collection of dots) at the same locations in the visual field. In this way, neurons involved in the processing of this numerosity (object location map, number-sensitive, and number-selective neurons) were neurally adapted. Occasionally, a deviant stimulus with a deviant number of dots and/or dots at deviant locations was presented. A factorial design was created in the deviant stimuli, which allowed us to calculate three independent contrasts, each sensitive to one of the three stages. The main result was that the three different stages, tested by three different contrasts, were indeed differently represented in different brain areas along the lines postulated by the model (see Figure 31.2). The activation of the object location map was present from the earliest parts of the occipito-parietal processing stream. The number-sensitive coding map exhibited a primarily nonlinear pattern of activation, with first increasing and then decreasing activation. The number-selective coding map became more pronounced further along the occipito-parietal processing stream. Such a posterior to anterior gradient along the intraparietal sulcus from number-sensitive to number-selective processing is consistent with the hypothesis that number-sensitive processing is a necessary intermediate processing step for non-symbolic number processing between early visual sensory analysis and a more abstract number-selective system. See Figure 31.2 for an illustration of the three stages as localized in the human brain.

Pathways for Symbolic and Non-Symbolic Number

The model above describes the processing of non-symbolic numbers. The Verguts and Fias (2004) model also concerned symbolic (e.g. Arabic) numbers. To simulate how initially arbitrary symbols can acquire numerical meaning by being associated with non-symbolic numerosities during development, the model was presented simultaneously with non-symbolic numerical input and the corresponding symbols. The latter were directly connected to the number-selective neurons because no nonlinearity is involved in the transformation from symbols to number-selective neurons. After training, it was observed that the number-selective neurons that were tuned to a specific numerosity also responded maximally to the corresponding symbolic input.

However, this representation is accessed through different pathways (Figure 31.1). For non-symbolic input, visual input is mapped to the object location map, is then followed by number-sensitive coding, and is finally converted in number-selective representations. For

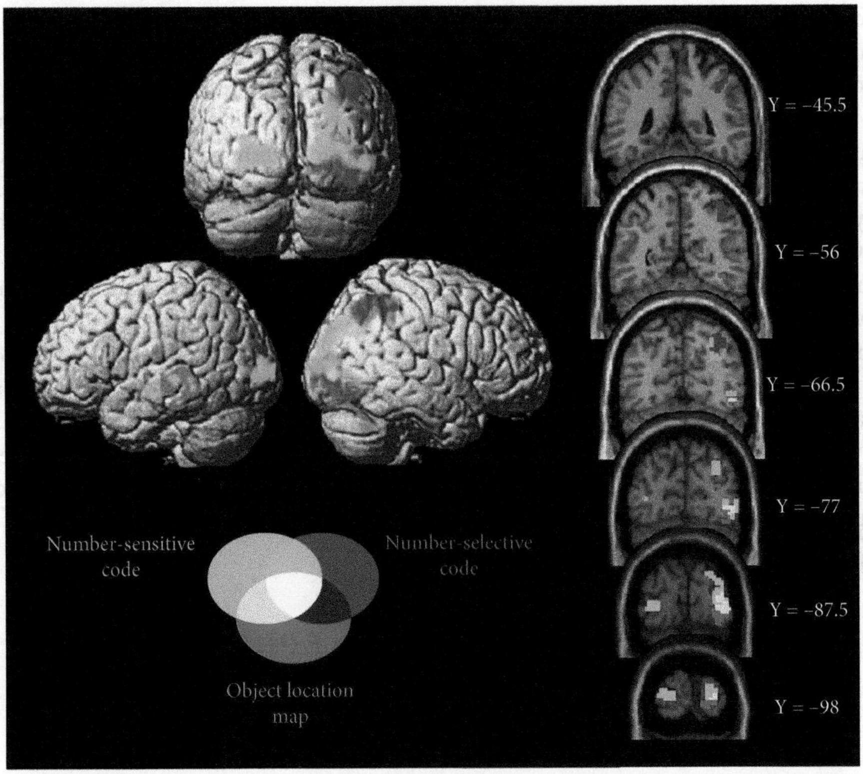

FIGURE 31.2 Activation of three different contrasts, each sensitive to one number processing stage: object location map (red), number-sensitive code (green), and number-selective code (blue).

symbolic input, in contrast, the mapping from symbol to number-selective code is linearly separable. For this reason, a direct pathway is possible from visual input to number-selective coding, without accessing the object location map and number-sensitive coding system as a necessary preprocessing step. In Santens et al. (2010, Experiment 2), we performed a connectivity analysis on separately acquired fMRI data to test this. We first localized areas that showed an increasing BOLD signal with increasing numerosity, while tightly controlling for different visual parameters. We identified an area in the posterior superior parietal cortex (human LIP) as the neural substrate for the number-sensitive coding system. As the neural substrate for a number-selective representation of number, we selected an area in the IPS which has been localized by both electrophysiological (Nieder and Miller 2004) and neuroimaging studies (Cantlon et al. 2006; Piazza et al. 2004, 2007). Piazza et al. (2007) showed that this representation is shared for symbolic and non-symbolic number (but see Cohen Kadosh et al. 2011; Cohen Kadosh and Walsh 2009). In our whole-brain analysis, we confirmed that this portion of the IPS is indeed activated by numerical stimuli, regardless of the input format. Subsequently, we investigated the functional connectivity of this area using structural equation modeling. It was confirmed that the area in the left IPS to which the number-selective representation of quantity was ascribed (Dehaene et al. 2003), shows a different functional connectivity with visual and number-sensitive areas for symbolic versus nonsymbolic quantities. In particular, the indirect pathway (visual input to number-sensitive to number-selective coding) was stronger for non-symbolic than for symbolic stimuli. In contrast, the direct pathway (visual input to number-selective coding) was stronger for symbolic than for non-symbolic numbers. In Roggeman et al. (2011) and Santens et al. (2010), non-symbolic number activation was bilateral but more pronounced in the right hemisphere. The symbolic stimuli used additionally by Santens et al. mainly activated a left-hemisphere network. This asymmetry may reflect gradients of processing symbolic (left hemisphere) versus non-symbolic (right hemisphere) materials more generally (Gevers et al. 2010; Kosslyn 2006), although this remains to be tested more thoroughly.

Our results revealed an anatomical distinction between number-sensitive and number-selective cortical regions. A number-sensitive processing area was shown in superior parietal cortex. In contrast, the IPS activation that was generated by both symbolic and non-symbolic quantities, was located more anteriorly, at a location that corresponds with activity observed in experiments that specifically investigated number-selective coding (Piazza et al. 2004, 2007). An analogous distinction between number-sensitive and number-selective processing has also been found in electrophysiological experiments in monkeys. Whereas number-sensitive neurons have been found in area LIP (Roitman et al. 2007), number-selective neurons are traditionally found more anteriorly in the IPS (Nieder and Miller 2004).

Beyond Small-Number Representation: Working Memory, Task Demands, and Multi-Digit Numbers

The Object Location Map as A Visual Short-Term Memory System

The model of Verguts and Fias (2004) takes an object location map as input. This is very similar to the concept of a saliency map known from the visual short-term memory literature. A saliency map is a two-dimensional topographic map in which neural activity represents the salient objects in the environment (e.g. de Brecht and Saiki 2006; Ipata et al. 2006; Itti and Koch 2000). The biological reality of such saliency maps has been shown by a number of studies. Bisley and Goldberg (2003) observed that monkey LIP represents attended locations in the visual field. In humans, Connolly et al. (2002) showed that a salience map in the human homologue of monkey area LIP (human LIP) holds a representation of the location of targets, similar to the monkey findings.

A salience map is typically modeled as consisting of a collection of nodes, with each node corresponding to a neuronal population coding for a given location in space. In working memory models, this map is equipped with an active maintenance mechanism implemented as recurrent self-excitation of the nodes (e.g. Grossberg 1980; Usher and Cohen 1999; Wong and Wang 2006). In addition, lateral inhibition between the nodes is implemented to reduce noise in randomly activated nodes. This leads to competitive interactions between the nodes in the map. With an appropriate balance between recurrent excitation and lateral inhibition, a stable encoding of spatial structure of the visual display can be created.

The combination of recurrent excitation and lateral inhibition leads to a maximal set size where all elements are retained. Beyond this set size, information in the map gets lost because of mutual competition. We can think of this maximum set size as the capacity limit of the map. Empirically, such a capacity limit has been observed both in fMRI (Todd and Marois 2004) and in EEG (Vogel and Machizawa 2004; Vogel et al. 2005) signatures. Moreover, different settings of the lateral inhibition parameter lead to different set sizes where the capacity limit is reached. Furthermore, the inhibition parameter functionally defines a threshold for neurons to become activated by input. If the threshold (inhibition parameter) is too low, neurons can be activated by random noise. If the inhibition parameter is too high, genuinely active neurons will be inhibited. Hence, a high level of lateral inhibition leads to precise representations (no noise), but also to a small capacity of the map. Lower levels of lateral inhibition lead to coarser representations (nodes activated by noise), but more items can be stored.

It is known that top down attention is able to bias competitive interactions in visual areas and beyond (e.g. Deco and Rolls 2005; Kastner et al. 1998). A method of controlling lateral inhibition by neurons upstream was recently described in stimulus-sensitive areas, and was suggested to be a principal method also in higher cortical areas (Arevian et al. 2008). Similarly, Edin et al. (2007) proposed that the amount of nonspecific input from upstream to downstream areas can determine working memory capacity.

Putting these pieces together, Roggeman et al. (2010) proposed that the task, and more exactly the representational precision required by the task, can top-down modulate the level of lateral inhibition in the object location / saliency map. The influence of task demands on the activation in a salience map was investigated as a function of set size. For this purpose, the working memory model of Usher and Cohen (1999) was implemented with different settings (high, medium, and low) of the lateral inhibition parameter. The model predictions were then compared with the BOLD activation in human LIP, where the saliency map is housed, in a series of fMRI experiments in which the attention to the items was manipulated. The activation in human LIP in a task that required high, medium or low attention, respectively, to the individual items, was found to be in perfect agreement with the predicted activation of the model with a high, medium or low inhibition parameter.

Symbolic Number and Working Memory

In the previous section, we discussed the object location map / saliency map for locating objects in space and as the basis for spatial working memory. We have shown the mutual dependency between non-symbolic number and the spatial saliency map. In the literature, also the link between symbolic number and space has been discussed extensively. A very robust link between number and space is found in the SNARC effect (Dehaene et al. 1993). In these experiments, it is found that small numbers are responded to faster with the left hand and large numbers with the right hand. Given that numbers are often represented in a left-to-right manner in Western cultures, this observation suggests that numbers are represented spatially, with the small numbers on the left side of space, and the large numbers on the right side of space. Zorzi et al. (2002) demonstrated that patients exhibiting left-side neglect, who neglect the left side of space, also overestimate the midpoints of numerical intervals (e.g. stating that the midpoint between 1 and 9 is 7). In the left-to-right representation of numbers, this observation suggests a corresponding neglect of the small (left) numbers. In addition, many other findings point toward an intimate relation between number and space (e.g., Dehaene et al. 1993; Fischer et al. 2003).

However, in recent years controversy has arisen, demonstrating that identification of number with space representations does not faithfully represent the rich empirical reality. Doricchi et al. (2005) demonstrated that a double dissociation can occur between bisection tasks in physical space and in 'numerical space' (cf. van Dijck et al. 2011, 2012; for review see Van Dijck et al. this volume). Recently, Chen and Verguts (2010) proposed that in order to reconcile these findings, it is necessary to assume that a link exists between symbolic number and space, but without assuming an identity relation. This particular model was built on the earlier Gevers et al. (2006) model, but added spatial representations, to which symbolic number representations were connected. Because of the association between number and space, right parietal damage led to number neglect in the model (as in Zorzi et al. 2002). However, because there was merely an association, not an identity relation, between numbers and space, dissociations between the two types of neglect were obtained in the model (as in Doricchi et al. 2005).

The nature of the associations between the spatial system and numbers remains relatively unspecified. One possibility is that it is exactly the same system that is also used for representing object locations, and is used as input for non-symbolic number representation (Section 5.1). Another possibility is that serial position in working memory is spatially coded and that it is this spatial code that is linked to number. This hypothesis is built on the observation that the SNARC effect did not depend on a number's magnitude but rather on the position of that number in working memory (van Dijck et al. 2011). Moreover, SNARC-like position-space associations can be established for non-numerical stimuli as well as for numbers, with the size of the numerical and non-numerical effects correlating (van Dijck et al. 2011). Further exploration of the validity of these and maybe other possibilities will need computational and imaging efforts in the future.

Adding Task Demands

Until now, we have only discussed how numbers are represented, and how these representations may be coupled to non-numerical representations. However, behavioral and neural signatures of number processing are almost always measured while subjects are engaged in a task. It is very well possible that also task-specific components leave their signature on the behavioral and neural data. This has been argued extensively in the developmental literature by Thelen, Smith, Schöner and others (e.g. Thelen et al. 2001). In particular, these authors have shown that claims about infant representational capacities must be treated with great caution, because infant performance is very heavily dependent on task demands.

By computationally modeling the different tasks used in the numerical cognition literature, we have come to the similar conclusion that behavioral and neural effects do not immediately inform us about underlying cognitive systems (e.g. Van Opstal and Verguts 2011; Verguts et al. 2005). For example, we have argued (Verguts et al. 2005) that there are different types of distance effects, depending on how the distance effect is measured. The priming distance effect was shown (see above) to emerge from overlap in number-selective representations (see Nieder et al. 2002, and Nieder, this volume, for direct evidence on the existence of such overlap). In contrast, the classic comparison distance effect (i.e. measured as a function of the distance between the relevant numbers in a comparison task; Moyer and Landauer 1967) was shown not to emerge from overlap in representations, but rather from the comparison process (at the decision or response levels). Although this is a rather isolated position in the numerical cognition literature, it connects numerical cognition to related domains, in which theories and models typically ascribe the distance effect as emerging from decision or response processes (e.g. for ordered sequences; Couvillon and Bitterman 1992; Frank et al. 2003; Leth-Steensen and Marley 2000). One prediction is that these two distance effects should in principle be dissociable. We tested this (Van Opstal et al. 2008) by comparing the distance effect in number comparison with the distance effect in letter comparison (for a similar view, see Cohen Kadosh et al. 2008). The classic distance effect was, as predicted, virtually indistinguishable across the two domains; this was predicted because a decision/response process is needed in both. In contrast, the priming distance effect appeared only for numbers, but not for letters.

At an applied level, this theoretical distinction may be important because the comparison distance effect is used increasingly often as a signature of elementary number processing, and correlated with more complex mathematical abilities (e.g. Halberda et al. 2008; Holloway and Ansari 2009; for review see Gebuis and Reynvoet, this volume). Despite the importance of this endeavor, many of the findings currently seem contradictory (but see Noël and Rousselle 2011). In our opinion, a research program looking at individual differences starting from a computationally motivated basis, may shed light on the current controversies. This research program currently remains to be carried out (but see Defever et al. 2011; Sasanguie et al. 2011).

Multi-Digit Numbers

Finally, a less-studied area of research, especially in computational terms, concerns how multi-digit numbers are represented. Verguts et al. (2005) argued that, because of the very low frequencies of multi-digit numbers, they could not have an explicit representation like small (high-frequency) numbers do. In this view, multi-digit numbers would be represented solely by their single-digit components (for a review see Nuerk et al. this volume); for example, the number 26 would be represented by its digit component 2 and its unit component 6. Moeller et al. (2011) implemented a neural network based on this concept and found that it could account for behavioral data on multi-digit numbers, better than models with a 'holistic' component where multi- (in this case, two-) digit numbers were represented as wholes. A related decomposed-multi-digit architecture was proposed by Grossberg and Repin (2003).

Concluding Remarks

The seminal article of Moyer and Landauer (1967) already ascribed both the distance and size effect for symbolic numbers to a semantic analogue magnitude system. Ever since this article, the concept of an analogue magnitude system for number has been extremely influential in the numerical cognition literature (e.g. Dehaene et al. 1993), up to the current day (e.g. Halberda et al. 2008). Although the concept certainly has its merits, it is our opinion that theories integrating psychology, neuroscience, and computational modeling, are now becoming indispensable. More detailed models need to be developed of how number is represented and processed in the brain. As a first step toward this ambitious goal, we have developed computational models of number processing with ensuing tests of the model predictions. The current chapter reviewed some of this work. In the introduction (Section 1), we described the number-selective coding system, which is the currently most accepted view of number representation. We described psychological and neural evidence for this representation. In Section 2, we discussed our models of core number representation. We showed that the number-selective coding system cannot exist in separation, and that a number-sensitive system is a necessary precursor step in the pathway leading up to a number-selective representation. In Section 3, we described empirical evidence for the existence of a number-sensitive system. In Section 4, we proposed how the different components may relate to one another in brain networks, and we illustrated the connection between the number-sensitive and the number-selective system, both in symbolic and non-symbolic number processing. Finally, we discussed a broader connection to other cognitive structures and processes, including working memory, response structures and multi-digit numbers (Section 5).

Despite our and other people's efforts, the end goal is far from reached. Besides connecting to biology, models need also to connect to complex numerical competencies, such as counting and understanding the natural number system (Ganor-Stern 2012; Widjaja et al. 2011). Such steps have recently been taken from a normative Bayesian point of view (Lee and Sarnecka 2009; Piantadosi et al. 2012). Eventually, the model should be able to explain how humans can work with concepts of number and its derivations, such as infinity, interest rates, and integral calculus (Rips et al. 2008). Moreover, models need to incorporate how number processing relates to other cognitive domains, such as working memory (Piazza et al. 2011), attention and language. Finally, models need to be defined and tested that describe the development of the number domain in children (Feigenson et al. 2002; Piantadosi et al. 2012; Spelke 2000).

Perhaps most importantly, such models should help finding out how number processing is impaired in dyscalculia (Piazza et al. 2010) and related afflictions. By way of comparison, computational models of the basal ganglia and their impairment in Parkinson's disease have helped understanding aspects of the disease and its remediation (Frank 2005; Frank et al. 2004). Ultimately, our goal should be to develop diagnostic tests to probe people and children for a failure in the system in very early stages, before it leads to observable defects. Early and directed intervention can then be developed for these people, in an attempt to remedy problems before they even arise.

References

Arevian, A.C., Kapoor, V., and Urban, N.N. (2008). Activity dependent gating of lateral inhibition in the mouse olfactory bulb. *Nature Neuroscience* 11(1): 80–87.

Awh, E. and Jonides, J. (2001). Overlapping mechanisms of attention and spatial working memory. *Trends in Cognitive Sciences* 5(3): 119–126.

Bisley, J.W. and Goldberg, M.E. (2003). Neuronal activity in the lateral intraparietal area and spatial attention. *Science* 299: 81–85.

Brannon, E.M. (2006). The representation of numerical magnitude. *Current Opinion in Neurobiology* 16: 222–229.

Buckley, P.B. and Gillman, C.B. (1974). Comparisons of digits and dot patterns. *Journal of Experimental Psychology* 103(6): 1131–1136.

Cantlon, J.F. and Brannon, E.M. (2006). Shared system for ordering small and large numbers in monkeys and humans. *Psychological Science* 17: 401–406.

Cantlon, J.F., Brannon, E.M., Carter, E.J., and Pelphrey, K.A. (2006). Functional imaging of numerical processing in adults and 4-y-old children. *PLOS Biology* 4(5): 844–854.

Chen, Q. and Verguts, T. (2010). Beyond the mental number line: A neural network model of number-space interactions. *Cognitive Psychology* 60: 218–240.

Cohen Kadosh, R. and Walsh, V. (2009). Numerical representation in the parietal lobes: Abstract or not abstract? *The Behavioral and Brain Sciences* 32(3–4): 313–328.

Cohen Kadosh, R., Brodsky, W., Levin, M., and Henik, A. (2008). Mental representation: What can pitch tell us about the distance effect? *Cortex* 44(4): 470–477.

Cohen Kadosh, R., Bahrami, B., Walsh, V., Butterworth, B., Popescu, T., and Price, C.J. (2011). Specialization in the human brain: the case of numbers. *Frontiers in Human Neuroscience* 5(July): 62.

Connolly, J.D., Goodale, M.A., Menon, R.S., and Munoz, D.P. (2002). Human fMRI evidence for the neural correlates of preparatory set. *Nature Neuroscience* 5(12): 1345–1351.

Couvillon, P.A. and Bitterman, M.E. (1992). A conventional conditioning analysis of 'transitive inference' in pigeons. *Journal of Experimental Psychology: Animal Behavior Processes* 18: 308–310.

de Brecht, M. and Saiki, J. (2006). A neural network implementation of a saliency map model. *Neural Networks* 19: 1467–1474.

Deco, G. and Rolls, E.T. (2005). Attention, short-term memory, and action selection: A unifying theory. *Progress in Neurobiology* 76: 236–256.

Dehaene, S. and Changeux, J.P. (1993). Development of elementary numerical abilities: A neuronal model. *Journal of Cognitive Neuroscience* 5: 390–407.

Dehaene, S., Dupoux, E., and Mehler, J. (1990). Is numerical comparison digital—analogical and symbolic effects in 2-digit number comparison. *Journal of Experimental Psychology—Human Perception and Performance* 16(3): 626–641.

Dehaene, S., Bossini, S., and Giraux, P. (1993). The mental representation of parity and magnitude. *Journal of Experimental Psychology: General* 122: 371–396.

Dehaene, S., Dehaene-Lambertz, G., and Cohen, L. (1998). Abstract representations of numbers in the animal and human brain. *Trends in Neurosciences* 21(8): 355–361.

Dehaene, S., Piazza, M., Pinel, P., and Cohen, L. (2003). Three parietal circuits for number processing. *Cognitive Neuropsychology* 20: 487–506.

Defever, E., Sasanguie, D., Gebuis, T., and Reynvoet, B. (2011). Children's representation of symbolic and nonsymbolic magnitude examined with the priming paradigm. *Journal of Experimental Child Psychology* 109(2): 174–186.

Doricchi, F., Guariglia, P., Gasparini, M., and Tomaiuolo, F. (2005). Dissociation between physical and mental number line bisection in right hemisphere brain damage. *Nature Neuroscience* 8: 1663–1665.

Edin, F., Macoveanu, J., Olesen, P., Tegner, J., and Klingberg, T. (2007). Stronger synaptic connectivity as a mechanism behind development of working memory-related brain activity during childhood. *Journal of Cognitive Neuroscience* 19(5): 750–760.

Feigenson, L., Carey, S., and Hauser, M. (2002). The representations underlying infants' choice of more: object files versus analog magnitudes. *Psychological Science* 13(2): 150–156.

Feigenson, L., Dehaene, S., and Spelke, E. (2004). Core systems of number. *Trends in Cognitive Sciences* 8(7): 307–314.

Fischer, M.H., Castel, A.D., Dodd, M.D., and Pratt, J. (2003). Perceiving numbers causes spatial shifts of attention. *Nature Neuroscience* 6(6): 555–556.

Frank, M.J. (2005). Dynamic dopamine modulation in the basal ganglia: A neurocomputational account of cognitive deficits in medicated and nonmedicated Parkinsonism. *Journal of Cognitive Neuroscience* 17: 51–72.

Frank, M.J., Rudy, J.W., and O'Reilly, R.C. (2003). Transitivity, flexibility, conjunctive representations, and the hippocampus, II. A computational analysis. *Hippocampus* 13: 299–312.

Frank, M.J., Seeberger, L.C., and O'Reilly, R.C. (2004). By carrot or by stick: Cognitive reinforcement learning in Parkinsonism. *Science* 306: 1940–1943.

Ganor-Stern, D. (2012). Fractions but not negative numbers are represented on the mental number line. *Acta Psychologica* 139: 350–357.

Gevers, W., Santens, S., Dhooge, E., Chen, Q., Van den Bossche, L., Fias, W., and Verguts, T. (2010). Verbal-spatial and visuospatial coding of number-space interactions. *Journal of Experimental Psychology: General* 139(1): 180–190.

Gevers, W., Verguts, T., Reynvoet, B., Caessens, B., and Fias, W. (2006). Numbers and space: A computational model of the SNARC effect. *Journal of Experimental Psychology, Human Perception and Performance* 32: 32-44.

Goldberg, M.E., Bisley, J., Powell, K.D., Gottlieb, J., and Kusunoki, M. (2002). The role of the lateral intraparietal area of the monkey in the generation of saccades and visuospatial attention. *Annual New York Academy of Science* 956: 205–215.

Grossberg, S. (1980). How does a brain build a cognitive code? *Psychological Review* 87(1): 1–51.

Grossberg, S. and Repin, D.V. (2003). A neural model of how the brain represents and compares multi-digit numbers: spatial and categorical processes. *Neural Networks* 16(8): 1107–1140.

Halberda, J., Mazzocco, M.M., and Feigenson, L. (2008). Individual differences in non-verbal number acuity correlate with maths achievement. *Nature* 455: 665–668.

Hinton, G.E (2007). Learning multiple layers of representation. *Trends in Cognitive Sciences* 11(10): 428–434.

Holloway, I. and Ansari, D. (2009). Mapping numerical magnitudes onto symbols. *Journal of Experimental Child Psychology* 103(1): 17–29.

Hubbard, E.M., Piazza, M., Pinel, P., and Dehaene, S. (2005). Interactions between number and space in parietal cortex. *Nature Reviews Neuroscience* 6: 435–48.

Ipata, A.E., Gee, A.L., Gottlieb, J., Bisley, J.W., and Goldberg, M.E. (2006). LIP responses to a popout stimulus are reduced if it is overtly ignored. *Nature Neuroscience* 9(8): 1071–1076.

Itti, L. and Koch, C. (2000). A saliency-based search mechanism for overt and covert shifts of visual attention. *Vision Research* 40: 1489–1506.

Kastner, S., De Weerd, P., Desimone, R., and Ungerleider, L.G. (1998). Mechanisms of directed attention in the human extrastriate cortex as revealed by functional MRI. *Science* 282: 108–111.

Koechlin, E., Naccache, L., Block, E., and Dehaene, S. (1999). Primed numbers: Exploring the modularity of numerical representations with masked and unmasked semantic priming. *Journal of Experimental Psychology-Human Perception and Performance* 25(6): 1882–1905.

Kosslyn, S.M. (2006). You can play 20 questions with nature and win: categorical versus coordinate spatial relations as a case study. *Neuropsychologia* 44(9): 1519–1523.

Lee, M.D. and Sarnecka, B.W. (2009). A model of knower-level behavior in number concept development. *Cognitive Science* 34: 51–67.

Leth-Steensen, C. and Marley, A.A. (2000). A model of response time effects in symbolic comparison. *Psychological Review* 107(1): 62–100.

Minsky, M.L. and Papert, S.A. (1969). *Perceptrons*. Cambridge: MIT Press.

Moeller, K., Huber, S., Nuerk, H.C., and Willmes, K. (2011). Two-digit number processing: holistic, decomposed or hybrid? A computational modelling approach. *Psychological Research* 75(4): 290–306.

Moyer, R.S. and Landauer, T.K. (1967). Time required for judgments of numerical inequality. *Nature* 215: 1519–1520.

Nieder, A. and Miller, E.K. (2004). A parieto-frontal network for visual numerical information in the monkey. *Proceedings of the National Academy of Sciences* 101: 7457–7462.

Nieder, A., Freedman, D.J., and Miller, E.K. (2002). Representation of the quantity of visual items in the primate prefrontal cortex. *Science* 297: 1708–1711.

Noël, M.P. and Rousselle, L. (2011). Developmental changes in the profiles of dyscalculia: an explanation based on a double exact-and-approximate number representation model. *Frontiers in Human Neuroscience* 5: 165.

Piantadosi, S.T., Tenenbaum, J.B., and Goodman, N.D. (2012). Bootstrapping in a language of thought: a formal model of numerical concept learning. *Cognition* 123(2): 199–217.

Piazza, M., Izard, V., Pinel, P., Le Bihan, D., and Dehaene, S. (2004). Tuning curves for approximate numerosity in the human intraparietal sulcus. *Neuron* 44: 547–55.

Piazza, M., Pinel, P., and Dehaene, S. (2007). A magnitude code common to numerosities and number symbols in human parietal cortex. *Neuron* 53: 293–305.

Piazza, M., Facoetti, A., Trussardi, A.N., Berteletti, I., Conte, S., Lucangeli, D., et al. (2010). Developmental trajectory of number acuity reveals a severe impairment in developmental dyscalculia. *Cognition* 116: 33–41.

Piazza, M., Fumarola, A., Chinello, A., and Melcher, D. (2011). Subitizing reflects visuo-spatial object individuation capacity. *Cognition* 121: 147–153.

Reynvoet, B. and Brysbaert, M. (2004). Cross-notation number priming investigated at different stimulus onset asynchronies in parity and naming tasks. *Experimental Psychology* 51(2): 81–90.

Reynvoet, B., Brysbaert, M., and Fias, W. (2002). Semantic priming in number naming. *Quarterly Journal of Experimental Psychology* 55: 1127–1139.

Rips, L.J., Bloomfield, A., and Asmuth, J. (2008). From numerical concepts to concepts of number. *Behavioral and Brain Sciences* 31: 623–687.

Roggeman, C., Verguts, T., and Fias, W. (2007). Priming reveals differential coding of symbolic and non-symbolic quantities. *Cognition* 105: 380–394.

Roggeman, C., Fias, W., and Verguts T. (2010). Salience maps in parietal cortex: Imaging and computational modeling. *Neuroimage* 52(3): 1005–1014.

Roggeman, C., Santens, S., Fias, W., and Verguts, T. (2011). Stages of nonsymbolic number processing in occipitoparietal cortex disentangled by fMRI adaptation. *Journal of Neuroscience* 31(19): 7168–7173.

Roitman, J.D., Brannon, E.M., and Platt, M.L. (2007). Monotonic coding of numerosity in macaque lateral intraparietal area. *PLOS Biology* 5(8): 1672–1682.

Sasanguie, D., Defever, E., Van den Bussche, E., and Reynvoet, B. (2011). The reliability of and the relation between non-symbolic numerical distance effects in comparison, same-different judgments and priming. *Acta Psychologica* 136(1): 73–80.

Santens, S., Roggeman, C., Fias, W., and Verguts, T. (2010). Number processing pathways in human parietal cortex. *Cerebral Cortex* 20: 77–88.

Spelke, E.S. (2000). Core knowledge. *American Psychologist* 55(11): 133–143.

Srimal, R. and Curtis, C.E. (2008). Persistent neural activity during the maintenance of spatial position in working memory. *Neuroimage* 39: 455–68.

Stoianov, I. and Zorzi, M. (2012). Emergence of a 'visual number sense' in hierarchical generative models. *Nature Neuroscience* 15: 194–196.

Thelen, E., Schöner, G., Scheier, C., and Smith, L.B. (2001). A field theory of infant perseverative reaching. *Behavioral and Brain Sciences* 24: 1–86.

Todd, J.J. and Marois, R. (2004). Capacity limit of visual short-term memory in human posterior parietal cortex. *Nature* 428: 751–754.

Usher, M. and Cohen, J.D. (1999). Short term memory and selection processes in a frontal-lobe model. In: D. Heinke, G.W. Humphryes, and A. Olsen. (eds), *Connectionist Models in Cognitive Neuroscience*, pp. 78–91. London: Springer-Verlag.

van Dijck, J.P., Gevers, W., Lafosse, C., Doricchi, F., and Fias, W. (2011). Non-spatial neglect for the mental number line. *Neuropsychologia* 49: 2570–2583.

van Dijck, J.P., Gevers, W., Lafosse, C., and Fias, W. (2012). The heterogeneous nature of number–space interactions. *Frontiers in Human Neuroscience* 5: 1–13.

Van Opstal, F. and Verguts, T. (2011). The origins of the numerical distance effect: The same-different task. *Journal of Cognitive Psychology* 23(1): 112–120.

Van Opstal, F., Gevers, W., De Moor, W., and Verguts, T. (2008). Dissecting the symbolic distance effect: Comparison and priming effects in numerical and non-numerical orders. *Psychonomic Bulletin and Review* 15: 419–425.

Verguts, T. and Fias, W. (2004). Representation of number in animals and humans: A neural model. *Journal of Cognitive Neuroscience* 16: 1493–1504.

Verguts, T., Fias, W., and Stevens, M. (2005). A model of exact small-number representation. *Psychonomic Bulletin and Review* 12: 66–80.

Vogel, E.K. and Machizawa, M.G. (2004). Neural activity predicts individual differences in visual working memory capacity. *Nature* 428: 745–751.

Vogel, E.K., McCollough, A.W., and Machizawa, M.G. (2005). Neural measures reveal individual differences in controlling access to working memory. *Nature* 438: 500–503.

Whalen, J., Gallistel, C.R., and Gelman, R. (1999). Nonverbal counting in humans: the psychophysics of number representation. *Psychological Science* 10(2): 130–137.

Widjaja, W., Stacey, K., and Steinle, V. (2011). Locating negative decimals on the number line: insights into the thinking of pre-service primary teachers. *Journal of Mathematical Behavior* 30: 80–91.

Wong, K.F. and Wang, X.J. (2006). A recurrent network mechanism of time integration in perceptual decisions. *Journal of Neuroscience* 26 (4): 1314–1328.

Xu, F. and Spelke, E.S. (2000). Large number discriminations in 6-months-old infants. *Cognition* 74: B1–11.

Xu, F., Spelke, E.S., and Goddard, S. (2005). Number sense in human infants. *Developmental Science* 8(1): 88–101.

Zorzi, M., Priftis, K., and Umiltà, C. (2002). Neglect disrupts the mental number line. *Nature* 417: 138–139.

CHAPTER 32

MAPPING THE BRAIN FOR MATH

reversible inactivation by direct cortical electrostimulation and transcranial magnetic stimulation

ELENA SALILLAS AND CARLO SEMENZA

Introduction

In the study of mathematical functions, as in other domains within human cognitive neuroscience, traditional anatomo-clinical correlation, electrophysiology, and various, more or less advanced, methods of neuroimaging have been complemented by different localization methods aimed at seeking for converging evidence. These methods include Transcranial Magnetic Stimulation (TMS) and intraoperative DCE. These two methods, while very different in several respects, share the property of deactivating brain functions safely and, crucially, in a quickly reversible way.

Intraoperative DCE is a very powerful method for investigating the location of brain functions (Duffau et al. 2008; Mandonnet et al. 2010). Brain mapping conducted with this method offers tremendous clinical value. Particularly when operating in the dominant hemisphere, it helps to limit the risk of personal and professional disturbances caused by acquired cognitive disorders in patients undergoing surgery for brain tumors or epilepsy. The purpose of this procedure is to gather precise information about the brain localization of functions that must be spared while removing tissues affected by pathology. Therefore, when possible (unfortunately, in a limited number of cases), neurosurgeons check the functions that they know might depend upon the area that is being operated. They do this by applying electrodes directly to the cortex, after removing part of the skull bone under local anesthesia. Because the brain lacks pain receptors, it is possible for the patient to be alert during the operation in order to interact with the operating team. He or she will commit errors in tasks sustained by stimulated areas, revealing the location of these certain functions. These areas will be spared in the operation, when possible. Despite several limitations

(Duffau 2011; Karnath and Steinbach, 2011; Shallice and Skrap, 2011; Borchers et al. and 2012), this methodology seems to provide surprisingly consistent results.

DCE is most often used in order to preserve sensory processing, motor functions, and speech. Only recently have mathematical skills received some attention in this respect. There are very good reasons to expect further research in this particular domain, however. The parietal areas have been singled out as critical for number and calculation skills since the first clinical studies on acquired calculation disorders (e.g., Henschen 1919; Hecaen et al. 1961). Mathematical functions in the parietal lobe have been further investigated and understood by means of neuroimaging, showing that numerical tasks involve a distributed network of areas, including the frontal cortex and left and right parietal lobes (e.g., Dehaene et al. 2003; Pinel et al. 2004; Eger et al. 2003; Cantlon et al. 2006; Ansari et al. 2006; Price et al. 2007; Emerson and Cantlon 2012). Neurosurgeons, therefore, need to take these findings into account in their intraoperative procedures. A safe surgery in the parietal areas must include proper testing of number processing and calculation. They must want, in fact, to limit damage to functions that are important in everyday life. Researchers in cognitive neuroscience can also benefit from findings with direct cortical stimulation. As argued in this chapter, these findings are not only interesting insofar as the method allows for direct interference with brain activity that can be quickly reversed. Additionally, a reasoned comparison with findings obtained using other techniques, such as TMS, may provide valuable insights, potentially leading to significant progress in the field.

Like DCE, TMS complements neuroimaging with causality. The question is whether all of the regions identified with functional neuroimaging are necessary for the performance of a given task (Robertson et al. 2003; Walsh and Cowey 2000). Essentially, fMRI findings reflect correlations between brain activations and a functional hypothesis in the experimental design. The fact that an fMRI study shows activations in a certain area does not imply that the area is crucial for a specific function. The basic idea of most TMS experiments is that if the temporary disruption of a brain area impacts accuracy and/or reaction times in a task or experimental condition, then this area is considered critical for the function behind the task. In other words, TMS is employed to test whether or not a region in which change in neural activity is associated with a given task is also *necessary* for the performance of this task. TMS has, thus, been used in the investigation of areas that have previously been linked with number processing, with the goal of determining their necessity for certain mathematical functions. Given the amount of neuroimaging evidence suggesting that core number representations reside in the intraparietal sulcus (IPS), the goal was to determine the essentiality of the IPS in core magnitude processing. Another important area of focus has been determining which brain areas are behind space-number relations, and just a few works have been dedicated to the mapping of calculation.

While TMS and DCE have been used for the temporary disruption of brain activity, a pair of recent studies employed another non-invasive stimulation technique, Transcranial Direct Current Stimulation (TDCS) (Utz et al. 2010), to improve numerical competence (Cohen Kadosh et al. 2010b; Iuculano and Cohen Kadosh 2013). TDCS consists of the delivery of a constant, low current (e.g. 1–2 mA). Saline-soaked surface electrodes supply the current through the anodal or cathodal end. Depending on the direction of the current, the effect is enhancing cortical functions (anodal) or temporarily disrupting cortical function (cathodal). The effects of TDCS are persistent due to modifications in postsynaptic connectivity. In the study by Cohen Kadosh et al. (2010b), anodal stimulation led to a specific improvement in the learning of artificial number symbols during anodal TDCS excitation of the right parietal lobe. The improvement was manifested in adult like numerical effects (linearization in quantity representation and numerical Stroop effects). In a recent study (Iuculano and Cohen Kadosh 2013), adult participants were administered TDCS on the posterior parietal cortex (PPC) and over the dorsolateral prefrontal cortex (DLPFC) areas, which are believed to be used in numerical understanding and learning, respectively. The authors showed a double dissociation: stimulation over PPC led to a better and faster overall learning of a new numerical notation, while automaticity, as measured by a numerical Stroop task, was impaired; stimulation over DLPFC enhanced automaticity while impairing numerical learning. Still another study (Snowball et al. 2013), combined near-infrared spectroscopy (NIRS) with transcranial random noise stimulation (TRNS). TRNS is a technique that alters cortical excitability in a polarity-dependent manner, in which an alternating current is applied over the cortex at random frequencies (see Terney et al. 2008 for an introduction to TRNS). Snowball and colleagues found that TRNS increased the learning rate for calculation and drill learning and decreased RTs for drill learning. Stimulation also had an effect on the amplitude and timing of the haemodynamic response. Behavioral and neurophysiological effects for calculation were still present in a follow-up test conducted six months later. This approach complements TMS and DCE, and, importantly, extends brain mapping to the treatment of numerical processing anomalies.

In what follows, we will report the (so far) relatively limited amount of work conducted with TMS and DCE methodologies in the mapping of brain for math, in an effort to understand the most robust findings and whether a direct comparison between the obtained results could lead to valuable conclusions and a significant advancement of research in this area.

Transcranial Magnetic Stimulation

TMS is based upon Faraday's principles of electromagnetic induction: an electrical current produces a magnetic field, and changing magnetic fields induce a secondary electric current in nearby conductors, such as human tissue. A TMS stimulator delivers a fast and large current that produces a magnetic field for about a millisecond. The magnetic field passes through the scalp and skull, inducing an electrical field sufficient to alter neuronal activity. A single pulse of TMS can modulate neurons up to a few hundred milliseconds. Different coil types exist, with changes in their focality and the depth of the effect. For example, a figure-of-eight coil targets an area about 1–2 cm² of cortex under the junction. It must be noted, however, that while TMS stimulates a particular cortical area, the effect is not restricted to that area. Even with the most focal coils, changes in activity can spread through several centimeters, and affect distant brain regions that are connected to the stimulated brain region (e.g. Ruff et al. 2006). Additional sources of variability in a TMS design come from the intensity, frequency, duration, and time point of stimulation, as well as the actual TMS paradigm and site localization method. When cognitive functions are under study, the location and timing of stimulation is determined using other brain imaging techniques (ERPs, MEG, PET, or fMRI). In a less precise approach, the researcher can use the 10–20 EEG positions, for example using P3 and P4 sites). Another way to carry out a functional search would be to apply TMS on proximal sites until maximum effects on related tasks (i.e. a visual search) are achieved, and then marking the relevant sites on MRI scans. But perhaps the most powerful way of localizing a site, for either an individual or a group, is to obtain an MRI for each subject and localize coordinates of interest. These coordinates can come from a previous fMRI study on the same participants, from similar fMRI studies on different participants or from a meta-analysis. During TMS application, frameless stereotaxic systems permit the on-line interactive navigation of the site of interest in the individual T1, allowing researchers to monitor the position of the coil during the experiment. When individual T1s are not available, a template image can also been used. Finally, the chosen intensity for stimulation can be fixed, or based on individual motor threshold (minimum intensity needed to generate visible hand movements or motor evoked potentials when stimulating primary motor cortex) or phosphene threshold (minimum intensity needed to generate stationary or moving phosphene perception when stimulating V1 or V5), albeit the correlation between these parameters is questionable (Stewart et al. 2001).

Magnetic stimulation can be applied in single pulse (spTMS) or repetitive pulse mode (rTMS). With spTMS, a single pulse is applied at a specific point in time. In rTMS, stimulation is delivered as a train of pulses of variable frequency, up to 50 Hz and for tens, hundreds, or thousands of ms. Thus, spTMS provides a good temporal resolution, which is useful when the experimenter knows when a certain process should occur. When used to disrupt performance, cognitive processes are better targeted with rTMS to produce an increase in errors and longer-lasting effects; however, when other dependent measures other than errors (i.e. RTs) are considered, spTMS can be successfully used in the study of cognition. Theta burst stimulation (TBS) is a form of rTMS in which brief trains of pulses are delivered at 5 Hz (theta frequency). One advantage of this technique is that the effect of 20 seconds of stimulation has been shown to last for 20 minutes. In turn, when TMS is used to disrupt the function of a targeted cortical site through a 'virtual brain lesion', information about the *where* and the *when* of that function can be obtained.

TMS designs require different means in order to assure specificity. Firstly, control tasks are required whose functions are not altered when a target area is stimulated, in contrast to the function of study. Secondly, control sites are required, the stimulation of which does not result in any change of the target function. Finally, a double dissociation can also be targeted: when Site 1 is stimulated, Task A is altered, but not Task B, whereas when Site 2 is stimulated, Task A is spared, and Task B is altered. Proximity can also be explored by stimulating proximal areas, thus possibly circumscribing effects to the target site. Dissociations in time are also possible: the role of functional areas can be explored through, for example, the inclusion of several Stimulus Onset Asynchronies (SOAs) for the pulse. This allows any differential effects at different time points to be detected. Finally, this technique can differentially affect functionally distinct neural populations, depending on their initial state of activity. This fact can be incorporated into the experimental design (state dependent TMS), greatly enhancing its functional specificity. As such, TMS is a very versatile technique, which has more to offer than simply 'disrupting site X alters function Y'. (More extensive reviews on the use of TMS in cognitive neuroscience can be found in Walsh and Rushworth 1999; Walsh and Cowey 2000; Pascual-Leone et al. 2000; Sandrini et al. 2011.)

TMS in The Study of Quantity

A number of studies have used TMS to investigate the crucial areas for number quantity processing, and almost all of them have focused on whether left and/or right parietal areas are crucial for quantity. While more convergence is being found among experiments into the link between space and numbers, lateralization for the processing of magnitude has not yielded such consistent results. There is now little doubt that the IPS is a locus for quantity processing: findings from fMRI (where bilateral IPS activations are found—see Dehaene et al. 2003), patient data (e.g. Cipolotti et al. 1991; Dehaene and Cohen 1991), physiological studies in monkeys (e.g. Nieder and Miller 2004), or studies in developmental dyscalculia (Price et al. 2007; Rotzer et al. 2008) all converge to support this. Divergence arises between (and within) TMS and fMRI research with regard to the differential involvement of the left and right IPS.

It is important to note that, apart from necessary differences in experimental paradigms, the above-mentioned TMS versatility is also found in the study of math cognition. Studies vary in site localization procedures, if participants' MRIs, or an initial fMRI, are not available, the use of a stereotactic system with a template image may not take into account the variability across subjects (Sack et al. 2009). On the other hand, the task used in a previous fMRI experiment should be measuring very similar processes to the process under study (i.e. distance or size effects and not just activations to a number comparison task). The selection of a stimulation pattern differs across studies (e.g. single pulse, repetitive—and Hz—triple pulse TMS, theta burst stimulation) and so does TMS effect. Selection of timing parameters often varies, as well: for example, SOAs are often used for single pulse or for the time locking of rTMS or TBS in relation to the stimuli. These are sources of variability that can explain discrepancies of findings. For these reasons, a correct approach might be to reconcile the positive results from the different studies.

It is important to observe in which of these works the distance effect (by which faster and more accurate responses occur when comparing two numbers, if the numerical distance separating them is relatively large) or other effects that stem from the numerical representations are impaired (i.e. disturbing a certain site of interest with a higher cost for closer numerosities would be the relevant outcome if one wants to show that magnitude processing has been targeted). Otherwise, a delay in RTs or decrease in accuracy could result from processes that occur earlier or later than the access to magnitude. In any case, the number comparison is generally taken as a task that activates the quantity code. However, only some of the studies below have based their choice of the pulse timing on the available ERP studies that had shown distance effects of around 200 ms (Libertus et al. 2007; Dehaene 1996; Pinel et al. 2001; Turconi et al, 2004). ERP effects may certainly occur later than the optimal time of choice in a TMS design (Walsh and Rushworth 1999), but when single neuron studies in monkeys, more accurate in timing (e.g. Ashbridge et al. 1997), are not available with comparable experimental designs, ERP effects can orient the TMS starting point and duration.

The first study that used magnitude comparison was the carried one out by Göbel et al. (2001). They showed delayed comparison times between a reference number (65) and two-digit numbers under rTMS stimulation of both the left and right angular gyrus (AG), but only a stronger effect for closer distances for numbers larger than the reference 65 in the left AG was shown.

Sandrini et al. (2004) used rTMS on the inferior parietal lobule (IPL) during a comparison between two single-digit numbers presented at the same time. Stimulation started during the presentation of digits and lasted 225 ms. They used sham stimulation as a comparison and the supramarginal gyrus as the control site. They found a general interference from the left IPL stimulation and no interaction with distance. The authors considered the possibility of having interfered before the magnitude was accessed—in turn, this would slow down the processing sequence from input to number semantics.

Andres et al. (2005) stimulated the posterior parietal cortex (PPC) using spTMS (150, 200, 250 ms after stimulus onset) on either one hemisphere or both (double pulse). They used sham stimulation as a comparison. Participants were required to compare one-digit numbers to the reference (5). Interestingly, a comparison of close distances was disrupted after bilateral and left PPC stimulation. A comparison on far distances was delayed only after bilateral PPC stimulation. The authors explain these effects as reflecting a fine discrimination in the left PPC, whereas each hemisphere could be able to perform the task by itself. There is, nevertheless, a stronger predominance of the left PPC in this study, as the sole stimulation of the right PPC produced no effect on far or close distances.

The only TMS study that uses non-symbolic stimuli (Cappelletti et al. 2007) explored the possible difference between symbolic and non-symbolic quantity processing using a comparison task between two-digit numerosities and a reference (65). They applied rTMS for 10 min before each task. TMS was delivered to the left and right IPS and the AG as a control site, in a between-subjects design for site vs. sham stimulation. They used a control task in which the orientation of ellipses had to be judged. The results showed left IPS involvement both in symbolic and non-symbolic comparisons. These effects were modulated by numerical distance with greater impairment for close numbers. Facilitation emerged after stimulation of the right IPS. The authors concluded that the left IPS is crucial both for symbolic and non-symbolic comparisons.

In a recent study, Sasanguie et al. (in press) investigated the role of right and left IPS in the processing of symbolic and non-symbolic magnitude using rTMS. They used sham stimulation as control and selected the targeted sites based on the average activation reflecting a distance effect in the same paradigm from a previous study (Notebaert et al. 2011). A priming paradigm was used in two experiments and stimulation was delivered on the presentation of the prime. Participants had to compare the second number of a pair to a reference 5. In a first experiment primes and targets were symbolic numerosities (Arabic or verbal). In a second experiment, primes and targets were symbolic and non-symbolic numerosities (dots or Arabic digits). The results showed that stimulation to right or left IPS was not enough to reduce the priming distance effect (PDE) between two symbolic numerosities. In contrast, left IPS stimulation tended to reduce the PDE between symbolic and non-symbolic numerosities. The authors conclude that the left hemisphere has a crucial role in the mapping between symbols and quantities. These data also reflect a differential brain basis in the processing of symbolic and non-symbolic magnitude.

A new approach to the study of magnitude with TMS was taken by Cohen Kadosh and collaborators (Cohen Kadosh et al. 2007; Cohen Kadosh et al. 2012). They used a size-congruity paradigm (numerical Stroop task) and asked subjects to attend to the physical size of a pair of digits while ignoring their numerical value, or in different blocks to attend the numerical value and ignore the physical size. Healthy participants, unlike dyscalculic participants, are faster in judging congruent trials (i.e. the numerical value matches the physical dimension (e.g. 2 4)), than in judging incongruent trials (e.g. 2 4). The authors used event-related triple pulse TMS to simulate dyscalculic-like behavior by disrupting right or left IPS in healthy participants. Another group of dyscalculics underwent the same paradigm without TMS. The results showed that healthy participants under sham stimulation and stimulation on right IPS or left IPS, as well as the dyscalculic group, all showed a cost in RTs for incongruent trials in comparison to neutral trials (e.g. 2 2) (interference). However, only after sham and left IPS stimulation, the facilitation effect (faster RTs for congruent compared to neutral) appeared. The facilitation effect was absent after right IPS stimulation and in the dyscalculic group. The authors concluded that the right IPS is crucial for the automatic, task-irrelevant processing of magnitude. More recently, Cohen Kadosh et al. (2012) provided the results from the other part of the task. They analyzed trials wherein healthy participants were requested to judge the numerical value of the number pairs, while ignoring the physical sizes of these pairs—that is, a number comparison task where the processing of magnitude was intentional. The results paralleled those of automatic processing: distance effects were significantly reduced after right IPS stimulation compared to sham and left IPS stimulation. No effect was found for physical size comparison in these sites. Thus, the study also implies some conclusions about the contrast between physical and a numerical magnitude: namely, that they might at least partially overlap in right IPS. The computation of congruency between physical and numerical magnitude appeared to rely on the right IPS, a site that was revealed to compute numerical distance during that size-congruity paradigm.

Numerical and Non-Numerical Quantities

Recent studies explicitly explore different types of non-numerical magnitude with the aim of dissociating or finding common neural substrates. It has been assumed that the adult brain has a *partially* shared magnitude system for time, space, and numbers (Walsh 2003). More generally, the parietal lobe could work within a broader network of areas that are involved in non-numerical magnitude representations (Fink et al. 2000; Pinel et al. 2004; Dormal et al. 2010; Holloway et al. 2010; see Cohen Kadosh et al. 2008 for a review). The few TMS studies addressing this issue—described below—have shown both overlap and dissociations between numeracy and non-numerical magnitude tasks (e.g. length, duration, time, categorical, and order information).

Dormal et al. (2008) tested this possible neural overlap in IPS by asking participants to compare the *duration* between flashed single dots in one task. In another task, participants judged the difference in numerosity of groups of flashing dots presented for a fixed interval. These tasks were performed prior to and after the application of 1 Hz rTMS for 15 minutes. The sites for stimulation were left or right IPS, and vertex was chosen as the control site, based on T1-weighted MRIs and using a frameless stereotactic system. The results showed a disruption of the numerosity task for close numerosities, after disrupting the left IPS. Specifically, RTs were always faster after rTMS application except after left IPS stimulation, for close numerosities, both compared to large distances and to vertex stimulation. Some facilitation was found (i.e. the fastest second block RTs) for the left IPS in duration processing, but this facilitation was not statistically significant. The authors concluded that there is at least one cerebral site wherein numerosity and duration processing dissociate.

Similarly, a recent study by Dormal et al. (2012) studied a possible overlap between the discrimination of *length* and numerosities in the IPS. The tasks for the TMS study were, again, comparable tasks that implied different functions: arrays with different numbers of dots constant in length, to which the participant had to give either a 'few' or 'many' response, compared with a continuous rectangle that varied in length, to which the participant had to give a 'short' or 'long' response. Therefore, a numerical vs. non-numerical size effect, was evaluated for the two dimensions, instead of a distance effect. rTMS was applied to the right or the left IPS. For localization, the study used another fMRI study (Dormal and Pesenti 2009). The IPS coordinates for stimulation were individually determined by subtracting the activations to a color detection task from those obtained by a numerosity task. Average coordinates were similar to the averages used in Dormal et al.'s study (2008), although standard deviations, especially in the left hemisphere, were twice as much. Vertex stimulation was used as the control site. The results showed significant disruption for numerosity *and* length processing after right IPS stimulation. Nevertheless, and although left IPS disruptive effect showed a tendency to differ from the vertex condition, this tendency was not significant. The authors concluded that the integrity of the right IPS is necessary both in numerosity and in length discrimination.

In an extensive study, Cappelletti et al. (2009) contrasted the crucial role of IPS in the processing of numerical quantity and *quantity processing with non-numerical stimuli*. Participants had to: (1) compare the magnitude of two numbers (e.g. 12.07 vs. 15.02) in a numerical quantitative task; (2) decide which of two dates referred to a summer month or to time (12.07 vs. 15.02 or time 11.01 vs. 12.06) in a numerical non-quantitative task; (3) decide which of two objects was larger (bikini vs. coat) in a categorical quantitative task; (4) decide whether an object was a summer object or not (bikini vs. coat) in a categorical non-quantitative task. As control tasks, the authors manipulated the color of the two types of stimuli, and a judgment of color was requested. A similar fMRI study identified the stimulation IPS sites as the common areas for the quantitative task on numbers and the quantitative task on object names (Cappelletti et al. 2010). A unique area in the left and right IPS was stimulated with 10 Hz rTMS for 500 ms at stimulus onset. Sham stimulation was used on the same areas as control sites. Results showed that quantitative and non-quantitative tasks on digits (e.g. numbers and months) were impaired after left or right IPS stimulation, with larger disruption after left IPS stimulation only during numerical comparison. Judgments on objects were disrupted after left or right IPS stimulation, but only when the task was quantitative (e.g. in determining which is the larger object). Again, disruption was larger after left IPS stimulation, in comparison to right IPS stimulation. No effects were found in the perceptual task on color. Importantly, only judgments of time implied larger impairment for close distances. A main effect of distance (but no site by distance interaction) appeared in the rest of conditions. No main effect or interaction implying distance was found in the color judgment task on objects or numbers. The authors concluded that bilateral IPS is crucial in performing quantitative or non-quantitative conceptual operations with numbers, as well as that these areas are crucial for quantity processing of numbers or objects. However, IPS is not critical in perceptual decisions based on numbers or in conceptual tasks not involving quantity in non-numerical stimuli.

In a recent study, Cheng et al. (2012) contrasted the role of IPS for *order* vs. quantity processing. They applied TBS for 20 seconds on right and left IPS and vertex for five minutes, then order and quantity tasks were performed. In the order task, a line of Xs and digits in either correct (XXX3XX4X) or incorrect order (XXX4XX3X) was presented, and a judgment on the correctness of the order was given. Numbers and Xs were always in two different colors; thus, a numerical task on the same stimuli consisted of judging which of the colors contained more Xs. After left IPS stimulation, a disruptive effect was shown for the quantity task, while the order task showed a facilitatory effect, with respect to sham stimulation. Neither the right IPS nor the vertex appeared as crucial for these tasks. The authors interpreted the results as evidence for, at least, partially different neuronal populations involved in order and quantity processing.

Use of Fingers during Early Arithmetic

The importance of fingers as an aid in the early stages of arithmetic is something evident. It is perhaps not a coincidence that the ten fingers have left us with a base ten system for math (see Andres and Pesenti, this volume). But have they left a trace in our brain? The skilled use of fingers has been found to be a predictor of math achievement (Fayol et al. 1998). The presence of Gerstmann syndrome (Gerstmann 1940) after a lesion in the left AG has been taken as evidence of an association between finger gnosis and numeracy (Butterworth 1999). Some TMS studies have examined this association in the posterior parietal lobe. Rusconi et al. (2005) used a magnitude matching task with related or unrelated primes. Five hundred ms of 10 Hz rTMS to the left AG impaired magnitude judgment in numbers with unrelated primes. Identification of hand finger after opposite finger stimulation also impaired after bilateral AG stimulation. Thus, the authors suggested that an association existed between finger and number tasks in the left AG, demonstrating a relationship in the left AG between numbers and body knowledge in skilled adults who no longer use their fingers for solving simple arithmetical tasks. Sato et al. (2007) showed increased amplitude in motor-evoked potentials for the right hand muscles specific for smaller numbers, suggesting a close relationship between hand and finger numerical representations.

In conclusion, all of the reported TMS studies have found sites surrounding the IPS as crucial for the representation or manipulation of magnitude. Many succeed in finding a larger disruption in the comparison between close numbers (i.e. disruption of the distance effect). Thus, taking into account all of the positive effects, the described research supports the idea that left and right IPS are crucial for magnitude processing. Figure 32.1 and Table 32.1 display brain sites that have been shown to be involved in the processing of quantity in the above described studies, with a focus on effects related to magnitude processing (i.e. distance or size effects disrupted). More recurrently, findings showed left IPS involvement in magnitude, which was often related to distance effects. On the other hand, an explanation of the divergences between studies should be found in a dissection of TMS parameters (including site localization), task and stimuli, overall design, etc. In both the studies of Cohen Kadosh et al. (2007) and Dormal et al. (2012), the only two studies that do not show left IPS as critical, size is used as a dimension in their tasks. The study of Cappelletti et al. (2009) implies a much more complex numerical decoding than the rest, and again, uses size, albeit representational.

More TMS studies are needed in order to understand the necessary areas for non-numerical quantities and their possible overlap with numerical quantities. They should use different tasks, implying the same non-numerical dimensions. Overall, it appears that

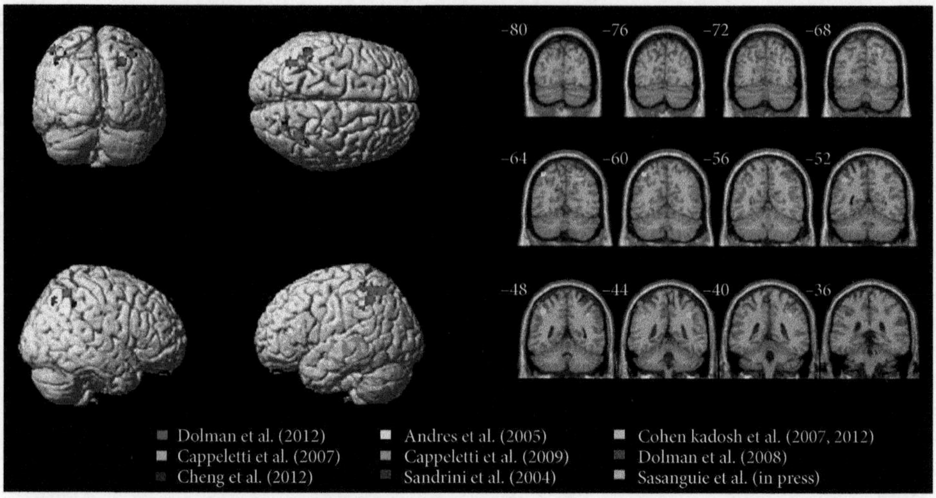

FIGURE 32.1 Reported brain sites whose stimulation through TMS has led to disruption in *quantity*. More overlap or sites with effect are found in left parietal areas. When Talairach coordinates had been reported, they were transformed to MNI coordinates. A sphere of a 5.25 mm radius and MNI coordinates as center was generated (Table 32.1). *Left*: General distribution of sites; yellow means overlap between the loci with effects in two or more experiments. *Right*: A different color is assigned to the locus/loci with effect for each experiment.

Table 32.1 TMS studies on quantity processing and the corresponding MNI coordinates of sites showing disruption. See Figure 32.1

		Left-Hemisphere			Right-Hemisphere		
PARIETAL	AREA	X	Y	Z	X	Y	Z
Sandrini et al. (2004)	IPL	−49.9	−43.31	58.82			
Andres et al. (2005)	PPC	−39.1	−61.87	54.87			
Cappeletti et al. (2007)	IPS	−41.5	−41.75	40.59			
Cohen Kadosh et al. (2007)	IPS				25.63	−66.73	45.28
Dormal et al. (2008)	IPS	−39	−52	55			
Cappeletti et al. (2009)	IPS	−42	−40	42	38	−44	40
Cheng et al. (2012)	IPS	−42	−40	42			
Dormal et al. (2012)	IPS				35	−54	53
Sasanguie et al. (in press)	IPS	−42	−50	46			

both overlaps or dissociations can be found, but why, in which tasks (i.e. processes), and why some dimensions show overlap while others do not, needs of further research. On the other hand, although being valuable in the study of crucial areas, the study of the overlapped vs. separate numerical and non-numerical dimensions using TMS is faced with the question of the inherent monolithic TMS effect. Because the hypothesis is about a partiality of overlap, no study showing dissociation between numerical and non-numerical areas will exclude the possibility that they can also overlap. In fact, as it is acknowledged in these studies (Dormal et al. 2008; Cheng et al. 2012), the most that can be stated is that that there is at least a site where those representations dissociate. In this sense, special care has to be taken with focality and with the variability in the targeted brain locus. As an example, the overlap between length and numerosity found by Dormal et al. (2012) might be explained by the variability in the exact site location across subjects. Ideally, studies should look for loci that double dissociate, showing overlap and separation in the same study, depending on the chosen brain locus. It is possible that the best way for the study of this issue would be the use of state-dependent TMS (Silvanto et al. 2008). As an example, putative neurons for one dimension could be adapted. If behavior on the deviant values of the same vs. a different dimension does not change after stimulation of Site A, then overlapped representations would have been detected. On the contrary, if after stimulating Site B, behavior to the deviant differed between the adapted vs. non-adapted dimension, then TMS would have detected an overlapped but distinct neural population for the two dimensions. Variations in proximity along parietal areas could be added to the design. Cohen Kadosh et al. (2010a) used this TMS adaptation paradigm for the study of format dependencies vs. number-specific parietal neurons. Segregated neurons for digits vs. verbal formats were found in the left IPS, but not in the right IPS. In turn, state-dependent use of TMS allows for the study of the functional role of distinct neural populations, an approach that should be fruitful in the study of the neural locus of numerical and non-numerical magnitude.

Space in Numbers

Many mathematicians report the intuition of using space in the manipulation of numerical concepts, and space is explicit in the use of math concepts many times. Spatial components in mathematics have, thus, a role in the curricula (e.g., using blocks or the mental number line (MNL) to introduce calculation or numerical concepts to first-learners). Cohen Kadosh and Gertner (2010) offer an explanation of this link between numbers and space (including time) in terms of synesthesia. The association is implicit for most people, as a sort of lower synesthetic proficiency, whereas number-space synesthetes vividly experience explicit associations between time, space and number in a sort of over-binding (Robertson 2003). Up to one out of five people experience this form of synesthesia consciously, but it is implicit for the rest of the population (Sagiv et al. 2006; Cohen Kadosh and Henik 2007). Nevertheless, the exact nature of the relationship between numbers and space, and the origin of this relationship are still unknown. It could be that it emerges over the course of education, giving rise to a MNL, but the representation of numbers could be intrinsically spatial (Dehaene 1997; Walsh 2003). In any event, a link between space and numbers is shown in many studies, even in congenitally blind participants (Castronovo and Seron 2007; Salillas et al. 2009, see Castronovo, this volume).

Empirically, the spatial characteristic of numbers is almost always addressed through the study of the so-called 'MNL'. In this MNL, numbers have been proposed to be represented along a horizontal left-to-right oriented mental line, with small numbers on its left and high numbers on its right portion (see van Dijck et al., this volume). A first index for this MNL was the Spatial Numerical Association of Response Codes (SNARC) effect, by which small numbers are responded to more quickly with a left lateralized response, and large numbers are responded to more quickly with a right lateralized response (Dehaene et al. 1993). However, other effects, which include a link between numbers and space in their explanation, have been reported since then. For example, Fischer and colleagues (Fischer et al. 2003) attained the orienting of spatial attention by the central presentation of numbers, and Calabria and Rossetti (2005) showed that a biased number strings bisection to the right or left, depending on the number repeated in a string. Even the visual field in which a number is presented determines the speed of response, with a preference for small numbers in the left visual field and large numbers in the right visual field (Lavidor et al. 2004).

Neuropsychological studies show a relation between number and space: Gerstmann's syndrome involves dyscalculia, spatial problems, finger agnosia, and dysgraphia, while hemispatial neglect patients show a bias in the bisection of number intervals, just as they show in line bisection (Zorzi et al. 2002; Vuilleumier et al.), which implies the use of a MNL in a representational side of neglect (but see Fias et al. 2010).

In a review about the space-number relationship, Hubbard et al. (2005) identified human homologues of spatial cognition areas in monkeys. They proposed that numerical-spatial interactions arise from common parietal circuits for attention to external space and internal numerical representations. Using TMS, the approach has been diverse: from findings that relate to MNL in number comparison tasks to direct tackling of the MNL through the study of the number bisection or spatial priming by numbers. Last, studies have extended their target areas to the frontal lobe, and others have incorporated motion perception as a part of the spatial quality of numbers. rTMS, spTMS, or triple pulse TMS has been used in the study of the MNL, but the exact timing for the studies was not inferred from previous ERPs studies.

As mentioned earlier, Göbel and collaborators (2001) used a number comparison to a reference number (65) while stimulating left and right AG with the supramarginal gyrus as the control site. Sites were studied as between-subjects variables. The parietal site at which the rTMS was delivered was identified through a visual search task. The authors found a greater disruption with rTMS over left AG for numbers greater than 65, which was even larger for numbers closer to the standard. rTMS over right AG disrupted the comparison of numbers larger and smaller than 65. The distance effect was not affected by stimulation. The authors concluded that within the left AG, the representation of numbers appears to be spatial in nature, differing from the right AG.

Attention over the MNL has been studied through biases in the bisection of lines occurring in pseudoneglect or hemispatial neglect (Zorzi et al. 2002, Rossetti et al. 2004). The anatomical basis of this approach was explored by Göbel et al. (2006a). The focus was on attention over a mental spatial representation—the MNL, rather than on the representation, per se. Given auditory number intervals (e.g. 117 166), participants were asked to provide the center of that interval without calculating. Left AG/adjacent posterior part of left and right IPS were stimulated with rTMS. The central occipital cortex was used as a control site, and trials without TMS were used as a control task. Interestingly, no significant disruption was found in the RTs, but bisection errors appeared: a systematic right displacement of the center (i.e. correction of pseudo-neglect in control trials) was found after right PPC stimulation. Although based, again, in a between-subjects comparison, this pseudo-neglect correction also seemed to appear after left PPC stimulation, compared to the control site. In any case, the effect after the right PPC stimulation was, indeed, larger. The authors concluded that the right PPC is crucially involved in the spatial representation of numbers.

The presentation of numbers has been shown to bias attention in space depending on their position in the MNL. Fischer and collaborators (2003) showed that the simple central presentation of a small number entails a faster detection of targets in the left visual field, and the presentation of a large number implies faster detection times of targets in the right visual field. This is something that could be taken as spatial priming by numbers. The critical brain basis of this effect was studied with a triple pulse TMS on the left and right AG by Cattaneo et al. (2009a). Cattaneo et al. (2009a) presented large or small numbers and subsequently asked participants to report the longer side of a bisected line. Trials with asterisks, instead of numbers and trials without TMS, served as controls. Without TMS, higher accuracy in 'left side is longer' responses after small numbers, and higher accuracy in 'right side is longer' responses after large numbers were reported. They stimulated during the interval between the prime and the line, and they found that when the number was small, a right AG stimulation made the higher rate of accuracy for 'left side is longer' disappear. For large number primes and after stimulation of the left AG, the higher rate of accuracy for 'right side is longer' disappeared. Right AG also had an effect with large number primes, with lower accuracy for 'right side is longer' responses. The authors explained the results by conventional attention accounts (e.g. right hemisphere processes information from both hemifield and left hemisphere processes information from the right hemifield only (Mesulam 1981)). As for the number domain, the authors explain the results as the AG being part of a network that mediates the impact of the MNL on visuospatial representations, rather than implying number processing, per se. The same network includes the visual cortex (V3, V4), as demonstrated by the effects of presentation of numbers from the low or high end of the MNL on visual cortex excitability. Namely, Cattaneo et al. (2009b) showed that small numbers increased the proportion of trials, on which phosphenes were induced from stimulating the right visual cortex. High numbers decreased the proportion of trials on which phosphenes were induced from stimulating the right visual cortex. The opposite pattern of effects was found during the stimulation of the left visual cortex. In turn, as the authors pointed out, the numbers magnitude modulates spatial attention; subsequently, these shifts induce changes in the excitability of both the left and right visual cortex.

The SNARC effect has been studied using TMS disruption, both within and outside of the parietal lobe. Rusconi et al. (2007) studied possible dissociations between the SNARC and Simon effects in the posterior parietal lobe (PPL), finding that the bilateral posterior part of PPL had a causal role in the SNARC effect. In a parity task to numbers 1 to 9, presented lateralized to fixation, they found that the SNARC effect was reduced after posterior, but not anterior, rTMS. The Simon effect was reduced after anterior and posterior PPL stimulation. More recently, Rusconi et al. (2011) compared two tasks that differed in the relevance of number magnitude. Areas in the frontal lobe responsible for visual scene analysis and visual conjunction (right frontal eye fields—FEF—and right inferior frontal gyrus—rIFG) were critical for the SNARC effect only for number comparison, where number magnitude is relevant, but not for a parity task, for which number magnitude is irrelevant. The authors explained the data as frontal regions, added to the parietal circuits as crucial for the representation and orienting to number space in humans. Besides, rFEF and rFIG rTMS effects were correlated, suggesting a common cognitive mechanism.

Another area in the PPC related to the perception of motion, but not space, has been studied using TMS (Salillas et al. 2009). Numerical processes have been found within the dorsal pathway (Fias et al. 2001; Izard et al. 2008). Salillas and collaborators (2009) hypothesized that the areas in the PPC with a role in motion perception could be sub-serving attentional processes over the MNL in a similar way that attention to motion operates over other visual images. In fact, a link between motion perception and approximate arithmetic processing had been proposed, perhaps with quantitative rules analogous to those characterizing movement in an internal continuum (McCrink et al. 2007). The motion area selected by Salillas and collaborators was the ventral IPS (VIPS), located in the occipital part of the IPS and reported as being connected to other motion responsive areas, such as MT or V3a (Orban et al. 2006; Vanduffel et al. 2002). The area was far enough to other areas classically implied as quantity responsive (Figure 32.2 and Table 32.2). In separate blocks, participants (a) compared numbers to a reference (5), responding only to numbers higher than the reference, (b) responded to numbers lower than the reference, (c) detected rightward coherent motion on random dot kinetograms (RDKs) with different proportion of dots moving coherently, (d) or detected leftward motion on RDKs with different proportion of dots moving coherently. Stimuli were presented lateralized to a central fixation point, in the left visual field or in the right visual field (RVF, LVF) while the contralateral VIPS was stimulated. spTMS was delivered at the right or left VIPS with varying SOAs (100, 150, 200 ms) from the presentation of the number or the RDK. A control task consisted of the presentation of shapes to which participants had to detect corners. The interhemispheric fissure, at the same Y and

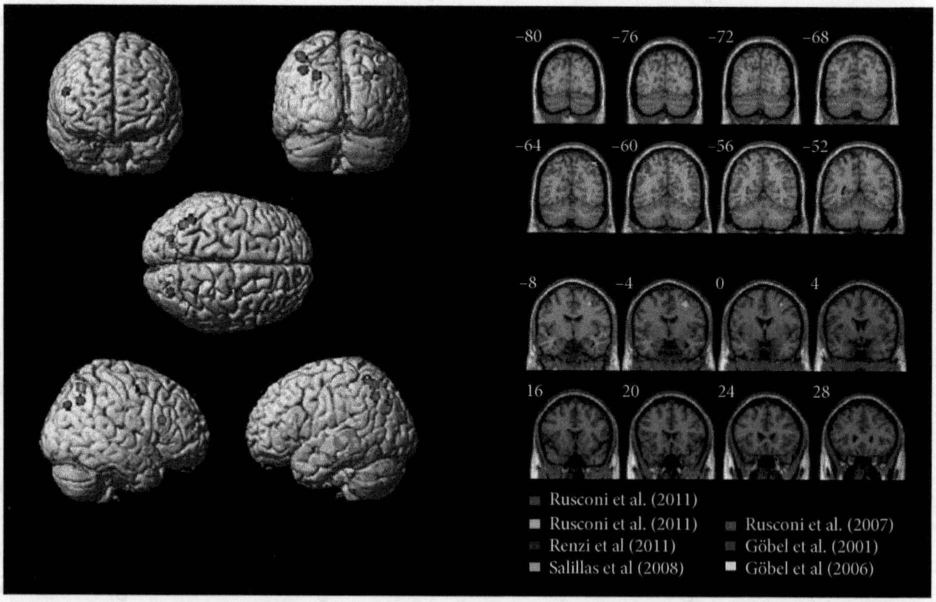

FIGURE 32.2 Reported brain sites whose stimulation through TMS has led to disruption in *spatial-numerical processes*. Similar number of sites with effect was found in left and right parietal areas. Frontal areas show to have a role in spatial-numerical processes when the task is number comparison. Figure 32.2 was built as Figure 32.1. See Table 32.2 with the MNI coordinates for each study.

Table 32.2 TMS studies on spatial-numerical processing and the corresponding MNI coordinates of sites showing disruption. See Figure 32.2

		Left-Hemisphere			Right-Hemisphere		
PARIETAL	AREA	X	Y	Z	X	Y	Z
Göbel et al. (2001)	AG	−43.38	−54.99	59.86			
Göbel et al. (2006a)	PPC				43.03	−62.54	54.66
Rusconi et al. (2007)	PPL	−37.09	−63.01	42.62	40.69	−62.73	41.27
Salillas et al. (2008)	VIPS	−24.16	−76.35	37	27.7	−76.17	36.1
Renzi et al. (2011)	PPC	−39.1	−61.87	54.87	43.01	−61.58	53.44
FRONTAL	AREA	X	Y	Z	X	Y	Z
Rusconi et al. (2011)	rFEF				31	−4.5	51
Rusconi et al. (2011)	rIFG				50	22	22

Z coordinates as VIPS, was used as control site. In the comparison task, results showed that at the shorter SOA, both right and left VIPS delayed RTs in responses to numbers higher than five. The effect remained at the 150 and 200 ms SOA, in the left VIPS/RVF only. Conversely, when responses to numbers lower than five were measured, a bigger disruption on the right VIPS/LVF was found at all SOAs, and only a tendency for the left VIPS/RVF was found at the shortest SOA. In the motion detection task, detection of coherent motion was slowed down when left or right VIPS was stimulated. Moreover, a displacement of the threshold (i.e. the proportion of coherent motion needed to clearly perceive it in more than 75% of the trials) was significantly increased when stimulating VIPS in comparison to central stimulation. The association between number comparison and sensitivity to motion coherence in random dot kinetograms was shown because of the impairment of both processes after the application of spTMS over VIPS. The authors concluded that bilateral VIPS was one of the possible areas within a functional network for the processing of quantity and that it was likely responsible for attention to motion over the MNL mental image continuum. Recently, another study by Renzi et al. (2011) used TMS-adaptation for the study of neuronal selectivity to quantity or motion in the PPC. In this study, adaptation was attained by the presentation of random dot kinetograms with coherent motion in rightward or leftward directions. Blocks of trials involving numerical magnitude judgment followed the adaptation blocks. The results showed that motion adaptation modulated in a direction-specific manner the initial state of neuronal representations causally involved in numerical magnitude judgments. This provided evidence for an overlap in neuronal representations of motion direction and numerical magnitude information. The exact nature of the link between attention to motion and number processes is not fully described to date. Nonetheless, and attending to the three described studies, commonalities between number manipulation and motion perception seems to occur in the space-in-time dimension. There might be processes entailing a moving focus of attention along space, which is what the perception of motion essentially entails from the perceiver. It also remains to be determined whether the same mechanisms are behind arithmetic operations and number comparison or, in fact, if there is another common factor between quantity and approximate calculation (or even during exact calculation). It is worth noting here that Salillas et al. (2012) found right VIPS stimulation effects in exact multiplication facts (see below).

Overall, TMS studies succeed in demonstrating a link between numbers and space. They suggest spatio-numerical representations within parietal and frontal areas: bilateral posterior parietal sites, bilateral sites in the AG and right FEF and IFG, or even VIPS (Figure 32.2 and Table 32.2). The described research has remarkably extended the IPS mapping to more posterior sites (VIPS) and to frontal areas.

The main role of TMS is identifying the need of certain brain loci for certain functions. Nonetheless, they must be put in a context of complex cognitive functions that imply functional segregation and integration of brain areas in time and possibly in hierarchies (e.g. Honey et al. 2007; Ioannides 2007; Varela et al. 2001). TMS studies have an important role in the study of the IPS within the mapping of numerical representations. Possibly, the use of TMS in math should also take connectivity into account. The goal would be to show double dissociations between processes, or even, ideally, within the same process at different time points. An example of such a functional connectivity fMRI study in the number domain is the recent work of Emerson and Cantlon (2012), which relates frontal and parietal areas during numerical tasks. spTMS at different time points in these two areas should distort behavior at different levels/times. As examples, two recent studies (Cohen Kadosh et al. 2011; Rusconi et al. 2013) have detailed functional numerical networks both within hemisphere, showing the contributions of rFEF and rIFG involvement in the SNARC effect using spTMS sampled between 0 and 400 ms (in Rusconi et al. 2013) and between the two parietal lobes, showing that the connectivity between both IPS is different depending on the numerical format (Cohen Kadosh et al. 2011).

TMS in the Study of Calculation

The few TMS research that has been dedicated to explore calculation (Göbel et al. 2006b; Andres et al. 2011; Salillas et al. 2012) has addressed the question of lateralization, as it was proposed by the Triple Code Model (Dehaene et al. 2003). According to this model, the bilateral horizontal portion of the IPS (hIPS) is proposed to be the brain basis for the core quantity system; is active in mental arithmetic, numerical comparison, or subtraction; and is domain specific. The posterior superior parietal lobe (PSPL) is responsible for the attentional processes non-specific to the number domain (i.e. spatial attention and orienting). Finally, the left AG is part of a left-lateralized network, including also perisylvian and subcortical areas, and it is responsible for memory retrieval and exact calculation (i.e. number facts retrieval). The model states elsewhere (Dehaene et al. 2004) that the left AG network includes the left HIPS, however.

The studied operations through TMS have been addition (Göbel et al. 2006b; Salillas et al. 2012), subtraction (Andres et al. 2011), and multiplication (Andres et al. 2011; Salillas et al. 2012). The triple code model would predict that subtraction, which differs from addition and multiplication, is sub-served bilaterally. Exact multiplication and addition would, similarly, imply the language-dependent left lateralized network. Although the model was based upon a compendium of some neuroimaging data, a discrepancy between the fMRI and TMS results emerged, similarly to what happens in quantity processing (see above). With the exception of Göbel et al. (2006a), the other two studies (Andres et al. 2011; Salillas et al. 2012) found the involvement of the bilateral hIPS in exact arithmetic.

Specifically, Göbel et al. (2006b) used additions that ranged from 21 + 22 to 49 + 47, presenting the two numbers simultaneously. Then, participants had to judge whether a given solution was correct or not. The presented solution was always one unit larger or smaller than the correct solution to enforce exact calculation. 10 Hz rTMS was delivered on the left and right AG at the onset of the inter-stimulus interval until the response. The control site was the anterior IPL (supramarginal gyrus). Sites were chosen by TMS disruption in a visual search task. Although stimulation in the two sites on the left hemisphere produced a disruption in response latencies, the left AG stimulation had a non-significant tendency for greater disruption. No TMS effects were found in the problem size effect. Left hemisphere predominance was suggested for exact addition.

Using fMRI-guided rTMS, Andres and collaborators (2011) investigated the role of left hIPS in simple multiplication vs. PSPL in subtraction. Participants multiplied a digit by three or four, or subtracted a number from 11 or 13 in different blocks. Sites were determined from regions showing increased activity in a subtraction and multiplication fMRI experiment by a conjunction analysis, obtaining coordinates in an individual basis. The vertex was used as a control site. Four 10 Hz rTMS pulse were delivered during 300 ms in each trial. Response latencies were slower after bilateral hIPS both in subtraction and multiplication in relation to the vertex condition. Multiplication errors in the verbal response increased after stimulation of hIPS, bilaterally. Interestingly, these effects were mainly operand-related errors (87%, i.e. the erroneous answer is a correct response for another problem which shares an operand, e.g. $4 \times 8 = 24$), and a few table (the erroneous answer is a correct response for another problem but there is no operand shared; e.g. $3 \times 8 = 28$) and non-table related errors (the erroneous answer is not a possible arithmetic response; e.g. $4 \times 6 = 26$) appeared, suggesting that retrieval had been affected by bilateral stimulation.

Salillas and collaborators (2012) used spTMS with different SOAs (150, 200, 250, 300 ms) time-locked to the presentation of easy or hard addition and multiplication problems. Stimulation was delivered to right/left hIPS or right/left VIPS (Salillas et al. 2009) contralateral to lateralized problems. The interhemispheric fissure was used as a control site for measuring disruption. In a second experiment, ipsilateral stimulation was contrasted with stimulation delivered contralateral to the problem. The results of both experiments showed that addition and multiplication differed in hard problems: Addition involved bilateral hIPS, whereas multiplication relied on left hIPS but involved right VIPS in both experiments. Moreover, right VIPS disruption predicted problem size effects (note that the right VIPS in Salillas et al. 2008, 2012 has a similar location to Cohen Kadosh et al. 2007, 2012). Finally, the contrast between ipsilateral and contralateral stimulation along the different SOAs allowed to describe the temporal course for the involvement of the different sites: the left hIPS disruption appeared late in the 300 ms SOA, and it was preceded by an involvement of the right VIPS from 150 to 250 ms SOAs. The study, thus, provided a precise description of the hemispheric involvement in exact multiplication and suggested that multiplication is more verbally mediated than addition. Moreover, similar to Salillas et al. (2008), the results suggested that more occipital areas are likely involved in the automatic solution search and, in turn, that verbal and visuospatial processes interact in exact calculation.

Direct Cortical Electrostimulation

DCE entails the application of electrical stimulation on the cortical surface (see Duffau 2007 or Szelenyi et al. 2010 or Borchers et al. 2012, for a more extended introduction to methods). The use of a bipolar probe has become the standard method. In bipolar stimulation, when both anode and cathode electrodes are active, the structures located between the two electrodes are stimulated, receiving the maximum current density. Bipolar electrodes are usually spaced 5–10 mm and deliver current (60 Hz) for a duration of 1–4 ms at different intensities, depending, for example, on whether the patient is on local or general anesthesia. Preoperative imaging is often used to plan the operation. The specific electrode position can be tailored (Figures 32.3 and 32.4), in which case, the precise number of electrodes can be determined in the operating room. Sterile tags with numbers or letters on stimulation-positive spots helps the visualization of the cortical areas involved in the function. Electrodes can also come in linear arrays or in grids. More specifically, the required intensity entails a trade-off with the duration of the pulse, depending also on the fiber speed. The spatial resolution of DCE is five mm, which is how far the actual DCE diffuses. There are, as in other stimulation techniques, diffusions caused by physiological propagation—that is,

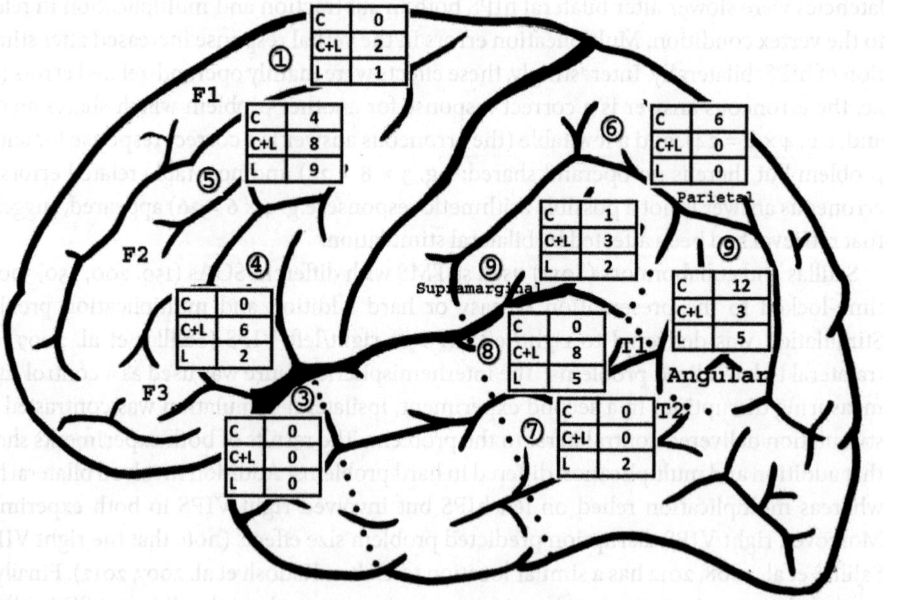

FIGURE 32.3 Scheme of the localization of calculation and language interference. Some of the areas overlapped and some were dissociated. C = number of specific calculation interferences found in the region tested; C + L = number of common calculation and language (naming and/or reading) interferences found; L = number of naming or reading interferences found (without calculation interference).

Reproduced from Franck-Emmanuel Roux, Leila Boukhatem, Louisa Draper, and Jean-François Démonet, Cortical calculation localization using electrostimulation: Clinical article, *Journal of Neurosurgery*, 110(6), pp. 1291–1299, figure 1, DOI: 10.3171/2008.8.JNS17649 © 2009, American Association of Neurological Surgeons.

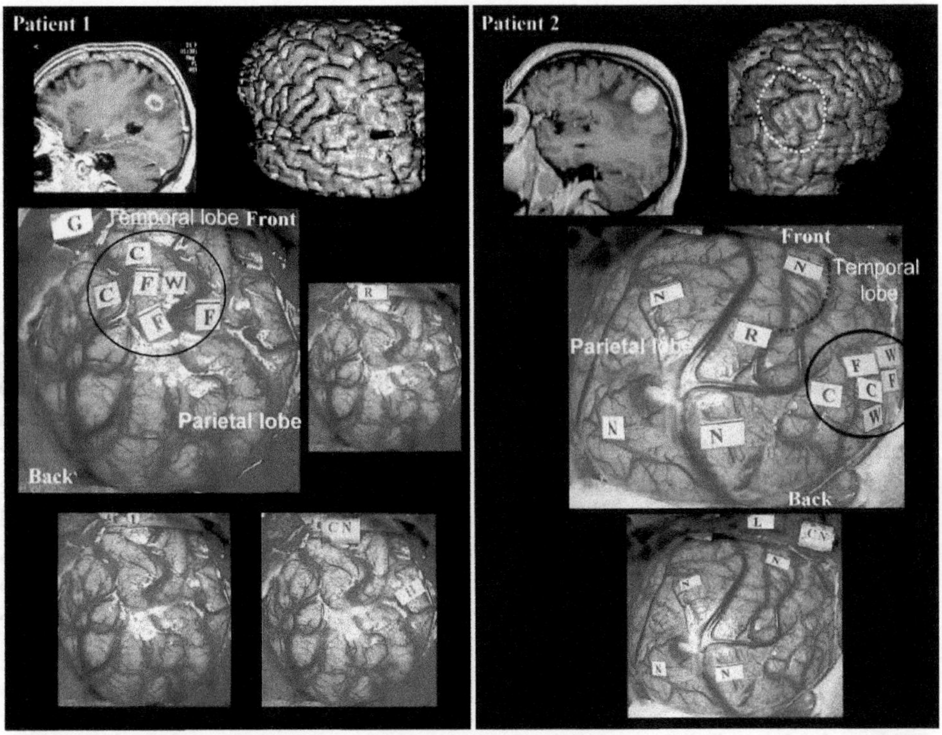

FIGURE 32.4 Examples of two patients reprduced from Figure 32.1 in Roux et al. (2003) showing the proximity of different sites in left AG where Gerstmann symptoms were observed C = acalculia site; CN = color-naming procedure; F = finger agnosia site; G = Gerstmann symptoms; H = hesitation during object-naming, color-naming, or reading procedure; L = language mapping: object naming- or color naming-interference site; N = areas negative for functions tested in this study; R = reading—interference site; W = agraphia site.

Reproduced from Franck-Emmanuel Roux, Sergio Boetto, Oumar Sacko, François Chollet, and Michel Trémoulet, Writing, calculating, and finger recognition in the region of the angular gyrus: a cortical stimulation study of Gerstmann syndrome, *Journal of Neurosurgery*, 99(4), pp. 716–727, Figure 1 © 2003, American Association of Neurological Surgeons.

the sub-circuits or networks to which the stimulated cortical tissue is connected. Ideally, concurrent electrophysiological recordings (through electrocorticography-ECoG) can measure the real physiological diffusion at the same time, which serves to monitor the excitability threshold in the avoidance of seizures. The topography of the tumor is delimited using a MRI, and functional neuroimaging can provide a first estimation of the location of the eloquent areas.

DCE has the advantage of not causing false negatives: eloquent structures with any role in a brain function that are disturbed by DCE will necessary induce a functional consequence. This is why, in the relative short time available for mapping, the assessed functions must be optimized. Nevertheless, a false negative would occur obviously, if the function was not included in the protocol or if the chosen intensity did not reach the excitability parameters of the tissue. DCE can cause false positives for different reasons: tiredness of the patient, partial seizures that look like positive effects, or by physiological spreading within a network that is wider than the tested area. Finally, DCE's positive effects cannot predict whether that function can be compensated, owing to brain plasticity. In fact, a tracking of postoperative outcomes is necessarily included in the protocols, with the aim of tracking the effects of surgery in a matching between intra-surgical positives and negatives and post-operative actual cognitive functioning.

DCE in The Study of Calculation

A limited amount of studies conducted with DCE are available so far, and they will be summarized hereafter, grouped by the different goals of the studies reported. They all concern calculation and have dealt with the differentiation and overlap between operations within the left parietal lobe, but they have also attempted the study of the differing right and left hemispheres' roles on calculation. Given the tradition of testing the dominant hemisphere for language before operation, most of the studies were only focused on the left parietal lobe, and only three cases of right hemisphere stimulation are reported in the literature. Association and dissociation of Gerstmann syndrome symptoms in the left AG have also been reported. Stimulated sites through DCE have covered the AG and the IPS, and one study so far has studied the middle frontal gyrus.

Dissociations Between Operations

In a first, single-patient investigation, Whalen et al. (1997) provided evidence as to how the production of one-digit multiplications solutions was disturbed by left anterior parietal stimulation in two consecutive sites in the AG (28% correct) to a much larger extent than single digit addition (87% correct responses). Twenty-seven trials were commission errors (vs. 19 trials without a response within the 7-second stimulation period), but the errors were not qualitatively analyzed. This pattern was not explained by the disruption of speech production or from impairment in attention. They ruled out possible differences between the difficulty of operations by comparing multiplication and addition problems that have the same answers (e.g., 9 + 7; 8 × 2); also, the two operations differed significantly in impairment. Stimulation on other sites (i.e. medial parietal, frontal) or other electrodes adjacent to the two positive anterior parietal sites did not impair any task. The authors explained the results as impairment in multiplication of arithmetic facts that are stored separately from addition facts, although an alternative explanation, in terms of over-learned additions, was considered.

Duffau et al. (2002) used both a multiplication and subtraction task in a single case of left parietal glioma. Multiplication consisted of simple single-digit problems with one operand < six. Subtraction consisted of subtracting seven from one- or two-digit numbers. Patients were requested to give an 'I don't know' response when they could not arrive to the solution to rule out speech arrest confound. Three positive sites were found in the AG: one site in the inferior part of the AG and similar in location to Whalen et al. (1997) showed disruption only in multiplication (100% errors); stimulation to one site in the superior part of AG and immediately below the IPS, disturbed only subtraction. In the area between these two sites, stimulation of another site disturbed both operations. The authors suggest a 'calculotopy' for calculation based on a well-ordered distribution of specific calculation sites, with overlapping permitting functional organization. The authors did not control for solution size.

Another, less detailed, single case study (Kurimoto et al. 2006) detected common addition and subtraction areas in the left AG.

Roux et al. (2009) combined a language and a calculation task to DCE, with the intention of sparing both language and calculation functions in their surgical procedures (Figure 32.3). Simple two-digit plus two-digit additions were used in the test of calculation functions. Interestingly, these authors argued that they chose complex addition (e.g. 26 + 43) instead of simple addition (2 + 2) or multiplication tables (7 × 3) because these latter tasks are generally solved in a completely automatic fashion. Discarding patients who presented preoperative acalculic symptoms, data were obtained from 16 patients. By stimulating brain areas possibly involved in calculation (the AG, intraparietal fissure, and dorsal premotor cortex-F2 in the left hemisphere), they caused various patterns of calculation interference. From 79 sites found with interference, 26 were naming- or reading-specific, 23 were calculation-specific, and 30 were common naming and calculation sites. Twelve out of 15 calculation specific areas were found in AG, six out of six were found in the intraparietal fissure, and four out of 12 were found in F2. Another was found in the supramarginal gyrus. From the 23 sites, 10 were 'single sites', in the sense than stimulating the surrounding 1 cm^2 area did not show calculation interference. The authors analyzed errors, as well, with 11 omissions, five wrong answers, two hesitations, and five mixed responses distributed across sites (Figure 32.3). Importantly, the proximity of the localized calculation areas to the corticectomy (removed or less than 2 cm) predicted postoperative acalculia.

Yu et al. (2011), described below, found a differentiation in the right parietal and temporal cortex between multiplication and subtraction. Stimulation of the right IPL and of the right AG impaired performance on simple subtraction but not on multiplication. Della Puppa et al. (2013) found different sites for addition and multiplication (see below). Thus, differentiation between operations can also be detected in the right hemisphere.

Right vs. Left IPS for Calculation

Pu et al. (2011) was the first DCE study contrasting the right and left hemispheres in different patients. They administered a numeric task to five patients: four were treated in the left hemisphere, and one was treated in the right hemisphere. Patients performed one-digit subtraction from a two-digit number and multiplications, wherein at least one of the operands was < six. While no interference was found in the right hemisphere, the authors found interferences with subtraction and multiplication in the left AG, in the left HIPS, and, with multiplication only, in the left supramarginal gyrus. Similar to the single case study by Duffau et al. (2002), the overlap between subtraction and multiplication was only partial, and sites were also detected where stimulation interfered with only one of the two types of operations. Solution size was not controlled in their operations. Rather than interpreting the results as a possible calculotopy, the author emphasized the differences between hemispheres, with the right hemisphere possibly implied in a more complex processing sequence or cognitive constituents of arithmetic, but not essential for it.

Yu et al. (2011) found in a single case study that multiplication was not affected by DCE of the right parietal and temporal cortex, while stimulation of the right IPL and of the right AG impaired performance on simple subtraction problems. They used subtractions where both of the operands were no more than 10 and single-digit multiplication with at least one operand < six. It is remarkable that the pre- and post-operative evaluation used fine measures, such as a numerical Stroop paradigm, apart from subtraction tasks. The authors explained the right parietal involvement in subtraction by the involvement of quantity processing, rather than verbal numerical processing, which was more relevant for multiplication. They also implied the role of spatial representation of numbers: spatial attention, visual imagery, and visuospatial working memory in the selected parietal sites.

Finally, Della Puppa et al. (2013) evaluated the clinical impact of simple calculation in DCE of right-handed patients affected by a right parietal brain tumor. Overall, numerical processing interferences were found in four parietal regions explored by electro-stimulation: three interferences (three multiplication) in the AG, three interferences (two multiplication and one addition) in the SMG, two interferences (two multiplication) in the HIPS, one interference (one multiplication) in the Superior lobule. This is the first DCE study that shows non-dominant right hemisphere involvement in multiplication and addition.

Gerstmann Syndrome and The Left AG

Roux et al. (2003) performed the first case series study with the double goal of characterizing the left AG sites behind the Gerstmann syndrome and possibly sparing the proposed as associated writing, finger gnosis and calculation, plus language functions. The authors found, in five patients, that different symptoms of Gerstmann syndrome can be found during DCE to the left AG, which may or may not be associated with language interference sites. According to the authors, the pattern of specific cortical stimulation findings explains the rarity of finding all of the Gertsmann syndrome symptoms at once, as only the distributed lesions within the AG would lead to manifestation of the complete syndrome (Figure 32.4).

Qualitative Analysis of Errors in DCE

We will briefly mention here some preliminary findings from our own laboratory that seem to add important information. The role of the left AG was explored through simple addition and multiplication in one patient suffering from low-grade left parietal glioma. While interference for simple addition was found in the posterior portion of the left AG, interference with simple multiplication was found in the anterior portion. Crucially, errors arising during simple multiplication were retrieval errors, while errors with a response close to the correct solution were given during addition, suggesting approximation errors. Thus, because different processes implied in different operations were disrupted, the effects cannot be described simply as a calculotopy. These results are exposed with the aim of highlighting the interest of the analysis of commission errors: in a possible case, if stimulating an area elicits an approximation error for addition, but stimulation in another area elicits retrieval errors (i.e. operand errors), the different contribution of the two stimulated sites could be inferred. In turn, the pattern of errors suggests different roles for each area. Similarly, the analysis of response latencies could follow a pattern, albeit a noisy one. While only two of the TMS studies studying calculation have reported effects in errors (Andres et al. 2011; Göbel et al. 2006b), the proportion of errors was the dependent variable mostly analyzed in DCE studies.

Overall, DCE succeeds in finding dissociations between operations. Interestingly, some studies, albeit not precisely compared, seem to converge in specific areas (e.g. left AG in Whalen et al. 1997 and Duffau et al. 2002 for multiplication). It is of interest that the pattern of overlap and separation between tasks appears in these two studies. Perhaps the hypothesis of calculotopy is too extreme, especially taking into account that size effect has not been fully controlled. They mainly show that addition and multiplication somehow differ, and they are not just two instances of arithmetic fact retrieval.

With regard to the differential involvement of right and left IPS, not much can be concluded so far from this technique, as only two patients with different results have been evaluated. Nonetheless, positive results have been shown for subtraction and even for arithmetic facts, in not always overlapped sites, in the right IPS. Also remarkable is the calculation impairment found in F2 for complex addition. Again, further characterization is needed for frontal areas using DCE in the study of math.

Discussion

So far, TMS studies in the research of math converge about the key role of parietal sites in the representation of numerical and some non-numerical quantities, in the link between space and numbers and in calculation. Bilateral posterior parietal sites are needed in spatio-numerical processes. IPS is essential during the access to magnitude, although stronger left IPS involvement appears. Conversely, calculation processes tend to be bilateral, contrary to the left lateralization that would be expected from fMRI studies or patient data. Consequently, TMS effects suggest that processes proposed to rely on a left lateralized verbal network could be also dependent on non-verbal components in the non-dominant hemisphere. Conversely, representations that are proposed to be bilateral may rely later on in the left hemisphere, as representations of numerosity link up with language (Rivera et al. 2005).

DCE studies show, on the one hand, that multiplication is specifically disrupted with respect to addition and subtraction in the left AG. They also show that areas in the left intraparietal fissure, AG, and medial prefrontal and supramarginal gyrus, all contribute to complex addition. Finally, in one occasion, subtraction disruption is found in the right hemisphere. Most of the DCE studies were not so detailed about distinct locations within the parietal lobe. With two exceptions, where the right hemisphere was stimulated, they all concerned the left hemisphere, and in no case was information about the nature of errors provided. An analysis of the type of errors after the surgery, although of no value for the surgeon, can inform the researcher about the part of the process that has been disrupted.

TMS and DCE in Numerical Cognition

What can be concluded at present from the convergent results in DCE and TMS, attending to the fact that DCE has only addressed calculation processes, is as follows: (1) The dissociation between arithmetic operations in different brain loci. DCE has shown also within-area dissociation and overlap. Note that this pattern of dissociation and overlap is precisely what is predicted between numerical and non-numerical quantities, thereby showing the sensitivity of DCE to this network pattern; (2) the common involvement of left parietal areas in subtraction and multiplication, associated in the left HIPS. Less can be concluded at this point for the right hemisphere by contrasting these two techniques; and (3) symptoms in Gerstmann syndrome are associated in the left AG.

Both TMS and DCE are techniques that imply a reversible disruption of functional brain areas allowing the mapping of brain functions, although both electrical stimulation and TMS are thought to silence all cortical areas that receive direct input from the fibers of the stimulated neurons (Borchers et al. 2012). Therefore, it is better to talk about networks, rather than areas. Despite the fact that they are similar, these techniques entail different experimental approaches. Given the pragmatic advantages of TMS (i.e. that there is no need of a between-subjects design, as well as that there are much lower restrictions in time or reaction time analysis), it can, and has, complemented these few DCE studies. This, on the other hand, allows for the avoidance of false negatives (Mandonnet et al. 2010) and allows for direct stimulation of the brain tissue. It is worth noting that DCE implies stimulation in patients with brain damage or epilepsy, and brain reorganization and compensation might have occurred previously to stimulation. Nonetheless, TMS could facilitate a more precise task design in DCE studies. On the other hand, DCE, in the study, as well as prevention of damage to math functions, needs of an 'omnibus' but simple task that really predicts relatively more complex future math performance in the patient, such as complex calculations, in order to assure more than an isolated preservation of arithmetic facts. Such a pertinent task—or, rather, group of tasks—still needs to be offered to the patient.

In conclusion, a truly meaningful comparison can hardly be made at the present time. However, this chapter is a first attempt to make a reasoned comparison between the results of two methods, implying temporary inactivation in the study of math processing. More is bound to follow.

References

Andres M, Seron X, and Olivier E (2005). Hemispheric lateralization of number comparison. *Brain Research Cognitive Brain Research* 25(1): 283–90.

Andres M, Pelgrims B, Michaux N, Olivier E, and Pesenti, M (2011). Role of distinct parietal areas in arithmetic: an fMRI-guided TMS study. *Neuroimage* 54(4): 3048–56.

Ansari D, Fugelsang JA, Dhital B, and Venkatraman V (2006). Dissociating response conflict from numerical magnitude processing in the brain: an event-related fMRI study. *Neuroimage* 32(2): 799–805.

Ashbridge E, Walsh V, and Cowey A (1997). Temporal aspects of visual search studied by transcranial magnetic stimulation. *Neuropsychologia* 35(8): 1121–31.

Borchers S, Himmelbach M, Logothetis N, and Karnath, HO (2012). Direct electrical stimulation of human cortex—the gold standard for mapping brain functions? *Nature Reviews Neuroscience* 13: 63–70.

Butterworth, B (1999). *The Mathematical Brain*. London: Macmillan.
Calabria M and Rossetti Y (2005). Interference between number processing and line bisection: a methodology. *Neuropsychologia* 43(5): 779–83.
Cantlon JF, Brannon EM, Carter EJ, and Pelphrey KA (2006). Functional imaging of numerical processing in adults and 4-year-old children. *PLoS Biology* 4(5): e125.
Cappelletti M, Barth H, Fregni F, Spelke ES, and Pascual-Leone A (2007). rTMS over the intraparietal sulcus disrupts numerosity processing. *Experimental Brain Research* 179(4): 631–42.
Cappelletti M, Muggleton N, and Walsh V (2009). Quantity without numbers and numbers without quantity in the parietal cortex. *Neuroimage* 46(2): 522–9.
Cappelletti M, Lee HL, Freeman ED, and Price CJ (2010). The role of right and left parietal lobes in the concpetual processing of numbers. *Journal of Cognitive Neuroscience* 22(2): 331–46.
Castronovo J and Seron X (2007). Semantic numerical representation in blind subjects: the role of vision in the spatial format of the mental number line. *Quarterly Journal of Experimental Psychology* 60(1): 101–19.
Cattaneo Z, Silvanto J, Pascual-Leone A, and Battelli L (2009a). The role of the angular gyrus in the modulation of visuospatial attention by the mental number line. *Neuroimage* 44(2): 563–8.
Cattaneo, Z, Silvanto J, Battelli L, and Pascual-Leone A (2009b). The mental number line modulates visual cortical excitability. *Neuroscience Letters* 462(3): 253–6.
Cheng GL, Tang J, Walsh V, Butterworth B, Cappelletti M (2013). Differential effects of left parietal theta-burst stimulation on order and quantity processing. *Brain Stimulation* 6(2): 160–5.
Cipolotti L, Butterworth B, and Denes G (1991). A specific deficit for numbers in a case of dense acalculia. *Brain* 114 (Pt. 6): 2619–37.
Cohen Kadosh R and Gertner L (2010). Synesthesia: Gluing together Time, Number, and Space. In: *Space, Time and Number in the brain*, edited by S Dehaene and E Brannon, pp. 123–32. London: Academic Press.
Cohen Kadosh R and Henik A (2007). Can synaesthesia research inform cognitive science? *Trends in Cognitive Sciences* 11: 177–84.
Cohen Kadosh R, Cohen Kadosh K, Schuhmann T, Kaas A, Goebel R, Henik A, et al. (2007). Virtual dyscalculia induced by parietal-lobe TMS impairs automatic magnitude processing. *Current Biology* 17(8): 689–93.
Cohen Kadosh R, Lammertyn J, and Izard V (2008). Are numbers special? An overview of chronometric, neuroimaging, developmental and comparative studies of magnitude representation. *Progress in Neurobiology* 84(2): 132–47.
Cohen Kadosh R, Muggleton N, Silvanto J, and Walsh V (2010a). Double dissociation of format-dependent and number-specific neurons in human parietal cortex. *Cerebral Cortex* 20: 2166–71.
Cohen Kadosh R, Soskic S, Iuculano T, Kanai R, and Walsh V (2010b). Modulating neuronal activity produces specific and long-lasting changes in numerical competence. *Current Biology* 20(22): 2016–20.
Cohen Kadosh R, Bahrami B, Walsh V, Butterworth B, Popescu T, and Price CJ (2011). Specialization in the human brain: the case of numbers. *Frontiers in Human Neuroscience* 5:62.
Cohen Kadosh R, Bien N, and Sack AT (2012). Automatic and intentional number processing both rely on intact right parietal cortex: a combined FMRI and neuronavigated TMS study. *Frontiers in Human Neuroscience* 6: 2.
Dehaene S (1996). The organization of brain activations in number comparison: Event-related potentials and the additive-factors method. *Journal of Cognitive Neuroscience* 8: 47–68.
Dehaene S (1997). *The Number Sense*. New York: Oxford University Press.
Dehaene S and Cohen L (1991). Two mental calculation systems: a case study of severe acalculia with preserved approximation. *Neuropsychologia* 29(11): 1045–54.
Dehaene S, Bossini S, and Giraux P (1993). The mental representation of parity and numerical magnitude. *Journal of Experimental Psychology: General* 122: 371–96.
Dehaene S, Piazza M, Pinel P, and Cohen L (2003). Three parietal circuits for number processing. *Cognitive Neuropsychology* 20: 487–506.

Dehaene S, Molko N, Cohen L, and Wilson AJ (2004). Arithmetic and the brain. *Current Opinion in Neurobiology* 14: 218–24.
Della Puppa A, De Pellegrin S, D'Avella E, Gioffré G, Munari M, Saladini M, et al. (2013). Right parietal cortex and calculation processing. Intra-operative functional mapping of multiplication and addition. *Journal of Neurosurgery* 119(5): 1107–11.
Dormal V and Pesenti M (2009). Common and specific contributions of the intraparietal sulci to numerosity and length processing. *Human Brain Mapping* 30(8): 2466–76.
Dormal V, Andres M, and Pesenti M (2008). Dissociation of numerosity and duration processing in the left intraparietal sulcus: a transcranial magnetic stimulation study. *Cortex* 44(4): 462–9.
Dormal V, Andres M, Dormal G, Pesenti M (2010). Mode-dependent and mode-independent representations of numerosity in the right intraparietal sulcus. *Neuroimage*, 52(4), 1677–1686.
Dormal V, Andres M, and Pesenti M (2012). Contribution of the right intraparietal sulcus to numerosity and length processing: an fMRI-guided TMS study. *Cortex* 48(5): 623–9.
Duffau H (2007). Contribution of cortical and subcortical electrostimulation in brain glioma surgery: methodological and functional considerations. *Neurophysiologie clinique = Clinical neurophysiology* 37: 373–82.
Duffau H (2011). Do brain tumours allow valid conclusions on the localization of human brain functions? *Cortex* 47: 1016–17.
Duffau H, Denvil D, Lopes M, Gasparini F, Cohen L, Capelle L, et al. (2002). Intraoperative mapping of the cortical areas involved in multiplication and subtraction: an electrostimulation study in a patient with a left parietal glioma. *Journal of Neurology, Neurosurgery and Psychiatry* 73(6): 733–8.
Duffau H, Gatignol P, Mandonnet E, Capelle L, and Taillandier L (2008). Intraoperative subcortical stimulation mapping of language pathways in a consecutive series of 115 patients with Grade II glioma in the left dominant hemisphere. *Journal of Neurosurgery* 109: 461–71.
Eger E, Sterzer P, Russ MO, Giraud AL, and Kleinschmidt A (2003). A supramodal number representation in human intraparietal cortex. *Neuron* 37(4): 719–25.
Emerson RW and Cantlon JF (2012). Early math achievement and functional connectivity in the fronto-parietal network. *Developmental Cognitive Neuroscience* 2(S1), S139–S151.
Fayol M, Barrouillet P, and Marinthe C (1998). Predicting arithmetical achievement from neuropsychological performance: A longitudinal study. *Cognition* 68: 63–70.
Fias W, Lauwereyns J, and Lammertyn J (2001). Irrelevant digits affect feature-based attention depending on the overlap of neural circuits. *Brain Research, Cognitive Brain Research* 12: 415–23.
Fias W, van Dijck JP, and Gevers W (2010). How is number associated with space? The role of working memory In: *Space, Time and Number in the brain*, edited by S Dehaene and E Brannon, Pp. 133–48. Academic Press.
Fink GR, Marshall JC, Shah NJ, Weiss PH, Halligan PW, Grosse-Ruyken M, et al. (2000). Line bisection judgments implicate right parietal cortex and cerebellum as assessed by fMRI. *Neurology* 54(6): 1324–31.
Fischer MH, Castel AD, Dodd, MD, and Pratt J (2003). Perceiving numbers causes spatial shifts of attention. *Nature Neuroscience* 6(6): 555–6.
Gerstmann J (1940). Syndrome of finger agnosia, disorientation for right and left, agraphia, acalculia. *Archives of Neurology and Psychology* 44: 398–408.
Göbel SM, Walsh V, and Rushworth MF (2001). The mental number line and the human angular gyrus. *Neuroimage* 14(6): 1278–89.
Göbel SM, Calabria M, Farnè A, and Rossetti Y (2006a). Parietal rTMS distorts the mental number line: simulating 'spatial' neglect in healthy subjects. *Neuropsychologia* 44(6): 860–8.
Göbel SM, Rushworth MF, and Walsh V (2006b). Inferior parietal rTMS affects performance in an addition task. *Cortex* 42(5): 774–81.
Hecaen H, Angelergues R, and Houllier S (1961). Les variétes cliniques de acalculies au cours de lesions retrorolandiques: approche statistique du problème. *Revue neurologique.* 105: 85–103.
Henschen SE (1919). Uber Sprach-, Musik- und Rechenmechanismen und ihre Lokalisation im Großhirn, *Zeitshrift für die gesamte neurologie und psychiatry* 52: 273–98.

Holloway ID, Price GR, and Ansari D (2010). Common and segregated pathways for the processing of symbolic and nonsymbolic numerical magnitude: An fMRI study. *NeuroImage* 49(1): 1006–17.

Honey CJ, Kötter R, Breakspear M, and Sporns O (2007). Network structure of cerebral cortex shapes functional connectivity on multiple time scales. *Proceedings of the National Academy of Sciences USA* 104(24): 10240–5.

Hubbard EM, Piazza M, Pinel P, and Dehaene S (2005). Interactions between number and space in parietal cortex. *Nature Reviews Neuroscience* 6(6): 435–48.

Ioannides AA (2007). Dynamic functional connectivity. *Current Opinion in Neurobiology* 17(2): 161–70.

Iuculano T and Cohen Kadosh R (2013) The mental cost of cognitive enhancement. *The Journal of Neuroscience* 33(10): 4482–6.

Izard V, Dehaene-Lambertz G, and Dehaene S (2008). Distinct cerebral pathways for object identity and number in human infants. *PLoS Biology* 6: e11.

Karnath HO and Steinbach JP (2011). Do brain tumours allow valid conclusions on the localisation of human brain functions? Objections. *Cortex* 47: 1004–6.

Kurimoto M, Asahi T, Shibata T, Takahashi C, Nagai S, Hayashi N, et al. (2006). Safe removal of glioblastoma near the angular gyrus by awake surgery preserving calculation ability. *Neurologia medico-chirurgica (Tokyo)* 46: 46–50.

Lavidor M, Brinksman V and Göbel SM (2004). Hemispheric asymmetry and the mental number line: comparison of double-digit numbers. *Neuropsychologia* 42(14):1927–33.

Libertus ME, Woldorff MG, and Brannon EM (2007). Electrophysiological evidence for notation independence in numerical processing. *Behavioral Brain Functions* 3, 1.

Mandonnet E, Winkler PA, and Duffau H (2010). Direct electrical stimulation as an input gate into brain functional networks: principles, advantages and limitations. *Acta Neurochirurgica (Wien)* 152(2): 185–93.

McCrink K, Dehaene S, and Dehaene-Lambertz G (2007). Moving along the number line: operational momentum in nonsymbolic arithmetic. *Perception and Psychophysics* 69(8): 1324–33.

Mesulam MM (1981). A cortical network for directed attention and unilateral neglect. *Annals in Neurology* 10(4): 309–25.

Nieder A, and Miller EK (2004). A parieto-frontal network for visual numerical information in the monkey. *Proceedings of the National Academy of Sciences USA* 101(19): 7457–62.

Notebaert K, Nelis S, and Reynboet B (2011). The magnitude representation of small and large symbolic numbers in the left and right hemisphere: an event-related fMRI study. *Journal of Cognitive Neuroscience* 23(3): 622–30.

Orban GA, Claeys K, Nelissen K, Smans R, Sunaert S, Todd JT, et al. (2006). Mapping the parietal cortex of human and non-human primates. *Neuropsychologia* 44(13): 2647–67.

Pascual-Leone A, Walsh V, and Rothwell J (2000). Transcranial magnetic stimulation in cognitive neuroscience—virtual lesion, chronometry, and functional connectivity. *Current opinion in neurobiology* 10: 232–7.

Pinel P, Dehaene S, Riviere D, and LeBihan D (2001). Modulation of parietal activation by semantic distance in a number comparison task. *Neuroimage* 14(5): 1013–26.

Pinel P, Piazza M, LeBihan D, and Dehaene S (2004). Distributed and overlapping cerebral representations of number size and luminance during comparative judgements. *Neuron* 41(6): 983–93.

Price GR, Holloway I, Räsänen P, Vesterinen M, and Ansari D (2007). Impaired parietal magnitude processing in developmental dyscalculia. *Current Biology* 18: 17(24).

Pu S, Li YN, Wu CX, Wang YZ, Zhou XL, and Jiang T (2011). Cortical areas involved in numerical processing: an intraoperative electrostimulation study. *Stereotactic and Functional Neurosurgery* 89: 42–7.

Renzi C, Vecchi T, Silvanto J, and Cattaneo, Z (2011). Overlapping representations of numerical magnitude and motion direction in the posterior parietal cortex: a TMS-adaptation study. *Neuroscience Letters* 490(2): 145–9.

Rivera SM, Reiss SM, Eckert V, and Menon V (2005). Developmental changes in mental arithmetic: evidence for increased functional specialization in the left inferior parietal cortex. *Cerebral Cortex* 15: 1779–90.

Robertson LC (2003). Binding, spatial attention and perceptual awareness. *Nature Reviews Neuroscience* 4: 93–102.

Robertson EM, Theoret H, and Pascual-Leone A (2003). Studies in cognition: the problems solved and created by transcranial magnetic stimulation. *Journal of Cognitive Neuroscience* 15: 948–60.

Rossetti Y, Jacquin-Courtois S, Rode G, Ota H, Michel C, and Boisson D (2004). Does action make the link between number and space representation? Visuo-manual adaptation improves number bisection in unilateral neglect. *Psychological Science* 15(6): 426–30.

Rotzer S, Kucian K, Martin E, von Aster M, Klaver P, and Loenneker T (2008). Optimized voxel-based morphometry in children with developmental dyscalculia. *Neuroimage* 39(1): 417–22.

Roux FE, Boetto S, Sacko O, Chollet F, and Trémoulet M (2003). Writing, calculating, and finger recognition in the region of the angular gyrus: a cortical stimulation study of Gerstmann syndrome. *Journal of Neurosurgery* 99: 716–27.

Roux FE, Boukhatem L, Draper L, Sacko O, and Démonet JF (2009). Cortical calculation localization using electrostimulation. *Journal of Neurosurgery* 110: 1291–9.

Ruff CC, Blankenburg F, Bjoertomt O, Bestmann S, Freeman E, Haynes JD, et al. (2006). Concurrent TMS-fMRI and psychophysics reveal frontal influences on human retinotopic visual cortex. *Current Biology* 16: 1479–88.

Rusconi E, Walsh V, and Butterworth B (2005). Dexterity with numbers: rTMS over left angular gyrus disrupts finger gnosis and number processing. *Neuropsychologia* 43(11): 1609–24.

Rusconi E, Turatto M, and Umilta C (2007). Two orienting mechanisms in posterior parietal lobule: an rTMS study of the Simon and SNARC effects. *Cognitive Neuropsychology* 24(4): 373–92.

Rusconi E, Bueti D, Walsh V, and Butterworth B (2011). Contribution of frontal cortex to the spatial representation of number. *Cortex* 47(1): 2–13.

Rusconi E, Dervinis M, Verbruggen F, and Chambers CD (2013). Critical Time Course of Right Frontoparietal Involvement in Mental Number Space. *Journal of Cognitive Neuroscience* 25: 465–83.

Sack, AT, Cohen Kadosh, R, Schuhmann, T, Moerel, M, Walsh, V, and Goebel, R (2009). Optimizing functional accuracy of TMS in cognitive studies: A comparison of methods. *Journal of Cognitive Neuroscience* 21: 207–21.

Sagiv N, Simner J, Collins J, Butterworth B, and Ward J (2006). What is the relationship between synaesthesia and visuo-spatial number forms? *Cognition* 101: 114–28.

Salillas E, Basso D, Baldi M, Semenza C, and Vecchi T (2009). Motion on numbers: transcranial magnetic stimulation on the ventral intraparietal sulcus alters both numerical and motion processes. *Journal of Cognitive Neuroscience* 21(11): 2129–38.

Salillas E, Graná A, El-Yagoubi R, and Semenza C (2009). Numbers in the blind's 'eye'. *PLoS One* 4(7): e6357.

Salillas E, Semenza C, Basso D, Vecchi T, and Siegal M (2012). Single pulse TMS induced disruption to right and left parietal cortex on addition and multiplication. *Neuroimage* 59(4): 3159–65.

Sandrini M, Rossini PM, and Miniussi C (2004). The differential involvement of inferior parietal lobule in number comparison: a rTMS study. *Neuropsychologia* 42(14): 1902–9.

Sandrini M, Umiltá C, and Rusconi E (2011). The use of transcranial magnetic stimulation in cognitive neuroscience: a new synthesis of methodological issues. *Neuroscience and Biobehavioral Reviews* 35: 516–36.

Sasanguie D, Göbel SM, and Reynboet B (2013). Left parietal TMS disturbs priming between symbolic and non-symbolic number representations. *Neuropsychologia* 51(8): 1528–33.

Sato M, Cattaneo L, Rizzolatti G, and Gallese V (2007). Numbers within our hands: modulation of corticospinal excitability of hand muscles during numerical judgment. *Journal of Cognitive Neuroscience* 19(4): 684–93.

Shallice T and Skrap M (2011). Localisation through operation for brain tumour: a reply to Karnath and Steinbach. *Cortex* 47: 1007–9.

Silvanto J, Muggleton N and Walsh V (2008). State-dependency in brain stimulation studies of perception and cognition. *Trends in Cognitive Sciences* 12(12): 447–54.

Snowball A, Tachtsidis I, Popescu T, Thompson J, Delazer M, Zamarian L, et al. (2013). Long-Term Enhancement of Brain Function and Cognition Using Cognitive Training and Brain Stimulation. *Current Biology* doi: 10.1016/j.cub.2013.04.045.

Stewart LM, Walsh V, and Rothwell JC (2001). Motor and phosphene thresholds: a transcranial magnetic stimulation correlation study. *Neuropsychologia* 39(4): 415–19.

Szelenyi A, Bello L, Duffau H, Fava E, Feigl GC, et al. (2010). Intraoperative electrical stimulation in awake craniotomy: methodological aspects of current practice. *Neurosurgical Focus* 28: E7.

Terney D, Chaieb L, Moliadze V, Antal A, and Paulus W (2008). Increasing human brain excitability by transcranial high-frequency random noise stimulation. *The Journal of neuroscience: the official journal of the Society for Neuroscience* 28 14147–55.

Turconi E, Jemel B, Rossion B, and Seron X (2004). Electrophysiological evidence for differential processing of numerical quantity and order in humans. *Brain Research Cognitive Brain Research* 21(1): 22–38.

Utz KS, Dimova, V, Oppenlander K, and Kerkhoff G (2010). Electrified minds: Transcranial direct current stimulation (tDCS) and galvanic vestibular stimulation (GVS) as methods of non-invasive brain stimulation in neuropsychology—a review of current data and future implications. *Neuropsychologia* 48: 2789–810.

Vanduffel W, Fize D, Peuskens H, Denys K, Sunaert S, Todd JT, et al. (2002). Extracting 3D from motion: differences in human and monkey intraparietal cortex. *Science* 298(5592): 413–15.

Varela F, Lachaux JP, Rodriguez E, and Martinerie J (2001). The brain web: phase synchronization and large-scale integration. *Nature Reviews Neuroscience* 2(4): 229–39.

Vuilleumier P, Ortigue S, and Brugger P (2004). The number space and neglect. *Cortex* 40(2): 399–410.

Walsh V (2003). A theory of magnitude: common cortical metrics of time, space and quantity. *Trends in Cognitive Science* 7(11): 483–8.

Walsh V and Cowey A (2000). Transcranial magnetic stimulation and cognitive neuroscience. *Nature Reviews Neuroscience* 1(1): 73–9.

Walsh V and Rushworth M (1999). A primer of magnetic stimulation as a tool for neuropsychology. *Neuropsychologia* 37(2): 125–35.

Whalen J, McCloskey M, Lesser RP and Gordon B (1997). Localizing Arithmetic Processes in the Brain: Evidence from a Transient Deficit During Cortical Stimulation. *Journal of Cognitive Neuroscience* 9(3): 409–17.

Yu X, Chen C, Pu S, Wu C, Li Y, Jiang T, and Zhou X (2011). Dissociation of subtraction and multiplication in the right parietal cortex: evidence from intraoperative cortical electrostimulation. *Neuropsychologia* 49: 2889–95.

Zorzi M, Priftis K, and Umiltà C (2002). Brain damage: neglect disrupts the mental number line. *Nature* 417(6885): 138–9.

CHAPTER 33

APPLICATIONS OF NEUROSCIENCE TO MATHEMATICS EDUCATION

BERT DE SMEDT AND ROLAND H. GRABNER

INTRODUCTION

THE last five years have witnessed an increasing interest and effort of forging connections between neuroscience and education (e.g., Blakemore & Frith, 2005; Geake, 2009; Howard-Jones, 2010; McCandliss, 2010), despite the initial criticism that such application may be "a bridge too far" (Bruer, 1997). In high-ranking scientific journals, as well as policy documents, there appears to be a momentum to think about applications of neuroscience to wider public domains, such as education. For example, the British Royal Society published a series of essays by leading neuroscientists in which the translation of neuroscientific knowledge into useful applications was discussed. One of these essays was entitled "Neuroscience: implications for education and lifelong learning" (The Royal Society, 2011). At the same time, review papers in *Science* have pointed to such applications in the fields of reading development (Gabrieli, 2009), mathematics achievement (Butterworth et al., 2011), and language learning (Meltzoff et al., 2009). Coupled with an increasing availability of non-invasive brain imaging methods that allow us to examine the neural correlates of academic skills and how these change throughout development and education, a new interdisciplinary and translational research field of Mind, Brain, and Education (e.g., Fischer, 2009), Neuroeducation (e.g., Howard-Jones, 2010), or Educational Neuroscience (e.g., Goswami & Szücs, 2011; McCandliss, 2010) has emerged, with the core aim to conduct research that connects the fields of (cognitive) neuroscience and education.

In this chapter, we focus on such connections in the field of mathematical cognition. In the past decade, there has been a tremendous increase in neuroscience research on mathematics learning (e.g. Ansari, 2008; Arsalidou & Taylor, 2011; Dehaene, 2011; Menon, 2010; Zamarian et al., 2009 for reviews; see also chapters by Zamarian & Delazer and Menon in this volume). In view of the importance of being numerate for daily life in our modern western society and the devastating effects of poor mathematical abilities on educational progress (Duncan et al., 2007), socioeconomic status (Ritchie & Bates, 2013), employment

(Bynner & Parsons, 1997), mortgage default (Gerardi et al., 2013), and medical decision making (Reyna et al., 2009), the field of mathematics learning has been proposed as an ideal workspace for making applications of neuroscience to education (Ansari et al., 2012; De Smedt et al., 2010; Goswami & Szucs, 2011).

This chapter explores three types of applications of neuroscience to education, i.e. neurounderstanding, neuroprediction, and neurointervention. *Neurounderstanding* refers to the idea that neuroscience is generating knowledge on how people acquire mathematical skills and how this learning is reflected at the biological level. The integration of this knowledge with psychological and educational theories has the potential to provide a better understanding of the typical and atypical development of school-taught mathematical competencies. *Neuroprediction* or neuroprognosis deals with the potential of neuroimaging data to predict future skill acquisition and response to educational interventions. In *neurointervention,* we discuss how brain imaging data have been used to ground educational interventions and how education shapes the neural circuitry that underlies complex school-taught skills, such as mathematics. In addition, we present very recently developed neurophysiological interventions that have been shown to affect mathematical performance or learning. While there is no doubt that these applications offer exciting opportunities for education, some potential caveats should be considered, which are discussed at the end of this chapter.

Neurounderstanding

Neuroscience data have the potential to enrich our understanding of the acquisition of mathematical skills and complement, as well as extend, the existing knowledge we have obtained on the basis of psychological and educational research (De Smedt et al., 2010). One of the first and most straightforward applications of this deals with understanding the causes of atypical mathematical development or dyscalculia (e.g., Butterworth et al., 2011). Notably, this research approach has already been successful in increasing our understanding of other *neuro*developmental disorders, such as dyslexia (e.g., Gabrieli, 2009) and autism (e.g., Just et al., 2012), where the origin of the difficulties in learning and development are situated at the neural level. Cognitive neuropsychological studies from adult patients with brain damage (e.g., Dehaene & Cohen, 1997) further suggested that mathematical cognition is not a unitary skill, but consists of different component skills (e.g., arithmetic, problem- solving, conceptual knowledge) that might be differentially impaired and that might have different developmental trajectories (see also Dowker, 2008). In addition, neuroscientific studies in healthy individuals have also helped us to understand the neural basis of various school-taught mathematical skills, such as arithmetic strategy use and mathematical problem-solving, and individual differences in the acquisition of these mathematical competencies.

The Cause of Poor Mathematics Achievement: Developmental Dyscalculia

Developmental dyscalculia (DD), or mathematics learning disability, is a learning disorder characterized by specific and persistent poor achievement in mathematics learning, despite otherwise normal intellectual skills and scholastic opportunities (e.g., Butterworth et al., 2011; Geary, 2013). When the term "developmental dyscalculia" was coined for the first time, it was already proposed that the origin of these difficulties was situated at the neural level: "a definition of developmental dyscalculia, stressing the hereditary or congenital affection of the brain substrate of mathematical functions is put forth" (Kosc, 1974, p. 164), although at that time it was not possible to directly examine this issue in children without manifest brain damage. With the advent of non-invasive imaging techniques about 20 years ago, neuroimaging studies are now unraveling the neural correlates of this specific learning disorder, pointing towards structural and functional abnormalities in the brain network that subserves numerical and mathematical processing (e.g., Butterworth et al., 2011; Kaufmann et al., 2011; and Kucian et al. (this volume) for a review). This research is converging on the conclusion that children with dyscalculia might have an abnormal development of the brain circuitry that supports the elementary processing of numerical magnitudes. There is wide consensus that this function is supported by the intraparietal sulcus (IPS) located in the posterior parietal cortex (Ansari, 2008; Dehaene et al., 2003). In line with this, magnetic resonance imaging (MRI) studies have revealed structural (Rotzer et al., 2008; Rykhlevskaia et al., 2009) and functional (De Smedt et al., 2011; Kucian et al., 2011; Mussolin et al., 2010; Price et al., 2007) differences in the IPS between children with DD and typically developing children. Similar parietal abnormalities have been observed by using electroencephalography or EEG (Heine et al., 2013; Soltesz et al., 2007). These differences have been interpreted to reflect the absence of an increasing parietal functional specialization in children with DD. Although there are some brain stimulation studies in healthy adults that reveal a causal link between parietal brain functioning and numerical processing (e.g., Göbel et al., 2001), it remains, however, to be empirically verified whether these parietal abnormalities in children with DD are truly the cause of DD, whether they arise as a consequence of DD, or both. It also should be noted that the existing number of imaging studies in children with DD is very small and that they typically included small samples, which limits the generalizability of these findings.

Nevertheless, these neuroimaging data have constrained our understanding of DD, by suggesting numerical magnitude processing as a potential source of the difficulties of children with DD. Meanwhile, this suggestion has fueled a large number of psychological and educational studies that have empirically confirmed this hypothesis at the behavioral level (see De Smedt et al., 2013, for a review), by consistently showing that individuals with DD have significant impairments in their ability to compare (symbolic) numbers.

The previously mentioned data on basic numerical magnitude processing in DD have been applied in two ways to education. Firstly, they have led to the construction of diagnostic instruments that can be used for screening and early identification of children at risk for developing DD, such as the computerized Dyscalculia Screener (Butterworth, 2003), a recently developed two-minute paper-and-pencil test (Nosworthy et al., 2013), as well as the standardized test battery TEDI-MATH (Grégoire et al., 2004). Secondly, these data have been used as background to develop interventions that foster the development of numerical magnitude processing (see the section 'Rationale for Behavioral Educational Interventions' in this chapter).

Numerical Magnitude Processing

Insights into numerical magnitude processing and its neural correlate in the IPS have also furthered our understanding of typical numerical development (e.g. De Smedt et al., 2013, for a review). One of the most robust findings in these studies is that children who are better in determining the larger of two symbolic numbers have higher achievement in mathematics, an association that even appears to be predictive (De Smedt et al., 2009). It has been hypothesized that the acquisition of more complex school-taught arithmetical skills builds on the previously mentioned elementary numerical magnitude processing abilities (e.g., Ansari, 2008), yet longitudinal data are needed to support this idea.

Recent data also indicate associations between the brain network that supports numerical magnitude processing in children and ecologically valid classroom measures of children's (Bugden et al., 2012; Cantlon & Li, 2013; Emerson & Cantlon, 2012) and adolescents' (Matjeko et al., 2013) mathematical performance. For example, individual differences in brain activity in the IPS during symbolic number comparison are correlated with standardized measures of mathematical achievement (Bugden et al., 2012). Cantlon & Li (2013) and Emerson & Cantlon (2012) recently showed that brain activity, more specifically the functional connectivity between frontal and parietal regions, in young children who viewed educational videos that had numerical content, was associated with children's performance on standardized tests of mathematics achievement. Matjeko et al. (2013) demonstrated that left parietal white matter was positively correlated with performance on the Preliminary Scholastic Aptitude Test (PSAT), a nationally administered entrance exam to college in the US. These studies provide critical bridges between neuroscience and education because they connect brain imaging data, which are acquired in the laboratory under very strict conditions, and ecologically valid measures of mathematics learning, i.e. in-classroom performance, which are also determined by interactions with teachers and learners, as well as other environmental factors, processes that are ubiquitous in educational practice. Without doubt, the brain network underlying these elementary numerical processing skills is, in turn, shaped by instructional and educational experiences, although the precise mechanisms behind this process need to be empirically investigated (see the section 'Effects of Education on Biology' in this chapter).

School-Taught Mathematical Skills

Although crucial for the successful acquisition of mathematics, individual differences in numerical magnitude processing cannot fully explain the success (or failure) in the acquisition of school-taught mathematical skills (e.g., Schneider et al., 2009). Indeed, the development of higher-order mathematical competence is characterized by a long-term learning process in which different types of knowledge are acquired and adaptively applied in problem-solving. Neuroscience data have also contributed to a better understanding of this long-term learning process.

Arithmetic

Developmental studies have shown that younger children predominantly rely on arduous and slow strategies that depend on procedural knowledge, such as counting, whereas older children and adults more frequently solve arithmetic problems by retrieving the correct solution from declarative memory (i.e., arithmetic fact retrieval; Jordan et al., 2003; Siegler et al., 1996). Thus, the successful acquisition of arithmetic skills is reflected in changes of problem-solving strategies based on acquired knowledge. Starting from these findings, cognitive neuroscience studies have recently begun to unravel the neural correlates of these arithmetic strategies. This research has mainly produced evidence on the brain regions as well as electrophysiological activation patterns related to procedural and retrieval strategies.

It has been shown that these two strategies differentially rely on two areas of the parietal cortex: the IPS and the angular gyrus (AG). The IPS was found to be particularly involved in procedural strategies, whereas the AG turned out to support the retrieval of arithmetic facts (Ansari, 2008; Dehaene et al., 2003; see also Zamarian et al. (this volume)). Although most of the evidence for this distinction is based on studies comparing brain activity while solving problems of different sizes, which are assumed to be accompanied by different strategies (e.g., small problems with sums ≤ 10 are typically solved by fact retrieval; Campbell & Xue, 2001; LeFevre et al., 1996), this distinction was also corroborated by a study which related strategy self-reports (i.e., participants are required to indicate the applied arithmetic strategy after solving a problem) with functional MRI (fMRI) data (Grabner et al., 2009a).

In addition, several EEG studies have provided important insights into the electrophysiological correlates of arithmetic strategies. These findings appear particularly noteworthy as EEG offers some advantages over fMRI in the educational context. For instance, it can be more easily applied in children and it is mobile, so it can also be administered outside the laboratory (Hüsing et al., 2006). Strategy-related differences in brain activation have been reported for two types of EEG activity: event-related potentials (ERPs; e.g., Jost et al., 2004a, 2004b; Ku et al., 2010; Nunez-Pena et al., 2006), and oscillatory brain activity (De Smedt et al., 2009; Grabner & De Smedt, 2011, 2012). While ERP studies reveal important information on the time course of arithmetic problem-solving, studies analyzing oscillatory brain activity provide insights into the dynamics of functional network formation related to different strategies. Specifically, a close link between changes in theta activity (around 3–6 Hz) and fact retrieval processes, on the one hand, and between changes in alpha activity (around 8–12 Hz) and procedural strategies, on the other hand, was found. This finding converged from three research approaches: comparison of different problem sizes (De Smedt et al., 2009), a fact retrieval training study (Grabner & De Smedt, 2012), and, most importantly, strategy self-reports (Grabner & De Smedt, 2011). One study also compared the typically used problem-size approach with the self-report approach and revealed that data from strategy self-reports are even more closely linked to brain processes than the experimental manipulation of problem size (Grabner & De Smedt, 2011).

The studies on arithmetic strategies illustrate two ways of how neuroscientific data can be applied in educational research. First, the sensitivity of this data to different mathematical problem-solving strategies paves the way to develop neuroscientific markers of strategy use which, in future, could complement verbal strategy reports or performance data in investigations of developmental and learning-related changes in mathematical strategies. Second, cognitive neuroscience data can be used to empirically validate, by means of methodological triangulation, behavioral measures such as strategy self-reports whose validity has been repeatedly questioned (e.g., Kirk & Ashcraft, 2001; Smith-Chant & LeFevre, 2003).

Mathematical Problem-Solving

Relatively little is known about other higher-order school-taught mathematical skills. This is mainly due to methodological restrictions of cognitive neuroscience techniques, which complicate the use of complex tasks eliciting multiple cognitive processes and neural activation patterns (see the section 'Cautionary Notes on Applications of Neuroscience to Education' in this chapter). Nonetheless, there are a few examples of studies which illustrate the value added of neuroscientific data for questions related to mathematics education.

Lee et al. (2007) focused on algebraic word problem-solving for which two different solution strategies are taught in schools in Singapore. In the model strategy, introduced to children in primary school, schematics are used to represent the quantitative and qualitative relationships given in the problem. Later on, in the secondary school, children are taught to use these problems using alphanumerical symbols (symbolic strategy). Using fMRI, the authors investigated the neural correlates of these problem-solving strategies in adults who were equally proficient in applying both strategies. Analyses revealed that the symbolic (compared to the model) strategy was accompanied by additional activation in brain areas supporting attentional processes (for similar evidence, see Lee et al., 2010). These results suggest that the two strategies differ in cognitive demands, specifically in mental effort, which were not apparent at the behavioral level. This finding can be considered when discussing questions of whether and when both methods are introduced in school.

Grabner et al. (2012) investigated bilingual mathematics learning and addressed the unresolved question of the sources of language-switching costs (i.e., poorer performance) that arise when arithmetic problems are solved in a language different from the language of instruction. Two general mechanisms have been proposed for these costs (Spelke & Tsivkin, 2001): they could either derive from the need to translate arithmetic knowledge from the language of instruction into the language of application or they could derive from additional domain-specific information processing. To test these hypotheses, bilingual adults underwent a four-day arithmetic training in one language, followed by an fMRI test session in which they were presented with the trained problems in both languages. The results revealed that language-switching costs were accompanied by increased activation in areas related with numerical information processing, such as the IPS, but not by increased activation in typical language areas. These findings favored the hypothesis that language-switching costs are due to additional domain-specific processes (such as magnitude processing) rather than mere translation of mathematical knowledge from one language into the other. Thus, this study can be regarded as another example of how neuroimaging data can provide incremental insights into school-relevant cognitive processes.

Individual Differences in Mathematical Competencies

Neuroimaging data can also add to our understanding of individual differences in mathematical competencies within the normal performance range. Such information is educationally relevant because it might help to tailor the instructional approach to the needs of a specific learner (Dowker, 2005).

One line of research dealt with individual differences in adults (e.g., Ansari et al., 2013; Grabner et al., 2007, 2009b, 2011; Menon et al., 2000; Núñez-Peña et al., 2011). In particular, several fMRI studies investigated how adults with higher mathematical competencies differ in brain activity from their less competent peers in various tasks of varying mathematical demands. In most of these studies, two groups were selected who differed in their mathematical competencies but not in other abilities (such as verbal or figural-spatial competencies). The first study of this type was conducted by Grabner et al. (2007) who presented these groups with easier and more complex arithmetic multiplication problems. Besides the expected performance differences, more competent individuals exhibited stronger activation of one brain region, (i.e., the left AG), whereas the less competent peers did not activate any brain region significantly stronger. In addition, the amount of left AG activation was linearly related to the individual's level of mathematical competence. Subsequent studies have corroborated this finding (Grabner et al., 2009b, 2011).

The stronger parietal brain activity in more competent adults is generally in line with developmental studies which show that with increasing age and competence, a functional specialization of the parietal cortex takes place (Kaufmann et al., 2011; Rivera et al., 2005). The critical question, however, is what cognitive processes underlie the differential activation of the left AG. Based on the presumed role of the AG in the retrieval of arithmetic facts, it was initially assumed that it may be due to a stronger reliance on fact retrieval processes in the more competent individuals (cf. Dehaene et al., 2003). At present, however, there is increasing evidence to suggest that the left AG supports a more fundamental function in mathematical cognition in which more and less competent adults may differ (e.g., Ansari, 2008; Grabner, 2009). Specifically, Ansari (2008) hypothesized that this brain region supports the automatic processing of mathematical symbols at various levels. For instance, in basic number processing, the presentation of a numerical symbol (e.g., an Arabic digit) would automatically activate its internal semantic representation (Holloway et al., 2010). Consequently, the differential activation of the left AG in mathematically more and less competent adults may reflect differences in their capability to process symbolic mathematical representations (see also Grabner, 2009). This assumption fits nicely with recent findings that the fast access of magnitude information from number symbols is crucial to the development of mathematical competence (e.g., De Smedt & Gilmore, 2011; Holloway & Ansari, 2009; Rousselle & Noel, 2007; Vanbinst et al., 2012). Together, these findings suggest that the mapping between numerical (or mathematical) symbols and their semantic knowledge seems to be an essential mechanism in the development of mathematical competencies that are taught in school.

Only a few neuroimaging studies have investigated the question of individual differences in mathematical competence in children (De Smedt et al., 2011; Price et al., 2013). De Smedt et al. (2011) showed that individual differences in arithmetical fluency in 10–12-year-olds modulated the brain activity in the right IPS during the solution of small and large addition and subtraction problems. More specifically, children with normal arithmetical fluency showed a problem-size effect in the IPS, with higher activity for large than for small problems, consistent with data in adults (e.g., Stanescu-Cosson et al., 2000). On the other hand, children with low arithmetical fluency did not show such problem-size effect, but recruited the IPS to the same extent for small and large problems, suggesting that these children also relied on (effortful) quantity-based arithmetic strategies in solving the small problems, which their more competent peers might have solved with memory retrieval. Most interestingly, this interaction was only apparent at the neural level and not in behavioral performance, providing another example of how neuroimaging data can capture a level of analysis and measurement that cannot be obtained by using behavioral measures alone. Price et al. (2013) recently extended these findings by revealing that brain activity during single-digit arithmetic was associated with ecologically valid measures of higher-order math competence (i.e., the PSAT) in 17-year-old high-school students. They reported that the more competent students displayed a stronger activation of the left supramarginal gyrus (adjacent to the AG) and the anterior cingulate cortex, whereas the less competent peers more strongly activated the right IPS, again suggesting that more competent students more strongly engaged systems associated with memory retrieval, whereas the less competent peers apparently relied on quantity-based strategies. Overall, these individual differences data suggest that individuals who fail to develop automatic mathematical information processing, either in terms of arithmetic strategies or symbol processing, achieve lower levels of mathematical competence.

Neuroprediction

Neuroprediction or neuroprognosis (Hoeft et al., 2011) refers to the idea that brain imaging measures may be used to predict future behavioral and educational outcomes as well as the emergence of symptoms in neurodevelopmental disorders. Although this approach has been mainly adopted in the context of psychiatric diseases, such as schizophrenia or depression (e.g., Kumari et al., 2009), it has been recently applied to the development of reading and dyslexia (e.g., Hoeft et al., 2011) and, to the best of our knowledge, to mathematics learning in only one study so far (Supekar et al., 2013).

Supekar et al. (2013) investigated which behavioral and neural measures could predict grade 3 children's response to an eight-week one-to-one math tutoring intervention that was specifically designed to foster children's arithmetical fluency. The intervention focused on conceptual knowledge of counting, the use of efficient counting strategies, and the development of arithmetic facts. Behavioral data (i.e. arithmetic skills, IQ, reading, working memory, math skills) as well as brain imaging data (i.e. structural MRI and resting-state fMRI, which measures the functional connectivity between different brain areas) were administered before the start of the intervention. Supekar et al. (2013) next investigated which of those measures was able to predict performance gains in arithmetical skills. Their data revealed that only the brain imaging measures (i.e. measures of gray matter volume and functional connectivity), and not the behavioral measures, were able to predict gains in arithmetical performance. More specifically, their data showed that children with larger gray matter volume in the right hippocampus and larger functional connectivity of the hippocampus had higher gains in performance across the intervention. It is important to emphasize that the intervention focused on only one aspect of mathematical performance, i.e. arithmetic fact retrieval skills. The observed association with the hippocampus is not unexpected, as this brain structure has been shown to play a role in the early stages of arithmetic fact retrieval in children (Cho et al., 2012; De Smedt et al., 2011). In all, the brain imaging measures were able to predict the response to an intervention specifically targeted at one particular mathematical skill, yet they did not predict overall mathematical achievement.

It has been suggested that neuroimaging data may be useful additional measures to identify children who are at risk for low academic performance at an early stage, i.e. before or at the start of formal instruction (e.g., Diamond & Amso, 2008; Gabrieli, 2009). A few attempts to discover such neural biomarkers have been made in the field of reading and dyslexia (e.g., Hoeft et al., 2011; Molfese, 2000), but, to our knowledge, such research has not been conducted in mathematical cognition. The discovery of such neural markers would, however, have substantial educational impact (Beddington et al., 2008), because this would allow us to understand how mathematical development goes awry before formal instruction and how different developmental trajectories as well as compensatory strategies emerge—information that can be exploited in new intervention approaches.

Neurointervention

Rationale for Behavioral Educational Interventions

One of the often-cited applications from neuroscience to education deals with the potential to use neuroscientific findings as an input to ameliorate existing educational interventions or to design new ones (e.g. Ansari et al., 2012; Butterworth et al., 2011). In the context of mathematics learning, this issue has been mainly explored in the case of atypical mathematical development such as dyscalculia and children with low socioeconomic status SES, a condition where children are at risk for developing mathematical difficulties due to environmental factors.

Neuroscientific research has pointed to brain networks that are important in processing numerical magnitudes and has highlighted that these networks are recruited during more complex mathematical processes (Ansari, 2008; see also chapters by Menon and by Kaufmann et al. in this volume), and, more recently, has shown that the quality of these networks relates to in-classroom performance (e.g., Bugden et al., 2012; Cantlon & Li, 2013; Emerson & Cantlon, 2013). Together with a large number of behavioral studies that have further refined our understanding of the association between numerical magnitude processing and mathematical development (see De Smedt et al., 2013, for a review), this has triggered the development of educational interventions that focus on these basic numerical skills. For example, such interventions, which include a wide variety of numerical activities (e.g., number recognition, counting, set comparison, board games), have been embedded in larger-scale kindergarten programs for children from low-income backgrounds (Dyson et al., 2013) or children at risk for dyscalculia (Toll & Van Luit, 2013). These interventions have been shown to have positive effects not only on specific measures of numerical magnitude processing, but also on more general tests of mathematical ability.

Smaller-scale interventions on numerical magnitude processing, such as board games (e.g., Ramani & Siegler, 2011) or computer games (e.g., Wilson et al., 2006), also seem to positively affect children's numerical magnitude processing (De Smedt et al., 2013; Ramani & Siegler, 2011). Notably, some of these effects seem to generalize to other mathematical skills (e.g., Ramani & Siegler, 2011), although not all studies are consistent with this (Rasanen et al., 2009). Nevertheless, such game-like interventions offer great promise for early intervention of at-risk children because they allow parents and teachers to stimulate numerical skills in a motivating environment and because they are easy to apply, given the availability of such board games as well as mobile devices that can be used to practise these skills at home.

Effects of Education On Biology

There is consistent evidence that the human brain is highly plastic and that its structure and function continue to undergo massive changes into late adolescence (Casey et al., 2005; Giedd & Rapoport, 2010). This development occurs not in isolation but requires the crucial input of the environment, a process referred to as experience-dependent plasticity (e.g., Greenough et al., 1987; Johnson & de Haan, 2011). Because children spend a large time at school, education is one of the most powerful factors affecting this type of plasticity, or, in other words, education is shaping our brain. Recent data by Rosenberg-Lee et al. (2011)

provided a first attempt to uncover such effects of education on brain development in mathematics, by examining the changes in brain activity during arithmetic problem-solving from second to third grade. Although their data showed that one year of schooling affected the function and connectivity of the brain, the precise mechanisms of how specific aspects of education are affecting brain development are not yet clear. This should be a focus of future research, which might ultimately lead to a characterization of the limits of brain plasticity of mathematical circuits that are shaped by education.

More progress has been made in unraveling the effects of specific remedial educational interventions on brain structure and function in atypical academic development, particularly in the field of reading and dyslexia (e.g., McCandliss, 2010). These studies have revealed that processes of normalisation and compensation take place after remediation in individuals with dyslexia (e.g., Temple et al., 2003). Such compensatory strategies might be difficult to detect by means of behavioral data alone. This information on how individuals with learning disorders compensate for their difficulties is extremely relevant for educators, because it can be useful to teach specific compensation strategies, provided that these strategies have beneficial effects on the learning progress (see also Ansari et al., 2012).

In the field of mathematical cognition, only one study has addressed neural effects of interventions that foster numerical magnitude processing. Kucian et al. (2011) investigated the effects of the Rescue calcularis game, a computerized intervention that focuses on numerical magnitude processing, on brain activity in children with and without DD. Their findings indicated that activity in frontal areas decreased after training, probably reflecting an increase in automatisation processes. Follow-up data at five weeks after the training showed a significant increase in activity in the posterior parietal cortex, which is known to play a crucial role in numerical representations. However, this is only one study and further replication is surely needed to draw firm conclusions.

Biological Interventions that Impact on Education

There is a small but increasing number of studies that use neurophysiological interventions, more specifically brain stimulation, to improve mathematical cognition (Cohen Kadosh et al., 2013). These studies applied the method of transcranial electric stimulation (TES), which has attracted much attention in diverse cognitive domains in recent years (Krause & Cohen Kadosh, 2013). In TES, a small electrical current (typically around 1–2 mA) is non-invasively applied to the brain through electrodes fixated at the scalp by straps or a cap. The applied current is assumed to alter the activation level of the cortical regions lying underneath the electrodes and may also influence long-term synaptic plasticity. Consequently, this method allows one to investigate causal relationships between activation in specific brain regions and behavior, by experimentally manipulating (increasing or decreasing) the neural activation level. In addition, and educationally more relevant, through the stimulation of task-specific brain regions, performance and/or learning can be enhanced. The latter issue is very appealing and may offer particular promise for the development of interventions for individuals suffering from learning disorders such as DD.

One type of TES is transcranial direct current stimulation (tDCS), which delivers a constant direct current to the scalp via two electrodes (Nitsche & Paulus, 2000). In an attempt to mimic the developmental process of learning symbols and their corresponding numerical meaning, Cohen Kadosh et al. (2010) examined whether parietal tDCS could beneficially influence the acquisition of new, artificial numerical symbols. To this end, 15 university students underwent six days' training in which they learned to associate the artificial symbols with magnitudes through a symbolic comparison task with feedback. It turned out that a specific stimulation of 20 minutes per day at the beginning of each learning session (right anodal, left cathodal), compared to sham (placebo) stimulation, resulted in increased automaticity of magnitude processing and more correct mappings of numerical magnitudes onto space. In addition, the students who received this stimulation also displayed long-term learning effects in a follow-up test after six months.

Could tDCS also have beneficial effects on more advanced mathematical skills, such as arithmetic? Recent data by Hauser et al. (2013) draw a promising picture. In two experiments with university students, they tested whether parietal tDCS improves performance in both basic number processing and mental arithmetic. Participants were presented with a number comparison task and a complex subtraction task before and after tDCS stimulation (or sham stimulation, respectively) at the parietal cortex for 25 minutes. Comparisons of the performance between these two measurements revealed that left anodal stimulation significantly improved performance in both tasks compared to sham. This suggests that tDCS over the parietal cortex has the potential to improve not only numerical but also mathematical performance (see also Clemens et al., 2013).

Another type of TES is transcranial random noise stimulation (TRNS), in which a current at quickly varying frequency bands is applied, so that there is no constant anode or cathode. Rather, each electrode is assumed to be capable of exciting the underlying brain region (Terney et al., 2008). Snowball et al. (2013) recently used TRNS to investigate whether it can improve arithmetic learning and performance. Specifically, university students participated in five days' training in which they had to learn rote artificial arithmetic facts (e.g., 4 # 12 = 17; drill training) or acquire a new calculation algorithm (e.g., to solve 5 § 18 by applying a certain arithmetic procedure; calculation training). TRNS was applied in half of the participants, for 20 minutes, at the bilateral lateral prefrontal cortex, at the beginning of the learning sessions, with the other half of participants receiving sham stimulation. In addition, frontal brain activity was assessed by means of near-infrared spectroscopy (NIRS). It turned out that the learning rates in both training types were higher in the TRNS group than in the sham group and that the amount of prefrontal brain activity in the TRNS group decreased more strongly throughout the training compared to sham, indicating the decreasing recruitment of attentional and working memory resources that are needed for effortful processing. A follow-up test of a sub-sample after six months revealed that the TRNS group solved calculation problems with the new algorithm still faster than the sham group, but no long-term effect was apparent for the artificial arithmetic facts. Thus, this study demonstrates that, also, TRNS can have beneficial and long-term effects on (specific types of) arithmetic learning and performance. Using a similar approach to study the improvement of numerosity discrimination in adults, Cappelletti et al. (2013) recently showed that the combination of parietal TRNS and cognitive training led to better improvements in numerosity discrimination, relative to cognitive training or to brain stimulation alone. The effects of the intervention also showed some transfer to time and space discrimination, but not to mental arithmetic.

With respect to the potential of TES for educational interventions, it needs to be emphasized that the stimulation needs to be paired with efficient cognitive training and that the timing of stimulation is important for task performance, indicating that brain stimulation alone is not sufficient to induce intervention effects. Furthermore, despite many advantages and promises of this method, there is only very limited work with children so far. In this population, questions of safety, efficacy, and potential side-effects (stimulating specific brain regions in a malleable, developing brain may be accompanied by cognitive side-effects, see Krause & Cohen Kadosh, 2013) need to be answered before applications can be developed.

Cautionary Notes on Applications of Neuroscience to Education

Although significant progress has been made in understanding the neural basis of mathematical cognition and even though there is a great enthusiasm in thinking about applications from neuroscience to education, such applications are facing many challenges that one needs to be aware of and that will have to be overcome in the next years. In the following, some of these challenges, which are not specific to the mathematical domain, are detailed (see also Ansari et al., 2012; De Smedt & Verschaffel, 2010; Grabner & Ansari, 2010).

Ecological Validity

One of the main methodological challenges in applying neuroscience research to education deals with overcoming the constraints of the available neuroscientific methods in order to achieve a high ecological validity (i.e., a generalization of findings obtained in the laboratory to learning in the classroom). Firstly, this concerns the participants tested in cognitive neuroscience studies. Although the number of studies on school-aged children is growing, most of the existing body of data is based on studies using adult populations. This is not surprising, because data acquisition in children is more frequently affected by motion artifacts, which have severe negative effects on the quality of neuroimaging data. Also, children are more difficult to recruit than adults and there might be considerations from ethics committees that make neuroscientific research in children more challenging. In light of the prominent developmental plasticity of the brain that continues into adulthood (Ansari, 2010; Kaufmann et al., 2011) and in view of the dynamic changes in the brain networks that support numerical and mathematical processing (e.g., Rivera et al., 2005), the generalization of neuroimaging findings from adult samples to children's and adolescents' scholastic learning, however, is questionable (see Ansari, 2010, for a discussion). Not considering these developmental changes may severely impair the potential of neurounderstanding, neuropredicition, and neurointervention.

Secondly, the mathematical tasks used in many cognitive neuroscience studies are very elementary and differ from the problems that children typically solve in the classroom. However, mathematical tasks that are frequent in the classroom cannot be easily administered in brain imaging studies for various reasons. For example, the signals in functional neuroimaging methods contain a large measurement error and require several trials of each task condition to be presented and averaged to obtain reliable data. This necessitates the use of homogeneous problems that can be solved in short time, such as arithmetic problems.

Problems in school are frequently not only characterized by a higher complexity and longer problem-solving time but also by heterogeneity that aims at facilitating the transfer of the acquired knowledge to different situations. In addition, the acquisition of brain imaging data limits the way of responding to the tasks, mainly due to the need to avoid movement artifacts. Thus, in typical neuroimaging studies, participants verify given response options by button press instead of actively producing problem solutions. However, the verification of solutions is accompanied by different neurocognitive processes than active production of solutions (Menon, 2010). Furthermore, mathematical problems taught at higher school levels require an elaborate and stepwise procedure, and usually involve a multitude of domain-general and domain-specific cognitive processes that can occur at various points during task processing and that exhibit intra- and inter-individual differences. This all represents a challenge to neuroimaging methods as (a) the more processes are involved, the more difficult it becomes to disentangle them at the neurophysiological level, and (b) the required aggregation over trials and participants reduces the sensitivity to detect individual differences. One way of overcoming this problem is to directly investigate how measures of brain activity acquired under laboratory conditions are correlated or predict ecologically valid and authentic measures of classroom learning that are affected by a wide range of environmental variables (see the section 'Numerical Magnitude Processing' in this chapter).

We Do Need Education

Most of the studies reviewed here have investigated mathematical performance in relative isolation from the educational context. Indeed, the participants' learning histories, as well as their educational environment, have been typically considered as nuisance variables that are ignored or controlled for. These variables are, however, crucial, as variability in these might have massive effects on brain structure and activity. Indeed, the development of mathematics cannot be studied in isolation from the learning context in which it is taught, as is documented by several decades of educational research. This is, for example, illustrated in the offer-use model by Helmke (2006), in which school learning is modeled as follows: a teacher makes a learning *offer* to students through instruction, which can be *used* by the students to acquire knowledge and to build up competencies. To what extent this offer is used, however, depends on students' learning prerequisites (e.g., intelligence, prior knowledge, motivation), the quality of instruction and material, teachers' competencies (e.g., content knowledge, pedagogical content knowledge, and psychological as well as pedagogical knowledge), other contextual factors (e.g., class climate), as well as interactions between each of these variables. Without knowledge on how mathematics is learned and taught at school, cognitive neuroscientists are at the risk of running naïve experiments with little or no relevance to educational practice (De Smedt & Verschaffel, 2010; see Cacioppo et al., 2008, for a similar rationale in applying neuroscience to psychology). At minimum, future neuroscience studies with application to education should, therefore, include careful descriptions of the educational environments and histories of their participants. More compelling, and surely a priority for future research, would be to understand how the characteristics of the educational environment affect the neural measures collected in neuroimaging research.

It has been suggested that, in order to make headway in making applications from neuroscience to the classroom, a two-way street scenario in which both neuroscience and educational science inform each other in a bidirectional and reciprocal way is needed (e.g., De Smedt et al., 2010). Moving back and forth between the two disciplines should create a productive cycle of true interdisciplinary research that yields a deep understanding of mathematical learning and how it can be fostered. More specifically, neuroscience data can deepen our understanding of the cognitive constraints in the learner and the learning process, but they do not directly determine how instruction should be designed to optimally foster this learning. As an example, some neuroscience-informed interventions to remediate DD were designed by cognitive neuroscientists who, based on their research findings, converted a task that is sensitive to DD (e.g., the number comparison task) into an intervention tool. It is, however, questionable whether the competencies captured by this task are simply trainable by repeating the task (which may also easily be boring) or whether a broader instructional approach would be more effective. Rather, educational research should combine the available neuroscientific data with domain-general (Van Merriënboer & Kirschner, 2007) and domain-specific (Nickerson & Whitacre, 2010) theories of instructional design into new learning environments. The effectiveness of these environments should then be evaluated in various phases of empirical research, ranging from small-scale interventions to larger randomised controlled trials, before such environments can be applied in the classroom (Sloane, 2008).

In addition to interdisciplinary collaborations, an interdisciplinary training of researchers that comprises both educational research and cognitive neuroscience, seems to be highly desirable (e.g., Ansari & Coch, 2006; Ansari et al., 2012). Notably, many universities have currently established such training programs. These programs have the potential to educate experts who are able to navigate between neuroscience and education because they speak both languages and they know critical research findings from both disciplines.

Unrealistic Expectations

The application of neuroscience to education faces a general problem that has repeatedly occurred in the history of applying empirical research to educational practice: the expectation that (this) research yields a panacea for the unresolved problems of institutional learning and teaching. More specifically, it is often expected that the possibility to look into the brains of learners will answer all questions related to how effective teaching should take place or how curricula should be designed and at which point certain abilities should be instructed. Such expectation, however, that the research findings can be directly applied to the classroom without translation seems indeed to be "a bridge too far" (Bruer, 1997).

There are several reasons for such high and unrealistic expectations of applications from neuroscience to education. One reason might lie in the prejudice that a biological explanation is more reliable, more convincing, or more informative than a non-biological explanation (for a review, see Beck, 2010). A second reason may be the lack of adequate communication of cognitive neuroscience findings (Beck, 2010). In light of the great public interest into brain research, there have been many publications in which authors, strongly differing in their expertise in cognitive neuroscience, describe how the brain learns and how brain-based teaching should take place. The messages that are conveyed in many of these publications are very simple (e.g., distributed learning is more effective than massed learning) and have long been known in psychology before the advent of brain imaging (Stern,

2005). Evidence from neuroscience (e.g., which supports well-established psychological knowledge) is often integrated in a way so that a non-expert may gain the impression that this knowledge has recently been explored through brain research. Such publications may also provide ground for the emergence of neuromyths (Dekker et al., 2012), which are (persistent) misconceptions that arise through a misunderstanding, misreading, or misquotation of neuroimaging findings in support of certain educational practices (Howard-Jones, 2010). An example of such a neuromyth is the idea that the brains of children with learning disorders cannot be changed through remedial interventions. Finally, a third reason could lie in the urgency to answer many open questions on teaching in order to improve the efficacy of the school system. If publications on brain-based teaching provide straightforward and easy answers (e.g., you have to present your material three times with gaps of ten minutes between; Kelley, 2008), they seem to be more readily accepted than the answers from educational researchers suggesting that there are no general recipes for effective teaching (Dumont et al., 2010).

One way to address such unrealistic expectations or misinterpretations deals, again, with interdisciplinary training (Ansari & Coch, 2006). Only if teachers have a certain level of (cognitive) neuroscientific literacy, in addition to their educational and psychological expertise, they can know about the chances and limitations of neuroscience to education and can critically evaluate pseudo-scientific recommendations on brain-based teaching. Such training should be very clear about the scope of a biological explanation for a certain educational phenomenon: such explanation does not mean that a particular behavior is hardwired or unchangeable. Rather, the human brain is characterized by remarkable plasticity and is shaped by education (see the section 'Effects of Education on Biology' in this chapter). On the other hand, there is also need for interdisciplinary researchers, as we have already outlined, who are educated in both neuroscience and education and who can translate neuroscience research findings into educational practice. In this vein, these "translators" should also play a critical role for disseminating scientific research to the public.

Conclusion

This chapter illustrated how neuroscience can be applied to education to enhance our understanding (neurounderstanding), to improve predictions (neuroprediction), and to develop effective interventions (neurointervention) within the domain of mathematical cognition. Mindful of the cautionary notes already outlined, we think that future translational research will strengthen the connection between neuroscience and education and will contribute to a more advanced knowledge base in which the findings from behavioral and neural levels of investigation are integrated and upon which effective mathematics education can be built.

Acknowledgments

Bert De Smedt is funded by grant GOA 2012/010 of the Research Fund KULeuven, Belgium.

References

Ansari, D. (2008). Effects of development and enculturation on number representation in the brain. *Nature Reviews Neuroscience*, 9(4), 278–291.

Ansari, D. (2010). Neurocognitive approaches to developmental disorders of numerical and mathematical cognition: the perils of neglecting the role of development. *Learning and Individual Differences*, 20(2), 123–129.

Ansari, D. & Coch, D. (2006). Bridges over troubled waters: education and cognitive neuroscience. *Trends in Cognitive Sciences*, 10, 146–151.

Ansari, D., De Smedt, B., & Grabner, R. (2012). Neuroeducation—a critical overview of an emerging field. *Neuroethics*, 5, 105–117.

Ansari, D., Grabner, R.H., Koschutnig, K., Reishofer, G., & Ebner, F. (2013). Individual differences in mathematical competence modulate brain responses to arithmetic errors: an fMRI study. *Learning and Individual Differences*, 21, 636–643.

Arsalidou, M. & Taylor, M.J. (2011). Is 2 + 2 = 4? Meta-analyses of brain areas needed for numbers and calculations. *Neuroimage*, 54, 2382–2393.

Beck, D.M. (2010). The appeal of the brain in the popular press. *Perspectives on Psychological Science*, 5(6), 762–766.

Beddington, J.C.C.L., Field, J., Goswami, U., Huppert, F.A., Jenkins, R., Jones, H.S., et al. (2008). The mental wealth of nations. *Nature*, 455, 1057–1060.

Blakemore, S.J. & Frith, U. (2005). *The Learning Brain: Lessons for Education*. Oxford: Blackwell.

Bruer, J. T. (1997). Education and the brain: a bridge too far. *Educational Researcher*, 26(8), 4–16.

Bugden, S., Price, G.R., Mclean, D.A., & Ansari, D. (2012). The role of the left intraparietal sulcus in the relationship between symbolic number processing and children's arithmetic competence. *Developmental Cognitive Neuroscience*, 2, 448–457.

Butterworth, B. (2003). *Dyscalculia Screener*. London: Nfer Nelson.

Butterworth, B., Varma, S., & Laurillard, D. (2011). Dyscalculia: from brain to education. *Science*, 332, 1049–1053.

Bynner, J. & Parsons, S. (1997). *Does Numeracy Matter*. London: Basic Skills Agency.

Cacioppo, J.T., Berntson, G.G., & Nusbaum, H.C. (2008). Neuroimaging as a new tool in the toolbox of psychological science. *Current Directions in Psychological Science*, 17, 62–67.

Campbell, J.I.D. & Xue, Q.L. (2001). Cognitive arithmetic across cultures. *Journal of Experimental Psychology-General*, 130(2), 299–315.

Cantlon, J.F. & Li, R. (2013). Neural activity during natural viewing of Sesame Street statistically predicts test scores in early childhood. *Plos Biology*, 11, e1001462.

Cappelletti, M., Gessaroli, E., Hithersay, R., Mitolo, M., Didino, D., Kanai, R., et al. (2013). Transfer of cognitive training across magnitude dimensions achieved with concurrent brain stimulation of the parietal lobe. *Journal of Neuroscience*, 33, 14899–14907.

Casey, B.J., Tottenham, N., Liston, C., & Durston, S. (2005). Imaging the developing brain: what have we learned about cognitive development? *Trends in Cognitive Sciences*, 9, 104–110.

Cho, S., Metcalfe, A.W.S., Young, C.B., Ryali, S., Geary, D.C., & Menon, V. (2012). Hippocampal-prefrontal engagement and dynamic causal interactions in the maturation of children's fact retrieval. *Journal of Cognitive Neuroscience*, 24, 1849–1866.

Clemens, B., Jung, S., Zvyagintsev, M., Domahs, F., & Willmes, K. (2013). Modulating arithmetic fact retrieval: a single-blind, sham-controlled tDCS study with repeated fMRI measurements. *Neuropsychologia*, 51(7), 1279–1286.

Cohen Kadosh, R., Dowker, A., Heine, A., Kaufmann, L., & Kucian, K. (2013). Interventions for improving numerical abilities: present and future. *Trends in Neuroscience and Education*, 2(2), 85–93.

Cohen Kadosh, R., Soskic, S., Iuculano, T., Kanai, R., & Walsh, V. (2010). Modulating neuronal activity produces specific and long-lasting changes in numerical competence. *Current Biology*, 20(22), 2016–2020.

De Smedt, B. & Gilmore, C.K. (2011). Defective number module or impaired access? Numerical magnitude processing in first graders with mathematical difficulties. *Journal of Experimental Child Psychology*, 108(2), 278–292.

De Smedt, B. & Verschaffel, L. (2010). Traveling down the road: from cognitive neuroscience to mathematics education … and back. *ZDM—The International Journal on Mathematics Education*, 42, 649–654.

De Smedt, B., Ansari, D., Grabner, R.H., Hannula, M.M., Schneider, M., & Verschaffel, L. (2010). Cognitive neuroscience meets mathematics education. *Educational Research Review*, 5, 97–105.

De Smedt, B., Grabner, R.H., & Studer, B. (2009). Oscillatory EEG correlates of arithmetic strategy use in addition and subtraction. *Experimental Brain Research*, 195(4), 635–642.

De Smedt, B., Holloway, I.D., & Ansari, D. (2011). Effects of problem size and arithmetic operation on brain activation during calculation in children with varying levels of arithmetical fluency. *Neuroimage*, 57(3), 771–781.

De Smedt, B., Noël, M.P., Gilmore, C., & Ansari, D. (2013). The relationship between symbolic and non-symbolic numerical magnitude processing skills and the typical and atypical development of mathematics: a review of evidence from brain and behavior. *Trends in Neuroscience and Education*, 2(2), 48–55.

Dehaene, S. (2011). *The Number Sense: How the Mind Creates Mathematics* (2nd edition). New York: Oxford University Press.

Dehaene, S. & Cohen, L. (1997). Cerebral pathways for calculation: double dissociation between rote verbal and quantitative knowledge of arithmetic. *Cortex*, 33, 219–250.

Dehaene, S., Piazza, M., Pinel, P., & Cohen, L. (2003). Three parietal circuits for number processing. *Cognitive Neuropsychology*, 20(3–6), 487–506.

Dekker, S., Lee, N.C., Howard-Jones, P.J., & Jolles, J. (2012). Neuromyths in education: prevalence and predictors of misconceptions among teachers. *Frontiers in Psychology*, 3, 429. doi: 10.3389/fpsyg.2012.00429.

Diamond, A. & Amso, D. (2008). Contributions of neuroscience to our understanding of cognitive development. *Current Directions in Psychological Science*, 17, 136–141.

Dowker, A. (2005). *Individual Differences in Arithmetic: Implications for Psychology, Neuroscience and Education*. New York: Psychology Press.

Dowker, A. (2008). Individual differences in numerical abilities in preschoolers. *Developmental Science*, 11, 650–654.

Dumont, H., Instance, D., & Benavides, F. (2010). *Educational Research and Innovation. The Nature of Learning Using Research to Inspire Practice*. OECD Publishing. http://www.oecd.org/edu/ceri/50300814.pdf

Duncan, G.J., Claessens, A.M.Y., Huston, A.C., Pagani, L.S., Engel, M., Sexton, H., et al. (2007). School readiness and later achievement. *Developmental Psychology*, 43, 1428–1446.

Dyson, N.I., Jordan, N.C., & Glutting, J. (2013). A number sense intervention for low-income kindergartners at risk for mathematics difficulties. *Journal of Learning Disabilities*, 46, 166–181.

Emerson, R.W. & Cantlon, J.F. (2012). Early math achievement and functional connectivity in the fronto-parietal network. *Developmental Cognitive Neuroscience*, 2, S139–S151.

Fischer, K.W. (2009). Mind, brain and education: building a scientific groundwork for teaching and learning. *Mind, Brain and Education*, 3, 3–16.

Gabrieli, J.E.D. (2009). Dyslexia—a new synergy between education and cognitive neuroscience. *Science*, 325, 280–283.

Geake, J. (2009). *The Brain at School: Educational Neuroscience in the Classroom*. Maidenhead: Open University Press.

Geary, D.C. (2013). Early foundations for mathematics learning and their relations to learning disabilities. *Current Directions in Psychological Science*, 22, 23–27.

Gerardi, K., Lorenz, G., & Meier, S. (2013). Numerical ability predicts mortgage default. *Proceedings of the National Academy of Sciences of the United States of America*; 110, 11267–11271.

Giedd, J.A.Y.N. & Rapoport, J.L. (2010). Structural MRI of pediatric brain development: what have we learned and where are we going? *Neuron*, 67, 728–734.

Göbel, S., Walsh, V., & Rushworth, M.F.S. (2001). The mental number line and the human angular gyrus. *Neuroimage*, 14, 1278–1289.
Goswami, U. & Szücs, D. (2011). Educational neuroscience: developmental mechanisms: towards a conceptual framework. *Neuroimage*, 57, 651–658.
Grabner, R.H. (2009). Expertise in symbol-referent mapping. *Behavioral and Brain Sciences*, 32(3–4), 338.
Grabner, R.H. & Ansari, D. (2010). Promises and potential pitfalls of a "cognitive neuroscience of mathematics learning." *ZDM—The International Journal on Mathematics Education*, 42(6), 655–660.
Grabner, R.H. & De Smedt, B. (2011). Neurophysiological evidence for the validity of verbal strategy reports in mental arithmetic. *Biological Psychology*, 87(1), 128–136.
Grabner, R.H. & De Smedt, B. (2012). Oscillatory EEG correlates of arithmetic strategies: a training study. *Frontiers in Psychology*, 3(426), 1-11.
Grabner, R.H., Ansari, D., Reishofer, G., Stern, E., Ebner, F., & Neuper, C. (2007). Individual differences in mathematical competence predict parietal brain activation during mental calculation. *Neuroimage*, 38(2), 346–356.
Grabner, R.H., Ansari, D., Koschutnig, K., Reishofer, G., Ebner, F., & Neuper, C. (2009a). To retrieve or to calculate? Left angular gyrus mediates the retrieval of arithmetic facts during problem solving. *Neuropsychologia*, 47(2), 604–608.
Grabner, R.H., Ischebeck, A., Reishofer, G., et al. (2009b). Fact learning in complex arithmetic and figural-spatial tasks: the role of the angular gyrus and its relation to mathematical competence. *Human Brain Mapping*, 30(9), 2936–2952.
Grabner, R.H., Reishofer, G., Koschutnig, K., & Ebner, F. (2011). Brain correlates of mathematical competence in processing mathematical representations. *Frontiers in Human Neuroscience*, 5(130), 1–11.
Grabner, R.H., Saalbach, H., & Eckstein, D. (2012). Language switching costs in bilingual mathematics learning. *Mind, Brain, and Education*, 6, 147–155.
Greenough, W.T., Black, J.E., & Wallace, C.S. (1987). Experience and brain development. *Child Development*, 58, 539–559.
Grégoire, J., Noël, M.P., & Van Nieuwenhoven, C. (2004). *TEDI-MATH*. Antwerpen: Harcourt.
Hauser, T.U., Rotzer, S., Grabner, R.H., Mérillat, S., & Jäncke, L. (2013). Enhancing performance in numerical magnitude processing and mental arithmetic using transcranial Direct Current Stimulation (tDCS). *Frontiers in Human Neuroscience*, 7(244), 1–9.
Heine, A., Wissmann, J., Tamm, S., De Smedt, B., Schneider, M., Stern, E., et al. (2013). An electrophysiological investigation of non-symbolic magnitude processing: numerical distance effects in children with and without mathematical learning disabilities. *Cortex*, 49(9), 2162-2177.
Helmke, A. (2007). *Unterrichtsqualität: erfassen, bewerten, verbessern*. Seelze: Kallmeyersche Verlagsbuchhandlung.
Hoeft, F., Mccandliss, B.D., Black, J.M., Gantman, A., Zakerani, N., Hulme, C., et al. (2011). Neural systems predicting long-term outcome in dyslexia. *Proceedings of the National Academy of Sciences of the United States of America*, 108, 361–366.
Holloway, I.D. & Ansari, D. (2009). Mapping numerical magnitudes onto symbols: the numerical distance effect and individual differences in children's mathematics achievement. *Journal of Experimental Child Psychology*, 103(1), 17–29.
Holloway, I.D., Price, G.R., & Ansari, D. (2010). Common and segregated neural pathways for the processing of symbolic and nonsymbolic numerical magnitude: an fMRI study. *Neuroimage*, 49, 1006–1017.
Howard-Jones, P. (2010). *Introducing Neuroeducational Research. Neuroscience, Education and the Brain from Contexts to Practice*. London: Routledge.
Hüsing, B., Jäncke, L., & Tag, B. (2006). *Impact Assessment of Neuroimaging*. Zurich: vdf Hochschulverlag.
Johnson, M.H. & de Haan, M. (2011). *Developmental Cognitive Neuroscience* (3rd edn). Malden, MA: Wiley-Blackwell.
Jordan, N.C., Hanich, L.B., & Kaplan, D. (2003). Arithmetic fact mastery in young children: a longitudinal investigation. *Journal of Experimental Child Psychology*, 85(2), 103–119.

Jost, K., Beinhoff, U., Hennighausen, E., & Rosler, F. (2004a). Facts, rules, and strategies in single-digit multiplication: evidence from event-related brain potentials. *Cognitive Brain Research*, 20(2), 183–193.

Jost, K., Hennighausen, E., & Rosler, F. (2004b). Comparing arithmetic and semantic fact retrieval: effects of problem size and sentence constraint on event-related brain potentials. *Psychophysiology*, 41(1), 46–59.

Just, M.A., Keller, T.A., Malave, V.L., Kana, R.K., & Varma, S. (2012). Autism as a neural systems disorder: a theory of frontal-posterior underconnectivity. *Neuroscience and Biobehavioral Reviews*, 36, 1292–1313.

Kaufmann, L., Wood, G., Rubinsten, O., & Henik, A. (2011). Meta-analyses of developmental fMRI studies investigating typical and atypical trajectories of number processing and calculation. *Developmental Neuropsychology*, 36(6), 763–787.

Kelley, P. (2008). *Making Minds: What's Wrong with Education—and What Should We Do About It?* New York: Routledge.

Kirk, E.P. & Ashcraft, M.H. (2001). Telling stories: the perils and promise of using verbal reports to study math strategies. *Journal of Experimental Psychology—Learning Memory and Cognition*, 27(1), 157–175.

Kosc, L. (1974). Developmental dyscalculia. *Journal of Learning Disabilities*, 7, 164–177.

Krause, B. & Cohen Kadosh, R. (2013). Can transcranial electrical stimulation improve learning difficulties in atypical brain development? A future possibility for cognitive training. *Developmental Cognitive Neuroscience*, 6, 176–194.

Ku, Y., Hong, B., Gao, X., & Gao, S. (2010). Spectra-temporal patterns underlying mental addition: an ERP and ERD/ERS study. *Neuroscience Letters*, 472(1), 5–10.

Kucian, K., Grond, U., Rotzer, S., Henzi, B., Schoenmann, C., Plangger, F., et al. (2011). Mental number line training in children with developmental dyscalculia. *Neuroimage*, 57, 782–795.

Kumari, V., Peters, E.R., Fannon, D., Antonova, E., Premkumar, P., Anilkumar, A.P., et al. (2009). Dorsolateral prefrontal cortex activity predicts responsiveness to cognitive-behavioral therapy in schizophrenia. *Biological Psychiatry*, 66, 594–602.

Lee, K., Lim, Z.Y., Yeong, S.H.M., Ng, S.F., Venkatraman, V., & Chee, M.W.L. (2007). Strategic differences in algebraic problem solving: neuroanatomical correlates. *Brain Research*, 1155, 163–171.

Lee, K., Yeong, S.H.M., Ng, S.F., Venkatraman, V., Graham, S., & Chee, M.W.L. (2010). Computing solutions to algebraic problems using a symbolic versus a schematic strategy. *The International Journal on Mathematics Education*, 42(6), 591–605.

LeFevre, J., Sadesky, G.S., & Bisanz, J. (1996). Selection of procedures in mental addition: reassessing the problem size effect in adults. *Journal of Experimental Psychology—Learning Memory and Cognition*, 22(1), 216–230.

Matjeko, A.A., Price, G.R., Mazzocco, M.M.M., & Ansari, D. (2013). Individual differences in left parietal white matter predict scores on the Preliminary Scholastic Aptitude Test. *Neuroimage*, 66, 604–610.

McCandliss, B.D. (2010). Educational neuroscience: the early years. *Proceedings of the National Academy of Sciences of the United States of America*, 107, 8049–8050.

Meltzoff, A.N., Kuhl, P.K., Movellan, J., & Sejnowski, T.J. (2009). Foundations for a new science of learning. *Science*, 325, 284–288.

Menon, V. (2010). Developmental cognitive neuroscience of arithmetic: implications for learning and education. *ZDM—The International Journal on Mathematics Education*, 42, 515–525.

Menon, V., Rivera, S.M., White, C.D., Eliez, S., Glover, G.H., & Reiss, A.L. (2000). Functional optimization of arithmetic processing in perfect performers. *Cognitive Brain Research*, 9(3), 343–345.

Molfese, D.L. (2000). Predicting dyslexia at 8 years of age using neonatal brain responses. *Brain and Language*, 72, 238–245.

Mussolin, C., De Volder, A., Grandin, C., Schlogel, X., Nassogne, M.C., & Noel, M.P. (2010). Neural correlates of symbolic number comparison in developmental dyscalculia. *Journal of Cognitive Neuroscience*, 22, 860–874.

Nickerson, S.D. & Whitacre, I. (2010). A local instruction theory for the development of number sense. *Mathematical Thinking and Learning*, 3, 227–252.

Nitsche, M.A., & Paulus, W. (2000). Excitability changes induced in the human motor cortex by weak transcranial direct current stimulation. *Journal of Physiology*, 527(3), 633–639.

Nosworthy, N., Bugden, S., Archibald, L., Evans, B., & Ansari, D. (2013). A two-minute paper-and-pencil test of symbolic and nonsymbolic numerical magnitude processing explains variability in primary school children's arithmetic competence. *Plos ONE*, 8(7), e67918.

Nunez-Pena, M.I., Cortinas, M., & Escera, C. (2006). Problem size effect and processing strategies in mental arithmetic. *Neuroreport*, 17(4), 357–360.

Núñez-Peña, M.I., Gracia-Bafalluy, M., & Tubau, E. (2011). Individual differences in arithmetic skill reflected in event-related brain potentials. *International Journal of Psychophysiology*, 80(2), 143–149.

Price, G.R., Mazzocco, M.M., & Ansari, D. (2013). Why mental arithmetic counts: brain activation during single digit arithmetic predicts high school math scores. *The Journal of Neuroscience*, 33(1), 156–163.

Price, G.R., Holloway, I., Rasanen, P., Vesterinen, M., & Ansari, D. (2007). Impaired parietal magnitude processing in developmental dyscalculia. *Current Biology*, 17, R1042–R1043.

Ramani, G.B. & Siegler, R.S. (2011). Reducing the gap in numerical knowledge between low- and middle-income preschoolers. *Journal of Applied Developmental Psychology*, 32, 146–159.

Rasanen, P., Salminen, J., Wilson, A.J., Aunio, P., & Dehaene, S. (2009). Computer-assisted intervention for children with low numeracy skills. *Cognitive Development*, 24, 450–472.

Reyna, V.F., Nelson, W.L., Han, P.K., & Dieckmann, N.F. (2009). How numeracy influences risk comprehension and medical decision making. *Psychological Bulletin*, 135, 943–973.

Ritchie, S.J. & Bates, T.C. (2013). Enduring links from childhood mathematics and reading achievement to adult socioeconomic status. *Psychological Science*, 24, 1301–1308.

Rivera, S.M., Reiss, A.L., Eckert, M.A., & Menon, V. (2005). Developmental changes in mental arithmetic: evidence for increased functional specialization in the left inferior parietal cortex. *Cerebral Cortex*, 15, 1779–1790.

Rosenberg-Lee, M., Barth, M., & Menon, V. (2011). What difference does a year of schooling make? Maturation of brain response and connectivity between 2nd and 3rd grades during arithmetic problem solving. *Neuroimage*, 57, 796–808.

Rotzer, S., Kucian, K., Martin, E., Von Aster, M., Klaver, P., & Loenneker, T. (2008). Optimized voxel-based morphometry in children with developmental dyscalculia. *Neuroimage*, 39, 417–422.

Rousselle, L. & Noel, M.P. (2007). Basic numerical skills in children with mathematics learning disabilities: a comparison of symbolic vs non-symbolic number magnitude processing. *Cognition*, 102(3), 361–395.

Rykhlevskaia, E., Uddin, L.Q., Kondos, L., & Menon, V. (2009). Neuroanatomical correlates of developmental dyscalculia: combined evidence from morphometry and tractography. *Frontiers in Human Neuroscience*, 3, 51.

Schneider, M., Grabner, R.H., & Paetsch, J. (2009). Mental number line, number line estimation, and mathematical achievement: their interrelations in grades 5 and 6. *Journal of Educational Psychology*, 101(2), 359–372.

Siegler, R.S., Adolph, K.E., & Lemaire, P. (1996). Strategy choices across the life span. In L.R. Reder (Ed.), *Implicit Memory and Metacognition* (pp. 79–121). Mahwah, NJ: Erlbaum.

Sloane, F.C. (2008). Randomized trials in mathematics education: recalibrating the proposed high watermark. *Educational Researcher*, 9, 624–630.

Smith-Chant, B.L. & LeFevre, J.A. (2003). Doing as they are told and telling it like it is: self-reports in mental arithmetic. *Memory & Cognition*, 31(4), 516–528.

Snowball, A., Tachtsidis, I., Popescu, T., et al. (2013). Long-term enhancement of brain function and cognition using cognitive training and brain stimulation. *Current Biology*, 23(11), 987–992.

Soltesz, F., Szucs, D., Dekany, J., Markus, A., & Csepe, V. (2007). A combined event-related potential and neuropsychological investigation of developmental dyscalculia. *Neuroscience Letters*, 417, 181–186.

Spelke, E.S. & Tsivkin, S. (2001). Language and number: a bilingual training study. *Cognition*, 78(1), 45–88.

Stanescu Cosson, R., Pinel, P., Moortele, P.F.v.d., Le Bihan, D., Cohen, L., & Dehaene, S. (2000). Understanding dissociations in dyscalculia. A brain imaging study of the impact of number size on the cerebral networks for exact and approximate calculation. *Brain*, 123, 2240–2255.

Stern, E. (2005). Pedagogy meets neuroscience. *Science*, 310, 745.

Supekar, K., Swigart, A.J., Tenison, C., et al. (2013). Neural predictors of individual differences in response to math tutoring in primary-grade school children. *Proceedings of the National Academy of Sciences*, 110, 8230–8235.

Temple, E., Deutsch, G.K., Poldrack, R.A., et al. (2003). Neural deficits in children with dyslexia ameliorated by behavioral remediation: evidence from functional MRI. *Proceedings of the National Academy of Sciences of the United States of America*, 100, 2860–2865.

Terney, D., Chaieb, L., Moliadze, V., Antal, A., & Paulus, W. (2008). Increasing human brain excitability by transcranial high-frequency random noise stimulation. *The Journal of Neuroscience*, 28(52), 14147–14155.

The Royal Society. (2011). *Brain Waves Module 2: Neuroscience. Implications for Education and Lifelong Learning*. London: The Royal Society.

Toll, S.W.M. & Van Luit, J.E.H. (2013). Accelerating the early numeracy development of kindergartners with limited working memory skills through remedial education. *Research in Developmental Disabilities*, 34, 745–755.

van Merriënboer, J.J.G. & Kirschner, P. (2007). *Ten Steps toComplex Learning. A Systematic Approach to Four-component Instructional Design*. New York: Lawrence Erlbaum Associates.

Vanbinst, K., Ghesquière, P., & De Smedt, B. (2012). Numerical magnitude representations and individual differences in children's arithmetic strategy use. *Mind, Brain, and Education*, 6(3), 129–136.

Wilson, A.J., Revkin, S.K., Cohen, D., Cohen, L., & Dehaene, S. (2006). An open trial assessment of "The Number Race," an adaptive computer game for remediation of dyscalculia. *Behavioral and Brain Functions*, 2, 20.

Zamarian, L., Ischebeck, A., & Delazer, M. (2009). Neuroscience of learning arithmetic—evidence from brain imaging studies. *Neuroscience and Biobehavioral Reviews*, 33, 909–925.

PART VI

NUMERICAL IMPAIRMENTS, CO-MORBIDITY, AND REHABILITATION

CHAPTER 34

WHEN NUMBER PROCESSING AND CALCULATION IS NOT YOUR CUP OF TEA

MARIE-PASCALE NOËL

Acquired Acalculia

Sometimes, people, who have previously had no problem with mathematics or who have even been good at the subject, may suffer a cerebral injury and, as a consequence of that, may present disorders of number or calculation skills called acquired acalculia.

The neuropsychology of acquired acalculia has developed over the last 35 years and is summarized in the chapter by Marinella Cappelletti (this volume). One of the first and more influential models of the number and calculation system was proposed by McCloskey et al. (1985). In support of the predictions of that model, many double dissociations have been reported. For instance, patients can suffer from a deficit in number comprehension or in number production, and within each of these, a specific deficit for the Arabic number format (e.g., Noël & Seron, 1993) or for the verbal format (Sokol et McCloskey, 1988) has been reported. Besides those number processing deficits, patients can also present difficulties in the processing of operation symbols (+, −, ×, ÷; e.g., Ferro & Botelho, 1980), arithmetic procedures (e.g., Girelli & Delazer, 1996), arithmetical facts (e.g., knowing that 3 × 5 = 15; e.g., Warrington, 1982), or conceptual knowledge about arithmetical operations (e.g., knowing that a × b = b × a; e.g., Delazer & Benke, 1997). More rarely, patients with a semantic deficit for numbers (e.g., Cipolotti et al., 1991) have also been reported. Sometimes, this deficit has been associated with other deficits of non-numerical magnitudes, such as time or space. These associations and dissociations between different magnitude dimensions suggest that they partly share common representations or decision mechanisms operating on those representations. These advances in the description and understanding of acquired acalculia have led to the development of assessment tools and of a few rehabilitation programs.

More recently, with the triple-code model of Dehaene and Cohen (1995), the neuropsychology of numerical cognition has moved into a brain-based approach looking more broadly at the relation between number and other cognitive domains. For instance, impaired visual–spatial abilities, as observed in hemineglect, have been shown to impact on number magnitude representation (e.g., Vuilleumier et al., 2004). Conversely, dissociations have been reported in the semantic domain between number magnitude representation (which is supported by the parietal lobe) and non-numerical semantics, such as the meaning of words or of objects (sustained by the left temporal lobe) (e.g., Cappelletti et al., 2001, 2005).

Brain Bases of Arithmetic

In the chapter by Laura Zamarian and Margarete Delazer (this volume), the brain bases of arithmetic are described. The brain activity underlying calculation has been shown to mainly recruit fronto-parietal circuits and to be influenced by many factors such as the complexity of the calculation, the type of arithmetical operation, the way the calculation has been learned, and the degree of expertise of the person. For instance, frontal regions are mostly involved in complex calculation, whereas the left parietal cortex is more involved in simple calculation (e.g., Delazer et al., 2006; Semenza et al., 1997). Among these parietal regions, the intraparietal sulcus is more activated in approximate than in exact calculation (Stanescu-Cosson et al., 2000) and the angular gyrus seems to be particularly important for the retrieval of a solution from long-term memory (e.g., Chochon et al., 1999).

Typically, learning arithmetic involves a shift from reliance on calculation procedures that require a lot of working memory resources and control, to direct memory retrieval and automatisation. In parallel, a decrease of brain activity can be measured in the frontal areas (working memory and control processes), together with a relative increase of activation in parietal brain areas, and, in particular, in the angular gyrus, as memory retrieval takes place. Some have argued that this involvement of the angular gyrus is evidence for the use of verbal strategies in the retrieval of arithmetical facts (Dehaene et al., 2003). However, more recent studies support the view that the angular gyrus would be engaged in general mapping processes (Ansari, 2008). For instance, an increase in angular gyrus activity was also seen after non-arithmetical training (i.e., training on figural–spatial problems led to a peak of activation in the right angular gyrus (Grabner et al., 2009)). Therefore, the angular gyrus may possibly be involved in general processes of fact retrieval from memory or in mapping to information stored in memory. The arithmetical operations which are typically learned through memorisation (e.g., multiplication tables) indeed lead to greater activation of the angular gyrus than do, for example, subtractions; examples of the latter are solved through manipulation of quantities and rely more on the intraparietal circuit (Chochon et al., 1999).

If the same operations are learned either by drill (emphasizing the retrieval from memory) or by emphasizing strategies, the former yields relative greater left angular gyrus activation (Delazer et al., 2005). As few as eight repetitions of the same problem suffice to lead to visible modifications of brain activation, with a decrease in fronto-parietal areas and an increase of activation in temporoparietal areas, including the left angular gyrus (Ischebeck et al., 2007). The degree of activation of this angular gyrus has also been related to individual differences in arithmetic skills. Moreover, transfer effects have also been reported. For instance, training in multiplication facts not only changes the brain activity in response to that operation but also has an impact on the corresponding division problems, which also evoke a greater activation of the left angular gyrus (Ischebeck et al., 2009).

Training not only leads to changes in brain activity, but can also impact on the brain structure itself. For instance, experience with mathematics is associated with different gray matter density in the parietal and frontal areas (Menon et al., 2000). Conversely, changes in brain activity can influence learning. Recent studies using brain stimulation techniques, such as the transcranial direct current stimulation, suggest that learning in the numerical domain could be significantly enhanced (Cohen Kadosh et al., 2010).

Dyscalculia

For a significant number of healthy children, learning mathematics is a real source of difficulty. Severe learning disability in arithmetic, despite normal intelligence and scholastic opportunities, is termed (developmental) dyscalculia. The prevalence of dyscalculia is estimated between 3% and 6%. These children are far behind their classmates in many different numerical tasks and present difficulties in retrieval of arithmetical facts, in using arithmetical procedures, in using immature problem-solving strategies, etc. Most individuals with mathematical learning difficulties in primary school continue to suffer from these years later (Shalev et al., 2005). These often lifelong mathematical difficulties lead to potentially debilitating outcomes, especially occupationally and economically.

Underlying Deficits and Brain Correlates

In recent years, research has focused on the very basic numerical deficits that may underlie dyscalculia. Brian Butterworth, Sashank Varma, and Diana Laurillard present those studies in their chapter in this volume. Neurobehavioral research has identified numerosity representations (the number of objects in a set) and symbolic numbers magnitude representations (Arabic or verbal numbers), foundational capacities in the development of arithmetic. These representations would be impaired in dyscalculic learners (e.g., Landerl et al., 2004; Noël & Rousselle, 2011; Rousselle & Noël, 2007). Indeed, numbers do not seem to be as meaningful for dyscalculics as they are for typically developing learners. As Butterworth et al. write in their chapter (this volume), "They do not intuitively grasp the size of a number or its value relative to other numbers." Brain imaging research has shown that the intraparietal sulcus supports the representation of number magnitude and is activated in almost all numerical and arithmetical processes. This brain area, among others, significantly differs between dyscalculic and typically developing children in terms of structure, connectivity, and activation.

Brain correlates of dyscalculia are reported in the chapter by Karin Kucian, Liane Kaufmann, and Michael von Aster (this volume). Although still scarce, the few studies that have investigated the brain correlates of numerical disabilities suggest that peculiarities are observed in core regions for number processing. In particular, dyscalculic children show deficits in gray matter in the intraparietal sulcus and adjacent regions, including the superior parietal lobe (Rotzer et al., 2008; Rykhlevskaia et al., 2009). Differences have also been found in other brain regions, but less consistently. They are also characterized by reduced fiber tracts connecting brain areas of the fronto-parietal network, which is important for number processing and calculation (Rykhlevskaia et al., 2009). Similarly, in typically developing children, integrity of the pathways connecting parietal and frontal areas correlates with skills in approximate arithmetic (Tsang et al., 2009). As regards brain activity in numerical tasks, dyscalculic participants show atypical activation in the parietal lobe, mainly close to the intraparietal sulcus. Deviant activation has also been reported in other cortical and subcortical areas and seems to be related to stronger recruitment of working memory, attention, or executive functions, as more effort and cognitive resources might be needed to solve the numerical tasks (see Kaufmann et al., 2011, for a review). Of course, these biological differences are not necessarily the causes of dyscalculia. Indeed, as Kucian et al. point out, in developmental disorders, one could imagine that the biological condition precedes and explains the cognitive (dys)function, but one could equally well imagine the reverse, as brain development is highly plastic and depends on the child's learning experiences and environment.

Interventions

In view of the basic numerical deficits that are observed in dyscalculia and the brain correlates of these numerical disabilities, remediation should first address this foundational number concept. Some adaptive software, informed by these neuroscientific findings, have been developed and are reviewed in Pekka Räsänen's chapter (this volume).

Pekka Räsänen first describes the historic evolution of computer-assisted interventions or technology-enhanced learning and reports the results of different meta-analytic reviews of the efficiency of these interventions. Although some studies had negative or null effects, in the main, positive effects were observed, though with a small effect size (around .3 in Cohen's d) which, moreover, tends to be lower in more recent studies than those published in the 90s. Studies with younger children have typically obtained better results than those with older students, and children with special needs in education seem to gain better benefit from those computer-assisted interventions than students in mainstream education, especially if their starting level is low due to social rather than cognitive reasons.

Recently, some computer-assisted interventions have been developed on the basis of the neuroscientific approach. For instance, the Number Race (Wilson et al., 2006) aimed at enhancing the approximate number system. A first study on children with mathematical learning difficulties (but without any control group) showed that an average of 12 hours of playing with this game led to significant improvement in the approximate number system but had no impact on addition. Another study with low socioeconomic children (5–6 years old) showed a positive impact of that intervention on the magnitude comparison of symbolic numbers (Arabic and verbal numbers) but, curiously, not on the magnitude comparison of dot sets (Wilson et al., 2009). The impact of this game was then compared with the impact of the Graphogame Math game, concentrating on the exact representation of numerical symbols. Both interventions proved to have positive impact on number comparison but not on arithmetic (Räsänen et al., 2009). Obersteiner et al. (2013) continued with this idea and made two versions of the Number Race: one for approximate and one for exact number representations. Compared to a language-learning group, both Number

Race groups showed numerical improvement, but in different ways. The approximate group improved more than the exact group on approximate tasks, such as estimation, whereas the exact group improved more than the other group in tasks requiring exact symbolic responses. Finally, both groups improved similarly in an arithmetic task, compared to the language-learning group.

Other interventions have been based on the idea that number magnitude representation is spatial and have used the metaphor of the number line. Thus, Kucian et al. (2011) developed Rescue Calcularis – a game in which children had to estimate, on a number line, the number corresponding to a given calculation. Both typical learners and children with dyscalculia showed improvement in positioning numbers on a number line after that intervention. Furthermore, a measure of brain functioning was also taken before and after the training and showed reduced recruitment of the brain regions that support number processing after the training. A full-scale educational application based on the same idea (Käser et al., 2011) showed greater improvement in an intervention group, compared with a waiting-list group, in number positioning on a number line, as well as in subtraction.

Using the same idea of a number line, Fischer et al. (2011) compared two forms of training. In the first one, children saw one collection or Arabic digit and had to compare it with the reference number "5" (presented either as a set or a digit) by pointing their finger towards the bigger one. In the second training, the reference number "5" was presented in the middle of a number line and the child had to compare numbers to that target by moving their body to the left or to the right if it was smaller or larger, respectively. They showed that children's estimates of the position of numbers on a 0–10 scale increased more in precision after the training involving full body movements and the number line than after the one involving only finger pointing. But here again, no transfer effect was obtained on calculation skills. More recently, Link et al. (2013) compared two types of number line training: a control one, in which children had to position a number on a number line placed in front of them, and an embodied condition, in which a much longer number line was laid on the floor and the child had to walk on that line to the estimated location of the number. They found more pronounced training effects after the embodied training than after the control training, as regards addition performance among others.

General Contributory Cognitive Factors

If very basic forms of number processing deficit have been repeatedly shown to characterize dyscalculia, other more general cognitive factors also seem to contribute to difficulties in mathematics. According to David Geary's chapter (this volume), most children with mathematical learning disability have poor working memory capacities, especially regarding the central executive component, which probably contributes to their slow numerical development independent of intelligence. In arithmetic, they tend to use less mature procedures and commit more procedural errors. With this development delay, it would take years to close the gap. Deficits in arithmetical fact retrieval, poor conceptual understanding (e.g., of the base-10 system), and low working memory capacities would contribute to this developmental delay. A common and persistent problem for these children is a difficulty in learning and retrieving basic arithmetical facts. A first possibility would be that these deficits are secondary to a more basic deficit in the number magnitude representation and the ability to approximate the result of an arithmetical operation (see Rousselle & Noël, 2008). Yet, as these children quite often produce errors corresponding to counting-string intrusions (e.g., $4 \times 5 = 6$), Geary et al. (2012) hypothesized that these difficulties could be due to an inability to inhibit irrelevant associations from entering working memory during the process of fact retrieval. More recently, however, De Visscher and Noël (2013, 2014) have proposed that such a difficulty might be associated with a greater sensitivity to interference in long-term memory which would prevent them from encoding properly these problem–response associations.

Different Profiles of Dyscalculia

Avishai Henik, Orly Rubinsten, and Sarit Ashknazi also underline in their chapter (this volume) the fact that arithmetic relies both on domain-specific and domain-general abilities such as working memory, attention, or visuospatial processing. Moreover, about 20 to 60% of children with dyscalculia also suffer from other learning problems such as dyslexia or ADHD (attention deficit disorder with hyperactivity). This, together with the fact that profiles of dyscalculia are heterogeneous, led Henik et al. to propose the existence of different profiles of dyscalculia. First, we might imagine that a deficit in the intraparietal sulcus, which is believed to support a spatial representation of the number magnitude, would lead to pure deficit in the numerical domain only, and result in pure dyscalculia. But deficit in the intraparietal sulcus could also lead to deficits both in processing numerical quantities and in attention, which would enhance the difficulties in arithmetic. Possibly, a deficit in frontal lobes could lead to deficits in executive functions (attention and working memory), which would lead to ADHD and to difficulties in arithmetic. Finally, as the left angular gyrus is involved in the integration of semantic information in an ongoing context and in associating symbols with non-symbolic events or, more generally, in learning facts and in retrieving facts (such as arithmetical facts, see the chapter by Zamarian and Delazer in this volume), a deficit at that level might lead to both reading and calculation difficulties. Currently, most studies have looked at dyscalculia as a whole, without distinguishing between different subtypes. This theoretical proposition of different subtypes of persons with developmental dyscalculia still needs to be fully tested, though a number of studies do indicate that mathematical difficulties involve a variety of components of arithmetic, and can take diverse forms (e.g., Desoete et al., 2003; Dowker, 2005a,b, this volume; Jordan & Hanich, 2000; Temple, 1991).

Reading Disability and Mathematical Outcomes

One path in this direction is the study of number processing and calculation in dyslexia. In her chapter in this volume, Silke Göbel presents this research as a way to see the impact of poor phonological abilities on mathematics. Earlier research on children and adults with reading disability had shown poor performance in mathematics. However, this might have been due to the high co-morbidity between reading and mathematics disability. More recently, some researchers have studied participants who are dyslexics (with a reading, decoding, or spelling difficulty) but who are not dyscalculics (i.e., who perform in the normal range on a standardized test of mathematics). Even in this very select population, difficulties with arithmetical fact retrieval (especially for small additions and multiplications) have been found (eg., Boets & De Smedt, 2010). To account for these difficulties, several hypotheses were proposed, such as a difficulty in paired associate learning, in acquiring automaticity, or in processing speed, but it seems that the phonological deficit hypothesis is the most prominent one. In typically developing children, retrieval of arithmetical facts is mostly related with phonological awareness (sensitivity to the sound structure in speech) and much less consistently with either phonological loop capacities or speed of lexical access (De Smedt et al., 2010). Currently, we do not know the neural bases of the fact retrieval deficits in dyslexics. However, according to Göbel (this volume), the left angular gyrus, which is involved in both phonological tasks and calculation, might be a good candidate. The genetic bases are unknown as well. However, Plomin and Kovas (2005) suggested that the same genetic risk factors may underlie mathematical and reading difficulties and, therefore, cause their high co-morbidity.

Visual Impairment and Mathematical Outcomes

Another way to address the interaction between number processing and other domain general functions is to examine its development in people who have grown up without one specific sensory modality. In recent years, several studies have examined the role of vision on numerical development. Indeed, most of the time, we access numerical information through vision, and a very large proportion of studies in the field of numerical cognition have used the visual modality to present the numbers. Vision permits the processing of a large number of items simultaneously and with a higher precision than with other senses such as touch or audition. We might thus ask whether vision plays a critical role in numerical cognition. Its role in the development of numerical representations and skills has been studied by examining the impact of early visual deprivation on numerical cognition. Julie Castronovo reports these results in her chapter (this volume). A series of studies tested the semantic representation of numbers using magnitude comparison of verbal numbers, number bisection, or parity judgment (e.g., Castronovo & Seron, 2007a; Szücs & Csépe, 2005). They clearly indicated that early visual deprivation did not prevent the development of a numerical representation similar to the one of sighted participants and the presence of similar spatial properties (i.e., a mental continuum oriented spatially). However, early visual deprivation does appear to have some impact on numerical cognition. First, whereas sighted individuals use an eye-centred coordinate frame of reference to map numbers onto space, individuals who are blind from an early stage use a body-centred coordinate frame of reference (Crollen et al., 2012). Second, early blinded individuals appear to have a higher capacity than sighted participants for numerical estimation. For instance, they are more precise when asked to estimate the number of auditory events (Castronovo & Seron, 2007b). Finally, vision has also been shown to play a critical role in the acquisition of finger–number associations. Indeed, early blinded children use finger counting less often and, when using it, do not show the canonical finger–number configurations (Crollen et al., 2011; 2014).

Genetic Syndromes and Mathematical Outcomes

The heterogeneity of mathematical learning difficulties can also be studied by examining the profile of children who share a known genotypic risk for poor mathematical outcomes. Michèle Mazzocco follows this in her chapter (this volume) by reporting the profile characteristics of two genetic syndromes related to the X chromosome: the fragile X syndrome and the Turner syndrome. High-functioning girls with fragile X show good memory-based learning but are more in trouble with conceptual numerical domains. For instance, they can recite the number sequence easily but show difficulties with counting principles; later on, they can name symbols for decimal values but are unsuccessful at ordering decimals or fractions. Difference from the general population can already be seen in kindergarten but is more obvious years later. Indeed, they show a slower rate of growth for mathematical calculation relative to their peers, possibly due to their weaker executive functions.

Turner syndrome leads to visual–spatial deficits but intact verbal skills. With respects to numerical cognition, difference from the general population is already evident in kindergarten but it accentuates over time, with up to 75% of these children meeting the criteria of mathematical learning disability by grade 3 (Murphy et al., 2006). In simple calculation, they show accurate but slow performance, despite numerical processing deficits. Their mathematical weaknesses do not seem obviously linked to their weak visuospatial abilities but might possibly be associated with their poor executive functions.

Although in both of these syndromes, the risk of mathematical learning disability is higher than in the normal population, not all the patients do show such difficulty. Moreover, these difficulties might well be secondary to other cognitive factors.

Williams' and Down syndromes are studied by Jo Van Herwegen and Annette Karmiloff-Smith in their chapter (this volume). They contrasted these two profiles as at the end state of their development, they are quite different. Indeed, patients with Williams' syndrome, who typically show relative strengths in the verbal area, are good at counting but show weaker number-line processing. On the contrary, in Down syndrome, the mental number line seems to be operating proficiently, while counting, which relies on their weaker verbal abilities, is impaired. In early infancy, those with Williams' syndrome are able to discriminate sets of two or three objects but not larger sets (8 vs. 16) that are supposedly coded on the number line, whereas those with Down syndrome do not react to changes in small numbers but can differentiate large collections (Paterson et al., 1999, 2006; Van Herwegen et al., 2008). Interestingly, in their chapter (this volume), Van Herwegen and Karmiloff-Smith look at even more basic peculiarities that could account for this difference and argue that these could be related to basic visuospatial scanning differences. Indeed, when confronted with large display, Down syndrome children are able to scan all items, while Williams' syndrome children fixate only on a few of the presented dots. In summary, the authors advocate for the need to trace full developmental trajectories from infancy onwards and to investigate basic-level abilities as well as the role of higher-level (numerical) abilities.

Summary

In summary, this section of this volume illustrates how varied the profiles of deficits in number processing or calculation are following brain damage. Valuable theoretical models have been developed to account for such a variety and have helped us in understanding how the normal brain functions and all the possible dissociations that can appear after a brain injury. More recently, research has been devoted to the cerebral bases of the numerical processes (see, for example, the section "The Neuroscience of Numbers"). Several brain structures are implicated in this, with a major role for the frontoparietal circuit. However, numerical cognition cannot be fully understood if one does not consider its interaction with other cognitive processes, such as visuospatial or language processing or the processing of other magnitudes such as time and space.

As regards developmental problems with mathematics, research has also suggested that this learning disability is related to both domain-specific and domain-general weaknesses. The former corresponds to a difficulty in processing the magnitude of numbers; the latter may include weaknesses in working memory capacities or phonological awareness or to a hyper sensitivity to interference in long-term memory. The differential roles of all these factors probably explain the heterogeneity of the profiles of dyscalculia and also the different associations that can be found with other learning disabilities such as dyslexia or ADHD.

The theoretical models coming from the neuropsychology of acalculia can only partially be used when considering dyscalculia. Indeed, these models are built on a static view of the cognitive system. In understanding dyscalculia, one has to take development, and therefore time, into account. Very few studies have depicted a developmental view of the difficulties encountered by dyscalculic children (Von Aster & Shalev, 2007) and even fewer have proposed a dynamic view of the causal factors that could account for such a learning disability (Noël & Rousselle, 2011). Furthermore, one should always consider the possibility that a deficit in basic number processing could, at least partly, be due to another impairment. This is, for example, the view that is adopted by Van Herwegen and Karmiloff-Smith who argue that Williams' syndrome patients' difficulty with number magnitude of large sets of items is due to a more basic impairment in visuospatial scanning. Such an approach should also be considered for children with dyscalculia. Much current research on dyscalculia treats participants as a homogeneous group, while actually, we know that this population is very heterogeneous (Dowker, 2005a,b). It is highly possible that different types, or even different combinations, of basic deficits may lead to mathematical learning disability. Henik et al. have made some suggestions of ways to differentiate profiles, but these hypotheses still need to be tested. Finally, the poorer domain investigated is, of course, interventions. Though some programs have been developed to enhance mathematical skills (for reviews, see also Butterworth et al., 2011; Cohen Kadosh et al., 2013), there has still been little work in this area as compared with, for example, interventions for reading disabilities. We hope that research in the future will consider this as a priority, especially in view of the devastating effects that such disabilities can have on both professional and financial domains.

Acknowledgment

The author is supported by the National Research Fund of Belgium.

References

Ansari, D. (2008). Effects of development and enculturation on number representation in the brain. *Nature Reviews Neuroscience*, 9, 278–291.

Boets, B. & De Smedt, B. (2010). Single-digit arithmetic in children with dyslexia. *Dyslexia, 16*, 183–191.

Butterworth, B., Sashank, V., & Laurillard, D. (2011). Dyscalculia: from brain to education. *Science, 332*, 1049–1053.

Cappelletti, M., Butterworth, B., & Kopelman, M. (2001). Spared numerical abilities in a case of semantic dementia. *Neuropsychologia, 39*, 1224–1239.

Cappelletti. M., Morton, J., Kopelman, M., & Butterworth, B. (2005). The progressive loss of numerical knowledge in a semantic dementia patient: a follow-up study. *Cognitive Neuropsychology, 22*, 771–793.

Castronovo, J. & Seron, X. (2007a). Semantic numerical representation in blind subjects: the role of vision in the spatial format of the mental number line. *The Quarterly Journal of Experimental Psychology, 60*(1), 101–119.

Castronovo, J. & Seron, X. (2007b). Numerical estimation in blind subjects: evidence of the impact of blindness and its following experience. *Journal of Experimental Psychology: Human, Perception and Performance, 33*(5), 1089–1106.

Chochon, F., Cohen, L., van De Moortele, P.F., & Dehaene, S. (1999). Differential contributions of the left and right inferior parietal lobules to number processing. *Journal of Cognitive Neuroscience, 11*, 617–630.

Cipolotti, L., Butterworth, B., & Denes, G. (1991). A specific deficit for numbers in case of dense acalculia. *Brain, 114*, 2619–2637.

Cohen Kadosh, R., Dowker, A., Heine, A., Kaufmann, L., & Kucian, K. (2013). Interventions for improving numerical abilities: present and future. *Trends in Neuroscience and Education*, 2, 85–93.

Cohen Kadosh, R., Soskic, S., Iuculano, T., Kanai, R., & Walsh, V. (2010). Modulating neuronal activity produces specific and long-lasting changes in numerical competence. *Current Biology, 20*, 2016–2020.

Crollen, V., Dormal, G., Seron, X., Lepore, F., & Collignon, O. (2012). The role of vision in the development of number-space interactions. *Cortex*. doi: 10.1016/j.cortex.2011.11.006

Crollen, V., Mahe, R., Collignon, O., & Seron, X. (2011). The role of vision in the development of finger-number interactions: finger-counting and finger-montring in blind children. *Journal of Experimental Child Psychology, 109*, 525–539.

Crollen, V., Noël, M-P., Seron, X., Mahau, P., Lepore, F. & Collignon, O. (2014) Visual experience influences the interactions between fingers and numbers. *Cognition*, 133, 91–96

De Smedt, B., Taylor, J., Archibald, L., & Ansari, D. (2010). How is phonological processing related to individual differences in children's arithmetic skills? *Developmental Science, 13*(3), 508–520.

De Visscher, A. & Noël, M.P. (2013). A case study of arithmetic facts dyscalculia caused by a hypersensitivity to interference in memory. *Cortex, 49*(1), 50–70.

De Visscher, A., & Noël, M. P. (2014). Arithmetic facts storage deficit: The hypothesis of hypersensitivity-to-interference in memory. *Developmental Science*, 17(3), 434–442.

Dehaene, S. & Cohen, L. (1995). Towards an anatomical and functional model of number processing. *Mathematical Cognition, 1*, 83–120.

Dehaene, S., Piazza, M., Pinel, P., & Cohen, L. (2003). Three parietal circuits for number processing. *Cognitive Neuropsychology, 20*, 487–506.

Delazer, M. & Benke, T. (1997). Arithmetic facts without meaning. *Cortex, 33*, 697–710.

Delazer, M., Ischebeck, A., Domahs, F., et al. (2005). Learning by strategies and learning by drill – evidence from an fMRI study. *NeuroImage, 25*, 838–849.

Delazer, M., Karner, E., Zamarian, L., Donnemiller, E., & Benke, T. (2006). Number processing in posterior cortical atrophy – a neuropsychological case study. *Neuropsychologia, 44*, 36–51.

Desoete, A., Roeyers, H., & DeClercq, A. (2003). Children with mathematical learning disabilities in Belgium. *Journal of Learning Disabilities, 37*, 32–41.

Dowker, A. (2005a). *Individual Differences in Arithmetic: Implications for Psychology, Neuroscience and Education*. Hove: Psychology Press.

Dowker, A. (2005b). Early identification and intervention for students with mathematics difficulties. *Journal of Learning Disabilities, 38*, 324–332.

Ferro, J.M. & Botelho, M. (1980). Alexia for arithmetical signs: a cause of disturbed calculation. *Cortex, 16*, 175–180.

Fischer, U., Moeller, K., Bientzle, M., Cress, U., & Nuerk, H.-C. (2011). Sensori-motor spatial training of number magnitude representation. *Psychonomic Bulletin & Review, 18*(1), 177–183.

Geary, D. C., Hoard, M. K., & Bailey, D. H. (2012). Fact retrieval deficits in low achieving children and children with mathematical learning disability. *Journal of Learning Disabilities*, 45(4), 291–307.

Girelli. L. & Delazer, M. (1996). Subtraction bugs in an acalculic patient. *Cortex, 32*, 547–555.

Grabner, R.H., Ischebeck, A., Reishofer, G., et al. (2009). Fact learning in complex arithmetic and figural-spatial tasks: the role of the angular gyrus and its relation to mathematical competence. *Human Brain Mapping, 30*, 2936–2952.

Ischebeck, A., Zamarian, L., Egger, K., Schocke, M., & Delazer, M. (2007). Imaging early practice effects in arithmetic. *NeuroImage, 36*, 993–1003.

Ischebeck, A., Zamarian, L., Schocke, M., & Delazer, M. (2009). Flexible transfer of knowledge in mental arithmetic – an fMRI study. *NeuroImage, 44*, 1103–1112.

Jordan, N.C. & Hanich, L. (2000). Mathematical thinking in second-grade children with different forms of LD. *Journal of Learning Disabilities, 33*, 567–578.

Käser, T., Kucian, K., Ringwald, M., Baschera, G.M., von Aster, M., & Gross, M. (2011). Therapy software for enhancing numerical cognition. In J. Özyurt et al. (Eds.), *Interdisciplinary Perspectives on Cognition, Education and the Brain* (pp. 207–216). Oldenburg: Hanse-Studies.

Kaufmann, L., Wood, G., Rubinsten, O., & Henik, A. (2011). Meta-analyses of developmental fMRI studies investigating typical and atypical trajectories of number processing and calculation. *Developmental Neuropsychology, 36*(6), 763–787.

Kucian, K., Grond, U., Rotzer, S., et al. (2011). Mental number line training in children with developmental dyscalculia. *NeuroImage, 57*(3), 782–795.

Landerl, K., Bevan, A., & Butterworth, B. (2004). Developmental dyscalculia and basic numerical capacities: a study of 8–9 year old students. *Cognition, 93*, 99–125.

Link, T., Moeller, K., Huber, S., Fischer, U., & Nuerk, H-C. (2013). Walk the number line – an embodied training of numerical concepts. *Trends in Neuroscience and Education*, 2, 74–84.

McCloskey, M., Caramazza, A., & Basili, A. (1985). Cognitive mechanisms in number processing and calculation: evidence from dyscalculia. *Brain and Cognition, 4*, 171–196.

Menon, V., Rivera, S.M., White, C.D., Eliez, S., Glover, G.H., & Reiss, A.L. (2000). Functional optimization of arithmetic processing in perfect performers. *Brain Research. Cognitive Brain Research, 9*, 343–345.

Murphy, M.M., Mazzocco, M.M.M., Gerner, G., & Henry, A.E. (2006). Mathematics learning disability in girls with Turner syndrome or fragile X syndrome. *Brain and Cognition, 6*, 195–210.

Noël, M-P. & Seron, X. (1993). Numeral number reading deficit: a single-case study or when 236 is read (2306) and judged superior to 1258. *Cognitive Neuropsychology, 10*, 317–339.

Noël, M.-P. & Rousselle, L. (2011). Developmental changes in the profiles of dyscalculia: an explanation based on a double exact-and-approximate number representation model. *Frontiers in Human Neuroscience, 5*(165), 1–4.

Obersteiner, A., Reiss, K., & Ufer, S. (2013). How training on exact or approximate mental representations of number can enhance first-grade students' basic number processing and arithmetic skills. *Learning and Instruction*, 23, 125–135.

Paterson, S.J., Brown, J.H., Gsödl, M.K., Johnson, M.H., & Karmiloff-Smith, A. (1999). Cognitive modularity and genetic disorders. *Science, 286*, 2355–2358.

Paterson, S.J., Girelli, L., Butterworth, B., & Karmiloff-Smith, A. (2006). Are numerical impairments syndrome specific? Evidence from Williams' syndrome and Down's syndrome. *Journal of Child Psychology and Psychiatry*, *47*(2), 190–204.

Plomin, R. & Kovas, Y. (2005). Generalist genes and learning disabilities. *Psychological Bulletin, 131*(4), 592–617.

Räsänen, P., Salminen, J., Wilson, A., Aunio, P., & Dehaene, S. (2009). Computer-assisted intervention for children with low numeracy skills. *Cognitive Development*, *24*(4), 450–472.

Rotzer, S., Kucian, K., Martin, E., von Aster, M., Klaver, P., & Loenneker, T. (2008). Optimized voxel-based morphometry in children with developmental dyscalculia. *Neuroimage, 39*(1), 417–422.
Rousselle, L. & Noël, M-P. (2008). Mental arithmetic in children with mathematics learning disabilities: the adaptive use of approximate calculation in an addition verification task. *Journal of Learning Disabilities, 41*(6), 498–513.
Rousselle, L. & Noël, M.-P. (2007). Basic numerical skills in children with mathematics learning disabilities: a comparison of symbolic vs non- symbolic number magnitude processing. *Cognition, 102*, 361–395.
Rykhlevskaia, E., Uddin, L.Q., Kondos, L., & Menon, V. (2009). Neuroanatomical correlates of developmental dyscalculia: combined evidence from morphometry and tractography. *Frontiers in Human Neuroscience, 3*, 51.
Semenza, C., Miceli, L., & Girelli, L. (1997). A deficit for arithmetical procedures: lack of knowledge or lack of monitoring? *Cortex, 33*, 483–498.
Shalev, R.S., Manor, O., & Gross-Tsur, V. (2005). Developmental dyscalculia: a prospective six-year follow-up. *Developmental Medicine & Child Neurology, 47*, 121–125.
Sokol, S.M. & McCloskey, M. (1988). Levels of representation in verbal number production. *Applied Psycholinguistics, 9*, 267–281.
Stanescu-Cosson, R., Pinel, P., van De Moortele, P.F., Le Bihan, D., Cohen, L., & Dehaene, S. (2000). Understanding dissociations in dyscalculia: a brain imaging study of the impact of number size on the cerebral networks for exact and approximate calculation. *Brain, 123*, 2240–2255.
Szücs, D. & Csépe, V. (2005). The parietal distance effect appears in both the congenitally blind and matched sighted controls in an acoustic number comparison task. *Neurosciences Letters, 384*, 11–16.
Temple, C.M. (1991). Procedural dyscalculia and number fact dyscalculia: double dissociation in developmental dyscalculia. *Cognitive Neuropsychology, 8*, 155–176.
Tsang, J. M., Dougherty, R. F., Deutsch, G. K., Wandell, B. A., and Ben-Shachar, M. (2009). Frontoparietal white matter diffusion properties predict mental arithmetic skills in children. *Proceedings of the National Academy of Sciences of the United States of America*, 106(52), 22546–22551.
Van Herwegen, J., Ansari, D., Xu, F., & Karmiloff-Smith, A. (2008). Small and large number processing in infants and toddlers with Williams syndrome. *Developmental Science, 11*(5), 637–643.
Von Aster, M.G. & Shalev, R.S. (2007). Number development and developmental dyscalculia. *Developmental Medicine and Child Neurology, 49*, 868–873.
Vuilleumier, P., Ortigue, S., & Brugger, P. (2004). The number space and neglect. *Cortex, 40*, 399–410.
Warrington, E.K. (1982). The fractionation of arithmetical skills: a single study. *Quarterly Journal of Experimental Psychology, 34*, 31–51.
Wilson, A.J., Dehaene, S., Dubois, O., & Fayol, M. (2009). Effects of an adaptive game intervention on accessing number sense in kindergarten children. *Mind, Brain and Education*, 3(4), 224–234.
Wilson, A.J., Revkin, S.K., Cohen, D., Cohen, L., & Dehaene, S. (2006). An open trial assessment of 'the number race', an adaptive computer game for remediation of dyscalculia. *Behavioural and Brain Functioning*, 2(1), 20.

CHAPTER 35

DYSCALCULIA

from brain to education

BRIAN BUTTERWORTH, SASHANK VARMA, AND DIANA LAURILLARD

The advancement and perfection of mathematics are ultimately connected with the prosperity of the state.

(Napoleon Bonaparte)

Introduction

Developmental dyscalculia is a mathematical disorder, with an estimated prevalence of about 5–7% (Shalev, 2007), roughly the same prevalence as developmental dyslexia (Gabrieli, 2009). A major report by the UK government concludes that: 'Developmental dyscalculia is currently the poor relation of dyslexia, with a much lower public profile. But the consequences of dyscalculia are at least as severe as those for dyslexia.' (Beddington, Cooper, Field, Goswami, Huppert, Jenkins, et al., 2008, p. 1060). The relative poverty of dyscalculia funding is clear from the figures – since 2000, NIH has spent $107.2m funding dyslexia research, but only $2.3m on dyscalculia (Bishop, 2010).

The classical understanding of dyscalculia as a clinical syndrome uses low achievement on mathematical achievement tests as the criterion without identifying the underlying cognitive phenotype (Gross-Tsur, Manor, & Shalev, 1996; Kosc, 1974; Shalev & Gross-Tsur, 2001). It has therefore been unable to inform pathways to remediation, whether in focused interventions, or in the larger, more complex context of the math classroom.

Why Is Mathematical Disability Important?

Low numeracy is a significant cost to nations, and improving standards could dramatically improve economic performance. In a recent analysis, the OECD demonstrated that an improvement of 'one-half standard deviation in mathematics and science performance at the individual level implies, by historical experience, an increase in annual growth rates of

Box 35.1 PISA question example

At Level 1 students can answer questions involving familiar contexts where all relevant information is present and the questions are clearly defined. They are able to identify information and carry out routine procedures according to direct instructions in explicit situations. They can perform actions that are obvious and follow immediately from the given stimuli. For example:

Mei-Ling found out that the exchange rate between Singapore dollars and South African rand was:

1 SGD = 4.2 ZAR

Mei-Ling changed 3000 Singapore dollars into South African rand at this exchange rate. How much money in South African rand did Mei-Ling get?

79% of 15-year-olds were able to answer this correctly.

GDP per capita of 0.87%' (OECD, 2010, p. 17). Time-lagged correlations show that improvements in educational performance contribute to increased GDP growth. A substantial long-term improvement in GDP growth (an added 0.68% per annum for all OECD countries) could be achieved just by raising the standard of the lowest-attaining students to the PISA minimum level. See Box 35.1 for an example of a PISA minimal level question.

In the USA, for example, this would mean bringing the lowest 19.4% up to the minimum level, with a corresponding 0.74% increase in GDP growth.

Besides reduced GDP growth, low numeracy is a significant financial cost to governments and personal cost to individuals. A large UK cohort study found that low numeracy was more of a handicap for the life chances of these individuals than low literacy. They earn less, spend less, are more likely to be sick, are more likely to be in trouble with the law, and need more help in school (Parsons & Bynner, 2005). It has been estimated that the annual cost to the UK of low numeracy is £2.4 billion (Gross, Hudson, & Price, 2009).

What is Dyscalculia?

Recent neurobehavioural and genetic research suggests that dyscalculia is a coherent syndrome that reflects a single core deficit. Although the literature is riddled with different terminologies, all seem to refer to the existence of a severe disability in learning arithmetic. The disability can be highly selective, affecting learners with normal intelligence and normal working memory (Landerl, Fussenegger, Moll, & Willburger, 2004), although it co-occurs with other developmental disorders, including reading disorders (Gross-Tsur et al., 1996) and attention deficit hyperactivity disorder (ADHD; Monuteaux, Faraone, Herzig, Navsaria, & Biederman, 2005) more often than would be expected by chance. There are high-functioning adults who are severely dyscalculic, but who are very good at geometry, using statistics packages, and doing degree-level computer programming (Butterworth, 1999).

There is evidence that mathematical abilities have high specific heritability. A multivariate genetic analysis of a sample of 1500 pairs of monozygotic and 1375 pairs of dizygotic 7-year-old twins found that about 30% of the genetic variance was specific to mathematics (Kovas, Haworth, Dale, & Plomin, 2007). Although there is a significant co-occurrence of dyscalculia with dyslexia, a study of first-degree relatives of dyslexic probands, revealed that numerical abilities constituted a separate factor, with reading-related and naming-related abilities being the two other principal components (Schulte-Körne, Ziegler, Deimel, Schumacher, Plume, Bachmann et al., 2007). These findings imply that arithmetical learning is at least partly based on a cognitive system that is distinct from those underpinning scholastic attainment more generally.

This genetic research is supported by neurobehavioural research that identifies the representation of numerosities – the number of objects in a set – as a foundational capacity in the development of arithmetic (Butterworth, 2010). This capacity is impaired in dyscalculic learners even in tasks as simple as enumerating small sets of objects (Landerl, Bevan, & Butterworth, 2004) or comparing the numerosities of two arrays of dots (Piazza, Facoetti, Trussardi, Berteletti, Conte, Lucangeli, et al., 2010). The ability to compare dot arrays has been correlated with more general arithmetical abilities both in children (Gilmore, McCarthy, & Spelke, 2010), and across the age range (Halberda, Mazzocco, & Feigenson, 2008; Piazza et al., 2010). This core deficit in processing numerosities is analogous to the core deficit in phonological awareness in dyslexia (Gabrieli, 2009). Examples of common indicators of dyscalculia are given in Box 35.2.

Although there is little longitudinal evidence, it seems that dyscalculia persists into adulthood (Shalev, Manor, & Gross-Tsur, 2005) even among individuals who are able in other cognitive domains (Butterworth, 1999). The effects of early and appropriate intervention with dyscalculia have yet to be investigated. This also leaves open the question as to whether there is a form of dyscalculia that is a delay, rather than a deficit, that will resolve, perhaps with appropriate educational support.

Converging evidence of dyscalculia as a distinct deficit comes from studies of impairments in the mental and neural representation of fingers. It has been known for many years, that fingers are used in acquiring arithmetical competence (Fuson, 1988). This involves understanding the mapping between the set of fingers and the set of objects to be enumerated. If the mental representation of fingers is weak, or if there is a deficit in understanding the numerosity of sets, then the child's cognitive development may fail to establish the link between fingers and numerosities. In fact, developmental weakness in finger representation ('finger agnosia') is a predictor of arithmetical ability (Noel, 2005). Gerstmann Syndrome, whose symptoms include dyscalculia and finger agnosia, is due to an abnormality in the parietal lobe, and in its developmental form is also associated with poor arithmetical attainment (Kinsbourne & Warrington, 1962).

Box 35.2 Dyscalculia observed

Examples of common indicators of dyscalculia are (i) carrying out simple number comparison and addition tasks by counting, often using fingers, well beyond the age when it is normal; and (ii) finding approximate estimation tasks difficult. Individuals identified as dyscalculic behave differently from their mainstream peers, for example:

To say which is the larger of two playing cards showing 5 and 8, they count all the symbols on each card.

To place a playing card of 8 in sequence between a 3 and a 9 they count up spaces between the two to identify where the 8 should be placed.

To count down from 10 they count up from 1 to 10, then 1 to 9, etc.

To count up from 70 in tens, they say '70, 80, 90, 100, 200, 300 ...'.

They estimate the height of a normal room as '200 feet?'.

To summarize, numbers do not seem to be meaningful for dyscalculics, at least, not meaningful in the way they are for typically developing learners. They do not intuitively grasp the size of a number and its value relative to other numbers. This basic understanding underpins all work with numbers and their relationships to one another.

What do we Know about The Brain and Mathematics?

The neural basis of arithmetical abilities in the parietal lobes, separate from language and domain-general cognitive capacities, has been broadly understood for nearly a hundred years from research on neurological patients (see Cipolotti & van Harskamp, 2001, for a review). One particularly interesting finding is that arithmetical concepts and laws can be preserved even when facts have been lost (Hittmair-Delazer, Semenza, & Denes, 1994) and, conversely, facts can be preserved even when an understanding of them has been lost (Delazer & Benke, 1997).

Neuroimaging experiments confirm this picture and show links from the parietal lobes to the left frontal lobe for more complex tasks (see Nieder & Dehaene, 2009; and Zamarian, Ischebeck, & Delazer, 2009 for reviews). One important new finding is that the neural organization of arithmetic is dynamic, shifting from one sub-network to another during the process of learning. Thus, learning new arithmetical facts primarily involves the frontal lobes and the intraparietal sulci (IPS), but using previously learned facts involves the left angular gyrus, which is also implicated in retrieving facts from memory (Ischebeck, Zamarian, Schocke, & Delazer, 2009). Figure 35.1 summarizes some of the principal links. Even prodigious calculators utilize this network, although supplementing it with additional brain areas (Pesenti, Zago, Crivello, Mellet, Samson, Duroux, et al., 2001) that appear to extend the capacity of working memory (Butterworth, 2006).

There is now extensive evidence that the IPS supports the representation of the magnitude of symbolic numbers (Dehaene, Piazza, Pinel, & Cohen, 2003; Pinel, Dehaene, Rivière, & Le Bihan, 2001), either as analogue magnitudes or as a discrete representation that codes cardinality, as evidenced by IPS activation when processing the numerosity of arrays of objects (Castelli, Glaser, & Butterworth, 2006). Moreover, when IPS functioning is disturbed by magnetic stimulation, the ability to estimate discrete magnitudes is affected (Cappelletti, Barth, Fregni, Spelke, & Pascual-Leone, 2007; Cohen Kadosh, Cohen Kadosh, Schuhmann, Kaas, Goebel, Henik, et al., 2007). The critical point is that almost all arithmetical and numerical processes implicate the parietal lobes, especially the IPS, suggesting that these are at the core of mathematical capacities.

Patterns of brain activity in 4-year-olds and adults show overlapping areas in the parietal lobes bilaterally when responding to changes in numerosity (Cantlon, Brannon, Carter, & Pelphrey, 2006). Nevertheless, there is a developmental trajectory in the organization of more complex arithmetical abilities. First, the organization of routine numerical activity changes with age, shifting from frontal areas, associated with executive function and working memory, and medial temporal areas, associated with declarative memory, to parietal areas, associated with magnitude processing and arithmetic fact retrieval, and occipito-temporal areas, associated with processing symbolic form (Ansari, 2008 for a review). These changes allow the brain to process numbers more efficiently and automatically, which enables it to carry out the more complex processing of arithmetical calculations. As A. N.

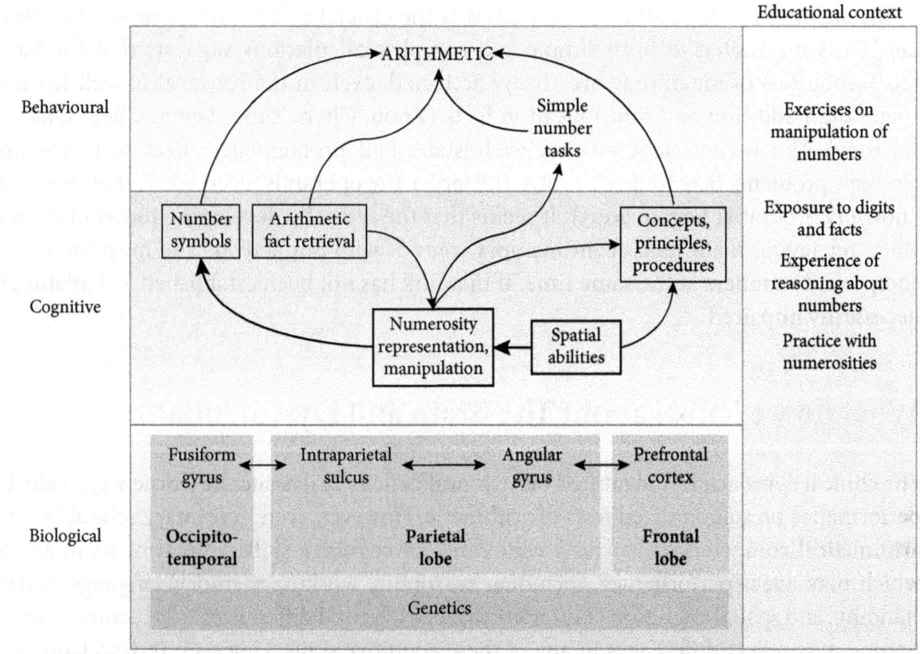

FIGURE 35.1 Causal model of possible interrelationships between biological, cognitive, and simple behavioural levels. Here, the only environmental factors we address are educational. If parietal areas, especially the IPS, fail to develop normally, there will be an impairment at the cognitive level in numerosity representation and consequential impairments for other relevant cognitive systems revealed in behavioural abnormalities. The link between the occipito-temporal and parietal cortex is required for mapping number symbols (digits and number words) to numerosity representations. Prefrontal cortex supports learning new facts and procedures. The multiple levels of the theory suggest the instructional interventions on which educational scientists should focus.

Whitehead observed, understanding symbolic notation relieves 'the brain of all unnecessary work … and sets it free to concentrate on more advanced problems' (Whitehead, 1948).

This suggests the possibility that the neural specialization for arithmetical processing may arise, at least in part, from a developmental interaction between the brain and experience (Ansari & Karmiloff-Smith, 2002; Johnson, 2001). Thus, one way of thinking about dyscalculia is that the typical school environment does not provide the right kind of experiences to enable the dyscalculic brain to develop normally to learn arithmetic.

Of course, mathematics is more than just simple number processing and retrieval of previously learned facts. In a numerate society we have to learn more complex mathematical concepts, such as place value, and more complex procedures, such as 'long' addition, subtraction, multiplication, and division. Recent research has revealed the neural correlates of learning to solve complex, multi-digit arithmetic problems (Cipolotti & van Harskamp, 2001, for a review) Again, this research shows that solving new problems requires more activation in the inferior frontal gyrus for reasoning and working memory, and the IPS for representing the magnitudes of the numbers involved, as compared with retrieval of previously learned facts (Delazer, Ischebeck, Domahs, Zamarian, Koppelstaetter, Siedentopf, et al., 2005).

The striking result in all of these studies is the crucial role of the parietal lobes. That the IPS is implicated in both simple and complex calculations suggests that the basic representations of magnitude are always activated, even in the retrieval of well-learned single-digit addition and multiplication facts (Zhou, Chen, Zang, Dong, Chen, Qiao, et al., 2007). This is consistent with the well-established 'problem-size effect', in which single-digit problems take longer to solve the larger the operands, even when they are well-known (Zbrodoff & Logan, 2005). It seems that the typically developing individual, even when retrieving math facts from memory, cannot help but activate the meaning of the component numbers at the same time. If that link has not been established, calculation is necessarily impaired.

What do we Know about The Brain and Dyscalculia?

The clinical approach has identified behavioural deficits in dyscalculic learners typically by performance on standardized tests of arithmetic. However, even in primary school (K1–5), arithmetical competence involves a wide range of cognitive skills, impairments in any of which may affect performance, including reasoning, working memory, language understanding, and spatial cognition (Butterworth, 2005). Neural differences in structure or functioning may reflect differences in any of these cognitive skills. However, if dyscalculia is a core deficit in processing numerosities, abnormalities should be found in the parietal network that supports the enumeration of small sets of objects (Landerl et al., 2004) and the comparison of numerosities of arrays of dots (Piazza et al., 2010).

The existence of a core deficit in processing numerosities is consistent with recent discoveries about dyscalculic brains:

- Reduced activation has been observed in children with dyscalculia during comparison of numerosities (Mussolin, De Volder, Grandin, Schlögel, Nassogne, & Noël, 2009; Price, Holloway, Räsänen, Vesterinen, & Ansari, 2007), comparison of number symbols (Mussolin et al., 2009), and arithmetic (Kucian, Loenneker, Dietrich, Dosch, Martin, & von Aster, 2006), i.e. these children are not using the IPS so much during these tasks;
- Reduced gray matter in dyscalculic learners has been observed in areas known to be involved in basic numerical processing, including the left IPS (Isaacs, Edmonds, Lucas, & Gadian, 2001), the right IPS (Rotzer, Kucian, Martin, von Aster, Klaver, & Loenneker, 2008), and the IPS bilaterally (Rykhlevskaia, Uddin, Kondos, & Menon, 2009), i.e. these learners have not developed these brain areas as much as typical learners (see Figure 35.2).
- Differences in connectivity among the relevant parietal regions, and between these parietal regions and occipito-temporal regions associated with processing symbolic number form (see Figure 35.2), are revealed by diffusion tensor imaging tractography (Rykhlevskaia et al., 2009), i.e. dyscalculic learners have not sufficiently developed the structures needed to coordinate the components needed for calculation.

Figure 35.1 suggests that the IPS is just one component in a large-scale cortical network that subserves mathematical cognition. This network may break down in multiple ways, leading to

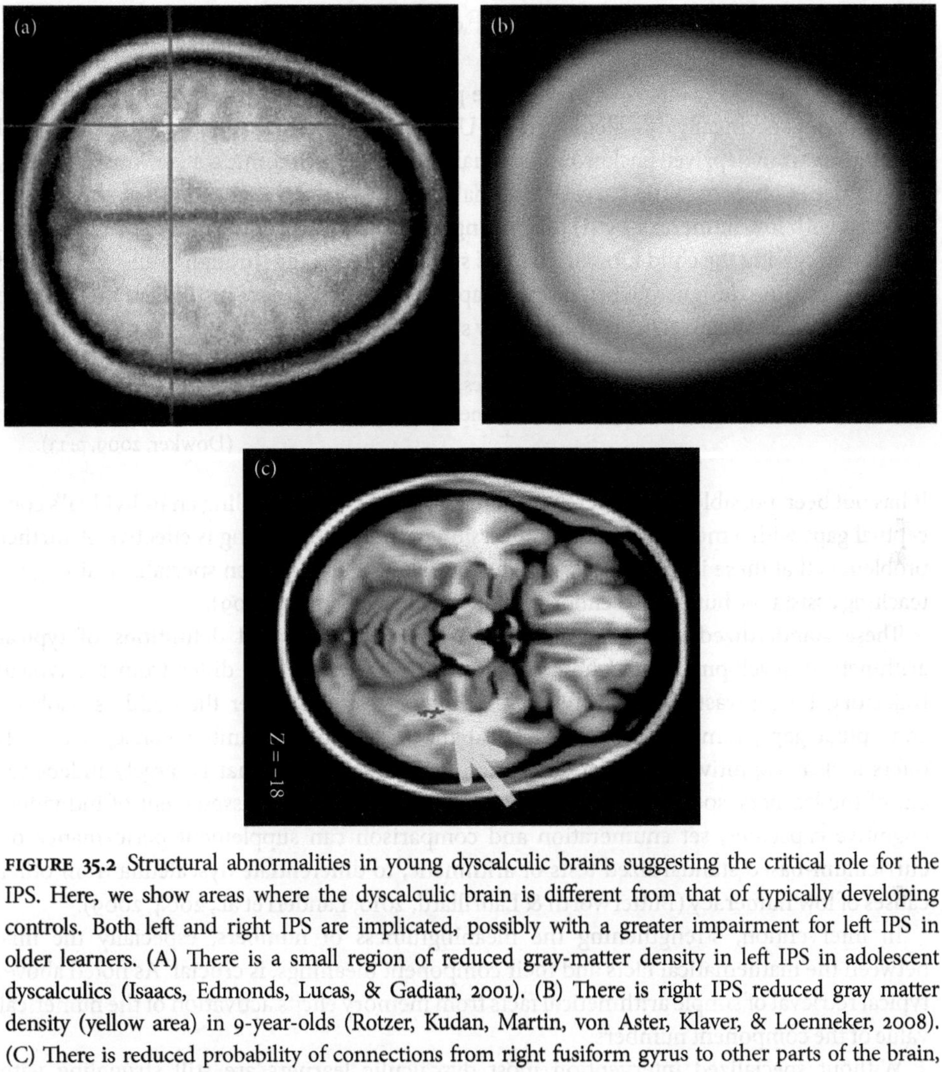

FIGURE 35.2 Structural abnormalities in young dyscalculic brains suggesting the critical role for the IPS. Here, we show areas where the dyscalculic brain is different from that of typically developing controls. Both left and right IPS are implicated, possibly with a greater impairment for left IPS in older learners. (A) There is a small region of reduced gray-matter density in left IPS in adolescent dyscalculics (Isaacs, Edmonds, Lucas, & Gadian, 2001). (B) There is right IPS reduced gray matter density (yellow area) in 9-year-olds (Rotzer, Kudan, Martin, von Aster, Klaver, & Loenneker, 2008). (C) There is reduced probability of connections from right fusiform gyrus to other parts of the brain, including the parietal lobes (*Rykhlevskala, Uddin, Kondos, & Menon, 2009*).

different dyscalculia patterns of presenting symptoms when the core deficit is combined with other cognitive deficits, including working memory, reasoning, or language (Rubinsten & Henik, 2009).

Moreover, the same presenting symptom, could reflect different impairments in the network. For example, abnormal comparison of symbolic numbers (Holloway & Ansari, 2009; Rousselle & Noël, 2007), could arise from an impairment in the fusiform gyrus associated with visual processing of number symbols, an impairment in IPS associated with the magnitude referents, analogue or numerosity, of the number symbols, or reduced connectivity between fusiform gyrus and IPS.

How is this Relevant to Maths Education?

There have been many attempts to raise the performance of children with low-numeracy, although not specifically dyscalculia. In the USA, evidence-based approaches have focused on children from deprived backgrounds, usually low socio-economic status (Mussolin et al., 2009; Price et al., 2007). The current National Strategy in the UK gives special attention to children with low numeracy by (i) diagnosing each child's conceptual gaps in understanding, and (ii) giving the child more individual support in working through visual, verbal, and physical activities designed to bridge each gap. Unfortunately, there is still little quantitative evaluation of the effectiveness of the strategy since it was first piloted in 2003:

> ... the evaluations used very diverse measures; and most did not include ratio gains or effect sizes or data from which these could be obtained.
>
> (Dowker, 2009, p. 13).

It has not been possible to tell, therefore, whether identifying and filling an individual's conceptual gaps with a more individualized version of the same teaching is effective. A further problem is that these interventions are effective when there has been specialist training for teaching assistants, but not all schools can provide this (Dowker, 2009).

These standardized approaches depend on curriculum-based definitions of typical arithmetical development, and how children with low numeracy differ from the typical trajectory. By contrast, neuroscience research suggests that, rather than address isolated conceptual gaps, remediation should build the foundational number concepts first. It offers a clear cognitive target for assessment and intervention that is largely independent of the learners' social and educational circumstances. In the assessment of individual cognitive capacities, set enumeration and comparison can supplement performance on curriculum-based standardized tests of arithmetic, to differentiate dyscalculia from other causes of low numeracy (Butterworth & Laurillard, 2010; Landerl et al., 2004, 2009).

In intervention, strengthening the meaningfulness of numbers, especially the link between the mathematical facts and their component meanings, is crucial. As noted above, typical retrieval of simple arithmetical facts from memory elicits activation of the numerical value of the component numbers.

Without specialized intervention most dyscalculic learners are still struggling with basic arithmetic in secondary school (Shalev et al., 2005). Effective early intervention may help to reduce the later impact on poor numeracy skills, as it does in dyslexia (Goswami, 2006). Although very expensive, it promises to repay 12–19 times the investment (Gross et al., 2009).

What Can be Done?

Although the neuroscience may suggest what should be taught, it does not specify how it should be taught. Concrete manipulation activities have been used for many decades in mathematical remediation because they provide tasks that make number concepts meaningful (Anning & Edwards, 1999), providing an intrinsic relationship between a goal, the learner's action, and the informational feedback on the action (Papert, 1980). Educators recognize that informational feedback provides intrinsic motivation in a task, and that is of greater value to the learner than the extrinsic motives and rewards provided by a supervising teacher (Bruner, 1961; Deci, Koestner, & Ryan, 2001; Dewey, 1938).

Experienced special educational needs (SEN) teachers employ these activities in the form of games with physical manipulables (such as Cuisenaire rods, number tracks, and playing cards) to give learners experience of the meaning of number. Through playing these games, learners can discover from their manipulations, for example, which rod fits with an 8-rod to match a 10-rod. However, these methods require specially-trained teachers working with a single learner or a small group of learners (Bird, 2007; Butterworth & Yeo, 2004; Yeo, 2003), and are allotted only limited time periods in the school schedule.

A promising approach, therefore, is to construct adaptive software informed by the neuroscience findings on the core deficit in dyscalculia (see chapter by Rasanen, this volume). Such software has the potential to reduce the demand on specially-trained teachers, and to transcend the limits of the school schedule. There have been two examples of adaptive games based on neuroscience findings.

The *Number Race* (Wilson, Revkin, Cohen, Cohen, & Dehaene, 2006) targets the inherited approximate numerosity system in the IPS (Feigenson, Dehaene, & Spelke, 2004) that may support early arithmetic (Gilmore et al., 2010). In dyscalculics this system is less precise (Piazza et al., 2010), and the training is designed to improve its precision. The task is to select the larger of two arrays of dots, and the software adapts to the learner, making the difference between the arrays smaller as their performance improves, and provides informational feedback as to which is correct.

Another adaptive game, *Graphogame-Maths,* targets the inherited system for representing and manipulating sets in the IPS, which is impaired in dyscalculia (Butterworth, 2005). Again, the basis of the game is the comparison of visual arrays of objects, but here the sets are small and can be counted, and the progressive tasks are to identify the link between the number of objects in the sets and their verbal numerical label, with informational feedback showing which is correct.

The effectiveness of the two games was compared in a carefully-controlled study of kindergarten children (aged 6–7), identified by their teachers as needing special support in early maths. After 10–15 minutes play per school-day for 3 weeks, there was a significant improvement in the task practiced in both games, namely number comparison, but the effect did not generalize to counting or arithmetic. *Graphogame-Maths* appeared to lead to slightly better and longer-lasting improvement in number comparison (Räsänen, Salminen, Wilson, Aunioa, & Dehaene, 2009).

Although the *Number Race* and *Graphogame-Maths* are adaptive games based on neuroscience findings, neither requires learners to manipulate numerical quantities. Manipulation is critical for providing an intrinsic relationship between task goals, a learner's actions, and informational feedback on those actions (Papert, 1980). When a learning environment provides informational feedback it enables the learner to work out how to adjust their actions in relation to the goal, and they can be their own 'critic,' not relying on the teacher to guide (Bruner, 1961; Deci et al., 2001; Dewey, 1938). This is analogous to the 'actor-critic' model of unsupervised reinforcement learning in neuroscience, which proposes a 'critic' element, internal to the learning mechanism, not a guide that is external to it, which evaluates the informational feedback in order to construct the next action (Dayan & Abbott, 2001).

Which number bonds make 10?

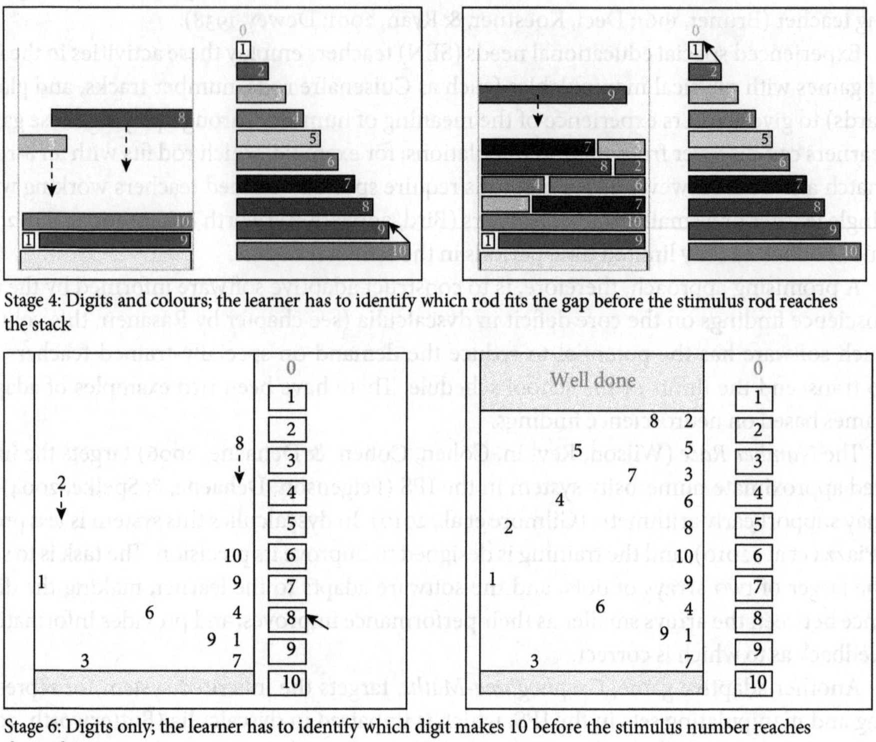

Stage 4: Digits and colours; the learner has to identify which rod fits the gap before the stimulus rod reaches the stack

Stage 6: Digits only; the learner has to identify which digit makes 10 before the stimulus number reaches the stack

FIGURE 35.3 Remediation using learning technology. The images are taken from an example of an interactive, adaptive game designed to help the learner make the link between digits and their meaning. The timed version of the number bonds game elicits many learner actions with informational feedback, and scaffolds abstraction, through stages 1, 5 colours + lengths, evens; 2, 5 colours + lengths, odds; 3, all 10 colours + lengths; 4, digits + colours + lengths; 5, digits + lengths; and 6, digits only. Each rod falls at a pace adapted to the learner's performance, and the learner has to click the corresponding rod or number to make 10 before it reaches the stack (initially 3 seconds); if there is a gap, or overlap, or they are too slow, the rods dissolve. When a stack is complete, the game moves to the next stage. The game is available from http://www.numbersense.co.uk/numberbonds/.

A different approach, one that emulates the manipulative tasks used by SEN teachers, has been taken in adaptive software that, driven by the neuroscience research on dyscalculia, focuses on numerosity processing (see Figure 3 for an example; Butterworth & Laurillard, 2010). The informational feedback here is not an external critic showing the correct answer. Rather, the visual representation of two rods that match a given distance, or not, enables the learner to interpret for themselves what the improved action should be – and can serve as their own internal critic.

An additional advantage of adaptive software is that learners can do more practice per unit time than with a teacher. It was found that for 'SEN learners (12-year-olds) using the

Number Bonds game (illustrated in Figure 35.3), 4–11 trials per minute were completed, while in an SEN class of three supervised learners only 1.4 trials per minute were completed during a 10-min observation.' In another SEN group of 11-year-olds, the game elicited on average 173 learner manipulations in 13 minutes (where a perfect performance, in which every answer is correct, is 88 in 5 minutes, since the software adapts the timing according to the response; Laurillard & Baajour, 2011). In this way, neuroscience research is informing what should be targeted in the next generation of adaptive software.

At present, it is not yet clear whether early and appropriately-targeted interventions can turn a dyscalculic into a typical calculator. Dyscalculia may be like dyslexia in that early intervention can improve practical effectiveness without making the cognitive processing like those of the typically developing.

What is The Outlook?

Recent research by cognitive and developmental scientists is providing a scientific characterization of dyscalculia as the reduced ability for understanding numerosities, and mapping number symbols to number magnitudes. Personalized learning applications developed by educational scientists can be targeted to remedy these deficits and can be implemented on hand-held devices for independent learning. Because there are also individual differences in numerosity processing in the normal range (Gilmore, McCarthy, & Spelke, 2007; Halberda et al., 2008; Piazza et al., 2010), the same programs can assist beginning mainstream learners, so one can envisage a future where all learners will benefit from these developments.

Although much progress has been made, a number of open questions remain:

- The possible existence of a variety of dyscalculia behavioural patterns of impairment raises the interesting question of whether one deficit – numerosity estimation – is a necessary or sufficient for a diagnosis of dyscalculia. This is critical, for it has implications for whether a single diagnostic assessment based on numerosity is sufficient, or whether multiple assessments are required. It also has implications for whether numerosity should be the focus of remediation, or whether other (perhaps more symbolic) activities should also be targeted.
- Does the sensitivity of neuroscience measures make it possible to identify learners at risk for dyscalculia earlier than is possible with behavioural assessments, as is the case with dyslexia (Lyytinen, Ahonen, Eklund, Guttorm, Laakso, Leinonen, et al., 2001)?
- Further research is needed on the neural consequences of intervention. Even where intervention improves performance, it may not be clear whether the learner's cognitive and neural functioning has become more typical or whether compensatory mechanisms have developed. This would require more extensive research, as in the case of dyslexia, where functional neuroimaging has revealed the effects of successful behavioural interventions on patterns of neural function (Gabrieli, 2009).
- Personalized learning applications enable fine-grain evaluation of theory-based instructional interventions. For example, the value of active manipulation versus associative learning on one hand, and of intrinsic versus extrinsic motivation and feedback on the other, can be assessed by orthogonally manipulating these features of learning environments. Their effects can be assessed behaviourally in terms of performance on

the tasks by target learners. Their effects can also be assessed by investigating neural changes over time in target learners, through structural and functional neuroimaging. Classroom-based but theory-based testing holds great potential for the development of educational theory, and could contribute in turn to testing hypotheses in neuroscience.

- At the moment, dyscalculia is not widely recognized by teachers or educational authorities, nor, it would seem, by research funding agencies. Recognition is likely to be the basis for improved prospects for dyscalculic sufferers.
- There is an urgent societal need to help failing learners achieve a level of numeracy where they can function adequately in the modern workplace. Contemporary research on dyscalculia promises a productive way forward, but it is still a 'poor relation' in terms of funding (Bishop, 2010), which means there is a serious lack of evidence-based approaches to dyscalculia intervention. An understanding of how the brain processes underlying number and arithmetic concepts will help focus teaching interventions on critical conceptual activities and will help focus neuroscience research on tracking the structural and functional changes that follow intervention. Learning more about how to help these learners is driving, and will continue to drive, where the science should go next.

Acknowledgment

This chapter was reproduced from Brian Butterworth, Sashank Varma, and Diana Laurillard, Dyscalculia: From Brain to Education, *Science*, 332 (6033) pp. 1049–1053, DOI: 10.1126/science.1201536

References

Anning, A., & Edwards, A. (1999). Promoting Children's Learning from Birth to Five: Developing the New Early Years Professional. Maidenhead: Open University Press.

Ansari, D. (2008). Effects of development and enculturation on number representation in the brain. Nature Reviews Neuroscience, 9, 278–291.

Ansari, D., & Karmiloff-Smith, A. (2002). Atypical trajectories of number development: a neuroconstructivist perspective. Trends in Cognitive Sciences, 6, 511–516.

Beddington, J., Cooper, C.L., Field, J., Goswami, U., Huppert, F.A., Jenkins, R., Jones, H.S., Kirkwood, T.B.L., Sahakian, B.J., & Thomas, S.M. (2008). The mental wealth of nations. Nature, 455, 1057–1060.

Bird, R. (2007). The Dyscalculia Toolkit. London: Paul Chapman Publishing.

Bishop, D.V.M. (2010). Which neurodevelopmental disorders get researched and why? PLoS One, 5: e15112.

Bruner, J.S. (1961). The act of discovery. Harvard Educational Review, 31, 21–32.

Butterworth, B. (1999). The Mathematical Brain. London: Macmillan.

Butterworth, B. (2005). Developmental dyscalculia. In: J. I. D. Campbell (ed.), Handbook of Mathematical Cognition. Hove: Psychology Press.

Butterworth, B. (2006). Mathematical expertise. In: K. A. Ericsson, N. Charness, P. J. Feltovich, & R. R. Hoffmann (eds.), Cambridge Handbook of Expertise and Expert Performance. Cambridge: Cambridge University Press.

Butterworth, B. (2010). Foundational numerical capacities and the origins of dyscalculia. Trends in Cognitive Sciences, 14, 534–541.

Butterworth, B., & Laurillard, D. (2010). Low numeracy and dyscalculia: identification and intervention. ZDM Mathematics Education, 42, 527–539.

Butterworth, B., & Yeo, D. (2004). Dyscalculia Guidance. London: NferNelson.

Cantlon, J.F., Brannon, E.M., Carter, E.J., & Pelphrey, K.A. (2006). Functional imaging of numerical processing in adults and 4-y-old children. PLoS, 4, e125.

Cappelletti, M., Barth, H., Fregni, F., Spelke, E.S., & Pascual-Leone, A. (2007). rTMS over the intraparietal sulcus disrupts numerosity processing. Experimental Brain Research, 179, 631–642.

Castelli, F., Glaser, D.E., & Butterworth, B. (2006). Discrete and analogue quantity processing in the parietal lobe: a functional MRI study. Proceedings of the National Academy of Science, 103, 4693–4698.

Cipolotti, L., & van Harskamp, N. (2001). Disturbances of number processing and calculation. In: R. S. Berndt (ed.), Handbook of Neuropsychology, 2nd edn. Amsterdam: Elsevier Science.

Cohen Kadosh, R., Cohen Kadosh, K., Schuhmann, T., Kaas, A., Goebel, R., Henik, A., & Sack, A. (2007). Virtual dyscalculia induced by parietal-lobe TMS impairs automatic magnitude processing. Current Biology, 17, 1–5.

Dayan, P., & Abbott, L. (2001). Theoretical Neuroscience: Computational and mathematical modeling of neural systems. Cambridge, MA. MIT Press.

Deci, E.L., Koestner, R., & Ryan, R.M. (2001). Extrinsic rewards and intrinsic motivation in education: reconsidered once again. Review of Educational Research, 71, 1–27.

Dehaene, S., Piazza, M., Pinel, P., & Cohen, L. (2003). Three parietal circuits for number processing. Cognitive Neuropsychology, 20, 487–506.

Delazer, M., & Benke, T. (1997). Arithmetic facts without meaning. Cortex, 33, 697–710.

Delazer, M., Ischebeck, A., Domahs, F., Zamarian, L., Koppelstaetter, F., Siedentopf, C., Kaufmann, L., Benke, T., & Felber, S. (2005). Learning by strategies and learning by drill – evidence from an fMRI study. Neuroimage, 25, 838–849.

Dewey, J. (1938). Experience and Education. New York: Kappa Delta Pi.

Dowker, A. (2009). What works for children with mathematical difficulties? The effectiveness of intervention schemes. The National Strategies: Primary. London: Department for Children, Schools and Families.

Feigenson, L., Dehaene, S., & Spelke, E. (2004). Core systems of number. Trends in Cognitive Sciences, 8, 307–314.

Fuson, K.C. (1988). Children's Counting and Concepts of Number. New York: Springer Verlag.

Gabrieli, J.D.E. (2009). Dyslexia: a new synergy between education and cognitive neuroscience. Science, 325, 280–283.

Gilmore, C.K., McCarthy, S.E., & Spelke, E.S. (2007). Symbolic arithmetic knowledge without instruction. Nature, 589–592.

Gilmore, C.K., McCarthy, S.E., & Spelke, E.S. (2010). Non-symbolic arithmetic abilities and mathematics achievement in the first year of formal schooling. Cognition, 115, 394–406.

Goswami, U.C. (2006). Neuroscience and education: from research to practice. Nature Reviews Neuroscience, 7, 406–413.

Gross, J., Hudson, C., & Price, D. (2009). The long term costs of numeracy difficulties. London: Every Child a Chance Trust, KPMG.

Gross-Tsur, V., Manor, O., & Shalev, R. S. (1996). Developmental dyscalculia: prevalence and demographic features. Developmental Medicine and Child Neurology, 38, 25–33.

Halberda, J., Mazzocco, M.M.M., & Feigenson, L. (2008). Individual differences in non-verbal number acuity correlate with maths achievement. Nature, 455, 665–668.

Hittmair-Delazer, M., Semenza, C., & Denes, G. (1994). Concepts and facts in calculation. Brain, 117, 715–728.

Holloway, I.D., & Ansari, D. (2009). Mapping numerical magnitudes onto symbols: the numerical distance effect and individual differences in children's mathematics achievement. Journal of Experimental Child Psychology, 103, 17–29.

Isaacs, E.B., Edmonds, C.J., Lucas, A., & Gadian, D.G. (2001). Calculation difficulties in children of very low birthweight: a neural correlate. Brain, 124, 1701–1707.

Ischebeck, A., Zamarian, L., Schocke, M., & Delazer, M. (2009). Flexible transfer of knowledge in mental arithmetic – an fMRI study. NeuroImage, 44, 1103–1112.

Johnson, M.H. (2001). Functional brain development in humans. Nature Reviews Neuroscience, 2, 475–483.

Kinsbourne, M., & Warrington, E.K. (1962). A study of finger agnosia. Brain, LXXXV, 47–66.

Kosc, L. (1974). Developmental dyscalculia. Journal of Learning Disabilities, 7, 159–162.

Kovas, Y., Haworth, C., Dale, P., & Plomin, R. (2007). The genetic and environmental origins of learning abilities and disabilities in the early school years. Monograph of the Society for Research in Child Development., 72, 1–144.

Kucian, K., Loenneker, T., Dietrich, T., Dosch, M., Martin, E., & von Aster, M. (2006). Impaired neural networks for approximate calculation in dyscalculic children: a functional MRI study. Behavioral and Brain Functions, 2, 1–17.

Landerl, K., Bevan, A., & Butterworth, B. (2004). Developmental dyscalculia and basic numerical capacities: a study of 8–9 year old students. Cognition, 93, 99–125.

Landerl, K., Fussenegger, B., Moll, K., & Willburger, E. (2009). Dyslexia and dyscalculia: two learning disorders with different cognitive profiles. Journal of Experimental Child Psychology, 103, 309–324.

Laurillard, D., & Baajour, H. (2011). Adaptive learning games for handhelds. Learning Without Frontiers Conference presentation, 09 January 2011, Barbican, London.

Lyytinen, H., Ahonen, T., Eklund, K., Guttorm, T.K., Laakso, M.-L., Leinonen, S., Leppanen, P.H.T., Lyytinen, P., Poikkeus, A.M., Puolakanaho, A., Richardson, U., & Viholainen, H. (2001). Developmental pathways of children with and without familial risk for dyslexia during the first years of life. Developmental Neuropsychology, 20, 535–554.

Monuteaux, M.C., Faraone, S.V., Herzig, K., Navsaria, N., & Biederman, J. (2005). ADHD and dyscalculia: evidence for independent familial transmission. Journal of Learning Disabilities, 38, 89–93.

Mussolin, C., de Volder, A., Grandin, C., Schlögel, X., Nassogne, M.-C., & Noël, M.-P. (2009). Neural correlates of symbolic number comparison in developmental dyscalculia. Journal of Cognitive Neuroscience, 22, 860–874.

Nieder, A., & Dehaene, S. (2009). Representation of number in the brain. Annual Review of Neuroscience, 32, 185–208.

Noël, M.-P. (2005). Finger gnosia: a predictor of numerical abilities in children? Child Neuropsychology, 11, 413–430.

OECD (2010). The High Cost of Low Educational Performance. The Long-run Economic Impact of Improving Educational Outcomes. London: OECD.

Papert, S. (1980). Mindstorms: Children, Computers, and Powerful Ideas. Brighton: Harvester Press.

Parsons, S., & Bynner, J. (2005). Does numeracy matter more? London: National Research and Development Centre for Adult Literacy and Numeracy, Institute of Education.

Pesenti, M., Zago, L., Crivello, F., Mellet, E., Samson, D., Duroux, B., Seron, X., Mazoyer, B., & Tzourio-Mazoyer, N. (2001). Mental calculation expertise in a prodigy is sustained by right prefrontal and medial-temporal areas. Nature Neuroscience, 4, 103–107.

Piazza, M., Facoetti, A., Trussardi, A.N., Berteletti, I., Conte, S., Lucangeli, D., Dehaene, S., & Zorzi, M. (2010). Developmental trajectory of number acuity reveals a severe impairment in developmental dyscalculia. Cognition, 116, 33–41.

Pinel, P., Dehaene, S., Rivière, D., & Le Bihan, D. (2001). Modulation of parietal activation by semantic distance in a number comparison task. NeuroImage, 14, 1013–1026.

Price, G.R., Holloway, I., Räsänen, P., Vesterinen, M., & Ansari, D. (2007). Impaired parietal magnitude processing in developmental dyscalculia. Current Biology, 17, R1042–R1043.

Rasanen, P. (this volume). Computer-assisted interventions on basic number skills. In: R. Cohen Kadosh, and A. Dowker, (eds), The Oxford Handbook of Numerical Cognition. Oxford: Oxford University Press.

Räsänen, P., Salminen, J., Wilson, A.J., Aunioa, P., & Dehaene, S. (2009). Computer-assisted intervention for children with low numeracy skills. Cognitive Development, 24, 450–472.

Rotzer, S., Kucian, K., Martin, E., von Aster, M., Klaver, P., & Loenneker, T. (2008). Optimized voxel-based morphometry in children with developmental dyscalculia. NeuroImage, 39, 417–422.

Rousselle, L., & Noël, M.-P. (2007). Basic numerical skills in children with mathematics learning disabilities: a comparison of symbolic vs non-symbolic number magnitude processing. Cognition, 102, 361–395.

Rubinsten, O., & Henik, A. (2009). Developmental dyscalculia: heterogeneity might not mean different mechanisms. Trends in Cognitive Sciences, 13, 92–99.

Rykhlevskaia, E., Uddin, L.Q., Kondos, L., & Menon, V. (2009). Neuroanatomical correlates of developmental dyscalculia: combined evidence from morphometry and tractography. Frontiers in Human Neuroscience, 3, 1–13.

Schulte-Körne, G., Ziegler, A., Deimel, W., Schumacher, J., Plume, E., Bachmann, C., Kleensang, A., Propping, P., Nöthen, M., Warnke, A., Remschmidt, H., & König, I.R. (2007). Interrelationship and familiality of dyslexia related quantitative measures. Annals of Human Genetics, 71, 160–175.

Shalev, R.S. (2007). Prevalence of developmental dyscalculia. In: D. B. Berch, & M. M. M. Mazzocco (eds), Why is Math so Hard for Some Children? The Nature and Origins of Mathematical Learning Difficulties and Disabilities. Baltimore, MD: Paul H Brookes Publishing Co.

Shalev, R.S., & Gross-Tsur, V. (2001). Developmental dyscalculia. Review article. Pediatric Neurology, 24, 337–342.

Shalev, R.S., Manor, O., & Gross-Tsur, V. (2005). Developmental dyscalculia: a prospective six year follow up. Developmental Medicine and Child Psychology, 47, 121–125.

Whitehead, A.N. (1948). An Introduction to Mathematics. London: Oxford University Press.

Wilson, A., Revkin, S., Cohen, D., Cohen, L., & Dehaene, S. (2006). An open trial assessment of 'The Number Race', an adaptive computer game for remediation of dyscalculia. Behavioral and Brain Functions, 2.

Yeo, D. (2003). A brief overview of some contemporary methodologies in primary maths. In: M. Johnson, & L. Peer (eds.), The Dyslexia Handbook. Reading: British Dyslexia Association.

Zamarian, L., Ischebeck, A., & Delazer, M. (2009). Neuroscience of learning arithmetic – evidence from brain imaging studies. Neuroscience & Biobehavioral Reviews, 33, 909–925.

Zbrodoff, N.J., & Logan, G.D. (2005). What everyone finds. The problem-size effect. In: J. I. D. Campbell (ed.), Handbook of Mathematical Cognition. Hove: Psychology Press.

Zhou, X., Chen, C.S., Zang, Y., Dong, Q., Chen, C.H., Qiao, S., & Gong, Q. (2007). Dissociated brain organization for single-digit addition and multiplication. NeuroImage, 35, 871–880.

CHAPTER 36

DEVELOPMENTAL DYSCALCULIA AS A HETEROGENEOUS DISABILITY

AVISHAI HENIK, ORLY RUBINSTEN, AND SARIT ASHKENAZI

NUMERICAL cognition is essential to many aspects of life and arithmetic abilities predict academic achievements better than reading (Estrada, Martin-Hryniewicz, Peek, Collins, & Byrd, 2004). Acquiring a solid sense of numbers and being able to mentally manipulate numbers are at the heart of this ability. Research suggests that infants already show a basic perception of quantities. However, a long, effortful, and sometimes painful developmental process is required until a child acquires the numerical representations in the adult sense (Butterworth, 2005). Studying arithmetic may be extremely difficult for those who suffer from specific learning disabilities in arithmetic, henceforth termed developmental dyscalculia (DD). Children with poor numeracy are at a disadvantage in both academic and everyday life situations (e.g. handling money). Adults with poor numeracy are more than twice as likely to be unemployed as those with competent numeracy (Parsons & Bynner, 1997; Rivera-Batiz, 1992). Poor numeracy often means low financial literacy with negative consequences for economic well-being. Estimates of prevalence of DD vary according to the definition of the disability, but for the relatively isolated difficulty, estimates are between 3 and 6% (Gross-Tsur, Manor, & Shalev, 1996; Lewis, Hitch, & Walker, 1994; Shalev & Gross-Tsur, 2001) and are comparable to estimates of learning disability in reading.

Research on DD is characterized by various trends: the study of high-level school-like concepts versus low-level processes (e.g. enumeration), or the study of domain-general (e.g. attention) versus domain-specific (e.g. number sense) factors. These research trends give rise to different notions about difficulties in numerical cognition and their roots. Whether DD might be an isolated deficiency or not (see Rubinsten & Henik, 2009), it is clear that similar to other learning disabilities, DD is rarely a very pure (e.g. Ashkenazi & Henik,

2010a,b) disorder. In most cases, children and adults who suffer from DD present difficulties in areas or mental processes different from pure numerical deficiencies.

The current chapter discusses heterogeneous aspects of DD both in terms of behaviour and cognitive operations, and in terms of the neural structures that underlie the deficiency.

Characteristics of Developmental Dyscalculia

Children with DD are far behind their classmates in a wide range of numerical tasks: they have difficulties in retrieval of arithmetical facts (Geary, 1993; Geary & Hoard, 2001; Ginsburg, 1997; Russell & Ginsburg, 1984), in using arithmetical procedures (Russell & Ginsburg, 1984; Shalev & Gross-Tsur, 2001), and use immature problem-solving strategies – for example, using finger counting (Jordan, Hanich, & Kaplan, 2003) – at an age when their classmates have already stopped using such strategies. Many studies have been directed at higher-level, school-like concepts. Hence, research has focused on general cognitive functions such as poor working memory (Geary, 1993), deficits in attention systems (Shalev, Auerbach, & Gross-Tsur, 1995), disorders of visuospatial functioning (Bull, Johnston, & Roy, 1999), and deficiencies in the retrieval of information (e.g. arithmetic facts) from memory (Kaufmann, Lochy, Drexler, & Semenza, 2004). In recent years, another trend toward identifying low-level deficits in DD has been advocated. Such research aims at revealing the building blocks of numerical cognition and the very basic deficiencies that underlie DD (Ansari & Karmiloff-Smith, 2002). Efforts in this direction revealed difficulties in subitizing (Ashkenazi, Mark-Zigdon, & Henik, 2013; Moeller, Neuburger, Kaufmann, Landerl, & Nuerk, 2009; Schleifer & Landerl, 2011) and counting (Geary, Bow-Thomas, & Yao, 1992; Geary, Hoard, & Hamson, 1999; Landerl, Bevan, & Butterworth, 2004), in comparative judgment of both symbolic (Ashkenazi, Mark-Zigdon, & Henik, 2009; Mussolin, Mejias, & Noël, 2010) and non-symbolic (Price, Holloway, Räsänen, Vesterinen, & Ansari, 2007) stimuli, and in automatic processing of numbers (Cohen Kadosh, Cohen Kadosh, Schuhmann, Kaas, Goebel, Henik, et al., 2007; Rubinsten & Henik, 2005). The latter led to a view of an innate, domain-specific foundation of arithmetic and to the suggestion that arithmetic disability involves a domain-specific deficit in the capacity to enumerate (Butterworth, 2010).

However, several findings suggest that this view needs to be examined carefully:

- arithmetic seems to rely on both domain-specific and domain-general abilities;
- in many cases behaviours, as well as cognition of DD could be characterized by deficits in other areas such as attention or memory and not only as a number sense deficiency;
- studies of the neural structures involved in DD reveal areas and mechanisms that hint toward heterogeneous damage.

Figure 36.1 presents several cognitive deficiencies, involving various brain mechanisms, which might contribute to different manifestations of DD.

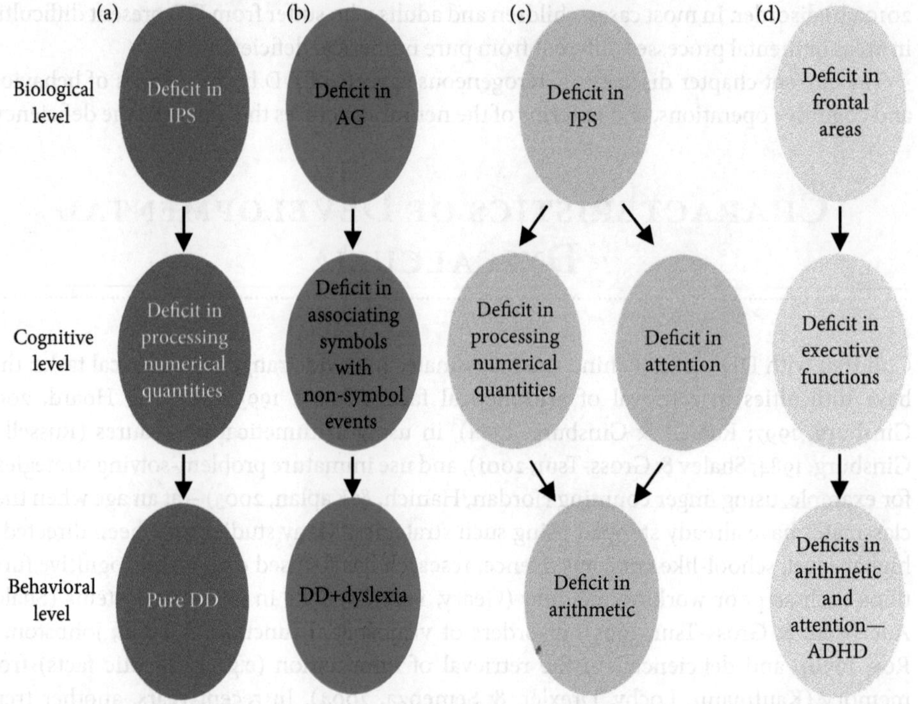

FIGURE 36.1 (a) Deficit in IPS leads to a pure deficit in numerical processing only, and results in pure DD. (b) Deficit to the AG leads to deficits in associating symbols with the events they symbolize. This, in turn, leads to difficulties both in language and DD. (c) Deficit in IPS leads to a deficit in processing numerical quantities and to a deficit in attention. Both numerical and attention deficits may create a deficit in arithmetic processing. The dashed line between the deficit in attention at the cognitive level and the deficit in arithmetic at the behavioural level suggests that the attention deficit might exist, but contribute only minimally to the arithmetic deficit. (d) Deficit in frontal lobe leads to deficient executive functions and, in turn, to a deficit in arithmetic and in attention. The latter is indicated in ADHD.

Symbolic versus Non-Symbolic Representations

There is a continuous debate in the scientific literature about the ability of people with DD to represent symbolic and non-symbolic numerical information. Specifically, symbolic numerical information involves number words such as 'four', 'ten', or 'plus'; or written numerical symbols, such as '4', '10', or '+'. Non-symbolic numerical information automatically extracts the understanding of the approximate quantity of concrete sets of objects (such as visual dots). Core knowledge of numbers is essentially non-symbolic. Symbolic numerical information, on the other hand, varies across cultures and is influenced by language. Numerical symbols, once acquired, become attached to their non-symbolic numerical representations. Accordingly, during development links between symbols and quantities should become automatic (for review see Ansari, 2008). There are disagreements regarding the question of whether DD is the result of deficits in non-symbolic numerical representations (e.g. Butterworth, 2010), in symbolic numerical representations (e.g. Mussolin, De Volder, Grandin, Schlögel, Nassogne, & Noël, 2009), in the ability to access numerical meaning (i.e. non-symbolic information) from symbols (e.g. Rousselle & Noël, 2007), or in the ability to automatically link symbolic and non-symbolic representations (e.g. Rubinsten & Henik, 2005).

The existent developmental brain imaging literature on DD, in regard to non-symbolic number processing (e.g. comparing the numerosity of two groups of dot patterns), is inconclusive. Researchers reported both group differences between children with and without DD in their ability to process non-symbolic numerical information (e.g. Price et al., 2007) and the absence of group differences (e.g. Kucian, Leoenneker, Dietrich, Dosch, Martin, & von Aster, 2006). However, even when using a symbolic comparison task, deficits in basic magnitude representation or quantity processing may appear. For example, Soltész and colleagues (Soltész, Szűcs, Dékány, Márkus, & Csépe, 2007), who examined symbolic comparison, found that adolescents with DD show no late event-related brain potentials (ERPs) distance effect between 400 and 440 ms on right parietal electrodes when comparing Arabic numerals. Such a finding may indicate that the processing of the magnitudes of numerical information is abnormal in DD. In addition, Mussolin and colleagues (Mussolin et al., 2009) found that compared with typically developing children, the numerical distance effect does not modulate intraparietal sulcus (IPS) activation in 9–11-year-old children with DD during a symbolic numerical comparison task. According to the authors, the IPS, which is typically involved with numerical representations, may be deficient in DD.

Only a few studies have investigated performance of DD participants in non-symbolic comparisons. Price and colleagues (Price et al., 2007), for example, found differences between DD participants and controls in non-symbolic comparisons in an fMRI study. Specifically, the difference was found in brain activation, but not in response time. Also, they found a weak IPS activation in DD children compared with controls when they compared non-symbolic numerical stimuli. Moreover, Landerl and colleagues (Landerl, Fussenegger, Moll, & Willburger, 2009) found that 8–10-year-old DD children were slower than controls in both symbolic and non-symbolic number comparisons. Mussolin et al. (2010) found that 10- and 11-year-old children with DD showed a larger distance effect in both symbolic and non-symbolic numerical comparisons, suggesting a deficit in the ability to process numerical magnitudes. Other studies with DD populations suggest that those with DD may be deficient in their ability to automatically associate between written digits (e.g. the symbol '3') and their corresponding quantities (e.g. the quantity of three items). Indeed, several studies support this weak-link theory by showing that there is actually an association between these two systems (Libertus, Waldorff, & Brannon, 2007; Notebaert, Nelis, & Reynvoet, 2011; Piazza, Pinel, Le Bihan, & Dehaene, 2007) and that it might be deficient in DD participants (Rousselle & Noël, 2007; Rubinsten & Henik, 2005) (although see Mussolin et al., 2009). Recently, in a review paper, Noël and Rousselle (2011) argued that the first deficit shown in those with DD appears in exact/symbolic numerical representations during the process of learning the symbolic numerical system. Deficiencies in non-symbolic numerical representations appear only later and are secondary to the first deficit.

Nonetheless, there is an increasing awareness that the core deficit approach, which implies a single-deficit view of DD (e.g. deficient non-symbolic representations), is not sufficient to account for the complex and often heterogeneous clinical picture of the disorder (Kaufmann & Nuerk, 2005; Rubinsten, 2009; Rubinsten & Henik, 2009). Such an argument fits with findings showing that 20–60% of children with DD have associated learning problems such as dyslexia (Dirks, Spyer, van Lieshout, & de Sonneville, 2008; Mayes & Calhoun, 2006) or attention deficit/hyperactivity disorder (Capano, Minden, Chen, Schachar, & Ickowicz, 2008; von Aster & Shalev, 2007).

Dyslexia Comorbidity and The Role of The Angular Gyrus

In a German sample of children ($n = 378$, reported by von Aster and Shalev, 2007), researchers found that although the prevalence of dyscalculia was about 6%, only 1.8% had pure dyscalculia, while 4.2% had co-morbid dyslexia. In conjunction with Noël and Rousselle's (2011) argument, it may be suggested that the ability to automatically associate written symbols with mental representations, such as quantities or phonemes may lead to both maths and reading difficulties, which are quite common. Such deficits are not conclusively proven scientifically, but they can serve as a basis for testable predictions at both the behavioural and the biological levels.

It is clear that the resolution of the debate concerning the extent to which arithmetic impairments are specific to DD or shared with dyslexia is challenged by the marked heterogeneity in behavioural symptoms in both dyslexia and DD (Simmons & Singleton, 2008; Tressoldi, Rosati, & Lucangeli, 2007; von Aster & Shalev, 2007). As can be seen in Figure 36.1, it is not clear if the deficit in maths, for example, is the result of a reading deficit (i.e. dyslexia) or is unique to the numerical faculty. In the last two decades, research has made impressive strides forward in studying numerical cognition and brain mechanisms involved in DD by using focused, low-level cognitive tasks (e.g. the numerical Stroop task), which resulted in a detailed description of the cognitive deficit; this was, instead, of or in addition to the paper and pencil tasks, which could not give an exact description of the deficit. Recently, for example, a double dissociation between DD and dyslexia was found in the ability to automatically associate quantities with written numbers versus automatically associating phonemes with written letters (Rubinsten & Henik, 2006). Only a few studies used similar low-level tasks to address co-morbidity issues. Landerl and colleagues (Landerl et al., 2004) argued that children with DD and DD + dyslexia do not have a different type of deficit, but suffer from the same number-processing deficit. Both groups of children, with DD only and DD + dyslexia, were slower on tests that included counting dots, comparing values of single digits, reciting number sequences, reading three-digit numbers, and writing numbers. Later, Landerl and colleagues (Landerl et al., 2009) examined a group of children with DD only, a group of children with dyslexia only, and a DD + dyslexia group. A phonological deficit was found in both dyslexia groups, regardless of numerical deficits, but not in the DD group. However, deficits in processing of symbolic and non-symbolic numerical information were found in the two DD groups, regardless of reading difficulties. Rousselle and Noël (2007) found no evidence for differential patterns of performance between children with DD and DD + dyslexia in tasks assessing basic numerical skills. However, others (Jordan et al., 2003;

Jordan, Kaplan, & Hanich, 2002) suggested that children with only DD have better numerical performance than children with DD + dyslexia in domains of arithmetic such as verbal problem solution and calculations.

Let us suggest a hypothesis for such a DD + dyslexia co-morbidity – one neural candidate that may be involved with both DD and dyslexia is the angular gyrus (and not necessarily the IPS). The angular gyrus (AG) is considered to be involved in integration of semantic information into an ongoing context (Humphreys, Binder, Medler, & Liebenthal, 2007), in associating symbols with non-symbolic events, such as written letters with sounds (i.e. phonemes; Booth, Burman, Meyer, Gitelman, Parrish, & Mesulam, 2004), and it was found to be recruited to a lesser extent in dyslexic children (Pugh, Mencl, Shaywitz, Shaywitz, Fulbright, Constable, et al., 2000) than in non-dyslexic children. This suggestion appears in Figure 36.1B. Note, that the AG is deactivated most of the time, rather than activated during numerical tasks (Wu, Chang, Majid, Caspers, Eickhoff, & Menon, 2009) and the activity level of the AG is modulated similarly after arithmetical and non-arithmetical training (i.e. figural-spatial task; Grabner, Ischebeck, Reishofer, Koschutnig, Delazer, Ebner et al., 2009). Our idea is in line with the suggestion that the left AG is involved in general processes of (fact) learning, skilled retrieval, and level of automatization (see Zamarian and Delazer, this volume).

Also, the left AG and/or surrounding perisylvian brain regions show larger activation during exact calculation that depends on instruction and often relies on verbal rote memorizing (Dehaene, Spelke, Pinel, Stanescu, & Tsivkin, 1999; Lee, 2000; Venkatraman, Siong, Chee, & Ansari, 2006; Zago, Turbelin, Petit, Vigneau, & Tzourio-Mazoyer, 2008; but see Rusconi, Walsh, & Butterworth, 2005, who found that the left AG is not essential for the storage of multiplication and addition facts). Specifically, the AG is involved in calculation (Grabner, Ansari, Koschutnig, Reishofer, Ebner, & Neuper, 2009; van Harskamp, Rudge, & Cipolotti, 2002), in retrieving simple arithmetic facts such as 2 × 3 = 4 (Dehaene, Molko, Cohen, & Wilson, 2004), in transfer of mental operations between multiplication and division (Ischebeck, Zamarian, Schocke, & Delazer, 2009), in mental representations of magnitudes or quantities (Göbel, Walsh, & Rushworth, 2001; Rusconi et al., 2005), and also in visuospatial attention that is induced by the mental number line (Cattaneo, Silvanto, Pascual-Leone, & Battelli, 2009). In addition, and with relevance to the current work, Kaufmann and colleagues (Kaufmann, Koppelstaetter, Siedentopf, Haala, Zimmerhackl, Zimmerhackl, et al., 2006) tested participants in a numerical Stroop task that required them to focus on one stimulus dimension (numerical value or physical size) and to ignore the other. Stimuli were classified into three categories:

- *Congruent:* physical and numerical comparison leads to the same response (e.g. 3 4).
- *Incongruent:* physical and numerical comparison leads to different responses (e.g. 3 4).
- *Neutral:* the stimuli differ only with regard to the task-relevant stimulus property (e.g. 3 4 for numerical comparison).

The results indicated that in the numerical comparisons, but not in physical comparisons, the left AG was activated (together with IPS and the supramarginal gyrus bilaterally). The activation of the AG was suggested to reflect the retrieval of numerical information that is associated with the number symbols, while this was not required in the physical task. This may be similar to the involvement of the AG with associating written letters with phonemes.

Hence, deficit in this structure may cause both reading impairments and calculation difficulties. Accordingly, it is reasonable to suggest that any lesion in the left AG could lead to calculation difficulties that result from either one or more of the following:

- deficits in the ability to associate mental representations (e.g. quantities or phonemes) with symbols (e.g. written numbers or letters);
- deficits in the ability to retrieve simple arithmetic facts from verbal memory.

As suggested by Rousselle and Noël (2007), such an AG developmental delay or deficit may lead to deficiencies in the symbolic system, but also to co-morbidities, such as DD and dyslexia.

Attention and Working Memory

Studies indicate that those with DD have attention deficits, such as impaired executive attention or visuospatial attention and alertness (Ashkenazi & Henik, 2010a,b). In recent years, various laboratories have been engaged in studying networks of attention by using the Attention Network Test – ANT (Fan, McCandliss, Sommer, Raz, & Posner, 2002) or various adaptations of this test (e.g. ANTI; Callejas, Lupiáñez, & Tudela, 2004). The ANT and similar tests were designed to examine three aspects of attention – executive function, alertness, and orientating. Executive control or selective attention is central to goal-directed behaviour. It is commonly studied by employing conflict situations like the Stroop or the flanker tasks. In such tasks, one stimulus or dimension is relevant and other stimuli, or dimensions are irrelevant. Participants are asked to pay attention to the relevant and ignore the irrelevant features of the stimuli. It has been suggested that these conflict situations involve the frontal lobe, mostly the midline frontal areas (anterior cingulate cortex) and the lateral prefrontal cortex (Fan, McCandliss, Fossella, Flombaum, & Posner, 2005). The alerting network is related to the state of awakeness. Its role is to activate and preserve attention. The alerting network is based on the distribution of the brain norepinephrine system, and sustains attention during long-lasting monotonic tasks (Posner & Petersen, 1990). The orienting network is involved in moving attention to a specific location in space (Posner & Petersen, 1990). Ashkenazi and Henik (2010a) examined participants suffering from pure DD (without any comorbidity and specifically without ADHD) and reported deficits in the alerting and executive networks. Specifically, DD participants showed a larger alerting effect and a larger congruity effect compared with matched controls. Furthermore, visuospatial attention deficits were found in the DD participants in a physical line bisection task. Non-deficit participants usually present a small leftward bias during the bisection of a physical line (pseudoneglect). Pseudoneglect is considered to be based on an asymmetry in visuospatial attention processing between the brain cerebral hemispheres, where the right hemisphere is strongly involved in attention processing. In contrast with non-deficit participants, those with DD had no consistent bias in the bisection of the physical line. This result implies a difficulty in visual and spatial attention (Ashkenazi & Henik, 2010a,b).

The results described above suggest that those with DD suffer from deficits in various aspects of attention. Moreover, numerical processing deficits in DD may involve attention or other domain-general mechanisms. For example, DD participants have shown an impaired behavioural facilitation component in the numerical Stroop task (Rubinsten & Henik,

2005). As mentioned above, the lack of facilitation in the size congruity effect in DD participants was explained by a weakness in the automatic connections between Arabic numeral and the internal magnitude representation. However, the size congruity effect, which is characteristic of the numerical Stroop task, is modulated by interactions between quantity and size (Cohen Kadosh, Henik, Rubinsten, Mohr, Dori, van de Ven, et al., 2005; Cohen Kadosh, Lammertyn, & Izard, 2008), and executive attention (Derrfuss, Brass, Neumann, & von Cramon, 2005). Importantly, a manipulation of cognitive load during the numerical Stroop task eliminated the facilitation component in typically developed adults. A similar pattern was presented by participants suffering from DD, in the task without the executive load (Ashkenazi, Rubinsten, & Henik, 2009). Furthermore, an event-related potential (ERP) study examined the numerical distance effect among those with DD. The results showed an early indication for a distance effect that was similar between DD and control participants (between 200 and 300 ms). However, a late right parietal distance effect, between 400 and 440 ms, was not evident in the DD group. This hints that DD impairments in number processing can be based on decelerated executive functioning weakness rather than a lack of automatic quantity activations (Szűcs & Soltész, 2007).

Another hypothesis is that pure DD is related to deficits in visuospatial attention (Ashkenazi & Henik, 2010a,b) or visuospatial working memory (Rotzer, Loenneker, Kucian, Martin, Klaver, & von Aster, 2009). The basis for this hypothesis is that the IPS is believed to support a spatial representation of a mental number line, which requires visuospatial working memory and attention (Simon, Mangin, Cohen, Le Bihan, & Dehaene, 2002). Furthermore, the IPS is strongly involved in working memory and specifically, in visuospatial working memory and spatial attention (Simon et al., 2002). Consistent with a visuospatial working memory deficit, it has been shown that those with DD have behavioural deficits in visuospatial working memory, as well as decreased activity in the IPS, right insula, and the right inferior frontal gyrus (IFG) relative to control participants during a visuospatial working memory task (Rotzer et al., 2009).

One fundamental question related to attentional deficits in DD is whether a domain- general weakness is the core of DD or just a co-occurrence? If a core of DD is attention weakness then targeted attention training should improve numerical processing in DD. However, attentional training improved most of the attentional deficits of those with DD, but it did not improve the abnormalities of the DD group in arithmetic or basic numerical processing. Thus, the deficits in attention among those with DD and the deficits in numerical processing may originate from different sources, or from the same source, at earlier developmental stages, that is going through cognitive and neural specialization as a function of development (Ashkenazi & Henik, 2012; see Figure 36.1C).

A major attention deficit is attention-deficit/hyperactivity disorder (ADHD). ADHD, which affects 4–10% of school-aged children (Skounti, Philalithis, & Galanakis, 2006) is associated with substantial academic underachievement in mathematics and reading (Spira & Fischel, 2005). For example, recent estimates suggest that 25% of children with ADHD have a comorbid disorder of mathematics (Mayes & Calhoun, 2006). The range and nature of mathematical difficulties associated with ADHD are unclear, primarily because of lack of data. A substantial proportion of individuals with ADHD manifest unexpected problems in mathematics that cause an impairment in academic achievement and daily functioning, with estimates ranging from 10 to 60% (Capano et al., 2008; Mayes, Calhoun, & Crowell, 2000). Specifically, existing studies indicate that children with ADHD exhibit problems in completing arithmetic calculations quickly and accurately (Barry, Lyman, & Klinger, 2002; Benedetto-Nasho & Tannock, 1999), and that these problems may persist into adulthood (Biederman, Mick, Fried, Aleardi, Potter, & Herziq, 2005; Biederman, Monuteaux, & Coyle, 2004). The rates of co-occurrence of mathematical difficulties and ADHD are greater than the rates of either math difficulties or ADHD alone in the general population (Shalev, 2004), but the underlying mechanisms for the overlap between ADHD and mathematical difficulties are unknown. Some researchers attribute the significant mathematical delays in children with ADHD to attention-based impairments (Lindsay, Tomazic, Levine, & Accardo, 2001) or working memory (Rosselli, Matute, Pinto, & Ardila, 2006).

These general cognitive impairments (i.e. not specific to mathematics) are considered to be integral features of both the DD and the ADHD syndrome, and hence, may cause mathematical difficulties in some of these children (i.e. DD + ADHD; Castellanos, Sonuga-Barke, Milham, & Tannock, 2006). An alternative proposition is that subgroups of children with ADHD and mathematical difficulties may exhibit different underlying mechanisms, including specific deficits in basic numerical processing (e.g. quantity processing), as is manifest in children with DD (i.e. two major cognitive deficits resulting in DD + ADHD). Rubinsten and colleagues (Rubinsten, Bedard, & Tannock, 2008) investigated effects of stimulant medication (methylphenidate; MPH) on arithmetic performance in children with ADHD. They identified three groups of children with ADHD from an existing large data base: one group with DD (DD + ADHD), one group with more general and less severe difficulties in arithmetic – mathematic learning difficulty MLD – (MLD + ADHD), and one group with good arithmetic abilities (ADHD). Children with DD + ADHD exhibited both general cognitive dysfunctions and specific deficits in understanding quantities. In contrast, arithmetic difficulties in children with MLD + ADHD were associated with deficits in executive function and working memory. Furthermore, MPH-enhanced performance in arithmetic problems dependent upon working memory (involving activation in the frontal lobes), but not upon processing numerical quantity (involving activation in the parietal lobes). These findings suggest the importance of distinguishing between DD + ADHD and MLD + ADHD. Namely, in some of the children, ADHD impairments in attention or working memory also caused mathematical difficulties resulting in co-occurrence of MLD. Figure 1D presents a possible case in which a deficient executive attention, due to a deficit to the frontal lobe, produces difficulties in arithmetic and in attention. The latter appears as ADHD.

In contrast, cases of DD + ADHD may be the result of double deficits in both quantity processing and executive functions. It should be noted that Kaufmann and colleagues have found that some 9–12-year-old children with ADHD (with no other developmental disorder) have deficient numerical processing skills, specifically in their ability to process numerical magnitude (Kaufmann & Nuerk, 2008).

Neuroanatomy of Developmental Dyscalculia

The previous discussions involved all levels of analyses – behavioural, cognitive, and neuroanatomical. Here, we aim to provide a holistic view on the neuroanatomical model of DD (see also Kucian, Kaufmann, & von Aster, this volume).

Converging evidence from infant, primates, and adults indicates that an a-modal, domain-specific representation of approximate quantities is supported by the IPS in the posterior parietal cortex (Cohen Kadosh et al., 2008; Dehaene, Piazza, Pinel, & Cohen, 2003). Hence, it is not surprising that the IPS has been suggested as the source of vulnerability in DD (Butterworth, Varma, & Laurillard, 2011). Most of the studies that investigated brain activity in those with DD concentrated on basic numerical processing tasks (Kaufmann, Vogel, Starke, Kremser, Schocke, & Wood, 2009; Kucian, Loenneker, Martin, & von Aster, 2011; Mussolin et al., 2009; Price, Holloway, Räsänen, Vesterinen, & Ansari, 2007). The majority of these studies have shown decreased activity in the IPS in participants that were diagnosed as having DD compared with control participants (Mussolin et al., 2009; Price et al., 2007). Specifically, a reduced numerical distance effect was found in the right IPS in non-symbolic (Price et al., 2007, but see reports by Kaufmann et al., 2009, and Kucian, Loenneker, Martin, & von Aster, 2011), and symbolic number comparison (Mussolin et al., 2009). Furthermore, a behavioural deficit in the retrieval of arithmetical facts is one of the most reported behavioural deficits in DD (Geary, 2004; Geary et al., 1992; Gross-Tsur et al., 1996). Hence, weakness in the right IPS activity was found also during high-level arithmetical problem solution in the DD population compared with matched controls (Ashkenazi, Rosenberg-Lee, Tenison, & Menon, 2012). Interestingly, structural neuroanatomical data of the DD population also points to the right IPS as a location of lower gray matter density in the DD group compared with controls (Rotzer, Kucian, Martin, von Aster, Klaver, & Loenneker, 2008; Rykhlevskaia, Uddin, Kondos, & Menon, 2009). Another support for the involvement of the right IPS in DD comes from a transcranial magnetic stimulation (TMS) study. Specifically, a magnetic pulse to the right IPS in typically developed adults resulted in DD-like size congruity effect (Cohen Kadosh et al., 2007).

However, the isolated involvement of the IPS in number processing is questionable. Moreover, it is widely accepted that networks of domain-general brain regions distributed across the cortex serve arithmetic and basic numerical processing. For example, a recent meta-analysis examining adult data indicated that visual areas, such as the fusiform gyrus, as well as multiple regions of the prefrontal cortex (such as the dorsolateral prefrontal cortex–DLPFC–and the IFG), are consistently active across different numerical and arithmetical tasks, and reflected in domain-general demands (Arsalidou & Taylor, 2011). Specifically, the IFG is thought to play a role in visual working memory necessary for arithmetic processing (Song & Jiang, 2006), while the bilateral DLPFC is believed to underlie attention and task difficulty during calculation tasks (Zago et al., 2008; Zhou, Zang, Dong, Qiao, & Gong, 2007). Moreover, both the right IFG and DLPFC are involved in tasks that require executive functions like stopping an incipient response and ignoring irrelevant distractors, respectively (Botvinick, Cohen, & Carter, 2004; Kalanthroff, Goldfarb, & Henik, 2013; Verbruggen & Logan, 2008).

In relation to the DD population, in addition to decreased activity and lower gray matter density in the IPS, individuals with DD have also shown abnormal activity patterns and abnormal structure in frontal, and visual brain regions (Ashkenazi et al., 2012; Mussolin et al., 2009; Price et al., 2007; Rotzer, Kucian, Martin, von Aster, Klaver, & Loenneker, 2008; Rykhlevskaia et al., 2009). Specifically, a reduced numerical distance effect was found in the DD population compared with the typical population, in frontal brain regions during symbolic and non-symbolic numerical comparisons (Mussolin et al., 2009; Price et al., 2007). Similarly, a reduced numerical distance effect was found in the left-hemisphere fusiform gyrus (Price et al., 2007, but see a recent report by Kucian et al., 2011). Moreover, children with DD showed lower arithmetical complexity task modulation than controls during arithmetical problem solution. Specifically, typical participants showed increased activity level in prefrontal and bilateral fusiform gyrus brain regions with increased arithmetical complexity. In contrast, the DD group showed the same level of activity regardless of the arithmetic complexity level (Ashkenazi et al., 2012). Studies of brain morphometry and tractography in those with DD have also provided evidence for abnormalities of white and gray matter in frontal regions (Rotzer et al., 2008), and the right fusiform and parahippocampal gyri, and cerebellum (Rykhlevskaia et al., 2009). Furthermore, deficiencies in right hemisphere micro-structure and long-range white matter projection fibers linking the right fusiform gyrus with the temporal-parietal region were also deficient in those with DD (Rykhlevskaia et al., 2009). Note that a recent meta-analysis of fMRI studies (Kaufmann, Wood, Rubinsten, & Henik, 2011) revealed that children recruit distributed networks encompassing parietal and frontal regions bilaterally. Namely, activation differences between children with and without dyscalculia were observed not only in number-relevant parietal regions, but also in the frontal and occipital cortex. Taken together, imaging studies on DD point to domain-specific weakness in designated numerical processing regions (such as the right IPS), as well as weakness in multiple frontal, visual, and middle temporal regions and fibre tracks connecting those regions. This fits in with the findings suggesting involvement of domain-general processes in the aetiology of DD, multiple behavioural symptoms, and potential comorbidity.

Conclusion

Parents, clinicians, and researchers are aware that homogeneity of symptoms is not as common as expected and heterogeneity of manifestations of a deficiency is not an exception. We (Henik, Rubinsten, & Ashkenazi, 2011; Rubinsten & Henik, 2009, 2010) have recently argued that core features of mental disorders are best understood in terms of deficits at the cognitive and the biological levels. Specifically, core ('common cause') deficits at the cognitive or brain level may show up as a network of symptoms, even when there is a single deficit (Karmiloff-Smith, 1998); and a single deficit at the behavioural or cognitive level may produce, through development, a cascade of difficulties that may end up as a network of symptoms at the behavioural level. Exploring the heterogeneity of DD and domain general weakness will potentially result in individual-level and targeted intervention programmes for the remediation of DD (Cohen Kadosh, Dowker, Heine, Kaufmann, & Kucian, 2013). Future research needs to examine heterogeneity in disability as part and parcel of the deficiency. Figure 36.1 gives several relevant examples.

Acknowledgment

This work was conducted as part of the research in the Center for the Study of the Neurocognitive Basis of Numerical Cognition, supported by the Israel Science Foundation (Grants 1799/12 and 1664/08) in the framework of their Centers of Excellence.

References

Ansari, D. (2008). Effects of development and enculturation on number representation in the brain. Nature Reviews Neuroscience, 9, 278–291.

Ansari, D., & Karmiloff-Smith, A. (2002). Atypical trajectories of number development: a neuroconstructivist perspective. Trends in Cognitive Sciences, 6, 511–516.

Arsalidou, M., & Taylor, M. J. (2011). Is 2 + 2 = 4? Meta-analyses of brain areas needed for numbers and calculations. NeuroImage, 54, 2382–2393.

Ashkenazi, S., & Henik, A. (2010a). Attentional networks in developmental dyscalculia. Behavioral and Brain Functions, 6, 2, 1–12.

Ashkenazi, S., & Henik, A. (2010b). A disassociation between physical and mental number bisection in developmental dyscalculia. Neuropsychologia, 48, 2861–2868.

Ashkenazi, S., & Henik, A. (2012). Does attentional training improve numerical processing in developmental dyscalculia? Neuropsychology, 26, 45–56.

Ashkenazi, S., Mark-Zigdon, N., & Henik, A. (2009). Numerical distance effect in developmental dyscalculia. Cognitive Development, 24, 387–400.

Ashkenazi, S., Mark-Zigdon, N., & Henik, A. (2013). Do subitizing deficits in developmental dyscalculia involve pattern recognition weakness? Developmental Science, 16, 35–46.

Ashkenazi, S., Rosenberg-Lee, M., Tenison, C., & Menon, V. (2012). Weak task-related modulation and stimulus representations during arithmetic problem solving in children with developmental dyscalculia. Developmental Cognitive Neuroscience, 2, S152–S166.

Ashkenazi, S., Rubinsten, O., & Henik, A. (2009). Attention, automaticity, and developmental dyscalculia. Neuropsychology, 23, 535–540.

Barry, T. D., Lyman, R. D., & Klinger, L. G. (2002). Academic underachievement and attention deficit/hyperactivity disorder: the negative impact of symptom severity on school performance. Journal of School Psychology, 40, 259–283.

Benedetto-Nasho, E., & Tannock, R. (1999). Math computation, error patterns and stimulant effects in children with Attention Deficit Hyperactivity Disorder. Journal of Attention Disorders, 3, 121–134.

Biederman, J., Mick, E., Fried, R., Aleardi, M., Potter, A., & Herziq, K. (2005). A stimulated work place experience for non-medicated adults with and without ADHD. Psychiatry Services, 56, 1617–1620.

Biederman, J., Monuteaux, M.C., & Coyle, A.E. (2004). Impact of executive function deficits and attention-deficit hyperactivity disorder (ADHD) on academic outcomes in children. Journal of Consulting and Clinical Psychology, 72, 757–766.

Booth, J.R., Burman, D.D., Meyer, J.R., Gitelman, D.R., Parrish, T.B., & Mesulam, M.M. (2004). Development of brain mechanisms for processing orthographic and phonologic representations. Journal of Cognitive Neuroscience, 16, 1234–1249.

Botvinick, M.M., Cohen, J.D., & Carter, C.S. (2004). Conflict monitoring and anterior cingulate cortex: an update. Trends in Cognitive Sciences, 8, 539–546.

Bull, R., Johnston, R.S., & Roy, J.A. (1999). Exploring the roles of the visual-spatial sketch pad and central executive in children's arithmetical skills: views from cognition and developmental neuropsychology. Developmental Neuropsychology, 15, 421–442.

Butterworth, B. (2005). The development of arithmetical abilities. Journal of Child Psychology and Psychiatry, 46, 3–18.

Butterworth, B. (2010). Foundational numerical capacities and the origins of dyscalculia. Trends in Cognitive Sciences, 14, 534–541.

Butterworth, B., Varma, S., & Laurillard, D. (2011). Dyscalculia: from brain to education. Science, 332, 1049–1053.

Callejas, A., Lupiáñez, J., & Tudela, P. (2004). The three attentional networks: on their independence and interactions. Brain and Cognition, 54, 225–227.

Capano, L., Minden, D., Chen, S.X., Schachar, R.J., & Ickowicz, A. (2008). Mathematical learning disorder in school-age children with attention-deficit/hyperactivity disorder. Canadian Journal of Psychiatry, 53, 392–399.

Castellanos, F.X., Sonuga-Barke, E.J.S., Milham, M.P., & Tannock, R. (2006). Characterizing cognition in ADHD: beyond executive dysfunction. Trends in Cognitive Sciences, 10, 117–123.

Cattaneo, Z., Silvanto, J., Pascual-Leone, A., & Battelli, L. (2009). The role of the angular gyrus in the modulation of visuospatial attention by the mental number line. NeuroImage, 15, 563–568.

Cohen Kadosh, R., Cohen Kadosh, K., Schuhmann, T., Kaas, A., Goebel, R., Henik, A., & Sack, A.T. (2007). Virtual dyscalculia induced by parietal-lobe TMS impairs automatic magnitude processing. Current Biology, 17, 689–693.

Cohen Kadosh, R., Dowker, A., Heine, A., Kaufmann, L., & Kucian, K. (2013). Interventions for improving numerical abilities: present and future. Trends in Neuroscience and Education, 2, 85–93.

Cohen Kadosh, R., Henik, A., Rubinsten, O., Mohr, H., Dori, H., van de Ven, V., Zorzi, M., Hendler, T., Goebel, R., & Linden, D.E.J. (2005). Are numbers special? The comparison systems of the human brain investigated by fMRI. Neuropsychologia, 43, 1238–1248.

Cohen Kadosh, R., Lammertyn, J., & Izard, V. (2008). Are numbers special? An overview of chronometric, neuroimaging, developmental and comparative studies of magnitude representation. Progress in Neurobiology, 84, 132–147.

Dehaene, S., Molko, N., Cohen, L., & Wilson, A.J. (2004). Arithmetic and the brain. Current Opinion in Neurobiology, 14, 218–224.

Dehaene, S., Piazza, M., Pinel, P., & Cohen, L. (2003). Three parietal circuits for number processing. Cognitive Neuropsychology, 20, 487–506.

Dehaene, S., Spelke, E., Pinel, P., Stanescu, R., & Tsivkin, S. (1999). Sources of mathematical thinking: behavioral and brain-imaging evidence. Science, *284*, 970–974.

Derrfuss, J., Brass, M., Neumann, J., & von Cramon, D.Y. (2005). Involvement of the inferior frontal junction in cognitive control: meta-analyses of switching and Stroop studies. Human Brain Mapping, 25, 22–34.

Dirks, E., Spyer, G., van Lieshout, E.C., & de Sonneville, L. (2008). Prevalence of combined reading and arithmetic disabilities. Journal of Learning Disabilities, 41, 460–473.

Estrada, C.A., Martin-Hryniewicz, M., Peek, B.T., Collins, C., & Byrd, J.C. (2004). Literacy and numeracy skills and anticoagulation control. American Journal of the Medical Sciences, *328*, 88–93.

Fan, J., McCandliss, B.D., Fossella, J., Flombaum, J.I., & Posner, M.I. (2005). The activation of attentional networks. NeuroImage, 26, 471–479.

Fan, J., McCandliss, B.D., Sommer, T., Raz, A., & Posner, M.I. (2002). Testing the efficiency and independence of attentional networks. Journal of Cognitive Neuroscience, 14, 340–347.

Geary, D.C. (1993). Mathematical disabilities: cognitive, neuropsychological, and genetic components. Psychological Bulletin, *114*, 345–362.

Geary, D.C. (2004). Mathematics and learning disabilities. Journal of Learning Disabilities, 37, 4–15.

Geary, D.C., Bow-Thomas, C.C., & Yao, Y. (1992). Counting knowledge and skill in cognitive addition: a comparison of normal and mathematically disabled children. Journal of Experimental Child Psychology, 54, 372–391.

Geary, D.C., & Hoard, M.K. (2001). Numerical and arithmetical deficits in learning-disabled children: relation to dyscalculia and dyslexia. Aphasiology, 15, 635–647.

Geary, D.C., Hoard, M.K., & Hamson, C.O. (1999). Numerical and arithmetical cognition: patterns of functions and deficits in children at risk for a mathematical disability. Journal of Experimental Child Psychology, 74, 213–239.

Ginsburg, H.P. (1997). Mathematics learning disabilities: a view from developmental psychology. Journal of Learning Disabilities, 30, 20–33.

Göbel, S., Walsh, V., & Rushworth, M.F.S. (2001). The mental number line and the human angular gyrus. NeuroImage, 14, 1278–1289.

Grabner, R.H., Ansari, D., Koschutnig, K., Reishofer, G., Ebner, F., & Neuper, C. (2009). To retrieve or to calculate? Left angular gyrus mediates the retrieval of arithmetic facts during problem solving. Neuropsychologia, 47, 604–608.

Grabner, R.H., Ischebeck, A., Reishofer, G., Koschutnig, K., Delazer, M., Ebner, F., & Neuper, C. (2009). Fact learning in complex arithmetic and figural-spatial tasks: the role of the angular gyrus and its relation to mathematical competence. Human Brain Mapping, 30, 2936–2952.

Gross-Tsur, V., Manor, O., & Shalev, R.S. (1996). Developmental dyscalculia: prevalence and demographic features. Developmental Medicine & Child Neurology, 38, 25–33.

Henik, A., Rubinsten, O., & Ashkenazi, S. (2011). The 'where' and 'what' in developmental dyscalculia. Clinical Neuropsychologist, 25, 989–1008.

Humphreys, C., Binder, J.R., Medler, D.A., & Liebenthal, E. (2007). Time course of semantic processes during sentence comprehension: an fMRI study. NeuroImage, 3, 924–932.

Ischebeck, A., Zamarian, L., Schocke, M., & Delazer, M. (2009). Flexible transfer of knowledge in mental arithmetic—an fMRI study. NeuroImage, 44, 1103–1112.

Jordan, N.C., Hanich, L.B., & Kaplan, D. (2003). A longitudinal study of mathematical competencies in children with specific mathematics difficulties versus children with co-morbid mathematics and reading difficulties. Child Development, 74, 834–850.

Jordan, N.C., Kaplan, D., & Hanich, L.B. (2002). Achievement growth in children with learning difficulties in mathematics: findings of a two-year longitudinal study. Journal of Educational Psychology, 94, 586–597.

Kalanthroff, E., Goldfarb, L., & Henik, A, (2013). Evidence for interaction between the stop-signal and the Stroop task conflict. *Journal of Experimental Psychology: Human Perception and Performance*, 39, 579–592.

Karmiloff-Smith, A. (1998). Development itself is the key to understanding developmental disorders. Trend in Cognitive Science, 2, 389–398.

Kaufmann, L., Koppelstaetter, F., Siedentopf, C., Haala, I., Haberlandt, E., Zimmerhackl, L.B., Felber, S., & Ischebeck, A. (2006). Neural correlates of the number-size interference task in children. NeuroReport, 17, 587–591.

Kaufmann, L., Lochy, A., Drexler, A., & Semenza, C. (2004). Deficient arithmetic fact retrieval - storage or access problem? A case study. Neuropsychologia, 42, 482–496.

Kaufmann, L., & Nuerk, H-C. (2005). Numerical development: current issues and future perspectives. Psychology Science, 47, 142–170.

Kaufmann, L., & Nuerk, H-C. (2008). Basic number processing deficits in ADHD: a broad examination of elementary and complex number processing skills in 9- to 12-year-old children with ADHD-C. Developmental Science, 11, 692–699.

Kaufmann, L., Vogel, S.E., Starke, M., Kremser, C., Schocke, M., & Wood, G. (2009). Developmental dyscalculia: compensatory mechanisms in left intraparietal regions in response to nonsymbolic magnitudes. Behavioral and Brain Functions, 5, 35.

Kaufmann, L., Wood, G., Rubinsten, O., & Henik, A. (2011). Meta-analysis of developmental fMRI studies investigating typical and atypical trajectories of number processing and calculation. Developmental Neuropsychology, 36, 763–787.

Kucian, K., Loenneker, T., Dietrich, T., Dosch, M., Martin, E., & von Aster, M. (2006). Impaired neural networks for approximate calculation in dyscalculic children: a functional MRI study. Behavior and Brain Functions, 5, 31.

Kucian, K., Loenneker, T., Martin, E., & von Aster, M. (2011). Non-Symbolic numerical distance effect in children with and without developmental dyscalculia: a parametric fMRI study. Developmental Neuropsychology, 36, 741–762.

Landerl, K., Bevan, A., & Butterworth, B. (2004). Developmental dyscalculia and basic numerical capacities: a study of 8–9-year-old students. Cognition, 93, 99–125.

Landerl, K., Fussenegger, B., Moll, K., & Willburger, E. (2009). Dyslexia and dyscalculia: two learning disorders with different cognitive profiles. Journal of Experimental Child Psychology, *103*, 309–324.

Lee, K.M. (2000). Cortical areas differentially involved in multiplication and subtraction: a functional magnetic imaging study with a case of selective acalculia. Annals of Neurology, 48, 657–661.

Lewis, C., Hitch, G., & Walker, P. (1994). The prevalence of specific arithmetic difficulties and specific reading difficulties in 9- and 10-year old boys and girls. Journal of Child Psychology and Psychiatry and Allied Disciplines, 35, 283–292.

Libertus, M.E., Waldorff, M.G., & Brannon, E.M. (2007). Electrophysiological evidence for notation intendance in numerical processing. Behavioral and Brain Functions, 3.

Lindsay, R., Tomazic, T., Levine, M., & Accardo, P. (2001). Attentional function as measured by a continuous performance task in children with dyscalculia. Journal of Developmental and Behavioral Pediatrics, 22, 287–292.

Mayes, S.D., & Calhoun, S.L. (2006). Frequency of reading, math and writing disabilities in children with clinical disorders. Learning and Individual Differences, 16, 145–157.

Mayes, S.D., Calhoun, S.L., & Crowell, E.W. (2000). Learning disabilities and ADHD: overlapping spectrum disorders. Journal of Learning Disabilities, 33, 417–424.

Moeller, K., Neuburger, S., Kaufmann, L., Landerl, K., & Nuerk, H-C. (2009). Basic number processing deficits in developmental dyscalculia: evidence from eye-tracking. Cognitive Development, 24, 371–386.

Mussolin, C., De Volder, A., Grandin, C., Schlögel, X., Nassogne, M-C., & Noël, M-P. (2009). Neural correlates of symbolic number comparison in developmental dyscalculia. Journal of Cognitive Neuroscience, 22, 860–874.

Mussolin, C., Mejias, S., & Noël, M-P. (2010). Symbolic and nonsymbolic number comparison in children with and without dyscalculia. Cognition, *115*, 10–25.

Noël, M-P., & Rousselle, L. (2011). Developmental changes in the profiles of dscalculia: an explanation based on a double exact-and-approximate number representation model. Frontiers in Human Neuroscience, 5, 165.

Notebaert, K., Nelis, S., & Reynvoet, B. (2011). The magnitude representation of small and large symbolic numbers in the left and right hemisphere: an event-related fMRI study. Journal of Cognitive Neuroscience, 23, 622–630.

Parsons, S., & Bynner, J. (1997). Numeracy and employment. Education & Training, 39, 43–51.

Piazza, M., Pinel, P., Le Bihan, D., & Dehaene, S. (2007). A magnitude code common to numerosities and number symbols in human intraparietal cortex. Neuron, 5, 293–305.

Posner, M.I., & Petersen, S.E. (1990). The attention system of the human brain. Annual Review of Neuroscience, 13, 25–42.

Price, G.R., Holloway, I., Räsänen, P., Vesterinen, M., & Ansari, D. (2007). Impaired parietal magnitude processing in developmental dyscalculia. Current Biology, 17, 1042–1043.

Pugh, K.R., Mencl, W.E., Shaywitz, B.A., Shaywitz, S.E., Fulbright, R.K., Constable, R. T., … Gore, J.C. (2000). The angular gyrus in developmental dyslexia: task-specific differences in functional connectivity within posterior cortex. Psychological Science, 11, 51–56.

Rivera-Batiz, F.L. (1992). Quantitative literacy and the likelihood of employment among young adults in the United States. Journal of Human Resources, 27, 313–328.

Rosselli, M., Matute, E., Pinto, N., & Ardila, A. (2006). Memory abilities in children with subtypes of dyscalculia. Developmental Neuropsychology, 30, 801–818.

Rotzer, S., Kucian, K., Martin, E., von Aster, M., Klaver, P., & Loenneker, T. (2008). Optimized voxel-based morphometry in children with developmental dyscalculia. NeuroImage, 39, 417–422.

Rotzer, S., Loenneker, T., Kucian, K., Martin, E., Klaver, P., & von Aster, M. (2009). Dysfunctional neural network of spatial working memory contributes to developmental dyscalculia. Neuropsychologia, 47, 2859–2865.

Rousselle, L., & Noël, M-P. (2007). Basic numerical skills in children with mathematics learning disabilities: a comparison of symbolic vs non-symbolic number magnitude processing. Cognition, *102*, 361–395.

Rubinsten, O. (2009). Co-occurrence of developmental disorders: the case of developmental dyscalculia. Cognitive Development, 24, 362–370.

Rubinsten, O., Bedard, A.C., & Tannock, R. (2008). Methylphenidate improves general but not core numerical abilities in ADHD children with co-morbid dyscalculia or mathematical difficulties. Journal of Open Psychology, 1, 11–17.

Rubinsten, O., & Henik, A. (2005). Automatic activation of internal magnitudes: a study of developmental dyscalculia. Neuropsychology, 19, 641–648.

Rubinsten, O., & Henik, A. (2006). Double dissociation of functions in developmental dyslexia and dyscalculia. Journal of Educational Psychology, 98, 854–867.

Rubinsten, O., & Henik, A. (2009). Developmental dyscalculia: heterogeneity may not mean different mechanisms. Trends in Cognitive Sciences, 13, 92–99.

Rubinsten, O., & Henik, A. (2010). Comorbidity: cognition and biology count! Behavioral and Brain Sciences, 33, 168–170.

Rusconi, E., Walsh, V., & Butterworth, B. (2005). Dexterity with numbers: rTMS over left anular gyrus disrupts finger anosis and number processing. Neuropsychlogia, 43, 1609–1624.

Russell, R.L., & Ginsburg, H.P. (1984). Cognitive analysis of children's mathematical difficulties. Cognition and Instruction, 1, 217–244.

Rykhlevskaia, E., Uddin, L.Q., Kondos, L., & Menon, V. (2009). Neuroanatomical correlates of developmental dyscalculia: combined evidence from morphometry and tractography. Frontiers in Human Neuroscience, 3:51.

Schleifer, P., & Landerl, K. (2011). Subitizing and counting in typical and atypical development. Developmental Science, 14, 280–291.

Shalev, R.S. (2004). Developmental dyscalculia. Journal of Child Neurology, 19, 765–771.

Shalev, R.S., Auerbach, J., & Gross-Tsur, V. (1995). Developmental dyscalculia behavioral and attentional aspects: a research note. Journal of Child Psychology and Psychiatry and Allied Disciplines, 36, 1261–1268.

Shalev, R.S., & Gross-Tsur, V. (2001). Developmental dyscalculia. Pediatric Neurology, 24, 337–342.

Simmons, F.R., & Singleton, C. (2008). Do weak phonological representations impact on arithmetic development? A review of research into arithmetic and dyslexia. Dyslexia, 14, 77–94.

Simon, O., Mangin, J-F., Cohen, L., Le Bihan, D., & Dehaene, S. (2002). Topographical layout of hand, eye, calculation, and language-related areas in the human parietal lobe. Neuron, 33, 475–487.

Skounti, M., Philalithis, A., & Galanakis, E. (2006). Variations in prevalence of attention deficit hyperactivity disorder worldwide. European Journal of Pediatrics, *166*, 117–123.

Soltész, F., Szűcs, D., Dékány, J., Márkus, A., & Csépe, V. (2007). A combined event-related potential and neuropsychological investigation of developmental dyscalculia. Neuroscience Letters, *417*, 181–186.

Song, J.H., & Jiang, Y. (2006). Visual working memory for simple and complex features: an fMRI study. NeuroImage, 30, 963–972.

Spira, E.G., & Fischel, J.E. (2005). The impact of preschool inattention, hyperactivity, and impulsivity on social and academic development: a review. Journal of Child Psychology and Psychiatry, 46, 755–773.

Szűcs, D., & Soltész, F. (2007). Event-related potentials dissociate facilitation and interference effects in the numerical Stroop paradigm. Neuropsychologia, 45, 3190–3202.

Tressoldi, P.E., Rosati, M., & Lucangeli, D. (2007). Patterns of developmental dyscalculia with or without dyslexia. Neurocase, 13, 217–225.

van Harskamp, N.J., Rudge, P., & Cipolotti, L. (2002). Are multiplication facts implemented by the left supramarginal and angular gyri? Neuropsychologia, 40, 1786–1793.

Venkatraman, V., Siong, S.C., & Chee, M.W., & Ansari, D. (2006). Effect of language switching on arithmetic: a bilingual fMRI study. Journal of Cognitive Neuroscience, 18, 64–74.

Verbruggen, F., & Logan, G.D. (2008). Response inhibition in the stop-signal paradigm. Trends in Cognitive Sciences, 12, 418–424.

von Aster, M., & Shalev, R.S. (2007). Number development and developmental dyscalculia. Developmental Medicine and Child Neurology, 49, 868–873.

Wu, S., Chang, T.T., Majid, A., Caspers, S., Eickhoff, S.B., & Menon, V. (2009). Functional heterogeneity of inferior parietal cortex during mathematical cognition assessed with cytoarchitectonic probability maps. Cerebral Cortex, 19, 2930–2945.

Zago, L., Petit, L., Turbelin, M.R., Andersson, F., Vigneau, M., & Tzourio-Mazoyer, N. (2008). How verbal and spatial manipulation networks contribute to calculation: an fMRI study. Neuropsychologia, 46, 2403–2414.

Zhou, X., Chen, C., Zang, Y., Dong, Q., Qiao, S., & Gong, Q. (2007). Dissociated brain organization for single-digit addition and multiplication. NeuroImage, 35, 871–880.

CHAPTER 37

THE CONTRIBUTIONS OF SYNDROME RESEARCH TO THE STUDY OF MLD

MICHÈLE M. M. MAZZOCCO

Introduction

There are many reasons why mathematics may be challenging for some children. This means that effective approaches to supporting those students will vary. Many children experience only occasional challenges with mathematics, such as when learning a new mathematics topic or solving a problem of unfamiliar complexity. For individuals with mathematics learning disability (MLD), however, mathematics difficulties may be a constant, long-term problem, even when faced with problems that others consider routine. In this chapter I focus on children whose MLD is associated with a known genetic disorder, and I illustrate how studies of genetic syndromes provide models of the kind of variation in MLD we might see in the general population. These models collectively demonstrate the heterogeneity of MLD, and they individually reveal distinct profiles of associated (or dissociated) mathematics and related skills. Together, these models have important implications for the roles of individual and developmental differences in mathematics achievement.

Syndrome Research Contributes to Knowledge of Individual Differences

The models of individual differences that emerge from studies of genetic phenotypes extend to some persons in the general population, because multiple pathways to mathematics achievement are not exclusive to genetic syndromes (see Jordan, this volume). Indeed, multiple pathways are implicated by the different relationships between mathematical and non-mathematical skills observed throughout the lifespan (as discussed by Desoete, this volume, and by the fact that predictors of mathematics performance vary depending on whether predicted outcomes are specific numerical skills (e.g. fact retrieval fluency, fractions knowledge) or broader measures of mathematics achievement (e.g. standardized test results). Multiple pathways are evident beginning in early childhood (e.g. Dowker 2008; Jordan, this volume; Kaufmann et al. 2013; LeFevre et al., 2010). This knowledge about individual differences in mathematical learning guides educational planning and instructional decision-making, and may guide the process of differential diagnosis when identifying or ruling out MLD.

Syndrome Research Contribute to Developing Models of MLD

At a minimum, phenotypic models of pathways to MLD contribute to building theoretical and practical guides to MLD classification and diagnosis, which in turn guide instruction. One advantage of genetic phenotype-based research is that it lacks the pervasive diagnostic and classification complications that plague general research on learning disabilities or MLD (Butterworth 2005; Mazzocco 2007), because it does not require a single widely accepted definition of MLD. Instead, in syndrome research, participant groups are often defined by presence of a well-established syndrome-specific biological marker. The overall mission to develop models of MLD is not accomplished by studying a single syndrome, because focusing on any single syndrome may yield inferences that are not relevant to many individuals with MLD. (Likewise, studies of MLD in the general population rarely describe all individuals with MLD, and often fail to capture important subgroup or subtype variations (Murphy et al. 2007)). Focusing on different genetic syndromes can, however, help to differentiate MLD phenotype models, and this is the approach used in a set of recently completed longitudinal studies described here.

Is there an MLD Phenotype? If we were charged with describing one MLD phenotype based on all of the characteristics reported across MLD studies to date, it would include quite a range of numerical, non-numerical, and non-cognitive features. The phenotype would be based on (but not limited to) the information that follows.

Relative to their peers, children with MLDs have difficulty processing non-symbolic quantities (Piazza et al. 2010; Mazzocco et al. 2011), mapping symbolic referents to non-symbolic quantities (Mazzocco et al. 2011), and accessing symbolic representations (Rousselle and Noël 2007). They rely on serial counting for quantities in the subitizing range (Koontz and Berch 1996), show persistent misconceptions of whole numbers (Mazzocco et al. 2013a), and make counting errors linked to poor short term memory (Geary et al. 2004). Young children with MLD are less likely to spontaneously focus on quantity as a feature of their environment (Hannula et al. 2010), make less accurate verbal magnitude comparisons, make more errors when reading or writing digits (Mazzocco and Thompson 2005), and have slower and less accurate computational skills. Computational errors occur for both exact and approximate arithmetic, although problems with exact arithmetic are more pronounced for children with co-morbid math and reading difficulties (Jordan et al. 2003). In middle school and beyond, children with MLD have persistent problems with rational numbers, and are less accurate when comparing magnitudes of two fractions values. Their rational-number-knowledge misconceptions persist through at least eighth grade (Mazzocco et al. 2013b) and into high school (Geary et al. 2013). Importantly, most children classified as MLD in primary school continue to meet those criteria several years later (Mazzocco and Myers 2003; Morgan et al. 2009).

It is unreasonable to expect that this collection of empirical findings describes a single phenotype, much less any single child with MLD. Collectively, studies of MLD reveal the three following overarching findings relevant to MLD phenotypes.

- Both mathematics-specific and non-mathematics cognitive skills play a role in mathematics outcomes (e.g. Cirino 2011; LeFevre et al. 2010). It follows that their relative contribution to mathematics outcomes will vary across individuals.
- Non-cognitive factors (e.g. motivational and environmental factors) also play a role in mathematics achievement outcomes (e.g. Ashcraft, this volume), even if it is not always clear when or whether these roles are causal or indirect, and uni- or bi-directional (e.g. Fuchs et al. 2010; LeFevre et al. 2010). Thus non-cognitive factors need to be considered in phenotype descriptions, and may be important for understanding consequences of MLD as well as causes of difficulties that incorrectly manifest as MLD.
- The nature of mathematics achievement outcomes can vary widely even among individuals with MLDs. That is, it is not only the underpinnings of MLD but also the manifestation of mathematics achievement that varies across individuals, and across development (e.g. Rubinsten and Henik, 2009).

Together, these studies of mathematics difficulties reveal the need to identify different underpinnings, associated features, and manifestations of MLD as well as developmental influences on the MLD phenotype, because one phenotype will not capture all children with MLD. These foci are reflected in our longitudinal studies of mathematics learning profiles in children with different genetic phenotypes.

Evidence of MLD Subtypes Shaped Longitudinal Studies of the MLD Phenotype

Questions about potential MLD subtypes emerged decades ago, when Rourke (Strang and Rourke 1985; Rourke 1993) proposed subtype models based on co-morbidities between mathematics and reading and/or spelling deficits, and Geary proposed theoretically and empirically guided subtype models (1993) based on neuropsychological correlates. Both proposals indicated an MLD + reading disability (RD) subtype, and other subtypes. Our longitudinal work was heavily influenced by Geary's models, which did not explicitly identify MLD + RD co-morbidity as a *defining* feature of a subtype, but indicated a higher rate of RD among children with the *semantic memory* subtype of MLD. Briefly, the semantic memory subtype is characterized by poor fact retrieval, slow and often inaccurate arithmetic solutions. These alleged fact retrieval deficits may be linked to lexical access difficulties (Geary et al. 2000) or competition from irrelevant associates (such as '8' as a possible solution for the problem, '3 × 5'). Geary describes this subtype as genetically linked and persistent over time. Geary also proposed a *visuospatial* subtype initially described as characterized by overt errors in alignment and failure to master place value concepts tied to relative positions of numerals; other researchers refer to spatial components involved in the representation of quantity and number line (Dehaene et al. 1999; Siegler and Opfer 2003) or spatial attention (e.g. Simon et al. 2008) also linked to basic numeracy skills. Thus contribution of spatial abilities, in children with this subtype, may pertain to the spatial, mental representation of quantity, rather than representation of visual space per se, as originally proposed. Finally, the *procedural* subtype proposed by Geary, which follows a delay rather than deficit trajectory, is characterized by immature strategy use and weak conceptual understanding of the arithmetic procedures. Executive function skills such as working memory are related to performance accuracy among children with this MLD subtype (Geary et al. 2004).

These three proposed subtypes, and later work emphasizing 'number sense' foundations in mathematics skills (Dehaene et al. 1999), informed prospective longitudinal study designs emerging in the late 1990s and at the turn of the century (e.g. Geary et al. 2007; Jordan et al. 2003; Mazzocco and Myers 2003; Passolunghi et al. 2008). Thus, a commonality across these early longitudinal studies was the inclusion of reading-related, visuospatial, executive function, and other measures in addition to measures of mathematical thinking and achievement. Almost in parallel, researchers' attention to specific numerical processing led to a temporary divide between research on dyscalculia (Butterworth 2005) and MLD (e.g. Geary and Hoard 2005) when, in fact, the degree to which these classifications overlapped varied across research labs. Some treated MLD and dyscalculia as synonymous with each other (e.g. Mazzocco et al. 2011) while others differentiated these constructs (e.g. Butterworth 2005). In reality, both approaches highlighted the heterogeneity of mathematics difficulties.

An obstacle to identifying MLD subtypes through normative prospective longitudinal studies lies in the reported prevalence of MLDs. Although MLD affects many individuals, the frequency with which strict criteria for MLD are met in the general population is too small to afford subtype analysis or confirmatory factor analysis unless normative sample sizes are very large, or unless children with MLD are oversampled. (This is in contrast to studies of children with a wide range of maths difficulties.) Another obstacle stems from the fact that subtypes may appear similar when outcome measures in research are limited to total composite scores on a broad measure of mathematics achievement, because phenotypic differences are likely to lie beneath this homogeneous surface. The study of children with genetic syndromes associated with mathematical learning difficulties affords the advantages of a prospective design with samples likely to over-represent subtypes MLDs, relative to studies of MLD in the general population.

Two specific syndromes have been the focus of longitudinal studies in our lab–fragile X and Turner syndromes. Although we focus on only these two syndromes in this chapter, we acknowledge the contributions to syndrome research made by many other labs devoted to studies of other biological conditions that provide models of pathways to mathematics, including (but not limited to) studies of spina bifida (Dennis et al. 2006), Williams syndrome (e.g. O'Hearn and Luna 2009), Barth syndrome (Mazzocco and Kelley 2001; Raches and Mazzocco 2012), and 22q deletion syndromes (Simon et al. 2008).

Fragile X syndrome

Brief Syndrome Overview. Fragile X syndrome is the leading known inherited cause of intellectual disability (incidence ~ 1:4000–8000 live births; Crawford et al. 2001), and is linked to a single gene mutation (Verkerk et al. 1991) that is now referred to as the fragile X mental retardation–1 gene (FMR1). This fragile X mutation, the first identified trinucleotide expansion, occurs in pre- and full-mutation forms. The syndrome is linked to the full mutation gene, which, in males, almost always leads to intellectual impairment. Protein expression linked to the typical FMR1 gene is decreased in persons with the mutation. Owing to the process of X-inactivation (Lyon 1972), females with fragile X syndrome have a much more variable cognitive phenotype, such that approximately 50% lack significant global intellectual impairment. Their impairment is negatively correlated with expression of FMR protein. Only these relatively high functioning females with fragile X were included in our longitudinal MLD studies, in order to focus on the specificity of mathematics difficulties that occur in the absence of global intellectual impairment. (For more detailed reviews of fragile X syndrome, see Cornish et al. 2007.)

The Fragile X MLD Phenotype. Fragile X syndrome provides a model of MLD marked by specific strengths and weaknesses in numerical skills and milestones. The phenotype demonstrates the need to avoid using limited performance strengths as evidence of mathematical mastery or to rule out mathematics learning difficulties. In other words, the fragile X phenotype makes a case for the need to evaluate multiple, comparable or related mathematics skills as indicators of either mastery or learning difficulties.

Low mathematics achievement (based on standardized testing) is observed in girls with fragile X syndrome as early as kindergarten, relative to their age- and FSIQ-matched kindergarten peers (Mazzocco 2001). At that age, comparable levels of low performance are *not* evident in tasks measuring vocabulary, verbal reasoning, verbal memory, letter and word identification, rapid naming, or motor-reduced visual perception, suggesting specific difficulties in maths. Moreover, girls with fragile X (as a group) are more likely to underperform in mathematics relative to their matched peers (and to other areas of performance), and are also more likely to perform in the impaired (vs. low average or mildly below average) range, thereby meeting our research criteria for MLD (i.e., scoring <86 on standardized testing). This group difference did not reach statistical significance in the kindergarten year of our study (56 vs. 20%, respectively, $p = .08$; Mazzocco 2001), but it became significant in later years.

Importantly, the relatively lower mathematics scores seen in kindergarten girls with fragile X are not evident across *all* mathematics measures, but *are* evident on global composite scores such as the Test of Early Mathematics Ability–Second Edition (TEMA-2) and on specific measures of counting principles. Group differences are not significant on subtests relating to use of measurement terminology and counting sequence knowledge (such as the measurement subtest of the Key-Math Revised test or the Quantitative Reasoning subtest of the Stanford Binet intelligence test (SBIV-QR)). This suggests an early profile of selective strengths and difficulties in mathematics. To date, to our knowledge, no similar studies have been carried out with younger girls with the full mutation.

The pattern of selective mathematics difficulties persists through later school years. By first grade, the frequency of scoring in the MLD range becomes significantly higher among girls with vs. without fragile X (but SBIQ-QR item performance still does not differ between groups). Where performance differences *are* most apparent are on items measuring conceptual principles, such as for counting sets (Murphy et al. 2006; see Table 37.1(a)). On rote items, first to third grade girls with fragile X tend to outperform their MLD counterparts (especially at kindergarten), and score in or above the level reported for children from the normative study who do not have MLD (Table 37.1). These item analyses demonstrate relative strengths in number knowledge skills that can be acquired by rote, and contrasting poor conceptual performance as compared with other children with MLD.

This performance profile is not limited to primary school. When tested during the transition to middle school (Grades four to six), girls with fragile X are as fast as their peers when completing a problem verification task (i.e. reporting if an arithmetic solution is correct or incorrect), but are significantly less accurate. Moreover, they are more likely to miss test items below their own basal levels (i.e. TEMA-2 items, which are normed for use with children through eight years of age), indicating inconsistent mastery of mathematics concepts or skills (Murphy and Mazzocco 2008a).

By middle school, the profile reappears for rational numbers. Girls with fragile X syndrome are as successful as children without MLD at correctly naming symbols for decimal values aloud (e.g. .7, .25), and at rank ordering fractions depicted by area models (which are visual aides commonly used to teach fractions). By contrast, they are quite unsuccessful at rank ordering decimals and fractions, despite having just named them correctly (Murphy and Mazzocco 2008a), performing at or below levels observed in their peers with MLD (Table 37.1)). When asked to identify tied values across rational number representations (e.g. 0.25 = ¼), girls with fragile X identify fewer correct pairs and generate more erroneous ties (e.g. 0.03 = ⅓) relative to their peers. In other words, their adeptness at naming decimals and rank ordering area models masks conceptual deficits that are only observed with further scrutiny.

Growth trajectories for mathematics calculation achievement (measured from Grade one to six) show slower rates of growth in girls with fragile X syndrome, relative to their peers (Murphy and Mazzocco 2009). What predicts rate of growth in mathematics and mathematics achievement at or before Grade six in this group? We tested candidate predictors drawn from normative studies of cognitive correlates or pathways to mathematics, such as working memory, cognitive flexibility, and visual spatial perceptual skills (see Desoete, this volume). Whereas working memory span (measured by digit span tests) and processing speed do not differentiate girls with vs. without fragile X syndrome, performance on a timed measure of cognitive shifting under low level working memory demands does (as measured by the Contingency Naming Test (CNT); Murphy and Mazzocco 2008b; Kirk et al. 2005). When measured at Grade three, variation in CNT performance accounts for differences in rate of growth in calculation skills up to Grade six (Murphy and Mazzocco 2009), and predicts differences in maths calculations scores at Grade six. Concurrent correlations between mathematics calculation and the CNT are also significant at Grades four to six for girls with fragile X (Spearman Rho ~ .73), but not for girls without fragile X (Rho ~ -.11), possibly because CNT performance levels plateau later in development for girls with fragile X syndrome, indicating lower thresholds for working memory demands that interfere with

Table 37.1 Percentage of children performing accurately on number knowledge items, among girls with fragile X syndrome and children without fragile X (with or without MLD).

	Participant Group			
Number Knowledge items	Girls with fragile X		Children with MLD	Children without MLD
(a) Primary school measures	Grade K n = 14	Grades K-2 n = 21	Grade K n = 23	Grades K-2 n = 226
Rote items				
Read one-digit numbers aloud	92	95	65	92
Read two-digit numbers aloud	75[a]	79	20	59
Write dictated one-digit numbers	86[a]	91	39	79
Count aloud to 41 (not sets)	62[a]	60	18	52
Count aloud by tens (not sets)	73[a]	82	30	59
What one-digit number comes next?	86[a]	76[b]	40	96
Use 1:1 correspondence when counting, examiner pointing	71[a]	71[b]	96	99
Conceptual items				
Verbal magnitude comparisons	-	48[b]	30	81
Identify *n*th position in a set	14	-	0	-
Use 1:1 correspondence counting scattered arrays	-	55[b*]	-	85
Number constancy	-	62[b]	-	90
		Grades 6/7		
b) Middle school measures		n = 9	n = 22	n = 169
Rote items				
Read decimals aloud		56[a]	12	77
Rank order 10 area model fractions		67	35	76
Conceptual items				
Rank order 10 decimal values		0[b]	6	58
Rank order decimals/fractions		0[b]	0	41

[a] Significantly *better* than MLD group and (nsd with the non-MLD group)

[b] Significantly *worse* than non-MLD group (nsd with MLD unless *no MLD data available).

Adapted from data reported by Murphy et al. 2006 (grades K-2) and Murphy and Mazzocco 2008a (Grades 6 and 7).

maths performance. Thus, there is evidence of an association between executive functions and mathematics achievement in this population.

These behavioral indices are further supported by fMRI studies supporting working memory deficits in this population (Kwon et al. 2001), in addition to impairments in sustained attention and inhibition (Cornish et al. 2004). Girls with fragile X have significantly less accurate performance on three-operand arithmetic problems (e.g. 2 + 7 + 4) vs. two-operand arithmetic (e.g. 2 + 7; Rivera et al. 2002), a finding aligned with the notion that girls with fragile X have a lower working memory threshold than their peers. This may prompt girls with fragile X syndrome to rely on different cognitive resources than their peers without fragile X, when faced with more taxing problems such as three-operand arithmetic. For instance, whereas females without fragile X show greater activation of prefrontal and parietal cortices on three- vs. two-operand addition, this increase is typically absent in girls with fragile X. These findings are expanded to biological markers of fragile X syndrome severity: In females with fragile X, brain activation during arithmetic processing, specifically the increase in prefrontal and parietal activation during three-operand arithmetic, is positively associated with levels of FMR protein expression (Rivera et al. 2002).

These behavioral and biological indicators of difficulty under working memory demands do not prove that executive functions are the causal pathway to maths ability in girls with fragile X, but they do support the notion that non-numerical factors are important. Moreover, several other cognitive measures also correlate significantly with maths calculations in this population. In fourth to sixth graders with fragile X, we observed associations not observed in their peers, between maths calculation performance and performance on measures of visual perception (matching spatial orientations; Rhos = .65 versus 03, respectively), phonological decoding (Rho = .45 versus .15), processing speed (as measured by rapid automatized naming; Rhos ~ –.58 versus ~.10), and reading fluency (Rho = –.54 versus -.08; Murphy and Mazzocco 2008a).

As illustrated here, girls with fragile X provide a compelling model of individual differences in mathematics performance and development. In this phenotype, performance on many early basic mathematics skills and procedures appears quite consistent with that of typically achieving children, and well above levels seen in children with MLD. We might be tempted, therefore, to report this phenotype as a model of early success on number awareness—but this would be an oversimplification, since the early indicators of number and counting knowledge do not appear to transfer to or otherwise support computational fluency. On more conceptual tasks, performance by individuals with fragile X no longer exceeds levels observed in their MLD peers, a dissociation suggestive of memory-based gains but numerical weaknesses that are further hindered by diminished executive function skills or resources. While rote performance may be viewed as a strength and an area to rely on to support individualized tutoring or intervention, its surface level success should not be interpreted as the kind of adaptive reasoning that accompanies understanding the rote skills and when to apply them to specific problem solving scenarios (NRC 2001).

Early numerical difficulties may have a complex influence on the associations observed between working memory and mathematics in females with fragile X syndrome. On the one hand, working memory deficits may exacerbate mathematical difficulties and interfere with mathematics learning, moderating the association between numerical and mathematics skills. On the other hand, if numerical skills are sufficiently weak, the working memory demands of 'routine' arithmetic tasks (such as two-operand arithmetic) may be quite demanding, regardless of foundational working memory ability; then, for harder tasks, the working memory demands that are typically viewed as achievable with effort (e.g. three-operand problems) for others are simply unsurpassable for the child with fragile X. This last scenario illustrates the lower thresholds of tolerance (for working memory demands) not due to working memory difficulties per se, but to weak numerical processing. Fragile X syndrome research may provide a model of this association (Murphy and Mazzocco 2009).

Turner syndrome

Brief Syndrome Overview

Turner syndrome results from the sporadic partial or total absence of the second X chromosome typically present in females (incidence ~ 1:200–5000 live female births; Nielsen and Wohlert 1991). The syndrome phenotype varies with karyotype characteristics (e.g. Ross et al. 2000; as reviewed by Davenport et al. 2007). Many studies of the phenotype focus on karyotypes with predominantly 45X, or mosaic (combination) karyotypes partially comprised of 45X cells. It is rare for females with Turner syndrome to have significant intellectual impairment, but specific cognitive and learning strengths and difficulties—including difficulties in mathematics—are widely reported. (For a detailed review of Turner syndrome, see Davenport et al. 2007.)

The Turner Syndrome MLD Phenotype. The history of research on the Turner syndrome cognitive phenotype illustrates how important it is to test assumptions about pathways to mathematics profiles that have not been tested for their face validity. That is, although the MLD phenotype in Turner syndrome appears to be aligned with MLD subtypes, early hypotheses concerning these pathways to MLD subsequently failed to be supported by empirical data. What led to these hypotheses was the consistently reported co-occurrence of mathematics difficulties and select impairments in spatial, executive function, and lexical retrieval skills in this population. But these co-occurrences alone were insufficient evidence that Turner syndrome fit the profile of visual-spatial, procedural, or semantic memory subtypes, respectively. This posed a challenge to accounting for MLD in the phenotype, if several subtypes were implicated. Moreover, the Turner syndrome phenotype is marked by accurate *untimed* calculation performance (e.g. Kesler et al. 2005; Murphy and Mazzocco 2008b) despite numerical processing deficits, leading to the intriguing possibility that it may be a model of compensatory mechanisms for some children with MLD.

If arithmetic accuracy is an aspect of the Turner syndrome phenotype, is MLD a justifiable feature of the syndrome? Yes, because the rate of poor mathematics *achievement* is four to five times greater among females with Turner syndrome relative to the general population. This group difference is apparent by kindergarten (when 43% vs. 10% of girls with or without Turner syndrome, respectively, met our research criteria for MLD; Mazzocco 2001), it increases by Grade three (when over 75% of girls with Turner syndrome meet these criteria; Murphy et al. 2006) and continues into the later school age years (where the reported difference is 55% vs. 7%; Rovet 1993). Although their mathematics achievement is, on average, two grades below age expectations during the early school years (Rovet et al. 1994), girls with Turner syndrome are not highly represented in special education, which Rovet proposed may be due to apparent strengths in other intellectual and language abilities. Rovet's notion is consistent with the notion of Turner syndrome as a model of compensatory mechanisms overcoming MLD.

Theoretically guided hypotheses about the routes to MLD in Turner syndrome were initially based on a well-established neuropsychological phenotype, the hallmark feature of which is a significant Verbal and Performance IQ score discrepancy (LaHood and Bacon 1985), averaging ~ 12–15 points (range = 0 to 40 points; as reviewed by Mazzocco 2006), in favor of intact verbal skills. Visual-spatial deficits were reported decades ago (e.g. Alexander et al. 1966; Rovet and Netley 1982; Shaffer 1962; Silbert et al. 1977; Waber 1979), which motivated early studies of spatial pathways to MLD in girls with Turner syndrome (Mazzocco 1998; Rovet et al. 1994; Temple and Marriott 1998). Despite clear evidence of spatial deficits (e.g. Temple and Carney 1995, 1996) these studies failed to yield significant correlations between spatial and mathematical abilities.

In our own work, many correlations between visuospatial and mathematics scores that did emerge were weaker in girls with Turner syndrome than in fragile X or typically developing comparison groups (Mazzocco et al. 2006), so they did not suggest pathways unique to Turner syndrome. We found inconsistent evidence of selective impairment in areas of math with a strong visuospatial component (Mazzocco 2001). For instance, third-graders with Turner syndrome were as capable as their peers at determining space-relationships (rank ordering items by length or size), but adolescents with Turner syndrome made more alignment errors on paper and pencil calculation tasks (such as '14 + 26 = 166') than their peers (Mazzocco 1998). Mathematics and visual-spatial skills are complex cognitive constructs, and their relation to each other may not be captured by broad standardized assessment tools. Perhaps such an association simply does not exist in girls with Turner syndrome, or it may be evident only when different or more fine-grained spatial skills are investigated (e.g. see Simon et al. 2008).

The evidence thus far has not provided unequivocal support for a visuospatial subtype of MLD in Turner syndrome. Moreover, the observed 'spatial errors' reported for girls with Turner syndrome may reflect weak executive function skills (Buchanan et al. 1998), which is another area in which impairments have been reported in the Turner syndrome cognitive phenotype, and which contribute to the procedural subtype of MLD. Rovet and colleagues' findings (1994) of frequently incomplete arithmetic procedures in girls with Turner syndrome implicated a role of executive functions.

The procedural MLD subtype is particularly evident on complex math problems, which may be more susceptible to poor working memory and other executive dysfunctions. When weak calculation skills are observed in studies of girls with Turner syndrome (relative to peers) these are most apparent on multi-digit versus single-digit arithmetic. This is aligned with the Turner syndrome neuropsychological phenotype that includes selective dysfunction in cognitive flexibility despite intact planning abilities (Temple et al. 1996). Cognitive flexibility and working memory limitations are also suggested by performance on the Stroop or Contingency Naming Test (CNT), both of which involve increasing working memory demands across trials. Even tasks with quite limited demands appear to lead to slower processing speed. For instance, on the CNT, a task that involves naming the color or shape of each stimulus while switching between two naming rules, third-graders with

Turner syndrome were less efficient than their peers as indicated by *slower* performance on trials that did not involve working memory demands, by *more errors* on the most demanding trials (which involved switching between three naming rules), and by the ratio of speed and accuracy on a trial of intermediate difficulty (which involved switching between two naming rules; Kirk et al. 2005).

However, a hallmark feature of procedural MLD is delayed acquisition of mathematics, rather than persistent difficulties over time. This is not consistent with the Turner syndrome phenotype. Through longitudinal studies we have shown that poor mathematics achievement in girls with Turner syndrome is evident in kindergarten (Mazzocco 2001), and continues throughout primary (Mazzocco et al. 2006; Murphy et al. 2006), and middle school (Mazzocco and Hanich 2010). Cross-sectional studies reveal math difficulties in adolescents (Mazzocco 1998; Kesler et al. 2005) and women with Turner syndrome (Bruandet et al. 2004; Molko et al. 2003). Thus, while aligned with the notion of executive functions influencing mathematics difficulties, the growth trajectories observed in Turner syndrome are not aligned with the procedural MLD subtype. Moreover, neuroimaging data are inconclusive on this point.

Findings from neuroanatomical, functional imaging, and ocular motor studies of Turner syndrome are mixed with regard to evidence of prefrontal involvement. For instance, functional imaging studies of performance on a visual working memory task (Haberecht et al. 2001) failed to reveal any activation deficits in the superior parietal region (which is believed to be involved with spatial location)—an unanticipated finding in view of reported behavioral deficits in visual working memory (as described earlier)—leading the authors to conclude that girls with Turner syndrome may have impairments in executive functions and storage/retrieval abilities. Yet findings from ocular motor studies reveal slower performance during visually-guided saccade tasks (used to measure reflective eye movements towards a rapidly-appearing stimulus) in girls with Turner syndrome relative to their peers, a performance profile associated with right posterior parietal dysfunction (Lasker et al. 2007). These are just a few of the competing pieces of evidence that suggest an MLD phenotype for Turner syndrome that is not wholly aligned with either the proposed visuospatial or procedural MLD subtype, despite impairments that may cross the boundaries of these proposed subtypes.

Geary's Semantic Memory subtype is not a likely aspect of the Turner syndrome phenotype, in view of intact to superior oral reading, visual reading, vocabulary, and decoding skills (Mazzocco 2001; Rovet 1993), despite *slow* addition fact responding (Temple and Marriott 1998) in this population. Strengths in reading and writing numerals and in accurate magnitude judgments in childhood (see Table 37.2), adolescence (Simon et al. 2008), and adulthood (Bruandet et al. 2004) further counter this notion. Slow fact retrieval and enumeration are aligned with the semantic memory subtype, but good accuracy on these tasks is not. The slower addition response time is noteworthy, because response speed is not slowed for all timed tasks measured in this population.

Non-cognitive contributors to the Turner syndrome MLD phenotype have also been explored. In our study of adolescent girls with Turner syndrome (and fragile X syndrome), we measured indicators of anxiety during mental arithmetic, risk taking, and divided attention tasks. Whereas girls with fragile X had heightened measure of arousal at baseline periods only, girls with Turner syndrome had the highest levels of arousal on all three tasks relative to both the fragile X and comparison groups—potentially because all three tasks were timed tasks—and not on arithmetic tasks specifically (Keysor et al. 2002). There was

Table 37.2 Percentage of children performing accurately on number knowledge items, among girls with Turner syndrome and children without Turner syndrome (with or without MLD).

	Participant Group			
Number Knowledge items	Girls with Turner Syndrome		Children with MLD	Children without MLD
(a) Primary school measures	Grade K n = 18	Grades K-2 n = 28	Grade K n = 23	Grades K-2 n = 226
Rote items				
Read one-digit numbers aloud	-	89	-	92
Read two-digit numbers aloud	56[a]	69	20	59
Write dictated one-digit numbers	-	79	39	79
Count aloud to 41 (not sets)	-	58	18	52
Count aloud by tens (not sets)	-	54	30	59
What one-digit number comes next?	67	86[b]	40	96
Use 1:1 correspondence when counting, examiner pointing	-	89[b]	96	99
Conceptual items				
Verbal magnitude comparisons	67[a]	79	30	81
Identify *n*th position in a set	33	-	0	-
Use 1:1 correspondence counting scattered arrays	-	70	-	85
Number constancy	-	86	-	90
		At Grades 6/7		
(b) Middle school measures		n = 13	n = 22	n = 169
Rote items				
Read decimals aloud		62	12	77
Rank order 10 area model fractions		62	35	76
Conceptual items				
Rank order 10 decimal values		31[b]	6	58
Rank order decimals/fractions		23[b]	0	41

[a] Significantly *better* than MLD group and (nsd with the non-MLD group).

[b] Significantly *worse* than non-MLD group.

Adapted from data reported by Murphy et al. 2006 (for grades K-2) and Murphy and Mazzocco 2008a (Grade 6 and 7), and unpublished Turner syndrome data.

no significant correlation between performance and arousal levels in the Turner syndrome group (Roberts et al. 2008).

The Turner syndrome phenotype reveals how a subtype-fitting exercise may pose challenges. It is tempting to consider that the Turner syndrome phenotype may model *more than one subtype*, depending on the type of mathematics with which girls are engaged; or that it represents an idiosyncratic sequence of mathematics-related development not captured by any of the proposed subtypes to date. The latter notion is implicated by imaging and behavioral data, such as a study using a two- versus three-operand arithmetic verification task. In this study, girls with Turner syndrome were as accurate as their typically developing peers on both types of problems. But the peer group showed greater prefrontal activity during three-operand vs. two-operand problems, whereas girls with Turner syndrome showed a shift from prefrontal to greater temporal activation during three- vs. two-operand problems (Kesler et al. 2005). The greater temporal lobe activation implicates recruiting verbal skills on moderately difficult tasks (which is a reasonable strategy for individuals characterized by superior verbal abilities). Diminished activation in posterior visuo-spatial networks also suggests different routes to complex problem solving during the more complex three-operand task in girls with Turner syndrome, relative to their peers.

The Turner syndrome MLD phenotype illustrates how subtype classifications can be inconclusive despite initial hypotheses based on strong but surface-level information. This, no doubt, occurs in classrooms throughout the world, when children's mathematics difficulties are attributed to 'obvious' associated deficits without benefit of differential diagnosis. Similarly, the Turner syndrome phenotype illustrates how problematic mathematics can be even when calculation accuracy is high, as is the effort with which that accuracy is achieved. Ironically, children whose difficulty is reflected by the long amount of time needed to accurately complete mathematics work may be viewed as putting forward insufficient effort, on the basis of completing less work in a prescribed allotted time, when in fact much effort is needed to maintain persistence in such cases, regardless of whether that persistence leads to task success.

The issue that remains is how to account for the source of difficulty—in this case, slowed and labored responses—in the Turner MLD phenotype. Evidence pointing to weak enumeration skill is remarkable in view of the consistent evidence of good number knowledge in young and older girls with Turner syndrome (Murphy et al. 2006; Temple and Sherwood 2002), reported earlier (Table 37.2). Moreover, Molko and colleagues (2003) found that the approximate and exact numerical tasks that activated different brain regions in typical adults did not lead to distinct activation patterns in women with Turner syndrome. Likewise, Bruandet et al. (2004) reported that women with Turner syndrome count, rather than subitize, very small exact sets. Both of these studies suggest a fundamental difference in rudimentary nonverbal number sense in females with Turner syndrome, and thus provide a foundation on which to explore potential models of compensatory skills that may account for why girls with Turner syndrome perform so accurately on number and arithmetic tasks. In our lab, we have found that girls with Turner syndrome show more accurate—albeit slower—numerical decomposition when working memory demands were imposed, compared to their own performance when no such demands were imposed. In contrast, children without Turner syndrome—regardless of whether they have MLD or typical mathematics achievement—show what we presumed to be the expected decrease in accuracy once such working memory demands are increased. We speculate that one consequence of deliberately slowing down performance (by demanding more of participants) prompted girls with Turner syndrome to rely on alternative (perhaps verbal mediation) skills that led to their performance improvement (defined, again, by accuracy and not speed), which is in stark contrast to the decreased accuracy seen in girls with fragile X when working memory demands increase (e.g., on three- vs. two-operand arithmetic tasks). Accepting that this is speculation, it is nonetheless an example of two very different outcomes to similar circumstances across two groups of individuals whose MLD is associated with very distinct genetic syndromes.

Conclusions

On the surface, one might be misled by the similarities between MLD seen in fragile X and Turner syndromes, on the basis of good number knowledge observed in both phenotypes, at least in early childhood, coupled with poor achievement on standardized school performance. But within and across subsets of rote and conceptual items, performance is more consistent among girls with Turner syndrome (Table 37.2), compared to girls with fragile X syndrome, who show a marked discrepancy across items (Table 37.1). Compared to children with MLD, a higher percentage of girls with fragile X achieve accuracy on the rote items—a pattern not observed in the Turner syndrome group—and yet significantly fewer girls with fragile X correctly solve conceptual items compared to girls with Turner syndrome or even children with MLD.

MLDs in children with fragile X meet the criteria of severity, but not of biologically primary numerical deficits (based on currently available data). Their pattern of relative difficulty across early mathematics skills differs from that observed in typically achieving kindergartners, and from kindergartners known to meet MLD criteria throughout primary school (in a retrospective analysis) or even from children with Turner syndrome: Whereas the frequency of performance accuracy is comparable for number constancy and reading numerals in these three groups, there is a large discrepancy in the frequency with which children with fragile X perform either of these types of item accurately. Strong verbal memory may facilitate performance, but not necessarily knowledge acquisition.

MLDs in girls with Turner syndrome meet the criteria of persistence, and there is some evidence of primary numerical deficits in this population, yet the criteria for severity in standardized test performance is not necessarily achieved or maintained. Despite deficits in several areas linked to poor mathematics performance—poor visual short-term memory, executive dysfunction, slower processing speed, and spatial attention deficits—girls with Turner syndrome's most consistent MLD profile reflects slow but accurate performance on simple arithmetic calculations, and remarkable enumeration deficits in adulthood despite much evidence of number knowledge from early childhood. Verbal skills may provide more than a mechanism to improve performance in this group—verbal skills may be recruited to engage in active problem solving—which is a testable hypothesis worthy of exploration.

This chapter is limited to just two genetic syndromes, and yet this limited comparison illustrates how syndrome research collectively reveals that MLDs vary. Phenotypes that share a heightened risk for MLD may not necessarily share common pathways. At issue then is when and how distinct pathways inform efforts to identify MLD and remediate or prevent the poor mathematics achievement that is a likely outcome of MLD.

References

Alexander, D., Ehrhardt, A.A., and Money, J. (1966). Defective figure drawing, geometric and human, in Turner's syndrome. *Journal of Nervous and Mental Disease* 142(2): 161–167.

Bruandet, M., Molko, N., Cohen, L., and Dehaene, S. (2004). A cognitive characterization of dyscalculia in Turner syndrome. *Neuropsychologia* 42: 288–298.

Buchanan, L., Pavlovic, J., and Rovet, J. (1998). A reexamination of the visuospatial deficit in Turner syndrome: Contributions of working memory. *Developmental Neuropsychology* 14: 341–367.

Butterworth, B. (2005). Developmental dyscalculia. In: J. Campbell (ed.), *Handbook of Mathematical Cognition*, pp. 455–467. Hove and New York: Psychology Press.

Cirino, P.T. (2011). The interrelationships of mathematical precursors in kindergarten. *Journal of experimental child psychology 108*(4): 713–733.

Cornish, K.M., Levitas, A., and Sudhalter, V. (2007). Fragile X syndrome: The journey from genes to behavior. In: M.M.M. Mazzocco and J.L. Ross (eds), *Neurogenetic developmental disorders: Manifestations and identification in childhood*, pp. 73–103. Cambridge, MA: MIT Press.

Cornish, K.M., Turk, J., Wilding, J., Sudhalter, V., Munir, F., Kooy, F., et al. (2004). Annotation: deconstructing the attention deficit in fragile X syndrome: a developmental neuropsychological approach. *Journal of Child Psychology and Psychiatry* 45(6): 1042–1053.

Crawford, D.C., Acuña, J.M., and Sherman, S.L. (2001). FMR1 and the fragile X syndrome: human genome epidemiology review. *Genetics in Medicine* 3(5): 359–371.

Davenport, M., Hooper, S., and Zeger, M. (2007). Turner syndrome throughout childhood. In: M.M.M. Mazzocco and J.L. Ross (eds), Neurogenetic developmental disorders: *Manifestations and identification in childhood*, pp. 3 – 45. Cambridge, MA: MIT Press.

Dennis, M., Landry, S.H., Barnes, M., and Fletcher, J.M. (2006). A model of neurocognitive function in spina bifida over the life span. *Journal of the International Neuropsychological Society* 12(02): 285–296.

Dehaene, S, Spelke, E, Pinel, P, et al. (1999). Sources of mathematical thinking: Behavioral and brain-imaging evidence. *Science* 284: 970–974.

Dowker, A. (2008). Individual differences in numerical abilities in preschoolers. *Developmental Science* 11(5): 650–654.

Fuchs, L.S., Geary, D.C., Compton, D.L., Fuchs, D., Hamlett, C.L., and Bryant, J.D. (2010). The contributions of numerosity and domain-general abilities to school readiness. *Child Development* 81(5): 1520–1533.

Geary D.C. (1993). Mathematical disabilities: Cognitive, neuropsychological, and genetic components. *Psychological Bulletin* 114: 345–362.

Geary D.C., and Hoard M.K. (2005). Learning disabilities in arithmetic and mathematics: theoretical and empirical perspectives. In: J.I.D. Campbell (ed.), *Handbook of Mathematical Cognition*, pp. 253–267. New York, NY: Psychology Press.

Geary, D.C., Hamson, C.O., and Hoard, M.K. (2000). Numerical and arithmetical cognition: A longitudinal study of process and concept deficits in children with learning disability. *Journal of Experimental Child Psychology* 77(3): 236–263.

Geary, D.C., Hoard, M.K., Byrd-Craven, J., and DeSoto, M.C. (2004). Strategy choices in simple and complex addition: Contributions of working memory and counting knowledge for children with mathematical disability. *Journal of Experimental Child Psychology* 88(2): 121–151.

Geary, D.C., Hoard, M.K., Nugent, L., and Byrd-Craven, J. (2007). Strategy use, long term memory, and working memory capacity. In: D. Berch and M.M.M. Mazzocco (Eds), *Why Is Math So Hard for Some Children? The Nature and Origins of Children's Mathematical Learning Difficulties and Disabilities*, pp. 83–105. Baltimore, MD: Paul H. Brookes Publishing Co.

Geary, D.C., Hoard, M.K., Nugent, L., and Bailey, D.H. (2013). Adolescents' functional numeracy is predicted by their school entry number system knowledge. *PLoS One* 8(1): e54651.

Haberecht, M.F., Menon, V., Warsofsky, I.S., et al. (2001). Functional neuroanatomy of visuo-spatial working memory in Turner syndrome. *Human Brain Mapping* 14: 96–107.

Hannula, M.M., Lepola, J., and Lehtinen, E. (2010). Spontaneous focusing on numerosity as a domain-specific predictor of arithmetical skills. *Journal of Experimental Child Psychology* 107(4): 394–406.

Jordan, N.C., Hanich, L.B., and Kaplan, D. (2003). A longitudinal study of mathematical competencies in children with specific mathematics difficulties versus children with comorbid mathematics and reading difficulties. *Child Development* 74(3): 834–850.

Kaufmann, L., Mazzocco, M.M., Dowker, A., von Aster, M., Göbel, S.M., Grabner, R. H., and Nuerk, H.C. (2013). Dyscalculia from a developmental and differential perspective. *Frontiers in Psychology* 4. doi: 10.3389/fpsycg.2013.00516

Kesler, S.R., Menon, V., Reiss, A.L. (2005). Neurofunctional differences associated with arithmetic processing in Turner syndrome. *Cerebral Cortex*16: 849–856.

Keysor C.S., Mazzocco, M.M.M., McLeod, D., and Hoehn-Saric, R. (2002). Physiological arousal in females with fragile X or Turner syndrome. *Developmental Psychobiology* 1: 133–146.

Kirk J., Mazzocco M.M.M., and Kover S.T. (2005). Assessing executive dysfunction in girls with fragile X or Turner syndrome using the Contingency Naming Test. *Developmental Neuropsychology* 28: 755–777.

Koontz, K.L. and Berch, D.B. (1996). Identifying simple numerical stimuli: Processing inefficiencies exhibited by arithmetic learning disabled children. *Mathematical Cognition* 2: 1–24.

Kwon, H., Menon, V., Eliez, S., Warsofsky, I.S., White, C.D., Dyer-Friedman, J., et al. (2001). Functional neuroanatomy of visuospatial working memory in fragile x syndrome: relation to behavior and molecular measures. *American Journal of Psychiatry* *158*: 1040–1051.

LaHood, B.J. and Bacon, G.E. (1985). Cognitive abilities of adolescent Turner's syndrome patients. *Journal of Adolescent Health Care* 6: 358–364.

Lasker, A.G., Mazzocco, M.M.M., and Zee, D.S. (2007). Ocular motor indicators of executive dysfunction in fragile X and Turner syndromes. *Brain and Cognition* 63: 203–220.

LeFevre, J.A., Fast, L., Skwarchuk, S.L., Smith-Chant, B.L., Bisanz, J., Kamawar, D., and Penner-Wilger, M. (2010). Pathways to mathematics: Longitudinal predictors of performance. *Child Development* 81(6): 1753–1767.

Lyon, M.F. (1972). X-chromosome inactivation and developmental patterns in mammals. *Biological Reviews* *47*: 1–35.

Mazzocco, M.M.M. (1998). A process approach to describing mathematics difficulties in girls with Turner syndrome. *Pediatrics* 102: 492–496.

Mazzocco, M.M.M. (2001). Math learning disability and math LD subtypes: Evidence from studies of Turner syndrome, fragile X syndrome, and Neurofibromatosis type 1. *Journal of Learning Disabilities* 34: 520–533.

Mazzocco, M.M.M. (2006). The cognitive phenotype of Turner syndrome: Specific learning disabilities. *International Congress SERIES* 1298: 83–92.

Mazzocco, M.M.M. (2007). Defining and differentiating mathematical learning disabilities and difficulties. In: D. Berch and M.M.M. Mazzocco (Eds), *Why Is Math So Hard for Some Children? The Nature and Origins of Children's Mathematical Learning Difficulties and Disabilities*, pp. 29–47. Baltimore, MD: Paul H. Brookes Publishing Co.

Mazzocco, M.M.M. and Hanich, L.B. (2010). Math achievement, numerical processing, and executive functions in girls with Turner syndrome: Do girls with Turner syndrome have math learning disability? *Learning and Individual Differences* 20: 70–81.

Mazzocco, M.M.M. and Kelley, R.I. (2001). Preliminary evidence for a cognitive phenotype in Barth syndrome. *American Journal of Medical Genetics* 102: 372–378.

Mazzocco, M.M.M., and Myers, G.F. (2003). Complexities in identifying and defining mathematics learning disability in the primary school age years. *Annals of Dyslexia* 53: 218–253.

Mazzocco, M.M.M. and Thompson, R.E. (2005). Kindergarten predictors of math learning disability. *Learning Disabilities Research and Practice* 20: 142–155.

Mazzocco, M.M.M., Singh Bhatia, N., and Lesniak-Karpiak, K. (2006). Visuospatial skills and their association with math performance in girls with fragile X or Turner Syndrome. *Child Neuropsychology* 12: 87–110.

Mazzocco, M.M.M., Feigenson L., and Halberda, J. (2011). Impaired acuity of the approximate number system underlies mathematical learning disability. *Child Development* 82: 1224–1237.

Mazzocco, M.M., Murphy, M.M., Brown, E.C., Rinne, L., and Herold, K.H. (2013a). Persistent consequences of atypical early number concepts. *Frontiers in Psychology* 4. doi: 10.3389/fpsyg.2013.00486

Mazzocco, M.M.M., Myers, G.F., Lewis, K.E., Hanich, L.B., and Murphy, M.M. (2013b). Limited knowledge of fraction representations differentiates middle school students with mathematics learning disability (dyscalculia) vs. low mathematics achievement. *Journal of Experimental Child Psychology* 115: 371–387.

Molko, N., Cachia, A., Rivière, D., Mangin, J.F., Bruandet, M., Le Bihan, D., et al. (2003). Functional and structural alterations of the intraparietal sulcus in a developmental dyscalculia of genetic origin. *Neuron* 40: 847–858.

Morgan, P.L., Farkas, G., and Wu, Q. (2009). Five-Year Growth Trajectories of Kindergarten Children With Learning Difficulties in Mathematics. *Journal of Learning Disabilities* 42(4): 306–321.

Murphy, M.M. and Mazzocco, M.M.M. (2008a). Rote numeric skills may mask underlying mathematical disabilities in girls with fragile X syndrome. *Developmental Neuropsychology* 33: 345–364.

Murphy, M.M. and Mazzocco, M.M.M. (2008b). Mathematics learning disability in girls with fragile X or Turner syndrome during late elementary school. *Journal of Learning Disabilities* 41: 29–46.

Murphy, M.M. and Mazzocco, M.M.M. (2009). The trajectory of mathematics skills and working memory thresholds in girls with fragile X syndrome. *Cognitive Development* 24: 430–449.

Murphy, M.M., Mazzocco, M.M.M., Gerner, G., and Henry, A.E. (2006). Mathematics learning disability in girls with Turner Syndrome or fragile X syndrome. *Brain and Cognition* 6: 195–210.

Murphy, M.M., Mazzocco, M.M.M., Hanich, L.B., and Early, M.C. (2007). Cognitive characteristics of children with mathematics learning disability (MLD) vary as a function of the cut-off criterion used to define MLD. *Journal of Learning Disabilities* 40: 458–478.

National Research Council (2001). *Adding it up: Helping children learn mathematics*, J. Kilpatrick, J. Swafford, and B. Findell (eds). Mathematics Learning Study Committee, Center for education, Division of Behavioral and Social Sciences and Education. Washington, D.C.: National Academy Press.

Nielsen, J. and Wohlert, M. (1991). Chromosome abnormalities found among 34,910 newborn children: results from a 13 year incidence study in Arhus, Denmark. *Human Genetics* 87: 81–83.

O'Hearn, K. and Luna, B. (2009). Mathematical skills in Williams syndrome: Insight into the importance of underlying representations. *Developmental Disabilities Research Reviews* 15(1): 11–20.

Passolunghi, M.C. and Cornoldi, C. (2008). Working memory failures in children with arithmetical difficulties. *Child Neuropsychology* 14(5): 387–400.

Piazza, M., Facoetti, A., Trussardi, A.N., Berteletti, I., Conte, S., Lucangeli, D., et al. (2010). Developmental trajectory of number acuity reveals a severe impairment in developmental dyscalculia. *Cognition 116*(1): 33 – 41.

Raches, D. and Mazzocco, M.M.M. (2012) Emergence and trajectory of mathematical difficulties in young children with Barth syndrome. *Journal of Developmental and Behavioral Pediatrics* 33: 328–335.

Rivera, S.M., Menon, V., White, C.D., Glaser, B., and Reiss, A.L. (2002). Functional brain activation during arithmetic processing in females with fragile X Syndrome is related to FMR1 protein expression. *Human Brain Mapping* 16: 206–218.

Roberts, J., Mazzocco, M.M.M., Murphy, M.M., and Hoehn-Saric, R. (2008). Arousal Modulation in Females with Fragile X or Turner Syndrome. *Journal of Autism and Developmental Disorders* 38: 20–27.

Ross, J.L., Roeltgen, D., Kushner, H., Wei, F., and Zinn, A.R. (2000). The Turner syndrome-associated neurocognitive phenotype maps to distal Xp. *American Journal of Human Genetics* 67: 672–681.

Rourke, B.P. (1993). Arithmetic Disabilities, Specific and Otherwise A Neuropsychological Perspective. *Journal of Learning Disabilities* 26(4): 214–226.

Rousselle, L. and Noël, M.P. (2007). Basic numerical skills in children with mathematics learning disabilities: A comparison of symbolic vs non-symbolic number magnitude processing. *Cognition* 102(3): 361–395.

Rovet, J.F. (1993). The psychoeducational characteristics of children with Turner syndrome. *Journal of Learning Disabilities* 26: 333–341.

Rovet, J. and Netley, C. (1982). Processing deficits in Turner's syndrome. *Developmental Psychology* 18: 77–94.

Rovet, J.F., Szekely, C., and Hockenberry, M.N. (1994). Specific arithmetic calculation deficits in children with Turner syndrome. *Journal of Clinical Experimental Neuropsychology* 16(6): 820–839.

Rubinsten, O. and Henik, A. (2009). Developmental dyscalculia: heterogeneity might not mean different mechanisms. *Trends in Cognitive Sciences* 13(2): 92–99.

Shaffer, J.W. (1962). A specific cognitive deficit observed in gonadal aplasia (Turner's syndrome). *Journal of Clinical Psychology* 18: 403–406.

Siegler, R.S. and Opfer, J.E. (2003). The development of numerical estimation: evidence for multiple representations of numerical quantity. *Psychological Science* 14(3): 237–243.

Silbert, A., Wolff, P.H., and Lilienthal, J. (1977). Spatial and temporal processing in patients with Turner syndrome. *Behavioral Genetics* 7: 11–21.

Simon, T.J., Takarae, Y., DeBoer, T., McDonald-McGinn, D.M., Zackai, E.H., and Ross, J.L. (2008). Overlapping numerical cognitive impairments in children with chromosome 22q11.2 deletion or Turner syndromes. *Neuropsychologia* 46: 82–94.

Strang, J. D., and Rourke, B. P. (1985). Arithmetic disability subtypes: The neuropsychological significance of specific arithmetical impairment in childhood. In: B. P. Rourke (ed.), *Neuropsychology of Learning Disabilities: Essentials of Subtype Analysis*, pp. 167–183, New York, NY: Guilford Press.

Temple, C.M. and Carney, R.A. (1995). Patterns of spatial functioning in Turner's syndrome. *Cortex* 31: 109–118.

Temple, C.M. and Carney, R.A. (1996). Reading skills in children with Turner's syndrome: An analysis of hyperlexia. *Cortex* 32: 335–345.

Temple C.M., Carney, R.A., and Mullarkey, S. (1996). Frontal lobe function and executive skills in children with Turner's syndrome. *Developmental Neuropsychology* 12: 343–363.

Temple, C.M. and Marriott, A.J. (1998). Arithmetical ability and disability in Turner's Syndrome: A cognitive neuropsychological analysis. *Developmental Neuropsychology* 14: 47–67.

Temple, C.M. and Sherwood, S. (2002). Representation and retrieval of arithmetial facts: Developmental difficulties. *The Quarterly Journal of Experimental Psychology: Human Experimental Psychology* 55(3): 733–752.

Verkerk, A.J., Pieretti, M., Sutcliffe, J.S., Fu, Y.H., Kuhl, D.P., Pizzuti, A., et al. (1991). Identification of a gene (FMR-1) containing a CGG repeat coincident with a breakpoint cluster region exhibiting length variation in fragile X syndrome. *Cell* 65: 905–914.

Waber, D.P. (1979). Neuropsychological aspects of Turner's syndrome. *Developmental Medicine and Child Neurology* 21: 58–70.

CHAPTER 38

NUMBER PROCESSING AND ARITHMETIC IN CHILDREN AND ADULTS WITH READING DIFFICULTIES

SILKE M. GÖBEL

INTRODUCTION

LITERACY and numeracy are key tools in the modern world. Both skills are cultural achievements yet neither develops on its own without extensive instruction. Literacy and numeracy play a pivotal role in education from primary school onwards well into higher and further education. Levels of literacy and numeracy within a population are related to economic growth, employment, prosperity, and adult well-being (OECD 2010) and poor literacy or numeracy are associated with higher rates of unemployment and poverty (Carpentieri 2008).

A major aim of successive governments in the developed world had been to boost literacy. In contrast, problems associated with numeracy have only come into focus in recent years. There are several reasons why individuals may struggle with numeracy *and* literacy: (1) they suffer from a reading difficulty and a separate math difficulty; (2) their difficulties with number processing and arithmetic could be a consequence of having a reading difficulty; (3) their difficulties with reading could be a consequence of having difficulties with numeracy; (4) reading and math difficulties follow from shared environmental risk factors including poor education; or (5) reading and math difficulties are a consequence of common genetic risk factors. The pattern of number and math difficulties displayed in those five cases might be different.

In this chapter I will review the literature on number processing and arithmetic difficulties in individuals with reading difficulties. First, reading difficulties will be defined and theories of reading difficulties will be presented briefly. Although reading comprehension difficulties will be mentioned, the main focus of this chapter will be on difficulties with decoding and/or spelling. Number processing and arithmetic difficulties in reading difficulties will be described in detail, followed by a discussion of potential cognitive and biological risk factors. Towards the end of the chapter the issue of co-morbidity will be described and a brief summary on implications for interventions will be given.

Reading Difficulties

According to the Diagnostic and Statistical Manual of Mental Disorders (DSM-IV; American Psychiatric Association 2004) individuals are affected by a reading disorder (RD) if their 'reading achievement, as measured by individually administered standardized tests of reading accuracy or comprehension, is substantially below that expected given the person's chronological age, measured intelligence, and age-appropriate education.' Reading disorder (sometimes called dyslexia) is a specific learning disorder that affects the development of reading and spelling. Children with RD have problems with decoding printed words and difficulties with word identification (Vellutino et al. 2004). Reading disorder is a disorder that, even if compensated, persists into adulthood with a relatively stable cognitive profile across the life-span (Bruck 1990). In recent years it has become clear that it is important to distinguish between difficulties in reading accuracy/fluency (RD, dyslexia) and in reading comprehension (reading comprehension impairment). Difficulties in decoding and in reading comprehension can occur independently of each other.

Children with a specific reading comprehension impairment (poor comprehenders) can read accurately and with fluency but show great difficulties in understanding the meaning of what they have read (Hulme and Snowling 2011; Cain 2010 for review). Despite poor oral language abilities and in clear contrast to poor decoders their phonological abilities are intact (Cain et al. 2000). To date research on the mathematical profile of poor comprehenders is surprisingly scarce and findings are inconsistent. However, there is evidence that mathematical difficulties associated with poor comprehension dissociate from those associated with RD. Pimperton and Nation (2010), for example, found age-appropriate arithmetic performance in poor comprehenders. In contrast, their mathematical reasoning, which places higher demands on verbal ability, semantic processing, and linguistic comprehension abilities, was impaired.

The focus of this chapter is on mathematical difficulties associated with poor reading accuracy/fluency (RD). Most children with RD show deficits in phonological awareness (segmentation and manipulation of speech sounds) and limitations of verbal short-term memory, lending strong support to the phonological processing account of RD. This account suggests that the cause of RD lies in poorly specified phonological representations hindering persons with RD to accurately map sound to meaning (Hulme and Snowling 2009 for a review).

Other theories of RD propose that the disorder arises at the level of sensory processing (Ramus 2004). Proponents of the rapid auditory processing theory view the phonological deficit as a consequence of a more basic auditory deficit: an impairment in the perception of short or rapidly varying sounds (e.g. Tallal 1980). The magnocellular deficit theory focuses on visual disturbance that affects the processing of text and proposes that in individuals with RD an impairment of the magnocellular pathway leads to visual processing difficulties

(Stein and Walsh 1997). The cerebellar theory of RD (Nicolson et al. 2001) proposes a deficit in the cerebellum as the cause of a reading deficit. A mild deficit in the functioning of the cerebellum could lead to difficulties in speech articulation (due to its involvement in motor control) and to difficulties in automatization of reading (due to its involvement in the automatization of overlearned tasks). It has been shown that in a group of adults with RD all participants showed deficits in phonological processing, but only small subsets of the adults showed deficits that would be predicted by the rapid auditory processing theory, the visual theories and the cerebellar theory (Ramus et al. 2003).

Task Demands of Learning to be Numerate

As a group, children with RD show poorer performance on mathematical tasks than expected given their cognitive abilities (Steeves 1983). Numeracy encompasses many different skills, e.g. non-symbolic and symbolic number comparison, counting, estimating, addition, subtraction, multiplication, and division. Number processing starts well before school entry (see McCrink & Birdsall; Hannula-Sormunen; Sarnecka; Gilmore; this volume). Preverbal infants can discriminate between two sets of objects if the ratio between the numbers in those sets is 1:2 (Lipton and Spelke 2003). This nonverbal sense of numerosity, the approximate number sense (ANS), might provide an early foundation for later arithmetic skills (Inglis et al 2011). Counting is acquired by most children before school entry. It typically takes children many months to move from beginning to utter count words to using them in a way that shows they have truly understood all principles of counting (Gelman and Gallistel 1978). Reading and writing Arabic digits is typically acquired next. Initially addition and subtraction tasks are solved with the help of counting strategies (e.g. counting on fingers). Over the course of development those strategies are replaced by fact retrieval, i.e. retrieving known answers (facts) from long-term memory. Procedural strategies are used for larger problems, i.e. the answer is calculated by decomposing the problem into smaller problems for which the answers can be retrieved and combined (De Smedt and Boets 2010). For multi-digit calculations understanding of the place-value system is essential. Place-value understanding is the knowledge that each digit in a multi-digit sequence depends on its location in the sequence and the total length of the sequence.

Number Processing and Arithmetic Difficulties in RD

According to educators, fact retrieval, counting, place-value understanding, and applied problem-solving are some of the most common mathematical weaknesses of children with RD (Simmons and Singleton 2009). First the behavioral deficits in number processing and arithmetic associated with RD will be described, then potential cognitive risk factors for arithmetic deficits will be discussed followed by a review of the literature on neural and genetic underpinnings of arithmetic difficulties in RD (Morton and Frith 1995).

Behavioral Level: Task-Specific Difficulties in Individuals With RD

Miles et al. (2001) compared the performance of ten-year-old children with RD to control children on a mathematical test. Test items covered a large range of topics (e.g. basic number and magnitude comparisons, geometry, single and double digit arithmetic, ratios, probability). Children with RD showed generally weaker performance in the mathematics test than controls. Importantly, however, only about six per cent of the variance in their mathematics test scores could be explained by reading comprehension scores. This is important for two reasons: first it points towards a dissociation between the effects of reading comprehension and decoding difficulties on mathematics performance. In addition, it shows that the difficulties of the RD group were not merely a consequence of difficulties with the actual reading and understanding of the questions. Some test items were much more difficult than others for the RD group, suggesting that the mathematical difficulties associated with RD might be quite specific.

Specific Difficulties of Children with RD

Fact retrieval. Most studies of RD report problems with fact retrieval (e.g. Steeves 1983). Turner Ellis et al. (1996) investigated multiplication performance in three groups of boys with at least average cognitive skills. Boys in the RD group had a spelling age of at least eighteen months below their chronological age, and at least four out of ten possible positive indicators on the Bangor Dyslexia Test (Miles 1997). Boys with RD took significantly longer than typically developing age-matched boys to respond to multiplication items. These differences were not driven by a general difference in response speed because there was no significant difference between the two groups in a simple reaction time task. There were no specific error patterns for children with RD. This is in line with Chinn's (1995) suggestion that errors made by children with RD in an untimed arithmetic test are not significantly different to those made by control children. However, he described one exception: the error of no attempt. Turner Ellis et al. (1996) suggest that children with RD possess a smaller repertoire of immediate number facts than their age-matched peers.

Further evidence for arithmetic difficulties in children with RD comes from Simmons and Singleton (2009). They found a significantly lower accuracy and longer reaction times (RT) on arithmetic facts for ten to eleven-year-old children with RD. This difference stayed significant after general differences in response speed were accounted for.

Studies on university students with RD have shown that these difficulties in fact retrieval persist into adulthood. Simmons and Singleton (2006) found significantly longer RTs for addition and subtraction items for students with RD. Göbel and Snowling (2010) showed that university students with RD were significantly slower in exact addition, but not on an approximate addition task. In contrast to Simmons and Singleton (2006) they found also a deficit in multiplication and no deficit in subtraction. Divergent results for multiplication and subtraction could be due to the low numbers of items per operation tested in both studies.

In summary, there is strong evidence to suggest that fact retrieval is an area of difficulty for children and adults with RD.

Counting. The evidence for a deficit in counting is less clear. Göbel and Snowling (2010) reported that adults with RD made more mistakes in counting than controls. Simmons and Singleton (2009) found that children with RD were significantly slower in counting without memory aides than controls, but the size of this effect was small. When age and RT differences were controlled, differences became non-significant.

Place-value understanding. Hanich et al. (2001) tested seven to eight-year-old children on their understanding of place value. The RD group performed worse than normally achieving children. In contrast Simmons and Singleton (2009) found no significant difference in place value understanding between ten to eleven-year old children with RD and age—and IQ-matched controls. Place-value understanding was strongly correlated with cognitive ability. This might explain the differences between the two studies, as Hanich et al. (2001) did not match groups for IQ. Further, in Hanich et al.'s study children were selected for the RD group if they had a reading composite score below the 35th percentile. This reading composite score included word reading as well as reading comprehension. Simmons and Singleton used a stricter selection criterion (single word reading standard score < 25th percentile). Hanich et al.'s RD group might have contained a mixture of children with RD and poor reading comprehension. Place-value understanding might be related to working memory (Fürst and Hitch 2000) and working memory deficits have been reported for poor comprehenders (Cain et al. 2004). Working memory predicted unique variance in reading comprehension after word reading ability, vocabulary, and verbal ability was controlled for (Cain et al. 2004), suggesting that working memory contributes towards reading comprehension more strongly than towards word reading. It is thus possible that difficulties in place-value understanding are more likely to be associated with poor reading comprehension than with RD.

Applied problem solving. Children with RD have sometimes also been reported to perform poorly in applied problem solving. For example, Vukovic et al. (2010) used an applied problems test with items that involved everyday activities such as counting money and telling the temperature. Children with RD performed significantly more poorly on the applied problems test than controls. When their performance on fact retrieval was controlled the difference between the two groups was still significant, the performance of the RD group, however, was now in the normal range. In this study no measures of cognitive ability were reported.

Methodological issues. In summary, the strongest support for mathematical difficulties in children with RD comes from studies on timed tests of single digit fact retrieval, but many of the studies are hampered by methodological problems. It is difficult to compare across studies because of varying selection criteria for RD. Across studies children with RD have had reading scores below the 35th (e.g. Hanich et al. 2001), 30th (e.g. Cirino et al. 2007), 25th (e.g. Simmons and Singleton 2009), 16th (e.g. Lewis et al. 1994), 15th (Vukovic et al. 2010), 10th (e.g. Boets and De Smedt 2010) or 7th (e.g. Andersson 2010) percentile. The stricter the criterion the more severe the deficit and the more likely it is to see deficits related to RD.

In addition, children have been identified as having RD by various measures: single word reading, spelling, or a composite reading score comprised of decoding and reading comprehension. Because it is possible to have poor comprehension despite good decoding skills, some groups of children with the RD label will include children who are good decoders but poor comprehenders. The underlying cognitive deficits of poor comprehension, however, are thought to be distinct from RD (Cain et al. 2000). Most of the studies reported have not measured phonological skills, thus it is theoretically also possible that only a subset of

RD suffers from phonological forms of RD. A few studies have used items from the Bangor Dyslexia Test (Miles 1997) to select children for the RD group, and to exclude children from the control group who score highly on those items. Those items include measures of left-right distinction, sequencing, working memory, and subtraction and multiplication abilities. Using those additional items might lead to two groups that not only differ on their ability to decode but also on other measures such as working memory or motor coordination. It is difficult to exclude that those factors affected mathematical performance. Clearly, the selection criteria for the control group are vital. Many studies do not report general cognitive abilities. This is unfortunate because IQ and mathematical performance are correlated (Dowker 2005). Only by matching the control and the RD group on cognitive abilities can we be certain that any differences are not driven by differences in IQ.

Pure Reading Difficulties (Pure RD) Versus Reading and Math Difficulties (MD+RD)

Measures of numerical and mathematical performance are typically continuous. Children with mathematical performance scores significantly below average or with scores that are out of line with what is expected given their general cognitive skills might suffer from a separate disorder: a math learning disorder (see Butterworth; Henik; Geary; this volume). In a sample of over 1000 nine to ten-year-olds Lewis et al. (1994) found that 1.3 per cent had math difficulties but no reading difficulties. Math difficulties (MD) can thus appear independently of RD. It is possible that the difficulties in number processing and arithmetic reported in the previous section in individuals with RD are related to the individual suffering from MD, in addition to but independently of RD, because most studies reviewed above have not tested for MD and thus might have included individuals with pure RD as well as individuals with RD *and* MD (MD+RD). Indeed, Lewis et al. (1994) found 3.9 per cent of children had pure reading difficulties (pure RD) while 2.3 per cent of children had reading difficulties *and* mathematical difficulties at the same time suggesting that a high percentage of children with reading difficulties also have mathematical difficulties (MD). So far no agreement on the cardinal features of MD has been reached (Cirino et al. 2007). This might be partially due to the relative independence of the literature on two instances on MD: dyscalculia and math learning difficulties.

Children and adults with dyscalculia have significant impairments in number processing and arithmetic (Landerl et al. 2004, Butterworth 2010), and show a significant discrepancy between their cognitive and arithmetic abilities. Landerl et al. (2004) for example used a speeded arithmetic test and only chose children for the dyscalculia group whose responses were at least three standard deviations slower than the mean of the distribution of the control group's response times (equivalent to scoring below the 0.13th percentile on an ability test). Children with math learning difficulties (MLD) score lower than expected on mathematical achievement tests, but typically have intact basic number processing abilities (Geary 2004). Research on MLD typically uses a lenient selection criterion (often < 35th percentile). Although studies on dyscalculia and MLD both use arithmetic performance as selection criterion, theoretically different underlying cognitive deficits have been proposed: MLD focuses on deficits in fact retrieval and arithmetic (Geary 2004) while for dyscalculia a deficit in the core number sense leading to difficulties in basic number processing is hypothesized (Landerl et al. 2004). The question whether dyscalculia is merely more severe than MLD or qualitatively different has not been fully answered empirically yet.

Based on neuropsychological and neuroimaging findings it has been suggested that numerical processing is independent from reading abilities (Tressoldi et al. 2007). In line with this, the pattern of numerical impairments for children with pure mathematical difficulties (pure MD) and for children with both mathematical and reading difficulties (MD + RD) is more similar than the pattern of numerical impairments in pure RD and MD + RD. Children with MD + RD and children with pure MD show impairments in basic number processing tasks such as number comparison, counting dots and reciting number sequences (Landerl et al. 2004). Several authors (e.g. Jordan et al. 2003, Landerl et al. 2009) have concluded that crucial features of dyscalculia are theoretically independent from RD and that in children with MD + RD the observed deficits in mathematics are solely consequences of their MD. This raises the question whether the evidence for number processing and arithmetic difficulties in children and adults with RD reported above has been driven by the subgroup of participants who suffer from MD in addition to RD.

Research on Pure RD

Recent research has addressed this issue by comparing mathematical performance of children with pure RD to pure MD, MD + RD and control groups. Typically participants are only selected for the pure RD group if their mathematical performance on a standardized test is within the normal range. Consequently any mathematical deficits found in the pure RD group will make a strong point because participants with below average overall mathematical performance have been excluded from this group.

In contrast to participants with pure MD or MD + RD, participants with pure RD show no impairment in most basic number processing tasks (Göbel and Snowling 2010; Landerl et al. 2004). Landerl et al. (2004) found a trend for children with pure RD to be slower at counting.

Turning to the performance of the group with pure RD in arithmetic, Cirino et al. (2007) tested nine to ten-year-old children on computerized arithmetic measures. Controls outperformed children with pure RD on small (e.g. 3 + 5) but not large (e.g. 9 + 4) addition problems. Hanich et al. (2001) asked children to solve simple addition and subtraction items in two different ways. When children could use any method they wanted to solve the items, they found no differences between the group with pure RD and the control group. However, when the children were told to respond immediately (fast retrieval instruction) children with pure RD performed significantly worse than controls. Thus the type of strategy used by children with pure RD might be crucial.

Multiplication items typically elicit more retrieval-based answers and subtraction items lead to more use of procedural strategies (Campbell and Xue 2001). Boets and De Smedt (2010) compared the performance of children on subtraction and multiplication items. Children with pure RD made more mistakes and were slower than controls on both multiplication and subtraction items. However, when digit naming speed was taken into account, there was no longer a group difference in RT for subtraction items. The group difference for multiplication items stayed significant. These findings suggest that children with pure RD either are less efficient at retrieval for multiplication items or use less retrieval than controls.

De Smedt and Boets (2010) further investigated this question by testing university students with pure RD. Strategy use was recorded after each multiplication and subtraction item. Adults with pure RD were significantly slower than controls on both subtraction and multiplication items with a significantly larger difference in multiplication. Analysis of strategy use first validated the assumption that retrieval was used more frequently for multiplication than subtraction items. Adults with pure RD used retrieval less often than controls. Interestingly, adults with pure RD were slower in executing both retrieval and procedural strategies. These studies provide strong evidence that a deficit in arithmetic, particularly with fact retrieval in addition and multiplication, also exists in pure RD groups.

Some studies, however, failed to find any arithmetic deficits in pure RD. Jordan et al. (2003) found no difference between pure RD and control children on forced retrieval of number facts. This null effect could be due to several factors. First, children were selected for the pure RD group if they had a reading composite score below the 35th compared to the 10th percentile used by Boets and De Smedt (2010). Second their reading score is a composite of decoding and comprehension. Consequently the pure RD group might include children with poor reading comprehension but good decoding (and intact phonological skills). Landerl et al. (2004) also found no significant differences between pure RD and control groups on addition, subtraction and multiplication items. This is probably a consequence of their selection criteria. Children were selected for the pure RD group if they scored below the 25th percentile on a word reading task, but they had to be within two standard deviations of the performance of the control group on the arithmetic tasks just described, i.e. they excluded all children with pure RD who had severe difficulties on this task from the pure RD group.

In conclusion there is clear evidence that fact retrieval, in particular in addition and multiplication, is also impaired in children and adults with pure RD.

Cognitive Level: Causal Explanations for Arithmetic Deficits

Several domain-general cognitive deficits underlying fact retrieval difficulties in RD have been proposed, for example difficulties with paired associated learning, a deficit in acquiring automaticity or slow processing speed. Within the domain-specific explanations, the phonological deficit hypothesis has been the most prominent.

Learning the association between a letter and a sound is essential for reading development. It has been suggested that children with RD have difficulties with paired associate learning (Ackerman and Dykman 1995). Forming and remembering an association between a sum and its answer is a form of paired associate learning. However, difficulties with paired associate learning would have a general effect on many aspects of learning and development and not just on arithmetic learning. This is in contrast to the relatively specific nature of the arithmetic deficit seen in RD (Robinson et al. 2002). The cerebellar theory of RD (Nicolson et al. 2001) proposes a failure to automatize at the heart of RD. Children with RD show some difficulties in automatization (Nicolson and Fawcett 1990), e.g. they are often impaired in rapid automatized naming tasks (Willburger et al. 2008). However, these difficulties are relatively specific and in most children with RD they do not extend to many other tasks that also require automatization, e.g. adults with RD have no difficulties with automatic activation of number magnitude (Rubinsten and Henik 2006). In addition performance on rapid automatized naming tasks does not explain unique variance in arithmetic performance (De Smedt and Boets 2010). Bull and Johnston (1997) suggested that number fact deficits might be related to general weaknesses in processing speed. In several studies difficulties with fact retrieval were still significant after controls for slower processing speed were applied (e.g. Simmons and Singleton 2009) and processing speed is not a significant contributor to individual differences in arithmetic for typical developing children (Berg 2008). This suggests that although some participants with RD may have difficulties in domain-general mechanisms such as paired associate learning, the acquisition of automaticity or slow processing speed, those difficulties are unlikely to be the core cognitive deficit underlying fact retrieval deficits in RD.

Phonological Processing Skills and Arithmetic Performance in Typically Developing Children

A deficit in phonological processing skills has been the most commonly proposed domain-specific core deficit underlying fact retrieval difficulties. As reviewed above most children and adults with RD have a phonological deficit. The role of phonological processing in arithmetic development for typically developing children has received limited attention in empirical research. Some researchers have proposed that phonological skills may be vital for the acquisition of number facts (e.g. Simmons and Singleton 2008, De Smedt et al. 2010), because items are often presented verbally and are often rehearsed orally. Individual differences in phonological skills predict later differences in arithmetic skills (Rasmussen and Bisanz 2005; for a review see Simmons and Singleton 2008). Most importantly, this relationship was still significant when individual differences in reading ability were controlled (Hecht et al. 2001; Simmons et al. 2008, Fuchs et al. 2005).

Wagner and Torgesen (1987) delineated three aspects of phonological processing skills that might be differentially related to reading development: phonological awareness, phonetic recoding in working memory, and phonological recoding in lexical access. This taxonomy will be used below to review studies of phonological skills and arithmetic.

Phonological awareness is the sensitivity to the sound structure in speech. This ability can be measured before reading acquisition begins and is typically assessed by tasks in which sounds have to be manipulated (e.g. phoneme deletion, phoneme blending). Landerl et al. (2009) are doubtful about the direct contribution of phonological awareness to arithmetic development, because in their view sub-lexical phonological manipulations as assessed in phonological awareness tasks are not required when solving arithmetic tasks. In a concurrent study of seven to ten-year-old children phoneme deletion was only a weak predictor of arithmetic skill and no longer contributed unique variance in arithmetic skill once verbal and non-verbal ability were controlled (Durand et al. 2005).

However, De Smedt et al. (2010) showed that, for nine to eleven-year-olds, phoneme awareness was a concurrent predictor of arithmetic when reading ability was controlled. Simmons et al. (2008) found that independently of reading, vocabulary and nonverbal reasoning, phonological awareness in five-year-olds was a significant predictor of their arithmetic performance one year later. Jordan et al. (2010) found that children with poor phonology at age five had significant impairments in some aspects of mathematics at age seven compared to control children. Phoneme awareness at age seven to eight was also a significant independent predictor of math ability at age eleven in a longitudinal study by

Hecht et al. (2001). They proposed that phonological awareness could be an indirect predictor of individual differences in mathematical skills because both phonological awareness and arithmetic skills draw on phonological memory and the central executive. However, this hypothesis has not received much empirical support (Rasmussen and Bisanz 2005). For example, a concurrent study in seven-year-old children reported a significant prediction of individual differences in arithmetic performance by phonological awareness over and above contributions of complex working memory tasks (Leather and Henry 1994). Phonological awareness has been conceptualized as an index of the quality of phonological representations in long-term memory (Snowling and Hulme 1994). During arithmetic learning verbal representations of quantities in the form of number words are commonly used and phonologically coded. Degraded phonological codes might thus lead to deficits in fact retrieval. Recent data from De Smedt et al. (2010) support this hypothesis. They report a highly specific concurrent association between phonological awareness and arithmetic that could explain some of the inconsistencies found in the literature. In nine to eleven-year-olds individual differences in phonological awareness were not significantly related to an untimed test of math achievement, but they were significantly related to both accuracy and RT on timed addition, subtraction, and multiplication items if their problem size was small (<=25). For items with large problems size (>25) the predictive power of phoneme awareness with reading ability controlled was not significant. In a further analysis the authors showed that phoneme awareness was a significant predictor of accuracy in retrieval problems, but not in procedural problems. Thus the quality of phonological representations in long-term memory seems to be related to individual differences in performance on those arithmetic items that are predominantly solved by retrieval.

Phonetic recoding in working memory is the ability to code, maintain, and manipulate sound-based representations in working memory, also known as the phonological loop (Baddeley 1986). It is typically assessed by verbal span tasks. Phonological working memory might be important for holding and manipulating information when doing mental arithmetic and therefore might affect the development of number fact knowledge (Landerl et al. 2009). Empirical evidence for a significant contribution of the phonological loop to arithmetic development so far is mixed. In a study by Durand et al. (2005) phonological memory was a weak concurrent predictor of arithmetic skill and no longer accounted for any unique variance once verbal and non-verbal ability were controlled. Gathercole and Baddeley (1993) found no significant predictive relationship between phonological memory measured at age four and arithmetic skills at age eight. Other studies however found a significant relationship between phonological loop functioning and arithmetic performance independent of reading, age, and general cognitive skills (Andersson 2010; Noël et al. 2004).

Results from several studies suggest that the relationship between the phonological loop and arithmetic performance might be age-dependent (Rasmussen and Bisanz 2005). De Smedt et al. (2009) found that phonological loop functioning in 6-year-olds predicts math achievement one year, but not four months later. Hecht et al. (2001) found phonological loop functioning at age seven to eight was a significant independent predictor of growth in mathematical skills to age eight and nine, but did not predict growth occurring between age seven to eight and ten or eleven. Together with findings of no independent contribution of the phonological loop to arithmetic performance in adults (De Smedt and Boets 2010) this suggest a time-limited, i.e. limited to a specific period in development, contribution of phonological loop functioning to arithmetic performance and growth between the age of seven to nine, the time period in which arithmetic facts are most commonly acquired.

Phonological recoding in lexical access is the speed of access to sound-based (segmental) representations in long-term memory and is typically measured by rapid automatized naming (RAN) tasks. This cognitive ability might influence the speed of retrieving phonological representations that are associated with the operators or answers of arithmetic problems from long-term memory. Bull and Johnston (1997) showed that individual differences in a RAN task for letters and digits were correlated with scores on a math achievement test. RAN Digits and Letters measured at age seven to eight was an independent predictor of growth in mathematical skills from the ages of seven to eight and eight to nine, but not from age seven to eight to ages ten or eleven (Hecht et al. 2001). In a study by Temple and Sherwood (2002) children with arithmetic deficits showed slow RAN performance. However, it is unclear whether some of the children in this study were also impaired in reading. Other studies failed to find a significant independent association between speed of lexical access and arithmetic performance (De Smedt and Boets 2010). Willburger et al. (2008) found no general RAN deficit in children with arithmetic deficits.

In summary, the majority of studies show that in typically developing children phonological awareness is related to arithmetic performance. The evidence for phonological loop functioning and speed of lexical access is less clear. Hecht et al. (2001) suggest that phonological loop functioning and speed of lexical access affect early arithmetic development, but later arithmetic development is influenced by central executive resources which are often also required in many tasks of phonological awareness.

Phonological Processing Skills in RD and Arithmetic Performance

Deficits in reading have been associated with all three aspects of phonological skills: phonological awareness (e.g. Griffiths and Snowling 2001), phonological loop functioning (e.g. de Jong 2006), and lexical access (e.g. Wimmer et al. 1998). There are currently only a few published studies investigating the relationship between phonological skills and the arithmetic deficit found in children and adults with RD.

Geary (1993) proposed three subtypes of mathematical difficulties. Of these subtypes, the semantic memory deficit subtype, has been hypothesized to occur often with phonetic forms of RD. Geary suggested two underlying difficulties that could cause fact retrieval deficits in RD. First, their ability to retrieve facts from a semantic-based long-term network could be impaired due to an impairment in phonological awareness or speed of lexical access. Alternatively, the retrieval process could be disturbed by difficulties in inhibiting the retrieval of irrelevant associations. The reduced capacity to inhibit irrelevant information would in turn lead to a reduced working memory capacity and increased error rate and RT (Barrouillet et al. 1997).

Geary et al. (2000)'s findings supported the second hypothesis. Children with RD had indeed difficulties to inhibit irrelevant associations when retrieving facts from long-term memory. Children with pure RD, with pure MD or with MD+RD made significantly more counting-string-associate errors than control children. In this type of error children give an answer that is the next number in the counting series after one of the operands, e.g. stating nine for three plus eight. However, Hanich et al. (2001) did not replicate this finding; in their study counting-string-associate errors were rare for all groups.

Landerl et al. (2009) investigated RAN digit deficits in children with pure RD, pure MD, MD+RD and control children. In their study RAN deficits in children with pure RD co-occurred with intact number fact knowledge. Intact performance on RAN occurred with number fact deficits in children with pure MD. They conclude that their findings do not support the idea that number fact deficits might be associated with difficulties in lexical access. However, they used a speeded arithmetic test for group selection and excluded children performing below the 25th percentile on the arithmetic test from the group with pure RD, i.e. those children who most likely were suffering from a fact retrieval deficit.

De Smedt and Boets (2010) investigated the separate contribution of the three aspects of phonological skills to subtraction and multiplication in adults. For subtraction items only a significant association between phonological awareness and fact retrieval emerged. In multiplication both phonological awareness and phonological loop functioning were related to frequency of fact retrieval, but only phonological awareness predicted unique variance in multiplication fact retrieval when entered after reading ability. The association between phonological awareness and multiplication fact retrieval was significant for both the control group and the RD group and stayed significant for both groups when reading ability was controlled for.

SUMMARY

Research directly investigating the relationship between phonological skills and arithmetic performance is surprisingly scarce. Existing results consistently show the importance of phonological awareness for number fact retrieval. It is possible that phonological awareness tasks tax phonological representations more than the other aspects of phonological skills and that phonological awareness tasks are therefore more sensitive measures of the quality of the underlying phonological representations. Alternatively, the task demands of number fact retrieval might be more similar to those of phonological awareness tasks. The results suggest that the poor quality of the underlying phonological representations impairs arithmetic fact retrieval in RD.

Critics might reject the idea of a number fact retrieval deficit in pure RD, because some participants with pure RD do not show any fact retrieval deficits (e.g. Jordan et al. 2003; Landerl et al. 2004) and if they do then participants with pure RD show typically less severe fact retrieval deficits than those with MD+RD. There are at least three possible explanations for these findings. The first explanation relates to sampling. In many studies individuals are excluded from the pure RD group if their performance on a test of arithmetic fact retrieval falls outside the normal range. By definition, this reduces the probability of finding fact retrieval deficits in the pure RD group. Second, learning to decode alphabetic scripts most certainly requires more sophisticated and higher developed phonological processing skills than those required for learning arithmetic facts (Robinson et al. 2002). It is thus feasible that participants with pure RD with milder deficits in phonological awareness will present with RD but that their deficit in phonological awareness will not be sufficient to impact the development of arithmetic, particularly if a lot of practice is given. Third, if indeed a deficit in phonological awareness is the underlying causal factor behind arithmetic deficits, a subgroup of individuals with RD stemming from problems other than phonological deficits, might be free of arithmetic difficulties.

Biological Level: Neural and Genetic Abnormalities in RD

During the last two decades structural and functional brain abnormalities have been documented in the brains of children and adults with RD (Eckert 2004). Many of these abnormalities could be as easily a consequence of having lived several years with RD as they could be causally related to RD (Castro-Caldas 2004). In any case, brain abnormalities clearly have the potential to affect number processing and arithmetic skills too. It is thus important to review this literature and to discuss potential implications for number processing and arithmetic. To date no studies have compared the neural networks involved in fact retrieval in participants with and without RD. However—in addition to the literature on structural and functional brain abnormalities that might be associated with RD—there is a growing literature on the brain networks involved in calculation (see Arsalidou and Taylor 2011 for review; and see Kaufmann; Menon; Holloway; Kucian; this volume). Thus it is possible to hypothesize about functional links between brain abnormalities in RD and their difficulties in fact retrieval. Theoretical proposals of neurological correlates of problems with fact retrieval in RD can be divided into three groups: proposals based on cognitive research on fact retrieval difficulties, on functional differences in brain activation in RD, and on structural brain differences in RD.

Cognitively Motivated Proposals

Non-verbal learning disability. Rourke (1993) proposed two subtypes of children with learning disabilities from clinical recruitment who showed distinct patterns of deficits as well as dissociable underlying neural bases of the deficit. One group, the non-verbal learning disability group, showed impaired performance in arithmetic but relatively normal reading and spelling and had a higher verbal than performance IQ. Children in this group showed deficits in visual-spatial organization, psychomotor, and tactile perceptual tasks. Those profound difficulties on nonverbal problem solving tasks led Rourke (1989) to propose a right hemisphere dysfunction as a 'final pathway' to non-verbal learning disabilities. Another group was severely impaired in reading and spelling with better arithmetic skills. However, the arithmetic skills were still below age expectation. In those children verbal IQ was typically lower than performance IQ. They had weak verbal and auditory perceptual skills, but showed normal performance on visual-spatial-organizational, psychomotor, and tactile-perceptual tasks. A dysfunction of the left hemisphere systems was proposed in those children. Although this proposal inspired much research recent neuroimaging studies have not backed up the left—vs. right-hemisphere dysfunction dichotomy (Richlan et al. 2011, Isaacs et al. 2001). In addition, behavioral studies (e.g. Dowker 1998) have not backed up the view that there is a sharp distinction between the types of mathematical characteristics and difficulties in children with higher verbal versus higher non-verbal IQ.

Semantic memory deficit. Geary (2004) proposed fact retrieval deficits might be associated with abnormal activations in association areas of the prefrontal cortex that support inhibitory mechanisms and the central executive. Inefficient inhibition results in the activation of irrelevant information that functionally lowers WM capacity. There is some evidence for frontal abnormalities in children with RD (Richlan et al. 2011), however so far functional abnormalities in children with RD during fact retrieval have only been reported in parietal areas (Evans et al., 2014).

Brain Function During Arithmetic in Typical Children and Adults

It is useful to consider functional brain abnormalities in the context of models and findings of typical brain function during number processing and arithmetic. Dehaene (2000) proposed a model of adult number processing and calculation, the Triple Code Model. The analogue magnitude code (or Approximate Number Sense, ANS) deals with non-symbolic stimuli and is involved in comparison, estimation, subitizing, and approximate calculation. As the core number semantics code it is essential for subtraction and division. The visual Arabic number form is activated when reading or writing Arabic numerals and is principally involved with multi-digit operations and parity tasks. The third proposed code, the verbal word frame, is activated by spoken or written number words and is responsible for counting, addition and multiplication tables. Dehaene suggested this code might be drawing upon general purpose modules for language processing. A reading impairment is expected to interfere most with the verbal word frame (Göbel and Snowling 2010).

Dehaene et al. (2003) have proposed three parietal brain circuits for number processing and calculation. In this conceptualization the bilateral horizontal segment of the intraparietal sulci (IPS) houses the core number semantics and is involved in all numerical tasks (comparison, calculation, approximation). The left angular gyrus (ANG) is activated in language-dependent tasks such as multiplication. This might be where rote arithmetic tables are stored in verbal form. The bilateral posterior superior parietal lobe is activated during subtraction and approximation. Based on the Triple Code Model, Arsalidou and Taylor (2011) described two main cortical networks for single digit calculation: a direct and an indirect route. The direct route is used for fact retrieval. It transcodes the operands into a verbal code, this code then elicits the rote memory of the operation and the associated answer. Arsalidou and Taylor suggest a left cortico-subcortical loop as neuroanatomical basis for the direct route. In the indirect route operands activate quantity representations on which calculation procedures will be applied. The inferior parietal and left perisylvian language networks are suggested to be involved in procedural solutions. Single digit addition and multiplication are proposed to mainly use the direct route while subtraction items will be solved through the indirect route.

Both neuroanatomical proposals are supported by findings from Grabner et al. (2009). Adults solving arithmetic problems reported after each trial which strategy they used. On fact retrieval trials the left ANG was more strongly activated than on procedural trials. Procedural strategies were associated with activations in a large frontoparietal network.

Functional Brain Abnormalities in Children and Adults with RD

The angular gyrus has received significant attention because it has traditionally been implied independently in both reading and calculation (e.g. Simmons and Singleton 2008). Consequently it has been suggested as a site of neural overlap between phonological processing and arithmetic fact retrieval (De Smedt and Boets 2010).

Early neurological conceptions saw the ANG as an essential structure for reading competence (Dejerine 1891). Currently the role of ANG in reading is debated (Price 2000). While some studies have reported left ANG activation during silent reading tasks (e.g. Bookheimer et al. 1995, Price et al. 1996), other haves not found any left ANG activation when unrelated words were read aloud (e.g. Rumsey et al. 1997). Two recent meta-analyses found little evidence for a dysfunction of ANG in RD (Richlan et al. 2009, 2011). Price (2000) suggests ANG could be part of a semantic memory system that is activated when word meaning is accessed. In several studies ANG activation was related to phonological awareness tasks (e.g. Gelfand and Bookheimer 2003; Pugh et al. 2001) and the ANG is underactivated in participants with RD during tasks that require phonological manipulations (e.g. Shaywitz et al. 1998). When demands on phonological awareness during print tasks were low, functional connectivity between ANG and occipital and temporal sites were normal in participants with RD (Pugh et al. 2000). In tasks that placed higher demands on phonological awareness the functional connectivity was significantly disrupted.

Lesions including the left ANG often lead to acalculia, an impairment of performing calculations (Cipolotti and van Harskamp 2001). Grabner et al. (2007) tested healthy adults on multiplication. Individuals with higher mathematical performance showed a significantly stronger activation of the left ANG. Hence a deficit in ANG could potentially cause both phonological deficits and calculation difficulties (Rubinsten and Henik 2009) and the fact retrieval deficit in RD might be functionally related to their abnormal patterns of ANG activation during phonological tasks (Simmons and Singleton 2008). To date no imaging studies on calculation of children or adults with RD have tested this hypothesis.

Fact retrieval. De Smedt et al. (2011) investigated brain activation during calculation in children with poor and normal fact retrieval. A frontoparietal network was activated for all children when they were solving subtraction and large addition problems, which are typically solved using procedural strategies. Solving small addition problems is thought to be more retrieval-based than subtraction. However, the left ANG was not more strongly activated during addition than subtraction. Children with poor arithmetic fact retrieval showed a significantly higher activation in the right IPS during problems with relatively small problem size than typical children. Thus in this study there was little support for the role of the left ANG during fact retrieval. Findings from this study are in contrast to findings from adults (Grabner et al. 2007) and suggest a possible difference between the networks underlying fact retrieval in children and adults. However, a recent neuroimaging study (Evans et al., 2014) found significant differences in brain activation during single-digit arithmetic between children with and without RD in an areas close to the left ANG: the left supramarginal gyrus (SMG). Children with RD showed significantly less activation than controls in this region for addition and subtraction. Furthermore, for controls, activation in the right SMG was significantly higher for subtraction than addition. In contrast, the children with RD did not show a significant modulation of right SMG activation by arithmetic operation. These finding provide preliminary evidence that children with RD might activate right parietal regions during single-digit arithmetic to compensate for underactivation of the left parietal areas.

It remains to be seen whether the activation patterns and the functional connectivity of the left ANG during fact retrieval differ between participants with and without RD. The prediction from both behavioral and neuroimaging studies suggest that it will be essential to record the strategies used during calculation because different networks are used for fact retrieval versus calculation.

Structural Brain Abnormalities of RD

Recent investigations into brain abnormalities in RD have produced complex and variable anatomical patterns (for a review see Eckert 2004). Differences in grey matter volume have been reported in many language-related brain structures, for example bilaterally in the planum temporale, inferior temporal cortex, cerebellar nuclei, and left superior and inferior temporal regions (Brambati et al. 2004; Brown et al. 2001). Silani et al. (2005) found reduced grey matter density in the left middle temporal gyrus. Posterior to this region, however, in the left middle temporal gyrus they also found an area with increased grey matter density. The higher the grey matter density in this area the worse the performance of participants with RD during reading tasks.

White matter differences between RD and typical readers have been found in left temporoparietal regions (e.g. Niogi and McCandliss 2006) and the left inferior frontal gyrus (e.g. Rimrodt et al. 2010), however, no agreement has been reached as to exactly from which white matter fiber tract these differences stem. Vandermosten et al. (2012) found decreased white matter in RD in a component of the left arcuate fasciculus, a white matter structure connecting part of the temporoparietal junction with the frontal cortex in the anterior-posterior direction. Decreases were significantly related to deficits in phoneme awareness. While Vandermosten et al. found no differences in white matter structure between adults with RD and good readers in another fiber tract, the left superior corona radiata, other studies did (Klingberg et al. 2000, Niogi and McCandliss 2006). The left superior corona radiata also passes through the temporoparietal region but in the inferior-superior direction. Niogi and McCandliss (2006) found that differences in the white matter microstructure of the superior corona radiata were significantly related to individual differences in word identification skills.

Direct studies of the association between arithmetic and white matter microstructure are currently missing. Van Eimeren et al. (2008) found two white fiber tracts differentially associated with performance of children aged seven to nine on standardized tests of mathematics. Variability in white matter microstructure in the left superior corona radiata was associated with performance on numerical operations and math reasoning, while variability in the left inferior longitudinal fasciculus, connecting temporal and occipital cortex, was only associated to performance on the numerical operations.

Summary

Clearly it is too early to propose any neural mechanisms for the co-morbidity of fact retrieval difficulties and RD. However, there are two candidates for future research: activation differences in the left ANG/SMG and structural differences in the left superior corona radiata. In adults the left ANG is involved in both phonological tasks and calculation. This makes it a clear candidate for the location of a functional abnormality during fact retrieval in adults with RD. Individual differences in white matter microstructure in the left superior corona radiata, connecting temporal and parietal regions, are associated with performance differences in skills underlying mathematics and reading in children. Putting these two candidates together, one might speculate that an abnormality in white matter connections in children with RD during fact learning leads to a lower functional ANG activation in adults with RD during fact retrieval.

It is important to keep in mind that nearly all studies so far have studied adults or school-aged children who already had several years of reading experience. Any of the functional and structural differences discussed between poor and normal readers could be a result of having RD (e.g. Castro-Caldas 2004). Experience and training change white and gray matter structures in children and adults (Eden et al. 2004; Scholz et al. 2009; Zatorre et al. 2012).

Genetics

Behavior genetics research (see Tosto, this volume) suggests that both RD and MD are moderately heritable, i.e. family members of adults or children with RD or MD are more likely to be affected themselves than are members of the general population (Geary 2004; Pennington and Olson 2005 for review). Several twin studies investigated the likelihood of one twin being affected by RD or MD when the other twin was affected by RD or MD, the so-called concordance rate. Monozygotic twins (MZ), who share 100 per cent of their genetic information, had significantly higher concordance rates for RD and MD than dizygotic twins (DZ) who share on average only 50 per cent of their genes (Kovas and Plomin 2007). None of these concordance rates were close to 100 per cent suggesting environmental factors also play a role. Genetic factors might put an individual at a higher risk of developing RD or MD.

Recent studies suggest that learning difficulties in mathematics and reading are caused by the same genetic factors that are associated with individual variation in reading and mathematical ability (Plomin and Kovas 2005). From a genetic perspective RD is the low end of normal variation in the normal distribution of reading ability and MD constitutes the low end of the normal distribution of mathematical ability. This is in line with behavioral findings from the MD and RD literature showing that mild difficulties in reading or mathematics did not appear to show qualitatively different patterns of impairment than severe difficulties (Shaywitz et al. 1992; Jordan and Hanich 2003).

Interestingly, recent studies suggest that RD and MD are also affected largely by the same genetic factors ('Generalist Genes', Kovas and Plomin 2007) and that the influence of non-shared environment is responsible for dissociations between RD and MD. This line of research implies that the causal risk factors, the genetic underpinnings, of developing RD and MD are to a large extent the same. For many recent accounts of RD and MD this hypothesis might be difficult to entertain because—based on the existence of children with RD without mathematics impairment and children with MD without reading impairment—many have argued for an independence of those learning difficulties (e.g. Tressoldi et al. 2007; Landerl et al. 2009). However, statistically those cases would occur even if MD and RD were highly correlated. The question is whether those cases occur more often than expected from a bivariate normal distribution (Plomin and Kovas 2005).

Plomin and Kovas (2005) propose that brain processes related to RD and MD will themselves be genetically correlated. Future studies in molecular genetics and genetic brain imaging might shed light onto whether one day generalist genes might serve as early warning systems for the development of MD or RD.

Co-Morbidity

Children and adults with RD are not only more likely to be affected by MD, they are also more likely to suffer from co-morbid language impairments (McArthur et al. 2000), attention difficulties (Tannock et al. 2000; Badian 1983; Gross-Tsur et al. 1996) and developmental co-ordination disorders (Rochelle and Talcott 2006). Any cognitive impairment present in those co-morbid disorders might influence mathematical performance. For example, ratings of attention, behavioral inattention in particular, are significantly related to mathematical performance (Fuchs et al. 2006). In addition, co-morbid disorders might limit otherwise available compensation mechanisms.

Intervention

While there is a large number of intervention studies for children with MD that are not specifically related to RD (e.g. see Butterworth et al. 2011; Dowker and Sigley 2010; Gersten et al. 2005; Jordan; Dowrick; this volume), the literature on effective interventions for arithmetic difficulties in RD is much smaller and existing studies have mainly taken the form of evaluations of educational programmes rather than theoretically motivated randomized controlled trials (e.g. Chinn and Ashcraft 1998; Miles and Miles 2004; Yeo 2003). Given our current knowledge of arithmetic difficulties in RD it is not surprising to see an underrepresentation of methodologically rigorous intervention studies in this field (Seethaler and Fuchs 2005). Reading intervention research has resulted in theoretically motivated intervention programmes to promote both reading accuracy and comprehension (Duff and Clarke 2011). However, it is not clear whether similar types of intervention would ameliorate difficulties in arithmetic fact learning (Robinson et al. 2002). It is somewhat surprising though that phonological intervention programmes have not been tried yet by researchers proposing a causal role of phonological awareness for fact retrieval. A positive effect of phonological awareness training on fact retrieval would provide strong evidence in favor of a causal role of phonological awareness (Wagner and Torgesen 1987).

Existing interventions have mainly focused on mathematical problem-solving interventions. Fuchs et al. (2004) evaluated the effectiveness of a 16-week mathematical problem-solving intervention for children at risk for pure MD, pure RD, MD+RD and for control children. Children at risk of pure RD improved more slowly than control children in their performance on arithmetic story problems. Children at risk of MD+RD showed the slowest progress. This finding highlights the issue of co-morbidity for intervention research (Fuchs and Fuchs 2005). It is currently unclear which cognitive deficits should be targeted in an intervention for arithmetic problems in children with RD. In a second step it will then be important to find out whether effective interventions for mathematical difficulties in children with RD also work for children with MD+RD or whether interventions will have to be specifically tailored to the set of deficits children present with.

Conclusions

Most early studies of children and adults with RD found poorer performance on mathematical tasks than expected given their cognitive abilities. In particular they were struggling with timed tests of single digit fact retrieval. However, those early studies are hampered by methodological issues such as varying selection criteria for RD and inadequate or no control groups. In addition, early studies have typically not tested for MD and might have included individuals with pure RD as well as individuals with RD *and* MD. The difficulties in number processing and arithmetic reported in early studies of individuals with RD could thus theoretically stem from a subset suffering from MD, in addition to but independently of RD. Recent research has addressed this issue by selecting participants for a group with pure RD only if their mathematical performance on a standardized test is within the normal range. Basic number processing skills of children and adults with pure RD are intact. However, they show also clear difficulties in arithmetic, in particular in tasks that require rapid fact retrieval (e.g. small addition problems and single digit multiplication problems). Fact retrieval deficits are associated with weaknesses in phonological awareness, the ability to segment and manipulate speech sounds, and possibly also with deficits in the phonological loop.

The left angular gyrus has been suggested as a site of neurological overlap between RD and fact retrieval deficits. While there is evidence for an involvement of the left ANG in fact retrieval in adults, the evidence for children is less clear. It is possible that during the acquisition of arithmetic facts ANG plays less of a role and only comes into play once these facts are consolidated within the semantic network. Research on structural brain changes associated with the process of learning to read and number fact acquisition is in its infancy. However, it is tempting to speculate that a structural abnormality in the left superior corona radiata, connecting temporal and parietal regions, might hamper reading and number fact acquisition in children. This structural difference might then over time lead to abnormalities in the functional connectivity and activation patterns of the ANG in adults. Alternatively, early difficulties in phonological awareness could lead to lower functional connectivity between temporal and parietal regions, in particular during language processing and reading tasks. This reduced connectivity could then lead to fact retrieval deficits. Currently it is impossible to decide whether the shared risk factor is at the level of poor phonological awareness (possibly leading to structural abnormalities) or at the level of structural brain abnormalities (leading to poor phonological awareness).

Behavioral genetic studies suggest that the same genetic risk factors may underlie difficulties in reading and mathematics and cause the high co-morbidity between RD and mathematical difficulties (and possibly other developmental disorders). This fits well with a dimensional view of developmental disorders (Snowling 2012). Fact retrieval difficulties would then be at the low end of a normal distribution of ability in fact retrieval and not a separate category. The dimensional view is supported by evidence from recent behavioral studies and has important implications for diagnosis and intervention. Given the high co-morbidity between RD and MD there might be scope for future theoretically motivated intervention programmes to target simultaneously poor literacy *and* numeracy in children and adults.

Acknowledgments

I would like to thank Maggie Snowling, Jelena Mirković, Graham Hitch and two anonymous reviewers for helpful comments on an earlier version of this chapter.

References

Ackerman, P. T. and Dykman, R. A. (1995). Reading-disabled students with and without comorbid arithmetic disability. *Developmental Neuropsychology* 11: 351–71.

American Psychiatric Association (2004). *Diagnostic and statistical manual of mental disorders* (DSM IV) (4th ed.). Washington, DC: American Psychiatric Association.
Andersson, U. (2010). Skill development in different components of arithmetic and basic cognitive functions: findings from a 3-year longitudinal study of children with different types of learning difficulties. *Journal of Educational Psychology 102*(1): 115–34.
Arsalidou, M. and Taylor, M. J. (2011). Is 2 + 2 = 4? Meta-analyses of brain areas needed for numbers and calculations. *NeuroImage* 54: 2382–93.
Baddeley, A. D. (1986). *Working memory*. Oxford, UK: OUP.
Badian, N. A. (1983). Dyscalculia and non-verbal disorders of learning. In: H. R. Myklebust (ed.), *Progress in learning disabilities*, Vol. 5, pp. 235–64. New York: Stratton.
Barrouillet, P., Fayol, M., and Lathulière, E. (1997). Difficulties in selecting between the competitors when solving elementary multiplication tasks: an explanation of the errors produced by adolescents with learning difficulties. *International Journal of Behavioural Development* 21: 253–75.
Berg, D. H. (2008). Working memory and arithmetic calculation in children: the contributory roles of processing speed, short-term memory and reading. *Journal of Experimental Child Psychology* 99(4): 288–308.
Boets, B. and De Smedt, B. (2010). Single-digit arithmetic in children with dyslexia. *Dyslexia* 16: 183–91.
Bookheimer, S. Y., Zeffiro, T. A., Blaxton, T., Gaillard, W., and Theodore, W. (1995). Regional cerebral blood flow during object naming and word reading. *Human Brain Mapping* 3: 93–106.
Brambati, S. M., Termine, C., Ruffino, M., et al. (2004). Regional reductions of grey matter volume in familial dyslexia. *Neurology* 63: 742–5.
Brown, W. E., Eliez, S., Menon, V., et al. (2001). Preliminary evidence of widespread morphological variations of the brain in dyslexia. *Neurology* 56: 781–3.
Bruck, M. (1990). Word recognition skills of adults with childhood diagnoses of dyslexia. *Developmental Psychology* 26: 439–54.
Bull, R. and Johnston, R. S. (1997). Children's arithmetic difficulties: contributions from processing speed, item identification, and short-term memory. *Journal of Experimental Child Psychology* 65: 1–24.
Butterworth, B. (2010). Foundational numerical capacities and the origins of dyscalculia. *Trends in Cognitive Sciences* 14: 534–41.
Butterworth, B., Sashank, V., and Laurillard, D. (2011). Dyscalculia: from brain to education, *Science* 332: 1049–53.
Cain, K. (2010). *Reading Development and Difficulties*. Oxford: Wiley-Blackwell.
Cain, K., Bryant, P. E., and Oakhill, J. (2004). Children's reading comprehension ability: Concurrent prediction by working memory, verbal ability, and component skills. *Journal of Educational Psychology* 96: 31–42.
Cain, K., Oakhill, J., and Bryant, P. E. (2000). Investigating the causes of reading comprehension failure: The comprehension-age match design. *Reading and Writing* 12(1–2): 31–40.
Campbell, J. D. and Xue, Q. L. (2001). Cognitive arithmetic across cultures. *Journal of Experimental Psychology General* 130: 299–315.
Carpentieri, J. D. (2008). *Research briefing: Numeracy*. London: NRDC.
Castro-Caldas, A. (2004). Targeting regions of interest for the study of the illiterate brain. *International Journal of Psychology* 39 5–17.
Chinn, S. J. (1995). A pilot a study to compare aspects of arithmetic skill. *Dyslexia Review* 4: 4–7.
Chinn, S. J. and Ashcroft, J. R. (1998). *Mathematics for Dyslexics: A Teaching Handbook*, 2nd edn. London: Whurr.
Cipolotti, L. and van Harskamp, N. (2001). Disturbances of number processing and calculation. In: R. S. Berndt (ed.), *Language and Aphasia*, Vol 3, pp. 305–31. Elsevier Science.
Cirino, P. T., Ewing-Cobbs, L., Barnes, M., Fuchs, L. S., and Fletcher, J. M. (2007). Cognitive arithmetic differences in learning disability groups and the role of behavioral inattention. *Learning Disabilities Research and Practice* 22: 25–35.
Dehaene, S. (2000). Cerebral bases of number processing and calculation. In: M. S. Gazzanigga (ed.), *The new cognitive neurosciences*, pp. 987–98). London: MIT Press.

Dehaene, S., Piazza, M., Pinel, P., and Cohen, L. (2003). Three parietal circuits for number processing. *Cognitive Neuropsychology* 20: 487–506.

Dejerine, J. (1891). Sur un cas de cécité verbale avec agraphie, suivi d'autopsie. *Mémoires de la Société Biologique* 3: 197–201.

De Jong, P. F. (2006). Understanding normal and impaired reading development: a working memory deficit. In: S. J. Pickering (ed.), *Working memory and education*, pp. 34–61. San Diego: Elsevier.

De Smedt, B. and Boets, B. (2010). Phonological processing and arithmetic fact retrieval: Evidence from developmental dyslexia. *Neuropsychologia* 48: 3973–81.

De Smedt, B., Janssen, R., Bouwens, K., Verschaffel, L., Boets, B., and Ghesquière, P. (2009). Working memory and individual differences in mathematics achievement: a longitudinal study from first to second grade. *Journal of Experimental Child Psychology* 103: 186–201.

De Smedt, B., Taylor, J., Archibald, L., and Ansari, D. (2010). How is phonological processing related to individual differences in children's arithmetic skills? *Developmental Science* 13(3): 508–20.

De Smedt, B., Holloway, I. D., and Ansari, D. (2011). Effect of problem size and arithmetic operation on brain activation during calculation with children with varying levels of arithmetical fluency. *NeuroImage* 57: 771–81.

Dowker, A. (1998). Individual differences in normal arithmetical development. In: C. Donlan (ed.), *The development of mathematical skills*, pp. 275–302. Hove: Psychology Press.

Dowker, A. (2005). *Individual differences in arithmetic*. Hove: Psychology Press.

Dowker, A. and Sigley, G. (2010). Targeted interventions for children with arithmetical difficulties. *British Journal of Educational Psychology, Monograph Series II* 7: 65–81.

Duff, F. J. and Clarke, P. J. (2011). Practitioner Review: Reading disorders: what are the effective interventions and how should they be implemented and evaluated? *Journal of Child Psychology and Psychiatry* 52(1): 3–12.

Durand, M., Hulme, C., Larkin, R., and Snowling, M. J. (2005). The cognitive foundations of reading and arithmetic skills in 7-to 10-year olds. *Journal of Experimental Child Psychology* 91: 113–36.

Eckert, M. (2004). Neuroanatomical markers for dyslexia: a review of dyslexia structural imaging studies. *Neuroscientist* 10(4): 362–71.

Eden, G. F., Jones, K. M, Cappell, K., et al. (2004). Neural changes following remediation in adult developmental dyslexia. *Neuron* 44: 411–22.

Evans, T.M., Flowers, D.L., Napoliello, E.M., Olulade, O.A., and Eden, G.F. (2014). The functional anatomy of single-digit arithmetic in children with developmental dyslexia. *Neuroimage*, 101, 644–652.

Fuchs, L. S. and Fuchs, D. (2005). Enhancing mathematical problem solving for students with disabilities. *Journal of Special Education* 39: 45–57.Fuchs, L. S., Fuchs, D., and Prentice, K. (2004). Responsiveness to mathematical problem-solving instruction among students with risk for mathematics disability with and without risk for reading disability. *Journal of Learning Disabilities* 37(4): 293–306.

Fuchs, L. S., Compton, D. L., Fuchs, D., Paulsen, K., Bryant, J. D., and Hamlett, C. L. (2005). The prevention, identification, and cognitive determinants of math difficulty. *Journal of Educational Psychology* 97: 493–513.

Fuchs, L. S., Fuchs, D., Compton, D. L., et al. (2006). The cognitive correlates of third-grade skill in arithmetic, algorithmic computation, and arithmetic word problems. *Journal of Educational Psychology* 98: 29–43.

Fürst, A. J. and Hitch, G. J. (2000). Separate roles for executive and phonological Components of working memory in mental arithmetic. *Memory & Cognition* 28: 774–82.

Gathercole, S. E. and Baddeley, A. D. (1993). Phonological working memory: A critical building block for reading development and vocabulary acquisition? *European Journal of Psychology of Education* 8: 259–72.

Geary, D. C. (1993). Mathematical disabilities: Cognitive, neuropsychological, and genetic components. *Psychological Bulletin* 114: 345–62.

Geary, D. C. (2004). Mathematics and Learning Disabilities. *Journal of Learning Disabilities* 37(1): 4–15.

Geary, D. C., Hamson, C. O., and Hoard, M. K. (2000). Numerical and arithmetical cognition: a longitudinal study of process and concept deficits in children with learning disability. *Journal of Experimental Child Psychology* 77: 236–63.

Gelfand, J. R. and Bookheimer, S. Y. (2003). Dissociating neural mechanisms of temporal sequencing and processing phonemes. *Neuron* 38: 831–42.

Gelman, R. and Gallistel, C. R. (1978). *The child's understanding of number.* Cambridge: Harvard University Press.

Gersten, R., Jordan, N., and Flojo, J. R. (2005). Early identification and interventions for students with mathematics difficulties. *Journal of Learning Disabilities* 38: 293–324.

Göbel, S. M. and Snowling, M. J. (2010). Number Processing Skills in Adults with Dyslexia. *The Quarterly Journal of Experimental Psychology* 63(7): 1361–73.

Grabner, R. H., Ansari, D., Koschutnig, K., Reishofer, G., Ebner, F., and Neuper, C. (2009). To retrieve or to calculate? Left angular gyrus mediates the retrieval of arithmetic facts during problem solving. *Neuropsychologia* 47: 604–8.

Grabner, R. H., Ansari, D., Reishofer, G., Stern, E., Ebner, F., and Neuper, C. (2007). Individual differences in mathematical competence predict parietal brain activation during mental calculation. *NeuroImage* 38: 346–56.

Griffiths, Y. M., and Snowling, M. J. (2001). Auditory word identification and phonological skills in dyslexic and average readers. *Applied Psycholinguistics* 22: 419–39.

Gross-Tsur, V., Manor, O., and Shalev, R. S. (1996). Developmental dyscalculia: prevalence and demographic features. *Developmental Medicine and Child Neurology* 38: 25–33.

Hanich, L. B., Jordan, N. C., Kaplan, D., and Dick, J. (2001). Performance across different areas of mathematical cognition in children with learning difficulties. *Journal of Educational Psychology* 93(3): 615–26.

Hecht, S. A., Torgesen, J. K., Wagner, R. K., and Rashotte, C. A. (2001). The relations between phonological processing abilities and emerging individual differences in mathematical computation skills: a longitudinal study from second to fifth grades. *Journal of Experimental Child Psychology* 79: 192–227.

Hulme, C. and Snowling, M. J. (2009). *Developmental disorders of language and cognition.* Oxford: Blackwell/Wiley.

Hulme, C. and Snowling, M. J. (2011). Children's reading comprehension difficulties: nature, causes and treatment. *Current Perspectives in Psychological Science* 20: 139–42.

Inglis, M., Attridge, N., Batchelor, S., and Gilmore, C. (2011). Non-verbal number acuity correlates with symbolic mathematics achievement: but only in children. *Psychonomic Bulletin and Review* 18(6): 1222–9.

Isaacs, E. B., Edmonds, C. J., Lucas, A., and Gadian, D. G. (2001). Calculation difficulties in children of very low birthweight: a neural correlate. *Brain* 124: 1701–7.

Jordan, N. and Hanich, L. B. (2003). Characteristics of children with moderate mathematics deficiencies: a longitudinal perspective. *Learning Disabilities Research & Practice* 18: 213–21.

Jordan, N. C., Hanich, L. B., and Kaplan, D. (2003). A longitudinal study of mathematical competencies in children with specific mathematics difficulties versus children with co-morbid mathematics and reading difficulties. *Child Development* 74(3): 834–50.

Jordan, J-A., Wylie J., and Mulhern, G. (2010). Phonological awareness and mathematical difficulty: A longitudinal perspective. *British Journal of Developmental Psychology* 28: 89–107.

Klingberg, T., Hedehus, M., Temple, E., et al. (2000). Microstructure of temporo-parietal white matter as a basis for reading ability: evidence from diffusion tensor magnetic resonance imaging. *Neuron* 25: 492–500.

Kovas, Y. and Plomin, R. (2007). Learning abilities and disabilities: generalist genes, specialist environments. *Current Directions in Psychological Science* 16(5): 284–8.

Landerl, K., Bevan, A., and Butterworth, B. (2004). Developmental dyscalculia and basic numerical capacities: A study of 8–9 year old students. *Cognition* 93: 99–125.

Landerl, K., Fussenegger, B., Moll, K., and Willburger, E. (2009). Dyslexia and dyscalculia: Two learning disorders with different cognitive profiles. *Journal of Experimental Child Psychology* 103: 309–24.

Leather, C. V. and Henry, L. A. (1994). Working memory span and phonological awareness tasks as predictors of early reading ability. *Journal of Experimental Child Psychology* 58: 88–111.

Lewis, C., Hitch, G. J., and Walker, P. (1994). The prevalence of specific arithmetic difficulties and specific reading difficulties in 9—to 10-year-old boys and girls. *Journal of Child Psychology and Psychiatry* 35(2): 283–92.

Lipton, J. S. and Spelke, E. S. (2003). Origins of number sense. Large-number discrimination in human infants. *Psychological Science* 14(5): 396–401.

McArthur, G. M., Hogben, J. H., Edwards, V. T., Heath, S. M., and Mengler, E. D. (2000). On the 'specifics' of specific reading disabilities and specific language impairment. *Journal of Child Psychology and Psychiatry* 41: 869–74.

Miles (1997). *The Bangor Dyslexia Test.* Wisbech, Cambridge: Learning Development Aids.

Miles, T. R. and Miles E. (2004). *Dyslexia and Mathematics*, 2nd edn. London: Routledge Falmer.

Miles, T. R., Haslum, M. M., and Wheeler, T. J. (2001). The mathematical abilities of dyslexic 10-year-olds. *Annals of Dyslexia* 51: 299–321.

Morton, J. and Frith, U. (1995). Causal modelling: a structural approach to developmental psychopathology. In: D. Cicchetti and D. J. Cohen (eds.), *Manual of developmental psychopathology.* New York: Wiley.

Nicolson, R. and Fawcett, A. J. (1990). Automaticity: a new framework for dyslexia research? *Cognition* 35(2): 159–82.

Nicolson, R., Fawcett, A. J., and Dean, P. (2001). Developmental dyslexia: the cerebellar deficit hypothesis. *Trends in Neurosciences* 24(9): 508–11.

Niogi, S. N. and McCandliss, B. D. (2006). Left lateralized whit matter microstructure accounts for individual differences in reading ability and disability. *Neuropsychologia* 44: 2178–88.

Noël, M.-P., Seron, X., and Trovarelli, F. (2004). Working memory as a predictor of addition skills and additions strategies in children. *Current Psychology of Cognition* 22: 3–25.

OECD (2010). *The high cost of low educational performance: the long-run economic impact of improving educational outcomes.* Paris: OECD.

Pennington, B. F. and Olson, R. K. (2005). Genetics of Dyslexia. In: M. J. Snowling and C. Hulme (eds.), *The Science of reading: a handbook*, pp. 453–72. Oxford: Blackwell.

Pimperton, H. and Nation, K. (2010). Understanding words, understanding numbers: an exploration of the mathematical profiles of poor comprehenders. *British Journal of Educational Psychology* 80(2): 255–68.

Plomin, R. and Kovas, Y. (2005). Generalist genes and learning disabilities. *Psychological Bulletin* 131(4): 592–617.

Price, C. J. (2000). The anatomy of language: contributions from functional neuroimaging. *Journal of Anatomy* 197: 335–59.

Price, C. J., Moore, C. J., and Frackowiak, R. S. J. (1996). The effect of varying stimulus rate and duration on brain activity during reading. *NeuroImage* 3: 40–52.

Pugh, K. R., Mencl, W. E., Jenner, A. R., et al. (2001). Neurobiological studies of reading and reading disability. *Journal of Communication Disorders* 34: 479–92.

Pugh, K. R., Mencl, W. E., Shaywitz, B. A., et al. (2000). The angular gyrus in developmental dyslexia: Task specific differences in functional connectivity in posterior cortex. *Psychological Science* 11: 51–6.

Ramus, F. (2004). Neurobiology of dyslexia: A reinterpretation of the data. *Trends in Neurosciences* 27(12): 720–6.

Ramus, F., Rosen, S., Dakin, S. C., et al. (2003). Theories of developmental dyslexia: insights from a multiple case study of dyslexic adults. *Brain* 126: 841–65.

Rasmussen, C. and Bisanz, J. (2005). Representation and working memory in early arithmetic. *Journal of Experimental Child Psychology* 91: 137–57.

Richlan, F., Kronbichler, M., and Wimmer, H. (2009). Functional abnormalities in the dyslexic brain: a quantitative meta-analysis of neuroimaging studies. *Human Brain Mapping* 30(10): 3299–308.

Richlan, F., Kronbichler, M., and Wimmer, H. (2011). Meta-analyzing brain dysfunction in dyslexic children and adults. *NeuroImage* 56 3: 1735–42.

Rimrodt, S. L., Peterson, D. J., Denckla, M. B., Kaufmann, W. E., and Cutting, L. E. (2010). White matter microstructural differences linked to left perisylvian language network in children with dyslexia. *Cortex* 46: 739–49.

Robinson, C. S., Menchetti, B. M., and Torgesen, J. K. (2002). Toward a two-factor theory of one type of mathematics disabilities. *Learning Disabilities Research and Practice* 17(2): 81–9.

Rochelle, K. and Talcott, J. (2006). Impaired balance in developmental dyslexia? A meta—analysis of contending evidence. *Journal of Child Psychology and Psychiatry* 47: 1159–66.

Rourke, B. P. (1989). *Nonverbal learning disabilities: the syndrome and the model.* New York: Guilford Press.

Rourke, B. P. (1993). Arithmetic disabilities, specific and otherwise: a neuropsychological perspective. *Journal of Learning Disabilities* 26(4): 214–26.

Rubinsten, O. and Henik, A. (2006). Double dissociation of functions in developmental dyslexia and dyscalculia. *Journal of Educational Psychology* 98: 854–67.

Rubinsten, O., and Henik, A. (2009). Developmental Dyscalculia: heterogeneity might not mean different mechanisms. *Trends in Cognitive Science* 13(2): 92–9.

Rumsey, J. M., Horwitz, B., Donohue, B. C., Nace, K. L., Maisog, J. M., and Andreason, P. (1997). Phonological and orthographic components of word recognition: a PET-rCBF study. *Brain* 120: 739–59.

Scholz, J., Klein, M. C., Behrens, T. E., and Johansen-Berg, H. (2009). Training induces changes in white-matter architecture. Nature Neuroscience 12: 1370–1.

Seethaler, P. M. and Fuchs, L. S. (2005). A drop in the bucket: Randomized controlled trials testing reading and math intervention. *Learning Disabilities Research and Practice* 20: 98–102.

Shaywitz, S. E., Escobar, M. D., Shaywitz, A. B., Fletcher, J. M., and Mahuch, K. (1992). Evidence that dyslexia may represent the lower tail of a normal distribution of reading ability. *New England Journal of Medicine* 326: 145–50.

Shaywitz, S. E., Shaywitz, B. A., Piugh, K. R., et al. (1998). Functional disruption in the organization of the brain for reading in dyslexia. *Proceedings of the National Academy of Sciences of the United States of America* 95: 2636–41.

Silani, G., Frith, U., Demonet, J.-F., et al. (2005). Brain abnormalities underlying altered activation in dyslexia: a voxel based morphometry study. *Brain* 128: 2453–61.

Simmons, F. R. and Singleton, C. (2006). The arithmetical abilities of adults with dyslexia. *Dyslexia* 12: 96–114.

Simmons, F. R. and Singleton, C. (2008). Do weak phonological representations impact on arithmetic development? A review of research into arithmetic and dyslexia. *Dyslexia* 14(2): 77–94

Simmons, F. R. and Singleton, C. (2009). The mathematical strengths and weaknesses of children with dyslexia. *Journal of Research in Special Educational Needs* 9: 154–63.

Simmons, F. R., Singleton, C., and Horne, J. K. (2008). Phonological awareness and visual spatial sketchpad functioning predict early arithmetic attainment: Evidence from a longitudinal study. *European Journal of Cognitive Psychology* 4: 711–22.

Snowling, M. J. (2012). Editorial: seeking a new characterisation of learning disorders. *Journal of Child Psychology and Psychiatry* 53(1): 1–2.

Snowling, M. J. and Hulme, C. (1994). The development of phonological skills. *Philosophical Transactions of the Royal Society B* 346: 21–8.

Steeves, K. J. (1983). Memory as a factor in the computational efficiency of dyslexic children with high abstract reasoning ability. *Annals of Dyslexia* 33: 141–52.

Stein, J. and Walsh, V. (1997). To see but not to read; the magnocellular theory of dyslexia. *Trends Neuroscience* 20: 147–52.

Tallal, P. (1980). Auditory-temporal perception, phonics and reading disabilities in children. *Brain and Language* 9: 182–98.

Tannock, R., Martinussen, R., and Frijters, J. (2000). Naming speed performance and stimulant effects indicate effortful, semantic processing deficits in attention-deficit/hyperactivity disorder. *Journal of Abnormal Child Psychology* 28: 237–52.

Temple, C. M. and Sherwood, S. (2002). Representation and retrieval of arithmetic facts: developmental difficulties. *Quarterly Journal of Experimental Psychology A* 55: 733–52.

Tressoldi, P. E., Rosati, M., and Lucangeli, D. (2007). Patterns of developmental dyscalculia with or without dyslexia. *Neurocase* 13(4): 217–225.

Turner Ellis, S. A., Miles, T. R., and Wheeler, T. J. (1996). Speed of multiplication in dyslexics and non-dyslexics. *Dyslexia* 2: 121–39.

Vandermosten, M., Boets, B., Poelmans, H., Sunaert, S., Wouters, J., and Ghesquière, P. (2012). A tractography study in dyslexia: neuroanatomic correlates of orthographic, phonological and speech processing. *Brain* 135: 935–48.

Van Eimeren, L., Niogi, S. N., McCandliss, B. D., Holloway, I. D., and Ansari, D. (2008). White matter microstructure underlying mathematical abilities in children. *NeuroReport* 19(11): 1117–21.

Vellutino, F. R., Fletcher, J. M., Snowling, M. J., and Scanlon, D. M. (2004). Specific reading disability (dyslexia): what have we learned in the past four decades? *Journal of Child Psychology and Psychiatry* 45(1): 2–40.

Vukovic, R. K, Lesaux, N. K., and Siegel, L. S. (2010). The mathematical skills of children with reading difficulties. *Learning and Individual Differences* 20: 639–43.

Wagner, R. K. and Torgesen, J. K. (1987). The nature of phonological processing and its causal role in the acquisition of reading skills. *Psychological Bulletin* 101: 192–212.

Willburger, E., Fussenegger, B., Moll, K., Wood, G., and Landerl, K. (2008). Naming speed in dyslexia and dyscalculia. *Learning and Individual Differences* 18: 224–36.

Wimmer, H., Mayringer, H., and Landerl, K. (1998). Poor reading: a deficit in skill automatization or a phonological deficit? *Scientific Studies of Reading* 2: 321–40.

Yeo, D. (2003). *Dyslexia, Dyspraxia and Mathematics*. London: Whurr.

Zatorre, R. J., Fields, R. D, and Johansen-Berg, H. (2012). Changes in brain structure during learning. *Nature Neuroscience* 15(4): 528–36.

CHAPTER 39

GENETIC DEVELOPMENTAL DISORDERS AND NUMERICAL COMPETENCE ACROSS THE LIFESPAN

JO VAN HERWEGEN AND
ANNETTE KARMILOFF-SMITH

Introduction

PREVIOUS chapters in this volume have already discussed the fact that animals, including monkeys (Hauser & Carey, 2003), salamanders (Uller, Jaeger, Guidry, & Martin, 2003), fish (Agrillo, Piffer, & Bisazza, 2010) and bees (Gross, Pahl, Si, Zhu, Tautz, & Zhang, 2009), are able to discriminate between different numerosities. However, the fact that other species have numerical abilities does not necessarily imply that such abilities are innately specified in the human case (Karmiloff-Smith, 2013). This holds even though studies investigating number abilities in typically developing (TD) infants have shown that, as young as 6 months of age, they can discriminate between both small and large numerosities (Xu, 2003; Xu, Spelke, & Goddard, 2005). However, it is still debated whether these discrimination abilities rely upon one or two distinct core systems, one for small or exact number and the other for large numerosities, i.e. an approximate magnitude system (Feigenson, Dehaene, & Spelke, 2004). The presence of double dissociations in adult brain-damaged patients has been used in favour of the argument that these sub-systems for number are neurally and cognitively distinct (Demeyenne, Lestor, & Humphreys, 2010; Lemer, Dehaene, Spelke, & Cohen, 2003). However, such adult studies cannot reveal how these systems have developed from infancy onwards or whether both core systems are pre-specified in the infant brain. In addition, as both core systems are already present in TD infants, studies of TD populations can, at best, only show correlations between the early discrimination abilities and number development later on in life. By contrast, genetic disorders often present with uneven cognitive profiles in which some abilities are better compared with others, despite overall cognitive delay. Therefore, genetic disorders provide a way to investigate the impact of impaired abilities present in infancy on the phenotypic outcome later in life.

Despite the fact that number skills are impaired in many developmental disorders, research is still relatively sparse. The current chapter will focus on two neurodevelopmental disorders, Williams syndrome (WS) and Down syndrome (DS), in order to evaluate the importance of the early number systems for subsequent number abilities. We will argue that the ability to discriminate between large numerosities, i.e. the magnitude system, is a better predictor for mathematical skills later in development. In addition, we will advocate that the cross-syndrome design is crucial to further our understanding of the relationship between basic-level processes and the development of number systems into adulthood. Throughout this chapter we will show that a truly developmental approach, or neuroconstructivist approach (Farran & Karmiloff-Smith, 2012; Karmiloff-Smith, 1998, 2009; Thomas, Purser, & Van Herwegen, 2012), is needed when evaluating cross-domain influences on cognitive abilities in different neurodevelopmental disorders.

The Need for a Developmental Perspective

With the advances in molecular genetics, the past decade has seen an increase in research in neurodevelopmental disorders, with studies attempting to link certain candidate genes to specific phenotypic outcomes. In addition, the advances and increased resolution of neuroimaging techniques have made it possible to investigate the links between brain and behaviour in both typical and atypical populations. More specifically, some cross-syndrome comparisons have tried to provide evidence about the innate specification of dissociated modules in the brain. For example, Pinker (1999) states: 'Overall, the genetic double dissociation is striking ... The genes of one group of children [specific language impairment] impair their grammar while sparing their intelligence: the genes of another group of children [WS] impair their intelligence while sparing their grammar' (p. 262). However, while double dissociations may be appropriate for the analysis of adult neuropsychological patients whose brains had developed normally and become specialized by the time of their brain insult, such notions are rather static and unsuitable for the analysis of neurodevelopmental disorders (Karmiloff-Smith, Brown, Grice, & Paterson, 2003). Indeed, the claim that the brain consists of innately specialized modules for specific cognitive functions (e.g. number, language, face processing, etc.) is based upon a static, adult neuropsychological interpretation of the brain in its mature state. Indeed, many examples of double dissociations can be found in adult patients with brain damage to specific areas (e.g. Brocca and Wernicke's aphasia). However, such a static view cannot and does not apply to genetic disorders as: 'brain volume, brain anatomy, brain chemistry, hemispheric asymmetry and temporal patterns of brain activity are all atypical [...]. How could the resulting system be described as a normal brain with part intact and parts impaired, as the popular view holds? Rather, the brains of infants with [neurodevelopment disorders] develop differently from the outset, which has subtle, widespread repercussions' (Karmiloff-Smith, 1998, p. 393). Thus, a truly developmental approach is required, because the start state in infants cannot be assumed from just examining phenotypic snapshots or the end state in adults (Karmiloff-Smith, 1998; Paterson, Brown, Gsödl, Johnson, & Karmiloff-Smith, 1999; Thomas & Karmiloff-Smith, 2002). To illustrate this point, let's take the example of language abilities in Down syndrome and Williams syndrome.

Language Abilities as an Example of the Neuroconstructivist Approach

Williams syndrome (WS) is a rare developmental disorder that affects about 1 in 15 000–20 000 live births (Morris, Demsey, Leonard, Dilts, & Blackburn, 1988) and is caused by a hemizygotic microdeletion on the long arm of chromosome 7, affecting some 28 genes (Donnai & Karmiloff-Smith, 2000; Tassabehji, 2003). Physically, individuals with WS can be recognized by their distinct facial morphology and unique disinhibited social behavioural profile. Generally, individuals with WS score within the moderate impaired range on IQ-tests with a mean score around 50–70 (Mervis, Morris, Bertrand, & Robinson, 1999). However, the cognitive profile in WS is uneven, characterized by proficient face-processing and verbal abilities, in contrast to the serious weaknesses in number skills, visuospatial abilities, route learning, and planning (Hudson & Farran, 2011; Mervis, Morris, Klein-Tasman, Bertrand, Kwitny, Appelbaum, et al., 2003).

Down syndrome (DS) is a very common genetic developmental disorder with a prevalence of about 1 in 800 live births (Dolk, Loane, Garne, De Walle, Queisser-Luft, De Vigan et al., 2005). The disorder is caused by trisomy of chromosome 21 (Silverman, 2007). Two rarer forms of DS, which occur only in 6% of all cases, are caused by translocations or mosaicism (Pangalos, Avramopoulos, Blouin, Raoul, deBlois, Prieur, et al., 1994). Although similarly to WS IQ scores are between 40 and 70 (Roizen & Patterson, 2003) and the cognitive profile in DS is also uneven, DS has different strengths and weaknesses compared with WS, with particular difficulties in language in contrast to their spatial abilities (Jarrold, Baddeley, & Hewes, 1999). Both neurodevelopmental disorders yield an overall cognitive impairment, but DS is often reported to show severe language impairment, while this is a relative strength in individuals with WS. However, when Singer-Harris and colleagues (Singer-Harris, Bellugi, Bates, Jones, & Rossen, 1997) compared the vocabulary production of 35 young children with WS (mean age 47 months) to that of 23 participants with DS (mean age 45 months), using the parental questionnaire MacArthur Communicative Development Inventory (CDI; Fenson, Dale, Reznick, Thal, Bates, Hartung, et al., 1993), they found that the participants with DS actually produced more words at this age compared with those with WS. This study suggests that the linguistic advantage of WS over DS, observed in later development, is not as clearly present in infancy. Thus, the advantage for language abilities in individuals WS over DS only *becomes* apparent with development (Paterson et al., 1999). In addition, detailed studies of infants have shown that vocabulary development is not only delayed in WS, but also atypical. For example, while in typically developing children pointing precedes the production of words, infants with WS start pointing after they have learned the names for objects (Laing, Butterworth, Ansari, Gsödl, Longhi, Panagiotaki, et al., 2002;

Mervis et al., 1999). These studies show that it is erroneous to presume that the phenotypic outcome that can be observed in adulthood is the same in infancy, or that the neurocognitive processes that underpin successful behaviour in developmental disorders are the same as those in individuals who follow a typical developmental trajectory. Thus, a truly developmental approach is needed when investigating neurodevelopmental disorders, which entails examining full developmental trajectories from infancy onwards.

Number Abilities in Children and Adults with Williams Syndrome and Down Syndrome

Studies that have investigated number abilities in older children and adults with WS have revealed that arithmetic skills are severely impaired in this population and show little development over time (O'Hearn & Landau, 2007; Udwin, Davies, & Howlin, 1996). However, as verbal abilities are generally better than non-verbal abilities in WS, further detailed research was carried out to investigate whether there were differences in those numerical abilities that rely upon verbal abilities compared with those that rely on the representation of a mental number line. Indeed, it has been shown that, although children with WS are proficient at counting sequences (Paterson, Girelli, Butterworth, & Karmiloff-Smith, 2006), they are impaired in their understanding of the meaning of counting or the cardinality principle (Ansari, Donlan, Thomas, Ewing, & Karmiloff-Smith, 2003; Paterson et al., 2006). In addition, verbal abilities predicted the variance within the WS group, whereas non-verbal spatial abilities predicted the variance in the TD controls, suggesting different developmental trajectories to reach an understanding of cardinality. Thus, although verbal abilities in WS may be compensating for the serious weaknesses in number development and thus enable proficient counting, verbal abilities are not able to compensate for all problems, such as the understanding of cardinality.

Another study has shown that children with WS are impaired on tasks that directly tap the representation of the mental number line, such as visual estimation tasks (Ansari, Donlan, & Karmiloff-Smith, 2007). In this study, performance of 31 children and adults with WS was much lower compared with 4- and 10-year old TD controls. However, it is interesting to note that the WS group performed better on small number items (1–3) than on larger number items (5, 7, 9, and 11). Moreover, evidence from a previous study by Paterson and collaborators (2006) found that, when reaction time was taken into account, older children and adults with WS participants were better at judging number pairs that were far apart (e.g. 2 versus 9) compared with those that were close to each other (e.g. 7 versus 9), as is the case in typical development. Finally, a recent study by Krajcsi, Lukacs, Igacs, Racsmany, and Pleh (2009) found that, compared with much younger TD controls, older children and adults with WS were especially slow on a number matching task, but not on a simple addition and subtraction task, which again indicates that the representation of the mental number line is impaired in this population.

In contrast, children with DS between the ages of 10 and 18 years of age have been found to produce more counting errors in comparison to younger TD controls and to those with general mental retardation of unknown etiologies (Stith & Fishbein, 1996). In addition, DS children had problems with establishing which set out of two options contained 'more' or a higher value. However, as verbal abilities are severely delayed in young children with DS, it is possible that the impairment in counting is in part a consequence of language difficulties, rather than solely a number problem in understanding cardinality. Indeed, studies that have investigated the Symbolic Distance Effect have shown that older participants and adults with DS show evidence of processing the mental number line. As in typical development, individuals with DS responded significantly faster to numerosities that were far apart than to those close to each other (Paterson et al., 2006). In addition, children with DS are able to discriminate between sets containing 8 versus 16 items (ratio 1:2) when these sets are presented rapidly, but not between sets including 8 versus 12 items (ratio 2:3; Camos, 2009). This performance did not differ from either chronological age or mental age-matched control groups and indicates that the approximate number system in older children with DS is relatively proficient.

These studies might suggest that there is a double dissociation for number abilities in older children and adults with DS, and those with WS. In DS the mental number line seems to be operating proficiently, while number abilities that rely upon verbal abilities, such as counting, are impaired. In contrast, individuals with WS show weaker number line processing and atypical development of many numerical abilities, but good counting. Yet, as discussed, terms such as 'double dissociations', 'intact', and 'sparing' are static notions (Karmiloff-Smith, 2009) borrowed from the adult neuropsychological literature. In addition, although these studies show how the two neurodevelopmental disorders differ in their end state, with individuals with DS outstripping participants with WS on a battery of number tasks (Paterson et al., 2006), we have already highlighted the fact that the start state in infancy cannot always be predicted from phenotypic outcomes in older children and adults. Therefore, without investigating how these abilities develop within and beyond infancy, it is impossible to ascertain whether the strengths in verbal abilities allow individuals with WS to compensate for their impaired approximate magnitude subsystem when counting and performing simple arithmetic tasks, and whether the impaired verbal abilities of individuals with DS hinder their number development. Alternatively, it is possible that other cognitive abilities, such as visuospatial cognition (Simon, 1997), attention, or general basic-level processes that influence the very early stages of number development (Karmiloff-Smith, D'Souza, Dekker, Van Herwegen, Xu, Rodic, et al., 2013), can provide a better explanation for the number difficulties and abilities observed in these neurodevelopmental disorders.

Number Abilities in Infants with Williams Syndrome and Down Syndrome

An initial study by Paterson and colleagues (1999) investigated the ability to discriminate between small number sets (2 versus 3 objects) in infants with WS and DS who were matched for both chronological and mental age. Infants were familiarized with pairs of objects depicting one number (e.g. pairs of two rabbits and two cars followed by pairs of two balls and two horses, etc.) and then their length of looking was assessed to a novel number (e.g. to one pair of two trees and one pair of three boats). In contrast to infants with WS, those with DS did not look longer when presented with the novel numerosity. This was the first study to reveal that although number abilities are severely impaired in older individuals with WS, early numerical abilities seemed to be proficient in WS infants, while the opposite was the case for infants with DS who perform better than WS in subsequent development. However, based upon discrimination abilities in TD infants, it has been argued that there are two distinct numerical systems, one for small or exact number and one for large approximate numerosities or magnitudes (see Feigenson et al., 2004, for an overview). The system for discriminating between small numerosities allows one-to-one mapping, i.e. the precise tracking of individual objects. By contrast, the system that enables discrimination between approximate quantities relates to subitizing abilities in adulthood. As Paterson et al. (1999) only investigated exact small number abilities in DS and WS, two further studies of WS and DS infants were carried out. These investigated both numerical sub-systems and included two experiments that have been replicated many times with TD infants. Based on the earlier study by Paterson et al. (1999, 2006) and the outcomes later in life, it was hypothesized that one of the core systems, namely the large magnitude discrimination system, would be predictive of later numerical abilities and that, therefore, DS infants would fare better than WS infants on the assessment of this system.

In, 2008, we started to investigate this hypothesis by examining looking behaviour to small and large numerosities in nine infants and toddlers with WS who had a mean chronological age of 35 months (range 13–53 months). In each of the experiments infants were presented with nine familiarization trails on a screen, displaying either a small (2 or 3 dots) or large number (8 or 16 dots). While the infants with WS looked longer at the novel number displayed when presented with small numerosities, their discrimination of large numerosities was significantly weaker (Van Herwegen, Ansari, Xu, & Karmiloff-Smith, 2008). A cross-syndrome comparison was required to test our hypothesis that large approximate number discrimination might be more predictive of subsequent mathematical abilities than small exact number discrimination. We therefore used the same paradigm to test nine infants and toddlers with DS, matched on chronological and mental age to the infants with WS in our previous study. The findings revealed that the infants with DS showed the opposite pattern to those with WS; while they did not discriminate between small numbers of dots as before, their looking time was longer to novel large numerosities (Karmiloff-Smith et al., 2013). One could argue that the ability to discriminate between the small numerosities with ratio 2:3 is harder compared with the large numerosities, which differed by ratio 1:2. Yet, the opposite pattern was observed in infants with WS who had a similar mental age, and studies of TD infants have shown that while ratio is relevant to large magnitude discrimination, it does not affect number discrimination abilities for small numerosities. It is therefore unlikely that a difference in ratio can explain the DS findings for small numerosities.

To summarize, these two studies lend support to our hypothesis that the ability to discriminate between large numerosities is a better predictor of subsequent number attainment. Infants with WS have difficulty computing approximate numerical magnitudes, and this impairment impacts their mathematic abilities in adulthood. In contrast, infants with DS are sensitive to differences in large approximate quantities and they thus outstrip those with WS on numerical tasks in adolescence and adulthood, even if they are impaired relative to TD controls. The delay in verbal abilities in DS, therefore, only initially hinders the development of number abilities and the ability to count, while number development in WS is not only delayed, but also atypical.

Pre-Specialization or Progressive Specialization Over Time?

Research in typically developing children has shown that the specialization of the parietal lobe for approximate numerical quantities only emerges over time as a result of gradual cortical specialization (Holloway & Ansari, 2010). Due to the fact that brain development in genetic disorders is atypical from the outset, it is unlikely that specific genetic mutations would impair only specific pre-specialized neural areas, while leaving others intact, e.g. an impaired small exact number system module, present at birth. Does development start out in terms of independently functioning modules? Or do regions interact initially, only to become specialized as a function of development? In order to address such questions, it is important to evaluate other abilities that might interact with the development of number, and contribute to an explanation of the atypicalities or impairments observed in developmental disorders.

One of the competencies that has been argued to be crucial for the successful development of numerical abilities is proficient spatial information processing (Simon, 2011). Counting and judging relationships between quantities requires the ability to mentally represent items spatially (e.g. 8 follows 7 and 8 is closer to 12 than to 16) and to keep track of them (i.e. if an individual loses track of which visually presented dots s/he has already counted, s/he ends up with the incorrect number). Previous studies of WS have pinpointed visuospatial impairments in a variety of tasks. For example, participants with WS generally have severe problems with pattern construction tasks, with performance close to floor and little increase in task performance over development in younger participants (Van Herwegen, Rundblad, Davelaar, & Annaz, 2011). In addition, individuals with WS have difficulties with mental rotation (Farran, Jarrold, & Gathercole, 2001; Stinton, Farran, & Courbois, 2008), and with recalling the location of targets in a static display (O'Hearn, Landau, & Hoffman, 2005). They also make more errors in visual search tasks compared with healthy controls (Scerif, Cornigh, Wilding, Driver, & Karmiloff-Smith, 2004). It has therefore been argued that their visuospatial difficulties impair the development of the mental number line in WS, thereby contributing to an explanation of the problems they experience in discriminating between large numerosities (Simon, 2011). However, although correlations do exist between visuospatial and number abilities, we argue that the deficits observed in both abilities might be rooted in basic-level visual and attention processes that interact with number processing from infancy onwards.

Studies by Brown and colleagues (Brown, Johnson, Paterson, Gilmore, Longhi, & Karmiloff-Smith, 2003) yielded impairments in infants with DS with respect to sustained attention, with fewer periods and less total time of sustained attention compared with control groups. In contrast, infants with WS performed similar to TD control groups for sustained attention, but they had problems with planning on a double-step saccade planning task. Based on this evidence, we suggested that the difficulties experienced by infants with

WS on large number discrimination might be due to poor visuospatial scanning abilities. Previous studies have made similar suggestions that WS is characterized by 'sticky fixation' (Paterson et al., 2006), especially towards faces and social stimuli (Mervis et al., 2003; Riby, Jones, Brown, Robinson, Langton, Bruce, et al., 2011). With recent methodological advances, such as infra red eye-tracking, researchers are now able to record very precise temporal and spatial data of when, for how long, and where participants look during stimuli presentations. Using a Tobii 1750 Eye Tracker (Tobii, 2003) we were able to obtain data from a few participants with DS and WS with respect to their processing of the large numerical displays. The scanning patterns of these infants revealed that those with DS scanned almost the entire array of dots in the number displays, while those with WS fixated only on a few of the dots within each display (Karmiloff-Smith et al., 2013). Thus, the fact that infants with WS have problems with planning eye movements gives rise to their disengagement difficulties and their focus on individual dots, rather than overall quantity. In contrast, as the infants with DS are able to scan the majority of the dots in large numerosity displays, they show no difficulties in discriminating between large numerosities. In addition, these problems in the basic-level visual and attention systems may also be used to explain the results for exact number discrimination in the two neurodevelopmental disorders. The WS inability to disengage from the stimuli actually might enable them to individualize objects for small sets (2 or 3 dots), while the problems with sustained attention in infants with DS may prevent them from staying focused on individual objects. Thus, the different deficits in basic-level visual attention systems – sustained attention for DS and impaired saccadic planning for WS – have cascading effects on the development of the two numerical sub-systems. We further predict that these deficits not only explain proficiencies and difficulties for the development of number abilities, but are likely to have an impact on other high-level cognitive abilities. This, in turn, allows us to explain emerging relationships between visuospatial abilities and number abilities. However, only through tracing full developmental trajectories from infancy onwards and investigating subtle differences in basic-level abilities between different neurodevelopmental disorders can a detailed pattern emerge of the underlying causes and relationships between different abilities (Karmiloff-Smith, 2009).

This approach is not only important for the development of theoretical explanations but, more importantly, should lead to better and more economically valid intervention programmes. First, our studies suggest that interventions should focus on low-level visual and attention abilities, rather than solely on specific number abilities. Secondly, interventions should be syndrome specific: infants with DS will need to be trained on sustained attention so that they can individuate objects and discriminate between small numerosities, while interventions for infants with WS will need to focus on saccade planning and disengagement from individual objects, so that entire displays will be scanned, leading to the successful discrimination of large numerosities.

Conclusions

Our findings have highlighted how important it is to take a truly developmental approach, tracing full trajectories from infancy onwards. However, longitudinal studies are expensive and can be challenging from a practical point. Yet, studies that only focus on a small age range or on the phenotypic outcome in adulthood may lead to incomplete or even inaccurate conclusions. In addition, our studies have shown that cross-syndrome comparisons are key in the investigation of how abilities relate to each other and which abilities predict others. The uneven cognitive profiles of our two neurodevelopmental disorders revealed that the system that is responsible for the discrimination of large approximate quantities allows older individuals with DS to outstrip those with WS on a battery of number tasks, even though both are impaired with respect to TD controls. However, one of the problems that is inherent to infancy studies in genetic disorders, especially for those where the disorder is rare and incidences are low, is that they often can only involve a small number of participants, which in turn does not allow for a full statistical exploration of differences. However, the fact that, in our studies with new infant participants with WS and DS, we were able to replicate the findings of Paterson et al. (2006), indicates that the samples are representative of the age groups in each syndrome. In addition, interestingly, for WS such a small sample represents nonetheless more than 1% of the entire WS population in the UK. Moreover, it has been shown that variability in WS groups is no greater than that of a TD population (Van Herwegen et al., 2011). This suggests that large numbers are not always essential, although they are, of course, preferable.

Numerous genetic disorders experience difficulties with numerical abilities (see Simon, 2011, for an overview), which may emerge from very different developmental trajectories. In this chapter, we have argued that, in order to understand atypical numerical development and plan successful intervention therapies, it is necessary to explore multiple low-level interacting processes that underpin both impaired and proficient abilities. Further research using the cross-syndrome design may elucidate the different basic-level processes involved. Not only can such studies lead to more specific interventions, they also allow for the early detection of basic deficits impacting on subsequent numerical cognition before numerical difficulties have become obvious in older children.

References

Agrillo, C., Piffer, L., & Bisazza, A. (2010). Large number discrimination in mosquitofish. PloS ONE, 5(12), e15232.

Ansari, D., Donlan, C., & Karmiloff-Smith, A. (2007). Typical and atypical development of visual estimation abilities. Cortex, 43, 758–768.

Ansari, D., Donlan, C., Thomas, M., Ewing, S., & Karmiloff-Smith, A. (2003). What makes counting count? Verbal and visuo-spatial contributions to typical and atypical number development. Journal of Experimental Child Psychology, 85, 50–62.

Brown, J.H., Johnson, M.H., Paterson, S.J., Gilmore, R., Longhi, E, & Karmiloff-Smith, A. (2003). Spatial representation and attention in toddlers with Williams syndrome and Down Syndrome. Neuropsychologia, 41, 1037–1046.

Camos, V. (2009). Numerosity discrimination in children with Down Syndrome. Developmental Neuropsychology, 34(4), 435–447.

Demeyenne, N., Lestor, V., & Humphreys, G. (2010). Neuropsychological evidence for a dissociation in counting and subitizing. Neurocase, 16, 219–237.

Dolk, H., Loane, M., Garne, E., De Walle, H., Queisser-Luft, A., De Vigan, C., Addor, M.C., Gener, B., Haeusler, M., Jordan, H., Tucker, D., Stoll, C., Feijoo, M., Lillis, D., & Bianchi, F. (2005). Trends and geographic inequalities in the prevalence of Down syndrome in Europe, 1980–1999. Revue d'épidémiologie et de Santé Publique, 53(2), 87–95.

Donnai, D., & Karmiloff-Smith, A. (2000). Williams syndrome: from genotype through to the cognitive phenotype. American Journal of Medical Genetics, 97, 164–171.

Gross, H.J., Pahl, M., Si, A., Zhu, H., Tautz, J., & Zhang, S. (2009). Number-based visual generalisation in the honeybee. PLoS ONE, 4(1), e4263.

Farran, E. K., Jarrold, C., & Gathercole, S.E. (2001). Block design performance in the Williams syndrome phenotype: a problem with mental imagery. Journal of Child Psychology and Psychiatry, 42, 719–728.

Farran, E.K., & Karmiloff-Smith, A. (Eds) (2012). Neurodevelopmental Disorders across the Lifespan: A Neuroconstructivist Approach. Oxford: Oxford University Press.

Feigenson, L., Dehaene, S.. & Spelke, E.S. (2004). Core systems of number. Trends in Cognitive Sciences, 8 (7), 307–314.

Fenson, L., Dale, P.S., Reznick, J.S., Thal, D., Bates, E., Hartung, J.P., Pethick, S., & Reilly, J.S. (1993). The MacArthur Communicative Development Inventories: User's Guide and Technical Manual. Baltimore: Paul H. Brokes Publishing Co.

Hauser, M.D., & Carey, S. (2003). Spontaneous representations of small numbers of objects by rhesus monkeys: examinations of content and format. Cognitive Psychology, 47(4), 367–401.

Holloway, I.D., & Ansari, D. (2010) Developmental specialization in the right intraparietal sulcus for the abstract representation of numerical magnitude. Journal of Cognitive Neuroscience, 22, 2627–2637.

Hudson, K.D., & Farran, E.K. (2011). Executive function and motor planning. In: Farran, E.K., & Karmiloff-Smith, A. (Eds), Neurodevelopmental Disorders Across the Lifespan: Lessons from Williams Syndrome. Oxford: Oxford University Press.

Jarrold, C., Baddeley, A.D., & Hewes, A.K. (1999). Genetically dissociated components of working memory: evidence from Down's and Williams syndrome. Neuropsychologia, 37, 637–651.

Karmiloff-Smith, A. (1998). Development itself is the key to understanding developmental disorders. Trends in Cognitive Sciences, 2, 389–398.

Karmiloff-Smith, A. (2009). Nativism versus neurconstructivism: rethinking the study of developmental disorders. Developmental Psychology, 45, 56–63.

Karmiloff-Smith, A. (2013). Challenging the use of adult neuropsychological models for explaining neurodevelopmental disorders: Developed versus developing brains. The Quarterly Journal of Experimental Psychology, 66, 1–14.

Karmiloff-Smith, A., Brown, J.H., Grice, S., & Paterson, S. (2003). Dethroning the myth: cognitive dissociations and innate modularity in Williams syndrome. Developmental Neuropsychology, 23, 227–242.

Karmiloff-Smith, A., D'Souza, D., Dekker, T., Van Herwegen, J., Xu, F., Rodic, M., & Ansari, D. (2013). Genetic and environmental vulnerabilities: the importance of cross-syndrome comparisons. PNAS *190* (2), 17261-17265.

Krajcsi, A., Lukacs, A., Igacs, J., Racsmany, M., & Pleh, C. (2009). Numerical abilities in Williams syndrome: dissociating the analogue magnitude system and verbal retrieval. Journal of Clinical and Experimental Neurospychology, 31(4), 439–446.

Laing, E., Butterworth, G., Ansari, D., Gsödl, M., Longhi, E., Panagiotaki, G., Paterson, S. & Karmiloff-Smith, A. (2002). Atypical development of language and social communication in toddlers with Williams syndrome. Developmental Science, 5, 233–246.

Lemer, C., Dehaene, S., Spelke, E., & Cohen, L. (2003). Approximate quantities and exact number words: dissociable systems. Neuropsychologia, 41, 1942–1958.

Mervis, C.B., Morris, C.A. Bertrand, J., & Robinson, B.F. (1999). Williams syndrome: findings from an integrated program of research. In: H. Tager-Flusberg (Ed.), Neurodevelopmental Disorders (pp. 65–110). Cambridge MA: MIT Press.

Mervis, C.B., Morris, C.A., Klein-Tasman, B.P., Bertrand, J., Kwitny, S., Appelbaum, L.G., & Rice, C.E. (2003). Attentional characteristics of infants and toddlers with Williams syndrome during triadic interactions. Developmental Neuropsychology, 23, 243–268.

Morris, C.A., Demsey, S.A., Leonard, C.O., Dilts, C., & Blackburn, B.L. (1988). Natural history of Williams syndrome: physical characteristics. Journal of Pediatrics, 113(2), 318–326.

O'Hearn, K., & Landau, B. (2007). Mathematical skill in individuals with Williams syndrome: evidence from a standardized mathematics battery. Brain & Cognition, 64, 238–246.

O'Hearn, K., Landau, B., & Hoffman, J.E. (2005). Multiple object tracking in people with Williams syndrome and in normally developing children. Psychological Science, 16(11), 905–912.

Pangalos, C., Avramopoulos, D., Blouin, J.L., Raoul, O., deBlois, M.C., Prieur, M., Schinzel, A.A., Gika, M., Abazis, D., & Antonarakis, S.E. (1994). Understanding the mechanism(s) of mosaic trisomy 21, by using DNA polymorphism analysis. American Journal of Human Genetics, 54, 473–481.

Paterson, S.J., Brown, J.H., Gsödl, M.K., Johnson, M.H., & Karmiloff-Smith, A. (1999). Cognitive modularity and Genetic disorders. Science, 286, 2355–2358.

Paterson, S.J., Girelli, L., Butterworth, B., and Karmiloff-Smith, A. (2006). Are numerical impairments syndrome specific? Evidence from Williams syndrome and Down's syndrome. Journal of Child Psychology and Psychiatry, 47(2), 190–204.

Pinker, S. (1999). Words and Rules. London: Weidenfeld & Nicolson.

Riby, D.M., Jones, N., Brown, P.H., Robinson, L.J., Langton, S.R., Bruce, V., & Riby, L.M. (2011). Attention to faces in Williams syndrome. Journal of Autism and Developmental Disorders, 41, 1228–1239.

Roizen, N.J., & Patterson, D. (2003). Down's syndrome. Lancet, 12(361), 1281–1289.

Scerif, G., Cornigh, K., Wilding, J., Driver, J., & Karmiloff-Smith, A. (2004). Visual search in typically developing toddlers and toddlers with Fragile X or Williams syndrome. Developmental Science, 7(1), 116–130.

Silverman, W. (2007). Down syndrome: cognitive phenotype. Mental Retardation and Developmental Disabilities, 13, 228–236.

Simon, T.J. (1997). Reconceptualizing the origins of number knowledge: a 'non-numerical' account. Cognitive Development, 12, 349–372.

Simon, T.J. (2011). Clues to the foundation of numerical cognitive impairments: evidence from genetic disorders. Developmental Neuropsychology, 36(6), 788–805.

Singer-Harris, N.G., Bellugi, U., Bates, E., Jones, W., & Rossen, M. (1997). Contrasting profiles of language development in children with Williams and Down syndromes. Developmental Neuropsychology, 13, 345–370.

Stinton, C., Farran, E., & Courbois, Y. (2008). Mental rotation in Williams syndrome: an impaired imagery ability. Developmental Neuropsychology, 33(5), 565–583.

Stith, L.E., & Fishbein, H.D. (1996). Basic money-counting skills of children with mental retardation. Research in Developmental Disabilities, 17(3), 185–201.

Tassabehji, M. (2003). Williams–Beuren syndrome: a challenge for genotype-phenotype correlations. Human Molecular Genetics, 12(2), 229–237.

Thomas, M.S.C., & Karmiloff-Smith, A. (2002). Are developmental disorders like cases of adult brain damage? Implications from connectionist modelling. Behavioral and Brain Sciences, 25, 727–788.

Thomas, M.S.C., Purser, H., & Van Herwegen, J. (2012). The developmental trajectories approach to cognition. In E. K. Farran & A. Karmiloff-Smith (Eds), Neurodevelopmental Disorders across the Lifespan: A Neuroconstructivist Approach. Oxford: Oxford University Press.

Tobii User Manual (2nd edn) (2003). Stockholm: Tobii.

Udwin, O., Davies, M., & Howlin, P. (1996). A longitudinal study of cognitive and education attainment in Williams syndrome. Developmental Medicine and Child Neurology, 38, 1020–1029.

Uller, C., Jaeger, R., Guidry, G., & Martin C. (2003). Salamanders (Plethodon cinereus) go for more: rudiments of number in an amphibian. Animal Cognition, 6, 105–112.

Van Herwegen, J., Ansari, D., Xu, F., & Karmiloff-Smith, A. (2008). Small and large number processing in infants and toddlers with Williams syndrome. Developmental Science, 11 (5), 637–643.

Van Herwegen, J., Rundblad, G., Davelaar, E.J., & Annaz, D. (2011). Variability and standardised test profiles in typically developing children and children with Williams syndrome. British Journal of Developmental Psychology, 29, 883–894.

Xu, F. (2003). Numerosity discrimination in infants: evidence for two systems of representations. Cognition, 89, B15–B25.

Xu, F., Spelke, E.S., & Goddard, S. (2005). Number sense in human infants. Developmental Science, 8(1), 88–101.

CHAPTER 40

BRAIN CORRELATES OF NUMERICAL DISABILITIES

KARIN KUCIAN, LIANE KAUFMANN, AND MICHAEL VON ASTER

INTRODUCTION

THE following chapter summarises the current state of research that relates the inability to mentally operate with numbers to corresponding brain mechanisms. Before looking at the sophistic details of different methods and results we would like to shortly address some core issues of how to define and conceptualise dyscalculia.

Dyscalculia can occur (i) as an acquired disorder, a loss of an established ability, or (ii) developmentally as an inability to acquire the ability. The former is easily explained by the circumstances that clinically link the different levels of ability before and after a damaging event. In contrast, the latter, which is called developmental dyscalculia (DD), can only be explained by influences on learning and development that are highly complex and far from being fully understood.

According to current classification manuals (ICD 10 by WHO, DSM IV by APA) DD is conceptualised as a domain specific learning disorder that emerges at a very early stage of development and cannot be explained by inappropriate schooling or deficient learning opportunities. Domain specificity commonly is defined by a discrepancy between general intelligence and specific ability. This concept implies a rather static and endogenous origin of the disorder, which is challenged by recent research and theory that addresses the nature of general intelligence (IQ) and the power of development.

Psychometric intelligence in diagnostic purposes serves as a stable component of personality that should be independent of the psychometrically discrepant domain. However, both assumptions (i.e. stability and independence) seem questionable. Arithmetic as a common component in IQ-testing is highly correlated with other verbal and non-verbal components of psychometric intelligence. At the brain level general intelligence seems to depend on intact white matter tracts linking frontal with parietal regions (Gläscher, Rudrauf, Colom, Paul, Tranel, Damasio et al., 2010). Similarly, proficient number processing and calculation relies on established fibre connections between frontal and parietal areas (Rykhlevskaia,

Uddin, Kondos, & Menon, 2009; Tsang, Dougherty, Deutsch, Wandell, & Ben-Shachar, 2009; van Eimeren, Niogi, McCandliss, Holloway, & Ansari, 2008). Hence, it is plausible that general intelligence and arithmetic alike involve partly overlapping white matter fibres connecting the frontal with the parietal lobes. Moreover, IQ-ability-discrepancy seems to make no difference at the behavioral and at the brain level – IQ-discrepant and IQ-non-discrepant poor readers did exhibit the same pattern of dysfunctional ability and correlating brain activity (Tanaka, Black, Hulme, Stanley, Kesler, Whitfrield-Gabrieliet al., 2011). The same seems to be true for dyscalculic children (Ehlert, Schroeders, & Fritz-Stratmann, 2012). IQ also seems much more variable as expected and can increase and decrease throughout the teenage years. These changes are correlated with changes in structural brain development and are specific for verbal and non-verbal IQ (Ramsden, Richardson, Josse, Thomas, Ellis, Shakeshaft et al., 2011).

When we correlate psychological symptoms to observable biological conditions, this often leads to implicit conclusions about causal relationships: The biological (mal-) condition precedes and explains the psychological (mal-) function. This obviously is true for acquired disorders, i.e. a loss of a particular mental function following stroke or brain injury. A well-known example is the so called Gerstmann Syndrome, first described in 1930, in which a circumscribed brain lesion in the left angular gyrus was followed by dyscalculia, dysgraphia, finger agnosia, and left-right-disorientation. The same four symptoms, however, have been described in children without any history of brain damage. Kinsbourne (1968) called that 'Developmental Gerstmann Syndrome'. Although we know from recent brain imaging research that in proficient adults, among other regions, the angular gyrus plays an important role in number processing and calculation (Grabner, Ansari, Koschutnig, Reishofer, & Ebner, 2011; Grabner, Ansari, Koschutnig, Reishofer, Ebner, & Neuper, 2009), the question of etiological causality remains open in developmental disorders. In contrast to acquired disorders a complete reverse direction of causal relationship should be kept in mind: the psychological (mal-) function precedes and explains the biological (mal-) condition. This paradigm reflects the fact that functional, as well as morphological brain development is highly plastic and depends on individual learning experiences and environments in which mental representations, factual knowledge, abilities, and attitudes are successively formed.

The development of number processing and calculation abilities is rather complex and dynamic. It parallels and is intertwined with the development of other cognitive domains such as language, as well as with domain general abilities like working memory, self-regulating and reflecting capacities at cognitive, emotional, and/or behavioral levels (e.g. see chapters by Geary, this volume; Jordan, this volume; Moore, Rudig &Ashcraft, this volume). A stepwise and hierarchical development of mental numerical representations, beginning with concrete magnitudes in infancy, followed by verbal and Arabic symbolization in the preschool years and a successively expanding spatially oriented mental number line during the primary school years has been proposed in a 'four-step-developmental-model' (von Aster & Shalev, 2007). It can account for different individual causes, affecting different contributing functions and neural circuits at different stages of development, and usually leading to an inability to create and automatise mental number line abilities at an age appropriate level. A more recent model, proposed by Kaufmann, Wood, Rubinsten, and Henik (2011) explicitly addresses the importance of different domain general factors that contribute to number processing and calculation development (see also Kaufmann, Kucian, and von Aster, this volume).

As children with DD very often show child psychiatric comorbidities on a clinical or even subclinical level, not only including symptoms from the ADHD and dyslexia spectrum, but also from the spectrum of socio-emotional and conduct disorders, DD should be conceptualised as one component within a broader neurodevelopmental spectrum disorder. According to issues of education, prevention, and therapy, concepts of the disorder that focus on only one single, presumably inherited core deficit seem to be a short cut and quite maladaptive. Narrow genetic concepts that focus on heritability alone obscure the view of familial and professional caregivers on relevant experience dependent environment-gene-interactions by epigenetic regulation mechanisms [Banaschewski, 2012; see also the 'generalist-genes' account for a differentiated view on the interplay between genetic and environmental factors on learning disabilities (Haworth, Kovas, Harlaar, Hayiou-Thomas, Petrill, Dale et al., 2009; Kovas & Plomin, 2006; see also the chapter by Tosto, Haworth, & Kovas, this volume)]. The early impact of epigenetic influences on the development of a vast variety of intellectual, as well as regulatory functions might be similar relevant as the experience of failure, anxiety, and avoidance in later school years.

Brain Correlates of Developmental Dyscalculia

Convergent evidence is growing that DD has a particular neural correlate. Recent work in the field of DD has emphasised the neural aspects of this learning disorder by means of contemporary brain imaging techniques such as electrophysiology and magnetic resonance imaging. Despite the relatively high prevalence of DD only very little imaging studies have addressed the question of neuronal correlates of this specific learning deficit. In the last few years, we have gained a clearer picture of functional processes in the typical adult brain during number processing and calculation. There is converging evidence that mainly the intraparietal sulcus and the superior and inferior parietal lobule display the core regions for numerical and mathematical processing. However, beside parietal areas, other regions of the brain are also important, depending on the task to be solved. Moreover, studies examining typically developing children showed that during schooling and learning the neuronal activation patterns are formed, adapted and fine-tuned (Kaufmann et al., 2011; Kucian & Kaufmann, 2009; Kucian, von Aster, Loenneker, Dietrich, & Martin, 2008; Rivera, Reiss, Eckert, & Menon, 2005). Not only the developmental stage and competence level, but also the type of numerical task has a significant impact on the brain activation pattern, which constitutes a special challenge in the comparison of different imaging studies. Hence, findings derived from mature brain systems are not readily applicable to developing brain systems (Karmiloff-Smith, 1998; Kaufmann et al., 2011), immature brain systems show higher inter-individual variance, which is even more pronounced in children with a specific developmental disorder, like dyscalculia. Furthermore, the examination of children by different imaging techniques requires special care, patience, and expertise. Despite these obstacles, few studies have met the challenge and provide first important insights in the neural correlates of DD. In the following sections, these studies are summarised regarding brain

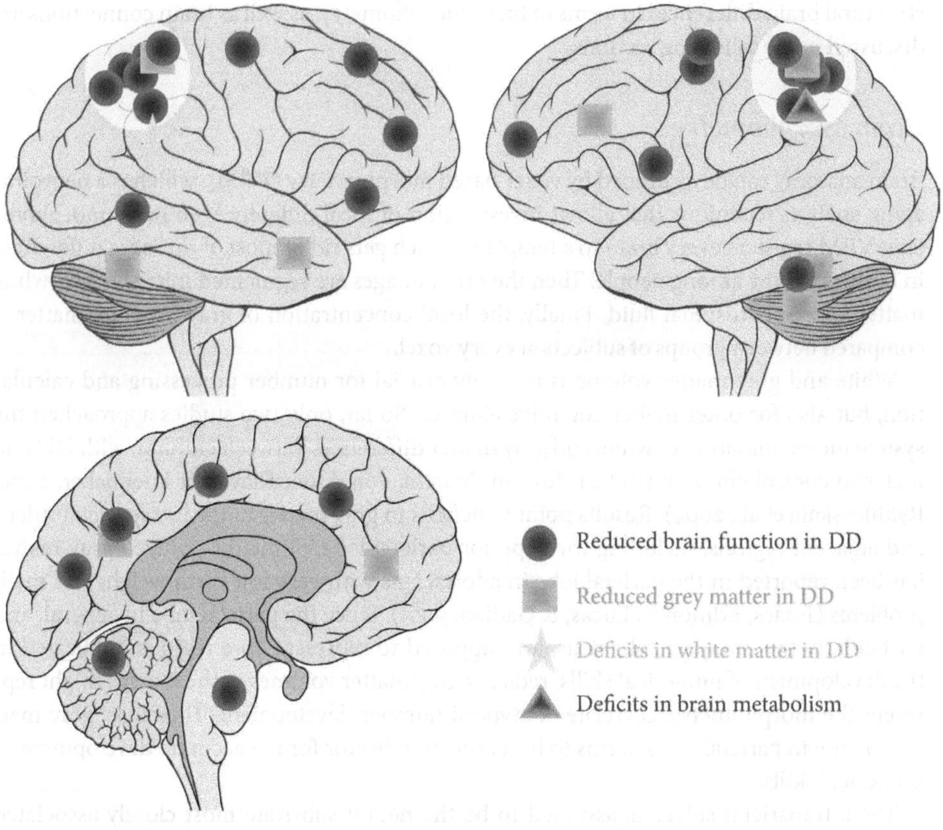

FIGURE 40.1 Summary of deficient brain function (pink circles), gray matter (green squares), white matter (yellow stars), and brain metabolism (blue triangle) in children with DD. Reported deficits include a variety of brain regions, however, there seems consistent evidence that DD is associated with deficits in the parietal lobes (marked in white) which host core regions for numerical understanding.

structure, metabolism, and function as well as the effects of intervention on neuronal systems is presented (please see also Figure 1).

Brain Structure

The acquisition of high resolution anatomical brain scans by means of magnetic resonance imaging allows the examination of morphometric differences in the brain between children with and without DD. In more detail, these images provide the chance to differentiate between gray and white brain matter, and to compare morphometric differences in these volumes between typically and atypically developing children.

Besides, magnetic resonance imaging also enables the measurement of connections between different brain regions using diffusion tensor imaging. Recent findings point to possible disconnections between brain areas in children with DD. Both findings of structural brain differences in terms of brain morphometry, as well as brain connections are discussed in the following sections.

Brain Morphometry

Brain anatomy can be examined by voxel-based morphometry (VBM), which is a neuroimaging analysis technique that allows investigation of focal differences in brain morphometry. VBM registers every brain to a template, which gets rid of most of the large differences in brain anatomy among people. Then the brain images are segmented into gray and white matter and cerebrospinal fluid. Finally, the local concentration of gray or white matter is compared between groups of subjects at every voxel.

White and gray matter volume is not only crucial for number processing and calculation, but also for other higher cognitive abilities. So far, only two studies approached the systematic examination of white and gray matter differences between children with DD and matched control children (Rotzer, Kucian, Martin, von Aster, Klaver, & Loenneker, 2008; Rykhlevskaia et al., 2009). Results point to deficits in gray matter in the intraparietal sulcus and adjacent regions, including the superior parietal lobe. Similarly, reduced gray matter has been reported in the parietal lobe in adolescents with very low birth weight and math problems (Isaacs, Edmonds, Lucas, & Gadian, 2001). Since the parietal lobe in general, and particularly the intraparietal sulcus, are supposed to represent core regions implicated in the development of numerical skills, reduced gray matter volumes in these areas might represent the morphometric correlate of atypical number development. Therefore, gray matter volume in parietal areas seems to be a crucial indicator for the accurate development of numerical skills.

The intraparietal sulcus is assumed to be the neural substrate most closely associated with DD. However, recent results also argue for a more distributed neuronal pattern. In line, reported gray matter deficits in children with DD emphasise widespread malformations in the brain. Reduced gray matter volume has also been reported in the anterior cingulate cortex, the left inferior frontal gyrus, and the dorsolateral prefrontal cortex (Rotzer et al. 2008). Based on the fact that an important component in the development of arithmetical skills is the growth of working memory for numerical information, the authors argue that deficits in these frontal areas may be of major importance in the development of dyscalculia.

In addition, the ventral visual stream seems to be affected as well (Rykhlevskaia et al., 2009). In particular, children with DD showed deficits in the fusiform gyrus, parahippocampal gyrus, and right anterior temporal cortex which might hinder the development of semantic memory representations important for rapid numerical fact retrieval (Rykhlevskaia et al., 2009).

Regarding white matter integrity, the results of existing studies report reduced white matter volume in children with DD, but regions are not consistent (Rotzer et al., 2008; Rykhlevskaia et al., 2009). Whereas, Rotzer et al. (2008) reported white matter deficits in left frontal lobe and in the right parahippocampal gyrus in dyscalculic children, Rykhlevskaia et al. (2009) found reduced white matter volume in the right temporoparietal region and the splenium of the corpus callosum. Deficits in these regions may have a negative influence on fact retrieval and spatial memory processing, as well as on visuospatial processes during the acquisition of mathematical skills (Rotzer et al., 2008; Rykhlevskaia et al., 2009).

Taken together, DD is characterised by reduced gray and white matter volume in areas that are supposed to play a key role for the domain specific development of numerical representations and abilities, as well as in regions important for the development of domain general abilities.

Brain Connections

In addition to the long neuronal fibre tracts that connect the brain to the rest of the body, there is a complicated network formed by short connections among different cortical and subcortical regions. Therefore, successful cognitive performance relies on the development and establishment of well-organised networks in the brain. Fast and accurate connections between the different brain regions are crucial for efficient transfer and adjustment of information. Magnetic resonance imaging provides a technique, called diffusion tensor imaging (DTI), to characterise the quality of brain connections. DTI is a magnetic resonance imaging (MRI) technique providing information about white matter microstructure non-invasively. Diffusion of water molecules is restricted by axons in white matter. This directional dependence of diffusion can be quantified as fractional anisotropy (FA), which reflects both size and number of myelinated axons, and the coherence of axonal orientation. Reduced FA is a marker of white matter integrity, and can point to a disruption or disorganisation of white matter tracts, which has been demonstrated in a variety of diseases [details about DTI can be found e.g. in the review article by (Le Bihan, Mangin, Poupon, Clark, Pappata, Molko et al., 2001)].

Thus far, only one DTI-study examined microstructural differences in white matter between children with DD and typically achieving children (Rykhlevskaia et al., 2009). The results emphasise the evidence of impaired fibre tracts connecting different brain areas of the frontoparietal network known to be responsible for number processing and calculation. In detail, the long-range white matter projection fibres linking the right fusiform gyrus with temporoparietal white matter are suggested to represent a specific source of vulnerability in children with DD (Rykhlevskaia et al., 2009). However, it has to be mentioned that Rykhlevskaia et al. (2009) only performed specific analysis of fibre tracts in regions that showed reduced white matter volume in a prior analysis of brain morphometry. Therefore, the present results display an important indication for the presence of deficient fibre connections as neurological correlate of dyscalculia, but presumably also additional fibre tracts might be affected.

Comparably, white matter integrity correlated significantly with mathematical abilities in typically developing children (Tsang et al., 2009; van Eimeren et al., 2008). However, in contrast to the results in children with DD, white matter microstructures in the left hemisphere seemed to be related to individual mathematical competencies in typically developing children (van Eimeren et al., 2008). Particularly, the posterior part of the left inferior longitudinal fasciculus correlated specifically with achievement in numerical operations test. White matter integrity in the inferior longitudinal fasciculus as part of the visual stream might be related to the efficiency with which children are able to process Arabic numerals presented in calculation tests (van Eimeren et al., 2008). Another study explicitly investigated the role of the anterior portion of the superior longitudinal fasciculus, a pathway connecting parietal and frontal areas (Tsang et al., 2009). The authors have chosen this tract based on consistent findings of brain activation in a frontoparietal network during number processing and calculation. DTI-analysis revealed a positive correlation of white matter integrity with arithmetic approximation skills in typically achieving children. Accordingly, lower approximation ability would be explained by deficiencies in one or several characteristics of efficient fibre tracts connecting parietal with frontal areas (Tsang et al., 2009).

Finally, a DTI-study examining children with velocardiofacial syndrome, a condition associated with low arithmetic skills, reported reduced white matter integrity adjacent to left inferior parietal lobe and intraparietal sulcus (Barnea-Goraly, Eliez, Menon, Bammer, & Reiss, 2005). FA-values in this area correlated significantly with arithmetic ability in this region suggesting an important role of fibre projections from and to the left inferior parietal lobe.

Taken together, to date, only very little knowledge is available regarding specific impairments of brain connections as possible neuronal correlates of DD. Moreover, results from typically-achieving children or children with math difficulties due to other neurodevelopmental disorders are difficult to consign to DD. Nevertheless, results on hand add a further explanation of DD probably caused by a disconnection or reduced connection between important areas for number processing.

Brain Metabolism

Magnetic resonance spectroscopy (MRS) provides a measure of brain chemistry. It can be used as a non-invasive technique to study changes in brain metabolites of different disorders. To our knowledge, only one single case study examined metabolic abnormalities in an 18-year-old man with DD (Levy, Reis, & Grafman, 1999). Magnetic resonance spectroscopy revealed specific and stable changes in a left temporoparietal region near the angular gyrus, with a significant decrease in choline and creatine levels, as well as a mild focal decrease in N-acetyl-aspartate. The authors conclude that changes in these brain metabolites are due to an alteration in cell density, impaired cellular energy metabolism in these neurons, and only little general neuronal loss (Levy et al., 1999). Since, only this single case study exists, it is difficult to draw strong conclusions, but these results point to significant abnormalities in brain chemistry near the angular gyrus, a region that has also been reported to play an important role in calculation and numerical fact retrieval (Grabner et al., 2009, 2011) and invites further research using magnetic resonance spectroscopy in children with DD.

Brain Function

Compared with brain imaging studies investigating brain structural or metabolic correlates of DD, differences in brain function between dyscalculic children and controls have received more attention from developmental neuroscientists. Nonetheless, functional magnetic resonance imaging (fMRI) or electrophysiological studies including children with DD are still scarce. In addition, study designs and the tested paradigms are quite different: symbolic (e.g. Arabic digits) or non-symbolic (e.g. dot patterns) presentations of numerical stimuli have been used, active tasks, where children were asked to give an answer, passive adaptation paradigms where no answer was necessary, basic number processing or arithmetical skills were tested in different difficulty levels. All these variables have a significant impact on the corresponding brain function. Another point that limits the comparability of obtained results is that the investigated groups of dyscalculic children differ between studies according to definitional criteria and assessment tools that have been used resulting in different neuropsychological profiles. Altogether it is still difficult to draw a clear picture of the functional brain correlates of DD. Nevertheless, current findings indicate consistently dysfunctions of the parietal lobe during number processing and calculation (Kaufmann, Vogel, Starke, Kremser, Schocke, & Wood, 2009; Kucian, Grond, Rotzer, Henzi, Schönmann, Plangger et al., 2011a; Kucian, Loenneker, Dietrich, Dosch, Martin, & von Aster, 2006; Kucian, Loenneker, Martin, & von Aster, 2011b; Mussolin, De Volder, Grandin, Schlogel, Nassogne, & Noel, 2010; Price, Holloway, Rasanen, Vesterinen, & Ansari, 2007; Rotzer, Loenneker, Kucian, Martin, Klaver, & von Aster, 2009; Soltész, Szucs, Dékány, Márkus, & Csépe 2007); for a review see the meta-analysis of Kaufmann et al. (2011). While some studies showed a general reduced brain activation in parietal areas (Kucian et al., 2006, 2011a; Rotzer et al., 2009), the findings of one study suggest enhanced – possibly compensatory activity – in parietal regions (Kaufmann et al., 2009) and still others point to a lack of modulation of parietal brain activation according to the numerical distance between two Arabic digits or sets of dots (Mussolin et al., 2010; Price et al., 2007; Soltész et al., 2007). Divergent results might derive from differences between studies regarding inclusion criteria of children with DD, conducted paradigms and fMRI-parameters. However, accordant and most prominent differences are evident in or in close vicinity to the intraparietal sulcus, but exact locations within the parietal lobes and their laterality varies also between studies. Taken together, there is consistent evidence that DD is associated with deviant parietal activity patterns indicating a deficient neuronal representation of numerosity. Corroborating evidence comes from simulated interruption of right parietal activity by transcranial magnetic stimulation (TMS), which evoked dyscalculic-like behavioral characteristics in adult volunteers without any problems in numerical understanding (Cohen Kadosh, Cohen Kadosh, Schuhmann, et al., 2007).

However, even the easiest arithmetic problem poses a high cognitive challenge in terms of neuronal processing and networks from different brain areas are necessary for a successful execution. Hence, it is not surprising that dysfunctions in areas other than the parietal lobes might contribute to the occurrence of DD. In line, there is growing evidence that the development of specific math problems might also be due to deficits in other or additional brain regions. In particular, several imaging studies reported impaired brain activation in distributed areas of the frontal lobe, including the superior, middle, and inferior frontal gyrus, medial prefrontal gyrus, the paracentral frontal lobe, the cingulate gyrus, or the insula (Kaufmann et al., 2011; Kucian et al., 2006, 2011a,b; Mussolin et al., 2010; Price et al., 2007; Rotzer et al., 2009). These frontal areas are attributed to domain general executive functions. A variety of cognitive control, regulation, or managing processes are included in the term executive functions, such as planning, working memory, attention, problem solving, inhibition, mental flexibility, or initiation and monitoring of actions. Many are mandatory for different components of number processing and calculation, including an accurate and automated access to mental number representations. In line, different studies reported deficits in various facets of executive functions in DD, like working memory (Kaufmann, Lochy, Drexler, & Semenza, 2004), recruiting attention (Askenazi & Henik, 2010), or problem solving (Osmon, Smerz, Braun, & Plambeck, 2006; Soltész et al., 2007). Moreover, there is growing knowledge of brain regions related to executive processes. For instance, the right inferior frontal gyrus, presupplementary motor area, and basal ganglia are necessary for successful inhibition and, whereas medial frontal brain regions are commonly associated with monitoring behavior (detection of errors, conflict processing), middle frontal regions are supposed to be related to adjusting behavior after conflicts or errors (Verbruggen & Logan, 2008). Future research will have to clarify the relationship between impairments of different areas within the frontal lobes and decelerated executive functioning in DD.

Besides these deficits in frontal regions, occipitotemporal areas and deep brain structures also seem to be affected in children with DD. For instance, the left fusiform gyrus has been reported to be deficient in dyscalculic children, presumably associated with problems of Arabic digit processing (Kaufmann et al., 2011; Price et al., 2007). Abnormal activation in subcortical areas, such as the thalamus, basal ganglia, or the parahippocampal gyrus also has been reported to be present in children with DD, which even broadens the impaired network to a disruption of cortico-subcortical loops (Delazer, Domahs, Lochy, Karner, Benke, & Poewe, 2004; Kucian et al., 2006, 2011a).

Regarding this wide range of reported abnormalities in neural networks for number processing and calculation, it might appear surprising that children with DD are still able to perform some numerical tasks quite well or even at the same level of their typically achieving peers. Different neuroimaging studies described equal behavioral performance, despite of differences in brain activation patterns between children with and without DD (Kaufmann et al., 2009; Kovas, Giampietro, Viding, Ng, Brammer, Happé et al., 2009; Kucian et al., 2006, 2011b; Mussolin et al., 2010; Soltész et al., 2007). Such results raise the question of possible compensatory mechanisms that children with DD might develop in order to cope with their learning disability. In fact, it has been reported that affected children show stronger activation of certain brain regions that might account for such compensatory mechanisms, including areas next to regions in the frontal lobes (supplementary motor area, post-central gyrus, superior frontal lobe, paracentral frontal lobe), also the left lingual gyrus, the thalamus and even inferior parietal areas (Kaufmann et al., 2011; Kovas et al., 2009; Kucian et al., 2011b; Mussolin et al., 2010). Kaufmann and colleagues (2011) argued that increased activation of more anterior parietal areas reaching into the post-central gyrus might be a result of the dependence on finger-based strategies upon solving simple arithmetical tasks. Moreover, stronger activation in frontal areas suggests that dyscalculic children compared with typically achieving peers need more effort to solve the same numerical problems. They seem to engage stronger demands on functions like visual working memory, attention or response selection (Kaufmann et al., 2011; Kucian et al., 2011b; Mussolin et al., 2010).

To summarize, based on existing literature, it can be concluded that the activation pattern of children with DD is less concise and main deficits are apparent in core regions for number processing, which mainly comprise parietal regions. However, also other cortical and subcortical regions that contribute to numerical cognition can be affected. The stronger recruitment of supporting areas associated with working memory, attention, monitoring, updating, or finger representation are supposed to reflect compensatory mechanisms in dyscalculic children. The increased need for bearing functions might be explained by a deficient development and automated access to adaptive number representations in children with DD.

Intervention

There is an urgent need to transform basic knowledge about neurocognitive development into evidence based practical applications for special needs education and therapy. Developmental cognitive scientists, teachers, and therapists are challenged to provide adequate and successful intervention techniques to failing learners. Moreover, research is needed on the neural consequences of intervention. Even where intervention improves performance, it may not be clear whether the child's cognitive and neural functioning has become more typical or whether compensatory mechanisms have been developed (Butterworth, Varma, & Laurillard, 2011).

Up to now, only one study tracked changes in brain function that follow specific intervention in children with DD (Kucian et al., 2011a). The authors developed a computer-based training program built on current neuroscientific concepts of DD. The aim of the 5-week intervention was to improve number representations and the close link of numbers to spatial properties on the internal mental number line in particular. Results were very promising and showed that children with and without DD enhanced their spatial number representations. In addition, a positive transfer effect has been observed demonstrated by improved arithmetical abilities, although children were never trained in exact arithmetic. This further highlights the importance of a precise mapping and automated access to the mental number line for calculation skills.

Next to behavioral improvements, the training was followed by a modulation of brain functions. Functional magnetic resonance imaging depicted a decrease in brain activation immediately after the training. Children showed a clear reduction in the recruitment of relevant brain regions after the training, including mainly frontal areas, bilateral intraparietal sulci and the left fusiform gyrus. Reduction of brain activation in these regions, and particularly of the frontal lobe, is argued to reflect automatization of cognitive processes necessary for mathematical reasoning (Ischebeck, Zamarian, Egger, Schocke, & Delazer, 2007; Ischebeck, Zamarian, Siedentopf, Koppelstätter, Benke, Felber et al., 2006; Pauli, Lutzenberger, Rau, Birbaumer, Rickard, Yaroush et al., 1994). Therefore, it seems that after completion of the training, the task puts less demand on quantity processing, working memory and requires less attentional effort. In a follow-up examination 5 weeks after training a significant increase of activity in parietal areas, including the intraparietal sulcus bilaterally was found in children with DD. Since the intraparietal sulcus is known to play a pivotal role in number representation, the authors speculate that time for consolidation after training was needed to establish neuronal representation.

In conclusion, domain specific training is associated with changes in functional circuitry within and between systems, and can shape corresponding brain activations.

Further studies are warranted to examine the effects and neural correlates of different intervention strategies in order to develop adaptive and differential indications for the particular needs of individual children. This may also include techniques like transcranial direct current stimulation (TDCS) that are supposed to facilitate learning of numerical abilities by specific stimulation of relevant brain regions during cognitive training (Cohen Kadosh, Soskic, Iuculano, Kanai, & Walsh, 2010).

General Conclusions

As mentioned in the beginning of that chapter, DD is far from being explained by correlating specific attributes in brain morphology and functional activity to certain ability patterns. To that point neuroscientific evidence should not be over-estimated. However, brain imaging up to now provides some important, but still quite vague answers to the questions if and where brains of children diagnosed with DD differ from those of typically achieving peers. Main deficits are apparent in core regions for number processing, which mainly comprise gray, as well as white matter in parietal lobes. Moreover, it already can be demonstrated that brains are changing according to development, learning, and intervention. However, regarding the question as to why children exhibit DD, up to now only few puzzle pieces really seem available. On the way from here to the whole picture prospective longitudinal and interdisciplinary research is necessary that includes cognitive developmental neuropsychology, as well as clinical and educational perspectives in order to address the huge variety of relevant factors that, not physically (as with acquired disorders), but psychologically 'damage' functional and structural brain development. The role of different comorbid symptoms like attentional, perceptual, emotional, and behavioral dysfunctions and dysregulations and their mutual effects on cognitive development seem as under-recognised as the impact of inappropriate learning environments, early school failure, conditioned anxieties, and avoiding behavior on future learning biographies.

References

APA. (2013). Diagnostic and statistical manual of mental disorders. Washington, DC: American Psychiatric Association.

Askenazi, S., & Henik, A. (2010). Attentional networks in developmental dyscalculia. Behavioral and Brain Functions, 6, 2.

Banaschewski, T. (2012). Editorial: can we dissect the interplay of genes and environment across development? Journal of Child Psychology and Psychiatry, and Allied Disciplines, 53(3), 217–218.

Barnea-Goraly, N., Eliez, S., Menon, V., Bammer, R., & Reiss, A.L. (2005). Arithmetic ability and parietal alterations: a diffusion tensor imaging study in velocardiofacial syndrome. Brain research. Cognitive Brain Research, 25(3), 735–740.

Butterworth, B., Varma, S., & Laurillard, D. (2011). Dyscalculia: from brain to education. Science, 332(6033), 1049–1053.

Cohen Kadosh, R., Cohen Kadosh, K., Schuhmann, T., Kaas, A., Goebel, R., Henik, A., Sack, AT. (2007). Virtual dyscalculia induced by parietal-lobe TMS impairs automatic magnitude processing. Current Biology, 17(8), 689–693.

Cohen Kadosh, R., Soskic, S., Iuculano, T., Kanai, R., & Walsh, V. (2010). Modulating neuronal activity produces specific and long-lasting changes in numerical competence. Current Biology, 20(22), 2016–2020.

Delazer, M., Domahs, F., Lochy, A., Karner, E., Benke, T., & Poewe, W. (2004). Number processing and basal ganglia dysfunction: a single case study. Neuropsychologia, 42(8), 1050–1062.

Ehlert, A., Schroeders, U., & Fritz-Stratmann, A. (2012). Criticism of the discrepancy criterion in the diagnosis of dyslexia and dyscalculia. Lernen und Lernstörungen, 1(3), 169–184.

Geary, D.C. (this volume). The classification and cognitive characteristics of mathematical disabilities in children. In The Oxford Handbook of Numerical Cognition, R. Cohen Kadosh, and A. Dowker, eds. (Oxford University Press).

Gläscher, J., Rudrauf, D., Colom, R., Paul L. K., Tranel D., Damasio H., Adolphs R.. (2010). Distributed neural system for general intelligence revealed by lesion mapping. Proceedings of the National Academy of Sciences of the United States of America, 107(10), 4705–4709.

Grabner, R. H., Ansari, D., Koschutnig, K., Reishofer, G., & Ebner, F. (2011). The function of the left angular gyrus in mental arithmetic: evidence from the associative confusion effect. Human Brain Mapping, 34(5), 1013-24.

Grabner, R.H., Ansari, D., Koschutnig, K., Reishofer, G., Ebner, F., & Neuper, C. (2009). To retrieve or to calculate? Left angular gyrus mediates the retrieval of arithmetic facts during problem solving. Neuropsychologia, 47(2), 604–608.

Haworth, C.M., Kovas, Y., Harlaar, N., Hayiou-Thomas, ME., Petrill, SA., Dale, PS., Plomin, R. (2009). Generalist genes and learning disabilities: a multivariate genetic analysis of low performance in reading, mathematics, language and general cognitive ability in a sample of 8000 12-year-old twins. Journal of Child Psychology and Psychiatry, and Allied Disciplines, 50(10), 1318–1325.

Isaacs, E.B., Edmonds, C.J., Lucas, A., & Gadian, D.G. (2001). Calculation difficulties in children of very low birthweight: a neural correlate. Brain, 124(Pt 9), 1701–1707.

Ischebeck, A., Zamarian, L., Egger, K., Schocke, M., & Delazer, M. (2007). Imaging early practice effects in arithmetic. NeuroImage, 36(3), 993–1003.

Ischebeck, A., Zamarian, L., Siedentopf, C., Koppelstätter, F., Benke, T., Felber, S., Delazer M. (2006). How specifically do we learn? Imaging the learning of multiplication and subtraction. NeuroImage, 30(4), 1365–1375.

Jordan, J.A. (this volume). Individual differences in children's paths to arithmetical development. In The Oxford Handbook of Numerical Cognition, R. Cohen Kadosh, and A. Dowker, eds. (Oxford University Press).

Karmiloff-Smith, A. (1998). Development itself is the key to understanding developmental disorders. Trends in Cognitive Sciences, 2(10), 389–398.

Kaufmann, L., Kucian, K., and von Aster, M. (this volume). Development of the numerical brain. In The Oxford Handbook of Numerical Cognition, R. Cohen Kadosh, and A. Dowker, eds. (Oxford: Oxford University Press).

Kaufmann, L., Lochy, A., Drexler, A., & Semenza, C. (2004). Deficient arithmetic fact retrieval – storage or access problem? A case study. Neuropsychologia, 42(4), 482–496.

Kaufmann, L., Vogel, S., Starke, M., Kremser, C., Schocke, M., & Wood, G. (2009). Developmental dyscalculia: compensatory mechanisms in left intraparietal regions in response to nonsymbolic magnitudes. Behavioral and Brain Functions, 5(1), 35.

Kaufmann, L., Wood, G., Rubinsten, O., & Henik, A. (2011). Meta-analyses of developmental fMRI studies investigating typical and atypical trajectories of number processing and calculation. Developmental Neuropsychology, 36(6), 763–787.

Kinsbourne, M. (1968). Developmental Gerstmann syndrome. Pediatric Clinics of North America, 15(3), 771–778.

Kovas, Y., Giampietro, V., Viding, E., Ng, V., Brammer, M., Barker, GJ., Happé, FG., Plomin, R. (2009). Brain correlates of non-symbolic numerosity estimation in low and high mathematical ability children. PLoS ONE, 4(2), e4587.

Kovas, Y., & Plomin, R. (2006). Generalist genes: implications for the cognitive sciences. Trends in Cognitive Sciences, 10(5), 198–203.

Kucian, K., Grond, U., Rotzer, S., Henzi, B., Schönmann, C., Plangger, F., Gälli, M., Martin, E., von Aster, M. (2011a). Mental number line training in children with developmental dyscalculia. NeuroImage, 57(3), 782–795.

Kucian, K., & Kaufmann, L. (2009). A developmental model of number representation. Behavioral and Brain Sciences, 32(3/4), 340–341.

Kucian, K., Loenneker, T., Dietrich, T., Dosch, M., Martin, E., & von Aster, M. (2006). Impaired neural networks for approximate calculation in dyscalculic children: a functional MRI study. Behavioral and Brain Functions, 2, 31.

Kucian, K., Loenneker, T., Martin, E., & von Aster, M. (2011b). Non-symbolic numerical distance effect in children with and without developmental dyscalculia: a parametric FMRI study. Developmental Neuropsychology, 36(6), 741–762.

Kucian, K., von Aster, M., Loenneker, T., Dietrich, T., & Martin, E. (2008). Development of neural networks for exact and approximate calculation: a FMRI study. Developmental Neuropsychology, 33(4), 447–473.

Le Bihan, D., Mangin, J.F., Poupon, C., Clark, CA., Pappata, S., Molko, N., Chabriat, H. (2001). Diffusion tensor imaging: concepts and applications. Journal of Magnetic Resonance Imaging, 13(4), 534–546.
Levy, L.M., Reis, I.L., & Grafman, J. (1999). Metabolic abnormalities detected by 1H-MRS in dyscalculia and dysgraphia. Neurology, 53(3), 639–641.
Moore, A.M., Rudig, N.O., and Ashcraft, M.H. (this volume). Affect, motivation, working memory, and mathematics. In The Oxford Handbook of Numerical Cognition, R. Cohen Kadosh, and A. Dowker, eds. (Oxford University Press).
Mussolin, C., De Volder, A., Grandin, C., Schlogel, X., Nassogne, M.C., & Noel, M.P. (2010). Neural correlates of symbolic number comparison in developmental dyscalculia. Journal of Cognitive Neuroscience, 22(5), 860–874.
Osmon, D.C., Smerz, J.M., Braun, M.M., & Plambeck, E. (2006). Processing abilities associated with math skills in adult learning disability. Journal of Clinical and Experimental Neuropsychology, 28(1), 84–95.
Pauli, P., Lutzenberger, W., Rau, H., Birbaumer, N., Rickard, TC., Yaroush, RA., Bourne, LE. Jr. (1994). Brain potentials during mental arithmetic: effects of extensive practice and problem difficulty. Cognitive Brain Research, 2, 21–29.
Price, G.R., Holloway, I., Rasanen, P., Vesterinen, M., & Ansari, D. (2007). Impaired parietal magnitude processing in developmental dyscalculia. Current Biology, 17(24), R1042–R1043.
Ramsden, S., Richardson, F.M., Josse, G., Thomas, MS., Ellis, C., Shakeshaft, C., Seghier, ML., Price, CJ. (2011). Verbal and non-verbal intelligence changes in the teenage brain. Nature, 479(7371), 113–116.
Rivera, S.M., Reiss, A.L., Eckert, M.A., & Menon, V. (2005). Developmental changes in mental arithmetic: evidence for increased functional specialization in the left inferior parietal cortex. Cerebral Cortex, 15(11), 1779–1790.
Rotzer, S., Kucian, K., Martin, E., von Aster, M., Klaver, P., & Loenneker, T. (2008). Optimized voxel-based morphometry in children with developmental dyscalculia. NeuroImage, 39(1), 417–422.
Rotzer, S., Loenneker, T., Kucian, K., Martin, E., Klaver, P., & von Aster, M. (2009). Dysfunctional neural network of spatial working memory contributes to developmental dyscalculia. Neuropsychologia, 47(13), 2859–2865.
Rykhlevskaia, E., Uddin, L.Q., Kondos, L., & Menon, V. (2009). Neuroanatomical correlates of developmental dyscalculia: combined evidence from morphometry and tractography. Frontiers in Human Neuroscience, 3, 51.
Soltész, F., Szucs, D., Dékány, J., Márkus, A., & Csépe, V. (2007). A combined event-related potential and neuropsychological investigation of developmental dyscalculia. Neuroscience Letters, 417(2), 181–186.
Tanaka, H., Black, J.M., Hulme, C., Stanley, LM., Kesler, SR., Whitfield-Gabrieli, S., Reiss, AL., Gabrieli, JD., Hoeft, F. (2011). The brain basis of the phonological deficit in dyslexia is independent of IQ. Psychological Science, 22(11), 1442–1451.
Tosto, M.G., Haworth, C.M.A., and Kovas, Y. (this volume). Behavioural genomics of mathematics. In The Oxford Handbook of Numerical Cognition, R. Cohen Kadosh, and A. Dowker, eds. (Oxford University Press).
Tsang, J.M., Dougherty, R.F., Deutsch, G.K., Wandell, B.A., & Ben-Shachar, M. (2009). Frontoparietal white matter diffusion properties predict mental arithmetic skills in children. Proceedings of the National Academy of Sciences of the United States of America, 106(52), 22546–22551.
van Eimeren, L., Niogi, S.N., McCandliss, B.D., Holloway, I.D., & Ansari, D. (2008). White matter microstructures underlying mathematical abilities in children. NeuroReport, 19(11), 1117–1121.
Verbruggen, F., & Logan, G.D. (2008). Response inhibition in the stop-signal paradigm. Trends in Cognitive Sciences, 12(11), 418–424.
von Aster, M., and Shalev, R. (2007). Number development and developmental dyscalculia. Developmental Medicine and Child Neurology, 49, 868–873.
WHO. (2005). ICD-10. International Statistical Classification of Diseases and Related Health Problems 10th Revision; Chapter V: Mental and behavioral disorders (F81.2). Geneva: World Health Organization.

CHAPTER 41

COMPUTER-ASSISTED INTERVENTIONS ON BASIC NUMBER SKILLS

PEKKA RÄSÄNEN

Introduction

> Lift from her [the teacher's] shoulders as much as possible of this burden and make her free for those inspirational and thought-stimulating activities which are, presumably, the real function of the teacher.
>
> Pressey (1926, p. 374)

> One can predict that in a few more years millions of schoolchildren will have access to what Philip of Macedonia's son Alexander enjoyed as a royal prerogative: the personal services of a tutor as well-informed and responsive as Aristotle.
>
> Suppes (1966, p. 207)

These two quotations describe contrasting early visions of how technology would change instruction. The first vision was given by one of the initial developers of mechanical teaching machines, professor of educational psychology, Sidney Pressey, who described "teaching machines" as tools for teachers who could use them as part of their educational work. In Pressey's vision, the machines would be suitable for rather boring, repetitive, drill-and-practice training, which he believed to be an inevitable and unavoidable part of any learning process. The same idea is often presented in the context of computer-assisted instruction (see e.g. Butterworth & Laurillard, 2010; Encyclopædia Britannica, 2011). Most of the small-scale educational computer and internet applications for numerical skills have followed this model and have only tried to put the drill-and-practice rehearsal into a more colorful and motivating package than textbooks. Nevertheless, the basic neuroscientific principle of repetition as a main route to successful learning and to strong long-term memory representations is exemplified in these applications.

Professor Patrick Suppes, the leader of one of the first large-scale projects on computer-assisted education in the 1960s, suggested computers as a potential tool to replace teachers in educational activities. Contrasting teacher-guided instruction for a control group with computer-assisted instruction for an intervention group is one of the most common research designs in computer-assisted intervention (CAI) studies. The more that computing power has increased, the more complex can be the modeling of human–computer interactions (HCI). The virtual worlds of entertainment games bring these visions closer.

Younger people are often more accustomed to computers, tablets, and mobile devices than to blackboards or paper and pencil. Children as young as three years old can control a computer mouse and click on their favorite website icon to play learning games. Typical two- to five-year-old pre-schoolers play on computers almost half an hour a day (Gentile, 2004). Rideout et al. (2003) found in their large-scale survey that children under seven years old in the United States spend an average of two hours a day with screen media (TV, DVD, video games, and computers) – about the same amount of time that they spend playing outside, and three times as much time as they spend reading books or being read to. This is a growing trend in all developed and developing countries (International Telecommunication Union, 2008).

This division of studies of CAI from other studies on educational effectiveness has been criticized. Clark (1983) and, later, Kozma (1994) stressed that the method or the content is more important than the medium in determining the learning effect. They argued that separating media from educational method is an unnecessary schism which does not produce real new insights in education. However, as discussed later in this chapter, the content and types of CAI applications have, from decade to decade, closely followed mainstream trends in educational methods and ideologies.

We are facing a similar revolution in society and education as we faced when textbooks were introduced as the primary teaching instrument in schools about two hundred years ago (Wakefield, 1998). Textbooks replaced verbal memory strategies and the rote learning of texts. The e-classroom will have blackboards replaced with computerized smart boards or smart walls, and textbooks replaced with e-books. Soon, when electronic schoolbooks largely replace traditional textbooks, there will be an increased demand for knowledge about CAI (e.g. Tolani-Brown et al., 2011; Wastiau et al., 2013).

Terminology

There are multiple ways of using computers and other technologies in education; therefore, we are faced with a variety of concepts and acronyms. Technology-enhanced learning (TEL) is commonly used as a broad umbrella term. This includes all different media and technical tools used for teaching: including radio, television, CD ROMs and DVDs, videos, graphic and other calculators, software solutions designed to deliver instructional content, integrated learning systems, computer applications for assessment of school achievement or for selecting educational contents, interactive whiteboards, and classroom or home computers used for school work. Even using web search engines in education can be included within TEL. Likewise, different ways of working with computers in learning are included: individually, in pairs collaborating or competing against each other in a learning game environment, or in groups where computers provide a platform for collaboration and communication (computer-supported collaborative learning, CSCL).

Computer-managed instruction (CMI) aids the teacher in instructional management without actually doing any teaching. In CMI, the computer is used to provide learning objectives, learning resources, or assessment of performance. Narrower terms like computer-based learning (CBL) or computer-based education (CBE) refer to a structured environment in which computers are used for teaching purposes as the only or as the main tool to provide the course content.

The acronym CAI has been used for computer-assisted instruction or interventions and for computer-aided instruction. The terms "assisted" and "aided" have been used interchangeably. The Association for Education Communications and Technology (1977), who use the term "computer-assisted instruction," has defined it as a method of training in which the computer is used to instruct the student, to teach, to guide, and to test the student until a desired level of proficiency is attained. One of the key features is that the computer program should be able to instruct the student without the involvement of a human teacher.

In their meta-analyses, Kulik and his colleagues (Kulik, Kulik, and Shwalb, 1986) used the term computer-enriched instruction (CEI) to describe a situation where educational computer applications are used as a supplement within conventional teaching or the computers are used to present simulations to illustrate mathematical contents like geometry or algebra. The distinction between CEI and CAI is far from clear in the literature, because the most common way of using educational computer programs is supplemental, not to replace the teacher-guided activities.

In the field of mathematics education, there are nowadays an uncountable number of computer applications. The most common programs designed to help children master basic mathematics are simple drill-and-practice tasks where the application presents a calculation task and the subject is required to reply by selecting from alternatives or by entering the answer. Some of these applications are designed to give immediate feedback about the correctness of the answer, and slightly more advanced programs also give a total score based on the number of items and the number of items answered correctly.

Whether a CAI application can be considered as real teaching machines depends on the definition. Typically, three requirements have been regarded as important: (1) the machine must present information in the form of a task, (2) it must provide some means to respond, and (3) it must provide feedback about the correctness of the response (Morrill, 1961). Possibilities afforded by automated computing enable us to include two other requirements: (4) the system should be able to adapt the task conditions online to maximize learning and (5) when an error occurs, the system should give feedback that minimizes the probability of failure next time. Most of the current educational applications for basic mathematics learning meet the first three requirements and very few have features matching the latter two.

Development of CAI on Numerical Skills

There has been a long history of teaching machines before digital computers. The idea of having mechanical machines to facilitate learning or even to carry out instruction is far from new.

From the beginning of numerical human life, there have been many assistive technologies for learning and doing calculations using finger systems, pebbles, tallies, and abacuses (Bogoshi et al., 1987). The first patent for a mechanical educational tool in arithmetic was given to George G. Altman (1897) who developed a machine which presented a problem, provided a means for the learner to respond, and provided feedback about the correctness of the response.

In the 1920s, Pressey developed mechanical machines for drill and practice of many school subjects, including mathematics (Pressey, 1926). His machines had two modes: test and teach. A large drum with paper attached to it exposed a narrow window with a multiple-choice question. There were four keys to be pressed by the subject. In the test mode, the subject chose one of the four keys, the machine recorded that response and advanced automatically to the next question where the subject again responded. In the teaching mode, a small lever on the back of the machine was raised. This action prevented the machine from advancing to the next question until the previous question had been answered correctly. Therefore, the subject could make multiple responses until the right answer was chosen. A counter on the back of the machine recorded the total number of items and the number of correct responses to them. Pressey believed that this procedure was effective for learning because it operated according to both the law of recency and the law of frequency (Benjamin, 1988). Similar attempts to develop teaching machines were undertaken in several countries (Hunka & Buck, 1996).

To the 1926 version of the teaching machine, Pressey added an additional technical attachment which he called a "reward dial." This allowed the teacher to set the number of correct responses required to receive a candy reward. Once the response criterion had been reached, the device automatically delivered a piece of candy to a container in front of the subject. The machine was able to provide both immediate and cumulative feedback. Still the same multiple-choice format with a similar feedback logic is used in many research-based educational applications (see e.g. Kyle et al., 2013). Within that historical context, Pressey's prototypes were revolutionary (Pagliaro, 1983). However, there was no commercial or public interest in his machines. Years later, B.F. Skinner (1958) argued that Pressey's machines failed only because the field of education was not ready for them.

Skinner was himself interested in teaching machines and built a prototype for a computerized arithmetic teaching machine (Skinner, 1958). Educationally, he objected to Pressey's multiple-choice format. In his mind, the student should learn to produce a response, instead of just recognizing a correct response. Secondly, Skinner thought that multiple-choice meant unnecessarily exposing the student to a number of wrong answers.

In the 1960s, when computers finally had sufficient computing power to provide a platform for programming languages, IBM developed "Coursewriter," the first computer language devoted to CAI programming. This made it easier to build drill-and-practice tasks based on "programmed instruction" covering almost all curriculum content. However, the computers required were mainframe computers, which were typically available only in universities and were the size of a room. Schools only had access to these computers via remote terminals and a telephone line. Two large-scale projects were conducted in the United States, one from Stanford (IMSSS) and another from Illinois University (PLATO). These two projects adopted different approaches. In the IMSSS project, the researchers developed a CAI program for teaching arithmetic from first to upper grades. The materials were carefully graded based on research findings to support a step-by-step learning process (Suppes et al., 1968; Suppes & Morningstar, 1972). The aim was to use this graded information to adapt the learning process individually, on the basis of each student's current level of skills. The PLATO project went technically further by offering an environment, "an authoring system", that allowed teachers to create their own drill-and-practice tasks, which led to a large collection of training tasks, from primary school to university level (Bitzer & Skaperdas, 1969).

The teaching machines and CAI received a lot of attention in the public media, even though the use of computers in the classrooms of the 1960s and 1970s was very limited. The biggest reasons for the low usage were the cost and the fact that teachers were not accustomed to using supplemental tools in education. Additionally, the cognitive and constructive theories of learning had started to challenge strongly the drill-and-practice approach in CAI (Dalgarno, 2001).

The new constructivist theories of learning opened the way to the next wave in technology-based approaches to mathematics teaching. Instead of giving a student questions for them to answer, the idea was to give them tools to enable them to build mathematical rules so that they might answer multiple questions. That educational tool was programming. Two programming languages are of particular interest: MS Basic and Logo. MS Basic (Beginner's All-purpose Symbolic Instruction Code, invented in 1964) was an easy-to-learn programming language to provide computer access to non-science students. It was the standard computing language for microcomputers when they entered the schools in late 1970s. A step further was taken with Logo, which was specifically designed to be used as an educational programming tool. The Logo programming language was developed by Wallace Feurzeig and Seymour Papert (1968; Papert, 1972). Papert, who had been a co-worker of Jean Piaget, developed the idea of microworlds in which children were able to construct geometrical and mathematical rules by means of simple programming.

The arrival of the microcomputer, in the latter half of the 1970s, not only accelerated the growing interest in programming activity but also led to the development of more specialized pieces of programming software, some developed for students as young as first graders (Shumway, 1984). The microcomputers' graphical interfaces also provided a way to create visual representations of mathematical objects, particularly for algebra and geometry (Campbell and Stewart, 1993). Real-time visualization as an educational tool for teaching algebra received an additional boost from the development of graphing calculators, which became available to schools in the middle 1980s (Ellington, 2003; Khoju, Jaciw, & Miller, 2005).

In the last two decades, the field of CAI for numerical skills has divided into several subfields, although all fall into one of two approaches to education: direct instruction with specific restricted content and response alternatives, and guided instruction with a wider focus on problem-solving and rule discovery. However, more and more eclectic ideas have been developed from both of these approaches for CAI applications.

Direct instruction has developed from the early drill-and-practice tasks to flash cards, small-scale applications, and games to train computational fluency in basic arithmetic skills or to teach early number skills such as comparison, ordering, and number knowledge. Many small modifications have been used to analyze the effects of different types of presentation formats. For example, Ted Hasselbring with his colleagues (Hasselbring, Goin, & Bransford, 1988) developed a calculation fluency program which, while introducing a new number combination to be solved, simultaneously reviewed already practiced facts. The idea was to support the strategy of using amalgamated knowledge of learned facts to derive new facts. Fuchs and her colleagues (2006) developed another variant of a similar task which also used flash cards to show simple addition tasks. However, in this approach, the duration that the stimulus remained on the screen corresponded to the student's earlier performance, with higher accuracy associated with shorter screen time, to maintain appropriate levels of challenge. The aim was to rewrite the item after the presentation. If the response was correct, reinforcement was provided; if not, corrective feedback was provided.

The constructivist criticisms of drill-and-practice applications led to two alternative approaches to the training of numerical problem-solving. Both approaches share the same features common to the ideas of discovery learning or guided instruction: the problem space is fully or partially open, and the student tries to construct the more general rules that underlie the problem. The first approach has been taken using the Logo programming language and its descendants (Healy and Kynigos, 2009). The second approach led to applications where the problem space itself was given, but the students had a control on the problem selection, solution strategies, and steps that they could take. For example, in Harskamp and Suhre's (2007) computer program, students were allowed to choose problems and were given prompts to support their problem-solving. These applications approach the principles of entertainment strategy games, where each step taken leads to a new situation to be evaluated and new problems to be solved.

The development of different types of CAI for numerical skills reflects the general changes in educational theories. However, there is no common ground or conclusive support for one or the other schools of educational approach.

Effectiveness of CAI on Learning Numerical Skills

Research on CAI for numerical skills has been very fragmented. The replication of earlier studies using the same computer applications with similar or different kinds of subject groups are rare, making it very difficult to build cumulative evidence. The speed by which computer technology develops is one reason for this. Today, an intervention study, from planning to published results, usually takes longer than the life cycle of a computer or mobile phone operating system. Updates may have new features, which may no longer be compatible with the CAI task developed for the study. In other words, learning technologies tend to become outdated during a research project. However, many commercial applications are now frequently updated, making it easier for researchers to replicate earlier studies.

The majority of CAI studies have been descriptive or quasi-experimental without offering enough measurable quantitative indicators about the improvement in learning. Very often, the studies have been undertaken without control groups, only comparing the performance levels at the start and at the end of the period of intervention, which does not give reliable information about the specific effects of the computer use compared with other factors. However, there is starting to be a reasonable database of different types of studies over several decades on the effectiveness of CAI on mathematical skills.

Summary of Effectiveness

Since the 1960s, several studies have claimed positive outcomes from using CAI, including pre-mathematical knowledge (Howard et al., 1994), counting skills (Clements & Nastasi,

1993; Hungate, 1982), recognizing numerals (McCollister et al., 1986), learning numerical concepts (Grover, 1986), as well as improvement in standardized tests of number skills (Elliot & Hall, 1997; Hughes & Macleod, 1986). The largest gains from the use of CAI have been achieved in primary-grade children's mathematics, especially when the CAI is used as additional practice (Lavin & Sanders, 1983; Niemiec & Walberg, 1984; Ragosta et al., 1981).

The first large-scale controlled IMSSS studies in the 1960s used basic drill-and-practice CAI tasks, with a simple "teletype" typewriter for responding and a system for giving feedback. It also contained a logic for individualizing the progress for each child (Suppes & Morningstar, 1969). The authors reported the SAT computation scores from their studies, which allows comparisons between the experimental and control groups. In the IMSSS studies, an extra five to ten minutes' daily CAI practice in calculation skills for eight months produced, for first- and third-grade students, a gain comparable to effect sizes[1] ranging from 0.05 to 0.73 in higher-performing schools and from 0.70 to 1.28 in more deprived areas with lower-performing children (Suppes et al., 1968).

These values are in accord with the early meta-analytic studies of CAI that were published prior to the appearance of microcomputers in schools. Dissertations by Hartley (1977) and Burns (1981) were the first reviews on CAI that used a meta-analytic approach. These studies reported significant improvements for students in elementary and secondary schools on computational arithmetic skills, with effect sizes ranging from 0.37 to 0.42. The first published reviews were undertaken by Kulik and colleagues (1985) and Bangert-Drowns and colleagues (1985). They both used stringent criteria for inclusion, in particular, that the studies had to have a control group that was taught in a conventionally instructed class. Therefore, the question was whether CAI produced a learning effect greater than when using only traditional teacher-guided classroom education. These studies found similar small to medium size (from 0.26 to 0.54) positive effects of computer-based instruction. However, many of the early studies showed the quite opposite favoring more traditional teaching methods (e.g. McKeown, 1965).

In the 1980s, both mainframe machines and microcomputers were used in CAI studies. However, the appearance of more advanced personal computer technology did not raise the average gains achieved by using CAI as part of mathematics education. For example, Kulik and Kulik (1991) calculated an average effect size of 0.47 for using CAI in elementary school mathematics education. In the 1990s, the studies were almost exclusively undertaken using microcomputers. Again, in general, positive outcomes were found for students with CAI experiences. The effect sizes for these meta-analyses ranged from 0.13 to 0.8, favoring CAI groups to non-CAI groups (Fletcher-Finn & Gravatt, 1995; Khalili & Shashaani, 1994; Kulik & Kulik, 1991).

However, more recent reviews show declining effects of CAI in mathematics education. For example, Li and Ma (2010) and Rakes and colleagues (2010) examined the impact of computer technology on mathematics achievement. However, they did not restrict their analysis only to CAI but also included other ways of using computers in mathematics education, including multimedia, web-based courseware or software, and distance education. In his review, Rakes also accepted additional tools such as graphic calculators. Again, the findings provided some evidence for the enhanced achievement in mathematics by the use of CAI in K-12 classrooms but with a reduced overall effect size of 0.28.

Two more recent reviews by Slavin and colleagues (Slavin & Lake, 2008; Slavin et al., 2009) applied very stringent inclusion criteria. They included a total of 38 educational technology studies in their review on elementary education and another 38 in a review on secondary education. They found a modest effect size of 0.19 for elementary schools and a small effect size of 0.10 for secondary schools. One reason for the modest effect sizes was that two recent, randomized, large-scale evaluations involving hundreds of schools found minimal, if non-existing, effects of various types of education technology applications (e.g., Cognitive Tutor, PLATO, and Larson Pre-Algebra) on mathematics achievement (Campuzano et al., 2009; Dynarski et al., 2007). These two studies are important because of their size and use of random assignment, and because they analyzed widely used forms of CAI.

Another example is Tucker (2009), who analyzed, in his dissertation, widely used commercial intelligent tutoring systems (ITS). The most popular of these was "Successmaker", which was based on the early IMSSS drill-and-practice tasks augmented by information processing and cognitive theories. ITS are programmed to analyze the types of errors students commit and then to adjust the tasks, and to give support or feedback according to individual learning patterns. However, despite the strong theoretical background, this study found no differences in mathematics achievement ($r = -0.04$) between schools which used "Successmaker" or other similar CAI tools and those that did not use CAI tools systematically. Tucker (2009) also reviewed other recent dissertation studies on CAI usage in the US. He found that in elementary education, only 38% of the studies showed significant gains, while in the upper grades, none of the studies showed large gains.

In summary, the majority of recent CAI reviews have concluded that, although there are always some studies with negative or non-existing effects, using educational technology has positive effects on mathematics learning and achievement. The overall study-weighted effect size of CAI would currently be close to 0.3 in Cohen's d, indicating a small effect (see Table 41.1).

However, there appears to be a small but significant trend that the average effect size is declining (approximately $r = -0.33$ between the publication year and the average effect size). This is especially notable because, given the increased use of computerized technologies in education, advances in technology, and the cumulating database of research on educational applications, a reverse trend might have been expected. In fact, this declining trend seems to be a rather new phenomenon. In the earlier studies, Kulik and Kulik (1987) reported that the average effect of computer-based instruction was improving. The average effect size for studies from 1966 to 1972 was only 0.24 compared to 0.36 for studies from 1974 to 1984. On the other hand, Fletcher-Finn and Gravatt (1995) and Liao (1998) did not find upward or downward patterns, while three more recent reviews have reported a declining trend. Christmann and Badgett (2003) found a 51% drop from the 1970s to the 1990s and Cheung and Slavin (2013) calculated a 35% drop in effect sizes from the 1980s to the 1990s. The latter also reported that the trend continued in the new millennium. Similarly, Li and Ma (2010) found a significant drop in effect sizes in studies published before 2000 than after it.

Educational Significance of CAI

The continuous increase in using and investing in computers for mathematics education allows us to ask whether the investments in CAI have been worthwhile. Should we continue investing in technologies for mathematics education or should we consider other factors more important than the medium of delivery?

Table 41.1 Meta-analytic studies of the effectiveness of CAI on mathematical skills in elementary education, 1985–2011

Authors	Publication Year	Years covered	n of studies	ES
Bangert-Drowns, Kulik, & Kulik	1985	1968–1982	22	0.26
Kulik, Kulik, & Bangert-Drowns	1985	1967–1982	17	0.54
Niemiec, Samson, Weistein, & Wallberg	1987	1986–1982	48	0.28
Kulik & Kulik	1991	1966–1986	9	0.39
Ryan	1991	1984–1989	8	0.30
Becker	1992	1984–1989	11	0.27
Khalili & Shashaani	1994	1988–1992	18	0.52
Fletcher-Flinn & Gravatt	1995	1987–1992	24	0.32
Liao	1998	1986–1997	5	0.13
Chrismann & Badgett	2003	1966–2001	12	0.34
Kulik	2003	1990–2001	12	0.38
Liao	2007	1983–2002	45	0.55
Rosen & Salomon	2007	1986–2002	32	0.46
Slavin & Lake	2008	1971–2006	38	0.19
Slavin, Lake, & Groff	2009	1971–2006	38	0.10
Li & Ma	2010	1990–2006	46	0.28
Rakes, Valentine, McGatha, & Ronau	2010	1968–2008	36	0.28
Cheung & Slavin	2011	1984–2009	45	0.17

Note: ES, effect size – Cohen's d

There is an interesting dilemma between those who have published studies on CAI and those who have reviewed those studies. While the former tend to describe the results as significant and, at the very least, promising, the latter are inclined to be more circumspect and sceptical. Those who have compared different types of interventions, including human-to-human interaction, have not found CAI as effective as other types of interventions. Kroesbergen and van Luit (2003) found that CAI did not appear to be effective for the domain of mathematical problem-solving and reported only low effect sizes for all other domains of mathematics. Similar conclusions were drawn by Slavin et al. (2009). Only Xin and Jitendra (1999) found that CAI was the most effective of all types of interventions on problem-solving – although this latter study found that CAI was more effective for the maintenance of skills than for generalization.

Wolf (1986) proposed an often-cited rule of thumb that educational methods with an effect size more than 0.25 should be considered educationally significant. More recently, Hattie (2008) has argued for an even stronger criterion, saying that educational interventions with consistent evidence of effect sizes above 0.40 are "worth having" and those below are not "educationally significant." Table 40.1 shows an overall estimate of effect sizes for different types of CAI in mathematics education, which is around 0.30, with a declining trend. It also offers some evidence for scepticism that the medium itself is not important. However, averaging studies without attention to their details may be deceptive. Within these average figures, there are methodological and conceptual details and subgroups of learners which demand more attention.

Trends in Effectiveness of CAI Studies on Mathematical Learning

The Effects of the Research Methods on the Results

The methodological details are especially important because an increasing number of CAI applications studied are commercial products. The popularity of CAI applications in schools has increased, and, therefore, the financial interests and investments in them have also risen. Cherry picking only those results or reporting results of only those subgroups that show positive outcomes is a risk, if the developers or vendors have supported the studies financially or conducted the studies themselves. For example, Becker (1992) re-analyzed the effectiveness of several vendor-conducted studies, finding cherry picking of results; and while recalculating the results, he found a drop in effect size from 0.80 to 0.35, which is much closer to the average of the other reviews.

In general, methodologically weaker studies, with or without commercial interests, have reported better results than methodologically strong studies (Niemiec et al., 1987). Therefore, it is important that independent researchers are granted funding to provide reliable studies. There are several clear trends in effect sizes, which are unrelated to the content of the studies but rather to the research methodologies and design:

- Non-experimental studies have reported twice as large effectiveness than randomized experimental studies.
- Studies with non-standardized tests have showed larger effects of technology on mathematics achievement than studies with standardized tests (Li & Ma, 2010).
- Studies with small sample sizes have produced, on average, twice the effect sizes of those with large sample sizes, irrespective of the general research design (Cheung & Slavin, 2013; Liao 1998). This kind of trend is very common in intervention studies, regardless of content (Moran et al., 2008; Slavin & Smith, 2008).
- The "file-drawer" effect, i.e. studies that are done but not published, is more likely in small-scale studies with null or negative effects than in large-scale studies requiring larger investments.

Small sample sizes do not always mean lower quality of research. Larger studies have often used rather general learning environments for non-selected samples. Therefore, these tend to have greater variances in initial skill levels and learning effects. Small-scale studies tend to have more selected groups and more focused target skills. Teaching multiplication tables with flash cards to a small group of second graders is easier to achieve than teaching problem-solving with a programming environment to a large group of randomly selected classes of seventh-grade adolescents.

Who Benefits from CAI?

Children in the primary grades or those who have special needs in mathematics education are the most likely to benefit from CAI as a supplementary education (Bangert-Drowns et al., 1985; Li & Ma, 2010; Niemiec et al., 1987; Slavin & Lake, 2008; Slavin et al., 2009). Studies on groups with younger students have typically produced better results than studies with older students, and similarly, children with special educational needs (SEN) seem to gain more from CAI than students in mainstream education. Kulik and colleagues suggested that older students " apparently have less need for highly structured, highly reactive instruction provided in computer drills and tutorials. They may be able to acquire basic textbook information with the cues and feedback that CAI systems provide" (Kulik et al., 1985, p. 71). Kulik et al.'s observation dates from the time when mainframe machines were used and drill-and-practice tasks were the most common format of CAI. However, this main trend has not changed, even though CAI applications are rarely now only drills and tutorials, often containing more interactive systems and being educational microworlds for discovering mathematical rules or for learning problem-solving strategies.

The factors that explain the greater success with younger children and children with SEN are connected to the type of training, the level of skills at the beginning of the intervention, and variables used in the studies. The groups in these studies have been smaller, especially when the focus has been on children with SEN. With smaller and more selective groups, the interventions can be more focused on the learning needs of the children, and the CAI applications can be better targeted to specific skills that these children are expected to master. In the case of mainstream students and older children, the variance in skill levels is usually larger and, therefore, the fit between the students' learning needs and the CAI application is inevitably smaller. Thus, it is to be expected that the effectiveness of the CAI tends to get smaller, the larger and the more unselected the group. With special groups, the measures have been more often tailored for the studies than for standardized tests, and a better match between the assessment and the training tends to produce a higher estimate of effectiveness.

The majority of published studies have, in fact, mostly been descriptive and have not included control groups. Moreover, in those studies with control groups, a poorer performing group has, typically, been selected as the intervention group and the rest of the children, the better performing children, have been the control group. This type of non-matching does not control for regression to the mean effects, producing an overestimation of the intervention's effectiveness.

Interestingly, socioeconomic factors or gender have not been found to play a significant role in the effectiveness of CAI. The question of the effects of gender has been especially inconclusive. Cultural (Else-Quest et al., 2010), cognitive (Geary et al., 2000), and affective (Devine et al., 2012) factors have been proposed to explain the differences between boys' and girls' performances and strategy selections in mathematical tasks. Computers are often thought to be "technical tools" which are, thus, somehow masculine. Accordingly, some studies have found that gender is related to children's motivation to use different types of CAI applications. For example, boys tend to view CAI more favorably than do girls (Vale &

Leder, 2004). This difference seems to increase with age, with pre-school-aged children showing no gender differences (Hattie, 1990; McLester, 1998). A common stereotypic finding is that girls prefer learning problems that require cooperation and communication, and prefer to work in larger groups cooperatively and interactively with the computer (Fiore, 1999; Hale, 2002), while boys benefit more from working individually or in smaller groups with competitive games. However, studies in which mathematics performance and the attitudes of girls and boys in cooperative and competitive games have been compared have not found any significant differences between genders (Ke & Grabowski, 2007; Nathan & Baron 1995). It appears that the impact of gender differences in studies of CAI fluctuates along the lines of the impact of gender differences in mathematics more generally. Depending on the perspective, the results vary. Currently, there are no results that show it would be beneficial to develop separate CAI applications for each gender.

Although neither socioeconomic status (SES) nor gender are indicators of an individual child's skills or educational needs, there does seem to be an indirect connection between SES and the effectiveness of CAI. From the first large-scale IMSSS studies in the 1950s, there was a clear difference in effectiveness between areas of different economic advantage. Similarly, the studies conducted in developing countries tend to produce higher effectiveness than those in developed countries. These results do not necessarily point to any educationally meaningful benefits but, possibly, to the fact that it seems to be easier to produce better results when the starting level is lower, especially if the reasons for lower starting level are SES and not cognitive. Children with cognitive deficits or mathematical learning disabilities have only recently become a focus of research, following neuroscientific discoveries of dyscalculia as a separate cognitive deficit.

Neuroscientific Approaches to CAI on Numerical Cognition

Since the 1990s, there has been an increasing interest in numerical cognition and disabilities in learning numerical skills. The number of published neuropsychological studies on mathematical disabilities has doubled during the last two decades (Räsänen & Koponen, 2011). Computer technology has played an important role in this development. Computers have allowed more precise presentation of stimuli and measurement of reaction times to that stimuli. Likewise, brain-imaging technologies require a huge amount of calculation power, impossible without advanced computer technology. Using computer-assisted tools to present stimuli (numbers, calculations, or clouds of dots of certain numerosity), to recode the response, and even to score online is nowadays a daily routine for researchers and an increasing practice in clinical work.

Moving from these experimental tasks to educationally effective neuroscience-based models of intervention tasks or games has required a theoretical jump. Discovering a consistent pattern of reaction time to numerical stimuli or brain activity at certain brain areas, or even a correlation between certain brain activity and performance in a specific test, does not yet tell us about the exact learning processes or requirements for learning a certain skill. However, thus far, three research groups have proposed models of the key basic numerical representations and tried to build computerized training tasks that mimic the basic numerical cognitions and representations that underlie how we process magnitudes and number symbols.

Approximate Number System in CAI

A French research group, led by Professor Stanislas Dehaene, developed a mathematics game, called Number Race, for remediating dyscalculia (Wilson, Dehaene, Pinel, Revkin, Cohen, & Cohen, 2006). The theoretical background of the game came from the triple code theory of numerical processing (Dehaene et al., 2003). The triple code theory argues that there are three modes of processing numerical and magnitude information: visual, verbal, and analog–magnitude. According to this theory, the analog–magnitude system is the heart of the semantic representation of approximate and relative quantities, while the visual and verbal symbolic representations are more concerned with exact quantities. The novel approach adopted in this CAI game was to concentrate on the approximate number system (ANS) instead of on exact quantities or number symbols. The Number Race game is a variant of traditional board games but with one exception: instead of throwing a dice, a child has to make a comparison between magnitudes. The game is adaptive: depending on the correctness of the child's selection (larger is better), the relative distance between the two magnitudes in the next item either decreases or increases (in the case of an incorrect response), pushing the child towards a more precise representation of magnitudes. After successful progress with non-symbolic clouds of dots, the game introduces comparisons presented as Arabic numbers or calculations.

There have been a few studies on the effectiveness of this game to support the development of children's numerical skills. For example, the developers (Wilson, Revkin, Cohen, Cohen, & Dehaene, 2006) assessed 7- to 9-year-old children with mathematical difficulties before and after five weeks of playing the game for 30 minutes a day. This study did not involve a comparison group. Nevertheless, children showed specific increases in performance on speed of subitizing and numerical comparison and in subtraction accuracy. However, their performance in addition or base-ten comprehension tasks did not improve over the period of the study. The developers also studied (Wilson et al., 2009) 5- to 6-year-old children with low SES using a crossover design with the Number Race game and a reading game. With a total of two hours playing time, children did show improvements in several numerical skills but surprisingly not in ANS.

In a third study, 6-year-old children with low number skills were randomly divided into two groups to play either Number Race or another game (GraphoGame-Math, GG-M) which concentrates on exact representations and numerical symbols (Räsänen, et al., 2009). In the GG-M game (version 1), a child selects a correct match between a spoken number and among two to four options of a visual representation, which can be a set of objects, an Arabic number, a calculation, or a combination of these. While the repetitions in Number Race aim to fine-tune the child's representations of magnitudes, the GG-M tries to build memory associations between sets of objects and symbols. In the study (Räsänen et al., 2009), three weeks of daily 15-minute playing sessions produced an improvement in both groups compared to typically performing children in number comparison speed, but not in other areas of numerical skills. A detailed analysis of the response time patterns in the number comparison task suggested that the two types of training produced slightly different types of learning. Oberstiener and colleagues (2013) continued with this idea with a more controlled study. They made two versions of the Number Race game, one for playing with approximate representations and one for playing with exact representations (i.e. the same game environment with different types of learning aims). They divided typically performing first-grade students into three groups: the first playing with the game using approximate representations, the second with exact representations, and a third group playing both games. The study also included a control group playing a language learning game. With five hours' total playing time over four weeks, the number game groups outperformed the language learning group. In addition, the approximate representations group improved more than the exact representations group in tasks requiring estimation, while the exact representation group improved more in tasks requiring exact symbolic responses, indicating task-specific learning effects. However, both groups improved similarly in an arithmetic task, gaining more than the control group and more than the group, which played both approximate representations and exact representations. This mixed model of training did not produce better results than either of the more specific types of training, and none of them produced strong transfer effects.

Representations of Number Line in CAI

An interesting addition to the triple code model is the idea of spatial representation of number magnitude, in the form of a mental number line that is activated whenever a number is encountered (Dehaene et al., 2003; Nuerk et al., 2006). The precision of locating numerical magnitude on this number line has been found to be associated with higher achievement in mathematics and to a better ability to learn arithmetic (Booth & Siegler, 2008). Playing linear, but not circular, board games is shown to improve these spatial representations of a mental number line (Siegler & Ramani, 2008; Whyte & Bull 2008). An Austrian group of neuroscientists developed a CAI game, Rescue Calcularis, based on the ideas of combining numerical and arithmetic learning with the estimation of numerical magnitudes on a visual number line. The effectiveness of this approach was studied with two groups of children, a group of typically performing children and a group of children with mathematical disabilities (Kucian et al., 2011). The children played the game for 15 minutes a day, five days a week, for five weeks. Both groups showed improvement in the spatial representation of numbers on a visual number line and in solving arithmetical problems. Of additional interest in this study were the brain imaging results. This was the first study to analyze the effects of a number game at a neural level. The researchers were able to find differences between the groups before and after the intervention, but no interaction effects, meaning that the intervention effects were similar in both groups. Children with mathematical disabilities showed less activation in their bilateral parietal regions, which reflects neuronal dysfunction in regions pivotal for number processing (Kucian et al., 2006; Price et al., 2008). In addition, after the training, both groups showed reduced activation of brain regions associated with number processing. The authors suggest that this could be attributed to the automatization of cognitive processes necessary for mathematical reasoning, attributable to the CAI training.

The same group has further developed their CAI model from simple number line estimation tasks to a full-scale educational application for arithmetic (Käser et al., 2011). The new model includes multiple games with different learning aims and an ITS to control the player's progress within and from one game to another. In the tasks, the properties of numbers are encoded with auditory and visual cues such as color, form, and 3D topology; while the two-dimensional visualization of the number line has been changed to a three-dimensional presentation. Adding depth was for educational reasons, to illustrate more easily the hierarchical structure of the base-ten system; there are no neuroscientific models that would suggest three-dimensionality in our spatial representation of the number system.

To date, the authors have published only preliminary results from their large-scale interventions. A group of 33 second- to fifth-grade students with difficulties in learning basic arithmetic participated in 20 minutes' daily training over six weeks. The authors used a waiting list design, to compare the trained group to those without training. Two types of data were collected and analyzed: the game log files and scores of two paper and pencil arithmetic tests. The ITS allowed children to go back and forth between easier and harder tasks depending on their performance. Analysis of the log files, which logged every action of the child in each task, showed that the possibility of going back to easier tasks produced substantially beneficial effects. The children not only immediately started reducing their rate of mistakes, but also learned faster. Secondly, the children's accuracy improved in tasks where an estimation of the position of a number on a two-dimensional number line was required. The improvement in the accuracy of positioning was a replication of the previous experiment. The training group also performed significantly better than the waiting group in paper and pencil subtraction tasks.

Fischer et al. (2011) took another approach towards improving children's skills in comparing magnitudes and numbers. They used similar number line visualizations to present collections of different sized squares (magnitude comparison) or Arabic numerals. However, in the intervention, the key element was not the visualization per se but *how* children provided their responses. Two groups of CAI game players were compared: one playing with a tablet and finger pointing, while another group used a digital dance mat and whole body movements to respond. Both training conditions used identical stimulus. The novelty in this study, therefore, was to include embodied cognition in the process of learning numbers. The idea being investigated was that the more sensoriperceptual features are shared by stimulus and bodily responses, the more holistic and faster is the learning. The study used a parallel randomized cross-over design. One half of the 5- and 6-year-old children were first trained on the dance mat and then on the tablet PC, while for the other half, the order of training conditions was reversed. Each child received only three 10–15 minutes' training sessions per condition. The increase in accuracy of the spatial position of children's estimates on the 0–10 scale was stronger after the dance mat training compared to the tablet PC control condition, but not on the 0–20 scale. In addition, a stronger transfer effect of the dance mat training compared to the finger pointing training was found for learning counting principles, but no effects were found for number knowledge or calculation skills.

The very small exposure time required to produce significant effects indicates that adding a physical–spatial aspect compared to only visualizing the spatial number line may produce stronger or more easily accessible representations. Currently, there are no controlled studies using movement recognition systems with similar tasks. These would allow us to analyze whether matching the symbolic distance with the amount of movement would produce even better results compared to only directional movement used in Fischer and colleagues' (2011) experiment. Another interesting question is whether using the finger to drag an item a certain length on the screen, rather than just pointing, would be sufficient to produce the same effects as moving the whole body through physical space. Current technology makes it possible to create such experiments.

Number Sets and Virtual Manipulatives in CAI

A third neuroscientific model of numerical cognition stresses the importance of the ability to recognize patterns of small quantities as the core feature of basic numerical cognition. At the heart of the theory is the so-called "number module", an innate brain system dedicated to domain-specific processes connected to quantities, allowing us to build culturally mediated, symbolic, numerical counting-based systems (Butterworth, 1999). Zorzi and Butterworth (1999; see also Stoianov & Zorzi, 2012) presented a computational model to show how the numerical system stems from the exact representations of small sets.

The majority of the non-computerized intervention programs follow this model by using manipulatives: small collections of objects to be ordered, categorized, compared, and counted (Clements & Samara, 2011; Samara & Clements, 2009). Investigations using computerized manipulatives for geometry and fractions show that these can lead to statistically significant gains in learning new concepts (Reimer & Moyer, 2005). Olson (1988) found that students who used both physical and software manipulatives demonstrated a greater sophistication in classification and logical thinking than did a control group that used physical manipulatives alone. Another benefit of digitalized versions of manipulatives is that compared to the use of physical materials, a computer environment offers students greater control and flexibility over the manipulatives, allowing them to, for example, duplicate and modify the computer bean sticks (Char, 1989; Moyer, Niezgoda, & Stanley, 2005).

The previously mentioned studies from Räsänen et al. (2009) and Oberstiener et al. (2013) are currently the only neuroscientifically motivated studies comparing the ANS and number module approaches in CAI. Both of them found training type specific effects on learning and small general learning effects for both types of training. However, in neither experiment were the children allowed to play freely with the small collections of objects and to manipulate the quantities "physically" on the screen; the experiments more resembled Pressey's teaching machine with multiple-choice drill-and-practice tasks. A true test of this approach would have allowed children to manipulate and create collections of objects to build rich representations of sets and the relations between sets (see Thomas & Laurillard, 2012). Actually, the same Austrian (Käser et al., 2011) and French (Dotan et al., 2011) brain research groups have moved to this type of activity in their new mathematics learning games.

Butterworth and Laurillard (2010) reviewed the benefits of digitalizing educational practices with computerized tasks. CAI provides children a learning environment where they can manipulate quantities to build their own microworlds, and discover the rules of numbers and calculations. Butterworth and Laurillard also presented case descriptions with a dot enumeration training task, although no quantitative results for the effectiveness of such training for children with mathematical disabilities. One of the benefits of using a digitalized version of the tasks was the increased number of tasks that the children were able to complete in any given time: the children managed to complete, on average, more than four times the number trials per minute working individually with a computer than completing the same tasks in a class with a specialist teacher. As the children were essentially unsupervised while the teacher dealt with another child, computerized tasks allowed all children to continue working and to receive constructive feedback from their activities. Computers with capabilities to recognize multi-touch interactions have just arrived in large scale to the markets. Hopefully that will encourage researchers to do more controlled experimental studies on the effectiveness of computerized manipulatives.

The Key Challenges of E-Learning

The arrival of computer technology is changing education just as the arrival of textbooks changed education from a verbal tradition to a literal tradition. Computers and other new technical tools based on automated computing will change the way information is searched, presented, and finally also learned. In particular, it will change again what kind of information is valuable and expected to be remembered. Already, a child can ask verbally his or her smart phone to answer almost any arithmetic or numerical task; while a web search database gives the answer with numbers and illustrations such as number lines and base-ten sticks. A key question will be how much and what type of knowledge will we need to have in our minds to be able to search for solutions and to apply it fluently in different contexts.

Biagi and Loi (2012) analyzed the latest OECD Pisa 2009 data to find correlations between the achievement scores and the different types of computer usage. To their surprise, the results were opposite to what would have expected: the intensity of school-related use of ICT tended to be negatively correlated with students' PISA test scores in mathematics, and gaming was the only activity with consistent positive correlations. The authors cited the OECD report as an explanation to this unexpected finding: "computer use can make the difference in educational performance if the student has the appropriate set of competences, skills and attitudes. Without these, no matter how intense the student's use of a computer, the expected benefits will not be realized" (OECD, 2010, p. 172). Learning to use ICT effectively in learning will be a new challenge for education.

The results discussed here also indicate that CAI is not yet integrated well into daily mathematics education, but is still mainly a separate act of a "computer class visit". E-books will change this, and the unseen technologies may challenge the whole idea of learning in classrooms. Wearable computing, voice recognition, and movement sensors will disengage children from computer screens, permitting new ways of interacting with CAI. However, as the history of CAI on numerical skills shows, although the technology does improve our understanding of how the learning brain works, it does not provide direct breakthroughs into how to teach the same brain to learn, whatever the HCI is. The models for educational applications come from the more general theories of learning and teaching. Inevitably, computer technology will give us better tools to analyze and to develop these theories and their educational applications.

Note

1. The effect sizes here are presented as Cohen's d values. Cohen's d is a widely used measure to indicate the standardized difference between two means. It uses standard deviation as a common scale to enable a comparison between effect sizes in different studies. Values below 0.2 are considered as small, below 0.5 as mediate but visible, and above 0.8 as large.

References

Altman, G.G. (1897). *Apparatus for Teaching Arithmetic*. Washington, DC: United States Patent Office; No. 588,371.

Association for Educational Communications and Technology (1977). *Educational Technology: Definition and Glossary of Terms* (Vol. 1). Washington, DC: Association for Educational Communications and Technology.

Bangert-Drowns, R.L., Kulik, J.A., & Kulik, C.L.C. (1985). Effectiveness of computer-based education in secondary schools. *Journal of Computer-Based Instruction, 12*, 59–68.

Becker, H.J. (1992). Computer-based integrated learning systems in the elementary and middle grades: a critical review and synthesis of evaluation reports. *Journal of Educational Computing Research, 8*(1), 1–41.

Benjamin, L.T., Jr (1988). A history of teaching machines. *American Psychologist, 43*, 703–712.

Biagi, F. & Loi, M. (2012). *ICT and Learning: Results from PISA 2009*. European Commission, JRC Scientific and Policy Reports. Luxembourg: Publications Office of the European Union.

Bitzer, D. & Skaperdas, D. (1969) *The Design of an Economically Viable Large-scale Computer Based Education*. Urbana: Computer-Based Education Research Lab.

Bogoshi, J., Naidoo, K., & Webb, J. (1987). The oldest mathematical artifact. *Mathematical Gazette, 71*(458), 294.

Booth, J.L. & Siegler, R.S. (2008). Numerical magnitude representations influence arithmetic learning. Child Development, *97*(4), 1016–1031.

Burns, P.K. (1981). A quantitative synthesis of research findings relative to the pedagogical effectiveness of computer-assisted mathematics instruction in elementary and secondary schools. *Dissertation Abstracts International, 42*, 2946A.

Butterworth, B. (1999). *The Mathematical Brain*. London: Macmillan.

Butterworth, B. & Laurillard, D. (2010). Low numeracy and dyscalculia: identification and intervention. *ZDM Mathematics Education, 42*, 527–539.

Campbell, P. & Stewart, E.L. (1993). Calculators and computers. In R. Jensen (Ed.), *Early Childhood Mathematics, NCTM Research Interpretation Project* (pp. 251–268). New York: Macmillan Publishing Company.

Campuzano, L., Dynarski, M., Agodini, R., & Rall, K. (2009). *Effectiveness of Reading and Mathematics Software Products: Findings from Two Student Cohorts*. Washington, DC: Institute of Education Sciences.

Char, C.A. (1989). *Computer Graphic Feltboards: New Software Approaches for Young Children's Mathematical Exploration*. San Francisco: American Educational Research Association.

Cheung, A.C.K. & Slavin, R.E. (2013). The effectiveness of educational technology applications for enhancing mathematics achievement in K-12 classrooms: a meta-analysis. *Educational Research Review, 9*, 88–113.

Clark, R.E. (1983). Reconsidering research on learning from media. *Review of Educational Research, 53*(4), 445–459.

Clements, D.H. & Nastasi, B.K. (1993). Electronic media and early childhood education. In B. Spodek (Ed.), *Handbook of Research on the Education of Young Children* (pp. 251–275). New York: Macmillan.

Clements, D.H. & Samara, J. (2011). Early childhood mathematics intervention. *Science, 333*, 968. Doi: 10.1126/science.1204537.

Christmann, E.P. & Badgett, J.L. (2003). A meta-analytic comparison of the effects of computer-assisted instruction on elementary students' academic achievement. *Information Technology in Childhood Education Annual, 15*, 91–104.

Dalgarno, B. (2001). Interpretations of constructivism and consequences for computer assisted learning. *British Journal of Educational Technology, 32*(2), 183–194.

Dehaene, S., Piazza, M., Pinel, P., & Cohen, L. (2003). Three parietal circuits for number processing. *Cognitive Neuropsychology, 20*(3–6), 487–506.

Devine, A., Fawcett, K., Szűcs, D., & Dowker, A. (2012). Gender differences in mathematics anxiety and the relation to mathematics performance while controlling for test anxiety. *Behavioral and Brain Functions, 8*(33), 1–9.

Dotan, D., Dehaene, S., Huron, C., & Piazza, M. (2011). *The Number Catcher Game*. INSERM-CEA Cognitive Neuroimaging Unit. Available at: http://www.thenumbercatcher.com/.

Dynarski, M., Agodini, R., Heaviside, S.N., et al. (2007). *Effectiveness of Reading and Mathematics Software Products: Findings from the First Student Cohort*. Washington, DC: Institute of Education Sciences.

Ellington, A.J. (2003). A meta-analysis of the effects of calculators on students' achievement and attitude levels in pre-college mathematics classes. Journal for Research in Mathematics Education, *34*(5), 433–463.

Elliot, A. & Hall, N. (1997). The impact of self-regulatory teaching strategies on "at-risk" preschoolers' mathematical learning in a computer mediated environment. *Journal of Computing in Childhood Education, 8*, 187–198.

Else-Quest, N.M., Hyde, J.S., & Linn, M.C. (2010). Cross-national patterns of gender differences in mathematics: a meta-analysis. *Psychological Bulletin, 136*(1), 103–127.

Encyclopædia Britannica. (2011). *Encyclopædia Britannica Online Academic Edition*. Encyclopædia BritannicaInc.Availableat:http://www.britannica.com/EBchecked/topic/130589/computer-assisted-instruction.

Feurzeig, W. & Papert, S. (1968). *Programming-Languages as a Conceptual Framework for Teaching Mathematics*. Nice, France: Proceedings of the NATO Conference on Computers and Learning. Available at: http://dl.acm.org/citation.cfm?id=965757.

Fiore, C. (1999). Awakening the tech bug in girls. *Learning and Leading with Technology, 26*(5), 10–17.

Fischer, U., Moeller, K., Bientzle, M., Cress, U., & Nuerk, H.-C. (2011). Sensori-motor spatial training of number magnitude representation. *Psychonomic Bulletin & Review, 18*(1), 177–183.

Fletcher-Finn, C. & Gravatt, B. (1995). The efficacy of computer-assisted instruction (CAI): a meta-analysis. *Journal of Educational Computing Research, 12*(3), 219–241.

Fuchs, L.S., Fuchs, D., Hamlet, C.L., Powell, S.R., Capizzi, A.M., & Seethaler, P.M. (2006). The effects of computer-assisted instruction on number combination skill in at-risk first graders. *Journal of Learning Disabilities*, 39(5), 467–475.

Gentile, D.D. (2004). The effects of video games on children: what parents need to know. *Pediatrics for Parents*, 21(6), 10. Available at: http://findarticles.com/p/articles/mi_mo816/is_6_21/ai_n9772319/.

Geary, D.C., Saults, S.J., Liu, F., & Hoard, M.K. (2000). Sex differences in spatial cognition, computational fluency, and arithmetical reasoning. *Journal of Experimental Child Psychology*, 77(4), 337–353.

Grover, S.C. (1986). Afield of the use of cognitive-developmental principles in microcomputer design for young children. *Journal of Educational Research, 79*, 325–332.

Hale, K.V. (2002). Gender differences in computer technology achievement. *Meridian: A Middle School Computer Technologies Journal, 5*, 2. Available at: http://www.ncsu.edu/meridian/sum2002/gender.

Harskamp, E. & Suhre, C. (2007). Schoenfeld's problem solving theory in a student controlled learning environment. *Computers & Education, 49*, 822–839.

Hartley, S.S. (1977). Meta-analysis of the Effects of Individually Paced Instruction in Mathematics. Unpublished doctoral dissertation, University of Colorado at Boulder.

Hasselbring, T.S., Goin, L.I., & Bransford, J.D. (1988). Developing math automaticity in learning handicapped children: the role of computerized drill and practice. *Focus on Exceptional Children*, *20*(6), 1–7.

Hattie, J. (1990). Performance indicators in education. *Australian Journal of Education*, 34, 249–276.

Hattie, J.A.C. (2008). *Visible Learning: A Synthesis of Over 800 Meta-Analyses Relating to Achievement*. London: Routledge.

Healy, L. & Kynigos, C. (2009). Charting the microworld territory over time: design and construction in mathematics education. *ZDM—The International Journal on Mathematics Education, 42*(1), 63–76.

Howard, J.R., Eatson, J.A., Brinkley, V.M., & Ingels-Young, G. (1994). Comprehension monitoring, stylistic differences, pre-math knowledge, and transfer: a comprehensive pre-math/spatial development computer-assisted instruction (CAI) and Logo curriculum designed to test their effects. *Journal of Educational Computer Research, 11*, 91–105.

Hughes, M. & Macleod, H. (1986). Using LOGO with very young children. In R.W. Lawler (Ed.), *Cognition and Computers: Studies in Learning* (pp. 179–219). Chichester, NY: Ellis Horwood.

Hungate, H. (1982). Computers in the kindergarten. *The Computing Teacher, 9*(5), 15–18.

Hunka, S. & Buck, G. (1996). The rise and fall of CAI at the University of Alberta's Faculty of Education. *Canadian Journal of Educational Communication, 21*(2), 153–170.

International Telecommunication Union (2008). *Use of Information and Communication Technology by the World's Children and Youth. A Statistical Compilation.* Available at: http://www.itu.int/ITU-D/ict/material/Youth_2008.pdf.

Käser, T., Kucian, K., Ringwald, M., Baschera, G.M., von Aster, M., & Gross, M (2011). Therapy software for enhancing numerical cognition. In J. Özyurt et al. (Eds.) *Interdisciplinary Perspectives on Cognition, Education and the Brain* (pp. 207–216). Oldenburg: Hanse-Studies.

Ke, F. & Grabowski, B. (2007). Gameplaying for maths learning: cooperative or not? *British Journal of Educational Technology, 38*(2), 249–259.

Khalili, A. & Shashaani, L. (1994). The effectiveness of computer applications: a meta-analysis. *Journal of Research on Computing in Education, 27*(1), 48–62.

Khoju, M., Jaciw, A., & Miller, G.I. (2005). *Effectiveness of Graphing Calculators in K-12 Mathematics Achievement: A Systematic Review.* Palo Alto, CA: Empirical Education.

Kozma, R.B. (1994). Will media influence learning? Reframing the debate. *Journal of Educational Technology Research and Development, 42*(2), 7–19.

Kroesbergen, E.H. & Van Luit, J.E.H. (2003). Mathematical interventions for children with special educational needs. *Remedial and Special Education, 24*, 97–114.

Kucian, K., Grond, U., Rotzer, S., et al. (2011). Mental number-line training in children with developmental dyscalculia. *NeuroImage, 57*(3), 782–95.

Kucian, K., Loenneker, T., Dietrich, T., Dosch, M., Martin, E., & von Aster, M. (2006). Impaired neural networks for approximate calculation in dyscalculic children: a functional MRI study. *Behavioral and Brain Functions, 2*, 31.

Kulik, J. A. (2003). *Effects of Using Instructional Technology in Elementary and Secondary Schools: What Controlled Evaluation Studies Say.* Arlington, VA: SRI International.

Kulik, J.A., Kulik, C.L.C., & Bangert-Browns, R.L. (1985). Effectiveness of computer-based education in elementary schools. *Computers in Human Behavior, 1*, 59–74.

Kulik, J. A., & Kulik, C. L. C. (1987). Review of recent research literature on computer-based instruction. *Contemporary Educational Psychology*, 12(3), 222–230.

Kulik, C.L.C. & Kulik, J.A. (1991). Effectiveness of computer-based instruction: an updated analysis. *Computers in Human Behavior, 7*(1–2), 75–94.

Kulik, C.L.C., Kulik, J.A., & Shwalb, B.J. (1986). The effectiveness of computer-based adult education: a meta-analysis. *Journal of Educational Computing Research, 2*(2), 235–252.

Kyle, F., Kujala, J., Richardson, U., Lyytinen, H., and Goswami, U. (2013). Theoretically motivated computer-assisted reading interventions in the United Kingdom : GG Rime and GG Phoneme. *Reading Research Quarterly, 48*(1), 61–76.

Lavin, R. & Sanders, J. (1983). *Longitudinal Evaluation of the C/A/I Computer Assisted Instruction Title 1 Project: 1979–1982.* Chelmsford, MA: Merrimack Education Center.

Li, Q. & Ma, X. (2010). A meta-analysis of the effects of computer technology on school students' mathematics learning. *Educational Psychology Review, 22*, 215–243.

Liao, Y.K. (1998). Effects of hypermedia versus traditional instruction on students' achievement: a meta-analysis. *Journal of Research on Computing in Education, 30*(4), 341–359.

McCollister, T.S., Burts, D.C., Wright, V.L., & Hildreth, G.J. (1986). Effects of computer-assisted instruction and teacher-assisted instruction on arithmetic task achievement scores of kindergarten children. *Journal of Educational Research, 80*, 121–125.

McKeown, E.N. (1965). A comparison of the teaching of arithmetic in grade four by teaching machine, programmed booklet, and traditional methods. *Ontario Journal of Educational Research, 7(3)*, 289–295.

McLester, S. (1998). Girls and technology: what's the story. *Technology and Learning, 19*(3), 18–26.

Moran, J., Ferdig, R., Pearson, P.D., Wardrop, J., & Blomeyer, R. (2008). Technology and reading performance in the middle-school grades: a meta-analysis with recommendations for policy and practice. *Journal of Literacy Research, 40*(1), 6–58.

Morrill, C.S. (1961). Teaching machines: a review. *Psychological Bulletin*, *58*, 363–375.

Moyer, P.S., Niezgoda, D., & Stanley, J. (2005). Young children's use of virtual manipulatives and other forms of mathematical representations. In W. Masalski & P.C. Elliott (Eds.), *Technology-supported Mathematics Learning Environments: 67th Yearbook* (pp. 17–34). Reston, VA: National Council of Teachers of Mathematics.

Nathan, R. & Baron, L.J. (1995). The effects of gender, program type, and content on elementary children's software preferences. *Journal of Research on Computing in Education*, *27*(3), 348.

Niemiec, R.P. &Walberg, H.J. (1984). Computers and achievement in the elementary schools. *Journal of Educational Computing Research*, *1*, 435–440.

Niemiec, R.P., Samson, G., Weinstein, T., & Walberg, H.J. (1987). The effects of computer based instruction in elementary schools: a quantitative synthesis. *Journal of Research on Computing in Education*, *20*(2), 85–103.

Nuerk, H.-C., Graf, M., & Willmes, K. (2006). Grundlagen der zahlenverarbeitung und des rechnens [Foundations of number processing and calculation]. *Sprache, Stimme, Gehör: Zeitschrift für Kommunikationsstörungen, Schwerpunktthema Dyskalkulie*, *30*, 147–153.

Obersteiner, A., Reiss, K., & Ufer, S. (2013). How training on exact or approximate mental representations of number can enhance first-grade students' basic number processing and arithmetic skills. *Learning and Instruction*, *23*, 125–135.

OECD (2010). *Are New Millennium Learners Making the Grade? Technology Use and Educational Performance in PISA*. Paris: OECD.

Olson, J.K. (1988). *Microcomputers Make Manipulatives Meaningful*. Budapest, Hungary: International Congress of Mathematics Education.

Pagliaro, L.A. (1983). The history and development of CAI: 1926–1981, an overview. *Alberta Journal of Educational Research*, *29*, 75–84.

Papert, S. (1972). Teaching children thinking. *Innovations in Education & Training International*, *9*(5), 245–255.

Pressey, S.L. (1926). A simple apparatus which gives tests and scores – and teaches. *School and Society*, *23*(586), 373–376.

Price, G., Holloway, I., Räsänen, P., Vesterinen, M., & Ansari, D. (2008). Impaired parietal magnitude processing in developmental dyscalculia. *Current Biology*, *17*(24), 1042–1043.

Ragosta, M., Holland, P., & Jamison, D.T. (1981). *Computer-assisted Instruction and Compensatory Education*. Princeton: Educational Testing Service.

Rakes, C.R., Valentine, J.C., McGatha, M.B., & Ronau, R.N. (2010). Methods of instructional improvement in algebra: a systematic review and meta-analysis. *Review of Educational Research*, *80*(3), 372–400.

Räsänen, P. & Koponen, T. (2011). Matemaattisten oppimisvaikeuksien neuropsykologisesta tutkimuksesta [About neuropsychological research on mathematical disorders]. *NMI-Bulletin*, *20*(3), 39–53.

Räsänen, P., Salminen, J., Wilson, A., Aunio, P., & Dehaene, S. (2009). Computer-assisted intervention for children with low numeracy skills. *Cognitive Development*, *24*(4), 450–472.

Reimer, K., & Moyer, P. (2005). Third graders learn about fractions using virtual manipulatives: A classroom study. *Journal of Computers in Mathematics and Science Teaching*, *24*(1), 5–25.

Rideout, V., Vandewater, E.A., & Wartella, E.A. (2003). *Zero to Six: Electronic Media in the Lives of Infants, Toddlers and Preschoolers*. Menlo Park, CA: The Henry J. Kaiser Foundation.

Rosen, Y. & Salomon, G. (2007). The differential learning achievements of constructivist technology-intensive learning environments as compared with traditional ones: a meta-analysis. *Journal of Educational Computing Research*, *36*(1), 1–14.

Ryan, A.W. (1991). Meta-analysis of achievement effects of microcomputer applications in elementary schools. *Educational Administration Quarterly*, *27*(2), 161–184.

Samara, J. & Clements, D.H. (2009). "Concrete" computer manipulatives in mathematics education. *Child Development Perspectives*, *3*(3), 145–150.

Shumway, R.J. (1984). Young children, programming, and mathematical thinking. In V.P. Hansen & M.J. Zweng (Eds.), *Computers in Mathematics Education* (pp. 127–134). Reston: NCTM.

Siegler, R.S. & Ramani, G.B. (2008). Playing linear numerical board games promotes low-income children's numerical development. *Developmental Science, 11*, 655–661.
Skinner, B.F. (1958). Teaching machines. *Science, 128*, 969–977.
Slavin, R.E. & Lake, C. (2008). Effective programs in elementary mathematics: a best evidence synthesis. *Review of Educational Research, 78*(3), 427–455.
Slavin, R.E., Lake, C., & Groff, C. (2009). Effective programs in middle and high school mathematics: a best evidence synthesis. *Review of Educational Research, 79*(2), 839–911.
Slavin, R.E. & Smith, D. (2008). Effects of sample size on effect size in systematic reviews in education. *Educational Evaluation and Policy Analysis, 31*(4), 500–506.
Stoianov, I. & Zorzi, M. (2012). Emergence of a 'visual number sense' in hierarchical generative models. *Nature Neuroscience, 15*(2), 194–196.
Suppes, P. (1966). The uses of computers in education. *Scientific American, 215*(3), 206–220.
Suppes, P. & Morningstar, M. (1969). Computer-assisted instruction: two computer-assisted instruction programs are evaluated. *Science, 166*, 343–350.
Suppes, P. & Morningstar, M. (1972), *Computer-assisted Instruction at Stanford, 1966–68: Data, Models, and Evaluation of the Arithmetic Programs*. New York: Academic Press.
Suppes, P., Jerman, M., & Brian, D. (1968). *Computer-assisted Instruction: The 1965–66 Stanford Arithmetic Program*. New York: Academic Press.
Thomas, M.S.C. & Laurillard, D. (2012). Computational modelling of learning and teaching. In D. Mareschal et al. (Eds.), *Handbook of Educational Neuroscience*. Oxford: Wiley-Blackwell.
Tolani-Brown, N., McCormac, M., & Zimmerman, R. (2011). An analysis of the research and impact of ICT in education in developing countries contexts. In J. Steyn & G. Johanson (Eds.), *ICTs and Sustainable Solutions for the Digital Divide: Theory and Perspectives* (pp. 218–243). Hershey, PA: Idea Group Inc.
Tucker, T.H. (2009). *The Relationships between Computer-assisted Instruction and Alternative Programs to Enhance Fifth-graders Mathematics Success on the Annual Texas Assessment of Knowledge and Skills*. Dissertation, Department of Education, University of Texas. Available at: http://digital.library.unt.edu/ark:/67531/metadc12208/.
Vale, C. & Leder, G. (2004). Student views of computer-based mathematics in the middle years: does gender make a difference? *Educational Studies in Mathematics, 56*, 287–312.
Wakefield, J.F. (1998). *A Brief History of Textbooks: Where Have We Been All These Years?* A paper presented at the Meeting of the Text and Academic Authors (St. Petersburg, FL, June 12–13). Available at: http://www.eric.ed.gov/PDFS/ED419246.pdf.
Wastiau, P., Blamire, R., Kearney, C., Quittre, V., Van de Gaer, E. & Monseur, C. (2013). The use of ICT in education: a survey of schools in Europe. *European Journal of Education, 48*(1), 11–27.
Wilson, A.J., Dehaene, S., Pinel, P., Revkin, S.K., Cohen, L., & Cohen, D. (2006). Principles underlying the design of "the number race", an adaptive computer game for remediation of dyscalculia. *Behavioural and Brain Functioning, 2*(1), 19.
Wilson, A.J., Revkin, S.K., Cohen, D., Cohen, L., & Dehaene, S. (2006). An open trial assessment of "the number race", an adaptive computer game for remediation of dyscalculia. *Behavioural and Brain Functioning, 2*(1), 20.
Wilson, A.J., Dehaene, S., Dubois, O., & Fayol, M. (2009). Effects of an adaptive game intervention on accessing number sense in kindergarten children. *Mind, Brain and Education, 3*(4), 224–234.
Whyte, J.C. & Bull, R. (2008). Number games, magnitude representation, and basic number skills in preschoolers. *Developmental psychology, 44*(2), 588–596.
Wolf, F.M. (1986). *Meta-analysis: Quantitative Methods for Research Synthesis*. Beverly Hills, CA: Sage.
Xin, Y.P. & Jitendra, A.K. (1999). The effects of instruction in solving mathematical word problems for students with learning problems: a meta-analysis. *Journal of Special Education, 32*(4), 207–225.
Zorzi, M. & Butterworth, B. (1999). A computational model of number comparison. In M. Hahn & S.C. Stoness (Eds.), *Proceedings of the Twenty First Annual Conference of the Cognitive Science Society* (pp. 778–783). Mahwah (NJ): Erlbaum.

CHAPTER 42

THE CLASSIFICATION AND COGNITIVE CHARACTERISTICS OF MATHEMATICAL DISABILITIES IN CHILDREN

DAVID C. GEARY

Introduction

Nationally representative studies conducted in the UK, USA, and Canada indicate that the costs associated with poorly developed mathematical skills are higher than those associated with poor reading skills, in part because more people have difficulty with mathematics than with reading and because the quantitative knowledge needed to function in modern economies has been increasing steadily (Bynner, 1997; Parsons & Bynner, 1997; Rivera-Batiz, 1992). The studies reveal that while poor reading skills reduce employment opportunities and wages once employed, poor mathematical skills result in even dimmer prospects (Every Child a Chance Trust, 2009). These skills include basic arithmetic, measurement, and simple algebra, and indicate that a substantial minority (23% in the UK, 22% in the USA, <http://nces.ed.gov/programs/coe/indicator_nal.asp>) of adults have not mastered the mathematics expected of an 8th grader, making them functionally innumerate (e.g. unable to interpret quantitative information in a news story, understand compound interest). Entry into high-paying science and technology fields requires an even deeper understanding of mathematics (Paglin & Rufolo, 1990). These and many other studies confirm the individual and society-level benefits of a workforce with strong mathematical abilities (National Mathematics Advisory Panel, 2008), and in doing so highlight the long-term costs to people who have difficulties learning mathematics.

In all, it appears that individuals falling in the bottom quartile of mathematics achievement are at risk of long-term underemployment and frequent unemployment (Bynner,

1997). Among these are the roughly 7% (ranging from 4% to 14% depending on classification methods) of students with a learning disability in mathematics (MLD; Barbaresi, Katusic, Colligan, Weaver, & Jacobsen, 2005; Lewis, Hitch, & Walker, 1994; Shalev, Manor, & Gross-Tsur, 2005), and another 10–15% of students who will experience mild but persistent learning difficulties in mathematics (Berch & Mazzocco, 2007) not attributable to intelligence (IQ), working memory, reading ability, or other known factors; hereafter low achieving (LA). I begin with an overview of the characteristics of MLD and LA children and then focus on their deficits in quantitative domains and in domain-general learning abilities.

Basic Characteristics of Children With MLD and LA Children

Diagnosis

There is no consistently used test, achievement cut-off score, or achievement-intelligence discrepancy for diagnosing MLD or LA (Gersten, Clarke, & Mazzocco, 2007; Mazzocco, 2007), although, as noted, children who are consistently in the bottom quartile of achievement on standardized mathematics tests should be considered at risk. Within this bottom quartile, researchers often distinguish MLD and LA groups (Geary, Hoard, Byrd-Craven, Nugent, & Numtee, 2007; Murphy, Mazzocco, Hanich, & Early, 2007). Children who score at or below the 10th percentile on standardized mathematics achievement tests for at least two consecutive academic years are categorized as MLD in many research studies, and children scoring between the 11th and the 25th percentiles, inclusive, across two consecutive years as LA. In an analysis of mathematics achievement from kindergarten through 5th grade, inclusive, Geary, Hoard, Nugent, and Bailey (2012a) found that both start point (kindergarten achievement) and grade-to-grade growth were normally distributed and showed no indication of distinct MLD and LA groups. Nevertheless, cut-points are necessary in many educational settings to determine eligibility for special education and because the deficits underlying the learning difficulties of children with MLD may differ to some extent from those contributing to LA. The latter indicates that the remediation programmes for children classified as MLD or LA may need to differ in some ways (see Geary, 2010, for discussion).

Geary et al. (2012a) used start point achievement less than the 25th percentile combined with achievement growth (i.e. rate of change in mathematics test raw scores) less than the 25th percentile to classify children as MLD and growth less than the 25th percentile to classify them as LA. The 1st to 5th grade standard mathematics achievement scores of the children in these groups were consistent with the criteria used by Geary et al. (2007) and Murphy et al. (2007), that is, the children in the MLD group had mean across-grade achievement scores between the 8th and 10th percentile, inclusive, and the children in the LA group had mean achievement scores between the 14th and 28th percentile, inclusive. At the end of 5th grade, the means were at the 8th and 19th percentiles for the MLD and LA groups, respectively. The implication is that children who start school in the bottom quartile on a standard mathematics achievement test are at high risk for MLD, and that children who start school in the average range but also show grade-to-grade growth that is in the bottom quartile are at high risk for long-term LA.

Prevalence

The lack of agreed-about criteria and diagnostic tests complicates attempts to estimating the prevalence of MLD and LA. As a result, estimates have ranged from 4% to 14% for MLD, with the best estimate that about 7% of children and adolescents will be diagnosable as MLD in at least one area of mathematics before graduating from high school (Barbaresi et al., 2005; Desoete, Roeyers, & DeClerq, 2004; Lewis et al., 1994; Shalev et al., 2005). Again, it is only an estimate but an additional 10–15% of children and adolescents will be identified as LA (Geary et al., 2007; Murphy et al., 2007). These percentages differ from the above-noted 10th percentile cut-off for MLD and 11th to 25th percentile for LA, because these cut-offs are based on performance across more than one academic year; low performance in one academic year alone is not sufficient for considering a child as MLD or LA (Geary, Brown, & Samaranayake, 1991, Geary, Hamson, & Hoard, 2000). At each grade level, 10% of children will necessarily score at or below the 10th percentile, but not all of them will score in this range across multiple years but roughly 7% will and are considered MLD. The same logic applies to LA.

Etiology

It is very likely that some mix of environmental and genetic factors contribute to MLD and LA, but the details of this mix and how it might vary from one student to the next are poorly understood (Kovas, Haworth, Dale, & Plomin, 2007; Light & DeFries, 1995; Shalev et al., 2001). Shalev and colleagues found that family members of children with MLD were 10 times more likely to be diagnosed with MLD than individuals in the general population. In a large twin study, Kovas et al. found genetic as well as shared (between the pair of twins) and unique environmental contributions to the classification as MLD, whether the latter was defined by cut-offs at the 5th or 15th mathematics achievement percentiles. The same genetic influences responsible for the low performance of children with MLD were responsible for individual differences at all levels of performance (Kovas et al., 2007; Oliver et al., 2004), consistent with Geary et al.'s (2012a) finding that MLD and LA groups represent different cut-offs along the normal distribution of mathematical achievement.

In other words, the genetic influences on MLD are the same as those that influence mathematics achievement across the continuum of scores, as contrasted with 'MLD genes'. Of the genetic effects on mathematics achievement, 1/3 was shared with intelligence. Kovas et al. (2007) did not administer working memory and speed of processing measures and, thus, given the correlations among intelligence, working memory, and speed of processing tests, it is possible that this shared genetic variation may be related to other domain-general abilities, not simply intelligence. In any case, another 1/3 of the genetic variation in mathematics achievement was shared with reading achievement independent of intelligence, and 1/3 was unique to mathematics.

The implication is that about 2/3 of the genetic influences on mathematics achievement and MLD and presumably LA are the same as those that influence learning in other academic areas, and 1/3 of these genetic influences only affect mathematics learning. Children with MLD tend to have low-average IQ scores and below-average reading achievement scores. One possibility then is that the genetic factors that are correlated with achievement across academic domains may explain why many children with MLD have reading disability (RD) or other difficulties that interfere with learning in school, such as attention deficit hyperactivity disorder (ADHD) (Barbaresi et al., 2005; Fletcher, 2005; Rubinsten, 2009). Barbaresi et al. found that between 57% and 64% of individuals with MLD also had RD, depending on the diagnostic criteria used for MLD. Because LA children tend to be average in terms of IQ and reading achievement (Geary et al., 2007, 2012a), any genetic influences on their slow mathematical development may be unique to mathematics. In any case, the genetic studies also indicate important unique environmental effects on mathematics achievement (i.e. effects independent of other academic domains) and it should be remembered that genetic influences on achievement do not mean that achievement levels cannot be improved with appropriate interventions.

Mathematical Deficits

The genetic studies suggest that there are competencies uniquely related to learning mathematics and to MLD and LA, above and beyond cognitive factors that affect learning across academic domains, such as IQ. In the search for these mathematics-specific competencies, scientists have focused primarily on number, knowledge of counting principles, and arithmetic (Butterworth & Reigosa, 2007; Butterworth, Varma, & Laurillard, 2011; Geary, 1993, 2004; Jordan & Montani, 1997; Rourke, 1993; Russell & Ginsburg, 1984; Temple, 1991). Overall, the studies to date have not consistently found differences in children with MLD and LA children's counting knowledge relative to their TA peers (see Geary, 2010, for review), and thus the focus below is on number and arithmetic.

Number

Infants and young children have potentially inherent brain and cognitive systems that support the representation and implicit (i.e. they respond to differences in quantity but cannot articulate them) understanding of the *exact quantity* of small collections of objects (e.g. ◊○□) and for representing the *approximate magnitude* of larger quantities. Debate continues as to whether these competencies are supported by single or independent brain and cognitive systems (Holloway & Ansari, 2008; Piazza et al., 2010), but either way the result for typically achieving (TA) children is the ability to (1) subitize, that is to apprehend the quantity of sets of 3–4 objects or actions without counting (Mandler & Shebo, 1982; Starkey & Cooper, 1980; Strauss & Curtis, 1984; Wynn, Bloom, & Chiang, 2002); (2) use non-verbal processes or counting to quantify small sets of objects and to add and subtract small quantities to and from these sets (Case & Okamoto, 1996; Levine, Jordan, & Huttenlocher,

1992; Starkey, 1992); and (3) estimate the relative magnitude of sets of objects beyond the subitizing range and estimate the results of simple numerical operations (Dehaene, 1997); for example, implicitly knowing that adding one item to a set of items results in 'more' (Wynn, 1992).

In theory, the exact representational system supports children's initial learning that Arabic numerals and number words represent distinct quantities (e.g. '◊○□' = '3' = 'three') and the approximate system appears to support learning in some other areas of basic mathematics, such as the ability to estimate quantity (e.g. determining the larger of 34 vs. 82) and map numbers onto a number line (Geary, 2011; Gilmore, McCarthy, & Spelke, 2007; Siegler & Opfer, 2003). It is a matter of debate, however, regarding exactly how culturally specific numerals and number words are mapped onto these potentially inherent systems and the extent to which performance on the associated tasks (e.g. number line) is a stronger or weaker measure of functioning of the proposed underlying system (e.g. approximate magnitude) (Barth & Paladino, 2011; Cohen & Blanc-Goldhammer, 2011; Feigenson, Dehaene, & Spelke, 2004; Núñez, 2009). Although these details remain to be resolved, it is nevertheless consistently found that performance on many basic measures of numerical representation and processing predict mathematics achievement (Booth & Siegler, 2006; Geary, 2011; Schneider, Rittle-Johnson, & Star, 2011; Siegler & Booth, 2004), and thus are potential contributors to MLD and LA (Butterworth et al., 2011; Geary, Hoard, Nugent, & Byrd-Craven, 2008; Halberda, Mazzocco, & Feigenson, 2008; Koontz & Berch, 1996; Landerl, Bevan, & Butterworth, 2003; Stock, Desoete, & Roeyers, 2010).

Koontz and Berch (1996) were the first to test the hypothesis that children with MLD had a deficit in the exact representational system. In this study, 3rd and 4th graders with MLD and their TA peers were asked to determine if combinations of Arabic numerals (e.g. 3-2), number sets (◊○-□☼), or numerals and sets were the same (2-◊○) or different (3-◊○). Confirming earlier findings (Mandler & Shebo, 1982), the TA children's reaction time (RT) patterns indicated fast and automatic subitizing for quantities of two and three, whether the code was an Arabic numeral or number set. The children with MLD showed fast access to representations of the quantity of two, but appeared to rely on counting to determine quantities of three. The results suggest that some children with MLD might not have an inherent representation for numerosities of three or the exact representational system does not reliably discriminate two and three.

Piazza et al. (2010) found evidence for a developmental delay in children with MLD's (termed dyscalculia in this study) ability to represent the approximate quantity of collections of items. The task involved asking the children (and a group of adults) to compare two collections of dots and determine the larger of the two. Ease of making such discriminations varies with the ratio of dots in the two sets, which is estimated by the Weber fraction, that is, the smallest difference between two set sizes that can be reliably discriminated. Three-year-olds can discriminate sets that differ by a 3 to 4 ratio (e.g. they can discriminate 6 from 8 items without counting but not 7 from 8 items); 6-year-olds have a Weber fraction of 5:6 and adults 10:11 (Halberda & Feigenson, 2008). Piazza et al. found that the Weber fraction of 10-year-old children with MLD was about the same as that of IQ-match TA 5-year-olds. Using a similar procedure, however, Rousselle and Noël (2007) did not find a deficit in the ability of 2nd graders with MLD to discriminate sets of objects, relative to same-age TA children, but the children with MLD were slower and made more errors when comparing symbolic representations (e.g. which is more, 16 vs. 24?). These results suggest that at least for some young children with MLD, the difficulty resides in the mapping of Arabic numerals onto an otherwise intact non-verbal representation of quantity.

Geary and colleagues developed the Number Sets Test to assess the speed and accuracy with which children process and add sets of objects and Arabic numerals to match a target number; e.g. whether the combination '●● 3' matches the target of '5' (Geary, Bailey, & Hoard, 2009; Geary et al., 2007). The items are similar to those used by Koontz and Berch (1996), but some involve magnitudes up to 9. Fluency should be aided by rapid subitizing, rapid mapping of Arabic numerals to the associated quantities, ease of estimating approximate quantity, and the ability to add and compare and contrast these quantities. In other words, the measure is more complex than the subitizing and magnitude discrimination tasks used by Koontz and Berch (1996), Piazza et al. (2010), and Rousselle and Noël (2007), but appears to tap the ability to dynamically use these basic number competencies. Performance on the Number Sets Test predicts mathematics but not reading achievement above and beyond the influence of intelligence, working memory, and speed of articulating number words (Geary, 2011; Geary et al., 2009). Figure 42.1 shows longitudinal performance on this test from 1st to 5th grade, inclusive, for groups of TA, MLD, and LA children

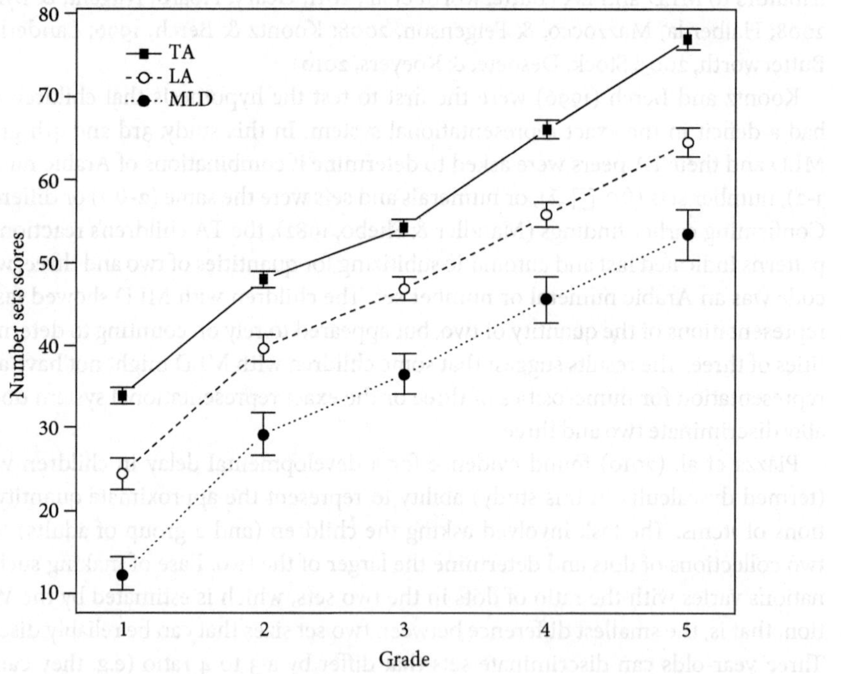

FIGURE 42.1 Fluency scores on the Number Sets Test comparing typically achieving (TA) children to children with mathematical learning disability (MLD) and persistent low achievement (LA) in mathematics.

Adapted from David C. Geary, Mary K. Hoard, Lara Nugent, and Drew. H. Bailey, Mathematical cognition deficits in children with learning disabilities and persistent low achievement: A five year prospective study, *Journal of Educational Psychology*, 104(1) pp. 206–223, doi: 110.1037/a0025398 © 2012, American Psychological Association.

(Geary et al., 2012a). In contrast to the catch-up in procedural skills described in arithmetic (below), the children in the MLD and LA groups started 1st grade behind their TA peers and showed no indication of catching up.

In all, the research to date is consistent with some very basic form of number processing deficit in children with MLD and LA children that cannot be attributed to below-average intelligence or poor working memory capacity. The exact nature of these deficits remains to be determined, with strong candidates including below-average subitizing capacity, deficits, or delays in the acuity of the system for representing approximate magnitudes, and difficulties mapping Arabic numerals and number words onto these representations of quantity. It is very likely that groups of children with MLD and LA children are heterogeneous, and that different children may have different forms or combinations of deficits or delays in these areas.

Arithmetic

The vast majority of the research on children with MLD's and LA children's competence in arithmetic has focused on whole number arithmetic, and much of this has focused on the sophistication of the procedures they use for problem solving and their difficulties in memorizing or retrieving basic facts from long-term memory (Geary, 1993), although several research groups have also examined aspects of their conceptual understanding of arithmetic and skill at solving word problems and these results are noted as well (Andersson, 2008, 2010; Jordan, Hanich, & Kaplan, 2003a). The basic results for whole number arithmetic are followed by a brief review of recent work in the area of rational numbers.

Whole Number Arithmetic

Procedural Development

Most kindergarten children can solve simple addition and subtraction problems and do so using a mix of problem-solving strategies (Geary & Burlingham-Dubree, 1989; Groen & Resnick, 1977; Siegler & Jenkins, 1989; Siegler & Shrager, 1984). The most common strategies involve counting sometimes using their fingers (finger counting strategy) and sometimes not using them (verbal counting strategy) (Siegler & Shrager, 1984), typically with the min or sum procedures (Groen & Parkman, 1972). The min procedure involves stating the larger-valued addend and then counting a number of times equal to the value of the smaller addend. The sum procedure involves counting both addends starting from 1.

The use of counting results in the development of memory representations of basic facts that in turn support the use of *direct retrieval* of arithmetic facts and *decomposition* (Siegler & Shrager, 1984). With direct retrieval, children state an answer that is associated in long-term memory with the presented problem, whereas with decomposition children reconstruct the answer based on the retrieval of a partial sum; for example, 6 + 7 might be solved by retrieving the answer to 6 + 6 and then adding 1 to this partial sum. Another decomposition process involves solving problems in one operation (e.g. subtraction, 10 − 7 =?) by retrieving the corresponding answer in the inverse operation (e.g. 7 + 3 = 10).

Developmental change does not simply involve switching from use of one strategy to another. Rather, for any given problem children can use one of the many strategies they know; they may retrieve the answer to 2 + 1 but count to solve 7 + 4. What changes is the mix of strategies, with sophisticated retrieval-based ones used more often and less sophisticated procedural-based ones less often (Siegler, 1996). The result is faster and more accurate solving of simple problems that in turn makes the solving of more complex, multi-digit problems easier (Andersson, 2008; Geary & Widaman, 1987, 1992).

One consistent finding is that children with MLD and LA children use the same types of problem-solving approaches as their TA peers, but commit more procedural errors when they solve simple arithmetic problems (4 + 3), simple word problems, and complex arithmetic problems (e.g. 745 – 198) (Andersson, 2008, 2010; Geary et al., 2007; Hanich, Jordan, Kaplan, & Dick, 2001; Jordan, Hanich, & Kaplan, 2003b; Russell & Ginsburg, 1984). Even when these children do not commit errors, they tend to use developmentally immature procedures, that is, procedures commonly used by younger TA children (Dowker, 2005, 2009; Geary, 1990; Hanich et al., 2001; Jordan et al., 2003b; Ostad, 1998; Raghubar, Cirino, Barnes, Ewing-Cobbs, Fletcher, & Fuchs, 2009). For the same groups shown in Figure 41.1, Geary et al. (2012a) documented the 1st to 5th grade procedural delays for groups of LA and MLD children (Figure 41.2, Panel A). The plot represents skilled use of min counting to solve relatively complex addition problems, such as 18 + 6; children who always used this approach when they counted (i.e. adjusting for proportional use of counting) and never committed a counting error received a score of 12. As shown in Figure 42.2, children with MLD started 1st grade far behind children in the LA and TA groups, and LA children started behind their TA peers. The children in the LA group closed the gap with the TA children by the beginning of 4th grade and the children with MLD by the beginning of 5th.

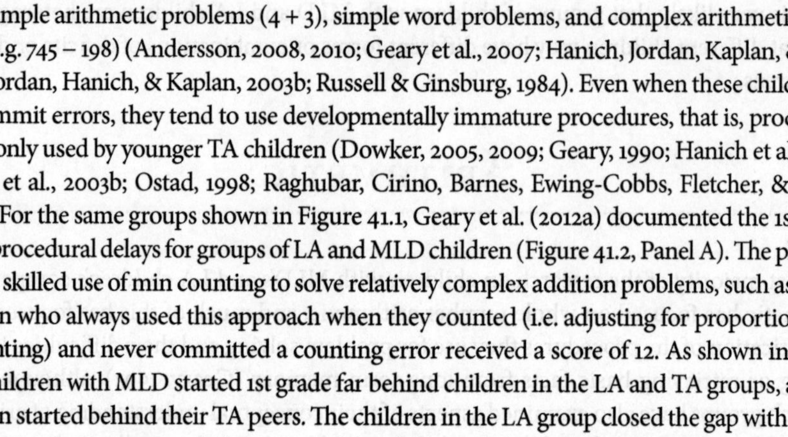

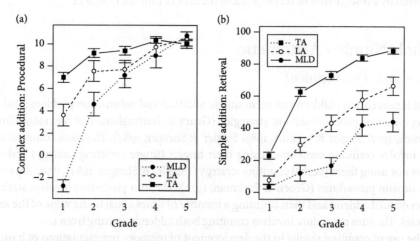

FIGURE 42.2 Panel A shows procedural competence scores for complex addition. A maximum score of 12 reflects use of the min procedure (e.g. 17 + 3 = 'seventeen, eighteen, nineteen, twenty') with either verbal or finger counting and without error; the score was adjusted to reflect the proportional use of counting strategies. Panel B shows the percentage of the simple addition problems that were correctly solved using retrieval. Brackets are standard errors.

Adapted from David C. Geary, Mary K. Hoard, Lara Nugent, and Drew. H. Bailey, Mathematical cognition deficits in children with learning disabilities and persistent low achievement: A five year prospective study, *Journal of Educational Psychology*, 104(1) pp. 206–223, doi: 110.1037/a0025398 © 2012, American Psychological Association.

The extent of these developmental delays appears to vary with the complexity of the procedure (Fuchs & Fuchs, 2002; Jordan & Hanich, 2000). During the solving of multi-step arithmetic problems (e.g. 45 × 12 or 126 + 537), Russell and Ginsburg (1984) found that 4th grade children with MLD committed more errors than their TA peers. The errors involved the misalignment of numbers while writing down partial answers or while carrying or borrowing from one column to the next. Raghubar et al. (2009) confirmed this finding and extended it to subtraction. Common subtraction errors included subtracting the larger number from the smaller one (e.g. 83 – 44 = 41), failing to decrement after borrowing from one column to the next (e.g. 92 – 14 = 88; the 90 was not decremented to 80), and borrowing across zeros (e.g. 900 – 111 = 899) (Geary, 1994). These patterns were found for children with MLD and LA children, regardless of their reading achievement.

The mechanisms underlying these developmental delays are not yet fully understood, but contributing factors include the fact retrieval deficits described below (Andersson, 2008), a poor conceptual understanding of the base-ten system and related concepts (e.g. place value; Andersson, 2010; Dowker, 2009; Hanich et al., 2001), and poor working memory (Geary, 2004). Whatever the combination of mechanisms, the closing of the gap shown in Figure 41.2 is not likely to occur until middle school or later for more complex arithmetic procedures.

Fact Retrieval

The best documented finding is that children with MLD and a subset of LA children have persistent difficulties committing basic arithmetic facts to long-term memory or retrieving them once they are committed (Andersson, 2008, 2010; Barrouillet, Fayol, & Lathuliére, 1997; Geary, 1990, 1993; Geary & Brown, 1991; Geary, Hamson, & Hoard, 2000; Geary et al., 2012b; Hanich et al., 2001; Jordan et al., 2003a; Jordan & Montani, 1997; Ostad, 1997). These children are able to memorize and retrieve some basic facts, but show persistent differences in the frequency with which they correctly retrieve them, as well as differences in the pattern of retrieval errors relative to TA children. An example of the persistence of this deficit is shown in Panel B of Figure 41.2 which depicts the 1st to 5th grade, inclusive, frequency with which the groups of TA, LA, and MLD children shown in Panel A correctly retrieved answers to simple (e.g. 5 + 4) addition problems (Geary et al., 2012a). Children in all of the groups were learning their basic facts, and after 2nd grade the rate of learning for the LA and TA groups was comparable, although the 5th grade LA children were correctly retrieving facts at a rate comparable to the TA children in 2nd grade and the children with MLD fared even worse.

Several mechanisms have been proposed as contributing to the retrieval deficit. The first is based on children's early reliance on counting to solve arithmetic problems. The gist is that any disruption in the ability to represent or retrieve information from the phonetic, such as number words, and semantic representational systems that support counting should, in theory, result in difficulties in forming problem–answer associations during the act of counting (Geary, 1993). Dehaene and colleagues' work with adults suggests that the retrieval of addition facts is indeed supported by a system of neural structures that appear to support phonetic and semantic representations and are engaged during incrementing processes, such as counting (Dehaene & Cohen, 1995, 1997). The results from neuropsychological and brain imaging studies of adults need to be applied cautiously to children's developing competences because the brain and cognitive systems that support early arithmetic learning probably differ in important ways from those that support the same competence in adulthood (Ansari, 2010; Cho, Ryali, Geary, & Menon, 2011). Although counting is frequently used to solve addition problems, it is used less frequently to solve subtraction, multiplication, and division problems, and thus the extent to which this mechanism is applicable across arithmetic operations is not clear.

The second mechanism is based on the work of Engle, Conway, Tuholski, and Shisler. (1995) and Engle, Kane, and Tuholski (1999) and involves a deficit in the ability to inhibit irrelevant associations from entering working memory during the process of fact retrieval (Barrouillet et al., 1997). For arithmetic, the mechanism is assessed by asking children to only solve problems by remembering the answer as quickly as possible (retrieval only task) and not to use counting or any other procedure (Jordan & Montani, 1997). If intrusions disrupt children's ability to retrieve the correct answer, then the corresponding retrieval errors should be common long-term memory associations between the numbers in the problem. As an example, for the problem 4×5, these would include 25 and 6. The first is a table-related error (i.e. a correct answer to a similar, 5×5, problem in the multiplication table; Campbell & Graham, 1985) and the second is a counting-string error (i.e. the retrieved answer follows one of the addends in the counting string; Siegler & Robinson, 1982).

Barrouillet et al. (1997) found that adolescents with MLD had frequent intrusions of table-related errors when they solved simple multiplication problems. Geary et al. (2000) found a similar pattern when a combined group of 2nd grade children with MLD/LA children solved simple addition problems. In a longitudinal assessment of retrieval errors from 2nd to 4th grade, inclusive, Geary et al. (2012b) found that LA children were composed of two subgroups. The first was defined based on the high percentage of retrieval errors on the retrieval only task; 85% of their retrieved answers were errors in all three grades, with little across-grade improvement. For the remaining LA children, 55% of their retrieved answers were errors in 2nd grade, and this dropped to 37% by 4th grade. For the children with MLD, 78% of their retrieved answers were errors in 2nd grade, dropping to 59% by 4th grade. The TA children had the fewest errors; 37% to 34% across grades. More important, counting-string intrusions were rare among the TA children (5% of retrieval errors in 2nd grade), but were especially frequent among the LA children with the high frequency of retrieval errors (21%) and among the children with MLD (21%). Unlike most other tasks in which LA children outperform children with MLD, the LA children with frequent retrieval errors showed no across-grade drop in the percentage of intrusion errors, but the percentage dropped from 21% in 2nd grade to 8% in 4th grade for the children with MLD.

The third mechanism is the earlier-described deficits or delays in the systems that support number representation and processing (Butterworth, 2005). Among the factors that facilitate children's early learning of arithmetic is their ability to estimate the approximate answer to simple problems (Gilmore et al., 2007) and, of course, their ability to eventually associate the numerals in arithmetic problems with specific quantities. In this view, retrieval deficits are secondary to a more basic deficit in the approximate representational system and in the ease with which numerals can be mapped onto this system with sufficient fidelity to allow discrimination among them. Empirical evaluation will require longitudinal studies to determine if there is a relation between poor discrimination of magnitudes, as represented by the Weber fraction, during the preschool years and retrieval deficits in the elementary school years. Geary et al.'s (2012b) study does not explicitly address this issue, but they did examine the relation between the overall frequency of retrieval and intrusion errors and performance on the Number Sets Test. The critical finding was that the two subgroups of LA children had similar delays in fluency on this test, despite very different patterns of retrieval errors. In other words, deficits in the number processing abilities assessed by the Number Sets Test were not related to the frequency of intrusion errors, contra Butterworth's (2005) prediction.

Overall, it is clear that difficulties in learning or retrieving basic arithmetic facts are a common and persistent problem for children with MLD and for a subset of LA children. Of the proposed mechanisms, the evidence is strongest for intrusion errors, that is, the retrieval deficits are related in part to the intrusion of related but task-irrelevant information into working memory when these children are attempting to remember arithmetic facts. Not all of their errors are due to intrusions, however, suggesting that multiple mechanisms may be involved and that different children may have retrieval deficits for different reasons. Whether these alternative mechanisms involve the language system and number processing deficits remains to be determined.

Rational Numbers

Achieving a conceptual understanding of rational numbers, especially fractions, decimals, and percentages, and gaining fluency in the use of associated problem-solving procedures is a critical step in children's mathematical development (National Mathematics Advisory Panel, 2008). The extent to which procedural skill facilitates concept learning, concept learning facilitates procedural skill, or whether they bi-directionally interact is debated. However, there is consensus that conceptual understanding is more difficult to acquire and more central than procedural skills to learning rational numbers (Byrnes & Wasik, 1991; Hallett, Nunes, & Bryant, 2010; Hecht, 1998; Mazzocco & Devlin, 2008; Rittle-Johnson, Siegler, & Alibali, 2001; Siegler, Thompson, & Schneider, 2011).

The few studies that have been conducted on rational number concepts and procedures indicate that children with MLD and LA children have particular difficulties in conceptually understanding the magnitudes represented by simple fractions or visual depictions of parts of a whole (e.g. 1/3 of a rectangle) (Hecht & Vagi, 2010; Mazzocco & Devlin, 2008). Mazzocco and Devlin found that 6th and 8th graders with MLD had difficulties reading decimals, identifying equivalent rational numbers (e.g. ½ = 0.5), and ordering fractions based on their magnitudes. Their LA peers had the most difficulty ordering rational numbers that were a mixture of fractions and decimals (e.g. 0.3, ½, 0.8, 8/9), again suggesting a poor understanding of the associated magnitudes.

There is clearly much that remains to be learned about the difficulties children with MLD and LA children have with learning rational numbers. Siegler et al. (2011) have proposed that the key to this learning is the insight that rational numbers represent magnitudes that can be ordered on the mathematical number line, which appears to be one area of difficulty for these children; i.e. they have difficulties with discriminating magnitudes and placing them on a number line (Geary et al., 2008; Piazza et al., 2010). More critically, Siegler et al.'s proposal leads to the prediction that poor acuity of the system for representing approximate magnitudes or difficulty mapping symbols onto these representations will make the conceptual understanding of rational numbers difficult to achieve, which in turn will result in later difficulties with algebra. In contrast, based on their delayed but otherwise intact procedural learning, children with MLD and LA children should be comparatively better at learning and using fractions procedures.

Domain-General Deficits

Children with learning disabilities by definition have below-average academic achievement, and thus factors that predict school achievement must be considered as potential sources of their academic delays; in particular fluid intelligence and working memory. Although performance on intelligence and working memory measures are correlated (Ackerman, Beier, & Boyle, 2005), they appear to tap unique competencies that can contribute to individual differences in mathematical learning, including MLD (Bull & Johnston, 1997; Deary, Strand, Smith, & Fernandes, 2007; Geary, 2011), and should be considered separately.

Intelligence

General intelligence is the best individual predictor of achievement across academic domains, including mathematics (e.g. Deary et al., 2007; Walberg, 1984). Intelligence is not a likely contributor to the persistently low mathematics achievement of LA children, because these children tend to be of average intelligence and reading ability. As a group, children with MLD tend to have intelligence scores in the low-average range (standard scores of 90–95; Geary et al., 2007; Murphy et al., 2007), but their achievement in mathematics and many of their specific mathematical cognition deficits are far below their intellectual potential.

In the Piazza et al. (2010) study, for instance, the children with MLD and their TA peers were of average intelligence and thus the poor acuity in the MLD group's system for representing approximate magnitudes could not be related to intelligence. Geary (2011) found that performance on the Number Sets Test, frequency of correct retrieval and use of decomposition to solve addition problems, and skilled placement of Arabic numerals on the mathematical number line predicted individual differences in school entry mathematics achievement or achievement growth through the end of 5th grade, above and beyond the influence of intelligence and working memory. Geary et al. (2012b) found that these mathematics cognition variables mediated the school entry mathematics achievement of children with MLD and contributed to the slow achievement growth of groups of children with MLD and LA children; i.e. these mathematical competencies were more important than intelligence in explaining the poor mathematics achievement of the children in these groups.

Working Memory

Working memory represents the ability to hold a mental representation in mind while simultaneously engaging in other mental processes. At the core is the central executive, which is expressed as attention-driven control of information that is active in two representational systems (Baddeley, 1986; Baddeley & Hitch, 1974; Cowan, 1995). These are a language-based phonological loop (Baddeley, Gathercole, & Papagno, 1998) and a visuospatial sketch pad (Logie, 1995). Children with higher capacity in the central executive score better than their lower-capacity peers on measures of mathematics achievement and cognition (Bull, Espy, & Wiebe, 2008; Mazzocco & Kover, 2007; Passolunghi, Vercelloni, & Schadee, 2007). The importance of the phonological loop and visuospatial sketch pad varies with the complexity and content of the mathematics being assessed (Bull et al., 2008; Geary et al., 2007). The phonological loop appears to be important for processes that involve the articulation of numbers, as in counting (Krajewski & Schneider, 2009) and the solving of mathematical word problems (Swanson & Sachse-Lee, 2001), and may be related to arithmetic fact retrieval (Fuchs et al., 2006; Geary, 1993). The visuospatial sketch pad appears to be involved in a broader number of mathematical domains, although many details remain to be determined (Andersson, 2010; De Smedt, Janssen, Bouwens, Verschaffel, Boets, & Ghesquière, 2009; Geary, 2011; Swanson, Jerman, & Zheng, 2008).

Most children with MLD have working memory deficits that contribute to their slow progress in learning mathematics independent of intelligence (Geary, Hoard, Byrd-Craven, & DeSoto, 2004; Geary et al., 2007; McLean & Hitch, 1999; Swanson, 1993; Swanson, Jerman, & Zheng, 2009; Swanson & Sachse-Lee, 2001). Although most of these children have deficits in all three working memory systems, their compromised central executive appears to be especially problematic (Bull, Johnston, & Roy, 1999; Geary et al., 2007; Swanson, 1993). The implications of the central executive deficit are not fully known because the three subcomponents of the central executive—maintaining information in working memory, task switching, and inhibiting the retrieval of irrelevant information—may affect mathematical learning in different ways (Bull & Scerif, 2001; Murphy et al., 2007; Passolunghi, Cornoldi, De Liberto, 1999; Passolunghi & Siegel, 2004) and the relation between these specific components of the central executive and the mathematical cognition deficits of children with MLD remain to be detailed.

Many LA children, in contrast, appear to have a normal phonological working memory, especially if reading achievement is average or better, and a normal ability to use the attentional control functions of the central executive to maintain information in working memory (Geary et al., 2012a). Many of these children also appear to have an intact visuospatial working memory system, but a subset of them may have more subtle deficits (McLean & Hitch, 1999). The most promising results suggest that LA children have subtle deficits in the inhibitory control and task-switching components of the central executive (Geary et al., 2007; Murphy et al., 2007), but we await confirmation.

Conclusion

Adults in the bottom quartile of mathematical competence are at high risk for long-term underemployment and frequent periods of unemployment (Bynner, 1997) and children in the bottom quartile of mathematics achievement are at high risk of becoming one of these functionally innumerate adults. Among these at-risk children are the roughly 7% with MLD and another roughly10% of children with persistent LA in mathematics that is unrelated to intelligence or reading ability. Although it is likely that MLD and LA represent different cut-offs along the normal distribution of mathematic achievement scores and not distinct groups with respect to achievement (Geary et al., 2012a; Kovas et al., 2007), selection for special education services or interventions must be based on explicit criteria, and thus the use of cut-off points has some practical value. Further, the study of the deficits and delays of these children and how to remediate them is also facilitated by using specific cut-off points; generally, a diagnostic cut-off for MLD at the 10th percentile on a mathematics achievement test for more than one grade (Geary et al., 2007; Murphy et al., 2007) and between the 11th and 25th percentiles, inclusive, for LA. Comparisons of students in these groups is scientifically important, because the cognitive and perhaps brain mechanisms that result in persistent achievement in the MLD or LA range of achievement scores differ to some extent (Geary et al., 2012a; Murphy et al., 2007; Rubinsten, 2009; Stock et al., 2010).

Much of the research in this area has focused on basic competencies in understanding number and arithmetic. Children with MLD and, to a lesser extent, their LA peers show a deficit or delay in their processing of numbers (e.g. accessing the quantity associated with '3', or sensitivity to differences among larger quantities), learning of arithmetic procedures, memorizing basic arithmetic facts, and their conceptual understanding of rational numbers (Berch & Mazzocco, 2007; Dowker, 2005, 2009; Geary, 1993; Mazzocco & Devlin, 2008; Piazza et al., 2010; Stock et al., 2010). These learning difficulties are related in part to low-average intelligence (i.e. 90–95) but more strongly to below-average working memory capacity for children with MLD but not for LA children, or at least not as substantially for these children. Children in LA groups are typically of average intelligence and working memory capacity—although subtle deficits cannot be ruled out at this time—but appear to have a below-average facility in dealing with numbers (e.g. adding □□◊ + 2 =?), show a delay in learning arithmetic procedures, and a subset of them have difficulty retrieving basic facts from long-term memory (Geary et al., 2012b; Jordan et al., 2003a). Children with MLD show deficits in all of these areas, above and beyond the influence of intelligence and working memory, most of which are more severe or longer term for developmental delays than those found in LA groups.

Further advances in our understanding of the cognitive mechanisms underlying the developmental delays and deficits of MLD and LA children will require not only fine-grain assessment of the targeted mathematical competence but also fine-grain assessment of potential mediators of these deficits. Of particular interest will be studies that simultaneously assess the multiple subsystems of the central executive component of working memory and that relate these to the mathematical deficits and delays of groups of children with MLD and LA children, such as their difficulties with arithmetic fact retrieval. Our understanding of these disorders will also require expansion of the mathematical content domains under study, with more extensive studies of conceptual understanding of rational numbers and procedural and conceptual competencies in algebra. The early identification of children at risk for later MLD and LA is also critical and some advances have been made in this regard but have focused on mathematical cognition in kindergarteners and 1st graders (Geary et al., 2009; Locuniak & Jordan, 2008). These studies need to be extended to the preschool years.

References

Ackerman, P. L., Beier, M. E., & Boyle, M. O. (2005). Working memory and intelligence: the same or different constructs? *Psychological Bulletin*, *131*, 30–60.

Andersson, U. (2008). Mathematical competencies in children with different types of learning difficulties. *Journal of Educational Psychology*, *100*, 48–66.

Andersson, U. (2010). Skill development in different components of arithmetic and basic cognitive functions: findings from a 3-year longitudinal study of children with different types of learning difficulties. *Journal of Educational Psychology, 102*, 115–134.

Ansari, D. (2010). Neurocognitive approaches to developmental disorders of numerical and mathematical cognition: the perils of neglecting development. *Learning and Individual Differences, 20*, 123–129.

Baddeley, A. D. (1986). *Working Memory*. Oxford: Oxford University Press.

Baddeley, A., Gathercole, S., & Papagno, C. (1998). The phonological loop as a language learning device. *Psychological Review, 105*, 158–173.

Baddeley, A. D. & Hitch, G. J. (1974). Working memory. In G. H. Bower (Ed.), *The Psychology of Learning and Motivation: Advances in Research and Theory* (Vol. 8, pp. 47–90). New York: Academic Press.

Barbaresi, W.J., Katusic, S.K., Colligan, R.C., Weaver, A.L., & Jacobsen, S.J. (2005). Math learning disorder: incidence in a population-based birth cohort, 1976–82, Rochester, Minn. *Ambulatory Pediatrics, 5*, 281–289.

Barrouillet, P., Fayol, M., & Lathuliére, E. (1997). Selecting between competitors in multiplication tasks: an explanation of the errors produced by adolescents with learning disabilities. *International Journal of Behavioral Development, 21*, 253–275.

Barth, H. C. & Paladino, A. M. (2011). The development of numerical estimation: evidence against a representational shift. *Developmental Science, 14*, 125–135.

Berch, D. B. & Mazzocco, M. M. M. (Eds) (2007). *Why Is Math So Hard for Some Children? The Nature and Origins of Mathematical Learning Difficulties and Disabilities*. Baltimore, MD: Paul H. Brookes.

Booth, J. L. & Siegler, R. S. (2006). Developmental and individual differences in pure numerical estimation. *Developmental Psychology, 41*, 189–201.

Bull, R., Espy, K. A., & Wiebe, S. A. (2008). Short-term memory, working memory, and executive functions in preschoolers: longitudinal predictors of mathematical achievement at age 7 years. *Developmental Neuropsychology, 33*, 205–228.

Bull, R. & Johnston, R. S. (1997). Children's arithmetical difficulties: contributions from processing speed, item identification, and short-term memory. *Journal of Experimental Child Psychology, 65*, 1–24.

Bull, R., Johnston, R. S., & Roy, J. A. (1999). Exploring the roles of the visual-spatial sketch pad and central executive in children's arithmetical skills: views from cognition and developmental neuropsychology. *Developmental Neuropsychology 15*, 421–442.

Bull, R. & Scerif, G. (2001). Executive functioning as a predictor of children's mathematical abilities: inhibition, switching, and working memory. *Developmental Neuropsychology, 19*, 273–293.

Butterworth, B. (2005). Developmental dyscalculia. In J. I. D. Campbell (Ed.), *Handbook of Mathematical Cognition* (pp. 455–467). New York: Psychology Press.

Butterworth, B. & Reigosa, V. (2007). Information processing deficits in dyscalculia. In D. B. Berch & M. M. M. Mazzocco (Eds), *Why Is Math So Hard for Some Children? The Nature and Origins of Mathematical Learning Difficulties and Disabilities* (pp. 65–81). Baltimore, MD: Paul H. Brookes.

Butterworth, B., Varma, S., & Laurillard, D. (27 May 2011). Dyscalculia: from brain to education. *Science, 332*, 1049–1053.

Bynner, J. (1997). Basic skills in adolescents' occupational preparation. *Career Development Quarterly, 45*, 305–321.

Byrnes, J. P. & Wasik, B. A. (1991). Role of conceptual knowledge in mathematical procedural learning. *Developmental Psychology, 27*, 777–786.

Campbell, J. I. D. & Graham, D. J. (1985). Mental multiplication skill: structure, process, and acquisition. *Canadian Journal of Psychology, 39*, 338–366.

Case, R. & Okamoto, Y. (1996). The role of central conceptual structures in the development of children's thought. *Monographs of the Society for Research in Child Development, 66* (1–2, Serial No. 246).

Cho, S., Ryali, S., Geary, D. C., & Menon, V. (2011). How do you solve 7+8?: decoding children's brain activity patterns during counting versus retrieval. *Developmental Science, 14*, 989–1001.

Cohen, D. & Blanc-Goldhammer, D. (2011). Numerical bias in bounded and unbounded number line tasks. *Psychonomic Bulletin & Review, 18*, 331–338.

Cowan, N. (1995). *Attention and Memory: An Integrated Framework*. New York: Oxford University Press.

Deary, I. J., Strand, S., Smith, P., & Fernandes, C. (2007). Intelligence and educational achievement. *Intelligence*, *35*, 13–21.

Dehaene, S. (1997). *The Number Sense: How the Mind Creates Mathematics*. New York: Oxford University Press.

Dehaene, S. & Cohen, L. (1995). Towards an anatomical and functional model of number processing. *Mathematical Cognition*, *1*, 83–120.

Dehaene, S. & Cohen, L. (1997). Cerebral pathways for calculation: double dissociation between rote verbal and quantitative knowledge of arithmetic. *Cortex*, *33*, 219–250.

De Smedt, B., Janssen, R., Bouwens, K., Verschaffel, L., Boets, B., & Ghesquière, P. (2009). Working memory and individual differences in mathematics achievement: a longitudinal study from first grade to second grade. *Journal of Experimental Child Psychology*, *103*, 186–201.

Desoete, A, Roeyers, H., & DeClerq, A. (2004). Children with mathematics learning disabilities in Belgium. *Journal of Learning Disabilities*, *37*, 50–61.

Dowker, A. (2005). *Individual Differences in Arithmetic: Psychology, Neuroscience and Education*. Hove: Psychology Press.

Dowker, A. (2009). Use of derived fact strategies by children with mathematical difficulties. *Cognitive Development*, *24*, 401–410.

Engle, R. W., Conway, A. R. A., Tuholski, S. W., & Shisler, R. J. (1995). A resource account of inhibition. *Psychological Science*, *6*, 122–125.

Engle, R. W., Kane, M. J., & Tuholski, S. W. (1999). Individual differences in working memory capacity and what they tell us about controlled attention, general fluid intelligence, and functions of the prefrontal cortex. In A. Miyake & P. Shah (Eds), *Models of Working Memory: Mechanisms of Active Maintenance and Executive Control* (pp. 102–134). Cambridge: Cambridge University Press.

Every Child a Chance Trust (2009). *The Long-Term Costs of Numeracy Difficulties*. <http://www.everychildachancetrust.org/counts/index.cfm>.

Feigenson, L., Dehaene, S., & Spelke, E. (2004). Core systems of number. *Trends in Cognitive Sciences*, *8*, 307–314.

Fletcher, J. M. (2005). Predicting math outcomes: reading predictors and comorbidity. *Journal of Learning Disabilities*, *38*, 308–312.

Fuchs, L. S. & Fuchs, D. (2002). Mathematical problem-solving profiles of students with mathematics disabilities with and without comorbid reading disabilities. *Journal of Learning Disabilities*, *35*, 573–573.

Fuchs, L.S., Fuchs, D., Compton, D.L., et al. (2006). The cognitive correlates of third-grade skill in arithmetic, algorithmic computation, and arithmetic word problems. *Journal of Educational Psychology*, *98*, 29–43.

Geary, D. C. (1990). A componential analysis of an early learning deficit in mathematics. *Journal of Experimental Child Psychology*, *49*, 363–383.

Geary, D. C. (1993). Mathematical disabilities: cognitive, neuropsychological, and genetic components. *Psychological Bulletin*, *114*, 345–362.

Geary, D. C. (1994). *Children's Mathematical Development: Research and Practical Applications*. Washington, DC: American Psychological Association.

Geary, D. C. (2004). Mathematics and learning disabilities. *Journal of Learning Disabilities*, *37*, 4–15.

Geary, D. C. (2010). Mathematical learning disabilities. In J. Holmes (Ed.), *Advances in Child Development and Behavior* (Vol. 38, pp. 45–77). San Diego, CA: Academic Press.

Geary, D. C. (2011). Cognitive predictors of individual differences in achievement growth in mathematics: a five year longitudinal study. *Developmental Psychology*, *47*, 1539–1552.

Geary, D. C., Bailey, D. H., & Hoard, M. K. (2009). Predicting mathematical achievement and mathematical learning disability with a simple screening tool: the number sets test. *Journal of Psychoeducational Assessment*, *27*, 265–279.

Geary, D. C. & Brown, S. C. (1991). Cognitive addition: strategy choice and speed-of- processing differences in gifted, normal, and mathematically disabled children. *Developmental Psychology*, *27*, 398–406.

Geary, D. C., Brown, S. C, & Samaranayake, V. A. (1991). Cognitive addition: a short longitudinal study of strategy choice and speed-of-processing differences in normal and mathematically disabled children. *Developmental Psychology, 27*, 787–797.

Geary, D. C. & Burlingham-Dubree, M. (1989). External validation of the strategy choice model for addition. *Journal of Experimental Child Psychology, 47*, 175–192.

Geary, D. C., Hamson, C. O., & Hoard, M. K. (2000). Numerical and arithmetical cognition: a longitudinal study of process and concept deficits in children with learning disability. *Journal of Experimental Child Psychology, 77*, 236–263.

Geary, D. C., Hoard, M. K., & Bailey, D. H. (2012b). Fact retrieval deficits in low achieving children and children with mathematical learning disability. *Journal of Learning Disabilities, 45*, 291–307.

Geary, D. C., Hoard, M. K., Byrd-Craven, J., & DeSoto, C. M. (2004). Strategy choices in simple and complex addition: contributions of working memory and counting knowledge for children with mathematical disability. *Journal of Experimental Child Psychology, 74*, 213–239.

Geary, D. C., Hoard, M. K., Byrd-Craven, J., Nugent, L., & Numtee, C. (2007). Cognitive mechanisms underlying achievement deficits in children with mathematical learning disability. *Child Development, 78*, 1343–1359.

Geary, D. C., Hoard, M. K., Nugent, L., & Bailey, D. H. (2012a). Mathematical cognition deficits in children with learning disabilities and persistent low achievement: a five year prospective study. *Journal of Educational Psychology, 104*, 206–223.

Geary, D. C., Hoard, M. K., Nugent, L., & Byrd-Craven, J. (2008). Development of number line representations in children with mathematical learning disability. *Developmental Neuropsychology, 33*, 277–299.

Geary, D. C. & Widaman, K. F. (1987). Individual differences in cognitive arithmetic. *Journal of Experimental Psychology: General, 116*, 154–171.

Geary, D. C. & Widaman, K. F. (1992). Numerical cognition: on the convergence of componential and psychometric models. *Intelligence, 16*, 47–80.

Gersten, R., Clarke, B., & Mazzocco, M. M. M. (2007). Historical and contemporary perspectives on mathematical learning disabilities. In D. B. Berch & M. M. M. Mazzocco (Eds), *Why Is Math So Hard for Some Children? The Nature and Origins of Mathematical Learning Difficulties and Disabilities* (pp. 7–28). Baltimore, MD: Paul H. Brookes.

Gilmore, C. K., McCarthy, S. E., & Spelke, E. S. (2007). Symbolic arithmetic knowledge without instruction. *Nature, 447*, 589–591.

Groen, G. J. & Parkman, J. M. (1972). A chronometric analysis of simple addition. *Psychological Review, 79*, 329–343.

Groen, G. & Resnick, L. B. (1977). Can preschool children invent addition algorithms? *Journal of Educational Psychology, 69*, 645–652.

Halberda, J. & Feigenson, L. (2008). Developmental change in the acuity of the 'number sense': the approximate number system in 3-, 4-, 5-, and 6-year-olds and adults. *Developmental Psychology, 44*, 1457–1465.

Halberda, J., Mazzocco, M. M. M., & Feigenson, L. (2008). Individual differences in non-verbal number acuity correlate with maths achievement. *Nature, 455*, 665–668.

Hallett, D., Nunes, T., & Bryant, P. (2010). Individual differences in conceptual and procedural knowledge when learning fractions. *Journal of Educational Psychology, 102*, 395–406.

Hanich, L. B., Jordan, N. C., Kaplan, D., & Dick, J. (2001). Performance across different areas of mathematical cognition in children with learning difficulties. *Journal of Educational Psychology, 93*, 615–626.

Hecht, S. A. (1998). Toward an information-processing account of individual differences in fraction skills. *Journal of Educational Psychology, 90*, 545–559.

Hecht, S. A. & Vagi, K. (2010). Sources of group and individual differences in emerging fraction skills. *Journal of Educational Psychology, 102*, 843–859.

Holloway, I. D. & Ansari, D. (2008). Domain-specific and domain-general changes in children's development of number comparison. *Developmental Science, 11*, 644–649.

Jordan, N. & Hanich, L. (2000). Mathematical thinking in second grade children with different forms of LD. *Journal of Learning Disabilities*, *33*, 567–578.

Jordan, N. C., Hanich, L. B., & Kaplan, D. (2003a). Arithmetic fact mastery in young children: a longitudinal investigation. *Journal of Experimental Child Psychology*, *85*, 103–119.

Jordan, N. C., Hanich, L. B., & Kaplan, D. (2003b). A longitudinal study of mathematical competencies in children with specific mathematics difficulties versus children with comorbid mathematics and reading difficulties. *Child Development*, *74*, 834–850.

Jordan, N. C. & Montani, T. O. (1997). Cognitive arithmetic and problem solving: a comparison of children with specific and general mathematics difficulties. *Journal of Learning Disabilities*, *30*, 624–634.

Koontz, K. L. & Berch, D. B. (1996). Identifying simple numerical stimuli: processing inefficiencies exhibited by arithmetic learning disabled children. *Mathematical Cognition*, *2*, 1–23.

Kovas, Y., Haworth, C. M. A., Dale, P. S., & Plomin, R. (2007). The genetic and environmental origins of learning abilities and disabilities in the early school years. *Monographs of the Society for Research in Child Development*, *72* (3, Serial No. 288).

Krajewski, K. & Schneider, W. (2009). Exploring the impact of phonological awareness, visual -spatial working memory, and preschool quantity-number competencies on mathematics achievement in elementary school: findings from a 3-year longitudinal study. *Journal of Experimental Child Psychology*, *103*, 516–531.

Landerl, K., Bevan, A., & Butterworth, B. (2003). Developmental dyscalculia and basic numerical capacities: a study of 8- to 9-year-old students. *Cognition*, *93*, 99–125.

Levine, S. C., Jordan, N. C., & Huttenlocher, J. (1992). Development of calculation abilities in young children. *Journal of Experimental Child Psychology*, *53*, 72–103.

Lewis, C., Hitch, G. J., & Walker, P. (1994). The prevalence of specific arithmetic difficulties and specific reading difficulties in 9-year-old to 10-year-old boys and girls. *Journal of Child Psychology and Psychiatry*, *35*, 283–292.

Light, J. G. & DeFries, J. C. (1995). Comorbidity of reading and mathematics disabilities: genetic and environmental etiologies. *Journal of Learning Disabilities*, *28*, 96–106.

Locuniak, M. N. & Jordan, N. C. (2008). Using kindergarten number sense to predict calculation fluency in second grade. *Journal of Learning Disabilities*, *41*, 451–459.

Logie, R. H. (1995). *Visuo-Spatial Working Memory*. Hove: Erlbaum.

McLean, J. F. & Hitch, G. J. (1999). Working memory impairments in children with specific arithmetic learning difficulties. *Journal of Experimental Child Psychology*, *74*, 240–260.

Mandler, G. & Shebo, B. J. (1982). Subitizing: an analysis of its component processes. *Journal of Experimental Psychology: General*, *111*, 1–22.

Mazzocco, M. M. M. (2007). Defining and differentiating mathematical learning disabilities and difficulties. In D. B. Berch & M. M. M. Mazzocco (Eds), *Why Is Math So Hard for Some Children? The Nature and Origins of Mathematical Learning Difficulties and Disabilities* (pp. 29–48). Baltimore, MD: Paul H. Brookes.

Mazzocco, M. M. M. & Devlin, K. T. (2008). Parts and 'holes': gaps in rational number sense in children with vs. without mathematical learning disability. *Developmental Science*, *11*, 681–691.

Mazzocco, M. M. M. & Kover, S. T. (2007). A longitudinal assessment of executive function skills and their association with math performance. *Child Neuropsychology*, *13*, 18–45.

Murphy, M. M., Mazzocco, M. M. M., Hanich, L. B., & Early, M. C. (2007). Cognitive characteristics of children with mathematics learning disability (MLD) vary as a function of the cutoff criterion used to define MLD. *Journal of Learning Disabilities*, *40*, 458–478.

National Mathematics Advisory Panel (2008). *Foundations for Success: Final Report of the National Mathematics Advisory Panel*. Washington, DC: United States Department of Education. <http://www.ed.gov/about/bdscomm/list/mathpanel/report/final-report.pdf>.

Núñez, R. (2009). Numbers and arithmetic: neither hardwired nor out there. *Biological Theory*, *4*, 68–83.

Oliver, B., Harlaar, N., Hayiou-Thomas, M. E., et al. (2004). A twin study of teacher-reported mathematics performance and low performance in 7-year-olds. *Journal of Educational Psychology*, *96*, 504–517.

Ostad, S. A. (1997). Developmental differences in addition strategies: a comparison of mathematically disabled and mathematically normal children. *British Journal of Educational Psychology*, *67*, 345–357.

Ostad, S. A. (1998). Developmental differences in solving simple arithmetic word problems and simple number-fact problems: a comparison of mathematically normal and mathematically disabled children. *Mathematical Cognition*, *4*, 1–19.

Paglin, M. & Rufolo, A. M. (1990). Heterogeneous human capital, occupational choice, and male–female earning differences. *Journal of Labor Economics*, *8*, 123–144.

Parsons, S. & Bynner, J. (1997). Numeracy and employment. *Education and Training*, *39*, 43–51.

Passolunghi, M. C., Cornoldi, C., & De Liberto, S. (1999). Working memory and intrusions of irrelevant information in a group of specific poor problem solvers. *Memory & Cognition*, *27*, 779–790.

Passolunghi, M. C. & Siegel, L. S. (2004). Working memory and access to numerical information in children with disability in mathematics. *Journal of Experimental Child Psychology*, *88*, 348–367.

Passolunghi, M. C., Vercelloni, B., & Schadee, H. (2007). The precursors of mathematics learning: working memory, phonological ability and numerical competence. *Cognitive Development*, *22*, 165–184.

Piazza, M., Facoetti, A., Trussardi, A. N., et al. (2010). Developmental trajectory of number acuity reveals a severe impairment in developmental dyscalculia. *Cognition*, *116*, 33–41.

Raghubar, K., Cirino, P., Barnes, M., Ewing-Cobbs, L., Fletcher, J., & Fuchs, L. (2009). Errors in multi-digit arithmetic and behavioral inattention in children with math difficulties. *Journal of Learning Disabilities*, *42*, 356–371.

Rittle-Johnson, B., Siegler, R. S., & Alibali, M. W. (2001). Developing conceptual understanding and procedural skills in mathematics: an iterative process. *Journal of Educational Psychology*, *93*, 346–362.

Rivera-Batiz, F. (1992). Quantitative literacy and the likelihood of employment among young adults in the United States. *Journal of Human Resources*, *27*, 313–328.

Rourke, B. P. (1993). Arithmetic disabilities, specific and otherwise: a neuropsychological perspective. *Journal of Learning Disabilities*, *26*, 214–226.

Rousselle, L. & Noël, M.-P. (2007). Basic numerical skills in children with mathematical learning disabilities: a comparison of symbolic vs non-symbolic number magnitude processing. *Cognition*, *102*, 361–395.

Rubinsten, O. (2009). Co-occurrence of developmental disorders: the case of developmental dyscalculia. *Cognitive Development*, *24*, 362–370.

Russell, R. L. & Ginsburg, H. P. (1984). Cognitive analysis of children's mathematical difficulties. *Cognition and Instruction*, *1*, 217–244.

Schneider, M., Rittle-Johnson, B., & Star, J. (2011). Relations among conceptual knowledge, procedural knowledge, and procedural flexibility in two samples differing in prior knowledge. *Developmental Psychology*, *47*, 1525–1538.

Shalev, R. S., Manor, O., & Gross-Tsur, V. (2005). Developmental dyscalculia: a prospective six-year follow-up. *Developmental Medicine & Child Neurology*, *47*, 121–125.

Shalev, R. S., Manor, O., Kerem, B., et al. (2001). Developmental dyscalculia is a familial learning disability. *Journal of Learning Disabilities*, *34*, 59–65.

Siegler, R. S. (1996). *Emerging Minds: The Process of Change in Children's Thinking*. New York: Oxford University Press.

Siegler, R. S. & Booth, J. L. (2004). Development of numerical estimation in young children. *Child Development*, *75*, 428–444.

Siegler, R. S. & Jenkins, E. (1989). *How Children Discover New Strategies*. Hillsdale, NJ: Erlbaum.

Siegler, R. S. & Opfer, J. (2003). The development of numerical estimation: evidence for multiple representations of numerical quantity. *Psychological Science*, *14*, 237–243.

Siegler, R. S. & Robinson, M. (1982). The development of numerical understandings. In H. Reese and L. P. Lipsitt (Eds), *Advances in Child Development and Behavior* (Vol. 16, pp. 241–312). New York: Academic Press.

Siegler, R. S. & Shrager, J. (1984). Strategy choice in addition and subtraction: how do children know what to do? In C. Sophian (Ed.), *Origins of Cognitive Skills* (pp. 229–293). Hillsdale, NJ: Erlbaum.

Siegler, R. S., Thompson, C. A., & Schneider, M. (2011). An integrated theory of whole number and fractions development. *Cognitive Psychology, 62*, 273–296.

Starkey, P. (1992). The early development of numerical reasoning. *Cognition, 43*, 93–126.

Starkey, P. & Cooper, R. G., Jr. (1980). Perception of numbers by human infants. *Science, 210*, 1033–1035.

Stock, P., Desoete, A., & Roeyers, H. (2010). Detecting children with arithmetic disabilities from kindergarten: evidence from a 3-year longitudinal study of the role of preparatory arithmetic abilities. *Journal of Learning Disabilities, 43*, 250–268

Strauss, M. S. & Curtis, L. E. (1984). Development of numerical concepts in infancy. In C. Sophian (Ed.), *Origins of Cognitive Skills: The Eighteenth Annual Carnegie Symposium on Cognition* (pp. 131–155). Hillsdale, NJ: Erlbaum.

Swanson, H. L. (1993). Working memory in learning disability subgroups. *Journal of Experimental Child Psychology, 56*, 87–114.

Swanson, H. L., Jerman, O., & Zheng, X. (2008). Growth in working memory and mathematical problem solving in children at risk and not at risk for serious math difficulties. *Journal of Educational Psychology, 100*, 343–379.

Swanson, H. L., Jerman, O., & Zheng, X. (2009). Math disabilities and reading disabilities: can they be separated? *Journal of Psychoeducational Assessment, 27*, 175–196.

Swanson, H. L. & Sachse-Lee, C. (2001). Mathematical problem solving and working memory in children with learning disabilities: both executive and phonological processes are important. *Journal of Experimental Child Psychology, 79*, 294–321.

Temple, C. M. (1991). Procedural dyscalculia and number fact dyscalculia: double dissociation in developmental dyscalculia. *Cognitive Neuropsychology, 8*, 155–176.

Walberg, H. J. (1984). Improving the productivity of America's schools. *Educational Leadership, 41*, 19–27.

Wynn, K. (1992). Addition and subtraction by human infants. *Nature, 358*, 749–750.

Wynn, K., Bloom, P., & Chiang, W.-C. (2002). Enumeration of collective entities by 5-month- old infants. *Cognition, 83*, B55–B62.

CHAPTER 43

NUMBERS IN THE DARK

numerical cognition and blindness

JULIE CASTRONOVO

Introduction

Numbers are an integral part of our daily life and have to be processed in a variety of contexts – to deal with money, to check what time it is, to use the right quantity of food when cooking, to use number pads on telephones, computers, etc. According to Butterworth (1999), we process numerical information at a rate of about 1000 times/hour and 16 000 times/day. These figures clearly demonstrate the importance and the pervasiveness of numerical cognition in humans. By taking a close look at Butterworth's (1999) anecdotal report of the different situations in which we have to process numerical information, a striking fact is that, most of the time, we access numerical information through the visual sensory modality. One may point out that this is not surprising as vision constitutes a predominant sensory modality. However, does this likely supremacy of vision in accessing numerical information imply that vision plays a critical role in numerical cognition?

This question is of particular importance as vision is a primary sensory modality, which might be at an advantage compared with other sensory modalities, especially in accessing numerical information. For example, in the simultaneous mode, vision allows a greater amount of information to be processed, as well as more precision, while auditory skills present limits of discrimination and touch limits in the amount of information that can be apprehended. Distant objects can easily be processed through vision, while touch and auditory skills are limited and/or less precise in accessing and exploring distant sources of information. Vision also allows great attentional modulations, as visual attention can be sharply focused and easily captured, while auditory skills can essentially be easily captured and touch easily focused. As a result, in the absence of external stimulations, blind people have to play an active role in order to maintain contact with their environment (Thinus-Blanc & Gaunet, 1997). For all these reasons, vision might be of greater efficiency in accessing numerical information.

Mix's (1999) study in children could be seen as an indicator of how vision could be more efficient compared with the other senses in accessing numerical information. In her study, Mix (1999) demonstrated that numerical equivalence for static sets is acquired earlier in children than numerical equivalence for sequential sets. Mix's (1999) results also showed that, when numerical information is sequentially presented, memory for the number of objects is acquired earlier than memory for the number of events. In the absence of vision, large numbers have to be sequentially accessed through touch and auditory skill. Moreover, they necessarily involve events in the auditory modality.

Vision is also predominant in research, as a great majority of the literature on numerical cognition has been based on studies conducted in the visual modality. Moreover, vision has for a long time been considered as central in the acquisition and development of numerical representations and skills (Klahr & Wallace, 1973; Simon, 1997, 1999; Trick & Pylyshyn, 1994). First, the 'object-file model' emphasized the importance of vision and the occipital cerebral regions in the acquisition and development of numerical representations and skills (Simon, 1997, 1999; Trick & Pylyshyn, 1994). In this model, subitizing is described as a visual pre-attentive process, accounting for numerosity discrimination abilities found in preverbal infants and constituting the foundations for the development of numerical cognition. Infants are assumed to be born with primary non-numerical abilities to individuate and create unique markers for up to four objects. These abilities are assumed to originate from the cerebral areas devoted to vision. According to this model, more advanced numerical abilities emerge with the development and the acquisition of counting abilities, which coincide with the development of visuospatial cerebral circuits. The critical role of visuospatial processes in the elaboration of numerical abilities has further been supported as many behavioural, neuropsychological, and neuroimaging data clearly demonstrate the existence of a strong link between numerical and visuospatial abilities (see De Hevia, Vallar and Girelli, 2008, for a review). It has also been hypothesized that a gradual specialization of neural populations initially involved in visuomotor functions could account for the involvement of the medial intraparietal sulcus (mIPS) in numerical processing (Cohen Kadosh, 2011; Cohen Kadosh, Bahrami, Walsh, Butterworth, Popescu, & Price, 2011; Johnson, 2001).

Recently, the number sense itself has been described as the 'visual number sense' (Burr & Ross, 2008; Ross & Burr, 2010; Stoianov & Zorzi, 2012). The number sense, also referred to as the approximate number system (ANS), corresponds to individuals' innate ability to approximately grasp numerical quantities (Dehaene, 2009; Halberda, Mazzocco & Feigenson, 2008). It presents ontogenetic (Izard, Sann, Spelke, & Streri, 2009) and phylogenetic (Hauser, Dehaene, Dehaene-Lambertz, & Patalano, 2002) origins. Burr and Ross (2008) defined the number sense as visual following the observation that the 'numerousness' of a set of objects (Butterworth, 2008) can be modified by visual adaptation, like other primary visual properties (e.g. colour). Their results demonstrate that the apparent numerosity of a set of dots is greater, when it is preceded by an adaptation phase to a small numerosity. On the contrary, the apparent numerosity of a set of dots is smaller, following an adaptation phase to a large numerosity (see Figure 43.1). According to Burr and Ross (2008), these findings indicate that the visual system in itself entails the capacity to approximate numerosities, following what the authors refer to as 'its extraordinary built-in computational powers.' In a second study, showing that numerosity and density judgements are independent from each other, Ross and Burr (2010) reaffirmed their conclusions, according to which vision presents the ability to automatically sense numerosity. The postulate according to which the number sense can be described as visual has also been supported by Stoianov and Zorzi (2012), as their hierarchical generative model showed that visual numerosity could be extracted and coded invariantly from other visual properties, such as cumulative area. Defining the number sense as visual, necessarily implies that our core innate intuition about quantities relies on vision and its corresponding occipital areas in the brain. Since the number sense

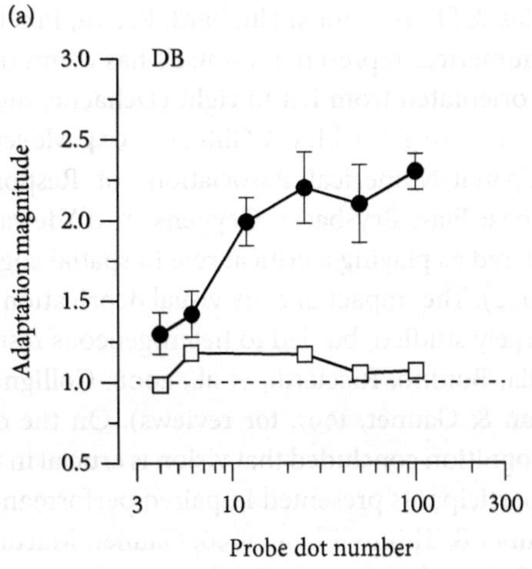

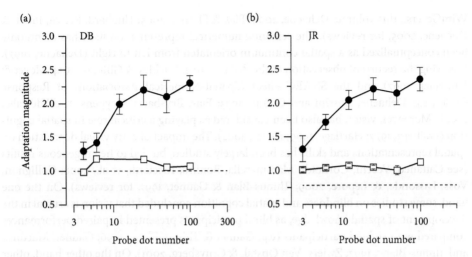

FIGURE 43.1 Illustration of the effect of adaptation on numerosity perception as a function of the number of dots in the probe. The adaptation magnitude is the ratio of test to probe dot number at the point of subjective equality (PSE) between test and probe. Filled circles correspond to the condition with adaptation. Open squares correspond to the condition without adaptation. The current figures are sample results from the authors DB and JR.

Adapted from *Current Biology*, 18(6), David Burr and John Ross, A Visual Sense of Number, pp. 425–8, Copyright (2008), with permission from Elsevier.

is considered as foundational in the elaboration of numerical representations and skills (see Piazza, 2010, for a review), defining it as visual sets vision back at a central place in the elaboration of numerical cognition.

To summarize, vision is a predominant sensory modality in everyday life. It has mainly been used in research. Moreover, its role has been implicitly and explicitly stressed in the acquisition and development of numerical representations and skills. This chapter reviews the different studies, which have directly addressed the issue of the role of vision in the elaboration of numerical representations and skills, by investigating the impact of early visual deprivation on numerical cognition. First, the studies on the role of vision in the elaboration of a semantic numerical representation with spatial properties are reviewed. Secondly, results on the impact of early visual deprivation on numerical processes, such as estimation and subitizing, will be presented. Finally, developmental results on the role of vision in the acquisition of finger numeral representation in children are presented. The chapter concludes by discussing the findings of this new exploratory field of research in numerical cognition.

Early Blindness and the Semantic Numerical Representation as a Spatial Continuum

A large set of behavioural, neuropsychological, and neuroimaging data demonstrates the existence of strong number-space interactions (see Van Dijck, Ginsburg, Girelli &

WimGevers., this volume; Dehaene, 2009; Fias & Fischer, 2005; Hubbard, Piazza, Pinel, & Dehaene, 2005, for reviews). The semantic numerical representation itself has commonly been conceptualized as a spatial continuum orientated from left to right (Dehaene, 1997), following the recurrent observation of the distance (e.g. Buckley & Gillman, 1974; Moyer & Landauer, 1967) and the SNARC effect (Spatial-Numerical Associations of Response Codes; e.g. Dehaene, Bossini and Giraux, 1993; Fias, Brysbaert, Geypens, & d'Ydewalle, 1996).[1] Moreover, vision has also been considered as playing a critical role in spatial cognition (Fraiberg, 1977; Hartlage, 1969; Heller, 2002). The impact of early visual deprivation on spatial representations and skills has been largely studied, but led to heterogeneous results (see Cattaneo, Vecchi, Cornoldi, Mammarella, Bonino, Ricciardi, et al., 2008; Collignon, Voss, Lassonde, & Lepore, 2009; Thinus-Blan & Gaunet, 1997, for reviews). On the one hand, some studies on blindness and spatial cognition concluded that vision is crucial in the development of spatial knowledge, as blind participants presented impaired performances compared with sighted participants (e.g. Gaunet & Thinus-Blanc, 1996; Gaunet, Martinez and Thinus-Blanc, 1997; Zwiers, Van Opstal, & Cruysberg, 2001). On the other hand, other studies demonstrated that blind people present similar, even enhanced spatial abilities compared with sighted people (e.g. Fieger, Roder, Teder-Salejarvi, Hillyard, & Neville, 2006; Lessard, Paré, Lepore, & Lassonde, 1998; Voss, Lassonde, Gougoux, Fortin, Guillemot, & Lepore, 2004). Visual sensory deprivation seems to entail both increased and altered spatial abilities, as blindness involves the development and use of particular compensatory mechanisms leading to particular patterns of performances depending on the task and the measures taken (Gaunet & Rossetti, 2006). Following these different observations, the main conclusion that can be drawn is that vision is not compulsory in the elaboration of spatial representations and skills. However, as numbers and space are intrinsically linked, what are the consequences of early visual deprivation on a semantic numerical representation with spatial properties?

Szucs and Csépe (2005) were the first to submit a group of congenitally blind participants and a group of sighted blindfolded participants to a comparison task to 5 in the auditory modality. Their aim was to study whether blind participants would present the classic distance effect, suggesting that numbers are represented on a mental continuum, using behavioural and electro-encephalography data. At a behavioural level, congenitally blind participants demonstrated similar distance and size effects compared with sighted participants. We have later replicated these findings and extended them to larger numbers, by submitting a group of congenitally and early blind participants, and a group of sighted blindfolded participants to a comparison task to 5, a parity judgement task, and a comparison task to 55 in the auditory modality (Castronovo & Seron, 2007a). Our results showed in both groups the occurrence of a distance effect in the comparison tasks, as well as of a SNARC effect in the comparison task to 5 and in the parity judgment task. Recently, Cattaneo and colleagues (Cattaneo, Fantino, Silvanto, & Vecchi, 2011) gave further support to these early findings. They submitted a group of congenitally blind participants and a group of sighted blindfolded participants to a numerical bisection task. In this task, participants were presented with pairs of 3-digit numbers varying in their numerical size. Without counting, participants had to judge the numerical midpoint of each number pair presented to them. Healthy adults submitted to a numerical bisection task usually present with pseudoneglect; they tend to misbisect numerical intervals to the left of their veridical centre, the same way they do when bisecting physical lines (e.g. Longo & Lourenco, 2007). Cattaneo et al.'s (2011) results demonstrated that blind and sighted participants presented similar leftward bias in bisecting numerical intervals.

These primary results clearly demonstrate that blind and sighted participants present similar behavioural effects, suggesting the existence of a semantic numerical representation with similar spatial properties to those generally postulated in sighted people – a mental continuum orientated from left to right. Therefore, it can be firmly concluded that vision is not necessary in the elaboration of a numerical representation having a spatial nature. However, although blind people seem to present a similar semantic numerical representation to sighted people, their use of this representation might present some peculiarity, as suggested by the electrophysiological data collected.

Szucs and Csépe's (2005) were the first to use EEG measures when submitting blind participants to numerical tasks. General electroencephalography (EEG) analyses showed that blind and sighted participants presented similar event-related brain potentials (ERPs) topographies and distance effect, suggesting that numerical comparison in blind people relies on a similar numerical representation and on similar neural circuits compared with sighted people. However, Szucs and Csépe's (2005) results also showed some dissimilarity between blind and sighted participants. First, preliminary EEG analyses showed that the P300 activity in the blind group was more pronounced than in the sighted group. Secondly, by conducting more detailed analyses, looking at the event-related spectral perturbation (ERSP), further differences were found between blind and sighted participants, particularly in the initial phase of numerical comparison. In this initial phase of the numerical comparison process, sighted participants presented left and right frontal and parietal ERSP distance effects. Blind participants failed to demonstrate similar ERSP effects compared with sighted participants, but they presented a left-parietal effect. In the following phase of the numerical comparison process, similar parietal effects were found in both groups of participants. These findings show that the distance effect's topography was different in blind and sighted participants' initial phase of numerical comparison process, while at a later stage similar expected parietal effects could be found in both groups (see Figure 42.2). According to the authors, these findings suggest that blind participants probably use a unique compensatory mechanism in the initial phase of the numerical process, while a similar parietal network to sighted participants is probably used at a later stage to represent numerical information. Therefore, following the observation of the classic behavioural distance and SNARC effects (Castronovo & Seron, 2007a; Szucs and Csépe, 2005), it appeared that early visual deprivation does not preclude the elaboration of semantic numerical representation with similar spatial properties to those usually assumed in sighted people: a mental continuum oriented from left to right. However, according to Szucs and Csépe (2005), the absence of vision involves the development of a 'partially normal number processing network.'

Likewise, Crollen, Dormal, Seron, Lepore, and Collignon (2013) found that, although vision is not necessary in the elaboration of a semantic numerical representation with similar spatial properties as in sighted people, it seems to play a critical role in the acquisition of an external coordinate system for the spatial representation of numbers. In their study, Crollen et al. (2013) submitted early and late blind, and blindfolded sighted participants to a comparison task to 5 and a control parity judgement task,[2] using two conditions. In the first condition, participants had to perform the tasks with their hands parallel, while in the second condition their hands were crossed over their body midline. In the comparison task with hands parallel, as well as in both conditions in the control parity judgement task, the three groups of participants presented a classic SNARC effect. However, in the crossed hands condition, early blind participants showed a reversed SNARC effect contrary to late blind and sighted participants. This reversed SNARC effect reflects the use in early blind participants of hand-based coordinates to perform the task, in opposition to the use of extracorporal coordinates in late blind and sighted individuals. These results indicate that, although early visual deprivation does not prevent the elaboration of a numerical representation with similar spatial properties as in sighted individuals, it involves a different body-centred coordinate frame of reference when mapping numbers onto space. According to the authors, these findings further support the hypothesis that the SNARC effect originates from our reading habits (Dehaene et al., 1993), as in sighted and early blind individuals, both coordinate systems used to map numbers onto space are related to their reading effector – the eye-centred coordinate system in sighted individuals versus hand-centred coordinate system in early blind individuals.

To investigate whether blind people similarly represent numerical information compared with sighed people, Salillas, Graná, El-Yagoubi, and Semenza (2009) used behavioural and electrophysiological measures. They submitted a group of congenitally blind and group of sighted participants to a detection task preceded by the presentation of a non-relevant number. Their aim was to study if they could replicate the attentional shift found in a detection task, when the presentation of a non-relevant number precedes the presentation of a non-numerical lateralized target. This congruency effect has been previously found by Fischer and colleagues (Fischer, Castel, Dodd, & Pratt, 2003). It results from the automatic activation of the semantic numerical representation, which induces a shift of attention in the corresponding visual field (left visual field for smaller numbers versus right visual field for larger numbers), leading to faster detection times when the lateralized target is presented in the congruent number-space location. Salillas and colleagues (2009) used a similar paradigm with early blind and sighted participants in the auditory modality, including ERP measurements. Smaller numbers ('one' and 'two'), as well as larger numbers ('eight' and 'nine') were binaurally presented with a fixed duration of 350 ms. Participants' task was to detect an auditory target, laterally presented with the use of dichotic listening (see Figure 43.2). The behavioural results showed the occurrence of a congruency effect in both groups of participants: congruent trials (small number/left-lateralized target, large number/right-lateralized target) were faster detected. Cattaneo and colleagues (Cattaneo, Fantino, Tinti, Silvanto, & Vecchi, 2010) also found a similar spatial shift of attention in blind and sighted participants following the presentation of numerical information. In Cattaneo and colleagues' (2010) study, early blind and blindfolded sighted participants had to explore rods having different lengths using their right index finger only and to indicate their midpoints, while a small number ('two'), a large number ('eight'), or the non-word 'blah' (control condition) was presented 10 consecutive times in the auditory modality. Results show that both groups of participants presented:

(1) pseudoneglect in the control condition;
(2) greater pseudoneglect when the small number was presented;
(3) reduced pseudoneglect when the large number was presented.

These data indicate that the automatic activation of the semantic numerical representation, as a spatial continuum orientated from left to right, can significantly and similarly affect spatial processing in a cross-modal paradigm in blind and sighted individuals.

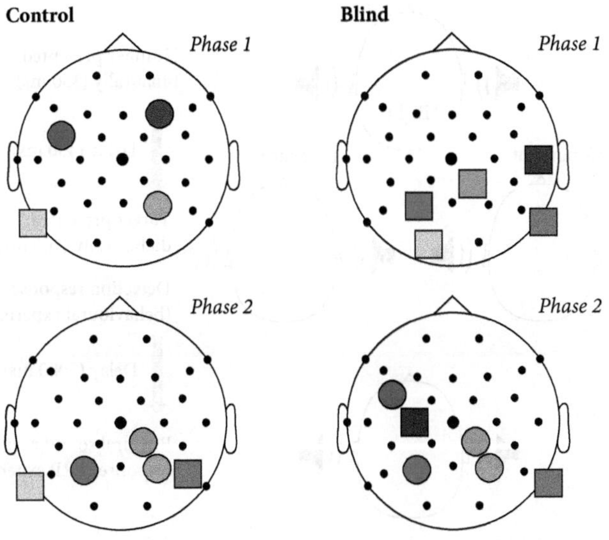

FIGURE 43.2 Szucs and Csépe's (2005) ERSP results illustrating the topography of electrode sites (each colour represents a different electrode site) showing the distance effect in phase 1 and in phase 2. Circles stand for similar, squares for different effects in the blind and control groups. Electrode Cz is marked by the black bold dot in the middle.

Reprinted from *Neuroscience Letters*, 384(1–2), Dénes Szűcs and Valéria Csépe, The parietal distance effect appears in both the congenitally blind and matched sighted controls in an acoustic number comparison task, pp. 11–16, Copyright (2005), with permission from Elsevier.

Like Szucs and Csépe (2005), Salillas et al. (2009) found that although blind and sighted participants present similar performances at a behavioural level, their ERPs results show some discrepancy. The two main ERPs components were differentially modulated by congruency in each group: congruency modulated the N100 component in the sighted group, while congruency modulated the P300 component in the blind group (see Figure 43.3). Moreover, blind participants presented larger ERPs amplitude in general compared with sighted participants. According to the authors, blind and sighted participants' results give further support to the idea that early visual deprivation does not prevent the elaboration of a semantic numerical representation with similar spatial properties to those of sighted individuals (Szucs & Csépe, 2005; Castronovo & Seron, 2007a; Crollen et al., 2013). However, the ERPs results would further suggest that early blindness involves a different use of this spatial numerical representation. On the one hand, the fact that congruency modulated the early sensory N100 component in the sighted group suggests an early amplification of auditory processes modulated by a higher order representation. On the other hand, the ERPs congruency effect found on the cognitive P300 component in the blind group would reflect higher cognitive processes, such as working memory processes. According to Salillas et al. (2009), the P300 congruency modulation in the blind group might result from the encoding and storage in working memory of the spatial location activated from the number presentation, while the absence of N100 congruency modulation might indicate that the target sensory process was not influence by the semantic numerical representation activation. Therefore, blind people might apply cognitive working memory resources to the computation of congruency, rather than a sensory process. Following their ERPs results, the authors suggest

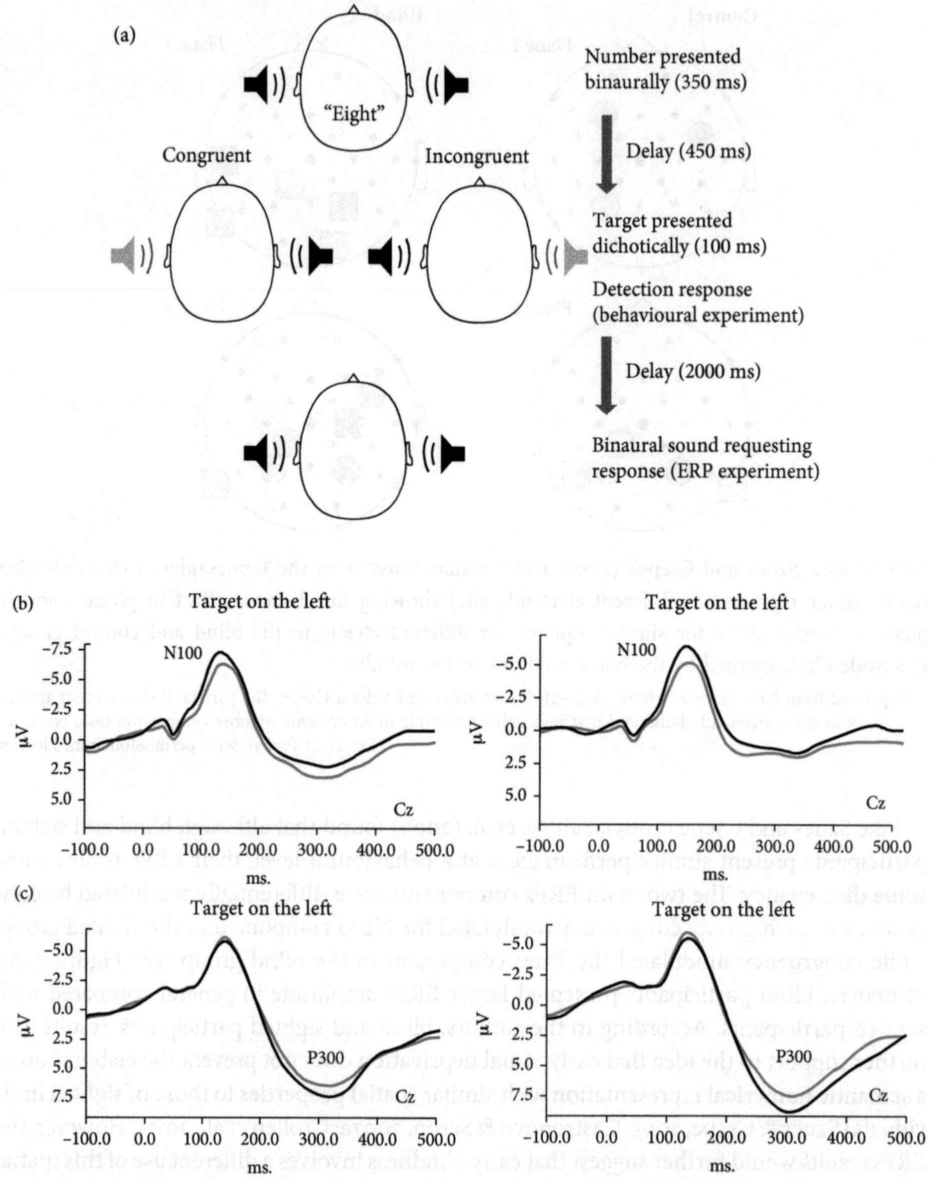

FIGURE 43.3 Illustration of the experimental design used in Salillas et al. (2009) (A); ERPs elicited by the target in the sighted group participants (B) and in the blind group participants (C), black lines represent congruent trials and red lines represent incongruent trials.

Adapted from Elena Salillas, Alessia Graná, Radouane El-Yagoubi, and Carlo Semenza , Numbers in the Blind's "Eye", *PLoS ONE*, 4(7), e6357, figure 1, doi: 10.1371/journal.pone.0006357. Copyright © 2009, Salillas, Graná, El-Yagoubi, and Semenza. Reproduced under the terms of the Creative Commons Attribution License.

that numerical processing in blind people might be more controlled, more dependent on working memory resources compared with numerical processing in sighted people.

This surprising finding that blind people might manipulate numerical representation in a different, more controlled, way compared with sighted people is supported by Collignon and colleagues' study (Collignon, Renier, Bruyer, Tranduy, & Veraart, 2006) demonstrating that blind people show enhanced spatial attention skills compared with sighted people. In an attempt to study blind people's spatial attentional abilities independently from their sensory skills in touch and auditory skills, Collignon et al. (2006) found that congenitally blind people present enhanced spatial attention performance in both tactile and auditory selective spatial attention task, as well as in bimodal divided spatial attention task. These results led the authors to conclude that following early visual deprivation blind people develop compensatory mechanisms involving high cognitive resources, such as attention, to deal with the environment. Therefore, Salillas et al.'s (2009) ERPs finding in a task involving a spatial attentional shift might, indeed, be accounted by blind people's greater higher-level cognitive resources, such as spatial attention, involving more controlled numerical process, which may account for their particular performances in numerical estimation.

Early Blindness and Numerical Estimation

The experience individuals have with numerical information has been assumed to play a significant role in the acquisition and development of numerical cognition. First, it has been assumed that experience with numerical information accounts for greater number sense acuity with age (Halberda & Feigenson, 2008). The core number sense is well-known for obeying Weber's law at a behavioural and neuronal level (Piazza, Izard, Pinel, Le Bihan, & Dehaene, 2004): the larger the numerosities, the more approximate their processing. Number sense acuity, measured with the Weber fraction (w), increases with age in children (i.e. the greater number sense acuity, the smaller w) and reaches in adults an estimated average between 0.11 and 0.15 (see Piazza & Izard, 2009, for a review). The observation that people from remote uneducated cultures present a slightly larger w (e.g. the Munduruku: $w = 0.17$), compared with the average w found in educated numerate Western cultures (e.g. French: $w = 0.12$) (Pica, Lemer, Izard, & Dehaene, 2004) is seen as reflecting the impact of experience with numerical information on the number sense acuity (Halberda & Feigenson, 2008). Secondly, it has been assumed that, with the acquisition of numerical and mathematical knowledge, the number sense is progressively refined into a formal, symbolic, linear exact number system (ENS), allowing the automatic access from symbolic numbers to their corresponding magnitudes (Dehaene, 2009). As a consequence, individuals would be equipped with an innate approximate number sense, useful to process non-symbolic quantities, and a later-acquired exact number system, useful to process symbolic numbers (see Gebuis & Reynvoet, this volume). Evidence of the role of experience with numerical information in the development of this later-acquired symbolic number system has been found, as the greater mathematics achievement, the better anchored and the more extended the ENS (see Ashcraft & Moore, 2012; Siegler & Opfer, 2003, for examples in children; see Castronovo & Göbel, 2012, for an example in adults). Thirdly, greater experience with numerical information has also been associated with greater mapping abilities between the symbolic and non-symbolic numerical representations (see Barth, Starr, & Sullivan, 2009; Mundy & Gilmore, 2009, for examples in children; see Castronovo and Göbel, 2012, for an example in adults).

Early visual deprivation necessarily involves a particular experience with numerical information. However, is this particular experience more efficient, less efficient, or as efficient as sighted people's experience with numerical information? We addressed this question, by investigating congenitally blind participants' numerical estimation skills (Castronovo & Seron, 2007b). We submitted a group of congenitally or early blind participants and a group of blindfolded sighted participants to two numerical estimation tasks: a key press estimation (KPE) task and an auditory event estimation (AEE) task. In the KPE task, participants had to actively and approximately produce numerosities corresponding to target numbers heard, by pressing a key at a fast rate to avoid counting. This task required a symbolic to non-symbolic mapping, as participants had to map symbolic verbal numerals ranging from 5 to 64 to their corresponding magnitude. On the contrary, the AEE task required a non-symbolic to symbolic mapping, as participants had to estimate how many auditory events were presented in unique non-periodic timed sequences with numerosities ranging from 8 to 62. In the AEE task, three different durations were used to create the non-periodic timed sequences: 2000 ms sequences for small numerosities [8–14]; 5400 ms sequences for medium numerosities [20–32]; and 8000 ms sequences for large numerosities [44–62]. The data collected in this first study on blind people's numerical estimation skills were particularly surprising and outstanding. In both tasks, blind and sighted participants presented the signature of Weber's law: constant coefficients of variation across target size, as expected (Cordes, Gelman, Gallistel, & Whalen, 2001; Whalen, Gallistel, & Gelman, 1999). Surprisingly, however, blind participants presented extraordinary good estimation skills, particularly in the KPE estimation task.

In the KPE task, blind participants presented less variability in their estimations (smaller coefficients of variation), but they also demonstrated much greater accuracy compared with sighted participants. Indeed, blind participants presented very good estimations of the number of key presses to produce in order to reach the target number heard, while sighted participants showed increasing overestimation of the number of key presses to produce with increasing target number (see Figure 43.4A). We used several control measures (verbal report, key press-to-key press interval patterns, and mean press-to-press interval) and a subvocal counting task to make sure that neither the blind, nor the sighted participants used a counting strategy to solve the task. Results of the control measures and analyses clearly indicated that no counting strategy was used in both groups. Two control tasks were also introduced to make sure that participants did not base their key press responses on duration, rather than on numerosity in the KPE task: the key press duration estimation (KPDE) task and the tone duration estimation (TDE) task. In the KPDE task, participants had to approximately estimate and reproduce the durations of their key press responses in the KPE task. In the TDE task, participants had to approximately replicate tone durations, which were in the range of their key press responses in the KPE task. In both duration tasks, both groups presented greater variability in their responses compared with their responses in the KPE task, suggesting that they did not base their estimations on duration in the KPE task. Interestingly, in both duration tasks, participants in the blind group showed a better performance compared with sighted participants: less variability and greater accuracy in their duration estimations. These results suggest that blind people might present greater estimation skills for quantitative information in general, rather than just for numerosity.

In the AEE task, differences between blind and sighted participants were less pronounced, as both groups presented similar variability in their estimations. However, blind participants still demonstrated better performances compared with sighted participants: they

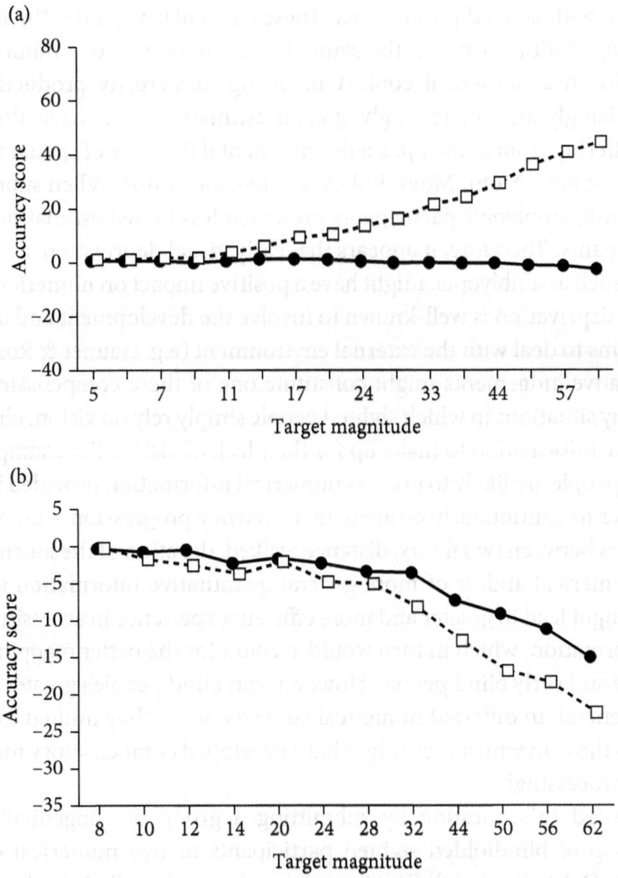

FIGURE 43.4 Castronovo and Seron's (2007b) results with blind (filled circles) and sighted (open squares) participants' accuracy scores (AS = response—target number) in the key press estimation task (A) and in the auditory events estimation tasks (B).

Adapted from Julie Castronovo and Xavuer Seron, Numerical estimation in blind subjects: Evidence of the impact of blindness and its following experience, *Journal of Experimental Psychology: Human Perception and Performance*, 33(5) pp. 1089–1106. doi: 10.1037/0096-1523.33.5.1089 © 2007, American Psychological Association.

show significantly less pronounced tendency to underestimate the target numerosities presented and greater accuracy in their estimations, but to a smaller extent (i.e. within the small range of numerosities [8–14]) (see Figure 43.4B). It is also worth noticing that blind participants showed in this particular task sensitivity to both targets' quantitative factors: numerosity and duration of the non-periodic auditory sequences; while sighted participants' performances were only predicted by the numerosity of the target sequences.

In this first study on early visual deprivation and estimation skills, results indicate that, although vision presents some advantages other the other sensory modalities, these advantages do not necessarily involved more efficient experience in accessing and processing numerical information. On the contrary, early visual deprivation, and its following experience in accessing and processing numerical information seem to have a positive impact on numerical cognition, as congenitally or early blind participants presented better estimation skills compared with sighted participants. These particularly good estimation skills reflect greater mapping abilities between the symbolic and non-symbolic numerical representations, especially in a numerical context involving numerosity production via proprioception. Surprisingly and interestingly, greater estimation skills have also been found in individuals suffering from amblyopia, a developmental disorder of spatial vision, leading to poor uniocular vision (Mohr, Mues, Robol, & Sireteanu, 2010). When submitted to a number bisection task, amblyopic participants presented less biased estimation compared with sighted participants. Therefore, it appears that early visual deprivation, as well as other visual disorders, such as amblyopia, might have a positive impact on numerical skills.

Early visual deprivation is well-known to involve the development and use of compensatory mechanisms to deal with the external environment (e.g. Gaunet & Rossetti, 2006). The use of quantitative judgements might constitute one of these compensatory mechanisms. Indeed, in many situations in which sighted people simply rely on vision, blind people might use quantitative information to make up for their lack of vision. For example, in their locomotion, blind people are likely to process numerical information provided by their proprioception in order to continuously estimate their journey progression – number of footsteps, number of stairs between two floors, distance walked, duration of the journey, etc. This possible use of numerical and/or of more general quantitative information as compensatory mechanisms might lead to greater and more efficient experience in accessing and processing numerical information, which in turn would account for the better mapping abilities found in congenitally and early blind people. However, can blind people's greater estimation skills be found, in general, in different numerical contexts or are they limited to numerical contexts similar to those in which they might have developed compensatory mechanisms based on numerical processing?

We investigated this question by submitting a group of congenitally blind participants and group of blindfolded sighted participants to two numerical estimation tasks (Castronovo & Delvenne, 2013). The first estimation task, called the footstep estimation task, was introduced for its similarity with blind people's daily situations in which they are likely the use compensatory mechanisms based on numerical and quantitative judgements. In the footstep estimation task, participants had to approximately reproduce the numerosity of target numbers heard, ranging from 5 to 64, by producing footsteps while holding on to a bar fixed on the wall. In order to avoid the use of a counting strategy, articulatory suppression was introduced. Participants had to continuously repeat the syllable 'da' at a quick rate, while producing their footstep estimation. The second estimation task, called the oral verbal estimation task, was introduced as a more general estimation task, disconnected from participants' daily experience in accessing and processing numerical information. In the oral verbal estimation task, participants had to produce the non-word 'bam' repeatedly until they felt they have approximately reached the quantity corresponding to the target number heard. The same target numbers as in the footstep estimation task were used. The oral verbal estimation task was performed under speed pressure, in order to avoid the use of a counting strategy. As blind individuals might present a greater use of short-term memory and working memory strategies involving numbers in their daily life (e.g. remembering lists of digits, such as Visa card number, remembering prices and mental computation while shopping; see Crollen, Mahe, Collignon, & Seron, 2011, for results in blind children supporting this hypothesis), working memory measures were also taken with the use of forwards and backwards digit-span tests, as well as a word-span test. We replicated the results found in

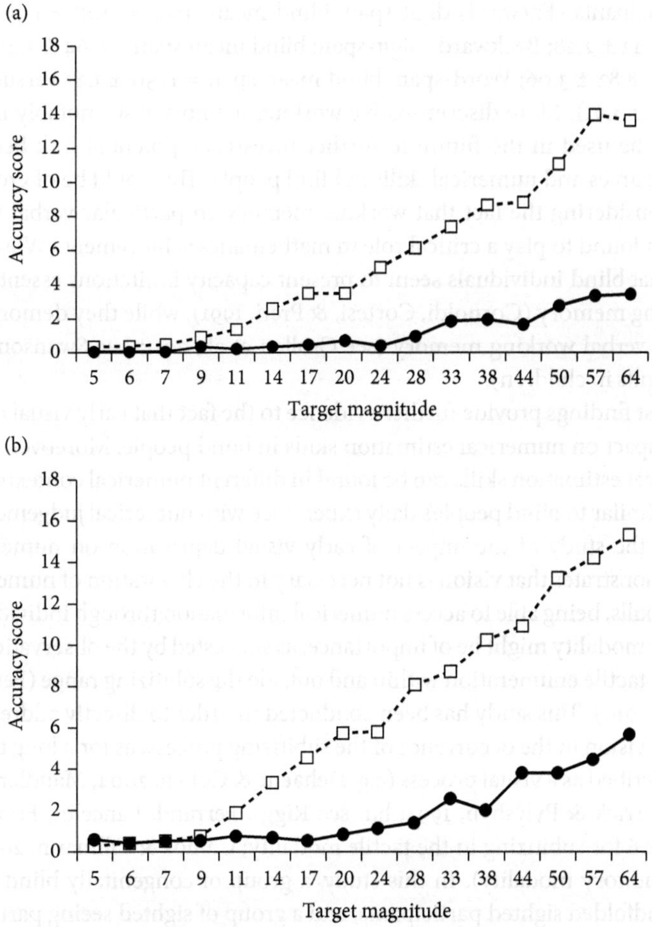

FIGURE 43.5 Castronovo and Delvenne's (2013) results with blind (filled circles) and sighted (open squares) participants' accuracy scores (|AS| = |response—target number| in the footstep estimation task (A) and in the oral verbal estimation tasks (B).

Adapted from *Cortex*, 5(6), Julie Castronovo and Jean-François Delvenne, Superior numerical abilities following early visual deprivation, pp. 1435–40, Copyright (2013), with permission from Elsevier.

Castronovo and Seron (2007b). In both the footstep and the oral verbal estimation tasks, blind and sighted participants presented constant coefficients of variation across target size (i.e. signature of Weber's law) (Cordes et al., 2001; Whalen et al., 1999). Again, participants in the blind group presented better estimation skills compared with sighted participants in both tasks; they exhibited smaller variability in their estimations (i.e. smaller coefficients of variation), less increasing tendency to misestimate the number of footsteps or non-words to produce to reach the target number heard, and greater accuracy scores (see Figure 42.5). No significant differences were found between the two tasks. On working memory, although previous studies found greater span scores (forward, backward digit spans and word spans) in blind children compared with sighted children (Hull & Mason, 1995; Lee Swanson & Luxenberg, 2009), we could not find significant differences between our two groups of participants. However, blind participants seem to show slightly better results compared with sighted participants (Forwards digit-span: blind mean span = 13.18 ± 1.99 versus sighted mean span = 12 ± 2.28; Backwards digit-span: blind mean span = 9.64 ± 3.41 versus sighted mean span = 8.82 ± 3.06; Word-span: blind mean span = 11.50 ± 1.27 versus sighted mean span = 10.10 ± 1.91). More discriminative working memory tests, notably involving numbers, should be used in the future to further investigate potential link between working memory resources and numerical skills in blind people. This could be of great importance, especially considering the fact that working memory, in particular verbal working memory, has been found to play a critical role in mathematics achievement (Wilson & Swanson, 2001) and that blind individuals seem to present capacity limitations essentially in visuospatial working memory (Cornoldi, Cortesi, & Preti, 1991), while they demonstrated greater resources in verbal working memory (see Crollen et al., 2011; Lee Swanson & Luxenberg, 2009, for results in children).

These latest findings provide further evidence to the fact that early visual deprivation has a positive impact on numerical estimation skills in blind people. Moreover, they also show that these great estimation skills can be found in different numerical contexts, which are not necessarily similar to blind people's daily experience with numerical judgements.

Although the study of the impact of early visual deprivation on numerical cognition strongly demonstrates that vision is not necessary in the elaboration of numerical representations and skills, being able to access numerical information through individuals' predominant sensory modality might be of importance, as suggested by the observations we made in our study on tactile enumeration within and outside the subitizing range (Ferrand, Riggs, & Castronovo, 2010). This study has been conducted in order to directly address the question of the role of vision in the occurrence of the subitizing process, as for a long time, subitizing has been described as a visual process (e.g. Dehaene & Cohen, 2004; Mandler & Shebo, 1982; Trick, 2008; Trick & Pylyshyn, 1994; but see Riggs, Ferrand, Lancelin, Fryziel, Dumur, & Simpson, 2006 for subitizing in the tactile modality; Camos & Tillmann, 2008, for subitizing in the auditory modality). In this study, a group of congenitally blind participants, a group of blindfolded sighted participants, and a group of sighted seeing participants had to perform an estimation task in the tactile modality. Participants' hands were placed in two boxes equipped with small metal rods, one for each finger. Their fingertips were stimulated by these metal rods. The task was to enumerate the number of fingers stimulated as quickly and as accurately as possible. One to ten fingers were stimulated. First, blind people's greater estimation skills were further supported, as blind participants presented a significantly decreasing tendency to misestimate the number of fingertips stimulated compared with the two groups of sighted participants. Secondly, analyses on error rates (i.e. [response—target magnitude]/target magnitude) and coefficients of variation indicated that the three groups presented similar performances within the subitizing range. However, outside the subitizing range congenitally blind participants, as well as sighted seeing participants showed better performances compared with blindfolded sighted participants, who presented larger variability in their estimations and larger error rates. These data provide strong evidence that subitizing is a general perceptual mechanism. They also further demonstrate that blind people show greater estimation skills outside the subitizing range, but mainly compared with sighted participants who were blindfolded to perform the task. When sighted participants were not blindfolded, they might have been able to rely on vision (i.e. their predominant sensory modality), which might account for the fact that their performances were similar to blind participants' performances. Therefore, although vision is clearly not necessary in the elaboration of numerical cognition, accessing numerical information through people's predominant sensory modality might facilitate its processing. It would be interesting in the future to further investigate the potential facilitative role a predominant sensory modality could have in accessing and processing numerical information.

Early Blindness and Finger-Counting

The association between number and space is not the only association that has been assumed in the literature on numerical cognition, as the association between numbers and fingers has also been largely documented (see Pesenti & Andres, this volume). It has notably been assumed that this particular association contributes, via finger-counting, to the acquisition of higher-level numerical and arithmetical knowledge (e.g. Butterworth, 1999, 2005). In order to study whether vision plays a critical role in the acquisition of the finger-number association, Crollen and colleagues (2011) studied the spontaneous use of finger-counting and finger-montring (i.e. activity of raising fingers to indicate a particular numerosity to someone else; Di Luca & Pesenti, 2008) in congenitally blind children compared with blindfolded sighted children, as well as the canonicity of their potential finger-counting and finger-montring habits. They first tested children's general cognitive skills, by submitting them to an IQ working memory test, a finger discrimination task, simple addition problems, and a task to test their knowledge of non-numerical ordered sequences. Results indicated that blind and sighted children presented similar performances; except that blind children showed greater results on working memory, especially when the task presented a higher level of complexity (pseudo-word repetition task). The authors also submitted the children to different numerical tasks in order to test their spontaneous use of finger-counting and finger-montring. Participants had, for example, to detect how many times a particular word appeared in a story without counting out loud, to count tactile stimulations, while an articulatory suppression was in place, to indicate numerosities using non-verbal strategy, etc. The results demonstrate that congenitally-blind children used finger-counting significantly less frequently compared with sighted children. However, they performed at the same level as sighted children, except in tasks involving interference with working memory, which led to poorer performance in blind children. These first results suggest that blind children might mainly rely on working memory to perform counting tasks, while sighted children rely on finger-counting. Finally, in a third set of tests, the authors investigate the explicit use of finger-counting and finger-montring in blind and sighted children. Their results indicate that both groups of children were able to efficiently use finger-counting and finger-montring. However, blind children appeared to use their fingers in a less canonical way than sighted children to count and to indicate numerosity.

These first developmental findings on blindness and numerical cognition demonstrate that vision plays a critical role in the acquisition of finger-number association. Indeed, blind children use finger-counting and finger-montring less spontaneously than sighted children. Moreover, when they use finger-counting and finger-montring, they do not show canonical finger-number configuration. The fact that blind children presented similar counting performances as sighted children suggests that even though finger-number association might facilitate the acquisition of counting and arithmetical skills, it is not necessary.

Conclusion

Although vision is a predominant sensory modality at advantage over the other senses (Thinus-Blanc & Gaunet, 1997), which has been mainly used in research on numerical cognition and has been considered as critical in the emergence of numerical representations and skills (Burr & Ross, 2008; Ross & Burr, 2010; Simon, 1997, 1999; Stoianov & Zorzi, 2012; Trick & Pylyshyn, 1994), the recent new line of research on blindness and numerical cognition clearly demonstrates that vision is not necessary in the elaboration of numerical representations and skills. Behavioural results demonstrate that early visual deprivation does not preclude the elaboration of a semantic numerical representation with similar spatial properties (Castronovo & Seron, 2007; Cattaneo, Fantino, Tinti, Silvanto, & Vecchi, 2010; Cattaneo, Fantino, Silvanto, & Vecchi, 2011; Crollen et al., 2013; Salillas et al., 2009; Szucs & Csépe, 2005) and obedience to Weber's law (Castronovo & Delvenne, 2013; Castronovo & Seron, 2007b; Ferrand et al., 2010), as in sighted people. However, early visual deprivation does appear to have an impact on numerical cognition. First, early blindness prevents the automatic use of an external coordinate system to map numbers onto physical space: early blindness is associated with the use of a body-centered coordinate system, involving a reversed SNARC effect in a comparison task, hands crossed (Crollen et al., 2013). Secondly, importantly and unexpectedly, early blindness seems to have a positive impact on numerical skills, with the observation in blind people of greater mapping abilities (Castronovo & Delvenne, 2013; Castronovo & Seron, 2007b; Ferrand et al., 2010) and the likely greater allocation of high-level cognitive resources, such as attention and working memory, to numerical tasks (Salillas et al., 2009; Szucs and Csépe, 2005), leading to outstanding estimation skills. Blind people's particular experience in accessing and processing numerical information probably accounts for their extraordinary results. Indeed, experience with numerical information (e.g. mathematical achievement) plays a critical role in the elaboration of numerical representations and skills, such as mapping abilities (e.g. Ashcraft & Moore, 2012; Castronovo & Göbel, 2012; Mundy & Gilmore, 2009). In order to compensate for their lack of vision, blind people might have developed compensatory strategies based on numerical judgements, such as in locomotion. Moreover, they are likely to present a greater use of numerical skills, such as mental calculation, compared with sighted people (e.g. when shopping – remembering products' prices, computing how much change to get back at the checkout, etc.). As a consequence, it is likely that early visual deprivation involves greater and more efficient experience in accessing and processing numerical information, which in turn accounts for better numerical skills. However, blind participants' greater estimation skills might also be accounted by enhanced high-level cognitive processes, such as working memory and attention.

Indeed, blind people's ERPs results suggest that they might use compensatory mechanisms involving high-level cognitive resources such as working memory and/or attention when performing numerical tasks (Salillas et al., 2009; Szucs and Csépe, 2005). These findings are in line with recent observations on blindness and spatial processing. Indeed, research on blindness and spatial processing demonstrates that blind people are able to show supra-normal auditory spatial skills when the level of difficulty of the task is increased. For example, blind participants presented better auditory spatial abilities compared with sighted participants when sounds to localize were presented in the peripheral space, in other words, when more subtle auditory cues had to be detected and exploited (e.g. Voss, Lassonde, Gougoux, Fortin, Guillemot, & Lepore, 2004; Fieger et al., 2006). It appeared that blind people show compensatory mechanisms and greater performances compared with sighted participants when more resources are needed to solve the task (Collignon et al., 2009). Obviously, numerical tasks involve high-level cognitive resources. Therefore, it is likely that blind people's particularly great numerical estimation skills outside the subitizing range (Castronovo & Delvenne, 2013; Castronovo & Seron, 2007b; Ferrand et al., 2010), as well as their ERPs results (i.e. greater modulation of the cognitive P300 component; Salillas et al., 2009; Szucs and Csépe, 2005), reflect the use of these high-level cognitive compensatory mechanisms, allowing greater resources to be allocated to the tasks and leading to better performances.

Dissimilarity, possibly reflecting greater resources allocated to the task, can also be found by taking a closer look at the behavioural tasks in which blind participants seemed to present similar effects compared with sighted people. In Szucs and Csépe (2005), although blind participants presented similar distance and SNARC effects compared with sighted participants, they also presented faster reaction times. In Castronovo and Seron (2007a), a size effect was only found in the blind group in the comparison task to 5. In the parity task, blind participants presented a more pronounced SNARC effect than sighted participants, while in the comparison task to 55, blind participants presented fewer errors than sighted participants. The observation of a size effect and a more pronounced SNARC effect in the blind group could be accounted by a different use of the semantic numerical representation (Salillas et al., 2009), involving high order cognitive resources, such as spatial attention and/or working memory, which has lately been suggested to be essential for the spatial coding of numbers (van Dijck & Fias, 2011). Blind participants' faster reaction times and fewer errors could also be explained by the use of greater attentional resources to solve the task, leading to a more controlled numerical process, which in turn involves greater performances. Crollen and colleagues' (2011) results are also in favour of the use in blind people of compensatory mechanisms, involving high-level cognitive resources, such as working memory, when performing numerical tasks. Indeed, although vision appears to be necessary in the acquisition of finger-number association, congenitally blind children presented similar counting and arithmetical performances as sighted children, except when working memory interference was introduced.

Therefore, early visual deprivation seems to involve the development of compensatory mechanisms involving high-level cognitive resources, such as attentional (Collignon, Renier, Bruyer, Tranduy, & Veraart, 2006) and working memory (Sallilas et al., 2009; Crollen et al., 2011) resources, which in combination with greater numerical mapping abilities, lead to greater numerical skills in blind people, compared with sighted people. Future research would be needed to try to disentangle the part play by blind people's potential greater attentional, working memory resources, and the part play by their potential greater mapping skills in their numerical performances. It would also be interesting to investigate by comparing early and late blind individuals whether the age of onset of blindness is critical in the acquisition of greater numerical skills.

Further research would also be needed in the future to explore whether early visual deprivation would have a positive impact on quantitative abilities in general, rather than essentially on numerical abilities. Primary results indicating that blind participants presented greater duration estimation skills, as well as sensitivity to numerical and duration information, tend to show that blind people greater skills might be generalized to different types of quantitative information. The observation in blind people of better quantitative skills, in general, would be in line with the hypothesis that their compensatory mechanisms involve different quantitative judgements for common situations, such as those based on locomotion – estimation of numerosity (e.g. number of footsteps between two landmarks; number of stairs to reach a floor), but also estimation of distance (e.g. between two landmarks) and duration (e.g. duration of the walk).

In the future, caution should be taken not to strictly categorize the number sense as visual, as this postulate implicitly suggests a foundational role of vision in the development of numerical cognition. However, this assumption is no longer sustainable, as early exploratory research on early blindness strongly demonstrates that vision is not crucial in numerical cognition. Nonetheless, we also need to be cautious in our conclusion, as although vision is not necessary in the elaboration of numerical representations and skills, being able to access numerical information via individuals' predominant sensory modality might be of importance and might play a facilitative role in numerical processing (Ferrand et al., 2010).

Notes

1. The distance effect corresponds to a decrease of performance (i.e. slower reaction times and less accuracy) the smaller the distance between numbers to process. The SNARC effect reflects the observation that relatively small numbers are responded faster when associated with the left side, compared with when associated with the right side, while the reverse is true for relatively large numbers (Dehaene et al., 1993).
2. The authors considered the parity judgement task as a control task, as it is assumed to rely on verbal-spatial associations with numbers, rather than on visuospatial associations as in the comparison task (van Dijck, Gevers, & Fias, 2009).

References

Andres, M., & Pesenti, M. (this volume). Finger-based representation of mental arithmetic. In R.C. Kadosh & A. Dowker (Eds), *The Oxford Handbook of Mathematical Cognition*. New York: Oxford University Press.

Ashcraft, M.H., & Moore, A.M. (2012). Cognitive processes of numerical estimation in children. *Journal of Experimental Child Psychology*, 111, 246–267.

Barth, H., Starr, A., & Sullivan, J. (2009). Children's mappings of large number words to numerosities. *Cognitive Development*, 24, 248–264.

Buckley, P.B., & Gillman, C.B. (1974). Comparisons of digits and dot patterns. *Journal of Experimental Psychology*, 103 (6), 1131–1136.

Burr, D.C., & Ross, J. (2008). A visual sense of number. *Current Biology*, 18, 425–428.

Butterworth, B. (1999). *The Mathematical Brain*. London: Macmillan.

Butterworth, B. (2005). The development of arithmetical abilities. *Journal of Child Psychology and Psychiatry*, 46, 3–18.

Butterworth, B. (2008). Numerosity perception: how many speckles on the hen? *Current Biology*, 18(9), R388-R389.

Camos, V., & Tillmann, B. (2008). Discontinuity in the enumeration of sequentially presented auditory and visual stimuli. *Cognition*, 107, 1135–1143.

Castronovo, J., & Delvenne, J.-F. (2013). Positive impact of early visual deprivation on numerical cognition: a general phenomenon? *Cortex*, 49(5), 1435–1440.

Castronovo, J., & Göbel, S. (2012). Impact of high mathematics education on the number sense. *PLoS One* 7(4):e33832.
Castronovo, J., & Seron, X. (2007a). Semantic numerical representation in blind subjects: the role of vision in the spatial format of the mental number line. *Quarterly Journal of Experimental Psychology*, 60(1), 101–119.
Castronovo, J., & Seron, X. (2007b). Numerical estimation in blind subjects: evidence of the impact of blindness and its following experience. *Journal of Experimental Psychology: Human, Perception and Performance*, 33(5), 1089–1106.
Cattaneo, Z., Fantino, M., Silvanto, C.T., & Vecchi, T. (2011). Blind individuals show pseudoneglect in bisecting numerical intervals. *Attention, Perception and Psychophysics*, 73, 1021–1028.
Cattaneo, Z., Fantino, M., Tinti, C., Silvanto, J., & Vecchi, T. (2010). Crossmodal interaction between the mental number line and peripersonal haptic space representation in sighted and blind individuals. *Attention, Perception and Psychophysics*, 72(4), 885–890.
Cattaneo, Z., Vecchi, T., Cornoldi, C., Mammarella, I., Bonino, D., Ricciardi, E., & Pietrini, P. (2008). Imagery and spatial processes in visual impairments. *Neuroscience and Biobehavioral Reviews*, 32, 1346–1360.
Cohen Kadosh, K. (2011). What can emerging cortical face networks tell us about mature brain organisation? *Developmental Cognitive Neuroscience*, 1, 246–255.
Cohen Kadosh, R., Bahrami, B., Walsh, V., Butterworth, B., Popescu, T., & Price C.J. (2011). Specialization in the human brain: the case of numbers. *Frontiers in Human Neuroscience*, 5, 62.
Collignon, O., Renier, L., Bruyer, R., Tranduy, D., & Veraart, C. (2006). Improved selective and divided spatial attention in early blind subjects. *Brain Research*, 1075, 175–182.
Collignon, O., Voss, P., Lassonde, M., & Lepore, F. (2009). Cross-modal plasticity for spatial processing of sounds in visually deprived subjects. *Experimental Brain Research*, 192, 343–358.
Cordes, S., Gelman, R., Gallistel, C.R., & Whalen, J. (2001). Variability signatures distinguish verbal from nonverbal counting for both large and small numbers. *Psychonomic Bulletin and Review*, 8, 698–707.
Cornoldi, C., Cortesi, A., & Preti, D. (1991). Individual differences in the capacity limitations of visuospatial short-term memory: Research on sighted and totally congenitally blind people. *Memory & Cognition*, 19(5), 459–468.
Crollen, V., Dormal, G., Seron, X., Lepore, F., & Collignon, Olivier. (2013). The role of vision in the development of number-space interactions. *Cortex*, 49(1), 276–283.
Crollen, V., Mahe, R., Collignon, O., & Seron, X. (2011). The role of vision in the development of finger-number interactions: finger-counting and finger-montring in blind children. *Journal of Experimental Child Psychology*, 109, 525–539.
De Hevia, M.D., Vallar, G., & Girelli, L. (2008). Visualizing numbers in the mind's eye: the role of visuo-spatial processes in numerical abilities. *Neuroscience and Biobehavioral Reviews*, 32, 1361–1372.
Dehaene, S. (1997). *The Number Sense*. New York: Oxford University Press.
Dehaene, S. (2009). Origins of mathematical intuitions: the case of arithmetic. *The Year in Cognitive Neuroscience: Annals of the New York Academy of Sciences*, 1156, 232–259.
Dehaene, S., Bossini, S., & Giraux, P. (1993). The mental representation of parity and number magnitude. *Journal of Experimental Psychology: General*, 122(3), 371–396.
Dehaene, S., & Cohen, L. (1994). Dissociable mechanisms of subitizing and counting: neuropsychological evidence from simultanagnosic patients. *Journal of Experimental Psychology: Human Perception and Performance*, 20, 958–975.
Di Luca, S., & Pesenti, M. (2008). Masked priming effect with canonical finger numerical configurations. *Experimental Brain Research*, 185, 27–39.
Ferrand, L., Riggs, K., & Castronovo, J. (2010). Subitizing in congenitally blind adults. *Psychonomic Bulletin and Review*, 17(6), 840–845.
Fias, W., Brysbaert, M., Geypens, F., & d'Ydewalle, G. (1996). The importance of magnitude information in numerical processing: evidence from the SNARC effect. *Mathematical Cognition*, 2(1), 95–110.
Fias, W., & Fischer, M. (2005). Spatial representation of numbers. In J. Campbell (Ed.), *Handbook of mathematical cognition* (pp. 43–54). New York: Psychology Press.

Fieger, A., Roder, B., Teder-Salejarvi, W., Hillyard, S.A., & Neville, H.J. (2006). Auditory spatial tuning in late-onset blindness in humans. *Journal of Cognitive Neuroscience*, 18, 149–157.
Fischer, M.H., Castel, A.D., Dodd M.D., & Pratt, J. (2003). Perceiving numbers causes spatial shifts of attention. *Nature Neuroscience*, 6, 555–556.
Fraiberg, S. (1977). *Insights from the blind*. New York: Basic Book.
Gaunet, F., & Rossetti, Y. (2006). Effects of visual deprivation on space representation: Immediate and delayed pointing toward memorised proprioceptive targets. *Perception*, 35, 107–124.
Gaunet, F., Martinez, J.-L., & Thinus-Blanc, C. (1997). Early-blind subjects spatial representation of manipulatory space: exploratory strategies and reaction to change. *Perception*, 26, 345–366.
Gaunet, F., & Thinus-Blanc, C. (1996). Early-blind subjects' spatial abilities in the locomotor space: exploratory strategies and reaction-to-change performance. *Perception*, 25, 967–981.
Gebuis, T., & Reynvoet, B. (this volume). Number Representations and Their Relation with Mathematical Ability. In R.C. Kadosh & A. Dowker (Eds), *The Oxford Handbook of Mathematical Cognition*. New York: Oxford University Press.
Halberda, J., & Feigenson, L. (2008). Developmental change in the acuity of the 'number sense': the approximate number system in 3-, 4-, 5-, and 6-year-olds and adults. *Developmental Psychology*, 44(5), 1457–1465.
Halberda, J., Mazzocco, M., & Feigenson, L. (2008). Individual differences in non-verbal number acuity correlate with maths achievement. *Nature*, 455 (2), 665–668.
Hartlage, L.C. (1969). Verbal tests of spatial conceptualization. *Journal of Experimental Psychology*, 80, 180–182.
Hauser, M.D., Dehaene, S., Dehaene-Lambertz, G., & Patalano, A. (2002). Spontaneous number discrimination of multi-format auditory stimuli in cotton-top tamarins. *Cognition*, 86, B23–B32.
Heller, M.A. (2002). Tactile picture perception in sighted and blind people. *Behavioral Brain Research*, 135, 65–68.
Hubbard, E.M., Piazza, M., Pinel, P., & Dehaene, S. (2005). Interactions between number and space in parietal cortex. *Nature Reviews Neuroscience*, 6, 435–448.
Hull, T., & Mason, H. (1995). Performance of blind children on digit-span tests. *Journal of Visual Impairment*, 89, 166–169.
Izard, V., Sann, C., Spelke, E.S., & Streri, A. (2009). Newborn infants perceive abstract numbers. *Proceedings of the National Academy of Sciences*, 106 (25), 10382–10385.
Johnson, M.H. (2001). Functional brain development in humans. *Nature Reviews Neuroscience*, 2, 475–483.
Klahr, D., & Wallace, J.G. (1973). The role of quantification operators in the development of conservation of quantity. *Cognitive Psychology*, 4, 301–327.
Lee Swanson, H., & Luxenberg, D. (2009). Short-term memory and working memory in children with blindness: support for a domain general or domain specific system? *Child Neuropsychology*, 15, 280–294.
Lessard, N., Paré, M., Lepore, F., & Lassonde, M. (1998). Early blind human subjects localize sound sources better than sighted subjects. *Nature*, 395, 278–280.
Longo, M.R., & Lourenco, S.F. (2007). Spatial attention and the mental number line: evidence for characteristic biases and compression. *Neuropsychologia*, 45, 1400–1407.
Mandler, G., & Shebo, B.J. (1982). Subitizing: an analysis of its component processes. *Journal of Experimental Psychology: General*, 111, 1–22.
Mix, K.S. (1999). Preschooler's recognition of numerical equivalence: sequential sets. *Journal of Experimental Child Psychology*, 74, 309–332.
Mohr, H.M., Mues, H.T., Robol, V., & Sireteanu, R. (2010). Altered mental number line in amblyopia—reduced pseudoneglect corresponds to a decreased bias in number estimation. *Neuropsychologia*, 48, 1775–1781.
Moyer, R.S., & Landauer, T.K. (1967). Time required for judgments of numerical equality. *Nature*, 215, 1519–1520.

Mundy, E., & Gilmore, C.K. (2009). Children's mapping between symbolic and nonsymbolic representations of number. *Journal of Experimental Child Psychology*, 103, 490–502.

Piazza, M. (2010). Neurocognitive start-up tools for symbolic number representations. *Trends in Cognitive Sciences*, 14(12), 542–551.

Piazza, M., & Izard, V. (2009). How humans count: numerosity and the parietal cortex. *Neuroscientist*, 15 (3), 262–273.

Piazza, M., Izard, V., Pinel, P., Le Bihan, D., & Dehaene, S. (2004). Tuning curves for approximate numerosity in the human intraparietal sulcus. *Neuron*, 44, 547–555.

Pica, P., Lemer, C., Izard, V., & Dehaene, S. (2004). Exact and approximate arithmetic in an Amazonian indigene group. *Science*, 306(5695), 499–503.

Riggs, K., Ferrand, L., Lancelin, D., Fryziel, L., Dumur, G., & Simpson, A. (2006). Subitizing in tactile perception. *Psychological Science*, 17, 271–272.

Ross, J., & Burr, D.C. (2010). Vision senses number directly. *Journal of Vision*, 10(2): 10, 1–8.

Salillas, E., Graná, A., El-Yagoudi, R., & Semenza, C. (2009). Numbers in the Blind's 'Eye'. *PLoS One*, 4(7), e6357.

Siegler, R.S., & Opfer, J.E. (2003). The development of numerical estimation: evidence for multiple representations of numerical quantity. *Psychological Science*, 14(3), 237–243.

Simon, T.J. (1997). Re-conceptualizing the origins of number knowledge: a 'non-numerical' account. *Cognitive Development*, 12, 349–372.

Simon, T.J. (1999). The foundations of numerical thinking in a brain without numbers. *Trends in Cognitive Sciences*, 3, 363–366.

Stoianov, I., & Zorzi, M. (2012). Emergence of a 'visual number sense' in hierarchical generative models. *Nature Neuroscience*, 15(2), 194–196.

Szucs, D., & Csépe, V. (2005). The parietal distance effect appears in both the congenitally blind and matched sighted controls in an acoustic number comparison task. *Neurosciences Letters*, 384, 11–16.

Thinus-Blanc, C., & Gaunet, F. (1997). Representation of space in blind persons: vision as a spatial sense? *Psychological Bulletin*, 121, 20–42.

Trick, L.M. (2008). More than superstition: differential effects of featural heterogeneity and change on subitizing and counting. *Perception and Psychophysics*, 70, 743–760.

Trick, L.M., & Pylyshynm Z.W. (1994). Why are small and large numbers enumerated differently? A limited-capacity pre-attentive stage in vision. *Psychological Sciences*, 3, 363–366.

Van Dijck, J.P., & Fias, W. (2011). A working memory account for spatial-numerical associations. *Cognition*, 119, 114–119.

Van Dijck, J.P., Gevers, W., & Fias, W. (2009). Numbers are associated with different types of spatial information depending on the task. *Cognition*, 113, 248–253.

Van Dijck, J.P., Ginsburg, V., Girelli, L., & Gevers, W. (this volume). Linking Numbers to Space: From the Mental Number Line towards a Hybrid Account. In R.C. Kadosh & A. Dowker (Eds), *The Oxford Handbook of Mathematical Cognition*. New York: Oxford University Press.

Voss, P., Lassonde, M., Gougoux, F., Fortin, M., Guillemot, J.-P., & Lepore, F. (2004). Early- and late-onset blind individuals show supra-normal auditory abilities in far-space. *Current Biology*, 14, 1734–1738.

Whalen, J., Gallistel, C.R., & Gelman, R. (1999). Nonverbal counting in humans: the psychophysics of number representation. *Psychological Science*, 10, 130–137.

Wilson, K., & Lee Swanson, H.L. (2001). Are mathematics disabilities due to a domain-general or a domain-specific working memory deficits? *Journal of Learning Disabilities*, 34(3), 237–248.

Zwiers, M.P., Van Opstal, A.J., & Cruysberg, J.R. (2001). A spatial hearing deficit in early-blind humans. *Journal of Neuroscience*, 21, RC142–RC145.

CHAPTER 44

THE NEUROPSYCHOLOGY OF ACQUIRED NUMBER AND CALCULATION DISORDERS

MARINELLA CAPPELLETTI

The Basic Components of The "Normal" Number System

The number system includes a *semantic system* which broadly involves the understanding and manipulation of number concepts; a *verbal system* consisting of numerical processes that are preferentially mediated by language, such as counting or reading and writing, and third, an *executive control/attention/memory system* which deals with numerical manipulations that require monitoring or attentional resources. Here, these three systems will be discussed separately, although some numerical processes can be supported by more than one system at a time. For instance, some arithmetical concepts are based on the semantic number system but also imply the retrieval of numerical information in a verbal format.

The Number Semantic System

Numbers can assume different meanings depending on the context in which they are used. For instance, number "4" indicates the numerosity of *4*, like the number of people in a room, the price of an item, or the result of an arithmetical operation ("2 × 2" or "3 + 1"). Numbers can refer to the way in which items are ordered, for example, houses along a street, and as such they do not imply any quantity; for instance, house number "4" is after but not larger than house number "3." Numbers can also assume a nominal or encyclopedic meaning, for example "4" can indicate a brand or the name of an item, like bus number "4" or "BBC4" in the UK (Butterworth, 1999; Cappelletti et al., 2008; Dehaene, 1997).

Symbolic Quantity

Among these meanings, quantity is the most frequently used and the one that has been more extensively studied. The representation of quantity is thought to be abstract in nature, i.e. to be the same irrespective of the format or the modality in which it is expressed (Dehaene & Cohen, 1997; McCloskey et al., 1990; but see Cohen Kadosh and Walsh, 2009), and to be shared with preverbal infants (e.g. Feigenson et al., 2004) and populations with minimal number words (e.g. Gordon, 2004; Pica et al., 2004). The quantity expressed by numbers is thought to be automatically processed in a variety of tasks, for instance, when comparing numbers to decide which of two products is the most expensive. In doing so, we are usually faster and more accurate when two numbers (or two prices) are more far apart, e.g. 1.20 vs 2.55, relative to when they are close to each other, e.g. 1.20 vs 1.15. This phenomenon is referred to as "distance effect" (Moyer & Landauer, 1967). Impairments in processing numerical quantity occasionally result in a "reverse" or "larger" distance effect, whereby the time needed to process numbers increases as the numerical distance between them increases or is particularly longer when numbers are close in distance (e.g. Delazer & Butterworth, 1997). This may be because the quantity associated with numbers cannot be accessed automatically and other strategies, for instance counting, have to be used instead. However, this pattern of performance is rare, and impairments in quantity processing more often result in unsystematic performance (Ashkenazi et al., 2008; Dehaene & Cohen, 1997; Lemer et al., 2003).

Non-Symbolic Quantity

This includes any quantifiable dimension that is not expressed by symbols, and can be distinguished into: discrete non-symbolic quantity, where individual items can be identified like the number of people in a room or of chairs in a cinema; and continuous quantity, namely any dimension that can be judged as a continuum, for instance, temporal or spatial durations, luminance, pitch, and others.

Some recent theoretical proposals have focused on the process of these continuous dimensions, and especially time and space, and suggested that these share some common processes with numbers (e.g. Cantlon et al., 2009; Walsh, 2003; see also Chapter by Walsh, this volume). The nature of these common processes is still a matter of debate as they have been defined in terms of magnitude processes (e.g. Cantlon et al., 2009; Feigenson et al., 2004; Walsh, 2003) or decision mechanisms (e.g. Cohen Kadosh et al., 2008; Santens & Verguts, 2011). Irrespective of their nature, the hypothesis of common processes is based on the observation that tasks requiring to estimate number, time, and space result in similar patterns of performance (Meck & Church, 1983) or similar rates of development (e.g. Brannon et al., 2006, 2007, 2008), and, to some extent, involve similar brain areas (e.g. Cohen Kadosh et al., 2008; Dehaene et al., 2003; Fias et al., 2003).

Arithmetical Conceptual Knowledge

This has been defined as "an understanding of arithmetical operations and laws pertaining to these operations" (Hittmair-Delazer et al., 1994). For instance, conceptual knowledge concerns the use of the commutativity principle, by which the result of an addition or multiplication problem does not change when swapping the position of the operands, like 4+3 is the same as 3 + 4. Another case is the rule by which a multiplication corresponds to a series of additions, for example 3 × 4 is the same as 3 + 3 + 3 + 3.

Number Encyclopedic Knowledge

Some numerical expressions acquire additional meanings which do not imply their quantity or ordinal connotations, and are termed "encyclopedic numbers." Typically, they have a *nominal* use, for example, when they identify objects like commercial brands (e.g. "Levis 501"), movies (e.g. "2001 Space Odyssey"), and TV channels (e.g. "BBC 1" in the UK). Encyclopedic numbers also refer to familiar (e.g. our birthday) and famous or historical dates (e.g. "September 11th" or "1789"), zip codes (e.g. "22063"), telephone numbers (e.g. "999"), and other general and autobiographical numbers (e.g. age of consent, PINs, wedding anniversary). Nominal number assignments are exclusively verbal and are found only in linguistic humans (Butterworth, 1999; Nieder, 2005).

The Verbal Number System

This includes language-based skills in the context of numerical processes.

Counting

One of these language-based number skills consists of the foundational ability to enumerate objects, which in numerically-normal people is typically mastered early in development. However, this has been relatively little studied in brain-damaged patients and it will, therefore, not be further discussed here.

Reading And Writing

These consist of converting numerals' semantic representation into an output format. The most commonly used formats are the Arabic, e.g. "4", and the verbal format, e.g. "four." The production of a number in one of these formats is based on input and output processes. For instance, reading Arabic numbers is based first on visually identifying an input consisting of a string of digits; second, on translating this string into a sequence of words according to a set of rules; and third, on producing an oral or written output. The transformation of numbers from one format or modality to another is usually referred to as "*transcoding*" (Dehaene & Cohen, 1995; Deloche & Seron, 1982; McCloskey et al., 1986; Seron & Deloche, 1984).

Simple Arithmetical Facts

Among arithmetical facts, simple addition and multiplication problems such as "3 + 3 = 6" or "3 × 3 = 9"are thought to be stored in verbal memory, whereas subtraction problems are thought to rely on quantity-based processes (Dehaene et al., 2003). Other authors instead suggest that each arithmetical operation is independent from others (Dagenbach & McCloskey, 1992; McCloskey et al., 1985; Van Harskamp, N.J., & Cipolotti, L. 2001). Correct processing of these simple facts typically implies the understanding of arithmetic symbols, e.g. +, x, and –.

Executive Function/Attention/Memory

These abilities are required when performing some numerical operations, for instance, attention and working memory are thought to be critical for navigating along a number line (e.g. Doricchi et al., 2005; van Dijck et al., 2011). Memory resources are also required to retrieve arithmetical facts, and executive functions are needed when performing more complex arithmetical operations. Specifically, these functions are critical for the correct use of calculation procedures, which consist of the specific algorithms required to solve multi-digit calculation. Examples of these are the carrying (e.g. in "234 + 159") and borrowing (e.g. in "234 – 159") procedures.

Neuronal Correlates of Number and Calculation Processes

A number of interconnected regions are consistently involved in numerical and calculation processes. These regions include the left and right parietal lobes, considered the most critical brain region for the core understanding of quantity (Dehaene et al., 2003) and that can be further subdivided into three regions: the intraparietal sulcus (IPS), important for quantity-related processes; the angular gyrus, involved in language-based number processes such as arithmetical facts and transcoding (which also involve additional left-hemisphere areas); and the superior parietal lobe for attention processes implicated in numerical manipulations. Besides these parietal regions, several other brain areas are involved in number processing, for instance, the left occipito-temporal cortex for the identification of number symbols (e.g. Cappelletti et al., 2010; Cohen & Dehaene, 1995, 2000), the frontal regions and, more specifically, the pre-frontal cortex for the monitoring and execution of arithmetical processes (e.g. Granà et al., 2006; McCloskey et al., 1991b), and some subcortical areas, for instance, the basal ganglia (Delazer et al., 2004; Hittmair-Delazer et al., 1994). For further discussion of this, please see the Chapter by Menon in this volume.

The Fractionation of The Number System

In the last three decades, various models have been proposed to characterize the architecture of number and calculation processes. One of the first accounts, McCloskey and colleagues' model (Macaruso et al., 1993; McCloskey et al., 1985), highlighted the modularity of the number system by assuming functionally independent number comprehension, number production, and calculation mechanisms which can be selectively spared or impaired. The modularity of the number and calculation system has been supported by several single-case studies showing selectively impaired or spared performance in tasks relying on the individual mechanisms hypothesized in McCloskey et al.'s model (1985).

Here, we will describe some of these selective impairments or preservations of (1) number semantic system, and, in particular, non-symbolic and symbolic quantity processing, arithmetical conceptual knowledge, and encyclopedic knowledge; (2) number verbal system and, specifically, reading and writing, and arithmetic facts; (3) executive functions specifically implicated in arithmetic procedures (see Table 44.1).

Table 44.1 Examples of impaired and spared number skills and their associated brain lesions within: (A) the number semantic system; (B) the verbal number system; and (C) the executive control/attention and memory systems. The examples, listed alphabetically, have been chosen because they refer to relatively selective impairment or preservation of various number skills. The list is not meant to be exhaustive. Unique entries have been used as much as possible

	Impaired	Lesion's location	Spared	Lesion's location
A. Number semantic system	**1. Symbolic quantity**			
	Askenazi et al. 2008	L IPS	Cappelletti et al. 2011	Various (group)
	Cipolotti et al. 1991	L P	Cohen & Dehaene 1995	L O T
	Dehaene & Cohen 1997	Pt1: LSC; Pt2: RP	Cohen & Dehaene 2000	L O T
	Delazer & Butterworth 1997	L F	Warrington 1982	L P
	Delazer et al. 2006	PCA+F		
	2. Non-symbolic quantity			
	Cappelletti et al. 2009	R FP	Dehaene & Cohen 1991	L T P O
	Lemer et al. 2003	L IPS		
	Revkin et al. 2008	L F		
	3. Conceptual knowledge			
	Delazer & Benke 1997	L P	Cappelletti et al. 2005	SD
	Delazer et al. 2004	L BG	Crutch & Warrington 2002	SD
			Hittmair-Delazer et al. 1994	L BG
B. Verbal number system	**4. Encyclopedic knowledge**			
	Cappelletti et al. 2007	L T	Cohen et al. 1994	L H
			Delazer & Girelli 1997	L H

(Continued)

Table 44.1 Continued

	Impaired	Lesion's location	Spared	Lesion's location
	5. Fact retrieval			
	Cohen & Dehaene 1994	L T P		
	Dehaene & Cohen 1991	L T P O		
	Dehaene & Cohen 1997	Pt1: LSC; Pt2: RP		
	Delazer & Benke 1997	R P; LSC		
	McCloskey et al. 1991b	NK		
	van Harskamp et al. 2005	L T		
	Warrington 1982	L P		
	6. Reading & writing			
	Cipolotti 1995	AD	Cohen et al. 1997	L T
	Cipolotti et al. 1995	L P	Domahs et al. 2006	PA
	Cohen & Dehaene 2000	L O T		
	Delazer & Denes 1998	AD		
	Deloche & Seron 1982	L T-P		
	Marangolo et al. 2005	C-SC atrophy		
	McCloskey et al. 1990	L H		
	Noel & Seron 1993	AD		
	Noel & Seron 1995	L P		
	Pesenti et al. 1994	NK		
	Sokol et al. 1991	L H		
	Thioux et al. 1999	AD		
C. Executive number system	**Arithmetical procedures**			
	Lucchelli & De Renzi 1993	R T P		
	McCloskey et al. 1985	L F		
	McCloskey et al. 1991a	NK		
	McNeil & Burgess 2002	AD		
	Semenza et al. 1997	AD		
	Sokol et al. 1991	R F P		
	Sokol & McCloskey 1991	R F P		
	Zamarian et al. 2006a	SD		

Legend: L = left; R = right; H = hemisphere; IPS = intra-parietal sulcus; P = parietal; T = temporal; F = frontal; O = occipital; SC = subcortical; NK = not known; AD = Alzheimer dementia; SD = semantic dementia; PCA = posterior cortical atrophy; PA = progressive aphasia; BG = basal ganglia; Pt = patient

Impairment or Preservation of The Semantic Number System

Symbolic Quantity

Some recent cases have shown that quantity processing can be maintained in the context of impairments in other aspects of number or calculation processing (e.g. Cappelletti et al., 2012a; Cohen & Dehaene, 1991, 1994, 1995, 1996, 2000; Grafman et al., 1989; Warrington, 1982). The opposite pattern, i.e. of disorders in processing number quantity, have also been reported (e.g. Ashkenazi et al., 2008; Cipolotti et al., 1991; Dehaene & Cohen, 1997; Delazer & Benke, 1997; Delazer & Butterworth, 1997; Delazer et al., 2006; Halpern et al., 2003, 2007; Lemer et al., 2003; Noël & Seron, 1993; Polk et al., 2001; Revkin et al., 2008; Rosselli & Ardila, 1989; Woods et al., 2006). For example, patient NR was impaired at understanding Arabic numerals such that she could no longer point to the larger of two Arabic numerals (e.g. "345 vs 785" or "265 vs 237"), and could not decide whether a number was larger in size than a reference number (Noël & Seron, 1993). One of the most profoundly acalculic patients so far reported (CG) presented with an unusual pattern of performance, as she had lost the meaning of all numbers above 4, although she retained the ability to process continuous quantities (Cipolotti et al., 1991). For example, CG was unable to say how many days there are in a week or whether 5 or 10 is the bigger number. Her deficit was so pervasive that it seriously limited her activities of daily living; for example, CG was unable to do her own shopping as she could no longer deal with money or check her change. She was also unable to make phone calls, use a calendar, or read the time, although her intellectual skills, language, memory, and visuo-spatial abilities were largely preserved.

Recently, a few case studies investigated whether quantity processing can be indirectly impaired when other non-numerical abilities are damaged. For instance, it has been shown that impaired visuo-spatial abilities, as observed in neglect patients, can impact on quantity processing (Cappelletti & Cipolotti, 2006; Sackur et al., 2008; Vuilleumier & Rafal, 1999; Vuilleumier et al., 2004). One study showed that neglect patients could accurately perform number quantity tasks; for example, they can correctly indicate the larger number among two centrally presented number stimuli from 1 to 9 with 5 being the middle number. However, although accurate, patients were slower to process numbers located to the left of the middle number 5, i.e. "4", relative to numbers located to its right, i.e. "6" (Vuilleumier et al., 2004). This performance in numerical tasks therefore mirrors neglect patients' impairment in processing the left side of space. In contrast, other studies showed that quantity can be maintained despite impaired visuo-spatial skills. For instance, neglect patients who typically are only aware of the stimuli presented in the right hemifield, were presented with pairs of Arabic numbers displayed bilaterally and simultaneously on the computer screen. Patients had to indicate whether each number stimulus presented on the right was larger or smaller than the reference number "5". Within each pair, numbers could be congruent if both the left and the right stimuli were either smaller or larger than 5, or incongruent if the right (or the left) was smaller and the left (or the right) was larger than 5. Although it was demonstrated that the left-hemifield number stimuli were not attended, the patients' response times were faster for the right presented stimuli that were congruent with the left unattended stimuli relative to incongruent stimuli. This suggests that patients with left unilateral neglect can still unconsciously process some semantic features of Arabic numerals, like their quantity (Cappelletti & Cipolotti, 2006).

Despite these clear cases, impairments in quantity processing are not often reported; this may be because quantity can be processed by more than one brain area (for instance, the left and right parietal) and, therefore, patients with either a left or right lesion can still show residual performance in quantity tasks. Impairments in quantity processing may also fail to be detected because patients' performance is often measured in terms of accuracy but not speed, although performance in several tasks based on quantity processing is more sensitive to speed. It is also possible that, among the various meanings that numbers can express, quantity is the most robust because it is suggested to be innate and does not depend on language and education, as other meanings do (Butterworth, 1999; Dehaene, 1997). This may be reflected in quantity processing being selectively preserved in the context of impaired calculation of transcoding.

Non-Symbolic Quantity

As previously mentioned, recent theoretical proposals have extended the study of quantity processing from number to other continuous dimensions like time and space (Cantlon et al., 2009; Walsh, 2003). Within this framework, a working hypothesis is that number, time, and space all share some common mechanisms, to process quantity itself or to make quantity-based decisions.

Neuropsychological cases have provided support for the idea of a shared magnitude or decision mechanism among dimensions by indicating that different magnitude dimensions can be equally impaired. One of the most elegant cases of such association concerns the link between number and space processing in neglect patients. These patients are equally biased at processing physical lines—a classical symptom of neglect—as well as number lines, a metaphor used to represent numbers as oriented from the smaller to the larger (Dehaene et al., 1993). A way to test the integrity of the mental number line is with a bisection task. This consists of orally presenting two numbers, e.g. "1" and "5", and asking patients which number falls in the middle. Neglect patients typically select a number which is biased towards the right of the middle one. For instance they state that "4" and not "3" is middle between "1" and "5," therefore resembling their performance in the line bisection task (Cappelletti et al., 2007a; Doricchi et al., 2005; Zorzi et al., 2002). Such association between physical and number line also depends on the orientation of the line, so that if patients neglect space along the horizontal but not the vertical dimension, they also neglect the number line along the horizontal but not an imaginary vertical dimension. For example, patients biased toward the left in the horizontal space would also say that the middle number between "1" and "5" is "4," when these numbers are meant to reflect a horizontal space such as, for example, house numbers. However, when the very same stimuli are imagined as indicating numbers along a vertical space, like floors when going up a lift, neglect patients with preserved vertical space are also able to bisect these "vertical" numbers correctly (Cappelletti et al., 2007a). Neglect patients' performance has also revealed other associations between magnitude dimensions, for example, between time and space, whereby neglect patients tend to underestimate the duration of stimuli presented in the neglected field but to overestimate the duration of stimuli presented in the preserved field (Basso et al., 1996).

A closer look at some of these associations, for instance, those in neglect patients, suggests that commonalities of impairments in space and time or space and numbers may sometimes be driven by factors which may not be magnitude-related. For instance, impairment in "auxiliary" skills that are needed when performing magnitude tasks such as, for example, working memory, could be reflected in impaired space processing (e.g. Doricchi et al., 2005).

These associations of symptoms—suggesting shared magnitude representations or decision mechanisms—have recently been challenged by evidence of dissociations between magnitude dimensions. Such dissociations would not be expected if these dimensions fully share magnitude or decision systems. However, some patients show selective impairment in time processing while number and space are maintained (e.g. Cappelletti et al., 2009a), or they present with number impairment whilst space and time are preserved (Cappelletti et al., 2011). In these cases, selectively impaired performance in time or number was nevertheless modulated by other task-irrelevant magnitude dimensions. For instance, time was severely underestimated in patient CB, who reported a veridical duration of 60 seconds as lasting about 18 seconds instead. Critically, this underestimation was influenced by the magnitude of the number stimuli used to test CB's performance. Hence, the duration of stimuli corresponding to small numbers (i.e. 1 to 4) was judged even shorter than the duration of large number stimuli (i.e. 6 to 9)(Cappelletti et al., 2009a). This suggests that although different magnitude dimensions can dissociate, they could still influence each other.

Together, these associations, dissociations, and interactions between magnitude dimensions suggest that they may only partly share common magnitude representations or decision mechanisms. A fully shared representation would not predict dissociations, and fully independent magnitude-specific mechanisms would not explain interactions. Instead, a combination of common and of magnitude-specific mechanisms may account for both dissociations and interactions between dimensions.

Arithmetical Conceptual Knowledge

One of the first reported cases of impaired arithmetical conceptual knowledge is patient DRC (Warrington, 1982), and more extensively studied cases have been recently described. For instance, patient JG could retrieve multiplication tables from memory despite severe problems in all tasks tapping conceptual knowledge; for example, the patient could correctly produce the result of operations such as "3×3" but she did not apply even very basic principles such as commutativity in multiplication (i.e. $4 \times 12 = 12 \times 4$), and she was unable to recognize that multiplication can be transformed into repeated addition ($4 \times 12 = 12 + 12 + 12 + 12$) (Delazer & Benke, 1997; see also Pesenti et al., 2000a; Venneri & Semenza, 2011).

The opposite side of the dissociation, namely, intact conceptual knowledge and impaired fact retrieval, was reported in two single-case studies (Hittmair-Delazer et al., 1994, 1995). In one case, a patient demonstrated excellent conceptual knowledge despite a severe acalculia. For example, when presented with arithmetical problems such as 8×6, he adopted the following strategy: $8 \times 10 = 80 \rightarrow 80{:}2=40 \rightarrow 40 + 8 = 48$ (see Figure 44.1a). Another patient, who was no longer able to solve simple elementary arithmetical facts such as 2+3, showed excellent understanding and use of abstract equations such as $(b \times a){:}(a \times b) = 1$ and $(cd + ed){:}d = c + e$ (Hittmair-Delazer et al., 1994). Similarly, a patient with a progressive disorder affecting the left temporal lobe, i.e. semantic dementia, showed well preserved understanding of arithmetical concepts despite the loss of arithmetical facts (Cappelletti et al., 2005). This patient (IH) spontaneously developed strategies that allowed him to solve arithmetical operations when he could not access simple facts or use canonical procedures (e.g. using multiple additions to solve a multiplication problem; see Figure 44.1b). This distinction

A) Performance of patient BE with impaired arithmetical fact retrieval but preserved conceptual knowledge. When asked to perform arithmetical problems such as 8×6, he adopted the strategy below. Data from M. Hittmair-Delazer, C. Semenza, and G. Denes, Concepts and facts in calculation, *Brain* 117(4), pp. 715–728, figure 4, doi:10.1093/brain/117.4.715, 1994.

$$8 \times 10 = 80; 80 : 2 = 40; 40 + 8 = 48$$

B) An example of IH's performance in solving a multidigit multiplication operation. Number 531 was decomposed into the subparts 1, 30, and 500. Numbers in squared brackets indicate the main steps in performing the operation. Step [1] is 27 × 1. Steps [2] to [4] show how 327×30 was obtained. Step [2] is 327 × 5; [3] is 327 × 10 and [4] is 327 × 30; [5] is 327×100 obtained through multiple additions; [6] is 327 × 500, again obtained through multiple additions. Finally, [7] is (327 × 500) + (327 × 30) + (327 × 1). The final result is correct.

Data from Dissociations in numerical abilities revealed by progressive cognitive decline in a patient with semantic dementia, Marinella Cappelletti, Michael D. Kopelman, John Morton, and Brian Butterworth, *Cognitive Neuropsychology*, 22(7), pp. 771–93, DOI: 10.1080/02643290442000293, 2007.

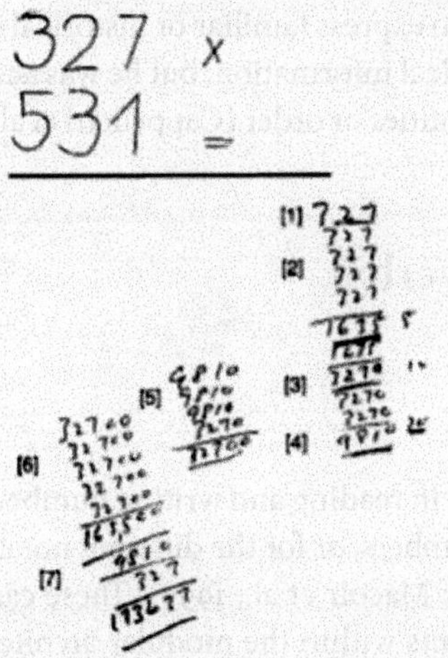

C) Performance of patient EB, with a selective impairment in processing arithmetical symbol such that he systematically 'plussed' the times. Data from *Neuropsychologia*, 35(3), M. Laiacona and A. Lunghi, A case of concomitant impairment of operational signs and punctuation marks, pp. 325–332, 1997.

$$\begin{array}{r} 48 \\ \times\ 67 \\ \hline 115 \end{array} \qquad \begin{array}{r} 59 \\ \times\ 29 \\ \hline 88 \end{array}$$

D) Performance of patient MT with impairment in using calculation procedures. He showed 'smaller from larger' subtraction bugs. Data from *Cortex*, 32(3), Luisa Girelli and Margarete Delazer, Subtraction Bugs in an Acalculic Patient, pp. 547–55, 1996.

$$\begin{array}{r} 923 \\ -\ 644 \\ \hline 321 \end{array} \qquad \begin{array}{r} 171 \\ -\ 48 \\ \hline 127 \end{array}$$

FIGURE 44.1 Examples of specific impairments or preservation of calculation skills.

between understanding numerical and arithmetical concepts and impaired simple facts and procedures has been observed in other semantic dementia patients (e.g. Cappelletti et al., 2006; Crutch & Warrington, 2002; Julien et al., 2010), and it suggests that conceptual knowledge is a functionally independent component within the number semantic system.

Number Encyclopedic Knowledge

A few studies have focused on the nominal features of numbers, for instance when used to indicate brands or objects. Both selective preservations and impairments have been reported. For instance, number stimuli with an encyclopedic meaning like "164" for Alfa Romeo (a car brand) or 1789 for the French revolution can facilitate verbal production in aphasic patients (Cohen & Dehaene, 1994; Delazer & Girelli, 1997). On the other hand, using numbers in an encyclopedic context could also be selectively impaired. For example, a patient with a left temporal dysfunction due to a traumatic accident could not process numbers when used as nominal labels, for instance to express familiar or historical dates, or to indicate other general or autobiographical numerical information, but he was accurate at processing the same numbers when referring to quantities or order (Cappelletti et al., 2006).

Impairment or Preservation of The Verbal Number System

Number Reading and Writing

Several studies have reported patients with disorders in reading and writing numbers which could be specific for understanding or producing numbers, or for the different notation systems, i.e. Arabic or verbal (e.g. Cipolotti et al., 1995; Macoir et al., 1999). These cases have been interpreted as the result of selective impairments within the modular architecture of number and calculation suggested by McCloskey and colleagues' model, which assumes distinct input and output modules operating independently but via an obligatory semantic representation of numbers (McCloskey et al., 1985).

Some of the specific mechanisms involved in transcoding have been identified by analyzing the errors made by patients in transcoding tasks (Seron & Deloche, 1984. For instance, two important cognitive mechanisms within number transcoding involve syntactic and lexical number processes. Syntactic processes involve the specification of the relationship between the elements of the number (e.g. number class). For example, reading aloud "600" requires retrieving the correct number class [hundred]. Syntactic errors are characterized by the selection of the wrong number class (Cipolotti et al., 1994; Furumoto, 2006). For example, in reading aloud Arabic numerals, patient SF made syntactic errors like "207" read as "two thousand and seven" and "80" read as "eight" (Cipolotti, 1995). Lexical processes involve manipulating the individual elements within the number; for instance, reading aloud number "600" requires retrieving the correct class [hundred] and then the correct element [6]. Lexical errors involve the incorrect production of one or more of the individual elements in a number, e.g. stimulus "29" read as "forty-nine." These errors have often been observed in patients; for instance, HY could access the correct lexical class of the number (ones, teens, tens) but not the correct position within that class, such that "17" was read as "thirteen" or "902" as "nine hundred six" (Sokol et al., 1991).

A frequent type of error in writing numbers to dictation or from a written input has been referred to as "intrusion errors." These consist of reproducing part of the input code within the output one; for instance, the Arabic number "75" written as "SEVENTY5" or "70" written as "7ty." These errors have been observed in patients with Alzheimer dementia as well as with focal brain lesions (Della Sala et al., 2000; Tegner & Nyback, 1990; Thioux et al., 1999), and have been explained in terms of the combination of an impaired transcoding mechanism on the one hand, and impaired inhibitory processes on the other (Thioux et al., 1999), or as originating from a unique impairment in attention capacities (Macoir et al., 2002).

Transcoding skills can operate independently from other numerical abilities, such as quantity processing and calculation. For example, some patients show well-preserved reading and writing skills but poor calculation (e.g. Cohen & Dehaene, 1994; Sokol et al., 1991), or, in contrast, difficulties in reading and writing numbers but maintained calculation (e.g. Cipolotti & Butterworth, 1995), or good number comprehension (Garcia-Orza et al., 2003; McCloskey et al., 1986; Noël & Seron, 1995). Some of these dissociations have challenged the idea that reading and writing necessarily operate via a semantic representation as assumed in McCloskey and colleagues' model (McCloskey et al., 1985). For instance, patients showing good understanding of the meaning of numbers despite making several errors in reading or in writing them (e.g. Cipolotti, 1995; Cipolotti et al., 1995; Noël & Seron, 1995) have prompted the hypothesis of an additional *asemantic route* for transcoding. This is supposed to parallel the semantic route proposed by McCloskey's model but with a direct link between input and output and no obligatory semantic mediation (Cipolotti & Butterworth, 1995; Deloche & Seron, 1982; Seron & Deloche, 1984).

Arithmetical Fact

Patients with a selective impairment of arithmetic fact retrieval typically present with severe difficulties in performing simple single-digit addition, subtraction, multiplication, or division problems. They may produce errors such as for example "5 + 7" = "13 roughly" (e.g. Dehaene and Cohen, 1991) and their response times can be abnormally slow (e.g. longer than 2 seconds per operation). A common analysis based on the patients' error (here, for instance, in performing multiplication problems) identified: (i) *operand errors*, if the incorrect answer is the correct answer to a problem that shares one of the operands (e.g. $6 \times 5 = 25$); (ii) *operation errors*, if the incorrect answer is the correct answer to another problem involving the same operands, but a different operation (e.g. $3 \times 4=7$); (iii) *table errors*, if the incorrect answer is an answer that is a product of two other single-digit numbers (e.g. $4 \times 4 = 25$); (iv) *non-table errors*, if the incorrect answer is not an operand, table, or operation error (e.g. $9 \times 8 = 52$); and (v) *errors specific to arithmetic rules*, i.e. in performing problems such as *n*x0 or *n*x1 (McCloskey et al., 1991c). Patients that make these errors could nevertheless show intact knowledge of arithmetical principles and procedures; for instance, they are typically able to retrieve and apply the appropriate arithmetical steps to solve complex arithmetical problems and they can define arithmetical operations adequately (e.g. Delazer & Benke, 1997; Sokol et al., 1991; Warrington, 1982). For this type of patient, everyday activities such as checking change or a bank statement poses great difficulties.

Arithmetical fact impairment can also manifest itself in a highly selective manner. Indeed, selectively preserved and selectively impaired arithmetical facts, according to the specific type of operation, have been reported; for example, in the case of multiplication (e.g. Dehaene & Cohen, 1997; Delazer & Benke, 1997; Grafman et al., 1989; McCloskey et al., 1991b; Whalen et al., 2002), subtraction (e.g. Dagenbach & McCloskey, 1992; Lampl et al., 1994; McNeil & Warrington, 1994; Pesenti et al., 1994; Sandrini et al., 2003), and addition and division facts (Cipolotti and de Lacy Costello, 1995; van Harskamp & Cipolotti, 2001). For example, Cipolotti and de Lacy Costello (1995) reported a patient who could no longer solve "4 ÷ 2," while being able to solve "27 × 26", and van Harskamp and Cipolotti (2001) described a patient with a selective addition impairment who could no longer solve 2 + 3 while still being able to solve 13 – 6 and 8 × 9.

These highly selective impairments of arithmetical facts have contributed to the characterization of the cognitive architecture of arithmetical operations in memory, which has led to the formulation of two influential proposals. One proposal, based on McCloskey and colleagues' view (Dagenbach & McCloskey, 1992; McCloskey et al., 1985), suggests that arithmetic facts are individually stored in memory, are segregated by arithmetical operation, and that these operations can dissociate arbitrary. An alternative view is based on Dehaene and Cohen's model (1997) and suggests two basic routes for arithmetic problems: an *asemantic route*, specific for maintaining and retrieving arithmetic problems stored as rote verbal knowledge, i.e. multiplication and simple addition problems; and a *semantic route*, specialised for arithmetical problems which are based on quantity processing, such as subtraction or complex addition. Within this framework, the two routes are suggested to be implemented by separate anatomical pathways, one relying on left hemispheric cortical and subcortical regions (frontal and superior/middle temporal gyrus as well as the basal ganglia and thalamic nuclei), the other based more strongly on the bilateral inferior parietal areas (Cohen et al., 2000). According to this second view, dissociations between operations are not arbitrary and, rather, they reflect the preservation of the underlying structure of the two main cerebral pathways for calculation.

As previously mentioned, processing arithmetical facts implies the understanding of arithmetical signs like + or –. The very few cases of selective impairments in understanding arithmetical signs suggest that this impairment can impact on processing arithmetical facts (e.g. Ferro & Botelho, 1980; Laiacona & Lunghi, 1997). For example, Laiacona and Lunghi (1997) investigated a patient who misnamed and misidentified arithmetical signs and performed written arithmetical operations according to their misidentification. Thus, for example, the patient systematically "plussed" the "times" (see Figure 43.1c). However, this deficit was not specific for arithmetical symbols, but was part of a more general impairment in processing relational symbols, for instance punctuation marks, which were also impaired (Laiacona & Lunghi, 1997).

Impairment or Preservation of The Executive/Attention and Memory Systems

Disorders in these systems can affect numerical processing in different ways. For instance, attentional/working memory disorders, as those occurring in neglect, can in some cases affect quantity processing, and, in other cases, leave this processing unimpaired (see for example Cappelletti & Cipolotti, 2006).

Another case is when impaired executive functions affect calculation procedures, which may manifest in various forms. For instance, difficulties in using these procedures may consist of errors with carrying processes in addition problems, or of not applying the problem-specific procedure in the correct order (e.g. Ardila & Rosselli, 1994; Benson & Weir, 1972; Granà et al., 2006; McNeil & Burgess, 2002; Semenza et al., 1997; Sokol et al., 1991; Sokol & McCloskey, 1988). Some systematic procedural errors have also been reported. For instance, Girelli and Delazer (1996) described patient MT, with a left hemisphere lesion, who consistently subtracted the smaller number from the larger one, irrespective of whether the larger digit was at the top or on the bottom line (see Figure 43.1d). His impairment, which was selective for multi-digit subtractions, was attributed to a defective knowledge of calculation procedures.

The Number System Relative to Other Cognitive Systems

In the last two decades, the study of numerical cognition has moved from the modular architecture of the number system, strongly supported by McCloskey and colleagues' model (1985), to a brain-based approach looking more broadly at the relation between number and other cognitive domains, and mainly expressed in the "Triple Code Model" (Dehaene & Cohen, 1995).

Within this more recent approach, several studies have established both dissociations and associations between numerical and calculation abilities and other cognitive skills (see Table 44.2). One of the most striking and clear case of double dissociation has been repeatedly reported between number and semantic knowledge (e.g. Cappelletti et al., 2001; Crutch & Warrington, 2002; Diesfeldt, 1993; Jefferies et al., 2004; Julien et al., 2008, 2010; Zamarian et al., 2006a). For instance, patients with semantic dementia who present with severe impairment in understanding the meaning of words and the use of objects, are usually able to understand numerical concepts and to perform arithmetical operations (e.g. Cappelletti et al., 2001; Crutch & Warrington, 2002; Julien et al., 2010), even at the late stages of illness (Cappelletti et al., 2005). This suggests that understanding numerical and arithmetical concepts can dissociate from understanding the meaning of words and objects. This pattern has been reported in several patients with semantic dementia, and recent cases have shown that precise numerical knowledge tends to decline with the severity of the semantic disorder, possibly because the atrophy of the left temporal lobe extends to brain areas more directly involved in numerical processing (e.g. Jefferies et al., 2005; Julien et al., 2010). Equally striking dissociations have been reported in the opposite direction whereby, for instance, patients may be severely incapacitated in processing numbers but maintain other cognitive skills well within the normal range. For instance, a patient with posterior cortical atrophy showed selective impairment in understanding numerical concepts but maintained comprehension of non-numerical concepts (Delazer et al., 2006). Together, these cases, which always consisted of dissociations rather than associations of impairments, strongly support the idea of a subdivision within semantic knowledge between numerical and non-numerical concepts.

Table 44.2 Examples of associations and dissociations between the number system and other cognitive systems. The examples, listed alphabetically, have been chosen because they refer to relatively specific dissociations or associations between the number system and other cognitive systems. The list is not meant to be exhaustive. Unique entries have been used as much as possible

A. Associations	Numbers preserved	Lesion/etiology	Numbers impaired	Lesion/etiology
1. Verbal skills Language	Cohen &	L H		
	Dehaene 1995	L H		
	Dehaene &	L H		
	Cohen 1997	R H		
	Delazer & Denes 1998	L T P F		
	Deloche & Seron 1982	L H		
	Garcia-Orza			
	et al. 2003			
	Seron & Deloche			
	1984			
2. Executive control/ attention/memory Visuo-spatial attention	Bonato et al. 2008	R H neglect		
	Cappelletti	R H neglect		
	et al. 2007a	L P		
	Pia et al. 2009	R H neglect		
	Vuilleumier	Parkinson		
	et al. 2004	R P		
	Zamarian et al. 2006b			
	Zorzi et al. 2002			
B. Dissociations				
1. Verbal skills Language	Basso et al. 2000	L H	Dehaene &	L post
	Cappelletti et al. 2005	SD	Cohen 1997	intra-cerebral
	Cohen et al. 1997	L T	Lucchelli &	R T P
	Cohen &	L O T	DeRenzi 1993	L T P
	Dehaene 2000	NK	Marangolo et al.	AD
	Delazer et al. 2002	PA	2004 Semenza	R H
	Domahs et al. 2006		et al. 1997	
	Rossor et al. 1995		Warrington 1982	
Reading and writing	Butterworth	SD	Cipolotti et al. 1995	AD
	et al. 2001	SD	Cipolotti 1995	L P
	Cappelletti et al. 2002	L T P	Noel & Seron 1995	AD
	Cohen & Dehaene			
	1991			
2. Executive control/ attention/memory Visuo-spatial attention	Cappelletti &	R P		
	Cipolotti 2006	R H		
	Doricchi et al. 2005	R H		
	Van Opstal et al. 2011			

(Continued)

Table 44.2 Continued

A. Associations	Numbers preserved	Lesion/etiology	Numbers impaired	Lesion/etiology
Semantic (number knowledge) Memory	Cappelletti et al. 2001	SD	Dehaene & Cohen 1997	LSC; R P
	Cappelletti et al. 2005	SD		RP; L SC
	Cappelletti et al. 2006	SD	Delazer & Benke 1997	PCA
	Crutch & Warrington 2004	SD	Delazer et al. 2006	
	Halpern et al. 2004	SD		
	Jefferies et al. 2005	SD		
	Julien et al. 2008	SD		
	Julien et al. 2010	L F T P		
	Thioux et al. 1998	SD		
	Zamarian et al. 2006a			
Short term	Butterworth et al. 1995	SD		
	Jefferies et al. 2005			

Legend: L = left; R = right; H = hemisphere; IPS = intra-parietal sulcus; P = parietal; T = temporal; F = frontal; O = occipital; SC = subcortical; NK = Not known; AD = Alzheimer dementia; SD = semantic dementia; PA = progressive aphasia; BG = basal ganglia; Pt = patient; PCA = progressive cortical atrophy

The relation between number and cognitive skills other than semantic knowledge is not always so unambiguous, as both associations and dissociations have been reported. For example, number skills have been shown to be largely independent from general intelligence (e.g. Remond-Besuchet et al., 1999), short-term memory (e.g. Butterworth et al., 1995), visuo-spatial attention (e.g. Cappelletti & Cipolotti, 2006; Rusconi et al., 2006; Sackur et al., 2008), and language (e.g. Cohen et al., 1997; Rossor et al., 1995; Thioux et al., 1998). In this latter case, for example, patients with both well-preserved language and severe quantity impairments or transcoding (e.g. Cipolotti et al., 1991; Dehaene & Cohen, 1997; Delazer & Benke, 1997; Denes & Signorini, 2001; Marangolo et al., 2005; Warrington, 1982) or impaired language and maintained number skills (e.g. Delazer et al., 2002; Domahs et al., 2006; Varley et al., 2005) have been reported. We discussed earlier that, despite the evidence of the independency of number and language, it is nevertheless indubitable that some numerical skills are mediated or preferentially supported by linguistic functions (e.g., counting, reading numbers, some arithmetical facts), which may explain the commonality of linguistic and numerical impairments often reported (e.g. Basso et al., 2000; Cohen & Dehaene, 1995; Dehaene & Cohen, 1997; Delazer et al., 1999a, b; Deloche & Seron, 1982; Garcia-Orza et al., 2003; Marangolo et al., 2004; Seron & Deloche, 1984). Likewise, some numerical skills are also intrinsically linked to working memory, executive functions, and spatial attention such that associations of impairments in number and in these other functions have frequently been reported (e.g. Bonato et al., 2008; Cappelletti et al., 2007a; Cohen & Dehaene, 1991; Cohen et al., 1994, 2007; Hoeckner et al., 2008; Mennemeier et al., 2005; Pesenti et al., 2000b; Pia et al., 2009; Priftis et al., 2008; Sathian et al., 1999; Vuilleumier et al., 2004; Zorzi et al., 2002). Whilst dissociations between numerical and other concepts support their independence at a semantic level, the pattern of associations and dissociations between number and other abilities suggest that some commonalities between these skills exist.

Localisation of Brain Lesions in Number and Calculation Disorders: The Parietal Lobes and Beyond Them

An overview of both group studies and single case studies has suggested that the majority of patients with number production and/or number comprehension impairments presented with left posterior lesions almost always involving the parietal lobe (Cappelletti & Cipolotti, 2010; Cipolotti & van Harskamp, 2001). Similarly, several patients with impairments in arithmetical fact retrieval had lesions mainly implicating the left parietal lobe (e.g. Dehaene & Cohen, 1997; Delazer & Benke, 1997; McNeil & Warrington, 1994; Pesenti et al., 1994; van Harskamp et al., 2002, 2005).

The neuropsychological evidence suggesting the involvement of the parietal lobe in numerical processing has recently been corroborated by neuroimaging studies (e.g. Cohen Kadosh et al., 2008; Dehaene et al., 1999, 2003; Pesenti et al., 2000c) which highlighted the involvement of these areas in normal number processing. Recently, the critical role of the parietal lobes, and not just their involvement in number and calculation, has been demonstrated in the context of reversible "lesions" obtained with transcranial magnetic stimulation (TMS). These temporary lesions showed that when the parietal lobes are suddenly unavailable, performance in some number tasks tends to be significantly slower (e.g. Andres et al., 2005; Cappelletti et al., 2007b, 2009b; see also Chapter by Salillas and Semenza in this volume).

If the parietal lobes are critical for number processing, number impairments might be expected to occur frequently following parietal lesions. However, the amount of patients with number impairments following parietal lesions is surprisingly small, and estimated to correspond to only about 20% of patients with left parietal damage (e.g. Jackson & Warrington, 1986). Moreover, most patients with right parietal lesions showed preserved comprehension of the core numerical concepts (e.g. Cappelletti & Cipolotti, 2006; Cappelletti et al., 2007a; Cohen et al., 2007; but see Langdon & Warrington, 1997; Rosselli & Ardila, 1989 for a different pattern of results). The relatively small number of patients described with acalculia may be because number and calculation processing are simply not diagnosed, leading to an underestimation of the extent of the disorder. Alternatively, as also envisaged by other authors (e.g. Cohen et al., 2007), number processing can rely on more than one brain region, for instance the left and right parietal or frontal lobes, following the idea of "degeneracy," i.e. that there may be more than one brain region capable to support a specific task (Price & Friston, 2002). In this case, lesions to a particular brain region may result in the controlateral area to "take over" such that residual number performance can be maintained. This possibility has recently been hypothesized to explain how patients with left or right parietal lesions could still perform number quantity tasks at a normal level (Cappelletti et al., 2012a).

Number impairments can also occur following lesions to other brain areas than the parietal. For instance, damage to the frontal and temporal lobes often leads to impaired calculation skills (e.g. Basso & Beschin, 2008; Cohen & Dehaene, 2000; Girelli & Delazer, 1996; Granà et al., 2001; Lucchelli & De Renzi, 1993; van Harskamp et al., 2005; Venneri & Semenza, 2011), but sometimes also to disorders in quantity processing (e.g. Delazer & Butterworth, 1997; Delazer et al., 2006; Polk et al., 2001). Likewise, damage to subcortical areas can lead to impairment in quantity processing, arithmetical fact retrieval, and conceptual knowledge (e.g. Dehaene & Cohen, 1997; Delazer & Benke, 1997; Delazer et al., 2004; Hittmair-Delazer et al., 1994; Koss et al., 2010; Lampl et al., 1994). More subtle number impairments can also occur following damage to posterior regions which are implicated in recognizing numerical symbols and words (Dehaene & Cohen, 1997). This was the case of patient JL who had damage to the left ventral occipito-temporal cortex (LvOT), an area activated during reading and object processing (e.g. Cohen & Dehaene, 2000), as well as during semantic tasks on Arabic numbers (Cappelletti et al., 2010). Following a LvOT lesion, JL had severely impaired written word recognition but he was remarkably accurate when making semantic decisions on visually presented numbers, albeit slower than normal (Cappelletti et al., 2012b). To account for JL's residual number performance following a lesion that is part of the "normal" number system, JL's brain activation was compared to healthy controls. This showed that JL had significantly increased activation in a left posterior middle temporal region which supported his residual semantic processing of numbers. These results suggest that although the left occipito-temporal cortex usually contributes to efficient number processing, activation in this region is not essential for accurate performance because other brain regions can take over, resulting in slow but accurate semantic decisions on numbers.

Overall, these neuropsychological cases suggest that number processing can be (1) most frequently impaired following lesions to the parietal lobes, which appear to be critical for accurate and efficient number performance; (2) compromised following damage to other brain regions, most notably frontal, occipital, and some subcortical areas; and (3) maintained, at least in part, when other brain regions are capable to take over the role of the damaged ones; this may be the reason why only a relatively small proportion of patients with parietal lobe lesions present with numerical impairments. This reinforces the idea that there is no one-to-one correspondence between number processing and parietal regions but, rather, that number skills – at least residually – can be supported by multiple brain areas.

Assessment and Rehabilitation of Acalculia

Diagnosis and Assessment Tools

The diagnosis of primary acalculia relies on establishing the presence of number processing and/or calculation impairment which is not a secondary consequence of other cognitive deficits (McCloskey et al., 1991a). This means excluding that a generalised impairment in language, attention, visuo-spatial functions, or other cognitive skills is underpinning the impairment in number processing and calculation.

A few paper-and-pencil testing batteries are available to evaluate number and calculation abilities in detail and to establish which specific components may be impaired (Delazer et al., 2003; Deloche et al., 1994, 1995, 1996). Another standardized test that evaluates mental calculation is the Graded Difficulty Arithmetic (GDA) Test, which comprises 12 timed, multi-digit addition and subtraction problems (Jackson and Warrington, 1986). See Table 44.3 for frequently used paper-and-pencil calculation tasks.

Error analyses, although usually not included in standardized tests, are very useful to corroborate the assessment as they can provide more detailed information about the nature of the impairment. Thus, when patients have problems in reading and writing numerals, a common analysis of errors is the distinction between lexical or syntactic errors. When patients have problems in the retrieval of arithmetical facts, a common analysis is the error classification proposed by McCloskey et al. (1991c) distinguishing between operand errors, operation errors, table errors, and non-table errors.

Table 44.3 Suggested numerical and calculation tasks for the diagnosis of acalculia

Function	Task	How?
Sequence	Counting	- Forwards or backwards from 1 to 20 or 20 to 1. - What comes next? Given a number, what comes before or after it?
Transcoding	Writing Reading	- Numbers to dictation, e.g. from spoken "two" to written TWO or 2. - Numbers from Arabic (2) into alphabetic format (TWO) or vice versa. - Numbers from Arabic (2) or alphabetic (TWO) format.
Quantity	Number comparison Comparison of non-symbolic discrete quantity Comparison of non-symbolic continuous quantity Number composition task	- Which is bigger: 6 or 9? Both accuracy and RTs to be collected. - Which display contains more dots? To be presented with either few fleshed dots or with many for unlimited time to test enumeration and counting respectively. - Space or "amount" discrimination, e.g. which line is longer or which container has more liquid? - Composing the value of a given number using poker chips ranging in value from 1 to 500.
Calculation	Arithmetical signs Simple facts Procedures Rules	- Read, point, and write the arithmetical signs. - Solve single-digit arithmetical problems mentally (e.g. 4 + 2, 3 × 4, or 5 − 2). - Solve multi-digit calculation (e.g. 294 + 12 = 306), either in written or oral modality. - Solve problems involving N × 0, N × 1, N + 1.

Natural Recovery of Acalculia

Relatively little is known about the prognosis of acalculia and only one study has investigated the natural recovery of acalculia without specific interventions in patients with left hemisphere vascular lesions (Caporali et al., 2000). This study indicates that some patients with severe calculation disorders completely recovered from acalculia in the first months post-stroke, even without specific rehabilitation (Caporali et al., 2000), and they continued to improve even several months after the stroke. However, it is not clear whether the recovery was specific for acalculia as most patients improved in other cognitive domains, such as language (Basso et al., 2005). This suggests that initial severity of acalculia may not predict the direction of recovery. Indeed, recovery might be influenced by the resolution of the diachisis in the first months after stroke (a period of up until 3–6 months), or by a more specialised functional reorganization that might have played a role several months post onset (>7 months) in these patients (Caporali et al., 2000).

Rehabilitation of Number Processing and Calculation Skills

The disabling consequences of acalculia underline the importance of developing remediation techniques and intervention for arithmetical disorders. So far, a very few studies have been devoted to the rehabilitation of numerical skills (for reviews, see Girelli & Seron, 2001; Lochy et al., 2004), and they concentrated on the rehabilitation of number transcoding deficits (e.g. Deloche et al., 1989), of arithmetical facts' retrieval (e.g. Domahs et al., 2003, 2007; Girelli et al., 1996, 2002; Hittmair-Delazer et al., 1994; Whetstone, 1998), and on the development of strategies for solving problems (Fasotti et al., 1992). A detailed assessment of a patient's number processing and calculation skills will constitute the basis for designing a suitable rehabilitation program (Girelli & Seron, 2001). For example, a patient may show a highly selective impairment in the retrieval of simple multiplication facts. However, he/she may still be able to perform simple addition and subtraction. The reacquisition of multiplication facts may then be facilitated by the use of back-up strategies based on addition (e.g. counting-on procedure: $3 \times 3 = 3 + 3 + 3$).

Rehabilitation of Transcoding Skills

Rehabilitating transcoding skills has focused on the ability to translate numerical stimuli into different codes (e.g. four → 4). As an example, an interesting rehabilitation program, implemented by Deloche et al. (1989), will be discussed. The authors treated a patient with selective difficulties in the production of written verbal numerals from Arabic numerals (e.g. 7001 → seven thousand zero one). This deficit was very severe as the patient showed a 45% error rate, making mostly syntactic errors (e.g. 114 → one hundred ten four). Deloche and colleagues' rehabilitation program consisted of re-teaching a set of explicit transcoding rules using a step-by-step procedure and several facilitation procedures (e.g. colour cues, vocabulary panels) to help the patient with his learning process. For example, using intensive and specific training, they rehabilitated the patient's ability to transcode a two-digit

Arabic numeral such as "*73*" into the corresponding written verbal numeral "*seventy-three*" by explicitly stating the rule "transcode the left digit by a ten name and the right digit by a unit name." During the training, a vocabulary panel with two coloured columns was placed in front of the patient, one panel containing the TENS names (ten, twenty, thirty … up to ninety), the other containing the UNIT names (one, two, three … up to nine). After 25 training sessions of 30 to 60 minutes each, the patient's performance was close to ceiling, and the improvement was maintained up to seven months post training.

Rehabilitation of Arithmetical Facts

Rehabilitating arithmetical fact retrieval aims primarily at re-teaching "lost" arithmetical knowledge. This can be achieved using either a *drill procedure* (i.e. extensive repetition of stimuli) or *conceptual training* (Domahs et al., 2003). In the first case, the underlying assumption is that practice may re-create and re-strengthen the lost associations between problem and answer. Thus, for example, two patients (TL and ZA) with a specific impairment of multiplication fact retrieval underwent twice-weekly training sessions over a period of eight weeks (Girelli et al., 1996). During the training sessions the arithmetical problems were presented in written form and simultaneously read aloud by the examiner. The patients were asked to answer either verbally whilst pointing to the number on a table or by writing the Arabic numeral. Errors were always corrected immediately. Patient TL relearned the answers as "labels" by reciting an operand table (e.g. $4 \times 3 = 4, 8, 12$). Patient ZA relearned the answers as serial addition (e.g. $4 \times 3 = 3 + 3 + 3 = 12$). At the end of the training, both patients improved considerably and showed a stable recovery as measured at one month follow-up. Strikingly, after treatment, their overall error rate dropped to 10% from pre-treatment error rates of 91% (TL) and 81% (ZA). Moreover, not only did the error rate decrease dramatically over the course of the remediation program for both patients, but the nature of the errors also changed, reflecting the different strategies used by the patients to relearn their facts: the recitation of Nx1 forward until the solution to the problem could be accessed (patient TL), and the repeated addition of the second operand (patient ZA). A similar training program for the rehabilitation of arithmetical fact impairment was adopted in other cases (e.g. Hittmair-Delazer et al., 1994; Whetstone, 1998). This conventional "drill" approach to the rehabilitation of arithmetical facts has recently been improved by presenting visual or auditory cues to the impaired problems which resulted in long-lasting improvements (Domahs et al., 2003, 2007).

Instead of continuous repetition of arithmetical facts, rehabilitation based on conceptual knowledge consists of inducing a meaningful reorganization of arithmetical facts according to the principles underlying them, e.g. the commutative law, or on decomposing them using known facts, e.g. 10s, ties, and 2s (Girelli et al., 2002). The use of back-up strategies based on the principles underlying arithmetical facts such as the order-irrelevant principle (e.g. $8 \times 6 = 6 \times 8 = 48$), of decomposition strategies (e.g. $4 \times 8 = (2 \times 8) + (2 \times 8) = 32$, or of repeated addition of the second operand ($3 \times 5 = 5 + 5 + 5 = 15$) seems important in rehabilitating arithmetical facts (Girelli and Seron, 2001).

Irrespective of the rehabilitation method adopted, a factor that contributes to improvement is minimising the opportunity to make mistakes, since repeated errors may strengthen the wrong associations between problems and answers (Girelli and Seron, 2001). Overall, these findings indicate that patients can significantly benefit from the training programs, and that they can often retain the improvement over time.

Conclusions

Acalculia is a heterogeneous disorder following cerebral damage. In the last three decades, the several single-case and group studies of impaired or spared number processing, calculation, or both have helped to highlight the multi-componential nature of the number system and how it can be fractionated. Moreover, these neuropsychological cases have contributed to the characterization of the number system relative to other cognitive systems, for instance language, memory, or general intelligence, from which number skills are at least partially independent. Acquired numerical impairments have also revealed the critical role of the parietal lobes in number and calculation skills, besides other brain areas such as frontal, temporal, and subcortical regions. More recent studies have expanded these findings, by showing that lesions to the parietal lobes may not necessarily lead to number impairments, possibly because of compensations with other brain regions, and that number impairments may occur following lesions to areas different from the parietal ones, suggesting that there is no one-to-one correspondence between number abilities and one specific brain region.

Acknowledgments

This work was supported by a Royal Society Dorothy Hodgkin fellowship. I thank Margarete Delazer and an anonymous reviewer for comments on an earlier version of this chapter.

References

Andres, M., Seron, X., & Olivier, E. (2005). Hemispheric lateralization of number comparison. *Cognitive Brain Research*, 25, 283–290.

Ardila, A., & Rosselli, M. (1994). Spatial acalculia. *International Journal of Neuroscience*, 78, 177–184.

Ashkenazi, S., Henik, A., Ifergane, G., &Shelef, I. (2008). Basic numerical processing in left intraparietal sulcus (IPS) acalculia. *Cortex*, 44, 439–448.

Basso, A., & Beschin, N. (2008). Number transcoding and number word spelling in a left-brain-damaged non-aphasic acalculic patient. *Neurocase*, 6, 129–139.

Basso, A., Burgio, F., & Caporali, A. (2000). Acalculia, aphasia and spatial disorders in left and right brain-damaged patients. *Cortex*, 36, 265–280.

Basso, A., Caporali, A., & Faglioni, P. (2005). Spontaneous recovery from acalculia. *Journal of the International Neuropsychological Society*, 11, 99–107.

Basso, A., Nichelli, P., & Frassinetti, F. (1996). Time perception in a neglected space. *Neuroreport*, 7, 2111–2114.

Benson, D.F., & Weir, W.F. (1972). Acalculia: Acquired anarithmetia. *Cortex*, 8, 465–472.

Brannon, E.M., Lutz, D., & Cordes, S. (2006). The development of area discrimination and its implications for number representation in infancy. *Developmental Science*, 9, F59–F64.

Brannon, E.M., Suanda, S., & Libertus, K. (2007). Temporal discrimination increases in precision over development and parallels the Magnitude Representation and Infants' development of numerosity discrimination. *Developmental Science*, 10, 770–777.

Brannon, E.M., Libertus, M.E., Meck, W.H., & Woldorff, M.G. (2008). Electrophysiological measures of time processing in infant and adult brains: Weber's Law holds. *Journal of Cognitive Neuroscience*, 20, 193–203.

Bonato, M., Priftis, K., Marenzi, R., & Zorzi, M. (2008). Modulation of hemispatial neglect by directional and numerical cues in the line bisection task. *Neuropsychologia*, 46, 426–433.

Butterworth, B. (1999). *The Mathematical Brain*. London: Macmillan.

Butterworth, B., Cipolotti, L., &Warrington, E.K. (1995). Short-term memory impairments and arithmetical ability. *Quarterly Journal of Experimental Psychology*, 49A, 251–262.

Cantlon, J., Platt, F., & Brannon, E.M. (2009). Beyond the number domain. *Trends in Cognitive Sciences*, 13, 83–91.

Caporali, A., Burgio, F., & Basso, A. (2000). The natural course of acalculia in left-brain-damaged patients. *Neurological Science*, 21, 143–149.

Cappelletti, M., & Cipolotti, L. (2006). Unconscious processing of Arabic numerals in unilateral neglect. *Neuropsychologia*, 44, 1999–2006.

Cappelletti, M., & Cipolotti, L. (2010). The neuropsychological assessment and treatment of calculation disorders. In P. Halligan et al. (Eds), *Handbook of Clinical Neuropsychology*. Oxford University Press, Oxford, UK.

Cappelletti, M., Butterworth, B., & Kopelman, M. (2001). Spared numerical abilities in a case of semantic dementia. *Neuropsychologia*, 39, 1224–1239.

Cappelletti, M., Butterworth, B., & Kopelman, M. (2006). The understanding of quantifiers in semantic dementia: A single-case study. *Neurocase* 12:136–145.

Cappelletti, M., Morton, J., Kopelman, M., & Butterworth, B. (2005). The progressive loss of numerical knowledge in a semantic dementia patient: A follow-up study. *Cognitive Neuropsychology*, 22, 771–793.

Cappelletti, M., Freeman, E.D., & Cipolotti, L. (2007a). The middle house or the middle floor: Bisecting horizontal and vertical mental number lines in neglect. *Neuropsychologia*, 45, 2989–3000.

Cappelletti, M., Barth, H., Fregni, F., Pascual Leone, A., & Spelke, E. (2007b). rTMS over the left and the right intraparietal sulcus disrupts discrete and continuous quantity processing. *Experimental Brain Research*, 179, 631–642.

Cappelletti, M., Jansari, A., Kopelman, M., & Butterworth, B. (2008). A case of selective impairment of encyclopaedic numerical knowledge or "*when December 25th is no longer Christmas day, but '20 + 5' is still 25.*" *Cortex*, 3, 325–336.

Cappelletti, M., Freeman, E., & Cipolotti, L. (2009a). Interactions and dissociations between time, numerosity and space processing. *Neuropsychologia*, 47, 2732–2748.

Cappelletti, M., Muggleton, N., & Walsh, V. (2009b). Quantity without numbers and numbers without quantity in the parietal cortex. *NeuroImage*, 46, 522–529.

Cappelletti, M., Lee, H.L., Freeman, E.D., & Price, C.J. (2010). The role of the left and right parietal lobe in the conceptual processing of numbers. *Journal of Cognitive Neuroscience*, 22, 331–346.

Cappelletti, M., Freeman, E., & Cipolotti, L. (2011). Number and time doubly dissociate. *Neuropsychologia*, 49, 3078–3092.

Cappelletti, M., Butterworth, B., & Kopelman, M. (2012a). Numerical abilities in patients with focal and progressive neurological disorders: A neuropsychological study. *Neuropsychology*, 26, 1–19.

Cappelletti, M., Leff, A., & Price, C. (2012b). How number processing survives left occipito-temporal damage. *Neurocase*, 18, 271–285.

Cipolotti, L. (1995). Multiple routes for reading words, why not numbers? Evidence from a case of Arabic numeral dyslexia. *Cognitive Neuropsychology*, 12, 313–342.

Cipolotti, L., & Butterworth, B. (1995). Toward a multiroute model of number processing: Impaired number transcoding with preserved calculation skills. *Journal of Experimental Psychology: General*, 124, 375–390.

Cipolotti, L., & De Lacy Costello, A. (1995). Selective impairment for simple division. *Cortex*, 31, 433–449.

Cipolotti, L., Butterworth, B., & Denes, G. (1991). A specific deficit for numbers in case of dense acalculia. *Brain*, 114, 2619–2637.

Cipolotti, L., Warrington, E., & Butterworth, B. (1994). From "one thousand nine hundred and forty-five" to 1000,945. *Neuropsychologia*, 32, 503–509.

Cipolotti, L., Warrington, E., & Butterworth, B. (1995). Selective impairment in manipulating Arabic numerals. *Cortex*, 31, 73–86.

Cipolotti, L., & van Harskamp, N. (2001). Disturbances of number processing and calculation. In *Handbook of Neuropsychology*, R.S. Bernt (Ed.), 2nd edition, 3, 305–331, Elsevier Science.

Cohen, L., & Dehaene, S. (1991). Neglect dyslexia for numbers? A case report. *Cognitive Neuropsychology*, 8, 39–58.

Cohen, L., & Dehaene, S. (1994). Amnesia for arithmetic facts: A single case study. *Brain and Language*, 47, 214–232.

Cohen, L., & Dehaene, S. (1995). Number processing in pure alexia: The effect of hemispheric asymmetries and task demands. *Neurocase*, 1, 121–137.

Cohen, L., Dehaene, S., & Verstichel, P. (1994). Number words and number non-words: A case of deep dyslexia extending to Arabic numerals. *Brain*, 117, 267–279.

Cohen, L., & Dehaene, S. (1996). Cerebral networks for number processing: Evidence from a case of posterior callosal lesion. *Neurocase*, 2, 155–174.

Cohen, L., Verstichel, P., & Dehaene, S. (1997). Neologistic jargon sparing numbers: A category-specific phonological impairment. *Cognitive Neuropsychology*, 14, 1029–1061.

Cohen, L., & Dehaene, S. (2000). Calculating without reading: Unsuspected residual abilities in pure alexia. *Cognitive Neuropsychology*, 17, 563–583.

Cohen, L., Dehaene, S., Chochon, F., Lehéricy, S., & Naccache, L. (2000). Language and calculation within the parietal lobe: A combined cognitive, anatomical and fMRI study. *Neurospychologia*, 138, 1426–1440.

Cohen, L., Wilson, A.J., Izard, V., & Dehaene, S. (2007). Acalculia and Gerstmann's Syndrome. In O. Godefroy & J. Bogousslavsky (Eds), *Cognitive and Behavioral Neurology of Stroke* (pp. 125–147). Cambridge, UK: Cambridge University Press.

Cohen Kadosh, R., & Walsh, V. (2009). Numerical representation in the parietal lobes: Abstract or not abstract? *Behavioral and Brain Sciences*, 32(3–4), 313–328.

Cohen Kadosh, R., Lammertyn, J., & Izard, V. (2008). Are numbers special? An overview of chronometric, neuroimaging, developmental and comparative studies of magnitude representation. *Progress in Neurobiology*, 84(2), 132–147.

Crutch, S., & Warrington, E.K. (2002). Preserved calculation skills in a case of semantic dementia. *Cortex*, 38, 389–399.

Dagenbach, D., & McCloskey, M. (1992). The organisation of arithmetical facts in memory: Evidence from a brain-damaged patient. *Brain and Cognition*, 20, 345–366.

Dehaene, S. (1997). *The Number Sense: How the Mind Creates Mathematics*. New York: Oxford University Press.

Dehaene, S., & Cohen, L. (1991). Two mental calculation systems: A case study of severe acalculia with preserved approximation. *Neuropsychologia*, 29, 1045–1074.

Dehaene, S., & Cohen, L. (1995). Towards an anatomical and functional model of number processing. *Mathematical Cognition*, 1, 83–120.

Dehaene, S., & Cohen, L. (1997). Cerebral pathways for calculation: Double dissociation between rote verbal and quantitative knowledge of arithmetic. *Cortex*, 33, 219–250.

Dehaene, S., Bossini, S., & Giraux, P. (1993). The mental representation of parity and number magnitude. *Journal of Experimental Psychology: Human, Memory, and Cognition*, 21, 314–326.

Dehaene, S., Spelke, E., Pinel, P., Stanescu, R., & Tsivkin, S. (1999). Sources of mathematical thinking: Behavioral and brain-imaging evidence. *Science*, 284, 970–974.

Dehaene, S., Piazza, M., Pinel, P., & Cohen, L. (2003). Three parietal circuits for number processing. *Cognitive Neuropsychology*, 20, 487–506.

Delazer, M., & Benke, T. (1997). Arithmetic facts without meaning. *Cortex*, 33, 697–710.

Delazer, M., & Butterworth, B. (1997). A dissociation of number meanings. *Cognitive Neuropsychology*, 14, 613–636.

Delazer, M., & Girelli, L. (1997). When 'Alfa Romeo' facilitate 164: Semantic effects in verbal number production. *Neurocase*, 3, 461–475.

Delazer, M., Girelli, L., & Benke, T. (1999a). Arithmetic reasoning and implicit memory: A neuropsychological study on amnesia. *Cortex*, 35, 615–627.

Delazer, M., Girelli, L., Semenza, C., & Denes, G. (1999b). Numerical skills and aphasia. *Journal of the International Neuropsychological Society*, 5, 213–221.

Delazer, M., Lochy, A., Jenner, C., Domahs, F., & Benke, T. (2002). When writing 0 (zero) is easier than writing O (o): A neuropsychological case study of agraphia. *Neuropsychologia*, 40, 2167–2177.

Delazer, M., Girelli, L., Grana, A., & Domahs, F. (2003). Number processing and calculation-normative data from healthy adults. *Clinical Neuropsychology*, 17, 331–350.

Delazer, M., Domahs, F., Lochy, A., Karner, E., Benke, T., & Poewe, W. (2004). Number processing and basal ganglia dysfunction: A single case study. *Neuropsychologia*, 42, 1050–1062.

Delazer, M., Benke, T., Trieb, T., Schocke, M., & Ischebeck, A. (2006). Isolated numerical skills in posterior cortical atrophy—an fMRI study. *Neuropsychologia*, 4, 1909–1913.

Della Sala, S., Gentileschi, V., Gray, C., & Spinnler, H. (2000). Intrusion errors in numerical transcoding by Alzheimer patients. *Neuropsychologia*, 38, 768–777.

Deloche, G., & Seron, X. (1982). From one to 1: An analysis of a transcoding process by means of neuropsychological data. *Cognition*, 12, 119–149.

Deloche, G., Seron, X., & Ferrand, I. (1989). Re-education of number transcoding mechanisms: A procedural approach. In X. Seron & S. Deloche (Eds), *Cognitive Approaches in Neuropsychological Rehabilitation* (pp. 249–287). Hillsdale, NJ: Lawrence Erlbaum.

Deloche, G., Seron, X., Larroque, C., et al. (1994). Calculation and number processing: Assessment battery; role of demographic factors. *Journal of Clinical and Experimental Neuropsychology*, 16, 195–208.

Deloche, G., Hannequin, D., Carlomagno, S., et al. (1995). Calculation and number processing in mild Alzheimer's disease. *Journal of Clinical and Experimental Neuropsychology*, 17, 634–639.

Deloche, G., Dellatolas, G., Vendrell, J., & Bergego, C. (1996). Calculation and number processing: Neuropsychological assessment and daily life difficulties. *Journal of the International Neuropsychological Society*, 2, 177–180.

Denes, G., & Signorini, M. (2001). Door but not four and 4. A category specific transcoding deficit in a pure acalculic patient. *Cortex*, 37, 267–277.

Diesfeldt, H.F.A. (1993). Progressive decline of semantic memory with preservation of number processing and calculation. *Behavioural Neurology*, 6, 239–242.

Domahs, F., Bartha, L., & Delazer, M. (2003). Rehabilitation of arithmetic abilities: Different intervention strategies for multiplication. *Brain and Language*, 87, 165–166.

Domahs, F., Bartha, L., Lochy, A., Benke, T., & Delazer, M. (2006). Number words are special: Evidence from a case of primary progressive aphasia. *Journal of Neurolinguistics*, 19, 1–3.

Domahs, F., Zamarian, L., & Delazer, M. (2007). Sound arithmetic: Auditory cues in the rehabilitation of impaired fact retrieval. *Neuropsychological Rehabilitation*, 1, 49–64.

Doricchi, F., Guariglia, P., Gasparini, M., & Tomaiuolo, F. (2005). Dissociation between physical and mental number line bisection in right hemisphere brain damage. *Nature Neuroscience*, 8, 1663–1666.

Fasotti, L., Bremer, J., & Eling, P. (1992). Influence of improved test encoding on arithmetical word problem solving after frontal lobe damage. *Neuropsychological Rehabilitation*, 2, 3–20.

Feigenson, L., Dehaene, S., & Spelke, E. (2004). Core systems of number. *Trends in Cognitive Science*, 8, 307–314.

Ferro, J.M., & Botelho, M. (1980). Alexia for arithmetical signs: A cause of disturbed calculation. *Cortex*, 16, 175–180.

Fias, W., Lammertyn, J., Reynvoert, B., Dupont, P., & Orban, G.A. (2003). Parietal representation of symbolic and nonsymbolic magnitude. *Journal of Cognitive Research*, 15, 47–56.

Furumoto, H. (2006). Pure misallocation of "0" in number transcoding: A new symptom of right cerebral dysfunction. *Brain and Cognition*, 60, 128–138.Garcia Orza, J., León-Carrlon, J., & Vega, O. (2003). Dissociating Arabic numeral reading and basic calculation: A case study. *Neurocase*, 9, 129–139.

Girelli, L., & Delazer, M. (1996). Subtraction bugs in an acalculic patient. *Cortex*, 32, 547–555.

Girelli, L., & Seron, X. (2001). Rehabilitation of number processing and calculation skills. *Aphasiology*, 15, 695–712.

Girelli, L., Delazer, M., Semenza, C., & Denes, G. (1996). The representation of arithmetical facts: Evidence from two rehabilitation studies. *Cortex*, 32, 49–66.

Girelli, L., Bartha, L., & Delazer, M. (2002). Strategic learning in the rehabilitation of semantic knowledge. *Neuropsychological Rehabilitation*, 12, 41–61.

Gordon, P. (2004). Numerical cognition without words: Evidence from Amazonia. *Science*, 306, 496–499

Grafman, J., Kampen, D., Rosenberg, J., Salazar, A., & Boller, F. (1989). Calculation abilities in a patient with a virtual left hemispherectomy. *Behavioural Neurology*, 2, 183–194.

Granà, A., Girelli, L., Gattinoni, F., & Semenza, C. (2001). Letter and number writing in agraphia: A single-case study. *Brain and Cognition*. 46, 149–153.

Granà, A., Hofer, R., & Semenza, C. (2006). Acalculia from a right hemisphere lesion. Dealing with "where" in multiplication procedures. *Neuropsychologia*, 44, 2972–2986.

Halpern, C.H., Glosser, G., Clark, R., et al. (2003). Dissociation of numbers and objects in corticobasal degeneration and semantic dementia. *Neurology*, 62, 1163–1169.

Halpern, C.H., Clark, R., Moore, P., Cross, K., & Grossman, M. (2007). Too much to count on: Impaired very small numbers in corticobasal degeneration. *Brain and Cognition*, 64, 144–149.

Henschen, S.E. (1919). Uber Sprach-Musik-und Rechenmechanismen und ihre Lokalisationen im Grosshirn. *Zeitschrift fur die gesamte Neurologie und Psychiatrie*, 52, 273–298.

Hittmair-Delazer, M., Semenza, C., & Denes, G. (1994). Concepts and facts in calculation. *Brain*, 117, 715–728.

Hittmair-Delazer, M., Sailer, U., & Benke, T. (1995). Impaired arithmetic facts but intact conceptual knowledge—A single case study of dyscalculia. *Cortex*, 31, 139–148.

Hoeckner, S.H., Moeller, K., Zauner, H., et al. (2008). Impairments of the mental number line for two-digit numbers in neglect. *Cortex*, 44(4), 429–438.

Jackson, M., & Warrington, E.K. (1986). Arithmetic skills in patients with unilateral cerebral lesions. *Cortex*, 22, 611–620.

Jefferies, E., Bateman, D., & Lambon-Ralph, M.A. (2005). The role of the temporal lobe semantic system in number knowledge: Evidence from late-stage semantic dementia. *Neuropsychologia*, 43, 887–905.

Jefferies, E., Patterson, K., Jones, R.W., Bateman, D., & Lambon-Ralph, M.A. (2004). A category-specific advantage for numbers in verbal short-term memory: Evidence from semantic dementia. *Neuropsychologia*, 42, 639–660.

Julien, C.L., Neary, D., & Snowden, J.S. (2010). Personal experience and arithmetic meaning in semantic dementia. *Neuropsychologia*, 48, 278–287.

Julien, C.L., Thompson, J.C., Neary, D., & Snowden, J.S. (2008). Arithmetic knowledge in semantic dementia: Is it invariably preserved? *Neuropsychologia*, 46, 2732–2744.

Koss, S., Clark, R., Vesely, L., et al. (2010). Numerosity impairment in corticobasal syndrome. *Neuropsychology*, 24, 476–492.

Laiacona, M., & Lunghi, A. (1997). A case of concomitant impairment of operational signs and punctuation marks. *Neuropsychologia*, 35, 325–332.

Lampl, Y., Eshel, Y., Gilad, R., & Sarova-Pinhas, I. (1994). Selective acalculia with sparing of the subtraction process in a patient with left parieto-temporal haemorrhage. *Neurology*, 44, 1759–1761.

Langdon, D.W., & Warrington, E.K. (1997). The abstraction of numerical relations: A role for the right hemisphere in arithmetic? *Journal of the International Neuropsychological Society*, 3, 260–268.

Lemer, C., Dehaene, S., Spelke, E., & Cohen, L. (2003). Approximate quantities and exact number words: Dissociable systems. *Neuropsychologia*, 41, 1942–1958.

Lochy, A., Domahs, F., & Delazer, M. (2004). Rehabilitation of acquired calculation and number processing disorders. In J. Campbell (Ed.), *The Handbook of Mathematical Cognition*, New York, Psychology Press.

Lucchelli, F., & De Renzi, E. (1993). Primary dyscalculia after a medial frontal lesion of the left hemisphere. *Journal of Neurology, Neurosurgery, and Psychiatry*, 56, 304–307.

Macaruso, P., McCloskey, M., & Aliminosa, D. (1993). The functional architecture of the cognitive numerical-processing system: Evidence from a patient with multiple impairments. *Cognitive Neuropsychology*, 10, 341–376.

Macoir, J., Audet, T., & Breton, M.F. (1999). Code-dependent pathways for number transcoding: Evidence from a case of selective impairment in written verbal numeral to Arabic transcoding. *Cortex*, 35, 629–645.

Macoir, J., Audet, T., Lecomte, S., & Delisle, J. (2002). From "Cinquante-Six" to "5quante-Six": The origin of intrusion errors in a patient with probable Alzheimer disease. *Cognitive Neuropsychology*, 19, 579–601.

Marangolo, P., Nasti, M., & Zorzi, M. (2004). Selective impairment for reading numbers and number words: A single case study. *Neuropsychologia*, 42, 997–1006.

Marangolo, P., Piras, F., & Fias, W. (2005). "I can write seven but I can't say it": A case of domain-specific phonological output deficit for numbers. *Neuropsychologia*, 43, 177–1188.

McCloskey, M., Caramazza, A., & Basili, A. (1985). Cognitive mechanisms in number processing and calculation: Evidence from dyscalculia. *Brain and Cognition*, 4, 171–196.

McCloskey, M., Sokol, S.M, & Goodman, R.A. (1986). Cognitive processes in verbal-number production: Inferences from the performance of brain-damaged subjects. *Journal of Experimental Psychology General*, 115, 307–330.

McCloskey, M., Sokol, S.M., Goodman-Schulman, R.A., & Caramazza, A. (1990). Cognitive representations and processes in number production: Evidence from cases of acquired dyscalculia. In A. Caramazza (Ed.), *Advances in Cognitive Neuropsychology and Neurolinguistics* (pp. 1–32). Hillsdale, NJ: Lawrence Erlbaum.

McCloskey, M., Aliminosa, D., & Macaruso, P. (1991a). Theory-based assessment of acquired dyscalculia. *Brain and Cognition*, 17, 285–308.

McCloskey, M., Aliminosa, D., & Sokol, S.M. (1991b). Facts, rules and procedures in normal calculation: Evidence from multiple single-patient studies of impaired arithmetic fact retrieval. *Brain and Cognition*, 17, 154–203.

McCloskey, M., Harley, W., & Sokol, S.M. (1991c). Models of arithmetic fact retrieval: An evaluation in light of findings from normal and brain-damaged subjects. *Journal of Experimental Psychology: Learning, Memory and Cognition*, 17, 377–397.

McNeil, J., & Warrington, E.K. (1994). A dissociation between addition and subtraction with written calculation. *Neuropsychologia*, 32, 717–728.

McNeil, J.E., & Burgess, P.W. (2002). The selective impairment of arithmetical procedures. *Cortex*, 38, 569–587.

Meck, W.H., & Church, R.M. (1983). A mode control model of counting and timing processes. *Journal of Experimental Psychology: Animal Behaviour Processes*, 9, 320–334.

Mennemeier, M., Pierce, C.A., Chatterjee, A., et al. (2005). Biases in attentional orientation and magnitude estimation explain crossover: Neglect is a disorder of both. *Journal of Cognitive Neuroscience*, 17, 1194–1211.

Moyer, R.S., & Landauer, T.K. (1967). Time required for judgements of numerical inequality. *Nature*, 215, 1519–1520.

Nieder, A. (2005). Counting on neurons: The neurobiology of numerical competence. *Nature Neuroscience Review*, 6, 177–190.

Noël, M.P., & Seron, X. (1993). Numeral number reading deficit: A single-case study or when 236 is read (2306) and judged superior to 1258. *Cognitive Neuropsychology*, 10, 317–339.

Noël, M.P., & Seron, X. (1995). Lexicalization errors in writing Arabic numerals. *Brain and Cognition*, 29, 151–179.

Pesenti, M., Seron, X., & Van Der Linden, M. (1994). Selective impairment as evidence for mental organisation of arithmetical facts: BB, a case of preserved subtraction? *Cortex*, 30, 661–671.

Pesenti, M., Depoorter, N., & Seron, X. (2000a). Noncommutability of the N + 0 arithmetical rule: A case study of dissociated impairment. *Cortex*, 36, 445–454.

Pesenti, M., Thioux, M., Samson, D., Bruyer, R., & Seron, X. (2000b). Number processing and calculation in a case of visual agnosia. *Cortex*, 36, 377–400.

Pesenti, M., Thioux, M., Seron, X., & De Volder, A. (2000c). Neuroanatomical substrates of Arabic number processing, numerical comparison, and simple addition: A PET study. *Journal of Cognitive Neuroscience*, 12, 461–479.

Pia, L., Latini Corazzini, L., Folegatti, A., Gindri, P., & Cauda, F. (2009). Mental number line disruption in a right-neglect patient after a left-hemisphere stroke. *Brain and Cognition*, 69, 81–88.

Pica, P., Lemer, C., & Izard, V. (2004). Exact and approximate calculation in an Amazonian indigene group with a reduced number lexicon. *Science*, 306, 499–503.

Polk, T., Reed, C., Keenan, J., Hogard, P., & Anderson, C.A. (2001). A dissociation between symbolic number knowledge and analogue magnitude information. *Brain and Cognition*, 47, 545–563.

Price, C.J., & Friston, K.J. (2002). Degeneracy and cognitive anatomy. *Trends in Cognitive Science*, 6, 416–421.

Priftis, K., Piccione, F., Giorgi, F., Meneghello, F., Umiltà, C., & Zorzi, M. (2008). Lost in number space after right brain damage: A neural signature of representational neglect. *Cortex*, 44, 449–453.

Remond-Besuchet, C., Noël, M.P., Seron, X., Thioux, M., Brun, M., & Aspe, X. (1999). Selective preservation of exceptional arithmetical knowledge in a demented patient. *Mathematical Cognition*, 5, 41–63.

Revkin, S.K., Piazza, M., Izard, V., Zamarian, L., Karner, E., & Delazer, M. (2008). Verbal numerosity estimation deficit in the context of spared semantic representation of numbers, a neuropsychological study of a patient with frontal lesions. *Neuropsychologia*, 46, 2463–2475.

Rosselli, M., & Ardila, A. (1989). Calculation deficits in patients with right and left hemisphere damage. *Neuropsychologia*, 27, 607–617.

Rossor, M.N., Warrington, E.K., & Cipolotti, L. (1995). The isolation of calculation skills. *Journal of Neurology*, 242, 78–81.

Rusconi, E., Priftis, K., Rusconi, M.L., & Umiltà, C. (2006). Arithmetic priming from neglected numbers. *Cognitive Neuropsychology*, 23, 227–239.

Sackur, J., Naccache, L., Pradat-Diehl, P., et al. (2008). Semantic processing of neglected numbers. *Cortex*, 44, 673–682.

Sandrini, M., Miozzo, A., Cotelli, M., & Cappa, S.F. (2003). The residual calculation abilities of a patient with severe aphasia: Evidence for a selective deficit of subtraction procedure. *Cortex*, 39, 85–96.

Santens, S., & Verguts, T. (2011). The size congruity effect: Is bigger always more? *Cognition*, 118, 97–113.

Sathian, K., Simon, T.J., Peterson, S., Patel, G.A., Hoffman, J.M., & Grafton, S.T. (1999). Neural evidence linking visual object enumeration and attention. *Journal of Cognitive Neuroscience*, 11, 36–51.

Semenza, C., Miceli, L., & Girelli, L. (1997). A deficit for arithmetical procedures: Lack of knowledge or lack of monitoring? *Cortex*, 33, 483–498.

Seron, X., & Deloche, (1984). From 2 to two: An analysis of a transcoding process by means of neuropsychological evidence. *Journal of Psycholinguistic Research*, **13**, 215–235.

Sokol, S.M., & McCloskey, M. (1988). Levels of representation in verbal number production. *Applied Psycholinguistics*, 9, 267–281.

Sokol, S.M., McCloskey, M., Cohen, N.J., & Aliminosa, D. (1991). Cognitive representations and processes in arithmetic: Inferences from the performance of brain-damaged subjects. *Journal of Experimental Psychology: Learning, Memory and Cognition*, 17, 355–376.

Tegner, R., & Nyback, H. (1990). "To hundred and twenty4our": A study of transcoding in dementia. *Acta Neurologica Scandinava*, 81, 177–178.

Thioux, M., Pillon, A., Samson, D., de Partz, M.P., Noël, M.P., & Seron, X. (1998). The isolation of numerals at the semantic level. *Neurocase*, 4, 371–389.

Thioux, M., Ivanoiu, A., Turconi, E., & Seron, X. (1999). Intrusion of the verbal code during the production of Arabic numerals: A single-case study in patient with probable Alzheimer disease. *Cognitive Neuropsychology*, 16, 749–773.

van Dijck, J.P., Gevers, W., Lafosse, C., Doricchi, F., & Fias, W. (2011). Non-spatial neglect for the mental number line. *Neuropsychologia*, 49, 2570–2583.

van Harskamp, N.J., & Cipolotti, L. (2001). Selective impairments in addition, subtraction and multiplication: Implications for the organisation of arithmetical facts. *Cortex*, 37, 363–388.

van Harskamp, N.J., Rudge, P., & Cipolotti, L. (2002). Are multiplication facts implemented by the left supramarginal and angular gyri? *Neuropsychologia*, 40, 1786–1793.

van Harskamp, N.J., Rudge, P., & Cipolotti, L. (2005). Does the left inferior parietal lobule contribute to multiplication facts? *Cortex*, 41, 742–752.

Varley, R.A., Klessinger, N.J.C., Romanowski, C.A.J., & Siegal, M. (2005). Agrammatic but numerate. *Proceedings of the National Academy of Sciences of the United States of America*, 102, 3519–3524.

Venneri, A., & Semenza, C. (2011). On the dependency of division on multiplication: Selective loss for conceptual knowledge of multiplication. *Neuropsychologia*, 49, 3629–3635.

Vuilleumier, P., & Rafal, R. (1999). "Both" means more than "two", localizing and counting in patients with visuospatial neglect. *Nature Neuroscience*, 2, 783–784.

Vuilleumier, P., Ortigue, S., & Brugger, P. (2004). The number space and neglect. *Cortex*, 40, 399–410.

Walsh, V. (2003). A theory of magnitude: Common cortical metrics of time, space and quantity. *Trends in Cognitive Sciences*, 7, 483–488.

Warrington, E.K. (1982). The fractionation of arithmetical skills: A single study. *Quarterly Journal of Experimental Psychology*, 34, 31–51.

Whalen, J., McCloskey, M., Lindemann, M., & Bouton, G. (2002). Representing arithmetic table facts in memory: Evidence from acquired impairments. *Cognitive Neuropsychology*, 19, 505–522.

Whetstone, T. (1998). The representation of arithmetic facts in memory: Results from retraining a brain-damaged patient. *Brain and Cognition*, 36, 290–309.

Woods, A.J., Mennemeier, M., Garcia-Rill, E., et al. (2006). Bias in magnitude estimation following left hemisphere injury. *Neuropsychologia*, 44, 1406–1412.

Zamarian, L., Karner, E., Benke, T., Donnemiller, E., & Delazer, M. (2006a). Knowing 7 x 8, but not the meaning of "elephant": Evidence for the dissociation between numerical and non-numerical semantic knowledge. *Neuropsychologia*, 44, 1708–1723.

Zamarian, L., Visani, P., Delazer, M., et al. (2006b). Parkinson's disease and arithmetics: The role of executive functions. *Journal of Neurological Sciences*, 25, 124–130.

Zorzi, M., Priftis, K., & Umiltà, C. (2002). Brain damage: Neglect disrupts the mental number line. *Nature*, 417, 138–139.

CHAPTER 45

ARITHMETIC LEARNING IN ADULTS

evidence from brain imaging

LAURA ZAMARIAN AND MARGARETE DELAZER

It has become well accepted that learning may lead to substantial changes in the functional and structural organization of the human brain. Learning may evolve after personal experience, informal practice, or targeted and repeated training. Here, we review recent neuroimaging studies on systematic arithmetic training in adults (for a review on neuroimaging studies with adults, see also Zamarian, Ischebeck, & Delazer, 2009b; for a review on neuroimaging studies with children, see Menon, 2010). Most of these studies are based on the grounds of cognitive psychology and on paradigms that are extensively used in learning experiments (e.g. Anderson, Fincham, & Douglass, 1999; Logan, 1988; Logan & Klapp, 1991; Rickard, 2004; Wenger, 1999). Modern neuroimaging methods offer a new tool to track the neural correlates underlying the acquisition of new expertise, as well as fascinating insights into the plasticity of the human brain.

In this article, we first summarize the neuropsychological findings about the cognitive architecture of arithmetic abilities and the cerebral networks supporting calculation. Then, we summarize the neuroimaging studies on the acquisition of new arithmetic expertise in healthy adults, as well as in a patient with acquired brain damage. Finally, we discuss the evidence about the relationship between arithmetic expertise, cerebral activation patterns, and brain structure (see Table 45.1 for a summary of the main results of learning studies).

Calculation in Adults

Over the past four decades, cognitive neuropsychological studies on adults have provided compelling evidence that arithmetic expertise requires the interplay of different types of knowledge and that number processing (e.g. reading number words or writing Arabic numerals) and calculation are subserved by different cognitive systems (for reviews, Dehaene, Piazza, Pinel, & Cohen, 2003; Domahs & Delazer, 2005; Zamarian, López Rolón, & Delazer, 2007). With regard to calculation processes, it has been shown that arithmetic fact

Table 45.1 Summary of the main results of learning studies

	Method	Behavioural results	Brain imaging results
Learning studies on normal users of calculation		(Behavioural results following training)	
Simple arithmetic			
Núñez-Peña (2008)	ERPs	Accuracy and speed improvements	Decrease of late positive slow wave with practice
		Reduction of problem-size effect	Increase of late positive slow wave with problem size
Pauli et al. (1994)	ERPs	Speed improvements	Reduction in frontocentral positivity
		Improvements greater for problems with larger operands than for problems with smaller operands	
Complex arithmetic			
Delazer et al. (2003)	fMRI	Accuracy and speed improvements with trained multiplication	Trained versus untrained multiplication: relatively stronger left AG activation
			Untrained versus trained multiplication: stronger IFG and IPS activation
Delazer et al. (2005)	fMRI	Comparable accuracy and speed improvements between drill and strategy conditions	Trained drill versus trained strategy: relatively stronger left AG activation
Grabner et al. (2009)	fMRI	Comparable accuracy and speed improvements between arithmetic and non-arithmetic conditions	Trained versus untrained multiplication: relatively stronger left AG activation
			Trained versus untrained figural-spatial problems: relatively stronger right AG activation
Ischebeck et al. (2006)	fMRI	Comparable accuracy and speed improvements between multiplication and subtraction	Trained versus untrained multiplication: relatively stronger left AG activation
			Trained versus untrained subtraction: no activation difference

(Continued)

Table 45.1 Continued

	Method	Behavioural results	Brain imaging results
Ischebeck et al. (2007)	fMRI	Accuracy and speed improvements for repeated problems relative to novel problems	Repeated versus novel problems: activation decrease in frontoparietal areas, activation increase in temporoparietal areas including left AG from the 8th repetition
Ischebeck et al. (2009)	fMRI	Accuracy and speed improvements with trained multiplication	Trained versus untrained multiplication: relatively stronger left AG activation
		Speed improvement with related division	Related versus unrelated division: relatively stronger left AG activation
			Positive correlation between behavioural and functional transfer effects
Studies on exceptional expertise			
Fehr et al. (2010)	fMRI	Greater accuracy and speed for CP relative to non-experts in simple and complex calculation	Complex versus simple calculation: bilateral activation of mid-frontal and posterior brain areas, largely similar in CP and non-experts
			Exponentiation versus complex/simple calculation for CP only: bilateral activation of mid-frontal and inferior parietal areas
Hanakawa et al. (2003)	fMRI	Abacus experts outperformed non-experts in one-, three-, and six-digit addition	Abacus experts versus non-experts: stronger activation of posterior superior parietal cortex/ precuneus
			Non-experts versus abacus experts: stronger activation of prefrontal and lateral parietal cortex
Pesenti et al. (2001)	PET		GR versus non-experts: increased activation of right frontal and parahippocampal areas
Wu et al. (2009b)	PET	Abacus experts outperformed non-experts in simple and complex addition	Non-experts versus experts: increased activation in frontal areas during complex calculation
			Experts versus non-experts: increased activation in left inferior parietal lobule during complex calculation

(Continued)

Table 45.1 Continued

	Method	Behavioural results	Brain imaging results
Studies on structural brain changes			
Aydin et al. (2007)	VBM		Mathematicians versus non-mathematicians: higher gray matter density in bilateral inferior parietal lobule and left IFG
			Positive correlation between time spent as mathematician and relative gray matter density increase in right parietal lobule
Takeuchi et al. (2011)	VBM	Performance improvements on verbal letter span, reverse Stroop interference, and complex arithmetic tasks for adaptive calculation training group relative to control groups	Gray matter density decreases in bilateral frontoparietal regions and left superior temporal region for adaptive calculation training group. Opposite tendency or no difference for controls
Study on rehabilitation			
Zaunmüller et al. (2008) (patient with left-sided brain lesion)	fMRI	Accuracy and speed improvements with trained multiplication following training	Trained versus untrained multiplication: relatively stronger right AG activation

knowledge (knowing that '2 + 2 = 4'), procedural knowledge (knowing how to multiply '34 × 67'), and the recognition of arithmetic signs (e.g. McCloskey, 1992; McCloskey, Caramazza, & Basili, 1985) are distinct components of arithmetic expertise. Conceptual knowledge, which entails a basic understanding of operations and arithmetic principles, is also an essential and independent component of meaningful and efficient arithmetic (e.g. Hittmair-Delazer, Sailer, & Benke, 1995; Hittmair-Delazer, Semenza, & Denes, 1994). While a number of influential neuropsychological studies have been focused on the investigation of the cognitive architecture of number processing and calculation (e.g. McCloskey, 1992), other studies have examined the relation between lesion localization and impaired cognitive components (for reviews, Dehaene et al., 2003; Domahs & Delazer, 2005; Zamarian et al., 2007). Results have pointed to the critical role of the left parietal cortex in simple calculation (e.g. Delazer, Karner, Zamarian, Donnemiller, & Benke, 2006; Lee, 2000; van Harskamp & Cipolotti, 2001; Warrington, 1982), whereas frontal regions have been found to be mostly involved in complex calculation processes (e.g. Lucchelli & De Renzi, 1993; Semenza, Miceli, & Girelli, 1997). Evidence has also been obtained that impairments in calculation do not only arise as a consequence of cortical lesions, but also in the case of subcortical lesions such as those affecting the basal ganglia and related structures (e.g. Corbett, McCusker, & Davidson,

1986; Dehaene & Cohen, 1997; Delazer, Domahs, Lochy, Karner, Benke, & Poewe, 2004; Whitaker, Habiger, & Ivers, 1985; Zamarian, Bodner, Revkin, Benke, Boesch, Donnemiller, et al., 2009a).

Modern neuroimaging techniques have offered the possibility of investigating the neuronal networks involved in arithmetic. In a pioneering study, Roland and Friberg (1985) used single photon emission computed tomography (SPECT) to assess regional cerebral blood flow changes in relation to performance on different cognitive tasks. They found activation in the angular gyrus (AG) bilaterally during performance on an arithmetic task (serial subtraction), which was interpreted as related to the retrieval of numerical information from memory. Several studies in the last 20 years have shown that numerical processing and calculation are mediated by a distributed network within the parietal lobes (e.g. Fias, Lammertyn, Reynvoet, Dupont, & Orban, 2003; Gruber, Indefrey, Steinmetz, & Kleinschmidt, 2001; Kong, Wang, Kwong, Vangel, Chua, & Gollub, 2005; Menon, Rivera, White, Glover, & Reiss, 2000b; Rickard, Romero, Basso, Wharton, Flitman, & Grafman, 2000; Venkatraman, Ansari, & Chee, 2005). According to a highly influential model (e.g. Dehaene et al., 2003; Dehaene & Cohen, 1997), three parietal circuits are involved in number processing. While the circuit relying on the horizontal intraparietal sulcus (IPS) bilaterally is thought to be domain-specific for number processing, the other two – the posterior parietal attention system and the left AG verbal system – are thought to be most probably shared with other cognitive domains. Evidence in support of the assumption that the IPS bilaterally is crucial for the representation of quantity has been gained by a large number of neuroimaging studies through different paradigms, among which number comparison tasks (e.g. Chochon, Cohen, van De Moortele, & Dehaene, 1999; Fias et al., 2003; Pesenti, Thioux, Seron, & De Volder, 2000; Pinel, Dehaene, Riviere, & LeBihan, 2001), discrete and analogue comparison tasks (Castelli, Glaser, & Butterworth, 2006), and adaptation to quantity tasks (Cohen Kadosh, Cohen Kadosh, Kaas, Henik, & Goebel, 2007; Piazza, Pinel, LeBihan, & Dehaene, 2007). In calculation paradigms, it has been found that the IPS is activated in approximate calculation more strongly than in exact calculation (e.g. Stanescu-Cosson, Pinel, van De Moortele, LeBihan, Cohen, & Dehaene, 2000) and in subtraction more strongly than in multiplication (e.g. Chochon et al., 1999; Kazui, Kitagaki, & Mori, 2000; Lee, 2000). The left AG system is thought to be involved in exact and highly automated calculation such as retrieval of multiplication facts from memory (e.g. '3 × 4 =?'; Chochon et al., 1999; Duffau, Denvil, Lopes, Gasparini, Cohen, Capelle, et al., 2002; Lee, 2000; but see Rickard et al., 2000), but not quantity-based operations, such as approximation, estimation, or number comparison. The posterior parietal system, including the superior parietal lobule is found to be activated during visuospatial processes, as well as attention and spatial working memory tasks related to numerical processing (Dehaene et al., 2003; Simon, Mangin, Cohen, LeBihan, & Dehaene, 2002). Although there is no doubt that the IPS, the AG, and the superior parietal lobule are involved in different aspects of numerical processing, some of the assumptions of this model have been controversially discussed. For example, several authors disagree that arithmetic fact retrieval is mediated by verbal processes (e.g. Butterworth & Walsh, 2011; Wu, Chang, Majid, Caspers, Eickhoff, & Menon, 2009a; Zago, Pesenti, Mellet, Crivello, Mazoyer, & Tzourio-Mazoyer, 2001; Zarnhofer, Braunstein, Ebner, Koschutnig, Neuper, Reishofer, et al., 2012). Recently, Wu et al. (2009a) have argued against the hypothesis that the left AG is primarily involved in verbally-mediated fact retrieval. In their study using probabilistic cytoarchitectonic maps of inferior parietal cortex subdivisions (IPS, AG, and supramarginal gyrus), participants performed two verification tasks, presented either in Arabic numerals or in Roman numerals. Both tasks showed activation in the IPS and deactivation in the AG regions. Compared with the more difficult Roman numerals task, the Arabic numerals task showed greater responses in the AG regions. This effect was driven by less deactivation for the Arabic numerals task. The AG deactivation overlapped with the default mode network, a set of brain areas that typically shows domain general reductions in brain responses during demanding cognitive tasks, in particular tasks in which attention is voluntarily directed. According to Wu et al. (2009a), the AG has a domain-general, rather than a domain-specific, role in mathematical cognition. A domain-general role has been attributed to the AG by other researchers. For example, Ansari (2008) suggested that the AG might be engaged in general mapping processes, such as the mapping between a symbol and a referent, or between a problem and its solution, independent of the problem domain. A domain-general role of the AG was also found in the study by Grabner, Ansari, Reishofer, Stern, Ebner, and Neuper (2007) comparing arithmetic learning to visuospatial learning. Since in both learning tasks training effects were found within the AG, it was suggested that relative activation increases in this brain region are not specific to the learning of arithmetic facts, but indicate more general processes of (fact) learning.

An update to the triple-code model by Dehaene et al. (2003) has also been proposed by Arsalidou and Taylor (2011) in a quantitative meta-analysis of arithmetic brain imaging studies. While the importance of the parietal lobes was confirmed in this analysis, additional areas – the cingulate gyri, the insula, and the cerebellum – were found to be activated in studies on number processing and calculation. Most importantly, working memory processes supported by the prefrontal cortices should be considered in a model on number processing and calculation. The meta-analysis provided evidence that activation of dorsolateral and frontopolar areas of the prefrontal cortices is modulated by the difficulty of the numerical tasks.

Brain Imaging Evidence of Arithmetic Learning

A signature phenomenon in the acquisition of arithmetic expertise is the shift from slow and effortful calculation to direct memory retrieval. Several developmental studies (e.g. Ashcraft, 1992; Barrouillet & Fayol, 1998; Fuson, 1982, 1988; Lemaire & Siegler, 1995; Siegler, 1988) and learning studies with adults (Anderson et al., 1999; Logan, 1988; Logan & Klapp, 1991; Rickard, 2004; Wenger, 1999) have confirmed this shift. Developmental studies have in particular investigated the acquisition of arithmetic fact knowledge (such as '3 × 5 = 15'); learning studies with adults have also explored the gain of arithmetic expertise by employing more complex paradigms. Several neuroimaging studies with adults have demonstrated how the shift from reliance on calculation procedures to direct memory retrieval after a short-term, but intensive arithmetic training, goes along with a significant change in brain activation patterns (e.g. Delazer, Domahs, Bartha, Brenneis, Lochy, Trieb, et al., 2003; Delazer, Ischebeck, Domahs, Zamarian, Koppelstätter, Siedentopf, et al.,

2005; Grabner, Ischebeck, Reishofer, Koschutnig, Delazer, Ebner, et al., 2009; Ischebeck, Zamarian, Siedentopf, Koppelstätter, Benke, Felber, et al., 2006; Ischebeck, Zamarian, Egger, Schocke, & Delazer, 2007; Ischebeck, Zamarian, Schocke, & Delazer, 2009; Pauli, Lutzenberger, Rau, Birbaumer, Rickard, Yaroush, et al., 1994; for a review, see also Zamarian et al., 2009b). Here below, we report the main findings of these neuroimaging studies.

Learning Simple Arithmetic

To our knowledge, the study by Pauli et al. (1994) was the first providing evidence that intensive arithmetic practice leads to significant modifications of brain activity. Pauli and colleagues (1994) monitored the evoked-related potential (ERP) responses of 14 young adults across four sessions of intensive practice on one-digit arithmetic problems. They found that response times decreased significantly with increasing practice and that this decrease was greater with problems with larger operands (e.g. '8 × 7') than with problems with smaller operands (e.g. '3 × 4'). Effects of practice in the ERPs were reflected in a reduction in frontocentral positivity. Parietal positivity remained relatively stable across time. In summary, arithmetic processing correlated with a stable level of parietal activity during all training, frontal activity diminished as a function of increasing practice. Results suggest that reliance on processes, such as working memory, monitoring, and attention, which are sustained by frontal areas, diminished with problem repetition, while participants relied more on memory retrieval.

A more recent study (Núñez-Peña, 2008) has also used ERPs to investigate learning effects in simple arithmetic. The main aim of this study was to assess whether training modulates the brain signature of the problem-size effect,[1] i.e. the late positive slow wave, which would suggest that training may affect the strength of arithmetic representations in memory. Participants performed an arithmetic fact verification task ('a + b = c?') during two intensive training sessions. Results showed a significant reduction of the problem-size effect following training in both behavioural measures and electrophysiological measures. Response times decreased and accuracy increased from the first session to the second session, indicating a significant training effect. The amplitude of the positive slow wave increased with problem size and decreased with practice. In summary, these results suggest that training has an impact on the problem-size effect, and may strengthen the association between problems and results. Training effects are more pronounced for larger, more difficult problems.

Learning Complex Arithmetic

Most of the recent studies on the effects of learning on brain activation patterns have used functional magnetic resonance imaging (fMRI; Delazer et al., 2003, 2005; Grabner et al., 2009; Ischebeck et al., 2006, 2007, 2009). These studies have principally employed more complex arithmetic problems in order to make sure that the problems to be trained were not already known by heart. Commonly, the solution of problems such as '23 × 8' or '6 × 17' is not part of our declarative memory and has to be calculated. This implies that one has to apply sometimes effortful, resource-demanding, and error-prone procedures, and that training effects should be reflected in accuracy and speed improvements at the behavioural level,

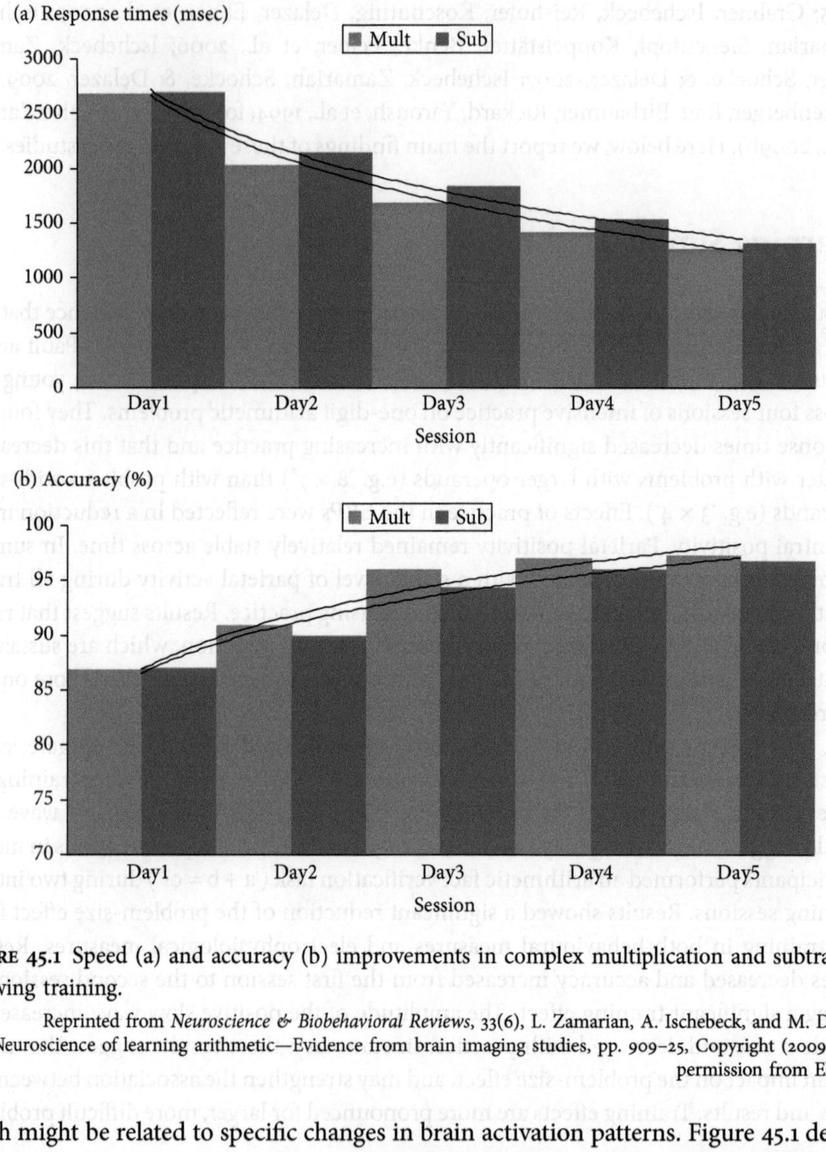

FIGURE 45.1 Speed (a) and accuracy (b) improvements in complex multiplication and subtraction following training.

Reprinted from *Neuroscience & Biobehavioral Reviews*, 33(6), L. Zamarian, A. Ischebeck, and M. Delazer, Neuroscience of learning arithmetic—Evidence from brain imaging studies, pp. 909–25, Copyright (2009), with permission from Elsevier.

which might be related to specific changes in brain activation patterns. Figure 45.1 depicts an example of accuracy and speed improvements in complex calculation after training.

In the first study of this series, Delazer and colleagues (2003) aimed at verifying whether expertise acquisition in complex calculation is actually reflected in specific brain activation changes. They found that, after an intensive training on a set of complex multiplication problems (e.g. '23 × 8'), young adults were able to solve the trained problems more accurately and faster than new, untrained problems of equal difficulty. Learning effects were also evident at the brain activation level. While the trained problems yielded relatively stronger activation compared with the untrained problems in the left AG,[2] the untrained problems yielded stronger activation compared with the trained problems in the IPS, the inferior parietal lobule, and the inferior frontal gyrus (IFG; all left lateralized). These results were interpreted as suggestive of a shift from quantity-based processing and complex procedures that rely on

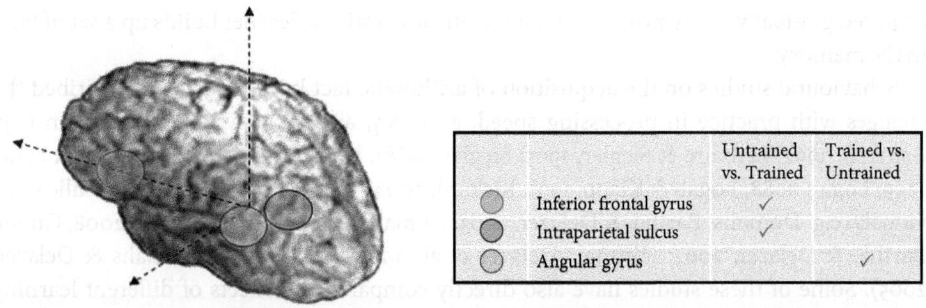

	Untrained vs. Trained	Trained vs. Untrained
○ Inferior frontal gyrus	✓	
○ Intraparietal sulcus	✓	
○ Angular gyrus		✓

FIGURE 45.2 Schematic illustration of brain activation changes in complex calculation following training. Extended activations in frontal and intraparietal areas (orange, blue) are typically found in the contrast untrained problems versus trained problems. A relative activation increase within the left angular gyrus (green) is found in the contrast trained problems versus untrained problems.

Reprinted from *Neuroscience & Biobehavioral Reviews*, 33(6), L. Zamarian, A. Ischebeck, and M. Delazer, Neuroscience of learning arithmetic—Evidence from brain imaging studies, pp. 909-25, Copyright (2009), with permission from Elsevier.

working memory and executive control (e.g. '23 × 8 = (20 × 8) + (3 × 8) = 160 + 24 = 184') to direct retrieval of correct solutions from memory (e.g. '23 × 8 = 184') after intensive training. Results agreed with the observation by Poldrack (2000) that learning is reflected in a shift from general purpose processes to task-specific processes (in this case, arithmetic fact retrieval). Figure 45.2 gives a schematic illustration of the brain activation changes in complex calculation following training.

More recent neuroimaging studies have tempted to explore these training-related effects on brain activation patterns in more detail. In particular, it has been investigated whether these changes are related to a specific learning method (Delazer et al., 2005) or learning content (Ischebeck et al., 2006), and when and how these changes take place (Ischebeck et al., 2007). Whether the adaptation of newly-acquired knowledge to new situations is reflected in specific brain activation changes (Ischebeck et al., 2009), and whether the activation changes found within the AG are specific to arithmetic learning or actually underlie more general processes of learning (Grabner et al., 2009) have also been examined.

Learning Effects Related to Different Learning Methods

The discussion about the effects of different learning methods on the acquisition of expertise has a long tradition in educational psychology as well as in cognitive psychology (Brownell, 1935; Thorndike, 1922). In the drill approach, instructions focus on the rote memorization of computational skills, i.e. arithmetic facts and procedures (Cowan, 2003). Children are taught bonds or associations between unrelated stimuli, and understanding is not seen as necessary for the formation of such associations (Baroody, 2003). In this view, direct instructions and drill learning are efficient methods of ensuring the acquisition of new arithmetic expertise. Alternative approaches emphasize the understanding of basic principles and arithmetic relations. They suggest that knowledge of arithmetic facts needs to be acquired gradually, using procedures and strategies, until the learner builds up a set of facts in the memory.

Behavioural studies on the acquisition of arithmetic fact knowledge have described the changes with practice in processing speed, accuracy, and strategy use for children (e.g. Ashcraft, 1992; Lemaire & Siegler, 1995; Siegler, 1988), healthy adults (e.g. Anderson et al., 1999; Logan, 1988; Logan & Klapp, 1991; Rickard, 2004; Wenger, 1999), and dyscalculic individuals (e.g. Domahs, Bartha, & Delazer, 2003; Domahs, Zamarian, & Delazer, 2008; Girelli, Bartha, & Delazer, 2002; Hittmair-Delazer et al., 1994; for a review, Domahs & Delazer, 2005). Some of these studies have also directly compared the effects of different learning methods on behavioural improvements (e.g. Logan & Klapp, 1991; Domahs et al., 2003). Both learning methods (by drill and by strategy) lead to skilled arithmetic performance and automated retrieval of answers from memory. To date, only one neuroimaging study (Delazer et al., 2005) has investigated whether the acquisition of new arithmetic expertise through different learning methods is reflected in the recruitment of the same cognitive processes and the activation of the same brain regions, or alternatively the brain activations still reflect the method of acquisition. In Delazer et al.'s (2005) study, young adults were trained on two new arithmetic operations. One operation – strategy condition – had to be learned by applying a new algorithm (e.g. '3 § 12 = {[(12 –3) + 1] + 12} = 22'); the other operation – drill condition – had to be learned by memorizing the operands-result association (e.g. '4 # 15 = 24'). Results indicated that the drill-trained problems yielded relatively greater left AG activation compared with the strategy-trained problems, which suggests that the solution by result retrieval was more dominant in the drill condition than in the strategy condition. The lesser AG activation for the strategy condition might be related to the participants sometimes, or simultaneously, applying the algorithm even in the case of the trained problems. The way problems are learned thus makes a great difference for the active involvement of brain areas that are important for problem processing, even when different training types yield similar behavioural improvements.

Learning Effects Related To Different Learning Contents

According to the triple-code model (e.g. Dehaene & Cohen, 1997; Dehaene et al., 2003), simple multiplication (e.g. '3 × 4') and simple subtraction (e.g. '9 – 4') depend on different cognitive processes, which are supported by partially different cortical substrates. Specifically, multiplication relies on skilled memory retrieval of arithmetic fact knowledge; subtraction is solved by manipulation of numerical representations. Consequently, multiplication most strongly relies on the left AG circuit, subtraction on the IPS circuit. One may hypothesize that these differences are related to the method how operations are taught at school. However, one may also assume that they are related to differences in efficiency. While procedures and back-up strategies might be the best and most simple way to solve addition and subtraction, memory retrieval might be the most efficient strategy with multiplication. The investigation of whether comparable training on two different operations leads to similar learning effects can help disentangle this matter of interest. In Ischebeck et al.'s (2006) study, identical training was provided for complex multiplication (e.g. '15 × 7') and complex subtraction (e.g. '43 – 26'). Behavioural results indicated a similar learning curve for both operations (see Figure 44.1), whereas imaging results pointed to systematic differences. Trained multiplication showed relatively stronger left AG activation compared with untrained multiplication. Differently, there were no significant differences in left AG activation between trained subtraction and untrained subtraction. These operation-specific brain activation changes after identical training may be interpreted as related to differences in efficiency between operations: well-known complex multiplication problems are possibly solved more efficiently through memory retrieval; fast and effortless computation may continue to play a role even in the case of well-known subtraction problems. Therefore, the effects of learning on brain activations seem to depend not only on the learning method, but also on the learning content.

Time Course of Learning Effects

The studies reviewed so far demonstrated adaptive changes in brain activation patterns once learning had taken place. However, when and how do these changes occur? Are there numerous or just a few repetitions needed? To investigate the ongoing process of learning itself, Ischebeck et al. (2007) provided no training prior to the fMRI session. In the scanner, participants were presented with two sets of complex multiplication problems – one set had high frequency of repetition ('repeated'); the other had low frequency of repetition ('novel'). As time went by, compared with the novel problems, the repeated problems yielded activation decrease in frontoparietal areas and activation increase in temporoparietal areas that included the left AG. These effects became significant after approximately 8 problem repetitions, remained stable over the course of the scanning session, and were, as a pattern, similar to the previously reported findings of experiments where participants had to train much more intensively (Delazer et al., 2003, 2005; Ischebeck et al., 2006). In summary, learning-related modifications of brain activation patterns appear very early during training and do not differ in principle from those observed after more intensive training.

Transfer Between Related Operations

Successful transfer between related operations is a key feature of efficient arithmetic. The investigation of transfer effects is an important issue because only the flexible use of acquired knowledge enables people to efficiently answer new situations. Even rehabilitation attempts cannot be considered as complete and successful if a patient learns to retrieve arithmetic facts from memory, but is unable to adapt this knowledge to new situations (for discussions, Delazer, 2003; Lochy, Domahs, & Delazer, 2005). Transfer effects between related operations have mostly been assessed in experimental psychology studies (e.g. Rickard, 2004, 2005) and neuropsychological case studies (Cipolotti & de Lacy Costello, 1995; Delazer et al., 2004; Hittmair-Delazer et al., 1994). In the study by Ischebeck et al. (2009), it was examined whether specific brain activation changes appear in association with transfer effects from newly acquired multiplication knowledge (e.g. '46 × 3 = 138') to related division (e.g. '138: 3 = 46'). If unknown related division problems are solved by using the multiplication knowledge acquired during training, then less activation should be found in frontal brain areas as the related division problems pose less demands on working memory and attention resources than unrelated division problems. Also, relatively less activation in IPS and more activation within the left AG should be expected for the related division problems solved by transfer. In their study, participants trained intensively on complex multiplication the day before the scanning session. In the scanner, participants had to solve the trained multiplication problems, as well as untrained multiplication problems and division problems that were either related or unrelated to the trained multiplication problems. Brain imaging results demonstrated significant training effects within the left AG for multiplication. With regard to transfer effects, a relatively greater left AG activation was found for related divisions relative to unrelated divisions, and this activation showed a positive correlation with the magnitude of transfer effects in the behavioural results. Overall, transfer effects were relatively weak, possibly due to strong inter-individual differences. In summary, the observation of a significant transfer effect within the left AG suggests that the newly-acquired multiplication knowledge was – at least to some extent – recruited for the solution of unknown related division problems. Inter-individual differences in behavioural measures of transfer were accompanied by brain activation differences within the left AG. Possibly, the conceptual understanding and/or the procedural knowledge required for successful transfer was not equally available to all participants, which might have yielded to these inter-individual differences in transfer effects. Inter-individual differences in mathematical competence will be further discussed under 'Inter-individual differences in arithmetic'.

Specificity of Brain Activation Changes Related to Learning

Previous studies showed that training-related activation changes within the left AG were more pronounced when the arithmetic operation being trained or the method of training emphasized arithmetic fact retrieval from memory. Grabner and colleagues (2009) investigated whether these changes are specific to arithmetic learning or can also emerge with non-arithmetic material, which would indicate that the AG is involved in more general processes of learning and automatization. In their study, young adults trained on complex multiplication and complex figural-spatial problems. During the multiplication training, they had to solve problems of the type '14 × 7 =?'; during the figural-spatial training, they were presented with drawings of 3D objects and had to determine the number of object faces. Behavioural improvements after training were comparable between conditions. Brain imaging results replicated the findings reported by previous studies for multiplication (Delazer et al., 2003; Ischebeck et al., 2006, 2007, 2009). Trained multiplication yielded relatively stronger activation compared with untrained multiplication within the AG bilaterally, with a peak of activation in the left AG. Untrained multiplication yielded stronger activation compared with trained multiplication within a frontoparietal network that included the IFG and the IPS. Results of the figural–spatial condition overlapped considerably with the pattern of brain activations found for the arithmetic condition. Trained figural–spatial problems yielded relatively stronger activation compared with untrained problems within the AG bilaterally, with a peak of activation in the right AG. Untrained problems showed stronger bilateral frontoparietal activations compared with trained problems. Similar training effects for multiplication and figural-spatial problems suggest that the relative activation increase within the AG after intensive training is not specific to the arithmetic domain. Although its function may slightly differ between hemispheres (left AG: verbal material;

right AG: figural material), the AG is possibly involved in general processes of fact retrieval and mapping to information stored in memory (e.g. mapping between arithmetic problems and their solutions, grapheme-phoneme mappings, or mapping between abstract numerical symbols and their magnitudes). Damage to the left AG has often been found in association with calculation deficits, finger agnosia, left-right disorientation, and writing or reading difficulties. Impairment to this mapping function of the AG might underlie all these disturbances (Ansari, 2008).

The Neural Basis of Exceptional Arithmetic Expertise

Exceptional arithmetic experts perform at extraordinary high levels after years of obsessive mental training and practicing on a task. Different explanations have been proposed to account for their exceptional skills (Jonides, 2004). Practice might lead to skill enhancements because the initial strategy is used more efficiently. In this case, skill enhancements should involve the same brain regions as found in normal processing. Exceptional skills might also be due to the application of new strategies, which would be reflected in the recruitment of additional cerebral networks (Fehr, Weber, Willmes, & Herrmann, 2010; Jonides, 2004). Insights into the behavioural and brain activation differences between experts and normal users of mathematics may further contribute to our understanding of the neuroplastic changes induced by intensive training.

Fehr and colleagues (2010) investigated the exceptional skills of a calculation prodigy (CP) in an fMRI study and reported evidence in favour of the first explanation. CP has trained his arithmetic abilities over 15 years for several hours a day. The brain activation patterns of CP were analysed in two steps. First, activation patterns during normal calculation tasks were compared between CP and controls without exceptional skills. Secondly, activation patterns of CP in exceptionally difficult tasks were compared with those shown in tasks of average difficulty. Difficult tasks were only administered to CP, and included exponentials with a base between 10 and 99, and exponents between 10 and 20. The comparison of CP to controls indicated largely similar activation patterns during average calculation tasks. In exceptionally difficult tasks, CP showed activation patterns located in regions near to those observed during normal calculation. As activations largely overlapped between CP and controls, Fehr and colleagues (2010) suggest that his exceptional skills are due to neuroplastic changes in initially recruited brain regions due to extensive training. Training extending over several years would thus lead to adaptation of neural networks that are usually used for normal working memory and calculation. According to Fehr et al. (2010), intensive training is a key aspect of exceptional performance. It is not necessary to develop new strategies and recruit new extraordinary brain regions to enhance performance dramatically. It seems sufficient to devote hours of daily training over several years. Prodigies may be better in their trained skill, but are not necessarily different from other persons.

Other studies with exceptionally performing individuals have pointed to activation of additional brain areas and use of strategies that are very different from those of average performing individuals. By using positron emission tomography (PET), Pesenti, Zago, Crivello, Mellet, Samson and Duroux (2001) compared the brain activation patterns related to complex calculation processes between a calculation expert (GR) and non-expert controls. GR has an impressive arithmetic knowledge acquired by memorizing arithmetic facts and complex calculation algorithms over several years for several hours a day. GR makes use of this extensive arithmetic knowledge and of specific episodic memory strategies to retrieve, for example, the square of two-digit numbers from memory (e.g. '76 × 76') or solve complex multiplication problems (e.g. '76 × 82'; a detailed description of the behavioural investigation is given in Pesenti, Seron, Samson, & Duroux, 1999). An analysis of the neuroimaging data revealed brain areas that were activated during calculation (as compared with memory retrieval) in both the expert and the non-expert controls, as well as areas that were more activated in the expert than in the non-experts. Computation-based problems (e.g. for the expert: '76 × 82'; for the non-experts: '14 × 25') were compared with memory-based problems (e.g. for the expert: '76 × 76'; for the non-experts: '3 × 9') in order to isolate the neural substrates specific to the calculation process. Both GR and non-experts showed bilateral activation (with a left-sided predominance) in the supramarginal gyrus, IPS, inferior occipital and middle occipital gyri, occipito-temporal junction, and frontal areas. GR, but not controls also showed activation in the left paracentral lobule, as well as in the right medial frontal gyrus, parahippocampal gyrus, anterior cingulate gyrus, and middle occipito-temporal junction. In summary, calculation performance of the expert and the non-expert controls relied on brain areas that are typically involved in visuospatial working memory, visual imagery, and numerical processing. However, GR also recruited areas that are associated with episodic memory processes, which might be related to his use of specific encoding-retrieval strategies during calculation.

A further example that expertise in calculation may be bound to the use of specific strategies relying on distinct cerebral circuits has been reported by Hanakawa, Honda, Okada, Fukuyama, and Shibasaki (2003). In their study, mental calculation of skilled abacus users and non-experts were assessed by means of fMRI. Abacus experts are able to solve complex mental calculations efficiently and exceptionally fast. After years of training, abacus experts can implement the abacus operations mentally on a virtual abacus, i.e. they can use an imaginary abacus. This technique is very efficient and allows experts to circumvent usual mental arithmetic strategies, which are relatively slow and limited in their capacity. Usual mental arithmetic strongly relies on working memory. Hanakawa and colleagues (2003) compared qualified Japanese abacus masters, with the highest degree for both mental calculation and abacus operations, to non-expert controls in a series of tasks requiring the manipulation of numerical, spatial, or verbal material. Abacus experts performed one-digit and three-digit additions with 100% accuracy and scored with six-digit additions as accurately as the non-experts in the one-digit condition. Brain imaging results indicated no significant differences between experts and non-experts with spatial and verbal material, whereas significant activation differences were found in the numerical condition. Both experts and non-experts showed significant activation in the cerebellum, fusiform gyrus, superior precentral sulcus, and posterior parietal cortex including the IPS. While this activation pattern was bilaterally symmetrical for the abacus experts, it was strongly left lateralized for the non-experts. Additional significant activation differences included stronger activation in the posterior superior parietal cortex/precuneus for the experts, and stronger activation in the prefrontal cortex and lateral parietal cortex for the non-experts. Abacus experts thus showed greater reliance compared with non-experts on brain areas that mediate visuospatial and visuomotor processing, a finding that might reflect their main strategy use during complex mental calculation.

The exceptional calculation abilities of abacus experts have also been investigated in a recent PET study by Wu, Chen, Huang, Liu, Hsieh, and Lee (2009b). Chinese abacus experts were compared with non-experts in three tasks: covert reading, simple addition, and complex addition. Abacus experts showed activation of a bilateral frontoparietal network in simple addition and complex addition. The two types of arithmetic conditions elicited similar activation patterns, suggesting that the abacus experts relied on the same procedure, with almost no increasing workload with the complex condition. For the non-experts, the activation patterns differed significantly between simple calculation and complex calculation. During simple calculation, activation was evident in the inferior frontal cortex (language processing) and in a left frontoparietal network (visuospatial processing). During complex calculation, activation was stronger in regions related to visuospatial processing. In summary, studies with high-performing individuals, who have trained extensively over several years, indicate that the extent of practice and the type of strategy play significant roles in modifying brain activation patterns.

Inter-Individual Differences in Arithmetic

Evidence that brain activation patterns are significantly modulated by the level of mathematical competence has been reported for normal users of arithmetic as well. Menon Rivera, White, Eliez, Glover, and Reiss (2000a) compared the brain activation patterns of participants performing at ceiling (perfect performers) with that of non-perfect performers (mean accuracy = 92%) in an arithmetic verification task (e.g. '4 + 2 − 1 = 4?'). Perfect performers showed less activation and less variability than non-perfect performers in the left AG, but not in other brain regions. Menon et al. (2000a) attributed the weaker activation (i.e. smaller percentage of activated voxels in the region of interest) to functional optimization within the left AG, which reflects skill mastery and long-term practice effects. This hypothesis was supported by faster reaction times in perfect performers relative to non-perfect performers. The same verification paradigm was used in a later study by Wu et al. (2009a) in two different formats, Arabic numerals and Roman numerals. In this study, an association was found between higher arithmetic performance and relatively greater activation (lesser deactivation) within the AG bilaterally. In both verification tasks (Arabic numerals, Roman numerals), stronger bilateral AG deactivation was associated with poorer performance. The more automated Arabic numerals task showed greater responses in both AG areas than the less automated – more difficult – Roman numerals task, but this effect was due to reduced deactivation for the Arabic numerals task. As Wu et al. (2009a) point out, the reported brain–behaviour correlations were associated with focal task-related deactivation, rather than with activation. These findings suggest that different aspects of subjective difficulty, i.e. automaticity, performance level, and general domain competence, have an impact on the responses of the AG during mathematical tasks.

Grabner and colleagues (2007) also explored the relationship between mathematical competence and brain activation patterns in average performing individuals. In order to ensure that results could be generalized beyond the specific demands of the experimental tasks, groups of low mathematical competent participants and high mathematical competent participants were made on the base of their performance on standardized psychometric tests of intelligence and arithmetic tasks. Groups differed in mathematical-numerical intelligence tests and experimental arithmetic tasks (simple and complex multiplication tasks), but not in verbal and figural-spatial intelligence measures. High-achieving participants outperformed the low-achieving participants in particular in the complex multiplication task. Compared with low-achieving participants, high-achieving participants showed stronger activation in the left AG, middle temporal gyrus, supplementary motor area, and medial superior frontal gyrus. The reverse contrast (low versus high) yielded no significant activation foci. Results also indicated that the relative activation within the left AG directly correlated with the individual level of mathematical competence. Performance of mathematically more competent individuals was thus supported by a neural network involving the left AG to a stronger extent than performance of less competent individuals.

In a second study by Grabner et al. (2009), more competent participants and less competent participants intensively practiced on complex multiplication problems. High-achieving participants outperformed the low-achieving participants during the entire training, as well as with untrained problems at post-training. Brain imaging results showed relatively stronger left AG activation with untrained problems for the high-achieving participants relative to the low-achieving participants. In line with results of Grabner et al. (2007), the left AG activation with untrained problems was directly correlated with the individual level of mathematical competence. There were no group differences with regard to activation with trained problems. Likely, the more competent individuals relied strongly on fact retrieval even in the case of untrained multiplication problems and the acquisition of mathematical expertise after training led to attenuation of competence-related activation differences within the left AG. This attenuation was due to stronger left AG activation for the less competent individuals following intensive training.

Future studies should also take into account individual cognitive styles, which may cause differences in learning. Zarnhofer et al. (2012) found that the verbal–visual cognitive style influences brain activations during multiplication and subtraction. Verbalizers, i.e. subjects who report repeating information during thinking verbally, showed higher activation in brain areas that are associated with language processing. The higher the self-reported tendency to verbalize, the higher the activation in the supramarginal gyrus, rolandic operculum, and Heschl's gyrus bilaterally. Significant correlations between cognitive style and activation patterns were found for verbalizers, but not for visualizers. Interestingly, no modulation of activation was observed in the AG. This finding questions the view that activation of the AG in mental arithmetic indicates verbal processing. As Zarnhofer et al. (2012) suggest, the AG seems to support number processing in a modality-general way.

All in all, these studies provided evidence that the relative activation of the AG during mental calculation is modulated by inter-individual differences in mathematical competence. In particular, they showed that individuals with higher mathematical competence have relatively higher left AG activation during multiplication, likely because they more strongly rely on automated retrieval of answers from memory. Differences in mathematical competence might have also an influence on transfer effects of newly-acquired multiplication knowledge to related division (Ischebeck et al., 2009). Findings on inter-individual differences in mathematical competence are in line with evidence from the learning studies reviewed so far, as well as from developmental studies showing that the acquisition of arithmetic competence is reflected in an increasing reliance on parietal cortices (for reviews, Ansari, 2008; Menon, 2010; Zamarian et al., 2009b).

Structural Brain Changes Due to Learning

Brain imaging studies have focused not only on changes in activation patterns induced by systematic training, but also on changes of the morphological structure of the brain. As shown by means of voxel-based morphometry (VBM) in several domains (e.g. juggling, Braille reading, music), experience and learning can significantly alter the anatomical structure (e.g. gray matter density, GMD) of the human brain (e.g. Draganski, Gaser, Busch, Schuierer, Bogdahn, & May, 2004; Draganski, Gaser, Kempermann, Kuhn, Winkler, Büchel, et al., 2006; Münte, Altenmüller, & Jäncke, 2002; Pantev, Oostenveld, Engelien, Ross, Roberts, & Hoke, 1998; Pascual-Leone & Torres, 1993). Recent neuroimaging studies have also investigated the structural brain effects of cognitive training (e.g. Aydin, Ucar, Oguz, Okur, Agayev, Unal, et al., 2007; Maguire, Gadian, Johnsrude, Good, Ashburner, Frackowiak, et al., 2000; Maguire, Spiers, Good, Hartley, Franckowiak, & Burgess, 2003; Mechelli, Crinion, Noppeney, O'Doherty, Ashburner, Frackowiak, et al., 2004; Takeuchi, Taki, Sassa, Hashizume, Sekiguchi, Fukushima, et al., 2011). However, only few VBM studies have been concerned with arithmetic training. Studies on the structural plasticity due to learning and expertise gained considerable interest in the past as they offer a promising view on the adaptivity of the human brain. This is particularly important in the field of rehabilitation after brain damage. Future studies integrating behavioural and brain imaging measures may contribute to the development of more efficient and targeted intervention programmes after acquired brain damage or in cases of atypical numerical development.

Aydin and colleagues (2007) compared academic mathematicians to non-mathematicians. Higher GMD was found in the bilateral inferior parietal lobule and left IFG for mathematicians relative to non-mathematicians. Also, the time spent as a mathematician showed to correlate significantly with the relative GMD increase in the right parietal lobule. The study thus suggests that extensive and persistent training in mathematical thinking over several years leads to structural brain changes in those areas that are involved with numerical processing, arithmetic calculation, and visuospatial processing. However, no behavioural testing was performed in this investigation, which limits the insight into the relationship between mathematical expertise and brain structure.

A recent study by Sader, Popescu, Thomas, Terhune, Dowker, Mars, et al. (submitted) combined the detailed evaluation of several mathematical skills, general cognitive status, and social abilities with the anatomical assessment of GMD and fractional anisotropy. Behavioural results indicated that mathematical expertise is grounded in domain-specific differences (found in basic and advanced mathematical tasks), as well as in domain-general differences. Contrary to what it might be expected on the basis of previous investigations showing an association between low numerical performance and low GMD in the right IPS (e.g. Molko, Cachia, Rivière, Mangin, Bruandet, LeBihan, et al., 2003), the study by Sader and colleagues (submitted) showed that mathematicians have lower GMD in the right IPS than non-mathematicians. This finding suggests that there is no linear relationship between GMD in the right IPS and mathematical abilities. Instead, an inverted U-shape may be assumed. Mathematicians showed greater GMD 20mm posterior to the right IPS. As the authors propose, the close vicinity between reduced GMD (right IPS) and increased GMD might indicate cortical specialization in a specific brain region, which affects the neighbouring structure. Exceptional mathematical abilities were associated with both increased and decreased GMD, depending on the specific brain regions. Importantly, the relationship between performance level and brain structure was found to be non-linear. Findings of this study also suggest that differences in brain structure are not only reflected in quantitative terms, but also in qualitative patterns. While higher GMD in the left superior frontal gyrus was associated with better performance in mathematicians, it was associated with worse performance in non-mathematicians. This double dissociation represents a qualitative difference in the brain–behaviour relationship as a function of expertise.

In a recent study, Takeuchi et al. (2011) showed that adaptive training on complex calculation significantly modifies the morphological structure of brain regions that are related to working memory. Average performing participants were divided into three groups: group 1 received intensive adaptive training of mental calculation, group 2 non-adaptive calculation training (placebo training), and group 3 no training at all. Training for the experimental group was adaptive, i.e. task difficulty was adapted to the participant performance. In the multiplication task, participants had to mentally solve problems of the type '37 × 45'. If they answered correctly, problems became increasingly more difficult (the task started from two-digit times two-digit multiplication, then became two-digit times three-digit multiplication, then three-digit times three-digit multiplication, and then three-digit times four-digit multiplication). In the addition task, participants were presented with two-digit numbers and were asked to add them. If they answered correctly, the ISI became shorter. Consecutive wrong answers made the task less difficult. Training was performed over 5 days for 4 hours a day. Participants underwent MRI and cognitive assessment before and after training. Results showed that group 1 profited from training in both tasks. For them, ISI in the addition task was significantly shorter and difficulty of mental multiplication was significantly higher at post-training compared with pre-training. Training effects for group 1 also extended to other cognitive tasks: performance on a verbal letter span task, a reverse Stroop interference task, and a complex arithmetic task increased; scores in a creativity task decreased. VBM analysis indicated regional GMD decreases for the adaptive calculation training group in regions that are associated with working memory (bilateral dorsolateral prefrontal cortex, bilateral parietal cortex, left superior temporal gyrus). The placebo training group did not show the same GMD changes. Takeuchi et al. (2011) speculate that increased synaptic elimination might have caused regional GMD decreases in group 1. This study suggests that extensive exercise causes plasticity of gray matter structure in frontal and language-related cortices, although it is assumed that the gray matter structure in these regions is genetically determined (Thompson, Cannon, Narr, van Erp, Poutanen, Huttunen, et al., 2001). It should, however, be noted that the processes mediating neuroplasticity are still poorly understood and that the cellular mechanisms underlying training-related gray matter changes in humans remain to be defined (Draganski & May, 2008; Taubert, Draganski, Anwander, Müller, Horstmann, Villringer, et al., 2010). Several different mechanisms may contribute to gray matter modifications, such as formation of new connections by dendritic spine growth, increase in the strength of existing connections (Draganski & May, 2008; Chklovskii, 2004), or glial hypertrophy and synaptogenesis (Anderson, Li, Alcantara, Issacs, Black, & Greenough, 1994). In a recent study on the effects of motor-balancing (Taubert et al., 2010), two training sessions sufficed to cause a rapid gray matter increase in sensorimotor-related brain areas. Taubert et al. (2010) suggest that rapid intracortical remodelling of dendritic spines and axonal terminals could underlie gray matter modifications in human learning. In conclusion, systematic training leads to significant modifications of gray matter structure. These modifications may result either in an increase or in a decrease of gray matter; in some cases, it results in a decrease of gray matter after a transient increase. Several factors influence the dynamics of training-related gray matter changes, including the duration of the training, the skill being trained, and the performance level reached (Driemeyer, Boyke, Gaser, Büchel, & May, 2008). As Sader et al. (submitted) point out, the developmental stage in which training-related changes occur, as well as the interaction between experience and pre-existing brain differences may also play a role. As they suggest, numerical expertise is reflected by quantitative changes in brain structures that are essential in numerical processing, as well as by qualitative differences in the anatomy–behaviour relationship as a function of expertise.

Brain Imaging and The Rehabilitation of Arithmetic Skills in Brain Damaged Patients

Learning studies with healthy adults reported reliable improvements in arithmetic skills, as well as significant changes in activation patterns and the morphological structure of the brain after relatively short periods of training. Better understanding of learning-related functional and structural brain modifications should also lead to better insight into the chances and limitations of rehabilitation approaches after brain damage. Deficits in arithmetic fact retrieval are often observed after left hemispheric lesions and in association with aphasic disorders (e.g. Delazer & Bartha, 2001; Delazer, Girelli, Semenza, & Denes, 1999; Domahs et al., 2008; Domahs & Delazer, 2005; Grafman, Passafiume, Faglioni, & Boller, 1982; McCloskey, Aliminosa, & Sokol, 1991). Targeted rehabilitation programmes are rare (for overviews, Girelli & Seron, 2001; Lochy et al., 2005), most of them focusing on a drill approach (i.e. extensive repetition of problems followed by immediate feedback; e.g. Lochy et al., 2005), some using a conceptual approach (Domahs et al., 2003; Girelli et al., 2002). All these rehabilitation studies providing targeted and intensive training reported good success. Whether behavioural improvements during rehabilitation are due to the recruitment of new brain regions, efficient use of preserved structures, and/or functional recovery of damaged areas is an open question. Thus far, only one study has investigated the brain activation changes that follow remediation of simple arithmetic skills. In their study, Zaunmüller, Domahs, Dressel, Lonnemann, Klein, Ischebeck, and colleagues (2008) assessed a left-sided brain damaged patient by means of fMRI before and after intensive training on multiplication facts. At post-training, brain imaging results showed relatively stronger activation increase in the right AG for the trained problems relative to the untrained problems, i.e. relatively stronger activation in the hemisphere that is contralateral to the side of the lesion and contralateral to the typical activation side reported for healthy adults. In summary, after rehabilitation, a homologue area in the right hemisphere, i.e. in the hemisphere contralateral to the lesion side, was recruited during arithmetic fact retrieval. Although very interesting, these findings can be interpreted only with caution. Results of single-case fMRI studies are difficult to replicate because of the large variability between single patients with regard to variables such as lesion location, lesion extension, aetiology, associated deficits, spontaneous recovery, and neurofunctional reorganization. Also, there is a considerable inter-individual variability in arithmetic performance even in healthy individuals and performance variability shows to significantly modulate the effects of practise on the brain (e.g. Grabner et al., 2007). Despite all these methodological difficulties, brain imaging studies with carefully selected patients could trace the time course of behavioural improvements in rehabilitation in conjunction with the repeated assessment of brain activation and/or brain structure changes. Possibly, such studies could lead to individualized rehabilitation approaches in the future, with targeted selection of training methods and a better prediction of the rehabilitation success.

A Way to Improve Learning Effects?

A very interesting recent study by Cohen Kadosh, Soskic, Iuculano, Kanai, & Walsh (2010) has used transcranial direct current stimulation (TDCS), a non-invasive brain stimulation technique, to facilitate basic numerical learning. During TDCS, a weak current is applied constantly over time to enhance (anodal stimulation) or reduce (cathodal stimulation) the excitation of neuronal populations. In the study by Cohen Kadosh et al. (2010), young adults were asked to learn the association between some artificial numerical digits over 6 days of training. TDCS was applied to the parietal lobes for 20 minutes from the beginning of each training session. Following the learning period, numerical proficiency with the newly-acquired numbers was tested by means of two standard tasks, a numerical Stroop task and a number-to-space mapping task with the learned symbols. The polarity of the TDCS showed to have specific facilitating or inhibiting effects on learning. Learning effects were better for the group of subjects who received anodal stimulation in the right parietal lobe and cathodal stimulation in the left parietal lobe compared with the control group (sham stimulation). Differently, application of anodal stimulation to the left parietal lobe and cathodal stimulation to the right parietal lobe led to underperformance in a second TDCS group. Importantly, the improvement was specific to the trained items and did not extend to Arabic digits. Also, it was still present 6 months after the learning experiment. These results have been replicated and extended by a study by Snowball, Tachtsidis, Popescu, Thompson, Delazer, Zamarian, et al. (2013) showing that transcranial random noise stimulation (TRNS) can induce long-lasting enhancement of learning effects in arithmetic. The outcome of these experiments is promising and suggests that non-invasive brain stimulation techniques such as TDCS and TRNS are possible tools for facilitating learning and/or rehabilitation.

Summary and Discussion

In this article, we have reviewed evidence that intensive practice of arithmetic skills not only modifies behavioural measures, such as reaction times, accuracy, and problem-size effect, but also has an impact on the functional and structural organization of cerebral networks. Several studies have shown that practice leads to a decrease of activation in frontal brain areas and to a relative increase of activation in parietal brain areas. This pattern has been found in learning studies with healthy young adults and mirrors a developmental trajectory in the first years of schooling (Kawashima, Taira, Okita, Inoue, Tajima, Yoshida, et al., 2004; Rivera, Reiss, Eckert, & Menon, 2005; for a review, see Menon, 2010). Apart from numerical processing, such as quantity manipulation, unknown calculation problems require planning, attentional control, working memory, and monitoring. Young children may be challenged by seemingly easy computations such a '10 – 3', healthy young adults by mental calculations such a '35 × 7', and calculation prodigies by the exponentiation of a two-digit number. Brain imaging studies are in line with behavioural studies showing that in difficult tasks heavy load is put on frontal lobe resources and executive functions. Practice may lead to a modification of strategies and to efficient retrieval of chunks of stored information from long-term memory. As automatization increases, the load on planning, attentional control, and working memory resources decreases, and this is reflected in a decrease of activation in frontal brain areas. Automatization also leads to a relative activation increase in the AG. In line with this general observation, it has been found that inter-individual differences in arithmetic skills are reflected in different activation levels within the AG. Although the AG seems to play an important role in arithmetic learning, activation changes within the AG are not specific to the number domain and probably reflect more general retrieval or mapping processes. It has also been proposed that the AG is part of the default mode network, a set of brain areas that typically shows domain general reductions in brain responses during demanding cognitive tasks.

What happens when we improve our arithmetic skills? One way to improve performance is to apply the initial strategy in a given problem type with greater efficacy. Another way is to acquire a new, more efficient strategy. As Jonides (2004) has lined out, it may be difficult to disentangle these possibilities by analysing behavioural data, and introspection gives only limited information. Brain imaging studies bring important insights into this matter of interest. When activation changes are found in a pre-defined network, i.e. in the network that has been initially devoted to solve the task, one may assume that the initial strategy has been performed more efficiently. When, on the other hand, new areas are recruited and the activation changes are observed outside the initial network, then the gain in competence is likely due to the adoption of a new, more efficient strategy. People with average arithmetic skills seem to use the same arithmetic network when exercising and improving arithmetic skills. Studies on calculation prodigies suggest that both strategies may lead to exceptional calculation performance.

Most of the studies reviewed here have been focused on the acquisition of simple or more complex fact knowledge in healthy young adults. These studies highlight a very limited aspect of arithmetic learning as regards the contents to be learned. Meaningful arithmetic requires not only the retrieval of facts or the application of procedures, but also insight and understanding. While increasing evidence is available on the functional and structural cerebral correlates of arithmetic fact learning, little is known about the acquisition of mathematical understanding. Similarly, there is limited evidence on how understanding of quantities is acquired during child development. A further limitation of the reviewed studies consists in the selected participant samples. Most of the studies have been performed with young adults, often with university students. It is yet to be explored whether learning in children and adults follows the same patterns and whether it results in comparable modifications of activation patterns. Likewise, it is not known which behavioural or functional changes occur over the later life span. In our modern society, learning and acquisition of new skills is also essential for people in advanced age. The investigation of arithmetic learning may be a good instance to study learning processes in old age. Comparative studies over different life spans (children, young adults, old adults) could be revealing about the specific learning processes in each group and the functional and structural correlates. Possibly, such a comparative investigation could also lead to a better understanding of the specific training needs and could give new hints for the development of successful training methods for different age groups. A further very important limitation of the studies reviewed here is that almost all have been performed with subjects with at least average arithmetic performance. Thus, they give us insight into the learning processes in healthy subjects, but not into the difficulties of dyscalculic children and adults. Deficits in the acquisition of quantity information and/or of numerical information may be one of the key issues in dyscalculia. Learning studies using behavioural and neuroimaging measures could deepen our understanding of dysfunctional cerebral networks in dyscalculia and possibly lead to promising ideas for intervention.

The studies reviewed here expand our understanding of the brain mechanisms mediating arithmetic learning. Ideally, they may also contribute to the development of ideas and approaches in teaching and remediation. However, one should be very cautious to generalize the results of neuroscience studies to the classroom context. Neuroimaging findings have often been overly simplified and misunderstood (Goswami, 2006). Although neuroscience has provided important insights into arithmetic processing and learning, it cannot provide educational programmes and classroom interventions. Expectations of teachers and educational services are often too high as regards the findings of neuroimaging studies. However, neuroscience can play an important role as a part of an inter-disciplinary work-force (Ansari & Coch, 2006; Goswami, 2006; Stern, 2005).

Acknowledgment

This work was supported by TWF-2010-1-993.

Notes

1. Typically, error rates and response latencies increase as problems get larger (measured as the sum of the operands or the size of the result; for a review, Zbrodoff & Logan, 2004).
2. Note that when we use 'relatively stronger activation' or 'relative activation increases' in this article we mean a lesser degree of deactivation in line with Wu et al. (2009a). For methodological

considerations of functional neuroimaging results that may be important for the interpretation of the arithmetic learning studies, we refer to the review by Zamarian et al. (2009b).

References

Anderson, J.R., Fincham, J.M., & Douglass, S. (1999). Practice and retention: a unifying analysis. *Journal of Experimental Psychology: Learning, Memory, and Cognition*, 25, 1120–1136.

Anderson, B.J., Li, X., Alcantara, A.A., Isaacs, K.R., Black, J.E., & Greenough, W.T. (1994). Glial hypertrophy is associated with synaptogenesis following motor-skill learning, but not with angiogenesis following exercise. *Glia*, 11, 73–80.

Ansari, D. (2008). Effects of development and enculturation on number representation in the brain. *Nature Reviews Neuroscience*, 9, 278–291.

Ansari, D., & Coch, D. (2006). Bridges over troubled waters: education and cognitive neuroscience. *Trends in Cognitive Sciences*, 10, 146–151.

Arsalidou, M., & Taylor, M.J. (2011). Is 2+2=4? Meta-analyses of brain areas needed for numbers and calculations. *NeuroImage*, 54, 2382–2393.

Ashcraft, M.H. (1992). Cognitive arithmetic: a review of data and theory. *Cognition*, 44, 75–106.

Aydin, K., Ucar, A., Oguz, K.K., Okur, O.O., Agayev, A., Unal, Z., et al. (2007). Increased gray matter density in the parietal cortex of mathematicians: a voxel-based morphometry study. *American Journal of Neuroradiology*, 28, 1859–1864.

Baroody, A.J. (2003). The development of adaptive expertise and flexibility: the integration of conceptual and procedural knowledge. In: A.J. Baroody and A. Dowker (Eds). *The Development of Arithmetic Concepts and Skills: Constructing Adaptive Expertise* (pp. 1–34). Mahwah, NJ: Lawrence Erlbaum Associates.

Barrouillet, P., & Fayol, M. (1998). From algorithmic computing to direct retrieval: evidence from number and alphabetic arithmetic in children and adults. *Memory and Cognition*, 26, 355–368.

Brownell, W.A. (1935). Psychological considerations in the learning and the teaching of arithmetic. *The teaching of arithmetic. The Tenth Yearbook of the National Council of Teachers of Mathematics*. Columbia University, New York: Teachers College.

Butterworth, B., & Walsh, V. (2011). Neural basis of mathematical cognition. *Current Biology*, 21, 618–621.

Castelli, F., Glaser, D.E., & Butterworth, B. (2006). Discrete and analogue quantity processing in the parietal lobe: a function MRI study. *Proceedings of the National Academy of Science USA*, 103, 4693–4698.

Chklovskii, D.B. (2004). Synaptic connectivity and neuronal morphology: two sides of the same coin. *Neuron*, 43, 609–617.

Chochon, F., Cohen, L., van De Moortele, P.F., & Dehaene, S. (1999). Differential contributions of the left and right inferior parietal lobules to number processing. *Journal of Cognitive Neuroscience*, 11, 617–630.

Cipolotti, L., & de Lacy Costello, A. (1995). Selective impairment for simple division. *Cortex*, 31, 433–449.

Cohen Kadosh, R., Cohen Kadosh, K., Kaas, A., Henik, A., & Goebel, R. (2007). Notation-dependent and -independent representations of numbers in the parietal lobes. *Neuron*, 53, 307–314.

Cohen Kadosh, R., Soskic, S., Iuculano, T., Kanai, R., & Walsh, V. (2010). Modulating neuronal activity produces specific and long-lasting changes in numerical competence. *Current Biology*, 20, 2016–2020.

Corbett, A.J., McCusker, E.A., & Davidson, O.R. (1986). Acalculia following a dominant-hemisphere subcortical infarct. *Archives of Neurology*, 43, 964–966.

Cowan, R. (2003). Does it all add up? Changes in children's knowledge of addition combinations, strategies, and principles. In: A.J. Baroody and A. Dowker (Eds). *The Development of Arithmetic Concepts and Skills: Constructing Adaptive Expertise*(pp. 35–74). Mahwah, NJ: Lawrence Erlbaum Associates.

Dehaene, S., & Cohen, L. (1997). Cerebral pathways for calculation: double dissociation between rote verbal and quantitative knowledge of arithmetic. *Cortex*, 33, 219–250.

Dehaene, S., Piazza, M., Pinel, P., & Cohen, L. (2003). Three parietal circuits for number processing. *Cognitive Neuropsychology*, 20, 487–506.

Delazer, M. (2003). Neuropsychological findings on conceptual knowledge of arithmetic. In: A.J. Baroody & A. Dowker (Eds). *The Development of Arithmetic Concepts and Skills: Constructing Adaptive Expertise*(pp. 385–407). Mahwah, NJ: Lawrence Erlbaum Associates.

Delazer, M., & Bartha, L. (2001). Transcoding and calculation in aphasia. *Aphasiology*, 15, 649–681.

Delazer, M., Domahs, F., Bartha, L., Brenneis, C., Lochy, A., Trieb, T., et al. (2003). Learning complex arithmetic-an fMRI study. *Brain Research. Cognitive Brain Research*, 18, 76–88.

Delazer, M., Domahs, F., Lochy, A., Karner, E., Benke, T., & Poewe, W. (2004). Number processing and basal ganglia dysfunction: a single case study. *Neuropsychologia*, 42, 1050–1062.

Delazer, M., Girelli, L., Semenza, C., & Denes, G. (1999). Numerical skills and aphasia. *Journal of the International Neuropsychological Society*, 5, 213–221.

Delazer, M., Ischebeck, A., Domahs, F., Zamarian, L., Koppelstätter, F., Siedentopf, C.M., et al. (2005). Learning by strategies and learning by drill—Evidence from an fMRI study. *NeuroImage*, 25, 838–849.

Delazer, M., Karner, E., Zamarian, L., Donnemiller, E., & Benke, T. (2006). Number processing in posterior cortical atrophy – a neuropsychological case study. *Neuropsychologia*, 44, 36–51.

Domahs, F., Bartha, L., & Delazer, M. (2003). Rehabilitation of arithmetic abilities: different intervention strategies for multiplication. *Brain and Language*, 87, 165–166.

Domahs, F., & Delazer, M. (2005). Some assumptions and facts about arithmetic facts. *Psychological Sciences*, 47, 96–111.

Domahs, F., Zamarian, L., & Delazer, M. (2008). Sound arithmetic: auditory cues in the rehabilitation of impaired fact retrieval. *Neuropsychological Rehabilitation*, 18, 160–181.

Draganski, B., Gaser, C., Busch, V., Schuierer, G., Bogdahn, U., & May, A. (2004). Neuroplasticity: changes in grey matter induced by training. *Nature*, 427, 311–312.

Draganski, B., Gaser, C., Kempermann, G., Kuhn, H.G., Winkler, J., Büchel, C., et al. (2006). Temporal and spatial dynamics of brain structure changes during extensive practice. *Journal of Neuroscience*, 26, 6314–6317.

Draganski, B., & May, A. (2008). Training-induced structural changes in the adult human brain. *Behavioural Brain Research*, 192, 137–142.

Driemeyer, J., Boyke, J., Gaser, C., Büchel, C., & May, A. (2008). Changes in gray matter induced by learning – revisited. *PLoS One*, 3, e2669.

Duffau, H., Denvil, D., Lopes, M., Gasparini, F., Cohen, L., Capelle, L., et al. (2002). Intraoperative mapping of the cortical areas involved in multiplication and subtraction: An electrostimulation study in a patient with a left parietal glioma. *Journal of Neurology, Neurosurgery, and Psychiatry*, 73, 733–738.

Fehr, T., Weber, J., Willmes, K., & Herrmann, M. (2010). Neural correlates in exceptional mental arithmetic – about the neural architecture of prodigious skills. *Neuropsychologia*, 48, 1407–1416.

Fias, W., Lammertyn, J., Reynvoet, B., Dupont, P., & Orban, G.A. (2003). Parietal representation of symbolic and nonsymbolic magnitude. *Journal of Cognitive Neuroscience*, 15, 47–56.

Fuson, K.C. (1982). An analysis of the counting-on procedure in addition. In: T.H. Carpenter, J.M. Moser, and T.H. Romberg (Eds). *Addition and Subtraction: A Cognitive Perspective*. Hillsdale, NJ: Lawrence Erlbaum, pp. 67–78.

Fuson, K.C. (1988). *Children's Counting and Concepts of Number*. New York: Springer.

Girelli, L., Bartha, L., & Delazer, M. (2002). Strategic learning in the rehabilitation of semantic knowledge. *Neuropsychological Rehabilitation*, 12, 41–61.

Girelli, L., & Seron, X. (2001). Rehabilitation of number processing and calculation skills. *Aphasiology*, 15, 695–712.

Goswami, U. (2006). Neuroscience and education: from research to practice? *Nature Reviews Neuroscience*, 7, 406–411.

Grabner, R.H., Ansari, D., Reishofer, G., Stern, E., Ebner, F., & Neuper, C. (2007). Individual differences in mathematical competence predict parietal brain activation during mental calculation. *NeuroImage*, 38, 346–356.

Grabner, R.H., Ischebeck, A., Reishofer, G., Koschutnig, K., Delazer, M., Ebner, F., et al. (2009). Fact learning in complex arithmetic and figural-spatial tasks: the role of the angular gyrus and its relation to mathematical competence. *Human Brain Mapping*, 30, 2936–2952.

Grafman, J., Passafiume, D., Faglioni, P., & Boller, F. (1982). Calculation disturbances in adults with focal hemispheric damage. *Cortex*, 18, 37–50.
Gruber, O., Indefrey, P., Steinmetz, P., & Kleinschmidt, A. (2001). Dissociating neural correlates of cognitive components in mental calculation. *Cerebral Cortex*, 11, 350–359.
Hanakawa, T., Honda, M., Okada, T., Fukuyama, H., & Shibasaki, H. (2003). Neural correlates underlying mental calculation in abacus experts: a functional magnetic resonance imaging study. *NeuroImage*, 19, 296–307.
Hittmair-Delazer, M., Sailer, U., & Benke, T. (1995). Impaired arithmetic facts but intact conceptual knowledge – single-case study of dyscalculia. *Cortex*, 31, 139–147.
Hittmair-Delazer, M., Semenza, C., & Denes, G. (1994). Concepts and facts in calculation. *Brain*, 117, 715–728.
Ischebeck, A., Zamarian, L., Egger, K., Schocke, M., & Delazer, M. (2007). Imaging early practice effects in arithmetic. *NeuroImage*, 36, 993–1003.
Ischebeck, A., Zamarian, L., Schocke, M., & Delazer, M. (2009). Flexible transfer of knowledge in mental arithmetic – an fMRI study. *NeuroImage*, 44, 1103–1112.
Ischebeck, A., Zamarian, L., Siedentopf, C., Koppelstätter, F., Benke, T., Felber, S., et al. (2006). How specifically do we learn? Imaging the learning of multiplication and subtraction. *NeuroImage*, 30, 1365–1375.
Jonides, J. (2004). How does practice makes perfect? *Nature Neuroscience*, 7, 10–11.
Kawashima, R., Taira, M., Okita, K., Inoue, K., Tajima, N., Yoshida, H., et al. (2004). A functional MRI study of simple arithmetic – a comparison between children and adults. *Brain Research. Cognitive Brain Research*, 18, 227–233.
Kazui, H., Kitagaki, H., & Mori, E. (2000). Cortical activation during retrieval of arithmetical facts and actual calculation: a functional magnetic resonance imaging study. *Psychiatry and Clinical Neurosciences*, 54, 479–485.
Kong, J., Wang, C., Kwong, K., Vangel, M., Chua, E., & Gollub, R. (2005). The neural substrate of arithmetic operations and procedure complexity. *Brain Research. Cognitive Brain Research*, 22, 397–405.
Lee, K.M. (2000). Cortical areas differentially involved in multiplication and subtraction: a functional magnetic resonance imaging study and correlation with a case of selective acalculia. *Annals of Neurology*, 48, 657–661.
Lemaire, P., & Siegler, R.S. (1995). Four aspects of strategic change: contributions to children's learning of multiplication. *Journal of Experimental Psychology: General*, 124, 83–97.
Lochy, A., Domahs, F., & Delazer, M. (2005). Assessment and rehabilitation of acquired calculation and number processing disorders. In: J.I.D. Campbell (Ed.). *Handbook of Mathematical Cognition*(pp. 469–485). NY: Psychology Press.
Logan, G.D. (1988). Toward an instance theory of automatization. *Psychological Review*, 95, 492–527.
Logan, G.D., & Klapp, S.T. (1991). Automatizing alphabet arithmetic: I. Is extended practice necessary to produce automaticity? *Journal of Experimental Psychology: Learning, Memory, and Cognition*, 17, 179–195.
Lucchelli, F., & De Renzi, E. (1993). Primary dyscalculia after a medial frontal lesion of the left hemisphere. *Journal of Neurology, Neurosurgery, and Psychiatry*, 56, 304–307.
Maguire, E.A., Gadian, D.G., Johnsrude, I.S., Good, C.D., Ashburner, J., Frackowiak, R.S., et al. (2000). Navigation-related structural change in the hippocampi of taxi drivers. *Proceedings of the National Academy of Science USA*, 97, 4398–403.
Maguire, E.A., Spiers, H.J., Good, C.D., Hartley, T., Franckowiak, R.S.J., & Burgess, N. (2003). Navigation expertise and the human hippocampus: a structural brain imaging analysis. *Hippocampus*, 13, 250–259.
McCloskey, M. (1992). Cognitive mechanisms in numerical processing: evidence from acquired dyscalculia. *Cognition*, 44, 107–157.
McCloskey, M., Aliminosa, D., & Sokol, S.M. (1991). Facts, rules, and procedures in normal calculation: evidence from multiple single-patient studies of impaired arithmetic fact retrieval. *Brain and Cognition*, 17, 154–203.
McCloskey, M., Caramazza, A., & Basili, A. (1985). Cognitive mechanisms in number processing and calculation: evidence from dyscalculia. *Brain and Cognition*, 4, 171–196.

Mechelli, A., Crinion, J.T., Noppeney, U., O'Doherty, J., Ashburner, J., Frackowiak, R.S., et al. (2004). Neurolinguistics: structural plasticity in the bilingual brain. *Nature*, 431, 757.

Menon, V. (2010). Developmental cognitive neuroscience of arithmetic: implications for learning and education. *ZDM Mathematics Education*, 42, 515–525.

Menon, V., Rivera, S.M., White, C.D., Eliez, S., Glover, G.H., & Reiss, A.L. (2000a). Functional optimization of arithmetic processing in perfect performers. *Brain Research. Cognitive Brain Research*, 9, 343–345.

Menon, V., Rivera, S.M., White, C.D., Glover, G.H., & Reiss, A.L. (2000b). Dissociating prefrontal and parietal cortex activation during arithmetic processing. *NeuroImage*, 12, 357–365.

Molko, N., Cachia, A., Rivière, D., Mangin, J.F., Bruandet, M., LeBihan, D., et al. (2003). Functional and structural alterations of the intraparietal sulcus in a developmental dyscalculia of genetic origin. *Neuron*, 40, 847–858.

Münte, T.F., Altenmüller, E., & Jäncke, L. (2002). The musician's brain as a model of neuroplasticity. *Nature Reviews Neuroscience*, 3, 473–478.

Núñez-Peña, M.I. (2008). Effects of training on the arithmetic problem-size effect: an event-related potential study. *Experimental Brain Research*, 190, 105–110.

Pantev, C., Oostenveld, R., Engelien, A., Ross, B., Roberts, L.E., & Hoke, M. (1998). Increased auditory cortical representation in musicians. *Nature*, 392, 811–814.

Pascual-Leone, A., & Torres, F. (1993). Plasticity of the sensorimotor cortex representation of the reading finger in Braille readers. *Brain*, 116, 39–52.

Pauli, P., Lutzenberger, W., Rau, H., Birbaumer, N., Rickard, T.C., Yaroush, R.A., et al. (1994). Brain potentials during mental arithmetic: effects of extensive practice and problem difficulty. *Brain Research. Cognitive Brain Research*, 2, 21–29.

Pesenti, M., Seron, X., Samson, D., & Duroux, B. (1999). Basic and exceptional calculation abilities in a calculating prodigy: a case study. *Mathematical Cognition*, 5, 97–148.

Pesenti, M., Thioux, M., Seron, X., & De Volder, A. (2000). Neuroanatomical substrates of Arabic number processing, numerical comparison, and simple addition: a PET study. *Journal of Cognitive Neuroscience*, 12, 461–479.

Pesenti, M., Zago, L., Crivello, F., Mellet, E., Samson, D., Duroux, B., et al. (2001). Mental calculation in a prodigy is sustained by right prefrontal and medial temporal areas. *Nature Neuroscience*, 4, 103–107.

Piazza, M., Pinel, P., LeBihan, D., & Dehaene, S. (2007). A magnitude code common to numerosities and number symbols in human intraparietal cortex. *Neuron*, 53, 293–305.

Pinel, P., Dehaene, S., Riviere, D., & LeBihan, D. (2001). Modulation of parietal activation by semantic distance in a number comparison task. *NeuroImage*, 14, 1013–1026.

Poldrack, R.A. (2000). Imaging brain plasticity: conceptual and methodological issues – a theoretical review. *NeuroImage*, 12, 1–13.

Rickard, T.C. (2004). Strategy execution in cognitive skill learning: an item-level test of candidate models. *Journal of Experimental Psychology: Learning, Memory, and Cognition*, 30, 65–82.

Rickard, T.C. (2005). A revised Identical Elements Model of arithmetic fact representation. *Journal of Experimental Psychology: Learning, Memory, and Cognition*, 31, 250–257.

Rickard, T.C., Romero, S.G., Basso, G., Wharton, C., Flitman, S., & Grafman, J. (2000). The calculating brain: an fMRI study. *Neuropsychologia*, 38, 325–335.

Rivera, S.M., Reiss, A.L., Eckert, M.A., & Menon, V. (2005). Developmental changes in mental arithmetic: evidence for increased functional specialization in the left inferior parietal cortex. *Cerebral Cortex*, 15, 1779–1790.

Roland, P.E., & Friberg, L. (1985). Localisation of cortical areas activated by thinking. *Journal of Neurophysiology*, 53, 1219–1243.

Sader, E., Popescu, T., Thomas, A., Terhune, D.B., Dowker, A., Mars, R.B., et al. (submitted). Examining cognitive and anatomical markers of the mathematical brain.

Semenza, C., Miceli, L., and and Girelli, L. (1997). A deficit for arithmetical procedures: lack of knowledge or lack of monitoring? *Cortex*, 33, 483–498.

Siegler, R.S. (1988). Strategy choice procedures and the development of multiplication skill. *Journal of Experimental Psychology: General*, 117, 258–275.

Simon, O., Mangin, J.F., Cohen, L., LeBihan, D., & Dehaene, S. (2002). Topographical layout of hand, eye, calculation, and language-related areas in the human parietal lobe. *Neuron*, 33, 475–487.

Snowball, A., Tachtsidis, I., Popescu, T., Thompson, J., Delazer, M., Zamarian, L., et al. (2013). Long-term enhancement of brain function and cognition using cognitive training and brain stimulation. *Current Biology*, 23, 987–993.

Stanescu-Cosson, R., Pinel, P., van De Moortele, P.F., LeBihan, D., Cohen, L., & Dehaene, S. (2000). Understanding dissociations in dyscalculia: a brain imaging study of the impact of number size on the cerebral networks for exact and approximate calculation. *Brain*, 123, 2240–2255.Stern, E. (2005). Pedagogy meets neuroscience. *Science*, 310, 745.

Takeuchi, H., Taki, Y., Sassa, Y., Hashizume, H., Sekiguchi, A., Fukushima, A., et al. (2011). Working memory training using mental calculation impacts regional gray matter of the frontal and parietal regions. *PLoS One*, 6, e23175.

Taubert, M., Draganski, B., Anwander, A., Müller, K., Horstmann, A., Villringer, A., et al. (2010). Dynamic properties of human brain structure: learning-related changes in cortical areas and associated fiber connections. *Journal of Neuroscience*, 30, 11670–11677.

Thompson, P.M., Cannon, T.D., Narr, K.L., van Erp, T., Poutanen, V.P., Huttunen, M., et al. (2001). Genetic influences on brain structure. *Nature Neuroscience*, 4, 1253–1258.

Thorndike, E.L. (1922). *The Psychology of Arithmetic*. New York: The Macmillan Co.

Van Harskamp, N.J., & Cipolotti, L. (2001). Selective impairments for addition, subtraction and multiplication: Implications for the organisation of arithmetical facts. *Cortex*, 37, 363–388.

Venkatraman, V., Ansari, D., & Chee, M.W.L. (2005). Neural correlates of symbolic and non-symbolic arithmetic. *Neuropsychologia*, 43, 744–753.

Warrington, E.K. (1982). The fractionation of arithmetical skills: a single case study. *Quarterly Journal of Experimental Psychology A*, 34, 31–51.

Wenger, M.J. (1999). On the whats and hows of retrieval in the acquisition of a simple skill. *Journal of Experimental Psychology: Learning, Memory, and Cognition*, 25, 1137–1160.

Whitaker, H.A., Habiger, T., & Ivers, R. (1985). Acalculia from a lenticular-caudate infarction. *Neurology*, 35, 161.

Wu, S.S., Chang, T.T., Majid, A., Caspers, S., Eickhoff, S.B., & Menon, V. (2009a). Functional heterogeneity of inferior parietal cortex during mathematical cognition assessed with cytoarchitectonic probability maps. *Cerebral Cortex*, 19, 2930–2945.

Wu, T.H., Chen, C.L., Huang, Y.H., Liu, R.S., Hsieh, J.C., & Lee, J.J. (2009b). Effects of long-term practice and task complexity on brain activities when performing abacus-based mental calculations: a PET study. *European Journal of Nuclear Medicine and Molecular Imaging*, 36, 436–445.

Zago, L., Pesenti, M., Mellet, E., Crivello, F., Mazoyer, B., & Tzourio-Mazoyer, N. (2001). Neural correlates of simple and complex mental calculation. *NeuroImage*, 13, 314–327.

Zamarian, L., Bodner, T., Revkin, S.K., Benke, T., Boesch, S., Donnemiller, E., et al. (2009a). Numerical deficits in a single case of basal ganglia dysfunction. *Neurocase*, 15, 390–404.

Zamarian, L., Ischebeck, A., & Delazer, M. (2009b). Neuroscience of learning arithmetic—evidence from brain imaging studies. *Neuroscience and Biobehavioral Reviews*, 33, 909–925.

Zamarian, L., López Rolón, A., & Delazer, M. (2007). Neuropsychological case studies on arithmetic processing. In: D.B. Berch and M.M.M. Mazzocco (Eds). *Why Is Math So Hard for Some Children? The Nature and Origins of Mathematical Learning Difficulties and Disabilities*(pp. 245–263). Baltimore: Brookes Publishing Co. Inc.

Zarnhofer, S., Braunstein, V., Ebner, F., Koschutnig, K., Neuper, C., Reishofer, G., et al. (2012). The influence of verbalization on the pattern of cortical activation during mental arithmetic. *Behavioral and Brain Functions*, 8, 13.

Zaunmüller, L., Domahs, F., Dressel, K., Lonnemann, J., Klein, E., Ischebeck, A., et al. (2008). Rehabilitation of arithmetic fact retrieval via extensive practice: a combined fMRI and behavioural case-study. *Neuropsychological Rehabilitation*, 19, 422–443.

Zbrodoff, N. J., & Logan, G. D. (2004). What everyone finds: the problem size effect. In: J. I. D. Campbell (Ed.). *Handbook of Mathematical Cognition* (pp. 331–346). New York: Psychology Press.

PART VII

INDIVIDUAL DIFFERENCES

CHAPTER 46

INDIVIDUAL DIFFERENCES

navigator chapter

CHRIS DONLAN

The study of individual differences has been a cornerstone of psychological science from its inception. Goldstein (2012) offers an insightful historical perspective, including a summary of the development of statistical methods for evaluating the strength of association between variables. Francis Galton's invention of the technique of correlation, and his use of regression to predict unknown measurements from known measurements (Galton 1890), provided the basis for the introduction of factor analysis by Spearman (1904). Subsequent refinements (e.g. Joreskog 1969) have produced the complex techniques of structural equation modelling in current use. This continuity of methodological approaches over more than century is remarkable. However, Goldstein highlights a further and less edifying continuity, based on Galton's 'use of evidence that does not *contradict* his theory as constituting strong evidence in its favour' (Goldstein 2012, p. 149). Despite his commitment to empirical approaches, Galton's enthusiasm for genetic explanations of individual differences made him, at times, blind to alternative (environmental) accounts (Galton 1890; Goldstein 2012, pp. 149–150). There is a lesson for us here: the weakness of correlational models as tools for experimental study has not altered as those models have become more sophisticated. In fact there is a risk that the complexity of current analyses may obscure limitations in design. As we explore the study of individual differences in mathematical skills and knowledge it will be important to bear in mind these fundamental principles: that correlational models are limited by the measurements they contain; that evidence consistent with a certain hypothesis does not necessarily constitute strong evidence in its favour.

The challenge of understanding individual differences in mathematics is considerable, defined in part by the numerous sources of variability involved, and the complex nature of their operation. The scene is set very effectively for us by Ann Dowker (Chapter 47). Focusing particularly on arithmetic, Dowker examines (a) the extraordinarily wide range of ability evidenced in cohort studies of both schoolchildren and adults, (b) the componential nature of arithmetic itself (comprising, e.g. counting, calculation, fact retrieval, word-problem solving) and the surprising extent of discrepancies between component skills within individuals (c) the varying effects of cultural factors on specific aspects of arithmetic operations, e.g. the effects of language difference on the development of counting and reading numerals, (d) the changing nature of individual differences at different stages of development. Under (d) Dowker raises important questions about the drivers of mathematical development, and the possibility that early emerging basic cognitive processes related to magnitude or number may be critical. This is an area of considerable current interest. Spelke and Kinzler (2007) provide a summary of evidence supporting the general proposal that 'core knowledge' in key domains of cognition (including number) are present in human infants, providing foundational capacities which continue to operate in adulthood. More specific proposals concerning core knowledge in number have been made, for example, by Feigenson et al. (2004) and Butterworth et al. (2011). Of particular interest for us, much of the work in this area examines individual differences in mathematical skills as a function of individual differences in core knowledge. Some key papers (cited by Ann Dowker in Chapter 47) merit discussion here by way of introduction to the field.

Holloway and Ansari (2009) looked at the extent to which children's mathematical attainment might be accounted for by their magnitude comparison skills. They used tasks derived from the seminal study of Moyer and Landauer (1967) which first reported a quasi-psychophysical function in the time taken to choose the greater of two single-digit Arabic numerals. Of particular interest was Moyer and Landauer's finding that response time increased as the numerical distance between stimuli decreased (e.g. participants took longer to judge 8 vs. 9 than to judge 2 vs. 9). This much replicated finding was intriguing insofar as it is the inverse of the function which would be expected if the comparison were made by rehearsal of the number word sequence; rather, it corresponds closely to the response time function found for judgements of physical phenomena (brightness, loudness, size), and was subsequently dubbed the 'symbolic distance effect' (Moyer and Bayer 1976). Sekuler and Mierkiewicz (1977) went on to demonstrate a developmental function whereby the slope of the distance effect for numeral judgments reduces substantially from kindergarten through to fifth grade.

Against this background, and in response to a raft of studies exploring the 'core knowledge' framework for number (see above), Holloway and Ansari (2009) pursued a specific research question concerning the cognitive processes underlying children's arithmetic skills. Might children's basic representations of magnitude (as indicated by the distance effect) underlie their arithmetic development? Furthermore, might there be differential effects for symbolic magnitudes (arabic numerals) vs. non-symbolic magnitudes (sets of items, in this case squares). Holloway and Ansari's study was timely. A paper published in Nature by Gilmore et al. (2007) had proposed that non-symbolic magnitude representations drive early arithmetic skills. This study was based on a small sample of kindergarten children, presenting an argument based primarily on the ability of these children to perform above chance levels on tasks requiring approximate addition and subtraction using arabic numerals. The tasks developed by Gilmore were ingenious, combining graphic presentation with spoken commentary, and requiring a response based on magnitude comparison. Participants had never received tuition in symbolic arithmetic; therefore, it was argued, they must be using non-symbolic magnitude representations in order to perform the symbolic operation. Holloway and Ansari took the argument further, recruiting a larger sample of older children (n=67, aged six to nine years) and using regression models to explore the relation between the distance effect for symbolic/non-symbolic stimuli and scores on a standardized test of mathematical attainment.

Holloway and Ansari took a number of wise precautions in their analyses of the data. They calculated the distance effect for each participant for each comparison task (symbolic and non-symbolic) as follows: mean RT for large distances was divided by mean RT for small distances (capturing the slope for each individual), then divided the result by mean RT for large distances (so that individual differences in RT were accounted for). The component scores for mathematical attainment (fluency and calculation), as well as composite scores, were included in the analyses. Also included was a standardized measure of reading, allowing evaluation of the specificity of any associations between basic processing (as indicated by distance effects) and mathematical skills. The resulting correlation matrix (Holloway and Ansari 2009, p. 22, Table 2) is extremely informative.

The distance effect for numeral comparisons was moderately but significantly correlated with Maths Fluency (–.339**) and Maths Calculation (–.222*). These negative correlations in were in line with expectation since Sekuler and Mierkiewicz (1977) had shown that the slope of the distance effect reduces over development. The distance effect for numeral comparisons was not correlated with Maths Composite scores, or with reading. The distance effect for non-symbolic judgments was not significantly correlated with any Maths attainment measure, but was significantly correlated with Reading (.249); note that this is a positive correlation.

Proceeding to examine specific hypotheses using regression techniques, Holloway and Ansari (Holloway and Ansari 2009, p. 23, Table 3) used a carefully considered hierarchical model which successively took account of associations between the dependent variable (Math Fluency) and age, speed of response, non-symbolic distance effect and reading, before evaluating further unique variance attributable to the distance effect for numerals. They did indeed find a significant contribution. Just under five per cent of variance in Math Fluency was uniquely accounted for by the distance effect for numerals. Applying the same regression model to Math Calculation produced very similar findings, though the unique contribution of the distance effect for numerals (accounting for three per cent of variance) did not reach significance. These findings, in particular the lack of influence of the non-symbolic distance effect on mathematical attainment, were confirmed through systematic further analysis.

We have seen how, through careful design and painstaking analysis, Holloway and Ansari were able to examine individual differences so as to elucidate the relation between basic number processing and attainment in primary school mathematics. The paper has been highly influential (at the time of writing it has 171 citations recorded by Web of Science). Of particular interest was the finding of association between symbolic comparison and mathematical attainment, in the absence of such association with non-symbolic comparison. The implications for core knowledge theories were clear. However, as we shall see, Holloway and Ansari's model was itself subject to limitations which only became clear in subsequent research.

Before examining recent research which calls for a re-examination of Holloway and Ansari's conclusions, we will consider the apparently contradictory findings of another influential paper. Mazzocco et al. (2011) used a non-symbolic comparison task, intriguingly adapted for pre-schoolers (n=17, age ranges 3; 5 to 4; 11 years) as an indicator of the child's Approximate Number System (ANS), a particular representation system proposed as a component of core knowledge of number (Feigenson et al. 2004). A strength of Mazzocco et al.'s study was its longitudinal design. They followed up each participant two years later, and administered standardized tests of mathematical attainment. Simple linear regression showed that individual differences in non-symbolic comparison predicted later mathematical attainment, accounting for 28% of variance. Taken at face value, these findings are strongly supportive of the core knowledge proposal, specifically suggesting that non-symbolic comparison (ANS) skills at pre-school are important drivers of later mathematical skills. As readers will appreciate, there are a number of reasons to be cautious about this conclusion, notably the low sample size and the study design, which is sparse in its range of measures, setting up substantial ambiguities and uncertainties in interpretation. A sample size as low as this runs the risk of producing unreliable results; a regression model, under this size restriction, may be distorted by a few exceptional scores. Most important in terms of design, is the extent to which the measurements, their timing, and the analyses to which they are subjected adequately address the developmental question under consideration. The great advantage of a longitudinal design, its aptness to address developmental change, is diminished here through paucity of measurement. The single measurement at first testing is non-symbolic comparison, leaving open the logical possibility that any correlation between this and measures at second testing may be accounted for by a further unmeasured factor. For example, if children's number knowledge were tested in the first test battery (we know from Mix et al. (2014) that even three year olds are capable to interpreting multi-digit numerals) this might correlate, via general cognitive factors, with non-symbolic comparison, and might therefore produce a spurious simple correlation with later mathematical attainment. Thus paucity of measurement at first testing imposes severe restriction on interpretation. Altogether, the issues of sample size and study design raised by close consideration of Mazzocco et al. (2011) are substantial. Nonetheless, the ingenuity of their non-symbolic comparison task, and the possibility that it is especially well adapted to the interests and skills of pre-school children, have attracted a substantial readership.

The above critique goes some way towards addressing the apparent conflict between the findings of Holloway and Ansari (2009) and Mazzocco et al. (2011). However, as indicated above, more recent work has extended the scope the 2009 study, and invited a re-examination of its findings. Goebel et al. (2014) tested 173 six year olds, re-testing 165 of these children a year later. A new measure was introduced: number knowledge was tested using a number identification task in which children were required match a spoken numeral (single-digit, two-digit, or three-digit), produced by the experimenter, to the corresponding arabic numeral target, presented along with a range of distractors. Multiple symbolic and non-symbolic comparison tasks were also used, corresponding almost exactly to those used by Holloway and Ansari (2009) but including a wider range of non-symbolic comparisons as well as letter comparisons ('which letter comes later in the alphabet?'). Standardized tests of non-verbal ability, vocabulary and arithmetic skills were conducted alongside the comparison tasks at Time 1; all the comparison tasks and the arithmetic task were repeated at Time 2. Structural equation modelling was used to construct latent variables based on the correlations between variables at Time 1. All magnitude comparison tasks (symbolic and non-symbolic) loaded on to a single latent variable, independent of letter comparison. Age, non-verbal ability, vocabulary, and arithmetic were entered as further independent latent variables. Most importantly, although magnitude comparison was a strong correlate (.60) of later arithmetic skills, its influence on these skills was completely accounted for in the structural model by number identification (which accounted for 32% of variance in Time

2 arithmetic) and Time 1 arithmetic skills (which accounted to 52% of variance in Time 2 arithmetic).

Goebel et al. (2014) used an extended set of measures, a more comprehensive design, and a more sophisticated analysis than earlier studies. We will give separate consideration to each of these important aspects of the research, then go on to examine the findings in relation to those of Holloway and Ansari (2009). First, the range of measurements used by Goebel and colleagues included a test of number knowledge. In this way they tapped in to an extensive but broadly separate strand of research in which children's knowledge of the spoken number sequence and the corresponding Arabic numerals, whatever its origins, is understood as a potent force in arithmetic development (Donlan et al. 2007; Jordan et al. 2010; Cowan et al. 2011). By including measures of number knowledge as well as comparison tasks (familiar in the context of core knowledge research) Goebel and colleagues brought together research measures which had rarely been combined before, and thereby enhanced the explanatory power of their study considerably. Furthermore, the inclusion of tests of language and non-verbal ability allowed more confident identification of specific effects of the theoretically important variables (number identification and comparison). The study design afforded several advantages. Longitudinal study of a large sample of children over a critical period in arithmetic development, with extensive repeated measurement, supported a comprehensive analytic procedure which addressed the question of developmental change directly. Several aspects of the analysis are important in this regard. The inclusion of a measure of individual differences in arithmetic at first testing is critical. Including this 'autocorrelate' allows identification of those further factors specifically associated with the growth of arithmetic skills during the period of measurement (recall that the structural model used by Goebel et al. identifies 32% of variance in later arithmetic skills accounted for by number identification, independent of the contribution of earlier arithmetic skills). Multiple measurements are also important. Several comparison tasks, symbolic and non-symbolic, were included in the study; the identification of a single latent variable comprising all but the letter comparison task is theoretically significant (this is a major point of contrast with Holloway and Ansari 2009) and affords an elegant and theoretically explicit evaluation of the drivers of change. Thus a single structural model captures the combined operation of numerous factors in the developmental process, and offers clear interpretation.

Goebel et al. (2014) set out to evaluate the factor structure underlying symbolic vs. non-symbolic magnitude comparisons and their influence on arithmetic skills in primary school. In this way they explicitly built on the findings of Holloway and Ansari (2009). The large sample size, multiple measurements and structural modelling they used produced an outcome at odds with Holloway and Ansari (2009) in the particular issue of the relation between symbolic vs. non-symbolic comparison. Goebel et al. found the two tasks to be part of a single factor; Holloway and Ansari argued that they were unrelated. What could explain this discrepancy? One possible account could be based on simple differences in measurement. Holloway and Ansari calculated distance effects for each task, based on response times for small and large distances, corrected for overall response time. It was these distance effects which failed to correlate (overall response time was highly correlated between tasks). Goebel et al. measured accuracy over a range of stimulus sets including small and large distances, but did not calculate distance effects per se. It is unclear how important this might be. More important perhaps is the strength of the correlations that were recorded; distance effects were moderately correlated with arithmetic (and showed an anomalous significant positive correlation with reading); correlations reported by Goebel et al. were extremely high (.60 to .77). Most importantly of all, the inclusion of a measure of number knowledge (matching spoken numerals to the corresponding Arabic forms) by Goebel et al. produced a radical departure from previous findings. It proved to be the sole (and substantial) unique predictor of later arithmetic (apart from the autocorrelate), suggesting that the children's ability to relate Arabic numerals to their spoken forms, rather than to representations of their magnitude, is critical to the developmental process. We await replication of this finding, but it appears to mark an important step forward in the field.

We have now completed a close examination of three studies which aim to identify the drivers of mathematical development. Each makes use of individual differences, and associations between them, in order to test specific hypotheses. The analyses used in these different papers vary considerably in their sophistication, from simple correlation and regression to multiple regression and structural equation modelling. However, our examination shows that the design of a study is crucial, defining and constraining the explanatory power of its analyses. We set out with two principles in mind: that correlational models are limited by the measurements they contain; that evidence consistent with a certain hypothesis does not necessarily constitute strong evidence in its favour. The studies we have evaluated so far provide strong support for these principles. The area of research they cover forms only a small part of the broader field represented in this section of the Companion, but the lessons we have learned are widely applicable, and should prepare us for the Chapters to come.

As indicated at the outset, Ann Dowker's Chapter (47) gives an informed overview of the range of topics covered under the general heading if Individual Differences in Mathematics. Dowker's own work has focussed particularly on the associations between 'components of arithmetic' (conceptual understanding, estimation and calculation) in children in primary school. This is highly original work of great practical importance, informed by Dowker's wide experience in intervention studies with teachers and children. She reports strong associations between the components of arithmetic, but also frequent dissociations. This finding prompts further investigation concerning the nature of mathematical difficulties. Children selected by their schools as having mathematical difficulties show greater independence between components, raising the possibility that a failure of integration may underlie some children's mathematical difficulties. This extension of findings in the typical population to apply to children with mathematical difficulties is thematic within the field.

Jo-Anne LeFevre, Emma Wells, and Carla Sowinski (Chapter 48) present the 'Pathways Model' of mathematical development, based on the interaction of cognitive capacities. Quantitative numerical capacities, linguistic capacities and Attentional capacities are sees as the drivers of mathematical development. Impressive here is the breadth of the framework. Under quantitative numerical capacities are included the basic non-symbolic processes referred to above within the context of 'core knowledge'. These are capacities which are early-emerging but specific to the domain of number. However, the Pathways model entails interaction with other capacities which operate across domains. Attentional capacities, including executive function and working memory play a central role in the model. The authors give an engaging account of the ways in which these domain-general capacities operate in the practice and development of arithmetic. Linguistic capacities provide the third central element of the model. This examined at a level of detail which is rare in mathematical research. Of great interest is the review of evidence linking grammatical development and children's growing knowledge of the count sequence. The comprehensive nature of the Pathways model is unparalleled. It draws on scholarship and research from viewpoints which are more often opposed (e.g. domain-general vs. domain-specific approaches) than interactive. The identification of linguistic capacities as a central element is distinctive, fitting well with the recent findings reported earlier in this Chapter (Goebel et al. 2014), and bringing to the forefront the significant ways in which cultural factors guide the development of mathematics.

It is salutary to observe that all the work so far discussed emanates from the USA or the UK. Research in English speaking communities has dominated. In Chapter 49, by Annemie Desoete from the University of Ghent, the balance is somewhat redressed. Desoete is particularly concerned with the identification and characteristics of children with mathematical difficulties. Drawing on the strength of European tradition she examines Piagetian tasks of classification and seriation as potentially discriminating children with mathematical difficulties. Much current research, as we have seen, is devoted to studying basic processes. The value of taking a broader view, encompassing both basic processes (as indicated by magnitude comparison) and conceptual understanding, (as indicated by conceptually oriented tasks) is underlined here. Desoete reports a study carried out with colleagues (Stock et al. 2010) in which a powerful longitudinal design is applied to problem of early identification. It is instructive to examine the ways in which this design builds on techniques outlined earlier, and extends them. A large sample of kindergarten children is tested on a battery including basic process and conceptual tasks, followed up at first and second grade with at Grade 1 and Grade 2 with standardized tests of arithmetic. The predictive value of the kindergarten tests is evaluated through regression analyses, indicating the extent to which individual differences in performance on the kindergarten battery are consistent with outcomes. These analyses confirmed significant influence of both basic and conceptual tasks. Now, based on arithmetic scores in Grades 1 and 2, children were classified according to their status as having persistent or fluctuating math difficulties, and different methods for predicting outcome status from kindergarten test performance are evaluated. The power of this prospective longitudinal design is clear, though the challenges of achieving successful prediction of outcome are considerable. In an important sense this study design represents the application of theoretical models and complex data analyses to real issues which can have profound effects on quality of life. The Chapter goes on to examine the importance of other factors, including language, motor, visual skills, as well as general cognitive factors, for mathematical difficulties. The complexity of possible influences, and the consequent variety of mathematical difficulties with which children present, sets very substantial challenges for applied research. However, significant progress has been made, and the Chapter concludes by summarizing practical evidence-based proposals for early identification.

The importance of individual differences in cognitive factors, general specific, basic or conceptual, tends to dominate research in the field. However, there is increasing recognition of the importance of emotional factors in the developmental process and in learning. Chapter 50, by Alex Moore, Nathan Rudig and Mark Ashcraft, provides a comprehensive survey of the research base in this area, They outline findings concerning interest, motivation, and self-efficacy mathematics, showing how the conjunction of positive self-report in these areas represent an 'approach constellation', associated with enhanced learning and high performance. Negative self-report conjunction constitutes 'avoidance constellation, associated with disengagement, low achievement and "math anxiety"'. Of interest is the fact that children in the early years of schooling (the stage at which cognitive systems studies of the type described earlier typically take place) very rarely show avoidance constellation, which emerges strongly in adolescence. The authors acknowledge that the correlational methodology used in studies of these phenomena. However, they point to a small but growing number of intervention studies, based on motivational theory, which test the hypothesis that reduction in math anxiety will enhance math performance. The Chapter goes on to examine, in some detail, the role of working memory in arithmetic performance, and increased influence of working memory on more complex problem solving (of the type encountered at later stages of schooling). Evidence is brought forward for specific effects of math anxiety on working memory, with particular costs to resource-demanding calculation. Recent work also shows effects of math anxiety on basic number processing. The Chapter closes with a proposal, that research across a range of areas in which math anxiety occurs (including transitory contexts such as 'choking under threat', stereotype threat) be brought together to build a broader base for theoretical and empirical study of anxiety reduction and enhancement of learning.

Chapter 51, by Lieven Verschaffel, Fien Depaepe, and Wim Van Dooren ends with a similar call to action. The chapter is based on a rich and insightful historical examination of research into children's arithmetic word-problem solving. Instructive in many ways, this account details successive approaches to understanding to a highly specific but particularly complex aspect of mathematical development. Studies, conducted over many decades, of the interface between linguistic comprehension (including grammatical and semantic elements), problem formulation, strategy selection and procedural solution have produced a fascinating evidence base which maps the development of approaches and methods in psychology more generally. The current growth of interest in language as a tool for mathematical learning is likely to motivate a careful re-examination of this literature. The authors draw particular attention to cognitive, metacognitive and affective factors as key areas of influence. Individual differences in word-problem solving have not been fully explored as a function of these factors, and should form the basis of future research.

Our examination of methodologies for evaluating individual differences is extended significantly in Chapter 52 by Julie-Ann Jordan. So far we have examined correlational techniques, elaborated through multiple regression and structural equation modelling, and looked at the ways in which developmental designs can address research questions concerning the drivers of development (e.g. through evaluation of the predictive value of factors measured at Time 1 for outcomes measured at Times 2 and 3). We have looked also at studies which examine the utility of predictive factors as early identifiers of low-performing individuals. In Chapter 52 the technique of growth curve modelling is introduced. This examines change over time in a different way. Here, instead of mapping individual differences concurrently (as in Holloway and Ansari 2009) or from one time point to another (as in Goebel et al. 2014), the progress (trajectory) of an individual across numerous time-points provides the basic measure, and individual differences in trajectories are the subject of analysis. The use of these techniques to mathematical development is relatively new. Jordan provides welcome and well-informed guidance to interpretation of findings. Several studies (many carried out, as it happens, by Julie-Ann's a namesake Nancy Jordan) examine topics familiar to us (e.g. basic processes and number knowledge, often grouped together under the heading 'Number Sense') using the new technique. Jordan's commentary underlines the importance of capturing variation in trajectories, especially where some individuals may show growth patterns which are highly discrepant. Some readers may be reminded here of Robert Siegler's influential cautionary advice concerning 'The perils of averaging data over strategies' (Siegler 1987). Jordan makes a similar point in this new context. In addition to this valuable methodological contribution, Chapter 52 review findings across a number of areas of great importance for individual differences, many of which are not examined in previous chapters. Sections are included on reading and language, socio-economic status, and teaching, curriculum and classroom experience. The latter is the particularly welcome since it is so frequently ignored or 'averaged out'. There is also a section which examines current contradictory findings in neuroscience, and identifies important design issues which may underlie them.

Our final Chapter (53) brings us full circle. We set out with reflections on Galton's work on individual differences, noting his (somewhat unscientific) devotion to explain them as wholly genetic in origin. Maria Tosto, Claire Haworth, and Yulia Kovas provide a review of current behavioural genetic research into mathematical development. The Chapter opens with a helpful guide to current methodologies (and an implicit assurance that behavioural genetic analyses are explicitly designed to be sensitive to environmental as well as genetic influence on individual differences!). The body of research reviewed here is extensive, primarily based on twin studies. The range of findings concerning the proportion of genetic vs. environmental influence is wide (though genetic factors are always significant), and a number of possible explanations are considered. Many studies have recognised the componential nature of children's mathematics, and used behavioural measurements of specific curricular areas, or of basic processes. Some of the study findings are counter-intuitive, motivating careful reconsideration of the theoretical basis for using particular measurements, and the compatibility of the methodologies involved. Particularly interesting are findings of correlated genetic influence on reading and mathematical development. Finally, the authors offer an insight into current research at a molecular level and the many challenges it presents.

As we have seen, research into Individual Differences in mathematics, particularly children's mathematics, is diverse, informative and continually evolving. The Chapters in this section provide an overview of current work in the field. Some important trends are evident. First, there is growing evidence of the importance of language in the developmental process. This marks a departure from modular approaches which have been popular in recent years. Neuropsychological research, based on studies of adults with brain damage specifically affecting mathematical function, has generated valuable models applicable to adult mathematical cognition (Butterworth 1999; Varley et al. 2005; Cipolotti and van Harskamp 2001). However, their extension to children's development (Gelman and Butterworth 2005) may have been premature. Many of our contributors support the notion that linguistic influence operates through the spoken number system and its relation to the Arabic numerals; recent evidence indicates further involvement of grammatical morphology (Almoammer et al. 2013). The developmental interaction between reading and language also emerges as an important area for future study.

Other important trends are evident. The need to include measures of affect within comprehensive models of mathematical development is indicated by several contributors. Reports of the extreme emotional reactions which can be elicited by mathematics in school raise the profile of this issue, especially as it relates to educational practice. The need to include take account of variation in input (teaching and curricular content) in any comprehensive model of development seems beyond argument, but this a pressing challenge which has been largely avoided by cognitive researchers. Current enthusiasm for Educational Neuroscience (Mareschal et al. 2013) may provide the just the motivation (and funding) that is needed to bring about the collaborations which will support such collaborative research in the future.

References

Almoammer, A., Sullivan, J., Donlan, C., Marusic, F., Zaucer, R., O'Donnell, T., and Barner, D. (2013). Grammatical morphology as a source of early number word meanings. Proceedings of the National Academy of Sciences of the United States of America. doi: 10.1073/pnas.1313652110.

Butterworth, B. (1999). *The Mathematical Brain*. Macmillan: London.

Butterworth, B., Varma, S., and Laurillard, D. (2011). Dyscalculia: From Brain to Education. *Science* 332 (6033): 1049–1053.

Cipolotti, L. and van Harskamp N. (2001). Disturbances of number processing and calculation. In R.S. Berndt, (ed.) *Handbook of Neuropsychology*, vol. 3, pp. 305–334. Elsevier Science: Amsterdam.

Cowan, R., Donlan, C., Shepherd, D.L., Cole-Fletcher, R., Saxton, M., and Hurry, J. (2011) Basic Calculation Proficiency and Mathematics Achievement in Elementary School Children. *Journal of Educational Psychology* 103 (4): 786–803.

Donlan, C., Cowan, R., Newton, E. J., and Lloyd, D. (2007). The role of language in mathematical development: Evidence from children with specific language impairments. *Cognition* 103: 23–33.

Feigenson, L., Dehaene S., and Spelke, E. (2004) Core systems of number. *Trends in Cognitive Sciences* 8 (7): 307–314.

Galton, F. 1890. Kinship and correlation. *North American Review* 150: 419–431.

Gelman, R. and Butterworth, B. (2005). Number and language: how are they related? *Trends in Cognitive Sciences* 9: 6–10.

Gilmore, C., McCarthy, S. and Spelke, E. (2007). Symbolic arithmetic knowledge without instruction. *Nature* 447 (7144): 589–590.

Goebel, S., Watson, S., Lervag, A., and Hulme, C. (2014). Children's arithmetic development: it is number knowledge, not the approximate number sense, that counts. *Psychological Science* 25 (3): 789–798.

Goldstein, H. (2012) Francis Galton, measurement, psychometrics and social progress. *Assessment in Education: Principles, Policy, and Practice* 19 (2): 147–154.

Holloway, D. and Ansari, D. (2009). Mapping numerical magnitudes onto symbols: The numerical distance effect and individual differences in children's mathematics achievement. *Journal of Experimental Child Psychology* 103: 17.

Jordan, N.C., Glutting, J., and Ramineni, C. (2010). The importance of number sense to mathematics achievement in first and third grades. *Learning and Individual Differences* 20 (2): 82–88.

Joreskog, K. 1969. A general approach to confirmatory maximum likelihood factor analysis. *Psychometrika* 34 (2): 183–202.

Mareschal, D., Butterworth, B., and Tolmie, A. (2013). *Educational Neuroscience*. Oxford: Wiley-Blackwell.

Mazzocco, M.M., Feigenson, L., and Halberda, J. (2011). Preschoolers' Precision of the Approximate Number System Predicts Later School Mathematics Performance *PLOS ONE* 6 (9), article number: e23749.

Mix, K.S., Prather, R.W., Smith, L.B., and Stockton, J.D. (2014). Young Children's Interpretation of Multidigit Number Names: From Emerging Competence to Mastery. *Child Development* 85 (3): 1306–1319.

Moyer R. and Bayer, R.H. (1976). Mental comparison and the symbolic distance effect. *Cognitive Psychology* 8 (2): 228–246.

Moyer, R.S. and Landauer, T.K. (1967). Time required for judgements of numerical inequality. *Nature* 215: 1519–1520.

Sekuler, R. and Mierkiewicz, D. (1977). Children's judgments of numerical inequality. *Child Development* 48: 630–633.

Siegler, R.S. (1987). The perils of averging over strategies—An example form children's addition. *Journal of Experimental Psychology-General* 116 (3): 250–264.

Spelke, E.S. and Kinzler, D. (2007). Core knowledge. *Developmental Science* 10 (1): 89–96.

Spearman, C. 1904. 'General intelligence' objectively determined and measured. *American Journal of Psychology* 15: 201–293.

Stock, P., Desoete, A., and Roeyers, H. (2010). Detecting children with arithmetic difficulties: evidence from a 3-year study on the role of preparatory arithmetic abilities. *Journal of Learning Disabilities* 43 (3): 250–268.

Varley, R.A., Klessinger, N.J.C., and Romanowski, C.A.J. (2005). Agrammatic but not numerate. *Proceedings of the National Academy of Sciences of the USA* 102 (9): 3519–3524.

CHAPTER 47

INDIVIDUAL DIFFERENCES IN ARITHMETICAL ABILITIES

the componential nature of arithmetic

ANN DOWKER

This chapter will discuss individual differences in arithmetic. It will deal first of all with the findings about the general large extent of such differences in both children and adults. This part of the chapter will be relatively brief, as it is not the main focus of the chapter. For a more detailed discussion of individual differences in general, readers are referred to Joanne LeFevre, Wells & Sowinski (this volume) and J.A. Jordan's (this volume).

The present chapter will then discuss findings that indicate that it is inadequate to speak of arithmetical ability as a single characteristic. Rather, it is made up of many components, which may correlate, but also show significant functional independence. Discrepancies between any two such components, in both directions, can be frequently observed. The chapter will conclude by discussing some important questions yet to be answered, regarding changes over time in the componential nature of arithmetical ability, and in the existence and nature of between-individual and within-individual differences. Is numerical ability multi-componential from the beginning, or does it start as a single ability and then differentiate? Indeed, does it make sense to speak of numerical ability as something that varies between individuals from the beginning, or do individual differences only begin at a later stage? To what extent are differences in specific components consistent over time and do they have specific predictive relationships to later-developing abilities? When might any such predictive relationships begin?

The Extent of Individual Difference in Arithmetic

There have been many studies of individual differences in arithmetic. It is clear that individual differences in this area are very great. For example, Cockcroft (1982) reported that an average British class of 11-year-olds is likely to contain the equivalent of a 7-year range in arithmetical ability. Despite many educational changes in the following 20 years, including the introduction of a standard national curriculum, almost identical results were obtained by Brown, Askew, Rhodes, Denvir, Ranson, and Wiliam (2002).

Although such individual differences are most easily studied in schoolchildren, who are undergoing regular assessments, it is clear that individual differences continue into adulthood. This is studied most often with regard to the significant proportion of the population, who experience persistent difficulties in arithmetic through adulthood. For example, Parsons & Bynner (2005) found that 22% of British 37-year-olds born in a particular week in 1958 had 'very low' numeracy skills that would make everyday tasks difficult to complete successfully, compared with only about 5% with similar difficulties in literacy. BIS (2011) found that only 22% of the working-age population in England (7.5 million adults) achieved the numeracy standard associated with a good grade in GCSE/matriculation examinations, whereas 57% of the population (19.3 million adults) achieved similar standards in literacy. Nearly half of the working population reached only the standard expected of those who complete primary school.

Although there are considerable international differences in mathematical performance, there seem to be a significant number of individuals with mathematical difficulties in most countries, including those with generally high standards of achievement in mathematics, such as Finland (Räsänen, Salminen, Wilson, Aunio, & Dehaene, 2009) and the Pacific Rim countries (Chan & Ho, 2010; Lee, Chang, & Lee, 2001).

While there is no doubt that mathematical difficulties are common (Butterworth, 2010; Geary, Hoard, Nugent, & Bailey, 2012), there are debates about the frequency of global, but specific deficits in mathematics, or even in arithmetic. Some studies suggest that it is rare for children to have difficulties with all aspects of arithmetic and at the same time no difficulties in other areas (Cowan, Donlan, Shepherd, Cole-Fletcher, Saxton, & Hurry, 2011; Dowker, 1998; Gifford & Rockliffe, 2008). For example, Gifford and Rockliffe (2008, p. 21) found that, in one group, 'no pure cases [of dyscalculia] were found, although the children presented complex patterns of learning difficulties and compensatory strategies. The range of contributory factors suggests the need for new theoretical perspectives to consider learning difficulties and the need to study individual mathematics learning trajectories.'

On the other hand, some researchers argue that 'pure' dyscalculia does exist, but only as a relatively small subset of mathematical difficulties. For example, Reigosa-Crespo, Valdes-Sosa, Butterworth, Estevez, Rodriguez, et al (2012) studied 11 652 children in Havana, Cuba from 2nd to 9th grades. At all ages, efficiency of numerical capacities (enumeration and numbers comparison) predicted over 25% of the variance of performance in a timed arithmetic test. Of the population, 3.4% had deficits in basic numerical capacities, and almost all showed calculation difficulties. An additional 9.35% showed poor calculation, without the basic numerical deficits. This indicates that there are several reasons for poor arithmetical ability, of which numerosity deficits are an important one, but hardly the only one. There were some qualitative differences between those with numerosity deficits, and other children with arithmetical difficulties. For example, the former problem was four times as common in boys as girls, while the latter showed no gender differences. Reigosa-Crespo et al. (2012) suggest that only those with numerosity deficits should be described as dyscalculic. Clearly, there is much more research to be done on the definition of dyscalculia and how it fits into the larger group of children with mathematical difficulties, and the broad range of arithmetical abilities.

Giftedness at Arithmetic

At the other end of the scale, some people are highly gifted at arithmetic. Extreme examples are the calculating prodigies who can, for example, multiply multidigit numbers in seconds or rapidly work out the day of the week on which a date will fall. These include some talented 'savant' calculators with autism and/or low IQs (Cowan, O'Connor, & Samella, 2003; Heavey, 2003; Hermelin & O'Connor, 1991), which sometimes leads to the mistaken impression that all or most outstanding calculators are autistic savants, but there are also many outstanding calculators who have no cognitive problems (Cowan, Stainthorp, Kapnogianni, & Anastasiou, 2004; Pesenti, Seron, Samson, & Duroux, 1999). There are a far larger number of people who are not super-calculators, but are gifted at arithmetic (Hope & Sherrill, 1987).

There have been some studies of mathematical giftedness (Benbow, 1992; Krutetskii, 1976), but they tend not to distinguish between ability at arithmetic and at other aspects of mathematics. Mathematics involves numerous areas – not only arithmetic, but also algebra, geometry, statistics, abstract reasoning (including proofs), etc. For example, most professional mathematicians are not outstanding calculators, although they do tend to be better at arithmetic than those with limited mathematical ability. Considerably more research is still needed on the nature of arithmetical giftedness and how it relates both to other forms of mathematical giftedness, and to arithmetical difficulties. Are arithmetical giftedness and weakness opposite ends of the same continuum, or should they be seen as qualitatively different?

Arithmetic is not Unitary: Individual Differences in Different Components of Arithmetic

Individual differences in arithmetic measured as a single unitary ability are very far from being the whole story. There is by now extensive evidence (e.g. Cowan et al, 2011; Dowker, 2005; Jordan, Mulhern, & Wylie, 2009) that arithmetical ability is not unitary, but includes a variety of components, ranging from counting to word problem solving to memory for arithmetical facts to understanding of arithmetical principles. Moreover, although the different components often correlate with one another, weaknesses in any one of them can occur relatively independently of weaknesses in the others. Weakness in even one component can ultimately take its toll on performance in other components – partly exclusively on another component, and failing to perceive and use relationships between different arithmetical processes and problems, and partly because when children fail at certain tasks, they may come to perceive themselves as 'no good at maths' and develop a negative attitude to the subject. However, the components of arithmetical thinking should not be seen as a hierarchy. An individual may perform well at an apparently difficult task (e.g. word problem solving), while performing poorly at an apparently easier component (e.g. remembering the counting word sequence.)

The evidence for the componential nature of arithmetic, and the existence of individual differences in the relative levels of functioning of the different components, comes from many converging sources. These include:

(1) Studies of patients who have become dyscalculic as the result of brain damage. For example patients can show selective impairments in knowledge of arithmetical facts; in carrying out procedures accurately; in understanding arithmetical principles; in reading or writing numbers; or in understanding and comparing the relative sizes of numbers. People may show impairments in any one of these components while showing little or no impairment in any of the other components (Dehaene & Cohen, 1997; Delazer, 2003; Demeyere, Lestou, & Humphreys, 2010; Demeyere, Rotshtein, & Humphreys, 2012; McCloskey, Alimosa, & Sokol, 1991).

Most such studies have involved patients who have experienced strokes, trauma, or other forms of selective injury to particular areas of the bran. Similar results have also been found for people who develop Alzheimer's disease and other degenerative brain disorders. For example, Kaufmann, Montanes, Jacquier, Matallana, Eibl, and Delazer (2002) found that people in the early stages of Alzheimer's disease showed a variety of dissociations between arithmetic facts, arithmetic procedures, and the ability to understand, and compare numbers. No component appeared to be a necessary prerequisite for other components.

One of the largest-scale studies of dissociations between components of numerical abilities was carried out by Cappelletti, Butterworth & Kopelman (2012). They investigated numeracy skills in 76 participants. Forty were healthy controls and 36 were patients with either neurodegenerative disorders or focal brain lesions. The participants were given a comprehensive battery of numerical and calculation tasks. In addition, patients with degenerative disorders and herpes simplex encephalitis patients also performed non-numerical semantic tasks. Results showed that all patients, including those with parietal lesions, had intact processing of number quantity. Patients with impaired semantic knowledge had much better preserved numerical than semantic knowledge. Most patients showed impaired calculation skills, but most with dementia did not. Patients also showed dissociations between different aspects of arithmetical ability.

(2) Since brain imaging techniques have become available, there is also increasing evidence that different components of arithmetic and of number representation can involve different areas and networks of the brain (e.g. Cappelletti, Lee, Freeman, & Price, 2010; Castelli, Glaser, & Butterworth, 2006; Gruber, Indefrey, Steinmetz, & Kleinschmidt, 2001; Stanescu-Cosson, Pinel, Van de Moortele, Le Bihan, Cohen, & Dehaene, 2000). For example, Castelli, Glaser & Butterworth (2006) found that continuous and discontinuous quantity tasks resulted in different patterns of brain activation. Chochon, Cohen, Van de Moortele, and Dehaene (1999) found that multiplication activated predominantly the left parietal lobe, number comparison mainly the right parietal lobe, and subtraction activated both equally. There have

been numerous brain imaging studies in recent years, which converge in showing a degree of functional independence of different numerical and arithmetical processes, but sometimes differ as to the details of the specific areas most strongly activated by different tasks. Results of brain imaging studies will not be discussed in great detail here, as there are several chapters of this book dedicated specifically to neuroscience. However, we may conclude that there is still some uncertainty and disagreement about the extent and nature of the associations and dissociations between specific functions and specific brain areas, but that findings from brain imaging studies concur with the findings from patients in supporting the view that arithmetical ability is not a single entity.

(3) Studies of typical adults also show that marked discrepancies can occur between different components of arithmetic. Indeed, typical adults often show significant individual differences even in what might appear to be very basic numerical abilities. For example, Deloche, Seron, Larroque, Magnien, Metz-Lutz, Riva, et al. (1994) gave 180 typical adult subjects the EC301, a standardized testing battery for the evaluation of brain-damaged adults in the area of calculation and number processing. Participants showed significant individual differences, even in components that might have intuitively been expected to result in ceiling effects, such as written and oral counting, and transcoding between digits and written and spoken number words. Moreover, discrepancies, akin to double dissociations, were found between participants for almost all pairs of components.

(4) There have been several factor analytic studies of arithmetic in both children and adults, based in part on the assumption that if several different factors are obtained, these may represent different components of arithmetic. It should, of course, be noted that the results of factor analytic studies are influenced by the items chosen for study in the first place, and by the factor analytic methods used. With this reservation noted, such studies concur with others in supporting componential theories of individual differences in arithmetic. Early studies of this nature (e.g. Thurstone & Thurstone, 1941; Very, 1967), supported by later work by Geary & Widaman (1992) indicate a distinction between two mathematical domains – numerical facility and mathematical reasoning. Some studies also indicated additional factors independent of the above two domains, e.g. dot counting (Thurstone & Thurstone, 1941) and arithmetical estimation (Very, 1967).

(5) *Cross-cultural studies* show that different aspects of arithmetic are affected to varying degrees by age and educational experience. For example, Dellatolas, Von Aster, Willadino-Braga, Meier and Deloche (2000) studied 460 children between the ages of 7 and 10 from three countries: France, Switzerland and Brazil. They found that age and educational level strongly predicted performance in some tasks: e.g. reading and writing numerals; number comparisons; mental calculation; and word problem solving. On the other hand, these characteristics had only a slight effect on certain other tasks: counting dots; counting backwards; and estimation. There are also a number of studies that indicate that people who have little or no schooling, or who are performing poorly at school, may perform extremely well at practical mental arithmetic in the workplace or marketplace, while performing poorly at formal written arithmetic, and the reverse can also be found. (e.g. Carraher, Carraher, & Schliemann, 1985).

More specific language characteristics can also have quite specific effects on specific components of arithmetic. For example, Dowker, Bala, & Lloyd (2008) compared children attending English – and Welsh-medium schools. All the children were studying the same curriculum in mathematics, but the Welsh counting system is more transparent than the English one, resembling in this respect such languages as Chinese and Japanese (see Okamoto, this volume). Children in Welsh medium schools did not outperform their English counterparts in all respects, but they did outperform them in reading and comparing two-digit numbers. Similarly, Zuber, Pixner, Moeller, and Nuerk (2009) studied German-speaking children, whose oral counting system has the property of inversion with respect to the written umber system (e.g. the number 23 is spoken as 'drei und zwanzig' or 'three and twenty,' rather than 'twenty three'). They did less well than speakers of other languages on some numerical tasks, notably transcoding, but not on all numerical tasks.

(6) *Studies of children with and without arithmetical disabilities:* as mentioned earlier in this chapter, several studies of children with arithmetical difficulties and disabilities (Desoete, Roeyers, & De Clercq, 2004; Gifford & Rockcliffe, 2008; Jordan, Hanich, & Uberti, 2003) also suggest that arithmetic is componential, to the point that, though this is controversial, some suggest that 'pure' dyscalculia rarely occurs.

For example, double dissociations have been reported between factual and procedural knowledge of arithmetic in children (Temple, 1991). Such studies also indicate the possibility of discrepancies between different aspects of arithmetical reasoning: for example, children with arithmetical disabilities are often impaired at arithmetical word problem solving, but not at other aspects of arithmetical reasoning such as derived fact strategy use and estimation (Russell & Ginsburg, 1984). Nor is it only among children with arithmetical difficulties that such discrepancies can be found, although these are the children most often studied from this from this point of view, and may show them to a particularly marked degree. For example, Jordan, Mulhern & Wylie (2009) studied 29 5–7-year-olds, testing them at four 6-month intervals on seven aspects of arithmetic: exact calculation, story problems, approximate arithmetic, place value, calculation principles (derived fact strategies), forced retrieval of facts, and place value. Some children showed marked variations in performance across the seven tasks, showing marked discrepancies in different directions between different tasks. Moreover, when development was investigated over time, there were significant differences in developmental trajectories both between children and within children across tasks.

Relationships between Calculation, Estimation, and Derived Fact Strategies in Primary School Children

The present author has investigated relationships between arithmetic performance, arithmetical estimation and derived fact strategies in primary school children (Dowker, 1998,

2005, 2009). I will here discuss these components specifically with regard to addition, although broadly similar results have been found with regard to subtraction (Dowker, 2005).

The main study included 291 children ranging from 5 years 2 months to 9 years 10 months from three state primary schools in Oxford. 130 were boys and 161 were girls. In order to evaluate the children's competence in addition calculations, a mental calculation task was given to each child. It consisted of a list of 20 addition sums graduated in difficulty from 4 + 5, 7 + 1, etc., to 235 + 349. These sums were simultaneously presented orally and visually in a horizontal format. The children's answers were oral. Testing continued with each child until (s)he had failed to give a correct response to six successive items.

The children were then divided into five levels according to their performance on the mental calculation task. With regard to *derived fact strategies*, children were given the Use of Principles Test (Dowker, 1998, 2005). In this test, they were given the answer to an addition sum, and were then asked to solve another problem that could be solved quickly by using this answer, together with the arithmetical principle under consideration. Problems preceded by answers to numerically unrelated problems were given as controls. The exact arithmetic problems given varied according to the previously assessed calculation ability of the child, and were selected to be just a little too difficult for the child to solve unaided. Such a set of problems is referred to here as the child's base corresponding set).

Each child was shown the arithmetic problems, while the experimenter simultaneously read them to him/her. Children were asked to respond orally. The children received three arithmetical problems per principle: on rare occasions, when there was serious ambiguity about the interpretation of their responses, they received a fourth problem.

The principles investigated were as follows, in order of their difficulty for the children:

(1) The identity principle (e.g. if one is told that 9 + 4 = 13, then one can automatically give the answer '13', without calculating, if asked 'What is 9 + 4?').
(2) The commutativity principle (e.g. if 68 + 21 = 89, 21 + 68 must also be 89).
(3) The $n + 1$ principle (e.g. if 8 + 6 = 14, 8 + 7 must be 14 = 1 OR 15.)
(4) The $n - 1$ principle (e.g. if 23 + 44 = 67, 23 + 43 must be 67–1 or 66).
(5) The addition/subtraction inverse principle (e.g. if 46 + 27 = 73, then 73 − 27 must be 46).

A child was deemed to be able to use a principle if (s)he could explain it and/or used it to derive at least two out of three unknown arithmetical facts, while being unable to calculate any sums of similar difficulty when there was no opportunity to use the principle.

For the estimation task (Dowker, 1997), each child was presented with a set of addition problems within their base correspondence as defined above. Each set included a group of nine sums to which a pair of imaginary characters ('Tom and Mary') estimated answers. Each set of 'Tom and Mary's' estimates included three good estimates (e.g. '7 + 2 = 10', '71 + 18 = 90'); three that were too small; and three that were too large. The children were asked to evaluate each guess on a five-point scale from 'very good' to 'very silly', represented by a set of schematic faces ranging from a broad smile to a deep frown. They were themselves asked to suggest 'good guesses' to the sums. The Estimation score was the number of reasonable estimates, out of a maximum score of 9, within the base correspondence. Reasonable estimates were defined as those that were within 30% of the correct answer, and were also larger than each of the addends.

Results showed significant independent effects of addition level on both the number of principles used in derived fact strategies) and the estimation score for, despite the fact that the difficulty of the sums presented to children for the estimation and derived fact strategy tasks was adapted to their assessed performance in calculation (the base correspondence). A strong independent relationship between derived fact strategy use and estimation was found, even after controlling for addition calculation level and for Verbal and Performance IQ scores. Nevertheless, individual children, at all achievement levels, showed marked discrepancies in all directions between their scores on all possible pairs of arithmetic tasks.

Thus, this study of unselected primary school children shows both a general association between different arithmetic tasks and, at the same time, frequent dissociations between them. It may be that some age or ability groups would show more or fewer dissociations than this group. Indeed, there is some evidence that the relationships between different aspects of arithmetic may be less strong in at least some groups of children with arithmetical difficulties.

Dowker (2009) studied 204 children who had been selected by their schools as having arithmetical difficulties and needing intervention (a subset of the children described in the intervention study of Dowker & Sigley, 2010) and compared them with 135 unselected children of similar age and school background studied at the same time. The unselected children gave very similar results to those in the earlier studies, despite some changes in mathematics teaching over the intervening period. Not surprisingly, the children with mathematical difficulties performed less well than the unselected children on all tasks: addition calculation, estimation and derived fact strategy use, as well as on standardized arithmetic tests. Moreover, they showed even greater functional independence between the different tasks: neither derived fact strategies nor estimation was independently related to standardized test performance in Arithmetic, British Abilities Scales Basic Numbers Skills test, or the WOND Numerical Operations test. In contrast with the unselected group, derived fact strategy use and estimation not only did not show an independent relationship in this group, but were not even correlated significantly *before* other factors were partialled out. This was not due to floor effects, as there was considerable variance in all scores.

It may be that in a group of children with arithmetical difficulties, there is even less relationship between different arithmetical components than in a typical sample. Perhaps in completely typical mathematical development, different components, though perhaps functionally separable, inform and reinforce one another in the course of development (as Baroody & Ginsburg, 1986) propose for the development of principles and procedures in younger children, in their 'mutual development' theory). In children with arithmetical difficulties, this integration may not occur to the same extent, either because it is impeded by marked weaknesses in individual components, or because of a failure in the integrative process itself.

How Early Do Individual Differences Begin?

An important, but difficult, question is when these within- and between-individual differences in numerical abilities begin. Numerical abilities can be assessed in infancy, but it is not entirely clear to what extent there are individual differences at this stage. It is still not clear whether numerical ability is componential from its inception or whether there is an initial unitary number concept that splits and becomes diversified with age. If the latter is the case, then the diversification would have to occur at a very young age. In order to investigate this question, it would be necessary to carry out longitudinal studies from infancy through the preschool period, and to investigate whether infants' responses in habituation studies involving number discriminations predict anything about individual differences in later preschool and school age.

There is still a lot of controversy about which numerical abilities emerge first in infancy, to what extent they elicit individual differences, and which are most predictive of later arithmetical cognition and performance. Moreover, many of the studies of such presumed foundational numerical abilities have, in fact, been carried out with children considerably beyond infancy, making it difficult to tell when the abilities, and the individual differences between them, had developed. Abilities that are emphasized by different researchers include:

(1) *The ability to subitize:* to recognize small numerical quantities without needing to count them. There is a lot of evidence that the ability to recognize quantities up to 3 is universal in infants and even in many animals, although there is a great deal of controversy about the ways in which this distinction is made, and the extent to which it is based on an abstract understanding of number versus perceptual factors.

Butterworth (1999, 2010) considers that deficits in this numerosity coding system are responsible for dyscalculia. However, he considers that individual differences in arithmetic as a whole are unlikely to be related to numerosity coding. In his view, individuals either do or do not have the numerosity coding module. If they do not, then dyscalculia results; if they do, then it is still possible for them to be low achievers in mathematics for a variety of reasons, but these are unrelated to dyscalculia.

(2) *The ability to estimate and compare quantities approximately:* according to Dehaene (1997) and others, this ability is universally present from infancy, before children have number words. There is a significant amount of evidence that approximate arithmetic and magnitude comparison are correlated with later mathematical performance. Of the basic numerical abilities proposed as foundations for later mathematical ability, this is perhaps the one that has so far received the largest amount of longitudinal study.

There is, indeed, some evidence that approximation and magnitude comparison are predictive of later arithmetical ability (Holloway & Ansari, 2009; also see Gilmore, this volume). Halberda, Mazzocco & Feigenson (2008) performed a correlational study, where they found large individual differences in the non-verbal approximation abilities of 14-year-old children, supporting the earlier findings that indicate that individual differences can be found in almost any numerical ability. They also found that such individual differences in approximation correlated with earlier scores on standardized maths achievement tests. Of course, such correlations do not indicate whether better approximation abilities are a result or cause of better arithmetical performance. A later study by Mazzocco, Feigenson & Halberda (2011) suggested that the differences in approximate mathematical abilities may be causal. They found that preschoolers' approximate numerical abilities predicted their performance on a school mathematics test at 6 years of age, but did not predict non-numerical cognitive abilities. Similarly, De Smedt, Verschaffel & Ghesquire (2009) gave 6-year-old school beginners a number comparison task and found that the size of the distance effect predicted mathematical achievement 1 year later.

(3) *Spontaneous focusing on numerosity:* attending to the numerical aspects of a scene, rather than other aspects, e.g. to the number of objects in a picture, rather than their colour or spatial arrangement. Hannula, Lepola and Lehtinen (2010) found that Finnish kindergarten children showed significant individual differences in spontaneous focusing on numerosity. Moreover, these individual differences predicted performance in arithmetic, but not in reading, 2 years later. It is still an open question whether such spontaneous focusing on numerosity is a foundational ability emerging in infancy, or whether it emerges in the preschool years with the development of counting.

(4) *Domain-general cognitive abilities:* Piaget (1952) and his followers considered logical development to be essential to an understanding of mathematics. Before the stage of 'concrete operations,' counting and arithmetic were just rote-learned procedures and did not imply understanding. This is not currently a commonly held view, at least in its purest form, but in the past meant that many researchers assumed that foundational numerical abilities simply did not exist.

Individual Differences in Infancy

Individual differences in infants' numerical abilities have proved difficult to study, because of the limitations in infants' attentional abilities; because most studies have involved pass/fail tasks where individual infants either responded or did not respond to a numerical change, rather than producing responses that could be scored on a scale; and because it is impossible to tell whether an individual infant's failure to respond is due to an inability to perceive or comprehend the numbers involved, or due to temporary mood or distraction factors. Very recently, such studies have become possible.

Libertus & Brannon (2010) carried out what was perhaps the first study of individual differences in approximate magnitude comparison in infancy. Previous studies had indicated that, as a group, 6-month-olds can discriminate numerical changes between numbers over 3 if the values differ by at least a 1:2 ratio (e.g. 8 versus 16), but only at 9 months can they discriminate between numbers that differ by a 2;3 ratio (e.g. 8 versus 12). Libertus & Brannon developed a technique for obtaining scores for numerical discrimination and studying individual differences in these scores, by comparing children's looking time for image streams involving alternations between different numbers of dots and image streams involving only images with the same numbers of dots). Using this technique, they studied individual differences in 6-month-olds' numerical perception and found that these were unlikely to be due to mood or random factors, as they reliably predicted the same infants' numerical discrimination abilities, but not their visual short-term memory, 3 months later at the age of 9 months.

Thus, individual differences in some aspects of numerical cognition do seem to be present in infancy. It would be interesting to find out whether numerical abilities are componential at their foundation, e.g. whether both subitization and approximation are present early on, and can show functional independence, as might be suggested by the fact that dissociations between the two can be found in developmental dyscalculia (Iuculano, Hall, & Butterworth, 2008). It would also be extremely interesting to investigate whether such early individual differences predicted any aspects of numerical abilities in the preschool and primary school years, and beyond.

Longitudinal studies from infancy onwards are difficult logistically, although of very great interest and importance if achievable. A probably somewhat easier form of longitudinal study would be to investigate whether specific patterns of strengths and weaknesses in *preschool* numerical abilities predict later patterns of strengths and weaknesses in arithmetic. There is already evidence that numerical abilities in the preschool years or at the time of school entrance do tend to predict achievement in primary school arithmetic (Aunola, Lehtinen, Lerkkanen, & Nurmi, 2004; Locuniak & Jordan, 2008; Muldoon, Lewis, & Berridge, 2007; Tymms, 1999). There is, however, little evidence as to whether *specific* components of number skills in the preschool years predict *specific* components of arithmetic later on.

Preschoolers and the Componential Nature of Early Numerical Abilities

Whatever may be the case in infancy, there is strong evidence that numerical abilities are already componential by 4 years. At first sight, it might have appeared possible that arithmetical cognition might start as a unitary ability and only later become componential because different arithmetical skills are taught separately. This was investigated by Dowker (2008), whose results suggested that numerical ability is already divisible into different components, with individual differences in these components and within-individual differences between the components, before formal instruction begins.

In this study, eighty 4-year-old children were tested on:

(1) *Accuracy of counting sets of 5, 8, 10, 12 and 21 objects:* for the purposes of the study, they were classed as proficient counters if they counted 10 objects accurately.

(2) *Understanding the cardinal word principle:* as assessed by Wynn's (1990) counting-versus-grabbing task, where children had to give an adult a given quantity of objects ('give me eight counters', etc.). It was noted whether they simply grabbed a set of items, or counted to establish the quantity.

(3) *Understanding the order-irrelevance principle:* i.e. that the result of counting a set of items will not change if the items are counted in a different order, whereas adding or subtracting items will change the result of the count (Cowan, Dowker, Christakis, & Bailey, 1996). The children watched an adult count a set of objects, and were then asked to predict the result of further counts
   (i) in the reverse order;
   (ii) after the addition of an object;
   (iii) after the subtraction of an object.

(4) *Repeated addition by 1:* they were shown a set of five items, and then shown one more item being added, and asked to say, without counting, how many there are now. This was repeated up to 12.

(5) *Repeated subtraction by 1:* they were shown a set of 10 items, and then shown one item being subtracted, and asked to say, without counting, how many there are now. This was repeated down to zero.

Individual differences were marked for most tasks. Most were reasonably proficient at counting – 62% were classed as proficient counters in that they counted 10 objects accurately, and a further 10% could count 8 objects accurately. 70% understood the cardinal word principle, but only 16% passed the order irrelevance task. As results repeated addition and subtraction by 1, the children could be divided into three approximately equal groups:

- those who were already able to use an internalized counting sequence for the simplest forms of addition and subtraction;
- those who relied on a repeated 'counting-all' procedure for such tasks;
- those who were as yet unable to cope with such tasks.

There were significant relationships between some, but not all, of the numerical tasks.

However, discrepancies in both directions could be found, within individuals, between almost any two components of numerical ability. Almost no component was an absolute prerequisite for any other. Even for tasks that showed a highly significant relationship to one another, there were almost always children who could perform one of the tasks without being able to perform the other. For example, despite a significant relationship between Counting Proficiency and the Cardinal Word task, 22% of proficient counters were 'grabbers' on the Cardinal Word task, and 41% of non-proficient counters counted on the Cardinal Word task. Despite a significant relationship between the Cardinal Word task and Addition Prediction, 32% of grabbers' made correct addition predictions, while 22% of 'counters' failed even to recount on the Addition Prediction task.

Thus, even if it is the case that there is an initial unitary number concept that splits and becomes diversified with age, this would have to occur at a very young age.

In longitudinal studies of the relationships between preschool and later abilities, it is, for example, important to investigate whether discrepancies between conceptual and procedural aspects of counting at 4 predict discrepancies between conceptual and procedural aspects of arithmetic at 6 or older? Finding the answer to this question is important from the point of view of understanding the components of numerical ability, and the extent to which they are continuous over time, in particular, whether the components found in older children's arithmetical cognition have direct roots in early specific numerical abilities

The Other End of the Scale: Components of Advanced Arithmetical Ability

Much of the research on the componential nature of arithmetic has dealt with people who have difficulties with the subject. Indeed, a lot of the original evidence came from the study of patients with acquired dyscalculia resulting from brain damage. There has been less study of the componential nature of arithmetic in people within the normal range, especially beyond the primary school years, and still less with regard to mathematically gifted individual and expert mathematicians As stated above, research on mathematical giftedness tends to lump different aspects of mathematics together, often not even separating such broad categories as arithmetic, algebra and geometry. Yet if we are to have a true understanding of the componential nature of arithmetical ability, across the entire range of ability and of development, we must study its nature in those who are gifted at mathematics as well as in those who experience difficulty with the subject; in adults who are studying mathematics at an advanced level, as well as in young children in the early stages of numerical development

Wei , Yuan, Chen, & Zhou (2011) found that, among Chinese university students of mathematics, advanced mathematical skills were independently correlated with both spatial and verbal abilities, but not with basic numerical skills or computation. By contrast, Dowker, Flood, Griffiths, Harriss, and Hook (1996) found that professional mathematicians were both more accurate, and used a wider range of strategies, in an arithmetical estimation task than other groups. However, much more research is needed on the nature and components of arithmetical ability in mathematicians and mathematically gifted individuals, before any strong conclusions can be drawn.

Conclusions

There is by now overwhelming evidence that, as stated by Dowker (2005, p. 26), 'there is no such thing as arithmetical ability; only arithmetical abilities.' Arithmetical cognition is componential at least from the preschool years onwards. The next stages of study should involve the longitudinal study of the development of the components over time; at how both the components themselves and the relationships between them may change with development, and predictive relationships between components found early in development and those found later in development. Studies should also look more at individual differences, and at the possible componential nature of numerical abilities, at the two rather neglected (from this point of view) extremes of development – infancy and advanced levels of mathematical ability in adults. More work also needs to be done on whether and how different components of arithmetic relate differentially to different aspects of non-numerical cognition (e.g. verbal ability, spatial ability, working memory; see, for example, Desoete, this volume). Finally, an important emerging area for research is the ways in which an understanding of the componential nature of arithmetic can contribute to the development of interventions that are targeted toward individuals' specific strengths and weaknesses (Cohen Kadosh, Dowker, Heine, Kaufmann, & Kucian, 2013; Dowker & Sigley, 2010; Holmes & Dowker, 2013). There is, indeed, scope for bi-directional relationships between research and intervention. Interventions may be inspired by research about the components of arithmetic, and may also serve to test theories about these components.

REFERENCES

Aunola, K., Lehtinen, E., Lerkkanen, M., & Nuomi, J. (2004). Developmental dynamics of math performance from preschool to grade 2. Journal of Educational Psychology, 96, 699–713.

Baroody, A.J. & Ginsburg, H.P. (1986). The relationship between initial meaningful and mechanical knowledge of arithmetic. In J. Hiebert (ed.) Conceptual and Procedural Knowledge: The Case of Mathematics (pp. 75–112). Hillsdale, N.J.: Lawrence Erlbaum Associates.

Benbow, C.P. (1992). Academic achievement in mathematics and science between the ages of 13 and 23: are there differences between students in the top 1% of mathematical ability. Journal of Educational Psychology, 84, 51–61.

BIS (2011). BIS Research Paper 57: Skills for Life Survey Headline Findings. London: Department of Business, Innovation and Skills.

Brown, M., Askew, M., Rhodes, V., Denvir, H., Ranson, E., & Wiliam, D. (2002). Characterizing individual and cohort progress in learning numeracy: results from the Leverhulme 5-year longitudinal study. Paper delivered at American Educational Research Association conference, Chicago, 21–25 April, 2002.

Butterworth, B. (1999). The Mathematical Brain. London: Allen Lane.

Butterworth, B. (2010). Foundational numerical capacities and the origins of dyscalculia. Trends in Cognitive Sciences, 14, 534–541

Cappelletti, M., Butterworth, B., & Kopelman, M. (2012). Numeracy skills in patients with degenerative disorders and focal brain lesions: a neuropsychological investigation. Neuropsychology, 26, 1–19.

Cappelletti, M., Lee, H.L., Freeman, E.D., & Price, C.J. (2010). The role of right and left parietal lobes in the conceptual processing of numbers. Journal of Cognitive Neuroscience, 22, 331–346.

Carraher, T.N., Carraher, D.W., & Schliemann, A.D. (1985). Mathematics in the streets and in the schools. British Journal of Developmental Psychology, 3, 21–29.

Castelli, F., Glaser, D.E., & Butterworth, B. (2006). Discrete and analogue quantity processing in the parietal lobe: a functional MRI study. Proceedings of the National Academy of Sciences, 103, 4693–4698.

Chan, B.M.Y., & Ho, C.S.H. (2010). The cognitive profile of Chinese children with mathematics difficulties. Journal of Experimental Child Psychology, 107, 260–279.

Chochon, F., Cohen, L., Van de Moortele, P., & Dehaene, S. (1999). Differential contributions of the left and right inferior parietal lobules to number processing. Journal of Cognitive Neuroscience, 11, 617–630.

Cockcroft, W.H. (1982). Mathematics Counts. London: HMSO.

Cohen Kadosh, R., Dowker, A., Heine, A., Kaufmann, L., & Kucian, K. (2013). Interventions for improving numerical abilities: present and future. Trends in Neuroscience and Education, 2, 85–93.

Cowan, R., Donlan, C., Shepherd, D.L., Cole-Fletcher, R., Saxton, M., & Hurry, J. (2011). Basic calculation proficiency and mathematics achievement in elementary school children. Journal of Educational Psychology, 103, 786–803.

Cowan, R., Dowker, A., Christakis, A., & Bailey, S. (1996). Even more precisely understanding children's understanding of the order irrelevance principle. Journal of Experimental Child Psychology, 62, 84–101.

Cowan, R., O'Connor, N., & Samella, K. (2003). The skills and methods of calendrical savants. Intelligence, 31, 51–65.

Cowan, R., Stainthorp, R., Kapnogianni, S., & Anastasiou, M. (2004). The development of calendrical skills. Cognitive Development, 19, 169–178.

De Smedt, B., Verschaffel, L., & Ghesquire, P. (2009). The predictive value of magnitude comparison for individual differences in mathematics achievement. Journal of Experimental Child Psychology, 103, 469–479.

Dehaene, S. (1997). The Number Sense. London: Macmillan.

Dehaene, S., & Cohen, L. (1997). Cerebral pathways for calculation: double dissociation between rote verbal and quantitative knowledge of arithmetic. Cortex, 33, 219–250.

Delazer, M. (2003). Neuropsychological findings on conceptual knowledge of arithmetic. In A. Baroody and A. Dowker (Eds), The Development of Arithmetic Concepts and Skills (pp. 385–407). Mahwah, NJ: Erlbaum.

Dellatolas, G., Von Aster, M., Willadino-Braga, L., Meier, M., & Deloche, G. (2000). Number processing and mental calculation in school children aged 7 to 10 years: a transcultural comparison. European Child and Adolescent Psychiatry, 9, 102–111.

Deloche, G., Seron, X., Larroque, C., Magnien, C., Metz-Lutz, M., Riva, I., Scils, J., Dordain, M., Ferrand, I., Baeta, E., Basso, A., Claros Salinas, D., Gaillard, E., Goldenberg, G., Howard, D., Mazzuchi, A., Tzavaras, A., Vendrell, J., Bergego, C. & Pradat-Diehl, P. (1994). Calculation and number processing: assessment battery: role of demographic factors. Journal of Clinical and Experimental Neuropsychology, 16, 195–208.

Demeyere, N., Lestou, V., & Humphreys, G.W. (2010). Neuropsychological evidence for a dissociation in counting and subitizing. Neurocase, 16, 219–237.

Demeyere, N., Rotshtein, P., & Humphreys, G. (2012). The neuroanatomy of visual enumeration: differentiating necessary neural correlates for subitizing versus counting in a neuropsychological voxel-based morphometry study. Journal of Cognitive Neuroscience, 24, 948–964.

Desoete, A. (2014, this volume). Cognitive predictors of mathematical abilities and disabilities.

Desoete, A., Roeyers, H., & De Clercq, A. (2004). Children with mathematics learning disabilities in Belgium. Journal of Learning Disabilities, 37, 32–41.

Dowker, A. (1997). Young children's addition estimates. Mathematical Cognition, 3, 141–154.

Dowker, A. (1998). Individual differences in arithmetical development. In C. Donlan (Ed.), The Development of Mathematical Skills (pp. 275–302). London: Taylor & Francis.

Dowker, A. (2005). Individual Differences in Arithmetic. Implications for Psychology, Neuroscience and Education. Hove: Psychology Press.

Dowker, A. (2008). Individual differences in numerical abilities in preschoolers. Developmental Science, 11, 650–654.

Dowker, A. (2009). Derived fact strategies in children with and without mathematical difficulties. Cognitive Development, 24, 401–410.

Dowker, A., Bala, S., & Lloyd, D. (2008). Linguistic influences on mathematical development: how important is the transparency of the counting system. Philosophical Psychology, 21, 523–538.

Dowker, A.D., Flood, A., Griffiths, H., Harriss, L., & Hook, L. (1996). Estimation strategies of four groups. Mathematical Cognition, 2, 113–135.

Dowker, A., & Sigley, G. (2010). Targeted interventions for children with mathematical difficulties. British Journal of Educational Psychology Monographs, II (7), 65–81.

Geary, D.C., Hoard, M.K., Nugent, L., & Bailey, D.H. (2012). Mathematical cognition deficits in children with learning disabilities and persistent low achievement: a five-year prospective study. Journal of Educational Psychology, 104, 206–223.

Geary, D.C., & Widaman, K.F. (1992). Numerical cognition: on the convergence of componential and psychometric models. Intelligence, 16, 47–80.

Gifford, S., & Rockliffe, F. (2008). In search of dyscalculia. *Proceedings of the British Society for Research into Learning Mathematics*, 28(1), 21–27.

Gruber, O., Indefrey, P., Steinmetz, H., & Kleinschmidt, A. (2001). Dissociating neural correlates of cognitive components in mental calculation, Cerebral Cortex, 11, 350–369.

Halberda, J., Mazzocco, M.M., & Feigenson, L. (2008). Individual differences in non-verbal number acuity correlate with maths achievement. Nature, 455, 665.

Hannula, M.M., Lepola, J., & Lehtinen, E. (2010). Spontaneous focusing on numerosity as a domain-specific predictor of arithmetical skills. Journal of Experimental Child Psychology, 107, 395–406.

Heavey, L. (2003). Arithmetical savants. In A. Baroody and A. Dowker (Eds), The Development of Arithmetic Concepts and Skills (pp. 409–433). Mahwah, NJ: Erlbaum.

Hermelin, B., & O'Connor, N. (1991). Factors and primes: a specific numerical ability. Psychological Medicine, 20, 163–169.
Holmes, W., & Dowker, A.D. (2013). Catch Up Numeracy: a targeted intervention for children who are low attaining in mathematics. Research in Mathematics Education, 15, 249–265.
Holloway, I., & Ansari, D. (2009). Mapping numerical magnitudes onto symbols: the numerical distance effect and individual differences in children's mathematics achievement. Journal of Experimental Child Psychology, 103, 17–29.
Hope, J.A., & Sherrill, J.M. (1987). Characteristics of unskilled and skilled mental calculators. Journal for Research in Mathematics Education, 18, 98–111.
Iuculano, T., Hall, C., & Butterworth, B, (2008). Core information processing deficits in dyscalculia and low numeracy. Developmental Science, 11, 669–680
Jordan, J.A. (2014, this volume). Individual differences in children's paths to mathematical development.
Jordan, J.A., Mulhern, G., & Wylie, J. (2009) Individual differences in trajectories of arithmetical development in typically achieving 5–7-year-olds. Journal of Experimental Child Psychology, 103, 455–468.
Jordan, N.C., Blanteno, L., & Uberti, H.Z. (2003). Mathematical thinking and learning difficulties. In A. Baroody and A. Dowker (Eds), The Development of Arithmetical Concepts and Skills (pp. 359–383). Mahwah, NJ: Erlbaum.
Kaufmann, L., Montanes, P., Jacquier, M., Matallana, D., Eibl, G., & Delazer, M. (2002). About the relationship between basic numerical processing and arithmetic in early Alzheimer's disease: a follow-up study. Brain and Cognition, 48, 398–405.
Krutetskii, V.A. (1976). The Psychology of Mathematical Abilities in Schoolchildren. London: University of Chicago Press.
Lee, N.H., Chang, S.C.A., & Lee, P.Y. (2001). The role of metacognition in the learning of mathematics among the low-achieving students. Teaching and Learning, 22, 18–30.
LeFevre. J.A. , Wells, E. & Sowinski, C. (2014, this volume). Individual differences in basic arithmetical processes in children and adults.
Libertus, M.W., & Brannon, E.M. (2010). Stable individual differences in number discrimination in infancy. Developmental Science, 13, 900–906.
Locuniak, M.N., & Jordan, N.C. (2008). Using kindergarten number sense to predict calculation fluency in second grade. Journal of Learning Disabilities, 41, 451–459.
Mazzocco, M.M.M., Feigenson, L., & Halberda, J. (2011). Preschoolers' precision of the approximate number system predicts later school mathematics performance Plos One, 6, No. e23749
McCloskey, M., Aliminosa, D., & Sokol, S. (1991). Facts, rules and procedures in normal calculation: evidence from multiple single-patient studies of impaired arithmetic fact retrieval. Brain and Cognition, 17, 154–203.
Muldoon, K.P., Lewis, C., & Berridge, D. (2007). Predictors of early numeracy: is there a place for mistakes when learning about number? British Journal of Developmental Psychology, 25, 543–558.
Okamoto, Y. (2014; this volume). Mathematics learning in the USA and East Asia: influences of language.
Parsons, S., & Bynner, J. (2005). Does Numeracy Matter More? London: NRDC.
Pesenti, M., Seron, X, Samson, D., & Duroux, B. (1999) Basic and exceptional calculation abilities in a calculating prodigy: a case study. Mathematical Cognition, 5, 97–148
Piaget, J. (1952). The Child's Conception of Number. London: Routledge.
Räsänen, O., Salminen, J., Wilson, A., Aunio, P., & Dehaene, S. (2009). Computer-assisted intervention for children with low numeracy skills. Cognitive Development, 24, 450–472.
Reigosa-Crespo, V., Valdes-Sosa, Butterworth, B., Estevez, N., Rodriguez, M., Santos, E., Torres, P., Suarez, R., & Lage, A. (2012). Basic numerical abilities and prevalence of developmental dyscalculia: the Havana survey. Developmental Psychology, 48, 123–135.
Russell, R., & Ginsburg, H.P. (1984). Cognitive analysis of children's mathematical difficulties. Cognition and Instruction, 1, 217–244.
Stanescu-Cosson, R., Pinel, P., Van de Moortele, P.F., Le Bihan, D., Cohen, L., & Dehaene, S. (2000). Understanding dissociations in dyscalculia: a brain-imaging study of the impact of number size on the cerebral networks for exact and approximate calculation. Brain, 123, 2240–2255.

Temple, C.M. (1991). Procedural dyscalculia and number fact dyscalculia: double dissociation in developmental dyscalculia. Cognitive Neuropsychology, 8, 155–176.

Thurstone, L.L., & Thurstone, T.G. (1941). Factorial Studies of Intelligence. Chicago: University of Chicago Press.

Tymms, P. (1999). Baseline assessment, value-added and the prediction of reading. Journal of Research in Reading, 22, 27–36.

Very, P.S. (1967). Differential factor structures in mathematical ability. Genetic Psychology Monographs, 75, 169–207.

Wei, W., Yuan, H., Chen, C., & Zhou, X. (2011). Cognitive correlates of performance in advanced mathematics. British Journal of Educational Psychology, 82, 157–181.

Zuber, J., Pixner, S., Moeller, K., & Nuerk, H.C. (2009). Transcoding in a language with inversion and relationship to working memory. Journal of Experimental Child Psychology, 102, 60–77.

CHAPTER 48

INDIVIDUAL DIFFERENCES IN BASIC ARITHMETICAL PROCESSES IN CHILDREN AND ADULTS

JO-ANNE LEFEVRE, EMMA WELLS, AND CARLA SOWINSKI

Acknowledgments

PREPARATION of this paper was supported by the Natural Sciences and Engineering Research Council of Canada through a grant to J. LeFevre, and by the Social Sciences and Humanities Research Council of Canada through a scholarship to C. Sowinski. We thank Kristina Dunbar for her assistance in preparing the manuscript and the Math Lab Group at Carleton University for many stimulating discussions.

Introduction

Why do some people find arithmetic difficult, whereas others seem to learn it easily? This question has motivated much research on arithmetic processes over the last 30 years. As a question, it has not lost much of its mystery in that time (see Ashcraft, 1995; Ashcraft & Guillaume, 2009). The range of individual differences in arithmetic skill is great. Some adults count on their fingers to solve 9 + 5 (Smith-Chant & LeFevre, 2003), whereas others can expand 41^5 in a few seconds (Pesenti, Seron, Samson, & Duroux, 1999). In this chapter, we summarize the commonly identified sources of individual differences. Knowing how children and adults manipulate numbers in arithmetic tasks is useful both for studying mathematical thinking and for learning how human memory works in general. This review is selective, with a goal of providing some guidance to the direction of research over the next 10 years.

Numeracy, with arithmetic as the central transformative process, is key to aspects of mathematics that involve manipulating quantities. Although we might not think about it as we retrieve the answer to 3×4, the written and spoken symbolic systems that allow us to recognize the problem, and to say 'twelve' or 'douze' or write '12', are tools that were developed to manage large quantities. Theorists have observed that the human brain did not evolve to do complex calculations (Butterworth, 1999; Dehaene, 1998; Geary, 1993). Despite this lack of fit between the tool and task, much of the advances in human civilization over the last few millennia have relied on increasingly sophisticated development of mathematical systems. In particular, the development of the Hindu–Arabic symbol system (see Butterworth, 1999; Dehaene, 1998) was hugely important for many scientific advances in areas such as astronomy, commerce, physics, and statistics. Of course, we have tools (e.g. calculators and computers) that can do the actual arithmetic for us, but in the past, the process of calculation was critical for scientist and merchant alike. Arguably, the tools are only useful to the extent that the user understands the underlying arithmetical procedures and can link his or her conceptual understanding to the output that is produced by those tools.

Arithmetic involves the manipulation of quantities, typically with symbolic representations of those quantities. Elementary arithmetic includes factual, procedural, and conceptual knowledge (Bisanz & LeFevre, 1990) as applied to integers, fractions, and decimals. Factual knowledge for arithmetic includes the addition, subtraction, multiplication, and division combinations (e.g. $3 + 2 = 5$; $3 \times 2 = 6$). Procedural knowledge includes the processes (mental or physical) involved in calculation (e.g. counting, carrying or borrowing) and the rules for combining those processes to solve single- and multi-digit problems. Conceptual knowledge includes the underlying principles of how the operations work, for example, that addition and multiplication are commutative (i.e. $3 + 4 = 4 + 3$), whereas subtraction and division are not. It also includes rules such as the precedence of different operations in equations with multiple operations (e.g. $6 \times 5 - 3 = 27$, not 12). Conceptual understanding also includes the knowledge that integer addition and multiplication both result in larger quantities, but with different underlying patterns. Arithmetic forms the basis of much further mathematical development and is a central learning task in the early grades. All of these types of knowledge potentially vary across individuals, but these variations presumably can be traced to differences in cognitive capacities or learning history. Because humans vary on many dimensions, it is not surprising that we can find persistent individual differences in the arithmetic performance of children and adults (see Dowker, 2005 for an extensive discussion).

The most important (or marker) variable in arithmetic performance is the problem-size effect – people perform worse on problems with larger numbers than on problems with smaller numbers (Ashcraft & Guillaume, 2009). The problem-size effect may occur because of variability in factual knowledge (i.e. retrieving the answer from memory is more efficient than using a calculation procedure, so retrieving $2 + 3$ will be faster than computing $8 + 9$ as $9 + 9 - 1$), procedural knowledge (e.g. counting procedures take longer for larger numbers, so $6 + 8$ will be slower than $6 + 2$), or conceptual knowledge (e.g. knowing the inversion principle will be a larger advantage on $8 + 27 - 27$ than on $2 + 3 - 3$). The problem-size effect persists among adults, even when they are retrieving answers to simple problems from memory (LeFevre, Sadesky, & Bisanz, 1996; LeFevre, Bisanz, et al., 1996). Essentially, individuals who are slower and less accurate at solving arithmetic problems have larger problem-size effects than individuals who are more skilled. The problem-size effect is an effect, therefore, not a cause of individual differences, and thus we ask: 'Why do some individuals achieve fast and accurate memory retrieval, thus showing smaller problem-size effects, whereas others do not?' and 'What cognitive capabilities influence acquisition of arithmetic?' In the present chapter, we review sources of individual differences in cognitive capabilities (quantitative, attentional, linguistic, and strategic), as well as experiential and cultural factors that influence acquisition of arithmetic knowledge, and that continue to be factors in variability in performance across individuals.

Cognitive Capacities as Sources of Individual Differences

Sources of individual differences in arithmetic can be linked to cognitive capacities. Although no single model has yet been proposed that captures all of these potential sources of individual differences, LeFevre, Fast, Skwarchuk, Smith-Chant, Bisanz, Kamawar, et al. (2010) proposed the 'Pathways' model of early numeracy acquisition in an attempt to outline a set of cognitive capacities that are important in early numeracy acquisition. In this model, individual differences in three cognitive capacities – quantitative, attentional, and linguistic – are independent sources of variability in children's numeracy acquisition. The three cognitive capacities proposed as most important in the Pathways model vary in how specific they are to learning numeracy. Quantitative knowledge is domain-specific, whereas linguistic knowledge applies to a wider range of complex tasks, and attentional abilities applies to most complex components of human cognition (see also Cirino, 2011; Geary, 2011; Jordan, Kaplan, Ramineni, & Locuniak, 2008; Kleemans, Segers, & Verhoeven, 2011a; Krajewski & Schneider, 2009a,b).

Numerical Quantity Knowledge

Numerical quantity knowledge is assumed to be an innate aspect of human and animal cognition (Ansari, 2008; Feigenson, Dehaene, & Spelke, 2004). It is based on the processing of objects, usually through comparison or discrimination. Several different types of numerical quantity processes have been described. First, the ability to make approximate discriminations between quantities appears to be a fundamental characteristic of species from amphibians to humans (Ansari, 2008). The accuracy of numerical comparisons among approximate quantities are a function of the ratio of the two amounts; thus, it is easier to discriminate 16 dots from 32 dots than it is to discriminate 16 dots from 20 dots. This ability is known as *approximate number acuity*. Approximate number acuity improves with development, such that infants can discriminate ratios of 1:2, whereas some adults can discriminate ratios as small as 7:8 (Halberda & Feigenson, 2008).

The second type of numerical quantity process that has been described is the ability to quickly identify small exact quantities without counting (1, 2, 3, and possibly 4), known as *subitizing* or *parallel non-verbal enumeration* (Feigenson, 2008). Subitizing has been shown by humans as early as infancy and by some animals. However, the ability to subitize may not be specifically numerical, but may depend on visual attention, which allows apprehension of only a few discrete objects or events (O'Hearn, Hoffman, & Landau, 2011). Feigenson (2008) has shown that the ability to apprehend three objects without counting can be extended to three sets of objects. Accordingly, Feigenson has suggested that this ability is a general cognitive capacity related to attention, and is not maths specific.

Butterworth (2010) has proposed a third type of quantity knowledge; he suggests that the essential quantity capacity is the ability to determine the numerosity of sets. He notes that 'although ... [numerosity coding] might seem like a simple extension of small number coding, no upper limit or a role for attention is assumed' (p. 538). Butterworth's view preserves the idea that the numerical quantity capacity involves exact representations of the numerosity of a set (as in subitizing), but assumes that this capacity is not limited to small sets, but allows representations of larger sets. From this point of view, it is possible that numerosity representations, through experience and development, extend the basic capacity related to subitizing through various cognitive processes, such as counting by amounts greater than one.

Finally, very recently it has been proposed that sensitivity to ordinal information, specifically, the ability to determine the order of symbolic quantities, may be a core cognitive capacity recruited for numerical tasks that is independent of numerical quantity knowledge (Lyons & Beilock, 2011; Rubinsten & Sury, 2011). However, no research has yet been done with young children showing that ordinal judgments are fundamentally distinct from quantity comparisons, so this possibility remains speculative.

Are there individual differences in some or all of these fundamental numerical quantity processes? Individuals with profound deficits in their ability to acquire arithmetical knowledge (i.e. dyscalculic) may have a deficit in one or all of these quantity processes (Ansari, 2008; Butterworth, 2010). Children with Williams Syndrome, a genetic defect associated with poor arithmetic abilities, have deficient subitizing (O'Hearn et al., 2011) and are not sensitive to differences in large quantities from an early age (Van Herwegen, Ansari, Xu, & Karmiloff-Smith, 2008). In typically achieving populations, however, researchers have only recently started to link basic quantitative processes to arithmetic skills in correlational studies (e.g. Bugden & Ansari, 2011; LeFevre et al., 2010; Lyons & Beilock, 2011; Reigosa-Crespo, Valdés-Sosa, & Butterworth, 2012).

To fully understand the role of quantity knowledge in individual differences in arithmetic, it is important to distinguish between symbolic and non-symbolic representations of numerosity. Children start out without any knowledge of symbolic representations for quantity, and have to acquire knowledge of both the symbols and their links to numerical quantities. Learning to count, which includes memorizing the counting words, the rules for generating numbers beyond the initial string, and the principles of counting (e.g. one-to-one correspondence, stable order, cardinality, that is, the number counted represents the quantity of the set) occupies children from late infancy to age 5 or 6 (Cordes & Gelman, 2005). Aspects of counting knowledge are developing as late as age 10 (Kamawar, LeFevre, Bisanz, Fast, Skwarchuk, Smith-Chant, et al., 2009). Children also learn the Arabic digits, and link them to the counting words and the quantities. In school, considerable time is spent learning the rules of the number system in relation to both Arabic digit representation (such as place value) and the verbal equivalents. One view is that symbolic knowledge is built upon the foundation provided by non-symbolic numerical quantity knowledge (e.g. Halberda, Mazzocco, & Feigenson, 2008). Alternatively, Noël and Rousselle (2011) proposed that children learn symbolic knowledge initially in relation to small exact quantities and then their developing symbolic knowledge is linked back to the approximate number system. Their claims are based on the findings that non-symbolic (approximate) and symbolic quantity knowledge are not correlated for children younger than age 8, but are correlated after age 10. This latter perspective is consistent with the findings that individual differences in arithmetic are more strongly linked (in older children and adults) to symbolic number comparisons than to non-symbolic comparisons (Bugden & Ansari, 2011; De Smedt & Gilmore, 2011; Holloway & Ansari, 2009; Lyons, Ansari, & Beilock, 2012).

Are there data that link individual differences in numerical quantity processing to arithmetic? Researchers have linked speed and accuracy of arithmetic performance to (a) subitizing latency (Landerl, Bevan, & Butterworth, 2004; LeFevre et al., 2010; Willburger, Fussenegger, Moll, Wood, & Landerl, 2008); (b) magnitude as indexed by symbolic comparison tasks, that is, deciding which of two digits or sets represents the large or smaller quantity (Bugden & Ansari, 2011; De Smedt, Verschaffel & Ghesquière, 2009; Reigosa-Crespo et al., 2012); (c) to symbolic priming tasks (Defever, Sasanguie, Gebuis, & Reynvoet, 2011); (d) to approximate number acuity (Halberda et al., 2008; Lyons & Beilock, 2011) and (e) to ordinal number knowledge (Lyons & Beilock, 2009, 2011). For example, Lyons and Beilock (2011) found a correlation between approximate number acuity and arithmetic knowledge with adults that was fully mediated by participants' efficiency of identifying ordered sequences of digits. Similarly, Cirino (2011) found that approximate quantity comparisons were linked to symbolic quantity comparisons (e.g. which is larger 4 or 7?), but symbolic quantity comparisons mediated the relationship between approximate acuity and kindergarten children's ability to solve symbolic arithmetic problems (e.g. 3 + 2). Wilson, Dehaene, Dubois, and Fayol (2009) found that training children on symbolic number comparison tasks increased their ability to link numbers to quantities, but did not appear to affect their ability to compare non-symbolic quantities. Thus, although the data are not comprehensive, it appears that children's knowledge of the *symbolic* number system will be the best predictor of individual differences in arithmetic skill.

Notably, we stress that there is presumably more than one source of variability in numerical quantity processes that are relevant to arithmetic. Furthermore, arithmetic is not a monolithic skill (Dowker, 2005). Arithmetic abilities can be grouped into domains such as procedural, factual, and conceptual abilities, and each of these domains include various subcomponents. For example, factual knowledge can include addition and multiplication facts; procedural knowledge can include procedures for multiplying fractions and performing long division. Thus, individuals can vary not only in their relative strengths and weaknesses between domains, but also in their relative strengths among the subcomponents within a domain (Dowker, 2005). In support of this view, Jordan, Mulhern, and Wylie (2009) found large individual differences in task performance in a longitudinal study with 5–7-year-olds completing various mathematical tasks. Correlations were extremely variable across four time points; some tasks showed consistently positive correlations (e.g. story problems), while others showed small or variable correlations (e.g. approximate arithmetic, retrieval fluency). Despite overall linear growth in skill, there were individual differences in growth patterns (e.g. quadratic or S-shaped growth on some tasks).

In summary, the various potential sources of differences in numerical quantitative knowledge (approximate number acuity, subitizing, numerosity coding, ordinality, counting) may all contribute to variability in individual outcomes, and these differences may be influenced by many factors (e.g. brain-based, cultural, social, and educational; Dowker, 2005). The state of research at this point does not support strong conclusions about exactly what individual differences in quantity knowledge will be most important, although the links between symbolic and non-symbolic representations are starting to emerge as crucial in understanding how symbolic number system knowledge and approximate non-symbolic quantity representations contribute to individual differences in arithmetic.

Attentional Processes

Over the last 10 years, it has become increasingly clear that attentional processes are crucial to arithmetical learning and to mathematics more generally (e.g. DeStefano & LeFevre, 2004; Geary, 2010; LeFevre, DeStefano, Coleman, & Shanahan, 2005; Raghubar, Barnes, & Hecht, 2010). Attentional processes include functions such as shifting, updating, inhibition, and response selection. General constructs that subsume attentional processing include executive functions and working memory. The focus in this section will be on delineating how certain aspects of attentional processes have been linked to specific individual differences in arithmetic.

Working memory is a system responsible for the concurrent processing and storage of information during cognitive tasks. There are a variety of models of working memory (see Miyake & Shah, 1999), but the one most often used to frame research in arithmetic is the one proposed by Baddeley (2001). According to this multi-component model, the working memory system is composed of a central executive processor, two short-term storage systems – known as the phonological loop and the visual-spatial sketchpad (VSSP) – and an episodic buffer, which integrates information between the two storage systems and long-term memory. The central executive is responsible for inhibition, planning, task switching, attention, monitoring information processing, and for controlling the phonological loop and the VSSP, hence, the activities of the central executive overlap with those processes included in the construct of executive functions (Best & Miller, 2010; Engle, 2002; McCabe, Roediger, McDaniel, Balota, & Hambrick, 2010). We use the term *executive attention* to capture both the central executive component of working memory and executive functions. The phonological loop is responsible for the maintenance and manipulation of verbal information, whereas the VSSP is responsible for storage and maintenance of visual and spatial information.

Correlational and dual task studies indicate that all aspects of working memory and executive functions influence performance and acquisition of arithmetic knowledge (e.g. Imbo, Duverne, & Lemaire, 2007; Imbo & Vandierendonck, 2007a,b). Different aspects of executive functions and working memory may be relevant for learning different skills, and may be important at different points in learning.

Executive Attention

The construct of executive attention is central to understanding complex arithmetic cognition because both adults and children presumably require executive attention to acquire arithmetic knowledge. Ongoing implementation of complex mental arithmetic is demanding of executive attention (Imbo & LeFevre, 2009; Imbo & Vandierendonck, 2007a,b, 2008a,b). Executive attention is presumably useful for updating, goal maintenance, inhibition of irrelevant information, strategic processes, and many other activities that are invoked in complex arithmetic (Imbo & Vandierendonck, 2007b, 2008b). Executive attention may be less critical for tasks that are well practiced, such as retrieval of basic facts (DeStefano & LeFevre, 2004; Geary, 2010; Imbo & Vandierendonck, 2008a), although the development of automatic retrieval may be linked to individual differences in executive attention (Geary, 2010; LeFevre, Berrigan, Vendetti, Kamawar, Smith-Chant, Bisanz, et al., 2013). When counting is used to solve simple problems, such as 6 + 8, even adults appear to require executive attention (Hecht, 2002). Imbo and LeFevre (2009) found cultural differences in the extent to which adults used executive attention and the phonological loop in solving multi-digit addition problems – Chinese students used less executive attention than Canadians or Belgians. Thus, the automaticity with which individuals access arithmetic knowledge from memory will have substantial effects of patterns of performance (e.g. Imbo & Vandierendonck, 2008a; Royer, Tronsky, Chan, Jackson, & Marchant, 1999).

Phonological Loop

Researchers have observed an age-related shift such that young children rely more on the VSSP to solve arithmetic problems, whereas older children and adults rely more on the phonological loop (Bull, Espy, & Wiebe, 2008; De Smedt, Janssen, Bouwens, Verschaffel, Boets, & Ghesquière, 2009a). In general, use of verbal strategies increases with age, and thus older children and adults often use verbal strategies when solving complex arithmetic problems (e.g. Kyttala & Lehto, 2008; Imbo & Vandierendonck, 2007b; Trbovich & LeFevre, 2003). Hecht (2002) showed that use of counting-based solution procedures on single-digit addition problems implicated the phonological loop, whereas retrieval from memory did not. The phonological loop is probably used to maintain intermediate solutions on complex problems and, if the operands are not available through the solution process, to maintain them in memory as well. Imbo and LeFevre (2009) found that Dutch-speaking Belgians required more phonological loop resources to solve complex addition problems than Chinese- or English-speaking solvers, possibly because of the requirement for verbally-produced answers to be reversed (i.e. 42 is named as two-and-forty in Dutch). Given the various roles in many aspects of arithmetic, phonological processes are clearly a source of individual differences in performance. Accordingly, correlational studies in which phonological loop capacity has been linked to arithmetic performance, show that the relationship is positive. The strongest links are found between arithmetic and verbal working memory measures (such as operation span) where both maintenance and manipulation of information is required, rather than with verbal short-term memory measures, such as digit span, where only maintenance is necessary (Fuchs, Fuchs, Compton, Powell, Seethaler, & Capizzi, 2006; Passolunghi, Vercelloni, & Schadee, 2007; Raghubar et al., 2010).

Visual-Spatial Sketch Pad

In several studies, researchers have found that the VSSP is more strongly related to arithmetic performance for younger children (up to about 8 years of age; Grade 2) than for older children where the phonological loop is more strongly related to performance (Bull et al., 2008; De Smedt et al., 2009a; Holmes & McGregor, 2007). LeFevre et al. (2010) showed that performance on a VSSP task was a predictor of arithmetic performance for children in Kindergarten and Grade 2. The VSSP may be involved, for adults, in representing the relations among numbers, at least under certain conditions (Trbovich & LeFevre, 2003). Older children and adults may rely on the VSSP for difficult arithmetic problems. Kyttala and Lehto (2008) found that mental rotation abilities were related to general maths and geometry, and that passive visuospatial processing was related to mental arithmetic in adults (see also Wei, Yuan, Chen, & Zhou, 2012). In general, there are no systematic investigations of how the VSSP is related to individual differences in arithmetic that are comprehensive in terms of age span or specific arithmetic tasks.

Attention Deficit Disorder

The importance of attentional resources in mathematics suggests that a deficit in one or more of these resources may lead to mathematical difficulties. Accordingly, children with attention deficit hyperactivity disorder (ADHD) are often assumed to have co-morbid problems with mathematics, even though the research supporting this view is limited (Semrud-Clikeman & Bledsoe, 2011). Kaufmann and Nuerk (2008) compared typically-developing children with children with ADHD (aged 9–12 years) on some basic tasks, but only found differences in performance on number comparisons not on two-digit mental arithmetic. Recent large scale studies support the view that the attentional processing differences related to working memory and executive functions are the source of differences in mathematics performance between typically-developing children and those with ADHD (Polderman, Huizink, Verhulst, van Beijsterveldt, Boomsma, & Bartels, 2011; Rogers, Hwang, Toplak, Weiss, & Tannock, 2011). Advancements will probably come from the development of detailed theories of how attentional processing is tied to arithmetical learning and development.

Linguistic Knowledge

A variety of linguistic skills have been linked to individual differences in arithmetic skill.

Phonological Properties of Number Words

One way that language can affect arithmetic processing is through the linguistic properties of the spoken number word system. For example, in Chinese, each number word is a single syllable. Stigler, Lee, & Stevenson (1986) showed that Chinese speakers can pronounce digits more quickly than English-speakers, giving them an advantage in maintaining digits in short-term memory. Because mental arithmetic can involve storing the results of intermediate calculations, some differences between speakers of different languages could be due to variability in demands the number words place on the memory system. Speed of digit naming could contribute to the development of arithmetic skills, possibly by reducing the working memory load at critical junctures during acquisition. To the extent that such language differences persist, and that individuals persist in 'doing' arithmetic in the language they learned it, comparisons of adults should take language factors into account when contrasting performance.

Phonological Awareness and Grammatical Knowledge

Research has shown that linguistic abilities in the form of phonological awareness and grammatical knowledge correlate with early numeracy skills, especially the ones that require verbal processes, such as counting and number knowledge (Kleemans et al., 2011a, Kleemans, Segers, & Verhoeven, 2001b; LeFevre et al., 2010; Simmons, Singleton, & Horne, 2008). Jordan, Wylie, & Mulhern (2010) found 7-year-olds with phonological and reading difficulty had significant impairments on formal mathematical measures (i.e. number facts, calculations, and number concepts) compared with typically achieving children; half of these children met the criteria for having mathematical difficulties. Krajewski and Schneider (2009b) found that the link between phonological awareness and numerical performance was mediated by basic numerical skills. Children (age 8) with specific language impairments (poor phonological and grammatical processing) were less skilled at counting and calculation in comparison to controls; furthermore, counting fully mediated the differences in calculation skills (Donlan, Cowan, Newton, & Lloyd, 2007). These findings suggest that *general* language deficits present a challenge to the acquisition of arithmetic procedures. Linguistic abilities appear to directly facilitate children's acquisition of the symbolic number system, separately from the contribution of non-symbolic quantity knowledge.

Strategic Knowledge

Strategic knowledge has received little direct attention as a source of individual differences in arithmetic. Bisanz and LeFevre (1990) defined a strategy as 'a procedure that is invoked in a flexible, goal-oriented manner and that influences the selection and implementation of subsequent procedures' (p. 236). Strategic processes have been included in models of arithmetic (Schunn, Reder, Nhouyvanisvong, Richards, & Stroffolino, 1997; Shrager & Siegler, 1998; Siegler & Shrager, 1984). Such models include at least four dimensions (Lemaire & Siegler, 1995):

1. *Repertoire:* that is, the collection of procedures used to solve arithmetic problems.
2. The frequency with which each available procedure is applied.
3. The efficiency with which a procedure is deployed.
4. The strategic component, that is, selection of the procedure used to solve an arithmetic problem.

Here, we focus on the fourth dimension – the strategic component. One approach to this issue is to explore individual differences in the use of strategic processing as a function of task and individual parameters. Lemaire and colleagues (e.g. Lemaire & Callies, 2009; Lemaire & Lecacheur, 2011) have described variations in strategic processing as a function of age and executive attention (see also Dubé & Robinson, 2010), however, more research is needed to provide a comprehensive view of how individual differences in strategic processes may influence arithmetic performance.

Strategic processing is often described in terms of whether or not it is adaptive (see Verschaffel, Luwel, Torbeyns, & Dooren, 2009); i.e. is the chosen strategy optimal for that problem or that person. Individual differences in strategic processing have been observed in research comparing arithmetic processes across individuals who have been educated in different cultures and, therefore, have had different learning experiences. For example, in a comparison of Belgian, Chinese, and Canadian participants who solved complex addition problems, such as 47 + 29 (Imbo & LeFevre, 2009), Chinese participants were more efficient than the Belgian and Canadian participants when told which procedure to use. Despite their superior efficiency, however, Chinese participants were less adaptive; they did not choose the strategy that maximized speed as often as individuals from the other two cultures. With computational estimation of problems such as 23 × 67 (Imbo & LeFevre, 2011), Chinese participants were also less adaptive than Belgian participants. These results were surprising because adaptivity and efficiency are often assumed to be tightly coupled in skilled problem solvers (Verschaffel et al., 2009). However, a disconnect between strategic choices and efficiency has also been observed in other domains (e.g. chess experts; Bilalic, McLeod, & Gobet, 2008, 2010) and high stress situations are shown to affect highly-skilled individuals more than less-skilled individuals, suggesting that strategic abilities may interact with attentional demands (e.g. Beilock & DeCaro, 2007). Thus, further research on the links among efficiency, adaptivity, and the processes that contribute to strategic performance will inform our understanding of individual differences in arithmetic skill.

Learning Experiences: Language, Culture, Practice

A variety of different experiential factors can cause differences in performance of arithmetic problems. These effects will show up as individual differences in diverse samples, e.g. dramatic differences in the problem-size effect and the solution strategies chosen by individuals educated in China versus Canada (Imbo & LeFevre, 2009; LeFevre & Liu, 1997). Although such findings are individual differences, we discuss them separately on the assumption that they are caused by experiential differences (e.g. the amount of practice) rather than by fundamental cognitive differences. Note, however, that these experiences may interact with cognitive and linguistic factors in interesting ways (cf. Cantlon & Brannon, 2007).

Number Word Complexity in Specific Languages

One link between linguistic knowledge, experience, and individual differences comes from the number language that children and adults use to access arithmetic knowledge. Most of the existing research compares Asian languages, especially Chinese, to English. In Chinese, the verbal labels for quantities correspond very closely to the base-10 structure used in symbolic (Arabic) numerals. Thus, the quantity 11 is named as the equivalent of 'ten-one', rather than with a novel word, such as 'eleven' in English or 'onze' in French; in Chinese, 20 is named as 'two-ten', 56 as 'five-ten-six' and so on. Thus, once Chinese children have learned nine basic number words, the word for ten (shi), and the rules for combining these words, they can easily learn to count to 99. English-speaking children need to learn the words for numbers to 19, the nine decade words, plus the rule for creating numbers within decades. French-speaking children also have to contend with compound words at 70, 80, and 90 (soixante-dix, quatre-vingt, and quatre-vingt-dix). Speakers of the Basque language combine multiples of 20 and the number words from one to 19 to create larger numbers (e.g. 35 is twenty-fifteen; Colome, Laka, & Sebastián-Gallés, 2010). These variations in the transparency of the links between the number language and the underlying quantities (and the transparency of links to the symbolic base-10 system) may account for some variations in early numeracy performance between Asian and English-speaking children (also French versus English; LeFevre, Clarke, & Stringer, 2002), but the degree to which number language differences have a persistent effect (that is, beyond learning to rote count) is not clear.

Consistency and transparency of the spoken number language could support the acquisition of arithmetic (especially addition and subtraction) beyond 10. Variation in number language has an effect on arithmetic in adults. Basque speakers were faster to solve addition problems where the units corresponded to those of the answer (e.g. 20 + 15 = twenty-fifteen) than when they did not (25 + 10; Colome et al., 2010). Chinese-speaking children (age 5 years) performed better than English-speaking children on similar tasks, presumably because they understood that 10 + 2 is equivalent to the quantity ten-two (twelve in English; Ho & Fuson, 1998). Thus, to the extent that arithmetic involves verbal codes, individual differences can arise through the language that the solver uses. Many researchers do not report the language of their participants or the information is available only implicitly. The language spoken should be included in research to facilitate comparisons among individuals who speak different languages.

Counting Skill

Understanding how to count and knowing why one counts are fundamental knowledge that children acquire in support of their learning of arithmetic. Counting appears to engage the phonological system, such that counting performance is correlated with measures such as digit span and verbal working memory (Aunola, Leskinen, Lerkkanen, & Nurmi, 2004; Kleemans et al., 2011b; Passolunghi et al., 2007). Early arithmetic procedures are counting-based (Groen & Parkman, 1972). Kindergarten children vary dramatically in their counting skill regardless of whether you consider rote counting, counting of objects, or knowledge of counting principles (Aunola et al., 2004; Kleemans et al., 2011a, b; Reigosa-Crespo et al., 2012). Speed and accuracy of counting improve with age (see Cordes & Gelman, 2005). Thus, counting performance and arithmetic performance are strongly related at least until children start to use other solution procedures, such as retrieval (e.g. Geary, 2011).

Counting skill is clearly important in initial acquisition of arithmetic knowledge, but is there an ongoing causal or even reciprocal relation between counting skill and arithmetic? It is possible that variations in counting speed continue to correlate with arithmetic performance in adults, but this relation might be based on past connections. At some point the most-skilled counters stop using counting to solve arithmetic problems and move to more sophisticated, memory-based procedures, such as retrieval or derived facts (e.g. 8 + 9 = 8 + 8 + 1; Geary, Hoard, Byrd-Craven, & DeSoto, 2004). Thus, the use of counting to solve arithmetic problems becomes negatively related to performance later in childhood (Geary, 2010). For example, Jordan et al. (2008) showed that, in kindergarten, the frequency with which children used their fingers to count while solving arithmetic problems was positively related to accuracy, whereas by the end of Grade 2, use of fingers was negatively correlated with accuracy. Furthermore, counting itself may evolve – a skilled counter may count sets of objects in units larger than 1, for example. This relation between counting in larger units and arithmetic skill has not been examined directly, although LeFevre and Bisanz (1986) found that adults' speed of identifying number sequences such as 3, 6, 9, 12 was correlated with their performance on a multi-digit arithmetic measure. The role for individual differences in counting via larger units is an interesting area for research with older children and adults. Charting the longitudinal patterns of how counting relates to arithmetic performance would help to develop a better model of how numerical quantity knowledge, counting processes, and arithmetic are inter-related (Kleemans et al., 2011a,b; Krajewski & Schneider, 2009a,b).

Automaticity and Practice

Arithmetic proficiency relies heavily on the automaticity with which people access stored knowledge from memory (Pesenti, Seron, Samson, & Duroux, 1999). Many researchers have shown that there are significant individual differences in the extent to which adults can retrieve answers to simple arithmetic facts from memory (LeFevre et al., 1996; LeFevre, Bisanz et al., 1996). In combination with the view that complex tasks require attentional resources, individuals with automatic fact retrieval presumably have more cognitive capacity to devote to other aspects of the task. Procedures tend to be demanding of attentional capacity (Hecht, 2002; Imbo & Vandierendonck, 2007b, 2008b) and so less use of procedures or automatic activation of those procedures is associated with higher levels of skill and smaller problem-size effects (Cumming & Elkins, 1999).

Children with maths difficulties seem to have particular difficulty memorizing or automating their memory for arithmetic 'facts' (Fuchs et al., 2006; Geary, 2010; Jordan et al., 2008). They also acquire counting-based procedures more slowly and use them less efficiently than their peers without maths difficulties. To the extent that accurate procedures help children acquire strong memory representations for those facts (Siegler & Shrager, 1984; cf. Pyke & LeFevre, 2011), difficulty in memorizing arithmetic facts may be a consequence of poor counting skills and deficient quantity representations. Donlan et al. (2007) found evidence in support of this link; counting fluency accounted for differences in arithmetic performance between normally achieving children and those with specific language impairment. Moreover, because rote memory is assumed to be a verbal skill, some researchers have proposed that maths difficulties stem from a similar deficit as for children with dyslexia, i.e. a particular inability to link sounds to symbols. An alternative explanation is that gaining automaticity is linked with attentional capacity, which is deficient in children with maths difficulties (Geary, 2010; LeFevre et al., 2012). Further research is needed to explore the source of the automaticity difficulties for these children.

Arithmetic fact knowledge is a very tight and interconnected subset of memory with incredible repetition and overlap (consider 3 + 4, 3 × 4, ¾, and 3 – 4). It may be that forming stable and accessible memory structures with these highly interconnected symbols is challenging. On this view, it may be those individuals who can successfully inhibit extraneous associations, while enhancing relevant ones who are most successful. Accordingly, Holmes and McGregor (2007) reported that adults who showed skill at memorizing arbitrary, but ordered sequences also had good arithmetic skills. Practice retrieving specific facts seems to play a role in developing automaticity (Pyke & LeFevre, 2011). Further work on the factors that contribute to successful development of memory structures will help to illuminate this issue.

The advantage shown by children and adults from Asian cultures over those from other cultures in terms of basic arithmetic skill is linked to differences in automaticity. In particular, individuals who spent more time practicing retrieval of arithmetic facts show continuing advantages in all forms of arithmetic that are speeded (which is probably due to greater practice retrieving basic facts and practice solving problems). Changes in the beliefs about the importance of memorization in many domains (not just arithmetic) in North American society are having a profound influence on the development of automaticity among children and young adults. To the extent that individual differences in arithmetic learning and performance are linked to automaticity, we might expect to see increasing diversity as the emphasis on practice and acquisition of memory structures continue to be de-emphasized (LeFevre et al., 2009a,b). Accordingly, the role of executive attention will be larger because more attention will be needed for processing basic facts and thus less will be available for conceptual and procedural processes.

Early Experiences

Children start 'formal' schooling (kindergarten, and/or Grade 1) showing substantial individual differences in early numeracy knowledge (Aunola et al., 2004; Jordan, Kaplan, Locuniak, & Ramineni, 2007). The assumption that these individual differences are related to preschool experiences is supported by the strong links between early numeracy skill and socio-economic status (SES). Children's performance on numerical tasks that involve language, such as counting or early arithmetic, and those that require knowledge of the symbolic system, are better in middle-class than in less-advantaged groups, whereas tasks that involve quantity manipulations that do not rely on language did not vary with SES (Jordan, Huttenlocher, & Levine, 1992). Kleemans et al. (2011b) note that differences between children learning mathematics in their first versus second language are not fully accounted for by language-related factors and suggest that differences in the home experiences of children are relevant.

LeFevre et al. (2009a,b; LeFevre, Polyzoi, Skwarchuk, Fast, & Sowinski, 2010) have shown that parents' reports of their home numeracy activities show modest, but consistent patterns of relations with children's performance, even when various other factors are controlled. Vandermaas-Peeler et al. (2009) showed that parents vary in the frequency of numeracy interactions, with high-income parents more likely than low-income parents to introduce counting, quantity, or size comparisons during joint storybook reading or free play. Vandermaas-Peeler, Ferretti, & Loving (2011) found that suggesting numeracy activities to parents increased the frequency of parents' numeracy-related guidance during a board game; children who received more guidance were more likely to produce correct responses (see also Bjorklund, Hubertz, & Reubens, 2004). A great deal of additional research is needed to explore how individual differences may influence children's environments and their amount of exposure to numeracy activities.

Summary and Questions for Further Research

In this chapter, we have described research that shows that many different factors can contribute to variability in arithmetic skill. Although an awareness of numerical quantity appears to be universal, variability in that awareness and in other cognitive precursor skills related to learning the symbolic number system differentiate children. Acquisition of symbolic numbers and connecting those with numerical quantity representations seem to present a significant challenge for children. Attentional capacities are critical to learning and implementing arithmetic skills. Finally, there is an important role of experiential factors (related to practice, education, culture, and language) in arithmetic performance. Here, we propose some questions for future research.

1. How are individual differences in the most basic quantitative knowledge – approximate number acuity, subitizing, numerosity coding, ordinality detection – related to individual differences in arithmetic? In an intriguing paper, Frank, Everett, Fedorenko, & Gibson (2008) showed that individuals from a society that has few (if any) exact number words can do tasks that show an ability to match exact quantities when the objects are visible, but when they are required to represent the objects mentally (i.e. use memory), their performance reflects the approximate system. Frank et al. suggest that the symbolic number system is a cognitive tool, developed to circumvent cognitive limitations, and not a reflection of a universal linguistic/cognitive capacity. This view is consistent with Noël and Rousselle's (2011) recent proposal that acquiring the symbolic number system is the critical task of early numeracy and that acquisition of the number system influences further development of the approximate number system. The question of which and how these fundamental quantity processes are involved in acquisition of numeracy skills is thus of critical importance for understanding individual differences in arithmetic.
2. How important are early experiences (e.g. differences across number languages, exposure to quantities in number games) in children's later acquisition of arithmetic? One possibility is that such experiences provide an early learning advantage, but that children with other languages or without early experience will easily catch up. Alternatively, those early experiences may change knowledge connections in the brain, resulting in differential learning potential. Cantlon and Brannon (2007) argue that experiences (including language-specific experiences) can change or influence brain structures – thus, understanding when, how, and why experience affects brain organization and processing is critical in understanding individual differences.
3. How do emotional factors link to individual differences that are caused by cognitive or experiential factors? Individuals with high maths anxiety perform more poorly on complex arithmetic tasks than those who are less anxious (Ashcraft & Ridley, 2005). An integrated model of how various individual differences influence arithmetic performance should include emotional factors.

4. Automaticity in basic arithmetic skill is an important source of individual differences in the speed and accuracy of performance and presumably will influence any task that requires arithmetic. Lack of automaticity, crucially, will increase working memory load. How will the greatly reduced level of automaticity of basic arithmetic knowledge affect current and future students' ability to acquire and understand mathematical tasks where some familiarity with the basic arithmetic facts is presupposed? This question emphasizes the importance of attentional processes and attentional limitations in mathematical learning.

References

Ansari, D. (2008). Effects of development and enculturation on number representation in the brain. Nature Reviews Neuroscience, 9(4), 278–291.

Ashcraft, M.H. (1995). Cognitive psychology and simple arithmetic: a review and summary of new directions. Mathematical Cognition, 1, 3–34.

Ashcraft, M.H., & Guillaume, M.M. (2009). Mathematical cognition and the problem size effect. In B. Ross (Ed.), Psychology of Learning and Motivation, Vol. 51 (pp. 121–151). Burlington: Academic Press.

Ashcraft, M.H., & Ridley, K.S. (2005). Math anxiety and its cognitive consequences: a Tutorial Review. In J. I. D. Campbell (Ed.) Handbook of Mathematical Cognition (pp. 315–327). New York: Psychology Press.

Aunola, K., Leskinen, E., Lerkkanen, M.K., & Nurmi, J.E. (2004). Developmental dynamics of math performance from preschool to Grade 2. Journal of Educational Psychology, 96(4), 699–713.

Baddeley, A.D. (2001). Is working memory still working? American Psychologist, 56(11), 851–864.

Beilock, S.L., & DeCaro, M.S. (2007). From poor performance to success under stress: working memory, strategy selection, and mathematical problem solving under pressure. Journal of Experimental Psychology: Learning, Memory, and Cognition, 33(6), 983–998.

Best, J.R., & Miller, P.H. (2010). A developmental perspective on executive function. Child Development, 81(6), 1641–1660.

Bilalic, M., McLeod, P., & Gobet, F. (2008). Inflexibility of experts – Reality or myth? Quantifying the Einstellung effect in chess masters. Cognitive Psychology, 56, 73–102.

Bilalic, M., McLeod, P., & Gobet, F. (2010). The mechanism of the Einstellung (set) effect: a pervasive source of cognitive bias. Current Directions in Psychological Science, 19(2), 111–115.

Bisanz, J., & LeFevre, J. (1990). Strategic and nonstrategic processing in the development of mathematical cognition. In D. Bjorklund (Ed.), Children's Strategies: Contemporary Views of Cognitive Development (pp. 213–244). Hillsdale, NJ: Erlbaum.

Bjorklund, D., Hubertz, M., & Reubens, A. (2004). Young children's arithmetic strategies in social context: how parents contribute to children's strategy development while playing games. International Journal of Behavioral Development, 28(4), 347–357.

Bugden, S., & Ansari, D. (2011). Individual differences in children's mathematical competence are related to the intentional but not automatic processing of Arabic numerals. Cognition, 118(1), 32–44.

Bull, R., Espy, K.A., & Wiebe, S.A. (2008). Short term memory, working memory, and executive functioning in preschoolers: longitudinal predictors of mathematical achievement at age 7 years. Developmental Neuropsychology, 33(3), 205–228.

Butterworth, B. (1999). The Mathematical Brain. London: Macmillan.

Butterworth, B. (2010). Foundational numerical capacities and the origins of dyscalculia. Trends in Cognitive Sciences, 14(12), 534–541.

Cantlon, J.F., & Brannon, E.M. (2007). Adding up the effects of cultural experience on the brain. Trends in Cognitive Sciences, 11(1), 1–4.

Cirino, P.T. (2011). The interrelationships of mathematical precursors in kindergarten. Journal of Experimental Child Psychology, 108(4), 713–733.

Colomé, A., Laka, I., & Sebastián-Gallés, N. (2010). Language effects in addition: how you say it counts. Quarterly Journal of Experimental Psychology, 63(5), 965–983.

Cordes, S., & Gelman, R. (2005). Cognitive representations for numbers and mathematics. In J. I. D. Campbell (Ed.), Handbook of Mathematical Cognition. New York: Psychology Press.

Cumming, J.J., & Elkins, J. (1999). Lack of automaticity in the basic addition facts as a characteristic of arithmetic learning. Mathematical Cognition, 5, 149–180.

De Smedt, B., & Gilmore, C. (2011). Defective number module or impaired access? Numerical magnitude processing in first graders with mathematical difficulties. Journal of Experimental Child Psychology, 108, 278–292.

De Smedt, B., Janssen, R., Bouwens, K., Verschaffel, L., Boets, B., & Ghesquière, P. (2009a). Working memory and individual differences in mathematics achievement: a longitudinal study from first grade to second grade. Journal of Experimental Child Psychology, 103(2), 186–201.

De Smedt, B., Verschaffel, L., & Ghesquière, P. (2009b). The predictive value of numerical magnitude comparison for individual differences in mathematics achievement. Journal of Experimental Child Psychology, 103(4), 469–479.

Defever, E., Sasanguie, D., Gebuis, T., & Reynvoet, B. (2011). Children's representation of symbolic and nonsymbolic magnitude examined with the priming paradigm. Journal of Experimental Child Psychology, 109(2), 174–186.

Dehaene, S. (1998). The Number Sense: How the Mind Creates Mathematics. New York: Oxford University Press.

DeStefano, D., & LeFevre, J. (2004). The role of working memory in mental arithmetic. European Journal of Cognitive Psychology, 16(3), 353–386.

Donlan, C., Cowan, R., Newton, E.J., & Lloyd, D. (2007). The role of language in mathematical development: Evidence from children with specific language impairments. Cognition, 103(1), 23–33.

Dowker, A. (2005). Individual Differences in Arithmetic. Hove: Psychology Press.

Dubé, A.K., & Robinson, K.M. (2010). Accounting for individual variability in inversion shortcut use. Learning and Individual Differences, 20(6), 687–693.

Engle, R.W. (2002). Working memory capacity as executive attention. Current Directions in Psychological Science, 11, 19–23.

Feigenson, L. (2008). Parallel non-verbal enumeration is constrained by a set-based limit. Cognition, 107, 1–18.

Feigenson, L., Dehaene, S., & Spelke, E. (2004). Core systems of number. Trends in Cognitive Sciences, 8(7), 307–314.

Frank, M.C., Everett, D.L., Fedorenko, E., & Gibson, E. (2008). Number as a cognitive technology: evidence from Pirahã language and cognition. Cognition, 108(3), 819–824.

Fuchs, L.S., Fuchs, D., Compton, D.L., Powell, S.R., Seethaler, P.M., & Capizzi, A.M. (2006). The cognitive correlates of third-grade skill in arithmetic, algorithmic computation, and arithmetic word problems. Journal of Educational Psychology, 98(1), 29–43.

Geary, D.C. (1993). Mathematical disabilities: cognitive, neuropsychological, and genetic components. Psychological Bulletin, 114, 345–362.

Geary, D.C. (2010). Mathematical disabilities: reflections on cognitive, neuropsychological, and genetic components. Learning and Individual Differences, 20(2), 130–133.

Geary, D.C. (2011). Cognitive predictors of achievement growth in mathematics: a 5-year longitudinal study. Developmental Psychology, 47(6), 1539–1552.

Geary, D.C., Hoard, M.K., Byrd-Craven, J., & DeSoto, M.C. (2004). Strategy choices in simple and complex addition: Contributions of working memory and counting knowledge for children with mathematical disability. Journal of Experimental Child Psychology, 88(2), 121–151.

Groen, G., & Parkman, J.M. (1972). A chronometric analysis of simple addition. Psychological Review, 79(4), 329–343.

Halberda, J., & Feigenson, L. (2008). Developmental change in the acuity of the 'number sense': the approximate number system in 3-, 4-, 5-, and 6-year-olds and adults. Developmental Psychology, 44(5), 1457–1465.

Halberda, J., Mazzocco, M.M., & Feigenson, L. (2008). Individual differences in non-verbal number acuity correlate with maths achievement. Nature, 455, 665–669.

Hecht, S.A. (2002). Counting on working memory in simple arithmetic when counting is used for problem solving. Memory & Cognition, 30(3), 447–455.

Ho, C.S., & Fuson, K.C. (1998). Children's knowledge of teen quantities as tens and ones: comparisons of Chinese, British, and American Kindergartners. Journal of Educational Psychology, 90(3), 536–544.

Holloway, I.D., & Ansari, D. (2009). Mapping numerical magnitudes onto symbols: The numerical distance effect and individual differences in children's mathematics achievement. Journal of Experimental Child Psychology, 103(1), 17–29.

Holmes, V.M., & McGregor, J.M. (2007). Rote memory and arithmetic fact processing. Memory & Cognition, 35(8), 2041–2051.

Imbo, I., Duverne, S., & Lemaire, P. (2007). Working memory, strategy execution, and strategy selection in mental arithmetic. Experimental Psychology, 60(9), 1246–1264.

Imbo, I., & LeFevre, J. (2009). Cultural differences in complex addition: efficient Chinese versus adaptive Belgians and Canadians. Journal of Experimental Psychology: Learning, Memory, and Cognition, 35(6), 1465–1476.

Imbo, I., & LeFevre, J. (2011). Cultural differences in strategic behavior: a study in computational estimation. Journal of Experimental Psychology: Learning, Memory, and Cognition, 37(5), 1294–1301.

Imbo, I., & Vandierendonck, A. (2007a). The development of strategy use in elementary school children: working memory and individual differences. Journal of Experimental Child Psychology, 96, 284–309.

Imbo, I., & Vandierendonck, A. (2007b). The role of phonological and executive working memory resources in simple arithmetic strategies. European Journal of Cognitive Psychology, 19(6), 910–933.

Imbo, I., & Vandierendonck, A. (2008a). Practice effects on strategy selection and strategy efficiency in simple mental arithmetic. Psychological Research, 72, 528–541.

Imbo, I., & Vandierendonck, A. (2008b). Effects of problem size, operation, and working-memory span on simple-arithmetic strategies: differences between children and adults? Psychological Research, 72, 331–346.

Jordan, N.C., Huttenlocher, J., & Levine, S.C. (1992). Differential calculation abilities in young children from middle- and low-income families. Developmental Psychology, 28(4), 644–653.

Jordan, N.C., Kaplan, D., Locuniak, M.N., & Ramineni, C. (2007). Predicting first-grade math achievement from developmental number sense trajectories. Learning Disabilities Research & Practice, 22(1), 36–46.

Jordan, N.C., Kaplan, D., Ramineni, C., & Locuniak, M.N. (2008). Development of number combination skill in the early school years: when do fingers help? Developmental Science, 11(5), 662–668.

Jordan, J.A., Mulhern, G., & Wylie, J. (2009). Individual differences in trajectories of arithmetical development in typically achieving 5- to 7-year-olds. Journal or Experimental Child Psychology, 103, 455–468.

Jordan, J.A., Wylie, J., & Mulhern, G. (2010). Phonological awareness and mathematical difficulty: a longitudinal perspective. British Journal of Developmental Psychology, 28, 89–107.

Kamawar, D., LeFevre, J., Bisanz, J., Fast, L., Skwarchuk, S.L., Smith-Chant, B., & Penner-Wilger, M. (2009). Knowledge of counting principles: how relevant is order irrelevance? Journal of Experimental Child Psychology, 105(1–2), 138–145.

Kaufmann, L., & Nuerk, H.C. (2008). Basic number processing deficits in ADHD: a broad examination of elementary and complex number processing skills in 9- to 12-year-old children with ADHD-C. Developmental Science, 11(5), 692–699.

Kleemans, T., Segers, E., & Verhoeven, L. (2011a). Precursors to numeracy in kindergartners with specific language impairment. Research in Developmental Disabilities, 32(6), 2901–2908.
Kleemans, T., Segers, E., & Verhoeven, L. (2011b). Cognitive and linguistic precursors to numeracy in kindergarten: evidence from first and second language learners. Learning and Individual Differences, 21(5), 555–561.
Krajewski, K., & Schneider, W. (2009a). Early development of quantity to number-word linkage as a precursor of mathematical school achievement and mathematical difficulties: findings from a four-year longitudinal study. Learning and Instruction, 19(6), 513–526.
Krajewski, K., & Schneider, W. (2009b). Exploring the impact of phonological awareness, visual-spatial working memory, and preschool quantity-number competencies on mathematics achievement in elementary school: findings from a 3-year longitudinal study. Journal of Experimental Child Psychology, 103(4), 516–531.
Kyttala, M., & Lehto, J. (2008). Some factors underlying mathematical performance: the role of visual-spatial working memory and non-verbal intelligence. European Journal of Psychology of Education, 23(1), 77–94.
Landerl, K., Bevan, A., & Butterworth, B. (2004). Developmental dyscalculia and basic numerical capacities: a study of 8 – 9-year-old students. Cognition, 93, 99–125.
LeFevre, J., Berrigan, L., Vendetti, C., Kamawar, D., Smith-Chant, B., Bisanz, J., & Skwarchuk, S.L. (2013). The role of executive attention in the acquisition of arithmetic skills for children in grades two through four. Journal of Experimental Child Psychology, 114, 243–261..
LeFevre, J., & Bisanz, J. (1986). A cognitive analysis of number-series problems: sources of individual differences in performance. Memory & Cognition, 14(4), 287–298.
LeFevre, J.-A., Bisanz, J., Daley, K. E., Buffone, L., Greenham, S. L., & Sadesky, G. S. (1996). Multiple routes to solution of single-digit multiplication problems. Journal of Experimental Psychology: General, 125(3), 284–306.
LeFevre, J., Clarke, T., & Stringer, A.P. (2002). Influences of language and parental involvement on the development of counting skills: comparisons of French- and English-speaking Canadian children. Early Child Development and Care, 172(3), 283–300.
LeFevre, J., DeStefano, D., Coleman, B., & Shanahan, T. (2005). Mathematical cognition and working memory. In J. I. D. Campbell (Ed.), Handbook of Mathematical Cognition (pp. 361–378). New York: Psychology Press.
LeFevre, J., Fast, L., Skwarchuk, S-L., Smith-Chant, B.L., Bisanz, J., Kamawar, D., et al. (2010). Pathways to mathematics: longitudinal predictors of performance. Child Development, 81(6), 1753–1767.
LeFevre, J., & Liu, J. (1997). The role of experience in numerical skill: multiplication performance in adults from Canada and China. Mathematical Cognition, 3(1), 31–62.
LeFevre, J., Penner-Wilger, M., Pyke, A.A., Shanahan, T., & Deslauriers, W.A. (2009a). Putting two and two together: declining arithmetic fluency among young adults. Unpublished manuscript.
LeFevre, J., Polyzoi, E., Skwarchuk, S., Fast, L., & Sowinski, C. (2010). Do home numeracy and literacy practices of Greek and Canadian parents predict the numeracy skills of kindergarten children? International Journal of Early Years Education, 18(1), 55–70.
LeFevre, J., Sadesky, G.S., & Bisanz, J. (1996). Selection of procedures in mental addition: reassessing the problem size effect in adults. Journal of Experimental Psychology: Learning, Memory, and Cognition, 22(1), 216–230.
LeFevre, J., Skwarchuk, S.L., Smith-Chant, B.L., Fast, L., Kamawar, D., & Bisanz, J. (2009b). Home numeracy experiences and children's math performance in the early school years. *Canadian Journal of Behavioural Science/Revue Canadienne des Sciences du Comportement*, 41(2), 55–66.
Lemaire, P., & Callies, S. (2009). Children's strategies in complex arithmetic. Journal of Experimental Child Psychology, 103(1), 49–65.
Lemaire, P., & Lecacheur, M. (2011). Cognitive development age-related changes in children's executive functions and strategy selection : a study in computational estimation. Cognitive Development, 26(3), 282–294.

Lemaire, P., & Siegler, R.S. (1995). Four aspects of strategic change : contributions to children's learning of multiplication. Journal of Experimental Psychology: General, 124(1), 83–97.

Lyons, I.M., Ansari, D., & Beilock, S.L. (2012). Symbolic estrangement: evidence against a strong association between numerical symbols and the quantities they represent. Journal of Experimental Psychology: General, 1–8.

Lyons, I.M., & Beilock, S.L. (2009). Beyond quantity: individual differences in working memory and the ordinal understanding of numerical symbols. Cognition, 113(2), 189–204.

Lyons, I.M., & Beilock, S.L. (2011). Numerical ordering ability mediates the relation between number-sense and arithmetic competence. Cognition, 121(2), 256–261.

McCabe, D.P., Roediger, H.L., McDaniel, M.A., Balota, D.A., & Hambrick, D.Z. (2010). The relationship between working memory capacity and executive functioning : evidence for a common executive attention construct. Neuropsychology, 24(2), 222–243.

Miyake, A., & Shah, P. (1999). Models of Working Memory: Mechanisms of Active Maintenance and Executive Control. Cambridge: Cambridge University Press.

Noël, M-P., & Rousselle, L. (2011). Developmental changes in the profiles of dyscalculia: an explanation based on a double exact-and-approximate number representation model. Frontiers in Human Neuroscience, 5,165.

O'Hearn, K., Hoffman, J.E., & Landau, B. (2011). Small subitizing range in people with Williams syndrome. Visual Cognition, 19(3), 289–312.

Passolunghi, M.C., Vercelloni, B., & Schadee, H. (2007). The precursors of mathematics learning: working memory, phonological ability and numerical competence. Cognitive Development, 22, 165–184.

Pesenti, M., Seron, X., Samson, D., & Duroux, B. (1999). Basic and exceptional calculation abilities in a calculating prodigy: a case study. Mathematical Cognition, 5(2), 97–148.

Polderman, T.J.C., Huizink, A.C., Verhulst, F.C., van Beijsterveldt, C.E.M., Boomsma, D.I., & Bartels, M. (2011). A genetic study on attention problems and academic skills: results of a longitudinal study in twins. *Journal of the Canadian Academy of Child and Adolescent Psychiatry/Journal de l'Académie Canadienne de Psychiatrie de L'enfant et de L'adolescent*, 20(1), 22–34.

Pyke, A.A., & LeFevre, J. (2011). Calculator use need not undermine direct-access ability: the roles of retrieval, calculation, and calculator use in the acquisition of arithmetic facts. Journal of Educational Psychology, 103(3), 607–616.

Raghubar, K.P., Barnes, M.A., & Hecht, S.A. (2010). Working memory and mathematics: a review of developmental, individual difference, and cognitive approaches. Learning and Individual Differences, 20(2), 110–122.

Reigosa-Crespo, V., Valdés-Sosa, M., Butterworth, B., et al. (2012). Basic numerical capacities and prevalence of developmental dyscalculia: the Havana survey. Developmental Psychology, 48(1), 123–135.

Rogers, M., Hwang, H., Toplak, M., Weiss, M., & Tannock, R. (2011). Inattention, working memory, and academic achievement in adolescents referred for attention deficit/hyperactivity disorder (ADHD). Child Neuropsychology : A Journal on Normal and Abnormal Development in Childhood and Adolescence, 17(5), 444–458.

Royer, J.M., Tronsky, L.N., Chan, Y., Jackson, S.J., & Marchant, H. (1999). Math-fact retrieval as the cognitive mechanism underlying gender differences in math test performance. Contemporary Educational Psychology, 266, 181–266.

Rubinsten, O., & Sury, D. (2011). Processing ordinality and quantity: the case of developmental dyscalculia. PloS one, 6(9), e24079.

Schunn, C.D., Reder, L.M., Nhouyvanisvong, A., Richards, D.R., & Stroffolino, P.J. (1997). To calculate or not to calculate: a source activation confusion model of problem familiarity's role in strategy selection. Journal of Experimental Psychology: Learning, Memory, and Cognition, 23(1), 3–29.

Semrud-Clikeman, M., & Bledsoe, J. (2011). Updates on attention-deficit/hyperactivity disorder and learning disorders. Current Psychiatry Reports, 13(5) 364–373.

Shrager, J., & Siegler, R.S. (1998). SCADS: A model of children's strategy choices and strategy discoveries. Psychological Science, 9(5), 405–410.

Siegler, R.S., & Shrager, J. (1984). A model of strategy choice. In C. Sophian (Ed.), Origins of Cognitive Skills (pp. 229–293). Hillsdale, NJ: Erlbaum.

Simmons, F., Singleton, C., & Horne, J. (2008). Phonological awareness and visual-spatial sketchpad functioning predict early arithmetic attainment: evidence from a longitudinal study. European Journal of Cognitive Psychology, 20(4), 711–722.

Smith-Chant, B.L., & LeFevre, J-A. (2003). Doing as they are told and telling it like it is: self-reports in mental arithmetic. Memory & Cognition, 31(4), 516–28.

Stigler, J.W., Lee, S.Y., & Stevenson, H.W. (1986). Digit memory in Chinese and English: evidence for a temporally limited store. Cognition, 23(1), 1–20.

Trbovich, P.L., & LeFevre, J. (2003). Phonological and visual working memory in mental addition. Memory & Cognition, 31(5), 738–745.

Van Herwegen, J., Ansari, D., Xu, F., & Karmiloff-Smith, A. (2008). Small and large number processing in infants and toddlers with Williams syndrome. Developmental Science, 11(5), 637–643.

Vandermaas-Peeler, M., Ferretti, L., & Loving, S. (2011). Playing the ladybug game: Parent guidance of young children's numeracy activities. Early Child Development and Care, 1–19.

Vandermaas-Peeler, M., Nelson, J., Bumpass, C., & Sassine, B. (2009). Numeracy-related exchanges in joint storybook reading and play. International Journal of Early Years Education, 17(1), 67–84.

Verschaffel, L., Luwel, K., Torbeyns, J., & Dooren, W. V. (2009). Conceptualizing, investigating, and enhancing adaptive expertise in elementary mathematics education. European Journal of Psychology of Education, 24(3) 335–359.

Wei, W., Yuan, H., Chen, C., & Zhou, X. (2012). Cognitive correlates of performance in advanced mathematics. British Journal of Educational Psychology, 82, 157–81.

Willburger, E., Fussenegger, B., Moll, K., Wood, G., & Landerl, K. (2008). Naming speed in dyslexia and dyscalculia. Learning and Individual Differences, 18(2), 224–236.

Wilson, A. J., Dehaene, S., Dubois, O., & Fayol, M. (2009). Effects of an adaptive game intervention on accessing number sense in low socio-economic status kindergarten children. Mind, Brain, and Education, 3(4), 224–234.Individual Differences -

CHAPTER 49

COGNITIVE PREDICTORS OF MATHEMATICAL ABILITIES AND DISABILITIES

ANNEMIE DESOETE

INTRODUCTION

IN everyday life situations, we need to be on time, pay bills, follow directions or use maps, look at bus or train timetables, or comprehend instruction leaflets and expiry dates. Even in music mathematical skills are needed to perceive rhythmic patterns (Desoete, 2012). Moreover, a lack of mathematical literacy was found to affect people's ability to gain full-time employment and often restricted employment options to manual and often low-paying jobs (Dowker, 2005).

The term mathematical abilities and disabilities/disorders (MD) refers to a significant degree of impairment in the mathematical skills (with substantially below performances). In addition, children with MD do not profit enough from (good) help. This is also referred to as a lack of responsiveness to intervention (RTI). Finally, the problems in MD cannot be totally explained by impairments in general intelligence or external factors that could provide sufficient evidence for scholastic failure.

There is some discussion about the percentile cut-off points to determine what 'substantially below means,' with cut-offs ranging from percentile 3 to even percentile 45 (see Stock, Desoete, & Roeyers, 2010). The reported prevalence rates for MD vary for that reason. Most practitioners and researchers currently report a prevalence between 2 and 14% of children (Barbaresi, Katuskic, Colligan, Weaver, & Jacobsen, 2005; Desoete, Roeyers, & De Clercq, 2004; Dowker, 2005; Geary, 2011; Shalev, Manor, & Gross-Tsur, 2005). The prevalence of MD in siblings even ranges from 40 to 64% (Shalev, Manor, Kerem, Ayali, Badichi, Friedlander, et al., 2001).

Persons with MD describe their problems as follows (Vanmeirhaeghe, 2012):

> *Charlotte (bachelor secondary education, choreographer):* The simple act of looking at numbers is blocking my mind immediately. I just don't see it … Numbers are stressing me. That is quite peculiar because all seems so logic and normal. But to me math is a complete abstract way of thinking.
>
> Mother of Bout (grade 4): It is quite confronting as a parent. For Bout it took until grade 2 … before he knew when it was his birthday. I didn't think it was normal. A child knows when it is his birthday.
>
> *Kristel (master in education):* Why was elementary school like hell? Because I felt a huge pressure on me. Open your manual on page 68. There we go again! Where is page 68? Other pupils already had taken down the title while I was still looking for page 68. It was a constant feeling of needing to exert myself. I have to take care that I can follow. That is what made it so hard for me. Everyone was faster than I was.
>
> *Yutta (bachelor student):* Once my mother asked me to go to the supermarket and get six hundred grams of spaghetti sauce … And I bought six kilos.
>
> *Sanne (bachelor in social work):* When I am filling up a shopping cart … I am never sure whether I have enough money with me and what the total amount will be. That's why I mostly pay by card. I never check the change.
>
> *Sara (bachelor in journalism):* Twice I arrived at an appointment at a wrong hour … just by swapping the numbers on the clock It says 18.00u for example. Then I think about the eight And then I think I should be there at eight.
>
> (Vanmeirhaeghe, 2012.)

Given that MD is associated with cost to society, family and the individual person, it is important to study the predictors of and causes of MD, so that treatments can be developed and targeted at the underlying causes. In addition, such research may have even wider implications to our understanding of mathematical development, and the factors that may facilitate it. In this chapter we will describe several studies aiming to detect cognitive predictors of mathematical abilities and disabilities.

Early Numeracy as Cognitive Predictors for Mathematical Abilities and Disabilities

Some of the most-described early numeracy abilities are the *Piagetian logical thinking abilities*, such as the classification and seriation skills. *Classification* is relevant to knowing the cardinality of a set (e.g. 'How many balls can you see on this picture?'); whereas *seriation* is needed when dealing with ordinal numbers (e.g. 'Circle the third ball from the beginning.'). Although a lot of criticism has been formulated on the theory of Piaget, the importance of the logical abilities is currently recognized. Even after controlling for differences in working memory, logical abilities in 6-year-old children remained a strong predictor for mathematical abilities 6 months later (Nunes, Bryant, Evans, Bell, Gardner, Gardner, et al., 2007). In addition, children who were successful in logical thinking tasks performed better on mathematical tests with mastery of seriation abilities having the strongest predictive power (Grégoire, 2005).

Besides the Piagetian operations, the neo-Piagetian area of *counting knowledge* also seems to be important in the development of adequate mathematical skills (Dowker, 2008; Sarnecka & Carey, 2008). It is obvious that early strategies for addition and subtraction involve counting. For example, in the 'count all' or 'sum' strategy, the child first counts each collection and then counts the combination of two collections starting from one (e.g. 2 + 5 = – 1, 2 ... 1, 2, 3, 4, 5 ... 1, 2, 3, 4, 5, 6, 7). Dowker (2005) suggested that counting knowledge includes both procedural and conceptual aspects.

Procedural counting knowledge is defined as children's ability to perform an arithmetic task, for example, when a child can successfully determine that there are five objects in an array (Le Fevre, Smith-Chant, Fast, Skwarchuk, Sargla, Arnup, et al., 2006). The knowledge of the sequence of counting words (the number row) is one of the most important procedural aspects of counting. This also includes the ability to count forward and backwards easily.

Besides the procedural aspect of counting knowledge, we can discern the conceptual counting knowledge. *Conceptual knowledge* reflects a child's understanding of why a procedure works or whether a procedure is legitimate (Le Fevre et al., 2006).

In what follows, we summarize the findings from a longitudinal study following up 455 children from kindergarten, at age 5 to 6 till grade 2, at age 7 to 8 (Desoete & Stock, 2010; Stock, Desoete, & Roeyers, 2010). The principal aim was to investigate whether we could predict typical and atypical mathematical abilities at age 7–8 from performances in kindergarten (on the TEDI-MATH; see Appendix A). Children were classified as having *MD* if they scored ≤ pc 10 on at least one of the mathematical tests, both in first (at age 6–7) and second (at age 7–8) grade. The *low achieving (LA) group* consisted of children who scored between the 10th and 25th percentiles on at least one of the mathematical tests, both at grade 1 and 2. Finally, *typical achievers (TA) scored* above the 25th percentile on all mathematical tests in both grades. Moreover, to study the importance of the RTI principle, the kindergarten skills were compared with the results of *children with fluctuating performances* (FL) and a severe underachievement during one grade and a typical performance during the other grade. Of the children, 64.23% appeared to have a stable arithmetic profile, whereas 35.77% had a profile with fluctuating arithmetic test scores (FL). The FL group performed significantly better than the MD group on procedural counting knowledge, conceptual counting knowledge, classification and magnitude comparison. A discriminant analysis revealed that 73.3% of the children were classified correctly into the MD, LA or TA group (see Table 49.1). In addition, only 8.2% of FL-children were classified correctly. In 83.5% of the cases FL-children were classified as TA children.

Generally, Table 49.1 showed that the prediction of children with weak mathematical abilities based on kindergarten numeracy abilities showed good specificity, but low sensitivity. It seemed to be easier to screen children who are not at risk than to detect the at-risk children with mathematical disabilities based on their early numeracy abilities based on a random dataset. However, based on clinical cut-off scores linked to a stratified normative group, it was possible to predict four out of five children with mathematical disabilities based on number comparison, conceptual counting and procedural counting screening in kindergarten (Stock et al., 2010).

Table 49.1 Percentage of observed and predicted group membership in children with stable mathematical abilities (LA and TA) and disabilities (MD)

		Predicted group membership		
		MD	LA	TA
Observed Group Membership	MD	***29.4 %***	5.9 %	64.7 %
	LA	13.1 %	***4.9 %***	82.0 %
	TA	2.3 %	1.4 %	***96.3 %***

Note. MD = mathematical disabilities group; LA = low achieving group; TA = typically achieving group.

Language as Cognitive Predictors for Mathematical Abilities

There is ample evidence for a significant relationship between language and formal or symbolic mathematics (Bull & Johnston, 1997; Dehaene, Spelke, Pinel, Stanescu, & Tsivkin, 1999; Geary, 1993; Jordan, Kaplan, Olah, & Locuniak, 2006; Jordan, Wylie, & Mulhern, 2010). Several researchers (Hooper, Roberts, Sideris, Burchinal, & Zeisel, 2010; Purpura, Hume, Sims, & Lonigan, 2011; Romano, Babchishin, Pagani, & Kohen, 2010; Sarnecka, Kamenskaya, Yamana, Ogura, & Ydovina, 2007) stressed the value of including language as predictors of mathematical abilities. However, it remains an open question to what extent mathematics is truly dependent on language. For example, some 'savant' calculators perform certain aspects of arithmetic very well despite language difficulties. Moreover, studies demonstrated that even preverbal infants already process numbers (Ceulemans, Loeys, Warreyn, Hoppenbrouwers, & Desoete, 2012).

To try to understand the role of language and numeracy, some tribes in Amazonia with an atypical language structure and an atypical way of dealing with numbers were studied. Gordon (2004) observed the Piraña, using a counting system of 'one-two-many' but managing to compare quantities despite their restricted numeric language. In addition, Pica, Lemer, Izar and Dehaene (2004) observed the Mundurukù, knowing, but not using number words up to five and making their selection based on estimation. Their number word 'five' can be translated by a handful, but they use that quantity also for 6, 7, 8, 9. In a comparison task, they performed as well as the French control group, but they failed on exact arithmetic. From this point of view Pica et al (2004) concluded that estimation is a basic skill independent of language. In handling mathematics, however, there is need of a well-developed lexicon of number words. An additional argument for the claim that language is associated with later mathematical abilities can be found in the model of adult problem solving with one (semantic), two (semantic and non-semantic), or three (semantic, visual and auditory-verbal word frame) supervariables (Cipolotti & Butterworth, 1995; Dehaene, 1992; McCloskey & Macaruso, 1995). Dehaene and Cohen assumed three such variables in their Triple Code model stating that numbers can be represented in three different ways: as a quantity system (a semantic representation of the size and distance relations between numbers), as a verbal system (where numerals are represented lexically, phonologically, and syntactically), and as a visual system (as strings of Arabic numerals).

In a recent study (Praet, Titeca, Ceulemans, & Desoete, 2013) we aimed to study the importance of language to different aspects of arithmetic. Participants (n = 132) were assessed at age 5–6 on receptive and expressive language. Our findings revealed that 10% of the children at that age had a language problem. All of the children with a language problem had additional problems with procedural counting and difficulties with the knowledge of the numerical system, even when intelligence as covariate was taken into account. Moreover, 7.31% of the 5–6-year-olds with counting problems had a lower receptive language index compared with their age-matched peers without counting problems. Finally, language has a value added of 21.6% to number naming and counting as predictors for early arithmetic abilities in kindergarten. Language in kindergarten especially predicted procedural counting abilities, knowledge of the numerical system and early calculation skills at age 5–6. When children were followed-up 1 year later in grade 1, kindergarten language still predicted arithmetic abilities even when kindergarten skills were taken into account. Especially expressive language at age 5–6 predicted prospectively number knowledge, mental arithmetic, number facts retrieval, and clock reading tasks at age 6–7.

Motor, Visual Perceptual, and Visuomotor Skills as Cognitive Predictors for Mathematical Abilities and Disabilities

Although the association between motor, visual perceptual, and visuomotor integration skills and mathematics is not fully understood, there are a number of important findings suggesting a link between those domains.

The relationship between *motor* and mathematical skills has been explained by the 'embodied cognition' literature or the theory that argues that cognitive processes are grounded in the interaction of the body with the world (e.g. Soylu, 2011). Moreover, the relationship between motor skills and mathematics was found in predictive studies in which fine motor skills were significantly related to mathematics scores (Luo, Jose, Huntsinger, & Pigott, 2007; Pagani, Fitzpatrick, Archambault, & Janosz, 2010; Vuijk, Hartman, Mombarg, Scherder, & Visscher, 2011).

In addition, Kulp, Earley, Mitchell, Timmerman, Frasco, and Geiger (2004) found that low scores for *visual perception* were related to poor mathematical abilities and Mazzocco and Myers (2003) described that poor performance for 'position in space' (i.e. the ability to match two figures according to their common features) occurred more frequently in the MD group than in the non-MD group. However, Vukovic and Siegel (2010) revealed, in line with Geary, Hamson, and Hoard (2000), and Morris, Stuebing, Fletcher, Shaywitz, Lyon, Shankweiler, et al. (1998), that a block rotation task could not reliably differentiate children with persistent mathematical difficulties from children with transient mathematical difficulties or control children, suggesting that visual perception is not a cognitive predictor for MD. Furthermore, Cirino, Morris, and Morris (2007) found that visual perception contributed to predicting the mathematical performance in college students referred for learning difficulties, although their previous study did not find this contribution (Cirino, Morris, & Morris, 2002). Finally, *visual-motor integration* has also been linked to mathematics. Kulp et al. (2004) found that visual-motor integration was significantly related to teachers' ratings of mathematical skills in children.

To conclude, research on motor, visual perception, and visual-motor integration skills as predictors for mathematics, and MD has given limited and inconclusive results. The main goal of this line of research has been to investigate co-morbidity of motor problems in various MD, and to gain insight into the relationships between motor, visual, and visual-motor skills and mathematical problems in elementary school children in three studies.

In a first study (Pieters, De Block, Scheiris, Eyssen, Desoete, Deboutte, Van Waelvelde, & Roeyers, 2012a), the profiles of children referred to rehabilitation centres for behavioural, developmental and sensorineural disorders in Flanders (the Dutch speaking part of Belgium) were studied (n = 3,608). An additional analysis revealed that 24.8% of the children with MD had motor problems.

Moreover, in a second study (Pieters, Desoete, Roeyers, Vanderswalmen, & Van Waelvelde, 2012b) it was investigated whether visual perception, motor skills, and visual-motor integration were related to mathematical scores for number fact retrieval and procedural calculation in a sample of 39 children with MD and 106 typically developing children belonging to three groups of different ages. All the measured domains (visual perception, motor skills, and visual-motor integration) explained a substantial proportion of the variance in either number fact retrieval (40%) or procedural calculation (38%). Furthermore, children with MD were found to have problems on all the measured domains in comparison with age-matched typically developing children. More specifically, we found a developmental delay of 1 year for visual perception, motor coordination (i.e. a tracing task) and visual-motor integration and a developmental delay of 2 years for motor skills. Finally, individual analyses showed that not all children with MD had problems with visual perception, motor skills, and visual-motor integration. The number of children with MD scoring between the first and the fifth percentile (i.e. having severe problems) for visual perception was much lower than that for motor skills and visual-motor integration (see Table 49.2).

Table 49.2 The prevalence of motor problems in children mathematical abilities (TA) and disabilities (MD)

	Children with mathematical disabilities (n = 39)					
	pc 1–5		pc 6–16		pc > 16	
Domain	n	%	n	%	n	%
Visual perception	3	7.7	11	28.2	25	64.1
Motor coordination	6	15.4	7	17.9	26	66.7
Motor skills	17	43.6	7	17.9	15	38.5
	Typically achieving children (n = 106)					
Visual perception	5	4.7	6	5.7	95	89.6
Motor coordination	1	0.9	5	4.7	100	94.4
Motor skills	1	0.9	8	7.5	97	91.6

In a third study, we examined whether we could identify subgroups of individuals with relatively homogeneous profiles on measures associated with mathematical skills. Therefore, we used a data-driven model-based clustering method in a sample of 410 children.

Based on the mathematical variables, three clusters were found, conforming to a subtype with number fact retrieval problems, a subtype with procedural calculation problems, and a cluster of children without mathematical problems. Whereas children with the procedural subtype of MD had no problems with number fact retrieval, children with the semantic memory subtype of MD had problems with both number fact retrieval, as well as procedural calculation. This latter cluster had the lowest scores on both tests and included the youngest children of all clusters.

Linear regressions (see Table 49.3) revealed a higher the value for belonging to cluster of the semantic memory MD the lower the score for reading of existing and pseudo-words,

Table 49.3 Prediction of the scores for reading, spelling, motor skills, and handwriting based on the probability of belonging to a cluster of mathematical abilities (cluster 1) or disabilities (cluster 2 and 3)

	Predictor	*B*	SD of *B*	*T*	β	Adj. R^2
Word reading	Cluster 1	1.65	0.49	3.41**	0.17	0.03
	Cluster 2	−1.06	0.54	−1.97*	−0.10	0.01
	Cluster 3	−1.32	0.64	−2.08*	−0.11	0.01
Non-word reading	Cluster 1	1.58	0.43	3.64***	0.18	0.03
	Cluster 2	−1.28	0.48	−2.66**	−0.13	0.02
	Cluster 3	−0.89	0.57	−1.56	−0.08	0.01
Spelling	Cluster 1	11.72	3.93	2.98**	0.15	0.02
	Cluster 2	−6.29	4.42	−1.42	−0.07	0.01
	Cluster 3	−11.14	5.11	−2.18*	−0.11	0.01
Motor skills	Cluster 1	3.30	0.40	8.26***	0.39	0.15
	Cluster 2	−2.83	0.46	−6.18***	−0.30	0.09
	Cluster 3	−1.64	0.55	−2.97**	−0.15	0.02
Handwriting quality	Cluster 1	−0.52	0.13	−3.87***	−0.20	0.04
	Cluster 2	1.00	0.14	7.00***	0.35	0.12
	Cluster 3	−0.45	0.17	−2.64**	−0.14	0.02
Handwriting speed difficulties	Cluster 1	0.60	0.13	4.59***	0.24	0.05
	Cluster 2	−1.62	0.12	−13.23***	−0.58	0.33
	Cluster 3	1.13	0.16	7.09***	0.16	0.12

Note. B = unstandardized regression coefficient; β = standardized regression coefficient; Cluster 1 = typical mathematical abilities; Cluster 2 = semantic memory disabilities; Cluster 3 = procedural calculation disabilities;

[a]The lower the quality of handwriting, the higher the score.

$^{*}p \leq 0.05$. $^{**}p < 0.01$. $^{***}p < 0.001$.

motor skills, handwriting quality, and handwriting speed. A higher probability of belonging to the cluster the procedural subtype of MD was significantly related to lower scores for reading of existing words, spelling and motor skills (see also Pieters, Roeyers, Rosseel, van Waelvelde, & Desoete, 2013).

Working Memory, Inhibition, Naming Speed, and Inference Control as Predictors for Mathematical Abilities and Disabilities

According to Baddeley (1986), *working memory* has to be seen as an active system that regulates complex cognitive behavior. Several studies have shown working memory problems in children with children with MD (e.g. Bull & Scerif, 2001; Swanson & Jerman, 2006; Temple & Sherwood, 2002).

According to Nigg (2000) *behavioural inhibition* is considered as the capacity to suppress a prepotent or dominant response, and entails the deliberate control of a primary motor response in compliance with changing context cues. Research revealed inhibition to be predictive for mathematical abilities (Bull & Scerif, 2001) and mathematical disabilities (Zhang & Wu, 2011) tested with a colour-word and a numerical Stroop paradigm. However, no differences were found on the numerical Stroop by Censabella and Noel (2005), and van der Sluis, van der Leij, & de Jong (2005) and also Passolunghi and colleagues revealed no differences between TA and MD children with a Go/no-go paradigm (Passolunghi, Marzocchi, & Fiorillo, 2005).

Naming speed is defined as the ability to quickly recognize and name a restricted set of serially or isolated presented high frequency symbols, objects or colors (Heikkila, Narhi, Aro, & Ahonen, 2009). Evidence for naming speed as cognitive predictor was provided by van der Sluis et al. (2004), who found that children with MD performed slower on naming speed of quantities and numbers, but not on naming speed of letters and objects. However, other studies revealed that this was a general naming speed problem (e.g., Geary, 2011; Temple & Sherwood, 2002). For instance, D'Amico and Passolunghi (2009) found slower naming speed on both numbers and letters in children with MD in comparison with age-matched peers. In addition, research also revealed a general relationship between math computation skills and naming speed (Hecht, Torgesen, Wagner, & Rashotte, 2001).

Finally, *interference control* or the capacity to prevent interference caused by resource or stimulus competition (Nigg, 2000) was impaired in MD according to some studies (e.g. Zhang & Wu, 2011), but not according to others (e.g. Censabella & Noel, 2005; van der Sluis et al., 2004), or only on some domains (Bull & Scerif, 2001).

To conclude, research to date on working memory, inhibition, naming speed, and inference control has given inconclusive results. Some of the differences might be explained by the tests used or by the fact that children had MD with or without RD, therefore, a study was set up with several tests compare performances of children with mathematical disabilities (MD and MD + RD) to typical achieving (TA) peers without learning disabilities.

Working memory was investigated by means of digit, word, and block recall tasks, listening and spatial span tasks (see Figure 49.1). The phonological loop was assessed with digit – and word list recall measures for the verbal recall of sequences. The visuospatial sketchpad was tested with block recall measures. Children were asked to repeat the sequence of the orange squares by clicking on the different blue squares. The central executive component was tested with a backward digit recall, backward word list recall and backward block recall task. Children had to recall sequences of digits, words or squares in the reverse order. In addition, two dual tasks were measured. In listening recall, children were presented with a sequence of spoken sentences (e.g. 'Lions have four legs'). The instructions were to verify the sentence by stating 'true' or 'false,' and to repeat the final word for each sentence. In the spatial span dual tasks, the child had to identify whether the shape on the right side was the same of the shape on the left, and to show the location of each red dot on the shape in the correct sequence.

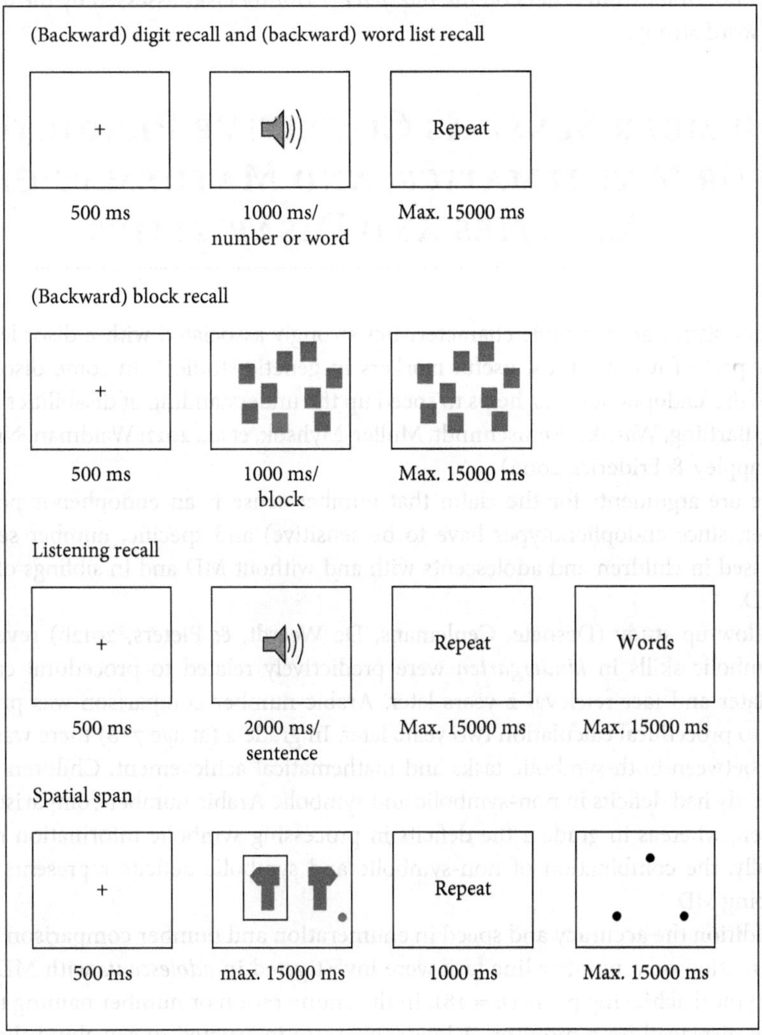

FIGURE 49.1 The working memory tasks.

The span scores of backward digit recall and block recall were the best predictors for mathematical disabilities. There were significant differences between the TA and MD group on both the phonological loop and the central executive and between the TA group and MD + RD group on every *working memory* component (De Weerdt, Roeyers, & Desoete, 2012).

In addition children's *behavioural inhibition* was tested with a go/no-go task in a non-symbolic picture modality, and symbolic letter and digit-modalities. Children with RD made significantly more commission errors on alphanumeric (letter and digit) modalities compared with the non-alphanumeric picture modality. As compared with children without MD, children with MD made as much commission errors on the picture modality as on the letter modality. No significant interaction-effect was found between RD and MD. These results can be considered as evidence for behavioral inhibition deficits related to alphanumeric stimuli in children with RD but not in children with MD (De Weerdt, Desoete, & Roeyers, 2013).

Finally, children with MD and MD+RD were slower on the quantity *naming speed*, but they did not differ from others on the *interference control* tasks assessed by the quantity or colour-word Stroop.

Number Sense as Cognitive Predictor for Mathematics, and Mathematical Abilities and Disabilities

Endophenotypes are heritable characteristics strongly associated with a disorder, but not actually part of it (and, thus, useful markers in genetic studies). In some disorders, the study of the 'endophenotypes' helps to speed up the understanding of disabilities (Neuhoff, Bruder, Bartling, Warnke, Remschmidt, Muller-Myhsok, et al., 2012; Waldman, Nigg, Gizer, Park, Rappley, & Friderici, 2006).

There are arguments for the claim that number sense is an endophenotype for MD. However, since endophenotypes have to be sensitive) and specific, number sense must be assessed in children and adolescents with and without MD and in siblings of children with MD.

A follow-up study (Desoete, Ceulemans, De Weerdt, & Pieters, 2012b) revealed that non-symbolic skills in *kindergarten* were predictively related to procedural calculation 1 year later and fact retrieval 2 years later. Arabic number comparison was predictively related to procedural calculation two years later. In grade 2 (at age 7–8) there was an association between both symbolic tasks and mathematical achievement. Children with MD had already had deficits in non-symbolic and symbolic Arabic number comparison in kindergarten, whereas in grade 2 the deficits in processing symbolic information remained. Especially, the combination of non-symbolic and symbolic deficits represents a risk of developing MD

In addition the accuracy and speed in enumeration and number comparison tasks and the estimation on a number line task were investigated in *adolescents* with MD (n = 23) and in typical achieving peers (n = 28). In the enumeration or number naming task stimuli were displayed on a monitor. Adolescents were instructed to say aloud the number of squares on the screen. All squares were black on a white background. The individual area, total area, and density of the squares were varied to ensure that participants could not use non-numerical cues to make a correct decision. The number comparison task was a task in which adolescents had to compare dot arrays and click on the side with the most dots, with the number of dots varying between 1 and 18. The dot patterns were controlled for perceptual variables. The Number Line Estimation task used a 0–100 interval. Adolescents were asked to put a single mark on the line to indicate the location of the number. The percentage absolute error (PAE) was calculated as a measure of the estimation accuracy. For example, if an adolescent was asked to estimate 25 on a 0–100 number line and placed the mark at the point on the line corresponding to 40, the PAE would be (40 – 25)/100 or 15%.

The Mann–Whitney *U*-test revealed that adolescents with MD or typical achieving peers did not significantly differ on the accurate enumeration of small (<4) numbers ($U = 238.50$, $p = 0.603$), but MD adolescents were significantly worse on the accurate enumeration of larger numbers ($U = 135.50$, $p = 0.009$), especially for the enumeration of five ($U = 122.50$, $p = 0.003$), six ($U = 123.50$, $p = 0.003$) and seven ($U = 164.00$, $p = 0.044$). In addition, also on the number comparison task, adolescents with MD were less accurate compared with TA peers ($U = 181.00$, $p = 0.007$). Finally the Mann–Whitney *U*-test revealed significant more PAE (error) on the number line estimation task ($U = 182.00$, $p = 0.029$) in MD. To conclude, our findings did support number sense as cognitive predictor for mathematical disabilities, especially with larger numbers. The findings did not support a limited capacity of subitizing in MD, since there were no significant differences between adolescents with and without MD on the enumeration of small numerical sets.

Finally, number sense was assessed in *siblings* of children with MD ($n = 9$) and age-matched children without family members with MD ($n = 63$). The results of the study confirmed that siblings were less proficient in number line placements compared with non-siblings, with a larger effect size for symbolic and especially number word estimation compared with the non-symbolic results. Siblings also differed from non-siblings on procedural and conceptual counting skills and logical thinking in kindergarten (Desoete, Praet, Titeca, & Ceulemans, 2013).

Conclusion

The chapter considered the cognitive predictors of MD.

Most findings suggested that *early numeracy* predict MD. Several studies revealed that typical achieving children perform significantly better than the MD group on seriation and classification as Piagetian logical thinking skills, on conceptual and procedural counting knowledge and on magnitude comparison assessed in kindergarten (Dowker, 2008; Nunes et al., 2007; Sarnecka & Carey, 2008; Stock et al., 2010). However, it was found easier to screen the children who are 'not at risk' than to detect the 'at-risk' children based on their early numeracy abilities (Stock et al., 2010). In addition, in line with Murphy, Mazzocco, Hanich and Early (2007), it was demonstrated that characteristics of children varied as a function of the criteria used to define the disabilities, with children not meeting the RTI criterion (or the FL-group) having less problems on counting, classification and comparison compared with MD children (Stock et al., 2010).

In addition, *language* in kindergarten was investigated as cognitive predictor for mathematical abilities. In line with previous studies revealing a significant relationship between language and mathematics (e.g. Hooper et al., 2010; Jordan et al., 2010; Purpura et al., 2011), a recent study demonstrated that children in kindergarten with a language problem had additional problems with procedural counting and knowledge of the numerical system. Moreover, language has a value added of 21.6% to number naming and counting as predictors for early arithmetic abilities in kindergarten. In addition, especially expressive language at age 5–6 predicted prospectively number knowledge, mental arithmetic, number facts retrieval, and clock reading tasks at age 6–7. These findings revealed that the co-occurrence of language and mathematical problems may be useful in answering parent's questions, anticipating challenges that children with language or counting problems might encounter, and planning services to address those.

Furthermore *motor, visual and visuomotor skills* were studied as predictors for mathematical abilities and disabilities. A comorbidity study revealed that about 25% of the MD children had additional motor problems (Pieters, De Block, Scheiris, Eyssen, Desoete, Deboutte, et al., 2012a) Visual perception, motor skills and visual-motor integration explained about 40% of the variance in number fact retrieval or procedural calculation (Pieters, et al., 2012b). In addition, children with MD had in general a developmental delay of 1 year for visual perception, motor coordination, and visual-motor integration, and a developmental delay of 2 years for motor skills. Moreover, the amount of children with MD failing on visual perception was remarkably lower than the number of children with MD failing on motor skills or visual-motor integration. Finally, in line with authors proposing a procedural and a semantic memory subtype within MD (Geary, 1993, 2004; Robinson, Menchetti, & Torgesen, 2002; Temple, 1991), these clusters were present in a recent a cluster analysis. In both clusters, motor skills were impaired. In the semantic memory cluster there were also problems with reading and handwriting, whereas in the procedural cluster especially spelling was impaired, indicating the importance of assessing *spelling* in MD.

In line with previous studies on *working memory* (e.g. Passolunghi & Siegel, 2004; Zhang & Wu, 2011), impairments were found on the phonological loop and the central executive in MD and on all components including the visuospatial sketchpad in MD + RD children. In addition, in line with van der Sluis et al. (2005), MD children experienced problems with the *naming speed* of non-symbolic numerical information. In contrast, with previous findings (Zhang & Wu, 2011) children with MD experienced no *behavioural inhibition* deficits related to alphanumeric stimuli (De Weerdt et al., 2013) and they had no *inference control* problems compared with TA peers. From these findings, one may conclude that tests of interference control or behavioural inhibition are not indicated for identifying children with MD.

Finally, in line with behavioural evidence (e.g. Mussolin, Mejias, & Noël, 2010; Piazza, Facoetti, Trussardi, Berteletti, Conte, Lucangeli, et al., 2010; von Aster & Shalev, 2007) number *sense* might be a good predictor for MD.

To conclude, these findings suggest that an assessment in young children at risk for MD should include early numeracy, number sense, and language measures. In both elementary school children and adolescents, motor and spelling skills, as well as number sense measures, seem important as predictors of mathematical abilities and disabilities, although, of course, spelling and motor deficits can occur in the absence any mathematical disabilities, and vice versa. In addition, different mathematical clusters were found, raising questions about the usefulness of placing children with MD into one single diagnostic category.

References

Baddeley, A. (1986). Working Memory. Oxford: Clarendon Press.

Barbaresi, W.J., Katusic, S.K., Colligan, R.C., Weaver, A.L., & Jacobsen, S.J. (2005). Learning disorder: incidence in a population-based birth cohort, 1976–82, Rochester, Minn. Ambulatory Pediatrics, 5, 281–289.

Bull, R., & Johnston, R. (1997). Children's arithmetical difficulties: contributions from processing speed, item identification and short term memory. Journal of Experimental Child Psychology, 65, 1–24.

Bull, R., & Scerif, G. (2001). Executive functioning as a predictor of children's mathematics ability: inhibition, switching, and working memory. Developmental Neuropsychology, 19, 273–293.

Censabella, S., & Noel, M.P. (2005). The inhibition of exogenous distracting information in children with learning disabilities. Journal of Learning Disabilities, 38, 400–410.

Ceulemans, A., Loeys, T., Warreyn, P., Hoppenbrouwers, K., & Desoete, A. (2012). Small number discrimination in early human development: the case of one versus three. Education Research International, 2012, Article ID 964052.

Cipolotti, L., & Butterworth, B. (1995). Toward a multiroute model of number processing: impaired number transcoding with preserved calculation skills. Journal of Experimental Psychology: General, 124, 375–390.

Cirino, P.T., Morris, M.K., & Morris, R.D. (2002). Neuropsychological concomitants of calculation skills in college students referred for learning difficulties. Developmental Neuropsychology, 21, 201–218.

Cirino, P.T., Morris, M.K., & Morris, R.D. (2007). Semantic, executive, and visuospatial abilities in mathematical reasoning of referred college students. Assessment, 14, 94–104.

D'Amico, A., & Passolunghi, M.C. (2009). Naming speed and effortful and automatic inhibition in children with arithmetic learning disabilities. Learning and Individual Differences, 19, 170–180.

Dehaene, S. (1992). Varieties of numerical abilities. Cognition & Emotion, 44, 1–42.

Dehaene, S. (1997). The Number Sense. New York: Oxford University Press.

Dehaene, S. (2001). Precis of the number sense. Mind & Language, 16, 16–36.

Dehaene, S., Spelke, E., Pinel, P., Stanescu, R., & Tsivkin, S. (1999) Sources of mathematical thinking: behavioral and brain-imaging evidence Science, 284, 970–974.

Desoete, A. (2012). How to support children with mathematical learning disabilities learning to play an instrument. Education Research International, Article ID 346858, 7 pages.

Desoete, A., Ceulemans, A., De Weerdt, F., & Pieters, S. (2012). Can we predict mathematical learning disabilities from symbolic and non-symbolic comparison tasks in kindergarten? British Journal of Educational Psychology, 82, 64–81.

Desoete, A., Praet, M., Titeca, D., & Ceulemans, A. (2013). Cognitive phenotype of mathematical learning disabilities: what can we learn from siblings? Research in Developmental Disabilities, 34, 404–412.

Desoete, A., & Roeyers, H. (2005). Cognitive skills in mathematical problem solving in grade 3. British Journal of Educational Psychology, 75, 119–138.

Desoete, A., Roeyers, H., & De Clercq, A. (2004). Children with mathematics learning disabilities in Belgium. Journal of Learning Disabilities, 37, 50–61.

Desoete, A., & Stock, P. (2010). Can we predict mathematical disabilities from abilities in kindergarten. In Spencer B. Thompson (Ed), Kindergartens: Programs, Functions and Outcomes (pp. 1–49). New York: Nova Science Publishers.

De Weerdt, F., Roeyers, H., & Desoete, A. (2012). Working memory in children with reading disabilities and/or mathematical disabilities, Journal of Learning Disabilities, 46(5), 461–472.

De Weerdt, F., Desoete, A., & Roeyers, H. (2013). Behavioral inhibition in children with learning disabilities. Research in Developmental Disabilities, 34, 1998–2007.

Dowker, A.D. (2005). Individual Differences in Arithmetic. Implications for Psychology, Neuroscience and Education. New York: Psychology Press.

Dowker, A. (2008). Individual differences in numerical abilities in preschoolers. Developmental Science, 11, 650–654.

Fuchs, L.S., & Fuchs, D. (2002). Mathematical problem-solving profiles of students with mathematics disabilities with and without comorbid reading disabilities. Journal of Learning Disabilities, 35, 563–573.

Fuchs, L.S., Fuchs, D., Compton, D.L., Bryant, J.D., Hamlett, C.L., & Seethaler, P.M. (2007). Mathematics screening and progress monitoring at first grade: implications for responsiveness to intervention. Exceptional Children, 73, 311–330.

Geary, D.C. (1993). Mathematical disabilities—cognitive, neuropsychological and genetic components. Psychological Bulletin, 114, 345–362.

Geary, D.C. (2004). Mathematics and learning disabilities. Journal of Learning Disabilities, 37, 4–15.

Geary, D.C. (2011). Consequences, characteristics, and causes of mathematical learning disabilities and persistent low achievement in mathematics. Journal of Developmental and Behavioral Pediatrics, 32, 250–263.

Geary, D.C., Hamson, C.O., & Hoard, M.K. (2000). Numerical and arithmetical cognition: a longitudinal study of process and concept deficits in children with learning disability. Journal of Experimental Child Psychology, 77, 236–263.

Gordon, P. (2004). Numerical cognition without words: evidence from Amazonia. Science, 306, 496–499.

Grégoire, J. (2005). Développement logique et compétences arithmétiques. Le modèle piagétien est-il toujours actuel? In M. Crahay, L. Verschaffel, E. De Corte, & J. Grégoire (Eds). Enseignement et apprentissage des mathématiques (pp. 57–77). Bruxelles: De Boeck.

Grégoire, J., Noël, M., & Van Nieuwenhoven, C. (2004). TEDI-MATH. Antwerpen: Harcourt.

Heikkila, R., Narhi, V., Aro, M., & Ahonen, T. (2009). Rapid automatized naming and learning disabilities: does ran have a specific connection to reading or not? Child Neuropsychology, 15, 343–358.

Hooper, S., Roberts, J., Sideris, J. Burchinal, M., & Zeisel, S. (2010). Longitudinal predictors of reading and math trajectories through middle school from African American versus Caucasian students across two samples. Development Psychology, 46, 1018–1029.

Jordan, N., Kaplan, D., Olah, L., & Locuniak, M. (2006). Number sense growth in kindergarten: a longitudinal investigation of children at risk for mathematics difficulties. Child Development, 77, 153–175.

Jordan, J., Wylie, J., & Mulhern, G. (2010). Phonological awareness and mathematical difficulty: A longitudinal perspective. Britisch Journal of Developmental Psychology, 28, 89–107.

Kulp, M.T., Earley, M.J., Mitchell, G.L., Timmerman, L.M., Frasco, C.S., & Geiger, M.E. (2004). Are visual perceptual skills related to mathematics ability in second through sixth grade children? Focus on Learning Problems in Mathematics, 26, 44–51.

Le Fevre, J-A., Smith-Chant, B.L., Fast, L., Skwarchuk, S-L., Sargla, E., Arnup, J.S., Penner-Wilger, M., Bisanz, J., & Kamawar, D. (2006). What counts as knowing? The development of conceptual and procedural knowledge of counting from kindergarten through grade 2. Journal of Experimental Child Psychology, 93, 285–303.

Luo, Z., Jose, P.E., Huntsinger, C.S., & Pigott, T.D. (2007). Fine motor skills and mathematics achievement in East Asian American and European American kindergartners and first graders. British Journal of Developmental Psychology, 25, 595–614.

Mazzocco, M.M.M., & Myers, G.F. (2003). Complexities in identifying and defining mathematics learning disability in the primary school-age years. Annals of Dyslexia, 53, 218–253.

McCloskey, M., & Macaruso, P. (1995). Representing and using numerical information. American Psychologist, 50, 351–363.

Morris, R. D., Stuebing, K. K., Fletcher, J. M., Shaywitz, S. E., Lyon, G. R., Shankweiler, D. P., Katz, L., Francis, D.J., & Shaywitz, B.A. (1998). Subtypes of reading disability: variability around a phonological core. Journal of Educational Psychology, 90, 347–373.

Murphy, M.M., Mazzocco, M.M.M., Hanich, L.B., & Early, M.C. (2007). Cognitive characteristics of children with mathematics learning disability (MLD) vary as a function of the cutoff criterion used to define MLD. Journal of Learning Disabilities, 40, 458–478.

Mussolin, C., Mejias, S., & Noël, M.P. (2010). Symbolic and nonsymbolic number comparison in children with and without dyscalculia. Cognition, 115, 10–25.

Neuhoff, N., Bruder, J., Bartling, J., Warnke, A., Remschmidt, H., Muller-Myhsok, B., & Schulte-Korne, G. (2012). Evidence for the late MMN as a neurophysiological endophenotype for dyslexia. Plos One, 7(5):e34909

Nigg, J.T. (2000). On inhibition/disinhibition in developmental psychopathology: views from cognitive and personality psychology and a working inhibition taxonomy. Psychological Bulletin, 126, 220–246.

Nunes, T., Bryant, P., Evans, D., Bell, D., Gardner, A., Gardner, A., & Carraher, J. (2007). The contribution of logical reasoning to the learning of mathematics in primary school. British Journal of Developmental Psychology, 25, 147-166.

Pagani, L.S., Fitzpatrick, C., Archambault, I., & Janosz, M. (2010). School readiness and later achievement: a French Canadian replication and extension. Developmental Psychology, 46, 984–994.

Passolunghi, M.C., Marzocchi, G.M., & Fiorillo, F. (2005). Selective effect of inhibition of literal or numerical irrelevant information in children with attention deficit hyperactivity disorder (ADHD) or arithmetic learning disorder (ALD). Developmental Neuropsychology, 28, 731–753.

Passolunghi, M.C., & Siegel, L.S. (2004). Working memory and access to numerical information in children with disability in mathematics. Journal of Experimental Child Psychology, 88, 348–367.

Piazza, M., Facoetti, A., Trussardi, A.N., Berteletti, I., Conte, S., Lucangeli, D., Dehaene, S., & Zorzi, M. (2010). Developmental trajectory of number acuity reveals a severe impairment in developmental dyscalculia. Cognition, 116, 33–41.

Pica, P., Lemer, C., Izard, V., & Dehaene, S. (2004) Exact and approximate arithmetic in an Amazonian indigene group. Science, 306, 499–503.

Pieters, S., De Block, K., Scheiris, J., Eyssen, M., Desoete, A., Deboutte, D., Van Waelvelde, H., & Roeyers, H. (2012a). How common are motor problems in children with a developmental disorder: rule or exception? Child: Care, Health and Development, 38, 139–145.

Pieters, S., Desoete, A., Roeyers, H., Vanderswalmen, R., & Van Waelvelde, H. (2012b). Behind mathematical learning disabilities: what about visual perception and motor skills? Learning and Individual Differences, 22, 498–504. 498–504.

Pieters, S. Roeyers, H., Rosseel, Y., Van Waelvelde, H., & Desoete, A. (2013). Identifying subtypes among children with developmental coordination disorder and mathematical learning disabilities, using model-based clustering. Journal of learning disabilities.

Praet, M., Titeca, D., Ceulemans, A., & Desoete, A. (2013). Language in the prediction of arithmetics in kindergarten and grade 1. Learning and Individual Differences, 27, 90–96.

Purpura, J., Hume, L.E., Sims, D,C., & Lonigan, C.J. (2011). Early literacy and early numeracy: the value of including early literacy skills in the prediction of numeracy development. Journal of Experimental Child Psychology, 110, 647–658.

Robinson, C.S., Menchetti, B.M., & Torgesen, J.K. (2002). Toward a two-factor theory of one type of mathematics disabilities. Learning Disabilities Research & Practice, 17, 81–89.

Romano, E., Babchishin, L. Pagani, L.S., & Kohen, D. (2010). School readiness and later achievement: Replication and extension using a nationwide Canadian survey. Developmental Psychology, 46, 995–1007.

Sarnecka, B.W., & Carey, S. (2008). How counting represents number: what children must learn and when they learn it. Cognition, 108, 662–674.

Sarnecka, B.W., Kamenskaya, V.G., Yamana, Y., Ogura, T., & Ydovina, Y.B. (2007). From grammatical number to exact numbers: early meanings of one, two and three in English, Russian and Japanese, Cognitive Psychology, 55, 136–168.

Shalev, R.S., Manor, O., & Gross-Tsur, V. (2005). Developmental dyscalculia: a prospective six-year follow-up. Developmental Medicine & Child Neurology, 47, 121–125.

Shalev, R.S., Manor, O., Kerem, B., Ayali, M., Badichi, N., Friedlander, Y., & Gross-Tsur, V. (2001). Developmental dyscalculia is a familial learning disability. Journal of Learning Disabilities, 34, 59–65.

Soylu, F. (2011). Mathematical cognition as embodied simulation. Paper presented at the 33rd Annual Conference of the Cognitive Science Society, 20-22 july. Austin, TX: Cognitive Science Society.

Stock, P., Desoete, A., & Roeyers, H. (2010). Detecting children with arithmetic disabilities from kindergarten: evidence form a three year longitudinal study on the role of preparatory arithmetic abilities. Journal of Learning Disabilities, 43, 250–268.

Swanson, H.L., & Jerman, O. (2006). Math disabilities: a selective meta-analysis of the literature. Review of Educational Research, 76, 249–274.

Temple, C.M. (1991). Procedural dyscalculia and number fact dyscalculia – double dissociation in developmental dyscalculia. Cognitive Neuropsychology, 8, 155–176.

Temple, C.M., & Sherwood, S. (2002). Representation and retrieval of arithmetical facts: developmental difficulties. Quarterly Journal of Experimental Psychology Section a-Human Experimental Psychology, 55, 733–752.

van der Sluis, S., van der Leij, A., & de Jong, P.F. (2005). Working memory in Dutch children with reading- and arithmetic-related LD. Journal of Learning Disabilities, 38, 207–221.

Vanmeirhaeghe, B. (2012). Divided by numbers. Studying with dyscalculia. Documentary: synopsis. Artevelde University College: Gent, Belgium. Available at: http://www.studerenmetdyscalculie.be/

von Aster, M.G., & Shalev, R.S. (2007). Number development and developmental dyscalculia. Developmental Medicine and Child Neurology, 49, 868–873.

Vuijk, P.J, Hartman, E., Mombarg, R., Scherder, E., & Visscher, C. (2011). Associations between the Academic and Motor Performance of Children with Learning Disabilities in Dutch Special Education. Journal of Learning Disabilities, 44, 276-282.

Vukovic, R.S., & Siegel, L.S. (2010). Academic and cognitive characteristics of persistent mathematics difficulty from first through fourth grade. Learning Disabilities Research & Practice, 25, 25–38.

Waldman, I.D., Nigg, J.T., Gizer, I.R., Park, L., Rappley, M.D., & Friderici, K. (2006). The adrenergic receptor alpha-2A gene (ADRA2A) and neuropsychological executive functions as putative endophenotypes for childhood ADHD. Cognitive Affective & Behavioral Neuroscience, 6, 18–30.

Zhang, H.Y., & Wu, H.R. (2011). Inhibitory ability of children with developmental dyscalculia. Journal of Huazhong University of Science and Technology-Medical Sciences, 31, 131–136.

Appendix A

Appendix A. Subtests and examples of test-items of the TEDI-MATH

Subtests	Items
1. Procedural knowledge of counting	Counting as far as possible Counting forward to an upper bound (e.g. 'up to 9') Counting forward from a lower bound (e.g. 'from 7') Counting forward from a lower bound to an upper bound (e.g. 'from 4 up to 8') Counting backward Counting by step (by 2 and by 10)
2. Conceptual knowledge of counting	Counting linear pattern of items Counting random pattern of items Counting a heterogeneous set of items Understanding of cardinality
3. Knowledge of the numerical system	*3.1. Arab numerical system* Judge if a written symbol is a number Which of two written numbers is the larger *3.2. Oral numerical system* Judge if a word is a number Judge if a number word is syntacticly correct Which of two numbers is the larger *3.3. Base-ten system* Representation of numbers with sticks Representation of numbers with coins Recognition of hundreds, tens and units in written numbers *3.4. Transcoding* Write a dictated number as an Arabic numeral Read a number written as an Arabic numeral
4. Logical thinking skills	*4.1. Seriation of numbers* Sort the cards from the one with fewer trees to the one with the most trees *4.2. Classification of numbers* Make groups with the cards that go together *4.3. Conservation of numbers* Do you have more counters than me? Do I have more counters than you? Or do we have the same number of counters? Why? *4.4. Inclusion of numbers* You put 6 counters in the envelope. Are there enough counters inside the envelope if you want to take out 8 of them? Why? *4.5. Additive decomposition of numbers* A shepherd had 6 sheep. He put 4 sheep in the first prairie, and 2 in the other one. In what other way could he put his sheep in the two prairies?

(*continued*)

Appendix A. Continued

Subtests	Items
5. Calculations	*5.1. Presented on pictures* There are 2 red balloons and 3 blue balloons. How many balloons are there in all? *5.2. Presented in arithmetical format* Addition (e.g. '6 + 3'; '5 + ... = 9,' ' ... + 3 = 6') Subtraction (e.g. '9 – 5', '9 – ... = 1', ' ... – 2 = 3') Multiplication (e.g. '2 × 4' '10 × 2') *5.3. Presented in verbal format* e.g. 'Denis had 2 marbles. He won two others. How many marbles had Denis in all?' *5.4. Understanding arithmetical operation properties (conditional knowledge)* e.g. commutativity of addition 'You know that 29 + 66 = 95. Would this information help you to compute 66 + 29? Why
6. Estimation of size	*6.1. Comparison of dot sets* *6.2. Estimation of number size* Comparison of distance between numbers e.g. target number is 5. What number is closed to this (3 or 9)?

CHAPTER 50

AFFECT, MOTIVATION, WORKING MEMORY, AND MATHEMATICS

ALEX M. MOORE, NATHAN O. RUDIG, AND MARK H. ASHCRAFT

> *If cognitive psychology aspires to an understanding of human thought and action, it can ill afford to leave out their emotional aspects.*
>
> (Mandler, 1989, p. 4.)

In a brief review of mathematics anxiety and its consequences, Ashcraft (2002) told of a female college student being tested in a laboratory experiment on subtraction (e.g. 34 − 19 = ?). Halfway through the session, she broke into tears. *Does this happen in sentence comprehension studies?* Baggett, Ehrenfeucht, and Main (2011) surveyed college students about their mathematics education experiences in grades K-12, and reported that 'many students were very emotional. ... Some even used obscenities in describing teachers whom they did not like.' *Do students describe their other teachers that way?* We are careful about how we describe our experiments to college students, saying, for example, 'a simple problem solving task,' rather than 'solving simple addition and subtraction problems.' *Do cognition researchers in other fields have to be this cautious when they recruit potential participants?*

This article is about affect, emotional states and reactions, and their impact on mathematics learning and performance. As suggested above, the study of mathematical cognition is rather unlike other topics studied in cognition, in that the knowledge being learned or tested often arouses affective reactions that can seriously disrupt learning and performance. Researchers who study memory or language generally do not have to be concerned with whether their participants have the base knowledge to perform adequately on their laboratory tasks, and are not concerned that reactions such as anxiety will cloud their participants' performance. However, such concerns are justified in the case of mathematical cognition. As such, the function of this article is to describe the literature on such affective states and reactions, describe the part of the cognitive system they impact, and portray the impact of those states and reactions on cognitive performance.

We begin this article by covering the several affective, emotional, and belief systems that have been investigated in the domain of mathematics learning and performance. We then turn to the component of the human cognitive system shown to be especially crucial to mathematics performance – working memory. Following this, we turn to the research that examines online influences of affect on performance, to see how emotion, anxiety, and pressure impact on actual arithmetic and mathematics problem solving. Given the critical role working memory plays in mathematics performance, it is not surprising that it is often the locus of disruption when emotions interfere with ongoing problem solving. The article concludes with a few observations about recent work aimed at reducing or eliminating these interfering effects.

Affect, Motivation, and Belief States

We begin with a discussion of the emotional states and beliefs shown to be related to learning and performance in mathematics. Specifically, we discuss the topics of interest, motivation, and self-efficacy in mathematics, as well as mathematics anxiety. All of the investigations rely heavily on correlations among these self-reported measures and various educational outcome measures like maths achievement. As such, firm causal inferences cannot be drawn. Nonetheless, the patterns of relatedness are very clear, and suggest that affect and belief are crucial in learning and understanding maths.

Interest and Motivation

Children begin formal education with a very positive view of mathematics. For example, Stevenson et al. (1990) reported that 72% of their first and fifth graders showed positive attitudes toward arithmetic and mathematics; in our own study, 87% of 1st grade children reported 'very good' feelings when doing arithmetic and mathematics, and 74% reported that their own abilities were 'very good' as well (Moore & Ashcraft, 2009). To be sure, both interest and motivation decline as children grow older, especially during adolescence, with at least part of this decline interpreted as reflecting the more difficult content of mathematics as the child advances in school.

Despite this overall decline, self-rated interest in mathematics still has a strong positive relationship with maths-related success. For example, a meta-analysis investigating the importance of attitudes in mathematics performance found that the strength of the maths attitude to achievement relationship grows with increasing age, and potentially grows to practical significance for students in high school (Ma & Kishor, 1997). Moreover, being interested in mathematics seems to play a large role in students' current and continued success in mathematics-related tasks and courses. Köller, Baumert, and Schnabel (2001), for example, examined ratings of interest in mathematics, scores on a standardized mathematics exam, and enrollment in mathematics courses from a longitudinal sample of 600 students, tested in 7th, 10th, and 12th grades. The results showed that students with higher levels of interest in mathematics were more successful in completing the higher level mathematics classes that were offered. Not surprisingly, those students who reported the highest levels of interest in mathematics also viewed the domain as being more important, took more mathematics courses, and obtained higher grades in maths courses, compared with those who showed little interest in the subject matter (Simpkins, Davis-Kean, & Eccles, 2006).

In a similar vein, high motivation shows the same positive relationships to performance, with continued engagement in maths classes and, subsequently, better achievement. High motivation was found to be positively related to mathematics achievement scores ($r = 0.31$) and negatively related to maths anxiety ($r = -0.72$; Zakaria & Nordin, 2008), to be predictive of college majors and career goals (Leuwerke, Robbins, Sawyer, & Hovland, 2004), and to supplement mathematics self-efficacy (Berger & Karabenick, 2011; Lopez, Lent, Brown, & Gore, 1997). Interestingly, Berger and Karabenick (2011) also found a relationship between high motivation, self-efficacy, and the sophistication of the learning strategies used to master new material; 9th graders with high self-efficacy found maths to be intrinsically valuable, and used more sophisticated strategies when learning new material (e.g. elaboration), whereas those with lower self-efficacy relied on simpler rehearsal and memorization strategies.

Needless to say, the positive effects of high motivation depend critically on whether the motivation is intrinsic or extrinsic. That is, intrinsically-motivated students generally express an interest in increasing their understanding and skills in maths, perceive maths to be important, and persevere in pursuing maths coursework. In contrast, those with extrinsic motivation wish to gain recognition such as good grades or praise from teachers and parents, and they tend to seek less help from teachers, thus avoiding situations in which they might experience negative judgments or consequences. When they experience such negative feedback, they tend to disengage or discontinue their pursuit of maths (e.g. National Mathematics Advisory Panel, 2008; Ryan & Pintrich, 1997).

Self-Efficacy

Ashcraft and Rudig (2012) adapted Bandura's (1977) definition of self-efficacy to the topic of mathematics, stating that '*self-efficacy* is an individual's confidence in his or her ability to perform mathematics and is thought to directly impact the choice to engage in, expend effort on, and persist in pursuing mathematics' (p. 249). As Ashcraft and Rudig noted, the definition combines elements of prior success, interest, and motivation to pursue mathematics. As such, it seems quite clear why research has shown that self-efficacy yields similar relationships to those found with interest and motivation.

As an example, Pietsch, Walker, and Chapman (2003) examined the relationship between self-efficacy and performance within mathematics in a study of 416 9th and 10th grade students. Self-efficacy was measured both at a general level (e.g. asking about confidence in a mathematics course), as well as for specific content areas (e.g. asking about confidence in computing percentages); a general mathematics self-concept measure was also collected. Although the general self-efficacy measure predicted end-of-year performance on a comprehensive maths exam more effectively than the self-concept score, the self-efficacy scores for specific content areas were even stronger predictors. Similar results are found in Lee (2009), who reported on 250,000 15-year-olds from 41 countries. The average correlation between self-efficacy and mathematics performance, across countries, was 0.42. The average correlation between performance and mathematics anxiety was −0.65 (see also Cooper & Robinson, 1991, for comparable results with incoming university students).

We suggest the literature can be summarized as follows. Those who show high interest and intrinsic motivation, as well as high self-efficacy, reliably show higher scores in mathematics achievement, and continue to engage in mathematics-related activities. These positive affective states and beliefs can be termed an *approach constellation* of factors, a constellation that accompanies enhanced learning and performance. Conversely, those who are low in interest, those espousing extrinsic motivation, and those with low self-efficacy, are predicted to be the students who show low persistence in the face of challenges, disengagement, and ultimately low maths achievement. Such negative states and beliefs can be grouped together as an *avoidance constellation* of factors. Although correlational, these patterns appear to be both substantial and influential.

Mathematics Anxiety

Mathematics anxiety, the apprehension or fear aroused when placed in a situation in which maths must be performed, has essentially the opposite effects to those of interest, motivation, and self-efficacy. Unlike those positive factors that lead people to approach mathematics, high mathematics anxiety can be characterized as an avoidance reaction; people who are high in mathematics anxiety are anxious or fearful when performing maths and, therefore, attempt to avoid maths when possible.

As Hembree's (1990) meta-analysis documents, people with high maths anxiety express a variety of poor attitudes about maths; the correlations with maths anxiety are −0.37 for usefulness of maths, −0.64 for motivation in maths, −0.82 for self-efficacy (grades 6–11; for college, $r = -0.65$), and −0.75 for enjoyment of maths (grades 5–12; for college, −0.47). At the college/university level, maths anxiety correlates −0.32 with the intent of taking more maths classes. The outcomes, in terms of performance, are equally dismal. According to data reported in Hembree, maths anxiety correlates −0.34 with mathematics achievement scores (grades 5–12; for college, $r = -0.31$), and −0.30 with course grades (grades 9–12; for college, $r = -0.27$). Highly maths-anxious students take fewer mathematics courses, gain lower grades in the courses they do take, and intend to avoid taking additional maths courses if possible. Consequently, they avoid mathematically-oriented college majors and career paths (for additional discussion, see Ashcraft, Krause, & Hopko, 2007; Ashcraft & Moore, 2009; Ashcraft & Rudig, 2012). Mathematics anxiety is far reaching; it is found in both western and eastern nations (Ho et al., 2000), and is associated with negative performance in both males and females, although the exact deficit in performance may be gender specific (Baloğlu & Koçak, 2006; Miller & Bichsel, 2004). Invariably, research has reported an inverse relationship between maths anxiety and the approach constellation factors discussed here, e.g. the Cooper and Robinson (1991) demonstration of an inverse relationship between self-efficacy and maths anxiety (see also Lee, 2009).

Theories of Intelligence

We propose that the variety of affective, emotional, and self-belief variables discussed here can be usefully grouped as constellations of approach or avoidance factors. A case can also be made that these factors are well captured by Dweck's social-cognitive theory of motivation (e.g. Dweck, 1986, 1999; Dweck & Leggett, 1988). Dweck argues that people adopt one of two sets of beliefs about intelligence and, therefore, one of two types of goals involving competence. These beliefs and goals then predict their activities when learning, in mathematics and other academic domains.

On the one hand, people may adopt an *incremental* theory of intelligence, a set of beliefs and goals that lines up closely with the approach constellation discussed here. In this theory, people believe that intelligence and abilities are malleable, that they measure effort, and can be increased through persistence. People with this belief system typically display *mastery goals* in learning situations, in which their motivation is to learn and master new material, even if that means receiving negative feedback during the learning process. The belief system, in other words, is that effort and persistence will increase skill and knowledge, and will ultimately lead to academic success. Failure, under such a belief, prods the individual to reassess how much effort has been expended, in order to adapt and overcome the challenge.

In contrast, people may adopt an *entity* theory of intelligence, believing that intelligence and abilities are largely fixed and do not change over time. People with such a belief system generally adopt *performance goals* when placed in a learning environment; they seek positive feedback about their competence (e.g. good test scores), and try to avoid negative feedback (poor grades). When confronted with negative feedback, they often withdraw their effort from the task. They do not persist when they face challenging new material, but instead retreat to material they have already mastered. The attitude is that, since intelligence is fixed, practice and effort are somewhat fruitless. The parallels to the avoidance constellation of attitudes and affective reactions are compelling, e.g. Burkley, Parker, Stermer, and Burkley (2010) found that female students who endorsed the fixed belief system were less likely to enjoy maths, to pursue a maths-related major, or to consider a career in maths. Although nothing in Dweck's approach speaks directly to maths anxiety, it seems clear that entity beliefs are consistent with higher levels of maths anxiety.

Although only a few studies have applied Dweck's ideas to the domain of mathematics, the results have been promising. In one report, Blackwell, Trzesniewski, and Dweck (2007) assessed 7th graders' beliefs, then recorded their maths grades over a 2-year span. Those espousing the malleable, incremental theory of beliefs improved their grades across the two years, favoured mastery over performance goals, and responded adaptively to difficult challenges, compared with those with a fixed theory of intelligence. In a follow-up study, randomly-selected students were taught to adopt the malleable theory of intelligence. After 1 year of the intervention, they earned significantly higher grades than the control group; they also increased their positive beliefs about effort, and were rated as having more adaptive classroom behaviour by their teachers. In a similar study, Burns and Isbell (2007) found that their intervention to promote malleable beliefs decreased levels of mathematics anxiety before a maths test and increased perceptions of improved performance (although performance did not improve significantly).

Summary

Affect and motivation are powerful factors that influence how students learn and master mathematics, an under-appreciated fact. Grouping these into the approach and avoidance constellations, and considering the motivational framework of Dweck's theories of intelligence, provides a way of organizing these effects and seeing their impact on learning.

Working Memory and Mental Arithmetic

We turn now to a seemingly unrelated topic, the role of working memory in mathematical calculation and performance. Focusing on working memory does not imply that other aspects of the human cognitive system are unimportant to overall maths learning and performance, of course; to state the obvious, the goal of mathematics education and learning is a rich, long-term memory representation of maths concepts, facts, and principles. Nonetheless, to understand how affect and emotion operate during maths performance, it is vital to understand the role played by working memory.

Take a moment to solve the problem (44 – 18)/4. Because the answers to 44–18 and 26/4 are surely not memorized, solving the problem correctly depends on executing a series of procedures, including borrowing, recalling mathematics facts from long-term memory, holding intermediate values in memory, performing the relatively slow division process, and keeping track of various intermediate steps and solutions. All of these steps demand a significant amount of cognitive effort, of course, not to mention time. By most accounts, the mechanism regulating this effort, the mental component responsible for this coordination of procedures, is working memory.

Working memory is commonly viewed as a limited capacity mechanism that allows the mind to integrate, compute, store, and manipulate information at the focus of a person's attention (Baddeley, 2000; Engle, 2002; Miyake & Shah, 1999). Working memory has been implicated in a wide range of cognitive domains, including attention, memory, language, and overall intelligence. As more research has accumulated, working memory has also been demonstrated to be an important component of arithmetic and mathematics performance (e.g. De Rammelaere, Stuyven, & Vandierendonck, 1999; Imbo & LeFevre, 2010; Imbo & Vandierendonck, 2007, 2008; Seyler, Kirk, & Ashcraft, 2003). What follows is a brief review of this evidence (see DeStefano & LeFevre, 2004; LeFevre, DeStefano, Coleman, & Shanahan, 2005; Raghubar, Barnes, & Hecht, 2010, for extensive reviews).

All existing models make the common claim that working memory has a limited capacity for processing and storing information; indeed, the limited capacity of immediate memory has been noted for over a century (e.g. James, 1890). Beyond that, the various models (e.g. Engle, 2002; Miyake & Shah, 1999) differ principally in whether working memory is viewed as an undifferentiated pool of resources or a system divided into multiple subcomponents. For the most part, researchers within mathematics cognition have explored the multi-component theory of working memory proposed by Baddeley and colleagues (Baddeley, 1996, 2000; Baddeley & Logie, 1999). This model describes working memory as the on-line coordination of three unique components of information processing: the central executive and two slave systems, the phonological loop and the visuospatial sketchpad (a recently proposed fourth component, the episodic buffer, has not yet been investigated to a significant degree). The central executive acts as the command centre of processing, fulfilling such activities as focusing and switching attentional scope, performing calculations, and coordinating the information temporarily maintained by the slave systems. The activities of the phonological loop include active rehearsal and storage of verbal and semantic information, while the visuospatial sketchpad is implicated in the production and storage of mental images that arise during processing the task at hand (Baddeley & Logie, 1999).

An attraction of the multi-component model for researchers is the possibility of testing the subcomponents separately, and relating those to possible verbal or visual characteristics of problem solving. This is typically done within a dual-task paradigm. That is, investigators design experimental situations to exploit the limited resources of one system while attempting to leave another system untouched, to see if this alters overall performance. In this paradigm, the participant completes two tasks concurrently, a primary task (a mathematics task) along with a secondary task selected to tax the processing or storage capacity of one particular component of working memory. The logic here is that if the secondary task interferes with accurate or efficient performance, it can be inferred that both tasks rely on the same type of processing or information. For example, if a verbal secondary task were to interfere with maths problem solving, we would infer that arriving at the correct solution requires the verbal processing resources consumed by the secondary task. (For an early demonstration of the role of working memory in maths, see Hitch, 1978, where decay and interference across time served to disrupt accuracy.)

Central Executive

There is clear evidence that the central executive plays an important role in addition and multiplication performance, whether participants are tested using the production (5 + 3 = ?) or verification (5 + 3 = 9, true or false?) task, using both single- and multiple-digit problems (e.g. DeRammelaere et al., 1999; DeRammelaere, Stuyven, & Vandierendonck, 2001; Imbo & Vandierendonck, 2007; Lemaire, Abdi, & Fayol, 1996; Seitz & Schumann-Hengsteler, 2000). In all of these studies, a secondary task involving the central executive (e.g. generating a random string of letters) was paired with the arithmetic task, showing interference on problem solving compared with control conditions.

The evidence of the central executive's role in arithmetic processing extends well beyond studies that tested the simple addition and multiplication facts, however. For example, Logie, Gilhooly, and Wynn (1994) had participants perform a central executive load task, while adding two-digit numbers across a 20-second span of time (they also tested several other secondary tasks). The results showed clear disruption of performance when the central executive was loaded. With auditory presentation of the numbers to be added, errors increased from 14% in the control condition to 38.5% in the dual-task condition; with visual presentation, errors increased from 3.5% to 44%. Similarly, Fürst and Hitch (2000) presented multiple-digit addition problems to their participants, and varied the number of carry operations required, while also loading the central executive with a secondary task (the Trails task, generating spoken sequences, alternating between the alphabet and months of the year, as in 'B, February, C, March'). Errors increased substantially (from 15 to 45%) when the central executive was loaded compared with the phonological load condition (see also Imbo, Vandierendonck, & DeRammelaere, 2007; the phonological loop has also been implicated in maintaining intermediate sums in multi-column addition problems; Heathcote, 1994).

Slave Systems

There is also substantial research showing the involvement of the phonological loop in arithmetic processing. As an example, Seyler et al. (2003) tested adults on simple subtraction facts, and used letter recall as the secondary task; participants held 2, 4, or 6 letters in working memory, then recalled them in order after solving the subtraction problem. Participants were also pre-tested on their working memory capacity, and separated into low, medium, and high capacity groups. The results showed clear working memory effects, with performance decrements on letter recall. Accuracy declined significantly with larger memory loads, especially on more difficult problems (e.g. 17 – 8), and also was lower for participants with lower baseline levels of working memory capacity. Importantly, memory load and working memory capacity also interacted significantly; those with lower working memory capacity were especially disrupted when confronted with a heavy working memory load. All of these disruptive effects, whether due to high load or low capacity, were particularly evident on the larger subtraction problems, those found to be solved more frequently by reconstructive procedures (e.g. counting, transformation).

Several studies have expanded the dual-task paradigm by using multiple secondary tasks, to see which components of working memory play a role in arithmetic processing. Hecht (2002), for example, tested the role of working memory in the selection and execution of calculation strategies, using an addition verification task and two secondary tasks. One secondary task involved articulatory suppression, requiring the participants to repeat a letter of the alphabet, thus loading the phonological loop, while the other was the random letter generation task, loading the central executive. After each trial, participants indicated the strategy they had used to solve the addition problem. The results showed that working memory resources were crucial for efficient strategy execution. The central executive load disrupted all strategies, but the articulatory suppression task also disrupted addition when participants were using a counting strategy. In other words, articulatory suppression specifically interfered with counting, suggesting that counting relies on phonological processing (see also Camos & Barrouillet, 2004).

Similarly, Trbovich and LeFevre (2003) investigated the roles of the two slave systems in complex addition. Problems like 47 + 5 were presented in conjunction with a secondary task that either taxed the phonological loop (remembering zero, one, or three non-words) or visuospatial sketchpad (remembering the location and patterns of one, four, or eight asterisks). In both secondary tasks, the participant had to verify if the probe, presented after the addition problem, was the same as the display presented before the problem. Furthermore, addition problems were presented in two formats, horizontally to one group, to be read from left to right, or vertically to the other group, to be read from top to bottom. Half of the participants in each format group completed the phonological load task, while the other half completed the visuospatial load task.

The results showed that interference from the secondary tasks depended on the format of the addition problems. When the problem was shown horizontally, performance (as measured by errors) suffered more under the phonological load than under the visuospatial load. The reverse pattern was observed for problems shown vertically; here, performance suffered more under the visuospatial load. These results led to an interpretation based on the activation of a specific code (phonological or visual) used for different procedural strategies in problem solving. That is, horizontal problems seemed to be represented and solved via a verbal code, as if participants were saying them sub-vocally during problem solving, whereas a visual representation of calculation seemed apparent in the vertical format condition, as if the problem were being written with pencil-and-paper (see also Imbo & LeFevre, 2010).

Although beyond the scope of this article, it should be noted that there are some inconsistencies in the literature regarding the influence of one or another of the components of working memory on mathematics performance. For example, some studies find phonological interference in subtraction (Seyler et al., 2003) while others do not (Lee & Kang, 2002). LeFevre et al. (2005) note that many of these inconsistencies may be attributable to differences in experimental methods and procedures, as well as educational and language backgrounds of participants (e.g. Lee & Kang's Korean participants (2002) solved only small subtraction facts). Nonetheless, the overall picture is clear; working memory plays an important role in arithmetic and mathematics processing, especially as problems become more difficult, and especially as processing relies on non-retrieval procedures. Note also that the bulk of the published research has been conducted on fairly simple arithmetic, going no higher than two-column addition or multiplication problems. Given the results on maintaining intermediate values, carrying, and the like, it is clear that more complex types of mathematics will be shown to rely extensively on the working memory system (e.g. see Beilock & Carr, 2005, below).

The Impact of Affect on Mathematics Performance

The first two sections of this article have yielded two firm conclusions. First, there is a clear pattern of influence between affect and attitudes, on the one hand, and mathematics achievement and learning on the other, positive in the case of the approach constellation factors, and negative for the avoidance constellation. Secondly, there is solid evidence that working memory plays a critical role in performing mathematics, even on the rather elementary procedures of carrying, borrowing, and counting, and certainly on problems that require multiple steps and keeping track of intermediate solutions.

This final section makes the explicit and important connection between these two points. That is, we now turn to the evidence that affective states and reactions indeed have an impact on performance, not just on global measures of performance as seen in achievement tests, but on local, online measures as seen in laboratory tasks. The evidence shows that the primary locus of this impact is in working memory. This evidence is clearest in three related domains, research on mathematics anxiety, on choking under pressure, and on stereotype threat.

Mathematics Anxiety

Our original research question was quite simple. How does mathematics anxiety affect mental processing, if it does at all? We tested this question using standard cognitive tasks that had been used in the mathematics cognition literature, looking to see if standard effects (e.g. the problem size effect; for a review, see Ashcraft, 1992) were obtained.

Our early exploratory studies (Ashcraft & Faust, 1994; Faust, 1988; Faust, Ashcraft, & Fleck, 1996) examined timed performance to simple and more complex arithmetic, e.g.

one- and two-column addition and multiplication, using the verification task. Three prominent effects were obtained. First, the low mathematics anxious group was invariably the fastest group to respond, and usually had the lowest error rates. In contrast, higher anxiety groups often showed dramatically higher errors, and frequently also showed rapid responses, in a classic speed-accuracy trade-off fashion. We interpreted this as a strategic form of responding on their part, speeding the session along to an early conclusion regardless of errors, and labelled it 'local avoidance' of the testing situation.

Secondly, we found that participants at higher levels of maths anxiety had difficulty rejecting incorrect problems, even when the stated answer was seriously incorrect (e.g. 8 + 7 = 38); their errors remained in the 7 to 13% range, whereas low anxious participants showed the typical (e.g. Ashcraft & Stazyk, 1981) decline in errors, from 8 to 2%, as answers deviated more and more from the correct value (Faust et al., 1996). We suggested that the quick and accurate rejection of 'unreasonable' answers reported earlier by Ashcraft and Stazyk reflected an aspect of 'number sense' (Dehaene, 1997), that is a sense that 38 seems too far away from a reasonable estimate for 8 + 7. If so, then finding high errors to such problems on the part of higher maths anxious participants suggests that those with higher levels of mathematics anxiety may, in fact, suffer from a more basic deficit in number sense than we originally appreciated.

Finally, there were prominent effects of mathematics anxiety when we examined performance to the more complex, two-column addition problems. Here, the higher anxiety groups took considerably longer to respond, and made considerably more errors, especially on problems requiring a carry. Although all groups slowed down on carry problems, and made more errors as well, these tendencies were especially pronounced for the high mathematics anxiety group.

The difficulty of carry problems for the high mathematics anxious participants, especially given the role of working memory, led us to hypothesize that working memory was the connection between mathematics anxiety and the processing disruption we observed for carry problems. Eysenck's processing efficiency theory provided a theoretical rationale for exactly such an explanation (Eysenck & Calvo, 1992; see Eysenck, Derakshan, Santos, & Calvo, 2007, for an updated version of this theory). In this theory, Eysenck claimed that general anxiety prompts mental rumination and negative, preoccupying thoughts about the anxiety itself. These thoughts consume an individual's working memory resources, and thus will have an effect on any cognitive process that normally relies on working memory. As such, those high in anxiety will show performance deficits whenever they perform a task that requires substantial working memory resources. We reasoned that mathematics anxiety should function much like generalized anxiety in terms of draining the working memory resources necessary for mathematics problem solving.

We tested this hypothesis directly in Ashcraft and Kirk (2001). We asked participants of low, medium, and high mathematics anxiety to perform a dual task experiment, using an addition task, with one and two-column addition problems, and a letter recall task, with either 2- or 6-letter sequences to be held in working memory. Our reasoning was that more difficult addition, especially when the problem required a carry, would start to burden working memory. The load should be even heavier when participants also had to hold six letters in working memory for later recall. High mathematics anxiety should then lead to an even greater drain on working memory due to the worries and ruminations that would be depleting additional resources.

In short, this is exactly what we found. As expected, there was interference (lower recall of letters) when carrying was required in the problems, and when working memory was loaded more heavily, and this was true for all groups. However, the most difficult combination of conditions—carrying, high working memory load, and high mathematics anxiety—led to a much larger deficit in letter recall than was true for the other conditions. Error rate in this condition was 39%, versus 14% for the high anxious group with a low memory load, and a 15% error rate in the control condition. In comparison, the low mathematics anxious group had only a 20% error rate in the most difficult condition. It was as if the high mathematics anxious participants were performing not just in a dual task, but in a *triple* task, since three separate factors—carrying, heavy letter load, and mathematics anxiety—were all operating to drain working memory of its critical resources.

We have found completely parallel results in a study on two-column subtraction with borrowing (Krause, Rudig, & Ashcraft, 2009), and comparable results using a novel mathematics task, modular arithmetic (Krause, 2008 see below for an explanation of this task). In comparison with their low mathematics anxious peers, high mathematics anxious groups committed far more errors when the task required considerable working memory resources (borrowing and heavy memory load), and required far more practice to master a novel mathematics task that relies heavily on working memory.

These studies indicate clearly that the locus of the mathematics anxiety deficit in performance is in working memory, and in particular in the way that mathematics anxiety robs working memory of the resources necessary for successful problem solving. While we are quite confident in this conclusion, there is now additional research showing mathematics anxiety effects on performance in more basic-level number tasks as well. First, Maloney, Risko, Ansari, and Fugelsang (2010) tested high and low mathematics anxious participants in the classic subitizing task, having them merely name how many simple objects (filled squares) were displayed on a screen. They obtained the familiar dog-leg pattern of latencies, with a flat pattern for displays up to size of 4, and an increasing function for display sizes from 4 through 9 attributable to counting (e.g. Trick & Pylyshyn, 1993). Critically, Maloney et al.'s data showed slower counting by high mathematics anxious participants. Consistent with the conclusion that counting is dependent on working memory capacity (e.g. Tuholski, Engle, & Baylis, 2001), when a measure of participants' working memory capacity was used as a covariate, the interaction of mathematics anxiety and set size was rendered non-significant.

In similar fashion, Maloney, Ansari, and Fugelsang (2011) had high and low mathematics anxious participants perform a number comparison task, deciding whether a presented digit was larger or smaller than a standard (5). In their second experiment, two digits were presented, and participants indicated which was larger. In both studies, high anxious participants showed a steeper numerical distance effect, i.e. were slower to judge numbers that were closer together in numerical magnitude (e.g. 4 and 5) than were low anxious participants. As with subitizing (Maloney et al., 2010), the number comparison task presumably taps a simpler, more basic level of numerical processing than the complex addition and subtraction studies reviewed above. Thus, the Maloney work is evidence that mathematics anxiety's effect on processing may be much more basic than was previously suspected. In particular, it suggests that rather elementary aspects of number processing, including the mental representation of numerosity and simple processes such as counting, may differ for high mathematics anxious individuals, possibly setting the stage for later learning difficulties.

Choking under Pressure

The term choking under pressure—or simply 'choking'—refers to the phenomenon in which an individual performs more poorly than expected, given the person's established level of skill in a particular domain, when placed in a high pressure situation. It occurs despite, or even because of, the individual's desire to perform well. Indeed, the desire to perform well is considered to be an integral part of choking under natural, everyday situations, e.g. the high-stakes testing situations students encounter in school.

Beilock and her colleagues (e.g. Beilock & Carr, 2005; Beilock, Kulp, Holt, & Carr, 2004; DeCaro, Rotar, Kendra, & Beilock, 2010) have pioneered the study of choking under pressure in the cognitive domain, using a novel mathematics task, modular arithmetic, as their preferred task. Modular arithmetic, introduced by Gauss (1801, as cited by Bogomolny, 1996) involves judging whether equations of the form $x \equiv y \pmod{z}$ are true or false; a mod arithmetic equation is true if $(x - y)/z$ leaves no remainder (or, alternatively, if x/z and y/z both leave the same remainder). Thus, $6 \equiv 3 \pmod{3}$ is true, because $(6 - 3)/3$ leaves no remainder, whereas $7 \equiv 3 \pmod{3}$ is false, because $4/3$ leaves a remainder of 1.

In their initial study, Beilock et al. (2004) asked participants to solve simple problems like the examples just provided, or more difficult problems, involving larger x and y values that often required borrowing, e.g. $44 \equiv 18 \pmod{4}$. Some participants were merely told to perform as well as they could. Others were placed in a high pressure situation involving speed demands, monetary rewards, and social pressure. In the face of this pressure situation, participants exhibited significant performance decrements on the modular arithmetic problems, but significantly, the performance decrements were only obtained on the difficult problems, those that placed demands on working memory. Problems like $6 \equiv 3 \pmod{3}$, where the subtraction and division were extremely simple, are considered to be performed in largely automatic fashion (e.g. Ashcraft, 1992). Thus, they demanded few, if any, working memory resources, and were unaffected by the pressure manipulation. However, difficult problems like $44 \equiv 18 \pmod{4}$, with large numbers and borrowing, were dramatically influenced by the pressure manipulation. In short, the participants choked under pressure. (Incidentally, this was essentially the problem that introduced working memory in this article, to illustrate its demands on working memory.)

Beilock and Carr (2005) explored this effect further, by categorizing participants as either low or high in working memory capacity, based on a pre-test. When tested under the low pressure condition, the high working memory group outperformed the low capacity group, not surprisingly, since difficult modular arithmetic is, in fact, fairly intensive in its reliance on working memory resources. However, when the groups were tested under high pressure, the high working memory group's performance fell to the level of the low capacity group—they were no longer able to demonstrate superior performance, given that their larger working memory capacities were now being disrupted by the pressure manipulation. A follow-up study, by Beilock and DeCaro (2007), determined the exact nature of their decline in performance. Under pressure, the high capacity group reverted to using more heuristic, short-cut strategies that minimized the need for working memory resources, essentially the same heuristics that the low capacity group routinely relied on in all circumstances (see Beilock & Ramirez, 2011, for a full account).

Stereotype Threat

Stereotype threat illustrates a third research domain in which individuals experience a performance decrement. Here, however, there is neither a personal anxiety at work nor a high-pressure situation that affects all individuals being tested. Instead, the operative factor is membership in a group to which some negative stereotype has been attached for the domain being tested. The phenomenon of stereotype threat, initially demonstrated by Steele and Aronson (1995), is actually quite straightforward; individuals who belong to a stigmatized group often experience anxiety in an evaluative situation when they expect that their own performance may falter and thus confirm the negative stereotype about their group. In the classic initial demonstration, Steele and Aronson tested African–American and European–American students on a verbal problem-solving task. When the task was described neutrally, there were no group differences, but when the task was described as diagnostic of intellectual ability, the description aroused the negative stereotype pertaining to African Americans concerning intellectual ability, and that group performed significantly worse than the European–American students.

The stereotype threat effect can apparently apply to anyone, to the degree that an existing and plausible stereotype exists for both the domain of knowledge or activity being tested, and the social, gender, or ethnic group to which the individual belongs. The effect is strengthened for individuals who value success more strongly in the domain being tested (domain identity, e.g. Spencer, Steele, & Quinn, 1999), and for those who identify more strongly with the stereotyped group (gender or ethnic identity, e.g. Schmader, 2002). Likewise, low domain identity, or low group identity, diminish the effects of stereotype threat (see Schmader, Johns, & Forbes, 2008, for an interesting discussion of stereotype lift, the situation in which a positive stereotype yields a small, but noticeable improvement in performance).

Several studies have used mathematics performance as a way of demonstrating the stereotype threat effect. For example, Aronson, Lustina, Good, Keough, Steele, and Brown (1999) exposed Caucasian men to the negative stereotype that Caucasians do more poorly than Asians on maths. Under this circumstance, the Caucasian men did worse than those not exposed to the stereotype. More analytically, Beilock, Rydell, and McConnell (2007) tested women on the modular arithmetic task, telling some that the task was being used to investigate why women do more poorly than men, and others that the researchers were merely studying problem solving. Under stereotype threat, accuracy dropped from 89 to 79% in the difficult mathematics condition; with no threat, performance actually improved from 86 to 92% from baseline to post-test, presumably due to practice. Critically, the decrement due to stereotype threat was only obtained on the difficult problems, those requiring the resources of working memory. As found in the choking studies, simple problems that did not rely on working memory were uniformly high in accuracy, and experienced no drop in accuracy due to stereotype threat. Thus, only when working memory needs to be actively engaged in problem solving does the decrement in its resources actually degrade performance.

Such results suggest strongly that depletion of working memory resources is the common thread that ties stereotype threat to mathematics anxiety (e.g. Beilock & Ramirez, 2011; Schmader et al., 2008). That is, stereotype threat appears to arouse one's fears and anxieties about confirming the negative stereotype. These fears and anxieties consume some of the resources of working memory that are necessary for problem-solving performance, leading to performance decrements. This is, of course, the same mechanism implicated in the 'affective drop' in performance seen in mathematics anxiety (Ashcraft & Moore, 2009) and in choking under pressure—inadequate working memory resources for the primary task because those resources are being devoted to an affective reaction, whether related to one's anxiety, to a situation of pressure, or to a personally relevant stereotype. As Schmader et al. (2008) note, a variety of processes can be triggered by these affective reactions, and all have the effect of draining resources from working memory, thus disrupting performance that depends on those resources. Importantly, the disruption is not only on performance, but has also now been shown to impair learning as well (Mangels, Good, Whiteman, Maniscalco, & Dweck, 2011).

Notice the interesting contrasts among mathematics anxiety, choking under pressure, and stereotype threat. All three phenomena show decrements in performance, which can be globally attributed to a reduction of working memory's resources due to the affective reaction during testing. However, the contrast in who is susceptible to the effect, and the source of the affective reaction, is fascinating. In the case of mathematics anxiety, we consider individuals who have an anxiety reaction to mathematics content, possibly stemming from elementary school years, and co-morbid difficulties that include attitudes and beliefs that yield both local and global avoidance, i.e. emotional reactions during the online performance of mathematics processing, and also long-term effects on learning and mastery. Little is known about the onset of mathematics anxiety, save for the fascinating study about children—especially girls—in early elementary school seeming to learn mathematics anxiety from their mathematics-anxious teachers (Beilock, Gunderson, Ramirez, & Levine, 2010).

For choking under pressure, however, no personal susceptibility to mathematics anxiety is indicated. Indeed, it is often those who have been shown to be especially likely to do well, and often those who intend (even strongly) to do well, who show performance decrements. The operative factor, instead, is the anxiety-inducing pressure of the testing situation. Finally, in stereotype threat, there is membership in a group that is the target of a negative stereotype, and identification with that group, with the attendant fear that one's poor performance will reflect poorly on oneself and the group as a whole. Schmader et al. (2008) discuss the several mechanisms, such as vigilance, self-regulation, and situation monitoring, that all drain working memory of the necessary resources for adequate performance. Importantly, these processes apparently lead to decrements in performance, regardless of the individual's actual levels of skill and expertise—in other words, even sufficiently high expertise does not prevent the performance decrement due to stereotype threat, provided sufficient group and domain identity. In both choking and stereotype threat, the decrement occurs despite the individual's intention to do well. Conversely, in the case of high maths anxiety, the evidence suggests that individuals have resigned themselves—possibly long ago—to doing poorly and, therefore, often lack high-level skill.

Reducing or Eliminating The Influence of Affect

We conclude with a brief discussion of the empirical evidence on efforts to reduce or eliminate the effects of emotion and affect on mathematics performance.

As noted in Hembree's (1990) meta-analysis, there is evidence that behavioural and cognitive behavioural therapies developed for treatments of test anxiety are effective at reducing or eliminating mathematics anxiety. The evidence is that, after therapeutic intervention, high maths-anxious individuals not only benefit from reduced anxiety levels, but also show increased mathematics achievement scores. That is, the treatment group showed significantly higher achievement scores than a control group with similarly high initial maths anxiety scores (Vance & Watson, 1994). A variety of other therapeutic interventions, however, including classroom curricular changes, relaxation therapy, and group counselling, were ineffective. Note, however, that the effective interventions are not brief (Richardson & Suinn, 1973; Schneider & Nevid, 1993).

Recent studies on the more transitory disruptions due to choking and stereotype threat have revealed several interesting and provocative findings. Ramirez and Beilock (2011) reported a series of studies on choking in which performance declined under pressure. However, an additional group that was exposed to the same pressure spent 10 minutes in an expressive writing intervention prior to the test of modular arithmetic, writing about their thoughts and feelings about taking the upcoming mathematics test (in two of the studies, a third group merely did unrelated writing). Interestingly, the expressive writing group showed a 5% improvement from pre- to post-test, compared with a 12% drop in accuracy for the choking group (7% drop in unrelated writing). Interestingly, the content of participants' expressive writing was heavily weighted toward negative thoughts and worries about the mathematics test, suggesting that expressing the worries frees the individual from considering them during actual testing.

Within the stereotype threat arena, several studies have shown the impact of two factors, first teaching participants about stereotype threat itself, and second guiding participants to reappraise the threatening situation. Concerning the first, Johns, Schmader, and Martens (2005) tested men and women on a mathematics test under one of three conditions, a regular problem solving condition with no threat, a conventional stereotype threat condition, and a 'threat plus teaching' condition; in the latter, participants were both exposed to the standard gender stereotype about mathematics and also given an explanation of stereotype threat, and how it could make women anxious while they did the mathematics test. There was no gender difference in the standard problem-solving condition, and a significant drop in performance for women in the conventional stereotype threat condition, as has been found repeatedly. However, in the 'threat plus teaching' condition, there was no gender difference at all; teaching the participants about the stereotype threat effect itself apparently immunized the women to the effect. This, of course, is an important lesson, given the concern that many if not most gender (and racial/ethnic) differences in grades and test scores across the years, according to Walton and Spencer (2009), may in fact have been due to subtle testing conditions that arouse stereotype threats.

Likewise, reappraisal of the fear and anxiety generated by stereotype threat has also been shown to eliminate the effects of stereotype threat. Johns, Inzlicht, and Schmader (2008) conducted several studies, always with a standard stereotype threat condition. In other conditions, however, they also instructed subjects to think of the upcoming mathematics test objectively and not as personally relevant, or they told them that previous research had shown that the anxiety they would experience would not hurt and might even help their performance. Consistently, the groups given a way to reappraise their anxieties and worries, were able to overcome the stereotype threat effect; the threat-induced depletion of working memory resources did not occur, having been circumvented by the reappraisal.

As discussed earlier, more theoretical and empirical work is needed to unify the topics of interest, motivation, self-efficacy, and mathematics anxiety with respect to mathematics learning and mastery, perhaps under the umbrella of entity versus incremental theories of intelligence. The approach- and avoidance-constellation ideas are convenient ways of grouping the effects, but still fall short of a full-fledged explanation of causality when it comes to understanding the onset of negative affect about mathematics. As the areas of choking and stereotype effect demonstrate, a clear understanding of the cognitive mechanisms—in this case disruption of working memory—can lead productively to demonstrations of how to alleviate the declines in performance. As yet, and with only a very few possible exceptions (e.g. Beilock et al., 2010; Blackwell et al., 2007), there is no clear evidence as to genuine causes for either the desirable approach-constellation attitudes and beliefs, or the undesirable avoidance-constellation ones. To the extent that we wish to furnish all students with adequate knowledge of important mathematics skills and adequately understand human cognitive processes in mathematics, efforts to determine the causes and effects of these affective influences need to be redoubled.

References

Aronson, J., Lustina, M.J., Good, C., Keough, K., Steele, C.M., & Brown, J. (1999). When White men can't do math: necessary and sufficient factors in stereotype threat. *Journal of Experimental Social Psychology*, 35, 29–46.

Ashcraft, M.H. (1992). Cognitive arithmetic: a review of data and theory. *Cognition*, 44, 75–106.

Ashcraft, M.H. (2002). Math anxiety: personal, educational, and cognitive consequences. *Current Directions in Psychological Science*, 11, 181–185.

Ashcraft, M.H., & Faust, M.W. (1994). Mathematics anxiety and mental arithmetic performance: an exploratory investigation. *Cognition and Emotion*, 8, 97–125.

Ashcraft, M.H., & Kirk, E.P. (2001). The relationships among working memory, math anxiety, and performance. *Journal of Experimental Psychology: General*, 130, 224–237.

Ashcraft, M.H., Krause, J.A., & Hopko, D.R. (2007). Is math anxiety a mathematical learning disability? In D. B. Berch & M. M. M. Mazzocco (Eds), *Why is Math so Hard for Some Children? The Nature and Origins of Mathematical Learning Difficulties and Disabilities* (pp. 329–348). Baltimore, MD: Paul H. Brookes.

Ashcraft, M.H., & Moore, A.M. (2009). Mathematics anxiety and the affective drop in performance. *Journal of Psychoeducational Assessment*, 27, 197–205.

Ashcraft, M.H., & Rudig, N.O. (2012). Higher cognition is altered by noncognitive factors: How affect enhances and disrupts mathematics performance in adolescence and young adulthood. In V. F. Reyna, S. B. Chapman, M. R. Dougherty, & J. Confrey (Eds), *The Adolescent Brain: Learning, Reasoning, and Decision Making* (pp. 243–263). Washington, D.C.: American Psychology Association.

Ashcraft, M.H., & Stazyk, E.H. (1981). Mental addition: a test of three verification models. *Memory & Cognition*, 9, 185–196.

Baddeley, A.D. (1996). Exploring the central executive. *Quarterly Journal of Experimental Psychology: Human Experimental Psychology*, 49A, 5–28.

Baddeley, A.D. (2000). The episodic buffer: a new component of working memory? *Trends in Cognitive Sciences*, 4, 417–423.

Baddeley, A.D., & Logie, R.H. (1999). Working memory: the multiple-component model. In A. Miyake & P. Shah (Eds), *Models of Working Memory: Mechanisms of Active Maintenance and Executive Control* (pp. 28–61). Cambridge: Cambridge University Press.

Baggett, P., Ehrenfeucht, A., & Main, M. (2011). University students' opinions about the mathematics they studied in grades K-12. Paper presented at the meetings of the Psychonomic Society, Seattle, November 2011.

Baloğlu, M., & Koçak, R. (2006). A multivariate investigation of the differences in mathematics anxiety. *Personality and Individual Differences*, 40, 1325–1335.
Bandura, A. (1977). Self-efficacy: toward a unifying theory of behavioral change. *Psychological Review*, 84, 191–215.
Beilock, S.L., & Carr, T.H. (2005). When high-powered people fail: working memory and 'choking under pressure' in math. *Psychological Science*, 16, 101–105.
Beilock, S.L., & DeCaro, M.S. (2007). From poor performance to success under stress: working memory, strategy selection, and mathematical problem solving under pressure. *Journal of Experimental Psychology: Learning, Memory, & Cognition*, 33, 983–998.
Beilock, S.L., Gunderson, E.A., Ramirez, G., & Levine, S.C. (2010). Female teachers' math anxiety affects girls' math achievement. *Proceedings of the National Academy of Sciences*, 107 (5), 1860–1863.
Beilock, S.L., Kulp, C.A., Holt, L.E., & Carr, T. H. (2004). More on the fragility of performance: Choking under pressure in mathematical problem solving. *Journal of Experimental Psychology: General*, 133, 584–600.
Beilock, S. L., & Ramirez, G. (2011). On the interplay of emotion and cognitive control: implications for enhancing academic achievement. In J. P. Mestre & B. H. Ross (Eds), *The Psychology of Learning and Motivation: Cognition in Education*, Vol. 55 (pp. 137–170). Oxford: Academic Press.
Beilock, S.L., Rydell, R.J., & McConnell, A.R. (2007). Stereotype threat and working memory: Mechanisms, alleviation, and spillover. *Journal of Experimental Psychology: General*, 136, 256–276.
Berger, J-L., & Karabenick, S.A. (2011). Motivation and students' use of learning strategies: evidence of unidirectional effects in mathematics classrooms. *Learning and Instruction*, 21, 416–428.
Blackwell, L., Trzesniewski, K., & Dweck, C.S. (2007). Implicit theories of intelligence predict achievement across an adolescent transition: a longitudinal study and an intervention. *Child Development*, 78, 246–263.
Bogomolny, A. (1996). Modular arithmetic. Retrieved from: http://www.cut-the-know.org/blue/Modoluo.shtml.
Burkley, M., Parker, J., Stermer, S.P., & Burkley, E. (2010). Trait beliefs that make women vulnerable to math disengagement. *Personality and Individual Differences*, 48, 234–238.
Burns, K.C., & Isbell, L.M., (2007). Promoting malleability is not one size fits all: priming implicit theories of intelligence as a function of self-theories. *Self and Identity*, 6, 51–63.
Camos, V., & Barrouillet, P. (2004). Adult counting is resource demanding. *British Journal of Psychology*, 95, 19–30.
Cooper, S.E., & Robinson, D.A. (1991). The relationship of mathematics self-efficacy beliefs to mathematics anxiety and performance. *Measurement and Evaluation in Counseling and Development*, 24, 4–11.
DeCaro, M.S., Rotar, K.E., Kendra, M.S., & Beilock, S.L. (2010). Diagnosing and alleviating the impact of performance pressure on mathematical problem solving. *Quarterly Journal of Experimental Psychology: Human Experimental Psychology*, 63, 1619–1630.
Dehaene, S. (1997). *The Number Sense*. New York: Oxford University Press.
De Rammelaere, Stuyven, E., & Vandierendonck, A. (1999). The contribution of working memory resources in the verification of simple mental arithmetic sums. *Psychological Research*, 62, 72–77.
DeRammelaere, S., Stuyven, E., & Vandierendonck, A. (2001). Verifying simple arithmetic sums and products: Are the phonological loop and the central executive involved? *Memory & Cognition*, 29, 267–273.
DeStefano, D., & LeFevre, J. (2004). The role of working memory in mental arithmetic. *European Journal of Cognitive Psychology*, 16, 353–386.
Dweck, C.S. (1986). Motivational processes affecting learning. *American Psychologist*, 41, 1040–1048.
Dweck, C.S. (1999). *Self-theories: Their Role in Motivation, Personality, and Development*. Philadelphia: Psychology Press.
Dweck, C.S., & Leggett, E.L. (1988). A social-cognitive approach to motivation and personality. *Psychological Review*, 95, 256–273.
Engle, R.W. (2002). Working memory capacity as executive attention. *Current Directions in Psychological Science*, 11, 19–23.

Eysenck, M.W., & Calvo, M.G. (1992). Anxiety and performance: the processing efficiency theory. *Cognition and Emotion*, 6, 409–434.

Eysenck, M.W., Derakshan, N., Santos, R., & Calvo, M.G. (2007). Anxiety and cognitive performance: attentional control theory. *Emotion*, 7, 336–353.

Faust, M.W. (1988). Arithmetic performance as a function of mathematics anxiety: an in-depth analysis of simple and complex addition problems. Unpublished M.A. thesis, Cleveland State University, Cleveland, Ohio.

Faust, M.W., Ashcraft, M. H., & Fleck, D.E. (1996). Mathematics anxiety effects in simple and complex addition. *Mathematical Cognition*, 2, 25–62.

Fürst, A.J., & Hitch, G.J. (2000). Separate roles for executive and phonological components of working memory in mental arithmetic. *Memory and Cognition*, 28, 774–782.

Heathcote, D. (1994). The role of visuo-spatial working memory in the mental addition of multi-digit addends. *Current Psychology of Cognition*, 13, 207–245.

Hecht, S.A. (2002). Counting on working memory in simple arithmetic when counting is used for problem solving. *Memory and Cognition*, 30, 447–455.

Hembree, R. (1990). The nature, effects, and relief of mathematics anxiety. *Journal for Research in Mathematics Education*, 21, 33–46.

Hitch, G.J. (1978). The role of short-term working memory in mental arithmetic. *Cognitive Psychology*, 10, 302–323.

Ho, H., Senturk, D., Lam, A.G., Zimmer, J.M., Hong., S., Okamoto, Y., Chiu, S., Nakazawa., Y., & Wang, C. (2000). The affective and cognitive dimensions of math anxiety: a cross-national study. *Journal for Research in Mathematics Education*, 31, 362–379.

Imbo, I., & LeFevre, J. (2010). The role of phonological and visual working memory in complex arithmetic for Chinese- and Canadian-educated adults. *Memory and Cognition*, 38, 176–185.

Imbo, I., & Vandierendonck, A. (2007). The role of phonological and executive working memory resources in simple arithmetic strategies. *European Journal of Cognitive Psychology*, 19, 910–933.

Imbo, I., & Vandierendonck, A. (2008). Effects of problem size, operation, and working-memory span on simple-arithmetic strategies: Differences between children and adults? *Psychological Research*, 72, 331–346.

Imbo, I., Vandierendonck, A., & De Rammelaere, S. (2007). The role of working memory in the carry operation of mental arithmetic: number and value of the carry. *Quarterly Journal of Experimental Psychology*, 60, 708–731.

James, W. (1890). *The Principles of Psychology*. Cambridge, MA: Harvard University Press.

Johns, M., Inzlicht, M., & Schmader, T. (2008). Stereotype threat and executive resource depletion: examining the influence of emotion regulation. *Journal of Experimental Psychology: General*, 137, 691–705.

Johns, M., Schmader, T., & Martens, A. (2005). Knowing is half the battle: teaching stereotype threat as a means of improving women's math performance. *Psychological Science*, 16, 175–179.

Krause, J.A. (2008). The effect of math anxiety on learning a novel math task. Unpublished M.A. thesis, University of Nevada Las Vegas.

Krause, J.A., Rudig, N.O., & Ashcraft, M.H. (2009, November). Math, working memory, and math anxiety effects. Poster presented at the meetings of the Psychonomic Society, Boston 2009.

Köller, O., Baumert, J., & Schnabel, K. (2001). Does interest matter? The relationship between academic interest and achievement in mathematics. *Journal for Research in Mathematics Education*, 32, 448–470.

Lee, J., (2009). Universals and specifics of math self-concept, math self-efficacy, and math anxiety across 41 PISA 2003 participating countries. *Learning and Individual Differences*, 19, 355–365.

Lee, K., & Kang, S. (2002). Arithmetic operation and working memory: differential suppression in dual tasks. *Cognition*, 83, B63–B68.

LeFevre, J., DeStefano, D., Coleman, B., & Shanahan, T. (2005). Mathematical cognition and working memory. In J. I. D. Campbell (Ed.), *Handbook of Mathematical Cognition* (pp. 361–377). New York: Psychology Press.

Lemaire, P., Abdi, H., & Fayol, M. (1996). The role of working memory resources in simple cognitive arithmetic. *European Journal of Cognitive Psychology*, 8, 73–103.

Leuwerke, W.C., Robbins, S., Sawyer, R., & Hovland, M. (2004). Predicting engineering major status from mathematics achievement and interest congruence. *Journal of Career Assessment*, 12, 135–149.

Logie, R.H., Gilhooly, K.J., & Wynn, V. (1994). Counting on working memory in mental arithmetic. *Memory & Cognition*, 22, 395–410.

Lopez, F.G., Lent, R.W., Brown, S.D., & Gore, P.A. (1997). Role of social-cognitive expectations in high school students' mathematics-related interest and performance. *Journal of Counseling Psychology*, 44, 44–52.

Ma, X., & Kishor, N. (1997). Assessing the relationship between attitude toward mathematics and achievement in mathematics: a meta-analysis. *Journal for Research in Mathematics Education*, 28, 26–47.

Maloney, E.A., Ansari, D., & Fugelsang, J.A. (2011). The effect of mathematics anxiety on the processing of numerical magnitude. *The Quarterly Journal of Experimental Psychology*, 64, 10–16.

Maloney, E.A., Risko, E.F., Ansari, D., & Fugelsang, J. (2010). Mathematics anxiety affects counting but not subitizing during visual enumeration. *Cognition*, 114, 293–297.

Mandler, G. (1989). Affect and learning: causes and consequences of emotional interactions. In D. B. McLeod & V. M. Adams (Eds), *Affect and Mathematical Problem Solving* (pp. 3–19). New York: Springer.

Mangels, J.A., Good, C., Whiteman, R.C., Maniscalco, B., & Dweck, C.S. (2011). Emotion blocks the path to learning under stereotype threat. *Social Cognitive and Affective Neuroscience*, 7, 230–241.

Miller, H., & Bichsel, J. (2004). Anxiety, working memory, gender, and math performance. *Personality and Individual Differences*, 37, 591–606.

Miyake, A., & Shah, P. (1999) *Models of working memory: mechanisms of active maintenance and executive control*. Cambridge: Cambridge University Press.

Moore, A.M., & Ashcraft, M.H. (2012). Relationships across mathematics tasks in elementary school children. Unpublished manuscript, University of Nevada, Las Vegas.

National Mathematics Advisory Panel (2008). Foundations for success. Reports of the Task Groups and Subcommittees. Washington, D.C.: U.S. Department of Education. Available at: www2.ed.gov/about/bdscomm/list/mathpanel/report/final-report.pdf.

Pietsch, J., Walker, R., & Chapman, E. (2003). The relationship among self-concept, self-efficacy, and performance in mathematics during secondary school. *Journal of Educational Psychology*, 95, 589–603.

Raghubar, K.P., Barnes, M.A., & Hecht, S.A. (2010). Working memory and mathematics: a review of developmental, individual difference, and cognitive approaches. *Learning and Individual Differences*, 20, 110–122.

Ramirez, G., & Beilock, S.L. (2011). Writing about testing worries boosts exam performance in the classroom. *Science*, 331, 211–213.

Richardson, F.C., & Suinn, R.M. (1973). A comparison of traditional systematic desensitization, accelerated massed desensitization, and anxiety management training in the treatment of mathematics anxiety. *Behavior Therapy*, 4, 212–218.

Ryan, A.M., & Pintrich, P.R. (1997). 'Should I ask for help?' The role of motivation and attitudes in adolescents' help seeking in math class. *Journal of Educational Psychology*, 89, 329–341.

Schmader, T. (2002). Gender identification moderates stereotype threat effects on women's math performance. *Journal of Experimental Social Psychology*, 38, 194–201.

Schmader, T., Johns, M., & Forbes, C. (2008). An integrated process model of stereotype threat effects on performance. *Psychological Review*, 115, 336–356.

Schneider, W.J., & Nevid, J.S. (1993). Overcoming math anxiety: a comparison of stress inoculation training and systematic desensitization. *Journal of College Student Development*, 34, 283–288.

Seitz, K., & Schumann-Hengsteler, R. (2000). Mental multiplication and working memory. *European Journal of Cognitive Psychology*, 12, 552–570.

Seyler, D.J., Kirk, E.P., & Ashcraft, M.H. (2003). Elementary subtraction. *Journal of Experimental Psychology. Learning, Memory, and Cognition*, 29, 1339–1352.

Simpkins, S.D., Davis-Kean, P.E., & Eccles, J.S. (2006). Math and science motivation: a longitudinal examination of the links between choices and beliefs. *Developmental Psychology*, 42, 70–83.

Spencer, S.J., Steele, C.M., & Quinn, D.M. (1999). Stereotype threat and women's math performance. *Journal of Experimental Social Psychology*, 35, 4–28.

Steele, C.M., & Aronson, J. (1995). Stereotype threat and the intellectual test performance of African-Americans. *Journal of Personality and Social Psychology*, 69, 797–811.

Stevenson, H.W., Lee, S., Chen, C., Lummis, M., Stigler, J., Fan, L., & Ge, F. (1990). Mathematics achievement of children in China and the United States. *Child Development*, 61, 1053–1066.

Trbovich, P.L., & LeFevre, J. (2003). Phonological and visual working memory in mental addition. *Memory & Cognition*, 31, 738–745.

Trick, L., & Pylyshyn, Z. (1993). What enumeration studies can show us about spatial attention: Evidence for limited capacity preattentive processing. *Journal of Experimental Psychology: Human Perception and Performance*, 19, 331–351.

Tuholski, S.W., Engle, R.W., & Baylis, G.C. (2001). Individual differences in working memory capacity and enumeration. *Memory & Cognition*, 29, 484–492.

Vance, W.R., & Watson, T.S. (1994). Comparing anxiety management training and systematic rational restructuring for reducing mathematics anxiety in college students. *Journal of College Student Development*, 35, 261–266.

Walton, G.M., & Spencer, S.J. (2009). Latent ability: grades and test scores systematically underestimate the intellectual ability of negatively stereotyped students. *Psychological Science*, 20, 1132–1139.

Zakaria, E., & Nordin, N.M. (2008). The effects of math anxiety on matriculation students as related to motivation and achievement. *Eurasia Journal of Mathematics, Science & Technology Education*, 4, 27–30.

CHAPTER 51

INDIVIDUAL DIFFERENCES IN WORD PROBLEM SOLVING

L. VERSCHAFFEL, F. DEPAEPE, AND W. VAN DOOREN

Introduction

Word problems are typically defined as essentially verbal descriptions of problem situations in which one or more questions are raised and the answers can be obtained by the application of mathematical operations to the numerical data available in the problem statement (Verschaffel, Greer, & De Corte, 2000, p. ix).

Importantly, the term 'word problem' does not necessarily imply that every task that meets the above definition represents a true *problem*, in the cognitive-psychological sense of the word, for a given student, i.e. a task for which no routine method of solution is available and which therefore requires the activation of (meta)cognitive strategies (Schoenfeld, 1992). Whether a word problem that a student encounters constitutes a genuine problem depends on his or her familiarity with the problem, his or her mastery of the various kinds of required knowledge and skills to solve it, the available tools, etc.

Word problems have attracted the attention of researchers in psychology and (mathematics) education for a very long time. Before the emergence of the information-processing approach, research on word problems focused mainly on the effects of various kinds of linguistic, computational, and/or presentational task features (e.g. number of words, grammatical complexity, presence of particular key words, number and nature of the required operations, nature and size of the given numbers) and subject features (e.g. age, gender, general intelligence, linguistic, and mathematical ability of the problem solver) on individuals' problem-solving performance (Goldin & McClintock, 1984). With the rise of the information-processing approach, researchers' attention shifted from learners' externally observable performance to the underlying cognitive schemes and processes. Accordingly, research methods changed as well; analyses of response accuracies were complemented with analyses of verbal protocols, individual interviews, reaction times, eye-movements, etc.

Whereas in the early days of information-processing psychology, solving a word problem was considered as a direct translation process from text to mathematical symbols, later it was thought of as a complex multi-phase process the 'heart' of which is formed by:

(1) The construction of an internal model of the problem situation, reflecting an understanding of the elements and relations in the problem situation.
(2) The transformation of this situation model into a mathematical model of the elements and relations that are essential for the solution.

In most current models of competent word problem solving, these two steps are then followed by:

(1) Working through the mathematical model to derive mathematical result(s).
(2) Interpreting the outcome of the computational work.
(3) Evaluating if the interpreted mathematical outcome is computationally correct and reasonable.
(4) Communicating the obtained solution.

This multi-phase model is not considered to be purely sequential; rather, individuals can go back and forth through the different phases of the model (Blum & Niss, 1991; Verschaffel et al. 2000).

Especially since the 1990s, insights from ethnomathematics and sociocultural theories have contributed to the insight that classical information-processing models are insufficient to grasp the complexity of learners' word problem-solving processes, and that they need to be enriched with the idea that word problem solving is a human activity situated in the particular microcosm of a mathematics classroom (Lave, 1992; Verschaffel et al., 2000). Therefore, students' word problem-solving behaviors can only be understood by also seriously taking into account the coping strategies, and the affects that they have built up along with their participation in the practice and culture of the mathematics classroom that they bring to the task at hand.

There is nowadays a rather broad consensus that the competencies that are required to solve word problems involve (De Corte, Greer, & Verschaffel, 1996; Schoenfeld, 1992):

- A well-organized and flexibly accessible knowledge base involving the relevant factual, conceptual, and procedural knowledge that is relevant for solving word problems.
- Heuristic methods, i.e. search strategies for problem analysis and transformation, which increase the probability of finding a solution.
- Metacognition, involving both metacognitive knowledge and metacognitive skills.
- Positive task-related affects, involving positive beliefs, attitudes, and emotions.
- Meta-affect, involving knowledge about one's affects and skills for regulating one's affective processes.

Given the theme of this section of the book – namely individual differences – we will document how individual differences in performance on word problems can be directly related to each of these components. Space restrictions do not allow us to include the extensive literature on the role of more general subject characteristics, such as age, gender, and socio-economic status (SES), but also general intelligence, verbal and mathematical ability, (non-verbal) problem-solving ability, working memory, cognitive inhibitory skills, and cognitive style, that may also – directly and/or indirectly – affect word problem-solving performance. Briefly, this literature reveals that older students (Morales, Shute, & Pellegrino, 1985), boys (Walch, Hicky, & Duffy, 1999), students with higher SES (Coley, 2002), more intelligent students (Endo, 2010), students with more language and/or mathematical skills (Fuchs, Fuchs, Steubing, & Fletcher, 2008; Muth, 1984), students with higher problem-solving skills (Fuchs et al., 2008), and children with better cognitive inhibitory skills (Walch et al., 1999), with better working memory (Passolunghi & Siegel, 2001), and with a more reflective cognitive style (Navarro, Aguilar, Alcalde, & Howell, 1999) typically perform better. Another restriction of this chapter is that we will only review research on word problems involving one or more basic arithmetic operations involving natural or rational numbers – so, essentially word problems encountered by elementary and lower secondary school students.

Differences in Students' Content Knowledge Base: Knowledge of Problem Structures and Arithmetic Operations

Knowledge of Problem Structures

Addition and Subtraction Problems

As explained above, throughout the 1980s and 1990s, concentrated research from an information-processing perspective was carried out on how students do arithmetic word problems. Although other types of word problems were investigated too, the most popular topic has been the solution of one-step addition and subtraction problems involving small whole numbers (typically less than 20) by elementary school children.

In the early 1980s a basic distinction emerged between three classes of problem situations modeled by addition and subtraction: situations involving a change from an initial state to a final state through the application of a transformation (change problems), situations involving the combination of two discrete sets or splitting of one set into two discrete sets (combine problems), and situations involving the quantified comparison of two discrete sets of objects (compare problems). Further distinctions were made depending on the nature of the unknown and the direction of the action or relation, resulting in 14 different types of one-step addition and subtraction problems (Riley, Greeno, & Heller, 1983; see also Fuson, 1992; Verschaffel & De Corte, 1997).

In line with the information-processing paradigm, researchers also developed computer simulation models of children operating at different levels of skill in solving these problems, the most influential one being Riley et al.'s (1983) model of how children learn to solve the 14 types of addition and subtraction word problems (by means of counting). The major feature of this model was the claim that children gradually develop and use cognitive schemes that reflect the above-mentioned semantic structures, as well as more abstract schemes (such as the part-whole schema underlying all addition and subtraction word problems) and that assist in the construction of a mental model of the given problem situation. So, understanding an additive word problem was considered as choosing and activating the proper semantic schema and filling the empty 'slots' of the activated schema with concrete information provided in the text. Some of these representations (such as change problems with the final set unknown or combine problems with the total set unknown) were claimed to link easily to counting or operation schemes available in the problem solver's cognitive repertoire (see 'Knowledge of arithmetic operations'). Other, more difficult, problems (such as change problems with an unknown start set or combine problems with one of the subsets unknown) were claimed to require additional *re*-representational steps (involving the use of the part-whole schema) before a link with a proper counting or computation scheme could be established.

Numerous empirical studies carried out during this period with children between 5 and 8 years old provided evidence for the validity of this classification scheme and its accompanying theory of word problem solving (Fuson, 1992; Verschaffel & De Corte, 1997; Verschaffel, Greer, & De Corte, 2007). First, word problems that can be solved by the same arithmetic operation (i.e. a direct addition or subtraction with the two given numbers in the problem), but that belong to different semantic problem types, were found to yield very different degrees of difficulty. For instance, the following two problems ' Pete has 8 marbles; he gives 5 marbles to Tom; how many marbles does he have now?' and 'Pete has 8 marbles; he has 5 more marbles than Tom; how many marbles does Tom have?' are both solved by the subtraction 8 – 5, but whereas the first one is correctly solved by almost all 6–8-year-olds, the second one is solved correctly by less than 1/3 of them (Verschaffel & De Corte, 1997). Secondly, problems requiring the same arithmetic operation but having different semantic structures, elicit different strategies for solving these problems (see 'Knowledge of arithmetic operations') and also different kinds of errors. Thirdly, as shown by Morales, Shute, and Pellegrino (1985), competent problem-solvers use these basic semantic schemes to categorize problems. Fourthly, in several investigations, such as the study by De Corte, Verschaffel, and De Win (1985), it was shown that reformulations of addition and subtraction word problems with a view to clarifying the underlying semantic structure were helpful for younger children (who are assumed not yet to have well-developed semantic schemata and are, therefore, more dependent on bottom-up processing), but not for older ones (who are assumed to have already developed these schemata and, therefore, are less dependent on bottom-up processing). Fifthly, several studies provided empirical support for the assumed role of part-whole knowledge in students' solutions of the most complex addition and subtraction word problems (e.g. Sophian & Vong, 1995). Finally, on the basis of the results of this research program, successful experimental programs for teaching elementary arithmetic word problems were developed, in which students were taught to materialize or schematize the addition and subtraction word problems in terms of the above-mentioned schemes before actually solving them (e.g. Fuson & Willis, 1989). While Fuson (1992, p. 269) stated that 'there is a fairly clear consensus about the development of the skill in solving one-step addition and subtraction word problems over time, both in terms of the underlying conceptual structures and the kind of solution strategies', there still remained several problematic issues and questions, which have continued to attract the researchers' attention (Verschaffel et al., 2007).

The first major remaining issue was that researchers had somehow neglected the details of the construction of the situation model (in the first phase of the solution process) and the influence of various specific textual variables on the construction of children's problem representations. In an attempt to resolve this problem, Kintsch and Greeno (1985) theorized that a dual representation is constructed when reading and trying to understand a word problem. This dual representation would include a propositional representation (called the text base) to capture the local meaning of the passage being read, and a representation of the situations and actions in the text (termed the situation model or problem model), which is inferred from the text base, but also from the reader's prior knowledge of arithmetic problems (i.e. his or her semantic schemes). However, several researchers (Munez Mendez, 2011; Thevenot 2010; Vicente, Orrantia, & Verschaffel, 2007) have suggested that the Kintsch and Greeno model still relies too excessively on the schema theory, 'jumping directly from the propositional text base to a problem schema that is activated by specific textual clues' (Munez Mendez, 2011, p. 75). Therefore, this model still does not successfully account for learners' comprehension processes during the initial stages of the word problem-solving process, especially when the learner is confronted (a) with word problems that do not fit nicely into one of the 14 problem types of Riley et al.'s (1983) classification schema (Verschaffel et al., 2007), (b) with word problems consisting of more than one step (Thevenot, 2010), or (c) with less restricted, more realistic word problems that also contain rich information of the temporal and functional structure of the situation and actions depicted in the problem text (Munez Mendez, 2011; Vicente et al., 2007). More recent models that represent attempts in this direction have been developed by Staub and Reusser (1995), and, later, by Thevenot (2010). Anyhow, whether word problems are solved via the mobilization of semantic problem schemes stored in long-term memory and triggered by propositions is still an unresolved issue. Possibly, this may (only) be the case for classically formulated one-step addition and subtraction word problems that are given to students who have a lot of experience with such problems.

A second major shortcoming of the initial (computer) models of addition and subtraction word problem solving is that they largely neglected that the solution of a word problem always takes place in a particular sociocultural context, typically a mathematics class, homework, or testing setting. The need to enrich the theoretical view with this socio-cultural perspective grew out of theoretical reflections about the gap between solving quantitative problems in and out of school (Resnick, 1987), as well as the observations of children's errors on arithmetic word problems that are due to the use of superficial coping strategies and/or inappropriate beliefs they have themselves developed within the particular culture of school mathematics, rather than to a purely cognitive failure. In this respect, we refer to the famous example of the captain's problem ('On a boat there are 20 sheep and 6 goats. How old is the captain?'). Several studies in France, Germany, and elsewhere found this problem was answered incorrectly by dramatically large numbers of elementary school children, who used an operation (e.g. an addition or a subtraction) on the two given numbers to produce an answer without any apparent sign of awareness that the problem or their solution was meaningless (see also 'Beliefs and word problem solving').

Multiplication and Division Problems

Of course, students are not only confronted with mathematical word problems involving additive relations between quantities. In the first years of elementary school, they are taught mathematical word problems that not only draw on additive *part-whole relations* between quantities, but also on multiplicative relations based on *correspondences (of different types) between quantities* (Nunes & Csapó, 2011).

Compared with the domain of addition and subtraction word problem solving, the field of multiplication and division word problems has developed somewhat differently, more slowly, and with less integration. Based on a conceptual/epistemological analysis, Greer (1992) proposed a synthesis of semantic types – which he termed 'models of situations' – for multiplication and division, involving types such as equal groups, equal measures, rate, Cartesian product, and rectangular area. He made a basic distinction between situations that are 'psychologically non-commutative' (or 'asymmetric') and 'psychologically commutative' (or 'symmetric'). In the asymmetric classes the multiplier and multiplicand can be distinguished – for example, in the class labeled 'equal groups,' the number of objects in each group is the multiplicand and the number of groups is the multiplier. By contrast, for Cartesian product (e.g. 'How many different combinations can be assembled if you have 3 shirts and 4 trousers?'), the numbers multiplied have symmetric roles. With respect to division for asymmetric cases, two types of division can be distinguished, namely division by the multiplier (partitive division) and division by the multiplicand (quotitive division), whereas for the symmetric situations, there is, according to Greer, no clear distinction between multiplier and multiplicand, and, consequently, only one type of division. For instance, for the above-mentioned Cartesian product multiplication problem, there is no substantial difference between the two possible corresponding division problems with, respectively, the three shirts and the four trousers in the role of unknown number.

Contrary to research on addition and subtraction, in which the categories of problem types also represent hypothetical cognitive structures that are assumed – at least by some theorists – to drive students' understanding and solving of particular problems, this does not seem to be so clearly the case for the different classes of multiplicative problems in Greer's classification scheme (Greer, 1992; Verschaffel & De Corte, 1997; Verschaffel et al., 2007). Although this line of research is more conceptual/epistemological in nature and has attracted less empirical research, some studies have yielded empirical evidence (in the form of accuracy, strategy, and/or error data) that supports the validity of the distinctions in this classification (Nunes & Csapó, 2011; Verschaffel et al. 2007).

Knowledge of Arithmetic Operations

Individual differences in students' ability to correctly and efficiently solve a word problem depends not only on the wealth and depth of their available 'models of (problem) situations,' but also on their conceptual and procedural knowledge of the arithmetic operations necessary to construct a mathematical model from the situation model and to work on that model in the later phases of the problem-solving process.

Addition and Subtraction

Again, most research has been done in the domain of one-step addition and subtraction word problems. Carpenter and Moser (1984) have carried out pioneering work about children's informal strategies for solving elementary addition and subtraction word problems. In a three-year longitudinal study they followed a large number of children from grade 1 through 3 by means of individual interviews wherein they confronted children with a subset of problems from Riley et al.'s (1983) classification scheme. Their results, first, demonstrated that early in their development children have a wide variety of 'material counting' strategies (based on the construction and manipulation of sets of concrete objects) and 'verbal counting' strategies (based on forward and backward counting without using concrete objects) for successfully solving addition and subtraction word problems, many of which are never taught explicitly and/or systematically at school. Gradually, these strategies develop into 'mental' strategies (i.e. 'known fact' strategies, such as knowing by heart that 3 and 7 makes 10, or 'derived fact' strategies, such as deriving that 5 and 7 equals 12 because 5 and 5 is 10 and 10 plus 2 equals 12). This finding was replicated in numerous other studies (for an overview see Fuson, 1992; Verschaffel et al., 2007).

Secondly, Carpenter and Moser (1984) found that the situational structure of a word problem significantly affects the nature of children's strategy choices. More specifically, children tended to solve each subtraction word problem with the type of strategy that corresponds most closely to its situation model. For instance, a problem like 'Pete had 6 apples. He gave 2 apples to Ann. How many apples does Pete have?' was typically solved by taking away 2 blocks from a group of 6 (material) or counting down 2 from 6 (verbal). In contrast, the problem 'First Pete had 2 apples. Now he has 6 apples. How many apples did he get?' was mainly solved by adding on blocks to a set of 2 until there were 6 (material), or counting up from 2 till 6 (verbal) strategies. In later studies, it was shown that this second finding also holds for material and verbal strategies for solving addition problems, and for children operating at the highest level of internalisation, using known and derived fact strategies (De Corte and Verschaffel, 1987).

Thirdly, research revealed that older and mathematically stronger students are more able to select and apply strategies that do not follow the semantic structure of the problem in an orderly fashion, allowing them to adapt their strategy to numerical features. For instance, consider the following problem: 'Pete has 4 apples. Ann gave him some more apples. Now he has 22 apples. How many apples did Ann give him?' Even if the semantic structure of this subtraction problem strongly evokes an indirect addition strategy, whereby one counts up or adds up from the smaller number to the larger one (e.g. $4 + 6 + 10 + 2 = 22$, so the answer is $6 + 10 + 2 = 18$), older and mathematically stronger students are more able to use the – for that problem computationally more efficient – direct subtraction strategy, whereby the solver finds the difference by counting down or subtracting the smaller number from the larger one (e.g. $22 - 2 - 2 = 18$, so the answer is 18; Kraemer, 2011).

To explain these individual and developmental differences in solution strategy use, researchers have pointed to students' conceptual knowledge of the operations of addition and subtraction, or – stated differently – to learners' 'models of operations'. According to Fischbein, Deri, Nello, and Marino's (1985) theory of primitive models of operations, each arithmetic operation is linked to a primitive, intuitive model, even long after that operation has acquired a formal status, and the identification of the operation that is needed to solve a word problem is mediated by these intuitive models (at least when the size of the numbers in the problem and/or the question format force them to operate at a formal level). As far as addition is concerned, distinction is made between a 'unary' and a 'binary' model of this operation (De Corte & Verschaffel, 1987). In the unary model, addition is conceived as 'adding a quantity to a given quantity' (implying that both quantities play asymmetric roles), whereas the binary model sees addition as 'two quantities being combined' (implying that their roles are symmetric). The unary model is assumed to be more primitive and thus develop earlier than the binary model (Baroody & Gannon, 1984). Absence of the latter model may hinder children's functional application of the commutativity principle for addition word problems in which the larger number is given second (De Corte & Verschaffel, 1987). As far as subtraction is concerned, two operational models are distinguished: the 'take-away' and the 'comparison' (or 'determine the difference') model (Selter, Prediger, Nührenbörger, & Hußmann, 2012). The interpretation of subtraction solely as taking away is too one-sided. According to Selter et al. (2012), the taking away model appears to be slightly more natural than the determine-the-difference model. However, it is important to interpret mathematical objects and operations in different ways, depending on the requirements of the problem. Learners who possess both models of subtraction will be more able to properly identify a variety of word problems as subtraction problems and to solve them by means of the most efficient strategy (Kraemer, 2011; Selter et al., 2012).

Multiplication and Division

As for addition and subtraction, a lot of research has unraveled the kinds of strategies that children apply to arrive at an answer of a multiplication or division word problem. Generally speaking, this research yielded similar outcomes as for addition and subtraction (Verschaffel & De Corte, 1997; Verschaffel et al., 2007). First, the results indicate that before students have been instructed in multiplication or division, many of them are able to solve simple word problems involving these operations by a variety of informal material and verbal counting-based strategies (e.g. Mulligan & Mitchelmore, 1997; Nunes & Bryant, 1995). Gradually, with increasing experience and increasing task demands, these strategies are replaced by more efficient mental strategies based on repeated addition and subtraction, repeated doubling and halving, partitioning one or both operands, etc., or the written computation algorithms (e.g. Baek, 1998; Verschaffel et al., 2007). Secondly, in the early stages of development, children's strategies tend to reflect the action or relationship described in the problem. For instance, they use different material strategies for solving partitive and quotitive division problems that tend to fit closely with the meaning of the problem. Thirdly, older and more able students are more likely to apply multiplicative strategies that no longer necessarily nicely match the semantic structure of the problem but are more efficient from a computational point of view (Verschaffel et al., 2007).

As was discussed for addition and subtraction, these observed shifts towards more efficient and more flexible strategy use depend on the learner's conceptual knowledge of these operations. According to Fischbein et al.'s (1985) above-mentioned theory of primitive models, the model affecting the meaning and use of multiplication is 'repeated addition', in which a number of collections of the same size are put together. The dominant intuitive model for division is equal sharing of a group of objects (partition), and a second less dominant interpretation relates division to finding the number of equal groups in a given total (quotation). Many research findings have been successfully interpreted within Fischbein et al.'s theory, such as:

- Children's reluctance to apply commutativity in their computations for certain types of (non-symmetric) multiplication problems (Nunes & Bryant 1995).
- The number and the kind of errors made when confronting learners with multiplicative problems involving different kinds of decimals in the role of multiplier or divisor (for an extensive overview see Greer, 1992).
- Children's verbalizations of their problem solving and reasoning processes around such problems (see also Greer, 1992).
- The kind of correct, as well as incorrect word problems that students generate when being asked to construct problems corresponding to particular multiplications and divisions (e.g. 1.5 × 4, 0.75 × 8, 4 ÷ 0.5; De Corte & Verschaffel, 1996).

Finally, there is some evidence that these primitive models of multiplication and division do not disappear in older and more expert problem solvers; rather, they seem to be more aware of the possible impact of these existent primitive models and more able to control them than younger and weaker subjects (Fischbein et al., 1985; see also 'The Role of Intuitions in Word Problem Solving').

Beyond the Purely Cognitive: Strategic Aspects of Word Problem Solving

Even in the early days of information-processing approach, various researchers pointed out that learners' problem-solving behavior could not be explained purely in terms of the above-mentioned cognitive task and subject factors, but also needed to draw on various kinds of 'strategic' aspects, such as students' use of superficial coping strategies, their use of heuristics and metacognitive strategies, and their oscillation between intuitive and analytic thinking.

Superficial Coping Strategies

A basic assumption in the chapter so far is that the word problem-solver *tries* to understand and represent the information in the problem statement, and build an integral representation of the problem situation before developing and working through a mathematical model. However, research has convincingly shown that many students follow a process whereby the initial comprehension and representation phases are completely bypassed and the problem text immediately triggers the mathematical operation(s) to be performed (Sowder, 1988; Verschaffel et al., 2000). We describe two such superficial coping strategies.

First, some students do not read or represent the problem, but simply perform the most familiar or recently taught operation(s) on the given numbers. This coping strategy has been clearly evidenced in an older study by Goodstein, Cawley, Gordon, and Helfgott (1971). Many mildly retarded elementary school children always added all numbers in a word problem, including irrelevant ones. Using eye movement data, De Corte and Verschaffel (1986) showed that weakly performing first graders who got a series of one-step addition and subtraction problems stopped reading at the second given number (without even glimpsing at the rest of the problem) and then simply always added the two given numbers.

Another superficial coping strategy is to rely on key words in the problem text to select an arithmetic operation. For instance, the word 'fewer' or 'lost' in the problem text automatically results in a choice for subtraction (Sowder, 1988). Particularly students with below-average reading skills in combination with above-average numeracy skills rely on such key word strategies, in order to compensate for poor reading comprehension (Nortvedt, 2011).

Several researchers have argued that these superficial strategies are typically not explicitly taught by teachers, but (implicitly) constructed by learners themselves from their daily classroom experiences. Several studies revealed that the way in which word problems are presented, formulated, and grouped in traditional mathematics textbooks and lessons, make the use of superficial coping strategies undeservedly successful (Verschaffel et al., 2000). For instance, Schoenfeld (1991) found that in some US textbooks up to 90% of the word problems could be solved correctly by means of key word strategies. The fact that these superficial coping strategies are typically observed among intellectually weaker suggests that it are especially these children who seem to be affected negatively by these aspects of the current classroom practice.

Heuristics

Heuristics are thinking or search strategies that can help a problem solver in transforming the initial problematic situation progressively into a routine task for which (s)he has the appropriate knowledge and skills to attain the solution of the problem. Although heuristics do not guarantee a correct solution, they significantly increase the probability because they induce a systematic approach to the task (De Corte, Depaepe, Op 't Eynde, & Verschaffel, 2011; Polya, 1945). Examples of such heuristics are making a sketch or a drawing, decomposing the problem into parts, and guessing-and-checking.

Some researchers have identified the (spontaneous) use of heuristics as a significant point of difference between students who are good and weak in word problem solving (De Bock, Verschaffel, & Janssens, 2002; Van Essen, 1991). For instance, De Bock et al. (2002) found that lower secondary school students massively gave erroneous (proportional) answers to problems about the effect of the enlargement of a square on its area, but students who spontaneously thought of making a drawing performed considerably better than those who did not.

Although the relevance and importance of heuristics for (word) problem solving is broadly recognized, there is only limited empirical evidence supporting their relationship with success in problem solving (e.g. Hegarty & Kozhevnikov, 1999). Moreover, initial attempts to teach heuristics strategies and enhance students' word problem-solving skills were not very successful, because learners often are unable to decide which strategy is appropriate, and/or do not know how to implement the chosen heuristic (De Corte et al., 1996, 2011; Schoenfeld, 1992).

Metacognitive Strategies

Students' superficial coping strategies and researchers' failure to establish strong links between heuristics and success in word problem solving, led researchers to study word problem solving from the perspective of metacognition (Stillman & Mevarech, 2010). Metacognition can be defined as knowledge and strategies that take the person's own cognitive activity as its object. Several studies have revealed that in many students' solution attempts of word problems, moments of metacognitive awareness or metacognitive activities such as analyzing the problem, monitoring the solution process, and evaluating the outcome, are completely absent (e.g. Garofalo & Lester, 1985).

However, research that shows that metacognition is a significant variable in success in word problem solving, is rather scarce. Several researchers reported differences in self-regulation between children who are more and less able in solving arithmetic word problems (e.g. Carr, Alexander, & Folds-Bennett, 1994). Moreover, research suggests that metacognitive skills develop through the early primary years (Stillman & Mevarech, 2010), and that enhancing them through instruction leads to better performance in word problems (e.g. Garofalo & Lester, 1985; Teong, 2003; Verschaffel, De Corte, Lasure, Van Vaerenbergh, Bogaerts, & Ratinckx, 1999).

The Role of Intuitions in Word Problem Solving

Recently, students' use of superficial coping strategies and the absence of valuable heuristics and (meta)cognitive skills have been investigated in terms of the dual process theories (e.g. Evans & Over, 1996; Sloman, 1996), which posit that people often rely on rapid, less effortful intuitive kinds of processing of tasks instead of slower, deliberative analytic processing that requires more working memory capacity. Intuitive processing relies on salient problem characteristics that immediately attract the problem solver's attention and direct his thinking. If these characteristics are relevant to the solution of the problem, this will be beneficial; if not, they will cause errors. For instance, young children often think that six spoons of sugar in two glasses of water taste sweeter than four spoons in one glass of water, because they intuitively focus on the amount of sugar per se (Stavy, Strauss, Orpaz, & Carmi, 1982).

In his theory of intuitive versus analytic reasoning, Fischbein (1987) distinguished primary and secondary intuitions. Primary intuitions develop independently of instruction (e.g. the idea that increasing the number of conditions imposed on an expected event diminishes its chances), whereas secondary intuitions are acquired through instruction. An example of a secondary intuition relates to the missing-value formulation of word problems. Consider the following 'runners' problem: 'Ellen and Kim are running around a track. They run equally fast but Ellen started later. When Ellen has run 5 laps, Kim has run 15 laps. When Ellen has run 30 laps, how many has Kim run?' Van Dooren, De Bock, Hessels, Janssens, and Verschaffel (2005) have shown that many elementary school students erroneously give the proportional answer 30 × 3 = 90 instead of the correct additive answer 30 + 10 = 40 to this word problem, and that the number of erroneous proportional answers *increases* from third to sixth grade, resulting in a *decrease* in the number of correct answers. It seems that students *learn to* intuitively associate the typical missing-value formulation with three known and one unknown value with proportional calculations (De Bock et al., 2002).

Fischbein (1987) has characterized intuitions by many descriptors, among others intrinsic certainty, self-evidence, immediacy, perseverance, coerciveness, and implicitness. These may explain why students often revert to intuitions. They come to mind first, and only when analytic processing takes place and manages to inhibit the result of intuitive processing, the original intuitions will be altered. The characteristics of rapidness and limited involvement of working memory that are often attributed to intuitive processing have been experimentally shown, with respect to word problem solving: Primary school students and adults more often gave incorrect proportional answers to the above-mentioned runners problem when working under time pressure or when their working memory was experimentally limited (Gillard, Van Dooren, Schaeken, & Verschaffel, 2009). Also, correlational studies have shown that individuals with a larger working memory capacity are better able to reason analytically and inhibit intuitive reasoning (Stanovich & West 2000). Finally, the existence of two different types of reasoning processes has also received support from neuro-imaging studies (e.g. Goel & Dolan, 2003), albeit not yet in the domain of word problems.

One might wonder how it is possible that in a mathematics classroom, where students are taught to think analytically about word problems, such intuitive processes occur so often (Leron & Hazzan, 2006). Many authors (e.g. Todd & Gigerenzer, 2000) argued that intuitive processing is inherently adaptive (or, in Todd and Gigerenzer's terms, ecologically rational). This may also apply to word problem solving. Intuitions as such are not wrong. After all, their frequent success is a major reason why intuitions originate and grow. Just like for superficial coping strategies (see 'Superficial coping strategies'), an intuitive approach to a school word problem, based on salient, but essentially irrelevant task characteristics, will frequently be successful. Moreover, in an environment where not only correctness, but also speed is valued, and where students are typically not fully cognitively committed to the tasks they receive (as in a traditional mathematics lesson), they might rely on intuitive processing, which is cognitively less demanding.

Affective Factors in Word Problem Solving

The interest in the interference of affective factors in students' word problem-solving behavior has mainly grown from the observation that students' negative affects can hamper or even impede their success in word problem solving, even in cases when the solver possesses the necessary conceptual and procedural content knowledge and (meta)cognitive skills. Since McLeod's (1992) influential analysis of the role of affect in mathematical thinking and learning, the domain is commonly conceptualized in terms of the following three subdomains – emotions (e.g. joy or frustration while solving a word problem), attitudes (e.g. (dis)interest in solving word problems), and beliefs (e.g. doing mathematics only involves memorizing rules and procedures). These three (deeply intertwined) subdomains vary in stability (with emotions being less stable than attitudes and beliefs), the time that they take to be developed (with emotions being evoked more quickly than attitudes and beliefs), intensity (varying from hot emotions to cool attitudes and cold beliefs), and the degree to which cognitions are involved (with emotions being less cognitive than attitudes and beliefs).

Emotions and Word Problem Solving

There is large agreement that emotions involve a complex interplay of cognitive processes (e.g. the way in which one interprets the situation, based on his prior knowledge, attitudes, and beliefs), physiological processes (e.g. starting to sweat in case of feeling frustration or having a high heart rate in case of feeling anxious), as well as motivational processes (e.g. whether one's personal goals are attained; (Hannula, 2002; Op 't Eynde, De Corte, & Verschaffel, 2006a,b). Typically, distinction is made between positive (e.g. happiness, proudness) and negative emotions (e.g. frustration, anxiety).

For the domain of mathematics as a whole, there exists a long-standing quantitative research line concerning the relationship between students' anxiety toward mathematics and their general mathematical achievement, mostly revealing a significant negative relationship between the two (Ma, 1999). Later, researchers started to apply qualitative methods (such as facial observations and interviews) as well, sometimes in combination with quantitative approaches (e.g. paper-and-pencil tests and questionnaires). A recent development is the increased and integrated use of measures of physiological changes in skin conductance, cardiovascular and respiratory activity, as well as of neurological changes in certain areas of the brain that are known to go along with changes in other expressions of people's emotional states, while doing mathematical tasks (Campbell, 2008). However, only few (recent) studies have explicitly addressed the relation between emotions and mathematical word problem solving.

McLeod, Metzger, and Craviotto (1989) asked, in the context of individual interviews, four experts (i.e. research-active professors of mathematics) and four novices (i.e. undergraduate students) in mathematics to solve word problems that were true problems for all of them. Based on thinking-aloud protocols and their 'on-line' verbal comments about their feelings during problem solving, McLeod et al. (1989) found that the novices, as well as the experts experienced rather intense negative emotions (e.g. frustration, aggravation, disappointment) during the solution of these non-routine problems and that both groups indicated considerable awareness of their emotions during problem solving. However, the experts tended to differ from novices in their causal attributions, with novices ascribing failure more to internal negative traits, such as being dumb and stupid. Moreover, experts and novices differed in the way they coped with negative emotions, whereas novices tended to keep on approaching the problem in the same (ineffective) way, without making any progress, the experts were more likely to switch to new approaches when getting stuck.

Similar conclusions were obtained by Op 't Eynde et al. (2006a), who described and interpreted 16 eighth-grade students' emotions when solving non-routine word problems through triangulation of data obtained from the observation of students' problem-solving physical behavior (e.g. facial and bodily actions, spontaneous emotional vocalizations), and their professed interpretations and appraisals in a video-based stimulated recall interview. All students were found to experience positive as well as negative emotions in the frame of a problem-solving process, even often in a particular pattern. Positive emotions (e.g. relief, happiness) were experienced at the moment that students were proceeding toward the realization of their personal goals, whereas negative emotions (e.g. nervousness, frustration) were experienced if they got stuck.

Different from the above-mentioned general research line on mathematics anxiety, these two latter studies recognize that negative emotions are inherently connected to mathematical problem solving and, even more importantly, that these negative emotions can be effective, for instance, in cases when they trigger the exploration of alternative solution paths and lead to moments of mathematical insight. Accordingly successful problem solvers differ from unsuccessful problem solvers not so much in the occurrence of negative emotions, but in the way they cope with them, or, in other words, in their meta-emotional skills. Meta-emotional skills then refer to their repertoire of strategies to regulate emotions and to their competence to consciously and effectively use them (De Corte, Verschaffel, & Van Dooren, 2011). In a study about meta-emotion in mathematics education, De Corte et al. (2011) investigated through a questionnaire the strategies that 393 eighth- and 10th-grade students of different educational tracks use to regulate their emotions in three different mathematical school settings (i.e. a difficult mathematics test, a difficult mathematics homework, and a difficult mathematics lesson). The results indicate that students of academically lower tracks tended to exercise less adequate meta-emotional strategies (e.g. abandoning and negotiation) than other students.

Attitudes and Word Problem Solving

Historically, students' mathematics-related attitudes have mostly been conceived and measured vis-à-vis mathematics in general, using questionnaires such as the Fennema and Sherman (1976) Mathematics Attitudes Scales. These studies revealed that attitude toward mathematics is positively related to mathematics achievement (Ma & Kishor, 1997), that students' attitude toward mathematics declines from elementary to secondary school (McLeod, 1994), and that there tends to be a small gender-difference in students' attitudes toward mathematics in favor of boys (Hyde, Fennema, Ryan, Frost, & Hopp, 1990).

To the best of our knowledge, systematic research on students' attitudes towards word problems is almost non-existent. It is generally accepted that one of the major reasons for including word problems in the mathematics curriculum is to enhance students' positive attitudes towards (learning) mathematics in general, by showing them, and having them experience, that they will really need the mathematics they are taught in school when they grow up and move into the world (Verschaffel et al., 2000). However, it is commonly accepted that after some years of traditional mathematics education, word problems have become one of the most disliked topics of the elementary school curriculum (as is suggested in Gary Larson's (2003) famous cartoon of a deceased man visiting hell's library, which is completely filled with books of ... word problems). The few available studies focus on students' interest in the specific topical content of the word problems. For instance, Hen-Yu and Sullivan (2000) investigated the impact of incorporating personal information and preferential topics provided by students into their mathematics word problems, on the performance of 72 fifth-grade students on and their interest in these word problems. The results showed that students performed significantly higher on and had more favorable attitudes for these personalized problems.

Beliefs and Word Problem Solving

Schoenfeld (1985) was one of the first scholars to emphasize the influence of beliefs on mathematical thinking and learning. Based on a review of the extensive literature on mathematical beliefs, Op 't Eynde et al. (2006b) distinguished between three broad categories:

(1) beliefs about mathematics education.
(2) beliefs about the self in relation to mathematics.
(3) beliefs about the social context.

Research has revealed that many primary and secondary school students hold inappropriate beliefs regarding the three categories, some of which relate directly to word problems. (For example, there is only one way to solve a mathematics problem – either one can learn mathematics or one cannot. In my class, solving mathematics word problems has nothing to do with real life; for a review, see Verschaffel et al., 2000.) However, as for the two previous affective subdomains, systematic empirical research on students' beliefs about word problems and how they relate to word problem solving is rare, except for intervention studies (which are beyond the scope of this chapter) that have involved the articulation, discussion, and alteration of students' inadequate beliefs about word problems and word problem solving as an essential part of the experimental program (see, for example, Verschaffel et al., 1999; Mason & Scrivani, 2004).

Some studies have focused on the role of mathematics-related self-efficacy beliefs in explaining students' failure or success in mathematics problem solving. Vermeer, Boekaerts, and Seegers (2000) investigated the relationship between 79 sixth-grade boys' and 79 girls' cognitive and affective variables, including a measure of their task-specific self-efficacy (before the task), as well as their causal attribution of their task-specific result (after the task), on the one hand, and their skill in solving a set of non-routine word problems, on the other hand. The results showed that task-specific subjective competence expressed before solving the word problems significantly contributed to task performance, after the influences of several objective measures of competence were taken into account. This was found for boys as well as girls, but additional effects of task-specific attributions on task performance were found for girls only.

Over the past 20 years, there is one specific line of word problem-solving research wherein beliefs have played an important role, namely the research on students' 'suspension of sense-making' (Schoenfeld, 1991) when solving word problems. This research line goes back to the above-mentioned famous example of the 'captain's problem' (see 'Knowledge of problem structures') and received a new impetus after two parallel studies by Greer (1993) and Verschaffel, De Corte, and Lasure (1994) showing that the vast majority of upper elementary and lower-secondary school students demonstrate a very strong tendency to respond in non-realistic ways to situationally problematic items (so-called P-items), presented in the context of a word problem-solving lesson or test. For instance, consider a P-problem like the following: 'A man wants to have a rope long enough to stretch between two poles 12 meters apart, but he only has pieces of rope 1.5 meters long. How many of these would he need to tie together to stretch between the poles?' Almost all of the students reacted by just doing 12:1.5 = 8, 'responding with '8 pieces' without any further comment or query about the fact that part of each 1.5-meter piece will be used to tie the knots together (whereas most of them would probably not demonstrate such 'lack of sense-making' behavior if they were to encounter a mathematically isomorphic problem out of the school context). With respect to individual differences, research evidence suggests that students' tendency to neglect realistic considerations about the problem context is associated with age, gender, social background, and school mathematics performances. Children with less years of experience with (traditional) schooling (Li & Silver, 2000), girls (Boaler, 1994), working-class children (Cooper & Dunne, 1998), and students with high school mathematics marks (Csaba, Kelemen, & Verschaffel, 2011) rely more on everyday knowledge when solving mathematical application problems.

Commenting on the initial studies of Greer (1993) and Verschaffel et al. (1994), as well as several replications (see Verschaffel et al., 2000, for a review), scholars have argued that students' tendency to neglect real-world knowledge and realistic considerations when confronted with this kind of problems is due to their beliefs about word problems and how to solve them, which they have gradually developed along with their participation in the practice and culture of the mathematics classroom (Lave, 1992; Reusser & Stebler, 1997; Schoenfeld, 1991; Verschaffel et al., 2000). Recently, Jiménez and Ramos (2011) investigated the impact of the following four specific misbeliefs about word problems that develop in students through their immersion in the culture of traditional mathematics classrooms, on their solutions to P-items:

(1) Every word problem is solvable.
(2) There is only one numerical and precise correct answer to every word problem.
(3) It is necessary to do calculations to solve a word problem.
(4) All numbers that are part of the word problem must be used in order to calculate the solution.

Specifically, 22 second and 22 third graders were asked in the context of an individual interview to solve four problem types of word problems that violate these four misbeliefs, i.e.:

(1) An unsolvable word problem.
(2) A word problem with multiple solutions.
(3) A word problem containing the solution in the problem statement.
(4) A word problem including irrelevant data.

All word problems were formulated with a change semantic structure (see 'Knowledge of problem structures') and the numbers that were used were comparable. Results revealed, first, that only one-third of all students responded correctly to the four problem types. Secondly, the percentage correct answers was higher for solution in the statement and irrelevant data problems (45.5 and 43.2%, respectively) than for unsolvable and multiple solution problems (20.5 and 23.9%, respectively). Thirdly, no differences were found between second and third graders. Fourthly, the vast majority of the errors originated from doing one or more arithmetic operations on all given numbers in the problem. Finally, many verbal explanations of erroneous responses contained spontaneous expressions of the above-mentioned misbeliefs about word problems. While this study further documents the strong connection between students' beliefs about word problems and their problem-solving behavior, it still does not yield convincing evidence on how these beliefs are actually established in the practice and culture of the mathematics class.

Conclusion

Research has convincingly shown that solving mathematical word problems requires more competencies than reading and computational skills (see also Fuchs et al., 2008; Nunes & Csapó, 2011). The heart of the solution process consists of building a model of the situation described in the problem statement and transforming that situation model into a mathematical one (that has to be computationally worked out afterwards). In these subsequent phases, knowledge of problem structures, as well as conceptual and procedural knowledge of the arithmetic operations is crucial, as has been shown in numerous studies over the past decades. Research has also documented how students' word problem solving is frequently affected by superficial coping strategies and intuitive approaches (that they have developed by participating in the cultural practice of the mathematics class), as well as the critical importance of heuristic, metacognitive and analytic strategies. Although less convincing, there is also growing evidence that students' emotions, attitudes, and beliefs and their awareness and control of these affects codetermine their success in solving word problems. While we have reviewed the research on the role of these various aspects in students word problem-solving processes and skills in separate sections, it should be clear that they are strongly interrelated and interdependent (see also De Corte et al., 1996; Vermeer et al., 2000). Finally, while this review has yielded a rather consistent and broadly accepted picture of the aspects that together may account for a students' success in solving a word problem, it should be acknowledged that the individual differences with respect to all these aspects of word problem solving are less clear, simply because the studies were not set up from the perspective of an individual difference. If attention was paid in these studies at individual differences, it involved the comparison of weak and strong word problem solvers on one or more of the cognitive, metacognitive, and/or affective aspects being addressed in this chapter, sometimes in relation to general subject factors, such as age, gender, or SES. Besides, there is another line of research -not the focus of this chapter - that links word problem solving skill to various kinds of other subject factors such as working memory, verbal and mathematical ability, (non-verbal) problem-solving ability, working memory, cognitive inhibitory skills, and cognitive style (see 'Introduction'). Future research on word problem solving would profit from studies in which students' individual scores on these subject factors would be linked not only to their word problem solving performance, but also to their competence with respect to the various cognitive, metacognitive, and affective factors discussed in this chapter.

References

Baek, J-M. (1998). Children's invented algorithms for multidigit multiplication problems. In L.J. Morrow & M.J. Kenney (Eds), *The Teaching and Learning of Algorithms in School Mathematics* (pp. 151–160). Reston: National Council of Teachers of Mathematics.

Baroody, A.J., & Gannon, K.E. (1984). The development of the commutativity principle and economical addition strategies. *Cognition and Instruction*, 1, 321–329.

Blum, W., & Niss, M. (1991). Applied mathematical problem solving, modelling, applications, and links to other subjects – state, trends, and issues in mathematics education. *Educational Studies in Mathematics*, 22, 37–68.

Boaler, J. (1994). When do girls prefer football to fashion? An analysis of female underachievement in relation to 'realistic' mathematical contexts. *British Educational Research Journal*, 20, 551–564.

Campbell, S.R. (2008). Mathematics educational neuroscience: origins, applications, and new opportunities. In P. Liljedahl (Ed.), *Proceedings of the 2007 Annual Meeting for the Canadian Mathematics Education Study Group* (pp. 71–78). Simon Fraser University, BC: CMESG/GCEDM.

Carpenter, T.P., & Moser, J.M. (1984). The acquisition of addition and subtraction concepts in grades one through three. *Journal for Research in Mathematics Education*, 15, 179–202.

Carr, M., Alexander, J., & Folds-Bennett, T. (1994). Metacognition and mathematics strategy use. *Applied Cognitive Psychology*, 8, 583–595.

Coley, R.J. (2002). *An Uneven Start: Indicator of Inequality in School Readiness*. Princeton, NJ: Educational Testing Service.

Cooper, B., & Dunne, M. (1998). Anyone for tennis? Social class differences in children's responses to national curriculum mathematics testing. *Sociological Review*, 46, 115–148.

Csaba, C., Kelemen, R., & Verschaffel, L. (2011). Fifth-grade students' approaches to and beliefs of mathematics word problem solving: a large sample Hungarian study. *ZDM Mathematics Education*, 43, 561–571.

De Bock, D., Verschaffel, L., & Janssens, D. (2002). The effects of different problem presentations and formulations on the illusion of linearity in secondary school students. *Mathematical Thinking and Learning*, 4, 65–89.

De Corte, E., Depaepe, F., Op 't Eynde, P., & Verschaffel, L. (2011). Students' self-regulation of emotions in mathematics: an analysis of meta-emotional knowledge and skills. *ZDM Mathematics Education*, 43, 483–495.

De Corte, E., Greer, B., & Verschaffel, L. (1996). Learning and teaching mathematics. In D. Berliner and R. Calfee (Eds), *Handbook of Educational Psychology* (pp. 491–549). New York: Macmillan.

De Corte, E., & Verschaffel, L. (1986). Eye movements of first graders during problem solving. In C. Hoyles & R. Noss (Eds), *Proceedings of the Tenth International Conference on the Psychology of Mathematics Education* (pp. 421–426). London: University of London, Institute of Education, Department of Mathematics, Statistics and Computing.

De Corte, E., & Verschaffel, L. (1987). The effect of semantic structure on first graders' solution strategies of elementary addition and subtraction word problems. *Journal for Research in Mathematics Education*, 18, 363–381.

De Corte, E., & Verschaffel, L. (1996). An empirical test of the impact of primitive intuitive models of operations on solving word problems with a multiplicative structure. *Learning and Instruction*, 6, 219–243.

De Corte, E., Verschaffel, L., & De Win, L. (1985). The influence of rewording verbal problems on children's problem representations and solutions. *Journal of Educational Psychology*, 77, 460–470.
De Corte, E., Verschaffel, L., & Van Dooren, W. (2011). Heuristics and problem solving. In N. Seel (Ed.), *Encyclopedia of the Sciences of Learning*, Vol. 3 (pp. 1421–1424). New York: Springer.
Endo, A. (2010). An intervention to improve arithmetic word-problem solving by a student with borderline intelligence: assessment of cognitive profile and errors in problem solving. *Japanese Journal for Educational Psychology*, 58, 224–235.
Evans, J.St.B.T., & Over, D.E. (1996). *Rationality and Reasoning*. Hove: Psychology Press.
Fennema, E., & Sherman, J.A. (1976). Fennema–Sherman Mathematics Attitudes Scales: instruments designed to measure attitudes toward the learning of mathematics by males and females. *Catalog of Selected Documents in Psychology*, 6(1), 31.
Fischbein, E. (1987). Intuition in Science and Mathematics: An Educational Approach. Dordrecht: Reidel.
Fischbein, E., Deri, M., Nello, M.S., & Marino, M.S. (1985). The role of implicit models in solving verbal problems in multiplication and division. *Journal for Research in Mathematics Education*, 16, 3–17.
Fuchs, L., Fuchs, D., Steubing, K., & Fletcher, J.M. (2008). Problem solving and computational skill: are they shared or distinct aspects of mathematical cognition? *Journal of Educational Psychology*, 100, 30–47.
Fuson, K.C. (1992). Research on whole number addition and subtraction. In D.A. Grouws (Ed.), *Handbook of Research on Mathematics Teaching and Learning: A Project of the National Council of Teachers of Mathematics* (pp. 243–275). New York: MacMillan.
Fuson, K.C., & Willis, G.B. (1989). Second graders' use of schematic drawings in solving addition and subtraction word problems. *Journal of Educational Psychology*, 81, 514–520.
Garofalo, J., & Lester, F.K. (1985). Metacognition, cognitive monitoring, and mathematical performance. *Journal for Research in Mathematics Education*, 16, 163–176.
Gillard, E., Van Dooren, W., Schaeken, W., & Verschaffel, L. (2009). Proportional reasoning as a heuristic-based process: time pressure and dual-task considerations. *Experimental Psychology*, 56, 92–99.
Goel, V., & Dolan, R.J. (2003). Explaining modulation of reasoning by belief. *Cognition*, 87, B11–B22.
Goldin, G.A., & McClintock, E. (Eds) (1984). Task Variables in Mathematical Problem Solving. Philadelphia: Franklin.
Goodstein, H.A., Cawley, J.F., Gordon, S., & Helfgott, J. (1971). Verbal problem solving among educable mentally retarded children. *American Journal of Mental Deficiency*, 76, 238–241.
Greer, B. (1992). Multiplication and division as models of situations. In D. A. Grouws (Ed.), *Handbook of Research on Mathematics Teaching and Learning: A Project of the National Council of Teachers of Mathematics* (pp. 276–295). New York: Macmillan.
Greer, B. (1993). The modelling perspective on wor(l)d problems. *Journal of Mathematical Behavior*, 12, 239–250.
Hannula, M. (2002). Attitude towards mathematics: emotions, expectations and values. *Educational Studies in Mathematics*, 49, 25–46.
Hegarty, M., & Kozhevnikov, M. (1999). Types of visual-spatial representations and mathematical problem solving. *Journal of Educational Psychology*, 91, 684–689.
Hen-Yu, K., & Sullivan, H.J. (2000). Personalization of mathematics word problems in Taiwan. *Educational Technology, Research and Development*, 48(3), 49–59.
Hyde, J.S., Fennema, E., Ryan, M., Frost, L.A., & Hopp, C. (1990). Gender comparisons of mathematics attitudes and affect. *Psychology of Women Quarterly*, 14, 299–324.
Jiménez, L., & Ramos, F.J. (2011). El impacto negativo del contracto didáctico en la resolución realista de problemas. Un estudio con alumnus de 2° y 3° de Educación Primaria [The negative impact of the didactic contract in realistic problems: a study with second- and third-grade students]. *Electronic Journal of Research in Educational Psychology*, 9, 1155–1182.

Kintsch, W., & Greeno, J. (1985). Understand and solving word arithmetic problems. *Psychological Review*, 92, 109–129.

Kraemer, J-M. (2011). Oplossingsmethoden voor Aftrekken tot 100 (Doctoral thesis). Arnhem: Cito.

Larson, G. (2003). The complete far side. Riverside, NJ: Andrews, & McMeel.

Lave, J. (1992). Word problems: a microcosm of theories of learning. In P. Light & G. Butterworth (Eds), *Context and Cognition: Ways of Learning and Knowing* (pp. 74–92). New York: Harvester Wheatsheaf.

Leron, U., & Hazzan, O. (2006). The rationality debate: application of cognitive psychology to mathematics education. *Educational Studies in Mathematics*, 62, 105–126.

Li, Y., & Silver, E. (2000). Can younger students succeed where older students fail? An examination of third graders' solutions of a division-with-remainder problem. *Journal of Mathematical Behavior*, 19, 233–246.

Ma, X. (1999). A meta-analysis of the relationship between anxiety toward mathematics and achievement in mathematics. *Journal for Research in Mathematics Education*, 30, 520–540.

Ma, X., & Kishor, N. (1997). Assessing the relationship between attitude toward mathematics and achievement in mathematics: meta-analyses. *Journal for Research in Mathematics Education*, 28, 26–47.

Mason, L., & Scrivani, L. (2004). Enhancing students' mathematical beliefs: an intervention study. *Learning and Instruction*, 14, 153–176.

Morales, R.V., Shute, V.J., & Pellegrino, J.W. (1985). Developmental differences in understanding and solving simple mathematics word problems. *Cognition and Instruction*, 2, 41–57.

McLeod, D.B. (1992). Research on affect in mathematics education: a reconceptualisation. In D. A. Grouws (Ed.), *Handbook of Research on Mathematics Teaching and Learning: A Project of the National Council of Teachers of Mathematics* (pp. 575–596). New York: Macmillan.

McLeod, D.B. (1994). Research on affect and mathematics learning in the JRME: 190 to the present. *Journal for Research in Mathematics Education*, 25, 637–647.

McLeod, D.B., Metzger, W., & Craviotto, C. (1989). Comparing experts' and novices' affective reactions to mathematical problem solving: an exploratory study. In G. Vergnaud (Ed.), *Proceedings of the Thirteenth International Conference for the Psychology of Mathematics Education*, Vol. 2 (pp. 296–303). Paris: Laboratoire de Psychologie et Développement de l'Education de l'Enfant.

Mulligan, J., & Mitchelmore, M. (1997). Young children's intuitive models of multiplication and division. *Journal for Research in Mathematics Education* 28, 309–330.

Munez Mendez, D. (2011). Representaciones Mentales en la Resolucion de Problemas Aritméticos (Doctoral thesis). Salamanca: Universidad di Salamanca.

Muth, K.D. (1984). Solving arithmetic word problems. Role of reading and computational skills. *Journal of Educational Psychology*, 76, 205–210.

Navarro, J.I., Aguilar, M., Alcalde, C., & Howell, R. (1999). Relationship of arithmetic problem solving and reflective-impulsive cognitive styles in third-grade students. *Psychological Reports*, 85, 179–186.

Nunes, T., & Bryant, P. (1995). Do problem situations influence children's understanding of the commutativity of multiplication? *Mathematical Cognition*, 1, 245–260.

Nunes, T., & Csapó, B. (2011). Developing and assessing mathematical reasoning. In B. Csapó and M. Szendrei (Eds), *Framework for Diagnostic Assessment of Mathematics* (pp. 57–93). Budapest: Nemzeti Tankönyvkiadó.

Nortvedt, G.A. (2011). Coping strategies applied to comprehend multistep arithmetic word problems by students with above-average numeracy skills and below-average reading skills. *Journal of Mathematical Behavior*, 30(1), 255–269.

Op 't Eynde, P., De Corte, E., & Verschaffel, L. (2006a). 'Accepting emotional complexity': a socio-constructivist perspective on the role of emotions in the mathematics classroom. *Educational Studies in Mathematics*, 63, 193–207.

Op 't Eynde, P., De Corte, E., & Verschaffel, L. (2006b). Framing students' mathematics-related beliefs. A quest for conceptual clarity and comprehensive categorization. In G. C. Leder, E. Pehkonen, & G.

Türner (Eds), Beliefs: *A Hidden Variable in Mathematics Education?* (pp. 13–37). Dordrecht: Kluwer Academic Publisher.

Passolunghi, M.C., & Siegel, L.S. (2001). Short term memory, working memory, and inhibitory control in children with specific arithmetic learning disabilities. *Journal of Experimental Child Psychology*, 80, 44–57.

Polya, G. (1945). *How to Solve it.* Princeton, NJ: Princeton University Press.

Resnick, L.B. (1987). Learning in school and out. *Educational Researcher* 16(9), 14–21.

Reusser, K., & Stebler, R. (1997). Every word problem has a solution: the social rationality of mathematical modeling in schools. *Learning and Instruction*, 7, 309–327.

Riley, M.S., Greeno, J.G., & Heller, J.I. (1983). Development of children's problem-solving ability in arithmetic. In H. P. Ginsburg (Ed.), *The Development of Mathematical Thinking* (pp. 153–196). New York: Academic Press.

Schoenfeld, A.H. (1985). *Mathematical Problem Solving.* New York: Academic Press.

Schoenfeld, A.H. (1991). On mathematics as sense-making: an informal attack on the unfortunate divorce of formal and informal mathematics. In J. F. Voss, D. N. Perkins, and J. W. Segal (Eds), *Informal Reasoning and Education* (pp. 311–343). Hillsdale, NJ: Lawrence Erlbaum Associates.

Schoenfeld, A.H. (1992). Learning to think mathematically. Problem solving, metacognition and sense-making in mathematics. In D. A. Grouws (Ed.), Handbook of Research on Mathematics Teaching and Learning (pp. 334–370). New York: Macmillan.

Selter, C., Prediger, S., Nührenbörger, M., & Hußmann, S. (2012). Taking away and determining the difference—a longitudinal perspective on two models of subtraction and its inverse relation to addition. *Educational Studies in Mathematics*, 79, 389–408.

Sloman, S.A. (1996). The empirical case for two systems of reasoning. *Psychological Bulletin*, 119, 3–22.

Sophian, C., & Vong, K.I. (1995). The parts and wholes of arithmetic story problems: developing knowledge in the preschool years. *Cognition and Instruction*, 13, 469–477.

Sowder, L. (1988). Children's solutions of story problems. *Journal of Mathematical Behavior*, 7(3), 227–238.

Stanovich, K.E., & West, R.F. (2000). Individual differences in reasoning: implications for the rationality debate. *Behavioral and Brain Sciences*, 23, 645–726.

Staub, F., & Reusser, K. (1995). The role of presentational structures in understanding and solving mathematical word problems. In C. A. Weaver III, S. Mannes, & C. R. Fletcher (Eds), *Discourse Comprehension: Essays in Honor of Walter Kintsch* (pp. 285–305). Hillsdale, NJ: Erlbaum.

Stavy, R., Strauss, S., Orpaz, N., & Carmi, G. (1982). U-shaped behavioral growth in ratio comparisons, or that's funny I would not have thought you were U-ish. In S. Strauss and R. Stavy (Eds), U-shaped Behavioral Growth (pp. 11–36). New York: Academic Press.

Stillman, G., & Mevarech, Z. (2010). Metacognition research in mathematics education: From hot topic to mature field. *ZDM Mathematics Education*, 42, 145–148.

Teong, S.K. (2003). The effect of metacognitive training on mathematical word-problem solving. *Journal of Computer-Assisted Learning*, 19(3), 46–55.

Thevenot, C. (2010). Arithmetic word problem solving: evidence for the construction of a mental model. *Acta Psychologica*, 133, 90–95.

Todd, P.M., & Gigerenzer, G. (2000). Simple heuristics that make us smart. *Behavioral and Brain Sciences*, 23(5), 727–741.

Van Dooren, W., De Bock, D., Hessels, A., Janssens, D., & Verschaffel, L. (2005). Not everything is proportional: effects of age and problem type on propensities for overgeneralization. *Cognition and Instruction*, 23, 57–86.

Van Essen, G (1991). Heuristics and Arithmetic World Problems, unpublished doctoral dissertation. Amsterdam: State University.

Vermeer, H.J., Boekaerts, M., & Seegers, G. (2000). Motivational and gender differences in sixth-grade students' mathematical problem-solving behavior. *Journal of Educational Psychology*, 92, 308–315.

Verschaffel, L., & De Corte, E. (1997). Word problems. A vehicle for promoting authentic mathematical understanding and problem solving in the primary school. In T. Nunes and P. Bryant (Eds), *Learning and Teaching Mathematics: An International Perspective* (pp. 69–97). Hove: Psychology Press.

Verschaffel, L., De Corte, E., & Lasure, S. (1994). Realistic considerations in mathematical modeling of school arithmetic word problems. *Learning and Instruction*, 4, 273–294.

Verschaffel, L., De Corte, E., Lasure, S., Van Vaerenbergh, G., Bogaerts, H., & Ratinckx, E. (1999). Design and evaluation of a learning environment for mathematical modeling and problem solving in upper elementary school children. *Mathematical Thinking and Learning*, 1, 195–230.

Verschaffel, L., Greer, B., & De Corte, E. (2000). *Making Sense of Word Problems*. Lisse: Swets and Zeitlinger.

Verschaffel, L., Greer, B., & De Corte, E. (2007). Whole number concepts and operations. In F.K. Lester (Ed.), *Second Handbook of Research on Mathematics Teaching and Learning* (pp. 557–628). Greenwich, CT: information Age Publishing.

Vicente, S., Orrantia, J., & Verschaffel, L. (2007). Influence of situational and conceptual rewording on word problem solving. *British Journal of Educational Psychology*, 77, 829–848.

Walch, M., Hicky, H., & Duffy, J. (1999). Influence of item content and stereotype situation on gender differences in mathematical problem solving. *Sex Roles*, 41(3–4), 219–240.

CHAPTER 52

INDIVIDUAL DIFFERENCES IN CHILDREN'S PATHS TO ARITHMETICAL DEVELOPMENT

JULIE ANN JORDAN, JUDITH WYLIE, AND GERRY MULHERN

INTRODUCTION

AT any point in time during the formal schooling period, a significant number of children are considered to have inadequate mathematical skills. In a UK context, this is evident from the Key Stage 1–3 teacher assessments for mathematics, which are designed to indicate how well children are progressing at ages 7, 11, and 14. The 2011 results of these assessments showed that, one in ten children aged 7 in England failed to meet the expected level, and at ages 11 and 14 the percentages of children failing to meet the expected level were 18% and 19% respectively (Department for Education, 2011a, b). The percentage point difference between the proportion of children who failed to meet the expected level at ages 7 (13%) and 14 (25%) is even greater in Wales (Statistical Directorate Welsh Assembly Government, 2010), and at ages 8 (5%) and 14 (23%) in Northern Ireland in 2009/10 (Department of Education, 2011). These figures suggest that many children, who initially have no apparent difficulty, start to struggle as the curriculum becomes more demanding. The proportion of children failing to achieve the expected level varies across curriculum area. For example, the Welsh 2009 data (Statistical Directorate Welsh Assembly Government, 2009) indicate that *Using and Applying Mathematics* at ages 7, 11, and 14 is the area in which most children fail to achieve the expected level, while the largest percentage point decrease over time in children achieving the expected level was in *Shape, Space and Measures*.

To some extent the differences between proportions of children meeting the expected level at ages 7 and 14 can be explained by cohort effects. For example, relative to the 2007 cohort the proportion of children meeting the expected level at ages 11 and 14 was 2–4% points greater for the 2011 cohort, although no change was found for 7-year-olds.

Additionally, as the data are cross-sectional, they do not tell us what proportions of children (1) have persistent difficulties; (2) are initially impaired but outgrow their difficulty; or (3) initially meet expected levels in mathematics but then develop difficulty. On the other hand, longitudinal data can provide a much more detailed and precise picture of children's development. The Department for Education (2011c) tracked English children from ages 7–11 and found that 18% failed to make the expected progress in mathematics. A comparable study that tracked children from 11 to 16 (Department for Education, 2011d) found that an even higher proportion (38%) of students from England failed to achieve their predicted score in the General Certificate of Secondary Education (GCSE) mathematics using predictions made based on mathematics performance at the end of primary school. However, as these studies only focus on two time points, they provide a one-dimensional picture of development and tell us nothing about the shape (e.g. linear, curvilinear) of children's trajectories. Information on the shape of growth could be obtained by linking the datasets for Key Stage 1, 2, and 3. Tracking students over even longer periods would be possible if the recommendation by The Royal Society (2011) to allow unique pupil numbers to be carried over from school and college into higher education is implemented. This would allow school and higher education datasets to be linked, thus enabling the association between children's trajectories in mathematics and their choice of subject in higher education to be studied.

Employers have raised concern over the numeracy levels of school leavers. For example, a survey by the Confederation of British Industry and Education Development International (CBI-EDI, 2011) revealed that two-fifths of employers have needed to provide remedial numeracy support to new employees who were school leavers. A more numerate workforce is likely to have a considerable economic impact, as those leaving school with inadequate numeracy skills are estimated to cost the UK taxpayer an estimated £2.4 billion per year (KPMG, 2009), and those with poor numeracy skills are five times more likely to be unemployed (OECD, 2006). The CBI (2010) proposed that more young people should complete some form of mathematics qualification post-16 to meet the need emphasized by employers for a more numerate workforce. These economic issues highlight that there is a strong need to understand what influences mathematics achievement, and, given that research indicates that children can develop mathematical difficulty at any stage during their schooling, this should be done in a developmental context. This chapter outlines research that has characterized children's trajectories on a range of mathematical tasks and identified several trajectory groups, including those with persistent difficulties from when they enter school, those who initially perform well but fall behind, and those who outgrow their difficulty. Furthermore, those factors that influence the characteristics of children's trajectories are considered.

Characteristics of Trajectories

Children's trajectories are characterized by their initial and final status, steepness, and shape (e.g. linear, uneven, and curvilinear). J Jordan, Mulhern, and Wylie (2009) examined the degree of heterogeneity amongst typically achieving children aged 5–7 years in terms of trajectory characteristics on seven mathematical tasks: exact calculation, story problems, approximate arithmetic, place value, calculation principles, forced retrieval, and written problems. Each child's development over four time points on the seven mathematical tasks was classified, using regression, as one of four trajectory types: linear, quadratic, s-curve, or flat. Considerable heterogeneity in trajectory characteristics between children was evident on all mathematical tasks. Specifically, linear, quadratic, and s-curve trajectories were found on all tasks, with flat trajectories also found on most tasks. Interestingly, although a linear trajectory best fitted the group data on most tasks, individual level analysis indicated that s-shaped trajectories were most prevalent on the majority of mathematical tasks. This shows that caution must be shown when documenting the shape of a group's trajectory, as the average trajectory may not adequately reflect the dominant trajectory type. This indicates that individual as opposed to average trajectory analysis may be a more reliable method for drawing conclusions about the shape of children's trajectories. Heterogeneity within children across tasks was evident; for example, linear development was the most common shape for place value and forced retrieval, while s-shaped was most common for the other five tasks. Furthermore, some children exhibited low performance and flat growth on some of the mathematical tasks (e.g. place value and calculation principles) despite achieving a standardized mathematics score above the 35th percentile.

Another study by N Jordan, Kaplan, Olah, and Locuniak (2006) focused on other characteristics of children's trajectories from ages 5–6; namely, steepness and performance at age 6. The children were examined at four time points on a battery of number sense measures: counting, enumeration, count sequence, counting principles, number knowledge, non-verbal calculation, story problems, estimation, number patterns, and number recognition. Using growth mixture modelling, they identified three trajectory types and labelled these according to the average growth rate and average performance at age 6 on the number sense battery: (1) low/flat—lower than average performance at age 6 and flat growth; (2) average/moderate—average performance at age 6 and moderate growth; (3) high/moderate—better than average performance at age 6 and moderate growth. Growth mixture modelling was also carried out on three of the number sense tasks, because there was evidence of divergent growth trajectories based on gender and income. When these number sense tasks, namely story problems, non-verbal calculation, and number combinations, were modelled, three trajectory types were found for each task. However, the nature of the three trajectory subtypes varied across task. For example, steep growth was evident for non-verbal calculation and number combinations but not for story problems. Individual differences within children and across these three tasks are evident from the variation in numbers of children falling into each trajectory classification. For example, while a large proportion of children had low/flat growth for story problems (70%) and number combinations (60%), a relatively smaller proportion of children had this type of trajectory for non-verbal combinations (44%). In contrast to the findings of J Jordan et al. (2009), no curvilinear trajectory types were identified. It is possible that a greater number of trajectory classifications would have been found had the tasks captured more growth. Indeed, very little growth was captured on the tasks as evidenced by the lack of significant growth rate and variation in growth rate.

Given that there is considerable variability amongst children with regard to cognitive and social variables (e.g. N Jordan et al., 2006; TIMSS, 2007), and that mathematical tasks vary considerably in terms of their cognitive demands and how they are perceived (e.g. Holmes, Adams, & Hamilton, 2008; Gregory, Snell, & Dowker, 1999), the degree to which cognitive and social variables explain trajectory characteristics should vary across mathematical tasks. It is therefore likely that a componential developmental approach which looks at development on a range of mathematical tasks is a more appropriate approach than non-componential methods.

This chapter documents many of the factors identified by longitudinal research as being associated with individual differences in development, including foundational numerical skills; reading and language; cognitive abilities; teaching, curriculum, and classroom experiences; social factors, socioeconomic status, and attitudes; and factors highlighted by neuroscience. Longitudinal studies employing a componential approach are also detailed; however, unfortunately, this type of approach is relatively uncommon and therefore in some sections the discussion is mainly limited to non-componential evidence.

Foundational Numerical Skills

N Jordan, Glutting, and Ramineni (2008, pp. 46–58) define number sense as abilities that involve numbers and operations during the 3- to 6-year-old period. On the whole, number sense tasks tend to require foundational number skills. For example, the number sense battery used by N Jordan et al. (2006) and N Jordan, Kaplan, Locuniak, and Ramineni (2007) included tasks such as counting and number knowledge (e.g. which number is bigger: 3 or 4?). Non-verbal number sense skills are diverse and while there is evidence to suggest that some non-verbal number sense skills are present at infancy (e.g. Mix, Huttenlocher, & Levine, 2002), there are debates about which abilities are present in that early time (e.g. Cohen and Marks, 2002) On the other hand, some number sense skills are generally considered to be more sensitive to preschool instruction (Dowker, 2005). N Jordan et al. (2007) tested children on a number sense battery at six time points from ages 5–6. Number sense performance and growth in number sense were found to explain 66% of the variation in mathematics performance as measured by the Woodcock Johnson test at age 7, even after controlling for gender, income, age, and reading ability. Given the similarities between the number sense battery and the items on the Woodcock Johnson test, the argument that number sense predicts later mathematics achievement could be viewed as circular. Arguably the real value of this research is that potentially children who will have weak mathematical development could be detected very early, as number sense tasks can be administered at the start of formal schooling. Furthermore, some non-verbal tasks (e.g. non-verbal magnitude comparison) could potentially be administered before formal schooling and used to predict later mathematics achievement.

Some aspects of number sense have emerged as better predictors of later mathematics achievement than others. Locuniak and N Jordan (2008) looked at how different aspects of number sense as measured at age 5 were related to later calculation fluency approximately 2 years later. Their number sense battery included the following tasks: counting, number knowledge, non-verbal calculation, story problems, and number combinations. When other variables such as memory, reading, and spatial ability were controlled for, number knowledge and, to a greater extent, number combinations emerged as the best predictors of later mathematics performance (calculation fluency), whereas counting, non-verbal calculation, and story problems did not.

Reading and Language

Verbal weaknesses can have a negative influence on mathematics development, as mathematics makes numerous demands on verbal ability; for example, the processing of speech sounds (Bull & Johnston, 1997; Geary, 1993; Hecht, Torgesen, Wagner, & Rashotte, 2001; Rourke & Conway, 1997), and retrieval and retention of verbal number codes (Robinson, Menchetti, & Torgesen, 2002). Additionally, there is neuropsychological evidence to suggest that the effect of language on mathematics varies across mathematical task, thus supporting the need for a componential approach. For example, exact calculation produces greater activation than approximate arithmetic in the angular gyrus, an area of the brain associated with language (Dehaene, Spelke, Pinel, Stanescu, & Tsivkin, 1999). There is also evidence to suggest that the effects of linguistic ability on addition are indirect, with linguistic effects mediated by skill on symbolic quantity tasks (Cirino, 2011).

In a longitudinal componential study by N Jordan et al. (2006) examining the performance of 5- to 6-year-old children, in which income status, gender, and age were controlled for, reading ability was positively associated with performance on a wide range of mathematics tasks. The results indicated that even those tasks that are considered to be relatively non-verbal, such as estimation and non-verbal calculation (e.g. Dehaene et al., 1999), have some language requirements; for example, understanding the examiner's instructions. When the analysis focused on growth rates rather than performance levels, better readers did not have stronger growth than poorer readers on any of the tasks, including story problems which are considered to be verbally demanding (N Jordan & Hanich, 2000; N Jordan & Montani, 1997). These results suggest that the poor readers started school with a disadvantage in many areas of mathematics, and they did not catch up with or fall further behind their typically achieving peers. However, as discussed previously, the lack of differences in growth rate may be attributed to the tasks being too difficult to capture sufficient growth and thus allow for meaningful statistical comparisons.

When sufficient levels of growth in mathematics are captured, the influence of phonological ability on mathematics achievement is evident for certain mathematical tasks (J Jordan, Wylie, & Mulhern, 2010). Children aged 5 years were classified by J Jordan et al. (2010) as one of the following subtypes: phonological difficulty (PD), co-morbid phonological difficulties and mathematical difficulties (PDMD), or typically achieving (TA). Children with PD exhibited weaker mathematical development than TA children, with approximately half of the PD subgroup meeting the criteria for PDMD at age 7 years. Further exploration indicated that those with PD tended not to progress as well as TA children on three out of four formal tasks—number facts, formal calculation, and concepts. On the other hand, similar development was found for most informal tasks—numbering, number comparison, and concepts. Informal mathematics is considered to be acquired through a child's interaction with their environment and some informal mathematical abilities are considered innate (Ginsburg & Baroody, 2003). In contrast, formal mathematical knowledge refers to the mathematics taught at school and tends to have greater language requirements than informal mathematics (Dowker, 2005). The weaker mathematical development of children with PD in this study appears to be associated with the shift from informal to formal mathematics in the curriculum, which was reflected in the items on the standardized Test of Early Mathematical Ability 3 (Ginsburg & Baroody, 2003) from age entry points 5–7. Interestingly, not all PD children struggled to cope with the increasing verbal demands of the mathematics test, as many showed similar growth to TA children. Although not explicitly investigated, J Jordan et al. (2010) speculated that these children may have been able to compensate for their weaknesses. For example, they may have been motivated to do extra work at home, or to employ alternative cognitive strategies.

Verbal weaknesses also appear to have a negative impact on mathematics performance in older children aged 7–9 years. Specifically, N Jordan, Kaplan, and Hanich (2002) found in a study using standardized tests of achievement that children with specific mathematical difficulties were more likely than those with co-morbid mathematics and reading difficulties to outgrow their mathematical difficulties. Yet in another study (N Jordan, Hanich, and Kaplan, 2003) using the same sample and design but a separate battery of mathematical tasks, no effects of reading ability on growth rate were found for any of the following mathematical subtasks: exact calculation, story problems, approximate arithmetic, place value, calculation principles, forced retrieval, and written computation. Unlike the studies of J Jordan et al. (2010) and N Jordan et al. (2002), in this study children were given the exact same items on each occasion. An effect of verbal ability on development may be absent simply because the actual verbal requirements of the test were static over time. None of these studies controlled for other cognitive variables such as working memory, and therefore caution must be shown when interpreting differences in performance at a single time point and in growth rate. In fact, in a study that did control for working memory (Locuniak and N Jordan, 2008), it was found that reading ability did not predict calculation fluency. As this study only examined one mathematical task, it is unclear if this finding would generalize to other mathematical tasks.

Cognitive Abilities

A case study from the Leverhulme study (Brown et al., 2008, pp. 85–108) highlights how memory problems, in general, can impact negatively on a child's mathematical development. The authors present a series of case studies, one of which describes a child with weak mathematical development in place value and fractions. Closer inspection revealed incorrect answers to questions that had been answered correctly at an earlier age. In addition, they made errors on simple questions, yet performed well in more difficult areas. Brown et al. suggest that this child's memory problems made it difficult for them to remember the procedural steps that they had been taught to answer these questions. In fact, they often tried to answer questions using procedural methods, but appeared to get answers correct only when they could remember the steps. This finding suggests that children with weak memory may have weaker growth or very uneven trajectories across some mathematical tasks compared to those with typical memory.

Much of the research looking at the influence of memory on mathematics has focused on the three components of the Baddeley and Hitch (1974) model of working memory: central executive (CE), phonological loop (PL), and visuospatial sketchpad (VSSP). The VSSP is considered to be important in areas of mathematics such as aligning columns of digits (Dowker, 2005), borrowing, and carrying (Bull, Espy, & Wiebe, 2008; Venneri, Cornoldi, & Garuti, 2003); number magnitude and estimation (Dehaene et al., 1999), and subtraction and written calculation (Venneri et al., 2003). On the other hand, neuroimaging and evidence from children with visuospatial learning difficulty indicates that spatial skills in general are less important in tasks such as exact calculation (Dehaene et al., 1999) and addition (Venneri et al., 2003). The effects of spatial ability may be indirect; for example, Cirino (2011) found that the effect of spatial ability on addition skill is mediated by performance on symbolic quantity measures. The CE function, storage and manipulation of information in long-term memory, plays an important role in arithmetic. For example, a child could use this function to solve an arithmetic problem such as 5 + 2, by recalling the answer to another fact such as 4 + 3 from long-term memory, then inferring the answer through the process of establishing equivalence (McLean & Hitch, 1999). There is also some evidence to suggest that the executive function of inhibition of irrelevant information is linked to mathematics performance (e.g. Espy, McDiarmid, Cwik, Stalets, Hamby, & Senn, 2004; Bull & Scerif, 2001), although not all studies have found evidence of a relationship (e.g. van der Sluis, de Jong, & van der Leij, 2004). Geary (1993) argued that the PL is important for solving arithmetic problems, as it is used to temporarily store both the problem and the solution, and the ability to do so will ultimately lead to the commitment of the problem and the solution to long-term memory. Furthermore, the better the ability to store a problem and solution in the PL, the stronger the long-term memory representation will become. In contrast, the PL is unlikely to be heavily involved in tasks such as estimation which tend not to produce strong activation in language areas of the brain (Dehaene et al., 1999).

Cross-sectional and longitudinal studies that have looked at how the influence of working memory on mathematics achievement varies over time have produced conflicting findings. For each component there have been reports of weakening (e.g. Bull et al., 2008; Holmes & Adams, 2006), stable (e.g. Geary, 2011), and increasing (e.g. Bull et al., 2008; De Smedt, Janssen, Bouwens, Verschaffel, Boets, & Ghesquiere, 2009; Geary, 2011) importance over time. There are a number of possible reasons for these conflicting findings, such as not controlling for the effects of other components of working memory (e.g. Holmes et al., 2008), use of cross-sectional design (e.g. Holmes et al., 2008), and not controlling for other key variables such as reading ability. For example, when Bull et al. (2008) controlled for reading ability, weaker relationships were found between mathematics and working memory.

In many studies the composite mathematics test given to each age group was comprised of different mathematical tasks. Yet detailed item or subtask analysis was not performed to see if changes in the influence of working memory over time were related to changes in the test. In some cases, each age group was given the same type of mathematical task; however, more difficult items were presented to the older children. For example, when testing number knowledge, De Smedt et al. (2009) gave younger children items requiring numbers from 1–10 but older children items with numbers 1–20. They acknowledged that the increasing importance of the PL that they observed in older children may in part be due to this task involving more knowledge of the number system and therefore greater language demands. Other studies have tested children on completely different mathematical subtasks at each time point. For example, Bull et al. (2008) tested children on graphical representation at age 7 (not at age 4 years), which may explain why the VSSP was significantly related to mathematics performance for this age group only. The use of different tasks at each time point makes it difficult to know if the relationship between working memory and mathematics achievement varies over time because the items/tasks used have different cognitive requirements or if children are solving the tasks via different cognitive routes as they get older. There is considerable evidence to suggest that mathematics is not a unitary ability, but rather it is composed of numerous mathematics abilities, and therefore greater insights can be gained by adopting a componential approach (Denvir & Brown, 1986; Dowker, 1998, 2005; Gifford & Rockliffe, 2008; Russell & Ginsburg, 1984). A study employing a componential approach suggests that the role of the VSSP varies across task, and over time, when subcomponents of the VSSP are considered separately. Specifically, Holmes et al. (2008) found that the VSSP explained similar amounts of variation in number and algebra skills at ages 7½ and 9½ years, yet did not explain significant variation in performance on shape, space, and measures, handling data, or mental arithmetic for either age group. Interestingly, they found that the spatial subcomponent of the VSSP was more important for younger children, whereas the visual subcomponent was more important for older children's mathematics. These results indicate the importance of considering both mathematical subtask and type of working memory subcomponent separately.

Teaching, Curriculum, and Classroom Experiences

Before children begin formal schooling, heterogeneity in mathematical abilities is already evident, and while some of this variability can be attributed to individual differences in aptitude, other factors such as learning experiences prior to attending school also explain unique variation (Crosnoe, 2006; Sadowski, 2006). Teaching has been highlighted as a limiting factor on children's mathematics development by Ofsted (2009) which found that many teachers needed help to enhance their own numeracy skills, and needed more guidance on teaching children the value of mathematics in everyday life.

Crosnoe et al. (2010) proposed that if a child has poor mathematics ability when they enter school, their trajectory will in part be influenced by the type of curriculum to which they are exposed and by their teacher–student relationships. Crosnoe et al. (2010) outlined how three different types of curriculum could influence a low-achieving child's development in mathematics: (1) *basic skills curriculum*—if those with weak mathematics skills are consistently exposed to a less challenging mathematics curriculum than their high-ability peers, they will not be able to catch up; (2) *common curriculum*—if given the same curriculum as their average-/high-achieving peers, although the material is more challenging, they have the potential to catch up. However, as many of these children enter school with poor foundation skills, they may find a more challenging curriculum too overwhelming; (3) *common curriculum with basic skills supplements*—to give low-achieving children the best chance of narrowing the achievement gap, they should be exposed to the same curriculum as their high-achieving peers, but should receive extra support from teachers (e.g. help with their basic skills). Crosnoe et al. (2010) found evidence to support their prediction that a *common curriculum with basic skills supplements* is more developmentally appropriate for those with weak mathematical ability. Their study revealed that children with low mathematics skills upon entry to school exposed to this curriculum were able to narrow their achievement gap as long as they had good relations with their teachers. However, the gap was only narrowed to a small extent, suggesting that the effect of this type of curriculum in comparison to other developmental influences on mathematics is relatively small.

The longitudinal Leverhulme study (Brown et al., 2008, pp. 85–108) of children aged 5–11 provides insights into classroom effects on development. Using data gained from researchers' observations and interviews with teachers, mathematics coordinators, and head teachers, inferences were made about the effects of classroom practice on mathematical development. Uneven trajectories were associated with changes in classroom practice, and these effects on trajectory depended on the ability level of the child. For example, in one year group, a small proportion of children struggled to understand their Year 5 teacher's explanations of place value, yet made very rapid progress when exposed to the more supportive and relaxed approach of their Year 6 teacher. In contrast, most children did not appear to find this teaching style challenging enough, because the class as a whole began to fall behind the rest of the sample during Year 6. Poor teaching in general, rather than a teaching style that favours some children, can lead to the whole class displaying uneven growth in mathematics. In one classroom, the observational data indicated that the Year 1 teacher had unclear objectives and undemanding teaching, yet the children progressed well under a different teacher in Year 2.

Social Factors: Socioeconomic Status and Attitudes

Higher income levels have been associated with better performance on a wide range of mathematical tasks in young children (N Jordan et al., 2006). National and international cross-sectional data from the Trends in International Mathematics and Science Study (TIMSS, 2007) provide some indication as to why children from low-income backgrounds do not perform as well in mathematics. At ages 10 and 14 years, those who have more books in their home tend to have better mathematics achievement. A computer at home with an Internet connection was also associated with better achievement. Growth curve modelling provides evidence to suggest that those from lower income backgrounds do not progress as well as those from higher income backgrounds on story problems (N Jordan et al., 2006) but progressed at a similar rate on seven other number sense tasks and number sense as whole. As the overall levels of growth on all tasks in the N Jordan et al. (2006) study were small, it is possible that the other mathematical tasks were not sensitive enough to allow the detection of growth rate differences.

N Jordan et al. (2006) suggested that these growth rate differences on the story problems task can be attributed to the high language and auditory attention requirements of the task (e.g. Levine, Jordan, & Huttenlocher, 1992). However, this seems unlikely given that reading ability did not predict growth in story problems. A more likely explanation provided by N Jordan et al. (2006) is that children are acquiring these skills outside the classroom.

Number sense as a whole was later modelled using six time points in the same group of children aged 5–6 (N Jordan et al., 2007). In contrast to when only four time points were used (N Jordan et al., 2006), income was found to be a significant predictor of growth rate in number sense. This may have been due to the greater accuracy in growth rate measurement achieved by using more time points or to greater growth at the last two time points.

The TIMSS (2007) data show that higher achievers tend to have more positive attitudes towards mathematics and greater self-confidence in learning mathematics at ages 10 and 14. Internationally, high levels of self-confidence in mathematics were more prevalent in 10-year-olds (57%) than in 14-year-olds (43%). This pattern was also found for high levels of positive affect towards mathematics in 10- (72%) and 14- (54%) year-olds. Declining levels of self-confidence throughout school towards mathematics were also found by the Scottish Survey of Achievement (2008), as well as lower levels of enjoyment and interest in older children. The lower levels of positive attitudes in older children mirror the lower levels of older children achieving the expected level. However, it is not clear if poor attitudes lead to lower performance, if lower performance leads to lower attitudes, or if the relationship is bi-directional (e.g. Dowker, 2005, p. 250). Measuring attitudes alongside mathematics performance longitudinally rather than relying on cross-sectional data will help to untangle cause and effect.

Neuroscience

Dehaene, Piazza, Pinel, and Cohen (2003) proposed that performance on complex arithmetical tasks depends on interactions between extraparietal areas and three regions within the parietal lobe. Each parietal region is considered to be involved in a different aspect of mathematics, namely numerical magnitude processing, language and phonologically mediated processing, and attentional and spatial orientation on the number line. LeFevre, Fast, and Skwarchuk (2010) have developed a model comprising three pathways (quantitative, linguistic, and spatial) based on the structure outlined by Dehaene et al. In the LeFevre et al. study, behavioural data indicated that at age 4, linguistic ability was related to number naming but not to non-linguistic arithmetic, while quantitative ability was related to non-linguistic arithmetic but not to number naming, indicating that quantitative and linguistic ability are two separate pathways. On the other hand, spatial ability was related to both non-linguistic arithmetic and number naming tasks, and this was interpreted as evidence to support a third spatial pathway. Similar to J Jordan et al. and N Jordan et al., the study found evidence to suggest that the linguistic pathway is more important in older children's mathematics. Specifically, linguistic ability uniquely predicted the performance of 7-year-olds on a wide variety of mathematical tasks including word reading, numeration, calculation, geometry, and measurement.

Unlike many of the other studies of mathematical development reviewed in this chapter, neuroscience studies tend to focus on mathematical subtasks separately. This approach is adopted for methodological reasons and because evidence suggests that different mathematical tasks engage different regions of the brain (e.g. Dehaene et al., 1999). Furthermore, the tasks tend to be very simple in nature, in order to permit researchers to develop an understanding of the basic skills that are required to perform complex tasks (Kaufmann,

2008, pp. 1–12). However, even studies of basic mathematical skills adopting a neuroscientific approach to development are uncommon, with most studies focusing solely on adults (Kaufmann, 2008, pp. 1–12).

Studies focusing on the same task have produced mixed evidence regarding developmental patterns of brain activation. Age-related shifts in brain activation from prefrontal to parietal areas were found for both symbolic (Ansari, Garcia, Lucas, Hamon, & Dhital, 2005) and non-symbolic (Ansari & Dhital, 2006) number processing. This has been interpreted as reflecting a developmental shift from using areas associated with working memory and attentional resources, towards more automatic processing. On the other hand, similar patterns of brain activation in tasks involving both symbolic and non-symbolic number processing were reported in children and adults by Cantlon, Brannon, Carter, and Pelphrey (2006) and Temple and Posner (1998). Considering that Temple and Posner (1998) used event-related potentials (ERP) source localization, a method with limited spatial resolution, overly strong conclusions cannot be drawn about the spatial activation patterns of both children and adults. Ansari, Holloway, Price, and Eimeren (2008, pp. 13–43) have suggested that the passive functional magnetic resonance imaging (fMRI) paradigm used by Cantlon et al. (2006) assessed low-level numerical processing rather than the ability to translate representations, and this may explain why age-related differences similar to those found in the studies by Ansari et al. (2005) and Ansari and Dhital (2006) were not observed.

Addition, subtraction, and multiplication have also been studied in both children and adults. When Kawashima et al. (2004) compared children aged 9–14 years with adults on these tasks they found broadly similar functional activation in both groups. By contrast, Rivera, Reiss, Eckert, and Menon (2005) found that between the ages of 8 and 19 when performing addition and subtraction, there was decreased activation in areas of the prefrontal and anterior cingulate cortex, areas that have been linked to working memory and attentional resources. The opposite pattern, increased activation with age, was observed for regions in the left parietal cortex shown by Dehaene et al. (1999) to be associated with arithmetic processing.

The inconsistencies between developmental imaging studies of mathematical abilities may be partly due to use of a cross-sectional design. With such a design, it is difficult to distinguish between group-specific differences and developmental differences (Ansari et al., 2008, pp. 13–43). The use of a cross-sectional design does not allow questions about individual differences in brain activation to be addressed, whereas a longitudinal approach would indicate if changes in activation are linear in some children and curvilinear for others, as was found with behavioural data (J Jordan et al., 2009).

Advances in Research Methods and Future Directions

The growing popularity of longitudinal statistical techniques has helped to build up a richer picture of children's mathematical development. Growth curve analysis is preferable to other techniques such as ANOVA for examining development, as it accounts for measurement error. This is clear when we compare the results from the same study when analysis was by ANOVA (Hanich, Jordan, Kaplan, & Dick, 2001) and when growth curve modelling was used (N Jordan et al. 2003). The ANOVA results indicate that at age 7, children with mathematical difficulty (MD) outperformed those with co-morbid mathematics and reading difficulty (MDRD) on exact calculation and story problems. However, the growth curve analysis shows that at age 9, MD had an advantage over MDRD on story problems but not exact calculation, and that they performed better on calculation principles. As there were no significant differences in growth rate in this study, this indicates that this pattern did not change between ages 7 and 9. Therefore the difference detected by the ANOVA at age 7 on exact calculation may have reflected measurement error rather than a true difference.

Another advantage of growth curve modelling is that it allows growth rate and shape to be compared statistically. For example, N Jordan and colleagues used growth curve modelling to compare growth rate and shape in mathematics of different mathematics and reading ability groups. However, as mentioned previously, even within a group of children with typical mathematics development, the mean trajectory for a group may bear little resemblance to the trajectories of the individuals within the group. Additionally, within MD subtypes there may be too much variation in mathematics performance within each subtype for meaningful comparisons to be made, as many factors other than reading are associated with MD (e.g. working memory, number sense, and attitudes). More recent studies have moved away from subtyping and have used growth curve modelling techniques which are able to explain greater amounts of variation in growth rate. For example, N Jordan et al. (2006) used growth curve modelling with multiple cognitive and demographic predictors to explain heterogeneity in development of number sense. The growth curve modelling methods being used to study mathematics development and individual differences in mathematical development are becoming progressively more advanced and more appropriate for examining development and heterogeneity. For example, rather than using one measurement of a variable to predict growth in mathematics, N Jordan, Kaplan, Ramineni, and Locuniak (2009) examined if growth in number sense predicted growth in mathematics. This is a more appropriate way of modelling predictors of mathematics development, such as reading ability, because they are not static in nature and individuals vary considerably in terms of rate and shape of growth on them. While growth curve modelling has been used to show how predictors affect the overall rate of development, multi-level modelling has been shown to be particularly useful at addressing questions about if and when the influence of predictors on mathematical development changes (e.g. Geary, 2011).

Growth mixture modelling and regression have been used to characterize the shape and/or steepness of each individual child's trajectory (e.g. J Jordan et al., 2009; N Jordan et al., 2006). Using regression to fit individual trajectories for each child to mathematical tasks can provide educationists with a very rich picture of a child's needs. This type of information would be particularly useful to those delivering interventions that are individually tailored to meet the needs of the child (e.g. Dowker & Sigley, 2010).

In cases where individually tailored intervention is not feasible, growth mixture modelling can provide a more simplified picture by grouping children with similar trajectories together. Furthermore, predictors can be added to these models to reveal the different factors associated with different patterns of development. Being able to identify which trajectory classification a child is likely to fall into enables predictions to be made about whether they are likely to outgrow their difficulty with little assistance or if intervention is likely to be needed (N Jordan et al., 2006).

In addition to using developmentally appropriate methods, future studies would benefit from a componential approach. In light of the findings of Holmes et al. (2008), it is quite likely that many of the findings from non-componential studies will not generalize to all mathematical abilities.

Using neuroscience in conjunction with other methods may provide further insights into the developmental issues discussed in this chapter. For example, such a combined approach may reveal if changes in strategy use as measured by behavioural data are mirrored by changes in activation patterns. Also as many of the studies of the influence of working memory over time are correlational, the neuroimaging evidence could indicate if changing relationships correspond to changes in brain activation. In addition, neuroimaging evidence may reveal if different trajectory types, or even the same trajectory type, are associated with different cognitive strategies.

Table 52.1 Key findings and concepts by section

Section	Key findings and concepts
Characteristics of trajectories	A child's trajectory can be described in terms of its initial and final status, growth rate, and shape of growth (e.g. linear or curvilinear). Componential and individual differences evidence indicates that there is considerable heterogeneity across mathematical tasks and within children in terms of trajectory characteristics.
Foundational numerical skills	Foundational numerical skills are required in number sense tasks such as counting and number knowledge. Number sense development is strongly associated with later mathematical performance.
Reading and language	Verbal skills are relatively more important in the formal mathematics taught at school than in informal mathematics. In studies using standardized mathematical tests, verbal skills play a greater role in older children's mathematics performance. This increase in importance has been associated with the shift from informal to formal mathematics in the curriculum and in some standardized mathematics tests.
Cognitive abilities	Evidence regarding the influence of working memory on mathematical development is inconsistent. Future studies may yield more consistent findings if a componential approach is adopted and variations in the composition of test administered to different age groups are considered.

(continued)

Table 52.1 Continued

Section	Key findings and concepts
Teaching, curriculum, and classroom experiences	Type of curriculum and teaching approach can influence a child's trajectory; for example, uneven trajectories may be associated with changes in teacher and teaching approach throughout schooling. Whether the effect of teaching approach on mathematics development is positive or negative depends on a child's learning style.
Social factors: socioeconomic status and attitudes	Income has been associated with growth on some tasks such as number sense and story problems. Younger children report more enjoyment in mathematics and are more likely to meet the expected level compared to older children; however, the direction of this relationship is unclear.
Neuroscience	Methodological issues such as not adopting a longitudinal approach may explain why evidence regarding age-related shifts in brain activation when performing mathematical tasks is mixed.
Advances in research methods and future directions	Growth curve modelling has identified some of the factors associated with individual differences in growth on mathematical tasks. Individual regression analyses identify the shape of each child's development on mathematical tasks, while growth mixture modelling groups similar trajectories together, thus providing useful information for individually tailored or group-based interventions.

Conclusion

The key findings and concepts outlined in each section are presented in Table 52.1. Cross-sectional and longitudinal data consistently indicate that mathematical difficulties are more prevalent in older than in younger children (e.g. Department of Education, 2011). Assessing children's mathematical performance longitudinally over two time points indicates if children perform at a satisfactory level or above, outgrow their difficulty, have persistent difficulty, or develop difficulties. On the other hand, by following children over a greater number of time points, it is possible to detect uneven developmental trajectories (J Jordan et al., 2009). The effects of a range of cognitive and social factors on mathematical development have been studied. However, for some of these variables (e.g. working memory and verbal ability) the effects on mathematical development are unclear. Such inconsistencies may be explained by the non-componential approach adopted by many studies. For example, Holmes et al. (2008) found that the influence of the VSSP on mathematical development varied across tasks. Studies examining individual differences in the shape of children's trajectories across mathematical tasks also indicate that there is a need to adopt a componential approach when studying children's mathematical development. This evidence suggests that interventions should be tailored to each child's needs rather than treating all children's MDs as homogenous. Indeed, there is now evidence to suggest that the most successful interventions are those that focus on the specific weaknesses of each individual child. For example, Dowker and Sigley (2010) compared the effectiveness of three levels of intervention: (1) individually administered and tailored to strengths and weaknesses; (2) individually administered but not tailored to strengths and weaknesses; and (3) no intervention. The individually tailored approach produced the greatest improvements in children's arithmetical development. Dowker and Sigley (2010) emphasized the need for more research that evaluates the relative effectiveness of individually tailored interventions for children with different levels/types of arithmetical difficulty. Furthermore, as the effectiveness of the intervention was only assessed at one point in time by Dowker and Sigley (2010), it is not known how the intervention affected the children's trajectories on the arithmetical tasks. For example, would the intervention have led to typical growth or even steeper than typical growth, allowing children with difficulty to catch up with their typically achieving peers? Future intervention and experimental studies adopting a componential approach and employing statistical methods appropriate for studying development are needed to build a clearer picture of how cognitive and social variables influence children's mathematical abilities over time.

References

Ansari, D. & Dhital, B. (2006). Age-related changes in the activation of the intraparietal sulcus during nonsymbolic magnitude processing: an event-related functional magnetic resonance imaging study. *Journal of Cognitive Neuroscience, 18* (11), 1820–1828.

Ansari, D., Garcia, N., Lucas, E., Hamon, K., & Dhital, B. (2005). Neural correlates of symbolic number processing in children and adults. *Neuroreport, 16* (16), 1769–1773.

Ansari, D., Holloway, I., Price, G., & Eimeren, L. (2008). Toward a developmental cognitive neuroscience approach to the study of the typical and atypical number development. In A Dowker (Ed.), *Mathematical Difficulties: Psychology and Intervention*, pp. 13–43. London: Elsevier.

Baddeley, A. & Hitch, G. (1974). Working memory. In G. H. Bower (Ed.) *The Psychology of Learning and Motivation* (pp. 47–90). New York: Academic Press.

Brown, M., Askew, M., & Hodgen, J., et al. (2008). Progression in numeracy ages 5-11: results from the Leverhulme longitudinal study. In: A Dowker (Ed.), *Mathematical Difficulties: Psychology and Intervention*, (pp. 85–108). London: Elsevier.

Bull, R., Espy, K., & Wiebe, S. (2008). Short-term memory, working memory, and executive functioning in preschoolers: longitudinal predictors of mathematical achievement at age 7 years. *Developmental Neuropsychology, 33* (3), 205–228.

Bull, R. & Johnston, R. (1997). Children's arithmetical difficulties: contributions from processing speed, item identification and short term memory. *Journal of Experimental Child Psychology, 65*, 1–24.

Bull, R. & Scerif, G. (2001). Executive functioning as a predictor of children's mathematics ability: inhibition, switching, and working memory. *Developmental Neuropsychology, 19* (3), 273–293.

Cantlon, J., Brannon, E., Carter, E., & Pelphrey, K. (2006). Functional imaging of numerical processing in adults and 4-year-old children. *PLOS Biology, 4* (5), e125.

CBI (2010). *Making It All Add Up: Business Priorities for Numeracy and Maths*. London: CBI.

CBI/EDI (2011). *Education and Skills Survey*. London: CBI.

Cirino, P. (2011). The interrelationships of mathematical precursors in kindergarten. *Journal of Experimental Child Psychology, 108*, 713–733.

Cohen, L. & Marks, K. (2002). How infants process addition and subtraction events. *Developmental Science*, 5 (2), 186–201.

Crosnoe, R. (2006). *Mexican Roots, American Schools: Helping Mexican Immigrant Children Succeed.* Palo Alto, CA: Stanford University Press.

Crosnoe, R., Morrison, F., Burchial, M., et al. (2010). Instruction, teacher–student relations, and math achievement trajectories in elementary school. *Journal of Educational Psychology, 102* (2), 407–417.

Dehaene, S., Piazza, M., Pinel, P., & Cohen, L. (2003). Three parietal circuits for number processing. *Cognitive Neuropsychology, 20* (3–6), 487–506.

Dehaene, S., Spelke, E., Pinel, P., Stanescu, R., & Tsivkin, S. (1999). Sources of mathematical thinking: behavioural and brain-imaging evidence. *Science, 284* (5416), 970–974.

Denvir, B. & Brown, M. (1986). Understanding of number concepts in low attaining 7–9 year olds: Part I. Development of descriptive framework and diagnostic instrument. *Educational Studies in Mathematics, 17*, 15–36.

Department for Education (2011a). *Interim Results for Key Stage 2 and 3 National Curriculum Assessments in England, 2010/11.* London: DfE.

Department for Education (2011b). *National Curriculum Assessments at Key Stage 1 in England, 2011.* London: DfE.

Department for Education (2011c). *Interim Percentage of Pupils Making Expected Progress in English and in Mathematics Between Key Stage 1 and Key Stage 2 in England.* London: DfE.

Department for Education (2011d). *Percentage of Pupils Making Expected Progress in English and Mathematics Between Key Stage 2 and Key Stage 4 in England.* London: DfE.

Department of Education (2011). *Percentages of Children in Northern Ireland Meeting the Expected Level in Mathematics at Key Stage One, Two and Three.* Data sourced from the DE.

De Smedt, B., Janssen, R., Bouwens, K., Verschaffel, L., Boets, B., & Ghesquiere, P. (2009). Working memory and individual differences in mathematics achievement: a longitudinal study from first grade to second grade. *Journal of Experimental Child Psychology, 103*, 186–201.

Dowker, A. (1998). Individual differences in arithmetical development. In C. Donlan (Ed.), *The Development of Mathematical Skills* (pp. 275–302). London: Taylor and Francis.

Dowker, A. (2005). *Individual Differences in Arithmetic.* Hove: Psychology Press.

Dowker, A. & Sigley, G. (2010). Targeted interventions for children with arithmetical difficulties. *BJEP Monograph Series II, 7.*

Espy, K., McDiarmid, M., Cwik, M., Stalets, M., Hamby, A., & Senn, T. (2004). The contribution of executive functions to emergent mathematical skills in preschool children. *Developmental Neuropsychology, 26*, 465–486.

Geary, D. (1993). Mathematical disabilities: cognitive, neuropsychological, and genetic components. *Psychological Bulletin, 114*, 345–362.

Geary, D. (2011). Cognitive predictors of individual differences in achievement growth in mathematics: a five year longitudinal study. *Developmental Psychology, 47* (6), 1539–1552.

Gifford, S. & Rockliffe, F. (2008). In search of dyscalculia. In M. Jourberr (Ed.), *Proceedings of the British Society for Research in Learning Mathematics, 28* (1), 21–27.

Ginsburg, H. & Baroody, A. (2003). *Test of Early Mathematics Ability 3.* Texas: Pro.Ed.

Gregory A., Snell, J., & Dowker, A. (1999). *Young children's attitudes to mathematics: a cross-cultural study.* Paper presented at the Conference on Language, Reasoning and Early Mathematical Development, University College London, September.

Hecht, S., Torgesen, J., Wagner, R., & Rashotte, C. (2001). The relations between phonological processing abilities and emerging individual differences in mathematical computation skills. A longitudinal study from second to fifth grades. *Journal of Experimental Child Psychology, 79*, 192–227.

Hanich, L., Jordan, N., Kaplan, D., & Dick, J. (2001). Performance across different areas of mathematical cognition in children with learning difficulties. *Journal of Educational Psychology, 93*, 615–626.

Holmes, J. & Adams, J. (2006). Working memory and children's mathematical skills: implications for mathematical development and mathematics curricula. *Educational Psychology, 26*, 339–366.
Holmes, J., Adams, J., & Hamilton, C. (2008). The relationship between visuospatial sketchpad capacity and children's mathematical skills. *European Journal of Cognitive Psychology, 20* (2), 272–289.
Jordan, J., Mulhern, G., & Wylie, J. (2009). Individual differences in trajectories of arithmetical development in typically achieving 5- to 7-year-olds. *Journal of Experimental Child Psychology, 103*, 455–468.
Jordan, J., Wylie, J., & Mulhern, G. (2010). Phonological awareness and mathematical difficulty: a longitudinal perspective. *British Journal of Developmental Psychology, 28*, 89–107.
Jordan, N., Glutting, J., & Ramineni, C. (2008). A number sense assessment tool for identifying children at risk of mathematical difficulties. In A Dowker (Ed.), *Mathematical Difficulties: Psychology and Intervention* (pp. 46–58). London: Elsevier.
Jordan, N. & Hanich, L. (2000). Mathematical thinking in second grade children with different types of learning difficulties. *Journal of Learning Disabilities, 33*, 567–578.
Jordan, N., Hanich, L., & Kaplan, D. (2003). A longitudinal study of mathematical competencies in children with specific mathematics difficulties versus children with co-morbid mathematics and reading difficulties. *Child Development, 74*, 834–850.
Jordan, N., Kaplan, D., & Hanich, L. (2002). Achievement growth in children with learning difficulties in mathematics: findings of a two year longitudinal study. *Journal of Educational Psychology, 94*, 586 597.
Jordan, N., Kaplan, D., Locuniak, M., & Ramineni, C. (2007). Predicting first-grade math achievement from developmental number sense trajectories. *Learning Disabilities Research and Practice*, 22 (1), 36–46.
Jordan, N., Kaplan, D., Olah, L., & Locuniak, M. (2006). Number sense growth in kindergarten: a longitudinal investigation of children at risk for mathematics difficulties. *Child Development, 77* (1), 153–175.
Jordan, N., Kaplan, D., Ramineni, C., & Locuniak, M. (2009) Early math matters: kindergarten number competence and later mathematics outcomes. *Developmental Psychology, 45*, 850–867.
Jordan, N. & Montani, T. (1997). Cognitive arithmetic and problem solving: a comparison of children with specific mathematical difficulties. *Journal of Learning Disabilities, 30*, 624–634.
Kaufmann, L. (2008). Neural correlates of number processing and calculation: developmental trajectories and educational implications. In A Dowker (Ed.), *Mathematical Difficulties: Psychology and Intervention* (pp. 1–12). London: Elsevier.
Kawashima, R., Taira, M., Okita, K., et al. (2004). A functional MRI study of simple arithmetic—a comparison between children and adults. *Brain Research Cognitive Brain Research, 18* (3), 227–233.
KPMG (2009). *Innumerate School Children Cost the Tax Payer up to £2.4bn a Year*. KPMG Press release.
LeFevre, J., Fast, L., & Skwarchuk, S., (2010). Pathways to mathematics: longitudinal predictors of performance. *Child Development, 81* (6), 1753–1767.
Levine, S., Jordan, N., & Huttenlocher, J. (1992). Development of calculation abilities in young children. *Journal of Experimental Child Psychology, 53*, 72–103.
Locuniak, M. & Jordan, N. (2008). Using kindergarten number sense to predict calculation fluency in second grade. *Journal of Learning Disabilities, 41* (5), 451–459.
McLean, J. & Hitch, G. (1999). Working memory impairments in children with specific arithmetic learning difficulties. *Journal of Experimental Child Psychology, 74*, 240–260.
Mix, K., Huttenlocher, J., & Levine, S. (2002). Multiple cues for quantification in infancy: is number one of them? *Psychological Bulletin, 128* (2), 278–294.
OECD (2006). *Education at a Glance*. Paris: OECD.
Ofsted (2009) *Implementation of 14–19 Reforms, Including the Introduction of Diplomas*. London: Ofsted.
Rivera, S., Reiss, A., Eckert, M., & Menon, V. (2005). Developmental changes in mental arithmetic: evidence for increased functional specialisation in the left inferior parietal cortex. *Cerebral Cortex, 15* (11), 1779–1790.

Robinson, C., Menchetti, B., & Torgesen, J. (2002). Toward a two-factor theory of one type of mathematics disabilities. *Learning Disabilities Research and Practice*, *17*, 81–89.

Rourke, B. & Conway, J. (1997). Disabilities of arithmetic and mathematical reasoning: perspectives from neurology and neuropsychology. *Journal of Learning Disabilities*, *30*, 34–46.

The Royal Society (2011). *Increasing the Size of the Pool. A Summary of the Key Issues from the Royal Society's 'State of the Nation' Report on Preparing for the Transfer from School and College Science and Mathematics Education to UK STEM Education*. London: Royal Society.

Russell, R. & Ginsburg, H. (1984). Cognitive analysis of children's mathematical difficulties. *Cognition and Instruction*, *1*, 217–244.

Sadowski, M. (2006). The school readiness gap. *Harvard Education Letter*, *22* (4), 1–2.

Scottish Survey of Achievement (2008). *Scottish Survey of Achievement: Mathematics and Core Skills*. Edinburgh: The Scottish Government.

Statistical Directorate Welsh Assembly Government (2009). *National Curriculum Assessments of 7, 11 and 14 Year Olds: Wales 2009*. Cardiff: SDWAG.

Statistical Directorate Welsh Assembly Government (2010). *National Curriculum Assessments of 7, 11 and 14 Year Olds: Wales 2010*. Cardiff: SDWAG.

Temple, E. & Posner, M. (1998). Brain mechanisms of quantity are similar in 5-year-old children and adults. *Proceedings of the National Academy of Sciences of the United States of America*, *95* (13), 7836–7841.

TIMSS (Trends in International Mathematics and Science Study) (2007). *TIMSS 2007 International Mathematics Report*. Chestnut Hill, MA: TIMSS.

Van der Sluis, S., de Jong, P., & van der Leij, A. (2004). Inhibition and shifting in children with learning deficits in arithmetic and reading. *Journal of Experimental Child Psychology*, *87*, 239–266.

Venneri, A. Cornoldi, C., & Garuti, M. (2003). Arithmetic difficulties in children with visuospatial learning disability (VLD). *Child Neuropsychology*, *9*, 175–183.

CHAPTER 53

BEHAVIOURAL GENOMICS OF MATHEMATICS

MARIA G. TOSTO, CLAIRE M. A. HAWORTH, AND YULIA KOVAS

The first complete DNA sequence of the human genome was outlined in 2001. This significant breakthrough occurred less than 50 years after James Watson and Francis Crick first proposed a model of the molecular structure of DNA. Since then, our understanding of human genetics has leapt forward exponentially, leading to the application of genetic knowledge and methodologies to unravelling the etiology of psychopathologies, and of individual differences in cognition, abilities, and personality.

This chapter addresses the contribution of behavioral genetic approaches, which includes both quantitative methods (e.g. twin, adoption, and other family studies) and molecular genetic methods (e.g. association studies) to the investigation of mathematical development. A brief overview of the quantitative genetic methodologies is provided at the beginning. The chapter then reviews the growing body of research into the relative contribution of genes and environments to the variation in mathematical ability at different ages and in different populations, including the examination of the etiology of any observed sex differences in mathematics. This review also includes a relatively small body of multivariate twin research into the etiological links between mathematics and other areas of cognition and achievement, as well as the links between mathematical ability and disability. The multivariate section reviews the latest results examining the genetic and environmental relationships between mathematical achievement and mathematical motivation. The molecular genetic section presents the few molecular genetic studies that have specifically explored mathematical abilities. The chapter concludes by outlining the future directions of the behavioral genetic research into mathematical learning and the potential implications of this research.

Quantitative Genetic Methodologies

Recent behavioral genetic research leaves no doubt that individual differences in behavior and cognition are a product of both genetic and environmental factors (e.g. Plomin, DeFries,

Knopik, & Neiderhiser, 2013). This research also suggests that the path leading from genes to behavior is intertwined with the environment. While molecular genetic research (reviewed in the next section) aims to detect and identify the genes implicated in the variation in different aspects of behavior and cognition, quantitative genetics aims to quantify the relative contribution of genes and environment to the variation in traits and co-variation among traits.

Genetic influences refer to the influence of multiple alleles – genetic markers that can differ in the population (rather than evolutionarily conserved invariant markers). Mostly, genetic influences are of the additive type, meaning that the variance of a trait that is attributed to genetic factors can be derived by adding the independent effects of all alleles at all loci that affect the trait. Some genetic influences may derive from interactions between genes at different loci. These epistatic processes, by which the effects of a gene on a specific trait depend on the influences of one or more other genes remain very poorly understood (e.g. Cordell, 2002).

From the behavioral genetic perspective, environmental influences are very broadly defined as effects on a trait produced by anything other than heritable DNA sequence variation. In twin and other family designs, any environmental influences that contribute to differences between family members are referred to as 'non-shared'; whereas any environmental influences that contribute to the similarity between family members are referred to as 'shared' (see Plomin & Daniels, 1987; Rijsdijk & Sham, 2002; Plomin et al., 2013).

Non-shared environments are defined as any event that is experienced or perceived differently by family members and contributing to dissimilarities between them. These may include perinatal events, accidents, surgical procedures, and different peers. Intuitively, environments that are objectively shared among individuals within a family seem more likely to increase their similarities. We can think of nutrition, parenting practices, or socio-economic status as shared experiences that may make family members more similar in a specific trait, if these factors affect the trait in question. For example, it is reasonable to think that family members would share similar eating habits that could increase similarity in weight among them. However, research shows that adult family members do not resemble each other in weight beyond genetically influenced similarity (e.g. Grilo & Pogue-Geile, 1991). Often, objectively shared environments lead to differences, rather than similarities, through differential perceptions and other poorly understood mechanisms (e.g. Dunn & Plomin 1990; Plomin & Daniels, 1987). Parental divorce, for instance, is a family event and, as such, is shared by siblings, but research shows that divorce often impacts on siblings' behavior in different ways (Amato, 2000; Hetherington & Clingempeel, 1992). The estimate of non-shared environment in quantitative genetic methodology also includes any measurement and procedural errors, as non-systematic error can only contribute to dissimilarity in assessed traits between twins or other family members.

One of the most prevalent methods used in quantitative genetic research is the twin design. As with any approach, the twin method has some limitations and relies on several assumptions (see Plomin et al., 2013, for additional details). One of these assumptions is that the same shared environmental influences will equally affect monozygotic or identical twins (MZ), and dizygotic or fraternal twins (DZ), that is to say that shared environment is the same for both MZ and DZ twins (e.g. Rijsdijk & Sham, 2002). Violation of the equal environments assumption would incorrectly include the environment together with genetic influences as causes for the observed MZ twins' similarity. Although MZ twins are more likely to be treated alike, research that has investigated the equal environments assumption has shown that this similar treatment is a reflection of their increased genetic similarities (e.g. Evans & Martin, 2000). Because twins are not a random sample, their representativeness has been questioned. For example, newborn twins on average weigh less than singletons; however, by middle childhood the weight differences disappear (MacGillivray, Campbell, & Thompson, 1988). Similarly, twins' achievement in adolescence has been shown to be the same as singletons' (Christensen, Petersen, Skytthe, Herskind, McGue, & Bingley, 2006), despite the slightly lower average IQ displayed by twins in early childhood (Ronalds, De Stavola, & Leon, 2005). Many studies tested the twin method, and concluded that findings from twin studies are valid and are applicable to the general population (e.g. Hart, Petrill, & Kamp Dush, 2010a; Kovas, Haworth, Dale, Plomin, Weinberg, Thomson, et al., 2007a; Plomin et al., 2013).

The twin method relies on the comparison of intraclass correlations between identical and fraternal twins. In the univariate genetic analysis, this comparison allows estimating the proportion of variance in a trait that can be attributed to genetic, shared and non-shared environmental influences (Plomin et al., 2013). MZ twins result from the division of one single zygote – one fertilised egg. It is generally assumed that MZ twins are genetically identical, although recent research suggests some genetic differences (e.g. Bruder, Piotrowski, Gijsbers, Andersson, Erickson, Diaz de Ståhl, et al., 2008). DZ twins occur when two eggs are fertilised at the same time. DZ twins, like any other pair of siblings, share on average 50% of the segregating genes. Twins brought up in the same family may be similar to each other because of the influence of their common environment, as well as because of their shared genes. If genes play an important role in a trait, identical (MZ) twins must be more similar on that trait, compared with fraternal (DZ) twins. The influence of genetic factors (heritability) is calculated as twice the difference between MZ and DZ twin correlations. Shared environmental factors are implicated if the DZ twin correlation is greater than half of the MZ twin correlation, and can be calculated as the difference between the MZ twin correlation and the heritability. Non-shared environmental influences are indicated by the extent to which the correlation between MZ twins is not 100%. In a practical example, if the MZ twins correlation for a trait is 0.8 and the DZ correlation is 0.5, heritability for that trait would be 60% [2 * (0.8 – 0.5) = 0.6], shared environment would be 20% (0.8 – 0.6 = 0.2) and non-shared environment would be 20% (1 – 0.8 = 0.2). The effects of genetic, shared and non-shared environmental influences are more accurately estimated using structural equation modelling, as the latent variables that are more likely to reproduce the observed MZ and DZ variance and covariance. The analysis also calculates confidence intervals around the estimates, which give an indication of their significance (see Plomin et al., 2013).

Multivariate genetic analysis allows the estimation of genetic and environmental sources of covariation among different traits (see Martin & Eaves, 1977). Comparison between MZ and DZ twins is conducted on the cross-trait twin correlations. For example, in the investigation of the etiology of the covariation between mathematics and reading the analyses would be conducted on the correlation between one of the twins in the pair in mathematics and the co-twin in reading. An MZ twin cross-trait correlation greater than the DZ twin cross-trait correlation indicates that common genetic factors contribute to the covariation between the two traits. DZ twin correlations exceeding half the MZ twin correlation indicate shared environmental influences. The absence of a significant cross-trait twin correlation implies that the common etiological influences are due to non-shared environmental factors. From multivariate genetic analysis it is also possible to determine the bivariate heritability, which indexes the extent to which the phenotypic correlation between two traits is genetically mediated. The remaining phenotypic correlation is explained by bivariate shared and bivariate non-shared environment. Another important statistic derived from multivariate analyses is the genetic correlation: the extent to which genetic influences on one trait correlate with the genetic influences on another trait – independently from univariate heritability of both traits. For example, it is possible that two traits have low heritability (genetic influences are very small), but have a high genetic correlation (the same genetic factors influence both traits) (for details, see Kovas et al., 2007a; Neale & Maes, 2003).

The multivariate method can be extended to longitudinal data, as an alternative to univariate analysis on cross-sectional data (see Loehlin, 1996). Longitudinal genetic analyses provide estimates of genetic (and environmental) stability and change. The analyses are conducted on the same trait (e.g. mathematical performance) assessed at different ages. For example, it is possible that mathematics is influenced by similar genetic (or environmental) factors at different ages, but these influences vary in their strength with time (quantitative differences). If different genetic (or environmental) factors affect mathematics across ages, qualitative changes take place.

Genetic methodologies are also well suited to unravel the etiology of the links between ability and disability. Low/high mathematical abilities may be qualitatively different from normal variation if different genetic or environmental factors influence variation at the extremes and in the normal range of mathematical performance. Alternatively, the same factors may drive individual differences in high/low and normal abilities, but to different extents (quantitative differences). One of the methods used to investigate the relationship between abilities and disabilities is the DeFries-Fulker extremes analysis (DeFries & Fulker, 1985). This method relies on identifying low or high ability groups by way of cut-offs, after which probandwise concordances are calculated. Probandwise concordances index the probability of both twins in a pair manifesting the same disorder/talent. The comparison of MZ and DZ concordances allows the calculation of the group heritability. This indexes the extent to which the mean differences between the proband group and the rest of the population are due to genetic or environmental factors. A concordance higher for MZ than DZ twins suggests that the mean differences between the proband group and the population (the unselected sample) are driven by genetic influences, to some extent. If identical and fraternal twins had the same concordance, the group heritability would be zero (no genetic influences). However, in the absence of genetic influences, the proband deficit could be due non-shared environmental factors (an injury or illness not shared by the co-twin) or shared environmental factors (family nutrition, for example) – or a combination of both. Group heritability helps to understand whether the proband group is 'special' or different from the rest of the population. If the etiology of individual differences in the proband sample is the same as in the whole distribution, the group heritability should not be significantly different from the heritability of the unselected sample (for details see Haworth, Kovas, Harlaar, Hayiou-Thomas, Petrill, Dale, et al., 2009a; Kovas et al., 2007a).

Finally, sex-limitation models can address whether the same genetic and environmental factors influence variation in males and females. Quantitative sex differences exist if the same genetic and environmental factors affect males and females but with different strength. If different environmental and/or genetic factors explain individual differences in males and females the differences are qualitative. It is also possible that there are no differences in the etiology of individual differences between males and females. In this case, the factors that drive individual differences in males are the same factors that drive individual differences in females (for details see Neale & Maes, 2003). It is important to note that the etiological sex differences related to variance do not necessarily translate into mean sex differences.

Genetic and Environmental Etiology of Individual Differences in Mathematics

Over the last few decades, the social desirability of good mathematic skills has increased due to the wide range of advantages associated with them. A recent survey reports a relationship between improvement in the numeracy skills of a population and the increase in productivity for that Nation (OECD, 2010). Similarly, social and economic disadvantages have been related to low numeracy (Gross, Hudson, & Price, 2009). The quest for understanding how different people acquire and can improve mathematical skills is therefore more important than ever. This importance is reflected in the increased number of quantitative genetic studies into the etiology of individual variation in mathematically-relevant traits.

One of the first twin studies that examined mathematical ability used 146 MZ and 132 DZ pairs of 6–12-year-old twins from the Western Reserve Twin Project. The study found that mathematics, measured by standardised tests of school achievement, was modestly heritable (0.20), with shared and non-shared environment explaining most of the variation (0.71 and 0.10, respectively; Thompson, Detterman, & Plomin, 1991). Another study used a sample of twins with age ranging between 8 and 20 years (Alarcón, Knopik, & DeFries, 2000). 570 twin pairs of this sample were controls, in 555 pairs one of the twins manifested reading or mathematical problems or both. The results showed an average heritability of 0.90 in both groups with virtually non-significant environmental influences. The wide range of univariate estimates illustrated in these studies deserves some methodological consideration. Twin correlations for a trait may be overestimated because of common factors unrelated to the trait. Twins in a pair are of the same age; MZ twins and around half of the DZ twins are of the same sex. In twin analyses it is common practice to correct for age and sex in order to avoid inflating correlations because of these factors (McGue & Bouchard, 1984). The estimates reported from Thompson et al. (1991) were, in fact, affected as the measures were not corrected. When the correction for age and sex was applied, the estimates revealed moderate genetic and shared environmental influences (~0.40 for both; in Kovas et al., 2007a).

Other factors also need to be considered. Quantitative genetic investigations of complex traits suggest somewhat different patterns of genetic and environmental influences on different traits. Reading abilities, for example, have shown consistently substantial genetic and shared environmental influences across ages and populations (e.g. Byrne, Samuelsson, Wadsworth, Hulslander, Corley, DeFries, et al., 2006; Byrne, Wadsworth, Corley, Samuelsson, Quain, DeFries, et al., 2005; Light, DeFries, & Olson 1998; Stromswold, 2001). Conversely, the heritability of *g* has been shown to increase consistently from early to middle childhood (Davis, Haworth, & Plomin, 2009a; Haworth, Wright, Luciano, Martin, de

Geus, van Beijsterveldt, et al., 2010). Although it is unclear whether the inconsistencies in the estimates of mathematical heritability can be explained by differences in participants' age, the findings suggest that age-homogeneous twin samples should be used in behavioral genetic investigations for at least two reasons. First, the trait itself changes across the school years – what we call 'mathematics' may involve very different cognitive and motivational processes at different ages, reflected in the changes in how mathematics is measured. Secondly, new genes and environments may become active or relevant during development, for example, reflecting changes in pubertal processes or in socialising. Estimates of genetic and environmental contributions are population-based as they explain the sources of individual differences within a particular population; these estimates may differ not only for different ages, but also for different countries or cultures. If a particular environment is uniform within a culture (e.g. a national curriculum or educational standards), this environment is unlikely to explain much of the inter-individual variation. In such a population, heritability of a trait may be higher. Much more research is needed in order to clarify the sources of inconsistencies among different studies, including careful examination of cultural norms and provisions. Furthermore, although Thompson et al. (1991) and Alarcón et al. (2000) used samples with a respectable number of twins, power calculations indicate that twin studies need samples as large as 600 twin pairs (Martin, Eaves, Kearsey, & Davies, 1978) to many thousand pairs (Marti, Eaves, & Kendler, 1994) in order to provide accurate estimates.

Much recent research into the etiology of variation in mathematical ability and achievement comes from the Twins Early Development Study (TEDS), a large-scale longitudinal study comprising of three cohorts of twins recruited at birth in 1994, 1995, and 1996 in the United Kingdom. TEDS twins have been regularly assessed from birth on multiple measures of behavior, cognition, and achievement – by different means: in person; with child, parent and teacher questionnaires; by telephone; and more recently with web-based test batteries. About 12,000 pairs are currently active in the study (Haworth, Davis, & Plomin, 2013; Oliver & Plomin, 2007). The large representative sample and the longitudinal nature of TEDS, have allowed researchers to address many of the methodological issues raised by previous studies.

The first large-scale assessment of mathematical achievement was conducted in over 2,000 TEDS twin pairs when they were 7 years of age (Oliver, Harlaar, Hayiou-Thomas, Kovas, Walker, Petrill, et al., 2004). Teacher-rated mathematics was found to be moderately heritable (0.66), with negligible shared environmental influences, and modest non-shared environmental influence (0.25). As the mathematical domain is multifaceted (e.g. Dowker, 2005), it is important to examine the relative contributions of genes and environments to the different aspects of mathematical ability and achievement. Oliver et al. (2004) estimated genetic and environmental influences on three different mathematical components, which are part of the UK National Curriculum for mathematics: 'Using and Applying Mathematics,' 'Numbers,' and 'Shapes, Space and Measures.' The heritability and environmental estimates were similar for the three components.

A very similar pattern of results emerged in the assessment of over 2,600 TEDS twin-pairs at 9 years (Kovas, et al., 2007a). Mathematics, rated by teachers, showed genetic influences of 0.72, almost non-existent shared environmental influences, and very modest (0.23) non-shared environment. The three mathematical components again yielded highly similar estimates:

- Genetic influences ranged between 0.63 (Shapes, Space, and Measures) and 0.73 (Using and Applying Mathematics).
- Shared environment was negligible for all three components.
- Estimates for non-shared environment ranged from 0.26 (Using and Applying Mathematics) to 0.28 (Shapes, Space, and Measures).

At 10 years, 2,674 twin pairs were assessed using an online battery of three mathematical sub-tests: 'Understanding Number,' 'Non-Numerical Processes,' and 'Computation and Knowledge' (Kovas, Haworth, Petrill, & Plomin, 2007b). Similarly to previous estimates, shared environment had very small effects on all three measures and the non-shared environmental influences were between 0.42 and 0.48. Although 'NON-NUMERICAL PROCESSES' showed a lower heritability (0.32) compared with the other two components (0.42 and 0.45, respectively), the differences in heritability among the three measures were not significant. Assessment of over 5,000 TEDS twin pairs at 12 years showed strong genetic influences on mathematical achievement (0.61), and small shared (0.18) and non-shared (0.21) environmental influences (Davis, Haworth, & Plomin, 2009b). The latest assessment of the TEDS twins at age 16, carried out on over 3000 twin pairs, again showed high genetic influences on web-tests of mathematical achievement (0.57), with small shared (0.21) and non-shared environmental influences (0.22) (results available from the authors. Overall, these results demonstrate consistent genetic and non-shared environmental influences on different aspects of mathematical ability across the school years.

Several other recent studies have addressed the etiology of different aspects of mathematical skills using different twin samples. The US-based Western Reserve Reading Project for Math (WRRPM) assessed 228 10-year-old twin pairs on four different mathematical components: 'Calculation,' 'Fluency,' 'Applied Problems,' and 'Quantitative Concepts' (Hart, Petrill, & Thompson, 2010b), reporting univariate heritability estimates of 0.35 and 0.34 for 'Calculation' and 'Fluency,' and a slightly higher heritability for 'Applied Problems' and 'Quantitative Concepts' (0.41 and 0.49, respectively). Interestingly, the shared environmental influences estimated in this study (0.32–0.46) were higher compared with the TEDS estimates for mathematical sub-components at the same age (0.07–0.23); whereas the non-shared influences (0.19–0.25) were lower than in TEDS (0.42–0.48). The discrepancy observed in the strength of environmental influences between the two samples could be due to different curricula or school environments in the two countries. It is possible that the UK educational system, with its standardised programmes across schools, ensures a homogeneous mathematical environment, which in turn reduces the proportion of variance in mathematics explained by shared environment. Alternatively, different estimates may reflect differences in the etiology of different facets of the mathematical domain assessed in the studies. Cross-cultural research using identical measures of mathematical ability and performance in large samples of the same age are needed in order to make meaningful comparisons and to establish the true sources of differences in estimates from different studies.

In summary, analyses of school-aged twins reveal substantial genetic influences on individual differences across the range of mathematical components, with environmental factors being primarily of the non-shared type, at least in the UK population. What specific environments could account for these influences? Teacher and school quality have been hypothesised as important in explaining differences in mathematical achievement (e.g. Darling-Hammond, 2000; Eide & Showalter, 1998; Saxe, Gearhart, & Nasir, 2001). If schools and teachers had equal effect on all the children within a class and school, children with the same genes (MZ twins) who have the same mathematics teacher should be more similar in mathematical skills than those taught by different teachers. However, this does not appear to be the case. At 10 years, twin pairs of the TEDS sample who were taught by the same teacher were compared with twins who had different teachers (Kovas et al., 2007a; Kovas, Petrill, & Plomin, 2007d). The results showed no significant differences in similarity in mathematical performance between twins taught by the same teacher versus different teachers, leading to the same estimates of heritability and environmental influences in both groups (Kovas et al. 2007a, 2007d). If being in the same class does not contribute to similarity among children in mathematics, the influence of a teacher and classroom may be hypothesised to have a non-shared effect, influencing mathematical abilities of different children in different ways. The absence of differences in twin similarity across classrooms may reflect the uniform National Curriculum and standardised teacher training applicable in the UK. The curriculum and schools contribute to invariance across the populations, whereas genetic and non-shared environmental effects explain individual differences.

The Etiology of the Relationships between Mathematics, and Other Cognitive and Motivational Traits

In recent years, behavioral genetic research has moved beyond simple evaluation of the relative contributions of genetic and environmental factors to a single phenotype (e.g. mathematical variation). Multivariate methodology examines the etiology of the relationships between mathematics, and other cognitive and motivational traits. For example, it has been suggested that motivational factors, such as self-perceived ability and enjoyment of mathematics may make unique contributions to mathematical development. The relationship between liking mathematics and achievement was investigated in the TEDS sample at age 9 (Spinath, Spinath, & Plomin, 2008). Liking mathematics was moderately influenced by genetic factors (0.40) and the rest of its variation came from non-shared environment (0.60). Moreover, the relationship between liking mathematics and mathematical achievement was very small, a regression analysis revealed that liking mathematics explained 0.5% of the total variance in mathematical scores, and only in the female sample.

Reciprocal relationships between academic achievement and self-evaluation have been previously reported (Marsh, Byrne, & Yeung, 1999; Marsh & Yeung, 1997), but the etiology of these relationships can only be examined in genetically sensitive studies. Self-evaluation of mathematical abilities has been assessed in TEDS at 9 and 12 years. At both ages this measure showed moderate genetic (~0.40) and non-shared environmental influences (~0.59), without any shared environmental influences (Spinath et al., 2008; Luo, Kovas, Haworth, Dale, & Plomin, 2011). A study that employed a cross-lagged design reported that self-evaluation at 9 predicted mathematical achievement at 12, and that achievement at 9 predicted self-evaluation at 12, although the effect size of these relationships was very small (r = ~0.10). These reciprocal relationships were found to be largely genetically mediated

(Luo et al., 2011). Genetic influences on self-evaluation at 9 years were also associated with the genetic influences on later mathematics at 12 years, independent of the genetic influences on mathematics at 9 years. This interesting phenomenon, requiring further investigation, suggests that the new genetic influences on mathematics that emerge at age 12, may be partially driven by the same genetic factors influencing self-evaluation in previous years.

The insight that self-perception of abilities and enjoyment of mathematics depend to a large extent on genetic and non-shared environmental factors, rather than shared (family-wide or school-wide) influences requires reconceptualisation of many hypothesised motivational mechanisms. It is possible that a better understanding of these mechanisms and their link with mathematics will lead to changes in the way we teach mathematics and perhaps other academic subjects. It is likely that the biggest progress in education will come when environments are individualized to fit the unique genetic profiles and unique needs of each child (Haworth, Asbury, Dale, & Plomin, 2011; Haworth & Plomin, 2011; Kovas & Plomin, 2012).

Self-evaluation is only one of the many domains that have been explored in terms of genetic and environmental sources of covariation with mathematics. One of the first studies for normal variation investigated the relationships between mathematics and English, and mathematics and vocabulary, in a sample of over 2,000 twin pairs in US High Schools (Martin, Jardine, & Eaves, 1984). The study reported genetic correlations of 0.52 and 0.39, respectively, suggesting that when genes associated with mathematical abilities are identified, many of the same genes will be associated with English and, to a lesser extent, with vocabulary. A later study (Thompson et al., 1991) reported strong genetic correlations between mathematics and reading (0.98), and mathematics and language (0.98), suggesting that largely the same genes contributed to variation in all of the examined traits. The same study showed that shared environments were also largely the same for the three traits (the shared environmental correlation was 0.93 on average). Non-shared environmental factors explained most of the differences among the traits (correlations ranged from 0.28 to 0.54). Multivariate analyses in 314 10-year-old twin pairs of the WRRPM study investigated the relationship between different components of mathematics, reading and general cognitive ability (*g*) (Hart, Petrill, Thompson, & Plomin, 2009). It was found that there was no significant genetic overlap between 'Reading Fluency', *g*, and the mathematical subcomponent of 'Calculation'. Conversely, the overlap in shared environmental influences on 'Reading Fluency' with those on 'Calculation' and *g* was significant, suggesting that the within-families environments that are important for 'Reading Fluency' and *g*, also influence the learning of 'Calculation'. 'Mathematical Fluency' shared genetic influences with 'Reading Fluency' (0.43) and *g* (0.20). However, some genetic influences on 'Mathematical Fluency' (0.59) were independent from reading and *g*. Furthermore, 'Mathematical Fluency' showed no significant environmental overlap with 'Reading Fluency' and *g*, indicating a degree of independence from these two abilities.

The assessment in TEDS at 7 years found a genetic correlation between mathematics and *g* of 0.67 and between mathematics and reading of 0.74 (Kovas, Harlaar, Petrill, & Plomin, 2005). In the same sample at 10 years, the genetic correlation between mathematics and *g* was 0.68, while the genetic correlation between mathematics and reading was 0.73 (Davis, Kovas, Harlaar, Busfield, McMillan, Frances, et al., 2008). Overall, the results of several TEDS studies suggest that, to a large extent, the same genes and the same shared environments contribute to mathematics and aspects of reading and general intelligence – explaining most of the observed correlations among these traits. On the contrary, non-shared environmental overlap was very small across the measures, indicating the contribution of non-shared environments to differences among the measures (Davis et al., 2008; Haworth, Kovas, Dale, & Plomin, 2008; Kovas et al., 2005).

The etiology of the relationship between mathematics and spatial abilities (e.g. Wai, Lubinski, & Benbow, 2009; Webb, Lubinski, & Benbow, 2007) has been also examined using data collected in TEDS at 12 years. The study investigated the relationship between three components of mathematics ('Understanding Numbers', 'Non-Numerical Processes', and 'Computation and Knowledge') and spatial abilities measured with two tests (Tosto, Hanscombe, Haworth, Davis, Petrill, Dale, et al., 2014a). Overall, spatial abilities showed modest (0.26) heritability, with no sex differences in the etiology of individual differences. The phenotypic correlation between spatial abilities and the three mathematical components was on average 0.43. Common genetic factors explained ~ 60% of the correlation between the three components of mathematics and spatial ability; shared and non-shared environmental factors explained 26% and 14% of the phenotypic correlation, respectively. The computational components of mathematics showed less genetic influence in common with spatial abilities (genetic correlation 0.66) compared with processes such as symmetry or rotation (genetic correlation of 0.91 between spatial ability and 'Non-Numerical Processes'). These results suggest that although mostly the same genes contribute to spatial and mathematical skills (genetic correlation between the composite of the three mathematical components and spatial ability was 0.75), individual differences in spatial abilities are largely driven by environmental influences, of the non-shared type. The extent to which mathematics and spatial abilities co-vary was explained by shared and non-shared environmental factors, as well as genetic factors.

In the TEDS sample multivariate genetic analyses have also been applied to investigate the etiology of the links among different aspects of mathematics. At age 10 years, phenotypic correlations among five different aspects of mathematics ('Mathematical Application,' 'Understanding Number,' 'Computation and Knowledge,' 'Mathematical Interpretation,' 'Non-Numerical Processes') ranged from 0.45 ('Computation and Knowledge' and 'Non-Numerical Processes') to 0.68 ('Mathematical Application' and 'Understanding Number'; Kovas et al., 2007d). On average, the genetic correlation among the five sub-tests was 0.91, indicating that the same genetic influences affect these different aspects of mathematics. For example, the genetic correlation between 'Understanding Number' and 'Mathematical Application' was 0.94, meaning that the genetic influences involved in these two mathematical components were almost the same. The examination of the bivariate heritability and enviromentalities suggested that the observed covariation among different aspects of mathematics is largely explained by genetic factors. For example, the bivariate heritability of 'Understanding Number' and 'Mathematical Application' was 0.49 indicating that half of the phenotypic correlation of 0.68 is genetically mediated. The bivariate shared environment explained 29% of the phenotypic correlation, while 22% was explained by non-shared environment. Overall, these results suggest that observed correlations among different components of the mathematical domain are largely explained by the same genetic factors affecting them. The dissociation in performance across the different components largely stems from component-specific influences of non-shared environments.

To summarise, multivariate results suggest that genetic influences on individual differences in mathematics are largely the same as those on a wide range of other cognitive and learning abilities, achievement, and motivation, supporting the Generalist Genes Hypothesis (Plomin & Kovas, 2005). According to this hypothesis, if a gene is involved in one domain (e.g. mathematics), the same gene is also likely to be associated with other abilities, such as spatial skills, language, *g*, and reading. On the contrary, most of the environmental effects on mathematics are not shared with other domains, suggesting that discrepancies in abilities largely stem from the influence of different environments (e.g. Davis et al., 2008).

Despite the large amount of genetic overlap, some genetic effects seem specific to mathematics, indicating that some genes are uniquely involved with mathematics (Kovas et al., 2005; Hart et al., 2009). It is possible that this genetic specificity is related to some specific aspect of mathematically-relevant cognition. One cognitive ability, assumed to be etiologically and behaviorally associated with mathematical skills, is estimation (e.g. Halberda, Mazzocco, & Feigenson, 2008; Siegler & Opfer, 2003). It is thought that estimation abilities may be evolutionarily conserved, as they are present in many animal species and in infants (e.g. Agrillo, Piffer, & Bisazza, 2010; Feigenson, Carey, & Spelke, 2002; Xu & Arriaga, 2007). In other words, something in the genetic code of humans and other animals enables the appreciation of approximate numerical information. For this reason, it has been hypothesised that individual differences in estimation abilities may also be largely driven by genetic factors. Moreover, it has been hypothesised that these genetic factors may be uniquely associated with mathematical variation, explaining the portion of genetic effects in mathematics that is not shared with other abilities (Tosto, Petrill, Halberda, Trzaskowski, Tikhomirova, Bogdanova, et al., 2014b). In order to test this hypothesis, over 3,400 twin pairs from TEDS were assessed on a large battery of mathematics-related tests at 16 years of age. The battery included measures categorised as *Number Sense:* estimation of numerical magnitude and estimation of numerosities (Tosto, et al., 2014b). Contrary to the predictions, these estimation measures (e.g. comparison of large numerosities) revealed modest genetic influences, indicating that most of the individual differences in estimation skills (~70%) are influenced by non-shared environmental factors.

It may seem puzzling that such skills, considered to be hardwired in the human biology, show such low heritability estimates. However, entirely different factors may explain the species-universal behavior (e.g. humans are capable of quantity discrimination) and individual differences in behavior (humans differ in precision of their quantity discrimination). From this first large-scale study examining genetic and environmental etiology of the individual variation in estimation, it appears that genes have only a modest effect on it. However, the multivariate genetic analysis showed that the phenotypic correlation between mathematics at 16 and two Number Sense measures (0.40 for number line and 0.29 for estimation of large numerosities) was largely explained by genetic factors (average bivariate heritability 0.73). This suggested that the modest genetic influences on Number Sense abilities were to a large extent the same as those on mathematical ability at this age (results available from the authors). The results also showed that there were no genetic influences shared between number sense and mathematics independently from *g*, suggesting that the relationship between number sense and mathematics is mediated by *g*. This first genetically sensitive investigation into the relationship between number sense and mathematics suggests that the mathematic-specific genetic influences do not refer to number sense; therefore, the question on the etiology of these mathematics specific influences is still open. More research is needed in order to examine whether these common number sense-mathematics genetic factor genes have more general effects on other cognitive traits.

Genetic and Environmental Influences on Stability and Change

Multivariate analyses have also been applied to the examination of the sources of phenotypic stability and change in mathematical ability and performance. As these analyses rely on large longitudinal samples, only a handful of studies to date have addressed the extent to which the same genetic and environmental factors drive variation across different ages (e.g. whether genetic and environmental influences on mathematics at one age are the same that influence mathematics at a different age). It is reasonable to expect that largely the same genes and environments will be contributing to mathematical learning across development. On the other hand, it is also reasonable to hypothesize that new genetic and environmental influences contribute to mathematical learning across development – reflecting the growing complexity of material, conceptual shifts, and maturational factors. For example, between the ages of 7 and 9, children in UK schools progress to a different stage in their formal mathematical education, leading to greater focus on children's ability to understand and manipulate mathematical concepts. In addition, genetic and environmental factors influencing individual variation in biological maturation, such as the onset of puberty, may also contribute to variation in cognition at different ages.

In order to investigate the extent of etiological change and continuity, one study examined mathematical performance in the same children at age 7 and 9, as part of the TEDS study (Haworth, Kovas, Petrill, & Plomin, 2007a). The results showed that genetic influences on mathematical variation at the two ages were largely the same, accounting for 80% of the 0.60 phenotypic correlation between mathematics at 7 and 9. In other words, genetic influences contributed to the stability of mathematical abilities at the two ages, although some genetic effects were specific to each age. Shared environmental influences were almost overlapping (0.87), but contributed very little (22%) to the total phenotypic correlation between mathematics at 7 and 9 years. The non-shared environments at 7 and 9 years were different and contributed to the changes in mathematical performance between the two ages. The extent to which children carried on in their mathematical learning from 7 to 9 was mostly due to genetic influences, as is the case for many other cognitive abilities (e.g. Bartels, Rietveld, van Baal, & Boomsma, 2002; Kovas et al., 2007a). A second investigation into the etiology of the stability and change in mathematical performance in TEDS looked at school ages of 7, 9, and 10 together (Kovas et al., 2007a). This study showed that almost half of the genetic variance at 10 years was shared with the genetic variance at 7 years, indicating that almost half of the genetic influences active at 7 were still influencing mathematics at 10. Some of the genetic influences were, however, time specific, meaning that some new genetic influences emerged respectively at 9 and 10 years. Some of this new genetic variance at 9 was shared with the genetic variance at 10, contributing to mathematics stability. Shared environment also contributed to continuity of mathematical performance across the three time points (shared environments influencing one age also influenced the other two ages to a large extent, although the overall influence of shared environment on all traits was very modest). Again, non-shared environments contributed mostly to change, meaning that they were specific to each age.

To summarise, longitudinal multivariate studies suggest that phenotypic stability in mathematical ability and other cognitive traits can largely be explained by the contribution of the same genes (e.g. Plomin, Owen, & McGuffin, 1994a; Bartels et al., 2002; Petrill, Lipton, Hewitt, Plomin, Cherny, Corley, et al., 2004). The continuity of genetic effects on mathematics at different ages implies that the same genetic factors are needed to support the complex cognitive functions required for mathematical reasoning across development. Many of these genetic influences also affect other cognitive and learning domains. At the same time, discontinuity of genetic effects may reflect changes in the phenotype (what we call mathematics and measure at different ages), as well as developmental changes in genetic expression – associated with hormonal and environmental changes.

The Nature of The Relationship between Mathematical Ability and Disability

So far this chapter has dealt with normal variation in mathematical ability. Another question to be addressed is whether mathematical disabilities and high abilities are influenced by the same genetic and environmental factors responsible for the normal variation in mathematics, or whether disabilities and exceptionally high performance are etiologically different from normal ability (qualitative differences). If the latter hypothesis is true, twin model fitting analyses should show qualitative differences in genetic and environmental factors that drive normal variation and extremely low or high performance.

Indeed, some mathematical impairment may stem from a particular genetic mutation or a particular case of severe environmental deprivation. For example, several single gene abnormalities, such as Prader–Willi Syndrome, Turner Syndrome, and others, are associated with disproportionate mathematical problems (e.g. Bertella, Girelli, Grugni, Marchi, Molinari, & Semenza, 2005; Mazzocco, 1998; Murphy & Mazzocco, 2008). The extent to which the normal variation in the genetic regions associated with these syndromes is also associated with normal variation in mathematics is currently unknown. However, quantitative genetic research, described in this section, suggests that most mathematical disability (and high performance) seem to lie on the same etiological continuum as the rest of the variation (Plomin, Haworth, & Davis, 2009).

Only a handful of studies have examined the etiology of low mathematical performance. One early study (Alarcón, DeFries, Light, & Pennington, 1997) investigated the etiology of low mathematical ability, finding a probandwise concordance of 0.73 for MZ, and 0.56 for DZ twins, suggesting moderate genetic influences for the low mathematical performance group. Similarly, in the TEDS sample at 7 years, low mathematical ability was found to be as heritable as normal mathematical ability; in fact, the heritability of normal and low ability was almost identical (group heritability for the low group was 0.65, and heritability was 0.66 in the unselected sample; Oliver et al., 2004). In TEDS at 9 years, the low ability children, selected with a 15% cut-off criteria, showed a group heritability very similar to the heritability of the whole distribution (0.75 low; 0.68 normal variation), again suggesting the same etiology for low and normal ability (Haworth et al., 2007a). In TEDS, low and normal ability groups at age 10 also showed similar heritability estimates in the three sub-components of mathematics 'Understanding Number,' 'Non-Numerical Processes,' 'Computation and Knowledge' (with moderate genetic influences and modest shared environment; Kovas et al., 2007b). However, all three measures suggested stronger genetic influences for low performance than for the normal range (normal ~0.40 vs low ~0.62), suggesting that the low end of the distribution may be more affected by genetic influences. It is possible that some genetic effects only operate in the low end, rather than contributing to the normal variation across all levels of ability.

Low mathematical performance, like mathematical performance in the normal range, has been shown to be affected by similarly stable genetic factors. In a longitudinal investigation of low mathematical ability in TEDS, low ability children were identified with scores below the 15th percentile (Haworth et al., 2007a). The low group heritability was 0.65 at the age of 7 and 0.75 at age 9. These estimates were very similar to the heritability in the whole sample (0.66 at age 7 and 0.68 at the age of 9). The genetic correlation of the low groups at the two ages was very high suggesting that the same genetic influences are involved in low abilities at the ages of 7 and 9.

The etiology of high mathematical ability also seems to be strongly related to the etiology of normal performance. At 10 years, high ability children in the TEDS sample were identified as scoring above the 85th percentile in web-administered mathematical tests (Petrill, Kovas, Hart, Thompson, & Plomin, 2009). Group heritability (0.53) in the high ability sample was similar to that in the normal distribution (0.49). Furthermore, no sex differences were found in the etiology of high mathematical abilities at this age. In other words, the genetic and environmental factors that contribute to individual differences in boys in the high ability group, also contribute to individual differences among high-mathematically achieving girls. The etiology of high mathematical abilities was again investigated in TEDS at age 12 (Haworth, Dale, & Plomin, 2009b). The results confirmed again that largely the same genetic influences affect high and normal performance, although some genetic influences are specific for high abilities only.

To summarise, research suggests that mathematical achievement in different ranges of ability is largely influenced by the same genetic factors (Plomin & Kovas, 2005; Kovas, Haworth, Harlaar, Petrill, Dale, & Plomin, 2007c; Haworth et al. 2009a,b; Petrill et al., 2009). More genetically sensitive research is needed into the etiology of mathematical giftedness, including measures that allow high levels of sensitivity – to detect and examine variation at the top end.

Several twin studies examined the relationship between mathematical disabilities and other cognitive disabilities. This research suggests that to some extent mathematics and other learning disabilities are driven by the same genetic and shared environmental factors. For example, the relationship between reading and mathematical disability was addressed in a study of 148 MZ and 111 DZ school-age twins, specifically selected for at least one of the twins in the pair displaying reading problems (Light & DeFries, 1995). The probandwise concordance for reading was 68% for MZ and 40% for DZ twins. Moreover, although the sample was not specifically selected for mathematical disability, it was found that 49% of the MZ and 32% of DZ co-twins displayed mathematical disability. Further analyses suggested that the observed cross-trait concordance was explained by both common genetic and common shared environmental influences on reading and mathematics. The authors concluded that approximately 26% of the proband reading deficits in this sample were due to genetic factors that also influence mathematics performance. Approximately 25% of the proband reading deficit was due to environmental factors also affecting mathematics.

Another study examined the nature of the covariation between reading and mathematical disability in a sample of 8–20-year-old twins, and found that the observed relationship between the two traits was substantially explained by genetic factors (Knopik, Alarcón, & DeFries, 1997). In a study of over 1,500 17–18-year-old twin pairs, unselected for disability, children with low reading and low mathematics were identified (Markowitz, Willemsen, Trumbetta, van Beijsterveldt, & Boomsma, 2005). Both mathematical and reading problems were highly heritable (~0.90), with a large degree of genetic overlap between the two traits: 64% of the genetic factors influencing mathematical disability also influenced reading disability. Moreover, the observed concordance between mathematical and reading disability in this sample was almost entirely explained by common genes.

In the TEDS sample, at 10 years, low achieving children in reading and mathematics were selected in the lowest 15% of the distribution (Kovas et al., 2007c). Reading and mathematical disability were moderately heritable and most of the covariation (63%) between reading and mathematics was genetically mediated. The shared environmental influences almost completely overlapped (0.96), although shared environment contributed only a very small amount to the observed covariation of the two disabilities as the influence of the shared environment on each trait was very small.

Very few studies examined the relationship between mathematical disabilities and other traits, such as low language, low general cognitive abilities (*g*) and psychopathologies. In the TEDS sample, over 4,000 twin pairs were assessed at 12 years, with the bottom 15% of the sample selected for low performance in mathematics, language, reading, and *g* (Haworth et al., 2009a). As expected, there was a significant amount of overlap among the measures (average observed comorbidity 0.58). The genetic correlations for the low performance groups in all measures were also high (0.67 average genetic correlation for mathematics, language, reading, and *g*). This genetic correlation was similar in magnitude to the average genetic correlation among the measures for the entire unselected sample (0.68) indicating similar etiology for disabilities and abilities in mathematics, language, reading, and *g*. The comorbidity between disabilities was largely mediated by genetic factors. A recent study that used TEDS data collected at 12 years, investigated the association between mathematics and attention-deficit/hyperactivity disorder (ADHD) separately for the hyperactive–impulsive and the inattentive dimensions. Mathematical abilities showed stronger phenotypic and genetic correlations with inattentiveness (phenotypic = -0.26, genetic= -0.41) compared with hyperactivity-impulsivity (phenotypic = -0.18, genetic = -0.22). Moreover, the genetic correlation between mathematics and inattentiveness was largely independent from hyperactivity-impulsivity suggesting that mathematical abilities show an association with ADHD because of a common genetic etiology with inattentiveness rather than with hyperactivity-impulsivity (Greven, Kovas, Willcutt, Petrill, & Plomin, 2014).

The relationship between high mathematical abilities and other cognitive abilities was investigated in TEDS at 12 years (Haworth et al., 2009b). High ability twins were identified as performing in the top 15% in mathematics, reading, language, and *g*. The observed overlap among the traits was high (0.59 on average). Similarly to low ability, there was substantial genetic and shared environmental overlap across all high abilities. As is the case for low abilities, the non-shared environment contributed to discrepancies in performance across abilities. A substantial portion of the phenotypic correlation among the high ability scores was explained by genetic and shared environmental influences. However, the contribution of the genetic effects to the phenotypic correlation in the high abilities was lower compared with low abilities (0.42 on average). For example, common genetic factors explained only 34% of the observed overlap between high mathematics and high *g*, as compared with explaining 58% of the observed overlap between low mathematics and *g*.

These studies suggest that the comorbidity between mathematics, and other low and high abilities is mediated by genetic and shared environmental factors, with some inconsistencies across the studies in the degree to which genes and shared environment contribute to this covariation. These inconsistencies may stem from the differences in samples, such as age or cultural differences, and need to be investigated further.

Overall, the research so far suggests that cognitive traits, such as mathematics, are polygenic – influenced by many genes of small effects (Plomin et al., 2013). These genes are called *Quantitative Trait Loci* (QTL; Plomin et al., 2013). For complex or quantitative traits, it appears that a disorder is also the result of the effect of many genes – each having only a small effect on the trait (Plomin & Kovas, 2005). This implies that genetic variants (the QTLs) contribute to the disorder or disability in a cumulative/quantitative way, producing a continuum of ability, rather than discrete ability groups. Therefore, for complex traits, disability can be defined as driven by an accumulation of risk factors, both genetic and environmental. In other words high ability and disability can be seen as the high and low end of the same distribution (Kovas & Plomin, 2006; Plomin & Kovas, 2005; Plomin, Pedersen, Lichtenstein, & McClearn, 1994b). Moreover, it is likely that a unique combination of polygenic and environmental effects leads to a particular observed performance profile in each person.

The Etiology of Sex Differences in Mathematics

Many psychological and educational studies have investigated sex differences in mathematical ability across development and across different aspects of mathematics (e.g. Benbow & Stanley, 1980; Lord & Clausen-May, 2003). In line with these studies, in the school years, males in the TEDS sample tend to have higher mathematical scores than females, although the effect size of these average differences is very small (e.g. Kovas et al., 2007a). Several reasons have been proposed for any observed average sex differences (Kovas et al., 2007a,b).

However, the sources of average group differences may differ from those driving individual variation within the groups. Quantitative genetic sex-limitation models examine the extent to which the same genetic and environmental factors contribute to variation within the sex groups (testing for qualitative differences); and whether these factors have the same effect on variation in each group (testing for quantitative differences). Because the genetic relatedness of DZ twins is 0.5 on average, if different genes influence males and females, the genetic correlation of DZ opposite sex pairs should be less than 0.5. If the genes influencing variation in males and females are the same, but the extent to which the genes influence variation is different, the estimates of genetic, shared and non-shared environment will be different for male and female pairs. Applying sex-limitation models to TEDS data at the age of 10, revealed no qualitative or quantitative sex differences in the etiology of mathematical abilities or disabilities (Kovas et al., 2007b). Furthermore, no sex differences were found in the three components of mathematics ('Understanding Number', 'Non-Numerical Processes', 'Computation and Knowledge'), both in the low ability group and in the unselected sample (Kovas et al., 2007b). At age 16, when the TEDS twins were assessed on a large battery of numerical tests, again no qualitative or quantitative sex differences emerged for mathematical performance (results available from the authors). Similarly, no etiological sex differences were found for parent-rated normal mathematical variation or low mathematical performance in a sample of 17-18-year-old Dutch twins (Markowitz et al., 2005). These findings suggest that genetic factors that make males better or worse at mathematics are the same genetic factors that make females better or worse at mathematics; and they exert the same amount of influence on males and females. The same studies also indicate equality of environmental effects on male and female variation.

Molecular Genetic Investigations into Mathematical Variation

Quantitative genetic investigations into the etiology of mathematical variation suggest that mathematics is highly heritable. Molecular genetic studies have begun to search for specific genes associated with mathematical abilities and disabilities. Approximately 99% of human DNA is invariant, meaning that out of the 3 billion base-pairs that form our DNA chain, only a small proportion differs across people. However, in a stretch of invariant DNA sequence, there may be a particular locus where some individuals may have, for example, the nucleotide adenine, while others have guanine. These variants are called single nucleotide polymorphisms (SNPs) and are the most common type of DNA variability. More than 3 million common SNPs have been discovered in human DNA, providing the basis for the quantitative trait locus (QTL) model described previously. Mapping phenotypic variation in complex traits, such as mathematical ability and disability, to QTLs is complicated by the fact that many DNA loci of very small effect are likely to be contributing to the variation.

In recent years, molecular genetic methods and technologies have advanced dramatically (Plomin & Davis, 2009; Plomin et al., 2013). The advent of microarray ('gene-chip') technology has revolutionised molecular genetic research, by way of tagging almost all of the common segregating polymorphic DNA sequence at the same time. Moreover, genetic variance not directly tagged by these chips can be reliably estimated through bio-informatic methods, such as imputation that considers the correlation between SNPs. Genome Wide Association Studies (GWAS) use this technology to comprehensively genotype large numbers of individuals for a large number of SNPs, and to look for associations between these genetic markers and variation in complex traits.

While a number of molecular genetic studies have been conducted on cognitive abilities, only one has focused on mathematics. This GWAS set out to identify genes associated with mathematical performance and achievement, comparing 10-year-old children selected from the TEDS sample for high and low mathematical performance (Docherty, Davis, Kovas, Meaburn, Dale, Petrill, et al., 2010a). The study adopted a cost-efficient method known as DNA pooling (Butcher, Meaburn, Liu, Fernandes, Hill, Al-Chalabi, et al., 2004; Davis, Plomin, & Schalkwyk, 2009c) to select the top-ranking 3000 SNPs with the highest probability of genetic association with mathematical performance. As a second stage, the top 43 SNPs were validated in an individual genotyped sample (2,356 TEDS individuals) and carried forward for association testing. Out of these 43 SNPs, 10 were found to be significantly associated with the phenotypic variance in this sample. The study revealed a number of functionally plausible candidate genes associated with mathematical performance. These include the *NRCAM* gene, which encodes a neuronal cell adhesion molecule and is implicated in memory processes (Hoffman, 1998). Taken individually, each of the 10 SNPs accounted for a very small amount of the variance in mathematics (0.13% to 0.58%); however, the 10 SNPs were found to work in an additive way, collectively explaining 2.9% of the total variance in individual differences in mathematics. These results are consistent with the QTL hypothesis, which suggests that the trait is influenced by many genes of small effect, leading to the observed continua of ability/performance. For a single molecular study, finding this amount of variance explained is very encouraging, considering that for cognitive traits, GWAS studies typically report considerably less variance explained. For example, a study (Davies, Tenesa, Payton, Yang, Harris, Liewald, et al., 2011) revealed that the SNPs found so far to be associated with intelligence explained only approximately 1% of individual variance in crystallised and fluid intelligence.

According to the Generalist Genes Hypothesis, genes implicated in one cognitive trait are likely to be implicated in other cognitive traits. Some support for this hypothesis has already been found. One study screened 100,000 SNPs in 4,258 TEDS and found that 10 SNPs were associated with early reading ability at age 7 (Haworth et al., 2007b). The same SNPs were also found to be associated with language, *g*, and 3 aspects of mathematical performance at the same age. The correlations between the genes found for reading and other cognitive abilities were very small (between 0.08 and 0.11), but statistically significant. To further test the generalist genes hypothesis, the 43 SNPs found to be associated with mathematics and the subset of 10 SNPs significantly associated with mathematics at 10 years were used in a follow-up behavioral genomic analysis and tested for association with three different components of mathematics, reading and *g* at age 10 (Docherty, Kovas, Petrill, & Plomin, 2010b). Both SNP-sets significantly correlated with all three mathematical components, with different degrees of magnitude. For example the web-test assessing 'Non-Numerical Processes' had the lowest correlation with both SNP-sets (0.11 with the 10 SNPs and 0.11 with 43 SNPs). The highest correlation was between the 10-SNP set and the web-test 'Understanding Number' (0.15). As predicted by the Generalist Genes Hypothesis (Kovas & Plomin, 2005), the sets also correlated with reading and *g* measured at the same age, with similar magnitude of effects (e.g. 0.11 correlation between 10-SNP set and reading). The same study found that the mathematics SNP sets identified at age 10 also explained variation in mathematics at ages 7, 9, and 12 years (correlations between 0.03 and 0.10). This last finding further supports the continuity of mathematical abilities throughout development, but also highlights some specificity of genetic influences at different ages, as predicted by the findings from quantitative genetic studies.

As described earlier, quantitative genetic findings are characterised by some degree of variability in heritability estimates, with differences in methods, measures, and samples potentially explaining the discrepancies in the literature. It is also possible that heritability estimates may change as a result of different environmental influences. In other words, the impact of genetic variation on a particular trait is dependent on the environmental background in which the genes operate. Heritability estimates may differ for mathematical abilities in populations exposed to different schooling, parenting, and socio-economic factors. For example, where the relevant environmental influences are more variable, heritability estimates may be lower. Once the mechanisms through which genes contribute to variation in complex traits are better understood, we will be able to use environmental interventions to modulate genetic propensities. To date, only one study examined gene-environment interactions in their effect on mathematical ability, testing the association between mathematics and the previously identified 10 SNPs, as a function of 10 environmental measures (Docherty, Kovas, & Plomin, 2011). The study used a sub-set of 1,888 TEDS children with complete genotype data for all 10 SNPs. Each SNP was assigned a value of 0 for the homozygote genotype (when the two genetic variants in the locus are of the same form) associated with low mathematical ability, a value of 1 was assigned for the heterozygote genotype and 2 for the homozygote genotype associated with higher mathematical ability. The values of the 10 SNPs were summed to obtain a global value for the entire 10 SNP set (ranging from 0 to 20). A SNP set with a low value indicated that the majority of the alleles across the 10 SNPs were associated with poorer mathematical performance. Environmental measures, assessed at the ages of 9 and 12 years, included socio-economic status, parental negativity, parental discipline, mathematical environment (mathematical relevant activities carried out in schools), home and classroom chaos (the level of environmental noise, confusion and disruption of routines), and children's perceptions of their teacher. Overall, the association between mathematics and the 10 SNPs was stronger in the condition of a greater home chaos and when the parents' negativity rate was high. However, the effect size of these interactions was small: after correcting for multiple testing, the interactions of the 10 SNP set with home chaos and parent negativity at the age of 12 explained only 0.49% each of the variance in mathematical scores. These results suggest that at the age of 12, the greatest genetic effects of the low mathematics SNP sets occurred when the parents' negativity and home chaos were high. In other words, the negative environments made the effects of the 'bad' genotype worse. In line with the diathesis-stress model (e.g. Asbury, Wachs, & Plomin, 2005), these results indicate the presence of a gene-environment interaction. Simply, taken in isolation, genetic risk factors may not be sufficient to fully account for poor mathematics. Genes are expressed in response to the environment. Chaos in the home environment and parents' negativity appear to trigger mathematical genetic risk response. Other types of gene-environment interplay, yet to be investigated, may also be important for mathematical learning.

These first genetic polymorphisms represent a promising beginning in the search for DNA influences on mathematics. The finding of several SNPs of additive small effect is in line with the predictions made by quantitative genetic studies of mathematical variation. The first 10 SNPs identified for mathematical ability need to be replicated in other samples, and further research is required in order to find many other SNPs and other types of DNA variation involved in the development of individual differences in mathematics.

Conclusions

The field of quantitative genetics has made important contributions to our understanding of the etiology of mathematical variation, its stability, and its relationship with other cognitive traits. Consistent with the QTL model, individual variation in mathematical ability depends on the influence of many genes, each of small effect. Genetic factors contributing to mathematical performance across the range of the distribution are mostly the same, suggesting that common mathematical disability is not a product of some abnormal genetic mutation, but rather stems from the accumulation of many ability-decreasing genetic variants that also operate throughout the ability range. The implication of these findings is that there are no genes for 'bad' mathematics or low abilities. From a genetic perspective, abilities and disabilities are not qualitatively different as they emerge from the additive effects of the same set of genes. Moreover, genetic variation is the foundation for the ability of humans to adapt to new environmental demands. Genetic variants that, under certain environmental conditions, today contribute to lower mathematical ability may become important for adaptation in other environments or with new societal demands. This means that molecular genetic research into complex traits does not aim to reduce genetic variance among people. On the contrary, the ultimate aim of genetic and genomic investigations is to optimise the current environments and to create new environments (e.g. individualised educational programmes) in order to capitalise on existing genetic variation.

Genes do not act in isolation, and although mathematics is a highly heritable trait, around 40% of individual differences in mathematics are due to the environment. Moreover, new evidence suggests that the effects of genes can be modulated by environments. The challenging task of molecular genetics will be to identify all the genes contributing to mathematical heritability at different ages. Quantitative genetic studies will continue to guide the molecular search for these genes. For example, the absence of sex differences in the etiology of individual differences in mathematics justifies combining the data from males and females in the search for genetic associations. New advances in molecular technologies, such as the ability to conduct complete genome sequencing on each individual, will lead to new discoveries, including the investigation of the role of rare genetic variants in mathematical variation. Better understanding of genetic effects on mathematical variation may help in the search for the relevant environments. Genetically-sensitive cross-cultural comparisons will also aid our understanding of the complex ways in which genetic and environmental effects interact. The exact applications of this future knowledge are currently difficult to predict. However, we believe that the way we conceptualise and influence mathematical development will be affected by continuous advances from the field of behavioral genetics.

Acknowledgments

The authors gratefully acknowledge the on-going contribution of the participants in the Twins Early Development Study (TEDS) and their families. TEDS is supported by a program grant [G0901245; and previously G0500079] from the UK Medical Research Council; our work on environments and academic achievement is also supported by grants from the US National Institutes of Health [HD44454, HD46167 and HD059215], YK's and MGT's research is supported by a grant from the Government of the Russian Federation [11. G34.31.0043]. CMAH is supported by a research fellowship from the British Academy.

References

Agrillo, C., Piffer, L., & Bisazza, A. (2010). Number versus continuous quantity in numerosity judgments by fish. *Cognition*, 119, 281–287.

Alarcón, M., DeFries, J.C., Light, J. G., & Pennington, B. F. (1997). A twin study of mathematics disability. *Journal of Learning Disabilities*, 30, 617–623.

Alarcón, M., Knopik, V.S., & DeFries, J.C. (2000). Covariation of mathematics achievement and general cognitive ability. *Journal of School Psychology*, 38, 63–77.

Amato, P.R. (2000). The consequences of divorce for adults and children. *Journal of Marriage and Family*, 62(4), 1269–1287.

Asbury, K., Wachs, T.D., & Plomin, R. (2005). Environmental moderators of genetic influence on verbal and nonverbal abilities in early childhood. *Intelligence*, 33(6), 643–661.

Bartels, M., Rietveld, M.J., van Baal, G.C., & Boomsma, D.I. (2002). Genetic and environmental influences on the development of intelligence. *Behavior Genetics*, 32, 237–249.

Benbow, C.P., & Stanley, J.C. (1980). Sex differences in mathematical ability: fact or artifact? *Science*, 210, 1262–1264.

Bertella, L., Girelli, L., Grugni, G., Marchi, S., Molinari, E., & Semenza, C. (2005). Mathematical skills in Prader–Willi Syndrome. *Journal of Intellectual Disability Research*, 49(2), 159–169.

Bruder, C.E., Piotrowski, A., Gijsbers, A.A.C.J., Andersson, R., Erickson, S., Diaz de Ståhl, T., Menze, U., Sandgren, J., von Tell, D., Poplawski, A., Crowley, M., Crasto, C., Partridge, E.C., Tiwari, H., Allison, D.B., Komorowski, J., van Ommen, G-J.B., Boomsma, D.I., Pedersen, N.L., den Dunnen, J.T., Wirdefeldt, K., & Dumanski, J.P. (2008). Phenotypically concordant and discordant monozygotic twins display different DNA copy-number-variation profiles. *American Journal of Human Genetics*, 82, 763–771.

Butcher, L.M., Meaburn, E., Liu, L., Fernandes, C., Hill, L., Al-Chalabi, A., Plomin, R., Schalkwyk, L.C., & Craig, I.W. (2004). Genotyping pooled DNA on microarrays: a systematic genome screen of thousands of SNPs in large samples to detect QTLs for complex traits. *Behavioral Genetics*, 34, 549–555.

Byrne, B., Samuelsson, S., Wadsworth, S., Hulslander, J., Corley, R., DeFries, J. C., Quain, P., Willcutt, E. G., & Olson, R. K. (2006). Longitudinal twin study of early literacy development: preschool through Grade 1. *Reading and Writing: An Interdisciplinary Journal*, 20, 77–102.

Byrne, B., Wadsworth, S., Corley, R., Samuelsson, S., Quain, P., DeFries, J. C., Willcutt, E. G. & Olson, R. K. (2005). Longitudinal twin study of early literacy development: preschool and kindergarten phases. *Scientific Studies of Reading*, 9, 219–235.

Christensen, K., Petersen, I., Skytthe, A., Herskind, A.M., McGue, M., & Bingley, P. (2006). Comparison of academic performance of twins and singletons in adolescence: follow-up study. *British Medical Journal*, 333, 1095.

Cordell, H.J. (2002). Epistasis: what it means, what it doesn't mean, and statistical methods to detect it in humans. *Human Molecular Genetics*, 11(20), 2463–2468.

Darling-Hammond, L. (2000). Teacher quality and student achievement: a review of State Policy evidence. *Education Policy Analysis Archives, North America*, 8, Available at: http://epaa.asu.edu/ojs/article/view/392 (Accessed 12 April 2012).

Davies, G., Tenesa, A., Payton, A., Yang, J., Harris, S. E., Liewald, D., Ke, X., Le Hellard, S., Christoforou, A., Luciano, M., McGhee, K., Lopez, L., Gow, A.J., Corley, J., Redmond, P., Fox, H.C., Haggarty, P., Whalley, L.J., McNeill, G., Goddard, M.E., Espeseth, T., Lundervold, A.J., Reinvang, I., Pickles, A., Steen, V.M., Ollier, W., Porteous, D.J., Horan, M., Starr, J.M., Pendleton, N., Visscher, P.M., & Deary,

I.J. (2011). Genome-wide association studies establish that human intelligence is highly heritable and polygenic. *Molecular Psychiatry*, 16, 996–1005.

Davis, O.S.P., Haworth, C.M.A., & Plomin, R. (2009a). Dramatic increase in heritability of cognitive development from early to middle childhood an 8-year longitudinal study of 8,700 pairs of twins. *Psychological Science*, 20(10), 1301–1308.

Davis, O.S.P., Haworth, C.M.A., & Plomin, R. (2009b). Learning abilities and disabilities: generalist genes in early adolescence. *Cognitive Neuropsychiatry*, 14(4–5), 312–331.

Davis, O.S.P., Kovas, Y., Harlaar, N., Busfield, P., McMillan, A., Frances, J., Petrill, S.A., Dale, P.S., & t Plomin, R (2008). Generalist genes and the Internet generation: etiology of learning abilities by web testing at age 10. *Genes, Brain and Behavior*, 7, 455–462.

Davis, O.S.P., Plomin, R., & Schalkwyk, L.C. (2009c). The SNPMaP package for R: a framework for genome-wide association using DNA pooling on microarrays. *Bioinformatics*, 25, 281–283.

DeFries, J.C., & Fulker, D.W. (1985). Multiple regression analysis of twin data. *Behavior Genetics*, 15, 467–473.

Docherty, S.J., Davis, O.S.P., Kovas, Y., Meaburn, E.L., Dale, P.S., Petrill, S.A., Schalkwyk, L.C., & Plomin, R. (2010a). A genome-wide association study identifies multiple loci associated with mathematics ability and disability. *Genes, Brain and Behavior*, 9, 234–247.

Docherty, S.J., Kovas, Y., Petrill, S.A., & Plomin, R. (2010b). Generalist genes analysis of DNA markers associated with mathematical ability and disability reveals shared influences across ages and abilities. *BMC Genetics*, 11(61).

Docherty, S.J., Kovas, Y., & Plomin, R. (2011). Gene-environment interaction in the etiology of mathematical ability using SNP sets. *Behavioral Genetics*, 41, 141–154.

Dowker, A. (2005). Early identification and intervention for students with mathematics difficulties. *Journal of Learning Disabilities*, 38 (4), 324–332.

Dunn, J.F., & Plomin, R. (1990). Separate Lives: Why Siblings Are So Different. New York: Basic Books.

Eide, E., & Showalter, M.H. (1998). The effect of school quality on student performance: a quantile regression approach. *Economics Letters*, 58, 345–350.

Evans, D.M., & Martin, N.G. (2000). The validity of twin studies. *GeneScreen*, 1(2), 77–79.

Feigenson, L., Carey, S., & Spelke, E.S. (2002). Infants' discrimination of number vs. continuous extent. *Cognitive Psychology*, 44, 33–66.

Greven, C.U., Kovas, Y., Willcutt, E.G., Petrill, S.A., & Plomin, R. (2014). Evidence for shared genetic risk between ADHD symptoms and reduced mathematics ability: a twin study. *Journal of Child Psychology and Psychiatry*, 55(1), 39-48.

Grilo, C.M., & Pogue-Geile, M.F. (1991). The nature of environmental influences on weight and obesity: a behavior genetic analysis. *Psychological Bulletin*, 10, 520–537.

Gross, J., Hudson, C., & Price, D. (2009). The Long Term Costs of Numeracy Difficulties. London: Every Child a Chance Trust and KPMG.

Halberda, J., Mazzocco, M.M.M., & Feigenson, L. (2008). Individual differences in nonverbal estimation ability predict maths achievement. *Nature*, 455, 665–668.

Hart, S.A., Petrill, S.A., & Kamp Dush, C.M. (2010a). Genetic influences on language, reading, and mathematics skills in a national sample: an analysis using the National Longitudinal Survey of Youth. *Language, Speech, and Hearing Services in Schools*, 41, 118–128.

Hart, S.A., Petrill, S.A., & Thompson, L.A. (2010b). A factorial analysis of timed and untimed measures of mathematics and reading abilities in school aged twins. *Learning and Individual Differences*, 20, 63–69.

Hart, S.A., Petrill S.A., Thompson. L.A., & Plomin, R. (2009). The ABCs of math: a genetic analysis of mathematics and its links with reading ability and general cognitive ability. *Journal of Educational Psychology*, 101(2), 388–402.

Haworth, C.M.A., Asbury, K., Dale, P.S., & Plomin, R. (2011). Added value measures in education show genetic as well as environmental influence. *PloS One*, 6(2), e16006.

Haworth, C.M.A., Dale, P.S., & Plomin, R. (2009b). Generalist genes and high cognitive abilities. *Behavior Genetics*, 39, 437–445.

Haworth, C.M.A., Davis, O.S.P., & Plomin, R. (2013). Twins Early Development Study (TEDS): a genetically sensitive investigation of cognitive and behavioural development from childhood to young adulthood. *Twin Research and Human Genetics*, 16(1), 117–125.

Haworth, C.M.A., Kovas, Y., Dale, P.S., & Plomin, R. (2008). Science in elementary school: generalist genes and school environments. *Intelligence*, 36(6), 694–701.

Haworth, C.M.A., Kovas, Y., Harlaar, N., Hayiou-Thomas, M. E., Petrill, S. A., Dale, P. S., & Plomin, R. (2009a). Generalist genes and learning disabilities: a multivariate genetic analysis of low performance in reading, mathematics, language and general cognitive ability in a sample of 8000 12-year-old twins. *Journal of Child Psychology and Psychiatry*, 50(10), 1318–1325.

Haworth, C.M.A., Kovas, Y., Petrill, S.A., & Plomin, R. (2007a). Developmental origins of low mathematics performance and normal variation in twins from 7 to 9 years. *Twin Research and Human Genetics*, 10, 106–117.

Haworth, C.M.A., Meaburn, E.L., Harlaar, N., & Plomin, R. (2007b). Reading and generalist genes. *Mind, Brain & Education*, 1, 173–180.

Haworth, C.M.A., & Plomin, R. (2011). Genetics and education: toward a genetically sensitive classroom. In: K. R. Harris, S. Graham, and T. Urdan (Eds), APA Educational Psychology Handbook, Vol. 1: Theories, Constructs, and Critical Issues. Washington, DC: American Psychological Association.

Haworth, C.M.A., Wright, M.J., Luciano, M., Martin, N.G., de Geus, E.J.C., van Beijsterveldt, C.E.M., Bartels, M., Posthuma, D., Boomsma, D.I., Davis, O.S.P. Kovas, Y., Corley, R.P., DeFries, J.C., Hewitt, J.K., Olson, R.K., Rhea, S-A., Wadsworth, S.J., Iacono, W.G., McGue, M., Thompson, L.A., Hart, S.A., Petrill, S.A., Lubinski, D., & Plomin, R. (2010). The heritability of general cognitive ability increases linearly from childhood to young adulthood. *Molecular Psychiatry*, 15, 1112–1120.

Hetherington, E.M., & Clingempeel, W.G. (1992). Coping with marital transitions: a family systems perspective. *Monographs of the Society for Research in Child Development*, 2–3, Serial No. 227.

Hoffman, K.B. (1998). The relationship between adhesion molecules and neuronal plasticity. *Cellular and Molecular Neurobiology*, 18, 461–475.

Kovas, Y., Harlaar, N., Petrill, S.A., & Plomin, R. (2005). 'Generalist genes' and mathematics in 7-year-old twins. *Intelligence*, 33, 473–489.

Kovas, Y., Haworth, C.M.A., Dale, P.S., Plomin, R., Weinberg, R. A., Thomson, J. M., & Fischer, K. W. (2007a). The genetic and environmental origins of learning abilities and disabilities in the early school years. *Monographs of the Society for Research in Child Development*, 72(3), vii, 1–160.

Kovas, Y., Haworth, C.M.A., Petrill, S.A., & Plomin, R. (2007b). Mathematical ability of 10-year-old boys and girls: genetic and environmental etiology of typical and low performance. *Journal of Learning Disabilities*, 40(6), 554–567.

Kovas, Y., Haworth, C.M.A., Harlaar, N., Petrill, S.A., Dale, P.S., & Plomin, R. (2007c). Overlap and specificity of genetic and environmental influences on mathematics and reading disability in 10-year-old twins. *Journal of Child Psychology and Psychiatry*, 48, 914–922.

Kovas, Y., Petrill, S.A., & Plomin, R. (2007d). The origins of diverse domains of mathematics: generalist genes but specialist environments. *Journal of Educational Psychology*, 99(1), 128–139.

Kovas, Y., & Plomin, R. (2006). Generalist genes: implications for cognitive sciences. *Trends in Cognitive Science*, 10, 198–203.

Kovas, Y., & Plomin, R. (2012). Genetics and genomics: good, bad and ugly. In: S. Della Salla and M. Anderson (Eds), Neuroscience in Education. The Good, The Bad and The Ugly (pp. 155–173). Oxford: Oxford University Press.

Knopik, V.S., Alarcón, M., & DeFries, J.C. (1997). Comorbidity of mathematics and reading deficits: evidence for a genetic etiology. *Behavior Genetics*, 27, 447–453.

Light, J.G., & DeFries, J.C. (1995). Comorbidity of reading and mathematics disabilities: genetic and environmental etiologies. *Journal of Learning Disabilities*, 28, 96–106.

Light, J.G., DeFries, J.C., & Olson, R.K. (1998). Multivariate behavioral genetic analysis of achievement and cognitive measures in reading-disabled and control twin pairs. *Human Biology*, 70, 215–237.

Loehlin, J.C. (1996). The Cholesky approach: a cautionary note. *Behavior Genetics*, 26(1), 65–69.

Lord, T.R., & Clausen-May, T. (2003). Comparing performance of pupils with high spatial-low numerical and high numerical-low spatial scores on a standardized mathematics test in the United Kingdom. *Perceptual and Motor Skills*, 97, 83–96.

Luo, Y.L.L., Kovas, Y., Haworth, C.M., Dale, P.S., & Plomin, R. (2011). The etiology of mathematical self-evaluation and mathematics achievement: understanding the relationship using a cross-lagged twin study from ages 9 to 12. *Learning and Individual Differences*, 21, 710–718.

MacGillivray, I., Campbell, D.M., & Thompson, B. (1988). Twinning and Twins. Chichester: John Wiley & Sons.

Markowitz, E.M., Willemsen, G., Trumbetta, S.L., van Beijsterveldt, T.C.E.M., & Boomsma, D.I. (2005). The etiology of mathematical and reading (dis)ability covariation in a sample of Dutch twins. *Twin Research and Human Genetics*, 8, 585–593.

Marsh, H.W., Byrne, B.M., & Yeung, A.S. (1999). Causal ordering of academic self-concept and achievement: reanalysis of a pioneering study and revised recommendations. *Educational Psychologist*, 34(3), 155–167.

Marsh, H.W., & Yeung, A.S. (1997). Causal effects of academic self-concept on academic achievement: structural equation models of longitudinal data. *Journal of Educational Psychology*, 89(1), 41–54.

Martin, N.G., & Eaves, L.J. (1977). The genetical analysis of covariance structure. *Heredity*, 38, 79–95.

Martin, N.G., Eaves, L.J. Kearsey, M.J., & Davies P. (1978). The power of the classical twin study. *Heredity*, 40(1), 97–116.

Martin, N.G., Eaves, L.J., & Kendler, K.S. (1994). The power of the classical twin study to resolve variation in threshold traits. *Behavior Genetics*, 24(3), 239–258.

Martin, N.G., Jardine, R., & Eaves, L.J. (1984). Is there only one set of genes for different abilities? A reanalysis of the National Merit Scholarship Qualifying Tests (NMSQT) data. *Behavior Genetics*, 14, 355–370.

Mazzocco, M.M.M. (1998). A process approach to describing mathematics difficulties in girls with Turner syndrome. *Pediatrics*, 102(2 Pt 3), 492–496.

McGue, M., & Bouchard, T.J. (1984). Adjustment of twin data for the effects of age and sex. *Behavior Genetics*, 14, 325–343.

Murphy, M.M., & Mazzocco, M.M.M. (2008). Mathematics learning disabilities in girls with fragile X or Turner syndrome during late elementary school. *Journal of Learning Disabilities*, 41(1), 29–46.

Neale, M.C., & Maes, H.H.M. (2003). Methodology for Genetic Studies of Twins and Families. Dordrecht, Netherlands: Kluwer Academic Publishers B.V.

OECD. (2010). The High Cost of Low Educational Performance: The Long-Run Economic Impact of Improving Educational Outcomes. Paris: OECD.

Oliver, B., Harlaar, N., Hayiou-Thomas, M.E., Kovas, Y., Walker, S.O., Petrill, S.A., Spinath, F.M., Dale, P.S., &, Plomin, R. (2004). A twin study of teacher-reported mathematics performance and low performance in 7-year-olds. *Journal of Educational Psychology*, 96, 504–517.

Oliver, B.R., & Plomin, R. (2007). Twins' Early Development Study (TEDS): a multivariate, longitudinal genetic investigation of language, cognition and behavior problems from childhood through adolescence. *Twin Research and Human Genetics*, 10(1), 96–105.

Petrill, S.A., Kovas, Y., Hart, S.A., Thompson, L.A. & Plomin, R. (2009). The genetic and environmental etiology of high math performance in 10-year-old twins. *Behavior Genetics*, 39(4), 371–379.

Petrill, S.A., Lipton, P.A., Hewitt, J.K., Plomin, R., Cherny, S.S., Corley, R., & DeFries, J.C. (2004). Genetic and environmental contributions to general cognitive ability through the first 16 years of life. *Developmental Psychology*, 40, 805–812.

Plomin, R., & Daniels, D. (1987). Why are children in the same family so different from each other? *Behavioral and Brain Sciences*, 10, 1–16.

Plomin, R., & Davis, O.S.P. (2009). The future of genetics in psychology and psychiatry: microarrays, genome-wide association, and non-coding RNA. *Journal of Child Psychology and Psychiatry*, 50, 63–71.

Plomin, R., DeFries, J.C., Knopik, V.S., & Neiderhiser, J.M. (2013). Behavioral Genetics, 6th edn. New York: Worth.

Plomin, R., Haworth, C.M.A., & Davis, O.S.P. (2009). Common disorders are quantitative traits. *Nature Reviews Genetics*, 10, 872–878.

Plomin, R., & Kovas, Y. (2005). Generalist genes and learning disabilities. *Psychological Bulletin*, 131(4), 592–617.

Plomin, R., Owen, M.J., & McGuffin, P. (1994a). The genetic basis of complex human behaviors. *Science*, 264, 1733–1739.

Plomin, R., Pedersen, N.L., Lichtenstein, P., & McClearn, G.E. (1994b). Variability and stability in cognitive abilities are largely genetic later in life. *Behavior Genetics*, 24, 207–215.

Rijsdijk, F.V., & Sham, P.C. (2002). Analytic approaches to twin data using structural equation models. *Briefings in Bioinformatics*, 3(2), 119–133.

Ronalds, G.A., De Stavola, B.L., & Leon, D.A. (2005). The cognitive cost of being a twin: evidence from comparisons within families in the Aberdeen children of the 1950s cohort study. *British Medical Journal*, 331(7528), 1306.

Saxe, G.B. Gearhart, M., & Nasir, N.S. (2001). Enhancing students' understanding of mathematics: a study of three contrasting approaches to professional support. *Journal of Mathematics Teacher Education*, 4(1) 55–79.

Siegler, R.S., & Opfer, J. (2003). The development of numerical estimation: evidence for multiple representations of numerical quantity. *Psychological Science*, 14, 237–243.

Spinath, F.M., Spinath, B., & Plomin, R. (2008). The nature and nurture of intelligence and motivation in the origins of sex differences in elementary school achievement. *European Journal of Personality*, 22(3), 211–229.

Stromswold, K. (2001). The heritability of language: a review and metaanalysis of twin, adoption and linkage studies. *Language*, 77, 647–723.

Thompson, L.A., Detterman, D.K., & Plomin, R. (1991). Associations between cognitive abilities and scholastic achievement: genetic overlap but environmental differences. *Psychological Science*, 2, 158–165.

Tosto, M.G., Hanscombe, K.B. Haworth, C.M.A., Davis, O.S.P., Petrill, S.A., Dale, P.S. Malykh, S., Plomin, R., & Kovas., Y. (2014a). Why do spatial abilities predict mathematical performance? *Development Science*, 17(3), 462–470.

Tosto, M.G., Petrill, S.A., Halberda, J., Trzaskowski, M., Tikhomirova, T.N., Bogdanova, O.Y., Ly, R., Wilmer, J.B., Naiman, D.Q., Germine, L., Plomin, R., & Kovas, Y. (2014b). Why do we differ in number sense? Evidence from a genetically sensitive investigation. *Intelligence*, 43, 35–46.

Wai, J., Lubinski, D., & Benbow, C.P. (2009). Spatial ability for STEM domains: aligning over 50 years of cumulative psychological knowledge solidifies its importance. *Journal of Educational Psychology*, 101(4), 817–835.

Webb, R.M., Lubinski, D., & Benbow, C.P. (2007). Spatial ability: a neglected dimension in talent searches for intellectually precocious youth. *Journal of Educational Psychology*, 99(2), 397–420.

Xu, F., & Arriaga, R. (2007). Number discrimination in 10-month-old infants. *British Journal of Developmental Psychology*, 25, 103–108.

PART VIII

EDUCATION

CHAPTER 54

EDUCATION

RICHARD COWAN

Education

RESEARCH in young children's mathematical cognition takes place in contexts of some lively and diverse debates. They include discussions about the importance of mathematics and why it should be taught in schools, how it should be taught, and what causes children to differ so much in their development. After sketching out these discussions, I shall locate the contributions in them.

The Importance of Learning Mathematics at School

Mathematics is a field of human endeavor with remarkable histories in different cultures and relevance to a wide range of human activities such as natural and social science, computing, and commerce. Some mathematics has developed during attempts to solve problems in other disciplines. Pure mathematics has developed in the contemplation of problems posed by mathematical objects themselves. Number theory is an area traditionally regarded as pure mathematics. Henry John Stephen Smith was a distinguished nineteenth century Oxford mathematics professor who worked in number theory. He is credited with proposing the toast 'Pure mathematics; may it never be of any use to anyone' as a challenge to the excessive emphasis of a utilitarian rationale for mathematics (Macfarlane 1916). The current importance of number theory in the development of cryptography, which is important for contemporary financial security, shows that it can be difficult to anticipate the future.

In the early development of the school curriculum the study of mathematics for its own sake had little place. The primary, or elementary, school curriculum focused on arithmetic, developing skills in the applications of mathematics in everyday life and commerce. Another justification for teaching it derived from a belief in its effects on the development of thinking. This was part of the doctrine of mental discipline that emphasized the distinct roles played by the study of particular subjects in the development of students' minds. Belief in mental discipline was dominant in education, as depicted by Hesse (1973, p. 8):

> In Greek, next to the irregular verbs, the main emphasis was laid on variety of sentence structure expressed through the use of particles, in Latin they were expected to concentrate on clear and precise statements, and become familiar with countless refinements of prosody, in mathematics pride of place was given to complicated problems of arithmetic. None of these things, as his teacher was never tired of repeating, had any apparent value for his later studies, but it was only 'apparently', for in point of fact, they were very important indeed, more important than many main subjects because they developed the logical faculties and formed a basis for all clear, sober and cogent reasoning.

The doctrine of mental discipline assumed that transfer of mental skills would be automatic, just as physical exercises to develop particular muscle groups would produce gains that automatically transferred to activities using those muscles. Early psychological research by Thorndike challenged the doctrine of mental discipline in several ways: his experiments on learning indicated that transfer was less common than expected, the limited amount of transfer that did occur could be explained without referring to general faculties, and gains in tests of thinking were not related to subjects studied as predicted by the leading version of mental discipline (Thorndike 1924a,b).

Subsequent research found that the amount and extent of transfer depended on the type of training (e.g. Woodrow 1927). This led some to refine mental discipline by allowing that whether a subject affects general intellectual development depends on how it is taught (Hamley 1939). Just as one can improve performance on intelligence or working memory tests without producing any benefits in educational achievement (Brody 1992; Melby-Lervåg and Hulme 2013) so may children's performance on school mathematics tests be improved without detectable benefits on their non-mathematical thinking and problem solving.

Holding schools accountable for children's performance on tests and publishing the results affects teaching in ways that increase the likelihood of improved test performance without transferable skills (National Research Council 2011). Teachers may focus more on the skills tested (Au 2007) or even more narrowly in coaching children on test items (Popham 2001). When targets specify a particular category of achievement, such as Level 4 at 11 years in English primary schools and Grade C or higher at GCSE in English secondary schools, some schools direct more effort towards those pupils whom they consider can reach the target with assistance, ignoring those who will meet the target anyway and those believed to be hopeless cases (Gillborn and Youdell 2000; Marks 2012).

It is still common to read assertions of the value of doing mathematics at school for general thinking skills: claims such as mathematical training 'disciplines the mind, develops logical and critical reasoning, and develops analytical and problem-solving skills to a high degree' (Smith 2004, p. 11) and mathematics 'nurtures both the faculties of logic and reasoning among learners' (Kaur and Vistro-Yu 2010, p. 453). Quite possibly the writers of these claims would acknowledge the dependence of these benefits on how it was taught and learnt. Certainly an impetus to changing how maths is taught is the perception that success in the current curriculum is not accompanied by the desired and intended benefits.

Changing the Teaching of Mathematics

Accounts of the history of curriculum reform emphasize the original 'Sputnik moment', the successful launching in 1957 of the Soviet Union's artificial satellite into orbit around the Earth, as prompting a critical reappraisal and reform of mathematics education in the West (Kaur and Vistro-Yu 2010). Although originally an attempt to reform US mathematics education for college-entry students, New Math came to be associated with changes in the teaching of mathematics at all levels with an emphasis of discovery methods, coverage of formal properties of number, and the introduction of unifying ideas such as set, variable, and function (Herrera and Owens 2001).

Coincidentally Piaget's (1952) investigations of children's number development had been prompted by an earlier debate in mathematical philosophy about whether the natural numbers had been constructed from more basic ideas and what these ideas might be (Brainerd 1979). Candidates included class, correspondence, order, and series. A natural number refers to a class and two sets are in the same number class if their members can be put in one to one correspondence with each other. This could be achieved without knowing any number words, for example by making a set of notches on a stick that is in one to one correspondence with a set of pebbles. Counting a set of objects can be seen as creating a one to one correspondence between number words and objects. The natural number sequence is an ordered series in which later numbers are greater than earlier numbers and adjacent numbers differ in magnitude by one. Piaget (1952) introduced several procedures to examine how children's grasp of these ideas developed, including number conservation as a test of understanding of the relation between one to one correspondence and number.

Early years mathematics in both the US and the UK came to include activities about sets, classes, and one to one correspondence. Curriculum developments in the 1960s such as the Nuffield Mathematics Project in the UK that attempted to develop a 'contemporary approach for students from 5 to 13' gratefully acknowledged their debts to Piaget. Number conservation became a curriculum topic (Evans 1977). Later primary school mathematics involved instruction in algebraic properties such as associativity and commutativity, and arithmetic with number bases other than ten. Unfortunately the rationale for these changes was lost on many teachers. Few primary teachers who had to teach the new curriculum had studied mathematics at university and many were baffled by their introduction to Piaget's work in training colleges. The provision of in-service training was limited. Parents complained that they could no longer support their children.

In the US New Math fell out of favor for a variety of reasons ranging from the critique by the mathematician Kline (1973), the outspoken criticisms by physicist Richard Feynman, and the perception of a fall in standards (Herrera and Owens 2001). New Math was even lampooned by Tom Lehrer. The continued availability of his song on the internet accompanied by animations is one of the most easily found legacies of New Math.

The importance of mathematics and the belief in its benefits for intellectual development may make for expectations that are unlikely to be fulfilled. The dissatisfaction with New Math is just one episode in the long history of dissatisfaction with the outcomes of mathematics education since the beginnings of compulsory schooling (McIntosh 1981). A procession of reports have suggested ways of improving learning and making school mathematics a better platform for understanding and enjoying mathematics (e.g. Cockcroft 1982; Great Britain Board of Education 1931; Williams 2008). These concerns and aspirations have been echoed in the United States (Ball et al. 2005; Baroody 2003).

A more recent stimulus to curriculum reform is provided by international surveys such as the Trends in International Mathematics and Science (TIMSS) and the Program for International Student Assessment (PISA). They come out every few years and their publication of rank orders of participating countries typically receives substantial media coverage. Their producers point out that providing a base for drawing policy recommendations, such as how teachers should teach, is beyond their scope (OECD 2009). Nevertheless some credit the surveys with a role in stimulating curriculum reform (Kaur and Vistro-Yu 2010) and suggest they exert a strong influence on politicians and their advisers (Gorard 2001; Jerrim 2013).

International league tables are hazardous for many reasons (Brown 1998). All such assessments are only a partial sample of what the children have learnt. Rank orders do not portray how much the items assessed differ: in a horse race there has to be a winner even if the difference between finishers is very small. There can be no control for the amount of time spent learning the assessed skills inside or outside school. There is also the problem that the activity of developing tests with desirable psychometric properties may have an undue influence on what aspects of mathematical attainment are assessed (Ridgway 1987): in discussing primary mathematics targets Schwarzenberger (1988) gave a detailed case for believing that the more accurately an attainment can be measured, the less likely it is to reflect genuine understanding. This provides a continuing challenge to researchers who seek to develop ways of assessing mathematical development as well as a reason for being cautious in holding schools or teachers to account on the basis of children's performance on tests.

The international surveys conducted to date commonly find that children in both England and the USA do less well on arithmetic tests than children from Asian countries such as Japan, Korea, Singapore, and Taiwan (Sturman et al. 2008). Differences favoring Japanese and Chinese children over American children have been found in more detailed and rigorous research (e.g. Stevenson et al. 1990) and been followed up with several investigations. So far the research has yielded many possible explanations with insufficient evidence to assess their relative merits (Cowan and Saxton 2010).

Imitating the curriculum of another country in the belief that it will make children more successful is magical thinking on a par with wearing the same brand of football boots as David Beckham in the belief that it will make one a better player. The problematic nature of such cultural borrowing has long been discussed in comparative education (Holmes 1981; Andrews 2010) emphasizes the importance of seeing mathematical education in its national cultural context. The study of other systems yields ideas to be discussed and hypotheses to be tested rather than 'lessons to be learned'.

Understanding Why Children Differ So Much

Within-School Variation is Greater Than between School or between Country

Another finding common to both international surveys and other research on children's mathematical skills is the extraordinary variation between children in the same country (Cockcroft 1982). This is consistently greater than the differences between countries: every country's distribution of fourth grade maths scores in TIMSS 2007 overlapped with the others (Mullis et al. 2009). All 36 participating countries had children who failed to achieve even the lowest benchmark and all except four had children who achieved the most advanced bench mark. Although some within-country variation is attributable to differences between schools it seems to be only a small part.

Most of the variation between children is within school. A vivid illustration of this is provided by data from our ESRC-funded project. In Year 4, when most children become nine years old, they were assessed using a standardized mathematics test, the Wechsler Individual Achievement Test (WIAT II^{UK}, Wechsler 2005). The WIAT consists of two subtests, Numerical Operations and Mathematical Reasoning. Numerical Operations largely assesses written computation skills initially with integers and later involving fractions, decimals, and percentages. In administering the Mathematical Reasoning subtest the tester reads aloud problems accompanied by illustrations and text. They require simple arithmetic skills and the ability to interpret charts. Table 53.1 shows the variation in standard scores, raw scores adjusted for chronological age with a mean of 100 and a standard deviation of 15, in each class. Standard scores above 76 are achieved by 95% of UK children that age according to the manual. Standard scores above 125 are achieved by less than 5%. All classes show substantial variation and if the children in those classes who are on the schools' Special Educational Needs registers are included the variation is greater. Although it can be misleading it is also possible to derive age equivalents from the raw scores (Wechsler 2005). Table 54.1 also shows the range in each class obtained by subtracting the lowest age equivalent from the highest. Given these children were to spend another two years in primary school and that differences between children increase with age rather than decrease, the results correspond well to the 'seven year difference' at 11 in mathematical skills noted by Cockcroft (1982).

Is Grouping a Solution or Part of the Problem?

How schools and teachers manage the differences between primary school children differs between schools. Many English primary schools have adopted some form of grouping of pupils (Hallam et al. 2003). There are three ways of grouping: streaming, where different

Table 54.1 Variation in standard scores and the age equivalent ranges on WIAT Numerical Operations and Mathematical Reasoning (Wechsler 2005) in each class when the children were in Year 4

	Numerical Operations		Mathematical Reasoning	
Class	Standard scores	Age equivalent ranges (years)	Standard scores	Age equivalent ranges (years)
A	74–140	4.7	82–138	5.5
B	89–133	2.7	91–130	7.7
C	78–139	4.7	76–124	6.7
D	82–149	8.7	91–128	5.7
E	85–147	6	79–137	8.7
F	78–149	5	68–138	6.7
G	81–147	4.7	91–138	8.7
H	65–142	7	76–130	6.3
I	81–130	6.7	70–124	4.3

Note. *N* = 212. Children on schools' registers of special education needs excluded. Number of children included in each class varies from 11 to 37.

ability groups are taught in separate classes; setting, where children from different classes are taught together in groups defined by their ability for a particular subject; and in-class grouping for numeracy, where children in the same class are given different tasks according to their group.

Although the use of particular grouping practices has varied over the last century, most children in English primary schools experience ability grouping when they are seven years old (Campbell 2013; Hallam and Parsons 2013). Some children are subject to more than one type which is why the following incidences of each type add up to more than 100%: streaming (17%), numeracy setting (36%), and in-class grouping (86%).

Advocates of grouping assert that it allows children to develop at their own pace, and thus maintaining interest and motivation, and that it makes teaching easier while opponents of grouping emphasize the inaccuracies in assignment, the demotivating effect of being in a lower group and the reification of ability (Whitburn 2001). Primary school children see some advantages and disadvantages to the grouping practices they experience (Hallam et al. 2004). Although some mathematics educators see the effects of ability grouping in primary school as inevitably harmful and increasing the differences between children (e.g. Boaler 2009; Marks 2012), the evidence is more equivocal (Hallam 2002). Both teaching quality and the nature of the activity may be more important (Hallam et al. 2004): for some mathematical tasks children may benefit from being in mixed ability groups; for other tasks, ability groups may have advantages.

Grouping need not reify ability if pupils, teachers, and parents understand that what ability (or attainment or aptitudes) measures assess are no more than sets of skills that develop.

Such an incremental view of ability needs reinforcing as the opposite view of maths ability as a fixed entity—something you either have or do not have and which is outside one's control—is common in the US and UK (Anastasi 1984). One does not have to have experienced ability grouping at school to acquire this pernicious belief which can harm children who believe they have it as well as those who believe they have not (Dweck 2008). Children's mathematical progress can be impeded by beliefs in maths ability as fixed whether these are held by their teachers, their parents, or themselves (Rattan et al. 2012).

The Relation between Early and Later Skills

The differences between children's number skills when they finish primary school at 11 are to some extent predictable from differences in their number skills when they start at age five (Duncan et al. 2007). This is a surprising finding given the rudimentary nature of the preschool skills and the more sophisticated number skills assessed at the end of primary school. Incidentally, there is predictability over even longer periods: a recent report produced analyses of cohort data that found differences in earnings by adults in their thirties were predictable to some extent from their number skills when they were 10 years old, even after controlling for reading, family background, and subsequent educational qualifications (Crawford and Cribb 2013). Understanding what is responsible for such long-term associations is limited. It certainly does not warrant confidence in believing that improving 10-year-olds' number skills will increase their subsequent earning power. That would assume the correlation implies a causal relationship, a common human error (Gould 1981).

Returning to the relation between early skills such as when children start primary school and later skills such as when they finish primary school, there is much work to be done to understand why differences in the earlier number skills matter, and what gives rise to the observed degree of predictability. One hypothesis is that the earlier number skills provide the foundation for later development and acquiring these skills is necessary to make progress. Other things being equal, by the time the children who have lower levels of skill improve, the children who started with higher levels will have progressed. So unless something is done the children who start behind other children will remain so. This view supports the belief that interventions which target children who start with lower entry skills and bring their skill levels up to their peers may suffice to bring about longer term improvement of their chances of later success.

Another hypothesis is that the differences in earlier skills are symptoms rather than causes of later differences in performance. In this view the predictability of later skills from early skills arises from overlap in the factors responsible for difference at both times: interventions that do not address these other factors may be limited in their long-term effects. The overlapping factors might be within the child such as numerical or general learning skills or in the child's environment such as access to learning opportunities and encouragement from parents. The involvement of other factors is strongly suggested by the finding that differences in preschool number skills predict differences in later reading skills as well as later number skills (Duncan et al. 2007). Although this is in part because individual differences in preschool number and reading skills are related this does not seem to be the whole explanation: statistical controls indicate that preschool number skills predict later reading skills independently of preschool reading skills.

As well as these different views, research on reading points to the need for longer term follow ups of early interventions. More research has been devoted to understanding the development of reading and how children with reading difficulties can be helped than understanding number and number difficulties. Nevertheless, even the benefits of interventions with a proven track record of short-term effectiveness and a theoretical grounding can sometimes disappear or 'wash out' over time (Hurry and Sylva 2007). Wash out is not inevitable (Holliman and Hurry 2013) and this raises the question of what factors affect it. It might be due to the expertise of those delivering the intervention, changes in the schools or families of participating children arising from the intervention, or differences between the children selected for the intervention.

Number Difficulties at School

An issue that divides many in mathematics education and mathematical cognition is the concept of mathematical learning disability, also known as specific arithmetic learning difficulty or developmental dyscalculia. Everyone acknowledges that some children make slower progress in learning than others (McDermott 1993) and knowledge is accumulating of the characteristics of children who perform less well on mathematical tests (Geary 2011) but there is less known and agreed about the roles played by biological, social, and environmental factors in the production of learning disabilities. Some follow Kosc's (1974) definition of developmental dyscalculia as an innate neural disorder that affects 6% of children (e.g. Butterworth et al. 2011) while others doubt a genuine neurological disorder affects so many children (Dehaene 1997).

Behavioral tests, such as the WIAT, rather than biological markers are used to identify children with mathematical learning disability. There is no agreed test, no agreement about how low the scores have to be, and no agreement on how long the difficulty has to have lasted, but some consensus is emerging (Cowan and Powell 2014). Even so, the reliance on purely statistical criteria such as scores in the lowest 10% and persistence over two years provide no evidence of the cause of difficulty.

Currently research is being stimulated by speculation that there may be numerosity systems present at birth which are common to other species and which can give rise to number difficulties if impaired: one is an approximate system assessable by discrimination tasks (Mazzocco et al. 2011), and the other an exact numerosity system assessable by single digit processing tasks (Butterworth 2010). These are interesting ideas but they are very far from being established. There are puzzling discrepancies between symbolic and non-symbolic tasks supposed to assess the same system (Gilmore et al. 2011; Rousselle and Noël 2007). No substantial relation between early infancy individual differences and later individual differences has yet been established though there are some indications of stability in approximate numerosity (Feigenson et al. 2013).

Some suggest infants' sensitivities to numerosity may have nothing to do with determining the acquisition of number skills or individual differences in them (Núñez 2009; Schneider et al. 2009). Any observed relations between individual differences in sensitivity to numerosity and in number skills may result from both being related to general cognitive skills (Anobile et al. 2013) or improved performance on numerosity tasks being the consequence of developing number skills. Research is needed to determine whether and how the functioning of these putative numerosity systems is causally related to number skills

(Feigenson et al. 2013). There is as yet no good evidence that improving numerosity sensitivity is what helps children with number difficulties as much or more than other interventions (see also the chapter by Räsänen in the section on Numerical Impairments, co-morbidity, and rehabilitation in this handbook).

What does seem clear from current research is that impaired functioning in the exact numerosity system is not necessary for poor mathematics progress: in a study with a large sample most of the children showing poor progress had unimpaired single digit processing skills (Reigosa-Crespo et al. 2012). Nor is poor single digit processing sufficient to bring about poor mathematical progress: the same study showed that about a quarter of the children who did show the poorest single digit processing skills were making average or superior progress in mathematics. Similar disconnection between single-digit processing tasks such as enumeration and magnitude comparison and more general arithmetic tests was found in our one-year study of Year 3 children (Cowan and Powell 2014). In our sample multi-digit skills and general cognitive skills were more important.

Evidence is emerging of the role of biological factors: behavioral genetic studies of twins do suggest a substantial genetic contribution to number skills in general once children are seven years old and number difficulties in particular. There is also substantial genetic overlap between number, reading and oral language (Kovas et al. 2007; Plomin and Kovas 2005). There is some genetic variance specific to number but whether this is manifested in particular number tasks is unknown: it might be manifested in some general learning skill that is more important for number development than reading or oral language. In common with other twin studies, studies of number disability do not support a simple genetic determinism: just because one twin has problems in maths does not guarantee their genetically identical twin does. Also, just because genes are implicated does not entail that the variability they account for is detectable at birth or even prior to instruction (see also the chapter by Tosto et al. in this handbook).

The reality of a neural basis for mathematical learning difficulties might be established by cognitive and behavioral studies in combination with brain-imaging studies. Conducting neural investigations of children with very poor number skills (e.g. Kovas et al. 2009; Rykhlevskaia et al. 2009) is interesting but insufficient. By themselves they cannot tell whether any observed neural abnormality is a cause or a consequence of number difficulties (Frith 2006).

There is much to be established about the factors responsible for the production and maintenance of mathematical learning difficulties. For example I was asked to assess a very low birthweight boy whose mother was concerned about his maths: she had seen a press report of calculation difficulties in very low birthweight children. He did indeed show a marked discrepancy between his reading skills which were very superior for his age and his number skills which were low average. He lived a sheltered life and enjoyed reading and talking about books with his mother. In contrast I saw no board games in the house and he did not mention any activity that might give him practice in arithmetic or stimulate his interest in numbers. Interests and skills show a reciprocal relationship even in young children (Fisher et al. 2012).

Children assigned to lower sets or remedial classes may receive less encouragement to develop independence in mathematical thinking by receiving more directive teaching or even less teaching. In this extract from a group interview of primary schoolchildren Muhi is talking with two other boys about what happened when he moved from the middle group where he was taught by Miss South to the lowest group where he is taught by Miss Middleton (Bibby et al. 2007, p. 40):

'Muhi: Yeah, I got to the lowest. And Miss Middleton hates maths, so guess what she does? She says 'right, get all your times tables done' and then she gets paper and we just have to color, like reception.

Mat: Miss Middleton is fun. I wish I was in her maths group

Muhi: But she hates maths. She doesn't even learn us maths. It's (very) boring, we just had to color like reception (...)

Emran: When Miss Middleton took us just for a bit she never taught us it and I think it was just a waste, I don't know, of teaching. Because I learned more from when we had to do sticking to make a collage'.

The Williams Review (Williams 2008) stressed the importance of having teachers who were competent in and enthusiastic about mathematics in the primary school, particularly in the delivery of remedial interventions. Whether they do not like maths or they believe that some children are incapable of learning maths (Rattan et al. 2012) the importance of the teacher in supporting children with difficulties is crucial. Unfortunately, but not inevitably, a belief in a biological basis to mathematical difficulties may discourage some parents, teachers, and pupils from making the efforts necessary to improve.

The Chapters in this Section

The following chapters present stimulating discussions of issues and research in primary mathematics learning by people who have made distinguished contributions to theory and practice in number development and mathematics teaching. The chapter by Rittle-Johnson and Schneider considers an important distinction in notions of mathematical competence, and identifies directions for future research. The chapters by Ramani and Siegler, Jordan et al., and Dowrick are all focused on interventions to support children at risk of developing mathematics difficulties. The chapters by Ginsburg et al. and Fuson et al. consider the whole curriculum.

As Rittle-Johnson and Schneider point out, the distinction between procedural and conceptual knowledge and the relations between them and their roles in mathematical competence have long exercised researchers in numerical cognition and mathematics education. Although some instructional contexts seem to favor the development of procedural knowledge over conceptual knowledge there are interventions that seem to enhance both. They argue that relations are often bi-directional and iterative and review studies which indicate this. In previous research the appearance of a sequence may reflect the measures used as much as the ways in which contexts privilege development of one aspect over the other. They therefore set out the issues researchers need to address in developing good measures.

Ramani and Siegler review a range of previous studies that suggest it is differences in experiences at home that play a major part in producing the variation between children in the number skills they start school with, including those that are associated with variation in family wealth. The theoretical and research effort devoted by Siegler and his colleagues has made a strong case for the importance of estimation of numerical magnitude in the development of number skills. Considering how estimation of relative magnitude might be enhanced in preschoolers led to the creation of a version of a familiar board game that provides children with experiences that support the development of number magnitude knowledge. Their most recent study (Ramani et al. 2012) shows the gains in this and related number skills are considerable even when the interventions are delivered to groups of children in schools by paraprofessionals with just one hour's training.

Estimation of numerical knowledge is a strong candidate for number sense in an educational context but it would be misleading to think that it is the only one: Berch (2005) lists 30 that vary considerably from each other. The chapter by Jordan et al. describes the work underlying the Number Sense Screener that combines assessments of knowledge of number magnitudes with tests of counting and calculation. It goes on to review rigorous interventions that are effective at various points in primary schooling.

The chapter by Dowrick reports and discusses a large scale follow-up study of Numbers Count, an intervention to support six- and seven-year-old children who were selected on the basis of low number skills. Debts are acknowledged to Reading Recovery in the design of sessions and there is similar careful attention to detail. The involvement of parents and class teachers seems a very worthwhile component as does the support provided to the Numbers Count teachers: currently in the UK some class teachers have no idea what their children are doing in remedial classes. Dowrick provides information that suggests children receiving Numbers Count show benefits in the skills that are the focus of the intervention and in their attitudes. Unfortunately I do not know what the change in attitude means. Like Reading Recovery, the selection on the basis of low achievement makes Numbers Count susceptible to criticism that the improvements are at least partly due to regression to the mean (Campbell and Kenny 1999). Whether there are generalized benefits in mathematics learning and whether these are sustained over a longer period seem to me important remaining questions. Also a comparison with the other interventions shown to be effective might help in determining whether the intervention which is admitted to be expensive is nevertheless cost-effective.

Ginsburg et al. advocate a very early start to maths education. They suggest that preschool maths education is very limited and primary maths education is not much better. It too readily drifts into repetitive practice of routines and procedures which at best help children develop some fluency. It should be more mathematical. It should involve more discussion and more development of understandings. This echoes the recommendations of many reports on mathematical education. What prevents change? Desforges and Cockburn (1987) investigated the primary classrooms of respected teachers to find out. They concluded that apart from the sheer number of demands teachers were dealing with moment to moment, children played an important part in determining the academic work of the classroom. In an interview following up responses to questionnaires some children explained that what they liked about maths was that it did not require thinking. Some children like maths the way it is while others want school maths to offer more scope for creativity and agency (Lange 2009).

Ginsburg et al. suggest that high quality software has a role to play in their new curriculum but they acknowledge much existing software falls short. They outline a series of steps for the development and evaluation of software. Certainly there is a need for some rigorous and impartial assessment of the many programs that are developed and websites offering instruction and games.

Fuson, Murata, and Abrahamson present a distillation of their thoughts on how the curriculum might be reformed to achieve the goals of making mathematics meaningful to children and supporting the development of fluency. These goals have at times been presented as mutually antagonistic in the fierce debates about mathematical education. By recognizing and planning for the development of both they justify their claim to offer a middle way. Also notable are the phases of instruction that draw on Murata's analysis of Japanese teaching, the emphasis on the role played by maths drawings, and the acknowledgment of the role of continuing professional development.

In setting out the context for the following chapters I have tried to represent the substantial differences in opinion on fundamental issues that range from the reasons for teaching mathematics to the factors that affect children's chances of success. Some of the issues are ideological and resolution is unlikely, similar to the people Sydney Smith saw arguing from houses on opposite sides of the street. He commented 'they will never agree: they are arguing from different premises'. Even the dialog between incompatible points of view can sharpen thinking. The value of these chapters is that they build on previous work and have found areas where it is possible to advance debate through research.

References

Anastasi, A. (1984). Aptitude and achievement tests: the curious case of the indestructible strawperson. In: B.S. Plake (ed.), *Social and Technical Issues in Testing: Implications for Test Construction and Usage*, pp. 129–140. Hillsdale, NJ: Lawrence Erlbaum Associates.

Andrews, P. (2010). *Acknowledging the cultural dimension in research into mathematics teaching and learning*. Paper presented at the British Congress for Mathematics Education 7 at Manchester University 2010. Available at: <http://www.bsrlm.org.uk/IPs/ip30-1/>.

Anobile, G., Stievano, P., and Burr, D.C. (2013). Visual sustained attention and numersity sensitivity correlate with math achievement in children. *Journal of Experimental Child Psychology 116*: 380–391.

Au, W. (2007). High-stakes testing and curricular control: A qualitative metasynthesis. *Educational Researcher 36*: 258–267.

Ball, D.L., Hill, H.C., and Bass, H. (2005, Fall). Knowing mathematics for teaching. *American Educator* 14–22.

Baroody, A.J. (2003). The development of adaptive expertise and flexibility: The integration of conceptual and procedural knowledge. In: A.J. Baroody and A. Dowker (eds), *The development of arithmetic concepts and skills: Constructing adaptive expertise*, pp. 1–33. Mahwah, NJ: Lawrence Erlbaum Associates.

Berch, D. (2005). Making sense of number sense: Implications for children with mathematical disabilities. *Journal of Learning Disabilities 38*: 333–339.

Bibby, T., Moore, A., Clark, S., and Haddon, A. (2007). *Children's learner-identities in mathematics at key stage 2: Full Research Report ESRC End of award report, RES-000-22-1272*. Swindon: ESRC.

Boaler, J. (2009). *The Elephant in the Classroom: Helping Children Learn and Love Maths*. London: Souvenir.

Brainerd, C. J. (1979). *The Origins of the Number Concept*. New York: Praeger.

Brody, N. (1992). *Intelligence*, 2nd edn. London: Academic Press.

Brown, M. (1998). The tyranny of the international horse race. In: R. Slee, G. Weiner, and S. Tomlinson (eds), *School Effectiveness for Whom? Challenges to the School Effectiveness and School Improvement Movements*, pp. 33–47. London: Falmer Press.

Butterworth, B. (2010). Foundational numerical capacities and the origins of dyscalculia. *Trends in Cognitive Sciences 14*: 534–541.

Butterworth, B., Varma, S., and Laurillard, D. (2011). Dyscalculia: from brain to education. *Science* 332: 1049–1053.

Campbell, T. (2013). *In-school ability grouping and the month of birth effect: Preliminary evidence from the Millennium Cohort Study*, CLS Cohort Studies Working Paper 2013/1. London: Institute of Education.

Campbell, D.T. and Kenny, D.A. (1999). *A Primer on Regression Artifacts*. New York: The Guilford Press.

Cockcroft, W. (1982). *Mathematics Counts*. London: HMSO.

Cowan, R. and Powell, D. (2014). The contributions of domain-general and numerical factors to third-grade arithmetic skills and mathematical learning disability *Journal of Educational Psychology* 106: 214–229. doi: 10.1037/a0034097

Cowan, R. and Saxton, M. (2010). Understanding diversity in number development. In: R. Cowan, M. Saxton, and A. Tolmie (eds), *Understanding Number Development and Difficulties. British Journal of Educational Psychology Monograph Series II: Psychological Aspects of Education—Current Trends.* Volume 7, pp. 1–14. Leicester: British Psychological Society.

Crawford, C. and Cribb, J. (2013). *Reading and Maths Skills at Age 10 and Earnings in Later Life: a Brief Analysis Using the British Cohort Study*. London: Centre for Analysis of Youth Transitions.

Dehaene, S. (1997). *The Number Sense*. Harmondsworth: Penguin.

Desforges, C. and Cockburn, A. (1987). *Understanding the Mathematics Teacher: A Study of Practice in First Schools*. London: Falmer Press.

Duncan, G.J., Dowsett, C.J., Claessens, A., Magnuson, K., Huston, A.C., Klebanov, P., et al. (2007). School readiness and later achievement. *Developmental Psychology 43*: 1428–1446.

Dweck, C.S. (2008). *Mindset: the New Psychology of Success*. New York: Ballantine Books.

Evans, D. (1977). *Mathematics: Friend or Foe? A Review of Some Ideas about Learning and Teaching Mathematics in the Primary School*. London: George Allen and Unwin.

Feigenson, L., Libertus, M.E., and Halberda, J. (2013). Links between the intuitive sense of number and formal mathematics ability. *Child Development Perspectives 7*: 74–79.

Fisher, P.H., Dobbs-Oates, J., Doctoroff, G.L., and Arnold, D.H. (2012). Early math interest and the development of math skills. *Journal of Educational Psychology 104*: 673–681.

Frith, C.D. (2006). The value of brain imaging in the study of development and its disorders. *Journal of Child Psychology and Psychiatry 47*: 979–982.

Geary, D.C. (2011). Consequences, characteristics, and causes of mathematical learning disabilities and persistent low achievement in mathematics. *Journal of Developmental and Behavioral Pediatrics 32*: 250–263.

Gillborn, D. and Youdell, D. (2000). *Rationing Education: Policy, Practice, Reform and Equality*. Buckingham: Open University Press.

Gilmore, C.K., Attridge, N., and Inglis, M. (2011). Measuring the approximate number system. *Quarterly Journal of Experimental Psychology 64*: 2099–2109.

Gorard, S. (2001). International comparisons of school effectiveness: The second component of the 'crisis account' in England? *Comparative Education 37*: 279–296.

Gould, S.J. (1981). *The Mismeasure of Man*. New York: Norton.

Great Britain Board of Education Consultative Committee (1931). *The primary school*. London: HMSO.

Hallam, S. (2002). *Ability Grouping in Schools: a Literature Review*. London: Institute of Education, University of London.

Hallam, S. and Parsons, S. (2013). Prevalence of streaming in UK primary schools: evidence from the Millennium Cohort Study. *British Educational Research Journal 39*: 514–544.

Hallam, S., Ireson, J., Lister, V., Andon Chaudhury, I., and Davies, J. (2003). Ability grouping in the primary school: a survey. *Educational Studies 29*: 69–83.

Hallam, S., Ireson, J., and Davies, J. (2004). Primary pupils' experiences of different types of grouping in school. *British Educational Research Journal 30*: 515–533.

Hamley, H.R. (1939). Memorandum on the cognitive aspects of transfer of training. In Great Britain Board of Education Consultative Committee (ed.), *Secondary Education*, pp. 439–452. London: HMSO.

Herrera, T.A. and Owens, D.T. (2001). The 'new new math'?: two reform movements in mathematics education. *Theory into Practice 40*: 84–92.

Hesse, H. (1973). *The Prodigy*. Harmondsworth: Penguin.

Holliman, A.J. and Hurry, J. (2013). The effects of Reading Recovery on children's literacy progress and special educational needs status: a three-year follow-up study. *Educational Psychology 33*: 719–733.

Holmes, B. (1981). *Comparative Education: Some Considerations of Method*. London: George Allen and Unwin.

Hurry, J. and Sylva, K. (2007). Long-term outcomes of early reading intervention. *Journal of Research in Reading 30*: 227–248.

Jerrim, J. (2013). The reliability of trends over time in International education test scores: is the performance of England's secondary school pupils really in relative decline? *Journal of Social Policy* 42: 259–279.

Kaur, B. and Vistro-Yu, C. P. (2010). Mathematics. In: P. Peterson, E. Baker, and B. McGaw (eds), *International Encyclopedia of Education*, 3rd edn., pp. 453–458. Oxford: Elsevier.

Kline, M. (1973). *Why Johnny Can't Add: The Failure of the New Math*. New York: St. Martin's Press.

Kosc, L. (1974). Developmental dyscalculia. *Journal of Learning Disabilities* 7: 164–177.

Kovas, Y., Haworth, C.M.A., Harlaar, N., Petrill, S.A., Dales, P.S., and Plomin, R. (2007). Overlap and specificity of genetic and environmental influences on mathematics and reading disability in 10-year-old twins. *Journal of Child Psychology and Psychiatry* 48: 914–922.

Kovas, Y., Giampietro, V., Viding, E., Ng, V., Brammer, M., Barker, G.J., et al. (2009). Brain correlates of non-symbolic numerosity estimation in low and high mathematical ability children. *PLoS ONE*, [Online] 24 Feb., 4, 1–8. Available at: <http://www.plosone.org/article/info%3Adoi%2F10.1371%2Fjournal.pone.0004587>.

Lange, T. (2009). *'Tell them that we like to decide for ourselves'- Children's agency in mathematics education*. Paper presented at the Sixth Congress of the European Society for Research in Mathematical Education, Lyon France. Available at: <http://ife.ens-lyon.fr/editions/editions-electroniques/cerme6/>.

Macfarlane, A. (1916). *Lectures on ten British Mathematicians of the Nineteenth Century*. London: John Wiley and Sons.

Marks, R. (2012). *'I get the feeling that it is really unfair': Educational triage in primary mathematics*. Paper presented at the Day Conference of the British Society for Research into Learning Mathematics, University of Sussex. Available at: <http://www.bsrlm.org.uk/IPs/ip32-2/BSRLM-IP-32-2-10.pdf>.

Mazzocco, M.M.M., Feigenson, L., and Halberda, J. (2011). Impaired acuity of the approximate number system underlies mathematical learning disability (dyscalculia). *Child Development* 82: 1224–1237.

McDermott, R.P. (1993). The acquisition of a child by a learning disability. In: S. Chaiklin and J. Lave (eds), *Understanding Practice: Perspectives on Activity and Context*, pp. 269–305. Cambridge: Cambridge University Press.

McIntosh, A. (1981). When will they ever learn? In A. Floyd (Ed.), *Developing mathematical thinking*, pp. 6–11. London: Addison-Wesley and Open University.

Melby-Lervåg, M. and Hulme, C. (2013). Is working memory training effective? A meta-analytic review. *Developmental Psychology* 49: 270–291.

Mullis, I.V. S., Martin, M.O., and Foy, P. (2009). *TIMSS 2007 International Mathematics Report: Findings from IEA's Trends in International Mathematics and Science Study at the Fourth and Eighth Grades*. Revised edn. Chestnut Hill, MA: TIMSS and PIRLS International Study Center, Boston College.

National Research Council. (2011). Incentives and test-based accountability in education. Committee on Incentives and Test-based Accountability in Public Education, M. Hout and S.W. Elliott (eds). Board on Testing and Assessment, Division of Behavioral and Social Sciences and Education. Washington, DC: The National Academies Press.

Núñez, R. (2009). Numbers and arithmetic: neither hardwired nor out there. *Biological Theory* 4: 68–83.

OECD (2009). The usefulness of PISA data for policy makers, researchers and experts on methodology. In OECD, *PISA Data Analysis Manual: SAS*, 2nd edn. Paris: OECD.

Piaget, J. (1952). *The Child's Conception of Number*. London: Routledge and Kegan Paul.

Plomin, R. and Kovas, Y. (2005). Generalist genes and learning disabilities. *Psychological Bulletin* 131: 592–617.

Popham, W.J. (2001). Teaching to the test? *Educational Leadership* 58: 16–20.

Ramani, G.B., Siegler, R.S., and Hitti, A. (2012). Taking it to the classroom: Number board games as a small group learning activity. *Journal of Educational Psychology* 104: 661–672.

Rattan, A., Good, C., and Dweck, C.S. (2012). 'It's ok—Not everyone can be good at math': Instructors with an entity theory comfort (and demotivate) students. *Journal of Experimental Social Psychology* 48: 731–737.

Reigosa-Crespo, V., Valdés-Sosa, M., Butterworth, B., Estévez, N., Rodríguez, M., Santos, E., et al. (2012). Basic numerical capacities and prevalence of developmental dyscalculia: The Havana study. *Developmental Psychology 48*: 123–135.

Ridgway, J. (1987). *A Review of Mathematics Tests*. Windsor: NFER-Nelson.

Rousselle, L. and Noel, M.P. (2007). Basic numerical skills in children with mathematics learning disabilities: A comparison of symbolic vs. non-symbolic number magnitude processing. *Cognition 102*: 361–395.

Rykhlevskaia, E., Uddin, L. Q., Kondos, L., and Menon, V. (2009). Neuroanatomical correlates of developmental dyscalculia: combined evidence from morphometry and tractography. *Frontiers in Human Neuroscience*. Available at: <http://www.frontiersin.org/human_neuroscience/10.3389/neuro.09.051.2009/abstract>.

Schneider, M., Grabner, R.H., and Paetsch, J. (2009). Mental number line, number line estimation, and mathematical achievement: Their interrelations in grades 5 and 6. *Journal of Educational Psychology 101*: 359–372.

Schwarzenberger, R. (1988). *Targets for Mathematics in Primary Education*. Stoke-on-Trent, England: Trentham Books.

Smith, A. (2004). *Making Mathematics Count: The Report of Professor Adrian Smith's Inquiry into Post-14 Mathematics Education*. London: The Stationery Office.

Stevenson, H.W., Lee, S.Y., Chen, C., Stigler, J.W., Hsu, C.C., and Kitamura, S. (1990). Contexts of achievement: A study of American, Chinese, and Japanese children. *Monographs of the Society for Research in Child Development*, *55*(Serial No. 221).

Sturman, L., Ruddock, G., Burge, B., Styles, B., Lin, Y., and Vappula, H. (2008). *England's Achievement in TIMSS 2007 National Report for England*. Slough: NFER.

Thorndike, E.L. (1924a). Mental discipline in high school studies. *Journal of Educational Psychology 15*: 1–22.

Thorndike, E.L. (1924b). Mental discipline in high school studies (continued from January). *Journal of Educational Psychology 15*: 83–98.

Wechsler, D. (2005). *Wechsler Individual Achievement Test—Second UK Edition (WIAT-IIUK)*. Oxford: Harcourt Assessment.

Whitburn, J. (2001). Effective classroom organization in primary schools: Mathematics. *Oxford Review of Education 27*: 411–428.

Williams, P. (2008). *Independent Review Of Mathematics Teaching in Early Years Settings and Primary Schools: Final Report*. London: Department for Children Schools and Families.

Woodrow, H. (1927). The effect of training upon transference. *Journal of Educational Psychology 18*: 159–172.

CHAPTER 55

USING LEARNING PATH RESEARCH TO BALANCE MATHEMATICS EDUCATION

teaching/learning for understanding and fluency

KAREN C. FUSON, AKI MURATA, AND DOR ABRAHAMSON

The past 40 years have seen an explosion of research on numerical cognition and some research on how best to teach various numerical topics. It is now clear that children develop different methods for solving numerical problems and that these move from simple and slow to more advanced and rapid. Early on children worldwide go through a progression of levels of counting, adding, and subtracting. However, as the numerical topics become more advanced, and cultural symbol systems become more central, children's methods increasingly depend on what they are taught in school.

Research on children's thinking and on methods they develop has often been carried out within a Piagetian perspective (e.g. 1965/1941). Research focused more on cultural symbol systems has often been carried out within a Vygotskiian perspective (Vygotsky, 1934/1986, 1978). Fuson (2009) discussed extreme views of teaching that are sometimes drawn from a Piagetian view ('learning without teaching' that values only children's invented methods) and from a Vygotskiian view ('teaching without learning' with a traditional emphasis on fluency, rather than on meaning making). A balanced learning–teaching approach that relates these views was summarized in that paper. In this approach, called learning path teaching, mathematically desirable methods that are accessible to children, and may have been invented by children, are linked to and explained using maths drawings or other visual referents to support meaning making. This approach enables all children to use general methods with understanding and move to fluency.

In this paper, we describe this balanced middle in more detail as the Class Learning Path Model, provide examples in three maths domains that show different aspects of such balanced learning-teaching, and then relate this view to past and present efforts to reform mathematics education in various countries. This approach allows us to exemplify major results on numerical cognition that affect mathematics education and to identify fruitful new directions for such research.

Teaching/Learning Within a Class Learning Path

The Class Learning Path Model integrates two theoretical foci – a Piagetian focus on learning and a Vygotskiian focus on teaching – and specifies phases in learning that reflect Vygotsky's assertion about children's move from spontaneous to scientific concepts (this model is discussed in more detail in Fuson & Murata 2007; Fuson, Murata, & Abrahamson, 2011; Murata & Fuson, 2006). Major aspects of the model were drawn from principles in two National Research Council reports on research on maths teaching and learning (Donovan & Bransford, 2005; Kilpatrick, Swafford, & Findell, 2001) and from the National Council of Teachers process standards (National Council of Teachers of Mathematics, 2000) in the United States. These reports and initiatives are based on intensive reviews of the research.

This model also draws from two other sources – research by the second author analysing Japanese approaches to teaching (Murata, 2008, 2013) and results of a 10-year research and curriculum development project, the Children's Math Worlds Project, directed by the first author. The project developed teaching materials, implemented them with teachers in a wide range of classrooms, and revised them in several cycles of revision. The materials sought to find and stimulate student methods in the middle that would relate to traditional methods, but be easier to understand and to carry out. This project drew from and contributed to on-going research and research reviews (e.g. Fuson, 1992, 2003). It was published as a Kindergarten through Grade 5 full maths programme *Math Expressions* (Fuson, 2006) and now includes Grade 6 (2012). This programme contains a coherent sequence of research-based visual models and methods in the middle that do connect children's invented methods with traditional methods along a learning path.

Mathematics education and cognitive research have many different terms. The lack of a shared language complicates communication efforts. However, considerable research has been reported as learning paths (trajectories, progressions) through which students move from basic and often informal understandings and methods to more formal, advanced, and fluent methods (e.g. Clements & Sarama, 2009, 2012). In Fuson et al. (2011), we sought to bring together various perspectives on understanding and fluency, provide a model of classroom teaching/learning that included this learning path research, and provide language that would communicate across different kinds of research literature. We found the word 'form' to be a central unifying term. We characterized Piaget's and Vygotsky's conceptual activity as involving three types of external maths forms: *situational (contextual), pedagogical,* and *cultural math forms* (Piaget, 1941/1965; Vygotsky, 1934/1986, 1978). Each learner continually forms and re-forms *individual internal forms* (IIFs) that are interpretations of the external forms. This parallel use of the word *forms* links the external and internal forms, but emphasizes that each individual internal form may vary from the external form because the internal form is an interpretation. Doing maths is using *individual-internal-forms-in-action* to form actions with *external forms.* Within a class learning path, each learner moves in a learning path from using informally-learned spontaneous forms to using explicitly-learned academic cultural maths forms (Vygotsky, 1934/1986). Such movement can be stimulated within the classroom by teaching/learning that is inter- or intra-semiotic mediation (re-forming forms) via instructional conversations within class learning zones. Teaching (by the Teacher and by all of the students at some times) leads learners' attention to aspects of the external forms and supports inter-forming them by:

Teaching: inscribing, speaking and gesturing about, and inter-forming external forms.

Learning: re-forming *individual internal forms* in response to teaching.

Such interactive cognizing over time leads to increasingly similar *individual internal forms* that can be taken-as-shared (Cobb & Bauersfeld, 1995). The individual forms become increasingly well-formed (correct and mathematically advanced), and they inter-form into networks of *individual internal forms* for the topic.

Vygotsky's *zone of proximal development* is what an individual can learn with assistance (1934/1986). We use the term *class learning zone* to mean what a given class can learn with the assistance of a teacher and of the pedagogical and situational external forms. Instructional conversations are possible because the external forms and the means of assistance illuminating and inter-forming the external forms direct and constrain the possible *individual internal forms* that individuals create and use within the classroom forms-in-action. Each student's *individual internal forms* evolve within the class learning path toward a *well-formed network* of individual internal forms that can inter-form with other students' networks, but still have idiosyncratic differences. This movement of individuals within the class learning path can be visualized as paths intertwining and coming closer within a corridor (Confrey, 2005) that overall looks more like a truncated cone as all class members inter-form their *individual internal forms* while cognizing interactively with assistance (Murata, 2013).

Notice that the whole learning path of methods can be elicited in Phase 1 or introduced early in Phase 2 (these methods are labelled as at Level 0, 1, and 2 in Table 55.1). The classroom instructional conversations support individualized instruction within whole-class activity as the methods of all students appear and are discussed. Diversity can be accepted and used to increase understanding by all, but the Class Learning Path model also assumes and makes possible the realization of high academic expectations by the early introduction and support of visual models and methods in the middle (Murata, 2013).

The four phases in the Class Learning Path model came from earlier work reported in Murata and Fuson (2006), Fuson and Murata (2007), and Murata (2008) that used research-based principles from NRC reports and the NCTM Process Standards to justify the parts of the model (see Box 54.1). The model itself was drawn from our on-going work describing common features of two kinds of classrooms that both used this model – those in the *Children's Math Worlds/Math Expressions* classrooms and in Japanese classrooms. Box 54.1 provides crucial detail that relates to student numerical cognizing, and provides a fuller view of classrooms in action that support and advance student cognizing.

An early version of these four phases was presented at a conference with Japanese maths educators (Lewis & Takahashi, 2006) to frame a requested discussion of the 'maths wars' in the United States. The response of the Japanese educators was that they use the same four phases in their elementary maths curricula. Japanese teachers' manuals that accompany elementary mathematics textbooks generally outline and describe

Table 55.1 Four Phases in the Class Learning Path to Well-Formed Networks of Individual Internal Forms

Solution Method or Situation Form by Level at Each Phase		
Phase 1 Guided Introducing: Eliciting Individual Internal Forms and Forming Initial Forms-in-Action		
Level 0	Level 1	Level 2
unformed	basic & slow	a) mathematically-desirable & accessible (MD&A)
	typical errors	b) mathematically-desirable & not accessible (often current common[a] methods)
Phase 2 Learning Unfolding: Forming Well-Formed Individual Nets-For-Action (Major Meaning-Making Phase)		
Level 0	Level 1	Level 2
		MD&A methods dominate and errors decrease
Phase 3 Kneading Knowledge: Major Fluency Phase for Fast-Forms-in-Action		
Level 0	Level 1	Level 2
		Each student fast-forms one Level 2 (mathematically desirable) method; many students inter-form ≥2 methods.
Phase 4 Maintaining Fluency and Relating to Later Topics: Remembering Fluent Methods and Re-Forming Individual Nets-For-Action		
Level 0	Level 1	Level 2
		Each student remembers and maintains Phase 3 performance.

a The current common method sometimes is mistermed "the standard algorithm" but should be considered as one variation of the standard algorithmic approach, which uses the major ideas of the method.

these phases. We use here translations of these terms by Murata (2008) for the first three phases in Table 55.1 and Box 55.1 – *guided introducing, learning unfolding, and kneading knowledge*.

Aspects of Table 55.1 and Box 55.1 obviously relate to many other theoretical and research perspectives. Some of these are discussed in the reports from which aspects were drawn, and some in the longer papers about this model (e.g. Gal'perin's instructional framework is discussed in Murata (2008), and the internalizing and abbreviating movement from 'speech for others' to 'speech to self' is discussed in the Fuson/Murata papers). We will mention here the importance of Case's neo-Piagetian framework as it undergirds the notion of learning paths as building another step onto a method that has become more fluent and of Case's educational work identifying bridging contexts as meaning-making supports (e.g. Case,

Box 55.1 NRC principles and NCTM Standards Summarize the Class Learning Path Model

Overall: Create The Year-Long Nurturing Meaning-Making Maths-Talk Community

- The teacher orchestrates collaborative instructional conversations focused on the mathematical thinking of classroom members (*How Students Learn Principle 1* and *NCTM Process Standards: Problem Solving, Reasoning & Proof, Communication).*
- Students and the teacher use responsive means of assistance that facilitate learning and teaching by all: engaging and involving, managing, and coaching: modelling, clarifying, instructing/explaining, questioning, and feedback.

For Each Maths Topic: Use Inquiry Learning Path Teaching–Learning

The teacher supports the meaning-making of all classroom members by using and assisting students to use and relate (inter-form) coherent mathematical situations, pedagogical forms, and cultural mathematical forms (*NCTM Process Standards: Connections & Representation*) as the class moves through four class learning zone teaching phases.

Phase 1 Guided Introducing

Supported by the coherent pedagogical forms, the teacher elicits and the class works with understandings that students bring to a topic (*How Students Learn Principle 1*).

(a) Teacher and students value and discuss student ideas and methods (they inter-form the individual internal forms-in-action using external forms).
(b) Teacher identifies different levels of solution methods used by students and typical errors and ensures that these are seen and discussed by the class.

Phase 2 Learning Unfolding (Major Meaning-Making Phase)

The Teacher helps students form emergent networks of forms-in-action (*How Students Learn Principle 2*):

(a) Explanations of methods and of mathematical issues continue to use maths drawings and other pedagogical supports (external forms) to stimulate correct relating (inter-forming) of the forms.
(b) Teacher focuses on or introduces mathematically-desirable and accessible method(s).
(c) Erroneous methods are analysed and repaired with explanations.
(d) Advantages and disadvantages of various methods including the current common method are discussed so that central mathematical aspects of the topic become explicit.

Phase 3 Kneading Knowledge (Fluency)

The Teacher helps students gain fluency with desired method(s):

- Students may choose a method.
- Fluency includes being able to explain the method.
- Some reflection and explaining still continue (kneading the individual internal forms)
- Students stop making maths drawings when they do not need them (*Adding It Up: fluency & understanding*).

Phase 4 Maintaining Fluency And Relating To Later Topics

The teacher assists remembering by giving occasional problems and initiates and orchestrates instructional discussions to assist re-forming individual internal forms to support (form-under) and stimulate new individual internal-nets-for-action for related topics.

Result: Together These Achieve The Overall High-Level Goal For All

Build resourceful self-regulating problem solvers (*How Students Learn Principle 3*) by continually intertwining the 5 strands of mathematical proficiency:

- Conceptual understanding.
- Procedural fluency.
- Strategic competence.
- Adaptive reasoning.
- Productive disposition (*Adding It Up*).

1991; Case & Okamoto, 1996). Our framework highlights the need for coherence within all maths programmes so that they can build systematically and the need for sufficient work to prepare children to have understandings required for simple Level 1 solution methods for Phase 1 before methods are elicited.

The Importance of Maths Drawings Linked to Mathematically-Desirable and Accessible Methods

MD&A Methods, and Extended Phase 2 Instructional Conversations

Phase 1 is emphasized in reform curricula – eliciting and discussing children's invented methods and focusing on understanding. Traditional curricula emphasize Phase 3 to focus on fluency. The new important part of the model is Phase 2 that connects Phase 1 and Phase 3, and provides deep and ambitious learning. As students compare, contrast, and analyse (inter-form) different methods in Phase 2, core maths concepts can be lifted up from the problem contexts or specific methods and connected (inter-formed within their *individual internal forms*). Coherent pedagogical forms (and especially the mathematically-desirable and accessible methods and maths drawings) enable students to inter-form their external forms to foster the growth of increasingly well-formed (culturally adapted and taken-as-shared) *individual internal forms*.

The word *inquiry* in *inquiry learning-path teaching–(learning)* is used here primarily in the sense of *problematizing* all topics (Freudenthal, 1983; Hiebert, Carpenter, Fennema, Fuson, Murray, Olivier, et al. 1996). Students are to approach all topics with inquiring minds, seeking to understand and to share their own thinking with classmates. Inquiry learning path teaching–learning for a given topic requires coherent mathematical situations, pedagogical forms, and cultural maths forms for a given topic to assist students in attuning their emergent individual networks-in-action through learning paths to well-formed networks of *individual internal forms*.

Phase 2 is the heart of this process, as the class focuses on and discusses mathematically-desirable and accessible (MD&A) methods with the help of some kind of visual supports for meaning-making. A major research result of the 10 years of classroom research underlying *Math Expressions* was the importance of maths drawings (visual models, diagrams) as pedagogical forms to support individual thinking and problem solving and instructional conversations (such Maths Talk is discussed in more detail in Fuson, Atler, Roedel, & Zaccariello, 2009; Hufferd-Ackles, Fuson, & Sherin, 2004). Maths drawings facilitate problem solving because students can relate steps in the maths drawing to steps with maths symbols (cultural maths external forms) and can label the drawing to relate to the problem situation or to maths concepts (e.g. tens). These drawings can help bridge problem situations with mathematical solutions through mathematizing (Murata & Kattubadi, 2012). Maths drawings assist instructional conversations because they can be put on the board for all to see and leave a trace of all steps in the thinking, so each step can be explained. They are inexpensive, easy to manage, can be used on homework, and remain after the problem is solved to support reflection and further explanation. Teachers can collect pages containing them and reflect on these windows into the minds of students outside of the class time. Many East Asian elementary maths programmes also have a history of using diagrams, as do some other countries around the world. Maths drawings initially can show all of the objects and later they can be diagrams with numbers in them. An initial phase of concrete objects may be helpful for very young children or for some special needs children, but for many maths topics this can be very short or non-existent.

We exemplify now coherent sets of visual models and the mathematically-desirable and accessible methods for two crucial domains of the elementary school content – problem-solving and multidigit computation. These are drawn from the extensive classroom-based design research of *Math Expressions*. These examples permit the reader to get some sense of how student cognizing has become the centre of new visions of teaching and learning. The research on these domains is vast and cannot be summarized here, nor can the methods or models be discussed in detail (for more details, see the Clements and Sarama, the Fuson, and the NRC reports referenced above and the Fuson, 2013, professional development webcasts for various topics listed under projects on https://www.sesp.northwestern.edu/profile/?p=61).

Diagrams For Representing Problem Situations

There is a large international research base about representing and solving word problem types as the bases for understanding of operations (+, –, ×, ÷). Learning paths of difficulty have been identified that depend on the problem types and the particular unknown. Algebraic problems are those where the *situation equation*, such as □ + 4 = 9, is not the same as the *solution equation*, 4 + □ = 9 or 9 – 4 = □. Students can also work in kindergarten with forms of equations with one number on the left (e.g. 5 = 2 + 3 and 5 = 4 + 1) as they decompose a given number (here, 5) and record each decomposition by a drawing or equation. Experience with these various forms of equations can eliminate the typical difficulty many students have with equations in algebra, where their limited experience with one form of equation leads them to expect only equations with one number on the right.

Figure 55.1 shows the final pedagogical forms used in *Math Expressions* to represent (form) the situations (for more details about these drawings, see Fuson & Abrahamson, 2012). The diagrams support a student with algebraic problem-solving – represent the situation by making a diagram and then use the numerical relationships in that diagram to find the solution. The diagrams are the Phase 2 MD&A methods in the middle for algebraic problem-solving. They have moved beyond students' Level 1 maths drawings that show all of the objects, and they are not yet algebra, which uses only an equation to represent the situation. These diagrams bridge these two levels and give students extensive experience with writing, understanding, solving, and explaining/discussing situation equations like □ – 538 = 286 or 5/7 = 2/7 + □.

Seeing the diagrams together shows their coherence, e.g. equal group situations arise from add to/take from or put together/take apart situations when an addend is a group that is added repeatedly, and additive comparisons likewise become specially restricted as multiplicative comparison situations. Figure 55.1 shows how different situations actually involve different meanings of the equals sign, indicated at the bottom of each cell. This single set of diagrams can be used for all of the quantities students experience through Grade 6 (from single-digit numbers through fractions and decimals) and for many multi-step problems. They thus also support Phase 4 connections among problem situations as students move through the grades.

Maths Drawings and MD&A Methods for Multidigit Computation

Understanding multidigit computation requires understanding the nature of the numbers involved. There is considerable research on ways to show the meanings of multidigit numbers and on different methods of computing. Continual reviewing of such research from around the world, and classroom-based research into supports and methods invented in our *Children's Math Worlds* classrooms, were carried out for many years. Mathematically-desirable methods were then tested with a range of students and teachers in other classrooms to find out how easy they were to understand and carry out. The final *Math Expressions* research-based maths drawings (in the left column) and MD&A methods for multidigit addition, subtraction, multiplication, and division are shown in the middle column of Figure 55.2. These methods were all invented in classrooms and are described in more detail in Fuson (2003), in National Council of Teachers of Mathematics (2010, 2011), and in the Number and Operations in Base Ten professional development webcasts listed under Projects on https://www.sesp.northwestern.

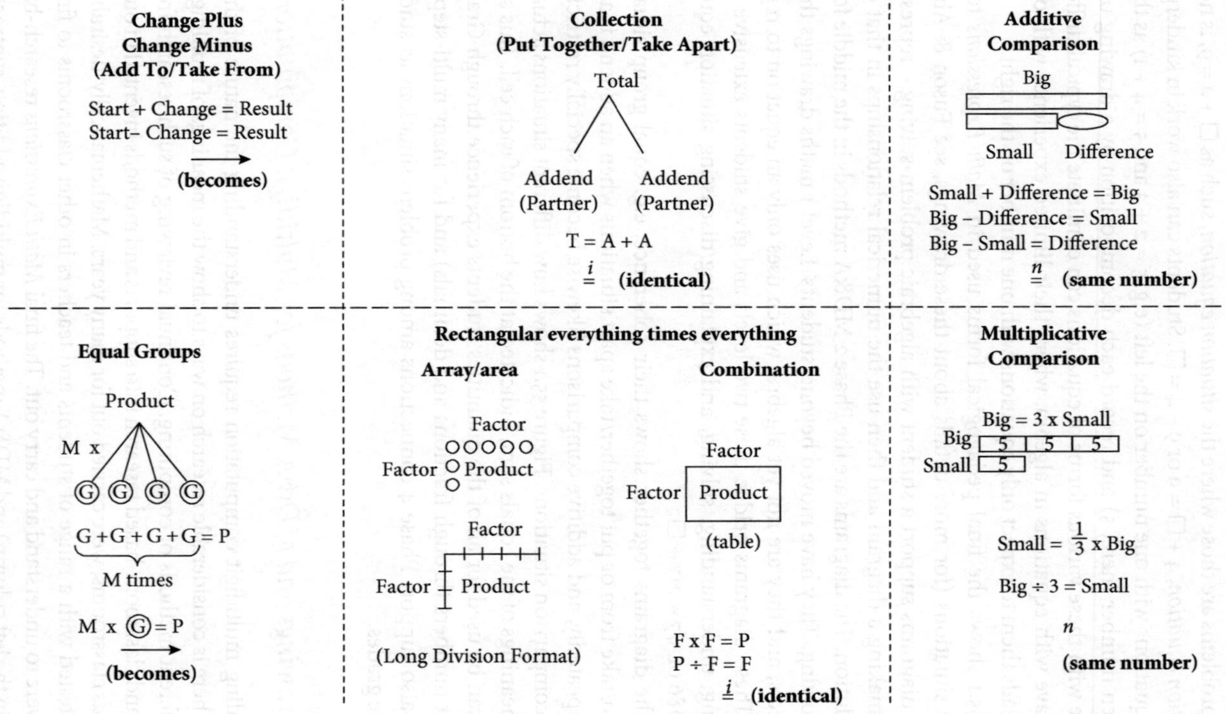

FIGURE 55.1 The related MD&A diagrams for addition (top row) and multiplication (bottom row) situations.

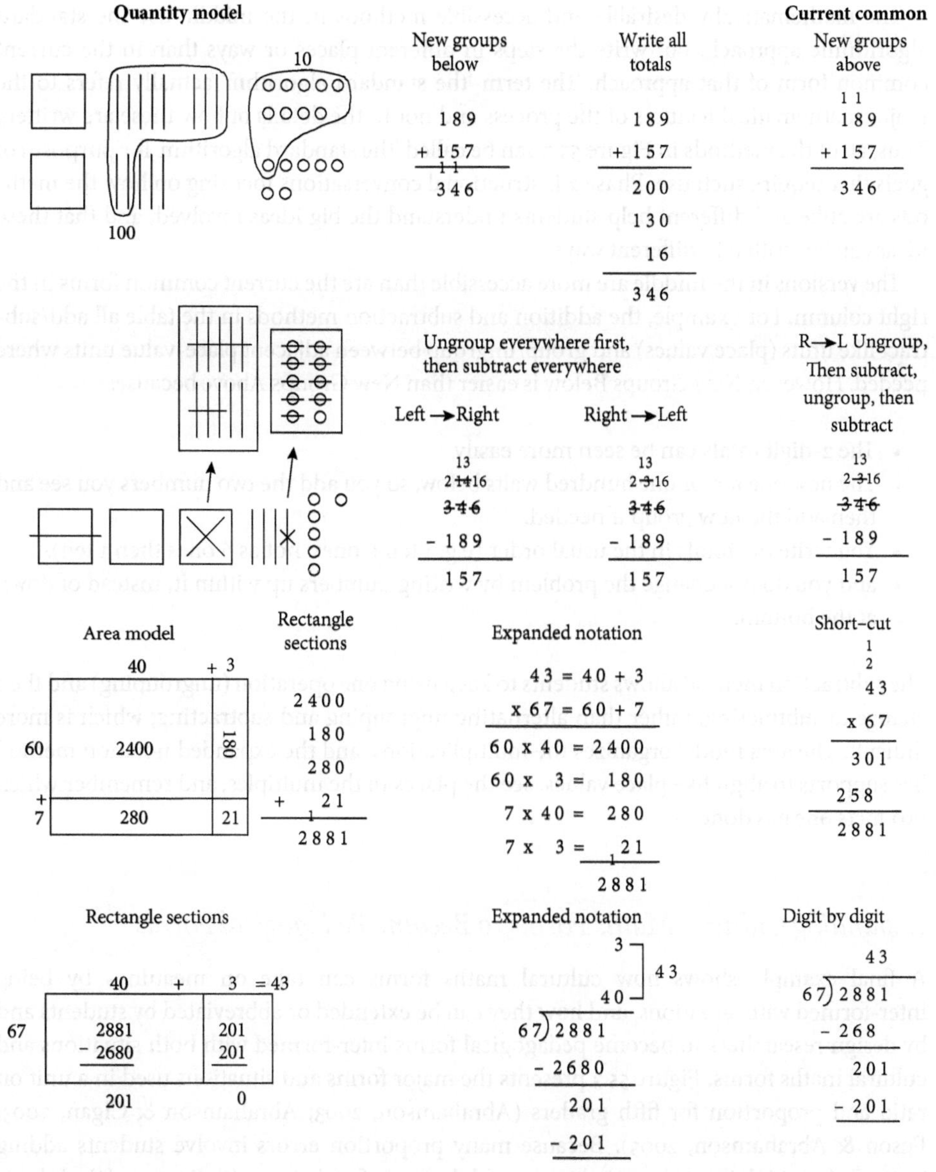

FIGURE 55.2 Mathematically-desirable and accessible algorithms and diagrams.

edu/profile/?p=61. Related pedagogical forms that support the meaningful development and use of the quick-hundreds, quick-tens, and ones and of the area models are also discussed in these resources. Students inter-form the drawings with the written methods (link them step-by-step) so that the cultural maths place-value symbols take on meanings as hundreds, tens, and ones. Students use such place-value language in their explanations to support this inter-forming. We found that some students want the extra support within a written method of the expanded notation forms that show the place values separately. Three of the MD&A methods show such expanded notation.

The mathematically-desirable and accessible methods in the middle use the standard algorithmic approach, but write the steps in different places or ways than in the current common form of that approach. The term 'the standard algorithm' actually refers to the major mathematical features of the process and not to the details of how these are written. Thus, all of the methods in Figure 55.2 can be called 'the standard algorithm' for purposes of goals that require such use. Phase 2 instructional conversations focusing on how the methods are alike and different help students understand the big ideas involved, and that these ideas can be written in different ways.

The versions in the middle are more accessible than are the current common forms in the right column. For example, the addition and subtraction methods in the table all add/subtract like units (place values) and group/ungroup between adjacent place-value units where needed. However, New Groups Below is easier than New Groups Above because:

- The 2-digit totals can be seen more easily.
- The new one ten or one hundred waits below, so you add the two numbers you see and then add the new group if needed.
- You write the totals in the usual order (e.g. 1 ten 6 ones, not as 6 ones then 1 ten).
- and you do not change the problem by writing numbers up within it, instead of down at the bottom.

The subtraction method allows students to keep using one operation (ungrouping) and then change to subtracting, rather than alternating ungrouping and subtracting, which is more difficult. The area model organizes the multiplications, and the expanded notation method has supports to align like place values, see the places in the multiples, and remember which products one has done.

Expanding Cultural Maths Forms to Become Pedagogical Forms

A final example shows how cultural maths forms can take on meanings by being inter-formed with situations, and how they can be extended or abbreviated by students and by design researchers to become pedagogical forms inter-formed with both situations and cultural maths forms. Figure 55.3 presents the major forms and situations used in a unit on ratio and proportion for fifth graders (Abrahamson, 2003; Abrahamson & Cigan, 2003; Fuson & Abrahamson, 2005). Because many proportion errors involve students adding instead of multiplying, we wanted to ground the topic firmly in multiplication. We did this by inter-forming ratio tables and 2×2 proportion forms with the multiplication table and with basic ratio incrementing stories, such as how the money in the banks of two siblings increased by \$3 (for Robin) and by \$7 (for Tim) a day. Students enacted this story by successive additions that they recorded in separate ratio tables and also in a joint ratio table with a connecting row on the left that showed the day (the unit that linked the successive ratios). All of these columns could be seen by students and were also discussed as columns from the multiplication table. A proportion was any two ratios from this story, which could be seen as (inter-formed as) two rows from the multiplication table and also from the linked ratio table. A Factor Puzzle was a proportion with one unknown value; factors of its rows and columns could be identified to find the unknown value.

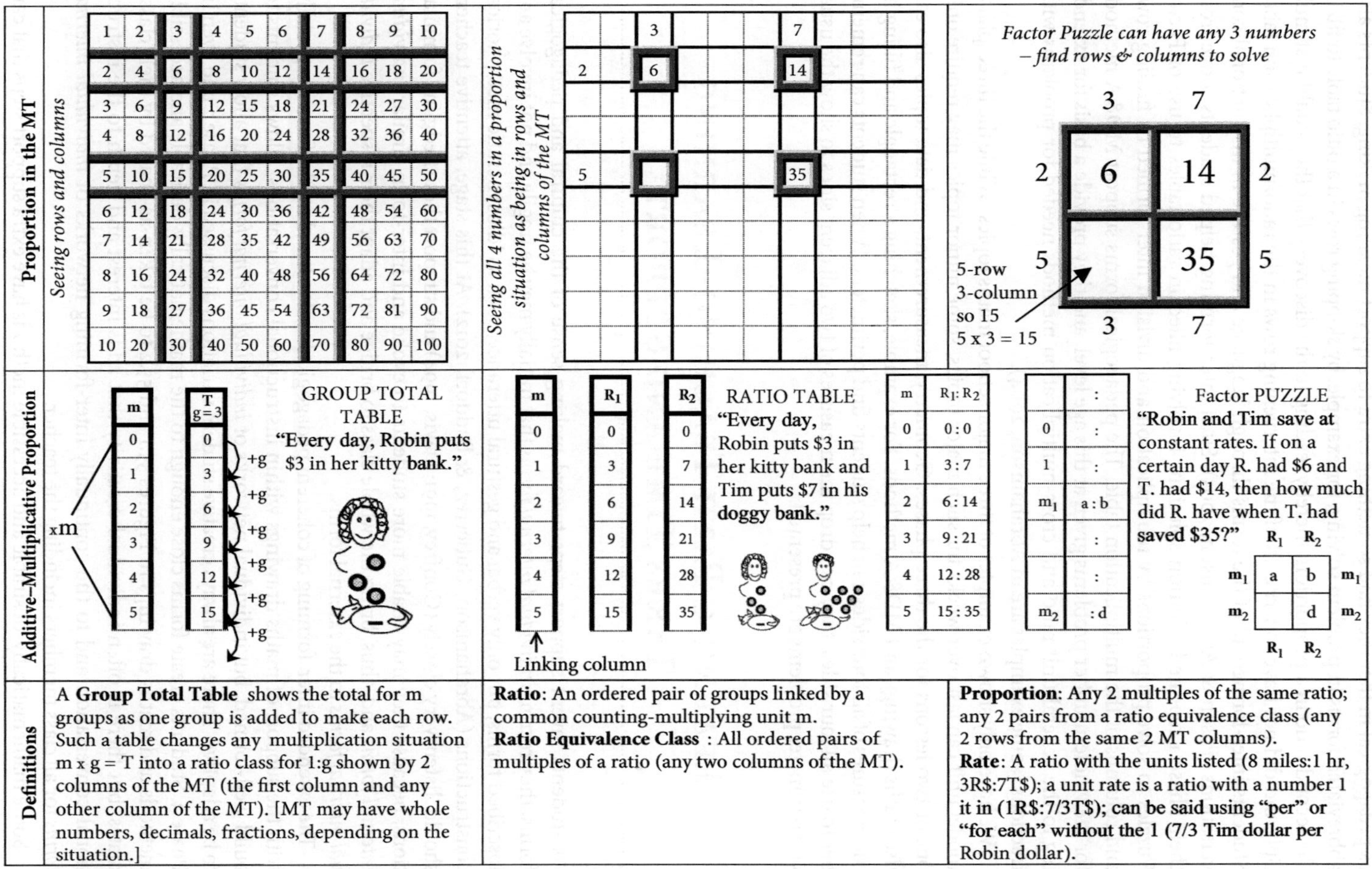

FIGURE 55.3 Using the multiplication table for teaching and learning ratio and proportion.

Most of these forms were very close to the cultural maths forms, but could be considered as pedagogical in that they had something extra to support meaning making. Students also abbreviated forms in problem solving, for example, by skipping rows in a ratio table to fill in the second ratio in a proportion, thereby leading to the discovery that they could just multiply and did not have to write all of the intervening rows in the ratio/multiplication table. Students inter-formed all of the forms through language and gesture in instructional conversations (Fuson & Abrahamson, 2005). Such inter-forming helped students move from their Phase 1 repeated-addition solutions involving filled-in ratio tables to use of Factor Puzzles to solve proportions by multiplication and division inter-formed by finding rows and columns of the multiplication table. The pedagogical forms support MD&A methods for the whole-number problems given at this age level, and they provide a basis for extending to the general unit ratio and cross-multiplication methods needed for problems with fractions (some examples are in Abrahamson, 2004).

We have only shown examples of three major school maths topics. While the three-phase model unit can last over weeks for such major topics, some minor topics may require only one or two lessons for the three phases. Geometry, measurement, and data topics also can use maths drawings and MD&A methods. For example, sketching a rectangle and filling in the measures of all four sides can help students find perimeters. When students can remember to use all four sides, they can drop the measures of two adjacent sides to show the usual way perimeter problems are presented.

Learning Path Teaching–Learning Takes Time and Support

As students invent support steps toward making sense of the cultural and pedagogical forms, they recruit *individual internal forms* that initially may be fragile and not involve any inscribed forms, but only verbal and gestural utterance, such as idiosyncratic metaphorical constructions (Abrahamson, Gutiérrez, & Baddorf, 2012). At this stage, attentive teachers should 'listen' very closely (Confrey, 1991; Davis, 1994) and support these fledgling formulations, because they may enable more students to evoke and inter-form similar *individual internal forms* and, thus, bring the whole class to bridge and adapt their respective *individual internal forms* to the external forms.

The extended inter-forming of coherent pedagogical and cultural maths forms, especially with the support of maths drawings within instructional conversations, allows students to build well-formed, but individual networks of *individual internal forms* that allow students to be adaptive. There are always creative student variations discussed in Phase 2 and even in Phase 3. Students create forms close enough to the mathematically-desirable and accessible methods and maths drawings in Figures 55.1 and 55.2 to be taken-as-shared by their classmates, but there is often individual creativity that adds interest and depth to the instructional conversation and to the continually inter-forming networks of *individual internal forms* for all class members including the teacher.

Some mathematically-desirable and accessible methods have extra support steps that can be dropped when they are no longer needed. For example, the partial products multiplication method in Figure 55.2 was invented by a class of low-achieving African–American students who wanted to put in all of the steps any student needed. Later many of them selectively dropped various supportive steps ('learner wheels'), just as students stop making maths drawings when they no longer need them. Pedagogical forms assist meaning-making for the cultural maths forms provided these are actively inter-formed within the student's network of *individual internal forms*. At some point for each student, the pedagogical form no longer needs to be used because the student's *individual internal forms* stimulate that meaning-in-action for the cultural maths forms (e.g. Abrahamson, 2002).

The *class* learning path simplifies the teacher's task to something that teachers perceive as do-able, especially if supported by a learning path programme. The teacher's task is not the commonly-perceived reform task of celebrating every student's methods and continually eliciting more ways. This can be overwhelming to teachers who think there are as many 'different' methods as the number of students (Murata, 2013). We instead use a big picture of the three levels of solution methods shown in Table 55.1: Level 1 basic and slow, Level 2a MD&A, and Level 2b mathematically-desirable and not accessible. These levels help teachers see certain methods as minor variations of each other and to place these within the phases of teaching a topic.

Learning Path Teaching–Learning requires Coherent External Forms

The teacher and (student) teachers 'tune' learners' *individual internal forms* toward the external forms and toward the more-advanced methods with the assistance of the pedagogical forms (similar to diSessa's tuning toward expertise in physics, 1993). Pedagogical forms need to be selected or designed to illuminate the central mathematical aspects of the cultural maths forms by their affordances and constraints (their *attunements*, Greeno, 1998). Such tuning takes time and much inter-forming by gesture and language by the teacher and students. Students' networks of *individual internal forms* have layers from less-advanced to more-advanced *individual-internal-forms-in-action*, and they may fold back (Martin, 2008; Pirie & Kieren, 1994) to a lower level to inter-form and make more meaningful a higher level. Because mathematics builds, the situational and pedagogical forms need to be coherent so that children can move among their layers of understanding easily. Our situational diagrams that work across all kinds of quantities is one such example. The use of the same quantity drawings for multidigit numbers in addition and in subtraction is another example. Methods in the middle that can extend from children's invented methods and relate to difficult formal maths methods require careful analysis and classroom research to reach coherence for teachers and for children. Pedagogical forms such as secret-code cards (layered cards that show 374, but with 300 under 70 under the 4) that compensate for difficult cultural forms such as English words for teens and tens are an important part of such analysis and research (see National Council of Teachers of Mathematics, 2011).

The extensive and excellent Dutch programme of research on Realistic Mathematics Education (e.g. Gravemeijer, 1994; Streefland, 1991) that extends Freudenthal's theory (e.g. Freudenthal, 1983) shares many features with the model proposed here. Our model does suggest a further examination of the coherence of that programme's situational and pedagogical 'models of' within and across topics, and raises the possibility that students might move to Phase 2 general methods in the middle more rapidly.

Using Research on Student Cognition for Mathematics Education

Initiatives have arisen in many countries to adapt mathematics education to reflect the research on student learning and to prepare students for the demands of lives infused with technology. Many of these initiatives have reflected the long-term perceived conflict between understanding and fluency, with emphases either on student inventing of methods or on traditional teaching of formal methods by teacher telling and showing (this conflict has been termed the 'maths wars' in the USA and some of these over-emphases are discussed in Fuson, 2009). We have seen here that there is a research-based middle ground: students can understand general methods and do not have to be limited to special methods that arise from particular numbers or from particular situations. This does raise questions about what seems to be a strong emphasis on such special methods in the National Numeracy Strategy in English primary schools (Askew & Brown, 2001; Vollaard, Rabinovich, Bowman, & van Stolk, 2008). A related emphasis in some countries on mental computation likewise seems too strong, especially if the methods cannot be written down by students. Such methods are often restricted to smaller numbers and do not generalize easily (e.g. methods of adding on or back for multidigit numbers; for more discussion of limitations of such methods, see National Council of Teachers of Mathematics, 2011; Fuson & Beckmann, 2012/2013). Our experience is that children are empowered by general methods they can understand and explain. Recording methods using place value notation is a core aspect of mathematics and can be done as early as age 7 if approached in the ways outlined in our model. However, it is also important to examine the form of the written methods that are to be taught because there may be variations that are easier for students to understand and/or to carry out.

A recent such initiative in the United States reflects this middle ground and heavily uses the research on cognition. There has been a special difficulty in the United States because different teaching/learning standards are adopted by each of the 50 states. There has been huge variation in the standards across states that led to characterizing the maths goals in the United States as 'a mile wide and an inch deep' (Schmidt, McKnight, & Raizen, 1997). Textbooks were enormous, and massive amounts of time and energy were spent on *what* to teach instead of *how* to teach it *well*. The new coherent teaching/learning standards, the Common Core State Standards (CCSSO/NGA, 2010), are based on research and were adopted by most states. The standards reflect research and curricula from around the world, and are the result of an intensive, prolonged feedback and revision period from many sources. Thus, they reflect a negotiated balance of views about how to fit together learning paths in various domains.

For example, the Common Core State Standards operations and algebraic thinking standards lay out an ambitious learning path with word problem types as the bases for understanding of operations (+, −, ×, ÷). The standards identify grade-appropriate levels at which students work with the various problem types and with unknowns for all three of the quantities. The standards appropriately specify that students use drawn models and equations with a symbol for the unknown number to represent the problem (*situation equations*, such as □ + 6 = 14). Thus, from grade 1 on students will have crucial experience with the more difficult algebraic problems (those in which the situation equation might vary from a solution equation, such as 6 + □ = 14 or □ + 6 = 14 for the situation equation 14 − 6 = □).

The Common Core State Standards drew from design-research and learning path research to include within standards the requirement that students are to use visual models, relate these to the problem situation or to the steps in a computation, and explain the reasoning used. For most numerical topics this meaning-making phase is one or two years ahead of the standard that calls for fluency, thus using phases that extend over years. These Phase 2 methods discussed above and shown in Figure 55.2 meet the more-advanced Common Core State Standards that students are to develop, discuss, and use efficient, accurate, and generalizable methods including the standard algorithm (for more see Fuson & Beckmann, 2012/2013). For more about features of these standards see Fuson (2012).

The Common Core State Standards emphasize the Maths Talk aspect of teaching-learning with eight Mathematical Practices that can be summarized in four pairs of practices:

- *Math Sense-Making (MP 1 and 6):* make sense and use appropriate precision.
- *Math Structure (MP 7 and 8):* see structure and generalize.
- *Math Drawings (MP 4 and 5):* model and use tools.
- *Math Talk (MP 2 and 3):* reason and explain.

This can be summarized as: Do math sense-making about math structure using math drawings to support math talk.

General Conclusion and Call for Future Research

In conclusion we assert that it is the responsibility of a research-based maths programme to provide:

(a) In Phase 1 the situations or pedagogical forms (especially maths drawings) to stimulate students' and Teacher's *individual internal forms* meaningfully toward the externals forms, and to have stimulated and practiced well-formed enough *individual internal forms* in earlier units before reaching Phase 1 for a topic.
(b) In Phase 2, the curriculum must provide to teachers research-based MD&A methods and pedagogical and/or situational forms to assist students to build well-formed *individual internal forms* that can assist them in using meaningful cultural maths forms.
(c) Pedagogical and formal maths external forms must be explained enough in the programme so that teachers can assist students to progress in their learning paths.

Point (c) is important because many teachers have not had sufficient opportunity to learn maths meaningfully themselves or to learn about student learning paths. They need the assistance of the coherent supports within Phase 1 and Phase 2 to form their own *individual internal forms* to teach with meaning. In our experience, teachers enthusiastically welcome the opportunity to learn meaningful maths and to use a 'teach while learning' approach. One group of pilot teachers articulated such feelings by calling the Children's Math Worlds programme 'maths therapy for teachers.'

Teachers' experiences vary considerably on the whole continuum of experiences from very constructivist to traditional. The Class Learning Path Model enables any Teacher to begin from initial strengths s/he has and to build new teaching–learning competencies as s/he moves along her/his own learning path. All teachers find phases within which they initially feel comfortable. They all gain confidence and knowledge from the learning supports in Phase 2. As they experience the Class Learning Path Model of teaching, they build competencies and understandings (*individual internal forms* about maths and about teaching) that enable them to use a more balanced approach to teaching in the following year.

The cultural maths forms for a topic are fairly well defined for a given culture, and there is relatively little variation in these around the world. What is and can be varied to affect learning are the situations, pedagogical forms including especially maths drawings and MD&A methods, and the sequence of problems and activities. More research and dialogue about the external teaching forms (situations, pedagogical, and cultural maths forms) would be beneficial. This dialogue needs to focus on the mathematical aspects of the learning paths (e.g. which methods are mathematically desirable?), as well as on data about them (e.g. How did the pedagogical forms work? How could they be made more coherent across topics and grades?). In many countries, an initiative to modify difficult forms of standard algorithms and other solution methods to mathematically-desirable forms more adapted to student cognizing would make mathematics education more successful and help students and teachers believe that mathematics is understandable.

References

Abrahamson, D. (2002). Optical illusions and computation formats: supporting Chandra's learning path to advanced ratio and proportion word problems. Unpublished manuscript, Northwestern University, Evanston, IL.

Abrahamson, D. (2003). Text talk, body talk, table talk: a design of ratio and proportion as classroom parallel events. In N. A. Pateman, B. J. Dougherty & J. Zilliox (eds.), Twenty Seventh Annual Meeting of the International Group for the Psychology of Mathematics Education. Eric Clearinghouse for Science, Mathematics, and Environmental Education, 2, 1–8.

Abrahamson, D. (2004). Keeping meaning in proportion: the multiplication table as a case of pedagogical bridging tools. Unpublished doctoral dissertation, Northwestern University, Evanston, IL. Available at: http://tinyurl.com/AbraDiss. ISBN 0-496-79720-0 (Accessed 20 Dec 2013).

Abrahamson, D., & Cigan, C. (2003). A design for ratio and proportion. Mathematics Teaching in the Middle School, 8(9), 493–501.

Abrahamson, D., Gutiérrez, J.F., & Baddorf, A.K. (2012). Try to see it my way: the discursive function of idiosyncratic mathematical metaphor. Mathematical Thinking and Learning, 14(1), 1–26.

Askew, M. & Brown, M. (eds) (2001). Teaching and Learning Primary Numeracy: Policy, Practice and Effectiveness. Southwell: British Educational Research Association.

CCSSO/NGA. (2010). Common Core State Standards for Mathematics. Washington, DC: Council of Chief State School Officers and the National Governors Association Center for Best Practices. Available at: http://corestandards.org/ (Accessed 1 July 2010).

Case, R. (1991). The Mind's Staircase: Exploring the Conceptual Underpinnings of Children's Thought and Knowledge. Hillsdale, N.J.: Lawrence Erlbaum Associates, Publishers.

Case, R. & Okamoto, Y. (1996). The role of central conceptual structures in the development of children's thought. Monographs of the Society for Research in Child Development, Series No. 246, Vol. 61, Nos. 1–2.

Clements, D.H., & Sarama, J. (2009). Learning and Teaching Early Math: The Learning Trajectories Approach. New York: Routledge.

Clements, D.H., & Sarama, J. (2012). Learning and teaching early and elementary mathematics. In J. Carlson, & J. Levin (eds). Instructional Strategies for Improving Students' Learning (pp. 107–162). Charlotte, NC: Information Age Publishing, 107–162.

Cobb, P., & Bauersfeld, H. (eds) (1995). The Emergence of Mathematical Meaning. Hillsdale, N.J.: Lawrence Erlbaum Associates, Publishers.

Confrey, J. (1991). Learning to listen: a student's understanding of powers of ten. In E. v. Glasersfeld (ed.), Radical Constructivism in Mathematics Education (pp. 111–138). Dordrecht, The Netherlands: Kluwer Academic Publishers.

Confrey, J. (2005). The evolution of design studies as methodology. In R. K. Sawyer (ed.), The Cambridge Handbook of the Learning Sciences (pp. 135–151). Cambridge, MA: Cambridge University Press.

Davis, B. (1994). Mathematics teaching: moving from telling to listening. Journal of Curriculum and Supervision, 9(3), 267–283.

diSessa, A.A. (1993). Towards an epistemology of physics. Cognition and Instruction, 10 (2–3), 105–225.

Donovan, M.S., & Bransford, J.D. (eds) (2005). How Students Learn: Mathematics in the Classroom. Washington, DC: National Academies Press.

Freudenthal, H. (1983). Didactical Phenomenology of Mathematical Structures. Dordrecht, Holland: D. Reidel Publishing Company.

Fuson, K.C. (1992). Research on whole number addition and subtraction. In D. Grouws (ed.), Handbook of Research on Mathematics Teaching and Learning (pp. 243–275). New York: Macmillan.

Fuson, K.C. (2003). Developing mathematical power in whole number operations. In J. Kilpatrick, W. G. Martin, & D. Schifter (eds), A Research Companion to Principles and Standards for School Mathematics (pp. 68–94). Reston, VA: National Council of Teachers of Mathematics.

Fuson, K.C. (2006, 2012, 2013). Math Expressions. Boston, MA: Houghton Mifflin.

Fuson, K. C. (2009). Avoiding misinterpretations of Piaget and Vygotsky: Mathematical teaching without learning, learning without teaching, or helpful learning-path teaching? Cognitive Development, 24, 343–361. doi:10.1016/j.cogdev.2009.09.009

Fuson, K.C. (2012). The Common Core Mathematics Standards as supports for learning and teaching early and elementary mathematics. In J. Carlson, & J. Levin (eds) Instructional Strategies for Improving Students' Learning (pp. 177–186). Charlotte, NC: Information Age Publishing.

Fuson, K. C. (2013). Professional development webcasts for Common Core State Standards. Links are listed under Projects on https://www.sesp.northwestern.edu/profile/?p=61).

Fuson, K.C., & Abrahamson, D. (2005). Understanding ratio and proportion as an example of the Apprehending Zone and Conceptual-Phase problem-solving models. In J. Campbell (ed.), Handbook of Mathematical Cognition (pp. 213–234). New York: Psychology Press.

Fuson, K.C., & Abrahamson, D. (2012). Word problem types, numerical situation drawings, and a Conceptual-Phase Model to implement an algebraic approach to problem-solving in elementary classrooms. Manuscript under revision.

Fuson, K.C., Atler, T., Roedel, S., & Zaccariello, J. (2009). Building a nurturing, visual, Math-Talk teaching-learning community to support learning by English Language Learners and students from backgrounds of poverty. New England Mathematics Journal, XLI (May), 6–16.

Fuson, K.C. & Beckmann, S. (2012/2013). Standard algorithms in the Common Core State Standards. National Council of Supervisors of Mathematics Journal of Mathematics Education Leadership, Fall/Winter, 14 (2), 14–30.

Fuson, K.C. & Murata, A. (2007). Integrating NRC principles and the NCTM Process Standards to form a Class Learning Path Model that individualizes within whole-class activities. National Council of Supervisors of Mathematics Journal of Mathematics Education Leadership, 10(1), 72–91.

Fuson, K.C., Murata, A., & Abrahamson, D. (2011). Forming minds to do math: ending the math wars through understanding and fluency for all. (Manuscript under revision.)

Gravemeijer, K.P.E (1994). Developing Realistic Mathematics Education. Utrecht: Freudenthal Institute.
Greeno, J. G. (1998). The situativity of knowing, learning, and research. American Psychologist, 53(1), 5–26.
Hiebert, J., Carpenter, T., Fennema, E., Fuson, K.C., Murray, H., Olivier, A., Human, P., & Wearne, D. (1996). Problem solving as a basis for reform in curriculum and instruction: the case of mathematics. Educational Researcher, 25(4), 12–21.
Hufferd-Ackles, K., Fuson, K., & Sherin, M.G. (2004). Describing levels and components of a math-talk community. Journal for Research in Mathematics Education, 35(2), 81–116.
Kilpatrick, J., Swafford, J., & Findell, B. (eds) (2001). Adding It Up: Helping Children Learn Mathematics. Mathematics Learning Study Committee. Center for Education, Division of Behavioral and Social Sciences and Education. Washington, DC: National Academy Press.
Lewis, C., & Takahashi, A. (eds) (2006). Learning Across Boundaries: US–Japan Collaboration in Mathematics, Science and Technology Education. Oakland, CA: Mills College School of Education.
Martin, L.C. (2008). Folding back and the dynamical growth of mathematical understanding: elaborating the Pirie–Kieren theory. Journal of Mathematical Behavior, 27(1), 64–85.
Murata, A., & Fuson, K.C. (2006). Teaching as assisting individual constructive paths within an interdependent class learning zone: Japanese first graders learning to add using ten. Journal for Research in Mathematics Education, 37(5), 421–456.
Murata, A. (2008). Mathematics teaching and learning as a mediating process: the case of tape diagrams. Mathematical Thinking and Learning, 10(4), 374–406.
Murata, A. (2013). Diversity and high academic expectations without tracking: Inclusively responsive instruction. *The Journal of Learning Sciences*, 21(2). 312-335. doi:10.1080/10508406.2012.682188
Murata, A., & Kattubadi, S. (2012). Grade 3 students' mathematization through modeling: situation models and solution models with multi-digit subtraction problem solving. Journal of Mathematical Behaviors, 31(1), 15–28.
National Council of Teachers of Mathematics (2000). Principles and Standards for School Mathematics. Reston, VA: Author.
National Council of Teachers of Mathematics (NCTM) (2010). Focus in Grade 1: Teaching with Curriculum Focal Points. Reston, VA: NCTM.
National Council of Teachers of Mathematics (NCTM) (2011). Focus in Grade 2: Teaching with Curriculum Focal Points. Reston, VA: NCTM.
Piaget, J. (1965/1941). The Child's Conception of Number. New York: Norton.
Pirie, S.E.B., & Kieren, T. (1994). Growth in mathematical understanding: how can we characterize it and how can we represent it? *Educational Studies in Mathematics*, 26, 165–190.
Schmidt, W.H., McKnight, C.C., & Raizen, S.A. (1997). A Splintered Vision: An Investigation of US Science and Mathematics Education. Dordrecht: Kluwer.
Streefland, L. (1991). Fractions in Realistic Mathematics Education. Dordrecht: Kluwer Academic Publishers.
Vollaard, B.A., Rabinovich, L., Bowman, R., & van Stolk, C. (2008). Ten Years of Reform in Primary Mathematics Education in England: A Review of Effectiveness. London: National Audit Office.
Vygotsky, L.S. (1934/1986). Thought and Language. Cambridge, MA: MIT Press.
Vygotsky, L.S. (1978). Mind in Society: The Development of Higher Psychological Processes. Cambridge, MA: Harvard University Press.

CHAPTER 56

NEW POSSIBILITIES FOR EARLY MATHEMATICS EDUCATION

cognitive guidelines for designing high-quality software to promote young children's meaningful mathematics learning

HERBERT P. GINSBURG, RACHAEL LABRECQUE, KARA CARPENTER, AND DANA PAGAR

Introduction

The thesis of this chapter is that the affordances of computer software, when guided by key principles of cognitive developmental psychology, can revolutionize early mathematics education. At present, only limited evidence supports our claim, but we hope to convince you that explorations of and experiments with software can result in extraordinary benefits for young children's mathematics learning.

The next sections present a brief account of the need for extensive early mathematics education, the current view of children's mathematical competence, the essential content of early mathematics education, and a basic principle of pedagogy. Then we offer a cognitive framework for the design, development, and evaluation of effective mathematics software. We illustrate the framework through an account of our own experiences in creating *MathemAntics*, comprehensive mathematics software for children from ages 3 to 8. We conclude with some lessons for early childhood educators, software developers, and researchers.

The Need for Transforming Early Mathematics Education

There are many compelling reasons for improving early childhood mathematics education. One is that, at least in the USA, many children perform poorly on standard tests of achievement as early as preschool (ages 3 and 4; Miller & Parades, 1996; Stevenson, Lee, & Stigler, 1986), but certainly by the 3rd or 4th grade (ages 8 and 9; Lemke & Gonzales, 2006). Even worse, within the USA, disadvantaged children's mathematics achievement is far below that of their more privileged counterparts (Arnold & Doctoroff, 2003).

Another reason is that even children who score well on standard achievement tests may not understand mathematics in a deep way (Ginsburg, Pappas, Lee, & Chiong, 2011). This should not be surprising when a widespread practice is teaching to uninspiring tests, which all too often produces students who can use procedures and memorize facts, but cannot think well.

A third reason is that, at least in the USA, preschools teach very little mathematics (Ginsburg, Lee, & Boyd, 2008). Most preschool teachers typically instruct children in a very narrow range of mathematical content. They often limit their focus to the names of the common shapes (Graham, Nash, & Paul, 1997) and the relatively small counting numbers, up to about 20. They generally do little to encourage addition or estimation, and seldom use proper mathematics terminology (Frede, Jung, Barnett, Lamy, & Figueras, 2007, p. 21). The situation does not improve after children leave for elementary school, where mathematics instruction, even for middle-SES students, is of poor quality (Pianta, Belsky, Houts, & Morrison, 2007).

In brief, many children (both high and low achievers as measured by test scores) are not learning mathematics as well as they should and the schools are not teaching it as effectively as they should, or are not teaching it at all, especially at the preschool level. The goal then must be to develop extensive and effective programs of early mathematics education designed to promote both skill and understanding.

Children's Mathematical Minds

Is the goal attainable? Yes. One reason is that young children are quite capable of learning mathematics when it is taught well. Over the past 35 or 40 years, researchers have produced a voluminous literature (Sarama & Clements, 2009) on children' mathematical thinking, beginning in infancy. We can touch only on a few major points here.

First, young children are surprisingly competent in many areas of mathematical thinking. They have a spontaneous and sometimes explicit interest in mathematical ideas. For example, as they play with blocks, 4-year-olds spend a good deal of time determining which tower is higher than another, creating and extending interesting patterns with blocks, exploring shapes, and creating symmetries (Seo & Ginsburg, 2004). Furthermore, young children understand basic ideas of addition and subtraction from an early age (Baroody, Lai, & Mix, 2006). They not only have the computational skill to add 3 toy dogs to 4 toy dogs to get the sum, but also have *ideas* about counting objects, including the abstraction principle (any discrete objects can be counted, from stones to unicorns; Gelman & Gallistel, 1986)

Secondly, young children generally begin schooling with positive motivation. They are self-confident and have positive attitudes towards school. However, their motivation decreases within the first few years of school, most likely because of educational factors such as boring and inappropriate teaching (Arnold & Doctoroff, 2003). Yet before schooling interferes, they are splendid little mathematicians.

At the same time, young children have a great deal to learn. They entertain various misconceptions about mathematics. For example, they fail to realize that an odd-looking isosceles triangle (for example, an extremely elongated, non-right-angle, 'skinny' triangle) is as legitimate a triangle as one with three sides the same length (Clements, Swaminathan, Hannibal, & Sarama, 1999). They often think that the set that looks bigger than another (perhaps because the former is spread out) is the more numerous (Piaget, 1952).

Perhaps most importantly, there is often a divide between what children already know and what schools teach them – namely isolated routines and surface manipulations (Hatano, 1988, p. 66). For example, children may use a sensible strategy to add three things to two and get five. However, at the same time, they may fail to understand the meaning of the expression 2 + 3 = 5 (Ginsburg et al., 2011). Although their everyday mathematics is relatively powerful, albeit with significant weaknesses, children have difficulty relating what they know to what is taught in school. For them, schooling in mathematics is merely 'academic,' in the sense of divorced from everyday life and meaningless.

The Content of Early Mathematics Education

The literature on mathematical thinking suggests a useful perspective on mathematics education. The subject matter of mathematics education is usually defined in terms of *topics*, including basic ideas of number, geometry, space, pattern, measurement, and more. But mathematics is more than topics.

Thinking

Early mathematics does not involve simply memorizing some number words or identifying plain plane shapes. It must include *abstract thought*. Children need to learn to *reason* about number (if 2 and 3 is 5, then 3 and 2 must also be 5), making *inferences* (if we add something other than 0 to 3, the sum must be bigger than 3) and developing a *mental number line* (100 is much further away from 2 than is 20). Understanding shape involves more than knowing a figure's name, although knowledge of correct mathematical vocabulary is certainly necessary. Children need to learn to analyse and construct shapes and to understand their defining features (Clements, 2004).

Metacognition

Various *metacognitive* functions also play a key role in mathematics learning. Children need to learn to be aware of and verbalize their mathematical strategies. They need to be able to put their thinking into words so that they can communicate it to others and thus make the first steps towards joining a community that values (or should value) discussion, argument, and proof. Learning mathematics is, in part, developing self-understanding and expressive language.

Mathematizing

Children also need to eventually conceive of problems in explicitly mathematical terms. For example, they need to learn that when you want to determine the combined number of two sets, you should 'add' and that the symbol + refers to the adding that they already understand and often can do. After all, the chief point of mathematics instruction is to help the child to think abstractly and symbolically and to understand the meaning of mathematical symbols and abstractions.

Positive Disposition

Learning mathematics also involves feelings and motives (Kilpatrick, Swafford, & Findell, 2001). Unfortunately, many adults, at least in the USA, have an aversion to the subject. Yet children begin school with good motivation to learn mathematics; unfortunately, the education system teaches them to dislike it, often beginning around the age of 8 or so. Therefore, it is important to promote a sense of competence, interest, initiative, persistence, and focused engagement in mathematical activities.

A Basic Pedagogical Principle: Building a Synthesis of the Everyday and the Formal

Given the gap between what children already know (their everyday mathematics) and what the schools teach (formal mathematics), a basic (and perhaps the primary) pedagogical principle is to begin with the child's existing knowledge, build on it, and integrate it with formal, school-based mathematics. The synthesis of the everyday and the scientific (explicitly organized knowledge) helps the child to understand the meaning of the formalization (Vygotsky, 1986). It is not enough simply to extend the child's everyday knowledge as such or to teach formal mathematics in isolation. Neither leads to genuine understanding.

What Computers Can Afford

Despite past false alarms, high quality software can serve as the basis for radical changes in mathematics learning, beginning with the 2-year-old exploring the iPad and extending throughout adulthood. Textbooks, desks, and teachers lecturing to a large number of cognitively empty seats are SO medieval (that is, 20th century and earlier). Although there is still a digital and bandwidth divide between the disadvantaged and advantaged, at least in the USA (Judge, Puckett, & Bell, 2006), relatively inexpensive computers seem to be in almost every pocket and pocketbook. We carry around our smart phones, touch pad tablets, and laptops. Many of us have at least one computer at home. Furthermore, everyday observation shows that even extremely young children can use smart phones, computers, and tablets, for better or worse: they employ the mouse to select and play games on the web, and manipulate (in the literal sense of using the fingers) icons and virtual objects on the iPad. The widespread (although far from universal) availability of relatively inexpensive computers and growing access to adequate bandwidth provides the opportunity to produce a transformative restructuring of mathematics education, just as it has produced major changes in the way we shop and read the news.

Software affordances can provide learning opportunities not possible in the typical classroom, limited as it is to the technology of the book, the white board, paper, and perhaps physical manipulatives. The world of the computer includes analogues of all those features, and in addition offers many affordances that the physical world, including manipulatives, cannot. Furthermore, software can enable children to develop 'situated abstraction' (Hoyles & Noss, 2009), the kind of synthesis between the everyday and the formal described above. Using tools, making connections between different types of objects, models, symbols, and concepts, and receiving immediate feedback, learners can develop a deep understanding of abstract concepts. Note that we do *not* claim that computers have magical powers that surpass everything that the white board, paper, and physical manipulatives can do. These traditional methods can contribute to education. Furthermore, as we shall see later, much available mathematics software is of poor quality. Our claim is that sound computer software has the *potential* to revolutionize mathematics education.

The computer can immerse and engage children in a 'microworld,' an environment artificially constructed to embody basic mathematical ideas and skills and to offer powerful tools to do significant mathematics (Hoyles & Noss, 2009; Papert, 1980).

Microworlds can include:

- *Goal-driven activities* (e.g. estimating the numerical value of a large set).
- *Virtual objects* to manipulate and explore (e.g. angry or happy birds).
- *Tools* (e.g. a device that organizes objects in straight lines);
- *Representations* (e.g. number line).
- *Scaffolds* (e.g. providing hints or an appropriate level of problem difficulty).
- *Pedagogical agents* (e.g. characters who teach).
- *Feedback* (e.g. providing information about accuracy).
- *Interaction, fantasy, challenge, and flow* (e.g. challenging and weird games).
- *Communication and collaboration* (e.g. sharing one's work with others).
- *Record keeping and reporting* (e.g. information about accuracy, levels, and types of achievement).

A microworld involving all of these features certainly appears to offer richer opportunities for learning than does a textbook!

Goal-Driven Activities

Software can provide goals, from the simple (e.g. 'Shoot down as many number 7s as you can') to the complex (e.g. 'Use these shapes to make a trapezoid'). The goals can be outcome-oriented (e.g. 'Try to get as many right as you can') to process-orientated (e.g. 'Figure out a good way to make squares out of triangles').

Virtual Objects

The objects in a microworld can be almost anything that can be conveyed visually, from the concrete to the abstract. They can be animals, shapes, written numerals, blocks like Unifix Cubes, images of the child using the software, and assorted fantasy characters, such as space aliens or monsters that can talk or dance. The unlimited nature and number of virtual objects afford the opportunity to engage the child in a meaningful narrative or to connect virtual objects of different types, like birds and number lines. The static illustrations offered by textbooks pale in comparison.

Tools

Technology can provide powerful tools. For example, the Building Blocks curriculum tools give children the power to rotate, flip, slide, duplicate, combine, and deconstruct shapes to solve increasingly difficult challenges (Sarama & Clements, 2002a). Most of these activities are easier to do with the computer than with real objects. The computer eliminates the need for difficult or impossible physical manipulation, and allows the child to focus on the mathematics of shape building rather than the physical constraints of adjustment and fit. For example, a geometric tool can allow manipulation of a triangle in such a way as to change its interior angles, stretch it out or squash it in.

Ironically, another tool that touch screen technology affords is the finger! Touch screens allow the child to move objects around the microworld. Not only does this eliminate the need for a mouse or touch pad, but it also can exploit gesture and 'embodied cognition' in learning (Goldin-Meadow, Cook, & Mitchell, 2009).

Representations

Technology also supports useful representational models not possible in traditional print media. For example, MotionMath™ provides children with a number line that zooms in to reveal more and more precise decimal notation and zooms back out to whole number notation. In order to locate the number 5, the child needs to be at the whole number level, but in order to locate the number .5, the child needs to zoom in to the tenths level between 0 and 1. The software allows the child to engage with a powerful visual model of a deep concept, the infinite divisibility of the number line.

Scaffolds

Software can offer supports designed to help children cope with problems at different levels of difficulty. For example, the microworld may offer different tools to solve different kinds of problems. It may speak hints, like 'Why don't you try to count?' In a sense, the software can provide a tutor attuned to each child's needs. In a real life class of 30 students, how can a teacher give appropriate and detailed scaffolding to individual students working at different levels simultaneously?

Pedagogical Agents

Another way of introducing scaffolding and teaching is through the use of pedagogical agents – characters who teach. Activities of this type can result in improved learning and interest (Lester, Converse, Kahler, Barlow, Stone, & Bhogal, 1997), especially when the agent provides interactive explanations (Moreno, Mayer, Spires, & Lester, 2001).

Feedback

Software can, of course, provide immediate feedback about accuracy. This alone can be useful. How often do students have to wait for the teacher's grading to learn if a solution is correct? Software can do much more: it can comment on the child's strategy of solution; it can provide visual information such as the impossibility of fitting two objects into a box designed to hold one; it can provide the opportunity for a child to feel that a circle he draws around 33 objects takes longer to make than a circle drawn around 3 and looks larger too.

Interaction, Fantasy, Challenge, and Flow

Because too many computer activities with an educational focus tend to fall flat with children, the educational arena has much to learn from video games (Gee, 2005). Successful computer games involve children as *interactive participants* whose choices affect ongoing play. Games often involve *fantasy* (Malone & Lepper, 1987). They tell a story that children can connect to. A successful fantasy does not need to involve sophisticated graphics; imagination fills in the gaps (Reeves & Nass, 1999). The early Pokémon computer characters were fuzzy, pixilated shapes, yet children fell in love with the characters and the stories (Schell, 2010). The danger in this case is that too many existing games involve violent and mindless adventures that should have no place in education.

Challenge is integrated into computer games by matching activities to a student's level, so that the activity is neither too dull nor too difficult. Ideally, educational games should assess a student's understanding and adjust the game accordingly.

Increasingly, game designers are striving to cultivate '*flow*' in gaming, the psychological state of being 'in the zone' (Csíkszentmihályi, 1991). Flow experiences are supported by clearly defined goals and rules as well as appropriate challenges and immediate feedback. Kiili (2005) builds on flow theory, creating an experiential gaming model that explores how to best integrate flow into educational games and discusses the importance of storytelling and considerations of cognitive load in game design.

Communication and Collaboration

Digital connectivity offers opportunities for sharing and communication. Elementary students can use digital tools to solve mathematics problems and then create digital solution sheets that are shared with classmates and teachers (Bottino & Chiappini, 2002). The solution sheets support metacognitive thinking by requiring students to select pertinent information from the workspace to copy into the solution space and to write notes to explain their work. Students can look at other students' solutions as an example, but may not copy and paste the others' work, and must instead recreate the solution themselves.

Connectivity also supports the development of communities of learners around particular tools. Scratch, a programming language designed by the MIT Media Lab to be simple enough for a kindergartner, is the focus of an active community in which young children, as well as older children and adults, share the interactive games and drawings they create, uploading more than 1500 interactive projects daily (Resnick, Maloney, Monroy-Hernández, Rusk, Eastmond, Brennan, et al. 2009). Children play each other's games and comment on each other's projects, sharing tips and learning from one another.

Through digital connectivity, children are also able to participate in knowledge building. Knowledge may not be transmitted or constructed only by the individual, but instead may be continually developed and refined by a community of learners, as in the case of Wikipedia (Scardamalia & Bereiter, 2006). Using Knowledge Forum®, students first create notes about a given topic or question using both text and graphics. Then, the students, teacher, or guests can organize the notes into different structures, creating different levels for summaries, main points, and details. Each iteration reflects the knowledge that the community is building.

Record Keeping and Reporting

Additionally, the recent trend towards stealth assessment in gaming offers the potential to capture information about what children learn while they are working with software and to send informative reports to teachers and parents (Shute, 2011). Increasingly software designers rely on data mining techniques that examine computer log files to identify important features of students' work, for example, the conditions under which they use certain tools or the patterns of errors that imply use of buggy procedures. Data mining methods help designers to create more effective educational software (Romero & Ventura, 2007) and can provide the basis for useful teacher reports.

Current Educational Software Misses the Mark

As we have seen, computers offer unique affordances and tools that can promote deep mathematics learning. Unfortunately, current educational software programs fail to exploit them or do it badly. Literally thousands of 'educational' software programs and apps designed for children of all ages, including infants, are commercially available; this is especially true for mathematics software, which ranges from free or cheap apps, to comprehensive academic tutoring systems.

Most available mathematics software can be placed into one of the following three categories: drill, edutainment, or free exploration environments (Clements, 2004). For example, drill-based software is little more than a glorified, but efficient, worksheet with emphasis on practice and repetition of procedural tasks. Instruction, scaffolds, and helpful feedback tend to be neglected. Edutainment software is often connected to popular children's television shows, using familiar, well-loved characters to initiate and maintain interest. Although perhaps fun and interactive, their limited content and crude pedagogy typically provide little to no educational value. Free exploration environments can offer opportunities for more profound learning and discovery of deep mathematical concepts. However, these environments tend to lead to superficial understanding (Sarama & Clements, 2002b). Additionally, children may lose interest unless given a specific task to complete (Clements & Battista, 2000).

Many publishers have recognized the importance of research as a marketing tool and, therefore, advertise their software as 'research-based,' and thus of high quality. This claim, however, is often questionable, as research is seldom used in meaningful and effective ways during the development process (Sarama & Clements, 2002b). Research may consist of small focus groups or limited beta-testing with a convenient sample, such as the developers' own children, and occur too late in the process to produce substantial changes (Clements & Battista, 2000). Many software developers now seem to believe that anyone can be an educational game developer; unfortunately, they are right. Individuals with little to no background in education or cognitive psychology, and with little understanding of children's developmental progressions, mathematical complexities, and meaningful research, are taking the lead in developing software for children.

Of course, there are some laudable exceptions. One is the *Building Blocks* Project (Sarama & Clements, 2002a), a mathematics curriculum for pre-kindergarten that utilizes both print- and computer-based activities developed from a rigorous research-based process. This curriculum has shown positive results on pre-kindergarten children's mathematics learning (Clements & Sarama, 2007). A second exception, *The Number Race,* specially developed to aid children with dyscalculia and based on cognitive neuroscience research (Wilson, Dehaene, Pinel, Revkin, Cohen, & Cohen, 2006), has also shown positive effects on a small sample of 7–9-year-olds with mathematical learning difficulties (Wilson, Revkin, Cohen, Cohen, & Dehaene, 2006). A third notable exception involves research-based computer software, *Dots2Track* (Butterworth & Laurillard, 2010), carefully designed to ameliorate core quantitative difficulties in children suffering from dyscalculia. In general, however, current software for young children fails to exploit the potential that the computer offers to guide meaningful and systematic mathematics learning (Bottino & Chiappini, 2002)

The Development Process

We next propose a framework for the development of high quality research-based educational software for young children. The five-stage development process consists of:

- Humble theory-based design.
- Formative research.
- Revision.
- Learning studies.
- Evaluation.

The development process is not simple or sequential; it is complex and iterative. For example, results of formative research might lead to revisions of fundamental designs, or to minor tweaks of a tool or of the appearance of a virtual object. Major and minor revisions then need to be evaluated through further formative research. Doing all this should precede a learning study, but the results of that study may send the software back to the drawing boards, resulting in new designs that again require formative research. We illustrate the process by means of our own work developing new mathematics software, *MathemAntics*.

Stage 1: Humble Theory-Based Design

The development of high quality mathematics software should draw upon a coherent theory of mathematical cognition and learning trajectories (Clements & Sarama, 2007). At the same time, software development is not totally a theory-based or rational process. For one thing, existing theory may be limited or even wrong, and hence should be applied with a grain of salty humility. For another, a good dose of creativity and intuition are also required. Software design is part humble science and part fanciful art.

Most importantly, to support learning, educational technology must be designed around the learner's needs and abilities. In *learner*-centred design, scaffolding tools guide children from novices to experts, as opposed to *user*-centred design, which assumes users already know how to do a task (e.g. write) and just need to learn a new digital tool (e.g. word processing) (Quintana, Shin, Norris, & Soloway, 2006). User-centred design tries to make tools as easy and intuitive as possible. In contrast, learner-centred design assumes that the users have varying levels of expertise and need different scaffolds to learn new information. The goal is not to make the tool easy, but to provide the appropriate level of challenge. Digital tools must be designed to provide scaffolding to guide learners and clear exit strategies to remove supports as learners become more capable.

Mathematics software should attend to both developmental progression and content expectations. *MathemAntics* consists of seven mathematics content areas that children can learn beginning at preschool and through the third grade:

- Enumeration.
- Equivalence.
- Estimation.
- Addition and subtraction.
- Written calculation and place value.
- Multiplication.
- Negative number.

Each content area is comprised of several activities designed to teach or assess specific concepts and skills. The content roughly reflects the Common Core Standards (National Governors Association Center for Best Practices and Council of Chief State School Officers, 2010), although some *MathemAntics* activities are more advanced than the Common Core. The *MathemAntics* age progression roughly reflects the developmental trajectories as portrayed by cognitive research. These, like the Standards, however, may be too conservative. Our informal observations and intuitions, and also some authorities (Papert, 1980) suggest that when provided with stimulating mathematics activities, children can learn more than the trajectories designate as age appropriate. One reason is that traditional research on trajectories generally does not involve teaching experiments, and instead, in the spirit of Piaget (but not Vygotsky), focuses on development in the absence of education. Of course, we recognize that our conjectures must be subjected to empirical test, and our research undertakes that task.

Next we show how *MathemAntics* employs each of the key computer affordances described earlier in the chapter to achieve its goals and demonstrate how research plays an integral role in software development.

Goal-Driven Activities and Virtual Objects

Software for young children often lacks helpful instructions or goals. Our review of existing mathematics software for young children suggests that the omission of instructions appears to be a growing trend. Some software has been designed so that children are able to immediately immerse themselves in the game, thus perhaps creating an instant flow state, but they are required to figure out the goal(s) of the activity on their own as they play. Although this approach works in some cases, it may jeopardize children's understanding of fundamental mathematical concepts, especially if goals are not learned or if learning the mathematics content is not directly a part of the activity's goals. For example, if collecting as many badges or stickers as possible as reward for completion of tasks is the child's primary goal, how much attention is he paying to the mathematical content of the task?

MathemAntics activities are designed so that each has a clearly defined mathematical goal. Some activity features encourage children to explore mathematical ideas in an unstructured microworld, while other activities promote the learning and practice of specific mathematical skills, for example, helping a farmer to figure out how many animals are in a field. Goals are clearly a part of initial instructions and reinforced during feedback on each activity. To achieve these goals, children are given a variety of key software features (objects, tools, scaffolds, characters, feedback, etc.) throughout the *MathemAntics* system.

Tools

Digital tools can provide powerful ways to demonstrate deep mathematical constructs and provide useful representational models (Hoyles & Noss, 2009) or mathematical metaphors (analogous to physical manipulatives) that are not possible in traditional print or tangible materials. A child cannot rotate, shrink, or enlarge a triangle on a piece of paper, but can on a computer. Similarly, a child cannot 'take away' a picture of a bear on a piece of paper, but can on a computer. Paper is static; computers are dynamic.

Mathematical tools are a vital component in *MathemAntics*. Designed to help children solve problems and explore ideas about mathematics, the tools are introduced to children as young as 3 years of age and used through the third grade.

Tools are used in *MathemAntics* in several different ways. They are used to directly manipulate objects, thereby allowing children to organize, enumerate, compare, and alter the physical arrangement of objects. Explicit strategy tools are specially designed to teach children

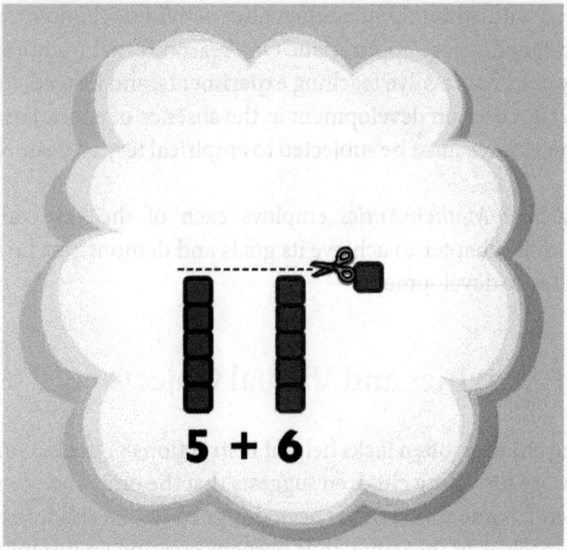

FIGURE 56.1 Scissors tool for creating doubles.

useful approaches to solving problems, as in the case of addition strategies for unknown number facts. For example, the scissors allows children to snip a block representation into already mastered doubles or tens facts. In the figure, the child snips off one block from the 6, which then turns into 5 + 5 with one left over, which the child easily determines is 11. The tool enables the child to use a known double fact (5 + 5) to solve the problem 5 + 6 (Figure 56.1).

Representations

In *MathemAntics,* tools are often linked to various mathematical representations. For example, imagine that a child used a blue 10-box and a white 3-box to organize 13 animals. After the child indicates her response by clicking on the number line, a thin continuous blue line spans from 0 to 10 on the number line followed by a continuous white line spanning from 10 to 13. On the screen, the child now sees the two grouping boxes (10- and-3-boxes) each filled with pigs, as well as the continuous lines showing 13 segmented by magnitude bars of 10 and 3 (each the same colour as the corresponding boxes). The child may also be shown the written number 13 and the number word, and may hear the spoken word 'thirteen.' (Figure 56.2).

Scaffolds

Scaffolds (or hints) are provided to assist the learning process. In *MathemAntics,* children are given various scaffolds when requested (by clicking on a help button), when inactive for a period of time, or when solving tasks inaccurately. As an example, one activity in *MathemAntics* asks children to locate or place numbers on a number line with only the endpoints labelled. The first time a child clicks on the help button or answers incorrectly, the

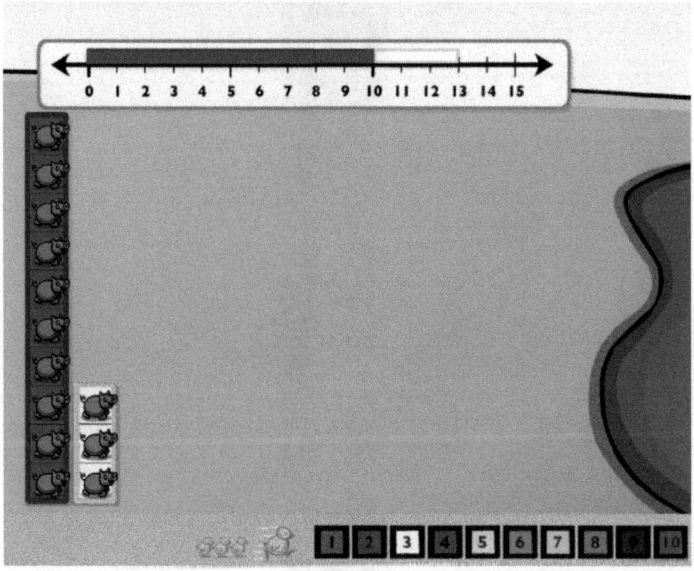

FIGURE 56.2 Grouping boxes and number line.

midpoint of the number line appears. With each subsequent request for help or error, the number line fills in more hatch marks – tens, and then units.

Various tools available in *MathemAntics* are deliberately designed to help scaffold learning. For example, a ghost-box is a grouping tool that has not been completely filled with animals. This occurs if there are fewer animals shown in the field than the size of the box created. So, if there are nine animals shown on a field and the child creates a 10-box, the nine animals fill the 10-box from the bottom-up, leaving the top box empty. This provides an immediate visual scaffold for the child. She knows there cannot be 10 animals as the 10-box has one empty box and may now begin to think of numbers in terms of their relation to the 10s. This is very helpful for dealing with all number fact problems involving 9 and even 8 (Figure 56.3).

Pedagogical Agents

MathemAntics utilizes a pedagogical agent (a child farmer) to guide children through activities by providing initial instruction and dynamic feedback. Pedagogical agents who guide children through their cognitive processing have been shown to increase learning outcomes (Baylor & Kim, 2005; Baylor & Ryu, 2003; Lester, et al., 1997; Moreno, et al., 2001). Furthermore, characters are used to provide short instructional demonstrations of how to complete an activity, and use various tools and strategies to solve problems. These instructional demos provide interactive examples to guide children in learning mathematical concepts and skills, and encourage strategy awareness and flexibility through a process of cognitive apprenticeship (Collins & Brown, 1991). Strategies used by the characters increase in difficulty according to their age; thus, older characters use more advanced strategies than younger characters. Importantly, these characters highlight that some strategies are more efficient than others in certain mathematical situations. Children are then able to use the tools to solve problems on their own (Figure 56.4).

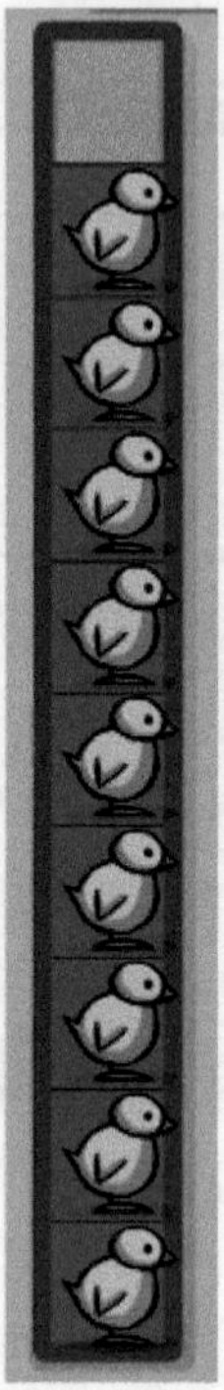

FIGURE 56.3 The Ghost Box grouping tool.

FIGURE 56.4 Four pedagogical agents.

Feedback

Our review of existing mathematics software suggests that feedback is often limited to positive sound effects or 'correct' for correct responses and negative sound effects or 'try again' for incorrect responses. This form of feedback provides no explanation of how or why answers are correct or incorrect. It does not help teach the child concepts or skills she lacks. Most drill-based software activities are specifically designed this way.

MathemAntics offers *dynamic* feedback. Characters or the pedagogical agent in *MathemAntics* explain the accuracy of a response through audio and visual explanations. For example, in a number facts activity, characters explicitly demonstrate and explain a useful strategy after incorrect responses. In equivalence activities, the pedagogical agent demonstrates how to use a *pair-up tool* to compare the number of objects when sets are aligned in a physical 1–1 correspondence. If there is nothing left over, the sets are the same number; any leftovers are highlighted to indicate which set has more than the other.

Interaction, Fantasy, Challenge, and Flow

MathemAntics incorporates fantasy through the creation of a farm environment. The child is asked to help the farmer complete a variety of activities, thereby becoming an active participant on the farm. As previously described, these activities include such things as helping the farmer feed the animals, counting how many animals are in the field, among many others.

Children are appropriately challenged through the use of automatic difficulty adjustments throughout the system. Activity goals and available tools are matched to meet each child's ability based on performance. This allows new tools to be introduced over time, increasing the length of activities and the possibilities for learning.

Flow is encouraged through the use of a variety of activities with which children can engage. Within activities, children are encouraged to complete several trials of the same activity, with objects or images changed each time (i.e. different colour barns, different animals, etc.). When ready to move on, children can choose another *MathemAntics* task. Importantly, we utilize formative research with children to help establish the threshold for maintaining attention and engagement in the activities, and create better flow.

Communication and Collaboration

We are in the early stages of incorporating communication and collaboration into *MathemAntics*. Currently, researchers create mini-lesson videos by recording an audio track as they play with the software. The children hear a child character (researchers' voice) walking them through how she uses the tools to solve the problem. We plan to work with classroom teachers to create their own video mini-lessons. Eventually, the tools for recording mini-lessons will be built into the software, giving teachers the ability to customize the software for their own classroom. We also envision students being able to record an audio track to accompany a particular problem, describing how they solved it. These solutions could be shared with teachers, parents, or other students.

Record Keeping and Reporting

A unique affordance of computers is the ability to capture large amounts of data. Textbooks and physical manipulatives may at times be helpful for classroom instruction, but they cannot record information about student performance. A customized logging system allows us to capture a wide range of information about each child's performance on every *MathemAntics* activity. This stealth assessment records such information as what level the child is on, whether the child increased or decreased levels during any given session, exact accuracy, near accuracy (e.g. how close the child's answer was to the correct answer), time elapsed before response, time elapsed between use of a tool and response, and usage of specific tools that promote specific strategies (such as the line-up tool for determining more, less, and equal).

Once data are gathered, the next challenge is to create informative reports for teachers. It is not enough for the software to adjust difficulty within the activities or to describe the child's frequent strategies: teachers must then be able to utilize information about student performance to inform classroom instruction. Reports can provide information about specific strategies that the child does *not* use and that can help her progress. Reports can provide suggestions about use of specific *MathemAntics* or classroom activities that might help the child develop specific skills or concepts. Furthermore, the reports may serve as a kind of professional development for teachers who could benefit from information about the kinds of strategies children use or the difficulties they face.

Stage 2: Formative Research: Usability and Learning

Formative research is a critical component in the design and development process. It is most often used to evaluate the usability of the software, whether it functions in the ways intended, or if any programming bugs and glitches are present. It allows the designer to learn through informal, preliminary observations whether an activity is engaging enough so that children enjoy it and want to keep playing. Usability research can be also used to evaluate the degree to which children understand the instructions and rules of the activity, and what additional scaffolds, if any, may be needed for children to understand or complete the activity successfully. Although usability research often involves informal observations, more structured methods may also be appropriate as in the case of observational rubrics used to assess attention and engagement.

Although usability research is necessary, it does not encompass all information critical for the development of software for children. In addition to usability, formative research that provides insight into children's *learning* and *understanding* is essential. To accomplish this, formative research often involves conducting exploratory case studies of individual or small groups of children. This method allows researchers to work flexibly with children over time, utilizing clinical interviews (Ginsburg, 1997) to gain initial information and insight about children's knowledge prior to and during use of the software, including their level of understanding, skills, and misconceptions of the content. Then the researcher can investigate how to improve the software so as to initiate *change* in understanding and performance. This is typically done through the careful introduction and use of software features, tools, and activities. Using these methods, the researcher is able to gain insight into the individual child's difficulties and then explore ways to optimize activities to be useful for many children, not only the one studied in depth. This method follows a learner-centred design approach as discussed earlier in the chapter. Results from case studies can thus inform design decisions and revisions (stage 3) and future learning studies (stage 4).

Working with children from the start of an activity idea through its final version is essential for the development of *MathemAntics*. Sometimes we start with physical materials or paper-and-pencil mock-ups of an activity to determine whether children understand its basic structure and task demands. The observations gathered during this process allow us to improve the activity design so as to make it more responsive to children's learning needs. For example, during formative case studies, we observed that when creating a number line second graders do not employ equal intervals and, instead, use a left-based clustering of numbers while ignoring the right-most side. Thus, numbers 1–9 tended to be clustered close towards the 0 end of a number line with only 0 marked and one end and 10 on the other. This finding then led us to devise a series of scaffolds (like the systematic introduction of hatch marks) to help children overcome the difficulty.

Stage 3: Revision

As shown with the number line example above, observations of children engaging in activities is a powerful method that allows researchers to inform and evaluate initial effectiveness of designs. The formative research stage emphasizes the importance of case studies to gather observations and initial data, while the revision stage stresses using those data to make informed decisions about the design of an activity. During the revision stage designers determine how to adjust activities to better teach mathematical content and understanding. Some form of revision is almost always necessary as a result of formative research. These revisions can include such features as a change in instructions, added scaffolding, a revised appearance of a character, new tools, etc. Revision and formative research should complement each other; thus the two stages should take place close in time. Information concerning the effectiveness of designs is informally gathered during formative research, which then allows the development of new design ideas during the revision phase, after which the designs are tested during formative research with the same children.

For example, using information gathered during the formative, paper and pencil phase of our work on the number line activity described above, we were able to design various prompts and scaffolding to support children's learning and then explore the initial effectiveness of these new prompts and scaffolds through continued formative research.

Stage 4: Learning Studies

After the convoluted process of the first three stages has reached what appears to be an effective design, focused learning studies designed to evaluate the effects of specific tools or activities should be conducted. Learning studies differ from formative research in that they use more formal research methods to evaluate learning. Learning studies also differ from large-scale evaluative studies in that the former tend to include a small-to-medium sized sample and are highly focused on specific learning objectives or particular aspects of the software. Unlike summative research, which is not appropriate at this point in the development of software (Butterworth & Laurillard, 2010), learning studies do not evaluate the general effects of the program on overall achievement.

A challenge of conducting research on learning effects involves implementing the most appropriate method to capture the phenomena of interest. Researchers have many empirical methods to choose from, but often, data gathered tends to be limited to traditional pre-/post-tests that may (or often do not) provide insight into the learning processes underlying the observed effects. Instead, we recommend using methods that allow researchers to examine the *process* of learning. These include microgenetic method and teaching experiments.

Microgenetic research is a valuable method for capturing change in understanding and performance in many areas, including mathematics (Siegler, 1995) and scientific reasoning (Kuhn & Phelps, 1982). The microgenetic method consists of multiple observations of specific skills, strategies and concepts as measured by task performance over a period of time (Siegler, 2006). Because this method requires multiple observations, large quantities of data can be generated. This data allows the researcher to evaluate five dimensions of change – the path (or process) of change, the rate at which change occurs, the breadth and application of change, the variability of change (within and between individuals), and the source of change. Microgenetic learning studies allow the designer to understand when, how, and why change occurs.

For example, to evaluate the effectiveness of the 10-box tool on children's learning, we recorded change in children's strategic use of the tool and accuracy on enumeration tasks over the course of several weeks. This study provided evidence for the effectiveness of the 10-box tool on use of advanced strategies (increased use of the count-on strategy as compared with count-all), improved accuracy over time, and the understanding of base-ten concepts.

Teaching experiments allow researchers to observe the *process* of learning under educationally-stimulating conditions. The goal is not to learn how the child develops normally, but to tap into the zone of proximal development so as to learn what the child can do under favourable conditions. Teaching experiments consist of a sequence of lessons and allow for flexibility of later lessons based on the conceptual analyses of students' performance and understanding of the material from previous lessons (Steffe & Thompson, 2000). Thus, tasks, scaffolds, and instructions may change from session to session depending on what children are capable of doing and learning. As suggested in the case of formative research, clinical interviews during teaching experiments can supplement observations of language and behaviour in order to capture children's thinking and understanding in depth.

Stage 5: Evaluation

The first four stages consist of an iterative cycle during which software designs are consistently evaluated and revised based on the results of formative research and small-scale learning studies. The fifth stage, evaluation, is designed to extend initial findings and includes three goals:

- Increase sample size to include a nationally representative sample.
- Evaluate effects of the software on standard measures of achievement.
- Investigate the potential for scaling up the software for typical classroom settings.

Representative Sample

Goals of the initial research stages are to inform design decisions and demonstrate potential learning effects. For these purposes, a nationally representative sample is unnecessary, but it is essential for a general evaluation when the goal is to understand if learning effects are stable across populations or to determine which groups of children benefit from software use and in what ways. These findings will ultimately influence how the software should be implemented in the classroom.

Standard Measures of Achievement

We argued previously that evaluation of learning should focus on the *process* by which learning occurs. To accomplish this, we proposed utilizing stealth assessment, and the microgenetic method to capture a variety of indicators of skill and understanding every time children use the software. This is a method not often utilized in education research and particularly absent from large-scale summative research, which typically employs standard measures of achievement. Unfortunately such tests are simply not very useful when it comes to providing insight into significant aspects of learning and thinking, particularly conceptual understanding. And the tests may not measure the complexities of thinking – for example judicious use of different strategies – made possible and revealed by comprehensive software. Large-scale evaluations require standard tests for some purposes. However, the computer software itself can provide stealth assessments of the processes of learning and thinking as they develop during children's engagement with the activities over time. Assessments of this type can be easily obtained from very large numbers of students working over short or long periods of time. Together, standard evaluation methods and software-derived stealth assessments can provide insights into learning, as well as overall achievement; this combination of methods can be much more powerful and informative than pre-and-post-test standardized data alone.

Scalability

The ultimate goal is not to develop educational software that is beneficial when used under unique and specialized conditions, but rather to offer an effective learning tool that can be used in traditional classroom settings. It is the scalability of the software that can make a substantial difference in education.

Of course it is clear that the mere presence of technology does not necessarily improve teaching or learning (Cuban, 2001). Good technology needs to be effectively implemented in the classroom. Yet many teachers at the preschool and elementary levels are uncomfortable with the use of software and some may even think it is not developmentally appropriate for young children to use computers (Lee, 2006). All this suggests that we cannot usefully evaluate large-scale learning effects unless teachers are properly trained to use the software in their classrooms. Even high quality software will not be maximally effective if teachers do not know how to use it and relate it to their classroom instruction.

A second problem in evaluating the scalability of software is the research method itself. Too often, large-scale summative research relies solely on traditional experimental designs that tend to compare treatment and control groups on limited assessments as previously described. This approach does not and cannot provide a comprehensive account of the software's effects. So what kind of large-scale research method(s) should be used to examine scalability? As suggested earlier, the desired research needed should include methods similar to microgenetic and teaching experiments recommended for small-scale learning studies and should involve stealth methods of assessment that provide richer information than do standard tests. Appropriate software makes possible rich, large scale, and rapid assessments that can enrich traditional pre- and post-test evaluations.

Conclusions

We have argued that computer software can make major contributions to mathematics education. What are the implications for early childhood educators, for software designers, and for researchers?

Early Childhood Education

Should early childhood education involve learning from computers? The answer is yes. Young children can and frequently do use computers, touch screen devices, and smart phones. Some good mathematics software is already available and more is on its way.

Unfortunately, the same is true of many poor quality software materials. As a result, early childhood educators (and parents) need to learn to separate the wheat from the chaff, and help children to engage in productive learning through sound software. To promote effective learning, teachers need to overcome their own fears of computers and mathematics. They need to understand the role of intentional teaching and curriculum in early education. They need to understand how to implement and supervise computer learning in the classroom.

We can help early childhood educators achieve these goals in several ways. One is by developing and implementing sound programmes of professional development, both in institutions of higher learning and the settings in which teachers work. Unfortunately such programmes are rare. A second is by providing sound analytical frameworks for evaluating the quality of software. These frameworks should be clear to teachers and parents so that both can help their children to engage in productive learning of mathematics. Unfortunately, few such frameworks are now available. A third is by developing and examining various methods for classroom use of mathematics software for young children. Computers cannot operate as completely stand-alone devices, particularly in early education. If so, what is the teacher's role? How can the teacher use the smart board to help children work with computers? To what extent should teachers intervene in the process of computer learning? Clearly, professional development needs to deal with these issues, and researchers can help to devise and explore productive possibilities.

Software Design and Analysis

There is a good deal of mindless computer design by people who do not understand children or mathematics education. Just as teachers need to understand computer learning, so do designers. They need to understand children's minds and the processes and goals of education. They should not be guided only by approaches useful for producing enjoyable (but often mindless and aggressive) games.

Research and Development

The relationships between the two are more complex than many realize. The research-based understanding of children's mathematical thinking and learning should guide development of software, as already mentioned. At the same time, current theories are not sufficient for the design process. Design involves art and taste, neither of which stems directly from research. Also, formative research should help guide the design process; the classical techniques of cognitive research are not sufficient. Designers must also accept that the existing research, while useful as a guide, may not be sufficient to understand children's computer learning, partly because computers offer affordances that other educational materials and everyday experience cannot. In particular, notions of developmental trajectories should be taken with a grain of salt. Formative research and teaching experiments may cause us to revise our notions of trajectories, and of what young children can learn. Designers should draw upon the very useful body of developmental and educational research, but should be flexible enough to entertain the possibility that effective software may require revising current expectations if young children's competence and developmental trajectories.

Finally, research can play an important role in evaluation of software, particularly if it employs rich stealth assessment that illuminates children's thinking and learning and does not rely only on standard achievement measures.

References

Arnold, D.H., & Doctoroff, G.L. (2003). The early education of socioeconomically disadvantaged children. Annual Review of Psychology, 54, 517–545.

Baroody, A.J., Lai, M., & Mix, K.S. (2006). The development of young children's early number and operation sense and its implications for early childhood education. In B. Spodek & O. Saracho (eds.), Handbook of Research on the Education of Young Children, Vol. 2 (pp. 187–221). Mahwah, NJ: Erlbaum.

Baylor, A.L., & Kim, Y. (2005). Simulating instructional roles through pedagogical agents. International Journal of Artificial Intelligence in Education, 15, 95–115.

Baylor, A.L., & Ryu, J. (2003). The effects of image and animation in enhancing pedagogical agent persona. Journal of Educational Computing Research, 28, 373–394.

Bottino, R.M., & Chiappini, G. (2002). Advanced technology and learning environments: their relationships within the arithmetic problem-solving domain. In L.D. English (Ed.), Handbook of International Research in Mathematics Education (pp. 757–785). Mahwah, NJ: Lawrence Erlbaum Associates.

Butterworth, B., & Laurillard, D. (2010). Low numeracy and dyscalculia: identification and intervention. ZDM Mathematics Education, 42, 527–539.

Clements, D.H. (2004). Geometric and spatial thinking in early childhood education. In D. H. Clements, J. Serama, & A.-M. DiBiase (eds), Engaging Young Children in Mathematics: Standards for Early Childhood Mathematics Education (pp. 267–297). Mahwah, NJ: Lawrence Earlbam Associates Publishers.

Clements, D.H., & Battista, M.T. (2000). Designing effective software. In K. R. A. Lesh & A. E. Lesh (eds), Handbook of Research Design in Mathematics and Science Education (pp. 761–776). Mahway, NJ: Lawrence Erlbaum Associates.

Clements, D.H., & Sarama, J. (2007). Effects of a preschool mathematics curriculum: Summative research on the Building Blocks project. Journal for Research in Mathematics Education, 38, 136–163.

Clements, D.H., Swaminathan, S., Hannibal, M.A.Z., & Sarama, J. (1999). Young children's concepts of shape. Journal for Research in Mathematics Education, 30(2), 192–212.

Collins, A., & Brown, J. (1991). Cognitive apprenticeship: making thinking visible. American Educator, 15, 6–11, 38–46.

Csíkszentmihályi, M. (1991). Flow: the Psychology of Optimal Experience. New York: HarperCollins.

Cuban, L. (2001). Oversold & Underused: Computers in the Classroom. Cambridge, MA: Harvard University Press.

Frede, E., Jung, K., Barnett, W.S., Lamy, C.E., & Figueras, A. (2007). The Abbott Preschool Program Longitudinal Effects Study (APPLES). Rutgers, NJ: National Institute for Early Education Research.

Gee, J.P. (2005). Learning by design: good video games as learning machines. E-Learning and Digital Media, 2, 5–16.

Gelman, R., & Gallistel, C.R. (1986). The Child's Understanding of Number. Cambridge, MA: Harvard University Press.

Ginsburg, H.P. (1997). Entering the Child's Mind: The Clinical Interview in Psychological Research and Practice. New York: Cambridge University Press.

Ginsburg, H.P., Lee, J.S., & Boyd, J.S. (2008). Mathematics education for young children: what it is and how to promote it. Social Policy Report.—Giving Child and Youth Development Knowledge Away: Society for Research in Child Development, 22(1), 1–24.

Ginsburg, H.P., Pappas, S., Lee, Y.-S., & Chiong, C. (2011). How did you get that answer? Computer assessments of young children's mathematical minds. In P. Noyce (ed.), New Frontiers in Formative Assessment. (pp. 49–67). Cambridge, MA: Harvard University Press.

Goldin-Meadow, S., Cook, S.W., & Mitchell, Z.A. (2009). Gesturing gives children new ideas about math. Psychological Science, 20, 267–272.

Graham, T.A., Nash, C., & Paul, K. (1997). Young children's exposure to mathematics: the child care context. Early Childhood Education Journal, 25(1), 31–38.

Hatano, G. (1988). Social and motivational bases for mathematical understanding. In G. Saxe & M. Gearhart (eds), Children's Mathematics. (pp. 55–70) San Francisco: Jossey-Bass.

Hoyles, C., & Noss, R. (2009). The technological mediation of mathematics and its learning. Human Development, 52, 129–147.

Judge, S., Puckett, K., & Bell, S.M. (2006). Closing the digital divide: update from the early childhood longitudinal study. Journal of Educational Research, 100(1), 52–60.

Kiili, K. (2005). Digital game-based learning: towards an experiential gaming model. Internet and Higher Education, 8, 13–24.

Kilpatrick, J., Swafford, J., & Findell, B. (eds) (2001). Adding It Up: Helping Children Learn Mathematics. Washington, DC: National Academy Press.

Kuhn, D., & Phelps, E. (1982). The development of problem-solving strategies. In W. R. Hayne (ed.), Advances in Child Development and Behavior, Volume 17 (pp. 1–44). London: Academic Press.

Lee, J.S. (2006). Preschool teachers' shared beliefs about appropriate pedagogy for 4-year-olds. Early Childhood Education Journal, 33(6), 433–441.

Lemke, M., & Gonzales, P. (2006). U.S. Student and Adult Performance on International Assessments of Educational Achievement: Findings from The Condition of Education 2006. U.S. Department of Education. Washington DC: National Center for Educational Statistics.

Lester, J.C., Converse, S.A., Kahler, S.E., Barlow, S.T., Stone, B.A., & Bhogal, R.S. (1997). The persona effect: affective impact of animated pedagogical agents. Paper presented at the Proceedings of the SIGCHI conference on human factors in computing systems.

Malone, T.W., & Lepper, M.R. (1987). Making learning fun: a taxonomy of intrinsic motivations for learning. In R. E. Snow & M. J. Farr (eds), Aptitude, Learning, and Instruction: Vol. 3. Conative and Affective Process Analysis (pp. 223–253). Hillsdale, NJ: Lawrence Erlbaum Associates.

Miller, K.F., & Parades, D.R. (1996). On the shoulders of giants: cultural tools and mathematical development. In R. J. Sternberg & T. Ben-Zeev (eds), The Nature of Mathematical Thinking (pp. 83–117). Mahwah, NJ: Lawrence Erlbaum Associates Publishers.

Moreno, R., Mayer, R.E., Spires, H.A., & Lester, J.C. (2001). The case for social agency in computer-based teaching: do students learn more deeply when they interact with animated pedagogical agents? Cognition and Instruction, 19, 177–213.

National Governors Association Center for Best Practices and Council of Chief State School Officers (2010). Common Core State Standards for Mathematics. Washington D.C.: National Governors Association Center for Best Practices, Council of Chief State School Officers.

Papert, S. (1980). Mindstorms: Children, Computers, and Powerful Ideas. New York: Basic Books.

Piaget, J. (1952). The Child's Conception of Number, transl. by C. Gattegno & F. M. Hodgson. London: Routledge & Kegan Paul Ltd.

Pianta, R.C., Belsky, J., Houts, R., & Morrison, F. (2007). Opportunities to learn in America's elementary classrooms. Science, 315(March), 1795–1796.

Quintana, C., Shin, N., Norris, C., & Soloway, E. (2006). Learner-centered design: reflections on the past and directions for the future. In R.K. Sawyer (ed.), The Cambridge Handbook of the Learning Sciences (pp. 119–134). Cambridge: Cambridge University Press.

Reeves, B., & Nass, C. (1999). The Media Equation: How People Treat Computers, Television, and New Media like Real People and Places. Cambridge: Cambridge University Press.

Resnick, M., Maloney, J., Monroy-Hernández, A., Rusk, N., Eastmond, E., Brennan, K., Millner, A., Rosenbaum, E., Silver, J., Silverman, B., & Kafai, Y. (2009). Scratch: programming for all. Communications of the ACM, 52(11), 60–67.

Romero, C., & Ventura, S. (2007). Educational data mining: a survey from 1995 to 2005. Expert Systems with Applications, 33(1), 135–146.

Sarama, J., & Clements, D.H. (2002a). Building Blocks for young children's mathematical development. Journal of Educational Computing Research, 27, 93–110.

Sarama, J., & Clements, D.H. (2002). Learning and teaching with computers in early childhood education. In O. N. Saracho & B. Spodek (Eds.), Contemporary perspectives in early childhood education (pp. 171–219). Greenwich, CT.: Information Age Publishing.

Sarama, J., & Clements, D.H. (2009). Early Childhood Mathematics Education Research: Learning Trajectories for Young Children. New York: Routledge.

Scardamalia, M., & Bereiter, C. (2006). Knowledge building: theory, pedagogy, and technology. In R.K. Sawyer (ed.), The Cambridge Handbook of the Learning Sciences (pp. 97–118). New York: Cambridge University Press.

Schell, J. (2010). Transmedia worlds. Paper presented at the Dust or Magic: Children's New Media Institute, Lambertsville, NJ, November 8, 2010.

Seo, K.-H., & Ginsburg, H.P. (2004). What is developmentally appropriate in early childhood mathematics education? Lessons from new research. In D.H. Clements, J. Sarama & A.-M. DiBiase (eds), Engaging Young Children in Mathematics: Standards For Early Childhood Mathematics Education (pp. 91–104). Hillsdale, NJ: Erlbaum.

Shute, V. (2011). Stealth Assessment In Computer-based Games to Support Learning Computer Games and Instruction. Charlotte, NC: Information Age Publishing.

Siegler, R. (2006). Microgenetic analyses of learning. In D. Kuhn & R.S. Siegler (eds), Handbook of Child Psychology: Volume 2 Cognition, Perception, and Language (pp. 464–510). Hoboken, NJ: John Wiley & Sons, Inc.

Siegler, R.S. (1995). How does change occur: a microgenetic study of number conservation. Cognitive Psychology, 28, 225–273.

Steffe, L.P., & Thompson, P.W. (2000). Teaching experiment methodology: underlying principles and essential elements. In R. Lesh & A. E. Kelly (eds), Research Design in Mathematics and Science Education (pp. 267–307). Hillsdale, NJ: Erlbaum.

Stevenson, H., Lee, S.S., & Stigler, J. (1986). The mathematics achievement of Chinese, Japanese, and American children. Science, 56, 693–699.

Vygotsky, L. S. (1986). Thought and Language, transl. by A. Kozulin. Cambridge, MA: MIT Press.

Wilson, A.J., Dehaene, S., Pinel, P., Revkin, S.K., Cohen, L., & Cohen, D. (2006). Principles underlying the design of 'The Number Race', an adaptive computer game for remediation of dyscalculia. Behavioral and Brain Functions, 2, 19.

Wilson, A.J., Revkin, S.K., Cohen, D., Cohen, L., & Dehaene, S. (2006). An open trial assessment of 'The Number Race', an adaptive computer game for remediation of dyscalculia. Behavioral and Brain Functions, 2, 20, 1–16.

CHAPTER 57

EARLY NUMBER COMPETENCIES AND MATHEMATICAL LEARNING

individual variation, screening, and intervention

NANCY C. JORDAN, LYNN S. FUCHS, AND NANCY DYSON

Poor achievement in mathematics is widespread and has serious educational and vocational consequences. If students do not develop foundational mathematics in elementary and middle school, they are less likely to graduate from college than higher-achieving students (National Mathematics Advisory Panel, 2008). Proficiency in advanced mathematics is important for success in science, technology, engineering, and mathematics (STEM) vocations (Sadler & Tai, 2007). Many students in US schools never develop essential skills for success in algebra, and low-income learners lag far behind their middle-income peers (Lee, Grigg, & Dion, 2007). Disturbingly, large mathematics disparities exist between middle- and low-income children before they enter primary school at about 6 years of age (Duncan, Dowsett, Claessens, Magnuson, Huston, Klebanov, et al., 2007; Jordan & Levine, 2009; Lee & Burkham, 2002; National Research Council, 2009).

Until recently, early interventions for children in mathematics have been overlooked, relative to the proliferation of early reading interventions in US schools (Berch & Mazzocco, 2007; Gersten, Jordan, & Flojo, 2005). In its recent report, *Mathematics Learning in Early Childhood: Paths toward Excellence and Equity* (National Research Council, 2009), the Committee on Early Childhood Mathematics identified number competence as of primary importance for success in school mathematics noting that 'developing an understanding of number, operations, and how to represent them is one of the major mathematical tasks for children during the early childhood years' (p. 22). Although mathematics includes content areas in addition to number, such as geometry and measurement, the focus of this article is on identifying and developing numerical competencies in young children. Our framework is motivated by the premises that:

1. Number competence in young children can be operationalized as knowledge of number, number relations, and number operations.
2. Number competencies can be assessed along a developmental progression
3. Number competence is highly predictive of future maths success.
4. Core deficiencies in number underlie maths difficulties/disabilities.
5. Deficiencies in number can be identified as early as pre-K
6. Targeted help in number competencies leads to improved maths achievement in school.

Individual Variation and the Need for Number Sense

Young children vary widely in their mathematical knowledge well before school entry (Levine, Suriyakham, Rowe, Huttenlocher, & Gunderson, 2010). Children's income status and their associated early home and preschool opportunities, in addition to their general cognitive capacity, all heavily influence their level of knowledge of numbers (Ginsburg, Lee, & Boyd, 2008; National Research Council, 2009). As early as 3 years of age, there is a significant gap between low- and middle-income children in mathematics-related skills (Clements, Sarama, & Gerber, 2005; Klibanoff, Levine, Huttenlocher, Vasilyeva, & Hedges, 2006). Parental social class and educational level predict mathematics achievement throughout the school years (Duncan et al., 2007). However, research findings are inconsistent with respect to the strength of these associations, the aspects of mathematics knowledge that are affected (e.g. Jordan, Huttenlocher, & Levine, 1992; Russell & Ginsburg, 1984), and the mechanisms that underlie the associations, including lack of opportunity for out-of-school learning and parenting characteristics (Blevins-Knabe & Musun Miller, 1996; Clements & Sarama, 2007).

Research has shown that young children are sensitive to number (Dehaene, 1999). Even infants 6 months of age can distinguish between quantities (Feigenson, Dehaene, & Spelke, 2004). These non-verbal abilities become connected to verbal representations as parents or caregivers engage their children in 'number talk' (Klibanoff et al., 2006), and encourage them to count and compare quantities in daily life. By the time most children come to school, they have developed a set of number competencies through these informal experiences (Ginsburg et al., 2008). Recent work has shown that the amount of 'number talk' provided in the home predicts young children's understanding of the meaning of number words (e.g. that three refers to collections of three items), over and above socioeconomic status (Levine et al., 2010). It also has been suggested that opportunities to practice number tasks with corrective feedback predict numerical competencies (National Research Council, 2009).

Early number competencies are important for establishing children's learning trajectories or paths in mathematics (Jordan, Kaplan, Ramineni, & Locuniak, 2009). These competencies also referred to as *number sense*, involve understanding of whole numbers, number relations, and number operations (Malofeeva, Day, Saco, Young, & Ciancio, 2004; National Research Council, 2009). Children must recognize:

1. That numbers represent quantities.
2. That counting is guided by principles related to one-to-one correspondence, stable order, and cardinality.
3. That children can use counting to compare the size of collections.
4. That the quantity of collections can be transformed through addition and subtraction (Gelman & Gallistel, 1978; Griffin, 2004).

Preschool number sense involves symbolic representations of numbers, verbal and written, in contrast to more fundamental preverbal number knowledge that appears to develop without much verbal input or instruction (Feigenson et al., 2004; Jordan & Levine, 2009). Children must be able to apply their basic understanding of numbers to a wide range of situations (Carpenter, Hiebert, & Moser, 1983; Gersten, et al., 2005; Hiebert, 1984). There can be four cookies, four fingers, four names, four turns taken, and so forth (National Research Council, 2009).

Early number competencies follow a developmental path in preschool (Clements & Sarama, 2007). Children move from acquiring basic knowledge of number (e.g. small number recognition and counting), to understanding numerical relations (e.g. finding the number after a given number, determining the larger of two numbers), to performing operations with number (e.g. mentally adding and subtracting with small numbers) with the progression repeating as larger numbers are learned. This progression represents the three core areas described by the National Research Council (2009): number, number relations, and number operations.

Number

Counting is of key importance to mathematical development because it expands quantitative understanding beyond very small numbers of four or less, which initially are recognized as whole units through subitization (Baroody, 1987; Baroody, Lai, & Mix. 2006; Ginsburg, 1989). Before formal schooling, most children master the count sequence beyond 10. They learn that each object in a collection is counted only once, that the count words are always used in the same sequence (i.e. one, two, three, etc.), and that the last number in the count always denotes the number of objects in the set (i.e. cardinality principle) (Gelman & Gallistel, 1978). Children come to understand that they can count any set presented in any configuration as long as they count each object only once in a stable order (Gelman & Gallistel, 1978). Children also must learn to read, write, and understand written number symbols (1, 3, 5, etc.) (National Research Council, 2009). Linking numerals to quantities through counting helps children see relations between and among numbers (e.g. more than, less than, equal to) (Griffin & Case, 1997).

Number Relations

Being able to think about the magnitudes of numbers is a key developmental accomplishment (Case & Griffin, 1990; Griffin, 2002, 2004). For example, young children learn that 4 is bigger than 3 or that 2 is smaller than 5 (Griffin, 2004). Case, Okamoto, Griffin, McKeough,

Bleiker, Henderson, et al. (1996) suggest the notion of central conceptual structures, which begin as separate understandings about the nature of numbers and quantities. For example, a child of four typically cannot answer the question, 'Which is bigger, four or five?' if the quantities are not present to observe. Yet, if the quantities are present, a four-year old can typically differentiate between quantities of four and five without counting and can correctly answer the question, 'Which is more?' This represents one structure. A second structure allows a typical four-year old to enumerate a set of four or five objects and answer the question, 'How many?' But answering the question, 'Which is bigger, four or five?' requires integrating these two structures, enabling children to make judgments about quantities without physical objects present. This integration typically takes place during kindergarten and allows children to make a 'mental counting line'& Deficits in number relations are associated with mathematics learning disabilities (Rousselle & Noël, 2007), emphasizing the importance of having these structures in place before first grade.

Number Operations

Competence with addition and subtraction combinations is important for success in elementary school mathematics (Jordan, et al., 2009). However, these abilities start to develop much earlier. Although preschoolers often have trouble solving verbal story problems ('Beth had 2 pennies. Jill gave her 1 more penny. How many pennies does Beth have now?') and number combinations ('How much is 2 and 1?'), many can solve nonverbal calculations with object representations (e.g. The child is shown 2 objects that are then hidden with a cover. One more object is slid under the cover by the examiner, and the child must calculate to determine that 3 objects are now under the cover) (Levine, Jordan, & Huttenlocher, 1992). Even children with limited counting facility can solve nonverbal problems with totals of 5 or less by visualizing set transformations in their heads (Huttenlocher, Jordan, & Levine, 1994). Early performance on nonverbal calculation is associated with later addition and subtraction skill (Levine et al., 1992).

Counting is an effective strategy for solving addition and subtraction problems. Knowing that the next number in the count sequence is always one more than the preceding number or one less than the following number allows children to solve problems in the form of n + 1 and n–1 (Baroody, Eiland, & Thompson, 2009). Counting also helps children find out how much less or how much more one number is from another (National Research Council, 2009). For example, for combinations with totals of 10 or less, children can count out each addend on their fingers and then count all of the fingers to get the total. By the end of kindergarten, many children can count on from the first number to get the total (e.g. for 4 + 3, the child counts 5, 6, 7 to get 7) a more efficient approach than counting out both addends (Baroody et al., 2006). Adaptive counting in kindergarten is associated with the development of calculation fluency in first and second grades (Jordan, Kaplan, Ramineni, & Locuniak, 2008).

Another important insight related to addition and subtraction is that numbers can be decomposed into smaller sets of numbers. For example, the number 4 can be broken into 1 and 3, or into 2 and 2. Fuson, Grandau, & Sugiyama (2001) refer to these combinations as 'partners' for particular totals. Thinking about different partners for the same sums and also relating them to subtraction (1 + 3 = 4 and 4 – 1 = 3) encourages flexibility when solving problems. Young children at risk for mathematics difficulties (i.e. low-income) do not make good use of counting and decomposition strategies to help them calculate with totals of 5 or more (Jordan, Kaplan, Olah, & Locuniak, 2006).

Predictability of Number Sense

Early number competencies matter for long-term achievement outcomes. Kindergarten (i.e. 5–6 years of age) number concepts are powerful predictors of adolescent learning outcomes across content areas (Duncan et al., 2007). Using longitudinal methods, Jordan and colleagues (Jordan et al., 2009; Jordan, Glutting, & Ramineni, 2009; Jordan, Glutting, Ramani, & Watkins, 2010; Locuniak & Jordan, 2008) found that core competencies in kindergarten (about 5 years of age) related to number, number relations, and number operations are highly predictive of later mathematics computation and problem solving proficiency in US third grade, when controlling for reading, age, and general cognitive factors. Similar findings have been reported in shorter-term studies (Clarke & Shinn, 2004; Lembke & Foegen, 2009; Methe, Hintze, & Floyd, 2008), and number competence in 3- and 4-year-old preschoolers, predicts performance on similar measures in kindergarten 5-year-olds (VanDerHeyden, Broussard, & Cooley, 2006).

Deficits in Number Sense and Learning Difficulties

Many children lacking in number competencies are from low-income populations (Jordan, Kaplan, Locuniak, & Ramineni 2007; Jordan et al., 2006). Low-income students enter kindergarten far behind middle-income children in number sense (Starkey, Klein, & Wakeley, 2004) and are four times more likely than their middle-income counterparts to show flat growth on numeracy tasks during kindergarten and first grade (Jordan et al., 2006, 2007). Jordan et al. (2007) found that *level* of performance in number competence in kindergarten, as well as *rate* of growth between kindergarten and first grade accounted for about 66% of the variance in mathematics learning at the end of first grade. Importantly, income status did not add explanatory variance over and above performance and growth in number competence. That is, the poor mathematics achievement of low-income learners is mediated by their weak number sense.

Young English language learners (ELL) may also experience difficulties developing number competencies if they are not instructed in their native language. Although little research has been conducted on the specific maths problems of young ELLs (National Research Council, 2009), achievement differences favoring native language learners are seen by third grade (Abedi, 2004). Bilingual children with little command of the language of instruction experience particular difficulties with mathematical word problems (Kempert, Saalbach, & Hardy, 2011).

Poor number sense may lead to reliance on rote memorization, which may lead to an inability to work meaningfully with numbers, which in turn may lead to poor problem solving skills (Robinson, Menchetti, & Torgesen, 2002). In reading, it has been found that those entering first grade with poor phonemic awareness (the ability to hear, identify, and reproduce the smallest meaningful units of sound) almost always become poor readers

(Juel, 1988). Without phonemic awareness, children cannot decode words and must rely on rote memorization. This, in turn, leads to poor comprehension. Trying to improve comprehension in later grades without going back to the root problem, poor phonemic awareness, was unsuccessful. It is reasonable to think that the same may be true for number sense and mathematics learning.

Deficits in number sense, especially those related to comparing symbolic magnitudes and enumerating sets of dots, also are associated with developmental dyscalculia, a congenital and persistent arithmetic disability that occurs across social class and language groups (Reigosa-Crespo, Valdes-Sosa, Butterworth, Estévez, Rodriguez, Santos, et al., 2011). Butterworth, Sashank, & Laurillard (2011) observe: 'Numbers do not seem to be meaningful for dyscalculics—at least, not meaningful in the way that they are for typically developing learners. They do not intuitively grasp the size of a number and its value relative to other numbers. This basic understanding underpins all work with numbers and their relationships to one another' (p. 1050).

Overall, mathematics learning difficulties/disabilities in young children are highly persistent. American children who perform in the lowest 10th percentile in kindergarten mathematics have a 70% chance of staying in that category after five years (Morgan, Farkus, & Wu, 2009). The persistence of mathematics difficulties has stimulated investment in early screening and intervention, similar to the widespread efforts that have been made in reading (Gersten, Clarke, Jordan, Newman-Gonchar, Hammond, & Wilkins, 2012)

Early Screening

Most early screening measures tap one or more aspects of the core content area of number, relations, or operations. Gersten et al. (2012) reviewed the predictive validity of single and multiple proficiency measures. All of the studies provided correlations between the screeners and later maths achievement outcomes. For example, low performance on a measure of magnitude comparison in kindergarten can be an indicator of later problems with calculation fluency (Locuniak & Jordan, 2008). The correlations were consistently positive and significant with tasks requiring children to name the larger of two numerals (e.g. Clarke, Baker, Smolkowski, & Shard 2008; Seethaler & Fuchs, 2010), to name the missing numeral from a string of numerals (e.g. Clarke et al., 2008; Lembke & Foegen, 2009), and to perform simple addition and subtraction tasks (e.g. Jordan et al., 2010; Seethaler & Fuchs, 2010), the last being especially predictive. Counting tasks (e.g. counting in 5's) were more predictive of maths outcomes in first grade than in kindergarten (Lembke & Foegen, 2009), suggesting that different types of screening items may be needed for first grade than in kindergarten (Gersten, et al., 2012). The number screening measures also demonstrate high classification accuracy (Clark, Nese, Alonzo, Mercier-Smith, Tindal, Kame'enui, et al., 2011; Geary, Bailey, & Hoard, 2009; Jordan et al., 2010; Seethhaler & Fuchs, 2010), i.e. the degree to which they provide correct classifications of children who will need additional help in mathematics (Gersten et al., 2012).

Based on the research on early predictors of mathematics outcomes, Jordan, Glutting, and Dyson (2012) developed the multiple proficiency Number Sense Screener (NSS), a measure that not only reliably identifies who will need support (Jordan et al., 2010), but also has strong treatment validity (Dyson, Jordan, & Glutting, 2011). That is, an intervention based on the NSS areas assessed (counting, number recognition, number comparisons, and arithmetic operations) led to meaningful gains not only on closely-aligned measures, but also ones tapping more general achievement. The content of the NSS is summarized in Table 57.1. To devise intervention plans, a close look at the child's performance in each sub-area is useful for planning intervention. For example, if a child has trouble on number combinations, but can solve non-verbal calculations, they may benefit from games and activities that connect number combinations to physical representations showing the part-part-whole relationship of number sets.

Seethaler and Fuchs (2010) assessed the predictive utility of three kindergarten measures for screening students for risk for mathematics difficulty with respect to:

1. Single- versus multiple-skill item screeners.
2. Fall versus spring screening.
3. Conceptual versus operational mathematics outcome.

The single-skill screener assessed students' ability to discriminate larger numbers from pairs of numbers ranging from 0–10 in 1 minute. The multi-skill screeners assessed computational fluency and various mathematical concepts central to early mathematical development (i.e. quantity discrimination, mental number lines, ordering numbers, estimation, patterns, counting backward, shape discrimination, number sentences, and one-to-one correspondence). Screening occurred in fall and spring of kindergarten; conceptual and operational maths outcomes were assessed at the end of first grade, with mathematics difficulties operationalized as performance below the 16th percentile on nationally norm-referenced tests. Logistic regression and receiver operating characteristics analyses indicated that the single- and multiple-skill screeners produced good and similar classification accuracy at the

Table 57.1 Number sense screener

Subarea	Examples of assessments
	The child is asked to:
Counting	• Count objects and indicate how many are in the set • Give a specified number of objects to the examiner • Count to 30
Number recognition	Name single- and double-digit numerals
Number comparisons	• Name numbers right after and right before a number • Choose the bigger/smaller of two numbers
Non-verbal calculation	Match a picture of dots to the number of tokens hidden under a box after a transformation (tokens have been added or removed) has taken place
Story problems	Solve addition or subtraction change story problems presented orally
Number combinations	Solve addition or subtraction presented orally

Data from N.C Jordan, J. Glutting, and N. Dyson, Number Sense Screener User's Guide. Baltimore: Paul H. Brookes, 2012.

fall and spring screening occasions in forecasting conceptual outcome. With respect to procedural outcome, the three screeners produced similar, but less accurate fits. Regardless of predictive model, however, high numbers of false positives (i.e. identifying children who do not go on to have mathematics difficulties) demonstrate the need for future work to increase the overall accuracy of screening methods.

Early Interventions

Number sense is malleable and can be taught to most young children (Berch, 2005, Berch & Mazzocco, 2007). Early identification and intervention promotes early success and confidence in mathematics, and prevents later failure (Geary, Hoard, Byrd-Craven, Nugent, & Numtee, 2005; Gersten, et al., 2005). The US Committee on Early Childhood Mathematics (National Research Council, 2009) urges all early childhood programmes to provide explicit instruction in mathematics. Although children learn mathematics partly through their experiences in the world, they also need instruction from adults to 'mathematize' (p. 334) or to focus on the mathematical aspects of familiar situations. Yet this is not so easy, as Clements and Sarama (2011) note:

> Early childhood teachers often believe they are doing 'mathematics' when they provide puzzles, blocks, and songs. Even when they teach mathematics, that content is often not the main focus, but is embedded in a fine-motor or reading activity. Unfortunately, evidence suggests that such an approach is ineffective, owing to a lack of explicit attention to mathematical concepts and procedures along with a lack of intentionality to engage in mathematical practices. (p. 968)

Although the following discussion primarily focuses on US intervention research, early intervention for those at risk for maths difficulties is an international concern (Dowker & Sigley, 2010; Dunn, Matthews, & Dowrick, 2010; Evans, 2007; Kroesbergen, & Van Luit, 2003; Opel, Zaman, Khanom, & Aboud, 2012).

Preschool Interventions (3–5 Years of Age)

There is evidence to suggest that number sense (and the ability to mathematize from everyday experience) can be developed in virtually every 3–5-year-old, or US preschooler, especially when taught along a developmental progression. However, relatively few studies have conducted causal evaluations. Clements and colleagues have conducted scientifically rigorous studies of early childhood mathematics interventions (see Clements & Sarama, 2011, for an overview of their work). At the preschool level, Clements and Sarama (2007, 2008) developed and tested the effectiveness of the *Building Blocks* mathematics curriculum with children from programmes serving low-income communities. Although the curriculum is comprehensive and covers a broad range of mathematical topics, core content number activities (e.g. counting, number comparisons, adding, and subtracting) are a focus of the 26-week instructional programme. Table 57.2 presents examples of activities aligned to the number concepts taught in the Building Blocks curriculum. The activities are structured

Table 57.2 Building Blocks preschool intervention activities (Clements & Sarama, 2009)

Number Concept	Activities
Subitizing	• Drawing the number of objects flashed on a card. • Matching a numeral to dot pattern on card; given a set of cards with dots—which card doesn't belong. • Using cards having different arrangements of dots for each number.
Verbal and object counting	• Counting up and down or by twos. • Touching objects while counting. • Finding desired number of objects in tins. • Touching head, shoulder knees, toes, etc. while counting. • Rolling a number cube with the numbers 1–5 and moving the corresponding spaces on a game board. • Jumping (or another motion) the number of times indicated on a card.
Comparing, ordering, and estimating	• Putting eggs in a carton (one-to-one correspondence). • Putting several cubes in a tin (many-to-one correspondence). • Comparing quantities of differing sizes without counting. • Comparing two sets by matching one-to-one: (e.g. one bone for each dog). • Estimating the number of dots on a card or page. • Estimating the value of a given location on a number line and placing a tick mark at an approximate place corresponding to a given number.
Arithmetic: early addition and subtraction and counting strategies	• Non-verbal +/−: showing a number of objects which are then covered, adding more objects under the cove, and matching the result to a physical model (sums up to 5). • Small number +/−: solving verbal problems with totals up to 5 using objects. • Showing a numeral and a number of dots and counting on the number of dots from the numeral amount, then moving the total number of spaces on a game board.

along a developmental sequence using learning trajectories for core areas of number (e.g. moving from small number recognition to counting to number relations and operations; non-verbal addition to small number word problems to counting on from an addend) in keeping with research on the development of early mathematics.

Using random assignment and a pre–post test design, children who received Building Blocks instruction increased significantly more than controls (with a business-as-usual curriculum) on a closely aligned Building Blocks assessment in general numeracy, with large and statistically meaningful effect sizes (ES = 1.06) (Clements & Sarama, 2008). Positive effects in numeracy also are seen in other randomized controlled studies with comprehensive preschool mathematics curricula (Dobbs, Doctoroff, Fisher, & Arnold, 2006; Klein, Starkey, Clements, Sarama, & Iyer, 2008)

Other researchers have developed targeted interventions to be used with small groups of preschoolers with established risk. Some of the most compelling work has come from Siegler and colleagues, who have identified specific number activities that can improve numerical understanding in high-risk preschoolers from low-income families. Playing board games that use linear number representations (i.e. numbered, equal-size spaces starting at 1) yields positive numeracy effects (Siegler, 2009). Using random assignment, Ramani and Siegler (2008) taught children from Head Start centres to play either a number-board game or a colour board game that did not involve numbers. Children played the games for 20 minutes on four occasions over a 2-week period. Children in the number board game condition became more accurate in their ability to estimate where numbers should be placed on a number line, relative to the children who played the colour board game without numbers. Relative to controls, the number board game also significantly improved the experimental children's ability to compare which of two numbers is bigger, identify written numbers by name, and count from 1 to 10, with the largest effects sizes for the number comparison task (ES = 1.40). The findings held at follow-up about 9 weeks later. Siegler (2009) suggests that because many low-income preschoolers have had little direct experiences with mathematics, number game interventions provide an effective environment for prompting development in a familiar context.

The importance of the game board representation is discussed in a companion study. Siegler and Ramani (2009) compared the effect of playing the game on a linear game board with playing a similar game constructed in a circle. The circular game board consisted of a circle with 10 equal wedges from two o'clock to ten o'clock, each containing one of the numbers 1 through 10, in order. There was also a third comparison group, which worked on numerical activities, such as oral counting, counting sets of objects, and number recognition. Young children who played the linear game performed considerably better than the other two groups on the measure related to numerical magnitudes, an important number sense competency.

There was another important result related to children's learning of arithmetic. After playing the game on four occasions, all the children received training in four story problems for which they had 0% success during the pretest. Immediately after this training, the children were given the posttest. The linear board game group again performed better than the other two groups (circular game and general numerical activities). Siegler and Ramani (2009) argue that understanding that numbers are represented linearly facilitates mathematics instruction. A mental number line helps children encode, store, and retrieve numerical information by providing a means for organizing the information around the numbers' magnitude.

Baroody, Eiland, and Thompson (2009) conducted a numeracy intervention study in which preschoolers between the ages of 4 and 5 years were instructed for 10 weeks, three times a week, in small groups. Manipulatives and games that focus on developmental prerequisites for mental arithmetic (basic number concepts, verbal counting, object counting, and numerical relations) were used. In a second phase, children were trained for another 10 weeks, one-on-one, in mental arithmetic related to the $n + 0$ and $n + 1$ rules. Children were randomly assigned to several conditions, including a semi-structured discovery learning condition and a haphazard practice of the number after condition. In terms of gains in arithmetic computation with 0 and 1, the structured discovery group outperformed the haphazard practice group, with small effect sizes. Lai, Baroody, & Johnson (2008) found that training in inverse principle (e.g. $3 + 1 - 1 = 3$) also significantly improved preschoolers' computational performance relative to controls (ES = 0.54).

Kindergarten Interventions (5–6 Years of Age)

Despite the importance of early number development in preschool, the National Research Council's Committee on Early Childhood Mathematics (2009) concludes, 'most early childhood programs spend little focused time on mathematics, and most of it is low in instructional quality. Many opportunities are missed for learning mathematics over the course of the preschool day' (p. 339). As a result, many children enter kindergarten with relatively few number experiences.

Chard, Baker, Clarke, Junghohann, Davis, and Smolkowski (2008) developed a kindergarten mathematics curriculum (Early Learning in Mathematics) that focused on number sense, among other areas of mathematics. Number lines were used to develop counting, numeral identification, quantity discrimination, and before and after activities. The impact of the curriculum was tested on a population of kindergartners, about half of whom came from low-income families. Children who received the curriculum in their classrooms tended to perform better than a comparison group on an early maths achievement test that was aligned with the curricula. Although the findings are suggestive, it is not possible to draw conclusions about causality because the study did not use random assignment for the experimental and comparison conditions, and there were substantial differences between the groups in mathematics at baseline.

Although many US kindergarten curricula are starting to provide more mathematics instruction into the school day, research has shown that children at risk for learning difficulties benefit most from intensive (at least 30 minutes per session) instruction in small-groups of three to six children (Gersten, R., Baker, S.K., Shanahan, T., Linan-Thompson, S., Collins, P., & Scarcella, R., 2007). Jordan and colleagues (Dyson, Jordan, & Glutting, 2011; Jordan, Glutting, Dyson, Hassinger-Das, & Irwin, 2012) developed a small group number sense intervention for at-risk kindergartners (i.e. from low-income communities and/or performed poorly on a number sense screener). The intervention targeted key number competencies that underlie mathematics difficulties, that is, number, number relations, and number operations. Examples of activities aligned to the number concepts taught in the kindergarten number sense intervention (Dyson et al., 2011; Jordan et al., 2012) are presented in Table 57.3. Intervention lessons were scripted and incorporated quantitative vocabulary (e.g. before, after, plus, minus, bigger, smaller, more, less, altogether, etc.). Consistent representations of quantities (primarily chips, black dots, and fingers) were used because previous work has shown that young children often focus on perceptual variables in tasks, rather than on relevant numerical information in mathematics-related activities (Rousselle, Palmers, & Noël, 2004). Activities were centred on a number list from one to ten (Ramani & Siegler, 2008), and adaptive strategy use was emphasized on addition and subtraction tasks.

The interventions were carried out in small groups of four children in 30-minute sessions three times per week (total of 24 sessions over 8 weeks). A contrasting language intervention was used to ensure that effects were due to the number sense intervention specifically and not simply because of additional focused attention. In several randomized trials, it was found that that the intervention group improved in number sense, as well as in general mathematics achievement relative to control groups (i.e. language intervention group and business-as-usual group) with moderate to large effect sizes (Dyson et al., 2011; Jordan et al., 2012). Controlling for initial mathematical knowledge at pretest, the number sense children performed better that controls on an immediate post-test and a delayed post-test about

Table 57.3 Kindergarten Number Sense Intervention Activities (Dyson et al., 2011)

Number concept	Activities
Number (subitization, counting, and number recognition)	• Saying the number of circles (1–5) flashed on a card. • Saying quickly how many fingers the instructor held up. • Counting quantities from 1 to 20. • Making quantities 10 to 100 using sticks of ten and single cubes. • Making numerals 10 to 100 using decade cards and unit overlays. • Using a hundreds chart to learn the pattern of base ten numerals. • Naming numerals flashed on a card (1 to 100) • Counting by tens to 100 to practice decade names.
Number relations (comparing quantities)	• Matching quantities shown as dot arrays, fingers, tally sticks, and numerals. • Using a cardinality chart, finding the bigger/smaller number. • Finding the bigger/smaller number given two numerals and no other representation of quantity. • Using a number list, children move a game piece to the number before/after. Visualizing movements on number list to find before/after numbers. • Using a number list, connect n +1 to the 'next number' or the number 'after.' • Using number flashcards, putting down the number 'after' or the number 'before.'
Number operations and place value (part-whole relationships, addition and subtraction)	• Showing part-part-whole relationships by using the fingers on two hands. • Using 10 frames to show numbers with respect to groups of 5 and 10. • 'Counting–on/down' 1 or 2 as they move their piece along a number list game board. • Solving combinations with totals up to 5 using objects, drawings, and fingers. • Solving combinations with totals up to 10 using objects, drawings, and fingers. • Solving written combinations in both horizontal and vertical format. • Separating dots in a horizontal line to show two smaller parts making the whole. • Matching story problems (e.g. Jane has three crayons, Rod gives her two more) to partners (e.g. ●●●| ●●) and number sentences (e.g. 3 + 2). • Making up stories that can be solved with a given number sentence.

2 months later. There were moderate to large effect sizes on both the closely aligned number sense assessment, as well as the more distal achievement test. Particularly noteworthy were the consistently large effect sizes on story problems (ES = 2.64 at post-test and 2.27 at delayed post-test). At delayed post-test, for example, 81% of children in the number sense group could solve the change plus story problem, 'Jose has 3 cookies. Sarah gives him 2 more cookies. How many cookies does Jose have now?' and 74% could solve the change minus problem, 'Paul has 5 oranges. Maria takes away 2 of his oranges. How many oranges does Paul have now?' versus 45 and 36% for the language group, and 43 and 36% for the control group. Without any intervention, an earlier study showed that low-income kindergartners are four times more likely

to fall into a low performance, flat growth group on story problems than are middle-income kindergartners (Jordan et al., 2006). The number sense children's performance on story problems in the present study were compared with that of middle-income children in a former study (i.e. Jordan et al., 2010) on a set of items that were common to both studies. In the spring of kindergarten, middle-income children on average were able to solve about 40% of the story problems vs. 80% for the number sense intervention children at delayed post-test.

The intervention group's gains on number combinations also are noteworthy, especially in light of the significant and unique relationship between knowledge of number combinations in kindergarten and the development of later number fact fluency (Locuniak & Jordan, 2008). Kindergarten knowledge of addition number combinations is a marker for mathematics learning disabilities (Mazzocco & Thompson, 2005). A closer look at children's performance on individual problems at delayed posttest reveals that the intervention group performed better, on average, than both the language and control groups on all items. The majority of the number sense children were able to add 1 to a larger number (e.g. 10 + 1) at post-test, suggesting they can construct the number-after rule for adding one (10 and 1 is the number after 10, that is, 11) (Baroody et al., 2009). Subtraction problems were hardest for all children, regardless of group membership.

Another study with kindergartners used adaptive computer games to train children directly in number comparisons, a core deficit for children with dyscalculia (Rasanen, Salminen, Wilson, Aunio, & Dehaene, 2009). Children identified as having difficulty with number were randomly assigned to one of two intervention conditions. One group played a computer game (The Number Race) that trained children in choosing (without counting) the larger of two arrays of dots; the other group played a computer game (Graphogame-Math) that trained children to match small sets of objects with the verbal numerical name. Playing either adaptive computer game helped children improve their accuracy on number comparisons, relative to a control group, but the findings did not transfer to more general maths tasks. The finding suggests that an approach that targets the interconnected number, relations, and operations areas may be more effective for promoting maths achievement in kindergarten. However, the computer-assisted game could be an effective component of a more comprehensive intervention.

First Grade Interventions (6–7 Years of Age)

By the time children enter first grade, most have a rudimentary understanding of addition and subtraction, and can count to problem solve. For addition, young children commonly count both addends; for subtraction, they represent the minuend with objects or fingers and sequentially separate objects or fold down fingers for the value of the subtrahend (Groen & Resnick, 1977; Siegler & Shrager, 1984). As understanding of cardinality and the counting sequence develops, children discover the number-after principle in counting, making +1/−1 problems among the easiest to learn (Baroody, 1999). They also understand the sum of 5 + 2 cannot be 6, but instead is two numbers beyond 5. In this way, children discover the efficiency of counting from the first addend and rely on more efficient counting procedures with greater frequency (Baroody, 1987; Siegler & Robinson, 1982).

For addition, counting on is the most efficient strategy and involves stating the cardinal value of the larger addend and counting the number of times equal to the smaller addend (e.g. 4 + 3 = 'four: five, six, seven'); for subtraction, counting up involves stating the subtrahend and counting to the minuend (e.g. 5 − 2 = 'two: three, four, five'; the number of counts is the answer). Frequent use of efficient counting procedures reliably produces the correct association between problem and answer, and results in the formation of long-term memories (Fuson & Kwon, 1992; Siegler & Robinson, 1982; Siegler & Shrager, 1984), which enables direct retrieval of answers.

Knowledge of mathematical principles also aids in fact memorization. The commutativity of addition facilitates retrieval of related addition problems (Rickard, 1994). Subtraction, which is not commutative, is more difficult, but can nonetheless be facilitated by retrieval of related addition facts (e.g. 8 − 5 = 3, based on 5 + 3 = 8; LeFevre & Morris, 1999), once children understand the inverse relation between addition and subtraction (Lai et al., 2008).

Not all children follow this developmental sequence with success, however. Some enter first grade with delays in the adoption of efficient counting procedures and make frequent counting errors during their execution. These types of difficulty are associated with failure to make the shift toward memory-based retrieval and can lead to long-term problems with mathematics (e.g. Geary, Bailey & Hoard, 2012; Goldman, Pellegrino, & Mertz, 1988). For these reasons, intervention research for at-risk first graders has placed a strong emphasis on simple arithmetic. Fuchs, Fuchs, Hamlett, Powell, Capizzi, & Seethaler (2006) conducted a randomized control study at first grade to assess the efficacy of practice designed to ensure quick, correct pairings of problems with answer. Correct addition or subtraction problems briefly flashed on a computer screen; then students generated the problem and answer from short-term memory. This is in line with Geary's (1993) proposal that practice strengthens retrieval only if problem stems and answers are simultaneously active in working memory. Compared with an analogous computer-assisted spelling practice condition, arithmetic practice (10 minutes, twice weekly for 18 weeks) produced significantly better performance on addition but not subtraction. Other mathematics outcomes were not assessed.

As mentioned, however, most researchers agree that a strong number knowledge is an important foundation for success with arithmetic. Two randomized control trials combined tutoring on number knowledge with arithmetic practice and addressed a broader range of outcomes. In Fuchs, Compton, Fuchs, Paulsen, Bryant, and Hamlett (2005), tutoring occurred 3 times per week for 16 weeks. Each session included 30 minutes of tutor-led instruction designed to build number knowledge plus 10 minutes of computer-mediated arithmetic practice, as just described. Results favoured tutoring over a no-tutoring control group on measures of concepts and applications (effect size (ES) in terms of Cohen's $d = 0.67$), procedural calculations (ES = 0.40 – 0.57), and word problems (ES = 0.48), but effects were not reliable on simple arithmetic (ESs = 0.15 – 0.40). Bryant, Bryant, Roberts, Vaughn, Pfannenstiel, Porterfield, et al. (2011) also integrated tutoring on number knowledge with practice (4 times per week for 19 weeks). In each session, 20 minutes were devoted to number knowledge; 4 minutes to practice, which focused on arithmetic problems, as well as reading numerals, counting on/back, writing dictated numerals, and writing three-number sequences. Effects were significantly stronger for tutoring compared to a no-tutoring control group on simple arithmetic (ES = 0.55), place value (ES = 0.39), and number sequences (ES = 0.47). However, tutoring did not enhance word-problem outcomes (ES = −0.05 and 0.07).

In the only randomized control trial to focus exclusively on number knowledge, Smith, Cobb, Farran, Cordray, and Munster (2011) evaluated Math Recovery (MR), in which tutors adapt lessons to meet student needs as reflected on MR assessments. Tutors introduce tasks and have students explain their reasoning, but practice is not provided. Tutoring was to occur 4–5 times per week, 30 minutes per session across 12 weeks, but the median number of sessions was 32. At the end of first grade, effects favoured MR over the control group on fluency with simple arithmetic (ES = 0.15), concepts and applications (ES = 0.28), quantitative concepts (ES = 0.24), and maths reasoning (ES = 0.30). Effects were stronger for students who began tutoring below the 25th percentile (ES = 0.31–0.40), but they are smaller than in other studies. Comparisons are, however, difficult because this effectiveness study allowed fidelity to vary, whereas other studies tried to ensure fidelity.

In perhaps the largest randomized control trial conducted at first grade, Fuchs, Geary, Compton, Fuchs, Schatschneider, Hamlett, et al. (2011) contrasted two forms of tutoring for improving simple addition and subtraction performance in first-grade children with risk for poor mathematics outcomes. The major focus in both conditions was number knowledge (e.g. cardinality) and relations (e.g. inverse relation between addition and subtraction) that facilitate the use of sophisticated counting procedures or retrieval-based processes (e.g. commutativity). Number knowledge comprised 25 of every 30-minute session and was identical in the two conditions. The conditions differed during the last 5 minutes of each session to contrast two forms of practice. Both practice conditions focused on the same content, provided immediate feedback, and encouraged strategic problem-solving behaviour. In the *non-speeded practice* condition, students played board games that reviewed key content from that day's number knowledge tutoring lesson. In the *speeded practice* condition, students had 90 seconds to answer flashcards. When the day's topic was arithmetic problems, for example, students answered as many flashcards as possible, 'Knowing the answer right off the bat' or using efficient counting strategies that were taught in both conditions. When an error occurred, tutors required students to use the counting strategies to derive the correct answer, as time continued to elapse. At the end of 90 seconds, children counted number correct and then had two opportunities to beat that score.

Number knowledge tutoring with non-speeded practice resulted in significantly better learning compared to at-risk control students who did not receive tutoring, with an ES of 0.38. At the same time, simple arithmetic learning among students in the non-speeded practice condition was comparable to that of their low-risk classmates, such that the gap in simple arithmetic did not narrow over the course of first grade. By contrast, incorporating speeded practice within number knowledge tutoring produced an effect size of 0.87 over at-risk control; was reliably more effective than number knowledge tutoring without practice (ES = 0.51); and produced superior improvement in simple arithmetic compared to low-risk classmates, with an effect size of 0.39. Thus, speeded practice students were narrowing the achievement gap. In this way, results indicate a substantial role for speeded practice in promoting simple arithmetic outcomes.

It should be noted that, in contrast to how practice is sometimes configured in schools (i.e. without sufficient scaffolding in number knowledge, in massed doses, and without support for correct responding), Fuchs et al. (2011) delivered speeded practice in the context of number knowledge instruction. Practice was formulated to help children generate many correct responses, support development of fluency with efficient counting strategies, require students to immediately correct errors with an efficient counting procedure, and encourage strategic meta-cognitive behaviour (i.e. requiring children to retrieve answers from memory, if confident; otherwise, use an efficient counting strategy). Therefore, findings should be generalized only to practice that incorporates similarly sound, theoretically motivated instructional design.

Summary and Conclusions

Poor achievement in mathematics is widespread with far-reaching consequences. Insufficient mathematics skills limit students' opportunities for higher education and their eventual employment in a global economy. The roots of mathematics difficulties appear in early childhood, with foundational number competencies being of primary importance. Number, number relations, and number operations can be reliably identified in young children, and weaknesses in these areas predict later mathematical learning. Income status, associated early home and preschool opportunities, and general cognitive capacity all influence children's level of numerical knowledge. Interventions based on a developmental progression and targeted to children's specific areas of need, such as the ability to count and sequence numbers, to compare numerical quantities, and to add and subtract small numbers have shown positive and lasting effects. Board game activities, where numerals are presented linearly, may be especially useful for preschool children who have less experience informal numerical abilities.

Guided practice for first graders is effective when configured to support efficient counting strategies, frequent correct responding, and meta-cognitive behavior, and when contextualized with a strong focus on number knowledge tutoring. Such practice helps children solve and develop fluency with number combinations. This is achieved via more correct procedural problem solving and stronger reliance on automatic retrieval, without inhibiting children's development of competence with number knowledge or word problems.

References

Abedi, J. (2004). The No Child Left Behind Act and English language learners: Assessment and accountability issues. Educational Researcher, 33, 4–14.

Baroody, A.J. (1987). The development of counting strategies for single-digit addition. Journal for Research in Mathematics Education, 18(2), 141–157.

Baroody, A.J. (1999). The development of basic counting, number, and arithmetic knowledge among children classified as mentally handicapped. In L. M. Glidden (Ed.), International Review of Research in Mental Retardation, Vol. 22 (pp. 51–103). New York: Academic Press.

Baroody, A.J., Eiland, M., & Thompson, B. (2009). Fostering at-risk preschoolers' number sense. Early Education and Development, 20, 80–128.

Baroody, A.J., Lai, M-L., & Mix, K.S. (2006). The development of young children's early number and operation sense and its implications for early childhood education. In B. Spodek & O. Saracho (Eds), Handbook of Research on the Education of Young Children (pp. 187–221). Mahwah, NJ: Lawrence Erlbaum Associates.

Berch, D.B. (2005). Making sense of number sense: implications for children with mathematical disabilities. Journal of Learning Disabilities, 38(4), 333–339.

Berch, D.B., & Mazzocco, M.M. (Eds). (2007). Why Is Math So Hard For Some Children? Baltimore: Paul H. Brookes Publishing Co.

Blevins-Knabe, B., & Musun-Miller, L., (1996). Number use at home by children and their parents and its relationship to early mathematical performance. Early Development and Parenting, 5(1), 35–45.

Bryant, D.P., Bryant, B.R., Roberts, G., Vaughn, S., Pfannenstiel, K.H., Porterfield, J., et al. (2011). Early numeracy intervention program for first-grade students with mathematics difficulties. Exceptional Children, 78, 7–23.

Butterworth, B., Sashank, V., & Laurillard, D. (2011). Dyscalculia: from brain to education. Science, 332, 1049–1053.

Carpenter, T.P., Hiebert, J., & Moser, J.M. (1983). The effect of instruction on children's solutions of addition and subtraction word problems. Educational Studies in Mathematics, 14(1), 55–72.

Case, R., & Griffin, S. (1990). Child cognitive development: the role of central conceptual structures in the development of scientific and social thoughts. In C. A. Hauert (Ed.), Advances in Psychology – Developmental Psychology: Cognitive, perception-Motor, and Neurological Perspectives. Amsterdam: North Holland.

Case, R., Okamoto, Y., Griffin, S., McKeough, A., Bleiker, C., Henderson, B., et al. (1996). The role of central conceptual structures in the development of children's thought [Monograph]. Monographs of the Society for Research in Child Development, 61(1/2), 1–295.

Chard, D.J., Baker, S.K., Clarke, B., Jungjohann, K., Davis, K., & Smolkowski, K. (2008). Preventing early mathematics difficulties: the feasibility of a rigorous kindergarten mathematics curriculum. Learning Disability Quarterly, 31(1), 11–20.

Clarke, B., Baker, S.K., Smolkowski, K., & Chard, D. (2008). An analysis of early numeracy curriculum-based measurement: examining the role of growth in student outcomes. Remedial and Special Education, 29(1), 46–57.

Clarke, B., Nese, J.F.T., Alonzo, J., Mercier-Smith, J., Tindal G., Kame'enui, E.J., & Baker, S. (2011). Classification accuracy of easy CBM first grade mathematics measures: Findings and implications for the field. Assessment for Effective Intervention, 36(4), 243–255.

Clarke, B. & Shinn, M. R. (2004). A preliminary investigation into the identification and development of early mathematics curriculum-based measurement. School Psychology Review, 33(2), 234–248.

Clements, D.H., & Sarama, J. (2007). Effects of a preschool mathematics curriculum: summative research on the Building Blocks project. Journal for Research in Mathematics Education, 38, 136–163.

Clements, D.H., & Sarama, J. (2008). Experimental evaluation of the effects of a research-based preschool mathematics curriculum. American Education Research Journal, 45(2), 443–494.

Clements, D.H., & Sarama, J. (2011). Early childhood mathematics intervention. Science, 333, 968–970.

Douglas H. Clements & Julie Sarama (2004) Learning Trajectories in Mathematics Education, Mathematical Thinking and Learning, 6:2, 81-89, DOI:10.1207/s15327833mtl0602_1

Clements, D.H., Sarama, J. and Gerber, S. (2005, April), Mathematics Knowledge of Entering Preschoolers. Paper presented at the Annual Meeting of the American Educational Research Association, Montreal, Canada.

Dehaene, S. (1999). The Number Sense: How the Mind Creates Mathematics. New York: Oxford University Press US.

Dobbs, J., Doctoroff, G.L., Fisher, P.H., & Arnold, D.H. (2006). The association between preschool children's socio-emotional functioning and their mathematic skills. Applied Developmental Psychology, 27, 97–108.

Dowker, A., & Sigley, G. (2010). Targeted interventions for children with arithmetical difficulties. British Journal of Educational Psychology, Monograph Series II, 7, 65–81.

Duncan, G.J., Dowsett, C.J., Claessens, A., Magnuson, K., Huston, A.C., Klebanov, P., et al. (2007). School readiness and later achievement. Developmental Psychology, 43(6), 1428–1446.

Dunn, S., Matthews, L. & Dowrick, N. (2010). Numbers count: developing a national approach to early intervention. In I. Thompson (Ed.) Issues in Teaching Numeracy in Primary Schools, 2nd edn. Maidenhead: Open University Press, 224–234.

Dyson, N.I., Jordan, N.C., & Glutting, J. (2013). A number sense intervention for kindergartners at risk for math difficulties. Journal of Learning Disabilities, 46(2), 186–181.

Evans, D. (2007). Developing mathematical proficiency in the Australian context: implications for children with learning disabilities. Journal of Learning Disabilities, 40, 420–426.

Feigenson, L., Dehaene, S., & Spelke, E.S. (2004). Core systems of number. Trends in Cognitive Sciences 8(7), 307–314.

Fuchs, L.S., Compton, D.L., Fuchs, D., Paulsen, K., Bryant, J.D., & Hamlett, C.L. (2005). The prevention, identification, and cognitive determinants of math difficulty. Journal of Educational Psychology, 97, 493–513.

Fuchs, L.S., Fuchs, D., Hamlett, C.L., Powell, S.R., Capizzi, A.M., & Seethaler, P.M. (2006). The effects of computer-assisted instruction on number combination skill in at-risk first graders. Journal of Learning Disabilities, 39, 467–475.

Fuchs, L.S., Geary, D.C., Compton, D.L., Fuchs, D., Schatschneider, C., Hamlett, C.L., et al. (2013). Effects of First-Grade Number Knowledge Tutoring with Contrasting Forms of Practice. Journal of Educational Psychology, *105* (1), pp. 58-77.

Fuson, K.C., Grandau, L., & Sugiyama, P.A. (2001). Achievable numerical understandings for all young children. Teaching Children Mathematics, 7(9), 522–526.

Fuson K.C., & Kwon Y. (1992). Learning addition and subtraction: effects of number word and other cultural tools. In J. Bideau, C. Meljac, & J. P. Fisher, (Eds), Pathways to Number (pp. 351–374). Hillsdale, NJ: Erlbaum.

Geary, D.C. (1993). Mathematical disabilities: cognitive, neuropsychological, and genetic components. Psychological Bulletin, 114, 345–362.

Geary, D.C., Bailey, D.H., & Hoard, M.K. (2009). Predicting mathematical achievement and mathematical learning disability with a simple screening tool: the Number Sets Test. Journal of Psychoeducational Assessment, 27, 265–279.

Geary, D.C., Hoard, M.K., & Bailey, D.H. (2012). Fact retrieval deficits in low achieving children and children with mathematical learning disability. Journal of Learning Disabilities, 45(4), 291–307.

Geary, D.C., Hoard, M.K., Byrd-Craven, J., Nugent, L., & Numtee, C. (2005). Early identification and intervention for students with mathematics difficulties. Journal of Learning Disabilities, 38(4), 324–332.

Gelman, R., & Gallistel, C.R. (1978). The Child's Understanding of Number. Cambridge, MA: Harvard University Press.

Gersten, R., Baker, S.K., Shanahan, T., Linan-Thompson, S., Collins, P., & Scarcella, R. (2007). Effective Literacy and English Language Instruction for English Learners in the Ele mentary Grades: A Practice Guide (NCEE 2007–4011). Washington, DC: National Center for Education Evaluation and Regional Assistance, Institute of Education Sciences, U.S. Department of Education.

Gersten, R., Clarke, B., Jordan, N.C., Newman-Gonchar, R., Hammond, K., & Wilkins, C. (2012). Universal screening in mathematics for the primary grades: Beginning of a research base. Exceptional Children, 78(4), 423–225.

Gersten, R., Jordan, N.C., & Flojo, J.R. (2005). Early identification and interventions for students with mathematics difficulties. Journal of Learning Disabilities, 38(4), 293–304.

Ginsburg, H.P. (1989). Children's Arithmetic, 2nd edn. Austin, TX: Pro-Ed.

Ginsburg, H.P., Lee, J.S., & Boyd, J.S. (2008). Mathematics education for young children: what it is and how to promote it. Social Policy Report, 22, Number 1.

Goldman, S.R., Pellegrino, J.W., & Mertz, D.L. (1988). Extended practice of addition facts: strategy changes in learning-disabled students. Cognition and Instruction, 5, 223–265.

Griffin, S. (2002). The development of math competence in the preschool and early school years: cognitive foundations and instructional strategies. In J. M. Roher (Ed.), Mathematical Cognition (pp. 1–32), Current Perspectives on Cognition, Learning, and Instruction Series. Greenwich, CT: Information Age Publishing, Inc.

Griffin, S. (2004). Building number sense with Number Worlds: a mathematics program for young children. Early Childhood Research Quarterly, 19, 173–180.

Griffin, S., & Case, R. (1997). Re-thinking the primary school math curriculum: an approach based on cognitive science. Issues in Education, 3(1), 1–49.

Groen, G. & Resnick, L. (1977). Can preschool children invent addition algorithms? Journal of Educational Psychology, 69, 645–652.

Hiebert, J. (1984). Children's mathematics learning: the struggle to link form and understanding. Elementary School Journal, 84(5), 496–513.

Huttenlocher, J., Jordan, N., & Levine, S. (1994). A mental model for early arithmetic. Journal of Experimental Psychology: General, 123 (3), 284–296.

Jordan, N.C., Glutting, J., & Dyson, N. (2012). Number Sense Screener User's Guide. Baltimore: Paul H. Brookes.

Jordan, N.C., Glutting, J., & Ramineni, C. (2009). The importance of number sense to mathematics achievement in first and third grades. *Learning and Individual Differences*, *20*, 81–88.
Jordan, N.C., Glutting, J., Dyson, N., Hassinger-Das, B., & Irwin, C. (2012). Building kindergartners' number sense: a randomized controlled study. Journal of Educational Psychology, *104* (3), 647–660.
Jordan, N.C., Glutting, J., Ramineni, C., & Watkins, M.W. (2010). Validating a number sense screening tool for use in kindergarten and first grade: prediction of mathematics proficiency in third grade. School Psychology Review, 39, 181–195.
Jordan, N.C., Huttenlocher, J., & Levine, S.C. (1992). Differential calculation abilities in young children from middle- and low-income families. Developmental Psychology, 28(4), 644–653.
Jordan, N.C., Kaplan, D., Olah, L., & Locuniak, M.N. (2006). Number sense growth in kindergarten: a longitudinal investigation of children at risk for mathematics difficulties. Child Development, 77, 153–175.
Jordan, N.C., Kaplan, D., Locuniak, M.N., Ramineni, C. (2007). Learning Disabilities Research & Practice, 22(1), 36–46
Jordan, N.C., Kaplan, D., Ramineni, C., & Locuniak, M.N. (2008). Development of number combination skill in the early school years: when do fingers help? Developmental Science, 11(5), 662–668.
Jordan, N.C., Kaplan, D., Ramineni, C., & Locuniak, M.N. (2009). Early math matters: kindergarten number competence and later mathematics outcomes. Developmental Psychology, 3(45), 850–867.
Jordan, N.C., & Levine, S.C. (2009). Socioeconomic variation, number competence, and mathematics learning difficulties in young children. Developmental Disabilities Research Reviews, 15, 60–68.
Juel, C. (1988). Learning to read and write: A longitudinal study of 54 children from first through fourth grades. Journal of Educational Psychology, *80*(4), 437–447.
Kempert, S., Saalbach, H., & Hardy, I. (2011). J Cognitive benefits and costs of bilingualism in elementary school students: The case of mathematical word problems. Journal of Educational Psychology, *103*(3), 547–561.
Klein, A., Starkey, P., Sarama, J., Clements, D.H., & Iyer, R. (2008). Effects of a pre-kindergarten mathematics intervention: a randomized experiment. Journal of Research on Educational Effectiveness, 1, 155–178.
Klibanoff, R.S., Levine, S.C., Huttenlocher, J., Vasilyeva, M., & Hedges, L.V. (2006). Preschool children's mathematical knowledge: the effect of teacher 'Math Talk'. Developmental Psychology, 42(1), 59–69.
Kroesbergen, E.H & Van Luit, J.E.H. (2003). Mathematics interventions with children with special educational needs: a meta-analysis. Remedial and Special Education, 24, 97–114.
Lai, M.-L., Baroody, A.J., & Johnson, A.R. (2008). Fostering Taiwanese preschoolers' understanding of the addition-subtraction inverse principle. Cognitive Development, 23, 216–235.
Lee, V.E., & Burkam, D.T. (2002). Inequality at the Starting Gate: Social Background Differences in Achievement as Children Begin School. Washington DC: Economic Policy Institute.
LeFevre, J., & Morris, J. (1999). More on the relation between division and multiplication in simple arithmetic: evidence for mediation of division solutions via multiplication. Memory & Cognition, 27, 803– 812.
Lembke, E., & Foegen, A. (2009). Identifying early numeracy indicators for kindergarten and first-grade students. Learning Disabilities Research & Practice, 24(1), 12–20.
Levine, S.C., Jordan, N., & Huttenlocher, J. (1992). Development of calculation abilities in young children. Journal of Experimental Child Psychology, 53 (1), 72–103.
Levine, S.C., Suriyakham, L.W., Rowe, M., Huttenlocher, J., & Gunderson, E.A. (2010). What counts in the development of children's number knowledge? Developmental Psychology, 46(5), 1309–1313.
Locuniak, M.N., & Jordan, N.C. (2008). Using kindergarten number sense to predict calculation fluency in second grade. Journal of Learning Disabilities, 41(5), 451–459.
Malofeeva, E., Day, J., Saco, X., Young L., & Ciancio, D. (2004). Construction and evaluation of a number sense test with head start children. Journal of Educational Psychology, 96(4), 648–659.
Mazzocco, M.M.M., & Thompson, R.E. (2005). Kindergarten predictors of math learning disability. Learning Disabilities Research and Practice, 20(3), 142–155.
Methe, S. A., Hintze, J.M., Floyd, R.G. (2008). Validation and decision accuracy of early numeracy skill indicators. School Psychology Review, *37*, 359–373.

Morgan, P.L., Farkas, G., & Wu, Q. (2009). Five-year growth trajectories of kindergarten children with learning difficulties in mathematics. Journal of Learning Disabilities, 42, 306–321.

Lee, J., Grigg, W., & Dion, G. (2007). The Nation's Report Card: Mathematics 2007 (NCES 2007–494). National Center for Education Statistics, Institute of Education Sciences, US Department of Education, Washington, D.C.

National Mathematics Advisory Panel. (2008). Foundations for Success: The Final Report of the National Mathematics Advisory Panel. Washington, D.C.: US Department of Education.

National Research Council. (2009). Mathematics Learning in Early Childhood: Paths Toward Excellence and Equity. Washington, DC: National Academies Press.

Opel, A., Zaman, S.S., Khanom, F., & Aboud, F.E. (2012). Evaluation of a mathematics program for preprimary children in rural Bangladesh. International Journal of Educational Development, 32, 104–110.

Ramani, G.B., & Siegler, R.S. (2008). Promoting broad and stable improvements in low-income children's numerical knowledge through playing number board games. Child Development, 79, 375–394.

Rasanen, P., Salminen, J., Wilson, A., Aunio, P., & Dehaene, S. (2009). Computer-assisted intervention for numerical skills in children with low numeracy: a review and a randomized controlled trial. Cognitive Development, 24, 450–472.

Reigosa-Crespo, V., Valdés-Sosa, M., Butterworth, B., Estévez, N., Rodríguez, M., Santos, E., et al. (2011). Basic numerical capacities and prevalence of developmental dyscalculia: the Havana survey. Developmental Psychology. 47, 1–13.

Rickard, J. (1994). Translation functors and equivalences of derived categories for blocks of algebraic groups. Proceedings, NATO ARW on Representations of Algebras and Related Topics, 255–264.

Robinson, C. S., Menchetti, B. M., & Torgesen, J. K. (2002). Toward a Two-Factor Theory of One Type of Mathematics Disabilities. Learning Disabilities Research & Practice, *17* (2), 81–89.

Rousselle, L., & Noël, M.-P. (2007). Basic numerical skills in children with mathematics learning disabilities: a comparison of symbolic vs non-symbolic number magnitude processing. Cognition, 102(3), 361–395.

Rousselle, L., Palmers, E., & Noël, M.-P. (2004). Magnitude comparison in preschoolers: what counts? Influence of perceptual variables. Journal of Experimental Child Psychology, 87 57–84.

Russell, R.L., & Ginsburg, H.P. (1984). Cognitive analysis of children's mathematical difficulties. Cognition and Instruction, 1, 217–244.

Sadler, P.M., & Tai, R.H. (2007) The two high-school pillars supporting college science. Science, 27, 457–458.

Seethaler, P.M., & Fuchs, L.S. (2010). The predictive utility of kindergarten screening for math difficulty: how, when, and with respect to what outcomes should it occur? Exceptional Children, 77, 37–60.

Siegler, R.S. (2009). Improving the numerical understanding of children from low-income families. Child Development Perspectives, 3(2), 118–124.

Siegler, R.S., & Ramani, G.B. (2009). Playing linear number board games—but not circular ones-improves low-income preschoolers' numerical understanding. Journal of Educational Psychology, 101(3), 545–560.

Siegler, R.S., & Robinson, M. (1982). The development of numerical understandings. In H. W. Reese & L. P. Lipsett (Eds), Advances in Child Development and Behavior, Vol. 16 (pp. 242–312). New York: Academic Press.

Siegler, R.S., & Shrager, J. (1984). Strategy choices in addition and subtraction: how do children know what to do? In C. Sophian (Ed.), The Origins of Cognitive Skills (pp. 229–293). Hillsdale, NJ: Erlbaum.

Smith, T.B., Cobb, P., Farran, D., Cordray, D., & Munster, C. (2013). Evaluating Math Recovery: assessing the causal impact of a diagnostic tutoring program on student achievement. American Educational Research Journal, 50 (2) 397–428.

VanDerHeyden, A. M., Broussard, C., Cooley, A. (2006). Further Development of Measures of Early Math Performance for Preschoolers. *Journal of School Psychology*, 44 (6), 533–553.

CHAPTER 58

NUMBERS COUNT

a large-scale intervention for young children who struggle with mathematics

NICK DOWRICK

Introduction

Children who find Mathematics Difficult

Some children find mathematics extremely difficult. In England, the proportion of 11-year-olds falling at least 4 years behind the expectations for their age in National Curriculum tests has remained around 6% every year since 1996 (Burr, 2008; Department for Education, 2014). While definitions of "mathematical difficulty" vary, there have been similar estimates in other countries of between 3% and 8% of primary-aged children needing remedial help in mathematics (Dowker, 2004; Geary, 2004).

Problems at a young age can have profound consequences. Children who fail to develop basic numerical understandings and strategies at ages 5–8 are at risk of being "stuck" with immature strategies throughout their schooling, falling further and further behind their classmates (Aubrey et al., 2006; Jordan et al., 2010). Haseler (2008) has suggested that the hierarchical nature of mathematics, for all that it is not fixed, makes recovery particularly difficult for children who have fallen behind. Low achievement in the high-stakes subject of mathematics can lead to a vicious circle of anxiety, loss of self-esteem, and further underachievement (Ashcraft and Moore, 2009; Krinzinger et al., 2009). Problems persist into adult life and often into the next generation (Gross et al., 2009; Moser, 1999).

Young children most often have difficulties in relation to four areas of mathematics: understanding the number system, retrieving number facts, using addition and subtraction strategies, and solving word problems (Dowker, 2004; Jordan and Levine, 2009). Within these areas, however, there is no consistency about the precise difficulties shown by each child (Geary, 2004; Gervasoni & Sullivan, 2007) or even on each day (Houssart, 2007). There is no such thing as a typical child with mathematical difficulties, or, to rephrase a remark by Dowker (2005, p. 26), there is no such thing as mathematical disability, only mathematical disabilities.

The causes of mathematical difficulties are equally varied and imperfectly understood (Geary, 2010). Explanations often involve language impairment with an effect on symbolic understanding, seriation, and number fact retrieval (Cowan et al., 2008; Turner Ellis et al., 2008) or neurological malfunctions associated with difficulties in subitising and number sense, linked to dyscalculia (Butterworth et al., 2011; Jordan et al., 2010). The most common diagnosis, however, is of children who appear to exhibit developmental delay: they follow the same path as younger peers without mathematical difficulties, but at a slower rate (Geary, 2004; Jordan & Levine, 2009). While these difficulties are often associated with overall delays in cognitive development, it is also common for children with mathematical difficulties to have at least average IQ scores (Jimenez & Garcia, 2002; Jordan & Montani, 1997) and to have no difficulties with literacy (Gross et al., 2009).

Dowker (2004) argued that we do not need to concern ourselves unduly with identifying the precise causes of children's arithmetical problems, because it is possible to help them to learn without this. Gross (2007, p. 152) also argued that it was unprofitable to search for the root causes of difficulties, and that "a medical–diagnostic model based on assumed pathology can obstruct the development of appropriate curricular responses." For her, it was more important to answer the question, "So what are we going to do about it?"

Mathematics Interventions

We can help many children to overcome most of their difficulties. Dowker's (2004, p. v) review of research concluded that "children's arithmetical difficulties are highly susceptible to intervention" and that individualised intervention programs achieve the greatest benefit for children with the greatest difficulties. She advised that they are most effective if they are implemented at an early age, individually or in very small groups, and if they begin with an individualised diagnostic assessment that establishes each child's precise needs in relation to distinct components of arithmetic. She also found that even a short intervention can be effective.

Fuchs et al. (2012, p. 258) summarized several research projects in the United States in which up to 108 first-grade and third-grade children received intervention support. They concluded that at-risk children's outcomes improved after small-group interventions that incorporated explicit instruction, attention to conceptual foundations, and well-chosen practice. Their study, however, was based on a broad definition of "at-risk" that encompassed the lowest achieving 20% to 30% of children in each age group, and they recommended that more intensive provision might be needed for about 5% of each age group. This chapter will focus on interventions that have been aimed more specifically at the lowest-achieving children and on those that have moved beyond the research phase to a wider implementation in schools in England.

A popular intervention has been Mathematics Recovery (Wright et al., 2006a, b), which was first developed in Australia and has since spread to the British Isles and other countries (Williams, 2008). It is designed for low-attaining 6-year-old children, who normally receive daily, individual, half-hour lessons for 12–15 weeks; their teachers begin with a diagnostic assessment linked to a framework of number development and then implement an individualised, problem-centred teaching program (Wright, 2008). There is relatively little hard evidence of the impact of Mathematics Recovery because children's progress is normally assessed in relation to the programme's internal objectives (e.g. Willey et al., 2007). However, a short study reported standardized score gains of 11–19 points for 88 children in England (Dowker, 2009).

Another successful program, popular in England, is Catch Up Numeracy (Dowker and Sigley, 2010). While it shares many features with Mathematics Recovery, differences include that it is designed for 6–10-year-olds with more moderate mathematical weaknesses, that children are normally taught by teaching assistants for about half an hour a week for up to 30 weeks, that it uses a different framework of number development, and that it takes a more instructional approach to teaching and learning. It also has more substantial evidence of progress. In a 30-week program, 146 children made standardized score gains of 3–4 points on mathematics tests. In a one-term program, 154 children achieved a mean ratio gain (number age gain divided by time spent on the program) of 2.2, whereas control groups gained little or nothing (Dowker, 2008; Dowker and Sigley, 2010). A more recent study (Education Endowment Foundation, 2014) found that 104 pupils made about 3 months of additional progress, but suggested that this was due to the amount of support that they received rather than to the intrinsic nature of the intervention.

Standardized test gains have additionally been reported by Numeracy Recovery in London, achieving 14–15 point standardized score gains for about 200 children, and by a Multisensory Maths program, which achieved an 18-point gain for 40 children (Dowker, 2009).

The above studies have all been on a relatively small scale, with the largest involving 200 children; Bell (2011) has reminded us that there is a danger that "the small-study effect" may not be replicated when a project is scaled up. While repeatedly demonstrating that interventions can help children who have difficulties in mathematics, the studies have reported little additional detail with, apart from Catch Up Numeracy, little evidence of statistical significance testing or the use of control groups. This chapter describes a new mathematics intervention and gives details of impact research that has attempted to address these issues.

Numbers Count

Overview

In 2008, the education ministry in England launched Every Child Counts, a nationally funded initiative aimed at tackling mathematical underachievement in primary schools (Every Child a Chance Trust, 2010). The core component of Every Child Counts was Numbers Count (Dunn et al., 2010), a new numeracy intervention that was developed specifically for the initiative. It was designed for 6- and 7-year-old children in primary schools who had the greatest difficulties with mathematics, particularly the lowest achieving 30 000 children (6% of each year group) who, without intervention, are on track to fall at least 4 years below national expectations at age 11, every year (Williams, 2008). Its aims were that

Table 58.1 The Numbers Count curriculum

Counting	The Number System
1 Thinking Mathematically about Counting 2 Oral Counting 3 Object Counting 4 Counting in Steps	1 Thinking Mathematically about the Number System 2 Reading and Writing Numbers 3 Comparing and Ordering Numbers 4 Place Value
Number Facts	**Calculating**
1 Thinking Mathematically about Number Facts 2 Recalling and Finding Number Pairs 3 Recalling and Finding Doubles and Halves	1 Thinking Mathematically about Calculating 2 Adding and Subtracting 3 Multiplying and Dividing

the children would make substantial progress while receiving intervention support and that they would continue to make good progress in class mathematics lessons afterwards. It focused on numeracy (see Table 58.1).

The children were taught by experienced teachers who were specially trained for Numbers Counts and who normally taught Numbers Count on a 0.5 timetable, every morning or every afternoon. Schools chose their own teachers to be trained for Numbers Count, often using existing members of staff but sometimes making new appointments; they were part-funded by the education ministry for their teachers' salaries and training costs.

The teachers' training included ten days of professional development, spread over a year and focusing on Numbers Count procedures, the mathematics curriculum, and pedagogy. Its aim was that teachers would understand the purpose and structure of Numbers Count and make independent decisions about what and how they would teach each child. They were given a bank of suggested learning activities for each element of the curriculum, but the emphasis of their training was on developing their ability to diagnose each child's individual needs and their understanding of the mathematics curriculum, common difficulties, and appropriate pedagogies so that they could adapt and develop their own activities to meet those needs.

Teachers were encouraged to collaborate by visiting each other in school and by sharing and analyzing videos of their lessons on professional development days. Group leaders visited each teacher in school up to five times to observe lessons and discuss individual children's needs. Teachers engaged in further professional development activities in their second and subsequent years.

Numbers Count teachers started to teach children after their first day of training, normally selecting the four children in Year 2 (6–7 years old) who had the greatest mathematical difficulties at the start of the school year to take part in Numbers Count. They aimed to help them to develop not just their skills but also their confidence, understanding, and learning behaviors so that they could carry on learning mathematics successfully in future years. The teachers' training emphasized the development of a "numerate child" who:

- is confident and enjoys mathematics
- has a sense of the size of a number and where it fits into the number system

- calculates accurately and efficiently
- makes sense of number problems, choosing and exploring strategies for solving them
- uses what s/he knows by heart to figure out answers mentally
- talks about mathematics using appropriate language and vocabulary
- has strong mental images of mathematics
- makes connections across mathematics and beyond
- is an active learner and is willing to "have a go"
- reflects on his/her learning.

Each child had a 30-minute, one-to-one lesson nearly every day in a well-resourced Numbers Count room, in addition to their normal class mathematics lessons. A program usually lasted for a term, in which time the child typically received between 40 and 45 lessons. While teachers could extend the program for a child who needed more time, they usually decided that it was more beneficial, overall, for four new children to take up places at the start of the next term; this meant that most teachers taught 11 or 12 children in a year. Children took the Sandwell Early Numeracy Test (Arnold et al., 2009) on entry to and exit from Numbers Count to assess their progress, and again 3 months and 6 months after exit to assess whether they had maintained this progress. Throughout the intervention, the Numbers Count teacher liaised closely with the children's class teachers and parents; both often came to observe Numbers Count lessons and discuss the children's development with the Numbers Count teacher.

The teacher began each child's intervention program with a 5-day diagnostic assessment phase. Drawing on Clay's (2005, p. 32) recommendation to begin a literacy intervention by "roaming around the known," they did not try to teach anything during this phase. Its purpose was to enable the teacher to find out about the child's strengths and weaknesses, and the child to develop confidence in their own knowledge and trust in the teacher. The teacher then used their diagnosis to identify priorities for the child within the Numbers Count curriculum (see Table 58.1) and planned an individualised program to address them.

A Numbers Count Lesson

Every Numbers Count lesson was planned from scratch, tailored to the child's needs. All lessons shared the same structure (see Table 58.2): a 30-minute sequence of episodes that ensured a good pace and progression of learning while giving the child daily opportunities to engage in a range of mathematical activities and to use and apply their learning.

The rest of this section describes and explains a typical lesson for Toby (whose name has been changed for this chapter).

Making a Positive Start

Toby begins the lesson by deciding whether he would like to play the "doubles game" or "snakes and ladders." He selects the "doubles game" as he is confident that he will beat his teacher, who forgot some of her doubles the last time they played. She asks him whether he wants to play it with the spinner or a die and he chooses the giant sponge die. He throws it onto the carpet, jumping in the air with excitement as he does so.

Table 58.2 30-minute lesson structure

Learning Episode	Approximate duration
Making a Positive Start	4 minutes
Counting	4 minutes
Current Learning Activity 1	10 minutes
Current Learning Activity 2	10 minutes
Ending on a Positive Note	2 minutes

TOBY. *6. Double 6 is 12* [immediately]*! A point for me!*
TEACHER. *Oh dear, it looks as if you're going to win again, Toby. How did you do that so quickly?*
TOBY. *I practised with my mum last night to make sure I would beat you.*

Like the warm-up at the start of a physical education lesson, Making a Positive Start prepares the child for the more strenuous learning activities that will follow. It sets a positive tone for the start of the lesson while building on the child's confidence and fluency. In this example, Toby is empowered to make a choice, which develops his independence and encourages him to take ownership and control of the learning situation. It also provides gentle opportunities for his teacher to strengthen and consolidate his mathematical knowledge and skills and to find out more about his learning.

Making a Positive Start often, but not always, takes the form of asking the child to choose a familiar mathematical activity with which to begin the lesson. Sometimes, Toby's teacher instead reviews with him some mathematics that he has done at home. At other times, she chats with him about the progress he has made in Numbers Count or about how he is getting on in class mathematics lessons.

Counting

Counting is the only element of mathematics that features in every Numbers Count lesson. With daily counting opportunities, children can develop the procedural skills of both oral and object counting and then proceed to more sophisticated skills such as counting forwards and backwards, counting in steps, and starting or finishing on a given number, all of which underpin their understanding of the number system and equip them for calculation. Toby's teacher uses a range of counting contexts and activities to enable him to build up his counting strategies and skills. Today, she goes with him to the shop that she has set up in the corner of the Numbers Count room.

TEACHER. *Toby, please could you help me to count these 5p coins while we put them back into the till? I need to buy some more toys because I've sold them all, but I'm not sure if I have enough money.*
TOBY. Picks up each 5p coin and places it into the money box. *5, 10, 15, 20....* Hesitates when he puts the next coin in. Looks around the room for something to help him. Goes to the 'hand number line' that he made with his teacher at the beginning of the intervention and which his teacher has put on the wall..... *I'll use this.... 25.... you have 25p!*
TEACHER. *Thank you Toby. I like the way you decided to use the number line to help you when you got stuck. That was a really useful idea.*

His teacher gives Toby specific praise for his strategy of using the number line because she wants him to be aware of and pleased about what he has done, so that he will do it again in the future. She has found out, by talking to his class teacher, that he rarely uses any resources to help him when he has a difficulty in his class mathematics lessons, and they have agreed they will both encourage him to do this. She does not praise him for getting the right answer, because this is less important and less under his control than the choice of an effective strategy.

The teacher now moves on to an activity that will help to develop Toby's conceptual knowledge of counting:

> *'Children who themselves count a set of objects correctly are credited with procedural knowledge while those who can detect errors in the counting of others are credited with having conceptual knowledge of counting.'*
>
> (Maclellan, 2008, p. 36)

She does this with Spotty, a soft toy dog.

TEACHER. *Toby can you please fetch Spotty? We need to see if he has learned to count backwards from 15 to 10.*
SPOTTY. *15, 14, 13, 11....*
TOBY. [Listening intently] *No, Spotty! You forgot to say 12. It's 15, 14, 13, 12, 11.*
SPOTTY. *15, 14, 13, 12, 11, 10.*
TOBY. *That's right.*

Toby's teacher uses this part of the lesson to enable him to practise and reinforce a wide range of existing counting skills. If she wants him to learn any new counting skills, on the other hand, she waits until the Current Learning Activities episode of the lesson.

Current Learning Activities

Most Numbers Count lessons contain two Current Learning Activities in order to maintain variety and pace of learning, although there are times when it is more appropriate to have only one, longer activity. Teachers may introduce a new concept, build on one that was introduced the day before, or give children an opportunity to practise and consolidate learning from the previous week.

After reflecting on the previous day's lesson, Toby's teacher has decided to develop counting-on strategies when adding two small numbers together as the main focus of today's Current Learning Activity. She sits at the table with Toby and uses an "I can" statement written on a star to share the learning objective with him.

TEACHER. *Toby, today we are going to work on counting on and at the end of our activity we will decide if we can place this star on your 'I Can' planet.*
BOTH. [reading together] *'I can count on when adding two small numbers.'*
TEACHER. *What do you think 'count on' means?*

She asks this question to activate and build on Toby's prior knowledge. Numbers Count is based upon a social constructivist understanding of learning. Teachers take to heart von Glasersfeld's (1995) observation that children do not passively receive and absorb knowledge from their teachers. They treat children instead as active learners who want to make sense of their world, who try to build on what they already know and believe when presented with new mathematical experiences, and who can be helped to do this through dialogue with others. In this instance, Toby's response indicates that he has constructed an understanding of counting on from playing with dominoes.

> TOBY. Goes to the resource trolley and comes back with a domino tile that has 4 spots and 2 spots. *Well, when we used these last week, I didn't have to count all the spots to find out how many. Look....* Points to the side with 4 spots..... *I know there are 4 spots on that side so I just counted 4 ....* Points to the individual spots on the other side.... *5, 6.* 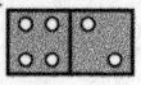

His teacher assesses that Toby understands the principle of counting on with small numbers in this context and decides that she now needs to help him to build on this by counting on from larger numbers and in other contexts. She uses a variety of short, linked activities during this single Current Learning Activity episode of the lesson, all with the same objective.

She begins with a "bucket count," dropping counters one at a time into a red bucket on the floor. Toby shuts his eyes and keeps count while listening to the sound of counters hitting the bottom of the bucket. When she stops, he announces that there are nine counters in the bucket. She shows him three more counters in her hand, and asks him, *"Can you work out how many counters will be in the bucket when I drop these three in?"* She continues the Current Learning Activity by going back to the shop with Toby and using coins to calculate an increase of 5p on each item. Finally, they kneel on the carpet with the "small world" garage scene that Toby made out of a shoe box during his diagnostic assessment. Toby drives eight cars that "need fixing" into the garage and closes the doors so that they cannot be seen; his teacher drives five more cars in front of the garage and asks him, *"Can you work out how many cars the mechanic has to fix now?"* They repeat this several times with different numbers, Toby's teacher encouraging him to count out loud *"so that I know what you are thinking"* and to check his calculations by counting all of the cars.

By the end of this Current Learning Activity, both Toby and his teacher are confident that he can add the star to his "I Can" planet. His teacher notes that he has been able to count on in a variety of contexts using concrete materials and she plans to develop this in problem-solving situations tomorrow. She then moves on to a second Current Learning Activity, focusing on place value, which is not described here.

During the Current Learning Activities, teachers illustrate mathematical concepts for children through a variety of situations. They give them an opportunity to develop their mathematical thinking through making connections, relating their mathematics to real-life scenarios, engaging in mathematical talk, reasoning, and solving problems in a range of different contexts.

Ending on a Positive Note

Numbers Count teachers encourage children to reflect on and assess their own learning throughout each lesson because this helps them to take responsibility for their own learning and to be actively involved in the learning process. Munns and Woodward (2006) found that pupils of all ages can learn to assess their learning and that doing so improves their engagement in learning and level of achievement. This is reinforced in the lesson's final episode.

In Ending on a Positive Note, the teacher initiates a dialogue with the child that celebrates the success of the lesson and helps the teacher to identify how the child feels he has progressed and to identify next steps. The dialogue also enables the child to identify what he can remember and understand, to know why his work is good, and to think about what he needs to do to improve. This self-assessment can and does take place at any time within a Numbers Count lesson: Toby has already decided in the Current Learning Activity that he deserves an "I Can" star for his counting on. So his teacher decides not to probe any further and to close today's lesson on a note of success.

TEACHER. Picks up a large sticker on which she has already printed 'Ask me about....' *Toby, I think you've done so well today that you deserve this sticker. Can you read it?*

TOBY. Smiles. He knows this sticker because he's had it twice before. *'Ask me about....'*

TEACHER. *What shall I write?*

TOBY. *Ask me about beating Mrs Kay at doubles!*

TEACHER. *I was afraid of that! OK.* Writes out the sticker for Toby, because she knows he is not a confident writer. Holds it up ready to stick on his chest and pauses – she has just seen another chance to highlight and reinforce an important learning strategy, which she wants him to be aware of and go on using even after he has finished his Numbers Count intervention. *Do you remember why you beat me, Toby?*

TOBY. *Because I practised with my mum.*

TEACHER. Puts the sticker on his chest. *Yes, practising with your mum was a really good idea. If you keep doing that I don't think I'll ever beat you. Do you think your mum deserves a sticker, too?*

TOBY. *Yes, coz she helped me.*

TEACHER. *Go on then, choose one for her.*

TOBY. Chooses a smaller sticker for his mother that says "Numbers Count winner".

Conclusion

Toby's teacher took him back to his classroom. Both were tired after 30 minutes of intensive thought and movement around the Numbers Count, but both were pleased with what they had achieved in the lesson.

IMPACT

Mathematical Achievement

A randomized control trial (Torgerson et al., 2011, 2013) compared the outcomes of 144 children who had taken part in Numbers Count with a matched waiting list control group of 274 children who had been selected to take part but not yet received any lessons. It found that the intervention group scored higher on a standardized test of general mathematics skills after taking part in Numbers Count than did the control group, with an effect size of .33 and $p < .0005$. The trial found no evidence that children's outcomes were related to pre-test score, age, gender, or entitlement to free school meals.

This chapter reports on a follow-up study over a 2-year period, 2009–2011, in which 9985 6- and 7-year-old children received a Numbers Count intervention. 8205 of them (82%) performed at least 6 months below the mean for their age when tested at entry; they are the subjects of this study, being defined as "low-attaining" and therefore matching the target

Table 58.3 Age and test scores

	n	Entry mean (SD)	Exit mean (SD)	Gain mean (SD)
Chronological Age*months*	7954	80.8 (4.53)	83.7 (4.48)	2.9 (0.65)
Test Number Age*months*	7948	65.5 (6.45)	79.5 (9.66)	14.0 (6.99)
Test Standardized Score*points*	7935	80.5 (7.85)	94.3 (12.34)	13.8 (9.99)

recipients of the intervention. They were taught by 667 teachers in 661 schools in 42 local authorities in England, 80% of them being taught by teachers in their first year of Numbers Count training. The study focused on the extent to which Numbers Count achieved the first of its two aims, that children would make good progress during the intervention.

7956 children (97% of the subjects) completed a program, receiving a mean of 42.6 lessons in 2.9 months. At entry, their mean chronological age was 6 years 9 months and their mean number age was 15.3 months lower, at 5 years 6 months (see Table 58.3). At exit 2.9 months later, they had made mean number age gains of 14.0 months.

The principal analysis of significance was a comparison of children's number age and chronological age gains between entry and exit, in an approach similar to that taken by Dowker and Sigley (2010). In effect, chronological age gain was used as a control for number age gain: the expectation was that, without an intervention, the two would be equal over any period for normally-achieving children and that number age gain would be lower for low-achieving children, so a larger number age gain in this study would indicate a positive effect of the intervention. A paired *t*-test indicated that the difference between number age gain (14.0 months) and chronological gain (2.9 months) was significant, at $p < .0001$, with a very large effect size ($r = .85$). The mean ratio gain (number age gain divided by chronological age gain) was 5.09; it was at least 1 for 97% of children and at least 4 for 64% of children. A further paired *t*-test indicated that the difference between exit and entry standardized scores was also significant, at $p < .0001$, with a very large effect size ($r = .81$).

The study also investigated the outcomes for children with different background characteristics. Dowker (2008) and Torgerson et al. (2011) had found no significant effects in relation to children's gender or free school meals status in exploratory investigations of these variables. With the large numbers of children involved, this study was able to explore, more thoroughly, the impact of the intervention on groups known to be vulnerable to underachievement: boys, children entitled to free school meals, children with special educational needs, children from minority ethnic backgrounds, children who spoke English as an additional language, and summer-born children, all of whom were more highly represented in Numbers Count than in the national primary school population. Table 58.4 shows that all groups achieved number age gains of at least 13.0 months and ratio gains of at least 4.7. Paired *t*-tests found that the difference between number age gain and chronological age gain was significant for all groups, at $p < .0001$, with very large effect sizes (r ranging from .83 to .87).

A comparison by independent *t*-tests of the difference between the number age gains made by each pair of sub-groups (e.g. boys and girls) found that children identified by their schools as having a special educational need made slightly lower number age gains

Table 58.4 Group outcomes

Group	*n*	Number Age Gain *mean*	Ratio Gain[1] *mean*	Paired *t*-test of difference between number age and chronological age gain *p*	*r*
Gender					
Boys	4246	14.15	5.04	<.0001	.849
Girls	3710	13.82	5.12	<.0001	.842
Free School Meals Entitlement					
Yes	3397	13.26	4.79	<.0001	.836
No	4559	14.54	5.31	<.0001	.854
Special Educational Needs					
Yes	4405	13.01	4.69	<.0001	.827
No	3551	15.22	5.58	<.0001	.870
Minority Ethnic Background					
Yes	3131	14.65	5.34	<.0001	.847
No	4825	13.58	4.92	<.0001	.846
English as an Additional Language					
Yes	2310	14.87	5.43	<.0001	.849
No	5646	13.64	4.95	<.0001	.845
Summer-born[2]					
Yes	3148	13.95	5.05	<.0001	.845
No	4808	14.03	5.11	<.0001	.846

[1] Number age gain divided by chronological age gain.
[2] Born in the summer months (April to July).

than children without such needs (small effect size, $r = .16$), while no other comparisons had an effect size larger than .1. In summary, these analyses showed that Numbers Count had a very large, positive impact on all vulnerable groups of children and that this impact was of a similar size for all groups but slightly lower for children with special educational needs.

A further investigation was made of whether children's gains were related to their entry age and attainment level. Analyses of correlations between children's entry chronological age, number age, and standardized score and their number age gains showed no significant relationships that had a reportable effect size. Figure 58.1 shows that children at all entry attainment levels made similar progress.

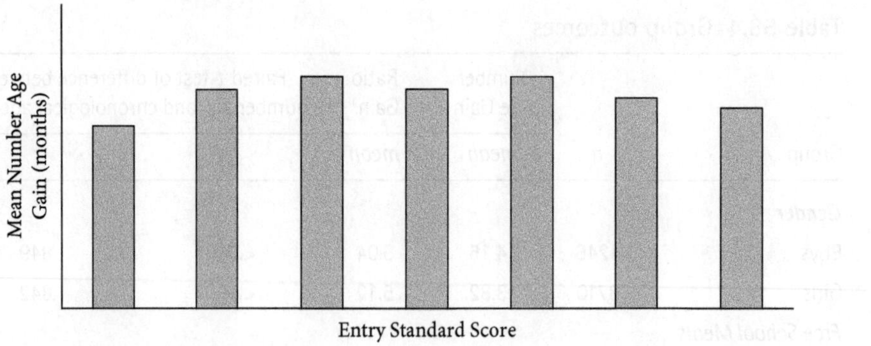

FIGURE 58.1 Entry standardized score and number age gain.

Attitudes towards Mathematics

Torgerson et al.'s (2011) study of Numbers Count found that the intervention group's self-reported attitudes to mathematics after exit were slightly higher than those of the control group (effect size .21; p <.03), but concluded that this could be the result of chance. An aim of this study was to investigate children's attitudes in more depth because of their importance for long-term mathematical achievement (Ashcraft and Moore, 2009; Nunes et al., 2009). It measured class teachers' perceptions of children's attitudes, rather than children's self-reports, because of the difficulty of obtaining accurate self-reports from very young children (Henry et al., 2007). No suitable standardized attitude measure was available, so a bespoke attitude survey was developed and administered before and after children took part in Numbers Count. Their class teachers used a five-point scale to indicate their agreement with four statements about children's willingness to engage in whole-class mathematics lessons and to explore problems, persevere, and talk about their mathematics. Teaching assistants separately responded to the same statements about the child's attitudes during small-group mathematics learning. All responses were scored from "5" as "very positive" to "1" as "very negative," and the eight scores were totalled to give a continuous attitude survey scale from 8 to 40 points. Children's mean entry and exit scores were 20.5 and 26.7 respectively (see Table 58.5). Their mean attitude gain of 6.2 points was significant on a paired t-test at $p < .0001$, with a very large effect size of $r = .72$.

Researchers interviewed teachers and parents to find out more about the development in children's attitudes. The almost universal response was that children had become much more confident and happy, both in Numbers Count lessons and in class mathematics lessons:

> *'He is so much more confident, absolutely loves to show off. We made a photo story of what he'd done about inverses and he showed it in class. He stood there beaming that this was all his work – something that he wouldn't have thought he could do before, and he was showing the children and explaining it to them.'*
>
> Numbers Count teacher

> *'She's got a completely different attitude. She seems happier.'*
>
> Class teacher

Table 58.5 Attitudes

	n	Entry mean (SD)	Exit mean (SD)	Gain mean (SD)
Attitude Survey Score *points*	7527	20.5 (5.79)	26.7 (6.12)	6.2 (6.05)

Further discussion and analysis revealed six overlapping factors which teachers repeatedly commented on as evidence of and contributions to children's increased overall confidence:

(1) greater enjoyment of mathematics
(2) more willing to 'having a go'
(3) longer concentration
(4) more resilient
(5) more independent
(6) more willing to talk about mathematics.

1. *Greater Enjoyment of Mathematics*

While some children professed a liking for mathematics at the start of their interventions, despite a relative lack of success in the classroom, many said that they disliked it.

> *'She said she couldn't do maths, she couldn't do any maths in class, so she didn't like maths.'*
>
> Numbers Count teacher

This quickly changed in the intervention. Numbers Count teachers reported that children responded positively to one-to-one interaction with a supportive adult, to the attractive environment of a well-resourced Numbers Count room, and to activities which initially were deliberately pitched at levels at which the children could be sure of success. Parents noticed the difference, too:

> *'She's forever talking about maths now. She wants to do it at home all the time, and she says she's better at it than me!'*
>
> Parent

2. *More Willing to "Have a Go"*

Many children were initially unwilling to try any activity at which they were not totally confident, or to respond to a simple question. Teachers started by ensuring that children could succeed, repeatedly making it clear to them that there was no penalty for failure, and tackling the main stumbling blocks identified in the initial diagnostic assessment.

> *'She learned to be more confident in her own ability and to have a go at things, as soon as we sorted out some basic things like how to add by counting on and how to subtract by counting back.'*
>
> Numbers Count teacher

3. *Longer Concentration*

Children were often described by their teachers as having a very short attention span at the start of their interventions, even when engaged in activities that they could do.

> *'He finds it very difficult to remain on task, he finds it very difficult to focus on one thing, he finds it hard to look at the teacher.'*
>
> Class teacher

Mathematical activities were repeatedly interrupted by questions about "What will we do next?" or by physical distractions and fidgeting, but these interruptions rapidly grew less as children became interested in the activities and teachers engaged them in dialogue.

4. *More Resilient*

Children also became more able to cope with difficulties. While they often became frustrated and gave up at the first hint of a problem when they started Numbers Count, they gradually began to keep trying for longer and to look for alternative ways to overcome a difficulty. Teachers attributed this to children's growing belief that they could succeed at mathematics, and to the fact that they often specifically praised children for "trying hard" and "thinking about how to do it," but rarely just for "getting it right."

> *'She still makes mistakes, but they don't worry her any more. She just keeps going, and she often gets it right in the end.'*
>
> Class teacher

5. *More Independent*

Children were commonly very dependent on teaching assistants in their classrooms at first, relying on them to tell them what resources to use and how to complete any activity. Numbers Count teachers encouraged children to make choices, starting with simple ones such as "Do you want to count the dinosaurs or the cats today?," and moving on to "Do you want to use the number line or the hundred square to help you with this calculation?" The effect in the classroom could be dramatic:

> *'Michael used to hate maths. He hid behind other children when I asked questions.... He wanted his teaching assistant to do everything for him, because he had no confidence. Now he's a different boy: always putting his hand up to answer questions.... Yesterday he told his teaching assistant, "I'll just get the number tiles and then I'll be alright on my own."'*
>
> Class teacher

Children also became more independent at choosing the strategies they would use to complete a task, after Numbers Count teachers had encouraged them to make their own decisions with questions such as, "Are you going to do this one by counting back or counting on?," followed later by "Why did you choose to count on? Would it have worked as well if you had counted back?"

6. *More Willing to Talk about Mathematics*

Some children were completely silent at the start of the intervention and many others fell silent when asked even the simplest of mathematical questions.

> *'He just answered the questions with a single word; he couldn't put any mathematical language into a sentence.'*
>
> Numbers Count teacher

Almost every child soon began to talk about the mathematics. Teachers attributed this, in part, to the child's enjoyment of it, partly to the child's need to engage in a duologue, and partly to the teachers' own specific modelling of mathematical language to give children the tools with which to express themselves.

By the end of the intervention, children were more willing and able to explain their mathematical strategies, not just to their Numbers Count teachers but also to their teachers and classmates in class mathematics lessons and to their parents at home.

> *'We were doing 'time' in the classroom, which isn't one of the things Omar has worked on in Numbers Count, but he was so confident in himself that he started actually teaching the other children in his group. He believed that he could do it.'*
>
> Class teacher

This not only indicated greater self-confidence but also a far greater desire, than at the start of the intervention, to engage in mathematical thinking and dialogue – in short, a more positive attitude towards learning mathematics.

Discussion

This has been the first large-scale study of a mathematics intervention in England. While previous studies showed impressive outcomes with small numbers of children, there is always the danger that the "small-study effect" (Bell 2011) will be diluted when an intervention is rolled out nationally to thousands of children in a wide range of contexts across a large number of schools, implemented by many different teachers who have learned about the intervention from a variety of trainers, none of whom were its authors. Numbers Count avoided such a dilution: the 7956 children's number age gain of 14.0 months after 3 months, with an effect size of .85, was at least as strong as the outcomes from previous small-scale studies of similar interventions.

The large numbers of children involved also enabled a more thorough investigation than previously possible of the impact of a mathematics intervention on children with different background characteristics. This study has confirmed Dowker's (2008) and Torgerson et al.'s (2011) initial findings of no significant differences. All groups, irrespective of gender, free school meals' status, special educational needs' status, ethnic background, home language, and season of birth, achieved number age gains in Numbers Count of at least 13.0 months, with effect sizes of at least .8. The greatest gains were made by children who spoke English as an additional language; feedback from participants suggested that this may be because one-to-one support enabled them to communicate more effectively with their teachers than in the classroom. While children with special educational needs made the least progress, indicating that their needs may have been more complex, their number ages still improved by an average of 13.0 months.

Children's attitudes towards mathematics also improved significantly, with an effect size of .72. Their parents and teachers reported a marked increase in enjoyment of mathematics, confidence, and positive learning behaviors that boded well for children's continued success in learning mathematics after leaving the Numbers Count intervention. Program data indicates that they continued to make good progress in class mathematics lessons over the next 6 months (Every Child a Chance Trust, 2010).

A qualitative analysis revealed six distinct but overlapping manifestations of children's increased confidence. Three of these (enjoyment, "having a go", and talking about mathematics) had been addressed in the teachers' training as part of the development of a "numerate child" outlined in the overview of Numbers Count at the start of this chapter. The other three (concentration, resilience, and independence), however, had not been explicitly targeted. Their emergence can be interpreted as a reminder that it is important for an intervention to help children to cope with the inevitable difficulties that are inherent in learning mathematics, and as an indication that Numbers Count succeeded in this.

Attributing causes to the success of Numbers Count must be relatively speculative, but three factors may be suggested. The first is that its design incorporates the findings of previous research into the essential features of successful interventions. Dowker's (2004) recommendations that interventions should be implemented individually and at an early age, should begin with a detailed diagnostic assessment, and should address distinct components of arithmetic are all implemented in Numbers Count. Torgerson et al. (2011) reported that observations of Numbers Count lessons found all the elements of teaching and learning that they expected to find in a successful intervention. Numbers Count is perhaps most different from other interventions in the specific emphasis that teachers place on fostering children's enjoyment of mathematics and effective and learning behaviors, reflected in the strong attitude gains that they made during the intervention.

Secondly, Numbers Count teachers engaged in a thorough professional development program to inform and support their teaching. Torgerson et al. (2011, para. 7.3.6) reported that "many head teachers took an interest in the training and reported that it was the best they had ever seen." The key contents of the program were a combination of practical training in the delivery of Numbers Count, study of early mathematics curriculum and pedagogy, and reflection on teachers' developing experience and understanding. Trainers aimed to develop a collaborative learning culture which would underpin teachers' continuous reflection and learning as they investigated the needs of each child and considered how best to meet them.

Finally, Numbers Count was underpinned by a rigorous quality assurance system. Each school's receipt of funding for its teacher was dependent on adherence to strict national standards for the selection of children, the deployment of the teacher, quality of teaching, and support by school management. Adherence was monitored by the training leaders through frequent school visits and through analysis of data about each child's progress. The training leaders gave individual support to any teachers and schools that fell short of expected standards and outcomes. At the same time, feedback from teachers and schools was used to refine and improve Numbers Count.

Success, however, came at a price. Numbers Count was an expensive intervention, both because it was delivered individually to children by a qualified teacher and because it was underpinned by extensive professional development and quality assurance. In a newly-prevailing climate of austerity, a national policy decision was taken to gradually withdraw state funding of the Every Child Counts initiative, which included the Numbers Count intervention; if it was to continue, it should be as a marketed program which schools could choose to purchase if they believed that it provided value for money (Department for Education, 2011). Torgerson et al. (2011) suggested that Numbers Count in its original form, for all its impact, was too expensive for most schools. Although analysts (Gross et al., 2009) had calculated that every £1 spent on an intervention such as Numbers Count would save the state between £12 and £18 in later spending, including special educational needs' support, welfare, and the criminal justice system, less than £1 of this saving was realised by the primary schools that made the spending decisions. While a decision to invest in support for young children who struggle in mathematics is not a purely monetary one, cost is nevertheless an important consideration.

Numbers Count was, therefore, modified after the period of this study, to make it more flexible and more affordable for schools. Following the success of a small-scale trial of teaching Numbers Count to children in groups of two and three (Torgerson et al., 2011) and in line with Dowker's (2004) finding that small groups can be effective, teachers were given the option of teaching up to three children together, once they had successfully learned to deliver Numbers Count to individual children; diagnostic assessment was still carried out individually, and teachers were trained and encouraged to make flexible decisions about group sizes to meet individual children's needs. The initial training program for teachers was reduced from 10 to 7 days, while still retaining its essential features, and the number of school visits made by group leaders was also reduced. Additionally, the quality assurance system that tied school funding to adherence to national standards was replaced by a system of accreditation for schools and teachers who voluntarily adhered to standards that allow more flexibility while focusing on the key features that ensured children's success in Numbers Count.

The early evidence is that the impact of Numbers Count actually increased during the first 3 years after these changes were introduced. Children's average number age gains between entry and exit rose from 14 months in this study to 16 months after the changes. While most children were still taught individually, equivalent progress was achieved by the 25% whose teachers taught them in pairs.

This chapter has reported on children's learning during the period of their Numbers Count interventions. The second aim of Numbers Count, that they would continue to make good progress subsequently, will be analyzed in separate studies. Initial results are encouraging, indicating that children who had completed Numbers Count in the first year of the intervention broadly kept pace with their peers over the next 4 years until they left primary school at age 11.

References

Arnold, C., Bowen, P., & Walden, B. (2009). *Sandwell Early Numeracy Test* (revised). West Bromwich: Sandwell Children & Young People's Services.

Ashcraft, M. & Moore, A. (2009). Mathematics anxiety and the affective drop in performance. *Journal of Psychoeducational Assessment*, 27(3), 197–205.

Aubrey, C., Dahl, S., & Godfrey, R. (2006). Early mathematics development and later achievement: further evidence. *Mathematics Education Research Journal*, 18(1), 27–46.
Bell, J. (2011). The small-study effect in educational trials. *Effective Education*, 3(1), 35–48.
Burr, T. (2008). *Mathematics Performance in Primary Schools: Getting the Best Results*. London: National Audit Office/Department for Children, Schools and Families.
Butterworth, B., Sashank, V., & Laurillard, D. (2011). Dyscalculia: from brain to education. *Science*, 332, 1049–1053.
Clay, M. (2005). *Literacy Lessons Designed for Individuals. Part One: Why: When? and How?* Auckland: Heinemann.
Cowan, R., Donlan, C., Newton, E., & Lloyd, D. (2008). Number development and children with specific language impairment. In A. Dowker (Ed.), *Mathematical Difficulties: Psychology and Intervention* (pp. 155–166). London: Elsevier.
Department for Education (2011). *Support and Aspiration: A New Approach to Special Educational Needs and Disability*. London: Department for Education.
Department for Education (2014). *National Curriculum Assessments at Key Stage 2 in England, 2014* (provisional). London: Department for Education.
Dowker, A. (2004). *What Works for Children with Mathematical Difficulties?* Research Report No. 554. Nottingham: Department for Education and Skills.
Dowker, A. (2005). *Individual Differences in Arithmetic: Implications for Psychology, Neuroscience and Education*. Hove: Psychology Press.
Dowker, A. (2008). Numeracy recovery with children with arithmetical difficulties: intervention and research. In A. Dowker (Ed.), *Mathematical Difficulties: Psychology and Intervention* (pp. 181–202). London: Elsevier.
Dowker, A. (2009). *What Works for Children with Mathematical Difficulties? The Effectiveness of Intervention Schemes*. London: Department for Children, Schools and Families.
Dowker, A. & Sigley, G. (2010). Targeted interventions for children with mathematical difficulties. *British Journal of Educational Psychology Monograph Series II, 7: Understanding Number Development and Difficulties*, 65–81.
Dunn, S., Matthews, L., & Dowrick, N. (2010). Numbers Count: developing a national approach to early intervention. In I. Thompson (Ed.), *Issues in Teaching Numeracy in Primary Schools* (2nd edn, pp. 223–234). Maidenhead: Open University Press.
Education Endowment Foundation (2014). *Catch Up Numeracy: evaluation report and executive summary*. London: Education Endowment Foundation.
Every Child a Chance Trust (2010). *Every Child a Chance Trust Impact Report 2010*. London: Every Child a Chance Trust.
Fuchs, L., Fuchs, D., & Compton, D. (2012). The early prevention of mathematics difficulty: its power and limitations. *Journal of Learning Disabilities*, 45(3), 257–269.
Geary, D. (2004). Mathematics and learning disabilities. *Journal of Learning Disabilities*, 37(1), 4–15.
Geary, D. (2010). Mathematical disabilities: reflections on cognitive, neuropsychological, and genetic components. *Learning and Individual Differences*, 20(2), 130–133.
Gervasoni, A. & Sullivan P. (2007). Assessing and teaching children who have difficulty learning arithmetic. *Educational and Child Psychology*, 24(2), 40–53.
Glasersfeld, E. von (1995). *Radical Constructivism: A Way of Knowing and Learning*. London: Falmer Press.
Gross, J. (2007). Supporting children with gaps in their mathematical understanding: the impact of the national numeracy strategy on children who find arithmetic difficult. *Educational and Child Psychology*, 24(2), 146–156.
Gross, J., Hudson, C., and Price, D. (2009). *The Long Term Costs of Numeracy Difficulties*. London: Every Child a Chance Trust.
Haseler, M. (2008). Making intervention in numeracy more effective in schools. In A. Dowker (Ed.), *Mathematical Difficulties: Psychology and Intervention* (pp. 225–241). London: Elsevier.
Henry, G., Mashbur, A., & Konold, T. (2007). Developing and evaluating a measure of young children's attitudes toward school and learning. *Journal of Psychoeducational Assessment*, 25(3), 271–284.

Houssart, J. (2007). Investigating variability in classroom performance amongst children exhibiting difficulties with early arithmetic. *Educational and Child Psychology*, 24(2), 83–97.

Jimenez, J. & Garcia, A. (2002). Strategy choice in solving problems: are there differences between students with learning difficulties, G-V performance, and typical achievement students? *Learning Disability Quarterly*, 25, 113–122.

Jordan, N. & Montani, T. (1997). Cognitive arithmetic and problem solving: a comparison of children with specific and general mathematical difficulties. *Journal of Learning Disabilities*, 28, 624–634.

Jordan, N., Glutting, J., & Ramineni, C. (2010). The importance of number sense to mathematics achievement in first and third grade. *Learning and Individual Differences*, 20(2), 82–88.

Jordan, N. & Levine, S. (2009). Socioeconomic variation, number competence, and mathematics learning difficulties in young children. *Developmental Disabilities Research Reviews*, 15, 60–68.

Krinzinger, H., Kaufmann, L., & Willmes, K. (2009). Math anxiety and math ability in early primary school years. *Journal of Psychoeducational Assessment*, 27(3), 206–225.

Maclellan, E. (2008). Counting: what it is and why it matters. In I. Thompson (Ed.), *Teaching and Learning Early Number* (2nd edn, pp. 72–81). Maidenhead: Open University Press.

Moser, C. (1999). *Improving Literacy and Numeracy: A Fresh Start. The Report of the Working Group Chaired by Sir Claus Moser*. London: Department for Education and Skills.

Munns, G. & Woodward, H. (2006). Student engagement and student self-assessment: the REAL framework. *Assessment in Education: Principles, Policy & Practices*, 13(2), 192–213.

Nunes, T., Bryant, T., Sylva, K., & Barros, R. (2009). *Development of Maths Capabilities and Confidence in Primary School; Research Report RB118*. London: Department for Children, Schools and Families.

Torgerson, C., Wiggins, A., Torgerson, D., et al. (2011). *Every Child Counts: The Independent Evaluation Technical Report; Research Report DFE-RR091a*. London: Department for Education.

Torgerson, C., Wiggins, A., Torgerson, D., Ainsworth, H., & Hewitt, C. (2013). Every Child Counts: testing policy effectiveness using a randomised control trial, designed, conducted and reported to CONSORT standards. *Research in Mathematics Education*, 15(2), 141–153.

Turner Ellis, S., Miles, T., & Wheeler, T. (2008). The performance of dyslexic and non-dyslexic boys at division sums. In A. Dowker (Ed.), *Mathematical Difficulties: Psychology and Intervention* (pp. 167–180). London: Elsevier.

Willey, R., Holliday, A., & Martland, J. (2007). Achieving new heights in Cumbria: raising standards in early numeracy through mathematics recovery. *Educational and Child Psychology*, 24(2), 108–118.

Williams, P. (2008). *Independent Review of Mathematics Teaching in Early Years Settings and Primary Schools: Final Report*. Nottingham: Department for Children, Schools and Families.

Wright, R. (2008). Interview-based assessment of early number knowledge. In I. Thompson (Ed.), *Teaching and Learning Early Number* (2nd edn, pp. 193-204). Maidenhead: Open University Press.

Wright, R., Martland, J., & Stafford, A. (2006a). *Early Numeracy: Assessment for Teaching and Intervention* (2nd edn). London: Sage.

Wright, R., Martland, J., Stafford, A., & Stanger, G. (2006b). *Teaching Number: Advancing Children's Skills and Strategies* (2nd edn). London: Chapman.

CHAPTER 59

DEVELOPING CONCEPTUAL AND PROCEDURAL KNOWLEDGE OF MATHEMATICS

BETHANY RITTLE-JOHNSON AND MICHAEL SCHNEIDER

Introduction

When children practise solving problems, does this also enhance their understanding of the underlying concepts? Under what circumstances do abstract concepts help children invent or implement correct procedures? These questions tap a central research topic in the fields of cognitive development and educational psychology: the relations between conceptual and procedural knowledge. Delineating how these two types of knowledge interact is fundamental to understanding how knowledge development occurs. It is also central to improving instruction.

The goals of the current paper were: (1) discuss prominent definitions and measures of each type of knowledge, (2) review recent research on the developmental relations between conceptual and procedural knowledge for learning mathematics, (3) highlight promising research on potential methods for improving both types of knowledge, and (4) discuss problematic issues and future directions. We consider each in turn.

Defining Conceptual and Procedural Knowledge

Although conceptual and procedural knowledge cannot always be separated, it is useful to distinguish between the two types of knowledge to better understand knowledge development.

First consider conceptual knowledge. A concept is 'an abstract or generic idea generalized from particular instances' (Merriam-Webster's Collegiate Dictionary, 2012). Knowledge of concepts is often referred to as *conceptual knowledge* (e.g. Byrnes & Wasik, 1991; Canobi, 2009; Rittle-Johnson, Siegler, & Alibali, 2001). This knowledge is usually not tied to particular problem types. It can be implicit or explicit, and thus does not have to be verbalizable (e.g. Goldin Meadow, Alibali, & Church, 1993). The National Research Council adopted a similar definition in its review of the mathematics education research literature, defining it as 'comprehension of mathematical concepts, operations, and relations' (Kilpatrick, Swafford, & Findell, 2001, p. 5). This type of knowledge is sometimes also called conceptual understanding or principled knowledge.

At times, mathematics education researchers have used a more constrained definition. Star (2005) noted that: 'The term *conceptual knowledge* has come to encompass not only what is known (knowledge of concepts) but also one way that concepts can be known (e.g. deeply and with rich connections)' (p. 408). This definition is based on Hiebert and LeFevre's definition in the seminal book edited by Hiebert (1986):

'Conceptual knowledge is characterized most clearly as knowledge that is rich in relationships. It can be thought of as a connected web of knowledge, a network in which the linking relationships are as prominent as the discrete pieces of information. Relationships pervade the individual facts and propositions so that all pieces of information are linked to some network' (pp. 3–4).

After interviewing a number of mathematics education researchers, Baroody and colleagues (Baroody, Feil, & Johnson, 2007) suggested that conceptual knowledge should be defined as 'knowledge about facts, [generalizations], and principles' (p. 107), without requiring that the knowledge be richly connected. Empirical support for this notion comes from research on conceptual change that shows that (1) novices' conceptual knowledge is often fragmented and needs to be integrated over the course of learning and (2) experts' conceptual knowledge continues to expand and become better organized (diSessa, Gillespie, & Esterly, 2004; Schneider & Stern, 2009). Thus, there is general consensus that conceptual knowledge should be defined as knowledge of concepts. A more constrained definition requiring that the knowledge be richly connected has sometimes been used in the past, but more recent thinking views the richness of connections as a feature of conceptual knowledge that increases with expertise.

Next, consider procedural knowledge. A procedure is a series of steps, or actions, done to accomplish a goal. Knowledge of procedures is often termed *procedural knowledge* (e.g. Canobi, 2009; Rittle-Johnson et al., 2001). For example, 'Procedural knowledge … is 'knowing how', or the knowledge of the steps required to attain various goals. Procedures have been characterized using such constructs as skills, strategies, productions, and interiorized actions' (Byrnes & Wasik, 1991, p. 777). The procedures can be (1) algorithms—a predetermined sequence of actions that will lead to the correct answer when executed correctly, or (2) possible actions that must be sequenced appropriately to solve a given problem (e.g. equation-solving steps). This knowledge develops through problem-solving practice, and thus is tied to particular problem types. Further, 'It is the clearly sequential nature of procedures that probably sets them most apart from other forms of knowledge' (Hiebert & LeFevre, 1986, p. 6).

As with conceptual knowledge, the definition of procedural knowledge has sometimes included additional constraints. Within mathematics education, Star (2005) noted that sometimes: 'the term procedural knowledge indicates not only what is known (knowledge of procedures) but also one way that procedures (algorithms) can be known (e.g. superficially and without rich connections)' (p. 408). Baroody and colleagues (Baroody et al., 2007) acknowledged that:

> 'some mathematics educators, including the first author of this commentary, have indeed been guilty of oversimplifying their claims and loosely or inadvertently equating "knowledge memorized by rote ... with computational skill or procedural knowledge" (Baroody, 2003, p. 4). Mathematics education researchers (MERs) usually define procedural knowledge, however, in terms of knowledge type—as sequential or "step-by-step [prescriptions for] how to complete tasks"
>
> (Hiebert & Lefevre, 1986, p. 6' (pp. 116–117).

Thus, historically, procedural knowledge has sometimes been defined more narrowly within mathematics education, but there appears to be agreement that it should not be.

Within psychology, particularly in computational models, there has sometimes been the additional constraint that procedural knowledge is implicit knowledge that cannot be verbalized directly. For example, John Anderson (1993) claimed: 'procedural knowledge is knowledge people can only manifest in their performance .... procedural knowledge is not reportable' (pp. 18, 21). Although later accounts of explicit and implicit knowledge in ACT-R (Adaptive Control of Thought—Rational) (Lebiere, Wallach, & Taatgen, 1998; Taatgen, 1999) do not repeat this claim, Sun, Merrill, and Peterson (2001) concluded that: 'The inaccessibility of procedural knowledge is accepted by most researchers and embodied in most computational models that capture procedural skills' (p. 206). In part, this is because the models are often of procedural knowledge that has been automatized through extensive practice. However, at least in mathematical problem solving, people often know and use procedures that are not automatized, but rather require conscious selection, reflection, and sequencing of steps (e.g. solving complex algebraic equations), and this knowledge of procedures can be verbalized (e.g. Star & Newton, 2009).

Overall, there is a general consensus that procedural knowledge is the ability to execute action sequences (i.e. procedures) to solve problems. Additional constraints on the definition have been used in some past research, but are typically not made in current research on mathematical cognition.

Measuring Conceptual and Procedural Knowledge

Ultimately, how each type of knowledge is *measured* is critical for interpreting evidence on the relations between conceptual and procedural knowledge. Conceptual knowledge has been assessed in a large variety of ways, whereas there is much less variability in how procedural knowledge is measured.

Measures of conceptual knowledge vary in whether tasks require implicit or explicit knowledge of the concepts, and common tasks are outlined in Table 59.1. Measures of implicit conceptual knowledge are often evaluation tasks on which children make a categorical choice (e.g. judge the correctness of an example procedure or answer) or make

Table 59.1 Range of tasks used to assess conceptual knowledge

Type of task Implicit measures	Sample task	Additional citations
a. Evaluate unfamiliar procedures	Decide whether ok for puppet to skip some items when counting (Gelman & Meck, 1983)	(Kamawar et al., 2010; LeFevre et al., 2006; Muldoon, Lewis, & Berridge, 2007; Rittle-Johnson & Alibali, 1999; Schneider et al., 2009; Schneider & Stern, 2010; Siegler & Crowley, 1994)
b. Evaluate examples of concept	a. Decide whether the number sentence 3 = 3 makes sense (Rittle-Johnson & Alibali, 1999); b. 45 + 39 = 84, Does puppet need to count to figure out 39 + 45? (Canobi et al., 1998)	(Canobi, 2005; Canobi & Bethune, 2008; Canobi, Reeve, & Pattison, 2003; Patel & Canobi, 2010; Rittle-Johnson et al., 2001; Rittle-Johnson et al., 2009; Schneider et al., 2011)
c. Evaluate quality of answers given by others	Evaluate how much someone knows based on the quality of their errors, which are or are not consistent with principles of arithmetic (Prather & Alibali, 2008)	(Dixon, Deets, & Bangert, 2001; Mabbott & Bisanz, 2003; Star & Rittle-Johnson, 2009)
d. Translate quantities between representational systems	a. Represent symbolic numbers with pictures (Hecht, 1998) b. Place symbolic numbers on number lines (Siegler & Booth, 2004; Siegler, Thompson, & Schneider, 2011)	(Byrnes & Wasik, 1991; Carpenter, Franke, Jacobs, Fennema, & Empson, 1998; Cobb et al., 1991; Hecht & Vagi, 2010; Hiebert & Wearne, 1996; Mabbott & Bisanz, 2003; Moss & Case, 1999; Prather & Alibali, 2008; Reimer & Moyer, 2005; Rittle-Johnson & Koedinger, 2009; Schneider et al., 2009; Schneider & Stern, 2010)
e. Compare quantities	Indicate which symbolic integer or fraction is larger (or smaller) (Hecht, 1998; Laski & Siegler, 2007)	(Durkin & Rittle-Johnson, 2012; Hallett et al., 2010; Hecht & Vagi, 2010; Laski & Siegler, 2007; Moss & Case, 1999; Murray & Mayer, 1988; Rittle-Johnson et al., 2001; Schneider et al., 2009; Schneider & Stern, 2010)
f. Invent principle-based shortcut procedures	On inversion problems such as 12 + 7–7, quickly stating the first number without computing (Rasmussen, Ho, & Bisanz, 2003)	(Canobi, 2009)
g. Encode key features	Success reconstructing examples from memory (e.g. a chess board or equations), with the assumption that greater conceptual knowledge helps people notice key features and chunk information, allowing for more accurate recall (Larkin, McDermott, Simon, & Simon, 1980)	(Matthews & Rittle-Johnson, 2009; McNeil & Alibali, 2004; Rittle-Johnson et al., 2001)

Table 59.1 Continued

Type of task Implicit measures	Sample task	Additional citations
h. Sort examples into categories	Sort 12 statistics problems based on how they best go together (Lavigne, 2005)	Mainly used in other domains, such as physics
Explicit measures		
a. Explain judgements	On evaluation task, provide correct explanation of choice (e.g. '29 + 35 has the same numbers as 35 + 29, so it equals 64, too.' (Canobi, 2009)	(Canobi, 2004, 2005; Canobi & Bethune, 2008; Canobi et al., 1998, 2003; Peled & Segalis, 2005; Rittle-Johnson & Star, 2009; Rittle-Johnson et al., 2009; Schneider et al., 2011; Schneider & Stern, 2010)
a. Generate or select definitions of concepts	Define the equal sign (Knuth, Stephens, McNeil, & Alibali, 2006; Rittle-Johnson & Alibali, 1999)	(Star & Rittle-Johnson, 2009; Vamvakoussi & Vosniadou, 2004) (Izsák, 2005)
b. Explain why procedures work	Explain why ok to borrow when subtract (Fuson & Kwon, 1992)	(Berthold & Renkl, 2009; Jacobs, Franke, Carpenter, Levi, & Battey, 2007; Reimer & Moyer, 2005; Stock, Desoete, & Roeyers, 2007)
c. Draw concept maps	Construct a map that identifies main concepts in introductory statistics, showing how the concepts are related to one another (Lavigne, 2005)	(Williams, 1998)

a quality rating (e.g. rate an example procedure as very-smart, kind-of-smart, or not-so-smart). Other common implicit measures are translating between representational formats (e.g. symbolic fractions into pie charts) and comparing quantities (see Table 58.1 for more measures).

Explicit measures of conceptual knowledge typically involve providing definitions and explanations. Examples include generating or selecting definitions for concepts and terms, explaining why a procedure works, or drawing a concept map (see Table 59.1). These tasks may be completed as paper-and-pencil assessment items or answered verbally during standardized or clinical interviews (Ginsburg, 1997). We do not know of a prior study on conceptual knowledge that quantitatively assessed how richly connected the knowledge was.

Clearly, there are a large variety of tasks that have been used to measure conceptual knowledge. A critical feature of conceptual tasks is that they be relatively unfamiliar to participants, so that participants have to derive an answer from their conceptual knowledge, rather than implement a known procedure for solving the task. For example, magnitude comparison problems are sometimes used to assess children's conceptual knowledge of number magnitude (e.g. Hecht, 1998; Schneider, Grabner, & Paetsch, 2009). However, children are sometimes taught procedures for comparing magnitudes or develop procedures with repeated practice; for these children, magnitude comparison problems are likely measuring their procedural knowledge, not their conceptual knowledge.

In addition, conceptual knowledge measures are stronger if they use multiple tasks. First, use of multiple tasks meant to assess the same concept reduces the influence of task-specific characteristics (Schneider & Stern, 2010). Second, conceptual knowledge in a domain often requires knowledge of many concepts, leading to a multi-dimensional construct. For example, for counting, key concepts include cardinality and order-irrelevance, and in arithmetic, key concepts include place value and the commutativity and inversion principles. Although knowledge of each is related, there are individual differences in these relationships, without a standard hierarchy of difficulty (Dowker, 2008; Jordan, Mulhern, & Wylie, 2009).

Measures of procedural knowledge are much less varied. The task is almost always to solve problems, and the outcome measure is usually accuracy of the answers or procedures. On occasion, researchers consider solution time as well (Canobi, Reeve, & Pattison, 1998; LeFevre et al., 2006; Schneider & Stern, 2010). Procedural tasks are familiar—they involve problem types people have solved before and thus should know procedures for solving. Sometimes the tasks include near transfer problems—problems with an unfamiliar problem feature that require either recognition that a known procedure is relevant or small adaptations of a known procedure to accommodate the unfamiliar problem feature (e.g. Renkl, Stark, Gruber, & Mandl, 1998; Rittle-Johnson, 2006).

There are additional measures that have been used to tap particular ways in which procedural knowledge can be known. When interested in how well automatized procedural knowledge is, researchers use dual-task paradigms (Ruthruff, Johnston, & van Selst, 2001; Schumacher, Seymour, Glass, Kieras, & Meyer, 2001) or quantify asymmetry of access, that is, the difference in reaction time for solving a practiced task versus a task that requires the same steps executed in the reverse order (Anderson & Fincham, 1994; Schneider & Stern, 2010). The execution of automatized procedural knowledge does not involve conscious reflection and is often independent of conceptual knowledge (Anderson, 1993). When interested in how flexible procedural knowledge is, researchers assess students' knowledge of multiple procedures and their ability to flexibly choose among them to solve problems efficiently (e.g. Blöte, Van der Burg, & Klein, 2001; Star & Rittle-Johnson, 2008; Verschaffel, Luwel, Torbeyns, & Van Dooren, 2009). Flexibility of procedural knowledge is positively related to conceptual knowledge, but this relationship is evaluated infrequently (see Schneider, Rittle-Johnson & Star, 2011, for one instance).

To study the relations between conceptual and procedural knowledge, it is important to assess the two independently. However, it is important to recognize that it is difficult for an item to measure one type of knowledge to the exclusion of the other. Rather, items are thought to predominantly measure one type of knowledge or the other. In addition, we believe that continuous knowledge measures are more appropriate than categorical measures. Such measures are able to capture the continually changing depths of knowledge, including the context in which knowledge is and is not being used. They are also able to capture variability in people's thinking, which appears to be a common feature of human cognition (Siegler, 1996).

Relations between Conceptual and Procedural Knowledge

Historically, there have been four different theoretical viewpoints on the causal relations between conceptual and procedural knowledge (cf. Baroody, 2003; Haapasalo & Kadijevich,

2000; Rittle-Johnson & Siegler, 1998). *Concepts-first views* posit that children initially acquire conceptual knowledge, for example, through parent explanations or guided by innate constraints, and then derive and build procedural knowledge from it through repeated practice solving problems (e.g. Gelman & Williams, 1998; Halford, 1993). *Procedures-first views* posit that children first learn procedures, for example, by means of explorative behaviour, and then gradually derive conceptual knowledge from them by abstraction processes, such as representational re-description (e.g. Karmiloff-Smith, 1992; Siegler & Stern, 1998). A third possibility, sometimes labelled *inactivation view* (Haapasalo & Kadijevich, 2000), is that conceptual and procedural knowledge develop independently (Resnick, 1982; Resnick & Omanson, 1987). A fourth possibility is an *iterative view*. The causal relations are said to be bi-directional, with increases in conceptual knowledge leading to subsequent increases in procedural knowledge and vice versa (Baroody, 2003; Rittle-Johnson & Siegler, 1998; Rittle-Johnson et al., 2001).

The iterative view is now the most well-accepted perspective. An iterative view accommodates gradual improvements in each type of knowledge over time. If knowledge is measured using continuous, rather than categorical, measures, it becomes clear that one type of knowledge is not well developed before the other emerges, arguing against a strict concepts- or procedures-first view. In addition, an iterative view accommodates evidence in support of concepts-first and procedures-first views, as initial knowledge can be conceptual or procedural, depending upon environmental input and relevant prior knowledge of other topics. An iterative view was not considered in early research on conceptual and procedural knowledge (see Rittle-Johnson & Siegler, 1998, for a review of this research in mathematics learning), but over the past 15 years there has been an accumulation of evidence in support of it.

First, positive correlations between the two types of knowledge have been found in a wide range of ages and domains. The domains include counting (Dowker, 2008; LeFevre et al., 2006), addition and subtraction (Canobi & Bethune, 2008; Canobi et al., 1998; Jordan et al., 2009; Patel & Canobi, 2010), fractions and decimals (Hallett, Nunes, & Bryant, 2010; Hecht, 1998; Hecht, Close, & Santisi, 2003; Reimer & Moyer, 2005), estimation (Dowker, 1998; Star & Rittle-Johnson, 2009), and equation solving (Durkin, Rittle-Johnson, & Star, 2011). In general, the strength of the relation is fairly high. For example, in a meta-analysis of a series of eight studies conducted by the first author and colleagues on equation solving and estimation, the mean effect size for the relation was 0.54 (Durkin, Rittle-Johnson, & Star, 2011). Further, longitudinal studies suggest that the strength of the relation between the two types of knowledge varies over time (Jordan et al., 2009; Schneider, Rittle-Johnson, & Star, 2011). The strength of the relation varies across studies and over time, but it is clear that the two types of knowledge are often related.

Second, evidence for predictive, bi-directional relations between conceptual and procedural knowledge has been found in mathematical domains ranging from fractions to equation solving. For example, in two samples differing in prior knowledge, middle-school students' conceptual and procedural knowledge for equation solving was measured before and after a 3-day classroom intervention in which students studied and explained worked examples with a partner (Schneider et al., 2011). Conceptual and procedural knowledge were modelled as latent variables to better account for the indirect relation between overt behaviour and the underlying knowledge structures. A cross-lagged panel design was used to directly test and compare the predictive relations from conceptual knowledge to procedural knowledge and vice versa. As expected, each type of knowledge predicted gains in the other type of knowledge, with standardized regressions coefficients of about 0.3, and the relations were symmetrical (i.e. they did not differ significantly in their strengths). Similar bi-directional relations have been found for elementary-school children learning about decimals (Rittle-Johnson & Koedinger, 2009; Rittle-Johnson et al., 2001). Overall, knowledge of one type is a good and reliable predictor of improvements in knowledge of the other type.

The predictive relations between conceptual and procedural knowledge are even present over several years (Cowan et al., 2011). For example, elementary-school children's knowledge of fractions was assessed in the winter of Grade 4 and again in the spring of Grade 5 (Hecht & Vagi, 2010). Conceptual knowledge in Grade 4 predicted about 5% of the variance in procedural knowledge in Grade 5 after controlling for other factors, and procedural knowledge in Grade 4 predicted about 2% of the variance in conceptual knowledge in Grade 5.

In addition to the predictive relations between conceptual and procedural knowledge, there is evidence that experimentally manipulating one type of knowledge can lead to increases in the other type of knowledge. First, direct instruction on one type of knowledge led to improvements in the other type of knowledge (Rittle-Johnson & Alibali, 1999). Elementary-school children were given a very brief lesson on a procedure for solving mathematical equivalence problems (e.g. $6 + 3 + 4 = 6 + \_\_$), the concept of mathematical equivalence, or were given no lesson. Children who received the procedure lesson gained a better understanding of the concept, and children who received the concept lesson generated correct procedures for solving the problems. Second, practice-solving problems can support improvements in conceptual knowledge when constructed appropriately (Canobi, 2009; McNeil et al., 2012). For example, elementary-school children solved packets of problems for 10 minutes on nine occasions during their school mathematics lessons. The problems were arithmetic problems sequenced based on conceptual principles (e.g. $6 + 3$ followed by $3 + 6$), the same arithmetic problems sequenced randomly, or non-mathematical problems (control group). Solving conceptually sequenced practice problems supported gains in conceptual knowledge, as well as procedural knowledge. Together, this evidence indicates that there are causal, bi-directional links between the two types of knowledge; improving procedural knowledge can lead to improved conceptual knowledge and vice versa, especially if potential links between the two are made salient (e.g. through conceptually sequencing problems).

An iterative view predicts that the bi-directional relations between conceptual and procedural knowledge persist over time, with increases in one supporting increases in the other, *which in turn supports increases in the first type of knowledge*. Indeed, prior conceptual knowledge of decimals predicted gains in procedural knowledge after a brief problem-solving intervention, which in turn predicted gains in conceptual knowledge (Rittle-Johnson et al., 2001). In addition, iterating between lessons on concepts and procedures on decimals supported greater procedural knowledge and equivalent conceptual knowledge compared to presenting concept lessons before procedure lessons (Rittle-Johnson & Koedinger, 2009). Both studies suggest that relations between the two types of knowledge are bi-directional over time (i.e. iterative).

Overall, there is extensive evidence from a variety of mathematical domains indicating that the development of conceptual and procedural knowledge of mathematics is often iterative, with one type of knowledge supporting gains in the other knowledge, which in turn supports gains in the other type of knowledge. Conceptual knowledge may help with the construction, selection, and appropriate execution of problem-solving procedures. At the same time, practice implementing procedures may help students develop and deepen understanding of concepts, especially if the practice is designed to make underlying concepts more apparent. Both kinds of knowledge are intertwined and can strengthen each other over time.

However, the relations between the two types of knowledge are not always symmetrical. In Schneider, Rittle-Johnson, and Star (2011), the relations were symmetrical—the strength of the relationship from prior conceptual knowledge to later procedural knowledge was the same as from prior procedural knowledge to later conceptual knowledge. However, in other studies, conceptual knowledge or conceptual instruction has had a stronger influence on procedural knowledge than vice versa (Hecht & Vagi, 2010; Matthews & Rittle-Johnson, 2009; Rittle-Johnson & Alibali, 1999). Furthermore, brief procedural instruction or practice solving problems does not always support growth in conceptual knowledge (Canobi, 2009; Perry, 1991; Rittle-Johnson, 2006), and increasing school experience is associated with gains in procedural knowledge for counting and arithmetic, but much less so with gains in conceptual knowledge (Canobi, 2004; LeFevre et al., 2006).

How much gains in procedural knowledge support gains in conceptual knowledge is influenced by the nature of the procedural instruction or practice. For example, in Canobi (2009) and McNeil et al. (2012), sequencing arithmetic practice problems so that conceptual relations were easier to notice supported conceptual knowledge, while random ordering of practice problems did not. In Peled and Segalis (2005), instruction that encouraged students to generalize procedural steps and connect subtraction procedures across whole numbers, decimals, and fractions led to greater conceptual knowledge than instruction on individual procedures. In general, it is best if procedural lessons are crafted to encourage noticing of underlying concepts.

The symmetry of the relations between conceptual and procedural knowledge also varies between individuals. Children in Grades 4 and 5 completed a measure of their conceptual and procedural knowledge of fractions (Hallett et al., 2010). A cluster analysis on the two measures suggested five different clusters of students, with clusters varying in the strength of conceptual and procedural knowledge. For example, one cluster had above-average conceptual knowledge and below-average procedural knowledge, another cluster was the opposite, and a third cluster was high on both measures. These cluster differences suggest that, although related in all clusters, the strength of the relations varied. Similar findings were reported for primary-school children's knowledge of addition and subtraction (Canobi, 2005), including a meta-analysis of over 14 studies (Gilmore & Papadatou-Pastou, 2009). At least in part, these individual differences may reflect different instructional histories between children.

Overall, the relations between conceptual and procedural knowledge are bi-directional, but sometimes they are not symmetrical. At times, conceptual knowledge more consistently and strongly supports procedural knowledge than the reverse. Crafting procedural lessons to encourage noticing of underlying concepts can promote a stronger link from improved procedural knowledge to gains in conceptual knowledge.

Promising Methods for Improving Both Types of Knowledge

Given the importance of developing both conceptual and procedural knowledge, instructional techniques that support both types of knowledge are critical. Here, we highlight examples of general instructional methods that are promising.

Promoting comparison of alternative solution procedures is one effective instructional approach. In a series of studies, students studied pairs of worked examples illustrating two different, correct procedures for solving the same problem and were prompted to compare them or studied the same examples one at a time and were prompted to reflect on them individually. For students who knew one of the solution procedures at pre-test, comparing procedures supported greater procedural knowledge (Rittle-Johnson & Star, 2007; Rittle-Johnson, Star, & Durkin, 2009) or greater conceptual knowledge (Rittle-Johnson & Star, 2009; Rittle-Johnson et al., 2009; Star & Rittle-Johnson, 2009). For novices, who did not know one of the solution procedures at pre-test, no benefits were found for conceptual or procedural knowledge (although comparison did improve procedural flexibility; see Rittle-Johnson et al., 2009; Rittle-Johnson, Star, & Durkin, 2011). In addition, having students compare incorrect procedures to correct ones aided conceptual and procedural knowledge and reduced misconceptions (Durkin & Rittle-Johnson, 2012). Overall, comparing procedures can help students gain conceptual and procedural knowledge, but its advantages are more substantial if students have sufficient prior knowledge.

A second approach is to encourage self-explanation when studying solution procedures. For example, prompting primary-school children to explain why solutions to mathematical equivalence problems were correct or incorrect supported greater procedural transfer (Rittle-Johnson, 2006). Similarly, prompting high-school students to self-explain when studying worked examples of probability problems supported greater conceptual knowledge of probability (although it seemed to hamper procedural knowledge; Berthold & Renkl, 2009).

A third approach is to offer opportunities for problem exploration *before* instruction (Schwartz, Chase, Chin, & Oppezzo, 2011). For example, primary-school children solved a set of unfamiliar mathematics problems and received a lesson on the concept of equivalence, and the order of problem solving and the lesson was manipulated (DeCaro & Rittle-Johnson, 2011). Children who solved the unfamiliar problems before the lesson made greater gains in conceptual knowledge, and comparable gains in procedural knowledge, compared to children who solved the problems after the lesson. Similarly, middle-school students who explored problems and invented their own formula for calculating density before instruction on density gained deeper conceptual and procedural knowledge of density than students who received the lessons first (Schwartz et al., 2011). Initial problem exploration fits with the recommendation from the mathematics education literature that students have opportunities to struggle—to figure out something that is not immediately apparent (Hiebert & Grouws, 2009).

Comparison, self-explanation, and exploration are all promising instructional methods for promoting conceptual and procedural knowledge, as are sequencing problems so that conceptual relations are more apparent (Canobi, 2009) and iterating between lessons on concepts and procedures (Rittle-Johnson & Koedinger, 2009). These are just some examples of effective methods; certainly there are numerous others (e.g. McNeil & Alibali, 2000) and more need to be identified.

Future Directions

Considerable progress has been made in understanding the development of conceptual and procedural knowledge of mathematics over the past 15 years. An important next step is to develop a more comprehensive model of the relations between conceptual and procedural knowledge. Some components that need to be considered in such a model are shown in Figure 59.1. To flesh out such a model, we will need a better understanding of numerous components. For example, are conceptual and procedural knowledge stored independently in long-term memory and does this change with expertise? How do age and individual differences impact the relations between conceptual and procedural knowledge and the effectiveness of different instructional methods? What additional instructional methods can be integrated into learning environments and what student behaviours and mental activities do they support? How do differences across topics impact the model (e.g. learning about counting vs. algebra)? What are alternative models for understanding the relations between conceptual and procedural knowledge?

However, before more progress can be made in understanding the relations between conceptual and procedural knowledge, we must pay more attention to the validity of measures of conceptual and procedural knowledge. Currently, no standardized approaches for assessing conceptual and procedural knowledge with proven validity, reliability, and objectivity have been developed. This is deeply problematic because knowledge is stored in memory and has to be inferred from overt behaviour. However, human behaviour arises from a complex interplay of a multitude of cognitive processes and usually does not reflect memory content in a pure and direct form. This makes it difficult to attribute learners' answers exclusively to one type of knowledge.

Each potential measure of conceptual or procedural knowledge has at least four different variance components (Schneider & Stern, 2010). First, if the measure has been developed carefully, it can be assumed to reflect the amount of the kind of knowledge it is supposed

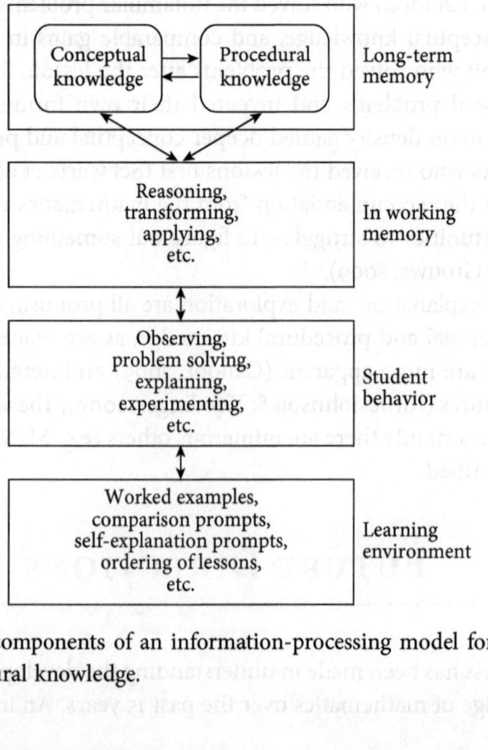

FIGURE 59.1 Potential components of an information-processing model for the relations between conceptual and procedural knowledge.

to assess. Second, each assessment task also requires task-specific knowledge. For example, when children answer interview questions, their answers reflect not only their conceptual knowledge, but also their vocabulary in the respective domain and more general verbal abilities. A diagram task designed to assess procedural knowledge about fractions reflects not only knowledge about fractions but also knowledge of and experience with the specific diagrams used in that task.

Third, under many circumstances, learners can derive new procedures from their conceptual knowledge (Gelman & Williams, 1998) and they can abstract new concepts from their procedural experience (Karmiloff-Smith, 1992). Thus, measures of conceptual knowledge often reflect some procedural knowledge and measures of procedural knowledge might also reflect conceptual knowledge to some degree.

Finally, random measurement error is present in virtually all psychological measures. This makes it hard to interpret findings about conceptual and procedural knowledge. For example, when a measure of conceptual knowledge and a measure of procedural knowledge show a low inter-correlation, is this due to a dissociation of conceptual and procedural knowledge, due to task-specific knowledge, or due to high measurement error?

A confirmatory factor analysis (Schneider & Stern, 2010) demonstrated that this problem is not just theoretical. Four commonly used hypothetical measures of conceptual knowledge and four commonly used hypothetical measures of procedural knowledge were completed by fifth and sixth graders. Conceptual and procedural knowledge were modelled as latent factors underlying these eight measures. However, each latent factor explained less than 50% of the variance of the measured variables, indicating that the measures reflected measure-specific variance components and random measurement error to a higher degree than the kind of knowledge they were supposed to assess.

Very little attention has been given to measurement validity in the literature on conceptual and procedural knowledge. Clearly, attention to validity is greatly needed. Future studies will have to validate tasks and measures to ensure that we are using good measures of conceptual and procedural knowledge. As noted by Hill and Shih (2009):

> 'Without conducting and reporting validation work on key independent and dependent variables, we cannot know the extent to which our instruments tap what they claim to. And without this knowledge, we cannot assess the validity of inferences drawn from studies' (p. 248).

Likely progress will require some mixture of traditional psychometric approaches, newer approaches based on item-response theory, and perhaps innovations in alternative ways to validate measures, especially of conceptual knowledge.

Conclusion

Mathematical competence rests on developing both conceptual and procedural knowledge. Although there is some variability in how these constructs are defined and measured, there is general consensus that the relations between conceptual and procedural knowledge are often bi-directional and iterative. Instructional methods for supporting both types of knowledge have emerged, such as promoting comparison of alternative solution methods, prompting for self-explanation, and providing opportunities for exploration before instruction. Future research needs to focus on more rigorous measurement of conceptual and procedural knowledge, providing evidence for the validity of the measures, and specify more comprehensive models for understanding how conceptual and procedural knowledge develop.

References

Anderson, J.R. (1993). *Rules of the Mind*. Hillsdale, NJ: Erlbaum.

Anderson, J.R. & Fincham, J. M. (1994). Acquisition of procedural skills from examples. *Journal of Experimental Psychology: Learning, Memory, and Cognition*, *20*, 1322–1340.

Baroody, A.J. (2003). *The development of adaptive expertise and flexibility: the integration of conceptual and procedural knowledge*. Mahwah, NJ: Erlbaum.

Baroody, A.J., Feil, Y., & Johnson, A.R. (2007). An alternative reconceptualization of procedural and conceptual knowledge. *Journal for Research in Mathematics Education*, *38*, 115–131.

Berthold, K. & Renkl, A. (2009). Instructional aids to support a conceptual understanding of multiple representations. *Journal of Educational Psychology*, *101*, 70–87. doi: 10.1037/a0013247.

Blöte, A.W., Van der Burg, E., & Klein, A.S. (2001). Students' flexibility in solving two-digit addition and subtraction problems: instruction effects. *Journal of Educational Psychology*, *93*, 627–638. doi: 10.1037//0022-0663.93.3.627.

Byrnes, J.P. & Wasik, B.A. (1991). Role of conceptual knowledge in mathematical procedural learning. *Developmental Psychology*, *27*, 777–786. doi: 10.1037//0012-1649.27.5.777.

Canobi, K.H. (2004). Individual differences in children's addition and subtraction knowledge. *Cognitive Development*, *19*, 81–93. doi: 10.1016/j.cogdev.2003.10.001.

Canobi, K.H. (2005). Children's profiles of addition and subtraction understanding. *Journal of Experimental Child Psychology*, *92*, 220–246. doi: 10.1016/j.jecp.2005.06.001.

Canobi, K.H. (2009). Concept-procedure interactions in children's addition and subtraction. *Journal of Experimental Child Psychology*, *102*, 131–149. doi: 10.1016/j.jecp.2008.07.008.

Canobi, K.H. & Bethune, N.E. (2008). Number words in young children's conceptual and procedural knowledge of addition, subtraction and inversion. *Cognition*, *108*, 675–686. doi: 10.1016/j.cognition.2008.05.011.

Canobi, K.H., Reeve, R.A., & Pattison, P.E. (1998). The role of conceptual understanding in children's addition problem solving. *Developmental Psychology*, *34*, 882–891. doi: doi:10.1037//0012-1649.34.5.882.

Canobi, K.H., Reeve, R.A., & Pattison, P.E. (2003). Patterns of knowledge in children's addition. *Developmental Psychology*, *39*, 521–534. doi: 10.1037/0012-1649.39.3.521.

Carpenter, T.P., Franke, M.L., Jacobs, V.R., Fennema, E., & Empson, S.B. (1998). A longitudinal study of invention and understanding in children's multidigit addition and subtraction. *Journal for Research in Mathematics Education*, *29*, 3–20. doi: 10.2307/749715.

Cobb, P., Wood, T., Yackel, E., Nicholls, J., Wheatley, G., Trigatti, B., & Perlwitz, M. (1991). Assessment of a problem-centered second-grade mathematics project. *Journal for Research in Mathematics Education*, *22*, 3–29. doi: doi:10.2307/749551.

Cowan, R., Donlan, C., Shepherd, D.-L., Cole-Fletcher, R., Saxton, M., & Hurry, J. (2011). Basic calculation proficiency and mathematics achievement in elementary school children. *Journal of Educational Psychology*, *103*, 786–803. doi: 10.1037/a0024556.

DeCaro, M. & Rittle-Johnson, B. (2011). *Preparing to learn from math instruction by solving problems first*. Paper presented at the Biennial Meeting of the Society for Research in Child Development, Montreal, QC.

diSessa, A.A., Gillespie, N.M., & Esterly, J.B. (2004). Coherence versus fragmentation in the development of the concept of force. *Cognitive Science*, *28*, 843–900.

Dixon, J.A., Deets, J.K., & Bangert, A. (2001). The representations of the arithmetic operations include functional relationships. *Memory and Cognition*, *29*, 462–477. doi: 10.3758/BF03196397.

Dowker, A. (1998). Individual differences in normal arithmetical development. In C. Donlan (Ed.), *The Development of Mathematical Skills* (pp. 275–301). Hove: Psychology Press.

Dowker, A. (2008). Individual differences in numerical abilities in preschoolers. *Developmental Science*, *11*, 650–654. doi: 10.1111/j.1467-7687.2008.00713.x.

Durkin, K. & Rittle-Johnson, B. (2012). The effectiveness of using incorrect examples to support learning about decimal magnitude. *Learning and Instruction*, *22* (3), 206–214.

Durkin, K., Rittle-Johnson, B., & Star, J.R. (2011). *Procedural flexibility matters for student achievement: how procedural flexibility relates to other outcomes*. Paper presented at the 14th Biennial Conference of the European Association for Research on Learning and Instruction, August, Exeter.

Fuson, K.C. & Kwon, Y. (1992). Korean children's understanding of multidigit addition and subtraction. *Child Development*, *63*, 491–506. doi: 10.1111/j.1467-8624.1992.tb01642.x.

Gelman, R. & Meck, E. (1983). Preschoolers' counting: principles before skill. *Cognition*, *13*, 343–359. doi: 10.1016/0010-0277(83)90014-8.

Gelman, R. & Williams, E.M. (1998). Enabling constraints for cognitive development and learning: domain specificity and epigenesis. In D. Kuhn & R. S. Siegler (Eds), *Handbook of Child Psychology: Cognition, Perception, and Language* (5th edn, Vol. 2, pp. 575–630). New York: John Wiley.

Gilmore, C.K. & Papadatou-Pastou, M. (2009). Patterns of individual differences in conceptual understanding and arithmetical skill: a meta-analysis. *Mathematical Thinking and Learning*, *11*, 25–40.

Ginsburg, H.P. (1997). *Entering the Child's Mind: The Clinical Interview in Psychological Research and Practice*. New York, NY: Cambridge University Press.

Goldin Meadow, S., Alibali, M.W., & Church, R.B. (1993). Transitions in concept acquisition: using the hand to read the mind. *Psychological Review*, *100*, 279–297. doi: 10.1037//0033-295X.100.2.279.

Haapasalo, L. & Kadijevich, D. (2000). Two types of mathematical knowledge and their relation. *JMD—Journal for Mathematic-Didaktik*, *21*, 139–157.

Halford, G.S. (1993). *Children's Understanding: The Development of Mental Models*. Hillsdale, NJ: Erlbaum.

Hallett, D., Nunes, T., & Bryant, P. (2010). Individual differences in conceptual and procedural knowledge when learning fractions. *Journal of Educational Psychology*, *102*, 395–406. doi: 10.1037/a0017486.

Hecht, S.A. (1998). Toward an information-processing account of individual differences in fraction skills. *Journal of Educational Psychology*, *90*, 545–559. doi: 10.1037/0022-0663.90.3.545.

Hecht, S.A., Close, L., & Santisi, M. (2003). Sources of individual differences in fraction skills. *Journal of Experimental Child Psychology*, *86*, 277–302. doi: 10.1016/j.jecp.2003.08.003.

Hecht, S.A. & Vagi, K.J. (2010). Sources of group and individual differences in emerging fraction skills. *Journal of Educational Psychology*, *102*, 843–859. doi: 10.1037/a0019824.

Hiebert, J. (1986). *Conceptual and Procedural Knowledge: The Case of Mathematics*. Hillsdale, NJ: Erlbaum.

Hiebert, J. & Grouws, D. (2009). Which teaching methods are most effective for maths? *Better: Evidence-Based Education*, *2*, 10–11.

Hiebert, J. & Lefevre, P. (1986). *Conceptual and Procedural Knowledge in Mathematics: An Introductory Analysis* (pp. 1–27). Hillsdale, NJ: Erlbaum.

Hiebert, J. & Wearne, D. (1996). Instruction, understanding, and skill in multidigit addition and subtraction. *Cognition and Instruction*, *14*, 251–283. doi: 10.1207/s1532690xci1403_1.

Hill, H.C. & Shih, J.C. (2009). Examining the quality of statistical mathematics education research. *Journal for Research in Mathematics Education*, *40*, 241–250.

Izsák, A. (2005). 'You have to count the squares': applying knowledge in pieces to learning rectangular area. *Journal of the Learning Sciences*, *14*, 361–403.

Jacobs, V.R., Franke, M.L., Carpenter, T., Levi, L., & Battey, D. (2007). Professional development focused on children's algebraic reasoning in elementary school. *Journal for Research in Mathematics Education*, *38*, 258–288.

Jordan, J.-A., Mulhern, G., & Wylie, J. (2009). Individual differences in trajectories of arithmetical development in typically achieving 5- to 7-year-olds. *Journal of Experimental Child Psychology, 103*, 455–468. doi: 10.1016/j.jecp.2009.01.011.

Kamawar, D., LeFevre, J.-A., Bisanz, J., et al. (2010). Knowledge of counting principles: how relevant is order irrelevance? *Journal of Experimental Child Psychology, 105*, 138–145. doi: 10.1016/j.jecp.2009.08.004.

Karmiloff-Smith, A. (1992). *Beyond Modularity: A Developmental Perspective on Cognitive Science.* Cambridge, MA: MIT Press.

Kilpatrick, J., Swafford, J.O., & Findell, B. (2001). *Adding it up: Helping Children Learn Mathematics.* Washington, DC: National Academy Press.

Knuth, E.J., Stephens, A.C., McNeil, N.M., & Alibali, M.W. (2006). Does understanding the equal sign matter? Evidence from solving equations. *Journal for Research in Mathematics Education, 37*, 297–312.

Larkin, J.H., McDermott, J., Simon, D.P., & Simon, H.A. (1980). Expert and novice performance in solving physics problems. *Science, 208*, 1335–1342. doi: 10.1207/s1532690xci0304_1.

Laski, E.V. & Siegler, R.S. (2007). Is 27 a big number? Correlational and causal connections among numerical categorization, number line estimation, and numerical magnitude comparison. *Child Development, 78*, 1723–1743.

Lavigne, N.C. (2005). Mutually informative measures of knowledge: concept maps plus problem sorts in statistics. *Educational Assessment, 101*, 39–71. doi: 10.1207/s15326977ea1001_3.

Lebiere, C., Wallach, D., & Taatgen, N.A. (1998). Implicit and explicit learning in ACT-R. In F. Ritter & R. Young (Eds), *Cognitive Modelling II* (pp. 183–193). Nottingham: Nottingham University Press.

LeFevre, J.-A., Smith-Chant, B.L., Fast, L., et al. (2006). What counts as knowing? The development of conceptual and procedural knowledge of counting from kindergarten through grade 2. *Journal of Experimental Child Psychology, 93*, 285–303. doi: 10.1016/j.jecp.2005.11.002.

Mabbott, D.J. & Bisanz, J. (2003). Developmental change and individual differences in children's multiplication. *Child Development, 74*, 1091–1107. doi: 10.1111/1467-8624.00594.

McNeil, N.M. & Alibali, M.W. (2000). Learning mathematics from procedural instruction: externally imposed goals influence what is learned. *Journal of Educational Psychology, 92*, 734–744. doi: 10.1037//0022-0663.92.4.734.

McNeil, N.M. & Alibali, M.W. (2004). You'll see what you mean: students encode equations based on their knowledge of arithmetic. *Cognitive Science, 28*, 451–466. doi: 10.1016/j.cogsci.2003.11.002.

McNeil, N.M., Chesney, D.L., Matthews, P.G., et al. (2012). It pays to be organized: organizing arithmetic practice around equivalent values facilitates understanding of math equivalence. *Journal of Educational Psychology, 104* (4), 1109–1121. doi: 10.1037/a0028997.

Matthews, P.G. & Rittle-Johnson, B. (2009). In pursuit of knowledge: comparing self-explanations, concepts, and procedures as pedagogical tools. *Journal of Experimental Child Psychology, 104*, 1–21. doi: 10.1016/j.jecp.2008.08.004.

Merriam-Webster's Collegiate Dictionary (2012). <http://www.m-w.com>.

Moss, J. & Case, R. (1999). Developing children's understanding of the rational numbers: a new model and an experimental curriculum. *Journal for Research in Mathematics Education, 30*, 122–147. doi: 10.2307/749607.

Muldoon, K.P., Lewis, C., & Berridge, D. (2007). Predictors of early numeracy: is there a place for mistakes when learning about number? *British Journal of Developmental Psychology, 25*, 543–558. doi: 10.1348/026151007x174501.

Murray, P.L. & Mayer, R.E. (1988). Preschool children's judgements of number magnitude. *Journal of Educational Psychology, 80*, 206–209.

Patel, P. & Canobi, K.H. (2010). The role of number words in preschoolers' addition concepts and problem-solving procedures. *Educational Psychology, 30*, 107–124. doi: 10.1080/01443410903473597.

Peled, I. & Segalis, B. (2005). It's not too late to conceptualize: constructing a generalized subtraction schema by abstracting and connecting procedures. *Mathematical Thinking and Learning, 7*, 207–230. doi: 10.1207/s15327833mtl0703_2.

Perry, M. (1991). Learning and transfer: instructional conditions and conceptual change. *Cognitive Development, 6*, 449–468. doi: 10.1016/0885-2014(91)90049-J.

Prather, R.W. & Alibali, M.W. (2008). Understanding and using principles of arithmetic: operations involving negative numbers. *Cognitive Science, 32*, 445–457. doi: 10.1080/03640210701864147.

Rasmussen, C., Ho, E., & Bisanz, J. (2003). Use of the mathematical principle of inversion in young children. *Journal of Experimental Child Psychology, 85*, 89–102. doi: 10.1016/s0022-0965(03)00031-6.

Reimer, K. & Moyer, P.S. (2005). Third-graders learn about fractions using virtual manipulates: a classroom study. *Journal of Computers in Mathematics and Science Teaching, 24*, 5–25.

Renkl, A., Stark, R., Gruber, H., & Mandl, H. (1998). Learning from worked-out examples: the effects of example variability and elicited self-explanations. *Contemporary Educational Psychology, 23*, 90–108. doi: 10.1006/ceps.1997.0959.

Resnick, L.B. (1982). Syntax and semantics in learning to subtract. In T.P. Carpenter, J. m. Moser, & T.A. Romberg (Eds), *Addition & Subtraction: A Cognitive Perspective* (pp. 136–155). Hillsdale, NJ: Erlbaum.

Resnick, L.B. & Omanson, S.F. (1987). Learning to understand arithmetic. In R. Glaser (Ed.), *Advances in Instructional Psychology* (Vol. 3, pp. 41–95). Hillsdale, NJ: Erlbaum.

Rittle-Johnson, B. (2006). Promoting transfer: effects of self-explanation and direct instruction. *Child Development, 77*, 1–15. doi: 10.1111/j.1467-8624.2006.00852.x.

Rittle-Johnson, B. & Alibali, M.W. (1999). Conceptual and procedural knowledge of mathematics: does one lead to the other? *Journal of Educational Psychology, 91*, 175–189. doi: 10.1037//0022-0663.91.1.175.

Rittle-Johnson, B. & Koedinger, K.R. (2009). Iterating between lessons concepts and procedures can improve mathematics knowledge. *British Journal of Educational Psychology, 79*, 483–500. doi: doi:10.1348/000709908X398106.

Rittle-Johnson, B. & Siegler, R.S. (1998). The relation between conceptual and procedural knowledge in learning mathematics: a review. In C. Donlan (Ed.), *The Development of Mathematical Skills* (pp. 75–110). London: Psychology Press.

Rittle-Johnson, B., Siegler, R.S., & Alibali, M.W. (2001). Developing conceptual understanding and procedural skill in mathematics: an iterative process. *Journal of Educational Psychology, 93*, 346–362. doi: 10.1037//0022-0663.93.2.346.

Rittle-Johnson, B. & Star, J.R. (2007). Does comparing solution methods facilitate conceptual and procedural knowledge? An experimental study on learning to solve equations. *Journal of Educational Psychology, 99*, 561–574. doi: 10.1037/0022-0663.99.3.561.

Rittle-Johnson, B. & Star, J.R. (2009). Compared with what? The effects of different comparisons on conceptual knowledge and procedural flexibility for equation solving. *Journal of Educational Psychology, 101*, 529–544. doi: 10.1037/a0014224.

Rittle-Johnson, B., Star, J.R., & Durkin, K. (2009). The importance of prior knowledge when comparing examples: influences on conceptual and procedural knowledge of equation solving. *Journal of Educational Psychology, 101*, 836–852. doi: 10.1037/a0016026.

Rittle-Johnson, B., Star, J.R., & Durkin, K. (2011). Developing procedural flexibility: are novices prepared to learn from comparing procedures? *British Journal of Educational Psychology, 82* (3), 436–455.doi: 10.1111/j.2044-8279.2011.02037.x.

Ruthruff, E., Johnston, J.C., & van Selst, M.A. (2001). Why practice reduces dual-task interference. *Journal of Experimental Psychology: Human Perception and Performance, 27*, 3–21.

Schneider, M., Grabner, R., & Paetsch, J. (2009). Mental number line, number line estimation, and mathematical achievement: their interrelations in Grades 5 and 6. *Journal of Educational Psychology, 101*, 359–372. doi: 10.1037/a0013840.

Schneider, M., Rittle-Johnson, B., & Star, J.R. (2011). Relations between conceptual knowledge, procedural knowledge, and procedural flexibility in two samples differing in prior knowledge. *Developmental Psychology, 47* (6), 1525–1538. doi: doi:10.1037/a0024997.

Schneider, M. & Stern, E. (2009). The inverse relation of addition and subtraction: a knowledge integration perspective. *Mathematical Thinking and Learning, 11*, 92–101. doi: 10.1080/10986060802584012.

Schneider, M. & Stern, E. (2010). The developmental relations between conceptual and procedural knowledge: a multimethod approach. *Developmental Psychology, 46*, 178–192. doi: 10.1037/a0016701.

Schumacher, E.H., Seymour, T.L., Glass, J.M., Kieras, D.E., & Meyer, D.E. (2001). Virtually perfect time sharing in dual-task performance: uncorking the central attentional bottleneck. *Psychological Science, 121*, 101–108.

Schwartz, D.L., Chase, C.C., Chin, D. B., & Oppezzo, M. (2011). Practicing versus inventing with contrasting cases: the effects of telling first on learning and transfer. *Journal of Educational Psychology, 103*, 759–775. doi: 10.1037/a0025140.

Siegler, R.S. (1996). *Emerging Minds: The Process of Change in Children's Thinking*. New York: Oxford University Press.

Siegler, R.S. & Booth, J.L. (2004). Development of numerical estimation in young children. *Child Development, 75*, 428–444. doi: 10.1111/j.1467-8624.2004.00684.x.

Siegler, R.S. & Crowley, K. (1994). Constraints on learning in nonprivileged domains. *Cognitive Psychology, 27*, 194–226. doi: 10.1006/cogp.1994.1016.

Siegler, R.S. & Stern, E. (1998). Conscious and unconscious strategy discoveries: a microgenetic analysis. *Journal of Experimental Psychology: General, 127*, 377–397. doi: 10.1037/0096-3445.127.4.377.

Siegler, R.S., Thompson, C.A., & Schneider, M. (2011). An integrated theory of whole number and fractions development. *Cognitive Psychology, 62*, 273–296. doi: 10.1016/j.cogpsych.2011.03.001.

Star, J.R. (2005). Reconceptualizing procedural knowledge. *Journal for Research in Mathematics Education, 36*, 404–411. http://www.jstor.org/stable/30034943

Star, J.R. & Newton, K.J. (2009). The nature and development of expert's strategy flexibility for solving equations. *ZDM Mathematics Education, 41*, 557–567. doi: 10.1007/s11858-009-0185-5.

Star, J.R. & Rittle-Johnson, B. (2008). Flexibility in problem solving: the case of equation solving. *Learning and Instruction, 18*, 565–579. doi: 10.1016/j.learninstruc.2007.09.018.

Star, J.R. & Rittle-Johnson, B. (2009). It pays to compare: an experimental study on computational estimation. *Journal of Experimental Child Psychology, 101*, 408–426. doi: 10.1016/j.jecp.2008.11.004.

Stock, P., Desoete, A., & Roeyers, H. (2007). Early markers for arithmetic difficulties. *Educational and Child Psychology. Special Issue: Arithmetical Difficulties: Developmental and Instructional Perspectives, 24*, 28–39.

Sun, R., Merrill, E., & Peterson, T. (2001). From implicit skill to explicit knowledge: a bottom-up model of skill learning. *Cognitive Science, 25*, 203–244.

Taatgen, N.A. (1999). Implicit versus explicit: an ACT-R learning perspective. *Behavioral and Brain Sciences, 22*, 785–786.

Vamvakoussi, X. & Vosniadou, S. (2004). Understanding the structure of the set of rational numbers: a conceptual change approach. *Learning and Instruction, 14*, 453–467. doi: 10.1016/j.learninstruc.2004.06.013.

Verschaffel, L., Luwel, K., Torbeyns, J., & Van Dooren, W. (2009). Conceptualizing, investigating, and enhancing adaptive expertise in elementary mathematics education. *European Journal of Psychology of Education, 24*, 335–359. doi: 10.1007/bf03174765.

Williams, C.G. (1998). Using concept maps to assess conceptual knowledge of function. *Journal for Research in Mathematics Education, 29*, 414–422.

CHAPTER 60

HOW INFORMAL LEARNING ACTIVITIES CAN PROMOTE CHILDREN'S NUMERICAL KNOWLEDGE

GEETHA B. RAMANI AND ROBERT S. SIEGLER

PROVIDING children with a strong foundation of mathematical knowledge is critical for success in school and beyond. Children's early mathematical knowledge predicts their rate of growth in mathematics (Aunola, Leskinen, Lerkkanen, & Nurmi, 2007; Jordan, Kaplan, Locuniak, & Ramineni, 2007), as well as mathematics achievement test scores in later elementary school and even into the high school years (Duncan et al., 2007; Jordan, Kaplan, Ramineni, & Locuniak, 2009; Locuniak & Jordan, 2008; Mazzocco & Thompson, 2005). Furthermore, mathematical achievement can impact children's performance in college and choice of careers (National Mathematics Advisory Panel, 2008).

Given the importance of maths education, the development of maths skills has been extensively researched over the past 20 years (Dehaene & Brannon, 2011; Geary, 2006). One specific area of interest is the wide range of individual variations in children's mathematical knowledge. Of particular concern to educators, policymakers, and parents is that many of these differences are evident prior to children beginning school. Individual differences in the numerical knowledge of preschool and kindergarten children have been demonstrated on a variety of foundational maths skills, such as counting, identifying written numerals, and simple arithmetic (see National Research Council of the National Academies, 2009, for a review). These early differences tend to increase the further children move through school (Alexander & Entwisle, 1988; Geary, 1994, 2006). Therefore, understanding the kinds of experiences that promote early numerical knowledge is of critical importance because it can impact children's long-term maths achievement in school and possibly their long-term career opportunities.

In this article, we review recent research on how children's early experiences in the home and, with informal learning activities, can influence their mathematical understanding and performance. First, we present a growing body of literature that demonstrates from a sociocultural perspective how contextual factors and early experiences, mainly with parents in the early home environment, can shape children's numerical development and can explain individual differences in numerical knowledge. Second, we review theories and empirical work regarding a specific aspect of numerical knowledge, the mental number line, which likely underlies children's number sense. Finally, we argue that by integrating this theoretical analysis of the mental number line with a sociocultural perspective can help inform the design of activities to promote children's numerical knowledge. To support our argument, we review research that we have conducted that shows that playing a linear number board game can promote the numerical knowledge of preschool-age children from low-income backgrounds.

Sociocultural Perspectives on the Development of Numerical Knowledge

The sociocultural perspective provides an impetus for examining how children's early home environment and interactions with adults can influence children's mathematical development. A central tenet of sociocultural theory is that social interactions with adults play a critical role in children's cognitive development (Gauvain, 2001; Rogoff, 1990; Vygotsky, 1976, 1978). Play and other informal activities are considered particularly important contexts in which adults provide children with new information, support their skill development, and extend their conceptual understanding. Everyday, informal activities can provide children with extensive numerical information in the home environment (Saxe, 2004). Furthermore, much of the development of mathematical understanding in early childhood is social in nature, occurring during activities with parents, such as meals, chores, and shopping. For example, a young child could learn about one-to-one correspondence, while setting the table, fractions while cooking, and arithmetic while at the grocery store.

The sociocultural perspective has also motivated research on how children from different cultural and socioeconomic status SES backgrounds can vary in numerical knowledge and how these differences are influenced by their early home experiences. Differences in mathematical knowledge between children from China and the United States have been found in preschoolers as young as age 3 years (Miller, Kelly, & Zhou, 2005). The advanced number skills of young Chinese children have been shown on familiar tasks, such as arithmetic problems, and also on novel tasks, such as a number line estimation task (Siegler & Mu, 2008). These differences are related to early home experiences – Chinese parents practice skills such as arithmetic at home much more than US parents do (Zhou, Huang, Wang, Wang, Zhao, Yang, L., et al., 2006) – and also to general societal differences, such as variations in number names (Miller et al., 2005).

Even within the United States, young children's maths achievement and their mathematical experience vary widely. Specifically, the numerical knowledge of young children from low-income backgrounds trails far behind that of their peers from middle-income backgrounds. These differences do not extend to all numerical tasks. On non-verbal numerical tasks, the performance of young children from low-income backgrounds is equivalent to that of age peers from wealthier backgrounds (Jordan, Huttenlocher, & Levine, 1992; Jordan, Levine, & Huttenlocher, 1994). However, the same studies find large differences on tasks with verbally stated numbers, story problems, written numerals, and higher-level maths problems. Specifically, differences have been found in areas such as knowing the cardinality principle when counting, identifying written numbers, and solving arithmetic problems (Lee & Burkam, 2002; Klibanoff, Levine, Huttenlocher, Vasilyeva, & Hedges, 2006; Saxe, Guberman, & Gearhart, 1987; Starkey, Klein, & Wakeley, 2004). These symbolic tasks are foundational for later mathematical concepts and, therefore, are the focus of the research presented in this article.

Numerical Activities in the Home

One source of individual differences in numerical knowledge is the early home environment. In this section, we focus on the kinds of informal learning activities children engage in at home, variations in exposure to these activities, and consequences of these variations for children's developing numerical knowledge.

Number-Related Activities and Parent–Child Interactions

The number-related activities and support for learning that parents provide to children have been examined through self-reports, naturalistic observations in the home, and structured observations in the home and the laboratory. Questionnaires and interviews suggest that parents engage their young children in both formal and informal numerically relevant activities. These activities include formal teaching through number-related activity books and worksheets, explanations of number concepts, and practicing number skills, such as identifying written numbers (Huntsinger, Jose, Larson, Balsink Krieg, & Shaligram, 2000; Skwarchuk, 2009). Parents also engage their children in informal activities involving numbers, such as playing board games and card games, singing songs and nursery rhymes, and measuring ingredients while cooking (Blevins-Knabe & Musun-Miller, 1996).

Naturalistic observations indicate that parents begin exposing their children to numbers at a young age. In a longitudinal study with primarily middle-income families (Durkin, Shire, Riem, Crowther, & Rutter, 1986), parents were observed using number words with children as young as 9 months. Over the next 2 years, parents more frequently used number words and engaged their children in number-related activities, such as counting and reading books involving numbers. A more recent longitudinal study of parental input (Levine, Suriyakham, Rowe, Huttenlocher, & Gunderson, 2010) found that variations in parent number talk is related to children's developing numerical knowledge. Families from diverse backgrounds were observed in their homes for five visits between the ages of 14 and 30 months. At 46 months, children were administered a measure of understanding of cardinality, in which children were shown two cards and asked to point to the one with a specific number (e.g. point to 6). The amount of number talk that the parents engaged in from 14 to 30 months predicted their children's understanding of cardinality at 46 months, even after controlling for SES. A follow-up study revealed that particular types of number talk seemed better at promoting children's cardinality understanding. Specifically, parents' number talk involving counting or labelling the cardinal value of visible objects and their talk about large sets of objects were the most predictive of children's later number knowledge (Gunderson & Levine, 2011).

Studies using structured observations have indicated that the kinds of activities, materials, and games in which parents engage their children influences parental talk and support about numbers and numerical concepts. For example, Vandermaas-Peeler, Nelson, Bumpass, and Sassine (2009) asked parents to read a book with their children about a shopping trip, and then provided them with materials related to the story, such as pretend money and food, as well as cooking materials. They found that providing parents with these everyday materials elicited talk about the materials' general properties, such as appropriate uses for money, but elicited little talk about their numerical properties, such as the value of the money. In contrast, other studies using such methods have found that many parents engage their preschoolers in counting and other numerical activities, while reading books, playing with blocks, and working on a mathematical workbook. Materials that had fewer applications to daily life elicited greater talk about numbers (Anderson, 1997; Anderson, Anderson, & Shapiro, 2004).

Others have also found that different materials and contexts elicit different kinds of talk and support from parents. For example, Bjorklund, Hubertz, and Reubens (2004) observed parents and their preschoolers while they played a modified version of *Chutes and Ladders* together, and while they solved maths problems together. Parents were more likely to give cognitive directives, such as modelling the correct answer and providing instruction on strategies, when solving the maths problems, whereas they were more likely to use simple prompts when playing the board game.

Together, these findings suggest that parents engage in a variety of activities and discussions with their children that could support early numerical knowledge. Furthermore, the type of material influences parents' mathematically relevant interactions with their children.

Relations between Familial Numerical Activities and Early Maths Knowledge

Parents vary considerably in the frequency with which they engage in mathematically relevant activities with their children, and these variations influence children's developing numerical knowledge. Observational studies of the early home environment suggest that the absolute frequency with which parents engage their children in mathematical activities in the home tends to be quite low. For example, Tudge and Doucet (2004) conducted an observational study of preschool children from both White and Black families from diverse SES backgrounds. Children were observed during their daily routines for 18 hours, distributed over many days, in places such as home, childcare centers, and parks. On average, children were observed engaging in a mathematical lesson or play activity in less than 1 out of 180 observations.

One reason for this low absolute frequency is that parents tend to place a greater emphasis on literacy development than on mathematics development for their young children (Barbarin, Early, Clifford, Bryant, Frome, Burchinal, et al., 2008; Cannon & Ginsburg, 2008). Observations of homes and preschools, as well as reports of teachers and parents, suggest that the home and preschool environments provide children with fewer mathematical than literacy-orientated experiences (LeFevre, Skwarchuk, Smith-Chant, Fast, Kamawar, & Bisanz, 2009; Plewis, Mooney, & Creeser, 1990; Tizard & Hughes, 1984; Tudge & Doucet, 2004).

Although the average amount of mathematical activity is low, there is considerable variation in children's experience with number-related activities. For example, Levine et al. (2010) found the quantity of number words that parents used varied from 4 to 257 words over the 7.5 hours of observation. As would be expected, children whose parents present them with more mathematical activities generally have greater mathematical knowledge and maths fluency. This relation is present for both amount of direct instruction and amount of informal learning activities involving numbers (Blevins-Knabe & Musun-Miller, 1996; Huntsinger et al., 2000; LeFevre et al., 2009). Frequency of engaging in non-mathematical informal learning activities at home also is predictive of subsequent maths achievement. Parents' reports of 3- and 4-year-olds' engagement in informal learning activities, such as rhyming and singing songs, as well as providing direct instruction about letters and numbers, predicts children's mathematical achievement at age 10 (Melhuish, Sylva, Sammons, Siraj-Blatchford, Taggart, Phan, M et al., 2008).

Differences in the mathematical knowledge of children from lower- and higher-income backgrounds also reflect differences in environmental support for maths learning. Lower SES families have numerous stressors, such as financial constraints and lower education, which can limit their ability to support their children's academic development (Rouse, Brooks-Gunn, & McLanahan, 2005). These factors likely influence SES-related differences in the types of resources and number-related activities parents report engaging in with their children at home. Interviews with parents of 2- and 4-year-olds revealed that middle-income parents reported engaging their children in numerical activities that were rated as higher in complexity than working class parents (Saxe et al., 1987). An example of a complex activity involved arithmetic operations and a simple activity involved songs involving numbers. Others also have found that parents of middle-class children report engaging their children in a wider range of number-related activities, and engaging in such activities more frequently, than families from lower-income backgrounds (Starkey et al., 2004; Starkey & Klein, 2008). However, even within lower-income families, the early home environment and the type of support they offer their children for preparing them for school vary greatly and in ways that influence their mathematics proficiency (Burchinal, McCartney, Steinberg, Crosnoe, Friedman, McLoyd, V. et al., 2011; Holloway, Rambaud, Fuller, & Eggers-Pierola, 1995). For example, within Head Start populations, children's numerical knowledge is positively related to the frequency of parents' engagement with them in both formal maths activities, such as practicing simple arithmetic, and informal numerical activities, such as board games and card games (Ramani, Rowe, Eason, & Leech, 2015).

Thus, informal learning activities and the early home environment appear to play a critical role in the development of children's number skills. Having more opportunities for practicing number skills, through both informal learning activities and direct instruction, is positively related to children's maths skills.

INTEGRATING SOCIOCULTURAL THEORY WITH A THEORETICAL ANALYSIS OF NUMERICAL SENSE

In this section, we discuss how we integrated the sociocultural orientation and research studies that followed from this orientation with a theoretical analysis of the mental number line, which is hypothesized to underlie children's number sense (e.g. Dehaene, 2011). Our goal was to understand individual differences in children's numerical magnitude knowledge and to identify ways of improving this understanding. We were interested in three specific questions:

1. How does numerical magnitude knowledge vary between children, especially children from different SES backgrounds.
2. What are the sources of these differences.
3. How can we promote young children's numerical magnitude knowledge?

Before discussing answers to these questions, however, we review research on the development of numerical magnitude knowledge and the importance of this knowledge for maths learning.

Representations of Numerical Magnitudes

Accurate estimation of numerical magnitudes is critical for mathematical achievement and is central to the concept of *number sense* (Berch, 2005; Siegler & Booth, 2005; Sowder, 1992). Estimation can involve approximating the answer to arithmetic problems (e.g. 124 + 272), the distance between two objects or places (e.g. about how many miles is it to school), or the number of objects in a set (e.g. about how many cookies are in the jar?). Children with more advanced numerical estimation skills in first grade show faster growth in maths skills over the elementary school years, even after controlling for other predictive factors, such as intelligence and working memory (Geary 2011). Having a strong understanding of numerical magnitudes can lay the foundation for learning later, more complex, mathematics.

The cognitive structure thought to underlie numerical magnitude knowledge is the *mental number line*, which is based on the hypothesis that numbers are represented spatially on a continuum. In cultures that use left-to-right orthographies, numerical magnitudes increase from left to right on the continuum (Dehaene, 2011). Both behavioural and neural research indicates the importance of the mental number line (Ansari, 2008; Hubbard, Piazza, Pinel, & Dehaene, 2005). One body of evidence comes from research on the numerical magnitude task. Specifically, people more quickly answer 'Which is *bigger*, 8 or 3' when correct responses require pressing a key on the right rather than on the left. However, people more quickly answer 'Which is *smaller*, 8 or 3' by pressing a key on the left rather than on the right (Dehaene, Bossini, & Giraux, 1993). This spatial-numerical association of response codes (SNARC) effect provides evidence of how representations of quantities are ordered horizontally in a left to right array.

The form of mental number line representations can be measured using a number line estimation task. The number line estimation task involves presenting people lines with a number at each end (e.g. 0 and 100) and a third number, printed above the line, in that range. The task is to estimate the location of the third number on the line (e. g., 'Mark where 36 would go on the line'). There are three major benefits of the task.

1. The task can be used with any range of numbers, because any two numbers can be used at the ends of the lines.
2. Any type of number – whole, fraction, decimal, percentage, positive, negative – can be located on the line.
3. The number line task parallels the ratio characteristics of the number system.

That is, just as 60 is twice as large as 30, the distance of the estimated position of 60 from 0 should be twice as great as the distance of the estimated position of 30 from 0.

Performance on the number line estimation task correlates strongly with mathematics achievement test scores at all grade levels from kindergarten through eighth grade (Booth & Siegler, 2006; Geary, Hoard, Nugent, & Byrd Craven, 2008; Holloway & Ansari, 2009; Schneider, Grabner, & Paetsch, 2009; Siegler & Booth, 2004; Siegler, Thompson, & Schneider, 2011). Causal connections between number line estimation and maths learning also are present. Improving the numerical magnitude knowledge of children improves their learning of arithmetic and other mathematical skills (Booth & Siegler, 2008)

Numerical magnitude knowledge develops over several years and involves knowledge of a range of numerical skills and concepts. Knowledge of the counting sequence likely contributes to the development of numerical magnitude knowledge, but is not sufficient. Children count correctly from 1 at least a year before they show much knowledge of numerical magnitudes in the same range or even know the rank order of those numbers (Le Corre, Brannon, Van de Walle, & Carey, 2006; Le Corre & Carey, 2007; Lipton & Spelke, 2005). For example, not until the age of 5 years do children's number line estimates become accurate and linear for the numbers 0–10, even though they can count to 10 at least a year earlier (Berteletti, Lucangeli, Piazza, Dehaene, & Zorzi, 2010). Similarly, not until second grade do number line estimates become accurate for the numbers between 0 and 100, despite children being able to count to 100 a year or two earlier (Ebersbach, Luwel, Frick, Onghena, & Verschaffel, 2008; Geary et al., 2008; Siegler & Booth, 2004).

Variations in Numerical Magnitude Knowledge

To answer our first question of whether numerical magnitude knowledge varies between children from different SES backgrounds, we compared the performance of middle- and lower income preschoolers on a 0–10 number line estimation task (Siegler & Ramani, 2008). Children from lower-income backgrounds tended to have poorer numerical magnitude knowledge than children from middle-income backgrounds. Analyses of individual children's estimates showed the best fitting linear function accounted for a mean of 60% of the variance in the estimates of the individual children from middle-income backgrounds, but only 15% of the variance among children from low-income backgrounds.

Children from lower-income background also had a poorer understanding of the order of numbers. Specifically, we compared each child's estimate of the magnitude of each number with the child's estimate for each of the other numbers, and calculated the percentage of estimates that were correctly ordered. Children from higher-income backgrounds correctly ordered 81% of the estimates compared with 61% correct from the children from lower-income backgrounds.

Sources of Ses-Related Variations in Numerical Magnitude Knowledge

Our second question concerned potential sources of these SES-related differences in numerical magnitude knowledge. Based on the above review of the literature on the role of home activities in preschoolers' maths learning, informal activities seem to be critical for promoting early numerical knowledge. One common activity that seems ideally designed for producing linear representations of the mental number line is playing linear, numerical, board games – that is, board games with linearly-arranged, consecutively numbered, equal-sized spaces (e.g. *Chutes and Ladders*.) As noted by Siegler and Booth (2004), linear board games provide multiple cues to both the order of numbers and the numbers' magnitudes. Specifically, the greater the number in a square, the greater:

1. The distance the token from the origin to its present location.
2. The number of discrete moves the child has made.
3. The number of number names the child has spoken.
4. The number of number names the child has heard.
5. The amount of time since the game began.

Thus, children playing the game have the opportunity to relate the number in each square to the time, distance, and number of manual and vocal actions required to reach that number. In other words, the linear relations between numerical magnitudes and these visuospatial, kinaesthetic, auditory, and temporal cues provide a broadly-based, multi-modal foundation for a linear representation of numerical magnitudes.

To test our hypothesis that such board games could account for SES-related differences in numerical knowledge, Ramani and Siegler (2008) asked young children from lower- and higher-income backgrounds directly about their experiences with board games at home – whether they play board games, card games, and video games at their own home and those of other friends and relatives, and which games they play. As predicted, more preschoolers from middle-income backgrounds reported playing board games and card games than children from low-income backgrounds. Specifically, 80% of the children from middle-income backgrounds reported playing board games at home or at other people's home, whereas only 47% of the children from low-income backgrounds did. A similar pattern was found for card games, but not for video games; 66% of the preschool children from lower-income backgrounds reported playing video games at home, whereas only 30% of the middle-income children did.

Children from low-income backgrounds who had more experience playing board games at their own and other people's homes exhibited greater numerical knowledge (Ramani & Siegler, 2008). The relationship was even present when only experience playing the single number board game *Chutes and Ladders* was considered. Reported experience playing card games and video games was not closely related to numerical knowledge, thus indicating that the correlations with board game experience were not due to numerically advanced children having better memory for their game playing experience or being more willing to report it.

Playing a Linear Number Board Game to Improve Numerical Magnitude Knowledge

To address our third question of how to improve the numerical knowledge of children from low-income backgrounds, Siegler and Ramani (2008) randomly assigned preschoolers from Head Start classrooms to either play a linear numerical board game with squares numbered from 1 to 10 or a colour board game that was identical, except for the squares varying in colours, rather than numbers (Figure 60.1).

At the beginning of each session, children were told that, on each turn, they would spin a spinner with a '1' or '2' on it and move their token that number of spaces; the first person to reach the end would win. Children in the colour board condition were given the identical instructions, except their spinner varied in colour. The children were told to say the number (colour) on each space as they moved. For example, children who played the number board game who were on 3 and spun a 2 would say, '4, 5' as they moved their token. Similarly, children who played the colour board game who were on a red space and spun a purple, would say, 'green, purple', as they moved. Children played one of the two games one-on-one with an experimenter for four 15–20-minute sessions distributed over a 2-week period. Each game lasted approximately 2–4 minutes, so that children played their game roughly 20 times over the four sessions. Children estimated the positions of the numbers 1–10 on a number line prior to Session 1 as a pretest and at the end of Session 4 as a post-test.

Children who played the number board game considerably improved their numerical magnitude knowledge. On the pretest, the best-fitting linear function accounted for an average of 15% of the variance in individual children's estimates; on the post-test, it accounted for 61%. In contrast, for children in the colour board game condition, the best fitting linear function accounted for 18% of the variance on both pretest and posttest. Children who had played the numerical board game also ordered correctly the magnitudes of more numbers on the posttest than on the pretest. Peers who played an identical game, except for the squares varying in colour, rather than number, did not show comparable improvements on the task.

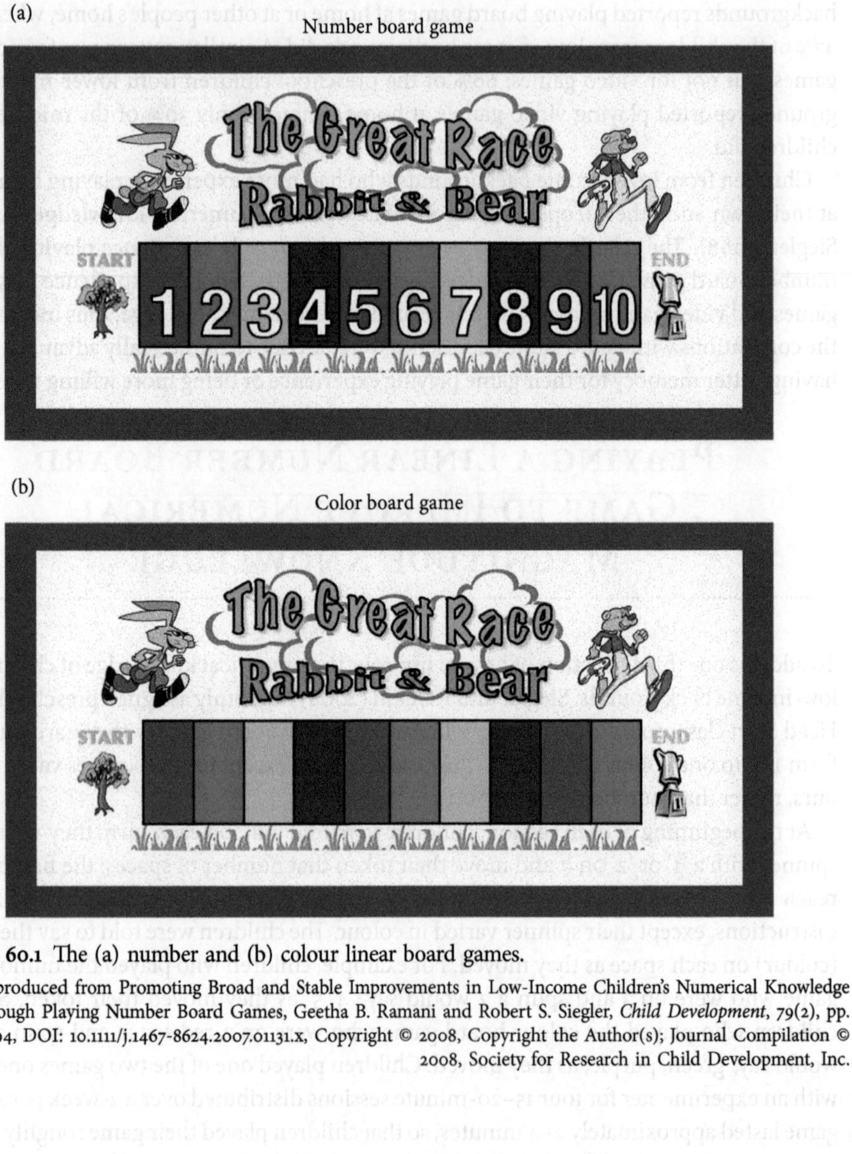

FIGURE 60.1 The (a) number and (b) colour linear board games.

Reproduced from Promoting Broad and Stable Improvements in Low-Income Children's Numerical Knowledge Through Playing Number Board Games, Geetha B. Ramani and Robert S. Siegler, *Child Development*, 79(2), pp. 375–394, DOI: 10.1111/j.1467-8624.2007.01131.x, Copyright © 2008, Copyright the Author(s); Journal Compilation © 2008, Society for Research in Child Development, Inc.

Improving Foundational Number Skills Over Time

We then tested the generality of the benefits of playing the number board game across various number tasks and over time (Ramani & Siegler, 2008). Playing the linear board game provides children with practice at counting and at numeral identification, because players are required to name the squares through which they move (e.g. saying '6, 7' after starting on the 5 and spinning a 2). Thus, playing such games would be expected to improve counting and numeral identification skills, as well as performance on tasks that require understanding of numerical magnitudes. We also wanted to examine whether children's learning remained apparent many weeks after the last game playing session.

Ramani and Siegler (2008) presented 124 Head Start children several measures of numerical knowledge of the numbers 1–10. Children were give the number line estimation task, a magnitude comparison task ('Which is bigger: N or M?'), a numeral identification task ('Can you tell me the number on this card?'), and rote counting ('Count from one to 10'). These tasks were presented on a pretest before the game playing began in Session 1, on a post-test immediately after the final game was played in Session 4, and in a follow-up session 9 weeks after the post-test.

After playing the number board game, children showed improvements on all four measures of numerical knowledge. These improvements were stable over time. After 9 weeks of not having played the board game, improvements on all four tasks remained significant, and they were at least as large on three of the four tasks as on the immediate post-test (Figure 60.2). As in the previous study, children who played an identical game, except for the squares varying in color, did not show comparable improvements.

Identifying Important Features of the Board Game's Design

Testing the effects of specific features of the board game on children's learning is critical for understanding why the game works and for creating future informal learning activities. One feature that seemed likely to be important was how the linearity of the game board influenced children's learning about numerical magnitudes. To test this feature, Siegler and

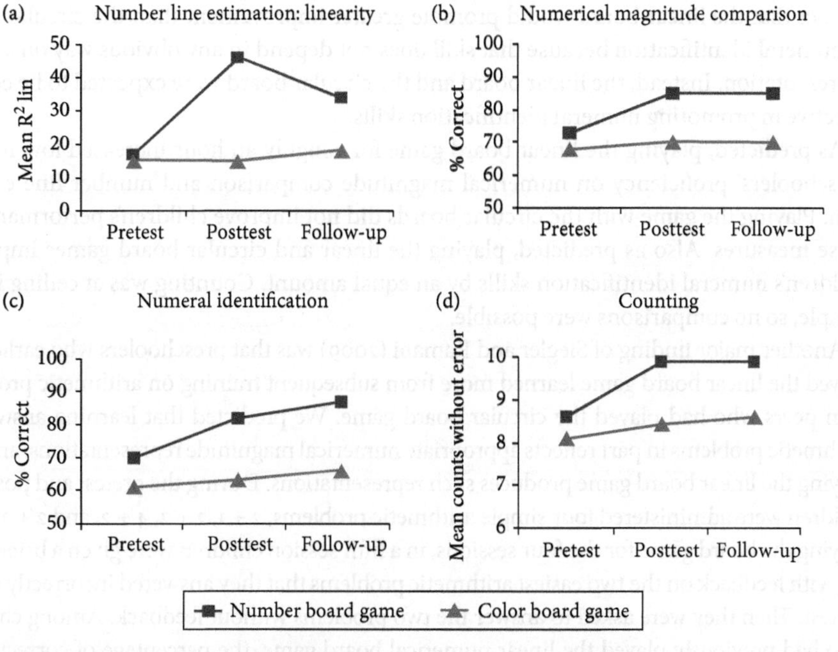

FIGURE 60.2 Performance of preschoolers from low-income backgrounds on four numerical tasks: (A) number line estimation; (B) magnitude comparison; (C) numerical identification; and (D) counting, before playing the number or colour board game (pretest), immediately after the fourth and final session of playing the game (post-test), and 9 weeks after the final game playing session (follow-up).

Reproduced from Promoting Broad and Stable Improvements in Low-Income Children's Numerical Knowledge Through Playing Number Board Games, Geetha B. Ramani and Robert S. Siegler, *Child Development*, 79(2), pp. 375–394, DOI: 10.1111/j.1467-8624.2007.01131.x, Copyright © 2008, Copyright Geetha B. Ramani and Robert S. Siegler; Journal Compilation © 2008, Society for Research in Child Development, Inc.

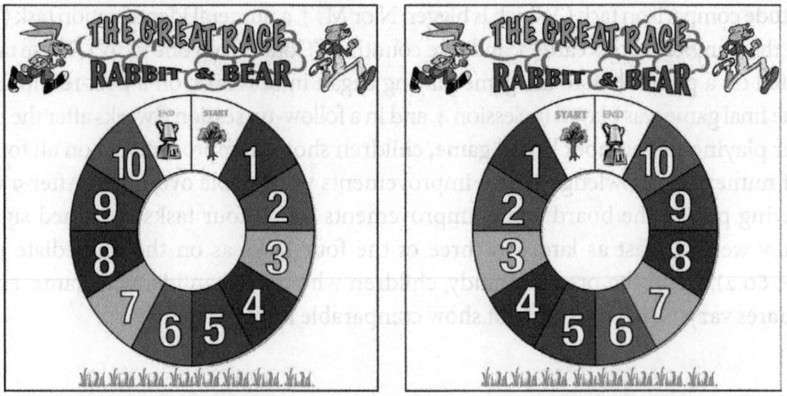

FIGURE 60.3 The circular number board games.

Adapted from Robert S. Siegler and Geetha B. Ramani, Playing linear number board games – but not circular ones – improves low-income preschoolers' numerical understanding, *Journal of Educational Psychology*, 101(3) pp. 545–560. doi: 10.1037/a0014239 © 2009, American Psychological Association.

Ramani (2009) assigned children to play either the linear board game or a circular board game (Figure 60.3). The linear game was expected to be more effective than the circular one on the magnitude tasks (number line and magnitude comparison), because the linear board is easier to translate into a linear mental number line. There was no obvious reason to predict that the linear board would promote greater improvement than the circular board in numeral identification because that skill does not depend in any obvious way on a linear representation. Instead, the linear board and the circular board were expected to be equally effective in promoting numeral identification skills.

As predicted, playing the linear board game for roughly an hour increased low-income preschoolers' proficiency on numerical magnitude comparison and number line estimation. Playing the game with the circular boards did not improve children's performance on these measures. Also as predicted, playing the linear and circular board games improved children's numeral identification skills by an equal amount. Counting was at ceiling in this sample, so no comparisons were possible.

Another major finding of Siegler and Ramani (2009) was that preschoolers who earlier had played the linear board game learned more from subsequent training on arithmetic problems than peers who had played the circular board game. We predicted that learning answers to arithmetic problems in part reflects appropriate numerical magnitude representations, and that playing the linear board game produces such representations. During the pretest and post-test, children were administered four simple arithmetic problems, 2 + 1, 2 + 2, 4 + 2, and 2 + 3. After playing the board game for the four sessions, in a fifth session children were given a brief training with feedback on the two easiest arithmetic problems that they answered incorrectly on the pretest. Then they were asked to answer the two problems without feedback. Among children who had previously played the linear numerical board game, the percentage of correct addition answers was higher than in the circular board game condition or in a control condition. Especially relevant to the idea that the gains came about through improving children's numerical magnitude representations, children's errors in the linear board condition, but not in the circular board condition, tended to become closer to the correct answer from pretest to post-test.

A second feature of the board game that we tested was whether the context of the game was important for promoting children's numerical knowledge. To determine whether playing numerical board games had effects above and beyond those of common number activities, Siegler and Ramani (2009) provided a third group of children practice with counting and numeral naming tasks, but not in a game context. We found that having children engage in a continuing cycle of the tasks – number string counting, numeral identification, and object counting – for the same length of time as children played the board game did not influence children's numerical knowledge. These children also did not improve their ability to learn the answers to novel arithmetic problems.

Scaling Up the Board Game Intervention

Determining whether an intervention can be used in everyday settings requires evidence beyond that the intervention is effective under controlled, laboratory conditions. Three types of evidence relevant to the present context are whether the board game is effective with different populations, when played with small groups of children, and when implemented by a paraprofessional from the children's classroom.

One variable that is important for scaling up the intervention is whether children from some groups benefit more than others. Preschoolers from low-income backgrounds at times show greater gains from mathematical interventions than preschoolers from middle-income backgrounds (Starkey et al., 2004). Because children from middle-income backgrounds have greater prior board game experience than children from low-income backgrounds, this greater experience might make playing the present board game redundant with the middle-income children's prior experience, and thus less effective for improving their numerical knowledge. We tested whether playing the linear board game would improve the numerical knowledge of preschoolers from middle-income backgrounds, or whether the benefits were unique to children from low-income backgrounds. Specifically, we compared the relative benefits of playing a linear number board game for two groups of children who before the experience had equal numerical knowledge: 3- and 4-year-olds from middle-income backgrounds and 4- and 5-year-olds from low-income backgrounds.

Children from low-income backgrounds learned at least as much, and on several measures more, than preschoolers from middle-income backgrounds with comparable initial knowledge. Within each group, children who initially knew less tended to learn more. Overall, the findings suggest that playing the linear board game is effective at promoting the numerical knowledge of children from both lower- and middle-SES backgrounds.

Interventions that are effective in lab settings are often ineffective in classrooms (Newcombe, Ambady, Eccles, Gomez, Klahr, Linn, M. et al., 2009; White, Frishkoff, & Bullock, 2008). Two likely reasons are that the 1:1 experimenter–child interactions that are typical of laboratory studies are often impossible in classrooms, and that the people executing the intervention in classrooms usually have fewer opportunities for training in executing the interventions than research personnel.

To better understand these challenges, we have begun to examine whether the number board game could be translated into a practical classroom activity that improves Head Start children's numerical knowledge. Participants within a condition were randomly divided into groups of three children and were presented six 20–25-min sessions within a 3- or 4-week period. Children played a similar linear number board game or colour board game to those used in previous studies. The procedures and game playing were also similar except the children played the game with each other, and the experimenter facilitated the game, instead of playing.

Playing the number board game as a small group learning activity promoted low-income children's number line estimation, magnitude comparison, numeral identification, and counting. Children who played the colour board game only improved in counting skills.

Improvements were also found when paraprofessionals from the children's classrooms played the game with small groups of children. The paraprofessionals were given roughly an hour of training prior to the start of the study. During this time, they were given the board game materials and an opportunity to practice with them. They were also given a short booklet that included the rules for the number and colour games, scripts for how to explain the games to the children, and standard prompts for correcting errors. The paraprofessionals also watched a demonstration video of a group of children playing the board games, and were told to correct errors or omissions by prompting children to say the required number or colour. If the errors continued, the paraprofessionals were to model the correct move and ask the child to repeat the move.

We found that playing the number board game as a small group activity supervised by a paraprofessional from the classroom improved children's numerical knowledge on four measures. Observations of the game playing sessions revealed that paraprofessionals adapted the feedback they provided to reflect individual children's improving numerical knowledge over the game playing sessions and that children remained engaged in the board game play even after multiple sessions (Ramani, Siegler, & Hitti, 2012). Thus, the linear number board game can be used to promote the numerical knowledge of children from a range of SES backgrounds and can be used effectively in preschool classrooms.

Preschool Mathematics Interventions and Curricula

Other targeted interventions also have been shown to be effective in improving young children's numerical understanding. An adaptive software program called 'The Number Race,' aimed at improving young children's number sense, has improved the skills of children with mathematical difficulties (Wilson, Revkin, Cohen, Cohen, & Dehaene, 2006; Wilson, Dehaene, Dubois, & Fayol, 2009). Other experimental interventions that have incorporated number board games have also produced improvements in young children's numerical knowledge (e.g. Malofeeva, Day, Saco, Young, & Cianco, 2004).

More comprehensive curricula for improving low-income preschoolers' and kindergartners' mathematical knowledge have also shown large positive effects. These programmes integrate informal learning activities with direct classroom instruction. One such curriculum is *Number Worlds,* which includes a wide range of numerical activities – songs about numbers, counting games, games involving money, and board games. The goal of the curriculum is to provide children with a strong foundation with numbers before teaching them more advanced concepts. Researchers have found that following participation in the Number Worlds curriculum, low-income kindergarteners had significantly better basic numerical skills than peers who did not receive the curriculum (Griffin, 2004).

Another early childhood curriculum, *Pre-K Mathematics,* combines school- and home-based activities to promote children's numerical knowledge. Children participate in small-group activities in the classrooms and parents are also provided with activities to do at home that link to the small-group activities in the school. Participation in the programme led to kindergartners from low-income backgrounds having mathematical knowledge at the end of the programme equivalent to that of middle-income peers who did not participate in it (Starkey et al., 2004).

Similarly, the *Building Blocks* curriculum (Clements & Sarama, 2007) includes classroom activities, with small group activities and computer games. Randomized control trials of preschoolers from low-income backgrounds indicated that children given the Building Blocks curriculum made much greater progress than a control group in number, geometry, measurement, and recognition of patterns. Overall, these curricula have found success by combining direct instruction, home involvement, and informal learning activities as ways to promote the numerical knowledge of young children.

Conclusions

The early home environment and children's experiences with informal learning activities play a critical role in mathematical development. Understanding that these experiences contribute to individual differences in numerical knowledge helped to inform the design of an intervention that can be used to promote the numerical knowledge of young children from lower-income backgrounds. A large body of evidence, including our research on board games, provides good reason to advocate that parents and teachers more frequently engage preschoolers in mathematical activities. This evidence includes the finding that early numerical knowledge lays the groundwork for later mathematical achievement, and that young children's number skills can be improved through a variety of informal and formal numerical activities. Many of these activities, including the linear number board game, are inexpensive or can be created at home, and require minimal time to play. Thus, it seems critical to play such games, engage in other informal maths activities, and make available mathematics curricula of proven effectiveness to a wider range of preschoolers, especially preschoolers from low-income backgrounds.

Acknowledgments

We would like to thank the Institute of Educational Sciences, which supported this research, Grants R305A080013 and R305H050035, and the Teresa Heinz Chair at Carnegie Mellon University.

References

Alexander, K.L., & Entwisle, D.R. (1988). Achievement in the first 2 years of school: patterns and processes. Monographs of the Society for Research in Child Development, 53(2, Serial No. 157).

Anderson, A. (1997). Families and mathematics: a study of parent–child interactions. Journal for Research in Mathematics Education, 28(4), 484–511.

Anderson, A., Anderson, A., & Shapiro, J. (2004). Mathematical discourse in shared storybook reading. Journal for Research in Mathematics Education, 35(1), 5–33

Ansari, D. (2008). Effect of development and enculturation on number representation in the brain. Nature Reviews Neuroscience, 9(4), 278–291.

Aunola, K., Leskinen, E., Lerkkanen, M.K., & Nurmi, J.E. (2007). Developmental dynamics of math performance from preschool to grade 2. Journal of Educational Psychology, 96, 699–713.

Barbarin, O.A., Early, D., Clifford, R., Bryant, D., Frome, P, Burchinal, M. et al. (2008). Parental conceptions of school readiness: relation to ethnicity, socioeconomic status, and children's skills. Early Education and Development, 19(5), 671–701.

Berch, D. B. (2005). Making sense of number sense: implications for children with mathematical disabilities. Journal of Learning Disabilities, 38, 333–339.

Berteletti, I., Lucangeli, D., Piazza, M., Dehaene, S., & Zorzi, M. (2010). Numerical estimation in preschoolers. Developmental Psychology, 46, 545–551.

Bjorklund, D.F., Hubertz, M.J., & Reubens, A.C. (2004). Young children's arithmetic strategies in social context: how parents contribute to children's strategy development while playing games. International Journal of Behavioral Development, 28(4), 347–357.

Blevins-Knabe, B., & Musun-Miller, L. (1996). Number use at home by children and their parents and its relationship to early mathematical performance. Early Development and Parenting, 5, 35–45.

Booth, J.L., & Siegler, R.S. (2006). Developmental and individual differences in pure numerical estimation. Developmental Psychology, 42, 189–201.

Booth, J.L., & Siegler, R.S. (2008). Numerical magnitude representations influence arithmetic learning. Child Development, 79, 1016–1031.

Burchinal, M., McCartney, K., Steinberg, L., Crosnoe, R., Friedman, S. L., McLoyd, V. et al. (2011). Examining the Black–White achievement gap among low-income children using the NICHD study of early child care and youth development. Child Development, 82(5), 1404–1420.

Cannon, J., & Ginsburg, H.P. (2008). 'Doing the math': maternal beliefs about early mathematics versus language learning. Early Education and Development, 19(2), 238–260.

Clements, D.H., & Sarama, J. (2007). Early childhood mathematics learning. In F. K. Lester (Ed.), Second Handbook of Research on Mathematics Teaching and Learning (pp. 461–555). New York: Information Age Publishing.

Dehaene, S. (2011). The Number Sense: How the Mind Creates Mathematics (revised edn). New York: Oxford University.

Dehaene, S., Bossini, S., & Giraux, P. (1993). The mental representation of parity and numerical magnitude. Journal of Experimental Psychology: General, 122, 371–396.

Dehaene, S., & Brannon, E. (Eds) (2011). Space, Time, and Number in the Brain: Searching for the Foundations of Mathematical Thought. London, Academic.

Durkin, K., Shire, B., Riem, R., Crowther, R. D., & Rutter, D. R. (1986). The social and linguistic context of early number word use. British Journal of Developmental Psychology, 4, 269–288.

Duncan, G. J., Dowsett, C. J., Claessens, A., Magnuson, K., Huston, A. C., Klebanov, P, et al. (2007). School readiness and later achievement. Developmental Psychology, 43, 1428–1446.

Ebersbach, M., Luwel, K., Frick, A., Onghena, P., & Verschaffel, L. (2008). The relationship between the shape of the mental number line and familiarity with numbers in 5- to 9-year-old children: evidence for a segmented linear model. Journal of Experimental Child Psychology, 99, 1–17.

Gauvain, M. (2001). The Social Context of Cognitive Development. New York, NY: Guilford.

Geary, D.C. (1994). Children's Mathematics Development: Research and Practical Applications. Washington, DC: American Psychological Association.

Geary, D.C. (2006). Development of mathematical understanding. In W. Damon & R. M. Lerner (Series Eds), and D. Kuhn & R. S. Siegler (Volume Eds), *Handbook of Child Psychology: Volume 2: Cognition, Perception, and Language* (6th edn, pp. 777–810). Hoboken, NJ: Wiley.

Geary, D.C. (2011). Cognitive predictors of individual differences in achievement growth in mathematics: a five year longitudinal study. *Developmental Psychology*. 47(6), 1539–1552.

Geary, D.C., Hoard, M.K., Nugent, L., & Byrd Craven, J. (2008). Development of number line representations in children with mathematical learning disability. Developmental Neuropsychology, 33, 277–299.

Griffin, S. (2004). Building number sense with Number Worlds: a mathematics program for young children. Early Childhood Research Quarterly, 19(1), 173–180.

Gunderson, E.A., & Levine, S.C. (2011). Some types of parent number talk count more than others: relations between parents' input and children's cardinal-number knowledge. Developmental Science, 14, 1021–1032.

Holloway, I.D. & Ansari, D. (2009). Mapping numerical magnitudes onto symbols: the numerical distance effect and individual differences in children's math achievement. Journal of Experimental Child Psychology, 103, 17–29.

Holloway, S.D., Rambaud, M.F., Fuller, B., & Eggers-Pierola, C. (1995). What is 'appropriate practice' at home and in child care? Low-income mothers' views on preparing their children for school. Early Childhood Research Quarterly, 10(4), 451–473.

Hubbard, E.M., Piazza, M., Pinel, P., & Dehaene, S. (2005). Interactions between numbers and space in parietal cortex. Nature Reviews Neuroscience, 6(6), 435–448.

Huntsinger, C.S., Jose, P.E., Larson, S.L., Balsink Krieg, D., & Shaligram, C. (2000). Mathematics, vocabulary, and reading development in Chinese American and European American children over the primary school years. Journal of Educational Psychology, 92, 745–760.

Jordan, N.C., Huttenlocher, J., & Levine, S.C. (1992). Differential calculation abilities in young children from middle- and low-income families. Developmental Psychology, 28, 644–653.

Jordan, N.C., Kaplan, D., Locuniak, M.N., & Ramineni, C. (2007). Predicting first-grade math achievement from developmental number sense trajectories. Learning Disabilities Research and Practice, 22, 36–46.

Jordan, N.C., Kaplan, D., Ramineni, C., & Locuniak, M.N. (2009). Early math matters: kindergarten number competence and later mathematics outcomes. Developmental Psychology, 45(3), 850–867.

Jordan, N.C., Levine, S.C., & Huttenlocher, J. (1994). Development of calculation abilities in middle- and low-income children after formal instruction in school. Journal of Applied Developmental Psychology, 15, 223–240.

Klibanoff, R.S., Levine, S.C., Huttenlocher, J., Vasilyeva, M., & Hedges, L.V. (2006). Preschool children's mathematical knowledge: the effect of teacher 'Math Talk'. Developmental Psychology, 42(1), 59–69.

Le Corre, M.L., Brannon, E.M, Van de Walle, G., & Carey, S. (2006). Re-visiting the competence/performance debate in the acquisition of the counting principles. Cognitive Psychology, 52, 130–169.

Le Corre, M.L. & Carey, S. (2007). One, two, three, four, nothing more: an investigation of the conceptual sources of the verbal counting principles. Cognition, 105, 395–438.

Lee, V.E., & Burkam, D. (2002). Inequality at the Starting Gate: Social Background Differences in Achievement as Children Begin School. Washington, DC: Economic Policy Institute.

LeFevre, J., Skwarchuk, S., Smith-Chant, B., Fast, L., Kamawar, D., & Bisanz, J. (2009). Home numeracy experiences and children's math performance in the early school years. Canadian Journal of Behavioural Science, 41, 55–66.

Levine, S.C., Suriyakham, L.W., Rowe, M.L., Huttenlocher, J., & Gunderson, E.A. (2010). What counts in the development of young children's number knowledge? Developmental Psychology, 46, 1309–1319.

Lipton, J.S., & Spelke, E.S. (2005). Preschool children's mapping of number words to nonsymbolic numerosities. Child Development, 76, 978–988.

Locuniak, M.N. & Jordan, N.C. (2008). Using kindergarten number sense to predict calculation fluency in second grade. Journal of Learning Disabilities, 41(5), 451–459.

Malofeeva, E., Day, J., Saco, X., Young, L., & Ciancio, D. (2004). Construction and evaluation of a number sense test with Head Start children. Journal of Educational Psychology, 96(4), 648–659.

Mazzocco, M.M.M., & Thompson, R.E. (2005). Kindergarten predictors of math learning disability. Learning Disabilities Research and Practice, 20(3), 142–155.

Melhuish, E., Sylva, K., Sammons, P., Siraj-Blatchford, I., Taggart, B., Phan, M. et al. (2008). Preschool influences on mathematics achievement. Science, 321(5893), 1161–1162.

Miller, K.F., Kelly, M.K., & Zhou, X. (2005). Learning mathematics in China and the United States: cross-cultural insights into the nature and course of mathematical development. In J. I. D. Campbell (Ed.), Handbook of Mathematical Cognition (pp. 163–173). New York: Psychology.

National Mathematics Advisory Panel (2008). Foundations for Success: The Final Report of the National Mathematics Advisory Panel. Washington, DC: US Department of Education.

National Research Council of the National Academies (2009). Mathematics learning in early childhood: paths toward excellence and equity. In C. T. Cross, T. A. Woods, & H. Schweingruber (Eds), Committee on Early Childhood Mathematics. Washington, DC: National Academies.

Newcombe, N.S., Ambady, N., Eccles, J., Gomez, L., Klahr, D., Linn, M. et al. (2009). Psychology's role in mathematics and science education. American Psychologist, 64(6), 538–550.

Plewis, I., Mooney, A., & Creeser, R. (1990). Time on educational activities at home and education progress in infant school. British Journal of Educational Psychology, 60, 330–337.

Ramani, G.B., Rowe, M.R., Easton, S., & Leech, K. (2015). Math talk during informal learning activities in Head Start families. *Cognitive Development*, 35, 15–33.

Ramani, G.B., & Siegler, R.S. (2008). Promoting broad and stable improvements in low-income children's numerical knowledge through playing number board games. Child Development, 79, 375–394.

Ramani, G.B., & Siegler, R.S. (2011). Reducing the gap in numerical knowledge between low- and middle-income preschoolers. Journal of Applied Developmental Psychology, 32, 146–159.

Ramani, G.B., Siegler, R.S. & Hitti, A. (2012). Taking it to the classroom: Number board games as a small group learning activity. *Journal of Educational Psychology*, 104, 661–672.

Rogoff, B. (1990). Apprenticeship in Thinking: Cognitive Development in Social Context. New York: Oxford University.

Rouse, C.E., Brooks-Gunn, J., & McLanahan, S. (2005). Introducing the issue. Future of Children, 15(1), 5–14.

Saxe, G.B. (2004). Practices of quantification from a sociocultural perspective. In K. A. Demetriou & A. Raftopoulos (Eds), Developmental Change: Theories, Models, and Measurement (pp. 241–263). New York, NY: Cambridge University.

Saxe, G.B., Guberman, S.R., & Gearhart, M. (1987). Social processes in early number development. Monographs of the Society for Research in Child Development, 52 (Serial No. 216)

Schneider, M., Grabner, R.H., & Paetsch, J. (2009). Mental number line, number line estimation, and mathematical school achievement: their interrelations in Grades 5 and 6. Journal of Educational Psychology, 101, 359–372.

Siegler, R.S., & Booth, J. (2004). Development of numerical estimation in young children. Child Development, 75, 428–444.

Siegler, R.S., & Booth, J. (2005). Development of numerical estimation: a review. In J. I. D. Campbell (Ed.), Handbook of Mathematical Cognition (pp. 197–212). New York, NY: Psychology.

Siegler, R.S., & Mu, Y. (2008). Chinese children excel on novel mathematics problems even before elementary school. Psychological Science, 19, 759–763.

Siegler, R.S., & Ramani, G.B. (2008). Playing board games promotes low-income children's numerical development. Developmental Science, Special Issue on Mathematical Cognition, 11, 655–661.

Siegler, R.S. & Ramani, G.B. (2009). Playing linear number board games – but not circular ones – improves low-income preschoolers' numerical understanding. Journal of Educational Psychology, 101, 545–560.

Siegler, R.S., Thompson, C.A., & Schneider, M. (2011). An integrated theory of whole number and fractions development. Cognitive Psychology, 62, 273–296.

Skwarchuk, S-L. (2009). How do parents support preschoolers' numeracy learning experiences at home? Early Childhood Education Journal, 37, 189–197.

Sowder, J.T. (1992). Estimation and number sense. In D. A. Grouws (Ed.), Handbook of Research on Mathematics Teaching and Learning (pp. 371–389). New York, NY: Macmillan.

Starkey, P., & Klein, A. (2008). Sociocultural Influences on Young Children's Mathematical Knowledge: Contemporary Perspectives on Mathematics in Early Childhood Education. Charlotte, NC: Information Age.

Starkey, P., Klein, A., & Wakeley, P. (2004). Enhancing young children's mathematical knowledge through a pre-kindergarten mathematics intervention. Early Childhood Research Quarterly, 19, 99–120.

Tizard, B., & Hughes, M. (1984). Children Learning at Home and in School. London: Fontana.

Tudge, J.R.H., & Doucet, F. (2004). Early mathematical experiences: observing young black and white children's everyday activities. Early Childhood Research Quarterly, 19, 21–39.

Vandermaas-Peeler, M., Nelson, J., Bumpass, C., & Sassine, B. (2009). Numeracy-related exchanges in joint storybook reading and play. International Journal of Early Years Education, 17(1), 67–84.

Vygotsky, L. S. (1976). Play and its role in the mental development of the child. In J. S. Bruner, A. Jolly, & K. Sylva (Eds), Play (pp. 537–559). New York, NY: Harper and Row.

Vygotsky, L.S. (1978). Mind and Society: The Development of Higher Psychological Processes. Cambridge, MA: Harvard University.

White, G., Frishkoff, G., & Bullock, M. (2008). Bridging the gap between psychological science and educational policy and practice. In S. Thurman, C. A. Fiorello, S. Thurman, C. A. Fiorello (Eds), Applied Cognitive Research in K–3 Classrooms (pp. 227–263). New York, NY: Routledge/Taylor and Francis Group.

Wilson, A.J., Dehaene, S., Dubois, O., & Fayol, M. (2009). Effects of an adaptive game intervention on accessing number sense in low-socioeconomic-status kindergarten children. Mind, Brain, and Education, 3, 224–234

Wilson, A.J., Revkin, S.K., Cohen, D., Cohen, L., & Dehaene, S. (2006). An open trial assessment of 'The Number Race', an adaptive computer game for remediation of dyscalculia. Behavioral and Brain Functions, 2, 20.

Zhou, X., Huang, J., Wang, Z., Wang, B., Zhao, Z., Yang, L., et al. (2006). Parent-child interaction and children's number learning. Early Child Development and Care, 176, 763–775.

Index

Note: Page numbers in *italics* refer to figures and tables. Footnotes are indicated by the suffix 'n' followed by the note number, for example 34n1.

E

F

H

I

Q

R